HANDBOOK OF SCIENTIFIC TABLES

HANDBOOK OF SCIENTIFIC TABLES

National Astronomical Observatory of Japan

MARUZEN PUBLISHING

World Scientific

NEW JERSEY • LONDON • SINGAPORE • BEIJING • SHANGHAI • HONG KONG • TAIPEI • CHENNAI • TOKYO

Published by

World Scientific Publishing Co. Pte. Ltd.
5 Toh Tuck Link, Singapore 596224
USA office: 27 Warren Street, Suite 401-402, Hackensack, NJ 07601
UK office: 57 Shelton Street, Covent Garden, London WC2H 9HE

British Library Cataloguing-in-Publication Data
A catalogue record for this book is available from the British Library.

HANDBOOK OF SCIENTIFIC TABLES

ISBN 978-981-3278-51-6 (hardcover)
ISBN 978-981-3278-52-3 (ebook for institutions)
ISBN 978-981-3278-53-0 (ebook for individuals)

For any available supplementary material, please visit
https://www.worldscientific.com/worldscibooks/10.1142/11218#t=suppl

Preface

As the title "Handbook of Scientific Tables" suggests, the tables found in this volume offer readers a comprehensive and contemporary collection of academic data and information in all areas of science. It presents areas such as astronomy, meteorology, physics and chemistry, earth science, life science, and environmental science in a style accessible to everyone seeking solid, authoritative information including the general public, students, and professional scientists.

Handbook of Scientific Tables is the first international version of the "Rika Nenpyo". First published in 1925 in Japanese by the then young Maruzen Publishing Co., Ltd., "Rika Nenpyo" has been republished every year since then, incorporating updates and completely new data to reflect rapid changes and advances in science.

While the Tables has continued to grow in quantity and scope since its early beginnings, it has established an exceptional style and structure in terms of accuracy, comprehensiveness, thoroughness, and accessibility during the revisions spanning almost one century. The purpose however remains the same as stated in the preface to the 1925 edition: to provide a complete collection of important and essential data in all-natural science disciplines and related areas. This has always been a fundamental and abiding principle of the Tables. This first English version is based on the 94th edition published in 2020.

All the individual numbers and information contained in the Tables have been through thorough examination and scrutiny. This rigorous process has earned the Tables a reputation as an authoritative but popular reference in science and related areas. Over the years, the book has gradually come to be well known and used by a much broader audience than just professional scientists and people interested in science.

Handbook of Scientific Tables is the result of the magnificent but painstaking labor of generations of solitary scientists with a deep sense of responsibility; they have provided information on the latest developments in various fields, innumerable and invaluable suggestion for additions and deletions, and expert advice on how to present the vast quantity of material in an attractive and accessible format. In this limited space, I can't properly acknowledge the strenuous efforts of all those who tirelessly contributed to the preparation of this international version and the preceding 94th editions of "Rika Nenpyo." The Tables is after all an invaluable encapsulation of the knowledge and wisdom that humanity has accumulated through persistent investigation of our world and Universe.

This groundbreaking edition is a reply to those who seek an authoritative source of scientific information and the keys to unlock even greater knowledge; it is a useful source of knowledge that can inspire new ideas in human endeavors far beyond natural sciences. We do hope that you consult the Tables frequently in a variety of situations, including your daily work, research activities, and school assignments.

March 2022

Saku Tsuneta
Director General
National Astronomical Observatory of Japan

Masahiko Hayashi
Professor emeritus (Former Director General)
National Astronomical Observatory of Japan

Toshio Fukushima
Professor emeritus (Former Director of Public Relation Center)
National Astronomical Observatory of Japan

Co-editors

Astronomy Section

National Astronomical Observatory of Japan (NAOJ)

Meteorology Section

Japan Meteorological Agency (JMA)

Physics and Chemistry Section

Takahiro Kuga (The University of Tokyo, Graduate School of Arts and Sciences)

Tatsuya Tsukuda (The University of Tokyo, Graduate School of Science)

Hiroshi Nishihara (Tokyo University of Science, Research institute for Science and Technology)

Earth Science Section

Toshihiko Sugai (The University of Tokyo, Graduate School of Frontier Sciences)

Kazuhito Ozawa (The University of Tokyo, Atmosphere and Ocean Research Institute)

Toshitsugu Fujii (Research Institute for Disaster Mitigation and Environmental Studies, Crisis & Environment Management Policy Institute)

Kazuki Koketsu (Keio University, Graduate School of Media and Governance)

Satoshi Taguchi (Kyoto University, Graduate School of Science)

Toshihiko Iyemori (Kyoto University, Professor emeritus)

Takuya Tsugawa (National Institute of Information and Communications Technology, Radio Research Institute)

Mamoru Ishii (National Institute of Information and Communications Technology, Radio Research Institute)

Life Science Section

Makoto Asashima (Teikyo University, Advanced Comprehensive Research Organization)

Environmental Science Section

Makoto Asashima (Teikyo University, Advanced Comprehensive Research Organization)

Toshihiko Sugai (The University of Tokyo, Graduate School of Frontier Sciences)

Yasuaki Hijioka (National Institute for Environmental Studies, Center for Social and Environmental Systems Research), Japan Meteorological Agency (JMA)

Authors

Astronomy Section Masaaki Hiramatsu (NAOJ, Public Relations Center), Jun-ichi Watanabe (NAOJ, Public Relations Center), Masatoshi Ohishi (NAOJ, Public Relations Center), Kumiko Usuda-Sato (NAOJ, Subaru Telescope), Masato Katayama (NAOJ, Public Relations Center), Ko Matsuda (NAOJ, Public Relations Center), Isao Endoh (NAOJ, Public Relations Center), Takashi Shibata (NAOJ, Public Relations Center), Yoichiro Hanaoka (NAOJ, Solar Science Observatory), Yoshiaki Tamura (NAOJ, Mizusawa VLBI Observatory), Tomoya Hirota (NAOJ, Mizusawa VLBI Observatory), Taihei Yano (NAOJ, JASMINE Project), Noriyuki Namiki (NAOJ, RISE Project), Tomonori Usuda (NAOJ, Thirty Meter Telescope (TMT)Project), Wako Aoki (NAOJ, Thirty Meter Telescope (TMT) Project), Takuya Yamashita (NAOJ, Thirty Meter Telescope (TMT) Project), Hirohisa Hara (NAOJ, SOLAR-C Project), Akiko Nakamura (Kobe University, Graduate School of Science), Masao Takata (The University of Tokyo, Graduate School of Science), Masashi Chiba (Tohoku University, Graduate School of Science), Tadayuki Takahashi (The University of Tokyo, Kavli Institute for the Physics and Mathematics of the Universe), Naoki Yasuda (The University of Tokyo, Kavli Institute for the Physics and Mathematics of the Universe), Nobuyuki Kawai (Tokyo Institute of Technology, Graduate School of Engineering), Makoto Yoshikawa (Japan Aerospace Exploration Agency (JAXA)), Tetsuharu Fuse (National Institute of Information and Communications, Technology Network Research Institute), Motohide Tamura (The University of Tokyo, Graduate School of Science), Tsutomu Takeuchi (Nagoya University, Graduate School of Science), Tadayuki Kodama (Tohoku University, Graduate School of Science), Fumihiko Usui (Japan Aerospace Exploration Agency (JAXA)), Yoshinori Suematsu (NAOJ, Solar Science Observatory), Nobuo Arimoto (NAOJ, Professor emeritus), Yoichi Takeda

Meteorology Section Japan Meteorological Business Support Center (JMBSC), Michihiko Tonouchi (Japan Meteorological Business Support Center (JMBSC)), Hirokazu Kumagai (Japan Meteorological Business Support Center (JMBSC))

Physics and Chemistry Section Atsutaka Maeda (The University of Tokyo, Graduate School of Arts and Sciences), Yoshihito Tanaka (University of Hyogo, Graduate School of Science), Yoshichika Otani (The University of Tokyo, The Institute for Solid State Physics), Masao Sugawara (Nihon University, Professor emeritus), Moritoshi Sato (The University of Tokyo, Graduate School of Arts and Sciences), Toshihiro Shimada (Hokkaido University, Graduate School of Engineering), Koichi Kato (National Institutes of Natural Science, Exploratory Research Center on Life and Living Systems), Norihiro Tokitoh (Kyoto University, Institute for

Chemical Research), Kei Goto (Tokyo Institute of Technology, School of Science), Tatsuya Tsukuda (The University of Tokyo, Graduate School of Science), Tomoji Ozeki (Nihon University, College of Humanities and Sciences), Makoto Sasaki (Tohoku University, Graduate School of Life Science), Seiichi Nishizawa (Tohoku University, Graduate School of Science), Kentaro Tanaka (Nagoya University, Graduate School of Science), Shoji Asai (The University of Tokyo, Graduate School of Science), Ken Yoshida (Tokushima University, Graduate School of Technology, Industrial and Social Sciences), Masaru Nakahara (Kyoto University, Professor emeritus), Toshiya Kohno (The University of Tokyo, Graduate School of Education), Takuji Okamoto (The University of Tokyo, Graduate School of Arts and Sciences), Shogo Higaki (The University of Tokyo, Isotope Science Center), Tetsuya Doi (Kobayasi Institute of Physical Research), Takashi Tanaka (RIKEN), Hiroshi Nishihara (Tokyo University of Science, Research institute for Science and Technology), Tadashi Watanabe (The University of Tokyo, Professor emeritus), Mitsuru Ebihara (Waseda University, Faculty of Education and Integrated Arts and Sciences)

Earth Science Section Geospatial Information Authority of Japan, Japan Coast Guard Hydrographic and Oceanographic Department/Japan Oceanographic Data Center, Ministry of Land, Infrastructure, Transport and Tourism, Water and Disaster Management Bureau, Yuji Esaki (Senshu University, School of Letters), Hiroko Nagahara (The University of Tokyo, Professor emeritus), Toshikatsu Yoshii (The University of Tokyo, Professor emeritus), Setsuya Nakada (The University of Tokyo, Professor emeritus), Takashi Nakajima (National Institute of Advanced Industrial Science and Technology), Takayuki Kaneko (The University of Tokyo, Earthquake Research Institute), Atsushi Yasuda (The University of Tokyo, Earthquake Research Institute), Tomoya Harada (The University of Tokyo, Earthquake Research Institute), Michi Nishioka (National Institute of Information and Communications Technology, Radio Research Institute), Kenji Nakayama (National Institute of Information and Communications Technology, Radio Research Institute), Kornyanat Hozumi (National Institute of Information and Communications Technology, Radio Research Institute), Septi Perwitasari (National Institute of Information and Communications Technology, Radio Research Institute), Mitsuhiro Yoshimoto (Mount Fuji Research Institute, Yamanashi Prefectural Government), Toshihiko Iyemori (Kyoto University, Professor emeritus), Yoko Odagi (Kyoto University, Graduate School of Science), Masahiko Takeda (Kyoto University, Graduate School of Science), Takashi Tsutsumi (Asama Jomon Museum), Tadahiro Hatakeyama (Okayama University of Science, Research Institute of Frontier Science and Technology), Japan Meteorological Agency (JMA), Seismological and Volcanological Department, Fire and Disaster Management Agency (FDMA), United States Geological Survey (USGS), The Japanese Alpine Club, Ministry of the Environment Government of Japan, Nature Conservation Bureau, Mount Fuji Research Institute, Yamanashi Prefectural Government

Life Science Section Shunsuke Mawatari (Hokkaido University, Professor emeritus), Masayoshi Kuwahara (The University of Tokyo, Graduate School of Agricultural and Life Sciences), Sadao Yasugi (Tokyo Metropolitan University, Professor emeritus), Haruki Hashimoto, Yoko Hirabayashi (National Institute of Health Sciences, Biological Safety Research Center), Kazuhiro Kaji (KOSÉ Research Laboratory), Takeo Kishimoto (Tokyo Institute of Technology, Professor emeritus), Michihiro Yoshida (Hokkaido University, Professor emeritus), Hiromichi Yonekawa (Tokyo Metropolitan Institute of Medical Science, Protein Metabolism Project), Junichi Yata (Tokyo Medical and Dental University, Professor emeritus), Nobuo Nara (Japan Accreditation Council for Medical Education), Wataru Yamori (The University of Tokyo, Graduate School of Agricultural and Life Sciences), Takeo Horiguchi (Hokkaido University, Faculty of Science), Hideki Takahashi (The Hokkaido University Museum), Satoshi Hanada (National Institute of Advanced Industrial Science and Technology), Yuji Tomaru (Japan Fisheries Research and Education Agency, Fisheries Technology Institute), Sachihiro Matsunaga (The University of Tokyo, Graduate School of Frontier Sciences), Takenori Sasaki (The University of Tokyo, The University Museum), Tsutomu Hikida (Kyoto University, Professor emeritus), Isao Nishiumi (National Museum of Nature and Science), Toshihiko Fujita (National Museum of Nature and Science), Tsuyoshi Hosoya (National Museum of Nature and Science), Yukimitsu Imahara (National Institute of Advanced Industrial Science and Technology), Takahito Suzuki (Lake Biwa Museum), Kensuke Yanagi (Coastal Branch Natural History Museum and Institute, Chiba), Mathew Hill Dick (Hokkaido University, Part-time lecturer), Hiroshi Kajihara (Hokkaido University, Faculty of Science), Keiichi Kakui (Hokkaido University, Faculty of Science), Daisuke Shimada (Hokkaido University, Faculty of Science), Masahiro Ohara (The Hokkaido University Museum), Masaharu Motokawa (The Kyoto University Museum), Hirofumi Yamasaki (Kyushu University, Faculty of Arts and Science), Kazuya Nagasawa (Hiroshima University, Professor emeritus), Teruaki Nishikawa (Nagoya University, Professor emeritus), Taichiro Goto (Mie University, Faculty of Education), Takuya Sato (Kobe University, Graduate School of Science), Keizo Nagasaki (Kochi University, Faculty of Science and Technology), Hidetaka Furuya (Osaka University, Graduate School of Science), Hiroaki Nakano (University of Tsukuba, Shimoda Marine Research Center), Tsutomu Tanabe (Kumamoto University, Faculty of Advanced Science and Technology), Hiroshi Namikawa (Showa Memorial Institute, National Museum of Nature and Science), Kazunori Yoshizawa (Hokkaido University, School of Agriculture), Masato Hirose (Kitasato University, School of Marine Biosciences), Masaatsu Tanaka (Keio University, Department of Biology), Keiichi Matsuura (National Museum of Nature and Science, Professor emeritus), Dongchon Kang (Kyushu University Hospital, Department of Clinical Chemistry and Laboratory Medicine), Fumiko Mashige, Kazuyoshi Endo (The University of Tokyo, Graduate School of Science), Yuji Ise (University of the Ryukyus, Tropical Biosphere Research Center),

Wataru Abe (Dokkyo Medical University, Premedical Sciences), Tohru Iseto (Japan Agency for Marine-Earth Science and Technology (JAMSTEC), Center for Earth Information Science and Technology), Nobuo Tsurusaki (Tottori University, Professor emeritus), Sakiko Shiga (Osaka University, Graduate School of Science), Masakatsu Ureshi (Saga University, Faculty of Education), Saori Yastomi (Scientific Illustrator), Ministry of Health, Labour and Welfare (MHLW), Statistics and Information Department, Minister's Secretariat, Kouji Yasuyama (Kawasaki University of Medical Welfare, Faculty of Rehabilitation), Masamichi Kanou (Ehime University, Professor emeritus)

Environmental Science Section Japan Meteorological Agency, Global Environment and Marine Department (JMA), Ministry of the Environment Government of Japan (MOE), Japan Wildlife Research Center, National Institute for Environmental Studies, Ministry of Agriculture, Forestry and Fisheries (MAFF), Statistics Department, Minister's Secretariat, Shinichi Sakai (Advanced Science, Technology & Management Research Institute of KYOTO), Yasuhiro Hirai (Kyoto University, Environment Preservation Research Center), Junya Yano (Kyoto University, Environment Preservation Research Center), Toru Hirawake (National Institute of Polar Research), Isao Koike (The University of Tokyo, Professor emeritus), Hidesuke Shimizu (The Jikei University School of Medicine, Professor emeritus), Hiroyuki Tsuda (Nagoya City University, Nanotoxicology Project), Hisanori Kohtsuka (The University of Tokyo, Misaki Marine Biological Station), Yoshiki Sohrin (Kyoto University, Institute for Chemical Research), Tsutomu Ikeda (Hokkaido University, Professor emeritus), Joji Ishizaka (Nagoya University, Institute for Space-Earth Environmental Research), Akihisa Takahashi (Gunma University, Heavy Ion Medical Center), Yukari Yoshida (Gunma University, Heavy Ion Medical Center), Yuichi Miyabara (Shinshu University, Faculty of Science), Hiroo Ohmori (The University of Tokyo, Professor emeritus), Taisen Iguchi (National Institute for Basic Biology, Professor emeritus), Motomi Ito (The University of Tokyo, Professor emeritus), Taku Kadoya (National Institute for Environmental Studies), Naoki Kumagai (National Institute for Environmental Studies, Center for Social and Environmental Systems Research), Takuya Kubo (Kyoto University, Graduate School of Engineering), Aiko Nagamatsu (Japan Aerospace Exploration Agency (JAXA)), Tatsuhiko Sato (Japan Atomic Energy Agency), Nobuyuki Hamada (Central Research Institute of Electric Power Industry)

Appendix Takuji Okamoto (The University of Tokyo, Graduate School of Arts and Sciences)

Contents

Astronomy Section

Meteorology Section

Physics and Chemistry Section

Units

Electric/Magnetic Properties

Sound

Light and Electromagnetic Waves

Optical Properties

Chemical Formulae and Reactions of Matter

Biological Material

Biologically Active Substances

Appendix

Earth Science Section

Geography

Geology and Mineralogy

Volcanoes

Earthquakes

Geomagnetism and Gravity

Ionosphere

Life Science Section

Shapes and Systems of Organisms

Reproduction, Development, Growth

Cell, Tissue, Organ

Genes, Immunity

Physiology

Metabolism and Biosynthesis System

Appendix

Environmental Science Section

Climate Change and Global Warming

Chemical Substances and Radiation

Appendix

Astronomy Section

The Size and Shape of the Earth The Earth's shape can be approximated by a spheroid and the shape of the meridian is an ellipse. The elements of geodetic reference system 1980 (GRS80) ellipsoid adopted in Japan, U.S.A., and many other countries in the world for current precise geodetic survey are as follows (a_E and f are fundamental parameters and all others are derived from them). Please refer to page **519** for varieties of Earth ellipsoids.

Equatorial radius	$a_E = 6\,378.137$ km
Polar radius	$a_P = 6\,356.752$ km
Flattening	$f = (a_E - a_P)/a_E = 1/298.257\,222\,101$
Meridian quadrant	$l = 10\,001.966$ km

System of Astronomical Constants This is based on the system of astronomical constants adopted at the 27th and 28th International Astronomical Union (IAU) General Assemblies (2009, 2012). It was determined at the 2012 General Assembly that the symbol "au" denoting the astronomical unit will be written in lowercase letters.

(Epoch J2000.0 = 2000 January 1.5 TT = JD 2 451 545.0 TT)

1) Defining Constants

Speed of light	$c = 2.997\,924\,58 \times 10^{8}$ m s^{-1}
Astronomical unit	au $= 1.495\,978\,707 \times 10^{11}$ m
1 − d(TT)/d(TCG)	$L_G = 6.969\,290\,134 \times 10^{-10}$
1 − d(TDB)/d(TCB)	$L_B = 1.550\,519\,768 \times 10^{-8}$
TDB − TCB at T_0	$\mathrm{TDB}_0 = -6.55 \times 10^{-5}$ s (T_0 = JD 2 443 144.500 372 5 TCB)
Earth rotation angle (2000 January 1.5 UT1)	$\theta_0 = 0.779\,057\,273\,264\,0$ revolutions
Rate of advance of Earth rotation angle	$d\theta/dt = 1.002\,737\,811\,911\,354\,48$ revolutions (UT1-day)$^{-1}$

2) Other Constants

Constant of gravitation	$G = 6.674\,28 \times 10^{-11}$ m^3 kg^{-1} s^{-2}
Average value of 1 − d(TCG)/d(TCB)	$L_C = 1.480\,826\,867\,41 \times 10^{-8}$
Equatorial radius of the Earth	$a_E = 6.378\,136\,6 \times 10^{6}$ m
Dynamical form factor of the Earth	$J_2 = 1.082\,635\,9 \times 10^{-3}$
Time rate of change in J_2	$dJ_2/dt = -3.0 \times 10^{-9}$ cy^{-1}
Solar mass parameter	$GM_S = 1.327\,124\,400\,41 \times 10^{20}$ m^3 s^{-2}
Terrestrial mass parameter	$GM_E = 3.986\,004\,356 \times 10^{14}$ m^3 s^{-2}
Potential of the geoid	$W_0 = 6.263\,685\,60 \times 10^{7}$ m^2 s^{-2}
Nominal mean angular velocity of the Earth	$\omega = 7.292\,115 \times 10^{-5}$ rad s^{-1}
Mean obliquity of the ecliptic (J2000.0)	$\varepsilon_{J2000} = 8.438\,140\,6 \times 10^{4}{}''$

The value of GM_S and GM_E (formerly known as Heliocentric and Geocentric gravitational constants, respectively) are based on TDB. The values for a_E and J_2 are "zero-tide system" values (see page **517**). The unit cy is 36 525 days.

Precession in right ascension (J2000.0) = $307^{s}\!.477$ cy^{-1}, Precession in declination (J2000.0) = $2\,004''\!.19$ cy^{-1}

Tropical year = 365.242 19 days, Sidereal year = 365.256 36 days, Anomalistic year = 365.259 64 days

1 mean sidereal day = 0.997 269 57 mean solar day = $23^{h}56^{m}04^{s}\!.090\,5$ mean solar time

1 mean solar day = 1.002 737 91 mean sidereal day = $24^{h}03^{m}56^{s}\!.555\,4$ mean sidereal time

The

Mercury, Venus, Mars, Jupiter, and Saturn were known since ancient times. Uranus was discovered in 1781, and Neptune was discovered in 1846. The table shows the orbital elements and related parameters of the planets. The Epoch is 2021 July 5.0 TT = JD 2 459 400.5 TT.

	Semimajor axis a	Eccentricity e	Inclination i		Longitude of perihelion ϖ	Longitude of ascending node Ω	Mean anomaly at the epoch M_0
			Ecliptic	Invariable plane			
	au		°	°	°	°	°
Mercury	0.387 1	0.205 6	7.004	6.343	77.490	48.304	282.128
Venus	0.723 3	0.006 8	3.394	2.196	131.565	76.620	35.951
Earth	1.000 0	0.016 7	0.003	1.578	103.007	174.821	179.912
Mars	1.523 7	0.093 4	1.848	1.680	336.156	49.495	175.817
Jupiter	5.202 6	0.048 5	1.303	0.328	14.378	100.502	312.697
Saturn	9.554 9	0.055 5	2.489	0.934	93.179	113.610	219.741
Uranus	19.218 4	0.046 4	0.773	1.028	173.024	74.022	233.182
Neptune	30.110 4	0.009 5	1.770	0.726	48.127	131.783	303.212

Parameters of the Sun,

	Amount of radiation received from the Sun (Earth = 1)	Apparent radius (At the average closest distance to Earth)	Equatorial radius	Flattening	Dynamical form factor J_2	Equatorial gravity (Earth = 1)	Volume (Earth = 1)	Number of satellites
		′ ″	km					
Sun	-	15 59.64	696 000	0	0	28.01	1 304 000	-
Mercury	6.67	5.49	2 439.4	0	0.050×10^{-3}	0.38	0.056	0
Venus	1.91	30.16	6 051.8	0	0.004×10^{-3}	0.91	0.857	0
Earth	1.00	-	6 378.1	0.003 35	1.083×10^{-3}	1.00	1.000	1 (1)
Mars	0.43	8.94	3 396.2	0.005 89	1.956×10^{-3}	0.38	0.151	2 (2)
Jupiter	0.037	23.46	71 492	0.064 87	14.70×10^{-3}	2.37	1 321	72 (79)
Saturn	0.011	9.71	60 268	0.097 96	16.32×10^{-3}	0.93	764	53 (85)
Uranus	0.0027	1.93	25 559	0.022 93	3.51×10^{-3}	0.89	63	27 (27)
Neptune	0.0011	1.17	24 764	0.017 08	3.54×10^{-3}	1.11	58	14 (14)
Moon	1.00	15 32.28	1 737.4	unequal 3 axes	0.202×10^{-3}	0.17	0.0203	-

The apparent radius for the Sun and the Moon is the value at mean distance from the Earth. The mass (Sun = 1) includes the satellites of the planet. The mass of the Sun is 1.988×10^{30} kg. The mass of the Earth is 5.972×10^{24} kg. The unit for mean density is 10^3 kg m^{-3} (= g cm^{-3}). The inclination of equator to the orbit is an angle between the equatorial plane and the orbital plane of each body. However actually, the inclination of equator to the orbit for the Sun is the angle of the equator to the ecliptic (average orbital plane of the Earth). The maximum magnitude is the magnitude as seen from the Earth when viewed at its brightest time.

The table on the right shows the names and positions of planetary rings (the mean distance from the parent planet's center in units of the equatorial radius of the parent planet).

The number of satellites shows the number of confirmed satellites, i.e. those numbered by IAU. The total number of satellites reported in the IAU Circulars, etc. by the end of August 2020 is shown in the parentheses.

Planets

The given values for the orbital elements are for the mean elements referred to the mean equinox and ecliptic of 2000 January 1.5 TT (JD 2 451 545.0 TT).

Distance from the Sun (10^8 km)			Radius of sphere of influence (10^6 km)	Sidereal daily mean motion μ	Sidereal orbit period P (Julian years)	Mean orbital velocity (km s^{-1})	Synodic period (days)	
Minimum $a(1-e)$	Mean a	Maximum $a(1+e)$						
				″				
0.460	0.579	0.698	0.11	14 732.42	0.240 85	47.36	115.9	Mercury
1.075	1.082	1.089	0.62	5 767.67	0.615 20	35.02	583.9	Venus
1.471	1.496	1.521	0.92	3 548.19	1.000 02	29.78	–	Earth
2.066	2.279	2.492	0.58	1 886.52	1.880 85	24.08	779.9	Mars
7.405	7.783	8.161	48.20	299.13	11.862 0	13.06	398.9	Jupiter
13.501	14.294	15.087	54.65	120.45	29.457 2	9.65	378.1	Saturn
27.417	28.750	30.084	51.84	42.23	84.020 5	6.81	369.7	Uranus
44.619	45.044	45.470	86.77	21.53	164.770 1	5.44	367.5	Neptune

the Planets and the Moon

Mass		Mean density	Escape velocity	Sidereal rotation period	Inclination of equator to the orbit	Albedo[1)]	Maximum magnitude	
(Sun = 1)	(Earth = 1)							
		10^3 kg m^{-3}	km s^{-1}	days	°			
1.000	332 946	1.41	617.5	25.3800	7.25	–	−26.8	Sun
1.6601×10^{-7}	0.05527	5.43	4.25	58.6461	0.03	0.08	−2.5	Mercury
2.4478×10^{-6}	0.8150	5.24	10.36	243.0185	177.36	0.76	−4.9	Venus
3.0404×10^{-6}	1.0000	5.51	11.18	0.9973	23.44	0.30	–	Earth
3.2272×10^{-7}	0.1074	3.93	5.02	1.0260	25.19	0.25	−2.9	Mars
9.5479×10^{-4}	317.83	1.33	59.53	0.4135	3.12	0.34	−2.9	Jupiter
2.8589×10^{-4}	95.16	0.69	35.48	0.4440	26.73	0.34	−0.5	Saturn
4.3662×10^{-5}	14.54	1.27	21.29	0.7183	97.77	0.30	+5.3	Uranus
5.1514×10^{-5}	17.15	1.64	23.50	0.6653	28.35	0.29	+7.8	Neptune
3.6943×10^{-8}	0.012300	3.34	2.38	27.3217	6.70	0.11	−12.9	Moon

1) Bond albedo: The ratio of all reflected energy to all incident energy for each planet.

Jovian rings		Saturnian rings		Uranian rings			
Halo	1.399–1.718	D	1.112–1.236	ζ	1.549	γ	1.864
Main	1.718–1.807	C	1.236–1.526	6	1.637	δ	1.889
Gossamer	1.807–2.996	B	1.526–1.950	5	1.652	λ	1.957
Neptunian rings		A	2.025–2.269	4	1.666	ε	2.001
Galle	1.69	F	2.327	α	1.750	ν	2.633
Leverrier	2.15	G	2.75–2.89	β	1.787	μ	3.823
Lassell	2.24	E	3.0–8.0	η	1.846		
Arago	2.33						
Adams	2.54						

The Major

The standard size of the orbit is the angular size of the semimajor axis of the satellite's orbit when viewed from 1 au apart; apparent value at an arbitrary time is obtained by dividing by the geocentric distance of the parent planet (in au) at that time.

Number		Satellite name	Discoverer	Discovery year	Magnitude	Standard size of the orbit	Semimajor axis of the orbit
						″	
Mars	I	Phobos	Hall	1877	11	12.93	2.76
	II	Deimos	Hall	1877	12	32.34	6.91
Jupiter	I	Io	Galileo	1610	5	581.6	5.90
	II	Europa	Galileo	1610	5	925.3	9.39
	III	Ganymede	Galileo	1610	5	1475.9	14.97
	IV	Callisto	Galileo	1610	6	2595.9	26.33
	V	Amalthea	Barnard	1892	14	250.1	2.54
	VI	Himalia	Perrine	1904	15	15802	160.3
	VII	Elara	Perrine	1905	16	16188	164.2
	VIII	Pasiphae	Melotte	1908	17	32573	330.4
	IX	Sinope	Nicholson	1914	18	33007	334.8
	X	Lysithea	Nicholson	1938	18	16155	163.9
	XI	Carme	Nicholson	1938	18	32269	327.4
	XII	Ananke	Nicholson	1951	19	29335	297.6
	XIII	Leda	Kowal	1974	19	15394	156.2
	XIV	Thebe	Synnott	1979	16	306	3.10
	XV	Adrastea	Voyager II	1979	19	178	1.80
	XVI	Metis	Voyager I	1979	18	176	1.79
Saturn	I	Mimas	Herschel	1789	13	255.82	3.08
	II	Enceladus	Herschel	1789	12	328.21	3.95
	III	Tethys	Cassini	1684	10	406.29	4.89
	IV	Dione	Cassini	1684	10	520.38	6.26
	V	Rhea	Cassini	1672	10	726.72	8.75
	VI	Titan	Huygens	1655	8	1684.70	20.27
	VII	Hyperion	Bond	1848	14	2069.5	24.90
	VIII	Iapetus	Cassini	1671	11	4909.7	59.08
	IX	Phoebe	Pickering	1898	17	17777	213.9
	X	Janus	Pascu	1980	14	208.8	2.51
	XI	Epimetheus	Cruikshank	1980	16	208.8	2.51
	XII	Helene	Laques	1980	18	520.36	6.26
	XIII	Telesto	Reitsema	1980	19	406.28	4.89
	XIV	Calypso	United States Naval Observatory	1980	19	406.28	4.89
	XV	Atlas	Voyager I	1980	19	189.8	2.28
	XVI	Prometheus	Voyager I	1980	16	192.2	2.31
	XVII	Pandora	Voyager I	1980	16	195.4	2.35
	XVIII	Pan	Voyager II	1981	19	184.2	2.22
Uranus	I	Ariel	Lassell	1851	13	263.2	7.47
	II	Umbriel	Lassell	1851	14	366.8	10.40
	III	Titania	Herschel	1787	13	601.6	17.07
	IV	Oberon	Herschel	1787	13	804.5	22.83
	V	Miranda	Kuiper	1948	15	179.1	5.08
Neptune	I	Triton	Lassell	1846	13	489.1	14.32
	II	Nereid	Kuiper	1949	20	7602.4	222.7

Many satellites besides the ones in the table have been discovered around Jupiter, Saturn, Uranus, and Neptune. Refer to the Table of "Parameters of the Sun, the Planets and the Moon" for the total number of the satellites.

Satellites

The semimajor axis of the orbit is expressed in units of the equatorial radius of the parent planet. The magnitude is the average magnitude at opposition of the parent planet. Mass is expressed in units of the mass of the parent planet.

Orbital period	Eccentricity	Orbital inclination	/reference plane	Radius	Mass	Number	
days		°		km			
0.3189	0.0151	1.08	Mars's Equator	13 × 11 × 9	1.67×10^{-8}	Mars	I
1.2624	0.0002	1.79	Mars's Equator	8 × 6 × 5	2.43×10^{-9}		II
1.7691	0.0041	0.036	Jupiter's Equator	1821	4.70×10^{-5}	Jupiter	I
3.5512	0.0094	0.466	Jupiter's Equator	1562	2.53×10^{-5}		II
7.1546	0.0013	0.178	Jupiter's Equator	2632	7.81×10^{-5}		III
16.6890	0.0074	0.192	Jupiter's Equator	2409	5.67×10^{-5}		IV
0.4982	0.0032	0.380	Jupiter's Equator	125 × 73 × 64	1.10×10^{-9}		V
250.56	0.1623	27.5	Jupiter's Orbit	85	2.2×10^{-9}		VI
259.64	0.2174	26.6	Jupiter's Orbit	40	4.58×10^{-10}		VII
743.63	0.4090	151.4	Jupiter's Orbit	18	1.58×10^{-10}		VIII
758.90	0.2495	158.1	Jupiter's Orbit	14	3.95×10^{-11}		IX
259.20	0.1124	28.3	Jupiter's Orbit	12	3.31×10^{-11}		X
734.17	0.2533	164.9	Jupiter's Orbit	15	6.94×10^{-11}		XI
629.77	0.2435	148.9	Jupiter's Orbit	10	1.58×10^{-11}		XII
240.92	0.1636	27.5	Jupiter's Orbit	5	5.76×10^{-12}		XIII
0.675	0.0176	1.080	Jupiter's Equator	58 × 49 × 42	7.89×10^{-10}		XIV
0.298	0.0018	0.054	Jupiter's Equator	10 × 8 × 7	3.95×10^{-12}		XV
0.295	0.0012	0.019	Jupiter's Equator	30 × 20 × 17	6.31×10^{-11}		XVI
0.9424	0.0196	1.574	Saturn's Equator	198	6.61×10^{-8}	Saturn	I
1.3702	0.0048	0.026	Saturn's Equator	252	1.90×10^{-7}		II
1.8878	0.0001	1.091	Saturn's Equator	531	1.09×10^{-6}		III
2.7369	0.0022	0.028	Saturn's Equator	561	1.93×10^{-6}		IV
4.5175	0.0002	0.333	Saturn's Equator	764	4.06×10^{-6}		V
15.9454	0.0288	0.306	Saturn's Equator	2575	2.37×10^{-4}		VI
21.2767	0.1035	0.644	Saturn's Equator	180 × 133 × 103	1.00×10^{-8}		VII
79.3311	0.0293	15.1	Saturn's Orbit	735	3.18×10^{-6}		VIII
546.41	0.1756	175.0	Saturn's Orbit	107	1.45×10^{-8}		IX
0.695	0.0068	0.163	Saturn's Equator	102 × 93 × 76	3.34×10^{-9}		X
0.694	0.0098	0.351	Saturn's Equator	65 × 57 × 53	9.26×10^{-10}		XI
2.74	0.000	0.212	Saturn's Equator	22 × 19 × 13	4.48×10^{-11}		XII
1.888	0.001	1.158	Saturn's Equator	16 × 12 × 10	1.27×10^{-11}		XIII
1.888	0.001	1.473	Saturn's Equator	15 × 12 × 7	6.33×10^{-12}		XIV
0.602	0.0012	0.003	Saturn's Equator	20 × 18 × 9	1.16×10^{-11}		XV
0.613	0.0022	0.008	Saturn's Equator	68 × 40 × 30	2.81×10^{-10}		XVI
0.629	0.0042	0.050	Saturn's Equator	52 × 41 × 32	2.41×10^{-10}		XVII
0.575	0.0000	0.000	Saturn's Equator	17 × 16 × 10	8.71×10^{-12}		XVIII
2.5204	0.0012	0.041	Uranus's Equator	579	1.56×10^{-5}	Uranus	I
4.1442	0.0039	0.128	Uranus's Equator	585	1.35×10^{-5}		II
8.7059	0.0011	0.079	Uranus's Equator	789	4.06×10^{-5}		III
13.4632	0.0014	0.068	Uranus's Equator	761	3.47×10^{-5}		IV
1.4135	0.0013	4.338	Uranus's Equator	236	0.8×10^{-6}		V
5.8769	0.0000	156.865	Neptune's Equator	1353	2.09×10^{-4}	Neptune	I
360.13	0.7507	7.1	Neptune's Orbit	170	3.01×10^{-7}		II

The Moon

Revolution The parameters related to the revolution of the Moon are as follows.

Synodic month 29.530 589 days Sidereal month 27.321 662 days Draconic month 27.212 221 days

Tropical month 27.321 582 days Anomalistic month 27.554 550 days

Period of sidereal precession of perigee 3 232.605 days (=about 8.85 years)

Period of sidereal regression of node 6 793.477 days (=about 18.6 years)

Sidereal mean motion 13°.176 358 day^{-1}

Orbit semimajor axis 384 399 km = about 60.268 2 × Earth's equatorial radius

Mean eccentricity 0.055 545 5 Mean inclination 5°09′24″.08 (to the ecliptic)

Sidereal secular acceleration of the Moon (Coefficients of the T^2 terms where T is the time measured in 36 525 days)

Mean longitude of the Moon	-6″.87 (-12″.93)	Contributions from tidal effects are shown in parentheses.
Mean longitude of perigee	-38″.26 (+0″.19)	
Mean longitude of node	+6″.36 (-0″.05)	

Main periodic term	Amplitude: Ecliptic longitude	Amplitude: Distance	Period
Equation of center	22 640″ (6°.29)	20 905 km	One anomalistic month
Evection	4 586 (1.27)	3 699	31.812 days
Variation	2 370 (0.66)	2 956	Half synodic month
Annual equation	666 (0.19)	49	One anomalistic year
Parallactic inequality	125 (0.03)	109	One synodic month

Cycle			
Metonic Cycle	19 tropical years = 6 939.602 days	235 synodic months = 6 939.688 days	235 = 19 × 12 + 7
Saros Cycle	19 eclipse years = 6 585.781 days 242 draconic months = 6 585.357 days 223 synodic months = 6 585.321 days 239 anomalistic months = 6 585.537 days		
Eclipse year	346.620 1 days		

Gravity coefficient of the Moon	Moment of inertia ratios and shape of the Moon
$J_2 = 20.22 \times 10^{-5}$ $C_{22} = 2.23 \times 10^{-5}$ $J_3 = 1.21 \times 10^{-5}$ $C_{31} = 3.07 \times 10^{-5}$ $S_{31} = 0.56 \times 10^{-5}$ $C_{41} = -0.72 \times 10^{-5}$	$(B-A)/C = 2.28 \times 10^{-4}$, $(C-A)/B = 6.32 \times 10^{-4}$, $C/MR^2 = 0.391$ Lunar ellipsoidal radius: 1738.4 km in the direction of Earth, 1736.7 km along the poles, 1737.5 km in direction perpendicular to both Observations such as by Japan's lunar explorer "KAGUYA" revealed that the offset of the center of figure from the center of mass is 1.93 km in direction of 202°.38E lunar longitude, 7°.12N lunar latitude. The lunar longitude and latitude are measured with respect to the mean direction toward the Earth and the polar axis.

Rotation The lunar equator is inclined by 1°32′33.″6 with respect to the ecliptic (Newhall and Williams 1997, CMDA, **66**). The ascending node of the Moon's equator with respect to the ecliptic is coincide with the descending node of the Moon's orbit with respect to the ecliptic. The lunar sidereal rotation period is exactly equal to 1 sidereal month. Because of this, the near side of the Moon never turns away from the Earth. However, by many inequalities in the orbital motion of the Moon, by inclination of lunar orbit which is about 5° with respect to the ecliptic, and by location of observer on the surface of the Earth, one can see the slight apparent motion of the lunar disk. This is called the optical libration of the Moon. Of the total lunar surface about 59 percent can be seen from the Earth.

Brightness Approximate visual magnitude of the Moon is listed below.

For more details, please refer to Rougier 1937, Ann. Obs. Strasbourg, **3**.

Apparent angular distance between the Sun and the Moon	180°	150°	120°	90°	60°	30°
Approximate age of the Moon (in days)	14.8	12.3 17.2	9.9 19.7	7.4 22.2	4.9 24.6	2.5 27.1
Visual magnitude	-12.6	-11.8	-11.0	-9.9	-8.7	-6.8

Solar Eclipses

Conjunction in right ascension (UT)					Central eclipse at sunrise		Central eclipse at noon		Central eclipse at sunset		Type (Maximum duration of totality)
					Longitude	Latitude	Longitude	Latitude	Longitude	Latitude	
year	month	day	hour	min	°	°	°	°	°	°	(min)
2017	2	26	14	38.8	-114	-43	-36	-37	+27	-11	Annular
	8	21	18	13.2	-172	+40	-93	+39	-27	+11	Total (2.7)
2018	2	15	20	15.1	-	-	-	-	-	-	Partial
	7	13	3	9.1	-	-	-	-	-	-	Partial
	8	11	9	20.1	-	-	-	-	-	-	Partial
2019	1	6	1	43.7	-	-	-	-	-	-	Partial
	7	2	19	21.7	-160	-38	-109	-17	-58	-36	Total (4.6)
	12	26	5	14.5	+48	+26	+101	+1	+157	+19	Annular
2020	6	21	6	41.4	+18	+1	+80	+31	+148	+11	Annular
	12	14	16	18.2	-133	-8	-66	-41	+11	-24	Total (2.2)
2021	6	10	11	1.1	-90	+50	(-165)	(+88)	+157	+64	Annular
	12	4	7	56.2	-51	-53	(-121)	(-79)	-134	-67	Total (2.0)
2022	4	30	19	40.8	-	-	-	-	-	-	Partial
	10	25	10	3.7	-	-	-	-	-	-	Partial
2023	4	20	3	55.5	+64	-48	+121	-15	-179	+3	Annular-total
	10	14	17	36.6	-147	+49	-88	+17	-29	-6	Annular
2024	4	8	18	36.1	-159	-8	-99	+31	-20	+48	Total (4.5)
	10	2	19	8.0	-166	+8	-110	-28	-37	-49	Annular
2025	3	29	11	46.2	-	-	-	-	-	-	Partial
	9	21	20	50.4	-	-	-	-	-	-	Partial
2026	2	17	11	18.8	(+137)	(-72)	-	-	+99	-50	Annular
	8	12	17	3.8	+113	+75	(+105)	(+85)	+5	+39	Total (2.4)
2027	2	6	15	44.4	-131	-39	-53	-35	+4	+6	Annular
	8	2	10	0.9	-44	+28	+31	+27	+90	-12	Total (6.4)
2028	1	26	15	24.7	-106	+3	-48	+6	+2	+41	Annular
	7	22	3	15.8	+76	-18	+133	-19	-179	-51	Total (5.2)
2029	1	14	17	46.9	-	-	-	-	-	-	Partial
	6	12	4	0.0	-	-	-	-	-	-	Partial
	7	11	16	14.4	-	-	-	-	-	-	Partial
	12	5	15	5.4	-	-	-	-	-	-	Partial
2030	6	1	6	30.7	+4	+29	+82	+57	+164	+34	Annular
	11	25	6	54.2	+2	-16	+73	-44	+151	-26	Total (3.8)
2031	5	21	7	12.2	+15	-17	+71	+9	+130	-5	Annular
	11	14	21	0.9	+164	+26	-139	-0	-79	+8	Annular-total
2032	5	9	13	7.1	-43	-70	-18	-55	+31	-56	Annular
	11	3	5	6.1	-	-	-	-	-	-	Partial
2033	3	30	18	33.2	+178	+61	-	-	(+143)	(+82)	Total (2.7)
	9	23	14	37.4	-	-	-	-	-	-	Partial
2034	3	20	10	27.1	-38	-1	+25	+18	+93	+35	Total (4.2)
	9	12	16	32.2	-129	-6	-69	-21	-5	-41	Annular
2035	3	9	22	49.6	+127	-43	-160	-32	-98	-9	Annular
	9	2	1	43.7	+80	+38	+154	+31	-143	+5	Total (3.0)

For commentary, refer to page 10.

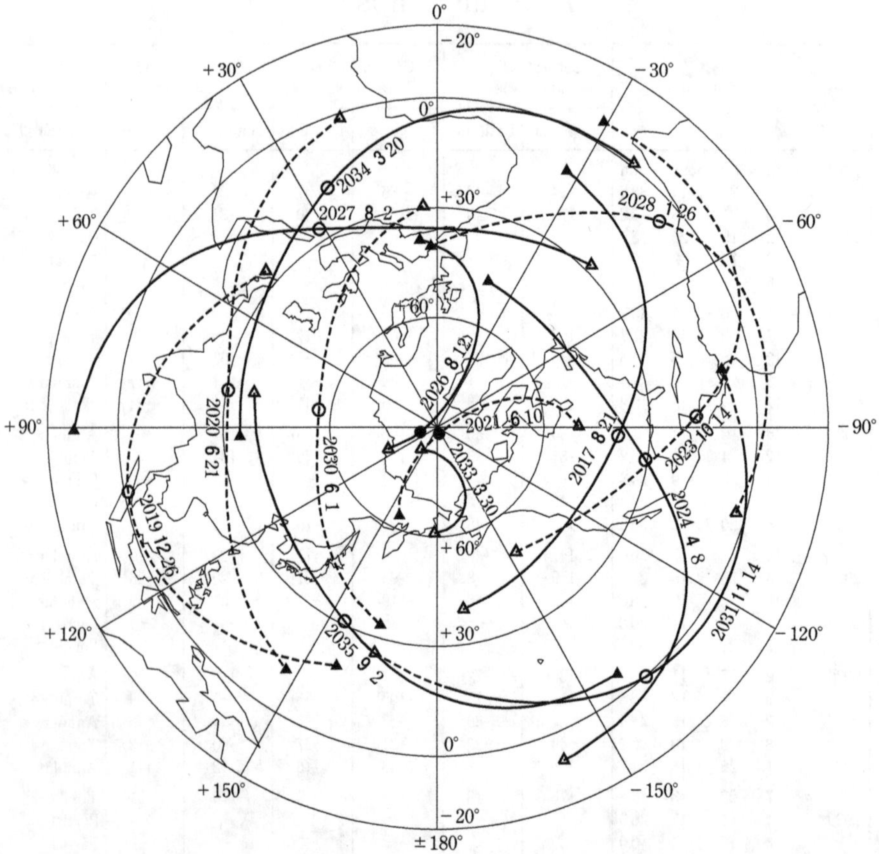

Figure 1A Map of solar eclipse central lines

This map shows the central lines of total, annular, and annular-total solar eclipses occurring between 2017 and 2035. The dates are given in UT. The symbols are:

△ : The location where the central eclipse occurs at sunrise.

○ : The location where the central eclipse occurs at local noon (i.e. when the sun crosses the meridian from east to west).

▲ : The location where the central eclipse occurs at sunset.

● : The location where the central eclipse occurs at local midnight (i.e. when the sun crosses the meridian from west to east).

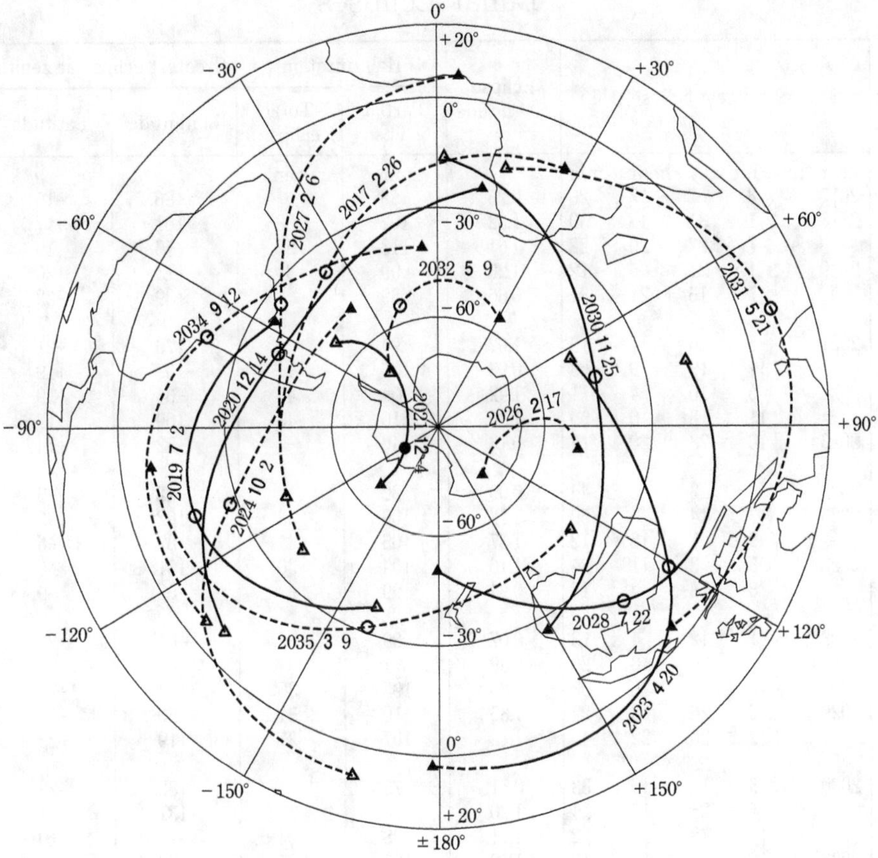

Figure 1B Map of solar eclipse central lines

Additionally, the lines indicate:

———————	Total Eclipse
- - - - - - - - - -	Annular Eclipse
- - - - ———— - - -	Annular-total Eclipse

A solid line is drawn for places where a total eclipse can be seen.

Lunar Eclipses

Time of greatest eclipse (UT)					Eclipse magnitude	Half duration		Greatest eclipse at zenith	
						Partial eclipse	Total eclipse	Longitude	Latitude
year	month	day	hour	min		min	min	°	°
2017	8	7	18	20	0.25	58	-	+86	-15
2018	1	31	13	30	1.32	102	38	+161	+17
	7	27	20	22	1.61	118	52	+56	-19
2019	1	21	5	12	1.20	99	31	-75	+20
	7	16	21	31	0.66	89	-	+39	-22
2021	5	26	11	19	1.02	94	9	-170	-21
	11	19	9	3	0.98	105	-	-139	+19
2022	5	16	4	11	1.42	104	43	-64	-19
	11	8	10	59	1.36	110	43	-169	+17
2023	10	28	20	14	0.13	40	-	+52	+14
2024	9	18	2	44	0.09	32	-	-42	-3
2025	3	14	6	59	1.18	109	33	-102	+3
	9	7	18	12	1.37	105	41	+87	-6
2026	3	3	11	34	1.16	104	30	-171	+6
	8	28	4	13	0.94	99	-	-63	-9
2028	1	12	4	13	0.07	29	-	-61	+23
	7	6	18	20	0.39	71	-	+86	-23
	12	31	16	52	1.25	105	36	+108	+23
2029	6	26	3	22	1.85	110	51	-50	-23
	12	20	22	42	1.12	107	27	+19	+23
2030	6	15	18	33	0.51	73	-	+82	-23
2032	4	25	15	14	1.20	106	33	+131	-14
	10	18	19	2	1.11	98	24	+71	+10
2033	4	14	19	13	1.10	108	25	+72	-9
	10	8	10	55	1.36	102	40	-167	+6
2034	9	28	2	46	0.02	16	-	-44	+1
2035	8	19	1	11	0.11	39	-	-17	-12

The solar eclipses in the years between 2017 and 2035 are given in the table on page 7 and the lunar eclipses in the same years are given in the above table. In these tables the longitudes are positive to the east and negative to the west, and the latitudes are positive to the north and negative to the south. In the solar eclipse table the longitudes and latitudes in the parentheses in the columns of "Central eclipse at sunrise", "Central eclipse at noon", and "Central eclipse at sunset" are actually those of the positions of central eclipse at sunset, central eclipse at midnight, and central eclipse at sunrise, respectively.

Atmospheric Compositions of Major Solar System Objects

The atmospheric composition of Solar System objects is obtained by spectral observations with ground based telescopes; remote observations by probes; and observations by atmospheric entry probes including spectroscopic observations, mass spectroscopy, gas analysis, and gas chromatography.

In the table, the atmospheric compositions of Venus, Earth, and Mars are actual measurement values in the vicinity of the surface. The composition of outer planets is determined from infrared/ultraviolet photometric and spectroscopic observations from Earth and probes.

The atmospheres of Mercury and the Moon are tenuous and show drastic time dependant variations. It is thought that the atmospheres on Triton and Pluto are also thin, and have changing seasons. Besides what is listed in the table it has been confirmed that the Jovian moon Ganymede has a thin atmosphere of oxygen and ozone that an oxygen atmosphere exists on Europa.

Material name	Mercury	Venus	Earth[4]	Mars	Jupiter	Saturn	Uranus	Neptune	Pluto	The Moon	Titan	Triton
	Percent volume	Percent volume	Percent volume	Percent volume	Percent volume	Percent volume	Percent volume	Percent volume	Percent volume	Percent volume	Percent volume	Percent volume
Hydrogen H_2					90	96	82.5	80		23	0.099	
Helium He	1.4	1.2×10^{-3}	5.2×10^{-4}		10	3.25	15.2	19		26		
Neon Ne		7×10^{-4}	1.8×10^{-3}	2.5×10^{-4}						29		
Argon Ar		7×10^{-3}	9.3×10^{-1}	1.6						19		
Sodium Na	86[2)]		—[2)]									
Potassium K	$<3.7\times10^{-1}$											
Nitrogen N_2		3.5	78	2.7					>90		94.2	~100
Oxygen O_2			21	1.3×10^{-1}								
Carbon dioxide CO_2		96.5	3.9×10^{-2} [1)]	95.3					5-7.5$\times10^{-2}$			
Sulfur dioxide SO_2		1.5×10^{-2}										
Water H_2O		2×10^{-3}	0–4[2), 4)]	2.1×10^{-2} [2)]	4×10^{-4}							
Carbon monoxide CO		1.7×10^{-3}	1.2×10^{-5}	7×10^{-2}	$\sim10^{-7}$			$\sim10^{-6}$	5×10^{-2}			
Ozone O_3			—[2)]	3×10^{-6}								
Methane CH_4					3×10^{-1}	4.5×10^{-1}	2.3	1.5	0.25		5.65	0.015
Ammonia NH_3					2.6×10^{-2}	1.25×10^{-2}		3×10^{-4}				
Acetylene C_2H_2					1.0×10^{-5}	3×10^{-5}	2×10^{-2}	10^{-4}	3×10^{-4}			
Ethane C_2H_6					5.8×10^{-4}	7.0×10^{-4}		1.5×10^{-4}				
Phosphorus hydride PH_3					$\sim10^{-4}$	$\sim10^{-4}$						
Surface atmospheric pressure (10^5 Pa)	$<5\times10^{-15}$ [2)]	92	1	0.006					1.3×10^{-5}	3×10^{-15} [2)]	1.5	1.9×10^{-5} [2)]
Atmospheric mass (kg)	<10000	4.8×10^{20}	5.1×10^{18}	2.5×10^{16}						25000		
〃 (km atm)[3)]					75±15	75±20	225±75	225±75				
Effective temperature (K)	437	227 (cloud top)	255 (sub-solar)	210	124.4±0.3	95.0±0.4	59.1±0.3	59.6±0.7	39	274	82	38
Surface temperature (K)	442	737	288	215					45±3	120–396	93.6	

Note 1) Value in 2012. It is gradually increasing. 2) Variable. 3) The equivalent depth (km) of hydrogen based on a 1 atmosphere, 0℃ reference. 4) Terrestrial atmospheric compositions are typically quoted in terms of dry air.

Dwarf Planets and Plutoids

In the Solar System, an object defined as a dwarf planet: (a) revolves around the Sun, (b) has enough mass to achieve an almost spherical shape (a shape in fluid mechanical equilibrium) through strong self gravity, (c) has not cleared its orbital region of other objects, (d) and is not a satellite. They have been called dwarf planets since a resolution at the August 2006 International Astronomical Union (IAU) General Assembly. At present, 5 objects are included in this category, but that number may increase. Those changes will be legislated by the IAU in the future.

A dwarf planet that exists in the outer Solar System is called a plutoid. This includes all the dwarf planets except Ceres. Orbital elements and various physical quantities of the dwarf planets at dynamic time 0 a.m. May 31, 2020 (JD 2459000.5) are shown in Table 1 and Table 2. m_0 is the mean visual magnitude at the opposition; M_0 is a mean anomaly; ω, Ω, and i are respectively the argument of perihelion, the longitude of ascending node, and the inclination; the equinox is J2000.0; e is the eccentricity; a is the semi-major axis shown in au (based primarily on Minor Planet Circulars). The visual absolute magnitude is the average magnitude assuming distances from the Sun and the Earth to the object to be one astronomical unit, and 0° phase angle. The albedo is the geometric albedo.

Table 1

Name	m_0	M_0	ω	Ω	i	e	a	Discovery year
	mag	°	°	°	°		au	
Ceres	6.8	162.7	73.7	80.3	10.6	0.078	2.768	1801
Pluto	15.6	42.9	115.3	110.3	17.1	0.253	39.861	1930
Haumea	16.5	217.8	238.8	122.2	28.2	0.195	43.182	2003
Makemake	16.4	165.5	294.8	79.6	29.0	0.161	45.429	2005
Eris	17.2	206.0	151.6	36.0	44.0	0.436	67.867	2003

Table 2

Name	Diameter	Average density	Visual absolute magnitude	Albedo	Rotation period	Orbital period
	km	10^3 kg m^{-3}	mag		Days	Years
Ceres	939	2.2	3.34	0.09	0.38	4.60
Pluto	2 377	1.85	-0.7	~0.6	6.4	248
Haumea	990 × 1540 × 1920	2.6	0.13	0.73	0.16	282
Makemake	1 400	1.7	-0.44	0.6	7.8	305
Eris	2 400	2.3	-1.2	0.86	~1.1 ?	561

Table 3 shows the orbital elements and the various physical quantities of the dwarf planets' satellites. The magnitudes of Pluto I, II, and III are the mean visual magnitudes at opposition, The visual magnitude at the time of discovery are given for Pluto IV and V, The magnitudes of Eris I, Haumea I, and II are rough values converted to visual magnitude from the near-infrared brightness at the time of discovery. The magnitude of Makemake I is the AB magnitude. The orbital radii of Pluto II and Pluto III are the values measured from the center of gravity of Pluto and Pluto I. The orbital inclinations of Pluto I, II, and III are based on the equator of Pluto. The orbital inclinations of Haumea I, II, and Makemake I are based on the average ecliptic coordinate system of J2000.0.

Table 3

Number	Name	Discoverer	Discovery year	Magnitude	Orbital radius	Orbital period	Eccentricity	Inclination	Radius	Mass (parent body =1)
				mag	km	Day		°	km	
Pluto I	Charon	Christy	1978	18.0	19 600	6.39	0.0	0.1	606	0.1
Pluto II	Nix	Weaver *et al.*	2005	24.4	48 694	24.9	0.0	0.1	25 × 18 × 17	4×10^{-5}
Pluto III	Hydra	Weaver *et al.*	2005	24.5	64 738	38.2	0.0	0.2	33 × 23 × 13	2×10^{-5}
Pluto IV	Kerberos	Showalter *et al.*	2011	26.1	57 783	32.2	0.0	0.4	10 × 5 × 5	-
Pluto V	Styx	Showalter *et al.*	2012	27.0	42 656	20.2	0.0	0.8	8 × 5 × 4	-
Haumea I	Hi'iaka	Brown *et al.*	2005	19	49 880	49.4	0.1	126	195	5×10^{-3}
Haumea II	Namaka	Brown *et al.*	2005	21	25 657	18.3	0.2	113	100	5×10^{-4}
Makemake I	S/2015(136472)1	Parker *et al.*	2015	25	≧21000	>12.4	0 ?	63-87?	88 ?	-
Eris I	Dysnomia	Brown *et al.*	2005	22	37 400	15.8	0.0	-	-	-

Small Solar System Bodies

All celestial bodies orbiting the Sun which are not planets, dwarf planets, or satellites are called Small Solar System bodies. In other words, Small Solar System bodies include the Trans-Neptunian objects (excluding plutoids), asteroids (excluding Ceres), comets and other small celestial bodies.

Trans-Neptunian Object (TNO) They have also been known as Kuiper Belt Objects or Edgeworth-Kuiper Belt Objects. About 3 400 objects have been discovered in the area beyond Neptune. The majority are distributed in a belt in the vicinity of the ecliptic plane. Some TNOs have a mean motion which has a simple interger ratio with Neptune's mean motion (21.5" day^{-1}), have been scattered by a close approach with the planets (Example of scattered object: 90377 Sedna) or have a satellite around them. Table 4 shows the orbital elements of typical TNOs at dynamic time 0 a.m. May 31, 2020 (JD 2459000.5). The meaning of the equinox and the denotations are the same as in the table of dwarf planets (based primarily on Minor Planet Circulars). A TNO also has an asteroid's designation.

Table 4

Designation	Name	m_0	M_0	ω	Ω	i	e	a	Mean motion resonance with Neptune
		mag	°	°	°	°		au	
79969	1999 CP133	22.8	58.1	158.4	334.4	3.2	0.081	34.909	4 : 5
15836	1995 DA2	23.2	65.2	333.3	127.6	6.6	0.071	36.303	3 : 4
15788	1993 SB	23.8	356.5	79.0	354.7	1.9	0.317	39.155	2 : 3
15809	1994 JS	24.0	356.1	237.3	56.3	14.0	0.225	42.664	3 : 5
50000	Quaoar	18.8	301.1	147.5	188.9	8.0	0.041	43.695	
60620	2000 FD8	23.0	328.6	81.0	184.9	19.5	0.226	44.022	4 : 7
15760	Albion	23.7	34.0	0.8	359.3	2.2	0.071	43.928	
20161	1996 TR66	24.3	57.6	309.6	342.9	12.4	0.401	47.957	1 : 2
69988	1998 WA31	24.8	47.1	309.2	20.7	9.5	0.431	55.102	2 : 5
	2012 VP113	28.2	3.6	293.4	90.7	24.1	0.693	262.109	
90377	Sedna	28.1	358.1	311.4	144.2	11.9	0.843	484.521	

Figure 2A shows the distribution of the eccentricity e versus the semimajor axis a for TNOs. (Only a part of both orbit semimajor axes and eccentricities are shown.). The numbers along the top indicate mean motion resonances with Neptune and the curve that stretches from the lower left to the upper right shows the minimum eccentricity that intersects with the Neptune's orbit for each semimajor axis.

Assuming the same reflectivity as a cometary nucleus, the diameter of the discovered objects range from tens of kilometers to hundreds of kilometers, with the largest exceeding 1000 km. Brightness variation observations have been performed for the bright objects, and the rotation periods have been found to be on the order of hours or tens of hours.

A TNO which changed to an orbit crossing planets' orbits through a collision, close approach, or other incident is called a "Centaur." Including the asteroid Chiron, about 270 Centaurs have been discovered.

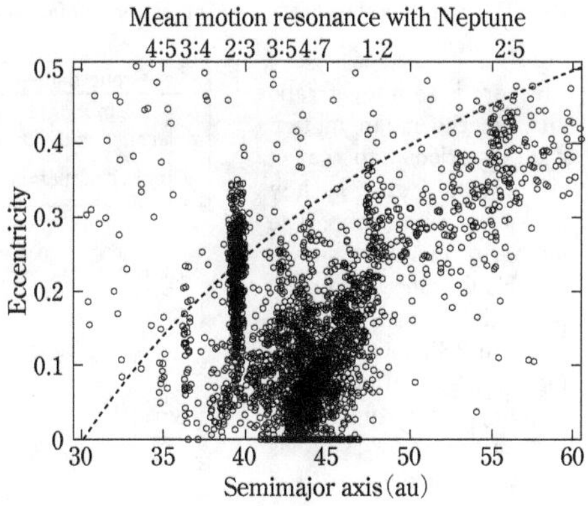

Figure 2A

Asteroids Among asteroids mainly between Mars and Jupiter, orbital elements of lager asteroids and ones with characteristic orbits at dynamical time 0 a.m. May 31, 2020 (JD 2459000.5) are listed in Table 5. The meaning of the equinox and the denotations are the same as in the table of dwarf planets (based primarily on Minor Planet Circulars). Earth crossing asteroids with small perihelion or asteroids with satellites are also discovered. Asteroids which had their orbits determined by June 3, 2020 have been assigned designations (numbers) up to 546077. There are also several hundred thousand asteroids which have not had their orbits determined.

Table 5

Designation	Name	m_0	M_0	ω	Ω	i	e	a	Discovery year
		mag	°	°	°	°		au	
2	Pallas	7.7	145.0	310.2	173.0	34.8	0.230	2.774	1802
3	Juno	8.4	125.4	248.1	169.9	13.0	0.257	2.668	1804
4	Vesta	5.5	204.3	150.9	103.8	7.1	0.089	2.362	1807
8	Flora	8.6	315.3	285.5	110.9	5.9	0.156	2.201	1847
9	Metis	8.9	23.9	6.3	68.9	5.6	0.123	2.386	1848
24	Themis	11.3	66.2	107.0	35.9	0.8	0.124	3.135	1853
25	Phocaea	10.4	278.9	90.3	214.1	21.6	0.255	2.400	1853
153	Hilda	13.0	225.0	39.1	228.1	7.8	0.140	3.979	1875
158	Koronis	12.8	295.3	146.4	277.7	1.0	0.052	2.868	1876
170	Maria	12.4	67.2	158.3	301.3	14.4	0.063	2.553	1877
221	Eos	11.6	331.9	192.3	141.7	10.9	0.104	3.013	1882
243	Ida	13.5	320.8	114.1	323.6	1.1	0.043	2.861	1884
253	Mathilde	13.5	130.1	157.7	179.5	6.7	0.263	2.649	1885
279	Thule	14.2	235.1	24.5	71.9	2.3	0.043	4.273	1888
401	Ottilia	13.7	33.3	299.5	36.1	6.0	0.032	3.343	1895
433	Eros	9.5	271.1	178.9	304.3	10.8	0.223	1.458	1898
434	Hungaria	12.5	10.0	123.9	175.3	22.5	0.074	1.944	1898
588	Achilles	15.0	271.7	133.3	316.5	10.3	0.147	5.209	1906
617	Patroclus	14.9	237.0	307.9	44.3	22.1	0.139	5.215	1906
887	Alinda	16.6	294.6	350.5	110.4	9.4	0.570	2.474	1918
903	Nealley	14.4	107.3	234.2	159.4	11.8	0.045	3.239	1918
944	Hidalgo	17.9	41.8	56.6	21.4	42.5	0.661	5.736	1920
951	Gaspra	13.6	172.9	129.9	253.1	4.1	0.174	2.209	1916
1221	Amor	18.7	38.5	26.7	171.3	11.9	0.435	1.919	1932
1373	Cincinnat	16.1	160.5	99.1	297.5	38.9	0.315	3.419	1935
1566	Icarus	11.5	192.1	31.4	88.0	22.8	0.827	1.078	1949
1862	Apollo	15.4	88.5	286.0	35.6	6.4	0.560	1.470	1932
2060	Chiron*	17.0	172.8	339.8	209.2	6.9	0.379	13.686	1977
2062	Aten	9.6	119.5	148.0	108.6	18.9	0.183	0.967	1976
5261	Eureka	15.6	263.8	95.5	245.0	20.3	0.065	1.524	1990
25143	Itokawa	17.4	187.6	162.8	69.1	1.6	0.280	1.324	1998
162173	Ryugu	16.1	193.6	211.5	251.6	5.9	0.190	1.190	1999

* Chiron physically shows the characteristics of a comet.

Table 6

Group name	Hilda group	Thule group	Trojan
Mean motion resonance with Jupiter	450″ Day^{-1} 3 : 2	400″ Day^{-1} 4 : 3	300″ Day^{-1} 1 : 1
Number of members	2447	12	3151 preceeding Jupiter 1834 trailing Jupiter

If there is an integer ratio between the mean motion of the asteriods and that of Jupiter (300″ day^{-1}), they are known as a group. The number of members shown in Table 6 reflects asteriods assigned designations.

Figure 2B shows the frequency distribution of the semimajor axis in 0.01 au bins out to 6 au for asteroids assigned designations. Near the 2 : 1, 7 : 3, 5 : 2, and 3 : 1 mean motion resonances of Jupiter, there are gaps with virtually no asteroids.

Though few in number, small bodies resembling Jupiter's Trojan have also been found for the Earth, Mars, and Neptune.

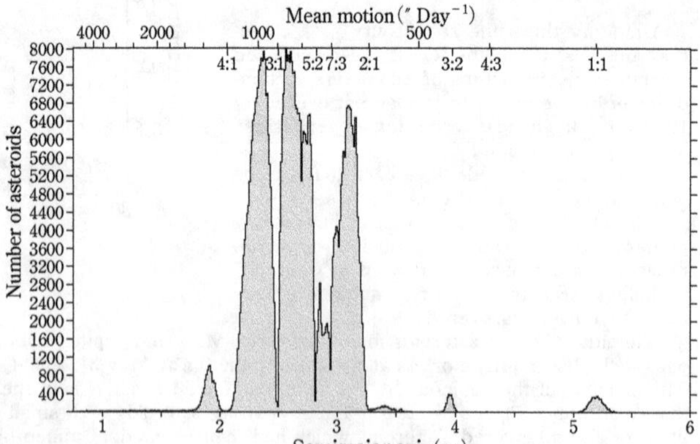

Figure 2B

Physical Properties of Asteroids and Meteorites

Physical properties of asteroids[1)]

The number of asteroids increases with decreasing size. Shapes of smaller sized asteroids are getting irregular. In general, the average density is smaller than those of the corresponding meteorite group, which is considered to be due to internal porosity.

Number	Asteroid name	Diameter (km)	Average density (10^3 kg m^{-3})	Absolute visual magnitude (mag)	Geometric albedo	Rotation period (h)	Taxonomy
2	Pallas	582×556×500*1	2.400±0.250*1	4.13	0.150*2	7.8132	B
3	Juno	282×249×220*3	3.32±0.40*3	5.33	0.246*2	7.210	S
4	Vesta	573×557×446*4	3.456±0.001*4	3.20	0.34±0.02*4	5.342	V
21	Lutetia	121×101×75*5	3.4±0.3*5	7.35	0.19±0.01*5	8.168270	Xc
243	Ida	59.8×25.4×18.6*6	2.6±0.5*6	9.94	0.206±0.032*6	4.633632	S
§	Dactyl	1.6×1.4×1.2*6	–	–*7	0.198±0.050*6	>8	S
253	Mathilde	66×48×46*8	1.3±0.2*8	10.3	0.047±0.005*8	417.7	Cb
433	Eros	38×15×14*8	2.67±0.03*8	11.16	0.23*8	5.270	S
951	Gaspra	18.2×10.5×8.9*6	–	11.46	0.23±0.06*6	7.042	S
2867	Steins	6.67×5.81×4.47*5	–	12.7	0.3*9	6.04679	Xe
5535	Annefrank	6.6×5.0×3.4*10	–	13.7	0.18–0.24*10	15.12	S
9969	Braille	2.1×1×1*11	–	15.9	0.34±0.03*11	226.4	Q
25143	Itokawa	0.535×0.294×0.209*12	1.9±0.1*12	19.2	0.24±0.02*12	12.1324*12	S
101955	Bennu	0.5647×0.5361×0.4985*13	1.190±0.013*13	20.9	0.044±0.002*13	4.296057*13	B
162173	Ryugu	1.004×1.004×0.875*14	1.19±0.02*14	19.3	0.045±0.002*14	7.63262*14	C

1) High precision measurements have been performed by in-situ observations with spacecrafts (*4 Dawn, *5 Rosetta, *6 Galileo, *8 NEAR Shoemaker, *10 Stardust, *11 Deep Space1, *12 Hayabusa, *13 OSIRIS-REx, *14 Hayabusa2).

§ Moon of Ida : 243 Ida I.

*1 Schmidt *et al.* 2009, Science, **326**, 275.

*2 Usui *et al.* 2011, PASJ, **63**, 1117.

*3 Viikinkoski *et al.* 2015, A&A, **581**, L3.

*7 Adopting the size and albedo measured by the spacecraft, it is approximately 7 magnitude fainter than Ida.

*9 Masiero *et al.* 2011, ApJ, **741**, 68.

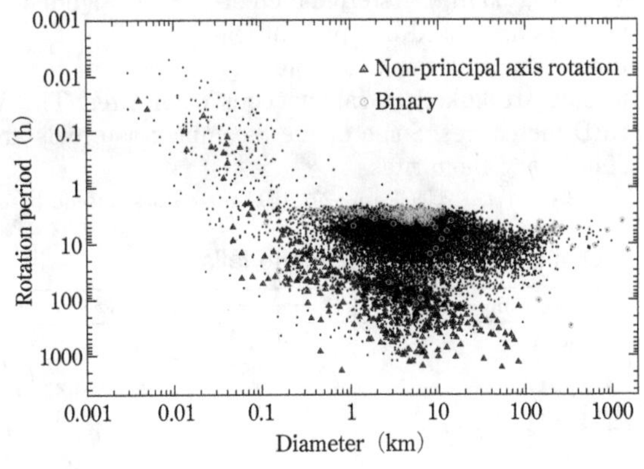

Figure 3 Rotation period versus diameter of asteroids[2)]

Rotation period

The rotation period of an asteroid larger than several km in diameter is 2.2 hours or more. Short rotation periods are seen for binary systems. Longer rotation periods are generally seen in asteroids of non-principal axis rotation (tumbling).

2) Warner *et al.* 2009, Icarus, **202**, 134.

Meteorite density[3)]

Classification	Type	Average Grain Density	Average Bulk Density
		(10^3 kg m^{-3})	
Carbonaceous chondrite	CI	2.5	1.6
	CM	2.9	2.3
	CO	3.4	3.0
	CV	3.3–3.5	2.8–3.1
	CK	3.6	2.9
Ordinary chondrite	H	3.7	3.4
	L	3.6	3.4
	LL	3.5	3.2
HED	Howardite	3.2	3.0
	Eucrite	3.2	2.9
	Diogenite	3.5	3.2

3) Based on Consolmagno, Britt, Macke, 2008. ChEG, **68**, 1.

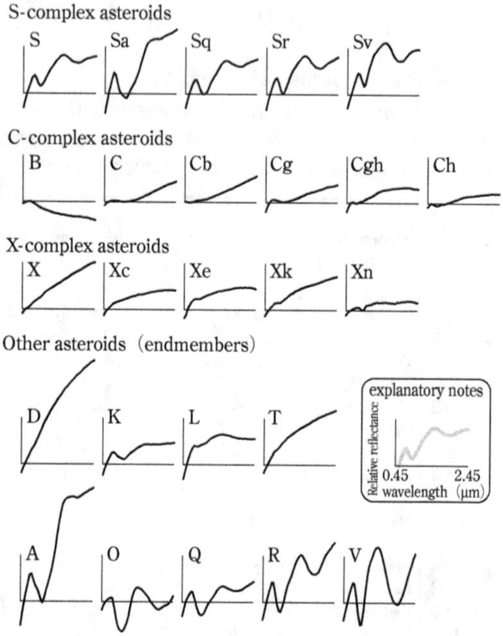

Figure 4 Asteroid reflectance spectra[4)]

Asteroid reflectance spectra

The reflectance spectra in the 0.45–2.45 μm wavelength range for each of the asteroid classes are shown. The vertical axis is reflectance spectrum normalized at the V band (0.55 μm). Asteroid-meteorite links are based on spectral features. Many of the C-complex asteroids are linked to cabonaceous chondrites. The V-type asteroids are linked to the HED meteorites. Some of the S-complex asteroids are thought to be the parent bodies of ordinary chondrites.

4) Demeo *et al.* 2009, Icarus, **202**, 160. (http://smass.mit.edu/busdemeoclass.html)

Orbital elements of meteorite falls[5)]

Meteorite (type)	Semimajor axis (au)	Eccentricity	Inclination (°)	Date of fall
Pribram (H5)[6)]	2.40	0.67	10.8	1959 **4** 7
Lostcity (H5)	1.66	0.42	12.0	1970 **1** 3
Innisfree (LL5)	1.87	0.47	12.3	1977 **2** 6
Peekskill (H6)[7)]	1.49	0.41	4.9	1992 **10** 9
Neuschwanstein (EL6)[6)]	2.40	0.67	11.4	2002 **4** 6
Almahata Sitta (Ureilite)[8)]	1.31	0.31	2.5	2008 **10** 7
Chelyabinsk (LL5)[9)]	1.8	0.58	4.9	2013 **2** 15

5) Values without notes are based on Schaifers and Voigt (Eds.). Landolt-Boörnstein. 1981.
6) Spurný *et al.* 2003, Nature, **423**, 151.
7) Brown *et al.* 1994, Nature, **367**, 624.
8) Jenniskens *et al.* 2009, Nature, **458**, 485. Asteroidal parent body is identified as 2008TC$_3$.
9) Popova *et al.* 2013, Science, **342**, 1069.

Comets

Among comets which have had their orbital characteristics determined and have been assigned a primary designation, comets brighter than 9.5 absolute magnitude are shown. T is a date of the perihelion passage. q is a perihelion distance shown in au. The remaining signs have the same meaning as on the dwarf planet table. Angle elements are given in J2000.0 equinox. Based primarily on data from JPL (NASA).

Prim. desig.	T	q	e	ω	Ω	i	Absolute magnitude	Orbital period	Name
	TT	au	°	°	°	°	mag	yrs.	
1P	1986 2 05	0.586	0.967	111.3	58.4	162.3	5.5	75.3	Halley
12P	1954 5 22	0.774	0.955	199.0	255.9	74.2	5.0	70.8	Pons-Brooks
13P	1956 6 19	1.178	0.930	64.6	86.1	44.6	5.0	69.5	Olbers
17P	2014 3 27	2.057	0.432	24.5	326.8	19.1	8.8	6.9	Holmes
19P	2001 9 14	1.360	0.623	353.4	75.4	30.3	8.2	6.9	Borrelly
23P	1989 9 11	0.479	0.972	129.6	311.6	19.3	7.8	70.5	Brorsen-Metcalf
29P	2004 7 21	5.724	0.045	49.7	312.6	9.4	8.5	14.7	Schwassmann-Wachmann 1
31P	2010 9 29	3.424	0.193	17.9	114.2	4.5	5.1	8.7	Schwassmann-Wachmann 2
35P	1939 8 9	0.748	0.974	29.3	356.0	64.2	8.3	155.0	Herschel-Rigollet
39P	1983 6 16	5.460	0.243	55.9	332.2	2.0	5.0	19.4	Oterma
43P	2016 8 19	1.357	0.595	191.6	249.9	16.0	8.6	6.1	Wolf-Harrington
48P	2011 9 28	2.303	0.368	207.8	117.3	13.7	9.3	7.0	Johnson
51P-D	2008 6 14	1.646	0.552	269.1	84.6	5.4	8.6	7.1	Harrington-D
52P	2014 3 9	1.776	0.540	139.7	336.9	10.2	7.9	7.6	Harrington-Abell
57P-A	2002 7 31	1.729	0.499	115.2	188.9	2.8	8.5	6.4	duToit-Neujmin-Delporte-A
58P	2004 1 10	1.388	0.661	200.5	160.6	13.5	7.7	8.3	Jackson-Neujmin
60P	2012 5 13	1.620	0.539	216.5	267.7	3.6	9.3	6.6	Tsuchinshan 2
62P	2011 6 30	1.384	0.598	30.1	90.4	9.7	7.6	6.4	Tsuchinshan 1
63P	2013 4 10	1.951	0.651	169.0	358.0	19.8	6.5	13.2	Wild 1
74P	2009 7 26	3.552	0.148	86.6	77.1	6.7	5.1	8.5	Smirnova-Chernykh
78P	2012 1 12	2.011	0.462	192.8	210.6	6.3	8.8	7.2	Gehrels 2
80P	2014 11 10	1.613	0.599	339.1	259.9	29.9	9.3	8.1	Peters-Hartley
81P	2010 2 22	1.598	0.537	41.8	136.1	3.2	8.8	6.4	Wild 2
82P	2001 8 31	3.626	0.124	227.7	239.6	1.1	7.5	8.4	Gehrels 3
90P	2017 6 19	2.975	0.510	29.3	13.3	9.6	6.5	15.0	Gehrels 1
91P	2005 6 27	2.606	0.330	354.9	247.9	14.1	7.3	7.7	Russell 3
99P	2007 1 16	4.718	0.228	172.9	28.4	4.3	6.8	15.1	Kowal 1
101P	2006 1 10	2.300	0.601	264.4	128.9	5.0	9.4	13.9	Chernykh
109P	1992 12 11	0.960	0.963	153.0	139.4	113.5	4.5	133.3	Swift-Tuttle
110P	2014 12 17	2.475	0.315	167.7	287.7	11.7	5.9	6.9	Hartley 3
111P	2013 1 30	3.704	0.110	3.4	89.8	4.2	8.4	8.5	Helin-Roman-Crockett
116P	2016 1 11	2.187	0.372	173.3	21.0	3.6	6.7	6.5	Wild 4
117P	2014 3 26	3.056	0.254	222.6	58.9	8.7	6.6	8.3	Helin-Roman-Alu 1
120P	2013 2 22	2.728	0.341	30.3	4.3	8.8	6.9	8.4	Mueller 1
121P	2013 9 2	3.752	0.184	11.8	94.2	20.2	6.2	9.9	Shoemaker-Holt 2
122P	1995 10 6	0.659	0.963	13.0	79.6	85.4	7.5	74.3	de Vico
123P	2019 2 4	2.126	0.449	102.9	46.5	15.4	8.2	7.6	West-Hartley
126P	2010 2 22	1.718	0.696	356.8	357.8	45.8	8.8	13.4	IRAS
128P	2017 1 10	3.054	0.322	210.5	214.3	4.4	7.1	9.6	Shoemaker-Holt 1
128P-A	2017 5 6	3.035	0.334	210.6	214.4	4.4	8.5	9.7	Shoemaker-Holt 1-A
128P-B	1997 11 20	3.047	0.321	210.2	214.5	4.4	5.6	9.5	Shoemaker-Holt 1-B
131P	2012 1 7	2.418	0.343	179.6	214.2	7.4	8.6	7.1	Mueller 2
134P	2014 5 21	2.571	0.588	18.6	202.1	4.3	5.3	15.6	Kowal-Vavrova
135P	1992 6 14	2.714	0.291	22.7	213.4	6.1	7.0	7.5	Shoemaker-Levy 8
136P	2007 10 23	2.964	0.293	225.1	137.6	9.4	5.0	8.6	Mueller 3
139P	2008 4 20	3.403	0.248	165.7	242.4	2.3	7.2	9.6	Vaisala-Oterma
147P	2008 9 22	2.756	0.276	346.8	93.8	2.4	4.5	7.4	Kushida-Muramatsu
151P	2015 10 7	2.496	0.570	215.5	143.6	4.7	8.4	14.0	Helin
153P	2002 3 18	0.507	0.990	34.7	93.4	28.1	4.0	365.5	Ikeya-Zhang

Comets Continued.

Prim. desig.	T	q	e	ω	Ω	i	Absolute magnitude	Orbital period	Name
	TT	au		°	°	°	mag	yrs.	
158P	2012 9 15	4.580	0.030	231.6	137.3	7.9	5.5	10.3	Kowal-LINEAR
160P	2012 9 18	2.067	0.479	18.2	337.0	17.3	6.8	7.9	LINEAR
165P	2000 6 15	6.830	0.622	126.2	0.6	15.9	6.3	76.7	LINEAR
166P	2002 5 20	8.564	0.383	321.9	64.5	15.4	5.8	51.7	NEAT
168P	2012 10 1	1.415	0.609	13.9	356.5	21.9	7.4	6.9	Hergenrother
179P	2007 12 4	4.082	0.307	295.7	115.9	19.9	5.3	14.3	Jedicke
180P	2008 5 25	2.481	0.356	95.3	84.7	16.9	5.6	7.6	NEAT
183P	2008 5 3	3.888	0.135	160.9	5.8	18.7	8.7	9.5	Korlevic-Juric
186P	2008 5 27	4.315	0.128	287.0	327.7	28.6	6.9	11.0	Garradd
187P	2008 10 21	3.713	0.173	134.6	111.8	13.7	6.0	9.5	LINEAR
193P	2008 2 16	2.122	0.403	8.0	336.1	10.8	7.2	6.7	LINEAR-NEAT
195P	2009 1 21	4.438	0.315	249.7	243.3	36.4	8.6	16.5	Hill
199P	2009 4 10	2.936	0.507	191.9	92.9	24.8	7.8	14.5	Shoemaker 4
213P-B	2011 6 16	2.123	0.380	3.3	312.7	10.2	7.3	6.3	Van Ness-B
216P	2001 2 12	2.154	0.445	151.6	359.9	9.0	4.9	7.7	LINEAR
231P	2011 5 16	3.032	0.247	42.4	133.1	12.3	7.2	8.1	LINEAR-NEAT
232P	2009 10 1	2.983	0.335	53.4	56.1	14.6	6.1	9.5	Hill
234P	2009 12 23	2.860	0.251	358.4	179.7	11.5	6.7	7.5	LINEAR
235P	2010 3 21	2.748	0.313	333.7	204.5	8.9	6.3	8.0	LINEAR
237P	2016 10 4	1.973	0.439	20.8	249.4	15.2	6.0	6.6	LINEAR
242P	2012 4 11	3.872	0.314	250.6	181.1	32.1	5.7	13.4	Spahr
244P	2012 1 5	3.937	0.204	90.3	354.3	2.3	6.5	11.0	Scotti
246P	2013 1 29	2.878	0.285	176.3	78.8	16.0	6.4	8.1	NEAT
261P	2012 9 29	2.187	0.390	58.9	298.4	6.3	7.6	6.8	Larson
269P	2014 11 26	4.077	0.435	223.9	249.0	6.6	5.6	19.4	Jedicke
271P	2013 7 5	4.252	0.390	35.0	9.6	6.9	6.4	18.4	van Houten-Lemmon
284P	2007 8 18	2.283	0.378	202.8	144.4	11.9	7.6	7.0	McNaught
287P	2014 12 27	3.052	0.269	188.9	139.0	16.3	4.6	8.5	Christensen
291P	2004 3 25	2.605	0.429	176.0	241.0	5.9	5.5	9.7	NEAT
297P	2008 3 21	2.409	0.309	131.8	98.3	10.3	6.1	6.5	Beshore
299P	2015 2 23	3.140	0.282	323.5	271.7	10.5	5.0	9.1	Catalina-PANSTARRS
324P	2015 11 29	2.620	0.154	58.6	270.7	21.4	7.0	5.4	La Sagra
328P	2007 5 11	1.880	0.552	30.3	341.8	17.7	7.1	8.6	LONEOS-Tucker
329P	2015 12 5	1.660	0.680	342.3	88.8	21.5	9.0	11.8	LINEAR-Catalina
332P	2010 10 13	1.579	0.489	152.4	3.8	9.4	5.2	5.4	Ikeya-Murakami
334P	2017 5 19	4.188	0.354	81.1	92.8	19.1	6.7	16.5	NEAT
336P	2006 8 18	2.633	0.453	314.0	299.2	18.6	8.0	10.6	McNaught
343P	2017 1 27	2.277	0.584	137.6	257.3	5.6	6.0	12.8	NEAT-LONEOS
349P	2017 8 27	2.496	0.301	256.3	331.8	5.5	6.9	6.7	Lemmon
353P	2009 6 23	2.209	0.469	230.5	121.6	28.4	8.5	8.5	McNaught
356P	2017 12 17	2.689	0.354	226.2	160.8	9.6	6.5	8.5	WISE
357P	2008 12 24	2.512	0.435	1.3	44.7	6.3	5.0	9.4	Hill
361P	2018 7 2	2.780	0.438	219.2	203.3	13.9	4.6	11.0	Spacewatch
370P	2002 1 25	2.503	0.613	356.9	55.3	19.4	6.2	16.4	NEAT
371P	2010 3 26	2.174	0.478	308.6	67.4	17.4	7.0	8.5	LINEAR-Skiff
372P	2009 4 21	3.804	0.154	27.5	325.9	9.5	7.1	9.5	McNaught
373P	2011 11 6	2.305	0.394	221.2	232.0	13.8	6.0	7.4	Rinner
374P	2007 12 9	2.671	0.463	51.4	8.2	10.8	6.8	11.1	Larson
376P	2005 8 16	2.838	0.515	285.5	315.1	1.2	9.4	14.2	LONEOS
377P	2020 7 14	5.037	0.251	354.6	226.0	9.0	7.1	17.4	Scotti
380P	2019 6 6	3.050	0.326	90.1	133.1	8.2	7.8	9.6	PANSTARRS
382P	2007 8 23	4.352	0.278	175.1	181.7	7.9	5.6	14.8	Larson
385P	2010 11 9	2.555	0.402	44.3	357.2	16.9	7.4	8.8	Hill

Various Physical Parameters of Comets

Number of comets As of June 2020, JPL (NASA) has published orbital elements for 3648 comets. 846 comets have orbital periods less than 200 years. Among those 468 have been given primary designations.

Nucli sizes and albedos Albedos and sizes for the nucli are based on results from space missions which have approached the nucli. The albedo is the geometrical albedo.

Table 7

Prim. desig.	Name	Size (km)	Albedo	Prim. desig.	Name	Size (km)	Albedo
1P	Halley	14.4×7.4×7.4 15.3×7.22×7.2	0.04	67P	Churyumov-Gerasimenko	4.1×3.3×1.8 and 2.6×2.3×1.8	0.06
9P	Tempel 1	7.6×4.9	0.04	81P	Wild 2	5.4×3.8×3.0	0.04
19P	Borrelly	4.0×1.58	0.029	103P	Hartley 2	0.69×2.33	0.04

Coma size When a comet approaches a heliocentric distance of 2.5–3 au, a coma is generated. The largest observed coma belonged to the Great Comet of 1811, with a 1.8 million km in diameter. The Humason VIII comet of 1962 achieved a diameter of 1.32 million km. There are also comets where the coma is hardly visible.

Comet tails A comet has a straight tail of expanding ionizing gas (type I) and a curved tail of dust (type II). A sodium tail was discovered in Comet Hale-Bopp in 1997. In Table 8, observing conditions have a large influcence on the apparent length of the tail. The real length is also nothing more than a rough standard.

Table 8

Name	Apparent length	Real length
1843 I	65°	3.2×10^8 km
1882 II	40	2.1
1811 I	25	1.6
Halley	125	1.2
Ikeya-Seki	25	1.2

The main spectral lines (photograph wavelength region)

Table 9

Head	Wavelength(nm)	Tail	Wavelength(nm)
OH	309	CO_2^+	351, 367
NH	336	CO^+	378, 380, 400, 402, 425, 427, 454, 457
CN	388, 422	N_2^+	391
C_2	(Band head)438, 474, 517, 564	H_2O^+	615, 620, 654, 658, 659, 669, 670
C_3	405	Na	589
[O]	630		

Major Meteor Showers

	Abbr.	Meteor shower	Appearance period	Maximum	Radiant Point α	Radiant Point δ	Number appearing	Parent object
Regular groups	QUA	Quadrantids	**12** 28 – **1** 12	**1** 4	230°	49°	Many	-
	LYR	April Lyrids	**4** 14 – **4** 30	**4** 22	271	34	Moderate	C/1861 G1
	ETA	η Aquariids	**4** 19 – **5** 28	**5** 6	338	−1	Many	1P/Halley
	SDA	Southern δ Aquariids	**7** 12 – **8** 23	**7** 30	340	−16	Moderate	-
	PER	Perseids	**7** 17 – **8** 24	**8** 13	48	58	Many	109P/Swift-Tuttle
	STA	Southern Taurids	**9** 10 – **11** 20	**10** 10	32	9	Few	2P/Encke
	ORI	Orionids	**10** 2 – **11** 7	**10** 21	95	16	Moderate	1P/Halley
	NTA	Northern Taurids	**10** 20 – **12** 10	**11** 12	58	22	Few	2P/Encke
	GEM	Geminids	**12** 4 – **12** 17	**12** 14	112	33	Many	(3200)Phaethon
Periodic groups	DRA	October Draconids	**10** 6 – **10** 10	**10** 8	262	54	(next appearance) 2024	21P/Giacobini-Zinner
	LEO	Leonids	**11** 6 – **11** 30	**11** 18	152	22	2034–2037	51P/Tempel-Tuttle
Daytime groups	ARI	Daytime Arietids	**5** 14 – **6** 24	**6** 7	42	25	Many	-
	ZPE	Daytime ξ Perseids	**5** 20 – **7** 5	**6** 9	62	23	Many	-
	BTA	Daytime β Taurids	**6** 5 – **7** 17	**6** 28	86	19	Moderate	-

The meteor shower name and its abbreviation are followed by IAU. Appearance period, maximum, radiant point are based on the International Meteor Organization. The maximum date could move 1–2 days depending on the year. Appearance period includes periods when the number of meteors is very low.

Characteristics of the Major Meteor Showers

Abbr.	H_B	H_E	V_E	V_H	q	e	Ω	ω	i
	km	km	km s^{-1}	km s^{-1}	au		°	°	°
QUA	103	91	40.5	38.2	0.975	0.614	283	171	71
LYR	107	88	47.1	41.8	0.920	0.966	32	214	80
ETA	116	100	65.3	40.7	0.571	0.940	44	96	164
SDA	100	88	38.6	34.1	0.069	0.958	306	155	28
PER	114	94	58.6	40.7	0.942	0.902	140	149	113
ORI	117	99	64.5	39.3	0.562	0.854	27	87	164
STA NTA	–	–	29.6	37.1	0.343	0.843	224	296	3
GEM	100	80	32.9	31.3	0.143	0.896	261	324	24
DRA	104	91	19.3	37.8	0.994	0.611	195	171	30
LEO	128	87	70.4	41.3	0.983	0.901	235	173	161
ARI	–	–	35.6	32.1	0.085	0.938	77	26	25
ZPE	–	–	25.1	33.8	0.365	0.755	81	61	7
BTA	–	–	27.2	34.5	0.274	0.834	102	52	0
*	99	85	33.5						

* Mean value of sporadic meteors.

The orbital elements are based primarily on Meteor Showers (G. W. Kronk, 1988). H_B and H_E are the average height of the origin and the disappearance point above the ground, respectively. V_E and V_H are geocentric and heliocentric speed. The meaning of other signs is the same as for the asteroids and the comets.

Abbr.	Years in which remarkable meteor showers were recorded
LYR	−686, −14, 582, 1093, 1094, 1095, 1096, 1122, 1123, 1136, 1803, 1922, 1945, 1982
PER	36, 714, 830, 833, 835, 841, 924, 925, 926, 933, 989, 1007, 1029, 1042, 1243, 1451, 1779, 1784, 1789, 1862, 1991, 1992, 1993, 1994
ORI	585, 930, 1436, 1439, 1465, 1623, 2006, 2007
DRA	1926, 1933, 1946, 1985, 1998, 2011
LEO	585, 902, 931, 934, 967, 1002, 1035, 1037, 1202, 1237, 1238, 1366, 1466, 1532, 1533, 1538, 1554, 1566, 1582, 1602, 1625, 1666, 1698, 1799, 1833, 1866, 1901, 1965, 1966, 1998, 1999, 2001, 2002

The Main Spectral Lines of Meteor Spectra

(Wavelength unit : nm)

Fast meteor		Slow meteor		Infrared spectral lines	
Ca II	393, 397	Na I	589, 590	Ca II	854, 866
Si II	635, 637	Fe I	401, 405, 406	O I	777, 778, 822,
Na I	589, 590		527, 533, 537		823, 845
Mg I	517, 518			NI	744, 747
Fe I	527, 533, 537				
[O I]	558				

Basic Constants of the Sun

Radius 6.960×10^8 m **Mass** 1.9884×10^{30} kg
Mean density 1.41×10^3 kg m^{-3} **Gravity at surface** 2.74×10^2 m s^{-2}
Escape velocity at solar surface 617.5 km s^{-1} **Spectral type** G2V
Total radiation 3.85×10^{26} W **Visual absolute magnitude** +4.82
Visual apparent magnitude −26.75 **Effective temperature** 5777 K
Color indices $B-V=+0.650$ $U-B=+0.195$
Synodic differential rotation period of the Sun (day)
$26.90 + 5.2 \sin^2\phi$ (ϕ is latitude on the surface of the Sun.)

Solar constant (flux of total radiation received per unit area at the mean Sun-Earth distance outside of the Earth's atmosphere, fluctuations are at the level of 0.1%): 1.37 kWm^{-2} (= 1.96 cal cm^{-2} min^{-1})

Limb darkening There is a rapid decrease in luminosity toward the limb. For instance at 0.5 μm, at 2% of the solar radius from the limb, the luminosity falls to 42% of the luminosity at disk center.

Various Phenomena on the Sun*

Atmospheric layer	Nomenclature	Size (Unit: 1000 km)	Lifetime	Temperature (K)	Remarks
Photosphere	Granulation	1.0	10 min.	$\pm 300^\dagger$	
	Oscillation pattern	3-Entire solar surface	5 min. (period)	the same as the photosphere	Solar p-mode oscillation
	Supergranulation	30	~20 hrs.	the same as the photosphere	Horizontal flow velocity 0.4 km s^{-1}
	Sunspot	10 (umbra) 30 (penumbra)	6 days-2 months	$-1600^\dagger$ (umbra)	Magnetic flux density 0.2-0.4 T
	Facular point	~0.2	~18 min.	A few hundred Kelvin higher†	Magnetic flux density 0.1 T
Chromosphere	Plage (facula)	150	1-6 months	$\geqq 6 \times 10^3$	Magnetic flux density on average 0.005 T
	Spicule	0.2-0.8	1-8 min.	$\sim 7 \times 10^3$	Electron density~10^{17} m^{-3}
	Flare	10-30	Several minutes ~ several hours	$\sim 10^4$	Electron density~10^{19} m^{-3}
Corona	Prominence	5 (width) × 30 (length)	Several minutes ~ several months	$\sim 7 \times 10^3$	Electron density~10^{17} m^{-3} Magnetic flux density 0.0005~0.01 T
	Coronal condensation	100	3 weeks	$\sim 2 \times 10^6$	Electron density~10^{15} m^{-3}
	Coronal streamer	300	Several months	$\sim 10^6$	Electron density~10^{14} m^{-3}
	Flare	10-30	A few minutes ~ several hours	$\sim 2 \times 10^7$	Electron density~10^{16} m^{-3}

* The numbers shown in this table are typical values: the values in reality vary around the shown value.
† This temperature shows differences from the photospheric temperature (6000 K).
Cox (ed.) 2000, Allen's Astrophysical Quantities, and ARA&A, 1963, **1**-1985, **23**

Spectral Distribution of Solar Radiation (Energy)

The amount of energy radiated per unit wavelength (1 μm) per second into a unit solid angle from a unit area (m^2) on the surface of the Sun is shown in Joules. To convert to the amount of radiation in Watts per 0.01 μm wavelength band reaching a 1 m^2 area outside of the Earth's atmosphere, multiply by 6.8×10^{-7}. According to Cox (ed.) 2000, Allen's, Astrophysical Quantities.

Wavelength (μm)	0.20	0.22	0.24	0.30	0.34	0.37	0.39	0.40	0.42	0.44
log (radiation energy)	5.00	5.85	5.90	6.92	7.13	7.22	7.18	7.34	7.40	7.43
Wavelength (μm)	0.46	0.50	0.60	0.8	1.0	1.4	2.0	5.0	10.0	20.0
log (radiation energy)	7.48	7.45	7.42	7.23	7.05	6.71	6.23	4.74	3.56	2.36

Structure of the Outer Solar Atmosphere

Name	Radial distance from the center of the Sun (solar radius = 1.0)	Height from the reference plane (km)	log P_g (Pa)	T (K)	log N (m^{-3})	log N_e (m^{-3})	τ_5
Photosphere	1.000	-50	4.19	7900	23.11	20.84	4.13
	1.000	0	4.07	6520	23.07	19.89	1
	1.000	125	3.68	5270	22.78	18.89	0.14
	1.000	250	3.24	4880	22.37	18.42	0.02
	1.001	400	2.66	4560	21.82	17.87	2.0×10^{-3}
	1.001	525	2.14	4400	21.32	17.38	2.4×10^{-4}
Chromosphere	1.001	855	0.93	5650	20.00	17.08	1.9×10^{-4}
	1.002	1278	-0.28	6390	18.73	16.90	4.0×10^{-6}
	1.002	1580	-0.96	6900	18.00	16.74	2.1×10^{-6}
	1.003	2017	-1.69	8400	17.08	16.63	5.4×10^{-7}
Corona-Interplanetary Space	1.01		-2.6	10^6	14.4	14.4	
	1.10		-2.9	10^6	14.0	14.0	
	1.4		-3.7	10^6	13.2	13.2	
	3		-5.4	10^6	11.5	11.5	
	20				9.2	9.2	
	215 (1 astronomical units)			2×10^5	6.7	6.7	

P_g : Gas pressure, N : Number of neutral and ionized atoms, N_e : Number of electrons, τ_5 : Optical depth at 0.5 μm. According to Cox (ed.) 2000, Allen's, Astrophysical Quantities.

Structure of the Solar Interior

The height with τ_5=1 is used as the reference plane.

Radius (Solar radius = 1.0)	Pressure (10^{14} Pa)	Temperature (10^6 K)	Density (10^3 kg m^{-3})	Fraction of mass (Mass of the Sun = 1.0)	Luminosity (Solar luminosity = 1.0)	Hydrogen content (mass ratio)
0.0	240	15.8	156	0.0	0.0	0.333
0.1	137	13.2	88	0.08	0.46	0.537
0.2	43	9.4	35	0.35	0.94	0.678
0.3	10.9	6.8	12.0	0.61	1.0	0.702
0.4	2.7	5.1	3.9	0.79	1.0	0.707
0.6	0.21	3.1	0.50	0.94	1.0	0.712
0.7	0.065	2.3	0.20	0.97	1.0	0.728
0.8	0.017	1.37	0.09	0.99	1.0	0.735
1.0	1.3×10^{-10}	0.0064	2.7×10^{-7}	1.00	1.0	0.735

Bahcall & Pinsonneault 1995, Rev. Mod. Phys., **67**, 781

Major Solar Absorption Lines

Strong lines, primarily from "The second revision of Roland's tables" are shown. Intensity is given by the equivalent width with a unit of 10^{-4} nm. * indicates that blends with other elements exist. * * indicates a line used for the magnetic field measurements. The symbols are based on Cox (ed.) 2000, Allen's, Astrophysical Quantities.

Symbol	Wavelength (nm)	Element	Intensity	Symbol	Wavelength (nm)	Element	Intensity	Symbol	Wavelength (nm)	Element	Intensity
	279.54	Mg$^+$			370.926	Fe*	573		407.772	Sr^{+*}	428
	280.23	Mg$^+$			371.995	Fe	1664	Hδ	410.175	H*	3133
	285.16	Mg		M	373.487	Fe	3027		413.207	Fe*	404
	288.11	Si			373.714	Fe	1071		414.388	Fe	466
	306.726	Fe*	663		374.557	Fe*	1202		416.728	Mg	200
	313.412	Ni*	414		374.827	Fe	497		420.204	Fe	326
	324.201	Ti$^+$	270		374.950	Fe	1907	g	422.674	Ca	1476
	324.757	Cu	246		375.825	Fe	1647		423.595	Fe*	385
	333.669	Mg	416		375.930	Ti$^+$	334		425.013	Fe*	342
	341.478	Ni	816		376.380	Fe	829		425.080	Fe*	400
	343.358	Ni*	492		376.720	Fe	820		425.435	Cr*	393
O {	344.063	Fe	1243		378.789	Fe	512		426.049	Fe	595
	344.102	Fe	634		379.501	Fe*	547		427.177	Fe	756
	344.388	Fe	655		380.672	Fe*	209		432.578	Fe*	793
	344.627	Ni	470		381.585	Fe	1272	Hγ	434.048	H	2855
	345.847	Ni	656	L	382.044	Fe	1712	d	438.356	Fe	1008
	346.167	Ni	758		382.589	Fe	1519		440.476	Fe	898
	347.546	Fe	622		382.783	Fe	897		441.514	Fe*	417
	347.671	Fe	465		382.937	Mg	874		452.863	Fe*	275
	349.059	Fe	830		383.231	Mg	1685		455.404	Ba$^+$	159
	349.298	Ni	826		383.423	Fe	624		470.300	Mg	326
	349.784	Fe	726		383.830	Mg	1920	Hβ	486.134	H	3680
	351.033	Ni	489		384.045	Fe	567		489.150	Fe	312
	351.507	Ni	718		384.106	Fe*	517		492.051	Fe*	471
	352.127	Fe	381		384.998	Fe	608		495.761	Fe*	696
	352.454	Ni	1271		385.638	Fe	648	b_4	516.733	Mg*	935
	355.494	Fe	404		385.992	Fe	1554	b_2	517.270	Mg	1259
	355.853	Fe*	485		387.803	Fe	555	b_1	518.362	Mg	1584
	356.540	Fe	990		388.629	Fe	920		525.022	Fe**	62
	356.638	Ni	458		389.972	Fe	436	E	526.955	Fe*	478
	357.013	Fe	1380		390.296	Fe*	530		532.805	Fe	375
	357.869	Cr	488		390.553	Si	816		552.842	Mg	293
N	358.121	Fe	2144		392.027	Fe	341	D_2	588.997	Na*	752
	358.699	Fe	532		392.292	Fe*	414	D_1	589.594	Na	564
	359.350	Cr	436		392.793	Fe	187		610.273	Ca	135
	360.887	Fe	1046		393.031	Fe	108		612.223	Ca	222
	361.878	Fe	1410	K	393.368	Ca^{+*}	20253		616.218	Ca	222
	361.940	Ni	568		394.402	Al	488		630.250	Fe**	83
	363.148	Fe*	1364		396.154	Al	621	Hα	656.281	H	4020
	364.785	Fe*	970	H	396.849	Ca^{+*}	15467		849.806	Ca$^+$	1470
	367.992	Fe*	448		404.583	Fe	1174		854.214	Ca$^+$	3670
	368.520	Ti$^+$	275		406.361	Fe*	787		866.217	Ca$^+$	2600
	370.558	Fe	562		407.175	Fe	723				

Major Solar UV Spectral Lines and Continua

Heroux & Hinteregger 1978, J. Geophy. Res., **83**, 5305

Wavelength(nm)	Element	Intensity	Wavelength(nm)	Element	Intensity	Wavelength(nm)	Element	Intensity
5–10		0.11	70–90		0.18	130.603	O I	0.019
10–40		1.2	70.336	O III	0.010	133.453	C II	0.027
25.63	He II, Si X	0.036	76.515	N IV	0.004	133.570	C II	0.037
28.415	Fe XV	0.015	77.041	Ne VIII	0.007	139.376	Si IV	0.019
30.331	Si XI	0.052	78.936	O IV	0.011	140.277	Si IV	0.013
30.378	HeII(Lyα)	0.45	90–110		0.39	150–194		33
36.807	MgIX	0.035	97.702	C III	0.089	154.819	C IV	0.049
40–70		0.22	102.572	H I(Lyβ)	0.068	155.077	C IV	0.025
46.522	Ne VII	0.012	103.191	O VI	0.040	156.10	C I	0.032
55.437	O IV	0.026	110–150		5.1	165.72	C I	0.10
58.433	He I	0.044	121.567	H I(Lyα)	4.1	180.801	Si II	0.10
60.976	Mg X	0.017	130.217	O I	0.017	181.645		0.060
62.973	O V	0.050	130.486	O I	0.017	181.693		0.15

The measured intensities outside of the Earth's atmosphere (10^{-3} W m^{-2} or mW m^{-2}) are summarized.

Major Chromospheric Spectral Lines

Dunn *et al.* 1968, ApJS, **15**, 275

Wavelength(nm)	Element	Intensity	Wavelength(nm)	Element	Intensity	Wavelength(nm)	Element	Intensity
368.519	Ti^+	90	383.829	Mg	60	518.360	Mg	65
369.156	H(H18)	29	388.905	$H_{H\zeta}$	381	587.565	He(D_3)	994
369.715	H(H17)	35	393.366	Ca^+	1100	656.282	H(Hα)	4740
370.386	H(H16)	43	396.847	Ca^+	1000	706.518	He	138
371.197	H(H15)	53	397.007	H(Hε)	306	777.196	O	91
372.194	H(H14)	73	402.636	He	24	777.418	O	75
373.437	H(H13)	99	407.771	Sr^+	75	777.540	O	53
375.015	H(H12)	108	410.174	H(Hδ)	459	849.802	Ca^+	380
375.929	Ti^+	90	421.552	Sr^+	51	854.209	Ca^+	1000
376.132	Ti^+	82	422.673	Ca	22	854.538	H(P15)	23
377.063	H(H11)	116	424.683	Sc^+	18	859.839	H(P14)	26
379.790	H(H10)	157	434.047	H(Hγ)	505	866.214	Ca^+	800
381.961	He	5	447.169	He	121	866.502	H(P13)	34
382.043	Fe	14	468.568	He^+	2	875.047	H(P12)	46
382.936	Mg	20	471.314	He	9	1082.91	He	
383.230	Mg	46	486.133	H(Hβ)	1630	1083.03	He	
383.539	H(H9)	228	501.567	He	6			

The intensity is the value integrated over a height higher than 1000 km above the photosphere. The unit is 10^6 W m^{-1} sr^{-1} or MW m^{-1} sr^{-1}.

Major Coronal Spectral Lines

Cox (ed.) 2000, Allen's Astrophysical Quantities

Wavelength(nm)	Element	Intensity	Wavelength(nm)	Element	Intensity	Wavelength(nm)	Element	Intensity
332.9	Ca XII	0.7	423.20	Ni XII	1.1	637.45	Fe X	5
338.82	Fe XIII	10	425.64	K XI	0.1	670.19	Ni XV	1.2
353.40	V X	1	435.10	Co XV	0.1	674.0	K XIV	0.1
360.09	Ni XVI	1.3	441.24	Ar XIV	0.3	705.96	Fe XV	0.8
364.28	Ni XIII	0.4	456.66	Cr IX	0.5	789.19	Fe XI	6
368.5	Mn XII	0.2	511.60	Ni XIII	0.8	802.42	Ni XV	0.3
380.07	Co XII	0.5	530.28	FeXIV	20	1074.68	Fe XIII	50
398.71	Fe XI	0.7	544.55	Ca XV	0.2	1079.79	Fe XIII	30
399.8	Cr XI	0.1	553.6	Ar X	0.3	*1252.49	S IX	–
408.65	Ca XIII	0.4	569.44	Ca XV	0.3	*1430.8	Si X	–

Intensity is in units for the equivalent width of 10^{-1} nm.

* indicates a line from Kuhn *et al.* 1996, ApJ, **456**, L67

Classification of Sunspot Groups

The Zurich Sunspot Classification (Left is to the west on the solar surface.)

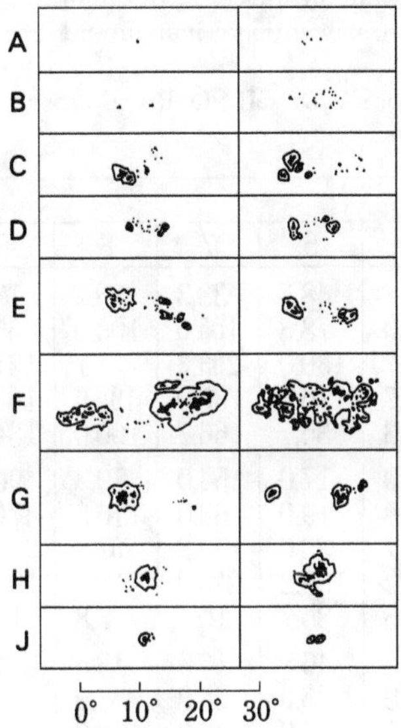

Figure 5 Classification of sunspot groups

A: Composed of a small single spot or a very small group of spots, mostly of short duration, concentrated in a region of 2–3 square degrees. No systematic structure of the group; spots without penumbra.

B: Bipolar group of spots without penumbra, the long axis of which is directed roughly east-west. Concentration of spots on east and west ends.

C: Bipolar group like B, but at least one main spot with penumbra.

D: Bipolar group, the largest spots showing penumbrae. Dimension of the group in longitude within 10°.

E: Large bipolar group showing a complicated structure, the two major spots each having a penumbra. Numerous small spots between the major spots. Dimension of the group in longitude at least 10°.

F: Very large bipolar or complex group. Dimension in longitude at least 15°.

G: Large bipolar group, without small spots between the two major spots. Dimension in longitude at least 10°.

H: Unipolar spot with penumbrae, sometimes with complicated structure. Diameter > 2.5°.

J: Unipolar spot with penumbra. Round shape, diameter < 2.5°.

(Based on Kiepenheuer's description (1953) of the classification by Waldmeier)

Latest Total Sunspot Numbers

This table shows the definitive sunspot numbers from WDC-SILSO, Royal Observatory of Belgium, Brussels.

Months	2013	2014	2015	2016	2017	2018	2019	2020
Jan	96.1	117.0	93.0	57.0	26.1	6.8	7.7	6.2
Feb	60.9	146.1	66.7	56.4	26.4	10.7	0.8	0.2
Mar	78.3	128.7	54.5	54.1	17.7	2.5	9.4	1.5
Apr	107.3	112.5	75.3	37.9	32.3	8.9	9.1	*5.4*
May	120.2	112.5	88.8	51.5	18.9	13.1	9.9	*0.2*
Jun	76.7	102.9	66.5	20.5	19.2	15.6	1.2	*5.8*
Jul	86.2	100.2	65.8	32.4	17.8	1.6	0.9	*6.3*
Aug	91.8	106.9	64.4	50.2	32.6	8.7	0.5	*7.6*
Sep	54.5	130.0	78.6	44.6	43.7	3.3	1.1	
Oct	114.4	90.0	63.6	33.4	13.2	4.9	0.4	(Provisional numbers are written in italics.)
Nov	113.9	103.6	62.2	21.4	5.7	4.9	0.5	
Dec	124.2	112.9	58.0	18.5	8.2	3.1	1.5	
Yearly mean	94.0	113.3	69.8	39.8	21.7	7.0	3.6	

Yearly mean = sum of the daily total sunspot number in the year/ number of days in the year

Yearly Mean Total Sunspot Numbers

The total sunspot number R, representing the amount of sunspots, is calculated by a formula, $R=k(10g+f)$, where g is the number of sunspot groups, f is the overall number of sunspots, and k is the factor depending on the observational instrument/observer/etc.

This table shows the definitive sunspot numbers from WDC-SILSO, Royal Observatory of Belgium, Brussels.

Decade	Year									
	0	1	2	3	4	5	6	7	8	9
1700	8.3	18.3	26.7	38.3	60.0	**96.7**	48.3	33.3	16.7	13.3
1710	5.0	0.0	0.0	3.3	18.3	45.0	78.3	**105.0**	100.0	65.0
1720	46.7	43.3	36.7	18.3	35.0	66.7	130.0	**203.3**	171.7	121.7
1730	78.3	58.3	18.3	8.3	26.7	56.7	116.7	135.0	**185.0**	168.3
1740	121.7	66.7	33.3	26.7	8.3	18.3	36.7	66.7	100.0	134.8
1750	**139.0**	79.5	79.7	51.2	20.3	16.0	17.0	54.0	79.3	90.0
1760	104.8	**143.2**	102.0	75.2	60.7	34.8	19.0	63.0	116.3	**176.8**
1770	168.0	136.0	110.8	58.0	51.0	11.7	33.0	154.2	**257.3**	209.8
1780	141.3	113.5	64.2	38.0	17.0	40.2	138.2	**220.0**	218.2	196.8
1790	149.8	111.0	100.0	78.2	68.3	35.5	26.7	10.7	6.8	11.3
1800	24.2	56.7	75.0	71.8	**79.2**	70.3	46.8	16.8	13.5	4.2
1810	0.0	2.3	8.3	20.3	23.2	59.0	**76.3**	68.3	52.9	38.5
1820	24.2	9.2	6.3	2.2	11.4	28.2	59.9	83.0	108.5	115.2
1830	**117.4**	80.8	44.3	13.4	19.5	85.8	192.7	**227.3**	168.7	143.0
1840	105.5	63.3	40.3	18.1	25.1	65.8	102.7	166.3	**208.3**	182.5
1850	126.3	122.0	102.7	74.1	39.0	12.7	8.2	43.4	104.4	178.3
1860	**182.2**	146.6	112.1	83.5	89.2	57.8	30.7	13.9	62.8	123.6
1870	**232.0**	185.3	169.2	110.1	74.5	28.3	18.9	20.7	5.7	10.0
1880	53.7	90.5	99.0	**106.1**	105.8	86.3	42.4	21.8	11.2	10.4
1890	11.8	59.5	121.7	**142.0**	130.0	106.6	69.4	43.8	44.4	20.2
1900	15.7	4.6	8.5	40.8	70.1	**105.5**	90.1	102.8	80.9	73.2
1910	30.9	9.5	6.0	2.4	16.1	79.0	95.0	**173.6**	134.6	105.7
1920	62.7	43.5	23.7	9.7	27.9	74.0	106.5	114.7	**129.7**	108.2
1930	59.4	35.1	18.6	9.2	14.6	60.2	132.8	**190.6**	182.6	148.0
1940	113.0	79.2	50.8	27.1	16.1	55.3	154.3	**214.7**	193.0	190.7
1950	118.9	98.3	45.0	20.1	6.6	54.2	200.7	**269.3**	261.7	225.1
1960	159.0	76.4	53.4	39.9	15.0	22.0	66.8	132.9	**150.0**	149.4
1970	148.0	94.4	97.6	54.1	49.2	22.5	18.4	39.3	131.0	**220.1**
1980	218.9	198.9	162.4	91.0	60.5	20.6	14.8	33.9	123.0	**211.1**
1990	191.8	203.3	133.0	76.1	44.9	25.1	11.6	28.9	88.3	136.3
2000	**173.9**	170.4	163.6	99.3	65.3	45.8	24.7	12.6	4.2	4.8
2010	24.9	80.8	84.5	94.0	**113.3**	69.8	39.8	21.7	7.0	3.6

The **bold** face shows the maxima.

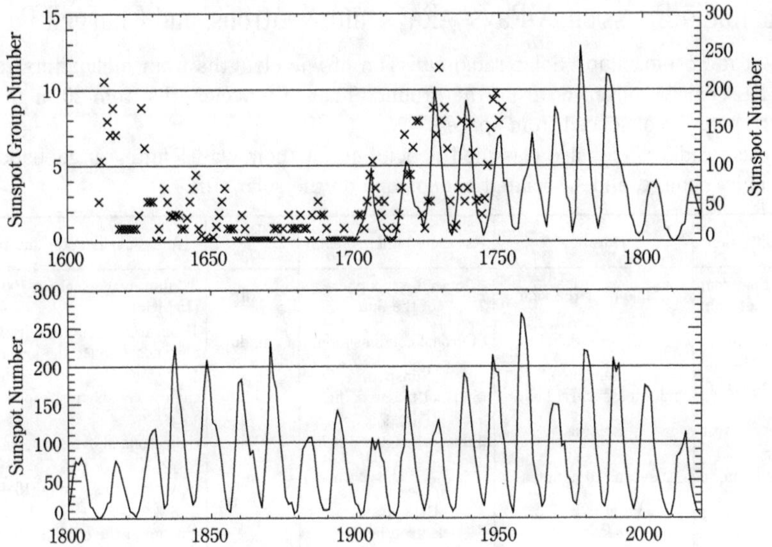

Figure 6 Sunspot number

The solid line shows the sunspot number, and the × symbols represent the number of sunspot groups. These are the definitive numbers from WDC-SILSO, Royal Observatory of Belgium, Brussels.

Latest Intense Flares

Date		Time (UT) h	m	Lat. °	Long. °	X-ray flux
2001						
4	2	22	03			X20.0
4	6	19	31	S 21	E 31	X 5.6
4	15	13	55	S 20	W85	X14.4
8	25	17	04	S 17	E 34	X 5.3
12	13	14	35	N 16	E 09	X 6.2
12	28	21	32			X 3.4
2002						
7	15	20	14	N 19	W01	X 3.0
7	20	21	54			X 3.3
7	23	00	47	S 13	E 72	X 4.8
8	24	01	31	S 02	W81	X 3.1
2003						
5	28	00	39			X 3.6
10	23	08	49	S 21	E 88	X 5.4
10	28	11	24	S 16	E 08	X17.2
10	29	21	01	S 15	W02	X10.0
11	2	17	39	S 14	W56	X 8.3
11	3	10	19	N 08	W77	X 3.9
11	4	20	06	S 19	W83	X28.0
2004						
7	16	14	01	S 10	E 35	X 3.6
2005						
1	17	10	07	N 15	W25	X 3.8
1	20	07	26	N 14	W61	X 7.1
9	7	18	03	S 11	E 77	X17.0
9	8	21	17	S 12	E 75	X 5.4
9	9	10	08	S 11	E 66	X 3.6
9	9	20	36	S 12	E 67	X 6.2
2006						
12	5	10	45	S 07	E 68	X 9.0
12	6	19	00	S 05	E 64	X 6.5
12	13	02	57	S 06	W23	X 3.4
2011						
8	9	08	08			X 6.9
2012						
3	7	00	40	N 17	E 27	X 5.4
2013						
5	14	01	20			X 3.2
11	5	22	15	S 13	E 44	X 3.3
2014						
2	25	01	03	S 12	E 82	X 4.9
10	24	22	13			X 3.1
2017						
9	6	12	02	S 08	W33	X 9.3
9	10	16	06			X 8.2

The flares of which flux is X3.0 or more after 2001 are listed above. (ftp://ftp.ngdc.noaa.gov/STP/space-weather/solar-data/solar-features/solar-flares/x-rays/goes/ and ftp://ftp.swpc.nasa.gov/pub/warehouse/)

Solar Radio Emission, X-Rays, γ-Rays and Neutrons, and Charged Particles

Solar radio emission Solar radio emission of wavelengths from millimeters to 20 m (15 MHz) can be observed from the ground. That of wavelengths from 20 m to 10 km (30 kHz) can be observed from space.

The solar radio waves are classified depending on their wavelength range, bandwidth, frequency change, and associated phenomena on the solar surface.

Classification	Duration	Associated phenomenon	Wavelength range	Apparent diameter and position
Radiation from the quiet Sun	Stable over the year	Ionized atmosphere of the Sun	All	32′ (mm waves)–about 100′ (15 MHz)
S-component	Several days	Coronal condensation	cm, dm	1′–several arcminutes, from the chromosphere to the lower corona
Noise Storm	tens of minutes–several hours	Sunspots (Sunspot magnetic field)	m	1′–4′, the corona above sunspots
Type I burst	Seconds	Sunspots (Sunspot magnetic field)	m	1′–4′, the corona above sunspots
Type II burst	Several–10 minutes	Magnetic shock wave	m	6′–12′, moving in the corona (10^3 km s^{-1})
Type III burst	several to tens of seconds	High-speed electron beam	m	3′–8′, moving in the corona (10^5 km s^{-1})
Type IV burst	Several to tens of minutes	flare	m, dm	4′–20′, the corona
Microwave burst	Several tens of seconds–several minutes	flare	cm	<1′–2′, the low corona

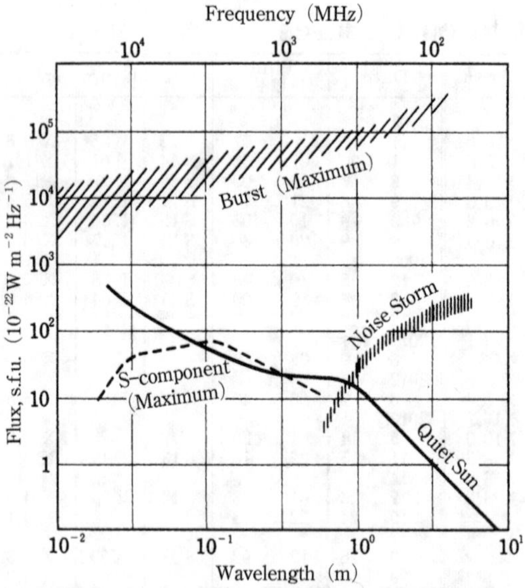

Figure 7 Spectra of solar radio emission

Radio flux density F is expressed as

$$F = \frac{2kT_b}{\lambda^2}\,\Omega\ (\mathrm{W\ m^{-2}\ Hz^{-1}})$$

where

k : The Boltzmann constant ($\mathrm{J\ K^{-1}}$),

T_b : Brightness temperature of the radio source (K),

λ : Wavelength (m),

Ω : Solid angle of the radio source (steradian)

T_b is calculated using Ω obtained with a radio interferometer etc. T_b reaches several million degrees at S-components, and exceeds 100 million degrees at bursts.

X-ray flares

Energy range	Spectrum	Flux (C-X class flares)	Duration	Size of X-ray source	Remarks
Soft X-rays (0.1–10 keV)	Exponential	10^{-6}–10^{-3} W m^{-2}	a few minutes–several hours	Several arcseconds–2′	Thermal emission[1)]
Hard X-rays (10 keV or more)	Power-law	10^{-10}–10^{-6} W m^{-2}	tens of seconds–several minutes[2)]	<1′	[3)]

Note: 1) This comes from the high temperature region formed in the low corona during flares. The temperature, the volume emission measure n_e^2V, the electron density n_e of this region are 10^7–3×10^7 K, 10^{51}–10^{53} m^{-3}, and 10^{16}–10^{17} m^{-3}, respectively.

2) This corresponds to the early phase of Hα flares and soft X-ray bursts, when the intensity increases most rapidly.

3) The spectrum in the range of 10–300 keV are the power-law type in many cases, but ones of the exponential type are also seen; there are significant differences among flares.

Solar γ-rays and neutrons High-energy protons and atomic nuclei accelerated in flares produce various kinds of excited atomic nuclei and neutrons through collisions with other nuclei. Prompt γ-rays are emitted from excited ^{12}C (4.44 MeV) and ^{16}O (6.13 MeV), and the neutron capture γ-rays are emitted from thermalized neutrons with relatively low energy through the process that

$$n+p \rightarrow d+\gamma \ (2.223 \text{ MeV}).$$

These γ-rays are observed along with the Bremsstrahlung γ-rays produced by high-energy electrons.

Neutrons with relatively high energy, on the other hand, survive until reaching the Earth, even though their lifetime is as short as about 900 s. They carry information on the original particle acceleration unperturbedly. Neutrons with the energy over 100 MeV are detected by ground-based monitoring instruments.

Solar charged particles

Particle	Flux (number m^{-2} s^{-1})	Time lag[1)]	Longitude of the source[2)]	Associated phenomena
Electron	10^5–10^8(>40 keV)	15 minutes–several hours	30°E–90°W	Type III bursts, early phase μ wave bursts, Hard X-ray flares
Ion	10–10^6(>10 MeV)	10 minutes–several hours	60°E–>90°W	Type II and Type IV bursts, Hard X-ray flares, Relativistic electrons

Note: 1) The time-lag between the start or the maximum of the corresponding Hα flare (on the solar surface) and the start of the particle increase near the Earth.

2) The longitude of the corresponding Hα flare or the microwave source on the solar surface.

Constellations

Constellation name	Abbr.	Meaning	Rough position		Culmination	
			R.A.	Dec.		
			h m	°	Month	Time of month
Andromeda (e)	And	the Chained Maiden	00 40	+38	11	Late
Antlia (e)	Ant	the Air Pump	10 00	-35	4	Mid
Apus (odis)	Aps	the Bird of Paradise	16 00	-76	7	Mid
*Aquarius (i)	Aqr	the Water Bearer	22 20	-13	10	Late
Aquila (e)	Aql	the Eagle	19 30	+02	9	Early
Ara (e)	Ara	the Altar	17 10	-55	8	Early
*Aries (tis)	Ari	the Ram	02 30	+20	12	Late
Auriga (e)	Aur	the Charioteer	06 00	+42	2	Mid
Boötes (is)	Boo	the Herdsman	14 35	+30	6	Late
Caelum (i)	Cae	the Engraving Tool	04 50	-38	1	Late
Camelopardalis	Cam	the Giraffe	05 40	+70	2	Mid
*Cancer (ri)	Cnc	the Crab	08 30	+20	3	Late
Canes (um) Venatici (orum)	CVn	the Hunting Dogs	13 00	+40	6	Early
Canis Major (is)	CMa	the Great Dog	06 40	-24	2	Late
Canis Minor (is)	CMi	the Lesser Dog	07 30	+06	3	Mid
*Capricornus (i)	Cap	the Sea Goat	20 50	-20	9	Late
Carina (e)	Car	the Keel	08 40	-62	3	Late
Cassiopeia (e)	Cas	the Seated Queen	01 00	+60	12	Early
Centaurus (i)	Cen	the Centaur	13 20	-47	6	Early
Cepheus (i)	Cep	the King	22 00	+70	10	Mid
Cetus (i)	Cet	the Sea Monster	01 45	-12	12	Mid
Chamaeleon (tis)	Cha	the Chameleon	10 40	-78	4	Late
Circinus (i)	Cir	the Compass	14 50	-63	6	Late
Columba (e)	Col	the Dove	05 40	-34	2	Early
Coma (e) Berenices	Com	the Bernice's Hair	12 40	+23	5	Late
Corona (e) Australis	CrA	the Southern Crown	18 30	-41	8	Late
Corona (e) Borealis	CrB	the Northern Crown	15 40	+30	7	Mid
Corvus (i)	Crv	the Crow	12 20	-18	5	Late
Crater (is)	Crt	the Cup	11 20	-15	5	Early
Crux (cis)	Cru	the Southern Cross	12 20	-60	5	Late
Cygnus (i)	Cyg	the Swan	20 30	+43	9	Late
Delphinus (i)	Del	the Dolphin	20 35	+12	9	Late
Dorado (us)	Dor	the Swordfish	05 00	-60	1	Late
Draco (nis)	Dra	the Dragon	17 00	+60	8	Early
Equuleus (i)	Equ	the Little Horse	21 10	+06	10	Early
Eridanus (i)	Eri	the River	03 50	-30	1	Mid
Fornax (cis)	For	the Furnace	02 25	-33	12	Late
*Gemini (orum)	Gem	the Twins	07 00	+22	3	Early
Grus (is)	Gru	the Crane	22 20	-47	10	Late
Hercules (is)	Her	the Hercules	17 10	+27	8	Early
Horologium (i)	Hor	the Clock	03 20	-52	1	Early
Hydra (e)	Hya	the Female Water Snake	10 30	-20	4	Late
Hydrus (i)	Hyi	the Male Water Snake	02 40	-72	12	Late
Indus (i)	Ind	the Indian	21 20	-58	10	Early
Lacerta (e)	Lac	the Lizard	22 25	+43	10	Late
*Leo (nis)	Leo	the Lion	10 30	+15	4	Late
Leo (nis) Minor (is)	LMi	the Lesser Lion	10 20	+33	4	Late
Lepus (oris)	Lep	the Hare	05 25	-20	2	Early
*Libra (e)	Lib	the Scales	15 10	-14	7	Early
Lupus (i)	Lup	the Wolf	15 00	-40	7	Early
Lynx (cis)	Lyn	the Lynx	07 50	+45	3	Mid
Lyra (e)	Lyr	the Lyre	18 45	+36	8	Late

Continued.

Constellation name	Abbr.	Meaning	Rough position R.A.	Rough position Dec.	culmination	
			h m	°	Month	Time of month
Mensa (e)	Men	the Table Mountain	05 40	-77	2	Early
Microscopium (i)	Mic	the Microscope	20 50	-37	9	Late
Monoceros (tis)	Mon	the Unicorn	07 00	-03	3	Early
Musca (e)	Mus	the Fly	12 30	-70	5	Late
Norma (e)	Nor	the Carpenter's Square	16 00	-50	7	Mid
Octans (tis)	Oct	the Octant	21 00	-87	10	Early
Ophiuchus (i)	Oph	the Serpent Bearer	17 10	-04	8	Early
Orion (is)	Ori	the Hunter	05 20	+03	2	Early
Pavo (nis)	Pav	the Peacock	19 10	-65	9	Early
Pegasus (i)	Peg	the Winged Horse	22 30	+17	10	Late
Perseus (i)	Per	the Hero	03 20	+42	1	Early
Phoenix (cis)	Phe	the Phoenix	01 00	-48	12	Early
Pictor (is)	Pic	the Painter's Easel	05 30	-52	2	Early
*Pisces (ium)	Psc	the Fishes	00 20	+10	11	Late
Piscis Austrinus (i)	PsA	the Southern Fish	22 00	-32	10	Mid
Puppis	Pup	the Stern	07 40	-32	3	Mid
Pyxis (dis)	Pyx	the Compass	08 50	-28	3	Late
Reticulum (i)	Ret	the Reticle	03 50	-63	1	Mid
Sagitta (e)	Sge	the Arrow	19 40	+18	9	Mid
*Sagittarius (i)	Sgr	the Archer	19 00	-25	9	Early
*Scorpius (i)	Sco	the Scorpion	16 20	-26	7	Late
Sculptor (is)	Scl	the Sculptor	00 30	-35	11	Late
Scutum (i)	Sct	the Shield	18 30	-10	8	Late
Serpens (tis)	Ser	the Serpent	15 35 18 00	+08 -05	7 8	Mid Mid
Sextans (tis)	Sex	the Sextant	10 10	-01	4	Mid
*Taurus (i)	Tau	the Bull	04 30	+18	1	Late
Telescopium (i)	Tel	the Telescope	19 00	-52	9	Early
Triangulum (i)	Tri	the Triangle	02 00	+32	12	Mid
Triangulum (i) Australe (is)	TrA	the Southern Triangle	15 40	-65	7	Mid
Tucana (e)	Tuc	the Toucan	23 45	-68	11	Mid
Ursa (e) Major (is)	UMa	the Great Bear	11 00	+58	5	Early
Ursa (e) Minor (is)	UMi	the Little Bear	15 40	+78	7	Mid
Vela (orum)	Vel	the Sails	09 30	-45	4	Early
*Virgo (inis)	Vir	the Maiden	13 20	-02	6	Early
Volans (tis)	Vol	the Flying Fish	07 40	-69	3	Mid
Vulpecula (e)	Vul	the Fox	20 10	+25	9	Mid

Out of the multitude of mutiple constellations used over the ages, the present constellations were fixed in the book 'DÉLIMITATION SCIENTIFIQUE DES CONSTELLATIONS' (written by E. Delporte) published by the International Astronomical Union (IAU) in 1930, which was based on discussions at the 3rd IAU general assembly held in 1928. The boundaries of the constellations use the right ascensions (R.A.) and declinations (Dec.) for the year 1875.0 equinox. A total of 88 constellations are provided.

Constellations marked with a * in the table are zodiacal constellations. They are closely assoicated with the 12 zodiac signs known from ancient times in astrology. The scientific name of the constellation is shown in Latin. When indicating individual stars, they are named with a Greek or Roman letter followed by the latin name with a suffix. In the table, suffixes are shown in ( ). The suffix is omitted when using the constellation name abbreviation. Most abbreviations consist of 3 letters.

Example Ursa Major's α star → α Ursae Majoris or α UMa (abbreviated form)
Corona Borealis's R star → R Coronae Borealis or R CrB (abbreviated form).

Moreover, the column "culmination" indicates the season when the outline of the constellation crosses over the meridian of Tokyo (UTC+09:00) at 8:00 pm.

Stars

Magnitude The apparent brightness of a star is expressed in magnitude. An increase by $\sqrt[5]{100}=2.512$ times in the amount of light corresponds to a decrease of 1 in the value of the magnitude. That is, the magnitude m of a star with luminosity L_m and the magnitude n of a star with luminosity L_n have the following relation:

$$n-m=2.5\log(L_m/L_n) \text{ (Pogson's equation)}$$

The zero-point of stellar magnitude is defined according to non-variable standard stars. The magnitudes given in the table are visual magnitudes. The data with a suffix 'd' denote the combined magnitude of double stars with angular separation of less than 1′ (almost corresponding to the resolving power of naked eye). In this case, the data of the brightest component are presented for proper motion, distance, and radial velocity. Regarding stars showing appreciable photometric variability, the brightest magnitude is given here with a suffix 'v'. Refer to page **47**, "Distance and Absolute Magnitude."

Color index The color of a star is expressed by the magnitude difference between two wavelength bands: $B-V$ is the difference between the B-band magnitude (blue region) and the V-band magnitude (visual/yellow region); $U-B$ is the difference between the magnitudes in the U-band (near-ultraviolet region) and the B-band. Color indices of blue/hot stars tend to be smaller and negative, while those of red/cool stars are larger and positive. The $B-V$ values of reddest stars can reach more than +1. Refer to page **46**, "Magnitude System and Interstellar Absorption" and page **48**, "Physical Properties of Stars".

Spectral type (Refer to page **38**, "Spectral Types of Stars")

Proper motion Proper motion is the systematic movement of a star on the celestial sphere, which stems from its intrinsic spatial motion relative to us in the solar system. In the table, proper motion is represented by $\mu_\alpha\cos\delta$ (components in right ascension) and μ_δ (component in declination). These data are taken from the Rivised Hipparcos Star Catalogue (2007).

Distance The distance of a star is usually determined by measuring its annual parallax (the angle subtended by the semi-major axis of the Earth's orbit as seen from the star) as long as it is not too far. Alternatively, the absolute magnitude (guessed from the luminosity class of stellar spectral classification) can be used to estimate the distance by comparing it with the apparent magnitude. The values given in the table are expressed in light-years (ly), which are based on the annual parallaxes and taken from the Revised Hipparcos Star Catalogue (2007). Data with colon (:) are uncertain values where the expected error exceeds 10%. Refer to page **47**, "Distance and Absolute Magnitude."

Radial velocity Radial velocity is the line-of-sight component of the relative velocity vector between the observer and the target, which is spectroscopically determinable by measuring the shift of spectral lines. Conventionally, the sign of radial velocity is taken positive (+) in cases where a target is receding from the observer, while negative (−) for approaching cases. In the table, "V" means that the star's radial velocity is variable (e.g., pulsating stars), while "B" denotes the case where the variability is known to originate from the orbital motion of the binary system. Time-averaged values are given for these radial velocity variables.

Note: The values for the right ascension (α), declination (δ), magnitude, color index, spectral type, and radial velocity are based on "The Bright Star Catalogue, 5th Revised Edition" (Hoffleit & Warren, 1991).

Stars

Star name	2000 Equinox		Visual magnitude	Color index		Spectral type	Proper motion		Distance	Radial velocity
	α	δ		$B-V$	$U-B$		$\mu_\alpha \cos\delta$	μ_δ		
	h m	° ′	mag	mag	mag		10^{-3}(″ year^{-1})	10^{-3}(″ year^{-1})	ly	km s^{-1}
α And	00 08.4	+29 05	2.1	-0.11	-0.46	B8IVp	+137	-163	97	-12B
β Cas	00 09.2	+59 09	2.3	+0.34	+0.11	F2III-IV	+524	-180	55	+12B
β Hyi	00 25.8	-77 15	2.8	+0.62	+0.11	G2IV	+2220	+324	24	+23
α Phe	00 26.3	-42 18	2.4	+1.09	+0.88	K0III	+233	-356	85	+75B
α Cas	00 40.5	+56 32	2.2	+1.17	+1.13	K0IIIa	+51	-32	228	-4V?
β Cet	00 43.6	-17 59	2.0	+1.02	+0.87	G9.5III	+233	+32	96	+13V?
γ Cas	00 56.7	+60 43	2.5	-0.15	-1.08	B0IVe	+25	-4	549	-7B
β And	01 09.7	+35 37	2.1	+1.58	+1.96	M0IIIa	+176	-112	197	+3V
δ Cas	01 25.8	+60 14	2.7	+0.13	+0.12	A5III-IV	+297	-49	99	+7B
α Eri[1]	01 37.7	-57 14	0.5	-0.16	-0.66	B3Vpe	+87	-38	139	+16V
β Ari	01 54.6	+20 48	2.6	+0.13	+0.10	A5V	+99	-110	59	-2B
α Hyi	01 58.8	-61 34	2.9	+0.28	+0.14	F0V	+264	+27	72	+1V
γ And	02 03.9	+42 20	2.2d	+1.17	+0.87	K3IIb+(B8V+A0V)	+42	-49	393:	-12B
α Ari	02 07.2	+23 28	2.0	+1.15	+1.12	K2III	+189	-148	66	-14B
β Tri	02 09.5	+34 59	3.0	+0.14	+0.10	A5III	+149	-39	127	+10B
o Cet[2]	02 19.3	-02 59	3.0v	+1.42	+1.09	M7IIIe+Bep	+9	-237	299:	+64V
α UMi[3]	02 31.8	+89 16	2.0	+0.60	+0.38	F7Ib-II	+44	-12	433	-17B
α Cet	03 02.3	+04 05	2.5	+1.64	+1.94	M1.5IIIa	-10	-77	249	-26V
β Per[4]	03 08.2	+40 57	2.1v	-0.05	-0.37	B8V+G	+3	-2	90	+4B
α Per	03 24.3	+49 52	1.8	+0.48	+0.37	F5Ib	+24	-26	506	-2V
α Tau[5]	04 35.9	+16 31	0.9	+1.54	+1.90	K5III	+63	-189	67	+54B
ι Aur	04 57.0	+33 10	2.7	+1.53	+1.78	K3II	+7	-15	493	+18V
β Eri	05 07.8	-05 05	2.8	+0.13	+0.10	A3III	-83	-75	89	-9
β Ori[6]	05 14.5	-08 12	0.1	-0.03	-0.66	B8Ia	+1	+1	863	+21B
α Aur[7]	05 16.7	+46 00	0.1d	+0.80	+0.44	G5IIIe+G0III	+75	-427	43	+30B
γ Ori	05 25.1	+06 21	1.6	-0.22	-0.87	B2III	-8	-13	252	+18B?
β Tau	05 26.3	+28 36	1.6	-0.13	-0.49	B7III	+23	-174	134	+9V
β Lep	05 28.2	-20 46	2.8	+0.82	+0.46	G5II	-5	-86	160	-14V?
δ Ori	05 32.0	-00 18	2.2d	-0.22	-1.05	O9.5II+B2V	+1	-1	692:	+16B
α Lep	05 32.7	-17 49	2.6	+0.21	+0.23	F0Ib	+4	+1	2219	+24
ε Ori	05 36.2	-01 12	1.7	-0.19	-1.04	B0Ia	+1	-1	1977:	+26B
α Col	05 39.6	-34 04	2.6	-0.12	-0.46	B7IVe	+2	-25	261	+35V?
ζ Ori	05 40.8	-01 57	2.0	-0.21	-1.07	O9.7Ib	+3	+2	736:	+18B
κ Ori	05 47.8	-09 40	2.1	-0.17	-1.03	B0.5Ia	+1	-1	647	+21V?
α Ori[8]	05 55.2	+07 24	0.5	+1.85	+2.06	M1-2Ia-Iab	+28	+11	498:	+21B
β Aur	05 59.5	+44 57	1.9	+0.03	+0.05	A2IV	-56	-1	81	-18B
θ Aur	05 59.7	+37 13	2.6	-0.08	-0.18	A0p	+44	-74	166	+30B
β CMa	06 22.7	-17 57	2.0	-0.23	-0.98	B1II-III	-3	-1	493	+34B
α Car[9]	06 24.0	-52 42	-0.7	+0.15	+0.10	F0II	+20	+23	309	+21
γ Gem	06 37.7	+16 24	1.9	+0.00	+0.04	A0IV	+14	-55	109	-13B
α CMa[10]	06 45.1	-16 43	-1.5	+0.00	-0.05	A1Vm	-546	-1223	8.6	-8B
ε CMa	06 58.6	-28 58	1.5	-0.21	-0.93	B2II	+3	+1	405	+27
δ CMa	07 08.4	-26 24	1.8	+0.68	+0.54	F8Ia	-3	+3	1607:	+34B
π Pup	07 17.1	-37 06	2.7	+1.62	+1.24	K3Ib	-10	+6	807	+16
η CMa	07 24.1	-29 18	2.5	-0.08	-0.72	B5Ia	-4	+6	1989:	+41V
β CMi	07 27.1	+08 17	2.9	-0.09	-0.28	B8Ve	-52	-38	162	+22B

1) Achernar, 2) Mira, 3) Polaris, 4) Algol, 5) Aldebaran, 6) Rigel, 7) Capella, 8) Betelgeuse, 9) Canopus, 10) Sirius.

Stars Continued.

Star name	2000 Equinox		Visual magnitude	Color index		Spectral type	Proper motion		Distance	Radial velocity
	α	δ		$B-V$	$U-B$		$\mu_\alpha \cos\delta$	μ_δ		
	h m	° ′	mag	mag	mag		10^{-3}(″ year^{-1})	10^{-3}(″ year^{-1})	ly	km s^{-1}
α Gem[1]	07 34.6	+31 53	1.6d	+0.03	+0.01	A1V+A2Vm	-191	-145	51	+6B
α CMi[2]	07 39.3	+05 14	0.4	+0.42	+0.02	F5IV-V	-715	-1037	11	-3B
β Gem[3]	07 45.3	+28 02	1.1	+1.00	+0.85	K0IIIb	-627	-46	34	+3V
ζ Pup	08 03.6	-40 00	2.2	-0.26	-1.11	O5f	-30	+17	1084	-24V?
γ Vel	08 09.5	-47 21	1.7d	-0.22	-0.98	(WC8+O9I)+B1IV	-6	+10	1117:	+35B
ε Car	08 22.5	-59 31	1.9d	+1.28	+0.19	K3III+B2V	-26	+22	605	+2
δ Vel	08 44.7	-54 43	2.0	+0.04	+0.07	A1V	+29	-103	81	+2V?
λ Vel	09 08.0	-43 26	2.2	+1.66	+1.81	K4.5Ib-II	-24	+14	545	+18
β Car	09 13.2	-69 43	1.7	+0.00	+0.03	A2IV	-156	+109	113	-5V?
ι Car	09 17.1	-59 17	2.2	+0.18	+0.16	A8Ib	-19	+12	766	+13
κ Vel	09 22.1	-55 01	2.5	-0.18	-0.75	B2IV-V	-11	+12	572	+22B
α Hya	09 27.6	-08 40	2.0	+1.44	+1.72	K3II-III	-15	+34	180	-4V?
α Leo[4]	10 08.4	+11 58	1.4	-0.11	-0.36	B7V	-249	+6	79	+6B
γ Leo	10 20.0	+19 51	2.1d	+1.15	+1.00	K1IIIb+G7III	+304	-154	130	-37B
μ Vel	10 46.8	-49 25	2.7d	+0.90	+0.57	G5III+G2V	+63	-54	117	+6B
β UMa	11 01.8	+56 23	2.4	-0.02	+0.01	A1V	+81	+33	80	-12B
α UMa	11 03.7	+61 45	1.8	+1.07	+0.92	K0IIIa	-134	-35	123	-9B
δ Leo	11 14.1	+20 31	2.6	+0.12	+0.12	A4V	+143	-130	58	-20V
β Leo	11 49.1	+14 34	2.1	+0.09	+0.07	A3V	-498	-115	36	+0V
γ UMa	11 53.8	+53 42	2.4	+0.00	+0.02	A0Ve	+108	+11	83	-13B
δ Cen	12 08.4	-50 43	2.6	-0.12	-0.90	B2IVne	-50	-7	415	+11V
γ Crv	12 15.8	-17 33	2.6	-0.11	-0.34	B8IIIp	-159	+22	154	-4B
α Cru	12 26.6	-63 06	0.8d	-0.25	-1.00	B0.5IV+B1V	-36	-15	322	-11B
γ Cru	12 31.2	-57 07	1.6	+1.59	+1.78	M3.5III	+28	-265	89	+21
β Crv	12 34.4	-23 24	2.7	+0.89	+0.60	G5II	+1	-57	146	-8V
α Mus	12 37.2	-69 08	2.7	-0.20	-0.83	B2IV-V	-40	-13	315	+13V
γ Cen	12 41.5	-48 58	2.2	-0.01	-0.01	A1IV	-186	+6	130	-6B
β Cru	12 47.7	-59 41	1.2	-0.23	-1.00	B0.5III	-43	-16	279	+16B
ε UMa	12 54.0	+55 58	1.8	-0.02	+0.02	A0p	+112	-8	83	-9B?
α CVn	12 56.0	+38 19	2.8d	-0.09	-0.31	A0p+F0V	-235	+54	115	-3V
ζ UMa	13 23.9	+54 56	2.1d	+0.04	+0.04	A1Vp+A1m	+119	-26	86	-6B
α Vir[5]	13 25.2	-11 10	1.0d	-0.23	-0.93	B1III-IV+B2V	-42	-31	250	+1B
ε Cen	13 39.9	-53 28	2.3	-0.22	-0.92	B1III	-15	-12	427	+3
η UMa	13 47.5	+49 19	1.9	-0.19	-0.67	B3V	-121	-15	104	-11B?
η Boo	13 54.7	+18 24	2.7	+0.58	+0.20	G0IV	-61	-356	37	+0B
ζ Cen	13 55.5	-47 17	2.5	-0.22	-0.92	B2.5IV	-57	-45	382	+7B
β Cen	14 03.8	-60 22	0.6	-0.23	-0.98	B1III	-33	-23	392	+6B
θ Cen	14 06.7	-36 22	2.1	+1.01	+0.87	K0IIIb	-521	-518	59	+1
α Boo[6]	14 15.7	+19 11	0.0	+1.23	+1.27	K1.5III	-1093	-2000	37	-5V
η Cen	14 35.5	-42 09	2.3	-0.19	-0.83	B1.5Vne	-35	-33	306	+0B
α Cen	14 39.6	-60 50	-0.3d	+0.75	+0.31	G2V+K1V	-3679	+474	4.3	-22B
α Lup	14 41.9	-47 23	2.3	-0.20	-0.89	B1.5III/Vn	-21	-24	465	+5B
ε Boo	14 45.0	+27 04	2.6d	+0.97	+0.73	K0II-III+A2V	-51	+21	203	-17V
β UMi	14 50.7	+74 09	2.1	+1.47	+1.78	K4III	-33	+11	131	+17V
α^2 Lib	14 50.9	-16 03	2.8	+0.15	+0.09	A3IV	-106	-68	76	-10B
β Lup	14 58.5	-43 08	2.7	-0.22	-0.87	B2III/IV	-36	-40	383	+0B

1) Castor, 2) Procyon, 3) Pollux, 4) Regulus, 5) Spica, 6) Arcturus.

Stars Continued.

Star name	2000 Equinox α	2000 Equinox δ	Visual magnitude	Color index $B-V$	Color index $U-B$	Spectral type	Proper motion $\mu_\alpha\cos\delta$	Proper motion μ_δ	Distance	Radial velocity
	h m	° ′	mag	mag	mag		10^{-3}(″ year^{-1})	10^{-3}(″ year^{-1})	ly	km s^{-1}
β Lib	15 17.0	-09 23	2.6	-0.11	-0.36	B8V	-98	-20	185	-35B
α CrB	15 34.7	+26 43	2.2d	-0.02	-0.02	A0V+G5V	+120	-90	75	+2B
α Ser	15 44.3	+06 26	2.7	+1.17	+1.24	K2IIIb	+134	+45	74	+3V?
β TrA	15 55.1	-63 26	2.8	+0.29	+0.05	F2III	-189	-402	40	+0
δ Sco	16 00.3	-22 37	2.3	-0.12	-0.91	B0.3IV	-10	-35	491:	-7B
β Sco	16 05.4	-19 48	2.5d	-0.06	-0.85	B1V+B2V	-5	-24	404	-1B
δ Oph	16 14.3	-03 42	2.7	+1.58	+1.96	M0.5III	-48	-143	171	-20V
η Dra	16 24.0	+61 31	2.7	+0.91	+0.70	G8IIIab	-17	+57	92	-14B?
α Sco[1]	16 29.4	-26 26	1.0d	+1.83	+1.34	M1.5 Iab-Ib+B4Ve	-12	-23	554:	-3B
β Her	16 30.2	+21 29	2.8	+0.94	+0.69	G7IIIa	-99	-15	139	-26B
ζ Oph	16 37.2	-10 34	2.6	+0.02	-0.86	O9.5Vn	+15	+25	366	-15V
α TrA	16 48.7	-69 02	1.9	+1.44	+1.56	K2IIb-IIIa	+18	-32	391	-3
ε Sco	16 50.2	-34 18	2.3	+1.15	+1.27	K2.5III	-615	-256	64	-3
η Oph	17 10.4	-15 43	2.4	+0.06	+0.09	A2V	+40	+99	88	-1B
β Ara	17 25.3	-55 32	2.8	+1.46	+1.56	K3Ib-IIa	-9	-25	646:	+0
β Dra	17 30.4	+52 18	2.8	+0.98	+0.64	G2Ib-IIa	-16	+12	380	-20V
υ Sco	17 30.8	-37 18	2.7	-0.22	-0.82	B2IV	-2	-30	576	+8B
α Ara	17 31.8	-49 53	3.0	-0.17	-0.69	B2Vne	-33	-67	267	+0B
λ Sco	17 33.6	-37 06	1.6d	-0.22	-0.89	B2IV+B	-9	-31	571:	-3B
α Oph	17 34.9	+12 34	2.1	+0.15	+0.10	A5III	+108	-222	49	+13B?
θ Sco	17 37.3	-43 00	1.9	+0.40	+0.22	F1II	+6	-3	300:	+1
κ Sco	17 42.5	-39 02	2.4	-0.22	-0.89	B1.5III	-6	-26	483	-14B
β Oph	17 43.5	+04 34	2.8	+1.16	+1.24	K2III	-41	+159	82	-12V
γ Dra	17 56.6	+51 29	2.2	+1.52	+1.87	K5III	-8	-23	154	-28
δ Sgr	18 21.0	-29 50	2.7	+1.38	+1.55	K3IIIa	+33	-26	348	-20V?
ε Sgr	18 24.2	-34 23	1.9	-0.03	-0.13	B9.5III	-39	-124	143	-15
α Lyr[2]	18 36.9	+38 47	0.0	+0.00	-0.01	A0Va	+201	+286	25	-14V
σ Sgr	18 55.3	-26 18	2.0	-0.22	-0.75	B2.5V	+15	-53	228	-11V
ζ Sgr	19 02.6	-29 53	2.6d	+0.08	+0.06	A2III+A4IV	+11	+21	88	+22B
β Cyg	19 30.7	+27 58	2.9d	+0.86	+0.22	(K3II+B9.5V)+B8Ve	-7	-6	434	-24V
γ Aql	19 46.3	+10 37	2.7	+1.52	+1.68	K3II	+17	-3	395	-2V
α Aql[3]	19 50.8	+08 52	0.8	+0.22	+0.08	A7V	+536	+385	17	-26
γ Cyg	20 22.2	+40 15	2.2	+0.68	+0.53	F8Ib	+2	-1	1832:	-8V
α Pav	20 25.6	-56 44	1.9	-0.20	-0.71	B2IV	+7	-86	179	+2B
α Cyg[4]	20 41.4	+45 17	1.2	+0.09	-0.24	A2Ia	+2	+2	1412:	-5B
ε Cyg	20 46.2	+33 58	2.5	+1.03	+0.87	K0III	+356	+331	73	-11B
α Cep	21 18.6	+62 35	2.4	+0.22	+0.11	A7V	+151	+49	49	-10V
β Aqr	21 31.6	-05 34	2.9	+0.83	+0.56	G0Ib	+19	-8	537	+7V?
ε Peg	21 44.2	+09 53	2.4	+1.53	+1.70	K2Ib	+27	+0	690	+5V
α Aqr	22 05.8	-00 19	3.0	+0.98	+0.74	G2Ib	+18	-9	524	+8V?
α Gru	22 08.2	-46 58	1.7	-0.13	-0.47	B7IV	+127	-147	101	+12
α Tuc	22 18.5	-60 16	2.9	+1.39	+1.54	K3III	-71	-39	200	+42B
β Gru	22 42.7	-46 53	2.1	+1.60	+1.67	M5III	+135	-4	177	+2
α PsA[5]	22 57.7	-29 37	1.2	+0.09	+0.08	A3V	+329	-165	25	+7
β Peg	23 03.8	+28 05	2.4	+1.67	+1.96	M2.5II-III	+188	+137	196	+9V
α Peg	23 04.8	+15 12	2.5	-0.04	-0.05	B9V	+60	-41	133	-4B

1) Antares, 2) Vega, 3) Altair, 4) Deneb, 5) Fomalhaut.

Nearby

Order	Star name	HIP	2000 Equinox α	δ	Parallax	Distance	Line-of-sight velocity
			h m	° ′	″	ly	km s^{-1}
1	α CenA	71 683	14 39.6	-60 50	0.755	4.3	-22
	α CenB	71 681					
	α CenC	70 890	14 29.7	-62 41	0.772	4.2	-16
2	Bernard star*	87 937	17 57.8	+04 42	0.548	5.9	-108
3§	Wolf 359		10 56.5	+07 01	0.421	7.7	+13
4	BD+36° 2147*	54 035	11 03.3	+35 58	0.393	8.3	-84
5§	Luyten 726-8 = A		01 39.0	-17 57	0.387	8.4	+29
	UV Cet = B						+32
6	Sirius A	32 349	06 45.1	-16 43	0.379	8.6	-8
	Sirius B						
7	Ross 154	92 403	18 49.8	-23 50	0.337	9.7	-4
8§	Ross 248		23 41.9	+44 11	0.314	10.4	-81
9	ε Eri	16 537	03 32.9	-09 27	0.311	10.5	+16
10	CD-36° 15693	114 046	23 05.9	-35 51	0.305	10.7	+10
11	Ross 128	57 548	11 47.7	+00 48	0.298	10.9	-13
12§	Luyten 789-6		22 38.6	-15 17	0.290	11.2	-60
13	61 Cyg A*	104 214	21 06.9	+38 45	0.287	11.4	-64
	61 Cyg B*	104 217			0.286	11.4	
14	Procyon A	37 279	07 39.3	+05 13	0.285	11.5	-3
	Procyon B						
15	BD+59° 1915A	91 768	18 42.8	+59 38	0.280	11.6	0
	BD+59° 1915B	91 772			0.289	11.3	+10
16	BD+43° 44A	1 475	00 18.4	+44 01	0.279	11.7	+13
	BD+43° 44B						+20
17§	G51-15		08 29.8	+26 47	0.278	11.7	
18	ε Ind	108 870	22 03.4	-56 47	0.276	11.8	-40
19	τ Cet	8 102	01 44.1	-15 56	0.274	11.9	-16
20	Luyten 725-32	5 643	01 12.5	-17 00	0.271	12.0	+28
21		82 724	16 54.5	-62 24	0.271	12.1	
22	BD+5° 1668*	36 208	07 27.4	+05 14	0.263	12.4	+26
23	Kapteyn star	24 186	05 11.7	-45 01	0.256	12.8	+245
24	CD-39° 14192	105 090	21 17.3	-38 52	0.253	12.9	+21
25	Krüger 60A	110 893	22 28.0	+57 42	0.250	13.0	-26
	Krüger 60B						
26		72 511	14 49.5	-26 06	0.248	13.2	
27	Ross 614A	30 920	06 29.4	-02 49	0.242	13.5	+24
	Ross 614B						
28		82 725	16 54.5	-62 24	0.241	13.5	
29	Van Maanen's Star	3 829	00 49.2	+05 23	0.235	13.9	+54

The right ascension (α), the celestial declination (δ), parallax, proper motion, and the color index are taken from the revised Hipparcos Star Catalogue (2007) referring to the Landolt-Born-stein Table (1982) by W. Gliese (Excluding stars with a § sign.). HIP is a number assigned to observed stars in Hipparcos Star Catalogue (1997). Spectral type is from Cox (ed.) 2000, Allen's Astrophysical Quantities. Magnitude is based on the Hipparcos Star Catalogue. However, magnitudes marked with # sign are from WDS (The Washington Double Star Catalog). Star names with A, B, and C are multiple star systems. A is the primary star, and the others are companion

Stars

Interstellar velocity	Proper motion		Spectral type	Visual magnitude	Absolute magnitude	Color index
	Size	Position angle				$B-V$
km s^{-1}	″ $year^{-1}$	°		mag	mag	mag
32	3.71	277	G2V	-0.01	4.38	0.71
			K1V	1.35	5.74	0.90
29	3.85	281	M5Ve	11.01	15.45	1.81
140	10.36	356	M5V	9.54	13.23	1.57
54	4.70	235	M8Ve	13.53	16.65	1.74
102	4.80	187	M2Ve	7.49	10.46	1.50
50	3.36	80	M6Ve	12.52	15.46	} 1.85
			M6Ve	13.02	15.96	
19	1.34	204	A1V	-1.44	1.45	0.01
			DA	8.58#	11.47	-0.12
10	0.67	107	M4.5Ve	10.37	13.01	1.51
85	1.60	176	M6Ve	12.29	14.77	1.48
22	0.98	271	K2V	3.72	6.18	0.88
108	6.90	79	M2Ve	7.35	9.77	1.48
25	1.36	154	M4.5Ve	11.12	13.49	1.75
80	3.26	46	M7Ve	12.18	14.49	1.96
108	5.30	52	K5Ve	5.20	7.49	1.07
	5.17	53	K7Ve	6.05	8.33	1.31
21	1.26	215	F5IV-V	0.40	2.67	0.43
			DF	10.8#	13.1	
37	2.25	324	M4V	8.94	11.18	1.50
38	2.25	324	M5V	9.70	12.00	1.56
51	2.92	82	M2V	8.09	10.32	1.56
			M6Ve	11.12#	13.35	1.80
	1.27	242		14.81	17.03	2.06
90	4.70	123	K5Ve	4.69	6.89	1.06
37	1.92	296	G8Vp	3.49	5.68	0.73
37	1.37	62	M5.5Ve	12.10	14.26	1.85
	0.39	56		11.96	14.12	1.73
72	3.74	171	M3.5	9.84	11.94	1.57
293	8.67	131	M0V	8.86	10.90	1.54
68	3.46	251	M0Ve	6.69	8.71	1.40
32	0.98	242	M2V	9.59	11.58	1.61
			M5Ve	11.41#	13.40	1.8
	1.38	277		11.72	13.69	1.48
30	0.93	131	M4.5Ve	11.12	13.04	} 1.69
				14.8#	16.7	
	0.34	56		11.72	13.63	0.00
81	2.98	155	DG	12.37	14.23	0.55

stars. * has an invisibile companion star or an unconfirmed companion star. The Sun's spectral type is G2V. Its visual magnitude is −26.75. Its absolute magnitude is +4.82. Its color index is $B-V$=+0.65. Note) α Cen C, Luyten 726-8B = UV Cet, Ross 154, and Kruger 60B are flare stars (Stars which normally show fairly constant brightness but, sometimes over the course of several minutes, the brightness will increase by several magnitude and then return to its original brightness.) Wolf 359 may also be a flare star. BD+43°44A is a spectral binary star.

Spectral Types of Stars

The spectral type of a star is determined by the physical condition and the chemical composition of its surface layer. Currently the most commonly used is the traditional Harvard Classification System, where stars are divided into 10 spectral types (O, B, A, F, G, K, M, R, N, and S) according to the appearance and strengths of spectral line. In addition, new spectral types (L, T, and Y) for objects with temperatures lower than M-type have been introduced recently been introduced. Each type is further subdivided by appending a numeral from 0 to 9 (even decimals or +/− are used for finer subdivision). Besides, a lower-case letter such as a, b, or c is sometimes attached to O, M, and N types. The characteristics for each spectral type are described below (given in the parenthesis is the typical star belonging to this type).

O (ζ Pup) Characterized by spectral lines of once-ionized helium or multiply-ionized oxygen, nitrogen and carbon. Those showing especially wide emission lines are called Wolf-Rayet stars, which are divided into WC-type (strong carbon lines) and WN-type (strong nitrogen lines).

B (γ Ori) Neutral helium lines are strongest in this spectral type. Hydrogen lines are getting strong.

A (α CMa) Hydrogen lines are strongest. Lines of singly-ionized metals are also of appreciable strength.

F (α CMi) Hydrogen lines become weaker, while calcium H, K lines and lines of neutral metals gradually strengthen.

G (Sun, α Aur) Calcium H and K lines are conspicuously strong, Hydrogen lines are not conspicuous any more, while the G-band due to CH molecule strengthens.

K (α Boo) CaI H, K lines are very strong and wide, and metallic lines of various species tend to overlap with each other. Continuum is depressed in the violet region.

M (α Ori, o Cet) Characterized by the strong absorption bands of TiO (titanium oxide) molecules.

R, N (19 Psc) Strong absorption bands of carbon-related molecules (such as CH, CN, and C_2) are the distinguished features. These R and N types are united into one class in the current C-type classification.

S (π^1 Gru) Characterized by the absorption bands of ZrO (zirconium oxide) molecules.

L (GD 165B) Conspicuous features are the absorptions due to alkali atoms, hydrides, H_2O and CO molecules. A part of brown dwarfs also belong to this class.

T (Gliese 229B) Notable spectral features are the absorptions of CH_4 (methane) and H_2O (water vapor). This is the typical class of brown dwarfs.

Y (WISEP J1541) Brown dwarfs with very low surface temperature (less than 400 K). Absorption of NH_3 (ammonia) is important in addition to H_2O and CH_4.

The sequence of Spectral Types is schematically expressed as follows, which starts from stars with highest surface temperature and further goes down toward those with lower temperature. Reflecting the difference of surface temperature, the color of stars changes as blue (O), bluish white (B), white (A), whity-yellow (F), yellow (G), orange (K), and eventually red (M). The bifurcation of the sequence, which occurs at G or K type, is due to the difference in the chemical composition.
Refer to "Physical Properties of Stars".

O—B—A—F—G—K—M—L—T—Y
(K) ╲ S
(G) ╲ R—N

In the Yerkes Spectral Classification (Morgan-Keenan or MK classification) which is most widely used nowadays, Roman numerals are appended after the spectral type in order to specify the absolute magnitude class: I — supergiants, II — bright giants, III — giants, IV — subgiants, V — dwarfs (main-sequence stars), and VI — subdwarfs. This class can be further subdivided by attaching a, ab, b (a is brighter, b is dimmer, ab is in-between) or by combining two classes with a hyphen (-). In the Mount Wilson classification system, luminosity class is indicated by lower case letters prefixed to the spectral type: c (supergiants), g (giants), sg (subgiants), d (dwarfs), sd (subdwarfs), and w or D (white dwarfs).

Besides, specific suffixes are attached to the spectral type in order to indicate notable characteristics of the spectra: "n" or "nn" — wide and blurred lines (usually due to large rotational velocities); "s" — sharp lines; "e" — emission lines; "v" — variable spectrum; "p" — peculiar spectrum; "m" — strong metallic lines. In particular, Am stars are "metallic-line A-type stars", which show peculiar spectra with weak calcium lines (as normal A-type stars) while strong lines of other metals (as normal F-type stars).

Pulsating Variable Stars

The pulsating variable stars are classified into several types shown in figure 8, based on the position on the Hertzsprung-Russell Diagram (HR diagram) and the properties of the light variations. The classical pulsating variables, represented by the δ Cephei-type stars (Cepheids) and the Mira variables, mainly show radial oscillations (or pulsations), which keep the stellar shape spherical during the oscillation, with a single or at most two eigenmode(s) (fundamental mode and/or a harmonic). On the other hand, many of the pulsating variables that were newly recognized after the latter half of 1960s show nonradial oscillations, which induce uneven variations of the stellar surface. There exist two types of nonradial-oscillation modes, acoustic (p) modes and (internal) gravity (g) modes, which are mainly restored by the gas pressure and the buovancy, respectively. The periods of the p (g) modes are shorter (longer) than that of the fundamental mode of the radial oscillation. In a star with an extremely high mass-to-luminosity ratio, there appears another type of mode, strange mode, which has a significantly different character from the modes in other stars, because the thermal energy changes a lot along with the oscillation.

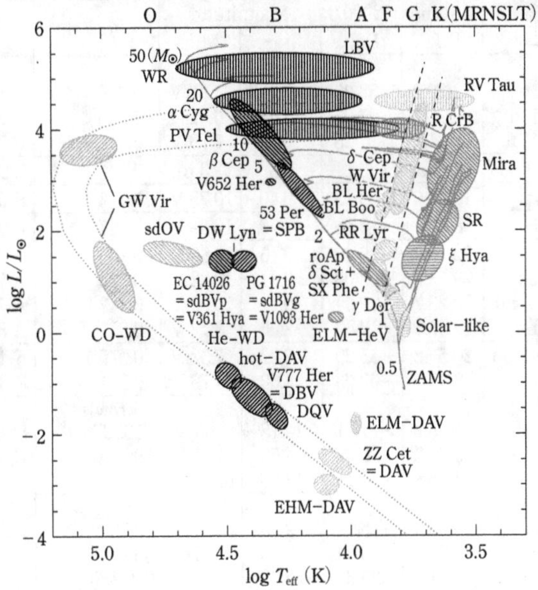

Figure 8 Positions of the various types of pulsating variable stars on the HR diagram (based on Fig 1 of C. S. Jeffery *et al.* 2015, MNRAS, **447**, 2836). The evolutionary tracks of the stars with the mass between half a solar mass ($M_\odot$) and 50 solar masses are shown, together with the main sequence and the sequence of white dwarfs. The regions hatched with lines from upper left to lower right indicate the positions of the stars with either the radial oscillations or the nonradial p-mode oscillations, while those with lines from upper right to lower left correspond to the stars with the nonradial g-mode oscillations. The stars in the regions shaded with vertical lines are supposed to show the strange modes, whereas those marked with horizontal lines are the solar-like oscillators in which the p modes are excited by the convection.

The properties of typical types of pulsating variable stars are given in the following table.

Type	Period range	Typical period	Light amplitude (Visual magnitude)	Spectral types	Absolute magnitude (M_V)
δ Cep	1–50 days	5–10 days	$\lesssim 2$	F5–K2	−0.5 to −6
RV Tau	20–150 days	75 days	$\lesssim 4$	F–G	−2 to −4
W Vir	8–35 days	12–20 days	$\lesssim 1.5$	A5–G8	−0.5 to −2
BL Her	0.8–8 days	1.5 days	$\lesssim 1.5$	A5–G2	+0.5 to −1.5
RR Lyr	1.5–24 hours	0.5 days	$\lesssim 2$	A2–F2	+3 to 0
δ Sct	1–4 hours	2 hours	$\lesssim 0.9$	A2–F5	+3 to +2
β Cep	4–6 hours	5 hours	$\lesssim 0.1$	B1–B2	−3.5 to −4.5
Mira variable	100–700 days	270 days	$\lesssim 11$	M_e, R_e, N_e, S_e	+1 to −2

Typical stars of various classical pulsating variables are listed. Symbols V, p and v in the column for magnitude mean that of the *UBV* system (page **46**), the photographic magnitude and the magnitude measured by eye, respectively. These data are based on Samus *et al.* 2013, General Catalogue of Variable Stars ; http://www.sai.msu.su/gcvs/gcvs/

Type	star name	Equinox J2000.0		Magnitude		Period	Spectral type
		R.A.	Dec.	Max.	Min.		
		h m	° ′	mag	mag	Days	
δ Cep	δ Cep	22 29.2	+58 25	3.48	4.37 V	5.366	F 5 Ib–G 1 Ib
	T Mon	06 25.2	+07 05	5.58	6.62 V	27.025	F 7 Iab–K 1 Iab
	SV Vul	19 51.5	+27 28	6.72	7.79 V	45.012	F 7 Iab–K 0 Iab
RV Tau	RV Tau	04 47.1	+26 11	9.8	13.3 p	79	G 2 Iae–M 2 Ia
W Vir	W Vir	13 26.0	−03 23	9.46	10.75 V	17.274	F 0I b–G 0 Ib
BL Her	BL Her	18 01.1	+19 15	9.70	10.62 V	1.307	F 0–F 6 II–III
RR Lyr	DH Peg	22 15.4	+06 49	9.15	9.80 V	0.256	A 5–F 0.5
	T Sex	09 53.5	+02 03	9.81	10.32 V	0.325	A 7 II–III–F 4 III
	SW And	00 23.7	+29 24	9.14	10.09 V	0.442	A 7 III–F 8 III
	RR Lyr	19 25.5	+42 47	7.06	8.12 V	0.567	A 5–F 7
δ Sct	DQ Cep	20 57.8	+55 29	7.22	7.32 V	0.079	F 4 III
	δ Sct	18 42.3	−09 03	4.60	4.79 V	0.194, 0.187	F 3 III p
						Seconds	
DAV	ZZ Cet	01 36.2	−11 21	14.13	(0.03)V	213, 274	DA
	VW Lyn	09 01.8	+36 07	14.55	(0.15)V	590, 260	DA
	ZZ Psc	23 28.8	+05 15	12.98	(0.23)V	612, 816	DA
						Days	
β Cep	γ Peg	00 13.2	+15 11	2.78	2.89 V	0.152	B 2 IV
	β Cep	21 28.7	+70 34	3.16	3.27 V	0.190	B 2 III ev
	σ Sco	16 21.2	−25 36	2.86	2.94 V	0.247, 0.240	B 2 III
	β CMa	06 22.7	−17 57	1.93	2.00 V	0.250, 0.239	B 1 II–III
Mira variable	R Leo	09 47.6	+11 26	4.4	11.3 v	310	M 6 e–M8 III e–M 9.5 e
	o Cet	02 19.3	−02 59	2.0	10.1 v	332	M 5 e–M 9 e
	χ Cyg	19 50.6	+32 55	3.3	14.2 v	408	S 6, 2 e–S10, 4 e(MSe)
	U Cyg	20 19.6	+47 54	5.9	12.1 v	463	C 7, 2 e–C 9, 2(Npe)
Semiregular variable	Z Aqr	23 52.2	−15 51	7.4	10.2 v	137	M 1e–M 7 III
	RR CrB	15 41.4	+38 33	8.4	10.1 p	61	M 5
	μ Cep	21 43.5	+58 47	3.43	5.1 v	730	M 2 Iae
	SX Her	16 07.5	+24 55	8.6	10.9 v	103	G 3 ep–K 0(M3)

Nova

Here we discuss stars which explosively become 9 to 13 magnitudes brighter over the course of several days, then slowly return to the state they were in before the explosion. From a spectroscopic observation immediately after the explosion, it is understood that material is discharged from the surface of the star. One explosion releases about 10^{38} J of energy, while the absolute magnitude reaches −6 to −7. In a close binary system that is composed of a white dwarf and a late-type G-M star (many are dwarf stars), hydrogen rich material coming from the companion accumulates on the surface of the white dwarf. When the temperature at the bottom of the hydrogen layer exceeds the critical temperature, an explosive nuclear reaction occurs. Each nova is categorized based on the dimming speed as a fast nova NA, slow nova NB, or very slow nova NC. The ones where explosions are recorded two or more times are called recurrent novae NR. The following table lists some examples of novae. The symbols M_1 and M_0 mean the maximum magnitude at time of explosion and the magnitude in quiescence, respectively.

Star name	Year	Equinox J2000.0		Type	Magnitude		Orbital period	Years with subsequent explosions
		R.A.	Dec.		M_1	M_0		
		h m s	° ′ ″		mag	mag	Days	
CK Vul	1670	19 47 38.0	+27 18 48	NB	2.7	20.7	−	
V841 Oph	1848	16 59 30.4	−12 53 27	NB	4.3	13.5	0.6014	
U Sco	1863	16 22 30.8	−17 52 43	NR	8.7	19.3	1.231	1906,1917,1936,1945,1969, 1979,1987,1999,2010
T CrB	1866	15 59 30.2	+25 55 13	NR	2.0	10.8	227.6	1946
AG Peg	1870	21 51 01.9	+12 37 32	NC	6.0	8.4	820	
Q Cyg	1876	21 41 43.9	+42 50 29	NA	3.0	15.6	0.4202	
T Pyx	1890	09 44 41.5	−32 22 47	NR	6.4	15.5	0.075	1890,1902,1920,1944,1966,2011
T Aur	1891	05 31 59.1	+30 26 46	NB	4.1	15.5	0.2044	
RS Oph	1898	17 50 13.2	−06 42 29	NR	4.3	12.5	455.72	1907?,1933,1958,1967,1985,2006
V1059 Sgr	1898	19 01 51.0	−13 09 39	NA	4.5	16.5	−	
GK Per	1901	03 31 12.0	+43 54 15	NA	0.2	14.0	1.9968	
DM Gem	1903	06 44 12.1	+29 56 42	NA	4.8	16.5	0.1228?	
DI Lac	1910	22 35 48.6	+52 42 59	NA	4.3	14.9	0.5438	
DN Gem	1912	06 54 54.4	+32 08 28	NB	3.6	15.8	0.1279	
GI Mon	1918	07 26 47.1	−06 40 30	NA	5.2	15.1	0.1802	
V603 Aql	1918	18 48 54.6	+00 35 03	NA	−1.4	12.0	0.1385	
V476 Cyg	1920	19 58 24.5	+53 37 07	NA	2.0	17.1	−	
RR Pic	1925	06 35 36.1	−62 38 24	NB	1.2	12.5	0.145	
DQ Her	1934	18 07 30.2	+45 51 32	NB	1.3	18.1	0.1936	
CP Lac	1936	22 15 41.1	+55 37 02	NA	2.1	16.6	0.127?	
V368 Aql	1936	19 26 34.4	+07 36 14	NA	5.0	15.7	0.3452	
V630 Sgr	1936	18 08 48.5	−34 20 21	NA	1.6	17.5	0.118	
CP Pup	1942	08 11 46.1	−35 21 05	NA	0.5	<17.0	0.06143	
DK Lac	1950	22 49 47.0	+53 17 20	NA	5.0	15.5	−	
RW UMi	1956	16 47 54.8	+77 02 12	NA	6.0	<21.0	0.0591	
V446 Her	1960	18 57 21.6	+13 14 29	NA	2.8	18.8	0.207	
V533 Her	1963	18 14 20.5	+41 51 23	NA	3.0	16.2	0.147	
QZ Aur	1964	05 28 34.1	+33 18 22	NA	6.0	18.0	0.3575	
HR Del	1967	20 42 20.3	+19 09 39	NB	3.7	12.4	0.2142	
LV Vul	1968	19 48 01.0	+27 10 19	NA	5.2	16.9	−	
FH Ser	1970	18 30 47.0	+02 36 52	NA	4.4	17.6	−	
V1500 Cyg	1975	21 11 36.6	+48 09 02	NA	1.7	<21.0	0.1396	
V1668 Cyg	1978	21 42 35.3	+44 01 55	NA	6.0	20.0	0.1384	
V1370 Aql	1982	19 23 21.1	+02 29 26	NA	6.0	20.0	−	
QU Vul	1984	20 26 47.0	+27 50 43	NA	5.2	17.5	0.1118	
V842 Cen	1986	14 35 52.6	−57 37 35	NA	4.6	18.6	−	
V838 Her	1991	18 46 31.5	+12 14 02	NA	5.0	20.0	0.2977	
V1974 Cyg	1992	20 30 31.7	+52 37 51	NA	4.2	17.5	0.08126	
V705 Cas	1993	23 41 47.2	+57 31 01	NA	5.8	<16.0	0.2280?	
V1494 Aql	1999	19 23 05.3	+04 57 20	NA	5.0	16.0	0.1346	
V382 Vel	1999	10 44 48.4	−52 25 31	NA	2.7	16.6	0.1462	
V339 Del	2013	20 23 30.7	+20 46 04		4.3	<13.0	−	

Supernova

There is a phenomenon where one day, one star suddenly starts to shine brightly, reaching a brightness comparable to an entire galaxy. This is a large explosion marking the last moments of a star's life. They are categorized into two groups: type I, where the hydrogen lines cannot be seen in the spectrum at the time of maximum, and type II, where hydrogen can be seen. Type I is further subdivided into type Ia with strong silicon absorption, type Ib where helium absorption is prominent, and type Ic, others. It is thought that type Ia explosions occur in a close binary system that includes a white dwarf, whereas the other types are thought to be stellar explosions originating in the graviational collapse of the iron core formed at the center during the final stages of the evolution of a massive star. Supernovae appear at a rate of a few per century in a given galaxy. Types Ib, Ic and II leave behind a neutron star or a black hole. Moreover these events expell gas rich in heavy elements which emits X-rays and radio waves. This is called a supernova remnant. It is thought that certain kinds of γ-ray burst are related to supernovae.

Designation	Equinox J2000.0		Max. mag.	Type	Host Galaxy (type)	Chronicler or discoverer	Corresponding Celestial Object Name
	R.A.	Dec.					
	h m s	° ′ ″	mag				
1006	15 02 08	-41 57	-8	Ia	Milky Way (SAB)	'Meigetsuki, Full Moons	The remnant is G 327.6+14.6
1054	05 34 05	+22 01	-6	II	Milky Way (SAB)	Diary', 'Songshi, History	CM Tau and the remnant is M 1
1181	02 05 06	+64 49	0	II	Milky Way (SAB)	of the Song' and others	The remnant is 3C 58
1572	00 25 03	+64 09	-4	Ia	Milky Way (SAB)	Schuler, Tycho	B Cas and the remnant is 3C 10
1604	17 30 06	-21 29	-3	Ia	Milky Way (SAB)	Brunowski, Kepler and others	V843 Oph and the remnant is 3C 358
1885A	00 42 43.07	+41 16 04.3	5.8	I	M 31 (SA)	Hartwig	
1895B	13 39 57.4	-31 38 12	8.0	Ia	NGC 5253 (I)	Fleming	
1920A	08 35 16.3	+28 28 30	11.7	II	NGC 2608 (SB)	Wolf	
1921C	10 18 23.1	+41 21 02	11.0	I	NGC 3184 (SAB)	Jones	
1937C	13 05 52.4	+37 37 08	8.5	Ia	IC 4182 (SAm)	Zwicky	
1954A	12 15 45.8	+36 16 14	9.8	Ib	NGC 4214 (I)	Wild	
1962M	03 18 12.2	-66 31 38	11.5	II	NGC 1313 (SB)	Sersic	
1966J	10 19 45.8	+45 30 10	11.3	Ib	NGC 3198 (SB)	Wild	
1968L	13 37 00.51	-29 51 59	11.9	II	M 83 (SAB)	Bennett	
1970A	12 32 41	+14 02 31	11.0	II?	IC 3476 (I)	Grizunova	
1970G	14 03 00.83	+54 14 32.8	11.4	II	NGC 5457 (SAB)	Lovas	
1972E	13 39 53.2	-31 40 15	8.4	Ia	NGC 5253 (I)	Kowal	
1979C	12 22 58.63	+15 47 51.7	11.6	II	M 100 (SAB)	G. Johnson	
1980K	20 35 30.07	+60 06 23.8	11.6	II	NGC 6946 (SAB)	Wild	
1983N	13 36 51.24	-29 54 02.7	11.4	Ib	M 83 (SAB)	Evans	
1987A	05 35 27.99	-69 16 11.5	2.9	II	LMC (SBm)	Shelton, *et al.*	
1991T	12 31 36.91	+02 56 28.3	11.5	Ia	NGC 4527 (SA)	Knight, *et al.*	
1993J	09 55 25	+69 01 13	10.5	II	M 81 (SA)	Garcia	
1994D	12 34 02.45	+07 42 04.7	11.8	Ia	NGC 4526 (SA)	UCB group	
1994I	13 29 54.01	+47 11 31.7	12.8	Ic	M 51 (SA)	Puckett, *et al.*	
1998S	11 46 06.18	+47 28 55.5	12.0	II	NGC 3877 (SA)	Beijing Astron. Obs.	
1998bw	19 35 03.31	-52 50 44.8	13.5	Ic	ESO 184-G82 (SB)	Galam, *et al.*	GRB 980425
2002ap	01 36 23.85	+15 45 13.2	12.3	Ic	M 74 (SA)	Y. Hirose	
2004dj	07 37 17.02	+65 35 57.8	11.2	II	NGC 2403 (SAB)	K. Itagaki	
2005cs	13 29 52.78	+47 10 35.7	13.9	II	M 51 (SA)	Kloehr	
2006aj	03 21 39.71	+16 52 02.6	17.3	Ic	Unnamed	Swift team	GRB 060218
2006gy	03 17 27.1	+41 24 19	14.2		NGC 1260	Quimby, Mondo	
2011fe	14 03 05.8	+54 16 25	9.9	Ia	M 101 (SAB)	Palomar Transient Factory	

Visual Binary Stars

Binary stars with known orbits and combined magnitudes equal to or brighter than 4.3 are shown. The symbols e, ρ and θ mean eccentricity, angular distance and position angle, respectively. (Based on Hartkopf, W. I., Mason, B. M., & Worley, C. E. 2012, Sixth Catalog of Orbits of Visual Binary Stars ; http://ad.usno.navy.mil/wds/orb6/orb6.html, etc.)

Star name	Equinox J2000.0		Visual magnitude		Spectral type		Period	e	Epoch 2000.0		Parallax
	R.A.	Dec.							ρ	θ	
	h m	° ′	mag	mag			Years		″	°	″
η Cas	00 49.1	+57 49	3.5	7.5	G0V	K7V	480	0.50	12.85	317 d	0.176
α Psc	02 02.0	+02 46	4.2	5.2	A0p	A3m	933	0.70	1.83	272 r	0.005
θ Per	02 44.2	+49 14	4.1	9.9	F7V	M1V	2720	0.13	20.04	304 d	0.082
α For	03 12.1	-28 59	3.9	6.5	G8IV	G7V	269	0.76	5.06	299 d	0.075
α Aur	05 16.7	+46 00	0.6	1.1	G8III	G0III	0.285	0.00	0.05	52 r	0.080
σ Ori	05 38.7	-02 36	4.1	5.3	O9V	B0V	157	0.07	0.24	115 r	0.007
μ Ori	06 02.4	+09 39	4.3	6.3	A1V	F2V	18.6	0.74	0.11	30 r	0.028
α CMa	06 45.1	-16 43	-1.5	8.5	A1V	DA	50.1	0.59	4.60	150 r	0.378
δ Gem	07 20.1	+21 59	3.6	8.2	F0IV	K3V	1200	0.11	5.77	225 d	0.061
α Gem	07 34.6	+31 53	1.9	3.0	A1V	A2Vm	467	0.34	3.98	67 r	0.067
α CMi	07 39.3	+05 14	0.4	10.8	F5IV-V	DA	40.7	0.40	4.64	76 d	0.292
ε Hya	08 46.8	+06 25	3.8	5.3	G5III	F5	15.1	0.66	0.25	203 d	0.027
10 UMa	09 00.6	+41 47	4.1	6.0	F3V	G5V	21.8	0.15	0.60	46 r	0.069
κ UMa	09 03.6	+47 09	4.2	4.4	A0IV-V	A0V	35.6	0.56	0.13	139 r	0.016
ψ Vel	09 30.7	-40 28	4.1	4.6	F3IV	F0IV	34.0	0.44	0.53	264 d	0.065
$\gamma^{1,2}$ Leo	10 20.0	+19 50	2.4	3.6	K1III	G7III	510	0.84	4.41	125 d	0.022
p Vel	10 37.3	-48 14	4.2	5.1	F4IV	A6	16.5	0.75	0.30	231 r	0.040
μ Vel	10 46.8	-49 25	2.7	5.6	G5III	G2V	116	0.79	2.50	58 d	0.022
α UMa	11 03.7	+61 45	2.0	5.0	K1II-III	K0V	44.4	0.44	0.36	162 r	0.038
ξ UMa	11 18.2	+31 32	4.4	4.9	G0V	F9V	59.8	0.40	1.77	273 r	0.137
ι Leo	11 23.9	+10 32	4.0	6.7	F1IV	G3V	186	0.53	1.72	116 r	0.052
γ Cen	12 41.5	-48 58	2.9	2.9	A0III	A0III	84.5	0.79	1.00	347 r	0.016
γ Vir	12 41.7	-01 27	3.5	3.5	F0V	F0V	169	0.88	1.84	267 r	0.099
β Mus	12 46.3	-68 06	3.7	4.0	B2V	B3V	194	0.60	1.29	43 d	0.015
d Cen	13 31.0	-39 24	4.5	4.7	G7III	G9III	78.7	0.46	0.21	75 r	0.012
$\alpha^{1,2}$ Cen	14 39.6	-60 50	0.0	1.3	G2V	K1V	79.9	0.52	14.14	222 d	0.750
ζ Boo	14 41.1	+13 44	4.5	4.8	A2III	A2III	123	0.99	0.80	300 r	0.009
γ Lup	15 35.1	-41 10	3.0	4.5	B2IV-V	B2IV-V	190	0.51	0.67	274 r	0.008
γ CrB	15 42.7	+26 18	4.0	5.6	B9IV	A3V	91.1	0.48	0.76	114 r	0.033
ξ Sco	16 04.4	-11 22	4.9	5.2	F5IV	F5IV	45.7	0.74	0.39	308 d	0.048
α Sco	16 29.4	-26 26	1.0 variable	5.5	M1I	B3Ve	1217	0.08	2.59	274 d	0.024
λ Oph	16 30.9	+01 59	4.2	5.2	A0V	A4V	129	0.61	1.54	30 d	0.010
ζ Her	16 41.3	+31 36	2.9	5.4	F9IV	G7V	34.5	0.46	0.75	13 r	0.102
η Oph	17 10.4	-15 43	3.0	3.5	A2V	A3V	87.6	0.95	0.60	244 r	0.052
70 Oph	18 05.5	+02 30	4.2	6.0	K0V	K4V	88.1	0.50	3.79	148 r	0.201
ζ Sgr	19 02.6	-29 53	3.2	3.4	A2III	A4IV	21.1	0.21	0.34	224 r	0.025
δ Cyg	19 45.0	+45 08	2.9	6.3	B9III	F1V	918	0.52	2.50	222 r	0.030
β Del	20 37.5	+14 36	3.6	4.9	F6III	F6IV	26.7	0.35	0.57	343 d	0.028
τ Cyg	21 14.8	+38 03	3.8	6.4	F2IV	G0V	49.8	0.24	0.77	306 r	0.055
$\zeta^{1,2}$ Aqr	22 28.8	-00 01	4.3	4.5	F3V	F6IV	487	0.34	2.10	192 r	0.022

Spectroscopic Binary Stars

Spectroscopic binary stars are known to be binary star systems through periodic shifts in the spectral lines or changes in the timing of pulses. Ones with visual magnitudes equal to or brighter than 2.9 and other important binary stars are shown. The symbols e, K, γ, ω and T mean eccentricity, radial velocity semi-amplitude, the relative center-of-mass radial velocity, longitude of periastron and time of periastron passage, respectively. Based on Pourbaix, D., Tokovinin, A. A, Batten, A. H., Fekel, F. C., Hartkopf, W. I., Levato, H., Morrell, N. I., Torres, G., & Udry, S. 2004, The Ninth Spectroscopic Binary Orbits; http://sb9.astro.ulb.ac.be.

Star name	Visual magnitude		Spectral type		Period	e	K		γ	ω	T
	Primary	Secondary	Primary	Secondary			Primary	Secondary			
	mag	mag			Days		km s^{-1}	km s^{-1}	km s^{-1}	°	JD2 400 000+
PSR 1913+16	22.5				0.323	0.62	199			179	42 321.43
Sco X-1	12.25		sdBe		0.787	0.0	58.2		-138.5	0	42 565.74
δ Cap	2.83		Am		1.023	0	75.3		-3.4	0	48 105.79
π Sco	2.89		B1V	B1V	1.570	0	124.1	196.1	-7.4	0	49 818.02
Her X-1	13.0		A		1.700	0	169.1			0	42 859.73
Cen X-3	13.4		O7		2.087	0.23	415.1	19	39	175	43 586.95
V711 Tau	5.71		K1V	G5IV	2.838	0	52.6	64.1	-15.9	0	42 767.40
β Per A	2.12		B8V	Am	2.867	0.02	44.0	201		62	28 482.74
α^1 Gem	1.58		Am	dM1e	2.928	0	31.9		-1.2	95	27 501.70
SMC X-1	13.15		B0I		3.892	0.0	299.5	19	180	0	42 838.63
β Aur	1.90	2.83	A2IV	A2IV	3.960	0	110.2	110.5	-15.8	139	43 915.70
α Vir	0.97		B1V	B3V	4.015	0.18	120	189	0	142	40 284.78
Cyg X-1	8.89		O9Iab		5.600	0.0	74.9		-2.7	0	43 045.07
δ Ori	2.14	2.26	O9II		5.733	0.09	97.5		20.4	117	45 139.84
β Sco	2.63		B0V		6.828	0.29	125.3	198.0	-1.7	17	18 501.53
ζ Cen	2.54		B2IV		8.024	0.5	110.7	159.4	6.5	290	29 798.46
Vel X-1	6.88		B0Ib		8.964	0.09	274.3	21.8		153	43 956.33
α^2 Gem			A1V	dM1e	9.213	0.50	12.9		5.2	266	27 543.94
α Pav	1.93		B3IV		11.753	0.0	7.2		2.0	0	17 547.68
PSR 2303+46					12.340	0.66	76.8			35	46 108.05
SS 433	13.0	15.13	peculiar		13.08	0.0	195		27	0	44 133.5
α CrB	2.23		B9IV	G	17.360	0.37	35.4	99.0	1.4	311	36 751.52
ζ^1 Uma	2.27		A2V		20.539	0.53	67.3	69.2	-6.3	285	38 085.7
ι Ori	2.76		O9III	B1III	29.134	0.76	111.9	195.7	31.3	130	51 121.66
α And	2.17		A0p		96.701	0.53	27.7	65.5	-10.0	77	47 374.6
α Aur	0.06		G5III	G0III	104.024	0.00	26.1	27.4	29.2	89	42 040.9
β Ari	2.60		A5V		106.994	0.88	34.1	66.6	-3.1	25	44 809.1
κ Vel	2.50		B2IV		116.65	0.19	46.5		21.9	96	16 459.00
β Her	2.78		G8III		413.1	0.59	12.3		-25.9	20	53 310.9
η Boo	2.69		G0IV		494.173	0.26	8.4		1.0	326	28 136.19
β Per AB	2.12		B8V	Am	680.081	0.23	12.0	31.6	3.7	313	34 042.13
η Peg	2.96		G2II	F0V	813	0.18	14.4		4.2	345	52 025
τ Pup	2.92		K0III		1 066.0	0.09	4.1		36.4	64	20 992.8
PSR 0820+02					1 710	0.0	5.2			0	42 910
α Phe	2.40		K0III		3 848.83	0.34	5.8		75.2	20	16 201.85
α Tuc	2.85		K3III		4 197.7	0.39	7.2		42.2	49	18 666.4
γ Per	2.94		G8III	A3V	5 329.89	0.79	13.7	18.6	3.1	350	32 288.2
ε Aur	2.98		F0Iap		9 890	0.20	15.0		-1.4	346	33 346
α UMi	2.02		F8Ib	(A5)	11 125.3	0.64	4.1		-16.4	307	25 422.41
α CMi	0.35		F5IV		14 847.1	0.36	1.7		-4.1	88	
α UMa	1.79		K0III	F	16 070.7	0.35	2.0		-8.7	174	18 636.2
α CMa	-1.47		A1V	A5wd	18 276.7	0.59	2.4		-7.6	146	
$\alpha^{1,2}$ Cen	-0.01	1.33	G2V	K1V	29 188.1	0.52	4.6	5.5	-22.4	52	35 319.2

Eclipsing Binary Stars

Typical eclipsing binary stars (also known as eclipsing variable stars) are shown in order of increasing period. The maximum magnitude, *M*, and depth of transit are based on visible light or blue light (in italic). The simbol *i* indicates orbital inclination. The type indicates the kind of the light curve (A: Algol type, B: β Lyrae type and W: W Ursa Major type). The unit of mass is Solar Mass $M_\odot$. They are classified depending on whether each component of the binary system fills the Roche lobe or not (D: detached type, SD: semi-detached type and C: contact type).

Star name	Period	Spectral type		Maximum magnitude	Depth of transit		Type	*i*	Mass		Class
		Primary	Companion		Primary	Secondary			Primary	Companion	
	Days			mag	mag	mag		°	$M_\odot$	$M_\odot$	
AM CVn	0.0122	DB		*13.86*	*0.03*	–	–				SD
WZ Sge	0.0567	sdBe		15.21*	0.3	0.2	–	76	0.8	0.09	SD
UdGem	0.177	sdBe	dM4	14.2*	0.8	–	–	67	1.18	0.56	SD
DQ Her	0.194	sdBe	dM3	14.17*	1.7	–	–	70	0.62	0.44	SD
i Boo	0.268	G2V	G2V	5.3	0.17	0.14	W	68.1	0.96	0.49	C
W UMa	0.334	F8V	F8	7.9	0.73	0.64	W	80.3	1.29	0.86	C
AH Vir	0.408	K0V	K0	9.0	0.61	0.52	W	80.8	1.38	0.58	C
V566 Oph	0.410	F4V	F4V	7.5	0.46	0.41	W	78.2	1.30	0.44	C
UU Sge	0.465	sdO	KV	14.18	1.41	0.23	–	86.6	1.1	0.6	D
V471 Tau	0.521	DA	K2V	9.40	0.31		–	79.5	0.79	0.73	D
ε CrA	0.591	F2V	F3	4.74	0.26	0.19	W	70	1.5	0.15	C
YY Gem	0.814	M1Ve	M1Ve	9.27*	0.50	0.49	A	86.1	0.56	0.56	D
δ Cap	1.023	A7IIIm		2.83	0.22		A				
μ^1 Sco	1.440	B1V	B6	*2.93*	*0.28*	*0.17*	B	60.0	14.0	9.24	SD
V Pup	1.454	B1V	B3V	*4.74*	*0.51*	*0.49*	B	74.8	19.1	11.3	SD
ζ Phe	1.670	B6V	A0V	3.92	0.50	0.30	A	84.7	6.10	3.00	D
Her X-1	1.700	B–A		13.2			–	85	1.9	0.9	SD
U Cep	2.493	B7V	G8III	6.80	2.30	0.09	A	86.4	3.19	1.53	SD
WW Aur	2.525	A3m	A4m	5.87	0.69	0.57	A	87.7	1.81	1.75	D
β Per	2.867	B8V	G8III	2.12	1.28	0.07	A	81.2	3.15	0.74	SD
AO Cas	3.523	O9III	O9III	5.9	0.16	0.15	B	55.5	23.0	18.0	~C
λ Tau	3.953	B3V	A4IV	*3.5*	*0.50*	*0.09*	A	86	8.77	2.07	SD
β Aur	3.960	A2IV	A2IV	1.90	0.07	0.08	A	78.5	2.34	2.25	D
V444 Cyg	4.212	O6	WN5	*8.3*	*0.29*	*0.14*	A	78.4	25.6	10.1	SD
RS CVn	4.798	F4IV-V	K0IV	7.93	1.21	0.26	A	87	1.34	1.40	D
δ Ori	5.732	B0III	O9V	*1.94*	*0.09*	*0.06*	A	67	26.9	10.2	D
η Ori	7.989	B0V	B0	*3.14*	*0.21*	*0.09*	B				
β Lyr	12.914	Bpe		3.34	1.00	0.50	B	80			
α CrB	17.360	A0V	G3	*2.21*	*0.11*	*0.02*	A	88.3	2.75	0.94	D
ε UMi	39.481	G5III		4.22	0.06	0.03	A	82.4	2.36	1.19	D
υ Sgr	137.939	B8p	F2p	*4.34*	*0.10*	*0.06*	B				
μ Sgr	180.45	B8Iap		3.79	0.13		A				
ζ Aur	972.164	K4Ib	B6V	3.75	0.15		A	90	8.30	5.60	D
VV Cep	7430.	M2Ia	B1V	*6.65*	*0.81*		A	90	84.4	41.3	D
ε Aur	9892.	F0Iap		2.94	0.89		A	90			

* These are explosive variable stars. The magnitude outside of eclipse during quiescence is shown.

Magnitude System and Interstellar Absorption

Effective wavelength of magnitude system When measuring the magnitude of a star in a certain system (wavelength band), the wavelength corresponding to the maximum efficiency is called the effective wavelength. Given the energy distribution of the star [$I(\lambda)$] and the efficiency of the observation [$S(\lambda)$] which depends on the detector sensitivity and the filter transmission function, the effective wavelength [$\lambda_{\rm eff}$] is defined as $\lambda_{\rm eff}=\int\lambda S(\lambda)I(\lambda)d\lambda/\int S(\lambda)I(\lambda)d\lambda$. The table shows the effective wavelength and the full width at half maximum (FWHM) of the efficiency function for the case of I=const. (wavelength is in unit of μm). Since I is not constant in real stars, the actual $\lambda_{\rm eff}$ should be slightly different from the values given here depending on the distribution of $I(\lambda)$. As for the details of the visible and infrared wavelength bands, refer to Fukugita *et al.* 1995, PASP, **107** and Cox (ed.) 2000, Allen's Astrophysical Quantities, respectively.

Interstellar absorption The light coming from a distant star gets more or less dimmer because of the absorption due to interstellar medium on the line of sight. Since the amount of interstellar extinction is almost inversely proportional to the wavelength in the visible region, blue light tends to suffer larger absorption than red light. Accordingly, a distant star appears generally redder compared to its actual color. The "color excess" is defined as the difference between the observed color index of a star and its actual color index, which can be estimated from its spectral type (based on the mean color of standard stars for the same type; cf. page **48**, "Physical Properties of Stars"). Accordingly, this color excess can be used as a quantitative measure of interstellar absorption. The values of interstellar absorption (A) given in the table are expressed relative to the tentatively assigned reference value of A_λ=0.27 mag at 1.03 μm (Whitford, 1948, ApJ, **107**, 102). Roughly speaking, the extent of A grows typically by ~0.4–2.0 mag as the distance of a star is increased by 1000 parsec. In reality, A depends strongly on the density distribution of the interstellar medium, and are much larger than the values in the table in some cases like the direction to the Galactic center.

	Band names	$\lambda_{\rm eff}$ (μm)	FWHM (μm)	A_λ (mag)	Methods and documentation
Old International System	*Pg* (photographic)	0.43	0.14	1.25	Photographic + Filter,
	Pv (photographic-visual)	0.54	0.052	0.97	Seares & Joyner 1943, ApJ, **98**, 302
UBV System (three-color system)	*U* (ultraviolet)	0.36	0.053	1.45	Photocell + filter; recently CCD +
	B (blue)	0.44	0.100	1.25	filter is frequently used.
	V (visual)	0.55	0.083	0.95	Johnson & Morgan 1953, ApJ, **117**, 313
Red	R_J (red)	0.69	0.21	0.62	Johnson 1965, ApJ, **141**, 923
	I_J (near infrared)	0.88	0.17	0.27	Photocell + filter; recently CCD
	R_C (red)	0.66	0.16		+ filter is frequently used.
	I_C (near infrared)	0.81	0.15		Cousins 1978, MNSSA, **37**, 8
					Kron 1980, ApJS, **43**, 305
Infrared	*J*	1.26	0.31		HgCdTe type (*I, J, H, K*),
	H	1.62	0.28		InSb type (*K, L, M, N*).
	K	2.15	0.35		SiAs type (*M, N, Q*)
	L	3.50	0.61		Johnson 1962, ApJ, **135**, 69
	M	4.85	0.62		Johnson 1965, ApJ, **141**, 923
	N	10.5	5.19		Low & Rieke 1974, Methods of
	Q	20.1	7.8		Experimental Physics, **12**, 415
Stromgren Intermediate-band	*u* (ultraviolet)	0.35	0.036		photoelectric tube + interference
	v (violet)	0.41	0.020		filter. (In recent days, CCD +
	b (blue)	0.47	0.018		interference filter is usually used.)
	y (yellow)	0.55	0.024		
SDSS System	*u′* (ultraviolet)	0.35	0.063		CCD + Filter, Fukugita *et al.* 1996,
	g′ (green)	0.48	0.14		AJ, and CCD + **111**, 1748
	r′ (red)	0.62	0.14		
	i′ (photographic infrared)	0.76	0.15		
	z′ (near infrared)	0.91	0.14		

Distance and Absolute Magnitude

Distance to celestial objects Distances of nearby objects (such as planets in the Solar System) are usually expressed in astronomical unit (au), which is the mean distance between the Sun and Earth (cf. page **1**, "Earth" section). Regarding stars and other objects outside of the Solar System, parsec (pc) or light-year (ly) is commonly used, where 1 parsec is a distance that corresponds to a parallax angle of 1 arcsecond (cf. page **32**, "Stars"). Accordingly, if an object shows a trigonometric parallax of π (arcsec), its distance is $1/\pi$ (pc). One light-year is a distance traveled by light in 1 year.

Absolute magnitude Absolute magnitude of an object is the fictitious magnitude which would be observed if it were placed at the standard distance of 10 parsec (or 32.6 light-years). This quantity is a measure of the intrinsic brightness. The following relation holds between the apparent magnitude m, the absolute magnitude M, and the parallax π (arcsec)

$$M = m + 5 + 5\log\pi$$

by which M is evaluated if m and π are known. Alternatively, this relation can be applied to estimate the distance a star spectroscopically (spectral parallax), since π can be determined from the spectroscopically guessed M (from the spectral classification) and the observed m. In this case, however, it may be necessary to correct for the effect of interstellar absorption (cf. page **46**, "Magnitude System and Interstellar Absorption"). The difference $m-M$ is called a distance modulus, which is another way of expressing the distance to a celestial object.

Various units of distance are compared with the distance modulus in the table.

pc	Light-year	au	km	Distance modulus
3.24×10^{-14}	1.06×10^{-13}	6.68×10^{-9}	1	
4.85×10^{-6}	1.58×10^{-5}	1	1.50×10^{8}	-31.57
0.307	1	6.32×10^{4}	9.46×10^{12}	-7.57
1	3.26	2.06×10^{5}	3.09×10^{13}	-5.00

pc	Light-year	Distance modulus	pc	Light-year	Distance modulus
1	3.26	-5.00	10^2	3.26×10^{2}	5
2	6.52	-3.49	10^3 = 1 kpc	3.26×10^{3}	10
5	16.3	-1.51	10^4	3.26×10^{4}	15
10	32.6	0.00	10^5	3.26×10^{5}	20
20	65.2	+1.51	10^6 = 1 Mpc	3.26×10^{6}	25
50	163	+3.49	10^9	3.26×10^{9}	40

Stellar Luminosity, Illuminance, and Radiant Energy

Illuminance of a star with apparent visual magnitude $m_v = 0$ observed at the Earth (just outside of the atmosphere) : 2.54×10^{-6} lux

Illuminance of the Sun ($m_v = -26.75$) observed at the Earth (just outside of the atmosphere) : 1.27×10^{5} lux

Intrinsic brightness of a star with absolute visible magnitude $M_v = 0$: 2.45×10^{29} cd

Intrinsic brightness of the Sun ($M_v = +4.82$) : 2.84×10^{27} cd

Radiation energy (per unit time, unit area, and unit wavelength) at 555.6 nm (human-eye sensitive wavelength) of a star with apparent visual magnitude $m_v = 0$ observed at the Earth (just outside of the atmosphere) : 3.56×10^{-11} W m^{-2} nm^{-1}

(cf. Gray, 2005. The Observation and Analysis of Stellar Photospheres)

Bolometric radiation flux (integrated over all wavelengths) of a star with apparent bolometric magnitude $m_{bol} = 0$ observed at the Earth (just outside of the atmosphere) : 2.48×10^{-8} W m^{-2}

Luminosity of a star with absolute bolometric magnitude $M_{bol} = 0$: 2.97×10^{28} W

Luminosity of the Sun ($M_{bol} = +4.74$) : 3.85×10^{26} W

(Based on Cox (ed.) 2000, Allen's Astrophysical Quantities)

Physical Properties of Stars

(1) Effective temperature Given that E is the radiation energy (Joule) emitted by a star per unit surface area (1 m^2) and per unit time (1 sec), its effective temperature (T_e) is defined as $E = 5.67 \times 10^{-8} T_e^4$ (Stefan-Boltzmann's Law).

(2) Bolometric correction Bolometric correction (B.C.) of a star is defined as B.C. = $m_{bol} - m_v$, which is the difference between the bolometric magnitude m_{bol} (corresponding to the total amount of radiation integrated over the whole spectral range) and the visual magnitude m_v (corresponding to the radiation observed in the visual band). While bolometric correction is generally negative, it approaches near to 0 around F-type stars. The B.C. of the Sun (G2V type) is -0.08.

(3) Color index The color indices $B-V$ and $U-B$ are the mutual differences between the magnitudes in the ultraviolet region (U), blue region (B), and the visual region (V). They are defined by requirement of $U-B = B-V = 0$ for the A0V type.

(4), (5), (6) Representative values of mass (in unit of solar mass), radius (in unit of solar radius), and absolute visual magnitude M_v (in mag) for main-sequence stars are given.

Spectral type (Main sequence)	(1) T_e (K)	(2) B.C. (mag)	(3) Color index $B-V$ (mag)	$U-B$ (mag)	(4) Mass ($M_\odot$)	(5) Radius ($R_\odot$)	(6) M_v (mag)
O 5	45000	-4	-0.3	-1.1	40	20	-5.5
B 0	29000	-2.8			15	8	-4
B 5	15000	-1.3	-0.16	-0.56	6	4	-1
A 0	9600	-0.2	0.00	0.00	3	2.5	+0.5
A 5	8300	0	+0.15	+0.11	2.0	1.7	+1.8
F 0	7200	0	+0.33	+0.03	1.7	1.4	+2.4
F 5	6600	0	+0.45	0.00	1.3	1.2	+3.2
G 0	6000	-0.1	+0.60	+0.12	1.1	1.0	+4.4
G 5	5600	-0.1	+0.68	+0.23	0.9	0.9	+5.1
K 0	5300	-0.2	+0.81	+0.46	0.8	0.8	+5.9
K 5	4400	-0.6	+1.15	+1.1	0.7	0.7	+7.2
M 0	3900	-1.2	+1.4	+1.2	0.5	0.6	+8.7
M 5	3300	-2.4	+1.6	+1.2	0.2	0.3	+12

(1), (2), (3) Schild *et al.* 1971, ApJ, **166**, 95 ; Davis & Webb 1970, IAU Symp., No.36 ; Johnson 1964, Bol. Tonantz. **3**. (4), (5) Allen 1973, Astrophysical Quantities. (6) Blaauw 1963, Basic Astronomical Data.

Main-sequence stars, giants, and supergiants When we plot the absolute magnitude (M) in the ordinate against the spectral type (O, B, A, F, G, K, and M) in the abscissa for a number of stars (so-called Hertzsprung-Russell diagram or HR diagram), many stars are found to locate in a narrow band running from the upper left (higher temperature and higher luminosity) to the lower right (lower temperature and lower luminosity), which is called "main sequence". Stars lying above the main-sequence on the HR diagram are called giants and supergiants. Supergiants are intrinsically very bright, the absolute magnitudes of which widely range from -3 mag to -8 mag. The luminosities of giants are between supergiants and main-sequence stars.

White dwarfs Intrinsically very faint stars with absolute magnitudes from +10 mag to +20 mag, which are thus located well below the main sequence on the HR diagram, are called "white dwarfs". It is believed that they are made up of degenerated matter of very high density. Their spatial density is quite large.

Examples of physical properties of stars The stellar mass can be directly measured only for the two cases: (1) visual binary system close to us and (2) eclipsing spectroscopic binaries. Regarding the stellar radius, it is directly determinable for the case (2) above. Alternatively, given that the distance is known, the radius can be evaluated from the angular diameter (θ_D), which can be determined by (3) stellar interferometry or (4) intensity interferometry (otherwise, the angular diameter can also be indirectly estimated from the absolute bolometric magnitude and the effective temperature). The parallax data (π) were adopted from the revised Hipparcos star catalog (2007) and Gaia (Italic) Data Release 2 (DR2 with (†) signs). Mass and radius are presented in unit of the solar values (mass = 1.988×10^{30} kg and radius = 6.960×10^8 m), and the mean density ($\bar{\rho}$) derived from mass and radius is also given. The values directly determined from observations or those indirectly derived based on observed data are shown in roman characters, while those guessed by using statistical techniques are in Italics. Asterisks (*) denote white dwarfs.

Note that the data with colon (:) indicate values with large uncertainties.

Star name	Spectral type	M_v (mag)	π (10^{-3}″)	Mass ($M_\odot$)	θ_D (10^{-3}″)	Radius ($R_\odot$)	T_e (K)	$\bar{\rho}$ (10^3 kg m^{-3})	Source
Sun	G2 V	-26.75		1.00		1.00	5777	1.41	
η Cas A	G0 V	3.44	168	0.87		*1.03*	*5940*	*1.14*	1
η Cas B	M0 V	7.22	168	0.54		*0.81*	*3800*	*1.41*	1
o^2 Eri B*	DA	9.6	201	0.59		*0.014*	*17100*	*3.3×10⁵*	6
α CMa A	A1 V	-1.46	379	2.14	6.0	1.7	10400	0.55	1, 3
α CMa B*	DA	8.44	379	0.98		*0.0085*	*26000*	*2.3×10⁶*	6
α CMi A	F5IV-V	0.38	285	1.78	5.5	2.1	6450	0.25	1, 3
70 Oph A	K0 V	4.03	197	0.89		*0.85*	*5290*	*2.0*	1
70 Oph B	K6 V	5.98	197	0.66		*0.80*	*4250*	*1.8*	1
Kr. 60 A	M4 V	9.77	250	0.26		*0.32*	*3150:*	*11:*	1
Kr. 60 B	M6 V	11.43	250	0.16		*0.25:*	*2950:*	*14:*	1
Y Cyg A	O9.5	8.3:	0.58†	17.4		5.9	*30000:*	0.12	2
YY Gem A	M1e	9.82	66.2†	0.64		0.62	3800	3.8	2
Z Her A	F4IV-V	7.3:	11.7†	1.22		1.6	*6800:*	0.42	2
Z Her B	K0IV	7.9:		1.10		2.6	*4900:*	0.09	2
ζ Phe A	B6 V	3.9:	14.7†	6.1		3.4	*15000:*	0.22	2
ζ Phe B	A0 V	5.8:		3.0		2.0	*11000:*	0.53	2
Z Vul A	B3-4 V	7.9:	1.8†	5.4		4.7	*18000:*	0.074	2
Z Vul B	A2-3 III	8.9:		2.3		4.7	*9100:*	0.031	2
ζ Aur A	K4 I b	4.0:	5.3†	8.3		160	*3700:*	2.4×10⁻⁶	2
32 Cyg A	K6 I	4.2:	2.1†	23		350	*3200:*	7.5×10⁻⁷	2
32 Cyg B	B3 V	5.6:		8.2		3.9	*19000:*	0.19	2
α Boo	K2 IIIp	-0.06	89		21	26	4200		3
α Sco	M1 I a	0.96	6:		40	720:	3500		3
β Peg	M2 II-III	2.42	17		18	110	3300		3
o Cet	M7e	8.1	11:		53	570:	2000		3
α Leo	B7 V	1.35	41		1.4	3.7	13000		4
α Lyr	A0 V	0.03	130		3.2	2.6	9500		3
α PsA	A3 V	1.16	130		2.2	1.8	9300		3
α Aql	A7 V	0.77	195		3.5	1.9	8250		3
α Ori	M2Iab	0.42	7:		45	690:	3600:		5

(1), (2) Strand (ed.) 1963, Basic Astr. Data. (3) Mozurkewich *et al.* 2003, AJ, **126**, 2502. (4) Brown *et al.* 1967, MNRAS, **137**, 393. (5) Dyck *et al.* 1996, AJ, **111**, 1705. (6) Giammichele *et al.* 2012, ApJS, **199**, 29.

The Number and Distribution of Stars

The number of stars in the full sky The number of stars brighter than magnitude 2 was counted from pages **33–35**, "Stars." The maximum magnitudes from pages **39–40**, "Pulsating variable stars," are used as the magnitudes of variable stars. The visual magnitude is rounded off to the nearest units. The star ratio shows the relative number compared to stars 1 magnitude brighter. The starlight is the brightness per square degree from this magnitude of stars. The unit is the number of 10 mag stars needed to produce equivalent brightness. (Refer to page **87**, "Night Sky Brightness", "Average Zenith Brightness of each Component of the Light", and "Brightness and Degree of Polarization Zodiacal Light"), where the full sky is about 41253 square degrees.

The number of stars in 1000 cubic parsecs The numbers of stars in the space near the Solar System against spectral type and absolute magnitude are shown, corresponding to a HR diagram (refer to page **39**, "Pulsating variable stars", and page **48**, "Physical Properties of Stars"). An empty column shows that there is no reliable value.

The frequency of observed star spectral types The number percentage for stars of each spectral type are shown down to 8.5 magnitude from the Henry Draper Catalogue.

The number of stars observed in one square degree The number of stars in each square degree decreases as galactic latitude increases. The concentration towards the galactic equator is stronger for fainter stars. The numbers in the table are the total numbers of stars brighter than the given photometeric magnitude. (Allen 1973, Astrophysical Quantities).

Number of stars in the full sky

Visual magnitude	Number of stars	Star ratio	Starlight
-1	2	-	1.2
0	7	(3.5)	1.7
1	12	(1.7)	1.2
2	67	(5.6)	2.6
3	190	2.8	2.9
4	710	3.7	4.3
5	2000	2.8	4.8
6	5600	2.8	5.4
7	1.6×10^4	2.8	6.1
8	4.3×10^4	2.7	6.5
9	1.2×10^5	2.8	7.2
10	3.5×10^5	3.0	8.4
11	8.7×10^5	2.5	8.4
12	2.3×10^6	2.7	9.0
13	5.6×10^6	2.4	8.6
14	1.3×10^7	2.4	8.2
15	3.2×10^7	2.4	7.9
16	6.9×10^7	2.1	6.7
17	1.4×10^8	2.0	5.3
18	2.8×10^8	2.0	4.2
19	4.2×10^8	1.5	2.5
20	7.1×10^8	1.7	1.7
21	1.3×10^9	1.8	1.2
>21.5	-	-	1.5
Total	2.9×10^9	-	118

Number of stars in 1000 cubic parsecs

M_v	O	B	A	F	G	K	M
-7	2×10^{-7}	5×10^{-7}	3×10^{-7}	3×10^{-7}	3×10^{-7}		
-6	5×10^{-7}	2.5×10^{-6}	1×10^{-6}	1×10^{-6}	1×10^{-6}	4×10^{-7}	4×10^{-7}
-5	1×10^{-6}	2.5×10^{-5}	1×10^{-5}	6×10^{-6}	8×10^{-6}	4×10^{-6}	1×10^{-5}
-4	3×10^{-6}	1.6×10^{-4}	1×10^{-5}	1.6×10^{-5}	2.5×10^{-5}	1.3×10^{-5}	1.3×10^{-5}
-3	1×10^{-5}	5×10^{-4}	5×10^{-5}	8×10^{-5}	8×10^{-5}	1×10^{-4}	6×10^{-5}
-2	1×10^{-5}	2.5×10^{-3}	8×10^{-5}	2×10^{-4}	3×10^{-4}	6×10^{-4}	4×10^{-4}
-1	1×10^{-5}	1.3×10^{-2}	1×10^{-3}	1.6×10^{-3}	1×10^{-3}	2.5×10^{-3}	3×10^{-3}
0	1×10^{-6}	2×10^{-2}	2×10^{-2}	2×10^{-3}	8×10^{-3}	2.5×10^{-2}	1×10^{-2}
1		3×10^{-2}	1×10^{-1}	3×10^{-2}	3×10^{-2}	1.2×10^{-1}	1×10^{-2}
2		2×10^{-2}	2×10^{-1}	1.6×10^{-1}	5×10^{-2}	1.1×10^{-1}	
3		1×10^{-2}	8×10^{-2}	7×10^{-1}	1.5×10^{-1}	1×10^{-1}	
4			3×10^{-2}	1.2	7×10^{-1}	1×10^{-1}	
5				6×10^{-1}	2	3×10^{-1}	
6				2×10^{-1}	1.5	1.5	1×10^{-2}
7				1×10^{-1}	8×10^{-1}	3	1×10^{-1}
8				1×10^{-2}	4×10^{-1}	2.5	1
9					2×10^{-1}	1.5	3
10		1×10^{-2}				4×10^{-1}	8
11		1×10^{-1}	3×10^{-2}	1×10^{-2}		2×10^{-1}	9
12		2×10^{-1}	4×10^{-1}	1×10^{-1}		1×10^{-1}	1×10
13		4×10^{-1}	6×10^{-1}	3×10^{-1}	1×10^{-1}	4×10^{-1}	1×10
14		8×10^{-1}	1	1	6×10^{-1}	8×10^{-1}	1×10
15		1.5	2	1	1.5	1.2	8
16		3	5	3	3		6

Frequency of observed star spectral type

Spectral type	O	B	A	F	G	K	M
% of stars	1	10	22	19	14	31	3

Number of stars observed in one square degree

Photographic magnitude	Galactic latitude =0°	±5°	±10°	±20°	±30°	±40°	±50°	±60°	±90°	Average
5	0.052	0.044	0.037	0.028	0.020	0.017	0.016	0.015	0.013	0.023
10	9.3	7.6	6.3	4.6	3.5	2.9	2.5	2.2	1.9	4.2
15	1200	1000	760	450	290	200	160	130	93	420
20	50000	50000	40000	16000	6000	4000	2500	2000	1300	15000

Stellar Dynamics

The solar motion The movement of individual stars in the solar neighborhood consists of the movement of the center of mass velocity for the entire cluster of stars (galactic rotation) and random motion relative to the center of mass velocity (peculiar motion). The peculiar motion of the Sun in particular is called the solar motion and its direction is called the solar apex. The value and the direction of the solar motion depends on the type of stars in the solar neighborhood used for reference and the analysis method but typical values are shown. The three components of the solar motion are written as radially outward from the galactic center $u_\odot$, the direction of galactic rotation $v_\odot$, and perpendicular to the Milky Way Disk $w_\odot$. (Table 1)

$u_\odot$ km s^{-1}	$v_\odot$ km s^{-1}	$w_\odot$ km s^{-1}	Total speed km s^{-1}	Solar apex		Reference
				Galactic longitude	Galactic latitude	
-9	12	7	16.5	53°	25°	Mihalas & Binney 1981, Galactic Astronomy
-11.10	12.24	7.25	18.0	48°	24°	Schönrich *et al.* 2010, MNRAS, **403**, 1829

Velocity distribution of stars The velocity distribution of the peculiar motion of stars in the solar neighborhood is described to a good approximation by an ellipsoidal body of three unequal axes (tri-axial velocity ellipsoid). This anisotropy of the velocity distributions shows that the statistical movement of the stellar group has not reached in the state of the thermal equilibrium yet. Orientation and velocity dispersion (σ_1, σ_2, σ_3) along the axis of the velocity ellipsoid are different for star clusters composed of different types of stars. The total velocity dispersions increase in proportion to roughly the 1/2-1/3th power of the age of the star cluster. (Table 2)

Star sample	Color index and spectral type	Velocity dispersion				Galactic longitude of major-axis orientation
		σ_1	σ_2	σ_3	Total velocity dispersion	
		km s^{-1}	km s^{-1}	km s^{-1}	km s^{-1}	°
Main Sequence	$0.14<B-V<0.31$	20	9	8	23	22.8
	$0.31<B-V<0.41$	22	12	9	27	19.8
	$0.41<B-V<0.47$	26	16	12	33	10.2
	$0.47<B-V<0.53$	30	18	13	37	6.9
	$0.53<B-V<0.58$	33	22	15	42	1.9
	$0.58<B-V<0.64$	38	23	21	49	10.2
	$0.64<B-V<0.72$	38	24	21	50	7.6
	$0.72<B-V<1.54$	37	26	18	49	13.1
Giant Star	A type	22	13	9	27	27
	F type	28	15	9	33	14
	G type	26	18	15	35	12
	M type	31	21	16	41	7
Subdwarf		100	75	50	135	-
RR Lyr		180	111	93	231	-

Dehnen & Binney 1998, MNRAS, **298**, 387; Delhaye 1965, Stars and Stellar Systems, ed. Blaauw & Schmidt, p. 61; Martin & Morrison 1998, AJ, **116**, 1724

Moving cluster Clusters consisting of dozens to hundreds of spatially grouped stars moving in a common direction are called a moving cluster. It is essentially equivalent to an open star cluster. The magnitude of the proper motion and the line-of-sight velocity of the member stars comprising the moving cluster seem to be almost constant, and the directions of the proper motions seem to converge towards a single point on the celestial globe. This one point is called the moving cluster's convergent point. (Table 3)

Name	Line-of-sight velocity	Proper motion		Rate of movement relative to the Sun	Convergent point (2000 Equinox)	
		$\mu_\alpha \cos\delta$	μ_δ		Right ascension	Celestial declination
	km s^{-1}	10^{-3} (" Year^{-1})	10^{-3} (" Year^{-1})	km s^{-1}	°	°
Hyades	38.4	106.0	-29.6	45.4	98	6
Coma Ber	-1.2	-11.8	-8.7	6.1	102	-38
Pleiades	5.7	20.1	-45.4	28.9	93	-48
Praesepe	33.6	-35.8	-12.9	47.0	89	1
α Per	-1.6	22.7	-26.5	28.6	103	-33

van Leeuwen 2009, A&A, **497**, 209

Exoplanetary Systems

Summary of exoplanetary systems

With the discovery of an object with some half of Jupiter mass orbiting a main-sequence star in 1995, exoplanet observations and theories have become one of the central topics of modern astronomy. As of June 2020, approximately 4 000 planets with masses ranging from 13 Jupiter masses down to approximately 1 Earth mass have been discovered. TESS mission has already discovered about 50 exoplanets. Behind the scenes of this discovery rush there were van de Kamp's observations of Barnard's star from the 1930's; the discovery of pulsar planets by Wolszczan *et al.* (1992), and Walker *et al.*'s trial to discover Jovian planets (1995) over the course of 12 years.

Major indirect observation methods

1) Doppler method A star is wobbled by the orbital motion of the planet so its speed and position fluctuate. This technique measures those velocity changes. Also known as the radial velocity method. Since the Mayor and Queloz's discovery of 51 Peg b in 1995, this has been one of the most successful methods. In the Solar System, the velocity variations of the Sun caused by orbital motion of Jupiter and Earth are 13 m s^{-1} and 0.1 m s^{-1}, respectively. A velocity measurement accuracy on the order of several m s^{-1} is required, even for the detection of a giant planet. A lower limit for the planetary mass, orbit semi-major axis, and eccentricity are obtained by this technique.

2) Astrometry method A method of detecting a planet by precise measurements of positional variations of the central star caused by the planetary orbital motion. This method has been tried for a long time, but the required positional measurement accuracy is much smaller than the Earth's atmospheric seeing. In the case of observing the Solar System from 10 pc away, the effect of Jupiter on the Sun would be a 0.5 milli-arcsecond wobble. Micro-arcsecond level accuracy is needed for the detection of Earth-like planets.

3) Transit method A technique for detecting small brightness changes caused by a planet passing in front of a star (occultation). The brightness changes caused by Jupiter and Earth are about 1% and 0.01%, respectively. In 2000, Charbonneau *et al.* reported a detection of brightness changes of HD 209458 in agreement with velocity variations found through the Doppler method for the first time, and confirmed the existence of the planet via two independent indirect methods.

4) Gravitational microlensing method Light from a distant celestial object is refracted and focused by the gravity of an intervening star (gravitational lensing). When the lensing object moves relative to the background object and the image is irresolvable, a characteristic brightening pattern is seen. In addition, when a planet exists around the lensing object, the brightening curve contains spike-like asymmetric variations. The existence of the planet can be detected indirectly by monitoring observations of this short brightening.

5) Pulsar timing method The periodical timing variations in the regular pulses from a neutron star due to the Doppler effect caused by the orbital motion of a planet are detected through radio observations. Pulsar PSR 1257+12 was the first celestial object found to be harboring exoplanets (but pulsars are not considered "normal stars" in this article.)

Direct observation methods

These are imaging observations of light directly from the planet. It is necessary to achieve high sensitivity, high resolution, and high contrast at the same time. The largest problem is contrast. At optical and near-infrared wavelengths the light from planets is the reflected light from the Sun and the brightness ratio reaches as much as 9–10 orders of magnitude. At mid-infrared and longer wavelengths, the brightness ratio is reduced due to the planet's own thermal emission, but still 7 orders of magnitude.

Exoplanet observation statistics as of June, 2020.

Technique	Number of planetary systems	Number of planets	Comment
Doppler method	654	888	
Transit method	2286	3052	Confirmed by other techniques.
Microlens method	111	122	
Pulsar timing method	37	43	Including objects other than pulsars
Direct imaging method	9	12	Less than 13 Jupiter masses, within 1000 au. Exduding objects around brown dwarfs.

Representative Exoplanets

Observation method	Name	Mass lower limit (Jupiter Masses)	Orbit semi-major axis (au)	Period (Days)	Distance (pc)	Remarks
Doppler method	Gliese 581b	0.05	0.041	5.4	6	
	c	0.017	0.073	12.9	6	5 Earth masses
	d	0.019	0.22	66.6	6	8 Earth masses
	e	0.006	0.03	3.1	6	2 Earth masses
	51 Pegasi b	0.47	0.05	4.2	15	First exoplanet discovered around a main sequence star
	55 Cancri b	0.82	0.12	14.7	13	Multiple planetary system and circumstellar dust exists around the primary star
	c	0.17	0.24	44.3	13	
	d	3.84	5.77	5 218	13	It resembles Jupiter
	e	0.027	0.016	0.7	13	
	f	0.14	0.78	260	13	
	HD20782b	1.9	1.4	592	36	The exoplanet with the greatest eccentricity (0.97) to date
	ε Eridani b	1.55	3.4	2 502	3	Asymmetric dust ring with a radius of 30–60 au
	Proxima Centauri b	0.004	0.049	11.2	1.3	1.3 Earth masses, nearest planet
Transit method + Doppler method	Kepler-11b	0.0027	0.011	1.51	12.1	4 Earth masses, multi-planet system discovered by Kepler
	c	0.0043	0.015	2.42	12.1	14 Earth masses
	d	0.0013	0.021	4.05	12.1	6 Earth masses
	e	0.0020	0.028	6.10	12.1	8 Earth masses
	f	0.0021	0.037	9.21	12.1	2 Earth masses
	g	0.0042	0.045	12.35	12.1	<302 Earth masses
	HAT-P-7b	1.8	0.038	2.2	320	Retrograde planet
	HD189733b	1.15	0.031	2.2	19.3	Sodium, water, and methane were discovered in the planet's atmosphere
	HD149026b	0.36	0.043	2.9	79	Planet is thought to have a huge core
	HD209458b	0.64	0.047	3.5	47	First exoplanet confirmed by both the transit method and the Doppler method
	Trappist-1b-h	0.4–1.4 Earth masses	0.02–0.06	1.5–2.0	12.1	7-planets system, mass measured by the transit timing variation method
Other techniques	PSR 1257+12b	0.020 Earth masses	0.19	25.3	300	Discovered by the pulsar method, 4-planets system
	c	4.3 Earth masses	0.36	66.5	300	
	d	3.9 Earth masses	0.46	98.2	300	
	OGLE-2013-BLG-0341 Lb	1.7 Earth masses	0.70	–	910	The smallest mass exoplanet discovered by the microlensing method
	S Orionis 70	3	–	–	440	An example of a sub-brown dwarf. Direct detection. There are several similar objects
	HR8799b	7	68	470 Years	39	A planet orbiting an A Type star (Direct imaging, assumed age 60 million years)
	c	10	38	190 Years	39	
	d	10	24	100 Years	39	
	e	9	14.5	50 Years	39	
	GJ 504 b	3	44	–	57	A planet orbiting a G Type star (direct imaging, assumed age 100 million years)
	κ Andromedae b	13	56	–	52	A planet orbiting a B Type star (direct imaging, assumed age 30 million years)

Candidate objects which have not been confirmed yet are also included in this table.

Note: 1) Brown dwarfs are defined as objects where stable hydrogen nuclear fusion does not occur, but deuterium ignites. Planets are defined as objects light-weight enough that even deuterium does not ignite which orbit a star. The boundary-mass is roughly 13 times the mass of Jupiter. An isolated celestial object with the mass of a planet which does not orbit a star is called a sub-brown dwarf. It is also called a rogue planet. It is possible to detect these by the microlensing method.

2) Because mass determined by the Doppler method and the pulsar method includes the uncertainty of the orbit inclination, this is a lower limit ($M \sin i$).

3) Because the orbit inclination is obtained from the transit method, the mass is not a lower limit but the exact value.

4) υ Andromeda is the first example of a multiple planet system orbiting a double star (distance 55″ = 750 au).

5) For the latest data, please refer to J. Schneider. http://exoplanet.eu/

6) No Doppler data yet for Trappist-1.

Brown Dwarf

A brown dwarf is a very low-mass celestial object that cannot maintain a stable hydrogen nuclear burning because its mass is too small. This corresponds celestial objects less than about 80 Jupiter masses (about 0.08 Solar masses). Such objects are relatively bright at infrared wavelengths immediately after they are formed due to gravitational energy release from contraction and deuterium nuclear fusion reactions; they cool with time, and become very faint objects at visible wavelengths. In terms of spectral classification, some M type stars, some of the even lower temperature L type stars, and all T type stars are brown dwarfs. Their existence was predicted theoretically by C. Hayashi and T. Nakano and independently by S. Kumar both in 1963. But, objects which could definitively be called brown dwarfs were not found until 1995 as members of open clusters or as companions to stars. The first example of what are now classified as L type brown dwarfs was discovered in 1988. Thanks to recent wide field surveys, many isolated brown dwarfs are found in the field as well as in open clusters and star forming regions. Likewise, among extremely low mass celestial bodies, objects with mass less than 13 Jupiter masses which circle a star are often called planets.

Representative Brown Dwarfs

Name	2000 Equinox		Infrared magnitude (mag)*	Distance (light-years)	Mass (Jupiter masses)	Spectral type	Discovery year	Remarks
	R.A.	Dec.						
	h m	° ′						
GD165 B	14 24	+09 17	14.1	103	~70	L4	1988	White dwarf companion
Gliese 229 B	06 11	-21 52	14.4	19	40	T6	1995	M star companion
Teide 1	03 47	+24 23	15.1	380	55	M8	1995	Open cluster member
OTS 44	11 10	-76 32	14.6	554	15	M9.5	1998	Star forming region member
Orion 107-453	05 35	-05 25	17.9	1500	~5	~M8	2005	Star forming region member, below 10 Jupiter masses
2M1207 A	12 08	-39 33	11.9	230	~25	M8	2002	Companion star has planetary mass
2M1207 B	12 08	-39 33	16.9	230	~6	M8.5-L4	2004	Binary brown dwarf
Luhman 16A	10 49	-53 19	8.8	6.5	~5	L7.5	2013	The nearest binary brown
Luhman 16B	10 49	-53 19	8.9	6.5	~4	T0.5	2013	dwarf to the Sun
WISEP J1541	15 41	-22 50	~21	27	~12	Y0	2011	One of the lowest temperature brown dwarfs (~350K)
UGPS0722	07 22	-05 40	17.1	13	~10	T10	2010	One of the lowest temperature brown dwarfs observed from the ground (~520K)

* Magnitude in the K or Ks band (wavelength approximately 2 μm).

The Milky Way Galaxy

Like the Andromeda Galaxy (NGC 224), the Milky Way Galaxy is a large spiral galaxy. It is composed of a disk, a spheroid, and a dark halo components. The disk consists of Population I objects rich in heavy elements, such as dG, dk, dM type stars, O, B type stars, open clusters, and interstellar gas, which are revolving at high speed associated with Galactic rotation. The interstellar gas and the open clusters are distributed on the disk in spiral patterns. Three spiral arms, the Sagittarius, Orion, and Pegasus arms can be clearly seen. The Sun is located on the outskirts of the inner side of the Orion Arm. The spheroid consists of Population II objects with limited amounts of heavy elements such as variable stars and globular clusters which hardly show rotation around the galaxy. The dark halo is an invisible part surrounding the disk and spheroid. Its existence is suggested by the movement of globular clusters and companion galaxies like the Magellanic Clouds in the halo of the Milky Way. The definite numbers for the extent and mass of the dark halo are unknown, but based on recent research using the movement of globular clusters and companion galaxies, it is inferred to reach 10 times the extend and mass of the visible portion of the galaxy.

Effective diameter of the Milky Way disk 100 000 light-years

Effective diameter of globular cluster system 150 000 light-years

Thickness of Milky Way disk 15 000 light-years (Central part) 2 000 light-years (in the Solar neighborhood)

Total Visible Mass of Milky Way $10^{11}\,M_\odot$ ($M_\odot$ is a solar mass)

Position of the Sun † The distance from the galactic center in the galactic plane 28 000 ± 3 000 light-years

Milky Way rotational speed in the solar neighborhood † 220 ± 20 km s^{-1}

Matter density in the solar neighborhood(stars + interstellar gas) $4.3 \times 10^{-3} M_\odot$ (light-year)$^{-3}$

Density of interstellar gas in the solar neighborhood $5.8 \times 10^{-4} M_\odot$ (light-year)$^{-3}$ = 0.8 hydrogen atoms cm^{-3}

† Value recommended by IAU commission 33 in 1985. The trend in recent observations, place the Sun's possition at 26 000–27 000 light-years and the galactic rotation velocity in the solar neighborhood at larger than 220 km s^{-1}. But the exact numbers haven't been decided.

Galactic coordinates On the celestial sphere Galactic coordinates (galactic longitude l and galactic latitude b) are defined using the great circle of maximum radio strength (corresponding roughly with the center line of the Milky Way) as the galactic equator. Galactic longitude l is measured based on the direction of radio source Sgr A in the constellation Sagittarius considered to be the center of the Milky Way. The following values are used for conversion from equatorial coordinates at the J2000.0 equinox to galactic coordinates.

The Milky Way Center	α = 192°.859 48	Equinox 2000.0
Right ascension of the Galactic North Pole	δ = +27°.128 25	
Celestial defination of the Galactic North Pole	l = 32°.931 92	

Galactic rotation The rotational speed distribution of the Milky Way galactic rotation is obtained from the analysis of the line-of-sight velocity and the proper motions of objects such as neutral hydrogen gas distributed in the galactic plane, H II regions, CO molecular clouds, open clusters, O and B type stars. The figure shows the Milky Way rotational speed distribution from the center of the Milky Way out to a distance of 15 kpc. According to recent radio VLBI observations, the rotation velocity can be said to be effectively constant in the region from 3 kpc to 15 kpc. The total mass of the Milky Way is strongly dependent on the state of the galactic rotation in the even outer region.

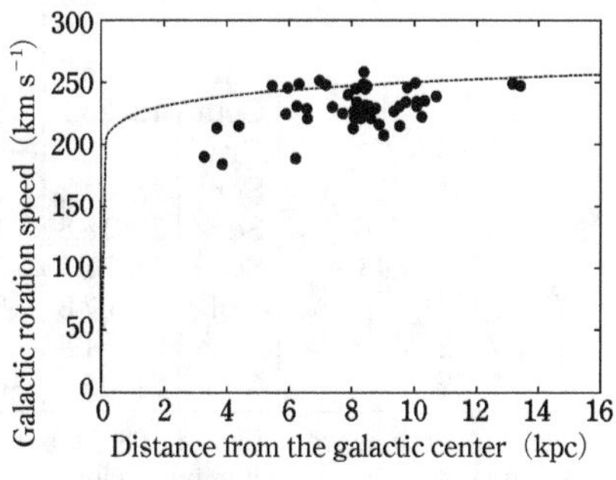

Figure 9 Galactic rotation speed
Honma *et al.* 2012, PASJ, **64**, 136

Star Clusters in

An object where many stars are gathered together on the celestial sphere is called a star cluster. There are two varieties: open clusters and globular clusters. The former is a comparatively incohesive group and in general the distances are short. In the latter, stars are more crowded closer to the center and in general the distances are long. There are up to 1 500 open clusters found near the galactic plane. About 150 globular clusters have been discovered distributed widely between the direction towards the ga-

Major Open Clusters

Name	Constellation	2000 Equinox R.A.	2000 Equinox Dec.	Galactic longitude l	Galactic latitude b	Apparent diameter	Brightest star	Distance	Color index $(B-V)$
		h m	° ′	°	°	′	mag	kly	
NGC 188	Cep	00 47.5	+85 15	122.8	+22.4	13	10	5.05	0.81
NGC 581, M103	Cas	01 33.4	+60 39	128.1	- 1.8	6	9	8.80	0.32
NGC 752	And	01 57.7	+37 47	137.1	-23.3	49	8	1.30	0.77
NGC 869, h Per[1)]	Per	02 19.0	+57 07	134.6	- 3.7	29	7	7.17	0.44
NGC 884, χ Per[1)]	Per	02 22.4	+57 07	135.1	- 3.6	29	7	7.50	0.48
NGC 1039, M34	Per	02 42.1	+42 45	143.7	-15.6	35	9	1.43	0.17
NGC 1245	Per	03 14.7	+47 14	146.6	- 8.9	10	12	7.50	0.76
Mel 20, α Per	Per	03 22.1	+48 37	146.6	- 5.9	184	3	0.55	
M45, Pleiades	Tau	03 47.0	+24 07	166.6	-23.5	109	3	0.41	
NGC 1528	Per	04 15.4	+51 12	152.1	+ 0.3	23	10	2.61	0.43
Mel 25, Hyades	Tau	04 26.9	+15 52	180.1	-22.3	329	4	0.16	0.40
NGC 1912, M38	Aur	05 28.7	+35 50	172.3	+ 0.7	21	8	4.30	0.29
NGC 1960, M36	Aur	05 36.3	+34 08	174.5	+ 1.1	12	9	4.14	0.09
NGC 2099, M37	Aur	05 52.3	+32 33	177.6	+ 3.1	23	11	4.40	0.59
NGC 2264, S Mon	Mon	06 41.0	+09 53	202.9	+ 2.2	20	5	2.45	
NGC 2324	Mon	07 04.1	+01 02	213.5	+ 3.3	7	12	9.45	0.43
NGC 2362	CMa	07 18.7	-24 57	238.2	- 5.5	8	8	5.05	
M 44, Praesepe	Cnc	08 40.0	+19 59	205.5	+32.5	95	6	0.59	
IC 2391	Vel	08 40.5	-53 02	270.4	- 6.8	49	4	0.46	
NGC 2682, M67	Cnc	08 51.3	+11 48	215.7	+31.9	29	9	2.35	0.70
IC 2602	Car	10 43.0	-64 24	298.6	- 4.9	49	3	0.51	
NGC 3532	Car	11 05.7	-58 45	289.6	+ 1.3	55	8	1.63	0.28
Mel 111, Coma Berenices	Com	12 25.1	+26 06	221.4	+84.0	275	5	0.28	0.46
NGC 4755, κ Cru	Cru	12 53.7	-60 21	303.2	+ 2.5	10	7	7.63	0.31
NGC 6531, M21	Sgr	18 04.2	-22 29	7.7	- 0.4	13	8	4.24	0.12
NGC 6611, M16	Ser	18 18.8	-13 48	17.0	+ 0.8	6	11	8.15	0.58
NGC 6705, M11	Sct	18 51.1	-06 16	27.3	- 2.8	13	11	5.61	0.52
NGC 6755	Aql	19 07.8	+04 16	38.6	- 1.7	14	11	4.89	1.10
NGC 7092, M39	Cyg	21 31.8	+48 26	92.4	- 2.2	31	7	0.88	0.06
NGC 7142	Cep	21 45.2	+65 46	105.3	+ 9.5	4	11	3.26	1.06

Data are taken from Lynga 1987, Open Cluster Data 5th Edition, and Dias *et al.* 2002, New catalogue of optically visible open clusters and candidates.

1) This pair cluster can be called h+χ Per.

the Milky Way

lactic center and the halo at high galactic latitudes surrounding it. The type (age and metallicity) of the stars comprising open clusters and globular clusters are systematically different. The former have young stars with many heavy elements and the latter have old stars with few heavy elements. The table shows the main examples. NGC, IC, and M are respectively New General Catalogue, Index Catalogue and Messier star catalogue numbers. Pal are the objects discovered by Abell from the Palomar chart star map.

Major Globular Clusters

Name	Constellation	2000 Equinox		Galactic longitude	Galactic latitude	Tidal diameter	Visual magnitude	Distance	Color index
		R.A.	Dec.	l	b				$(B-V)$
		h m	° ′	°	°	′	mag	kly	
NGC 104, 47Tuc	Tuc	00 24.1	-72 05	305.9	-44.9	94	4.0	14.7	0.88
NGC 2419	Lyn	07 38.1	+38 53	180.4	+25.2	17	10.4	274.5	0.66
Pal 4	UMa	11 29.3	+28 58	202.3	+71.8	7	14.2	356.0	
NGC 4147	Com	12 10.1	+18 33	252.9	+77.2	13	10.3	62.9	0.59
NGC 4590, M68	Hya	12 39.5	-26 45	299.6	+36.1	61	7.8	33.3	0.63
NGC 5024, M53	Com	13 12.9	+18 10	333.0	+79.8	44	7.6	58.0	0.64
NGC 5053	Com	13 16.5	+17 42	335.7	+78.9	27	9.5	53.5	0.65
NGC 5139, ω Cen	Cen	13 26.8	-47 29	309.1	+15.0	114	3.7	17.3	0.78
NGC 5272, M3	CVn	13 42.2	+28 23	42.2	+78.7	76	6.2	33.9	0.69
NGC 5466	Boo	14 05.5	+28 32	42.2	+73.6	68	9.0	51.8	0.67
Pal 5	Ser	15 16.1	-00 07	0.9	+45.9	33	11.8	75.6	
NGC 5904, M5	Ser	15 18.6	+02 05	3.9	+46.8	57	5.7	24.5	0.72
NGC 6121, M4	Sco	16 23.6	-26 32	351.0	+16.0	65	5.6	7.2	1.03
NGC 6171, M107	Oph	16 32.5	-13 03	3.4	+23.0	35	7.9	20.9	1.10
NGC 6205, M13	Her	16 41.7	+36 28	59.0	+40.9	50	5.8	25.1	0.68
NGC 6218, M12	Oph	16 47.2	-01 57	15.7	+26.3	35	6.7	16.0	0.83
NGC 6254, M10	Oph	16 57.1	-04 06	15.1	+23.1	43	6.6	14.3	0.90
NGC 6266, M62	Oph	17 01.2	-30 07	353.6	+ 7.3	18	6.5	22.5	1.19
NGC 6341, M92	Her	17 17.1	+43 08	68.3	+34.9	30	6.4	26.7	0.63
NGC 6356	Oph	17 23.6	-17 49	6.7	+10.2	16	8.3	49.6	1.13
NGC 6397	Ara	17 40.7	-53 40	338.2	-12.0	32	5.7	7.5	0.73
NGC 6402, M14	Oph	17 37.6	-03 15	21.3	+14.8	66	7.6	30.3	1.25
NGC 6522	Sgr	18 03.6	-30 02	1.0	- 3.9	33	8.3	25.4	1.21
NGC 6656, M22	Sgr	18 36.4	-23 54	9.9	- 7.6	58	5.1	10.4	0.98
NGC 6712	Sct	18 53.1	-08 42	25.4	- 4.3	15	8.1	22.5	1.17
NGC 6779, M56	Lyr	19 16.6	+30 11	62.7	+ 8.3	17	8.3	32.9	0.86
NGC 6838, M71	Sge	19 53.8	+18 47	56.7	- 4.6	18	8.2	13.0	1.09
NGC 7078, M15	Peg	21 30.0	+12 10	65.0	-27.3	43	6.2	33.6	0.68
NGC 7089, M2	Aqr	21 33.5	-00 49	53.4	-35.8	43	6.5	37.5	0.66
NGC 7492	Aqr	23 08.4	-15 37	53.4	-63.5	17	11.3	84.1	0.42

Data are taken from Harris 2003, Catalog of Parameters for Milky Way Globular Clusters : Harris 1996, AJ, **112**, 1487.

Nebulae in the Milky Way

A nebula is a region in interstellar space where the interstellar medium density is high. A nebula shines either with light emitted by ionized gas, or light reflected and scattered by solid particles.

Diffuse nebula An emission nebula (abbreviated as E) is a shining gas cloud ionized by nearby hot stars. A reflection nebula (abbreviated as C) reflects light from nearby stars and bright in continuum. They can reach diameters of tens of light-years. They often accompany a young open cluster.

Planetary nebula (abbreviation : PN) An object of illuminated gas emitted from a central, high temperature dead star. The diameter is on the order of 1 light-year. The longevity is on the order of tens of thousands of years.

Supernova remnant (abbreviation : SNR) Shining gas compressed by a shockwave traveling through the interstellar medium. The diameter is on the order of 100 light-years. The longevity is on the order of hundreds of thousands of years.

Name	Constellation	Equinox J2000		Galactic longitude l	Galactic latitude b	Apparent diameter	Distance	Classification
		R.A.	Dec.					
		h m	° ′	°	°	′	ly	
NGC 7822, S171, W1	Cep	00 04	+68 37	119	+ 6	170	5.0	E
NGC 246	Cet	00 47	-11 53	119	-75	4	1.3	PN
NGC 281	Cas	00 52	+56 36	123	- 6	12	5.5	E
IC 1795, W3	Cas	02 26	+61 51	133	+ 1	20	6.4	E
IC 1805, W4	Cas	02 33	+61 26	135	+ 1	50	6.2	E
IC 1848, W5	Cas	02 51	+60 25	137	+ 1	50	4.9	E
NGC 1432	Tau	03 45	+24 22	167	-24	40	0.41	C
NGC 1499, California Nebula	Per	04 00	+36 37	161	-12	140	2.3	E
NGC 1952, M1, Crab Nebula	Tau	05 34	+22 01	185	- 6	5	7.2	SNR
NGC 1976-7, M42, Orion Nebula, W10	Ori	05 35	-05 27	209	-19	35	1.4	E
IC 434, W12	Ori	05 41	-02 24	207	-17	30	1.1	CE
NGC 2068, M78	Ori	05 46	+00 03	205	-14	4	1.6	C
NGC 2174-5, W13	Ori	06 09	+20 30	190	0	15	5.2	E
NGC 2237-38-44-46, Rosette Nebula, W16	Mon	06 32	+05 03	206	- 2	60	4.6	E
NGC 2261, Hubble's Variable Nebula	Mon	06 39	+08 44	204	+ 1	0.5	4.9	CE
NGC 2264	Mon	06 40	+09 54	203	+ 2	60	2.6	E
NGC 3132, Eight-Burst Nebula	Ant	10 07	-40 26	272	+12	0.8	3.8	PN
NGC 3587, M97, Owl Nebula	UMa	11 14	+55 01	148	+57	3	1.8	PN
NGC 6514, M20, Trifid Nebula	Sgr	18 02	-23 02	7	0	15	5.6	E
NGC 6523, M8, Lagoon Nebula, W29	Sgr	18 03	-24 23	6	- 1	25	3.9	E
NGC 6611, M16, Eagle Nebula, W37	Ser	18 18	-13 47	17	+ 1	12	5.5	E
NGC 6618, M17, Omega Nebula, W38	Sgr	18 20	-16 11	15	- 1	20	6.5	E
NGC 6720, M57, Ring Nebula	Lyr	18 53	+33 02	63	+14	1	2.6	PN
NGC 6853, M27, Dumbbell Nebula	Vul	19 59	+22 43	61	- 4	7	0.82	PN
NGC 6960-92-95, Veil Nebula	Cyg	20 45	+30 43	75	- 9	150	1.8	SNR
IC 5067-68-70, Pelican Nebula, W80	Cyg	20 48	+44 22	85	0	60	2.0	CE
NGC 7000, North America Nebula, W80	Cyg	20 58	+44 20	86	- 1	100	2.0	CE
NGC 7009, Saturn Nebula	Aqr	21 04	-11 22	37	-34	0.7	4.1	PN
NGC 7027	Cyg	21 07	+42 14	84	- 3	0.3	4.4	PN
NGC 7293, Helix Nebula	Aqr	22 29	-20 48	36	-57	13	0.49	PN

Bečvář 1964 ; Allen 1973, Astrophysical Quantities ; Cahn & Kaler 1971, ApJS, **22**, 319 ; Sky Catalog 2000.0 ; Stanghellini *et al.* 2008, ApJ, **689**, 194. Refer to pages **69, 71, 76.**

Galaxies

Outside the Milky Way, celestial objects with the similar scales and structures as the Milky Way are called galaxies. They are classified broadly into elliptical (E), lenticular (S0), spiral (S), and the irregular (I) type according to their morphologies. E and S are further subdivided into sub-classes. E is appended with numbers ranging from 0 to 7 from round to elongated galaxies and S is appended with a letter a,b,c,d,m from tight to open spirals. This system is called the Hubble Sequence. The left (E type side) are called early type and the right (I type side) are called late type. However, this nomenclature is unrelated to the actual evolution of the galaxies. The presence of a bar-like structure is shown with A (no bar), B (with bar), and AB (intermediate). The suffix p indicates peculiar and the prefix d indicates dwarf.

The group of galaxies within a radius of about three million light-years centered on the Milky Way and Andromeda galaxies is called as the Local Group. The major member galaxies, excluding the Milky Way, are shown in the table below. The table in the next page shows other bright galaxies. The distances are based on Tully *et al.* 2016, AJ, **152**, 50. The average values and standard deviations calculated from the distances derived by various distance measurement methods are listed as the distance and uncertainty, respectively.

Galaxies of the Local Group

Name	Equinox J2000		Galactic longitude l	Galactic latitude b	Type	Magnitude B	Color index $B-V$	Apparent diameter	Distance
	R.A.	Dec.							
	h m	° ′	°	°		mag	mag	′ ′	10 kly
NGC 147	00 33.2	+48 30	119.8	-14.3	E5p	10.4	0.94	13× 8	218(15)
NGC 185	00 39.0	+48 20	120.8	-14.5	E3p	10.1	0.90	11× 10	206(13)
NGC 205, M110	00 40.4	+41 41	120.7	-21.1	E5p	8.9	0.84	17× 10	264(19)
NGC 221, M32	00 42.7	+40 52	121.2	-22.0	E2	9.2	0.94	8× 6	253(19)
NGC 224, M31	00 42.7	+41 16	121.2	-21.6	SAb	4.4	0.91	180× 63	250(14)
SMC	00 52.7	-72 50	302.8	-44.3	SBmp	2.8	0.50	280×160	20(1)
Sculptor	01 00.0	-33 42	287.8	-83.2	dE3p	9	-	20× 20:	27(2)
IC 1613	01 04.8	+02 07	129.7	-60.6	IABm	10.0	0.60	12× 11	243(11)
NGC 598, M33	01 33.9	+30 39	133.6	-31.3	SAcd	6.3	0.55	62× 39	296(14)
Fornax	02 39.9	-34 31	237.3	-65.7	dE0p	9.0	-	20× 14	48(4)
LMC	05 23.6	-69 45	280.5	-32.9	SBm	0.6	0.55	650×550	16(1)
Carina	06 41.6	-50 58	260.1	-22.2	dE	10.4:	-	24× 16:	35(3)
Leo I	10 08.5	+12 18	226.0	+49.1	dE3	11.1	0.97	11× 8	84(5)
Leo II	11 13.5	+22 10	220.2	+67.2	dE0p	12.4:	0.90	14× 13	78(6)
Ursa Minor	15 08.8	+67 12	105.0	+44.8	dE4	11.6:	-	32× 21	22(3)
Draco	17 20.0	+57 55	86.4	+34.8	dE0p	11.9:	-	40× 25	26(3)
NGC 6822	19 44.9	-14 48	25.3	-18.4	IBm	9.4	-	10× 10	157(9)

In addition to the galaxies listed in the table, the Local Group includes about 90 galaxies if very faint galaxies found in recent years are included. In the table M31 is the Andromeda Galaxy, SMC is the Small Magellanic Could and LMC is the Large Magellanic Cloud. (:) indicates a very uncertain value.

Bright Galaxies Outside the Local Group

Name	Constellation	Equinox J2000 R.A.	Equinox J2000 Dec.	Galactic longitude l	Galactic latitude b	Type	Magnitude B	Color index $B-V$	Apparent diameter	Distance
		h m	° ′	°	°		mag	mag	′ ′	10 kly
NGC 55	Scl	00 14.9	-39 11	333	-76	SBm	7.9	-	32×6	650(40)
NGC 247	Cet	00 47.1	-20 46	114	-84	SABd	9.4	0.65	20×7	1 150(60)
NGC 253	Scl	00 47.6	-25 17	98	-88	SABc	8.0	0.97	25×7	1 160(90)
NGC 300	Scl	00 54.9	-37 41	299	-79	SAd	8.7	-	20×15	650(40)
NGC 628, M74	Psc	01 36.7	+15 47	139	-46	SAc	9.8	0.58	10×10	3 190(290)
NGC 1068, M77	Cet	02 42.7	-00 01	172	-52	SAb	9.5	0.70	7×6	5 140(520)
NGC 1291	Eri	03 17.3	-41 08	248	-57	SB0/a	9.4	0.93	10×9	2 960(270)
NGC 1313	Ret	03 18.3	-66 30	283	-45	SBd	9.4	-	9×7	1 380(100)
NGC 1316	For	03 22.7	-37 12	240	-57	SAB0p	9.7	0.90	7×5	5 690(390)
IC 342	Cam	03 46.8	+68 06	138	+11	SABcd	9.1	-	18×17	1 100(70)
NGC 2403	Cam	07 36.9	+65 36	151	+29	SABcd	8.9	0.50	18×11	1 040(60)
NGC 2903	Leo	09 32.2	+21 30	209	+45	SABbc	9.5	0.64	13×7	3 040(420)
NGC 3031, M81	UMa	09 55.6	+69 04	142	+41	SAab	7.8	0.93	26×14	1 180(70)
NGC 3034, M82	UMa	09 55.8	+69 41	141	+41	I0	9.3	0.87	11×5	1 150(80)
NGC 3521	Leo	11 05.8	-00 02	256	+53	SABbc	9.7	0.84	10×5	4 320(600)
NGC 3627, M66	Leo	11 20.2	+12 59	242	+64	SABb	9.7	0.70	9×4	2 930(180)
NGC 4236	Dra	12 16.7	+69 28	127	+47	SBdm	10.0	0.40	19×7	1 440(130)
NGC 4258, M106	CVn	12 19.0	+47 18	138	+69	SABbc	9.0	0.68	18×8	2 380(110)
NGC 4449	CVn	12 28.2	+44 06	137	+72	IBm	9.9	0.41	5×4	1 390(130)
NGC 4472, M49	Vir	12 29.8	+08 00	287	+70	E2	9.3	0.94	9×7	5 240(580)
NGC 4486, M87	Vir	12 30.8	+12 24	284	+74	E0-1p	9.6	0.94	7×7	5 390(700)
NGC 4594, M104	Vir	12 40.0	-11 37	298	+51	SAa	9.3	0.97	9×4	3 680(410)
NGC 4631	CVn	12 42.1	+32 32	143	+84	SBd	9.8	0.54	15×3	2 400(220)
NGC 4649, M60	Vir	12 43.7	+11 33	296	+74	E2	9.8	1.00	7×6	5 670(630)
NGC 4725	Com	12 50.4	+25 30	295	+88	SABabp	10.0	0.74	11×8	4 200(290)
NGC 4736, M94	CVn	12 50.9	+41 07	123	+76	SAab	8.9	0.75	11×9	1 500(110)
NGC 4826, M64	Com	12 56.7	+21 41	316	+84	SAab	9.4	0.84	9×5	1 730(120)
NGC 4945	Cen	13 05.4	-49 28	305	+13	SBcd	9.5	-	20×4	1 210(90)
NGC 5055, M63	CVn	13 15.8	+42 02	106	+74	SAbc	9.3	0.73	12×8	2 930(230)
NGC 5128	Cen	13 25.5	-43 01	310	+19	S0p	8.0	0.98	18×14	1 190(70)
NGC 5194, M51	CVn	13 29.9	+47 12	105	+69	SAbcp	9.0	0.60	11×8	2 800(260)
NGC 5236, M83	Hya	13 37.0	-29 52	315	+32	SABc	8.2	-	11×10	1 520(100)
NGC 5457, M101	UMa	14 03.2	+54 21	102	+60	SABcd	8.2	0.46	27×26	2 270(130)
NGC 6744	Pav	19 09.8	-63 51	332	-26	SABbc	9.0	-	15×10	2 920(230)
NGC 6946	Cep	20 34.8	+60 09	96	+12	SABcd	9.6	0.80	11×10	2 510(90)
NGC 7793	Scl	23 57.8	-32 35	5	-77	SAdm	9.7	0.59	9×7	1 170(70)

Galaxy Group

A group of between 3 and a few tens of galaxies is called a galaxy group. Typical galaxy groups have five members, a diameter of 1.5 million light-years, and a velocity dispersion of 100 km s^{-1}. The average density is over 20 times higher than the background density (Tully 1987, ApJ, **321**, 280; Geller & Huchra 1983, ApJS, **52**, 61). Nearby galaxy groups with galaxies brighter than 10th magnitude, except for the members of the Local Group, are shown in the table below.

Nearby Galaxy Groups

Galaxy group name	Galactic longitude l	Galactic latitude b	Apparent diameter	Line-of-sight velocity	Distance	Major member galaxies' NGC numbers (in order of brightness)
	°	°	° °	km s^{-1}	10 kly	
Sculptor group	5	−80	25×20	245	700[b]	253, 55, 300, 247, 7793, 45
M81 group	142	+41	40×20	220	1 200[b]	3031, 2403, 3034, 4236, 2976
Canes Venatici I group	162	+80	28×14	361	1 600	4736, 4258, 4826, 4449, 4214
NGC 5128 group	310	+20	30× −	315	1 400	5236, 5128, 4945, 5102, 5068
M101 group	102	+60	23×16	511	2 200	5457, 5194, 5055, 5195, 5585
M66 group	241	+64	7× 4	615	2 700	3627, 3628, 3623, 3489, 3593
Canes Venatici II group	138	+75	22×12	687	3 000	4631, 4490, 3675, 4656, 4051
Comae Berenices I group	198	+86	11× 5	1 031	4 500	4725, 4559, 4565, 4414, 4494
Cetus I group	178	−56	12× 9	1 350	5 900	1068, 936, 1084, 1087, 1055

The tables in "Galaxies and the Universe" (1975) by G. de Vaucouleurs were used as a reference. The line-of-sight velocity is taken from the catalog by Sandage & Tammann (1981) and the catalog by Tully (1987). An affixed 'b' indicates uncertainty on the level of 30%. Other than those, the uncertainty is about 50%.

Galaxy Cluster

A group where more than 50 galaxies are gathering together in a region of about 10 million light-years is called a galaxy cluster. The velocity dispersion of a galaxy cluster is from a few hundreds to 1000 km s^{-1}. Based on the survey with the Schmidt telescopes, Abell *et al.* made an all sky catalog with about 4100 galaxy clusters and Zwicky *et al.* made a catalog of about 9600 galaxy clusters north of declination of −3°. The apparent shape of galaxy clusters are arranged in a series from nearly round regular ones to irregular ones. The former is thought to be more dynamically evolved than the latter. In the BM classification by Bautz and Morgan, galaxy clusters are classified into 5 types based on a comparison between the brightest galaxy in the cluster and the other galaxies: Type I (the cD galaxy at the center dominates all the others), Type II (the brightest galaxy is dimmer than a cD galaxy and brighter than a usual elliptical galaxy), Type III (no galaxy particularly dominates the others) and the intermediate types Type I–II and Type II–III. In contrast, the RS classification system by Rood and Struble, classifies them into cD (supergiant), B (binary), L (line), C (core), F (flat), and I (irregular) based on the appearance of the 10 brightest galaxies. Nearby famous galaxy clusters, galaxy clusters with comparatively large amounts of data, as well as distant galaxy clusters are shown in the next page.

Galaxy Clusters

Name	Abell number	Equinox J2000		BM type	RS type	Magnitude[1]	Number of galaxies[2]	Redshift z[3]	Distance[4]
		R.A.	Dec.						
		h m	° ′						100 Mly
Virgo	-	12 30.8	+12 23	III	-	9.4:	45	0.0039	0.59
Fornax	S373	03 38.5	-35 27	I	-	10.3	-	0.0046	0.63
Antlia	S636	10 30.0	-35 19	I - II	-	13.4	1	0.0087	1.2
Centaurus	A3526	12 48.9	-41 18	I - II:	-	13.2	33	0.0110	1.5
Hydra I	A1060	10 36.9	-27 31	III	C	12.7	50	0.0114	1.6
Pavo II	S805	18 47.2	-63 19	I	-	14.7	8	0.0139	1.9
Cancer	-	08 20.6	+21 04	-	-	13.4:	-	0.0160	2.2
Perseus	A426	03 18.6	+41 30	II - III	L	12.5	88	0.0183	2.5
-	A1367	11 44.5	+19 50	II - III:	F	13.5	117	0.0215	3.0
Coma Berenices	A1656	12 59.8	+27 58	II	B	13.5	106	0.0232	3.2
-	A2199	16 28.6	+39 31	I	cD	13.9	88	0.0309	4.2
Hercules	A2151	16 05.2	+17 44	III	F	13.8	87	0.0371	5.1
-	A85	00 41.6	-09 20	I	cD	15.7	59	0.0518	7.0
Corona Borealis	A2065	15 22.7	+27 43	III	C	15.6	109	0.0721	9.6
-	A1132	10 58.3	+56 46	III	B	17.0	74	0.136	17
-	A520	04 54.3	+02 56	III	I	17.4	186	0.203	25
-	A370	02 39.8	-01 35	II - III	B	17.8	40	0.373	41
Cl0024+1654	-	00 26.6	+17 10	-	-	-	-	0.392	43
Cl0016+1609	-	00 18.6	+16 27	-	-	-	-	0.55	54
MS 1054-0321	-	10 57.0	-03 37	-	-	-	-	0.82	70
RDCS J0910+5422	-	09 10.0	+54 22	-	-	-	-	1.11	83
XMMXCS J2215	-	22 15.9	-17 38	-	-	-	-	1.46	94
Cl J1449+0856	-	14 49.2	+08 56	-	-	-	-	2.00	105

1) Magnitude: Magnitude in the red wavelength region of the 10th brightest galaxy. A colon indicates a visual magnitude.
2) Number of Galaxies: The number of galaxies between the magnitude of the 3rd brightest galaxy (m_3) and m_3+2.
3) The redshift: Though the redshift ($z = \Delta\lambda/\lambda_0$) is caused by the cosmic expansion, if this is considered to be the recession velocity of galaxies (v km s^{-1}), then the relation between z and v is given by $v/c=\{(1+z)^2-1\}/\{(1+z)^2+1\}$ (where c is the speed of light in units of km s^{-1}). When $z<0.05$, the approximate expression $v=cz$ can be used with negligible error.
4) Distance: Except for the Virgo Cluster, the distances were calculated as products of lookback time (the time showing how long ago the light we are observing was emitted by that galaxy cluster) and the speed of light based on Hubble's Constant $H_0=70$ km s^{-1} Mpc^{-1} and a cosmological model where the ratio of the contributions of matter and the Cosmological Constant to the energy density is 3 : 7 ($\Omega_0=\Omega_M+\Omega_\Lambda=1$; $\Omega_M=0.27$, $\Omega_\Lambda=0.73$; Age of the Universe 13.7 billion years).
5) For distant galaxy clusters, a part of them observed by X-ray and their redshifts are measured spectroscopically are shown. Detailed list can be seen at NASA/IPAC EXTRAGALACTIC DATA BASE.

Reference literature: Abell *et al.* 1989, ApJS, **70**, 1; Strubble & Rood 1987a, b, ApJS, **63**, 543, 555; Bruzual & Spinrad 1978, ApJ, **220**, 1; Allen 1973, Astrophysical Quantities; Stanford *et al.* 2002, AJ, **123**, 619; Hilton *et al.* 2009, ApJ, **697**, 436; Gobat *et al.* 2013, ApJ, **776**, 9.

Galaxy Superclusters and the Large Scale Structure of the Universe

When multiple galaxy clusters and galaxy groups are joined together forming a structure larger than about 100 million light-years, this is called a supercluster of galaxies. The galaxies within about 100 million light-years from the Milky Way are centered on the Virgo Cluster and, as shown in Figure 10 A, forming the Local Supercluster with a thin disk and a halo surrounding it. Besides this, other known superclusters are Coma-A1367, Hercules, Perseus-Pisces, Hydra-Centaurus, etc. In contrast, regions with the scale of 100 million light-years where there are virtually no galaxies are known as voids. The large scale structure of the Universe is formed as a mixture of superclusters, the filaments which connect them and the voids between (Figures 10 B, C). It is known that this large scale structure exists universally over about 2.5 billion light-years of space.

Color and Number of Galaxies

The color of a galaxy reflects differences in the type of stars within it and changes systematically according to the galaxy type. Early types (E, S0) are red; galaxies become increasingly blue moving towards the late types containing many young stars. Average colors for several types of galaxies are shown in the table below. The filter sets are Johnson (U, B, V), Cousins (R_C, I_C) and SDSS (u', g', r', i', z').

Type	$U-B$	$B-V$	$V-R_C$	R_C-I_C	$u'-g'$	$g'-r'$	$r'-i'$	$i'-z'$
E	0.64	0.96	0.61	0.70	1.99	0.77	0.43	0.36
S0	0.42	0.85	0.54	0.61	1.70	0.68	0.34	0.29
Sab	0.33	0.78	0.56	0.65	1.60	0.66	0.38	0.32
Sbc	0.00	0.57	0.52	0.62	1.16	0.52	0.33	0.32
Scd	-0.08	0.50	0.50	0.57	1.04	0.48	0.28	0.29
Im	-0.35	0.27	0.31	0.33	0.64	0.20	0.04	0.11

Fukugita *et al.* 1995, PASP, **107**, 945

Taking $N(m)$ to be the number of galaxies observed in one square degree on the celestial sphere between apparent magnitude m-0.25 and m+0.25, the value of log $N(m)$ for several bands are summarized in the following table. For each band, the left-hand column is the observed value (including Milky Way stars) and the right-hand column is the estimated value corrected for incompleteness of the observation after removing Milky Way stars.

m	B		V		R_C		i'		z'	
20.25	2.48	2.16	2.81	2.57	2.99	2.81	3.15	3.00	3.28	3.19
21.25	2.85	2.65	3.12	3.02	3.36	3.28	3.53	3.47	3.64	3.60
22.25	3.27	3.20	3.54	3.48	3.74	3.70	3.88	3.85	3.96	3.93
23.25	3.77	3.75	3.98	3.96	4.09	4.07	4.18	4.18	4.26	4.26
24.25	4.29	4.30	4.41	4.43	4.48	4.50	4.52	4.55	4.56	4.60
25.25	4.65	4.68	4.74	4.78	4.77	4.81	4.78	4.85	4.79	4.97
26.25	4.88	4.96	4.93	5.13	4.94	5.09	4.93	5.18	4.89	5.33
27.25	4.99	5.29	4.95	5.42	4.97	5.36				

Kashikawa *et al.* 2004, PASJ, **56**, 1011

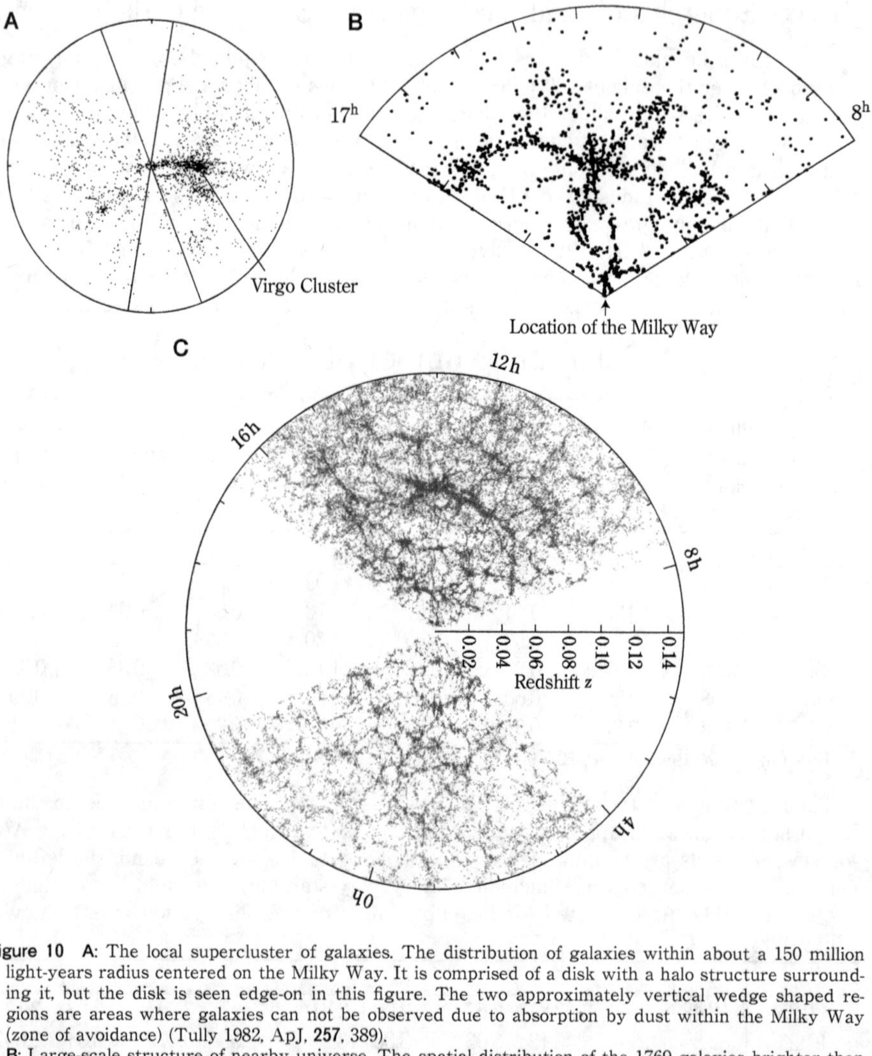

Figure 10 A: The local supercluster of galaxies. The distribution of galaxies within about a 150 million light-years radius centered on the Milky Way. It is comprised of a disk with a halo structure surrounding it, but the disk is seen edge-on in this figure. The two approximately vertical wedge shaped regions are areas where galaxies can not be observed due to absorption by dust within the Milky Way (zone of avoidance) (Tully 1982, ApJ, **257**, 389).
B: Large-scale structure of nearby universe. The spatial distribution of the 1769 galaxies brighter than B=15.5 mag in the region of right ascension 8^h–17^h and declination +26.5° to +38.5° The 12° width of declination is projected onto a single plane. The radius of the arch is about 700 million light-years (Geller *et al.* 1987, IAU Symp., **124**, 301).
C: Wide-area Large-scale Structure of the Universe. The spatial distribution of the approximately 50 000 galaxies that are brighter than r' = 17.8 magnitude in the toroidal celestial region of celestial declination 2°. 5 width centered on the celestial equator. The Milky Way is at the center. The scale positioned to the right of the center shows the redshift. The position of the outer circle corresponds to 0.15 (about 1.9 billion light-years).
(Based on the Sloan Digital Sky Survey; http://sdss.physics.nyu.edu/pie/main_all_z0.150.jpg)

Chemical Composition of the Universe

Following the nucleosysnthesis in the Big Bang the elemental abundances of the material in the Universe have been changing through a series of physical processes like star formation, stellar mass-loss, supernova explosions, etc. The chemical composition that is best understood among astronomical objects is that of the Solar System. It is obtained from the spectroscopic analysis of light from the solar photosphere and chemical analysis of meteorites known as C1 chondrites and of the solar wind. The accuracy is on the order of 3–10% in general. The table shows the chemical composition (i.e., ratio of number density) of the Solar System normalized to the silicon (Si) abundance. The abundances of volatile elements like hydrogen, carbon, and oxygen, which are not well preserved in meteorites, are determined by analysis of the solar spectrum. The abundance of noble gases is not well constrained by spectroscopic analysis of light from the solar photosphere. The helium abundance is determined by solar seismology, whereas the abundances of neon, argon, etc. are estimated from corona spectra and solar wind, in which the accuracy is not as good as that of other methods.

The mass ratios of hydrogen, helium, and heavier elements are represented by X, Y, and Z, respectively. The values are $X=0.7381$; $Y=0.2485$; $Z=0.0134$ $(X+Y+Z=1)$. The mass ratio of element A is obtained using the elemental abundances given in the table by
(atomic weight of A/atomic weight of hydrogen) × (abundance of A/abundance of hydrogen) × (mass ratio of hydrogen X). For instance, the mass ratio of silicon is 0.00067.

Chemical composition of the solar system

Atomic number, Element	Number density ratio	Atomic number, Element	Number density ratio	Atomic number, Element	Number density ratio
1 H	3.09×10^{10}	31 Ga	37.2	62 Sm	0.269
2 He	2.63×10^{9}	32 Ge	117	63 Eu	0.100
3 Li	56.2	33 As	6.17	64 Gd	0.347
4 Be	0.617	34 Se	67.6	65 Tb	0.0646
5 B	19.1	35 Br	10.7	66 Dy	0.417
6 C	8.32×10^{6}	36 Kr	55.0	67 Ho	0.0912
7 N	2.09×10^{6}	37 Rb	7.08	68 Er	0.257
8 O	1.51×10^{7}	38 Sr	23.4	69 Tm	0.0407
9 F	8.13×10^{2}	39 Y	4.57	70 Yb	0.257
10 Ne	2.63×10^{6}	40 Zr	10.5	71 Lu	0.0380
11 Na	5.75×10^{4}	41 Nb	0.794	72 Hf	0.158
12 Mg	1.05×10^{6}	42 Mo	2.69	73 Ta	0.0234
13 Al	8.32×10^{4}	44 Ru	1.78	74 W	0.138
14 Si	1.00×10^{6}	45 Rh	0.355	75 Re	0.0562
15 P	8.32×10^{3}	46 Pd	1.38	76 Os	0.692
16 S	4.37×10^{5}	47 Ag	0.490	77 Ir	0.646
17 Cl	5.25×10^{3}	48 Cd	1.58	78 Pt	1.29
18 Ar	7.76×10^{4}	49 In	0.178	79 Au	0.195
19 K	3.72×10^{3}	50 Sn	3.63	80 Hg	0.457
20 Ca	6.03×10^{4}	51 Sb	0.316	81 Tl	0.182
21 Sc	34.7	52 Te	4.68	82 Pb	3.39
22 Ti	2.51×10^{3}	53 I	1.10	83 Bi	0.138
23 V	282	54 Xe	5.37	90 Th	0.0335
24 Cr	1.35×10^{4}	55 Cs	0.372	92 U	0.00891
25 Mn	9.33×10^{3}	56 Ba	4.68		
26 Fe	8.71×10^{5}	57 La	0.457		
27 Co	2.29×10^{3}	58 Ce	1.17		
28 Ni	4.90×10^{4}	59 Pr	0.178		
29 Cu	550	60 Nd	0.871		
30 Zn	1.32×10^{3}				

Asplund *et al.* 2009, ARA&A, **47**, 481

Chemical compositions of objects outside the Solar System are based entirely on the spectroscopic analysis of stellar light. Observations for a variety of objects are useful to reveal the individual nucleosynthesis processes in the Universe and the chemical evolution of galaxies. Whereas nearby stars have similar metallicities and chemical compositions to those of the Sun, a wide distribution of elemental abundances are observed in stars in the Milky Way, reflecting the differences in the era and place in which individual stars were formed. Moreover, analysis of absorption lines appearing in the spectra of high redshift quasars enables measurements of the elemental abundances of gas clouds from the early to the present Universe.

The abundances of ^{2}H, ^{3}He, ^{4}He, and ^{7}Li are measured for quasar absorption systems, nearby stars, and star forming regions, providing a constraint on the Big Bang nucleosynthesis models. Spectroscopic studies for ultraviolet, infrared, X-ray, and gamma-ray, in addition to visible light, have been identifying many elements, and uncovering the composition of material ejected from supernova explosions and planetary nebulae. Chemical compositions of early generation stars, which have only a small amount of heavy elements and reflect the products from individual stars and supernova explosions, provide useful constraints on models of nucleosynthesis for stars and supernovae. Such measurements sometimes successfully obtain abundances of elements heavier than iron, which are useful to understand the nucleosynthesis processes of heavy elements like rapid (r-) and slow (s-) neutron-capture processes.

Cosmic Rays

Cosmic rays and meteorites are both matter flying down to the Earth from space. Meteorites from within and around the Solar System have macroscopic size, but only minute particles can achieve the speeds near the speed of light to cross interstellar space and arrive from far away. High-energy elementary particles and atomic nuclei are detected raining down to the Earth from space. These are called cosmic rays. The propagation of low energy cosmic rays (on the order of less than several hundred MeV), emitted from the Sun, is affected by the 11 year solar activity cycle and the magnetic fields. High energy cosmic rays from outside of the Solar System has a spectrum of a power law $E^{-\alpha}$ (α~2.7, where E is the energy.) It is thought that most of them are originated from acceleration within our galaxy. On the other hand, Cosmic rays attaining the highest energies observed to date, ~10^{20} eV, are believed to come from outside of our galaxy. In terms of chemical composition, about 90% of cosmic rays are protons (nucleus of hydrogen), and the remainder is α particles (nucleus of helium) and heavier nuclei. Though this composition closely resembles the elementary composition of the universe, compositions of light elements such as Li, Be, and B are relatively larger. It is thought that these are produced when heavy elements in cosmic rays are destroyed by collisions with matter in the interstellar medium during propagation. These cosmic rays from the outside of the Earth are called primary cosmic rays.

Cosmic rays plunging into Earth's atmosphere collide with the air atoms and generate secondary particles. Cosmic rays above several 100 GeV produce numerous particles to be generated one after another, creating an extensive air shower phenomena as a swarm of particles flies through the air. Secondary cosmic rays generated in the atmosphere such as electrons, positrons, γ-rays, pions, muons, neutrons, and protons have strength of ~10^{4} particles$\cdot m^{-2} \cdot min^{-1}$ at the Earth's surface. Muons account for about 3/4 of the charged particles in secondary cosmic rays. Secondary cosmic rays have been used to study high energy reactions of elementary particles through the detection of positrons, pions, muons and particles with strangeness. Muons and neutrinos which interact weakly with matter, penetrate deep underground. Study of neutrino interactions, such as neutrino oscillations, and study of solar neutrinos is on going.

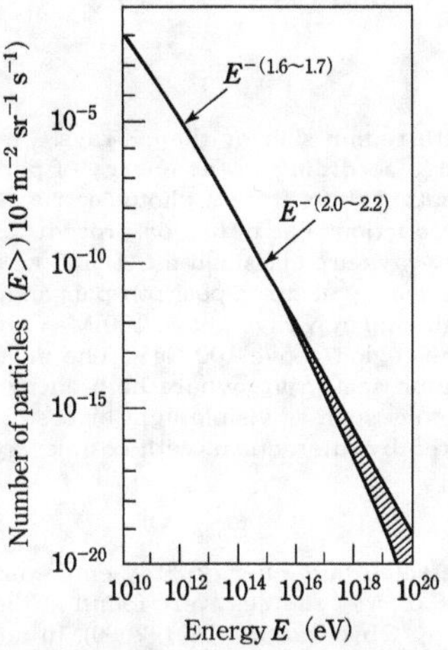

Figure 11 Integral energy spectrum of cosmic ray particles

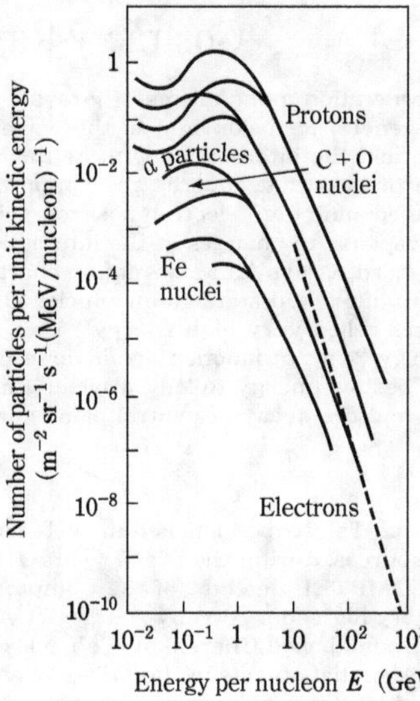

Figure 12 The differential energy spectrum of cosmic rays as measured from above the Earth's atmosphere. The low energy side is influenced by the solar cycle. In each case, the lower curve corresponds to the solar maximum. The dashed line corresponds to the energy spectrum of the primary electrons (m^{-2} sr^{-1} s^{-1} $(MeV)^{-1}$). It is measured to have a linear relation up to 2×10^3 GeV.

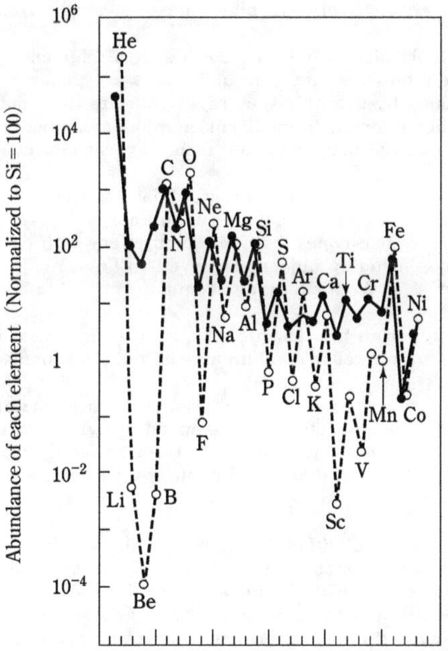

Figure 13 The cosmic ray elemental abundances measured at Earth (● is an element with an energy of 70–280 MeV for each nucleon. ○ shows the solar system abundances.) (From Simpson 1983, Ann. Rev. Nucl. Part. Sci. **33**, 323)

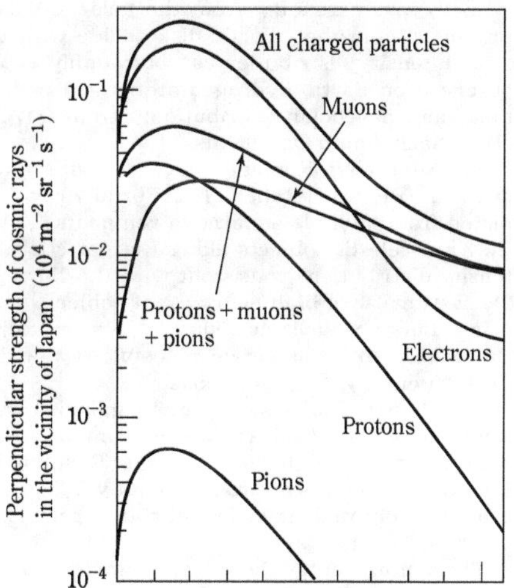

Figure 14 Flux of cosmic rays that come from the vertical direction. Near the surface of the Earth, the flux is proportional to $\cos^2\theta$ (θ is the zenithal angle.) It becomes isotropic near the maximum. Above that, the number of horizontal incidences becomes large. The horizontal axis is in units of the air mass per each unit area normalized to the atmospheric zenith (equivalent to atmospheric pressure).

Cosmic γ-Rays

Classification and generation mechanisms of γ-rays

γ (gamma) ray is a general term for light in the wavelength region shorter than X-rays. The production mechanism and the interactions with matter change according to the energy of photons. For instance, the process that governs absorption by matter shifts from a photoelectric effect, to the Compton effect, and onto electron-positron pair production. The detection procedures are also different. In response to changes in the interactions, γ-rays are classified as: (1) 100 keV regime also known as hard X-rays (2) MeV regime characterized by electron-positron pair annihilation, the line spectra of excited state atomic nuclei. (3) high energy γ-rays above 100 MeV. In addition, there are cases called very high energy γ-rays in the region above 100 GeV. The main processes of high energy γ-ray production are inverse Compton scattering, where high energy electrons or positrons bestow energy to long-wavelength photons such as visible light or cosmic background radiation, and the decay of neutral pions generated by interactions with cosmic ray protons.

Typical γ-ray sources

The Fermi γ-ray Space Telescope launched in 2008, has detected high energy γ-ray emissions from more than 3000 sources, during the first 4 years. About 30 γ-ray sources were found in the MeV regime by the COMPTEL detector of the Compton γ-ray Observatory (1991–2000). In addition, more than 50 very-high-energy γ-ray sources have been discovered recently through the developement of the ground-based Cherenkov TeV telescope.

1) High energy γ-ray celestial objects in the Milky Way

(a) Pulsars Over 60 γ-ray pulsars have been discovered in high energy γ-rays and one has been discovered in the MeV regime.

(b) γ-rays extending from the galactic disk, molecular clouds Cosmic rays (high-energy protons etc.) in the galactic disk collide with the interstellar medium and γ-rays are generated. γ-ray intensity distribution can be roughly explained by flux of cosmic rays similar to the values observed on Earth, hydrogen atoms and molecules inferred from 21 cm or millimeter observations, and molecular distribution. Stronger γ-rays have been detected from several molecular clouds including ρ-Ophiuchus.

(c) Supernova Remnants etc. Supernova remnants are a prime candidate as the source of cosmic-ray acceleration. The Fermi γ-ray Space Telescope discovered high energy γ-rays radiated from multiple supernova remnants. TeV γ-ray telescopes discovers many very-high energy γ-ray celestial objects along the galactic plane, including objects observed to have γ-ray intensity distributions consistent with the X-ray emission from supernova remnants.

2) Extragalactic high energy γ-ray objects

(a) Large Magellanic Cloud Similar to γ-rays from the galactic disk, it is thought that in this galaxy, collisions between cosmic rays and the interstellar medium are the main mechanism for high energy γ-ray emissions.

(b) Active galaxies Active galaxies called blazars with characteristics such as strong time variabilities, hard radio spectrum, or the existence of jets near the speed of light, are the main sources of high-energy γ-rays. Relativistic electrons in jets seems to form γ-rays through Inverse Compton Scattering. Recently, it has been observed that the high energy γ-rays are also radiated from radio galaxies, starburst galaxies, etc.

3) γ-ray burst

There are events where γ-rays greater than tens of keV (there are cases of observations exceeding 10 GeV) are emitted in times on the order 10 seconds. These are the brightest phenomena known in the Universe. After their discovery since 1970, the distances to the sources and their true natures remained a mystery. Around 1997, it became possible to observe corresponding objects in X-rays, visible light and radio waves through afterglow. Based on this, it was proven that they are phenomenon taking place in cosmologically distant galaxies with redshifts on the order of 1, the total amount of radiant energy from a γ-ray burst is equivalent to the energy released by an object of about the mass of the Sun all at once. It can be approximately explained with so-called a "Fire Ball" model, pushing out with speeds very near to the speed of light. However, many unsettled problems remain about the origin of such an explosion and explanations for the details of the observed results.

Major High-Energy γ-Ray Sources

Name[1)]	2000 Equinox		γ line[2)] intensity	Remarks
	Right ascension	Celestial declination		
	h m s	° ′ ″		
3FGL J0023.9-7203	00 23 55.7	-72 03 64	0.4	Globular cluster 47 Tucanae
3FGL J0047.5-2516	00 47 32.9	-25 16 30	0.1	Starburst galaxy NGC 253
3FGL J0240.5+6113	02 40 33.1	+61 13 40.8	4.6	X-ray binary star LSI+61 303
3FGL J0534.5+2201	05 34 32.9	+22 01 26.4	15.7	Crab pulsar
3FGL J0617.2+2234	06 17 14.4	+22 34 48	6.3	Supernova remnant IC443[4)]
3FGL J0835.3-4510	08 35 21.1	-45 10 40.8	130	Vela pulsar
3FGL J0955.4+6940	09 55 28.8	+69 40 22.8	0.1	Starburst galaxy M82[3)]
3FGL J1104.4+3812	11 04 28.3	+38 12 36	3.0	Active galaxy Mrk 421 ($z=0.030$)[3)]
3FGL J1229.1+0202	12 29 08.9	+02 02 27.6	0.9	Active galaxy 3C 273 ($z=0.0158$)[4)]
3FGL J1230.9+1224	12 30 54.7	+12 24 21.6	0.1	Radio galaxy M87
3FGL J1256.1-0547	12 56 09.8	-05 47 27.6	2.1	Active galaxy 3C 279 ($z=0.536$)[4)]
3FGL J1325.4-4301	13 25 28.1	-43 01 48	0.3	Radio galaxy M87 Cen A[3)]
3FGL J1653.9+3945	16 53 54.7	+39 45 14.4	1.0	Active galaxy Mrk 501 ($z=0.034$)
3FGL J1826.2-1450	18 26 16.3	-14 50 49.2	1.9	X-ray binary star LS5039
3FGL J1855.9+0121	18 55 57.6	+01 21 18	7.1	Supernova remnant W 44
3FGL J2254.0+1608	22 54 00.5	+16 18 45.6	10.6	Active galaxy 3C 454.3[4)] ($z=0.859$)
			%	
HESS J0047-253	00 47 33.6	-25 18 08	0.3	Starburst galaxy NGC 253
HESS J0534+220	05 34 32.0	+22 00 52	100	Crab nebula[3)]
HESS J0852-463	08 52 00.13	+46 22 00	100	Supernova remnant Vela Junior
HESS J1104+382	11 04 27.3	+38 12 32	300	Active galaxy Mrk 421 ($z=0.030$)
HESS J1514-591	15 14 07	-59 09 27	15	Pulsar wind nebula MSH15-52
HESS J1713-397	17 13 57.6	-39 45 00	66	Supernova remnant RXJ 1713.7-3946
HESS J1826-148	18 26 13.8	-14 50 54	3	X-ray binary star LS5039
HESS J2009-488	20 09 29.3	-48 49 19	2.5	Active galaxy PKS 2005-489 ($z=0.071$)
HESS J2158-302	21 58 52.7	-30 13 18	20	Active galaxy PKS 2155-304 ($z=0.116$)

1) Selected γ-ray sources, which have been identified by observations at other wavelength (including candidate sources) in the energy range above 100 MeV. Prefix 3GFL and HESS denote the third fermi satellite catalog (https://fermi.sfc.nasa.gov/ssc/data/access/lat/4yr_catalog/3FGL-table) and the H. E. S. S. telescope catalog (https://www.mpihd.mpg.de/hfm/HESS/pages/home/sources/), resepctively. Coordinates which are determined by γ-ray observations are used in the table (also see SIMBAD database for coordinates by radio or other observations). 3033 γ-ray sources are listed on the third Fermi catalog, and approximately two-thirds of objects are identified or associated with pulsars, active galactic nuclei and others, however the rest is unidentified.

2) For 3 FGL sources, fluxes in 1 GeV to 100 GeV with the units of 10^{-4} photon m^{-2} s^{-1} are used. Flux of TeV sources is an approximate ratio to a γ-ray flux of the Crab nebula above 1 TeV.
It should be noted that rapid time variations are often observed in AGN.

3) X-ray sources (Refer to page **71**, "Typical X-Ray Sources").

4) Radio sources (Refer to page **77**, "Radio Galaxy and Quasar").

Cosmic X-Rays

Types of cosmic X-ray sources

1) X-ray sources in the Milky Way

(a) X-ray binaries In close binary systems composed of a normal star and a compact object (i.e. a neutron star, black hole, or white dwarf), gravitational energy is released by the accretion of the gas from the normal star onto the compact object to form the high temperature (10^7–10^8 K) gas that emits X-rays. Binary systems including a rotating neutron star with a strong magnetic fields become accreting X-ray pulsars. An X-ray burster is an X-ray source that produces thermo-nuclear explosions (bursts) of the helium or hydrogen atmosphere on the surface of a neutron star with a weak magnetic field that has a low mass companion. The X-ray luminosities of binary systems containing white dwarfs (cataclysmic variables) are 2–3 orders of magnitudes smaller than those of neutron star binary systems. The stellar-mass black holes are formed through the gravitational collapse of massive stars or the coallescence of binary neutron stars, and have masses from several to more than ten times that of the Sun. Most X-ray novae are binary star systems consisting of a late-type star and a black hole or a neutron star. Some X-ray novae containing a black hole emanate jets with almost the speed of light.

(b) Supernova Remnants X-rays are emitted from high temperature gas and high-energy particles formed by the shock wave of a supernova explosion. As a supernova remnant ages, its temperature becomes lower, and it becomes a soft X-ray source. Some supernova remnants contain a neutron star, sometimes as a rotation-powered pulsar with a strong magnetic field.

(c) Normal stars A late-type star emits primarily soft X-rays from its high temperature corona (10^6–10^7 K). The early-type stars radiate X-rays via shock waves formed by their stellar wind.

2) Extragalactic X-ray sources

(a) Active galaxies Active galactic nuclei such as quasars, Seyfert galaxies, and BL Lac objects emit strong X-rays. It is believed that a super-massive black hole with a mass of millions to hundreds of million million times that of the Sun exists at the centers of these active galaxies.

(b) Galaxy clusters In a galaxy cluster many galaxies are gravitationally bound in an area where dark matter is concentrated. High temperature gas with a mass several times that of the sum of all the galaxies is distributed in this region and radiates X-rays.

(c) Nearby galaxies X-rays are observed from the supernova remnats and X-ray binaries in galaxies near the Milky Way, such as the Andromeda Galaxy and the Large Magellanic Cloud. Ultra luminous X-ray sources (ULXs) emit more than ten times the energy of the brightest X-ray sources in the Milky Way Galaxy.

3) X-ray background radiation

The majority of X-ray background above 2 keV is distributed almost uniformly over the entire sky. It originates from countless distant active galaxies and galactic clusters which are not spatially resolved. The X-ray surface brightness in 2–10 keV is 5.1×10^{-11} W m^{-2} sr^{-1} (=7 counts cm^{-2} s^{-1} sr^{-1}). In the 0.1–2 keV energy band, contributions of the high temperature regions (about 10^6 K) surrounding the solar system and the charge-exchange reaction between the solar wind and the earth's atmosphere are greater than those from distant X-ray sources. The intensity of these foreground emission depends on the direction and the latter varies in time.

X-ray generation mechanisms

The elementary processes are (a) Bremsstrahlung generated when a high-energy electron that approaches an atomic nucleus is decelerated or deflected by the electric field. (b) line emission generated when an electron bound to an ion is excited by a collision with a high-energy electron, and then falls to a lower energy level. (c) synchrotron emissions by highly energetic electrons being deflected by magnetic fields, (d) Inverse Compton Scattering where a photon is scattered by a high energy electron and gains energy. The radiation from electrons in thermal equilibrium is called thermal emission. (a) and (b) are the major component of thermal emission, contributing the continum and line spectra respectively. The spectral components depend primarily on the temperature of the gas. When the density is sufficiently high, the spectrum is modified to approach a blackbody radiation.

Representative X-Ray Sources

Name	2000 Equinox[1)]		X-ray flux		Remarks
	R.A.	Dec.	max.[2)]	[variability][3)]	
	h m	° ′			
AM Her	18 16.2	+49 52.1	1.4×10^{-10}	[3]	Cataclysmic variable, 3.1 hour period
Algol	03 08.2	+40 57.3	1.8×10^{-10}	[4]	β Per, 2.9 d period eclipsing binary
ρ Oph	16 25.6	-23 26.7	5×10^{-11}		Dark nebula, star forming region, distance 160 pc
SS Cyg	21 42.7	+43 35.6	4×10^{-10}	[>20]	Dwarf nova, 9 s, 6.6 hour period
ζ Pup	08 03.6	-40 00.2	1×10^{-11}	(0.1-2 keV)	Early type star (spectral type O 4 If)
π^1 UMa	08 39.2	+65 01.2	5×10^{-12}	(0.15-4 keV)	Late type star (spectral type G 1.5 Vb)
HZ 43	13 16.4	+29 06.4	3×10^{-10}	(0.15-4 keV)	White dwarf
Cyg X-1	19 58.4	+35 12.0	2.6×10^{-8}	[5]	Black hole candidate, 5.6 d period binary
SS 433	19 11.8	+04 59.0	2×10^{-10}	[5]	0.26 c jet, 13.1 d period binary
Her X-1	16 57.8	+35 20.5	3.2×10^{-9}	[>100]	1.24 s accretion powered pulsar, 1.7 d period binary
Cyg X-3	20 32.4	+40 57.6	8.6×10^{-9}	[5]	4.8 d period binary radio outbursts
Sco X-1	16 19.9	-15 38.4	3.8×10^{-7}	[3]	Low mass neutron star binary system, 0.79 d period
Vela X-1	09 02.1	-40 33.2	5.6×10^{-9}	[>10]	283 s, stellar wind accretion powered pulsar, 8.96 d period binary
X 1636-536	16 40.9	-53 45.0	7.8×10^{-9}	[3]	X-ray burster, 1.7 ms period during burst
GRS 1915+105	19 15.2	+10 56.7	6×10^{-8}	[>10 000]	0.92 c radio jet, 33 d period black hole binary
A 0620-00	06 22.8	-00 20.7	1×10^{-6}	[>10 000]	7.75 hour period, black hole binary
Crab Nebula	05 34.5	+22 01.0	2×10^{-8}		M1, centrally concentrated supernova remnant, 33 ms pulsar
Cas A	23 23.5	+58 50	1.1×10^{-9}		Shell type supernova remnant
Cygnus loop	20 48.8	+32 05	3×10^{-8}	(0.1-4 keV)	Veil Nebula, shell type supernova remnant
Vela SNR	08 35.4	-45 10.6	1.3×10^{-8}	(0.15-2 keV)	Gum Nebula, composite type supernova remnant, 89 ms pulsar
SN 1987 A	05 35.5	-69 16.2	2×10^{-12}	(0.5-2 keV)	Supernova in Large Magellanic Cloud
M 31	00 42.7	+41 16.4	4×10^{-11}		Andromeda Galaxy, primary components are acreeting binary stars, etc.
M 51	13 29.9	+47 11.5	3.5×10^{-12}		Spiral galaxy with a weak active nucleus, distance, 7 Mpc
M 82	09 55.9	+69 40.8	6×10^{-11}		Starburst galaxy, contains intermediate mass black hole candidate
NGC 4636	12 42.8	+02 41.2	6.6×10^{-12}	(0.5-4.5 keV)	Elliptical galaxy
NGC 4151	12 10.5	+39 24.4	4×10^{-10}	[5]	Type 1 Seyfert galaxy, z = 0.00332
NGC 1068	02 42.7	-00 00.8	5×10^{-12}		Type 2 Seyfert galaxy, z = 0.00379
3C 273	12 29.1	+02 03.1	2×10^{-10}	[3]	Radio-loud QSO, z = 0.15834
Cen A	13 25.5	-43 01.1	6×10^{-10}	[3]	Radio galaxy NGC 5128, z = 0.00183
Mrk 421	11 04.5	+38 12.5	2×10^{-9}	[>100]	BL Lac object, z = 0.03002, TeV γ-ray source
Perseus cluster	03 19.8	+41 30.8	1×10^{-9}		Radio galaxy NGC 1275, Per A
Virgo cluster	12 30.8	+12 23.5	4.6×10^{-10}		M87 (Vir A) in the center
Coma cluster	12 59.6	+27 57.7	4.4×10^{-10}		A1656

Examples of the various types of X-ray sources inside and outside of the Galaxy have been selected. (4U Catalog: Forman *et al.* 1978, ApJS, **38**, 357; HEAO A-1 Catalog: Wood *et al.* 1984, ApJS, 56, 507, etc) In many cases these objects have also been observed in other wavelength bands. (Refer to pages **69**, **73**, **76**, **77**).

1) Position: Most probable position, uncertainty less than 0.1 degrees.

2) Maximum flux: The unit is 10^{-3} W m^{-2} (= erg cm^{-2} s^{-1}). The integration energy range is shown in ( ). The energy range is 2-10 keV if not indicated.

3) Variability: For objects with large variability, approximate ratios of the highest and lowest fluxes are shown.

Cosmic Infrared Radiation

Wavelengths of infrared radiation

The wavelength of infrared radiation is defined as from 1 μm to 1 000 μm: 1-5 μm for near infrared; 5-25 μm for mid-infrared; 25-1 000 μm for far-infrared. (The 100-1 000 μm region is also called submillimeter waves.) Most of celestial objects primarily emit thermal radiation in the infrared region corresponding to their temperature. The various types of infrared sources emit infrared radiation with characteristic radiation mechanisms and wavelengths, as listed below.

Typical infrared sources

1) Planets, satellites, comets, and other solar system objects In addition to the reflection and scattering of sunlight in the visible and infrared, in the mid to far-infrared regions these objects exhibit thermal emission (approximately blackbody radiation) corresponding to the surface temperature of each.
2) Red giants/supergiants and infrared stars All stars are infrared sources. In particular, red giants/supergiants with low temperatures are the strongest (near) infrared sources. Among those, the stars surrounded by dust cloud radiate strongly in the near to mid infrared region with obstructed visible light, which are known as infrared stars.
3) Protostars (Young Stellar Object; YSO) and T Tauri stars Stars are formed in molecular clouds in the results of the processes of fragmentation and collapse of interstellar gas. The newly born protostars and T Tauri stars radiate in wide infrared wavelength bands. It is thought that the primary energy source of their radiation is the gravitational energy of the contracting gas. The observed infrared radiation is primarily thermal radiation from the dust particles and spectral line emissions from molecules in the gas.
4) Molecular clouds and globules Stars are formed in molecular clouds due to density variations in the interstellar matter through relatively high density phases known as molecular cores and globules. These areas contain low temperature gas and dust, and observed particularly in the far-infrared region.
5) H II regions, planetary nebulae, and Galactic nuclei These objects consist of high temperature, ionized regions. They radiate thermal radiation from dust heated and excited by Lyman α and infrared light, and spectral lines of various ions and atoms in a wide wavelength range from near-infrared to far-infrared.
6) External galaxies, Seyfert galaxies, and Quasars A galaxy's infrared radiation represents a composite of a stellar component (mainly red giants), an interstellar medium component (dust thermal radiation and gas spectral lines), and a nucleus component. In a quasar, the nucleus component shines overpoweringly strong due to some immense energy emission mechanism in the nucleus.
7) Diffuse light In addition to the above representative infrared sources, there is widely distributed infrared radiation: thermal emissions from interplanetary dust (mid-infrared), and thermal emissions from interstellar dust concentrated in the galactic plane and the isotropic infrared cosmic background radiation (far-infrared).

Major Infrared Sources

Name	2000 Equinox		Infrared radiation intensity			Remarks
	R.A.	Dec.	2.2 μm	10 μm	100 μm	
	h m s	° ′ ″	Mag	Mag	Jy	
Red giant/supergiant, Infrared star						
NML Tau	03 53 28.9	+11 24 22	-1.1	-4.6	101 (IRAS)	Infrared star, M type star
α Ori	05 55 10.3	+07 24 25	-4.1	-5.2	95 (IRAS)	Brightest star at 2 μm
VY CMa	07 22 58.3	-25 46 03	-0.7	-5.9	331 (IRAS)	OH/IR star, M type star
IRC+10216	09 47 57.4	+13 16 44	0.6 (10″)	-7.8 (10″)	2 100 (54″)	Infrared star, C type star
η Car	10 45 03.6	-59 41 04	1.2	-7.9 (16″)	5 200 (32″)	Brightest star at 20 μm
χ Cyg	19 50 33.9	+32 54 51	-1.8	-3.4	17 (IRAS)	Long-period variable, S type star
NML Cyg	20 46 25.5	+40 06 59	0.4	-5.2		Infrared star, M type star
AFGL2688	21 02 18.3	+36 41 37	8.0	-2.3 (8″)	–	Egg Nebula
Protostar, T Tauri star						
T Tau	04 21 59.4	+19 32 06	5.4 (15″)	1.0	63 (37″)	T Tau star
L1551-IRS5	04 31 34.2	+18 08 05	8.9 (30″)	3.0 (3″.8)	470 (54″)	Bipolar molecular ouflow object
BN Object	05 35 14.1	-05 22 23	4.5 (15″)	-1.2 (13″)	–	OMC-1, protostar
R Mon	06 39 10.0	+08 44 10	5.2	0.4	42 (37″)	T Tau star, nebular variable nebula
M17-IRS1	18 20 19.6	-16 13 31	6.8 (17″)	0.7 (17″)	–	Protostar in an H II region
AFGL2591	20 29 24.9	+40 11 19	5.5 (26″)	-1.5 (26″)	5 700 (IRAS)	Protostar in an molecular cloud
S140-IRS1	22 19 18.3	+63 18 46	5.5 (32″)	-1.4 (3″.5)	14 000 (IRAS)	Protostar in an molecular cloud
Molecular cloud, H II region, Planetary nebula, Galactic nucleus						
LkHα 101	04 30 14.4	+35 16 24	3.1	-2.4 (26″)	510 (37″)	H II region
KL Nebula	05 35 14.5	-05 22 29	–	-2.2 (26″)	90 000 (1′)	OMC-1, infrared nebula
GC-IRS16	17 45 40.0	-29 00 28	8.3 (3″.8)	–	2 600 (28″)	Galactic nucleus, SgrA
BD+30 3639	19 34 45.2	+30 30 59	8.1 (20″)	0.0 (11″)	55 (55″)	Planetary nebula
B335	19 37 01.0	+07 34 11	–	–	41 (IRAS)	Globule
S106	20 27 26.5	+37 22 42	9.6 (6″)	4.9 (5″)	13 000 (IRAS)	AFGL2584 molecular cloud
MWC 349	20 32 45.5	+40 39 37	3.3 (10″)	-1.6 (11″)	8.5 (40″)	H II region
NGC 7027	21 07 01.6	+42 14 10	7.1 (11″)	-1.1 (11″)	206 (55″)	Planetary nebula
Galaxy, Seyfert galaxy, Quasar						
NGC 253	00 47 33.1	-25 17 20	6.9 (25″)	2.1 (6″)	1 000 (2′.2)	Spiral galaxy
NGC 1068	02 42 40.8	-00 00 48	7.5 (12″)	0.6 (5″.7)	239 (IRAS)	Seyfert galaxy
M82	09 55 52.4	+69 40 47	4.9 (7″.8)	0.5 (25″)	1 400 (2′.2)	Peculiar galaxy
NGC 4151	12 10 32.6	+39 24 21	8.7 (22″)	3.6 (6″)	8.0 (2′)	Seyfert galaxy
3C 273	12 29 06.7	+02 03 09	9.8	5.1 (17″)	2.8 (IRAS)	Quasar
Arp220	15 34 57.2	+23 30 12	11.4 (5″)	5.6 (6″)	149 (2′)	Interacting galaxies
BL Lac	22 02 43.3	+42 16 40	10.4	6.4	0.4 (40″)	BL Lac object

The majority of the stars in the table are variable stars. In the column of infrared radiation intensity, magnitude and Jy ($=10^{-26}$ W Hz^{-1} m^{-2}) are used. The 0 magnitudes for 2.2 μm and 10 μm correspond to roughly 630 Jy, and 43 Jy, respectively. The brightness of extended celestial objects depends on the size of the observational field of view. The numbers in ( ) show the size of the field of view for the observed infrared intensity. (IRAS) indicates infrared intensity measured by the IRAS satellite.

Cosmic Radio Waves

Radiation mechanisms of cosmic radio waves

Continuum radiation originates when a free electron moves under acceleration. It is divided into nonthermal radiation, such as synchrotron emissions radiated when electrons accelerated to high energies rotate around magnetic field lines due to the Lorentz force, and thermal radiation, such as the Bremsstrahlung (or free-free radiation) when thermal electrons interact with ions through the Coulomb Force in ionized interstellar matters. Spectral lines are radiated by transitions between energy levels such as electron quantum levels in atoms, and vibrational and rotational states in molecules.

Celestial objects observed in radio wavelength

1) Supernova remnant Within the Galaxy, there are over 290 confirmred continuum radio sources which are the remnants after supernova explosion of high mass stars (over 8–10 solar masses) or white dwarfs which exceeded the Chandraskar limit through accreting mass from a companion star. They are broadly classified into the shell type, illustrated by Cas A (Cassiopeia A) which are shell shaped with spectral energy distribution that weakens steeply on the high frequency side, and the Crab-like type, illustrated by the Crab Nebula with a centralized continuum source with a flat spectral energy distribution.

2) Planetary nebula Celestial objects formed by mass ejection from intermediate mass stars below 8 solar masses. Thermal continuum emissions from ionized gas and molecular spectral lines from surrounding media are emitted.

3) Pulsar Neutron stars formed by the gravitational collapse in supernova explosion emit regular pulsed non-thermal continuum emissions due to its rotation; these can be observed in the visible light and X-rays as well. Over 2 500 are confirmed in the Milky Way. The existence of gravitational waves is suggested from the change in the orbital period of binary pulsars.

4) H II region Interstellar gas ionized by the ultraviolet photons radiated from high temperature B1 or earlier type stars (Refer to page **38**, "Special Types of Stars" and page **48**, "Physical Properties of Stars"), lies next to high-mass star-forming regions in molecular clouds. In addition to thermal continuum emissions, recombination lines emitted by atom such as H, He, C, S are also observed.

5) H I gas This is warm, low density interstellar gas consisting primarily of neutral hydrogen atoms (HI). It is observed in the 21 cm line (1.420 46 GHz) that is the transition between hyperfine structures of the hydrogen atom. It is used to study the structure and rotation of the Milky Way.

6) Molecular cloud Hydrogen molecule (H_2) is a principal ingredient in this relatively high density interstellar gas. These clouds have masses of 10^2–10^6 solar masses. Although there is no radio spectral line of H_2 itself, it is possible to observe the emission and absorption lines of interstellar molecules like CO (115.271 GHz etc.) and H_2O (22.235 GHz). These clouds become the sites for star formation. (Refer to Figure 17).

7) Star Continuum emission is extremely weak, and not observed, except in peculiar examples, but strong maser emissions occur in molecular lines before a planetary nebular is formed by an intermediate mass evolved star. Masers such as OH (1.667 GHz etc.), H_2O, and SiO (43.122 GHz etc.) have been discovered.

8) Normal galaxy and galactic background radiation In normal galaxies, broad continuum emissions are observed as a result of synchrotron emissions from interactions between interstellar magnetic fields and high energy electrons in the galaxy. Within the Milky Way, this is observed as a galactic background radiation. In the case of disk galaxies, it is concentrated in the disk and shows central concentration. The 21 cm line radiated by HI gas is observed primarily in disk galaxies, and in many cases has a ring shaped distribution. The molecular spectral lines radiated by interstellar molecules are also observed primarily in disk galaxies. There are galaxies where the distribution is concentrated in the center; and those, like M31 for example, with a ring shape distribution.

9) Quasar and radio galaxy Quasars are explosive phenomenon in the nucleus of galaxies. High-energy particles are ejected in the form of jets. Radio galaxies have more extended lobe-like structures ahead of the jets. Currently, over 600 000 quasars have been identified. The number of radio loud quasars is less than 1/10 of radio quiet quasars. The radiation is non-thermal and extremely bright; the emission energy reaches 10^{40} J s^{-1} in some cases. Quasars have high redshifts and some exceeds z=7. It is presumed that the energy source is the release of gravitational energy, but the details like the acceleration mechanism of the jet, etc. are not yet well understood.

10) Cosmic background radiation There is a radio wave distributed with uniform intensity regardless of the direction on the celestial sphere. It is a thermal radiation emitted when the Universe changed from an ionized state to a neutral state and redshifted as the Universe expanded. It has a 2.73 K blackbody radiation spectrum. In addition to a dipole anisotropic distribution caused by the motion of our Galaxy, anisotropies at the level of 3×10^{-5} K have been discovered, which is caused by fluctuations in the gravitational field at the beginning of the Universe.

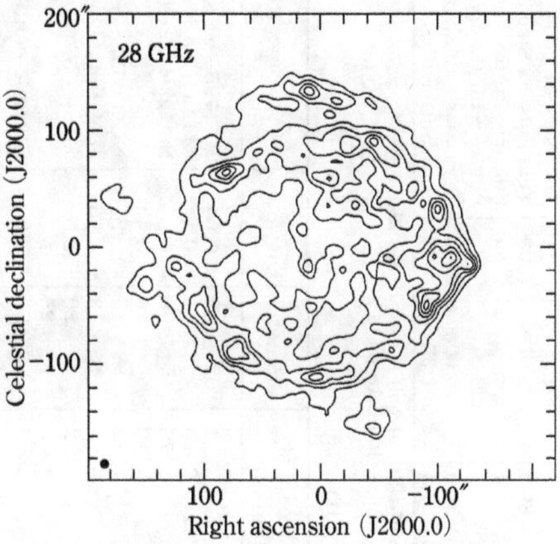

Center coordinates: right ascension (J2000.0) = $23^h23^m25^s$, Celestial declination (J2000.0) = 58°49′00″

Figure 15 Radio intensity map of supernova remnant Cas A (Wright *et al.* 1999, ApJ, **518**, 284)

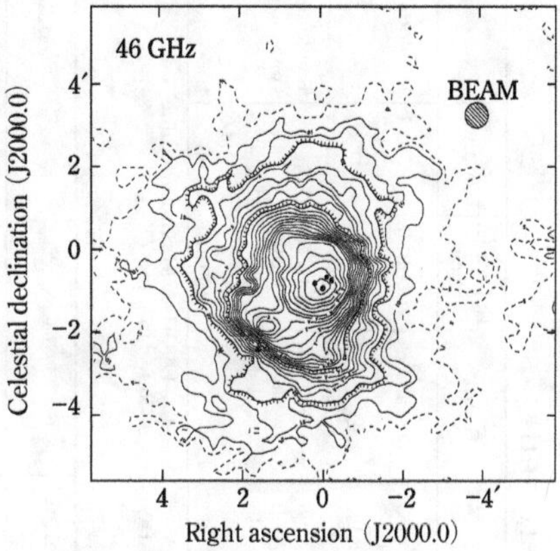

Center coordinates: right ascension (J2000.0) = $5^h35^m15^s$, Celestial declination (J2000.0) = −5°22′30″

Figure 16 Radio intensity map of HII region Ori A (M42) (Akabane *et al.* 1985, PASJ, **37**, 123)

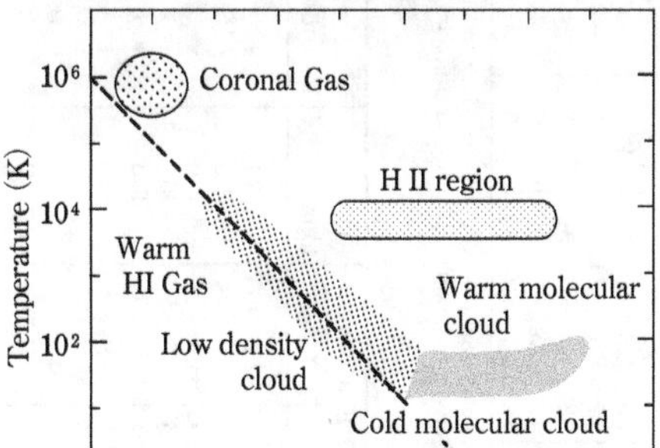

Figure 17 Temperature and density of interstellar gas

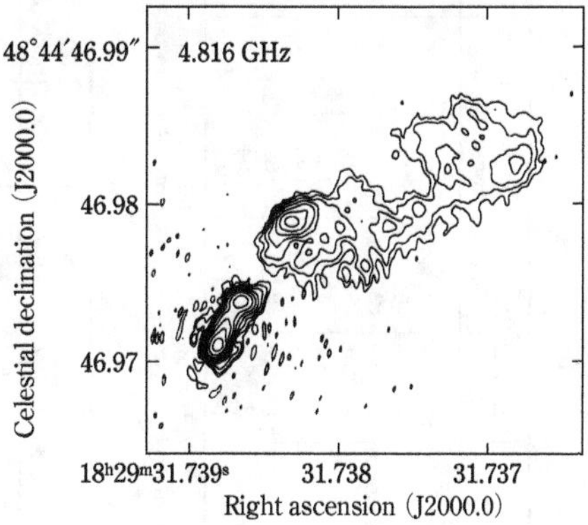

Figure 18 Radio intensity map of quasar 3C 380 (Kameno *et al.* 2000, PASJ, **52**, 1045)

Supernova Remnant

Name	2000 Equinox R.A.	2000 Equinox Dec.	Galactic longitude l	Galactic latitude b	Radio flux density (1 GHz) S[1]	Spectral index α[2]	Apparent diameter	Type	Expansion velocity	Distance	Remarks
	h m	° ′	°	°	Jy		′		km s^{-1}	10 kly	
Kepler	17 30.7	−21 28	5	+ 7	19	0.64	3	Spherical shell	200–300	1.4	Exploded in 1604
W49B	19 11.1	+09 06	43	− 0	38	0.48	4×3	Spherical shell		3.3	
Cygnus Loop	20 50.0	+30 34	74	− 9	210	—[3]	230×160	Spherical shell		0.18	Veil Nebula, NGC 6992/95
§Cas A	23 23.4	+58 49	112	− 2	2 720	0.77	5	Spherical shell	7 400	0.9	
Tycho	00 25.4	+64 09	120	+ 1	56	0.61	8	Spherical shell	4 700	0.7–1.6	Exploded in 1572
△§Tau A	05 34.5	+22 01	185	− 6	1 040	0.30	7×5	Crab-like	1 500	0.7	Crab Nebula M1 Crab Nebula, exploeded in 1054
IC 443	06 16.9	+22 47	189	+ 3	160	0.36	45	Spherical shell		0.2–0.5	
△§Vela XYZ	08 33.9	−45 11	264	− 3	1 750	—[3]	255	Mixed Morphology		0.16	
§SN1987A	05 35.5	−69 16	214	−34	0.14[4]	−0.2[4]	<1″		30 000	16	In the Large Magellanic Cloud, exploded in 1987.

1) 1 Jy = 10^{-26} W Hz^{-1} m^{-2}. 2) The Exponent when written as $S \propto \nu^{-\alpha}$ 3) Variable 4) 1.4 GHz. Maximum value 3 days after the explosion.
△ : Primerily a high-energy γ-ray source (Refer to page **69**) § : Typical X-ray source (Refer to page **71**)

Pulsar

Name	2000 Equinox R.A.	2000 Equinox Dec.	Galactic longitude l	Galactic latitude b	Radio flux density[1] S	FWHM[2]	Period P	Change in period dP/dt	Distance	Remarks
	h m	° ′	°	°	mJy	ms	ms	10^{-15}s s^{-1}	10 kly	
△§PSR B0531+21	05 34.5	+22 01	185	−6	646	3.0	33.1	422.77	0.7	Crab pulsar at the center of the Crab Nebula SNR (M1)
△§PSR B0833−45	08 35.3	−45 11	264	−3	5 000	2.1	89.3	125.01	0.09	Vela pulsar . Strongest radio signal
PSR J1748−2446ad	17 48.1	−24 47	4	+2	–	–	1.4	–	2.84	Pulsar with the shortest period
PSR J1841−0456	18 41.3	−04 56	27	−0	–	–	11 765.7	41 330	–	Pulsar with the longest period
PSR B1913+16	19 15.5	+16 06	50	+2	4	7.0	59.0	0.0086	2.32	Binary star system with a 465 minutes orbital period
PSR B1919+21	19 21.7	+21 53	56	+4	57	30.9	1 337.3	1.35	0.22	First pulsar discovered

1) Radio flux density: Time average at 400 MHz in one cycle. 2) FWHM: Full-width at half max, pulse width at 50% of peak strength
△ : Major high-energy γ-ray source (Refer to page **69**) § : Typical X-ray source (Refer to page **71**)

H II Region

Name	2000 Equinox R.A.	2000 Equinox Dec.	Galactic longitude l	Galactic latitude b	Radio flux density (2.7 GHz) S	Apparent diameter	Diameter	Distance	Remarks
	h m	° ′	°	°	Jy	′	ly	10 kly	
W3	02 25.7	+62 06	134	+ 1	110	1	2	0.64	IC 1795
Ori A	05 35.4	−05 27	209	−19	420	(Core) 3×6	1×2	0.14	M42 Orion Nebula
W16	06 32.0	+04 55	206	− 2	250	75	130	0.5	NGC 2244 Rosette Nebula
NGC 6334	17 20.6	−35 49	351	+ 1	176	6×10	4×6	0.44	
W22	17 25.6	−34 21	353	+ 1	269	13	13	0.3	NGC 6357
M17	18 20.4	−16 12	15	− 1	552	5×7	10×13	0.65	ω Nebula
W49 A	19 10.2	+09 06	43	− 0	53	2	20	3.6	
W51 A	19 23.7	+14 31	50	− 0	119	3×5	15×25	1.7	
Sgr A	17 45.7	−29 00	0	− 0	250	Halo 2×3 Core<0.001″	20×30 <0.0001	2.8	Galactic center nucleus, thermal and nonthermal radio sources

Interstellar Molecular Cloud

Name	2000 Equinox R.A.	2000 Equinox Dec.	Galactic longitude l	Galactic latitude b	Radio strength T_b (CO)[1]	Apparent diameter	Diameter	Distance	Remarks
	h m	° ′	°	°	K		ly	ly	
Taurus Molecular Cloud -1	04 41.4	+25 47	172	−13	10	15′ × 5′	2 × 0.7	430	Cold molecular cloud. It is located in a large dark cloud complex.
Orion Molecular Cloud -1	05 35.2	−05 23	209	−19	10	7° × 1.5°	170 × 40	1 400	Giant molecular cloud. Massive stars are being formed.
					(Core) 60	10′ × 3′	4 × 1.2		There is a compact core with number densities over 10^{13} m^{-3}.
L183 (L134N)	15 54.1	−02 53	6	37	10	30′	5	520	Cold molecular cloud.
ρ Ophiuchus Molecular Cloud	16 26.3	−24 26	353	17	10–25	2° × 1°	18 × 9	450	There is a core containing many young stars.
Sgr A	17 45.7	−29 00	0	0	40	25′	250	28 000	Giant molecular cloud in galactic center. Violent activities are observed.
Sgr B2	17 47.3	−28 23	1	0	20	15′	150	28 000	Giant molecular cloud in the vicinity of galactic center. Formation site of high mass stars.
M17	18 20.4	−16 14	15	−1	10–50	3° × 0.5°	340 × 60	6 500	Giant molecular cloud. Formation site of high mass stars.

1) Intensity of the CO molecular line at 115.271 GHz (J=1–0, at the wavelength of 2.6 mm) in unit of the temperature. It is roughly equal to the temperature of the molecular gas.

Radio Galaxy and Quasar

Name	2000 Equinox R.A.	2000 Equinox Dec.	Redshift z[1]	Type of galaxy	Visual magnitude	Radio strength (2.7 GHz)	Extent of radio structure	Remarks
Radio galaxy	h m	° ′			mag	Jy[2]		
3C 84	03 19.8	+41 31	0.0176	ED 2	12.5	9.9*	Halo (26′)	Per A, NGC 1275
							Core (10″, 0.001″)	Seyfert Galaxy
PKS 0320−37	03 22.6	−37 14	0.0044	D 3–4	10.5	94	Bipolar structure (29′)	For A, NGC 1316
§3C 274	12 30.8	+12 23	0.0041	E 2	10.9	118.3	Halo (12′ × 16′)	Vir A, M87
							Core (50″)	
△§PKS 1322−42	13 25.3	−43 01	0.0009	DE 3	6.1	890	Bipolar structure (200′ and 7.′2)	Cen A, NGC 5128
3C 348	16 51.1	+05 00	0.154	cD 4	18.0	22.4	Bipolar structure (31″)	Her A
3C 405	19 59.5	+40 44	0.056	cD 3	15.1	785	Bipolar structure (106″)	Cyg A
BL Lac	22 02.7	+42 17	0.069	–	14.7*	4.5*	0.007″	BL Lac object
Quasar								
3C 48	01 37.7	+33 10	0.367	–	16.2*	9.0		First quasar identified
PKS 0237−23	02 40.1	−23 09	2.225	–	16.6	5.3		There are many absorption lines.
PKS 0458−02	05 01.3	−02 00	2.286	–	19.5	2.0		There are 21cm absorption lines at of z = 2.
0957+561A, B	10 01.3	+55 54	1.413	–	17.0	0.2		First gravitationally lensed double quasar discovered
BR 1202−0725	12 05.4	−07 43	4.694	–	18.7‡	–		Protogalaxy candidate harboring a large amount of molecular gas and interstellar dust.
△§3C 273	12 29.1	+02 03	0.158	–	12.9*	41.4*		First quasar to have z measured
△3C 279	12 56.2	−05 47	0.538	–	17.8*	12.0*		
H 1413+117	14 15.8	+11 30	2.546	–	17.0	–		Cloverleaf quasar. Quadrupal image caused by gravitational lens
△3C 454.3	22 54.0	+16 09	0.859	–	16.1*	10.7*		

1) For the case that z is significantly smaller than 1, the distance to the object is calculated approximately by $(cz/H_0)\{1(q_0+1)z/2\}$ (where H_0 = 75 km s^{-1} Mpc^{-1} is Hubble's Constant, q_0 is the deceleration parameter; for instance in the Einstein-de Sitter Universe model, q_0 = 0.5).
2) 1 Jy = 10^{-26} W Hz^{-1} m^{-2}.
* Variable ‡ R band magnitude. △ : Major high-energy γ-ray source (Refer to page **69**) § : Major X-ray source (Refer to page **71**)

Interstellar Molecules Observed to Date

(208 species confirmed out of 217, as of July, 2020)

Simple hydrides, oxides, sulfides, halides, etc.				
H_2(IR)	CO	H_2O	CS	$NaCl^*$
HF	SiO	NH_3	SiS	$AlCl^*$
HCl	SO_2	CH_4^*(IR)	H_2S	KCl^*
N_2(UV)	O_2	CH_3Cl	OCS	AlF^*
N_2O	CO_2	SiH_4^*(IR)	PH_3	AlO^*
H_2O_2	TiO	CH_3SiH_3	PN	AlOH
	TiO_2	SiH_3CN	HCP	
Nitriles, acetylene derivatives, etc.				
C_2^* (IR)	HCN	CH_3CN	HNC	HC_2H^* (IR)
C_3^* (IR)	HC_3N	CH_2CHCN	HNCO	HC_4H^* (IR)
C_5^* (IR)	HC_5N	CH_3CH_2CN	HCNO	HC_6H^* (IR)
C_3O	HC_7N	CH_3C_3N	HNCS	CH_3C_2H
C_3S	HC_9N	CH_3C_5N	HNCCC	CH_3C_4H
C_4Si^*	$HC_{11}N$?	CH_2CCHCN	HCCNC	CH_3C_6H
$C_2H_4^*$ (IR)	KCN^*	*n*-C_3H_7CN	CH_3NC	HOCN
	HCOCN	*i*-C_3H_7CN	CH_3NCO	HSCN
	CNCN	$HOCH_2CN$	HONO	
Aldehydes, alcohols, ethers, ketones, amids, etc.				
H_2CO	CH_3OH	HCOOH	CH_2NH	H_2C_3
H_2CS	C_2H_5OH	$HCOOCH_3$	CH_2CNH ?	H_2C_4
CH_3CHO	CH_2CHOH	CH_3COOH	HNCNH	H_2C_6
NH_2CHO	CH_3OCH_2OH	H_2CCO	CH_3CHNH	
HC_2CHO	CH_3SH	CH_2CHCH_3	*e*-HNCHCN	
CH_2OHCHO	CH_3CH_2SH ?	C_2H_5OCHO	CH_3NH_2	
CH_2CHCHO ?	$(CH_3)_2O$	$C_2H_5OCH_3$?	NH_2CN	
CH_3NHCHO ?	$(CH_3)_2CO$	CH_3CHCH_2O	CH_3CONH_2	
C_2H_5CHO	$(CH_2OH)_2$	CH_3COOCH_3	NH_2CH_2CN	
			NH_2CONH_2	
Cyclic molecules				
c-C_3H	*c*-SiC_2	*c*-C_2H_4O	*c*-$C_6H_6^*$ (IR)	C_{60}^* (IR)
c-C_3H_2	*c*-Si_2C^*	*c*-H_2C_3O	*c*-C_6H_5CN	C_{60}^+(IR)
	c-SiC_3^*			C_{70}^* (IR)
Molecular ions				
H_3^+	HCO^+	$HCNH^+$	H_2O^+	C_4H^-
CH^+(OPT)	HOC^+	H_2Cl^+	SO^+	C_6H^-
OH^+	HCS^+	H_3O^+	NO^+ ?	C_8H^-
ClH^+	HN_2^+	H_2COH^+	NS^+	CN^-
ArH^+	$HOCO^+$	HC_3NH^+	NH_2CO^+	C_3N^-
SH^+	CO^+	*l*-C_3H^+	NH_4^+	C_5N^-
CF^+		HeH^+	$NCCNH^+$	
Radicals				
OH	C_2H	CN	NO	C_2O
CH	C_3H	C_2N	HNO	C_2S
CH_2	C_4H	C_3N	NS	CP^*
CH_3	C_5H	C_5N	SO	C_2P^*
NH(UV)	C_6H	HCCN	PO	SiC^*
NH_2	C_7H	HC_4N^*	FeO ?	SiN^*
SH(IR)	C_8H	CH_2CN	MgNC	$SiCN^*$
SiH ?	HCO	CH_2N	$MgCN^*$	$SiNC^*$
HO_2	HCCO	CH_3O	$HMgNC^*$	$AlNC^*$
HS_2	HC_5O	HCS	$NaCN^*$	$CaNC^*$
NCO	HC_7O	HSC	$FeCN^*$	
		MgC_2H^*	MgC_3N^*	MgC_4H^*

For molecules observed other than radio waves, the wavelength range is shown in ( ). IR is infrared, OPT is optical, and UV is ultraviolet.

* indicates it is found only in red giants.

? indicates that it has been reported, but not yet confirmed.

Time System

The definitions of the time system and the polar motion changed drastically as a consequence of a series of the IAU recommendations between 1991 through 2006 that, along with the adoption the general theory of relativity as a fundamental frame, revised the precession-nutation theory and the definitions of the celestial and terrestrial frames.

Universal Time (UT)

The relation between the Geocentric Celestial Reference System (GCRS) and the International Terrestrial Reference System (ITRS) at time t is written as: [GCRS] = $Q(t)R(t)W(t)$ [ITRS]. Here $Q(t)$, $R(t)$ and $W(t)$ respectively stand for the rotation of the celestial pole in the celestial system, that of the Earth around the axis of the pole and the polar motion. The pole of celestial system realized by $R(t)$ $W(t)$ [ITRS] is called CIP, the Celestial Intermediate Pole. The Earth rotates around CIP. The origin of the celestial system fixed over the true equator in the instantaneous CIP axis is called CIO, Celestial Intermediate Origin. The direction of the first axis of the rotating Earth coordinate system (≈ longitude 0°) is called TIO, the Terrestrial Intermediate Origin. The angle measured from CIO to TIO is ERA, the Earth Rotation Angle. Universal time UT1 is linear in ERA: $\mathrm{ERA}(T_u) = 2\pi\,(0.779\,057\,273\,264\,0 + 1.002\,737\,811\,911\,354\,48 T_u)$. Here T_u = Julian UT1 date − 2 451 545.0 and UT1 = UTC + (UT1 − UTC). UT1 − UTC is determined from such as VLBI observations. The relation between ERA and Greenwich Sidereal Time is defined by $\mathrm{GST}(T_u, t) = \mathrm{ERA}(T_u) +$ polynomials of $t + \mathrm{EE}(t)$. The right side terms up to the second correspond to GMST, the Greenwich Mean Sidereal Time given by $\mathrm{GMST}(T_u, t) = \mathrm{ERA}(T_u) + 0''.014\,506 + 4\,612''.156\,534t + 1''.391\,581\,7t^2 - 4''.4 \times 10^{-7}t^3 - 2''.9956 \times 10^{-5}t^4 - 3''.68 \times 10^{-8}t^5$. $\mathrm{EE}(t)$ called the Equation of Equinox is caused primarily by the nutation and is the sum of the periodic and $t \times$ periodic terms. Here t is Terrestrial Time TT (Refer to Dynamic Time). Polar motion (x, y) is the movement of CIP in ITRF. Universal time UT0 obtained from position astronomy observations, besides UT1, contains the longitude variation due to the polar motion as UT0 = UT1 + $\Delta\lambda$. $\Delta\lambda$ is given by $\Delta\lambda = 1000(x \sin\lambda + y \cos\lambda) \tan\phi \,/\, 15$ (ms). Here, λ and ϕ are the longitude and latitude of an observation point. The final values of x, y, UT1 − UTC and UTC − TAI are calculated on the basis of observations and published by the central bureau of the International Earth Rotation and Reference System Service (IERS). In addition the predicted values of UT1 − UTC known as DUT1 can be obtained with accuracy of ±0.1 seconds from standard time and frequency services of individual countries.

Dynamical Time

Dynamical time is the time system used for astrodynamics and ephemerides based on the general theory of relativity. It is classified into Barycentric Coordinate Time (TCB) and Geocentric Coordinate Time (TCG). TCB is used to express the motion of the solar system bodies while TCG is used to express the motion around the Earth's center of mass. TCB − TCG is calculated by using astronomical ephemerides and the relation in seconds of time is given by TCB − TCG $= L_c \times (\mathrm{JD_{TAI}} - 2\,443\,144.5) \times 86\,400 \,/\, (1 - L_B)$ + terms depending on the Earth's speed and the position of the solar system bodies. Here $L_B = 1.550\,519\,768 \times 10^{-8}$ and $L_c = 1.480\,826\,867\,41 \times 10^{-8}$. L_B is a defining constant. Since TCG is a time system at the geocenter, TCG as it is cannot be in observations on the Earth's surface. Terrestrial Time (TT) the rate of which is uniform and coincides with that of the SI second on the surface of the geoid. However, since the geoid surface cannot be realized with sufficient accuracy, TT is defined by the following relation to TCG,

$\mathrm{TCG} - \mathrm{TT} = L_G \,/\, (1 - L_G) \times (\mathrm{JD_{TAI}} - 2\,443\,144.5) \times 86\,400$ (s), $L_G = 6.969\,290\,134 \times 10^{-10}$,
L_G being a defining constant and a transformation coefficient and the Julian Day JD in international atomic time TAI. The relation between TT and TAI is TT = TAI + 32.184 seconds.

The difference of TT from UT1, ΔT is expressed by

ΔT = TT − UT1 = −(UT1 − UTC) − (UTC − TAI) + 32.184 seconds and predicted to be about 70 seconds in 2020. For UT1 − UTC and UTC − TAI refer to Coordinated Universal Time. Conventional Barycentric Dynamical Time (TDB) in seconds is redefined by the following linear expression of TCB as time close to TT in the geocentric system.

$\mathrm{TCB} - \mathrm{TDB} = L_B \times (\mathrm{JD_{TCB}} - 2\,443\,144.500\,372\,5) \times 86\,400 + \mathrm{TDB_0}$(s).

Here, $\mathrm{TDB_0} = -6.55 \times 10^{-5}$ is a defining constant.

International Atomic Time (TAI)

The unit of time interval, "second", was defined at the 13th General Conference of Weights and Measures in Paris in October, 1967 as "The second is the duration of 9 192 631 770 periods of the radiation corresponding to the transition between the two hyperfine levels of the ground state of the cesium 133-atom.". Making a clock that beats once per every f cycles by measuring with an oscillator which is referred to this frequency, f = 9 192 631 770 Hz, the time indicated by such clock is called atomic time.

The definitive time calculated and synthesized at the International Bureau of Weights and Measures (BIPM, France) from an ensemble of atomic clocks distributed worldwide is called International Atomic Time (TAI). Its initial epoch was matched as 00:00:00 January 1st, 1958 at the instant 00:00:00 UT2 on that day.

Coordinated Universal Time (UTC)

Coordinated Universal Time (UTC) is an artificially coordinated atomic time whose difference from the time based on the Earth rotation is maintained to fall in a certain range. In the old UTC used up to the end of 1971, the time difference (UT2 − UTC) was maintained to be within ±0.1 seconds by shifting the frequency from the defining frequency and adjusting interval of 1 second as well as by 0.1 second step-adjustments. In the new UTC started on January 1st, 1972, frequency offsets were abolished and the time difference (UT1 − UTC) is maintained so as not to exceed the range of ±0.9 seconds (revised on January 1, 1975) through 1 step-second adjustments. This means that the interval of one second is not adjusted at all as defined. Now, the time difference between UTC and TAI is always an integral number of seconds.

1 second step-adjustments are made by inserting or extracting 1 second after the last second of the day (UTC) on the last day of December or June (1st preference), or the last day of March or September (2nd preference), or on the last day of any month if needed. The central bureau of IERS determines dates of the 1 second step-adjustments. The inserted (extracted) second is called a positive (negative) leap second. DUT1 is a term showing the predicted value of UT1 − UTC to accuracy of 0.1 seconds. This numerical value (from −0.8 to +0.8 seconds) is notified every minute by a time code sent out in standard time and frequency services. (Refer to "Time Comparison")

Dates and amount of UTC−TAI after 1972 are shown in the tables.

Y M D	UTC-TAI	Y M D	UTC-TAI	Y M D	UTC-TAI
	s		s		s
1972 1 1	-10	1981 7 1	-20	1996 1 1	-30
1972 7 1	-11	1982 7 1	-21	1997 7 1	-31
1973 1 1	-12	1983 7 1	-22	1999 1 1	-32
1974 1 1	-13	1985 7 1	-23	2006 1 1	-33
1975 1 1	-14	1988 1 1	-24	2009 1 1	-34
1976 1 1	-15	1990 1 1	-25	2012 7 1	-35
1977 1 1	-16	1991 1 1	-26	2015 7 1	-36
1978 1 1	-17	1992 7 1	-27	2017 1 1	-37
1979 1 1	-18	1993 7 1	-28		
1980 1 1	-19	1994 7 1	-29		

Time Comparison

Time and frequency dissemination services can be used as a means for comparison and calibration of time and/or frequency standards. Radio time and frequency services use shortwave, low frequency and very low frequencies. The services emit carrier waves of standard frequencies, indicating time by audio or pulse signal modulations or intermittence of carrier waves. The time scale of the signal is that of the Coordinated Universal Time (UTC) .

A typical time code for a shortwave time and frequency service is embodied by emitting a continuous carrier wave generated by a frequency standard and by modulating the carrier wave by 200 × *n* Hz audio signal with 5 millisecond duration. The audio modulation starts at integral seconds. This is the method recommended by the International Telecommunication Union Radio communication sector (UTC−R). For instance, HLA, WWV and WWVH, respectively adopt 9, 5 and 6 as *n* values.

DUT1 of WWVH, for example, is expressed as *m*-signals between 1 and 8 seconds every minute in the case DUT1 is +0.1 × *m* seconds. On the other hand in the case that DUT1 is −0.1 × *n* seconds it is expressed by *m* signals between subsequent 9 and 16 seconds. *m* and *n* take 0, 2, ⋯, 8 depending on the value of DUT1. DUT1 = 0, if no signal.

Because of high precision of ground waves, standard radio waves with low and very low frequencies have been used as high precision frequency standards mainly for calibrating clock rate and frequency oscillators. In addition, the low frequency standard time and frequency (JJY) operated by National Institute of Information and Communications Technology is superimposed on the time code (Refer to the chart. At 15th minutes and 45th minutes of each hour, after the 40 second, the format is replaced by the call sign and advanced notification of service interruption) and is used for disseminating Japan Standard Time.

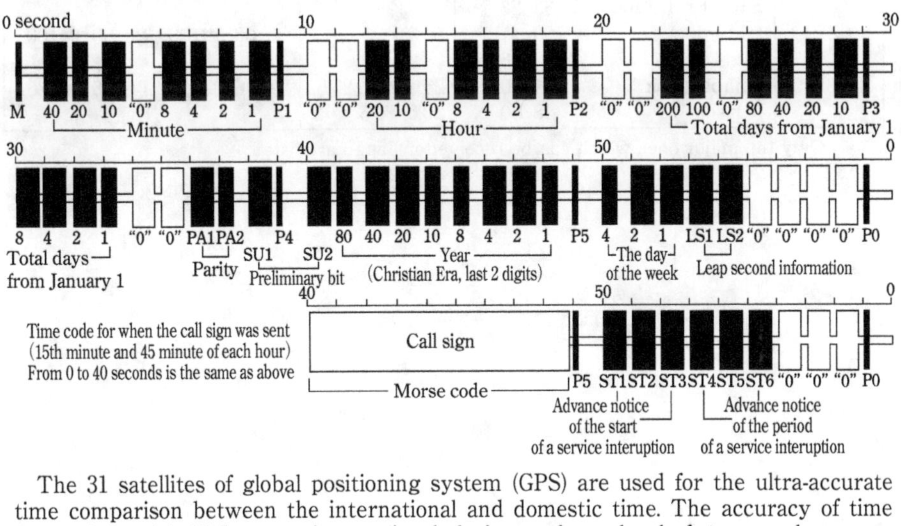

The 31 satellites of global positioning system (GPS) are used for the ultra-accurate time comparison between the international and domestic time. The accuracy of time comparison with GPS among international clocks reaches a level of ± several nanoseconds.

Shortwave Standard Time and Frequency Services

Station			Standard time and frequency emission	
Call sign	Location	Longitude/Latitude	Frequency (MHz)	Transmission time (UTC)
BPM*	Pucheng China	109°31′ E 35 00 N	2.5, 5, 10, 15	(2.5)07:30-01:00, (15)01:00-09:00 5, 10 MHz are continuous.
CHU	Ottawa Canada	75 45 W 45 18 N	3.330, 7.850, 14.670	Continuous
HLA	Daejeon Rep. of Korea	127 22 E 36 23 N	5	Continuous
WWV	Fort-Collins CO, USA	105 03 W 40 41 N	2.5, 5, 10, 15, 20	Continuous
WWVH	Kauai HI, USA	159 46 W 21 59 N	2.5, 5, 10, 15	Continuous

* BPM emits UT1 time between 25–29 minutes and 55–59 minutes of each hour and UTC time advanced by 20 milliseconds in the other periods.

Low Frequency Standard Time and Frequency Services*

Station			Standard time and frequency emission	
Call sign	Location	Longitude/Latitude	Frequency (kHz)	Transmission time (UTC)
JJY	Ootakadoyayama Fukushima Pref. Japan	140°51′ E 37 22 N	40.0	Continuous
JJY	Haganeyama Saga Pref. Japan	130 11 E 33 28 N	60.0	Continuous
RTZ	Irkutsk Russia	103 41 E 52 26 N	50.0	00:00–19:00, 20:00–24:00
RAB-99	Khabarovsk Russia	134 50 E 48 30 N	25.0, 25.1, 25.5, 23.0, 20.5	02:06–02:36, 06:06–06:36

* Only the major ones which can be received in Japan are listed.

Polar Motion and Rotation Speed Variation

Figure 19 shows the position of North Pole of Earth rotation axis (polar motion) since January 1, 2014 in reference to the IERS Reference Pole (IRP). The x-axis is taken as the Greenwich meridian, and the y-axis directs toward longitude 90 degrees west. The marks in the figure show the positions at the beginning of each month of each year. These are taken from EOP14 C04 (IAU2000), the final values published by the IERS Central Bureau. The relations between the average longitude and latitude of a site (λ_0, φ_0) referred to the IRP and the latitude and longitude (λ, φ) obtained by astronomical observation at some moment are given with setting east longitude positive:

$$\lambda = \lambda_0 + 1000(x \sin \lambda_0 + y \cos \lambda_0) \tan \varphi_0 / 15 \text{ (ms)},$$

$$\varphi = \varphi_0 + (x \cos \lambda_0 - y \sin \lambda_0) \text{ (arcsecond)}$$

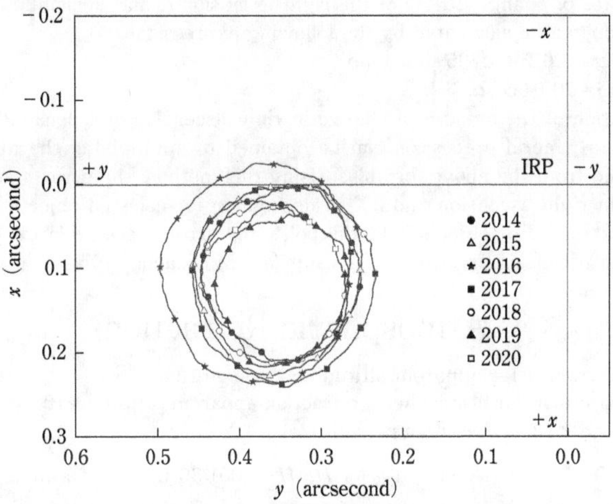

Figure 19 Polar motion (2014.1.1-2020.5.24)

Earth rotation speed variation: Figure 20 shows the final daily values of the excess over 86400 seconds (24 hours) of the Length of Day (LOD) based on EOP14 C04 (IAU2000) published by the IERS Central Bureau. The fluctuations caused by the long period tides with periods shorter than 35 days have been removed (LODR).

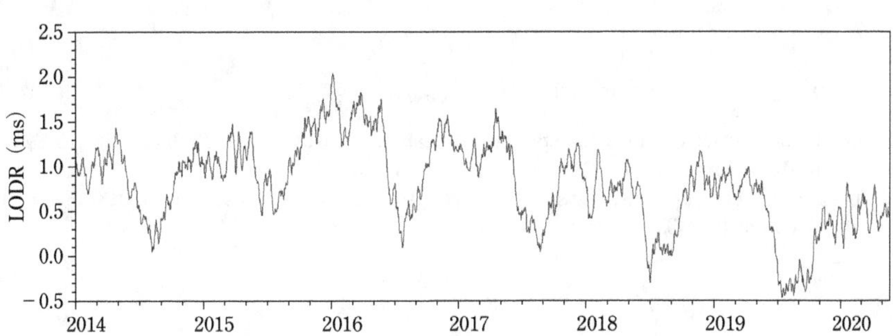

Figure 20 The excess of length of day over 24 hours (2014.1.1-2020.5.24)

Positional Changes due to Precession

The equatorial plane of the Earth is constantly moving due to the effects of the gravity from the Moon and Sun on the equatorial bulge of the Earth. Therefore, the Celestial North Pole also moves on the Celestial Sphere. That movement is divided into the rotation around the ecliptic pole with a period of approximately 26 000 years and with an almost constant obliquity, and periodic motion with amplitudes of about 9″ and less. The former is called the precession of the equator and the latter is called the nutation. Also, the ecliptic moves because of the gravitation of planets affecting the Earth and the Moon. This movement is called the precession of the ecliptic. The combination of the precession of the equator and the precession of the ecliptic is called the general precession.

The annual rates of change $\Delta\alpha$, $\Delta\delta$ of the right ascension α and declination δ of a star due to general precession can be calculated by the following expressions.

$$\Delta\alpha = 3\overset{s}{.}075 + 1\overset{s}{.}336 \sin\alpha \tan\delta$$

$$\Delta\delta = 20\overset{''}{.}04 \cos\alpha$$

An approximation of the amount of change in right ascension and declination over long periods of time due to general precession can be obtained by multiplying the number of years by the value obtained from the above formulas. Using this method, the error margin after 50 years is less than 10^{s} in right ascension and 20″ in declination for celestial objects between the declination of -80° and +80°. For objects near the poles, the errors become large. For instance, after 50 years the right ascension error of the Polestar (α UMi) reaches 10^{m}.

Atmospheric Refraction

atmospheric refraction (R) = apparent altitude − true altitude

Approximate expression for atmospheric refraction(apparent altitude>10°) :

$R = R_0 \tan Z + R_1 \tan^3 Z$ ($Z = 90° -$ apparent altitude)

$R_0 = (n_0 - 1)(1 - H)$, $R_1 = \frac{1}{2}(n_0 - 1)^2 - (n_0 - 1)H$, $H \approx 0.001\,30$ (H is the atmosphere scale height in units of the Earth's radius, R_0 and R_1 are in radians).

n_0 is the refractive index of the air at the observation location, and it is given by the following expression for 0.03% carbon dioxide content air, air temperature T (K), pressure P (hPa), water vapor pressure F (hPa), and observation wavelength λ (μm) (Owens 1967, Applied Optics, **6**, 51):

$$(n_0 - 1) \times 10^8 = C(\lambda) \times \frac{P}{T}\left[1 + P\left\{57.90 \times 10^{-8} - 9.325\,0 \times 10^{-4} \times \frac{1}{T} + 0.258\,44 \times \left(\frac{1}{T}\right)^2\right\}\right] \times \left(1 - 0.16\frac{F}{P}\right)$$

Here, $C(\lambda) = 2\,371.34 + 683\,939.7\left(130 - \frac{1}{\lambda^2}\right)^{-1} + 4\,547.3\left(38.9 - \frac{1}{\lambda^2}\right)^{-1}$

(Example) The refractive index for standard dry air ($P = 1\,013.25$ hPa, $F = 0$, $T = 288.15$ K) at wavelength 0.575 μm is, $n_0 = 1.000\,277\,4$. Therefore $R_0 = 57.14''$, $R_1 = -0.07''$.

Atmospheric Refraction : Radau's calculations (Visible Ray, Air Pressure 1 013.25 hPa, Temperature 10℃, Latitude 35°)

Apparent altitude	Atmospheric refraction	Apparent altitude	Atmospheric refraction	Apparent altitude	Atmospheric refraction	Apparent altitude	Atmospheric refraction	Apparent altitude	Atmospheric refraction
°	′ ″	°	′ ″	°	′ ″	°	′ ″	°	′ ″
0.0	34 24	1.5	20 53	5.0	9 49	12	4 26	30	1 40
0.2	31 54	2.0	18 13	6.0	8 25	14	3 48	40	1 09
0.4	29 41	2.5	16 04	7.0	7 22	16	3 19	50	0 49
0.6	27 41	3.0	14 20	8.0	6 31	18	2 56	60	0 33
0.8	25 54	3.5	12 54	9.0	5 51	20	2 38	70	0 21
1.0	24 17	4.0	11 41	10.0	5 17	25	2 04	80	0 10

Radau 1889, Ann. Obs. Paris, Mémoires, **19**, G.

Atmospheric Extinction

The difference between the magnitude of a star observed from the ground and that observed outside of the atmosphere (Δm) is called the atmospheric extinction usually expressed as $\Delta m = a\,F(z)$, where a is the extinction coefficient and $F(z)$ is the air-mass (a function of the zenith distance z).

The extinction coefficient (a) depends on the wavelength as well as the condition of the atmosphere. The typical values of a corresponding to the standard condition of the atmosphere are given in the table below, where the atmospheric transmittancy (calculated as $10^{-0.4a}$) is also shown. (Allen 1973, Astrophysical Quantities)

Wavelength (μm)	0.30	0.32	0.36	0.40	0.5	0.6	0.8	1.0	2.0
a (molecules)	1.21	0.92	0.56	0.361	0.144	0.068	0.0215	0.0087	0.0005
a (ozone absorption)	3.2	0.24	0.00	0.000	0.012	0.044	0.001		
a (solid particles)	0.143	0.132	0.113	0.099	0.074	0.058	0.040	0.030	0.012
Total	4.5	1.30	0.68	0.46	0.23	0.170	0.062	0.039	0.013
Transmittancy	0.011	0.273	0.51	0.63	0.79	0.84	0.939	0.962	0.987

Air-mass function $F(z)$ is a function of the zenith distance (z), which represents the amount of atmosphere along the path of light coming from outside the atmosphere to the surface of the Earth. $F(z)$ is normalized to be unity at the zenith ($z = 0$), and is approximately equal to sec z as long as z is not so large (e.g., $z < 60°$). (Hdb. d. Ap., II, **1**, 1930)

z	$F(z)$	z	$F(z)$	z	$F(z)$	z	$F(z)$	z	$F(z)$
0°	1.000	20°	1.064	40°	1.30	60°	1.99	80°	5.60
5	1.004	25	1.103	45	1.41	65	2.36	85	10.4
10	1.015	30	1.154	50	1.56	70	2.90	87	15.4
15	1.035	35	1.220	55	1.74	75	3.82	89	27

The following table shows the representative values of atmospheric extinction corresponding to clear and moonless nights, which were measured at the National Astronomical Observatory of Japan (Mitaka Campus). (Dermawan & Fukushima 2002, unpublished). B, V, R_C, and I_C are the magnitudes in the *UBVRI* system.

z	B	V	R_C	I_C
0°	0.44	0.27	0.24	0.12
20	0.47	0.28	0.25	0.13
40	0.57	0.35	0.31	0.16
60	0.87	0.53	0.47	0.25
80	2.45	1.49	1.32	0.69

Wavelength Dependence of Atmospheric Absorption

These three figures (corresponding to visible, near-infrared, and mid-infrared band, respectively) show how the optical thickness (τ_λ) of the Earth's atmosphere in the vertical direction depends on the wavelength and which species of molecular lines make major contributions to the atmospheric absorption. The upper line (larger τ_λ) corresponds to the typical location on the surface of Earth, while the lower line (smaller τ_λ) to the typical high-elevation site (such as the summit of Mauna Kea, where the altitude is ~4200 m and the atmospheric pressure is as low as ~60% of the sea level).

In each of the figures, the position of $\tau_\lambda = 1$ is shown by a horizontal dashed line. As a rule of thumb, those wavelength bands satisfying the condition $\tau_\lambda < 1$ are called "atmospheric windows," where the extinction of the Earth's atmosphere is not very significant and thus observations from the ground are possible.

The most representative window is the 0.3–1 μm region, which includes the visible region (0.4–0.7 μm) perceived by human eye. In addition, several windows separately existing in the near-infrared region (~1–5 μm) are important for near infrared observations, which are called J, H, K, L, L′ and M bands, respectively. Similarly, in the mid-infrared region, two bands N (~10 μm) and Q (~20 μm) are defined and widely used. These graphs indicate that molecular absorptions are especially important, such as due to H_2O (water vapor), O_2 (oxygen), O_3 (ozone), CO_2 (carbon dioxide), or CH_4 (methane).

The Earth's atmosphere is opaque in the UV (shorter than about 0.3 μm) and X-ray regions, where ground-based observations are impossible. On the other hand, while Earth's atmosphere is still opaque in the far-infrared region (at wavelengths longer than mid-infrared), a window appears above several hundred μm (submillimeter band).

In the radio domain (wavelengths from millimeter to meter), the atmosphere is generally transparent and observations are possible from the ground. (Note that absorption by the Earth's ionosphere begins at wavelengths of longer than several tens of meters.)

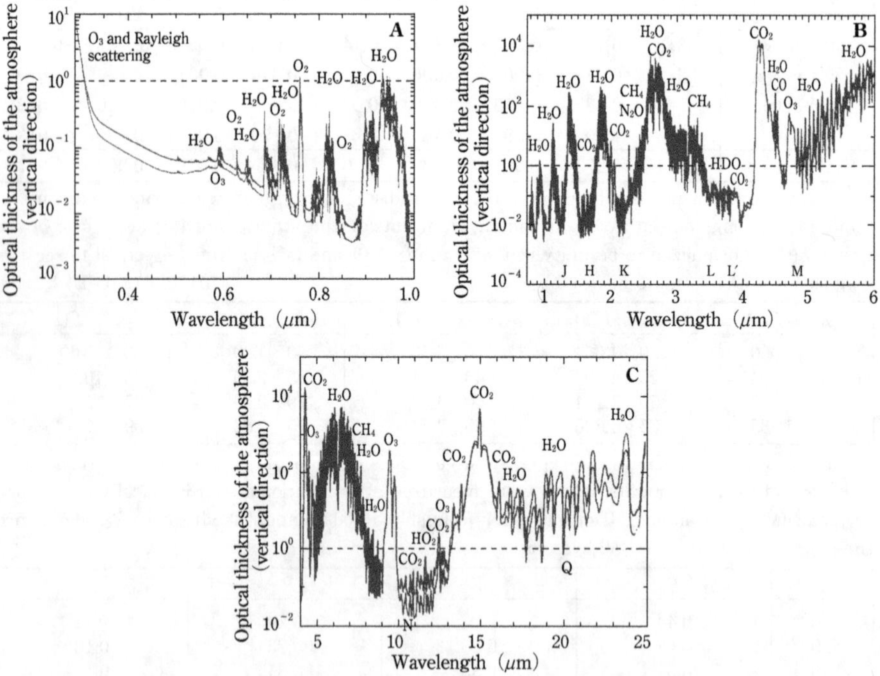

Figure 21 Wavelength-dependent absorption of electromagnetic radiation due to Earth's atmosphere (optical thickness), and representative molecules significantly contributing to the absorption. **A**: Around the Visible region, **B**: Near-infrared region, **C**: Mid-infrared region. (From Allen's Astrophysical Quantities, ed. A. N. Cox, 2000)

Night Sky Brightness

The light that comes from the entire night sky is called night sky brightness. It is primarily light from the following 3 sources.

(1) Airglow: Light emitted by molecules and atoms in the upper layers of the Earth's atmosphere

(2) The zodiacal light: Light from sunlight scattered by fine dust in the Solar System.

(3) Scattered Starlight: Accumulated light from faint stars and nebulae

The brightness of these types of light varies depending on the position on the celestial sphere and the time. The following tables show the rough averages, the units are:

S_{10}(vis) = the number of Magnitude 10 (visual magnitude) stars needed produce the same level of brightness. 1 S_{10} corresponds to approximately Magnitude 27.78 per square second.

Average Zenith Brightness for each Component of the Light

(wavelength 530 nm)

Light component	Brightness		
	S_{10} (vis)	10^{-11} J m^{-2} s^{-1} sr^{-1} nm^{-1}	
Airglow	55	66	Continumn spectrum only
Zodiacal light	135	162	
Scattered starlight	130	156	

1 S_{10}(vis) = 1.20 × 10^{-11} J m^{-2} s^{-1} sr^{-1} nm^{-1} at 530 nm

Major Airglow Lines

(The strengths are average values for mid-latitudes on the Earth's Surface)

Source of luminescence	Wavelength (nm)	Strength (R)	Source of luminescence	Wavelength (nm)	Strength (R)
〔O I〕	557.7	260	O_2	300–400	4.3×10^2
〔O I〕	630.0, 636.4	80		(Herzberg band)	
Na I	589.0, 589.6	Summer 30	O_2	864.5 (0–1) band	1.5×10^3
		Winter 200	OH	1580 (0–4) band	1.75×10^5
H I	121.6(Lyα)	2500–5000	OH	Visible	1.3×10^2
	656.3(Hα)	10?		(5–0) (7–1) (8–2) (9–3)	
〔N I〕	519.9	3		band	
			OH	All wavelength bands	4×10^6

1 R (rayleigh) = 10^{10} photon m^{-2} (column) s^{-1}

Brightness and Degree of Polarization of Zodiacal Light

(along the Ecliptic, wavelength 530 nm)

Elongation from the Sun	Brightness S_{10} (vis)	Polarization degree (%)	Elonga-tion	Brightness S_{10} (vis)	Polarization degree (%)	Elonga-tion	Brightness S_{10} (vis)	Polarization degree (%)
1°	4.1×10^6	0	50°	570	19	130°	137	8
2	9.0×10^5	0	60	394	20	140	136	6
5	1.3×10^5	1	70	296	20	150	137	3
10	2.5×10^4	7	80	239	19	160	142	0
15	9600	14	90	202	17	170	158	-2
20	4990	15	100	174	14	180	180	0
30	1940	17	110	154	12			
40	920	18	120	142	10			

Leinert 1975, SSRv, **18**, 281

Observatory Sites of Major Groundbased Telescopes

1. Island of La Palma, Spain

Observatorio del Roque de los Muchachos (17°43′ West, 28°46′ North, altitude 2430 m)
Optical-Infrared telescopes: WHT 4.2 m, TNG 3.6 m, INT 2.5 m, NOT 2.5 m, Liverpool 2 m, GTC 10 m; 1 m solar telescope, 17 m high energy γ-ray telescope (MAGIC), etc.

2. Chile

Chajnantor region (67.5° West, 23° South, altitude 5000 m)
Optical-Infrared telescopes: The University of Tokyo Atacama 6.5 m (TAO), 1 m (miniTAO)
Radio telescopes: ASTE 10 m, CBI 2 1.37 m × 13, APEX 12 m, NANTEN2 4 m (Nagoya University), ALMA 12 m × 54 and 7 m × 12

Las Campanas Observatory (70°42′ West, 29°1′ South, altitude 2300 m)
Optical-Infrared telescopes (Carnegie Observatories): du Pont 2.5 m, Magellan 6.5 m × 2, solar telescope, etc.

European Southern Observatory

La Silla (70°44′ West, 29°15′ South, altitude 2400 m)
Optical-Infrared telescopes: 3.6 m, 2.2 m, NTT 3.5 m, 1.5 m Danish Telescope, 1.2 m Swiss Telescope, etc.

Cerro Paranal (70°24′ West, 24°38′ South, altitude 2635 m)
Optical-Infrared telescopes: VLT 8.1 m × 4, VISTA 4 m, VST 2.5 m, and AT 1.8 m × 4

Cerro Tololo Inter-American Observatory
Cerro Tololo summit region (70°49′ West, 30°10′ South, altitude 2200 m)
Optical-Infrared telescopes: Blanco 4 m, 1.5 m, YALO 1 m, 0.9 m, etc.
Cerro Pachon (70°43′ west, 30°14′ south, altitude 2700 m)
Optical-Infrared telescopes: Gemini South 8 m, SOAR 4.1 m, and Vera Rubin 6.7 m

3. United States of America

Maunakea Observatories (Maunakea summit area, 155°28′ West, 19°50′ North, altitude 4200 m)
Optical-Infrared telescopes: W. M. Keck Observatory 10 m × 2, Subaru Telescope 8.2 m, Gemini North 8 m, UKIRT 3.8 m, CFHT 3.6 m, IRTF 3 m, and University of Hawai'i 2.2 m
Radio telescopes: JCMT 15 m, SMA 6 m × 8, VLBA (Very Long Baseline Array) 25 m

Mount Graham International Observatory (109°53′ West, 32°42′ North, altitude 3000 m)
Optical-Infrared telescopes: LBT 8.4 m × 2, VATT 1.8 m
Radio telescope: SMT 10 m

Kitt Peak National Observatory (111°36′ West, 31°58′ North, altitude 2060 m)
Optical-Infrared telescopes: Mayall 3.8 m, WIYN 3.5 m & 0.9 m, KPNO 2.1 m, etc.
Radio telescope: KP 12 m

Palomar Observatory (116°52′ West, 33°21′ North, altitude 1900 m)
Optical-Infrared telescopes: Hale 5.1 m, 1.5 m, 1.2 m, etc.

4. Sutherland, South Africa

South African Astronomical Observatory (20°49′ East, 32°23′ South, altitude 1750 m)
Optical-Infrared telescopes: SALT 9.2 m, Radcliffe 1.9 m, IRSF 1.4 m (Nagoya University), etc.

5. Asia

Russian Academy of Sciences Special Astrophysical Observatory (41°26′ East, 43°39′ North, altitude 2070 m)
Optical-Infrared telescopes: BTA 6 m, 1 m, 0.6 m, etc.

Okayama Astrophysical Observatory (133°36′ East, 34°26′ North, altitude 370 m)
Optical-Infrared telescopes: Seimei 3.8 m, 1.88 m, 0.91 m, 0.5 m, etc.

Nobeyama Radio Observatory (138°29′ East, 35°56′ North, altitude 1350 m)
Radio telescopes: 45 m, 10 m × 6, 0.8 m × 84

Bohyunsan Optical Astronomy Observatory, South Korea (128°58′ East, 36°10′ North, altitude 1160 m)
Optical-infrared telescopes: 1.8 m, solar flare telescope 0.2 m × 2 & 0.15 m × 2

Xinglong Observing Station, China (117°35′ East, 40°24′ North, altitude 950 m)
Optical-Infrared telescopes: 2.16 m, 1.26 m, 0.85 m, 0.8 m, Schmidt telescope, solar telescope, etc.

Devasthal Observatory, India (79°41′ East, 29°22′ North, altitude 2540 m)
Optical-infrared telescopes: 3.6 m, 1.3 m.

6. Australia

Siding Spring Observatory (149°4′ East, 31°17′ South, altitude 1160 m)
Optical-Infrared telescopes: AAT 3.9 m, ATT 2.3 m, 1.2 m United Kingdom Schmidt Telescope, etc.

7. Antarctica

The South Pole (90°South, altitude 2835 m)
Radio telescopes: AST/RO 1.7 m, SPT 10 m
Optical-infrared telescope: SPIREX/Abu 0.6 m

Dome A (80°22′ East, 77°21′ South, altitude 4090 m)
Optical-infrared telescopes: CSTAR 0.14 m × 4

Dome C (123°20′ East, 75°6′ South, altitude 3230 m)
Optical-infrared telescope: A STEP 0.4 m, IRAIT 0.8 m

Dome Fuji (39°42′ East, 77°19′ South, altitude 3810 m)

Major Radio Telescopes

Single dish telescope

Name	Location	Diameter (m)	Resolution (Frequency)[1]
Effelsberg	Germany	100	40″ (22 GHz)
GBT	USA	100	34″ (22 GHz)
Tianma (SHAO)	China	65	22″ (43 GHz)
SRT	Italy	64	50″ (22 GHz)
Parkes	Australia	64	4.4′ (5 GHz)
LMT	Mexico	50	15″ (100 GHz)
Nobeyama (NAOJ)	Japan	45	15″ (115 GHz)
Haystack	USA	37	49″ (43 GHz)
Yamaguchi Univ./NAOJ	Japan	32	4′ (8 GHz)
Ibaraki Univ./NAOJ	Japan	32×2	4′ (8 GHz)
IRAM (Pico Veleta)	Spain	30	13″ (230 GHz)
Onsala	Sweden	20	33″ (115 GHz)
JCMT	USA	15	14″ (345 GHz)
Taeduk Radio Astronomy Observatory	South Korea	13.7	46″ (115 GHz)
Purple Mountian Observatory	China	13.7	46″ (115 GHz)
Ishioka (GSI)	Japan	13.2	10′ (8 GHz)

Name	Location	Diameter (m)	Resolution (frequency)[1]
APEX	Chile	12	18″ (345 GHz)
Kitt Peak	USA	12	30″ (230 GHz)
Gifu Univ.	Japan	11	5′ (22 GHz)
ASTE (NAOJ)	Chile	10	22″ (345 GHz)
HHT	USA	10	22″ (345 GHz)
Mizusawa (NAOJ)	Japan	10	5′ (22 GHz)
Misato Observatory/Wakayama Univ.	Japan	8	1.5° (1.4 GHz)
Nagoya Univ. NANTEN2	Chile	4	38″ (490 GHz)
Nagoya Univ.	Japan	4	2.7′ (115 GHz)
CfA	USA	1.2	8′ (115 GHz)
The Univ. of Tokyo VST-1	Japan	0.6	8′ (230 GHz)
FAST	China	500[2]	2.9′ (1.4 GHz)
Arecibo	Puerto Rico	305[2]	3′ (1.4 GHz)
RATAN-600	Russia	600[3]	17″ (15 GHz)
Ootacamund	India	529×30[4]	2.3°×3.3′ (330 MHz)
Nançay	France	300×35[4]	3.5×19′ (1.6 GHz)

1) Only the typical frequency is shown. In many cases, it is possible to use other frequencies as well. 2) Fixed spherical mirror 3) Circle of fixed mirrors 4) Semi fixed mirror

Radio interferometer

Name	Location	Maximum baseline length	Antenna	Resolution (Frequency)[1]
VLBA	USA	8 000 km	25 m × 10 antennas	0.0003″ (22 GHz)
VERA	Japan	2 300	20 m × 4 antennas	0.001″ (22 GHz)
JVN	Japan	2 300	64–11 m, total 10 antennas	0.001″ (22 GHz)
KVN	South Korea	500	21 m × 3 antennas	0.005″ (22 GHz)
e-MERLIN	UK	217	76–15 m, total 7 antennas	0.35″ (1.6 GHz)
GMRT	India	25 (Y shape)	45 m × 30 antennas	9″ (330 MHz)
JVLA	USA	21 (Y shape)	25 m × 27 antennas	0.4″ (5 GHz)
ALMA	Chile	16	12 m × 54 antennas, 7 m × 12 antennas	0.01″ (345 GHz)
ASKAP	Australia	6	12 m × 36 antennas	10″ (1.4 GHz)
MeerKAT	South Africa	8	13.5m × 64 antennas	6″ (1.6 GHz)
ATCA	Australia	6	22 m × 6 antennas	1″ (10 GHz)
Cambridge	UK	5	13 m × 8 antennas	4″ (2.7 GHz)

Name	Location	Maximum baseline length	Antenna	Resolution (frequency)[1]
Westerbork	Netherlands	3.2	25 m × 14 antennas	3.9″ (5 GHz)
SMA	USA	508 m (ring)	6 m × 8 antennas	0.3″ (345 GHz)
NMA (NAOJ)	Japan	570 m×545 m (T shape)	10 m × 6 antennas	1″ (115 GHz)
IRAM (Plateau de Bure)	France	160 m×288 m (T shape)	15 m × 6 antennas	1″ (115 GHz)
Waseda Univ.	Japan	20 m×20 m	2.4 m × 64 antennas	6′ (10.6 GHz)
Nagoya Univ. (Nobeyama)[2]	Japan	489 m×220 m (T shape)	0.8 m × 84 antennas	10″ (17 GHz)
CBI	Chile	5.5 m (single platfrom design)	0.9 m × 13 antennas	5′ (31 GHz)
SSRT[2]	Russia	622 m×622 m	2.5 m × 256 antennas	15″ (5.73 GHz)
Owens Valley Solar Array[2]	USA	670 m×366 m	27 m × 2 antennas, 2 m × 3 antennas	5.1″ (18 GHz)

1) Resolution indicated is the highest value at the typical observation frequency (in parentheses). The resolution of radio interferometer varies according to the array configuration, the observation frequency, and the celestial declination of the object. 2) For solar activity observations.

Major Optical-Infrared Telescopes

Diameter (m)	Name [Type*1]	Foci & Focal Ratio*2	Affiliation; Location	Altitude (m)	Completion Year*3
16.2 (8.1×4)	Very Large Telescope, VLT [IF]	Cs 13, Ns 15, Cd	ESO; Cerro Paranal, Chile	2 635	1998, 2000
14.1 (10×2)	Keck I & II [IF, M]*4	Keck I (Cs/Ns: 15 & 25), Keck II (Cs/Ns: 15 & 40)	WMKO; Maunakea, USA	4 123	1993, 1996
11.8 (8.3×2)	Large Binocular Telescope, LBT [IF]	P 1.4, Ns 15	Arizona University, etc.; Mt. Graham, USA	3 050	2005, 2006
10.4	Gran Telescopio CANARIAS, GTC [M]*4	Cs/Ns 15	IAC; ORM, Spain	2 400	2009
9.2	South African Large Telescope, SALT [M]*5	P 4.2	South African Astronomical Observatory; Sutherland, South Africa	1 767	2004
9.2	Hobby-Eberly Telescope HET [M]*5	P 4.7	MacDonald Obs.; Mt. Fowlkes, USA	2 026	1996
8.2	Subaru Telescope	P 1.87*6, Cs 12.2, Ns 12.6 & 13.6	NAOJ; Maunakea, USA	4 139	1999
8.0	Gemini North*7	Cs 16	NOIRLab; Maunakea, USA	4 200	1999
8.0	Gemini South*7	Cs 16	NOIRLab; Cerro Pachon, Chile	2 737	2002
6.7	Vera Rubin*8	P 1.234	LSST; Cerro Pachon, Chile	2 123	Under construction
6.5	Baade (Magellan 1)	Cs 15, Ns 11	OCIW; Cerro Manqui, Chile	2 282	2000
6.5	Clay (Magellan 2)	Cs 15, Ns 11	OCIW; Cerro Manqui, Chile	2 282	2002
6.5	MMT*9	Cs 5.4 & 9 & 15	Steward Obs.; Mt. Hopkins, USA	2 606	2000
6.5	The Univ. of Tokyo Atacama 6.5 m Telescope	Ns 12.2	Univ. of Tokyo; Cerro Chajnantor	5 640	Under construction
6.0	Bolshoi Teleskop Azimutalnyi (Large Altazimuth Telescope), BTA	P 4, Ns 30	Special Ap. Obs., Russia	2 070	1976
6.0	LZT (Large Zenith Telescope)	P 1.5	Liquid Mirror Observatory, Canada	395	2003
5.08	Hale	P 3.3, Cs 9 & 16 & 30	Palomar Obs.; Mt Palomar, USA	1 900	1948
4.2	Discovery Channel Telescope	P 2.3, Cs 6.1	Lowell; Happy Jack, USA	2 361	2012
4.2	William Herschel Telescope, WHT	P 2.8, Cs 10.9, Ns 11.1	ING; ORM, Spain	2 370	1987
4.1	Southern Astrophysical Research (SOAR) Telescope	Cs 16, Ns 16	NOIRLab; Cerro Pachon, Chile	2 738	2004
4.0	Visible Infrared Survey Telescope for Astronomy, VISTA	Cs 3	ESO; Cerro Paranal, Chile	2 600	2008
4.0	Blanco	P 2.87, Cs 8 & 14.5 & 30	NOIRLab; Cerro Tololo, Chile	2 399	1974
4.0	LAMOST	F 5	National Astronomical Observatories, Chinese Academy of Sciences; Xinglong Station, China.	950	2010
4.0	DKIST	Ns 13, Cd 20 & 60	NSO; Hale Akala, USA	3 058	Under construction
3.89	Anglo-Australian Telescope, AAT	P 3.3, Cs 8 & 15, Cd 36	AAO; Siding Springs, Australia	1 164	1974
3.81	Mayall	P 2.7, Cs 8 & 28.5, Cd 160	NOIRLab; Kitt Peak, USA	2 064	1973
3.8	United Kingdom Infrared Telescope, UKIRT [IR]	Cs 9 & 35, Cd 20	NASA; Maunakea, USA	4 194	1979
3.78	Seimei (Okayama 3.8 m Telescope)	Ns 6	Kyoto Univ.; Okayama, Japan	370	2018
3.67	AEOS	Cd 200, Ns 32	Air Force Research Laboratory; Hale Akala, USA	3 058	1997
3.6	Devasthal Optical Telescope	Cs 9	ARIES; Devasthal Obs., India	2 540	2017
3.58	Telescopio Nazionale Galileo, TNG	Ns 11	Italy; ORM, Spain	2 370	1998
3.58	New Technology Telescope, NTT	Ns 11	ESO; La Silla, Chile	2 375	1989
3.58	Canada France Hawaii Telescope, CFHT	P 3.8, Cs 8 & 36, Cd 110	CFH Corp.; Maunakea, USA	4 200	1979
3.57	ESO 3.6 m	Cs 8 & 35	ESO; La Silla, Chile	2 400	1977
3.5	WIYN	Cs 13.7, Ns 6.3	NOIRLab; Kitt Peak, USA	2 096	1994
3.5	Apache Point 3.5 m	Ns 10	ARC; Apache Point, USA	2 788	1993
3.5	Max Planck Institute, MPI 3.5 m	P 3.48 & 3.93, Cs 10 & 35, Cd 35	MPIfA; Calar Alto, Spain	2 168	1985
3.5	Starfire 3.5 m	Cd 90, Ns 5.6	Kirtland; Air Force Research Laboratory, USA	91	1997
3.05	Shane	P 5, Cs 17, Cd 36	Lick Obs.; Mt. Hamilton, USA	1 283	1959
3.0	NASA Infrared Telescope Facility, IRTF	Cs 37, Cd 120	NASA; Maunakea, USA	4 160	1979

* 1 Type abbreviations: M is a segmented mirror or multi mirror, IF is an Interferometer, IR indicates dedicated infrared observations.
* 2 Abbreviations for reflecting telescope foci: P prime focus, N newtonian, Cs Cassegrain, Ns Nasmyth, Cd Coude.
* 3 The year of completion is based on the year of the first light.
* 4 GCT, Keck I, Keck II: Primary mirror is composed of 36 hexagonal 1.8 m mirror segments.
* 5 SALT, HET: Primary mirror is spherical which is composed of 91 hexagonal 1 m mirror segments. Fixed Altitude.
* 6 P1.87 does not include the corrective lens.
* 7 Gemini Telescopes: The following six countries have participated: USA, Canada, Chile, Australia, Argentina, and Brazil.
* 8 Since the 8.4 m primary and 5.1 m tertiary mirrors form a continuous surface on a single glass, the effective diameter is 6.7 m.
* 9 MMT: Currently it is a single piece 6.5 m mirror, however it had previously six 1.8 m mirrors which provided the equivalent gathering area of 4.5 m telescope.

· Abbreviations for observatories generally used in the table, and those not listed in page **88**, "Observatory Sites of Major Groundbased Telescopes"
NOIRLab : National Optical-Infrared Astronomy Research Laboratory, USA
OCIW : Observatories of the Carnegie Institution of Washington
NSO : National Solar Observatory, USA
ARIES : Aryabhatta Research Institute of Observational Sciences

Major Astronomical Observation Satellites and Solar System Probes

Name	Launch year	Countries	Remarks	Name	Launch year	Countries	Remarks
Astronomical satellite				**Solar system Probe**			
Einstein Observatory (HEAO 2)	1978	USA	X-rays	Apollo 11	1969	USA	Manned Moon landing
Infrared Astronomical Satellite (IRAS)	1983	USA, NL, UK	Infrared	Pioneer 11	1973	USA	Jupiter and Saturn flyby
Cosmic Background Explorer (COBE)	1989	USA	Infrared-radio	Viking 1, 2	1975	USA	Mars orbiter/lander
Hubble Space Telescope (HST)	1990	USA	2.4 m telescope	Voyager 2	1977	USA	Jupiter, Saturn, Uranus, Neptune flyby
Compton Gamma Ray Observatory (CGRO)	1991	USA	γ-rays	Giotto	1985	ESA	Halley's comet flyby
Infrared Space Observatory (ISO)	1995	ESA	Infrared	Magellan	1989	USA	Venus orbiter
Solar and Heliospheric Observatory (SOHO)	1995	ESA	Sun	Galileo	1989	USA	Jupiter orbiter
Far Ultraviolet Spectroscopic Explorer (FUSE)	1999	USA	Far-ultraviolet	Cassini	1997	USA	Saturn orbiter
				Stardust	1999	USA	Comet sample return
Chandra	1999	USA	X-rays				
XMM-Newton	1999	ESA	X-rays	Mars Express	2003	ESA, Russia	Mars orbiter
Wilkinson Microwave Anisotropy Probe (WMAP)	2001	USA	Microwaves	Rosetta	2004	ESA	Comet rendezvous
				Messenger	2004	USA	Mercury orbiter
Spitzer Space Telescope	2003	USA	Infrared				
Swift	2004	USA	γ-rays	Deep Impact	2005	USA	Comet flyby and impact
COROT	2006	France	Stellar brightness measurements	Mars Reconnaissance Orbiter	2005	USA	Mars orbiter
Fermi (Gamma-Ray Large Area Space Telescope: GLAST)	2008	USA	γ-rays	Venus Express	2005	ESA	Venus orbiter
				New Horizons	2006	USA	Pluto flyby
Kepler	2009	USA	Extrasolar planets				
Herschel Space Observatory	2009	ESA	Infrared	Dawn	2007	USA	Vesta Ceres orbiter
Planck	2009	ESA	Microwaves	Chang'e 1	2007	China	Moon orbiter
Wide-field Infrared Survey Explorer (WISE)	2009	USA	Infrared	Chandrayaan 1	2008	India	Moon orbiter
				Juno	2011	USA	Jupiter orbiter
Solar Dynamics Observatory	2010	USA	Sun				
Gaia	2013	ESA	Astrometry of stars	Mars Science Laboratory	2011	USA	Mars lander

In the Country column, "ESA" indicates the European Space Agency, and NL indicates Netherlands.

Recently Launched Major Astronomical Observation Satellites and Solar System Probes

International designator	Name	Launch date	Countries	Orbit	Remarks
2016-017A	ExoMars 2016	2016.03.14	ESA	Mars orbiter	Mars orbiter and a lander
2016-055A	OSIRIS-REx	2016.09.08	USA	Asteroid rendezvous	Sample return from Bennu
2017-034A	HXTM	2017.06.15	China	Earth orbiter	High energy X-ray
2018-038A	TESS	2018.04.18	USA	Earth orbiter	Extrasolar planets
2018-042A	InSight	2018.05.05	USA	Mars lander	Mars lander with a seismometer
2018-080A	BepiColombo	2018.10.20	ESA, Japan	Mercury orbiter	Two orbiters around Mercury
2018-103A	Chang'e-4	2018.12.07	China	Lunar lander	Lunar lander with a rover
2019-040A	Spektr-RG	2019.07.13	Russia	L2	X-ray observation
2019-042A	Chandrayaan 2	2019.07.22	India	Lunar orbiter	Lunar orbiter and a lander
2020-010A	Solar Orbiter	2020.02.08	ESA	Sun orbiter	Observation in the vicinity of the Sun

Satellites or probes launched between March 2016 and June 2020 are shown. For earlier missions, please refer previous issues of the Rikanenpyo. The dates are given in Universal Time. L2 refers the Lagrangian point that locates anti-solar direction from the Earth.

Astronomical Discoveries and Inventions

Christian era	Event	Inventor/Discoverer/Involved parties (country)
circa 2400–2000 BC	Construction of Stonehenge	(UK)
circa 1300 BC	Astronomical events recorded in Yinxu inscriptions on oracle bones and tortoise shells	(China)
circa 700–100 BC	Astronomical tables and astronomical events in cuneiform clay tablets	(Babylonia)
circa 800 BC	Establishment of the 28 lunar-lodge constellations	(China)
circa 600 BC	Discovery of the saros series in solar eclipses	(Babylonia)
548 BC	Discovery of the inclination of the ecliptic	Anaximander (Greece)
circa 500 BC	Establishment of the 19-year cycle of leap months	(Babylonia)
circa 500 BC	Establishment of the 12 zodiacal constellations	(Persia)
433 BC	Discovery of the Metonic cycle	Meton (Greece)
circa 350 BC	Establishment of the quarter remainder calendars and discovery of the 19-year cycle of leap months (zhangfa rule)	(China)
330 BC	Discovery of the Callippic cycle	Callippus (Greece)
circa 270 BC	Measurement of the Earth's size	Eratosthenes (Greece)
225 BC	Eccentric center and epicycle theory of planetary motions	Apollonius of Perga (Greece)
circa150 BC	Discovery of precession	Hipparchus (Greece)
129 BC	Completion of the Hipparchus Catalog	Hipparchus (Greece)
circa 100–150BC	Invention of the astronomical computer "Antikythera Mechanism"	(Greece)
45 BC	Introduction of the Julian calendar	Sosigenes, Julius Ceasar (Italy)
78	Establishment of 24 seasons (*Qis*) in the luni-solar calendars	(China)
circa 150	Completion of the "Almagest", and description of atmospheric refraction	Ptolemy (Greece)
335-342	Independent discovery of precession in China	Yu Xi (China)
circa 300–900	Use of the Mayan calendar	(Mexico, Guatemala)
circa the mid-11th century	Publication of Arabic astronomical tables "Toledan Tables"	al-Zarqali [Arzahel] (Spain)
1011–21	Writing of "Book of Optics"	Alhazen [Ibn al-Haitham] (Persia)
1092	Publication of the oldest printed star chart "Xin Yi Xiang Fa Yao" (New essential methods of astronomical instruments)	Su Song (China)
circa 1252–70	Publication of astronomical tables "Alfonsine Tables"	King Alfonso X (Spain)
1281	Introduction of the Shoushi Calendar	Wang Xun, Guo Shoujing (China)
1420	Construction of the astronomical observatory in Samarkand	Ulugh Beg (Uzbekistan)
1543	Publication of "On the Revolutions of Celestial Spheres" advocating the heliocentrism	Copernicus [Kopernik] (Poland)
1551	Ephemerides "Prutenicae Tabulae"	Reinhold (Germany)
1582	Use [or Introduction] of the Gregorian Calendar	Clavius, Pope Gregory XIII (Italy)
1596	Discovery of the light variation of the star Mira	Fabricius (Germany)
1603	First modern star chart "Uranometria"	Bayer (Germany)
1608	Invention of the telescope	Lipperhey (Netherlands), and others
1609–10	Various discoveries on celestial bodies using telescopes	Galilei (Italy), Fabricius (Germany), and others
1609–19	Publication on the new laws of planetary motions	Kepler (Germany)
1610	Publication of "Sudereus Nuncius (Sidereal Messenger)"	Galilei (Italy)
1627	Ephemerides "Rudolphine Tables"	Kepler (Germany)
1656	Recognition of Saturn's Ring	Huygens (Netherlands)
1668	Production of a reflecting telescope	Newton (UK)
1672	Measurement of solar parallax	Cassini (Italy)
1672	Completion of Paris Observatory	(France)
1675	Foundation of the Royal Greenwich Observatory	(UK)
1676	Confirmation of the finite speed of light via obervations of eclipses of Jupiter's satellites	Rømer (Denmark)
1678	Proposal of the wave theory of light	Huygens (Netherlands)
1687	Publication of "Principia", law of universal gravitation	Newton (UK)

Continued.

Christian era	Event	Inventor/Discoverer/Involved parties (country)
1704	Proposal of the particle theory of light	Newton (UK)
1705	Discovery of the first periodic comet (Halley's comet)	Halley (UK)
1718	Discovery of the proper motion of fixed stars	Halley (UK)
1728	Discovery of the aberration of light	Bradley (UK)
1735	Production of Chronometer H1 for navigation	Harrison (UK)
1736	"The Figure of the Earth", confirmation of the Oblate Shape of the Earth	Maupertuis, Clairaut (France)
1747	Discovery of nutation	Bradley (UK)
1755	Island univserse theory and nebular hypothesis of solar system origin	Kant (Germany)
1758	Patent for the achromatic lens	Dollond (Britain)
1772	Discovery of the Titius-Bode law	Titius, Bode (Germany)
1781	Discovery of Uranus	W. Herschel (UK)
1781-84	Messier Catalog of nebulosity	Messier (France)
1783	Measurement of the solar motion	W. Herschel (UK)
1785	Observational determination of the size and the shape of the Universe (the Milky Way)	W. Herschel (UK)
1796	Nebular hypothesis of Solar System origin	Laplace (France)
1798	Theory of the formation process of the Solar System "Konton Bumpan Zusetsu"	T. Shitsuki (Japan)
1800	Discovery of infrared light	Herschel (UK)
1801	Experiment of the interference of light	Young (UK)
1801	Discovery of ultraviolet light	Ritter (Germany)
1801	Discovery of asteroid Ceres	Piazzi (Italy)
1802	Discovery of binary star systems	W. Herschel (UK)
1814-15	Discovery of absorption lines in the solar spectrum (Fraunhofer lines)	Fraunhofer (Germany)
1816-19	Theory and experiment of the diffraction and polarization of light	Fresnel (France)
1838-39	Measurement of stellar annual parallax	Bessel (Germany), Henderson (UK), Struve (Russia)
1842	Prediction of the Doppler effect in stars	Doppler (Austria)
1843	Discovery of the sunspot activity cycle (Solar cycle)	Schwabe (Germany)
1846	Discovery of Neptune	Le Verrier (France), Adams (UK), and Galle (Germany)
circa 1850	Establishment of astronomical photography	Bond (USA) and de la Rue (UK)
1851	Foucault pendulum experiment, proof of the Earth rotation	Foucault (France)
1856	Definition of the relation between stellar flux and magnitude	Pogson (UK)
1859	Establishment of the fundamentals of spectral analysis	Kirchhoff (Germany) and Bunsen (Germany)
1861	Electromagnetic radiation theory of light and the equations of electromagnetic fields	Maxwell (UK)
1863-66	Classification of stellar spectra	Huggins (UK), Rutherfurd (USA), and Secchi (Italy)
1866	Discovery of the interrelation between comets and meteor showers	Schiaparelli (Italy)
1868	Discovery of Helium in the Sun	Lockyer, Frankland (UK)
1887	Michelson-Morley experiment	Michelson (USA), Morley (USA)
1888	New General Catalog of Nebulae and Clusters of Stars (NGC)	Dreyer (Ireland)
1888-90	Measurement of the line-of-sight velocity of celestial objects (Doppler effect)	Vogel (Germany), Scheiner (Germany), Keeler (USA)
1888-91	Discovery of latitude variations	Küstner (Germany) and Chandler (USA)
1889	Discovery of a spectroscopic binary star	Pickering (USA)
1901	Harvard spectral classification scheme for stars	Cannon Inc. (USA) and Pickering (USA)
1902	Introduction of the new Z-term in the latitude variation analysis	H. Kimura (Japan)
1902	Theory of gravitational instability of interstellar gas	Jeans (UK)
1905	Distinction between giant and dwarf stars	Hertzsprung (Denmark)

Astronomical Discoveries and Inventions Continued.

Christian era	Event	Inventor/Discoverer/Involved parties (country)
1905	Special relativity	Einstein (Germany, Switzerland)
1905	Photon hypothesis of light	Einstein (Germany, Switzerland)
1908–12	Period-luminosity relation of cepheid variables	Leavitt (USA)
1908	Discovery of sunspot magnetic fields	Hale (USA)
1911–12	Discovery of cosmic rays	Hess (Austria)
1911–14	Relation between spectral type and absolute magnitude of stars (H-R diagram)	Russell (USA)
1915	Spectrum of Sirius B (white dwarf)	Adams (USA)
1915–16	Theory of general relativity	Einstein (Germany, Switzerland)
1916	Method of spectroscopic parallax	Adams (USA)
1918	Discovery of asteroid families (Hirayama family)	K. Hirayama (Japan)
1919	Verification of general relativity by astrometric observation at the total solar eclipse	Dyson (UK), Eddington (UK) *et al.*
1919	Foundation of the International Astronomical Union (IAU)	–
1920–21	Measurement of diameter of a fixed star with an interferometer	Anderson (USA), Michelson (USA), and Pease (USA)
1921	Physical theory of stellar spectra (Saha's ionization equation)	Saha (India, UK)
1922	Adoption of the Latin designations and abbreviations for constellations	International Astronomical Union (IAU)
1924	Theory of wave-particle duality	de Broglie (France)
1924	Mass-luminosity relation for stars	Eddington (UK)
1924	True nature of spiral nebulae (discovery of Cepheid variable stars)	Hubble (USA)
1924–25	Confirmation of high density of Sirius B (white dwarf)	Eddington (UK), Adams (USA)
1925	Discovery that hydrogen is the principal ingredient of stars	Payne (Payne-Gaposchkin) (UK, USA)
1927	Rotation of the Milky Way	Oort (Netherlands), Lindblad (Sweden)
1927, 1929	Velocity-distance relation for galaxies	Lemaître (Belguim), Hubble (USA)
1928	Identification of nebular lines	Bowen (USA)
1928–30	88 constellations and their boundaries	International Astronomical Union (IAU)
1930	Discovery of Pluto	Tombaugh (USA)
1930	Design and production of a Schmidt camera	Schmidt (Germany)
1930	Corona observations at non-solar eclipse (invention of the coronagraph)	Lyot (France)
1930	Confirmation of interstellar extinction	Trumpler (USA)
1931	Proposal of the Chandrasekhar mass limit	Chandrasekhar (USA, India)
1931	Discovery of cosmic radio waves	Jansky (USA)
1933	Invention of Lyot filters	Lyot (France)
1934	Clarification of the character of supernovae	Baade (USA), Zwicky (Switzerland)
1934–35	Theoretical prediction of neutron stars	Baade (USA), Zwicky (Switzerland), and Eddington (UK)
1937	Indication of the missing mass problem in clusters of galaxies	Zwicky (Switzerland)
1938–39	Explanation of solar energy by nuclear reactions	von Weizsäcker (Germany), Bethe (USA)
1939	Concept for hydrogen ionization regions	Strömgren (Sweden and Denmark)
1939	Theoretical prediction of stellar-mass black holes	Oppenheimer (USA), Volkoff (USA), and Snyder (USA)
1939	Identification of hydrogen anions as the source of continuous photo-absorption in stars	Wildt (USA)
1939–40	Identification of coronal lines in the solar spectrum	Grotrian (Germany), Edlen (Sweden)
1940	Discovery of interstellar molecules	McKeller (Canada)
1942	Discovery of solar radio waves	Hay (UK), Southworth (USA)
1942–47	Discovery of T Tauri type stars	Joy (USA) *et al.*
1944	Distinction between population I and population II stars	Baade (USA)

Continued.

Christian era	Event	Inventor/Discoverer/Involved parties (country)
1945	Prediction of radio emissions from interstellar neutral hydrogen	van de Hulst (Netherlands)
1946	Proposal of the Big Bang theory	Gamow (Russia, USA)
1947	Observation of magnetic fields in stars	Babcock (USA)
1948	Completion of the Palomar Observatory 200-inch telescope	(USA)
1948	Proposal of the steady state cosmology	Hoyle (UK), Bondi (UK) and Gold (UK)
1950	"Dirty Snow Ball" theory of cometary nuclei	Whipple (USA)
1951	Observation of radio emissions from the interstellar neutral hydrogen	Ewen (USA), Purcell (USA)
1952	Difference in the period-luminosity relation for Cepheid variables of Population I and Population II	Baade (USA)
1953	Principle of adaptive optics	Babcock (USA)
1953	Triple alpha reaction	Hoyle (UK)
1955	Dynamo theory	Parker (USA)
1955	Stellar initial mass function	Salpeter (USA)
1957	The first artificial satellite, Sputnik 1	(Soviet Union)
1957	Proposal of stellar interiors as the site of element synthesis	Burbidge, Burbidge (UK, USA), Fowler (USA), and Hoyle (UK)
1959	The first artificial satellite to reach the Moon, Luna 1	(Soviet Union)
1961	Convection equilibrium solution for protostars (Hayashi phase)	C. Hayashi (Japan)
1961–63	Discovery of quasars (quasi-stellar radio sources)	Schmidt (USA), Matthew (USA), and Sandage (USA)
1962	Discovery of 5-minute oscillations in the Sun	Leighton (USA), Noyes (USA), Simon (USA)
1962	Discovery of an X-ray star	Giacconi (USA), Gursky (USA), Paolini (USA), and Rossi (USA)
1965	Indication of the missing mass problem in the solar neighborhood	Oort (Netherlands)
1965	Prediction of cosmic microwave background radiation	Dicke (USA), Peebles (USA), Roll (USA), Wilkinson (USA)
1965	Discovery of cosmic microwave background radiation	Penzias (USA) and Wilson (USA)
1967	Discovery of a pulsar (neutron star)	Bell (UK), Hewish (UK) *et al.*
1967	Discovery of Orion BN Object (protostar)/KL nebula	Becklin (USA), Neugebauer (USA), Kleinman (USA), Low (USA)
1967	Electro weak unification theory	Weinberg (USA), Salam (Pakistan), and Glashow (USA)
1969	First human landing on the Moon (Apollo 11)	(USA)
1971–72	Discovery of X-ray radiation from clusters of galaxies	Gursky (USA) and the UHURU satellite team, Meekins (USA), *et al.*
1971–72	Identification of a secure black-hole candidate (Cyg X-1)	Webster (UK) *et al.*, Pringle (UK) *et al.*, M. Oda (Japan) *et al.*
1973	Discovery of gamma-ray bursts	Klebesabel (USA), Strong (USA), and Olson (USA)
1975	Discovery of a binary pulsar system	Hulse (USA), Taylor (USA)
1976–77	Detection of iron spectral lines in the X-ray spectrum of clusters of galaxies	Michell (USA) *et al.*, Serlemitsos (USA) *et al.*
1978–86	Discovery of the large-scale structure of the Universe;	Gregory (USA), Kirshner (USA), Davis (USA), Huchra (USA), Geller (USA) and others
1979	Detection of the twin image of a quasar formed by a gravitational lens	Walsh (UK), Carswell (UK), and Weymann (USA)
1979	Indirect confirmation of gravitational wave emissions by a decrease in the orbital period of a binary pulsar	Taylor (USA), Fowler (USA), McCulloch (USA)
1979–84	Observation of X-ray radiation from elliptical galaxies	Forman (USA) *et al.*, Nulsen (UK) *et al.*
1980	Discovery of bipolar molecular outflows	Snell (USA) *et al.*

Astronomical Discoveries and Inventions Continued.

Christian era	Event	Inventor/Discoverer/Involved parties (country)
1980–82	Observation of the flat rotation curve of spiral galaxies	Rubin (USA), Ford (USA), Thonnard (USA) *et al.*
1987	Detection of neutrinos from supernova 1987A	M. Koshiba and the Kamiokande group (Japan)
1988	Discovery of L-type brown dwarfs	Boecklin, Zuckerman (USA)
1990	Launch of the Hubble Space Telescope	(USA and Europe)
1990	Precision measurement of the cosmic microwave background radiation spectrum	Mather and the COBE satellite team
1992	Discovery of fluctuations in the cosmic microwave background radiation	Smoot and the COBE satellite team
1992	Discovery of a Kuiper Belt object (1992QB1)	Jewitt (USA), Luu (USA)
1992	Discovery of planets orbiting around a pulsar	Wolszczan (Poland) Frail (USA)
1993	Discovery of massive compact halo objects (MACHO)	Alcock (USA) *et al.*, Aubourg (France) *et al.*, Udalski (Poland) *et al.*
1993	Direct detection of a protoplanetary disk	O'Dell (USA) *et al.*
1995–96	Discovery of an exoplanet orbiting a normal star (Dopper Method)	Mayor (Switzerland), Queloz (Switzerland), Marcy (USA), and Butler (USA)
1995	Discovery of T-type brown dwarf	T. Nakajima (USA, Japan), Oppenheimer (USA), and Kulkarni (USA) *et al.*
1995	Discovery of a young brown dwarf	Rebolo (Spain) *et al.*
1996	Observation of X-ray radiation from a protostar	K. Koyama (Japan) *et al.*
1997–98	Discovery of the afterglow of a gamma-ray burst and confirmation of their extra galactic origin	Costa (Italy) *et al.*, van Paradijs (Netherlands) and the rest of the Beppo-SAX satellite team
1998	Termination of the Royal Greenwich Observatory	(UK)
1998	Discovery of evidence for a black hole existing in the Galactic center	Ghetz (USA), Kleine (USA), Morris (USA), Becklin (USA)
1998-99	Discovery of the accelerated expansion of the Universe	Perlmutter (USA) and the SCP team, Schmidt (Australia), Riess (USA) and the Hi-z SS team
2000	Beginning of Subaru Telescope operation	(Japan)
2000	Detection of an exoplanet by the transit method	Henry (USA) *et al.*, Charbonneau (USA) *et al.*
2001	Detection of a Gunn-Peterson trough (neutral hydrogen absorption band in the reionization epoch)	Becker (USA) and the SDSS team
2001	High accuracy determination of Hubble constant	Freedman (USA) and the Hubble Space Telescope key project team
2003	Precise determination of the cosmological parameters	Spergel (USA) and the WMAP satellite team
2004	Detection of an exoplanet by gravitational microlensing	Bond (UK), Udalski (Poland) *et al.*
2006	Adoption of the definition of a planet and the names for the varieties of Solar System objects	International Astronomical Union (IAU)
2008	Adoption of the term Plutoid	International Astronomical Union (IAU)
2008-09	Direct imaging of exoplanets	Kalas (USA) *et al.*, Marois (Canada) *et al.*, Lagrange (France) *et al.*, M. Tamura and the rest of the HiCIAO/AO/SEEDS team (USA, Germany, and others)
2009	Discovery of an exoplanet with a retrograde orbit	N. Nartia (Japan) *et al.*, Winn (USA) *et al.*
2009	Discovery of a rocky exoplanet	Leger (France), Rouen (France) *et al.*
2013	Start of regular ALMA observations	Japan, East-Asia, North America, Europe
2016	Observation of gravitational waves from a binary black hole merger	LIGO team (USA)
2017	Multi-wavelength and gravitational wave detections of a neutron star merger (kilonova)	LIGO/Virgo and various observation teams around the world (Japan)
2019	Imaging of the black hole shadow at the center of M87	EHT team (USA, Japan, Germany, France Canada, and others)
2020	Start of science operation of a gravitational wave detector KAGRA	KAGRA team (Japan)

Meteorology Section

Climatological Normals, Records, etc. for Weather Observed in Japan

Climatological normals, maximum records, etc. of meterological observation data observed at 82 stations out of 157 stations (as of December 31, 2017) including meteorological offices and weather stations in Japan are listed in pages **99** to **189**.

Climatological normal The average of observation values over a continuous 30-year period starting from a Western calendar year in which the final digit is "1." The normal value is updated every 10 years by the Japan Meteorological Agency. This report lists average values from 1981 to 2010 as normal values. In consideration for the impact of replacing/relocating measurement equipment or reallocation of measurement stations, an average value for a period shorter than 30 years is used as the normal value, or a normal value is calculated which has corrected for the impact of the aforementioned factors. Meanwhile, a normal value is not calculated if the observation site has been recently located or if there were a large number of incorrect measurements. From among normal values for 5-day periods, the value for the sixth 5-day period in February is listed for both normal years and leap years.

Maximum records, etc. Maximum records, etc. which have been observed from the start of keeping statistics until 2017 are listed in this report. In some cases, the start of keeping statistics is the start of observation. In other cases, it is the time at which measurement devices were relocated, observation stations were reallocated, etc.

Sea-level pressure Pressure value at a sea-level which is adjusted from local pressure value.

Wind From 1971, the average value for a 10-minute period prior to the observation time has been used for wind speed (excluding maximum instantaneous wind speed) and wind direction. Prior to 1970, calculation methods for the wind speed and wind direction varied depending on the year. Wind direction is expressed in 16 directions using English symbols (E: East, S: South, W: West, N: North). The normal value for the prevailing wind direction is the most frequently observed wind direction over the 30-year period. If calm conditions (wind speed of 0.2m/s or lower) were most frequently observed, then the 2nd most frequent wind direction is used.

Top 10 rankings for temperature, precipitation, and wind speed From among 155 stations of meteorological offices and weather stations etc. in Japan (excludes Mount Fuji and Syowa Station (Antarctica) from the 157 stations of Japan), the top 10 rankings for maximum records, etc. from the start of keeping statistics until 2017 are listed in this report.

Cloud amount Uses a scale of 0 to 10 to express the ratio of clouds to the entire sky.

Number of days for atmospheric phenomena The number of days without sunlight is the number of days for with sunlight hours of less than 0.1 hours. The number of snow (snowfall) days is the number of days during which at least one of the following phenomena such as snow, shower snow, blizzards, sleet, snow grain, or ice crystals occurred. The number of thunder days is the number of days on which lightning or thunder occurred; this excludes the number of days on which only light thunder occurred.

First day/last day of frost/snow (snowfall) From autumn until spring, the first day is the day on which frost/snow (snowfall) was first observed. The last day is the day on which frost/snow (snowfall) was last observed. Snow (snowfall) also includes shower snow, blizzards, sleet, snow grain, and ice crystals.

Solar radiation amount The amount of direct incident solar radiation from the surface of the sun which is received by a vertical surface placed in the ray of incident light. The atmospheric transmission rate is calculated by converting the ratio of the direct solar irradiation amount to the extraterrestrial solar radiation amount (amount of solar radiation received by a vertical surface placed in the rays of the sun at the top of the atmosphere) for cases in which the sun is at its zenith. The atmospheric turbidity coefficient is the Feussner-Dubois turbidity coefficient. The 12:00 pm values indicate observation values at 12:00 pm for local true solar time. The global solar radiation amount is the amount of incident solar radiation from all directions of the sky which is received on a horizontal surface.

Phenology The flowering date is the first day on which several blossoms of the flower opened. However, the flower head of dandelions is counted as a single bloom. The flowering date for Japanese silver grass is the first day on which the number of ears protruding from the sheath reaches approximately 20% of the total forecasted number. The full bloom date is the first day on which approximately 80% of all flowers which will bloom have opened. The red (yellow) leaf date is the first day on which the majority of leaves have changed to a red (yellow) color and almost no green color can be observed when viewing the overall plant. The date of first sight is the day on which the bird is first seen. The date of first chirping is the day on which chirping is first heard.

Japanese Weather Stations (Meteorological Offices) Referenced in Meteorology Section

(As of December 31, 2019)

Station	Latitude (N)		Lognitude (E)		Altitude*	Station	Latitude (N)		Lognitude (E)		Altitude*
	°	′	°	′	m		°	′	°	′	m
Sapporo	43	3.6	141	19.7	17.4	Iida	35	31.4	137	49.3	516.4
Hakodate	41	49.0	140	45.2	35.0	Karuizawa	36	20.5	138	32.8	999.1
Asahikawa	43	45.4	142	22.3	119.8	Gifu	35	24.0	136	45.7	12.7
Kushiro	42	59.1	144	22.6	4.5	Takayama	36	9.3	137	15.2	560.0
Obihiro	42	55.3	143	12.7	38.4	Shizuoka	34	58.5	138	24.2	14.1
Abashiri	44	1.0	144	16.7	37.6	Hamamatsu	34	45.2	137	42.7	45.9
Rumoi	43	56.7	141	37.9	23.6	Nagoya	35	10.0	136	57.9	51.1
Wakkanai	45	24.9	141	40.7	2.8	Tsu	34	44.0	136	31.1	2.7
Nemuro	43	19.8	145	35.1	25.2	Owase	34	4.1	136	11.6	15.3
Suttu	42	47.7	140	13.4	33.4	Hikone	35	16.5	136	14.6	87.3
Urakawa	42	9.7	142	46.6	36.7	Kyoto	35	0.8	135	43.9	40.8
Aomori	40	49.3	140	46.1	2.8	Osaka	34	40.9	135	31.1	23.0
Morioka	39	41.9	141	9.9	155.2	Kobe	34	41.8	135	12.7	5.3
Miyako	39	38.8	141	57.9	42.5	Nara	34	40.4	135	50.2	102.1
Sendai	38	15.7	140	53.8	38.9	Wakayama	34	13.7	135	9.8	13.9
Akita	39	43.0	140	5.9	6.3	Shionomisaki	33	27.0	135	45.4	67.5
Yamagata	38	15.3	140	20.7	152.5	Tottori	35	29.2	134	14.3	7.1
Sakata	38	54.5	139	50.6	3.1	Matsue	35	27.4	133	3.9	16.9
Fukushima	37	45.5	140	28.2	67.4	Hamada	34	53.8	132	4.2	19.0
Onahama	36	56.8	140	54.2	3.3	Saigo	36	12.2	133	20.0	26.5
Mito	36	22.8	140	28.0	29.0	Okayama	34	41.4	133	55.5	5.3
Utsunomiya	36	32.9	139	52.1	119.4	Hiroshima	34	23.9	132	27.7	3.6
Maebashi	36	24.3	139	3.6	112.1	Shimonoseki	33	56.9	130	55.5	3.3
Kumagaya	36	9.0	139	22.8	30.0	Tokushima	34	4.0	134	34.4	1.6
Choshi	35	44.3	140	51.4	20.1	Takamatsu	34	19.1	134	3.2	9.4
Tokyo	35	41.5	139	45.0	25.2	Matsuyama	33	50.6	132	46.6	32.2
Oshima	34	44.9	139	21.7	74.0	Kochi	33	34.0	133	32.9	0.5
Hachijojima	33	7.3	139	46.7	151.2	Murotomisaki	33	15.1	134	10.6	185.0
Yokohama	35	26.3	139	39.1	39.1	Shimizu	32	43.3	133	0.6	31.0
Niigata	37	53.6	139	1.1	4.1	Fukuoka	33	34.9	130	22.5	2.5
Takada	37	6.4	138	14.8	12.9	Saga	33	15.9	130	18.3	5.5
Aikawa	38	1.7	138	14.4	5.5	Nagasaki	32	44.0	129	52.0	26.9
Toyama	36	42.5	137	12.1	8.6	Izuhara	34	11.8	129	17.5	3.7
Kanazawa	36	35.3	136	38.0	5.7	Fukue	32	41.6	128	49.6	25.1
Wajima	37	23.4	136	53.7	5.2	Kumamoto	32	48.8	130	42.4	37.7
Fukui	36	3.3	136	13.3	8.8	Oita	33	14.1	131	37.1	4.6
Tsuruga	35	39.2	136	3.7	1.6	Miyazaki	31	56.3	131	24.8	9.2
Kofu	35	40.0	138	33.2	272.8	Kagoshima	31	33.3	130	32.8	3.9
Nagano	36	39.7	138	11.5	418.2	Naze	28	22.7	129	29.7	2.8
Matsumoto	36	14.8	137	58.2	610.0	Naha	26	12.4	127	41.2	28.1
Mt. Fuji	35	21.6	138	43.6	3775.1	Syowa station (Antarctica)	69(S)	0.3	39	34.8	18.4

* Height above sea-level of the survey makers placed by meteorological offices.

Map of Japanese Weather Stations (Meteorological Offices) Referenced in Meteorology Section

Figure 1

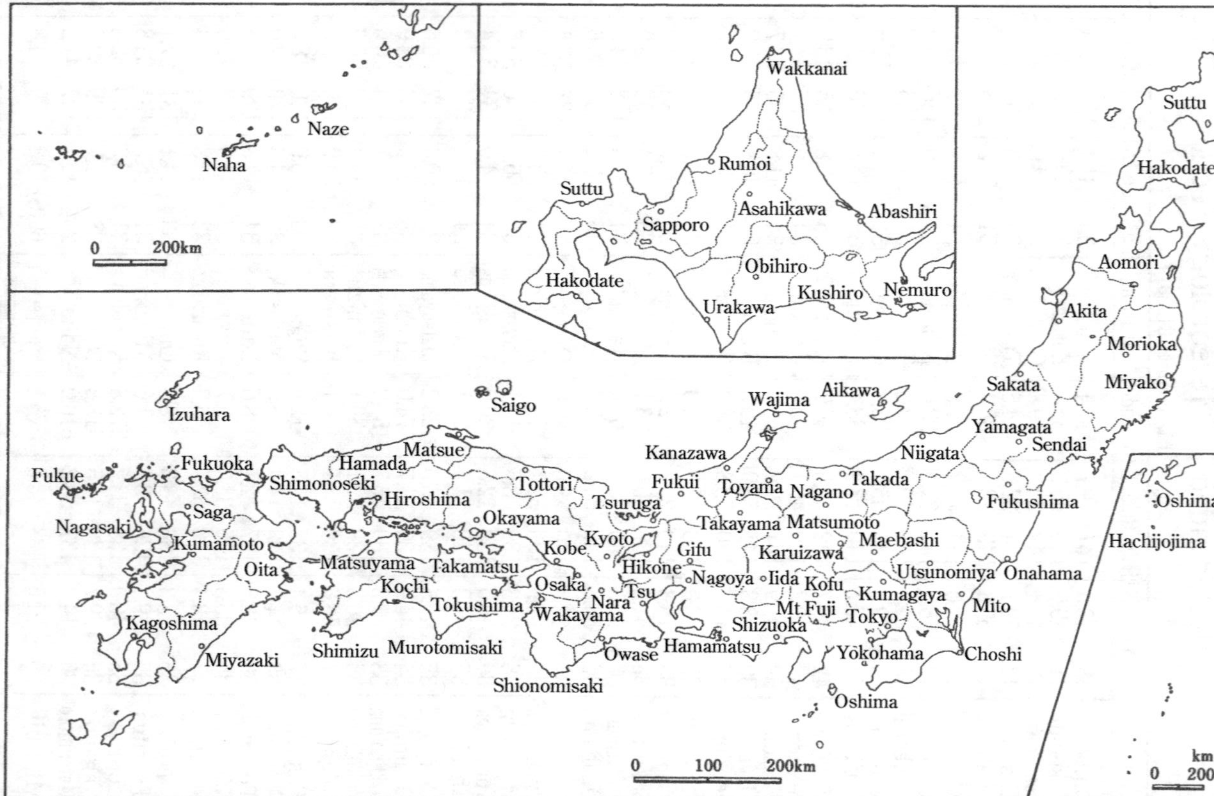

For the annual averages and records highs of the temperature, wind-speed, hours of sunshine, precipitation, and snowfall from meteorological offices and Automated Meteorological Data Acquisition System points not listed on this map, please refer to the Japan Meteorological Agency homepage (http://www.jma.go.jp/jma/indexe.html).

Climatological Normal Monthly Sea-Level Atmospheric Pressure (hPa)

(1981–2010 Averages)

Location	Jan.	Feb.	Mar.	Apr.	May	Jun.	Jul.	Aug.	Sep.	Oct.	Nov.	Dec.	Annual
Sapporo	1013.1	1013.5	1013.6	1012.6	1010.7	1009.4	1008.5	1010.0	1013.3	1015.4	1015.6	1014.1	1012.5
Hakodate	1013.9	1014.2	1014.4	1013.4	1011.4	1009.8	1008.9	1010.2	1013.4	1016.0	1016.5	1015.0	1013.1
Asahikawa	1013.3	1013.7	1013.7	1012.7	1010.7	1009.4	1008.5	1010.2	1013.6	1015.6	1015.5	1014.0	1012.6
Kushiro	1012.0	1012.6	1013.1	1013.0	1011.8	1010.9	1010.2	1011.4	1014.0	1015.3	1014.7	1012.8	1012.7
Obihiro	1012.7	1013.1	1013.3	1012.8	1011.4	1010.4	1009.7	1011.1	1014.0	1015.6	1015.1	1013.4	1012.7
Abashiri	1012.0	1012.8	1013.0	1012.4	1011.0	1010.2	1009.4	1010.7	1013.6	1014.8	1014.2	1012.5	1012.2
Rumoi	1012.6	1013.1	1013.3	1012.4	1010.7	1009.5	1008.5	1010.0	1013.2	1015.0	1014.9	1013.4	1012.2
Wakkanai	1012.2	1012.7	1012.7	1011.8	1010.4	1009.8	1008.9	1010.0	1012.8	1013.8	1013.5	1012.4	1011.8
Nemuro	1011.2	1012.0	1012.7	1012.9	1011.8	1011.1	1010.4	1011.6	1014.0	1015.2	1014.3	1012.1	1012.4
Suttu	1013.6	1014.0	1014.0	1012.8	1010.8	1009.4	1008.4	1009.9	1013.2	1015.6	1016.0	1014.7	1012.7
Urakawa	1012.7	1013.1	1013.7	1013.4	1011.9	1010.5	1009.7	1010.9	1013.7	1015.7	1015.7	1013.7	1012.9
Aomori	1014.9	1015.0	1015.1	1013.7	1011.5	1009.5	1008.6	1010.0	1013.4	1016.5	1017.3	1015.9	1013.5
Morioka	1015.7	1015.6	1015.8	1014.5	1012.1	1009.7	1009.0	1010.4	1013.9	1017.3	1018.3	1016.7	1014.1
Miyako	1014.4	1014.5	1014.9	1014.0	1011.9	1010.0	1009.2	1010.6	1013.8	1016.7	1017.2	1015.4	1013.6
Sendai	1015.7	1015.7	1015.9	1014.7	1012.4	1010.0	1009.3	1010.6	1013.8	1017.1	1018.2	1016.7	1014.2
Akita	1016.2	1016.1	1016.2	1014.6	1012.0	1009.3	1008.5	1009.9	1013.3	1016.9	1018.3	1017.1	1014.0
Yamagata	1016.8	1016.7	1016.6	1014.9	1012.1	1009.4	1008.7	1010.1	1013.6	1017.5	1019.0	1017.8	1014.5
Sakata	1016.6	1016.6	1016.5	1014.8	1012.1	1009.3	1008.5	1009.8	1013.2	1017.0	1018.6	1017.4	1014.2
Fukushima	1016.3	1016.2	1016.3	1014.8	1012.2	1009.7	1009.1	1010.4	1013.8	1017.3	1018.7	1017.3	1014.4
Onahama	1015.9	1015.8	1015.9	1014.8	1012.6	1010.0	1009.3	1010.7	1013.6	1016.9	1018.2	1016.8	1014.2
Mito	1015.9	1015.7	1015.7	1014.6	1012.4	1009.8	1009.2	1010.6	1013.4	1016.8	1018.2	1017.0	1014.1
Utsunomiya	1016.0	1015.8	1015.6	1014.4	1012.1	1009.5	1008.9	1010.2	1013.3	1016.8	1018.3	1017.1	1014.0
Maebashi	1016.4	1016.2	1015.9	1014.4	1011.9	1009.2	1008.6	1010.0	1013.1	1016.9	1018.5	1017.4	1014.0
Kumagaya	1016.2	1016.0	1015.7	1014.4	1012.0	1009.3	1008.7	1010.1	1013.2	1016.8	1018.4	1017.3	1014.0
Choshi	1015.4	1015.3	1015.3	1014.4	1012.3	1009.7	1009.1	1010.5	1013.0	1016.2	1017.7	1016.5	1013.8
Tokyo	1016.0	1015.7	1015.5	1014.3	1012.0	1009.4	1008.7	1010.0	1012.9	1016.5	1018.1	1017.0	1013.9
Oshima	1015.6	1015.3	1015.1	1014.2	1012.1	1009.4	1009.0	1010.3	1012.6	1015.9	1017.5	1016.7	1013.6
Hachijojima	1016.1	1015.6	1015.1	1014.4	1012.3	1009.6	1009.6	1010.3	1011.9	1014.8	1017.3	1017.1	1013.7
Yokohama	1015.8	1015.6	1015.4	1014.3	1012.0	1009.4	1008.8	1010.1	1012.9	1016.4	1018.0	1016.9	1013.8
Niigata	1017.6	1017.5	1017.2	1015.1	1012.2	1009.2	1008.5	1009.7	1013.1	1017.2	1019.1	1018.3	1014.6
Takada	1018.2	1018.1	1017.5	1015.0	1012.1	1009.1	1008.5	1009.6	1013.0	1017.2	1019.4	1018.9	1014.7
Aikawa	1017.6	1017.5	1017.0	1014.9	1012.0	1009.1	1008.4	1009.5	1012.9	1016.9	1018.9	1018.2	1014.4
Toyama	1019.1	1018.8	1017.8	1015.2	1012.1	1009.1	1008.5	1009.5	1012.9	1017.3	1019.7	1019.7	1015.0
Kanazawa	1019.0	1018.7	1017.7	1015.1	1012.1	1009.0	1008.4	1009.4	1012.8	1017.2	1019.6	1019.5	1014.9
Wajima	1018.5	1018.3	1017.5	1015.0	1012.0	1009.1	1008.3	1009.5	1013.1	1017.2	1019.3	1019.0	1014.8
Fukui	1019.4	1019.0	1017.9	1015.4	1012.3	1009.1	1008.5	1009.5	1012.8	1017.3	1019.9	1020.0	1015.1
Tsuruga	1019.4	1018.9	1017.8	1015.5	1012.4	1009.2	1008.7	1009.6	1012.8	1017.2	1019.8	1020.0	1015.1
Kofu	1016.6	1015.8	1015.1	1013.6	1011.3	1008.6	1008.1	1009.5	1012.4	1016.4	1018.5	1017.8	1013.6
Nagano	1018.5	1018.1	1017.2	1014.7	1011.6	1008.6	1008.0	1009.3	1012.9	1017.4	1019.7	1019.3	1014.6
Matsumoto	1018.0	1017.5	1016.5	1014.1	1011.1	1008.1	1007.6	1008.9	1012.5	1017.1	1019.4	1018.9	1014.2
Mt. Fuji													

Climatological Normal Monthly Sea-Level Atmospheric Pressure (1981–2010 Averages) Continued.

Location	Jan.	Feb.	Mar.	Apr.	May	Jun.	Jul.	Aug.	Sep.	Oct.	Nov.	Dec.	Annual
Iida	1018.2	1017.3	1016.2	1014.3	1011.6	1008.5	1008.2	1009.4	1012.4	1016.9	1019.6	1019.3	1014.3
Karuizawa													
Gifu	1018.7	1018.1	1017.0	1015.0	1012.2	1009.1	1008.7	1009.8	1012.5	1016.6	1019.4	1019.6	1014.7
Takayama	1020.0	1019.4	1018.2	1015.6	1012.2	1008.8	1008.5	1009.6	1013.1	1017.9	1020.8	1020.8	1015.4
Shizuoka	1015.7	1015.3	1014.9	1014.0	1011.9	1009.2	1008.8	1010.0	1012.4	1015.7	1017.5	1016.8	1013.5
Hamamatsu	1017.2	1016.5	1015.7	1014.4	1012.0	1009.2	1008.9	1010.0	1012.3	1015.8	1018.2	1018.1	1014.0
Nagoya	1018.6	1018.0	1016.8	1014.9	1012.2	1009.1	1008.8	1009.8	1012.4	1016.5	1019.3	1019.5	1014.7
Tsu	1018.9	1018.3	1017.2	1015.1	1012.3	1009.2	1008.8	1009.8	1012.6	1016.8	1019.5	1019.7	1014.9
Owase	1018.2	1017.5	1016.4	1014.7	1012.1	1009.2	1008.8	1009.8	1012.3	1016.2	1018.8	1019.0	1014.4
Hikone	1019.5	1018.9	1017.8	1015.6	1012.6	1009.3	1008.8	1009.7	1012.8	1017.1	1019.9	1020.2	1015.2
Kyoto	1019.7	1019.1	1017.7	1015.3	1012.3	1009.0	1008.6	1009.5	1012.6	1017.1	1020.0	1020.5	1015.1
Osaka	1019.8	1019.1	1017.7	1015.2	1012.1	1008.9	1008.5	1009.3	1012.4	1016.9	1019.9	1020.5	1015.0
Kobe	1019.8	1019.1	1017.6	1015.1	1012.1	1008.9	1008.5	1009.4	1012.3	1016.8	1019.8	1020.5	1015.0
Nara	1019.8	1019.1	1017.7	1015.2	1012.1	1008.9	1008.5	1009.4	1012.4	1017.0	1020.0	1020.5	1015.1
Wakayama	1020.0	1019.2	1017.6	1015.1	1012.1	1008.9	1008.6	1009.4	1012.3	1016.8	1019.8	1020.6	1015.0
Shionomisaki	1018.3	1017.5	1016.2	1014.5	1011.9	1009.0	1008.9	1009.6	1011.9	1015.7	1018.6	1019.1	1014.3
Tottori	1020.2	1019.6	1018.1	1015.2	1012.0	1008.7	1008.1	1009.2	1012.9	1017.5	1020.2	1020.7	1015.2
Matsue	1020.4	1019.8	1018.1	1015.1	1011.8	1008.6	1008.0	1009.0	1012.8	1017.5	1020.3	1020.8	1015.2
Hamada	1020.9	1020.0	1018.1	1015.1	1011.8	1008.4	1007.9	1008.7	1012.4	1017.3	1020.3	1021.2	1015.2
Saigo	1019.9	1019.5	1018.1	1015.1	1011.8	1008.8	1008.0	1009.2	1013.0	1017.6	1020.0	1020.2	1015.1
Okayama	1020.6	1019.8	1018.0	1015.2	1012.0	1008.7	1008.2	1009.1	1012.5	1017.3	1020.4	1021.2	1015.3
Hiroshima	1021.2	1020.2	1018.2	1015.3	1012.0	1008.6	1008.2	1009.0	1012.4	1017.4	1020.6	1021.7	1015.4
Shimonoseki	1021.6	1020.5	1018.3	1015.3	1011.9	1008.5	1008.1	1008.6	1012.2	1017.3	1020.7	1022.0	1015.4
Tokushima	1020.1	1019.3	1017.7	1015.1	1012.0	1008.8	1008.5	1009.3	1012.4	1017.0	1020.0	1020.7	1015.1
Takamatsu	1020.5	1019.7	1017.9	1015.1	1011.9	1008.6	1008.2	1009.0	1012.3	1017.1	1020.2	1021.1	1015.2
Matsuyama	1021.0	1019.9	1017.9	1015.1	1011.8	1008.5	1008.2	1008.9	1012.0	1016.9	1020.2	1021.5	1015.2
Kochi	1019.7	1018.7	1017.0	1014.9	1011.9	1008.9	1008.7	1009.4	1011.9	1016.2	1019.5	1020.4	1014.8
Murotomisaki	1018.9	1018.1	1016.5	1014.5	1011.7	1008.8	1008.6	1009.3	1011.7	1015.8	1018.9	1019.6	1014.4
Shimizu	1019.9	1018.7	1016.9	1014.7	1011.8	1008.7	1008.6	1009.1	1011.6	1015.9	1019.3	1020.5	1014.7
Fukuoka	1022.0	1020.6	1018.3	1015.3	1011.8	1008.3	1007.9	1008.4	1012.1	1017.3	1020.8	1022.4	1015.4
Saga	1022.1	1020.7	1018.3	1015.2	1011.7	1008.2	1008.0	1008.3	1011.8	1017.1	1020.8	1022.6	1015.4
Nagasaki	1022.0	1020.6	1018.1	1015.1	1011.7	1008.2	1008.2	1008.3	1011.5	1016.8	1020.6	1022.4	1015.3
Izuhara	1022.2	1020.9	1018.5	1015.3	1011.8	1008.4	1007.7	1008.5	1012.6	1017.7	1021.0	1022.5	1015.6
Fukue	1022.3	1020.8	1018.3	1015.2	1011.7	1008.1	1007.8	1008.1	1011.7	1017.1	1020.8	1022.6	1015.4
Kumamoto	1022.1	1020.6	1018.1	1015.1	1011.6	1008.2	1008.2	1008.3	1011.6	1016.8	1020.7	1022.5	1015.3
Oita	1021.4	1020.3	1018.2	1015.2	1011.9	1008.5	1008.1	1008.7	1012.1	1017.1	1020.5	1021.9	1015.3
Miyazaki	1020.6	1019.3	1017.3	1014.8	1011.6	1008.5	1008.3	1008.6	1011.5	1016.2	1019.8	1021.2	1014.8
Kagoshima	1021.3	1019.8	1017.5	1014.9	1011.7	1008.5	1008.5	1008.3	1011.1	1016.0	1019.9	1021.7	1015.0
Naze	1020.6	1019.4	1017.3	1014.8	1011.6	1008.6	1008.8	1007.6	1010.3	1014.9	1018.8	1020.9	1014.5
Naha	1020.5	1019.4	1017.2	1014.6	1011.2	1008.7	1008.6	1006.9	1009.5	1014.1	1018.0	1020.5	1014.1
Syowa station (Antarctica)	989.8	987.0	984.7	984.9	987.4	989.9	985.7	984.9	984.1	983.1	985.4	987.9	986.2

Sea-level pressures at Mt. Fuji and Karuizawa are not calculated because those stations are located at high altitude.

Climatological Normal Monthly Air Temperature (°C)

(1981-2010 Averages)

Station	Jan.	Feb.	Mar.	Apr.	May	Jun.	Jul.	Aug.	Sep.	Oct.	Nov.	Dec.	Annual
Sapporo	-3.6	-3.1	0.6	7.1	12.4	16.7	20.5	22.3	18.1	11.8	4.9	-0.9	8.9
Hakodate	-2.6	-2.1	1.4	7.2	11.9	15.8	19.7	22.0	18.3	12.2	5.7	0.0	9.1
Asahikawa	-7.5	-6.5	-1.8	5.6	11.8	16.5	20.2	21.1	15.9	9.2	1.9	-4.3	6.9
Kushiro	-5.4	-4.7	-0.9	3.7	8.1	11.7	15.3	18.0	16.0	10.6	4.3	-1.9	6.2
Obihiro	-7.5	-6.2	-1.0	5.8	11.1	14.8	18.3	20.2	16.3	10.0	3.2	-3.7	6.8
Abashiri	-5.5	-6.0	-1.9	4.4	9.4	13.1	17.1	19.6	16.3	10.6	3.7	-2.4	6.5
Rumoi	-4.4	-4.1	-0.4	5.5	10.6	15.0	19.2	20.9	16.8	10.9	4.3	-1.5	7.7
Wakkanai	-4.7	-4.7	-1.0	4.4	8.8	12.7	16.8	19.6	16.8	11.1	3.6	-2.0	6.8
Nemuro	-3.7	-4.3	-1.3	3.4	7.3	10.6	14.2	17.3	15.7	11.3	5.3	-0.5	6.3
Suttu	-2.4	-2.1	1.0	6.4	11.0	14.9	18.9	21.1	17.8	11.9	5.3	-0.1	8.6
Urakawa	-2.5	-2.4	0.5	5.0	9.3	13.1	17.2	19.9	17.3	12.0	5.9	0.2	7.9
Aomori	-1.2	-0.7	2.4	8.3	13.3	17.2	21.1	23.3	19.3	13.1	6.8	1.5	10.4
Morioka	-1.9	-1.2	2.2	8.6	14.0	18.3	21.8	23.4	18.7	12.1	5.9	1.0	10.2
Miyako	0.3	0.4	3.3	8.7	13.0	16.0	19.8	22.2	18.8	13.3	7.8	3.1	10.6
Sendai	1.6	2.0	4.9	10.3	15.0	18.5	22.2	24.2	20.7	15.2	9.4	4.5	12.4
Akita	0.1	0.5	3.6	9.6	14.6	19.2	22.9	24.9	20.4	14.0	7.9	2.9	11.7
Yamagata	-0.4	0.1	3.5	10.1	15.7	19.8	23.3	24.9	20.1	13.6	7.4	2.6	11.7
Sakata	1.7	1.9	4.6	10.2	15.3	19.6	23.3	25.3	21.1	15.1	9.3	4.5	12.7
Fukushima	1.6	2.2	5.3	11.5	16.6	20.1	23.6	25.4	21.1	15.1	9.2	4.4	13.0
Onahama	3.8	4.0	6.6	11.3	15.2	18.4	22.0	24.2	21.5	16.4	11.1	6.4	13.4
Mito	3.0	3.6	6.7	12.0	16.4	19.7	23.5	25.2	21.7	16.0	10.4	5.4	13.6
Utsunomiya	2.5	3.3	6.8	12.5	17.2	20.6	24.2	25.6	21.9	16.1	10.1	4.9	13.8
Maebashi	3.5	4.0	7.3	13.2	18.0	21.5	25.1	26.4	22.4	16.5	10.8	6.0	14.6
Kumagaya	4.0	4.7	7.9	13.6	18.2	21.7	25.3	26.8	22.8	17.0	11.2	6.3	15.0
Choshi	6.4	6.6	9.1	13.3	16.9	19.5	22.9	25.2	23.0	18.7	14.0	9.2	15.4
Tokyo	5.2	5.7	8.7	13.9	18.2	21.4	25.0	26.4	22.8	17.5	12.1	7.6	15.4
Oshima	7.3	7.4	9.9	14.2	17.9	20.8	24.1	25.7	23.0	18.5	14.2	9.9	16.1
Hachijojima	10.1	10.2	12.2	15.6	18.3	20.9	24.9	26.3	24.5	20.7	16.7	12.7	17.8
Yokohama	5.9	6.2	9.1	14.2	18.3	21.3	25.0	26.7	23.3	18.0	13.0	8.5	15.8
Niigata	2.4	2.7	5.7	11.0	16.4	20.5	24.3	26.4	22.5	16.3	10.3	5.2	13.6
Takada	2.4	2.4	5.4	11.5	16.6	20.6	24.6	26.3	22.0	16.0	10.2	5.3	13.6
Aikawa	3.9	3.8	6.2	11.2	15.5	19.5	23.6	26.0	22.1	16.9	11.6	6.9	13.9
Toyama	2.7	3.0	6.3	12.1	17.0	20.9	24.9	26.6	22.3	16.4	10.8	5.7	14.1
Kanazawa	3.8	3.9	6.9	12.5	17.1	21.2	25.3	27.0	22.7	17.1	11.5	6.7	14.6
Wajima	3.1	3.1	5.7	11.0	15.7	19.6	23.9	25.7	21.6	15.9	10.5	6.0	13.5
Fukui	3.0	3.4	6.8	12.8	17.7	21.6	25.6	27.2	22.7	16.6	11.0	5.9	14.5
Tsuruga	4.5	4.7	7.8	13.2	17.8	21.7	25.8	27.4	23.4	17.6	12.3	7.4	15.3
Kofu	2.8	4.3	8.0	13.8	18.3	21.9	25.5	26.6	22.8	16.5	10.4	5.0	14.7
Nagano	-0.6	0.1	3.8	10.6	16.0	20.1	23.8	25.2	20.6	13.9	7.5	2.1	11.9
Matsumoto	-0.4	0.2	3.9	10.6	16.0	19.9	23.6	24.7	20.0	13.2	7.4	2.3	11.8
Mt. Fuji	-18.4	-17.8	-14.2	-8.7	-3.4	1.1	4.9	6.2	3.2	-2.8	-9.2	-15.1	-6.2

Climatological Normal Monthly Air Temperature (1981–2010 Averages) Continued.

Station	Jan.	Feb.	Mar.	Apr.	May	Jun.	Jul.	Aug.	Sep.	Oct.	Nov.	Dec.	Annual
Iida	0.8	2.1	5.6	11.7	16.4	20.3	23.9	25.1	21.2	14.4	8.2	3.2	12.8
Karuizawa	-3.5	-3.1	0.5	6.8	11.8	15.6	19.5	20.5	16.3	10.0	4.4	-0.7	8.2
Gifu	4.4	5.1	8.6	14.4	19.0	22.8	26.5	28.0	24.1	18.1	12.2	6.9	15.8
Takayama	-1.4	-0.9	2.9	9.6	15.1	19.4	23.0	24.1	19.7	12.9	6.6	1.4	11.0
Shizuoka	6.7	7.3	10.3	14.9	18.8	22.0	25.7	27.0	24.1	18.9	13.9	9.0	16.5
Hamamatsu	8.2	8.6	11.4	15.6	19.2	22.1	25.4	27.0	24.3	20.2	15.3	10.6	17.3
Nagoya	4.5	5.2	8.7	14.4	18.9	22.7	26.4	27.8	24.1	18.1	12.2	7.0	15.8
Tsu	5.3	5.6	8.5	14.0	18.6	22.4	26.3	27.5	24.0	18.3	12.7	7.8	15.9
Owase	6.3	6.9	9.9	14.6	18.4	21.7	25.4	26.4	23.6	18.3	13.4	8.6	16.1
Hikone	3.7	3.9	6.9	12.3	17.2	21.4	25.6	27.1	23.2	17.1	11.4	6.3	14.7
Kyoto	4.6	5.1	8.4	14.2	19.0	23.0	26.8	28.2	24.1	17.8	12.1	7.0	15.9
Osaka	6.0	6.3	9.4	15.1	19.7	23.5	27.4	28.8	25.0	19.0	13.6	8.6	16.9
Kobe	5.8	6.1	9.3	14.9	19.4	23.2	26.8	28.3	25.2	19.3	13.9	8.7	16.7
Nara	3.9	4.4	7.6	13.4	18.0	21.9	25.8	26.9	22.9	16.6	11.1	6.2	14.9
Wakayama	6.0	6.4	9.5	14.9	19.3	23.0	27.0	28.1	24.7	18.8	13.5	8.5	16.7
Shionomisaki	6.0	6.5	9.9	14.6	18.7	22.1	25.8	27.5	24.2	18.9	13.7	8.5	16.4
Tottori	4.0	4.4	7.5	13.0	17.7	21.7	25.7	27.0	22.6	16.7	11.6	6.8	14.9
Matsue	4.3	4.7	7.6	12.9	17.5	21.3	25.3	26.8	22.6	16.8	11.6	6.9	14.9
Hamada	6.0	6.2	8.7	13.3	17.4	21.1	25.2	26.5	22.6	17.4	12.8	8.6	15.5
Saigo	4.2	4.3	6.9	11.9	16.3	20.1	24.2	26.0	22.0	16.5	11.6	7.1	14.3
Okayama	4.9	5.5	8.8	14.5	19.3	23.3	27.2	28.3	24.4	18.1	12.3	7.3	16.2
Hiroshima	5.2	6.0	9.1	14.7	19.3	23.0	27.1	28.2	24.4	18.3	12.5	7.5	16.3
Shimonoseki	6.9	7.2	9.9	14.5	18.6	22.3	26.3	27.6	24.4	19.4	14.2	9.4	16.7
Tokushima	6.1	6.5	9.6	14.8	19.2	22.7	26.6	27.8	24.5	18.9	13.5	8.5	16.6
Takamatsu	5.5	5.9	8.9	14.4	19.1	23.0	27.0	28.1	24.3	18.4	12.8	7.9	16.3
Matsuyama	6.0	6.5	9.5	14.6	19.0	22.7	26.9	27.8	24.3	18.7	13.3	8.4	16.5
Kochi	6.3	7.5	10.8	15.6	19.7	22.9	26.7	27.5	24.7	19.3	13.8	8.5	17.0
Murotomisaki	7.5	7.9	10.6	15.0	18.5	21.4	24.8	26.1	23.8	19.4	14.9	10.1	16.7
Shimizu	8.7	9.5	12.4	16.7	20.2	22.9	26.4	27.5	25.4	21.0	16.2	11.3	18.2
Fukuoka	6.6	7.4	10.4	15.1	19.4	23.0	27.2	28.1	24.4	19.2	13.8	8.9	17.0
Saga	5.4	6.7	9.9	15.0	19.5	23.3	26.8	27.8	24.2	18.6	12.9	7.6	16.5
Nagasaki	7.0	7.9	10.9	15.4	19.4	22.8	26.8	27.9	24.8	19.7	14.3	9.4	17.2
Izuhara	5.7	6.7	9.7	14.0	17.8	21.1	25.1	26.4	23.2	18.4	13.0	8.0	15.8
Fukue	7.4	8.1	10.7	14.7	18.6	22.0	26.1	27.0	24.0	19.3	14.3	9.6	16.8
Kumamoto	5.7	7.1	10.6	15.7	20.2	23.6	27.3	28.2	24.9	19.1	13.1	7.8	16.9
Oita	6.2	6.9	9.7	14.5	18.8	22.4	26.5	27.3	23.9	18.6	13.4	8.5	16.4
Miyazaki	7.5	8.6	11.9	16.1	19.9	23.1	27.3	27.2	24.4	19.4	14.3	9.6	17.4
Kagoshima	8.5	9.8	12.5	16.9	20.8	24.0	28.1	28.5	26.1	21.2	15.9	10.6	18.6
Naze	14.8	15.2	17.1	19.8	22.7	26.0	28.7	28.4	26.8	23.7	20.2	16.5	21.6
Naha	17.0	17.1	18.9	21.4	24.0	26.8	28.9	28.7	27.6	25.2	22.1	18.7	23.1
Syowa station (Antarctica)	-0.7	-2.9	-6.5	-10.1	-13.5	-15.2	-17.3	-19.4	-18.1	-13.5	-6.8	-1.6	-10.4

Climatological Normal Air Temperature (≈ 5 Day Intervals) (℃)

(1981-2010 Averages)

Mo.	Dates \ Station	Sapporo	Sendai	Tokyo	Niigata	Nagoya	Osaka	Hiroshima	Takamatsu	Fukuoka	Kagoshima	Naha
1	1- 5	-2.7	2.4	5.6	3.2	5.0	6.6	5.8	6.2	7.3	9.0	17.4
	6-10	-3.2	2.0	5.4	2.8	4.7	6.4	5.5	6.0	7.0	8.8	17.2
	11-15	-3.5	1.7	5.2	2.5	4.6	6.2	5.4	5.7	6.8	8.6	17.1
	16-20	-3.8	1.5	5.0	2.2	4.5	5.9	5.2	5.4	6.5	8.4	17.0
	21-25	-4.0	1.3	4.9	2.0	4.3	5.6	4.9	5.1	6.2	8.2	16.7
	26-31	-4.1	1.3	4.9	1.8	4.1	5.4	4.8	5.0	6.1	8.1	16.4
2	1- 5	-3.9	1.4	5.1	1.9	4.2	5.6	5.0	5.2	6.4	8.4	16.5
	6-10	-3.6	1.7	5.5	2.2	4.7	6.0	5.6	5.6	7.0	9.1	16.8
	11-15	-3.3	2.0	5.7	2.6	5.2	6.4	6.0	6.0	7.6	9.8	17.2
	16-20	-2.9	2.2	5.9	2.9	5.5	6.6	6.3	6.2	7.8	10.3	17.5
	21-25	-2.3	2.5	6.1	3.3	5.9	6.9	6.5	6.4	8.0	10.6	17.5
	26-28	-1.9	2.9	6.5	3.6	6.3	7.2	6.9	6.7	8.3	10.7	17.6
	26-29	-1.9	2.9	6.5	3.6	6.3	7.2	6.9	6.7	8.3	10.7	17.6
3	1- 5	-1.5	3.3	7.0	4.0	6.8	7.6	7.3	7.2	8.6	10.9	17.7
	6-10	-0.9	3.8	7.7	4.6	7.4	8.3	7.9	7.8	9.2	11.3	18.0
	11-15	0.1	4.5	8.4	5.3	8.2	9.1	8.7	8.6	10.0	12.2	18.7
	16-20	1.1	5.2	9.0	6.1	9.0	9.8	9.5	9.3	10.8	13.0	19.3
	21-25	1.9	5.8	9.5	6.7	9.7	10.4	10.2	9.8	11.3	13.6	19.7
	26-31	2.9	6.7	10.3	7.4	10.6	11.3	11.0	10.6	12.0	14.1	20.0
4	1- 5	4.4	7.9	11.5	8.5	11.8	12.5	12.2	11.8	13.0	15.0	20.3
	6-10	5.6	9.0	12.6	9.7	13.0	13.7	13.4	13.1	13.9	15.9	20.7
	11-15	6.7	9.9	13.5	10.7	14.1	14.7	14.4	14.2	14.8	16.4	21.1
	16-20	7.6	10.7	14.4	11.5	14.9	15.6	15.2	15.0	15.6	17.1	21.6
	21-25	8.6	11.6	15.3	12.5	15.8	16.5	16.0	15.7	16.4	18.0	22.2
	26-30	9.7	12.6	16.1	13.5	16.6	17.4	16.9	16.6	17.2	18.8	22.7
5	1- 5	10.6	13.5	16.9	14.6	17.5	18.3	17.8	17.6	18.1	19.5	23.2
	6-10	11.3	14.1	17.5	15.3	18.2	18.9	18.5	18.3	18.7	20.2	23.6
	11-15	12.0	14.5	17.8	15.8	18.5	19.3	18.9	18.7	19.1	20.6	23.9
	16-20	12.7	15.1	18.3	16.5	19.0	19.7	19.3	19.2	19.5	20.8	24.1
	21-25	13.4	15.8	18.9	17.3	19.7	20.4	20.0	19.9	20.1	21.3	24.3
	26-31	14.1	16.5	19.6	18.1	20.5	21.3	20.8	20.7	20.8	21.9	24.7
6	1- 5	14.9	17.2	20.3	19.0	21.3	22.1	21.6	21.5	21.5	22.5	25.2
	6-10	15.7	17.8	20.8	19.8	21.9	22.7	22.3	22.2	22.1	23.0	25.7
	11-15	16.4	18.2	21.1	20.2	22.3	23.2	22.8	22.7	22.7	23.5	26.4
	16-20	17.0	18.7	21.5	20.7	22.8	23.7	23.3	23.3	23.3	24.1	27.1
	21-25	17.7	19.2	22.0	21.2	23.4	24.2	23.7	23.9	23.9	25.0	27.8
	26-30	18.4	19.7	22.5	21.7	24.1	24.9	24.3	24.6	24.6	25.9	28.4

Climatological Normal Air Temperature (1981-2010 Averages) Continued.

Mo.	Dates	Sapporo	Sendai	Tokyo	Niigata	Nagoya	Osaka	Hiroshima	Takamatsu	Fukuoka	Kagoshima	Naha
7	1- 5	18.9	20.3	23.3	22.3	24.8	25.6	25.2	25.4	25.4	26.7	28.7
	6-10	19.4	21.0	24.0	23.0	25.5	26.4	26.0	26.2	26.2	27.5	28.8
	11-15	19.9	21.6	24.7	23.7	26.1	27.1	26.7	26.8	27.0	28.0	28.9
	16-20	20.6	22.3	25.3	24.5	26.6	27.7	27.3	27.3	27.6	28.4	29.0
	21-25	21.5	23.2	25.8	25.5	27.2	28.3	28.0	27.8	28.0	28.7	29.0
	26-31	22.3	23.9	26.4	26.4	27.8	28.8	28.5	28.3	28.3	28.8	29.0
8	1- 5	22.7	24.4	26.6	26.8	28.0	29.0	28.7	28.5	28.5	28.8	28.8
	6-10	22.8	24.5	26.7	26.8	28.0	29.0	28.6	28.4	28.5	28.7	28.8
	11-15	22.6	24.4	26.6	26.7	28.0	28.9	28.4	28.2	28.3	28.6	28.7
	16-20	22.2	24.1	26.4	26.4	27.9	28.8	28.2	28.0	28.0	28.5	28.7
	21-25	21.8	23.9	26.2	26.0	27.6	28.5	27.9	27.7	27.7	28.3	28.7
	26-31	21.4	23.6	25.9	25.6	27.2	28.1	27.4	27.3	27.3	28.0	28.5
9	1- 5	20.6	22.9	25.3	24.8	26.5	27.4	26.8	26.7	26.6	27.6	28.3
	6-10	19.6	22.1	24.4	23.9	25.6	26.5	25.9	25.8	25.7	27.1	28.1
	11-15	18.7	21.1	23.4	22.9	24.7	25.5	24.9	24.9	24.9	26.5	27.9
	16-20	17.7	20.2	22.3	22.0	23.7	24.6	24.0	23.9	24.1	25.8	27.5
	21-25	16.6	19.3	21.3	21.0	22.6	23.5	23.0	22.8	23.2	25.1	27.1
	26-30	15.5	18.4	20.4	19.9	21.5	22.4	21.9	21.8	22.2	24.3	26.8
10	1- 5	14.5	17.6	19.6	18.9	20.5	21.4	20.9	20.8	21.4	23.5	26.4
	6-10	13.5	16.7	18.8	17.9	19.7	20.6	20.0	19.9	20.7	22.7	26.0
	11-15	12.5	15.8	18.0	17.0	18.8	19.7	19.0	19.1	19.9	21.9	25.6
	16-20	11.3	14.8	17.2	15.9	17.7	18.6	17.9	18.0	18.9	20.9	25.0
	21-25	10.2	13.7	16.2	14.8	16.5	17.6	16.8	16.9	17.8	19.9	24.5
	26-31	9.2	12.7	15.3	13.7	15.5	16.7	15.8	15.9	16.9	18.9	24.0
11	1- 5	8.2	11.9	14.5	12.8	14.6	15.9	15.0	15.1	16.1	18.1	23.4
	6-10	6.9	11.0	13.6	12.0	13.8	15.2	14.2	14.4	15.4	17.4	23.0
	11-15	5.4	9.9	12.7	10.8	12.7	14.1	13.1	13.4	14.4	16.4	22.4
	16-20	4.0	8.7	11.6	9.6	11.5	12.9	11.9	12.2	13.2	15.2	21.8
	21-25	3.1	7.8	10.6	8.6	10.6	12.0	10.9	11.2	12.3	14.3	21.3
	26-30	2.1	7.0	9.9	7.9	9.8	11.2	10.1	10.5	11.4	13.5	20.7
12	1- 5	1.1	6.3	9.3	7.1	8.9	10.3	9.2	9.6	10.6	12.5	20.0
	6-10	0.1	5.5	8.6	6.3	8.0	9.5	8.5	8.8	9.8	11.6	19.5
	11-15	-0.9	4.7	7.9	5.4	7.2	8.8	7.7	8.1	9.1	10.8	19.0
	16-20	-1.6	4.1	7.2	4.7	6.6	8.2	7.1	7.6	8.6	10.3	18.4
	21-25	-1.8	3.7	6.7	4.3	6.1	7.7	6.7	7.1	8.2	9.9	18.0
	26-31	-2.2	3.1	6.2	3.8	5.5	7.2	6.3	6.7	7.8	9.4	17.7

Climatological Normal Daily High Temperature by Month (℃)

(1981-2010 Averages)

Station	Jan.	Feb.	Mar.	Apr.	May	Jun.	Jul.	Aug.	Sep.	Oct.	Nov.	Dec.	Annual
Sapporo	-0.6	0.1	4.0	11.5	17.3	21.5	24.9	26.4	22.4	16.2	8.5	2.1	12.9
Hakodate	0.7	1.5	5.3	11.8	16.5	19.9	23.4	25.8	22.7	16.8	9.7	3.3	13.1
Asahikawa	-3.5	-2.1	2.6	11.7	17.7	22.9	25.8	26.3	21.6	14.8	5.8	-0.8	11.9
Kushiro	-0.6	-0.4	2.7	7.7	12.0	15.2	18.6	21.2	19.7	14.8	8.7	2.5	10.2
Obihiro	-1.9	-0.6	4.0	11.9	17.6	20.8	23.5	25.2	21.5	15.6	8.0	1.1	12.2
Abashiri	-2.4	-2.5	1.6	8.9	14.2	17.2	20.8	23.4	20.2	14.8	7.4	0.7	10.4
Rumoi	-1.3	-0.8	2.8	9.4	14.8	18.8	22.7	24.6	20.9	14.9	7.6	1.4	11.3
Wakkanai	-2.7	-2.5	1.2	7.2	12.0	15.7	19.7	22.3	19.7	13.7	6.1	0.1	9.4
Nemuro	-1.1	-1.6	1.6	7.2	11.5	14.4	17.9	20.8	18.8	14.4	8.3	2.1	9.5
Suttu	-0.3	0.1	3.5	10.0	15.1	18.8	22.3	24.5	21.3	15.4	8.3	2.2	11.8
Urakawa	0.7	0.9	3.9	8.8	13.0	16.4	20.2	23.0	20.9	15.9	9.5	3.6	11.4
Aomori	1.6	2.3	6.3	13.5	18.4	21.7	25.4	27.7	24.0	18.0	10.9	4.6	14.5
Morioka	1.8	2.9	7.0	14.4	19.7	23.5	26.4	28.3	23.6	17.6	10.6	4.6	15.0
Miyako	4.9	5.1	8.3	14.3	18.4	20.5	23.9	26.4	23.1	18.5	13.2	7.8	15.4
Sendai	5.3	5.9	9.2	15.0	19.4	22.3	25.7	27.9	24.4	19.4	13.7	8.4	16.4
Akita	2.8	3.5	7.4	14.0	19.0	23.2	26.5	29.0	24.7	18.6	11.9	5.9	15.5
Yamagata	3.1	4.0	8.4	16.2	22.0	25.4	28.4	30.4	25.2	19.0	12.2	6.4	16.7
Sakata	4.3	4.8	8.3	14.8	19.7	23.6	27.1	29.6	25.3	19.5	13.2	7.7	16.5
Fukushima	5.5	6.5	10.4	17.4	22.5	25.2	28.3	30.4	25.6	20.0	14.1	8.7	17.9
Onahama	8.4	8.5	10.9	15.5	18.9	21.8	25.2	27.5	25.0	20.5	15.7	11.1	17.4
Mito	9.0	9.4	12.2	17.5	21.3	23.8	27.6	29.6	25.8	20.8	16.0	11.4	18.7
Utsunomiya	8.3	9.1	12.6	18.5	22.5	25.2	28.7	30.5	26.4	20.9	15.5	10.7	19.1
Maebashi	8.8	9.4	12.9	19.0	23.5	26.2	29.7	31.3	26.7	21.2	16.0	11.4	19.7
Kumagaya	9.4	10.2	13.5	19.5	23.9	26.4	30.1	31.9	27.2	21.7	16.4	11.8	20.2
Choshi	9.9	9.8	12.2	16.4	19.9	22.3	25.9	28.1	25.4	21.1	16.9	12.5	18.4
Tokyo	9.6	10.4	13.6	19.0	22.9	25.5	29.2	30.8	26.9	21.5	16.3	11.9	19.8
Oshima	10.7	11.2	13.5	18.0	21.5	24.0	27.1	29.2	26.2	21.5	17.4	13.3	19.5
Hachijojima	12.7	13.3	15.5	18.7	21.2	23.4	27.2	29.3	27.5	23.6	19.7	15.7	20.7
Yokohama	9.9	10.3	13.2	18.5	22.4	24.9	28.7	30.6	26.7	21.5	16.7	12.4	19.7
Niigata	4.9	5.8	9.4	15.9	20.4	24.3	28.0	30.6	25.5	20.3	14.0	8.5	17.3
Takada	5.9	6.3	10.3	17.5	22.2	25.4	29.1	31.3	26.8	21.2	15.3	9.5	18.4
Aikawa	6.4	6.3	9.2	14.7	19.5	22.9	26.7	29.4	25.5	20.2	14.7	9.7	17.1
Toyama	6.0	6.8	10.9	17.3	21.9	25.1	29.0	30.9	26.5	21.1	15.3	9.6	18.4
Kanazawa	6.8	7.3	11.0	16.9	21.6	25.0	28.8	30.9	26.6	21.3	15.5	10.2	18.5
Wajima	6.1	6.5	9.9	15.9	20.3	23.5	27.5	29.8	25.8	20.5	14.8	9.4	17.5
Fukui	6.5	7.4	11.5	18.1	22.7	26.0	29.7	31.8	27.3	21.6	15.7	9.9	19.0
Tsuruga	7.5	8.0	11.6	17.6	22.0	25.3	29.4	31.5	27.3	21.7	16.2	10.7	19.1
Kofu	8.8	10.3	14.2	20.4	24.6	27.3	30.9	32.5	28.0	21.9	16.4	11.2	20.5
Nagano	3.5	4.7	9.5	17.3	22.5	25.7	29.1	31.0	25.6	19.2	13.0	6.8	17.3
Matsumoto	5.0	6.0	10.5	17.8	22.9	26.0	29.4	31.1	25.7	19.3	13.6	8.0	17.9
Mt. Fuji	-15.4	-14.7	-10.9	-5.7	-0.8	3.6	7.5	9.3	6.1	-0.1	-6.4	-12.2	-3.4

Climatological Normal Daily High Temperature by Month (1981-2010 Averages) Continued.

Station	Jan.	Feb.	Mar.	Apr.	May	Jun.	Jul.	Aug.	Sep.	Oct.	Nov.	Dec.	Annual
Iida	6.5	7.5	11.4	18.5	22.9	26.1	29.4	31.1	26.6	20.4	14.3	8.7	18.6
Karuizawa	2.0	2.8	6.9	13.9	18.6	21.1	24.7	25.9	21.2	15.7	10.7	5.2	14.1
Gifu	8.8	10.0	13.7	19.8	24.2	27.4	31.0	33.0	28.8	23.1	17.2	11.6	20.7
Takayama	3.0	4.2	9.1	16.9	22.3	25.7	29.0	30.7	25.6	19.2	12.7	6.2	17.0
Shizuoka	11.5	12.0	14.8	19.5	23.0	25.7	29.5	30.8	27.9	23.1	18.4	14.0	20.9
Hamamatsu	10.5	11.1	14.6	19.2	22.9	26.2	29.8	31.3	28.5	22.9	18.2	13.0	20.7
Nagoya	9.0	10.1	13.9	19.9	24.1	27.2	30.8	32.8	28.6	22.8	17.0	11.6	20.7
Tsu	9.2	9.7	12.9	18.4	22.6	26.0	30.0	31.2	27.7	22.2	16.9	11.9	19.9
Owase	11.3	12.0	14.9	19.4	22.8	25.5	29.2	30.4	27.7	23.0	18.5	13.9	20.7
Hikone	6.8	7.3	11.0	17.3	22.1	25.7	29.7	31.6	27.3	21.3	15.3	9.8	18.8
Kyoto	8.9	9.7	13.4	19.9	24.6	27.8	31.5	33.3	28.8	22.9	17.0	11.6	20.8
Osaka	9.5	10.2	13.7	19.9	24.5	27.8	31.6	33.4	29.3	23.3	17.6	12.3	21.1
Kobe	9.0	9.6	12.8	18.7	23.2	26.6	30.0	31.8	28.5	22.7	17.3	11.9	20.2
Nara	8.7	9.6	13.4	19.8	24.1	27.2	30.8	32.6	28.2	22.2	16.5	11.4	20.4
Wakayama	9.7	10.4	13.8	19.6	23.8	26.9	30.8	32.4	28.8	23.0	17.7	12.5	20.8
Shionomisaki	11.4	12.3	14.8	18.7	22.1	24.9	27.8	29.6	27.3	23.1	18.6	13.9	20.4
Tottori	7.7	8.5	12.4	18.7	23.3	26.6	30.4	32.2	27.4	22.0	16.4	11.0	19.7
Matsue	8.0	8.9	12.6	18.5	22.7	25.9	29.3	31.3	26.8	21.7	16.2	11.0	19.4
Hamada	9.2	9.9	12.9	17.8	21.8	25.0	28.6	30.4	26.6	21.9	17.0	12.0	19.4
Saigo	7.8	8.2	11.3	16.6	20.8	24.1	27.6	29.8	25.9	21.1	15.9	10.9	18.3
Okayama	9.0	9.8	13.3	19.6	24.4	27.7	31.4	32.7	28.4	22.5	16.8	11.6	20.6
Hiroshima	9.7	10.6	14.0	19.7	24.1	27.2	30.8	32.5	29.0	23.4	17.4	12.3	20.9
Shimonoseki	9.4	10.1	13.1	18.0	22.1	25.5	29.4	30.9	27.5	22.6	17.2	12.2	19.8
Tokushima	9.8	10.5	13.8	19.4	23.6	26.6	30.3	31.9	28.3	22.8	17.5	12.5	20.6
Takamatsu	9.4	10.1	13.4	19.5	24.1	27.3	31.2	32.4	28.4	22.8	17.2	12.1	20.7
Matsuyama	9.8	10.6	13.9	19.4	23.6	26.8	30.9	32.1	28.6	23.3	17.8	12.6	20.8
Kochi	11.9	12.9	15.9	20.8	24.4	27.0	30.7	31.9	29.3	24.5	19.3	14.3	21.9
Murotomisaki	10.5	11.1	13.8	18.1	21.3	23.7	27.3	28.7	26.3	21.9	17.5	12.9	19.4
Shimizu	12.2	13.0	15.7	19.7	22.9	25.2	28.7	29.9	28.0	23.8	19.3	14.6	21.1
Fukuoka	9.9	11.1	14.4	19.5	23.7	26.9	30.9	32.1	28.3	23.4	17.8	12.6	20.9
Saga	9.8	11.4	14.6	20.4	25.1	28.0	31.5	32.5	29.1	23.9	17.8	12.3	21.4
Nagasaki	10.4	11.7	14.8	19.7	23.5	26.4	30.1	31.7	28.6	23.8	18.3	13.1	21.0
Izuhara	8.9	10.2	13.0	17.8	21.5	24.3	27.8	29.5	26.3	22.0	16.8	11.6	19.1
Fukue	10.6	11.5	14.4	18.8	22.7	25.6	29.3	30.6	27.7	23.2	18.2	13.3	20.5
Kumamoto	10.5	12.1	15.7	21.3	25.6	28.2	31.7	33.2	29.9	24.6	18.5	13.0	22.0
Oita	10.5	11.1	14.1	19.3	23.5	26.5	30.6	31.8	28.0	22.9	17.9	13.0	20.8
Miyazaki	12.7	13.8	16.7	20.7	24.1	26.8	31.4	31.0	28.1	24.3	19.5	15.1	22.0
Kagoshima	12.8	14.3	17.0	21.6	25.2	27.6	31.9	32.5	30.1	25.4	20.3	15.3	22.8
Naze	17.5	18.0	20.2	23.1	26.1	29.2	32.3	31.8	30.2	26.9	23.3	19.5	24.8
Naha	19.5	19.8	21.7	24.1	26.7	29.4	31.8	31.5	30.4	27.9	24.6	21.2	25.7
Syowa station (Antarctica)	2.0	-0.5	-4.3	-7.6	-10.7	-12.1	-14.1	-15.8	-14.9	-10.8	-4.0	1.1	-7.6

Climatological Normal Daily Low Temperature by Month (°C)

(1981-2010 Averages)

Station	Jan.	Feb.	Mar.	Apr.	May	Jun.	Jul.	Aug.	Sep.	Oct.	Nov.	Dec.	Annual
Sapporo	-7.0	-6.6	-2.9	3.2	8.3	12.9	17.3	19.1	14.2	7.5	1.3	-4.1	5.3
Hakodate	-6.2	-5.9	-2.6	2.6	7.5	12.1	16.6	18.7	14.1	7.4	1.4	-3.5	5.2
Asahikawa	-12.3	-12.7	-6.3	0.0	5.4	11.6	15.9	16.8	11.2	3.9	-2.0	-7.9	2.0
Kushiro	-10.4	-9.9	-4.9	0.3	5.0	9.0	12.8	15.5	12.3	5.5	-0.8	-7.1	2.3
Obihiro	-13.7	-12.6	-6.0	0.6	5.7	10.3	14.5	16.4	12.1	4.8	-1.5	-8.9	1.8
Abashiri	-9.4	-10.1	-5.5	0.4	5.4	9.8	14.0	16.6	12.9	6.6	0.1	-5.9	2.9
Rumoi	-7.9	-7.8	-4.0	1.6	6.6	11.8	16.3	17.8	12.7	6.6	0.9	-4.4	4.2
Wakkanai	-6.8	-7.1	-3.5	1.8	6.0	10.1	14.5	17.3	14.0	8.1	1.0	-4.2	4.3
Nemuro	-6.9	-7.6	-4.3	0.4	4.1	7.7	11.5	14.7	13.1	8.2	1.9	-3.6	3.3
Suttu	-4.9	-4.6	-1.8	2.8	7.3	11.8	16.4	18.4	14.3	8.2	2.1	-2.7	5.6
Urakawa	-5.9	-5.9	-2.8	1.6	5.9	10.3	14.8	17.4	13.9	8.0	2.2	-3.0	4.7
Aomori	-3.9	-3.7	-1.3	3.7	8.9	13.5	18.0	19.8	15.1	8.6	3.0	-1.4	6.7
Morioka	-5.6	-5.2	-2.2	3.0	8.5	13.8	18.1	19.6	14.6	7.3	1.5	-2.4	5.9
Miyako	-3.8	-3.8	-1.1	3.8	8.5	12.7	17.0	19.2	15.2	8.7	2.8	-1.2	6.5
Sendai	-1.7	-1.5	0.9	6.1	11.1	15.5	19.5	21.4	17.6	11.2	5.2	0.9	8.9
Akita	-2.5	-2.3	-0.1	5.1	10.5	15.5	19.8	21.3	16.5	9.8	4.1	0.0	8.2
Yamagata	-3.4	-3.3	-0.7	4.5	10.1	15.2	19.4	20.7	16.2	9.2	3.2	-0.7	7.5
Sakata	-1.0	-1.1	1.0	5.7	11.1	16.1	20.2	21.7	17.3	11.0	5.5	1.6	9.1
Fukushima	-1.8	-1.5	0.9	6.2	11.5	16.1	20.1	21.8	17.6	11.0	4.8	0.7	8.9
Onahama	-0.5	-0.2	2.3	7.1	11.7	15.7	19.6	21.7	18.6	12.5	6.6	1.9	9.8
Mito	-2.2	-1.5	1.6	6.7	12.0	16.3	20.3	21.9	18.3	11.8	5.4	0.2	9.2
Utsunomiya	-2.7	-1.9	1.5	7.0	12.5	16.9	20.8	22.2	18.4	11.8	5.0	-0.3	9.3
Maebashi	-0.8	-0.4	2.6	8.0	13.1	17.5	21.4	22.6	18.9	12.5	6.5	1.7	10.3
Kumagaya	-0.7	0.0	3.1	8.4	13.4	17.8	21.7	23.0	19.3	13.0	6.7	1.6	10.6
Choshi	2.7	3.0	5.9	10.3	14.2	17.2	20.7	23.0	21.0	16.3	10.7	5.4	12.5
Tokyo	0.9	1.7	4.4	9.4	14.0	18.0	21.8	23.0	19.7	14.2	8.3	3.5	11.6
Oshima	3.7	3.4	6.2	10.3	14.5	18.1	21.8	23.1	20.5	15.7	11.0	6.2	12.9
Hachijojima	7.5	7.5	9.0	12.9	16.0	19.0	22.9	23.9	22.2	18.3	13.9	9.7	15.2
Yokohama	2.3	2.6	5.3	10.4	15.0	18.6	22.4	24.0	20.6	15.0	9.6	4.9	12.5
Niigata	-0.2	-0.1	2.1	6.6	12.5	17.6	21.3	23.0	18.8	12.7	6.7	2.3	10.3
Takada	-0.6	-1.0	1.0	5.8	11.2	16.5	21.0	22.4	18.2	11.6	5.9	1.8	9.5
Aikawa	1.2	0.9	2.8	7.4	11.5	16.2	20.8	22.8	18.9	13.5	8.0	3.9	10.6
Toyama	-0.1	-0.3	2.2	7.2	12.6	17.4	21.5	22.9	18.8	12.4	6.8	2.4	10.3
Kanazawa	0.9	0.7	3.0	8.2	13.1	18.0	22.3	23.7	19.5	13.3	7.7	3.4	11.2
Wajima	0.2	-0.2	1.5	5.9	11.1	16.0	20.6	22.0	17.8	11.5	6.4	2.5	9.6
Fukui	0.3	0.1	2.5	7.8	13.1	17.9	22.2	23.4	19.1	12.5	6.9	2.6	10.7
Tsuruga	1.6	1.5	3.9	8.9	13.7	18.3	22.7	24.2	20.0	13.7	8.5	4.1	11.8
Kofu	-2.4	-1.0	2.7	8.3	13.3	17.9	21.8	22.8	19.1	12.3	5.5	-0.2	10.0
Nagano	-4.1	-3.8	-0.8	4.9	10.5	15.8	20.0	21.3	16.9	9.7	3.1	-1.6	7.7
Matsumoto	-5.2	-4.8	-1.5	4.1	9.9	14.9	19.2	20.2	15.9	8.4	2.1	-2.7	6.7
Mt. Fuji	-21.7	-21.5	-17.8	-12.1	-6.5	-1.6	2.4	3.6	0.4	-5.8	-12.2	-18.3	-9.3

Climatological Normal Daily Low Temperature by Month (1981–2010 Averages) Continued.

Station	Jan.	Feb.	Mar.	Apr.	May	Jun.	Jul.	Aug.	Sep.	Oct.	Nov.	Dec.	Annual
Iida	-3.8	-2.7	0.3	5.9	10.7	15.7	19.9	20.6	17.1	10.1	3.7	-1.5	8.0
Karuizawa	-8.7	-8.5	-5.1	0.3	5.8	11.3	15.9	16.7	12.8	5.5	-0.7	-5.7	3.3
Gifu	0.5	0.9	3.9	9.3	14.2	19.0	23.0	24.3	20.4	13.8	7.7	2.7	11.6
Takayama	-5.1	-5.2	-2.0	3.2	9.0	14.6	18.9	19.7	15.7	8.5	2.4	-2.1	6.5
Shizuoka	1.8	2.5	5.7	10.4	14.7	18.8	22.7	23.8	20.8	15.0	9.4	4.1	12.5
Hamamatsu	2.2	2.4	5.4	10.5	14.9	19.2	23.0	24.4	21.2	15.8	9.8	4.4	12.8
Nagoya	0.8	1.1	4.2	9.6	14.5	19.0	23.0	24.3	20.7	14.1	8.1	3.1	11.9
Tsu	1.9	2.0	4.7	9.9	14.9	19.3	23.4	24.4	21.0	14.8	9.0	4.2	12.5
Owase	1.6	2.1	4.9	9.8	14.1	18.4	22.4	23.1	20.3	14.3	8.8	3.8	12.0
Hikone	0.7	0.8	3.3	8.0	13.1	18.0	22.4	23.6	19.9	13.4	7.6	2.9	11.1
Kyoto	1.2	1.4	4.0	9.0	14.0	18.8	23.2	24.3	20.3	13.6	7.8	3.2	11.7
Osaka	2.8	2.9	5.6	10.7	15.6	20.0	24.3	25.4	21.7	15.5	9.9	5.1	13.3
Kobe	2.7	3.0	6.0	11.3	16.2	20.4	24.4	25.8	22.5	16.1	10.6	5.4	13.7
Nara	-0.2	-0.1	2.3	7.4	12.5	17.5	21.8	22.6	18.8	12.1	6.4	1.9	10.3
Wakayama	2.6	2.8	5.4	10.4	15.2	19.7	23.9	24.6	21.2	15.0	9.5	4.8	12.9
Shionomisaki	5.2	5.1	7.9	12.2	16.4	20.1	23.6	24.7	22.1	17.3	12.2	7.5	14.5
Tottori	0.8	0.7	2.8	7.5	12.5	17.6	22.1	22.9	18.7	12.3	7.3	3.1	10.7
Matsue	1.1	1.0	3.2	8.0	13.0	17.8	22.3	23.4	19.2	12.7	7.6	3.4	11.1
Hamada	2.8	2.6	4.3	8.7	13.1	17.7	22.3	23.2	19.1	13.3	8.9	5.2	11.8
Saigo	0.9	0.6	2.4	6.9	11.8	16.5	21.4	22.8	18.4	12.1	7.4	3.4	10.4
Okayama	1.1	1.4	4.3	9.6	14.6	19.4	23.7	24.7	20.7	14.0	8.2	3.3	12.1
Hiroshima	1.7	2.1	4.8	9.9	14.7	19.4	23.8	24.8	20.8	14.2	8.5	3.7	12.4
Shimonoseki	4.5	4.6	7.0	11.4	15.8	19.9	24.0	25.4	22.0	16.6	11.5	6.9	14.1
Tokushima	2.7	2.8	5.6	10.5	15.2	19.6	23.6	24.6	21.4	15.4	9.8	4.9	13.0
Takamatsu	1.6	1.8	4.4	9.4	14.4	19.3	23.6	24.4	20.7	14.2	8.5	3.7	12.2
Matsuyama	2.3	2.5	5.2	10.0	14.7	19.1	23.5	24.2	20.8	14.5	9.2	4.5	12.5
Kochi	1.6	2.7	6.0	10.7	15.2	19.4	23.5	24.0	21.0	14.9	9.2	3.8	12.7
Murotomisaki	4.8	5.1	7.8	12.4	16.3	19.5	23.0	24.1	21.8	17.3	12.5	7.6	14.3
Shimizu	5.3	6.0	8.9	13.5	17.4	20.8	24.5	25.4	23.1	18.2	13.1	7.9	15.3
Fukuoka	3.5	4.1	6.7	11.2	15.6	19.9	24.3	25.0	21.3	15.4	10.2	5.6	13.6
Saga	1.3	2.3	5.3	9.8	14.7	19.6	23.6	24.1	20.2	13.9	8.3	3.1	12.2
Nagasaki	3.8	4.4	7.3	11.6	15.8	20.0	24.3	25.1	21.8	16.1	10.8	5.9	13.9
Izuhara	2.2	3.1	6.1	10.0	14.0	18.0	22.9	23.8	20.5	14.9	9.3	4.4	12.4
Fukue	3.9	4.2	6.5	10.3	14.5	18.7	23.4	24.0	20.7	15.3	10.2	5.7	13.1
Kumamoto	1.2	2.3	5.6	10.3	15.2	19.8	24.0	24.4	20.8	14.2	8.3	3.1	12.5
Oita	2.2	2.7	5.4	9.9	14.5	18.9	23.2	23.8	20.5	14.5	9.1	4.1	12.4
Miyazaki	2.6	3.4	7.2	11.5	15.9	19.7	23.9	24.1	21.1	15.1	9.6	4.7	13.2
Kagoshima	4.6	5.7	8.4	12.7	17.1	21.0	25.3	25.6	22.8	17.5	11.9	6.7	14.9
Naze	12.0	12.3	14.0	16.7	19.7	23.3	25.9	25.8	24.1	20.9	17.3	13.7	18.8
Naha	14.6	14.8	16.5	19.0	21.8	24.8	26.8	26.6	25.5	23.1	19.9	16.3	20.8
Syowa station (Antarctica)	-3.7	-5.5	-9.2	-13.0	-16.6	-18.7	-20.8	-23.3	-22.0	-17.2	-10.4	-4.6	-13.7

Climatological Normal Number of Days with Daily High/Low Temperatures Exceeding Reference Values

(1981–2010 Averages)

Station	Daily highes over 30.0℃	Daily highes over 25.0℃	Daily lows under 0.0℃	Daily highs under 0.0℃
Sapporo	8.0	49.1	124.8	45.0
Hakodate	3.4	38.7	124.0	29.7
Asahikawa	9.8	58.9	160.3	76.0
Kushiro	0.1	5.5	150.2	44.7
Obihiro	10.5	44.5	155.9	55.7
Abashiri	3.9	25.0	148.3	73.7
Rumoi	2.2	24.0	134.0	56.2
Wakkanai	0.1	8.0	131.5	74.9
Nemuro	0.5	7.4	137.4	58.3
Suttu	1.4	23.9	116.7	42.6
Urakawa	0.1	9.8	125.3	29.0
Aomori	12.5	60.0	106.2	20.4
Morioka	19.1	70.8	123.3	14.9
Miyako	12.6	49.7	107.0	2.4
Sendai	17.9	66.0	70.3	1.7
Akita	18.3	73.8	85.2	10.2
Yamagata	37.1	96.4	99.5	9.4
Sakata	24.2	80.6	58.4	3.7
Fukushima	42.2	97.4	71.6	2.0
Onahama	8.6	65.1	50.3	0.0
Mito	32.6	87.8	74.0	0.0
Utsunomiya	43.6	101.3	79.7	0.1
Maebashi	52.3	111.5	50.5	0.0
Kumagaya	56.7	117.2	48.2	0.0
Choshi	13.5	69.8	9.2	0.0
Tokyo	46.4	108.7	20.5	0.0
Oshima	20.2	90.1	5.9	0.0
Hachijojima	14.7	99.1	0.1	0.0
Yokohama	43.3	103.8	7.9	0.0
Niigata	31.8	89.4	43.8	1.4
Takada	42.6	108.5	56.2	1.0
Aikawa	20.7	77.9	26.3	0.7
Toyama	40.3	103.2	44.0	1.2
Kanazawa	41.1	103.3	26.2	0.3
Wajima	27.3	85.0	47.1	0.8
Fukui	49.4	116.3	37.2	0.3
Tsuruga	45.6	109.1	16.8	0.0
Kofu	65.5	130.6	68.5	0.1
Nagano	43.5	103.3	104.6	7.1
Matsumoto	46.3	108.0	116.8	2.9
Mt. Fuji	0.0	0.0	276.7	213.6

Days with high temperatures above 30.0°C are referred to as Midsummer Days. Days with high temperatures above 25.0°C are referred to as Summer Days. Days with low temperatures below 0.0°C are referred to as Winter Days. Days with high temperatures below 0.0°C are referred to as Midwinter days.

Climatological Normal Number of Days with Daily High/Low Temperatures Exceeding Reference Values (1981–2010 Averages) Continued.

Station	Daily highes over 30.0℃	Daily highes over 25.0℃	Daily lows under 0.0℃	Daily highs under 0.0℃
Iida	47.3	112.9	94.4	1.5
Karuizawa	5.1	50.4	155.2	20.2
Gifu	67.7	133.0	32.9	0.1
Takayama	40.7	104.3	117.7	10.0
Shizuoka	47.7	119.2	19.6	0.0
Hamamatsu	53.4	122.3	13.8	0.0
Nagoya	64.3	131.4	28.5	0.0
Tsu	49.7	115.3	15.5	0.0
Owase	41.0	115.3	20.6	0.0
Hikone	48.3	110.4	29.7	0.1
Kyoto	71.3	136.8	22.9	0.0
Osaka	73.2	139.0	6.8	0.0
Kobe	54.9	124.4	7.8	0.0
Nara	63.9	130.8	52.9	0.0
Wakayama	63.6	131.9	8.2	0.0
Shionomisaki	21.6	108.0	1.0	0.0
Tottori	54.5	121.2	29.7	0.2
Matsue	43.9	111.6	28.0	0.2
Hamada	33.8	103.8	8.8	0.0
Saigo	26.9	87.4	39.1	0.3
Okayama	66.7	133.3	24.0	0.0
Hiroshima	63.1	136.3	17.0	0.0
Shimonoseki	44.1	114.2	2.6	0.0
Tokushima	57.2	127.2	8.3	0.0
Takamatsu	65.0	131.8	21.0	0.0
Matsuyama	61.5	132.6	12.8	0.0
Kochi	64.0	144.9	21.6	0.0
Murotomisaki	11.3	89.3	2.7	0.0
Shimizu	28.9	120.3	2.6	0.0
Fukuoka	57.1	132.4	4.3	0.0
Saga	70.6	145.8	24.6	0.0
Nagasaki	54.2	132.4	4.5	0.0
Izuhara	25.9	95.8	18.2	0.2
Fukue	40.6	119.4	6.1	0.0
Kumamoto	78.6	154.5	29.4	0.0
Oita	55.7	126.0	16.6	0.0
Miyazaki	59.5	137.9	16.0	0.0
Kagoshima	76.8	157.3	3.0	0.0
Naze	90.0	183.9	0.0	0.0
Naha	96.0	207.4	0.0	0.0
Syowa station (Antarctica)	0.0	0.0	363.3	299.9

For Major Cities, the (yearly) changes in the number of days with high temperatures above 30.0°, low temperatures above 25.0°C and low temperatures below 0.0° can be found in pages **874–876**.

Air Temperature Record Highs and Lows

(From the start of record keeping until 2019)

Station	Record high temperature ℃	Yr.	Mo.	Day	1st year on record	Record low temperature ℃	Yr.	Mo.	Day	1st year on record
Sapporo	36.2	1994	8	7	1876	-28.5	1929	2	1	1876
Hakodate	33.6	1999	8	4	1872	-21.7	1891	1	29	1872
Asahikawa	36.0	1989	8	7	1888	-41.0	1902	1	25	1888
Kushiro	32.4	2010	6	26	1910	-28.3	1922	1	28	1910
Obihiro	38.8	2019	5	26	1892	-38.2	1902	1	26	1892
Abashiri	37.6	1994	8	7	1889	-29.2	1902	1	25	1889
Rumoi	35.0	1978	8	3	1943	-23.4	1985	1	25	1943
Wakkanai	31.3	1946	8	22	1938	-19.4	1944	1	30	1938
Nemuro	34.0	2019	5	26	1879	-22.9	1931	2	18	1879
Suttu	34.0	1904	8	20	1884	-15.7	1912	1	3	1884
Urakawa	31.2	1989	8	6	1927	-15.5	1979	1	29	1927
Aomori	36.7	1994	8	12	1882	-24.7	1931	2	23	1882
Morioka	37.2	1924	7	12	1923	-20.6	1945	1	26	1923
Miyako	37.3	1933	7	23	1883	-17.3	1908	1	23	1883
Sendai	37.3	2018	8	1	1926	-11.7	1945	1	26	1926
Akita	38.2	1978	8	3	1882	-24.6	1888	2	5	1882
Yamagata	40.8	1933	7	25	1889	-20.0	1891	1	29	1889
Sakata	40.1	1978	8	3	1937	-16.9	1940	1	22	1937
Fukushima	39.1	1942	8	15	1889	-18.5	1891	2	4	1889
Onahama	37.7	1994	8	3	1910	-10.7	1952	2	5	1910
Mito	38.4	1997	7	5	1897	-12.7	1952	2	5	1897
Utsunomiya	38.7	1997	7	5	1890	-14.8	1902	1	24	1890
Maebashi	40.0	2001	7	24	1896	-11.8	1923	1	3	1896
Kumagaya	41.1	2018	7	23	1896	-11.6	1919	2	9	1896
Choshi	35.3	1962	8	4	1887	-7.3	1893	2	13	1887
Tokyo	39.5	2004	7	20	1875	-9.2	1876	1	13	1875
Oshima	34.3	2004	7	21	1938	-4.0	1996	2	3	1938
Hachijojima	34.8	1942	8	2	1906	-2.0	1981	2	27	1906
Yokohama	37.4	2016	8	9	1896	-8.2	1927	1	24	1896
Niigata	39.9	2018	8	23	1881	-13.0	1942	2	12	1881
Takada	40.3	2019	8	14	1922	-13.2	1942	2	12	1922
Aikawa	38.1	2019	8	15	1911	-7.5	1915	1	13	1911
Toyama	39.5	2018	8	22	1939	-11.9	1947	1	29	1939
Kanazawa	38.5	1902	9	8	1882	-9.7	1904	1	27	1882
Wajima	38.2	2000	7	31	1929	-10.4	1943	1	30	1929
Fukui	38.6	1942	7	19	1897	-15.1	1904	1	27	1897
Tsuruga	37.6	1918	8	13	1897	-10.9	1904	1	27	1897
Kofu	40.7	2013	8	10	1894	-19.5	1921	1	16	1894
Nagano	38.7	1994	8	16	1889	-17.0	1934	1	24	1889
Matsumoto	38.5	1942	8	2	1898	-24.8	1900	1	27	1898
Mt. Fuji	17.8	1942	8	13	1932	-38.0	1981	2	27	1932

Air Temperature Record Highs and Lows
(From the start of record keeping until 2019)

Continued.

Station	Record high temperature ℃	Yr.	Mo.	Day	1st year on record	Record low temperature ℃	Yr.	Mo.	Day	1st year on record
Iida	37.7	2018	8	6	1897	-16.5	1954	1	27	1897
Karuizawa	34.2	1946	7	16	1925	-21.0	1936	3	1	1925
Gifu	39.8	2007	8	16	1883	-14.3	1927	1	24	1883
Takayama	37.7	2019	8	13	1899	-25.5	1939	2	11	1899
Shizuoka	38.7	1995	8	28	1940	-6.8	1960	1	25	1940
Hamamatsu	39.8	2013	8	11	1882	-6.0	1923	1	2	1882
Nagoya	40.3	2018	8	3	1890	-10.3	1927	1	24	1890
Tsu	39.5	1994	8	5	1889	-7.8	1904	1	27	1889
Owase	38.6	2016	7	3	1938	-6.9	1963	1	24	1938
Hikone	37.7	2014	7	26	1893	-11.3	1904	1	27	1893
Kyoto	39.8	2018	7	19	1880	-11.9	1891	1	16	1880
Osaka	39.1	1994	8	8	1883	-7.5	1945	1	28	1883
Kobe	38.8	1994	8	8	1896	-7.2	1981	2	27	1896
Nara	39.3	1994	8	8	1953	-7.8	1977	2	16	1953
Wakayama	38.5	2013	8	11	1879	-6.0	1945	1	28	1879
Shionomisaki	35.6	1915	7	14	1913	-5.0	1981	2	26	1913
Tottori	39.1	1994	7	23	1943	-7.4	1981	2	26	1943
Matsue	38.5	1994	8	1	1940	-8.7	1977	2	19	1940
Hamada	38.5	2017	8	6	1893	-7.6	1981	2	26	1893
Saigo	35.8	1994	8	14	1939	-8.9	1981	2	26	1939
Okayama	39.3	1994	8	7	1891	-9.1	1981	2	27	1891
Hiroshima	38.7	1994	7	17	1879	-8.6	1917	12	28	1879
Shimonoseki	37.0	1960	8	10	1883	-6.5	1901	2	3	1883
Tokushima	38.4	1994	7	15	1891	-6.0	1945	2	9	1891
Takamatsu	38.6	2013	8	11	1941	-7.7	1945	1	28	1941
Matsuyama	37.4	2018	8	7	1890	-8.3	1913	2	12	1890
Kochi	38.4	1965	8	22	1886	-7.9	1977	2	17	1886
Murotomisaki	35.0	1942	7	30	1920	-6.6	1981	2	26	1920
Shimizu	35.5	1942	7	30	1940	-5.0	1981	2	26	1940
Fukuoka	38.3	2018	7	20	1890	-8.2	1919	2	5	1890
Saga	39.6	1994	7	16	1890	-6.9	1943	1	13	1890
Nagasaki	37.7	2013	8	18	1878	-5.6	1915	1	14	1878
Izuhara	36.9	2018	7	26	1886	-8.6	1895	2	22	1886
Fukue	35.9	2013	8	19	1962	-5.4	1977	2	19	1962
Kumamoto	38.8	1994	7	17	1890	-9.2	1929	2	11	1890
Oita	37.8	2013	7	24	1887	-7.8	1918	2	19	1887
Miyazaki	38.0	2013	8	1	1886	-7.5	1904	1	26	1886
Kagoshima	37.4	2016	8	22	1883	-6.7	1923	2	28	1883
Naze	37.3	1960	7	9	1896	3.1	1901	2	12	1896
Naha	35.6	2001	8	9	1890	4.9	1918	2	20	1890
Syowa station (Antarctica)	10.0	1977	1	21	1957	-45.3	1982	9	4	1957

Climatological Normal Monthly Precipitation (mm)

(1981-2010 Averages)

Station	Jan.	Feb.	Mar.	Apr.	May	Jun.	Jul.	Aug.	Sep.	Oct.	Nov.	Dec.	Annual
Sapporo	113.6	94.0	77.8	56.8	53.1	46.8	81.0	123.8	135.2	108.7	104.1	111.7	1106.5
Hakodate	77.2	59.3	59.3	70.1	83.6	72.9	130.3	153.8	152.5	100.0	108.2	84.7	1151.7
Asahikawa	69.6	51.3	54.0	47.6	64.8	63.6	108.7	133.5	130.9	104.3	117.2	96.6	1042.0
Kushiro	43.2	22.6	58.2	75.8	111.9	107.7	127.7	130.8	155.6	94.6	64.0	50.8	1042.9
Obihiro	42.8	24.9	42.4	58.9	81.0	75.5	106.4	139.1	138.1	75.0	57.6	46.1	887.8
Abashiri	54.5	36.0	43.5	52.1	61.6	53.5	87.4	101.0	108.2	70.3	60.0	59.4	787.6
Rumoi	100.0	70.2	52.9	43.3	58.8	50.8	97.1	121.3	139.6	131.3	144.4	117.5	1127.0
Wakkanai	84.3	60.7	50.3	49.0	67.6	53.0	90.6	116.0	123.5	134.1	120.9	112.8	1062.7
Nemuro	35.5	22.6	52.5	66.5	102.1	90.9	121.7	120.8	167.0	106.3	84.5	50.4	1020.8
Suttu	112.1	78.0	58.9	57.0	68.4	51.9	88.4	125.7	137.7	136.2	142.3	120.6	1177.1
Urakawa	36.9	24.4	49.1	75.6	121.2	88.3	145.5	159.9	140.9	99.3	80.7	50.5	1072.3
Aomori	144.9	111.0	69.9	63.4	80.6	75.6	117.0	122.7	122.7	103.9	137.7	150.8	1300.1
Morioka	53.1	48.7	80.5	87.5	102.7	110.1	185.5	183.8	160.3	93.0	90.2	70.8	1266.0
Miyako	60.6	60.1	82.1	100.6	93.9	116.4	159.0	171.3	213.7	125.7	80.1	64.8	1328.0
Sendai	37.0	38.4	68.2	97.6	109.9	145.6	179.4	166.9	187.5	122.0	65.1	36.6	1254.1
Akita	119.2	89.1	96.5	112.8	122.8	117.7	188.2	176.9	160.3	157.2	185.8	160.1	1686.2
Yamagata	83.0	62.7	68.6	68.4	75.4	110.5	157.0	150.8	127.2	92.4	84.5	82.7	1163.0
Sakata	168.1	114.0	106.7	102.4	121.4	120.7	209.0	178.5	162.1	180.5	225.0	204.0	1892.4
Fukushima	49.4	44.3	75.6	81.0	92.6	122.1	160.4	154.0	160.3	119.1	65.5	41.8	1166.0
Onahama	52.8	58.0	107.5	125.3	142.0	148.7	150.4	135.5	188.2	173.8	82.4	44.4	1408.9
Mito	51.0	59.4	107.6	119.5	133.3	143.2	134.0	131.8	181.3	167.5	79.1	46.1	1353.8
Utsunomiya	33.9	42.9	88.4	120.5	146.6	174.7	205.8	209.8	220.4	146.5	68.1	35.5	1493.1
Maebashi	26.2	32.1	61.5	78.1	101.9	145.2	197.3	202.3	220.6	115.5	44.7	23.1	1248.5
Kumagaya	32.6	34.6	70.5	92.9	111.8	145.4	161.6	192.6	208.3	146.1	59.0	31.0	1286.3
Choshi	91.6	88.9	158.0	126.7	132.8	168.7	118.9	109.6	220.7	234.6	129.6	79.9	1659.8
Tokyo	52.3	56.1	117.5	124.5	137.8	167.7	153.5	168.2	209.9	197.8	92.5	51.0	1528.8
Oshima	130.5	146.9	258.2	238.7	259.8	337.8	246.5	231.0	353.1	329.0	194.9	100.8	2827.1
Hachijojima	190.0	202.9	308.6	226.8	251.2	380.5	224.6	179.3	338.9	465.9	250.7	182.9	3202.4
Yokohama	58.9	67.5	140.7	144.1	152.2	190.4	168.9	165.0	233.8	205.5	107.0	54.8	1688.6
Niigata	186.0	122.4	112.6	91.7	104.1	127.9	192.1	140.6	155.1	160.3	210.8	217.4	1821.0
Takada	419.1	262.0	194.2	96.1	95.7	145.3	210.6	150.4	206.2	210.8	342.0	423.1	2755.3
Aikawa	127.3	91.6	91.9	88.4	106.8	128.5	172.3	125.4	142.2	125.2	157.0	150.0	1506.4
Toyama	259.5	172.1	158.5	122.2	134.2	182.6	240.4	168.3	220.2	160.7	234.4	247.0	2300.0
Kanazawa	269.6	171.9	159.2	136.9	155.2	185.1	231.9	139.2	225.5	177.4	264.9	282.1	2398.9
Wajima	212.3	141.7	133.3	113.2	127.6	163.4	201.8	155.8	213.5	156.4	227.9	253.6	2100.4
Fukui	284.8	169.7	156.8	127.3	146.2	166.5	233.4	127.6	202.3	144.9	205.3	272.9	2237.6
Tsuruga	269.5	166.9	150.2	118.7	142.2	165.7	195.8	125.5	188.2	135.2	185.0	282.4	2136.4
Kofu	40.2	46.1	87.9	77.7	86.3	122.5	132.6	149.5	180.3	125.2	54.9	32.1	1135.2
Nagano	51.1	49.8	59.4	53.9	75.1	109.2	134.4	97.8	129.4	82.8	44.3	45.5	932.7
Matsumoto	35.9	43.5	79.6	75.3	100.0	125.7	138.4	92.1	155.6	101.9	54.9	28.1	1031.0
Mt. Fuji													

The amount of precipitation have not been measured at Mt. Fuji wasn't.

Climatological Normal Monthly Precipitation (1981–2010 Averages) Continued.

Station	Jan.	Feb.	Mar.	Apr.	May	Jun.	Jul.	Aug.	Sep.	Oct.	Nov.	Dec.	Annual
Iida	62.6	77.4	136.6	127.3	158.4	203.1	216.0	138.8	218.5	133.6	86.8	52.4	1611.5
Karuizawa	33.0	40.8	67.8	78.0	110.7	155.6	189.4	159.3	206.5	119.9	53.6	27.2	1241.7
Gifu	67.0	82.1	143.0	161.2	204.7	245.3	261.6	148.9	237.3	125.5	93.0	58.0	1827.5
Takayama	97.2	99.4	122.9	118.9	136.9	172.1	230.9	165.1	235.5	133.5	99.3	87.8	1699.5
Shizuoka	75.0	102.6	216.8	209.9	213.0	292.8	277.6	250.9	292.0	199.9	131.5	63.0	2324.9
Hamamatsu	57.0	78.3	149.4	167.5	190.5	241.3	190.0	150.8	248.9	164.5	118.8	52.3	1809.1
Nagoya	48.4	65.6	121.8	124.8	156.5	201.0	203.6	126.3	234.4	128.3	79.7	45.0	1535.3
Tsu	43.9	59.0	109.9	127.9	177.1	200.4	180.3	137.0	273.1	150.7	83.5	38.5	1581.4
Owase	100.7	118.8	253.1	289.4	371.8	405.7	397.2	468.2	691.9	395.7	249.8	106.5	3848.8
Hikone	106.9	102.2	120.1	114.4	150.2	190.5	217.7	109.0	168.8	115.5	84.5	91.1	1570.9
Kyoto	50.3	68.3	113.3	115.7	160.8	214.0	220.4	132.1	176.2	120.9	71.3	48.0	1491.3
Osaka	45.4	61.7	104.2	103.8	145.5	184.5	157.0	90.9	160.7	112.3	69.3	43.8	1279.0
Kobe	37.8	56.9	98.5	101.6	149.7	181.6	152.1	90.9	144.6	98.3	63.4	40.9	1216.2
Nara	49.6	63.3	103.2	97.7	143.5	188.8	165.1	111.8	163.3	111.1	71.4	47.3	1316.0
Wakayama	44.4	61.0	96.5	100.3	150.0	188.6	144.9	86.0	183.8	121.5	90.5	49.5	1316.9
Shionomisaki	99.7	105.0	183.3	212.7	249.0	351.9	290.6	233.2	304.8	243.8	160.2	84.7	2519.0
Tottori	202.0	159.8	141.9	108.6	130.6	152.1	200.9	116.6	204.0	144.1	159.4	194.0	1914.0
Matsue	147.2	121.9	132.6	109.4	134.6	189.8	252.4	113.7	197.9	119.5	130.6	137.6	1787.2
Hamada	101.3	85.1	122.4	116.5	144.9	197.3	276.5	122.7	180.8	103.0	109.0	104.4	1663.8
Saigo	159.9	117.3	119.2	110.1	139.8	171.7	219.7	121.1	224.7	114.2	137.7	159.5	1794.8
Okayama	34.2	50.5	86.7	92.3	125.0	171.5	160.9	87.4	134.4	81.1	51.2	31.0	1105.9
Hiroshima	44.6	66.6	123.9	141.7	177.6	247.0	258.6	110.8	169.5	87.9	68.2	41.2	1537.6
Shimonoseki	75.5	81.2	128.4	135.5	165.5	274.8	287.1	153.3	173.9	70.3	78.8	60.2	1684.3
Tokushima	38.9	52.8	94.5	108.2	148.4	190.8	148.8	172.9	210.0	146.2	97.2	45.2	1453.8
Takamatsu	38.2	47.7	82.5	76.4	107.7	150.6	144.1	85.8	147.6	104.2	60.3	37.3	1082.3
Matsuyama	51.9	65.6	102.3	107.8	141.5	223.6	191.6	89.6	130.3	96.7	68.0	46.0	1314.9
Kochi	58.6	106.3	190.0	244.3	292.0	346.4	328.3	282.5	350.0	165.7	125.1	58.4	2547.5
Murotomisaki	88.6	111.6	177.8	200.7	247.0	300.6	256.2	205.6	297.1	202.3	167.8	70.8	2326.1
Shimizu	93.2	124.9	201.1	224.2	235.8	330.3	211.9	246.7	366.1	230.0	141.5	72.9	2478.5
Fukuoka	68.0	71.5	112.5	116.6	142.5	254.8	277.9	172.0	178.4	73.7	84.8	59.8	1612.3
Saga	56.7	77.5	128.6	156.2	198.2	339.0	338.5	196.9	179.5	75.5	75.9	47.7	1870.1
Nagasaki	64.0	85.7	132.0	151.3	179.3	314.6	314.4	195.4	188.8	85.8	85.6	60.8	1857.7
Izuhara	77.4	93.4	159.2	193.3	231.9	331.5	367.4	301.7	235.1	97.8	93.6	53.0	2235.2
Fukue	98.9	105.9	184.1	236.7	243.9	317.5	314.2	234.5	284.7	108.8	119.7	86.9	2335.8
Kumamoto	60.1	83.3	137.9	145.9	195.5	404.9	400.8	173.5	170.4	79.4	80.6	53.6	1985.8
Oita	45.4	65.2	112.1	129.3	150.3	273.8	252.5	172.2	219.5	120.9	69.1	34.4	1644.6
Miyazaki	63.8	90.8	182.1	212.5	239.3	429.2	309.4	290.2	354.6	181.8	95.0	60.0	2508.5
Kagoshima	77.5	112.1	179.7	204.6	221.2	452.3	318.9	223.0	210.8	101.9	92.4	71.3	2265.7
Naze	200.0	162.0	233.2	229.0	258.5	410.3	202.4	268.2	302.7	234.5	180.0	156.9	2837.7
Naha	107.0	119.7	161.4	165.7	231.6	247.2	141.4	240.5	260.5	152.9	110.2	102.8	2040.8
Syowa station (Antarctica)													

The amount of precipitation have not been measured at Syowa station (Antarctica).

Climatological Normal Precipitation (≈5 Day Intervals) (mm)

(1981-2010 Averages)

Mo.	Dates \ Station	Sapporo	Sendai	Tokyo	Niigata	Nagoya	Osaka	Hiroshima	Takamatsu	Fukuoka	Kagoshima	Naha
1	1- 5	19.7	6.0	6.2	31.3	6.6	6.0	6.4	5.2	8.9	11.6	19.9
	6-10	18.9	6.0	7.2	31.1	7.1	6.1	7.4	5.7	10.2	12.6	18.4
	11-15	18.1	5.6	8.6	29.1	7.7	6.9	7.9	6.5	12.2	13.6	17.2
	16-20	18.2	5.9	9.4	28.2	8.4	8.3	8.3	7.0	13.3	14.0	16.3
	21-25	18.7	6.3	10.0	28.8	8.5	8.5	7.6	6.9	12.0	13.1	15.7
	26-31	21.3	6.9	11.3	33.8	10.1	9.8	8.0	7.7	11.5	14.6	19.7
2	1- 5	16.7	5.1	7.9	25.5	7.7	7.3	6.8	5.8	8.3	13.0	18.3
	6-10	16.5	5.4	7.2	23.9	7.7	7.6	8.2	5.9	9.2	15.8	19.5
	11-15	16.8	7.0	9.4	22.6	10.0	9.7	10.9	8.0	11.9	21.0	21.5
	16-20	17.5	8.7	12.3	20.9	13.4	12.5	13.5	9.9	14.4	24.1	23.4
	21-25	16.4	8.7	13.6	20.0	15.7	14.6	15.1	10.9	15.8	24.0	22.0
	26-28	8.8	4.8	8.5	11.8	9.6	9.3	9.5	6.6	10.1	14.4	12.1
	26-29	11.7	6.4	11.3	15.7	12.8	12.4	12.7	8.8	13.5	19.2	16.1
3	1- 5	13.4	8.6	14.5	19.2	15.6	15.3	16.2	10.5	17.5	24.2	20.5
	6-10	13.0	9.9	14.5	18.2	15.8	15.0	17.6	10.8	17.5	25.5	24.0
	11-15	12.7	10.8	16.0	18.4	18.1	16.0	20.4	12.5	18.4	27.5	28.4
	16-20	12.1	11.5	19.0	18.8	21.6	17.6	22.7	14.9	19.3	30.7	28.7
	21-25	12.0	12.0	21.5	17.5	23.6	18.4	22.6	15.6	18.9	32.5	26.9
	26-31	13.6	13.8	26.8	18.9	26.6	21.2	25.2	16.9	22.7	36.8	32.4
4	1- 5	9.6	12.2	22.0	14.7	21.6	18.1	21.1	13.5	19.5	30.2	26.4
	6-10	8.5	14.3	22.0	14.7	23.3	19.4	24.0	14.0	18.9	33.6	26.8
	11-15	8.9	16.7	22.0	15.1	23.4	19.3	25.9	13.9	18.0	36.8	29.5
	16-20	10.4	18.6	22.0	15.7	21.8	18.0	25.3	13.0	18.6	37.8	30.5
	21-25	10.9	18.6	20.4	16.0	19.6	16.6	23.1	11.9	19.6	36.3	28.6
	26-30	10.1	16.5	17.1	16.3	19.7	17.2	22.8	12.3	21.0	35.8	27.6
5	1- 5	9.5	15.0	15.8	18.0	24.6	21.3	26.3	14.9	23.0	36.9	28.2
	6-10	9.5	16.0	18.9	19.4	28.8	25.6	31.3	18.3	25.6	33.5	30.1
	11-15	9.2	18.5	23.1	18.5	28.5	27.2	33.8	19.9	27.3	30.8	33.4
	16-20	8.2	20.1	25.3	17.6	25.4	24.7	30.6	18.2	24.5	32.8	35.2
	21-25	8.0	19.8	25.1	15.4	21.7	20.7	25.1	15.7	19.3	35.3	39.2
	26-31	9.0	20.8	26.7	14.4	22.7	21.9	26.7	18.1	21.4	48.0	58.2
6	1- 5	7.0	16.4	19.4	10.5	18.7	18.3	21.8	15.7	19.5	46.2	53.5
	6-10	7.1	18.7	21.7	11.9	22.9	20.3	24.5	16.7	22.8	57.1	56.7
	11-15	7.8	21.5	27.4	15.7	30.1	26.0	31.7	19.1	31.4	78.0	55.9
	16-20	8.0	24.9	31.9	22.3	38.0	34.7	44.6	26.1	47.0	95.9	43.1
	21-25	8.4	29.2	33.8	30.1	43.7	41.0	60.3	34.9	62.6	94.0	26.4
	26-30	9.8	32.9	33.8	34.3	44.5	41.9	68.2	38.0	68.1	80.3	17.1

Climatological Normal Precipitation (1981–2010 Averages) Continued.

Mo.	Dates	Sapporo	Sendai	Tokyo	Niigata	Nagoya	Osaka	Hiroshima	Takamatsu	Fukuoka	Kagoshima	Naha
7	1- 5	11.7	33.5	32.8	33.4	40.5	38.1	61.4	33.2	63.0	68.2	16.2
	6-10	13.4	34.4	30.2	36.8	38.1	32.1	49.2	26.9	55.5	56.8	20.7
	11-15	14.5	34.8	26.6	41.9	38.3	29.5	44.3	25.4	50.7	48.3	24.7
	16-20	13.8	29.9	23.5	36.6	33.9	26.6	40.7	23.7	43.5	43.0	23.0
	21-25	12.4	24.8	22.1	25.5	27.1	19.1	32.2	19.0	34.4	40.4	21.9
	26-31	16.0	25.9	27.2	22.9	27.0	16.3	28.2	17.2	31.0	47.6	33.7
8	1- 5	17.1	22.5	25.5	19.2	18.6	13.1	17.9	12.5	21.3	35.3	34.7
	6-10	19.0	24.7	27.6	20.8	17.7	13.6	17.2	13.1	25.6	35.5	41.4
	11-15	17.8	24.3	26.3	22.2	19.1	14.5	19.0	14.3	31.1	39.4	43.4
	16-20	19.6	24.3	23.8	23.0	20.4	15.0	19.1	14.5	31.4	40.9	36.3
	21-25	23.6	26.7	24.8	23.0	22.0	14.6	17.6	14.1	29.2	39.9	32.5
	26-31	28.8	35.7	33.4	28.8	29.9	19.1	21.6	17.3	34.5	45.1	46.7
9	1- 5	23.4	29.5	30.0	25.6	28.7	20.1	23.0	16.5	29.9	35.3	43.7
	6-10	23.1	31.2	32.5	28.4	35.3	24.9	27.9	19.3	30.0	31.6	42.0
	11-15	21.4	34.3	35.0	29.2	42.6	27.4	28.8	22.2	30.0	30.4	39.0
	16-20	20.2	35.0	37.0	26.2	43.9	26.7	28.2	25.5	29.7	33.6	42.2
	21-25	21.1	33.8	38.8	24.5	41.2	27.2	28.4	28.7	28.0	35.7	48.6
	26-30	20.8	30.2	38.6	24.7	37.3	28.4	26.2	29.2	24.9	33.1	43.8
10	1- 5	18.3	27.4	40.7	25.7	31.7	26.1	21.8	25.5	20.5	26.4	34.8
	6-10	16.4	25.7	41.6	26.2	26.1	22.5	17.9	20.8	15.9	19.4	29.9
	11-15	16.6	20.3	35.0	25.7	21.1	20.0	15.0	18.9	12.6	15.9	23.3
	16-20	19.0	16.2	28.5	25.5	18.3	17.7	13.4	17.9	11.1	16.0	20.5
	21-25	19.7	16.4	26.2	26.4	16.9	15.0	12.3	14.6	10.3	16.7	22.2
	26-31	21.0	18.3	25.8	33.5	18.2	15.7	13.9	13.8	13.2	19.9	24.9
11	1- 5	16.4	12.3	16.7	28.7	13.8	12.1	11.7	11.0	12.9	16.2	18.6
	6-10	17.3	11.2	15.2	31.0	13.8	11.3	12.2	10.7	13.9	16.9	18.8
	11-15	17.5	10.8	14.5	34.9	13.3	10.9	11.6	9.6	13.6	15.9	18.8
	16-20	16.8	9.6	14.0	36.1	12.1	10.5	9.9	8.2	12.2	14.0	18.5
	21-25	16.7	9.1	14.3	36.9	12.4	10.8	9.6	8.2	12.4	13.5	18.3
	26-30	17.4	9.4	14.9	37.9	12.9	11.0	10.0	9.0	14.1	14.3	17.2
12	1- 5	17.3	8.3	13.3	37.3	11.1	10.1	9.3	8.4	13.8	15.0	15.5
	6-10	17.2	6.5	10.3	37.4	8.5	8.6	8.1	7.1	11.5	13.9	14.7
	11-15	17.7	5.2	8.2	38.7	7.1	7.4	7.3	6.2	9.7	11.7	16.5
	16-20	17.6	5.2	7.2	37.3	7.0	6.6	6.7	5.9	9.0	9.8	18.4
	21-25	18.4	5.8	7.5	32.7	7.0	6.0	5.7	5.4	8.2	9.0	18.5
	26-31	23.8	7.2	8.1	36.4	7.9	7.2	6.2	6.0	9.5	11.8	23.0

Climatological Normal Number of Days with Precipitation Exceeding 1 mm by Month

(1981–2010 Averages)

Station	Jan.	Feb.	Mar.	Apr.	May	Jun.	Jul.	Aug.	Sep.	Oct.	Nov.	Dec.	Annual
Sapporo	18.1	16.0	14.2	9.0	8.5	6.5	8.0	8.5	9.7	11.7	13.9	15.4	139.5
Hakodate	15.7	12.1	12.9	8.9	9.3	7.3	9.2	8.7	10.0	10.7	13.7	14.6	133.1
Asahikawa	17.6	13.8	13.6	9.4	9.7	8.1	10.3	9.6	11.8	14.0	18.1	20.6	156.6
Kushiro	5.0	4.0	6.4	7.4	8.7	8.1	10.2	9.5	9.6	6.9	6.7	5.8	88.3
Obihiro	5.0	4.4	6.1	6.9	8.0	7.7	9.6	9.4	9.4	6.8	6.5	5.9	85.6
Abashiri	12.9	8.6	8.9	8.3	9.4	8.7	9.9	9.3	10.4	9.1	9.9	12.0	117.3
Rumoi	20.9	16.9	13.3	8.9	8.9	7.6	8.6	9.2	12.4	15.4	18.9	21.5	162.5
Wakkanai	20.3	16.1	11.7	8.4	8.6	7.4	7.9	8.6	10.8	14.0	17.0	20.8	151.7
Nemuro	7.4	4.6	6.9	7.5	8.8	8.1	9.4	8.9	9.7	8.5	8.4	7.5	95.8
Suttu	21.6	17.6	13.4	9.1	9.2	7.7	7.7	9.1	11.0	13.3	17.6	20.5	158.1
Urakawa	8.4	5.9	7.8	9.1	9.9	8.6	10.2	9.7	10.0	10.0	12.3	10.8	112.7
Aomori	22.3	18.8	13.9	9.4	9.7	8.2	9.0	9.7	10.3	12.6	16.9	20.7	161.6
Morioka	9.9	8.2	11.2	10.5	10.3	9.3	12.8	10.5	11.4	10.2	11.9	11.2	127.4
Miyako	4.6	4.9	7.4	7.5	8.6	9.3	11.5	9.8	10.7	7.2	5.8	4.5	91.8
Sendai	5.3	5.0	7.2	8.0	9.0	10.7	13.5	10.6	11.2	7.8	6.2	4.7	99.2
Akita	21.2	16.8	14.7	11.3	11.0	9.9	11.8	10.0	12.3	13.8	18.1	21.4	172.3
Yamagata	15.1	12.4	11.7	9.1	9.0	9.7	12.2	9.6	10.8	9.8	11.9	14.1	135.3
Sakata	23.1	18.7	17.0	12.1	11.4	10.5	13.0	10.5	13.0	15.0	19.4	22.9	186.5
Fukushima	8.1	7.1	8.2	7.4	8.1	10.3	12.9	9.7	10.5	7.6	6.6	6.9	103.3
Onahama	4.4	5.2	9.3	9.5	10.6	11.2	11.1	7.8	11.2	10.0	6.9	4.5	101.7
Mito	4.7	5.6	9.7	10.1	11.0	11.4	10.8	7.3	10.5	9.9	6.4	4.8	102.3
Utsunomiya	3.5	5.0	8.8	9.7	11.2	13.2	13.9	11.2	12.8	9.6	6.2	3.4	108.5
Maebashi	2.9	4.2	7.7	8.7	9.8	12.8	14.9	11.8	12.9	8.6	5.0	2.8	102.0
Kumagaya	3.2	4.3	8.2	8.6	9.6	11.8	12.2	9.3	11.8	8.8	5.5	2.9	96.2
Choshi	7.2	7.6	12.4	10.9	10.3	11.2	8.8	5.9	10.5	11.3	9.0	7.0	112.2
Tokyo	4.5	5.5	9.9	9.9	10.3	11.4	10.3	7.7	11.0	9.8	6.8	4.2	101.4
Oshima	7.4	7.8	12.4	11.1	11.2	12.4	10.4	8.3	12.2	12.1	9.2	6.8	121.4
Hachijojima	13.5	13.0	16.8	12.7	13.1	14.0	11.2	11.0	14.1	15.9	12.9	13.2	161.4
Yokohama	5.1	5.8	10.6	10.0	10.2	12.1	10.2	7.5	11.4	10.1	7.4	4.8	105.3
Niigata	21.5	17.3	16.0	11.1	10.1	10.1	12.1	8.5	11.7	14.2	18.0	21.7	172.2
Takada	24.7	20.8	19.1	11.4	10.5	11.5	12.9	10.3	14.1	15.0	18.1	22.2	190.5
Aikawa	19.6	15.2	14.3	9.8	9.6	9.0	10.8	8.3	10.7	12.6	16.7	19.8	156.5
Toyama	22.3	18.1	16.7	12.0	10.8	10.9	13.5	9.8	12.6	12.7	16.5	20.7	176.5
Kanazawa	23.5	19.0	16.4	11.6	10.4	10.7	12.9	8.8	11.6	13.1	16.8	22.0	176.8
Wajima	22.8	17.9	16.2	10.7	10.0	9.9	11.5	8.7	12.5	13.1	17.1	22.1	172.6
Fukui	22.3	18.4	16.3	11.6	11.1	10.7	12.2	8.0	11.6	11.7	15.7	20.7	170.2
Tsuruga	21.6	17.6	15.2	11.5	11.0	10.9	12.2	8.0	11.5	11.5	14.2	19.5	164.7
Kofu	4.1	4.9	8.9	7.7	8.4	10.6	10.8	9.0	10.1	7.9	5.4	3.4	91.2
Nagano	9.7	9.2	9.7	8.0	8.7	9.9	12.0	8.8	10.0	7.5	6.8	8.1	108.4
Matsumoto	4.5	5.3	8.5	7.9	8.6	9.9	11.5	8.2	9.9	7.6	5.4	4.3	91.7
Mt. Fuji													

The amount of precipitation have not been measured at Mt. Fuji wasn't.

Climatological Normal Number of Days with Precipitation Exceeding 1 mm by Month (1981–2010 Averages) Continued.

Station	Jan.	Feb.	Mar.	Apr.	May	Jun.	Jul.	Aug.	Sep.	Oct.	Nov.	Dec.	Annual
Iida	6.7	6.8	10.5	9.6	10.6	12.4	12.8	9.2	11.5	9.1	7.9	6.8	113.8
Karuizawa	4.9	5.8	8.9	9.1	10.2	12.4	14.8	10.9	11.7	9.1	5.9	4.5	108.1
Gifu	8.0	7.8	9.8	9.8	11.0	11.6	12.6	8.5	11.2	8.2	6.9	7.5	112.9
Takayama	14.7	12.5	13.6	10.8	10.9	11.6	14.1	10.8	12.2	10.3	10.6	12.9	145.0
Shizuoka	5.5	5.7	10.3	9.7	10.4	12.1	11.5	9.5	11.8	9.3	6.8	4.8	107.3
Hamamatsu	5.5	5.8	10.0	9.9	10.0	11.4	10.1	7.7	11.2	10.0	6.5	5.3	103.3
Nagoya	5.3	6.3	9.3	9.2	10.2	11.7	12.2	7.7	10.5	8.7	6.2	5.5	102.8
Tsu	5.2	5.9	9.8	8.8	10.3	11.3	11.2	8.5	10.3	8.9	6.0	5.0	101.2
Owase	6.0	6.5	10.7	10.1	11.5	14.4	13.1	12.0	13.8	11.1	7.8	5.2	122.2
Hikone	14.1	12.3	12.4	10.3	10.4	11.8	12.3	7.5	10.4	8.9	8.9	12.1	131.4
Kyoto	6.3	7.3	10.3	9.3	10.2	11.4	11.6	7.6	9.6	7.8	6.4	6.3	104.0
Osaka	5.6	6.3	9.9	9.3	10.0	11.2	9.9	6.9	9.4	7.9	6.2	5.5	98.2
Kobe	4.7	6.0	9.7	9.1	9.5	10.9	10.2	6.0	8.8	7.9	5.6	5.1	93.4
Nara	6.1	6.8	10.6	9.4	10.4	11.7	10.7	7.7	10.2	8.7	7.0	6.4	105.7
Wakayama	6.0	6.5	9.7	9.4	9.3	11.0	9.2	6.0	8.9	8.2	6.5	5.2	96.0
Shionomisaki	6.6	7.6	11.3	10.3	10.5	13.0	11.2	11.0	11.5	10.6	8.4	5.9	118.1
Tottori	20.2	16.4	15.0	11.1	10.8	10.7	12.2	9.3	12.0	10.9	13.5	17.1	159.1
Matsue	17.9	15.8	13.9	10.2	10.3	10.8	11.7	9.1	11.4	10.5	13.1	15.9	150.5
Hamada	13.0	11.1	12.5	9.8	10.0	10.3	11.0	8.3	10.6	8.9	10.5	12.0	128.1
Saigo	19.6	15.3	12.7	9.6	8.3	8.9	10.0	7.3	10.3	9.9	13.4	17.2	142.7
Okayama	4.9	5.8	9.0	8.8	9.2	10.4	9.2	6.0	8.8	6.3	5.5	4.2	88.2
Hiroshima	5.7	7.0	9.6	9.3	9.2	10.7	10.1	6.8	9.1	6.0	6.1	4.9	94.5
Shimonoseki	9.7	8.7	11.0	9.5	9.1	11.1	10.3	8.3	8.9	6.2	8.1	8.4	109.4
Tokushima	5.4	6.2	9.6	9.2	9.0	11.8	9.8	7.9	9.8	7.8	6.3	4.6	97.5
Takamatsu	6.5	6.9	9.8	9.0	8.8	10.3	9.4	6.8	9.1	7.8	6.6	6.2	97.2
Matsuyama	7.4	7.0	10.4	9.3	9.5	11.5	9.1	6.8	8.9	6.9	6.8	6.4	99.8
Kochi	5.4	6.4	10.3	10.0	10.6	12.9	12.1	11.4	11.3	7.7	6.1	4.5	108.8
Murotomisaki	6.9	7.3	11.6	10.5	11.3	12.9	10.8	10.7	11.7	9.6	7.6	5.6	116.6
Shimizu	6.3	7.1	11.6	10.8	11.2	13.5	10.3	10.6	11.7	9.2	7.6	5.3	115.1
Fukuoka	9.1	8.3	11.1	9.8	9.3	11.2	10.6	8.9	9.9	6.2	8.3	8.5	111.3
Saga	7.5	7.4	10.7	9.7	9.2	12.0	11.7	9.4	8.9	5.4	6.7	6.2	104.8
Nagasaki	9.4	8.3	11.3	9.7	9.7	12.1	10.3	8.9	8.8	5.5	7.8	7.8	109.7
Izuhara	6.5	7.1	9.6	9.1	9.2	10.5	11.6	10.6	9.5	5.4	6.2	5.5	101.0
Fukue	11.0	9.3	11.4	10.1	9.6	11.8	10.6	10.1	9.8	5.9	8.4	9.8	117.7
Kumamoto	7.2	7.7	11.4	9.6	9.8	13.2	12.4	9.7	9.5	6.2	7.0	6.6	110.3
Oita	5.2	6.3	9.7	9.2	9.5	12.0	11.2	9.0	9.7	6.4	5.3	3.8	97.4
Miyazaki	5.7	6.9	10.9	10.6	10.9	15.2	11.5	12.7	12.2	7.9	6.3	4.7	115.5
Kagoshima	8.9	8.8	12.9	10.4	10.0	14.6	11.2	10.5	10.2	7.2	7.3	7.7	119.8
Naze	16.0	14.1	15.8	13.6	13.8	15.1	10.0	13.7	13.9	11.6	10.7	13.0	161.3
Naha	10.5	10.2	11.8	10.5	11.5	10.6	8.8	11.8	11.2	8.3	8.5	8.1	121.9
Syowa station (Antarctica)													

The amount of precipitation have not been measured at Syowa station (Antarctica).

Climatological Normal Number of Days with Precipitation Exceeding 10 mm by Month

(1981-2010 Averages)

Station	Jan.	Feb.	Mar.	Apr.	May	Jun.	Jul.	Aug.	Sep.	Oct.	Nov.	Dec.	Annual
Sapporo	3.7	2.6	2.3	1.6	1.7	1.5	2.5	3.5	3.8	3.1	3.2	3.7	33.1
Hakodate	1.4	1.3	1.4	2.1	2.8	2.3	4.3	4.6	4.8	3.0	3.5	2.4	33.9
Asahikawa	1.0	0.6	0.9	1.2	1.9	2.1	3.5	4.2	3.9	3.6	4.2	2.1	29.2
Kushiro	1.4	0.8	1.8	2.2	3.5	3.2	4.0	3.8	4.6	2.8	2.0	1.6	31.7
Obihiro	1.4	0.7	1.2	1.7	2.6	2.6	3.8	4.0	4.2	2.4	1.7	1.6	27.9
Abashiri	1.2	0.7	0.9	1.8	1.9	1.7	3.1	3.0	3.3	2.1	1.8	1.1	22.5
Rumoi	2.4	1.0	0.8	1.0	2.0	1.7	3.1	3.7	4.5	5.0	4.9	2.9	32.9
Wakkanai	1.4	0.9	1.1	1.4	2.2	1.8	2.9	3.4	3.8	4.6	3.9	3.0	30.4
Nemuro	0.9	0.6	1.5	2.1	3.7	2.9	3.7	3.5	4.6	3.1	2.8	1.5	30.9
Suttu	2.7	1.5	1.3	1.7	2.2	1.7	2.6	3.9	4.5	4.8	5.1	3.4	35.5
Urakawa	0.7	0.6	1.3	2.5	3.7	3.1	4.6	4.6	4.7	3.1	2.2	1.2	32.3
Aomori	4.8	3.2	1.5	2.0	2.5	2.6	3.9	3.8	3.6	3.3	5.0	4.6	40.8
Morioka	1.4	1.4	2.8	3.2	3.4	3.7	5.9	5.1	4.6	3.1	2.9	2.2	39.9
Miyako	1.5	1.8	2.7	2.9	2.7	3.7	4.0	4.3	4.8	2.8	2.3	1.4	35.0
Sendai	1.0	1.1	2.3	3.5	3.4	4.4	5.3	4.6	4.9	3.3	1.8	0.9	36.4
Akita	3.4	2.4	3.1	4.5	4.2	3.9	5.9	5.2	5.2	5.8	7.1	5.5	56.1
Yamagata	2.2	1.4	2.0	2.1	2.9	3.8	5.2	4.2	3.8	2.5	3.0	2.4	35.3
Sakata	5.4	3.3	3.2	3.8	4.0	3.9	6.4	5.2	5.7	6.2	8.5	7.5	63.1
Fukushima	1.4	1.5	2.5	2.9	2.8	3.9	5.0	4.1	4.2	3.3	2.2	1.3	35.0
Onahama	1.5	2.2	3.7	3.9	4.5	4.3	4.3	3.0	5.2	4.4	2.7	1.4	41.1
Mito	1.7	2.0	3.7	4.1	4.6	4.5	3.9	3.1	5.1	4.4	2.9	1.4	41.4
Utsunomiya	1.3	1.5	3.4	4.2	5.0	5.5	5.9	5.2	6.4	4.4	2.3	1.2	46.3
Maebashi	0.8	1.1	2.0	2.7	3.5	5.0	6.2	5.3	6.1	3.3	1.5	0.8	38.2
Kumagaya	1.1	1.0	2.5	3.1	3.7	5.0	4.6	4.1	5.8	3.9	1.8	1.0	37.4
Choshi	2.7	2.9	5.2	4.2	4.6	5.0	3.0	2.4	5.7	5.8	3.7	2.7	48.0
Tokyo	1.8	2.0	4.2	4.2	4.8	5.5	4.2	3.6	5.4	5.1	2.6	1.6	45.1
Oshima	3.5	3.8	7.3	6.5	6.5	7.8	5.3	4.2	6.8	6.7	4.7	2.9	66.0
Hachijojima	6.1	5.6	8.7	6.8	6.8	7.7	5.5	4.5	7.2	9.0	6.4	5.3	79.5
Yokohama	1.9	2.6	4.9	4.7	5.0	5.9	4.5	3.5	6.0	5.1	2.9	1.9	48.9
Niigata	7.0	4.1	3.6	3.6	3.2	4.2	5.9	3.7	5.0	5.9	8.2	8.2	62.5
Takada	14.9	10.2	7.4	3.5	3.3	4.5	6.5	4.4	6.2	6.8	10.8	13.4	92.0
Aikawa	4.5	2.6	2.9	3.3	3.9	4.2	5.3	3.4	4.6	4.5	6.0	5.5	50.8
Toyama	10.5	6.6	6.1	4.6	4.7	5.4	6.7	4.7	6.0	5.7	8.2	9.8	79.1
Kanazawa	10.9	5.7	6.0	5.0	5.2	5.5	6.4	4.3	6.0	6.3	8.7	10.6	80.6
Wajima	7.6	4.9	4.6	4.1	4.1	4.6	5.4	3.8	5.7	5.4	7.9	10.1	68.2
Fukui	11.5	6.6	5.9	4.3	4.9	5.0	5.9	4.0	5.6	5.3	7.4	10.6	77.0
Tsuruga	9.7	6.4	5.6	4.4	5.0	5.3	5.5	3.4	5.5	4.6	6.2	9.6	71.4
Kofu	1.5	1.7	3.4	2.7	3.1	4.0	4.1	4.0	4.6	3.6	2.0	1.2	35.8
Nagano	1.3	1.4	1.8	2.1	2.5	3.8	4.5	3.4	4.3	2.7	1.6	1.2	30.6
Matsumoto	1.1	1.4	3.1	2.7	3.6	4.4	4.5	3.0	4.3	3.0	1.9	0.8	33.7
Mt. Fuji													

The amount of precipitation have not been measured at Mt. Fuji wasn't.

Climatological Normal Number of Days with Precipitation Exceeding 10 mm by Month (1981–2010 Averages) Continued.

Station	Jan.	Feb.	Mar.	Apr.	May	Jun.	Jul.	Aug.	Sep.	Oct.	Nov.	Dec.	Annual
Iida	2.2	2.9	5.1	4.5	4.9	5.9	5.9	4.2	5.8	4.3	3.0	1.7	50.5
Karuizawa	1.0	1.3	2.6	3.0	4.4	5.6	5.9	4.3	5.4	3.6	1.8	0.8	39.8
Gifu	2.3	2.9	4.8	4.9	6.1	6.8	7.4	4.2	5.8	3.9	2.9	1.9	53.9
Takayama	2.9	3.4	4.4	4.3	4.4	5.3	6.8	4.8	6.0	4.1	3.5	2.6	52.5
Shizuoka	2.6	3.0	5.9	5.5	5.6	6.7	6.3	5.1	6.6	4.8	3.4	2.1	57.4
Hamamatsu	2.0	2.7	4.8	4.8	5.5	6.4	5.0	3.3	6.0	4.7	3.0	1.6	49.7
Nagoya	1.8	2.5	4.6	4.6	5.4	6.1	5.8	3.4	5.3	3.7	2.5	1.5	47.2
Tsu	1.1	2.3	3.8	4.1	5.3	6.2	5.4	3.5	5.8	4.2	2.4	1.0	44.9
Owase	2.8	3.3	6.2	6.4	6.8	8.3	7.6	7.6	8.8	6.6	4.0	2.4	70.8
Hikone	3.8	4.0	4.5	4.2	5.2	5.9	6.3	3.3	4.7	3.9	2.9	3.0	51.6
Kyoto	1.8	2.6	4.2	4.1	4.9	6.1	5.9	3.6	4.6	3.6	2.4	1.5	45.3
Osaka	1.5	2.3	4.2	3.8	4.8	5.6	4.8	2.7	4.3	3.6	2.3	1.6	41.7
Kobe	1.4	2.1	3.7	3.8	4.5	5.6	4.5	2.7	3.9	3.1	2.2	1.4	38.9
Nara	1.6	2.4	3.8	3.4	4.4	5.9	5.1	3.4	4.7	4.0	2.4	1.7	43.0
Wakayama	1.4	2.1	3.3	3.8	4.3	5.7	4.3	2.5	4.5	3.6	2.5	1.7	39.6
Shionomisaki	3.0	3.0	5.6	5.8	6.1	7.9	6.3	5.4	6.6	5.9	4.1	2.2	61.9
Tottori	7.1	6.5	5.6	3.7	4.7	4.8	5.8	3.7	5.6	4.3	4.9	7.0	63.7
Matsue	5.3	4.1	4.9	4.0	4.4	5.0	6.1	3.6	5.9	3.9	4.4	5.1	56.6
Hamada	3.5	2.9	4.0	4.0	4.6	5.2	6.3	3.5	5.3	3.3	3.6	3.5	49.7
Saigo	5.6	3.8	4.2	3.7	4.1	4.1	5.0	3.2	5.4	3.5	4.5	6.1	53.4
Okayama	1.3	2.0	3.5	3.3	4.3	5.7	4.7	2.6	3.6	2.5	1.9	1.0	36.2
Hiroshima	1.8	2.5	4.3	4.7	5.0	6.0	5.9	2.9	4.8	2.5	2.3	1.4	44.1
Shimonoseki	2.3	2.9	4.7	4.6	4.8	6.3	6.0	4.3	4.4	2.2	2.8	1.9	47.2
Tokushima	1.2	1.7	3.2	3.3	4.3	5.9	4.2	3.7	4.6	3.4	2.7	1.2	39.4
Takamatsu	1.0	1.7	3.1	2.5	3.8	4.9	4.2	2.2	4.1	3.1	2.2	1.3	33.8
Matsuyama	1.8	2.6	4.1	4.0	4.7	6.3	4.8	2.6	3.8	3.1	2.3	1.4	41.5
Kochi	2.2	3.1	5.4	6.1	6.4	7.8	6.4	5.6	6.9	3.7	3.0	1.7	58.3
Murotomisaki	2.7	3.3	5.8	6.1	6.2	7.8	6.1	5.2	6.5	4.8	4.0	2.3	60.7
Shimizu	2.9	3.8	6.1	6.1	6.5	8.3	5.5	5.3	7.0	5.0	3.7	2.1	62.3
Fukuoka	2.2	2.4	4.3	3.8	4.1	6.1	5.8	4.4	5.0	2.4	2.8	1.8	45.2
Saga	1.9	2.8	4.8	4.8	5.1	7.9	7.1	4.7	4.6	2.3	2.1	1.5	49.6
Nagasaki	2.0	3.2	4.6	4.9	4.9	7.0	5.6	4.4	4.6	2.7	2.4	1.9	48.1
Izuhara	2.4	2.9	4.9	5.1	5.3	6.0	6.8	6.1	5.3	2.7	2.4	1.6	51.6
Fukue	3.1	3.3	5.7	5.7	6.0	6.9	6.0	5.9	5.2	2.6	2.9	2.5	55.8
Kumamoto	2.1	3.0	4.9	5.0	4.9	8.4	7.6	4.7	4.5	2.5	2.4	1.8	51.9
Oita	1.3	2.3	4.1	4.3	4.3	6.8	5.7	4.2	5.2	2.7	2.2	1.1	44.1
Miyazaki	2.0	3.0	6.0	5.8	6.4	9.8	6.2	6.5	6.5	3.9	2.6	1.8	60.6
Kagoshima	2.8	3.3	5.8	5.6	5.7	9.2	6.3	5.3	4.7	2.6	3.0	2.5	56.8
Naze	5.8	5.5	7.8	6.7	6.8	9.0	4.1	6.1	6.7	5.1	4.8	5.2	73.5
Naha	3.8	3.6	4.9	4.5	6.1	5.4	3.0	4.6	4.9	3.4	3.0	2.9	50.1
Syowa station (Antarctica)													

The amount of precipitation have not been measured at Syowa station (Antarctica).

Climatological Normal Number of Days with Precipitation Exceeding 30 mm by Month

(1981-2010 Averages)

Station	Jan.	Feb.	Mar.	Apr.	May	Jun.	Jul.	Aug.	Sep.	Oct.	Nov.	Dec.	Annual
Sapporo	0.1	0.2	0.1	0.2	0.2	0.1	0.5	1.3	1.2	0.7	0.5	0.5	5.6
Hakodate	0.3	0.1	0.0	0.3	0.4	0.5	1.4	1.9	1.5	0.6	0.5	0.2	7.8
Asahikawa	0.0	0.0	0.0	0.1	0.1	0.1	0.7	0.9	0.8	0.2	0.1	0.0	3.1
Kushiro	0.2	0.1	0.4	0.4	1.0	0.9	1.1	1.4	1.5	1.0	0.5	0.3	8.8
Obihiro	0.3	0.1	0.2	0.4	0.4	0.3	0.6	1.2	1.1	0.7	0.2	0.2	5.7
Abashiri	0.1	0.1	0.1	0.1	0.2	0.1	0.4	0.8	0.6	0.3	0.1	0.1	3.0
Rumoi	0.0	0.0	0.0	0.1	0.2	0.1	0.7	0.9	1.1	0.3	0.2	0.0	3.7
Wakkanai	0.0	0.0	0.0	0.1	0.2	0.1	0.7	1.1	1.0	0.9	0.4	0.0	4.5
Nemuro	0.1	0.0	0.3	0.3	0.6	0.7	1.1	1.3	1.5	1.0	0.6	0.3	7.7
Suttu	0.1	0.1	0.0	0.2	0.1	0.2	0.7	1.3	1.2	0.6	0.6	0.1	5.1
Urakawa	0.1	0.0	0.2	0.5	1.2	0.7	1.3	1.7	1.4	0.8	0.4	0.1	8.4
Aomori	0.1	0.0	0.0	0.1	0.4	0.3	1.0	1.1	1.0	0.4	0.2	0.3	5.0
Morioka	0.1	0.1	0.2	0.4	0.6	0.8	1.9	2.0	1.3	0.6	0.3	0.1	8.4
Miyako	0.6	0.6	0.6	0.8	0.6	0.9	1.3	1.8	2.0	1.3	0.7	0.4	11.7
Sendai	0.2	0.2	0.4	0.7	0.9	1.3	1.6	1.6	1.9	1.2	0.6	0.2	10.8
Akita	0.1	0.0	0.1	0.5	1.0	1.0	2.1	1.9	1.4	1.1	0.9	0.4	10.6
Yamagata	0.1	0.0	0.1	0.3	0.3	0.8	1.3	1.6	0.9	0.7	0.3	0.1	6.5
Sakata	0.3	0.2	0.1	0.5	0.8	0.8	2.0	1.7	1.4	1.3	1.4	0.8	11.3
Fukushima	0.2	0.1	0.5	0.5	0.7	0.9	1.1	1.5	1.6	1.3	0.5	0.1	9.0
Onahama	0.4	0.4	0.9	1.0	1.1	1.5	1.4	1.3	1.9	1.5	0.7	0.3	12.4
Mito	0.3	0.4	0.7	0.7	1.0	1.3	1.3	1.3	1.6	1.6	0.6	0.2	11.0
Utsunomiya	0.2	0.1	0.5	1.0	1.3	1.5	2.1	2.1	2.0	1.2	0.5	0.3	12.7
Maebashi	0.2	0.1	0.2	0.5	0.8	1.1	1.9	2.2	2.3	0.9	0.2	0.0	10.7
Kumagaya	0.3	0.2	0.2	0.7	0.8	1.3	1.6	1.9	2.1	1.3	0.3	0.2	10.9
Choshi	0.7	0.8	1.3	0.8	1.2	1.7	1.3	1.2	2.2	2.2	1.2	0.6	15.1
Tokyo	0.5	0.3	0.8	1.0	1.0	1.6	1.5	1.5	1.7	1.9	0.8	0.3	12.9
Oshima	1.3	1.8	3.2	2.7	3.0	3.9	2.8	2.1	3.6	3.5	1.9	1.0	30.8
Hachijojima	1.8	2.1	3.0	2.3	2.7	4.4	2.3	1.7	3.2	4.6	3.0	1.6	32.6
Yokohama	0.5	0.5	1.1	1.1	1.3	1.9	1.6	1.5	2.4	2.1	0.9	0.4	15.3
Niigata	0.6	0.2	0.2	0.3	0.7	1.1	1.9	1.3	1.2	1.1	1.3	1.0	11.0
Takada	4.0	1.7	0.7	0.2	0.5	1.4	2.1	1.5	1.6	2.1	4.0	4.9	24.6
Aikawa	0.2	0.2	0.1	0.3	0.6	1.2	1.7	1.0	1.2	0.6	0.4	0.3	7.8
Toyama	1.5	0.7	0.8	0.8	1.1	1.9	2.4	1.8	2.5	1.2	2.4	1.6	18.7
Kanazawa	1.4	0.5	0.8	1.2	1.5	2.0	2.6	1.4	2.3	1.6	2.9	2.1	20.2
Wajima	1.0	0.4	0.4	0.7	1.1	1.4	2.2	1.6	2.2	1.0	1.9	1.6	15.6
Fukui	1.7	0.3	0.7	0.8	1.3	1.8	2.4	1.3	2.3	1.0	1.8	1.9	17.3
Tsuruga	1.8	0.7	0.5	0.6	1.0	1.7	2.4	1.4	2.1	1.1	1.6	2.5	17.5
Kofu	0.3	0.2	0.3	0.4	0.4	1.1	1.0	1.3	1.7	1.1	0.4	0.2	8.2
Nagano	0.1	0.0	0.0	0.1	0.2	0.7	1.0	0.8	1.0	0.6	0.1	0.0	4.6
Matsumoto	0.2	0.2	0.4	0.5	0.8	1.1	1.1	0.8	1.6	1.0	0.3	0.1	8.1
Mt. Fuji													

The amount of precipitation have not been measured at Mt. Fuji wasn't.

Climatological Normal Number of Days with Precipitation Exceeding 30 mm by Month (1981-2010 Averages) Continued.

Station	Jan.	Feb.	Mar.	Apr.	May	Jun.	Jul.	Aug.	Sep.	Oct.	Nov.	Dec.	Annual
Iida	0.4	0.4	1.0	1.1	1.4	2.2	2.4	1.4	2.3	1.2	0.6	0.2	14.7
Karuizawa	0.0	0.1	0.1	0.2	0.6	1.2	1.6	1.2	1.9	1.1	0.2	0.1	8.2
Gifu	0.3	0.4	1.3	1.7	2.4	2.9	3.0	1.7	2.7	1.2	0.8	0.2	18.6
Takayama	0.3	0.4	0.6	0.6	1.1	1.5	2.5	1.7	2.7	1.1	0.7	0.2	13.6
Shizuoka	0.5	1.1	2.4	2.5	2.3	3.3	3.0	2.6	2.8	1.8	1.3	0.4	24.2
Hamamatsu	0.4	0.7	1.5	1.8	2.1	2.7	2.0	1.6	2.6	1.5	1.2	0.3	18.4
Nagoya	0.2	0.3	0.9	1.0	1.5	2.2	2.2	1.4	2.4	1.2	0.7	0.2	14.2
Tsu	0.3	0.3	0.8	1.2	1.9	2.1	1.7	1.2	2.5	1.5	0.5	0.2	14.1
Owase	0.7	1.1	2.6	3.1	4.0	4.7	3.8	4.3	5.3	3.5	1.9	0.7	35.8
Hikone	0.4	0.2	0.6	0.7	1.4	1.9	2.5	1.1	1.7	1.1	0.4	0.4	12.4
Kyoto	0.1	0.3	0.7	1.0	1.8	2.4	2.6	1.4	1.8	1.2	0.5	0.3	14.0
Osaka	0.2	0.3	0.4	0.7	1.5	1.8	1.7	0.8	1.5	1.0	0.6	0.2	10.7
Kobe	0.1	0.3	0.4	0.7	1.5	1.7	1.7	0.9	1.3	0.8	0.4	0.2	10.0
Nara	0.1	0.2	0.5	0.6	1.4	1.8	1.8	1.1	1.6	1.0	0.4	0.1	10.6
Wakayama	0.2	0.3	0.4	0.6	1.6	1.9	1.2	0.9	1.7	1.2	0.7	0.3	11.1
Shionomisaki	1.0	0.9	1.7	2.5	2.9	4.2	2.8	2.4	3.1	2.7	1.7	0.8	26.8
Tottori	1.0	0.7	0.7	0.5	0.9	1.4	2.1	1.0	2.0	1.2	1.3	1.2	13.9
Matsue	0.7	0.4	0.7	0.6	1.2	1.8	2.8	1.0	1.9	0.8	0.9	0.5	13.3
Hamada	0.4	0.4	0.9	0.8	1.5	1.9	3.1	1.2	2.0	0.8	0.8	0.5	14.1
Saigo	0.4	0.3	0.4	0.9	1.3	1.7	2.1	1.2	2.3	0.9	1.0	0.6	13.2
Okayama	0.1	0.1	0.4	0.6	1.1	1.5	1.8	1.0	1.3	0.7	0.2	0.1	8.7
Hiroshima	0.0	0.2	1.0	1.5	2.0	2.7	3.1	1.2	1.9	0.8	0.6	0.2	15.3
Shimonoseki	0.3	0.4	1.0	1.4	1.9	3.0	3.1	1.7	2.0	0.6	0.4	0.2	16.1
Tokushima	0.1	0.2	0.6	0.9	1.5	1.4	1.5	1.6	2.1	1.2	0.9	0.4	12.5
Takamatsu	0.0	0.0	0.2	0.3	0.8	1.2	1.5	0.8	1.2	0.7	0.3	0.1	7.1
Matsuyama	0.1	0.1	0.4	0.6	1.3	2.5	2.2	0.8	1.4	0.7	0.4	0.1	10.6
Kochi	0.4	0.8	1.8	2.8	3.1	3.6	3.4	2.6	3.5	1.6	1.1	0.4	25.0
Murotomisaki	0.8	0.9	1.7	2.1	2.7	3.6	2.6	2.0	3.5	1.9	1.7	0.5	23.8
Shimizu	0.7	1.2	1.8	2.8	2.9	3.8	2.1	2.6	3.5	2.2	1.5	0.6	25.7
Fukuoka	0.2	0.4	0.5	0.9	1.4	2.8	2.9	2.1	1.9	0.5	0.6	0.1	14.2
Saga	0.2	0.4	0.8	1.6	2.4	3.9	3.6	2.3	1.9	0.7	0.6	0.3	18.8
Nagasaki	0.1	0.5	0.9	1.6	1.9	3.6	3.1	1.8	2.0	0.6	0.8	0.2	17.2
Izuhara	0.7	0.9	1.7	2.2	2.4	3.3	3.8	3.5	2.6	1.0	1.0	0.4	23.3
Fukue	0.6	0.9	1.9	3.0	2.9	3.8	3.5	2.7	2.7	1.1	1.0	0.5	24.6
Kumamoto	0.3	0.4	0.9	1.3	2.1	4.4	3.9	1.9	1.9	0.7	0.5	0.2	18.3
Oita	0.3	0.5	0.8	1.2	1.6	3.3	2.6	1.7	2.2	1.0	0.6	0.2	15.7
Miyazaki	0.5	0.8	2.1	2.5	2.8	5.2	3.5	2.9	3.2	1.7	0.8	0.4	26.4
Kagoshima	0.3	0.9	1.7	2.3	2.7	5.1	3.2	2.3	2.2	0.9	0.8	0.4	22.7
Naze	1.5	0.8	2.4	2.3	2.6	4.5	2.0	2.1	2.7	2.1	1.7	1.1	25.8
Naha	0.7	1.0	1.2	1.9	2.5	2.7	1.3	2.1	2.2	1.5	0.9	0.8	18.7
Syowa station (Antarctica)													

The amount of precipitation have not been measured at Syowa station (Antarctica).

Climatological Normal Number of Days with Precipitation Exceeding 50 mm by Month

(1981–2010 Averages)

Station	Jan.	Feb.	Mar.	Apr.	May	Jun.	Jul.	Aug.	Sep.	Oct.	Nov.	Dec.	Annual
Sapporo	0.0	0.0	0.0	0.1	0.0	0.0	0.1	0.4	0.6	0.2	0.0	0.0	1.6
Hakodate	0.0	0.0	0.0	0.0	0.0	0.1	0.2	0.6	0.6	0.3	0.1	0.0	2.1
Asahikawa	0.0	0.0	0.0	0.0	0.0	0.0	0.1	0.4	0.2	0.0	0.0	0.0	0.8
Kushiro	0.0	0.0	0.1	0.2	0.3	0.4	0.5	0.4	0.7	0.4	0.1	0.1	3.2
Obihiro	0.1	0.0	0.1	0.1	0.1	0.1	0.1	0.6	0.4	0.2	0.1	0.1	1.9
Abashiri	0.0	0.0	0.0	0.0	0.0	0.0	0.1	0.2	0.2	0.1	0.0	0.0	0.8
Rumoi	0.0	0.0	0.0	0.0	0.0	0.0	0.3	0.5	0.3	0.0	0.0	0.0	1.2
Wakkanai	0.0	0.0	0.0	0.0	0.0	0.0	0.2	0.4	0.3	0.2	0.0	0.0	1.1
Nemuro	0.0	0.0	0.0	0.1	0.2	0.2	0.5	0.4	0.7	0.4	0.2	0.0	2.7
Suttu	0.0	0.0	0.0	0.0	0.1	0.0	0.2	0.5	0.5	0.1	0.0	0.0	1.4
Urakawa	0.0	0.0	0.0	0.1	0.3	0.1	0.6	0.6	0.3	0.1	0.2	0.0	2.2
Aomori	0.0	0.0	0.0	0.0	0.1	0.0	0.1	0.4	0.4	0.2	0.1	0.0	1.2
Morioka	0.0	0.0	0.0	0.0	0.2	0.1	0.5	0.9	0.5	0.1	0.0	0.1	2.3
Miyako	0.2	0.1	0.2	0.4	0.3	0.3	0.4	0.9	1.1	0.7	0.3	0.3	5.2
Sendai	0.1	0.1	0.1	0.3	0.3	0.5	0.8	0.7	0.8	0.7	0.2	0.1	4.6
Akita	0.0	0.0	0.0	0.1	0.2	0.3	0.6	0.9	0.5	0.1	0.0	0.0	2.7
Yamagata	0.0	0.0	0.0	0.1	0.0	0.2	0.4	0.7	0.4	0.3	0.0	0.1	2.1
Sakata	0.0	0.0	0.0	0.0	0.2	0.3	0.6	0.8	0.3	0.3	0.3	0.1	3.0
Fukushima	0.0	0.0	0.1	0.1	0.2	0.2	0.5	0.6	0.7	0.5	0.1	0.1	3.1
Onahama	0.2	0.1	0.2	0.3	0.5	0.5	0.5	0.6	0.8	0.8	0.1	0.1	4.6
Mito	0.1	0.1	0.1	0.3	0.4	0.3	0.3	0.6	0.7	0.6	0.1	0.1	3.7
Utsunomiya	0.0	0.1	0.0	0.3	0.4	0.3	0.5	1.0	0.9	0.6	0.1	0.1	4.2
Maebashi	0.0	0.0	0.0	0.0	0.1	0.2	0.6	1.1	0.9	0.5	0.1	0.0	3.5
Kumagaya	0.0	0.0	0.0	0.1	0.1	0.3	0.4	1.1	0.8	0.6	0.1	0.0	3.7
Choshi	0.3	0.2	0.4	0.2	0.2	0.7	0.5	0.7	1.0	1.2	0.5	0.1	6.2
Tokyo	0.1	0.1	0.2	0.4	0.4	0.5	0.6	0.9	1.0	0.9	0.3	0.1	5.4
Oshima	0.5	0.7	1.0	1.1	1.5	2.2	1.3	1.3	2.1	2.2	0.9	0.5	15.2
Hachijojima	0.6	0.7	1.2	0.9	1.4	2.6	1.2	0.7	1.7	2.9	1.3	0.7	16.0
Yokohama	0.2	0.1	0.2	0.5	0.5	0.7	0.7	0.8	1.2	0.9	0.3	0.1	6.2
Niigata	0.0	0.0	0.0	0.0	0.1	0.4	0.6	0.7	0.3	0.2	0.1	0.1	2.6
Takada	1.1	0.1	0.0	0.0	0.1	0.3	0.8	0.7	0.7	0.4	1.3	1.5	7.1
Aikawa	0.0	0.0	0.0	0.0	0.2	0.4	0.5	0.5	0.4	0.2	0.0	0.0	2.3
Toyama	0.1	0.0	0.0	0.0	0.3	0.6	1.2	0.7	1.1	0.3	0.4	0.3	5.0
Kanazawa	0.1	0.1	0.0	0.0	0.5	0.6	1.2	0.6	1.1	0.3	0.7	0.4	5.6
Wajima	0.1	0.0	0.1	0.1	0.4	0.6	1.0	0.8	1.1	0.4	0.3	0.1	4.8
Fukui	0.1	0.0	0.0	0.1	0.3	0.5	1.1	0.5	0.8	0.1	0.3	0.1	4.1
Tsuruga	0.5	0.0	0.1	0.0	0.2	0.5	0.6	0.4	0.8	0.3	0.3	0.5	4.2
Kofu	0.0	0.0	0.0	0.0	0.1	0.2	0.4	0.8	1.0	0.5	0.1	0.0	3.1
Nagano	0.0	0.0	0.0	0.0	0.0	0.2	0.3	0.2	0.4	0.2	0.0	0.0	1.2
Matsumoto	0.0	0.0	0.0	0.0	0.1	0.2	0.3	0.2	0.7	0.3	0.1	0.0	1.9
Mt. Fuji													

The amount of precipitation have not been measured at Mt. Fuji wasn't.

Climatological Normal Number of Days with Precipitation Exceeding 50 mm by Month (1981–2010 Averages) Continued.

Station	Jan.	Feb.	Mar.	Apr.	May	Jun.	Jul.	Aug.	Sep.	Oct.	Nov.	Dec.	Annual
Iida	0.0	0.1	0.2	0.2	0.5	0.8	0.8	0.5	0.9	0.5	0.2	0.1	4.8
Karuizawa	0.0	0.0	0.0	0.0	0.0	0.3	0.5	0.7	0.8	0.4	0.0	0.0	2.7
Gifu	0.0	0.1	0.3	0.6	0.8	1.4	1.4	0.8	1.4	0.4	0.4	0.1	7.7
Takayama	0.0	0.0	0.0	0.1	0.2	0.7	0.8	0.6	1.2	0.3	0.2	0.0	4.2
Shizuoka	0.2	0.4	1.1	1.0	0.9	1.7	1.5	1.5	1.5	1.0	0.6	0.2	11.8
Hamamatsu	0.1	0.2	0.3	0.8	0.7	1.2	1.0	0.9	1.4	0.8	0.7	0.1	8.3
Nagoya	0.1	0.1	0.1	0.1	0.4	0.8	1.0	0.7	1.0	0.4	0.2	0.0	5.0
Tsu	0.1	0.1	0.1	0.4	0.7	0.7	0.7	0.6	1.5	0.7	0.3	0.0	6.0
Owase	0.4	0.3	1.4	1.8	2.1	2.4	2.7	2.7	4.1	2.3	1.2	0.5	21.9
Hikone	0.0	0.0	0.0	0.1	0.3	0.8	1.0	0.5	0.7	0.3	0.1	0.0	3.7
Kyoto	0.0	0.1	0.1	0.2	0.4	0.9	1.1	0.6	0.8	0.4	0.2	0.0	4.7
Osaka	0.0	0.0	0.0	0.1	0.4	0.8	0.5	0.3	0.7	0.3	0.1	0.0	3.3
Kobe	0.0	0.0	0.0	0.1	0.6	0.6	0.7	0.4	0.5	0.3	0.2	0.0	3.5
Nara	0.0	0.0	0.0	0.1	0.4	0.8	0.6	0.3	0.6	0.2	0.1	0.0	3.3
Wakayama	0.0	0.1	0.1	0.1	0.4	0.8	0.5	0.3	0.9	0.4	0.3	0.1	3.9
Shionomisaki	0.4	0.3	0.5	1.0	1.5	2.2	1.6	1.2	1.8	1.2	0.9	0.3	13.0
Tottori	0.3	0.1	0.1	0.1	0.2	0.6	0.7	0.4	0.7	0.5	0.3	0.2	4.1
Matsue	0.0	0.1	0.1	0.1	0.4	0.8	1.4	0.4	0.9	0.3	0.1	0.1	4.6
Hamada	0.0	0.0	0.1	0.1	0.4	1.0	1.7	0.5	0.7	0.3	0.1	0.0	5.0
Saigo	0.1	0.0	0.1	0.2	0.6	0.9	1.4	0.6	1.3	0.4	0.2	0.1	5.9
Okayama	0.0	0.0	0.0	0.1	0.3	0.5	0.6	0.3	0.5	0.2	0.0	0.0	2.6
Hiroshima	0.0	0.1	0.2	0.3	0.8	1.5	1.6	0.4	0.7	0.4	0.1	0.0	6.2
Shimonoseki	0.1	0.1	0.2	0.3	0.7	1.7	2.1	0.8	0.9	0.3	0.1	0.0	7.1
Tokushima	0.0	0.1	0.1	0.3	0.4	0.6	0.6	0.9	1.3	0.7	0.3	0.1	5.4
Takamatsu	0.0	0.0	0.0	0.0	0.1	0.5	0.6	0.3	0.6	0.3	0.1	0.0	2.5
Matsuyama	0.0	0.0	0.0	0.1	0.3	1.2	1.1	0.4	0.4	0.3	0.1	0.0	4.0
Kochi	0.1	0.3	1.0	1.3	1.5	1.8	1.9	1.7	2.0	0.8	0.6	0.2	13.3
Murotomisaki	0.2	0.5	0.5	0.9	1.4	1.8	1.1	1.0	1.8	1.2	0.7	0.2	11.3
Shimizu	0.3	0.6	0.7	1.0	1.1	1.9	1.1	1.3	2.4	1.4	0.7	0.2	12.6
Fukuoka	0.0	0.0	0.1	0.2	0.4	1.4	1.7	0.9	0.9	0.2	0.2	0.0	6.0
Saga	0.0	0.1	0.2	0.5	1.1	2.2	2.2	1.4	0.8	0.2	0.3	0.0	9.1
Nagasaki	0.0	0.1	0.2	0.4	1.0	2.0	1.9	1.1	1.1	0.3	0.2	0.2	8.4
Izuhara	0.2	0.3	0.6	0.8	1.5	2.4	2.4	2.1	1.4	0.5	0.5	0.1	12.5
Fukue	0.1	0.3	0.9	1.5	1.6	1.9	2.2	1.3	1.7	0.6	0.5	0.2	12.7
Kumamoto	0.0	0.1	0.3	0.4	1.1	2.8	2.4	0.8	0.7	0.2	0.3	0.1	9.2
Oita	0.1	0.0	0.0	0.3	0.5	1.5	1.5	1.0	0.9	0.5	0.2	0.1	6.7
Miyazaki	0.1	0.2	0.3	1.1	1.3	2.8	1.9	1.7	2.0	1.0	0.3	0.2	12.9
Kagoshima	0.0	0.3	0.5	1.0	1.2	3.0	2.0	0.9	1.1	0.5	0.3	0.1	11.0
Naze	0.6	0.3	0.6	1.0	1.3	2.7	1.3	1.1	1.5	1.1	0.8	0.4	12.7
Naha	0.2	0.3	0.5	0.8	1.2	1.3	0.8	1.1	1.5	0.9	0.3	0.3	9.3
Syowa station (Antarctica)													

The amount of precipitation have not been measured at Syowa station (Antarctica).

Climatological Normal Number of Days with Precipitation Exceeding 100 mm by Month

(1981–2010 Averages)

Station	Jan.	Feb.	Mar.	Apr.	May	Jun.	Jul.	Aug.	Sep.	Oct.	Nov.	Dec.	Annual
Sapporo	0.0	0.0	0.0	0.0	0.0	0.0	0.0	0.1	0.1	0.0	0.0	0.0	0.3
Hakodate	0.0	0.0	0.0	0.0	0.0	0.0	0.0	0.0	0.0	0.0	0.0	0.0	0.1
Asahikawa	0.0	0.0	0.0	0.0	0.0	0.0	0.0	0.1	0.0	0.0	0.0	0.0	0.2
Kushiro	0.0	0.0	0.0	0.0	0.0	0.0	0.0	0.0	0.1	0.0	0.0	0.0	0.1
Obihiro	0.0	0.0	0.0	0.0	0.0	0.1	0.0	0.1	0.1	0.0	0.0	0.0	0.3
Abashiri	0.0	0.0	0.0	0.0	0.0	0.0	0.0	0.0	0.1	0.0	0.0	0.0	0.1
Rumoi	0.0	0.0	0.0	0.0	0.0	0.0	0.0	0.1	0.0	0.0	0.0	0.0	0.1
Wakkanai	0.0	0.0	0.0	0.0	0.0	0.0	0.0	0.0	0.0	0.0	0.0	0.0	0.0
Nemuro	0.0	0.0	0.0	0.0	0.0	0.0	0.0	0.1	0.2	0.0	0.0	0.0	0.4
Suttu	0.0	0.0	0.0	0.0	0.0	0.0	0.1	0.0	0.0	0.0	0.0	0.0	0.2
Urakawa	0.0	0.0	0.0	0.0	0.0	0.0	0.0	0.1	0.0	0.0	0.0	0.0	0.2
Aomori	0.0	0.0	0.0	0.0	0.0	0.0	0.0	0.0	0.0	0.1	0.0	0.0	0.2
Morioka	0.0	0.0	0.0	0.0	0.0	0.0	0.0	0.1	0.1	0.0	0.0	0.0	0.2
Miyako	0.0	0.0	0.0	0.1	0.1	0.1	0.1	0.1	0.4	0.2	0.0	0.1	1.2
Sendai	0.0	0.0	0.0	0.0	0.0	0.0	0.1	0.1	0.2	0.1	0.0	0.0	0.6
Akita	0.0	0.0	0.0	0.0	0.0	0.0	0.0	0.1	0.0	0.0	0.0	0.0	0.1
Yamagata	0.0	0.0	0.0	0.0	0.0	0.0	0.0	0.1	0.1	0.0	0.0	0.0	0.2
Sakata	0.0	0.0	0.0	0.0	0.0	0.0	0.1	0.1	0.0	0.0	0.0	0.0	0.3
Fukushima	0.0	0.0	0.0	0.0	0.0	0.1	0.1	0.1	0.1	0.1	0.0	0.0	0.4
Onahama	0.0	0.0	0.0	0.1	0.1	0.1	0.1	0.2	0.1	0.3	0.0	0.0	1.0
Mito	0.0	0.0	0.0	0.0	0.0	0.0	0.1	0.2	0.2	0.3	0.0	0.0	0.8
Utsunomiya	0.0	0.0	0.0	0.0	0.0	0.1	0.1	0.3	0.2	0.1	0.0	0.0	0.8
Maebashi	0.0	0.0	0.0	0.0	0.0	0.0	0.1	0.1	0.2	0.1	0.0	0.0	0.5
Kumagaya	0.0	0.0	0.0	0.0	0.0	0.0	0.1	0.3	0.2	0.2	0.0	0.0	0.8
Choshi	0.0	0.0	0.0	0.0	0.0	0.0	0.0	0.2	0.2	0.2	0.0	0.0	0.8
Tokyo	0.0	0.0	0.0	0.0	0.0	0.0	0.1	0.3	0.2	0.2	0.1	0.0	1.0
Oshima	0.1	0.2	0.2	0.2	0.4	0.4	0.4	0.5	0.6	0.6	0.2	0.1	4.1
Hachijojima	0.1	0.1	0.3	0.1	0.1	0.5	0.3	0.2	0.6	1.0	0.2	0.1	3.4
Yokohama	0.0	0.0	0.0	0.0	0.0	0.0	0.2	0.3	0.2	0.3	0.1	0.0	1.3
Niigata	0.0	0.0	0.0	0.0	0.0	0.0	0.1	0.1	0.1	0.0	0.0	0.0	0.3
Takada	0.0	0.0	0.0	0.0	0.0	0.1	0.1	0.0	0.1	0.1	0.1	0.1	0.6
Aikawa	0.0	0.0	0.0	0.0	0.0	0.1	0.1	0.1	0.0	0.0	0.0	0.0	0.3
Toyama	0.0	0.0	0.0	0.0	0.0	0.2	0.1	0.1	0.1	0.0	0.0	0.0	0.5
Kanazawa	0.0	0.0	0.0	0.0	0.0	0.1	0.0	0.0	0.3	0.0	0.0	0.0	0.5
Wajima	0.0	0.0	0.0	0.0	0.0	0.1	0.1	0.2	0.2	0.0	0.0	0.0	0.7
Fukui	0.0	0.0	0.0	0.0	0.1	0.0	0.2	0.0	0.2	0.0	0.0	0.0	0.5
Tsuruga	0.0	0.0	0.0	0.0	0.0	0.0	0.0	0.1	0.1	0.0	0.0	0.0	0.3
Kofu	0.0	0.0	0.0	0.0	0.0	0.0	0.0	0.2	0.2	0.1	0.0	0.0	0.5
Nagano	0.0	0.0	0.0	0.0	0.0	0.0	0.0	0.0	0.1	0.0	0.0	0.0	0.1
Matsumoto	0.0	0.0	0.0	0.0	0.0	0.0	0.0	0.0	0.1	0.1	0.0	0.0	0.2
Mt. Fuji													

The amount of precipitation have not been measured at Mt. Fuji wasn't.

Climatological Normal Number of Days with Precipitation Exceeding 100 mm by Month (1981–2010 Averages)

Continued.

Station	Jan.	Feb.	Mar.	Apr.	May	Jun.	Jul.	Aug.	Sep.	Oct.	Nov.	Dec.	Annual
Iida	0.0	0.0	0.0	0.0	0.0	0.1	0.1	0.0	0.2	0.0	0.0	0.0	0.5
Karuizawa	0.0	0.0	0.0	0.0	0.0	0.0	0.1	0.1	0.2	0.0	0.0	0.0	0.5
Gifu	0.0	0.0	0.0	0.0	0.1	0.1	0.1	0.1	0.3	0.1	0.0	0.0	0.7
Takayama	0.0	0.0	0.0	0.0	0.0	0.0	0.1	0.0	0.2	0.0	0.0	0.0	0.4
Shizuoka	0.0	0.1	0.2	0.2	0.2	0.4	0.4	0.6	0.5	0.3	0.1	0.0	2.9
Hamamatsu	0.0	0.0	0.0	0.1	0.2	0.2	0.2	0.2	0.4	0.1	0.1	0.0	1.5
Nagoya	0.0	0.0	0.0	0.0	0.0	0.0	0.1	0.1	0.4	0.1	0.1	0.0	0.7
Tsu	0.0	0.0	0.0	0.0	0.1	0.1	0.2	0.1	0.6	0.1	0.1	0.0	1.3
Owase	0.2	0.1	0.4	0.5	0.9	0.7	0.8	1.3	2.0	1.0	0.7	0.3	8.8
Hikone	0.0	0.0	0.0	0.0	0.0	0.0	0.1	0.1	0.1	0.0	0.0	0.0	0.3
Kyoto	0.0	0.0	0.0	0.0	0.1	0.2	0.2	0.1	0.2	0.1	0.0	0.0	0.8
Osaka	0.0	0.0	0.0	0.0	0.0	0.1	0.0	0.1	0.1	0.0	0.0	0.0	0.3
Kobe	0.0	0.0	0.0	0.0	0.0	0.1	0.0	0.1	0.2	0.0	0.0	0.0	0.5
Nara	0.0	0.0	0.0	0.0	0.0	0.0	0.0	0.1	0.1	0.0	0.0	0.0	0.3
Wakayama	0.0	0.0	0.0	0.0	0.1	0.1	0.1	0.1	0.2	0.1	0.1	0.0	0.8
Shionomisaki	0.0	0.1	0.0	0.2	0.2	0.6	0.5	0.4	0.6	0.3	0.2	0.1	3.3
Tottori	0.0	0.0	0.0	0.0	0.0	0.0	0.2	0.0	0.2	0.0	0.0	0.0	0.5
Matsue	0.0	0.0	0.0	0.0	0.0	0.3	0.3	0.0	0.1	0.1	0.0	0.0	0.8
Hamada	0.0	0.0	0.0	0.0	0.0	0.3	0.3	0.1	0.1	0.0	0.0	0.0	0.7
Saigo	0.0	0.0	0.0	0.0	0.1	0.1	0.3	0.1	0.3	0.0	0.0	0.0	1.0
Okayama	0.0	0.0	0.0	0.0	0.0	0.0	0.1	0.0	0.1	0.1	0.0	0.0	0.3
Hiroshima	0.0	0.0	0.0	0.0	0.1	0.3	0.4	0.1	0.2	0.0	0.0	0.0	1.1
Shimonoseki	0.0	0.0	0.0	0.0	0.1	0.3	0.5	0.1	0.2	0.0	0.0	0.0	1.2
Tokushima	0.0	0.0	0.0	0.1	0.1	0.1	0.1	0.4	0.5	0.2	0.2	0.0	1.8
Takamatsu	0.0	0.0	0.0	0.0	0.0	0.0	0.0	0.1	0.2	0.1	0.0	0.0	0.4
Matsuyama	0.0	0.0	0.0	0.0	0.0	0.1	0.2	0.0	0.0	0.0	0.0	0.0	0.4
Kochi	0.0	0.2	0.2	0.4	0.5	0.6	0.5	0.7	0.6	0.3	0.2	0.1	4.4
Murotomisaki	0.1	0.0	0.1	0.1	0.3	0.2	0.4	0.2	0.5	0.3	0.3	0.0	2.4
Shimizu	0.0	0.1	0.1	0.2	0.3	0.5	0.3	0.4	0.9	0.5	0.1	0.0	3.4
Fukuoka	0.0	0.0	0.0	0.0	0.1	0.4	0.7	0.1	0.2	0.0	0.0	0.0	1.4
Saga	0.0	0.0	0.0	0.0	0.2	0.5	0.7	0.2	0.2	0.0	0.0	0.0	1.8
Nagasaki	0.0	0.0	0.0	0.1	0.0	0.3	0.7	0.4	0.2	0.1	0.0	0.0	1.9
Izuhara	0.0	0.0	0.0	0.2	0.4	0.6	0.8	0.6	0.3	0.1	0.1	0.0	3.0
Fukue	0.0	0.0	0.0	0.2	0.2	0.5	0.6	0.3	0.5	0.2	0.1	0.0	2.6
Kumamoto	0.0	0.0	0.0	0.0	0.1	0.8	0.9	0.1	0.1	0.0	0.0	0.0	2.1
Oita	0.0	0.0	0.0	0.0	0.0	0.3	0.4	0.3	0.3	0.2	0.0	0.0	1.5
Miyazaki	0.0	0.0	0.0	0.1	0.1	0.6	0.6	0.4	0.7	0.3	0.1	0.1	3.0
Kagoshima	0.0	0.0	0.0	0.2	0.2	0.9	0.7	0.4	0.3	0.1	0.0	0.0	3.0
Naze	0.2	0.1	0.1	0.1	0.3	0.7	0.4	0.5	0.5	0.3	0.2	0.0	3.4
Naha	0.0	0.0	0.1	0.0	0.2	0.4	0.2	0.6	0.7	0.2	0.1	0.1	2.7
Syowa station (Antarctica)													

The amount of precipitation have not been measured at Syowa station (Antarctica).

Record Precipitation in 1 Day, 1 Hour, 10 Minutes

(From the start of record keeping until 2019)

Station	Precipitation in 1 day					Precipitation in 1 hour					Precipitation in 10 minutes				
	mm	Yr.	Mo.	Day	1st year on record	mm	Yr.	Mo.	Day	1st year on record	mm	Yr.	Mo.	Day	1st year on record
Sapporo	207.0	1981	8	23	1876	50.2	1913	8	28	1889	19.4	1953	8	14	1937
Hakodate	176.0	1939	8	25	1872	63.2	1939	8	25	1889	21.3	1959	9	11	1937
Asahikawa	184.2	1955	8	17	1888	57.3	1912	8	14	1908	29.0	2000	7	25	1937
Kushiro	182.4	1941	9	6	1910	55.9	1947	8	26	1937	21.8	1952	6	20	1937
Obihiro	174.0	1988	11	24	1892	56.5	1975	7	17	1919	26.1	1943	8	9	1938
Abashiri	163.0	1992	9	11	1889	38.5	2009	9	16	1919	28.0	2009	9	16	1937
Rumoi	147.5	1973	8	18	1943	57.5	1988	8	25	1943	15.6	1953	7	31	1943
Wakkanai	192.0	2016	9	6	1938	64.0	1938	9	1	1938	21.0	1995	8	31	1938
Nemuro	211.5	1992	9	11	1879	53.5	2015	8	10	1889	19.0	2015	8	10	1937
Suttu	206.3	1962	8	3	1884	57.5	1990	7	25	1938	18.0	2010	8	24	1938
Urakawa	190.0	1981	8	5	1927	60.0	2012	9	9	1939	21.0	2017	9	24	1939
Aomori	208.0	2007	11	12	1882	67.5	2000	7	25	1937	20.5	2000	7	25	1937
Morioka	198.0	2007	9	17	1923	62.7	1938	8	15	1923	22.0	1953	8	1	1940
Miyako	319.0	2000	7	8	1883	84.5	2019	10	13	1937	24.5	2016	8	30	1940
Sendai	312.7	1948	9	16	1926	94.3	1948	9	16	1937	30.0	1950	7	19	1937
Akita	186.8	1937	8	31	1882	72.4	1964	8	13	1938	27.0	1964	8	13	1942
Yamagata	217.6	1913	8	27	1889	74.5	1981	8	3	1931	29.0	1958	8	2	1937
Sakata	171.0	2011	6	23	1937	77.8	1949	8	24	1937	23.7	1965	9	5	1937
Fukushima	233.5	2019	10	12	1889	71.0	2017	7	28	1937	26.8	1966	8	12	1937
Onahama	227.2	1966	6	28	1910	69.5	2007	8	22	1937	31.5	2007	8	22	1937
Mito	276.6	1938	6	29	1897	81.7	1947	9	15	1906	36.3	1959	7	7	1937
Utsunomiya	219.4	1957	8	7	1890	100.5	1957	8	7	1930	35.5	1982	6	21	1938
Maebashi	357.4	1947	9	15	1896	114.5	1997	9	11	1912	32.0	2001	7	25	1937
Kumagaya	301.5	1982	9	12	1896	88.5	1943	9	3	1915	35.8	1943	9	3	1937
Choshi	311.6	1947	8	28	1887	140.0	1947	8	28	1912	31.2	1957	10	6	1937
Tokyo	371.9	1958	9	26	1875	88.7	1939	7	31	1886	35.0	1966	6	7	1940
Oshima	525.5	2013	10	16	1938	122.5	2013	10	16	1938	29.0	2003	7	24	1938
Hachijojima	438.9	1941	9	19	1906	129.5	1999	9	4	1937	32.5	1999	9	4	1937
Yokohama	287.2	1958	9	26	1896	92.0	1998	7	30	1937	39.0	1995	6	20	1937
Niigata	265.0	1998	8	4	1881	97.0	1998	8	4	1914	24.0	1967	8	28	1937
Takada	176.0	1985	7	8	1922	91.0	2006	10	29	1922	33.0	2006	10	29	1937
Aikawa	240.0	2002	7	15	1911	79.8	1961	8	4	1925	26.5	2010	9	12	1937
Toyama	207.7	1948	7	25	1939	75.0	1970	8	23	1939	33.0	1970	8	23	1939
Kanazawa	234.4	1964	7	18	1882	77.3	1950	9	18	1937	29.0	1953	8	24	1937
Wajima	218.8	1966	7	12	1929	73.7	1936	9	15	1929	24.9	1967	8	24	1930
Fukui	201.4	1933	7	26	1897	75.0	2004	7	18	1940	23.0	2013	9	3	1940
Tsuruga	211.2	1965	9	17	1897	58.5	2014	6	12	1937	23.5	2014	6	12	1937
Kofu	244.5	1945	10	5	1894	78.0	2004	8	7	1937	28.0	2016	8	1	1937
Nagano	132.0	2019	10	12	1889	63.0	1933	8	13	1903	26.5	1947	8	17	1937
Matsumoto	155.9	1911	8	4	1898	59.0	1981	7	18	1936	24.3	1947	8	28	1937
Mt. Fuji															

The amount of precipitation have not been measured at Mt. Fuji wasn't.

Record Precipitation in 1 Day, 1 Hour, 10 Minutes
(From the start of record keeping until 2019)

Continued.

Station	Precipitation in 1 day					Precipitation in 1 hour					Precipitation in 10 minutes				
	mm	Yr.	Mo.	Day	1st year on record	mm	Yr.	Mo.	Day	1st year on record	mm	Yr.	Mo.	Day	1st year on record
Iida	325.3	1961	6	27	1897	79.7	1960	8	5	1929	22.0	1973	8	4	1937
Karuizawa	318.8	1949	8	31	1925	69.4	1960	8	2	1931	38.5	1960	8	2	1937
Gifu	260.2	1961	6	26	1883	99.6	1914	7	24	1903	30.5	2012	4	3	1937
Takayama	266.1	1910	9	7	1899	62.0	2018	7	4	1914	24.5	1975	6	15	1937
Shizuoka	401.0	2019	10	12	1940	113.0	2003	7	4	1940	29.0	2003	7	4	1940
Hamamatsu	344.1	1910	8	9	1882	87.5	1982	11	30	1933	31.5	1982	11	30	1934
Nagoya	428.0	2000	9	11	1890	97.0	2000	9	11	1890	30.0	2013	7	25	1937
Tsu	427.0	2004	9	29	1889	118.0	1999	9	4	1916	30.0	1946	10	12	1913
Owase	806.0	1968	9	26	1938	139.0	1972	9	14	1938	36.1	1960	10	7	1940
Hikone	596.9	1896	9	7	1893	63.5	2001	7	17	1894	27.5	2001	7	17	1937
Kyoto	288.6	1959	8	13	1880	88.0	1980	8	26	1906	26.5	2019	8	19	1937
Osaka	250.7	1957	6	26	1883	77.5	2011	8	27	1889	27.5	2013	8	25	1937
Kobe	319.4	1967	7	9	1896	87.7	1939	8	1	1897	36.5	2012	4	3	1937
Nara	196.5	2017	10	22	1953	79.0	2000	5	13	1953	27.0	2013	8	5	1953
Wakayama	353.5	2000	9	11	1879	122.5	2009	11	11	1933	34.5	1950	4	5	1937
Shionomisaki	420.7	1939	10	17	1913	145.0	1972	11	14	1937	38.0	1972	11	14	1938
Tottori	187.5	1976	9	10	1943	68.0	1981	7	3	1943	28.0	2016	8	16	1943
Matsue	263.8	1964	7	18	1940	77.9	1944	8	25	1940	25.6	1958	8	1	1940
Hamada	394.5	1988	7	15	1893	91.0	1983	7	23	1912	27.4	1963	8	30	1937
Saigo	236.0	1991	9	14	1939	93.0	1988	9	27	1939	29.0	2007	10	17	1939
Okayama	187.0	2011	9	3	1891	73.5	1994	7	7	1933	30.5	2014	7	20	1939
Hiroshima	339.6	1926	9	11	1879	79.2	1926	9	11	1888	26.0	1987	8	13	1937
Shimonoseki	336.7	1904	6	25	1883	77.4	1953	6	28	1908	32.5	2004	9	16	1937
Tokushima	471.5	1891	8	2	1891	90.5	2009	8	10	1901	32.0	1983	9	7	1937
Takamatsu	210.5	2004	10	20	1941	68.5	1998	9	22	1941	23.5	2017	8	21	1941
Matsuyama	215.1	1943	7	23	1890	60.5	1992	8	2	1890	24.0	2012	8	19	1937
Kochi	628.5	1998	9	24	1886	129.5	1998	9	24	1937	28.5	1998	9	24	1938
Murotomisaki	446.3	1949	7	5	1920	149.0	2006	11	26	1925	38.0	1942	9	17	1940
Shimizu	421.0	1980	8	4	1940	150.0	1944	10	17	1940	49.0	1946	9	13	1940
Fukuoka	307.8	1953	6	25	1890	96.5	1997	7	28	1896	23.5	2007	7	12	1937
Saga	366.5	1953	6	25	1890	110.0	2019	8	28	1926	26.9	1950	8	6	1926
Nagasaki	448.0	1982	7	23	1878	127.5	1982	7	23	1897	36.0	1959	7	8	1937
Izuhara	392.5	1916	9	24	1886	116.0	2003	7	23	1904	29.4	1927	9	2	1904
Fukue	432.5	2005	9	10	1962	113.5	1967	7	9	1962	28.5	1989	9	21	1962
Kumamoto	480.5	1957	7	25	1890	94.0	2016	6	20	1890	27.0	1991	6	30	1937
Oita	443.7	1908	8	10	1887	81.5	1993	9	3	1937	29.0	1948	8	16	1941
Miyazaki	587.2	1939	10	16	1886	139.5	1995	9	30	1925	38.5	1995	9	30	1937
Kagoshima	324.0	1995	8	11	1883	104.5	1995	8	11	1902	33.0	1998	10	7	1938
Naze	622.0	2010	10	20	1896	116.4	1949	10	21	1896	28.0	1968	9	23	1937
Naha	468.9	1959	10	16	1890	110.5	1998	7	17	1900	29.5	1979	6	11	1941
Syowa station (Antarctica)															

The amount of precipitation have not been measured at Syowa station (Antarctica).

Precipitation (mm) for Wettest/Driest Years on Record (Top 3)

(From the start of record keeping until 2019)

Station	High values						1st year on record	Low values						1st year on record
	1st mm	Year	2nd mm	Year	3rd mm	Year		1st mm	Year	2nd mm	Year	3rd mm	Year	
Sapporo	1671.5	1981	1559.0	1972	1444.5	2000	1876	724.6	1897	725.0	1984	840.7	1910	1876
Hakodate	1745.5	1873	1589.2	1932	1578.0	2018	1872	673.0	1984	704.1	1886	797.8	1883	1872
Asahikawa	1741.2	1955	1556.1	1932	1538.0	2000	1888	728.0	1984	779.0	2008	793.2	1921	1888
Kushiro	1703.9	1920	1577.0	2009	1498.0	2016	1910	704.5	1984	790.3	1940	797.4	1924	1910
Obihiro	1489.0	1975	1410.7	1920	1366.4	1955	1892	476.5	2008	540.9	1900	542.0	1984	1892
Abashiri	1231.4	1912	1206.0	2016	1152.4	1966	1889	544.8	1905	578.0	1984	583.5	1988	1889
Rumoi	1899.1	1955	1815.5	1947	1660.5	1946	1943	819.0	1984	858.0	2007	899.5	2008	1943
Wakkanai	1753.7	1962	1619.7	1966	1571.8	1965	1938	749.0	2019	776.5	1986	795.5	2008	1938
Nemuro	1617.5	2009	1459.6	1920	1439.4	1966	1879	639.3	1905	654.6	1900	658.8	1898	1879
Suttu	1745.4	1936	1702.6	1932	1674.5	2010	1884	654.0	1984	828.6	1906	874.5	2019	1884
Urakawa	1579.8	1955	1552.8	1935	1521.0	1936	1927	642.5	1984	767.5	1944	782.5	1982	1927
Aomori	1972.8	1947	1929.8	1922	1916.5	1966	1882	943.5	1887	1003.1	1928	1003.5	2015	1882
Morioka	1702.0	1990	1650.8	1947	1643.0	2013	1923	736.0	1924	792.4	1929	827.5	1994	1923
Miyako	2049.4	1890	2036.9	1920	1919.8	1911	1883	818.0	1978	823.0	1973	860.0	1985	1883
Sendai	1892.3	1950	1796.5	1991	1647.6	1948	1926	813.5	1973	819.5	1984	824.9	1943	1926
Akita	2439.4	1922	2373.0	2013	2357.4	1947	1882	1230.5	1994	1256.0	2008	1327.0	1970	1882
Yamagata	1551.4	1937	1541.1	1958	1531.3	1938	1889	760.0	1975	810.0	1970	837.5	1984	1889
Sakata	2727.0	2013	2549.0	1937	2400.5	2018	1937	1340.0	1994	1459.0	1974	1468.5	1988	1937
Fukushima	1621.3	1890	1612.5	1991	1589.2	1910	1889	652.5	1973	700.5	1970	708.0	1984	1889
Onahama	1989.5	2006	1965.9	1920	1869.5	1911	1910	813.0	1984	946.1	1926	954.0	1978	1910
Mito	2096.8	1920	2030.9	1938	1954.5	1991	1897	760.5	1984	893.9	1926	901.5	1978	1897
Utsunomiya	2214.0	1920	1990.4	1948	1984.0	1989	1890	864.0	1984	989.0	1978	1012.0	1926	1890
Maebashi	1776.6	1955	1733.0	1910	1657.5	1989	1896	799.0	1963	815.5	1996	816.5	1984	1896
Kumagaya	1870.0	1998	1832.5	1991	1803.3	1950	1896	713.0	1984	738.3	1933	805.5	1973	1896
Choshi	2352.0	1989	2293.5	1991	2288.0	1954	1887	1052.5	1926	1066.0	1984	1090.9	1900	1887
Tokyo	2229.6	1938	2193.7	1920	2155.2	1941	1875	879.5	1984	1011.3	1933	1023.3	1967	1875
Oshima	4384.4	1941	4007.8	1954	3890.5	2019	1938	1769.5	1997	1977.6	1947	2085.5	1994	1938
Hachijojima	4889.3	1910	4809.6	1908	4603.0	1998	1906	2080.6	1926	2191.0	2013	2316.0	1984	1906
Yokohama	2535.2	1941	2334.3	1938	2317.4	1921	1896	996.0	1984	1064.4	1967	1118.2	1947	1896
Niigata	2397.0	1998	2351.2	1958	2327.0	2013	1881	1273.1	1928	1331.2	1882	1352.0	2019	1881
Takada	3748.4	1944	3722.5	1945	3656.4	1927	1922	1793.5	1987	2075.2	1954	2185.0	1977	1922
Aikawa	2109.3	1961	2102.5	2013	2083.1	1945	1911	988.0	1994	1125.0	1984	1158.0	2000	1911
Toyama	3123.0	1985	3017.0	1961	2925.0	1998	1939	1562.5	1994	1643.5	1982	1697.0	1987	1939
Kanazawa	3476.2	1917	3318.0	2013	3307.0	1985	1882	1600.5	1994	1820.5	2007	1859.0	2008	1882
Wajima	3106.8	1956	2907.0	2013	2885.7	1958	1929	1575.5	1994	1619.5	2001	1666.0	2008	1929
Fukui	3388.3	1917	3097.6	1956	3042.0	1980	1897	1528.0	1994	1644.5	1987	1817.0	1978	1897
Tsuruga	3381.3	1917	3300.2	1956	3275.0	1945	1897	1535.0	1994	1611.9	1898	1658.5	2000	1897
Kofu	1876.3	1938	1653.8	1910	1652.5	1991	1894	705.6	1940	725.0	1987	765.0	1984	1894
Nagano	1296.9	1903	1274.2	1897	1268.5	1953	1889	555.5	1994	556.0	1987	643.0	1986	1889
Matsumoto	1537.3	1923	1530.1	1911	1488.3	1903	1898	578.2	1926	636.5	1987	642.5	1994	1898
Mt. Fuji														

The amount of precipitation have not been measured at Mt. Fuji wasn't.

Precipitation for Wettest/Driest Years on Record
(From the start of record keeping until 2019) Continued.

Station	High values							Low values						
	1st mm	Year	2nd mm	Year	3rd mm	Year	1st year on record	1st mm	Year	2nd mm	Year	3rd mm	Year	1st year on record
Iida	2254.5	2010	2244.0	1991	2214.1	1903	1897	951.5	1984	1089.5	1994	1142.0	2005	1897
Karuizawa	1838.2	1950	1815.7	1959	1734.2	1928	1925	820.0	1994	856.5	1978	864.9	1926	1925
Gifu	3448.9	1896	2811.3	1903	2792.0	1976	1883	1208.0	1994	1225.6	1947	1302.7	1940	1883
Takayama	2385.0	1945	2374.9	1953	2363.8	1916	1899	1181.0	1994	1264.7	1939	1296.0	1987	1899
Shizuoka	3731.2	1941	3399.0	1998	3391.5	2004	1940	1348.0	1984	1471.0	1940	1548.5	1994	1940
Hamamatsu	2765.5	1938	2666.5	1941	2635.2	1910	1882	1118.5	1984	1200.3	1940	1212.5	2005	1882
Nagoya	2323.6	1896	2128.1	1903	2115.7	1905	1890	900.5	2005	1061.0	1994	1082.5	2002	1890
Tsu	2332.3	1959	2264.6	1902	2220.0	1915	1889	928.0	2005	1008.0	1987	1047.0	1994	1889
Owase	6174.5	1954	5762.2	1950	5699.0	1998	1938	2317.0	2005	2413.0	1986	2437.5	1987	1938
Hikone	3065.5	1896	2165.2	1921	2152.0	1923	1893	1098.5	1939	1137.5	1994	1173.3	1924	1893
Kyoto	2150.6	1921	2134.6	1923	2061.0	2010	1880	880.5	1994	954.5	2005	983.5	1924	1880
Osaka	1858.0	1903	1765.9	1896	1750.9	1959	1883	744.0	1994	818.5	1947	884.0	1978	1883
Kobe	1759.9	1935	1743.5	1903	1737.2	1897	1896	599.5	1994	687.0	2005	787.6	1947	1896
Nara	1790.2	1959	1693.0	1998	1646.8	1965	1953	715.5	1994	911.0	2005	911.5	1978	1953
Wakayama	2030.9	1905	2012.9	1890	2009.4	1952	1879	617.0	1994	895.1	1964	908.3	1947	1879
Shionomisaki	3620.8	1966	3533.4	1954	3514.0	1998	1913	1406.0	1940	1586.0	1934	1733.5	1994	1913
Tottori	2689.7	1945	2689.4	1953	2506.3	1965	1943	1343.5	1973	1405.5	1978	1509.0	1994	1943
Matsue	2683.0	1953	2500.0	1964	2485.0	1972	1940	1105.5	1973	1338.5	1984	1473.0	2005	1940
Hamada	2677.5	1972	2372.0	1997	2349.8	1923	1893	947.4	1939	1125.0	1994	1205.5	1984	1893
Saigo	2450.7	1953	2430.5	1993	2299.7	1958	1939	979.0	1973	1297.0	1994	1341.0	1974	1939
Okayama	1660.1	1923	1646.5	1993	1620.4	1965	1891	593.2	1939	691.8	1924	732.5	2005	1891
Hiroshima	2540.9	1923	2390.5	1993	2223.0	1972	1879	739.5	1978	785.3	1939	917.9	1883	1879
Shimonoseki	2836.5	1980	2585.4	1923	2444.5	1972	1883	957.2	1939	970.0	1978	1091.5	1994	1883
Tokushima	2694.5	1899	2628.5	2004	2562.5	2011	1891	860.5	2007	941.5	1981	966.5	1984	1891
Takamatsu	1618.5	1993	1604.5	2004	1604.0	2011	1941	737.5	1978	765.5	2002	768.0	1984	1941
Matsuyama	2040.4	1943	2029.2	1923	1933.0	1993	1890	696.0	1994	759.5	1978	792.9	1939	1890
Kochi	4383.0	1998	4156.9	1890	3668.1	1954	1886	1543.6	1930	1605.4	1894	1732.5	1996	1886
Murotomisaki	3537.0	2016	3507.6	1949	3321.4	1954	1920	1584.5	2005	1593.5	1995	1675.5	1987	1920
Shimizu	3674.0	1990	3583.0	2001	3386.8	1954	1940	1498.0	1995	1668.5	1994	1703.0	1986	1940
Fukuoka	2976.5	1980	2440.5	1953	2420.5	2016	1890	891.0	1994	999.8	1939	1020.0	2005	1890
Saga	2643.7	1953	2627.5	1980	2586.0	2016	1890	1013.5	1994	1065.5	1978	1129.1	1894	1890
Nagasaki	2842.0	1993	2826.0	1980	2701.3	1927	1878	922.0	1994	1028.6	1894	1243.0	1978	1878
Izuhara	3483.5	1985	3345.5	1972	3254.0	1891	1886	1176.5	1978	1411.2	1944	1429.5	1984	1886
Fukue	3491.5	1972	3133.0	1963	3124.0	1999	1962	1585.0	1996	1601.0	2007	1623.5	1994	1962
Kumamoto	3369.0	1993	2848.2	1963	2800.5	2006	1890	861.7	1894	920.5	1994	1129.0	1967	1890
Oita	2859.0	1993	2527.0	1980	2434.8	1954	1887	986.5	1978	1040.4	1926	1072.5	1994	1887
Miyazaki	4174.5	1993	3832.7	1954	3544.5	1943	1886	1498.9	1904	1622.0	1926	1695.0	1944	1886
Kagoshima	4022.0	1993	3663.5	2015	3550.6	1905	1883	1397.8	1894	1478.8	1904	1519.5	1974	1883
Naze	4429.5	1959	4403.5	1998	4280.1	1954	1896	1708.8	1963	1839.0	1986	1914.7	1940	1896
Naha	3322.0	1998	3190.9	1941	3176.2	1966	1890	969.8	1963	1072.9	1904	1330.5	1993	1890
Syowa station (Antarctica)														

The amount of precipitation have not been measured at Syowa station (Antarctica).

Climatological Normal Number of Days per Month with Maximum Sustained Winds above 10 m/s

(1981–2010 Averages)

Station	Jan.	Feb.	Mar.	Apr.	May	Jun.	Jul.	Aug.	Sep.	Oct.	Nov.	Dec.	Annual
Sapporo	1.8	2.0	2.5	3.6	3.3	1.7	1.4	1.3	1.6	2.2	2.1	1.6	25.1
Hakodate	4.0	4.2	6.7	5.5	3.9	1.5	1.0	1.6	2.6	4.6	5.7	4.9	46.1
Asahikawa	0.7	1.1	1.7	1.5	1.5	0.2	0.2	0.4	0.8	1.5	1.7	1.1	12.3
Kushiro	7.1	6.6	9.5	7.3	6.3	2.7	1.7	2.5	4.8	8.7	10.6	9.5	77.5
Obihiro	0.8	1.3	1.3	1.8	0.8	0.1	0.1	0.1	0.5	0.6	1.5	0.9	9.8
Abashiri	4.4	2.6	3.8	3.9	2.7	0.5	0.3	0.7	1.7	3.0	3.1	4.3	31.0
Rumoi	14.8	10.6	10.6	8.2	6.3	2.2	1.6	2.2	5.3	11.1	15.9	17.5	106.4
Wakkanai	9.8	7.2	8.0	8.3	8.1	4.6	2.3	3.3	4.8	8.3	9.4	9.9	83.8
Nemuro	13.1	9.2	11.2	9.7	7.8	3.9	2.5	2.8	5.5	10.4	13.4	13.8	103.4
Suttu	8.4	6.5	7.5	6.5	6.7	5.5	5.1	4.1	2.9	4.8	6.0	6.7	70.7
Urakawa	15.7	13.7	15.2	9.9	7.0	2.9	2.3	3.2	6.5	12.6	16.5	15.9	121.3
Aomori	4.4	4.0	6.6	6.9	5.1	2.0	1.9	1.3	2.3	3.5	4.8	5.0	47.8
Morioka	1.3	1.3	3.3	4.0	1.8	0.4	0.3	0.3	0.9	0.7	1.9	1.9	18.0
Miyako	0.4	0.4	0.3	0.6	0.2	0.1	0.1	0.2	0.6	0.6	0.4	0.4	4.4
Sendai	6.5	6.6	8.8	7.4	4.5	1.5	1.1	1.6	2.0	3.0	4.4	6.8	54.2
Akita	13.5	11.1	10.4	7.7	4.4	1.8	1.7	2.0	2.8	5.8	8.5	12.9	82.7
Yamagata	0.0	0.0	0.0	0.0	0.0	0.0	0.0	0.1	0.1	0.0	0.0	0.1	0.3
Sakata	14.5	12.3	10.0	7.1	4.2	2.3	2.1	1.8	3.3	5.1	9.1	14.2	86.0
Fukushima	0.7	1.2	1.8	1.9	1.0	0.2	0.1	0.4	0.3	0.6	0.7	1.3	10.3
Onahama	2.0	2.1	3.1	1.7	0.4	0.3	0.3	0.7	0.8	1.3	1.4	2.1	16.0
Mito	0.5	0.6	0.9	1.0	0.2	0.2	0.3	0.3	0.5	0.6	0.1	0.2	5.3
Utsunomiya	2.6	3.2	3.9	2.9	1.6	0.7	1.2	1.4	1.6	1.0	1.2	2.1	23.3
Maebashi	1.5	2.1	2.4	1.5	0.5	0.1	0.2	0.3	0.3	0.4	0.8	1.3	11.4
Kumagaya	1.1	2.6	2.9	1.4	0.4	0.2	0.1	0.3	0.4	0.3	0.5	1.4	11.5
Choshi	14.4	14.3	17.2	14.8	11.3	7.5	8.1	7.2	9.7	12.6	12.8	13.3	143.3
Tokyo	2.2	3.4	4.0	2.8	1.5	0.9	0.5	0.9	1.3	1.4	1.4	1.9	22.1
Oshima	14.0	11.3	13.3	11.4	9.9	7.8	7.9	5.1	7.7	9.2	10.6	13.9	122.1
Hachijojima	16.4	15.3	15.5	12.2	9.2	8.7	5.3	4.4	9.2	11.7	11.8	14.5	134.3
Yokohama	3.4	3.6	5.1	4.1	2.5	1.4	1.3	1.2	2.0	1.6	2.5	2.9	31.5
Niigata	8.7	6.7	4.5	4.1	2.2	1.0	1.1	1.1	1.7	2.4	6.0	8.9	48.5
Takada	1.1	1.2	1.3	2.1	1.0	0.3	0.1	0.3	0.5	0.7	1.1	1.7	11.4
Aikawa	23.3	18.2	14.2	7.8	3.5	1.6	1.5	2.0	3.9	9.1	15.5	22.1	122.6
Toyama	2.2	1.9	3.3	4.2	2.7	1.7	1.0	0.9	1.4	1.5	1.7	2.4	24.8
Kanazawa	9.2	7.2	7.4	5.3	3.3	1.5	2.0	1.9	2.4	3.1	6.2	9.2	58.8
Wajima	5.8	5.2	4.9	5.0	3.9	1.9	2.7	2.0	2.8	2.4	4.9	6.6	48.2
Fukui	1.1	1.4	1.8	2.2	1.7	1.0	0.7	1.0	1.3	0.9	0.9	1.3	15.3
Tsuruga	9.3	9.5	8.8	6.7	4.5	2.2	2.2	1.9	2.9	3.9	6.7	9.2	67.8
Kofu	4.2	4.2	4.8	3.5	1.3	0.3	0.5	0.6	0.4	1.0	1.9	3.7	26.4
Nagano	0.3	0.8	1.8	2.3	1.8	0.9	0.2	0.5	0.7	0.9	0.7	0.4	11.4
Matsumoto	0.2	0.2	0.5	0.5	0.5	0.1	0.1	0.2	0.3	0.1	0.1	0.1	3.0
Mt. Fuji													

Winds have not been observed at Mt. Fuji.

Climatological Normal Number of Days per Month with Maximum Sustained Winds above 10 m/s (1981–2010 Averages) Continued.

Station	Jan.	Feb.	Mar.	Apr.	May	Jun.	Jul.	Aug.	Sep.	Oct.	Nov.	Dec.	Annual
Iida	0.5	0.9	1.6	1.4	0.5	0.1	0.2	0.2	0.1	0.2	0.4	0.9	7.0
Karuizawa	0.0	0.0	0.0	0.0	0.0	0.0	0.0	0.0	0.0	0.0	0.0	0.0	0.0
Gifu	0.8	0.8	1.8	1.7	1.0	0.2	0.4	0.8	0.8	0.5	0.6	0.7	10.3
Takayama	0.0	0.0	0.1	0.1	0.0	0.0	0.0	0.1	0.2	0.1	0.0	0.0	0.7
Shizuoka	0.1	0.3	0.1	0.2	0.2	0.1	0.1	0.3	0.2	0.2	0.2	0.2	2.1
Hamamatsu	2.6	3.6	4.1	2.3	1.0	0.4	0.3	0.4	0.3	0.6	1.2	2.3	19.0
Nagoya	1.8	2.1	4.1	2.8	1.3	0.4	0.2	0.8	1.1	1.1	1.2	1.0	17.9
Tsu	10.7	10.0	12.1	9.0	8.2	5.0	3.7	5.3	5.2	5.3	6.9	8.8	90.2
Owase	1.9	1.5	1.2	0.9	0.4	0.2	0.4	1.2	1.0	0.5	0.7	2.1	12.2
Hikone	3.9	4.6	4.1	2.0	1.0	0.2	0.3	0.7	1.4	1.9	2.5	4.0	26.6
Kyoto	0.0	0.0	0.0	0.0	0.0	0.0	0.1	0.0	0.2	0.1	0.0	0.0	0.4
Osaka	3.0	3.2	2.7	2.3	1.8	1.4	1.6	2.2	1.3	1.2	1.8	2.4	24.9
Kobe	5.2	3.9	4.5	3.9	3.1	2.0	1.3	2.3	2.1	3.2	3.4	4.6	39.5
Nara	0.0	0.0	0.0	0.0	0.0	0.0	0.0	0.1	0.2	0.1	0.0	0.0	0.6
Wakayama	3.8	3.1	4.6	3.6	3.2	2.7	3.2	2.4	2.0	1.7	2.2	3.7	36.4
Shionomisaki	5.6	5.1	7.1	5.4	4.3	4.1	2.7	2.8	3.3	3.4	3.4	4.2	51.3
Tottori	4.2	3.9	4.3	3.7	2.4	1.2	0.8	1.2	1.3	1.4	2.4	3.5	30.3
Matsue	8.7	7.6	6.8	7.1	5.9	3.9	5.6	2.6	2.2	2.5	4.4	8.6	65.9
Hamada	9.8	7.9	8.6	6.0	3.4	2.7	4.4	2.1	2.7	3.4	6.0	9.4	66.4
Saigo	4.6	3.2	4.0	3.9	2.7	2.0	3.6	2.1	2.5	1.9	3.1	5.0	38.6
Okayama	4.0	3.5	3.4	2.6	1.9	0.8	1.3	1.3	1.2	1.2	2.3	3.7	27.2
Hiroshima	2.8	3.6	4.8	4.0	2.4	1.0	1.7	2.1	2.6	2.8	2.7	2.9	33.4
Shimonoseki	4.8	4.4	4.2	4.1	2.4	1.8	1.3	2.6	1.6	1.4	3.1	5.2	36.9
Tokushima	1.7	1.6	2.4	2.8	2.2	1.3	1.0	1.4	1.3	0.7	1.0	1.3	18.7
Takamatsu	1.4	1.7	1.3	1.1	0.4	0.2	0.3	0.7	0.6	0.4	0.8	1.6	10.6
Matsuyama	0.0	0.1	0.2	0.1	0.0	0.0	0.1	0.3	0.3	0.0	0.0	0.0	1.1
Kochi	0.0	0.0	0.0	0.0	0.0	0.0	0.1	0.4	0.3	0.2	0.1	0.0	1.2
Murotomisaki	25.2	22.2	24.9	21.0	20.5	19.4	18.2	15.5	19.2	22.7	21.7	23.1	253.7
Shimizu	2.9	3.2	5.0	3.5	3.0	3.7	2.7	1.9	2.4	1.9	2.1	2.1	34.5
Fukuoka	1.0	1.3	1.8	1.3	0.7	0.4	0.6	1.1	1.6	1.2	1.1	1.2	13.3
Saga	2.1	2.0	3.2	2.5	1.1	2.0	2.5	2.1	2.0	1.2	1.3	1.5	23.5
Nagasaki	1.3	1.6	1.2	0.9	0.3	0.7	0.6	0.6	0.6	0.2	0.7	1.2	10.0
Izuhara	3.2	3.4	3.9	4.2	3.4	2.9	3.5	2.4	1.3	1.1	2.3	3.2	34.9
Fukue	2.3	2.9	3.3	2.9	1.1	1.7	1.7	1.8	1.8	1.1	1.3	2.2	24.3
Kumamoto	0.4	0.4	0.6	0.2	0.4	0.4	0.5	1.2	0.6	0.2	0.2	0.4	5.4
Oita	0.3	0.3	0.3	0.3	0.1	0.2	0.4	0.6	0.6	0.1	0.1	0.2	3.7
Miyazaki	2.9	2.5	2.7	1.5	0.9	0.9	1.3	1.4	1.1	0.4	0.7	2.0	18.3
Kagoshima	1.1	2.1	2.4	1.4	0.7	1.4	1.5	1.7	1.4	1.2	0.7	1.2	16.8
Naze	0.1	0.1	0.2	0.1	0.1	0.1	0.3	0.6	0.6	0.6	0.2	0.2	3.2
Naha	7.6	7.0	8.4	6.4	5.8	8.5	6.2	6.8	5.8	6.8	7.3	7.0	83.5
Syowa station (Antarctica)	12.5	16.9	21.9	21.5	19.6	18.5	18.6	16.7	16.6	17.6	18.7	14.2	213.2

Climatological Normal Number of Days per Month with Maximum Sustained Winds above 15 m/s

(1981-2010 Averages)

Station	Jan.	Feb.	Mar.	Apr.	May	Jun.	Jul.	Aug.	Sep.	Oct.	Nov.	Dec.	Annual
Sapporo	0.1	0.2	0.2	0.2	0.3	0.1	0.0	0.1	0.1	0.0	0.1	0.2	1.5
Hakodate	0.2	0.1	0.5	0.3	0.1	0.2	0.0	0.1	0.3	0.1	0.2	0.3	2.5
Asahikawa	0.0	0.1	0.2	0.1	0.0	0.0	0.0	0.0	0.0	0.1	0.2	0.0	0.8
Kushiro	1.5	1.3	2.1	1.3	0.8	0.3	0.2	0.3	0.7	1.5	2.3	2.0	14.2
Obihiro	0.0	0.0	0.0	0.0	0.0	0.0	0.0	0.0	0.0	0.0	0.0	0.0	0.1
Abashiri	0.4	0.3	0.1	0.1	0.0	0.0	0.0	0.1	0.2	0.3	0.2	0.3	2.1
Rumoi	1.9	1.0	1.5	0.7	0.4	0.1	0.0	0.1	0.4	1.5	2.6	2.1	12.3
Wakkanai	1.1	1.0	1.1	0.4	0.4	0.1	0.0	0.1	0.6	0.9	1.1	1.2	7.9
Nemuro	2.2	1.8	2.2	1.6	0.7	0.2	0.1	0.3	0.9	1.7	2.1	2.1	15.8
Suttu	1.1	0.8	1.4	1.3	1.4	1.1	0.9	0.8	0.7	0.9	0.5	1.1	11.8
Urakawa	4.0	2.8	3.6	1.6	0.7	0.3	0.1	0.3	1.1	2.4	5.0	5.2	27.1
Aomori	0.1	0.3	0.3	0.3	0.1	0.0	0.0	0.1	0.2	0.2	0.1	0.2	1.9
Morioka	0.0	0.0	0.0	0.0	0.0	0.0	0.0	0.1	0.1	0.0	0.0	0.0	0.4
Miyako	0.1	0.0	0.0	0.0	0.0	0.0	0.0	0.1	0.1	0.2	0.0	0.1	0.5
Sendai	0.8	0.5	0.8	0.7	0.3	0.0	0.1	0.2	0.3	0.2	0.4	0.7	5.0
Akita	2.0	1.4	0.8	0.4	0.0	0.1	0.1	0.2	0.6	0.4	1.7	2.2	9.7
Yamagata	0.0	0.0	0.0	0.0	0.0	0.0	0.0	0.0	0.0	0.0	0.0	0.0	0.0
Sakata	1.2	1.1	0.6	0.3	0.2	0.1	0.0	0.2	0.4	0.3	1.0	1.4	6.7
Fukushima	0.0	0.0	0.0	0.1	0.0	0.0	0.0	0.0	0.0	0.0	0.0	0.0	0.1
Onahama	0.0	0.0	0.1	0.0	0.0	0.0	0.0	0.1	0.1	0.1	0.0	0.0	0.5
Mito	0.0	0.0	0.0	0.0	0.0	0.0	0.0	0.0	0.0	0.0	0.0	0.0	0.1
Utsunomiya	0.1	0.3	0.3	0.1	0.1	0.1	0.1	0.0	0.2	0.2	0.1	0.0	1.7
Maebashi	0.0	0.0	0.0	0.0	0.0	0.0	0.0	0.0	0.0	0.0	0.0	0.0	0.1
Kumagaya	0.0	0.0	0.0	0.0	0.0	0.0	0.0	0.0	0.0	0.0	0.0	0.0	0.1
Choshi	2.8	3.1	4.0	3.1	1.2	0.6	0.8	0.8	2.0	2.5	2.1	2.5	25.4
Tokyo	0.0	0.0	0.1	0.0	0.0	0.0	0.0	0.0	0.1	0.0	0.0	0.1	0.4
Oshima	3.1	3.0	2.5	1.7	1.0	1.0	0.8	0.8	1.1	1.6	1.9	2.6	20.9
Hachijojima	2.7	2.9	3.7	1.7	1.7	1.3	0.7	0.5	1.4	2.0	1.3	2.5	22.4
Yokohama	0.0	0.0	0.0	0.2	0.0	0.0	0.0	0.1	0.2	0.1	0.1	0.0	0.8
Niigata	0.6	0.3	0.2	0.2	0.1	0.0	0.0	0.1	0.2	0.2	0.4	0.8	3.2
Takada	0.0	0.0	0.2	0.1	0.0	0.0	0.0	0.0	0.1	0.0	0.0	0.1	0.5
Aikawa	9.5	6.6	3.3	1.0	0.3	0.1	0.1	0.2	0.7	1.8	5.0	9.0	37.6
Toyama	0.0	0.0	0.2	0.3	0.0	0.1	0.0	0.0	0.2	0.1	0.1	0.0	1.1
Kanazawa	2.0	1.5	0.6	0.5	0.1	0.1	0.1	0.2	0.4	0.3	0.9	2.3	9.0
Wajima	0.2	0.2	0.1	0.2	0.1	0.0	0.1	0.2	0.4	0.2	0.3	0.4	2.5
Fukui	0.0	0.0	0.0	0.1	0.0	0.1	0.0	0.1	0.2	0.1	0.1	0.0	0.8
Tsuruga	0.7	1.2	0.7	0.5	0.4	0.2	0.0	0.2	0.5	0.4	0.4	0.9	6.2
Kofu	0.0	0.2	0.2	0.3	0.0	0.0	0.0	0.0	0.0	0.0	0.1	0.1	1.0
Nagano	0.0	0.0	0.0	0.0	0.0	0.0	0.0	0.0	0.0	0.0	0.0	0.0	0.1
Matsumoto	0.0	0.0	0.0	0.0	0.0	0.0	0.0	0.0	0.0	0.0	0.0	0.0	0.0
Mt. Fuji													

Winds have not been observed at Mt. Fuji.

Climatological Normal Number of Days per Month with Maximum Sustained Winds above 15 m/s (1981–2010 Averages) Continued.

Station	Jan.	Feb.	Mar.	Apr.	May	Jun.	Jul.	Aug.	Sep.	Oct.	Nov.	Dec.	Annual
Iida	0.0	0.1	0.0	0.0	0.0	0.0	0.0	0.0	0.0	0.0	0.0	0.0	0.1
Karuizawa	0.0	0.0	0.0	0.0	0.0	0.0	0.0	0.0	0.0	0.0	0.0	0.0	0.0
Gifu	0.0	0.0	0.0	0.0	0.0	0.0	0.0	0.1	0.2	0.0	0.0	0.0	0.3
Takayama	0.0	0.0	0.0	0.0	0.0	0.0	0.0	0.0	0.0	0.0	0.0	0.0	0.0
Shizuoka	0.0	0.0	0.0	0.0	0.0	0.0	0.0	0.0	0.0	0.0	0.0	0.0	0.0
Hamamatsu	0.0	0.0	0.0	0.0	0.0	0.0	0.0	0.1	0.0	0.0	0.0	0.0	0.1
Nagoya	0.0	0.0	0.0	0.0	0.0	0.0	0.0	0.1	0.2	0.1	0.0	0.0	0.5
Tsu	0.6	0.5	1.1	0.8	0.4	0.4	0.4	0.8	1.2	0.6	0.5	0.4	7.8
Owase	0.0	0.0	0.0	0.1	0.0	0.1	0.1	0.1	0.2	0.1	0.1	0.0	0.8
Hikone	0.0	0.0	0.0	0.0	0.0	0.0	0.0	0.0	0.1	0.1	0.0	0.0	0.2
Kyoto	0.0	0.0	0.0	0.0	0.0	0.0	0.0	0.0	0.0	0.0	0.0	0.0	0.0
Osaka	0.1	0.1	0.1	0.0	0.1	0.1	0.0	0.2	0.3	0.1	0.0	0.1	1.2
Kobe	0.1	0.3	0.3	0.1	0.0	0.1	0.1	0.3	0.4	0.2	0.2	0.1	2.3
Nara	0.0	0.0	0.0	0.0	0.0	0.0	0.0	0.0	0.0	0.0	0.0	0.0	0.0
Wakayama	0.2	0.3	0.7	0.4	0.4	0.5	0.3	0.4	0.5	0.2	0.3	0.4	4.7
Shionomisaki	0.4	0.6	0.8	0.3	0.1	0.4	0.3	0.4	0.5	0.3	0.2	0.2	4.6
Tottori	0.1	0.1	0.3	0.4	0.2	0.2	0.0	0.1	0.3	0.0	0.1	0.1	1.8
Matsue	0.8	0.6	0.6	0.7	0.2	0.2	0.2	0.3	0.4	0.3	0.4	0.9	5.6
Hamada	1.7	0.9	1.5	1.3	0.4	0.1	0.3	0.2	0.4	0.4	1.0	1.4	9.7
Saigo	0.1	0.2	0.0	0.2	0.1	0.2	0.1	0.3	0.4	0.1	0.0	0.1	1.9
Okayama	0.3	0.1	0.2	0.2	0.1	0.1	0.1	0.2	0.3	0.0	0.1	0.2	1.8
Hiroshima	0.1	0.1	0.1	0.1	0.0	0.1	0.2	0.5	0.5	0.3	0.1	0.1	2.1
Shimonoseki	0.1	0.1	0.1	0.2	0.0	0.3	0.1	0.5	0.3	0.1	0.0	0.0	1.9
Tokushima	0.0	0.0	0.0	0.0	0.0	0.1	0.0	0.4	0.4	0.1	0.0	0.0	1.1
Takamatsu	0.0	0.0	0.0	0.0	0.0	0.0	0.0	0.0	0.0	0.0	0.0	0.0	0.2
Matsuyama	0.0	0.0	0.0	0.0	0.0	0.0	0.0	0.0	0.0	0.0	0.0	0.0	0.0
Kochi	0.0	0.0	0.0	0.0	0.0	0.0	0.0	0.0	0.1	0.0	0.0	0.0	0.1
Murotomisaki	11.4	11.3	12.8	9.3	8.4	8.4	7.0	4.9	7.4	8.9	8.4	10.0	108.2
Shimizu	0.2	0.2	0.3	0.2	0.0	0.1	0.1	0.4	0.4	0.1	0.1	0.1	2.3
Fukuoka	0.0	0.0	0.0	0.0	0.0	0.1	0.1	0.2	0.3	0.1	0.0	0.0	0.8
Saga	0.0	0.0	0.1	0.0	0.0	0.1	0.2	0.5	0.4	0.1	0.0	0.0	1.4
Nagasaki	0.0	0.0	0.0	0.0	0.0	0.0	0.0	0.1	0.2	0.0	0.0	0.0	0.3
Izuhara	0.0	0.0	0.3	0.1	0.1	0.1	0.3	0.2	0.3	0.0	0.1	0.0	1.5
Fukue	0.0	0.0	0.0	0.0	0.0	0.1	0.1	0.4	0.4	0.2	0.0	0.0	1.2
Kumamoto	0.0	0.0	0.0	0.0	0.0	0.0	0.0	0.1	0.3	0.0	0.0	0.0	0.4
Oita	0.0	0.0	0.0	0.0	0.0	0.0	0.0	0.2	0.2	0.0	0.0	0.0	0.4
Miyazaki	0.0	0.1	0.1	0.0	0.0	0.0	0.2	0.4	0.4	0.0	0.0	0.1	1.3
Kagoshima	0.0	0.0	0.0	0.0	0.0	0.1	0.3	0.5	0.5	0.0	0.0	0.0	1.4
Naze	0.0	0.0	0.0	0.0	0.0	0.0	0.0	0.2	0.3	0.0	0.0	0.0	0.5
Naha	0.0	0.3	0.1	0.0	0.0	0.4	0.7	2.0	1.7	0.7	0.2	0.3	6.2
Syowa station (Antarctica)	5.1	9.1	12.9	13.6	12.3	11.4	11.6	9.3	8.9	9.8	9.2	6.0	119.2

Record Wind Speeds

(From the start of record keeping until 2019)

Station	Maximum wind speed						Maximum gusts					
	m/s	Wind direction	Yr.	Mo.	Day	1st year on record	m/s	Wind direction	Yr.	Mo.	Day	1st year on record
Sapporo	28.8	N N W	1912	3	19	1876	50.2	S W	2004	9	8	1943
Hakodate	27.9	W N W	1928	2	7	1872	46.5	W S W	1999	9	25	1940
Asahikawa	24.6	WSW	2010	3	21	1888	34.1	W S W	2010	3	21	1942
Kushiro	31.8	S	2016	8	17	1910	43.2	S	2016	8	17	1942
Obihiro	20.3	W N W	1936	4	27	1892	32.3	S E	2002	10	2	1943
Abashiri	29.8	W N W	1950	11	28	1890	37.5	S S W	2004	9	8	1952
Rumoi	36.7	S W	1951	2	22	1943	43.9	S S W	2004	9	8	1957
Wakkanai	27.0	N	1955	2	21	1938	44.9	W S W	1995	11	8	1940
Nemuro	30.7	N W	1910	2	11	1890	42.2	N N E	2006	10	8	1939
Suttu	49.8	S S E	1952	4	15	1884	53.2	S W	1954	9	26	1942
Urakawa	39.6	W N W	1958	1	10	1927	48.5	W N W	1958	1	10	1949
Aomori	29.0	S W	1991	9	28	1882	53.9	S W	1991	9	28	1937
Morioka	22.2	W N W	1951	4	10	1923	38.6	S W	2004	11	27	1941
Miyako	31.4	W S W	1912	9	23	1883	43.5	S S E	2002	10	2	1941
Sendai	24.0	W N W	1997	3	11	1926	41.2	W N W	1997	3	11	1937
Akita	30.7	S W	1954	9	26	1882	51.4	S S W	1991	9	28	1937
Yamagata	21.4	S W	1957	12	13	1889	32.6	S E	1959	9	27	1941
Sakata	37.7	W S W	1961	9	16	1937	49.0	W S W	1961	9	16	1942
Fukushima	22.9	W	1959	4	10	1889	32.2	W	1979	3	31	1947
Onahama	28.8	S S E	2002	10	1	1910	48.1	S E	2002	10	1	1940
Mito	28.3	N	1961	10	10	1897	44.2	N N E	1939	8	5	1937
Utsunomiya	24.2	N	1938	10	21	1890	42.7	S E	1966	9	25	1937
Maebashi	29.9	N	1900	9	28	1896	40.2	E S E	1966	9	25	1940
Kumagaya	31.7	W	1900	9	28	1896	41.0	S E	1966	9	25	1940
Choshi	48.0	S S E	1948	9	16	1887	52.2	S	2002	10	1	1937
Tokyo	31.0	S	1938	9	1	1875	46.7	S	1938	9	1	1937
Oshima	39.0	S W	1948	9	16	1938	57.0	S	2005	8	25	1940
Hachijojima	44.2	W	1938	10	21	1906	67.8	S	1975	10	5	1937
Yokohama	37.4	N E	1938	9	1	1896	48.7	N E	1938	9	1	1938
Niigata	40.1	S W	1929	4	21	1886	45.5	W S W	1991	9	28	1937
Takada	23.1	S	1959	4	5	1922	42.0	S W	1998	9	22	1937
Aikawa	31.3	N W	1945	9	18	1911	46.2	N W	1961	9	16	1940
Toyama	26.0	S S E	1947	4	1	1939	42.7	S	2004	9	7	1939
Kanazawa	32.8	S S W	1950	9	3	1882	44.3	S S W	2018	9	4	1937
Wajima	31.3	S S W	1991	9	28	1929	57.3	S S W	1991	9	28	1929
Fukui	30.9	S	1950	9	3	1897	48.8	S S E	1991	9	27	1940
Tsuruga	30.4	S E	1950	9	3	1897	47.9	E S E	2018	9	4	1909
Kofu	33.9	E S E	1959	8	14	1894	43.2	E S E	1959	8	14	1937
Nagano	25.8	N W	1916	9	26	1889	31.4	N W	1948	8	23	1937
Matsumoto	24.7	S	1959	9	27	1898	37.6	S	1998	9	22	1939
Mt. Fuji	72.5	W S W	1942	4	5	1932	91.0	S S W	1966	9	25	1965

] : Indicates that the actual wind-speed/gusts exceeded the maximum measureable value shown.
For Mt. Fuji the maximum as of August 2004 is listed.

Record Wind Speeds (From the start of record keeping until 2019) Continued.

Station	Maximum wind speed m/s	Wind direction	Yr.	Mo.	Day	1st year on record	Maximum gusts m/s	Wind direction	Yr.	Mo.	Day	1st year on record
Iida	21.8	N N E	1932	11	14	1897	37.0	S	1959	9	26	1940
Karuizawa	24.5	W	1929	4	21	1925	36.3	N E	1959	8	14	1925
Gifu	32.5	S S E	1959	9	26	1886	44.2	E S E	1959	9	26	1918
Takayama	20.9	S	1921	9	26	1899	36.0	S W	1998	9	22	1942
Shizuoka	24.1	W S W	1959	8	14	1940	40.0	S E	1966	9	25	1940
Hamamatsu	37.0	S	1926	9	4	1887	42.0	S S E	1959	9	26	1941
Nagoya	37.0	S S E	1959	9	26	1890	45.7	S S E	1959	9	26	1937
Tsu	36.8	E S E	1959	9	26	1889	51.3	E S E	1959	9	26	1937
Owase	28.1	S E	1959	9	26	1938	56.1	S E	1990	9	19	1938
Hikone	31.2	S S E	1934	9	21	1893	42.5	S E	1950	9	3	1920
Kyoto	28.0	S	1934	9	21	1880	42.1	S	1934	9	21	1915
Osaka	33.3	S S E	1961	9	16	1883	60.0]	S	1934	9	21	1934
Kobe	33.4	N E	1950	9	3	1897	48.5	S S E	1965	9	10	1937
Nara	25.0]	S S E	1961	9	16	1953	47.2	S	1979	9	30	1953
Wakayama	39.7	S S W	2018	9	4	1879	56.7	S	1961	9	16	1940
Shionomisaki	33.6	W	1921	9	26	1913	59.5	S S E	1990	9	19	1941
Tottori	29.2	N W	1961	9	16	1943	48.6	S	1991	9	27	1943
Matsue	28.5	W	1991	9	27	1940	56.5	W N W	1991	9	27	1940
Hamada	29.6	S S W	1922	3	23	1893	48.9	W S W	1991	9	27	1937
Saigo	26.9	S S W	2004	9	7	1939	55.8	S W	2004	9	7	1939
Okayama	25.8	S E	1896	8	18	1891	41.4	N E	2004	10	20	1940
Hiroshima	36.0	S	1991	9	27	1879	60.2	S	2004	9	7	1937
Shimonoseki	34.2	E	1942	8	27	1883	45.3	E S E	1991	9	27	1937
Tokushima	37.8	S E	1941	8	15	1891	67.0]	S S E	1965	9	10	1940
Takamatsu	24.4	S W	1954	9	26	1941	39.5	N E	1965	9	10	1941
Matsuyama	25.4	S S E	1945	9	17	1890	42.1	S S E	1945	9	17	1937
Kochi	29.2	E	1970	8	21	1886	54.3	E	1970	8	21	1940
Murotomisaki	69.8	W S W	1965	9	10	1920	84.5]	W S W	1961	9	16	1921
Shimizu	35.8	S W	1970	8	21	1941	52.1	E	1975	8	17	1941
Fukuoka	32.5	N	1951	10	14	1890	49.3	S	1987	8	31	1937
Saga	32.7	S	1930	7	18	1890	54.3	S E	1991	9	14	1941
Nagasaki	43.5	S S E	1900	8	24	1878	54.3	S W	1991	9	27	1951
Izuhara	27.1	S S E	2004	8	19	1886	52.1	S E	1987	8	31	1918
Fukue	31.3	S	1987	8	31	1962	55.6	S	1987	8	31	1962
Kumamoto	38.7	E	1902	8	10	1890	52.6	S	1991	9	27	1937
Oita	25.0	W N W	1945	9	18	1887	44.3	S S E	1999	9	24	1940
Miyazaki	39.2	S S E	1945	9	17	1886	57.9	S E	1993	9	3	1937
Kagoshima	39.3	S S E	1942	8	27	1883	58.5	S S E	1996	8	14	1940
Naze	33.7	N	1964	9	24	1896	78.9	E S E	1970	8	13	1937
Naha	49.5	E N E	1949	6	20	1927	73.6	S	1956	9	8	1953
Syowa station (Antarctica)	47.4	E N E	2009	2	20	1957	61.2	N E	1996	5	27	1957

] : Indicates that the actual wind-speed/gusts exceeded the maximum measureable value shown.

Climatological Normal Most Frequent Wind Direction by Month (16 Compass Points, Frequency %)

(1990–2010 Averages)

Station	Jan.	Feb.	Mar.	Apr.	May	Jun.	Jul.	Aug.	Sep.	Oct.	Nov.	Dec.	Annual
Sapporo	NW 14	NW 15	NW 16	NW 16	SE 18	SE 20	SE 23	SE 23	SE 18	SSE 16	SSE 14	NW 14	SE 16
Hakodate	WNW 25	WNW 24	WNW 19	WNW 12	ESE 13	ESE 17	ESE 19	ESE 18	ESE 13	WNW 13	WNW 21	WNW 27	WNW 14
Asahikawa	SSE 13	SSE 13	SSE 12	WNW 12	WNW 13	WNW 17	WNW 15	WNW 12	WNW 9	SSE 12	SSE 15	SSE 16	SSE 11
Kushiro	NNE 23	NNE 23	NNE 16	NNE 14	S 14	S 18	S 17	S 17	NNE 17	NNE 17	NNE 16	NNE 21	NNE 16
Obihiro	WNW 18	WNW 18	WNW 17	WNW 12	E 16	E 19	E 19	E 15	E 10	WNW 12	WNW 18	WNW 19	WNW 11
Abashiri	SW 13	NW 12	NW 10	N 11	SSE 11	N 13	N 13	S 14	S 14	SW 15	SW 15	SW 16	SW 10
Rumoi	ESE 34	ESE 32	ESE 29	ESE 33	ESE 31	ESE 30	ESE 34	ESE 40	ESE 42	ESE 40	ESE 29	ESE 29	ESE 34
Wakkanai	W 18	W 15	W 15	SSW 15	SSW 17	SSW 15	E 16	SW 14	WSW 12	WSW 17	W 25	W 26	W 12
Nemuro	NW 16	NNW 15	NNW 13	SW 10	S 12	S 13	SE 14	S 14	S 11	SSW 12	SW 12	NW 15	S 9
Suttu	NW 28	NW 27	NW 21	SSE 22	SSE 26	SSE 32	SSE 36	SSE 32	SSE 21	SSE 15	NW 22	NW 29	SSE 20
Urakawa	WNW 23	WNW 27	WNW 28	WNW 22	WNW 19	WNW 16	ESE 16	ESE 15	NE 13	WNW 15	WNW 21	NW 25	WNW 19
Aomori	SW 26	SW 24	SW 21	SW 17	SW 13	NNW 11	N 11	SW 10	SW 14	SW 19	SW 24	SW 25	SW 18
Morioka	S 11	S 12	S 14	S 17	S 22	S 24	S 25	S 23	S 17	S 13	S 13	S 11	S 17
Miyako	WSW 27	WSW 24	WSW 21	WSW 18	WSW 15	NNE 18	NNE 19	WSW 14	WSW 21	WSW 26	WSW 29	WSW 29	WSW 20
Sendai	NNW 17	WNW 16	WNW 15	SE 12	SE 17	SE 21	SE 21	SE 20	NNW 16	NNW 19	NNW 17	NNW 16	NNW 13
Akita	NW 16	WNW 15	SE 17	SE 22	SE 24	SE 25	SE 26	SE 29	SE 28	SE 27	SE 24	SE 18	SE 23
Yamagata	SSW 15	SSW 15	SSW 13	SSW 9	N 10	N 11	N 11	ESE 10	N 10	SSW 9	SSW 13	SSW 15	SSW 10
Sakata	WNW 25	WNW 23	WNW 19	ESE 16	ESE 18	ESE 19	ESE 18	ESE 22	ESE 23	SE 22	SE 21	WNW 21	SE 16
Fukushima	WNW 18	WNW 18	WNW 17	NE 12	NE 16	NE 23	NE 22	NE 19	NE 14	NE 8	WNW 13	WNW 16	NE 12
Onahama	NNW 25	NNW 22	NNW 19	N 16	S 16	S 16	S 20	S 18	N 21	N 25	NNW 27	NNW 27	N 17
Mito	NNW 24	NNW 20	N 19	N 14	E 13	E 19	E 20	ENE 17	ENE 14	NNW 21	NNW 26	NNW 27	NNW 15
Utsunomiya	NNE 17	NNE 18	NNE 18	NNE 17	NNE 14	NNE 14	NNE 14	NNE 14	NNE 22	NNE 23	NNE 20	NNE 19	NNE 18
Maebashi	NNW 35	NNW 34	NNW 33	NNW 26	ESE 20	ESE 23	ESE 23	ESE 24	NW 20	NW 26	NW 34	NNW 34	NNW 24
Kumagaya	NW 27	NW 26	NW 23	NW 15	SE 11	E 13	E 15	E 15	E 12	NW 19	WNW 23	WNW 25	NW 16
Choshi	WNW 19	WNW 15	NNE 15	SSW 16	SSW 21	SSW 19	SSW 27	SSW 24	NNE 22	NNE 24	NNE 17	WNW 18	NNE 15

Tokyo	NNW 39	NNW 33	NNW 26	NNW 15	SW 14	SW 13	SW 15	SW 15	NNW 13	NNW 25	NNW 33	NNW 36	NNW 20
Oshima	NE 21	NE 21	NE 23	NE 17	SSW 18	SSW 19	SSW 27	SSW 22	NE 26	NE 35	NE 29	NE 21	NE 20
Hachijojima	W 40	W 34	W 28	W 22	SW 17	SW 25	SW 29	ENE 17	ENE 23	ENE 29	W 26	W 39	W 22
Yokohama	N 44	N 39	N 32	N 21	N 14	SW 14	SW 19	SW 17	N 24	N 37	N 42	N 43	N 27
Niigata	WNW 16	WNW 13	W 10	WSW 11	NNE 10	NNE 12	NNE 10	S 12	SSE 11	S 14	S 16	S 15	S 12
Takada	S 20	S 18	S 18	S 16	S 13	N 13	S 12	S 17	S 18	S 24	S 25	S 22	S 18
Aikawa	NW 28	NW 23	NW 17	SE 13	NNW 15	NNW 16	NNW 14	SE 14	SE 14	ESE 13	NW 13	NW 21	NW 14
Toyama	SSW 25	SW 23	SW 18	SW 15	NNE 17	NNE 20	NNE 16	NNE 18	NNE 16	SW 22	SW 25	SW 26	SW 18
Kanazawa	SSW 13	ENE 11	ENE 14	ENE 13	ENE 14	ENE 15	SW 14	ENE 15	ENE 21	ENE 18	ENE 13	SSW 14	ENE 14
Wajima	SSW 21	SSW 24	SSW 27	SSW 30	SSW 28	SSW 27	SSW 29	SSW 31	SSW 31	SSW 37	SSW 32	SSW 26	SSW 28
Fukui	S 17	S 15	NNW 15	S 14	S 14	NNW 13	S 15	S 17	S 15	S 18	S 19	S 18	S 16
Tsuruga	SSE 17	N 18	SSE 19	SSE 25	SSE 31	SSE 30	SSE 33	SSE 36	SSE 28	SSE 24	SSE 23	SSE 21	SSE 25
Kofu	NNW 11	NNW 13	NW 12	SW 13	SW 17	SW 19	SW 19	SW 19	SW 13	SW 9	WNW 9	NNW 10	SW 12
Nagano	E 18	E 18	E 14	WSW 11	WSW 15	WSW 14	WSW 14	WSW 16	WSW 15	WSW 13	E 12	E 15	E 12
Matsumoto	N 14	N 16	N 16	N 15	N 14	NNW 12	S 12	N 11	NNW 13	N 16	N 15	N 14	N 14
Iida	WSW 13	WSW 12	W 12	W 13	S 15	S 16	S 15	S 15	S 11	WSW 9	WSW 10	WSW 13	WSW 12
Karuizawa	WSW 23	WSW 23	WSW 19	WSW 13	ENE 17	NE 19	NE 17	NE 20	NE 21	NE 16	WSW 15	WSW 22	WSW 14
Gifu	NW 20	NW 22	NW 21	NW 15	WNW 12	W 11	W 11	SSW 11	WNW 12	NW 17	NW 19	NW 18	NW 14
Takayama	NW 13	NNW 15	NNW 16	NNW 14	NNW 10	NNW 9	NNW 9	NW 10	NW 11	NW 11	NNW 12	NW 12	NNW 12
Shizuoka	WNW 12	WNW 11	NE 12	NE 12	S 13	S 14	S 15	S 15	NE 13	NE 13	NE 11	WNW 14	NE 11
Hamamatsu	WNW 42	WNW 40	WNW 31	WNW 19	NE 12	WSW 14	WSW 16	E 11	NE 14	NE 19	WNW 28	WNW 41	WNW 21
Nagoya	NNW 25	NNW 26	NNW 25	NNW 18	NNW 13	SSE 16	SSE 18	SSE 19	NNW 18	NNW 25	NNW 27	NNW 25	NNW 19
Tsu	NW 25	NW 30	NW 30	NW 21	NW 16	ESE 14	SE 14	SE 16	NW 19	NW 24	NW 24	NW 23	NW 21
Owase	W 28	W 25	W 23	W 22	W 20	W 17	ENE 19	W 19	W 23	W 27	W 28	W 30	W 23
Hikone	NW 15	NW 22	NW 24	NW 19	NW 18	NW 16	NW 14	NW 14	NW 18	NW 19	NW 17	SSE 16	NW 17
Kyoto	W 10	N 11	N 13	N 11	NNE 11	NNE 11	NNE 10	NNE 11	N 13	N 14	N 12	N 10	N 11
Osaka	W 16	NNE 13	NNE 17	NNE 16	NNE 15	NNE 15	WSW 18	WSW 15	NNE 21	NNE 23	NNE 18	W 15	NNE 15
Kobe	W 19	W 15	N 14	ENE 11	ENE 12	WSW 14	WSW 17	SW 12	ENE 15	N 18	N 15	W 18	N 12
Nara	S 12	NNW 13	NNW 16	NNW 13	NNE 15	NNE 17	NNE 15	NE 17	NNE 18	NNE 16	NNE 12	S 12	NNE 13
Wakayama	ENE 19	ENE 17	ENE 15	ENE 15	ENE 15	WSW 13	WSW 14	WSW 13	ENE 18	ENE 26	ENE 27	ENE 24	ENE 18

Shionomisaki	NW 27	NW 22	NE 15	NE 14	NE 19	W 19	W 28	W 17	NE 25	NE 28	NE 21	NW 25	NE 17
Tottori	ESE 27	ESE 24	ESE 23	ESE 23	ESE 21	ESE 18	ESE 17	ESE 20	ESE 24	ESE 32	ESE 32	ESE 28	ESE 24
Matsue	W 22	W 17	W 15	W 17	W 19	E 21	W 22	E 17	E 18	E 14	W 16	W 21	W 16
Hamada	W 18	ENE 18	ENE 22	ENE 22	ENE 20	ENE 18	SW 21	ENE 22	ENE 30	ENE 32	ENE 25	W 17	ENE 21
Saigo	NW 24	NW 24	NW 21	NW 16	NW 12	ENE 13	WSW 17	NE 14	NE 17	NW 25	NW 24	NW 24	NW 18
Okayama	W 19	W 14	N 10	ENE 11	ENE 14	ENE 13	ENE 13	ENE 13	N 13	N 15	N 12	W 17	ENE 10
Hiroshima	N 30	NNE 30	N 29	N 25	N 22	SW 18	SSW 21	NNE 19	NNE 37	NNE 42	NNE 38	N 32	NNE 28
Shimonoseki	E 14	E 16	E 17	E 19	E 21	E 23	E 20	ESE 23	E 18	E 18	E 19	E 16	E 18
Tokushima	WNW 39	WNW 31	WNW 21	WNW 15	SSE 15	SSE 15	SSE 16	SSE 19	WNW 18	WNW 30	WNW 34	WNW 40	WNW 22
Takamatsu	W 18	W 16	WSW 12	WSW 12	WSW 11	ENE 12	WSW 11	SW 10	SW 10	SW 15	SW 15	W 16	WSW 12
Matsuyama	WNW 15	WNW 13	WNW 10	WNW 11	E 11	E 12	E 13	E 16	E 14	ESE 14	ESE 13	WNW 14	ESE 11
Kochi	W 27	W 25	W 24	W 22	W 22	W 19	W 15	W 19	W 23	W 28	W 28	W 28	W 23
Murotomisaki	WNW 24	WNW 20	NE 23	NE 20	NE 21	NE 19	W 22	NE 15	NE 29	NE 35	NE 27	WNW 24	NE 21
Shimizu	N 17	N 15	NNE 14	NNE 14	NNE 14	W 17	W 24	E 12	NNE 18	NNE 21	N 19	N 18	NNE 14
Fukuoka	SE 18	SE 16	N 15	N 18	N 17	N 18	N 16	N 17	N 24	N 18	SE 19	SE 21	SE 15
Saga	NNW 12	NNW 13	NNW 12	NNW 12	NNW 10	S 13	S 20	S 12	NE 14	NE 15	NNW 15	NNW 14	NNW 11
Nagasaki	NNE 14	NNE 14	NNE 15	SW 12	SW 17	SW 24	SW 28	SW 17	NNE 19	NNE 17	NNE 13	NNE 12	NNE 12
Izuhara	NNW 34	NNW 32	NNW 26	NNW 21	NNW 19	NNW 20	NNW 15	NNW 20	NNW 36	NNW 36	NNW 36	NNW 34	NNW 27
Fukue	NW 26	NW 21	NW 15	N 12	N 10	SSW 15	SSW 24	SSW 14	N 18	N 20	NNW 16	NW 21	N 12
Kumamoto	NW 15	NNW 14	NNW 15	NNW 12	SW 16	SW 19	SW 22	SW 16	NNW 16	NNW 18	NNW 16	N 14	NNW 13
Oita	S 17	S 17	NNW 16	S 17	S 18	S 16	S 15	S 19	S 18	S 21	S 24	S 22	S 18
Miyazaki	W 30	W 25	W 20	WNW 18	WNW 16	WSW 14	WSW 18	WNW 14	WNW 19	WNW 25	WNW 30	WNW 30	WNW 20
Kagoshima	NNW 34	NNW 27	NNW 21	NNW 14	NNW 12	SSE 9	SSE 11	NE 10	NNW 13	NNW 21	NNW 29	NNW 35	NNW 19
Naze	NNW 24	NNW 22	S 17	SSE 17	SSE 19	S 24	S 19	SSE 18	SSE 18	SSE 17	NNW 19	NNW 23	SSE 16
Naha	NNE 21	N 20	N 15	ESE 12	E 11	SSW 23	SE 15	SE 15	ESE 13	NNE 30	NNE 30	NNE 26	NNE 15
Mt. Fuji													
Syowa station (Antarctica)	NE 26	NE 30	ENE 26	ENE 26	NE 23	NE 21	NE 22	NE 22	NE 19	NE 25	NE 25	NE 26	NE 24

Winds have not been observed at Mt. Fuji.

Top 10 Air Temperature, Precipitation, Wind Speed

(As of December 31, 2019)

Rank	Station	Record	Yr.	Mo.	Day	1st year on record	Rank	Station	Record	Yr.	Mo.	Day	1st year on record
High temperature							**Low temperature**						
1	Kumagaya	41.1℃	2018	7	23	1896	1	Asahikawa	-41.0℃	1902	1	25	1888
2	Yamagata	40.8	1933	7	25	1889	2	Obihiro	-38.2	1902	1	26	1892
3	Kofu	40.7	2013	8	10	1894	3	Kutchan	-35.7	1945	1	27	1944
4	Takada	40.3	2019	8	14	1922	4	Abashiri	-29.2	1902	1	25	1889
4	Nagoya	40.3	2018	8	3	1890	5	Sapporo	-28.5	1929	2	1	1876
6	Uwajima	40.2	1927	7	22	1922	6	Kushiro	-28.3	1922	1	28	1910
7	Sakata	40.1	1978	8	3	1937	7	Omu	-27.5	1978	2	18	1942
8	Maebashi	40.0	2001	7	24	1896	8	Kitamiesashi	-26.4	1947	2	12	1942
9	Niigata	39.9	2018	8	23	1881	8	Haboro	-26.4	1923	1	27	1921
9	Hita	39.9	2018	8	13	1942	10	Takayama	-25.5	1939	2	11	1899
Daily precipitation							**Hourly precipitation**						
1	Owase	806.0mm	1968	9	26	1938	1	Shimizu	150.0mm	1944	10	17	1940
2	Yonagunijima	765.0	2008	9	13	1956	2	Murotomisaki	149.0	2006	11	26	1925
3	Kochi	628.5	1998	9	24	1886	3	Shionomisaki	145.0	1972	11	14	1937
4	Naze	622.0	2010	10	20	1896	4	Yamaguchi	143.0	2013	7	28	1966
5	Hikone	596.9	1896	9	7	1893	5	Choshi	140.0	1947	8	28	1912
6	Miyazaki	587.2	1939	10	16	1886	6	Miyazaki	139.5	1995	9	30	1925
7	Kumejima	577.5	2001	9	12	1958	7	Owase	139.0	1972	9	14	1938
8	Yakushima	557.3	1942	8	27	1937	8	Miyakojima	138.0	1970	4	19	1938
9	Akune	555.5	1971	7	23	1939	9	Unzendake (Mt. Unzen)	134.5	2015	8	25	1941
10	Oshima	525.5	2013	10	16	1938	10	Hachijojima	129.5	1999	9	4	1937
							10	Kochi	129.5	1998	9	24	1937
Maximum wind speed and direction							**Maximum gust speed and direction**						
1	Murotomisaki	69.8m/s WSW	1965	9	10	1920	1	Miyakojima	85.3m/s NE	1966	9	5	1938
2	Miyakojima	60.8 NE	1966	9	5	1938	2	Murotomisaki	84.5] WSW	1961	9	16	1921
3	Unzendake (Mt. Unzen)	60.0 ESE	1942	8	27	1924	3	Yonagunijima	81.1 SE	2015	9	28	1957
4	Yonagunijima	54.6 SE	2015	9	28	1956	4	Naze	78.9 ESE	1970	8	13	1937
5	Ishigakishima	53.0 SE	1977	7	31	1897	5	Naha	73.6 S	1956	9	8	1953
6	Yakushima	50.2 ENE	1964	9	24	1937	6	Uwajima	72.3 W	1964	9	25	1939
7	Suttu	49.8 SSE	1952	4	15	1884	7	Ishigakishima	71.0 SSW	2015	8	23	1941
8	Naha	49.5 ENE	1949	6	20	1927	8	Iriomoteshima	69.9 NE	2006	9	16	1972
9	Irouzaki	48.8 E	1959	8	14	1939	9	Yakushima	68.5 ENE	1964	9	24	1937
10	Choshi	48.0 SSE	1948	9	16	1887	10	Hachijojima	67.8 S	1975	10	5	1937

Compiled based on the maximum values at each location from among the meteorological/weather observation stations throughout Japan. "] " attached to a value indicates that the actual value exceeded the maximum observable value listed. "×" indicates no data available.

Climatological Normal Monthly Relative Humidity (%)

(1981-2010 Averages)

Station	Jan.	Feb.	Mar.	Apr.	May	Jun.	Jul.	Aug.	Sep.	Oct.	Nov.	Dec.	Annual
Sapporo	70	69	66	62	66	72	76	75	71	67	67	69	69
Hakodate	73	71	68	67	73	79	82	81	76	72	71	72	74
Asahikawa	80	79	69	67	66	74	78	79	77	78	81	82	76
Kushiro	69	67	68	77	79	87	88	88	82	75	69	67	76
Obihiro	70	68	66	66	69	79	83	82	79	73	68	68	73
Abashiri	73	74	72	69	73	80	83	81	77	70	68	70	74
Rumoi	77	75	72	71	75	81	84	82	78	73	73	75	76
Wakkanai	72	72	71	75	79	85	86	84	75	68	67	70	75
Nemuro	71	73	75	78	82	89	91	89	84	75	69	68	79
Suttu	70	69	67	69	74	82	85	84	78	71	68	69	74
Urakawa	65	67	71	77	83	89	91	89	83	74	68	64	77
Aomori	78	76	69	66	70	78	80	79	76	73	72	77	74
Morioka	73	70	67	65	69	75	80	79	80	77	75	74	74
Miyako	59	62	64	66	74	84	87	85	83	76	67	62	72
Sendai	66	64	62	64	71	80	83	81	78	72	68	66	71
Akita	73	71	67	67	72	75	79	76	75	72	72	73	73
Yamagata	81	77	69	62	65	72	77	75	77	77	78	80	74
Sakata	71	70	67	68	71	75	79	76	75	72	71	71	72
Fukushima	68	64	61	59	63	72	77	75	76	72	69	68	69
Onahama	58	59	63	69	77	83	86	84	80	74	68	62	72
Mito	64	64	67	71	75	81	83	81	81	79	75	69	74
Utsunomiya	62	60	61	64	70	77	80	78	77	74	71	66	70
Maebashi	54	53	54	56	62	71	74	73	74	68	62	57	63
Kumagaya	54	54	57	61	66	73	76	75	76	70	65	59	66
Choshi	62	63	68	75	81	88	90	87	84	76	71	65	76
Tokyo	52	53	56	62	69	75	77	73	75	68	65	56	65
Oshima	64	65	69	75	79	84	87	86	83	77	73	67	76
Hachijojima	71	68	69	80	86	96	92	86	86	82	73	70	80
Yokohama	53	54	60	65	70	78	78	76	76	71	64	56	67
Niigata	71	74	70	66	68	73	78	73	73	72	72	75	72
Takada	78	76	72	67	71	77	80	77	79	77	77	76	75
Aikawa	69	67	66	67	72	78	81	77	73	69	68	69	71
Toyama	82	79	73	69	72	79	81	77	79	77	77	80	77
Kanazawa	75	72	67	65	69	75	77	73	74	71	71	72	72
Wajima	74	73	71	70	73	79	81	79	79	76	75	75	76
Fukui	81	78	72	67	69	74	76	72	76	76	78	80	75
Tsuruga	74	72	68	66	69	75	76	72	75	72	71	72	72
Kofu	57	54	56	58	64	71	73	71	72	71	68	61	65
Nagano	78	74	68	61	64	70	74	72	74	75	76	77	72
Matsumoto	67	66	64	59	62	70	72	71	75	75	71	69	68
Mt. Fuji			58	60	61	70	79	73	68	53	50	47	

Climatological Normal Monthly Relative Humidity (1981-2010 Averages) Continued.

Station	Jan.	Feb.	Mar.	Apr.	May	Jun.	Jul.	Aug.	Sep.	Oct.	Nov.	Dec.	Annual
Iida	66	62	58	61	66	71	75	71	76	77	75	71	69
Karuizawa	75	74	72	70	76	85	87	87	89	85	79	76	80
Gifu	67	63	60	60	65	71	74	70	71	67	67	68	67
Takayama	81	78	73	68	69	74	78	76	79	80	81	82	77
Shizuoka	57	57	63	66	71	78	79	77	75	70	66	60	68
Hamamatsu	56	56	61	64	70	78	75	76	71	70	64	60	67
Nagoya	64	61	59	60	65	71	74	70	71	68	66	65	66
Tsu	62	62	63	65	70	75	77	75	73	69	66	64	69
Owase	60	60	64	68	74	81	82	80	79	75	70	64	71
Hikone	75	75	72	70	72	76	78	73	75	74	74	75	74
Kyoto	66	65	62	59	62	67	70	66	68	68	68	68	66
Osaka	61	60	59	59	62	68	70	66	67	65	64	62	64
Kobe	62	63	61	62	66	72	75	71	70	64	63	61	66
Nara	69	69	68	65	69	75	77	74	77	77	76	72	72
Wakayama	61	61	61	61	66	72	73	70	70	67	65	63	66
Shionomisaki	57	58	62	70	75	83	87	83	77	71	64	60	71
Tottori	76	74	70	67	69	74	77	74	78	76	75	74	74
Matsue	75	74	72	70	72	78	81	77	79	76	76	75	75
Hamada	65	66	68	69	74	81	82	79	80	74	69	65	73
Saigo	73	72	71	72	75	80	83	80	79	75	73	72	75
Okayama	65	64	62	60	64	69	73	69	70	69	69	67	67
Hiroshima	68	67	64	63	66	72	74	71	70	68	69	69	68
Shimonoseki	62	63	65	68	71	77	78	75	73	67	66	63	69
Tokushima	60	60	61	63	67	74	76	72	71	67	65	62	67
Takamatsu	63	63	64	63	66	72	74	72	73	71	69	66	68
Matsuyama	63	63	64	63	66	72	72	69	71	68	67	64	67
Kochi	60	59	62	64	70	77	78	75	73	68	67	63	68
Murotomisaki	60	61	66	71	78	86	89	85	80	71	67	61	73
Shimizu	57	58	62	67	74	82	83	81	76	67	63	58	69
Fukuoka	63	63	65	65	68	74	75	72	73	67	67	64	68
Saga	69	66	66	66	67	73	77	72	71	69	70	69	70
Nagasaki	66	64	66	68	72	79	80	75	73	67	67	66	70
Izuhara	59	61	65	67	74	82	83	81	77	69	67	60	71
Fukue	66	66	68	72	76	82	84	81	77	70	68	67	73
Kumamoto	70	67	67	66	68	75	77	73	72	69	72	71	71
Oita	62	63	66	67	70	76	77	75	75	71	69	64	69
Miyazaki	65	66	69	70	73	82	76	79	80	75	73	70	73
Kagoshima	65	65	66	68	71	76	75	73	71	67	67	67	69
Naze	69	69	71	73	76	79	77	78	78	74	71	69	74
Naha	67	70	73	76	79	83	78	78	76	71	69	66	74
Syowa station (Antarctica)	67	68	71	72	67	65	66	64	64	69	68	68	67

Climatological Normal Monthly Hours of Sunshine (h)

(1981–2010 Averages)

Station	Jan.	Feb.	Mar.	Apr.	May	Jun.	Jul.	Aug.	Sep.	Oct.	Nov.	Dec.	Annual
Sapporo	92.5	104.0	146.6	176.5	198.4	187.8	164.9	171.0	160.5	152.3	100.0	85.9	1740.4
Hakodate	103.4	119.3	157.6	187.7	193.5	173.3	135.6	149.5	158.1	167.5	109.7	92.9	1748.0
Asahikawa	73.8	105.8	149.6	167.1	197.6	189.3	161.8	147.3	142.2	132.0	64.0	60.3	1590.9
Kushiro	182.0	181.9	200.6	181.9	188.3	129.3	107.4	127.1	149.7	180.9	166.6	173.6	1969.5
Obihiro	183.4	190.1	217.8	194.5	192.3	152.8	117.6	128.9	143.0	175.0	166.7	171.3	2033.2
Abashiri	114.3	139.4	172.4	177.8	189.0	174.0	168.7	172.1	165.2	160.1	121.3	115.0	1869.3
Rumoi	51.3	72.6	131.5	171.0	199.4	186.2	169.5	174.3	171.2	128.8	52.6	32.4	1540.9
Wakkanai	45.6	80.3	138.6	171.8	185.6	166.3	146.8	148.5	177.1	135.9	57.6	30.1	1484.4
Nemuro	152.8	164.8	190.8	177.4	176.2	135.6	112.6	127.7	145.5	167.7	146.5	146.0	1843.6
Suttu	29.7	46.6	112.6	170.7	192.5	181.4	155.9	162.8	156.6	127.7	55.4	29.4	1420.7
Urakawa	139.5	162.3	194.2	185.7	189.3	145.6	109.5	137.5	159.5	174.9	122.1	114.1	1834.2
Aomori	51.3	69.8	130.5	182.3	201.0	179.6	159.5	180.3	158.4	149.7	87.6	52.8	1602.7
Morioka	116.9	127.5	160.4	173.7	185.4	154.7	128.5	149.1	123.7	145.8	116.9	101.6	1684.1
Miyako	161.0	152.9	178.6	189.3	181.2	149.4	133.8	160.6	128.0	155.2	147.3	147.4	1882.6
Sendai	148.1	151.8	177.0	188.5	185.2	133.8	119.5	144.4	121.2	148.6	139.6	138.6	1796.1
Akita	39.9	62.5	124.7	170.4	182.0	176.2	150.3	193.0	153.8	145.4	82.7	45.1	1526.0
Yamagata	84.8	98.9	140.3	176.1	191.5	158.8	143.7	178.4	128.7	132.1	99.2	80.7	1613.3
Sakata	39.4	59.2	117.2	172.4	191.2	178.6	164.0	208.2	150.7	141.5	81.9	43.9	1552.1
Fukushima	132.0	142.3	174.2	186.4	187.5	136.6	123.6	152.5	114.2	135.8	128.3	125.2	1738.8
Onahama	189.8	177.9	185.5	188.8	188.6	142.1	147.9	185.7	139.5	152.7	160.5	183.6	2042.5
Mito	186.3	167.8	173.9	176.6	176.4	129.4	140.9	175.6	127.9	141.5	148.4	177.2	1921.7
Utsunomiya	204.8	186.2	187.9	179.5	166.9	112.1	114.1	138.9	112.2	145.0	164.5	199.1	1911.3
Maebashi	210.1	194.3	206.2	199.7	192.6	132.2	139.0	165.1	126.2	161.0	179.3	205.1	2110.9
Kumagaya	210.6	192.2	196.0	190.2	182.0	125.5	136.9	166.5	120.8	148.2	169.9	203.2	2042.1
Choshi	173.5	154.4	161.2	176.9	178.6	135.8	165.0	220.6	150.3	140.5	138.3	165.0	1959.9
Tokyo	184.5	165.8	163.1	176.9	167.8	125.4	146.4	169.0	120.9	131.0	147.9	178.0	1876.7
Oshima	151.7	142.8	148.6	168.5	171.6	127.3	142.6	190.6	135.9	137.2	138.7	149.3	1804.9
Hachijojima	85.7	83.8	122.3	133.5	135.6	91.8	118.5	170.0	134.2	106.8	108.1	108.2	1398.5
Yokohama	186.4	164.0	159.5	175.2	177.1	131.7	162.9	206.3	130.7	141.0	149.3	180.4	1964.4
Niigata	58.2	78.6	133.2	169.8	202.1	168.5	160.1	211.1	162.8	140.1	89.9	60.5	1631.9
Takada	65.4	79.6	120.7	181.1	196.3	150.9	153.8	195.0	129.4	134.5	104.1	80.0	1591.7
Aikawa	49.9	68.3	132.0	178.9	195.0	171.4	164.2	215.7	152.4	152.4	93.9	53.7	1631.2
Toyama	68.1	86.3	131.3	174.9	191.1	150.2	147.1	201.3	133.1	142.7	102.8	75.8	1612.1
Kanazawa	63.5	84.1	141.3	185.5	202.3	152.6	158.9	221.5	144.1	150.4	104.1	72.5	1680.8
Wajima	43.3	64.5	127.4	187.5	201.9	157.2	156.1	206.8	138.2	142.0	88.4	51.6	1564.9
Fukui	64.2	85.4	129.9	174.6	185.0	142.8	149.9	206.8	142.0	151.8	110.0	77.0	1619.4
Tsuruga	62.3	76.2	124.4	167.1	176.6	136.4	146.5	201.6	139.5	145.8	106.5	77.3	1560.1
Kofu	204.8	189.9	198.7	202.0	196.3	148.9	164.1	197.3	142.2	160.9	176.6	201.3	2183.0
Nagano	127.2	131.3	168.5	198.3	207.3	165.2	168.8	204.3	141.7	152.4	139.1	135.6	1939.6
Matsumoto	170.7	163.5	185.0	202.1	209.0	163.6	171.3	205.4	141.8	159.9	159.2	166.0	2097.5
Mt. Fuji													

The amount of sunshine had not been observed at Mt. Fuji before 2003.

Climatological Normal Monthly Hours of Sunshine (1981-2010 Averages) Continued.

Station	Jan.	Feb.	Mar.	Apr.	May	Jun.	Jul.	Aug.	Sep.	Oct.	Nov.	Dec.	Annual
Iida	176.7	164.1	178.4	186.6	193.8	151.4	168.4	200.1	140.2	151.8	144.8	162.0	2018.2
Karuizawa	173.3	164.8	186.0	195.1	191.3	134.0	136.9	165.3	118.9	142.1	157.5	173.5	1940.0
Gifu	160.3	163.6	188.3	196.0	199.0	159.4	167.0	202.2	157.8	174.2	157.3	160.2	2085.1
Takayama	95.6	112.8	150.9	174.6	181.3	143.0	146.4	180.5	124.1	125.8	98.9	89.0	1623.7
Shizuoka	201.6	181.0	179.1	185.1	183.3	132.1	154.2	201.4	148.9	160.9	170.3	201.1	2099.0
Hamamatsu	200.4	182.4	198.6	193.6	191.9	143.8	179.3	198.1	162.6	162.6	168.3	197.5	2179.2
Nagoya	170.1	170.0	189.1	196.6	197.5	149.9	164.3	200.4	151.0	169.0	162.7	172.2	2091.6
Tsu	166.5	146.0	179.4	189.0	185.3	142.6	188.8	214.8	168.6	164.8	163.3	179.9	2089.0
Owase	179.4	168.3	184.0	183.9	173.7	129.5	155.7	175.3	130.5	142.4	151.9	177.9	1946.9
Hikone	98.5	110.9	153.3	182.1	191.6	153.7	167.6	209.1	156.5	163.3	130.1	109.3	1825.8
Kyoto	123.2	117.4	146.8	175.4	180.9	138.3	142.3	182.7	136.8	157.4	138.1	135.8	1775.1
Osaka	142.6	135.4	159.5	188.6	194.3	156.2	182.1	216.9	156.7	163.9	148.5	151.6	1996.4
Kobe	154.9	141.9	164.0	194.7	190.4	170.1	194.1	228.3	159.6	170.0	142.7	162.0	2072.6
Nara	116.7	115.5	147.4	180.3	184.8	143.5	162.7	205.4	150.3	154.5	134.5	127.3	1823.0
Wakayama	134.8	141.0	171.4	195.4	202.3	163.8	207.4	237.9	169.6	171.0	145.4	142.2	2088.8
Shionomisaki	191.0	188.3	185.9	191.9	181.5	135.9	207.4	234.6	172.0	162.8	187.4	200.7	2240.1
Tottori	70.2	79.5	124.3	177.3	197.4	158.2	163.0	206.8	139.9	148.5	108.8	89.5	1663.2
Matsue	68.2	84.7	132.8	180.6	202.2	161.3	166.7	202.1	142.9	158.0	112.7	84.0	1696.2
Hamada	63.6	84.7	139.3	181.7	201.5	161.5	177.4	215.5	159.1	167.6	116.0	79.3	1747.2
Saigo	72.5	88.6	139.1	191.3	212.1	170.8	160.8	210.9	147.8	160.2	109.7	81.9	1745.9
Okayama	150.6	142.3	169.3	190.3	200.7	160.0	171.9	207.0	156.6	173.5	151.9	156.7	2030.7
Hiroshima	137.2	139.7	169.0	190.1	206.2	161.4	179.5	211.2	165.3	181.8	151.6	149.4	2042.3
Shimonoseki	96.6	114.2	154.7	185.9	200.3	154.6	175.1	209.5	162.2	177.7	134.8	110.0	1880.5
Tokushima	157.5	150.2	171.2	192.9	196.8	157.9	195.2	230.4	159.9	166.7	150.8	163.3	2092.9
Takamatsu	141.2	141.6	168.2	192.5	203.3	165.8	195.0	225.2	159.6	169.3	145.2	148.6	2053.9
Matsuyama	125.8	138.9	166.7	189.0	198.5	160.2	192.9	221.9	165.0	177.3	144.4	136.5	2017.1
Kochi	188.4	173.1	184.1	191.7	185.6	142.4	175.7	205.8	162.0	182.4	170.3	192.7	2154.2
Murotomisaki	174.9	166.1	184.4	192.4	190.7	144.5	186.2	230.5	173.5	178.7	167.9	181.9	2174.2
Shimizu	180.6	171.2	181.2	191.1	186.6	143.9	205.5	236.3	177.8	183.9	166.7	183.1	2210.1
Fukuoka	102.1	121.0	149.8	181.6	194.6	149.4	173.5	202.1	162.8	177.1	136.3	116.7	1867.0
Saga	124.3	138.0	156.3	183.2	191.1	140.1	170.2	206.7	176.3	189.9	150.3	142.7	1969.0
Nagasaki	102.8	119.7	148.5	174.7	184.4	135.3	178.7	210.7	172.8	181.4	137.9	119.1	1866.1
Izuhara	146.1	144.0	152.7	180.3	191.6	143.7	133.9	166.8	135.4	163.6	148.3	156.1	1860.8
Fukue	81.2	103.2	141.9	172.6	184.5	133.7	160.3	199.2	166.6	179.3	130.5	103.1	1756.1
Kumamoto	132.6	139.5	158.3	181.4	187.2	141.0	184.5	211.0	175.9	189.7	153.0	147.5	2001.6
Oita	150.1	148.9	164.8	186.0	187.3	146.2	183.6	207.3	154.2	168.0	149.0	156.6	2001.8
Miyazaki	182.8	167.3	175.8	179.0	173.3	133.6	205.5	208.6	155.5	176.9	168.1	189.6	2116.1
Kagoshima	132.7	135.1	148.8	167.5	174.2	121.8	190.9	206.2	176.7	186.7	155.2	149.8	1935.6
Naze	60.3	58.6	82.4	107.5	126.7	121.7	209.7	177.1	143.0	112.8	86.3	73.7	1359.9
Naha	94.2	87.1	108.3	123.8	145.8	163.6	238.8	215.0	188.9	169.6	123.0	115.6	1774.0
Syowa station (Antarctica)	375.7	203.2	120.1	58.0	17.7	0.0	4.8	64.1	136.5	191.0	316.0	434.6	1925.9

Climatological Normal Number of Days per Month with less than 0.1 Hours of Sunshine

(1981–2010 Averages)

Station	Jan.	Feb.	Mar.	Apr.	May	Jun.	Jul.	Aug.	Sep.	Oct.	Nov.	Dec.	Annual
Sapporo	3.7	2.8	2.5	2.8	3.5	3.5	4.0	3.7	3.3	2.2	3.3	4.1	39.6
Hakodate	3.1	2.5	3.2	3.7	5.0	5.8	7.0	6.0	4.7	3.1	3.2	4.0	51.3
Asahikawa	4.3	2.8	3.1	3.2	3.5	3.3	4.1	4.1	3.3	2.6	5.5	5.5	45.3
Kushiro	2.9	2.6	3.3	5.2	6.9	8.8	9.8	9.0	6.6	3.8	3.6	3.2	65.6
Obihiro	3.0	2.4	2.7	4.1	5.9	7.9	9.4	8.5	7.1	3.5	3.5	3.5	61.4
Abashiri	4.7	3.3	4.0	3.8	5.0	5.6	6.0	5.0	4.0	3.1	3.8	3.9	52.2
Rumoi	9.0	4.3	3.3	3.5	3.8	4.3	5.2	3.7	2.8	2.7	7.6	11.5	61.7
Wakkanai	8.4	4.3	4.5	4.3	5.2	6.2	7.1	6.0	3.4	3.4	6.6	10.2	69.6
Nemuro	3.5	3.0	4.4	4.8	6.6	8.1	9.6	7.8	6.7	3.7	3.7	3.7	65.7
Suttu	10.4	5.6	3.8	3.8	4.4	4.6	4.3	4.5	3.6	2.8	7.0	11.8	67.0
Urakawa	3.0	2.5	3.4	4.3	5.3	6.3	8.3	6.5	4.6	2.8	3.6	4.1	54.8
Aomori	6.8	3.7	3.0	3.0	3.6	3.9	3.6	2.7	3.3	2.8	4.4	6.4	47.1
Morioka	3.3	2.6	3.9	4.0	4.6	5.4	6.3	4.2	6.1	3.9	4.2	4.0	52.3
Miyako	2.3	2.6	3.5	3.9	5.2	6.3	7.3	5.5	6.5	4.3	3.0	2.4	53.0
Sendai	2.2	2.3	4.0	4.4	5.4	8.4	9.0	6.0	7.4	4.3	3.6	2.2	59.2
Akita	6.7	4.1	4.0	4.3	5.3	4.5	6.8	3.3	4.0	3.5	5.5	7.3	59.5
Yamagata	4.6	3.4	3.7	3.0	3.5	3.9	4.3	2.3	5.1	3.7	3.9	4.6	45.9
Sakata	8.2	5.0	3.9	4.2	4.2	4.4	5.4	2.3	4.5	3.6	6.1	7.9	59.3
Fukushima	2.8	2.6	4.1	3.5	4.8	6.7	6.9	4.7	7.3	4.7	3.9	2.5	54.5
Onahama	2.8	3.2	4.7	4.7	5.0	7.0	6.0	3.2	5.2	5.1	3.8	2.8	53.6
Mito	3.1	3.8	5.6	5.2	5.2	7.4	6.3	3.4	5.8	5.8	4.6	3.2	59.4
Utsunomiya	2.8	3.1	5.0	4.7	5.9	8.0	6.6	4.2	6.8	5.8	4.6	2.7	60.1
Maebashi	2.2	2.6	4.3	4.1	5.4	7.6	6.5	3.6	7.4	5.9	3.9	1.8	55.3
Kumagaya	2.9	3.1	4.9	4.7	5.8	8.1	6.8	3.9	7.6	6.2	5.0	2.6	61.6
Choshi	4.4	5.0	6.1	5.2	5.2	7.1	5.6	2.5	5.6	6.1	5.3	3.9	62.0
Tokyo	3.3	4.1	6.3	5.4	5.7	8.0	6.0	3.3	6.9	6.4	5.5	3.3	64.2
Oshima	3.5	3.6	6.1	5.0	5.7	8.0	5.8	2.4	5.4	6.7	4.8	2.9	59.9
Hachijojima	3.8	3.9	6.1	5.0	6.1	8.7	5.7	1.4	2.9	5.3	4.6	3.2	56.9
Yokohama	3.4	4.0	6.1	4.9	5.3	7.4	5.0	2.9	6.1	6.2	5.1	3.2	59.6
Niigata	7.7	5.0	4.2	4.0	3.9	4.2	4.5	2.1	4.8	4.0	5.9	7.2	57.6
Takada	9.2	6.5	5.5	4.1	4.3	4.8	4.7	2.3	5.3	4.2	5.9	7.5	64.5
Aikawa	6.3	4.2	4.0	3.9	4.5	4.8	5.2	2.6	4.3	3.9	4.7	6.7	55.0
Toyama	7.6	5.4	5.2	5.0	4.9	5.6	5.2	2.5	5.2	5.0	5.6	6.9	64.0
Kanazawa	6.6	4.5	4.8	4.4	4.8	5.5	4.5	2.1	5.0	4.3	4.8	5.3	56.7
Wajima	7.7	5.2	4.5	3.9	4.5	4.4	5.0	2.2	4.9	3.7	5.0	7.5	58.5
Fukui	5.6	4.1	5.2	4.6	5.0	5.8	4.7	2.0	4.8	4.5	4.0	4.7	55.0
Tsuruga	6.8	5.0	5.9	5.2	5.2	6.7	4.3	2.1	4.7	4.8	4.4	6.1	61.3
Kofu	2.5	2.5	4.1	3.5	3.9	5.1	3.8	1.7	5.1	5.2	3.4	2.1	42.8
Nagano	3.3	2.9	3.3	2.5	3.0	3.7	2.6	1.3	4.1	3.9	2.8	3.0	36.4
Matsumoto	2.6	2.6	3.9	3.3	3.2	4.1	2.6	1.3	4.5	4.5	3.0	2.0	37.6
Mt. Fuji													

The number of days with less than 0.1 hour of sunshine is referred to as the Number of Sunless Days.
The amount of sunshine had not been observed at Mt. Fuji before 2003.

Climatological Normal Number of Days per Month with less than 0.1 Hours of Sunshine (1981–2010 Averages)

Continued.

Station	Jan.	Feb.	Mar.	Apr.	May	Jun.	Jul.	Aug.	Sep.	Oct.	Nov.	Dec.	Annual
Iida	3.2	3.2	4.9	4.0	4.4	4.9	3.7	1.3	4.8	5.1	3.6	3.0	46.1
Karuizawa	2.4	2.8	4.2	3.7	4.4	5.7	4.6	2.2	6.4	5.4	3.5	2.0	47.4
Gifu	2.9	2.6	5.0	4.7	5.4	6.6	5.4	2.1	5.6	4.7	3.8	2.6	51.1
Takayama	3.8	3.4	4.8	4.2	4.3	4.7	3.6	1.7	4.9	5.0	4.2	4.0	48.6
Shizuoka	2.9	3.2	5.2	4.7	5.3	7.0	4.2	2.3	4.5	5.3	3.8	2.4	50.8
Hamamatsu	2.7	3.0	4.8	4.5	5.4	6.1	4.2	1.4	4.1	5.1	3.9	2.2	47.4
Nagoya	3.1	2.4	5.1	4.7	5.1	6.3	4.4	1.4	4.7	4.9	3.8	2.6	48.4
Tsu	2.9	2.6	5.1	4.7	5.7	6.8	4.3	2.0	5.6	5.8	4.1	2.5	52.1
Owase	3.2	2.8	5.1	5.3	6.9	7.9	5.3	3.7	6.4	7.0	4.9	2.7	61.8
Hikone	4.8	3.3	5.2	4.5	5.0	5.6	3.9	1.4	4.6	4.3	3.8	3.3	49.8
Kyoto	3.3	3.2	4.9	4.6	5.1	6.2	4.0	1.4	4.4	4.6	3.3	2.8	47.9
Osaka	3.1	3.5	4.7	3.9	4.1	5.5	2.7	0.9	3.8	4.0	3.3	2.7	42.1
Kobe	2.9	3.5	4.8	3.9	4.6	6.2	3.3	1.3	4.2	4.0	3.2	2.3	44.4
Nara	3.2	3.2	4.7	3.8	4.1	5.2	3.2	0.9	3.7	4.4	3.4	2.9	42.6
Wakayama	3.2	3.1	4.7	3.6	4.2	5.3	2.4	0.8	3.6	3.9	3.3	2.3	40.3
Shionomisaki	3.1	3.1	5.1	4.3	4.8	6.4	3.4	1.4	3.3	4.2	4.1	2.3	45.8
Tottori	5.9	5.0	5.8	4.3	4.5	5.2	4.1	2.1	4.7	4.3	4.5	5.1	55.5
Matsue	4.9	4.1	5.2	4.5	4.4	5.1	4.8	2.2	4.6	3.4	4.3	3.7	51.2
Hamada	5.3	4.1	5.3	4.3	4.3	5.3	4.7	2.2	4.5	3.4	3.7	4.2	51.2
Saigo	3.9	3.6	4.5	3.7	4.3	4.2	4.5	2.2	4.5	2.7	3.3	3.1	44.4
Okayama	3.0	3.3	4.6	4.2	4.4	5.5	3.3	1.4	4.1	3.9	3.4	2.2	43.3
Hiroshima	3.1	3.3	4.8	4.9	4.4	6.1	4.2	2.0	4.2	3.5	3.3	2.3	46.0
Shimonoseki	4.3	3.8	5.3	4.8	4.6	5.8	4.3	1.9	4.2	3.0	3.6	3.2	48.9
Tokushima	3.1	3.2	4.9	3.8	4.4	5.9	2.9	1.4	4.6	4.5	3.8	2.4	44.9
Takamatsu	2.8	3.6	4.5	3.8	3.9	4.9	2.7	1.3	4.4	4.2	3.5	2.6	42.2
Matsuyama	3.5	3.5	5.2	4.1	4.2	5.8	3.3	1.4	3.7	3.4	3.4	2.5	44.1
Kochi	3.0	3.1	5.1	5.0	5.9	6.9	3.8	1.9	4.4	4.1	3.2	2.4	48.9
Murotomisaki	3.0	3.2	5.2	4.5	5.2	6.8	3.3	1.6	3.3	4.1	3.6	2.1	45.9
Shimizu	3.3	3.2	5.6	4.7	5.2	6.6	3.0	1.5	3.4	4.3	3.7	1.9	46.2
Fukuoka	4.8	4.5	5.6	4.8	5.1	5.8	4.1	1.6	3.8	3.2	3.9	3.2	50.5
Saga	3.6	3.7	4.9	4.6	4.9	6.9	4.4	1.7	3.2	2.9	3.4	2.5	46.8
Nagasaki	4.5	3.6	5.5	4.9	5.1	7.2	4.5	1.7	3.1	2.8	3.4	3.3	49.8
Izuhara	3.8	4.5	5.7	5.2	5.8	7.5	7.3	4.0	5.4	3.3	3.5	2.2	58.1
Fukue	5.0	4.2	5.9	5.4	5.8	6.8	4.8	1.8	3.5	2.7	3.8	3.3	52.8
Kumamoto	3.7	3.8	5.8	4.5	4.6	6.8	3.9	1.4	3.0	3.0	3.4	2.8	46.7
Oita	3.2	3.7	5.6	5.2	4.9	6.4	3.8	2.0	4.6	3.9	3.9	2.3	49.6
Miyazaki	3.2	3.6	6.1	5.0	5.8	7.4	2.7	2.0	3.9	4.6	4.0	2.3	50.6
Kagoshima	3.8	3.9	6.0	5.1	5.2	7.0	3.0	1.5	2.8	2.8	3.5	2.8	47.8
Naze	7.6	7.2	6.9	5.9	5.8	6.0	1.6	2.2	2.3	4.6	5.6	5.7	61.4
Naha	5.7	6.3	5.6	4.3	4.4	3.1	1.1	1.7	1.5	2.1	4.2	4.8	44.7
Syowa station (Antarctica)	1.7	3.8	8.2	14.5	24.2	30.0	28.1	14.8	8.3	6.6	2.2	1.2	143.6

The number of days with less than 0.1 hour of sunshine is referred to as the Number of Sunless Days.

Climatological Normal Number of Days per Month with Average Daily Cloud Amount less than 1.5

(1981–2010 Averages)

Station	Jan.	Feb.	Mar.	Apr.	May	Jun.	Jul.	Aug.	Sep.	Oct.	Nov.	Dec.	Annual
Sapporo	0.2	0.5	0.9	1.7	1.7	1.6	0.9	1.3	1.5	2.0	1.0	0.2	13.5
Hakodate	0.2	0.2	1.2	2.2	1.8	1.5	1.1	1.3	1.7	2.6	1.3	0.5	15.6
Asahikawa	0.1	0.6	1.2	1.7	1.6	1.7	0.9	0.7	0.9	1.2	0.4	0.1	11.1
Kushiro	7.2	5.0	3.7	2.1	1.3	0.5	0.3	0.6	1.6	4.6	6.7	8.5	42.1
Obihiro	6.8	5.2	4.4	2.2	1.6	0.8	0.5	0.8	1.3	3.7	4.4	7.2	38.9
Abashiri	1.0	1.4	1.6	1.5	1.8	1.9	1.2	1.5	1.9	2.6	1.6	1.4	19.4
Wakkanai	0.0	0.4	1.2	1.3	1.5	1.5	0.8	1.1	2.1	1.5	0.5	0.0	11.9
Aomori	0.0	0.1	0.8	2.1	1.9	1.1	0.9	1.2	1.3	2.0	0.9	0.2	12.6
Morioka	0.2	0.4	0.8	1.8	1.4	0.7	0.3	0.6	0.7	1.9	1.6	0.7	11.1
Sendai	1.0	1.0	1.3	2.7	1.8	0.8	0.4	0.7	0.7	1.9	2.2	1.6	16.1
Akita	0.0	0.1	0.4	1.3	1.5	1.1	1.5	2.1	1.2	1.5	0.9	0.2	11.9
Yamagata	0.1	0.2	0.7	1.9	1.8	0.9	0.8	1.1	0.9	1.3	1.2	0.3	11.1
Fukushima	0.6	0.9	1.2	2.8	2.1	0.9	0.1	0.6	0.6	2.0	1.8	1.3	14.9
Mito	8.7	5.0	3.4	2.9	1.5	0.9	0.6	0.9	1.2	2.7	5.1	8.9	41.9
Utsunomiya	7.5	4.7	3.6	2.7	1.5	0.6	0.3	0.3	0.5	2.5	4.8	8.6	37.6
Maebashi	7.9	5.2	3.5	3.1	2.1	0.8	0.3	0.5	0.4	2.7	6.0	8.5	41.2
Kumagaya	11.7	7.2	5.1	4.2	2.5	1.2	0.7	1.2	1.1	3.8	7.5	12.3	58.6
Choshi	6.7	4.6	2.6	2.7	1.9	1.0	0.7	2.4	1.3	2.1	3.8	6.4	36.3
Tokyo	8.1	4.9	3.2	2.7	1.9	0.9	0.7	1.0	0.8	2.7	5.2	8.3	40.3
Yokohama	6.3	4.0	2.5	2.3	1.5	0.9	0.9	1.5	1.1	2.4	4.6	7.0	35.0
Niigata	0.1	0.2	0.4	1.7	1.6	0.9	1.3	1.9	1.1	1.6	0.9	0.2	11.9
Toyama	0.5	0.4	1.0	2.4	1.9	0.6	0.7	1.8	0.9	2.1	2.0	0.8	15.2
Kanazawa	0.5	0.5	1.1	2.7	2.5	1.1	1.3	2.3	1.4	2.3	2.1	0.8	18.5
Fukui	0.4	0.8	1.7	3.3	2.6	1.3	1.1	2.0	1.6	2.9	2.8	1.2	21.7
Kofu	9.7	6.2	3.8	2.7	1.5	0.5	0.5	0.5	0.8	2.4	6.5	9.6	44.8
Nagano	1.7	1.3	1.7	2.8	2.0	0.7	0.5	0.5	0.7	2.1	2.8	2.5	19.4
Gifu	3.1	3.2	3.7	4.7	3.0	1.5	1.3	1.8	2.1	5.0	5.5	4.5	39.4
Shizuoka	10.1	6.6	4.4	3.5	2.0	0.9	0.8	1.2	1.1	3.7	6.9	11.2	52.4

Climatological Normal Number of Days per Month with Average Daily Cloud Amount less than 1.5 (1981–2010 Averages) Continued.

Station	Jan.	Feb.	Mar.	Apr.	May	Jun.	Jul.	Aug.	Sep.	Oct.	Nov.	Dec.	Annual
Nagoya	3.0	3.2	3.2	3.5	2.1	1.3	0.6	0.9	1.4	3.5	4.5	4.8	31.9
Tsu	3.0	2.3	2.8	3.6	2.6	1.3	0.8	1.8	2.1	4.1	5.4	5.1	34.8
Hikone	1.0	1.2	2.0	3.4	3.0	1.4	1.0	1.8	1.9	3.6	3.4	2.1	25.7
Kyoto	0.9	0.7	1.6	2.4	1.9	1.1	0.4	1.0	1.0	2.5	2.8	2.1	18.6
Osaka	1.3	1.1	1.8	2.6	2.1	0.9	0.6	0.9	1.0	2.8	3.3	2.9	21.5
Kobe	2.2	1.5	2.2	2.8	2.1	0.9	0.8	1.2	0.9	3.0	3.8	3.7	25.1
Nara	1.5	1.5	2.1	3.2	2.3	1.2	0.8	1.3	1.3	3.1	3.9	3.3	25.4
Wakayama	1.2	1.7	2.4	3.2	2.2	1.0	1.4	2.0	1.1	3.0	3.1	2.9	25.2
Tottori	0.2	0.6	1.0	2.4	2.8	1.2	1.0	1.9	0.8	1.8	1.8	1.1	16.6
Matsue	0.2	0.6	1.3	3.3	2.7	1.2	1.2	1.9	1.2	2.3	1.8	0.9	18.6
Okayama	2.6	2.4	3.0	4.0	3.0	1.6	1.1	1.8	2.0	4.2	4.6	4.2	34.5
Hiroshima	1.4	1.5	2.8	3.1	2.7	1.2	1.2	1.9	2.2	4.4	4.1	3.2	29.8
Shimonoseki	0.5	1.1	2.1	3.0	3.0	1.3	1.5	1.9	2.2	4.3	3.1	2.1	26.3
Tokushima	2.5	2.2	2.7	3.3	2.0	0.9	1.3	2.5	1.3	2.9	3.4	4.1	29.1
Takamatsu	1.6	1.7	2.8	3.7	2.7	1.5	1.4	2.2	1.4	3.3	3.9	3.5	29.6
Matsuyama	1.3	1.6	2.4	3.2	2.6	1.1	1.9	2.4	2.1	4.2	4.0	3.1	29.9
Kochi	5.8	3.9	3.4	3.4	2.3	0.9	1.2	1.9	2.0	4.9	5.5	7.1	42.4
Fukuoka	1.0	1.2	2.0	3.1	3.2	0.9	1.3	1.5	2.2	4.4	3.3	2.2	26.3
Saga	2.7	2.9	3.8	3.8	3.6	1.4	1.4	2.0	3.5	6.2	5.7	4.7	41.8
Nagasaki	2.1	1.7	2.2	2.9	3.1	0.9	1.3	1.7	3.6	6.0	4.7	3.3	33.6
Kumamoto	2.8	2.7	2.6	2.7	2.4	0.8	0.8	1.4	2.6	5.8	4.5	4.1	33.3
Oita	3.1	2.8	2.9	3.1	2.7	1.0	1.7	1.7	2.1	4.0	4.7	4.4	34.2
Miyazaki	8.2	6.1	3.9	3.5	2.4	0.5	1.9	2.0	2.5	4.8	7.2	9.9	52.7
Kagoshima	2.9	2.7	2.2	2.6	1.8	0.5	1.1	0.9	2.0	5.4	5.5	4.7	32.3
Naze	0.5	0.3	0.6	0.5	0.9	0.3	0.7	0.5	0.5	0.7	0.7	0.7	7.0
Naha	0.9	1.0	0.6	0.5	0.8	0.2	0.2	0.3	0.7	1.4	1.1	1.1	8.9
Syowa station (Antarctica)	4.2	2.5	1.8	2.0	3.4	4.0	3.9	4.3	4.2	2.8	3.9	4.7	41.6

Climatological Normal Number of Days per Month with Average Daily Cloud Amount greater than 8.5

(1981-2010 Averages)

Station	Jan.	Feb.	Mar.	Apr.	May	Jun.	Jul.	Aug.	Sep.	Oct.	Nov.	Dec.	Annual
Sapporo	15.7	13.7	13.7	11.3	13.1	13.8	16.9	14.7	10.5	9.2	12.2	15.0	159.8
Hakodate	12.9	11.4	11.0	10.0	12.9	15.2	19.5	16.8	12.1	8.4	11.5	13.5	155.2
Asahikawa	17.8	14.2	14.2	12.1	13.1	15.1	17.5	16.4	12.3	12.0	18.8	20.3	183.8
Kushiro	4.5	5.1	8.0	11.3	14.9	19.3	22.8	20.3	15.1	7.9	5.4	4.4	139.0
Obihiro	4.8	5.0	7.3	9.8	13.6	17.2	21.6	19.2	14.7	8.4	5.2	4.6	131.2
Abashiri	11.9	9.6	10.9	11.0	13.7	16.0	17.4	15.6	11.7	7.7	8.5	9.1	143.0
Wakkanai	24.7	18.3	14.4	12.4	14.6	16.8	19.8	17.4	9.3	11.9	20.5	25.8	205.8
Aomori	24.0	20.7	15.8	11.8	12.8	15.8	17.8	14.8	13.3	11.4	15.6	22.1	196.0
Morioka	9.9	9.8	10.7	11.6	13.5	16.9	19.7	17.3	16.6	11.2	9.8	11.0	158.0
Sendai	5.9	6.7	8.1	10.0	13.6	18.6	20.9	17.2	16.3	10.4	6.9	6.0	140.7
Akita	25.5	21.7	17.7	12.7	14.3	16.0	18.5	13.4	13.9	12.3	17.2	23.3	206.5
Yamagata	18.2	15.2	13.5	11.5	14.2	17.4	18.6	13.6	15.5	12.8	13.3	16.8	180.6
Fukushima	8.7	7.7	9.4	10.7	13.4	18.8	20.3	15.4	16.6	11.8	8.4	8.6	149.8
Mito	4.8	6.4	9.9	11.3	14.7	19.7	18.9	13.1	16.0	12.4	8.1	4.6	139.8
Utsunomiya	4.3	5.4	9.6	11.5	16.2	21.1	22.0	17.4	17.9	13.7	7.9	3.7	150.8
Maebashi	3.4	4.5	7.5	9.9	14.0	19.6	20.4	14.7	17.2	12.0	6.7	3.0	133.0
Kumagaya	5.0	6.4	9.9	12.9	16.0	20.4	19.5	14.3	18.1	13.7	8.8	4.8	149.7
Choshi	6.8	8.5	12.1	11.8	15.6	20.2	18.4	11.3	14.3	12.9	9.6	5.8	147.3
Tokyo	5.4	7.3	10.8	11.7	15.3	19.5	18.1	13.5	16.0	13.2	9.2	4.7	144.7
Yokohama	5.8	7.6	12.2	12.4	15.8	19.8	18.3	12.5	16.2	13.5	9.6	5.4	149.1
Niigata	23.8	19.7	16.7	11.8	13.4	16.5	17.3	11.5	15.1	12.7	16.2	21.4	196.3
Toyama	21.7	17.0	15.0	11.5	13.0	18.4	18.2	11.5	15.2	12.8	14.7	18.0	187.1
Kanazawa	23.0	17.8	15.0	11.0	12.8	17.1	16.7	10.2	14.2	11.7	14.0	18.9	182.3
Fukui	22.8	17.7	16.0	12.8	15.5	19.4	18.6	11.7	14.9	12.4	14.9	19.6	196.1
Kofu	4.1	6.0	8.9	10.0	14.5	19.7	18.5	12.4	15.7	12.6	7.7	3.8	134.0
Nagano	11.4	10.6	11.2	9.5	12.3	17.8	17.2	11.1	15.5	11.6	7.9	8.9	144.9
Gifu	8.2	7.2	10.1	11.7	14.0	17.9	16.2	11.7	14.0	10.7	8.9	7.4	138.0
Shizuoka	6.6	7.9	12.6	12.6	16.7	20.4	19.0	13.5	15.5	12.9	9.1	5.0	151.9

Climatological Normal Number of Days per Month with Average Daily Cloud Amount greater than 8.5 (1981-2010 Averages) Continued.

Station	Jan.	Feb.	Mar.	Apr.	May	Jun.	Jul.	Aug.	Sep.	Oct.	Nov.	Dec.	Annual
Nagoya	5.4	5.2	8.8	9.7	12.9	18.2	16.1	11.4	13.7	10.2	7.0	4.8	123.3
Tsu	6.8	7.2	11.1	11.8	15.8	18.7	16.0	11.6	14.7	11.5	8.7	5.6	139.5
Hikone	15.6	13.5	13.7	13.2	16.3	19.2	17.7	11.9	14.7	11.8	11.6	12.9	172.1
Kyoto	8.1	8.6	11.1	10.6	14.2	18.3	16.5	11.0	13.5	9.8	7.7	6.8	136.4
Osaka	6.4	7.6	10.3	9.8	13.6	17.9	15.0	10.0	12.4	9.5	7.7	5.5	125.7
Kobe	5.8	6.4	9.7	9.4	13.1	17.5	16.1	10.1	12.9	9.2	6.9	4.7	121.9
Nara	10.2	9.6	12.6	12.6	16.0	19.6	16.9	11.4	14.5	11.5	9.7	8.1	152.8
Wakayama	7.0	7.3	10.1	10.0	13.3	17.4	14.6	8.3	12.8	9.8	7.6	6.3	124.5
Tottori	20.3	17.6	15.8	10.9	13.2	17.6	17.0	11.3	14.7	11.2	13.6	16.2	179.4
Matsue	20.7	16.6	14.0	10.9	12.7	16.9	16.3	11.4	14.3	10.1	12.5	16.6	172.9
Okayama	7.4	7.5	10.9	11.5	14.4	17.4	15.7	11.3	13.7	9.9	8.4	6.1	134.1
Hiroshima	7.7	7.6	9.6	9.9	11.8	16.0	15.1	9.9	12.1	8.3	7.3	7.1	122.3
Shimonoseki	13.9	11.2	11.1	10.5	12.4	16.2	14.0	9.5	11.8	7.9	8.9	11.5	139.0
Tokushima	6.1	6.9	9.5	9.6	12.6	17.0	14.2	8.6	12.3	9.9	7.6	4.9	119.1
Takamatsu	6.6	6.9	9.5	9.3	11.9	15.6	13.4	8.2	12.3	8.8	7.2	5.9	115.5
Matsuyama	7.6	7.3	9.9	9.7	12.2	15.2	13.2	8.7	11.8	8.1	8.1	7.4	119.1
Kochi	4.6	5.7	9.2	9.9	13.5	18.1	15.5	11.1	13.3	8.4	6.7	3.9	119.9
Fukuoka	12.7	10.3	11.7	10.5	12.5	16.7	14.4	9.6	11.5	8.2	8.7	10.3	137.0
Saga	11.3	9.8	12.3	11.9	14.1	18.5	15.2	11.0	12.3	8.9	9.7	9.1	144.0
Nagasaki	11.6	9.5	11.8	10.4	13.2	17.5	14.9	9.8	10.8	7.7	7.9	10.0	135.0
Kumamoto	8.6	7.7	10.9	10.8	13.3	18.4	15.0	9.3	11.5	8.2	7.8	6.2	127.8
Oita	6.5	7.1	10.4	10.3	13.1	16.7	13.5	9.2	12.8	8.3	7.5	5.4	120.7
Miyazaki	5.2	6.6	10.8	11.5	13.9	17.9	13.1	9.8	11.7	9.2	7.7	4.3	121.7
Kagoshima	9.0	9.0	12.5	11.6	14.7	19.1	12.7	9.4	10.9	8.4	7.7	7.4	132.6
Naze	18.6	17.4	18.4	16.0	16.5	18.4	9.7	9.9	10.2	11.7	13.3	15.5	175.6
Naha	15.7	15.8	16.0	15.4	16.7	16.5	7.9	7.9	8.1	8.6	12.3	12.8	153.8
Syowa station (Antarctica)	13.2	15.3	17.4	16.8	15.5	13.2	13.9	13.1	13.1	17.1	14.5	12.2	175.3

Climatological Normal Number of Days per Month with Fog

(1981-2010 Averages)

Station	Jan.	Feb.	Mar.	Apr.	May	Jun.	Jul.	Aug.	Sep.	Oct.	Nov.	Dec.	Annual
Sapporo	0.0	0.0	0.1	0.5	0.6	0.4	0.2	0.4	0.1	0.1	0.2	0.0	2.7
Hakodate	0.0	0.1	0.1	0.9	1.5	2.3	2.3	0.9	0.1	0.0	0.1	0.0	8.4
Asahikawa	2.2	1.2	1.0	1.1	0.6	0.5	0.4	0.9	4.8	6.0	2.6	1.7	23.1
Kushiro	1.4	2.1	4.1	9.3	12.4	16.1	17.0	16.5	10.7	7.2	2.6	2.0	101.4
Obihiro	1.2	1.1	1.9	3.8	4.8	7.1	7.3	7.4	6.5	6.0	3.1	1.8	52.0
Abashiri	0.2	0.6	1.3	2.6	3.9	5.0	6.0	2.6	0.7	0.2	0.3	0.1	23.7
Wakkanai	0.0	0.0	0.2	0.9	2.4	3.5	3.4	2.2	0.3	0.2	0.0	0.0	13.1
Aomori	0.1	0.2	0.3	1.0	1.5	1.9	2.5	1.7	0.5	0.2	0.1	0.1	10.0
Morioka	0.9	0.4	0.9	0.5	0.3	0.3	0.6	0.7	1.1	1.8	1.6	1.5	10.7
Sendai	0.2	0.1	0.5	2.2	3.3	4.9	7.0	3.6	1.6	0.7	0.4	0.3	24.7
Akita	0.5	0.3	0.8	1.2	1.3	1.4	1.6	0.8	0.7	1.1	0.8	0.6	11.0
Yamagata	2.5	1.6	2.2	0.9	0.5	0.4	0.9	0.4	1.4	4.0	4.2	4.1	22.9
Fukushima	0.7	0.4	0.7	0.7	0.2	0.5	0.8	0.2	0.4	1.4	1.7	0.9	8.7
Mito	1.3	1.4	2.4	3.4	3.6	4.4	5.1	3.4	2.7	3.2	3.5	2.0	36.4
Utsunomiya	1.1	1.2	1.2	1.8	1.7	1.4	1.5	0.7	1.2	1.2	2.0	1.4	16.4
Maebashi	0.2	0.2	0.4	0.4	0.4	1.1	0.6	0.2	0.4	0.2	0.3	0.1	4.5
Kumagaya	0.5	0.6	0.9	1.0	0.8	1.2	1.1	0.3	0.7	0.8	1.1	0.6	9.7
Choshi	0.4	0.5	1.3	2.8	5.2	9.8	11.8	6.8	2.3	0.9	1.1	0.6	43.7
Tokyo	0.1	0.1	0.1	0.6	0.2	0.4	0.3	0.1	0.2	0.3	0.4	0.1	2.7
Yokohama	0.2	0.4	1.0	1.4	0.5	1.0	0.8	0.2	0.4	0.6	1.1	0.2	8.0
Niigata	0.3	0.4	0.2	0.7	0.9	0.7	0.3	0.2	0.1	0.3	0.2	0.4	4.7
Toyama	0.3	0.5	0.3	1.1	1.0	1.3	0.8	0.2	0.3	0.1	0.1	0.1	6.1
Kanazawa	0.0	0.0	0.1	0.3	0.2	0.4	0.4	0.0	0.0	0.0	0.0	0.0	1.3
Fukui	2.8	3.2	2.4	1.1	1.0	0.8	0.4	0.3	1.7	3.7	3.6	3.5	24.4
Kofu	2.1	1.0	0.7	0.3	0.1	0.1	0.0	0.0	0.1	0.5	1.3	2.0	8.4
Nagano	1.4	1.0	0.6	0.5	0.1	0.4	0.5	0.1	0.4	1.4	2.8	1.9	11.0
Gifu	0.4	0.1	0.4	0.5	0.5	0.5	0.4	0.1	0.3	0.4	0.6	0.8	4.9
Shizuoka	0.2	0.1	0.5	0.5	0.3	0.6	0.4	0.1	0.1	0.1	0.2	0.2	3.1

Climatological Normal Number of Days per Month with Fog (1981–2010 Averages) Continued.

Station	Jan.	Feb.	Mar.	Apr.	May	Jun.	Jul.	Aug.	Sep.	Oct.	Nov.	Dec.	Annual
Nagoya	0.6	0.5	1.3	0.7	0.7	0.5	0.3	0.1	0.4	0.7	0.8	0.6	7.3
Tsu	0.2	0.3	0.6	1.0	0.8	0.9	0.9	0.2	0.1	0.2	0.3	0.2	5.5
Hikone	0.4	0.5	0.9	1.2	0.5	0.3	0.1	0.0	0.0	0.1	1.1	0.8	6.0
Kyoto	0.1	0.1	0.1	0.0	0.0	0.0	0.0	0.0	0.0	0.1	0.2	0.4	1.0
Osaka	0.2	0.3	0.3	0.3	0.2	0.3	0.1	0.1	0.2	0.2	0.3	0.4	2.9
Kobe	0.1	0.2	0.1	0.3	0.2	0.1	0.1	0.0	0.0	0.1	0.0	0.0	1.3
Nara	1.0	1.1	0.8	0.5	0.5	0.1	0.1	0.1	0.4	1.3	1.5	1.2	8.6
Wakayama	0.1	0.2	0.2	0.2	0.2	0.1	0.1	0.0	0.0	0.2	0.1	0.2	1.7
Tottori	0.2	0.1	0.3	0.4	0.2	0.2	0.0	0.0	0.0	0.2	0.2	0.2	2.1
Matsue	1.1	0.9	1.7	1.3	0.8	0.6	0.2	0.3	1.0	1.9	2.0	1.6	13.3
Okayama	0.6	0.3	0.7	0.6	0.5	0.3	0.1	0.1	0.1	0.6	1.0	1.2	6.2
Hiroshima	0.7	0.6	0.8	0.8	0.5	0.7	0.3	0.1	0.1	0.1	0.6	0.6	5.8
Shimonoseki	0.1	0.2	0.4	0.6	0.5	0.6	0.4	0.0	0.0	0.0	0.1	0.2	3.2
Tokushima	0.3	0.1	0.3	0.7	0.8	1.1	1.0	0.2	0.1	0.0	0.1	0.1	4.9
Takamatsu	0.3	0.3	0.5	0.8	0.6	0.7	0.5	0.1	0.1	0.1	0.4	0.3	4.7
Matsuyama	0.3	0.3	0.9	1.6	1.1	1.2	0.3	0.0	0.1	0.1	0.1	0.2	6.2
Kochi	0.2	0.0	0.1	0.1	0.2	0.2	0.1	0.0	0.0	0.1	0.1	0.1	1.2
Fukuoka	0.1	0.1	0.1	0.2	0.4	0.4	0.0	0.1	0.0	0.0	0.0	0.0	1.4
Saga	0.7	0.7	0.7	0.4	0.2	0.3	0.2	0.0	0.1	0.5	0.8	1.2	5.9
Nagasaki	0.1	0.2	0.2	0.6	1.0	0.9	0.3	0.0	0.0	0.1	0.2	0.2	3.7
Kumamoto	1.2	1.1	1.2	0.9	0.9	0.3	0.1	0.2	0.4	0.6	1.6	2.2	10.7
Oita	0.2	0.2	0.6	0.9	0.7	1.0	0.9	0.0	0.1	0.1	0.3	0.3	5.3
Miyazaki	0.6	0.9	0.6	0.7	0.9	0.7	0.5	0.3	0.7	0.7	1.0	0.9	8.3
Kagoshima	0.0	0.2	0.1	0.0	0.2	0.2	0.1	0.0	0.0	0.0	0.0	0.1	1.0
Naze	0.0	0.0	0.1	0.0	0.0	0.0	0.0	0.0	0.0	0.0	0.0	0.0	0.2
Naha	0.0	0.0	0.2	0.5	0.5	0.1	0.0	0.0	0.0	0.0	0.0	0.0	1.3
Syowa station (Antarctica)	2.2	0.4	0.2	0.2	0.4	0.4	0.4	0.8	0.7	0.7	0.6	1.5	8.4

Climatological Normal Number of Days per Month with Thunder

(1981–2010 Averages)

Station	Jan.	Feb.	Mar.	Apr.	May	Jun.	Jul.	Aug.	Sep.	Oct.	Nov.	Dec.	Annual
Sapporo	0.3	0.2	0.5	0.1	0.5	1.0	1.0	1.6	0.8	1.7	0.8	0.4	8.8
Hakodate	0.1	0.2	0.4	0.7	0.8	0.9	1.0	1.8	1.6	2.4	1.8	0.6	12.2
Asahikawa	0.0	0.0	0.0	0.0	1.0	1.7	1.3	1.9	2.1	1.7	0.5	0.1	10.2
Kushiro	0.0	0.0	0.0	0.1	0.3	0.6	0.8	0.7	0.9	0.7	0.6	0.2	4.9
Obihiro	0.0	0.0	0.0	0.1	0.3	0.7	1.0	1.0	0.7	0.5	0.4	0.1	4.9
Abashiri	0.1	0.0	0.0	0.1	0.7	1.0	1.4	1.5	1.0	0.8	0.2	0.0	6.8
Wakkanai	0.1	0.2	0.1	0.0	0.4	0.7	0.5	0.9	3.2	3.1	1.4	0.6	11.2
Aomori	0.5	0.5	0.5	0.8	1.2	1.4	1.6	2.2	1.7	2.3	1.9	0.9	15.5
Morioka	0.0	0.2	0.2	0.7	1.8	1.7	2.3	3.3	1.3	1.0	0.8	0.3	13.7
Sendai	0.1	0.0	0.1	0.2	1.2	1.1	2.3	2.5	1.1	0.4	0.2	0.1	9.3
Akita	2.6	1.7	1.2	1.5	1.5	1.5	1.7	2.6	2.1	4.8	5.7	4.5	31.4
Yamagata	0.2	0.2	0.2	0.6	1.4	2.0	2.9	3.9	1.2	0.6	0.7	0.7	14.7
Fukushima	0.0	0.0	0.0	0.5	1.4	1.7	3.4	3.9	1.6	0.3	0.2	0.2	13.3
Mito	0.1	0.2	0.5	1.5	2.3	1.9	3.4	3.5	2.0	0.8	0.2	0.3	16.7
Utsunomiya	0.0	0.2	0.5	1.7	3.2	3.2	5.3	6.4	2.8	1.0	0.3	0.2	24.8
Maebashi	0.0	0.0	0.1	0.6	2.4	2.6	5.2	6.3	2.4	0.4	0.1	0.1	20.4
Kumagaya	0.0	0.2	0.3	0.9	2.4	2.2	4.8	5.8	2.1	0.5	0.3	0.2	19.7
Choshi	0.9	0.7	1.2	1.2	1.2	1.1	1.2	1.8	1.4	1.0	1.1	0.9	13.8
Tokyo	0.1	0.2	0.5	1.1	1.4	1.1	2.3	3.0	1.9	0.6	0.5	0.2	12.9
Yokohama	0.2	0.3	0.9	1.1	1.2	0.8	1.9	2.4	1.9	0.8	0.6	0.5	12.6
Niigata	4.3	3.0	1.4	1.2	1.3	1.7	2.4	3.0	2.1	3.1	5.3	6.1	34.8
Toyama	3.1	1.5	1.1	1.4	1.7	1.8	4.6	4.8	2.2	2.2	3.5	4.2	32.2
Kanazawa	7.0	4.6	2.1	1.5	1.3	1.6	3.1	3.0	1.8	2.5	5.6	8.2	42.4
Fukui	5.4	3.3	1.4	1.3	1.1	1.2	3.5	3.1	2.0	2.0	4.0	6.8	35.0
Kofu	0.1	0.2	0.3	0.7	1.6	1.4	3.3	4.6	2.0	0.4	0.5	0.2	15.4
Nagano	0.0	0.1	0.2	0.7	1.9	2.6	4.6	5.2	1.9	0.4	0.5	0.5	18.6
Gifu	0.2	0.3	0.3	0.9	1.6	1.8	5.0	4.9	3.3	0.7	0.6	0.4	19.9
Shizuoka	0.5	0.6	0.5	1.3	1.4	1.2	2.9	3.4	2.7	1.0	0.9	0.6	17.0

Climatological Normal Number of Days per Month with Thunder
(1981–2010 Averages)

Continued.

Station	Jan.	Feb.	Mar.	Apr.	May	Jun.	Jul.	Aug.	Sep.	Oct.	Nov.	Dec.	Annual
Nagoya	0.1	0.1	0.3	0.8	1.0	1.8	4.2	4.3	2.6	0.7	0.4	0.3	16.6
Tsu	0.2	0.2	0.4	0.5	1.0	1.3	3.1	3.2	2.5	0.7	0.3	0.2	13.6
Hikone	0.2	0.4	0.4	0.8	1.4	2.0	4.3	4.4	2.1	0.8	0.4	0.8	18.1
Kyoto	0.2	0.2	0.4	0.9	1.5	1.8	5.7	5.0	2.7	1.1	0.4	0.5	20.3
Osaka	0.4	0.4	0.5	0.7	1.3	1.1	3.0	4.0	2.4	1.0	0.7	0.8	16.2
Kobe	0.1	0.2	0.3	0.5	1.2	1.3	2.6	3.0	2.2	0.8	0.6	0.6	13.5
Nara	0.4	0.2	0.7	0.9	1.7	1.8	4.6	5.6	3.2	1.4	0.9	0.7	22.2
Wakayama	0.1	0.1	0.4	0.5	1.3	1.0	2.0	2.4	1.9	1.0	0.6	0.3	11.8
Tottori	2.6	2.3	1.3	1.0	1.3	1.3	3.4	3.4	2.2	1.2	2.2	4.2	26.4
Matsue	2.4	1.7	1.4	0.9	1.6	1.3	3.5	3.2	2.1	1.3	3.0	3.0	25.4
Okayama	0.0	0.0	0.4	0.6	1.4	1.5	2.9	2.6	1.5	0.5	0.3	0.1	11.9
Hiroshima	0.0	0.1	0.4	0.5	1.1	1.7	4.3	3.4	2.0	0.6	0.5	0.2	14.9
Shimonoseki	0.5	0.7	1.3	1.1	1.1	1.7	4.0	3.8	2.3	0.5	1.0	0.8	18.7
Tokushima	0.2	0.2	0.4	0.4	1.1	1.1	3.3	3.7	2.6	1.0	0.5	0.1	14.6
Takamatsu	0.2	0.2	0.5	0.4	0.8	1.4	2.6	3.0	1.9	0.6	0.4	0.4	12.3
Matsuyama	0.4	0.2	0.5	0.5	0.9	1.4	2.7	2.5	1.4	0.8	1.1	0.7	13.2
Kochi	0.2	0.2	0.3	0.7	1.2	1.4	3.2	3.5	3.0	0.8	0.5	0.2	15.2
Fukuoka	1.0	1.0	1.4	1.3	1.5	1.9	4.9	5.7	2.8	0.7	1.4	1.1	24.7
Saga	0.3	0.6	1.0	1.2	1.4	1.9	5.0	6.4	2.8	0.6	1.0	0.3	22.6
Nagasaki	0.8	0.6	1.2	1.6	1.3	1.6	3.6	4.5	2.5	0.8	1.6	1.1	21.1
Kumamoto	0.4	0.7	1.2	1.4	1.5	2.6	6.6	6.8	3.1	0.7	1.0	0.5	26.6
Oita	0.1	0.2	0.4	1.0	1.3	2.3	5.9	5.2	2.9	0.4	0.4	0.0	20.0
Miyazaki	0.2	0.5	0.9	1.1	1.6	3.7	5.3	5.3	3.8	0.9	0.5	0.4	24.1
Kagoshima	0.7	1.0	1.8	2.0	1.6	2.7	4.6	4.8	3.1	0.9	1.2	0.7	25.1
Naze	0.4	0.4	2.2	2.6	1.5	3.5	3.1	3.0	3.3	0.7	0.8	0.5	21.9
Naha	0.4	0.8	2.1	2.0	2.2	3.1	2.8	3.6	2.7	1.2	0.4	0.3	21.6
Syowa station (Antarctica)	0.0	0.0	0.0	0.0	0.0	0.0	0.0	0.0	0.0	0.0	0.0	0.0	0.0

First and Last Days with Frost; Climatological Normals and Records

(Sample Period: Climatological normals are calculated for autumn 1980 to spring 2010.
The earliest and latest dates are taken from the start of observations through the spring of 2019.)

Station	First day					Last day					1st year on record
	Normal date		Earliest on record			Normal date		Latest on record			
Sapporo	10	25	1888	9	9	4	24	1908	6	28	1877
Hakodate	10	19	2001	9	22	5	5	1917	6	7	1874
Asahikawa	10	8	1913	9	14	5	15	1890	7	7	1889
Kushiro	10	18	1935	9	16	5	9	1922	6	27	1911
Obihiro	10	9	1913	9	14	5	15	1922	6	27	1893
Abashiri	10	25	1913	9	15	5	9	1908	7	1	1890
Wakkanai	11	7	1969	10	1	5	4	1963	6	25	1939
Aomori	10	29	1905	10	1	4	27	1919	5	27	1887
Morioka	10	22	1984	9	27	5	3	1985	6	15	1924
Sendai	11	10	1944	10	3	4	9	1928	5	20	1927
Akita	11	11	1906	9	27	4	16	1912	5	24	1883
Yamagata	10	30	1941	9	28	4	28	1942	6	2	1891
Fukushima	11	9	1938	10	9	4	10	1921	6	4	1890
Mito	11	6	1926	10	14	4	14	1908	5	16	1897
Utsunomiya	11	2	1973	10	10	4	21	1981	5	31	1891
Maebashi	11	16	1936	10	23	3	31	1902	5	13	1898
Kumagaya	11	17	1901	10	21	3	31	1917	5	16	1898
Choshi	12	11	1904	11	1	3	8	1934	5	4	1888
Tokyo	12	20	1937	10	21	2	20	1926	5	16	1877
Yokohama	12	10	1945	10	29	3	1	1926	4	27	1897
Niigata	11	25	1890	10	25	3	30	1911	5	16	1886
Toyama	11	20	1945	10	28	4	6	1957	5	4	1940
Kanazawa	12	1	1945	10	28	4	1	1927	5	12	1883
Fukui	11	24	1897	10	22	4	5	1927	5	12	1898
Kofu	11	3	1922	10	12	4	14	1898	5	16	1895
Nagano	10	28	1963	10	4	4	28	1981	5	31	1890
Gifu	11	20	1901	10	2	4	3	1893	5	19	1884
Shizuoka	11	28	1976	10	30	3	23	1956	4	30	1941

\- No instances.

First and Last Days with Frost; Climatological Normals and Records
(Sample Period: Climatological normals are calculated for autumn 1980 to spring 2010.
The earliest and latest dates are taken from the start of observations through the spring of 2019.) Continued.

Station	First day					Last day					1st year on record
	Normal date		Earliest on record			Normal date		Latest on record			
Nagoya	11	27	1899	10	13	3	25	1902	5	13	1891
Tsu	12	2	1936	10	24	3	13	1956	4	30	1890
Hikone	11	20	1986	10	20	4	3	1908	5	16	1894
Kyoto	11	18	1892	10	2	4	6	1928	5	19	1881
Osaka	12	5	1888	10	23	3	17	1940	5	6	1883
Kobe	12	21	1899	10	24	3	12	1915	4	25	1897
Nara	11	12	1986	10	20	4	12	1969	5	7	1954
Wakayama	12	17	1889	11	13	3	6	1887	4	24	1880
Tottori	12	4	1986	10	31	3	31	1974	5	6	1944
Matsue	11	22	1942	10	26	4	9	1974	5	6	1941
Okayama	12	5	1957	10	19	3	12	1952	5	9	1892
Hiroshima	12	14	1942	10	25	3	13	1893	5	5	1880
Shimonoseki	1	10	1913	11	1	2	23	1885	5	12	1884
Tokushima	12	10	1913	11	4	3	18	1911	5	3	1892
Takamatsu	11	24	1986	10	31	4	1	1980	5	2	1942
Matsuyama	12	1	1918	10	26	3	26	1919	5	16	1891
Kochi	11	25	1949	10	31	3	16	1947	4	23	1886
Fukuoka	12	12	1903	10	21	3	10	1913	5	11	1891
Saga	11	27	1899	10	12	3	27	1904	5	7	1891
Nagasaki	12	10	1941	10	28	3	14	1947	4	23	1889
Kumamoto	11	19	1927	10	15	4	1	1940	5	6	1891
Oita	12	4	1888	10	17	3	21	1926	4	30	1888
Miyazaki	11	27	1888	10	22	3	18	1894	5	2	1886
Kagoshima	12	10	1926	10	20	3	1	1929	4	22	1887
Naze			-	-	-			-	-	-	1898
Naha			1918	2	20			1918	2	20	1891
Syowa station (Antarctica)											

- No instances.

Frost has not been observed at Syowa station (Antaractica).

Climatological Normal Number of Days per Month with Snowfall

(1981–2010 Averages)

Station	Jan.	Feb.	Mar.	Apr.	May	Jun.	Jul.	Aug.	Sep.	Oct.	Nov.	Dec.	Annual
Sapporo	28.8	25.4	23.5	6.4	0.1	0.0	0.0	0.0	0.0	1.2	13.9	26.5	125.9
Hakodate	27.8	24.9	21.7	6.0	0.0	0.0	0.0	0.0	0.0	1.5	12.1	25.1	119.1
Asahikawa	30.4	26.7	27.1	12.3	1.7	0.0	0.0	0.0	0.0	3.9	20.7	30.0	152.7
Kushiro	15.7	15.8	16.5	8.5	0.5	0.0	0.0	0.0	0.0	1.1	5.7	12.1	79.3
Obihiro	15.7	15.8	16.5	8.5	0.5	0.0	0.0	0.0	0.0	1.1	10.0	13.7	81.7
Abashiri	27.0	24.1	23.4	15.1	4.4	0.0	0.0	0.0	0.0	2.4	15.6	24.4	136.9
Wakkanai	30.1	26.2	25.1	12.1	2.1	0.0	0.0	0.0	0.0	4.3	20.7	28.2	148.7
Aomori	29.3	26.0	22.6	5.4	0.1	0.0	0.0	0.0	0.0	0.3	10.4	25.1	119.1
Morioka	27.5	23.5	21.2	6.1	0.1	0.0	0.0	0.0	0.0	0.4	9.5	24.7	112.9
Sendai	20.8	17.7	11.3	1.5	0.1	0.0	0.0	0.0	0.0	0.0	2.5	12.7	66.5
Akita	28.6	25.2	19.0	3.3	0.0	0.0	0.0	0.0	0.0	0.1	8.0	24.0	108.3
Yamagata	27.6	24.1	20.5	4.5	0.0	0.0	0.0	0.0	0.0	0.0	6.2	21.3	104.4
Fukushima	25.2	21.8	16.5	0.6	0.0	0.0	0.0	0.0	0.0	0.0	3.9	17.8	88.1
Mito	6.7	7.2	5.7	0.2	0.0	0.0	0.0	0.0	0.0	0.0	0.1	3.3	23.2
Utsunomiya	6.6	8.0	6.5	0.6	0.0	0.0	0.0	0.0	0.0	0.0	0.1	3.0	24.8
Maebashi	8.1	8.9	6.3	0.5	0.0	0.0	0.0	0.0	0.0	0.0	0.1	4.3	28.3
Kumagaya	6.5	5.8	4.2	0.3	0.0	0.0	0.0	0.0	0.0	0.0	0.0	2.9	19.7
Choshi	5.1	6.5	2.1	0.0	0.0	0.0	0.0	0.0	0.0	0.0	0.0	0.1	13.9
Tokyo	2.8	3.7	2.2	0.2	0.0	0.0	0.0	0.0	0.0	0.0	0.0	0.8	9.7
Yokohama	5.1	7.8	4.3	0.2	0.0	0.0	0.0	0.0	0.0	0.0	0.0	2.5	19.9
Niigata	22.9	20.2	11.4	0.9	0.0	0.0	0.0	0.0	0.0	0.0	2.1	13.3	70.8
Toyama	24.0	19.1	13.0	1.7	0.0	0.0	0.0	0.0	0.0	0.0	1.1	12.5	71.7
Kanazawa	23.0	19.9	12.3	2.7	0.0	0.0	0.0	0.0	0.0	0.0	1.9	13.7	73.9
Fukui	24.2	19.6	11.4	1.5	0.0	0.0	0.0	0.0	0.0	0.0	1.0	13.0	70.9
Kofu	6.2	5.9	4.0	0.3	0.0	0.0	0.0	0.0	0.0	0.0	0.0	3.6	20.0
Nagano	23.6	20.8	16.1	2.9	0.0	0.0	0.0	0.0	0.0	0.1	4.1	18.7	86.5
Gifu	13.4	10.1	4.9	0.2	0.0	0.0	0.0	0.0	0.0	0.0	0.1	6.7	35.5
Shizuoka	2.0	2.0	0.3	0.0	0.0	0.0	0.0	0.0	0.0	0.0	0.0	0.3	4.5

Climatological Normal Number of Days per Month with Snowfall
(1981–2010 Averages) Continued.

Station	Jan.	Feb.	Mar.	Apr.	May	Jun.	Jul.	Aug.	Sep.	Oct.	Nov.	Dec.	Annual
Nagoya	6.4	5.4	2.0	0.0	0.0	0.0	0.0	0.0	0.0	0.0	0.0	2.6	16.6
Tsu	8.7	10.1	4.6	0.1	0.0	0.0	0.0	0.0	0.0	0.0	0.0	2.9	26.4
Hikone	18.6	16.1	8.2	0.3	0.0	0.0	0.0	0.0	0.0	0.0	0.2	8.0	51.6
Kyoto	17.0	14.6	7.7	0.3	0.0	0.0	0.0	0.0	0.0	0.0	0.1	6.4	46.3
Osaka	5.0	6.3	2.3	0.0	0.0	0.0	0.0	0.0	0.0	0.0	0.0	1.9	15.5
Kobe	9.2	10.8	3.8	0.1	0.0	0.0	0.0	0.0	0.0	0.0	0.0	4.1	28.1
Nara	11.8	12.7	5.2	0.1	0.0	0.0	0.0	0.0	0.0	0.0	0.0	5.7	35.6
Wakayama	9.3	9.0	2.4	0.0	0.0	0.0	0.0	0.0	0.0	0.0	0.0	5.9	26.9
Tottori	19.2	16.4	8.9	0.3	0.0	0.0	0.0	0.0	0.0	0.0	0.6	9.8	55.5
Matsue	18.9	14.1	6.8	0.3	0.0	0.0	0.0	0.0	0.0	0.0	0.7	8.7	49.8
Okayama	8.5	9.5	3.0	0.1	0.0	0.0	0.0	0.0	0.0	0.0	0.0	4.1	25.2
Hiroshima	8.7	7.1	2.6	0.0	0.0	0.0	0.0	0.0	0.0	0.0	0.2	4.6	23.3
Shimonoseki	10.7	7.4	3.1	0.0	0.0	0.0	0.0	0.0	0.0	0.0	1.1	7.3	29.7
Tokushima	8.0	5.5	2.4	0.0	0.0	0.0	0.0	0.0	0.0	0.0	0.0	3.1	19.1
Takamatsu	5.5	5.0	1.4	0.0	0.0	0.0	0.0	0.0	0.0	0.0	0.0	2.3	14.3
Matsuyama	7.1	5.9	2.1	0.0	0.0	0.0	0.0	0.0	0.0	0.0	0.0	4.7	19.9
Kochi	4.2	3.3	1.5	0.0	0.0	0.0	0.0	0.0	0.0	0.0	0.0	2.2	11.3
Fukuoka	6.9	4.3	1.9	0.0	0.0	0.0	0.0	0.0	0.0	0.0	0.1	3.8	17.1
Saga	9.3	6.0	2.4	0.0	0.0	0.0	0.0	0.0	0.0	0.0	0.0	6.4	24.3
Nagasaki	7.7	5.4	1.3	0.0	0.0	0.0	0.0	0.0	0.0	0.0	0.0	6.0	20.5
Kumamoto	7.3	4.4	2.2	0.0	0.0	0.0	0.0	0.0	0.0	0.0	0.0	5.3	19.3
Oita	5.7	4.9	1.9	0.0	0.0	0.0	0.0	0.0	0.0	0.0	0.0	5.0	17.7
Miyazaki	1.5	1.3	0.2	0.0	0.0	0.0	0.0	0.0	0.0	0.0	0.0	1.2	4.2
Kagoshima	2.3	1.8	0.5	0.0	0.0	0.0	0.0	0.0	0.0	0.0	0.0	0.8	5.5
Naze	0.0	0.0	0.0	0.0	0.0	0.0	0.0	0.0	0.0	0.0	0.0	0.0	0.0
Naha	0.0	0.0	0.0	0.0	0.0	0.0	0.0	0.0	0.0	0.0	0.0	0.0	0.0
Syowa station (Antarctica)	11.0	14.2	19.3	20.1	18.5	17.2	18.9	19.3	18.2	20.2	13.9	10.7	200.6

First and Last Days with Snowfall; Climatological Normals and Records

(Sample Period: Climatological normals are calculated for autumn 1980 to spring 2010.
The earliest and latest dates are taken from the start of observations through the spring of 2019.)

Station	First day					Last day					1st year on record
	Normal date		Earliest on record			Normal date		Latest on record			
Sapporo	10	28	1880	10	5	4	19	1941	5	25	1877
Hakodate	10	29	1878	10	11	4	16	1898	5	22	1873
Asahikawa	10	17	1898	10	2	4	30	1981	5	30	1889
Kushiro	11	7	1923	10	11	5	5	1912	5	21	1911
Obihiro	11	2	1945	10	11	4	28	1941	5	26	1893
Abashiri	10	28	1954	10	7	5	13	1941	6	8	1890
Wakkanai	10	18	1978	10	6	5	14	1981	5	29	1939
Aomori	11	5	1986	10	17	4	16	1983	5	17	1887
Morioka	11	6	1986	10	18	4	23	1983	5	17	1924
Sendai	11	24	1995	11	8	4	7	1991	5	3	1927
Akita	11	13	1986	10	17	4	8	1890	5	14	1883
Yamagata	11	16	1918	10	25	4	11	1892	5	2	1891
Fukushima	11	16	1904	10	31	4	6	1892	5	2	1890
Mito	12	20				3	21				2020
Utsunomiya	12	15				4	3				2020
Maebashi	12	10				3	30				2020
Kumagaya	12	19				3	15				2020
Choshi	1	6				3	11				2020
Tokyo	1	3	1900	11	17	3	11	2010	4	17	1877
Yokohama	12	13				3	21				2020
Niigata	11	24	2009	11	3	3	30	1928	4	23	1886
Toyama	11	30	2002	11	4	4	5	1947	4	23	1940
Kanazawa	11	21	2002	11	5	4	6	1926	4	26	1883
Fukui	11	30	2002	11	5	3	30	1947	4	23	1898
Kofu	12	10				3	20				2020
Nagano	11	17				4	6				2020
Gifu	12	10	1904	11	6	3	27	1928	4	24	1884
Shizuoka	1	5	1973	12	7	2	11	2010	3	29	1941

- No instances.

First and Last Days with Snowfall; Climatological Normals and Records
(Sample Period: Climatological normals are calculated for autumn 1980 to spring 2010.
The earliest and latest dates are taken from the start of observations through the spring of 2019.) Continued.

Station	First day					Last day					1st year on record
	Normal date		Earliest on record			Normal date		Latest on record			
Nagoya	12	20	1904	11	7	3	7	1902	4	11	1891
Tsu	12	22	1922	11	27	3	21	1929	4	14	1890
Hikone	12	6	1904	11	6	3	30	1947	4	22	1894
Kyoto	12	9	1904	11	6	3	25	1996	4	13	1881
Osaka	12	22	1938	11	12	3	11	1996	4	12	1883
Kobe	12	14	1938	11	12	4	1	1996	4	12	1897
Nara	12	12	2013	11	29	3	24	1996	4	13	1954
Wakayama	12	17	1938	11	12	3	14	1902	4	11	1880
Tottori	12	4	1981	11	8	3	28	2010	4	15	1944
Matsue	12	6	1990	11	10	3	28	1996	4	12	1941
Okayama	12	9	1938	11	12	3	18	1996	4	12	1892
Hiroshima	12	11	1921	11	8	3	11	1902	4	12	1880
Shimonoseki	12	8	1950	11	14	3	14	1963	4	9	1884
Tokushima	12	17	1924	11	9	3	7	1909	4	6	1892
Takamatsu	12	23	1966	11	21	3	1	1972	4	1	1942
Matsuyama	12	17	1938	11	12	3	7	1902	4	11	1891
Kochi	12	18	1924	11	9	2	22	1958	3	30	1886
Fukuoka	12	15	1938	11	12	3	5	1962	4	4	1891
Saga	12	11	1924	11	9	3	16	1962	4	4	1891
Nagasaki	12	7	1924	11	9	3	4	1982	4	9	1889
Kumamoto	12	12	1938	11	12	3	10	1962	4	4	1891
Oita	12	5	1938	11	12	3	5	1887	4	4	1888
Miyazaki	1	21	1988	12	16	2	7	1952	3	24	1886
Kagoshima	1	2	1987	12	2	2	16	1958	3	29	1887
Naze			2016	1	24			1901	2	12	1898
Naha			-	-	-			-	-	-	1891
Syowa station (Antarctica)											

- No instances.
The first/last days are not be determined at Syowa station (Antarctica).

Climatological Normal Number of Days per Month with Daily Maximum Snow Depth over 0 cm

(1981–2010 Averages)

Station	Jan.	Feb.	Mar.	Apr.	May	Jun.	Jul.	Aug.	Sep.	Oct.	Nov.	Dec.	Annual
Sapporo	31.0	28.2	29.6	6.4	0.0	0.0	0.0	0.0	0.0	0.6	8.7	27.9	132.4
Hakodate	30.0	26.7	19.6	1.5	0.0	0.0	0.0	0.0	0.0	0.0	7.2	23.1	108.4
Asahikawa	31.0	28.2	30.4	10.8	0.2	0.0	0.0	0.0	0.0	1.0	15.8	30.2	147.7
Kushiro	26.6	25.2	19.7	4.3	0.2	0.0	0.0	0.0	0.0	0.0	2.1	13.2	91.1
Obihiro	30.7	27.7	23.8	4.0	0.3	0.0	0.0	0.0	0.0	0.0	4.2	23.0	113.9
Abashiri	31.0	28.2	30.1	9.8	0.6	0.0	0.0	0.0	0.0	0.3	8.3	27.1	135.3
Rumoi	30.9	28.2	28.4	4.5	0.0	0.0	0.0	0.0	0.0	0.2	9.5	27.4	129.2
Wakkanai	31.0	28.2	29.7	8.7	0.1	0.0	0.0	0.0	0.0	0.7	14.2	29.3	142.0
Nemuro	25.6	24.9	19.8	3.3	0.1	0.0	0.0	0.0	0.0	0.0	1.2	11.6	86.7
Suttu	30.7	28.2	26.1	2.3	0.0	0.0	0.0	0.0	0.0	0.0	7.5	25.5	120.6
Urakawa	23.8	19.9	7.7	0.7	0.0	0.0	0.0	0.0	0.0	0.0	1.6	12.9	66.6
Aomori	30.8	28.1	25.0	3.4	0.0	0.0	0.0	0.0	0.0	0.0	6.6	24.3	118.4
Morioka	26.8	25.3	15.3	1.4	0.0	0.0	0.0	0.0	0.0	0.0	2.8	16.3	88.2
Miyako	9.8	12.9	9.4	0.6	0.0	0.0	0.0	0.0	0.0	0.0	0.1	4.0	36.9
Sendai	13.0	11.0	5.6	0.5	0.0	0.0	0.0	0.0	0.0	0.0	0.4	5.2	36.0
Akita	27.3	24.4	13.5	0.5	0.0	0.0	0.0	0.0	0.0	0.0	3.0	16.6	85.5
Yamagata	27.8	26.0	14.4	1.0	0.0	0.0	0.0	0.0	0.0	0.0	2.4	14.8	86.8
Sakata	22.3	19.5	7.1	0.1	0.0	0.0	0.0	0.0	0.0	0.0	1.2	9.4	59.6
Fukushima	17.9	13.5	6.3	0.4	0.0	0.0	0.0	0.0	0.0	0.0	0.4	6.8	45.9
Mito	2.6	3.1	1.1	0.1	0.0	0.0	0.0	0.0	0.0	0.0	0.0	0.5	7.3
Utsunomiya	3.6	3.8	1.9	0.2	0.0	0.0	0.0	0.0	0.0	0.0	0.0	1.1	10.6
Maebashi	2.9	3.5	1.7	0.2	0.0	0.0	0.0	0.0	0.0	0.0	0.0	1.0	9.4
Kumagaya	2.5	2.8	1.2	0.2	0.0	0.0	0.0	0.0	0.0	0.0	0.0	0.5	7.2
Choshi	0.3	0.4	0.2	0.0	0.0	0.0	0.0	0.0	0.0	0.0	0.0	0.0	1.0
Tokyo	1.6	2.0	0.8	0.1	0.0	0.0	0.0	0.0	0.0	0.0	0.0	0.4	4.8
Yokohama	1.6	2.3	0.7	0.0	0.0	0.0	0.0	0.0	0.0	0.0	0.0	0.3	4.9
Niigata	19.8	18.4	6.5	0.2	0.0	0.0	0.0	0.0	0.0	0.0	0.7	8.6	54.4
Takada	26.1	25.4	17.3	2.7	0.0	0.0	0.0	0.0	0.0	0.0	0.6	12.4	84.7
Aikawa	10.3	9.2	2.4	0.0	0.0	0.0	0.0	0.0	0.0	0.0	0.1	2.9	25.0
Toyama	21.7	19.4	7.9	0.3	0.0	0.0	0.0	0.0	0.0	0.0	0.5	8.6	58.9
Kanazawa	18.7	16.9	6.2	0.2	0.0	0.0	0.0	0.0	0.0	0.0	0.6	7.1	50.0
Wajima	15.7	14.7	4.1	0.1	0.0	0.0	0.0	0.0	0.0	0.0	0.0	5.9	40.5
Fukui	19.9	17.5	5.9	0.1	0.0	0.0	0.0	0.0	0.0	0.0	0.3	7.6	51.5
Tsuruga	13.7	11.2	3.5	0.0	0.0	0.0	0.0	0.0	0.0	0.0	0.1	4.4	33.0
Kofu	4.5	3.1	0.9	0.1	0.0	0.0	0.0	0.0	0.0	0.0	0.0	0.6	9.4

Climatological Normal Number of Days per Month with Daily Maximum Snow Depth over 0 cm (1981–2010 Averages)

Continued.

Station	Jan.	Feb.	Mar.	Apr.	May	Jun.	Jul.	Aug.	Sep.	Oct.	Nov.	Dec.	Annual
Nagano	22.9	19.6	8.2	0.6	0.0	0.0	0.0	0.0	0.0	0.0	0.9	10.8	63.4
Matsumoto	11.2	9.8	4.9	0.4	0.0	0.0	0.0	0.0	0.0	0.0	0.1	3.7	30.4
Iida	8.3	7.3	1.6	0.1	0.0	0.0	0.0	0.0	0.0	0.0	0.0	2.8	20.4
Karuizawa	20.0	21.2	12.7	1.4	0.0	0.0	0.0	0.0	0.0	0.1	0.8	8.5	64.4
Gifu	5.0	4.8	0.9	0.0	0.0	0.0	0.0	0.0	0.0	0.0	0.0	2.4	13.2
Takayama	25.3	24.1	10.5	0.7	0.0	0.0	0.0	0.0	0.0	0.0	1.2	12.4	74.4
Shizuoka	0.0	0.2	0.0	0.0	0.0	0.0	0.0	0.0	0.0	0.0	0.0	0.0	0.2
Nagoya	2.1	2.8	0.4	0.0	0.0	0.0	0.0	0.0	0.0	0.0	0.0	0.9	6.2
Tsu	1.5	2.2	0.4	0.0	0.0	0.0	0.0	0.0	0.0	0.0	0.0	0.4	4.6
Hikone	10.1	9.2	2.3	0.1	0.0	0.0	0.0	0.0	0.0	0.0	0.0	3.5	25.3
Kyoto	3.1	3.9	1.0	0.0	0.0	0.0	0.0	0.0	0.0	0.0	0.0	1.2	9.3
Osaka	0.3	0.9	0.2	0.0	0.0	0.0	0.0	0.0	0.0	0.0	0.0	0.1	1.5
Kobe	0.5	1.2	0.5	0.0	0.0	0.0	0.0	0.0	0.0	0.0	0.0	0.4	2.6
Nara	1.4	1.9	0.8	0.0	0.0	0.0	0.0	0.0	0.0	0.0	0.0	0.7	4.9
Wakayama	0.9	1.0	0.1	0.0	0.0	0.0	0.0	0.0	0.0	0.0	0.0	0.2	2.1
Tottori	15.5	13.2	4.0	0.2	0.0	0.0	0.0	0.0	0.0	0.0	0.1	6.7	39.9
Matsue	11.2	9.1	2.1	0.1	0.0	0.0	0.0	0.0	0.0	0.0	0.2	4.3	27.1
Saigo	12.2	9.3	2.0	0.0	0.0	0.0	0.0	0.0	0.0	0.0	0.0	5.1	29.5
Okayama	0.9	1.1	0.4	0.0	0.0	0.0	0.0	0.0	0.0	0.0	0.0	0.1	2.5
Hiroshima	2.7	2.3	0.7	0.0	0.0	0.0	0.0	0.0	0.0	0.0	0.0	1.2	7.1
Shimonoseki	1.8	1.3	0.4	0.0	0.0	0.0	0.0	0.0	0.0	0.0	0.0	0.5	4.2
Tokushima	1.2	1.2	0.2	0.0	0.0	0.0	0.0	0.0	0.0	0.0	0.0	0.2	2.8
Takamatsu	0.8	1.1	0.3	0.0	0.0	0.0	0.0	0.0	0.0	0.0	0.0	0.1	2.3
Matsuyama	0.8	1.1	0.2	0.0	0.0	0.0	0.0	0.0	0.0	0.0	0.0	0.2	2.3
Kochi	0.4	0.2	0.1	0.0	0.0	0.0	0.0	0.0	0.0	0.0	0.0	0.2	0.9
Fukuoka	1.7	0.9	0.3	0.0	0.0	0.0	0.0	0.0	0.0	0.0	0.1	0.8	3.7
Saga	1.8	1.3	0.3	0.0	0.0	0.0	0.0	0.0	0.0	0.0	0.0	0.6	3.9
Nagasaki	1.2	0.9	0.2	0.0	0.0	0.0	0.0	0.0	0.0	0.0	0.0	0.2	2.5
Kumamoto	1.0	0.8	0.1	0.0	0.0	0.0	0.0	0.0	0.0	0.0	0.0	0.3	2.2
Oita	0.5	0.4	0.1	0.0	0.0	0.0	0.0	0.0	0.0	0.0	0.0	0.1	1.1
Miyazaki	0.1	0.0	0.0	0.0	0.0	0.0	0.0	0.0	0.0	0.0	0.0	0.0	0.2
Kagoshima	0.6	0.6	0.0	0.0	0.0	0.0	0.0	0.0	0.0	0.0	0.0	0.1	1.4
Naze	0.0	0.0	0.0	0.0	0.0	0.0	0.0	0.0	0.0	0.0	0.0	0.0	0.0
Naha	0.0	0.0	0.0	0.0	0.0	0.0	0.0	0.0	0.0	0.0	0.0	0.0	0.0
Syowa station (Antarctica)	13.3	17.5	29.6	29.5	30.8	29.3	30.3	30.6	29.5	30.5	30.0	28.5	

Climatological Normal Number of Days per Month with Daily Maximum Snow Depth over 10 cm

(1981-2010 Averages)

Station	Jan.	Feb.	Mar.	Apr.	May	Jun.	Jul.	Aug.	Sep.	Oct.	Nov.	Dec.	Annual
Sapporo	30.8	28.2	27.9	3.5	0.0	0.0	0.0	0.0	0.0	0.0	1.7	19.5	111.9
Hakodate	22.2	23.4	10.3	0.2	0.0	0.0	0.0	0.0	0.0	0.0	1.1	8.7	66.2
Asahikawa	31.0	28.2	29.5	6.9	0.0	0.0	0.0	0.0	0.0	0.1	8.4	27.2	131.6
Kushiro	15.0	17.4	8.9	0.5	0.0	0.0	0.0	0.0	0.0	0.0	0.2	3.4	45.3
Obihiro	27.6	26.2	18.4	0.9	0.0	0.0	0.0	0.0	0.0	0.0	1.1	14.8	89.5
Abashiri	29.2	27.6	25.4	2.9	0.0	0.0	0.0	0.0	0.0	0.0	0.6	14.6	100.6
Rumoi	30.2	28.2	26.9	2.9	0.0	0.0	0.0	0.0	0.0	0.0	2.0	19.8	110.5
Wakkanai	30.4	28.2	27.2	4.2	0.0	0.0	0.0	0.0	0.0	0.0	2.4	20.5	113.1
Nemuro	11.3	15.6	10.6	0.7	0.0	0.0	0.0	0.0	0.0	0.0	0.0	2.6	40.7
Suttu	28.1	27.3	23.1	1.2	0.0	0.0	0.0	0.0	0.0	0.0	1.2	15.3	96.6
Urakawa	9.4	8.6	1.4	0.0	0.0	0.0	0.0	0.0	0.0	0.0	0.1	2.0	21.4
Aomori	29.8	27.9	21.7	1.9	0.0	0.0	0.0	0.0	0.0	0.0	2.1	17.4	101.0
Morioka	17.4	18.6	6.6	0.1	0.0	0.0	0.0	0.0	0.0	0.0	0.3	4.6	47.9
Miyako	3.1	7.1	4.3	0.1	0.0	0.0	0.0	0.0	0.0	0.0	0.0	1.0	15.6
Sendai	1.5	1.5	0.8	0.0	0.0	0.0	0.0	0.0	0.0	0.0	0.0	0.4	4.6
Akita	15.8	17.0	3.8	0.0	0.0	0.0	0.0	0.0	0.0	0.0	0.4	4.7	41.9
Yamagata	20.3	21.0	7.2	0.2	0.0	0.0	0.0	0.0	0.0	0.0	0.2	5.5	54.7
Sakata	12.7	12.0	2.1	0.0	0.0	0.0	0.0	0.0	0.0	0.0	0.3	3.7	30.7
Fukushima	6.5	3.5	1.0	0.1	0.0	0.0	0.0	0.0	0.0	0.0	0.0	1.5	13.0
Mito	0.3	0.5	0.1	0.0	0.0	0.0	0.0	0.0	0.0	0.0	0.0	0.0	0.9
Utsunomiya	0.6	0.4	0.2	0.0	0.0	0.0	0.0	0.0	0.0	0.0	0.0	0.1	1.3
Maebashi	0.7	0.5	0.2	0.0	0.0	0.0	0.0	0.0	0.0	0.0	0.0	0.1	1.4
Kumagaya	0.5	0.5	0.1	0.0	0.0	0.0	0.0	0.0	0.0	0.0	0.0	0.0	1.1
Choshi	0.0	0.0	0.0	0.0	0.0	0.0	0.0	0.0	0.0	0.0	0.0	0.0	0.0
Tokyo	0.3	0.4	0.0	0.0	0.0	0.0	0.0	0.0	0.0	0.0	0.0	0.0	0.7
Yokohama	0.4	0.6	0.0	0.0	0.0	0.0	0.0	0.0	0.0	0.0	0.0	0.0	1.1
Niigata	9.0	8.8	1.9	0.0	0.0	0.0	0.0	0.0	0.0	0.0	0.0	2.5	22.2
Takada	22.2	24.6	14.8	2.1	0.0	0.0	0.0	0.0	0.0	0.0	0.1	7.6	71.6
Aikawa	2.2	2.3	0.5	0.0	0.0	0.0	0.0	0.0	0.0	0.0	0.0	0.9	5.8
Toyama	15.3	13.5	3.8	0.0	0.0	0.0	0.0	0.0	0.0	0.0	0.0	4.2	37.0
Kanazawa	9.8	9.1	2.5	0.0	0.0	0.0	0.0	0.0	0.0	0.0	0.1	2.6	24.1
Wajima	8.2	7.2	1.4	0.0	0.0	0.0	0.0	0.0	0.0	0.0	0.0	1.9	18.9
Fukui	13.6	11.3	3.0	0.0	0.0	0.0	0.0	0.0	0.0	0.0	0.0	3.3	31.3
Tsuruga	8.2	7.3	2.1	0.0	0.0	0.0	0.0	0.0	0.0	0.0	0.0	2.2	20.0
Kofu	1.5	0.9	0.3	0.0	0.0	0.0	0.0	0.0	0.0	0.0	0.0	0.0	2.7

Climatological Normal Number of Days per Month with Daily Maximum Snow Depth over 10 cm (1981–2010 Averages)

Continued.

Station	Jan.	Feb.	Mar.	Apr.	May	Jun.	Jul.	Aug.	Sep.	Oct.	Nov.	Dec.	Annual
Nagano	11.7	9.1	1.6	0.0	0.0	0.0	0.0	0.0	0.0	0.0	0.0	3.1	26.2
Matsumoto	3.2	4.0	1.3	0.1	0.0	0.0	0.0	0.0	0.0	0.0	0.0	0.5	9.4
Iida	2.2	2.2	0.3	0.0	0.0	0.0	0.0	0.0	0.0	0.0	0.0	0.5	5.2
Karuizawa	9.6	12.4	7.3	0.6	0.0	0.0	0.0	0.0	0.0	0.1	0.0	2.6	32.9
Gifu	1.3	0.7	0.0	0.0	0.0	0.0	0.0	0.0	0.0	0.0	0.0	0.5	2.5
Takayama	18.4	19.7	5.4	0.2	0.0	0.0	0.0	0.0	0.0	0.0	0.2	4.4	48.9
Shizuoka	0.0	0.0	0.0	0.0	0.0	0.0	0.0	0.0	0.0	0.0	0.0	0.0	0.0
Nagoya	0.2	0.5	0.0	0.0	0.0	0.0	0.0	0.0	0.0	0.0	0.0	0.1	0.8
Tsu	0.1	0.1	0.0	0.0	0.0	0.0	0.0	0.0	0.0	0.0	0.0	0.0	0.1
Hikone	3.0	3.1	0.2	0.0	0.0	0.0	0.0	0.0	0.0	0.0	0.0	1.2	7.6
Kyoto	0.1	0.1	0.1	0.0	0.0	0.0	0.0	0.0	0.0	0.0	0.0	0.1	0.4
Osaka	0.0	0.0	0.0	0.0	0.0	0.0	0.0	0.0	0.0	0.0	0.0	0.0	0.1
Kobe	0.0	0.0	0.0	0.0	0.0	0.0	0.0	0.0	0.0	0.0	0.0	0.0	0.0
Nara	0.0	0.2	0.1	0.0	0.0	0.0	0.0	0.0	0.0	0.0	0.0	0.0	0.4
Wakayama	0.0	0.0	0.0	0.0	0.0	0.0	0.0	0.0	0.0	0.0	0.0	0.0	0.0
Tottori	8.8	7.2	1.3	0.0	0.0	0.0	0.0	0.0	0.0	0.0	0.0	2.6	20.0
Matsue	3.2	1.5	0.3	0.0	0.0	0.0	0.0	0.0	0.0	0.0	0.0	1.0	6.1
Saigo	4.6	3.2	0.5	0.0	0.0	0.0	0.0	0.0	0.0	0.0	0.0	2.0	10.9
Okayama	0.0	0.0	0.0	0.0	0.0	0.0	0.0	0.0	0.0	0.0	0.0	0.0	0.0
Hiroshima	0.1	0.1	0.0	0.0	0.0	0.0	0.0	0.0	0.0	0.0	0.0	0.1	0.3
Shimonoseki	0.0	0.0	0.0	0.0	0.0	0.0	0.0	0.0	0.0	0.0	0.0	0.0	0.1
Tokushima	0.0	0.0	0.0	0.0	0.0	0.0	0.0	0.0	0.0	0.0	0.0	0.0	0.1
Takamatsu	0.1	0.0	0.0	0.0	0.0	0.0	0.0	0.0	0.0	0.0	0.0	0.0	0.1
Matsuyama	0.1	0.0	0.0	0.0	0.0	0.0	0.0	0.0	0.0	0.0	0.0	0.0	0.1
Kochi	0.0	0.0	0.0	0.0	0.0	0.0	0.0	0.0	0.0	0.0	0.0	0.0	0.0
Fukuoka	0.0	0.0	0.0	0.0	0.0	0.0	0.0	0.0	0.0	0.0	0.0	0.0	0.1
Saga	0.1	0.0	0.0	0.0	0.0	0.0	0.0	0.0	0.0	0.0	0.0	0.0	0.1
Nagasaki	0.1	0.0	0.0	0.0	0.0	0.0	0.0	0.0	0.0	0.0	0.0	0.0	0.1
Kumamoto	0.0	0.0	0.0	0.0	0.0	0.0	0.0	0.0	0.0	0.0	0.0	0.0	0.0
Oita	0.1	0.0	0.0	0.0	0.0	0.0	0.0	0.0	0.0	0.0	0.0	0.0	0.1
Miyazaki	0.0	0.0	0.0	0.0	0.0	0.0	0.0	0.0	0.0	0.0	0.0	0.0	0.0
Kagoshima	0.1	0.0	0.0	0.0	0.0	0.0	0.0	0.0	0.0	0.0	0.0	0.1	0.1
Naze	0.0	0.0	0.0	0.0	0.0	0.0	0.0	0.0	0.0	0.0	0.0	0.0	0.0
Naha	0.0	0.0	0.0	0.0	0.0	0.0	0.0	0.0	0.0	0.0	0.0	0.0	0.0
Syowa station (Antarctica)	11.4	10.9	16.7	23.0	29.8	29.3	30.3	30.6	29.5	30.5	30.0	27.3	

Climatological Normal Number of Days per Month with Daily Maximum Snow Depth over 20 cm

(1981–2010 Averages)

Station	Jan.	Feb.	Mar.	Apr.	May	Jun.	Jul.	Aug.	Sep.	Oct.	Nov.	Dec.	Annual
Sapporo	29.5	28.2	26.5	2.5	0.0	0.0	0.0	0.0	0.0	0.0	0.5	12.9	100.5
Hakodate	13.5	19.3	6.5	0.0	0.0	0.0	0.0	0.0	0.0	0.0	0.2	2.7	42.2
Asahikawa	30.9	28.2	28.4	5.4	0.0	0.0	0.0	0.0	0.0	0.0	4.0	23.1	120.5
Kushiro	8.2	9.8	4.7	0.1	0.0	0.0	0.0	0.0	0.0	0.0	0.0	0.8	23.6
Obihiro	22.0	23.3	14.9	0.2	0.0	0.0	0.0	0.0	0.0	0.0	0.3	8.1	68.4
Abashiri	20.7	22.9	20.0	1.7	0.0	0.0	0.0	0.0	0.0	0.0	0.0	6.3	71.8
Rumoi	28.7	28.1	25.1	2.0	0.0	0.0	0.0	0.0	0.0	0.0	0.5	12.1	96.9
Wakkanai	26.7	27.8	25.3	3.0	0.0	0.0	0.0	0.0	0.0	0.0	0.9	13.4	97.4
Nemuro	5.0	9.0	5.6	0.3	0.0	0.0	0.0	0.0	0.0	0.0	0.0	0.8	20.8
Suttu	22.3	26.2	19.5	0.6	0.0	0.0	0.0	0.0	0.0	0.0	0.1	7.0	75.7
Urakawa	1.5	2.3	0.3	0.0	0.0	0.0	0.0	0.0	0.0	0.0	0.0	0.4	4.4
Aomori	26.4	27.0	20.1	1.5	0.0	0.0	0.0	0.0	0.0	0.0	0.8	11.5	87.6
Morioka	9.0	12.4	3.9	0.0	0.0	0.0	0.0	0.0	0.0	0.0	0.0	1.2	26.6
Miyako	1.1	3.7	1.6	0.0	0.0	0.0	0.0	0.0	0.0	0.0	0.0	0.0	6.4
Sendai	0.3	0.4	0.2	0.0	0.0	0.0	0.0	0.0	0.0	0.0	0.0	0.0	1.0
Akita	9.0	10.2	1.8	0.0	0.0	0.0	0.0	0.0	0.0	0.0	0.1	1.5	22.7
Yamagata	12.5	15.5	4.4	0.0	0.0	0.0	0.0	0.0	0.0	0.0	0.1	2.6	35.5
Sakata	6.1	5.2	0.9	0.0	0.0	0.0	0.0	0.0	0.0	0.0	0.2	1.7	14.1
Fukushima	1.6	1.1	0.2	0.0	0.0	0.0	0.0	0.0	0.0	0.0	0.0	0.2	3.3
Mito	0.0	0.1	0.0	0.0	0.0	0.0	0.0	0.0	0.0	0.0	0.0	0.0	0.1
Utsunomiya	0.1	0.1	0.0	0.0	0.0	0.0	0.0	0.0	0.0	0.0	0.0	0.0	0.3
Maebashi	0.3	0.1	0.0	0.0	0.0	0.0	0.0	0.0	0.0	0.0	0.0	0.0	0.4
Kumagaya	0.1	0.1	0.0	0.0	0.0	0.0	0.0	0.0	0.0	0.0	0.0	0.0	0.2
Choshi	0.0	0.0	0.0	0.0	0.0	0.0	0.0	0.0	0.0	0.0	0.0	0.0	0.0
Tokyo	0.0	0.1	0.0	0.0	0.0	0.0	0.0	0.0	0.0	0.0	0.0	0.0	0.1
Yokohama	0.1	0.2	0.0	0.0	0.0	0.0	0.0	0.0	0.0	0.0	0.0	0.0	0.3
Niigata	5.2	5.4	0.9	0.0	0.0	0.0	0.0	0.0	0.0	0.0	0.0	1.1	12.6
Takada	18.5	23.3	13.1	1.9	0.0	0.0	0.0	0.0	0.0	0.0	0.0	4.5	61.6
Aikawa	0.5	0.6	0.1	0.0	0.0	0.0	0.0	0.0	0.0	0.0	0.0	0.2	1.4
Toyama	11.0	9.8	2.8	0.0	0.0	0.0	0.0	0.0	0.0	0.0	0.0	2.7	26.4
Kanazawa	6.6	5.6	1.4	0.0	0.0	0.0	0.0	0.0	0.0	0.0	0.0	1.6	15.2
Wajima	4.6	3.9	1.0	0.0	0.0	0.0	0.0	0.0	0.0	0.0	0.0	0.8	10.3
Fukui	8.7	7.8	2.1	0.0	0.0	0.0	0.0	0.0	0.0	0.0	0.0	2.0	20.7
Tsuruga	5.2	5.1	1.5	0.0	0.0	0.0	0.0	0.0	0.0	0.0	0.0	1.5	13.5
Kofu	0.4	0.2	0.0	0.0	0.0	0.0	0.0	0.0	0.0	0.0	0.0	0.0	0.7

Climatological Normal Number of Days per Month with Daily Maximum Snow Depth over 20 cm (1981–2010 Averages) Continued.

Station	Jan.	Feb.	Mar.	Apr.	May	Jun.	Jul.	Aug.	Sep.	Oct.	Nov.	Dec.	Annual
Nagano	5.1	3.6	0.7	0.0	0.0	0.0	0.0	0.0	0.0	0.0	0.0	1.0	10.8
Matsumoto	1.6	1.2	0.4	0.0	0.0	0.0	0.0	0.0	0.0	0.0	0.0	0.2	3.6
Iida	0.7	0.8	0.0	0.0	0.0	0.0	0.0	0.0	0.0	0.0	0.0	0.1	1.6
Karuizawa	4.2	7.9	4.3	0.1	0.0	0.0	0.0	0.0	0.0	0.0	0.0	0.7	17.7
Gifu	0.3	0.2	0.0	0.0	0.0	0.0	0.0	0.0	0.0	0.0	0.0	0.3	0.8
Takayama	11.4	13.0	3.3	0.0	0.0	0.0	0.0	0.0	0.0	0.0	0.0	2.1	30.2
Shizuoka	0.0	0.0	0.0	0.0	0.0	0.0	0.0	0.0	0.0	0.0	0.0	0.0	0.0
Nagoya	0.0	0.0	0.0	0.0	0.0	0.0	0.0	0.0	0.0	0.0	0.0	0.0	0.0
Tsu	0.0	0.0	0.0	0.0	0.0	0.0	0.0	0.0	0.0	0.0	0.0	0.0	0.0
Hikone	1.2	1.5	0.1	0.0	0.0	0.0	0.0	0.0	0.0	0.0	0.0	0.4	3.2
Kyoto	0.0	0.0	0.0	0.0	0.0	0.0	0.0	0.0	0.0	0.0	0.0	0.0	0.0
Osaka	0.0	0.0	0.0	0.0	0.0	0.0	0.0	0.0	0.0	0.0	0.0	0.0	0.0
Kobe	0.0	0.0	0.0	0.0	0.0	0.0	0.0	0.0	0.0	0.0	0.0	0.0	0.0
Nara	0.0	0.0	0.0	0.0	0.0	0.0	0.0	0.0	0.0	0.0	0.0	0.0	0.0
Wakayama	0.0	0.0	0.0	0.0	0.0	0.0	0.0	0.0	0.0	0.0	0.0	0.0	0.0
Tottori	5.9	4.3	0.7	0.0	0.0	0.0	0.0	0.0	0.0	0.0	0.0	1.3	12.2
Matsue	1.0	0.3	0.1	0.0	0.0	0.0	0.0	0.0	0.0	0.0	0.0	0.4	1.8
Saigo	1.9	1.5	0.1	0.0	0.0	0.0	0.0	0.0	0.0	0.0	0.0	0.9	4.7
Okayama	0.0	0.0	0.0	0.0	0.0	0.0	0.0	0.0	0.0	0.0	0.0	0.0	0.0
Hiroshima	0.0	0.0	0.0	0.0	0.0	0.0	0.0	0.0	0.0	0.0	0.0	0.0	0.0
Shimonoseki	0.0	0.0	0.0	0.0	0.0	0.0	0.0	0.0	0.0	0.0	0.0	0.0	0.0
Tokushima	0.0	0.0	0.0	0.0	0.0	0.0	0.0	0.0	0.0	0.0	0.0	0.0	0.0
Takamatsu	0.0	0.0	0.0	0.0	0.0	0.0	0.0	0.0	0.0	0.0	0.0	0.0	0.0
Matsuyama	0.0	0.0	0.0	0.0	0.0	0.0	0.0	0.0	0.0	0.0	0.0	0.0	0.0
Kochi	0.0	0.0	0.0	0.0	0.0	0.0	0.0	0.0	0.0	0.0	0.0	0.0	0.0
Fukuoka	0.0	0.0	0.0	0.0	0.0	0.0	0.0	0.0	0.0	0.0	0.0	0.0	0.0
Saga	0.0	0.0	0.0	0.0	0.0	0.0	0.0	0.0	0.0	0.0	0.0	0.0	0.0
Nagasaki	0.0	0.0	0.0	0.0	0.0	0.0	0.0	0.0	0.0	0.0	0.0	0.0	0.0
Kumamoto	0.0	0.0	0.0	0.0	0.0	0.0	0.0	0.0	0.0	0.0	0.0	0.0	0.0
Oita	0.0	0.0	0.0	0.0	0.0	0.0	0.0	0.0	0.0	0.0	0.0	0.0	0.0
Miyazaki	0.0	0.0	0.0	0.0	0.0	0.0	0.0	0.0	0.0	0.0	0.0	0.0	0.0
Kagoshima	0.0	0.0	0.0	0.0	0.0	0.0	0.0	0.0	0.0	0.0	0.0	0.0	0.0
Naze	0.0	0.0	0.0	0.0	0.0	0.0	0.0	0.0	0.0	0.0	0.0	0.0	0.0
Naha	0.0	0.0	0.0	0.0	0.0	0.0	0.0	0.0	0.0	0.0	0.0	0.0	0.0
Syowa station (Antarctica)	9.2	6.6	11.6	15.3	21.0	21.6	26.0	26.9	27.2	28.5	29.8	24.0	

Climatological Normal Number of Days per Month with Daily Maximum Snow Depth over 50 cm

(1981–2010 Averages)

Station	Jan.	Feb.	Mar.	Apr.	May	Jun.	Jul.	Aug.	Sep.	Oct.	Nov.	Dec.	Annual
Sapporo	15.6	24.7	18.6	1.0	0.0	0.0	0.0	0.0	0.0	0.0	0.0	2.0	62.0
Hakodate	1.0	2.9	1.6	0.0	0.0	0.0	0.0	0.0	0.0	0.0	0.0	0.0	5.5
Asahikawa	21.3	25.8	21.7	1.7	0.0	0.0	0.0	0.0	0.0	0.0	0.2	5.2	75.9
Kushiro	0.3	0.5	0.4	0.0	0.0	0.0	0.0	0.0	0.0	0.0	0.0	0.0	1.2
Obihiro	6.6	7.3	4.1	0.0	0.0	0.0	0.0	0.0	0.0	0.0	0.0	0.7	18.7
Abashiri	3.0	6.1	5.7	0.0	0.0	0.0	0.0	0.0	0.0	0.0	0.0	0.1	13.4
Rumoi	12.5	19.8	14.6	0.6	0.0	0.0	0.0	0.0	0.0	0.0	0.0	2.3	50.1
Wakkanai	10.1	18.8	14.9	0.4	0.0	0.0	0.0	0.0	0.0	0.0	0.1	2.3	46.6
Nemuro	0.1	0.3	0.2	0.0	0.0	0.0	0.0	0.0	0.0	0.0	0.0	0.0	0.7
Suttu	6.9	15.5	8.6	0.1	0.0	0.0	0.0	0.0	0.0	0.0	0.0	0.5	32.5
Urakawa	0.0	0.0	0.0	0.0	0.0	0.0	0.0	0.0	0.0	0.0	0.0	0.0	0.0
Aomori	14.9	22.1	13.3	0.8	0.0	0.0	0.0	0.0	0.0	0.0	0.1	3.1	54.2
Morioka	0.6	1.0	0.1	0.0	0.0	0.0	0.0	0.0	0.0	0.0	0.0	0.3	1.9
Miyako	0.0	0.3	0.1	0.0	0.0	0.0	0.0	0.0	0.0	0.0	0.0	0.0	0.4
Sendai	0.0	0.0	0.0	0.0	0.0	0.0	0.0	0.0	0.0	0.0	0.0	0.0	0.0
Akita	0.6	0.9	0.2	0.0	0.0	0.0	0.0	0.0	0.0	0.0	0.0	0.3	2.0
Yamagata	2.5	3.9	1.3	0.0	0.0	0.0	0.0	0.0	0.0	0.0	0.0	0.1	8.1
Sakata	0.1	0.5	0.1	0.0	0.0	0.0	0.0	0.0	0.0	0.0	0.0	0.0	0.7
Fukushima	0.0	0.0	0.0	0.0	0.0	0.0	0.0	0.0	0.0	0.0	0.0	0.0	0.0
Mito	0.0	0.0	0.0	0.0	0.0	0.0	0.0	0.0	0.0	0.0	0.0	0.0	0.0
Utsunomiya	0.0	0.0	0.0	0.0	0.0	0.0	0.0	0.0	0.0	0.0	0.0	0.0	0.0
Maebashi	0.0	0.0	0.0	0.0	0.0	0.0	0.0	0.0	0.0	0.0	0.0	0.0	0.0
Kumagaya	0.0	0.0	0.0	0.0	0.0	0.0	0.0	0.0	0.0	0.0	0.0	0.0	0.0
Choshi	0.0	0.0	0.0	0.0	0.0	0.0	0.0	0.0	0.0	0.0	0.0	0.0	0.0
Tokyo	0.0	0.0	0.0	0.0	0.0	0.0	0.0	0.0	0.0	0.0	0.0	0.0	0.0
Yokohama	0.0	0.0	0.0	0.0	0.0	0.0	0.0	0.0	0.0	0.0	0.0	0.0	0.0
Niigata	0.5	1.4	0.3	0.0	0.0	0.0	0.0	0.0	0.0	0.0	0.0	0.1	2.3
Takada	10.0	17.6	8.8	1.4	0.0	0.0	0.0	0.0	0.0	0.0	0.0	1.9	39.8
Aikawa	0.0	0.1	0.0	0.0	0.0	0.0	0.0	0.0	0.0	0.0	0.0	0.0	0.1
Toyama	3.9	4.7	1.5	0.0	0.0	0.0	0.0	0.0	0.0	0.0	0.0	0.8	11.0
Kanazawa	2.5	2.7	0.4	0.0	0.0	0.0	0.0	0.0	0.0	0.0	0.0	0.2	5.9
Wajima	0.6	1.1	0.5	0.0	0.0	0.0	0.0	0.0	0.0	0.0	0.0	0.0	2.2
Fukui	3.9	3.0	1.0	0.0	0.0	0.0	0.0	0.0	0.0	0.0	0.0	0.8	8.8
Tsuruga	2.2	2.8	0.7	0.0	0.0	0.0	0.0	0.0	0.0	0.0	0.0	0.5	6.3
Kofu	0.0	0.0	0.0	0.0	0.0	0.0	0.0	0.0	0.0	0.0	0.0	0.0	0.0

Climatological Normal Number of Days per Month with Daily Maximum Snow Depth over 50 cm (1981–2010 Averages) Continued.

Station	Jan.	Feb.	Mar.	Apr.	May	Jun.	Jul.	Aug.	Sep.	Oct.	Nov.	Dec.	Annual
Nagano	0.0	0.2	0.0	0.0	0.0	0.0	0.0	0.0	0.0	0.0	0.0	0.0	0.2
Matsumoto	0.2	0.0	0.0	0.0	0.0	0.0	0.0	0.0	0.0	0.0	0.0	0.0	0.2
Iida	0.1	0.0	0.0	0.0	0.0	0.0	0.0	0.0	0.0	0.0	0.0	0.0	0.1
Karuizawa	0.5	0.2	0.3	0.0	0.0	0.0	0.0	0.0	0.0	0.0	0.0	0.0	1.1
Gifu	0.0	0.0	0.0	0.0	0.0	0.0	0.0	0.0	0.0	0.0	0.0	0.0	0.0
Takayama	2.6	3.5	1.1	0.0	0.0	0.0	0.0	0.0	0.0	0.0	0.0	0.6	8.0
Shizuoka	0.0	0.0	0.0	0.0	0.0	0.0	0.0	0.0	0.0	0.0	0.0	0.0	0.0
Nagoya	0.0	0.0	0.0	0.0	0.0	0.0	0.0	0.0	0.0	0.0	0.0	0.0	0.0
Tsu	0.0	0.0	0.0	0.0	0.0	0.0	0.0	0.0	0.0	0.0	0.0	0.0	0.0
Hikone	0.0	0.2	0.0	0.0	0.0	0.0	0.0	0.0	0.0	0.0	0.0	0.0	0.2
Kyoto	0.0	0.0	0.0	0.0	0.0	0.0	0.0	0.0	0.0	0.0	0.0	0.0	0.0
Osaka	0.0	0.0	0.0	0.0	0.0	0.0	0.0	0.0	0.0	0.0	0.0	0.0	0.0
Kobe	0.0	0.0	0.0	0.0	0.0	0.0	0.0	0.0	0.0	0.0	0.0	0.0	0.0
Nara	0.0	0.0	0.0	0.0	0.0	0.0	0.0	0.0	0.0	0.0	0.0	0.0	0.0
Wakayama	0.0	0.0	0.0	0.0	0.0	0.0	0.0	0.0	0.0	0.0	0.0	0.0	0.0
Tottori	0.9	1.1	0.1	0.0	0.0	0.0	0.0	0.0	0.0	0.0	0.0	0.3	2.5
Matsue	0.0	0.0	0.0	0.0	0.0	0.0	0.0	0.0	0.0	0.0	0.0	0.1	0.1
Saigo	0.0	0.0	0.0	0.0	0.0	0.0	0.0	0.0	0.0	0.0	0.0	0.1	0.2
Okayama	0.0	0.0	0.0	0.0	0.0	0.0	0.0	0.0	0.0	0.0	0.0	0.0	0.0
Hiroshima	0.0	0.0	0.0	0.0	0.0	0.0	0.0	0.0	0.0	0.0	0.0	0.0	0.0
Shimonoseki	0.0	0.0	0.0	0.0	0.0	0.0	0.0	0.0	0.0	0.0	0.0	0.0	0.0
Tokushima	0.0	0.0	0.0	0.0	0.0	0.0	0.0	0.0	0.0	0.0	0.0	0.0	0.0
Takamatsu	0.0	0.0	0.0	0.0	0.0	0.0	0.0	0.0	0.0	0.0	0.0	0.0	0.0
Matsuyama	0.0	0.0	0.0	0.0	0.0	0.0	0.0	0.0	0.0	0.0	0.0	0.0	0.0
Kochi	0.0	0.0	0.0	0.0	0.0	0.0	0.0	0.0	0.0	0.0	0.0	0.0	0.0
Fukuoka	0.0	0.0	0.0	0.0	0.0	0.0	0.0	0.0	0.0	0.0	0.0	0.0	0.0
Saga	0.0	0.0	0.0	0.0	0.0	0.0	0.0	0.0	0.0	0.0	0.0	0.0	0.0
Nagasaki	0.0	0.0	0.0	0.0	0.0	0.0	0.0	0.0	0.0	0.0	0.0	0.0	0.0
Kumamoto	0.0	0.0	0.0	0.0	0.0	0.0	0.0	0.0	0.0	0.0	0.0	0.0	0.0
Oita	0.0	0.0	0.0	0.0	0.0	0.0	0.0	0.0	0.0	0.0	0.0	0.0	0.0
Miyazaki	0.0	0.0	0.0	0.0	0.0	0.0	0.0	0.0	0.0	0.0	0.0	0.0	0.0
Kagoshima	0.0	0.0	0.0	0.0	0.0	0.0	0.0	0.0	0.0	0.0	0.0	0.0	0.0
Naze	0.0	0.0	0.0	0.0	0.0	0.0	0.0	0.0	0.0	0.0	0.0	0.0	0.0
Naha	0.0	0.0	0.0	0.0	0.0	0.0	0.0	0.0	0.0	0.0	0.0	0.0	0.0
Syowa station (Antarctica)	3.4	2.2	2.3	7.3	5.8					18.7	14.1	12.7	

Climatological Normal Number of Days per Month with Daily Maximum Snow Depth over 100 cm

(1981-2010 Averages)

Station	Jan.	Feb.	Mar.	Apr.	May	Jun.	Jul.	Aug.	Sep.	Oct.	Nov.	Dec.	Annual
Sapporo	0.5	3.3	1.1	0.0	0.0	0.0	0.0	0.0	0.0	0.0	0.0	0.0	4.9
Hakodate	0.0	0.0	0.0	0.0	0.0	0.0	0.0	0.0	0.0	0.0	0.0	0.0	0.0
Asahikawa	0.6	2.7	3.2	0.0	0.0	0.0	0.0	0.0	0.0	0.0	0.0	0.0	6.5
Kushiro	0.0	0.0	0.0	0.0	0.0	0.0	0.0	0.0	0.0	0.0	0.0	0.0	0.0
Obihiro	0.0	0.1	0.0	0.0	0.0	0.0	0.0	0.0	0.0	0.0	0.0	0.0	0.1
Abashiri	0.0	0.2	0.1	0.0	0.0	0.0	0.0	0.0	0.0	0.0	0.0	0.0	0.2
Rumoi	0.4	2.1	1.4	0.0	0.0	0.0	0.0	0.0	0.0	0.0	0.0	0.0	3.9
Wakkanai	0.0	0.4	0.5	0.0	0.0	0.0	0.0	0.0	0.0	0.0	0.0	0.0	1.0
Nemuro	0.0	0.0	0.0	0.0	0.0	0.0	0.0	0.0	0.0	0.0	0.0	0.0	0.0
Suttu	0.0	0.2	0.0	0.0	0.0	0.0	0.0	0.0	0.0	0.0	0.0	0.0	0.2
Urakawa	0.0	0.0	0.0	0.0	0.0	0.0	0.0	0.0	0.0	0.0	0.0	0.0	0.0
Aomori	3.4	8.1	4.7	0.1	0.0	0.0	0.0	0.0	0.0	0.0	0.0	0.1	16.5
Morioka	0.0	0.0	0.0	0.0	0.0	0.0	0.0	0.0	0.0	0.0	0.0	0.0	0.0
Miyako	0.0	0.0	0.0	0.0	0.0	0.0	0.0	0.0	0.0	0.0	0.0	0.0	0.0
Sendai	0.0	0.0	0.0	0.0	0.0	0.0	0.0	0.0	0.0	0.0	0.0	0.0	0.0
Akita	0.0	0.0	0.0	0.0	0.0	0.0	0.0	0.0	0.0	0.0	0.0	0.0	0.0
Yamagata	0.1	0.0	0.0	0.0	0.0	0.0	0.0	0.0	0.0	0.0	0.0	0.0	0.1
Sakata	0.0	0.0	0.0	0.0	0.0	0.0	0.0	0.0	0.0	0.0	0.0	0.0	0.0
Fukushima	0.0	0.0	0.0	0.0	0.0	0.0	0.0	0.0	0.0	0.0	0.0	0.0	0.0
Mito	0.0	0.0	0.0	0.0	0.0	0.0	0.0	0.0	0.0	0.0	0.0	0.0	0.0
Utsunomiya	0.0	0.0	0.0	0.0	0.0	0.0	0.0	0.0	0.0	0.0	0.0	0.0	0.0
Maebashi	0.0	0.0	0.0	0.0	0.0	0.0	0.0	0.0	0.0	0.0	0.0	0.0	0.0
Kumagaya	0.0	0.0	0.0	0.0	0.0	0.0	0.0	0.0	0.0	0.0	0.0	0.0	0.0
Choshi	0.0	0.0	0.0	0.0	0.0	0.0	0.0	0.0	0.0	0.0	0.0	0.0	0.0
Tokyo	0.0	0.0	0.0	0.0	0.0	0.0	0.0	0.0	0.0	0.0	0.0	0.0	0.0
Yokohama	0.0	0.0	0.0	0.0	0.0	0.0	0.0	0.0	0.0	0.0	0.0	0.0	0.0
Niigata	0.0	0.0	0.0	0.0	0.0	0.0	0.0	0.0	0.0	0.0	0.0	0.0	0.0
Takada	5.1	7.0	4.2	0.5	0.0	0.0	0.0	0.0	0.0	0.0	0.0	0.3	17.1
Aikawa	0.0	0.0	0.0	0.0	0.0	0.0	0.0	0.0	0.0	0.0	0.0	0.0	0.0
Toyama	1.1	0.8	0.0	0.0	0.0	0.0	0.0	0.0	0.0	0.0	0.0	0.0	1.9
Kanazawa	0.7	0.0	0.0	0.0	0.0	0.0	0.0	0.0	0.0	0.0	0.0	0.0	0.8
Wajima	0.0	0.0	0.0	0.0	0.0	0.0	0.0	0.0	0.0	0.0	0.0	0.0	0.0
Fukui	1.0	0.7	0.1	0.0	0.0	0.0	0.0	0.0	0.0	0.0	0.0	0.0	2.0
Tsuruga	0.8	0.6	0.0	0.0	0.0	0.0	0.0	0.0	0.0	0.0	0.0	0.0	1.3
Kofu	0.0	0.0	0.0	0.0	0.0	0.0	0.0	0.0	0.0	0.0	0.0	0.0	0.0

Climatological Normal Number of Days per Month with Daily Maximum Snow Depth over 100 cm (1981–2010 Averages) Continued.

Station	Jan.	Feb.	Mar.	Apr.	May	Jun.	Jul.	Aug.	Sep.	Oct.	Nov.	Dec.	Annual
Nagano	0.0	0.0	0.0	0.0	0.0	0.0	0.0	0.0	0.0	0.0	0.0	0.0	0.0
Matsumoto	0.0	0.0	0.0	0.0	0.0	0.0	0.0	0.0	0.0	0.0	0.0	0.0	0.0
Iida	0.0	0.0	0.0	0.0	0.0	0.0	0.0	0.0	0.0	0.0	0.0	0.0	0.0
Karuizawa	0.0	0.0	0.0	0.0	0.0	0.0	0.0	0.0	0.0	0.0	0.0	0.0	0.0
Gifu	0.0	0.0	0.0	0.0	0.0	0.0	0.0	0.0	0.0	0.0	0.0	0.0	0.0
Takayama	0.4	0.1	0.0	0.0	0.0	0.0	0.0	0.0	0.0	0.0	0.0	0.0	0.5
Shizuoka	0.0	0.0	0.0	0.0	0.0	0.0	0.0	0.0	0.0	0.0	0.0	0.0	0.0
Nagoya	0.0	0.0	0.0	0.0	0.0	0.0	0.0	0.0	0.0	0.0	0.0	0.0	0.0
Tsu	0.0	0.0	0.0	0.0	0.0	0.0	0.0	0.0	0.0	0.0	0.0	0.0	0.0
Hikone	0.0	0.0	0.0	0.0	0.0	0.0	0.0	0.0	0.0	0.0	0.0	0.0	0.0
Kyoto	0.0	0.0	0.0	0.0	0.0	0.0	0.0	0.0	0.0	0.0	0.0	0.0	0.0
Osaka	0.0	0.0	0.0	0.0	0.0	0.0	0.0	0.0	0.0	0.0	0.0	0.0	0.0
Kobe	0.0	0.0	0.0	0.0	0.0	0.0	0.0	0.0	0.0	0.0	0.0	0.0	0.0
Nara	0.0	0.0	0.0	0.0	0.0	0.0	0.0	0.0	0.0	0.0	0.0	0.0	0.0
Wakayama	0.0	0.0	0.0	0.0	0.0	0.0	0.0	0.0	0.0	0.0	0.0	0.0	0.0
Tottori	0.0	0.0	0.0	0.0	0.0	0.0	0.0	0.0	0.0	0.0	0.0	0.0	0.0
Matsue	0.0	0.0	0.0	0.0	0.0	0.0	0.0	0.0	0.0	0.0	0.0	0.0	0.0
Saigo	0.0	0.0	0.0	0.0	0.0	0.0	0.0	0.0	0.0	0.0	0.0	0.0	0.0
Okayama	0.0	0.0	0.0	0.0	0.0	0.0	0.0	0.0	0.0	0.0	0.0	0.0	0.0
Hiroshima	0.0	0.0	0.0	0.0	0.0	0.0	0.0	0.0	0.0	0.0	0.0	0.0	0.0
Shimonoseki	0.0	0.0	0.0	0.0	0.0	0.0	0.0	0.0	0.0	0.0	0.0	0.0	0.0
Tokushima	0.0	0.0	0.0	0.0	0.0	0.0	0.0	0.0	0.0	0.0	0.0	0.0	0.0
Takamatsu	0.0	0.0	0.0	0.0	0.0	0.0	0.0	0.0	0.0	0.0	0.0	0.0	0.0
Matsuyama	0.0	0.0	0.0	0.0	0.0	0.0	0.0	0.0	0.0	0.0	0.0	0.0	0.0
Kochi	0.0	0.0	0.0	0.0	0.0	0.0	0.0	0.0	0.0	0.0	0.0	0.0	0.0
Fukuoka	0.0	0.0	0.0	0.0	0.0	0.0	0.0	0.0	0.0	0.0	0.0	0.0	0.0
Saga	0.0	0.0	0.0	0.0	0.0	0.0	0.0	0.0	0.0	0.0	0.0	0.0	0.0
Nagasaki	0.0	0.0	0.0	0.0	0.0	0.0	0.0	0.0	0.0	0.0	0.0	0.0	0.0
Kumamoto	0.0	0.0	0.0	0.0	0.0	0.0	0.0	0.0	0.0	0.0	0.0	0.0	0.0
Oita	0.0	0.0	0.0	0.0	0.0	0.0	0.0	0.0	0.0	0.0	0.0	0.0	0.0
Miyazaki	0.0	0.0	0.0	0.0	0.0	0.0	0.0	0.0	0.0	0.0	0.0	0.0	0.0
Kagoshima	0.0	0.0	0.0	0.0	0.0	0.0	0.0	0.0	0.0	0.0	0.0	0.0	0.0
Naze	0.0	0.0	0.0	0.0	0.0	0.0	0.0	0.0	0.0	0.0	0.0	0.0	0.0
Naha	0.0	0.0	0.0	0.0	0.0	0.0	0.0	0.0	0.0	0.0	0.0	0.0	0.0
Syowa station (Antarctica)	0.0	0.0	0.0	0.0	0.0					7.4	7.5	3.5	

Deepest Snow Depth on Record

(From the start of record keeping until spring 2019)

Station	Deepest snow depth cm	Yr.	Mo.	Day	1st year on record	Station	Deepest snow depth cm	Yr.	Mo.	Day	1st year on record
Sapporo	169	1939	2	13	1890	Iida	81	2014	2	15	1897
Hakodate	91	2012	2	27	1872	Karuizawa	99	2014	2	15	1925
Asahikawa	138	1987	3	4	1893	Gifu	58	1936	2	1	1891
Kushiro	123	1939	3	9	1910	Takayama	128	1981	1	8	1899
Obihiro	177	1970	3	17	1892	Shizuoka	10	1945	2	25	1940
Abashiri	143	2004	2	23	1892	Hamamatsu	27	1907	2	11	1906
Rumoi	204	1946	3	17	1943	Nagoya	49	1945	12	19	1890
Wakkanai	199	1970	2	9	1938	Tsu	26	1951	2	14	1889
Nemuro	115	2014	3	21	1879	Owase	5	2005	2	1	1938
Suttu	189	1945	3	17	1884	Hikone	93	1918	1	9	1893
Urakawa	52	1928	1	7	1927	Kyoto	41	1954	1	26	1886
Aomori	209	1945	2	21	1894	Osaka	18	1907	2	11	1901
Morioka	81	1938	2	19	1924	Kobe	17	1945	2	25	1914
Miyako	101	1944	3	12	1883	Nara	21	1990	2	1	1953
Sendai	41	1936	2	9	1926	Wakayama	40	1883	2	8	1880
Akita	117	1974	2	10	1890	Shionomisaki	5	1948	1	16	1916
Yamagata	113	1981	1	8	1893	Tottori	129	1947	2	22	1943
Sakata	100	1940	2	3	1938	Matsue	100	1971	2	4	1940
Fukushima	80	1936	2	9	1901	Hamada	53	1982	1	17	1893
Onahama	28	1945	2	26	1916	Saigo	107	1962	1	27	1939
Mito	32	1945	2	26	1897	Okayama	26	1945	2	25	1891
Utsunomiya	32	2014	2	15	1890	Hiroshima	31	1893	1	5	1883
Maebashi	73	2014	2	15	1896	Shimonoseki	39	1900	1	26	1883
Kumagaya	62	2014	2	15	1896	Tokushima	42	1907	2	11	1891
Choshi	17	1936	3	2	1887	Takamatsu	19	1984	1	31	1941
Tokyo	46	1883	2	8	1875	Matsuyama	34	1907	2	11	1890
Oshima	32	1945	2	22	1939	Kochi	10	1987	1	13	1912
Hachijojima	3	2006	2	4	1906	Murotomisaki	4	1986	2	11	1920
Yokohama	45	1945	2	26	1896	Shimizu	4	1968	1	15	1941
Niigata	120	1961	1	18	1890	Fukuoka	30	1917	12	30	1894
Takada	377	1945	2	26	1922	Saga	21	1959	1	17	1893
Aikawa	65	1936	2	6	1912	Nagasaki	15	1967	1	17	1906
Toyama	208	1940	1	30	1939	Izuhara	9	1901	2	21	1886
Kanazawa	181	1963	1	27	1882	Fukue	43	1963	1	26	1962
Wajima	110	1945	1	18	1929	Kumamoto	13	1945	2	7	1890
Fukui	213	1963	1	31	1897	Oita	15	1997	1	22	1916
Tsuruga	196	1981	1	15	1897	Miyazaki	3	1945	1	24	1886
Kofu	114	2014	2	15	1894	Kagoshima	29	1959	1	17	1892
Nagano	80	1946	12	11	1892	Naze	-	-	-	-	1896
Matsumoto	78	1946	3	3	1898	Naha	-	-	-	-	1891
Mt. Fuji	338	1989	4	27	1951	Syowa station (Antarctica)	195	2016	11	30	1999

- No instances.
The deepest snow depth at Mt. Fuji is as of August 2004.

Climatological Normal Daily Accumulated Total Solar Radiation by Month (MJ/m^2)

(1981–2010 Averages)

Station	Jan.	Feb.	Mar.	Apr.	May	Jun.	Jul.	Aug.	Sep.	Oct.	Nov.	Dec.	Annual
Sapporo	5.9	8.7	12.5	15.8	17.9	18.8	16.9	15.6	13.0	9.6	6.0	4.9	12.1
Hakodate	6.4	9.5	12.6	15.9	17.4	17.7	15.3	14.4	12.8	10.2	6.3	5.2	12.0
Asahikawa	5.8	9.0	12.8	15.4	17.8	18.8	17.0	15.3	12.5	8.6	5.0	4.5	11.9
Obihiro	7.8	11.4	15.2	16.6	17.7	17.1	14.8	13.8	12.2	10.4	7.5	6.4	12.5
Abashiri	6.1	9.9	13.8	16.2	18.0	18.9	17.8	16.0	13.3	9.7	6.3	5.2	12.6
Wakkanai	4.0	7.3	11.5	15.2	17.5	18.0	16.3	14.5	13.2	8.4	4.1	2.9	11.1
Aomori	5.0	7.8	11.7	15.8	17.9	18.5	17.0	16.1	13.0	9.7	5.6	4.1	11.9
Morioka	6.9	9.9	13.0	15.6	17.3	17.1	15.1	15.0	11.8	9.8	6.8	5.6	12.0
Sendai	8.1	10.8	13.7	16.4	17.2	15.2	13.8	14.3	11.5	10.4	8.1	7.1	12.2
Akita	4.7	7.4	11.7	15.8	17.2	18.1	16.0	17.1	13.3	10.2	5.9	4.0	11.8
Tateno (Tsukuba)	9.6	11.8	13.5	16.3	17.5	15.6	16.2	17.0	12.6	10.5	8.7	8.3	13.1
Maebashi	9.6	12.1	14.6	16.8	17.5	14.9	14.8	15.2	11.7	10.8	9.2	8.6	13.0
Tokyo	8.9	10.9	12.5	15.3	16.2	14.0	14.6	15.2	11.1	9.6	8.1	7.8	12.0
Niigata	5.2	7.8	11.5	16.0	18.1	17.8	16.8	17.6	13.1	10.0	6.1	4.5	12.0
Fukui	5.9	8.6	11.8	16.0	17.7	16.3	16.1	17.9	13.3	10.9	7.4	5.4	12.3
Nagoya	9.1	11.8	14.2	16.9	17.7	16.0	16.1	17.2	13.4	11.5	9.4	8.4	13.5
Hikone	6.9	9.5	12.7	16.1	17.6	16.5	16.8	17.8	13.7	11.3	8.2	6.5	12.8
Osaka	7.8	9.8	12.5	16.1	17.4	16.3	17.1	17.5	13.4	11.0	8.5	7.3	12.9
Hiroshima	8.5	10.6	13.3	16.6	18.2	16.8	17.2	18.1	14.4	12.4	9.4	8.0	13.6
Takamatsu	8.1	10.5	13.2	16.8	18.5	17.3	18.3	18.6	13.9	11.6	8.8	7.6	13.6
Kochi	10.1	12.3	14.3	16.9	17.6	16.0	17.6	18.3	14.4	12.6	10.1	9.5	14.1
Fukuoka	7.4	10.2	12.9	16.5	17.9	16.2	16.9	17.6	14.4	12.5	9.0	7.1	13.2
Kumamoto	8.3	10.8	13.1	16.5	17.6	15.6	17.5	18.2	15.1	13.0	9.5	8.1	13.6
Miyazaki	10.5	12.4	13.7	16.4	17.4	15.2	18.7	18.5	14.6	12.7	10.3	9.9	14.2
Kagoshima	9.0	11.3	13.2	16.2	17.4	14.9	18.3	18.5	15.7	13.6	10.3	9.0	13.9
Naze	6.4	7.7	10.0	13.0	14.9	14.9	19.3	16.8	14.3	11.0	8.3	6.7	11.9
Naha	8.7	10.0	12.2	14.9	16.6	17.9	20.8	19.0	17.0	14.0	10.7	9.1	14.3
Ishigakishima	8.8	10.2	12.8	15.4	17.9	20.3	22.3	20.3	17.7	14.3	10.9	9.1	15.0
Chichijima	10.9	13.1	15.7	17.1	18.2	20.7	22.3	20.0	18.4	15.2	11.6	9.8	16.1
Marcus	12.5	15.6	19.1	21.5	23.3	24.6	22.3	20.8	20.3	17.8	14.7	12.1	18.7
Syowa station (Antarctica)	26.3	16.4	7.7	2.3	0.3	0.0	0.1	1.5	6.4	14.6	24.8	30.3	10.9

Direct Solar Irradiance (12:00 PM) Climatological Normals by Month (kW/m^2)

(1981-2010 Averages)

Station	Jan.	Feb.	Mar.	Apr.	May	Jun.	Jul.	Aug.	Sep.	Oct.	Nov.	Dec.
Sapporo	0.80	0.83	0.82	0.83	0.82	0.85	0.82	0.83	0.85	0.83	0.77	0.76
Tateno (Tsukuba)	0.87	0.87	0.85	0.83	0.81	0.81	0.81	0.80	0.83	0.83	0.83	0.83
Fukuoka	0.79	0.80	0.78	0.79	0.78	0.77	0.78	0.79	0.79	0.80	0.77	0.76
Ishigakishima	0.83	0.83	0.81	0.82	0.83	0.85	0.87	0.87	0.87	0.85	0.83	0.82

Atmospheric Transmittance (12:00 PM) Climatological Normal Maximum Values by Month

(1981-2010 Averages)

Station	Jan.	Feb.	Mar.	Apr.	May	Jun.	Jul.	Aug.	Sep.	Oct.	Nov.	Dec.
Sapporo	0.80	0.77	0.72	0.71	0.70	0.70	0.68	0.70	0.74	0.76	0.78	0.80
Tateno (Tsukuba)	0.81	0.79	0.75	0.71	0.68	0.65	0.67	0.67	0.70	0.76	0.79	0.81
Fukuoka	0.77	0.73	0.70	0.67	0.66	0.63	0.64	0.66	0.69	0.73	0.75	0.76
Ishigakishima	0.72	0.69	0.67	0.64	0.65	0.67	0.68	0.68	0.69	0.69	0.71	0.72

Atmospheric Transmittance (12:00 PM) Climatological Normals by Month

(1981-2010 Averages)

Station	Jan.	Feb.	Mar.	Apr.	May	Jun.	Jul.	Aug.	Sep.	Oct.	Nov.	Dec.
Sapporo	0.78	0.74	0.69	0.66	0.64	0.66	0.64	0.66	0.70	0.73	0.75	0.78
Tateno (Tsukuba)	0.77	0.73	0.69	0.64	0.62	0.62	0.62	0.63	0.66	0.70	0.74	0.76
Fukuoka	0.71	0.68	0.63	0.61	0.60	0.58	0.60	0.61	0.63	0.67	0.69	0.71
Ishigakishima	0.69	0.66	0.62	0.61	0.62	0.64	0.66	0.65	0.67	0.67	0.68	0.69

Atmospheric Turbidity Coefficient (12:00 PM) Climatological Normal Values by Month

(1981–2010 Averages)

Station	Jan.	Feb.	Mar.	Apr.	May	Jun.	Jul.	Aug.	Sep.	Oct.	Nov.	Dec.
Sapporo	3.1	3.5	4.1	4.5	4.8	4.4	4.7	4.4	3.9	3.6	3.5	3.3
Tateno (Tsukuba)	3.1	3.5	4.1	4.7	5.0	5.1	5.0	4.9	4.4	4.0	3.5	3.3
Fukuoka	3.9	4.3	5.0	5.2	5.4	5.6	5.4	5.2	4.9	4.5	4.3	4.0
Ishigakishima	4.2	4.5	5.1	5.1	5.0	4.6	4.4	4.4	4.3	4.3	4.2	4.1

Refer to page **873** "Atmospheric turbidity coefficient".

Total Daily Direct Solar Radiation Climatological Normals by Month (MJ/m^2)

(1981–2010 Averages)

Station	Jan.	Feb.	Mar.	Apr.	May	Jun.	Jul.	Aug.	Sep.	Oct.	Nov.	Dec.
Sapporo	5.46	6.81	8.69	11.44	12.51	12.88	10.34	11.21	11.57	10.04	6.32	5.01
Tateno (Tsukuba)	15.27	14.82	11.96	11.71	10.59	7.59	8.36	10.76	8.31	9.52	11.07	13.61
Fukuoka	6.55	8.56	9.29	11.62	11.73	8.69	10.25	12.36	10.63	11.77	9.11	7.42
Ishigakishima	5.84	6.03	7.01	7.83	10.29	14.32	18.84	16.35	13.64	10.68	8.02	6.72

Total Daily Direct Solar Radiation Maximum Values by Month (MJ/m^2)

(1981–2010 Averages)

Station	Jan.	Feb.	Mar.	Apr.	May	Jun.	Jul.	Aug.	Sep.	Oct.	Nov.	Dec.
Sapporo	18.14	21.28	24.70	30.01	33.62	35.39	31.09	31.42	29.89	24.81	19.68	16.94
Tateno (Tsukuba)	27.01	28.81	29.81	30.87	31.28	28.79	27.08	27.64	27.61	27.43	26.23	25.39
Fukuoka	21.49	24.64	25.58	28.88	30.98	27.62	28.09	28.77	27.44	26.74	23.70	22.03
Ishigakishima	22.48	22.06	24.74	24.93	28.27	31.29	33.13	31.28	28.21	25.89	23.05	22.74

Monthly Average Air Temperature by Year (℃)

Sapporo

Year	Jan.	Feb.	Mar.	Apr.	May	Jun.	Jul.	Aug.	Sep.	Oct.	Nov.	Dec.	Annual
1980	-3.9	-5.7	-0.5	4.1	12.3	17.6	19.4	19.0	16.9	10.8	5.1	-0.3	7.9
1981	-4.9	-3.2	-0.5	6.6	10.3	14.9	21.0	20.9	16.2	11.0	1.8	-0.2	7.8
1982	-4.4	-4.6	0.7	6.1	13.0	15.8	20.4	22.8	17.8	12.5	5.9	0.3	8.9
1983	-2.9	-4.5	0.1	9.5	12.4	13.6	18.2	22.3	17.7	9.7	5.1	-1.8	8.3
1984	-4.7	-5.6	-2.2	5.0	11.4	18.2	22.1	23.6	17.7	10.3	3.7	-2.1	8.1
1985	-6.9	-2.2	-0.1	7.6	12.4	15.5	20.3	24.6	17.1	10.9	4.1	-3.5	8.3
1986	-5.8	-5.5	-0.4	6.9	11.4	16.1	18.2	22.3	18.1	8.9	3.5	-1.5	7.7
1987	-4.9	-3.8	-0.1	6.0	12.0	17.3	20.6	20.6	17.8	11.9	3.6	-2.3	8.2
1988	-3.8	-5.2	-0.1	7.1	11.4	16.6	18.3	22.9	17.6	10.9	2.9	-0.4	8.2
1989	-1.9	-1.4	2.4	7.3	11.4	15.2	21.0	23.0	18.4	11.8	6.6	-0.6	9.4
1990	-4.9	-0.7	2.9	7.9	13.4	17.7	20.8	23.1	18.7	13.0	7.5	1.9	10.1
1991	-1.2	-2.8	0.3	8.0	13.8	18.8	20.2	21.5	18.4	12.7	4.7	-0.9	9.5
1992	-2.4	-2.5	1.0	6.9	11.7	16.4	20.7	21.1	16.2	11.6	4.8	0.1	8.8
1993	-1.7	-2.1	1.3	6.2	11.9	15.4	19.1	20.0	17.5	11.5	5.8	-0.1	8.7
1994	-4.7	-1.3	-0.5	7.0	13.5	16.9	22.2	24.8	19.8	12.4	5.4	-1.4	9.5
1995	-3.3	-2.3	0.9	7.3	13.3	15.6	21.5	21.3	17.4	13.1	5.5	0.0	9.2
1996	-3.7	-3.4	0.5	5.8	10.5	15.7	20.4	20.9	18.2	11.5	4.1	-0.7	8.3
1997	-2.5	-1.8	0.2	7.2	11.9	15.8	22.0	20.5	17.1	10.8	7.1	0.4	9.1
1998	-5.9	-3.6	1.6	9.0	13.5	15.5	20.3	21.1	19.2	13.2	3.2	-1.7	8.8
1999	-3.2	-3.6	-0.5	6.9	11.9	17.8	22.1	24.9	19.7	11.9	5.5	-1.1	9.4
2000	-3.1	-3.8	0.2	6.1	14.0	16.7	22.2	23.9	18.6	11.6	3.9	-2.5	9.0
2001	-5.7	-5.5	0.2	8.3	13.4	16.7	20.7	21.1	17.1	12.1	5.0	-3.4	8.3
2002	-2.5	-0.6	2.5	9.6	13.6	15.9	20.5	20.1	17.7	12.0	2.8	-3.0	9.1
2003	-3.6	-3.5	0.7	8.2	12.9	17.1	17.7	20.7	17.5	11.6	6.2	0.0	8.8
2004	-2.5	-1.3	0.7	6.6	13.8	18.5	21.3	21.9	18.4	12.5	7.2	-0.8	9.7
2005	-3.5	-3.9	0.1	6.2	10.7	18.3	20.1	23.5	18.8	13.2	5.5	-2.6	8.9
2006	-4.1	-2.7	1.3	5.2	12.9	15.7	20.6	24.3	18.5	11.7	6.2	-0.5	9.1
2007	-1.8	-1.5	0.9	6.3	12.5	18.8	19.6	23.5	19.1	11.7	3.9	-0.6	9.4
2008	-4.3	-3.4	3.3	9.4	12.4	17.0	21.4	21.2	19.2	12.9	4.6	0.8	9.5
2009	-1.3	-2.2	1.5	7.7	13.9	17.5	19.7	21.5	17.8	12.5	5.1	-0.7	9.4
2010	-2.0	-3.2	-0.1	5.5	12.2	19.2	22.1	24.8	20.0	12.2	5.9	0.6	9.8
2011	-3.8	-1.1	0.7)	6.9	11.1	17.3	21.8	23.6	19.2	12.1	6.0	-2.0	9.3
2012	-4.5	-4.4	0.1	7.0	13.0	17.1	21.8	23.4	22.4	13.0	5.5	-2.3	9.3
2013	-4.7	-4.0	0.0	6.3	11.3	17.6	22.5	23.1	18.8	12.9	6.3	0.8	9.2
2014	-4.1	-3.5	0.5	7.3	14.0	18.7	22.5	22.4	18.1	11.3	6.1	-1.3	9.3
2015	-1.5	-0.8	3.8	8.7	14.2	16.7	21.3	22.4	18.4	10.8	5.4	0.8	10.0
2016	-3.5	-2.3	2.1	7.8	14.9	16.3	20.7	23.9	19.4	10.6	2.1	-1.0	9.3
2017	-3.9	-2.0	1.4	7.7	14.4	16.0	22.9	21.7	17.7	11.3	4.3	-2.0	9.1
2018	-2.6	-4.2	2.4	8.2	13.4	16.6	21.4	21.2	18.9	13.0	6.4	-1.0	9.5
2019	-3.0	-2.6	2.5	8.0	15.7	17.4	21.7	22.5	19.3	13.3	3.9	-0.8	9.8

) indicates a Quasi-normal value

Sendai (Monthly Average Air Temperature by Year) Continued.

Year	Jan.	Feb.	Mar.	Apr.	May	Jun.	Jul.	Aug.	Sep.	Oct.	Nov.	Dec.	Annual
1980	1.4	-0.2	3.8	9.2	15.6	20.1	19.9	20.1	19.6	14.3	9.6	3.6	11.4
1981	0.0	1.4	4.0	10.0	13.3	15.5	23.0	23.3	18.5	14.2	6.5	4.1	11.2
1982	1.1	0.9	5.1	9.6	16.3	18.0	20.3	23.7	19.7	15.1	10.1	5.6	12.1
1983	2.2	0.6	4.3	12.4	15.3	16.4	19.9	24.0	20.4	13.7	8.1	2.8	11.7
1984	-0.5	-1.0	1.5	6.9	12.4	17.8	22.3	25.7	19.9	13.5	8.0	3.1	10.8
1985	-1.1	2.4	3.7	9.9	15.0	17.4	23.3	26.2	20.2	14.4	9.2	3.0	12.0
1986	0.1	-0.3	3.8	9.4	14.5	17.4	20.1	23.6	21.1	13.1	7.8	5.0	11.3
1987	1.0	1.9	4.5	9.3	14.9	18.8	23.1	23.3	20.3	15.3	9.0	4.8	12.2
1988	2.5	-0.1	4.1	10.2	14.3	18.2	18.6	24.5	19.8	14.2	7.2	4.4	11.5
1989	3.1	3.3	6.2	11.3	14.5	17.0	21.1	24.5	20.9	14.5	10.9	5.0	12.7
1990	0.8	4.3	6.2	10.4	16.1	20.1	22.2	25.9	22.1	16.2	11.5	7.2	13.6
1991	2.4	1.7	5.1	11.1	15.6	20.6	21.9	22.2	20.6	15.5	8.9	4.9	12.5
1992	2.7	2.4	5.0	10.1	13.3	17.3	22.5	23.4	19.9	14.7	9.7	5.0	12.2
1993	2.8	3.4	4.6	9.5	14.4	18.1	18.5	21.6	19.8	14.1	10.5	4.6	11.8
1994	1.8	2.8	4.0	11.0	15.5	18.6	24.2	26.6	22.2	16.6	9.6	4.4	13.1
1995	1.1	2.4	4.7	11.3	15.6	17.0	23.7	25.4	20.3	16.4	8.8	3.9	12.6
1996	1.3	1.8	4.3	8.5	13.9	18.0	22.5	22.9	20.0	15.1	8.9	5.4	11.9
1997	2.4	2.6	5.7	11.1	15.0	18.6	23.1	23.9	19.7	14.4	10.5	5.2	12.7
1998	0.5	2.4	6.1	11.6	16.5	17.6	21.6	23.1	21.9	16.8	9.0	5.2	12.7
1999	2.1	2.4	5.3	11.0	16.5	19.7	23.4	25.4	22.3	15.8	10.7	4.3	13.2
2000	3.6	1.8	4.6	10.0	15.9	19.6	24.3	25.3	22.0	15.5	9.1	3.6	12.9
2001	0.0	1.1	4.9	11.6	15.4	19.3	24.7	21.8	20.1	15.3	9.4	2.7	12.2
2002	3.2	3.6	7.5	11.9	14.6	17.9	23.3	24.4	20.5	15.6	7.0	3.2	12.7
2003	1.5	2.5	4.8	11.1	14.6	19.1	18.4	22.2	20.2	14.5	10.3	5.9	12.1
2004	1.7	3.1	5.4	10.9	15.3	19.9	23.8	23.6	21.2	14.5	11.8	5.4	13.1
2005	1.6	0.8	4.1	11.0	13.4	19.5	21.4	25.0	21.5	16.3	9.5	1.8	12.2
2006	0.7	2.2	5.0	9.0	15.2	18.9	21.5	24.5	20.4	15.5	10.1	4.8	12.3
2007	3.8	3.9	5.3	9.5	15.4	19.8	20.9	25.6	22.3	16.0	9.2	4.9	13.1
2008	1.3	1.6	6.6	11.1	14.8	18.5	22.9	23.1	21.0	16.2	9.4	5.5	12.7
2009	2.9	3.2	5.5	11.5	16.5	19.2	22.7	22.9	19.9	15.6	10.4	4.9	12.9
2010	2.8	2.1	4.4	8.2	14.7	20.4	25.3	27.2	21.7	16.2	10.1	5.7	13.2
2011	0.5	3.2	3.8)	10.0	15.6	20.6	24.8	24.9	22.1	15.9	10.5	3.4	12.9
2012	0.4	0.3	4.5	9.8	15.9	18.2	22.8	26.2	23.9	16.6	9.7	3.3	12.6
2013	0.7	1.1	5.8	10.2	14.4	19.0	22.2	25.6	21.9	16.7	9.6	4.7	12.7
2014	1.9	1.4	5.5	10.9	16.5	20.6	23.7	24.6	20.5	15.3	10.0	2.8	12.8
2015	2.6	3.0	6.8	11.7	18.0	20.0	24.8	24.3	20.5	15.5	10.7	5.9	13.7
2016	2.4	3.5	7.0	11.9	17.0	19.8	23.0	25.7	22.1	15.7	8.6	5.7	13.5
2017	2.5	3.2	5.4	11.5	17.0	18.6	25.1	23.0	21.1	14.9	9.1	3.5	12.9
2018	1.4	1.4	7.5	12.5	17.0	20.3	25.5	24.9	20.8	16.5	10.7	4.3	13.6
2019	2.4	3.7	7.0	10.2	17.4	19.0	22.4	26.2	22.4	16.9	10.0	5.4	13.6

) indicates a Quasi-normal value

Tokyo

Continued.

Year	Jan.	Feb.	Mar.	Apr.	May	Jun.	Jul.	Aug.	Sep.	Oct.	Nov.	Dec.	Annual
1980	5.6	5.2	8.2	13.6	19.2	23.6	23.8	23.4	23.0	18.2	13.0	7.7	15.4
1981	4.4	5.3	9.0	13.9	17.5	20.2	26.3	26.2	21.8	17.6	10.4	7.6	15.0
1982	5.8	5.5	9.9	14.0	20.7	21.4	23.1	27.1	22.3	18.0	14.3	9.5	16.0
1983	6.2	6.1	8.6	15.9	19.7	20.5	23.8	27.5	23.1	17.7	12.3	7.1	15.7
1984	3.7	3.0	5.9	11.6	17.2	21.8	26.2	28.6	23.5	17.7	12.2	7.7	14.9
1985	4.1	6.5	7.8	14.2	19.1	20.2	26.3	27.9	23.1	17.9	13.3	7.4	15.7
1986	4.5	4.3	7.8	13.9	17.9	21.1	23.9	26.8	23.7	17.1	12.3	8.5	15.2
1987	5.8	6.8	9.3	14.4	19.3	22.1	27.0	27.3	23.3	18.9	12.8	8.1	16.3
1988	7.7	4.9	8.4	14.3	18.2	22.3	22.4	27.0	22.8	17.5	11.4	8.4	15.4
1989	8.1	7.5	9.6	15.6	17.7	20.7	24.1	27.1	25.2	17.5	14.2	9.2	16.4
1990	5.0	7.8	10.6	14.7	19.2	23.5	25.7	28.6	24.8	19.2	15.1	10.0	17.0
1991	6.3	6.5	9.5	15.4	18.8	23.6	26.7	25.5	23.9	18.1	13.0	9.2	16.4
1992	6.8	6.9	9.7	15.1	17.3	20.6	25.5	27.0	23.3	17.3	13.0	9.4	16.0
1993	6.2	7.7	8.7	13.4	18.1	21.7	22.5	24.8	22.9	17.5	14.1	8.5	15.5
1994	5.5	6.6	8.1	15.8	19.5	22.4	28.3	28.9	24.8	20.2	13.4	9.0	16.9
1995	6.3	6.5	8.9	15.0	19.1	20.4	26.4	29.4	23.7	19.5	12.7	7.7	16.3
1996	6.6	5.4	9.2	12.7	18.1	22.6	26.2	26.0	22.4	18.0	13.2	9.3	15.8
1997	6.8	7.0	10.5	15.2	19.2	22.7	26.6	27.0	22.9	18.7	14.3	9.2	16.7
1998	5.3	7.0	10.1	16.3	20.5	21.5	25.3	27.2	24.4	20.1	13.9	9.0	16.7
1999	6.6	6.7	10.1	15.0	19.9	22.8	25.9	28.5	26.2	19.5	14.2	9.0	17.0
2000	7.6	6.0	9.4	14.5	19.8	22.5	27.7	28.3	25.6	18.8	13.3	8.8	16.9
2001	4.9	6.6	9.8	15.7	19.5	23.1	28.5	26.4	23.2	18.7	13.1	8.4	16.5
2002	7.4	7.9	12.2	16.1	18.4	21.6	28.0	28.0	23.1	19.0	11.6	7.2	16.7
2003	5.5	6.4	8.7	15.1	18.8	23.2	22.8	26.0	24.2	17.8	14.4	9.2	16.0
2004	6.3	8.5	9.8	16.4	19.6	23.7	28.5	27.2	25.1	17.5	15.6	9.9	17.3
2005	6.1	6.2	9.0	15.1	17.7	23.2	25.6	28.1	24.7	19.2	13.3	6.4	16.2
2006	5.1	6.7	9.8	13.6	19.0	22.5	25.6	27.5	23.5	19.5	14.4	9.5	16.4
2007	7.6	8.6	10.8	13.7	19.8	23.2	24.4	29.0	25.2	19.0	13.3	9.0	17.0
2008	5.9	5.5	10.7	14.7	18.5	21.3	27.0	26.8	24.4	19.4	13.1	9.8	16.4
2009	6.8	7.8	10.0	15.7	20.1	22.5	26.3	26.6	23.0	19.0	13.5	9.0	16.7
2010	7.0	6.5	9.1	12.4	19.0	23.6	28.0	29.6	25.1	18.9	13.5	9.9	16.9
2011	5.1	7.0	8.1	14.5	18.5	22.8	27.3	27.5	25.1	19.5	14.9	7.5	16.5
2012	4.8	5.4	8.8	14.5	19.6	21.4	26.4	29.1	26.2	19.4	12.7	7.3	16.3
2013	5.5	6.2	12.1	15.2	19.8	22.9	27.3	29.2	25.2	19.8	13.5	8.3	17.1
2014	6.3	5.9	10.4	15.0	20.3	23.4	26.8	27.7	23.2	19.1	14.2	6.7	16.6
2015	5.8	5.7	10.3	14.5	21.1	22.1	26.2	26.7	22.6	18.4	13.9	9.3	16.4
2016	6.1	7.2	10.1	15.4	20.2	22.4	25.4	27.1	24.4	18.7	11.4	8.9	16.4
2017	5.8	6.9	8.5	14.7	20.0	22.0	27.3	26.4	22.8	16.8	11.9	6.6	15.8
2018	4.7	5.4	11.5	17.0	19.8	22.4	28.3	28.1	22.9	19.1	14.0	8.3	16.8
2019	5.6	7.2	10.6	13.6	20.0	21.8	24.1	28.4	25.1	19.4	13.1	8.5	16.5

Note) Values before May 2012 are values before repositioning.

Niigata (Monthly Average Air Temperature by Year) Continued.

Year	Jan.	Feb.	Mar.	Apr.	May	Jun.	Jul.	Aug.	Sep.	Oct.	Nov.	Dec.	Annual
1980	2.6	1.4	4.3	10.0	16.5	21.6	22.6	23.9	21.2	15.6	10.4	4.7	12.9
1981	1.2	1.9	5.2	10.5	14.9	19.0	25.7	25.1	20.2	15.1	7.8	5.1	12.6
1982	2.5	2.0	6.0	11.3	17.5	19.8	23.6	26.0	20.9	16.3	11.4	6.3	13.6
1983	2.9	1.5	5.5	13.4	17.0	19.5	23.0	26.6	22.8	15.0	9.3	3.7	13.4
1984	0.5	-0.2	2.4	9.2	15.2	21.9	25.5	27.6	21.6	15.2	9.3	4.4	12.7
1985	0.1	3.2	5.0	11.9	16.9	20.0	24.2	29.2	21.5	15.5	10.0	3.5	13.4
1986	1.0	0.7	5.3	11.0	15.6	19.4	22.3	26.4	22.5	14.6	9.3	6.1	12.9
1987	2.3	3.1	5.5	10.7	16.6	21.5	25.2	26.0	22.8	16.9	10.5	5.9	13.9
1988	3.7	0.8	4.9	10.7	15.2	20.3	22.6	27.2	22.5	15.1	8.0	5.1	13.0
1989	4.8	4.4	6.9	11.7	16.1	19.8	24.4	26.6	21.9	15.7	11.7	6.1	14.2
1990	1.9	5.6	7.1	12.1	16.5	21.9	25.3	27.4	23.4	17.1	12.6	7.5	14.9
1991	3.0	2.4	5.7	12.0	16.4	22.1	24.0	24.9	22.7	16.8	9.8	6.4	13.9
1992	3.9	2.9	6.5	11.8	14.8	19.7	24.2	26.2	21.9	16.1	10.9	6.2	13.8
1993	4.1	4.5	5.6	10.2	15.2	19.4	22.8	23.3	20.9	15.2	11.6	5.6	13.2
1994	2.5	3.2	4.5	11.9	17.6	20.4	26.3	28.9	23.9	18.0	10.8	5.6	14.5
1995	2.4	3.2	6.1	12.0	16.6	19.6	24.2	26.8	21.0	17.6	9.6	4.7	13.7
1996	2.5	1.9	5.3	9.7	15.1	20.3	24.9	25.9	21.3	16.2	10.4	5.6	13.3
1997	3.5	3.5	6.3	11.2	17.0	20.7	24.8	26.0	21.4	15.3	11.6	6.3	14.0
1998	2.2	3.7	6.7	13.6	18.3	20.4	25.5	25.1	23.8	18.3	10.3	6.3	14.5
1999	3.1	2.6	6.7	12.0	16.8	20.8	25.2	28.3	24.1	17.1	11.3	5.3	14.4
2000	4.4	2.2	5.4	11.4	17.2	20.8	26.0	28.2	23.7	16.5	10.9	5.1	14.3
2001	1.5	2.1	5.9	12.0	18.3	20.8	26.5	26.5	22.1	16.9	10.3	4.7	14.0
2002	3.8	4.3	7.7	13.9	16.3	20.9	25.6	26.9	22.3	16.5	7.4	4.2	14.2
2003	2.3	3.5	5.5	11.8	17.4	21.4	22.0	24.9	22.3	15.8	12.1	6.8	13.8
2004	3.0	4.5	6.4	12.0	17.5	21.5	26.0	26.0	23.6	16.5	13.0	6.8	14.7
2005	3.1	1.9	5.0	11.7	15.4	22.4	24.3	27.1	23.6	17.5	10.3	2.7	13.8
2006	1.9	3.4	5.6	10.2	16.5	20.8	23.3	27.8	22.1	17.2	11.3	6.1	13.9
2007	4.9	5.1	6.0	10.8	16.8	21.3	22.9	27.0	24.7	16.7	10.2	6.3	14.4
2008	3.0	2.1	7.5	12.6	17.0	20.3	25.4	25.8	22.7	17.4	10.3	6.8	14.2
2009	3.7	4.2	6.3	11.8	17.6	21.4	23.7	24.9	21.6	16.7	11.5	5.7	14.1
2010	3.4	2.9	5.6	9.7	15.7	21.6	26.3	29.0	23.7	17.6	10.8	6.7	14.4
2011	1.3	3.7	4.2	10.2	16.3	21.2	26.5	27.0	23.7	16.7	11.8	4.4	13.9
2012	1.6	1.0	5.3	11.2	16.3	20.6	25.4	27.9	25.2	17.3	9.9	3.9	13.8
2013	1.8	1.5	6.0	10.1	16.3	22.1	25.1	26.9	22.9	18.3	9.8	5.2	13.8
2014	2.5	2.5	6.1	11.1	17.1	22.0	24.9	26.1	21.6	16.0	11.1	3.4	13.7
2015	3.1	3.9	7.0	12.1	18.6	21.2	25.2	25.8	21.1	15.8	12.1	7.0	14.4
2016	3.1	3.6	6.9	12.4	18.4	21.4	24.6	27.0	23.4	16.4	9.9	6.4	14.5
2017	3.1	3.3	6.1	11.6	17.7	19.0	25.9	26.2	21.7	16.4	9.1	3.9	13.7
2018	1.7	1.4	7.5	12.7	17.0	21.1	27.4	26.6	21.8	17.2	11.6	5.9	14.3
2019	3.0	4.0	7.2	10.8	18.0	20.8	25.2	27.5	23.4	17.7	11.0	6.6	14.6

Note) Values before May 2012 are values before repositioning.

Nagoya Continued.

Year	Jan.	Feb.	Mar.	Apr.	May	Jun.	Jul.	Aug.	Sep.	Oct.	Nov.	Dec.	Annual
1980	4.1	3.4	7.9	12.9	18.4	23.1	24.8	25.1	22.5	17.2	12.1	5.1	14.7
1981	1.8	3.6	8.2	13.4	17.5	21.7	26.8	26.1	21.5	16.0	9.2	5.6	14.3
1982	3.9	4.3	8.9	13.5	20.2	21.9	23.4	26.0	22.1	17.2	13.4	7.1	15.2
1983	4.8	4.3	8.2	15.7	19.3	21.8	25.4	28.0	23.9	17.2	11.2	5.0	15.4
1984	2.1	2.1	4.8	12.9	18.4	22.8	26.6	28.5	23.3	17.2	12.1	6.4	14.8
1985	2.8	5.3	8.9	14.7	19.4	21.6	26.7	27.7	23.9	17.6	11.3	5.5	15.5
1986	2.7	2.6	7.8	13.9	17.9	22.1	24.9	27.3	23.9	16.2	11.4	7.6	14.9
1987	4.5	5.4	8.3	14.3	18.6	23.0	26.8	27.8	23.6	18.7	12.4	7.4	15.9
1988	5.6	3.4	7.8	13.8	18.2	22.3	24.8	27.2	24.2	16.6	9.6	5.9	15.0
1989	6.6	6.9	9.1	14.9	18.0	21.6	25.2	27.6	24.5	17.4	13.0	7.1	16.0
1990	4.2	8.5	9.6	14.4	18.7	23.7	26.9	28.4	24.6	18.4	14.1	7.8	16.6
1991	4.6	4.2	9.7	15.2	18.6	23.8	26.6	26.8	24.8	17.9	12.2	8.2	16.1
1992	6.0	5.3	10.1	14.6	17.4	21.7	26.6	27.4	23.8	17.8	12.0	7.7	15.9
1993	5.9	5.9	8.0	13.3	18.0	21.6	23.6	25.2	22.2	16.8	13.0	6.9	15.0
1994	4.7	4.6	7.4	15.4	19.8	22.9	28.9	29.3	24.9	19.8	13.3	7.7	16.6
1995	4.0	5.3	8.9	13.5	18.3	21.5	26.5	30.1	23.0	18.7	10.1	5.2	15.4
1996	4.5	3.8	7.8	12.0	18.4	22.7	26.9	27.0	22.5	17.3	12.0	6.9	15.2
1997	4.3	4.8	9.6	14.7	18.9	22.7	25.8	27.4	23.6	16.9	12.8	7.7	15.8
1998	4.5	6.8	9.8	17.2	20.3	22.4	26.8	28.5	24.7	19.9	12.1	8.2	16.8
1999	4.7	4.8	9.9	14.4	19.5	22.8	26.0	27.9	25.8	19.2	12.8	6.8	16.2
2000	6.1	3.5	7.6	13.8	19.9	22.6	28.0	28.8	24.9	18.9	13.7	7.0	16.2
2001	3.6	5.5	8.5	14.8	19.9	23.3	28.5	27.4	23.7	18.4	11.6	6.7	16.0
2002	4.9	6.1	10.4	15.8	19.0	22.8	27.8	28.5	24.0	18.0	9.4	6.7	16.1
2003	3.8	5.9	7.9	14.8	19.1	22.6	23.7	26.9	24.9	16.9	14.5	7.1	15.7
2004	4.3	6.2	9.1	15.6	20.0	24.0	28.6	27.5	25.0	18.3	14.1	8.6	16.8
2005	4.6	4.8	7.8	15.0	18.3	24.0	26.7	28.1	25.5	19.0	11.6	3.4	15.7
2006	3.8	5.5	7.8	13.0	18.7	23.3	26.2	28.5	23.9	19.5	13.2	7.6	15.9
2007	6.1	7.7	9.0	14.0	19.0	23.1	25.2	29.1	26.1	19.1	12.5	8.0	16.6
2008	5.1	4.0	10.4	15.3	19.6	22.4	28.2	28.1	24.3	19.0	12.2	8.0	16.4
2009	5.3	7.3	9.5	15.4	19.9	23.3	26.4	27.3	24.1	18.5	12.9	7.6	16.5
2010	4.6	7.0	9.1	13.3	18.7	23.9	27.8	29.4	26.1	19.4	12.1	7.9	16.6
2011	2.8	6.6	7.5)	13.3	19.0	23.8	27.5	28.3	25.1	18.8	13.9	6.7	16.1
2012	4.2	4.1	8.3	14.2	19.2	22.3	26.9	28.4	25.8	19.0	11.3	5.3	15.8
2013	4.0	4.6	10.5	13.8	19.4	23.6	28.1	29.3	24.9	20.2	11.5	6.4	16.4
2014	4.6	5.3	9.3	14.6	19.5	24.0	27.4	27.1	23.4	18.9	13.2	5.4	16.1
2015	4.9	5.7	9.7	15.2	21.3	22.3	26.5	28.1	23.1	18.4	14.3	9.3	16.6
2016	5.8	6.5	10.5	15.9	20.6	22.9	27.0	28.6	25.2	19.7	12.6	8.1	17.0
2017	4.8	5.2	8.4	14.7	20.5	22.4	28.1	28.1	23.6	17.9	11.5	5.7	15.9
2018	3.8	4.7	11.2	16.5	19.8	23.4	29.3	29.7	23.6	18.9	13.8	8.1	16.9
2019	5.1	7.2	10.1	14.1	20.4	23.1	25.9	28.9	26.7	20.3	13.4	8.8	17.0

) indicates a Quasi-normal value

Osaka (Monthly Average Air Temperature by Year) Continued.

Year	Jan.	Feb.	Mar.	Apr.	May	Jun.	Jul.	Aug.	Sep.	Oct.	Nov.	Dec.	Annual
1980	5.7	4.8	8.5	13.9	19.5	23.9	25.8	26.6	23.3	18.7	13.7	6.8	15.9
1981	3.7	5.1	9.3	14.4	18.5	23.2	28.3	27.4	23.1	17.6	11.1	7.6	15.8
1982	5.3	5.5	9.7	14.0	20.5	22.3	24.7	27.2	22.9	18.4	14.5	8.6	16.1
1983	6.1	5.6	8.7	16.5	20.1	22.5	26.7	29.4	24.9	17.9	12.6	6.9	16.5
1984	3.6	3.3	6.0	13.7	18.9	24.0	27.5	29.4	24.1	17.8	13.6	7.9	15.8
1985	4.3	6.6	9.3	15.5	20.1	22.5	27.6	29.2	25.3	19.0	12.8	6.6	16.6
1986	4.2	3.8	8.3	14.8	18.6	23.0	26.4	28.2	24.5	17.0	12.4	8.9	15.8
1987	6.2	6.5	8.9	14.2	19.3	23.8	27.6	28.7	24.4	19.6	13.4	8.7	16.8
1988	7.2	5.0	8.4	14.3	19.0	23.3	26.0	27.9	24.8	17.8	11.5	7.6	16.1
1989	7.8	7.6	9.5	15.5	18.8	22.6	26.7	28.0	24.9	18.2	14.1	8.5	16.9
1990	5.7	8.7	10.4	14.9	19.1	24.5	28.0	29.5	25.4	18.9	14.8	9.4	17.4
1991	6.4	5.8	10.2	15.9	19.2	24.4	28.1	28.1	25.6	18.7	13.3	9.8	17.1
1992	7.4	6.5	10.5	15.3	18.3	22.4	27.0	28.2	24.5	19.2	13.4	9.6	16.9
1993	6.9	7.5	8.7	14.4	18.9	22.7	25.4	26.6	23.3	17.9	14.7	8.4	16.3
1994	6.1	6.2	8.3	16.5	20.7	24.2	29.9	30.2	26.0	20.8	14.6	9.3	17.7
1995	6.1	6.2	9.8	14.6	18.9	22.1	27.5	30.3	24.2	19.9	12.1	7.2	16.6
1996	6.3	5.2	8.9	12.7	19.6	23.4	27.5	28.3	23.3	18.5	13.7	8.7	16.3
1997	6.0	5.8	10.1	15.1	19.9	23.7	26.7	28.5	24.3	18.2	14.1	8.9	16.8
1998	5.9	7.8	10.6	17.7	21.6	23.3	27.8	29.4	25.6	20.8	13.7	9.8	17.8
1999	6.4	5.9	10.6	14.7	20.5	23.5	26.8	29.1	27.2	20.1	14.1	8.7	17.3
2000	7.0	4.9	8.9	14.6	20.7	23.3	28.7	29.6	25.8	19.7	14.6	8.8	17.2
2001	5.2	6.6	9.7	15.5	20.6	24.0	29.2	28.8	24.4	19.5	13.2	8.3	17.1
2002	7.2	7.3	11.6	16.6	20.0	23.6	29.0	29.0	25.1	18.9	11.1	8.2	17.3
2003	5.1	6.8	8.4	15.9	20.2	23.7	25.3	28.3	25.9	18.1	15.5	9.1	16.9
2004	5.8	7.9	10.2	16.4	21.1	24.8	29.5	28.4	26.2	19.0	15.2	10.2	17.9
2005	6.2	6.1	9.2	16.2	19.5	24.9	27.5	28.7	26.1	19.8	13.7	5.9	17.0
2006	5.5	6.7	8.6	13.6	19.7	24.3	27.2	29.8	24.6	20.4	14.8	9.1	17.0
2007	7.5	8.7	10.1	14.6	19.8	23.6	25.9	29.9	27.2	20.0	13.7	9.6	17.6
2008	5.8	5.1	10.8	15.4	20.0	23.1	28.7	28.4	24.5	19.6	13.4	9.1	17.0
2009	6.5	7.9	9.7	15.5	19.7	24.0	27.3	28.0	24.5	19.2	13.6	8.7	17.1
2010	6.1	7.8	9.6	13.6	18.8	23.9	27.9	30.5	26.7	19.9	13.2	9.0	17.3
2011	4.4	7.4	8.1	13.8	19.6	24.2	27.8	28.9	25.2	19.5	15.2	8.1	16.9
2012	5.6	5.1	9.1	15.2	19.6	23.0	27.8	29.4	26.0	19.3	12.4	6.6	16.6
2013	5.2	5.6	10.7	14.3	19.8	24.3	28.5	30.0	25.1	20.8	12.9	7.8	17.1
2014	5.9	5.8	9.9	14.8	19.8	23.9	27.8	27.8	24.0	19.5	14.2	6.8	16.7
2015	6.1	6.9	10.2	15.9	21.5	22.9	27.0	28.6	23.2	19.0	15.2	10.1	17.2
2016	6.8	7.4	10.8	16.6	21.2	23.3	28.0	29.5	25.8	20.3	13.4	9.4	17.7
2017	6.2	6.3	9.2	15.7	21.1	22.7	28.8	29.2	24.4	18.4	12.6	7.0	16.8
2018	5.0	5.3	11.5	16.9	20.1	23.4	29.5	29.7	24.1	19.7	14.6	9.4	17.4
2019	6.5	7.8	10.6	14.6	21.0	23.7	26.5	29.1	26.6	20.7	14.2	9.5	17.6

Hiroshima Continued.

Year	Jan.	Feb.	Mar.	Apr.	May	Jun.	Jul.	Aug.	Sep.	Oct.	Nov.	Dec.	Annual
1980	4.5	3.5	7.8	12.3	17.7	21.8	23.6	24.3	21.9	16.8	12.0	4.8	14.3
1981	2.1	3.8	8.1	12.6	16.6	21.5	26.6	25.9	21.4	16.3	10.0	6.2	14.3
1982	3.9	4.6	8.9	13.1	19.1	21.6	23.8	26.5	21.6	17.4	13.2	7.3	15.1
1983	5.1	4.5	8.2	15.2	18.7	21.7	25.0	28.1	23.6	16.9	10.9	5.8	15.3
1984	2.8	2.4	5.9	13.1	18.0	22.5	26.7	28.0	22.7	16.8	12.4	6.8	14.8
1985	3.7	5.6	9.0	13.8	18.8	21.0	26.2	28.1	24.2	17.9	11.6	5.2	15.4
1986	3.0	3.1	7.8	13.7	17.6	22.0	24.9	27.1	22.8	15.7	11.5	7.9	14.8
1987	5.4	6.1	8.2	13.0	18.2	22.4	25.4	26.5	22.4	18.7	12.6	7.5	15.5
1988	6.0	5.0	8.1	13.6	18.3	22.8	26.8	27.4	24.2	17.5	10.3	6.5	15.5
1989	7.4	7.1	9.0	15.3	18.5	22.1	26.8	27.4	24.1	17.2	13.0	8.0	16.3
1990	4.8	8.6	10.3	14.4	19.1	24.0	28.1	29.2	24.9	18.2	14.2	7.9	17.0
1991	5.6	5.0	10.2	15.2	18.3	23.6	26.9	27.7	25.2	18.4	12.2	8.9	16.4
1992	6.5	6.2	10.1	15.3	18.2	22.2	26.4	27.7	24.1	18.2	12.3	8.2	16.3
1993	6.5	6.8	8.4	14.2	18.5	22.4	25.2	25.7	22.9	16.9	13.2	7.8	15.7
1994	5.5	6.0	7.9	15.9	20.0	23.2	30.1	29.8	25.4	19.4	14.0	8.7	17.2
1995	5.2	6.0	9.6	13.9	18.5	21.7	27.2	29.7	23.3	18.7	10.7	6.4	15.9
1996	5.4	4.5	8.4	11.8	19.5	23.2	27.6	28.2	23.7	18.0	13.1	7.1	15.9
1997	5.3	5.7	10.4	14.8	19.6	23.7	26.3	28.0	23.6	17.6	13.7	8.6	16.4
1998	5.5	7.9	10.3	17.4	21.1	23.2	27.5	29.0	25.7	20.5	13.0	9.5	17.6
1999	5.8	5.6	10.4	14.7	19.9	23.3	26.0	28.0	26.2	19.6	12.9	7.6	16.7
2000	6.7	4.7	8.8	14.1	19.5	23.0	27.9	28.6	24.4	19.1	13.6	7.9	16.5
2001	4.2	6.1	9.3	15.1	20.0	23.3	28.2	28.4	23.8	18.7	11.6	6.8	16.3
2002	5.9	6.6	11.0	15.7	19.5	23.5	27.9	28.3	24.7	17.9	9.7	7.4	16.5
2003	4.2	6.3	8.3	15.2	19.7	22.8	24.7	27.3	25.0	17.6	14.6	7.4	16.1
2004	4.7	7.3	9.7	15.5	20.0	24.0	28.9	28.0	24.7	18.1	13.9	8.8	17.0
2005	5.1	4.9	8.1	15.6	19.2	24.5	26.9	27.9	25.6	19.3	12.5	4.0	16.1
2006	5.3	6.1	8.1	13.2	19.2	23.4	26.6	29.0	23.4	20.1	13.6	7.9	16.3
2007	6.2	8.2	9.6	14.0	19.4	23.4	25.7	28.8	27.0	20.0	12.8	8.3	17.0
2008	5.4	4.4	9.8	14.9	19.4	22.7	28.5	27.9	24.9	19.1	12.0	7.8	16.4
2009	5.2	7.8	9.7	15.1	19.8	23.3	25.8	27.5	24.2	18.5	12.7	7.2	16.4
2010	5.2	7.6	9.1	13.0	18.5	23.3	27.2	30.3	26.2	19.2	12.0	7.3	16.6
2011	2.9	6.6	7.2	13.4	19.5	23.6	27.6	28.2	24.9	18.5	14.7	6.9	16.2
2012	4.7	4.3	8.7)	15.0	19.6	23.2	27.4	29.5	25.6	18.9	11.7	5.5	16.2
2013	4.4	6.0	10.7	13.5	19.7	24.0	28.3	29.5	24.6	19.9	11.9	6.5	16.6
2014	5.7	6.2	10.0	14.3	19.6	23.2	26.9	26.9	23.9	18.7	13.4	5.5	16.2
2015	5.8	6.1	10.0	15.8	20.5	22.5	26.5	27.5	23.1	18.0	14.6	9.3	16.6
2016	5.6	6.5	10.4	16.2	20.3	23.3	27.7	29.3	25.1	20.2	13.1	8.9	17.2
2017	5.5	6.1	8.8	15.6	20.6	22.5	28.4	29.0	23.4	18.4	11.9	5.8	16.3
2018	4.3	4.7	10.9	16.2	19.8	23.1	29.1	29.8	23.7	18.5	13.3	8.5	16.8
2019	6.4	7.6	10.6	14.8	20.5	23.2	26.4	28.5	26.3	20.3	13.5	8.6	17.2

) indicates a Quasi-normal value
Note) Values before 1987 are values before repositioning.

Takamatsu (Monthly Average Air Temperature by Year) Continued.

Year	Jan.	Feb.	Mar.	Apr.	May	Jun.	Jul.	Aug.	Sep.	Oct.	Nov.	Dec.	Annual
1980	4.6	4.2	7.7	12.7	17.9	22.8	24.8	24.8	22.1	17.2	11.8	5.6	14.7
1981	3.1	4.1	8.2	13.2	17.2	21.9	27.1	26.0	21.8	16.5	10.4	6.6	14.7
1982	4.6	4.8	9.0	13.4	19.4	22.2	24.5	26.7	22.0	17.3	13.7	7.8	15.5
1983	5.7	4.9	8.3	15.8	19.2	22.5	26.5	28.7	24.5	17.5	11.6	6.7	16.0
1984	3.4	2.7	5.6	12.5	18.1	23.1	26.9	28.4	23.3	17.1	12.7	7.3	15.1
1985	4.0	5.8	8.7	14.1	19.0	21.5	27.0	27.8	24.9	18.2	12.1	5.7	15.7
1986	3.7	3.6	7.7	13.9	18.0	22.9	25.8	27.5	23.6	16.1	11.6	8.3	15.2
1987	6.1	6.0	8.5	13.4	18.4	22.8	26.8	27.4	23.0	18.6	13.0	7.8	16.0
1988	6.2	4.8	7.6	13.5	18.0	22.6	25.8	26.7	23.7	17.2	11.1	6.5	15.3
1989	7.0	6.6	8.8	14.7	17.9	22.1	25.6	26.6	23.4	16.6	12.8	7.9	15.8
1990	4.9	8.0	9.9	14.2	18.5	24.1	27.8	28.4	24.5	18.4	14.2	8.5	16.8
1991	5.8	5.3	9.3	14.6	18.4	23.5	27.4	27.0	24.5	18.1	12.3	8.8	16.3
1992	6.8	6.1	9.5	14.6	18.1	21.7	26.6	27.0	23.8	18.2	12.3	8.8	16.1
1993	6.4	7.0	8.0	14.1	18.4	22.7	25.0	25.7	22.4	16.9	13.6	8.0	15.7
1994	5.3	5.9	7.8	15.7	20.2	23.4	29.6	29.6	25.1	19.3	13.8	8.9	17.1
1995	5.8	5.9	9.4	13.5	18.4	21.7	27.4	29.8	23.5	18.6	11.3	6.6	16.0
1996	5.4	4.7	8.3	12.2	19.4	23.3	27.2	28.0	23.2	17.6	13.1	7.5	15.8
1997	5.7	5.5	10.0	14.8)	20.0	23.3	26.6	28.0	23.9	17.6	13.6	8.6	16.5
1998	5.9	7.2	10.2	16.9	21.0	23.0	27.8	29.2	25.2	20.6	13.4	9.6	17.5
1999	6.2	5.9	9.7	14.1	20.2	23.0	25.9	27.9	26.3	19.5	13.2	7.7	16.6
2000	6.3	5.0	8.8	14.3	19.8	22.7	28.4	29.2	25.1	19.5	13.7	8.0	16.7
2001	5.1	6.0	9.5	14.7	20.0	23.9	28.3	28.4	23.8	19.0	12.2	7.6	16.5
2002	6.9	6.9	11.4	16.0	19.8	23.5	28.3	28.5	24.9	18.6	10.3	7.8	16.9
2003	5.0	6.4	8.2	15.2	19.4	23.1	25.4	27.7	25.4	17.8	14.9	9.0	16.5
2004	5.5	7.5	9.5	16.0	20.5	23.9	29.1	27.7	25.3	18.6	14.4	9.1	17.3
2005	5.8	5.5	8.8	15.7	19.5	25.2	27.3	28.2	25.6	19.4	13.1	5.3	16.6
2006	5.2	6.2	8.2	13.2	19.2	23.6	27.0	29.5	23.7	20.0	14.1	8.5	16.5
2007	6.7	8.2	9.6	14.3	19.9	23.7	26.1	29.3	27.0	19.9	13.2	9.1	17.3
2008	5.7	5.0	10.1	14.9	19.5	22.6	29.1	28.5	24.8	19.5	13.0	8.3	16.8
2009	6.1	7.8	9.9	15.6	19.8	24.0	26.7	27.8	24.3	19.1	13.4	8.2	16.9
2010	5.9	7.4	9.3	13.2	18.8	23.9	27.8	30.4	26.7	19.8	12.7	8.3	17.0
2011	4.1	6.6	7.9	13.6	19.6	24.0	27.3	28.6	25.1	19.2	15.0	7.9)	16.6
2012	5.2	4.7	8.9	15.0	19.4	22.8	27.7	29.3	25.2	18.9	12.3	6.3	16.3
2013	4.7	5.8	10.4	13.6	19.9	24.2	29.0	29.8	24.5	20.3	12.5	7.4	16.8
2014	5.8	5.7	9.8	14.3	19.8	23.6	27.6	26.9	24.0	19.1	13.6	6.5	16.4
2015	6.3	6.5	9.5	15.4	21.0	22.5	26.7	28.1	23.1	18.4	15.0	9.9	16.9
2016	6.6	6.9	10.3	16.1	20.8	23.1	28.1	29.5	25.2	20.5	13.6	9.3	17.5
2017	6.2	6.4	9.0	15.7	20.8	22.8	28.7	29.4	23.9	18.4	11.9)	6.4	16.6
2018	4.7	4.8	10.5	16.1	19.7	22.9	29.1	29.7	24.1	19.1	13.5	9.2	17.0
2019	6.8	7.5	10.2	14.5	20.6	23.6	26.5	28.4	26.4	20.7	13.5	9.1	17.3

) indicates a Quasi-normal value

Fukuoka Continued.

Year	Jan.	Feb.	Mar.	Apr.	May	Jun.	Jul.	Aug.	Sep.	Oct.	Nov.	Dec.	Annual
1980	6.2	5.1	9.6	14.0	18.7	23.1	24.9	24.5	22.4	18.1	13.5	6.4	15.5
1981	4.0	6.1	10.2	14.6	18.6	22.8	28.5	27.4	23.1	18.1	11.7	8.2	16.1
1982	5.9	6.7	10.7	14.2	20.3	22.1	24.8	27.1	22.4	18.7	14.9	8.6	16.4
1983	6.7	6.0	10.1	16.3	19.4	22.1	26.9	28.1	24.8	19.1	12.4	7.4	16.6
1984	4.1	4.4	7.6	14.5	18.0	23.8	27.8	28.6	23.7	18.1	14.4	8.3	16.1
1985	4.7	7.0	10.0	14.5	19.5	22.0	28.0	29.3	25.4	19.3	12.7	6.7	16.6
1986	4.5	4.7	9.4	15.2	18.9	22.9	25.6	27.8	23.7	17.0	12.6	9.6	16.0
1987	7.3	7.9	9.7	14.4	19.0	22.5	26.8	27.2	22.8	19.9	14.4	9.0	16.7
1988	7.6	6.4	8.9	14.4	19.2	22.5	26.5	26.8	23.8	18.5	12.2	8.5	16.3
1989	9.1	8.7	10.3	16.2	18.4	21.8	26.3	27.3	24.3	18.3	13.7	9.6	17.0
1990	6.0	10.2	11.4	14.8	19.3	24.2	28.3	29.6	25.4	18.7	15.2	8.7	17.7
1991	7.0	6.1	10.4	14.6	18.3	23.5	27.4	26.2	24.4	18.3	13.1	10.0	16.6
1992	7.8	7.6	10.8	15.4	18.9	21.4	26.5	26.7	24.6	18.5	13.2	10.0	16.8
1993	7.5	8.3	9.8	14.6	18.8	22.9	25.1	25.2	22.6	17.7	14.4	8.9	16.3
1994	7.1	7.2	8.9	15.8	20.5	22.3	29.6	29.8	24.6	19.7	15.2	10.1	17.6
1995	6.9	7.1	10.1	14.4	18.4	21.7	27.4	29.3	23.5	19.2	12.3	7.4	16.5
1996	6.5	6.1	9.9	12.5	19.4	23.9	27.1	28.2	23.8	19.0	14.2	8.6	16.6
1997	6.2	7.7	11.3	15.3	20.4	23.6	26.6	28.0	23.3	18.6	14.9	9.5	17.1
1998	6.6	9.4	10.9	17.6	21.1	22.9	27.6	29.2	25.7	20.8	14.3	10.6	18.1
1999	7.3	7.2	11.3	14.8	19.7	23.2	25.3	27.6	26.2	19.9	13.8	8.6	17.1
2000	7.7	6.1	10.4	15.1	19.4	23.0	28.2	28.6	24.4	19.7	14.5	9.6	17.2
2001	6.2	7.7	10.9	15.3	20.1	23.6	27.8	28.5	24.0	19.8	13.2	8.3	17.1
2002	7.9	8.4	12.5	16.6	19.3	23.6	27.9	27.8	24.4	18.8	11.5	9.1	17.3
2003	5.9	8.3	9.9	16.1	20.0	23.2	25.5	27.2	25.6	18.8	16.1	9.3	17.2
2004	6.1	9.0	10.9	16.2	20.5	24.0	28.7	28.6	24.6	19.0	15.1	10.7	17.8
2005	6.4	6.1	9.6	16.7	19.4	24.8	27.6	28.4	26.0	20.5	14.4	6.0	17.2
2006	6.9	7.7	10.1	14.6	19.2	23.2	27.3	29.0	23.3	20.6	15.0	9.5	17.2
2007	7.6	9.8	11.3	15.1	20.4	23.8	26.3	29.4	27.0	20.9	14.1	9.8	18.0
2008	7.5	6.3	10.7	15.0	19.4	22.2	29.0	27.6	25.0	20.3	13.4	9.1	17.1
2009	6.4	9.8	11.7	15.6	19.9	23.6	26.8	27.6	24.4	19.7	13.7	8.9	17.3
2010	6.6	9.4)	10.9	13.8	19.2	23.5	27.7	30.3	26.3	20.0	13.2	8.8	17.5
2011	3.8	8.2	8.8)	14.7	19.8	23.9	27.9	28.5	25.2	19.7	16.3	8.5	17.1
2012	6.3	5.7	10.7	16.2	20.1	23.1	28.0	29.1	24.5	19.2	12.9	7.6	17.0
2013	6.1	7.8	12.3	14.7	20.3	23.7	30.0	30.0	25.2	20.7	13.4	8.1	17.7
2014	7.5	7.6	11.5	15.6	20.5	22.6	27.1	26.5	24.2	19.7	14.7	7.6	17.1
2015	7.9	7.6	11.1	16.2	20.7	22.6	26.0	27.4	23.2	18.9	16.0	10.3	17.3
2016	7.0	7.9	11.5	16.8	20.8	23.6	28.3	29.3	25.1	21.3	14.5	10.5	18.1
2017	7.4	8.3	10.5	16.5	21.0	23.1	29.4	29.5	24.3	19.8	13.6	7.4	17.6
2018	5.7	6.2	11.9	17.1	20.8	23.7	28.7	30.0	24.8	19.1	14.3	10.2	17.7
2019	8.0	9.4	11.9	15.4	21.1	23.4	26.4	28.0	25.9	20.5	14.9	10.3	17.9

) indicates a Quasi-normal value

Kagoshima (Monthly Average Air Temperature by Year) Continued.

Year	Jan.	Feb.	Mar.	Apr.	May	Jun.	Jul.	Aug.	Sep.	Oct.	Nov.	Dec.	Annual
1980	8.0	6.6	12.1	16.0	20.3	24.2	26.9	27.1	24.4	20.1	15.4	7.1	17.4
1981	5.3	8.5	12.1	16.3	18.9	24.1	28.6	28.2	24.7	19.9	13.3	9.0	17.4
1982	6.9	8.6	13.4	16.1	20.9	23.1	26.4	27.3	24.0	20.7	16.3	9.3	17.8
1983	7.8	8.2	12.5	17.6	20.7	23.0	27.4	28.5	25.8	20.7	13.2	8.3	17.8
1984	5.6	6.6	9.6	16.9	20.0	24.9	28.0	28.4	24.4	19.3	17.0	10.3	17.6
1985	6.2	9.4	13.3	16.9	21.5	24.1	27.7	28.4	26.7	21.0	13.7	8.5	18.1
1986	6.0	6.7	11.5	17.1	20.4	23.9	27.7	28.0	25.3	18.2	14.5	10.9	17.5
1987	8.3	9.3	12.3	16.4	20.7	23.9	27.6	28.3	24.2	21.7	16.2	9.8	18.2
1988	9.4	8.6	11.3	15.7	21.0	23.8	28.1	27.4	25.4	20.5	12.5	9.3	17.8
1989	11.1	11.1	12.0	17.3	19.8	23.0	27.1	27.8	25.9	19.8	15.0	10.4	18.4
1990	8.1	12.6	13.1	15.7	19.7	24.5	28.8	29.0	26.5	20.7	16.5	9.8	18.8
1991	8.3	8.1	13.8	17.4	19.4	24.7	28.6	27.8	26.1	19.9	14.6	11.6	18.4
1992	9.4	8.3	14.0	16.9	19.9	21.9	26.5	27.6	25.9	19.9	15.2	11.2	18.1
1993	9.0	9.5	11.3	15.5	19.9	23.8	26.2	27.1	23.9	19.1	16.7	10.2	17.7
1994	8.2	9.4	10.9	17.6	21.0	23.8	29.2	28.8	25.5	21.4	17.2	11.9	18.7
1995	8.1	9.0	11.8	15.8	20.0	22.5	28.0	29.1	25.5	21.7	13.8	8.7	17.8
1996	8.3	7.8	12.1	14.1	20.4	25.0	27.9	28.5	26.4	21.3	17.0	10.5	18.3
1997	8.1	9.4	13.8	17.3	21.4	23.8	27.9	28.5	25.1	19.6	17.0	12.0	18.7
1998	9.2	12.3	12.5	19.7	22.7	24.4	28.7	29.6	27.1	23.0	16.5	12.1	19.8
1999	9.1	9.1	14.5	17.4	20.3	24.4	27.0	27.9	27.4	22.3	15.8	9.9	18.8
2000	9.9	7.8	12.6	16.1	20.6	24.0	28.3	28.4	26.1	22.5	17.6	11.6	18.8
2001	8.7	10.8	12.2	17.0	21.1	24.9	29.0	29.1	25.7	22.3	14.7	10.8	18.9
2002	9.7	10.1	14.0	17.7	21.6	24.3	28.0	28.5	26.3	20.3	13.1	11.7	18.8
2003	7.4	10.4	12.2	18.1	21.6	23.7	28.1	28.6	27.0	20.2	18.9	10.5	18.9
2004	7.9	10.2	13.2	17.7	21.4	25.0	28.8	29.3	26.6	21.2	16.6	12.9	19.2
2005	8.0	9.2	10.9	17.5	21.0	24.7	28.4	28.8	27.3	22.3	15.9	7.5	18.5
2006	9.1	11.1	11.8	16.5	21.1	24.2	28.9	29.3	26.1	23.2	17.0	11.6	19.2
2007	9.5	11.6	13.3	16.3	20.8	24.5	27.9	29.1	28.0	23.1	15.9	12.0	19.3
2008	9.9	7.8	13.1	17.0	21.1	23.6	29.2	28.7	26.4	22.1	15.3	10.2	18.7
2009	8.7	12.5	13.1	16.9	21.1	24.3	27.8	29.2	26.9	20.9	15.9	10.6	19.0
2010	8.3	11.9	13.3	16.2	21.1	23.8	27.8	29.6	27.2	21.6	15.1	10.5	18.9
2011	5.2	10.8	10.6	16.0	20.8	24.1	28.2	28.7	26.1	21.4	18.2	10.1	18.4
2012	8.0	9.1	12.8	16.9	21.2	23.7	28.0	28.9	25.2	21.1	14.3	9.7	18.2
2013	7.9	10.3	14.1	16.3	21.4	24.6	29.4	30.0	26.8	22.5	14.6	9.3	18.9
2014	9.2	10.5	13.1	16.9	20.4	23.1	27.5	27.7	25.3	22.0	16.6	9.1	18.5
2015	9.1	9.1	12.9	18.8	21.2	22.7	26.7	27.9	25.1	20.8	18.3	12.6	18.8
2016	9.0	9.6	13.3	18.4	21.8	24.7	28.6	29.8	27.3	23.8	16.7	12.6	19.6
2017	9.0	9.2	11.6	17.5	21.1	23.3	29.2	29.7	25.7	22.4	15.3	8.7	18.6
2018	7.6	8.2	14.5	18.5	21.7	24.7	28.6	29.6	26.4	20.1	16.0	12.2	19.0
2019	9.6	11.3	13.6	17.5	21.3	24.0	27.2	28.8	27.8	23.0	17.0	12.1	19.4

Note) Values before January 1994 are values before repositioning.

Naha Continued.

Year	Jan.	Feb.	Mar.	Apr.	May	Jun.	Jul.	Aug.	Sep.	Oct.	Nov.	Dec.	Annual
1980	16.3	15.1	19.1	20.5	23.9	27.6	28.7	28.8	27.3	24.2	21.8	16.8	22.5
1981	15.0	16.3	18.9	21.6	22.6	26.1	28.1	28.6	26.8	24.5	21.1	17.3	22.2
1982	15.6	16.9	19.8	20.2	24.7	25.5	28.6	28.0	26.7	24.2	22.5	18.1	22.6
1983	17.1	15.9	18.4	23.0	24.3	26.3	28.7	28.6	28.2	26.2	21.0	17.0	22.9
1984	15.2	15.6	17.5	20.7	23.2	27.0	28.5	28.2	27.3	24.3	22.3	18.3	22.3
1985	16.2	17.3	19.4	20.2	24.6	25.9	28.0	27.5	27.3	25.4	20.5	17.7	22.5
1986	15.0	14.6	17.1	21.3	23.4	26.5	28.8	28.3	27.3	23.8	21.4	18.4	22.2
1987	16.1	16.6	19.3	21.8	24.4	25.9	28.7	28.7	27.2	26.1	23.1	18.7	23.1
1988	18.5	17.6	19.2	20.6	23.6	27.4	29.7	28.3	27.7	25.0	20.1	17.9	23.0
1989	17.8	17.7	17.8	21.3	23.7	26.6	28.6	28.1	27.5	24.4	21.3	17.9	22.7
1990	16.5	18.5	19.0	20.2	23.8	27.2	29.3	29.0	27.4	24.2	22.4	18.9	23.0
1991	17.3	16.4	20.5	22.1	24.6	28.8	29.5	28.9	27.9	24.4	21.0	19.3	23.4
1992	17.2	16.0	20.8	21.6	23.7	26.0	28.2	28.6	27.4	24.6	21.3	19.1	22.9
1993	17.4	17.0	18.3	20.7	24.6	27.0	29.1	28.9	27.5	24.9	22.8	18.8	23.1
1994	17.3	17.1	17.1	22.4	23.4	26.9	29.4	28.7	26.8	24.4	22.5	20.2	23.0
1995	16.4	15.7	17.6	22.1	23.3	26.3	28.5	29.0	27.8	26.1	21.0	17.8	22.6
1996	17.1	16.3	19.0	19.5	22.4	27.5	29.4	28.1	27.9	24.6	23.2	18.1	22.8
1997	16.5	17.1	19.9	22.0	24.6	26.1	28.4	27.8	26.8	24.3	22.6	19.7	23.0
1998	18.7	18.5	20.2	23.5	26.1	27.4	29.5	30.1	28.4	26.7	23.0	20.2	24.4
1999	18.0	17.4	20.9	21.7	23.7	27.5	28.3	28.4	28.2	26.3	22.6	18.4	23.5
2000	17.9	16.2	18.6	20.7	23.1	27.2	28.0	27.9	26.3	26.2	23.4	19.9	23.0
2001	17.7	18.7	18.3	21.2	23.5	27.8	29.9	29.6	27.5	25.6	21.8	19.2	23.4
2002	17.1	17.4	19.6	22.4	25.1	27.0	28.2	28.7	27.5	24.7	21.1	19.4	23.2
2003	15.7	18.1	18.4	22.6	24.4	26.6	29.9	29.6	28.5	24.7	23.8	18.7	23.4
2004	16.8	17.1	19.2	21.9	25.5	26.7	28.8	29.0	27.6	24.9	22.6	20.2	23.4
2005	16.6	17.9	17.2	21.5	24.2	26.6	29.2	29.0	28.2	26.2	22.8	17.2	23.1
2006	18.1	17.9	18.4	21.2	24.8	26.8	29.1	29.2	27.8	26.1	22.9	19.7	23.5
2007	17.8	18.2	19.6	20.7	23.8	26.7	29.6	28.8	28.2	26.6	22.2	19.9	23.5
2008	18.5	16.1	18.7	21.4	24.1	27.6	29.4	29.0	28.2	26.4	22.5	18.7	23.4
2009	16.7	19.9	19.6	20.5	23.7	26.4	29.2	29.5	29.0	25.3	22.7	18.3	23.4
2010	16.8	18.3	19.9	21.2	23.8	26.7	28.7	28.9	28.0	25.7	21.4	18.1	23.1
2011	14.9	17.6	17.1)	20.4	23.9	27.9	28.9	28.3	27.9	25.2	23.7	18.6	22.9
2012	17.0	17.5	19.6	21.7	24.4	26.9	29.1	28.5	27.2	24.6	21.0	18.5	23.0
2013	17.0	18.6	20.4	20.6	23.7	27.9	29.4	29.6	28.3	25.3	21.3	17.3	23.3
2014	16.8	17.9	18.4	20.9	23.6	26.9	29.3	28.7	28.8	25.4	22.6	17.6	23.1
2015	16.6	16.8	19.0	22.2	24.9	28.7	29.0	28.7	27.8	25.5	23.8	20.1	23.6
2016	17.4	16.9	18.7	23.0	25.7	28.4	29.8	29.5	28.4	27.7	23.2	20.5	24.1
2017	18.4	17.1	18.3	21.6	24.2	26.6	29.9	30.4	28.9	27.0	22.8	18.0	23.6
2018	17.2	16.9	19.9	21.6	25.6	27.8	28.3	28.5	28.4	23.9	23.1	20.4	23.5
2019	18.1	20.0	19.9	22.3	24.2	26.5	28.9	29.2	28.0	26.0	23.1	20.0	23.9

) indicates a Quasi-normal value

Phenological Normals

(1981–2010 Averages)

Phenomenon / Station	Plum blooming	Dandelion blooming	Someiyoshi-no cherry blooming	Someiyoshio cherries full bloom	Rhodo-dendron blooming	Wistera floribunda blooming	Crape myrtle blooming	Miscanthus blooming	Ginkgo tree yellow leaves	Japanese maple autumn colors	First squeal of the nightingale	First sight of the swallow	First sight of the cabbage butterfly	First sight of the firefly	First squeal of the cicada (*Graptopsaltria nigrofuscata*)	First squeal of the shrike
Sapporo	501	429[a]	503	507	519				1104	1025[a]			502		728	
Hakodate	429	421[a]	430	504	518	526		821	1031	1029[b]	419	425	428		729	
Asahikawa		505[a]	505[a]	507[a]	523					1023[a]			501			
Kushiro		510[a]	517[a]	520[a]	506[a]					1015[c]			510			
Obihiro		501[a]	504[a]	507[a]	525				1025	1018[b]			424			
Abashiri		505[a]	511[a]	514[a]	530						505		505			
Wakkanai		509[a]	514[a]	517[a]	604						429					
Aomori	421[a]	430	424	429	511	518		827	1102	1112	417	422	425	723	731	
Morioka	412	415	421	425	507	515	803	827	1027	1108	401	421	422	705	727	928
Sendai	227	403	411	416	428	508	811	812	1118	1118	309	411	408	701	720	923
Akita	409	425	418	422	501	514	815	822	1102	1108	414	418	416	629	719	1013
Yamagata	405	418	415	419	507	510	801	815	1104	1120	409	409	409	712	720	924
Fukushima	307	330	409	413	424	504	730	826	1030	1113	310	410	409	630	723	927
Mito	130	313	402	408	426	430	804	904	1118	1120	309	403	331	606	727	910
Utsunomiya	214	326	401	408	422	427	731	824	1124	1116	309	404	325	621	722	907
Maebashi	210	317	331	406	423	426	727	903	1123	1205	305	331	328		801	1008
Kumagaya	211	312	329	405	418	421	720	913	1122	1126	306	401	328		721	930
Choshi	120	309	331	408	426	423	808	928	1128	1212	224	401	320	705	729	919
Tokyo	126		326	403	421	421	716	909	1120	1127	306	407			724	
Yokohama	204	122	326	403	419	429	804	922	1126	1214	317	408	407		725	1006
Niigata	315	413	409	414	504	504	802	829	1109	1112	327	410	408	623	717	921
Toyama	302	407	405	410	430	426	805	828	1115	1120	316	331	404	613	716	919
Kanazawa	226	405	404	410	429	429	803	809	1112	1121	324	330	407	617	716	1005
Fukui	227	407	403	409	505	501	803	905	1115	1122	317	402	404	618	715	930
Kofu	223	314	327	403	419	419	717	825	1112	1126	310	401	321	605	722	918
Nagano	317	404	413	417	430	430	804	816	1108	1107	321	410	406	627	717	
Gifu	214	309	326	404	414	417	716	905	1129	1126	306	330	321	602	716	921

Shizuoka	120	204	325	403	413	413	731	914	1121	1205	309	402	318		721	1004
Nagoya	202	223	326	403	414	419	723	916	1120	1126	311	327	323		713	920
Tsu	201	318	330	405	416	424	802	909	1117	1126	309	326	325	606	719	929
Hikone	217	308	402	409	425	426	802	903	1117	1125	303	329	327	603	719	929
Kyoto	220	228	328	405	422	421	723	830	1124	1203	301	326	326	602	717	1001
Osaka	210	321	328	405	413	417	712	908	1120	1202		404	331		716	
Kobe	214	321	328	405	427	424	716	916	1114	1203	312	329	407		717	1003
Nara	205	319	329	405	420	422	710	913	1113	1118	301	327	327	608	712	920
Wakayama	212	305	326	404	420	417	724	922	1121	1204	313	323	320	527	720	924
Tottori	210	312	331	407	423	424	722	908	1122	1126	311	326	327	602	714	921
Matsue	124	304	331	408	429	428	801	910	1117	1122	301	324	328	603	716	926
Okayama	209	317	329	406	413	420	725	912	1120	1129	305	322	322	523	711	914
Hiroshima	206	305	327	404	407b	418	724	919	1117	1120	310	326	326		712	1006
Shimonoseki	130	314	327	405	507	423	802	1001	1121	1204	308	325	324	524	709	928
Tokushima	206	303	328	405	419	419	727	916	1208	1129		324	320		716	928
Takamatsu	120	215	328	405	504	420	726	923	1125	1118		319	316		723	917
Matsuyama	106	220	325	404	404c	418	719	913	1123	1205	305	321	318	524	713	917
Kochi	125	215	322	330	414	411	716	822	1112	1128	228	320	309	516	712	914
Fukuoka	202	303	323	401	517	412	727	923	1112	1125	304	324	317	528	708	915
Saga	129	304	324	403	505	415	727	921	1118	1126	225	315	312	523	711	1005
Nagasaki	124	222	324	403	419	417	724	927	1203	1203	306	320	312	520	713	1003
Kumamoto	131		323	401	510	411	711	916	1123	1202	226	312	309	511	711	911
Oita	128		324	403	502	420	804	917	1130	1126	214	312	305	517	714	908
Miyazaki	122		324	402	425	408	730	927	1202	1205	225		302	516	709	910
Kagoshima	131	124	326	404	414	409	723	925	1124	1211	228	306	303	512	716	916
Naze			119b	130b		401	723	1023			307		313			
Naha	115		118b	204b	222d		620	1010			222	314		504a	608a	

Plum blooming: a) Bungo plum
Dandelion blooming: a) Common dandelion (Taraxacum officinale)
Someiyoshino cherry blooming: a) Sargent's cherry (Prunus sargentii), b) Wild Himalayan cherry (Cerasus cerasoides)
Someiyoshio cherries full bloom: a) Sargent's cherry (Prunus sargentii), b) Wild Himalayan cherry (Cerasus cerasoides)
Rhododendron blooming: a) Bidens parvifora, b) Japanese hornwort azalea, c) Rhododendron reticulatum, d) Rhododendron scabrum
Japanese maple autumn colors: a) Japanese mountain maple (Acer palmatum var.matsumurae), b) Japanese large maple (Acer palmatum var.amoenum), c) Painted maple (acer pictum)
First sight of the firefly: a) Luciola kuroiwae
First squeal of the cicada (*Graptopsaltria nigrofuscata*): a) Ryukyu cicada
In cases where the climatological normals could not be determined because there are many years that the phenomena did not occur, the columns are left blank.

Climatological Normals Observed for Global Weather

Normal values for weather observed at 240 sites throughout the world are listed in pages **191** to **239**.

List of global observation sites listed by the weather division

Based on the locations, countries, regions, latitude, longitude and altitude (elevation or official elevation of airfields; if such elevation is unknown, the reference surface altitude used for reporting atmospheric pressure) listed in the WMO (World Meteorological Organization) Publication No.9-Volume A "Observing Stations" (issued in December 2010; older versions used for some locations). In the case of subsequent changes to country names, etc., the latest version is used.

Note that, in accordance with the source listed above, Taipei is generally listed as Taiwan in this report.

Temperature/precipitation amount

Average values for the 30-year period from 1981 to 2010 which were created by the Japan Meteorological Agency based on Global Historical Climatology Network data distributed by the National Climactic Data Center (USA) and data of monthly climate reports issued by meteorological organizations in countries throughout the world. The Japan Meteorological Agency updates climatological normals once every 10 years. If data for a 30-year period was not available, normal values were calculated when data for at least an 8-year period existed for each month.

Relative humidity

Average values for the 30-year period from 1961 to 1990 which are listed in the Climatological Normals (CLINO) for the period 1961–1990 (WMO/OMM-No. 847, Secretariat of the World Meteorological Organization, Geneva – Switzerland, 1996), a WMO summary of weather observation data collected from countries throughout the world. The WMO summarizes climatological normals every 30 years. For cases in which it was not possible to obtain sufficient data, some climatological normals are for periods shorter than 30 years.

The 56 locations marked with an asterisk (*) have average values for 4-year periods to 19-year periods cited from the JMA (Japan Meteorological Agency) Observation Technology Report No. 51 (JMA, 1987) as reference values. This values were calculated by the JMA based on data from the National Climactic Data Center (USA).

World Weather Observation Stations Referenced in Meteorology Section

#	Station	Country or region	Latitude (deg. min.)	Longitude (deg. min.)	Altitude (m)
1	OSLO/GARDERMOEN	Norway	60 12 N	11 04 E	202
2	STOCKHOLM/BROMMA	Sweden	59 21 N	17 53 E	15
3	HELSINKI-VANTAA	Finland	60 19 N	24 58 E	51
4	HEATHROW	England	51 28 N	00 27 W	24
5	DUBLIN AIRPORT	Ireland	53 26 N	06 15 W	68
6	REYKJAVIK	Iceland	64 08 N	21 54 W	54
7	NUUK (GODTHAAB)	Greenland	64 10 N	51 45 W	80
8	KOEBENHAVN/LANDBOHOEJSKOLEN	Denmark	55 41 N	12 32 E	7
9	DE BILT AWS	Netherlands	52 05 N	05 10 E	1
10	UCCLE	Belgium	50 48 N	04 21 E	99
11	LUXEMBOURG/LUXEMBOURG	Luxembourg	49 37 N	06 13 E	376
12	ZUERICH/FLUNTERN	Switzerland	47 22 N	08 33 E	555
13	LE BOURGET	France	48 58 N	02 25 E	66
14	LYON-BRON	France	45 43 N	04 56 E	197
15	MARIGNANE	France	43 26 N	05 13 E	21
16	BARCELONA/AEROPUERTO	Spain	41 17 N	02 04 E	4
17	MADRID, RETIRO	Spain	40 24 N	03 40 W	667
18	GIBRALTAR	Gibraltar	36 09 N	05 20 W	5
19	LISBOA/GEOF	Portugal	38 43 N	09 09 W	77
20	SAL	Cabo Verde	16 44 N	22 56 W	54
21	BERLIN-TEMPELHOF	Germany	52 28 N	13 24 E	48
22	FRANKFURT/MAIN	Germany	50 02 N	08 35 E	112
23	MUENCHEN-FLUGHAFEN	Germany	48 21 N	11 48 E	443
24	WIEN/HOHE WARTE	Austria	48 14 N	16 21 E	198
25	PRAHA/RUZYNE	Czech Republic	50 06 N	14 15 E	380
26	SLIAC	Slovakia	48 39 N	19 09 E	314
27	WARSZAWA-OKECIE	Poland	52 09 N	20 57 E	106
28	BUDAPEST/PESTSZENTLORINC	Hungary	47 26 N	19 11 E	138
29	BEOGRAD	Serbia	44 48 N	20 28 E	132
30	PODGORICA-GRAD	Montenegro	42 26 N	19 17 E	49
31	SKOPJE PETROVEC	Macedonia	41 58 N	21 39 E	238
32	VLORE	Albania	40 28 N	19 29 E	1
33	LJUBLJANA/BEZIGRAD	Slovenia	46 03 N	14 30 E	299
34	ZAGREB/GRIC	Croatia	45 48 N	15 58 E	157
35	SARAJEVO-BJELAVE	Bosnia-Herzegovina	43 52 N	18 26 E	630
36	BUCURESTI BANEASA	Romania	44 30 N	26 04 E	90
37	SOFIA (OBSERV.)	Bulgaria	42 39 N	23 23 E	595
38	VERONA/VILLAFRANCA	Italy	45 23 N	10 52 E	73
39	ROMA/FIUMICINO	Italy	41 48 N	12 14 E	4
40	LUQA	Malta	35 51 N	14 29 E	91

World Weather Observation Stations Referenced in Meteorology Section Continued.

#	Station	Country or region	Latitude (deg. min.)	Longitude (deg. min.)	Altitude (m)
41	ATHINAI AP HELLINIKON	Greece	37 44 N	23 44 E	28
42	ISTANBUL/GOZTEPE	Turkey	40 54 N	29 09 E	18
43	ANKARA/CENTRAL	Turkey	39 57 N	32 53 E	891
44	LARNACA AIRPORT	Cyprus	34 52 N	33 37 E	9
45	OSTROV DIKSON	Russia	73 30 N	80 24 E	47
46	OJMJAKON	Russia	63 15 N	143 09 E	741
47	TALLINN-HARKU	Estonia	59 23 N	24 36 E	32
48	ST.PETERSBURG (VOEJKOVO)	Russia	59 58 N	30 18 E	6
49	JELGAVA	Latvia	56 40 N	23 44 E	4
50	KAUNAS	Lithuania	54 53 N	23 50 E	76
51	MINSK	Belarus	53 55 N	27 38 E	223
52	MOSKVA	Russia	55 50 N	37 37 E	156
53	OMSK	Russia	55 01 N	73 23 E	122
54	IRKUTSK	Russia	52 16 N	104 19 E	469
55	VLADIVOSTOK	Russia	43 07 N	131 56 E	183
56	KIEV	Ukraine	50 24 N	30 34 E	166
57	CHISINAU	Moldova	47 01 N	28 59 E	173
58	KARAGANDA	Kazakhstan	49 48 N	73 09 E	553
59	GORI	Georgia	41 59 N	44 06 E	609
60	TASHIR	Armenia	41 07 N	44 16 E	1507
61	AGSTAPHA AIRPORT	Azerbaijan	41 08 N	45 25 E	331
62	BISHKEK	Kirghiz	42 51 N	74 32 E	756
63	TASHKENT	Uzbekistan	41 20 N	69 18 E	488
64	DUSHANBE	Tajikistan	38 33 N	68 47 E	800
65	ASHGABAT KESHI	Turkmenistan	37 55 N	58 08 E	312
66	DAMASCUS INT. AIRPORT	Syria	33 25 N	36 31 E	608
67	JERUSALEM	Israel	31 52 N	35 13 E	757
68	AMMAN AIRPORT	Jordan	31 59 N	35 59 E	778
69	RIYADH OBS. (O.A.P.)	Saudi Arabia	24 42 N	46 44 E	635
70	TEHRAN-MEHRABAD	Iran	35 41 N	51 19 E	1204
71	KABUL AIRPORT	Afghanistan	34 33 N	69 13 E	1791
72	BAHRAIN (INT. AIRPORT)	Bahrain	26 16 N	50 39 E	2
73	DOHA INTERNATIONAL AIRPORT	Qatar	25 15 N	51 34 E	11
74	ABU DHABI BATEEN AIRPORT	United Arab Emirates	24 26 N	54 28 E	5
75	SEEB, INT'L AIRPORT	Oman	23 35 N	58 17 E	15
76	TAIZ	Yemen	13 41 N	44 08 E	1402
77	PESHAWAR	Pakistan	34 01 N	71 35 E	359
78	KARACHI AIRPORT	Pakistan	24 54 N	67 08 E	21
79	CHANDPUR	Bangladesh	23 16 N	90 42 E	6
80	NEW DELHI/SAFDARJUNG	India	28 35 N	77 12 E	211

Continued.

#	Station	Country or region	Latitude (deg. min.)	Longitude (deg. min.)	Altitude (m)
81	CHERRAPUNJI	India	25 15 N	91 44 E	1313
82	KOLKATA/ALIPORE	India	22 32 N	88 20 E	6
83	BOMBAY/COLABA	India	18 54 N	72 49 E	9
84	CHENNAI/MINAMBAKKAM	India	13 00 N	80 11 E	13
85	COLOMBO	Sri Lanka	06 54 N	79 52 E	7
86	ULAANBAATAR	Mongolia	47 55 N	106 52 E	1307
87	KATHMANDU AIRPORT	Nepal	27 42 N	85 22 E	1337
88	HONG KONG OBSERVATORY	Hong Kong, China	22 18 N	114 10 E	31
89	PYONGYANG	North Korea	39 02 N	125 47 E	36
90	SEOUL	Korea	37 34 N	126 57 E	86
91	BUSAN	Korea	35 06 N	129 02 E	69
92	YANGON	Myanmar	16 46 N	96 10 E	14
93	BANGKOK METROPOLIS	Thailand	13 43 N	100 33 E	3
94	KUALA LUMPUR/SUBANG	Malaysia	03 07 N	101 33 E	27
95	SINGAPORE/CHANGI AIRPORT	Singapore	01 22 N	103 59 E	5
96	HA NOI	Vietnam	21 02 N	105 48 E	5
97	VIENTIANE	Laos	17 57 N	102 34 E	171
98	PHNOM-PENH/POCHENTONG	Cambodia	11 33 N	104 51 E	10
99	Urumqi	China	43 47 N	87 39 E	936
100	Kashgar	China	39 28 N	75 59 E	1291
101	Changchun	China	43 54 N	125 13 E	238
102	Shenyang	China	41 44 N	123 31 E	49
103	Beijing	China	39 56 N	116 17 E	55
104	Dalian	China	38 54 N	121 38 E	97
105	Lhasa	China	29 40 N	91 08 E	3650
106	Kunming	China	25 01 N	102 41 E	1892
107	Xian	China	34 18 N	108 56 E	398
108	Wuhan	China	30 36 N	114 03 E	24
109	Shanghai	China	31 25 N	121 27 E	9
110	Fuzhou	China	26 05 N	119 17 E	85
111	Taipei	China	25 02 N	121 31 E	9
112	LAS PALMAS DE GRAN CANARIA/GANDO	Canary Islands	27 55 N	15 23 W	24
113	DAKHLA	Western Sahara	23 43 N	15 56 W	12
114	RABAT-SALE	Morocco	34 02 N	06 45 W	74
115	DAR-EL-BEIDA	Algeria	36 41 N	03 13 E	25
116	TUNIS-CARTHAGE	Tunisia	36 50 N	10 14 E	4
117	NIAMEY-AERO	Niger	13 29 N	02 10 E	223
118	BAMAKO/SENOU	Mali	12 32 N	07 57 W	380
119	NOUAKCHOTT	Mauritania	18 06 N	15 57 W	2
120	DAKAR/YOFF	Senegal	14 43 N	17 30 W	24

World Weather Observation Stations Referenced in Meteorology Section Continued.

#	Station	Country or region	Latitude (deg. min.)	Longitude (deg. min.)	Altitude (m)
121	BOLAMA	Guinea-Bissau	11 35 N	15 29 W	18
122	LUNGI	Sierra Leone	08 37 N	13 12 W	25
123	PLAISANCE (MAURITIUS)	Mauritius	20 26 S	57 41 E	55
124	TRIPOLI INTERNATIONAL AIRPORT	Libya	32 42 N	13 05 E	63
125	CAIRO	Egypt	30 06 N	31 24 E	116
126	ASSWAN	Egypt	23 57 N	32 49 E	201
127	KHARTOUM	Sudan	15 36 N	32 33 E	382
128	DJIBOUTI	Djibouti	11 33 N	43 09 E	13
129	ADDIS ABABA-BOLE	Ethiopia	09 02 N	38 45 E	2354
130	KAMPALA	Uganda	00 19 N	32 37 E	1144
131	JOMO KENYATTA INTERNATIONAL AIRPORT	Kenya	01 19 S	36 54 E	1624
132	DAR ES SALAAM INT	Tanzania	06 52 S	39 12 E	55
133	SEYCHELLES INTERNATIONAL AIRPORT	Seychelles	04 40 S	55 31 E	3
134	LISALA	DR Congo	02 19 N	21 34 E	463
135	KIGALI	Rwanda	01 57 S	30 07 E	1491
136	BUJUMBURA	Burundi	03 19 S	29 19 E	782
137	POINTE-NOIRE	Republic of Congo	04 49 S	11 54 E	17
138	BITAM	Gabon	02 05 N	11 29 E	600
139	BOSSEMBELE	Central Africa	05 16 N	17 38 E	673
140	NDJAMENA	Chad	12 08 N	15 02 E	295
141	DOUALA OBS.	Cameroon	04 00 N	09 44 E	10
142	BENIN CITY	Nigeria	06 19 N	05 36 E	79
143	COTONOU	Benin	06 21 N	02 23 E	5
144	LOME	Togo	06 10 N	01 15 E	20
145	KUMASI	Ghana	06 43 N	01 36 W	286
146	OUAGADOUGOU	Burkina Faso	12 21 N	01 31 W	303
147	ABIDJAN	Côte d'Ivoire	05 15 N	03 56 W	7
148	ROBERTS FIELD	Liberia	06 15 N	10 21 W	8
149	ANTANANARIVO/IVATO	Madagascar	18 48 S	47 29 E	1279
150	MAPUTO/MAVALANE	Mozambique	25 55 S	32 34 E	39
151	KABWE	Zambia	14 27 S	28 28 E	1206
152	CHILEKA	Malawi	15 41 S	34 58 E	766
153	HARARE (KUTSAGA)	Zimbabwe	17 55 S	31 08 E	1479
154	WINDHOEK	Namibia	22 34 S	17 06 E	1725
155	MAHALAPYE	Botswana	23 05 S	26 48 E	991
156	PRETORIA EENDRACHT	South Africa	25 44 S	28 11 E	1308
157	MASERU-MIA	Lesotho	29 27 S	27 33 E	1631
158	CAPE TOWN INTNL. AIRPORT	South Africa	33 58 S	18 36 E	46
159	BARROW/W. POST W. ROGERS	Alaska, USA	71 17 N	156 47 W	12
160	ANCHORAGE/INT., AK	Alaska, USA	61 09 N	149 59 W	52

Continued.

#	Station	Country or region	Latitude (deg. min.)	Longitude (deg. min.)	Altitude (m)
161	MONTREAL/PIERRE ELLIOTT TRUDEAU INT'L A, QUE	Canada	45 28 N	73 45 W	35
162	WINNIPEG RICHARDSON INT'L A, MAN	Canada	49 55 N	97 14 W	238
163	EDMONTON CITY CENTRE AWOS, ALTA	Canada	53 34 N	113 31 W	671
164	VANCOUVER INT'L A, BC	Canada	49 11 N	123 10 W	4
165	EUREKA, NU	Canada	79 59 N	85 56 W	10
166	MIAMI, FL	USA	25 45 N	80 23 W	4
167	ATLANTA/MUN., GA.	USA	33 39 N	84 25 W	312
168	NEW ORLEANS/MOISANT INT., LA.	USA	29 59 N	90 15 W	1
169	DALLAS-FORT WORTH/FORT WORTH REG.AIRPORT, TX.	USA	32 54 N	97 02 W	182
170	LOS ANGELES /INT., CA.	USA	33 56 N	118 24 W	38
171	LAS VEGAS/MCCARRAN, NV.	USA	36 05 N	115 10 W	662
172	WASHINGTON/NAT., VA.	USA	38 51 N	77 02 W	5
173	DENVER/STAPLETON INT., CO.	USA	39 46 N	104 52 W	1611
174	SAN FRANCISCO/INT., CA.	USA	37 37 N	122 23 W	6
175	NEW YORK/LA GUARDIA, NY.	USA	40 46 N	73 54 W	7
176	BOSTON/LOGAN INT., MA.	USA	42 22 N	71 02 W	6
177	CHICAGO/O'HARE, IL.	USA	41 59 N	87 54 W	203
178	DETROIT/METROPOLITAN, MI.	USA	42 14 N	83 20 W	195
179	MINNEAPOLIS/ST.PAUL INT., MN.	USA	44 53 N	93 13 W	256
180	SEATTLE/S.-TACOMA, WA.	USA	47 27 N	122 18 W	130
181	MEXICO (CENTRAL), D.F.	Mexico	19 24 N	99 11 W	2309
182	NASSAU AIRPORT NEW PROVIDENCE	Bahamas	25 03 N	77 28 W	7
183	CASA BLANCA, LA HABANA	Cuba	23 10 N	82 21 W	50
184	KINGSTON/NORMAN MANLEY	Jamaica	17 56 N	76 47 W	3
185	SANTO DOMINGO	Dominica	18 26 N	69 53 W	14
186	SAN JUAN/INT., PUERTO RICO	Puerto Rico	18 25 N	65 59 W	4
187	BELIZE/PHILLIP GOLDSTON INTL. AIRPORT	Belize	17 32 N	88 18 W	5
188	GUATEMALA (OBSERVATORIO NACIONAL)	Guatemala	14 35 N	90 31 W	1502
189	TEGUCIGALPA	Honduras	14 03 N	87 13 W	1007
190	MANAGUA A.C.SANDINO	Nicaragua	12 09 N	86 10 W	56
191	JUAN SANTAMARIA INT. AIRPORT	Costa Rica	09 59 N	84 11 W	908
192	HEWANORRA INT'L AIRPORT	Saint Lucia	13 45 N	60 57 W	3
193	GRANTLEY ADAMS	Barbados	13 04 N	59 29 W	50
194	PIARCO INT. AIRPORT, TRINIDAD	Trinidad and Tobago	10 37 N	61 21 W	12
195	BOGOTA/ELDORADO	Columbia	04 42 N	74 09 W	2547
196	CARACAS/MAIQUETIA AEROP. INTL. SIMON BOLIVAR	Venezuela	10 36 N	66 59 W	43
197	GEORGETOWN	Guyana	06 48 N	58 09 W	1
198	ZANDERIJ	Suriname	05 27 N	55 12 W	15
199	MANAUS	Brazil	03 08 S	60 01 W	72
200	SALVADOR	Brazil	13 01 S	38 31 W	52

World Weather Observation Stations Referenced in Meteorology Section Continued.

#	Station	Country or region	Latitude (deg. min.)	Longitude (deg. min.)	Altitude (m)
201	BRASILIA	Brazil	15 47 S	47 56 W	1159
202	RIO DE JANEIRO	Brazil	22 55 S	43 10 W	5
203	SAO PAULO	Brazil	23 30 S	46 37 W	792
204	ESMERALDAS (TACHINA) AEROPUERTO	Ecuador	00 58 N	79 37 W	10
205	LIMA/CALLAO	Peru	12 01 S	77 07 W	12
206	LA PAZ/ALTO	Bolivia	16 31 S	68 11 W	4058
207	PUDAHUEL	Chile	33 23 S	70 47 W	481
208	AEROPUERTO SILVIO PETTIROSSI, LUQUE	Paraguay	25 16 S	57 38 W	83
209	ROCHA	Uruguay	34 29 S	54 18 W	18
210	BUENOS AIRES OBSERVATORIO	Argentina	34 35 S	58 29 W	25
211	USHUAIA AERO	Argentina	54 48 S	68 19 W	28
212	VOSTOK	Antarctic	78 27 S	106 52 E	3488
213	HONOLULU, OAHU, HAWAII	Hawaii, USA	21 21 N	157 56 W	2
214	WEATHER FORECAST OFFICE, GUAM, MARIANA IS.	Guam Mariana Islands	13 28 N	144 47 E	75
215	MAJURO/MARSHALL IS. INTNL.	Marshall Islands	07 05 N	171 23 E	3
216	WEATHER SERVICE OFFICE, KOROR, PALAU WCI.	Palau	07 20 N	134 29 E	30
217	HONIARA/HENDERSON	Solomon Islands	09 25 S	160 03 E	8
218	PEKOA AIRPORT (SANTO)	Vanuatu	15 31 S	167 13 E	41
219	NOUMEA (NLLE-CALEDONIE)	New Caledonia	22 16 S	166 27 E	69
220	TARAWA	Kiribati	01 21 N	172 55 E	2
221	FUNAFUTI	Tuvalu	08 31 S	179 13 E	1
222	NADI AIRPORT	Fiji	17 45 S	177 27 E	13
223	APIA	Samoa	13 48 S	171 47 W	2
224	FUA'AMOTU	Tonga	21 14 S	175 09 W	38
225	TAHITI-FAAA	Polynesia, France	17 33 S	149 36 W	2
226	MADANG W.O.	Papua New Guinea	05 13 S	145 48 E	3
227	AUCKLAND AIRPORT	New Zealand	37 01 S	174 48 E	7
228	CHRISTCHURCH	New Zealand	43 29 S	172 33 E	38
229	DARWIN AIRPORT	Australia	12 25 S	130 53 E	31
230	CAIRNS AERO	Australia	16 52 S	145 44 E	3
231	ALICE SPRINGS AIRPORT	Australia	23 47 S	133 53 E	545
232	BRISBANE AERO	Australia	27 23 S	153 07 E	4
233	PERTH AIRPORT	Australia	31 55 S	115 58 E	20
234	SYDNEY AIRPORT AMO	Australia	33 56 S	151 10 E	6
235	MELBOURNE AIRPORT	Australia	37 39 S	144 49 E	132
236	CANBERRA AIRPORT	Australia	35 18 S	149 12 E	575
237	BRUNEI AIRPORT	Brunei Darussalam	04 56 N	114 56 E	22
238	BALIKPAPAN/SEPINGGAN	Indonesia	01 16 S	116 54 E	3
239	JAKARTA/OBSERVATORY	Indonesia	06 11 S	106 50 E	8
240	NINOY AQUINO INTERNATIONAL AIRPORT	Philippines	14 31 N	121 00 E	14

Map of World Weather Observation Stations Referenced in Meteorology Section

Figure 2 Ⅰ Europe

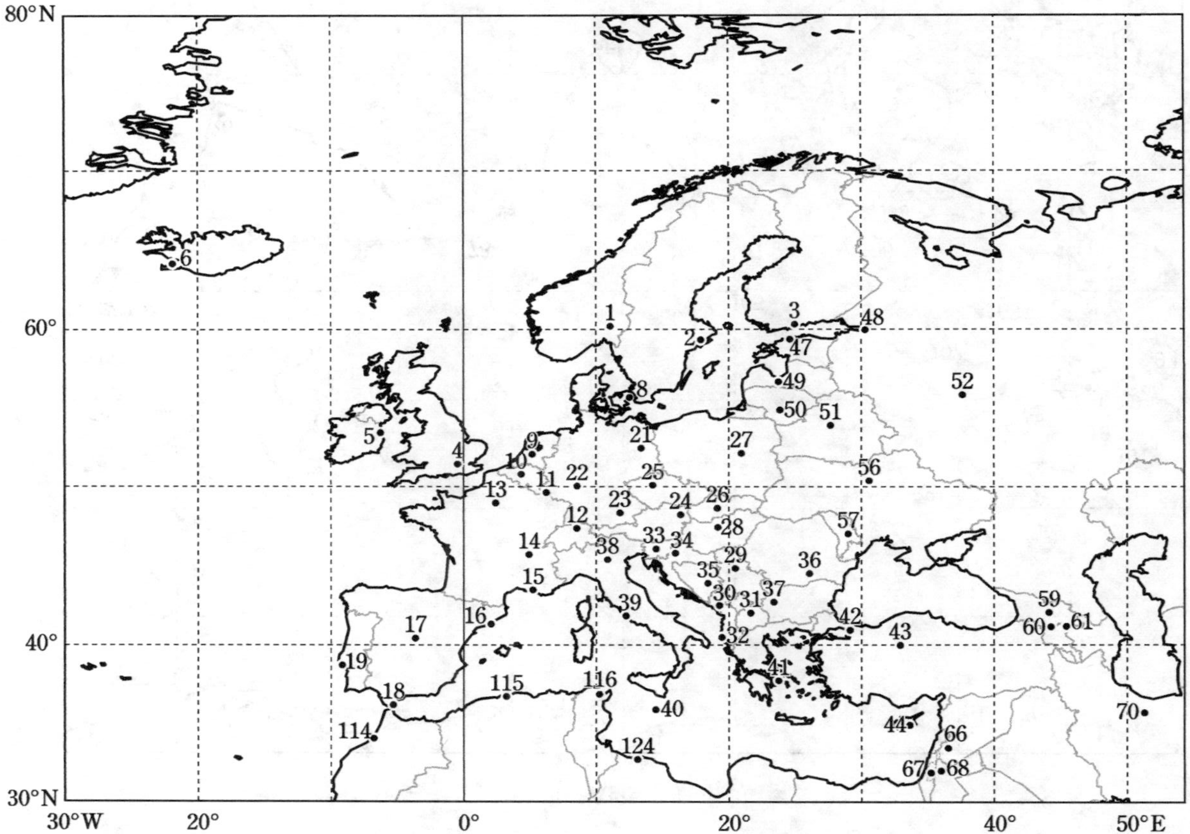

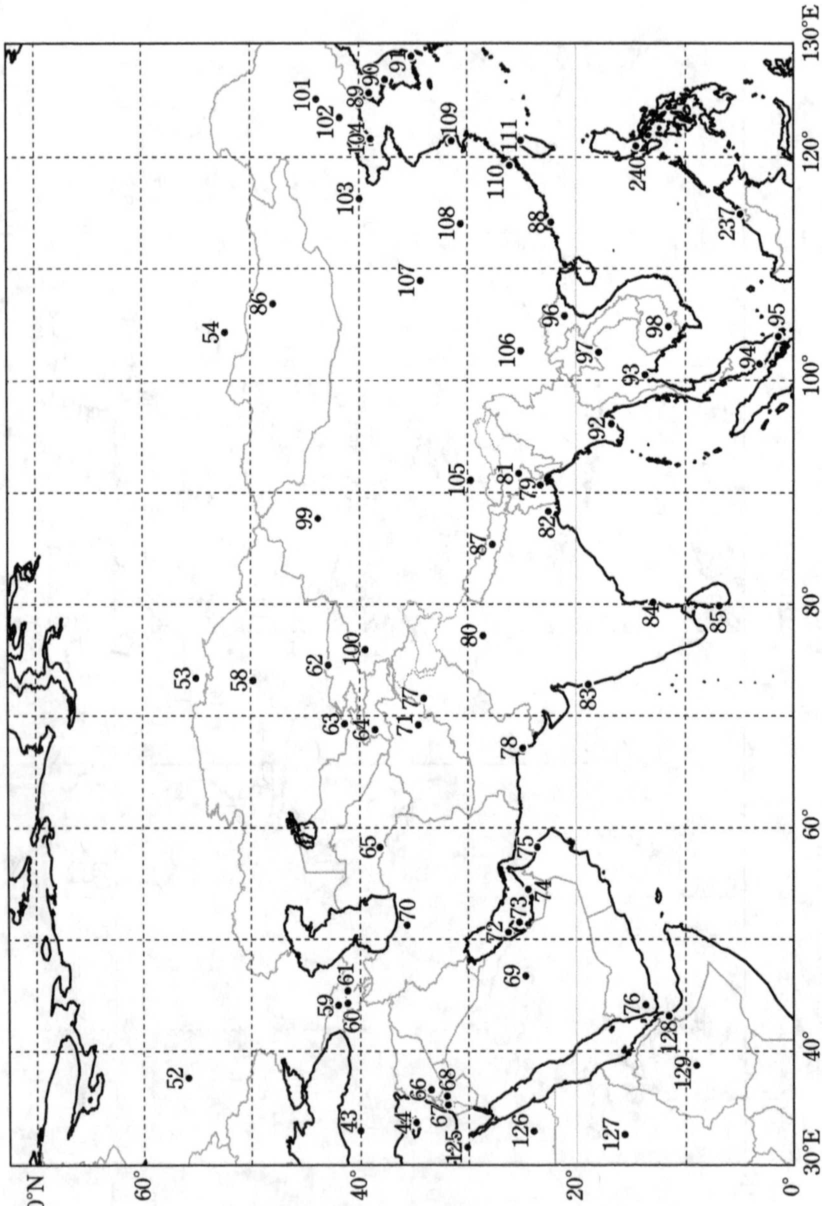

II Asia (1)

II Asia (2)

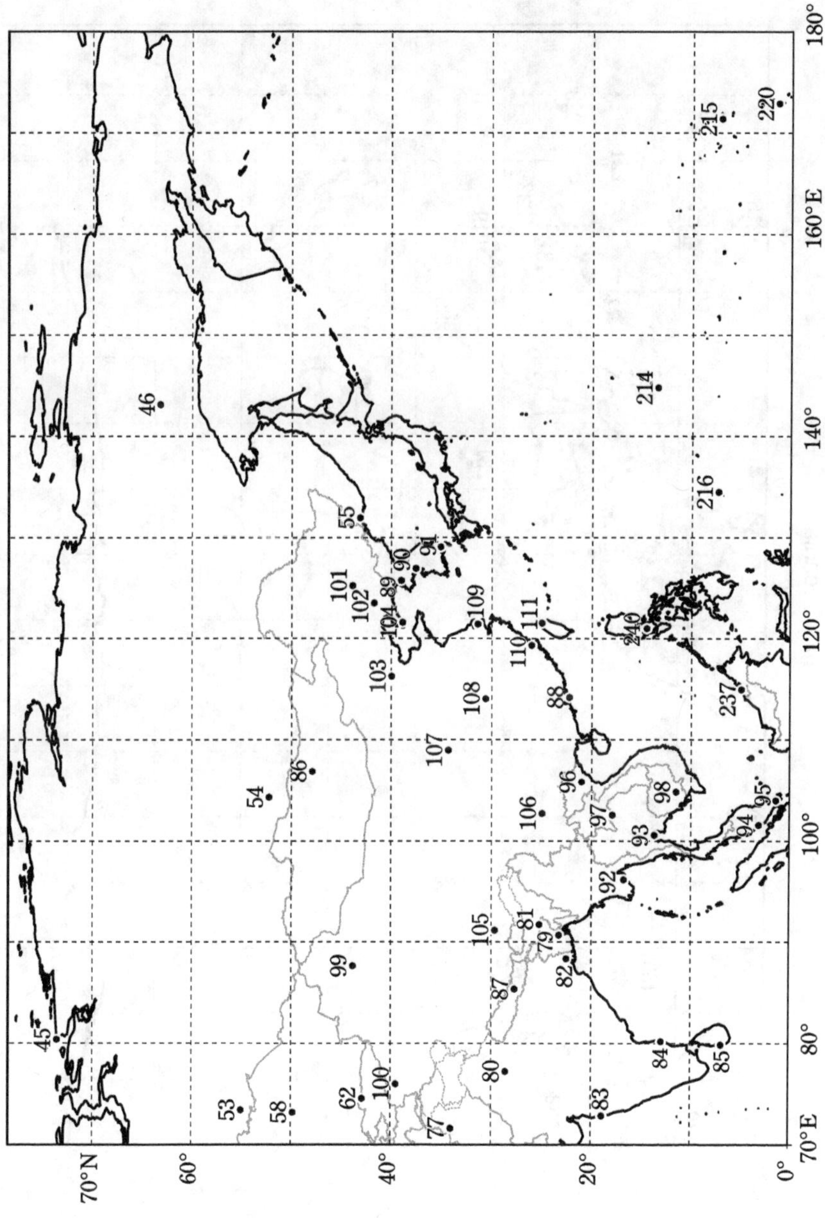

III Oceania

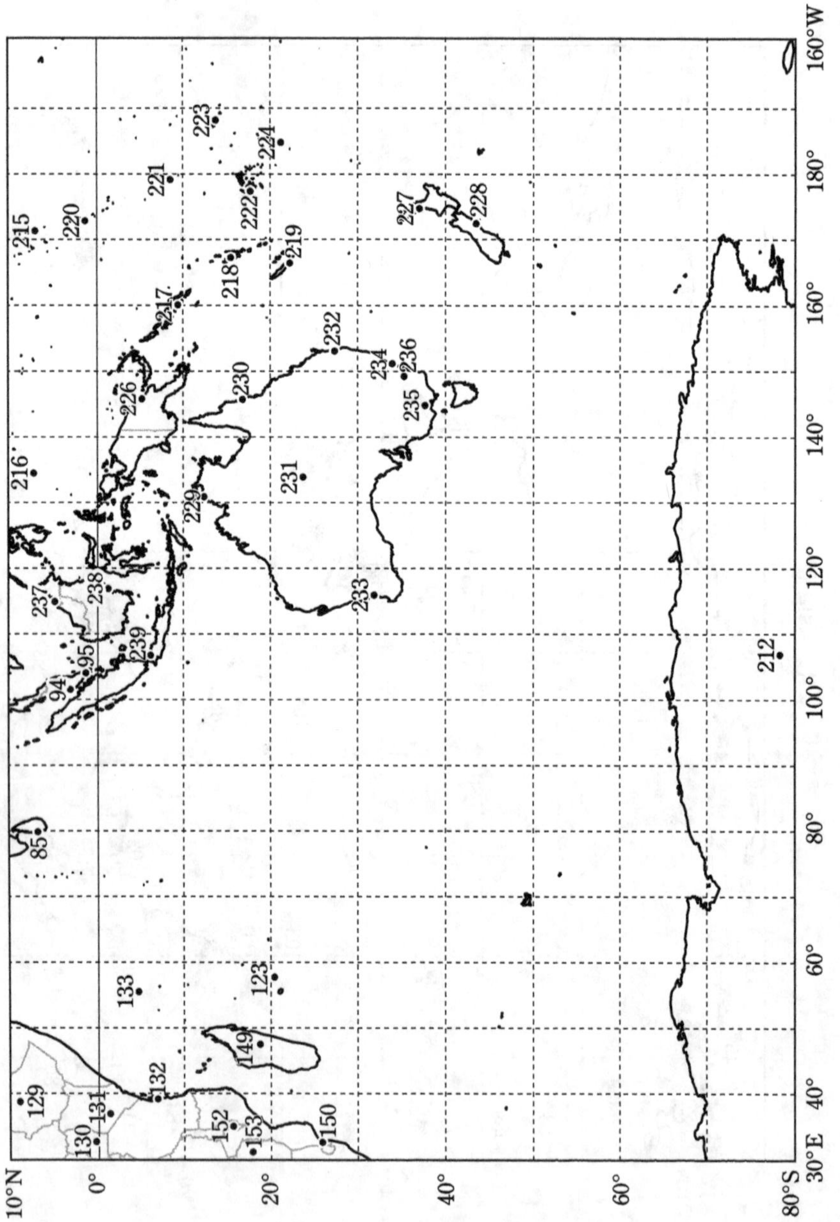
10°N
0°
20°
40°
60°
80°S
30°E
40°
60°
80°
100°
120°
140°
160°
180°
160°W
85
94
95
123
129
130
131
132
133
149
150
152
153
212
215
216
217
218
219
220
221
222
223
224
226
227
228
229
230
231
232
233
234
235
236
237
238
239

IV North America

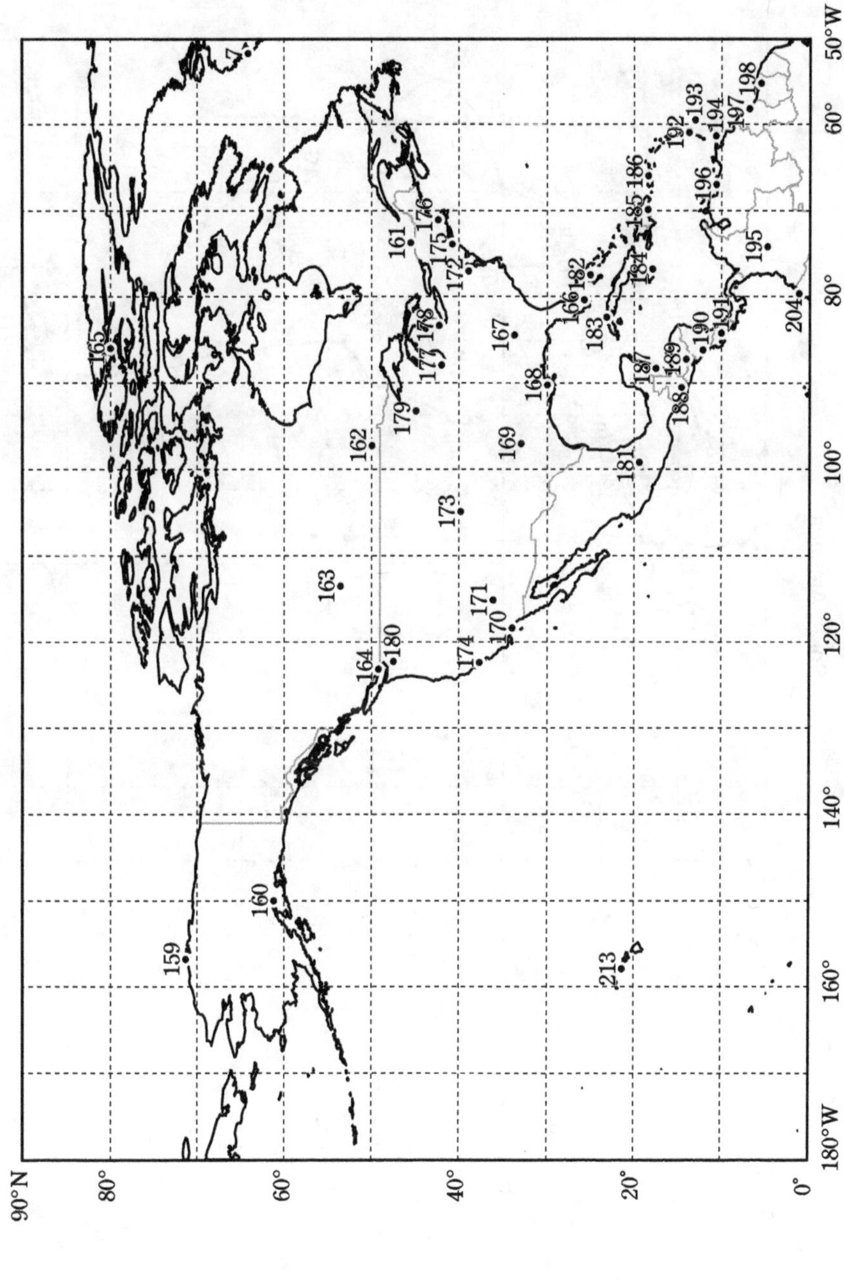

V South America

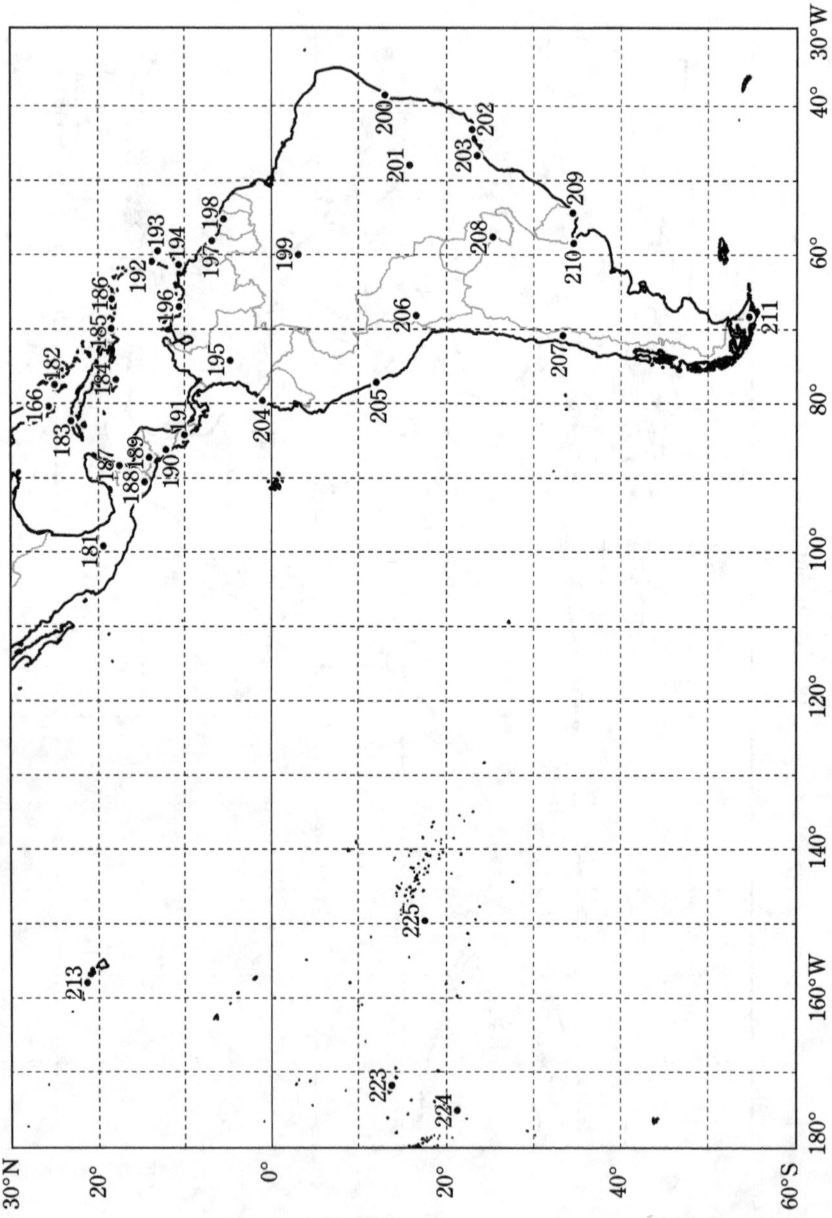

VI Africa

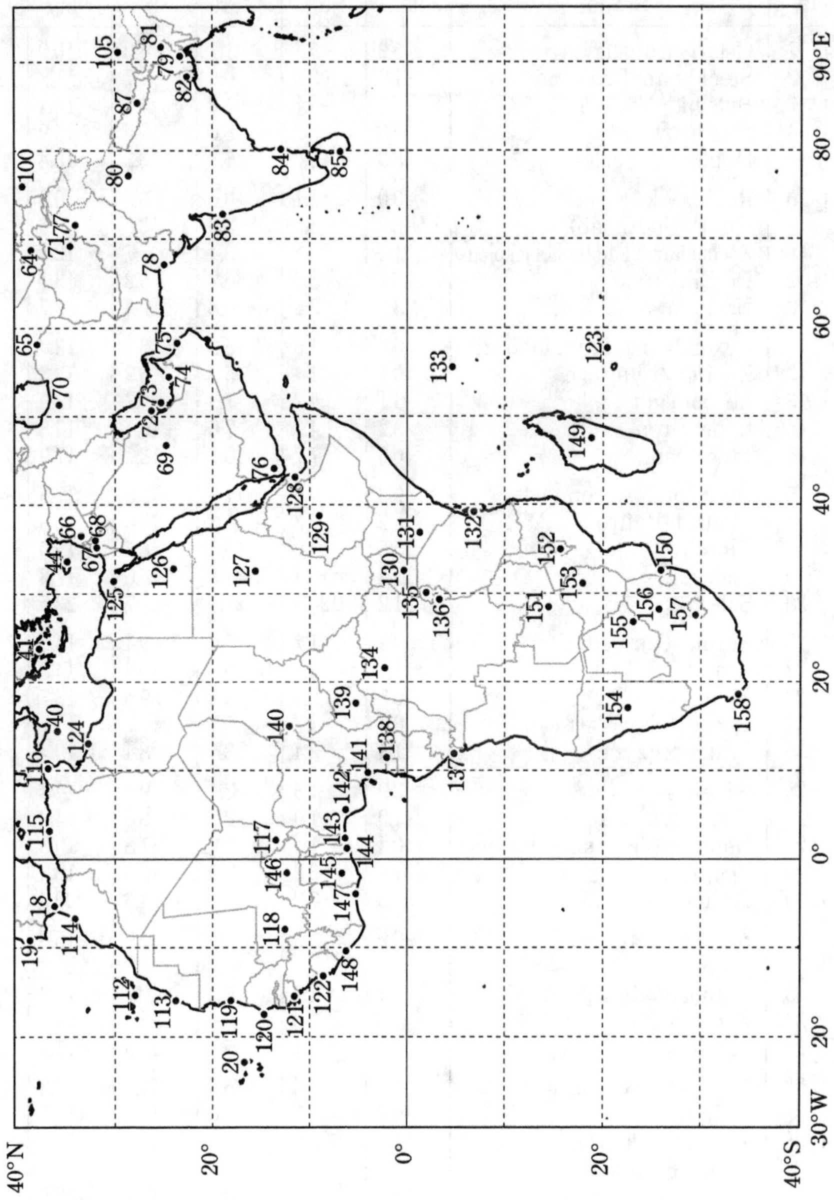
40°N
20°
0°
20°
40°S
30°W
20°
0°
20°
40°
60°
80°
90°E
19 18 114 115 116 40 124 41 44 66 67 68 125 126 127 128 129 76 69 72 73 74 75 70 65 64 71 77 100 78 83 80 84 85 87 105 81 79 82
112 113 119 120 20 121 122 148 118 147 146 145 117 143 144 142 141 140 139 138 134 130 131 135 136 137 132 133 123 149 150 151 152 153 154 155 156 157 158

Climatological Normal Monthly

#	Station	Jan.	Feb.	Mar.	Apr.	May	Jun.
1	Oslo/Gardermoen	-5.2	-5.1	-1.1	4.0	10.0	13.9
2	Stockholm/Bromma	-1.0	-3.2	0.6	4.7	11.2	14.7
3	Helsinki-Vantaa	-5.0	-5.7	-1.9	4.1	10.4	14.6
4	Heathrow	5.8	6.2	8.0	10.5	13.9	17.0
5	Dublin Airport	5.3	5.4	6.7	8.2	10.8	13.4
6	Reykjavik	0.0	0.1	0.6	3.0	6.6	9.5
7	Nuuk (Godthaab)	-8.3	-8.9	-8.3	-3.8	0.6	4.3
8	Koebenhavn/Landbohoejskolen	1.4	1.4	3.4	7.7	12.4	15.6
9	De Bilt Aws	3.1	3.3	6.2	9.2	13.1	15.6
10	Uccle	3.3	3.6	6.8	9.8	13.7	16.3
11	Luxembourg/Luxembourg	0.8	1.6	5.2	8.7	13.0	15.9
12	Zuerich/Fluntern	0.4	1.4	5.2	8.9	13.3	16.4
13	Le Bourget	4.1	4.2	7.3	9.7	13.9	16.5
14	Lyon-Bron	3.2	4.2	7.6	10.4	14.8	18.3
15	Marignane	6.6	7.7	10.7	13.7	17.9	21.7
16	Barcelona/Aeropuerto	9.3	9.6	11.8	13.6	17.5	21.0
17	Madrid, Retiro	6.2	7.8	11.3	13.0	16.8	22.1
18	Gibraltar	13.6	14.0	15.5	16.6	19.0	21.8
19	Lisboa/Geof	11.4	12.4	14.5	15.6	17.8	20.7
20	Sal	21.2	20.5	21.4	21.7	22.5	23.5
21	Berlin-Tempelhof	0.9	1.7	5.0	9.4	14.6	17.4
22	Frankfurt/Main	1.4	2.5	6.3	10.3	14.8	17.9
23	Muenchen-Flughafen	-0.5	0.4	4.2	9.5	13.7	17.4
24	Wien/Hohe Warte	0.4	1.7	5.7	10.7	15.7	18.8
25	Praha/Ruzyne	-1.4	-0.4	3.6	8.4	13.4	16.1
26	Sliac	-2.9	-1.3	3.4	9.3	14.4	17.4
27	Warszawa-Okecie	-1.9	-1.0	2.7	8.6	14.2	16.8
28	Budapest/Pestszentlorinc	-0.4	1.4	6.0	11.8	16.8	19.8
29	Beograd	1.4	3.2	7.6	12.8	18.3	20.9
30	Podgorica-Grad	6.5	7.3	10.9	14.6	20.6	25.0
31	Skopje Petrovec	-0.9	2.3	7.9	11.9	16.8	21.0
32	Vlore						
33	Ljubljana/Bezigrad	-0.1	1.7	6.3	10.7	15.9	19.1
34	Zagreb/Gric	1.3	3.3	7.8	12.5	17.2	20.1
35	Sarajevo-Bjelave	0.0	2.1	5.8	10.3	15.1	18.2
36	Bucuresti Baneasa	-1.7	0.6	5.3	11.2	16.8	20.9
37	Sofia (Observ.)	-0.7	0.7	5.0	10.3	15.1	18.7
38	Verona/Villafranca	1.9	3.9	8.7	12.3	17.9	21.7
39	Roma/Fiumicino						
40	Luqa	12.7	12.5	13.9	16.1	19.8	23.9

Air Temperatures (°C)

Jul.	Aug.	Sep.	Oct.	Nov.	Dec.	Annual	Statistics Period	#
16.3	14.9	10.0	4.9	-0.3	-4.6	4.8	1981-2010	1
17.6	16.1	11.3	7.1	2.4	-0.8	6.7	1982-1994	2
17.7	15.8	10.7	5.6	0.4	-3.1	5.3	1981-2010	3
18.7	18.5	16.2	12.4	8.5	5.7	11.8	1997-2010	4
15.5	15.2	13.2	10.3	7.4	5.6	9.8	1981-2010	5
11.2	10.7	8.0	4.4	1.9	0.6	4.7	1981-2010	6
6.6	6.3	3.8	-0.3	-3.4	-5.7	-1.4	1981-2010	7
18.1	17.7	14.0	9.8	5.6	2.5	9.1	1981-2010	8
17.9	17.5	14.5	10.7	6.7	3.7	10.1	1981-2010	9
18.4	18.0	14.9	11.1	6.8	4.1	10.6	1981-2010	10
18.2	17.7	14.0	9.5	4.7	1.8	9.3	1981-2010	11
18.6	18.1	14.2	9.9	4.4	1.5	9.4	1981-2010	12
19.3	18.8	15.4	11.7	7.1	5.3	11.1	1982-1995	13
22.2	21.2	17.1	12.7	7.1	4.5	11.9	1982-1995	14
24.7	24.1	20.2	16.2	10.9	7.6	15.2	1982-2010	15
24.1	24.5	21.6	17.8	12.8	9.8	16.1	1986-2010	16
25.5	25.1	20.8	15.1	9.8	6.8	15.0	1982-2010	17
24.2	24.6	22.9	19.8	16.5	14.6	18.6	1981-2010	18
22.7	23.1	21.7	18.5	15.1	12.3	17.2	1981-2010	19
24.5	25.8	26.4	25.5	24.2	22.3	23.3	1981-2005	20
19.8	19.2	14.9	9.9	5.1	1.6	10.0	1981-2010	21
19.8	19.4	15.2	10.7	5.6	2.4	10.5	1981-2010	22
19.5	18.1	13.8	8.8	4.7	0.5	9.2	1993-2010	23
20.9	20.2	15.5	10.3	5.1	1.1	10.5	1981-2010	24
18.2	17.8	13.5	8.5	3.1	-0.3	8.4	1981-2010	25
19.4	18.6	13.9	8.6	3.3	-1.9	8.5	1981-2010	26
19.1	18.3	13.4	8.5	3.3	-0.7	8.4	1981-2010	27
22.0	21.5	16.6	11.1	5.1	0.5	11.0	1981-2010	28
22.9	22.8	18.1	12.9	7.1	2.5	12.5	1981-2010	29
27.7	27.1	21.1	17.0	11.5	7.4	16.4	1994-2010	30
22.7	22.9	18.8	12.8	6.1	1.8	12.0	1981-1992	31
								32
21.2	20.3	15.9	11.2	5.5	1.1	10.7	1981-2010	33
22.2	21.9	17.2	12.3	6.7	2.6	12.1	1981-2010	34
20.3	20.3	14.5	10.7	6.1	1.1	10.4	2001-2010	35
22.8	22.2	16.3	10.5	5.0	-0.8	10.8	1983-2010	36
21.0	20.7	16.2	10.9	4.9	0.5	10.3	1981-2010	37
24.3	24.1	19.3	13.6	7.4	3.2	13.2	1981-2005	38
								39
26.7	27.1	24.7	21.5	17.6	14.3	19.2	1981-2010	40

Climatological Normal Monthly Air Temperatures

#	Station	Jan.	Feb.	Mar.	Apr.	May	Jun.
41	Athinai Ap Hellinikon	10.3	10.2	12.8	16.3	21.5	26.2
42	Istanbul/Goztepe	6.4	6.2	8.1	12.4	17.1	21.8
43	Ankara/Central	0.8	2.1	6.1	11.5	16.1	20.4
44	Larnaca Airport	12.0	12.0	14.2	17.5	21.6	25.5
45	Ostrov Dikson	-24.8	-25.7	-22.3	-17.4	-7.7	0.5
46	Ojmjakon	-46.4	-42.2	-31.1	-13.5	2.7	12.5
47	Tallinn-Harku	-3.3	-4.0	-1.0	4.6	10.2	14.3
48	St.Petersburg (Voejkovo)	-5.5	-5.8	-1.3	5.1	11.3	15.7
49	Jelgava						
50	Kaunas	-3.4	-2.9	0.6	7.1	13.0	15.6
51	Minsk	-4.4	-4.5	0.0	7.2	13.3	16.4
52	Moskva	-6.5	-6.7	-1.0	6.7	13.2	17.0
53	Omsk	-16.2	-15.0	-7.3	3.7	12.5	17.9
54	Irkutsk	-17.7	-14.4	-6.4	2.4	10.1	15.4
55	Vladivostok	-12.3	-8.4	-1.9	5.1	9.8	13.6
56	Kiev	-3.5	-2.8	1.9	9.3	15.6	18.6
57	Chisinau	-2.1	-0.9	3.9	10.4	16.7	19.9
58	Karaganda	-12.9	-12.6	-6.0	5.7	13.3	18.8
59	Gori						
60	Tashir						
61	Agstapha Airport						
62	Bishkek	-2.2	0.2	7.5	12.5	17.6	23.0
63	Tashkent	1.9	4.0	9.3	15.5	20.4	25.9
64	Dushanbe	3.9	4.5	10.1	15.8	20.6	25.5
65	Ashgabat Keshi	3.2	4.6	10.2	17.3	23.2	29.0
66	Damascus Int. Airport	6.0	7.8	11.2	16.1	20.9	25.0
67	Jerusalem	7.6	7.6	10.5	15.4	18.8	21.4
68	Amman Airport	8.1	8.5	11.5	16.4	20.7	24.0
69	Riyadh Obs. (O.A.P.)	14.5	16.8	21.4	26.5	32.6	35.3
70	Tehran-Mehrabad	5.1	7.5	12.1	17.5	23.0	28.4
71	Kabul Airport	-0.8	0.2	6.5	13.5	18.8	23.3
72	Bahrain (Int. Airport)	16.9	18.0	20.7	25.5	30.6	33.2
73	Doha International Airport	16.8	18.2	21.5	26.2	31.8	34.3
74	Abu Dhabi Bateen Airport						
75	Seeb, Int'l Airport	21.1	22.0	24.9	29.4	34.2	34.9
76	Taiz						
77	Peshawar	11.6	13.5	18.3	23.8	29.4	32.6
78	Karachi Airport	19.0	21.2	25.6	29.0	31.0	31.7
79	Chandpur						
80	New Delhi/Safdarjung	14.1	17.4	22.7	28.9	32.7	33.2

Continued.

Jul.	Aug.	Sep.	Oct.	Nov.	Dec.	Annual	Statistics Period	#
28.7	28.8	24.4	19.6	15.2	11.8	18.8	1982-2010	41
24.2	24.2	20.5	15.9	11.4	8.2	14.7	1982-2010	42
23.7	23.5	18.8	13.0	6.9	2.6	12.1	1982-2010	43
27.8	28.1	25.8	22.6	17.7	13.7	19.9	1982-2010	44
5.0	5.5	1.7	-7.5	-17.5	-22.7	-11.1	1981-2010	45
14.8	10.4	2.2	-14.5	-35.0	-45.4	-15.5	1981-2010	46
17.2	16.2	11.3	6.6	1.9	-1.5	6.0	1982-2010	47
18.8	16.9	11.6	6.2	0.1	-3.9	5.8	1981-2010	48
								49
18.1	17.1	12.3	7.2	2.2	-1.9	7.1	1981-2010	50
18.5	17.5	12.1	6.6	0.6	-3.3	6.7	1981-2010	51
19.2	17.0	11.3	5.6	-1.2	-5.2	5.8	1981-2010	52
19.6	16.9	10.4	3.6	-7.0	-13.9	2.1	1981-2010	53
18.3	15.9	9.1	1.8	-7.9	-15.3	0.9	1981-2010	54
17.6	19.9	16.1	9.0	-0.9	-9.0	4.9	1981-2010	55
20.5	19.7	14.2	8.3	1.7	-2.2	8.4	1981-2010	56
22.0	21.8	16.3	10.6	4.4	-0.3	10.2	1981-2010	57
20.4	18.6	12.5	4.5	-4.9	-11.0	3.9	1981-2010	58
								59
								60
								61
25.0	24.3	18.7	12.1	5.8	-1.6	11.9	2001-2010	62
27.8	26.3	20.6	13.8	8.7	3.4	14.8	1982-2010	63
27.7	26.4	21.0	14.7	9.9	5.4	15.5	1981-2010	64
31.4	29.3	23.2	16.2	9.8	4.7	16.8	1982-2010	65
27.0	26.9	23.9	18.7	11.8	7.5	16.9	1981-2010	66
22.6	23.2	21.9	18.9	13.8	9.2	15.9	1982-1995	67
25.5	25.7	24.3	20.5	14.1	9.6	17.4	1981-1998	68
36.6	36.5	33.4	28.2	21.3	16.1	26.6	1981-2010	69
31.2	30.2	26.2	19.8	13.1	6.8	18.4	1984-2010	70
25.7	25.1	20.9	14.3	7.7	1.9	13.1	1981-2009	71
34.5	34.5	32.6	29.4	24.4	19.3	26.6	1982-2010	72
35.2	34.7	32.7	28.9	24.3	19.6	27.0	1981-2010	73
								74
33.5	31.6	30.8	29.3	25.6	22.5	28.3	1982-2010	75
								76
32.1	31.0	29.0	23.6	17.7	12.8	23.0	1982-2010	77
30.4	29.2	29.4	28.6	24.6	20.5	26.7	1982-2010	78
								79
31.4	30.3	29.6	26.0	20.5	15.5	25.2	1981-2010	80

Climatological Normal Monthly Air Temperatures

#	Station	Jan.	Feb.	Mar.	Apr.	May	Jun.
81	Cherrapunji	11.5	13.3	16.2	18.3	19.3	20.3
82	Kolkata/Alipore	20.0	23.6	28.0	30.4	30.9	30.4
83	Bombay/Colaba	24.9	25.3	27.1	28.9	30.5	29.3
84	Chennai/Minambakkam	25.2	26.6	28.7	30.9	32.9	32.4
85	Colombo	27.1	27.3	28.1	28.6	28.7	28.0
86	Ulaanbaatar	-21.7	-16.1	-7.0	1.8	10.0	16.0
87	Kathmandu Airport	10.9	12.9	16.9	20.2	22.3	24.1
88	Hong Kong Observatory	16.1	16.1	18.6	22.0	25.5	27.8
89	Pyongyang	-5.9	-2.7	3.3	10.9	16.7	21.3
90	Seoul	-2.4	0.6	5.7	12.5	17.8	22.2
91	Busan	3.2	4.9	8.7	13.6	17.5	20.7
92	Yangon	24.8	26.5	28.6	31.0	29.2	27.4
93	Bangkok Metropolis	27.3	28.6	29.8	30.9	30.1	29.7
94	Kuala Lumpur/Subang	26.9	27.3	27.6	27.7	28.0	27.9
95	Singapore/Changi Airport	26.6	27.2	27.6	28.0	28.4	28.4
96	Ha Noi						
97	Vientiane	22.8	25.0	27.5	29.3	28.7	28.5
98	Phnom-Penh/Pochentong						
99	Urumqi	-12.0	-9.3	-0.5	10.3	17.0	21.8
100	Kashgar	-4.8	0.4	8.5	15.7	20.1	23.7
101	Changchun	-14.8	-9.8	-1.5	8.5	15.8	20.9
102	Shenyang	-11.1	-6.4	1.4	10.6	17.6	22.3
103	Beijing	-3.1	0.2	6.7	14.8	20.8	24.9
104	Dalian	-3.7	-1.4	3.6	10.6	16.4	20.6
105	Lhasa	-0.8	2.2	5.7	8.9	12.7	16.4
106	Kunming	8.9	10.9	14.1	17.2	19.3	20.2
107	Xian	0.3	3.4	8.7	15.4	20.5	25.2
108	Wuhan	4.0	6.7	10.8	17.5	22.7	26.1
109	Shanghai	4.8	6.6	10.2	15.5	20.7	24.5
110	Fuzhou	11.2	11.6	14.0	18.5	22.7	26.2
111	Taipei	16.3	16.5	18.3	22.0	25.3	27.8
112	Las Palmas De Gran Canaria/Gando	17.8	18.0	18.7	19.4	20.4	22.1
113	Dakhla	18.2	18.6	19.5	19.2	19.9	21.2
114	Rabat-Sale	12.0	12.8	14.8	15.7	18.2	20.8
115	Dar-El-Beida	10.5	11.2	13.1	15.1	18.6	22.4
116	Tunis-Carthage	11.8	12.2	14.0	16.6	20.4	24.5
117	Niamey-Aero	24.0	27.2	31.5	34.3	34.4	32.0
118	Bamako/Senou	25.1	28.3	31.2	32.5	31.5	28.8
119	Nouakchott	21.3	22.6	24.1	24.4	25.6	26.8
120	Dakar/Yoff	20.8	20.7	21.1	21.5	23.0	25.6

Continued.

Jul.	Aug.	Sep.	Oct.	Nov.	Dec.	Annual	Statistics Period	#
20.2	20.8	20.5	19.7	16.7	13.2	17.5	1981-2010	81
29.4	29.3	29.2	28.1	25.0	21.2	27.1	1981-2010	82
27.8	27.4	27.8	28.9	28.4	26.5	27.7	1981-2010	83
30.9	30.3	29.8	28.4	26.5	25.3	29.0	1981-2010	84
27.9	27.9	27.8	27.3	27.1	27.0	27.7	1981-2010	85
18.5	16.0	9.5	0.9	-10.6	-19.0	-0.1	1983-2010	86
24.3	24.4	23.4	20.3	15.8	11.9	19.0	1981-2010	87
28.8	28.7	27.7	25.3	21.4	17.4	23.0	1981-2004	88
24.0	24.5	19.3	12.4	4.5	-2.9	10.5	1981-2010	89
24.9	25.7	21.2	14.8	7.2	0.4	12.6	1981-2010	90
24.1	26.0	22.3	17.6	11.6	5.8	14.7	1981-2010	91
26.8	26.9	27.5	27.6	27.3	25.0	27.4	1981-2005	92
29.3	29.1	28.7	28.4	27.9	26.6	28.9	1981-2010	93
27.5	27.4	27.2	27.1	26.8	26.7	27.3	1981-2010	94
27.9	27.8	27.7	27.7	27.0	26.6	27.6	1981-2010	95
								96
28.2	27.8	27.7	27.1	25.1	22.4	26.7	1981-2010	97
								98
23.8	22.7	16.9	8.4	-1.5	-9.3	7.4	1981-2010	99
25.6	24.5	19.6	12.5	4.4	-2.9	12.3	1981-2010	100
23.2	22.1	16.0	7.7	-3.2	-11.6	6.1	1981-2010	101
24.6	23.8	17.9	10.0	0.2	-7.9	8.6	1982-2010	102
26.7	25.5	20.7	13.7	5.0	-0.9	12.9	1981-2010	103
23.7	24.4	20.8	14.2	6.1	-0.4	11.2	1981-2010	104
16.3	15.5	13.6	9.0	3.3	-0.4	8.5	1981-2010	105
20.2	20.0	18.3	16.0	12.1	9.0	15.5	1981-2010	106
26.9	25.1	20.3	14.1	7.2	1.4	14.0	1981-2005	107
29.0	28.5	24.1	18.4	12.1	6.2	17.2	1982-2010	108
28.6	28.3	24.8	19.7	13.8	7.6	17.1	1991-2010	109
29.2	28.8	26.2	22.4	18.2	13.4	20.2	1981-2010	110
29.7	29.2	27.5	24.5	21.5	18.0	23.1	1982-2010	111
23.9	24.5	24.2	23.0	21.0	19.2	21.0	1982-2010	112
21.5	22.7	23.0	22.5	21.3	19.2	20.6	1988-2010	113
22.7	23.0	21.6	19.0	15.5	13.3	17.5	1988-2010	114
25.4	26.1	23.4	19.7	15.2	12.0	17.7	1982-2010	115
27.6	28.2	25.2	21.6	16.7	13.1	19.3	1981-2010	116
29.4	28.1	29.4	31.1	28.5	25.2	29.6	1981-2010	117
26.4	25.7	26.2	27.5	27.4	25.2	28.0	1981-2010	118
27.3	28.4	29.5	28.6	25.8	22.6	25.6	1981-2010	119
27.2	27.6	27.7	27.7	25.9	23.3	24.3	1981-2010	120

Climatological Normal Monthly Air Temperatures

#	Station	Jan.	Feb.	Mar.	Apr.	May	Jun.
121	Bolama						
122	Lungi	26.2	27.1	27.6	27.8	27.4	26.6
123	Plaisance (Mauritius)	26.5	26.5	26.1	25.2	23.6	22.1
124	Tripoli International Airport	12.0	13.3	15.2	19.0	23.1	26.9
125	Cairo	14.1	14.8	17.3	21.6	24.5	27.4
126	Asswan	16.2	18.1	22.4	27.6	31.8	33.9
127	Khartoum	23.2	25.7	28.1	32.8	34.9	35.1
128	Djibouti	25.0	25.8	27.4	28.8	31.0	34.0
129	Addis Ababa-Bole	15.7	17.0	17.9	18.2	18.3	16.9
130	Kampala						
131	Jomo Kenyatta International Airport	19.9	21.1	21.3	20.6	19.6	18.4
132	Dar Es Salaam Int	28.0	28.1	27.6	26.6	25.7	24.4
133	Seychelles International Airport	27.1	27.7	28.1	28.3	28.0	26.9
134	Lisala						
135	Kigali	20.4	20.9	20.1	20.1	20.4	20.6
136	Bujumbura						
137	Pointe-Noire	26.6	27.0	27.2	27.1	26.2	23.2
138	Bitam	24.3	24.7	24.4	24.2	23.9	23.3
139	Bossembele						
140	Ndjamena	22.9	26.2	30.4	33.7	33.4	31.0
141	Douala Obs.	27.2	27.9	27.7	27.2	26.8	26.0
142	Benin City						
143	Cotonou	27.4	28.8	29.2	28.9	28.2	26.9
144	Lome	27.1	28.5	28.7	28.5	27.8	26.6
145	Kumasi	26.6	28.1	28.0	27.2	26.7	25.7
146	Ouagadougou	24.8	27.9	31.6	33.4	32.2	29.8
147	Abidjan	26.9	27.9	28.3	28.3	27.8	26.5
148	Roberts Field	25.9	26.7	27.2	27.2	26.9	25.8
149	Antananarivo/Ivato	20.7	20.8	20.5	19.4	17.2	15.0
150	Maputo/Mavalane	26.6	26.6	25.8	23.9	22.0	19.6
151	Kabwe	21.6	21.8	21.5	20.4	19.0	16.8
152	Chileka	24.1	23.5	23.4	22.5	20.9	19.4
153	Harare (Kutsaga)	20.7	20.2	19.8	18.1	15.5	13.2
154	Windhoek						
155	Mahalapye						
156	Pretoria Eendracht	23.0	22.9	21.6	18.8	15.3	12.3
157	Maseru-Mia						
158	Cape Town Intnl. Airport	21.0	21.1	19.8	17.3	15.0	12.8
159	Barrow/W. Post W. Rogers	-25.3	-25.8	-24.8	-16.6	-6.0	2.1
160	Anchorage/Int., Ak	-8.2	-6.5	-3.0	2.7	8.8	12.9

Continued.

Jul.	Aug.	Sep.	Oct.	Nov.	Dec.	Annual	Statistics Period	#
								121
25.5	25.3	25.9	26.3	27.0	26.5	26.6	1981-1998	122
21.3	21.3	21.8	22.8	24.2	25.6	23.9	1981-2010	123
27.9	28.3	27.2	23.5	17.9	13.6	20.7	1982-2010	124
28.0	28.2	26.6	24.0	19.2	15.1	21.7	1981-1995	125
34.8	34.4	32.3	28.6	22.4	17.4	26.7	1982-2010	126
32.7	31.7	32.6	32.9	28.7	24.8	30.3	1983-2010	127
35.3	35.1	32.3	29.5	27.2	25.4	29.7	1981-1999	128
16.2	16.1	16.3	16.1	15.2	14.9	16.6	1983-2010	129
								130
17.2	17.9	19.4	20.4	19.7	19.6	19.6	1981-2010	131
23.7	23.8	24.5	25.6	26.6	27.5	26.0	1981-2010	132
26.1	26.1	26.6	27.1	27.1	27.1	27.2	1981-2010	133
								134
20.9	21.7	21.3	20.3	19.8	20.0	20.5	1982-2010	135
								136
22.2	22.2	23.9	25.3	26.1	26.2	25.3	1981-2010	137
22.7	22.8	23.1	23.4	23.5	24.0	23.7	1981-2010	138
								139
27.9	26.6	27.9	28.8	27.2	24.1	28.3	1987-2010	140
25.0	24.8	25.2	25.5	26.5	26.9	26.4	1981-2010	141
								142
25.8	25.5	26.1	27.0	28.3	28.0	27.5	1981-2010	143
25.4	25.1	25.7	26.8	27.9	27.6	27.1	1981-2010	144
24.8	24.4	25.0	25.7	26.5	26.5	26.3	1984-2010	145
27.7	26.6	27.5	29.2	28.0	25.4	28.7	1981-2010	146
25.0	24.3	24.9	26.6	27.9	27.4	26.8	1981-2010	147
24.7	24.5	25.2	26.0	26.6	25.6	26.0	1981-1996	148
14.0	14.9	16.6	18.6	19.7	20.5	18.2	1981-2010	149
19.5	20.4	21.9	22.7	24.0	25.6	23.2	1981-2010	150
16.5	19.0	22.4	24.7	24.2	22.1	20.8	1981-2010	151
18.7	20.7	23.4	25.3	25.9	24.4	22.7	1981-2010	152
13.1	15.1	18.8	20.3	21.4	20.5	18.1	1981-2010	153
								154
								155
12.2	15.4	19.3	20.8	21.8	22.4	18.8	1981-2010	156
								157
12.2	12.7	14.4	16.3	18.3	20.1	16.8	1982-2010	158
5.0	4.0	0.1	-8.2	-17.0	-22.1	-11.2	1981-2010	159
14.9	13.7	9.3	1.6	-5.4	-7.2	2.8	1981-2010	160

Climatological Normal Monthly Air Temperatures

#	Station	Jan.	Feb.	Mar.	Apr.	May	Jun.
161	Montreal/Pierre Elliott Trudeau int'l A, Que	-10.1	-7.8	-2.3	6.1	13.5	18.4
162	Winnipeg Richardson Int'l A, Man	-16.5	-12.7	-5.3	4.2	12.0	17.0
163	Edmonton City Centre Awos, Alta	-10.3	-7.8	-1.9	5.7	11.7	15.3
164	Vancouver Int'l A, Bc	4.0	4.8	7.0	9.4	12.7	15.5
165	Eureka, Nu	-36.6	-37.3	-36.7	-26.5	-10.5	3.0
166	Miami, Fl	20.1	21.0	22.4	24.1	26.5	28.0
167	Atlanta/Mun., Ga.	6.3	8.5	12.5	16.7	21.3	25.2
168	New Orleans/Moisant Int., La.	11.6	13.5	16.7	20.3	24.6	27.2
169	Dallas-Fort Worth/Fort Worth Reg.Airport, Tx.	8.2	10.4	14.3	18.7	23.4	27.5
170	Los Angeles/Int., Ca.	14.1	14.3	14.8	16.1	17.4	19.0
171	Las Vegas/Mccarran, Nv.	8.8	11.2	15.0	19.2	24.8	29.9
172	Washington/Nat., Va.	2.3	3.9	8.2	13.8	18.9	24.1
173	Denver/Stapleton Int., Co.	-0.3	1.1	4.9	9.0	14.3	19.7
174	San Francisco/Int., Ca.	10.1	11.6	12.7	13.8	15.3	16.8
175	New York/La Guardia, Ny.	1.0	2.0	5.9	11.6	17.1	22.4
176	Boston/Logan Int., Ma.	-1.5	0.0	3.6	9.1	14.5	20.0
177	Chicago/O'hare, Il.	-4.6	-2.4	3.2	9.3	15.0	20.5
178	Detroit/Metropolitan, Mi.	-3.2	-2.3	2.9	9.6	15.4	20.8
179	Minneapolis/St. Paul Int., Mn.	-8.8	-6.3	0.5	8.7	15.1	20.7
180	Seattle/S.-Tacoma, Wa.	5.6	6.3	8.0	10.2	13.4	16.1
181	Mexico (Central), D.F.	14.0	15.3	17.6	19.0	19.3	18.3
182	Nassau Airport New Providence	21.6	22.1	22.6	24.0	25.8	27.6
183	Casa Blanca, La Habana						
184	Kingston/Norman Manley	26.3	26.1	26.6	27.3	28.2	28.9
185	Santo Domingo	25.0	24.9	25.4	26.1	26.8	27.2
186	San Juan/Int., Puerto Rico	25.1	25.3	25.8	26.6	27.5	28.2
187	Belize/Phillip Goldston Intl. Airport	24.1	24.9	26.0	27.3	28.2	28.5
188	Guatemala (Observatorio Nacional)						
189	Tegucigalpa	19.7	20.9	22.4	23.7	23.8	23.0
190	Managua A.C.Sandino	26.5	27.0	28.0	29.2	29.0	27.8
191	Juan Santamaria Int. Airport	22.6	23.0	23.5	23.8	23.0	22.6
192	Hewanorra Int'l Airport						
193	Grantley Adams	25.8	25.8	26.1	26.9	27.6	27.6
194	Piarco Int. Airport, Trinidad	25.6	25.6	26.4	27.2	27.4	26.9
195	Bogota/Eldorado	13.1	13.4	13.6	13.9	13.9	13.7
196	Caracas/Maiquetia Aerop. Intl. Simon Bolivar	25.2	25.2	25.7	26.4	27.2	27.4
197	Georgetown	26.5	26.8	27.1	27.5	27.2	26.9
198	Zanderij	25.4	25.5	25.8	26.2	26.1	25.9
199	Manaus	26.4	26.4	26.6	26.4	26.7	26.6
200	Salvador	26.9	27.2	27.1	26.3	25.5	24.5

Continued.

Jul.	Aug.	Sep.	Oct.	Nov.	Dec.	Annual	Statistics Period	#
21.0	19.8	15.0	8.4	1.8	-5.9	6.5	1981-2001	161
19.6	18.9	12.5	5.1	-5.3	-14.2	2.9	1981-2001	162
17.5	17.1	11.5	5.4	-3.9	-9.0	4.3	1981-2001	163
17.6	17.9	14.9	10.2	6.3	3.4	10.3	1981-2001	164
6.1	3.2	-6.4	-20.7	-29.6	-33.2	-18.8	1981-2010	165
28.8	28.9	28.2	26.4	23.7	21.3	25.0	1982-2010	166
26.8	26.4	23.0	17.5	12.2	7.4	17.0	1981-2010	167
28.2	28.2	26.2	21.6	16.7	13.0	20.7	1982-2010	168
29.8	29.8	25.7	20.0	13.7	8.7	19.2	1986-2010	169
20.7	21.0	20.5	19.1	16.7	14.0	17.3	1986-2010	170
33.2	32.2	27.7	20.4	13.1	8.3	20.3	1981-2010	171
26.6	25.6	21.7	15.3	9.8	4.3	14.5	1981-2010	172
23.0	22.3	17.3	10.5	3.6	-1.1	10.4	1981-2008	173
17.5	18.0	18.2	16.4	13.2	10.3	14.5	1981-2010	174
25.3	24.8	20.8	14.7	9.2	3.7	13.2	1982-2010	175
23.1	22.4	18.4	12.3	7.2	1.6	10.9	1981-2010	176
23.3	22.4	18.2	11.4	4.6	-2.3	9.9	1981-2010	177
23.1	22.3	18.0	11.2	5.1	-1.0	10.2	1986-2010	178
23.2	21.7	16.9	9.4	1.1	-6.2	8.0	1986-2010	179
18.7	18.9	16.3	11.7	7.6	4.9	11.5	1986-2010	180
17.0	17.2	16.9	16.2	15.1	14.3	16.7	1981-2010	181
28.5	28.5	28.0	26.6	24.5	22.7	25.2	1981-2010	182
								183
29.2	28.9	28.4	28.1	27.4	26.7	27.7	1981-2010	184
27.5	27.5	27.6	27.2	26.6	25.7	26.5	1981-2004	185
28.4	28.6	28.3	27.9	26.7	25.8	27.0	1981-2010	186
28.2	28.2	28.1	27.0	25.5	24.5	26.7	1981-2010	187
								188
22.3	22.5	22.2	21.8	21.0	20.1	22.0	1981-2010	189
27.0	27.1	27.2	26.6	26.4	26.4	27.4	1982-2010	190
22.7	22.6	22.1	21.9	22.2	22.5	22.7	1981-2010	191
								192
27.5	27.8	27.7	27.5	27.1	26.5	27.0	1981-2010	193
26.9	27.0	27.1	27.0	26.4	26.0	26.6	1981-2010	194
13.3	13.2	13.2	13.3	13.3	13.1	13.4	1981-2010	195
27.2	28.0	28.2	27.9	27.2	26.0	26.8	1986-2010	196
26.9	27.5	27.9	28.0	27.6	27.0	27.2	1981-2007	197
26.1	26.7	27.3	27.3	27.0	25.8	26.3	1981-2010	198
26.9	27.5	27.6	28.0	27.6	26.8	27.0	1981-2010	199
23.7	23.8	24.3	25.4	26.0	26.4	25.6	1981-2010	200

Climatological Normal Monthly Air Temperatures

#	Station	Jan.	Feb.	Mar.	Apr.	May	Jun.
201	Brasilia	21.8	21.8	21.7	21.5	20.3	19.0
202	Rio De Janeiro	26.6	26.7	26.2	25.0	23.2	21.6
203	Sao Paulo	22.8	23.2	22.4	20.8	18.3	17.0
204	Esmeraldas (Tachina) Aeropuerto						
205	Lima/Callao	22.6	23.1	22.8	20.4	18.4	17.1
206	La Paz/Alto	9.4	9.5	9.4	8.6	8.0	6.3
207	Pudahuel	20.4	19.5	17.5	13.9	10.5	8.3
208	Aeropuerto Silvio Pettirossi, Luque	27.3	26.7	26.0	23.0	19.4	17.5
209	Rocha	22.1	21.9	20.3	17.2	13.8	11.2
210	Buenos Aires Observatorio	24.8	23.4	21.8	17.8	14.6	11.8
211	Ushuaia Aero	9.7	9.5	8.0	6.0	3.8	1.7
212	Vostok	-31.8	-43.9	-57.6	-64.5	-66.3	-65.8
213	Honolulu, Oahu, Hawaii	23.0	23.0	23.7	24.6	25.5	26.9
214	Weather Forecast Office, Guam, Mariana Is.	27.1	27.0	27.4	28.1	28.4	28.5
215	Majuro/Marshall Is. Intnl.	27.5	27.6	27.7	27.7	27.8	27.7
216	Weather Service Office, Koror, Palau Wci.	27.5	27.5	27.7	28.1	28.1	27.8
217	Honiara/Henderson	27.0	26.9	26.9	26.8	26.6	26.3
218	Pekoa Airport (Santo)	26.5	26.3	26.5	26.1	25.5	24.8
219	Noumea (Nlle-Caledonie)	26.0	26.2	25.5	24.3	22.5	21.0
220	Tarawa	28.1	28.0	27.9	27.9	28.0	28.0
221	Funafuti	28.2	28.2	28.1	28.3	28.5	28.3
222	Nadi Airport	26.8	27.0	26.8	25.9	24.8	23.8
223	Apia	26.9	26.9	27.1	26.9	26.5	26.4
224	Fua'amotu						
225	Tahiti-Faaa	27.4	27.5	27.7	27.4	26.5	25.5
226	Madang W.O.						
227	Auckland Airport	19.5	19.9	18.4	16.0	13.5	11.7
228	Christchurch	17.0	16.6	14.7	11.9	9.0	6.4
229	Darwin Airport	28.2	28.0	28.1	28.2	27.0	25.1
230	Cairns Aero	27.5	27.4	26.6	25.3	23.7	21.9
231	Alice Springs Airport	29.5	28.4	25.4	20.7	15.8	12.1
232	Brisbane Aero	25.0	24.8	23.5	21.1	18.3	15.6
233	Perth Airport	24.4	24.6	22.8	19.5	16.3	13.7
234	Sydney Airport Amo	22.9	22.9	21.5	18.9	16.1	13.4
235	Melbourne Airport	19.7	19.8	18.0	15.1	12.1	10.0
236	Canberra Airport	20.8	20.2	17.8	13.5	9.8	6.7
237	Brunei Airport	26.9	27.1	27.6	28.1	27.9	27.8
238	Balikpapan/Sepinggan	27.0	26.9	27.2	27.2	27.6	27.1
239	Jakarta/Observatory	27.0	27.1	27.6	28.3	28.6	28.3
240	Ninoy Aquino International Airport	26.1	26.5	27.8	29.3	29.5	28.9

Continued.

Jul.	Aug.	Sep.	Oct.	Nov.	Dec.	Annual	Statistics Period	#
19.3	20.6	22.0	22.5	21.7	21.5	21.1	1982-2010	201
21.2	21.6	21.6	22.8	24.5	25.4	23.9	1981-1991	202
16.5	17.8	18.3	19.9	21.3	22.1	20.0	1981-2010	203
								204
17.0	16.5	16.8	17.5	18.8	20.9	19.3	1981-2010	205
6.3	7.3	8.8	9.8	10.3	10.0	8.6	1981-2010	206
7.5	8.9	11.1	14.1	16.8	19.3	14.0	1981-2010	207
17.0	19.1	20.5	23.1	25.2	26.6	22.6	1981-2010	208
10.5	11.7	12.7	15.5	18.0	20.4	16.3	1981-2010	209
11.0	12.9	14.6	17.7	20.5	23.2	17.8	1981-2006	210
2.4	2.9	4.2	6.3	7.7	8.8	5.9	1981-2010	211
-66.4	-68.5	-66.6	-57.2	-42.5	-31.4	-55.2	1982-2010	212
27.4	27.9	27.6	26.8	25.4	23.8	25.5	1981-2010	213
28.1	27.9	27.9	28.0	28.0	27.6	27.8	1981-2010	214
27.7	27.8	27.8	27.8	27.8	27.5	27.7	1981-2009	215
27.6	27.8	27.9	28.0	28.1	27.9	27.8	1981-2009	216
26.0	25.9	26.3	26.6	26.7	27.0	26.6	1981-2000	217
24.3	24.2	24.6	25.0	25.9	26.3	25.5	1981-2010	218
20.0	19.9	21.0	22.4	23.6	25.1	23.1	1981-2010	219
28.0	28.0	28.1	28.1	28.2	27.9	28.0	1981-1998	220
28.1	28.1	28.2	28.3	28.4	28.4	28.3	1981-2010	221
23.1	23.2	24.0	25.0	25.9	26.6	25.2	1981-2010	222
25.8	25.9	26.0	26.5	26.6	26.7	26.5	1981-1994	223
								224
25.0	25.0	25.5	26.0	26.7	27.0	26.4	1981-2010	225
								226
10.8	11.6	12.9	14.2	16.3	18.3	15.3	1981-2010	227
5.8	7.2	9.3	11.2	13.4	15.5	11.5	1981-2010	228
24.7	25.6	27.7	29.0	29.2	28.8	27.5	1981-2010	229
21.2	21.7	23.3	25.0	26.3	27.3	24.8	1981-2010	230
11.9	14.6	19.7	23.3	26.1	28.1	21.3	1981-2010	231
14.8	15.6	18.2	20.5	22.3	24.0	20.3	1981-2010	232
12.7	13.2	14.6	16.6	19.7	22.1	18.4	1981-2010	233
12.5	13.7	16.2	18.2	19.8	21.8	18.2	1981-2010	234
9.3	10.0	12.0	13.7	16.4	17.8	14.5	1999-2010	235
5.8	7.2	10.2	13.0	16.2	18.8	13.3	1981-2010	236
27.6	27.8	27.5	27.1	27.1	27.1	27.5	1987-2010	237
26.9	27.1	27.4	27.6	27.2	26.8	27.2	1981-2010	238
28.1	28.3	28.3	28.6	28.3	27.5	28.0	1981-2010	239
27.9	27.5	27.7	27.6	27.2	26.2	27.7	1981-2008	240

Climatological Normal Monthly

#	Station	Jan.	Feb.	Mar.	Apr.	May	Jun.
1	Oslo/Gardermoen	89.0	86.0	80.0	71.0	67.0	66.0
2	Stockholm/Bromma	88.0	83.0	74.0	71.0	65.0	65.0
3	Helsinki-Vantaa	91.0	90.0	82.0	76.0	68.0	66.0
4	Heathrow						
5	Dublin Airport						
6	Reykjavik	78.2	78.6	77.4	78.0	75.7	79.0
7	Nuuk (Godthaab)	78.0	79.0	81.0	81.0	84.0	84.0
8	Koebenhavn/Landbohoejskolen						
9	De Bilt Aws	89.0	84.0	81.0	77.0	75.0	76.0
10	Uccle	87.4	82.9	80.2	75.8	75.2	76.8
11	Luxembourg/Luxembourg	90.0	83.0	77.0	71.0	70.0	71.0
12	Zuerich/Fluntern						
13	Le Bourget						
14	Lyon-Bron						
15	Marignane						
16	Barcelona/Aeropuerto						
17	Madrid, Retiro						
18	Gibraltar						
19	Lisboa/Geof	80.0	78.0	71.0	69.0	66.0	66.0
20	Sal	70.0	71.0	71.0	71.0	73.0	75.0
21	Berlin-Tempelhof	89.0	83.0	76.0	68.0	64.0	61.0
22	Frankfurt/Main						
23	Muenchen-Flughafen						
24	Wien/Hohe Warte	79.0	76.0	69.0	64.0	66.0	66.0
25	Praha/Ruzyne	85.0	82.0	76.0	70.0	70.0	71.0
26	Sliac	85.0	81.0	76.0	69.0	71.0	72.0
27	Warszawa-Okecie	92.0	90.0	83.0	74.0	73.0	69.0
28	Budapest/Pestszentlorinc	84.0	80.0	71.0	63.0	65.0	66.0
29	Beograd						
30	Podgorica-Grad						
31	Skopje Petrovec	83.0	77.0	70.0	65.0	67.0	61.0
32	Vlore						
33	Ljubljana/Bezigrad						
34	Zagreb/Gric						
35	Sarajevo-Bjelave						
36	Bucuresti Baneasa						
37	Sofia (Observ.)	77.0	74.0	69.0	63.0	67.0	67.0
38	Verona/Villafranca						
39	Roma/Fiumicino	75.0	75.0	75.0	75.0	75.0	73.0
40	Luqa	79.0	79.0	79.0	77.0	74.0	71.0

Note) Sample periods indicated with a * are the values calculated by the JMA.

Relative Humidity (%)

Jul.	Aug.	Sep.	Oct.	Nov.	Dec.	Annual	Statistics Period	#
70.0	75.0	79.0	84.0	88.0	89.0	79.0	1961-1990	1
69.0	75.0	78.0	83.0	86.0	88.0	76.0	1961-1967*	2
73.0	81.0	85.0	89.0	93.0	93.0	82.0	1961-1967*	3
								4
								5
80.9	80.9	79.0	79.6	79.0	78.5	78.7	1961-1990	6
87.0	87.0	83.0	78.0	76.0	77.0	81.0	1961-1990	7
								8
77.0	79.0	83.0	86.0	88.0	89.0	82.0	1961-1990	9
77.5	78.2	82.5	85.8	88.0	88.4	81.6	1961-1990	10
70.0	74.0	79.0	85.0	88.0	90.0	79.0	1961-1990	11
								12
								13
								14
								15
								16
								17
								18
63.0	61.0	67.0	72.0	77.0	79.0	70.0	1961-1990	19
75.0	75.0	77.0	75.0	73.0	71.0	73.0	1961-1990	20
65.0	69.0	73.0	79.0	87.0	89.0	75.0	1961-1967*	21
								22
								23
64.0	68.0	74.0	78.0	80.0	80.0	72.0	1961-1990	24
70.0	72.0	77.0	81.0	85.0	85.0	77.0	1961-1990	25
70.0	73.0	77.0	80.0	84.0	87.0	77.1	1961-1990	26
70.0	75.0	79.0	84.0	91.0	93.0	81.0	1961-1967*	27
63.0	65.0	70.0	75.0	83.0	86.0	72.6	1961-1990	28
								29
								30
57.0	58.0	65.0	72.0	81.0	84.0	70.0	1971-1990	31
								32
								33
								34
								35
								36
63.0	62.0	65.0	70.0	77.0	79.0	68.0	1961-1990	37
								38
72.0	73.0	75.0	76.0	77.0	77.0	74.8	1961-1990	39
69.0	73.0	77.0	78.0	77.0	79.0	76.0	1961-1990	40

Climatological Normal Monthly Relative Humidity

#	Station	Jan.	Feb.	Mar.	Apr.	May	Jun.
41	Athinai Ap Hellinikon	69.1	68.8	67.2	63.1	59.6	53.9
42	Istanbul/Goztepe						
43	Ankara/Central						
44	Larnaca Airport						
45	Ostrov Dikson	83.0	82.0	85.0	84.0	87.0	91.0
46	Ojmjakon	75.0	76.0	73.0	69.0	63.0	60.0
47	Tallinn-Harku	86.0	87.0	81.0	76.0	74.0	75.0
48	St.Petersburg (Voejkovo)	87.0	85.0	78.0	72.0	67.0	68.0
49	Jelgava						
50	Kaunas	85.0	84.0	81.0	74.0	74.0	70.0
51	Minsk	87.0	85.0	81.0	76.0	68.0	65.0
52	Moskva	80.0	76.0	73.0	67.0	64.0	63.0
53	Omsk	79.0	77.0	78.0	63.0	52.0	58.0
54	Irkutsk	91.0	85.0	71.0	58.0	51.0	62.0
55	Vladivostok	61.0	57.0	61.0	66.0	73.0	87.0
56	Kiev	86.0	85.0	80.0	69.0	59.0	63.0
57	Chisinau						
58	Karaganda	76.0	76.0	79.0	64.0	53.0	52.0
59	Gori						
60	Tashir						
61	Agstapha Airport						
62	Bishkek						
63	Tashkent	70.0	68.0	63.0	60.0	53.0	41.0
64	Dushanbe						
65	Ashgabat Keshi	76.0	72.0	67.0	59.0	48.0	37.0
66	Damascus Int. Airport						
67	Jerusalem						
68	Amman Airport	73.0	70.0	65.0	54.0	45.0	36.0
69	Riyadh Obs. (O.A.P.)	50.0	40.0	35.0	33.0	22.0	14.0
70	Tehran-Mehrabad	65.0	57.0	47.0	39.0	33.0	24.0
71	Kabul Airport						
72	Bahrain (Int. Airport)						
73	Doha International Airport						
74	Abu Dhabi Bateen Airport	68.0	67.0	63.0	58.0	55.0	60.0
75	Seeb, Int'l Airport	63.0	64.0	58.0	45.0	42.0	49.0
76	Taiz						
77	Peshawar	55.0	60.0	59.0	59.0	38.0	36.0
78	Karachi Airport	49.0	59.0	61.0	68.0	70.0	76.0
79	Chandpur						
80	New Delhi/Safdarjung	63.0	55.0	47.0	34.0	33.0	46.0

Note) Sample periods indicated with a * are the values calculated by the JMA.

Continued.

Jul.	Aug.	Sep.	Oct.	Nov.	Dec.	Annual	Statistics Period	#
48.1	47.8	53.9	62.1	69.2	70.1	61.1	1961-1990	41
								42
								43
								44
90.0	90.0	90.0	88.0	84.0	87.0	86.0	1961-1967*	45
69.0	72.0	73.0	80.0	76.0	77.0	73.0	1961-1967*	46
79.0	84.0	84.0	85.0	87.0	86.0	82.0	1961-1967*	47
72.0	78.0	82.0	85.0	87.0	87.0	79.0	1961-1967*	48
								49
74.0	79.0	81.0	85.0	88.0	89.0	80.0	1961-1967*	50
68.0	73.0	77.0	84.0	88.0	89.0	78.0	1961-1967*	51
69.0	74.0	78.0	79.0	82.0	82.0	74.0	1961-1967*	52
64.0	72.0	68.0	77.0	83.0	79.0	71.0	1961-1967*	53
73.0	80.0	75.0	76.0	86.0	90.0	75.0	1961-1967*	54
91.0	90.0	76.0	65.0	62.0	63.0	73.0	1961-1967*	55
65.0	67.0	68.0	77.0	87.0	88.0	74.0	1961-1967*	56
								57
56.0	61.0	58.0	74.0	76.0	76.0	64.0	1961-1967*	58
								59
								60
								61
								62
40.0	43.0	46.0	58.0	66.0	71.0	57.0	1961-1990	63
								64
35.0	35.0	41.0	57.0	66.0	77.0	55.8	1961-1990	65
								66
								67
41.0	43.0	50.0	49.0	57.0	70.0	57.0	1961-1967*	68
15.0	14.0	18.0	24.0	37.0	46.0	29.0	1961-1990	69
24.0	24.0	25.0	36.0	47.0	61.0	40.0	1961-1990	70
								71
								72
								73
61.0	63.0	64.0	65.0	65.0	68.0	63.0	1971-1991	74
60.0	67.0	63.0	55.0	60.0	65.0	58.0	1979-1990	75
								76
53.0	59.0	63.0	52.0	57.0	60.0	54.0	1961-1967*	77
79.0	80.0	78.0	66.0	58.0	47.0	68.0	1963-1967*	78
								79
70.0	73.0	62.0	52.0	55.0	62.0	54.0	1971-1990	80

Climatological Normal Monthly Relative Humidity

#	Station	Jan.	Feb.	Mar.	Apr.	May	Jun.
81	Cherrapunji	70.0	69.0	70.0	82.0	86.0	92.0
82	Kolkata/Alipore	66.0	58.0	58.0	66.0	70.0	77.0
83	Bombay/Colaba	69.0	67.0	69.0	71.0	70.0	80.0
84	Chennai/Minambakkam	73.0	72.0	70.0	69.0	62.0	57.0
85	Colombo						
86	Ulaanbaatar						
87	Kathmandu Airport						
88	Hong Kong Observatory						
89	Pyongyang						
90	Seoul	64.4	64.0	63.1	61.3	64.5	72.2
91	Busan	51.3	55.0	58.7	66.5	69.8	78.7
92	Yangon						
93	Bangkok Metropolis						
94	Kuala Lumpur/Subang	79.0	77.0	80.0	83.0	82.0	82.0
95	Singapore/Changi Airport						
96	Ha Noi						
97	Vientiane	70.0	68.0	66.0	69.0	78.0	82.0
98	Phnom-Penh/Pochentong						
99	Urumqi						
100	Kashgar	66.0	61.0	47.0	38.0	39.0	37.0
101	Changchun						
102	Shenyang	62.0	56.0	52.0	52.0	54.0	67.0
103	Beijing	44.0	49.0	51.0	51.0	54.0	62.0
104	Dalian						
105	Lhasa	27.0	25.0	29.0	37.0	44.0	55.0
106	Kunming	69.0	64.0	60.0	60.0	67.0	76.0
107	Xian	66.0	67.0	67.0	72.0	72.0	58.0
108	Wuhan						
109	Shanghai						
110	Fuzhou						
111	Taipei	83.0	84.0	83.0	82.0	81.0	82.0
112	Las Palmas De gran Canaria/Gando						
113	Dakhla						
114	Rabat-Sale						
115	Dar-El-Beida	80.0	77.0	78.0	76.0	76.0	75.0
116	Tunis-Carthage	76.0	74.0	73.0	71.0	68.0	64.0
117	Niamey-Aero	22.0	17.0	18.0	28.0	43.0	55.0
118	Bamako/Senou						
119	Nouakchott	31.0	35.0	39.0	49.0	55.0	58.0
120	Dakar/Yoff	68.0	72.0	73.0	77.0	78.0	76.0

Note) Sample periods indicated with a * are the values calculated by the JMA.

Continued.

Jul.	Aug.	Sep.	Oct.	Nov.	Dec.	Annual	Statistics Period	#
95.0	92.0	90.0	81.0	73.0	72.0	81.0	1971–1990	81
83.0	83.0	81.0	73.0	67.0	68.0	71.0	1971–1990	82
86.0	86.0	83.0	78.0	71.0	69.0	75.0	1971–1990	83
64.0	66.0	72.0	77.0	78.0	77.0	70.0	1971–1990	84
								85
								86
								87
								88
								89
81.3	78.9	72.5	67.4	66.3	65.9	68.5	1961–1990	90
85.1	80.5	74.8	65.4	59.3	53.6	66.6	1961–1990	91
								92
								93
81.0	82.0	83.0	84.0	85.0	83.0	81.0	1961–1967*	94
								95
								96
82.0	84.0	83.0	78.0	72.0	70.0	75.2	1961–1990	97
								98
								99
38.0	43.0	46.0	51.0	61.0	70.0	50.0	1961–1979*	100
								101
78.0	78.0	71.0	66.0	64.0	62.0	64.0	1961–1979*	102
78.0	81.0	72.0	66.0	61.0	51.0	60.0	1961–1979*	103
								104
65.0	68.0	65.0	51.0	40.0	34.0	45.0	1961–1979*	105
84.0	85.0	84.0	81.0	78.0	74.0	74.0	1961–1979*	106
71.0	73.0	81.0	80.0	75.0	70.0	71.0	1961–1979*	107
								108
								109
								110
77.0	77.0	79.0	79.0	83.0	84.0	81.0	1961–1967*	111
								112
								113
								114
73.0	74.0	72.0	75.0	76.0	81.0	76.0	1961–1967*	115
62.0	64.0	68.0	72.0	74.0	77.0	70.0	1961–1990	116
67.0	77.0	73.0	52.0	35.0	26.0	43.0	1961–1967*	117
								118
69.0	70.0	68.0	56.0	40.0	31.0	49.0	1961–1967*	119
75.0	79.0	80.0	80.0	75.0	67.0	75.0	1961–1967*	120

Climatological Normal Monthly Relative Humidity

#	Station	Jan.	Feb.	Mar.	Apr.	May	Jun.
121	Bolama						
122	Lungi	77.0	76.0	75.0	75.0	80.0	85.0
123	Plaisance (Mauritius)	82.0	84.0	84.0	83.0	81.0	79.0
124	Tripoli International Airport	66.0	64.0	62.0	55.0	50.0	47.0
125	Cairo						
126	Asswan	40.0	32.0	24.0	19.0	17.0	16.0
127	Khartoum	27.0	22.0	17.0	16.0	19.0	28.0
128	Djibouti	74.0	75.0	73.0	76.0	71.0	59.0
129	Addis Ababa-Bole						
130	Kampala						
131	Jomo Kenyatta International Airport						
132	Dar Es Salaam Int						
133	Seychelles International Airport	82.0	80.0	79.0	80.0	79.0	79.0
134	Lisala						
135	Kigali						
136	Bujumbura						
137	Pointe-Noire	85.0	85.0	84.0	86.0	87.0	85.0
138	Bitam						
139	Bossembele						
140	Ndjamena	29.0	23.0	21.0	28.0	39.0	52.0
141	Douala Obs.						
142	Benin City						
143	Cotonou	81.0	81.0	80.0	81.0	82.0	86.0
144	Lome	82.0	81.0	82.0	83.0	84.0	88.0
145	Kumasi						
146	Ouagadougou	25.0	21.0	22.0	36.0	50.0	64.0
147	Abidjan	84.0	83.0	82.0	83.0	83.0	85.0
148	Roberts Field	85.0	82.0	82.0	83.0	87.0	90.0
149	Antananarivo/Ivato						
150	Maputo/Mavalane	77.0	77.0	76.0	78.0	75.0	75.0
151	Kabwe	81.3	81.3	78.6	72.6	64.8	60.2
152	Chileka	79.0	80.0	78.0	75.0	69.0	66.0
153	Harare (Kutsaga)	77.0	78.0	71.0	67.0	61.0	57.0
154	Windhoek						
155	Mahalapye						
156	Pretoria Eendracht	62.0	63.0	63.0	63.0	56.0	54.0
157	Maseru-Mia						
158	Cape Town Intnl. Airport	71.0	72.0	74.0	78.0	81.0	81.0
159	Barrow/W. Post W. Rogers	72.7	70.0	70.9	76.8	87.0	88.5
160	Anchorage/Int., Ak	73.4	71.4	66.1	64.3	61.6	65.6

Note) Sample periods indicated with a * are the values calculated by the JMA.

Continued.

Jul.	Aug.	Sep.	Oct.	Nov.	Dec.	Annual	Statistics Period	#
								121
87.0	88.0	87.0	85.0	84.0	80.0	82.0	1961-1967*	122
78.0	78.0	78.0	78.0	79.0	81.0	80.4	1961-1990	123
50.0	53.0	59.0	60.0	59.0	64.0	57.0	1961-1967*	124
								125
18.0	21.0	22.0	27.0	36.0	42.0	26.2	1961-1990	126
43.0	49.0	40.0	28.0	27.0	30.0	29.0	1961-1990	127
42.0	46.0	62.0	68.0	73.0	71.0	67.0	1964-1967*	128
								129
								130
								131
								132
80.0	79.0	78.0	79.0	80.0	82.0	79.8	1971-1990	133
								134
								135
								136
84.0	84.0	84.0	82.0	85.0	86.0	85.0	1961-1967*	137
								138
								139
68.0	76.0	72.0	49.0	33.0	31.0	43.0	1961-1990	140
								141
								142
86.0	85.0	85.0	85.0	84.0	83.0	84.0	1961-1967*	143
88.0	86.0	86.0	85.0	83.0	82.0	84.0	1961-1967*	144
								145
72.0	80.0	77.0	60.0	38.0	29.0	48.0	1961-1967*	146
85.0	86.0	89.0	87.0	83.0	84.0	85.0	1961-1967*	147
91.0	90.0	90.0	90.0	88.0	87.0	87.0	1961-1967*	148
								149
74.0	73.0	74.0	75.0	76.0	76.0	75.0	1961-1967*	150
56.4	48.1	40.7	44.0	60.9	78.0	63.9	1962-1991	151
64.0	50.0	52.0	53.0	61.0	75.0	66.8	1961-1990	152
56.0	51.0	44.0	45.0	61.0	71.0	61.0	1961-1967*	153
								154
								155
50.0	45.0	44.0	52.0	59.0	61.0	56.0	1961-1990	156
								157
81.0	80.0	77.0	74.0	71.0	71.0	76.0	1961-1990	158
87.9	91.1	90.6	85.6	79.4	74.0	81.2	1961-1990	159
71.4	75.1	75.9	74.5	77.1	77.1	71.2	1961-1990	160

Climatological Normal Monthly Relative Humidity

#	Station	Jan.	Feb.	Mar.	Apr.	May	Jun.
161	Montreal/Pierre Elliott Trudeau Int'l A, Que	73.0	71.0	69.0	65.0	64.0	68.0
162	Winnipeg Richardson Int'l A, Man						
163	Edmonton City Centre Awos, Alta	73.0	72.0	71.0	58.0	53.0	60.0
164	Vancouver Int'l A, Bc						
165	Eureka, Nu					79.0	78.0
166	Miami, Fl	72.7	70.9	69.5	67.3	71.6	76.2
167	Atlanta/Mun., Ga.	67.6	63.4	62.4	61.0	67.2	69.8
168	New Orleans/Moisant Int., La.						
169	Dallas-Fort Worth/Fort Worth Reg. Airport, Tx.	67.5	66.4	63.7	65.3	69.7	65.8
170	Los Angeles/Int., Ca.	63.4	67.9	70.5	71.0	74.0	75.9
171	Las Vegas/Mccarran, Nv.	45.1	39.6	33.1	25.0	21.3	16.5
172	Washington/Nat., Va.	62.1	60.5	58.6	58.0	64.5	65.8
173	Denver/Stapleton Int., Co.	55.2	55.8	53.7	49.6	51.7	49.3
174	San Francisco/Int., Ca.	78.3	75.8	73.2	70.9	71.2	71.4
175	New York/La Guardia, Ny.	61.0	60.2	59.5	59.3	63.8	64.6
176	Boston/Logan Int., Ma.	62.3	62.0	63.1	63.0	66.7	68.5
177	Chicago/O'hare, Il.	72.2	71.6	69.7	64.9	64.1	65.6
178	Detroit/Metropolitan, Mi.	74.7	72.5	70.0	66.0	65.3	67.3
179	Minneapolis/St. Paul Int., Mn.						
180	Seattle/S.-Tacoma, Wa.	78.0	75.2	73.6	71.4	68.9	67.1
181	Mexico (Central), D.F.						
182	Nassau Airport New Providence	78.0	78.0	76.0	74.0	77.0	79.0
183	Casa Blanca, La Habana	75.0	74.0	73.0	72.0	75.0	79.0
184	Kingston/Norman Manley	70.0	69.0	72.0	71.0	74.0	74.0
185	Santo Domingo	82.5	80.8	79.3	78.8	82.5	83.7
186	San Juan/Int., Puerto Rico						
187	Belize/Phillip Goldston Intl. Airport						
188	Guatemala (Observatorio Nacional)	68.0	68.0	68.0	70.0	74.0	79.0
189	Tegucigalpa						
190	Managua A.C.Sandino						
191	Juan Santamaria Int. Airport						
192	Hewanorra Int'l Airport						
193	Grantley Adams	76.0	73.0	72.0	73.0	73.0	77.0
194	Piarco Int. Airport, Trinidad	81.0	80.0	77.0	77.0	79.0	84.0
195	Bogota/Eldorado						
196	Caracas/Maiquetia Aerop. Intl. Simon Bolivar	82.0	81.0	81.0	83.0	84.0	83.0
197	Georgetown						
198	Zanderij	82.0	79.0	77.0	77.0	82.0	81.0
199	Manaus	86.0	87.0	88.0	87.0	87.0	83.0
200	Salvador	79.4	79.0	79.8	82.2	83.1	82.3

Note) Sample periods indicated with a * are the values calculated by the JMA.

Continued.

Jul.	Aug.	Sep.	Oct.	Nov.	Dec.	Annual	Statistics Period	#
70.0	73.0	75.0	74.0	77.0	76.0	71.0	1953-1991	161
								162
66.0	68.0	69.0	64.0	74.0	75.0	67.0	1953-1991	163
								164
76.0	81.0	85.0					1954-1990	165
74.8	76.2	77.8	74.9	73.8	72.5	73.2	1961-1990	166
74.4	74.8	73.9	68.5	68.1	68.4	68.3	1961-1990	167
								168
59.8	59.5	66.5	65.7	67.4	67.5	65.4	1961-1990	169
76.6	76.6	74.2	70.5	65.5	62.9	70.8	1961-1990	170
21.1	25.6	25.0	28.8	37.2	45.0	30.2	1961-1990	171
66.9	69.3	69.7	67.4	64.7	64.1	64.3	1961-1990	172
47.8	49.3	50.1	49.2	56.3	56.6	52.0	1961-1990	173
73.2	74.5	71.9	71.9	74.8	77.5	73.7	1961-1990	174
64.7	67.0	67.2	65.2	64.2	63.5	63.4	1961-1990	175
68.4	70.8	71.8	68.5	67.5	65.4	66.5	1961-1990	176
68.5	70.7	71.1	68.6	72.5	75.5	69.6	1961-1990	177
68.5	71.5	73.4	71.6	74.6	76.7	71.0	1961-1990	178
								179
65.4	68.2	73.2	78.6	79.8	80.1	73.3	1961-1990	180
								181
77.0	79.0	81.0	80.0	78.0	78.0	77.9	1961-1990	182
78.0	78.0	79.0	80.0	77.0	75.0	76.2	1961-1990	183
69.0	72.0	74.0	76.0	72.0	72.0	72.0	1961-1967*	184
83.8	84.6	85.1	85.7	84.2	83.3	82.9	1961-1990	185
								186
								187
75.0	75.0	80.0	77.0	75.0	70.0	78.0	1961-1967*	188
								189
								190
								191
								192
81.0	80.0	80.0	82.0	80.0	77.0	78.0	1961-1967*	193
84.0	84.0	84.0	85.0	86.0	84.0	82.1	1960-1990	194
								195
83.0	82.0	81.0	82.0	82.0	83.0	82.0	1961-1990	196
								197
79.0	78.0	74.0	75.0	80.0	83.0	80.0	1961-1967*	198
80.0	77.0	77.0	79.0	81.0	85.0	83.0	1961-1990	199
81.5	80.0	79.6	80.7	81.5	81.1	80.8	1961-1990	200

Climatological Normal Monthly Relative Humidity

#	Station	Jan.	Feb.	Mar.	Apr.	May	Jun.
201	Brasilia	76.0	77.0	76.0	75.0	68.0	61.0
202	Rio De Janeiro	79.0	79.0	80.0	80.0	80.0	79.0
203	Sao Paulo	80.0	79.0	80.0	80.0	79.0	78.0
204	Esmeraldas (Tachina) Aeropuerto						
205	Lima/Callao	84.0	83.0	84.0	83.0	85.0	84.0
206	La Paz/Alto	64.0	68.0	73.0	61.0	49.0	34.0
207	Pudahuel						
208	Aeropuerto Silvio Pettirossi, Luque	68.0	71.0	72.0	75.0	76.0	76.0
209	Rocha						
210	Buenos Aires Observatorio	64.0	68.0	72.0	76.0	77.0	79.0
211	Ushuaia Aero	75.0	76.0	78.0	80.0	81.0	82.0
212	Vostok						
213	Honolulu, Oahu, Hawaii	73.3	70.8	68.8	67.3	66.1	64.4
214	Weather Forecast Office, Guam, Mariana Is.						
215	Majuro/Marshall Is. Intnl.	77.7	77.1	79.0	80.7	81.9	81.1
216	Weather Service Office, Koror, Palau Wci.	84.3	83.7	83.8	83.3	85.4	86.2
217	Honiara/Henderson						
218	Pekoa Airport (Santo)						
219	Noumea (Nlle-Caledonie)						
220	Tarawa	80.0	80.0	81.0	81.0	79.0	79.0
221	Funafuti	82.0	80.0	81.0	82.0	81.0	83.0
222	Nadi Airport						
223	Apia	83.0	84.0	86.0	85.0	84.0	83.0
224	Fua'amotu						
225	Tahiti-Faaa						
226	Madang W.O.						
227	Auckland Airport						
228	Christchurch						
229	Darwin Airport	79.0	80.0	78.0	70.0	57.0	51.0
230	Cairns Aero						
231	Alice Springs Airport	28.0	27.0	29.0	34.0	44.0	48.0
232	Brisbane Aero						
233	Perth Airport	48.0	50.0	51.0	63.0	71.0	75.0
234	Sydney Airport Amo						
235	Melbourne Airport						
236	Canberra Airport	58.0	60.0	64.0	68.0	74.0	76.0
237	Brunei Airport						
238	Balikpapan/Sepinggan						
239	Jakarta/Observatory	86.0	84.0	83.0	82.0	80.0	78.0
240	Ninoy Aquino International Airport	74.0	71.0	67.0	65.0	69.0	78.0

Note) Sample periods indicated with a * are the values calculated by the JMA.

Continued.

Jul.	Aug.	Sep.	Oct.	Nov.	Dec.	Annual	Statistics Period	#
56.0	49.0	53.0	66.0	75.0	79.0	67.0	1961-1990	201
77.0	77.0	79.0	80.0	79.0	80.0	79.0	1961-1990	202
77.0	74.0	77.0	79.0	78.0	80.0	78.0	1961-1990	203
								204
86.0	87.0	86.0	84.0	84.0	83.0	84.0	1961-1967*	205
36.0	38.0	49.0	52.0	52.0	58.0	53.0	1961-1967*	206
								207
70.0	70.0	66.0	67.0	67.0	68.0	70.0	1961-1990	208
								209
79.0	74.0	70.0	69.0	66.0	63.0	78.0	1961-1990	210
82.0	80.0	76.0	73.0	72.0	74.0	78.0	1961-1988	211
								212
64.6	64.1	65.5	67.5	70.4	72.4	67.9	1961-1990	213
								214
80.5	79.3	79.4	79.4	79.9	79.7	79.7	1961-1990	215
85.3	84.9	83.7	84.8	85.1	85.0	84.7	1961-1990	216
								217
								218
								219
79.0	77.0	77.0	77.0	77.0	80.0	80.0	1961-1967*	220
82.0	83.0	83.0	81.0	81.0	82.0	82.0	1961-1967*	221
								222
81.0	81.0	81.0	83.0	82.0	83.0	83.0	1961-1967*	223
								224
								225
								226
								227
								228
50.0	59.0	65.0	64.0	69.0	74.0	67.0	1961-1967*	229
								230
40.0	37.0	28.0	24.0	26.0	25.0	34.0	1961-1967*	231
								232
77.0	72.0	72.0	62.0	56.0	50.0	63.0	1963-1967*	233
								234
								235
75.0	72.0	70.0	69.0	61.0	60.0	68.0	1961-1967*	236
								237
								238
75.0	74.0	73.0	75.0	78.0	82.0	77.0	1961-1965*	239
81.0	82.0	82.0	80.0	78.0	77.0	75.0	1961-1990	240

Climatological Normal

#	Station	Jan.	Feb.	Mar.	Apr.	May	Jun.
1	Oslo/Gardermoen	61.7	46.8	56.5	50.5	58.5	75.4
2	Stockholm/Bromma	37.2	26.1	30.3	33.2	31.7	52.3
3	Helsinki-Vantaa	54.4	38.8	36.9	31.4	39.7	59.3
4	Heathrow	55.0	46.8	41.9	46.4	49.1	46.8
5	Dublin Airport	65.3	48.2	52.5	55.8	61.6	65.1
6	Reykjavik	82.5	83.9	82.5	54.9	51.2	43.4
7	Nuuk (Godthaab)	40.4	40.9	36.8	35.5	40.7	40.4
8	Koebenhavn/Landbohoejskolen	41.8	27.8	33.2	32.7	46.0	60.6
9	De Bilt Aws	68.9	58.3	64.6	42.8	61.9	66.1
10	Uccle	76.5	63.7	70.0	51.3	67.2	72.5
11	Luxembourg/Luxembourg	80.4	65.8	71.4	57.4	79.2	76.1
12	Zuerich/Fluntern	58.9	65.9	78.8	86.9	124.7	130.4
13	Le Bourget	58.2	44.4	55.8	59.1	57.3	60.4
14	Lyon-Bron	46.1	50.1	51.5	90.2	102.6	74.8
15	Marignane	49.4	32.0	29.6	57.0	42.4	23.9
16	Barcelona/Aeropuerto	40.5	37.2	34.6	39.7	46.0	30.6
17	Madrid, Retiro	34.4	35.1	25.1	43.1	51.6	21.3
18	Gibraltar	114.6	110.5	72.9	60.7	26.4	8.5
19	Lisboa/Geof	92.9	90.5	58.1	69.8	48.4	18.3
20	Sal						
21	Berlin-Tempelhof	49.0	38.6	44.5	34.2	54.7	59.7
22	Frankfurt/Main	39.4	44.9	43.0	35.3	66.7	53.1
23	Muenchen-Flughafen	46.8	33.0	49.1	56.7	78.3	103.7
24	Wien/Hohe Warte	37.2	42.6	53.6	46.0	70.5	70.9
25	Praha/Ruzyne	21.3	19.2	28.0	27.7	70.0	66.6
26	Sliac	42.8	39.8	42.0	50.3	69.4	77.6
27	Warszawa-Okecie	26.4	28.4	30.7	33.4	54.3	66.5
28	Budapest/Pestszentlorinc	30.5	27.1	31.0	39.4	63.4	63.5
29	Beograd	45.0	39.4	51.3	56.0	56.4	103.6
30	Podgorica-Grad	130.0	143.7	146.0	130.7	83.3	55.0
31	Skopje Petrovec						
32	Vlore	163.3	165.5	149.3	92.0	81.7	70.0
33	Ljubljana/Bezigrad	72.4	71.8	92.3	114.6	101.2	121.9
34	Zagreb/Gric	49.3	45.2	59.0	60.0	73.9	95.2
35	Sarajevo-Bjelave	76.1	62.6	74.1	74.1	74.2	98.1
36	Bucuresti Baneasa	33.8	34.1	38.6	52.4	56.4	79.2
37	Sofia (Observ.)	33.9	33.8	38.0	50.1	66.2	74.2
38	Verona/Villafranca	39.5	35.5	46.9	70.8	73.0	85.3
39	Roma/Fiumicino	68.7	68.7	50.8	62.2	40.9	23.8
40	Luqa	86.6	49.1	35.3	16.0	6.9	4.9

Monthly Precipitation (mm)

Jul.	Aug.	Sep.	Oct.	Nov.	Dec.	Annual	Statistics Period	#
76.7	92.7	82.2	96.0	90.5	62.2	849.7	1981-2010	1
65.8	62.2	64.7	44.1	43.8	44.4	535.8	1982-1994	2
65.8	77.6	65.3	81.0	71.5	56.9	678.6	1982-2010	3
46.8	57.8	50.8	70.6	72.4	55.9	640.3	1997-2010	4
63.5	73.2	60.9	78.7	76.9	73.5	775.2	1982-2010	5
53.6	67.4	75.0	75.4	80.3	97.0	847.1	1982-2010	6
55.4	90.9	72.7	60.0	57.8	40.1	611.6	1982-2010	7
53.8	70.7	58.8	57.3	50.8	48.6	582.1	1982-2010	8
80.1	74.6	79.1	79.2	80.2	72.7	828.5	1982-2010	9
73.6	79.3	69.1	74.5	76.5	80.9	855.1	1981-2010	10
74.8	73.4	74.0	87.8	81.8	85.0	907.1	1982-2010	11
122.3	124.6	95.8	82.8	78.7	79.8	1129.6	1982-2010	12
50.9	39.1	58.7	63.5	46.6	58.8	652.8	1982-1995	13
54.6	60.7	98.3	108.8	68.4	49.2	855.3	1982-1995	14
8.4	28.1	80.5	67.6	58.4	44.6	521.9	1982-2010	15
20.4	59.0	84.5	91.5	61.8	43.9	589.7	1981-2010	16
11.9	8.9	21.6	65.1	63.2	55.6	436.9	1982-2010	17
1.0	6.6	26.1	88.1	128.0	190.3	833.7	1982-2010	18
4.3	7.5	32.1	84.3	130.8	116.1	753.1	1982-2010	19
								20
58.9	58.8	46.6	38.7	46.4	48.2	578.3	1981-2010	21
78.0	61.6	50.1	52.4	50.6	47.6	622.7	1995-2010	22
98.8	84.3	53.8	49.3	43.7	52.5	750.0	1993-2010	23
70.0	76.3	59.6	37.7	48.3	47.3	660.0	1982-2010	24
78.0	65.7	37.7	27.0	30.1	27.3	498.6	1981-2010	25
78.0	60.3	59.4	51.4	61.3	54.2	686.5	1982-2010	26
71.3	62.9	49.9	31.4	39.0	35.4	529.6	1982-2010	27
56.8	56.2	45.2	34.6	49.2	39.3	536.2	1981-2010	28
61.6	57.5	48.6	50.1	53.4	58.7	681.6	1981-2010	29
47.3	61.8	147.1	161.1	249.0	222.9	1577.9	1981-2010	30
								31
16.5	47.4	103.5	164.0	217.2	189.0	1459.4	1981-1990	32
141.5	133.6	160.6	127.4	136.4	127.8	1401.5	1997-2010	33
79.1	92.1	92.8	84.9	74.1	67.0	872.6	1981-2010	34
92.2	59.1	109.3	105.0	97.4	85.5	1007.7	2001-2010	35
63.1	51.4	54.5	49.5	43.0	44.2	600.2	1981-2010	36
53.9	63.2	46.4	43.5	39.5	43.4	586.1	1982-2010	37
61.9	84.0	75.0	88.2	66.6	52.4	779.1	1981-2005	38
18.8	27.7	73.3	91.7	88.5	91.5	706.6	1981-1996	39
0.2	2.1	46.4	76.6	112.3	107.8	544.2	1982-2010	40

Climatological Normal Monthly Precipitation

#	Station	Jan.	Feb.	Mar.	Apr.	May	Jun.
41	Athinai Ap Hellinikon	44.6	38.2	37.9	39.8	12.7	6.5
42	Istanbul/Goztepe	80.6	71.8	62.4	38.0	29.0	30.9
43	Ankara/Central	36.0	34.7	38.5	48.9	48.0	38.8
44	Larnaca Airport	76.0	48.0	29.3	16.3	10.5	2.3
45	Ostrov Dikson	35.7	29.0	25.5	20.0	21.2	32.5
46	Ojmjakon	6.6	6.2	4.9	6.0	12.7	34.9
47	Tallinn-Harku	54.9	37.8	37.4	32.7	31.3	53.1
48	St.Petersburg (Voejkovo)	44.2	33.3	35.3	31.6	46.4	72.0
49	Jelgava	41.2	25.4	42.6	40.3	64.1	70.2
50	Kaunas	43.7	32.9	38.8	36.0	50.4	75.9
51	Minsk	45.7	39.2	46.0	42.3	62.1	86.9
52	Moskva	51.6	43.1	35.2	36.3	50.3	80.4
53	Omsk	23.4	17.8	16.9	19.9	35.2	50.8
54	Irkutsk	14.1	8.1	11.3	18.6	35.8	78.5
55	Vladivostok	13.4	15.2	24.0	48.0	82.6	107.6
56	Kiev	36.4	37.1	36.4	44.8	58.4	83.9
57	Chisinau	50.9	36.2	39.6	35.1	45.8	56.8
58	Karaganda	23.4	21.0	23.3	24.2	40.6	33.9
59	Gori	43.5	36.7	33.7	41.0	64.7	74.7
60	Tashir	16.3	36.3	42.4	59.9	134.5	133.8
61	Agstapha Airport	13.6	26.1	26.9	33.6	60.0	56.1
62	Bishkek	27.1	34.6	53.5	73.4	66.3	36.2
63	Tashkent	55.0	65.3	69.5	61.9	42.2	13.7
64	Dushanbe						
65	Ashgabat Keshi	20.6	27.9	40.4	32.8	22.3	5.9
66	Damascus Int. Airport	27.7	33.0	20.1	12.2	8.4	0.9
67	Jerusalem	134.3	157.9	77.7	15.9	5.0	2.5
68	Amman Airport	52.0	69.6	48.8	4.3	2.4	0.0
69	Riyadh Obs. (O.A.P.)	14.3	17.6	27.8	34.4	11.1	0.0
70	Tehran-Mehrabad						
71	Kabul Airport						
72	Bahrain (Int. Airport)	16.8	16.0	21.8	3.7	0.8	0.0
73	Doha International Airport						
74	Abu Dhabi Bateen Airport	12.0	16.1	22.0	9.7	0.0	0.0
75	Seeb, Int'l Airport	11.6	16.0	15.6	8.8	0.0	3.6
76	Taiz	0.6	8.0	48.5	46.9	84.5	57.7
77	Peshawar	35.3	58.0	87.2	63.0	22.6	17.9
78	Karachi Airport	7.5	6.4	3.6	3.9	0.2	11.4
79	Chandpur	4.9	16.6	50.6	151.4	355.3	456.3
80	New Delhi/Safdarjung	20.8	25.4	16.2	13.1	31.8	87.9

Continued.

Jul.	Aug.	Sep.	Oct.	Nov.	Dec.	Annual	Statistics Period	#
10.8	2.2	18.1	40.1	63.0	61.4	375.3	1982-2010	41
22.9	27.6	36.1	88.1	94.6	98.0	680.0	1982-2010	42
17.7	14.0	17.3	33.8	40.0	40.1	407.8	1982-2010	43
0.6	0.3	6.5	13.8	43.1	80.7	327.4	1982-2010	44
33.5	40.8	43.2	36.4	27.8	38.0	383.6	1982-2010	45
43.4	39.6	22.3	14.0	12.9	6.9	210.4	1982-2010	46
82.5	79.3	70.3	71.4	73.8	54.7	679.2	1982-2010	47
80.1	83.3	64.1	66.9	55.6	49.2	662.0	1982-2010	48
94.8	77.9	80.1	51.4	62.0	52.0	702.0	1981-1990	49
82.6	72.3	53.8	54.7	47.9	42.8	631.8	1982-2010	50
90.2	66.6	62.3	53.2	48.8	47.1	690.4	1982-2010	51
84.3	82.0	66.8	71.3	54.9	50.3	706.5	1982-2010	52
65.4	55.9	38.1	30.7	34.8	32.4	421.3	1982-2010	53
109.2	93.1	52.0	21.2	20.6	16.0	478.5	1982-2010	54
168.8	159.0	112.7	55.6	29.0	21.1	837.0	1982-2010	55
70.1	58.7	59.0	36.6	48.4	41.3	611.1	1982-2010	56
68.5	49.0	53.8	43.8	48.1	44.4	572.0	1999-2010	57
50.0	27.1	20.8	28.6	29.9	27.2	350.0	1982-2010	58
49.9	41.6	21.0	41.7	53.5	29.3	531.3	1981-1990	59
80.9	62.9	42.6	52.5	39.5	16.0	717.6	1981-1991	60
25.4	27.8	11.0	40.8	30.7	10.4	362.4	1981-1990	61
18.4	12.4	18.2	42.6	45.1	34.3	462.1	1981-2010	62
4.1	1.3	6.1	27.0	43.5	61.8	451.4	1982-2010	63
								64
2.8	1.4	3.6	10.3	16.8	22.1	206.9	1982-2010	65
0.0	0.0	0.4	11.1	30.6	31.7	176.1	1981-2010	66
0.0	0.0	0.4	18.3	85.5	134.5	632.0	1982-1995	67
0.0	0.0	0.2	9.5	33.8	47.9	268.5	1982-1998	68
0.0	0.8	0.0	1.9	12.6	19.0	139.5	1981-2010	69
								70
								71
0.2	0.0	0.4	0.5	11.9	16.8	88.9	1982-2010	72
								73
1.0	0.1	0.0	0.0	0.3	10.4	71.6	1983-1997	74
2.7	0.3	0.2	0.8	1.6	14.4	75.6	1981-2010	75
57.9	83.1	58.6	8.9	0.2	0.0	454.9	1981-1990	76
49.4	65.3	24.4	15.7	12.1	19.2	470.1	1982-2010	77
67.7	64.0	10.3	2.6	0.4	5.4	183.4	1982-2010	78
472.7	460.6	285.7	129.3	48.5	12.4	2444.3	1981-1991	79
187.0	232.4	117.6	16.3	6.6	12.6	767.7	1981-2010	80

Climatological Normal Monthly Precipitation

#	Station	Jan.	Feb.	Mar.	Apr.	May	Jun.
81	Cherrapunji						
82	Kolkata/Alipore	12.6	19.7	35.2	58.8	137.4	303.8
83	Bombay/Colaba	0.9	0.3	0.9	0.6	15.6	518.0
84	Chennai/Minambakkam	30.9	23.8	11.2	19.1	53.3	80.7
85	Colombo	96.5	70.9	116.4	222.1	309.1	244.4
86	Ulaanbaatar	2.3	2.9	3.3	9.1	21.1	45.6
87	Kathmandu Airport	16.6	25.0	34.4	45.3	118.5	231.5
88	Hong Kong Observatory	21.3	68.3	97.9	190.4	361.4	328.6
89	Pyongyang	11.7	14.7	25.9	58.3	83.6	95.4
90	Seoul	20.8	24.9	47.3	63.3	105.9	133.3
91	Busan	37.1	52.7	83.3	126.8	158.8	214.6
92	Yangon	0.2	1.1	17.3	24.3	237.4	490.9
93	Bangkok Metropolis	15.1	18.3	39.3	86.6	245.8	162.0
94	Kuala Lumpur/Subang	210.3	203.6	273.7	286.7	198.1	139.7
95	Singapore/Changi Airport	246.3	114.1	173.8	151.5	167.4	136.1
96	Ha Noi	20.6	32.6	53.0	98.5	210.7	235.5
97	Vientiane	10.4	16.3	46.1	117.4	218.2	222.5
98	Phnom-Penh/Pochentong	10.0	4.0	30.2	70.7	123.6	149.3
99	Urumqi	11.0	12.6	19.4	36.8	41.1	32.2
100	Kashgar	2.5	3.8	7.1	5.1	11.2	9.1
101	Changchun	4.2	4.3	12.4	24.9	51.3	94.2
102	Shenyang	7.2	8.8	20.4	41.7	53.2	95.7
103	Beijing	2.5	4.4	9.8	24.8	37.3	72.1
104	Dalian	8.2	6.6	13.6	29.5	51.8	77.3
105	Lhasa	0.9	1.2	3.0	8.0	28.2	71.3
106	Kunming	15.9	14.4	18.4	27.0	79.4	177.2
107	Xian	6.7	8.6	27.0	37.2	54.8	64.7
108	Wuhan	46.7	71.5	89.8	135.8	169.3	224.7
109	Shanghai	59.1	58.8	90.6	83.8	91.6	161.0
110	Fuzhou	50.8	85.5	142.8	152.4	189.0	199.9
111	Taipei	103.5	180.0	189.5	198.3	233.9	322.6
112	Las Palmas De Gran Canaria/Gando	32.8	22.2	10.8	5.4	0.8	0.3
113	Dakhla	5.0	4.5	2.3	0.7	5.1	0.0
114	Rabat-Sale	86.9	61.0	52.7	39.2	14.7	3.4
115	Dar-El-Beida	86.0	67.9	62.3	49.5	40.7	9.0
116	Tunis-Carthage	58.2	50.2	41.3	36.7	25.7	12.5
117	Niamey-Aero	0.0	0.0	2.0	7.3	27.3	74.9
118	Bamako/Senou	0.0	0.1	3.8	13.9	68.0	113.0
119	Nouakchott	0.9	0.1	0.1	0.0	0.0	2.7
120	Dakar/Yoff	1.8	1.0	0.0	0.0	0.3	11.4

Continued.

Jul.	Aug.	Sep.	Oct.	Nov.	Dec.	Annual	Statistics Period	#
								81
409.4	336.4	318.2	165.1	36.1	9.0	1841.7	1981-2010	82
729.3	505.1	315.1	82.2	12.1	1.8	2181.9	1981-2010	83
105.9	136.1	151.9	298.4	382.0	158.0	1451.3	1981-2010	84
124.6	111.5	227.7	343.2	301.7	153.8	2321.9	1981-2010	85
75.2	76.5	27.0	9.6	5.2	3.6	281.4	1983-2010	86
374.3	324.4	202.6	65.6	12.2	25.8	1476.2	1982-2010	87
362.2	333.9	300.8	103.8	45.6	31.9	2246.1	1981-2004	88
310.4	226.5	122.4	46.8	39.2	14.4	1049.3	1982-2010	89
373.4	364.2	169.3	51.5	52.6	22.5	1429.0	1981-2010	90
323.6	256.0	134.9	58.2	45.3	26.2	1517.5	1982-2010	91
459.8	468.0	259.9	114.0	32.5	2.3	2107.7	1981-2005	92
171.4	207.9	349.2	302.2	47.9	7.4	1653.1	1982-2010	93
150.3	164.0	200.2	256.6	333.5	255.6	2672.3	1982-2010	94
155.8	154.0	163.1	156.2	265.9	314.8	2199.0	1982-2010	95
279.1	279.1	222.9	146.4	56.3	10.2	1644.9	1981-2010	96
293.1	352.7	268.3	117.2	17.5	2.0	1681.7	1982-2010	97
159.8	193.0	279.3	244.0	117.0	26.1	1407.0	1985-2010	98
38.8	26.3	24.8	23.2	19.6	19.4	305.2	1982-2010	99
9.4	7.8	6.4	5.5	2.0	1.6	71.5	1981-2010	100
170.1	134.0	43.5	22.1	13.1	6.3	580.4	1982-2010	101
174.1	167.8	63.8	39.6	20.0	11.0	703.3	1982-2010	102
160.5	138.9	48.8	23.1	9.8	2.3	534.3	1981-2010	103
130.3	144.9	60.0	34.3	18.2	9.7	584.4	1981-2010	104
121.3	123.5	64.3	7.7	0.7	0.5	430.6	1981-2010	105
206.0	198.4	114.5	80.5	35.1	13.2	980.0	1981-2010	106
100.0	81.3	94.2	61.8	21.5	7.4	565.2	1981-2005	107
236.0	126.7	77.4	81.5	63.0	28.9	1351.3	1981-2010	108
148.9	198.0	112.8	60.3	55.7	36.4	1157.0	1981-2010	109
124.6	167.7	154.9	46.8	40.9	34.1	1389.4	1981-2010	110
251.1	349.1	372.4	163.6	98.0	72.0	2534.0	1982-2010	111
0.0	0.3	9.5	18.9	20.7	40.5	162.2	1981-2010	112
2.5	0.9	3.2	2.5	0.3	8.2	35.2	1982-2010	113
0.6	0.8	13.7	50.6	87.2	102.9	513.7	1988-2010	114
3.9	6.7	34.1	57.2	89.9	90.7	597.9	1982-2010	115
5.1	8.0	46.2	50.9	57.6	74.5	466.9	1982-2010	116
136.9	161.2	85.6	13.3	0.0	0.0	508.5	1982-2010	117
216.4	266.0	177.7	48.9	1.4	0.0	909.2	1982-2010	118
7.0	38.2	52.5	7.9	2.1	6.4	117.9	1982-2010	119
55.4	152.4	133.5	23.6	0.4	0.0	379.8	1981-2010	120

Climatological Normal Monthly Precipitation

#	Station	Jan.	Feb.	Mar.	Apr.	May	Jun.
121	Bolama	0.2	1.6	0.5	0.5	21.7	189.2
122	Lungi						
123	Plaisance (Mauritius)	221.4	250.9	245.7	180.2	121.2	84.9
124	Tripoli International Airport	59.5	22.7	26.8	13.6	6.8	0.1
125	Cairo	7.1	4.3	6.9	1.2	0.4	0.0
126	Asswan	0.7	0.0	0.8	0.4	0.3	0.0
127	Khartoum						
128	Djibouti						
129	Addis Ababa-Bole	15.6	40.9	67.4	87.4	95.6	132.1
130	Kampala	56.8	30.4	99.7	168.8	126.9	49.7
131	Jomo Kenyatta International Airport	74.1	40.3	79.6	137.0	101.0	26.5
132	Dar Es Salaam Int	66.3	60.8	143.3	246.7	169.4	25.3
133	Seychelles International Airport	380.9	236.6	192.7	182.9	120.0	94.8
134	Lisala	72.3	115.3	172.8	184.5	149.4	188.6
135	Kigali	84.1	103.3	119.5	158.1	95.9	20.5
136	Bujumbura	93.0	72.8	134.9	110.2	47.6	9.0
137	Pointe-Noire	158.2	223.7	207.3	114.8	36.5	1.5
138	Bitam	33.4	71.9	152.6	184.1	177.9	122.6
139	Bossembele	13.0	20.1	78.5	91.9	144.3	198.8
140	Ndjamena	0.0	0.0	0.0	4.2	23.9	51.4
141	Douala Obs.						
142	Benin City	9.5	41.5	108.9	165.3	211.5	230.1
143	Cotonou	14.4	24.0	80.4	122.5	197.0	341.5
144	Lome	5.5	13.8	58.1	104.3	149.5	187.4
145	Kumasi	17.5	51.6	102.2	147.6	159.5	192.6
146	Ouagadougou	0.0	0.1	2.0	23.6	52.8	86.5
147	Abidjan	20.7	32.6	81.4	186.3	277.1	444.4
148	Roberts Field						
149	Antananarivo/Ivato	404.6	291.1	219.8	53.9	27.0	3.4
150	Maputo/Mavalane						
151	Kabwe						
152	Chileka						
153	Harare (Kutsaga)						
154	Windhoek	74.3	96.2	47.0	29.2	4.2	1.0
155	Mahalapye	78.9	89.5	67.7	20.4	0.8	3.5
156	Pretoria Eendracht	111.3	104.5	86.4	29.1	17.5	7.5
157	Maseru-Mia	96.0	96.5	81.1	49.9	17.4	12.9
158	Cape Town Intnl. Airport	10.1	15.0	13.5	47.4	80.7	93.4
159	Barrow/W. Post W. Rogers	3.4	3.2	3.4	3.9	5.0	8.0
160	Anchorage/Int., Ak	18.3	18.8	15.2	12.2	18.0	24.8

Continued.

Jul.	Aug.	Sep.	Oct.	Nov.	Dec.	Annual	Statistics Period	#
499.5	590.8	401.2	189.1	17.3	0.2	1911.8	1981-2010	121
								122
87.3	75.2	70.4	55.5	69.2	137.2	1599.1	1982-2010	123
0.0	0.0	5.0	29.4	45.1	52.1	261.1	1982-2010	124
0.0	0.3	0.0	0.1	6.4	7.9	34.6	1981-1998	125
0.0	0.0	0.0	0.6	0.2	0.1	3.1	1982-2010	126
								127
								128
238.2	254.0	144.0	46.9	16.8	6.3	1145.2	1983-2010	129
54.4	80.3	99.7	115.6	149.4	98.9	1130.6	1981-1992	130
19.8	13.9	21.8	47.4	108.9	98.6	768.9	1981-2010	131
22.5	19.2	15.1	74.0	117.9	111.5	1072.0	1982-2010	132
95.2	131.6	182.4	190.6	193.0	299.7	2300.4	1982-2010	133
180.8	117.9	144.8	156.5	99.0	69.4	1651.3	1981-1988	134
12.5	42.8	71.1	112.3	164.2	75.2	1059.5	1982-2010	135
1.4	7.7	33.0	68.9	101.2	101.9	781.6	1982-2000	136
0.2	4.2	15.4	88.7	200.1	160.3	1210.9	1981-2010	137
53.3	59.9	229.1	287.5	168.0	59.9	1600.2	1981-2010	138
255.4	263.4	214.7	180.6	44.2	5.2	1510.1	1981-2010	139
165.0	181.1	82.2	21.0	0.0	0.0	528.8	1981-2010	140
								141
315.6	321.6	317.7	223.6	52.5	24.5	2022.3	1981-1998	142
107.0	64.6	135.3	156.7	36.7	21.2	1301.3	1982-2010	143
65.8	27.2	77.7	79.9	21.5	6.9	797.6	1982-2010	144
140.4	85.4	148.9	151.0	44.9	31.8	1273.4	1981-2010	145
174.5	195.3	133.0	30.5	0.6	0.1	699.0	1982-2010	146
92.4	40.0	57.0	157.0	127.5	84.8	1601.2	1982-2010	147
								148
8.0	6.0	14.0	70.1	124.8	294.4	1517.1	1981-2010	149
								150
								151
								152
								153
0.0	0.0	0.7	11.4	19.6	25.7	309.3	1981-2010	154
1.1	0.9	5.2	29.7	67.7	73.2	438.6	1981-1997	155
2.7	4.4	16.6	72.2	100.2	113.8	666.2	1982-2010	156
7.3	21.9	20.9	82.1	87.0	76.6	649.6	1981-1996	157
91.5	78.2	44.6	35.3	23.1	13.0	545.8	1982-2010	158
24.0	26.6	18.8	10.9	4.9	3.8	115.9	1982-2010	159
43.2	80.3	76.2	50.2	28.1	27.8	413.1	1982-2010	160

Climatological Normal Monthly Precipitation

#	Station	Jan.	Feb.	Mar.	Apr.	May	Jun.
161	Montreal/Pierre Elliott Trudeau int'l A, Que	83.5	63.9	66.5	81.3	80.6	85.2
162	Winnipeg Richardson Int'l A, Man	16.6	14.6	23.8	31.6	51.9	93.7
163	Edmonton City Centre Awos, Alta	15.1	11.2	12.1	27.8	47.6	86.7
164	Vancouver Int'l A, Bc	163.4	109.9	114.8	93.0	74.5	56.7
165	Eureka, Nu	2.5	3.3	2.3	3.7	3.0	8.4
166	Miami, Fl	43.0	55.9	77.3	76.3	130.3	248.3
167	Atlanta/Mun., Ga.	108.3	116.3	123.8	85.2	93.5	103.1
168	New Orleans/Moisant Int., La.	139.3	122.0	118.0	116.0	119.4	201.1
169	Dallas-Fort Worth/Fort Worth Reg.Airport, Tx.	54.2	68.6	84.2	80.0	115.9	95.4
170	Los Angeles/Int., Ca.	72.1	90.6	42.4	15.7	6.6	2.8
171	Las Vegas/Mccarran, Nv.	13.6	19.3	11.9	4.2	3.0	2.1
172	Washington/Nat., Va.	74.5	63.8	90.9	78.3	101.2	97.0
173	Denver/Stapleton Int., Co.	12.4	13.2	32.6	45.8	59.0	43.4
174	San Francisco/Int., Ca.	100.7	105.8	70.3	32.3	14.4	3.1
175	New York/La Guardia, Ny.	82.5	67.8	105.1	102.1	97.3	101.8
176	Boston/Logan Int., Ma.	85.8	78.8	113.9	94.7	91.5	97.1
177	Chicago/O'hare, Il.	44.3	44.9	63.8	81.3	92.9	86.7
178	Detroit/Metropolitan, Mi.	50.6	50.0	56.6	74.0	84.4	89.9
179	Minneapolis/St. Paul Int., Mn.	22.0	17.1	45.6	67.9	82.7	108.2
180	Seattle/S.-Tacoma, Wa.	150.5	84.2	97.2	73.9	52.0	36.9
181	Mexico (Central), D.F.	7.6	7.2	13.0	67.1	118.9	268.3
182	Nassau Airport New Providence	49.1	53.9	66.9	66.4	103.5	236.5
183	Casa Blanca, La Habana	76.7	74.3	61.0	76.1	87.3	207.6
184	Kingston/Norman Manley	19.1	17.1	20.3	31.8	95.2	59.9
185	Santo Domingo	83.1	79.1	57.8	63.2	192.1	132.6
186	San Juan/Int., Puerto Rico	91.1	53.2	42.5	114.4	135.6	100.0
187	Belize/Phillip Goldston Intl. Airport	160.4	79.4	54.5	40.0	121.2	233.0
188	Guatemala (Observatorio Nacional)	3.2	3.1	18.4	14.6	109.6	242.2
189	Tegucigalpa						
190	Managua A.C.Sandino						
191	Juan Santamaria Int. Airport	7.6	14.2	21.8	77.5	271.2	251.4
192	Hewanorra Int'l Airport	122.8	87.7	93.3	97.4	119.2	118.0
193	Grantley Adams	69.1	37.5	39.8	52.1	100.5	126.5
194	Piarco Int. Airport, Trinidad	94.7	41.9	37.5	53.7	87.0	241.3
195	Bogota/Eldorado						
196	Caracas/Maiquetia Aerop. Intl. Simon Bolivar	29.2	24.4	15.9	27.8	40.0	44.0
197	Georgetown						
198	Zanderij						
199	Manaus	292.5	296.8	315.7	311.5	244.1	115.5
200	Salvador	79.7	99.1	147.7	337.8	297.8	258.2

Continued.

Jul.	Aug.	Sep.	Oct.	Nov.	Dec.	Annual	Statistics Period	#
83.2	90.0	79.2	71.0	96.4	77.1	957.9	1982-2001	161
85.0	72.9	42.7	41.0	25.1	17.2	516.1	1982-2001	162
89.5	60.3	46.9	20.1	16.8	13.5	447.6	1982-2001	163
41.3	37.2	47.6	109.1	193.5	156.6	1197.6	1982-2001	164
14.4	16.1	9.3	7.6	3.9	3.6	78.1	1982-2010	165
169.3	221.2	246.2	163.2	84.2	53.4	1568.6	1982-2010	166
136.0	100.0	113.8	83.7	106.8	96.3	1266.8	1982-2010	167
149.0	155.6	133.7	92.1	117.2	134.7	1598.1	1982-2010	168
56.3	47.7	71.2	89.2	68.4	67.9	899.0	1986-2010	169
1.1	0.3	3.5	14.8	18.7	53.4	322.0	1986-2010	170
9.9	8.4	6.8	7.1	9.8	12.1	108.2	1981-2010	171
92.7	74.2	95.3	86.2	83.1	76.8	1014.0	1982-2010	172
56.8	45.4	27.9	27.8	25.1	17.1	406.5	1982-2001	173
0.1	1.1	4.4	22.8	58.6	103.5	517.1	1982-2010	174
111.4	107.9	94.5	96.9	87.8	90.3	1145.4	1982-2010	175
86.6	85.6	88.2	99.5	100.7	94.1	1116.5	1982-2010	176
91.7	123.2	82.5	80.8	79.9	55.5	927.5	1982-2010	177
79.8	78.4	83.0	62.8	65.2	60.1	834.8	1986-2010	178
109.7	106.6	80.3	55.6	41.2	25.1	762.0	1986-2010	179
16.7	23.3	34.5	86.0	169.2	142.2	966.6	1986-2010	180
276.9	201.1	141.8	71.2	5.1	11.8	1190.0	1982-2010	181
135.7	215.0	182.2	148.1	71.5	51.3	1380.1	1982-2010	182
93.6	92.3	160.0	168.1	123.5	57.1	1277.6	1981-2010	183
48.6	99.4	159.4	132.1	105.8	40.4	829.1	1981-2010	184
139.6	194.0	194.7	178.7	152.8	101.6	1569.3	1981-2004	185
124.7	132.0	137.6	136.0	150.4	117.0	1334.5	1982-2010	186
216.4	204.3	252.4	278.4	225.0	145.8	2010.8	1981-2010	187
185.7	175.3	244.4	112.1	15.4	5.4	1129.4	1981-1989	188
								189
								190
176.4	240.6	339.2	338.0	146.4	34.7	1919.0	1981-2010	191
234.5	204.9	242.4	226.2	230.6	118.9	1895.9	1981-1990	192
135.3	142.1	162.8	200.9	153.5	121.2	1341.3	1982-2010	193
280.9	261.4	173.5	217.9	237.7	179.0	1906.5	1982-2010	194
								195
60.7	62.4	57.2	59.2	72.2	45.1	538.1	1981-2010	196
								197
								198
78.7	66.4	75.3	101.6	172.6	252.9	2323.6	1981-2010	199
199.3	131.1	105.0	105.2	98.7	83.5	1943.1	1981-2010	200

Climatological Normal Monthly Precipitation

#	Station	Jan.	Feb.	Mar.	Apr.	May	Jun.
201	Brasilia	210.8	180.0	210.4	136.0	39.8	5.5
202	Rio De Janeiro	121.3	135.3	140.5	145.5	75.0	70.7
203	Sao Paulo	289.7	236.0	198.4	87.6	88.4	52.1
204	Esmeraldas (Tachina) Aeropuerto	124.0	168.8	146.3	105.8	92.1	43.7
205	Lima/Callao	0.0	0.4	0.3	0.0	0.1	0.3
206	La Paz/Alto						
207	Pudahuel	0.7	1.2	3.6	7.4	47.1	81.4
208	Aeropuerto Silvio Pettirossi, Luque	142.8	147.0	111.4	166.2	117.6	67.7
209	Rocha	100.5	109.1	88.3	110.9	112.0	102.9
210	Buenos Aires Observatorio	144.7	120.5	144.2	136.0	93.8	60.8
211	Ushuaia Aero	45.7	43.5	51.2	54.2	43.2	56.6
212	Vostok	1.0	0.3	1.1	1.1	2.5	1.3
213	Honolulu, Oahu, Hawaii	49.6	52.3	49.9	14.9	16.9	6.2
214	Weather Forecast Office, Guam, Mariana Is.	98.2	89.5	62.0	72.9	111.7	158.3
215	Majuro/Marshall Is. Intnl.	198.4	203.1	149.7	239.1	238.3	272.2
216	Weather Service Office, Koror, Palau Wci.	268.6	243.3	192.6	200.4	319.5	450.8
217	Honiara/Henderson						
218	Pekoa Airport (Santo)	353.2	344.2	273.5	256.8	174.5	281.5
219	Noumea (Nlle-Caledonie)	104.3	129.1	161.0	118.0	94.8	104.4
220	Tarawa	217.8	156.0	94.9	196.4	173.0	136.4
221	Funafuti	420.7	319.8	390.1	295.3	234.1	212.1
222	Nadi Airport	327.3	308.4	313.1	177.8	100.4	56.1
223	Apia	424.3	357.2	330.9	202.6	193.6	138.9
224	Fua'amotu	202.0	221.9	184.9	154.0	105.2	99.3
225	Tahiti-Faaa	243.8	204.9	182.0	111.0	116.5	76.7
226	Madang W.O.	324.7	303.9	340.6	405.9	379.9	211.4
227	Auckland Airport	71.8	71.0	77.7	87.9	94.4	98.6
228	Christchurch	36.5	41.2	46.4	42.0	61.4	50.2
229	Darwin Airport	449.6	386.5	311.3	112.8	22.6	0.6
230	Cairns Aero	355.7	453.9	363.7	213.0	88.6	41.8
231	Alice Springs Airport	39.5	36.6	23.4	22.2	19.3	13.9
232	Brisbane Aero	109.4	139.8	109.7	102.1	115.3	64.7
233	Perth Airport	11.2	25.9	19.0	36.1	84.9	138.9
234	Sydney Airport Amo	79.7	121.1	87.4	123.1	109.7	100.1
235	Melbourne Airport	24.5	43.9	40.2	39.2	27.3	28.0
236	Canberra Airport	52.4	48.3	41.7	44.4	37.0	43.2
237	Brunei Airport	319.5	190.6	128.6	220.8	249.6	206.7
238	Balikpapan/Sepinggan	238.7	254.9	246.5	270.3	233.1	312.1
239	Jakarta/Observatory						
240	Ninoy Aquino International Airport	6.8	4.9	4.2	12.4	90.7	182.9

Continued.

Jul.	Aug.	Sep.	Oct.	Nov.	Dec.	Annual	Statistics Period	#
7.4	23.6	48.7	158.3	219.9	247.1	1487.5	1981-2010	201
62.8	50.3	92.1	83.3	77.6	168.2	1222.6	1981-1991	202
53.7	44.1	91.8	115.7	131.2	229.8	1618.5	1982-2010	203
28.1	15.6	24.8	16.7	3.6	41.8	811.3	1981-2000	204
0.1	0.3	0.5	0.0	0.1	0.1	2.2	1982-2010	205
								206
60.4	34.1	17.4	6.5	5.2	1.3	266.3	1983-2010	207
48.9	55.0	78.5	152.5	147.0	181.7	1416.3	1981-2010	208
108.5	93.9	102.0	111.8	99.3	85.3	1224.5	1981-2010	209
59.9	76.2	71.6	127.1	127.4	110.6	1272.8	1982-2006	210
59.4	40.0	44.5	34.2	37.8	41.2	551.5	1981-2010	211
1.9	1.9	1.3	1.5	1.2	0.9	16.0	1982-2010	212
11.3	12.8	17.9	42.6	62.2	74.3	410.9	1982-2010	213
282.5	395.7	345.3	293.8	209.1	144.4	2263.4	1981-2010	214
331.0	275.5	320.6	358.3	325.0	281.9	3193.1	1982-2010	215
440.1	325.0	318.7	314.2	301.5	290.5	3665.2	1982-2010	216
								217
112.1	147.1	127.8	142.8	191.5	192.7	2597.7	1982-2010	218
74.7	74.8	41.2	48.3	50.8	74.1	1075.5	1982-2010	219
176.4	173.5	128.3	146.4	98.6	228.6	1926.3	1982-1998	220
273.6	230.0	254.8	264.1	263.6	418.5	3576.7	1981-2010	221
54.3	70.3	71.2	70.9	129.4	249.8	1929.0	1982-2010	222
102.1	99.9	126.9	138.6	230.9	365.8	2711.7	1981-2002	223
106.6	122.5	110.7	90.7	90.0	149.8	1637.6	1981-2002	224
65.0	55.9	59.9	101.5	129.7	325.9	1672.8	1982-2010	225
143.0	83.6	136.3	237.5	286.6	382.6	3236.0	1982-2010	226
119.2	124.2	82.2	76.9	78.2	94.8	1076.9	1981-2010	227
65.7	63.4	40.6	48.6	46.8	50.0	592.8	1981-2010	228
0.2	4.6	13.6	67.8	138.0	281.8	1789.4	1982-2010	229
30.2	29.9	33.2	48.5	111.4	180.5	1950.4	1982-2010	230
16.7	5.3	7.1	21.1	31.6	40.7	277.4	1982-2010	231
46.0	41.2	34.2	83.2	87.0	132.9	1065.5	1982-2010	232
147.4	114.2	77.2	35.8	28.2	7.6	726.4	1982-2010	233
70.1	81.2	60.9	55.4	72.4	71.4	1032.5	1982-2010	234
32.1	41.5	33.9	50.0	65.0	55.2	480.8	1999-2010	235
46.6	45.0	53.2	50.9	67.2	56.9	586.8	1982-2010	236
252.3	240.2	264.8	325.3	340.2	384.3	3122.9	1981-2010	237
277.0	202.6	242.2	262.0	229.6	306.1	3075.1	1981-2010	238
								239
273.4	427.9	277.1	323.3	138.8	45.1	1787.5	1982-1999	240

Climatological Normals for Aerological Observation

The JMA (Japan Meteorological Agency) conducts aerological observation at 16 sites of meteorological offices and weather stations, etc. in Japan. The Climatological Normals listed here are average values for a 30-year period from 1981 to 2010. However, in some cases, average values for a period shorter than 30 years are used for the normals in consideration of the impact of replacing measuring equipment, etc. In some cases, the normals were not calculated when sufficient data were not available, etc. The JMA updates the normals every 10 years.

Geopotential altitude During aerological observation using radiosonde, altitude is indirectly calculated from atmospheric pressure, temperature, humidity and acceleration of gravity. This value is called "geopotential altitude" and is expressed in units of meters (m). Geopotential altitude is calculated by dividing the geopotential (energy required to move an object of unit mass from average sea level to a certain height) by the standard acceleration of gravity. At altitudes of approximately 30,000 m or less, the difference with geometric altitude is within 0.6%.

Synthetic wind The result of synthesizing the 30-year average value for the east-west component and north-south component of wind. Wind direction is expressed in the following angles: 90° for east, 180° for south, 270° for west, and 360° for north.

Normal Values for Aerological Observation

Station	Latitude (deg. min.)		Longitude (deg. min.)		Altitude* (m)
Wakkanai	45	25	141	41	11.7
Sapporo	43	4	141	20	26.3
Nemuro	43	20	145	35	27.2
Akita	39	43	140	6	6.5
Wajima	37	24	136	54	6.8
Tateno	36	3	140	8	31.0
Hachijojima/Omure	33	7	139	47	156.2
Yonago	35	26	133	20	7.8
Shionomisaki	33	27	135	45	75.5
Fukuoka	33	35	130	23	15.1
Kagoshima	31	33	130	33	31.7
Naze/Honchyatoge	28	24	129	33	294.9
Ishigakijima	24	20	124	10	14.9
Minamidaitojima	25	50	131	14	20.7
Chichijima	27	6	142	11	8.3
Minamitorishima	24	17	153	59	8.5

* Height above sea-level calculated from atmospheric pressure.

For the Nemuro aerological observations, the observation point was reallocated to Kushiro in March 2010.

For the Yonago aerological observations, the observation point was reallocated to Matsue in March 2010.

Aerological Observation Climatological Normal Relative Humidity (%) at 21:00 by Month

(1981–2010 Averages)

Month	Air pressure (hPa)	Wak-kanai	Sap-poro	Ne-muro	Akita	Wa-jima	Tate-no	Hachi-jojima	Yona-go	Shio-nomi-saki	Fuku-oka	Kago-shima	Naze	Ishi-gaki-jima	Mina-mida-itojima	Chi-chi-jima	Mina-mitori-shima
Jan.	300												18	17	18	18	17
	500		40	40	38	34	31	34	32	33	34	35	30	20	24	26	24
	700	55	49	47	55	52	42	37	43	33	34	37	49	56	47	36	36
	850	79	75	67	81	82	57	72	79	56	66	57	79	83	76	74	76
Feb.	300													18	19	19	15
	500			40	40	36	34	35	33	35	34	36	36	24	27	29	22
	700	55	49	48	52	48	45	39	42	35	36	38	50	58	49	40	35
	850	75	74	66	78	75	59	71	74	57	60	55	77	83	74	73	74
Mar.	300												33	25	26	27	22
	500	44	41	42	38	38	37	40	36	39	37	40	42	31	33	33	23
	700	52	48	51	49	45	49	41	40	40	39	43	55	56	51	43	34
	850	64	70	66	69	65	65	67	64	58	56	56	69	78	69	69	68
Apr.	300												44	37	39	38	32
	500	43	42	43	39	38	36	39	35	37	36	40	47	44	44	41	33
	700	54	47	50	48	43	51	38	41	38	38	42	53	61	55	48	44
	850	56	58	61	58	54	64	57	57	60	54	56	61	75	66	63	71
May	300							49		48	43	45	45	41	43	40	32
	500	44	43	43	41	39	40	42	37	41	36	42	51	53	51	47	36
	700	56	52	53	53	49	56	44	44	45	41	43	53	62	57	51	45
	850	58	62	62	62	57	67	59	61	63	58	61	63	73	68	67	70
Jun.	300					48	52	52	49	52	49	50	47	40	40	36	31
	500	42	42	42	38	38	44	49	38	46	43	50	57	55	53	47	38
	700	59	57	59	56	55	65	55	53	57	53	56	62	61	60	55	47
	850	63	70	67	69	65	75	67	72	72	71	75	74	74	74	72	70
Jul.	300	43	43	44	45	46	46	42	44	44	43	42	37	36	36	34	40
	500	42	43	43	44	45	50	48	47	49	47	47	47	48	46	44	49
	700	60	59	60	62	62	68	58	62	61	60	58	55	56	55	53	58
	850	68	76	71	78	73	76	68	78	76	76	77	71	72	73	70	74
Aug.	300	44	44	45	44	41	41	34	42	37	40	37	35	36	36	35	41
	500	39	42	42	44	43	43	41	46	43	46	46	45	49	47	45	51
	700	55	54	55	57	59	65	55	59	60	59	58	58	61	58	57	60
	850	67	73	70	74	71	76	69	76	75	76	79	74	75	76	75	76
Sep.	300				44	44	43	38	42	40	41	39	37	35	33	32	33
	500	35	38	39	39	43	46	44	43	46	45	47	46	48	43	41	40
	700	46	45	48	48	50	60	54	50	55	49	54	54	61	55	53	53
	850	61	68	68	68	70	78	71	75	75	70	76	73	76	75	74	75
Oct.	300							37		36	34	32	30	29	30	30	29
	500	39	34	36	34	34	39	42	35	41	36	40	42	41	40	38	30
	700	49	42	42	42	40	44	45	36	41	33	38	49	55	51	48	44
	850	63	65	64	65	65	73	69	67	66	57	62	73	77	75	74	72
Nov.	300											30	29	24	27	28	25
	500	40	38	38	37	34	32	38	30	34	32	34	39	35	39	37	27
	700	51	46	44	49	44	39	39	39	36	33	35	45	56	49	45	36
	850	73	74	66	73	73	66	67	71	58	59	55	74	81	74	72	72
Dec.	300												20	20	21	22	21
	500	41	38	40	37	32	28	29	28	28	29	28	27	25	29	31	26
	700	52	47	45	54	50	36	34	39	30	32	33	42	52	43	41	39
	850	80	77	66	81	80	58	68	76	55	63	56	78	80	74	72	74

Aerological Observation Climatological Normal Geopotential Height (m) at 21:00 by Month

(1981-2010 Averages)

Station	Air pressure (hPa)	Jan.	Feb.	Mar.	Apr.	May	Jun.	Jul.	Aug.	Sep.	Oct.	Nov.	Dec.
Wakkanai	30	23603	23550	23615	23731	23927	24135	24285	24264	24091	23850	23669	23577
	100	15727	15722	15806	15995	16217	16416	16604	16634	16431	16161	15921	15767
	300	8574	8591	8702	8928	9133	9321	9496	9534	9335	9085	8845	8655
	500	5189	5202	5277	5426	5546	5656	5742	5768	5668	5527	5369	5241
	700	2792	2801	2843	2920	2977	3029	3068	3088	3051	2982	2893	2820
	850	1348	1354	1374	1405	1423	1442	1456	1474	1473	1445	1401	1362
Sapporo	30	23624	23585	23633	23750	23922	24115	24258	24246	24086	23870	23709	23624
	100	15804	15801	15876	16048	16255	16441	16624	16662	16484	16234	16007	15856
	300	8648	8667	8777	8996	9188	9360	9537	9591	9415	9173	8941	8744
	500	5236	5250	5324	5469	5579	5676	5765	5801	5711	5580	5434	5303
	700	2822	2829	2870	2946	2997	3040	3081	3107	3074	3014	2933	2857
	850	1366	1372	1391	1421	1436	1450	1464	1485	1485	1466	1428	1386
Nemuro	30	23640	23602	23648	23757	23926	24116	24258	24246	24089	23881	23721	23647
	100	15814	15806	15883	16054	16255	16434	16609	16653	16491	16249	16025	15871
	300	8643	8654	8772	8995	9185	9347	9521	9582	9428	9187	8953	8745
	500	5227	5235	5315	5462	5575	5666	5756	5795	5714	5583	5437	5297
	700	2810	2814	2859	2938	2993	3034	3076	3103	3074	3012	2931	2848
	850	1351	1356	1379	1415	1435	1449	1465	1485	1486	1462	1421	1373
Akita	30	23646	23611	23664	23771	23931	24103	24229	24230	24091	23901	23743	23657
	100	15918	15919	15984	16138	16324	16495	16659	16705	16561	16341	16127	15979
	300	8790	8808	8918	9114	9286	9439	9602	9663	9530	9308	9092	8895
	500	5325	5337	5410	5539	5636	5715	5800	5842	5772	5657	5531	5400
	700	2882	2886	2923	2988	3029	3060	3100	3130	3105	3058	2995	2923
	850	1408	1410	1426	1450	1458	1460	1476	1497	1501	1493	1469	1431
Wajima	30	23656	23626	23676	23783	23930	24084	24203	24208	24086	23911	23760	23671
	100	16020	16024	16076	16209	16377	16534	16687	16725	16600	16406	16213	16078
	300	8908	8929	9031	9199	9354	9499	9649	9694	9585	9383	9188	9006
	500	5406	5417	5483	5591	5674	5744	5826	5860	5802	5702	5592	5474
	700	2931	2934	2964	3018	3051	3075	3115	3139	3120	3082	3033	2970
	850	1442	1441	1451	1466	1468	1465	1482	1501	1507	1507	1493	1462
Tateno	30	23679	23645	23690	23789	23926	24071	24183	24194	24079	23910	23776	23699
	100	16085	16081	16129	16249	16407	16551	16684	16723	16624	16457	16281	16152
	300	8993	9003	9102	9261	9412	9544	9670	9712	9635	9461	9279	9102
	500	5452	5457	5522	5625	5704	5767	5843	5875	5834	5743	5641	5525
	700	2944	2945	2979	3033	3066	3087	3129	3151	3136	3099	3054	2987
	850	1436	1435	1449	1471	1476	1474	1491	1510	1515	1511	1496	1460
Hachijojima	30	23692	23662	23711	23804	23927	24057	24154	24165	24072	23920	23791	23717
	100	16214	16210	16246	16340	16474	16595	16696	16725	16659	16532	16389	16281
	300	9171	9171	9251	9380	9510	9622	9707	9729	9687	9568	9422	9275
	500	5554	5556	5611	5693	5758	5812	5869	5888	5868	5803	5720	5619
	700	2997	2995	3023	3065	3091	3109	3146	3159	3153	3126	3093	3036
	850	1463	1460	1470	1486	1487	1484	1503	1514	1519	1518	1514	1486
Yonago	30	23669	23639	23686	23790	23930	24072	24180	24190	24079	23915	23774	23691
	100	16101	16109	16149	16266	16423	16568	16705	16732	16622	16449	16277	16157
	300	9025	9049	9134	9278	9417	9546	9680	9707	9617	9442	9272	9114
	500	5478	5491	5546	5636	5708	5767	5843	5865	5820	5735	5642	5540
	700	2968	2971	2996	3039	3066	3083	3122	3139	3126	3098	3059	3006
	850	1466	1464	1468	1475	1474	1467	1483	1497	1508	1515	1508	1487

(1981-2010 Averages) Continued.

Station	Air pressure (hPa)	Jan.	Feb.	Mar.	Apr.	May	Jun.	Jul.	Aug.	Sep.	Oct.	Nov.	Dec.
Shionomisaki	30	23689	23659	23708	23803	23931	24060	24161	24170	24076	23921	23792	23715
	100	16196	16196	16231	16328	16468	16595	16709	16734	16654	16512	16363	16255
	300	9143	9156	9234	9360	9489	9606	9708	9727	9671	9534	9383	9236
	500	5544	5552	5604	5682	5746	5800	5867	5882	5855	5784	5701	5604
	700	2997	2997	3021	3060	3085	3102	3141	3152	3144	3119	3086	3033
	850	1469	1466	1472	1485	1484	1478	1498	1508	1514	1518	1514	1490
Fukuoka	30	23679	23650	23697	23794	23931	24063	24163	24173	24073	23920	23785	23700
	100	16176	16186	16220	16321	16468	16601	16722	16742	16646	16492	16335	16225
	300	9131	9155	9227	9351	9476	9592	9707	9722	9651	9504	9348	9207
	500	5545	5559	5605	5677	5739	5790	5860	5873	5839	5770	5687	5596
	700	3004	3006	3025	3058	3079	3093	3132	3141	3134	3115	3084	3036
	850	1486	1480	1479	1483	1478	1470	1488	1496	1509	1521	1520	1504
Kagoshima	30	23691	23667	23710	23798	23928	24045	24139	24147	24063	23923	23797	23718
	100	16266	16273	16297	16382	16512	16625	16723	16738	16664	16541	16410	16313
	300	9248	9265	9319	9425	9538	9638	9723	9729	9682	9573	9444	9321
	500	5609	5619	5658	5719	5771	5815	5873	5877	5858	5807	5736	5656
	700	3035	3034	3050	3076	3094	3105	3142	3143	3143	3130	3106	3064
	850	1495	1489	1487	1490	1484	1476	1495	1496	1509	1523	1526	1513
Naze	30	23694	23678	23724	23807	23920	24020	24100	24101	24032	23916	23798	23727
	100	16385	16389	16405	16465	16568	16657	16716	16722	16674	16591	16496	16424
	300	9417	9423	9452	9525	9614	9689	9735	9727	9706	9649	9565	9476
	500	5697	5703	5733	5777	5815	5850	5885	5875	5872	5849	5801	5738
	700	3081	3080	3091	3108	3116	3128	3152	3139	3147	3149	3135	3104
	850	1519	1511	1506	1503	1494	1488	1502	1490	1508	1527	1538	1532
Ishigakijima	30	23710	23694	23732	23806	23905	23999	24054	24058	24005	23905	23810	23740
	100	16497	16500	16509	16551	16628	16686	16706	16706	16675	16627	16566	16519
	300	9572	9573	9583	9624	9682	9725	9734	9724	9717	9699	9659	9607
	500	5789	5792	5809	5836	5855	5874	5882	5869	5875	5877	5856	5818
	700	3126	3125	3132	3140	3138	3142	3147	3132	3146	3161	3159	3141
	850	1538	1530	1523	1515	1501	1492	1494	1480	1501	1529	1544	1546
Minamidaitojima	30	23708	23693	23732	23807	23906	24002	24063	24069	24013	23909	23803	23732
	100	16469	16470	16478	16520	16599	16664	16695	16703	16671	16621	16552	16499
	300	9537	9536	9550	9597	9662	9716	9733	9724	9716	9690	9644	9583
	500	5765	5767	5790	5822	5848	5875	5890	5876	5882	5874	5848	5802
	700	3111	3110	3122	3135	3137	3150	3159	3141	3153	3159	3155	3129
	850	1529	1523	1520	1516	1505	1506	1510	1492	1509	1525	1540	1537
Chichijima	30	23713	23693	23739	23812	23913	24011	24082	24090	24029	23912	23803	23733
	100	16457	16449	16459	16499	16577	16645	16678	16693	16672	16626	16557	16501
	300	9515	9502	9522	9580	9655	9715	9730	9728	9721	9698	9646	9580
	500	5748	5742	5774	5817	5853	5886	5896	5890	5894	5886	5855	5800
	700	3096	3093	3114	3136	3148	3167	3171	3162	3170	3170	3160	3125
	850	1515	1511	1518	1522	1520	1526	1525	1516	1526	1534	1540	1528
Minamitorishima	30	23712	23703	23740	23808	23902	23989	24044	24046	23998	23901	23807	23741
	100	16542	16531	16532	16547	16590	16627	16635	16648	16646	16633	16604	16576
	300	9624	9605	9609	9641	9686	9720	9717	9715	9716	9716	9698	9669
	500	5825	5811	5833	5862	5885	5904	5897	5891	5899	5904	5895	5865
	700	3133	3126	3147	3166	3176	3187	3180	3171	3178	3183	3180	3160
	850	1525	1522	1535	1543	1544	1547	1539	1530	1537	1544	1548	1537

Aerological Observation Climatological Normal Temperature (℃) at 21:00 by Month

(1981–2010 Averages)

Station	Air pressure (hPa)	Jan.	Feb.	Mar.	Apr.	May	Jun.	Jul.	Aug.	Sep.	Oct.	Nov.	Dec.
Wakkanai	30	-49.5	-50.4	-50.3	-51.2	-51.6	-50.1	-49.9	-50.7	-52.0	-52.8	-51.4	-49.5
	100	-49.8	-50.9	-52.1	-54.4	-55.2	-56.3	-59.0	-61.3	-58.6	-56.0	-54.2	-52.3
	300	-53.3	-53.5	-52.5	-49.6	-45.6	-41.4	-35.5	-34.1	-40.0	-46.3	-50.3	-52.5
	500	-37.1	-36.6	-33.6	-26.8	-20.5	-14.7	-9.4	-9.0	-15.5	-22.8	-29.3	-34.6
	700	-23.0	-22.6	-19.0	-11.2	-4.8	0.8	5.2	5.7	-0.2	-7.3	-14.7	-20.5
	850	-15.0	-14.4	-10.4	-2.3	4.2	9.4	13.3	13.7	8.0	1.1	-6.8	-12.6
Sapporo	30	-50.4	-51.3	-51.2	-51.6	-52.0	-50.7	-50.6	-51.2	-52.5	-53.4	-52.0	-50.4
	100	-51.3	-52.3	-53.7	-55.9	-56.9	-58.2	-61.3	-63.9	-61.8	-58.4	-56.0	-53.8
	300	-51.8	-52.0	-50.9	-48.3	-44.2	-39.9	-34.1	-32.4	-37.4	-44.1	-48.5	-51.1
	500	-35.3	-34.7	-31.5	-25.1	-19.1	-13.8	-8.4	-7.6	-13.5	-20.7	-27.0	-32.3
	700	-21.0	-20.5	-16.9	-9.4	-3.3	1.7	6.1	7.1	1.9	-5.2	-12.1	-18.0
	850	-12.9	-12.6	-8.9	-1.0	5.1	9.9	13.8	14.7	9.3	2.6	-4.8	-10.5
Nemuro	30	-50.3	-51.0	-51.0	-51.8	-52.0	-50.5	-50.3	-51.1	-52.4	-53.5	-52.4	-50.2
	100	-50.7	-51.4	-53.0	-55.9	-57.0	-57.9	-60.7	-63.5	-62.0	-58.5	-55.6	-53.1
	300	-51.5	-51.7	-50.8	-47.9	-43.9	-39.9	-34.7	-32.8	-36.8	-43.5	-47.9	-50.6
	500	-35.1	-34.7	-31.2	-24.9	-19.1	-14.0	-8.9	-7.7	-13.0	-20.1	-26.5	-32.1
	700	-20.5	-20.4	-16.9	-9.4	-3.6	1.1	5.4	6.7	1.9	-4.8	-11.5	-17.4
	850	-12.5	-12.6	-8.9	-1.5	4.5	8.9	12.8	14.0	9.1	2.9	-3.9	-9.9
Akita	30	-52.1	-52.9	-52.5	-52.3	-52.5	-51.5	-51.3	-51.9	-52.9	-54.2	-53.2	-52.3
	100	-54.8	-55.4	-56.4	-58.3	-59.6	-61.4	-64.5	-67.1	-65.8	-62.3	-59.6	-57.3
	300	-48.9	-49.0	-48.2	-45.6	-41.4	-36.8	-31.9	-30.7	-34.2	-40.7	-45.3	-47.8
	500	-32.0	-31.3	-27.8	-22.0	-16.5	-11.7	-6.8	-5.7	-10.1	-17.1	-23.5	-29.0
	700	-18.3	-17.6	-13.9	-7.0	-1.4	3.3	7.4	8.8	4.4	-2.2	-8.7	-14.8
	850	-9.6	-9.3	-5.9	1.4	7.2	11.8	15.4	16.9	12.2	5.5	-1.1	-6.6
Wajima	30	-53.5	-54.3	-53.4	-52.9	-52.8	-52.1	-52.1	-52.6	-53.3	-54.6	-54.0	-53.6
	100	-57.6	-58.4	-59.0	-60.7	-62.4	-64.2	-67.3	-69.0	-68.1	-65.0	-61.9	-59.6
	300	-47.6	-47.5	-46.5	-43.8	-39.4	-34.6	-30.4	-29.8	-32.8	-38.8	-43.5	-46.2
	500	-28.8	-27.9	-24.8	-19.7	-14.9	-10.0	-5.6	-4.8	-8.5	-15.1	-21.1	-26.3
	700	-15.4	-14.5	-10.8	-4.7	0.5	4.8	8.6	9.5	5.7	-0.3	-6.2	-12.0
	850	-6.9	-6.5	-3.2	4.0	9.2	13.4	17.0	18.1	13.5	7.4	1.4	-3.9
Tateno	30	-53.7	-54.5	-53.7	-53.4	-53.0	-52.3	-52.3	-52.6	-53.5	-55.1	-54.8	-54.0
	100	-59.4	-59.6	-60.4	-62.3	-64.2	-66.2	-68.6	-70.1	-70.3	-67.8	-64.5	-61.8
	300	-45.6	-45.9	-45.1	-42.4	-37.8	-33.2	-30.2	-29.8	-31.9	-36.6	-41.0	-43.5
	500	-25.5	-25.0	-22.1	-17.7	-13.2	-8.8	-5.5	-4.5	-7.1	-12.5	-18.3	-23.0
	700	-12.2	-11.8	-8.3	-3.1	1.5	5.4	8.9	9.8	7.1	2.0	-3.4	-8.6
	850	-3.5	-3.4	-0.3	5.5	10.1	13.7	17.7	18.3	14.6	9.1	4.3	-0.4
Hachijojima	30	-55.2	-55.5	-54.9	-53.9	-53.2	-52.8	-52.7	-53.2	-53.8	-55.3	-55.7	-55.5
	100	-64.8	-65.1	-64.9	-65.9	-67.8	-69.9	-71.0	-71.5	-72.6	-71.9	-69.4	-67.0
	300	-40.1	-40.9	-41.8	-39.5	-35.2	-31.2	-29.5	-29.6	-31.0	-34.0	-37.1	-38.3
	500	-20.7	-20.3	-17.7	-14.2	-10.2	-6.6	-4.7	-4.2	-5.6	-9.3	-14.1	-18.2
	700	-6.9	-6.7	-3.7	0.6	4.5	7.7	10.2	10.7	9.2	5.4	0.6	-4.0
	850	0.2	0.4	3.3	7.9	12.0	15.7	18.8	19.0	16.9	12.5	7.7	2.8
Yonago	30	-54.2	-55.2	-54.0	-53.1	-52.8	-52.4	-52.5	-52.8	-53.4	-54.6	-54.6	-54.2
	100	-60.5	-61.4	-61.6	-62.8	-64.5	-66.6	-69.4	-70.4	-69.8	-67.2	-64.6	-62.2
	300	-45.5	-45.4	-44.9	-42.1	-37.8	-32.9	-29.5	-29.3	-32.0	-37.2	-41.5	-43.9
	500	-25.0	-24.1	-21.5	-17.2	-12.8	-8.5	-4.8	-4.3	-7.4	-13.3	-18.6	-23.0
	700	-12.4	-11.4	-7.9	-2.5	2.3	6.1	9.7	10.2	6.9	1.5	-3.9	-9.5
	850	-5.3	-4.5	-0.9	5.6	10.5	14.3	17.7	18.3	14.1	8.3	2.8	-2.5

(1981-2010 Averages) Continued.

Station	Air pressure (hPa)	Jan.	Feb.	Mar.	Apr.	May	Jun.	Jul.	Aug.	Sep.	Oct.	Nov.	Dec.
Shionomisaki	30	-54.9	-55.5	-54.5	-53.6	-53.0	-52.7	-52.7	-53.1	-53.7	-55.0	-55.2	-55.1
	100	-63.9	-64.4	-64.3	-65.2	-67.1	-69.3	-71.0	-71.6	-72.1	-70.8	-68.2	-65.8
	300	-41.5	-42.0	-42.5	-40.0	-35.6	-31.4	-29.2	-29.2	-31.0	-34.7	-38.0	-39.7
	500	-21.7	-20.9	-18.3	-14.7	-10.8	-6.9	-4.5	-4.0	-5.9	-10.4	-15.4	-19.4
	700	-7.9	-7.3	-4.4	-0.1	4.0	7.4	10.2	10.6	8.6	4.3	-0.6	-5.4
	850	-1.3	-0.8	2.6	7.3	11.6	15.2	18.6	18.8	16.2	11.2	6.4	1.3
Fukuoka	30	-54.8	-55.7	-54.4	-53.5	-53.0	-52.8	-52.9	-53.2	-53.7	-54.8	-54.9	-54.8
	100	-64.0	-64.8	-64.5	-65.1	-66.7	-68.7	-71.1	-71.6	-71.4	-69.6	-67.3	-65.4
	300	-42.7	-42.8	-43.0	-40.3	-36.0	-31.6	-28.8	-28.8	-31.2	-35.6	-39.1	-41.3
	500	-22.3	-21.1	-18.7	-14.9	-11.0	-7.1	-4.1	-3.9	-6.3	-11.5	-16.6	-20.5
	700	-8.9	-7.8	-4.8	-0.3	3.9	7.3	10.5	10.7	8.0	3.4	-1.7	-6.7
	850	-3.4	-2.1	1.7	7.2	11.8	15.4	18.6	18.7	15.4	10.0	4.6	-0.8
Kagoshima	30	-55.7	-56.2	-55.2	-54.0	-53.4	-53.1	-53.2	-53.5	-53.9	-55.0	-55.4	-55.7
	100	-67.5	-68.1	-67.5	-67.6	-69.2	-71.2	-72.7	-72.7	-73.1	-72.5	-70.5	-68.6
	300	-38.9	-39.2	-40.4	-38.4	-34.3	-30.4	-28.7	-28.7	-30.6	-33.7	-36.3	-37.9
	500	-19.1	-18.1	-16.0	-12.7	-9.1	-5.9	-4.0	-3.7	-5.4	-9.2	-13.8	-17.2
	700	-5.4	-4.5	-1.8	1.9	5.5	8.5	11.0	11.1	9.2	5.5	0.8	-3.4
	850	0.0	1.1	4.4	8.8	12.7	15.8	18.8	18.7	16.5	12.0	7.4	2.4
Naze	30	-56.9	-57.2	-55.8	-54.5	-53.6	-53.5	-53.6	-54.0	-54.3	-55.1	-56.1	-56.8
	100	-72.9	-73.2	-72.3	-71.9	-72.9	-74.4	-74.5	-74.2	-75.1	-76.0	-75.0	-73.6
	300	-34.2	-34.5	-36.3	-35.6	-32.2	-29.3	-28.6	-28.6	-30.0	-32.0	-33.4	-34.0
	500	-14.3	-13.9	-12.6	-10.2	-7.1	-4.9	-4.1	-3.6	-4.6	-6.9	-9.9	-12.5
	700	-1.7	-0.6	1.7	4.8	7.5	9.8	11.5	11.4	10.2	7.8	3.9	0.0
	850	3.9	4.6	7.8	11.5	14.8	17.8	19.6	19.2	17.6	14.0	9.9	5.6
Ishigakijima	30	-57.9	-58.1	-56.6	-55.0	-53.8	-53.5	-53.5	-54.1	-54.4	-55.5	-56.3	-57.5
	100	-77.2	-77.3	-76.5	-76.1	-76.5	-77.2	-76.5	-76.4	-77.1	-78.6	-78.7	-77.9
	300	-32.1	-32.1	-33.0	-32.8	-30.3	-28.5	-28.4	-28.3	-29.5	-30.7	-31.5	-32.0
	500	-9.2	-9.0	-9.0	-7.6	-5.4	-4.2	-4.0	-3.6	-4.2	-5.4	-6.9	-8.3
	700	2.3	3.3	5.5	7.8	9.5	11.3	11.9	11.7	10.7	9.2	6.8	3.9
	850	8.3	9.2	11.8	14.8	17.2	19.2	20.0	19.7	18.3	15.8	12.8	9.4
Minamidaitojima	30	-57.5	-57.6	-56.4	-54.8	-53.7	-53.5	-53.8	-54.2	-54.4	-55.2	-56.2	-57.5
	100	-76.4	-76.6	-75.5	-74.9	-75.1	-76.1	-75.5	-75.1	-76.2	-77.8	-77.9	-77.2
	300	-32.4	-32.5	-33.8	-33.8	-31.3	-29.3	-29.1	-28.9	-30.1	-31.1	-32.1	-32.4
	500	-10.2	-10.2	-9.9	-8.4	-6.0	-4.7	-4.4	-3.9	-4.5	-5.7	-7.3	-9.0
	700	1.8	2.5	4.5	6.8	8.9	10.5	11.3	11.4	10.9	9.5	6.8	3.8
	850	7.1	7.9	10.5	13.7	16.2	18.5	19.3	19.3	18.3	16.1	12.8	8.8
Chichijima	30	-57.4	-57.7	-56.3	-55.0	-53.8	-53.5	-53.6	-54.1	-54.4	-55.4	-56.6	-57.7
	100	-75.5	-75.4	-74.1	-73.4	-73.8	-75.0	-74.0	-73.6	-75.1	-77.2	-77.1	-76.6
	300	-32.5	-32.7	-34.6	-34.8	-32.3	-30.1	-29.9	-29.8	-30.6	-31.5	-32.6	-32.6
	500	-10.6	-11.0	-10.7	-9.0	-6.6	-5.2	-4.8	-4.4	-4.8	-5.7	-7.4	-8.9
	700	2.0	2.0	3.8	6.3	8.4	10.0	10.7	10.9	10.6	9.7	7.4	4.6
	850	7.0	7.0	9.4	12.8	15.5	17.9	18.7	18.6	18.3	16.8	13.7	9.7
Minamitorishima	30	-58.3	-58.4	-57.2	-55.1	-54.0	-53.6	-53.9	-54.4	-54.6	-55.3	-56.5	-57.7
	100	-78.8	-78.2	-76.7	-75.6	-75.0	-74.9	-73.0	-72.8	-74.9	-77.7	-78.6	-79.3
	300	-31.6	-31.7	-33.4	-34.1	-32.7	-31.4	-31.1	-30.8	-31.4	-31.8	-32.3	-31.8
	500	-7.3	-7.8	-8.2	-7.6	-6.2	-5.5	-5.6	-5.3	-5.3	-5.3	-5.8	-6.4
	700	6.4	5.7	6.5	7.8	9.1	10.0	9.9	10.1	10.3	10.3	9.6	8.3
	850	11.3	10.5	12.1	14.5	16.1	17.6	18.0	18.1	18.0	17.5	15.9	13.9

Aerological Observation Climatological Normal for Magnitude and Direction of Resultant Wind at 21:00 by Month

(1981–2010 Averages)

Station	Air pressure (hPa)	January		February		March		April		May		June	
		Magni-tude	Wind direction	Magni-tude	Wind direction	Magni-tude	Wind direction	Magni-tude	Wind direction	Magni-tude	Wind direction	Magni-tude	Wind direction
Wakkanai	30	11.1	238	10.8	240	9.9	242	6.4	246	0.4	270	5.5	81
	100	24.9	262	25.1	265	23.2	262	18.5	263	13.3	267	9.0	278
	300	22.0	270	23.1	275	23.6	269	22.0	269	17.9	270	11.6	279
	500	13.9	277	14.5	281	14.1	277	13.1	275	9.6	273	5.9	284
	700	9.0	278	8.9	286	8.3	284	8.2	278	6.2	275	3.1	285
	850	5.2	295	5.8	299	5.5	290	5.8	266	5.0	255	2.7	237
Sapporo	30	8.8	241	9.7	243	8.6	243	6.3	250	0.5	270	5.5	81
	100	29.7	264	29.3	267	27.2	264	21.6	265	16.0	268	11.6	276
	300	30.3	269	30.3	274	30.9	269	27.2	268	22.4	266	16.2	272
	500	19.3	273	18.9	278	18.5	274	15.7	271	12.4	268	8.0	274
	700	11.7	283	11.6	288	11.0	284	9.7	279	7.2	273	4.1	274
	850	7.7	292	7.6	293	6.9	283	6.0	268	4.5	250	2.5	222
Nemuro	30	8.1	238	8.9	240	7.9	241	5.5	252	0.5	270	5.3	82
	100	27.7	263	27.8	264	25.6	262	20.9	263	16.0	266	11.9	275
	300	30.3	267	29.8	270	30.1	266	27.0	266	23.7	266	17.6	271
	500	18.6	270	18.2	275	17.8	270	15.9	270	13.0	266	8.5	275
	700	11.1	279	10.6	286	10.0	282	9.1	276	7.2	269	4.1	280
	850	7.0	294	7.1	299	6.4	292	5.8	281	4.4	269	1.9	270
Akita	30	7.4	247	9.7	251	8.4	253	5.9	256	0.6	301	6.0	80
	100	37.0	266	36.4	269	33.2	267	26.4	267	20.0	269	14.6	274
	300	42.1	269	40.6	273	40.5	269	32.2	267	28.2	264	24.5	267
	500	27.4	273	26.0	276	25.1	273	19.3	270	15.3	267	11.1	269
	700	16.8	271	16.0	275	15.0	273	12.1	269	10.1	264	6.8	266
	850	10.2	278	9.4	277	8.7	270	7.3	257	6.3	247	4.5	233
Wajima	30	8.0	254	9.3	254	8.5	254	5.4	259	0.6	321	6.3	80
	100	38.6	265	39.0	268	35.2	267	28.2	267	22.1	269	16.0	273
	300	49.6	268	48.5	271	45.7	269	36.1	267	30.9	264	27.2	264
	500	28.9	273	28.0	276	25.9	274	19.8	271	15.7	269	12.5	265
	700	15.2	271	14.4	276	13.5	273	10.7	266	8.7	261	6.6	256
	850	8.5	277	8.0	276	7.0	267	7.2	251	6.6	247	5.2	242
Tateno	30	5.9	252	8.5	253	7.4	255	3.9	264	1.0	45	7.1	81
	100	42.1	265	41.0	268	36.9	266	29.4	266	22.5	269	15.3	274
	300	55.4	266	52.5	269	47.9	267	38.3	265	33.0	261	28.1	263
	500	31.7	269	30.4	272	27.7	269	20.9	266	16.7	263	14.3	264
	700	14.6	274	13.9	276	12.7	271	9.5	267	7.4	263	6.2	265
	850	7.0	273	6.4	276	4.8	259	3.5	243	2.7	231	1.8	223
Hachijojima	30	5.3	253	7.8	254	6.5	257	2.6	274	2.3	65	9.1	81
	100	45.7	265	44.6	267	39.6	266	30.8	266	21.9	270	13.5	279
	300	66.4	265	62.1	269	51.1	267	39.8	265	31.4	261	24.4	262
	500	36.1	269	34.9	271	30.8	269	23.2	266	18.2	261	15.5	260
	700	18.6	272	17.9	274	15.8	270	11.2	264	8.7	257	9.7	258
	850	10.2	274	9.9	276	8.1	267	6.3	262	4.8	250	6.4	252
Yonago	30	8.1	258	9.5	258	8.6	259	5.0	265	0.7	34	7.3	81
	100	42.4	266	42.0	268	37.7	268	29.4	269	22.9	272	15.8	277
	300	53.5	268	52.0	270	46.2	270	37.3	269	31.2	265	27.1	263
	500	31.2	275	30.4	277	27.2	274	20.3	273	15.8	270	12.5	265
	700	15.2	285	14.7	287	13.0	285	9.9	278	7.6	274	6.3	265
	850	8.0	285	7.0	285	5.7	277	5.4	256	4.6	244	4.2	236

(1981–2010 Averages) Continued.

Station	Air pressure (hPa)	January		February		March		April		May		June	
		Magni-tude	Wind direction	Magni-tude	Wind direction	Magni-tude	Wind direction	Magni-tude	Wind direction	Magni-tude	Wind direction	Magni-tude	Wind direction
Shionomisaki	30	6.0	258	8.4	257	7.0	259	3.4	270	2.1	61	8.6	81
	100	44.8	265	44.2	268	39.6	267	30.6	268	22.4	271	14.2	278
	300	62.6	266	59.6	269	49.1	268	39.4	266	31.7	263	25.4	261
	500	34.3	270	33.1	272	29.4	270	22.2	268	16.9	266	14.5	260
	700	17.5	276	17.0	278	14.2	274	10.1	269	7.7	263	7.9	257
	850	9.2	284	8.3	287	5.9	283	3.7	270	2.4	255	3.1	248
Fukuoka	30	7.2	257	8.3	258	7.6	261	4.4	269	1.7	57	8.4	78
	100	45.5	265	44.5	267	39.9	267	31.5	269	23.4	273	15.3	281
	300	58.9	266	56.9	268	47.8	268	37.9	268	31.3	266	25.5	263
	500	32.4	273	31.5	275	27.6	273	21.1	272	16.1	271	13.5	264
	700	16.6	284	15.5	285	13.0	280	9.7	275	7.1	272	6.6	257
	850	7.1	295	6.0	291	4.5	284	3.9	253	3.2	238	4.1	223
Kagoshima	30	6.7	258	8.3	257	7.1	261	3.9	271	2.4	66	9.5	80
	100	45.3	265	44.4	267	39.2	267	30.8	269	21.5	274	13.0	285
	300	61.8	265	59.6	267	48.4	267	37.1	267	29.2	266	22.2	262
	500	32.9	270	31.6	271	27.8	270	21.8	269	16.3	268	13.7	260
	700	17.5	280	16.7	280	14.7	275	10.9	272	7.6	268	8.3	254
	850	8.8	297	7.2	295	5.1	285	3.1	266	2.1	245	3.9	230
Naze	30	4.9	257			5.3	261	2.4	285				
	100	40.1	263	38.9	265	35.3	266	27.3	269	17.2	278	7.7	306
	300	58.2	263	55.9	265	47.2	266	34.5	267	23.2	267	13.8	262
	500	32.7	267	30.8	268	27.1	267	21.0	266	15.1	265	11.4	257
	700	15.6	274	15.9	274	14.9	270	11.7	267	8.1	263	9.3	248
	850	6.5	305	6.6	299	5.9	282	4.7	270	3.4	253	6.7	238
Ishigakijima	30	1.9	235	3.4	243	1.8	248	1.2	48				
	100	29.9	261	28.6	263	26.7	265	20.4	269	9.6	291	7.0	28
	300	43.0	262	42.0	263	38.5	265	28.5	268	15.1	270	4.5	271
	500	27.7	264	26.5	265	23.0	264	17.7	263	10.4	261	5.9	243
	700	10.4	264	10.5	263	10.0	261	8.4	260	5.9	258	5.6	230
	850	1.6	11	0.9	288	2.0	243	2.8	235	2.2	216	5.4	208
Minamidaitojima	30	2.2	248	3.5	246	2.5	256	0.8	7	7.1	76	14.4	81
	100	33.7	262	32.4	264	30.3	265	23.3	269	12.4	284	5.0	359
	300	48.8	262	47.6	265	42.8	266	30.8	268	17.7	266	7.3	261
	500	31.7	264	29.9	267	25.9	266	19.3	265	12.3	261	8.0	248
	700	13.9	272	14.1	274	13.4	269	10.9	265	7.9	255	7.2	238
	850	4.3	309	4.6	293	5.2	274	4.6	264	3.8	245	6.0	229
Chichijima	30	1.7	245	3.7	244	2.2	254	0.2	0	6.4	77	13.4	82
	100	38.2	262	38.4	265	33.9	266	25.1	270	13.2	280	4.3	345
	300	54.2	264	54.6	267	45.7	269	30.4	268	17.1	264	7.2	267
	500	35.0	265	33.9	268	27.3	266	18.9	263	12.6	255	7.9	251
	700	16.7	270	17.0	272	14.8	267	11.4	260	8.4	249	6.6	246
	850	7.2	281	7.8	278	7.1	268	5.1	259	5.1	240	5.6	239
Minamitorishima	30	0.5	68	0.5	158	0.8	120	2.8	72	9.1	79	15.6	83
	100	25.1	263	26.3	265	23.4	269	16.9	274	6.8	297	5.9	44
	300	34.9	269	37.3	271	31.6	274	19.5	276	7.3	280	2.5	16
	500	25.0	266	25.3	269	18.5	269	10.2	266	4.3	258	0.1	90
	700	13.3	265	13.2	267	8.6	265	4.3	254	1.9	234	0.9	122
	850	4.4	273	5.0	275	2.7	262	1.3	193	1.5	160	1.6	128

Aerological Observation Climatological Normal for Magnitude and Direction of Resultant Wind at 21:00 by Month Continued.

Station	Air pressure (hPa)	July		August		September		October		November		December	
		Magni-tude	Wind direction	Magni-tude	Wind direction	Magni-tude	Wind direction	Magni-tude	Wind direction	Magni-tude	Wind direction	Magni-tude	Wind direction
Wakkanai	30	8.5	83	4.9	82	1.8	270	9.2	254	15.5	246	14.9	242
	100	8.4	294	12.1	274	20.8	258	26.3	256	29.1	257	28.5	259
	300	13.5	282	21.3	266	28.3	255	29.2	258	31.2	262	27.4	266
	500	6.9	279	11.3	264	13.9	261	17.1	263	20.9	265	18.5	270
	700	3.5	275	6.0	263	7.3	265	10.7	264	12.9	264	11.2	271
	850	3.3	239	4.0	250	4.5	261	7.7	263	8.8	265	7.1	278
Sapporo	30	9.3	83	6.3	84	0.5	281	7.4	257	13.4	250	12.7	244
	100	8.9	295	10.9	277	21.1	258	28.7	256	32.3	258	32.8	260
	300	15.5	277	20.8	265	32.1	255	34.1	257	36.8	261	34.6	265
	500	8.4	273	11.6	262	16.0	258	19.0	262	23.8	264	23.2	268
	700	4.6	269	6.4	261	8.0	261	10.7	268	14.5	270	14.1	277
	850	3.0	224	3.6	231	3.8	247	6.5	266	8.9	273	8.6	283
Nemuro	30	9.2	83	5.9	83	0.9	276	7.1	255	11.6	248	11.6	241
	100	8.8	295	10.7	282	21.1	260	28.9	257	31.6	259	32.0	260
	300	15.9	278	20.1	271	33.2	256	36.4	255	38.2	259	35.8	263
	500	8.7	274	11.7	266	17.2	258	19.9	259	24.3	262	23.4	266
	700	4.6	269	5.9	263	8.6	259	10.9	264	14.3	266	13.5	274
	850	2.4	251	3.0	249	3.9	261	6.1	264	8.2	265	7.9	282
Akita	30	10.6	84	8.4	84	1.5	79	5.2	262	9.6	254	10.3	249
	100	8.4	298	7.5	284	19.3	258	30.5	257	36.9	260	39.5	262
	300	17.8	274	16.4	265	31.3	253	37.8	254	42.4	260	43.6	264
	500	10.7	271	10.5	260	16.5	255	20.7	260	25.6	264	28.1	269
	700	7.6	262	7.2	255	9.0	256	11.7	263	16.1	264	17.6	268
	850	5.4	234	4.6	229	4.9	238	6.6	256	9.5	266	10.5	274
Wajima	30	11.6	84	10.1	84	3.1	84	4.0	264	8.8	257	10.2	255
	100	7.3	302	5.0	290	17.0	258	29.3	257	36.9	259	40.0	262
	300	17.0	273	12.2	261	27.8	253	38.5	255	45.4	259	49.3	264
	500	10.9	268	8.4	255	15.6	254	19.9	260	25.0	264	27.8	270
	700	7.9	257	6.2	241	7.3	248	9.2	261	13.0	263	14.9	268
	850	6.8	241	4.8	233	3.7	239	5.0	253	7.3	260	8.6	271
Tateno	30	12.5	84	11.3	85	4.8	83	2.0	270	6.2	256	7.8	253
	100	6.2	314	3.1	327	13.5	258	27.8	256	38.0	260	43.3	261
	300	14.8	275	8.1	270	22.3	254	39.0	254	48.7	258	55.4	261
	500	9.9	269	5.7	257	14.5	250	21.1	255	26.2	261	29.9	265
	700	5.8	269	3.5	248	6.4	247	8.5	255	12.0	263	14.3	272
	850	1.9	235	1.6	194	1.2	204	1.8	251	5.0	254	6.8	267
Hachijojima	30	14.5	84	14.1	85	8.0	85	1.1	49	3.5	258	5.7	254
	100	4.6	355	4.5	42	7.3	257	23.1	256	36.5	260	44.6	261
	300	9.2	278	2.4	287	13.1	253	33.2	255	49.8	259	63.0	261
	500	7.8	269	2.2	252	9.7	248	19.7	253	27.6	260	34.0	264
	700	6.3	264	1.8	223	5.1	239	8.8	250	12.8	262	16.4	268
	850	5.2	257	1.9	198	2.3	218	2.2	248	5.5	265	9.1	273
Yonago	30	12.6	83	11.8	85	4.9	83	2.9	266	8.2	258	10.0	257
	100	5.7	313	2.6	302	14.6	259	27.5	259	37.6	261	42.9	263
	300	14.3	269	9.3	261	23.6	255	37.2	259	46.2	260	52.0	263
	500	9.9	265	6.3	250	13.7	255	19.4	262	24.9	266	28.6	271
	700	7.4	260	4.4	237	5.8	254	8.1	274	12.2	275	14.5	282
	850	6.1	237	3.7	214	1.6	225	2.9	276	5.6	276	8.0	281

(1981–2010 Averages) Continued.

Station	Air pressure (hPa)	July		August		September		October		November		December	
		Magni-tude	Wind direction	Magni-tude	Wind direction	Magni-tude	Wind direction	Magni-tude	Wind direction	Magni-tude	Wind direction	Magni-tude	Wind direction
Shionomisaki	30	14.1	84	13.6	84	7.4	84	0.4	34	4.5	261	6.9	258
	100	4.5	346	3.1	36	9.3	257	24.8	256	37.4	260	44.2	261
	300	9.9	275	3.8	264	16.3	253	35.1	256	50.3	259	60.3	261
	500	8.0	267	3.2	240	11.0	250	19.9	256	26.5	262	31.9	267
	700	6.2	260	2.1	209	5.2	242	7.9	261	12.5	268	15.7	274
	850	3.7	251	1.5	152	1.0	169	0.5	292	4.7	277	8.2	283
Fukuoka	30	14.2	84	13.7	84	7.1	84	0.7	278	6.5	259	8.5	258
	100	4.4	337	1.6	7	11.6	261	26.2	260	38.6	261	45.2	262
	300	11.3	268	6.5	259	19.7	258	36.2	261	48.9	260	56.5	262
	500	8.7	261	4.5	243	11.8	255	19.1	264	25.2	266	29.9	270
	700	7.3	253	3.4	225	5.0	254	7.5	277	11.9	276	14.8	281
	850	5.9	227	3.3	191	1.0	204	1.8	308	4.5	288	6.6	291
Kagoshima	30	15.5	83	15.5	84	9.2	84	1.3	77	4.5	261	7.3	259
	100	4.7	20	4.1	61	6.9	260	22.1	258	35.5	261	43.6	262
	300	6.9	272	2.4	253	13.2	258	31.6	261	47.8	260	58.1	261
	500	6.5	260	2.1	229	8.8	251	18.2	262	25.2	263	31.0	267
	700	6.0	251	2.2	185	4.1	244	7.0	272	11.2	273	15.2	278
	850	3.9	230	2.9	136	1.5	157	0.8	315	4.2	292	7.7	296
Naze	30	17.6	84	18.3	85	12.4	85	4.1	84	2.2	257	4.9	259
	100	8.4	56	8.4	73	0.2	207	14.6	256	28.2	259	36.9	260
	300	0.9	302	1.8	87	5.6	261	21.6	263	39.5	261	52.2	259
	500	2.4	248	1.4	124	4.3	240	13.2	261	23.2	260	30.1	263
	700	2.9	221	2.2	129	1.7	211	5.3	270	8.9	269	12.9	274
	850	3.3	209	3.3	133	1.7	121	2.5	14	2.9	338	5.2	318
Ishigakijima	30	19.9	84	20.5	85			7.8	87	1.4	102	1.7	242
	100	13.9	71	12.6	78	6.8	82	5.3	249	17.7	256	26.3	257
	300	3.5	82	2.8	63	1.1	10	10.4	271	24.9	261	36.9	260
	500	2.5	127	2.1	90	0.8	90	5.4	265	15.3	257	23.6	261
	700	2.7	158	1.7	111	1.1	80	1.2	312	4.5	256	8.3	264
	850	4.1	167	2.3	130	2.8	73	4.6	49	4.3	54	3.3	35
Minamidaitojima	30	20.0	84	20.2	85	14.8	85	6.8	85	0.3	135	2.4	253
	100	11.6	67	11.0	76	5.1	86	7.4	248	20.4	256	29.2	258
	300	2.5	81	3.1	84	0.6	231	12.5	262	27.9	261	41.3	260
	500	1.2	125	2.6	99	1.4	176	7.3	255	18.0	257	26.9	260
	700	2.1	151	2.9	112	1.9	137	2.4	251	6.9	260	11.2	268
	850	2.5	164	3.4	118	2.8	111	2.2	63	2.1	31	3.1	341
Chichijima	30	18.8	85	19.4	86	14.1	86			0.8	83	1.6	255
	100	9.8	61	10.3	67	5.3	80	7.1	249	21.2	259	31.5	259
	300	1.8	34	2.4	68	1.1	131	11.1	254	27.5	262	43.2	262
	500	0.1	270	2.6	99	2.4	150	7.5	241	18.4	255	28.9	260
	700	0.9	180	2.8	122	2.7	146	4.0	227	8.6	251	14.1	261
	850	1.5	188	3.5	130	3.3	135	2.1	166	1.1	243	4.3	283
Minamitorishima	30	20.5	86	20.7	86	16.4	86	9.1	84	2.4	81	1.7	65
	100	9.7	61	9.5	64	8.7	76	3.1	68	8.1	271	17.7	261
	300	3.4	52	2.6	36	3.8	81	0.4	56	11.5	276	24.0	270
	500	2.9	88	2.3	85	4.4	95	2.6	119	5.8	262	16.4	263
	700	3.1	115	2.7	118	4.5	100	3.6	109	1.0	233	8.1	259
	850	3.6	126	3.5	125	5.1	105	4.9	108	3.2	103	1.2	239

Aerological Observation Climatological Normals; Annual Average Values Observed at 21:00 (JST)

(1981–2010 Averages)

Station	Air pressure (hPa)	Geopotential height (m)	Air temperature (℃)	Relative humidity (%)	Resultant wind direction (degrees)	Resultant wind magnitude (m)
Wakkanai	30	23856	-50.8			
	100	16117	-55.0		263	19.7
	300	9017	-46.2		267	22.4
	500	5468	-24.2		271	13.2
	700	2939	-9.3	54	273	7.9
	850	1413	-1.0	67	269	5.2
Sapporo	30	23869	-51.6			
	100	16174	-56.9		264	22.3
	300	9087	-44.6		266	27.5
	500	5511	-22.4		269	16.1
	700	2964	-7.5	50	276	9.3
	850	1429	0.4	70	270	5.4
Nemuro	30	23878	-51.5			
	100	16179	-56.6		264	21.7
	300	9084	-44.3		264	28.0
	500	5505	-22.3		267	16.3
	700	2958	-7.4	50	273	9.1
	850	1423	0.2	66	277	5.1
Akita	30	23880	-52.5			
	100	16263	-60.2		266	25.6
	300	9204	-41.7		265	32.9
	500	5580	-19.5	39	268	19.6
	700	3007	-5.0	52	267	12.1
	850	1457	3.1	71	260	7.0
Wajima	30	23879	-53.3			
	100	16329	-62.8		265	26.1
	300	9286	-40.1		264	35.5
	500	5631	-17.3	38	268	19.8
	700	3036	-2.9	50	265	10.5
	850	1474	5.3	69	257	6.4
Tateno	30	23885	-53.6		225	0.7
	100	16369	-64.6		266	26.2
	300	9348	-38.6		263	36.8
	500	5666	-15.3	38	264	20.7
	700	3051	-1.1	52	267	9.5
	850	1477	7.1	68	255	3.5
Hachijojima	30	23888	-54.3		94	1.3
	100	16447	-68.5		266	25.3
	300	9458	-35.7		263	37.1
	500	5729	-12.1	40	264	21.5
	700	3083	2.3	45	264	11.0
	850	1492	9.8	67	263	5.7
Yonago	30	23880	-53.8			
	100	16380	-65.1		267	26.4
	300	9357	-38.5		265	35.7
	500	5672	-15.0	37	269	19.9
	700	3056	-0.9	46	276	9.8
	850	1484	6.5	71	264	4.8

(1981-2010 Averages) Continued.

Station	Air pressure (hPa)	Geopotential height (m)	Air temperature (℃)	Relative humidity (%)	Resultant wind direction (degrees)	Resultant wind magnitude (m)
Shionomisaki	30	23887	-54.2		90	0.6
	100	16437	-67.8		266	25.8
	300	9438	-36.2		263	36.8
	500	5718	-12.8	39	266	20.8
	700	3078	1.6	44	268	10.2
	850	1491	8.9	64	274	3.9
Fukuoka	30	23885	-54.1		135	0.1
	100	16428	-67.5		267	26.8
	300	9422	-36.8		264	36.4
	500	5712	-13.2	38	269	20.0
	700	3076	1.1	42	274	9.6
	850	1493	8.1	64	267	3.6
Kagoshima	30	23882	-54.6		90	1.3
	100	16479	-70.1		267	24.6
	300	9492	-34.8		264	34.8
	500	5750	-11.2	40	266	19.5
	700	3094	3.2	45	271	9.8
	850	1499	9.9	64	277	3.2
Naze	30					
	100	16541	-73.8		267	18.9
	300	9582	-32.4		264	29.2
	500	5800	-8.7	43	264	17.4
	700	3119	5.5	52	266	8.6
	850	1510	12.2	72	283	2.8
Ishigakijima	30					
	100	16597	-77.2		267	10.4
	300	9658	-30.8	30	264	19.7
	500	5844	-6.4	40	261	12.6
	700	3141	7.8	58	256	5.0
	850	1516	14.7	77	121	0.6
Minamidaitojima	30	23869	-55.4			
	100	16578	-76.2		267	13.3
	300	9641	-31.4	31	263	22.6
	500	5837	-7.0	40	261	14.6
	700	3138	7.4	53	262	6.9
	850	1518	14.0	73	270	1.4
Chichijima	30					
	100	16568	-75.0		267	15.3
	300	9633	-32.0	30	265	23.9
	500	5837	-7.4	38	261	15.5
	700	3143	7.2	48	258	8.1
	850	1524	13.8	71	252	3.2
Minamitorishima	30	23865	-55.8			
	100	16593	-76.3		277	7.7
	300	9676	-32.0	28	275	13.2
	500	5872	-6.4	33	265	7.8
	700	3166	8.7	44	252	3.2
	850	1538	15.3	73	146	1.1

Aeronomy

Balloons, wind profilers, satellites and lidar are used to observe data such as temperature, atmospheric pressure, and density in atmospheres which reach several hundred kilometers (km) from the earth. The observed data are then used to calculate the temperature distribution by altitude. Based on the distribution, the atmosphere is divided into several different layers which are named as shown in Figure 3. Here, a table and figure are shown regarding the U.S. Standard Atmosphere 1976 which was determined in America based on these observed values.

The table uses the standard atmosphere listed above to assign values of 0to 1,000 km to temperature (K), atmospheric pressure (hPa), and density (kg/m^3) for the geometric altitude (km). Singular points marked with an asterisk (*) in the table represent densities which are a border where the variation rate differs according to the altitude of temperature. Figure 4 shows the altitude distribution for temperature and density.

Note that significant changes in aeronomy are caused by latitude and season.

† World Meteorological Organization, CAe-III (3rd Session of the Commission for Aerology) Recommendation 17, 1961, Rome

Figure 3 Atmosphere designations

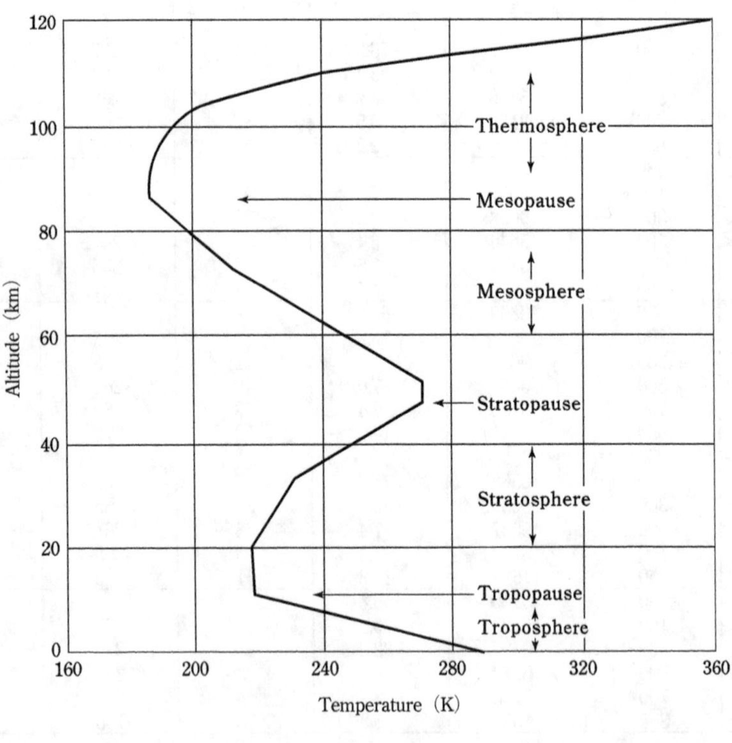

Air Temperature, Atmospheric Pressure, Density Height Distribution

(1976, U.S. Standard Atmosphere)

Altitude Z (km)	Air temperature T (K)	Pressure P (hPa)**	Density ρ (kg/m³)**	Altitude Z (km)	Air temperature T (K)	Pressure P (hPa)**	Density ρ (kg/m³)**
0	288.150	1.01325 +3	1.2250 +0	*47.4	270.650	1.1022 +0	1.4187 -3
1	281.651	8.9876 +2	1.1117	50	270.650	7.9779 -1	1.0269
2	275.154	7.9501	1.0066	*51.0	270.650	7.0458	9.0690 -4
3	268.659	7.0121	9.0925 -1	50	260.771	4.2525	5.6810
4	262.166	6.1660	8.1935	60	247.021	2.1958	3.0968
5	255.676	5.4048	7.3643	65	233.292	1.0929	1.6321
6	249.187	4.7217	6.6011	70	219.585	5.2209 -2	8.2829 -5
7	242.700	4.1105	5.9002	*72.0	214.263	3.8362	6.2374
8	236.215	3.5651	5.2579	75	208.399	2.3881	3.9921
9	229.733	3.0800	4.6706	80	198.639	1.0524	1.8458
10	223.252	2.6499	4.1351	*86.0	186.87	3.7338 -3	6.958 -6
11	216.774	2.2699	3.6480	90	186.87	1.8359	3.416
*11.1	216.650	2.2346	3.5932	*91.0	186.87	1.5381	2.860
12	216.650	1.9399	3.1194	100	195.08	3.2011 -4	5.604 -7
13	216.650	1.6579	2.6660	110	240.00	7.1042 -5	9.708 -8
14	216.650	1.4170	2.2786	120	360.00	2.5382	2.222
15	216.650	1.2111	1.9476	130	469.27	1.2505	8.152 -9
16	216.650	1.0352	1.6647	140	559.63	7.2028 -6	3.831
17	216.650	8.8497 +1	1.4230	160	696.29	3.0395	1.233
18	216.650	7.5652	1.2165	180	790.07	1.5271	5.194 -10
19	216.650	6.4674	1.0400	200	854.56	8.4736 -7	2.541
*20.0	216.650	5.5293	8.8910 -2	250	941.33	2.4767	6.073 -11
21	217.581	4.7289	7.5715	300	976.01	8.7704 -8	1.916
22	218.574	4.0475	6.4510	350	990.06	3.4498	7.014 -12
23	219.567	3.4668	5.5006	400	995.83	1.4518	2.803
24	220.560	2.9717	4.6938	450	998.22	6.4468 -9	1.184
25	221.552	2.5492	4.0084	500	999.24	3.0236	5.215 -13
26	222.544	2.1883	3.4257	550	999.67	1.5137	2.384
27	223.536	1.8799	2.9298	600	999.85	8.2130 -10	1.137
28	224.527	1.6161	2.5076	650	999.93	4.8865	5.712 -14
29	225.518	1.3904	2.1478	700	999.97	3.1908	3.070
30	226.509	1.1970	1.8410	750	999.99	2.2599	1.788
*32.2	228.756	8.6314 +0	1.3145	800	999.99	1.7036	1.136
35	236.513	5.7459	8.4634 -3	850	1000.00	1.3415	7.824 -15
40	250.350	2.8714	3.9957	900	1000.00	1.0873	5.759
45	264.164	1.4910	1.9663	1000	1000.00	7.5138 -11	3.561

* Transition point.

** The values +3, +2, ⋯⋯, −15, ⋯⋯ are powers of 10.

Figure 4 Temperature, density altitudinal distribution

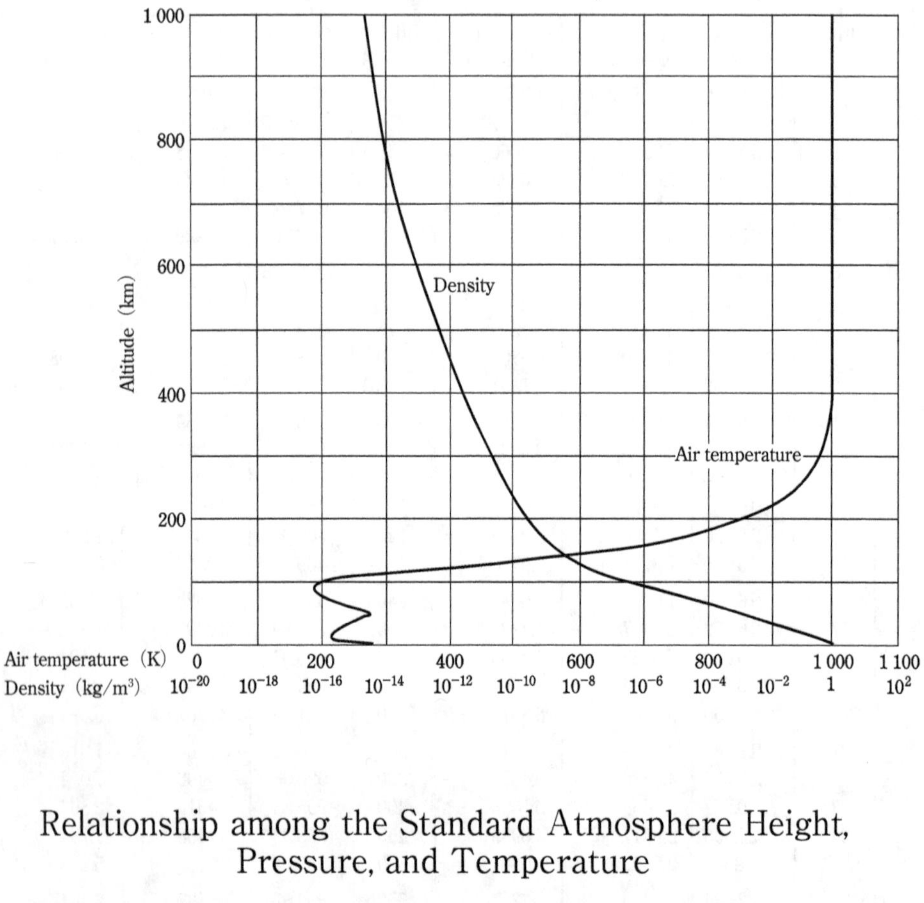

Relationship among the Standard Atmosphere Height, Pressure, and Temperature

Standard atmosphere is one type of atmospheric model. A well-known international standard atmosphere is the ICAO Standard Atmosphere which was adopted by the International Civil Aviation Organization (ICAO) in 1964. This is the same as the lower area (up to an altitude of 32 km) of the U.S. Standard Atmosphere 1976. The ICAO Standard Atmosphere is represented by three straight temperature lines in relation to geopotential altitude. In other words, the following values are used:

Ground pressure, temperature: 1013.25 hPa, 15.0℃

Temperature change rate until an altitude of 11 km: −6.5℃/km

Tropopause altitude, pressure, temperature: 11 km, 226.32 hPa, -56.5℃
Temperature change rate until an altitude of 20 km: 0.0℃/km
Temperature change rate until an altitude of 32 km: +1.0℃/km

The height used is the geopotential altitude determined at the 11th WMO World Meteorological Congress (1991.5). At locations with standard gravity (9.80665 m/s^2), this value matches the geometric altitude value. At other locations, it is nearly equivalent to the geometric altitude value. The minus (-) symbol in the m column (geopotential altitude) refers to altitudes below sea level.

(m)	Air temperature (℃)	Pressure (hPa)	(m)	Air temperature (℃)	Pressure (hPa)	(m)	Air temperature (℃)	Pressure (hPa)
-400	17.6	1062.2	4800	-16.2	554.8	10000	-50.0	264.4
-200	16.3	1037.5	5000	-17.5	540.2	10200	-51.3	256.4
0	15.0	1013.3	5200	-18.8	525.9	10400	-52.6	248.6
200	13.7	989.5	5400	-20.1	511.9	10600	-53.9	241.0
400	12.4	966.1	5600	-21.4	498.3	10800	-55.2	233.5
600	11.1	943.2	5800	-22.7	484.9	11000	-56.5	226.3
800	9.8	920.8	6000	-24.0	471.8	11500	-56.5	209.2
1000	8.5	898.7	6200	-25.3	459.0	12000	-56.5	193.3
1200	7.2	877.2	6400	-26.6	446.5	12500	-56.5	178.6
1400	5.9	856.0	6600	-27.9	434.3	13000	-56.5	165.1
1600	4.6	835.2	6800	-29.2	422.3	13500	-56.5	152.6
1800	3.3	814.9	7000	-30.5	410.6	14000	-56.5	141.0
2000	2.0	795.0	7200	-31.8	399.2	14500	-56.5	130.3
2200	0.7	775.4	7400	-33.1	388.0	15000	-56.5	120.4
2400	-0.6	756.3	7600	-34.4	377.1	15500	-56.5	111.3
2600	-1.9	737.5	7800	-35.7	366.4	16000	-56.5	102.9
2800	-3.2	719.1	8000	-37.0	356.0	17000	-56.5	87.9
3000	-4.5	701.1	8200	-38.3	345.8	18000	-56.5	75.0
3200	-5.8	683.4	8400	-39.6	335.9	19000	-56.5	64.1
3400	-7.1	666.2	8600	-40.9	326.2	20000	-56.5	54.7
3600	-8.4	649.2	8800	-42.2	316.7	22000	-54.5	40.0
3800	-9.7	632.6	9000	-43.5	307.4	24000	-52.5	29.3
4000	-11.0	616.4	9200	-44.8	298.4	26000	-50.5	21.5
4200	-12.3	600.5	9400	-46.1	289.6	28000	-48.5	15.9
4400	-13.6	584.9	9600	-47.4	281.0	30000	-46.5	11.7
4600	-14.9	569.7	9800	-48.7	272.6	32000	-44.5	8.7

Reduction of Atmospheric Pressure to Sea Level

The observed value for atmospheric pressure is dependent on the altitude of the observation station. Therefore, in order to compare the observed value at one station with the value at another station, it is necessary to use values at the same altitude. Consequently, in Japan, altitudes are converted to the average sea level value of Tokyo Bay. This is referred to as "reduction to sea level."

The following explanation of reduction to sea level assumes a temperature decrease rate of 0.5 (℃/100 m) and ignores the effects of humidity. When the observed value for atmospheric pressure is P (hPa), the barometer altitude based on the average sea level of Tokyo Bay is h (m), and the temperature at the sites is (℃), the average temperature from the sea level to the observation point is calculated as follows: $t = th + 0.0025h$ (℃). At this time, sea level reduction value is calculated as $\Delta P = P(\exp(gh/R(t+273.15))-1)$. (gravitational acceleration g = 9.80665 m/s^2, gas constant of dry air R = 287.05 (Jkg^{-1}K^{-1})) The table shows the sea level reduction value ΔP when $P = 1000$ (hPa).

h (m) \ t (℃)	-10	-5	0	5	10	15	20	25	30
	hPa	hPa	hPa	hPa	hPa	hPa	hPa	hPa	hPa
10	1.3	1.3	1.3	1.2	1.2	1.2	1.2	1.1	1.1
20	2.6	2.6	2.5	2.5	2.4	2.4	2.3	2.3	2.3
30	3.9	3.8	3.8	3.7	3.6	3.6	3.5	3.4	3.4
40	5.2	5.1	5.0	4.9	4.8	4.8	4.7	4.6	4.5
50	6.5	6.4	6.3	6.2	6.0	5.9	5.8	5.7	5.6
60	7.8	7.7	7.5	7.4	7.3	7.1	7.0	6.9	6.8
70	9.1	9.0	8.8	8.6	8.5	8.3	8.2	8.0	7.9
80	10.4	10.2	10.0	9.9	9.7	9.5	9.4	9.2	9.1
90	11.7	11.5	11.3	11.1	10.9	10.7	10.5	10.4	10.2
100	13.1	12.8	12.6	12.3	12.1	11.9	11.7	11.5	11.3

mmHg to hPa Conversion Chart

Atmospheric pressure in mmHg can be converted to hPa by using the following equation.

$$1 \text{ mmHg} = 13.5951 \times 980.665 \times 10^{-4} \text{ hPa}$$

mmHg	0	1	2	3	4	5	6	7	8	9
	hPa	hPa	hPa	hPa	hPa	hPa	hPa	hPa	hPa	hPa
690	919.9	921.3	922.6	923.9	925.3	926.6	927.9	929.3	930.6	931.9
700	933.3	934.6	935.9	937.3	938.6	939.9	941.3	942.6	943.9	945.3
710	946.6	947.9	949.3	950.6	951.9	953.3	954.6	955.9	957.3	958.6
720	959.9	961.3	962.6	963.9	965.3	966.6	967.9	969.3	970.6	971.9
730	973.3	974.6	975.9	977.3	978.6	979.9	981.3	982.6	983.9	985.3
740	986.6	987.9	989.3	990.6	991.9	993.3	994.6	995.9	997.3	998.6
750	999.9	1001.3	1002.6	1003.9	1005.3	1006.6	1007.9	1009.3	1010.6	1011.9
760	1013.3	1014.6	1015.9	1017.2	1018.6	1019.9	1021.2	1022.6	1023.9	1025.2
770	1026.6	1027.9	1029.2	1030.6	1031.9	1033.2	1034.6	1035.9	1037.2	1038.6
780	1039.9	1041.2	1042.6	1043.9	1045.2	1046.6	1047.9	1049.2	1050.6	1051.9
790	1053.2	1054.6	1055.9	1057.2	1058.6	1059.9	1061.2	1062.6	1063.9	1065.2
800	1066.6	1067.9	1069.2	1070.6	1071.9	1073.2	1074.6	1075.9	1077.2	1078.6

mmHg 10 fractions	1	2	3	4	5	6	7	8	9
hPa	0.1	0.3	0.4	0.5	0.7	0.8	0.9	1.1	1.2

Humidity Determined by Ventilated Psychrometers

When the wet-bulb is not frozen

Dry-bulb (t)	Difference between dry-bulb and wet-bulb (t–t')																				
	0.0	0.5	1.0	1.5	2.0	2.5	3.0	3.5	4.0	4.5	5.0	5.5	6.0	6.5	7.0	7.5	8.0	8.5	9.0	9.5	10.0
40	100	97	94	91	88	85	82	80	77	74	72	69	67	64	62	59	57	55	53	51	48
35	100	97	93	90	87	84	81	78	75	72	69	67	64	61	59	56	54	51	49	47	44
30	100	96	93	89	86	83	79	76	73	70	67	64	61	58	55	52	50	47	44	42	39
25	100	96	92	88	84	81	77	74	70	67	63	60	57	53	50	47	44	41	38	35	33
20	100	96	91	87	83	78	74	70	66	62	59	55	51	48	44	40	37	34	30	27	24
15	100	95	90	85	80	75	70	66	61	57	52	48	44	39	35	31	27	23	19	16	12
10	100	94	88	82	76	71	65	60	54	49	44	39	33	28	23	19	14	9	4		
5	100	93	86	78	71	65	58	51	45	38	32	25	19	13	7	1					
0	100	91	82	73	64	56	47	39	30	22	14	6									
-5	100	88	77	65	54	43	32	21	10												
-10	100	84	69	54	38	23	8														

When the wet-bulb is frozen

Dry-bulb (t)	Difference between dry-bulb and wet-bulb (t–t')												
	0.0	0.5	1.0	1.5	2.0	2.5	3.0	3.5	4.0	4.5	5.0	5.5	6.0
0	100	91	82	74	65	57	49	41	33	25	17	10	2
-5	95	84	73	63	52	42	31	21	11	1			
-10	91	76	62	48	35	21	7						
-15	86	67	48	29	10								
-20	82	55	28	1									

Note) The values in this chart were calculated using an atmospheric pressure of 1013.25 hPa in the formula for relative humidity.

Humidity Determined by Psychrometers

When the wet-bulb is not frozen

Dry-bulb (t)	Difference between dry-bulb and wet-bulb (t–t')																				
	0.0	0.5	1.0	1.5	2.0	2.5	3.0	3.5	4.0	4.5	5.0	5.5	6.0	6.5	7.0	7.5	8.0	8.5	9.0	9.5	10.0
40	100	97	94	91	88	85	82	79	76	73	71	68	66	63	60	58	56	53	51	49	47
35	100	97	93	90	87	83	80	77	74	71	68	65	63	60	57	54	52	49	47	44	42
30	100	96	92	89	85	82	78	75	72	68	65	62	59	56	53	50	47	44	41	39	36
25	100	96	92	88	84	80	76	72	68	65	61	57	54	51	47	44	41	37	34	31	28
20	100	95	91	86	81	77	72	68	64	60	56	52	48	44	40	36	32	29	25	21	18
15	100	94	89	84	78	73	68	63	58	53	48	43	39	34	30	25	21	16	12	8	4
10	100	93	87	80	74	68	62	56	50	44	38	32	27	21	15	10	5				
5	100	92	84	76	68	61	53	46	38	31	24	16	9	2							
0	100	90	80	70	60	50	40	31	21	12	3										
-5	100	87	73	60	47	34	22	9													
-10	100	82	64	46	29	11															

When the wet-bulb is frozen

Dry-bulb (t)	Difference between dry-bulb and wet-bulb (t–t')										
	0.0	0.5	1.0	1.5	2.0	2.5	3.0	3.5	4.0	4.5	5.0
0	100	90	80	71	61	52	43	34	25	16	8
-5	95	83	71	58	47	35	23	11			
-10	91	74	58	42	26	11					
-15	86	64	42	20							
-20	82	50	18								

Note) The values in this chart were calculated using an atmospheric pressure of 1013.25 hPa in the formula for relative humidity.

Figure 5 Tracks of Historic Typhoons and Major Recent Typhoons

Number	Typhoon name	Date of landfall/ closest approach	Dead or Missing people
①	Muroto	1934. 9. 21	3 036
②	Makurazaki	1945. 9. 17	3 756
③	Kathleen	1947. 9. 15	1 930
④	Toya Maru	1954. 9. 26	1 761
⑤	Isewan	1959. 9. 26	5 098

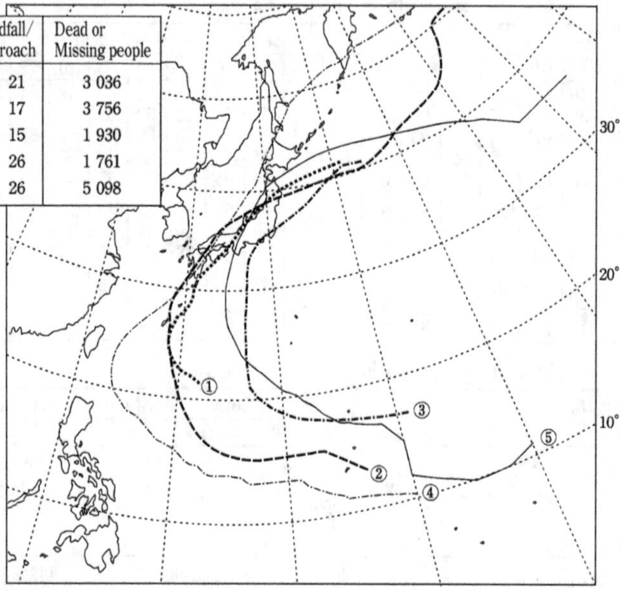

Number	Typhoon name	Date of landfall/ closest approach	Dead or Missing people
⑥	1982 No.10	1982. 8. 2	95
⑦	1983 No.10	1983. 8. 28	44
⑧	1990 No.19	1990. 9. 19	40
⑨	1991 No.19	1991. 9. 27	62
⑩	1993 No.13	1993. 9. 3	48
⑪	2004 No.18	2004. 9. 7	47
⑫	2004 No.23	2004. 10. 20	99
⑬	2011 No.12	2011. 9. 3	98
⑭	2013 No.26	2013. 10. 16	43
⑮	2019 Eastern Japan typhoon	2019. 10. 12	107*

* Include the damage caused by the heavy rain from October 25.

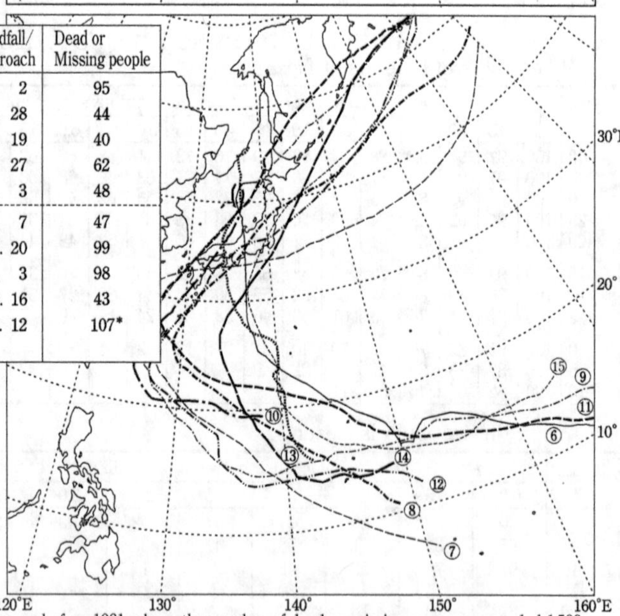

The top chart is the tracks of typhoons before 1981 where the number of dead or missing persons exceeded 1,500. The bottom chart is the tracks of typhoons after 1982 where the number of dead or missing persons exceeded 40.

Typhoons

The term typhoon is applied to tropical cyclone where the maximum sustained wind speed exceeds 17.2 m/s in the northwestern Pacific Ocean (including the norther Pacific Ocean west of 180° East Longitude and the South China Sea with its attached seas).

Number of typhoons that form/pass-close/make landfall each month (1981-2010 Averages)

	Jan.	Feb.	Mar.	Apr.	May	Jun.	Jul.	Aug.	Sep.	Oct.	Nov.	Dec.	Annual
Number that form	0.3	0.1	0.3	0.6	1.1	1.7	3.6	5.9	4.8	3.6	2.3	1.2	25.6
Number of approach	-	-	-	0.2	0.6	0.8	2.1	3.4	2.9	1.5	0.6	0.1	11.4
Number of landfall	-	-	-	-	0.0	0.2	0.5	0.9	0.8	0.2	0.0	-	2.7

Note) Number of landfall: Number of typhoons where the center reached the coastline of Hokkaido, Honshu, Shikoku, and Kyushu
Number of approach: Number of typhoons where the center came within 300 km of the Japanese coastline.
In cases where a single typhoon extended into two different months, it is counted as 1 typhoon in each month.

Number of typhoons that approach annually by region (1981-2010 Averages)

Region	Number of approach (per year)	Region	Number of approach (per year)	Region	Number of approach (per year)
Hokkaido	1.8	Kinki	3.2	Amami	3.8
Tohoku	2.6	Chugoku	2.6	Okinawa	7.4
Hokuriku	2.5	Shikoku	3.1	Izu, Ogasawara	5.4
Kanto Koshin	3.1	Northern Kyushu	3.2		
Tokai	3.3	Southern Kyushu	3.3		

Note) Yamaguchi Prefecture is included in Northern Kyushu, not in Chugoku.
Southern Kyushu does not include Amamigunto.
Kanto-Koshin does not include Izu or the Ogasawara Islands.

Categories of wind strength and size (Japan Meteorolgical Agency)

(1) Strength categories

Category	Maximum wind speed (m/s)
(No designation)	< 33
Strong	33 - 44
Very strong	44 - 54
Violent	54 <

(2) Size Categories

Category	Radius (km) with wind speeds over 15 m/s
(No designation)	< 500
Large	500 - 800
Very large	800 <

Normal Value for Start/End of Rainy Season

The climatological normals for the start of the rainy season and the end of the rainy season as determined by the JMA are listed with consideration given to average weather development in regions throughout Japan. On average, the start/end of the rainy season vary by about 5 days. Therefore, the expression "around (date)" is used. The climatological Normal is an average value for the 30-year period from 1981 to 2010. However, these normal values do not include statistics for years in which it was not possible to identify the start/end of the rainy season.

Local constituency min	Northern Tohoku	Southern Tohoku	Kanto Koshin	Hokuriku	Tokai	Kinki
Beginning of rainy season	June 14	June 12	June 8	June 12	June 8	June 7
End of rainy season	July 28	July 25	July 21	July 24	July 21	July 21
Local constituency min	Chugoku	Shikoku	Northern Kyushu	Southern Kyushu	Amami	Okinawa
Beginning of rainy season	June 7	June 5	June 5	May 31	May 11	May 9
End of rainy season	July 21	July 18	July 19	July 14	June 29	June 23

Weather-Related

This table is a summary from 1998 onward, which is based on the in-
Research on the Epidemiology of Disasters (CRED) or reports by govern-

Year	Asia		
	East Asia and Eastern Siberia	Southeast Asia	South Asia to Middle East
1998	**Heavy rain, Flood:** China, Korea (May - Aug.) **Wildfire:** Sakhalin (Jul. - Oct.)	**Drought:** Philippines (Apr. - Dec.) **Flood:** Indonesia (Aug.) **Tropical Storm:** Philippines (Oct.), Vietnam (Nov.)	**Cold wave:** Bangladesh (Jan.), India (Dec.) **Heavy rain, Flood:** India (Mar., Aug. - Sep.), Pakistan (Mar.), Turkey (May), Nepal (Jul. - Aug.), Bangladesh (Jul. - Sep.), Sri Lanka (Jul., Dec. - Jan. 1999) **Heat wave:** India (May) **Cyclone:** India (Jun.), Bangladesh (Nov.) **Tropical Storm:** India (Oct.)
1999	**Drought:** Northern China (Mar. - Dec.) **Heavy rain, Flood:** Southern China (Jun. - Sep.) **Typhoon, Tropical Storm:** Southeastern China (Sep.)	**Drought:** Thailand, Vietnam (All year round) **Heavy rain, Flood:** Southeast Asia (Jul. - Aug., Oct. - Dec.)	**Heat wave:** India (Apr.) **Drought:** Iran (Apr. - Dec.), Pakistan (Nov. - Dec.) **Heavy rain, Flood:** Sri Lanka (Apr.), Nepal (Jun.), Bangladesh (Jul. - Aug.), India (Jul. - Sep.) **Cyclone:** Pakistan (May), India (May, Oct.)
2000	**Heavy snow, Storm:** Mongolia (Jan. - Apr., Dec.) **Drought:** Northen and Eastern China (Mar. - Dec.) **Heavy rain, Flood:** Central to Southern China (Jun. - Aug., Oct.), Japan (Sep.)	**Heavy rain, Flood, Landslide:** Philippines (Jan. - Feb.), Indochina Peninsula (Jul. - Nov.), Indonesia (Nov. - Dec.)	**Cold wave:** India to Bangladesh (Jan.) **Drought:** Pakistan to Iran (All year round), Northern India, Tajikistan (Apr. - Dec.) **Heavy rain, Flood:** Bangladesh (Jun., Sep. - Oct.), India (Jul. - Oct.), Sri Lanka (Sep., Nov.)
2001	**Heavy snow, Storm:** China (Jan. - Feb.) **Drought:** Southen China (Mar. - Dec.) **Heavy rain:** Southeastern China (Jun.) **Typhoon:** Southeastern China (Jul.) **Flood:** Siberia (Jul.)	**Heavy rain:** Indonesia (Feb., Jul.) **Typhoon:** Philippines (Jul., Nov.) **Flood:** Indochina Peninsula (Aug. - Nov.)	**Cold wave:** India (Jan.), Afghanistan (Jan. - Feb., Nov.) **Drought:** Northern India to Iran (All year round), Sri Lanka (Sep. - Dec.) **Heavy rain, Flood:** Bangladesh (Jun., Aug.), India (Jun. - Sep.), Pakistan (Jul.), Iran (Aug.)
2002	**Drought:** China (Apr. - Oct.) **Heavy rain, Flood:** Southern China (May - Jun., Aug.) **Heavy snow, Storm:** Mongolia (Dec. - Jan. 2003), China (Dec.)	**Heavy rain, Flood:** Indonesia (Jan. - Feb.), Indochina Peninsula (Aug. - Nov.) **Drought:** Thailand (Feb. - Dec.), Vietnam (May - Dec.)	**Drought:** Pakistan to Afghanistan, Sri Lanka (All year round), India (Jul. - Aug.) **Heat wave:** India, Pakistan (May) **Heavy rain, Flood, Landslide:** Nepal, India, Bangladesh (Jun. - Aug.), Iran (Aug.), Sri Lanka (Dec.) **Cold wave:** In and around Northern India (Dec.)
2003	**Drought:** Northern China (Jan.), Southeastern China (Jul. - Aug.) **Heavy rain, Flood:** Eastern and Southeastern China (May - Oct.)	**Heavy rain, Flood:** Indonesia (Mar., Nov. - Dec.), Indochina Peninsula (Oct. - Dec.) **Heavy rain, Flood, Landslide:** Philippines (Dec.)	**Drought:** Pakistan (All year round) **Cold wave:** Bangladesh (Jan.), In and around Northern India (Dec.) **Heat wave:** India, Pakistan (May - Jun.) **Heavy rain, Flood:** Sri Lanka (May), India, Nepal, Pakistan (Jun. - Oct.)

* EM-DAT: The Emergency Events Database - Université catholique de Louvain (UCL) - CRED,

Disasters

ternational disaster database (EM-DAT)* maintained by the Centre for ments or UN agencies.

Western Russia and Europe	Africa	Nothern and Central America	South America	Oceania
Cold wave: Western Russia, Europe (Nov.)	**Heavy rain, Flood:** Zambia (Feb.), Senegal (Jul.), Benin (Aug.), Sudan (Aug. - Sep.), Nigeria (Oct.)	**Heat wave:** USA (May - Jul.) **Hurricane, Heavy rain:** Central America, Caribbean countries (Sep. - Nov.) **Cold wave:** Mexico (Dec.)	**Drought, Wildfire:** Nothern part of South America (- Apr.) **Flood:** Argentina (Apr. - Jun.)	**Drought:** Fiji (All year round)
Heavy rain, Avalanche: Europe (Feb.) **Melting snow flood:** Central Europe (May - Jun.) **Cold wave:** Poland (Oct.), Western Russia (Dec.) **Heavy rain, Storm, Cold wave:** Europe (Nov. - Dec.)	**Heavy rain, Flood:** Northern part of Eastern Africa to Western Africa (Aug. - Oct.) **Drought:** Norhern part of Eastern Africa (Sep. -)	**Heat wave, Drought:** Eastern to Central USA (Jul. - Oct.) **Heavy rain, Flood:** Central America (Sep. - Oct.)	**Heavy rain, Flood:** Peru (Mar., May), Colombia (Oct.), Venezuela (Dec.)	**Flood:** Papua New Guinea (Apr.)
Cold wave: Russia (Oct.)	**Heavy rain:** In and around the Southern part of Eastern Africa (Jan. - Apr.) **Drought:** Northen part of Eastern Africa (All year round)		**Heavy rain, Flood:** Chile (Jun.)	
Cold wave: Russia (Jan. - Feb., Oct.), Eastern Europe (Oct., Dec.) **Melting snow flood:** Eastern Europe (Mar.) **Heat wave:** Western Russia (Jul.)	**Drought:** Northen part of Eastern Africa (All year round), Zimbabwe (Mar. - Dec.), Northern part of Western Africa (Apr. - Dec.) **Heavy rain, Flood:** Southern part of Eastern Africa (Jan. - Apr., Jul.), Western Africa (Jun., Aug. -.Oct.), Algeria (Nov.)		**Heavy rain, Flood:** Bolivia (Jan. - Mar.), Argentina (Oct. - Nov.) **Drought:** Brazil (Jun. - Dec.)	
Heavy rain, Flood: Western Russia (Jun., Aug.), Eastern Europe (Aug.) **Cold wave:** Poland to Western Russia (Oct.)	**Drought:** Eritrea, Kenya, Malawi, Zimbabwe, Northern part of Western Africa (All year round) **Flood:** Senegal (Jan.), Malawi (Jan. - Mar., Dec.), Kenya (Apr. - May)	**Flood:** Southern USA (Jul.)	**Flood:** Chile (May - Jun.)	**Drought:** Australia (Apr. - Nov.) **Typhoon:** Federated States of Micronesia (Jul.)
Heat wave, Drought, Wildfire: Europe (Jun. - Aug.)	**Flood:** Eastern Africa (Jan. - Feb., Apr. - May, Jul. - Aug.), Nigeria (Sep. - Oct.) **Drought:** Eastern Africa, Mauritania (All year round)	**Flood:** Haiti (Dec.)	**Flood:** Argentina (May), Brazil (Dec. - Feb. 2004) **Cold wave:** Peru (Jul. - Aug.)	

D. Guha-Sapir - www.emdat.be, Brussels, Belgium.

Weather-Related Disasters

Year	Asia		
	East Asia and Eastern Siberia	Southeast Asia	South Asia to Middle East
2004	**Heavy rain, Flood:** Eastern and Southern China (Jun. - Jul., Sep.), D.P.R.Korea (Jul.)	**Cyclone:** Myanmar (May) **Flood:** Thailand (Aug.) **Typhoon:** Philippines (Nov.)	**Cold wave:** In and around Northern India (Jan.) **Heavy rain, Flood:** India, Bangladesh, Nepal (Jun. - Oct.), Tajikistan (Jul.), Sri Lanka (Dec.)
2005	**Heavy rain, Flood:** China (Jun. - Oct.), D.P.R.Korea (Jun. - Jul.) **Drought:** Northern and Southern China (Jul. - Oct.) **Typhoon:** Eastern and Southeastern China (Sep.) **Heavy snow:** Japan (Dec. - Jan. 2006)	**Drought:** Vietnam to Cambodia (Apr. - May) **Heavy rain, Flood:** Thailand (Aug., Nov. - Dec.), Philippines (Dec.)	**Heavy rain, Flood, Avalanche:** Afghanistan, Pakistan, India (Jan. - Mar.) **Heat wave:** India, Pakistan (Jun.) **Heavy rain, Flood:** India (Jun. - Dec.), Pakistan (Jun. - Aug.), Bangladesh (Jul.), Sri Lanka (Nov.) **Cold wave, Dense fog:** India, Bangladesh (Dec.)
2006	**Heavy rain, Flood:** China (May - Jul.) **Drought:** China (May - Sep.) **Heavy rain:** Korean Peninsula (Jul.) **Typhoon:** Southeastern China (Jul. - Aug.)	**Heavy rain, Flood:** Indonesia (Jan., Jun., Dec.), Philippines (Jan. - Feb., Dec.), Thailand (May - Jun., Aug. - Dec.), Malaysia (Dec.) **Typhoon:** Vietnam (May), Philippines (Sep. - Oct., Nov. - Dec.)	**Cold wave:** India, Pakistan (Jan.) **Heat wave:** India, Pakistan (May) **Heavy rain, Flood:** India (Jun. - Nov.), Pakistan (Aug.), Bangladesh (Aug. - Sep.), Sri Lanka (Oct. - Nov.), Afghanistan (Nov.) **Drought:** Afghanistan (Jul. - Dec.) **Tropical Depression:** India, Bangladesh (Sep. - Oct.)
2007	**Heavy rain, Flood:** Eastern to Southern China (May - Aug.), D.P.R.Korea (Aug.)	**Heavy rain, Flood:** Indochina Peninsula (Jan., Jul. - Aug., Oct. - Nov.), Indonesia (Feb., Dec.)	**Cold wave, Dense fog:** Bangladesh (Jan.) **Avalanche, Landslide:** Pakistan (Mar.) **Heavy rain, Flood:** Sri Lanka (May, Dec.), Pakistan (Jun.), India (Jun. - Oct.), Afghanistan (Jun. - Oct.), Bangladesh (Jul. - Aug.), Nepal (Jul. - Oct.) **Tropical Storm:** Pakistan (Jun. - Jul.) **Cyclone:** Bangladesh (Nov.)
2008	**Cold wave, Storm:** China to Central Asia (Jan. - Feb.) **Heavy rain:** Central to Southern China (May - Aug., Nov.), Central China (Sep.) **Drought:** China (Nov. - Feb. 2009)	**Drought:** Thailand (Apr. - Dec.) **Cyclone:** Myanmar (May) **Heavy rain, Flood:** Philippines (Feb. - Mar., Aug. - Sep., Nov.), Indochina Peninsula (May, Aug. - Dec.), Indonesia (Sep.) **Typhoon:** Philippines (Jun.)	**Cold wave, Storm:** Afghanistan (Jan. - Feb.) **Heavy rain, Flood:** Sri Lanka (May - Jun., Nov. - Dec.), In and around Northern India (Jun. - Sep.), Southeastern India (Nov. - Dec.)
2009	**Drought:** Kyrgyzstan (All year round), Southeastern China (Oct. - May 2010) **Heavy rain, Flood:** Eastern to Southern China (Apr. - Sep.) **Typhoon:** Taiwan (Aug.)	**Flood:** Philippines (Jan., Jul. - Aug.), Vietnam (Jul.), Thailand (Nov.) **Typhoon:** Philippines, Vietnam (Sep. - Oct.)	**Heat wave:** India (Apr. - Jun.) **Cyclone:** Bangladesh, India, Bhutan (May) **Heavy rain, Flood:** India (Jul. - Oct.), Bangladesh (Jul.), Nepal (Oct.), Sri Lanka (Dec.) **Cold wave:** Bangladesh (Dec. - Jan. 2010)

Continued.

Western Russia and Europe	Africa	Nothern and Central America	South America	Oceania
Flood: Bosnia and Herzegovina (Apr.), Macedonia (Jun.)	**Drought:** Northen part of Eastern Africa, South Africa (All year round) **Flood:** Angola (Jan. - Feb.), Zambia (Feb. - May) **Cyclone:** Madagascar (Mar.)	**Heavy rain:** Haiti, Dominican Republic (May) **Wildfire:** Southwestern USA (May) **Hurricane:** Haiti (Sep.)	**Flood:** Colombia (Jan. - Jun., Oct.)	
Cold wave, storm: Poland (Oct. - Dec.)	**Drought:** Mali (All year round), Eastern Africa, Niger (Latter half of the year) **Heavy rain:** Ethiopia (Apr.) **Flood:** Sudan (Aug. - Sep.)	**Hurricane:** Southern USA (Aug. - Sep.), Central America (Oct.)	**Heavy rain, Flood:** Guyana (Jan. - Feb.), Colombia (Sep. - Nov.)	
Cold wave, Storm: Western Russia to Eastern Europe (Jan.) **Heat wave:** Europe (Jun. - Jul.)	**Drought:** Eastern Africa, Mali (All year round), Niger (- Jul.) **Heavy rain:** In and around the Northern part of Eastern Africa (Aug. - Dec.)	**Heat wave:** Southwestern USA (Jul. - Aug.)	**Heavy rain, Flood:** Colombia, Bolivia (Jan. - Apr.), Brazil (Mar. - Apr.)	**Drought:** Australia (Oct. - Dec.)
Heat wave, Wildfire: Southeastern Europe (Jun. - Aug.) **Flood:** UK (Jul.)	**Heavy rain, Flood:** Southern part of Eastern Africa (Jan. - May, Dec.), Tropical Africa (Jul. - Oct.) **Drought:** Rwanda, Zimbabwe (All year round), Southern part of Eastern Africa (Latter half of the year)	**Heavy rain:** Central America (Sep. - Nov.) **Wildfire:** Southwestern USA (Oct.)	**Flood:** Bolivia (Jan. - May, Dec. - Feb. 2008), Brazil (Jan.), Colombia (Mar. - Oct.), Uruguay (May) **Wildfire:** Paraguay (Sep.) **Drought:** Northeastern Brazil (Oct. - Dec.)	**Cyclone:** Papua New Guinea (Nov.)
Heavy rain, Flood: In and around Ukraine (Jul.)	**Drought:** Syria, Zimbabwe, Nortern part of Eastern Africa (All year round) **Heavy rain, Flood:** Angola to Namibia (Jan. - Mar.), Eastern part of Western Africa (Jul. - Aug.)	**Heavy rain, Flood:** In and around the Midwest of USA (Jun.), Central America (Oct. - Nov.) **Hurricane:** Haiti (Aug. - Sep.)	**Heavy rain, Flood:** Northwestern part of South America (Jan. - Mar., Sep. - Dec.), Brazil (Mar. - Apr.), Chile (Aug. - Sep.)	**Heavy rain, Flood:** Papua New Guinea (Dec.)
Cold wave: Eastern to Central Europe (Nov. - Jan. 2010)	**Drought:** Somalia (- Apr.), Syria, Sudan to Kenya, Zimbabwe (All year round), Niger (Sep. - Mar. 2010) **Heavy rain, Flood:** South part of Eastern Africa to Southern Africa (Mar. - Apr.), Western Africa (Aug. - Sep.), Sudan (Aug.), Saudi Arabia (Nov.)	**Hurricane:** Central America (Nov.) **Flood:** Mexico (Nov.)	**Drought:** Guatemala (Mar. - Dec.) **Flood:** Brazil (Apr. - May, Nov.), Peru (Dec. - Mar. 2010) **Cold wave:** Peru (May - Aug.)	**Heat wave, Wildfire:** Southeastern Australia (Jan. - Feb.)

Weather-Related Disasters

Year	Asia		
	East Asia and Eastern Siberia	Southeast Asia	South Asia to Middle East
2010	**Heavy rain:** China (May - Aug.), Southern China (Oct.) **Heat wave:** Japan (Jul. - Sep.) **Drought:** Central China (Dec. - May 2011)	**Drought:** Thailand (Mar. - Mar. 2011) **Heavy rain, Flood:** Philippines (Jun., Dec. - Jan. 2011), Indochina Peninsula (Oct. - Dec.), Indonesia (Oct.)	**Cold wave:** India (Jan.) **Avalanche:** Afghanistan (Feb.) **Heat wave:** India (Mar. - May) **Heavy rain, Flood:** Sri Lanka (May), Pakistan (Jul. - Aug.), India (Jul. - Sep.), Bangladesh (Oct.), Southern India to Sri Lanka (Nov. - Dec.)
2011	**Heavy rain:** Central China (Jun.) **Flood:** China (Jul. - Sep.)	**Flood:** Philippines (Jan. - Mar., Jun., Dec.), Thailand (Mar. - Apr.), Indochina Peninsula (Aug. - Dec.) **Tropical Storm:** Philippines (Dec.)	**Cold wave:** India, Nepal, Bangladesh (Jan., Dec.) **Flood:** Sri Lanka (Jan., Feb.), Bangladesh (Jul.) **Heavy rain:** Pakistan, India (Aug. - Oct.) **Drought:** Afghanistan (Jan. - Aug.)
2012	**Heavy snow:** Japan (Jan.) **Heavy rain, Flood:** China (Apr. - Aug.) **Drought:** D.P.R.Korea (Apr. - Dec.)	**Flood:** Thailand (Aug. - Sep.), Philippines (Aug.) **Typhoon:** Philippines (Dec.)	**Drought:** Sri Lanka (Jan. - Nov.) **Avalanche:** Pakistan (Apr.) **Flood:** Bangladesh (Jun. - Jul., Sep.), India (Jun. - Jul., Sep.), Sri Lanka (Dec.) **Heavy rain:** Pakistan (Aug. - Oct.) **Cold wave:** India, Bangladesh (Dec. - Feb. 2013)
2013	**Drought:** Central and Western China (Jan. - Jul.) **Heat wave:** Japan (May - Sep.) **Heavy rain, Flood:** China (May - Sep.), D.P.R.Korea (Jul.)	**Heavy rain, Flood:** Indonesia (Jan., Apr.), Philippines (Jan., Aug. - Oct.), Indochina Peninsula (Jun. - Nov.) **Typhoon:** Philippines (Nov.)	**Heat wave:** India (Apr. - Jun.) **Heavy rain, Flood:** India (Jun. - Jul., Oct.), Pakistan (Aug.)
2014	**Flood:** Central to Southern China (May - Sep.) **Drought:** Northern to Northeastern China (Jun. - Oct.)	**Flood:** Indonesia (Jan.), Cambodia (Jul.), Malaysia (Dec.) **Drought:** Malaysia (Mar.)	**Drought:** Sri Lanka (Jan. - Aug.) **Heavy rain, Flood, Landslide:** Afghanistan (Apr. - Jun.), Iran (Jun.), Sri Lanka (Jun., Oct., Dec.), India (Jul. - Oct.), Bangladesh (Aug. - Sep.), Nepal (Aug. - Sep.), Pakistan (Sep.)
2015	**Flood:** Southern China (May, Jul. - Aug.) **Heavy snow:** China (Feb.) **Drought:** D.P.R.Korea (May - Jun.)	**Heavy rain, Flood:** Indonesia (Jan.), Myanmar (Jun. - Aug.) **Drought:** Indonesia (Aug. - Sep.), Vietnam (All year round), Thailand (All year round) **Tropical Storm:** Myanmar (Aug.) **Typhoon:** Philippines (Dec.) **Wildfire:** Indonesia (Oct. - Nov.)	**Heavy rain, Flood:** Sri Lanka (Jan.), Bangladesh (Jun. - Aug.) **Heavy rain, Flood, Landslide:** India (Jun. - Sep., Nov. - Dec.), Pakistan (Jul. - Sep.) **Avalanche, Flood, Landslide:** Afghanistan (Feb. - Apr.) **Heat wave:** India (May), Pakistan (Jun.) **Cyclone:** India (Aug.)
2016	**Cold wave:** China, Korea, Japan (Jan.) **Flood:** Southern China (Mar.) **Heavy rain, Flood:** China (Apr. - Jul.), D.P.R.Korea (Aug. - Sep.) **Tornado:** Eastern China (Jun.) **Drought:** Northern to Northeastern China (Jun. - Aug.)	**Drought:** Southeast Asia (Jan. - May) **Flood:** Myanmar (Jun. - Aug.), Thailand (Oct., Dec.), Philippines (Aug.), Vietnam (Oct. - Dec.), Indonesia (Nov. - Dec.)	**Heat wave:** India (Mar. - May) **Heavy rain:** Afghanistan to Pakistan (Mar. - Apr.), Nepal (Jul. - Aug.) **Heavy rain, Flood:** India (Apr., Jul. - Oct.), Sri Lanka (May) **Flood:** Bangladesh (Jul. - Aug.)

Continued.

Western Russia and Europe	Africa	Nothern and Central America	South America	Oceania
Heat wave, Wildfire: Western Russia (Jun. - Aug.) **Cold wave:** Eastern Europe to Western Russia (Nov. - Feb. 2011)	**Drought:** Syria, Somalia, Sudan (All year round), Ethiopia, Chad (Former half of the year), Zimbabwe (May -) **Heavy rain, Landslide:** Uganda (Feb.) **Heavy rain, Flood:** Kenya (Mar. - May), Angola to Namibia (Mar.), in and around the Eastern part of Western Africa (Jul. - Nov.), Sudan (Aug.)	**Flood:** Mexico (Sep.)	**Heavy rain, Flood:** Bolivia (Jan.), Brazil (Apr., Jun.) **Flood, Landslide:** Colombia (Apr. - Mar. 2011) **Cold wave:** In and around Peru (Jul.)	**Flood:** Eastern Australia (Dec. - Feb. 2011)
	Drought: Eastern Africa, Northern part of Western Africa (All year round) **Flood:** South Africa (Jan. - Feb.), Namibia (Mar. - Apr.)	**Drought:** Southern USA to Northern Mexico (Jan. - Nov.) **Tornado:** Southeastern USA, Central USA (Apr. - May) **Flood:** Mexico (Jul.), Central America (Sep. - Oct.)	**Heavy rain, Flood:** Southeastern Brazil (Jan., Sep.), Colombia (Apr. - Jul., Sep. - Dec.), Peru (Dec. - May 2012)	**Drought:** Tuvalu (Mar. - Oct.)
Cold wave: Russia, Eastern to Southern Europe (Jan. - Feb., Dec.) **Drought:** Ukraine (Apr. - Jul.), Russia (Jun. - Sep.) **Flood:** Georgia (May) **Heavy rain:** Southwestern Russia (Jul.)	**Drought:** Sudan to Eastern Africa, Northern part of Western Africa (All year round), Chad, Malawi, Angola (Latter half of the year) **Flood:** Kenya (Apr. - May), Sudan to Niger (Jul. - Sep.), South Africa (Oct.) **Heavy rain:** Nigeria (Jul. - Oct.)	**Heat wave, Drought:** Eastern to Central USA (Jun. - Sep.)	**Drought:** Paraguay (Jan.) **Flood:** Peru (Feb. - Apr.), Brazil (May) **Cold wave:** Peru (Jun.)	**Flood:** Papua New Guinea (Sep.)
Flood: Czechia (Jun.) **Heat wave:** UK (Jul.)	**Heavy rain:** Mozambique to Zimbabwe (Jan.), Kenya (Mar. - Apr.), Niger (Jul. - Aug.), Sudan, South Sudan (Aug. - Oct.), Senegal (Sep.) **Drought:** Chad, Malawi, Zimbabwe (All year round)	**Flood:** Canada (Jun.)	**Flood:** Brazil (Jan.), Peru (Jan. - Feb.), Bolivia (Jan. - Feb., Oct. - Feb. 2014), Argentina (Apr.)	**Flood:** Papua New Guinea (Jan.)
Flood: Southeastern Europe (May)	**Drought:** Kenya (Jan. - Aug.), Burkina Faso (May - Dec.) **Flood:** Niger (Jun. - Aug.), Sudan (Jul. - Sep.), Morocco (Nov.)	**Drought:** Southwestern USA (All year round), Guatemala (Aug. - Dec.)	**Cold wave:** Peru (Apr. - Sep.) **Flood:** Paraguay, Southern Brazil (Jun.), Bolivia (Oct. - Feb. 2015)	**Heat wave:** Australia (Jan.) **Heavy rain, Flood:** Solomon Islands (Apr.), Papua New Guinea (Jun. - Jul.)
Flood: Southern Europe (Feb.), France (Oct.) **Heat wave:** France, Belgium (Jul.)	**Heavy rain, Flood:** Malawi (Jan. - Feb.), Mozambique (Jan.), Ethiopia (Apr.), Nigeria (Sep. - Oct.), Somalia (Oct. - Nov.) **Drought:** Ethiopia (Feb. - Mar., Jun. - Sep.), Somalia (Jun. - Aug.), Southern Africa (All year round) **Heat wave:** Egypt (Aug.)	**Drought:** California, the USA (All year round) **Landslide:** Guatemala (Oct.)	**Heavy rain, Flood:** Brazil, Argentina, Paraguay (Mar., Dec.), Chile (Mar., Aug.) **Drought:** Brazil, Bolivia, Peru (All year round)	**Heavy rain, Flood:** Papua New Guinea (Mar.), Australia (Apr.) **Drought:** Papua New Guinea (Apr. - Dec.) **Cyclone:** Vanuatu (Mar.)
	Drought: Madagascar, Mozambique (All year round), South Sudan (Feb. - Nov.), Lesotho, Swaziland (Feb. - Dec.), Kenya (Jun. - Dec.) **Heavy rain, Flood:** Ethiopia (May), Sudan (Jun. - Aug.) **Flood:** Niger (Jun. - Oct.), Mozambique (Oct. - Dec.)	**Drought:** Haiti, California, the USA (All year round) **Heavy snow:** Eastern USA (Jan.) **Hurricane:** Haiti, Southeastern USA (Oct.) **Flood:** Honduras (Oct.), Dominican Republic (Nov.)		

Weather-Related Disasters

Year	Asia		
	East Asia and Eastern Siberia	Southeast Asia	South Asia to Middle East
2017	**Drought:** Northeastern China (Apr. - Jun.) **Heavy rain, Typhoon, Tropical Strom:** Southern China (Jun. - Aug.)	**Flood:** Philippines (Jan. - Feb.), Thailand (Jan., Jul. - Aug., Oct. - Dec.), Indonesia (Jun.) **Typhoon, Tropical Strom, Heavy rain:** Vietnam (Sep. - Nov.) **Typhoon, Tropical Strom:** Philippines (Dec.)	**Heat wave:** India (Apr. - Jun.) **Heavy rain:** Sri Lanka (May), South Asia to Afghanistan (Jun. - Sep.) **Cyclone:** Southern India to Sri Lanka (Nov. - Dec.)
2018	**Flood:** Central to Southeastern China (May - Jul.), D.P.R.Korea (Aug. - Sep.) **Heavy rain, Flood:** Japan (Jun. - Jul.) **Heat wave:** Japan (Jul.)	**Flood:** Philippines (Jan.), Indonesia (Feb.), Myanmar (Jul. - Aug.), Vietnam (Dec.)	**Drought:** Afghanistan (Apr. - Jul. 2019), Southwestern India (Sep. - Jun. 2019) **Heat wave:** Pakistan (May) **Dust storm, Thunderstorm:** Northern India (May) **Flood:** Sri Lanka (May) **Heavy rain, Flood:** India (Jun. - Sep.)
2019	**Drought:** D.P.R.Korea (Jan. - Feb.) **Heavy rain, Flood:** Eastern to Southern China (Jun. - Jul.) **Heat wave:** Japan (Jul. - Aug.) **Typhoon:** Japan (Sep. - Oct.)	**Heavy rain, Flood:** Indonesia (Mar., Dec.), Laos, Thailand, Cambodia (Sep.)	**Drought:** Pakistan (Jan. - Feb.) **Heavy rain, Flood:** Northern Middle East to India (Mar. - Apr.), Pakistan (Jul. - Sep.), India (Jul. - Oct.), Sri Lanka (Sep. - Dec.)

Continued.

Western Russia and Europe	Africa	Nothern and Central America	South America	Oceania
	Cyclone: Zimbabwe (Feb.) **Drought:** Chad (Mar. - May), Mauritania (Apr. - Aug.), Angola, Niger (May - Jun.) **Flood:** Ghana (Apr. - Jul.), Niger (Aug. - Sep.) **Landslide:** Sierra Leone, DR Congo (Aug.)	**Hurricane:** Southeastern USA to Caribbean countries (Aug. - Sep.)	**Heavy rain:** Peru (Jan. - Mar.), Colombia (Feb. - Apr.) **Flood:** Brazil (May - Jun.)	
Cold wave: Hungary (Dec. - Feb. 2019)	**Heavy rain, Flood:** Northern to Central part of Eastern Africa (Mar. - May), Niger (Jul. - Aug.), Nigeria (Jul. - Sep.) **Drought:** Madagascar (Apr. - Dec.) **Flood:** Ghana (Sep.)	**Drought:** Southern part of Central America (Jun. - Mar. 2019) **Flood:** Costa Rica, Trinidad and Tobago (Oct.) **Wildfire:** Western USA (Nov.)	**Drought:** In and around Northern Argentina (Jan. - Mar.)	**Drought:** Southeastern Australia (Jan. - Sep.)
Heat wave: Northern to Central Europe (Jun. - Jul.) **Flood:** Spain (Sep.)	**Drought:** Kenya (Jan. - Sep.), Somalia (Feb. - Oct.), Zimbabwe (Feb. - Mar. 2020), Zambia (In 2019 - Dec. 2020) **Cyclone:** Southern part of Eastern Africa (Mar. - Apr.) **Heavy rain, Flood:** South Sudan (Jun.), Sudan (Jul. - Sep.), Niger (Sep.), Northern to Western part of Eastern Africa (Oct. - Dec.), DR Congo (Nov.)	**Hurricane:** Eastern USA to Bahamas (Sep.)	**Heavy rain, Flood:** Bolivia (Feb. - Apr.), Paraguay (Apr. - May)	**Wildfire:** Australia (Sep. - Feb. 2020)

Major Meteorological Disasters in Japan

Data up through 1970 are taken from the "Japan Meteorological Agency Weather Handbook," data since 1971 are the totals released each February by the Japan Meteorological Agency based on materials related to disaster prevention. The collection standards are roughly as follows.

① This list focuses on disasters caused by typhoons, heavy rain, heavy snow, strong wind, etc., with a large number of dead or missing (100 people until 1950, 50 people until 1965, 25 people until 1975, 15 people since 1985, and 5 people since then). Other damage is also considered.

② Cases of large scale damage from drought, cold, or frost.

③ Fires are cases where more than 1000 homes were destroyed.

Date	Disaster	Damaged areas (region)	Damage
1927. 9.11- 9.24	Typhoon	Kyushu-Tohoku	D 373, M 66, I 181, H 2,211, F 3493
1932.11.14-11.15	Typhoon	Pacific Coast (Tokai-Tohoku)	D 146, M 111, I 345, H 13,672, F 65,081, S 2,230
1934. 3.21	Fire	Hakodate City	D 2,015, B 11,102
1934. Jul.-Aug.	Cold weather damege	Tohoku-Hokkaido	Tohoku Index of paddy field rice production: 61
1934. 7.10- 7.11	Heavy rainfall	Hokuriku	D 119, M 26, I 370, H 465, F 17,129, L 11,257
1934. 9.20- 9.21	Typhoon "Muroto"	Kyushu-Tohoku (In particular, Osaka pref.)	D 2,702, M 334, I 14,994, H 92,740, F 401,157, D 27,594
1935. 6.26- 6.30	Heavy rainfall	Western Japan	D 147, M 9, I 283, H 2,041, F 232,202
1935. 8.21- 8.25	Heavy rainfall [Low]	Northern Tohoku	D 20, M 181, I 4, H 284, F 17,799, L 25,854
1935. 8.27- 8.30	Typhoon	All over Japan	D 48, M 25, I 98, H 1,451, F 60,550, L 399
1935. 9.23- 9.26	Typhoon	All over Japan (In particular, Gunma pref.)	D 317, M 60, I 276, H 3,423, F 110,153, L 15,613 S 389
1936. Jan.-Feb.	Heavy snowfall	Hokuriku, Tohoku	Extensive damage in Fukui, Niigata, Fukushima and Yamagata pref.
1938. 1. 4- 1.14	Avalanche	Niigata pref.	D 70
1938. 6.28- 7. 5	Heavy rainfall [Front]	Kinki-Tohoku (In particular, Hyogo pref.)	D 708, M 217, I 3,393, H 9,123, F 501,201
1938. 9. 1	Typhoon	Chubu-Tohoku	D 201, M 44, I 137, H 13,223, F 158,536, S 378
1938. 9. 5	Typhoon	Shikoku, Kinki (In particular, Tokushima pref.)	D 81, M 23, I 45, H 1,130, F 31,388, S 145
1938.10.14	Typhoon	Southern Kyushu	D 289, M 178, I 594, H 2,161, F 6,897, L 605, S 24
1939. Jun.-Aug.	Drought	Kyushu-Kinki	A 28

Continued.

Date	Disaster	Damaged areas (region)	Damage
1939.10.15-10.17	Typhoon	Kyushu-Shikoku (In particular, Miyazaki pref.)	D 74, M 25, I 31, H 2,580, F 13,798, L 1,849, S 321
1940. 1.15	Fire	Shizuoka City	B 5,121
1941. Mid-Jun.-Late-Jun.	Heavy rainfall [Front]	Western Japan	D 100, M 12, I 50, H 534, F 48,556, L 44,623
1941. 7.22- 7.23	Typhoon	Tokai-Tohoku	D 66, M 32, I 15, H 1,044, F 213,767, L 202,180
1941. 9.30-10. 1	Typhoon	Kyushu-Kinki	D 108, M 102, I 169, H 5,492, F 46,525, L 828,224, S 320
1942. 8.27- 8.28	Typhoon	Kyushu-Kinki (In particular, Yamaguchi pref.)	D 891, M 267, I 1,438, H 102,374, F 132,204, L 26,846, S 3,936
1943. 7.22- 7.25	Typhoon	Northern Kyushu-Kinki	D 211, M 29, I 231, H 4,531, F 33,440, L 58,092
1943. 9.18- 9.20	Typhoon	Kyushu-Chugoku (In particular, Shimane pref.)	D 768, M 202, I 491, H 21,587, F 76,323, L 114,566, S 830
1944. 7.19- 7.22	Heavy rainfall [Front]	Hokuriku-Tohoku	D 59, M 29, I 18, H 575, F 35,734, L 92,048
1944.10. 7-10. 8	Typhoon	Shikoku-Hokkaido	D 58, M 45, I 47, H 1,995, F 29,418, L 12,562, S 2,969
1945. Jul.-Aug.	Cold weather damege	Hokuriku-Hokkaido	Hokkaido Index of paddy field rice production: 42
1945. 9.17- 9.18	Typhoon "Makurazaki"	Western Japan (In particular, Hiroshima pref.)	D 2,473, M 1,283, I 2,452, H 89,839, F 273,888, I 128,403
1945.10. 9-10.13	Typhoon "Akune"	Western Japan (In particular, Hyogo pref.)	D 377, M 74, I 202, H 6,181, F 174,146, L 158,893
1947. 4.20	Fire	Iida city (Nagano pref.)	B 3,984
1947. Jul.-Aug.	Drought	Western Japan, Chubu	No Harvest: Miyazaki pref. 12,014 ha, Mie pref. 6,520 ha etc.
1947. 9.14- 9.15	Typhoon "Kathleen"	Eastern Japan	D 1,077, M 853, I 1,547, H 9,298, F 384,743, L 12,927
1948. 9.11- 9.12	Heavy rainfall [Low]	Northern Kyushu	D 121, M 126, I 317, H 1,263, F 2,290, L 739, S 45
1948. 9.15- 9.17	Typhoon "Ione"	Shikoku-Tohoku (In particular, Iwate pref.)	D 512, M 326, I 1,956, H 18,017, F 120,035, L 113,427, S 435
1949. 2.20	Fire	Noshiro City (Akita pref.)	D 3, I 265, H 1,414
1949. 6.20- 6.23	Typhoon "Della"	Kyushu-Tohoku (In particular, Ehime pref.)	D 252, M 216, I 367, H 5,398, F 57,553, L 80,300, S 4,242
1949. 8.15- 8.19	Typhoon "Judith"	Kyushu, Shikoku	D 154, M 25, I 213, H 2,561, F 101,994, L 104,973, S 123
1949. 8.31- 9. 1	Typhoon "Kitty"	Chubu-Hokkaido	D 135, M 25, I 479, H 17,203, F 144,060, L 48,598, S 2,907
1950. 1. 9- 1.15	Strong wind	Kyushu-Kanto	D-M 120, H 810, S 17

Major Meteorological Disasters in Japan Continued.

Date	Disaster	Damaged areas (region)	Damage
1950. 8. 3- 8. 6	Tropical cyclone	Chubu-Tohoku	D 40, M 59, I 764, H 376, F 32,293, L 47,722
1950. 9. 2- 9. 4	Typhoon "Jane"	Shikoku, Chugoku, Kinki, and East of Japan (In particular, Osaka pref.)	D 336, M 172, I 10,930, H 56,131, F 166,605, L 85,018, S 2,752
1950. 9.12- 9.14	Typhoon "Kezia"	Kyushu-Chugoku	D 35, M 8, I 75, H 4,836, F 121,924, L 90,215, S 845
1951. 7. 7- 7.17	Heavy rainfall [Front]	Western Japan, Chubu (In particular, Kyoto pref.)	D 162, M 144, I 358, H 1,585, F 103,298, L 139,821, S 98
1951. Jul.-Aug.	Drought	All over Japan	Rice paddy damage : 139,885 ha, Dry land rice damage: 91,824 ha
1951.10.13-10.15	Typhoon (5115) "Ruth"	All over Japan (In particular, Yamaguchi pref.)	D 572, M 371, I 2,644, H 221,118, F 138,273, L 128,517, S 9,596
1952. 4.17	Fire	Tottori city	D 2, I 362, H 5,228 50 ha forest burned down
1952. 6.22- 6.25	Typhoon (5202) "Dinah"	Western Japan, Chubu, Kanto	D 65, M 70, I 28, H 425, F 39,712, L 40,924, S 178
1952. 7. 7- 7.18	Heavy rainfall [Front]	Chugoku-Tokai	D 67, M 73, I 101, H 664, F 161,027, L 50,184, S 23
1953. 6. 4- 6. 8	Typhoon (5302), Front	Kyushu-Chubu	D 37, M 17, I 56, H 1,802, F 33,640, L 74,353, S 139
1953. 6.25- 6.29	Heavy rainfall [Front]	Kyushu-Chugoku (In particular, Kumamoto pref.)	D 748, M 265, I 2,720, H 34,655, F 454,643, L 269,813, S 618
1953. 7.16- 7.24	Heavy rainfall	All over Japan	D 713, M 411, I 5,819, H 10,889, F 86,479, L 98,046, S 112
1953. 8.14- 8.15	Heavy rainfall [Front]	Eastern Kinki	D 290, M 139, I 994, H 1,777, F 21,517, L 11,876
1983. Aug.-Sep.	Cold weather damege	Eastern Japan	Hokkaido Grains declined Over 30%
1953. 9.24- 9.26	Typhoon (5313)	All over Japan (In particular, Kinki)	D 393, M 85, I 2,559, H 86,398, F 495,875, L 318,657, S 5,582
1954. 5. 9- 5.10	Strong wind [Low]	Tohoku-Hokkaido	D 31, M 330, H 12,359, F 23, L 73, S 348
1954. 7. 4- 7. 6	Heavy rainfall [Front]	Chugoku-Kinki	D 33, M 12, I 65, H 317, F 32,645, L 36,790, S 8
1954. 8.17- 8.20	Typhoon (5405)	Kyushu, Shikoku	D 28, M 33, I 69, H 5,442, F 32,265, L 30,476, S 191
1954. Jul.-Aug.	Cold weather damege	All over Japan	Hokkaido Index of paddy field rice production: 61
1954. 9.10- 9.14	Typhoon (5412)	Western Japan, Chubu, Kanto	D 107, M 39, I 311, H 39,855, F 181,380, L 61,722, S 688
1954. 9.17- 9.19	Typhoon (5414)	Shikoku-Tohoku	D 34, M 20, I 41, H 422, F 43,762, L 66,645, S 126

Continued.

Date	Disaster	Damaged areas (region)	Damage
1954. 9.25- 9.27	Typhoon (5415) "Touyamaru"	All over Japan	D 1,361, M 400, I 1,601, H 207,542, F 103,533, L 82,963, S 5,581
1955. 2.19- 2.21	Strong wind [Low]	All over Japan	D 16, M 107, I 18, H 573, F 296, L 13, S 84
1955. 4.14- 4.18	Heavy rainfall [Front]	Kyushu-Chugoku	D 91, M 4, I 34, H 110, F 18,533, L 15,861, S 13
1955. 5.11	Dense fog	Areas around the Seto Inland Sea	D 166, S 2, "Shiun-Maru" sank
1955. 7. 3- 7. 8	Heavy rainfall [Front]	Kyushu, Hokkaido	D 17, M 32, I 40, H 216, F 30,218, L 47,068, S 9
1955. 9.29-10. 1	Typhoon (5522), Fire	All over Japan (except Kanto)	D 54, M 14, I 537, H 85,554, F 51,294, L 30,271, S 1,483, Niigata city conflagration
1956. 4.16- 4.18	Heavy rainfall [Low]	Tohoku-Hokkaido	D-M 58, H 35, F 2,407, L 13,104, S 12
1956. 7.14- 7.17	Heavy rainfall [Front]	Hokuriku, Tohoku	D 50, M 10, I 37, H 708, F 31,066, L 42,608, S 22
1956. 8.16- 8.19	Typhoon (5609), Fire	All over Japan	D 33, M 3, I 213, H 37,341, F 10,431, L 11,285, S 2,700, Odate city conflagration
1956. 9. 7- 9.10	Typhoon (5612), Fire	Okinawa-Chubu	D 41, M 2, I 313, H 32,044, F 11,489, L 6,729, S 994, Uozu city conflagration
1956. 9.25- 9.27	Typhonn (5615)	Okinawa-Kanto	D 20, M 11, I 41, H 4,170, F 47,520, L 20,378, S 109
1956.10.28-10.31	Strong wind, Heavy rainfall [Low]	Kyushu-Hokkaido	D 24, M 48, I 22, H 146, F 5,373, L 583, S 7
1957. 6.27- 6.28	Typhoon (5705), Front	Kyushu-Kanto	D 30, M 23, I 33, H 396, F 129,673, L 21,939, S 14
1957. 7.25- 7.28	Heavy rainfall "Isahaya Gouu"	Kyushu (In particular, Nagasaki pref.)	D 856, M 136, I 3,860, H 6,811, F 72,565, I 43,566, S 222
1957.12.12-12.13	Rainstorm [Front]	All over Japan	D 14, M 29, I 156, H 15,913, F 2,076, S 122
1958. 1.26- 1.27	Strong wind [Low]	Southern shore of Honshu	D 7, M 194, I 8, H 3, F 6, S 8, "Nankai-maru" sank
1958. 3.29- 3.31	Frost damage	All over Japan	A 230
1958. 7.22- 7.23	Typhoon (5811)	Kinki, Eastern Japan	D 26, M 14, I 64, H 1,089, F 46,243, L 27,673, S 21
1958. 7.23- 7.29	Heavy rainfall [Front]	Shikoku-Tohoku	D 22, M 8, I 51, H 423, F 40,874, L 45,916, S 48
1958. 8.24- 8.26	Typhoon (5817)	Kinki-Chubu	D 15, M 30, I 39, H 1,996, F 17,641, L 11,679, S 33
1958. 9.15- 9.18	Typhoon (5821)	All over Japan	D 25, M 47, I 111, H 5,648, F 48,700, L 25,530, S 226

Major Meteorological Disasters in Japan Continued.

Date	Disaster	Damaged areas (region)	Damage
1958. 9.26- 9.28	Typhoon (5822) "Kanogawa"	Kinki, East of Japan (In particular, Shizuoka pref.)	D 888, M 381, I 1,138, H 16,743, F 521,715, L 89,236, S 260
1959. 4. 6- 4. 7	Strong wind [Low]	Hokkaido	M 85, I 3, H 189, S 20
1959. 7.13- 7.15	Typhoon (5905), Front	Western Japan, Chubu	D 44, M 16, I 77, H 603, F 77,288, L 32,452, S 3
1959. 8.13- 8.14	Typhoon (5907), Front	Kinki-Tohoku (In particular, Nagano and Yamanashi pref.)	D 188, M 47, I 1,528, H 76,199, F 148,607, L 74,169, S 111
1959. 9.15- 9.18	Typhoon (5914) "Miyakojima"	All over Japan (except Kanto)	D 47, M 52, I 509, H 16,632, F 14,360, L 3,566, S 778
1959. 9.26- 9.27	Typhoon (5915) "Isewan"	All over Japan (except Kyushu)	D 4,697, M 401, I 38,921, S 7,576, H 833,965, F 363,611, L 210,859
1960. 1.16- 1.18	Snowstorm [Low]	Tohoku-Hokkaido	D 14, M 70, I 14, H 167, F 7, S 87
1960. 8. 2- 8. 3	Heavy rainfall [Front]	Northern Tohoku, Hokkaido	D 21, M 9, I 50, H 866, F 9,577, L 5,274, S 3
1960. 8.12- 8.13	Typhoon (6012)	Kinki, Chubu	D 28, M 19, I 154, H 449, F 21,144, L 7,764
1960. 8.28- 8.30	Typhoon (6016)	Western Japan, Chubu	D 50, M 11, I 145, H 2,265, F 45,009, L 17,195, S 120
1961. 5.28- 5.29	Typhoon (6104), Fire	Tohoku-Hokkaido	D 14, I 225, H 6,764, F 260, L 937, S 31, 25,077 ha forest burned down, Sanriku conflagration
1961. 6.24- 7.10	Heavy rainfall ["Baiu" Front]	All over Japan (except Hokkaido)	D 302, M 55, I 1,320, H 8,464, F 414,362, L 340,449, S 21
1961. 9.15- 9.17	Typhoon (6418) "2nd Muroto"	All over Japan (In particular, Kinki)	D 194, M 8, I 4,972, H 499,444, F 384,120, L 82,850, S 2,540
1961.10.25-10.29	Heavy rainfall [Low]	Kyushu-Chubu	D 78, M 36, I 86, H 819, F 60,748, L 32,190, S 186
1962. 7. 1- 7. 9	Heavy rainfall [Front]	Kyushu, Tokai	D 92, M 10, I 82, H 395, F 91,604, L 66,113, S 5
1963. Jan.	Heavy snowfall	All over Japan	D 228, M 3, I 356, H 6,005, F 7,028
1963. 6.29- 7. 5	Heavy rainfall [Low]	Northern Kyushu	D 36, M 1, I 42, H 567, F 44,929, L 20,458
1963. 8. 7- 8.11	Typhoon (6309)	Kyushu-Kinki	D 23, M 6, I 46, H 2,064, F 25,166, L 23,009, S 94
1964. 7.17- 7.19	Heavy rainfall	Sanin-Hokuriku (In particular, Shimane pref.)	D 123, M 5, I 291, H 2,048, F 67,517, L 46,042, S 15
1964. 9.24- 9.25	Typhoon (6420)	Kyushu-Tohoku	D 47, M 9, I 530, H 71,269, F 44,751, L 16,326, S 594
1964. Apr.-Oct.	Cold weather damege	Aomori pref., Hokkaido	A 522

Continued.

Date	Disaster	Damaged areas (region)	Damage
1965. 1. 8- 1.10	High waves [Low]	Tohoku-Hokkaido	D 1, I 44, H 320, F 1,594, L 109, S 1,759
1965. 6.30- 7. 3	Heavy rainfall [Front]	Middle of Kyushu	D 20, M 16, I 1, H 880, F 21,659, L 14,829, S 7
1965. 7.21- 7.23	Heavy rainfall [Front]	Chugoku, Chubu	D 32, M 1, I 35, H 451, F 21,761, L 16,265, S 5
1965. 8. 4- 8. 6	Typhoon (6515)	Kyushu-Chugoku	D 28, I 368, H 58,951, F 5,716, L 5,418, S 260
1965. 9. 9- 9.11	Typhoon (6523)	All over Japan	D 67, M 6, I 883, H 63,436, F 49,626, L 12,353, S 619
1965. 9.13- 9.18	Typhoon (6524), Front	All over Japan	D 98, M 9, I 330, H 8,105, F 251,820, L 81,649, S 110
1965.10. 6-10. 7	Typhoon (6529)	Mariana Sea Area	D 1, M 208, Fishing boat was distressed
1966. 3. 5	Air turbulence	Sky over Mt. Fuji	D 124, BOAC flight crash
1966. 6.27- 6.29	Typhoon (6604)	Chubu-Hokkaido	D 64, M 19, I 91, H 433, F 128,041, L 129,195, S 12
1966. 8.14- 8.16	Typhoon (6613), Front	Pacific Coast (Kyushu-Kinki)	D 36, M 3, I 22, H 74, F 19,142, L 9,261
1966. 9. 5	Typhoon (6618) "2nd Miyakojima"	Miyakojima Island, Ishigakijima Island	I 41, H 7,765, F 30, S 56
1966. 9.24- 9.25	Typhoon (6624, 6626)	All over Japan (In particular, Yamanashi pref.)	D 275, M 43, I 976, H 73,166, F 53,601, L 34,159, S 107
1966.10.12-10.17	Heavy rainfall [Low]	Eastern Japan, Northern Japan	D 27, M 3, I 8, H 245, F 25,420, L 6,611, S 79
1966. Jun.-Oct.	Cold weather damege	Northern Japan	Hokkaido Index of paddy field rice production: 73
1967. 7. 7- 7.10	Heavy rainfall	Northern Kyushu-Kanto	D 365, M 6, I 618, H 3,756, F 301,445, L 44,444, S 5
1967. 8.26- 8.29	Heavy rainfall	Niigata pref., Yamagata pref.	D 113, M 33, I 190, H 2,594, F 69,424, L 62,678
1967.10.27-10.29	Typhoon (6734)	Kyushu-Tohoku	D 37, M 10, I 41, H 2,959, F 26,842, L 2,481, S 181
1967. Jul.-Oct.	Drought	Western Japan	A 682
1968. 8.15- 8.18	Typhoon (6807), Front	Western Japan	D 112, M 21, I 63, H 443, F 14,662, L 1,946, S 88, Bus fell into Hida River
1968. 8.25- 8.31	Typhoon (6810), Front	All over Japan	D 25, M 2, I 68, H 263, F 24,386, L 6,550, S 13
1968. 9.22- 9.27	Typhoon (6816) "3rd Miyakojima"	Okinawa, Western Japan	D 11, I 80, H 5,715, F 15,322, L 2,747, S 85
1969. 2. 4- 2. 7	Snowstorm	Hokuriku-Hokkaido	D 31, I 82, H 117, F 99, S 1,085
1969. 6.24- 7. 1	Heavy rainfall [Front]	West of Japan, Chubu, Kanto	D 81, M 8, I 184, H 976, F 64,390, L 55,064, S 9

Major Meteorological Disasters in Japan Continued.

Date	Disaster	Damaged areas (region)	Damage
1969. 8. 7- 8.12	Heavy rainfall [Front]	Hokuriku, Tohoku	D 27, M 14, I 83, H 608, F 34,360, L 20,564, S 36
1970. 1.30- 2. 2	Low	Chubu, Hokuriku-Hokkaido	D 14, M 11, I 45, H 916, F 4,422, L 271, S 293
1970. 7. 1	Heavy rainfall [Front]	Southern Kanto	D 22, M 2, I 42, H 1,758, F 14,424, L 26,771, S 7
1970. 8.20- 8.22	Typhoon (7010)	Shioku, Chugoku	D 23, M 4, I 556, H 48,652, F 59,961, L 14,329, S 1,403
1971. 8. 1- 8. 6	Typhoon (7119)	Kyushu-Chugoku	D 62, M 7, I 209, H 1,691, F 18,113, L 10,588, S 47
1971. Jul.-Aug.	Cold weather damege	Northern Japan	Hokkaido Index of paddy field rice production: 66
1971. 8.28- 9. 1	Typhoon (7123)	Western Japan, Chubu, Kanto	D 37, M 7, I 103, H 1,427, F 122,290, L 46,720, S 57
1971. 9. 6- 9. 7	Typhoon (7125)	Kinki-Kanto	D 56, M 28, I 1, H 202, F 11,504, L 1,652
1971. 9. 9- 9.10	Heavy rainfall [Front]	South of Mie pref.	D 41, M 1, I 39, H 79, F 1,200, L 158
1971. 9.26- 9.27	Typhoon (7129)	Kinki-Kanto	D 9, M 11, I 11, H 30, F 59,665, L 1,373, S 1
1972. 7. 3- 7.13	Heavy rainfall	All over Japan	D 410, M 32, I 534, H 4,339, F 194,691, L 84,794, S 2
1972. 9. 6- 9.10	Heavy rainfall [Tropical cyclone]	Kyushu-Chubu	D-M 16, I 30, H 137, F 36,925, L 1,567
1972. 9.13- 9.20	Typhoon (7220), Front	All over Japan	D-M 85, I 157, H 4,213, F 146,547, L 43,342, S 322
1973. 5. 7- 5. 9	Heavy rainfall [Front]	Kyushu-Hokuriku	D 14, M 10, I 10, H 28, F 1,809, L 217, S 5
1973. 7.30- 7.31	Typhoon (7306), Low	Kyushu	D 14, M 15, I 10, H 110, F 37,783, L 209
1973. Jun.-Aug.	Drought	All over Japan (except Hokkaido)	A 807
1973.11.17-11.24	Strong wind, Heavy rainfall	Chubu-Tohoku	D 9, M 17, I 15, H 12, S 6
1974. 4.20- 4.22	Heavy rainfall [Low]	Kyushu-Hokkaido	D 17, M 6, I 55, F 218, S 29
1974. 4.26	Landslide [Snowmelt]	Yamagata pref.	D 17, I 13, H 20
1974. 7. 3- 7.11	Typhoon (7408), Front	Okinawa-Chubu	D 111, I 171, H 1,448, F 148,934, L 16,230, S 19
1975. 8. 5- 8. 8	Heavy rainfall [Front]	Northern Kyushu-Tohoku	D-M 35, I 63, H 536, F 12,622, L 3,627
1975. 8.17- 8.20	Typhoon (7505)	Shikoku-Hokkaido	D-M 77, I 209, H 2,419, F 50,222, L 12,712, S 12
1975. 8.21- 8.24	Typhoon (7506)	Shikoku-Hokkaido	D-M 33, I 51, H 711, F 48,832, L 80,033, S 28, A 340

Continued.

Date	Disaster	Damaged areas (region)	Damage
1976. 6.21- 6.26	Heavy rainfall [Front]	Kyushu-Chubu	D-M 43, I 30, H 164, F 3,474, L 4,564, A 151
1976. 9. 8- 9.17	Typhoon (7617), Front	All over Japan	D-M 169, I 435, H 11,193, F 442,317, L 80,304, S 237, A 2,080
1976. Jun.-Sep.	Cold weather damege	Koshin, Tohoku, Hokkaido	A 4,093
1976.10.28-10.30	Strong Wind, Fire	The sea of Japan side (Hokuriku-Tohoku)	D-M 2, I 981, H 3,767, F 663, S 302, Sakata city conflagration
1976.12.26-1977. Mar.	Snow damege	All over Japan	D-M 84, I 320, H 826, F 770, A 458
1977. 9. 8- 9.10	Typhoon (7709) "Okinoerabu"	Pacific side	D 1, H 5,119, F 3,207, L 105, S 22, A 61
1978. May-Sep.	Drought	All over Japan	A 1,448
1979. 3.29- 4. 4	Rainstorm, Snowfall [Low]	All over Japan (except Okinawa)	D-M 20, I 153, H 4,725, F 184, L 616, A 193
1979. 6.25- 7. 4	Heavy rainfall [Front]	All over Japan (except Kanto)	D-M 29, I 52, H 273, F 48,208, L 35,991, S 3
1979. 9.24-10. 2	Typhoon (7916)	All over Japan	D-M 12, I 83, H 1,503, F 68,216, L 7,042, S 133, A 1,074
1979.10.14-10.20	Typhoon (7920)	All over Japan	D-M 111, I 478, H 7,523, F 37,450, L 25,451, A 1,057
1980. 8.26- 8.31	Heavy rainfall [Low]	All over Japan	D-M 26, I 50, H 405, F 39,141, L 10,069, S 9, A 404
1980. Jul.-Sep.	Cold weather damege	All over Japan (except Okinawa)	A 6,919
1980.10.25-10.28	Strong Wind, Waves	All over Japan	D-M 14, I 17, H 133, F 976, L 130, S 541
1980. Late-Dec.-1981. Feb.	Heavy snowfall	All over Japan	D-M 103, I 1,305, H 5,819, F 5,553, L 25, S 1,269, A 1,203
1981. 8.20- 8.27	Typhoon (8115)	Kinki, Eastern Japan, Northern Japan	D-M 43, I 173, H 4,401, F 31,082, L 65,821, S 264, A 2,272
1981. Aug.-Mid-Sep.	Cold weather damege	Tohoku-Hokkaido	A 2,622
1982. 7.10- 7.26	Heavy rainfall	Western Japan, Chubu, Kanto	D 337, M 8, I 661, H 851, F 52,165, L 15,354, S 30, A 474
1982. 8. 1- 8. 3	Typhoon (8210), Front	Chugoku-Tohoku	D-M 95, I 174, H 5,312, F 113,902, L 28,311, S 12, A 5,916
1982. 9. 8- 9.14	Typhoon (8218), Front	Chugoku, Kinki, Eastern Japan, Northern Japan	D-M 38, I 174, H 651, F 136,308, L 20,012, S 3, A 1,258
1982. Late-Jun.-Mid-Sep.	Cold weather damege	Koshin, Kanto-Northern Japan	A 1,165
1982.11.28-11.30	Rainstorm [Low]	Western Japan-Kanto	D-M 15, I 19, H 76, F 8,700, S 12
1983. 4.24- 4.28	Forest Fire	Tohoku	D 1, I 21, H 118, S 66, 8685 ha forest burned down

Major Meteorological Disasters in Japan Continued.

Date	Disaster	Damaged areas (region)	Damage
1983. 7.20- 7.27	Heavy rainfall	Kyushu-Tohoku	D-M 117, I 166, H 3,669, F 17,141, L 7,796, A 1,302
1983. Jun.-Jul.	Cold weather damege	Koshin, Tohoku, Hokkaido	A 2,095
1983. 9.24- 9.30	Typhoon (8310), Front	Western Japan, Chubu	D-M 44, I 118, H 640, F 56,267, L 5,651, S 26, A 805
1983. Late-Dec.- 1984. Apr.	Heavy snowfall	All over Japan	D-M 96, H 939, F 515
1984. 6.22- 6.29	Heavy rainfall [Front]	Western Japan, Chubu	D-M 16, I 6, H 21, F 2,967, L 492
1984. Mid-Jun.- Sep.	Drought	All over Japan	A 289
1984. Mid-Dec.- 1985. Jan.	Heavy snowfall	All over Japan	D-M 46, H 139
1985. 2.15	Landslide	Niigata pref.	D-M 10, I 5, H 7, F 72
1985. 4.11- 4.15	Heavy rainfall [Low]	Kyushu-Kinki	D 7, M 4, I 8, S 1, 11 dead or missing in fishing boat sinking
1985. 6.18- 7. 6	Typhoon (8506), Front	Kyushu-Tohoku	D-M 16, I 49, H 811, F 12,691, L 31,617, S 1, A 1,303
1985. 7. 3- 7.15	Heavy rainfall [Front]	The sea of Japan side (Chugoku-Chubu)	D-M 13, I 42, H 59, F 3,609, L 3,135
1985. 7.26	Landslide	Nagano city	D-M 26, I 4, H 69, L 3
1985. 8.29- 9. 2	Typhoon (8512, 8513, 8514)	Kyushu-Hokkaido	D-M 31, I 232, H 7,805, F 2,858, L 2,112, S 1,144, A 714
1985. Mid-Jul.- Mid-Sep.	Drought	Chugoku-Tohoku	A 132
1986. 3.20- 3.26	Heavy snowfall	Shikoku-Hokkaido	D-M 17, I 148, H 38, F 7, S 2, A 159
1986. 7. 4- 7.17	Heavy rainfall [Front]	Western Japan, Chubu	D-M 23, I 24, H 175, F 3,638, L 809, S 1, A 121
1986. 8. 3- 8. 9	Typhoon (8610)	Tokai-Tohoku	D-M 21, I 106, H 2,683, F 105,072, L 85,119, S 9, A 1,055
1987. 2. 2- 2. 4	Heavy snowfall, Strong wind [Low]	Kyushu-Tohoku	D-M 22, I 46, H 2, S 6
1987. 7.13- 7.21	Typhoon (8705), Front	Okinawa-Kanto	D-M 9, I 24, H 51, F 3,426, L 3,778, S 7, A 155
1987. 8. 4- 8. 7	Thunderstorm	Chugoku, Shikoku-Tohoku	D 6, I 6, H 12, F 1,725, L 113
1987. 8.28- 9. 1	Typhoon (8712)	Okinawa-Hokkaido	D-M 8, I 189, H 37,078, F 611, L 78, S 791, A 874
1987.10.13-10.18	Typhoon (8719)	Chugoku-Kinki	D-M 9, I 17, H 216, F 24,044, L 6,802, S 99, A 529
1988. 3.22	Strong wind [Low]	Off the coast of Shiogama	D-M 21, S 1
1988. 7. 9- 7.29	Heavy rainfall [Front]	Ksyuhu-Tohoku	D-M 27, I 61, H 613, F 10,083, L 3,021, S 22
1988. Jul.-Oct.	Cold weather damege	Chubu-Tohoku	A 3,245

Continued.

Date	Disaster	Damaged areas (region)	Damage
1989. 7.15- 7.17	Heavy rainfall	Hokuriku-Tohoku	D 15, H 436
1989. 7.28- 8. 4	Topical cyclone, Typhoon (8911, 8912)	Kyushu-Kanto	D-M 11, H 76, F 10,664, L 30
1989. 8. 5- 8. 7	Typhoon (8913)	Chubu-Tohoku	D-M 15, I 26, H 159, F 5,063, L 1,384
1989. 9. 1- 9.15	Heavy rainfall	Kyushu-Hokkaido	D-M 17, I 44, H 50, F 21,581, L 2,883
1989. 9.17- 9.20	Typhoon (8922)	Kyushu-Tohoku	D-M 8, I 8, H 247, F 5,824, L 584, S 39
1990. 1.21- 1.28	Heavy snowfall, Strong wind [Low]	Kyushu-Kanto	D 9, I 15, S 2
1990. 2.11	Heavy rainfall, Strong wind	Kinki, Kanto	D 5, I 3, H 1
1990. 3.23- 3.25	Heavy rainfall	Kyushu	D 7, S 1
1990. 4.20- 4.24	Heavy rainfall [Low]	Kanto-Hokkaido	D-M 9, H 2, F 183, L 2,496, S 1, A 14
1990. 6.25- 7. 4	Heavy rainfall [Low, Front]	Kyushu, Kinki	D 27, I 81, H 592, F 42,141, L 20,765, S 11
1990. 7.14- 7.20	Heavy rainfall	Kyushu-Tohoku	D 5, I 3, H 5, F 239, L 1,126, A 17
1990. 8.21- 8.24	Typhoon (9014)	Kyushu-Hokkaido (except Kanto, Tohoku)	D 6, I 24, H 67, F 420, L 1,845, S 16, A 151
1990. 9.16- 9.20	Typhoon (9016)	Okinawa-Tohoku	D 40, I 131, H 16,541, F 18,183, L 41,954, S 413, A 1,322
1990. 9.25-10. 1	Typhoon (9020)	Kyushu-Kanto	D-M 6, I 20, H 210, F 13,318, L 4,273, S 31, A 229
1990.11. 3-11. 5	Heavy rainfall	Shikoku-Hokkaido	D-M 12, I 12, H 102, F 5,283, L 1,620, S 37, A 161
1990.11.27-12. 3	Typhoon (9028)	Okinawa-Hokkaido	D-M 4, I 12, H 162, F 1,544, L 33, A 48
1990.12.11-12.12	Heavy rainfall, Tornado	Shikoku-Kanto (In particular, Chiba pref.)	D-M 6, I 81, H 2,099, F 5, A 17
1991. 2.13- 2.19	Haevy rainfall, Strong wind	Kinki-Hokkaido	D-M 26, I 14, H 178, F 54, S 316, A 161
1991. 7. 7- 7. 8	Strong wind [Front]	Shimane pref., Hyogo pref., Kyoto pref.	D-M 5, I 2
1991. 5. 1- 7.18	Long spell of rain, Poor sunshine	Kyushu	A 177
1991. 7.25- 7.31	Typhoon (9109)	Okinawa-Chubu	D-M 6, I 39, H 64, F 1,472, L 263, S 120, A 101
1991. 8.19- 8.24	Typhoon (9112)	Kyushu-Southern Tohoku	D-M 16, I 11, H 53, F 4,162, L 4,142, S 4, A 119
1991. 9. 7- 9. 9	Typhoon (9115)	Miyazaki pref., Tokushima pref., Kanto	D 6, I 4, H 77, F 1,505, A 16
1991. 9.12- 9.15	Typhoon (9117)	Okinawa-Chubu	D 11, I 94, H 382, F 2,586, L 875, S 69, A 511

Major Meteorological Disasters in Japan Continued.

Date	Disaster	Damaged areas (region)	Damage
1991. 9.16- 9.21	Typhoon (9118)	Kinki-Tohoku	D-M 12, I 23, H 225, F 52,662, L 4,973, S 2, A 242
1991. 9.24-10. 1	Typhoon (9119)	All over Japan	D 62, I 1,499, H 170,447, F 23,965, L 362, S 930, A 5,735
1992. 1.31- 2. 1	Heavy snowfall, Strong wind [Low]	Kyushu-Tohoku	D-M 8, I 311, H 6, S 6, A 30
1992. 5.20- 5.30	Hail, Thunder, Tornado	Kyushu-Southern Tohoku	D 7, I 3, H 108, A 66
1992. 8. 6- 8.10	Typhoon (9210)	Western Japan, Hokkaido	D-M 7, I 67, H 1,561, F 1,508, L 4,802, S 47, A 740
1992. 8.17- 8.20	Typhoon (9211)	Western Japan	D-M 8, I 2, H 13, F 379, S 2, A 17
1992. 9. 9- 9.12	Typhoon (9217)	Hokkaido	D 1, H 9, F 3,128, L 24,549, A 139
1993. 2.20- 2.25	Strong wind [Low]	Nagasaki pref., Hyogo pref., Osaka pref.	D 2, M 22, I 1, H 18, F 53, S 2
1993. 3.23- 3.25	Strong wind [Low]	Tokushima pref.	D 17, M 12, S 1
1993. 6.28- 7. 8	Heavy rainfall [Front]	Kyushu-Kanto	D 20, M 1, I 18, H 84, F 1,392, L 3,632, A 174
1993. 7.24- 8. 1	Typhoon (9304, 9305, 9306)	All over Japan (except Okinawa)	D-M 18, I 20, H 143, F 4,316, L 4,527, S 2, A 178
1993. 7.31- 8. 7	Heavy rainfall	Western Japan (In particular, Southern Kyushu)	D 74, M 5, I 154, H 824, F 21,987, A 746
1993. 8.31- 9. 5	Typhoon (9313)	All over Japan (except Okinawa)	D-M 48, I 266, H 1,892, F 10,447, L 7,905, A 1,755
1993. Jun.-Oct.	Cold weather damege	All over Japan (except Okinawa)	A 9,791
1994. 2.11- 2.15	Strong wind, Heavy snowfall [Low]	Chugoku-Tohoku	D 12, I 1,462, H2
1994. 2.20- 2.27	Strong wind, Heavy rainfall [Low]	Chubu, Kanto, Tohoku, Hokkaido	D 5, I 54, H 372, F 92, S 79
1994. 9. 8	Gust, Thunderstorm	Saitama pref.	I 73, H 79
1994. 4. 4-Oct.	Drought, Intense heat	All over Japan	D 14, I 662
1995. 1. 4	Avalanche, Strong wind [Low]	Nagano pref., Chiba pref.	D 6, H 64
1995. 1.10- 1.20	Heavy snowfall, Strong wind [Low]	Chugoku-Tohoku (except Shikoku)	D 6, I 51, S 2, A 12
1995. 4. 1- 4. 2	Strong wind [Low]	Chiba pref., Tochigi pref., Akita pref.	D 5
1995. 5.11- 5.16	Heavy rainfall, Strong wind [Low]	Kyushu-Southern Tohoku	D 3, M 2, I 7, H 7, F 506, L 1,238, A 5
1995. 6.30- 7.15	Heavy rainfall [Low]	All over Japan (except Okinawa)	D-M 9, I 12, H 235, F 15,520, L 1,573, A 344

Continued.

Date	Disaster	Damaged areas (region)	Damage
1995. 7. 1- 9.30	Drought, Intense heat	Okinawa-Kanto	D 8, I 704, A 28
1995. 7.15- 7.24	Typhoon (9503), Front	Okinawa-Hokuriku	D 3, M 4, I 10, S 1, H 6, F 1,303, A 28
1995. 7.30- 8.11	Red tide	Kagawa pref.	A 4
1996. 1.24- 2. 7	Heavy snowfall [Low]	Kyushu-Hokuriku	D 7, I 104
1996. 7. 1- 7. 3	Hail, Lightning	Kanto-Hokkaido	D 2, I 16, H 6, F 254, A 110
1996. 7.17- 7.21	Typhoon (9606)	Kyushu-Kanto	D 1, I 27, H 2,361, F 933, S 3, L 7, A 788.2
1996. 8.10- 8.16	Typhoon (9612)	All over Japan	D 4, I 77, H 2,989, F 1,675, S 20, L 44, A 205.7
1996. 9.20- 9.24	Typhoon (9617)	Kinki-Hokkaido	D 11, I 73, H 898, F 12,226, S 61, L 309, A 167.1
1996.12. 6	Debris Flow	Nagano pref.	D 14, I 9, H 2
1997. 1. 1- 1. 7	Strong wind, High wave	The sea of Japan side (Kyushu-Hokkaido), Shikoku	D 5, S 3
1997. 7. 3- 7.19	Heavy rainfall [Front]	All over Japan (except Okinawa)	D 26, I 16, H 89, F 7,681, L 1,089, A 84.7
1997. 7.25- 7.31	Typhoon (9709)	Kyushu-Tohoku	D 3, I 47, H 78, F 2,057, L 112, A 110.5, S 14
1997. 8. 3- 8. 8	Heavy rainfall [Front]	All over Japan (except Okinawa)	D 5, I 2, H 19, F 14,563, L 629, A 43.3
1997. 9.12- 9.20	Typhoon (9719)	All over Japan (except Okinawa)	D 12, I 25, H 216, F 16,016, L 557, A 402.6, S 34
1998. 1.15- 1.19	Heavy snowfall, Strong wind [Low]	Kyushu-Tohoku	D-M 7, I 396, H 542, S 11
1998. 2. 7- 2.10	Strong wind [Low]	Shikoku-Tohoku	D-M 6, I 1, S 1
1998. 3. 5- 3. 7	Strong wind [Low]	Kinki-Hokkaido	D-M 5, I 7
1998. 5.15- 5.19	Heavy rainfall [Low]	Okinawa-Kinki	D 6, I 1, H 4, F 1,924
1998. 7. 3- 7. 5	Intense heat [High]	Kanto	D 6, I 52
1998. 7.27- 7.31	Heavy rainfall, Thunderstorm	Tokai-Tohoku	I 3, H 10, F 1,260
1998. 8. 2- 8. 9	Heavy rainfall	Chugoku-Tohoku	D 2, I 5, H 45, F 18,207
1998. 8.11- 8.19	Heavy rainfall [Front]	All over Japan	D 3, H 9, F 2,836
1998. 8.25- 9. 1	Heravy rainfall [Typhoon, Front]	All over Japan (except Okinawa)	D-M 25, I 55, H 486, F 13,927
1998. 9.14- 9.17	Typhoon (9805), Front	Shikoku-Hokkaido	D 6, I 50, H 601, F 5,840
1998. 9.21- 9.24	Typhoon (9807, 9808), Front	Shikoku-Hokkaido	D-M 18, I 569, H 21,165, F 8,692
1998. 9.23-10. 1	Typhoon (9809), Front	Okinawa-Tokai	D 9, I 14, H 143, F 17,806

Major Meteorological Disasters in Japan Continued.

Date	Disaster	Damaged areas (region)	Damage
1998.10.13-10.20	Typhoon (9810), Front	All over Japan	D-M 14, I 67, H 770, F 12,548
1999. 1. 7- 1.15	Heavy snowfall, Strong wind [Low]	Shikoku-Hokkaido	D-M 7, I 46, H 4, S 1, A 1
1999. 6.22- 7. 4	Heavy rainfall, Strong wind [Low]	Kyushu-Tohoku	D-M 40, I 64, H 615, F 12,453, L 570, S 4, A 374
1999. 7.10- 7.15	Heavy rainfall [Tropical cyclone]	Chubu-Tohoku	D-M 2, I 1, H 36, F 2,033, L 2,033, A 63
1999. 7.21- 7.22	Heavy rainfall [Front, Thunderstorm]	Koshin-Tohoku	D 3, I 4, H 9, F 5,289, L 81
1999. 8. 9- 8.17	Heavy rainfall [Tropical cyclone]	Shikoku-Tohoku	D-M 17, I 10, H 63, F 7,524, L 2,625, A 138
1999. 9.13- 9.16	Typhoon (9916), Front	Kyushu-Tohoku	D 2, I 6, H 112, F 1,663, L 278, A 21
1999. 9.16- 9.25	Typhoon (9918), Front	All over Japan	D-M 36, I 1,077, H 47,150, F 23,218, S 552, A 1,631
1999.10.27-10.28	Heavy rainfall, Strong wind [Low]	Kinki-Hokkaido	D-M 5, I 6, H 217, F 4,705, L 50, S 45, A 262
2000. 2. 8- 2.10	Heavy snowfall, Strong wind [Low]	Kyushu-Kanto	D-M 6, I 55, H 6, S 3
2000. 7. 1- 7. 5	Heavy rainfall [Thunderstorm]	Kyushu-Hokkaido	D-M 3, I 3, H 42, F 1,673, A 61
2000. 7. 6- 7. 9	Typhoon (0003)	Shikoku-Hokkaido	D-M 1, I 7, H 36, F 2,713, L 521, S 2, A 38
2000. 7.13- 7.20	Heavy rainfall [Front, Thunderstorm]	Kyushu-Tohoku	D-M 2, I 2, H 1, F 1,711, L467, A 5
2000. 9. 8- 9.17	Typhoon (0014), Front	Okinawa-Tohoku	D-M 11, I 103, H 609, F 70,017, L 5,562, S 1, A 248
2000.12.24-12.29	Heavy snowfall, Strong wind [Low]	Shikoku-Hokkaido	D-M 5, I 5, H 4
2001. 1. 1- 1. 6	Heavy snowfall, Strong wind [Low]	Kinki-Hokkaido	D-M 8, I 4, H 1, F 15
2001. 1. 7- 1.11	Heavy snowfall, Strong wind [Low]	Shikoku-Hokkaido	D-M 9, I 54, H 6, S 1, A 7
2001. 1.12- 1.22	Heavy snowfall, Strong wind [Low]	Kyushu-Tohoku	D-M 14, I 413, H 21, F 19, S 2, A 8
2001. 1.25- 1.28	Heavy snowfall, Strong wind [Low]	Kinki-Kanto	D-M 9, I 949, H 7, F 8, A 5
2001. 6.18- 6.23	Heavy rainfall [Low, Front]	Kyushu-Chubu	D-M 1, I 16, H 262, F 1,037, L 111, A 86
2001. 8.18- 8.24	Typhoon (0111), Front	Kyushu-Hokkaido	D-M 8, I 141, H 154, F 1,052, L 917, S 1, A 63
2001. 9. 8- 9.13	Typhoon (0115), Front	Shikoku-Hokkaido	D-M 8, I 48, H 140, F 1,158, L 7,348, S 9, A 111
2002. 7. 8- 7.12	Typhoon (0206), Front	All over Japan	D-M 6, I 29, H 212, F 10,302, L 25,861, A 354

Continued.

Date	Disaster	Damaged areas (region)	Damage
2002. 9.30-10. 3	Typhoon (0221), Front	Tokai-Hokkaido	D-M 5, I 96, H 1,076, F 1,950, L 58, S 25, A 183
2003. 1. 3- 1. 8	Heavy snowfall, Strong wind [Low]	All over Japan	D 6, I 400
2003. 1.25- 1.31	Heavy snowfall, Heavy rainfall, Strong wind [Low]	All over Japan	D-M 6, I 95, H 11, A 2
2003. 3.16- 3.20	Strong wind, High wave [Low]	Okinawa, Kagoshima pref., Hokkaido	D 5, I 1, S 1
2003. 7. 3- 7. 9	Heavy rainfall ["Baiu" Front]	Kyushu-Tohoku	H 4, F 1,998, L 8, A 9
2003. 7.18- 7.20	Heavy rainfall ["Baiu" Front]	Kyushu-Chubu	D 23, I 67, H 265, F 7,845, L 1,183, S 4, A 104
2003. 8. 6- 8.10	Typhoon (0310), Front	All over Japan	D-M 20, I 93, H 713, F 2,263, L 295, S 25, A 353
2003. 9.10- 9.14	Typhoon (0314)	All over Japan (In particular, Miyakojima-island)	D 3, I 107, H 1,498, F467, L 9, S 262, A 113
2003. 9.18- 9.25	Typhoon (0315)	Okinawa-Kanto	M 11, I 9, H 194, F 1, S 8, A 3
2003. May-Oct.	Low temperature, Poor sunshine	Chubu-Hokkaido	A 2,338
2003.12.25-12.27	Strong wind, High Wave [Monsoon]	Kyushu-Hokkaido	D-M 6, H 1, F 21
2004. 4.18- 4.21	Heavy rainfall, Strong wind [Low]	All over Japan	D-M 5, I 12, H 129, S 4, A 0.4
2004. 5.12- 5.13	Heavy rainfall, Strong wind [Low]	Kyushu-Kinki	H 2, F 1,636
2004. 6.18- 6.25	Typhoon (0406), Front	All over Japan	D-M 7, I 116, H 180, F 202, S 1, A 92
2004. 7.12- 7.20	Heavy rainfall	Niigata pref., Fukushima pref.	D-M 16, I 4, H 5,518, F 8,402, L 13,662, A 400
2004. 7.17- 7.21	Heavy rainfall	Gifu pref., Hokuriu, Tohoku	D-M 5, I 20, H 409, F 13,950, L 3,946, A 219
2004. 7.28- 8. 2	Typhoon (0410), Front	Western Japan, Kanto	D-M 3, I 17, H 113, F 2,215, L 2, S 4, A 103
2004. 8.16- 8.20	Typhoon (0415)	All over Japan	D-M 12, I 24, H 513, F 2,724, L 71, S 90, A 397
2004. 7. 1- 8.22	Intense heat	Kyushu, Chugoku, Kinki-Hokkaido	D 11, I 1,570
2004. 8.22- 8.27	Typhoon (0417)	Okinawa, Kagoshima pref., Oita pref., Yamaguchi pref., Ehime pref., Gifu pref.	D 4, I 28, H 3,719, F 1,256, L 101, S 847, A 67
2004. 8.26- 9. 2	Typhoon (0416)	All over Japan	D-M 18, I 285, H 8,627, F 46,581, L 102, S 995, A 1,054
2004. 9. 4- 9. 8	Typhoon (0418)	All over Japan	D-M 47, I 1,364, H 57,466, F 10,026 L 104, S 1,592, A 1,262

Major Meteorological Disasters in Japan Continued.

Date	Disaster	Damaged areas (region)	Damage
2004. 9.24- 9.30	Typhoon (0421)	Okinawa-Tohoku	D-M 27, I 95, H 3,068, F 19,153, L 1,213, S 74, A 210
2004.10. 7-10.10	Typhoon (0422)	Okinawa, Kagoshima pref., Shimane pref., Kinki-Tohoku	D-M 8, I 169, H 5,553, F 7,843, L 5,099, S 95, A 58
2004.10.17-10.21	Typhoon (0423)	Okinawa-Tohoku	D-M 99, I 704, H 19,235, F 54,850, L 12,329, S 494, A 934
2004.11.11-11.12	Heavy rainfall, Strong wind [Low]	Kagoshima pref., Shikoku-Kanto	D 1, H 7, F 1,084, L 10, A 1
2004.12. 4-12. 6	Heavy rainfall, Heavy snowfall, Strong wind [Low]	Kyushu-Hokkaido	D-M 6, I 46, H 966, F 293, L 77, S 31, A 13
2005. 1. 8- 1.13	Heavy snowfall, Strong wind [Low]	Chugoku-Tohoku	D-M 6, I 24
2005. 1.29- 2. 9	Heavy snowfall, Strong wind [Low]	All over Japan (except Okinawa)	D-M 13, I 377, H 2, F 16, A 0.5
2005. 7. 1- 7. 6	Heavy rainfall	Shikoku, Chugoku-Kanto	D-M 5, I 3, H 17, F 2,956, L 330, A 47
2005. 7. 9- 7.16	Heavy rainfall	Kyushu-Kanto, Hokkaido	D-M 6, I 3, H 8, F 584, A 5
2005. 7.22	Poor visibility	Chiba pref.	D-M 9, S 2
2005. 8. 3- 8. 5	Intense heat	Hokuriku, Kanto, Tohoku	D-M 5, I 93
2005. 8.10- 8.13	Heavy rainfall	Hokuriku	D-M 1, I 4, H 27, F 1,094
2005. 9. 3- 9. 8	Typhoon (0514), Front	All over Japan	D-M 29, I 179, H 7,452, F 21,160, L 4, S 81, A 949
2005. Dec.- 2006. Mar.	Heavy snowfall	Shiokoku-Hokkaido	D-M 152, I 2,136, H 4,713, F 113
2006. 4. 8- 4. 9	Strong wind, Avalanche	Gifu pref., Nagano pref.	D-M 10, I 5
2006. 7.13- 7.15	Intense heat	Kyushu, Tokai, Hokuriku, Kanto	D-M 5, I 143
2006. 7.15- 7.24	Heavy rainfall [Front]	Kyushu-Tohoku	D-M 30, I 46, H 1,708, F 6,996, L 562, S 50, A 151
2006. 8.22- 8.24	Heavy rainfall	Kyushu-Tohoku	D-M 2, I 1, H 2, L 1,160
2006. 9.15- 9.20	Typhoon (0613)	Okinawa, Kyushu, Shikoku, Chugoku, Hokkaido	D-M 11, I 556, H 9,251, F 934, S 276, A 174
2006.10. 4-10. 9	Heavy rainfall, Strong wind, High wave [Low, Front]	Shikoku-Hokkaido	D-M 50, I 57, H 1,154, F 1,206, L 2,600, S 1,038, A 293
2006.11. 7-11. 8	Tornado, Strong wind, High wave [Low, Front]	Shikoku-Hokkaido	D-M 9, I 42, H 64, F 2, L 14, S 1, A 1
2007. 2.13- 2.16	Strong wind, High wave, Avalanche	All over Japan	D-M 11, I 20, H 116, F 1, S 4, A 1
2007. 7. 1- 7.17	Front, Typhoon (0704)	Okinawa-Tohoku	D-M 7, I 83, H 295, F 3,993, L 55, S 4, A 253

Continued.

Date	Disaster	Damaged areas (region)	Damage
2007. Jun.-Sep.	Intense heat	All over Japan	D-M 66, I 5,371
2007. 9. 5- 9.12	Typhoon (0709), Front	Kinki-Hokkaido	D=M 3, I 87, H 672, F 1,345, L 1,603, S 19, A 166
2007. 9.15- 9.18	Heavy rainfall	Tohoku	D-M 4, I 5, H 238, F 1,396, L 13,685, S 3, A 90
2007.11.19-11.23	Avalanche, Strong wind, High wave, Heavy snowfall	Kyushu-Hokkaido	D-M 5, I 2, H 1
2007.12.29- 2008. 1. 3	Avalanche, Heavy snowfall, Strong wind	Kyushu-Tohoku	D-M 5, I 2, H 41
2008. 6.19- 6.25	Heavy rainfall	Kyushu-Tohoku	D-M 1, I 2, H 14, F 1,172, L 33, S 2, A 3
2008. 7.27- 7.29	Heavy rainfall, Lightning strike, Gust	Shikoku, Chugoku-Tohoku	D-M 8, I 32, H 72, F 2,797, L 244, S 1, A 3
2008. 8. 4- 8. 9	Heavy rainfall, Lightning strike	Kyushu-Kanto	D-M 8, I 3, H 21, F 3,819, A 4
2008. 8.26- 8.31	Heavy rainfall	Kyushu-Hokkaido	D-M 2, I 7, H 52, F 21,844, L 994, A 16
2009. 7. 1- 8.31	Long spell of rain, Low temperature, Poor sunshine	Hokkaido	A 595
2009. 7.19- 7.26	Heavy rainfall	Kyushu-Kanto	D-M 39, I 34, H 378, F 11,541, L 590, A 102
2009. 8. 8- 8.11	Typhoon (0909)	Kyushu-Tohoku	D-M 28, I 29, H 1,173, F 5,217, L 446, A 71
2009.10. 6-10. 9	Typhoon (0918)	All over Japan	D-M 6, I 133, H 2,387, F 3,310, I 886, S 91, A 218
2010. Jun.-Sep.	Intense heat, Heavy rainfall	All over Japan	D-M 271, I 20,998, A 495
2010. 7. 1- 7. 6	Heavy rainfall, Strong wind, Lightning strike	All over Japan	D-M 5, I 7, H 14, F 1,186, L 167, S 2, A 18
2010. 7.10- 7.16	Heavy rainfall	Kyushu-Tohoku	D-M 14, I 15, H 251, F 5,380, L 112, A 167
2011. 7.27- 7.30	Heavy rainfall	Niigata pref., Fukushima pref.	D-M 6, I 13, H 1,107, F 9,025
2011. 8.30- 9. 5	Typhoon (1112)	Shikoku-Hokkaido	D-M 98, I 113, H 4,008, F 22,094
2011. 9.15- 9.22	Typhoon (1115)	All over Japan	D-M 19, I 337, H 3,739, D 7,840
2012. 7.11- 7.14	Heavy rainfall	Northern Kyushu	D-M 32, I 27, H 2,176, F 12,606
2013. 7.22- 8. 1	Heavy rainfall	Kyushu-Hokkaido	D-M 5, I 17, H 84, F 3,586
2013. 8. 9- 8.10	Heavy rainfall	Tohoku	D-M 8, I 12, H 131, F 1,941
2013. 9.15- 9.16	Typhoon (1318)	All over Japan	D-M 7, I 143, H 1,650, F 10,089

Major Meteorological Disasters in Japan Continued.

Date	Disaster	Damaged areas (region)	Damage
2013.10.14-10.16	Typhoon (1326)	Kanto	D-M 43, I 130, H 1,094, F 6,142
2014. 7.30- 8.26	Heavy rainfall	All over Japan	D-M 91, I 167, H 4,817, F 16,517
2014.10. 4-10. 6	Typhoon (1418)	Tokai-Kanto	D-M 7, I 72, H 257, F 2,540
2015. 9. 7- 9.11	Heavy rainfall [Typhoon (1518)]	Shikoku-Tohoku	D-M 20, I 82, H 7,555, F 15,782
2016. 6. 6- 7.15	Heavy rainfall	All over Japan	D-M 7, I 12, H 391, F 2,535
2016. 8.16- 8.31	Typhoon (1607, 1609, 1610, 1611)	All over Japan	D-M 31, I 90, H 4,575, F 5,283
2016. 9.17- 9.20	Typhoon (1616)	All over Japan	D-M 1, I 47, H 2,279, F 2,455
2017. 6. 7- 7.27	"Baiu" Front, Typhoon (1703)	All over Japan	D-M 44, I 39, H 1,578, F 4,525, A 1,124
2017. 9.13- 9.19	Typhoon (1718), Front	All over Japan	D-M 5, I 73, H 1,424, F 7,475, A 343
2017.10.20-10.23	Typhoon (1721)	All over Japan	D-M 8, I 245, H 3,132, F 8,183, A 660
2018. 1.22- 1.27	Heavy snowfall Snowstorm	Kanto-Tohoku	D 5, I 976, H4
2018. 2. 3- 2. 8	Heavy snowfall	Hokuriku, Tohoku, Sea of Japan side	D 22, I 320, H 39, F 11
2018. 6.28- 7. 8	Heavy rainfall [Front] Typhoon (1807)	All over Japan (Especially western Japan)	D-M 271, I 484, H 22,491, F 28,619, L 21,168, S 86
2018. 9. 3- 9. 5	Typhoon (1821)	Kansai-Tohoku	D 14, I 980, H 97,910, F 707, L 30,996, S 131
2018. 9.28-10. 1	Typhoon (1824)	All over Japan	D 4, I 231, H 10,407, F 2,163, L 64,590, S 433
2019. 8.26- 8.29	Heavy rainfall [Front]	Northern Kyushu	D 4, I 4, H 1,040, F 5,678, L 9,425, S 7
2019. 9. 8- 9. 9	Typhoon (1915)	Kanto	D 9, I 160, H 93,096, F 276, L 14,841, S 395
2019.10.11-10.13	Typhoon (1919)	Eastern Japan	D-M 107, I 384, H 70,652, F 31,021, L 22,949, S 307

●Explanatory note

D: Dead, D-M: Dead or Missing, M: Missing, I: Injured,
B: Number of houses burned down
H: Homes completely or partially destroyed by fire,
F: Homes flood to above/below floor level,
L: Arable land washed away, buried, and flooded (ha),
S: Ships sunk, washed way, and damaged,
A: Damage to agricultural, forestry, fishery industry (units 100 Million Yen)

Physics and Chemistry Section

Units

The values of a physical quantity are expressed as the product of a numerical number and a unit[*1]. The unit is a particular example of the physical quantity and used as a reference of that quantity. The number is the ratio of the physical quantity value to the unit. Accordingly, it is possible to express a physical quantity value with several combinations of a number and a unit. However, in order to facilitate social activities, it is vital to select a single internationally accepted unit for each physical quantity. The selected units *must be readily available to any person*, *must be constant throughout time and space*, and *must be easily realized with high accuracy*.

Furthermore, the equations (physical laws) between the physical quantities determine the relations between the units. A coherent system of units is a system of units that consists of a small number of units (**base units**) and units derived by physical laws or definitions of physical quantities (**derived units**): the derived units should be expressed by simple products of powers of the base units, no conversion factor between units is required.

International System of Units (SI)

At the General Conference on Weights and Measures in 1960, the decision was made to adopt an expanded version of the metric system as **the International System of Units (Système International d'Unités, SI)**[*2]. The SI is a system of units that is widely used throughout the world in all applicable fields.

Before 2019, the SI has defined the seven base units (s, m, kg, A, K, mol, cd) corresponding to the seven base quantities (time, length, mass, electric current, temperature, amount of substance, luminous intensity) by specific artificial objects (International Prototype of the Kilogram), idealized experimental prescriptions (zero thickness conductor) and a specific physical state (triple point of water), *etc.*

The SI revision, which took effect in 2019 (the resolution was adopted in 2018), first defines the numerical values and units of seven **defining constants**. These defining constants are the unperturbed ground state hyperfine transition frequency of the caesium 133 atom $\Delta\nu_{Cs}$, the speed of light in vacuum c, the Planck constant h, the elementary charge e, the Boltzmann constant k, the Avogadro constant N_A, and the luminous efficancy of monochromatic radiation of frequency 540×10^{12} Hz, K_{cd}. The seven base units are determined from the defining constants by dimensional analysis using multiplication and division of the defining constants.

The SI units are composed of these seven base units and the derived units formed as products of powers of the base units. Special names have been given to some of the derived units.

---

*1 As subsequently explained, a numerical number and a unit are equivalent. Therefore, a unit must not be placed in parentheses when representing a value of physical quantity. For example, the value for gravitational acceleration g on the surface of the earth is written as $g = 9.8$ m s^{-2}.

*2 For historical details, refer to page **290** "History of Units and Standards".

*3 The source is the SI brochure, the Le Système International d'Unités, 9e édition (9th edition of the International System of Units) issued by the Bureau International des Poids et Measures (the International Bureau of Weights and Measures) in 2019.

Previous SI Base Units (Before 2019)

The SI base units are defined as follows.

Time: The **second** (s) is the SI unit of time. It is defined by taking the fixed numerical value of the caesium frequency, $\Delta\nu_{Cs}$, the unperturbed ground-state hyperfine transition frequency of the caesium 133 atom, to be 9 192 631 770 when expressed in the unit Hz, which is equal to s^{-1}.

Electric current: The **metre** (m) is the SI unit of length. It is defined by taking the fixed numerical value of the speed of light in vacuum, c, to be 299 792 458 when expressed in the unit m s^{-1}, where the second is defined in terms of the caesium frequency $\Delta\nu_{Cs}$.

Mass: The **kilogram** (kg) is the SI unit of mass. It is defined by taking the fixed numerical value of the Planck constant, h, to be $6.626\,070\,15\times10^{-34}$ when expressed in the unit J s, which is equal to kg m^2 s^{-1}, where the metre and the second are defined in terms of c and $\Delta\nu_{Cs}$.

Electric current: The **ampere** (A) is the SI unit of electric current. It is defined by taking the fixed numerical value of the elementary charge, e, to be $1.602\,176\,634\times10^{-19}$ when expressed in the unit C, which is equal to A s, where the second is defined in terms of $\Delta\nu_{Cs}$.

Temperature: The **kelvin** (K) is the SI unit of thermodynamic temperature. It is defined by taking the fixed numerical value of the Boltzmann constant, k, to be $1.380\,649\times10^{-23}$ when expressed in the unit J K^{-1}, which is equal to kg m^2 s^{-2} K^{-1}, where the kilogram, metre and second are defined in terms of h, c and $\Delta\nu_{Cs}$.

Amount of substance: The **mole** (mol) is the SI unit of amount of substance. One mole contains exactly $6.022\,140\,76\times10^{23}$ elementary entities. This number is the fixed numerical value of the Avogadro constant, N_A, when expressed in the unit mol^{-1} and is called the Avogadro number. The amount of substance, symbol n, of a system is a measure of the number of specified elementary entities. An elementary entity may be an atom, a molecule, an ion, an electron, any other particle or specified group of particles.

Luminous intensity: The **candela** (cd) is the SI unit of luminous intensity in a given direction. It is defined by taking the fixed numerical value of the luminous efficacy of monochromatic radiation of frequency 540×10^{12} Hz, K_{cd}, to be 683 when expressed in the unit lm W^{-1}, which is equal to cd sr W^{-1}, or cd sr kg^{-1} m^{-2} s^3, where the kilogram, metre and second are defined in terms of h, c and $\Delta\nu_{Cs}$.

---

1) Vienna Standard Mean Ocean Water.

SI Derived Units with Special Names and Symbols

An SI assembly unit that has a unique name and is expressed as a multiplication or division of a basic unit.

Derived quantity	Name	Symbol	Expressed in terms of other SI units	Expressed in terms of SI base units
plane angle	radian[1)]	rad		m/m
solid angle	steradian[1)]	sr		m^2/m^2
frequency	hertz[2)]	Hz		s^{-1}
force	newton	N		$kg\ m\ s^{-2}$
pressure, stress	pascal	Pa	N/m^2	$kg\ m^{-1}\ s^{-2}$
energy, work, amount of heat	joule	J	N m	$kg\ m^2\ s^{-2}$
power, radiant flux	watt	W	J/s	$kg\ m^2\ s^{-3}$
electric charge, amount of electricity	coulomb	C		A s
electric potential difference, electromotive force	volt	V	W/A	$kg\ m^2\ s^{-3}\ A^{-1}$
capacitance	farad	F	C/V	$kg^{-1}\ m^{-2}\ s^4\ A^2$
electric resistance	ohm	Ω	V/A	$kg\ m^2\ s^{-3}\ A^{-2}$
electric conductance	siemens	S	A/V	$kg^{-1}\ m^{-2}\ s^3\ A^2$
magnetic flux	weber	Wb	V s	$kg\ m^2\ s^{-2}\ A^{-1}$
magnetic flux dencity	tesla	T	Wb/m^2	$kg\ s^{-2}\ A^{-1}$
inductance	henry	H	Wb/A	$kg\ m^2\ s^{-2}\ A^{-2}$
Celsius temperature	degree Celsius[3)]	℃		K
luminous flux	lumen	lm	cd sr	cd
illuminance	lux	lx	lm/m^2	$cd\ m^{-2}$
activity reffered to a radionuclide	becquerel[2)]	Bq		s^{-1}
absorbed dose, specific energy (imparted), kerma	gray	Gy	J/kg	$m^2\ s^{-2}$
dose equivalent, ambient dose equivalent, directional dose equivalent, personal dose equivalent	sievert	Sv	J/kg	$m^2\ s^{-2}$
catalytic activity	katal	kat		$mol\ s^{-1}$

1) The radian and steradian are special names for the number one that may be used to convey information about the quantity concerned. In practice the symbol rad and sr are used where appropriate, but the symbol for the derived unit one is generally omitted in specifying the values of dimensionless quantities.

2) A steradian is a coherent unit of solid angle.1 sr is a solid angle cut on the surface of a sphere with an area equal to a square whose vertex is the center of the sphere and whose side is the radius of the sphere.

3) The heltz is used only for periodic phenomena and the becquerel is used only for stochastic processes in activity reffered to a radionuclide.

4) The Celsius temperature (θ) is defined in terms of the thermodynamical temperature (T) by the following equation: $\theta/℃ = T/K - 273.15$.

Examples of SI Derived Units

Derived quantity	Name	Symbol	Expressed in terms of SI base units
area	square meter	m^2	
volume	cubic meter	m^3	
density	kilogram per cubic meter	kg/m^3	
speed, velocity	meter per second	m/s	
acceleration	meter per square second	m/s^2	
angular velocity	radian per second	rad/s	$m\ m^{-1}\ s^{-1} = s^{-1}$
moment of force	newton meter	N m	$kg\ m^2\ s^{-2}$
surface tension	newton per meter	N/m	$kg\ s^{-2}$
dynamic viscosity[1)]	pascal second	Pa s	$kg\ m^{-1}\ s^{-1}$
kinematic viscosity[1)]	square meter per second	m^2/s	
heat flux density, irradiance	watt per square meter	W/m^2	$kg\ s^{-3}$
heat capacity, entropy	joule per kelvin	J/K	$kg\ m^2\ s^{-2}\ K^{-1}$
specific heat capacity, specific entropy	joule per kilogram kelvin	$J\ kg^{-1}\ K^{-1}$	$m^2\ s^{-2}\ K^{-1}$
thermal conductivity	watt per meter kelvin	$W\ m^{-1}\ K^{-1}$	$kg\ m\ s^{-3}\ K^{-1}$
electric field strength	volt per meter	V/m	$kg\ m\ s^{-3}\ A^{-1}$
electric flux density, electric dsiplacement	coulomb per square meter	C/m^2	$A\ m^{-2}\ s$
permittivity	farad per meter	F/m	$kg^{-1}\ m^{-3}\ s^4\ A^2$
current density	ampere per square meter	A/m^2	
magnetic field strength	ampere per meter	A/m	
permeability	henry per meter newton per (ampere)^2	H/m N/A^2	$kg\ m\ s^{-2}\ A^{-2}$
magnetomotive force, magnetic potential difference	ampere	A	
amount concentration	mole per cubic meter	mol/m^3	
luminance	candela per square meter	cd/m^2	
wavenumber	reciprocal meter	m^{-1}	
exposure	coulomb per kilogram	c/kg	$A\ s\ kg^{-1}$

SI Prefixes for Expressing the Powers of 10

Name	Symbol	Factor	Name	Symbol	Factor
yotta	Y	10^{24}	deci	d	10^{-1}
zetta	Z	10^{21}	centi	c	10^{-2}
exa	E	10^{18}	milli	m	10^{-3}
peta	P	10^{15}	micro	μ	10^{-6}
tera	T	10^{12}	nano	n	10^{-9}
giga	G	10^{9}	pico	p	10^{-12}
mega	M	10^{6}	femto	f	10^{-15}
kilo	k	10^{3}	atto	a	10^{-18}
hecto	h	10^{2}	zepto	z	10^{-21}
deca	da	10	yocto	y	10^{-24}

Note: The above prefixes can not be combined to create new prefixes.

Symbols of physical quantities and units, and notation of values of physical quantities

The symbols for physical quantities and units recommended by the SI and the notation for the values of physical quantities are as follows The SI recommended symbols for physical quantities and units, and the notation for the values of physical quantities are as follows.

Symbols for physical quantities

- The symbol for a physical quantity is usually a single letter of the alphabet in italics. The symbol for a physical quantity is usually a single letter of the alphabet in italics.
- Additional information should be written in subscript or superscript, or in parentheses. If the additional information is derived from a physical quantity, it should be written in italics; otherwise, it should be written in Roman (solid) type.

  Example: velocity υ A of object A (A is a solid), constant pressure molar specific heat CP (P is an italic)

Unit symbol

- The unit symbol is written in solid letters. Normally lowercase letters are used, but special names derived from human names should have the first letter capitalized. The exception is the unit symbol for liters. The exception is the unit symbol for liters.

  The exception is the unit symbol for liters, for which an uppercase L is allowed to be used to avoid confusion between the lowercase l and the numeral 1. The exception is the unit symbol for liters.

- The unit symbol is a letter symbol in mathematics, not an abbreviation. Therefore, no periods should be added. The unit symbol and the name of the unit should not be used together when describing a physical quantity, nor should they be pluralized.
  Bad example: pressure p = 101 325 newton m^{-2} (correct: p = 101 325 N m^{-2})
- The unit symbol should not have any special meaning. It is the symbols for physical quantities that can have special meanings.
  Bad example: maximum voltage drop U = 1000 V_{max} (correct: U_{max} = 1000 V)
- For the product and quotient of unit symbols, the usual mathematical rules apply. For multiplication, use one-byte space or one-byte black (.). For division, use the usual fraction form, a slash (/), or a negative power exponent. When combining several unit symbols, care should be taken not to mislead. In particular, it is recommended that the slash (/) be used only once in a notation. Or use parentheses appropriately to avoid misunderstandings.

Value of physical quantity

- When writing the value of a physical quantity, the numerical value, single-byte space, and unit should be written in this order. The exceptions are the unit symbols for plane angles, °, ′, and ″, which are not preceded by a space, as in ϕ = 30° 22′8″. A similar notation is used for the unit℃ for temperature in degrees Celsius, but this unit is preceded by a space, as in t = 30.2℃.
- The value of a physical quantity is expressed as a product of a number and a unit. Therefore, the usual mathematical rules apply between the value of the physical quantity, the number, and the unit.
  Example: The pressure p = 101 325 Pa can also be written as p/Pa = 101 325.
- When writing numerical values in a table or on the axes of a graph, use the name of the entry in the table or the axis label in the graph that means the symbol of the physical quantity divided by the unit, i.e., a numerical value, such as p/Pa.
- Use a dot (.) or comma (,) for the decimal point of a number.) or a comma (,).
- If a number has five or more digits starting from the decimal point, a single-byte space should be inserted to separate each of the three digits.
  Example: f = 384 227.981 9 GHz or f = 384 227.9819 GHz.
- To express the value of a physical quantity including uncertainty, write m_n = 1.674 927 471 (21) × 10^{-27} kg. where m_n is the symbol of the physical quantity (in this case, the mass of the neutron). The number in parentheses represents the standard uncertainty, which means that the last two digits of the previous number have this uncertainty. If the number in parentheses is a single digit, the uncertainty is in the last digit.

History of Units and Standard

There have been numerous units serving as a reference of a physical quantity since ancient times. These units have changed with the passage of time. The following table lists major changes in a definition of unit, particularly the SI base unit, and its standard, a thing or method to practically realize a corresponding definition of unit, in chronological order. The following abbreviations are used in the table: CGPM (General Conference on Weights and Measures, Conférence Générale des Poids et Mesures: conference at which final decisions are made), and CIPM (International Committee for Weights and Measures, Comité International des Poids et Mesures: administrative committee which makes proposals to CGPM).

Era	Details
Medieval Period	《1 second is 1/86 400 of a mean solar day (mean solar second)》 A solar day is the time required for a imaginary sun (mean sun) moving at the average speed of the sun during one year along the celestial equator to pass the meridian and then rotate completely around to pass the same meridian again.
1789 and later	At the time of the French Revolution, **a decimal metric system** was established.
1791	《1 meter is 1/10 000 000 of the meridian between the North Pole and the Equator》 《1 kilogram is defined as, at 1 atmosphere, the mass of 1000 cm^3 of pure water at its maximum density》
1792 to 1798	Triangulation between Dunkirk (France) and Barcelona (Spain) (for the creation of a standard to realize the definition of 1 meter)
1799	Two **platinum standards** representing the meter and the kilogram were stored in the collection of the National Archives (France). This standard meter bar was known as **the mètre des Archives**. The French metric system was promulgated. 《1 meter is defined as the length of the meter prototype》, 《1 kilogram is defined as the mass of the kilogram prototype》
1832	C. F. Gauss began investigating the terrestrial magnetism by using the decimal system based on the 3 mechanical units of millimeter, gram and second, which were used for the quantities of length, mass and time, respectively. Afterwards, Gauss also cooperated with W. E. Weber to investigate electromagnetic phenomena.
1860s	J. C. Maxwell and W. Thomson advocated the need for **a coherent system of units** with base units and derived units.
1874	The British Association for the Advancement of Science implemented **the CGS system of units** as a three-dimensional coherent system of units based on the 3 mechanical units of centimeter, gram and second.
1875	**Metre Convention** was signed on May 20.
1880s	The British Association for the Advancement of Science and the International Electrical Congress (precursor to the International Electrotechnical Commission) approved a mutually coherent set of practical units. The approved units included the ohm for electrical resistance, the volt for electromotive force and the ampere for electrical current (in the fields of electricity and magnetism, it had become clear that the size of the coherent CGS units was inconvenient to use).
1889	1st CGPM · **Sanction of the international prototypes of the meter and the kilogram**: 《1 meter is the distance between the two lines carved on the international prototype of the meter》, 《1 kilogram is the mass of the international prototype of the kilogram》 Together with the astronomical second (mean solar second) as the unit of time, these units constituted a three-dimensional mechanical unit system with the base units of meter, kilogram, and second, **the MKS system**, similar to the CGS system.
1901	3rd CGPM · Declaration on the unit of mass (the mass of the international prototype of the kilogram is 1 kg) and on the definition of weight (weight is the same physical quantity as force).

Era	Details
1901	G. Giorgi showed the possibility to constitute a single coherent four-dimensional system by adding a fourth unit of electromagnetism (e.g. ampere, ohm) to the three mechanical units (meter, kilogram, second).
1927	7th CGPM · **Definition of the meter by the international Prototype**: Reconfirmed the length of the meter as the distance between the two lines carved on the international prototype of the meter, and specified the method to realize the definition of the meter practically.
1939	International Bureau of Weights and Measures, Advisory Committee on Electricity · Proposal for **the MKSA system**, a four-dimensional system based on the meter, kilogram, second, and ampere.
1946	CIPM · Definition of photometric units (bougie nouvelle, precursor to the candela, etc.). · Definition of the electric units (ampere, volt, ohm, etc.).
1948	9th CGPM · Thermodynamic temperature scale with a single fixed point (triple point of water). · Proposal for establishing a practical system of units of measurement (establishment of the SI) based on the definition from the CIPM held in 1946.
1954	10th CGPM · **Definition of the thermodynamic temperature scale**: 《the thermodynamic temperature for the triple point of water is 273.16 K, exactly》 · Practical system of units: Decided to adopt the following six units as base units: length (meter), mass (kilogram), time (second), electric current (ampere), thermodynamic temperature (kelvin), and luminous intensity (candela).
1960	11th CGPM · **New definition of the meter**: 《The meter is the length equal to 1 650 763.73 wavelengths in vacuum of the radiation corresponding to the transition between the levels $2p_{10}$ and $5d_5$ of the ^{86}Kr atom》 · **New definition of the unit of time (second)**: 《The second is the fraction 1/31 556 925.9747 of the tropical year for 1900 January 0 at 12 hours ephemeris time》 (Ratification of definition established at the CIPM in 1956) · **Establishment of the International System of Units (Système International d'Unités; "SI")**: Based on resolution recommended by the CIPM in 1956, six base units, two supplementary units (rad and sr) and many derived units were adopted.
1964	12th CGPM · Atomic and molecular frequency standards: Empowered the CIPM to name the atomic or molecular frequency standards. In the same year, the CIPM declared that the standard to be employed is the transition between the hyperfine levels of the ground state of ^{133}Cs atom.
1967/ 1968	13th CGPM · **SI unit of time (second)**: Abrogated previous definitions and ratified the previous definition (page **286**). · **SI unit of thermodynamic temperature (kelvin)**: Restricted the use of the names "degree kelvin" and "degree", and the symbols "°K" and "deg". · **Definition of the SI unit of thermodynamic temperature (kelvin)**: Ratified the previous definition (page **286**). (Further clarification of the existing definition)
1967/ 1968	· **SI unit of luminous intensity (candela)**: 《The candela is the luminous intensity, in the perpendicular direction, of a surface of 1/600 000 m^2 of a black body at the temperature of freezing platinum under a pressure of 101 325 N/m^2》. Abrogated the unit name of "bougie nouvelle".
1970	Definition of International Atomic Time (Temps Atomique International, TAI)
1971	14th CGPM · **SI unit of amount of substance (mole)**: The mole was added as a base unit of the SI and the previous definition was established (page **286**).
1975	15th CGPM · Recommended value for the speed of light: The use of the value for the speed of propagation of electromagnetic waves in vacuum $c = 299\,792\,458$ m/s. · Recommended the use of Coordinated Universal Time (UTC)
1979	16th CGPM · **SI unit of luminous intensity (candela)**: Abrogated previous definitions and ratified the previous definition (page **286**).

Era	Details
1980	CIPM ratified **the previous remark for the unit of amount of substance** (page **286**).
1983	17th CGPM · **Definition of the meter**: Abrogated previous definitions and ratified the previous definition (page **286**). · On the realization of the definition of the meter: Issued a request to CIPM for selection of radiation to serve as standards of wavelength for the interference measurement of length. In the same year, CIPM recommended **a list of radiation** and methods to practically realize the definition of the meter.
1987	18th CGPM · Forthcoming adjustment to the representations of the volt and of the ohm: Instructed CIPM to recommend the adjustments.
1988	CIPM · Representation of the volt by means of the Josephson effect (**voltage standard**): As the agreed value for the Josephson constant, recommended the value $K_{J\text{-}90} = 483\,597.9$ GHz/V. · Representation of the ohm by means of the quantum Hall effect (**resistance standard**): As the agreed value for the von Klitzing constant, recommended the value $R_{K\text{-}90} = 25\,812.807\ \Omega$.
1989	CIPM · International Temperature Scale of 1990 (ITS-90): Abrogated the International Practical Temperature Scale of 1968 (IPTS-68) and the 1976 Provisional 0.5 K to 30 K Temperature Scale (EPT-76), and recommended that the ITS-90 came into force on January 1, 1990.
1995	20th CGPM · Elimination of the class of supplementary units in the SI: Decided to interpret the supplementary units in the SI, namely the radian and the steradian, as dimensionless derived units (provided by CIPM in 1980).
1997	CIPM confirmed **the previous note** (page **286**) **for the unit of time (second)**.
1999	21st CGPM · The definition of the kilogram: Recommended that national laboratories continue their efforts to refine experiments that link the unit of mass to fundamental or atomic constants with a view to a future redefinition of the kilogram.
2002	CIPM · **Revision of the practical realization of the definition of the meter**[1]: Considering advancements in technology for measuring wavelengths in optical domains, recommended that the list of recommended radiation be replaced by the list of radiation which includes radiations from trapped cold atoms and ions and wavelengths of stable lasers.
2007	23rd CGPM · **Clarification of the definition of the kelvin, unit of thermodynamic temperature**: Decided that the definition of the kelvin refers to water of a specified isotopic composition (previous note (page **286**)) (ratification of the decision made by the CIPM in 2005).
2011	24th CGPM · On the possible future revision of the International System of Units (SI): While maintaining the present seven base units, proposed a revision of the SI that links the definition of base units to exact **numerical values of the invariant of nature** (fundamental constants and atomic constants; specifically, Planck constant, elementary charge, Boltzmann constant, Avogadro constant, etc.).
2014	25th CGPM · Revision of the International System of Units (SI) in the future: All necessary efforts will be made to define and resolve the new SI at the next CGPM (2018). For the future revision of the International System of Units (SI): requested the National Standards Institutes, University Institutes, BIPM and CIPM to make all necessary efforts to define and resolve the new SI at the next CGPM (2018). For future revision of the International System of Units (SI).
2018	26th CGPM · Revision of the International System of Units (SI): The SI is the system of units described in page **285**. The previous definitions of the base units (page **286**) are abrogated. Also the agreed value of the voltage and resistance standards ($K_{J\text{-}90}$ and $R_{K\text{-}90}$) are abrogated.

1) After this, it was performed once every few years. The latest version of the list of recommended radiations is found at http://www.bipm.org/en/publications/mises-en-pratique/

System of Units for Electricity and Magnetism

International System of Units (SI)

Based on the MKSA system of units, the four base units of m, kg, s, and A are used for electricity and magnetism.

CGS Electrostatic System of Units (CGS-ESU)

Using the three base units (cm, g, s) and assuming that the vacuum permittivity is the dimensionless quantity of value 1, 1 esu is defined as the electric quantity when the equivalent force exerted on the electric quantity interval at a distance of 1 cm in vacuum is 1 dyn.

CGS Electromagnetic System of Units (CGS-EMU)

Assuming that the vacuum permeability is the dimensionless quantity of value 1, 1 emu is defined as the strength of the magnetic pole when the equivalent force exerted on the magnetic pole at a distance of 1 cm in vacuum is 1 dyn.

The CGS-Gauss System of Units (CGS Symmetrical System of Units)

Assuming that the vacuum permittivity and permeability is the dimensionless quantity of value 1, esu is used for electric quantity and emu is used for magnetic quantity. In relational expressions that include electric quantity and magnetic quantity, the speed of light in vacuum appears as a proportional constant having the velocity dimension.

Comparison of System of Units for Electromagnetism

(The numerical value of 1 unit value in the SI when expressed in the CGS units system.)

Quantity and symbol	SI	CGS	
		esu	emu
amount of electricity Q	C	$c\cdot10^{-1}$	10^{-1}
electric flux density D	C/m^2	$4\pi c\cdot10^{-5}$	$4\pi\cdot10^{-5}$
polarization P	C/m^2	$c\cdot10^{-5}$	10^{-5}
current I	A	$c\cdot10^{-1}$	10^{-1}
electric potential V	V	$\frac{1}{c}\cdot10^{8}$	10^{8}
electric field E	V/m	$\frac{1}{c}\cdot10^{6}$	10^{6}
electric resistance R	Ω	$\frac{1}{c^2}\cdot10^{9}$	10^{9}
capacitance C	F	$c^2\cdot10^{-9}$	10^{-9}
permittivity ε	F/m	$4\pi c^2\cdot10^{-11}$	$4\pi\cdot10^{-11}$
magnetic pole strength Q_m	Wb	$\frac{1}{4\pi c}\cdot10^{8}$	$\frac{1}{4\pi}\cdot10^{8}$
magnetic flux φ	Wb	$\frac{1}{c}\cdot10^{8}$	10^{8} 1)
magnetic flux density B	T	$\frac{1}{c}\cdot10^{4}$	10^{4} 2)
magnetization M	A/m	$c\cdot10^{-3}$	10^{-3} 2)
magnetomotive force, magnetic potential F_m	A	$4\pi c\cdot10^{-1}$	$4\pi\cdot10^{-1}$ 3)
strength of magnetic field H	A/m	$4\pi c\cdot10^{-3}$	$4\pi\cdot10^{-3}$ 4)

Continued.

Quantity and symbol	SI	CGS	
		esu	emu
inductance L	H	$\frac{1}{c^2}\cdot 10^9$	10^9
permeability μ	H/m N/A^2	$\frac{1}{4\pi c^2}\cdot 10^7$	$\frac{1}{4\pi}\cdot 10^7$

In this chart, c is the numecial value of the speed of light when expressed in cm/s, c =2.997924588 × 10^{10} (dimensionless), not the actual speed of light.

In the SI $D=\varepsilon_0 E+P$, $H=\frac{B}{\mu_0}-M$, where ε_0: electric constant μ_0: magnetic constant. (Refer to page **298**)

In the CGS-Gauss units systems, the quantities related to electricity (quantities ε and above in this chart) use the fundamental units in the CGSesu, and the quantities related to magnatism (Q_m and below in the chart) use the fundamental units in the CGSemu.

1)~4) have proper names as fundamental units of the CGS-emu

1) maxwell, Mx; 2) gauss, G; 3) gilbert, Gi; 4) oersted, Oe

Non-SI Units

The non-SI units used in various fields are shown here.

(underline) Units which are not recommended for use in the SI
(Bold-face type) Units which can be used together with the SI
* The value in the SI units has been obtained experimentally; can be used in combination with the SI units.
** A unit used provisionally
† CGS unit with a proper name

Time

minute, min : 1 min = 60 s.
hour, h : 1 h = 60 min = 3600 s.
day, d : 1 d = 24 h = 86 400 s.

Length

fermi : 1 fermi = 1 fm = 10^{-15} m.
ångström**, Å : 1 Å = 0.1 nm = 10^{-10} m.
micron, μ : Because of the possiblity to confuse this with the SI prefix μ, it is not used. Instead, μm should be used since 1 μ = 10^{-3} m = 1 μm.
astronomical unit*, au : 1 au = 149 597 870 700 m (defined value)[1)]
parsec, pc : 1 pc ≒ 3.085 678 × 10^{16} m.
The distance at which 1 au subtends an angle of 1 arcsecond. (Refer to page **47**)
light-year, ly : 1 ly = 9 460 730 472 580 800 m.
The distance that light travels in vacuum in one Julian year (365.25 days). (Refer to page **47**)
(international) nautical mile** : 1 nautical mile = 1 852 m (defined value).
Nautical miles are used in aviation.

Area

barn**, b : 1 b = 10^{-28} m^2 = 100 fm^2.
used for effective cross secions in nuclear physics.
are, a : 1 a = 100 m^2.
hectare, ha : 1 ha = 10^4 m^2 = 1 hm^2.

Volume

litre, L[2) : 1 L = 1 dm^3 = 10^{-3} m^3.

Planar Angle

degree, ° : 1° = 1/90 of a right angle = (π/180) rad.
1 rad ≒ 57.295 78° ≒ 57°17′44″.

minute, ′ : 1′ = 1/60 degree = (π/10 800) rad.

second, ″ : 1″ = 1/60 minute = (π/648 000) rad.

Mass

unified atomic mass unit, u : 1 u = 1.660 539 040(20) × 10^{-27} kg.

1/12 the mass of 1 atom of the ^{12}C isotope at rest.

dalton, Da : 1 Da = 1 u.

tonne, t : 1 t = 1 000 kg.

Speed

knot** : 1 knot = 1 nautical mile/h = 1.852 km/h (defined value) ≒ 0.5144 m/s.

Acceleration

gal**, Gal : 1 Gal = 1 cm/s^2 = 10^{-2} m/s^2.

Unit of acceleration used in gravimetry and geophysics[3).

Force (weight)

dyne†, dyn : 1 dyn = 1 g cm/s^2 = 10^{-5} N.

kilogram-force, kgf : 1 kgf = 9.806 65 N (defined value)[4).

Pressure

bar, bar : 1 bar = 10^6 dyn/cm^2 = 10^5 N/m^2 = 10^5 Pa.

Milimeters of mercury, mmHg = torr. Torr : 1 Torr ≒ 133.322 Pa.

standard atmosphere, atm : 1 atm = 101 325 Pa = 760 mmHg (defined value).

kilogram weight per square centimeter, kgf/cm^2 : 1 kgf/cm^2 = 1 kp/cm^2 = 98 066.5 Pa (defined value).

Work, Energy

erg†, erg : 1 erg = 1 dyn cm = 10^{-7} J.

electron volt, eV : 1 eV = 1.602 176 620 8(98) × 10^{-19} J, Kinetic energy gained by an electron moving across an electric potential difference of one volt.

Caloric Value[5)

calorie, cal : When the temperature is not specified, 1 cal = 4.184 J (defined value).

kilocalorie, also known as a large calorie, or food calorie : 1 kcal = 1000 cal.

Power

horse-power (metric), PS : 1 PS = 75 m kgf/s = 735.5 W.

horse-power (imperial/mechanical), hp : 1 hp = 550 ft lbf/s = 745.7 W.

Viscosity

poise†, P : 1 P = 1 dyn s/cm^2 = 0.1 Pa s.

Kinematic Viscosity

stokes†, St : 1 St = 1 cm^2/s = 10^{-4} m^2/s.

Magnatism

gauss†, G : 1 G = 10^{-4} T (CGS-emu unit of magnetic induction).

oersted†, Oe : 1 Oe = $(1/4\pi)10^3$ A/m (CGS-emu unit of magnetic field strength).

maxwell†, Mx : 1 Mx = 10^{-8} Wb (CGS-emu unit of magnetic flux).

gamma, γ : 1 $\gamma = 10^{-9}$ T.

Light

stilb†, sb : 1 sb = 1 cd/cm^2 = 10^4 cd/m^2 (CGS unit of luminance).

photo, ph : 1 ph = 10^4 lx (CGS unit of illuminance).

Qunatities Related to Radiation

Radioactivity

curie†, Ci : 1 Ci = 3.7×10^{10} Bq (defined value)[7)].

Absorbed Dose

rad, rad : 1 rad = 10^{-2} Gy.

Dose Equivalent

rem, rem : 1 rem = 10^{-2} Sv.

Quantities Related to Irradiation

coulomb per kilogram, C/kg : 1 C/kg = the amount of irradiation required to ionize 1 kg of air to 1 C.

röntgen†, R : 1 R = 2.58 × 10^{-4} C/kg (the amount of irradiation required to produce a charge of 1 esu in 0.001 293 g of air (1 cm^3 of dry air at 0° C, 760 mmHg).

For information about acoustic units, refer to page **369**.

---

1) At first, 1 au was defined as a length of the semi-major axis of the elliptic orbit of Earth around the Sun. In 1976, a revision was made that 1 au should be computed by substituting the Gravitational constant and mass of the Sun into Kepler's third law of planetary motion. The value of 1 au listed in the SI brochure (8th edition, 2006) is 1.495 978 706 91(6) × 10^{11} m, where the number in parentheses is the numerical value of the combined standard uncertainty referred to the corresponding last digits of the quoted result. The present definition of au (= 1.495 978 707 00 × 10^{11} m, exact) was adopted by the International Astronomical Union in 2012 for keeping the relationship with the SI unit of length, and the SI changed the definition of au in 2014.

2) Until 1964, the definition of liter was 1 L = 1000.028 cm^3, the volume of water of 1 kg at its maximum density at 1 atmosphere.

3) The gravitational acceleration is usually denoted by g and the value of the standard gravity g_0 was defined as 9.806 65 m/s^2 = 980.665 Gal at CGPM in 1901.

4) There is a system of units different from the SI called the gravitational unit system, which uses force based on standard gravitational acceleration (9.806 65 m/s^2) instead of mass. The unit of force used in this system is the gravitational kilogram kgf. The unit of force in this system is the gravitational kilogram kgf.

5) The unit of calorific value is the same as that of work or energy. The unit of heat is the same as work or energy, and the International Conference on Weights and Measures in 1948 resolved that the calorie should be used as little as possible, and that if it is used, the value in joules, which is equivalent to one calorie, should be added.

6) French horse power can be used for the time being, but the use of British horse power has been prohibited in Japan since 1959.

7) Originally, it was defined as the radioactivity per gram of radium.

Fundamental Constants

Quantity and symbol		Numerical value	Unit
Universal and electromagnetic constants			
speed of light in vacuum[3]	$c,\ c_0$	2.99 792 458	m s^{-1}
vacuum magnetic permittivity	μ_0	$1.256\ 637\ 062\ 12(19) \times 10^{-6}$	N A^{-2}
	$\mu_0/(4\pi \times 10^{-7})$	1.000 000 000 55(15)	N A^{-2}
vacuum electric permittivity	$\varepsilon_0 = 1/\mu_0 c^2$	$8.854\ 187\ 8128(13) \times 10^{-12}$	F m^{-1}
Newtonian constant of gravitation	G	$6.674\ 30(15) \times 10^{-11}$	N m^2 kg^{-2}
Planck constant[3]	h	$6.626\ 070\ 15 \times 10^{-34}$	J s
	$\hbar = h/2\pi$	$1.054\ 571\ 817\cdots \times 10^{-34}$	J s
elementary charge[3]	e	$1.602\ 176\ 634 \times 10^{-19}$	C
magnetic flux quantum[4]	$\Phi_0 = h/2e$	$2.067\ 833\ 848\cdots \times 10^{-15}$	Wb
condctance quantum[4]	$G_0 = 2e^2/h$	$7.748\ 091\ 729\cdots \times 10^{-5}$	S
Josephson constant[5]	$K_J = 2e/h$	$483\ 597.848\ 416\ 984\cdots \times 10^{9}$	Hz V^{-1}
von Klitzing constant[5]	$R_K = h/e^2$	25 812.807 459 3045⋯	Ω
Bohr magneton	$\mu_B = e\hbar/2m_e$	$9.274\ 010\ 0783(28) \times 10^{-24}$	J T^{-1}
nuclear magneton	$\mu_N = e\hbar/2m_p$	$5.050\ 783\ 7461(15) \times 10^{-27}$	J T^{-1}
Atomic and nuclear constant			
fine structure constant	$\alpha = e^2/4\pi\varepsilon_0\hbar c$	$7.297\ 352\ 5693(11) \times 10^{-3}$	
	$1/\alpha$	137.035 999 084(21)	
Rydberg constant	$R_\infty = \alpha^2 m_e c/2h$	$1.097\ 373\ 156\ 8160(21) \times 10^{7}$	m^{-1}
Bohr radius	$a_0 = \alpha/4\pi R_\infty$	$5.291\ 772\ 109\ 03(80) \times 10^{-11}$	m
Hartree energy	$E_h = e^2/4\pi\varepsilon_0 a_0$	$4.359\ 744\ 722\ 2071(85) \times 10^{-18}$	J
quantum of circulation	$h/2m_e$	$3.636\ 947\ 5516(11) \times 10^{-4}$	m^2 s^{-1}

1) CODATA (Committee on Data for Science and Technology) recomended 2018 value.
2) The number in parentheses is the numerical value of the conbined standard uncertainty referred to the corresponding last digit of quoted result. For example, the numerical value of G is $(6.674\ 08 \pm 0.000\ 31) \times 10^{-11}$.
3) The defining constant.
4) The calculated value from the defining constant.
5) 15 digit calculated value in Appendix 2 of the SI Brochure 9th edition.

of Physics and Chemistry[1),2)]

Quantity and symbol		Numerical value	Unit
electron mass	m_e	$9.109\,383\,7015(28) \times 10^{-31}$	kg
muon mass	m_μ	$1.883\,531\,627(42) \times 10^{-28}$	kg
tau mass	m_τ	$3.167\,54(21) \times 10^{-27}$	kg
proton mass	m_p	$1.672\,621\,923\,69(51) \times 10^{-27}$	kg
neutron mass	m_n	$1.674\,927\,498\,04(95) \times 10^{-27}$	kg
electron magnetic moment	μ_e	$-9.284\,764\,7043(28) \times 10^{-24}$	J T^{-1}
electron g-factor	$2\mu_e/\mu_B$	$-2.002\,319\,304\,362\,56(35)$	
muon magnetic moment	μ_μ	$-4.490\,448\,30(10) \times 10^{-26}$	J T^{-1}
proton magnetic moment	μ_p	$1.410\,606\,797\,36(60) \times 10^{-26}$	J T^{-1}
proton g-factor	$2\mu_p/\mu_N$	$5.585\,694\,6893(16)$	
neutron magnetic moment	μ_n	$-9.662\,3651(23) \times 10^{-27}$	J T^{-1}
(electron) Compton wavelength	$\lambda_C = h/m_e c$	$2.426\,310\,238\,67(73) \times 10^{-12}$	m
proton Compton wavelength	$\lambda_{C,p} = h/m_p c$	$1.321\,409\,855\,39(40) \times 10^{-15}$	m
electron charge to mass quotient	$-e/m_e$	$-1.758\,820\,010\,76(53) \times 10^{11}$	C kg^{-1}
classical electron radius	$r_e = e^2/4\pi\varepsilon_0 m_e c^2$	$2.817\,940\,3262(13) \times 10^{-15}$	m

Physicochemical constant

Quantity and symbol		Numerical value	Unit
atomic mass constant[6)]	m_u, u	$1.660\,539\,066\,60(50) \times 10^{-27}$	kg
Avogadro constant[3)]	N_A, L	$6.022\,140\,76 \times 10^{23}$	mol^{-1}
Boltzmann constant[3)]	k	$1.380\,649 \times 10^{-23}$	J K^{-1}
Faraday constant[4)]	$F = N_A e$	$96\,485.332\,12\cdots$	C mol^{-1}
molar gas constant[4)]	$R = N_A k$	$8.314\,462\,618\cdots$	J mol^{-1} K^{-1}
molar volume of ideal gas[4)] (0℃, 1 atm[7)])	V_m	$22.413\,969\,54\cdots \times 10^{-3}$	m^3 mol^{-1}
Stefan-Boltzmann constant[4)]	$\sigma = \pi^2 k^4/60\hbar^3 c^2$	$5.670\,374\,419\cdots \times 10^{-8}$	W m^{-2} K^{-4}

6) 1/12 mass of the ^{12}C isotope. Equal to the unified atmic mass unit 1 u.
7) 0℃ = 273.15 k, 1 atm = 101.325 kPa.

Energy

	J	eV	kg	Hz
1 J (energy)	(1 J) = 1 J	(1 J) = 6.241 509 074··· $\times 10^{18}$ eV	(1 J)/c^2 = 1.112 650 056··· $\times 10^{-17}$ kg	(1 J)/h = 1.509 190 179··· $\times 10^{33}$ Hz
1 eV (energy)	(1 eV) = 1.602 176 634 $\times 10^{-19}$ J	(1 eV) = 1 eV	(1 eV)/c^2 = 1.782 661 921··· $\times 10^{-36}$ kg	(1 eV)/h = 2.417 989 242··· $\times 10^{14}$ Hz
1 kg (mass)	(1 kg)c^2 = 8.987 551 787··· $\times 10^{16}$ J	(1 kg)c^2 = 5.609 588 603··· $\times 10^{35}$ eV	(1 kg) = 1 kg	(1 kg)c^2/h = 1.356 392 489··· $\times 10^{50}$ Hz
1 Hz (frequency)	(1 Hz)h = 6.626 070 15 $\times 10^{-34}$ J	(1 Hz)h = 4.135 667 696··· $\times 10^{-15}$ eV	(1 Hz)h/c^2 = 7.372 497 323··· $\times 10^{-51}$ kg	(1 Hz) = 1 Hz
1 m^{-2} (wavenumber)	(1 cm^{-1})hc = 1.986 445 857··· $\times 10^{-25}$ J	(1 cm^{-1})hc = 1.239 841 984··· $\times 10^{-6}$ eV	(1 cm^{-1})h/c = 2.210 219 094··· $\times 10^{-42}$ kg	(1 cm^{-1})c = 299 792 458 Hz
1 K (absolute temperature)	(1 K)k = 1.380 649 $\times 10^{-23}$ J	(1 K)k = 8.617 333 262··· $\times 10^{-5}$ eV	(1 K)k/c^2 = 1.536 179 187··· $\times 10^{-40}$ kg	(1 K)k/h = 2.083 661 912··· $\times 10^{10}$ Hz
1 T (magnetic induction)	(1 T)μ_B = 9.274 010 0783(28) $\times 10^{-24}$ J	(1 T)μ_B = 5.788 381 8060(17) $\times 10^{-5}$ eV	(1 T)μ_B/c^2 = 1.031 872 783 35(31) $\times 10^{-40}$ kg	(1 T)μ_B/h = 1.399 624 493 61(42) $\times 10^{10}$ Hz

1) CODATA 2018.

Conversion Table[1)]

$cm^{-1} = 10^2\ m^{-1}$	K	T	Remarks
$(1\ J)/hc =$ 5.034 116 567··· $\times 10^{24}\ m^{-1}$	$(1\ J)/k =$ 7.242 970 516··· $\times 10^{22}$ K	$(1\ J)/\mu_B =$ 1.078 282 201 07(33) $\times 10^{23}$ T	
$(1\ eV)/hc =$ 8.065 543 937··· $\times 10^{5}\ m^{-1}$	$(1\ eV)/k =$ 1.160 451 812··· $\times 10^{4}$ K	$(1\ eV)/\mu_B =$ 1.727 598 547 42(52) $\times 10^{4}$ T	
$(1\ kg)c/h =$ 4.524 438 335··· $\times 10^{41}\ m^{-1}$	$(1\ kg)c^2/k =$ 6.509 657 260··· $\times 10^{39}$ K	$(1\ kg)c^2/\mu_B =$ 9.691 117 1235(29) $\times 10^{39}$ T	$c =$ 2.997 924 58 $\times 10^{8}\ m\ s^{-1}$ (speed of light in vacuum)
$(1\ Hz)/c =$ 3.335 640 951··· $\times 10^{-9}\ m^{-1}$	$(1\ Hz)h/k =$ 4.799 243 073··· $\times 10^{-11}$ K	$(1\ Hz)h/\mu_B =$ 7.144 773 5058(22) $\times 10^{-11}$ T	$h =$ 6.626 070 15 $\times 10^{-34}$ J s (Planck constant)
$(1\ m^{-1}) =$ 1 m^{-1}	$(1\ m^{-1})hc/k =$ 1.438 776 877··· $\times 10^{-2}$ K	$(1\ m^{-1})hc/\mu_B =$ 2.141 949 211 16(65) $\times 10^{-2}$ T	
$(1\ K)k/hc =$ 6.950 348 004··· $\times 10^{-1}\ cm^{-1}$	(1 K) = 1 K	$(1\ K)k/\mu_B =$ 1.488 729 242 63(45) T	$k =$ 1.380 649 $\times 10^{-23}\ J\ K^{-1}$ (Boltzmann constant)
$(1\ T)\mu_B/hc =$ 4.668 644 7783(14) m^{-1}	$(1\ T)\mu_B/k =$ 6.717 138 1563(20) $\times 10^{-1}$ K	(1 T) = 1 T	$\mu_B =$ 9.274 010 0783(28) $\times 10^{-24}\ J\ T^{-1}$ (Bohr magneton)

Element

Standard Atomic Weights

The atomic weight is defined as the relative mass of the element when $^{12}C = 12$. This table shows the atomic weight recommended in 2015 by the Commission on Isotopic Abundances and Atomic Weights (CIAAW) of the International Union of Pure and Applied Chemistry (IUPAC). It contains the latest information at the present time. In this table, the atomic weight and uncertainty (applies to the final digit of significant figures for atomic weights) shown in parenthesis are applied to elements of matter which originates from the earth. For elements which may show fluctuations that exceed the atomic weight uncertainty shown in this table, the fluctuation formula is given in (Notes) using the symbols g, m, and r. From 2009, the IUPAC decided to indicate the atomic weight of several elements as a fluctuation range instead of a single number. In this table, the fluctuation range is represented by [a, b]. This means that the atomic weight is equal to or greater than a and equal to or less than b.

Elements marked with an asterisk (*) are elements without a stable isotope. An example of the mass number for the radioactive isotope of the element is shown in square brackets ([ ]). However, the radioactive elements bismuth, thorium, protoactinium and uranium possess specific isotopic compositions on the earth, so an atomic weight is assigned.

(Notes) g : The isotopic composition of some elements fluctuates greatly among geological samples. The atomic weight of some of these elements may exceed the uncertainty listed in this table. m : Due to unknown or inappropriate isotope fractionation, some commercial products contain substances with significant fluctuation of the isotopic composition for applicable elements. In some cases, the atomic weight of those elements may exceed the uncertainty listed in this table. r : Due to relatively significant fluctuations in the isotopic composition of applicable elements in normal substances found on the earth, there is a limit to the accuracy of the atomic weight for those elements. In some cases, the atomic weight may exceed the uncertainty listed in this table.

Atomic number	Element	Symbol	Atomic weight	Note
1	Hydrogen	H	[1.007 84, 1.008 11]	m
2	Helium	He	4.002 602(2)	g r
3	Lithium	Li	[6.938, 6.997]	m
4	Beryllium	Be	9.012 183 1(5)	
5	Boron	B	[10.806, 10.821]	m
6	Carbon	C	[12.009 6, 12.011 6]	
7	Nitrogen	N	[14.006 43, 14.007 28]	m
8	Oxygen	O	[15.999 03, 15.999 77]	m
9	Fluorine	F	18.998 403 163(6)	
10	Neon	Ne	20.179 7(6)	g m

Continued.

Atomic number	Element	Symbol	Atomic weight	Note
11	Sodium	Na	22.989 769 28(2)	
12	Magnesium	Mg	[24.304, 24.307]	
13	Aluminium[1)]	Al	26.981 538 5(7)	
14	Silicon	Si	[28.084, 28.086]	
15	Phosphorus	P	30.973 761 998(5)	
16	Sulfur	S	[32.059, 32.076]	
17	Chlorine	Cl	[35.446, 35.457]	m
18	Argon	Ar	[39.792, 39.963]	g r
19	Potassium	K	39.098 3(1)	
20	Calcium	Ca	40.078(4)	g
21	Scandium	Sc	44.955 908(5)	
22	Titanium	Ti	47.867(1)	
23	Vanadium	V	50.941 5(1)	
24	Chromium	Cr	51.996 1(6)	
25	Manganese	Mn	54.938 043(2)	
26	Iron	Fe	55.845(2)	
27	Cobalt	Co	58.933 194(3)	
28	Nickel	Ni	58.693 4(4)	r
29	Copper	Cu	63.546(3)	r
30	Zinc	Zn	65.38(2)	r
31	Gallium	Ga	69.723(1)	
32	Germanium	Ge	72.630(8)	
33	Arsenic	As	74.921 595(6)	
34	Selenium	Se	78.971(8)	r
35	Bromine	Br	[79.901, 79.907]	
36	Krypton	Kr	83.798(2)	g m
37	Rubidium	Rb	85.467 8(3)	g
38	Strontium	Sr	87.62(1)	g r
39	Yttrium	Y	88.905 84(1)	
40	Zirconium	Zr	91.224(2)	g
41	Niobium	Nb	92.906 37(1)	
42	Molybdenum	Mo	95.95(1)	g
43	Technetium*	Tc	[99]	
44	Ruthenium	Ru	101.07(2)	g
45	Rhodium	Rh	102.905 49(2)	
46	Palladium	Pd	106.42(1)	g
47	Silver	Ag	107.868 2(2)	g
48	Cadmium	Cd	112.414(4)	g
49	Indium	In	114.818(1)	
50	Tin	Sn	118.710(7)	g
51	Antimony	Sb	121.760(1)	g
52	Tellurium	Te	127.60(3)	g
53	Iodine	I	126.904 47(3)	
54	Xenon	Xe	131.293(6)	g m
55	Caesium[2)]	Cs	132.905 451 96(6)	
56	Barium	Ba	137.327(7)	
57	Lanthanum	La	138.905 47(7)	g
58	Cerium	Ce	140.116(1)	g
59	Praseodymium	Pr	140.907 66(1)	
60	Neodymium	Nd	144.242(3)	g
61	Promethium*	Pm	[145]	
62	Samarium	Sm	150.36(2)	g
63	Europium	Eu	151.964(1)	g
64	Gadolinium	Gd	157.25(3)	g
65	Terbium	Tb	158.925 354(8)	

1) Aluminum (an alternative spelling listed in parentheses on the IUPAC atomic weight table)
2) Cesium

Standard Atomic Weights Continued.

Atomic number	Element	Symbol	Atomic weight	Note
66	Dysprosium	Dy	162.500(1)	g
67	Holmium	Ho	164.930 328(7)	
68	Erbium	Er	167.259(3)	g
69	Thulium	Tm	168.934 218(6)	
70	Ytterbium	Yb	173.045(10)	g
71	Lutetium	Lu	174.966 8(1)	g
72	Hafnium	Hf	178.486(6)	
73	Tantalum	Ta	180.947 88(2)	
74	Tungsten	W	183.84(1)	
75	Rhenium	Re	186.207(1)	
76	Osmium	Os	190.23(3)	g
77	Iridium	Ir	192.217(3)	
78	Platinum	Pt	195.084(9)	
79	Gold	Au	196.966 570(4)	
80	Mercury	Hg	200.592(3)	
81	Thallium	Tl	[204.382, 204.385]	
82	Lead	Pb	207.2(1)	g r
83	Bismuth*	Bi	208.980 40(1)	
84	Polonium*	Po	[210]	
85	Astatine*	At	[210]	
86	Radon*	Rn	[222]	
87	Francium*	Fr	[223]	
88	Radium*	Ra	[226]	
89	Actinium*	Ac	[227]	
90	Thorium*	Th	232.037 7(4)	g
91	Protactinium*	Pa	231.035 88(1)	
92	Uranium*	U	238.028 91(3)	g m
93	Neptunium*	Np	[237]	
94	Plutonium*	Pu	[239]	
95	Americium*	Am	[243]	
96	Curium*	Cm	[247]	
97	Berkelium*	Bk	[247]	
98	Californium*	Cf	[252]	
99	Einsteinium*	Es	[252]	
100	Fermium*	Fm	[257]	
101	Mendelevium*	Md	[258]	
102	Nobelium*	No	[259]	
103	Lawrencium*	Lr	[262]	
104	Rutherfordium*	Rf	[267]	
105	Dubnium*	Db	[268]	
106	Seaborgium*	Sg	[271]	
107	Bohrium*	Bh	[272]	
108	Hassium*	Hs	[277]	
109	Meitnerium*	Mt	[276]	
110	Darmstadtium*	Ds	[281]	
111	Roentgenium*	Rg	[280]	
112	Copernicium*	Cn	[285]	
113	Nihonium*	Nh	[278]	
114	Flerovium*	Fl	[289]	
115	Moscovium*	Mc	[289]	
116	Livermorium*	Lv	[293]	
117	Tennessine*	Ts	[293]	
118	Oganesson*	Og	[294]	

(Source: M. Ebihara)

Periodic Table of the Elements

Period \ Group	1	2	3	4	5	6	7	8	9	10	11	12	13	14	15	16	17	18
1	1 H [1.00784, 1.00811]																	2 He 4.002602
2	3 Li [6.938, 6.997]	4 Be 9.0121831											5 B [10.806, 10.821]	6 C [12.0096, 12.0116]	7 N [14.00643, 14.00728]	8 O [15.99903, 15.99977]	9 F 18.998403163	10 Ne 20.1797
3	11 Na 22.98976928	12 Mg [24.304, 24.307]											13 Al 26.9815385	14 Si [28.084, 28.086]	15 P 30.973761998	16 S [32.059, 32.076]	17 Cl [35.446, 35.457]	18 Ar [39.792, 39.963]
4	19 K 39.0983	20 Ca 40.078	21 Sc 44.955908	22 Ti 47.867	23 V 50.9415	24 Cr 51.9961	25 Mn 54.938043	26 Fe 55.845	27 Co 58.933194	28 Ni 58.6934	29 Cu 63.546	30 Zn 65.38	31 Ga 69.723	32 Ge 72.630	33 As 74.921595	34 Se 78.971	35 Br [79.901, 79.907]	36 Kr 83.798
5	37 Rb 85.4678	38 Sr 87.62	39 Y 88.90584	40 Zr 91.224	41 Nb 92.90637	42 Mo 95.95	43 Tc [99]	44 Ru 101.07	45 Rh 102.90550	46 Pd 106.42	47 Ag 107.8682	48 Cd 112.414	49 In 114.818	50 Sn 118.710	51 Sb 121.760	52 Te 127.60	53 I 126.90447	54 Xe 131.293
6	55 Cs 132.90545196	56 Ba 137.327	57~71 ※	72 Hf 178.486	73 Ta 180.94788	74 W 183.84	75 Re 186.207	76 Os 190.23	77 Ir 192.217	78 Pt 195.084	79 Au 196.966570	80 Hg 200.592	81 Tl [204.382, 204.385]	82 Pb 207.2	83 Bi 208.98040	84 Po [210]	85 At [210]	86 Rn [222]
7	87 Fr [223]	88 Ra [226]	89~103 ※※	104 Rf [267]	105 Db [268]	106 Sg [271]	107 Bh [272]	108 Hs [277]	109 Mt [276]	110 Ds [281]	111 Rg [280]	112 Cn [285]	113 Nh [278]	114 Fl [289]	115 Mc [289]	116 Lv [293]	117 Ts [293]	118 Og [294]

※	57 La 138.90547	58 Ce 140.116	59 Pr 140.90766	60 Nd 144.242	61 Pm [145]	62 Sm 150.36	63 Eu 151.964	64 Gd 157.25	65 Tb 158.925354	66 Dy 162.500	67 Ho 164.930328	68 Er 167.259	69 Tm 168.934218	70 Yb 173.045	71 Lu 174.9668
※※	89 Ac [227]	90 Th 232.0377	91 Pa 231.03588	92 U 238.02891	93 Np [237]	94 Pu [239]	95 Am [243]	96 Cm [247]	97 Bk [247]	98 Cf [252]	99 Es [252]	100 Fm [257]	101 Md [258]	102 No [259]	103 Lr [262]

※ Lanthanoids ※※ Actinoids

The number listed above the chemical symbol is the atomic number. The number listed under the symbol is the atomic weight. Refer to page **302** for information on atomic weight shown in the range. For natural elements which do not indicate a certain isotopic composition and do not possess a stable isotope, an example of the mass number for the radioactive isotope of those elements is shown in square brackets ([ ]). Group numbers (1 to 18) are according to the revised IUPAC Nomenclature of Inorganic Chemistry (1989). Elements with an atomic number of 104 and higher are provisionally positioned in the periodic table.

Electronic Structures of Atoms and Ions

In the following table, [He], [Ne], [Ar], etc. represent inert noble gases. If no notes are provided, the electronic configurations for ions are those without one electron in bold in the ground state configuration.

Principal quantum number	1	2	3	4	5	6	7
Shell name	K	L	M	N	O	P	Q

Azimuthal quantum number	0	1	2	3	4	5	6	7
Electron orbit name	s	p	d	f	g	h	i	k
Resultant orbital angular momentum	S	P	D	F	G	H	I	K

LS coupling notation is used for most of the elements. The notation is written as $^{2S+1}L_J^p$, where *S* represents the total spin angular momentum, *L* represents the total orbital angular momentum, *J* represents the total angular momentum. For *p*, there is no symbol for even parity. Odd parity is denoted by an "o". For some heavy elements, *jj* coupling notation is used (refer to Reference Literature 1 for details on notation).

The value for first (second) ionization potential is the minimum energy required to infinitely separate one electron from the ground state of the atom (monovalent ion).

The electron affinity is the energy value released during the process in which electrons attached to atoms, molecules, etc. The degree of electronegativity proposed by Pauling is used in order to indicate the tendency of atoms to attract electrons. The uncertainty for the smallest digit is indicated in parenthesis.

Blank columns indicate that the electronic configuration or value is not reported.

Atomic number	Element	Atomic ground state electron configuration	Atomic ground state	1st ionization energy E_{i1} (eV)	Ground state of ion	2nd ionization energy E_{i2} (eV)	Electron affinity (eV)	Electro-negativity
1	H	**1s**	$^2S_{1/2}$	13.598 434	–	–	0.754 195(19)	2.20
2	He	**1s**^{2}	$^1S_0$	24.587 388	$^2S_{1/2}$	54.417 765	<0	
3	Li	[He]**2s**	$^2S_{1/2}$	5.391 715	$^1S_0$	75.640 09	0.618 049(20)	0.98
4	Be	-**2s**^{2}	$^1S_0$	9.322 70	$^2S_{1/2}$	18.211 1	<0	1.57
5	B	-2s^2 **2p**	$^2P^o_{1/2}$	8.298 02	$^1S_0$	25.154 8	0.279 273(25)	2.04
6	C	-2s^2 **2p**^{2}	$^3P_0$	11.260 288(1)	$^2P^o_{1/2}$	24.383 15(2)	1.262 119(20)	2.55
7	N	-2s^2 **2p**^{3}	$^4S^o_{3/2}$	14.534 1	$^3P_0$	29.601 3	<0	3.04
8	O	-2s^2 **2p**^{4}	$^3P_2$	13.618 05	$^4S^o_{3/2}$	35.121 1	1.461 113 5(12)	3.44
9	F	-2s^2 **2p**^{5}	$^2P^o_{3/2}$	17.422 8	$^3P_2$	34.970 8	3.401 189 5(25)	3.98
10	Ne	-2s^2 **2p**^{6}	$^1S_0$	21.564 54	$^2P^o_{3/2}$	40.963 0	<0	

Continued.

Atomic number	Element	Atomic ground state electron configuration	Atomic ground state	1st ionization energy E_{i1} (eV)	Ground state of ion	2nd ionization energy E_{i2} (eV)	Electron affinity (eV)	Electro-negativity
11	Na	[Ne]$\mathbf{3s}$	$^2S_{1/2}$	5.139 077	$^1S_0$	47.286 4(3)	0.547 926(25)	0.93
12	Mg	$\mathbf{-3s^2}$	$^1S_0$	7.646 23	$^2S_{1/2}$	15.035 27	<0	1.31
13	Al	$-3s^2\ \mathbf{3p}$	$^2P^o_{1/2}$	5.985 77	$^1S_0$	18.828 6	0.432 83(5)	1.61
14	Si	$-3s^2\ \mathbf{3p^2}$	$^3P_0$	8.151 68	$^2P^o_{1/2}$	16.345 9	1.389 521 3(13)	1.90
15	P	$-3s^2\ \mathbf{3p^3}$	$^4S^o_{3/2}$	10.486 69	$^3P_0$	19.769 5	0.746 5(3)	2.19
16	S	$-3s^2\ \mathbf{3p^4}$	$^3P_2$	10.360 00(1)	$^4S^o_{3/2}$	23.337 89(3)	2.077 104 18(71)	2.58
17	Cl	$-3s^2\ \mathbf{3p^5}$	$^2P^o_{3/2}$	12.967 63	$^3P_2$	23.813 6	3.612 724(27)	3.16
18	Ar	$-3s^2\ \mathbf{3p^6}$	$^1S_0$	15.759 611	$^2P^o_{3/2}$	27.629 7	<0	
19	K	[Ar]$\mathbf{4s}$	$^2S_{1/2}$	4.340 6637	$^1S_0$	31.625 0	0.501 47(10)	0.82
20	Ca	$\mathbf{-4s^2}$	$^1S_0$	6.113 155	$^2S_{1/2}$	11.871 72	0.024 55(10)	1.00
21	Sc	$-3d\ \mathbf{4s^2}$	$^2D_{3/2}$	6.561 5	$^3D_1$	12.799 8(3)	0.188(20)	1.36
22	Ti	$-3d^2\ \mathbf{4s^2}$	$^3F_2$	6.828 12	$^4F_{3/2}$	13.576(3)	0.079(14)	1.54
23	V	$-3d^3\ \mathbf{4s^2}$	$^4F_{3/2}$	6.746 19	$^5D_0$ 1)	14.634(7)	0.525(12)	1.63
24	Cr	$-3d^5\ \mathbf{4s}$	$^7S_3$	6.766 5	$^6S_{5/2}$	16.486 3	0.666(12)	1.66
25	Mn	$-3d^5\ \mathbf{4s^2}$	$^6S_{5/2}$	7.434 038	$^7S_3$	15.640 0	<0	1.55
26	Fe	$-3d^6\ \mathbf{4s^2}$	$^5D_4$	7.902 468	$^6D_{9/2}$	16.199 2	0.151(3)	1.83
27	Co	$-3d^7\ \mathbf{4s^2}$	$^4F_{9/2}$	7.881 0	$^3F_4$ 2)	17.084	0.662(3)	1.88
28	Ni	$-3d^8\ \mathbf{4s^2}$	$^3F_4$	7.639 88	$^2D_{5/2}$ 3)	18.168 84	1.156(10)	1.91
29	Cu	$-3d^{10}\mathbf{4s}$	$^2S_{1/2}$	7.726 38	$^1S_0$	20.292 4	1.235(5)	1.90
30	Zn	$-3d^{10}\mathbf{4s^2}$	$^1S_0$	9.394 20	$^2S_{1/2}$	17.964 4(2)	<0	1.65
31	Ga	$-3d^{10}4s^2\mathbf{4p}$	$^2P^o_{1/2}$	5.999 302	$^1S_0$	20.515 14	0.43(3)	1.81
32	Ge	$-3d^{10}4s^2\mathbf{4p^2}$	$^3P_0$	7.899 44	$^2P^o_{1/2}$	15.934 61(3)	1.232 712(15)	2.01
33	As	$-3d^{10}4s^2\mathbf{4p^3}$	$^4S^o_{3/2}$	9.788 6(3)	$^3P_0$	18.589	0.804(8)	2.18
34	Se	$-3d^{10}4s^2\mathbf{4p^4}$	$^3P_2$	9.752 39	$^4S^o_{3/2}$	21.20	2.020 670(25)	2.55
35	Br	$-3d^{10}4s^2\mathbf{4p^5}$	$^2P^o_{3/2}$	11.813 8	$^3P_2$	21.591	3.363 588(2)	2.96
36	Kr	$-3d^{10}4s^2\mathbf{4p^6}$	$^1S_0$	13.999 605(2)	$^2P^o_{3/2}$	24.359 8	<0	3.04
37	Rb	[Kr]$\mathbf{5s}$	$^2S_{1/2}$	4.177 128	$^1S_0$	27.289 5	0.485 92(2)	0.82
38	Sr	$\mathbf{-5s^2}$	$^1S_0$	5.694 8674	$^2S_{1/2}$	11.030 276	0.048(6)	0.95
39	Y	$-4d\ \mathbf{5s^2}$	$^2D_{3/2}$	6.217 3	$^1S_0$	12.224	0.307(12)	1.22
40	Zr	$-4d^2\ \mathbf{5s^2}$	$^3F_2$	6.634 13	$^4F_{3/2}$	13.13	0.426(14)	1.33
41	Nb	$-4d^4\ \mathbf{5s}$	$^6D_{1/2}$	6.758 85(4)	$^5D_0$	14.32	0.916(5)	1.6
42	Mo	$-4d^5\ \mathbf{5s}$	$^7S_3$	7.092 43(4)	$^6S_{5/2}$	16.16(12)	0.748(2)	2.16
43	Tc	$-4d^5\ \mathbf{5s^2}$	$^6S_{5/2}$	7.119 4	$^7S_3$	15.26	0.55(20)	2.10
44	Ru	$-4d^7\ \mathbf{5s}$	$^5F_5$	7.360 5	$^4F_{9/2}$	16.76(6)	1.05(15)	2.2
45	Rh	$-4d^8\ \mathbf{5s}$	$^4F_{9/2}$	7.458 9	$^3F_4$	18.08	1.137(8)	2.28
46	Pd	$\mathbf{-4d^{10}}$	$^1S_0$	8.336 84	$^2D_{5/2}$	19.43(12)	0.562(5)	2.20
47	Ag	$-4d^{10}\mathbf{5s}$	$^2S_{1/2}$	7.576 23(3)	$^1S_0$	21.484	1.302(7)	1.93
48	Cd	$-4d^{10}\mathbf{5s^2}$	$^1S_0$	8.993 82	$^2S_{1/2}$	16.908 31	<0	1.69
49	In	$-4d^{10}5s^2\mathbf{5p}$	$^2P^o_{1/2}$	5.786 355	$^1S_0$	18.870 4	0.3(2)	1.78
50	Sn	$-4d^{10}5s^2\mathbf{5p^2}$	$^3P_0$	7.343 92	$^2P^o_{1/2}$	14.633 1	1.112 067(15)	1.96

Electronic Structure of Atoms and Ions Continued.

Atomic number	Element	Atomic ground state electron configuration	Atomic ground state	1st ionization energy E_{i1} (eV)	Ground state of ion	2nd ionization energy E_{i2} (eV)	Electron affinity (eV)	Electro-negativity
51	Sb	$-4d^{10}5s^2\mathbf{5p}^3$	$^4S^o_{3/2}$	8.608 39	$^3P_0$	16.626(25)	1.046(5)	2.05
52	Te	$-4d^{10}5s^2\mathbf{5p}^4$	$^3P_2$	9.009 81	$^4S^o_{3/2}$	18.6(4)	1.970 876(7)	2.1
53	I	$-4d^{10}5s^2\mathbf{5p}^5$	$^2P^o_{3/2}$	10.451 26(3)	$^3P_2$	19.131 3	3.059 046 3(38)	2.66
54	Xe	$-4d^{10}5s^2\mathbf{5p}^6$	$^1S_0$	12.129 844	$^2P^o_{3/2}$	20.975(4)	<0	2.60
55	Cs	[Xe]$\mathbf{6s}$	$^2S_{1/2}$	3.893 905 69(2)	$^1S_0$	23.157 5	0.471 626(25)	0.79
56	Ba	$-\mathbf{6s}^2$	$^1S_0$	5.211 665	$^2S_{1/2}$	10.003 83	0.144 62(6)	0.89
57	La	$-5d\mathbf{6s}^2$	$^2D_{3/2}$	5.577	$^3F_2$ [4]	11.185 0	0.47(2)	1.10
58	Ce	$-4f5d\mathbf{6s}^2$	$^1G^o_4$	5.538 6(4)	$^4H^o_{7/2}$ [5]	10.96(2)	0.65(3)	1.12
59	Pr	$-4f^3\mathbf{6s}^2$	$^4I^o_{9/2}$	5.470	$(9/2, 1/2)^o_4$	10.63(2)	0.962(24)	1.13
60	Nd	$-4f^4\mathbf{6s}^2$	$^5I_4$	5.525	$^6I^o_{7/2}$	10.78(2)	>1.916	1.14
61	Pm	$-4f^5\mathbf{6s}^2$	$^6H^o_{5/2}$	5.581 87(4)	$^7H^o_2$	10.94(2)		
62	Sm	$-4f^6\mathbf{6s}^2$	$^7F_0$	5.643 7	$^8F_{1/2}$	11.08(2)		1.17
63	Eu	$-4f^7\mathbf{6s}^2$	$^8S^o_{7/2}$	5.670 39	$^9S^o_4$	11.240(6)	0.864(24)	
64	Gd	$-4f^7 5d\mathbf{6s}^2$	$^9D^o_2$	6.149 80(4)	$^{10}D_{5/2}$	12.08(2)		1.20
65	Tb	$-4f^9\mathbf{6s}^2$	$^6H^o_{15/2}$	5.864	$(15/2, 1/2)^o_8$	11.51(2)	>1.165	
66	Dy	$-4f^{10}\mathbf{6s}^2$	$^5I_8$	5.939 1	$(8, 1/2)_{17/2}$	11.65(2)	>0	1.22
67	Ho	$-4f^{11}\mathbf{6s}^2$	$^4I^o_{15/2}$	6.022	$(15/2, 1/2)^o_8$	11.78(2)		1.23
68	Er	$-4f^{12}\mathbf{6s}^2$	$^3H_6$	6.108	$(6, 1/2)_{13/2}$	11.92(2)		1.24
69	Tm	$-4f^{13}\mathbf{6s}^2$	$^2F^o_{7/2}$	6.184 3	$(7/2, 1/2)^o_4$	12.07(2)	1.029(22)	1.25
70	Yb	$-4f^{14}\mathbf{6s}^2$	$^1S_0$	6.254 16	$^2S_{1/2}$	12.179 19(3)	-0.020	
71	Lu	$-4f^{14}5d\ \mathbf{6s}^2$	$^2D_{3/2}$	5.425 87	$^1S_0$	13.13(5)	0.34(1)	1.0
72	Hf	$-4f^{14}5d^2\ \mathbf{6s}^2$	$^3F_2$	6.825 07	$^2D_{3/2}$	14.61(4)	0.017	1.3
73	Ta	$-4f^{14}5d^3\ \mathbf{6s}^2$	$^4F_{3/2}$	7.549 57(3)	$^5F_1$	16.2(5)	0.322(12)	1.5
74	W	$-4f^{14}5d^4\ \mathbf{6s}^2$	$^5D_0$	7.864 0	$^6D_{1/2}$	16.37(15)	0.816 26(7)	1.7
75	Re	$-4f^{14}5d^5\ \mathbf{6s}^2$	$^6S_{5/2}$	7.833 5	$^7S_3$	16.6(5)	0.15(15)	1.9
76	Os	$-4f^{14}5d^6\ \mathbf{6s}^2$	$^5D_4$	8.438 2(2)	$^6D_{9/2}$	17.0(10)	1.1(2)	2.2
77	Ir	$-4f^{14}5d^7\ \mathbf{6s}^2$	$^4F_{9/2}$	8.967 0(2)	$^5F_5$	17.0(3)	1.563 8(5)	2.2
78	Pt	$-4f^{14}5d^9\ \mathbf{6s}$	$^3D_3$	8.958 8	$^2D_{5/2}$	18.56(12)	2.128(2)	2.2
79	Au	$-4f^{14}5d^{10}\mathbf{6s}$	$^2S_{1/2}$	9.225 554(4)	$^1S_0$	20.20(3)	2.308 63(3)	2.4
80	Hg	$-4f^{14}5d^{10}\mathbf{6s}^2$	$^1S_0$	10.437 50	$^2S_{1/2}$	18.756 9	<0	1.9
81	Tl	$-4f^{14}5d^{10}6s^2\mathbf{6p}$	$^2P^o_{1/2}$	6.108 287	$^1S_0$	20.428 3(6)	0.377(13)	1.8
82	Pb	$-4f^{14}5d^{10}6s^2\mathbf{6p}^2$	$(1/2, 1/2)_0$	7.416 680	$^2P^o_{1/2}$	15.032 50	0.364(8)	1.8
83	Bi	$-4f^{14}5d^{10}6s^2\mathbf{6p}^3$	$^4S^o_{3/2}$	7.285 52	$^3P_0$	16.703(4)	0.942 362(13)	1.9
84	Po	$-4f^{14}5d^{10}6s^2\mathbf{6p}^4$	$^3P_2$	8.418 070(4)	$^4S^o_{3/2}$	19.3(17)	1.9(3)	2.0
85	At	$-4f^{14}5d^{10}6s^2\mathbf{6p}^5$	$^2P^o_{3/2}$	9.317 5	$^3P_2$	17.88(2)	2.8(2)	2.2
86	Rn	$-4f^{14}5d^{10}6s^2\mathbf{6p}^6$	$^1S_0$	10.748 50	$^2P^o_{3/2}$	21.4(19)	<0	
87	Fr	[Rn]$\mathbf{7s}$	$^2S_{1/2}$	4.072 741	$^1S_0$	22.4(19)	0.486(2)	0.7
88	Ra	$-\mathbf{7s}^2$	$^1S_0$	5.278 424	$^2S_{1/2}$	10.147 2	0.10	0.9
89	Ac	$-6d\ \mathbf{7s}^2$	$^2D_{3/2}$	5.380 23(2)	$^1S_0$	11.75(3)	0.35	1.1
90	Th	$-6d^2\mathbf{7s}^2$	$^3F_2$	6.306 7(3)	$^2D_{3/2}$	12.1(2)		1.3

Continued.

Atomic number	Element	Atomic ground state electron configuration	Atomic ground state	1st ionization energy E_{i1} (eV)	Ground state of ion	2nd ionization energy E_{i2} (eV)	Electron affinity (eV)	Electro-negativity
91	Pa	$-5f^{2}(^{3}H_{4})6d7s^{2}$	$^{4}K_{11/2}$	5.89(12)	$^{3}H_{4}$	11.9(4)		1.5
92	U	$-5f^{3}(^{4}I^{o}_{9/2})6d7s^{2}$	$^{5}L^{o}_{6}$	6.194 1	$^{4}I^{o}_{9/2}$	11.6(4)		1.7
93	Np	$-5f^{4}(^{5}I_{4})6d7s^{2}$	$^{6}L_{11/2}$	6.265 5(3)	$^{7}L_{5}$	11.5(4)		1.3
94	Pu	$-5f^{6}7s^{2}$	$^{7}F_{0}$	6.025 8(3)	$^{8}F_{1/2}$	11.5(4)		1.3
95	Am	$-5f^{7}7s^{2}$	$^{8}S^{o}_{7/2}$	5.973 8(3)	$^{9}S^{o}_{4}$	11.7(4)		
96	Cm	$-5f^{7}6d7s^{2}$	$^{9}D^{o}_{2}$	5.991 4(3)	$^{8}S^{o}_{7/2}$	12.4(4)		
97	Bk	$-5f^{9}\ 7s^{2}$	$^{6}H^{o}_{15/2}$	6.197 8(3)	$^{7}H_{8}$	11.9(4)		
98	Cf	$-5f^{10}7s^{2}$	$^{5}I_{8}$	6.281 7(3)	$^{6}I^{o}_{17/2}$	12.0(4)		
99	Es	$-5f^{11}7s^{2}$	$^{4}I^{o}_{15/2}$	6.367 6(3)	$^{5}I^{o}_{8}$	12.2(4)		
100	Fm	$-5f^{12}7s^{2}$	$^{3}H_{6}$	6.50(7)	$^{4}H_{13/2}$	12.4(4)		
101	Md	$-5f^{13}7s^{2}$	$^{2}F^{o}_{7/2}$	6.58(7)	$^{3}F^{o}_{4}$	12.4(4)		
102	No	$-5f^{14}7s^{2}$	$^{1}S_{0}$	6.626 21(5)	$^{2}S_{1/2}$	12.5(4)		
103	Lr	$-5f^{14}7s^{2}7p$?	$^{2}P^{o}_{1/2}$?	4.96(8)	$^{1}S_{0}$	14.3(4)		
104	Rf	$-5f^{14}6d^{2}7s^{2}$?	$^{3}F_{2}$?	6.0(1)	$^{2}D_{3/2}$	14.5(2)		
105	Db	$-5f^{14}6d^{3}7s^{2}$?	$^{4}F_{3/2}$?	6.8(5)	$^{3}F_{2}$?	14.0(16)		
106	Sg	$-5f^{14}6d^{4}7s^{2}$?	$?_{0}$?	7.8(5)	$^{4}F_{3/2}$?	17.1(5)		
107	Bh	$-5f^{14}6d^{5}7s^{2}$?	$?_{5/2}$?	7.7(5)	$?_{0}$?	17.5(5)		
108	Hs	$-5f^{14}6d^{6}7s^{2}$?	$?_{4}$?	7.6(5)	$?_{5/2}$?	18.2(5)		
109	Mt							
110	Ds							
111	Rg	[Rn]$5f^{14}6d^{9}7s^{2}$	$^{2}D_{5/2}$	10.6?	$^{3}D_{4}$			
112	Cn	$-5f^{14}6d^{10}7s^{2}$	$^{1}S_{0}$	11.7(3)	$^{2}D_{5/2}$	22.2(3)		
113	Nh	$-5f^{14}6d^{10}7s^{2}7p$	$^{2}P^{o}_{1/2}$	7.2(2)	$^{1}S_{0}$	23.6?		
114	Fl	$-5f^{14}6d^{10}7s^{2}7p^{2}$	$^{1}S_{0}$	8.5(1)	$^{2}P^{o}_{1/2}$	17.0(2)		
115	Mc			5.5(1)		18.4?	0.31?	
116	Lv			6.9?		13.6?	0.78?	
117	Ts			7.5(2)		15.2?	1.6?	
118	Og			8.6?				

electron configuration of ions 1) $-3d^{4}$ 2) $-3d^{8}$ 3) $-3d^{9}$ 4) $-5d^{2}$ 5) $-4f5d^{2}$

References:

1 : Springer Handbook of Atomic, Molecular and Optical Physics, 2nd Ed. Chap. 10 (2006).
2 : CRC Handbook of Chemistry and Physics 92 ed. (2011).
3 : Amer. Inst. Phys. Handbook 3rd ed. (1972).
4 : NIST Atomic Spectra Database (ver. 5.7) (2019).
5 : Phys. Rev. A94, 042503 (2016).
6 : Eur. J. Phys. D44, 51 (2007).
7 : Nucl. Phys. A944, 518 (2015).
8 : J. Phys. Chem. A117, 8555 (2013).
9 : Hyperfine interact 237, 160 (2016).
10 : J. Chem. Phys. 141, 1641044 (2014).
11 : Phys. Rev. A 91, 020501 (R) (2015).
12 : J. Chem. Phys. 124, 064305 (2006).
13 : Phys. Rev. Lett. 120, 263003 (2018).
14 : J. Am. Chem. Soc. 140, 14609 (2018).

Mechanical Properties

Density

Density of the elements

The state of matter is solid not otherwise specified. Temperature not indicated in a column is room temperature. The unit of density is 10^3 kg m^{-3} = g cm^{-3} for solid (s) and liquid (l) and kg m^{-3} = g L^{-1} for gas (g) under the pressure of 101.325 kPa.

Element (Symbol)	State	t/℃	Density ρ	Element (Symbol)	State	t/℃	Density ρ
Hydrogen (H)	(g)	0	0.08988	Strontium (Sr)			2.64
	(l)	-252.879	0.07099	Yttrium (Y)			4.472
	(s)	-259.16	0.0763	Zirconium (Zr)			6.52
Helium (He)	(g)	0	0.1786	Niobium (Nb)			8.57
	(l)	-268.928	0.125	Molybdenum (Mo)			10.28
	(s)	-272.20	0.19*	Technetium (Tc)			11
Lithium (Li)			0.534	Ruthenium (Ru)			12.45
Beryllium (Be)			1.85	Rhodium (Rh)			12.41
Boron (B)			2.34	Palladium (Pd)			12.023
Carbon (C)	Diamond		3.513	Silver (Ag)			10.49
	Graphite		2.267	Cadmium (Cd)			8.65
Nitrogen (N)	(g)	0	1.251	Indium (In)			7.31
	(l)	-195.795	0.808	Tin (Sn)	White/tetragonal		7.265
Oxygen (O)	(g)	0	1.429		Gray/cubical crystal		5.769
	(l)	-182.962	1.141	Antimony (Sb)			6.697
Fluorine (F)	(g)	0	1.696	Tellurium (Te)			6.24
	(l)	-188.11	1.505	Iodine (I)			4.933
Neon (Ne)	(g)	0	0.9002	Xenon (Xe)	(g)	0	5.894
	(l)	-246.046	1.207		(l)	-108.099	2.942
Sodium (Na)			0.968	Cesium (Cs)	(l)	28.44	1.843
Magnesium (Mg)			1.738		(s)		1.93
Aluminum (Al)			2.70	Barium (Ba)			3.51
Silicon (Si)			2.3290	Lantern (La)			6.162
Phosphorus (P)	White(Yellow)		1.823	Cerium (Ce)			6.770
	Red		2.2-2.34	Praseodymium (Pr)			6.77
	Black		2.69	Neodymium (Nd)			7.01
Sulfur (S)	α(Rhombic crystal)		2.07	Promethium (Pm)			7.26
	β(Monoclinic)		1.96	Samarium (Sm)			7.52
	γ(Monoclinic)		1.92	Europium (Eu)			5.244
Chlorine (Cl)	(g)	0	3.2	Gadolinium (Gd)			7.90
	(l)	-34.04	1.5625	Terbium (Tb)			8.23
Argon (Ar)	(g)	0	1.784	Dysprosium (Dy)			8.540
	(l)	-185.848	1.3954	Holmium (Ho)			8.79
Potassium (K)			0.862	Erbium (Er)			9.066
Calcium (Ca)			1.55	Thulium (Tm)			9.32
Scandium (Sc)			2.985	Ytterbium (Yb)			6.90
Titanium (Ti)			4.506	Lutetium (Lu)			9.841
Vanadium (V)			6.0	Hafnium (Hf)			13.31
Chrome (Cr)			7.19	Tantalum (Ta)			16.69
Manganese (Mn)			7.21	Tungsten (W)			19.25
Iron (Fe)			7.874	Rhenium (Re)			21.02
Cobalt (Co)			8.90	Osmium (Os)			22.59
Nickel (Ni)			8.908	Iridium (Ir)			22.56
Copper (Cu)			8.96	Platinum (Pt)			21.45
Zinc (Zn)			7.14	Gold (Au)			19.30
Gallium (Ga)			5.91	Mercury (Hg)	(l)		13.534
Germanium (Ge)			5.323	Thallous (Tl)			11.85
Arsenic (As)			5.727	Lead (Pb)			11.34
Selenium (Se)	Gray		4.81	Bismuth (Bi)			9.78
	α		4.39	Polonium (Po)	α		9.196
	Amorphous		4.28		β		9.398
Bromine (Br)	(l)		3.1028	Radon (Rn)	(g)	0	9.73
Krypton (Kr)	(g)	0	3.749	Radium (Ra)			5.5
	(l)	-153.415	2.413	Thorium (Th)			11.724
Rubidium (Rb)	(l)	39.30	1.46	Uranium (U)			19.1
	(s)		1.532	Plutonium (Pu)			19.816

* 2.5 MPa

Source: https://en.wikipedia.org/wiki/(Element name), https://en.wikipedia.org/wiki/Densities_of_the_elements_(data_page). The latter collects data from several handbook such as CRC.

Density of various substances

For the density of porous or powdery substances, values marked with an asterisk (*) are the actual density of the substance. Other values are the apparent density determined by the volume of the external surface. Temperatures are room temperature except otherwise specified. For minerals, refer to page **592** and later in the Geology tables.

(unit : 10^3 kg m^{-3} = g cm^{-3})

Material	Density ρ
Liquid † indicates density at 20°C	
Acetone	0.791†
Aniline	1.022†
Linseed oil	0.91-0.94
Ammonia (−40℃)	0.690
Ethyl alcohol	0.789†
Seawater	1.01-1.05
Hydrogen peroxide	1.442†
Gasoline	0.66-0.75
Milk	1.03-1.04
Air[1)] (−194℃)	0.92
Glycerin	1.264†
Chloroform	1.489†
Whale oil	0.88
Acetic acid (purity)	1.049†
Heavy water (purity)	1.105†
Crude petroleum	0.85-0.90
Nitric acid (purity)	1.502
Oil (crude oil from Japan)	0.80-0.98
Oil (kerosene)	0.80-0.83
Turpentine	0.87
Rapeseed oil	0.91-0.92
Carbon disulfide	1.263†
Paraffin oil	-0.8
Ricinus	0.96-0.97
Benzene	0.879†
Methyl alcohol	0.793†
Sulfuric acid (purity)	1.834†
Wood Dried, in air	
Japanese evergreen oak (Quercus acuta)	0.85
Princess tree (Paulownia tomentose)	0.31
Chestnut	0.60

Material	Density ρ
Zelkova	0.70
Ebony	1.1-1.3
Passing	0.40
Bamboo	0.31-0.40
Teak	0.58-0.78
Hemlock-spruce	0.53
Japanese cypress	0.49
Pine	0.52
Mahogany	0.45-1.06
Yamazakura	0.67
Solids	
Asphalt	1.04-1.40
Ebonite	1.1-1.4
Granite	2.6-2.7
Paper (western-style paper)	0.7-1.1
Glass (common)	2.4-2.6
Glass (crown)	2.2-3.6
Glass (flint)	2.8-6.3
Glass (pyrex)	2.32
Tuff	1.4-2.6
Solid carbon dioxide (dry ice) (−80℃)	1.565
Rubber (India rubber)	0.91-0.96
Ice (0℃)	0.917
Cork	0.22-0.26
Concrete[2)]	2.4
Acetylcellulose	1.15-1.25
Sugar	1.59*
Porcelain (normal)	2.0-2.6
Porcelain (insulator)	2.3-2.5
Salt	2.17*
Camphor (10℃)	0.99
Crystal	2.65
Sand (dry)	1.4-1.7
Slate	2.7-2.9

Material	Density ρ
Fused silica (transparency)	2.22
Fused silica (opacity)	2.07
Quicklime	2.3*-3.2*
Slaked lime	1.15-1.25
Coal	1.2-1.5
Coal (smokeless coal)	1.4-1.7
Asbestos	2.0*-3.0*
Cement	3.0*-3.15*
Celluloid	1.35-1.60
Fiber hemp	1.50*-1.52*
Silk	1.30*-1.37*
Artificial silk	1.51*-1.52*
Wool	1.28*-1.33*
Cotton	1.50*-1.55*
Ivory	1.8-1.9
Marble	1.52-2.86
Soil (nomral conditions)	-2
Nylon	1.12
Naphthalene	1.16
Leather	0.86-1.02
Nitrocellulose	1.35-1.50
Paraffin	0.87-0.94
Bakelite (pure)	1.20-1.29
Bakelite (cellulose filled)	1.32-1.40
Calcite	2.71
Bone	1.7-2.0
Polyethylene	0.92-0.97
Polyvinyl chloride	1.2-1.6
Polystyrene	1.056
Beeswax	0.96
Agate	2.5-2.8
Charcoal	0.3-0.6
Charcoal	1.4*-1.9*
Polymethyl methacrylate	1.16-1.20
Fresh snow	-0.12
Brick	1.2-2.2

1) 20.9% oxygen, 2) compounding ratio 1 part cement, 2 parts sand, and 4 parts gravel (by volume).

Density of water (single phase for liquid phase only)

Density of water at 1 atm = 101325 Pa. The unit is kg m^{-3} = 10^{-3}g cm^{-3}. The maximum value is 999.97 kg m^{-3} at 3.98℃. The boiling point is approximately 99.974℃.

t/℃	0	1	2	3	4	5	6	7	8	9
0	999.84	999.90	999.94	999.96	999.97	999.96	999.94	999.90	999.85	999.78
10	999.70	999.60	999.50	999.38	999.24	999.10	998.94	998.77	998.59	998.40
20	998.20	997.99	997.77	997.54	997.30	997.04	996.78	996.51	996.23	995.94
30	995.65	995.34	995.02	994.70	994.37	994.03	993.68	993.33	992.96	992.59
40	992.21	991.83	991.43	991.03	990.62	990.21	989.79	989.36	988.92	988.48
50	988.03	987.57	987.11	986.64	986.17	985.69	985.20	984.71	984.21	983.70
60	983.19	982.67	982.15	981.62	981.09	980.55	980.00	979.45	978.89	978.33
70	977.76	977.19	976.61	976.02	975.43	974.84	974.24	973.63	973.02	972.41
80	971.78	971.16	970.53	969.89	969.25	968.61	967.96	967.30	966.64	965.97
90	965.30	964.63	963.95	963.27	962.58	961.88	961.18	960.48	959.77	959.06

Calculated by the relational expression in Kaye & Laby (web version 2008).

Density of water and water vapor in a saturated state (boundary of liquid/gas phase)

p : saturation pressure, ρ_L : density of water in a saturated state, ρ_V : density of saturated water vapor.

T_{90}/K	p/MPa	ρ_L/ kg m^{-3}	ρ_V/ kg m^{-3}	T_{90}/K	p/MPa	ρ_L/ kg m^{-3}	ρ_V/ kg m^{-3}
273.16[1)]	0.000 612	999.793	0.004 85	500	2.639 2	831.313	13.199
280	0.000 992	999.862	0.007 68	510	3.165 5	817.772	15.833
290	0.001 920	998.758	0.014 36	520	3.769 0	803.535	18.900
300	0.003 537	996.513	0.025 59	530	4.456 9	788.527	22.470
310	0.006 231	993.342	0.043 66	540	5.236 9	772.657	26.627
320	0.010 546	989.387	0.071 66	550	6.117 2	755.808	31.474
330	0.017 213	984.750	0.113 57	560	7.106 2	737.831	37.147
340	0.027 188	979.503	0.174 40	570	8.213 2	718.530	43.822
350	0.041 682	973.702	0.260 29	580	9.448 0	697.638	51.739
360	0.062 194	967.386	0.378 58	590	10.821	674.781	61.242
370	0.090 535	960.587	0.537 92	600	12.345	649.411	72.842
380	0.128 85	953.327	0.748 30	610	14.033	620.65	87.369
390	0.179 64	945.624	1.021 2	620	15.901	586.88	106.31
400	0.245 77	937.486	1.369 4	630	17.969	544.25	132.84
410	0.330 45	928.921	1.807 6	640	20.265	481.53	177.15
420	0.437 30	919.929	2.351 8	641	20.509	473.01	183.90
430	0.570 26	910.507	3.020 2	642	20.756	463.67	191.53
440	0.733 67	900.649	3.832 9	643	21.006	453.14	200.38
450	0.932 20	890.341	4.812 0	644	21.259	440.73	210.99
460	1.170 9	879.569	5.982 6	645	21.515	425.05	224.45
470	1.455 1	868.310	7.372 7	646	21.775	402.96	243.46
480	1.790 5	856.537	9.013 9	647	22.038	357.34	286.51
490	2.183 1	844.219	10.942	647.096[2)]	22.064	322	

W. Wagner and A. Pruss, J. Phys. Chem. Ref. Data, 31, 387 (2002).
1) triple point 2) critical point

Aqueous solution density of various substances (1)

The unit of concentration is wt% and the unit of density ρ is 10^3 kg m^{-3} = g cm^{-3}. ρ_{20}, ρ_{18}, ρ_{15} are the density at 20 ℃, 18 ℃, and 15 ℃ respectively. Δ is the ratio of density change per temperature increase of 1℃.

Concentration (%)	Hydrochloric acid		Nitric acid		Ammonia	
	ρ_{20}	Δ	ρ_{20}	Δ	ρ_{20}	Δ
		$\times 10^{-4}$		$\times 10^{-4}$		$\times 10^{-4}$
1	1.0032	−2.1	1.0036	−2.2	0.9939	−2.0
6	1.0279	2.8	1.0312	3.1	0.9730	2.7
10	1.0474	3.2	1.0543	3.8	0.9575	3.4
16	1.0776	4.0	1.0903	4.8	0.9362	4.3
20	1.0980	4.5	1.1150	5.5	0.9229	5.0
26	1.1290	5.3	1.1534	6.7	0.9040	6.0
30	1.1493	5.8	1.1800	7.5	0.8920	6.3
35	1.1789(36%)		1.2140	8.5		
40	1.1980		1.2463	9.4		
45			1.2783	10.4		
50			1.3100	11.4		

Concentration (%)	Sodium hydroxide		Potassium hydroxide	Sulfuric acid	Hydrogen peroxide
	ρ_{20}	Δ	ρ_{15}	ρ_{20}	ρ_{18}
		$\times 10^{-4}$			
1	1.0095	−2.5	1.0083	1.0051	1.0022
6	1.0648	3.8	1.0544	1.0384	1.0204
10	1.1089	4.4	1.0918	1.0661	1.0351
16	1.1751	5.1	1.1493	1.1094	1.0574
20	1.2191	5.5	1.1884	1.1394	1.0725
26	1.2848	5.9	1.2489	1.1863	1.0959
30	1.3279	6.1	1.2905	1.2185	1.1122
35	1.3798	6.5	1.3440	1.2599	1.1327
40	1.4300	6.8	1.3991	1.3028	1.1536
45	1.4779	7.1	1.4558	1.3476	1.1749
50	1.5253	7.3	1.5143	1.3951	1.1966

Aqueous solution density of various substances (20℃) (2)

Material	4%	10%	20%	30%	40%	50%
KCl	1.023 91	1.063 29	1.132 77			
KNO_3	1.023 41	1.062 66	1.132 58			
K_2SO_4	1.031 0	1.081 7				
NaBr	1.029 81	1.080 30	1.174 46	1.284 10	1.413 81	
NaI	1.029 78	1.080 42	1.176 88	1.290 64	1.427 08	1.594 15
$CaCl_2$	1.031 6	1.083 5	1.177.5	1.281 6	1.395 7	

Aqueous solution density of various substances (20℃) (3)

Material	4%	10%	20%	30%	40%	50%
NaCl	1.026 77	1.070 65	1.147 76			
$NaNO_3$	1.025 4	1.067 4	1.142 9	1.225 6	1.317 5	
Na(Ac)	1.018 6	1.049 5	1.102 1			
$CuSO_4$	1.040 1	1.084 0				
$ZnSO_4$	1.040 3	1.107 1				
$MgCl_2$	1.031 1	1.081 6	1.170 6	1.268 8		
$FeCl_3$	1.032 4	1.085 1	1.182 0	1.291 0	1.417 5	1.551 0
$SrCl_2$	1.034 4	1.092 5	1.201 0	1.325(24%)		
$MgSO_4$	1.039 2	1.103 4	1.219 8	1.270 1		
$BaCl_2$	1.034 1	1.092 1	1.203 1	1.2531(24%)		
NH_4Cl	1.010 7	1.028 6	1.056 7			
KBr	1.027 44	1.073 96	1.160 02	1.259 24	1.374 51	
KI	1.028 08	1.076 07	1.165 94	1.271 15	1.395 87	1.545 72
K_2CO_3	1.034 5	1.090 4	1.189 8	1.297 9	1.414 1	1.540 4
LiCl	1.021 45	1.055 91	1.115 01	1.179 14		
$AgNO_3$	1.032 7	1.088 2	1.171 5			
$CdCl_2$	1.033 9	1.091 2	1.199 2	1.327 3	1.483 3	1.676 2
NH_4NO_3	1.014 7	1.039 7	1.082 8	1.127 7	1.175 4	1.225 8

Density of mercury

At 1 atm = 101 325 Pa. 10^3 kg m^{-3} = g cm^{-3}.

Temp. t/℃	0	-5	-10	-15	-20	-25	-30	-35	-38.83	
0	13.595 08	.607 43	.619 8	.632 2	.644 61	.657 04	.669 49	.681 96	.691 53	
t/℃	0	1	2	3	4	5	6	7	8	9
0	13.595 08	.592 61	.5901 4	.587 68	.585 21	.582 75	.580 28	.577 82	.575 35	.572 89
10	13.570 43	.567 97	.5655 1	.563 05	.560 59	.558 13	.555 67	.553 22	.550 76	.548 31
20	13.545 85	.543 40	.5409 4	.538 49	.536 04	.533 59	.531 14	.528 69	.526 24	.523 79
30	13.521 34	.518 89	.5164 5	.514 00	.511 56	.509 11	.506 67	.504 22	.501 78	.499 34
40	13.496 90	.494 46	.4920 2	.489 58	.487 14	.484 70	.482 26	.479 82	.477 39	.474 95
50	13.472 51	.470 08	.4676 5	.465 21	.462 78	.460 35	.457 91	.455 48	.453 05	.450 62
60	13.448 19	.445 76	.4433 3	.440 90	.438 48	.436 05	.433 62	.431 20	.428 77	.426 35
70	13.423 92	.421 50	.4190 8	.416 65	.414 23	.411 81	.409 39	.406 97	.404 55	.402 13
80	13.399 71	.397 29	.3948 7	.392 45	.390 03	.387 62	.385 20	.382 78	.380 37	.377 95
90	13.375 54	.373 13	.3707 1	.368 30	.365 89	.363 47	.361 06	.358 65	.356 24	.353 83

t/℃	Density ρ	t/℃	Density ρ	t/℃	Density ρ	t/℃	Density ρ	t/℃	Density ρ
100	13.351 42	150	13.231 4	200	13.112 0	250	12.992 9	300	12.873 7
110	.327 3	160	.207 5	210	.088 2	260	.969 1	310	.849 8
120	.303 3	170	.183 6	220	.064 4	270	.945 3	320	.825 8
130	.279 3	180	.159 7	230	.040 6	280	.921 4	330	.801 9
140	.255 3	190	.135 9	240	.016 7	290	.897 5	356.73	.737 7

Calculated by the relational expression in Kaye & Laby (web version 2008).

Density of air

The following table shows values for the density of dried air including 0.04% of CO_2 and correction amounts due to relative humidity. The unit is kg $m^{-3} = 10^{-3}$ g cm^{-3}. The correction amount hardly depends on a pressure.

t/℃ \ P/kPa	92	93	94	95	96	97	98	99	100	101	102	103	104
0	1.174	1.187	1.199	1.212	1.225	1.238	1.251	1.263	1.276	1.289	1.302	1.314	1.327
5	1.153	1.165	1.178	1.19	1.203	1.215	1.228	1.240	1.253	1.266	1.278	1.291	1.303
10	1.132	1.145	1.157	1.169	1.182	1.194	1.206	1.219	1.231	1.243	1.255	1.268	1.280
15	1.113	1.125	1.137	1.149	1.161	1.173	1.185	1.197	1.209	1.222	1.234	1.246	1.258
20	1.094	1.105	1.117	1.129	1.141	1.153	1.165	1.177	1.189	1.201	1.212	1.224	1.236
25	1.075	1.087	1.099	1.110	1.122	1.134	1.145	1.157	1.169	1.180	1.192	1.204	1.215
30	1.057	1.069	1.080	1.092	1.103	1.115	1.126	1.138	1.149	1.161	1.172	1.184	1.195

t/℃ \ Humidity %	0	10	20	30	40	50	60	70	80	90	100
0	0.000	0.000	-0.001	-0.001	-0.001	-0.001	-0.002	-0.002	-0.002	-0.003	-0.003
5	0.000	0.000	-0.001	-0.001	-0.002	-0.002	-0.002	-0.003	-0.003	-0.004	-0.004
10	0.000	-0.001	-0.001	-0.002	-0.002	-0.003	-0.003	-0.004	-0.005	-0.005	-0.006
15	0.000	-0.001	-0.002	-0.002	-0.003	-0.004	-0.005	-0.005	-0.006	-0.007	-0.008
20	0.000	-0.001	-0.002	-0.003	-0.004	-0.005	-0.006	-0.007	-0.008	-0.009	-0.011
25	0.000	-0.001	-0.003	-0.004	-0.006	-0.007	-0.008	-0.010	-0.011	-0.013	-0.014
30	0.000	-0.002	-0.004	-0.006	-0.007	-0.009	-0.011	-0.013	-0.015	-0.017	-0.018

Calculated by the relational expression in Kaye & Laby (web version 2005).

Density of gas

The following table shows the density of various gases in standard conditions (0℃, 101 325 Pa), and the relative density in relation to air in the same condition. (Unit of density : kg $m^{-3} = 10^{-3}$g cm^{-3})

Gas	Density	Relative density	Gas	Density	Relative density
Acetylene	1.173	0.907	Hydrogen iodide	5.789	4.477
Air	1.293	1	Hydrogen sulfide	1.539	1.190
Ammonia	0.771	0.597	Krypton	3.739	2.891
Argon	1.784	1.380	Methane	0.717	0.555
Arsenous hydride (AsH_3)	3.50	2.71	Neon	0.900	0.696
Carbon dioxide	1.977	1.529	Nitric oxide (NO)	1.340	1.036
Carbon monoxide	1.250	0.967	Nitrogen	1.250	0.967
Chlorine	3.214	2.486	Nitrous oxide	1.978	1.530
Cyanogen	2.34	1.81	Oxygen	1.429	1.105
Dimethyl ether	2.108	1.630	Ozone	2.14	1.66
Ethane	1.356	1.049	Phosphine	1.531	1.184
Ethylene	1.260	0.974	Propane	2.02	1.56
Fluorine	1.696	1.312	Radon	9.73	7.53
Freon 12	5.083	3.931	Steam (100℃)	0.598	0.463
Helium	0.1785	0.138	Sulfur dioxide (sulfurous acid gas)	2.926	2.264
Hydrogen	0.0899	0.0695			
Hydrogen bromide	3.644	2.818	Uranium fluoride (Ⅵ)	4.68	3.62
Hydrogen chloride	1.639	1.268	Xenon	5.887	4.553

Fixed Pressure Point

Similar to the fixed temperature point, it is possible to use the transformation point of pure substances for the fixed point of pressure measurement. The following table shows the main fixed pressure points used as practical criteria for high-pressure measurement.

Material	Transformation type	Detection method*	p/GPa	t/℃
Carbon dioxide	liquid/vapor equilibrium		{ 3.4856×10^{-3} / 3.4847×10^{-3}	0.01 (triple point of water) / 0
Water	ice_I– ice_{III}– liquid water (triple point)		0.2091	−22.2
Ammonium fluoride	Polymorphic transformation (NH_4F I-II)	V	0.3605	25
Mercury	Solid/Liquid Equilibrium	V, E	0.7569	0
Ammonium fluoride	Polymorphic transformation (NH_4F II-III)	V	1.1531	25
Potassium bromide	Polymorphic transformation (KBr I-II)	V	1.74	About 25
Potassium chloride	Polymorphic transformation (KCl I-II)	V	1.92	〃
Bismuth	Polymorphic transformation (Bi I-II)	V, E	2.550	〃
Thallous	Polymorphic transformation (Tl II-III)	V, E	3.67	〃
Barium	Polymorphic transformation (Ba I-II)	V, E	5.5	〃
Bismuth	Polymorphic transformation (Bi III-V)	V, E	7.7	〃
Tin	Polymorphic transformation (Sn I-II)	V, E	9.2– 9.6	〃
Iron	Polymorphic transformation (Fe α-ε)	E	11.0–11.3	〃
Barium	Polymorphic transformation (Ba II-III)	E	11.8–12.2	〃
Lead	Polymorphic transformation (Pb I-II)	E	12.8–13.2	〃
Rubidium	Polymorphic transformation (Rb II-III)	E	14.2–15.3	〃

* "V" indicates a change in volume, "E" indicates a change in electrical resistance.

Viscosity (Viscosity Coefficient)

When a fluid is flowing in a laminar fashion, the internal frictional forces acting on both sides of the fluid with respect to a plane parallel to the flow line in the fluid are proportional to the product of the area of this plane and the velocity gradient in the direction perpendicular to the plane. The proportionality constant η is called the **viscosity** of the fluid. The dimension of η is $[ML^{-1} T^{-1}]$ and its unit is pascal-second (Pa s) or newton-second/square meter (N s m^{-2}), or poise (poise, P) in the CGS unit system. 1 Pa s = 1 N s m^{-2} = 10 P = 10^3 cP.

The amount of viscosity divided by the density of the fluid in its viscous state is called **kinematic viscosity.** Kinematic viscosity the dimension of v is $[L^2 T^{-1}]$ and the unit is square meters per second ($m^2 s^{-1}$), or Stokes (St) in the CGS unit system. 1 $m^2 s^{-1}$ = 10^4 St = 10^6 cSt.

Dynamic viscosity (η) and kinematic viscosity (v) of water

(Pressure : 1 atm = 101 325 Pa)

t/℃	$\eta/10^{-3}$ Pa s	$v/10^{-6}$ m^2 s^{-1}	t/℃	$\eta/10^{-3}$ Pa s	$v/10^{-6}$ m^2 s^{-1}
0	1.7906	1.7909	40	0.6524	0.6576
5	1.5185	1.5186	50	0.5469	0.5535
10	1.3064	1.3068	60	0.4668	0.4748
15	1.1378	1.1388	70	0.4045	0.4137
20	1.0016	1.0034	80	0.3550	0.3653
25	0.8899	0.8925	90	0.3150	0.3263
30	0.7970	0.8005	100	0.2821	0.2943

JIS Z 8803 : 2011.

Dynamic viscosity of liquid

(Pressure : 1 atm = 101 325 Pa, Unit : 10^{-3} Pa s)

Material	-100℃	-50℃	0℃	25℃	50℃	75℃	100℃
Acetone	–	–	0.402	0.310	0.247	0.200	0.165
Aniline	–	–	9.450	3.822	1.982	1.201	0.808
Benzene	–	–	–	0.603	0.436	0.332	0.263
Carbon tertrachloride	–	–	1.341	0.912	0.662	0.503	0.395
Diethyl ether	1.545	0.544	0.288	0.224	0.179	0.146	0.119
Ethyl alcohol	98.96	8.318	1.873	1.084	0.684	0.459	0.323
Mercury	–	–	1.616	1.528	1.401	1.322	1.255
Methyl alcohol	–	2.258	0.797	0.543	0.392	0.294	0.227
Ricinus	–	–	–	700	125	42.0	16.9
Sulfuric acid	–	–	–	23.8	11.7	6.6	4.1

Kaye & Laby (web version 2005).

Dynamic viscosity of gas[1)]

(Unit : 10^{-6} Pa s)

Material	0℃	20℃	50℃	100℃	200℃	300℃	400℃	500℃	600℃
Air	17.3	18.2	19.6	22.0	26.1	29.8	33.2	36.4	39.4
Ammonia	9.2	9.9	11.0	13.0	16.8	20.6	24.4	28.2	31.9
Argon	21.0	22.3	24.2	27.3	32.8	37.7	42.2	46.4	50.4
Benzene	7.0	7.5	8.1	9.4	12.0	–	–	–	–
Carbon dioxide	13.7	14.7	16.1	18.5	23.0	27.1	30.8	34.2	37.4
Carbon monoxide	16.6	17.4	18.8	21.0	25.2	29.0	32.5	35.6	38.6
Chlorine	12.3	13.2	14.5	16.9	21.0	25.0	–	–	–
Chloroform	9.4	10.1	11.1	12.8	16.2	19.5	–	–	
Ethylene	9.7	10.3	11.2	12.8	15.4	17.9	–	–	–
Helium	18.7	19.6	21.0	23.2	27.3	31.2	34.8	38.4	41.8
Hydrogen	8.4	8.8	9.4	10.4	12.1	13.7	15.3	16.9	18.4
Krypton	23.4	25.0	27.4	31.2	38.0	44.2	49.9	55.2	60.2
Methane	10.3	11.0	11.9	13.5	16.3	18.8	21.1	23.3	25.3
Neon	29.8	31.3	33.6	37.0	43.2	48.9	54.3	59.4	64.4
Nitrogen	16.6	17.6	18.9	21.2	25.1	28.6	31.9	34.9	37.8
Nitrous oxide	13.7	14.7	16.1	18.4	22.9	27.0	30.7	34.0	37.0
Oxygen	19.5	20.4	21.8	24.4	29.3	33.7	37.6	41.3	44.7
Steam	9.2	9.7	10.6	12.4	16.2	20.3	24.5	28.6	32.6
Sulfur dioxide	11.6	12.6	14.0	16.4	20.9	25.1	29.0	32.6	36.1
Xenon	21.2	22.8	25.1	28.8	35.7	42.0	47.9	53.4	58.6

1) The viscosity of gas is almost completely unrelated to pressure in a wide range from several dozen Pa to several atmospheres.
Kaye & Laby (web version 2005).

Surface Tension

Surface tension of water (γ) (Surface tension (unit of γ) : 10^{-3} N m^{-1})

t/℃	0	1	2	3	4	5	6	7	8	9
0	75.646*	75.508	75.367	75.226	75.084	74.942	74.799	74.655	74.511	74.366
10	74.221	74.075	73.929	73.782	73.634	73.486	73.337	73.188	73.038	72.887
20	72.736	72.584	72.432	72.279	72.126	71.972	71.818	71.663	71.507	71.351
30	71.194	71.037	70.879	70.721	70.562	70.402	70.242	70.081	69.920	69.759
40	69.596	69.434	69.270	69.106	68.942	68.777	68.611	68.445	68.279	68.112
50	67.944	67.776	67.607	67.438	67.268	67.098	66.927	66.755	66.584	66.411
60	66.238	66.065	65.891	65.716	65.541	65.366	65.190	65.013	64.836	64.659
70	64.481	64.302	64.123	63.944	63.764	63.583	63.402	63.220	63.038	62.856
80	62.673	62.489	62.305	62.121	61.936	61.750	61.565	61.378	61.191	61.004
90	60.816	60.628	60.439	60.250	60.060	59.870	59.679	59.488	59.296	59.104
100	58.912	58.719	58.525	58.332	58.137	57.943	57.747	57.552	57.356	57.159
110	56.962	56.765	56.567	56.368	56.170	55.970	55.771	55.571	55.370	55.169
120	54.968									

* Value at 0.01℃
Calculated by the relational expression in NIST web book (2017) http://webbook.nist.gov/chemistry/fluid/.

Surface tension of various substances (γ) (Unit of γ : 10^{-3} N m^{-1})

Material	Contacting gas	t/℃	γ	Material	Contacting gas	t/℃	γ
Helium (liquid)	Its own vapor	−268.9	0.098	Hexane	Air	20	18.42
Helium (liquid)	Its own vapor	−271.6	0.354	Methyl alcohol	Nitrogen	20	22.55
Hydrogen (liquid)	Its own vapor	−253.1	1.98	Nitrobenzene	Air	20	43.35
Nitrogen (liquid)	Its own vapor	−203.1	10.53	Oil	Air	18	26
Oxygen (liquid)	Its own vapor	−183.6	13.55	Olive oil[1)]	Air	20	32
Acetic acid	Air	20	27.7	Paraffin oil[2)]	Air	25	26.4
Benzene	Air	20	28.86	Sal volatile (20%)	Air	18	59.3
Carbon dioxide	Air	20	35.3	Sulfuric acid (98.5%)	Air	20	55.1
Carbon tertrachloride	Air	20	27.63	Toluene	Its own vapor	20	28.53
Chloroform	Air	20	27.28	Gold	Hydrogen	1 200	1 120
Diethyl ether	Its own vapor	20	16.96	Iron	Helium	1 570	1 720
Dioxane	Its own vapor	20	33.55	Lead	Hydrogen	350	442
Ethyl alcohol	Nitrogen	20	22.27	Mercury	Nitrogen	25	482.1
Glycerin	Air	20	63.4	Sodium chloride	Air	803	117.6

Elastic Modulus

In following table, E denotes Young's modulus, G shear modulus, σ Poisson's ratio, k bulk modulus, for substances of uniform isotropy, the following relation is satified between the quantities.

$$E = 2\,G(1+\sigma) = 3\,k(1-2\sigma)$$

Material (20℃)	E/GPa	G/GPa	σ	k/GPa
Aluminum	70.3	26.1	0.345	75.5
Bismuth	31.9	12.0	0.330	31.3
Brass[2)]	100.6	37.3	0.350	111.8
Cadmium	49.9	19.2	0.300	41.6
Constantan	162.4	61.2	0.327	156.4
Copper	129.8	48.3	0.343	137.8
German silver[5)]	132.5	49.7	0.333	132.0
Gold	78.0	27.0	0.44	217.0
Invar[1)]	144.0	57.2	0.259	99.4
Iron (cast)[4)]	152.3	60.0	0.27	109.5
Iron (soft)	211.4	81.6	0.293	169.8
Lead[4)]	16.1	5.59	0.44	45.8
Nickel (hard)[4)]	219.2	83.9	0.306	187.6
Nickel (soft)[4)]	199.5	76.0	0.312	177.3
Nylon-6, 6	1.2-2.9	–	–	–
Platinum	168.0	61.0	0.377	228.0
Polyethylene	0.4-1.3	0.26	0.458	–
Polystyrene	2.7-4.2	1.43	0.340	4.0
Quartz (fused silica)	73.1	31.2	0.17	36.9
Silver	82.7	30.3	0.367	103.6
Stainless steel[3)]	215.3	83.9	0.293	166.0
The glass (crown)	71.3	29.2	0.22	41.2
The glass (Flint)	80.1	31.5	0.27	57.6
Tin	49.9	18.4	0.357	58.2
Titanium	115.7	43.8	0.321	107.7
Tungsten carbide[4)]	534.4	219.0	0.22	319.0
Wood (teak)	13.0	–	–	–
Zinc	108.4	43.4	0.249	72.0

1) 36Ni, 63.8Fe, 0.2C 2) 70Cu, 30Zn 3) Approximately : 0.02C, 0.5Si, 0.7Mn, 2Ni, 18Cr, residual Fe 4) Typical values 5) 55Cu, 18Ni, 27Zn

Kaye & Laby (web version 2005).

Compressibility of Liquid

Compressibility of liquid is expresed as $\kappa = \frac{1}{k} = -\frac{1}{v}\frac{dv}{dp}$ at constant temperature.

Liquid	t/℃	p/atm	κ/(GPa)$^{-1}$	Liquid	t/℃	p/atm	κ/(GPa)$^{-1}$
Acetone	20	1	1.26	Glycerin	20	5 000	0.12
Acetone	20	5 000	0.21	Mercury	21.9	1 000	0.0388
Benzene	20	1	0.95	Mercury	21.9	10 000	0.0300
Carbon disulfide	20	1	0.93	Methyl alcohol	20	1	1.23
Chloroform	20	1	1.01	Methyl alcohol	20	5 000	0.21
Ether	20	1	1.87	Toluene	20	1	0.91
Ether	20	5 000	0.22	Water	20	1	0.45
Ethyl alcohol	20	1	1.11	Water	20	1 000	0.36
Ethyl alcohol	20	5 000	0.22	Water	20	5 000	0.18
Glycerin	20	1	0.21	Water	60	10 000	0.12

Amer. primarily Inst. Phys. Handbook. By using 3rd ed. 1 atm = 101 325 Pa

Diffusion Coefficient

The diffusion coefficient inside of a solid changes significantly depending on temperature. It follows the Arrhenius equation $D = D_0 \exp(-U/RT)$. D_0 is the pre-exponential factor depending on the frequency of collisions. U is the activation energy, R is the molar gas constant, and T is the absolute temperature. At certain temperatures, the diffusion coefficient of ionic crystals is significantly impacted by impurities.

(Unit of D_0 : 10^{-4} m^2 s^{-1}, Unit of U : kJ mol^{-1})

Diffusion atom	Solid	D_0	U	Diffusion atom	Solid	D_0	U	Diffusion atom	Solid	D_0	U
Na	Na	0.145	42.2	Ni	Cu	2.7	236	Na^+	NaCl	0.5	155
Cu	Cu	0.62	207.4	Zn	Cu	0.34	191	Cl^-	NaCl	110	215
Ag	Ag	0.44	185.3	Ga	Cu	0.55	192	K^+	KCl	0.04	142
Au	Au	0.091	174.5	Ag	Cu	0.63	195	Cl^-	KCl	10	193

Impurities D_0, U in Germanium and Silicon

(Unit of D_0 : 10^{-4} m^2 s^{-1}, Unit of U : kJ mol^{-1})

Impurities	Semiconductor			
	Germanium		Silicon	
	D_0	U	D_0	U
B	1.8×10^9	439	16.0	356
Al	-	-	4.8	335
Ga	20.0	297	3.6	339
In	20	259	16.5	377
Tl	0.06	(259)	16.5	377
P	12.3	247	10.5	356
As	3.0	234	69.6	405
Sb	1.3	218	12.9	381
Bi	3.3	239	1 033	448
Li	25×10^{-4}	49.3	23×10^{-4}	63.6
Fe	-	-	6.2×10^{-3}	84
Ge	10.8	307	-	-
Au	12.6	218	2.75×10^{-3}	196

Tensile Strength

The following table shows the maximum stress T generated in a material when it is subjected to a tensile force and fractures. The unit is $Pa = N\ m^{-2}$. In general, the strength of a wire increases when it is stretched to become a wire.

Substance	T/Pa	Substance	T/Pa	Substance	T/Pa
	$\times 10^8$		$\times 10^8$		$\times 10^8$
Zinc (rolled)	1.1–1.5	Copper (rolled)	2.0–4.0	Brass (brass)	3.5–5.5
Aluminium (cast)	0.9–1.0	Nylon - 6,6	0.62–0.83	Tungsten	15–35
Aluminum (rolled)	0.9–1.5	Lead (cast)	0.12–0.17	Tantalum	8–11
Glass	0.3–0.9	Solder	0.55–0.75	Duralumin	4.0–5.5
Silk thread	2.6	Gunmetal[3]	1.9–2.6	Zirconium (soft)	2.6–3.9
Spider Thread	1.8	Polyethylene	0.21–0.35	Zirconium (hard)	10
Leather belt	0.3–0.5	Polystyrene	0.34–0.52	Iron (steel)	11–15.5
Brass (cast)[1]	1.5–1.9	Methyl methacrylate	0.50–0.76	Iron (piano wire)	18.6–23.3
Brass (rolled)[1]	2.3–2.7	Wood (normal)[4]	0.2–0.7	Copper (hard)	4.0–4.6
Tin (cast)	0.2–0.35	Wood (hardwood)[4, 5]	0.6–1.1	Copper (soft)	2.8–3.1
Quartz yarn	About 10	Phosphor bronze(cast)	1.8–2.8	Nickel	5.0–9.0
Intestine wire (gut)	4.2	Wire		Platinum	3.3–3.7
Iron (cast steel)	4.0–6.0			Platinum Rhodium[6]	6.3
Iron (steel)	7.0–10.8	Aluminum	2.0–4.5	Phosphor Bronze (Hard)	6.9–10.8
Iron (nickel steel)[2]	8.0–10.0	Gold	2.0–2.5	Western Silver	4.6
Copper (cast)	1.2–1.7	Silver	2.9		

1) 66% Cu, 34% Zn 2) 5% Ni 3) 90% Cu, 10% Sn 4) Parallel to grain in timber form 5) Oak (ash), teak, beech, mahogany, etc. 6) 10% rhodium According to Kaye & Laby (web version 2005).

Mechanical Strength of Metallic Materials

The mechanical strength of a metallic material (including alloys) is generally defined by its characteristic value on the "stress-strain curve" as shown in the figure below. The mechanical strength of metallic materials (including alloys) is generally defined by the characteristic value of the "stress-strain curve" as shown in the figure below. The stress-strain curve is shown in Figure (a) for a material with a yield point and in Figure (b) for a material without a yield point. The stress-strain curve is shown in Figure (a) for a material with a yield point and Figure (b) for a material without a yield point. The Young's modulus E is the proportional limit of the relationship of stress σ is within the proportional limit, the Young's modulus E can be obtained from the relation $\sigma = E\varepsilon$. The shear modulus corresponds to the Young's modulus during shear deformation and is also called the stiffness modulus. The yield strength is the stress at the point where the yield stress shown in the figure is generated in case (a), and the stress at the point where the strain of 0.2% is generated in case (b). In the case of (b), the yield strength is defined as the stress at the point where 0.2% strain occurs (called the 0.2% proof stress). The tensile strength is the maximum stress value at increasing strain. The tensile strength is the maximum stress value when the strain is increased. The following table shows the mechanical strength at room temperature of typical metallic materials used in practical applications. The following table shows the mechanical strength at room temperature of typical metallic materials used in practical applications. The following table shows the mechanical strength at room temperature of typical metallic materials used in practical applications. The composition, for example, Fe–0.45 C–0.25 Si–0.8 In the composition, for example, Fe–0.45 C–0.25 Si–0.8 Mn indicates iron (Fe) alloyed with 0.45% C, 0.25% Si, and 0.8% Mn.

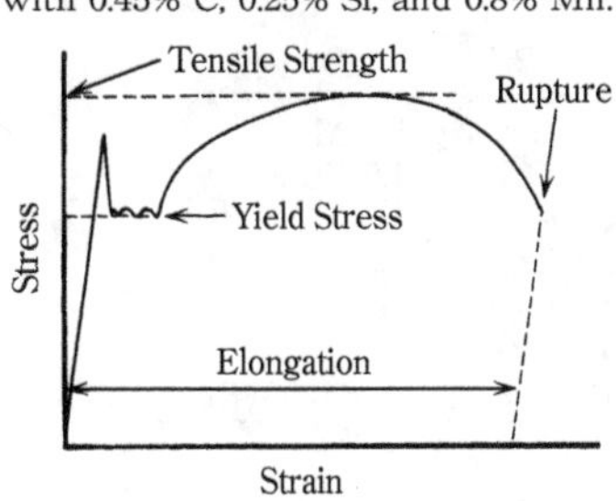

(a)

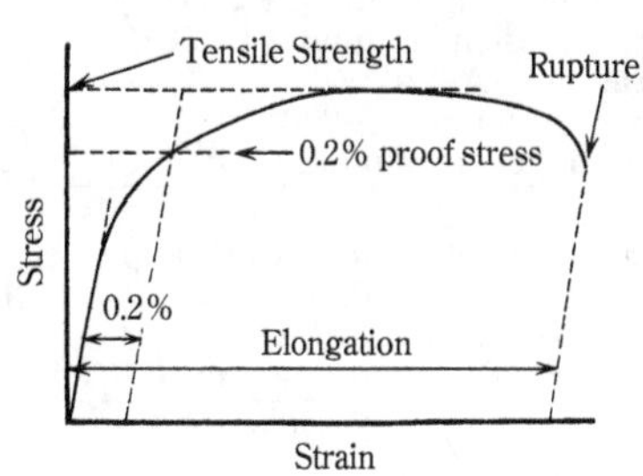

(b)

Mechanical

Material	Composition (wt %)
Industrial pure iron	99.96 Fe
Rolled steels for ordinary structures (SS 400)	Fe-0.1 C
Carbon steels for mechanical structures (S 45 C)	Fe-0.45 C-0.25 Si-0.8 Mn
High-tensile steel (HT 80)	Fe-0.12 C-0.8 Mn-1.0 Ni-0.5 Cr-0.4 Mo
Chromium-molybdenum steel (SCM 440)	Fe-0.4 C-0.7 Mn-1.0 Cr-0.25 Mo
Nickel-chromium-molybdenum steel (SNCM 439)	Fe-0.40 C-0.30 Si-0.70 Mn-1.85 Ni-0.80 Cr-0.25 Mo
Hot work tool steel (SKD 6)	Fe-0.37 C-1.0 Si-5.0 Cr-1.25 Mo-0.4 V
Spring steel (SUP 7)	Fe-0.6 C-2.0 Si-0.85 Mn
9% Ni steel for cryogenic vessels (SL 9 N 590)	Fe-0.05 C-0.3 Si-0.9Mn-9.0 Ni
Maraging steel (350)	Fe-17.5 Ni-12.5 Co-3.75 Mo-1.8 Ti-0.15 Al
Extraction stiffening type stainless steel (SUS 631)	Fe-0.06 C-0.4 Si-0.6 Mn-7.0 Ni-17.0 Cr-1.2 Al
System martensite stainless steel (SUS 410)	Fe-0.15>C-1.0>Si-1.0>Mn-12.5 Cr
Ferritic stainless steel (SUS 430)	Fe-0.12>C-0.75>Si-1.0>Mn-17 Cr
Austenitic stainless steel (SUS 304)	Fe-0.08>C-1.0>Si-2.0>Mn-9 Ni-19 Cr
Parakeet Roy 800 (NCF 800)	Fe-32.5 Ni-21 Cr-0.4 Al-0.4 Ti
Gray iron	Fe-3.3 C-2 Si-0.5 Mn
Spheroidal graphite cast iron (FCD 370)	Fe-2.5 C-2 Si
Austempering spheroidal graphite cast iron (FCD 900 A)	Fe-3.5 C-3 Si-0.2 Mn
Blackheart malleable iron (FCMB 360)	Fe-2.5 C-1 Si-0.4 Mn
Nickel	99.99 Ni
Inconel 600 (NCF 600)	72 Ni-15.5 Cr-8 Fe
Hastelloy X	Ni-22 Cr-9 Mo-0.6 W-18.5 Fe-1.5 Co-0.6 W
Monel metal	Ni-30 Cu-4 Si-2 Fe-1.0 Mn
Nichrome (GNC 108)	80 Ni-20 Cr
Oxygen-free copper (C 1020)	Cu>99.96
7/3 brasss (C 2600)	70 Cu-30 Zn
6/4 brasss (C 2801)	60 Cu-40 Zn
Naval brass (C 4640 P)	Cu-40 Zn-0.8 Sn
Rin bronze (C 5212 P)	Cu-8 Sn-0.2 P
Nickel silver (C 7521 P)	65 Cu-18 Ni-Zn(remainder)
Beryllium copper (C 1720)	Cu-1.9 Be-0.2 Ni
Yellow brass (YBsC 2)	68 Cu-2 Pb-Zn(remainder)
Yellow brass (BC 2 C)	Cu-8 Sn-4 Zn
Phosphor bronze castings (PBC 2 C)	Cu-11 Sn-0.3 P
Manganin (CMW)	84 Cu-12 Mn-4 Ni
Industrial aluminum (A 1085 P)	Al>99.85
Anti-corrosion aluminum (A 5083 P)	Al-4.5 Mg-0.5 Mn
Duralumin (A 2017 P)	Al-4 Cu-0.6 Mg-0.5 Si-0.6 Mn
Super-duralumin (A 2024 P)	Al-4.5 Cu-1.5 Mg-0.6 Mn
Extra super duralumin (A 7075 P)	Al-5.6 Zn-2.5 Mg-1.6 Cu
Silumin (AC 3A)	Al-12 Si
Magnesium alloys (board) (MP 5)	Mg-3.5 Zn-0.6 Zr
Magnesium alloys (bar) (MB 1)	Mg-3 Al-1 Zn
Magnesium alloy castings (MC 1)	Mg-6 Al-3 Zn-0.3 Mn
Commercially pure titanium (C.P.Ti)	H<0.013-O<0.20-N<0.05-Fe<0.20
Titanium 6Al-4V alloy (Grade 5)	Ti-6 Al-4 V
Tatanium 5-2.5 alloy	Ti-5 Al-2.5 Sn
Alloies for zinc die casting (ZDC1)	Zn-4.0 Al-1.0 Cu-0.04 Mg

Strength of Metals

(The JIS designation is given in ( ) after the material name. Compositions are representative.)

Heat-treatment	Density ρ/kg m^{-3}	Young's modulus E/GPa	Shear modulus G/GPa	Yield strength Y/MPa	Tensile strength T/MPa	Expansion (%)
Annealing	7.9×10^3	205	81	98	196	60
Annealing	7.9	206	79	240	450	21
Quenching, Tempering	7.8	205	82	727	828	22
Quenching, Tempering	–	203	73	834	865	26
Quenching, Tempering	7.8	–	–	833	980	12
Quenching, Tempering	7.8	204	–	1 471	1 765	8
Quenching, Tempering	7.8	206	82	–	1 550	–
Quenching, Tempering	–	–	–	1 080	1 230	9
Quenching, Tempering	–	193	74	588	760	23
Annealing, Aging	8.0	186	71	–	2 403	10
Tempering, Aging	7.8	204	–	1 029	1 225	4
Quenching, Tempering	7.8	200	–	345	540	25
Annealing	7.8	200	–	205	450	22
Solution treatment	8.0	197	74	205	520	40
Annealing	8.02	196	73	205	520	30
As cast	7.2	100	40		450	2
As cast	7.1	176	69	230	370	17
Austempering	–	–	–	600	900	8
Normalizing	7.4	172	68	215	360	14
Annealing	8.9	204	81	58	335	28
Annealing	8.4	214	76	245	550	30
Annealing	8.2	197	75	384	775	43
Annealing	8.8	179	66	515	775	10
As manufactured	8.4	214	–	–	690–930	20
Full annealing	8.9	117	–	–	195	35
Full annealing	8.5	110	41	–	280	50
Full annealing	8.4	103	38	–	330	40
As manufactured	8.4	103	–	–	375	25
Complete hardening	8.8	110	43	–	600	12
Complete hardening	8.7	120	47	–	540	3
Complete hardening	8.2	130	–	–	900	–
–	8.5	78	–	–	195	20
–	8.7	96	–	–	275	15
–	8.8	–	–	165	295	10
–	8.2	123	46	–	340–590	10
Annealing	2.7	69	27	15	55	30
Annealing	2.7	72	–	195	345	16
Normal temperature prescription (T 4)	2.8	69	–	195	355	15
Normal temperature prescription (T 4)	2.8	74	29	323	430	15
Quenching, Tempering(T 6)	2.8	72	28	505	573	11
As cast	2.7	71	–	–	140	2
As manufactured	1.8	40	17	160	250	6
As manufactured	1.8	40	17	140	230	6
Normal temperature prescription (T 4)	1.8	45	16	70	240	7
Annealing	4.6	106	45	170	320	27
Annealing	4.4	106	41	920	980	14
Annealing	–	118	48.0	800	860	16
–	6.6	89	–	–	325	7

Heat and Temperature

Temperature Standard

The SI determines the thermodynamic temperature scale introduced by Kelvin using the triple point of water as the fundamental fixed point. There is an implicit agreement that the other fixed point is absolute zero (0 K). However, since no one can reach 0 K, it is impossible to realize the SI temperature scale. The CIPM specifies several fixed temperature points which are easy to replicate and defines methods for interpolating intervals of fixed temperature points. Since 1948, the optimal thermodynamic temperature at that point in time has been assigned to each fixed temperature point. However, a major revision was performed in 1968 (IPTS-68). In 1989, the CIPM particularly proposed revision at high temperatures and a new temperature standard system (the International Temperature Scale of 1990, ITS-90) was established.

<table>
<tr><th></th><th>Quantity symbol</th><th>Unit name</th><th>Unit symbol</th><th>Remarks</th></tr>
<tr><td>Thermodynamics temperature scale</td><td>T</td><td>kelvin</td><td>K</td><td>One of the basic units of the International System of Units (SI) (Refer to page 288).</td></tr>
<tr><td>Celsius temperature scale</td><td>t</td><td>degree Celsius</td><td>℃</td><td>Defined by $t/℃ = T/\mathrm{K} - 273.15$. The scale intervals 1℃ and 1 K are equal.</td></tr>
<tr><td>International Kelvin Temperatures</td><td>T_{90}</td><td>kelvin</td><td>K</td><td rowspan="2">Measured with the International Temperature Scale of 1990.
$t_{90}/℃ = T_{90}/\mathrm{K} - 273.15$</td></tr>
<tr><td>International Celsius Temperatures</td><td>t_{90}</td><td>degree Celsius</td><td>℃</td></tr>
</table>

Temperature differences are expressed in kelvins, or degrees Celsius. Values expressed in the old system are denoted with T_{68} or t_{68}.

The International Temperature Scale of 1990 (ITS-90)[1)]

ITS-90 is defined by reproducible temperatures values (defining fixed points, see the following table) and measurement instruments (thermometers) which are calibrated at those temperatures. Temperatures other than fixed points are calculated using the prescribed formula.

Number	Defining fixed point	T_{90}/K	t_{90}/℃
1	evaporation point of helium[2)]	3-5	-270.15 --268.15
2	triple point of equilibrium hydrogen	13.8033	-259.3467
3, 4	evaportaion/boiling point of equilibrium hydrogen[2)]	about 17 and 20.3	about -256.15 and -252.85
5	triple point of neon	24.5561	-248.5939
6	triple point of oxygen	54.3584	-218.7916
7	triple point of argon	83.8058	-189.3442
8	triple point of mercury	234.3156	-38.8344
9	triple point of water	273.16	0.01
10	melting point of gallium[3)]	302.9146	29.7646
11	freezing point of indium	429.7485	156.5985
12	freezing point of tin	505.078	231.928
13	freezing point of zinc	692.677	419.527
14	freezing point of aluminum	933.473	660.323
15	freezing point of silver	1234.93	961.78
16	freezing point of gold	1337.33	1 064.18
17	freezing point of copper	1357.77	1 084.62

1) The International Temperature Scale of 1990, based on the resolution of the 1989 International Committee for Weights and Measures (CIPM) (Refer to page **293**).
2) In this chart, an approximate range is indicated. For detail, see next subsection.
3) The freezing points and melting points are where the solid phase and liquid phase can coexist under standard pressure, 101 325 Pa.

Note: The boiling point of water is no longer a defining fixed point. In the new temperature scale, t_{90}, its value under standard pressure is about 99.974°C.

a. Range 0.65 K to 5.0 K

T_{90} is defined in terms of the vapor pressure p of ^{3}He and ^{4}He using equations of the form :

$$T_{90}/\mathrm{K} = A_0 + \sum_{i=1}^{9} A_i \left\{[\ln(p/\mathrm{Pa}) - B]/C\right\}^i$$

The values of A, A_i, B and C are given in the table below for ^{3}He and ^{4}He.

Constants of helium vapor-pressure temperature equations

	^{3}He 0.65 K−3.2 K	^{4}He 1.25 K−2.1768 K	^{4}He 2.1768 K−5.0 K
A_0	1.053 447	1.392 408	3.146 631
A_1	0.980 106	0.527 153	1.357 655
A_2	0.676 380	0.166 756	0.413 923
A_3	0.372 692	0.050 988	0.091 159
A_4	0.151 656	0.026 514	0.016 349
A_5	−0.002 263	0.001 975	0.001 826
A_6	0.006 596	−0.017 976	−0.004 325
A_7	0.088 966	0.005 409	−0.004 973
A_8	−0.004 770	0.013 259	0
A_9	−0.054 943	0	0
B	7.3	5.6	10.3
C	4.3	2.9	1.9

b. Range 3.0 K to 24.5561 K

T_{90} is defined in terms of a constant volume gas thermometer using ^{3}He or ^{4}He that has been calibrated at three defining fixed points. These are the triple point of neon (24.5561 K), the triple point of equilibrium hydrogen (13.8033 K) and a temperature between 3 K to 5 K. The last temperature is determined using a ^{3}He or a ^{4}He vapor pressure thermometer as specified in a.

From 4.2 K to 24.5561 K, T_{90} is defined by the relation :

$$T_{90} = a + bp + cp^2,$$

where p is the pressure in the ^{4}He gas thermometer and a, b and c are coefficients, the numerical values of which are obtained from measurements made at the three defining points given above, but the lowest one lies between 4.2 K to 5.0 K.

From 3.0 K to 24.5561 K, T_{90} is defined by the relation :

$$T_{90}=\frac{a+bp+cp^2}{1+B_x(T_{90})N/V},$$

where N/V is the gas density with N being the quantity of gas and V the volume.

$B_x(T_{90})$ is the second virial coefficient calculated as follows. ^{x}He($x=3, 4$)

$$B_3(T_{90})/\mathrm{m^3\,mol^{-1}}=\{16.69-336.98(T_{90}/\mathrm{K})^{-1}+91.04(T_{90}/\mathrm{K})^{-2}-13.82(T_{90}/\mathrm{K})^{-3}\}\times 10^{-6}$$

$$B_4(T_{90})/\mathrm{m^3\,mol^{-1}}=\{16.708-374.05(T_{90}/\mathrm{K})^{-1}-383.53(T_{90}/\mathrm{K})^{-2}+1799.2(T_{90}/\mathrm{K})^{-3}-4033.2(T_{90}/\mathrm{K})^{-4}+3252.8(T_{90}/\mathrm{K})^{-5}\}\times 10^{-6}$$

c. Range 13.8033 K to 961.78°C

T_{90} is defined in terms of a platinum resistance thermometer calibrated at specified sets of defining fixed points. A platinum resistance thermometer in use must satisfy at least one of following conditions: W(29.7646℃) ≧ 1.118 07 or W(−38.8344℃) ≦ 0.844 235, where $W(T_{90})$ = $R(T_{90})/R$(273.16 K), $R(T_{90})$ resistance at temperature T_{90}. When using up to temperatures of 961.78℃, the resistor must also satisfy the following condition: W(961.78℃) ≧ 4.2844.

The temperature T_{90} is calculated using one of the following formulas which include the resistance ratio function $W(T_{90})$ and the reference function $W_r(T_{90})$. The coefficients a, b, $c_{(i)}$ and d in the formula are obtained from measurements at the defining fixed points.

(1) $$W(T_{90})-W_r(T_{90})=a[W(T_{90})-1]+b[W(T_{90})-1]^2+\sum_{i=1}^{5} c_i[\ln(W(T_{90}))]^{i+n}$$

(2) $$W(T_{90})-W_r(T_{90})=a[W(T_{90})-1]+b[W(T_{90})-1]^2+c[W(T_{90})-1]^3+d[W(T_{90})-W(660.323℃)]^2$$

(3) $$W(T_{90})-W_r(T_{90})=a[W(T_{90})-1]+b[W(T_{90})-1]\ln(W(T_{90}))$$

The table shows the formula and related conditions used according to the temperature range.

Defining fixed points and conditions to determine coefficients in formulas

Temperature range	Formula	Defining fixed point[*1]	Condition
13.803 3 K–273.16 K	(1)	2–9	$n=2$
24.556 1 K–273.16 K	(1)	2, 5–9	$n=0$ $c_4=c_5=0$
54.358 4 K–273.16 K	(1)	6–9	$n=1$ $c_2=c_3=0$ $c_4=c_5=0$
83.805 8 K–273.16 K	(3)	7–9	
234.315 6 K–302.914 6 K	(2)	8–10	$c=d=0$
0℃–29.764 6℃	(2)	9, 10	$b=c=d=0$
0℃–156.598 5℃	(2)	9, 11	$b=c=d=0$
0℃–231.928℃	(2)	9, 11, 12	$c=d=0$
0℃–419.527℃	(2)	9, 12, 13	$c=d=0$
0℃–660.323℃	(2)	9, 12–14	$d=0$
0℃–961.78℃	(2)	9, 12–15	*2

*1 The numbers for the fixed points correspond to the numbers on the left side of the table on page **325**.

*2 a, b and c are determined by the measurements at the fixed points lying between 0℃ and 660.323℃ ; d is defined by the measurement at the melting point of silver (961.78℃).

One of the following reference functions $W_r(T_{90})$ is used depending on the temperature range. Representative values for $W_r(T_{90})$ are shown on page **330**.

(1) $\ln[W_r(T_{90})] =$

$$A_0+\sum_{i=1}^{12} A_i\{[\ln(T_{90}/273.16\ \mathrm{K})+1.5]/1.5\}^i$$

$$(13.803\ 3\ \mathrm{K} \leqq T_{90} \leqq 273.16\ \mathrm{K})$$

(2) $W_r(T_{90}) =$

$$C_0 + \sum_{i=1}^{9} C_i\{(T_{90}/K - 754.15)/481\}^i$$

$$(273.15\ K \leqq T_{90} \leqq 1\,234.93\ K)$$

Constant of the reference function

A_0	-2.135 347 29	C_0	2.781 572 54
A_1	3.183 247 20	C_1	1.646 509 16
A_2	-1.801 435 97	C_2	-0.137 143 90
A_3	0.717 272 04	C_3	-0.006 497 67
A_4	0.503 440 27	C_4	-0.002 344 44
A_5	-0.618 993 95	C_5	0.005 118 68
A_6	-0.053 323 22	C_6	0.001 879 82
A_7	0.280 213 62	C_7	-0.002 044 72
A_8	0.107 152 24	C_8	-0.000 461 22
A_9	-0.293 028 65	C_9	0.000 457 24
A_{10}	0.044 598 72		
A_{11}	0.118 686 32		
A_{12}	-0.052 481 34		

d. Range 961.78℃ or higher

T_{90} is defined by the equation derived from Planck radiation law :

$$\frac{L_\lambda(T_{90})}{L_\lambda(T_{90}(x))} = \frac{\exp(c_2[\lambda T_{90}(x)]^{-1}) - 1}{\exp(c_2[\lambda T_{90}]^{-1}) - 1}$$

where $T_{90}(x)$ refers to any one of the freezing points of silver (1234.93 K, x= Ag), gold (1337.33 K, x= Au) or copper (1357.77 K, x=Cu). $L_\lambda(T_{90})$ and $L_\lambda(T_{90}(x))$ is the spectral concentrations of the radiance of a blackbody at wavelength λ in vacuum at T_{90} and $T_{90}(x)$ respectively, and c_2=0.0143 88 m K.

Representative values of the reference function $W_r(T_{90})$

T_{90}/K	$W_r(T_{90})$	T_{90}/K	$W_r(T_{90})$	T_{90}/K	$W_r(T_{90})$
13.803 3	0.001 190 07	100	0.286 074 10	550	2.058 408 07
14	0.001 238 46	120	0.371 891 42	600	2.239 991 28
16	0.001 863 82	140	0.456 497 71	650	2.418 685 80
18	0.002 776 70	160	0.540 064 72	700	2.594 482 57
20	0.004 035 94	180	0.622 788 86	750	2.767 356 49
22	0.005 692 22	200	0.704 809 73	800	2.937 269 58
24	0.007 785 78	220	0.786 216 45	850	3.104 176 96
26	0.010 345 17	240	0.867 067 92	900	3.268 034 65
28	0.013 387 05	260	0.947 405 65	950	3.428 807 65
30	0.016 916 85	280	1.027 253 01	1 000	3.586 477 09
40	0.041 464 85	300	1.106 614 06	1 050	3.741 045 07
50	0.075 134 00	340	1.263 887 11	1 100	3.892 536 57
60	0.114 304 23	380	1.419 241 76	1 150	4.040 998 20
70	0.156 248 44	420	1.572 691 45	1 200	4.186 494 59
80	0.199 347 21	460	1.724 249 46	1 234.93	4.286 420 53
90	0.242 755 51	500	1.873 929 46		

Reference Electromotive Force of Thermocouple

The reference electromotive forces of thermocouples broadly used in temperature measurement are defined by the reference functions (IEC 60584-1:2013). The reference function is given by Eq. (1). For type-K thermocouples above 0℃ Eq. (2) has to be used.

$$E/\mu\text{V} = \sum_{i=0}^{n} a_i \,(t_{90}/℃)^i \quad (1)$$

$$E/\mu\text{V} = \sum_{i=0}^{n} b_i (t_{90}/℃)^i + c_0 \exp[c_1 (t_{90}/℃ - 126.9686)^2] \quad (2)$$

The following tables show coefficients of the reference functions and the calculated reference electromotive forces for several thermocouples. Electromotive force is expressed in units of mV when the reference junction is maintained at 0℃ and the measurement junction at t_{90}/℃. These are reference values. In precise temperature measurements, it is necessary to calibrate each thermocouple at the defining fixed points of the ITS-90.

Copper — constantan thermocouple (Type T)

(unit: mV)

Temperature range ℃	-270–0				0–400
Index n	14				8
a_0	0.000 000 000 0	a_8	$3.849\,393\,988\,3 \times 10^{-12}$	a_0	0.000 000 000 0
a_1	$3.874\,810\,636\,4 \times 10^{1}$	a_9	$2.821\,352\,192\,5 \times 10^{-14}$	a_1	$3.874\,810\,636\,4 \times 10^{1}$
a_2	$4.419\,443\,434\,7 \times 10^{-2}$	a_{10}	$1.425\,159\,477\,9 \times 10^{-16}$	a_2	$3.329\,222\,788\,0 \times 10^{-2}$
a_3	$1.184\,432\,310\,5 \times 10^{-4}$	a_{11}	$4.876\,866\,228\,6 \times 10^{-19}$	a_3	$2.061\,824\,340\,4 \times 10^{-4}$
a_4	$2.003\,297\,355\,4 \times 10^{-5}$	a_{12}	$1.079\,553\,927\,0 \times 10^{-21}$	a_4	$-2.188\,225\,684\,6 \times 10^{-6}$
a_5	$9.013\,801\,955\,9 \times 10^{-7}$	a_{13}	$1.394\,502\,706\,2 \times 10^{-24}$	a_5	$1.099\,688\,092\,8 \times 10^{-8}$
a_6	$2.265\,115\,659\,3 \times 10^{-8}$	a_{14}	$7.979\,515\,392\,7 \times 10^{-28}$	a_6	$-3.081\,575\,877\,2 \times 10^{-11}$
a_7	$3.607\,115\,420\,5 \times 10^{-10}$			a_7	$4.547\,913\,529\,0 \times 10^{-14}$
				a_8	$-2.751\,290\,167\,3 \times 10^{-17}$

t_{90}/℃	0	-10	-20	-30	-40	-50	-60	-70	-80	-90
-200	-5.603	-5.753	-5.888	-6.007	-6.105	-6.180	-6.232	-6.258		
-100	-3.379	-3.657	-3.923	-4.177	-4.419	-4.648	-4.865	-5.070	-5.261	-5.439
0	0.000	-0.383	-0.757	-1.121	-1.475	-1.819	-2.153	-2.476	-2.788	-3.089
t_{90}/℃	0	10	20	30	40	50	60	70	80	90
0	0.000	0.391	0.790	1.196	1.612	2.036	2.468	2.909	3.358	3.814
100	4.279	4.750	5.228	5.714	6.206	6.704	7.209	7.720	8.237	8.759
200	9.288	9.822	10.362	10.907	11.458	12.013	12.574	13.139	13.709	14.283
300	14.862	15.445	16.032	16.624	17.219	17.819	18.422	19.030	19.641	20.255
400	20.872									

Platinum — platinum 10% rhodium thermocouple (Type S)

(unit: mV)

Temperature range ℃	-50–1 064.18	1 064.18–1 664.5	1 664.5–1 768.1
Index n	8	4	4
a_0	0.000 000 000 00	$1.329\,004\,440\,85\times10^{3}$	$1.466\,282\,326\,36\times10^{5}$
a_1	5.403 133 086 31	3.345 093 113 44	$-2.584\,305\,167\,52\times10^{2}$
a_2	$1.259\,342\,897\,40\times10^{-2}$	$6.548\,051\,928\,18\times10^{-3}$	$1.636\,935\,746\,41\times10^{-1}$
a_3	$-2.324\,779\,686\,89\times10^{-5}$	$-1.648\,562\,592\,09\times10^{-6}$	$-3.304\,390\,469\,87\times10^{-5}$
a_4	$3.220\,288\,230\,36\times10^{-8}$	$1.299\,896\,051\,74\times10^{-11}$	$-9.432\,236\,906\,12\times10^{-12}$
a_5	$-3.314\,651\,963\,89\times10^{-11}$		
a_6	$2.557\,442\,517\,86\times10^{-14}$		
a_7	$-1.250\,688\,713\,93\times10^{-17}$		
a_8	$2.714\,431\,761\,45\times10^{-21}$		

t_{90}/℃	0	-10	-20	-30	-40	-50	-60	-70	-80	-90
0	0.000	-0.053	-0.103	-0.150	-0.194	-0.236				
t_{90}/℃	0	10	20	30	40	50	60	70	80	90
0	0.000	0.055	0.113	0.173	0.235	0.299	0.365	0.433	0.502	0.573
100	0.646	0.720	0.795	0.872	0.950	1.029	1.110	1.191	1.273	1.357
200	1.441	1.526	1.612	1.698	1.786	1.874	1.962	2.052	2.141	2.232
300	2.323	2.415	2.507	2.599	2.692	2.786	2.880	2.974	3.069	3.164
400	3.259	3.355	3.451	3.548	3.645	3.742	3.840	3.938	4.036	4.134
500	4.233	4.332	4.432	4.532	4.632	4.732	4.833	4.934	5.035	5.137
600	5.239	5.341	5.443	5.546	5.649	5.753	5.857	5.961	6.065	6.170
700	6.275	6.381	6.486	6.593	6.699	6.806	6.913	7.020	7.128	7.236
800	7.345	7.454	7.563	7.673	7.783	7.893	8.003	8.114	8.226	8.337
900	8.449	8.562	8.674	8.787	8.900	9.014	9.128	9.242	9.357	9.472
1 000	9.587	9.703	9.819	9.935	10.051	10.168	10.285	10.403	10.520	10.638
1 100	10.757	10.875	10.994	11.113	11.232	11.351	11.471	11.590	11.710	11.830
1 200	11.951	12.071	12.191	12.312	12.433	12.554	12.675	12.796	12.917	13.038
1 300	13.159	13.280	13.402	13.523	13.644	13.766	13.887	14.009	14.130	14.251
1 400	14.373	14.494	14.615	14.736	14.857	14.978	15.099	15.220	15.341	15.461
1 500	15.582	15.702	15.822	15.942	16.062	16.182	16.301	16.420	16.539	16.658
1 600	16.777	16.895	17.013	17.131	17.249	17.366	17.483	17.600	17.717	17.832
1 700	17.947	18.061	18.174	18.285	18.395	18.503	18.609			

Platinum — platinum 13% rhodium thermocouple (Type R)

(unit: mV)

Temperature range ℃	-50–1 064.18	1 064.18–1 664.5	1 664.5–1 768.1
Index n	9	5	4
a_0	0.000 000 000 00	$2.951\,579\,253\,16\times10^{3}$	$1.522\,321\,182\,09\times10^{5}$
a_1	5.289 617 297 65	-2.520 612 513 32	$-2.688\,198\,885\,45\times10^{2}$
a_2	$1.391\,665\,897\,82\times10^{-2}$	$1.595\,645\,018\,65\times10^{-2}$	$1.712\,802\,804\,71\times10^{-1}$
a_3	$-2.388\,556\,930\,17\times10^{-5}$	$-7.640\,859\,475\,76\times10^{-6}$	$-3.458\,957\,064\,53\times10^{-5}$
a_4	$3.569\,160\,010\,63\times10^{-8}$	$2.053\,052\,910\,24\times10^{-9}$	$-9.346\,339\,710\,46\times10^{-12}$
a_5	$-4.623\,476\,662\,98\times10^{-11}$	$-2.933\,596\,681\,73\times10^{-13}$	
a_6	$5.007\,774\,410\,34\times10^{-14}$		
a_7	$-3.731\,058\,861\,91\times10^{-17}$		
a_8	$1.577\,164\,823\,67\times10^{-20}$		
a_9	$-2.810\,386\,252\,51\times10^{-24}$		

t_{90}/℃	0	-10	-20	-30	-40	-50	-60	-70	-80	-90
0	0.000	-0.051	-0.100	-0.145	-0.188	-0.226				
t_{90}/℃	0	10	20	30	40	50	60	70	80	90
0	0.000	0.054	0.111	0.171	0.232	0.296	0.363	0.431	0.501	0.573
100	0.647	0.723	0.800	0.879	0.959	1.041	1.124	1.208	1.294	1.381
200	1.469	1.558	1.648	1.739	1.831	1.923	2.017	2.112	2.207	2.304
300	2.401	2.498	2.597	2.696	2.796	2.896	2.997	3.099	3.201	3.304
400	3.408	3.512	3.616	3.721	3.827	3.933	4.040	4.147	4.255	4.363
500	4.471	4.580	4.690	4.800	4.910	5.021	5.133	5.245	5.357	5.470
600	5.583	5.697	5.812	5.926	6.041	6.157	6.273	6.390	6.507	6.625
700	6.743	6.861	6.980	7.100	7.220	7.340	7.461	7.583	7.705	7.827
800	7.950	8.073	8.197	8.321	8.446	8.571	8.697	8.823	8.950	9.077
900	9.205	9.333	9.461	9.590	9.720	9.850	9.980	10.111	10.242	10.374
1 000	10.506	10.638	10.771	10.905	11.039	11.173	11.307	11.442	11.578	11.714
1 100	11.850	11.986	12.123	12.260	12.397	12.535	12.673	12.812	12.950	13.089
1 200	13.228	13.367	13.507	13.646	13.786	13.926	14.066	14.207	14.347	14.488
1 300	14.629	14.770	14.911	15.052	15.193	15.334	15.475	15.616	15.758	15.899
1 400	16.040	16.181	16.323	16.464	16.605	16.746	16.887	17.028	17.169	17.310
1 500	17.451	17.591	17.732	17.872	18.012	18.152	18.292	18.431	18.571	18.710
1 600	18.849	18.988	19.126	19.264	19.402	19.540	19.677	19.814	19.951	20.087
1 700	20.222	20.356	20.488	20.620	20.749	20.877	21.003			

Chromel — alumel thermocouple (Type K)

(unit: mV)

Temperature range ℃	-270-0		0-1 372		
Index n	10		9		
a_0	0.000 000 000 0	b_0	$-1.760\,041\,368\,6\times10^{1}$	c_0	$1.185\,976\times10^{2}$
a_1	$3.945\,012\,802\,5\times10^{1}$	b_1	$3.892\,120\,497\,5\times10^{1}$	c_1	$-1.183\,432\times10^{-4}$
a_2	$2.362\,237\,359\,8\times10^{-2}$	b_2	$1.855\,877\,003\,2\times10^{-2}$		
a_3	$-3.285\,890\,678\,4\times10^{-4}$	b_3	$-9.945\,759\,287\,4\times10^{-5}$		
a_4	$-4.990\,482\,877\,7\times10^{-6}$	b_4	$3.184\,094\,571\,9\times10^{-7}$		
a_5	$-6.750\,905\,917\,3\times10^{-8}$	b_5	$-5.607\,284\,488\,9\times10^{-10}$		
a_6	$-5.741\,032\,742\,8\times10^{-10}$	b_6	$5.607\,505\,905\,9\times10^{-13}$		
a_7	$-3.108\,887\,289\,4\times10^{-12}$	b_7	$-3.202\,072\,000\,3\times10^{-16}$		
a_8	$-1.045\,160\,936\,5\times10^{-14}$	b_8	$9.715\,114\,715\,2\times10^{-20}$		
a_9	$-1.988\,926\,687\,8\times10^{-17}$	b_9	$-1.210\,472\,127\,5\times10^{-23}$		
a_{10}	$-1.632\,269\,748\,6\times10^{-20}$				

t_{90}/℃	0	-10	-20	-30	-40	-50	-60	-70	-80	-90
-200	-5.891	-6.035	-6.158	-6.262	-6.344	-6.404	-6.441	-6.458		
-100	-3.554	-3.852	-4.138	-4.411	-4.669	-4.913	-5.141	-5.354	-5.550	-5.730
0	0.000	-0.392	-0.778	-1.156	-1.527	-1.889	-2.243	-2.587	-2.920	-3.243
t_{90}/℃	0	10	20	30	40	50	60	70	80	90
0	0.000	0.397	0.798	1.203	1.612	2.023	2.436	2.851	3.267	3.682
100	4.096	4.509	4.920	5.328	5.735	6.138	6.540	6.941	7.340	7.739
200	8.138	8.539	8.940	9.343	9.747	10.153	10.561	10.971	11.382	11.795
300	12.209	12.624	13.040	13.457	13.874	14.293	14.713	15.133	15.554	15.975
400	16.397	16.820	17.243	17.667	18.091	18.516	18.941	19.366	19.792	20.218
500	20.644	21.071	21.497	21.924	22.350	22.776	23.203	23.629	24.055	24.480
600	24.905	25.330	25.755	26.179	26.602	27.025	27.447	27.869	28.289	28.710
700	29.129	29.548	29.965	30.382	30.798	31.213	31.628	32.041	32.453	32.865
800	33.275	33.685	34.093	34.501	34.908	35.313	35.718	36.121	36.524	36.925
900	37.326	37.725	38.124	38.522	38.918	39.314	39.708	40.101	40.494	40.885
1 000	41.276	41.665	42.053	42.440	42.826	43.211	43.595	43.978	44.359	44.740
1 100	45.119	45.497	45.873	46.249	46.623	46.995	47.367	47.737	48.105	48.473
1 200	48.838	49.202	49.565	49.926	50.286	50.644	51.000	51.355	51.708	52.060
1 300	52.410	52.759	53.106	53.451	53.795	54.138	54.479	54.819		

Electromotive Forces against Platinum for Various Metals

The plus (+) sign for electromotive force means that electric current flows toward platinum through a 0℃ junction. The minus (–) sign indicates a flow in the opposite direction. The unit is mV.

Metal	t/℃		Metal	t/℃	
	–100	+100		–100	+ 100
Alumel	–	– 1.29	Manganin[5]	–	+ 0.61
Aluminum	– 0.06	+ 0.42	Mercury	–	+ 0.045
Antimony	–	+ 4.89	Nichrome[2]	–	+ 1.14
Bismuth	+ 7.54	– 7.34	Nichrome[3]	–	+ 0.85
Brass[1]	–	+ 0.60	Nickel	+ 1.22	– 1.48
Cadmium	– 0.31	+ 0.90	Palladium	+ 0.48	– 0.57
Calcium	– 0.13	– 0.51	Potassium	+ 0.78	–
Carbon	–	+ 0.70	Rhodium	– 0.34	+ 0.70
Chromel	–	+ 2.81	Silicon	+37.17	–41.56
Cobalt	–	– 1.33	Silver	– 0.39	+ 0.74
Constantan	–	– 3.51	Sodium	+ 0.29	–
Copper	– 0.37	+ 0.76	Solder[4]	–	+ 0.46
Germanium	–26.62	+33.90	Stainless steel (18-8)	–	+ 0.44
Gold	– 0.39	+ 0.78	Tantalum	– 0.10	+ 0.33
Indium	–	+ 0.69	Tin	– 0.12	+ 0.42
Iron	– 1.94	+ 1.98	Tungsten	– 0.15	+ 1.12
Lead	– 0.13	+ 0.44	Zinc	– 0.33	+ 0.76
Magnesium	– 0.09	+ 0.44			

1) 70Cu+30Zn 2) 80Ni+20Cr 3) 60Ni+24Fe+16Cr
4) 50Sn+50Pb 5) 84Cu+4Ni+12Mn
Source: Kaye & Laby (web version 2005).

Temperature of Flame

flame	t/℃
Wax candle	1 400
Alcohol (ethyl)	1 700
Bunsen burner (with sufficent air)	1 800
Hydrogen	1 900
Acetylene	2 500
Carbon monoxide and oxygen	2 600
Hydrogen and oxygen (oxyhydrogen flame)	2 800
Acetylene and oxygen	3 800

Temperature and Color

color	t/℃
First red flame	500
Dark red flame	700
Cherry red flame	900
Cherry red flame turns translucent	1 000
Bitter orange/yellow flame	1 100
Bitter orange/yellow flame turns translucent	1 200
White hot	1 300
Dazzling white heat	>1 500

Temperature and Brightness Temperature of Thermal Radiators

The temperature measured using a conventional optical pyrometer is the brightness temperature of that object; in other words, the temperature of the perfect black body which is assumed to release radiation that is equivalent to the radiation released by that object. The following chart contrasts the 'real' temperatures T_r and the brightness temperature T_{eu} of that object as deduced for monochromatic light with a wavelength of 0.655 μm.

T_r/K	Brightness temperature T_{eu}/K of various substances							
	C[2)]	W	Fe	Mo	Ni	Pt	Cu	Au
1 000	995	966	-	958	956	950	-	908
1 100	1 092	1 058	-	1 049	1 047	1 037	-	990
1 200	1 189	1 149	-	1 139	1 137	1 124	-	1 071
1 300	1 286	1 240	-	1 228	1 226	1 211	-	1 151
1 400	1 382	1 330	-	1 316	1 315	1 296	1 255*	-
1 500	1 478	1 420	1 412	1 403	1 403	1 381	1 333*	-
1 600	1 574	1 509	1 499	1 489	-	1 466	1 413*	-
1 700	1 670	1 597	1 586	1 574	-	1 551	1 490*	-
1 800	1 766	1 684	1 673	1 658	-	1 634	1 566*	-
1 900	1 862	1 771	1 759*	1 741	-	1 717	-	-
2 000	1 958	1 857	1 844*	1 824	-	1 800	-	-
2 200	2 150	2 026	2 016*	1 986				
2 400	2 340	2 192	-	2 143				
2 600	-	2 356	-	2 297				
2 800	-	2 516	-	2 448				

1) When material surface is sufficently clean.
2) Unprocessed carbon filament
* molten

Source: Intenational Critical Tables, 1929.

Thermal Properties

Melting point and boiling point of simple substances

The following table shows the melting point (t_m) and boiling point (t_b) in Celsius temperature for simple substances at 1 atmosphere.

Element (Symbol)	Melting point t_m/℃	Boiling point t_b/℃	Element (Symbol)	Melting point t_m/℃	Boiling point t_b/℃
Hydrogen (H)	-259.16	-252.879	Ruthenium (Ru)	2334	4150
Helium (He)	-272.20[1)]	-268.928	Rhodium (Rh)	1964	3695
Lithium (Li)	180.50	1330	Palladium (Pd)	1554.9	2963
Beryllium (Be)	1287	2469	Silver (Ag)	961.78[*2]	2162[*1]
Boron (B)	2076	3927	Cadmium (Cd)	321.07	767
Carbon (graphite) (C)		3825[2)]	Indium (In)	156.60[*2]	2072
Nitrogen (N)	-210.00	-195.795	Tin (Sn)	231.928[*2]	2602
Oxygen (O)	-218.79	-182.962	Antimony (Sb)	630.63	1587[*1]
Fluorine (F)	-219.67	-188.11	Tellurium (Te)	449.51	988
Neon (Ne)	-248.59	-246.046	Iodine (I)	113.7	184.3
Sodium (Na)	97.794	882.940	Xenon (Xe)	-111.75	-108.099
Magnesium (Mg)	650	1090	Cesium (Cs)	28.5	671
Aluminum (Al)	660.323[*2]	2519[*1]	Barium (Ba)	727	1845[*1]
Silicon (Si)	1414	3265	Lanthanum (La)	920	3464
Phosphorus (white)[3)] (P)	44.15	277	Cerium (Ce)	795	3443
Phosphorus (red)		431[2)]	Praseodymium (Pr)	935	3520[*1]
Sulfur α (Rhombic crystal) (S)	95.3[4)]	444.6	Neodymium (Nd)	1024	3074
Sulfur β (Monoclinic)	115.21	444.6	Promethium (Pm)	1042	3000
Sulfur γ (Monoclinic)	106.8	444.72	Samarium (Sm)	1072	1794[*1]
Chlorine (Cl)	-101.5	-34.04	Europium (Eu)	826	1529
Argon (Ar)	-189.34	-185.848	Gadolinium (Gd)	1312	3273[*1]
Potassium (K)	63.5	759	Terbium (Tb)	1356	3123
Calcium (Ca)	842	1484	Dysprosium (Dy)	1407	2567
Scandium (Sc)	1541	2836	Holmium (Ho)	1461	2700[*1]
Titanium (Ti)	1668	3287	Erbium (Er)	1529	2868
Vanadium (V)	1910	3407	Thulium (Tm)	1545	1950
Chrome (Cr)	1907	2671	Ytterbium (Yb)	824	1196[*1]
Manganese (Mn)	1246	2061	Lutetium (Lu)	1652	3402
Iron (Fe)	1538	2862	Hafnium (Hf)	2233	4603
Cobalt (Co)	1495	2927	Tantalum (Ta)	3017	5458
Nickel (Ni)	1455	2913[*1]	Tungsten (W)	3422	5555[*1]
Copper (Cu)	1084.62[*2]	2562	Rhenium (Re)	3186	5596
Zinc (Zn)	419.527[*2]	907	Osmium (Os)	3033	5012
Gallium (Ga)	29.7646[*3]	2204[*1]	Iridium (Ir)	2446	4428[*1]
Germanium (Ge)	938.25	2833	Platinum (Pt)	1768.3	3825
Arsenic (hexagonal structure) (As)		614[2)]	Gold (Au)	1064.18[*2]	2856[*1]
Selenium (Se)	221	685	Mercury (Hg)	-38.8290	356.73
Bromine (Br)	-7.2	58.8	Thallous (Tl)	304	1473
Krypton (Kr)	-157.37	-153.415	Lead (Pb)	327.46	1749
Rubidium (Rb)	39.30	688	Bismuth (Bi)	271.5	1564
Strontium (Sr)	777	1377[*1]	Polonium (Po)	254	962
Yttrium (Y)	1526	3345[*1]	Radon (Rn)	-71	-61.7
Zirconium (Zr)	1855	4377	Radium (Ra)	700	1737
Niobium (Nb)	2477	4744	Thorium (Th)	1750[*1]	4788
Molybdenum (Mo)	2623	4639	Uranium (U)	1132.2	4131
Technetium (Tc)	2157	4265	Plutonium (Pu)	639.4	3228

1) 2.5 MPa, 2) sublimation, 3) sometimes called yellow, 4) transition to β

Source: https://en.wikipedia.org/wiki/(element name) not otherwise specified.

*1 CRC Handbook of Chemistry and Physics (2011).

*2 Freezing point, The defining fixed points of the International Temperature Scale of 1990.

*3 The defining fixed points of the International Temperature Scale of 1990.

Melting point and boiling point of chemical compounds

( ) indicates an approximate value.

Material	t_m/℃	t_b/℃	Material	t_m/℃	t_b/℃
Acetic acid	16.6	117.9	Hydrogen sulfide	-82.9	-60.19
Acetone	-94.82	56.5	Magnesium oxide	2 800	3 600
Acetylene	-80.8[1]	-84.0[2]	Methane	-182.6	-161.5
Aluminum oxide (α)	2 054	2 974	Methyl alcohol	-97.78	64.65
Ammonia	-77.7	-33.48	Mineral fiber	(700-1 500)*	
Aniline	-6.0	184.55	Naphthalene	80.5	217.9
Benzene	5.5	80.1	Nitric acid	-42	83.8
Calcium chloride	772	2 008	Nitric oxide (NO)	-161	-151.8
Calcium oxide	2 572	2 850	Nitroglycerine	13.0	160(15mm)
Camphor	178	209	Nitrous oxide (N_2O)	-91	-88.57
Carbon dioxide	-56.6[1]	-78.5[2]	Olive oil	(20)	(300)
Carbon monoxide	-205	-191.6	Ozone	-193	-112
Carbon tertrachloride	-23.8	76.74	Porcelain	(1 100-1 400)*	
Carbonization tantalum	3 880	5 500	Propane	-188	-42.1
Carborundum	>2700	-	Refractory brick	(1 550-1 900)*	
Chloroform	-63.5	61.2	Seawater	(-1.9)	(103.7)
Crystal	1 610	2 227	Sodium chloride	801	1 485
Diethyl ether	-116.3	34.6	Sulfur dioxide (SO_2)	-75.5	-10.1
Dimethyl ether	-141.5	-24.8	Sulfuric acid	10.35	340 (decomposition)
Ethane	-183.6	-88.6			
Ethyl alcohol	-114.5	78.32	The glass (lead)	(500)*	-
Ethylene	-169	-103.9	The glass (Pyrex)	(800)*	-
Formic acid	8.4	100.6	The glass (quartz)	(1600)*	-
Glycerin	17.8	167(10mm) 290 (decomposition)	The glass (soda)	(550)*	-
			Toluene	-94.99	110.6
Hydrogen chloride	-114.2	-85.1	*o*-Xylene	-25.2	144.4
Hydrogen cyanide	-13.3	25.64	*m*-Xylene	-47.9	139.0
Hydrogen peroxide	-0.9	150.0	*p*-Xylene	13.3	138.3

* indicates softening temperature, 1) triple point, 2) sublimation point

Boiling point of water[3]

p/kPa	T_{90}/K	p/kPa	T_{90}/K	p/kPa	T_{90}/K	p/kPa	T_{90}/K	p/kPa	T_{90}/K
88.0	369.22	92.0	370.44	96.0	371.62	100.0	372.76	104.0	373.86
88.5	369.38	92.5	370.59	96.5	371.76	100.5	372.90	104.5	373.99
89.0	369.53	93.0	370.74	97.0	371.91	101.0	373.03	105.0	374.13
89.5	369.68	93.5	370.89	97.5	372.05	101.5	373.17	105.5	374.25
90.0	369.84	94.0	371.04	98.0	372.19	102.0	373.31	106.0	374.39
90.5	369.99	94.5	371.18	98.5	372.33	102.5	373.45	106.5	374.53
91.0	370.14	95.0	371.33	99.0	372.48	103.0	373.58	107.0	374.66
91.5	370.29	95.5	371.47	99.5	372.62	103.5	373.72	107.5	374.79

At 1 atmoshpere (1 atm = 101.325 kPa) this is approximately 373.124 K = 99.974°C.

3) The pressure/temperature relation from W. Wagner and A. Pruss, J. Phys. Chem, Ref. Data, 31, 387 (2002) has been reverse engineered Please refer to page **339**, note 2).

Vapor Pressure

The next two tables show the vapor pressures (p) of H_2O on the vapor-solid (ice) and the vapor-liquid (water) phase boundary as a function of temperatures (t or T).

Vapor pressure of ice[1]

(Unit: Pa)

T_{90}/K	0	2	4	6	8	10	12	14	16	18
170	0.0007	0.0011	0.0017	0.0025	0.0037	0.0054	0.0078	0.0113	0.0161	0.0229
190	0.0323	0.0452	0.0629	0.0868	0.119	0.162	0.220	0.297	0.397	0.529
210	0.7012	0.9241	1.212	1.581	2.053	2.653	3.412	4.370	5.572	7.074
230	8.944	11.26	14.13	17.65	21.98	27.26	33.69	41.50	50.94	62.33
250	76.01	92.40	112.0	135.3	163.0	195.8	234.5	280.2	333.8	396.6
270	470.1	555.7								

Vapor pressure at the ice point (273.15 K, 0°C) is 611.153 Pa.

1) Kaye & Laby, 1995.

Vapor pressure of water[2]

(Unit: kPa)

T_{90}/K	0	2	4	6	8	10	12	14	16	18
270			0.650	0.750	0.863	0.992	1.137	1.300	1.483	1.689
290	1.920	2.178	2.465	2.786	3.142	3.537	3.975	4.459	4.993	5.582
310	6.231	6.944	7.726	8.583	9.521	10.546	11.664	12.882	14.208	15.649
330	17.213	18.909	20.744	22.730	24.874	27.188	29.681	32.366	35.253	38.354
350	41.682	45.249	49.070	53.158	57.527	62.194	67.172	72.478	78.129	84.142
370	90.535	97.326	104.53	112.18	120.28	128.85	137.93	147.52	157.66	168.36
390	179.64	191.54	204.08	217.28	231.17	245.77	261.11	277.22	294.13	311.87
410	330.45	349.93	370.32	391.66	413.97	437.30	461.67	487.11	513.67	541.38
430	570.26	600.36	631.72	664.36	698.33	733.67	770.41	808.59	848.26	889.45
450	932.20	976.56	1022.6	1070.3	1119.7	1170.9	1223.9	1278.8	1335.6	1394.3
470	1455.1	1517.9	1582.8	1649.8	1719.0	1790.5	1864.2	1940.3	2018.8	2099.7
490	2183.1	2269.0	2357.5	2448.7	2542.6	2639.2	2738.6	2840.9	2946.1	3054.3
510	3165.5	3279.8	3397.2	3517.9	3641.7	3769.0	3899.5	4033.6	4171.1	4312.2
530	4456.9	4605.2	4757.4	4913.3	5073.1	5236.9	5404.7	5576.5	5752.5	5932.7
550	6117.2	6306.0	6499.3	6697.0	6899.3	7106.2	7317.9	7534.4	7755.7	7981.9
570	8213.2	8449.6	8691.2	8938.1	9190.3	9448.0	9711.2	9980.0	10255	10535
590	10821	11113	11412	11716	12027	12345	12669	12999	13337	13681
610	14033	14391	14757	15131	15512	15901	16297	16703	17116	17538
630	17969	18409	18858	19317	19786	20265	20756	21259	21775	

Vapor pressure is 0.611657 kPa at the triple point (273.16 K) and 22.06 MPa at the critical point (647.1 K).

2) W. Wagner and A. Pruss, J. Phys. Chem. Ref. Data, 31, 387 (2002). The isotropic ratio of water is specified as follows: $^2H/^1H = 155.76 \pm 0.1$ ppm, $^3H/^1H = (1.85 \pm 0.36) \times 10^{-11}$ ppm, $^{18}O/^{16}O = 2005.20 \pm 0.43$ ppm, $^{17}O/^{16}O = 379.9 \pm 1.6$ ppm, known as Vienna Standard Mean Ocean Water.

Vapor pressure of simple substances

Absolute temperatures are listed for pressures from 0.2 to 5000 kPa. (Unit: K)

Element	Pressure p/kPa												
	0.2	1	2	5	10	20	50	101.325	200	500	1000	2000	5000
Aluminum	1885	2060	2140	2260	2360	2430	2640	2790					
Antimony	1065	1220	1295	1405	1495	1595	1740	1890					
Argon	55.0	60.7	63.6	67.8	71.4	75.4	81.4	87.3	94.3	105.8	116.6	129.8	
Arsenic	656	701	723	754	780	808	849	883					
Barium	1165	1290	1360	1455	1540	1635	1780	1910					
Bromine	227	244.1	252.1	263.6	275.7	290.0	312.0	332.0	354	389	421	459	523
Cadmium	683	748	781	829	870	915	983	1045					
Calcium	1105	1220	1280	1365	1440	1525	1650	1765					
Cesium	559	624	657	708	752	802	883	959					
Chlorine	158	170	176.3	187.4	197.0	207.8	224.3	239.2	256	283	308	338	387
Chrome	2020	2180	2260	2370	2470	2580	2730	2870					
Copper	1940	2110	2200	2320	2420	2540	2710	2860					
Fluorine	53	58	61	65.3	68.9	73.0	79.3	85.0					
Gallium	1665	1820	1900	2000	2100	2200	2350	2480					
Gold	2100	2290	2390	2520	2640	2770	2960	3120					
Helium	1.3	1.66	1.85	2.17	2.48	2.87	3.54	4.22					
Hydrogen	10	11.4	12.2	13.4	14.5	16.0	18.2	20.3	22.8	27.1	31.2		
Iodine	318	341.8	353.1	369.3	382.7	400.8	430.6	457.5					
Iron	2110	2290	2380	2500	2610	2720	2890	3030					
Krypton	77	84.3	88.1	93.8	98.6	103.9	112.0	119.7	129.2	144.7	159.2	176.8	206.0
Lead	1285	1420	1485	1585	1670	1760	1905	2030					
Lithium	1040	1145	1200	1275	1345	1415	1530	1630					
Magnesium	891	979	1025	1085	1140	1205	1295	1375					
Manganese	1560	1710	1780	1890	1980	2080	2240	2390					
Mercury	408	448.8	468.8	498.4	523.4	551.2	592.9	629.8					
Molybdenum	3420	3720	3860	4060	4220	4410	4680	4900					
Neon	16.3	18.1	19.0	20.4	21.6	22.9	24.9	27.1	29.6	33.7	37.5	42.2	
Nickel	2230	2410	2500	2630	2740	2860	3030	3180					
Nitrogen	48.1	53.0	55.4	59.0	62.1	65.8	71.8	77.4	83.6	94.0	103.8	115.6	
Oxygen	55.4	61.3	64.3	68.8	72.7	77.1	83.9	90.2	97.2	108.8	119.6	132.7	154.4
Phosphorus (black)	572	605	620	642	660	678	705	727					
Phosphorus (yellow)	365	398	414	439	460	484	520.5	553.6					
Platinum	2910	3150	3260	3420	3560	3700	3910	4090					
Potassium	633	704	740	794	841	894	975	1050					
Radon	139	152	158	168	176	184	198	211					
Rubidium	583	649	684	735	779	829	907	978					
Selenium	636	695	724	767	803	844	904	958					
Silicon	2380	2590	2690	2840	2970	3100	3310	3490					
Silver	1640	1790	1865	1970	2060	2160	2310	2440					
Sodium	729	807	846	904	954	1010	1095	1175					
Strontium	1040	1150	1205	1285	1355	1430	1550	1660					
Sulfur	462	505.7	528.0	561.3	590.1	622.5	672.4	717.8					
Tellurium	806	888	929	990	1040	1100	1190	1270					
Thallous	1125	1235	1290	1370	1440	1515	1630	1730					
Tin	1930	2120	2210	2350	2470	2600	2800	2990					
Tungsten	4300	4630	4790	5020	5200	5400	5690	5940					
Xenon	107	117.3	122.5	130.1	136.6	143.8	154.7	165.0	177.8	198.9	218.6	242.5	282.2
Zinc	780	854	891	945	991	1040	1120	1185					

Kaye & Laby (web version 2005).

Vapor pressure of chemical compounds

Ethanol C_2H_5OH									
t/℃	p/Torr	t/℃	p/Torr	t/℃	p/Torr	t/℃	p/Torr	t/℃	p/Torr
0	12.73	8	21.31	16	34.6	24	55.7	50	219.8
1	13.65	9	22.66	17	36.8	25	59.0	60	350.2
2	14.60	10	24.08	18	39.0	26	62.5	70	541
3	15.59	11	25.59	19	41.5	27	66.2	80	812
4	16.62	12	27.19	20	44.0	28	70.1	100	1692
5	17.70	13	28.9	21	46.7	29	74.1	120	3220
6	18.84	14	30.7	22	49.5	30	78.4	140	5670
7	20.04	15	32.6	23	52.5	40	133.4	160	9370

Acetylene C_2H_2		Ammonia NH_3		Sulfur dioxide SO_2		Ether $(C_2H_5)_2O$		Chloroform $CHCl_3$	
t/℃	p/atm	t/℃	p/atm	t/℃	p/atm	t/℃	p/Torr	t/℃	p/Torr
-90Gas	0.69	-77.7	0.060	-30	0.39	-10	112.3	20	160.5
-85 〃	1.00	-60	0.216	-20	0.63	0	184.9	30	248
-81	1.25	-33.4	1.000	-10	1.00	10	290.8	40	369
-70	2.22	-20	1.877	0	1.53	20	439.8	50	535
-50	5.3	0	4.238	10	2.26	40	921	60	755
-23.8	13.2	20	8.459	20	3.24	60	1734	70	1042
0	26.05	40	15.34	30	4.52	80	2974	80	1408
20.2	42.8	80	40.90	40	6.15	100	4855	90	1865
36.5	61.6	132.4	112.3	50	8.19	193.8	27060	100	2429

Acetic acid CH_3CO_2H		Carbon dioxide CO_2		Carbon disulfide CS_2		Benzene C_6H_6	
t/℃	p/Torr	t/℃	p/atm	t/℃	p/Torr	t/℃	p/Torr
17	9.8	-100 (solid)	0.137	-20	47.3	-10	14.8
30	20.6	-78.5 (sublime point)	1.000	-10	79.4	0	26.5
50	56.2	-56.6 (triple point)	5.112	0	128	10	45.4
70	133	-40	9.93	10	198	20	74.7
90	288	-20	19.44	20	298	40	181.1
110	582	0	34.40	40	618	60	389
130	1068	10	44.4	60	1164	80	754
150	1847	20	56.5	80	2033	100	1344
200	5905	31.9 (critical point)	72.8	100	3325	120	2238

The pressure conversion is 1 atm = 101 325 Pa, 1 Torr ≒ 133.322 Pa

Constants at the Critical Point of Substance

The table shows the critical temperature (T_c), critical pressure (p_c), and critical molar volume (V_m^c).

Substance	T_c/K	p_c/MPa	V_m^c/cm^3 mol^{-1}
Acetic acid	592.7	5.79	171
Acetylene	308.3	6.14	113
Ammonia	405.5	11.35	72.5
Argon	150.9	4.90	74.6
Benzene	562.0	4.90	256
Bromine	588	10.3	127
Butane	425.1	3.80	255
Carbon dioxide	304.1	7.38	94.0
Carbon disulfide	552	7.9	173
Carbon monoxide	132.9	3.50	93.1
Chlorine	416.9	7.98	124
Chloroform	536.4	5.47	239
Diethyl ether	466.7	3.64	280
Ethyl alcohol	514.0	6.14	168
Helium	5.19	0.227	57.2
Hydrogen	33.2	1.297	65.0
Hydrogen sulfide	373.2	8.94	98.5
Methane	190.6	4.60	98.6
Methanol	512.5	8.08	117
Neon	44.4	2.76	41.7
Nitrogen	126.2	3.39	89.5
Oxygen	154.6	5.04	73.4
Propane	369.8	4.25	200
Toluene	591.8	4.11	316
Water	647.1	22.06	56.0

Kaye & Laby (web version 2005).

Coefficients of Expansion

Coefficients of linear expansion of solids

Coefficients of linear expansion is given by $\alpha=\frac{1}{l}\frac{dl}{dt}$, where l is the length at that temperature.

Material	$\alpha/10^{-6}$ K^{-1}			
	100 K	293 K (20℃)	500 K	800 K
Elements				
Aluminum	12.2	23.1	26.4	34.0
Antimony	9.1	11.0	11.7	11.7
Barium	18.1–21.0 (0–300℃)			
Beryllium	1.3	11.3	15.1	19.1
Bismuth	12.3	13.4	12.7	–
Boron	–	4.7	5.4	6.2
Cadmium	26.9	30.8	36.0	–
Calcium	22 (0–300℃)			
Carbon (diamond)	0.05	1.0	2.3	3.7
Carbon (graphite)	–	3.1	3.3	3.6
Chrome	2.3	4.9	8.8	11.8
Cobalt	6.8	13.0	15.0	15.2
Copper	10.3	16.5	18.3	20.3
Germanium	2.4	5.7	6.5	7.2
Gold	11.8	14.2	15.4	17.0
Indium	25.4	32.1	–	–
Iridium	4.4	6.4	7.2	8.1
Iron	5.6	11.8	14.4	16.2
Lead	25.6	28.9	33.3	–
Lithium	56 (0–100℃)			
Magnesium	14.6	24.8	29.1	35.4
Manganese α	22.3 (0–20℃)			
Manganese β	18.7–24.9 (0–20℃)			
Manganese γ	14.8 (0–20℃)			
Molybdenum	3.7–5.3 (20–100℃)			
Nickel	6.6	13.4	15.3	16.8
Osmium	–	4.7	–	–
Palladium	8.0	11.8	13.2	14.5
Platinum	6.6	8.8	9.6	10.3

Based primarily on Kaye & Laby, 1986.

Coefficients of Expansion Continued.

Material	α/10^{-6} K^{-1}			
	100 K	293 K (20℃)	500 K	800 K
Potassium	-	85	-	-
Rhodium	5.0	8.2	9.3	10.8
Selenium (amorphism)	48.7 (0–21℃)			
Selenium (polycrystal)	20.3 (-78–19℃)			
Silicon	-0.4	2.6	3.5	4.1
Silver	14.2	18.9	20.6	23.7
Sodium	70 (0–50℃)			
Tantalum	4.8	6.3	6.8	7.2
Tellurium	-	16.8	-	-
Thorium	11.3 (20–100℃)			
Tin	16.5	22.0	27.2	-
Titanium	4.5	8.6	9.9	11.1
Tungsten	2.6	4.5	4.6	5.0
Vanadium	5.1	8.4	9.9	10.9
Zinc	24.5	30.2	32.8	-
Zirconium	5.4 (20–200℃)			
Alloy				
Aluminum bronze (90Cu, 5Al, 4.5Ni)	12–14	15.9	18.1	20.3
Brass (67Cu, 33Zn)	-	17.5	20.0	22.5
Bronze (85Cu, 15Sn)	-	17.3	19.3	21.9
Cadmium sulfide (∥ to axis)	-	4	-	-
Cadmium sulfide (⊥ to axis)	-	6	-	-
Carbon steel	6.7	10.7	13.7	16.2
Constantan (65Cu, 35Ni)	11.2	15.0	17.4	19.2
Duralumin	13.1	21.6	27.5	30.1
Fernico (54Fe, 31Ni, 15Co)	5.0 (25–300℃)			
Gunmetal (80Cu, 20Sn)	-	17–18	-	-
Lead selenide	-	20	-	-
Lead sulphide	19 (40℃)			
Lead telluride	-	27	-	-
Magnalium (90Al, 10Mg)	-	about 23	-	-
Manganin	18.1 (20–100℃)			
Monel metal (63Ni, 30Cu, Fe, Mn, Pb)	15.9–16.7 (25–600℃)			
Nickel steel (50Fe, 50Ni)	-	9.4	9.6	12.5
Nickel steel (64Fe, 36Ni)[1]	1.4	0.13	5.1	17.1

1) Invar

Coefficients of Expansion Continued.

Material	$\alpha/10^{-6}\,K^{-1}$			
	100 K	293 K (20℃)	500 K	800 K
Phosphorus bronze	–	17.0	20.0	–
Platiniridium (90Pt, 10Ir)	–	8.7	–	–
Speculum metal Alloy	16 (20–100℃)			
Stainless steels (18Cr, 8Ni)	11.4	14.7	17.5	20.2
Stellite (65Co, 25Cr, 10W)	6.9	11.2	14.6	17.2
Tin (gray)	5.3 (–163–18℃)			
Titania (TiO_2)	–	9	–	–
Uranium oxide (UO_2)	11.5 (20–720℃)			
Y alloy	–	22	–	–
Others				
Bakelite	–	21–33	–	–
Brick	–	3–10	–	–
Cadmium sulfide (⊥ to axis)	9.1	12.2	19.5	37.6
Cadmium sulfide (⊥ to axis)	5.44 (0–80℃)			
Calcite (It is ∥ in the axis)	26.3 (0–80℃)			
Celluloid	–	90–160	–	–
Concrete and cement	–	7–14	–	–
Crystal (∥ to axis)	4.0	6.8	11.4	31.4
Ebonite	–	50–80	–	–
Fluorspar	–	19	–	–
Fused silica	–	0.4–0.55	–	–
Granite	–	4–10	–	–
Ice	0.8 (–200℃), 33.9 (–100℃), 45.6 (–50℃), 52.7 (0℃)			
Marble	–	3–15	–	–
Paraffin	106.6 (0–16℃), 130.3 (16–38℃), 477.1 (38–49℃)			
Poly methyl methacrylate	–	80	–	–
Polyethylene	–	100–200	–	–
Polystyrene	–	34–210	–	–
Porcelain (insulation)	–	2–6	–	–
Rock salt	40.4 (40℃)			
Rubber (elasticity)	77 (16.7–25.3℃)			
Slate and sandstone	–	5–12	–	–
The glass (average)	8–10 (0–300℃)			
The glass (Flint)	–	8–9	–	–
The glass (Pyrex)	–	2.8	–	–
Wood (∥ to the fiber)	–	3–6	–	–
Wood (⊥ to the fiber)	–	35–60	–	–

Coefficients of cubic expansion of solids

Coefficient of the cubic expansion is given by $\beta = \frac{1}{v}\frac{dv}{dt}$, where v is volume at that temperature.
For isotropic solids, the coefficient of cubic expansion β is almost equal to three times the coefficient of linear expansion α.

Unit	t/℃	β/K^{-1}	Unit	t/℃	β/K^{-1}
		$\times 10^{-6}$			$\times 10^{-6}$
Cesium	0-23	291	Potassium	0-55	240
Cobalt	100	35.6	Rubidium	0-38	270
Diamond	25-650	9.1	Selenium	0-100	175
Gallium	0-30	55	Sodium	0-53	207
Lithium	0-100	162	Sulfur (orthorhombic)	-273-18	139
Nickel	100	38.2	Tin	80	68
	300	46.5	Zinc	50	89
Phosphorus	-79-19	362			
	0-44	372			

Coefficients of cubic expansion of liquid (20℃)

The table shows the values of $\beta = \frac{1}{v}\frac{dv}{dt}$ at $t = 20$℃. The coefficient of cubic expansion of liquids increases slightly as temperature rises.

Material	β_{20}/K^{-1}	Material	β_{20}/K^{-1}
	$\times 10^{-3}$		$\times 10^{-3}$
Acetone	1.43	Glycerin	0.47
Aniline	0.85	Mercury[1)]	0.181
Benzene	1.22	Methyl alcohol	1.19
Carbon disulfide	1.19	Pentane	1.55
Carbon tertrachloride	1.22	Phenol	0.79
Chloroform	1.27	Sulfuric acid (100%)	0.56
Diethyl ether	1.63	Toluene	1.07
Ethyl alcohol	1.08	Water[2)]	0.21
Ethylene glycol	0.64	*m*-Xylene	0.99

For the densities of 1), 2) refer to pages **314** and **312** respectively.

Thermal Conductivity

The thermal conductivity k of a substance is defined as the quantity of heat transmitted in 1 s in a direction normal to a surface of 1 m^2 area, due to temperature gradient of 1 K/m.

Thermal conductivities of materials

Material	t/℃	k / W m^{-1} K^{-1}	Material	t/℃	k / W m^{-1} K^{-1}
Acrylic	RT*	0.17–0.25	Glass (soda)	RT*	0.55–0.75
Alumina	RT*	21	Glass wool	RT*	0.04
	800	7	Gypsum	RT*	0.13
Asbestos (cement board)	RT*	0.3	Ice	0	2.2
Asbestos (cloth)	RT*	0.1	Linoleum	20	0.08
Asbestos (cotton)	RT*	0.06	Mica	100–600	0.55–0.79
Ash	20	0.03	Nylon	RT*	0.27
Asphalt	RT*	1.1–1.5	Paper	RT*	0.06
Blanket	30	0.04	Paraffin	RT*	0.24
Brick (porous)	20	0.2	Plaster	20	0.8
Brick (red)	RT*	0.5–0.6	Polyethylene	RT*	0.25–0.34
Calcite (∥ to axis)	0	5.39	Polystyrene	RT*	0.08–0.12
Calcite (⊥ to axis)	0	4.51	Porcelain	RT*	1.5
Carbon (amorphous)	0	1.5	Pumice (density 0.6)	20	0.2
	300	2.2	Quartz glass	0	1.4
	700	2.5		100	1.9
Carbon (graphite)	0	80–230	Refractory bricks	600	1.1
	300	50–130		1000	1.3
	700	35–70	Rubber (hard)	0	0.2
Cardboard	RT*	0.2	Rubber (molle)	RT*	0.1–0.2
Concrete	RT*	1	Rubber (sponge)	25	0.04
Cork	RT*	0.04–0.05	Sand	20	0.3
Cotton	RT*	0.03	Selenium (amorphous)	0	0.43
Cotton cloth	40	0.08	Silicon	0	168
Crystal (∥ to axis)	70	9.3	Silk	40	0.05
Crystal (⊥ to axis)	70	5.4	Snow (density 0.11)	0	0.11
Diatomaceous earth	25–650	0.07–0.1	Snow (density 0.45)	0	0.57
Felt	RT*	0.04	Soil (dry)	20	0.14
Fiber	50	0.2–0.3	Sulfur (amorphous)	0	0.2
Fluorspar	0	10.3	Sulfur (monoclinic)	100	0.16–0.17
Germanium	0	67	Sulfur (oblique)	20	0.27
Glass (lead)	15	0.6	Wood (dry)	18–25	0.14–0.18
Glass (pyrex)	30–75	1.1	Wool	RT*	0.04

* RT denotes room temperature.

Thermal conductivities (k) of metal and alloy (Unit: W m^{-1} K^{-1})

Material	t/℃				
	-100	0	100	300	700
Aluminum	241	236	240	233	92
Antimony	33	25.5	22	19	27
Beryllium	367	218	168	129	93
Bismuth	11	8.2	7.2	13	17
Brass[1)]	89	106	128	146	-
Cadmium	100	97	95	89	45
Chrome	120	96.5	92	82	66
Cobalt	130	105	89	69	53
Constantan[6)]	19	22	24	27	-
Copper	420	403	395	381	354
Gold	324	319	313	299	272
Gunmetal (bronze)[10)]	-	53	60	80	-
Indium	92	84	76	42	-
Iridium	156	147	145	139	-
Iron	99	83.5	72	56	34
Lead	37	36	34	32	21
Magnesium	160	157	154	150	-
Manganese	7	8	-	-	-
Mercury	29.5	7.8	9.4	11.7	-
Molybdenum	145	139	135	127	113
Monel metal[11)]	19	21	24	30	-
Nichrome[7)]	11	13	14	17	-
Nickel	113	94	83	67	71
Palladium	72	72	73	79	93
Platiniridium[8)]	-	31	-	-	-
Platinum	73	72	72	73	78
Platinum rhodium[9)]	-	46	51	58	-
Potassium	105	104	53	45	32
Silver	432	428	422	407	377
Sodium	141	142	88	78	60
Steel (18-8 stainless steels)[5)]	12	15	16.5	19	-
Steel (carbon)[2)]	48	50	48.5	41.5	-
Steel (Ni-Cr)[3)]	31	33	36	37.5	-
Steel (silicon)[4)]	-	25	28.5	31	-
Tantalum	58	57	58	58.5	60
Thallous	51	47	44	-	-
Tin	76	68	63	32	40
Titanium	26	22	21	19	21
Tungsten	188	177	163	139	119
Zinc	117	117	112	104	66

1) Cu 70% Zn 30, 2) C 0.8 Mn 0.3, 3) Ni 3.6 Cr 0.8 Mn 0.6 C 0.4, 4) Si 2.0 Mn 0.9 Cu 0.6 C 0.5, 5) Cr 17.9 Ni 8.0 Mn 0.3, 6) Cu 60, 7) Ni 77 Cr 21, 8) Pt 90 Ir 10, 9) Pt 60 Rh 40, 10) Cu 90 Sn 10, 11) Ni 67 Cu 29. Based primarily on Kaye & Laby, 1986.

Thermal conductivity of liquid

The following table shows the thermal conductivity (k) of the liquid at the temperature t/℃ shown in parenthesis. Among the temperatures shown, dk/dt may be considered as almost constant.

Material	$k(t)$/W m^{-1} K^{-1}	Material	$k(t)$/W m^{-1} K^{-1}
Acetone	0.198(−80), 0.146(60)	Glycerin	0.28670), 0.292(60)
Aniline	0.172(20)	Methyl alcohol	0.223(−40), 0.186(60)
Benzene	0.147(20), 0.137(50)	Toluene	0.159(−80), 0.119(80)
Carbon tertrachloride	0.115(−20), 0.102(60)	Transfomer oil	0.136(0), 0.127(100)
Ethyl alcohol	0.189(−40), 0.150(80)	Water	0.561(0), 0.673(80)

Thermal conductivity of gas

The thermal conductivity k of gas is almost completely unrelated to pressure with a wide range of pressure from several hundred Pa to several MPa.

(Unit : 10^{-2}W m^{-1} K^{-1})

Gas	t/℃					Gas	t/℃				
	-200	-100	0	100	1000		-200	-100	0	100	1000
Air		1.58	2.41	3.17	7.6	Hydrogen sulfide			1.2		
Ammonia			2.18	3.38		Krypton		0.57	0.87	1.15	2.9
Argon		1.09	1.63	2.12	5.0	Methane		1.88	3.02		
Carbon dioxide			1.45	2.23	7.9	Neon	1.74	3.37	4.65	5.66	12.8
Carbon monoxide		1.51	2.32	3.04		Nitric monoxide			1.51		
Chlorine			0.79	1.15		Nitrogen		1.59	2.40	3.09	7.4
Ethane			1.80			Nitrous oxide		1.54	2.38		
Ethylene			1.64			Oxygen		1.59	2.45	3.23	8.6
Freon			0.85	1.35		Steam			1.58	2.35	
Helium	5.95	10.45	14.22	17.77	41.9	Sulfur dioxide			0.77		
Hydrogen	5.09	11.24	16.82	21.18		Xenon		0.34	0.52	0.70	1.9

Source: Kaye & Laby, 1986.

Flash Points of Substances

When flammable substances (mainly liquids) are heated and flame is brought near the substances, flash point (t_{fl}) is the lowest temperature at which substance give off a sufficient amount of vapor to ignite instantaneously.

Material	t_{fl}/℃	Material	t_{fl}/℃
Diethyl ether	−45	Xylene	27
Gasoline	−43 or under	Kerosene	40−60
Petroleum benzine	−40 or under	Light crude petroleum	50−70
Carbon disulfide	−30	Crude petroleum	60−100
Acetone	−20	Aniline	70
Benzene	−11	Naphthalene	79
Thinners	− 9	Nitrobenzene	88
Toluene	4	Machine oil	106−270
Methyl alcohol	11	Sesame oil	289−304*
Ethyl alcohol	13	Rapeseed oil	313−320*

Based primarily on the Chemcial Society of Japan, Chemical Accident Prevetion manual I, Maruzen (1979).
*Japan Inspection Institute of Fats & Oils, Heisei 4 (1992) Ranking Results Report, (1993).

Ignition Point of Substances

When heating substances in the air, ignition point (t_{ig}) is the lowest temperature at which the substance catches fire without the presence of a fire source. The following values differ significantly depending on the sample shape and measurement method.

Material	t_{ig}/℃	Material	t_{ig}/℃
Hydrogen	500	Polypropylene	201*
Methane	537	Polystyrene	282*
Propane	432	Melamine	380*
Ethylene	450	Teflon	492*
Acetylene	305	Old tire	150−200*
Carbon monoxide	609	Wood	250−260*
Hydrogen sulfide	260	Newspaper	291
Carbon disulfide	90	Charcoal	250−300
Diesel fuel oil	225	Peat	225−280
Compressor oil (mineral oil)	250−280*	Cocoa	180*
Acetone	469	Coffee	398*
Benzene	498	Starch (corn)	381*
Aniline	615		
Yellow phosphorus	30		
Red phosphorus	260		
Sulfur	232		
Naphthalene	526		

* Komamiya, Morisaki, Wakakura, Chemical Danger Forecast Data, Japan Policy Institute (1984).

Electric/Magnetic Properties

Electric Resistance of Metal

Resistance of a uniform material with the length l m, the cross section a m² is $R=\rho(l/a)$, where $\rho(\Omega\ \mathrm{m})$ is the volume resistivity of the material*.

When the volume resistivity at 0℃ is ρ_0 and the volume resistivity at 100℃ is ρ_{100}, the average temperature coefficient for the volume resistivity between 0℃ and 100℃ is calculated by $\alpha_{0,100}=(\rho_{100}-\rho_0)/100\rho_0$.

Metal	$\rho/10^{-8}\Omega$ m					
	-195℃	0℃	100℃	300℃	700℃	1 200℃ (others)
Alumel	–	28.1	34.8	43.8	53.2	65.1
Aluminum	0.21	2.50	3.55	5.9	24.7	32.1
Antimony	8	39	59	–	114	123.5 (1000℃)
Arsenic	5.5	26	–	–	–	–
Beryllium	–	2.8	5.3	11.1	26	–
Bismuth	35	107	156	129	155	172 (1000℃)
Brass	–	6.3	–	–	–	–
Bronze	–	13.6	–	–	–	–
Cadmium	1.6	6.8	9.8	–	–	36.3 (600℃)
Calcium	0.7	3.2	4.75	7.8	20	–
Cesium			21 (20℃)	($\alpha_{0,100}/10^{-3}=4.8$)		
ChromelP	–	70.0	72.8	79.3	89.3	100.1
Chromium	0.5	12.7	16.1	25.2	47.2	80
Cobalt	0.9	5.6	9.5	19.7	48	88.5
Constantan	–	49	–	–	–	–
Copper	0.2	1.55	2.23	3.6	6.7	21.3 (1083℃)
Duralumin (soft)			3.4 (Room temperature)			
German silver	–	40	–	–	–	–
Gold	0.5	2.05	2.88	4.63	8.6	31 (1063℃)
Indium	1.8	8.0	12.1	36.7	47	55 (1000℃)
Invar		75		($\alpha_{0,100}/10^{-3}=2$)		
Iridium	0.9	4.7	6.8	10.8	22	33.5
Iron (cast)			57–114 (Room temperature)			
Iron (pure)	0.7	8.9	14.7	31.5	85.5	122；139 (1550℃)
Iron (steel)			10–20 (Room temperature)			($\alpha_{0,100}/10^{-3}=1.5$–$5$)
Lead	4.7	19.2	27	50	108	126 (1000℃)
Lithium	1.04	8.55	12.4	30	40.5	53
Magnesium	0.62	3.94	5.6	10.0	27.7	28.7 (900℃)
Manganin	–	41.5	–	–	–	–
Mercury	5.8	94.1	103.5	128	214	630
Molybdenum	0.7	5.0	7.6	12.7	23.3	37.2
Nichrome	–	107.3	108.3	110.0	110.3	–
Nickel	0.55	6.2	10.3	22.5	40	109 (1500℃)
Nickelin			27–45 (Room temperature)		($\alpha_{0,100}/10^{-3}=0.2$–$0.34$)	
Osmium	–	8.1	11.4	17.8	30.4	46
Palladium	1.73	10.0	13.8	21	33	42
Phosphor bronze			2–6 (Room temperature)			
Platinoid			34–41 (Room temperature)		($\alpha_{0,100}/10^{-3}=0.25$–$0.32$)	
Platinum	1.96	9.81	13.6	21.0	34.3	48.3
Platinum rhodium[1)]	–	18.7	21.8	–	–	–
Potassium	1.38	6.1	17.5	28.2	66.4	160
Rhodium	0.46	4.3	6.2	10.2	20	33
Rubidium	2.2	11.0	27.5	48	99	260
Silver	0.3	1.47	2.08	3.34	6.1	19.4
Sodium	0.8	4.2	9.7	16.8	39.2	89
Strontium	5	20	30	52.5	94.5	–
Tantalum	2.5	12.3	16.7	25.5	43.0	61.5
Thallous	3.7	15	22.8	38	85	88 (800℃)
Thorium	3.9	14.7	20.8	32.5	53.6	68
Tin	2.1	11.5	15.8	50	60	72
Tungsten	0.6	4.9	7.3	12.4	24	39
Zinc	1.1	5.5	7.8	13.0	–	37 (500℃)
Zirconium	7.3	40	58	88	125	110 (1000℃)

1) Platinum 90, Rhodium 10 alloy

* volume resistivity is shown in units of Ω cm, the values in this tables should be multiplied by 100.

Superconductors

Critical temperature and critical magnetic field of superconducting materials

Material	Critical temperature (T_c/K)	Critical magnetic field[1)] ($B_0(0)$/G)
Element		
Al	1.196	99
Cd	0.56	30
Ga	1.091	51
Ga (β)	6.2	-
Ga (γ)	7.62	-
Hf	0.165	-
Hg (α)	4.154	411
Hg (β)	3.949	339
In	3.4035	293
Ir	0.14	19
La (α)	4.9	-
La (β)	6.06	1600
Mo	0.92	98
Nb	9.23	1980
Os	0.655	65
Pa	1.4	-
Pb	7.193	803
Re	1.699	198
Ru	0.49	66
Sn	3.722	305.5
Ta	4.39	780
Tc	7.92	1410
Th	1.368	162
Ti	0.39	100
Tl (α)	2.39	171
U (α)	0.68	-300
V	5.3	1020
W	0.012	1.07
Zn	0.852	53
Zr	0.546	47
Li	20 (48GPa)	
Ba	5 (high pressure)	
Bi	3.93, 7.25 (high pressure)	
Cs	1.5 (high pressure)	
La	12 (high pressure)	
Bi	6 (thin film)	
Compound		
MgB_2	39	
Nb_3Al	18.8	3.20×10^5 (4.2 K)
Nb_3Au	11.5	
Nb_3Ge	-18	
Nb_3Sn	18.3	2.45×10^5
$Nb_{0.79}(Al_{0.73}Ge_{0.27})_{0.21}$	21.05	4.2×10^5 (4.2 K)
V_3Ga	16.5	2.08×10^5
V_3Si	17.1	2.35×10^5
$NbTc_3$	10.5	
Tc_6Zr	9.7	
HfN	6.2	

Material	Critical temperature (T_c/K)	Critical magnetic field[2)] ($B_{c2}(0)$/G)
β-MNX (M = Ti, Zr, Hf, X = Cl, Br, I)	25.5 (M-Hf, X = Cl, intercalation)	
NbC	6	
$NbN_{0.91}$	15.72	1.8×10^5
$NbN_{0.72}C_{0.28}$	17.9	
NbB	8.25	
BiNi	4.25	
$CoSi_2$	1.22	
$PbTe_2$	1.53	
YOs_2	4.7	
ZrV_2	8.8	
$Hf_{0.5}Zr_{0.5}V_2$	10.1	2.1×10^5 (4.2 K)
FeU_6	3.86	
$InLa_3$	10.4	
MoN	12	
YM_2B_2C (M = Ni, Pd)	23 (M = Pd)	
Alloy[3)]		
Pb_xI_y	3.39-7.26	
$Nb_{0.75}Zr_{0.25}$	10.8	-9.1×10^4
$Nb_{0.40}Ti_{0.60}$	9.3	-1.2×10^5 (4.2 K)
$Nb_{0.70}Ta_{0.30}$	6.9	9.5×10^3
$Ta_{0.53}Ti_{0.47}$	8.1	8.8×10^4 (4.2 K)
$Ti_{0.50}V_{0.50}$	7.4	-1.2×10^5
$Nb_{0.36}Ti_{0.60}Ta_{0.04}$	9.9	1.24×10^5 (4.2 K)
$Rh_{0-0.15}Zr_{1-0.85}$	9.8-12.2	
$Mo_{0.3}Tc_{0.7}$	12	
$PuCoGa_5$	18.5 (heavy electron system)	7.40×10^5
Oxide*1		
$La_{2-x}(Sr, Ba)_xCuO_4$ (x = 0.15)	38	-10^6 *2
$RBa_2Cu_3O_{7-\delta}$ (R = Y, La, Nd, Sm, Eu, Gd, Dy, Ho, Er, Tm, Yb)	90-95 *3	-10^6 *2
$Bi_2Sr_2Ca_{n-1}Cu_nO_y$ *4	40 (n = 1) 80 (n = 2) 110 (n = 3)	-10^6 *2
$HgBa_2Ca_2Cu_3O_y$	153 (high pressure the under)	-10^6 *2
$SrTiO_3$	0.38 ($n = 1 \times 10^{20}$ cm^{-3})[4)]	
$Li_{1+x}Ti_{2-x}O_4$	13	3.7×10^5 (x = 0.1)
(Ba, K)BiO_3	30	
AOs_2O_6 (A = Cs, Rb, K)	9.6 (A = K)	
Organic compound		
$(TMTSF)_2ClO_4$	1.4 †1	2.1×10^4 (0.5 K) †2
β-$(BEDT\text{-}TTF)_2Cu(NCS)_2$ †3	10.4	-2.0×10^5 (2 K) †2
C_8K (graphite)	0.39	
$C_{60}K_3$ (fullerene)	19.7	2.6×10^5

$1G = 10^{-4}T$

Material	Critical temperature (T_c/K)	Critical magnetic field[2] ($B_{c2}(0)$/G)	Material	Critical temperature (T_c/K)	Critical magnetic field[2] ($B_{c2}(0)$/G)
Organic compound (Continued.)			$Gd_{0.09}La_{0.91}Os_2$	-6 (ferromagnetic)	
			$Ce_{0.92}Gd_{0.08}Ru_2$	3.1 (ferromagnetic)	
$C_{60}Cs$ (fullerene)	38 (high pressure the under)		$R(O_{1-x}F_x)FeAs$ (R = La, Cr, Pr, Nd, Sm *etc.*)	50 (R = Sm)	
Others					
B doped C (diamond)	6		H_3S	203†4 (high pressure the under)	
GeTe	0.30 ($n = 1.5 \times 10^{21}$ cm^{-3})		LaH_{10}	260†4 (high pressure the under)	
SnTe	0.21 ($n = 2 \times 10^{21}$ cm^{-3})				

1) The critical magnetic field shows the thermodynamical critical field at 0 K for each element. For all elements except Nb and V, this is equal to the critical magnetic field observed in type I superconductors. Because Nb and V are type II superconductors, the observed lower critical magnetic field H_{C1}, and upper critical magentic field H_{C2} are different.
2) Type II superconductor (hard supercondutor) alloys show the upper bound critical magnetic field H_{C2} for each compound.
3) For alloys with a wide range of mixing ratios, the specific composition used for the measurements is shown.
4) For semiconductors which exhibit superconductivity, the critical temperature changes drastically based on the carrier concentration n.
* 1 The temperature where the resistance goes to 0 is taken as the critical temperature. Please understand that vaules shown here are rough estimates which can vary based on the composistion and prodution methods.
* 2 The critical magnetic field for superconducting oxides have large anisotropies. And because the absolute values are very large, there is no sufficently reliable data to put into the table.
* 3 x indicates values which have not been determined at this time.
* 4 electron-doped type copper oxide high-temperature supercondutor
† 1 T_c is the midpoint of the resistive transition.
† 2 Anisotropies exist, the largest values are shown.
† 3 T_c for BEDT-TTE deureride is 11.0 to 11.2 K.
† 4 An onset temperature of the resistance decrease and the Meissrer effect.

TMTSF 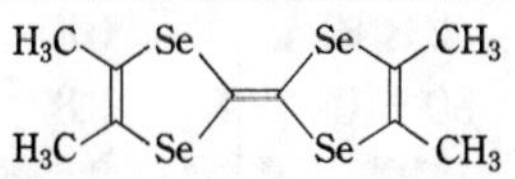 Tetramethyltetraselenafulvalene

BEDT-TTF 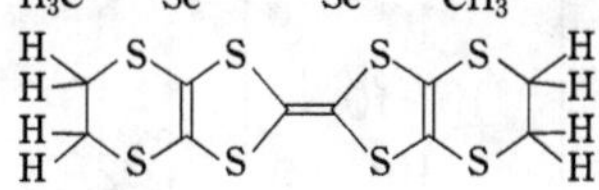 *bis*(ethylenedithiolo)tetrathiafulvalene

Characteristic Quantity of Semiconductors

Material	Melting point t/℃	Width of prohibition belt (0 K) E_G/eV	$\beta = -\frac{\partial E_G}{\partial T}$ $10^4\beta$/eV K^{-1}	Electron mobility (300 K) $\mu_n/10^{-4}m^2\ V^{-1}\ s^{-1}$	Hole mobility (300 K) $\mu_p/10^{-4}m^2\ V^{-1}\ s^{-1}$
Ge	959	0.785	+4.0	3 600	1 800
Si	1 410	1.206	+4.2	1 500	500
ZnSb	544	0.56			150
CdSb	456	0.48		660	362
Cu_2O	1 236	2.172			50
AlP		-3.0			
AlAs	>1 600	2.2			
AlSb	>1 060	1.55	+3.5	200	200
GaP	1 350	2.32	+5.4	175	75
GaAs	1 280	1.53	+4.9	9 700	420
GaSb	728	0.81	+3.5	4 000	700
InP	1 055	1.29	+4.6	3 400	50
InAs	942	0.35	+3.5	33 000	460
InSb	525	0.17	+2.7	82 000	-750
ZnO	1 975	3.436			
ZnS(hex)	1 850	3.91			
ZnSe	-1 500	2.83		-100	
ZnTe	1 239	2.39		-100	
CdS(hex)	1 750	2.582	+5.5	210	
CdSe	1 350	1.84			
CdTe	1 098	1.607		650	60
InSe	660	1.2		-5	
PbS	1 114	0.29	-5.0	640	800
PbSe	1 065	0.17	-5.0	1 200	850
PbTe	905	0.19	-5.0	2 100	840
Bi_2Te_3	580	0.2	+0.9	800	400

The specific electron conductivity σ is given by $\sigma = e(n\mu_n + p\mu_p)$
n and p are the number of electrons or holes, respectively, in the valence/conduction band within a unit volume

Dielectric

The dielectric constant $k_\varepsilon(=\varepsilon/\varepsilon_0)$ changes according to the temperature t and the frequency f. When using complex notation for alternating current, k_ε is expressed as a complex number, and the loss angle δ of dielectric loss is given by $\tan\delta = \mathrm{Im}(k_\varepsilon)/\mathrm{Re}(k_\varepsilon)$.

Solids

Material	t/℃	f/Hz	k_ε	$\tan\delta/10^{-4}$
Alumina	20–100	$50\text{–}10^6$	8.5	20–5
Cardboard	20	50	3.2	80
Crystal (⊥ to axis)	20	10^3	4.5	2
Diamond	20	500–3 000	5.68	
Flint glass	20	$10^3\text{–}10^6$	6.9	17–13
Fluorite	20	2×10^6	6.8	
Fused silica	20–150	$50\text{–}10^8$	3.8	10–1
Granite	20	10^6	8	
KCl	20	$10^6\text{–}10^{10}$	4.8	
Kraft paper	20	10^3	2.9	45
Marble	20	10^6	8	400
Mica	20–100	$50\text{–}10^8$	7.0	10–2
NaCl	25	$10^3\text{–}10^{10}$	5.9	5–1
Paraffin	20	$10^6\text{–}10^9$	2.2	2
Rubber (nature)	20–80	$10^6\text{–}10^7$	2.4	15–100
Rubber (neoprene)	20	$10^3\text{–}10^6$	6.5–5.7	300–900
Rubber (silicone)	20	$50\text{–}10^8$	8.6–8.5	50–10
Sand (dry)	20	10^6	2.5	
Sapphire (⊥ to axis)	20	$50\text{–}10^9$	9.4	2
Soda glass	20	$10^6\text{–}10^8$	7.5	100–80
Soil (dry)	20	10^6	3	
$SrTiO_3$	25	10^3	332	
Steatite	20	$10^6\text{–}10^9$	6	2–20
Umber	20	$10^6\text{–}3\times10^9$	2.8–2.6	2–90

Liquids k_ε is the dielectric constant for a sufficiently low frequency. The temperature coefficient a for k_ε is $a=(dk_\varepsilon/dt)/k_\varepsilon$. For polar molecular liquid (Polar, abbreviated as "P"), k_ε and δ change dramatically according to the frequency.

Material	t/℃	k_ε	$a/10^{-5}$
Acetone (P)	25	20.7	-470
Benzene	20	2.284	- 85
Carbon disulfide	20	2.64	-100
Carbon tertrachloride	20	2.24	- 90
Ethyl alcohol (P)	25	24.3	
Liquid argon	82 K	1.53	-220
Liquid helium	4.19 K	1.048	
Liquid hydrogen	20.4 K	1.23	-280
Liquid nitrogen	70 K	1.45	-200
Liquid oxygen	80 K	1.51	-160
Methyl alcohol (P)	25	32.6	-600
Nitrobenzene (P)	25	34.8	-520
Paraffin oil	20	2.2	
Silicon oil	20	2.2	
Toluene	20	2.39	-100
Transformer oil	20	2.2	
Water (P)	t(℃)	$k_\varepsilon=88.15-0.414\,t+0.131\times10^{-2}t^2-0.046\times10^{-4}t^3$	

Gas The dielectric constant k_ε is unrelated to the frequency until the bottom of the infrared region. $k_\varepsilon-1$) for non-polar molecular gas is proportional to density. For polar molecular gas (Polar, abbreviated as "P"), approximation consists of the formula $(k_\varepsilon-1)\propto$ (pressure)·(absolute temperature)$^{-2}$.

Material (1 atm)	t/℃	$(k_\varepsilon-1)/10^{-4}$	Material (1 atm)	t/℃	$(k_\varepsilon-1)/10^{-4}$
Air (dry)	20	5.36	Helium	0	0.7
Ammonia (P)	1	71	Hydrogen	0	2.72
Argon	20	5.17	Methyl alcohol (P)	100	57
Benzene	100	32.7	Nitrogen	20	5.47
Carbon dioxide	20	9.22	Oxygen	20	4.94
Carbon disulfide	29	29.0	Steam (P)	100	60
Ethyl alcohol (P)	100	78	Sulfur dioxide (P)	22	82

1 atm = 101 325 Pa

Ferroelectrics and Antiferroelectrics

Ferroelectrics are substances which possess spontaneous polarization and which are capable of reversing the orientation of poles according to an external magnetic field. Antiferroelectrics are substances which possess anti-parallel sub-lattice polarization of equivalent size and which are capable of making the pole parallel according to an external magnetic field. In many cases, these characteristics appear below a certain temperature (ferroelectric/anti-ferroelectric phase transition point). The ferroelectric phase contains a hysteresis curve for polarization and external magnetic field. However, in the case of the anti-ferroelectric phase, a double hysteresis curve is observed, and an abnormal dielectric constant is shown at each phase transition point.

Ferroelectric crystals

Material	Chemical formula	Curie point θ/K	Maximum value of spontaneous polarization $P/10^{-2}$C m^{-2}	Relative permittivity to small alternating voltage (Room temperature) $\varepsilon_a/\varepsilon_0$	$\varepsilon_b/\varepsilon_0$	$\varepsilon_c/\varepsilon_0$
Rochelle salt	$KNaC_4H_4O_6 \cdot 4H_2O$	First 297 Last 255	0.24 (276 K)	4000 (θ above)	10.0	9.6
Potassium phosphate monobasic	KH_2PO_4	123	4.95 ($T \ll \theta$)	42	42	21
Phosphoric acid double hydrogen potassium	KD_2PO_4	213	4.8 ($T \ll \theta$)	88	88	90
Barium titanate	$BaTiO_3$	393	26 (296 K)	-5000	-5000	-160
Potassium niobate	$KNbO_3$	707	3.78 (683 K)	-	-	-500
Titanic acid bismuth	$Bi_4Ti_3O_{12}$	916-948	3.5		112	
Lithium niobate	$LiNbO_3$	1468	71	84	84	30
Niobate sodium barium (banana)	$Ba_2NaNb_5O_{15}$	833-858	40	242	242	51
Glycine sulfate	$(CH_2NH_2COOH)_3H_2SO_4$	320	3.2 ($T \ll \theta$)	9.2	45	6.4
Ammonium sulfate	$(NH_4)_2SO_4$	223.5	0.465 ($T < \theta$)	10.0	9.3	9.2
Sodium nitrite	$NaNO_2$	433	8.6 ($T < \theta$)	6.0	8.0	8.0
Hexacyanoferrate (II) acid Potassium (yellow prussiate of potash)	$K_4[Fe(CN)_6] \cdot 3H_2O$	248.5	1.3 (218.5 K)	$\varepsilon[101]/\varepsilon_0$ = 11* (* 218.5 K)	6*	$\varepsilon[101]/\varepsilon_0$ = 25*
Antimony sulfide iodide	SbSI	294	25 (273 K)	25	25	5×10^4 (RT)

Antiferroelectric bodies

Material	Chemical formula	Transition point T_c/K	Relative permittivity to small alternating voltage (Room temperature) $\varepsilon_a/\varepsilon_0$	$\varepsilon_b/\varepsilon_0$	$\varepsilon_c/\varepsilon_0$
Ammonium dihydrogenphosphate	$NH_4H_2PO_4$	148	56	56	15.5
Lead zirconate (porcelain)	$PbZrO_3$	506		80	
Sodium niobate	$NaNbO_3$	911	76	76	670
Formic acid copper	$Cu(HCOO)_2 \cdot 4H_2O$	235.5	-30	-420	

Piezoelectric Substances

When force (stress) is applied to a certain type of substance (crystal), an electric charge (voltage between surfaces) appears due to dielectric polarization. Conversely, when an electric charge is applied, deformation (warping) occurs. The former phenomena is called the "piezoelectric effect," and the latter is called the "inverse piezoelectric effect." Constraints between the direction of applied stress and the direction of the generated voltage are determined by crystal symmetry.

Piezoelectric crystal

Material	Chemical formula	A piezo-electric rate $d/10^{-12}$C N^{-1} or $d/10^{-12}$m V^{-1}		
Crystal (left-handed quartz)	SiO_2	$d_{11}=2.31$ $d_{14}=-0.727$		
Selenium	Se	$d_{11}=63$ $d_{14}=$		
Tellurium	Te	$d_{11}=37$ $d_{14}=$		
Cadmium sulphide	CdS	$d_{15}=-13.98$		$d_{31}=-5.18$ $d_{33}=10.32$
Cadmium selenide	CdSe	$d_{15}=-10.51$		$d_{31}=-3.92$ $d_{33}=7.84$
Zinc oxide	ZnO	$d_{15}=-12$		$d_{31}=-4.7$ $d_{33}=12$
Barium titanate	$BaTiO_3$	$d_{15}=392$		$d_{31}=-34.5$ $d_{33}=85.6$
Ammonium dihydrogenphosphate (ADP)	$NH_4H_2PO_4$	$d_{14}=-1.5$		$d_{36}=48$
Potassium phosphate monobasic (KDP)	KH_2PO_4	$d_{14}=1.3$		$d_{36}=21$
Rochelle salt	$KNaC_4H_4O_6 \cdot 4H_2O$	$d_{14}=345$	$d_{25}=54$	$d_{36}=12$
Antimony sulfide iodide	SbSI			$d_{33}=4\times10^3$ (19℃) $d_{33}=0.33\times10^3$ (-100℃)
Lead zirconate titanate (PZT)	$PbZr_xTi_{1-x}O_3$ $x=0.52$			$d_{31}=-93.5$ $d_{33}=223$
Gallium arsenide	GaAs	$d_{14}=2.6$		
Lithium niobate	$LiNbO_3$	$d_{15}=74$	$d_{22}=20.8$	$d_{31}=-0.86$ $d_{35}=16.2$
Polyvinylidene difluoride (PVDF)	$-(CF_2-CH_2)_n-$			$d_{31}=25$

Insulators

Electric current which flows through insulators is composed of internal current and surface current. Internal resistance changes according to the purity of the substance, while surface resistance is influenced by the surface condition. An increase in temperature causes both internal resistance and surface resistance to decrease. The following table shows the internal resistivity and surface resistivity (both at ambient temperature) which were measured 1 minute after the start of applying voltage. In addition to the conduction current, the electric current at this time includes the electric charge stored inside the insulator. The resistivity value calculated from an electric current in equilibrium is significantly greater than the values shown in this table. The strength of breakdown is the value obtained at commercial frequency for plate samples of 1 to 3 mm.

Material	Internal resistivity ρ/Ω m	Surface resistivity (50–60% in humidity) R/Ω	Strength of dielectric breakdown $F/10^6$V m^{-1}
Ceramics			
Alumina	10^9–10^{12}	-	10–16
Feldspar porcelain (glazes)	10^{10}–10^{12}	10^{11}	-
Feldspar porcelain (substrate)	10^{10}–10^{12}	10^9	10–15
Steatite	10^{11}–10^{13}	-	8–14
Glass (Pyrex)	10^{12}	-	-
Glass (quartz)	$>10^{16}$	3×10^{12}	20–40
Glass (soda)	10^9–10^{11}	10^{10}–10^{12}	-
Insulation (mineral) oil	10^{11}–10^{15}	-	-
Marble	10^7–10^9	10^9	-
Mica (molded)	10^{13}	5×10^{13}	-
Paraffin	10^{13}–10^{17}	10^{15}	8–12
Plastic			
Acrylic fiber	$>10^{13}$	$>10^{14}$	
Epoxy	10^{12}–10^{13}	3×10^7–$>10^{14}$	16–22
Nylon	10^8–10^{13}	10^{11}–10^{15}	15–20
Polyethylene	$>10^{14}$	-	18–28
Polystyrene	10^{15}–10^{19}	$>10^{14}$	20–30
Polyvinyl chloride (hard)	5×10^{12}–10^{13}	10^{12}–10^{15}	17–50
Polyvinyl chloride (soft)	5×10^6–5×10^{12}	$>10^{14}$	10–30
Teflon	10^{15}–10^{19}	4×10^{12}–10^{17}	20
Rubber (chloroprene)	10^{10}–10^{11}	-	10–15
Rubber (nature)	10^{13}–10^{15}	-	20–30
Rubber (silicone)	10^{12}–10^{13}	-	5–25
Sulfur	10^{14}–10^{15}	7×10^{15}	-

Magnetic Properties

The quantity expressing the magnetization strength of a substance is the magnetic moment M (unit: A/m) per unit volume, or $P_M = \mu_0 M$ (unit: Wb/m^2 or T (tesla)). These are referred to as magnetization strength and magnetic polarization, respectively. P_M is sometimes called magnetization strength, and symbols such as M, I, or J are sometimes used instead of P_M. $B = \mu_0(H+M) = \mu_0 H + P_M$ is used to express the relationship between the two vector quantities expressing the magnetic field, namely B (magnetic flux density, unit: Wb/m^2 or T) and H (strength of magnetic field, unit: A/m). In the case of the CGS electromagnetic system of units which is used in the field of magnetism, the relationship among B, H, and the magnetization strength M_{CGS} is expressed by $B = H + 4\pi M_{CGS}$. The unit of quantity for these variables is calculated the same [cm$^{-1/2}$g$^{1/2}$s^{-1}]. However, the units are referred to as G (Gauss), Oe (Oersted), and G (Gauss), respectively. When converting to MKSA units, 1G of magnetic flux density is 10^{-4}T, and 1 Oe of magnetic field is $10^3/4\pi(79.577\text{–}80)$A/m. For the quantity expressing magnetization, 1G is equivalent to 10^3A/m of M, or to $4\pi \times 10^{-4}$ T of P_M.

Customarily, the strength of magnetization is often expressed as the magnetic moment $\sigma_g(= M_{CGS}/$(density in the CGS system of units)) per unit quantity (1 g) in the CGS electromagnetic system of units, or as the magnetic moment σ_{mol} per 1 mole of substance. These units of quantity are often expressed as G/g and emu/g, or as G/mol and emu/mol.

In the case of diamagnetic substances and paramagnetic substances, M is proportional to H, unless H is extremely strong. This proportional constant $\chi = M/H = \mathrm{P}_M/\mu_0 H$ is referred to as the "magnetic susceptibility" (if magnetization strength is expressed as P_M, the constant is sometimes referred to as the "magnetic susceptibility ratio"). The value for magnetic susceptibility $\chi_{CGS} = M_{CGS}/H$ in the CGS system of units is χ $(1/4\pi)$ times the SI system. Practically speaking, the formulas $\chi_{CGS,g} = \sigma_g/H$ and $\chi_{CGS,mol} = \sigma_{mol}/H$ are often used for magnetic susceptibility per 1 g and 1 mole. These quantity units are cm^3/g and cm^3/mol, respectively. The units are often written as emu/g and emu/mol. In the case of many paramagnetic substances, the magnetic susceptibility undergoes temperature changes and conforms with the Curie-Weiss law of $\chi_{CGS,mol} = C_{mol}/(T - \Theta_p)$. C_{mol} and Θ_p are referred to as the Curie constant and the asymptotic Curie temperature, respectively. On the other hand, temperature normally has almost no effect on the magnetic susceptibility of metal paramagnetic substances and diamagnetic substances.

The magnetization M of ferromagnetics shows the hysteresis (history) phenomenon. Therefore, M and B are dependent on the previous method of application for H, as well as on the size of H. This condition is shown by the M–H magnetization curve which shows the relationship between M and H, and by the B–H magnetization curve which shows the relationship between B and H. The state in which $M=0$ at $H=0$ is referred to as the "demagnetized state." The magnetization curve occurring when the magnetic field is increased from the demagnetized state is called the "initial magnetization curve." The $1/\mu_0$ factor for the angle near the origin of the B–H initial magnetization curve is written as μ_i and referred to as the "initial permeability." The $1/\mu_0$ factor for the angle of the tangent drawn from the origin of the B–H magnetization curve is written as μ_m and referred to as the "maximum permeability." As the magnetic field becomes larger, the strength of magnetization nears the fixed saturation magnetization M_S. Generally speaking, ferromagnetics are divided into categories called "magnetic domains." Within a single magnetic domain, the magnetic moment of atoms is aligned in the same direction. The magnetization strength created by this phenomenon is called "spontaneous magnetization." In actuality, saturation magnetization is equivalent to spontaneous magnetization. The strength of spontaneous magnetization is the substance-specific quantity. The strength decreases as temperature increases and becomes zero upon reaching the Curie temperature T_c. Both conceptually and in actuality, T_c is a quantity that differs from Θ_p.

When a magnetic field equal to or greater than the magnetic field (saturation magnetic field) required to saturate the magnetization of ferromagnetics is sent back and forth between the positive value and negative value of the magnetic field, the magnetization curve changes to a closed curve. This is referred to as the "major hysteresis loop." The major hysteresis loop of the B–H magnetization curve is essential for application of ferromagnetics. When the magnetic field is decreased after saturation of the magnetization for ferromagnetics, the magnetization changes through a different route than when the magnetic field is increased. Even if the magnetic field reaches zero, it possesses the limited magnetic flux density B_r (can also be described as the limited magnetization M_r). B_r is called the "residual magnetic flux density," and M_r is called the "residual magnetization." Furthermore, when the magnetic field is increased in the opposite direction, the magnetic flux density becomes zero at $H=-{}_BH_c$. This ${}_BH_c$ is called "coercive force" (coercive force is also defined as the magnetic field ${}_MH_c$ where magnetization is zero on the M–H magnetization curve). When the force of the magnetic field is increased in the opposite direction, saturation is reached in the opposite direction. If the direction of change for the magnetic field is reversed again, the (B, H) and (M, Hwhich show the state of ferromagnetics make a complete circuit of the major hysteresis loop. The area $W_h=\oint H\mathrm{d}B$ ($(1/4\pi)\oint H\mathrm{d}B$ for the CGS electromagnetic system of units) surrounding the major hysteresis loop of the B–H magnetization curve is the energy lost due to magnetic loss during 1 cycle of alternating magnetization within the unit volume of the ferromagnetic. This is called "hysteresis loss." The B–H magnetization curve and variables B_r, ${}_BH_c$ (or ${}_MH_c$), W_h, μ_i, μ_m, etc. which characterize the curve have different properties according to the material, and change significantly depending on additives, thermal processing, fabrication, and other procedures. The tables "Permanent magnet material" and "High permeability material" show a list of representative values. Generally speaking, it is preferable for B_r and ${}_BH_c$ to be large in the case of permanent magnet material. In the case of high permeability material used in the magnetic core of transformers or other applications, it is preferable for μ_i and μ_m to be large and for ${}_BH_c$ and W_h to be small.

I. Paramagnetic ions

The Curie constant for paramagnetic substances which include paramagnetic ions is given by $C_{mol}=Np^2\mu_B{}^2/3k_B$. N is the number of paramagnetic ions within 1 mole. p is the number of effective Bohr magnetons. It is determined by the electric state of magnetic ions.

Iron group ions

Ion	3d electron number	Free ion term	p (calculated) $=2\sqrt{S(S+1)}$	p (experimental)
Ti^{3+}, V^{4+}	1	${}^2D_{3/2}$	1.73	1.7–1.9
V^{3+}	2	${}^3F_2$	2.83	2.7–2.9
Cr^{3+}, V^{2+}	3	${}^4F_{3/2}$	3.87	3.8–3.9
Mn^{3+}	4	${}^5D_0$	4.90	4.8–4.9
Mn^{2+}, Fe^{3+}	5	${}^6S_{5/2}$	5.92	5.8–5.9
Fe^{2+}	6	${}^5D_4$	4.90	5.2–5.5
Co^{2+}	7	${}^4F_{9/2}$	3.87	4.8–5.1
Ni^{2+}	8	${}^3F_4$	2.83	2.8–3.3
Cu^{2+}	9	${}^2D_{5/2}$	1.73	1.8–2.0

Rare-earth ions

Ion	4f electron number	Free ion term	p (calculated) $=g\sqrt{J(J+1)}$	p (experimental)
Ce^{3+}	1	${}^2F_{5/2}$	2.54	2.4
Pr^{3+}	2	${}^3H_4$	3.58	3.5
Nd^{3+}	3	${}^4I_{9/2}$	3.62	3.5
Pm^{3+}	4	${}^5I_4$	2.68	
Sm^{3+}	5	${}^6H_{5/2}$	1.55*	1.5
Eu^{3+}	6	${}^7F_0$	3.40*	3.4
Gd^{3+}	7	${}^8S_{7/2}$	7.94	8.0
Tb^{3+}	8	${}^7F_6$	9.72	9.5
Dy^{3+}	9	${}^6H_{15/2}$	10.65	10.6
Ho^{3+}	10	${}^5I_8$	10.61	10.4
Er^{3+}	11	${}^4I_{15/2}$	9.58	9.5
Tm^{3+}	12	${}^3H_6$	7.56	7.3
Yb^{3+}	13	${}^2F_{7/2}$	4.54	4.5

* Value by theory of van Vleck-Frank

$$g=\frac{3}{2}+\frac{S(S+1)-L(L+1)}{2J(J+1)}$$

For other symbols, refer to page **306**.

II. Magnetic susceptibility of paramagnetic substances and diamagnetic substances

Material	Temperature (K)	$\chi_{CGS,g}/10^{-6}$ cm^3 g^{-1} *2
Al	293	0.61
C_6H_6	Room temperature	−0.702
C(graphite)	289	−3.0
CO_2	298	−0.454
Cu	296	−0.086
$CuSO_4 \cdot 5H_2O$ *1	293	5.85
GaAs	300	−0.230
Ge	298	−0.106
H_2	293	−1.987
H_2O	273	−0.720
N_2	293	−0.43
$(NH_4)_2Fe(SO_4)_2 \cdot 6H_2O$ *1	290.3	32.57
NaCl	Room temperature	−0.517
O_2	90	240.6
O_2	293	107.8
Pt	298	0.983

*1 The magnetic susceptibilities of these materials change with temperature according to the Curie-Weiss law.
*2 The electromagnetic CGS system magnetic susceptibility χ_{CGS} can be obtained by multipling this value by the density (in units of g/cm). SI magnetic susceptibility χ can be obtained by multiplying that value by an additional 4π.

III. Curie temperature and magnetization of ferromagnetic elements

Element	T_C/K	Room temperature		0 K(extrapolation value)	
		σ_g/G cm^3 g^{-1} *1	$M/10^3$A m^{-1}	σ_g/G cm^3 g^{-1}	μ/μ_B *2
Fe	1 044	217.6(293 K)	1 713	221.7	2.216
Co *3	1 390	161.9(293 K)	1 440	163.76	1.728
Ni	631.0	55.07(298 K)	490.2	58.57	0.616
Gd	291.2			268.4	7.56

*1 The saturation magnetic moment for 1 gram, 1 G cm^3 = $4\pi \times 10^{-10}$ Wb m.
*2 The value indicating the saturation magnetic moment for 1 atom in units of Bohr magnetons (Refer to page **298**.)
*3 The crystal structure changes at 700 K.

IV. Curie temperature and magnetization main ferromagnetics

Material	T_C/K	σ_g/G cm^3 g^{-1} (Room temperature)	σ_g/G cm^3 g^{-1} (4.2 K)
CoPt *1	850	22	
Cu_2MnAl	603		96
70Fe−30Co *2	>1 250	242	247
FeCo *3	>1 000	230	235
$FeNi_3$ *3	>775	110	115
Ni_3Mn *1	820	90	98
$Nd_2Fe_{14}B$	588	167	194
$SmCo_5$	1 015	91	99
$BaFe_{12}O_{19}$ *4	725	70	100
Fe_3O_4 *4	860	92	98
$NiFe_2O_4$ *4	860	50	56
$Y_3Fe_5O_{12}$ *4	560	27	38

*1 ordered alloy *2 disorderd alloy *3 ordered alloy. Below T_C an order-disorder transistion occurs.
*4 Ferrimagnetic substance

V. Permanent magnet materials (mainly quoted from "Permanent Magnets" (AGNE Gijutsu Center, 2007))

Name	The main composition *2	Processing	$_MH_c$/kA m^{-1}	B_r/T	$(BH)_{max}$/kJ m^{-3} *3	Classification
KS steel	Fe–0.9C, 35Co, 5Cr, 4W	Forging, Hardening, Tempering	20 *1	0.9	7.6	Martensite Steel magnet
Alnico 5	Fe–8Al, 24Co, 14Ni, 3Cu	Casting, Magnetic field inside cooling, Ageing treatment	50–62	1.25–1.35	40–64	Alnico magnet
Iron chrome cobalt	Fe–26Cr, 10Co, 1.5Ti	Casting, Magnetic field inside annealing, Ageing treatment	47	1.44	54	Iron chrome cobalt magnet
Samarium cobalt	Sm_2Co_{17}+Fe, Cu, Zr	Magnetic field inside molding, Sintering	533–1 030	1.02–1.20	176–264	Rare earth permanent magnet
Iron nedymium boron	$Nd_2Fe_{14}B$	Magnetic field inside molding, Sintering	875–2 626	1.05–1.51	206–437	
Barium ferrite	BaO·6 Fe_2O_3	In a magnetic field at the water-based wet-pressing stage, Sintering	175	0.41	29.8	Oxide magnet M type ferrite sintering magnet
Strontium ferrite	SrO·6 Fe_2O_3+La, Co	In a magnetic field at the water-based wet-pressing stage, Sintering	151–354 *1	0.36–0.47	23.8–42.0	Oxide magnet M type ferrite sintering magnet

*1 The value for $_BH_C$/kA m^{-1} is shown. (Normally $_MH_C > {}_BH_C$.)
*2 On lines 1–3, the major atomic constituants and the % by weight for compoents other than iron are shown. On the other lines, the chemical formula for the compound is given.
*3 The ferromagnet performance index, largest absolute value for the $B·H$ area which occurs at the 2nd quadrant of the large B–H hysteresis loop.
1 kJ m^{-3} corresponds to $4\pi \times 10^{-2}$ MG·Oe ($\approx$1/8 MG·Oe).

VI. High permeability materials

Name	The main composition *1	Heat-treatment *2	μ_i	μ_m	$_BH_c$/A m^{-1}	W_h/10^{-1} J m^{-3} s^{-1}	Kind
Pure iron		950℃	200–300	6 000–8 000	50–90		Iron and iron alloy
Silicon steel	Fe–4Si	800℃	500	7 000	40	3 500	
Directionality silicon steel	Fe–3Si	800℃	1 500	40 000	10	700	
Alperm	Fe–16Al	Cooling quickly from 600℃	3 000	55 000	3		
Permendur	Fe–2V, 50Co	800℃	650	6 000	160		
Sendust	Fe–5.5Al, 9.5Si		30 000	120 000	2	100	
78 permalloy	Fe–78.5Ni	1 050℃, Cooling quickly from 600℃	8 000	100 000	4	580	Permalloy
Supermalloy	Fe–79Ni, 5Mo	1 300℃ (H_2)	100 000	6 000 000	0.16		
Mu metal	Fe–77Ni, 2Cr, 5Cu	1 175℃ (H_2)	20 000	1 000 000	4		
Meta glass (2605 S 2)	Fe–3B, 5Si		5 000	500 000	3.2		Amorphous alloys
Manganese zinc ferrite	$Mn_{0.5}Zn_{0.4}Fe_{2.1}O_4$		10 000		7		Ferrite
Nickel zinc ferrite	$Ni_{0.35}Zn_{0.65}Fe_2O_4$		1 500		24		

*1 On lines 1–10, the major atomic constituants and the % by weight for compoents other than iron are shown.
*2 A defining characteristic of high permeability material metallic systems is that they depend strongly on the heat treatment conditions. Here, the heat treatment conditions corresponding to the characteristics given to the columns to the right.

Sound

Speed of Sound Inside of Various Substances

Speed of sound inside of gases

(1) The speed of sound inside of gases is almost completely unrelated to pressure. The speed increases in proportion to the square root of the absolute temperature.
(2) Generally speaking, in the case of monoatomic molecule gases, the absorption of sound is proportional to two times the frequency, and is determined by the viscosity and thermal conductivity of the gas. However, for other than monoatomic molecules, the internal degree of freedom such as oscillation and rotation inside of molecules is related to absorption. This causes internal dispersion and abnormal absorption of the speed of sound to occur.
(3) The acoustic impedance is the ratio of particle velocity and sound pressure in the plane wave which propagates inside an infinitely wide medium. It is equivalent to the product of density and speed of sound. The reflectance R and the transmittance T of the wave energy injected vertically into the boundary surface of two mediums which are adjacent on a level surface are given by the formula $R=(z_1-z_2)^2/(z_1+z_2)^2$, $T=1-R$. However, z_1 and z_2 respectively are the acoustic impedance for each medium.
(4) By using the following formula, the speed of sound c' when a water vapor with pressure p exists within air of pressure H can be determined from the speed of sound in dry air at the same temperature.

$$c'=c/\sqrt{1-\frac{p}{H}\left(\frac{\gamma_w}{\gamma}-0.622\right)}$$

γ_w and γ are the ratio of specific heat at constant pressure and the specific heat at constant volume for water vapor and dry air, respectively.

Material	Density ρ/kg m^{-3} (0℃, 1 atm)	Speed of sound c/m s^{-1} (0℃, 1 atm)	Temperature coefficient of c Δ/m s^{-1} ℃$^{-1}$ (0℃)	Acoustic impedance ρc/N s m^{-3}
Air (dryness)	1.2929	331.45	0.607	428.6
Ammonia	0.7710	415	0.73	319.9
Argon	1.7837	319	-	569
Carbon dioxide	1.9769	258 (low frequency)→ 268.6 (high frequency)	0.87	508
Carbon monoxide	1.2504	337	0.604	421
Chlorine	3.214	205.3	-	660
Deuterium	0.1784	890	1.58	158.7
Ethane	1.3566	308 (10℃)	-	418
Ethylene	1.2604	314	0.56	396
Helium	0.17847	970	1.55	173
Hydrogen	0.08988	1269.5	2.00	114.1
Hydrogen sulfide	1.539	289	-	445
Methane	0.7168	430	0.62	308
Neon	0.90035	435	0.78	385
Nitric oxide	1.9778	258	-	518
Nitrogen	1.25055	337	0.85	421
Nitrous oxide	1.3402	325	-	435
Oxygen	1.4290	317.2	0.57	453
Steam (100℃)	0.5980	473	-	242
Sulfur dioxide	2.9269	211	-	617.6

1 atm = 101 325 Pa

Speed of sound inside of liquids

(1) With the exception of very few special cases, the speed of sound in liquids is in the range of 1000 to 1500 m s^{-1}. In many cases, the speed decreases by 2 to 5 m s^{-1} for every 1°C change in temperature. The speed of sound in water increases together with temperature and reaches a maximum at 74℃.

(2) The absorption constant α is expressed by neper m^{-1}. When the dm destination decreases to I_d, the strength I_0 of the initial sound is given by $\alpha = (1/2\ d)\log_e(I_0/I_d)$neper m^{-1}. When expressed as a decibel unit, the value is 8.686 α dB m^{-1}. In the case of normal liquids, α is proportional to the square of the frequency f.

Material	t/℃	Density $\rho/10^3$kg m^{-3}	Speed of sound c/m s^{-1}	$\alpha f^{-2}/10^{-13}$ neper m^{-1} Hz^{-2}	Acoustic impedance $\rho c/10^6$N s m^{-3}
Benzene	23-27	0.87	1 295	8.3	1.13
Carbon disulfide	23-27	1.26	1 149	74	1.45
Carbon tertrachloride	23-27	1.59	930	4.80	1.48
Chloroform	25	1.49	995	3.80	1.48
Diethyl ether	23-27	0.71	985	1.40	0.70
Ethyl alcohol	23-27	0.786	1 207	0.9	0.95
Ethyl bromide	28	1.428	892	0.62	1.27
Glycerin	23-27	1.26	1 986	26	1.90
Heavy water	20	1.1053	1 381	0.32	1.53
Mercury	23-27	13.6	1 450	0.66	19.8
Pentane	18	0.632	1 052	1.0	0.66
Water (distillation)	23-27	1.00	1 500	0.33	1.50
Water (seawater and salinity 30/1000)	20	1.021	1 513	3	1.54

Transmission Loss for Solid Walls

Transmission loss shows the decrease in decibels for the intensity of transmitted sound waves I_t in relation to the intensity of the incident sound I_i when sound waves enter a solid wall. Transmission loss is given by the following formula. Transmission loss $= 10\ \log_{10}(I_i/I_t)$dB

Material	Surface density ρs/kg m^{-2}	Penetration loss (dB)					
		Frequency f/Hz					
		125	250	500	1 000	2 000	4 000
Lauan plywood (6mm thickness)	3.0	11	13	16	21	25	22
Plaster board (9 mm thick)	8.7	20	22	25	28	34	23
Plate glass (6mm thickness)	15.0	20	27	31	31	26	37
Iron plate (1mm thickness)	8.2	17	21	25	28	34	38
Iron plate (4.5mm thickness)	36.9	28	33	37	41	40	38
Lead sheeting (1mm thickness)	11.3	28	26	29	33	38	43
Aerated concrete (75 mm thickness, 6 mm of plaster painted on both sides)	60	27	31	31	41	48	54
Concrete block (100 mm thickness, 15 mm of plaster painted on both sides)	160	33	37	42	49	56	60
6 mm double walled plywood (wooden base, 100 mm air layer)	-	11	20	29	38	45	42

Speed of sound

(1) Even for the same substances, the speed of sound inside of solids changes dramatically according to the crystallization state and orientation for metal, and according to the frequency and mixture ratio for rubber, etc. Generally speaking, the speed of sound decreases as temperature increases.

Material	Density $\rho/10^3$ kg m^{-3}	c_1/m s^{-1}	c_2/m s^{-1}
ADP (Z direction)	1.80	4 300	-
Aluminum	2.69	6 420	3 040
Beryllium	1.82	12 890	8 880
Brass (70Cu, 30Zn)	8.6	4 700	2 100
Cadmium	8.56	2 780	-
Chromium	7.193	6 200	3 860
Concrete	-	4 250-5 250	-
Copper	8.96	5 010	2 270
Crystal (X direction)	2.65	5 720	-
Duralumin (17 S)	2.79	6 320	3 130
Ebonite	1.2	2 500	-
Germanium	5.322	5 944	3 555
Glass (crown)	2.4-2.6	5 100	2 840
Glass (Flint)	2.9-5.9	3 980	2 380
Glass (pane)	2.42	5 440	-
Gold	19.32	3 240	1 220
Ice	0.917	3 230	1 600
Iron	7.86	5 950	3 240
Lead	11.34	1 960	690
Magnesium	1.54	5 770	3 050
Marble	2.65	6 100	2 900
Melted crystal	2.2	5 968	3 764
Monel metal	8.90	5 350	2 720
Nickel	8.90	6 040	3 000
Nylon-6,6	1.11	2 620	1 070
Platinum	21.62	3 260	1 730
Polyethylene (soft)	0.90	1 950	540
Polystyrene	1.056	2 350	1 120
Pyrex glass (702)	2.32	5 640	3 280
Rubber (nature)	0.97	1 500 (1 MHz)	120 (1 MHz)
Rubber (styrene butadiene rubber)	1.00	1 760 (1 MHz)	530 (1 MHz)
Silicon (100 plane)	2.33	8 433	5 843
Silver	10.49	3 650	1 660
Stainless steels (347)	7.91	5 790	3 100
Tin	7.3	3 320	1 670
Titanium	4.58	5 990	2 960
Tungsten	19.2	5 410	2 640
Uranium	18.7	3 370	1 940
Zinc	7.18	4 210	2 440
Zirconium	6.44	4 650	2 250

inside of solids

(2) The speed of longitudinal waves in free solids c_1 is given by the formula $\sqrt{(K+\frac{4}{3}G)/\rho}$, the speed of transverse in free solids c_2 is given by $\sqrt{G/\rho}$, and the speed of longitudinal vibration in rods c_3 is given by $\sqrt{E/\rho}$. ρ is the density, K is the bulk modulus, G is the shear modulus, and E is Young's modulus. A free solid is a solid which is of sufficient size compared to the wavelength. A bar is a solid which is of sufficient thinness compared to the wavelength.

c_3/m s^{-1}	impedance $\rho c_1/10^6$N s m^{-3}	impedance $\rho c_2/10^6$N s m^{-3}
3 500	7.77	–
5 000	17.3	8.2
12 870	24.1	16.6
3 480	40.6	18.3
2 400	23.8	–
5 900	44.6	27.3
–	–	–
3 750	44.6	20.2
5 440	15.16	–
5 150	17.1	8.5
1 570	3.0	–
–	31.63	18.91
4 540	11.4	6.35
3 720	15.4	9.22
–	13.2	–
2 030	62.5	23.5
–	2.96	1.47
5 120	46.4	25.3
1 210	22.4	7.85
4 940	10.0	5.3
3 810	16.2	7.7
5 760	13.1	8.29
4 400	47.5	24.2
4 900	53.5	26.6
1 800	2.86	1.18
2 800	69.7	37.0
920	1.75	0.48
2 240	2.48	1.18
5 170	13.1	7.6
210 (1 MHz)	1.5	0.12
–	1.76	0.53
–	19.64	13.61
2 680	38.0	17.5
5 000	45.7	24.5
2 730	24.6	11.8
–	27.4	13.5
4 320	103	50.5
–	63.02	36.3
3 850	30	17.3
–	29.95	14.49

Sound Absorption Coefficient

When the intensity of sound entering an object is I_i and the intensity of the reflected sound is I_r, the sound absorption coefficient is given by the following formula. Sound absorption coefficient $\alpha_a = 1 - (I_r/I_i)$.

Material	Thickness t/mm	Specific gravity	α_a Frequency f/Hz					
			125	250	500	1 000	2 000	4 000
Glass wool board (Affixed to rigid wall)	25	0.016–0.024	0.08–0.16	0.22–0.34	0.54–0.71	0.71–0.88	0.69–0.88	0.77–0.94
Glass wool board (Affixed to rigid wall)	50	0.016–0.024	0.13–0.27	0.55–0.75	0.87–0.98	0.78–0.93	0.78–0.93	0.85–0.98
Glass wool board (100 mm in back air layer)	25	0.016–0.024	0.18–0.33	0.53–0.75	0.84–0.98	0.78–0.92	0.70–0.86	0.77–0.94
Rock wool (Affixed to rigid wall)	25	-0.12	0.04–0.09	0.23–0.34	0.58–0.80	0.78–0.91	0.75–0.84	0.75–0.90
Rock wool (Affixed to rigid wall)	25	0.12-	0.05–0.16	0.26–0.50	0.70–0.92	0.80–0.97	0.72–0.93	0.77–0.96
Flexible urethane foam (Affixed to rigid wall)	20	0.01–0.02	0.06–0.11	0.16–0.24	0.32–0.42	0.47–0.58	0.54–0.67	0.56–0.70
Wood-fiber reinforced cement board	25	0.43	0.01	0.10	0.23	0.64	0.69	0.73
Insulation board for absorbing sound (wooden foundation, back air layer 300 mm)								
A (Pore size 5 mm, ϕ-pitch 12.7 mm)	9	-	0.27–0.39	0.24–0.38	0.38–0.52	0.55–0.59	0.67–0.68	0.77–0.78
A (5 ϕ-12.7)	12	-	0.35	0.28	0.45	0.64	0.75	0.83
G (pin hole)	9	-	0.33–0.40	0.30–0.33	0.38–0.41	0.30–0.48	0.27–0.47	0.30–0.47
Rock wool ceiling sound insulation board (metal frame, gypsum board under coating, back air layer 300 mm)	12	-	0.28–0.32	0.22–0.32	0.42–0.48	0.52–0.74	0.64–0.79	0.67–0.87
Perforated gypsum board (6 ϕ-22, board hand paper nailed to back, 300 mm air layer in rear)	7	-	0.63	0.51	0.29	0.21	0.25	0.31
Apertured board (5 ϕ-15, glasswool board 25 mm substrates, back air layer 150 mm)	4	-	0.25	0.74	0.86	0.46	0.45	0.28
Apertured board (9 ϕ-15, glasswool board 25 mm substrates, back air layer 500 mm)	5	-	0.83	0.72	0.80	0.90	0.87	0.70

Acoustic Terminology and Units

Sound intensity (W m^{-2}) The intensity of sound in a specific direction and at a certain point is the energy per second which passes through a vertical unit surface area in that direction.

Sound pressure (Pa) 1 Pa = 1 N m^{-2}, pressure changes shown by sound. This is normally expressed as an effective value.

Sound intensity level (dB) 10 times the common logarithm for the ratio of sound intensity I and standard sound intensity I_0; in other words, 10 $\log_{10}(I/I_0)$. Here, $I_0 = 10^{-12}$ W m^{-2}

Sound pressure level (dB) 20 times the common logarithm for the ratio of sound pressure P and standard sound pressure P_0; in other words, 20 $\log_{10}(P/P_0)$. Here, $P_0 = 20\ \mu$Pa. In the case of plane waves, can be considered to be the same as the sound intensity level for practical purpose.

Sound loudness level (phon) The sound pressure level (dB) of 1000 Hz pure tone which is heard at the same loudness by the listener. For example, if a sound is heard at the same loudness as 1000 Hz pure tone at 60 dB, the sound is 60 phon.

Reverberation time (s) After stopping the sound source, the time required for the sound energy density in a room to return to 10^{-6} of the original value.

Noise level (dB) Value obtained by measuring sound using the normal sound level meter defined by JIS C 1502 or the precision sound level meter defined by JIS C 1505.

Sound power level (dB) 10 times the common logarithm for the ratio of a certain sound output and the standard sound output. The standard sound output is 1 pW. The quantity symbol is L_w or L_p. The unit symbol is dB.

Sound transmission loss (dB) A quantity indicating the level of reciprocal sound transmissivity in order to express the sound insulation performance of sheet material. Calculated using the following formula.

$$\text{Sound transmission loss}: TL = 10 \log_{10}\left(\frac{1}{\tau}\right)$$

However, τ: sound transmissivity of material.

Diffraction effect Effect caused by the diffraction of waves. In other words, the effect resulting from a change in the traveling direction of waves due to the presence of obstacles, different mediums, etc.

The diffraction effect is prominent when obstacles are about the same size as the wavelength.

Sound absorption When sound waves enter the surface of a substance which differs from the propagation medium (generally speaking, the air), the phenomenon in which a portion of sound wave energy is converted into heat energy inside that substance and appears to have been absorbed.

Distance attenuation When sound waves emitted from a sound source propagate through a medium, the strength attenuates due to a variety of factors. The most basic of these factors is distance attenuation (also known as "geometric attenuation"). This is the phenomenon in which the area of the sound wave surface widens as it travels further from the sound source, causing the sound intensity to decrease in reverse proportion to distance.

Ultra low frequency sound (Infrasound) Sound waves from 1 Hz to 20 Hz. Defined in the international standards ISO-7196.

Low frequency sound (Low frequency noise) Sound waves with a low frequency. In Japan, this refers to low frequency sound with an approximate range from 1 Hz to less than 100 Hz. There are no international standards for a low frequency range. The low frequency range differs depending on the country.

Low frequency air vibration General name for low frequency sound including that outside the visible and audible range. Divided into two categories: under 20 Hz, and 20 Hz to around 100 Hz. These categories are referred to as infrasound and low frequency sound, respectively.

Particle velocity (m/s) Time derivative for instantaneous values of particle displacement. Normally expressed as an effective value. The quantity symbol is u or v.

Fundamental Frequency of Music

The equal temperament scale as based on the international standards of $a^1 = 440$ Hz.*

(Unit: Hz)

	C_2	C_1	C	c	c^1	c^2	c^3	c^4	c^5
C	16.352	32.703	65.406	130.81	261.63	523.25	1046.5	2093.0	4186.0
C#	17.324	34.648	69.296	138.59	277.18	554.37	1108.7	2217.5	4434.9
D	18.354	36.708	73.416	146.83	293.66	587.33	1174.7	2349.3	4698.6
D#	19.445	38.891	77.782	155.56	311.13	622.25	1244.5	2489.0	4978.0
E	20.602	41.203	82.407	164.81	329.63	659.26	1318.5	2637.0	5274.0
F	21.827	43.654	87.307	174.61	349.23	698.46	1396.9	2793.8	5587.7
F#	23.125	46.249	92.499	185.00	369.99	739.99	1480.0	2960.0	5919.9
G	24.500	48.999	97.999	196.00	392.00	783.99	1568.0	3136.0	6271.9
G#	25.957	51.913	103.83	207.65	415.30	830.61	1661.2	3322.4	6644.9
A	27.500	55.000	110.00	220.00	**440.00**	880.00	1760.0	3520.0	7040.0
A#	29.135	58.270	116.54	233.08	466.16	932.33	1864.7	3729.3	7458.6
H	30.868	61.735	123.47	246.94	493.88	987.77	1975.5	3951.1	7902.1

* At an international conference held in London in May 1939, the equal temperament was defined as a^1, and a proposal was made for this value to be used in all musical performances including solos, choruses, orchestral music, etc. However, $a^1 = 442$ Hz is mainly used in music today. The time signal broadcast by NHK is at 440 and 880 Hz. In some physics experiments, an equal temperament based on $c^1 = 2^8 = 256$ (also known as "physics tone") was used. For the sake of convenience, the latter expressed the frequency of the acoustic signal C in the power of two. The following is an explanation of sound symbols and musical notes used in the table above.

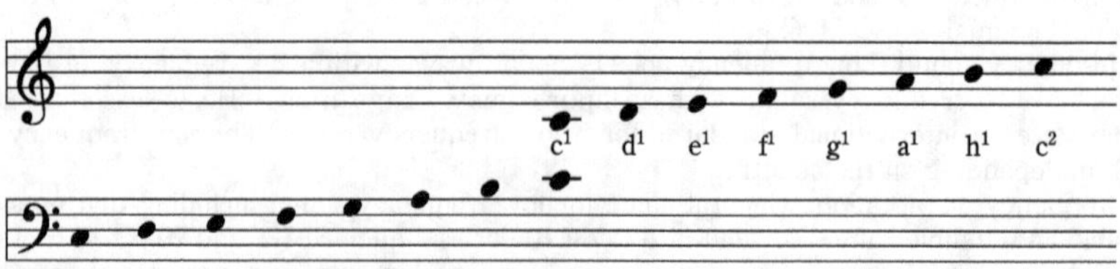

Range of Frequency for Singing and Musical Instruments

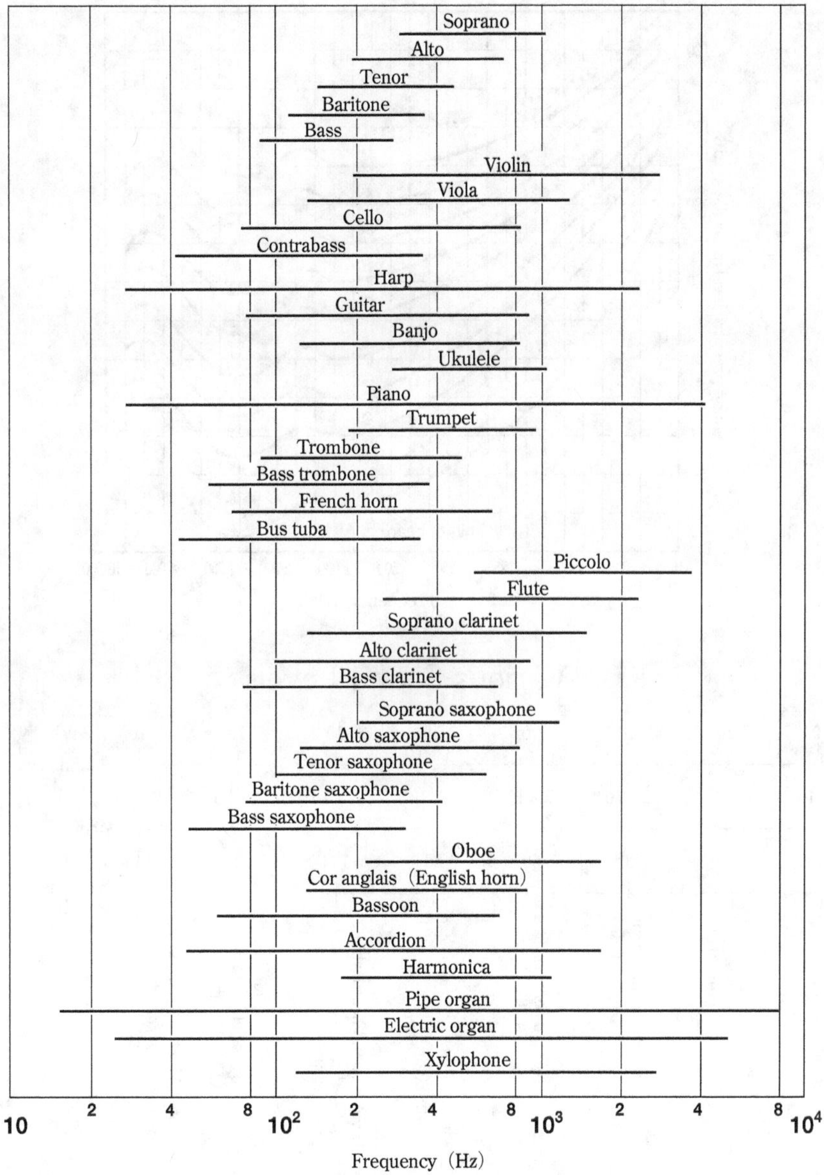

Equal Loudness Curve (ISO226 (2003))

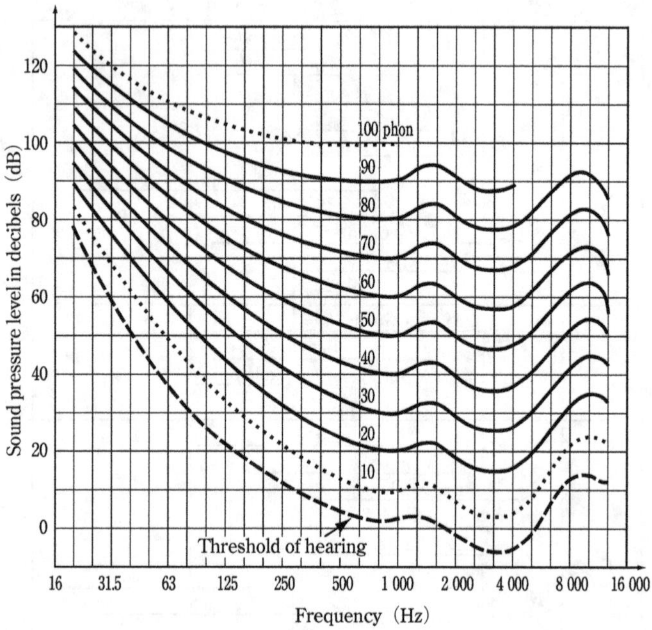

Various Sound Levels

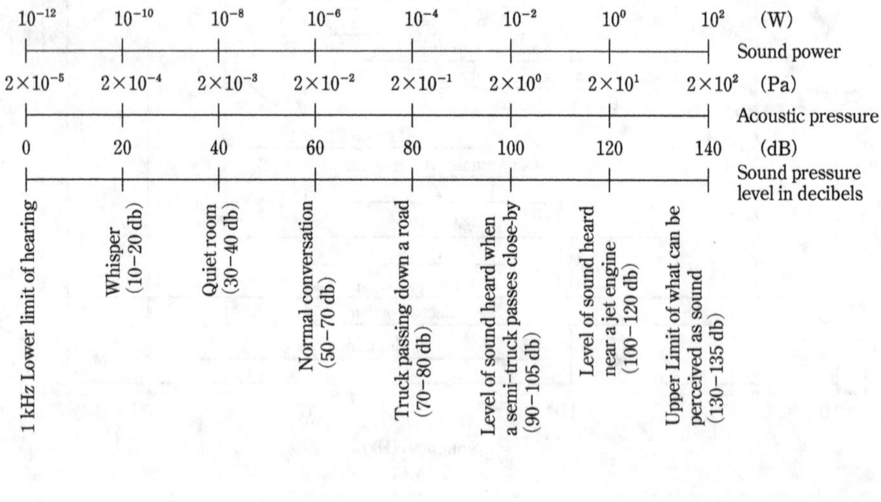

Light and Electromagnetic Waves

Wavelength and Frequency of Electromagnetic Waves

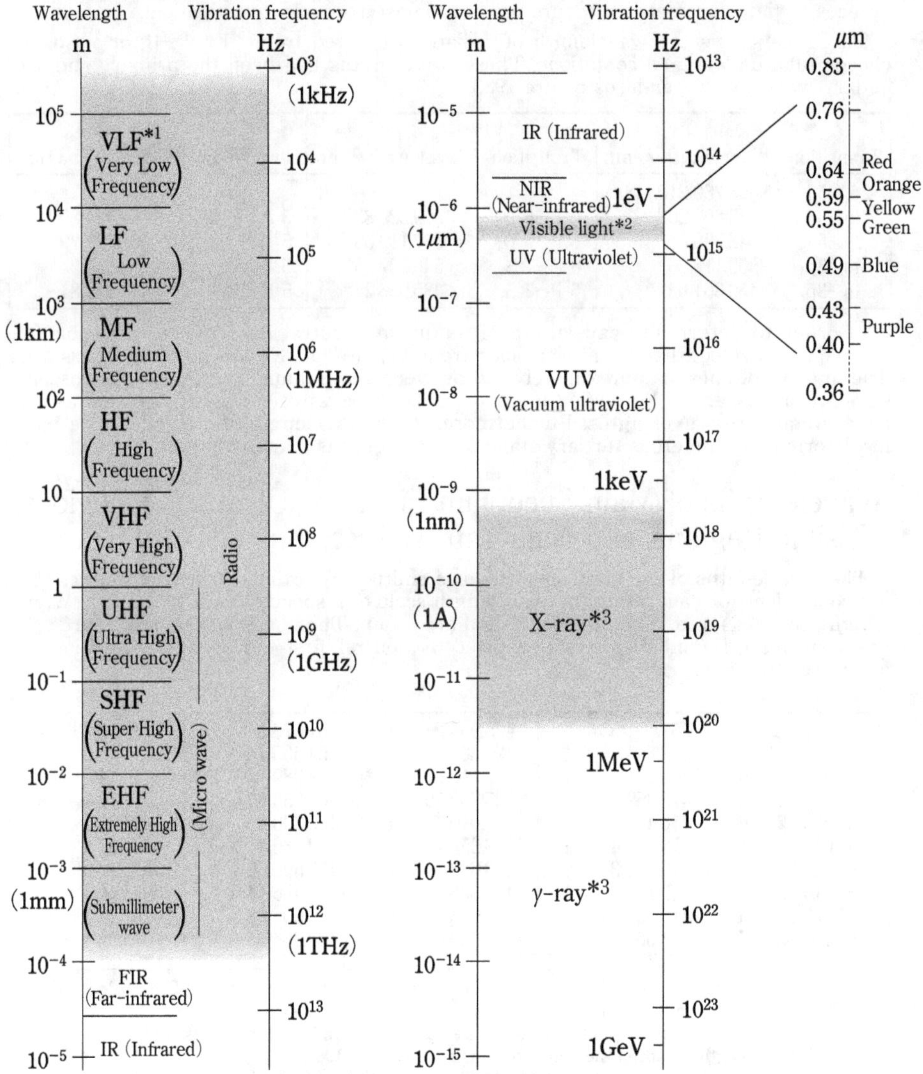

*1 The letter designations of the radio frequency bands are from the International Telecommunication Convention wireless rules.

*2 JIS Z 8120

*3 The distinction between X-rays and γ-rays is not based on wavelength or frequency : an X-ray is a radiation produced by electron state transistion; a γ-ray is a radiation produced by nuclear state transition.

Standard Wavelength[1)]

The meter, one of the SI base unit, is defined by the propagation time of light propagating in vacuum. However, for the practical realization (*mise en practique*, MeP) of the definition of the meter, several radiations are recommended by the CIPM as stable references for interferometric measurements and spectroscopic investigations[2)].

The table shows the wavelength of radiations emitted from ^{86}Kr, ^{198}Hg or ^{114}Cd discharge tube under fixed conditions. These wavelengths had been the primary and secondary wavelength standards before 1983.

^{86}Kr		^{198}Hg		^{114}Cd	
Transition	Wavelength λ/nm	Transition	Wavelength λ/nm	Transition	Wavelength λ/nm
$2p_{10}$-$5d_5$	605.780 2103 (8)[3)]				
$2p_9$-$5d'_4$	645.807 20[4)]	6^1P_1-6^1D_2	579.226 83[4)]	5^1P_1-5^1D_2	644.024 80[4)]
$2p_8$-$5d_4$	642.280 06[4)]	6^1P_1-6^3D_2	577.119 83[4)]	5^3P_2-6^3S_1	508.723 79[4)]
$1s_3$-$3p_{10}$	565.112 86[4)]	6^3P_2-7^3S_1	546.227 05[4)]	5^3P_1-6^3S_1	480.125 21[4)]
$1s_4$-$3p_8$	450.361 62[4)]	6^3P_1-7^3S_1	435.956 24[4)]	5^3P_0-6^3S_1	467.945 81[4)]

In order to determine wavelength of unknown spectral lines by the interpolation technique, wavelengths of emission lines from Fe, Ne and Kr are precisely measured. The number of lines is approximately 340 between 240 nm and 700 nm in spectroscopic standard air (refer to "Refraction index of air"). These wavelengths are defined as secondary standard wavelengths. Furthermore, these wavelengths are used as the basis for determining numerous tertiary standard wavelengths in spectroscopy.

Wavelengths of Main Spectrum Lines for Ultraviolet Regions, Visible Regions, and Near Infrared Regions (200–2200 nm)[6)]

The wavelengths of the main spectral lines emitted by ordinary spectral sources that are convenient for calibrating the wavelength scale of a spectrograph are shown. Wavelengths are measured in standard air[5)] and are in nm. The lines marked with * are relatively strong. I, II, and III are lines emitted by neutral, first-ionized, and second-ionized atoms, respectively.

(Units : nm)

Argon		357.661 53	II	425.936 2	I	451.073 3	I	501.716 26	II
		358.844 03	II	426.628 6	I	454.505 16	II	591.208 5	I
231.371 91	II	404.441 8	I	427.216 9	I	458.989 76*	II	603.212 7	I
231.629 73	II	415.859 0	I	427.752 79	II	460.956 69	II	641.630 7	I
294.289 32	II	416.418 0	I	430.010 1	I	465.790 09	II	667.728 2	I
348.050 20	III	418.188 4	I	433.356 1	I	472.686 81*	II	675.283 4	I
349.153 56	II	419.071 3	I	433.533 8	I	473.590 55	II	687.128 9	I
351.114 85	III	419.102 9	I	434.806 35	II	476.486 44	II	693.766 4	I
355.950 79	II	419.831 7	I	440.098 60	II	480.602 02	II	696.543 1	I
356.103 03	II	420.067 4	I	442.600 08	II	487.986 34	II	703.025 1	I

1) On the historical reasons, the title of this section is "Standard Wavelength". The SI does not use "Standard Wavelength" no longer.
2) http://www.bipm.org/en/publications/mises-en-pratique/standard-frequencies.html
3) The wavelength had been the definition of the meter until 1983.
4) Relative expanded uncertainty, $U = ku_c$ ($k=3$): ^{86}Kr: 2×10^{-8}, ^{198}Hg: 5×10^{-8}, ^{114}Cd: 7×10^{-8}.
5) The refractive index of air is refer to page **395**
6) A. Kramida, Yu. Ralchenko, J. Reader, and NIST ASD Team (2014). NIST Atomic Spectra Database (ver. 5.2), [Online]. http://physics.nist.gov/asd

Continued.

706.721 8	I	1 300.826 4	I	292.927	II	464.187 6	I	484.329 34	I
706.873 6	I	1 321.399 0	I	298.062 04*	I	464.237 3	I	484.433*	II
714.704 2	I	1 322.810 7	I	298.136 25	I	496.503 1	I	486.245	II
720.698 0	I	1 323.090	I	298.185 38	I	508.422 6	I	488.353	II
727.293 6	I	1 327.264	I	308.082 24	I	509.717 1	I	491.650 7	I
735.329 3	I	1 331.321 0	I	313.316 66	I	509.920 0	I	492.148	II
737.211 8	I	1 336.711 1	I	325.252 40	I	511.224 9	I	492.315 2	I
738.398 0	I	1 350.419 1	I	326.105 48	I	532.327 6	I	502.827 94	I
750.386 9*	I	1 362.265 9	I	340.365 21	I	533.968 8	I	508.062	II
751.465 2	I	1 367.855 0	I	346.619 96*	I	534.297 0	I	529.222	II
763.510 6*	I	1 371.857 7	I	346.765 47	I	535.957 4	I	531.387	II
772.376 1	I	1 409.364 0	I	361.050 77*	I	578.238 4	I	533.933	II
772.420 7	I	1 504.650	I	361.287 29	I	580.175 2	I	541.915*	II
794.817 6*	I	1 694.058 0	I	361.445 29	I	581.214 8	I	547.261	II
800.615 7*	I	2 061.623 0	I	413.477	II	583.188 7	I	582.389 0	I
801.478 6*	I			441.563*	II	691.108 4	I	582.480 0	I
810.369 3*	I	**Cadmium**		467.814 93	I 1)	693.628 4	I	597.646	II
810.369 31	II	200.060	III	479.991 23	I 1)	693.876 7*	I	603.620	II
811.531 1*	I	203.983	III	508.582 17*	I 1)	696.467 2	I	605.115	II
826.452 2	I	204.561	III	533.748*	II	766.489 913*	I	609.759	II
840.821 0	I	208.791	III	537.813*	II	769.896 456*	I	617.830 3	I
842.464 8*	I	211.160	III	538.189	II	850.345	I	617.966 5	I
852.144 2	I	214.441*	II	609.914 21	I	850.511	I	618.242 0	I
866.794 4	I	219.456*	II	611.149 5	I	890.219	I	631.806 2	I
884.991	I	226.502*	II	632.516 61	I	890.402	I	635.635	II
912.296 7*	I	228.802 25*	I	635.472	II	959.570	I	646.970 5	I
919.463 8	I	231.277*	II	635.998	II	959.783	I	647.284 1	I
922.449 9	I	232.107	II	643.846 95*	I 1)	1 101.987	I	648.776 5	I
929.153 1	I	249.981	III	646.494	II	1 102.267	I	650.418 0	I
935.422 0	I	254.461 30	I	672.578	II	1 169.021 9	I	659.501	II
965.778 6*	I	257.293	II	734.567 04*	I	1 176.963 7	I	666.892 0	I
978.450 3	I	258.010 62	I	1 571.2	I	1 177.283 8	I	672.800 8	I
1 005.206	I	262.897 86	I	1 912.5	I	1 243.227 4*	I	680.574	II
1 047.005 4	I	263.941 96	I			1 252.214 1*	I	682.731 5	I
1 050.650	I	266.032 53	I	**Potassium**				688.215 5	I
1 067.356 5	I	267.753 97	I	203.540 8	II	**Xenon**		699.088*	II
1 148.810 9	I	271.250 49	I	203.700 5	II	271.735 0	III	711.959 8	I
1 166.871 0	I	273.381 99	I	204.528 3	II	378.100	III	716.483	II
1 211.232 6	I	274.854*	II	207.679 8	II	395.092 4	I	739.379 3	I
1 213.973 8	I	276.389 35	I	207.869 4	II	396.754 11	I	746.082	III
1 234.339 3	I	276.423 05	I	208.493 3	II	418.010	II	758.468 0	I
1 240.282 7	I	276.699	III	208.985 5	II	419.352 8	I	764.202 4	I
1 243.932 1	I	277.495 85	I	212.919 0	II	433.052	II	788.739 3	I
1 245.612	I	280.559	III	213.068 7	II	446.219	II	796.734 2	I
1 248.766 3	I	283.689 99	I	214.469 9	II	460.303	II	805.725 8	I
1 270.228 1	I	286.817 99	I	344.637 6	I	473.415 18	I	806.133 9	I
1 280.273 9	I	288.076 70	I	344.737 6	I	479.261 9	I	820.633 6	I
1 293.319 5	I	288.122 45	I	404.413 6	I	480.701 90	I	823.163 36*	I
1 295.665 9	I	291.467	II	404.720 8	I	482.970 8	I	826.652 0	I

1) The wavelength of this line in vacuum is the standard wavelength (meter present wavelength) (refer to page **374**).

Wavelengths of main spectrum lines — Continued.

828.011 62*	I	431.855 24	I	821.840	I	1 818.507	I	410.805 4	I
834.682 17*	I	431.957 95*	I	826.324 26	I	1 858.089	I	412.207	III
840.918 94*	I	435.135 969	I	828.105 22	I	1 869.630 6	I	421.674	III
857.601 0	I	435.547 73*	II	829.810 99	I	1 878.548	I	434.749 45	I
864.854 0	I	436.264 157	I	850.887 28	I	1 879.771	I	435.833 50*	I [1)]
869.686 0	I	437.612 159	I	877.675 05	I	2 020.989	I	496.010	I
873.937 2	I	439.996 634	I	892.869 33	I	2 042.399	I	497.357	III
881.941 06*	I	442.519 007	I	935.223	I	2 116.552	I	512.063 7	I
886.232 0	I	443.681 22	II	985.624	I	2 190.253 23*	I	535.403 4	I
890.873 0	I	445.391 749	I	1 022.152*	II	**Mercury**		542.525 3*	II
893.083 0	I	446.369 000	I	1 029.693	I			546.075 00*	I [1)]
895.225 09	I	447.501 41	II	1 036.037	I	234.543	I	567.581	I
898.757 0	I	450.235 427	I [1)]	1 087.490	I	237.832 4	I	567.710 5*	II
904.544 66	I	457.720 87	II	1 118.711	I	239.935	I	576.961 00	I [1)]
916.265 20	I	461.916 58*	II	1 125.770 4	I	248.200 1	I	579.067 00	I [1)]
951.337 7	I	463.388 50	II	1 145.747 7	I	248.271 6	I	580.378 2	I
968.532	I	465.887 61*	II	1 179.241	I	248.381 5	I	585.925 4	I
979.969 7*	I	473.900 19*	II	1 181.937 85*	I	253.477 2	I	587.127 9*	II
992.319 8*	I	476.574 41*	II	1 199.710 2	I	253.652 10*	I	588.893 9*	II
1 052.785 7	I	483.207 73	II	1 207.721	I	257.628 5	I	614.643 5*	II
1 070.678	I	484.661 15	II	1 211.779	I	265.203 9	I	614.947 50*	II
1 083.834	I	556.222 534	I	1 220.453 57	I	265.369 0*	I	671.634	I
1 089.532 4	I	557.028 944*	I	1 278.252	I	265.513 4	I	690.746	I
1 108.523 7	I	558.038 729	I	1 278.252	I	269.882 8	I	708.190 1	I
1 112.718 9	I	564.956 177	I [1)]	1 287.874	I	269.937 6	I	709.186 0	I
1 114.114 5	I	583.285 661	I	1 317.741 10	I	272.443	III	794.455 5*	II
1 174.223 6	I	587.091 599*	I	1 362.241 64	I	275.277 7	I	1 013.975	I
1 365.648	I	599.385 020	I	1 363.422 06	I	280.346 6	I	1 128.71	I
1 473.238	I	605.612 628	I [1)]	1 442.679 27	I	284.767 50*	II	1 320.995	I
1 555.71	I	642.102 700	I [1)]	1 473.444 15	I	289.360 10	I	1 342.657	I
1 597.95	I	645.628 888	I [1)]	1 496.189 1	I	292.541 4	I	1 346.838	I
1 655.45	I	669.922 960	I	1 523.962 23	I	296.728 30	I	1 350.558	I
1 672.815 8*	I	690.467 87	I	1 533.496 72	I	302.150 40	I	1 357.021	I
1 732.579 8	I	722.410 3	I	1 537.204 1	I	302.347 1	I	1 367.351	I
1 878.814 6	I	728.726 2	I	1 547.403 3	I	302.749 0	I	1 395.055	I
2 026.224 3*	I	742.554	I	1 582.009 0	I	312.567 40	I	1 529.582	I
2 147.010 1	I	748.686 2	I	1 672.653	I	313.155 50	I	1 688.148	I
Krypton		758.741 36*	I	1 678.513 29	I	313.184 40	I	1 692.016	I
		760.154 57*	I	1 685.349 81	I	334.148 40	I	1 707.279	I
324.569	III	768.524 59	I	1 689.045 38	I	354.346	I	1 732.941	I
332.575	III	769.454 01	I	1 689.676 47	I	365.015 80	I	1 813.038	I
350.742	III	785.482 33	I	1 693.581 34	I	365.484 20	I	1 970.017	I
366.532 54	I	805.950 48	I	1 709.877 93	I	366.288 70	I	**Hydrogen**	
367.956 09	I	810.402	I	1 736.761 40	I	366.328 40	I		
367.961 11	I	810.436 60*	I	1 784.274 4	I	379.000	I	379.790 9	I [2)]
427.396 943*	I	811.290 12	I	1 800.223 00*	I	398.393 12*	II	383.539 7	I [2)]
428.296 734	I	813.298	I	1 809.944	I	404.656 50*	I	388.906 4	I [2)]
429.292 33	II	819.005 66	I	1 816.732 73	I	407.783 70	I	397.007 5	I [2)]

1) The wavelength of this line in vacuum is the standard wavelength (wavelength in meters) (refer to page **374**).

2) Balmer series

Continued.

410.173 4	I [2]	418.022	I	819.482 37	I	585.248 78	I	953.416 40	I
434.047 2	I [2]	418.55	I	864.992	I	588.189 50	I	966.542 00	I
486.128 694 9	I [2]	419.598	I	865.089	I	597.553 43	I	1 056.240 89	I
486.129 776 1	I [2]	420.216	I	894.296	I	602.999 68	I	1 079.804 30	I
656.270 970	I [2]	420.49	I	915.388	I	607.433 76	I	1 084.447 74	I
656.272 483	I [2]	421.613	I	946.594	I	614.306 27	I	1 114.302 00	I
656.277 153	I [2]	422.381	I	996.128	I	616.359 37	I	1 117.752 46	I
656.285 175	I [2]	424.208 2	I	1 056.600	I	621.728 12	I	1 139.043 33	I
1 004.98	I [3]	425.252 0	I	1 057.228	I	626.649 52	I	1 140.913 38	I
1 093.817	I [3]	425.38	I	1 074.644	I	633.442 76	I	1 152.274 50	I
1 281.807 2	I [3]	427.364 2	I	1 074.929	I	638.299 14	I	1 152.502 03	I
1 875.13	I [3]	427.39	I	1 083.487	I	640.224 80	I	1 153.634 46	I
Sodium		427.678 7	I	1 119.019	I	650.652 77	I	1 161.408 05	I
		428.67	I	1 119.721	I	659.895 28	I	1 176.679 29	I
223.032 8	III	429.100 6	I	1 138.145	I	692.946 72*	I	1 178.904 44	I
224.670 6	III	432.140 0	I	1 140.378	I	702.405 00	I	1 198.491 39	I
245.930 6	III	432.461 5	I	1 267.917	I	703.241 28*	I	1 206.633 43	I
247.473 2	III	434.148 9	I	1 476.754	I	705.910 79	I	1 268.920 32	I
249.702 2	III	434.473 6	I	1 477.975	I	717.393 80*	I	1 291.201 41	I
268.033 1	I	439.002 9	I	1 637.387	I	724.516 65*	I	1 742.006 18	II
268.042 1	I	439.334 0	I	1 846.539	I	743.889 81	I	1 764.366 08	II
285.281 1	I	441.988 5	I	**Neon**		748.887 12	I	1 827.664 15	I
285.301 3	I	442.324 6	I			753.577 39	I	1 828.261 40	I
313.548 3*	II	443.23	I	258.999 71	III	754.404 39	I	1 830.396 74	I
314.928 3*	II	449.417 7	I	259.355 52	III	794.318 05	I	1 838.482 56	I
316.373 6*	II	449.765 8	I	267.790 47	III	813.640 61	I	1 838.993 66	I
330.236 9	I	454.163 3	I	277.762 88	III	826.607 69	I	1 842.240 16	I
330.297 9	I	454.518 6	I	286.671 87	III	830.032 48	I	1 859.154 1	I
341.62	I	466.481 07	I	331.972 2	II	837.635 90	I	1 859.769 8	I
342.686 2	I	466.855 95	I	332.373 505	II	837.760 70*	I	2 098.614 46	II
342.73	I	474.794 10	I	334.545 437	II	841.842 65	I	**Helium**	
348.90	I	475.182 18	I	336.990 69	I	849.535 91	I		
350.25	I	497.854 14	I	337.821 931	II	859.125 83	I	257.76	I
351.10	I	498.281 34	I	339.280 061	II	863.464 72	I	281.82	I
384.80	I	507.12	I	352.047 14	I	864.704 12	I	294.510 6	I
386.55	I	514.883 81	I	453.775 45	I	865.438 28	I	301.37	I
387.29	I	515.340 24	I	470.439 49	I	865.552 20	I	318.774 5	I
388.18	I	516.25	I	470.885 94	I	867.949 36	I	381.960 74	I
388.57	I	525.64	I	471.006 50	I	868.192 16	I	388.864 8*	I
390.04	I	562.10	I	471.206 33	I	877.165 75	I	396.472 91	I
391.79	I	568.263 33	I	471.534 40	I	878.062 23	I	402.619 14	I
392.56	I	568.819 34	I	478.892 58	I	878.375 39	I	412.081 54	I
393.06	I	568.820 46	I	482.733 80	I	885.386 69	I	438.792 96	I
394.26	I	574.42	I	488.491 70	I	886.575 62	I	447.148 02	I
398.03	I	588.995 095*	I [4]	495.703 35	I	891.950 07	I	471.314 57	I
399.77	I	589.592 424*	I [5]	533.077 75	I	914.867 20	I	492.193 13	I
400.88	I	615.422 53	I	534.109 38	I	920.175 88	I	501.567 83	I
415.740	I	616.074 70	I	534.328 34	I	922.005 98	I	504.773 8	I
416.767	I	818.325 56	I	540.056 16	I	930.085 32	I	587.562 1*	I
417.67	I	819.479 05	I	576.441 88	I	932.650 72	I	667.815 1	I

2) Balmer series, 3) Paschen series, 4) D_2, 5) D_1

Wavelengths of main spectrum lines Continued.

706.519 0	I	307.571 94	I	368.305 45	I	387.857 30*	I	446.655 15	I
706.571	I	309.996 79	I	368.410 74	I	387.867 07	I	452.861 39	I
728.134 9	I	319.322 57	I	368.745 64	I	388.628 20	I	487.131 79	I
1 091.292	I	322.206 69	I	370.556 58*	I	388.704 80	I	489.075 48	I
1 196.912	I	322.578 70	I	370.782 18	I	388.851 32	I	489.149 21	I
1 252.752	I	323.622 21	I	370.791 96	I	389.565 62	I	491.899 37	I
1 278.479	I	328.675 28	I	370.924 61	I	389.788 96	I	492.050 28	I
1 278.499	I	340.745 94	I	372.256 27*	I	389.970 71*	I	495.729 83	I
1 279.057	I	341.313 21	I	372.437 67	I	390.294 55	I	495.759 65	I
1 296.845	I	342.711 93	I	372.761 88	I	390.647 95	I	500.611 88	I
1 508.364	I	344.060 57*	I	373.239 61	I	392.025 78	I	501.206 81	I
1 700.247	I	344.098 85*	I	373.331 73*	I	392.291 15*	I	504.175 57	I
1 868.534*	I	344.387 63	I	374.336 19	I	392.792 0	I	511.041 28	I
1 869.723	I	344.514 89	I	374.556 10*	I	393.029 64*	I	513.946 25	I
1 908.938	I	346.586 04	I	374.589 93*	I	395.667 7	I	516.748 81	I
1 954.308	I	347.545 00*	I	374.826 19*	I	396.925 70	I	517.159 60	I
2 058.130*	I	347.670 16	I	374.948 51*	I	399.739 19	I	519.234 39	I
2 112.007	I	349.057 38*	I	375.823 27	I	400.524 17	I	519.494 14	I
2 112.143	I	349.784 03	I	376.004 94	I	404.581 22*	I	521.627 37	I
2 113.203	I	351.381 77	I	376.378 88	I	406.359 39	I	522.718 91	I
		352.126 1	I	376.553 87	I	407.173 77	I	523.294 00	I
Iron		352.604 06	I	376.719 15	I	411.854 47	I	526.655 50	I
234.349 480	II	352.616 56	I	378.594 81	I	413.205 79	I	526.953 70*	I
238.203 733	II	354.108 31	I	378.667 66	I	414.341 43	I	527.035 60	I
239.562 504	II	354.207 54	I	378.787 99	I	414.386 78	I	532.417 87	I
240.488 550	II	355.492 43	I	379.009 27	I	418.175 44	I	532.803 83	I
259.939 515	II	355.851 48	I	379.433 95	I	418.703 87	I	532.853 13	I
273.954 721	II	356.537 87	I	379.500 19	I	418.779 52	I	534.102 37	I
275.573 620	II	357.025 39	I	379.751 46	I	419.909 5	I	537.148 93	I
293.690 32	I	358.465 92	I	379.851 11	I	420.202 89	I	539.712 76	I
294.787 58	I	358.531 87	I	379.954 73	I	421.618 35	I	540.577 49	I
295.393 98	I	358.570 51	I	380.534 2	I	422.742 63	I	542.969 64	I
295.736 43	I	358.611 24	I	381.296 43	I	423.360 25	I	543.452 35	I
296.689 84	I	358.698 45	I	381.584 00	I	423.593 67	I	544.691 64	I
297.009 92	I	360.320 33	I	382.117 76	I	425.011 92	I	545.560 92	I
297.010 58	I	360.545 25	I	382.444 35*	I	425.078 66	I	633.533 04	I
297.313 22	I	360.667 91	I	382.588 09	I	426.047 41	I	633.682 39	I
297.323 52	I	360.885 91	I	382.782 24	I	427.115 35	I	639.360 09	I
298.144 48	I	361.015 88	I	383.422 22	I	427.176 02	I	640.000 08	I
299.951 16	I	361.876 77	I	383.925 56	I	428.240 26	I	640.801 79	I
300.094 76	I	362.146 04	I	384.043 72	I	429.412 45	I	641.164 89	I
300.728 22	I	362.200 31	I	384.104 77	I	429.923 5	I	642.135 04	I
300.813 80	I	362.318 58	I	384.325 65	I	430.790 20	I	643.084 60	I
300.956 91	I	363.146 29*	I	384.996 64	I	431.508 43	I	649.498 00	I
302.049 05	I	364.038 83	I	385.081 76	I	432.576 16	I	654.623 90	I
302.403 24	I	364.784 3	I	385.637 13*	I	437.592 98	I	656.921 51	I
302.584 22	I	364.950 61	I	385.921 23	I	438.354 47	I	659.291 34	I
303.738 85	I	365.146 67	I	386.552 28	I	440.475 01	I	659.387 01	I
305.744 56	I	366.952 09	I	387.250 09	I	441.512 22	I	666.344 17	I
305.908 56	I	367.762 74	I	387.376 03	I	442.730 96	I	667.798 65	I
306.724 37	I	367.991 31	I	387.801 80	I	446.165 25	I	675.015 20	I

Wavelengths (in Vacuum) of Spectral Lines in Vacuum Ultraviolet Regions

The notes I, II, III and IV for the element symbols indicate spectrums emitted by neutral, primary ionized, secondary ionized, and tertiary ionized-atoms, respectively.

Wave-length λ/nm	Spectrum	Wave-length λ/nm	Spectrum	Wave-length λ/nm	Spectrum	Wave-length λ/nm	Spectrum
193.0900	C I	108.5701	N II	77.5965	N II	53.8312	C III
174.5249	N I	108.4580	N II	76.9152	Ar III	52.4189	Ar V
171.852	N IV	108.3990	N II	76.4357	N III	47.9372	Ar VII
165.8120	C I	102.5722	H, L_β, I	74.5322	Ar II	47.5656	Ar VII
165.7905	C I	99.1579	N III	74.4925	Ar II	47.3938	Ar VII
165.7377	C I	98.9790	N III	74.6270	Ar II	46.2007	Ar VI
165.7007	C I	97.7020	C III	73.0129	Ar II	45.9633	C III
165.6265	C I	93.2053	Ar II	72.3361	Ar II	45.9521	C III
156.1437	C I	92.3211	N IV	70.0278	Ar IV	44.5997	Ar V
156.0682	C I	91.9782	Ar II	68.9007	Ar IV	41.9525	C IV
155.077	C IV	91.6703	N II	68.7345	C II	41.9525	C IV
154.820	C IV	91.6015	N II	68.6335	N III	38.4178	C IV
149.4668	N I	90.1162	Ar IV	68.5816	N III	38.4032	C IV
149.2615	N I	90.0362	Ar IV	68.5513	N III	31.2453	C IV
133.5708	C II	88.7404	Ar III	68.4996	N III	31.2422	C IV
133.4532	C I	88.3179	Ar III	68.3278	Ar III	28.3579	N IV
132.9600	C I	87.9622	Ar III	67.9400	Ar II	28.3420	N IV
132.9577	C I	87.8728	Ar III	67.7951	Ar II	24.7710	N V
121.5670	H, L_α, I	87.6057	Ar I	66.0286	N II	24.7563	N V
120.0710	N I	87.5534	Ar III	64.5178	N II	20.9270	N V
120.0224	N I	87.1099	Ar III	64.1808	Ar III	4.0270	C V
119.9550	N I	85.0602	Ar IV	63.7282	Ar III	3.4973	C V
117.7694	N I	84.3772	Ar IV	59.6694	Ar VI	3.3736	C VI
117.6508	N I	84.0029	Ar IV	58.8921	Ar VI	2.8787	N VI
117.5711	C II	80.1913	Ar IV	55.5754	Ar VII	2.4898	N VI
113.4981	N I	80.1409	Ar IV	57.4381	C III	2.4781	N VII
113.4415	N I	80.1086	Ar IV	55.1378	Ar VI		
113.4166	N I	80.0573	Ar IV	54.9905	Ar VI		
				54.4731	Ar VI		

Based primarily on B. Edlén, Progress in Physics, 23 (1963).

Wavenumbers of Spectral Lines in Infrared Regions

The wavenumbers of spectral lines shown here are useful for the wavelength calibration of infrared spectrometers with a moderate resolution ($\Delta\tilde{\nu} = 0.5$ to 1.0 cm^{-1}). Values for argon are for atomic emission spectra in vacuum. All others are molecular absorption spectra.

(Units: cm^{-1})

Argon	3663.92	3191.52	2683.76	Indene	1856.9	Indene :	1871
4920.64	3654.46	3151.86	2637.00	3927.2	1797.7	Camphor :	1802
4849.22	3597.96	3142.14	2623.25	3901.6	1661.8	Cyclohexane	1601
4841.97	3545.80	3141.90	2608.84	3798.9	1609.8	(1 : 1 : 1)	1583
4821.77	3540.33	3102.19	2576.58	3660.6	1587.5	592.1	1542
4821.37	3530.85	3100.17	2566.67	3297.8	1553.2	551.7	1493
4803.83	3516.79	3095.41	2552.32	3139.5	1457.3	521.4	1452
4763.76	3508.07	3092.73	2542.60	3110.2	1393.5	490.2	1373
4686.32	3504.05	3040.57	2531.75	3025.4	1361.1	420.5	1328
4642.51	3494.03	3023.09	2521.16	2887.6	1312.4	393.1	1312
4536.06	3484.58	3016.73	2520.75	2673.3	1288.0	381.6	1181
4528.33	3482.77	3003.60	2515.80	2622.3	1226.2	301.4	1154
4321.61	3474.28	2860.37	2505.69	2598.4	1205.1		1069
4192.60	3467.04	2838.59	2499.49	2525.5	1166.1		1028
4171.35	3449.56	2838.39	2494.99	2305.1	1122.4		1003
3978.97	3435.42	2760.84	2494.25	2271.4	1067.7	Polystyrene	980
3922.40	3432.41	2753.88	2493.90	2258.7	1018.5	3103	964
3919.70	3417.30	2747.97	2489.74	2172.8	947.2	3082	942
3895.90	3415.22	2747.63	2489.44	2135.8	942.4	3060	907
3810.72	3382.23	2740.33	2473.22	2113.2	914.7	3027	841
3766.44	3356.07	2701.71	2469.70	2090.2	861.3	3000	757
3725.36	3282.77	2696.44	2445.51	1943.1	830.5	2924	699
3715.12	3273.02	2692.26		1915.3	765.3	2851	
3672.01	3226.20	2689.17		1885.1	692.6	1944	

The following references show the standard absorption lines used for wavelength calibration in more detail.

International Union of Pure and Applied Chemistry, Commission on Molecular Structure and Spectroscopy, A. R. H. Cole, Chapter: "Tables of Wavenumbers for the Calibration of Infrared Spectrometers", Pergamon Press, 1977.

Wavelength of Characteristic X-Rays

K series (1)

(Units: nm)

Element and atomic number	Characteristic X-rays and the strength ratio					
	$\alpha_{1,2}$	α_1	α_2	β_1	β_3	β_2
	150	100	50	15		5
Li 3	23.0					
Be 4	11.3					
B 5	6.7					
C 6	4.4					
N 7	3.1603					
O 8	2.3707					
F 9	1.8307					
Ne 10	1.4615			1.4460		
Na 11	1.1909			1.1574	1.1726	
Mg 12	0.9889			0.9559	0.9667	
Al 13	0.8339	0.8338	0.8341	0.7960	0.8059	
Si 14	0.7126	0.7125	0.7127	0.6778		
P 15	0.6155	0.6154	0.6157	0.5804		
S 16	0.5373	0.5372	0.5375	0.5032		
Cl 17	0.4729	0.4728	0.4731	0.4403		
Ar 18	0.4192	0.4191	0.4194	0.3886		
K 19	0.3744	0.3742	0.3745	0.3454		
Ca 20	0.3360	0.3359	0.3362	0.3089		
Sc 21	0.3032	0.3031	0.3034	0.2780		
Ti 22	0.2750	0.2749	0.2753	0.2514		
V 23	0.2505	0.2503	0.2507	0.2285		
Cr 24	0.2291	0.2290	0.2294	0.2085		
Mn 25	0.2103	0.2102	0.2105	0.1910		
Fe 26	0.1937	0.1936	0.1940	0.1757		
Co 27	0.1791	0.1789	0.1793	0.1621		
Ni 28	0.1659	0.1658	0.1661	0.1500		0.1489
Cu 29	0.1542	0.1540	0.1544	0.1392	0.1393	0.1381
Zn 30	0.1437	0.1435	0.1439	0.1296		0.1284
Ga 31	0.1341	0.1340	0.1344	0.1207	0.1208	0.1196
Ge 32	0.1256	0.1255	0.1258	0.1129	0.1129	0.1117
As 33	0.1177	0.1175	0.1179	0.1057	0.1058	0.1045
Se 34	0.1106	0.1105	0.1109	0.0992	0.0993	0.0980
Br 35	0.1041	0.1040	0.1044	0.0933	0.0933	0.0921
Kr 36	0.0981	0.0980	0.0984	0.0879	0.0879	0.0866
Rb 37	0.0927	0.0926	0.0930	0.0829	0.0830	0.0817
Sr 38	0.0877	0.0875	0.0880	0.0783	0.0784	0.0771
Y 39	0.0831	0.0829	0.0833	0.0740	0.0741	0.0728
Zr 40	0.0788	0.0786	0.0791	0.0701	0.0702	0.0690
Nb 41	0.0748	0.0747	0.0751	0.0665	0.0666	0.0654
Mo 42	0.0710	0.0709	0.0713	0.0632	0.0633	0.0621

K series (2)

(Units: nm)

Element and atomic number	Characteristic X-rays and the strength ratio					
	$\alpha_{1,2}$	α_1	α_2	β_1	β_3	β_2
	150	100	50	15		5
Tc 43	0.0676	0.0675	0.0679	0.0601		0.0590
Ru 44	0.0644	0.0643	0.0647	0.0572	0.0573	0.0562
Rh 45	0.0614	0.0613	0.0617	0.0546	0.0546	0.0535
Pd 46	0.0587	0.0585	0.0590	0.0521	0.0521	0.0510
Ag 47	0.0561	0.0559	0.0564	0.0497	0.0498	0.0487
Cd 48	0.0536	0.0535	0.0539	0.0475	0.0476	0.0465
In 49	0.0514	0.0512	0.0517	0.0455	0.0455	0.0445
Sn 50	0.0492	0.0491	0.0495	0.0435	0.0436	0.0426
Sb 51	0.0472	0.0470	0.0475	0.0417	0.0418	0.0408
Te 52	0.0453	0.0451	0.0456	0.0400	0.0401	0.0391
I 53	0.0435	0.0433	0.0438	0.0384	0.0385	0.0376
Xe 54	0.0418	0.0416	0.0421	0.0369		0.0360
Cs 55	0.0402	0.0401	0.0405	0.0355	0.0355	0.0346
Ba 56	0.0387	0.0385	0.0390	0.0341	0.0342	0.0333
La 57	0.0373	0.0371	0.0376	0.0328	0.0329	0.0320
Ce 58	0.0359	0.0357	0.0362	0.0316	0.0317	0.0309
Pr 59	0.0346	0.0344	0.0349	0.0305	0.0305	0.0297
Nd 60	0.0334	0.0332	0.0337	0.0294	0.0294	0.0287
Pm 61	0.0322	0.0321	0.0325	0.0283		
Sm 62	0.0311	0.0309	0.0314	0.0274	0.0274	0.0267
Eu 63	0.0301	0.0299	0.0304	0.0264	0.0265	0.0258
Gd 64	0.0291	0.0289	0.0294	0.0255	0.0256	0.0249
Tb 65	0.0281	0.0279	0.0284	0.0246	0.0246	0.0239
Dy 66	0.0272	0.0270	0.0275	0.0237	0.0238	0.0231
Ho 67	0.0263	0.0261	0.0266			
Er 68	0.0255	0.0253	0.0258	0.0222	0.0223	0.0217
Tm 69	0.0246	0.0244	0.0250	0.0215	0.0216	
Yb 70	0.0238	0.0236	0.0241	0.0208	0.0209	0.0203
Lu 71	0.0231	0.0229	0.0234	0.0202	0.0203	0.0197
Hf 72	0.0224	0.0222	0.0227	0.0195	0.0196	0.0190
Ta 73	0.0217	0.0215	0.0220	0.0190	0.0191	0.0185
W 74	0.0211	0.0209	0.0213	0.0184	0.0185	0.0179
Re 75	0.0204	0.0202	0.0207	0.0179	0.0179	0.0174
Os 76	0.0198	0.0196	0.0201	0.0173	0.0174	0.0169
Ir 77	0.0193	0.0191	0.0196	0.0168	0.0169	0.0164
Pt 78	0.0187	0.0185	0.0190	0.0163	0.0164	0.0159
Au 79	0.0182	0.0180	0.0185	0.0159	0.0160	0.0155
Hg 80	0.0177	0.0175	0.0180	0.0154	0.0155	0.0150
Tl 81	0.0172	0.0170	0.0175	0.0150	0.0151	0.0146
Pb 82	0.0167	0.0165	0.0170	0.0146	0.0147	0.0147
Bi 83	0.0162	0.0161	0.0165	0.0142	0.0143	0.0138
Po 84	0.0158	0.0156	0.0161	0.0138		0.0133

K series (3)

(Units: nm)

Element and atomic number	Characteristic X-rays and the strength ratio					
	$\alpha_{1,2}$	α_1	α_2	β_1	β_3	β_2
	150	100	50	15		5
At 85						
Rn 86						
Fr 87						
Ra 88						
Ac 89						
Th 90	0.0135	0.0133	0.0138	0.0117	0.0118	0.0114
Pa 91						
U 92	0.0128	0.0126	0.0131	0.0111	0.0112	0.0108

L series (1)

(Units: nm)

Element and atomic number	Characteristic X-rays and the strength ratio					Element and atomic number	Characteristic X-rays and the strength ratio				
	α	α_1	α_2	β_1	β_2		α_1	α_2	β_1	β_2	γ_1
	110	100	10	50	20		100	10	50	20	10
Na 11						Y 39	0.6449	0.6456	0.6211		
Mg 12						Zr 40	0.6070	0.6077	0.5836	0.5586	0.5384
Al 13						Nb 41	0.5725	0.5732	0.5492	0.5238	0.5036
Si 14						Mo 42	0.5406	0.5414	0.5176	0.4923	0.4726
P 15						Tc 43					
S 16						Ru 44	0.4846	0.4854	0.4620	0.4372	0.4182
Cl 17						Rh 45	0.4597	0.4605	0.4374	0.4130	0.3944
Ar 18						Pd 46	0.4368	0.4376	0.4146	0.3909	0.3725
K 19						Ag 47	0.4154	0.4162	0.3935	0.3703	0.3523
Ca 20	3.6393			3.6022		Cd 48	0.3956	0.3965	0.3739	0.3514	0.3336
Sc 21	3.1393			3.1072		In 49	0.3752	0.3781	0.3555	0.3339	0.3162
Ti 22	2.7445			2.7074		Sn 50	0.3600	0.3609	0.3385	0.3175	0.3001
V 23	2.4309			2.3898		Sb 51	0.3439	0.3448	0.3226	0.3023	0.2852
Cr 24	2.1713			2.1323		Te 52	0.3290	0.3299	0.3077	0.2882	0.2712
Mn 25	1.9489			1.9158		I 53	0.3148	0.3157	0.2937	0.2751	0.2582
Fe 26	1.7602			1.7290		Xe 54					
Co 27	1.6000			1.5698		Cs 55	0.2892	0.2902	0.2683	0.2511	0.2348
Ni 28	1.4595			1.4308		Ba 56	0.2776	0.2785	0.2567	0.2404	0.2242
Cu 29	1.3357			1.3079		La 57	0.2665	0.2674	0.2458	0.2303	0.2141
Zn 30		1.2282		1.2009		Ce 58	0.2561	0.2570	0.2356	0.2208	0.2048
Ga 31		1.1313		1.1045		Pr 59	0.2463	0.2473	0.2259	0.2119	0.1961
Ge 32		1.0456		1.0194		Nd 60	0.2370	0.2382	0.2166	0.2035	0.1878
As 33		0.9671		0.9414		Pm 61	0.2283		0.2081		
Se 34		0.8990		0.8735		Sm 62	0.2199	0.2210	0.1998	0.1882	0.1726
Br 35		0.8375		0.8126		Eu 63	0.2120	0.2131	0.1920	0.1812	0.1657
Kr 36						Gd 64	0.2046	0.2057	0.1847	0.1746	0.1592
Rb 37		0.7318	0.7325	0.7075		Tb 65	0.1976	0.1986	0.1777	0.1682	0.1530
Sr 38		0.6863	0.6870	0.6623		Dy 66	0.1909	0.1920	0.1710	0.1623	0.1473

L series (2)

(Units: nm)

Element and atomic number	Characteristic X-rays and the strength ratio					Element and atomic number	Characteristic X-rays and the strength ratio				
	α_1	α_2	β_1	β_2	γ_1		α_1	α_2	β_1	β_2	γ_1
	100	10	50	20	10		100	10	50	20	10
Ho 67	0.1845	0.1856	0.1647	0.1567	0.1417	Pb 82	0.1175	0.1186	0.0982	0.0983	0.0840
Er 68	0.1785	0.1796	0.1587	0.1514	0.1364	Bi 83	0.1144	0.1155	0.0952	0.0955	0.0814
Tm 69	0.1726	0.1738	0.1530	0.1463	0.1316	Po 84	0.1114	0.1126	0.0921	0.0929	0.0786
Yb 70	0.1672	0.1682	0.1476	0.1416	0.1268	At 85					
Lu 71	0.1619	0.1630	0.1424	0.1370	0.1222	Rn 86					
Hf 72	0.1569	0.1580	0.1374	0.1327	0.1179	Fr 87	0.1030		0.0840	0.0858	0.0716
Ta 73	0.1522	0.1533	0.1327	0.1285	0.1138	Ra 88	0.1005	0.1017	0.0814	0.0836	0.0694
W 74	0.1476	0.1487	0.1282	0.1245	0.1098	Ac 89					
Re 75	0.1433	0.1444	0.1238	0.1206	0.1061	Th 90	0.0956	0.0968	0.0766	0.0794	0.0653
Os 76	0.1391	0.1402	0.1197	0.1169	0.1025	Pa 91	0.0933	0.0945	0.0742	0.0774	0.0634
Ir 77	0.1352	0.1363	0.1158	0.1135	0.0991	U 92	0.0911	0.0923	0.0720	0.0755	0.0615
Pt 78	0.1313	0.1325	0.1120	0.1102	0.0958	Np 93	0.0890	0.0901	0.0698	0.0735	0.0597
Au 79	0.1277	0.1288	0.1083	0.1070	0.0927	Pu 94	0.0868	0.0880	0.0678	0.0719	0.0579
Hg 80	0.1242	0.1253	0.1049	0.1040	0.0897	Am 95	0.0849	0.0860	0.0658	0.0701	0.0562
Tl 81	0.1207	0.1218	0.1015	0.1010	0.0868						

M series

(Units: nm)

Element and atomic number	Characteristic X-rays and the strength ratio					Element and atomic number	Characteristic X-rays and the strength ratio				
	α_1	α_2	β	γ	1		α_1	α_2	β	γ	1
K 19					68.0	Dy 66	band		0.9364	0.8144	1.243
Cu 29					17.0	Ho 67	band		0.8965	0.7865	1.186
Ru 44				2.685		Er 68	band		0.8593	0.7545	1.137
Rh 45				2.500		Tm 69	0.8460		0.8246		
Pd 46						Yb 70	0.8139	0.8155	0.7909	0.7023	1.048
Ag 47				2.180		Lu 71	0.7840		0.7600	0.6761	1.007
Cd 48				2.046		Hf 72	0.7539	0.7546	0.7304	0.6543	0.969
In 49						Ta 73	0.7251	0.7258	0.7022	0.6312	0.932
Sn 50				1.794		W 74	0.6983	0.6990	0.6756	0.6088	0.896
Sb 51				1.692		Re 75	0.6528		0.6504	0.5887	0.863
Te 52				1.593		Os 76	0.6490		0.6267	0.5681	
Ba 56				1.2700		Ir 77	0.6261	0.6275	0.6037	0.5501	0.802
La 57	1.488		1.451	1.2064		Pt 78	0.6046	0.6057	0.5828	0.5320	0.774
Ce 58	1.406		1.378	1.1534	1.838	Au 79	0.5840	0.5854	0.5623	0.5145	0.747
Pr 59				1.0997		Hg 80	0.5666		0.5452		
Nd 60	1.2675		band	1.0504		Tl 81	0.5461	0.5472	0.5250	0.4825	0.697
Pm 61						Pb 82	0.5285	0.5299	0.5075	0.4674	0.674
Sm 62	band		band	0.9599		Bi 83	0.5118	0.5129	0.4909	0.4531	0.652
Eu 63	band		1.0744	0.9211	1.422	Th 90	0.4138	0.4151	0.3942	0.3679	0.524
Gd 64	band		1.0253	0.8844	1.357	Pa 91	0.4022	0.4035	0.3827	0.3577	0.508
Tb 65	band		0.9792	0.8485	1.298	U 92	0.3910	0.3924	0.3715	0.3480	0.495

Source: Encyclopedia of X-Rays and gamma Rays. For more details, refer to ASTM DataSeries, DS37(X-Ray Emission Line Wavelength and Two-Theta Tables).

X-ray absorption edges (1)

(Unit: eV)

Element and atomic number	K	L III	L II	L I	M V	M IV	M III	M II	M I
H 1	13.6								
He 2	24.6								
Li 3	54.8								
Be 4	111.0								
B 5	188.0	4.7	4.7						
C 6	283.8	6.4	6.4						
N 7	401.6	9.2	9.2						
O 8	532.0	7.1	7.1	23.7					
F 9	685.4	8.6	8.6	31.0					
Ne 10	866.9	18.3	18.3	45.0					
Na 11	1072.1	31.1	31.1	63.3					
Mg 12	1305.0	51.4	51.4	89.4					
Al 13	1559.6	73.1	73.1	117.7					
Si 14	1838.9	99.2	99.2	148.7					
P 15	2145.5	132.2	132.2	189.3					
S 16	2472.0	164.8	164.8	229.2					
Cl 17	2822.4	200.0	201.6	270.2			6.8		17.5
Ar 18	3202.9	245.2	247.3	320.0			12.4		25.3
K 19	3607.4	293.6	296.3	377.1			17.8		33.9
Ca 20	4038.1	346.4	350.0	437.8			25.4		43.7
Sc 21	4492.8	402.2	406.7	500.4	6.6		32.3		53.8
Ti 22	4966.4	455.5	461.5	563.7	3.7		34.6		60.3
V 23	5465.1	512.9	520.5	628.2	2.2		37.8		66.5
Cr 24	5989.2	574.5	583.7	694.6	2.3		42.5		74.1
Mn 25	6539.0	640.3	651.4	769.0	3.3		48.6		83.9
Fe 26	7112.0	708.1	721.1	846.1	3.6		54.0		92.9
Co 27	7708.9	778.6	793.8	925.6	2.9		59.5		100.7
Ni 28	8332.8	854.7	871.9	1008.1	3.6		68.1		111.8
Cu 29	8978.9	931.1	951.0	1096.6	1.6		73.6		119.8
Zn 30	9658.6	1019.7	1042.8	1193.6	8.1		86.6		135.9
Ga 31	10367.1	1115.4	1142.3	1297.7	17.4		102.9	106.8	158.1
Ge 32	11103.1	1216.7	1247.8	1414.3	28.7		120.8	127.9	180.0
As 33	11866.7	1323.1	1358.6	1526.5	41.2		140.5	146.4	203.5
Se 34	12657.8	1435.8	1476.2	1653.9	56.7		161.9	168.2	231.5
Br 35	13473.7	1549.9	1596.0	1782.0	69.0	70.1	181.5	189.3	256.5
Kr 36	14325.6	1674.9	1727.2	1921.0	88.9		213.8	222.7	
Rb 37	15199.7	1804.4	1863.9	2065.1	110.3	111.8	238.5	247.4	322.1
Sr 38	16104.6	1939.6	2006.8	2216.3	133.1	135.0	269.1	279.8	357.5
Y 39	17038.4	2080.0	2155.5	2372.5	157.4	159.6	300.3	312.4	393.6
Zr 40	17997.6	2222.3	2306.7	2531.6	180.0	182.4	330.5	344.2	430.3
Nb 41	18985.6	2370.5	2464.7	2697.7	204.6	207.4	363.0	378.4	468.4
Mo 42	19999.5	2520.2	2625.1	2865.5	227.0	230.3	392.3	409.7	504.6
Tc 43	21044.0	2676.9	2793.2	3042.5	252.9	256.4	425.0	444.9	
Ru 44	22117.2	2837.9	2966.9	3224.0	279.4	283.6	460.6	482.8	585.0
RH 45	23219.9	3003.8	3146.1	3411.9	307.0	311.7	496.2	521.0	627.1
Pd 46	24350.3	3173.3	3330.3	3604.3	334.7	340.0	531.5	559.1	669.9

X-ray absorption edges (2)

(Unit: eV)

Element and atomic number	K	L III	L II	L I	M V	M IV	M III	M II	M I
Ag 47	25514.0	3351.1	3523.7	3805.8	366.7	372.8	571.4	602.4	717.5
Cd 48	26711.2	3537.5	3727.0	4018.0	403.7	410.5	616.5	650.7	770.2
In 49	27939.9	3730.1	3938.0	4237.5	443.1	450.8	664.3	702.2	825.6
Sn 50	29200.1	3928.8	4156.1	4464.7	484.8	493.3	714.4	756.4	883.8
Sb 51	30491.2	4132.2	4380.4	4698.3	527.5	536.9	765.6	811.9	943.7
Te 52	31813.8	4341.4	4612.0	4939.2	572.1	582.5	818.7	869.7	1006.0
I 53	33169.4	4557.1	4852.1	5188.1	619.4	631.3	874.6	930.5	1072.1
Xe 54	34561.4	4782.2	5103.7	5452.8	672.3		937.0	990.0	
Cs 55	35984.6	5011.9	5359.4	5714.3	725.5	739.5	997.6	1065.0	1217.1
Ba 56	37440.6	5247.0	5623.6	5988.8	780.7	796.1	1062.2	1136.7	1292.8
La 57	38924.6	5482.7	5890.6	6266.3	831.7	848.5	1123.4	1204.4	1361.3
Ce 58	40443.0	5723.4	6164.2	6548.8	883.3	901.3	1185.4	1272.8	1434.6
Pr 59	41990.6	5964.3	6440.4	6834.8	931.0	951.1	1242.2	1337.4	1511.0
Nd 60	43568.9	6207.9	6721.5	7126.0	977.7	999.9	1297.4	1402.8	1575.3
Pm 61	45184.0	6459.3	7012.8	7427.9	1026.9	1051.5	1356.9	1471.4	
Sm 62	46834.2	6716.2	7311.8	7736.8	1080.2	1106.0	1419.8	1540.7	1722.8
Eu 63	48519.0	6976.9	7617.1	8052.0	1130.9	1160.6	1480.6	1613.9	1800.0
Gd 64	50239.1	7242.8	7930.3	8375.6	1185.2	1217.2	1544.0	1688.3	1880.8
Tb 65	51995.7	7514.0	8251.6	8708.0	1241.2	1275.0	1611.3	1767.7	1967.5
Dy 66	53788.5	7790.1	8580.6	9045.8	1294.9	1332.5	1675.6	1841.8	2046.8
Ho 67	55617.7	8071.1	8917.8	9394.2	1351.4	1391.5	1741.2	1922.8	2128.3
Er 68	57485.5	8357.9	9264.3	9751.3	1409.3	1453.3	1811.8	2005.8	2206.5
Tm 69	59389.6	8648.0	9616.9	10115.7	1467.7	1514.6	1884.5	2089.8	2306.8
Yb 70	61332.3	8943.6	9978.2	10486.4	1527.8	1576.3	1949.8	2173.0	2398.1
Lu 71	63313.8	9244.1	10348.6	10870.4	1588.5	1639.4	2023.6	2263.5	2491.2
Hf 72	65350.8	9560.7	10739.4	11270.7	1661.7	1716.4	2107.6	2365.4	2600.9
Ta 73	67416.4	9881.1	11136.1	11681.5	1735.1	1793.2	2194.0	2468.7	2708.0
W 74	69525.0	10206.8	11544.0	12099.8	1809.2	1871.6	2281.0	2574.9	2819.6
Re 75	71676.4	10535.3	11958.7	12526.7	1882.9	1948.9	2367.3	2681.6	2931.7
Os 76	73870.8	10870.9	12385.0	12968.0	1960.1	2030.8	2457.2	2792.2	3048.5
Ir 77	76111.0	11215.2	12824.1	13418.5	2040.4	2116.1	2550.7	2908.7	3173.7
Pt 78	78394.8	11563.7	13272.6	13879.9	2121.6	2201.9	2645.4	3026.5	3296.0
Au 79	80724.9	11918.7	13733.6	14352.8	2205.7	2291.1	2743.0	3147.8	3424.9
Hg 80	83102.3	12283.9	14208.7	14839.3	2294.9	2384.9	2847.1	3278.5	3561.6
Tl 81	85530.4	12657.5	14697.9	15346.7	2389.3	2485.1	2956.6	3415.7	3704.1
Pb 82	88004.5	13035.2	15200.0	15860.8	2484.0	2585.6	3066.4	3554.2	3850.7
Bi 83	90525.9	13418.6	15711.1	16387.5	2579.6	2687.6	3176.9	3696.3	3999.1
Po 84	93105.0	13813.8	16244.3	16939.3	2683.0	2798.0	3301.9	3854.1	4149.4
At 85	95729.9	14213.5	16784.7	17493.0	2786.7	2908.7	3426.0	4008.0	4317.0
Rn 86	98404.0	14619.4	17337.1	18049.0	2892.4	3021.5	3538.0	4159.0	4482.0
Fr 87	101137.0	15031.2	17906.5	18639.0	2999.9	3136.2	3663.0	4327.0	4652.0
Ra 88	103921.9	15444.4	18484.3	19236.7	3104.9	3248.4	3791.8	4489.5	4822.0
Ac 89	106755.3	15871.0	19083.2	19840.0	3219.0	3370.2	3909.0	4656.0	5002.0
TH 90	109650.9	16300.3	19693.2	20472.1	3332.0	3490.8	4046.1	4830.4	5182.3
Pa 91	112601.4	16733.1	20313.7	21104.6	3441.8	3611.2	4173.8	5000.9	5366.9
U 92	115606.1	17166.3	20947.6	21757.4	3551.7	3727.6	4303.4	5182.2	5548.0

Source: J. A. Eearden et al., Reviews of Modern Physics **39**, 125 (1967).

X-Ray Mass Absorption Coefficient for Materials

The absorption coefficient is determined using the following expression.

$I = I_0 \exp(-\lambda d)$. In this expression, I_0 is the intensity of the incident X-rays, I is the intensity after traveling distance d in the material, and λ is the absorption coefficient.

The following table shows the absorption coefficient divided by the density of the material; i. e. the mass absorption coefficient. (Unit: cm^2/g)

Wave-length nm	Absorbing element and its atomic number											
	H 1	C 6	N 7	O 8	Al 13	Fe 26	Cu 29	Ag 47	Sn 50	Pt 78	Au 79	Pb 82
0.030	0.348	0.208	0.230	0.260	0.552	3.35	4.51	16.5	19.0	11.1	11.5	12.8
0.040	0.359	0.258	0.311	0.382	1.08	7.54	10.2	35.0	39.5	24.6	25.5	28.1
0.050	0.367	0.336	0.438	0.577	1.93	14.2	19.1	9.45	11.5	45.2	46.6	50.6
0.060	0.374	0.449	0.625	0.865	3.20	23.9	31.9	15.9	19.2	73.3	75.2	80.1
0.070	0.379	0.605	0.884	1.26	4.95	36.9	48.9	24.7	29.8	109	111	116
0.080	0.384	0.811	1.23	1.79	7.25	53.6	70.4	36.0	43.4	151	153	148
0.090	0.389	1.07	1.66	2.47	10.2	74.2	96.6	50.2	60.3	179	183	141
0.100	0.394	1.40	2.21	3.31	13.8	99.1	128	67.3	80.7	158	165	74.2
0.110	0.400	1.80	2.87	4.33	18.2	128	163	87.6	105	82.1	85.4	96.0
0.120	0.407	2.28	3.67	5.55	23.5	162	204	111	132	104	108	121
0.130	0.414	2.84	4.60	6.99	29.6	199	248	138	164	129	134	150
0.140	0.422	3.50	5.69	8.66	36.8	242	40.0	169	199	156	162	181
0.150	0.431	4.25	6.95	10.6	44.9	288	48.9	203	239	187	194	216
0.160	0.441	5.12	8.38	12.8	54.1	338	58.9	240	282	220	228	254
0.170	0.452	6.09	9.99	15.3	64.5	392	70.3	281	328	257	266	295
0.180	0.465	7.19	11.8	18.0	76.0	53.7	82.9	325	378	295	306	338
0.190	0.478	8.41	13.8	21.1	88.8	62.8	97.0	372	431	337	348	384
0.200	0.493	9.76	16.1	24.6	103	72.9	112	423	487	380	393	432
0.210	0.510	11.3	18.5	28.4	118	83.9	129	476	545	426	439	482
0.220	0.527	12.9	21.3	32.5	135	96.0	148	532	606	473	487	532
0.230	0.547	14.7	24.2	37.0	154	109	168	591	668	521	537	584
0.240	0.568	16.6	27.5	42.0	173	123	190	652	732	571	587	636
0.250	0.592	18.8	31.0	47.3	195	139	213	714	796	621	637	687
0.260	0.617	21.1	34.7	53.1	217	155	239	778	861	671	687	738
0.270	0.643	23.5	38.8	59.3	242	173	266	843	926	720	737	787

Source: International Tables for X-ray Crystallography, vol. III (1968)

Standard Crystals for X-Ray Diffractometers

Prece measurement of the X-ray wavelengths and the lattice constant of crystals is performed with respect to the lattice parameter of Si crystal. The lattice parameter is defined as the unit cell edge length of an ideal single Si crystal with natural abundance free of impurities and imperfections in vacuum at 22.5℃. The value is evaluated in which the carbon and oxygen concentration included in the crystal are extrapolated to zero.

The Si (220) lattice spacing (in vacuum, 22.5℃) is $d_{220} = 192.015\,5714 \times 10^{-12}$ m and its relative uncertainty is 0.016 ppm. The lattice parameter of Si (in vacuum, 22.5℃) is then $a = 543.102\,0504\ (89) \times 10^{-12}$ m. The lattice parameter of Si at 1 atm (101 325 Pa) is calculated to be $a = 543.101\,863 \times 10^{-12}$ m by using its linear expansion coefficient $\alpha = 2.5$ ppm (± 0.05 ppm)/K and elastic constants $c_{11} = 166$ GPa and $c_{12} = 64$ GPa.

Lattice Parameter and Linear Expansion Coefficient of Standard Crystal

A single lattice parameter and a linear expansion coefficient are listed for cubic crystals. Two values are listed for hexagonal crystals such as those with crystal a axis and c axis. 1 pm $= 10^{-12}$ m.

Material	Lattice parameter a, c/pm	Measurement temperature t/℃	Linear coefficient of expansion $\alpha/10^{-6}\ K^{-1}$
W (tungsten)	316.524	25	4.5
Ag	408.651	25	18.9
Al	404.94	25	23.1
C (diamond)	356.703	18	1.0
C (graphite)	$a = 246.1$	20	-1.2
	$c = 670.8$	20	25.9
LiF	402.6		35
α-quartz	$a = 490.3$	20	14.3
	$c = 539.4$	20	8.7
NaCl	564.0	20	40.0

Lattice Parameter used for Powder X-Ray Diffractometry

At room temperature (20°C). Please refer to the NIST Standard Reference Materials.

Material	Lattice parameter a, c/pm	Material	Lattice parameter a, c/pm
α-Al_2O_3 (corundum)	a 475.9397	CeO_2	a 541.1102
	c 1 299.237	Si	a 543.0940
ZnO (wurtzite)	a 324.9074	SiN_4 (α phase)	a 775.2630
	c 520.6535		c 561.9372
TiO_2 (rutile)	a 459.3939	SiN_4 (β phase)	a 760.2293
	c 295.8862		c 290.6827
Cr_2O_3	a 495.9610		
	c 1 358.747		

Oscillation Wavelength of Laser

Solid laser

Laser, (alias etc.)	Oscillation wavelength (in air) $\lambda/\mu m$	Excitation method*	Feature etc.
Cr^{3+} : Al_2O_3, (ruby)	0.6943(R_1, 300 K), 0.6929(R_2, 300 K)	FL	The first solid laser
Nd^{3+} : $Y_3Al_5O_{12}$, (YAG)	0.946(low temperature), 1.05205($R_2 \to Y_1$), 1.06152($R_1 \to Y_1$) 1.06414($R_2 \to Y_3$ and strength), 1.0646($R_1 \to Y_2$) 1.0738($R_1 \to Y_3$), 1.0780($R_1 \to Y_4$) 1.1121($R_2 \to Y_6$), 1.1159($R_1 \to Y_5$) 1.12267($R_1 \to Y_6$), 1.3188($R_2 \to X_1$) 1.3382($R_2 \to X_3$), 1.3564($R_1 \to X_4$) 1.839(low temperature)	FL, AL, L	Highly effective, High power, Multi wavelength oscillation
Nd^{3+} : Phosphate glass, (glass laser)	1.054	FL	Large-scale, High power
Nd^{3+} : The Kei acid glass (glass laser)	1.062	FL	Large-scale, High power
Nd^{3+} : Cr^{3+} : $[Gd_{3-x}Nd_x][Cr_xSc_{2-x}]Ga_3O_{12}$, (GSGG) Nd^{3+} : Cr^{3+} : $[Gd_{3-x}Nd_x][Cr_xSc_{2-x}]Al_3O_{12}$, (GSAG)	1.06	FL	High power, High efficiency due to Cr-sensitization
Nd^{3+} : $LiYF_4$, (YLF)	1.047(π), 1.053(σ), 1.321(π), 1.313(σ)	FL	birefingent, It can be amplification with glass laser. Q switch characteristic is excellent.
Nd^{3+} : $Nd_xY_{1-x}Al_3(BO_3)_4$, (NYAB)	1.062 (fundmental wave), 531 (twice wave)	L (SC)	Green light by second harmonic generation, All solid state
Er^{3+} : Glass, (Fiber-optic laser)	1.522–1.575, 1.590–1.603 (Variable)	L (SC)	Amplification for optical communication >900 km, 1.2 Gbit/s
Ho^{3+} : $Y_3Al_5O_{12}$ (Ho-YAG)	2.098 (2.090–2.100 Variable, <200 K)	FL, L (SC)	Strong absorption by H_2O
KCl : Li(F_A II) (color center laser)	2.2–3.2 (low temperature)	L, FL (SC)	Variable wavelength
$NaCl(F_2^+)_H$ (color center laser)	1.4–1.8	L	Variable wavelength
Cr^{3+} : $BeAl_2O_3$ (alexandrite)	0.701–0.818	FL, L, L (SC)	Variable wavelength, High power, long fluorescence life
Ti^{3+} : Al_2O_3 (Titanium sapphire)	0.653–1.150	FL, L	Variable wavelength, Wideband, High power
Co^{2+} : MgF_2	1.63–2.50	L	Variable wavelength, Wideband, Long wavelength

* The letters indicate how the lines can be excited:
FL = flash lamp, AL = arc lamp, L= laser, L(SC) = semiconductor laser

Gas laser

Laser	Oscillation wavelength (in air) $\lambda/\mu m$
H_2, HD, D_2	0.1098–0.1268 ($C^1\Pi_u$–$X^1\Sigma_g^+$; Werner belt), 0.1246–0.1646 ($B^1\Sigma_u^+$–$X^1\Sigma_g^+$; Lyman belt)

Gas laser (Continued.)

Laser		Oscillation wavelength (in air) $\lambda/\mu m$	
Ar_2	Excimer laser (pulsed oscillation)	0.126	
ArCl		0.175	
ArF		0.193	
KrCl		0.222	
KrF		0.248, 0.486	
XeBr		0.282	
XeCl		0.308	
XeF		0.353	
ArO		0.558	
Cd^+ : He		0.325, 0.442	
N_2		0.337 ($C^3\Pi_u$–$B^3\Pi_g$; second positive band)	
Au	Metallic vapour Laser (pulsed oscillation)	0.3122, 0.6278	
Pb		0.4057, 0.7229	
Bi		0.4722	
Cu		0.5105, 0.5782	
Tl		0.5350	
Ca		0.8662	
Ba		1.13	
Mn		1.30	
Ar^+	Rare gas*1 ion laser (continuous wave oscillation)	0.3511, 0.3638, 0.4545, 0.4579, 0.4658, 0.4727, 0.4765, 0.4880,	
Laser		0.4965, 0.5017, 0.5145(strongest), (0.514673466368 (4)*2)), 0.5287, 1.090	
Kr^+		0.3374, 0.3507, 0.3564, 0.4131, 0.4680, 0.4762, 0.4825, 0.5208,	
		0.5309, 0.5682, 0.6471(strongest), 0.6764, 0.7525, 0.7993	
Ne : He		0.63282(0.632991212579 (13)*2)), 1.1523, 3.39(3.392231397327 (10)*3))	
CO : He, N_2, Xe		5.1–7.1 ($X_1\Sigma^+$, 5-4 band–26-25 band ; About 1200 lines)	
CO_2 : He, N_2		9.1–11.5 (00^01—$[10^00,\ 02^00]_{I,\ II}$ band, About 600 lines using $^{12}C\,^{16}O_2$, $^{13}C\,^{16}O_2$, $^{12}C\,^{18}O_2$, and $^{12}C\,^{16}O\,^{18}O$)	
N_2O : He, N_2		10.3–11.0 (001-100 band. About 80 lines)	
I		1.315 (CH_3I, CF_3I, Atomic I oscillations produced by photodissociation of alkyl halide molecules or haloalkane molecules such as ... C_3F_7I, etc., Significant power)	
HF	Chemical laser	2.4–3.4	Excitation by chemcial reactions of H_2, F_2, etc.
DF		3.5–4.7	
HCl		3.6–4.1	
DCl		5.0–5.5	
HBr		4.0–4.6	
DBr		5.8–6.3	
H_2O	Gas discharge excited far infrared lasers	28.0, 33.0, 35.8, 39.7, 47.5, 78.4, 118.6, 220.3	
D_2O		36.3, 72.7, 84.3, 107, 172, 218	
HCN		310, 337, 373, 538, 774	
DCN		190, 195	
NH_3	CO_2 laser excited far infrared lasers	12.1, 12.8, 81.5, 153	
CF_4		16.3	
CH_3OH		41.7, 70.5, 119	
D_2O		50.3, 66, 114, 119, 385	
CH_2F_2		166, 185, 214	
HCOOH		394, 433, 513	
$^{13}CH_3F$		1222	
$CH_3\,^{81}Br$		311, 545, 586, 659, 831, 2650	

*1 IUPAC's official notation is nobel gas.

*2 The wavelength (in vacuum) of a laser locked to the absorption line of $^{127}I_2$ (molecular iodine).

*3 The wavelength (in vacuum) of a laser locked to the absorption line of CH_4(υ_3, P(7), $F_2^{(2)}$ transition). The absolute frequency of this oscillation is measured by the optical heterodyne method by several groups, and its relative uncertainty is reached to 10^{-10} level. Before the new defintion of the meter (Refer to page **286**), the frequency and wavelength of this oscillation (in terms of the 1960 defintion of the meter) were used to determine the speed of light as 299 792 458 ± 1.2 $m \cdot s^{-1}$. The new defintion of the meter (1983) was determined to match this value.

Dye laser

Laser	Oscillation wavelength (in air) $\lambda/\mu m$
BM-terphenyl; 2,2′-Dimethyl-*p*-terphenyl	0.3085-0.360
p-Terphenyl	0.322 -0.365
Butyl-PBD; 2-(4-*tert*-Butylphenylyl)-1,3,4-oxadiazole	0.351 -0.395
BBQ; 4,4‴-*Bis*-butyloctyloxy-*p*-quaterphenyl	0.359 -0.410
Polyphenyl 1; Sulphonated *p*-quaterphenyl*	0.360 -0.415
DPS; 4,4′-Diphenyl stilbene	0.393 -0.420
POPOP; *p*-*Bis*(5-phenyloxazolyl) benzene	0.390 -0.451
Stilbene 3; 4,4′-*Bis*(2-sulfostyryl) biphenyl*	0.403 -0.493
Coumarin 120; 7-Amino-4-methyl coumarin*	0.419 -0.477
4-MU; Coumarin 4; 4-Methylumbelliferone*	0.385 -0.574
DAMC; Coumarin 1; 7-Diethylamino-4-methylcoumarin*	0.438 -0.505
Coumarin 6; 3-(2-Benzothiazolyl)-7-diethylamino-coumarin*	0.507 -0.585
Brilliant sulfoflavine	0.465 -0.620
Fluorescein (Na salt)*	0.522 -0.600
Rhodamine 110*	0.525 -0.603
Rhodamine 6G*	0.550 -0.655
Rhodamine B*	0.570 -0.675
DCM; 4-Dicyanomethylene-2-methyl-6-*p*-dimethyl-aminostyryl-4H-pyran*	0.610 -0.710
Cresyl violet*	0.612 -0.709
Oxazine 1*	0.685 -0.825
DOTC; 3,3′-Dimethyloxatricarbocyanine iodide*	0.720 -0.820
HITC; 1,3,3,1′,3′,3′-Hexamethylindotricarbocyanine iodide*	0.775 -0.940
IR-140; 5,5′-Dichloro-11-diphenylamino-3,3′-diethyl-10,12-ethylene-thiatricarbocyanine perchlorate*	0.858 -1.03
1,1′-Diethyl-4,4′-quinocarbocyanine iodide	0.970 -1.145
DNXTPC; 3,3′-Diethyl-9,11,15,17-dineopentylene-5,6,5′,6′-tetramethoxy-thiapentacarbocyanine perchlorate	1.107 -1.285

For the excitation of dye lasers, Flash lamps, N_2 lasers, excimer lasers, Cu vapor lasers, and harmonics of Nd: YAG lasers (pulse emission), and argon/krypton lasers (continuous oscillation; marked by an asterisk (*)) are used.

Semiconductor laser (injection diode laser)*

Laser	Oscillation wavelength (in air) $\lambda/\mu m$	Laser	Oscillation wavelength (in air) $\lambda/\mu m$
GaN/AlGaN	0.35-0.36	$In_xGa_{1-x}As$	1.0- 3.5
$Al_xIn_yGa_{1-x-y}N$	0.35-0.39	$InAs_{1-x}P_x$	1.0- 3.5
$In_xGa_{1-x}N$	0.39-0.53	$Ga_xIn_{1-x}N_yAs_{1-y}$	1.2- 1.4
$Cd_xZn_{1-x}Se$	0.47-0.51	$InAs_{1-x}Sb_x$	3.1- 5.4
$(Al_xGa_{1-x})_{0.5}In_{0.5}P$	0.58-0.69	$Pb_{1-x}Cd_xS$	2.5- 4
$In_{1-x}Ga_xP$	0.65-1.0	$PbS_{1-x}Se_x$	4 - 8.5
$Al_xGa_{1-x}As$	0.7 -1.0	$Pb_{1-x}Sn_xTe$	6.3-32
$GaAs_{1-x}P_x$	0.7 -1.0	$Pb_{1-x}Sn_xSe$	8.5-34

* By controling the stoichiometry, it is possible to make laser diodes which oscillate in the above ranges.

Synchrotron Radiation

Synchrotron radiation (SR) is an electromagnetic wave emitted by a high energy electron deflected by a magnetic field. As a light source, SR is utilized in various applications for dynamical and structural analyses of matter, because it is polarized, tunable in wavelength, and highly directional. Generally speaking, SR generated in a circular accelerator called a storage ring is used as a light source for the above applications, in a wavelength region shorter than the vacuum ultraviolet. To generate the magnetic field to deflect the electron beam, two types of magnetic devices are used: bending magnets (BMs) and insertion devices (IDs). The former generates a uniform magnetic field, while the latter generates a sinusoidal one, whose polarity alternates periodically along the path of the electron beam.

Bending Magnet SR generated in a BM is broadband light, whose photon intensity peaks near the wavelength called the critical wavelength (λ_c). The electron beam moves along a circular orbit; SR is isotropic in the orbital plane, while it is highly collimated in the direction perpendicular to it, with a typical angular divergence of γ^{-1}. Here, γ is the energy of the electron beam normalized by the electron rest energy ($m_e c^2 = 0.511$ MeV), and is called the Lorentz factor. The polarization property depends on the observation condition; SR emitted in the orbital plane is linearly polarized, while that emitted outside the orbital plane is elliptically polarized, with a helicity depending on the observation position (above or below the orbital plane).

BM SR

	Formula	Practical Unit
Critical Wavelength	$\lambda_c = \dfrac{4\pi m_e c}{3\gamma^2 eB}$	$\lambda_c = \dfrac{1.864}{BE^2}$
Angular Power Density	$\dfrac{dP}{d\Omega} = \dfrac{7\gamma^4}{64\pi}\dfrac{e^2}{\varepsilon_0}\dfrac{eB}{m_e c}\dfrac{I}{e}$	$\dfrac{dP}{d\Omega} = 5.421 \times 10^{-3} BE^4 I$
Total Power	$P = \dfrac{\gamma^3}{3}\dfrac{e^2}{\varepsilon_0}\dfrac{eB}{m_e c}\dfrac{I}{e}$	$P = 26.52BE^3 I$
Angular Flux Density	$\dfrac{dF}{d\Omega} = \dfrac{3\alpha\gamma^2}{4\pi^2} f\left(\dfrac{\lambda_c}{\lambda}\right)\dfrac{I}{e}$	$\dfrac{dF}{d\Omega} = 1.325 \times 10^{13} f\left(\dfrac{\lambda_c}{\lambda}\right)E^2 I$
Total Flux	$F = \sqrt{3}\,\alpha\gamma g\left(\dfrac{\lambda_c}{\lambda}\right)\dfrac{I}{e}$	$F = 1.544 \times 10^{17} g\left(\dfrac{\lambda_c}{\lambda}\right)EI$

Undulator SR

	Formula	Practical Unit
Deflection Parameter	$K=\frac{eB\lambda_u}{2\pi m_e c}$	$K=93.37B\lambda_u$
Fundamental Wavelength	$\lambda_1=\frac{\lambda_u(1+K^2/2)}{2\gamma^2}$	$\lambda_1=130.6\lambda_u(1+K^2/2)/E^2$
Angular Power Density	$\frac{dP}{d\Omega}=\frac{7\gamma^4}{16}\frac{e^2}{\varepsilon_0 c}\frac{c}{\lambda_u}NKp(K)\frac{I}{e}$	$\frac{dP}{d\Omega}=1.161\times10^{-4}p(K)E^4IKN/\lambda_u$
Total Power	$P=\frac{\pi\gamma^2 N}{3}\frac{e^2}{\varepsilon_0}\frac{K^2}{\lambda_u}\frac{I}{e}$	$P=7.257\times10^{-5}E^2K^2NI/\lambda_u$
Angular Flux Density	$\frac{dF}{d\Omega}=\alpha\gamma^2N^2q_n(K)\frac{I}{e}$	$\frac{dF}{d\Omega}=1.744\times10^{14}q_n(K)E^2N^2I$
Total Flux	$F=\frac{\pi\alpha}{2}Nr_n(K)\frac{I}{e}$	$F=7.154\times10^{13}r_n(K)NI$

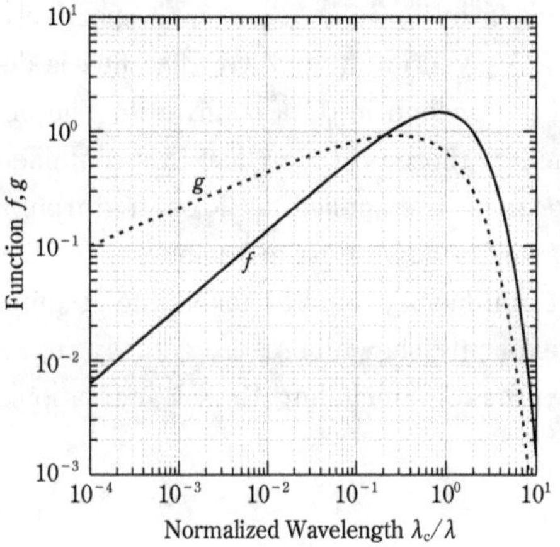

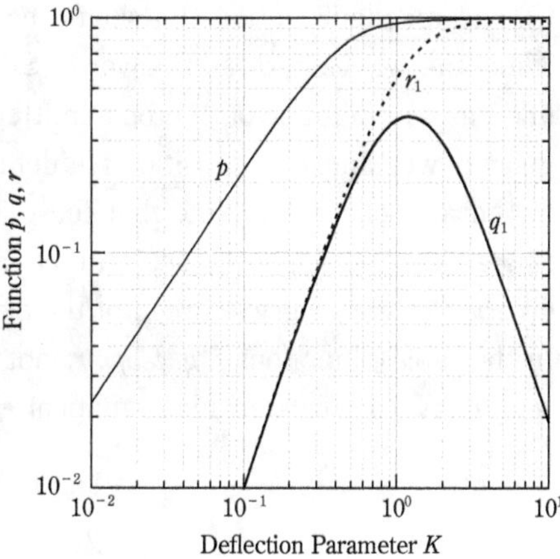

Functions f, g, p, q_n, r_n obtained by numerical calculations

Insertion Device SR generated in an ID is quasi-monochromatic light, whose photon intensity peaks at several wavelengths defined as λ_1/n, where λ_1 is called the fundamental wavelength, and n is an odd integer called a harmonic order, with a bandwidth inversely proportional to the number of magnetic periods (N). Because λ_1 depends on the magnetic field strength (B), magnetic period (λ_u), and Lorentz factor (γ), an arbitrary wavelength can be selected by adjusting these parameters. SR generated in an ID is horizontally polarized almost perfectly, and is highly collimated with a typical angular divergence of $\sqrt{\lambda_1/(2nN\lambda_u)}$, both in the horizontal and vertical directions. The ID is further classified into two types according to the harmonic order n to be utilized: undulator (lower order) and wiggler (higher order). Note that the spectral and directional properties explained above are lost in higher-harmonic orders because of various factors, and thus the characteristics of wiggler SR is similar to those of BM SR, except for the photon intensity. SR generated in an ID is characterized by a dimensionless parameter usually denoted by K, which is called the deflection parameter.

Formulas to calculate the light source performances of SR[1)]

Mathematical expressions to calculate the power (P), flux (F), and their angular densities ($dP/d\Omega$, $dF/d\Omega$) of SR are summarized in the following tables. In practical-unit representations, the magnetic field (B), electron energy (E), current (I), and magnetic period (λ_u) should be given in T, GeV, A, and m, respectively. Then, the wavelength and power of SR are given in nm and kW. The flux is defined as the number of photons emitted per second in 0.1% bandwidth. The angular power and flux densities are defined as those on the optical axis contained in 1-mrad^2 solid angle. Note that these formulas are applied to SR emitted from a single electron. For more accurate characterization, the finite beam size, angular divergence and energy spread of the electron beam should be taken into account. In the tables, functions f, g, p, q_n, r_n not explicitly shown here, are actually represented by complicated mathematical expressions including Bessel and rational functions.

Free Electron Laser

The free electron laser (FEL) is a laser source based on the undulator SR, which drastically enhances the intensity and improves the coherence. The amplification process in the FEL is based on the interaction between the electron beam and light (seed light) in the undulator; when the wavelength of the seed light is the same as the fundamental wavelength (λ_1) of undulator SR, the electrons exchange energy with the seed light, and are modulated in energy with the pitch of λ_1. The energy modulation is converted to a density modulation while the electron beam travels along the undulator, which emits SR with the wavelength of λ_1. Because the SR emitted by the density-modulated electron beam is coherent both in time and space, the seed light is exponentially amplified, which eventually results in lasing. Even in the spectral regions where no seed light is available, the spontaneous SR emitted by the electron beam works as the seed light, which is referred to as a self-amplified spontaneous emission (SASE) FEL, and thus lasing in the SASE FEL is in principle achieved at any wavelengths. In short-wavelength regions, however, the density of the electron beam to achieve lasing in SASE FELs should be much higher than that available in the storage ring, and thus linear accelerators are mainly used for this purpose.

Although rigorous numerical simulations are necessary to describe the amplification process in the FEL and evaluate its light source performances, qualitative discussions are approximately possible using a dimensionless physical quantity called the FEL parameter defined by the following formula[2),]

$$\rho = \left[\frac{\gamma q_1(K)}{16\pi^2} \frac{\lambda_1^2}{\sigma_x \sigma_y} \frac{I_p}{4\pi\varepsilon_0 m_e c^3/e} \right]^{1/3} = 0.08992 \left[E I_p q_1(K) \frac{\lambda_1^2}{\sigma_x \sigma_y} \right]^{1/3},$$

where E, I_p, σ_x, σ_y denote the electron energy (GeV), peak current (A), horizontal and vertical beam sizes, and λ_1 is the undulator fundamental wavelength (= lasing wavelength). In the amplification process, the FEL power (P) exponentially increases with the undulator length (L), namely, $P \propto e^{L/L_g}$, where $L_g = \lambda_u/(4\pi\sqrt{3}\rho)$ is called the gain length. The exponential amplification continues until the FEL power reaches the saturation power P_s, defined as $P_s = \rho\gamma m_e c^2 I_p/e$. In the SASE FELs, the undulator length needed to reach saturation is roughly given by $L_s = \lambda_u/\rho$.

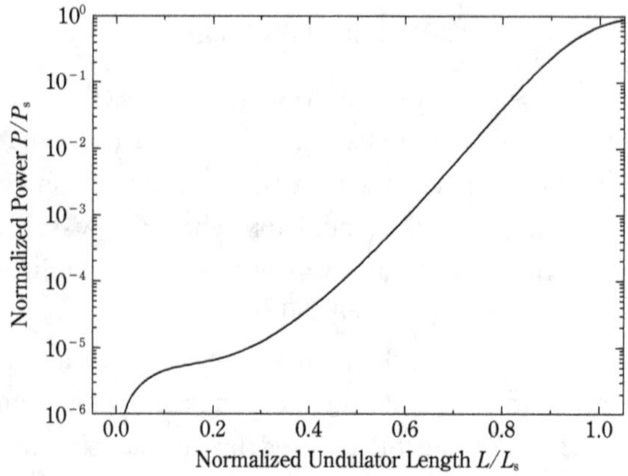

Amplification process in the SASE FEL evaluated by a numerical simulation[3)]

1) K. J. Kim, "Characteristics of Synchrotron Radiation," in Physics of Particle Accelerators, AIP Conf. Proc. **184** (Am. Inst. Phys., New York, 1989), pp. 565–632

2) E. Saldin, E.V. Schneidmiller and M.V. Yurkov, "The Physics of Free Electron Lasers" (Springer-Verlag Berlin Heidelberg, 2000)

3) T. Tanaka, "SIMPLEX: simulator and postprocessor for free-electron laser experiments", Journal of Synchrotron Radiation **22**, 1319–1326 (2015)

Optical Properties

Refractive Index of Air

The refractive index n_s of standard air (dry air with 0.045% carbon dioxide gas at 15℃ and at 1 atm = 0.101325 MPa) is given by the following formula in a range of λ_{vac}=0.23 to 1.69 μm where λ_{vac} is wavelength in vacuum.

$$(n_s-1)\times 10^8=\frac{5792105}{238.0185-(\lambda_{vac}/\mu m)^{-2}}+\frac{167917}{57.362-(\lambda_{vac}/\mu m)^{-2}}$$

The table shows the value n_s determined using this formula.

$\lambda_{vac}/\mu m$	n_s	$\lambda_{vac}/\mu m$	n_s	$\lambda_{vac}/\mu m$	n_s	$\lambda_{vac}/\mu m$	n_s
0.23	1.000 308 0	0.42	1.000 281 8	0.62	1.000 276 7	1.00	1.000 274 2
0.24	1.000 304 5	0.44	1.000 280 9	0.64	1.000 276 4	1.05	1.000 274 0
0.26	1.000 298 9	0.46	1.000 280 2	0.66	1.000 276 2	1.10	1.000 273 9
0.28	1.000 294 8	0.48	1.000 279 5	0.68	1.000 276 0	1.15	1.000 273 8
0.30	1.000 291 6	0.50	1.000 279 0	0.70	1.000 275 8	1.20	1.000 273 7
0.32	1.000 289 0	0.52	1.000 278 5	0.75	1.000 275 4	1.30	1.000 273 5
0.34	1.000 287 0	0.54	1.000 278 0	0.80	1.000 275 0	1.40	1.000 273 4
0.36	1.000 285 3	0.56	1.000 277 6	0.85	1.000 274 8	1.50	1.000 273 3
0.38	1.000 283 9	0.58	1.000 277 3	0.90	1.000 274 5	1.65	1.000 273 2
0.40	1.000 282 8	0.60	1.000 277 0	0.95	1.000 274 3	1.69	1.000 273 2

* P. E. Ciddor, Appl. Opt. **35**, 1566 (1996).

Refractive Index of Various Substances

The following table shows the relative refractive index to air at indicated temperature for various wavelength of light and various substances. The Δ listed in the bottom row is the temperature coefficient of refractive index at the sodium D line (589.3 nm).

Wave-length λ/nm	Calcite 18℃		Quartz crystal 18℃		Fused silica 18℃	Fluorite (CaF_2) 18℃	Rock salt 18℃	Sylvius (KCl) 18℃	Water 20℃
	Ordinary ray	Extraordinary ray	Ordinary ray	Extraordinary ray					
9429.0	-	-	-	-	-	1.3161	1.4983	1.4587	-
4200.0	-	-	1.4569	-	-	4078	5213	4720	-
2172.0	1.6210	1.4746	5180	1.5261	-	4230	5262	4750	-
1256.0	6388	4782	5316	5402	-	4275	5297	4778	1.3210
656.3	1.6544	1.4846	1.5419	1.5509	1.4564	1.4325	1.5407	1.4872	1.3311
589.3	6584	4864	5443	5534	4585	4339	5443	4904	3330
546.1	6616	4879	5462	5553	4602	4350	5475	4931	3345
404.7	6813	4969	5572	5667	4697	4415	5665	5097	3428
303.4	1.7196	1.5136	1.5770	1.5872	1.4869	1.4534	1.6085	1.5440	1.3581
214.4	8459	5600	6305	6427	5339	4846	7322	6618	4032
185.2	-	-	6759	6901	5743	5099	8933	8270	-
Δ/k	$+5\times10^{-6}$	$+14\times10^{-6}$	-5×10^{-6}	-6×10^{-6}	-3×10^{-6}	-1×10^{-5}	-4×10^{-5}	-4×10^{-5}	-8×10^{-5}

Source: Kaye

The following tables show the refractive index at the sodium D line (589.3 nm). Solids and liquids are the relative refractive index to air, while gases are the absolute refractive index.

Gases and liquids

Material	Refractive index	Material	Refractive index	Material	Refractive index
Gas (0℃, 1 atm)		Helium	0035	Benzene	1.5012
		Hydrogen	0138	Carbon tertrachloride	1.4607
Air	0292	Mercury	1.000933	Diethyl ether	1.3538
Argon	1.000284	Neon	0067	Diiodomethane	1.737
Benzene	1762	Nitrogen	0297	Ethyl alcohol	1.3618
Bromine	1125	Oxygen	0272	Glycerin	1.4730
Cadmium	2675	Steam	0252	Methyl alcohol	1.3290
Carbon dioxide	0450	Sulfur	1111	Paraffin oil	1.48
Carbon monoxide	0334	**Liquid** (20℃)		Seda oil	1.516
Chlorine	0768			Water	1.3330
Chloroform	1455	Aniline	1.586	α-bromonaphthalene	1.660

Optically isotropic solids

(20℃)

Material	Refractive index	Material	Refractive index
Canada balsam	1.542	Polychroickisholmetacrylat (15℃)	1.507
Diamond	2.4195	Polymethyl methacrylate	1.491
Germanium	4.092	Polystyrene (15℃)	1.592
Iodide of potash	1.6666	Potassium bromide	1.5599
KRS-5 (Thallium Bromoiodide)	2.395	Silicon	3.448
Lithium Fluoride (21℃)	1.3921	Silver chloride	2.09
Magnesium oxide	1.7373	Zincblende (ZnS)	2.370

Uniaxial crystals

(20℃)

Material	Refractive index	
	Ordinary ray	Extraordinary ray
Ammonium dihydrogenphosphate (ADP)	1.5242	1.4787
Cadmium sulphide	2.506	2.529
Chile saltpeter ($NaNO_3$)	1.5854	1.3369
Corundum (Sapphire and Al_2O_3)	1.768	1.760
Hematite (Fe_2O_3) (670.8 nm)	2.940	3.220
Ice (0℃)	1.309	1.313
Potassium phosphate monobasic (KDP)	1.5095	1.4684
Rutile (TiO_2)	2.616	2.903
Tourmaline	1.669	1.638
Wurtzite (ZnS)	2.356	2.378

Biaxial crystals

(20℃)

Material	Refractive index		
	n_1	n_2	n_3
Aragonite ($CaCO_3$)	1.6862	1.6810	1.5309
Gypsum ($CaSO_4 \cdot 2H_2O$)	1.5298	1.5228	1.5208
Mica	1.5993	1.5944	1.5612
Niter (KNO_3)	1.5064	1.5056	1.3346
Stibnite (Sb_2S_3)	4.460	4.303	3.194

Refractive Index of Optical Glass

Type	Wavelength λ/nm								
	1014.0	768.2	656.3	587.6	546.1	486.1	435.8	404.7	365.0
FK1	1.4623	1.4660	1.4685	1.4707	1.4724	1.4755	1.4793	1.4823	1.4875
BK7	1.5073	1.5115	1.5143	1.5168	1.5187	1.5224	1.5267	1.5302	1.5363
K3	1.5083	1.5125	1.5155	1.5182	1.5203	1.5243	1.5291	1.5331	1.5399
BaK4	1.5576	1.5623	1.5658	1.5688	1.5713	1.5759	1.5815	1.5861	1.5941
SK5	1.5781	1.5828	1.5862	1.5891	1.5914	1.5958	1.6010	1.6053	1.6126
SSK1	1.6048	1.6099	1.6137	1.6172	1.6199	1.6252	1.6315	1.6368	1.6459
LaK3	1.6794	1.6852	1.6896	1.6935	1.6966	1.7025	1.7097	1.7156	1.7259
KzF2	1.5180	1.5228	1.5263	1.5294	1.5319	1.5366	1.5422	1.5469	1.5551
BaF10	1.6550	1.6611	1.6658	1.6700	1.6734	1.6800	1.6880	1.6948	1.7068
LaF2	1.7262	1.7335	1.7390	1.7440	1.7479	1.7556	1.7649	1.7728	1.7867
LF5	1.5667	1.5726	1.5772	1.5814	1.5848	1.5915	1.5996	1.6067	1.6192
F2	1.6028	1.6096	1.6150	1.6200	1.6241	1.6321	1.6421	1.6507	1.6663
SF2	1.6286	1.6361	1.6421	1.6477	1.6522	1.6612	1.6725	1.6823	1.7003
SF13	1.7151	1.7248	1.7331	1.7408	1.7471	1.7598	1.7761	1.7907	–
SFS1	1.8781	1.8927	1.9054	1.9176	1.9277	1.9484	1.9753	1.9997	–

Source: Jenaer Glas für die Optik

Spectral Reflectivity of Metal Surfaces

The following table shows the reflectivity (%) when the light of various wavelengths (μm) is irradiated vertically on a fresh surface made by the vacuum deposition of metal. Superior deposition surfaces are obtained by performing rapid deposition in a high vacuum. Normally, the reflectivity of these deposition surfaces is higher than the reflectivity of surfaces which were made by polishing or spattering the same metal.

Wave-length	Alumi-num	Silver	Gold	Cop-per	Rho-dium	Wave-length	Alumi-num	Silver	Gold	Cop-per	Rho-dium
0.220	91.5	28.0	27.5	40.4	58.5	0.700	89.9	98.5	97.0	97.5	80.4
0.240	91.9	29.5	31.6	39.0	61.3	0.750	88.0	98.6	97.4	97.9	81.2
0.250	92.1	30.4	33.2	37.0	63.0	0.800	86.3	98.6	97.7	98.1	82.0
0.260	92.2	29.2	35.6	35.5	65.0	0.850	85.8	98.7	97.8	98.3	82.8
0.280	92.3	25.2	37.8	33.0	68.5	0.900	88.9	98.7	98.0	98.4	83.5
0.300	92.3	17.6	37.7	33.6	71.2	0.950	91.8	98.8	98.1	98.4	84.2
0.315	92.4	5.5	37.3	35.5	73.0	1.0	93.9	98.9	98.2	98.5	85.0
0.320	92.4	8.9	37.1	36.3	73.6	1.5	96.8	98.9	98.2	98.5	88.2
0.340	92.5	72.9	36.1	38.5	75.5	2.0	97.2	98.9	98.3	98.6	90.5
0.360	92.5	88.2	36.3	41.5	77.0	3.0	97.5	98.9	98.3	98.6	92.5
0.380	92.5	92.8	37.8	44.5	77.4	4.0	97.6	98.9	98.3	98.7	94.0
0.400	92.4	94.8	38.7	47.5	77.6	5.0	97.7	98.9	98.3	98.7	94.5
0.450	92.2	96.6	38.7	55.2	77.2	6.0	97.7	98.9	98.3	98.7	94.8
0.500	91.8	97.7	47.7	60.0	77.4	7.0	97.8	98.9	98.4	98.7	95.2
0.550	91.6	97.9	81.7	66.9	78.0	8.0	97.9	98.9	98.4	98.7	95.5
0.600	91.1	98.1	91.9	93.3	79.1	9.0	97.9	98.9	98.4	98.8	95.8
0.650	90.3	98.3	95.5	96.6	79.9	10.0	98.0	98.9	98.4	98.8	96.0

Transmittance of Optical Crystals

When light penetrates a crystal, the light intensity decreases due to a partial reflection on the two surfaces and absorption inside of the crystal. The ratio of intensity to the incident light of transmitted light is called "transmittance." Transmittance decreases in shorter wavelength due to absorption by electrons. It decreases in longer wavelength due to absorption caused by oscillation of atoms. The figure shows transmittance of light which vertically entered a sheet-shaped sample. (KRS-5 in the figure idicates thallium bromo-iodide)

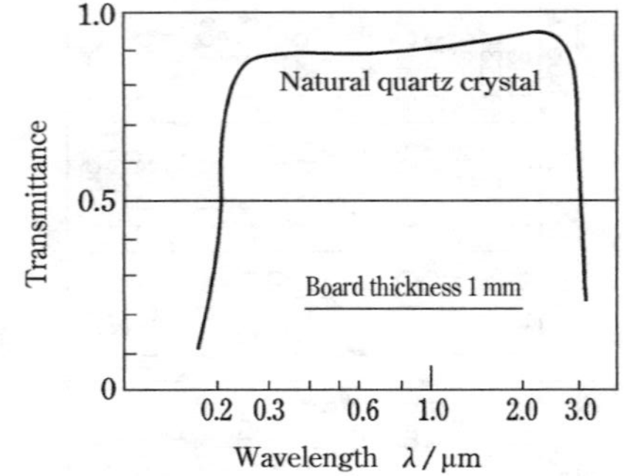

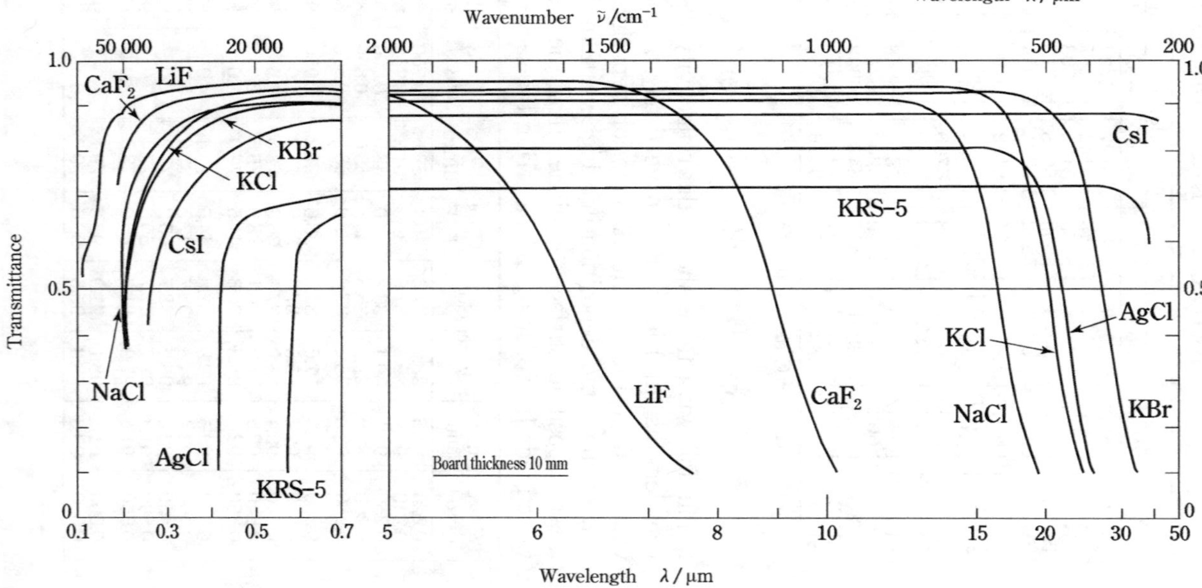

Polarization Rotation due to Magnetic Field

When linearly polarized monochromatic light propagates the length l/m in a medium located in the magnetic field H/A m^{-1} parallel to the direction of propagation, the polarization plane rotates at the angle α/min in a clockwise direction relative to the direction of propagation (Faraday effect). The rotation angle is expressed as $\alpha = VHl$, where V is Verdet constant of the material. V changes according to temperature and is inversely proportional to the square of wavelength λ approximately. For all materials in the following table, the sign of V is "+." 1 min = $(1/60)^\circ = (\pi/10\,800)$ rad.

Gas (0℃, 1 atm)	$V/10^{-6}$ min A^{-1}	Liquid (20℃)	$V/10^{-2}$ min A^{-1}
Oxygen	7.598	Water	1.645
Nitrogen	8.861	Acetone	1.42
Helium	0.667	Methyl alcohol	1.17
Hydrogen	8.867	Ethyl alcohol	1.41
Carbon dioxide	13.22	Chloroform	2.06
Methane	24.15		
Propane	50.05		
Iso-butane	68.12		

$\lambda = 589.3$ nm   1 atm = 101 325 Pa

Solid (20℃)	$V/10^{-2}$ min A^{-1}	Solid (20℃)	$V/10^{-2}$ min A^{-1}
Potassium chloride(KCl)	4.12*	Crystal (∥ to crystal axis)	2.091
Sodium chloride(NaCl)	5.15*	Fused Silica(25℃)	2.175*
Fluorite(CaF_2)	1.127	The glass	
Cupric chloride(I)(CuCl)	25*	Crown ($n_d = 1.519$)	2.4
Diamond(C)	2.93	Flint ($n_d = 1.623$)	4.85
Zinc sulphide(ZnS)	28.4	Heavy Flint ($n_d = 1.920$)	13.3

* $\lambda = 546.1$ nm, others $\lambda = 589.3$ nm.
Based primarily on Kaye & Laby, 1986.

Birefringence of Liquids in Electric Field

When an electric field is applied to isotropic gases and liquids, they behave as uniaxial crystals and have different indices of refraction for light polarized parallel ($n_{\parallel}$) to or perpendicular ($n_{\perp}$) to the applied field. Therefore, when light of wavelength λ propagates l/m perpendicular to the electric field E/V m^{-1}, the phase difference $\Gamma = l(n_{\parallel} - n_{\perp})/\lambda = KlE^2$ occurs between two polarization components which are parallel and perpendicular to the applied electric field. K is the Kerr constant and changes depending on wavelength and temperature. The table shows value for K when $\lambda = 589$ nm and temperature is 20℃.

Material	$K/10^{-12}$ m V^{-2}	Material	$K/10^{-12}$ m V^{-2}
Paraldehyde	−0.256	Benzenesulfonyl chloride	+1.00
Carbon disulfide	+0.036	*o*-Nitro toluene	+1.93
Water	+0.044	*m*-Nitro toluene	+1.97
Acetone	+0.181	*p*-Nitro toluene	+2.47
Benzaldehyde	+0.898	Nitrobenzene	+3.62

Optical Rotation of Quartz

The table shows the angle α of optical rotation when the linearly polarized light of wavelength λ propagates in quartz of 1 mm thickness along the optical axis at temperature of 20℃.

λ/nm	α/(°)	λ/nm	α/(°)	λ/nm	α/(°)	λ/nm	α/(°)
226.50	201.9	348.53	68.59	508.58	29.73	794.8	11.589
231.29	190.5	396.85	51.12	546.07	25.54	1014.1	6.976
250.33	153.9	410.17	47.50	589.29	21.72	1200.0	4.889
274.87	121.10	435.83	41.55	643.85	18.02	1529.6	2.930
303.41	95.02	467.82	35.60	670.79	16.54	2058.2	1.527
340.37	72.45	486.13	32.76	728.14	13.92	2500.0	0.972

Source: International Critical Tables

Optical Rotation Substances

$$[\alpha]^{D}_{20} = \left(\begin{array}{l}\text{rotation angle (°) of a polarization plane when the} \\ \text{sodium D line propagates through a solution of 10} \\ \text{cm-long and 20℃ in temperature.}\end{array}\right) \Big/ \left(\begin{array}{l}\text{amount (g) of the optical} \\ \text{rotation substance within} \\ \text{a solution of 1 cm}^3\end{array}\right)$$

In the table, c is the amount of the optical rotation substance (measured in g) within a solution of 100 cm^3, p is the amount of the optical rotation substance (measured in g) within a solution of 100 g (in other words, wt%) , and all solvent is water.

Material	Condition	$[\alpha]^{D}_{20}$/(°)
Fructose	$c = 10$	−104 (6 min after dissolution), −90 (33 min after dissolution)
Fructose	$p = 2-31$	$-91.9 - 0.11\,p$ (After reaching equilibrium.)
Saccharose	$c = 0-65$	$+66.462 + 0.00870\,c - 0.000235\,c^2$
Invert sugar	$p = 9-68$	$-19.447 - 0.06068\,p + 0.000221\,p^2$
Glucose (dextrorotation)	$c = 9.1$	+105.2 (5.5 min after dissolution), +52.5 (6 h after dissolution)
Glucose (dextrorotation)	$p = 1-18$	$+52.50 + 0.0188\,p + 0.000517\,p^2$ (After reaching equilibrium.)
Glucose (levorotation)	$p = 4$	−94.4 (7 min after dissolution), −51.4 (7 h after dissolution)
Tartaric acid	$p = 4.7-18$	$+14.83 - 0.149\,p$
Tartaric acid	$p = 18-36$	$+14.849 - 0.144\,p$
Potassium tartrate	$p = 9-55$	$+27.62 + 0.1064\,p - 0.00108\,p^2$
Rochelle salt	−	$+29.73 - 0.0078\,c$

Photometry and Colorimetry

Photometry The reciprocal of the energy of monochromatic light (wavelength λ) required to produce a certain sense of brightness for a standard observer is called the visual sensitivity at that wavelength λ. The monochromatic light with the greatest visual sensitivity is green light with a wavelength of 0.555 μm. The reciprocal of the energy of the wavelength λ is called the visual sensitivity of that wavelength. The ratio of the visual sensitivity of wavelength λ to the maximum visual sensitivity (K_m) is called the specific visual sensitivity (V_λ) of wavelength λ. This value is equal to the value of y- in the following table.

Spectral energy distribution of a radiant flux passing through a surface in unit time, In other words, if the energy for a unit wavelength interval at wavelength If the energy for a unit wavelength interval at wavelength λ is E (λ), then this is the standard The luminous flux F as seen by the observer is

$$F=K_m\int_{\lambda_1}^{\lambda_2} V_\lambda E(\lambda)\,d\lambda \text{ (Unit lm)}$$

This is calculated by $\lambda_1=0.380\ \mu$m, $\lambda_2=0.760\ \mu$m, and $K_m=683$ lm W^{-1}.

$1/K_m=0.001\ 464$ W lm^{-1} is also called the work equivalent of light.

Color measurement The chromaticity coordinates x, y, and z of the radiation with spectral energy distribution E (λ) in the International Commission on Illumination (CIE) XYZ color system are calculated by the following equation.

$$x=\frac{a}{(a+b+c)},\quad y=\frac{b}{(a+b+c)},\quad z=\frac{c}{(a+b+c)}$$

provided that

$$a=\int_{\lambda_1}^{\lambda_2}\bar{x}E(\lambda)\,d\lambda,\quad b=\int_{\lambda_1}^{\lambda_2}\bar{y}E(\lambda)\,d\lambda,\quad c=\int_{\lambda_1}^{\lambda_2}\bar{z}E(\lambda)\,d\lambda$$

where λ_1 and λ_2 are the same as above, and the values of $\bar{x}$, $\bar{y}$,, and $\bar{z}$, are given in the following table.

Chromaticity Coordinates of Monochromatic Light and Tristimulus Values of Isoenergetic Monochromatic Radiation (1)

Chromaticity coordinates of monochromatic light			Wavelength λ/μm	Tristimulus values for isoenergetic monochromatic radiation		
x	y	z		$\bar{x}$	$\bar{y}$*	$\bar{z}$
0.174 1	0.005 0	0.820 9	0.380	0.001 4	0.000 0	0.006 5
0.173 8	0.004 9	0.821 3	390	0.004 2	0.000 1	0.020 1
0.173 3	0.004 8	0.821 9	400	0.014 3	0.000 4	0.067 9
0.172 6	0.004 8	0.822 6	0.410	0.043 5	0.001 2	0.207 4
0.171 4	0.005 1	0.823 5	420	0.134 4	0.004 0	0.645 6
0.168 9	0.006 9	0.824 2	430	0.283 9	0.011 6	1.385 6
0.164 4	0.010 9	0.824 7	440	0.348 3	0.023 0	1.747 1
0.156 6	0.017 7	0.825 7	450	0.336 2	0.038 0	1.772 1

Chromaticity Coordinates of Monochromatic Light and Tristimulus Values of Isoenergetic Monochromatic Radiation (2)

Chromaticity coordinates of monochromatic light			Wavelength $\lambda/\mu m$	Tristimulus values for isoenergetic monochromatic radiation		
x	y	z		$\bar{x}$	$\bar{y}$*	$\bar{z}$
0.144 0	0.029 7	0.826 3	0.460	0.290 8	0.060 0	1.669 2
0.124 1	0.057 8	0.818 1	470	0.195 4	0.091 0	1.287 6
0.091 3	0.132 7	0.776 0	480	0.095 6	0.139 0	0.813 0
0.045 4	0.295 0	0.659 6	490	0.032 0	0.208 0	0.465 2
0.008 2	0.538 4	0.453 4	500	0.004 9	0.323 0	0.272 0
0.013 9	0.750 2	0.235 9	0.510	0.009 3	0.503 0	0.158 2
0.074 3	0.833 8	0.091 9	520	0.063 3	0.710 0	0.078 2
0.154 7	0.805 9	0.039 4	530	0.165 5	0.862 0	0.042 2
0.229 6	0.754 3	0.016 1	540	0.290 4	0.954 0	0.020 3
0.301 6	0.692 3	0.006 1	550	0.433 4	0.995 0	0.008 7
0.373 1	0.624 5	0.002 4	0.560	0.594 5	0.995 0	0.003 9
0.444 1	0.554 7	0.001 2	570	0.762 1	0.952 0	0.002 1
0.512 5	0.486 6	0.000 9	580	0.916 3	0.870 0	0.001 7
0.575 2	0.424 2	0.000 6	590	1.026 3	0.757 0	0.001 1
0.627 0	0.372 5	0.000 5	600	1.062 2	0.631 0	0.000 8
0.665 8	0.334 0	0.000 2	0.610	1.002 6	0.503 0	0.000 3
0.691 5	0.308 3	0.000 2	620	0.854 4	0.381 0	0.000 2
0.707 9	0.292 0	0.000 1	630	0.642 4	0.265 0	0.000 0
0.719 0	0.280 9	0.000 1	640	0.447 9	0.175 0	0.000 0
0.726 0	0.274 0	0.000 0	650	0.283 5	0.107 0	0.000 0
0.730 0	0.270 0	0.000 0	0.660	0.164 9	0.061 0	0.000 0
0.732 0	0.268 0	0.000 0	670	0.087 4	0.032 0	0.000 0
0.733 4	0.266 6	0.000 0	680	0.046 8	0.017 0	0.000 0
0.734 4	0.265 6	0.000 0	690	0.022 7	0.008 2	0.000 0
0.734 7	0.265 3	0.000 0	700	0.011 4	0.004 1	0.000 0
0.734 7	0.265 3	0.000 0	0.710	0.005 8	0.002 1	0.000 0
0.734 7	0.265 3	0.000 0	720	0.002 9	0.001 0	0.000 0
0.734 7	0.265 3	0.000 0	730	0.001 4	0.000 5	0.000 0
0.734 7	0.265 3	0.000 0	740	0.000 7	0.000 3	0.000 0
0.734 7	0.265 3	0.000 0	750	0.000 3	0.000 1	0.000 0
0.734 7	0.265 3	0.000 0	0.760	0.000 2	0.000 1	0.000 0
0.734 7	0.265 3	0.000 0	770	0.000 1	0.000 0	0.000 0
0.734 7	0.265 3	0.000 0	780	0.000 0	0.000 0	0.000 0

* It is equal to the specific visual sensitivity V_λ.

Standard light for colorimetry

The International Commission on Illumination (CIE) used to specify three types of light as standards for colorimetry A, B, and C (1931), but B, which was not widely used, was excluded and D_{65} was newly introduced (1964).

Standard light A Represents the spectral energy distribution corresponding to Planck's radiation formula at 2 856 K. This is achieved as radiation from a standard light source A, a gas-filled tungsten bulb with a color temperature of 2 856 K (T_{68}).

Standard light C It has a spectral energy distribution almost identical to that of sunlight, including blue sky light, and a color temperature equivalent to about 6774 K. This is obtained by passing the light from the standard light source A through the following two solution filters C_1 and C_2, each 10 mm thick.

C_1 Solution	
Copper sulfate $CuSO_4 \cdot 5H_2O$	3.412 g
Mannitol $C_6H_8(OH)_6$	3.412 g
Pyridine C_5H_5N	30.0 mL
Add distilled water to make the total volume	1 L
C_2 Solution	
Ammonium cobalt sulfate $CoSO_4(NH_4)_2SO_4 \cdot 6H_2O$	30.580 g
Copper sulfate $CuSO_4 \cdot 5H_2O$	22.520 g
Sulfuric acid (specific gravity 1.835)	10.0 mL
Add distilled water to make the total volume	1 L

Standard light D_{65} This corresponds to the spectral energy distribution of daylight with a color temperature of about 6 504 K (T_{68}). However, a light source that can achieve this with sufficient accuracy has not yet been specified.

Atoms, Atomic Nuclei, Elementary Particles

Stable Isotopes

Isotopes are expressed in the format "$^{A}_{Z}$X," where Z is the atomic number, A is the mass number, and X is the chemical symbol. The mass in this table is expressed in atomic mass unit which uses "12" for the mass of neutral ^{12}C atom. This is in accordance with M. Wang et al., Chinese Physics, 41 (2017) 030003. Isotopic abundance is in accordance with J. Meija et al., Pure and Applied Chemistry 88 (2016) 293. The number listed in parenthesis for the abundance is the uncertainty. This applies to the final number. This table is a summary of stable isotopes. However, it also displays values for radioactive isotopes which have a long half-life (marked with a †) and which show certain natural isotopic composition. Note that the values in this table may not correspond with the atomic weight.

Isotope	Atomic Mass	Abundance (atomic percent)
$^{1}_{1}$H	1.007 825 032 24	[99.972, 99.999]
$^{2}_{1}$H($^{2}_{1}$D)	2.014 101 778 11	[0.001, 0.028]
$^{3}_{2}$He	3.016 029 322 65	0.0002(2)*1
$^{4}_{2}$He	4.002 603 254 13	99.9998(2)
$^{6}_{3}$Li	6.015 122 887 4	[1.9, 7.8]*2
$^{7}_{3}$Li	7.016 003 437	[92.2, 98.1]
$^{9}_{4}$Be	9.012 183 07	100
$^{10}_{5}$B	10.012 936 862	[18.9, 20.4]
$^{11}_{5}$B	11.009 305 167	[79.6, 81.1]
$^{12}_{6}$C	12.000 000 0	[98.84, 99.04]
$^{13}_{6}$C	13.003 354 835 21	[0.96, 1.16]
$^{14}_{7}$N	14.003 074 004 46	[99.578, 99.663]*1
$^{15}_{7}$N	15.000 108 898 9	[0.337, 0.422]
$^{16}_{8}$O	15.994 914 619 60	[99.738, 99.776]
$^{17}_{8}$O	16.999 131 756 6	[0.0367, 0.0400]
$^{18}_{8}$O	17.999 159 612 8	[0.187, 0.222]
$^{19}_{9}$F	18.998 403 162 9	100
$^{20}_{10}$Ne	19.992 440 176 2	90.48(3)*1
$^{21}_{10}$Ne	20.993 846 69	0.27(1)
$^{22}_{10}$Ne	21.991 385 110	9.25(3)
$^{23}_{11}$Na	22.989 769 282 0	100
$^{24}_{12}$Mg	23.985 041 697	[78.88, 79.05]
$^{25}_{12}$Mg	24.985 836 96	[9.988, 10.034]
$^{26}_{12}$Mg	25.982 592 97	[10.96, 11.09]
$^{27}_{13}$Al	26.981 538 41	100
$^{28}_{14}$Si	27.976 926 535 0	[92.191, 92.318]
$^{29}_{14}$Si	28.976 494 665 3	[4.645, 4.699]
$^{30}_{14}$Si	29.973 770 137	[3.037, 3.110]
$^{31}_{15}$P	30.973 761 998 6	100
$^{32}_{16}$S	31.972 071 174 4	[94.41, 95.29]
$^{33}_{16}$S	32.971 458 909 9	[0.729, 0.797]
$^{34}_{16}$S	33.967 867 01	[3.96, 4.77]
$^{36}_{16}$S	35.967 080 70	[0.0129, 0.0187]
$^{35}_{17}$Cl	34.968 852 69	[75.5, 76.1]
$^{37}_{17}$Cl	36.965 902 58	[23.9, 24.5]

*1 Value of the gasious elements in air (applying to all isotopes of the element concerned)

*2 Care is required because lithium and uranium which have been processed to remove ^{6}Li and ^{235}U, respectively, are commercially available.

Continued.

Isotope	Atomic Mass	Abundance (atomic percent)
$^{36}_{18}\mathrm{Ar}$	35.967 545 105	0.3336(210) *1
$^{38}_{18}\mathrm{Ar}$	37.962 732 10	0.0629(70)
$^{40}_{18}\mathrm{Ar}$	39.962 383 1238	99.6035(250)
$^{39}_{19}\mathrm{K}$	38.963 706 487	93.2581(44)
$^{40}_{19}\mathrm{K}^{\dagger}$	39.963 998 17	0.0117(1)
$^{41}_{19}\mathrm{K}$	40.961 825 258	6.7302(44)
$^{40}_{20}\mathrm{Ca}$	39.962 590 866	96.941(156) *3
$^{42}_{20}\mathrm{Ca}$	41.958 617 83	0.647(23)
$^{43}_{20}\mathrm{Ca}$	42.958 766 43	0.135(10)
$^{44}_{20}\mathrm{Ca}$	43.955 481 5	2.086(110)
$^{46}_{20}\mathrm{Ca}$	45.953 688 0	0.004(3)
$^{48}_{20}\mathrm{Ca}$	47.952 522 90	0.187(21)
$^{45}_{21}\mathrm{Sc}$	44.955 907 5	100
$^{46}_{22}\mathrm{Ti}$	45.952 626 86	8.25(3)
$^{47}_{22}\mathrm{Ti}$	46.951 757 75	7.44(2)
$^{48}_{22}\mathrm{Ti}$	47.947 940 93	73.72(3)
$^{49}_{22}\mathrm{Ti}$	48.947 864 63	5.41(2)
$^{50}_{22}\mathrm{Ti}$	49.944 785 84	5.18(2)
$^{50}_{23}\mathrm{V}$	49.947 155 8	0.250(10)
$^{51}_{23}\mathrm{V}$	50.943 956 9	99.750(10)
$^{50}_{24}\mathrm{Cr}$	49.946 041 4	4.345(13)
$^{52}_{24}\mathrm{Cr}$	51.940 505 0	83.789(18)
$^{53}_{24}\mathrm{Cr}$	52.940 647 0	9.501(17)
$^{54}_{24}\mathrm{Cr}$	53.938 878 0	2.365(7)
$^{55}_{25}\mathrm{Mn}$	54.938 043 2	100
$^{54}_{26}\mathrm{Fe}$	53.939 608 3	5.845(105)
$^{56}_{26}\mathrm{Fe}$	55.934 935 6	91.754(106)
$^{57}_{26}\mathrm{Fe}$	56.935 392 1	2.119(29)
$^{58}_{26}\mathrm{Fe}$	57.933 273 7	0.282(12)
$^{59}_{27}\mathrm{Co}$	58.933 193 7	100
$^{58}_{28}\mathrm{Ni}$	57.935 341 8	68.0769(190)

Isotope	Atomic Mass	Abundance (atomic percent)
$^{60}_{28}\mathrm{Ni}$	59.930 785 3	26.2231(150)
$^{61}_{28}\mathrm{Ni}$	60.931 054 9	1.1399(13)
$^{62}_{28}\mathrm{Ni}$	61.928 344 9	3.6345(40)
$^{64}_{28}\mathrm{Ni}$	63.927 966 3	0.9256(19)
$^{63}_{29}\mathrm{Cu}$	62.929 597 2	69.15(15)
$^{65}_{29}\mathrm{Cu}$	64.927 789 5	30.85(15)
$^{64}_{30}\mathrm{Zn}$	63.929 141 8	49.17(75)
$^{66}_{30}\mathrm{Zn}$	65.926 033 7	27.73(98)
$^{67}_{30}\mathrm{Zn}$	66.927 127 5	4.04(16)
$^{68}_{30}\mathrm{Zn}$	67.924 844 3	18.45(63)
$^{70}_{30}\mathrm{Zn}$	69.925 319 2	0.61(10)
$^{69}_{31}\mathrm{Ga}$	68.925 573 5	60.108(50)
$^{71}_{31}\mathrm{Ga}$	70.924 702 5	39.892(50)
$^{70}_{32}\mathrm{Ge}$	69.924 248 7	20.52(19)
$^{72}_{32}\mathrm{Ge}$	71.922 075 83	27.45(15)
$^{73}_{32}\mathrm{Ge}$	72.923 458 96	7.76(8)
$^{74}_{32}\mathrm{Ge}$	73.921 177 762	36.52(12)
$^{76}_{32}\mathrm{Ge}$	75.921 402 727	7.75(12)
$^{75}_{33}\mathrm{As}$	74.92 1594 6	100
$^{74}_{34}\mathrm{Se}$	73.922 475 935	0.86(3)
$^{76}_{34}\mathrm{Se}$	75.919 213 704	9.23(7)
$^{77}_{34}\mathrm{Se}$	76.919 914 15	7.60(7)
$^{78}_{34}\mathrm{Se}$	77.917 309 24	23.69(22)
$^{80}_{34}\mathrm{Se}$	79.916 521 8	49.80(36)
$^{82}_{34}\mathrm{Se}$	81.916 699 5	8.82(15)
$^{79}_{35}\mathrm{Br}$	78.918 337 6	[50.5, 50.8]
$^{81}_{35}\mathrm{Br}$	80.916 288 2	[49.2, 49.5]
$^{78}_{36}\mathrm{Kr}$	77.920 366 3	0.355(3) *1
$^{80}_{36}\mathrm{Kr}$	79.916 378 0	2.286(10)
$^{82}_{36}\mathrm{Kr}$	81.913 481 155	11.593(31)
$^{83}_{36}\mathrm{Kr}$	82.914 126 518	11.500(19)
$^{84}_{36}\mathrm{Kr}$	83.911 497 729	56.987(15)
$^{86}_{36}\mathrm{Kr}$	85.910 610 626	17.279(41)

*3 Because they increase through radioactive decay, there are cases where the amounts for the isotopes considered exceed the ranges listed in this table.

Stable Isotopes Continued.

Isotope	Atomic Mass	Abundance (atomic percent)
$^{85}_{37}$Rb	84.911 789 738	72.17(2)
$^{87}_{37}$Rb†	86.909 180 531	27.83(2)
$^{84}_{38}$Sr	83.913 419 1	0.56(2)
$^{86}_{38}$Sr	85.909 260 726	9.86(20)
$^{87}_{38}$Sr	86.908 877 496	7.00(20)^{*3}
$^{88}_{38}$Sr	87.905 612 256	82.58(35)
$^{89}_{39}$Y	88.905 841 2	100
$^{90}_{40}$Zr	89.904 698 76	51.45(4)
$^{91}_{40}$Zr	90.905 640 22	11.22(5)
$^{92}_{40}$Zr	91.905 035 32	17.15(3)
$^{94}_{40}$Zr	93.906 312 52	17.38(4)
$^{96}_{40}$Zr	95.908 277 62	2.80(2)
$^{93}_{41}$Nb	92.906 373 2	100
$^{92}_{42}$Mo	91.906 807 16	14.649(106)
$^{94}_{42}$Mo	93.905 083 59	9.187(33)
$^{95}_{42}$Mo	94.905 837 44	15.873(30)
$^{96}_{42}$Mo	95.904 674 77	16.673(8)
$^{97}_{42}$Mo	96.906 016 90	9.582(15)
$^{98}_{42}$Mo	97.905 403 61	24.292(80)
$^{100}_{42}$Mo	99.907 468 0	9.744(65)
$^{96}_{44}$Ru	95.907 588 91	5.54(14)
$^{98}_{44}$Ru	97.905 287	1.87(3)
$^{99}_{44}$Ru	98.905 930 3	12.76(14)
$^{100}_{44}$Ru	99.904 210 5	12.60(7)
$^{101}_{44}$Ru	100.905 573 1	17.06(2)
$^{102}_{44}$Ru	101.904 340 3	31.55(14)
$^{104}_{44}$Ru	103.905 425 4	18.62(27)
$^{103}_{45}$Rh	102.905 494 1	100
$^{102}_{46}$Pd	101.905 632 1	1.02(1)
$^{104}_{46}$Pd	103.904 030 4	11.14(8)
$^{105}_{46}$Pd	104.905 079 5	22.33(8)
$^{106}_{46}$Pd	105.903 480 3	27.33(3)
$^{108}_{46}$Pd	107.903 891 8	26.46(9)
$^{110}_{46}$Pd	109.905 172 9	11.72(9)
$^{107}_{47}$Ag	106.905 091 5	51.839(8)
$^{109}_{47}$Ag	108.904 755 8	48.161(8)
$^{106}_{48}$Cd	105.906 459 8	1.245(22)
$^{108}_{48}$Cd	107.904 183 6	0.888(11)
$^{110}_{48}$Cd	109.903 007 5	12.470(61)
$^{111}_{48}$Cd	110.904 183 8	12.795(12)
$^{112}_{48}$Cd	111.902 763 88	24.109(7)
$^{113}_{48}$Cd	112.904 408 10	12.227(7)
$^{114}_{48}$Cd	113.903 364 99	28.754(81)
$^{116}_{48}$Cd	115.904 763 23	7.512(54)
$^{113}_{49}$In	112.904 060 45	4.281(52)
$^{115}_{49}$In†	114.903 878 774	95.719(52)
$^{112}_{50}$Sn	111.904 824 9	0.97(1)
$^{114}_{50}$Sn	113.902 780 13	0.66(1)
$^{115}_{50}$Sn	114.903 344 697	0.34(1)
$^{116}_{50}$Sn	115.901 742 82	14.54(9)
$^{117}_{50}$Sn	116.902 954 0	7.68(7)
$^{118}_{50}$Sn	117.901 606 6	24.22(9)
$^{119}_{50}$Sn	118.903 311 2	8.59(4)
$^{120}_{50}$Sn	119.902 201 9	32.58(9)
$^{122}_{50}$Sn	121.903 444 0	4.63(3)
$^{124}_{50}$Sn	123.905 276 7	5.79(5)
$^{121}_{51}$Sb	120.903 810 1	57.21(5)
$^{123}_{51}$Sb	122.904 214 0	42.79(5)
$^{120}_{52}$Te	119.904 060	0.09(1)
$^{122}_{52}$Te	121.903 043 4	2.55(12)
$^{123}_{52}$Te†	122.904 269 7	0.89(3)
$^{124}_{52}$Te	123.902 817 1	4.74(14)
$^{125}_{52}$Te	124.904 429 9	7.07(15)
$^{126}_{52}$Te	125.903 310 9	18.84(25)
$^{128}_{52}$Te	127.904 461 3	31.74(8)
$^{130}_{52}$Te	129.906 222 747	34.08(62)

Continued.

Isotope	Atomic Mass	Abundance (atomic percent)
$^{127}_{53}$I	126.904 472	100
$^{124}_{54}$Xe	123.905 891 6	0.095(5)[*1]
$^{126}_{54}$Xe	125.904 297	0.089(3)
$^{128}_{54}$Xe	127.903 531 0	1.910(13)
$^{129}_{54}$Xe	128.904 780 859	26.401(138)
$^{130}_{54}$Xe	129.903 509 349	4.071(22)
$^{131}_{54}$Xe	130.905 084 136	21.232(51)
$^{132}_{54}$Xe	131.904 155 087	26.909(55)
$^{134}_{54}$Xe	133.905 393 034	10.436(35)
$^{136}_{54}$Xe	135.907 214 476	8.857(72)
$^{133}_{55}$Cs	132.905 451 961	100
$^{130}_{56}$Ba	129.906 320 9	0.11(1)
$^{132}_{56}$Ba	131.905 061 1	0.10(1)
$^{134}_{56}$Ba	133.904 508 4	2.42(15)
$^{135}_{56}$Ba	134.905 688 6	6.59(10)
$^{136}_{56}$Ba	135.904 576 0	7.85(24)
$^{137}_{56}$Ba	136.905 827 4	11.23(23)
$^{138}_{56}$Ba	137.905 247 2	71.70(29)
$^{138}_{57}$La†	137.907 118	0.08881(71)
$^{139}_{57}$La	138.906 358 8	99.91119(71)
$^{136}_{58}$Ce	135.907 129 4	0.186(2)
$^{138}_{58}$Ce	137.905 989	0.251(2)[*3]
$^{140}_{58}$Ce	139.905 446 4	88.449(51)
$^{142}_{58}$Ce	141.909 249 9	11.114(51)
$^{141}_{59}$Pr	140.907 658 4	100
$^{142}_{60}$Nd	141.907 728 9	27.153(40)
$^{143}_{60}$Nd	142.909 819 9	12.173(26)[*3]
$^{144}_{60}$Nd†	143.910 092 9	23.798(19)
$^{145}_{60}$Nd	144.912 579 2	8.293(12)
$^{146}_{60}$Nd	145.913 122 5	17.189(32)
$^{148}_{60}$Nd	147.916 899 1	5.756(21)
$^{150}_{60}$Nd	149.920 901 5	5.638(28)

Isotope	Atomic Mass	Abundance (atomic percent)
$^{144}_{62}$Sm	143.912 006 4	3.08(4)
$^{147}_{62}$Sm†	146.914 904 1	15.00(14)
$^{148}_{62}$Sm†	147.914 829 0	11.25(9)
$^{149}_{62}$Sm	148.917 191 4	13.82(10)
$^{150}_{62}$Sm	149.917 282 2	7.37(9)
$^{152}_{62}$Sm	151.919 739 0	26.74(9)
$^{154}_{62}$Sm	153.922 216 2	22.74(14)
$^{151}_{63}$Eu	150.919 856 9	47.81(6)
$^{153}_{63}$Eu	152.921 237 0	52.19(6)
$^{152}_{64}$Gd†	151.919 798 8	0.20(3)
$^{154}_{64}$Gd	153.920 873 4	2.18(2)
$^{155}_{64}$Gd	154.922 629 8	14.80(9)
$^{156}_{64}$Gd	155.922 130 6	20.47(3)
$^{157}_{64}$Gd	156.923 967 9	15.65(4)
$^{158}_{64}$Gd	157.924 111 6	24.84(8)
$^{160}_{64}$Gd	159.927 061 5	21.86(3)
$^{159}_{65}$Tb	158.925 353 9	100
$^{156}_{66}$Dy	155.924 284 0	0.056(3)
$^{158}_{66}$Dy	157.924 414 6	0.095(3)
$^{160}_{66}$Dy	159.925 203 2	2.329(18)
$^{161}_{66}$Dy	160.926 939 1	18.889(42)
$^{162}_{66}$Dy	161.926 804 2	25.475(36)
$^{163}_{66}$Dy	162.928 736 9	24.896(42)
$^{164}_{66}$Dy	163.929 180 5	28.260(54)
$^{165}_{67}$Ho	164.930 328 0	100
$^{162}_{68}$Er	161.928 787 0	0.139(5)
$^{164}_{68}$Er	163.929 207 4	1.601(3)
$^{166}_{68}$Er	165.930 299 0	33.503(36)
$^{167}_{68}$Er	166.932 054 1	22.869(9)
$^{168}_{68}$Er	167.932 376 2	26.978(18)
$^{170}_{68}$Er	169.935 470 7	14.910(36)
$^{169}_{69}$Tm	168.934 218 4	100

Stable Isotopes Continued.

Isotope	Atomic Mass	Abundance (atomic percent)
$^{168}_{70}$Yb	167.933 889 1	0.123(3)
$^{170}_{70}$Yb	169.934 767 246	2.982(39)
$^{171}_{70}$Yb	170.936 331 517	14.086(140)
$^{172}_{70}$Yb	171.936 386 659	21.686(130)
$^{173}_{70}$Yb	172.938 216 215	16.103(63)
$^{174}_{70}$Yb	173.938 867 548	32.026(80)
$^{176}_{70}$Yb	175.942 574 709	12.995(83)
$^{175}_{71}$Lu	174.940 777 3	97.401(13)
$^{176}_{71}$Lu†	175.942 691 8	2.599(13)
$^{174}_{72}$Hf	173.940 048 5	0.16(12)
$^{176}_{72}$Hf	175.941 409 9	5.26(70)*3
$^{177}_{72}$Hf	176.943 230 3	18.60(16)
$^{178}_{72}$Hf	177.943 708 5	27.28(28)
$^{179}_{72}$Hf	178.945 825 8	13.62(11)
$^{180}_{72}$Hf	179.946 559 7	35.08(33)
$^{180}_{73}$Ta	179.947 468 4	0.01201(32)
$^{181}_{73}$Ta	180.947 999 3	99.98799(32)
$^{180}_{74}$W	179.946 713 4	0.12(1)
$^{182}_{74}$W	181.948 205 7	26.50(16)
$^{183}_{74}$W	182.950 224 5	14.31(4)
$^{184}_{74}$W	183.950 933 3	30.64(2)
$^{186}_{74}$W	185.954 365 2	28.43(19)
$^{185}_{75}$Re	184.952 958 3	37.40(5)
$^{187}_{75}$Re†	186.955 752 3	62.60(5)
$^{184}_{76}$Os	183.952 492 9	0.02(2)
$^{186}_{76}$Os†	185.953 837 7	1.59(64)
$^{187}_{76}$Os	186.955 749 6	1.96(17)*3
$^{188}_{76}$Os	187.955 837 4	13.24(27)
$^{189}_{76}$Os	188.958 146 0	16.15(23)
$^{190}_{76}$Os	189.958 445 5	26.26(20)
$^{192}_{76}$Os	191.961 478 9	40.78(32)
$^{191}_{77}$Ir	190.960 591 5	37.3(2)
$^{193}_{77}$Ir	192.962 923 8	62.7(2)
$^{190}_{78}$Pt	189.959 949 9	0.012(2)
$^{192}_{78}$Pt	191.961 042 7	0.782(24)
$^{194}_{78}$Pt	193.962 683 5	32.864(410)
$^{195}_{78}$Pt	194.964 794 4	33.775(240)
$^{196}_{78}$Pt	195.964 954 7	25.211(340)
$^{198}_{78}$Pt	197.967 896 7	7.356(130)
$^{197}_{79}$Au	196.966 570 1	100
$^{196}_{80}$Hg	195.965 833	0.15(1)
$^{198}_{80}$Hg	197.966 769 2	10.04(3)
$^{199}_{80}$Hg	198.968 281 0	16.94(12)
$^{200}_{80}$Hg	199.968 326 9	23.14(9)
$^{201}_{80}$Hg	200.970 303 0	13.17(9)
$^{202}_{80}$Hg	201.970 643 6	29.74(13)
$^{204}_{80}$Hg	203.973 494 0	6.82(4)
$^{203}_{81}$Tl	202.972 344 0	[29.44, 29.59]
$^{205}_{81}$Tl	204.974 427 2	[70.41, 70.56]
$^{204}_{82}$Pb	203.973 043 4	1.4(6)
$^{206}_{82}$Pb	205.974 465 1	24.1(30)*3
$^{207}_{82}$Pb	206.975 896 7	22.1(50)*3
$^{208}_{82}$Pb	207.976 651 9	52.4(70)*3
$^{209}_{83}$Bi†	208.980 398 5	100
$^{230}_{90}$Th†	230.033 132 4	0.02(2)
$^{232}_{90}$Th†	232.038 053 7	99.98(2)
$^{231}_{91}$Pa†	231.035 882 6	100
$^{234}_{92}$U†	234.040 950 4	0.0054(5)
$^{235}_{92}$U†	235.043 928 2	0.7204(6)*2
$^{238}_{92}$U†	238.050 787 0	99.2742(10)

(Source: Mitsuru Ebihara)

Main Radionuclides (Radioisotopes)

From among the currently known radionuclides (radioisotopes), this table shows a selection of about 420 main radionuclides, and list their half-lives and decay modes.

The number shown at the upper left of the nuclide chemical symbol is the mass number. *m* indicates the metastable state.

The symbols used in the section for half-life and decay modes are as follows.

y: years (=365.242 d); d: days; h: hours; m: minutes; s: seconds;

α: α Decay; β^-: β^- Decay; β^+: β^+ Decay; EC: Orbital electron capture; IT: Nuclear isomer transition, SF: Spontaneous fission.

Nuclide	Half-life	Decay mode	Nuclide	Half-life	Decay mode
^{3}H	12.32 y	β^-	^{42}K	12.355 h	β^-
^{7}Be	53.22 d	EC	^{43}K	22.3 h	β^-
^{10}Be	1.51×10^6 y	β^-	^{45}Ca	162.61 d	β^-
^{11}C	20.364 m	β^+, EC	^{47}Ca	4.536 d	β^-
^{14}C	5.70×10^3 y	β^-	^{44m}Sc	58.61 h	IT, EC
^{13}N	9.965 m	β^+, EC	^{44}Sc	3.97 h	β^+, EC
^{15}O	122.24 s	β^+, EC	^{46}Sc	83.79 d	β^-
^{18}F	109.77 m	β^+, EC	^{47}Sc	3.3492 d	β^-
^{22}Na	2.6018 y	β^+, EC	^{48}Sc	43.67 h	β^-
^{24}Na	14.997 h	β^-	^{49}Sc	57.18 m	β^-
^{27}Mg	9.458 m	β^-	^{44}Ti	59.1 y	EC
^{28}Mg	20.915 h	β^-	^{45}Ti	184.8 m	β^+, EC
^{26}Al	7.17×10^5 y	β^+, EC	^{51}Ti	5.76 m	β^-
^{28}Al	2.245 m	β^-	^{48}V	15.9735 d	EC, β^+
^{31}Si	157.36 m	β^-	^{49}V	330 d	EC
^{30}P	2.498 m	β^+, EC	^{52}V	3.743 m	β^-
^{32}P	14.268 d	β^-	^{51}Cr	27.7010 d	EC
^{33}P	25.35 d	β^-	^{52m}Mn	21.1 m	β^+, EC, IT
^{35}S	87.37 d	β^-	^{52}Mn	5.591 d	EC, β^+
^{36}Cl	3.013×10^5 y	β^-, EC, β^+	^{53}Mn	3.74×10^6 y	EC
^{38}Cl	37.230 m	β^-	^{54}Mn	312.20 d	EC
^{37}Ar	35.011 d	EC	^{56}Mn	2.5789 h	β^-
^{41}Ar	109.61 m	β^-	^{52}Fe	8.275 h	β^+, EC
^{42}Ar	32.9 y	β^-	^{55}Fe	2.744 y	EC
^{40}K	1.248×10^9 y	β^-, EC	^{59}Fe	44.490 d	β^-

The half-lives and decay modes are based on National Nuclear Data Center, Brookhaven National Laboratory, Evaluated Nuclear Structure Data File (ENSDF) (2020-06), http://www.nndc.bnl.gov/ensdf/index.jsp

Main Radionuclides Continued.

Nuclide	Half-life	Decay mode	Nuclide	Half-life	Decay mode
^{55}Co	17.53 h	EC, β^+	^{73}As	80.30 d	EC
^{56}Co	77.236 d	EC, β^+	^{74}As	17.77 d	EC, β^+, β^-
^{57}Co	271.74 d	EC	^{76}As	26.24 h	β^-
^{58m}Co	9.10 h	IT	^{77}As	38.79 h	β^-
^{58}Co	70.86 d	EC, β^+	^{72}Se	8.40 d	EC
^{60m}Co	10.467 m	IT, β^-	^{75}Se	119.78 d	EC
^{60}Co	5.2712 y	β^-	^{77m}Se	17.36 s	IT
^{56}Ni	6.075 d	EC	^{79}Se	3.27×10^5 y	β^-
^{57}Ni	35.60 h	EC, β^+	^{81m}Se	57.28 m	IT, β^-
^{59}Ni	7.6×10^4 y	EC	^{81}Se	18.45 m	β^-
^{63}Ni	101.2 y	β^-	^{76}Br	16.2 h	EC, β^+
^{65}Ni	2.5175 h	β^-	^{77}Br	57.04 h	EC, β^+
^{66}Ni	54.6 h	β^-	^{80m}Br	4.4205 h	IT
^{61}Cu	3.339 h	β^+, EC	^{80}Br	17.68 m	β^-, EC, β^+
^{62}Cu	9.67 m	β^+, EC	^{82}Br	35.282 h	β^-
^{64}Cu	12.701 h	EC, β^+, β^-	^{83}Br	2.374 h	β^-
^{66}Cu	5.120 m	β^-	^{79}Kr	35.04 h	EC, β^+
^{67}Cu	61.83 h	β^-	^{81m}Kr	13.10 s	IT, EC
^{62}Zn	9.193 h	EC, β^+	^{81}Kr	2.29×10^5 y	EC
^{63}Zn	38.47 m	β^+, EC	^{83m}Kr	1.83 h	IT
^{65}Zn	243.93 d	EC, β^+	^{85m}Kr	4.480 h	β^-, IT
^{69m}Zn	13.756 h	IT, β^-	^{85}Kr	10.739 y	β^-
^{69}Zn	56.4 m	β^-	^{81m}Rb	30.5 m	IT, EC
^{72}Zn	46.5 h	β^-	^{81}Rb	4.572 h	EC, β^+
^{66}Ga	9.49 h	β^+, EC	^{82}Rb	1.2575 m	β^+, EC
^{67}Ga	3.2617 d	EC	^{83}Rb	86.2 d	EC
^{68}Ga	67.71 m	β^+, EC	^{84}Rb	32.82 d	EC, β^+, β^-
^{70}Ga	21.14 m	β^-, EC	^{86}Rb	18.642 d	β^-, EC
^{72}Ga	14.10 h	β^-	^{87}Rb	4.81×10^{10} y	β^-
^{68}Ge	270.93 d	EC	^{88}Rb	17.773 m	β^-
^{69}Ge	39.05 h	EC	^{82}Sr	25.35 d	EC
^{71}Ge	11.43 d	EC	^{83}Sr	32.41 h	β^+, EC
^{75}Ge	82.78 m	β^-	^{85}Sr	64.849 d	EC
^{77m}Ge	53.7 s	β^-, IT	^{87m}Sr	2.815 h	IT, EC
^{77}Ge	11.211 h	β^-	^{89}Sr	50.563 d	β^-
^{71}As	65.30 h	β^+, EC	^{90}Sr	28.91 y	β^-
^{72}As	26.0 h	β^+, EC	^{91}Sr	9.65 h	β^-

Continued.

Nuclide	Half-life	Decay mode	Nuclide	Half-life	Decay mode
^{86m}Y	47.4 m	IT, β^+, EC	^{109}Pd	13.59 h	β^-
^{86}Y	14.74 h	EC, β^+	^{111}Pd	23.4 m	β^-
^{87}Y	79.8 h	EC, β^+	^{112}Pd	21.04 h	β^-
^{88}Y	106.626 d	EC, β^+	^{105}Ag	41.29 d	EC
^{90}Y	64.05 h	β^-	^{107m}Ag	44.3 s	IT
^{91m}Y	49.71 m	IT	^{108}Ag	2.382 m	β^-, EC, β^+
^{91}Y	58.51 d	β^-	^{109m}Ag	39.79 s	IT
^{88}Zr	83.4 d	EC	^{110m}Ag	249.83 d	β^-, IT
^{89m}Zr	4.161 m	IT, EC, β^+	^{110}Ag	24.56 s	β^-, EC
^{89}Zr	78.41 h	β^+, EC	^{111m}Ag	64.8 s	IT, β^-
^{93}Zr	1.61×10^6 y	β^-	^{111}Ag	7.45 d	β^-
^{95}Zr	64.032 d	β^-	^{112}Ag	3.130 h	β^-
^{97}Zr	16.749 h	β^-	^{107}Cd	6.50 h	EC, β^+
^{90}Nb	14.60 h	β^+, EC	^{109}Cd	461.9 d	EC
^{92m}Nb	10.15 d	EC, β^+	^{111m}Cd	48.50 m	IT
^{93m}Nb	16.12 y	IT	^{115m}Cd	44.56 d	β^-
^{94}Nb	2.03×10^4 y	β^-	^{115}Cd	53.46 h	β^-
^{95m}Nb	3.61 d	IT, β^-	^{117m}Cd	3.36 h	β^-
^{95}Nb	34.991 d	β^-	^{117}Cd	2.49 h	β^-
^{97m}Nb	58.7 s	IT	^{109}In	4.159 h	EC, β^+
^{97}Nb	72.1 m	β^-	^{110}In	4.92 h	EC, β^+
^{93}Mo	4.0×10^3 y	EC	^{111}In	2.8047 d	EC
^{99}Mo	65.976 h	β^-	^{112}In	14.88 m	EC, β^-, β^+
^{92}Tc	4.25 m	β^+, EC	^{113m}In	1.6579 h	IT
^{95m}Tc	61 d	EC, IT, β^+	^{114m}In	49.51 d	IT, EC, β^+
^{95}Tc	20.0 h	EC	^{114}In	71.9 s	β^-, EC, β^+
^{99m}Tc	6.0067 h	IT, β^-	^{115m}In	4.486 h	IT, β^-
^{99}Tc	2.111×10^5 y	β^-	^{115}In	4.41×10^{14} y	β^-
^{103}Ru	39.247 d	β^-	^{116m}In	54.29 m	β^-, EC
^{105}Ru	4.439 h	β^-	^{117m}In	116.2 m	β^-, IT
^{106}Ru	371.8 d	β^-	^{117}In	43.2 m	β^-
^{99}Rh	16.1 d	EC, β^+	^{119m}In	18.0 m	β^-, IT
^{103m}Rh	56.114 m	IT	^{119}In	2.4 m	β^-
^{105m}Rh	42.9 s	IT	^{113}Sn	115.09 d	EC, β^+
^{105}Rh	35.341 h	β^-	^{117m}Sn	14.00 d	IT
^{106}Rh	30.07 s	β^-	^{119m}Sn	293.1 d	IT
^{103}Pd	16.991 d	EC	^{121m}Sn	43.9 y	IT, β^-

Main Radionuclides Continued.

Nuclide	Half-life	Decay mode	Nuclide	Half-life	Decay mode
^{121}Sn	27.03 h	β^-	^{130}Cs	29.21 m	EC, β^+, β^-
^{123m}Sn	40.06 m	β^-	^{131}Cs	9.689 d	EC
^{123}Sn	129.2 d	β^-	^{132}Cs	6.480 d	EC, β^+, β^-
^{125}Sn	9.64 d	β^-	^{134m}Cs	2.912 h	IT
^{122}Sb	2.7238 d	β^-, EC, β^+	^{134}Cs	2.0652 y	β^-, EC
^{124}Sb	60.20 d	β^-	^{135}Cs	2.3×10^6 y	β^-
^{125}Sb	2.75856 y	β^-	^{137}Cs	30.08 y	β^-
^{127}Sb	3.85 d	β^-	^{131}Ba	11.50 d	EC
^{121m}Te	164.2 d	IT, EC	^{133m}Ba	38.93 h	IT, EC
^{121}Te	19.17 d	EC	^{133}Ba	10.551 y	EC
^{123m}Te	119.2 d		^{137m}Ba	2.552 m	IT
^{123}Te	$>9.2\times10^{16}$ y	EC	^{139}Ba	82.93 m	β^-
^{125m}Te	57.40 d	IT	^{140}Ba	12.751 d	β^-
^{127m}Te	106.1 d	IT, β^-	^{138}La	1.03×10^{11} y	EC, β^-
^{127}Te	9.35 h	β^-	^{140}La	1.67858 d	β^-
^{129m}Te	33.6 d	IT, β^-	^{139}Ce	137.63 d	EC
^{129}Te	69.6 m	β^-	^{141}Ce	32.511 d	β^-
^{132}Te	3.204 d	β^-	^{143}Ce	33.039 h	β^-
^{121}I	2.12 h	EC, β^+	^{144}Ce	284.91 d	β^-
^{123}I	13.2235 h	EC	^{142}Pr	19.12 h	β^-, EC
^{124}I	4.1760 d	EC, β^+	^{143}Pr	13.57 d	β^-
^{125}I	59.407 d	EC	^{144m}Pr	7.2 m	IT, β^-
^{126}I	12.93 d	EC, β^+, β^-	^{144}Pr	17.28 m	β^-
^{128}I	24.99 m	β^-, EC, β^+	^{144}Nd	2.29×10^{15} y	α
^{129}I	1.57×10^7 y	β^-	^{147}Nd	10.98 d	β^-
^{130}I	12.36 h	β^-	^{149}Nd	1.728 h	β^-
^{131}I	8.0252 d	β^-	^{151}Nd	12.44 m	β^-
^{132}I	2.295 h	β^-	^{147}Pm	2.6234 y	β^-
^{133}I	20.83 h	β^-	^{149}Pm	53.08 h	β^-
^{134}I	52.5 m	β^-	^{151}Pm	28.40 h	β^-
^{135}I	6.58 h	β^-	^{147}Sm	1.060×10^{11} y	α
^{131m}Xe	11.84 d	IT	^{148}Sm	7×10^{15} y	α
^{133m}Xe	2.198 d	IT	^{151}Sm	90 y	β^-
^{133}Xe	5.2475 d	β^-	^{153}Sm	46.284 h	β^-
^{135m}Xe	15.29 m	IT, β^-	^{155}Sm	22.18 m	β^-
^{135}Xe	9.14 h	β^-	^{152m}Eu	9.3116 h	β^-, EC, β^+
^{129}Cs	32.06 h	EC, β^+	^{152}Eu	13.517 y	EC, β^+, β^-

Continued.

Nuclide	Half-life	Decay mode	Nuclide	Half-life	Decay mode
^{154}Eu	8.601 y	β^-, EC, β^+	^{185}Os	93.6 d	EC
^{155}Eu	4.753 y	β^-	^{186}Os	2.0×10^{15} y	α
^{156}Eu	15.19 d	β^-	^{191m}Os	13.10 h	IT
^{152}Gd	1.08×10^{14} y	α	^{191}Os	15.4 d	β^-
^{153}Gd	240.4 d	EC	^{193}Os	29.830 h	β^-
^{159}Gd	18.479 h	β^-	^{191m}Ir	4.899 s	IT
^{157}Tb	71 y	EC	^{192}Ir	73.829 d	β^-, EC
^{160}Tb	72.3 d	β^-	^{193m}Ir	10.53 d	IT
^{161}Tb	6.89 d	β^-	^{194}Ir	19.28 h	β^-
^{157}Dy	8.14 h	EC, β^+	^{193m}Pt	4.33 d	IT
^{165}Dy	2.334 h	β^-	^{193}Pt	50 y	EC
^{166}Dy	81.6 h	β^-	^{197}Pt	19.8915 h	β^-
^{166m}Ho	1.20×10^{3} y	β^-	^{199}Pt	30.80 m	β^-
^{166}Ho	26.824 h	β^-	^{195}Au	186.01 d	EC
^{169}Er	9.392 d	β^-	^{197m}Au	7.73 s	IT
^{171}Er	7.516 h	β^-	^{198}Au	2.6941 d	β^-
^{170}Tm	128.6 d	β^-, EC	^{199}Au	3.139 d	β^-
^{171}Tm	1.92 y	β^-	^{197m}Hg	23.8 h	IT, EC
^{169}Yb	32.018 d	EC	^{197}Hg	64.14 h	EC
^{175}Yb	4.185 d	β^-	^{203}Hg	46.594 d	β^-
^{177}Yb	1.911 h	β^-	^{206}Hg	8.32 m	β^-
^{176m}Lu	3.664 h	β^-, EC	^{200}Tl	26.1 h	EC, β^+
^{176}Lu	3.76×10^{10} y	β^-	^{201}Tl	3.0421 d	EC
^{177}Lu	6.6443 d	β^-	^{202}Tl	12.31 d	EC
^{175}Hf	70 d	EC	^{204}Tl	3.783 y	β^-, EC, β^+
^{180m}Hf	5.53 h	IT, β^-	^{206}Tl	4.202 m	β^-
^{181}Hf	42.39 d	β^-	^{207}Tl	4.77 m	β^-
^{180}Ta	8.154 h	EC, β^-	^{208}Tl	3.053 m	β^-
^{182}Ta	114.74 d	β^-	^{209}Tl	2.162 m	β^-
^{181}W	121.2 d	EC	^{210}Tl	1.30 m	β^-
^{185}W	75.1 d	β^-	^{200}Pb	21.5 h	EC
^{187}W	24.000 h	β^-	^{201}Pb	9.33 h	EC, β^+
^{188}W	69.78 d	β^-	^{202m}Pb	3.53 h	IT, EC
^{183}Re	70.0 d	EC	^{202}Pb	5.25×10^{4} y	EC, α
^{186}Re	3.7183 d	β^-, EC	^{203}Pb	51.92 h	EC
^{187}Re	4.33×10^{10} y	β^-	^{207m}Pb	0.806 s	IT
^{188}Re	17.005 h	β^-	^{209}Pb	3.234 h	β^-

Main Radionuclides Continued.

Nuclide	Half-life	Decay mode	Nuclide	Half-life	Decay mode
^{210}Pb	22.20 y	β^-, α	^{229}Th	7.88×10^3 y	α
^{211}Pb	36.1 m	β^-	^{230}Th	7.54×10^4 y	α
^{212}Pb	10.64 h	β^-	^{231}Th	25.52 h	β^-
^{214}Pb	27.06 m	β^-	^{232}Th	1.40×10^{10} y	α
^{206}Bi	6.243 d	EC, β^+	^{233}Th	21.83 m	β^-
^{207}Bi	31.55 y	EC, β^+	^{234}Th	24.10 d	β^-
^{208}Bi	3.68×10^5 y	EC	^{231}Pa	3.276×10^4 y	α
^{209}Bi	2.01×10^{19} y	α	^{233}Pa	26.975 d	β^-
^{210}Bi	5.012 d	β^-, α	^{234m}Pa	1.159 m	β^-, IT
^{211}Bi	2.14 m	α, β^-	^{234}Pa	6.70 h	β^-
^{212}Bi	60.55 m	β^-, α	^{232}U	68.9 y	α
^{213}Bi	45.59 m	β^-, α	^{233}U	1.592×10^5 y	α, SF
^{214}Bi	19.9 m	β^-, α	^{234}U	2.455×10^5 y	α, SF
^{215}Bi	7.6 m	β^-	^{235m}U	26 m	IT
^{208}Po	2.898 y	α, EC, β^+	^{235}U	7.04×10^8 y	α, SF
^{210}Po	138.376 d	α	^{236}U	2.342×10^7 y	α, SF
^{211}Po	0.516 s	α	^{237}U	6.752 d	β^-
^{213}Po	3.72×10^{-6} s	α	^{238}U	4.468×10^9 y	α, SF
^{214}Po	1.643×10^{-4} s	α	^{239}U	23.45 m	β^-
^{215}Po	1.781×10^{-3} s	α	^{237}Np	2.144×10^6 y	α, SF
^{216}Po	0.145 s	α	^{238}Np	2.099 d	β^-
^{218}Po	3.097 m	α, β^-	^{239}Np	2.356 d	β^-
^{211}At	7.214 h	EC, α	^{238}Pu	87.7 y	α, SF
^{220}Rn	55.6 s	α	^{239}Pu	2.4110×10^4 y	α, SF
^{222}Rn	3.8235 d	α	^{240}Pu	6.561×10^3 y	α, SF
^{221}Fr	4.9 m	α	^{241}Pu	14.329 y	β^-, α
^{223}Fr	22.00 m	β^-, α	^{242}Pu	3.75×10^5 y	α, SF
^{223}Ra	11.43 d	α	^{241}Am	432.6 y	α, SF
^{224}Ra	3.6319 d	α	^{242}Am	16.02 h	β^-, EC
^{225}Ra	14.9 d	β^-	^{243}Am	7.364×10^3 y	α
^{226}Ra	1.600×10^3 y	α	^{242}Cm	162.8 d	α, SF
^{228}Ra	5.75 y	β^-	^{244}Cm	18.1 y	α, SF
^{225}Ac	9.920 d	α	^{246}Cm	4.706×10^3 y	α, SF
^{227}Ac	21.772 y	β^-, α	^{248}Cm	3.48×10^5 y	α, SF
^{228}Ac	6.15 h	β^-	^{247}Bk	1.38×10^3 y	α
^{227}Th	18.697 d	α	^{252}Cf	2.645 y	α, SF
^{228}Th	1.9116 y	α			

(Compiled by Yoshihiro Makide)

Natural Radionuclides

Radionuclides (radioisotopes) found in nature are categorized into the following three classes.

1) Radionuclides belonging to radioactive decay chains with long-life parent radioactive elements (nuclides) such as uranium (^{235}U and ^{238}U) and thorium. These nuclides are shown in the decay chain diagram which appears later (refer to "Radio active decay series").

2) ^{40}K and other long-life nuclides which do not belong to any radioactive decay chain. The main nuclides for 2) are shown in the following table.

3) ^{3}H, ^{14}C and other nuclides which are produced by nuclear reactions with cosmic rays. In the upper atmosphere, cosmic rays collide with atomic nuclei of the elements in the atmosphere such as nitrogen, oxygen and argon. This collision causes nuclear reactions (spallation reaction). The neutrons emitted during these reactions cause secondary nuclear reactions which generate radionuclides such as ^{3}H, ^{7}Be, ^{10}Be, ^{14}C, ^{22}Na, ^{32}P, ^{35}S, and ^{36}Cl.

Long-lived naturally-ocurring radionuclides which do not belong to any radioactive decay chain

Nuclide	Half-life	Decay mode	Nuclide	Half-life	Decay mode
^{40}K	1.248×10^{9} y	β^-, EC	^{144}Nd	2.29×10^{15} y	α
			^{147}Sm	1.060×10^{11} y	α
^{87}Rb	4.81×10^{10} y	β^-	^{148}Sm	7×10^{15} y	α
^{113}Cd	8.04×10^{15} y	β^-	^{152}Gd	1.08×10^{14} y	α
^{115}In	4.41×10^{14} y	β^-	^{176}Lu	3.76×10^{10} y	β^-
^{123}Te	$>9.2\times10^{16}$ y	EC	^{174}Hf	2.0×10^{15} y	α
			^{187}Re	4.33×10^{10} y	β^-
^{138}La	1.03×10^{11} y	EC, β^-	^{186}Os	2.0×10^{15} y	α
			^{190}Pt	6.5×10^{11} y	α

The half-lives and decay modes are the same as for the table, "Main radionuclides (radioisotopes)" in page 411.

(Compiled by Shogo Higaki)

Radio Active

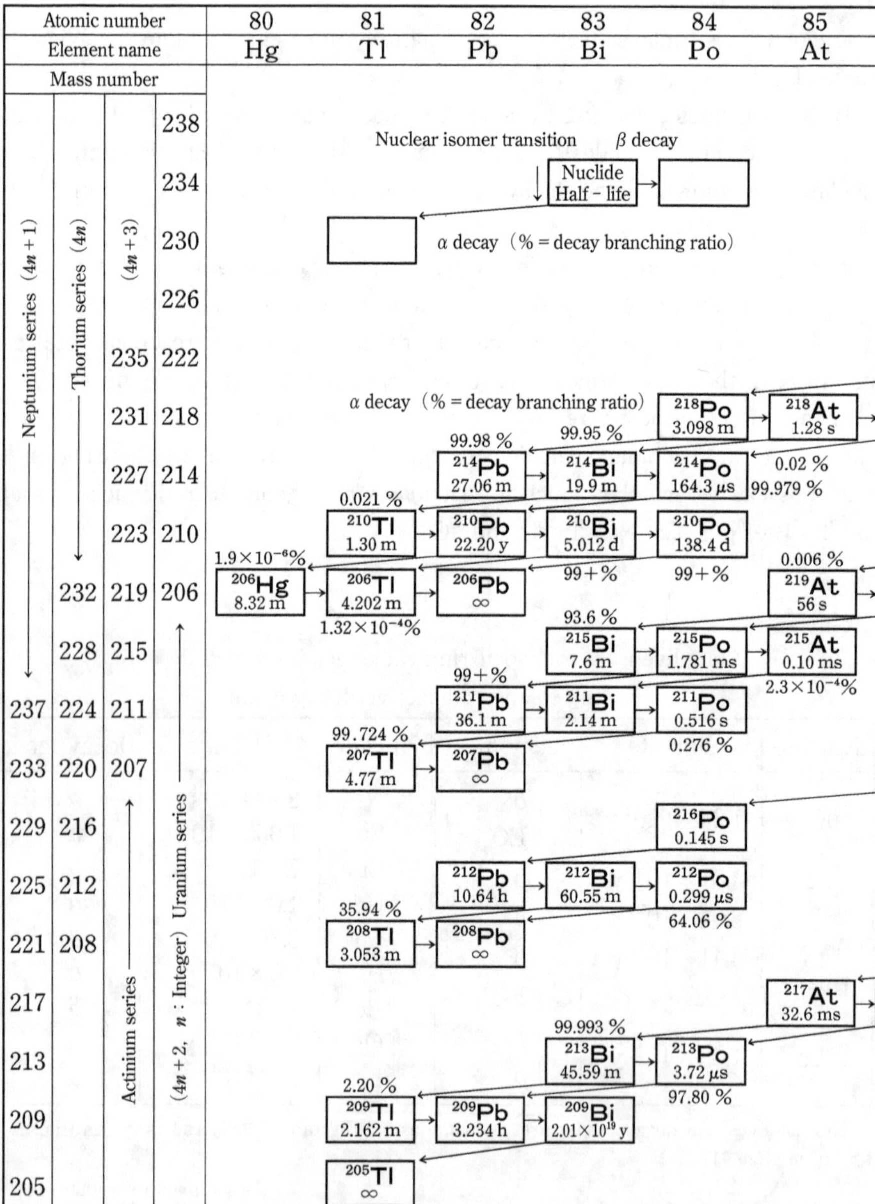

* The sources are the same as the table in page 411.

Decay Series

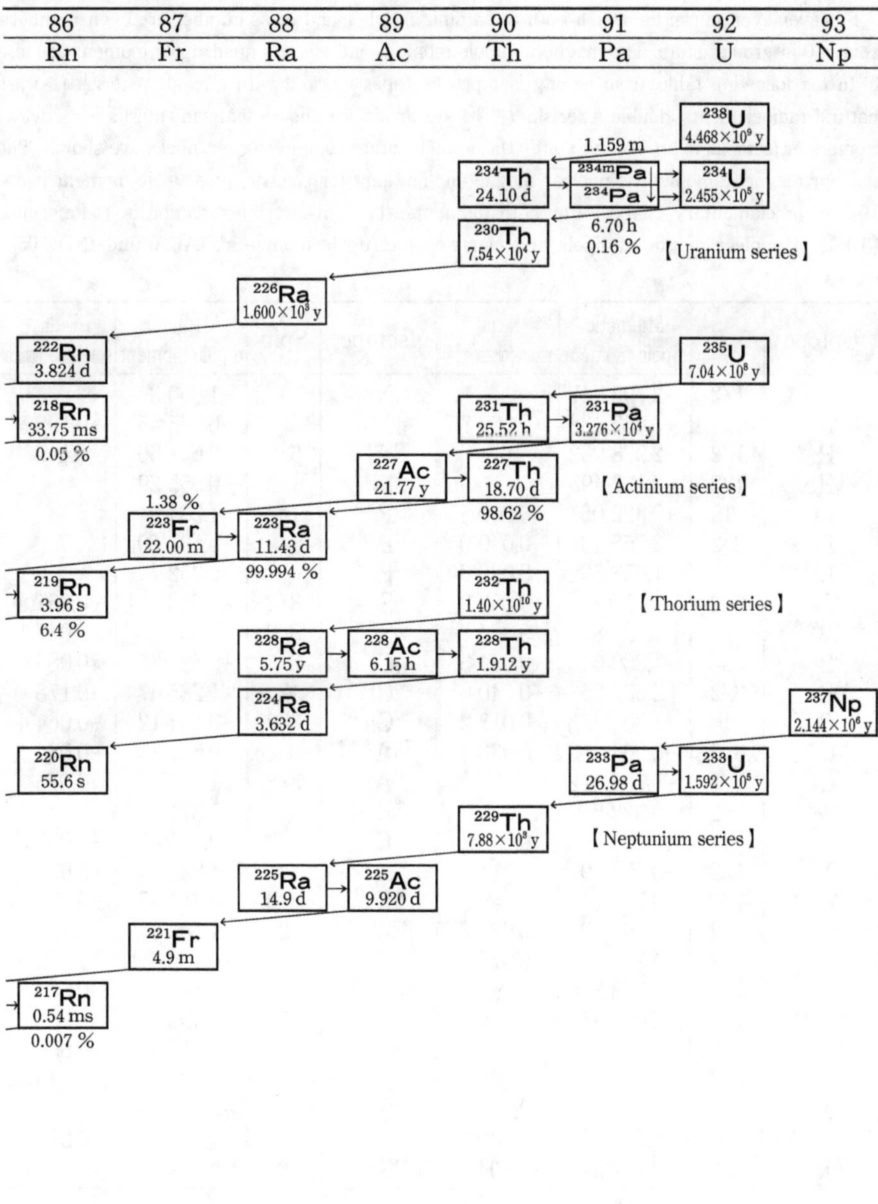

Key: y (years), d (days), h (hours), m (minutes), s (seconds), ms (10^{-3}s), μs (10^{-6}s) (Compiled by Shogo Higaki)

Magnetic Moment and Electric Quadrupole Moment of Nucleus

For even-even nuclei for which both the atomic number and mass number are even numbers, spin is 0 in ground state. The magnetic dipole moment and electric quadrupole moment are also 0. In the following table, from among isotopes (isotopes marked with a single asterisk (*) are natural radioelements, double asterisks (* *) are artificial radioelements, m marks are nuclear isomers or isotopes in an excited state), those nuclei other than even-even nuclei are shown. The unit for magnetic moment is nuclear magneton. The unit for electric quadrupole moment is $e\times 10^{-28}\ m^2$ (e: elementary charge). For both moments, the sign for spin direction is +. Reference "Table of nuclear magnetic dipole and electric quadrupole moments", IAEA, and INDC-0658 (2014).

Isotope	Spin	Magnetic dipole moment	Electric quadrupole moment	Isotope	Spin	Magnetic dipole moment	Electric quadrupole moment
$^{1}_{1}H$	1/2	+2.792 847		^{24}Na**	4	+1.690 3	
^{2}H	1	+0.857 438	+0.002 86	$^{25}_{12}Mg$	5/2	−0.855 45	+0.199
^{3}H**	1/2	+2.978 962		$^{27}_{13}Al$	5/2	+3.641 50	+0.147
$^{3}_{2}He$	1/2	−2.127 498		$^{29}_{14}Si$	1/2	−0.555 29	
$^{6}_{3}Li$	1	+0.822 05	−0.000 81	$^{29}_{15}P$**	1/2	1.234 6	
^{7}Li	3/2	+3.256 43	−0.040 0	^{31}P	1/2	+1.131 60	
^{8}Li**	2	+1.653 56	0.032 6	^{32}P**	1	−0.252 4	
$^{9}_{4}Be$	3/2	−1.177 43	+0.052 9	$^{33}_{16}S$	3/2	+0.643 821	−0.067 8
$^{8}_{5}B$**	2	1.035 8	+0.064 3	^{35}S**	3/2	+1.00	+0.047 1
^{10}B	3	+1.800 64	+0.084 5	$^{35}_{17}Cl$	3/2	+0.821 87	0.085 0
^{11}B	3/2	+2.688 65	+0.040 6	^{36}Cl**	2	+1.285 47	−0.178
^{12}B**	1	+1.00	0.013 2	^{37}Cl	3/2	+0.684 12	−0.064 4
$^{11}_{6}C$**	3/2	−0.964	0.033 3	$^{35}_{18}Ar$**	3/2	+0.632 2	−0.084
^{13}C	1/2	+0.702 412		^{37}Ar**	3/2	+1.145	+0.076
$^{13}_{7}N$**	1/2	0.322 2		$^{38}_{19}K$**	3	+1.371	
^{14}N	1	+0.403 76	+0.020 44	^{39}K	3/2	+0.391 47	+0.058 5
^{15}N	1/2	−0.283 19		^{40}K*	4	−1.298 10	−0.073
$^{15}_{8}O$**	1/2	0.719 5		^{41}K	3/2	+0.214 87	+0.071 1
^{17}O	5/2	−1.893 79	−0.025 6	^{42}K**	2	−1.142 5	
$^{17}_{9}F$**	5/2	+4.721 3	0.076	^{43}K**	3/2	+0.163 3	
^{19}F	1/2	+2.628 868	0.094 2	$^{41}_{20}Ca$**	7/2	−1.594 78	−0.665
^{20}F**	2	+2.093 35	0.056	^{42m}Ca	6	−2.49	
$^{19}_{10}Ne$**	1/2	−1.884 6		^{43}Ca	7/2	−1.317 3	−0.055
^{21}Ne	3/2	−0.661 797	+0.102	$^{41}_{21}Sc$**	7/2	+5.431	−0.145
$^{21}_{11}Na$**	3/2	+2.386 30	+0.138	^{43m}Sc	19/2	+3.122	0.199
^{22}Na*	3	+1.746	+0.180	^{43}Sc**	7/2	+4.528	−0.27
^{23}Na	3/2	+2.217 52	+0.104	^{44}Sc**	2	+2.499	+0.10

$^{1}_{0}n$ 1/2 −1.913 043 (neutron)

Continued.

Isotope	Spin	Magnetic dipole moment	Electric quadrupole moment	Isotope	Spin	Magnetic dipole moment	Electric quadrupole moment
${}^{44m}\mathrm{Sc}^{**}$	6	+3.833	−0.19	${}^{75}_{33}\mathrm{As}$	3/2	+1.439 48	0.314
${}^{45}\mathrm{Sc}$	7/2	+4.756 487	−0.220	${}^{76}\mathrm{As}^{**}$	2	−0.902	
${}^{46}\mathrm{Sc}^{**}$	4	+3.042	+0.12	${}^{75}_{34}\mathrm{Se}^{**}$	5/2	0.683	1.1
${}^{47}\mathrm{Sc}^{**}$	7/2	+5.34	−0.22	${}^{77}\mathrm{Se}$	1/2	+0.535 04	
${}^{45}_{22}\mathrm{Ti}^{**}$	7/2	0.095	0.015	${}^{79}\mathrm{Se}^{**}$	7/2	−1.018	+0.8
${}^{47}\mathrm{Ti}$	5/2	−0.788 48	+0.302	${}^{76}_{35}\mathrm{Br}^{**}$	1	0.54 821	0.255
${}^{49}\mathrm{Ti}$	7/2	−1.104 2	0.247	${}^{79}\mathrm{Br}$	3/2	+2.106 4	+0.313
${}^{50}_{23}\mathrm{V}^{*}$	6	+3.345 69	+0.21	${}^{80}\mathrm{Br}^{**}$	1	0.514 0	+0.185
${}^{51}\mathrm{V}$	7/2	+5.148 71	−0.043	${}^{80m}\mathrm{Br}^{**}$	5	+1.317 7	+0.71
${}^{49}_{24}\mathrm{Cr}^{**}$	5/2	0.476		${}^{81}\mathrm{Br}$	3/2	+2.270 56	+0.262
${}^{53}\mathrm{Cr}$	3/2	−0.474 54	−0.15	${}^{82}\mathrm{Br}^{**}$	5	+1.627 0	+0.707
${}^{51}_{25}\mathrm{Mn}$	5/2	+3.568 3	0.42	${}^{83}_{36}\mathrm{Kr}$	9/2	−0.970 67	+0.259
${}^{51}\mathrm{Mn}^{**}$	6	+3.062 2	+0.50	${}^{85}\mathrm{Kr}^{**}$	9/2	−1.005	+0.443
${}^{53}\mathrm{Mn}^{**}$	7/2	5.035	+0.17	${}^{81}_{37}\mathrm{Rb}^{**}$	3/2	+2.059 5	+0.48
${}^{54}\mathrm{Mn}^{**}$	3	3.306	+0.37	${}^{82m}\mathrm{Rb}^{**}$	5	+1.510 008	+1.2
${}^{55}\mathrm{Mn}$	5/2	+3.453 2	+0.330	${}^{83}\mathrm{Rb}^{**}$	5/2	+1.424 9	+0.24
${}^{56}\mathrm{Mn}^{**}$	3	+3.226 6	+0.48	${}^{84}\mathrm{Rb}^{**}$	2	−1.324 12	−0.02
${}^{57}_{26}\mathrm{Fe}$	1/2	+0.090 44		${}^{85}\mathrm{Rb}$	5/2	+1.352 98	+0.276
${}^{57m}\mathrm{Fe}$	3/2	−0.154 9	+0.160	${}^{86}\mathrm{Rb}^{**}$	2	−1.692 0	+0.23
${}^{56}_{27}\mathrm{Co}^{**}$	4	3.85	+0.25	${}^{87}\mathrm{Rb}^{*}$	3/2	+2.751 31	+0.133 5
${}^{57}\mathrm{Co}^{**}$	7/2	+4.720	+0.54	${}^{87}_{38}\mathrm{Sr}$	9/2	−1.092 8	+0.305
${}^{58}\mathrm{Co}^{**}$	2	+4.044	+0.23	${}^{89}_{39}\mathrm{Y}$	1/2	−0.137 41	−0.43
${}^{59}\mathrm{Co}$	7/2	+4.627	+0.42	${}^{90}\mathrm{Y}^{**}$	2	−1.630	−0.125
${}^{60}\mathrm{Co}^{**}$	5	+3.799	+0.46	${}^{91}\mathrm{Y}^{**}$	1/2	0.164 1	
${}^{61}_{28}\mathrm{Ni}$	3/2	−0.750 02	+0.162	${}^{91}_{40}\mathrm{Zr}$	5/2	−1.303 6	−0.176
${}^{61m}\mathrm{Ni}$	5/2	+0.480	−0.20	${}^{93}_{41}\mathrm{Nb}$	9/2	+6.170 5	−0.32
${}^{61}_{29}\mathrm{Cu}^{**}$	3/2	+2.108 3	−0.221	${}^{95}_{42}\mathrm{Mo}$	5/2	−0.914 2	−0.022
${}^{63}\mathrm{Cu}$	3/2	+2.223 6	−0.211 70	${}^{97}\mathrm{Mo}$	5/2	−0.933 5	+0.255
${}^{64}\mathrm{Cu}^{**}$	1	−0.216 4	+0.75	${}^{99}_{43}\mathrm{Tc}^{**}$	9/2	+5.684 7	−0.129
${}^{65}\mathrm{Cu}$	3/2	+2.381 7	−0.204	${}^{99}_{44}\mathrm{Ru}$	5/2	−0.641	+0.079
${}^{66}\mathrm{Cu}^{**}$	1	−0.282 3	+0.059	${}^{99m}\mathrm{Ru}$	3/2	−0.284	+0.231
${}^{65}_{30}\mathrm{Zn}^{**}$	5/2	+0.769 0	−0.023	${}^{101}\mathrm{Ru}$	5/2	−0.719	+0.46
${}^{67}\mathrm{Zn}$	5/2	+0.875 48	+0.15	${}^{103}_{45}\mathrm{Rh}$	1/2	−0.884 0	
${}^{67}_{31}\mathrm{Ga}^{**}$	3/2	+1.850 7	+0.197	${}^{105}_{46}\mathrm{Pd}$	5/2	−0.642	0.660
${}^{68}\mathrm{Ga}^{*}$	1	0.011 75	−0.027 7	${}^{104}_{47}\mathrm{Ag}^{**}$	5	3.916	+1.06
${}^{69}\mathrm{Ga}$	3/2	+2.016 59	+0.171	${}^{104m}\mathrm{Ag}^{**}$	2	3.691	
${}^{71}\mathrm{Ga}$	3/2	+2.562 27	+0.107	${}^{105}\mathrm{Ag}^{**}$	1/2	0.101 4	
${}^{72}\mathrm{Ga}^{**}$	3	−0.132 24	+0.53	${}^{107}\mathrm{Ag}$	1/2	−0.113 57	
${}^{71}_{32}\mathrm{Ge}^{**}$	1/2	+0.547		${}^{108}\mathrm{Ag}^{**}$	1	2.688 4	
${}^{73}\mathrm{Ge}$	9/2	−0.879 46	−0.196	${}^{109}\mathrm{Ag}$	1/2	−0.130 69	

Magnetic Moment and Electric Quadrupole Moment of Nucleus Continued.

Isotope	Spin	Magnetic dipole moment	Electric quadruple moment
$^{111}Ag^{**}$	1/2	-0.146	
$^{112}Ag^{**}$	2	0.054 7	
$^{113}Ag^{**}$	1/2	0.159	
$^{107}_{48}Cd^{**}$	5/2	-0.615 0	+0.60
$^{109}Cd^{**}$	5/2	-0.827 8	+0.60
^{111}Cd	1/2	-0.594 89	
$^{113}Cd^{*}$	1/2	-0.622 30	
$^{113m}Cd^{**}$	11/2	-1.087 78	-0.61
$^{115}Cd^{**}$	1/2	-0.648 4	
$^{115m}Cd^{**}$	11/2	-1.041 0	-0.48
$^{109}_{49}In^{**}$	9/2	+5.538	+0.80
$^{111}In^{**}$	9/2	+5.503	+0.76
^{113}In	9/2	+5.528 9	0.759
$^{113m}In^{**}$	1/2	-0.210 7	
$^{114m}In^{**}$	5	+4.653	+0.703
$^{115}In^{*}$	9/2	+5.540 8	+0.770
$^{115m}In^{**}$	1/2	-0.243 98	
$^{116m}In^{**}$	5	+4.435	+0.762
$^{117m}In^{**}$	1/2	-0.251 74	
$^{115}_{50}Sn$	1/2	-0.918 83	
^{117}Sn	1/2	-1.001 0	
^{119}Sn	1/2	-1.047 8	
^{119m}Sn	3/2	+0.633	-0.132
$^{121}_{51}Sb$	5/2	+3.363 4	-0.543
$^{122}Sb^{**}$	2	-1.90	+1.28
^{123}Sb	7/2	+2.549 8	-0.692
$^{123}_{52}Te^{*}$	1/2	-0.736 94	
^{125}Te	1/2	-0.888 50	
^{125m}Te	3/2	+0.605	-0.31
$^{125}_{53}I^{**}$	5/2	2.821	-0.761
^{127}I	5/2	+2.813 3	-0.696
$^{129}I^{**}$	7/2	+2.621	-0.488
^{129m}I	5/2	+2.805	-0.604
$^{131}I^{**}$	7/2	+2.742	-0.34
$^{132}I^{**}$	4	3.088	0.08
$^{133}I^{**}$	7/2	+2.856	-0.23
$^{129}_{54}Xe$	1/2	-0.777 97	
^{129m}Xe	3/2	+0.58	-0.393
^{131}Xe	3/2	+0.691 5	-0.114
$^{127}_{55}Cs^{**}$	1/2	+1.459	
$^{129}Cs^{**}$	1/2	+1.491	
$^{130}Cs^{**}$	1	+1.460	-0.056
$^{131}Cs^{**}$	5/2	+3.53	+0.59
$^{132}Cs^{**}$	2	+2.222	+0.48
^{133}Cs	7/2	+2.582 0	-0.003 4
$^{134}Cs^{**}$	4	+2.993 7	+0.37
$^{134m}Cs^{**}$	8	+1.097 8	+0.92
$^{135}Cs^{**}$	7/2	+2.732 4	+0.048
$^{137}Cs^{**}$	7/2	+2.851 3	+0.048
$^{135}_{56}Ba$	3/2	+0.837 9	+0.16
^{137}Ba	3/2	+0.937 4	+0.245
$^{138}_{57}La^{*}$	5	+3.713 6	+0.39
^{139}La	7/2	+2.783 0	+0.20
$^{141}_{58}Ce^{**}$	7/2	1.09	
$^{141}_{59}Pr$	5/2	+4.275 4	-0.077
$^{142}Pr^{**}$	2	+0.234	+0.039
$^{143}_{60}Nd$	7/2	-1.065	-0.61
^{145}Nd	7/2	-0.656	-0.31
$^{147}Nd^{**}$	5/2	0.578	0.9
$^{147}_{61}Pm^{**}$	7/2	+2.58	+0.74
$^{151}Pm^{**}$	5/2	1.8	2.2
$^{147}_{62}Sm^{*}$	7/2	-0.812	-0.26
^{149}Sm	7/2	-0.668	+0.078
^{149m}Sm	5/2	-0.623 8	+1.01
$^{153}Sm^{**}$	3/2	-0.021	+1.3
$^{151}_{63}Eu$	5/2	+3.471 7	+0.903
^{151m}Eu	7/2	+2.591	1.28
$^{152}Eu^{**}$	3	-1.940	+2.72
^{153}Eu	5/2	+1.532 4	+2.41
$^{154}Eu^{**}$	3	-2.005	+2.85
$^{155}_{64}Gd$	3/2	-0.257 2	+1.27
^{157}Gd	3/2	-0.339 8	+1.35
$^{159}_{65}Tb$	3/2	+2.014	+1.432
$^{161}_{66}Dy$	5/2	-0.480	+2.51
^{161m}Dy	5/2	+0.594	+2.51
^{163}Dy	5/2	+0.673	+2.65

Continued.

Isotope	Spin	Magnetic dipole moment	Electric quadrupole moment	Isotope	Spin	Magnetic dipole moment	Electric quadrupole moment
$^{165}_{67}\mathrm{Ho}$	7/2	+4.17	3.58	$^{199}\mathrm{Au}$**	3/2	+0.261	+0.510
$^{166m}_{68}\mathrm{Er}$	2	+0.649	−1.9	$^{195}_{80}\mathrm{Hg}$**	1/2	+0.541 47	
$^{167}\mathrm{Er}$	7/2	−0.563 8	+3.57	$^{195m}\mathrm{Hg}$**	13/2	−1.044 65	+1.08
$^{169}\mathrm{Er}$**	1/2	+0.52	-	$^{197}\mathrm{Hg}$**	1/2	+0.527 37	
$^{171}\mathrm{Er}$**	5/2	0.659	2.86	$^{197m}\mathrm{Hg}$**	13/2	−1.027 7	+1.25
$^{166}_{69}\mathrm{Tm}$**	2	+0.092	+2.14	$^{199}\mathrm{Hg}$	1/2	+0.505 88	
$^{169}\mathrm{Tm}$	1/2	−0.231 0		$^{201}\mathrm{Hg}$	3/2	−0.560 22	+0.387
$^{169m}\mathrm{Tm}$	3/2	+0.515	−1.2	$^{203}\mathrm{Hg}$**	5/2	+0.849 0	+0.34
$^{170}\mathrm{Tm}$**	1	+0.246	+0.74	$^{199}_{81}\mathrm{Tl}$**	1/2	+1.60	
$^{171}_{70}\mathrm{Yb}$	1/2	+0.493 6		$^{201}\mathrm{Tl}$**	1/2	+1.605	
$^{173}\mathrm{Yb}$	5/2	−0.648	+2.80	$^{203}\mathrm{Tl}$	1/2	+1.622 257	
$^{175}\mathrm{Lu}$	7/2	+2.232 3	+3.49	$^{204}\mathrm{Tl}$**	2	0.09	
$^{176}\mathrm{Lu}$*	7	+3.16	+4.92	$^{205}\mathrm{Tl}$	1/2	+1.638 2	
$^{176m}\mathrm{Lu}$**	1	+0.311	−1.45	$^{207}_{82}\mathrm{Pb}$	1/2	+0.592 58	
$^{177}\mathrm{Lu}$**	7/2	+2.239	+3.39	$^{203}_{83}\mathrm{Bi}$**	9/2	+4.02	0.93
$^{177}_{72}\mathrm{Hf}$	7/2	+0.793 5	+3.37	$^{204}\mathrm{Bi}$**	6	+4.32	−0.7
$^{179}\mathrm{Hf}$	9/2	−0.640 9	+3.79	$^{205}\mathrm{Bi}$**	9/2	+4.065	−0.81
$^{181}_{73}\mathrm{Ta}$	7/2	+2.370 5	+3.17	$^{206}\mathrm{Bi}$**	6	+4.361	−0.54
$^{183}_{74}\mathrm{W}$	1/2	+0.117 8		$^{209}\mathrm{Bi}$	9/2	+4.110 3	−0.516
$^{185}_{75}\mathrm{Re}$	5/2	+3.187 1	+2.18	$^{210}\mathrm{Bi}$**	1	−0.044 5	+0.190
$^{186}\mathrm{Re}$**	1	+1.739	+0.618	$^{205}_{84}\mathrm{Po}$**	5/2	+0.76	
$^{187}\mathrm{Re}$*	5/2	+3.219 7	+2.07	$^{207}\mathrm{Po}$**	5/2	+0.79	
$^{188}\mathrm{Re}$**	1	+1.788	+0.572	$^{227}_{89}\mathrm{Ac}$*	3/2	+1.1	+1.7
$^{187}_{76}\mathrm{Os}$	1/2	+0.064 65		$^{229}_{90}\mathrm{Th}$**	5/2	+0.46	+4.3
$^{189}\mathrm{Os}$	3/2	+0.659 933	+0.98	$^{231}_{91}\mathrm{Pa}$*	3/2	2.01	−1.72
$^{191}_{77}\mathrm{Ir}$	3/2	+0.150 7	+0.816	$^{233}\mathrm{Pa}$**	3/2	4.0	−3.0
$^{193}\mathrm{Ir}$	3/2	+0.163 7	+0.751	$^{233}_{92}\mathrm{U}$**	5/2	0.59	3.663
$^{195}_{78}\mathrm{Pt}$	1/2	+0.609 52		$^{235}\mathrm{U}$*	7/2	−0.38	4.936
$^{190}_{79}\mathrm{Au}$**	1	−0.065		$^{237}_{93}\mathrm{Np}$**	5/2	+3.14	+3.866
$^{191}\mathrm{Au}$**	3/2	+0.136 9	+0.72	$^{239}_{94}\mathrm{Pu}$	1/2	+0.203	
$^{192}\mathrm{Au}$**	1	−0.010 7	−0.228	$^{241}\mathrm{Pu}$**	5/2	−0.683	+6
$^{193}\mathrm{Au}$**	3/2	+0.139	+0.66	$^{241}_{95}\mathrm{Am}$**	5/2	+1.58	+3.8
$^{194}\mathrm{Au}$**	1	+0.076	−0.24	$^{242}\mathrm{Am}$**	1	+0.387 9	−2.4
$^{195}\mathrm{Au}$**	3/2	+0.148	+0.61	$^{243}\mathrm{Am}$**	5/2	+1.50	+2.86
$^{196}\mathrm{Au}$**	2	+0.58	0.81				
$^{197}\mathrm{Au}$	3/2	+0.145 75	+0.55				
$^{197m}\mathrm{Au}$	1/2	+0.420					
$^{198}\mathrm{Au}$**	2	+0.64	+0.64				

Range of Charged Particles (in Lead)

When a group of charged particles which are heavier than electrons pass through matter, the number of passing particles decreases dramatically at a certain distance. The distance, where number of incident particles becomes 1/2, is defined as the range. Below, the range of μ, π, K,p,d and α particles (10 kg m^{-2}) and the energy loss (10^{-1} MeV kg^{-1} m^2) are shown as the function of the momentum (GeV/c). The ionization potential of lead is 823 eV for these calculations. Among the same matter, the range R_b for particles with mass M_b, charge z_b, and momentum p_b is given as $R_b(M_b, z_b, p_b) = ((M_b/M_a)/(z_b^2/z_a^2)) \times R_a(M_a, z_a, p_a)$, when the range R_a of particles is mass M_a, charge z_a, and momentum $p_a = p_b M_a / M_b$. However, this excludes electrons.

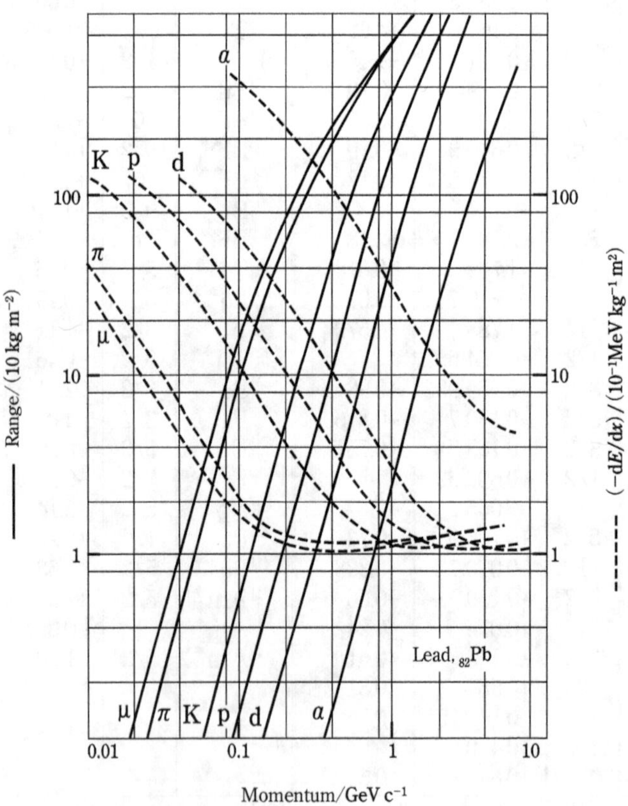

Practical Range of Electrons (in Aluminum)

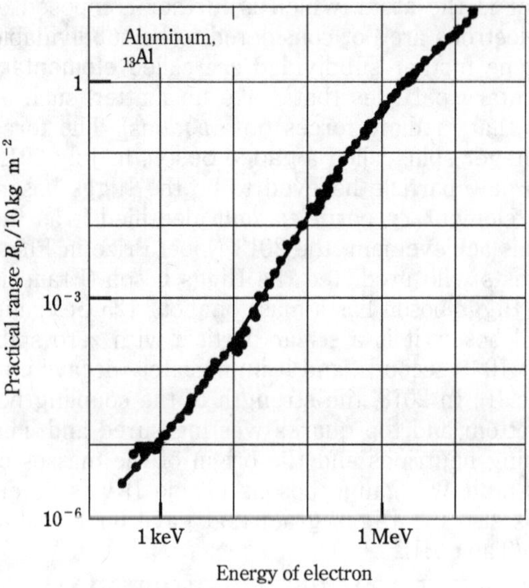

When electrons pass through matter, the absorption curve is smooth. It is difficult to determine the range as in the case of heavily charged particles. The practical range R_p is defined as the distance at which intensity becomes 0. The absorption line is extapolated with a straight line. The figure shows R_p of electrons in aluminum.

Cross-Sections of Photonic Processes (in Lead)

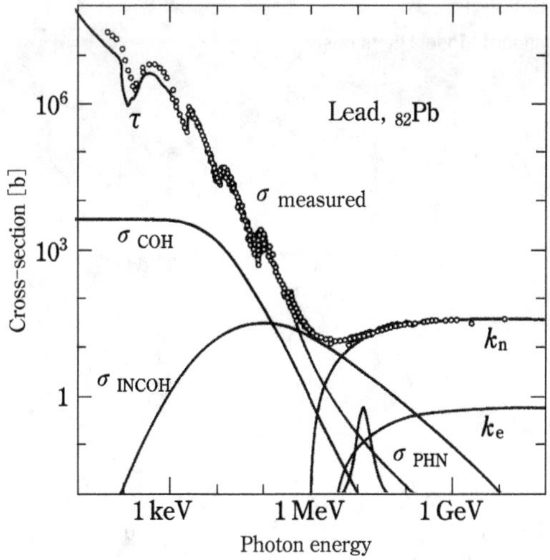

The cross-sections of photon in matter is given as the sum of various processes. The size and contribution processes depend on the photon energy. The figure shows the cross-sections of various processes in Lead [unit: b (barn) = 10^{-28} m^2].

τ: photoelectric effect, σ_{COH}: coherent scatter (Rayleigh scatter) σ_{INCOH}: Compton effect, k_n: electron pair production (by nucleus field), k_e: electron pair production (by electron field), σ_{PHN}: photonuclear reaction.

Elementary Particle

The unit that constitutes matter is the atom, which is further composed of electrons and nuclei. At present, electrons are not considered to be subdividable any further. Particles that cannot be further subdivided are called **elementary particles.** In addition to the elementary particles that make up matter, such as electrons, there are also particles that mediate forces (interactions). The force mediator is a Bose particle with integer spin, called a **gauge boson.** In July 2012, the LHC (Table VII) discovered a new particle believed to be the Higgs boson, which is thought to give mass to elementary particles, and identified it as the Higgs boson in March 2013. For this achievement, the 2013 Nobel Prize in Physics was awarded to the two scientists who predicted the Higgs boson (François Angoulême and Peter Higgs). The Higgs boson has a mass of about 125 GeV, and unlike the gauge boson and Fermi boson, it is a scalar particle with zero spin. The lifetime is thought to be about 10^{-21} seconds, and it immediately decays into known elementary particles (Table II). In 2018, the strength of the coupling between the Higgs boson and the bottom and top quarks was measured and The third generation of quarks, excluding neutrinos and the origin of the masses of the leptons (Table IV) and the Z^0 and $W^{\pm}$ gauge bosons (Table I) was determined to be the Higgs boson. It is also the Higgs grain that gave birth to the generation of elementary particles (Table III).

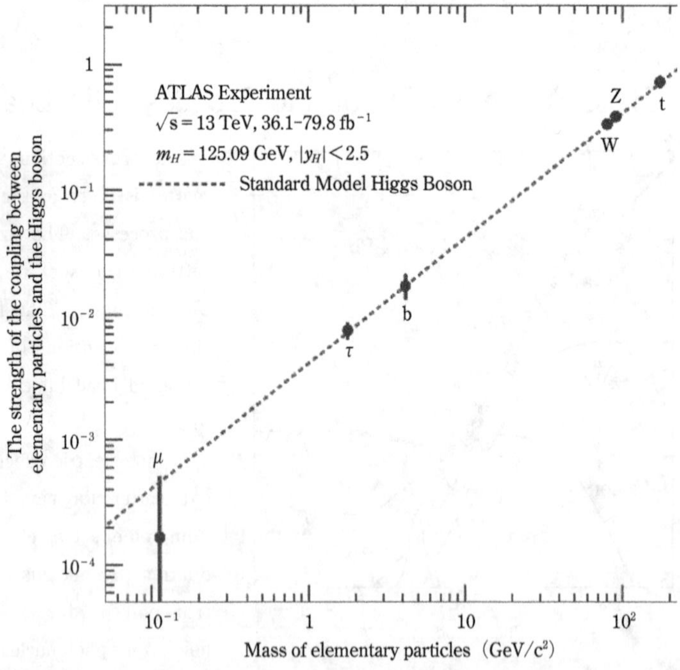

The electron family is collectively called **leptons.** There are two types of leptons: charged leptons, with a charge of -1, and electrically neutral neutrinos (neutral neutrinos, ν). Leptons are charged leptons with a charge of -1, and electrically neutral **neutrinos** (neutrinos, denoted by ν), where the charge is in units of the elementary charge e. Neutrinos were thought to have zero mass, but an experiment called K2K, conducted in collaboration between KEK and Super-Kamiokande, has shown that they do have a small mass.

An atomic nucleus is made up of several nucleons. Nucleons include protons and neutrons, which have a further structure, each consisting of three particles called **quarks.** A quark has a fractional charge of $+2/3$ or $-1/3$.

Leptons and quarks are both spin 1/2 Fermi particles. These are the ultimate particles that make up matter. Leptons and quarks are divided into four types of "generations" as shown in Table III. The existence of the first three generations has been confirmed so far. There is experimental evidence that, under certain assumptions, leptons and quarks can have no more than three generations, and no more than four.

Each generation of charged leptons is called an electron (e), a muon (μ), or a tau (τ), and there are corresponding electron neutrinos (ν_e), muon neutrinos (ν_μ), and tau neutrinos (ν_τ).

Quarks with charge $+2/3$ are called up (u), charm (c), and top (t), while quarks with charge $-1/3$ are called down (d), strange (s), and bottom (b). Table IV shows the properties of leptons and quarks. Quarks have not been observed alone, and are thought to exist only as **mesons** in composite particles consisting of quarks and antiquarks, or as **baryons** in composite particles of three quarks. Such composite particles are also sometimes called elementary particles for historical reasons. The mesons and baryons are collectively called **hadrons.** Tables V and VI show the main properties of mesons and baryons.

For all elementary particles, there is an **antiparticle** with the same mass, lifetime, and spin, but with the opposite sign of charge. For neutral elementary particles, there are both equal and different cases of particle and antiparticle. A hadronic antiparticle is one in which all the constituent quarks are replaced by their respective antiparticles (antiquarks).

In the following table, charge is denoted by the elementary charge e, spin is denoted by the unit $\hbar$, and mass M is denoted by the unit of energy eV according to the relation $E = Mc^2$. The average lifetime is the time it takes for the number of particles to decrease to the original $1/e$ (e: the bottom of the natural logarithm). For short-lived materials, the lifetime may be expressed as the resonance width $\Gamma = (\hbar/\text{average lifetime})$.

Recent experimental studies of elementary particles have been carried out mainly using colliders (collider particle accelerators). Table VI lists the major e^+e^- colliders and pp, $\bar{p}$p, and ep colliders that are currently in operation or scheduled for construction.

The following table is based primarily on the Particle Data Group's Review of Particle Physics, Prog. Prog. Theor. Exp. Phys. 083 C01 (2020).

I. Properties of gauge bosons

Particle	Charge	Antiparticle	Spin	Mass (GeV)	Lifetime	Mediated interaction
γ (photon)	0	γ	1	0	Stable	Electromagnetic interaction
g (gluon)	0	g	1	0*		Strong interaction
W^+	+1	W^-	1	80.379 ± 0.012	$\Gamma = 2.085 \pm 0.042$ GeV	Weak interaction
Z^0	0	Z^0	1	91.188 ± 0.002	$\Gamma = 2.495 \pm 0.002$ GeV	Weak interaction

$\Gamma = (\hbar/\text{lifetime})$. * estimated value in nucleons

II. Properties of Higgs bosons

Symbol	Charge	Spin	Mass (GeV)	Lifetime	Major decay products
H	0	0^+ (Spin 0 Positive parity)	125.10 ± 0.24	Unstable theoretical prediction of lifetime $\sim 10^{-21}$ sec	$\gamma\gamma$, ZZ, W^+W^-, $b\bar{b}$, $\tau^+\tau^-$, these 5 have been confirmed.

III. Generations of leptons and quarks

	Charge	1st	2nd	3rd
Lepton	0	ν_e	ν_μ	ν_τ
	−1	e	μ	τ
Quark	$+\frac{2}{3}$	u	c	t
	$-\frac{1}{3}$	d	s	b

IV. Properties of leptons and quarks

	Particle	Charge	Antiparticle	Spin	Mass	Average life span
Lepton	ν_e	0	$(\bar{\nu}_e)^\dagger$	$\frac{1}{2}$	<2 eV‡	Stable
	ν_μ	0	$(\bar{\nu}_\mu)^\dagger$	$\frac{1}{2}$	<0.19 MeV‡	Stable
	ν_τ	0	$(\bar{\nu}_\tau)^\dagger$	$\frac{1}{2}$	<18.2 MeV‡	Stable
	e^-	−1	e^+	$\frac{1}{2}$	0.510 998 946 ± 0.000 000 03 MeV	Stable
	μ^-	−1	μ^+	$\frac{1}{2}$	105.658 372 ± 0.000 002 MeV	$(2.196\,981 \pm 0.000\,02) \times 10^{-6}$ s
	τ^-	−1	τ^+	$\frac{1}{2}$	1776.9 ± 0.12 MeV	$(290.3 \pm 0.5) \times 10^{-15}$ s
Quark	u	$+\frac{2}{3}$	$\bar{u}$	$\frac{1}{2}$	$2.2^{+0.5}_{-0.3}$ MeV	Stable
	d	$-\frac{1}{3}$	$\bar{d}$	$\frac{1}{2}$	$4.7^{+0.5}_{-0.2}$ MeV	Stable
	s	$-\frac{1}{3}$	$\bar{s}$	$\frac{1}{2}$	93^{+11}_{-5} MeV	Unstable $\sim 10^{-8}$ s
	c	$+\frac{2}{3}$	$\bar{c}$	$\frac{1}{2}$	1.27 ± 0.02 GeV	Unstable $\sim 10^{-13}$ s
	b	$-\frac{1}{3}$	$\bar{b}$	$\frac{1}{2}$	$4.18^{+0.03}_{-0.02}$ GeV	Unstable $\sim 10^{-12}$ s
	t	$+\frac{2}{3}$	$\bar{t}$	$\frac{1}{2}$	172.8 ± 0.3 GeV	$\Gamma = 1.42^{+0.19}_{-0.15}$ GeV

† ν is considered as Majorana particle. Antineutrio $\bar{\nu}$ is the same as ν.

† In addition to the absolute values, Δm^2 (the square of the mass difference) are obtained from the ν oscillation.

V. Properties of major mesons

Particle	Charge	Composition	Antiparticle	Spin	Mass (MeV)	Lifetime
π^+	+1	$u\bar{d}$	π^-	0	139.57039 ± 0.00018	$(2.6033 \pm 0.0005) \times 10^{-8}$ s
π^0	0	$\frac{1}{\sqrt{2}}(u\bar{u}-d\bar{d})$	π^0	0	134.9768 ± 0.0005	$(8.52 \pm 0.18) \times 10^{-17}$ s
η	0	$c_1(u\bar{u}+d\bar{d})+c_2 s\bar{s}$	η	0	574.86 ± 0.017	$\Gamma = 1.31 \pm 0.05$ keV
$\rho(770)^+$	+1	$u\bar{d}$	ρ^-	1	775.26 ± 0.25	$\Gamma = 149.1 \pm 0.8$ MeV
$\rho(770)^0$	0	$\frac{1}{\sqrt{2}}(u\bar{u}-d\bar{d})$	ρ^0	1	775.26 ± 0.25	$\Gamma = 149.1 \pm 0.8$ MeV
$\omega(782)$	0	$c_1(u\bar{u}+d\bar{d})+c_2 s\bar{s}$	ω	1	782.65 ± 0.12	$\Gamma = 8.49 \pm 0.08$ MeV
$\eta'(958)$	0	$c_1(u\bar{u}+d\bar{d})+c_2 s\bar{s}$	η'	0	957.78 ± 0.06	$\Gamma = 0.188 \pm 0.006$ MeV
$\phi(1020)$	0	$c_1(u\bar{u}+d\bar{d})+c_2 s\bar{s}$	ϕ	1	1019.461 ± 0.02	$\Gamma = 4.249 \pm 0.013$ MeV
K^+	+1	$u\bar{s}$	K^-	0	493.677 ± 0.016	$(1.2380 \pm 0.0020) \times 10^{-8}$ s
K^0	0	$d\bar{s}$	$\bar{K}^0$	0	497.611 ± 0.013	K^0_S : $(0.8956 \pm 0.0003) \times 10^{-10}$ s K^0_L : $(5.116 \pm 0.021) \times 10^{-8}$ s
D^+	+1	$c\bar{d}$	D^-	0	1869.7 ± 0.05	$(1.040 \pm 0.007) \times 10^{-12}$ s
D^0	0	$c\bar{u}$	$\bar{D}^0$	0	1864.83 ± 0.05	$(0.410 \pm 0.002) \times 10^{-12}$ s
D^+_s	+1	$c\bar{s}$	D^-_s	0	1968.34 ± 0.07	$(0.504 \pm 0.004) \times 10^{-12}$ s
B^+	+1	$u\bar{b}$	B^-	0	5279.3 ± 0.1	$(1.638 \pm 0.004) \times 10^{-12}$ s
B^0	0	$d\bar{b}$	$\bar{B}^0$	0	5279.7 ± 0.1	$(1.519 \pm 0.004) \times 10^{-12}$ s
B^0_s	0	$s\bar{b}$	$\bar{B}^0_s$	0	5366.9 ± 0.1	$(1.515 \pm 0.004) \times 10^{-12}$ s
$\eta_c(1s)$	0	$c\bar{c}$	$\eta_c(1s)$	0	2983.9 ± 0.5	$\Gamma = 32.0 \pm 0.8$ MeV
$J/\psi(1s)$	0	$c\bar{c}$	$J/\psi(1s)$	1	3096.90 ± 0.006	$\Gamma = 92.9 \pm 2.8$ keV
$\chi_{c0}(1P)$	0	$c\bar{c}$	$\chi(1P)$	0	3414.71 ± 0.30	$\Gamma = 10.8 \pm 0.6$ MeV
$\chi_{c1}(1P)$	0	$c\bar{c}$	$\chi(1P)$	1	3510.67 ± 0.05	$\Gamma = 0.84 \pm 0.04$ MeV
$\chi_{c2}(1P)$	0	$c\bar{c}$	$\chi(1P)$	2	3556.17 ± 0.07	$\Gamma = 1.97 \pm 0.09$ MeV
$\psi(2s)$	0	$c\bar{c}$	$\psi(2s)$	1	3686.10 ± 0.06	$\Gamma = 294 \pm 8$ keV
$\Upsilon(1s)$	0	$b\bar{b}$	$\Upsilon(1s)$	1	9460.30 ± 0.26	$\Gamma = 54.02 \pm 1.25$ keV
$\Upsilon(2s)$	0	$b\bar{b}$	$\Upsilon(2s)$	1	10023.26 ± 0.31	$\Gamma = 31.98 \pm 2.63$ keV

The values of the quark composition coefficients c_1 and c_2 depend on the particle. $\omega(782)$ has a composition close to $(u\bar{u}+d\bar{d})/\sqrt{2}$, on the other hand $\phi(1620)$ has a composition close to $s\bar{s}$. $\Gamma = (\hbar/\text{lifetime})$.

Ⅵ. Properties of major baryons

Particle	Charge	Composition	Antiparticle	Spin	Mass (MeV)	Lifetime
p (proton)	+1	uud	$\bar{p}$	$\frac{1}{2}$	938.272081 ± 0.000006	Stable
n (neutron)	0	udd	$\bar{n}$	$\frac{1}{2}$	939.565413 ± 0.000006	879.4 ± 0.4 s
Δ^{++} (1232)	+2	uuu		$\frac{3}{2}$	1209–1211	Γ = 114–120 MeV
Δ^{+} (1232)	+1	uud		$\frac{3}{2}$	1209–1211	Γ = 114–120 MeV
Δ^{0} (1232)	0	udd		$\frac{3}{2}$	1209–1211	Γ = 114–120 MeV
Δ^{-} (1232)	-1	ddd		$\frac{3}{2}$	1209–1211	Γ = 114–120 MeV
Λ	0	uds		$\frac{1}{2}$	1115.683 ± 0.006	$(2.632 \pm 0.020) \times 10^{-10}$ s
Σ^{+}	+1	uus		$\frac{1}{2}$	1189.37 ± 0.07	$(0.8018 \pm 0.0026) \times 10^{-10}$ s
Σ^{0}	0	uds		$\frac{1}{2}$	1192.64 ± 0.02	$(7.4 \pm 0.7) \times 10^{-20}$ s
Σ^{-}	-1	dds		$\frac{1}{2}$	1197.449 ± 0.030	$(1.479 \pm 0.011) \times 10^{-10}$ s
Ξ^{0}	0	uss		$\frac{1}{2}$	1314.9 ± 0.2	$(2.90 \pm 0.09) \times 10^{-10}$ s
Ξ^{-}	-1	dss		$\frac{1}{2}$	1321.71 ± 0.07	$(1.639 \pm 0.015) \times 10^{-10}$ s
Ω^{-}	-1	sss		$\frac{3}{2}$	1672.45 ± 0.29	$(0.821 \pm 0.011) \times 10^{-10}$ s
Λ_c^{+}	+1	udc		$\frac{1}{2}$	2286.46 ± 0.14	$(0.200 \pm 0.006) \times 10^{-12}$ s

Among baryon antiparticles, other than $\bar{p}$ and $\bar{n}$ are not described here.

$\Gamma = (\hbar/\text{lifetime})$.

Only lowest mass (ground state) particles are given in Ⅴ and Ⅳ. In addtion to these, many high-mass (excited state) hadrons have been discovered. Please refer to Review of Particle Properties for more details.

Ⅶ. High-energy colliders which are currently in operation or scheduled for construction

e^+e^- collider

Laboratory (place)	Abbreviation of collider	Experiment first year	The highest beam energy (GeV)	Circumference (km)
Institute of High Energy Physics of the Chinese Academy of Sciences (Beijing)	BEPC-II	2008	2.3	0.2375
Budker Institute of Nuclear Physics (Novosibirsk)	VEPP-4M	1994	5.5	0.366
Budker Institute of Nuclear Physics (Novosibirsk)	VEPP-2000	2008	1.0	0.024
Frascati National Institute (Frascati)	DAΦNE	1999	0.51	0.032
National Laboratory for High Energy Physics (KEK) (Tsukuba)	Super-KEKB	2017	7/4*	3.016

* Two numbers indicate the maximum energy e^- and e^+ in asymmetrical colliders where their energies aren't equal.

pp, $\bar{p}$p, and ep collider

Laboratory (place)	Abbreviation of collider	Experiment first year	The highest beam energy (TeV)	Circumference (km)	Remarks
European Organization for Nuclear Research (CERN) (Geneva)	LHC	2008	$4\text{–}7^{\dagger}$	26.659	pp

Structural Chemical Properties and Molecular Spectroscopic Properties

Wavenumber for Main Infrared Absorption Bands

Vibration type	Wavenumber $\tilde{\nu}/cm^{-1}$	Shape of compound
$-OH$ expansion and contraction	3600–3200	H_2O, ROH
$-NH$ 〃	3400–3100	Amine, Amid
$\equiv C-H$ expansion and contraction	3300–3270	$R-C\equiv CH$
$=C-H$ 〃	3100–3000	Aromatic series, Olefin compound
$-C-H$ 〃	2960–2850	Saturated hydrocarbons
$-C\equiv N$ 〃	2250–2200	RCN, ArCN
$>C=O$ 〃	1820–1650	Carbonyl compound
$>C=C<$ 〃	1680–1640	Unsaturated hydrocarbon
$-NH_2$ scissoring	1640–1560	$R-NH_2$
Bending	1610–1590	Benzene derivative
	1500–1480	〃
$-CH_2$ scissoring	-1450	Saturated hydrocarbon
$-CH_3$ degenerate bending	-1450	〃
$-CH_3$ symmetric bending	1380	〃
$C-O$ expansion and contraction	1080–1050	Alcohol, Ester
$>C=C-H$ out-of-plane bending	990– 965	$RCH=CH_2$, $RCH=CHR$
CH out-of-plane bending	880– 720	Benzene substitution product
$C-Cl$ expansion and contraction	780– 615	Group $RC-Cl$

R: alkyl group, Ar: aryl group

Wavenumber for Infrared Absorption of Gas $\tilde{\nu}/cm^{-1}$

H_2O (D_2O) steam	Vibration type
3756 (2788)	OH_2 (OD_2) antisymmetrical stretching
3657 (2671)	OH_2 (OD_2) symmetry expansion and contraction
1595 (1178)	HOH (DOD) bending

SO_2	Vibration type
1362	SO_2 antisymmetrical stretching
1151	SO_2 symmetryexpansion and contraction
518	OSO bending

H_2S	
2626	SH_2 antisymmetrical stretching
2615	SH_2 symmetry expansion and contraction
1183	HSH bending

NO_2	
1618	NO_2 antisymmetrical stretching
1318	NO_2 symmetry expansion and contraction
750	ONO bending

NO	
1876	NO expansion and contraction

CO_2	
2349	OCO antisymmetrical stretching
667	OCO bending

CO	
2143	CO expansion and contraction

NH_3	
3444	NH_3 antisymeteric stretching mode
3336	NH_3 symmetry expansion and contraction
1626	NH_3 degenerate bending
950	NH_3 symmetric bending

CH_4	Vibration type
3019	CH_4 antisymeteric stretching mode
1306	CH_4 degenerate bending

C_2H_6	
2986	CH_3 antisymeteric stretching mode
2896	CH_3 symmetry expansion and contraction
1469	CH_3 degenerate bending
1379	CH_3 symmetric bending
822	CH_3 roll

C_2H_4	
3106	CH_2 antisymmetrical stretching
2989	CH_2 symmetry expansion and contraction
1444	CH_2 scissoring
949	CH_2 pitch
826	CH_2 roll

C_6H_6	
3063	CH expansion and contraction
1486	Bending
1038	CH in-plane bending
673	CH out-of-plane bending

Wavenumber for Rotation Spectrum Lines of Water Vapor (Vacuum)

The rotation spectrum caused by water vapor in the air indicates absorption into the far infrared range (550cm^{-1} to 10cm^{-1}). It is used in wavelength calibration for far infrared spectrographs.

(Unit: cm^{-1})

550.00	461.45	394.28	327.60	208.48	80.98
547.81	457.96	385.54	325.78	197.45	79.77
546.30	457.73	384.88	323.92	195.83	78.97
545.29	456.87	383.82	323.67	188.21	78.21
541.07	452.89	380.98	315.08	183.46	75.55
536.25	451.72	378.55	314.74	181.40	74.09
525.97	446.93	376.27	311.72	177.55	73.24
519.60	446.36	375.38	304.87	176.05	72.21
517.79	443.70	374.54	304.55	161.79	69.20
516.81	442.09	373.88	303.02	160.17	68.13
515.89	441.75	370.02	301.87	158.89	64.01
515.05	436.46	369.61	298.42	153.46	62.29
514.28	434.82	369.36	290.73	149.06	59.95
510.54	431.16	362.75	289.46	144.96	58.87
506.94	428.85	358.49	285.57	140.70	57.29
504.41	426.33	357.27	284.78	135.89	53.44
502.26	425.34	354.60	282.72	135.06	51.42
501.57	423.03	354.19	282.00	128.66	50.87
492.05	419.88	351.98	280.34	127.02	47.05
491.63	419.13	351.80	278.47	121.88	44.13
489.56	418.52	351.21	277.94	117.92	42.56
486.12	417.90	350.50	276.15	116.64	32.93
483.97	417.70	349.77	271.86	111.11	30.48
481.05	400.49	345.84	266.21	107.79	25.11
476.38	400.25	343.21	247.94	107.15	18.58
472.75	399.51	341.27	227.83	104.55	16.24
472.39	398.97	340.55	226.27	101.55	15.60
472.20	397.69	335.68	223.72	100.53	12.67
470.52	397.34	335.16	221.67	99.07	
468.75	396.44	334.63	213.95	92.54	
467.93	394.63	328.16	212.59	82.11	

(Source: Tatsuya Tsukuda)

Nuclear Magnetic Resonance

Representative nuclides and their resonance frequencies and relative sensitivities

Nuclide	Spin	Resonant frequency* /MHz	Relative sensititivity to ^{1}H	Nuclide	Spin	Resonant frequency* /MHz	Relative sensititivity to ^{1}H
Nuclides related to biological sciences				**Nuclides related to materials and environmental sciences**			
^{1}H	1/2	100.000	1.00	^{3}He	1/2	76.178	4.42×10^{-1}
^{2}H	1	15.351	9.65×10^{-3}	^{7}Li	3/2	38.863	2.94×10^{-1}
^{13}C	1/2	25.144	1.59×10^{-2}	^{11}B	3/2	32.084	1.59×10^{-2}
^{14}N	1	7.222	1.01×10^{-3}	^{27}Al	5/2	26.057	2.07×10^{-1}
^{15}N	1/2	10.133	1.04×10^{-3}	^{29}Si	1/2	19.865	7.86×10^{-3}
^{17}O	5/2	13.557	2.91×10^{-2}	^{53}Cr	3/2	5.651	9.08×10^{-4}
^{19}F	1/2	94.077	8.32×10^{-1}	^{59}Co	7/2	23.614	2.78×10^{-1}
^{23}Na	3/2	26.451	9.27×10^{-2}	^{63}Cu	3/2	26.505	9.39×10^{-2}
^{31}P	1/2	40.481	6.65×10^{-2}	^{89}Y	1/2	4.917	1.19×10^{-4}
^{33}S	3/2	7.67	2.27×10^{-3}	^{113}Cd	1/2	22.193	1.11×10^{-2}
^{35}Cl	3/2	9.798	4.72×10^{-3}	^{119}Sn	1/2	37.291	5.27×10^{-2}
^{39}K	3/2	4.667	5.10×10^{-4}	^{171}Yb	1/2	17.61	5.52×10^{-3}
^{67}Zn	5/2	6.254	2.87×10^{-3}	^{195}Pt	1/2	24.14	1.04×10^{-2}
^{81}Br	3/2	27.006	9.95×10^{-2}	^{195}Hg	1/2	17.827	5.94×10^{-3}
^{127}I	5/2	20.007	9.54×10^{-2}	^{205}Tl	1/2	57.634	2.02×10^{-1}
^{129}Xe	1/2	27.66	2.16×10^{-2}	^{207}Pb	1/2	20.921	9.06×10^{-3}

* at 2.3488 T

Chemical shifts of ^{1}H nuclei in organic compounds (Reference: TMS)

Type of proton	Shift/ppm	Type of proton	Shift/ppm
CH_3-X X = -N<	2.3-3.4	Alcohol -O-CH_3	3.3-4.0
-O-	3.3-4.0	>C = CH_2	4.5-6.5
-S-	2.2-2.8	-CH = C<	4.5-8.0
-ϕ	2.3-2.6	-C ≡ CH	2.4-3.1
-CO-	1.8-2.7	Aromatic ring ϕ-H	6.5-8.6
-C =	1.6-2.0	Heteroaromatic ring-H	6.5-8.8
-CY<	0.8-2.0	Aldehyde-C(O)H	9.4-10.6
-M(Si, Li, Ge, etc.)	<0	Carboxyl group-OH	9-12
-CH_2-X X = -O-	3.6-4.6	Amine-NH_2	3.6-4.7
-CO-	2.2-2.6	Thioalcohol-SH	3.5-4.0
-C-	1.2-1.7	Alcohol-OH	1.0-5.2
Alcohol -O-CH<	3.8-4.3	Phenol-OH	4.0-10.0
-O-CH_2-	3.6-4.6		

Chemical shifts of ^{13}C nuclei in organic compounds (Reference: TMS)

Type of carbon		Shift/ppm	Type of carbon		Shift/ppm
>CO	Ketone	202–227	⋗C–X	X = –N<	61–76
	Aldehyde	183–207		–S–	52–71
	Carboxylic acid	171–183		–Halogen	34–77
	Ester and amide	160–175	⋗C(tertiary)–CH<		32–78
–C=N	Nitrile	117–124	$-CH_2-$		2–71
>C=C<	Alkene	104–143	H_3C–X	X = –C⋖	4–29
>C=C<	Aromatic ring	112–136		–O–	47–59
–C≡C–	Alkyne	74–93		–N<	12–43
⋗C–X	X = –C⋖	28–51		–S–	10–19
	–O–	73–84		–Halogen	2–27

Chemical shifts of other nuclei (without correction for second-order quadrupolar effects)

Nuclide	Reference substance	Range of chemical shift	
^{11}B	$BF_3O(C_2H_5)_2$	12–19	B(trigonal-planar) : BO_3
		-3.3–2.0	B(tetrahedral) : BO_4
^{14}N	NO_3^-	315–385	Amine
		220–285	Amide
		-14–48	Nitro
		-296–420	Inorganic complex
^{15}N	NH_3(25℃)	0–100	Amine
		100–160	Amide
		150–200	-N<
		330–380	NO_2, =N–, –N*=N
		150–450	>N=N=N
		220–520	Pyridine and related compound
		500–560	Azo
^{17}O	$H_2{}^{17}O$	37–259	-O-
		13–595	=O
^{19}F	$CFCl_3$	-150–270	Inorganic compound
		-15–280	Organic compound
		-429	Gas : F_2
^{27}Al	Al^{3+}aqueous acid (0.2–1 M)	-170–46	Al^{3+}(l, Solvated)
		-100–30	Halogeno Al acid complexes
		51–61	Al(tetrahedral, s) : Q^4 Silicate
		66–75	Q^3 Sheet silicate
		70–77	Oxide
		1–22	Al(octahedral, s, l), $[Alacac]^{3+}$
^{29}Si	$(CH_3O)_4Si$	-120–25	Organic Si compound
		-110–-75	Silicate(SiO_4–nAl):
		-114–-102	Q^4(nAl) : $n=0$
		-107–-96	$n=1$
		-102–-93	$n=2$
		-97–-88	$n=3$
		-89–-80	$n=4$
		-99–-93	Q^3(nAl) : $n=0$
		-99–-85	$n=1$
		-86–-84	$n=2$
		–-76	$n=3$
^{31}P	85% H_3PO_4 solution	-50–260	: P<(linear)
		-220–460	: P⋖(trigonal-planar): PX_3
		40–100	: PH_2R
		120–165	: PHR_2
		-90–135	: PR_3
		-210–-100	: P(Y_2Z), P(YZ_2), Y, Z = R, NR, OR, SR, X
		-130–50	–P⋖(tetrahedral)
		-100–100	>P⋖(trigonal-bipyramidal)
		140–280	⋗P⋖(octahedral)
	PCr (creatine phosphate)	pH dependent	PO_4^{3-} : $\delta_{acid}=3.02$, $\delta_{base}=5.75$, $pH=pK_a+\log[(\delta_{obs}-\delta_{acid})/(\delta_{base}-\delta_{obs})]$ $pK_a=6.78$, $\delta_{PCr}=0.00$

Primary and secondary standards frequently used for chemical shift referencing (s: for solid-state NMR)

Nuclide	Standard	Secondary standards and their chemical shifts relative to the primary reference
^{1}H	Tetramethylsilane(TMS)	C_6H_6(6.771), $CHCl_3$(7.392), H_2O(4.877, s), Adamantane(1.91, s)
^{2}H	D_2O	
^{7}Li	1.0 M LiCl aq	LiCl(−1.19, s), LiBr(−2.04, s)
^{11}B	$(C_2H_5)_2O \cdot BF_3$	H_3BO_3sat.(19.49), BPO_4(−3.60), $NaBH_3$(−42.06)
^{13}C	TMS	C_6H_6(128.475), $CHCl_3$(77.966), Adamantane(38.520, 29.472, s)
^{15}N	Nitromethane NO_3^-, NH_4^+ (4.5 mol ammonium sulfate/3M hydrochloric acid)	$HCONH_3$(−266.712), $^{15}NH_4Cl$(−341.168)
^{17}O	$H_2{}^{17}O$	
^{23}Na	1.0 M NaCl aq	NaCl(7.21, s), NaBr(5.04, s), $NaBH_4$(−8.16, s)
^{27}Al	1.0 M $Al(NO_3)_3$ aq	1.0 M $AlCl_3$ aq(−0.10), $AlNH_4(SO_4)_2 \cdot 12\,H_2O$(−0.54)
^{29}Si	TMS	Hexamethyldisiloxane(6.679, s), $Si(Si(CH_3)_3)_4$ (−9.843, −135.402, s)
^{31}P	85% H_3PO_4 aq	$(NH_4)_2HPO_4$(1.33, s), $NH_4H_3PO_4$(1.0, s)
^{35}Cl	1.0 M NaCl aq	KCl(3.90, s), NaCl(−45.83, s),
^{39}K	1.0 M KCl aq	KCl(47.8, s), KBr(55.1, s), KI(59.3, s)
^{79}Br	1.0 M KBr aq	KBr(42.7, s), NaBr(−10.19, s)

Chemical shifts and spin coupling constants in hydrogen bonds

Compound	Shift /ppm	O···H length /Å
O–H···O	4–20	2–1.2
KOH	−4.4(s)	2.776
$NaClO_4 \cdot H_2O$	4.8(s)	2.152
α-$(COOH)_2 \cdot 2\,H_2O$	6.0(s)	1.9818
$K_2C_2O_4 \cdot H_2O$	7.3(s)	1.801
$CaHPO_4 \cdot 2\,H_2O$	6.4(H_2O, s)	1.811
	10.4(H, s)	1.694
$BeSO_4 \cdot 4\,H_2O$	10.4(s)	1.656
KHsuccinate	19.3(H(3), s)	1.222
KHmalonate	20.5(H(2), s)	1.234
H_2O(l)	5.11(1℃), 4.86(23.3℃), 4.68(48.5℃)	
H_3CCOCH_3	2.075	
$H_2O \cdots H_3CCOCH_3$	3(xw = 0.05), 4.9(xw = 0.96), 23℃	
$H_2O \cdots H_3CCOCH_3$	2.06(xw = 0.05), 2.32(xw = 0.96), 23℃	
H_2O	1.2(g)	without hydrogen bonding

Hydrogen bond	Shift /ppm	Remarks
–NH	12–14	Oligonucleotide, tRNA
$-NH_2$	8–10	(Purine-pyrimidine)
	Magnitude of high-frequency shift /ppm	
$>C^{17}O \cdots H$	50–60	
$>^{17}O \cdots H$	2–5	
$^{17}O-H \cdots X$	5–20	(e.g., Alcohol and phenol)
$^{15}NH \cdots O$	13	
$^{13}O=O \cdots H$	2–40	
	Spin coupling constant /Hz	
$J(^{14}N-H) =$	$-0.7129J(^{15}N-H)$	without hydrogen bonding
	6–7	$^{h}J(^{15}N-H \cdots {}^{15}N)$
	2–3.6	$^{h}J(^{15}N-{}^{1}H \cdots {}^{15}N)$

(Source: Tokuko Watanabe)

Metallic Radii

Metallic radius of an element is defined by the half of the shortest interatomic distance in the metallic crystal of the element. (Unit: pm)

Element	Radius/pm	Element	Radius/pm	Element	Radius/pm	Element	Radius/pm
Li	152	La	187	Co	125	Au	144
Na	186	Cr	125	Ni	125	Zn	133
K	230	Mo	136	Pd	138	Hg	150
Mg	160	W	137	Pt	139	Al	143
Ca	198	Mn	112	Cu	128	Sn	151
Sr	215	Fe	124	Ag	144	Pb	175
Ba	217						

The value for Mn was calculated using data of Oberteuffer, J. A. ; Ibers, J. A. *Acta Crystallogr., Sect. B* 26, 1499-1504 (1970) (ICSD 42743). Those for other elements were calculated using lattice constants shown in CRC Handbook of Chemistry and Physics 85 th ed (2004-2005).

Ionic Radii (In case of octahedral)

The following values are taken from those listed as crystal radii of ions (CR) by R. D. Shannon, *Acta Cryst.*, **A32**, 751-767 (1976). Note that he also lists effective radii of ions (IR), which are also widely used. CR and IR have simple relationships: CR = IR + 14 pm for cations and CR = IR − 14 pm for anions. Although CR and IR give similar results in many cases, care must be taken to ensure consistency. (Unit: pm)

Ion	Radius/pm	Ion	Radius/pm	Ion	Radius/pm	Ion	Radius/pm
Li^+	90	Cr^{3+}	76	Ni^{2+}	83	Pb^{2+}	133
Na^+	116	Mn^{2+} (HS)	97	Cu^{2+}	87	P^{5+} (IV)	31
K^+	152	Mn^{3+} (HS)	79	Zn^{2+}	88	O^{2-}	126
Mg^{2+}	86	Fe^{2+} (LS)	75	Hg^{2+}	116	S^{2-}	170
Ca^{2+}	114	Fe^{2+} (HS)	92	B^{3+} (IV)	25	F^-	119
Ba^{2+}	149	Fe^{3+} (LS)	69	Al^{3+}	68	Cl^-	167
La^{3+}	117	Fe^{3+} (HS)	79	Si^{4+} (IV)	40	Br^-	182
Ti^{4+}	75	Co^{2+} (HS)	89	Sn^{4+}	83	I^-	206

Coordination numbers are 4 for the ions marked with (IV) and 6 otherwise. HS and LS denote high spin and low spin electron configurations, respectively.

van der Waals Radii

(Unit: pm)

H	120 [1.2×10^2]							He	140
		N	155 [1.5×10^2]	O	152 [140]	F	147 [135]	Ne	154
		P	180 [1.9×10^2]	S	180 [185]	Cl	175 [180]	Ar	188
		As	185 [2.0×10^2]	Se	190 [200]	Br	185 [195]	Kr	202
		Sb	[2.2×10^2]	Te	206 [220]	I	198 [215]	Xe	216

Radius of the methyl group, $-CH_3$ [2.0×10^2]
Half thickness of aromatic molecules [1.7×10^2]

Source: values in square brackets: L. Pauling, elsewhere: A. Bondi, *J. Phys. Chem.*, **68**, 441-451 (1964)

Interatomic Distances for Typical Covalent Bonds

Bond	Compound	Interatomic distance/pm	Bond	Compound	Interatomic distance/pm
H-H	H_2	74.14	Br-Br	Br_2	228.1
O-O	O_2	120.7	I-I	I_2	266.6
O-H	H_2O	95.75	C-C	C_2H_6	153.5
N-N	N_2	109.8	C=C	C_2H_4	133.9
N-H	NH_3	101.2	C⩦C	C_6H_6	139.9
N-O	NO_2	119.3	C≡C	C_2H_2	120.3
N=O	NO	115.1	C-H	C_2H_6	109.4
S-H	H_2S	133.6	C-Cl	C_2H_5Cl	180.2
S-O	SO_2	143.1	Si-H	SiH_4	148.0
S-S	S_8	207	Si-Si	Silicon crystal	235.2
Cl-Cl	Cl_2	198.8	Si-O	SiO_2 (Quartz)	161

The value for Si-O is the average of six high-precision data retrieved from ICSD (Levien et al., 1980; Ogata et al., 1987; Hazen et al., 1989; Kihara et al., 1990; Glinnemann et al., 1992; Dusek et al). The value for Si-Si is calculated by multiplying the lattice constant of α Si of 0.543 06 nm in the CRC Handbook of Chemistry and Physics 85th ed. (2004-2005) by $\sqrt{3}/4$.

Bond Angles

It is the angle between two atomic bonds in a molecule.

Compound	Bonds	Angle/°	Compound	Bonds	Angle/°
H_2O	H-O-H	104.5	C_2H_2	C-C-H	180
CO_2	O-C-O	180	C_6H_6	C-C-C	120
H_2S	H-S-H	92.1	CH_4	H-C-H	109.47
NH_3	H-N-H	106.7	C_3H_8	C-C-C	112
O_3	O-O-O	117.5			

Source: CRC Handbook of Chemistry and Physics 85th ed (2004-2005)

Dipole Moment of Molecules

Material	Chemical formula	Electric dipole moment $\mu_e/3.3356\times10^{-30}$ C m	Material	Chemical formula	Electric dipole moment $\mu_e/3.3356\times10^{-30}$ C m
Iodide of potash	KI	11.05	Nitrobenzene	$C_6H_5NO_2$	4.21
Carbon monoxide	CO	0.110	Chlorobenzene	C_6H_5Cl	1.78
Hydrogen chloride	HCl	1.12	Acetic acid	CH_3CO_2H	1.70
Hydrogen bromide	HBr	0.827	Ethanol	C_2H_5OH	1.69
Hydrogen cyanide	HCN	2.985	Methanol	CH_3OH	1.66
Water	H_2O	1.94	Acetone	CH_3COCH_3	2.90
Ammonia	NH_3	1.468	Ethyl ether	$C_2H_5OC_2H_5$	1.16

(Source: Tatsuya Tsukuda)

Thermochemistry

Heat Capacity

Isobaric heat capacity C_p/J K^{-1} mol^{-1} of simple substances at the constant pressure (p)

The symbols S, L, and G indicate, respectively, the solid, liquid, and gas states.

Simple substance	Temperature/K				
	100	200	298.15	400	600
Aluminum (Al)	(S) 13.0	(S) 21.6	(S) 24.3	(S) 25.6	(S) 28.1
Antimony (Sb)	(S) 20.6	(S) 24.4	(S) 25.4	(S) 25.9	(S) 27.4
Boron (B)	(S) 1.1	(S) 6.1	(S) 11.1	(S) 13.8	(S) 17.5
Bromine (Br_2)[1]	(S) 43.6	(S) 53.8	(L) 75.7	(G) 36.7	(G) 37.3
Cadmium (Cd)	(S) 22.1	(S) 24.9	(S) 26.0	(S) 27.2	(L) 29.7
Calcium (Ca)	(S) 19.5	(S) 24.7	(S) 26.3	(S) 27.4	(S) 30.6
Chlorine (Cl_2)[1]	(S) 42.3	(L) 66.2	(G) 33.8	(G) 35.3	(G) 36.6
Chromium (Cr)	(S) 9.7	(S) 20.1	(S) 23.4	(S) 26.1	(S) 29.4
Cobalt (Co)	(S) 14.0	(S) 22.3	(S) 25.0	(S) 26.5	(S) 29.7
Copper (Cu)	(S) 16.0	(S) 22.7	(S) 24.5	(S) 25.3	(S) 26.8
Diamond (C)[2]	(S) 0.2	(S) 2.3	(S) 6.1	(S) 10.2	(S) 16.1
Fullerene (C_{60})[2]	(S) 1.6	(S) 5.0	(S) 8.8	(S) 12.0	-
Gold (Au)	(S) 21.4	(S) 24.4	(S) 25.4	(S) 25.8	(S) 26.8
Graphite (C)[2]	(S) 1.7	(S) 4.9	(S) 8.5	(S) 11.9	(S) 16.9
Iodine (I_2)[1]	(S) 45.7	(S) 51.6	(S) 54.4	(L) 37.2	(G) 37.6
Iron (Fe, α)	(S) 12.1	(S) 21.5	(S) 25.0	(S) 27.4	(S) 32.1
Lead (Pb)	(S) 24.4	(S) 25.9	(S) 26.8	(S) 27.5	(S) 29.4
Lithium (Li)	(S) 12.8	(S) 20.6	(S) 23.6	(S) 27.6	(L) 29.6
Magnesium (Mg)	(S) 15.7	(S) 22.7	(S) 24.8	(S) 26.3	(S) 28.6
Mercury (Hg)	(S) 24.3	(S) 27.3	(L) 28.0	(L) 27.4	(L) 27.2
Nickel (Ni, α)	(S) 13.6	(S) 22.5	(S) 26.1	(S) 28.8	(S) 34.6
Palladium (Pd)	(S) 17.8	(S) 24.2	(S) 26.2	(S) 26.6	(S) 27.7
Phosphorus (yellow) (P)	(S) 13.7	(S) 21.1	(S) 23.8	-	-
Platinum (Pt)	(S) 19.7	(S) 24.8	(S) 26.8	(S) 26.3	(S) 27.4
Potassium (K)	(S) 24.6	(S) 27.0	(S) 29.5	(L) 31.5	(L) 30.1
Rhodium (Rh)	(S) 15.1	(S) 22.6	(S) 24.5	(S) 26.4	(S) 28.2
Silicon (Si)	(S) 7.3	(S) 15.6	(S) 20.0	(S) 22.1	(S) 24.2
Silver (Ag)	(S) 20.2	(S) 24.3	(S) 25.5	(S) 25.9	(S) 27.1
Sodium (Na)	(S) 22.4	(S) 26.0	(S) 28.2	(L) 31.5	(L) 29.7
Sulfur (orthorhombic) (S)	(S) 12.8	(S) 19.4	(S) 22.6	-	-
Tin (white) (Sn)	(S) 22.4	(S) 25.4	(S) 26.4	(S) 29.0	(L) 30.5
Uranium (U)	(S) 22.4	(S) 26.0	(S) 27.5	(S) 29.6	(S) 34.5
Zinc (Zn)	(S) 19.3	(S) 24.0	(S) 25.5	(S) 26.4	(S) 28.4

1) Value per one mole of molecules.
2) Value per one mole of carbon atoms.

The C_p/J K^{-1} mol^{-1} of inorganic materials

Inorganic	Temperature/K				
	100	200	298.15	400	600
Aluminium oxide (alumina) (Al_2O_3)	(S) 12.9	(S) 51.1	(S) 79.0	(S) 96.1	(S)112.6
Aluminium sulfate ($Al_2(SO_4)_3$)	(S) 91.5	(S)191.0	(S)259.4	(S)322	(S)373
Ammonia(NH_3)	(S) 26.1	(L) 73.5	(G) 35.5	(G) 38.5	(G) 44.7
Ammonium chloride (NH_4Cl)	(S) 37.9	(S) 69.5	(S) 85.4	(S)107.1	-
Calcite ($CaCO_3$)	(S) 39.2	(S) 66.5	(S) 83.5	(S) 97	(S)111
Calcium chloride ($CaCl_2$)	(S) 48.8	(S) 67.3	(S) 72.6	(S) 75.4	(S) 78.8
Calcium oxide (CaO)	(S) 16.2	(S) 34.8	(S) 42.8	(S) 46.6	(S) 50.4
Carbon disulfide (CS_2)	(S) 46.1	(L) 75.1	(L) 75.7	(G) 49.6	(G) 54.4
Carbon tetrachloride (CCl_4)	(S) 66.9	(L)103.3	(L)131.7	(G) 91.7	(G) 99.7
Copper oxide (CuO)	(S) 16.5	(S) 34.8	(S) 42.3	(S) 47.0	(S) 52.3
Diborane (B_2H_6)	(S) 54.0	-	(G) 56.4	(G) 71.7	(G) 98.9
Ferrous chloride ($FeCl_2$)	(S) 50.9	(S) 70.7	(S) 76.3	(S) 79.7	(S) 83.1
Hematite (Fe_2O_3)	(S) 31.5	(S) 76.6	(S)103.7	(S)120	(S)141
Hydrogen chloride (HCl)	(S) 40.0	-	(G) 29.1	(G) 29.2	(G) 29.7
Hydrogen iodide (HI)	(S) 43.7	(L) 47.9	(G) 29.1	(G) 29.3	(G) 30.1
Hydrogen sulfide (H_2S)	(S) 39.2	(L) 68.0	(G) 34.2	(G) 35.6	(G) 38.7
Magnetite (Fe_3O_4, α)	(S) 54.7	(S)120.8	(S)150.8	(S)172	(S)213
Nitric acid (HNO_3)	(S) 42.1	(S) 61.5	(L)109.8	(G) 63.6	-
Nitric monoxide (NO)	(S) 35.7	-	(G) 29.8	(G) 30.0	(G) 31.3
Phosphine (PH_3)	(S) 46.9	-	(G) 37.1	(G) 41.8	(G) 50.9
Potassium chloride (KCl)	(S) 39.1	(S) 48.5	(S) 51.5	(S) 52.1	(S) 55.3
Potassium sulfate (K_2SO_4, α)	(S) 79.1	(S)110.0	(S)129.9	(S)149	(S)175
Silane (SiH_4)	(L) 60.7	-	(G) 42.8	(G) 51.4	(G) 65.8
Sodium chloride (NaCl)	(S) 35.1	(S) 46.9	(S) 49.7	(S) 52.5	(S) 55.7
Sodium hydroxide (NaOH)	(S) 27.7	(S) 49.6	(S) 59.5	(S) 65.1	(S) 86.1
Sulfur dioxide (SO_2)	(S) 48.1	(S) 87.7	(G) 39.9	(G) 43.5	(G) 49.0
Zinc carbonate ($ZnCO_3$)	(S) 34.4	(S) 63.8	(S) 80.5	(S) 94.1	(S)122

The C_p/J K^{-1} g^{-1} of water at the saturated vapor pressure

t/℃	0	1	2	3	4	5	6	7	8	9
0	4.22	4.22	4.21	4.21	4.21	4.21	4.20	4.20	4.20	4.20
10	4.20	4.19	4.19	4.19	4.19	4.19	4.19	4.19	4.19	4.19
20	4.18	4.18	4.18	4.18	4.18	4.18	4.18	4.18	4.18	4.18
30	4.18	4.18	4.18	4.18	4.18	4.18	4.18	4.18	4.18	4.18
40	4.18	4.18	4.18	4.18	4.18	4.18	4.18	4.18	4.18	4.18
50	4.18	4.18	4.18	4.18	4.18	4.18	4.18	4.18	4.18	4.18
60	4.19	4.19	4.19	4.19	4.19	4.19	4.19	4.19	4.19	4.19
70	4.19	4.19	4.19	4.19	4.19	4.19	4.19	4.19	4.20	4.20
80	4.20	4.20	4.20	4.20	4.20	4.20	4.20	4.20	4.20	4.20
90	4.21	4.21	4.21	4.21	4.21	4.21	4.21	4.21	4.21	4.21

A quantity per g is called "specific". The molar value, in the unit of K^{-1} mol^{-1}, is obtained when the specific one is multiplied by the molecular weight of water, 18.02.

The C_p/J K^{-1} g^{-1} of water at the saturated vapor pressure beyond the boiling point

t/℃	C_p	t/℃	C_p	t/℃	C_p	t/℃	C_p	t/℃	C_p
100	4.22	160	4.34	220	4.61	280	5.29	340	8.21
110	4.23	170	4.37	230	4.69	290	5.49	350	10.1
120	4.24	180	4.41	240	4.77	300	5.75	360	15.0
130	4.26	190	4.45	250	4.87	310	6.08	370	45.2
140	4.28	200	4.50	260	4.99	320	6.54	373	244
150	4.31	210	4.55	270	5.12	330	7.19	373.9	10000

E. W. Lemmon, M. O. McLinden and D. G. Friend, "Thermophysical Properties of Fluid Systems" in NIST Chemistry WebBook, NIST Standard Reference Database Number 69, Eds. P. J. Linstrom and W. G. Mallard, National Institute of Standards and Technology, Gaithersburg MD, 20899, http://webbook.nist.gov, (retrieved June 25, 2021).

The C_p/J K^{-1} g^{-1} of organic substances at the atmospheric pressure

Symbols S, L, and G indicate the solid, liquid, and gaseous states, respectively.

Organic	C_p/J K^{-1} mol^{-1}				
	100 K	200 K	298.15 K	400 K	600 K
Acetic acid (CH_3COOH)	(S) 50.3	(S) 67.2	(L)123.4	(G) 81.7	(G)105.2
Acetone ($(CH_3)_2CO$)	(S) 65.6	(L)117	(L)125	(G) 92.1	(G)122.8
Aniline ($C_6H_5NH_2$)	(S) 52.3	(S) 95.1	(L)192.0	–	–
Benzene (C_6H_6)	(S) 50.4	(S) 83.8	(L)136.1	(G)111.9	(G)157.9
Benzoic acid (C_6H_5COOH)	(S) 63.9	(S)102.9	(S)146.8	(L)253.3	–
Cyclohexane (c-C_6H_{12})	(S) 58.6	(S)109.5	(L)156.5	(G)149.9	(G)225.2
Dimethylamine ($(CH_3)_2NH$)	(S) 53.3	(L)126.2	(G) 69.0	(G) 87.4	(G)118.9
Ethanol (C_2H_5OH)	(S) 47.0	(L) 91.9	(L)111.4	(G) 87.9	(G)112.2
Methane (CH_4)	(L) 54.8	–	(G) 35.8	(G) 40.7	(G) 52.5
Methanol (CH_3OH)	(S) 43.6	(L) 70.7	(L) 81.6	(G) 51.4	(G) 67.0
Octane (C_8H_{18})	(S)100.8	(S)164.0	(L)254.1	(G)244.0	(G)327.5
Toluene ($C_6H_5CH_3$)	(S) 61.9	(L)134	(L)162	(G)139.1	(G)194.9
Urea ($(NH_2)_2CO$)	(S) 41.3	(S) 67.0	(S) 93.1	–	–

Rough heat capacity of useful materials

(mixtures, multicomponent systems): C_p/J K^{-1} g^{-1}

Material	t/℃	$\frac{C_P}{\text{J K}^{-1}\text{ g}^{-1}}$	Material	t/℃	$\frac{C_P}{\text{J K}^{-1}\text{ g}^{-1}}$
Alloy			Chalk	0	0.75
Brass	0	0.39	Ebonite	20–100	1.38
Constantan		-0.4	Glass	10–50	-0.7
Solder	0	0.18	Natural rubber	0	1.80
Stainless steels			Nylon 6	0	1.31
18Cr/10Ni	0	0.52	PET	0	1.03
18Cr/12Ni	0	0.47	Polyethylene	0	1.55
24Cr/20Ni	0	0.46	Polystyrene	0	1.11
Organic and inorganic substance					
Basalt	20–100	0.84–1.0			

Source: Tables of Physical & Chemical Constants (16th edition 1995). 2.3.6 Specific heat capacities. Kaye & Lady Online. Version 1.0 (2005), achived version is available at https://web.archive.org/web/20190506031327/http://www.kayelaby.npl.co.uk/

Lattice Energy U (kJ mol^{-1}) of Metal Halide and Metal Oxide (298.15 K)

The lattice energy of a metal salt is the energy required for the infinite separation in the gas phase of the component species confined to the solid lattice space at the closest distances. In the case of a salt composed of univalent cations and anions, for example, the process is expressed as MX (S) → M^+ (G) + X^- (G).

For example, the lattice energy is larger for the smaller ion with the larger surface charge density, cf., the smaller Li^+ and the larger K^+.

LiF	1019 ± 3	NaCl	771 ± 2	RbCl	679 ± 2	MgO	3760 ± 20
LiCl	839 ± 2	NaBr	733 ± 2	RbBr	650 ± 2	CaF_2	2596 ± 5
LiBr	793 ± 2	KCl	701 ± 2	CsCl	646 ± 5	CaO	3371 ± 21
LiI	750 ± 2	KBr	670 ± 2	CsBr	610 ± 6	BaO	3019 ± 21

Enthalpy of Fusion (Heat of Fusion) $\Delta_{\text{fus}}H$, at Saturated Vapor Pressure

Material	t_{fus}/℃	$\frac{\Delta_{\text{fus}}H}{\text{kJ mol}^{-1}}$	Material	t_{fus}/℃	$\frac{\Delta_{\text{fus}}H}{\text{kJ mol}^{-1}}$
Acetic acid (CH_3COOH)	16.6	11.7	Ethanol (C_2H_5OH)	-114.5	5.02
Aluminum (Al)	660.1	10.7	Gold (Au)	1064.2	12.7
Ammonia (NH_3)	-77	5.66	Heavy ice (D_2O)	3.8	6.28
Antimony (Sb)	630.7	20	Hydorgen (ortho, para equilibrium) (H_2)	-259.3	0.12
Benzene (C_6H_6)	5.5	9.84			
Bromine (Br_2)	-7.2	10.5	Hydrogen chloride (HCl)	-114.2	1.97
Caesium bromide (CsBr)	636.9	23.6	Ice (H_2O)	0.0	6.01
Caesium chloride (CsCl)	645.9	20.4	Iron (Fe)	1535	15.1
Carbon dioxide (CO_2)[1)]	-56.2	8.33	Lead (Pb)	327.5	4.77
Carbon monoxide (CO)	-205.1	0.84	Lithium bromide (LiBr)	549.9	17.7
Chlorine (Cl_2)	-101	6.41	Lithium chloride (LiCl)	609.9	19.8
Copper (Cu)	1084.6	13.3	Lithium fluoride (LiF)	848.2	27.1

Continued.

Material	t_{fus}/℃	$\frac{\Delta_{fus}H}{kJ\ mol^{-1}}$	Material	t_{fus}/℃	$\frac{\Delta_{fus}H}{kJ\ mol^{-1}}$
Lithium iodide (LiI)	468.9	14.6	Potassium (K)	63.5	2.4
Magnesium (Mg)	651	9.2	Potassium bromide (KBr)	733.9	25.5
Mercury (Hg)	−38.8	2.33	Potassium chloride (KCl)	770.9	26.3
Methane (CH_4)	−182.6	0.94	Silver (Ag)	961.8	11.3
Methanol (CH_3OH)	−97.8	3.17	Sodium bromide (NaBr)	746.9	26.2
Nickel (Ni)	1455	17.6	Sodium chloride (NaCl)	800.7	28.2
Nitrogen (N_2)	−209.9	0.72	Sodium (Na)	97.8	2.63
Oxygen (O_2)	−218.8	0.44	Tin (Sn)	231.9	7.07
Phosphorus (yellow) (P)	44	0.63	Zinc (Zn)	419.6	6.57
Platinum (Pt)	1772	21.7			

1) The value at 5.180×10^5 Pa. Those without notes are the valuse at 10^5 Pa (1 bar).

Enthalpy of vaporization (heat of vaporization) $\Delta_{vap}H$, at saturated vapor pressure

Material	t_{vap}/℃	$\frac{\Delta_{vap}H}{kJ\ mol^{-1}}$	Material	t_{vap}/℃	$\frac{\Delta_{vap}H}{kJ\ mol^{-1}}$
Simple substance			Benzoic acid (C_6H_5COOH)	249	61.5
Argon (Ar)	−185.9	6.5	Benzyl alcohol ($C_6H_5CH_2OH$)	208.8	53.6
Bromine (Br_2) ($p = 2.85 \times 10^4$ Pa)	25	30.7	Butane (C_4H_{10})	−1.1	22.4
Chlorine (Cl_2)	−34.1	20.4	Chloroform ($CHCl_3$)	61.2	29.4
Fluorine (F_2)	−187.9	6.3	*cis*-2-butene (C_4H_8)	3.7	23.3
Helium (He)	−268.9	0.084	Cyclohexane (C_6H_{12})	25	33.3
Hydrogen (ortho, para body equilibrium) (H_2)	−252.8	0.904	Diethyl ether ($(C_2H_5)_2O$)	34.6	26.5
Iodine (I_2) (sublimation)	25	62.3	Dimethyl ether ($(CH_3)_2O$)	−24.8	21.5
Mercury (Hg)	356.7	58.1	Ethane (C_2H_6)	−88.6	14.7
Nitrogen (N_2)	−195.8	5.58	Ethanol (C_2H_5OH)	78.3	38.6
Oxygen (O_2)	−183	6.8	Ethene (C_2H_4)	−103.8	13.5
Silver (Ag)	2193	254	Ethyl bromide (C_2H_5Br)	38.4	30.5
Sodium (Na)	890	89.1	Formaldehyde (HCHO)	−19.3	23.3
Sulfur (S)	444.6	9.6	Formic acid (HCOOH)	100.7	22.7
Xenon (Xe)	−108.1	12.6	Glycerin ($C_3H_8O_3$)	289.9	59.8
Inorganic substance			Hexane (C_6H_{14})	68.8	28.9
Ammonia (NH_3)	−33.5	23.4	Methane (CH_4)	−161.5	8.2
Carbon dioxide (CO_2) (sublimation)	−78.5	25.2	Methanol (CH_3OH)	64.7	35.3
Carbon disulfide (CS_2) (sublimation)	46.3	26.8	*m*-Xylene (*m*-$C_6H_4(CH_3)_2$)	139.1	36.4
Carbon monoxide (CO)	−191.6	6	Naphthalene ($C_{10}H_8$)	217.9	49.4
Hydrogen chloride (HCl)	−85.1	16.2	*n*-Propanol (C_3H_7OH)	97.2	41
Nitrogen monoxide (N_2O)	−88.5	16.6	Octane ($CH_3(CH_2)_6CH_3$)	125.7	35
Sulfur dioxide (SO_2)	−10	24.9	o-Xylene (o-$C_6H_4(CH_3)_2$)	144.4	36.8
Water (H_2O) ($p = 1.25 \times 10^4$ Pa)	25	44.0	Pentane ($CH_3(CH_2)_3CH_3$)	36.1	25.8
Water (value at 100℃)	100	40.7	Phenol (C_6H_5OH)	182	48.5
Organic compound			Propyne (C_3H_4)	−23.2	23.3
1-Butyne (C_4H_6)	8.1	24.5	Propane (C_3H_8)	−42.1	18.8
1-Butene (C_4H_8)	−6.3	21.8	Propene (C_3H_6)	−47.8	18.4
2-Butyne (C_4H_6)	17.9	26.1	Propionic acid (CH_3CH_2COOH)	141.1	28.5
Acetaldehyde (CH_3CHO)	20	27.2	*p*-Xylene (*p*-$C_6H_4(CH_3)_2$)	138.3	36.1
Acetic acid (CH_3COOH)	118	24.4	Toluene ($C_6H_5CH_3$)	110.6	33.5
Acetone ($(CH_3)_2CO$)	56.5	29	*trans*-2-butene (C_4H_8)	0.9	22.8
Benzene (C_6H_6) ($p = 1.25 \times 10^4$ Pa)	25	31.7			

Combustion enthalpy (Heat of Combustion) $\Delta_c H$, ordinary pressure

(Value at 25 ℃)

Material		$\Delta_c H$	Material		$\Delta_c H$
Benzene (C_6H_6)	(L)	-3268 kJ mol^{-1}	Charcoal	(S)	-28–32 kJ g^{-1}
Cyclohexane (c-C_6H_{12})	(L)	-3920 kJ mol^{-1}	Coal	(S)	-20–30 kJ g^{-1}
Diethyl ether ($(C_2H_5)_2O$)	(G)	-2751 kJ mol^{-1}	Coke	(S)	-24–32 kJ g^{-1}
Dimethyl ether ($(CH_3)_2O$)	(G)	-1461 kJ mol^{-1}	Crude petroleum	(L)	-38–47 kJ g^{-1}
Ethane (C_2H_6)	(G)	-1560 kJ mol^{-1}	Jet fuel	(L)	below -44 kJ g^{-1}
Ethanol (C_2H_5OH)	(L)	-1368 kJ mol^{-1}	Kerosene	(L)	-44–47 kJ g^{-1}
Ethylene (C_2H_4)	(G)	-1411 kJ mol^{-1}	Coal gas	(G)	-16–24 kJ dm^{-3}
Methanol (CH_3OH)	(L)	-726 kJ mol^{-1}	LP gas	(G)	~ -100 kJ dm^{-3}
Naphthalene ($C_{10}H_8$)	(S)	-5156 kJ mol^{-1}	Natural gas	(G)	-29–42 kJ dm^{-3}
Octane (C_8H_{18})	(L)	-5470 kJ mol^{-1}	Producer gas	(G)	-4.6–6.3 kJ dm^{-3}
Propane (C_3H_8)	(G)	-2220 kJ mol^{-1}	Utility gas	(G)	~ -45 kJ dm^{-3}
Toluene ($C_6H_5CH_3$)	(L)	-3910 kJ mol^{-1}	Water gas	(G)	-11–12 kJ dm^{-3}

G=gas; L=liquid; S=solid. The values listed in the right column are in the unit of g for solids and liquids and in the unit of dm^3 under standard pressure for gases.

Standard Formation Enthalpy $\Delta_f H^\circ$, Standard Gibbs Formation Energy $\Delta_f G^\circ$, and Standard Entropy S°

G, L, and S indicate the gas, liquid, and solid starts, respectively.

Material		$\Delta_f H^{\circ 1)}$ kJ mol^{-1}	$\Delta_f G^{\circ 1)}$ kJ mol^{-1}	$S^{\circ 1)}$ J K^{-1} mol^{-1}
Inorganic				
Alumina (Al_2O_3)	(S)	−1675.7	−1582.3	50.9
Ammonia (NH_3)	(G)	−45.9	−16.4	192.8
Ammonium chloride (NH_4Cl)	(S)	−314.4	−202.9	94.6
Ammonium sulfate ($(NH_4)_2SO_4$)	(S)	−1180.9	−901.7	220.1
Aragonite ($CaCO_3$)	(S)	−1207.1	−1127.8	88.7
Cadmium sulfide (CdS)	(S)	−161.9	−156.5	64.9
Calcite ($CaCO_3$)	(S)	−1206.9	−1128.8	92.9
Calcium chloride ($CaCl_2$)	(S)	−795.8	−748.1	104.6
Calcium oxide (CaO)	(S)	−635.1	−604.1	39.8
Carbon dioxide (CO_2)	(G)	−393.5	−394.4	213.6
Carbon disulfide (CS_2)	(L)	89.7	65.3	151.3
Carbon monoxide (CO)	(G)	−110.5	−137.2	197.6
Copper (II) sulfate anhydrous ($CuSO_4$)	(S)	−771.4	−662.2	109.2
Copper (II) sulfate pentahydrate ($CuSO_4 \cdot 5H_2O$)	(S)	−2279.7	−1880.1	300.4
Diamond (C)	(S)	1.895	2.9	2.377
Diborane (B_2H_6)	(G)	36.4	87.6	232.1
Dinitrogen monoxide (N_2O)	(G)	82.1	104.2	219.7
Fullerene C_{60} (C)	(S)	2327.0	2302.0	426.0

1) The enthalpy change $\Delta_f H^\circ$ and the Gibbs energy change $\Delta_f G^\circ$ when generating the material listed below from stable simple materials at standard pressure. The entropy S° for the materials at standard conditions (value at 25 ℃).

The values of $\Delta_f H^\circ$, $\Delta_f G^\circ$, and S°

Continued.

Material		$\Delta_f H^{\circ 1)}$ / kJ mol^{-1}	$\Delta_f G^{\circ 1)}$ / kJ mol^{-1}	$S^{\circ 1)}$ / J K^{-1} mol^{-1}
Gallium arsenide (GaAs)	(S)	−71.0	−67.8	64.2
Hematite (Fe_2O_3)	(S)	−824.2	−742.2	87.4
Hydrogen chloride (HCl)	(G)	−92.3	−95.3	186.8
Hydrogen iodide (HI)	(G)	26.5	1.7	206.6
Hydrogen sulfide (H_2S)	(G)	−20.6	−33.4	205.8
Iron (II) sulfate ($FeSO_4$)	(S)	−928.4	−820.8	107.5
Lead (II) sulfide (PbS)	(S)	−100.4	−98.7	91.2
Magnesium hydroxide ($Mg(OH)_2$)	(S)	−924.5	−833.5	63.2
Magnesium sulfate ($MgSO_4$)	(S)	-1284.9	-1170.6	91.6
Magnetite (Fe_3O_4, α)	(S)	-1118.4	-1015.4	146.4
Nitric acid (HNO_3)	(L)	−174.1	−80.7	155.6
Nitrogen dioxide (NO_2)	(G)	33.2	51.3	240.1
Nitrogen monoxide (NO)	(G)	91.3	87.6	210.8
Ozone (O_3)	(G)	142.7	163.2	238.8
Phosphorous trichloride (PCl_3)	(L)	−319.7	−272.3	217.1
Potassium chloride (KCl)	(S)	−436.5	−408.5	82.6
Potassium hydroxide (KOH)	(S)	−424.6	−379.4	81.2
Potassium iodide (KI)	(S)	−327.9	−324.9	106.3
Pyrite (FeS_2)	(S)	−178.2	−166.9	52.9
Silane (SiH_4)	(G)	34.3	56.9	204.6
Sodium chloride (NaCl)	(S)	−411.2	−384.1	72.1
Sodium hydroxide (NaOH)	(S)	−425.8	−379.7	64.4
Sulfur dioxide (SO_2)	(G)	−296.8	−300.2	248.1
Sulfuric acid (H_2SO_4)	(L)	−814.0	−690.0	156.9
Water (H_2O)	(L)	−285.8	−237.2	69.9
Organic				
Acetic acid (CH_3COOH)	(L)	−484.3	−389.9	159.8
Acetone (CH_3COCH_3)	(L)	−248.1	−155.4	199.8
Benzene (C_6H_6)	(L)	49.1	124.5	173.3
Chloroform ($CHCl_3$)	(L)	−134.1	−73.7	201.7
Cyclohexane (C_6H_{12})	(L)	−156.4	26.7	204.4
1,2-dichloroethane ($ClCH_2CH_2Cl$)	(L)	−166.8	−79.5	208.5
Ethane (C_2H_6)	(G)	−84.0	−32.0	229.2
Ethanol (C_2H_5OH)	(L)	−277.0	−174.8	160.7
Ethene (ethylene) (C_2H_4)	(G)	52.4	68.4	219.3
Ethyl acetate ($CH_3CO_2C_2H_5$)	(L)	−479.3	−332.7	257.7
Ethyne (acetylene) (C_2H_2)	(G)	226.7	209.9	200.9
Formic acid (HCOOH)	(L)	−425.0	−361.4	129.0
Hexane (C_6H_{14})	(L)	20.0	−4.4	296.1
Methane (CH_4)	(G)	−74.6	−50.5	186.3
Methanol (CH_3OH)	(L)	−239.2	−166.6	126.8
Octane ($CH_3(CH_2)_6CH_3$)	(L)	−250.1	6.5	361.2
Propene (propylene) (C_3H_6)	(G)	12.4	62.7	266.6
Toluene (C_7H_8)	(L)	−166.8	113.8	221.0
ortho-Xylene (o-$C_6H_4(CH_3)_2$)	(L)	−24.4	110.5	246.0
meta-Xylene (m-$C_6H_4(CH_3)_2$)	(L)	−25.4	107.7	253.8
para-Xylene (p-$C_6H_4(CH_3)_2$)	(L)	−24.4	110	243.5

2) The pressure of 10^5 Pa (=1 bar) is taken as the standard pressure.

Standard formation enthalpy $\Delta_f H^\circ$, standard Gibbs formation energy $\Delta_f G^\circ$, and standard entropy S° for ions in aqueous solution [1)]

Ion	$\Delta_f H^\circ$/kJ mol^{-1}	$\Delta_f G^\circ$/kJ mol^{-1}	S°/J K^{-1} mol^{-1}
Cation			
Calcium (Ca^{2+})	-543.0	-533.0	-53.1
Copper (Cu^{2+})	64.8	65.5	-99.6
Copper (Cu^{+})	71.7	50.0	40.6
Iron (Fe^{3+})	-48.5	-4.5	-293.3
Iron (Fe^{2+})	-89.1	-78.9	-113
Potassium (K^{+})	-254.1	-283.3	102.5
Proton (H^{+})	0	0	0
Sodium (Na^{+})	-240.1	-260.9	59.0
Anion			
Carbonate (CO_3^{2-})	-677.1	-527.9	-56.9
Chloride (Cl^{-})	-167.2	-131.3	56.5
Hydrogencarbonate (HCO_3^{-})	-692.0	-586.8	91.2
Hydroxide (OH^{-})	-230.0	-157.3	-10.7
Iodide (I^{-})	-55.2	-51.6	111.3
Nitrate (NO_3^{-})	-207.4	-111.3	146.4
Sulfate (SO_4^{2-})	-909.3	-744.6	20.1

1) The standard here is the condition in which the activity is equal to the unity, that is, the condition in which the effective concentration in an aquious solution is equal to 1 mol dm^{-3}. As a rule, the values of $\Delta_f H^\circ$, $\Delta_f G^\circ$, S° for H^+ (aq) are all set to 0 as a reference.

Standard generation enthalpy $\Delta_f H^\circ$ of free radicals in vapor

Free radical	$\Delta_f H^\circ$ kJ mol^{-1}	Free radical	$\Delta_f H^\circ$ kJ mol^{-1}	Free radical	$\Delta_f H^\circ$ kJ mol^{-1}	Free radical	$\Delta_f H^\circ$ kJ mol^{-1}
AlCl	-52 ± 4	CF	255 ± 8	SH	145 ± 20	C_6H_5	325 ± 10
AlH	259 ± 20	CN	423 ± 4	SO	6.8 ± 1.3	CHO	33 ± 6
BH	443 ± 8	NH	339 ± 4	CH	594 ± 4	CH_3COO	-208 ± 8
C_2	838 ± 4	NH_2	168 ± 13	CH_2	385 ± 4	CH_3NH	190 ± 4
CCl_3	80 ± 4	OH	39 ± 1	CH_3	142 ± 5	CH_3NH_2	155 ± 8

Lower and Upper Explosive Limits (LEL/UEL) of Gases in the Mixture with Air

Material	LEL	UEL	Material	LEL	UEL
Ammonia (NH_3)	16	25	Ethyne (acetylene)(C_2H_2)	2.5	81
Benzene (C_6H_6)	1.4	7.1	Gasoline	1.2	7.1
Carbon disulfide (CS_2)	1.3	44	Hydrogen (H_2)	4.0	75
Carbon monoxide (CO)	12.5	74	Methane (CH_4)	5.3	14.0
Diethyl ether ($(C_2H_5)_2O$)	1.9	48	Methanol (CH_3OH)	7.3	36
Dimethyl ether ($(CH_3)_2O$)	3.4	27	Toluene ($C_6H_5CH_3$)	1.4	6.7
Ethanol (C_2H_5OH)	4.3	19			

Explosive limits are shown in volume percentage.

Electrochemistry and Solution Chemistry

Standard Electrode Potential E°

(standard hydrogen electrode reference in an aqueous solution of 25℃, pH = 0)

The majority of E° are values calculated based on the standard molar Gibbs formation energy $\Delta_f G^\circ$ of the matter.

Electronic transfer equilibrium		E° (V *vs.* SHE)
$Li^+ + e^-$	$= Li$	−3.045
$K^+ + e^-$	$= K$	−2.925
$Rb^+ + e^-$	$= Rb$	−2.924
$Ba^{2+} + 2e^-$	$= Ba$	−2.92
$Sr^{2+} + 2e^-$	$= Sr$	−2.89
$Ca^{2+} + 2e^-$	$= Ca$	−2.84
$Na^+ + e^-$	$= Na$	−2.714
$Mg^{2+} + 2e^-$	$= Mg$	−2.356
$Be^{2+} + 2e^-$	$= Be$	−1.97
$Al^{3+} + 3e^-$	$= Al$	−1.676
$U^{3+} + 3e^-$	$= U$	−1.66
$Ti^{2+} + 2e^-$	$= Ti$	−1.63
$Zr^{4+} + 4e^-$	$= Zr$	−1.55
$Mn^{2+} + 2e^-$	$= Mn$	−1.18
$Zn^{2+} + 2e^-$	$= Zn$	−0.763
$Cr^{3+} + 3e^-$	$= Cr$	−0.74
$Ag_2S + 2e^-$	$= 2Ag + S^{2-}$	−0.691
$S + 2e^-$	$= S^{2-}$	−0.447
$Fe^{2+} + 2e^-$	$= Fe$	−0.44
$Cr^{3+} + e^-$	$= Cr^{2+}$	−0.424
$Cd^{2+} + 2e^-$	$= Cd$	−0.403
$PbSO_4 + 2e^-$	$= Pb + SO_4^{2-}$	−0.351
$O_2 + e^-$	$= O_2^-$	−0.284
$Co^{2+} + 2e^-$	$= Co$	−0.277
$PbCl_2 + 2e^-$	$= Pb + 2Cl^-$	−0.268
$Ni^{2+} + 2e^-$	$= Ni$	−0.257
$V^{3+} + e^-$	$= V^{2+}$	−0.255
$Mo^{3+} + 3e^-$	$= Mo$	−0.2
$CO_2 + 2H^+ + 2e^-$	$= HCOOH$	−0.199
$CuI + e^-$	$= Cu + I^-$	−0.182
$AgI + e^-$	$= Ag + I^-$	−0.152
$Sn^{2+} + 2e^-$	$= Sn$	−0.138
$Pb^{2+} + 2e^-$	$= Pb$	−0.126
$2H^+ + 2e^-$	$= H_2$ (standard)	**0.000**
$AgBr + e^-$	$= Ag + Br^-$	+0.071
$CuCl + e^-$	$= Cu + Cl^-$	+0.121
$Sn^{4+} + 2e^-$	$= Sn^{2+}$	+0.15

Electronic transfer equilibrium		E° (V *vs.* SHE)
$Cu^{2+} + e^-$	$= Cu^+$	+0.159
$S + 2H^+ + 2e^-$	$= H_2S$	+0.174
$CO_3^{2-} + 6H^+ + 4e^-$	$= HCHO + 2H_2O$	+0.197
$AgCl + e^-$	$= Ag + Cl^-$	+0.222
$Hg_2Cl_2 + 2e^-$	$= 2Hg + 2Cl^-$	+0.268
$Cu^{2+} + 2e^-$	$= Cu$	+0.337
$Fe(CN)_6^{3-} + e^-$	$= Fe(CN)_6^{4-}$	+0.361
$Cu^+ + e^-$	$= Cu$	+0.520
$O_2 + 2H^+ + 2e^-$	$= H_2O_2$	+0.695
$Fe^{3+} + e^-$	$= Fe^{2+}$	+0.771
$Hg_2^{2+} + 2e^-$	$= 2Hg$	+0.796
$Ag^+ + e^-$	$= Ag$	+0.799
$NO_3^- + 2H^+ + 2e^-$	$= NO_2^- + H_2O$	+0.835
$Hg^{2+} + 2e^-$	$= Hg$	+0.85
$Pd^{2+} + 2e^-$	$= Pd$	+0.915
$NO_3^- + 4H^+ + 3e^-$	$= NO + 2H_2O$	+0.957
$Br_2 + 2e^-$	$= 2Br^-$	+1.065
$Pt^{2+} + 2e^-$	$= Pt$	+1.188
$O_2 + 4H^+ + 4e^-$	$= 2H_2O$	+1.229
$MnO_2 + 4H^+ + 2e^-$	$= Mn^{2+} + 2H_2O$	+1.23
$Cl_2 + 2e^-$	$= 2Cl^-$	+1.358
$Cr_2O_7^{2-} + 14H^+ + 6e^-$	$= 2Cr^{3+} + 7H_2O$	+1.36
$MnO_4^- + 8H^+ + 5e^-$	$= Mn^{2+} + 4H_2O$	+1.51
$Mn^{3+} + e^-$	$= Mn^{2+}$	+1.51
$Au^{3+} + 3e^-$	$= Au$	+1.52
$HClO + 2H^+ + 2e^-$	$= Cl_2 + 2H_2O$	+1.630
$PbO_2 + SO_4^{2-} + 4H^+ + 2e^-$	$= PbSO_4 + 2H_2O$	+1.698
$Ce^{4+} + e^-$	$= Ce^{3+}$	+1.71
$H_2O_2 + 2H^+ + 2e^-$	$= 2H_2O$	+1.763
$Au^+ + e^-$	$= Au$	+1.83
$S_2O_8^{2-} + 2e^-$	$= 2SO_4^{2-}$	+1.96
$O_3 + 2H^+ + 2e^-$	$= O_2 + H_2O$	+2.705
$F_2 + 2e^-$	$= 2F^-$	+2.87

Reference Electrode and Potential (25℃)

The reference electrode normally uses SHE as the origin. For example, silver-silver chloride electrodes have a measurement value of −0.500 V *vs.* Ag-AgCl. In SHE scale, this value is shown as −0.301 V *vs.* SHE.

Reference electrode	Electrode system	Electronic transfer equilibrium	Potential
Standard hydrogen electrode (SHE)	Pt/H^+ (a=1), H_2 (a=1)	$2H^+ + 2e^- = H_2$	0.000
Silver-silver chloride (Ag-AgCl) electrode	Ag/AgCl/KCl (Saturated)	$AgCl + e^- = Ag + Cl^-$	+0.199
Saturated calomel electrode (SCE)	$Hg/Hg_2Cl_2/KCl$ (Saturated)	$Hg_2Cl_2 + 2e^- = 2Hg + 2Cl^-$	+0.241

Example of Primary Batteries

Generally speaking, discharge reactions are quite complex and there are many aspects which have not yet been clarified. Therefore, we have shown only those reactions which are considered to be main reactions.

Battery name	Electrical discharge reaction type	Electrolyte	Nominal voltage
Daniell cell	$Zn + CuSO_4 \rightarrow ZnSO_4 + Cu$	$ZnSO_4 \parallel CuSO_4$	1.10 V
Manganese Dry battery			
[NH_4Cl Type]	$Zn + 2MnO_2 + 2NH_4Cl$ $\rightarrow Zn(NH_3)_2Cl_2 + 2MnOOH$	NH_4Cl	1.50 V
[$ZnCl_2$ Type]	$4Zn + 8MnO_2 + ZnCl_2 + 8H_2O$ $\rightarrow ZnCl_2 \cdot 4Zn(OH)_2 + 8MnOOH$	$ZnCl_2$	1.50 V
Alkaline-manganese cell	$Zn + 2MnO_2 + 2H_2O + 2OH^-$ $\rightarrow Zn(OH)_4^{2-} + 2MnOOH$	KOH	1.50 V
Mercury cell	$Zn + HgO + H_2O + 2OH^-$ $\rightarrow Zn(OH)_4^{2-} + Hg$	KOH	1.35 V
Silver oxide cell	$Zn + Ag_2O + H_2O + 2OH^-$ $\rightarrow Zn(OH)_4^{2-} + 2Ag$	KOH	1.55 V
Lithium battery	$4Li + 2SOCl_2 \rightarrow 4LiCl + S + SO_2$	$LiAlCl_4(SOCl_2)$	3.66 V
Metal-air battery	$2Zn + O_2 + 2H_2O \rightarrow 2Zn(OH)_2$	KOH	1.65 V
Hydrogen-oxygen fuel cell	$2H_2 + O_2 \rightarrow 2H_2O$	H_3PO_4 Others	0.7-1.0 V

Example of Secondary Batteries (Storage Batteries)

Battery name	Electrical charge and discharge reaction type (The right is discharged)	Electrolyte	Nominal voltage
Lead-acid storage battery	$Pb + PbO_2 + 2H_2SO_4 \rightleftarrows 2PbSO_4 + 2H_2O$	H_2SO_4	2.04 V
Nickel-Cadmium battery	$Cd + 2NiOOH + 2H_2O$ $\rightleftarrows Cd(OH)_2 + 2Ni(OH)_2$	KOH	1.33 V
Nickel-hydrogen battery	$MH + NiOOH \rightleftarrows M^{(*)} + Ni(OH)_2$	KOH	1.33 V
Lithium-ion battery	$Li_xC + Li_{1-x}CoO_2 \rightleftarrows C + LiCoO_2$	$LiClO_4$ Others	4.10 V

* M indicates solid metal hydrides (alloys of La, Nd, Co, Ni, Al, etc.).

Electric Conductivity κ (25℃) of Aqueous Electrolyte Solution

κ possesses the dimensions of S m^{-1} ($\Omega^{-1}m^{-1}$). Solution with a thickness of l (m) enclosed between two polar plates with an area of A (m^2) shows a resistance R (Ω) which is almost equivalent to $R = l/(\kappa \cdot A)$. The following table shows the κ value at 0.1 mol dm^{-3}.

Electrolyte	HCl	HNO_3	H_2SO_4	NaOH	NH_3	KCl	NaCl	$CuSO_4$
κ/S m^{-1}	3.898	3.842	2.343	2.215	0.036	1.290	1.067	0.436

Electric Conductivity κ (25℃) of KCl Standard Aqueous Solution

Density(g-KCl/kg-H_2O)		76.5829	7.47458	0.745819
κ/S m^{-1}	0℃	6.5144	0.7134	0.07733
	18℃	9.782	1.1164	0.12202
	25℃	11.132	1.2853	0.14085

Limit Molar Conductivity λ^∞ (in 25℃ Water) for Ions

The value for molar conductivity $\lambda = \kappa_l/c$ (c: $molm^{-3}$ unit concentration) of ions extrapolated to$c \to 0$.

Ion	λ^∞ (S m^2 mol^{-1})	Ion	λ^∞ (S m^2 mol^{-1})	Ion	λ^∞ (S m^2 mol^{-1})	Ion	λ^∞ (S m^2 mol^{-1})
H^+	0.03498	NH_4^+	0.00735	OH^-	0.01986	NO_3^-	0.00714
Li^+	0.003869	1/2 Mg^{2+}	0.005306	Cl^-	0.007635	CH_3COO^-	0.00409
Na^+	0.005011	1/2 Ca^{2+}	0.00595	Br^-	0.00781	1/2 CO_3^{2-}	0.0072
K^+	0.00735	1/3 Al^{3+}	0.0061	I^-	0.00768	1/2 SO_4^{2-}	0.00800

Cation Transport Number t^+ (25℃) in Aqueous Solution

Value at a molarity of 0.1 mol dm^{-3} for electrolytes. Generally speaking, the transport number value does not change significantly depending on concentration. The anion transport number t_- is $1-t_+$.

Electrolyte	$AgNO_3$	$CaCl_2$	HCl	HNO_3	KCl	KNO_3	NaCl	Na_2SO_4
t_+	0.468	0.395	0.831	0.840	0.490	0.510	0.385	0.383

Mean Activity Coefficient $\gamma_\pm$ (in 25℃ Water) for Strong Electrolytes

The mean activity is calculated by multiplying the mass molarity m (mol kg^{-1}) of electrolyte solution by $\gamma_\pm$.

m/mol kg^{-1} \ Electrolyte	HCl	H_2SO_4	NaOH	KOH	NaCl	KCl	KNO_3	$NaClO_4$
0.10	0.803	0.265	0.766	0.798	0.778	0.769	0.739	0.775
0.50	0.757	0.154	0.693	0.728	0.679	0.651	0.548	0.670
1.0	0.809	0.130	0.679	0.756	0.656	0.604	0.443	0.629
2.0	1.009	0.124	0.698	0.888	0.670	0.576	0.333	0.609

Dissociation Constant K_a (in 25℃ Water) of Weak Acid and Weak Base

When the dissociation constant of equilibrium ionization $HA \rightleftarrows H^+ + A^-$ is $K_a = [H^+][A^-]/[HA]$, the table shows the values for $pK_a = -\log_{10}K_a$. Multistage ionization is categorized into (1), (2) and (3). For the base, HA is used for the conjugate acid. Acid becomes stronger as pK_a becomes smaller. The base becomes stronger as pK_a becomes larger. Furthermore, when $pH = pK_a$, the formula $[HA] = [A^-]$ holds true.

Material	HA	pK_a	Material	HA	pK_a
Inorganic compound			**Organic compound**		
Sulphurous acid (1)	H_2SO_3	1.86	Aniline	$C_6H_5NH_3^+$	4.65
Sulphurous acid (2)	HSO_3^-	7.19	Benzoic Acid	C_6H_5COOH	4.00
Ammonia	NH_4^+	9.24	Formic Acid	HCOOH	3.55
Sodium hypochlorite solution	HClO	7.53	Glycine (1)	$H_3N^+CH_2COOH$	2.36
Hydrocyanic acid	HCN	9.21	Glycine (2)	$H_3N^+CH_2COO^-$	9.57
Carbonic acid (1)	$H_2CO_3 (CO_2 + H_2O)$	6.35	Acetic acid	CH_3COOH	4.76
Carbonic acid (2)	HCO_3^-	10.33	Salicylic acid (1)	$C_6H_4(OH)COOH$	2.81
Hydrofluoric acid	HF	3.17	Salicylic acid (2)	$C_6H_4(OH)COO^-$	13.4
Hydrogen sulfide acid (1)	H_2S	7.02	Oxalate (1)	HOOC-COOH	1.04
Hydrogen sulfide acid (2)	HS	13.9	Oxalate (2)	$HOOC\text{-}COO^-$	3.82
Sulfuric acid (2)	HSO_4^-	1.99	Lactic acid	$CH_3CH(OH)COOH$	3.66
Phosphoric acid (1)	H_3PO_4	2.15	Pyrdine	$C_5H_5NH^+$	5.42
Phosphoric acid (2)	$H_2PO_4^-$	7.20	Phenol	C_6H_5OH	9.82
Phosphoric acid (3)	HPO_4^{2-}	12.35	Butyric acid	$CH_3(CH_2)_2COOH$	4.63

Example of Ion-Selective Electrodes

Type of electrode	Measurement ion	Composition of film	Measurement range/ mol dm^{-3}	Interfering elements (selection coefficients)
Glass film	H^+	$Li_2O\text{-}Cs_2O\text{-}La_2O_3\text{-}SiO_2$	pH 0-14	$Na^+(-10^{-15})$
Solid film	F^-	LaF_3	$1\text{-}10^{-6}$	$OH^-(-0.1)$
	I^-	AgI	$1\text{-}10^{-5}$	S^{2-}(Coexistence not possible)
	Cu^{2+}	CuS-AgS	$1\text{-}10^{-6}$	Ag^+, Hg_2^{2+}(Coexistence not possible)
Liquid membrane	Li^+	TTD-14-Crown 4/TPB*1 Conductor/membrane material*2	$10^{-1}\text{-}10^{-5}$	$Na^+(-10^{-3})$, $K^+(-2\times10^{-4})$
	Na^+	Bis(12-crown 4)/TBP Conductor/membrane material*2	$10^{-1}\text{-}10^{-5}$	$K^+(-10^{-2})$, $Mg^{2+}(-10^{-3})$, $Ca^{2+}(-10^{-3})$
	K^+	Valinomycin/membrane material*2	$10^{-1}\text{-}10^{-6}$	$Na^+(-10^{-4})$, $Rb^+(-3)$, $Cs^+(-0.4)$
	Ca^{2+}	Ionophore-K23E1/TBP Conductor/membrane material*2	$10^{-1}\text{-}10^{-5}$	$Na^+(-10^{-4})$, $Mg^{2+}(-10^{-4})$

*1 tetraphenylborate. *2 Polyvinyl chloride + film solvent (Source: Tadashi Watanabe)

Solubility

Solubility of inorganic materials in water

The values listed below are the amount in grams of material (anhydride) dissolved in 100 grams of saturated solution at each temperature. Each number is shown as the weight of the anhydride calculated using the chemical formula. For systems marked with asterisks, the solubility is shown as the amount of the solute dissolved in 100 grams of the solvent water.

Material	Formula	t/℃					
		0	20	40	60	80	100
Aluminum sulfate	$Al_2(SO_4)_3$	27.50	27.82(25°)	28.8	31.0	37.4(80.6°)	43.90(99.2°)
Alums*	$AlK(SO_4)_2$	3.0	5.9	11.70	24.75	71.0	119.0(92.5°)
Ammonium alums	$Al(NH_4)(SO_4)_2$	2.97	5.91	14.41(45°)	22.0(55°)	47.3(85°)	-
Ammonium chloride*	NH_4Cl	29.4	37.2	45.8	55.2	65.6	77.3
Ammonium nitrate	NH_4NO_3	54.2	65.5	71.0	80.7	86.9	90.3
Ammonium sulfate	$(NH_4)_2SO_4$	41.35	42.85	44.7	46.64	48.47	50.42
Aresenite (tetraarsenic oxide)	As_4O_6	1.20	1.81	2.85	4.45(62°)	5.62(75°)	8.18(98.5°)
Barium chloride	$BaCl_2$	23.8	26.30	28.9	31.6	34.3	37.5
Barium hydroxide	$Ba(OH)_2$	1.65	3.74	7.60	17.32	50.35	-
Barium nitrate	$Ba(NO_3)_2$	4.72	8.27	12.35	16.9	21.4	25.6
Borax	$Na_2B_4O_7$	1.18	2.58	6.00	14.82	19.88	28.22
Boric acid	H_3BO_3	2.70	4.65	8.17	12.96	19.06	27.53
Cadmium sulfate*	$CdSO_4$	75.5	76.4	78.4	83.7	67.5(79°)	58.4(99°)
Caesium chloride	$CsCl$	61.7	65.1	67.5	69.7	71.4	73.0
Calcium chloride	$CaCl_2$	37.3	42.7	53.4	57.8	59.5	61.4
Calcium hydroxide	$Ca(OH)_2$	0.143	0.129(25°)	0.107	0.0917	0.0800(70°)	0.0523(99°)
Calcium sulfate*	$CaSO_4$	0.1759	0.2080(25°)	0.2097	0.2009(55°)	0.1847(75°)	0.1619
Chrome alum*	$CrK(SO_4)_2$	-	12.51(25°)	-	-	-	-
Cobalt (Ⅱ) chloride	$CoCl_2$	30.3	34.6	46.0	48.4	49.4	51.5
Copper (Ⅱ) nitrate	$Cu(NO_3)_2$	45.5	55.5	62.0	64.5	67.5	71.2
Copper (Ⅱ) sulfate	$CuSO_4$	12.3	16.8	22.3	28.5	35.9	43.4
Disodium hydrogen phosphate*	Na_2HPO_4	1.6	7.7	52.7	82.8	93.5	103.3
Iron (Ⅱ) chloride	$FeCl_2$	33.2	38.5	40.7	43.9	47.4	48.7
Iron (Ⅱ) sulfate	$FeSO_4$	13.6	20.8	28.6	35.5	35.6	30.5
Iron (Ⅲ) chloride	$FeCl_3$	42.66	47.88	60.01(37°)	78.86	84.03	84.26
Lead (Ⅱ) bromide*	$PbBr_2$	0.4554	0.9744(25°)	1.7457(45°)	2.574(65°)	3.343	4.751
Lead (Ⅲ) nitrate*	$Pb(NO_3)_2$	38.8	56.5	75	95	115	138.8
Magnesium chloride	$MgCl_2$	34.6	35.3	36.5	37.9	39.8	42.3
Magnesium sulfate	$MgSO_4$	18.0	25.2	30.8	35.3	35.8	33.5
Mercury (Ⅰ) chloride	Hg_2Cl_2	0.000140	0.000038	-	-	-	-
Mercury (Ⅱ) chloride	$HgCl_2$	3.5	6.1	9.3	14.4	23.4	36.5

1) When the temperature differs from the value at the top of the table, the temperature is shown in (-°) in the unit of ℃.

Continued.

Material	Formula	t/℃					
		0	20	40	60	80	100
Nickel (Ⅱ) sulfate*	$NiSO_4$	27.6	38.0	47.9	56.4	66	78
Nickel chloride	$NiCl_2$	34.8	38.29	42.3	46.1	45.96	46.7
Oversodium chlorate	$NaClO_4$	62.87	67.82(25°)	70.38(38°)	-	-	76.75
Potassium bromide	KBr	34.9	39.4	43.2	46.1	48.7	51.0
Potassium carbonate	K_2CO_3	51.25	52.5	53.9	55.9	58.3	60.9
Potassium chlorate	$KClO_3$	3.2	6.8	12.2	19.2	27.3	36.0
Potassium chloride	KCl	21.92	25.5	28.6	31.4	33.9	36.0
Potassium chromate*	K_2CrO_4	58.8	63.9	68.0	72.2	76.3	80.1
Potassium dichromate*	$K_2Cr_2O_7$	4.6	12.2	26.0	46.5	70	97
Potassium ferrocyanide	$K_4[Fe(CN)_6]$	12.5	22.0	27.8(35°)	36.8(65°)	40.1	42.63
Potassium hexacyanoferrate (III)	$K_3[Fe(CN)_6]$	23.22(0.1°)	32.80(25°)	37.22	41.10(58°)	45.27	47.60(99°)
Potassium hydrogen carbonate	$KHCO_3$	18.6	25.0	31.3	37.5	-	-
Potassium hydroxide	KOH	49.2	52.8	58.03	-	61.73	65.15
Potassium iodide	KI	56.0	59.0	61.6	63.8	65.7	67.4
Potassium nitrate	KNO_3	11.7	24.0	39.0	52.2	62.8	71.0
Potassium permanganate	$KMnO_4$	2.75	5.96	11.13	18.15	20.2	-
Potassium sulfate	K_2SO_4	7.20	10.0	12.9	15.4	17.6	19.4
Rubidium chloride	RbCl	43.5	47.7	50.9	53.6	56.0	58.9
Silver chloride	AgCl	0.000070	0.000155	0.00036	-	-	0.0021
Silver nitrate	$AgNO_3$	54.8	68.4	75.7	81.5	85.4	88.0
Sodium bromide	NaBr	44.47	-	-	54.10	-	-
Sodium carbonate	Na_2CO_3	6.54	18.1	33.1	31.6	31.1	30.9
Sodium chloride	NaCl	26.28	26.38	26.65	27.05	27.54	28.2
Sodium dihydrogenphosphate	NaH_2PO_4	36.5	45.5	57.0	65	68	71
Sodium hydrogen carbonate	$NaHCO_3$	6.48	8.72	11.29	14.10	-	19.1
Sodium hydroxide	NaOH	29.6	52.2	56.3	63.5	75.8	78.5(110°)
Sodium iodide	NaI	61.54	64.1	67.2	72.0	74.7	75.14
Sodium nitrate	$NaNO_3$	41.9	46.0	-	-	-	62.39(94°)
Sodium sulfate*	Na_2SO_4	4.5	19.0	48.1	45.2	43.2	42.2
Sodium thiosulfate	$Na_2S_2O_3$	33.40	41.20	55.33(45°)	-	-	-
Strontium chloride	$SrCl_2$	30.3	35.8(25°)	39.5	45.0	47.5	50.2
Strontium nitrate	$Sr(NO_3)_2$	28.2(0.1°)	40.7	47.2(35°)	48.3	49.2	51.2(105°)
Sub-sodium sulphide*	Na_2SO_3	13.3	30.7(25°)	35.7	31.7	28.0	26.3
Thallous chloride	TlCl	0.16	0.33	-	-	1.5	-
Thallous nitrate	$TlNO_3$	3.76	8.72	17.33	31.55	52.6	80.54
Trisomium phosphate*	Na_3PO_4	5.38	14.53(25°)	23.3	46.2	68.0	94.6
Zinc nitrate	$Zn(NO_3)_2$	48.3(0.4°)	58.1(30°)	70.2(45°)	87.2(59°)	-	-
Zinc sulfate (orthorhombic)	$ZnSO_4$	28.58	36.65(25°)	41.2	42.98	39.2(85°)	37.7

Solubility of organic compounds in water (1)
Polar organic compounds

Organic	Chemical formula	t/℃					
		0	20	40	60	80	100
Benzoic acid	$C_6H_5CO_2H$	0.17	0.29	0.56	1.16	2.71	5.88
Succinic acid	$(CH_2CO_2H)_2$	2.8	6.9	16.2	35.8	70.8	120.9
Oxalic acid	$(COOH)_2$	3.50	9.52	21.5	44.3	84.5	–
Tartaric acid (dextrorotatory)	$(CHOH\cdot CO_2H)_2$	115	139	176	218	273	343
Racemic acid(*dl*)	$(CHOH\cdot CO_2H)_2$	9.23	20.6	43.3	78.3	125	185
Picric acid	$C_6H_2(NO_2)_3OH$	0.68	1.11	1.78	2.81	4.41	7.24
Phenol	C_6H_5OH	–	8.5	9.7	17.5	–	–
Sucrose	$C_{12}H_{22}O_{11}$	179.2	203.9	238.1	287.3	362.1	485.2
Sodium acetate	$Na(CH_3CO_2)$	36.3	46.5	65.5	–	–	–
Calcium acetate	$Ca(CH_3CO_2)_2$	37.4	34.73	33.22	32.70	33.50	29.65

Solubility of organic compounds in water (2)
Aromatic hydrocarbons

(Vicinity of room temperature: 25℃ and unit of concentration: mM=10^{-3}mol dm^{-3})

Organic	Chemical formula	Solubility/mM
Benzene	C_6H_6	23
Toluene	$C_6H_5CH_3$	6.5
1,4-dimethylbenzene	$C_6H_4(CH_3)_2$	1.9
Ethyl benzene	$C_6H_5C_2H_5$	1.6
Propyl benzene	$C_6H_5C_3H_7$	0.71
Butylbenzene	$C_6H_5C_4H_9$	0.37
Biphenyl	$(C_6H_5)_2$	0.047
Acenaphthene	$C_{10}H_{10}$	0.025
Naphthalene	$C_{10}H_8$	0.098
Phenanthrene	$C_{14}H_{10}$	0.0090
Anthracene	$C_{14}H_{10}$	0.00042
Pyrene	$C_{16}H_{10}$	0.00087

(Source: Masaru Nakahara and Ken Yoshida)

Solubility in water of organochloride compounds (3)
Organic chlorine compounds

(Temperature: 25℃ and unit of concentration: mM=10^{-3} mol dm^{-3})

Organic	Chemical formula	Solubility/mM
Carbon tertrachloride	CCl_4	5.2
Chloroform	$CHCl_3$	66
Methylene chloride	CH_2Cl_2	231
1,1-dichloroethane	$C_2H_4Cl_2$	52
1,2-dichloroethane	$C_2H_4Cl_2$	87
1,1,1-trichloroethane	$C_2H_3Cl_3$	9.5
1,1,2-trichloroethane	$C_2H_3Cl_3$	33
Chlorobenzene	C_6H_5Cl	3.9
Ortho-dichlorobenzene	$C_6H_4Cl_2$	0.98
Meta-dichlorobenzene	$C_6H_4Cl_2$	0.86
Paradichlorobenzene	$C_6H_4Cl_2$	0.51

Source : S. H. Yalkowsky, Y. He, P. Jain, Handbook of Aqueous Solubility Data, 2nd ed.

Solubility of gases in water

At each temperature, the following table shows the volume when 1 atm = 101325 Pa of gas is dissolved in 1 cm^3 of water, and is then converted to the volume for 0℃, 1 atm. The unit is cm^3.

Material	Formulae	t/℃					
		0	20	40	60	80	100
Air	–	0.029	0.019	0.014	0.012	0.011	0.011
Ammonia*	NH_3	1176	702	–	–	–	–
Argon	Ar	0.053	0.035	0.027	0.021	0.019	0.019
Carbon dioxide	CO_2	1.71	0.88	0.53	0.36	–	–
Carbon monoxide	CO	0.035	0.023	0.018	0.015	0.014	0.014
Chlorine	Cl_2	4.61	2.30	1.44	1.02	0.68	0.00
Ethene (ethylene)	C_2H_4	0.226	0.122	0.081	0.063	0.053	0.049
Ethyne (acetylene)	C_2H_2	1.73	1.03	0.71	0.56	0.48	0.46
Helium	He	0.0093	0.0088	0.0084	0.0092	0.0101	0.0114
Hydrogen	H_2	0.022	0.018	0.016	0.016	0.016	0.016
Hydrogen chloride	HCl	507	442	386	339	–	–
Hydrogen sulfide	H_2S	4.67	2.58	1.66	1.19	0.92	0.81
Methane	CH_4	0.056	0.033	0.024	0.020	0.018	0.017
Neon	Ne	0.013	0.0104	0.0095	0.0094	0.0103	0.0115
Nitrogen monoxide	NO	0.074	0.047	0.035	0.030	0.027	0.026
Nitrogen	N_2	0.024	0.016	0.012	0.010	0.0096	0.0095
Oxygen	O_2	0.049	0.031	0.023	0.019	0.018	0.017
Sulfur dioxide (sulfurous acid gas)	SO_2	80	39	19	–	–	–

* reference value

Solubility Product of Slightly Soluble Salt

The solubility product K_S is the product of the concentrations of dissolved ions in a saturated aqueous solution. For example, when the salt is composed of A_mB_n dissociates in water as $A_mB_n \rightleftarrows mA+nB$, the value for $[A]^m[B]^n$ is the solubility product. The number in the square brackets [ ] is the ion concentration expressed in the unit of mol dm^{-3}.

Material	Ion product concentration	t/℃	K_S
Barium sulfate	$[Ba^{2+}][SO_4^{2-}]$	25	1.0×10^{-10}
Calcium carbonate	$[Ca^{2+}][CO_3^{2-}]$	25	3.6×10^{-9}
Calcium oxalate	$[Ca^{2+}][C_2O_4^{2-}]$	25	3.0×10^{-9}
Iron (II) hydroxide	$[Fe^{2+}][OH^-]^2$	25	8×10^{-16}
Iron (III) hydroxide	$[Fe^{3+}][OH^-]^3$	25	2.5×10^{-39}
Lead carbonate	$[Pb^{2+}][CO_3^{2-}]$	25	8×10^{-14}
Lead (II) iodide	$[Pb^{2+}][I^-]^2$	25	6.4×10^{-9}
Lead (II) sulfate	$[Pb^{2+}][SO_4^{2-}]$	25	1.6×10^{-8}
Magnesium carbonate	$[Mg^{2+}][CO_3^{2-}]$	12	1×10^{-5}
Magnesium hydroxide	$[Mg^{2+}][OH^-]^2$	25	1.8×10^{-11}
Mercury (I) chloride	$[Hg_2^{2+}][Cl^-]^2$	25	1.3×10^{-18}
Silver (I) hydroxide	$[Ag^+][OH^-]$	25	1.9×10^{-8}
Silver bromide	$[Ag^+][Br^-]$	25	4.9×10^{-13}
Silver chloride	$[Ag^+][Cl^-]$	25	1.7×10^{-10}
Silver iodide	$[Ag^+][I^-]$	25	8.3×10^{-17}

Buffer Solution

If these is almost no change in the hydrogen ion concentration (activity) when adding a small amount of acid or base to a solution, or when diluting a solution, the solution is said to possess a buffer effect. There are various types of buffer solutions which maintain a fixed hydrogen ion concentration (activity) of solution. Examples of frequently used buffer solutions are listed below.

1) Hydrochloric acid-potassium chloride buffer solution (25℃)

Add x cm^3 of 0.2 mol dm^{-3} hydrochloric acid solution to 25 cm^3 of 0.2 mol dm^{-3} potassium chloride solution, and diulte to 100 cm^3 with water.

x	67.0	42.5	26.6	16.2	10.2	6.5	3.9
pH	1.00	1.20	1.40	1.60	1.80	2.00	2.20

2) Potassium hydrogenphthalate-hydrochloric acid buffer solution (25℃)

Add x cm^3 of 0.1 mol dm^{-3} hydrochloric acid solution to 50 cm^3 of 0.1 mol dm^{-3} potassium hydrogenphthalate solution, and dilute to 100 cm^3 with water.

x	49.5	42.2	35.4	28.9	22.3	15.7	10.4	6.3	2.9	0.1
pH	2.20	2.40	2.60	2.80	3.00	3.20	3.40	3.60	3.80	4.00

3) Potassium hydrogenphthalate-sodium hydroxide buffer solution (25℃)

Add x cm^3 of 0.1 mol dm^{-3} sodium hydroxide solution to 50 cm^3 of 0.1 mol dm^{-3} potassium hydrogenphthalate solution, and dilute to 100 cm^3 with water.

x	3.0	6.6	11.1	16.5	22.6	28.8	34.1	38.8	42.3
pH	4.20	4.40	4.60	4.80	5.00	5.20	5.40	5.60	5.80

4) Potassium hydrogenphosphate sodium hydroxide buffer solution (25℃)

Add x cm^3 of 0.1 mol dm^{-3} sodium hydroxide solution to 50 cm^3 of 0.1 mol dm^{-3} potassium dihydrogenphosphate solution, and dilute to 100 cm^3 with water.

x	3.6	5.6	8.1	11.6	16.4	22.4	29.1	34.7	39.1	42.4	44.5	46.1
pH	5.80	6.00	6.20	6.40	6.60	6.80	7.00	7.20	7.40	7.60	7.80	8.00

5) Boric acid-sodium hydroxide buffer solution (25℃)

Add x cm^3 of 0.1 mol dm^{-3} sodium hydroxide solution to 50 cm^3 of soultion containing 0.1 mol dm^{-3} of potassium chloride and 0.1 mol dm^{-3} of sodium hydroxide solution, and dilute to 100 cm^3 with water.

x	3.9	6.0	8.6	11.8	15.8	20.8	26.4	32.1	36.9	40.6	43.7	46.2
pH	8.00	8.20	8.40	8.60	8.80	9.00	9.20	9.40	9.60	9.80	10.0	10.20

* Dissolve 6.184 g H_3BO_3 and 7.455 g KCl into 1 dm^3.

6) Sodium hydrogencarbonate-sodium hydroxide buffer solution (25℃)

Add x cm^3 of 0.1 mol dm^{-3} sodium hydroxide soultion to 50 cm^3 of 0.05 mol dm^{-3} sodium hydrogencarbonate solution, and dilute to 100 cm^3 with water.

x	5.0	7.6	10.7	13.8	16.5	19.1	21.2	22.7
pH	9.60	9.80	10.00	10.20	10.40	10.60	10.80	11.00

7) Disodium hydrogenphosphate-sodium hydroxide buffer solution (25℃)

Add x cm^3 of 0.1 mol dm^{-3} sodium hydroxide solution to 50 cm^3 of 0.05 mol dm^{-3} potassium dihydrogenphosphate solution, and dilute to 100 cm^3 with water.

x	4.1	6.3	9.1	13.5	19.4	23.0	26.9
pH	11.00	11.20	11.40	11.60	11.80	11.90	12.00

8) Sodium hydroxide-potassium chloride buffre solution (25℃)

Add x cm^3 of 0.1 mol dm^{-3} sodium hydroxide solution to 25 cm^3 of 0.2 mol dm^{-3} potassium chloride solution, and dilute to 100 cm^3 with water.

x	6.0	10.2	16.2	25.6	41.2	66.0
pH	12.00	12.20	12.40	12.60	12.80	13.00

9) Tris buffer solution (25℃)

Add x cm^3 of 0.1 mol dm^{-3} hydrochloric acid solution to 50 cm^3 of 0.1 mol dm^{-3} Tris solution, and dilute to 100 cm^3 with water.

x	46.6	44.7	42.0	38.5	34.5	29.2	22.9	17.2	12.4	8.5	5.7
pH	7.00	7.20	7.40	7.60	7.80	8.00	8.20	8.40	8.60	8.80	9.00

Tris: Tris (hydroxymethyl) aminomethane

10) Acetic acid-sodium acetate buffer solution (23℃)

Add x cm^3 of 0.2 mol dm^{-3} acetic acid solution to 50$-x$ cm^3 of 0.2 mol dm^{-3} sodium acetate solution, and dilute to 100 cm^3 with water.

x	46.3	44.0	41.0	36.8	30.5	25.5	20.0	14.8	10.5	8.8
pH	3.6	3.8	4.0	4.2	4.4	4.6	4.8	5.0	5.2	5.4

Gomori (1955), Methods in Enzymology, 1, 141.

11) collidine (2,4,6-Trimethylpyridine)-hydrochloric acid buffer

Add x cm^3 of 0.2 mol dm^{-3} hydrochloric acid solution to 25 cm^3 of 0.2 mol dm^{-3} collidine solution, and dilute to 100 cm^3 with water.

x	22.5	20.0	17.5	15	12.5	10	7.5	5.0	2.5
pH	6.45	6.80	7.03	7.22	7.40	7.57	7.77	8.00	8.35

Gomori (1946), *Proc. Soc. Exp. Biol. Med.*, 62, 33.

12) Ammonia-ammonium chloride buffer solution (20℃)

Add x cm^3 of 2 mol dm^{-3} ammonia solution to 10$-x$ cm^3 of 2 mol dm^{-3} ammonium chloride solution, and dilute to 100 cm^3 with water.

x	0.50	1.0	2.0	3.0	5.0	7.0	8.0	9.0	9.5
pH	8.25	8.61	8.96	9.21	9.58	9.94	10.18	10.51	10.82

Gottschalk (1959), *Zeit. Anal. Chem.*, 167, 342. (Source: Masao Sugawara)

Acid-Base Indicators

Acid-base indicators are pigment which changes in conjunction with changes in the hydrogen ion concentration (activity) of solution. The change in color is used to determine the pH of the solution. These indicators can also be used to determine the equivalence point (terminus) of neutralization caused by the acid-base. Normally, acid-base indicators are prepared as water of approximately 0.1 – 0.2 wt % , ethyl alcohol, or a solution which is a mixture of both.

Indicator	Symbol	Indicator range pH	Color variation Acidity → Basicity
Thymor blue (acid side)	TB	1.2– 2.8	Red → Yellow
Bromophenol Blue	BPB	3.0– 4.6	Yellow → Blue purple
Methyl orange	MO	3.1– 4.4	Red → Light yellow
Bromcresol green	BCG	3.8– 5.4	Yellow → Blue
Methylic red	MR	4.2– 6.3	Red → Yellow
Bromocresol purple	BCP	5.2– 6.8	Yellow → Purple
Bromothymol Blue	BTB	6.0– 7.6	Yellow → Blue
Phenol red	PR	6.8– 8.4	Yellow → Red
Cresol red	CR	7.2– 8.8	Yellow → Red
Thymor blue (alkali side)	TB	8.0– 9.6	Yellow → Blue
Phenolphthalein	PP	8.3–10.0	No → Red purple
Thymolphthalein	TP	9.3–10.5	No color → Blue

Analytical Reagent Main types of compounds for ion/molecular recognition

In an aqueous solution or organic solvent, this reagent recognizes the structure of the target ion/molecule, and to form a complex. In general, the binding proceeds with high selectivity. Typical examples are given for each type of compound. The ions in parentheses represent the target ion/molecule by each reagent.

1. Chelating reagent

Mainly used in aqueous solutions, one or multiple molecules form a chelate complex with a metal ion. It has been used as a separation/analytical reagent.

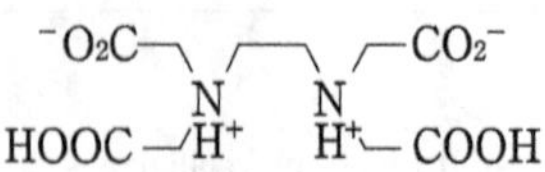

Ethylenediaminetetraacetic acid (EDTA) (various divalent metal ions)

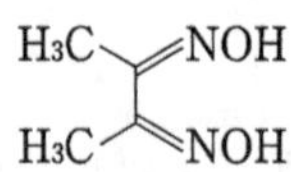

Dimethylglyoxime (Ni^{2+})

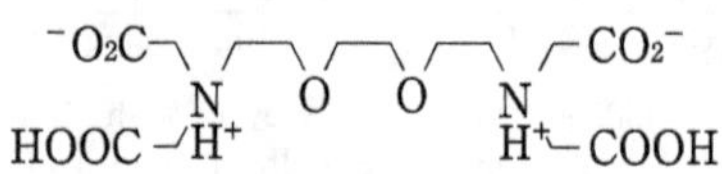

Ethylene glycol-bis-(β-amino-ethyl ether) ethylene glycol tetraacetic acid (EGTA) (Ca^{2+})

2. Crown ether and related compounds

Mainly used in organic solvents, crown ether selectively binds to alkaline-earth metal ions and alkaline metal ions which fit the cavity size. Many analogous compounds are used as separation/analytical reagents.

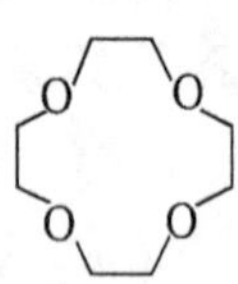

12-crown-4 (Li^+)

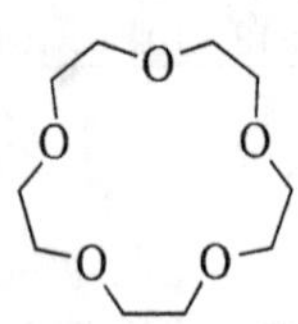

15-crown-5 (Na^+)

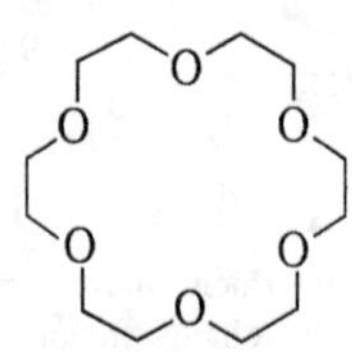

18-crown-6 (K^+)

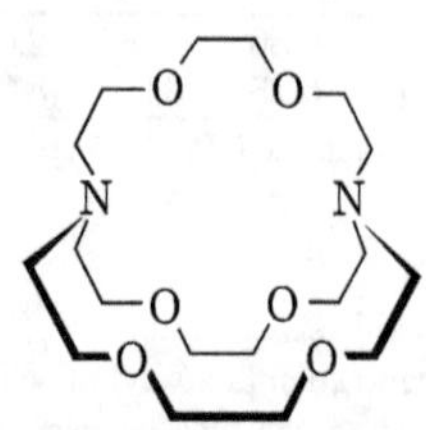

2.2.2-Cryptand (K^+)

3. Macrocyclic polyamine

Mainly used in aqueous solutions, macrocyclic polyamine binds to transition metal ions and heavy metal ions. In the protonated form, on the other hand, it complexes multivalent organic anions such as carboxylate anions and nucleotides. It is applied to separation/analytical reagents.

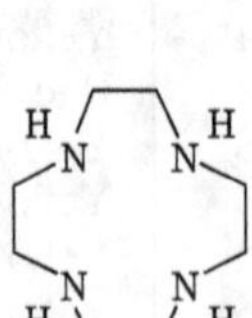

Cyclen

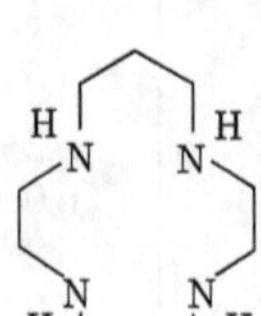

Cyclam

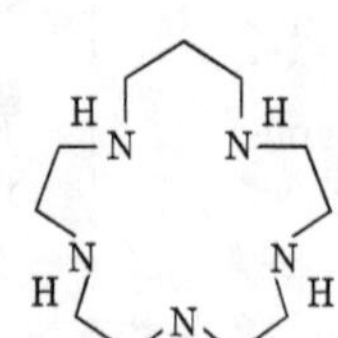

4. Calixarene and related compounds

A macrocyclic compound which is derived by the condensation of phenols and aldehydes. Possesses an intramolecular cavity with a well-defined structure. Mainly in organic solvents, captures organic/inorganic substance which fits the shape and size. It is applied to separation/analytical reagents.

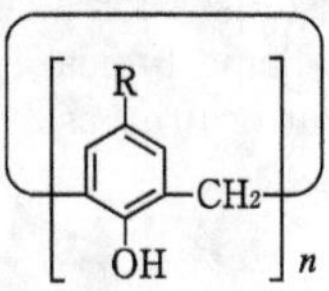

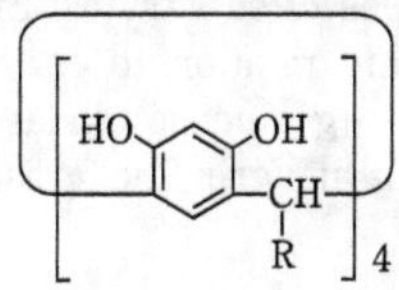

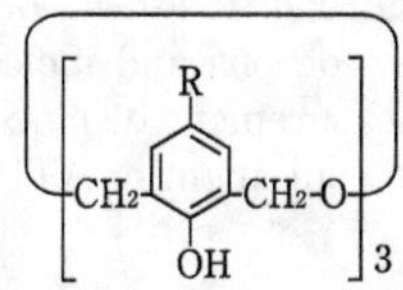

Calix[4]arene ($n=4$)
Calix[6]arene ($n=6$)
Calix[8]arene ($n=8$)

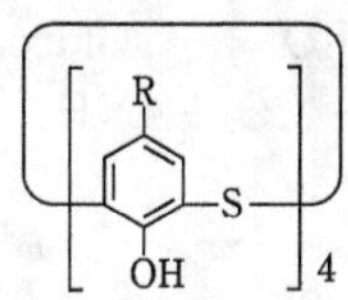

5. Cyclodextrin

Cyclodextrin is a macrocyclic compound made of the cyclic α(1→4)-linked D-glucose. It is obtained as a degradation product of starch. Possesses an intramolecular cavity with a well-defined structure and size dependenting on the number of glucose units. Mainly in aqueous solutions, captures a variety of organic/inorganic substances which fit the shape and size. Cyclodextrin has many applications including use in separation/analytical reagents, nonvolatile aromatic preparations, and stabilizing pharmaceuticals.

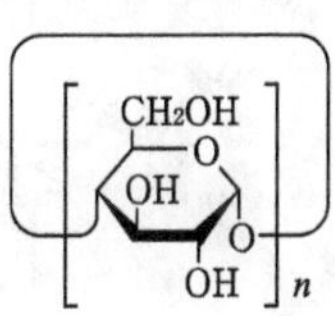

α-cyclodextrin ($n=6$)
β-cyclodextrin ($n=7$)
γ-cyclodextrin ($n=8$)

6. Cyclophane

Cyclophane can be used to design and synthesize intramolecular cavities in a variety of shapes and sizes for capturing organic and inorganic substance in aqueous solutions or organic solvents. Application to separation/analytical reagents is expected.

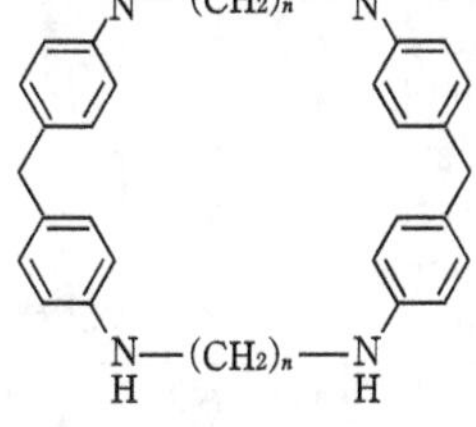

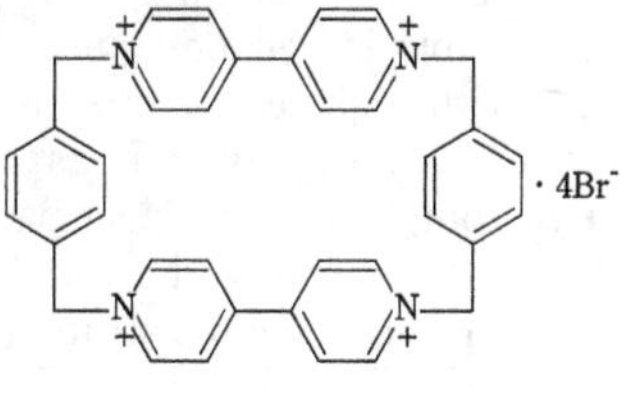

7. Biological material

1) Enzymes

An enzyme is a biopolymer which is composed mainly of protein. Enzymes are capable of acting as substrate-specific catalysts for biochemical reactions and are used as analytical reagent for various types of substrates. Enzymes are also used for quantitative analysis of ions and molecules in relation to enzyme reactions. Some typical examples are measurement of glucose using glucose oxidase, measurement of urea using urease, and measurement of ATP using luciferin/luciferase reaction.

$$\text{Glucose} + O_2 + H_2O \longrightarrow \text{Gluconic acid} + H_2O_2$$

$$\text{Urea} + H_2O \longrightarrow CO_2 + 2\,NH_3$$

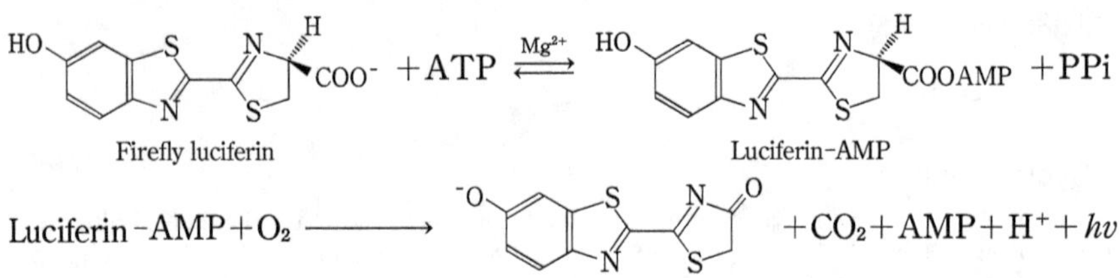

2) Antibodies

Antibodies are a protein which recognizes specific biopolymers (antigens), and are used as an analytical reagent for antigens. Antibodies have a Y-shaped structure composed of four polypeptide chains. Antigens are bonded between the ends of two of the peptide chains. By realizing the production of antibodies in an organism, it is possible to create analytical reagents for a variety of target biopolymers.

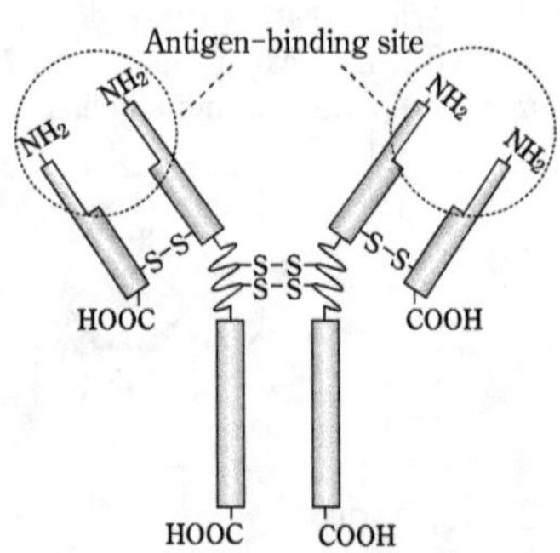

Basic structure of antibody

3) Others

Antibiotics are used as an analytical reagent. A representative example is valinomycin which binds to a potassium ion with high selectivity. Also, single-strand DNAs are used as an analytical reagent for nucleic acid. Through complementary base pairing, they recognize the base sequence of the target single-strand DNA or RNA, forming a double strand (hybridization).

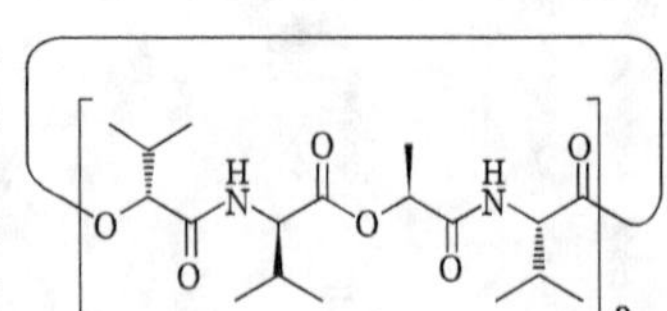

Valinomycin

(Source: Seiichi Nishizawa)

Chemical Formulae and Reactions of Matter

Representative Structures and Reactions of Inorganic Substances

I. Structures

1. Nonmetallic elements

1) Monoatomic molecules
   He, Ne, Ar, Kr, Xe
2) Diatomic molecules
   H_2, N_2, O_2, F_2, Cl_2, Br_2, I_2
3) Other examples

C: graphite (a layered planar structure consisting of sp^2 carbon atoms arranged in a hexagonal honeycomb lattice)
Diamond (sp^3 carbon atoms arranged in a variation of the face-centered cubic crystal)
Amorphous carbon (noncrystalline structure where the nanometer scale net-like planes of carbon hexagons are assembled irregularly)
Fullerene (spherical and spheroidal clusters of carbon, C_{60} has a soccer ball shape)
Carbon nanotubes (tubes with nanometer diameter made from hexagonal sheets)
Graphene (monolayer of hexagonal lattice made of sp^2 carbon atoms)

P: Yellow phosphorus (white phosphorus)(tetrahedral), Purple phosphorus (layered structure, monoclinic),
Black phosphorus (layered structure, orthorhombic), red phosphorus (amorphous solid)

O: Ozone (V shape)

S: S_8 (eight-membered ring with crown conformation) plastic sulfur (noncrystalline solid)

Se: Red selenium (ring) metallic selenium (chain)

2. Metals

(a) Body-centered cubic lattice Li, Na, K, Ca, V, Cr, Mo, W, Fe
(b) Hexagonal lattice Be, Mg, Sc, Y, La, Ti, Zr, Zn, Cd
(c) Face-centered cubic lattice Co, Ni, Pt, Cu, Ag, Au, Al, Pb

3. Small molecular compounds and ions

1) Linear $BeCl_2$, $HgCl_2$, CO_2, XeF_2, N_3^-, NO_2^+, I_3^-
2) Bent $SnCl_2$, SO_2, NO_2^-, ONCl, H_2O, NH_2^-, ClO_2^-
3) Trigonal planar BF_3, BO_3^{3-}, NO_3^-, CO_3^{2-}, SO_3
4) T-shaped ClF_3, BrF_3
5) Trigonal pyramidal NH_3, NF_3, PH_3, PCl_3, AsO_3^{3-}, SO_3^{2-}, $SOCl_2$, ClO_3^-
6) Tetrahedral NH_4^+, BH_4^-, BF_4^-, PO_4^{3-}, SO_4^{2-}, SO_2Cl_2, ClO_4^-, MnO_4^-
7) Square planar XeF_4, IF_4^-
8) Seesaw type $TeCl_4$, SF_4
9) Trigonal bipyramidal PF_5, PCl_5
10) Octahedral SF_6, SiF_6^{2-}

4. Covalent macromolecules and ions (capable of forming large molecules and ions)

1) Polyacids Polysilicate ions (polyacid of ortho silicate ions)
   SiO_3^{2-} (single chain), $Si_4O_{11}^{6-}$ (double chain), $Si_2O_5^{2-}$ (network structure), SiO_2 (three dimension)
   Polyphosphoric acid, Polyboric acid
   Isopoly acid and heteropoly acid of molybdate and tungstate

Continued.

2) Nitride BN (graphite and diamond structures), AlN (wurtzite structure), Si_4N_4 (hexagonal)

3) Carbide SiC, B_4C

4) Sulfide $(NS)_x$ (single chain)

5) Boron hydride (boran) $B_nH_n^{2-}$(closo structure), B_nH_{n+4} (nido structure), B_nH_{n+6} (arachno structure)

6) Silicon-containing polymers Polysilylene$(SiR_2)_x$ (R = Me, Et...), Polysiloxane $(SiR_2O)_x$

5. Ionic crystal

1) Rock salt (NaCl) structure	NaH, LiF, AgBr, MgO, CaS, ScN
2) Cesium chloride (CsCl) structure	CsBr, CsI, TlCl,
3) Zincblende (ZnS) structure	CuCl
4) Wurtzite (ZnS) structure	AlN, GaN, ZnO, CdS
5) Nickel arsenide (NiAs) structure	FeS, CoS, NiS
6) Fluorite (CaF_2) structure	SrF_2, CdF_2, Li_2O, K_2O, CeO_2, ThO_2, Na_2S
7) Cadmium chloride ($CdCl_2$) structure	$CdBr_2$
8) Rutile (TiO_2) structure	MnF_2, FeF_2, NiF_2, SnO_2, PbO_2, MnO_2
9) α-alumina (Al_2O_3) structure	Cr_2O_3, Fe_2O_3
10) Spinel ($MgAl_2O_4$) structure	$CoAl_2O_4$, $MnAl_2O_4$, Fe_3O_4
11) Ilmenite ($FeTiO_3$) structure	$MnTiO_3$, $CoTiO_3$, $NiTiO_3$
12) Perovskite ($CaTiO_3$) structure	$SrTiO_3$, $LaAlO_3$, $LiTaO_3$, $LiNbO_3$

6. Complexes

(A) Mononuclear complex (CN = coordination number)

1) CN = 2 Linear $[CuCl_2]^-$, $[Ag(NH_3)_2]^+$, $[Au(CN)_2]^-$, $[Hg(CH_3)_2]$

2) CN = 3 Trigonal planar $[Pd(PPh_3)_3]$, $[HgI_3]^-$

3) CN = 4 (a) Square planar $[NiCl_4]^{2-}$, $[Cu(NH_3)_4]^{2+}$, $[AuCl_4]^-$

(b) Tetrahedral $[MnO_4]^-$, $[FeCl_4]^-$, $[CoBr_4]^{2-}$, $[Ni(CO)_4]$

4) CN = 5 (a) Square pyramidal $[VO(acac)_2]$(acac: 2, 4-pentanedionato), $[InCl^5]^{2-}$

(b) Trigonal bipyramidal $[CuCl_5]^{3-}$, $[CdCl_5]^{3-}$, $[HgCl_5]^{3-}$

5) CN = 6 (a) Octahedral $[TiF_6]^{2-}$, $[V(H_2O)_6]^{2+}$, $[Cr(NH_3)_6]^{3+}$, $[Mn(CN)_6]^{3-}$, $[Fe(CN)_6]^{4-}$, $[Co(NH_3)_6]^{3+}$, $[NiF_6]^{3-}$, $[PtCl_5(NO)]$, $[Cu(en)_3]^{2+}$(en: ethylenediamine), $[CdCl_6]^{4-}$, $[UF_6]$

(b) Trigonal prismatic $[Mo(S_2C_2H_2)_3]$, $[V(S_2C_2Ph_2)_3]$

Structures Continued.

6) CN=7 (a) Pentagonal bipyramidal $[UF_5O_2]^{3-}$, $[V(CN)_7]^{4-}$
(b) Face-sharing octahedral $[NbCl_4(PMe_3)_3]$, $[TaCl_4(PMe_3)_3]$
(c) Face-sharing trigonal prismatic $[Mo(CNMe)_7]^{2+}$
7) CN=8 (a) Cubic $[UF_8]^{3-}$
(b) Square antiprismatic $[TaF_8]^{3-}$
(c) Dodecahedral $[Ti(NO_3)_4]$

(B) Binuclear and trinuclear complexes
1) Binuclear complexes
(a) No metal-metal bonds $[Cu(CH_3COO)_2 \cdot H_2O]_2$
(b) Metal-metal single bond $[Mn_2(CO)_{10}]$
(c) Metal-metal double bond $[Nb_2Cl_6(SMe_2)_3]$
(d) Metal-metal triple bond $[Mo_2(NMe_2)_6]$, $[W_2Cl_2(NMe_2)_4]$
(e) Metal-metal quadruple bond $[Re_2Cl_8]^{2-}$, $[Mo_2(O_2CMe)_4]$
2) Trinuclear complexes
(a) Linear metal alignment $[Mo_3S_9]$(no metal-metal bond)
(b) Triangle $[Ru_3(CO)_{12}]$, $[M_3S_4(H_2O)_9]^{4+}$ (including metal-metal bonds)
3) Tetranuclear and higher multinuclear complexes
(a) Tetrahedral $[Ir_4(CO)_{12}]$
(b) Cubane type (metal and ligand molecues alternately occupy vertexes of the cubic structure) $[\{Mn(CO)_3\}_4(SEt)_4]$, $[\{Fe(SEt)_4\}S_4]$, $[Mo_3FeS_4(H_2O)_{10}]^{4+}$
(c) Octahedral $[Mo_6Cl_8]^{4+}$, $[Mo_6S_8(PEt_3)_6]$
(d) Linear polymers $[Pt(en)_2][PtCl_2(en)_2]Cl_2$ (one-dimensional chain of -Pt-Cl-Pt-Cl-)
(e) Three-dimensional polymers $[Ni(CN)_2NH_3 \cdot C_6H_6]$(clathrate), $[CuCl_2]$

7. **Interstitial compounds** (possess a structure in which small atoms such as H or N locate in the interstitial hole in the metal lattice of the close-packed structure)
1) Hydride Pd_2H, TiH_2
2) Nitride TiN, Fe_4N, Mn_4N, Mo_2N
3) Carbide WC, TiC, TaC
4) Boride Mn_4B (monatomic inclusion type), V_3B_2 (B_2 inclusin type), FeB (B-B chain bonding), Ta_3B_4 (double chain), MgB_2 (layered structure), CaB_6 (three dimensional structure)

8. **Noncrystalline solids**
1) Inorganic gel (solid where material dispered in sol (a solution containing colloid) has congealed)
Iron hydroxide (III) gel, silica gel, and silica-alumina gel
2) Glass (solidified supercooled liquid)
Silicate glass (fused silica, soda-lime glass, and borosilicate glass), chalcogenide glass

II. Reactions

Continued.

1. Oxidation-reduction

Reaction which increases or decreases the oxidation number of atoms

1) Oxidation with oxygen (combustion)
   a) $C + O_2 \rightarrow CO_2$
   b) $2\,NO + O_2 \rightarrow 2\,NO_2$
   c) $2\,Cu + O_2 \rightarrow 2\,CuO$
2) Oxidation with oxidizing agent such as halogens
   a) $H_2O + Cl_2 \rightarrow HCl + HClO$
   b) $Ca(OH)_2 + Cl_2 \rightarrow CaCl(ClO) \cdot H_2O$ (chlorinated lime)
   c) $NaSO_3 + HClO \rightarrow Na_2SO_4 + HCl$
   d) $2\,HI + H_2O_2 \rightarrow I_2 + 2\,H_2O$
3) Reduction with hydrogen
   a) $CuO + H_2 \rightarrow Cu + H_2O$
   b) $H_2 + Cl_2 \rightarrow 2\,HCl$
4) Metal dissolution and deposition reaction
   a) $Fe + 2\,HCl \rightarrow FeCl_2 + H_2$
   b) $Zn + CuSO_4 \rightarrow Cu + ZnSO_4$
   c) $2\,Na + 2\,H_2O \rightarrow 2\,NaOH + H_2$
5) Electron transfer reaction in metal complexes
   5-1) Outer-sphere mechanism (where chemical processes are not involved)
      a) $[Fe(CN)_6]^{4-} + [IrCl_6]^{2-} \rightarrow [Fe(CN)_6]^{3-} + [IrCl_6]^{3-}$
      b) $[CoCl(NH_3)_5]^{2+} + [Ru(NH_3)_6]^{2+} \rightarrow [CoCl(NH_3)_5]^{+} + [Ru(NH_3)_6]^{3+}$
   5-2) Inner-sphere mechanism (where chemical processes such as ligand exchange are involved)
      $[CoCl(NH_3)_5]^{2+} + [Cr(H_2O)_6]^{2+} \rightarrow [Co(NH_3)_5(H_2O)]^{3+} + [CrCl(H_2O)^5]^{2+}$
6) Electrochemical reaction
   a) $Cu^{2+} + 2\,e^- \rightarrow Cu$ (electrodeposition)
   b) $[Fe(\eta^5\text{-}C_5H_5)_2] \rightleftarrows [Fe(\eta_5\text{-}C_5H_5)_2]^+ + e^-$
      (ferrocene) (ferrocenium ion)

2. Neutralization

Reaction between acid and base

a) $HCl + NaOH \rightarrow NaCl + H_2O$
b) $H_2SO_4 + NaOH \rightarrow NaHSO_4 + H_2O$
c) $H_2SO_4 + 2\,NaOH \rightarrow Na_2SO_4 + 2\,H_2O$
d) $CO_2 + Ca(OH)_2 \rightarrow CaCO_3 + H_2O$

Continued.

3. Substitution

Reaction which substitutes ligands without changing the oxidation number or coordination number of central atoms

a) $Ag^+ + Cl^- \rightarrow AgCl$ (precipitation)
b) $Ba^{2+} + SO_4^{2-} \rightarrow BaSO_4$ (precipitation)
c) $FeS + H_2SO_4 \rightarrow FeSO_4 + H_2S$
d) $[PtCl_4]^{2-} + NH_3 \rightarrow [PtCl_3(NH_3)]^- + Cl^-$
e) $[Cr(CO)_6] + py \rightarrow [Cr(CO)_5(py)] + CO$ (py: pyridine)
f) $FeCl_2 + 2\,C_5H_5Na \rightarrow [Fe(\eta^5\text{-}C_5H_5)_2] + 2\,NaCl$

4. Addition-dissociation

Reaction which adds or dissociates the atomic group (ligands in the case of metal complexes) without changing the oxidation number of central atoms

a) $Ni + 4\,CO \rightarrow [Ni(CO)_4]$
b) $BF_3 + F^- \rightarrow [BF_4]^-$
c) $TiCl_4 + 2\,Cl^- \rightarrow [TiCl_6]^{2-}$
d) $[Cr(CO)_5(CS)] \rightarrow [Cr(CO)_4(CS)] + CO$

5. Oxidative addition / reductive elimination

Reaction which increases/decreases the oxidation number and increases/decreases the coordination number of central atoms

1) Oxidative addition reaction
   a) $2[Co(CN)_5]^{3-} + H_2 \rightarrow 2[Co(CN)_5H]^{3-}$
   b) $[IrCl(CO)(PPh_3)_2]$(Vaska's complex) $+ H_2 \rightarrow [IrClH_2(CO)(PPh_3)_2]$
   c) $PCl_3 + Cl_2 \rightarrow PCl_5$
2) Reductive elimination reaction
   a) $[IrCl(CO)(PPh_3)_2O_2] \rightarrow [IrCl(CO)(PPh_3)_2] + O_2$
   b) $[PdCl_6]^{2-} \rightarrow [PdCl_4]^{2-} + Cl_2$

6. Free radical reaction

Reaction which includes radical species generated by photochemical reactions, etc.

a) $2[Cr(H_2O)_6]^{2+} + CHCl_3 \rightarrow [CrCl(H_2O)_5]^{2+} + [Cr(CHCl_2)(H_2O)_5]^{2+}$
   $+ 2\,H_2O$($CHCl_3 \rightarrow CHCl_2^{\bullet} + Cl^{\bullet}$ occurs)
b) $[Mn_2(CO)_{10}] \xrightarrow{h\nu} [Mn(CO)_5]^{\bullet}$, $[Mn(CO)_5]^{\bullet} + PPh_3 \rightarrow$
   $[Mn(CO)_4(PPh_3)] + CO$

7. Insertion and abstraction (deinsertion) reaction

Reaction in which an atomic group inserts between the central atom and the ligands, and the corresponding reverse reaction

a) $[CH_3Mn(CO)_5] + PPh_3 \rightarrow [(CH_3CO)Mn(CO)_4(PPh_3)]$
b) $[PtH(OCMe_2)(PEt_3)_2]^+ + C_2H_4 \rightarrow [Pt(C_2H_5)(OCMe_2)(PEt_3)_2]^+$

8. Isomerization

Reaction which changes the structure without changing the chemical composition

$$\textit{trans}\text{-}[CoCl_2(en)_2] \underset{HCl}{\overset{NH_3}{\rightleftarrows}} \textit{cis}\text{-}[CoCl_2(en)_2]$$

(Source: Hiroshi Nishihara)

Main Organic Chemical Reactions

1. Hydrolysis

Decomposition reaction caused by water

a) $C_6H_5COOCH_2CH_3 + H_2O \xrightarrow{\text{Acid or alkali}} C_6H_5COOH + CH_3CH_2OH$

b) $CH_3CONHC_6H_5 + H_2O \xrightarrow{\text{Acid or alkali}} CH_3COOH + C_6H_5NH_2$

2. Condensation reaction

Reaction which forms a covalent bond in conjunction with the elimination of simple molecules such as water, hydrogen halide, and alcohol from two or more organic molecules

1) Esterification

$$CH_3COOH + CH_3CH_2OH \xrightarrow[-H_2O]{\text{Acid}} CH_3COOCH_2CH_3$$

2) Amidation

$$C_6H_5NH_2 + CH_3COCl \xrightarrow[-HCl]{} C_6H_5NHCOCH_3$$

3) Aldol condensation

$$2\,CH_3CHO \xrightarrow{\text{NaOH}} CH_3-\underset{\displaystyle OH}{\underset{|}{C}}HCH_2CHO \xrightarrow[-H_2O]{} CH_3CH=CHCHO$$

4) Generation of oxime and hydrazone

a) $(CH_3)_2C=O + NH_2OH \xrightarrow[-H_2O]{\text{Acid}} (CH_3)_2C=NOH$

b) $C_6H_5CHO + C_6H_5NHNH_2 \xrightarrow[-H_2O]{\text{Acid}} C_6H_5CH=NNHC_6H_5$

5) Wittig reaction

$$(C_6H_5)_2C=O + (C_6H_5)_3P=CH_2 \longrightarrow (C_6H_5)_2C=CH_2 + (C_6H_5)_3P=O$$

3. Substitution reaction

Reaction in which the atoms or atomic group included in the molecule are replaced by other atoms or another atomic group

1) Aliphatic substitution reaction

a) $CH_3CH_2CH_2ONa + CH_3CH_2I \longrightarrow CH_3CH_2CH_2OCH_2CH_3 + NaI$
(Williamson's ether synthesis method)

b) $CH_3CH_2CH_2OH + HBr \longrightarrow CH_3CH_2CH_2Br + H_2O$
(Alkyl halide generated from alcohol)

2) Aroma family substitution reaction

a) Halogenation : $C_6H_6 + Br_2 \xrightarrow{Fe} C_6H_5Br + HBr$

b) Nitration : $C_6H_6 + HNO_3 \xrightarrow{H_2SO_4} C_6H_5NO_2 + H_2O$

c) Sulfonation : $C_6H_6 + H_2SO_4 \longrightarrow C_6H_5SO_3H + H_2O$

Continued.

d) Friedel-Crafts reaction：

Acylation $C_6H_6 + CH_3COCl \xrightarrow{AlCl_3} C_6H_5COCH_3 + HCl$

Alkylation $C_6H_6 + 3\ C_2H_5Cl \xrightarrow{AlCl_3}$ 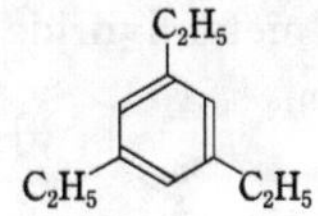

e) Diazonium salt reaction (sandmeyer' reaction)

$C_6H_5NH_2 \xrightarrow{NaNO_2-HCl} C_6H_5N_2^+Cl^- \xrightarrow{CuCl} C_6H_5Cl$

4. Oxidation reaction

Reaction in which oxygen is obtained or hydrogen is lost

1) Dehydration 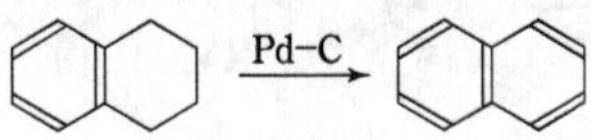

2) Epoxidation

a) $C_6H_5CH{=}CH_2 + m\text{-}ClC_6H_4C({=}O)OOH \longrightarrow C_6H_5CH{-}CH_2$ (with O bridging, epoxide) $+ m\text{-}ClC_6H_4COOH$

b) 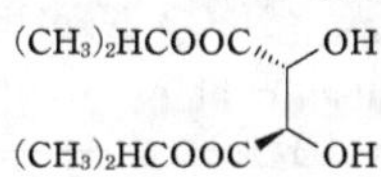

$CH_2{=}CHCH_2OH + (CH_3)_3CCOOH \xrightarrow{\text{Titanium catalyst}}$ 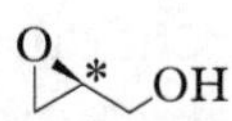

(Sharpress asymmetric epoxidation)

3) Wacker oxidation $H_2C{=}CH_2 + O_2 \xrightarrow{PdCl_2,\ CuCl_2} CH_3CHO$

4) Baeyer-Villiger oxidation

$$(C_6H_5)(C_6H_5)C{=}O + C_6H_5C({=}O)OOH \longrightarrow (C_6H_5)(C_6H_5O)C{=}O + C_6H_5COOH$$

5) Ozone oxidation 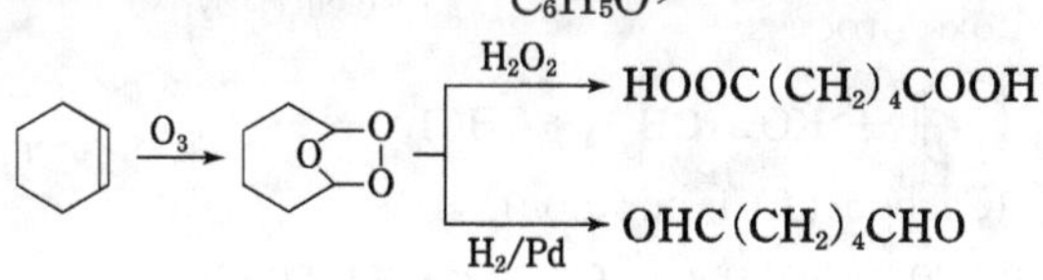

6) Oxidation with metallic oxide

a) Hydrocarbon oxidation：$C_6H_5CH_3 \xrightarrow{KMnO_4} C_6H_5COOK \xrightarrow{H_2SO_4} C_6H_5COOH$

b) Oxidation of alcohol： 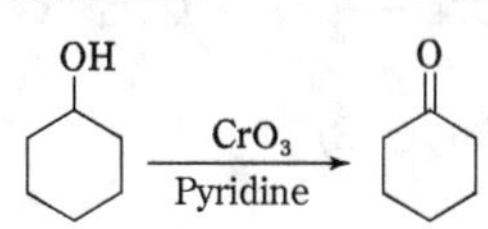

5. Reduction reaction

Reaction in which oxygen is lost or hydrogen is obtained

Main Organic Chemical Reactions Continued.

1) Catalytic hydrogenation 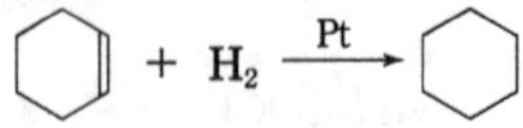

2) Reduction via metal hydride $C_6H_5COCH_3 \xrightarrow{LiAlH_4} C_6H_5CH(OH)CH_3$

3) Birch reduction 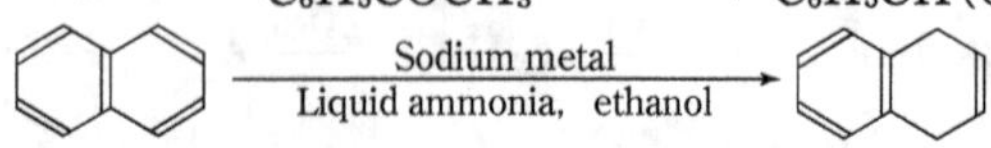

4) Sodium metal $C_6H_5NO_2 \xrightarrow{Sn/HCl} C_6H_5NH_2$

5) Noyori asymmetric hydrogenation

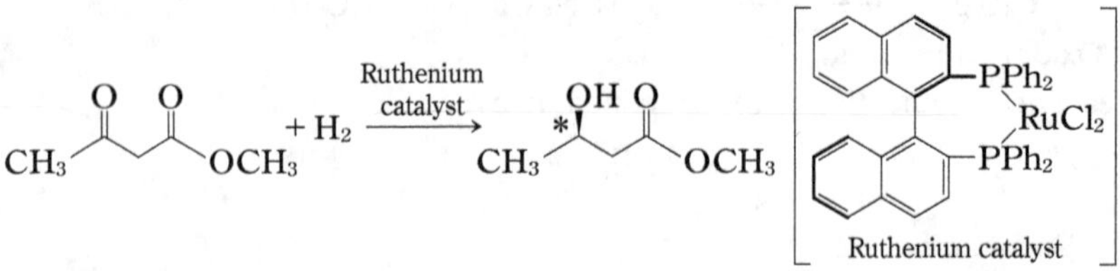

6. Addition reaction

Reaction in which hydrogen halide, halogen, water, alcohol, metal hydride, organometallic compounds, or other substances are added to an unsaturated bond ($C=C$, $C\equiv C$, $C=O$, $C=N$, $C\equiv N$, etc.)

1) Additional carbon-carbon double bond recations

a) $(CH_3)_2C=CH_2+HCl \longrightarrow (CH_3)_3CCl$

b) $C_6H_5CH=CH_2+Br_2 \longrightarrow C_6H_5CHBrCH_2Br$

c) $H_2C=CH_2+H_2O \xrightarrow{H_2SO_4} CH_3CH_2OH$

d) $n\text{-}C_4H_9CH=CH_2 \xrightarrow{B_2H_6} (n\text{-}C_4H_9CH_2CH_2)_3B \xrightarrow[NaOH]{H_2O_2} n\text{-}C_4H_9CH_2CH_2OH$

(Hydroboration)

e) $CH_3CH=CH_2+CO+H_2 \xrightarrow[\text{rhodium catalyst}]{\text{Cobalt or}} CH_3CH_2CH_2CHO$

(oxo process)

f) 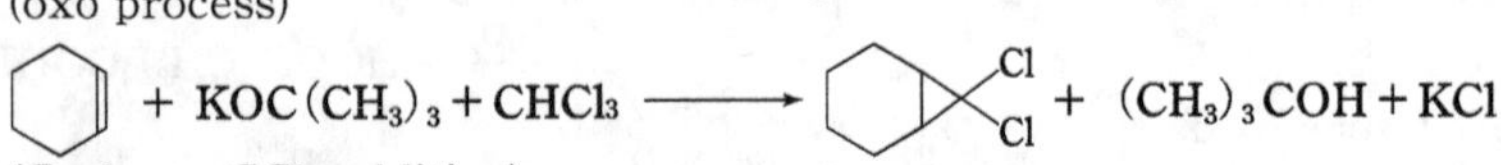

(Carbene: CCl_2 addition)

2) Addition reactions of carbonyl compounds

a) $CH_3-\underset{\underset{O}{\|}}{C}-CH_3+HCN \xrightarrow{\text{Alkali}} CH_3-\underset{\underset{OH}{|}}{\overset{\overset{CN}{|}}{C}}-CH_3$

(generation of cyanohydrin)

b) $C_6H_5-\underset{\underset{O}{\|}}{C}-CH_3+C_2H_5MgBr \xrightarrow{\text{Addition}} \xrightarrow{H_2O} C_6H_5-\underset{\underset{OH}{|}}{\overset{\overset{C_2H_5}{|}}{C}}-CH_3$

(Grignard reaction)

Continued.

3) Conjugate addition reaction

$$C_6H_5CH{=}CHCOOC_2H_5 + CH_2(COOC_2H_5)_2 \xrightarrow{NaOC_2H_5} C_6H_5{-}\underset{\displaystyle CH(COOC_2H_5)_2}{\underset{|}{C}}HCH_2COOC_2H_5$$

(Michael reaction)

4) Cycloaddition reaction

a)

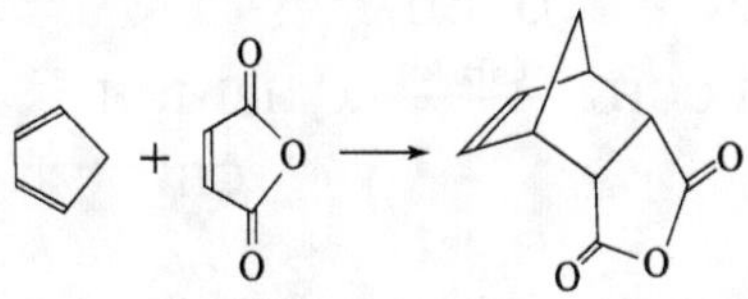

(Diels-Alder reaction)

b)

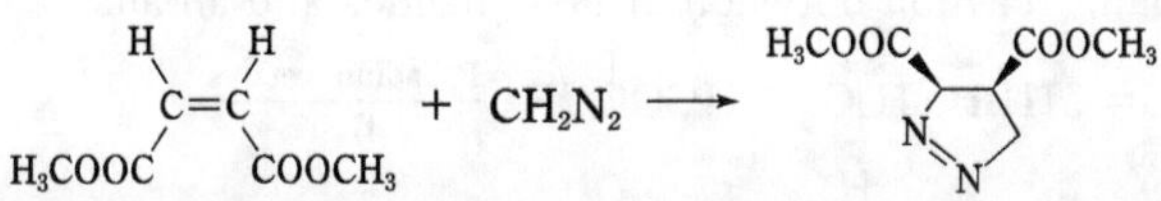

(1 and three-dipole addition reaction)

7. Separation reaction

Reaction in which molecules such as water, hydrogen halide, halogen, or amines are separated and an unsaturated bond is formed.

1) $C_6H_5CH_2CH_2Br \xrightarrow{NaOC_2H_5} C_6H_5CH{=}CH_2$

2)

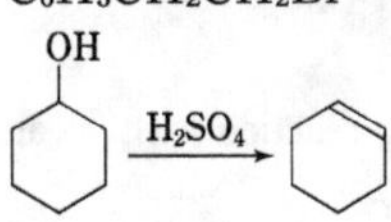

3) $CH_3CH_2CH_2\overset{+}{N}(CH_3)_3OH^- \xrightarrow{Heating} CH_3CH{=}CH_2 + (CH_3)_3N + H_2O$

(Hofmann escape revolt respondent)

8. Rearrangement reaction

Reaction in which the position of atoms or atomic groups inside of molecules is changed and bonds are reconfigured

1) Beckmann rearrangement $\begin{matrix} C_6H_5 \\ C_6H_5 \end{matrix}\!\!>\!C{=}N\diagdown_{OH} \xrightarrow{Acid} C_6H_5{-}\underset{\displaystyle O}{\underset{\|}{C}}{-}NHC_6H_5$

2) Curtius rearrangement $C_6H_5{-}\underset{\displaystyle O}{\underset{\|}{C}}{-}\bar{N}{-}\overset{+}{N}{\equiv}N \xrightarrow{Heating} C_6H_5{-}N{=}C{=}O + N_2$

3) Cope rearrangement

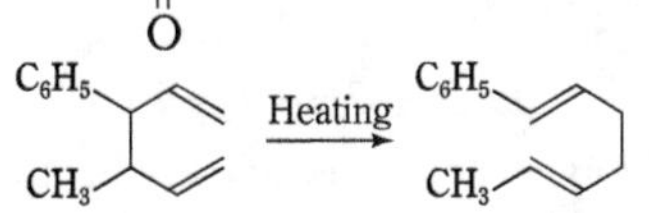

4) claisen rearrangement

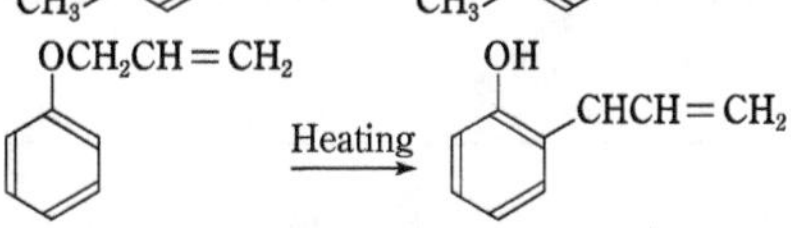

Main Organic Chemical Reactions Continued.

5) Pinacol rearrangement (A dislocation reaction accoplised through carbon-ation is normally known as a Wagner-Meerwein Rearrangement. Pinacol Re-arrangement is one of type.)

$$CH_3-\underset{OH}{\overset{CH_3}{C}}-\underset{OH}{\overset{CH_3}{C}}-CH_3 \xrightarrow{H_2SO_4} CH_3-\underset{\|\atop O}{C}-\underset{CH_3}{\overset{CH_3}{C}}-CH_3$$

6) Wittig rearrangement $C_6H_5CH_2OCH_3 \xrightarrow{C_6H_5Li} C_6H_5\underset{CH_3}{CH}OH$

9. Coupling reaction

1) Cross-coupling reaction between organic halides and an alkene or alkyne

a) $(CH_3)_2C=CHBr+H_2C=CHCOOCH_3 \xrightarrow[\text{Base}]{\text{Palladium catalyst}}$

$$(CH_3)_2C=\underset{H}{C}-\overset{H}{\underset{H}{C}}=CCOOCH_3$$

(Mizoroki-Heck reaction)

b) $C_6H_5I+HC\equiv CC_6H_5 \xrightarrow[\text{Base}]{\text{The catalyst palladium/both of copper}} C_6H_5C\equiv CC_6H_5$

(Sonogashira cross-coupling)

2) Organometallic, reactive medicine and intersection coupling reaction with organic halid

a) $C_6H_5MgBr+ClCH=CH_2 \xrightarrow{\text{Nickel catalyst}} C_6H_5CH=CH_2$

(Kumada-Tamao-Corriu cross-coupling)

b) o-$CH_3C_6H_4ZnCl+p$-$BrC_6H_4NO_2 \xrightarrow{\text{Palladium catalyst}}$ 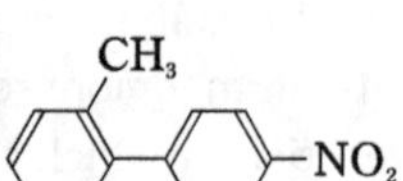

(Negishi reaction)

c) $C_6H_5B(OH)_2+p$-$BrC_6H_4CHO \xrightarrow[\text{Base}]{\text{Palladium catalyst}}$ 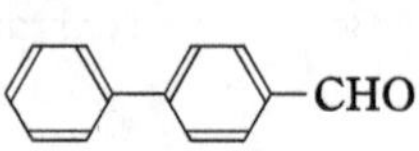

(Suzuki-Miyaura cross coupling)

10. Free radical reaction

1) Optical chlorination

$$CH_4+Cl_2 \xrightarrow{\text{Light}} CH_3Cl+CH_2Cl_2+CHCl_3+CCl_4$$

2) bromination of allylic positions with *N*-bromosuccinimide (Wohl-Ziegler reaction)

Continued.

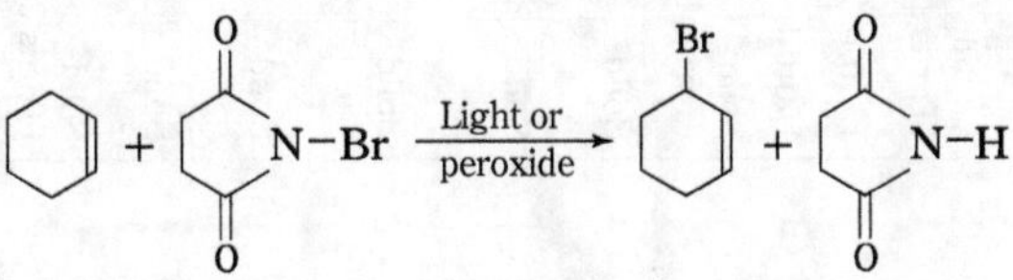

3) Annomolous addition for hydrogen bromide (the addition reaction of hydrogen bromide to an alkene in the presence of oxygen or peroxide. Also known as the oxygen effect. Different from an ion reaction, addition occurs according to the inverse Markovnikov rule

$$CH_2{=}CHCH_2Br + HBr \xrightarrow[\text{peroxide}]{\text{Oxygen or}} BrCH_2CH_2CH_2Br$$

4) Autoxidation (mild oxidation not accompanied by combustion of molecular oxygen)

$$C_6H_5CH(CH_3)_2 + O_2 \longrightarrow C_6H_5\overset{\overset{\large OOH}{|}}{C}(CH_3)_2$$

5) Conversion from an alcohol nitrite ester to oxime alcohol (Barton reaction)

$$2CH_3(CH_2)_4COONa \xrightarrow{\text{Electroanalysis}} CH_3(CH_2)_8CH_3$$

6) Conversion from alcohol nitrite ester to oxime alcohol (Barton reaction)

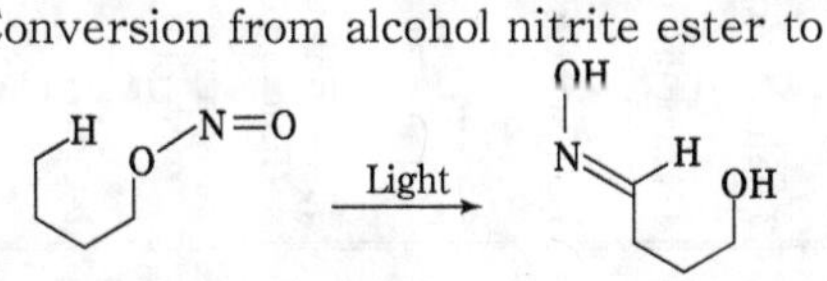

11. Polymerization reaction

Reaction where many small molecules bind together to form a large molecule (polymer)

1) Addition polymerization (Chain polymerization by additional reactions. There are three kinds (the radical polymerization, the cation polymerization, and the anion polymerization) depending on polymerization initiator's kind).

$$nCH_2{=}CH_2 \longrightarrow [-CH_2-CH_2-]_n \quad \text{(polyethylene)}$$

2) Condensation polymerization (Successive polymerization by condensation reactions)

$$nH_2N(CH_2)_6NH_2 + nHOOC(CH_2)_4COOH \xrightarrow[-2nH_2O]{}$$

$$\left[-NH(CH_2)_6-NH-\underset{\underset{O}{\|}}{C}(CH_2)_4-\underset{\underset{O}{\|}}{C}-\right]_n \quad \text{(Nylon 66)}$$

3) Polyaddition (Successive polymerization by additional reactions)

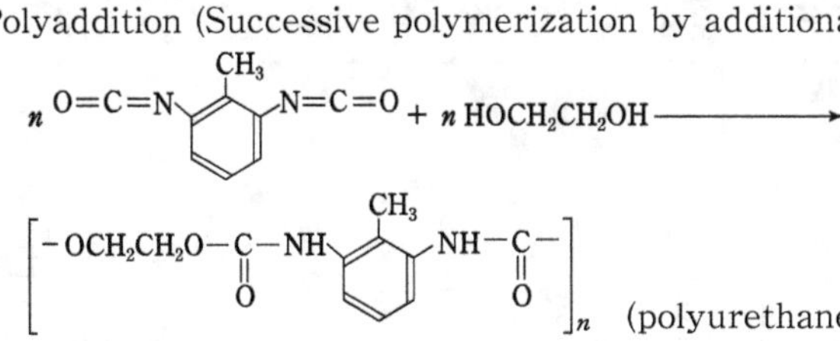

(polyurethane)

(Source: Kei Goto)

Organic Matter

Compound name	Chemical formula	Shape	Compound name	Chemical formula	Shape
Hydrocarbons			**Alcohol, phenols** (Continued.)		
Methane	CH_4	Gas	2-Propanol (Isopropyl alcohol)	$(CH_3)_2CHOH$	Liquid
Ethane	CH_3CH_3	Gas	2-Methyl-2-propanol (*t*-Butyl alcohol)	$(CH_3)_3COH$	Liquid (Torus 26℃)
Propane	$CH_3CH_2CH_3$	Gas	1-Pentanol (Pentyl alcohol)	$CH_3(CH_2)_4OH$	Liquid
Butane	$CH_3CH_2CH_2CH_3$	Gas	Glycerin (Glycerol)	H_2C-OH $HC-OH$ H_2C-OH	Liquid (Torus 18℃)
Hexane	$CH_3(CH_2)_4CH_3$	Liquid	Propargyl alcohol	$HC\equiv C-CH_2OH$	Liquid
Paraffin	C_nH_{2n+2}	Liquid or solid	Phenethyl alcohol	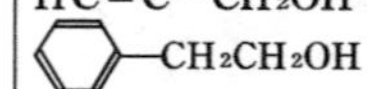	Liquid
Ethylene	$CH_2=CH_2$	Gas	Phenol (Carbolic acid)	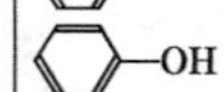	Solid (Torus 41℃)
Acetylene	$HC\equiv CH$	Gas	Hydroquinone	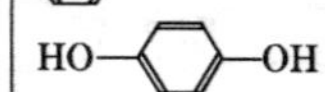	Solid
Cyclohexane	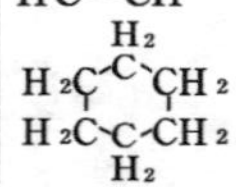	Liquid	Catechol (Pyrocatechol)	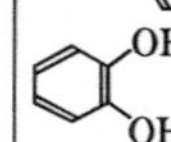	Solid
Decahydronaphthalene (Decalin)	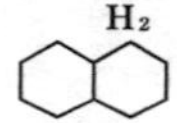 *Cis-trans* isomer	Liquid	1,2,3-Trihydroxybenzene (Pyrogallol)	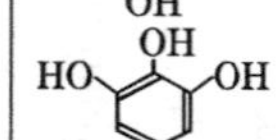	Solid
Benzene		Liquid	1,3,5-Trihydroxybenzene (Phloroglucinol)	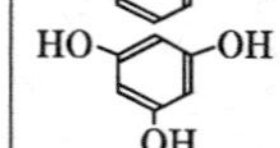	Solid
Naphthalene	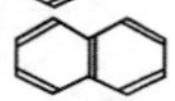	Solid	2-Naphthol (*β*-Naphthol)	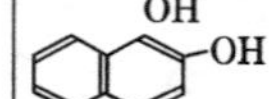	Solid
Anthracene	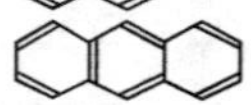	Solid	*o*-Cresol	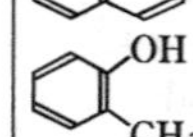	Solid (Torus 31℃)
Toluene	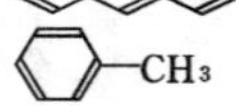	Liquid	1,2-Dihydroxy anthraquinone (Alizarine)	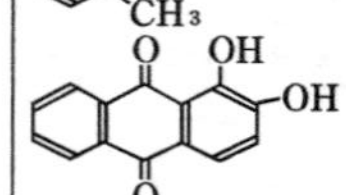	Solid
p-Xylene	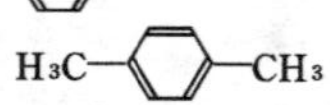	Liquid (Torus 13℃)			
Cumene	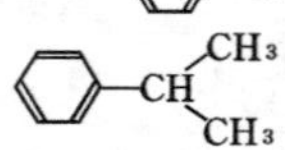	Liquid			
Biphenyl	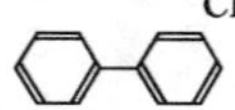	Solid			
Alcohol, phenols					
Methyl alcohol (Methanol)	CH_3OH	Liquid			
Ethyl alcohol (Ethanol)	CH_3CH_2OH	Liquid			

Continued.

Compound name	Chemical formula	Shape
Alcohol, phenols (Continued.)		
(+) Borneol	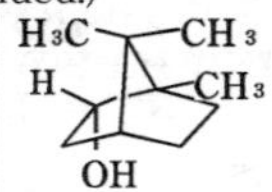	Solid
Ethers		
Dimethyl ether	CH_3OCH_3	Gas
Diethyl ether (Ether)	$(CH_3CH_2)_2O$	Liquid
Anisole	$C_6H_5OCH_3$	Liquid
Ethylene oxide		Gas (aqua 10.7℃)
Tetrahydrofuran (THF)	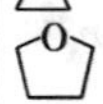	Liquid
Furan		Liquid
1,4-Dioxane	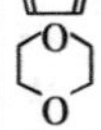	Liquid
1,2-Dimethoxyethane (Ethylene glycol dimethyl ether)	$(CH_3OCH_2)_2$	Liquid
Organohalides		
Methyl bromide	CH_3Br	Gas
Methyl iodide	CH_3I	Liquid
Dichloromethane (Methylene chloride)	CH_2Cl_2	Liquid
Chloroform	$CHCl_3$	Liquid
Iodoform	CHI_3	Solid
Hexachlorocyclohexane (Benzene hexachloride; BHC)	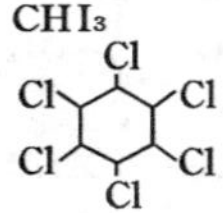	Solid
Aldehydes		
Formaldehyde	$HCH=O$	Gas
Acetaldehyde	$CH_3CH=O$	Liquid
Acrylic aldehyde (Acrolein)	$H_2C=CH-CH=O$	Liquid
Benzaldehyde	$C_6H_5CH=O$	Liquid
Cinnamic aldehyde	$C_6H_5CH=CH-CH=O$	Liquid
Aldehydes (Continued.)		
Glyoxal	$O=CH-CH=O$	Liquid (Torus 15℃)
o-1,2-Benzenedicarboxylic aldehyde	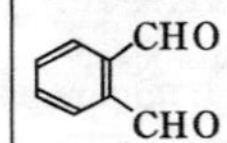	Solid
Ketones		
Acetone	$(CH_3)_2C=O$	Liquid
Benzophenone	$(C_6H_5)_2C=O$	Solid
Methylvinylketone (3-butene-2-on)	$H_2C=CHC(=O)CH_3$	Liquid
Cyclohexanone	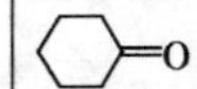	Liquid
p-benzoquinone (Quinones)	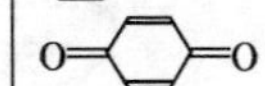	Solid
Anthraquinone	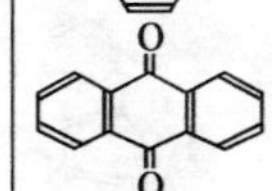	Solid
Camphor	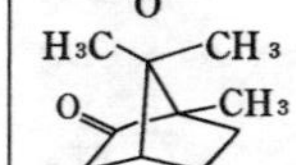	Solid
Carboxylic acid		
Formate	HCO_2H	Liquid
Acetate	CH_3CO_2H	Liquid
Oxalate	$(CO_2H)_2$	Solid
Butyrate (Butanoic acid)	$CH_3(CH_2)_2CO_2H$	Liquid
Eicosapentaenoic acid (EPA)	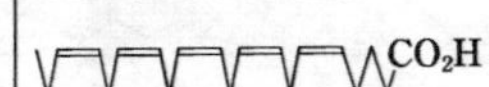	Liquid
Docosahexaenoic acid (DHA)		Liquid
Benzoic acid	$C_6H_5CO_2H$	Solid
Salicylate	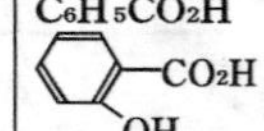	Solid

Organic Matter — Continued.

Compound name	Chemical formula	Shape	Compound name	Chemical formula	Shape
Carboxylic acids			**Amines** (Continued.)		
Lactate	$CH_3CH(OH)CO_2H$	Solid (L torus 25.8 ℃)	Morpholine	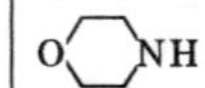	Liquid
Citrate	H_2C-CO_2H $HO-C-CO_2H$ H_2C-CO_2H	Solid	Quinoline	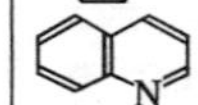	Liquid
			Quinoxalines (Benzopyrazine)	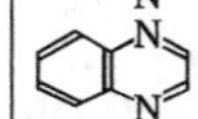	Liquid (Torus 29℃)
Gallic acid	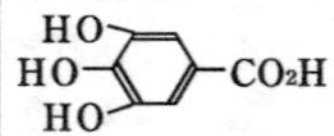	Solid	*o*-Toluidine	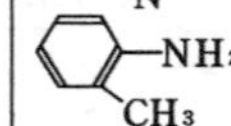	Liquid
Stearate	$CH_3(CH_2)_{16}CO_2H$	Solid			
Malate	$HO-CH-CO_2H$ H_2C-CO_2H	Solid	*p*-Anisidine	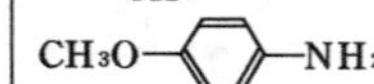	Solid
			Imidazole	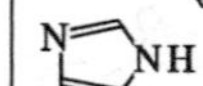	Solid
Tartrate	$HO-CH-CO_2H$ $HO-CH-H$ CO_2H	Solid	Indole	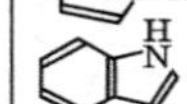	Solid
Esters			Pyrazine	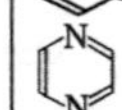	Solid (Torus 52℃)
Ethyl chloroformate	$ClCO_2CH_2CH_3$	Liquid			
Ethyl acetate	$CH_3CO_2CH_2CH_3$	Liquid	Pyrimidine	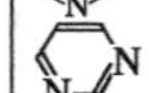	Liquid (Torus 20℃)
Phenyl acetate	$CH_3CO_2C_6H_5$	Solid			
Dioctylphthalate (Bis(2-ethylhexyl) phthalate)	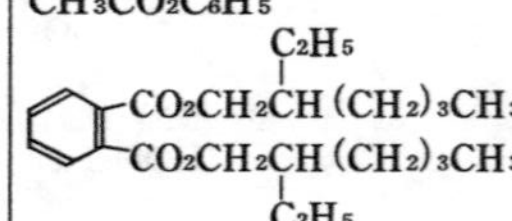	Solid	Pyridazines	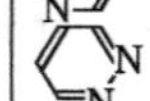	Liquid
			Quinuclidines		Solid
Amines			**Combination with other including nitrogen compounds**		
Dimethylamine	$(CH_3)_2NH$	Gas	Azobenzene	$C_6H_5N=NC_6H_5$	Solid
Triethylamine	$(CH_3CH_2)_3N$	Liquid	Hydrazobenzene (*N,N'*-Diphenylhydrazine)	$C_6H_5NHNHC_6H_5$	Solid
Aniline	$C_6H_5NH_2$	Liquid			
Pyrrolidine	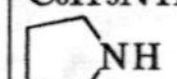	Liquid	Nitrobenzene	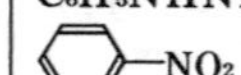	Liquid
Piperidines	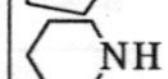	Liquid	2,4,6-Trinitrotoluene (TNT)	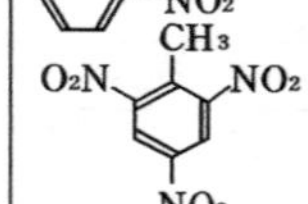	Solid
Pyridine	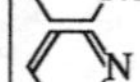	Liquid			

Continued.

Compound name	Chemical formula	Shape
Combination with other including Nitrogen compounds (Continued.)		
Nitroglycerin	H_2C-ONO_2 / $HC-ONO_2$ / H_2C-ONO_2	Solid
Auramine O		Solid
Urea	$(NH_2)_2C=O$	Solid
Thiourea	$(NH_2)_2C=S$	Solid
Dye indicators group		
Indigo (Indigo blue)	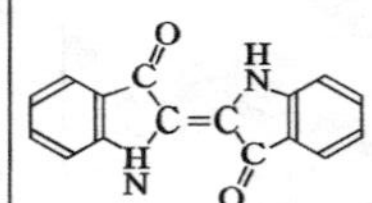	Solid
Methylene blue (Methylene blue)	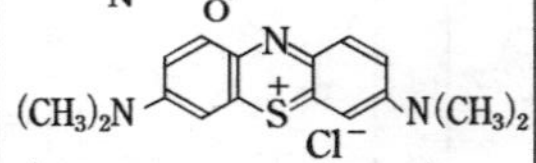	Solid
Congo Red	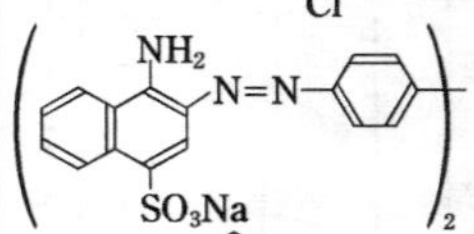	Solid
Eosin B	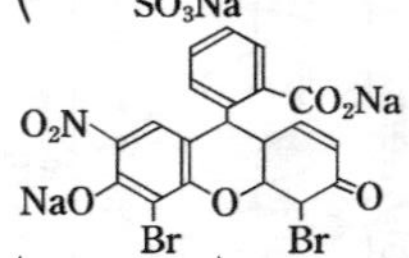	Solid
Phenolphthalein	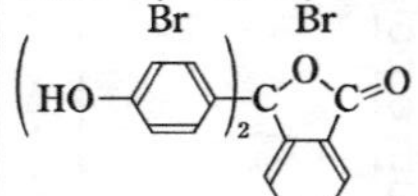	Solid
Amino acid, etc		
Alanine (α-Alanine)	$CH_3CH(NH_2)CO_2H$	Solid
Amino acid, etc (Continued.)		
Glycine	$(H_2N)CH_2CO_2H$	Solid
Serine	$HOCH_2-C(NH_2)(H)-CO_2H$	Solid
Cysteine	$HSCH_2-C(NH_2)(H)-CO_2H$	Solid
Nicotinic	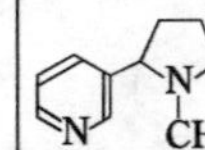	Liquid
Quinine (Kinin)	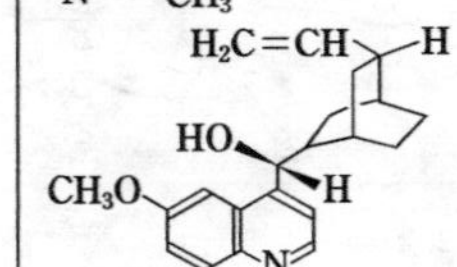	Solid
Strychnine	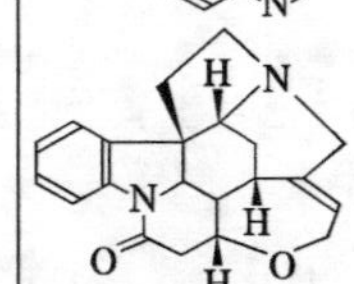	Solid
Aspirin (Acetylsalicylic acid)	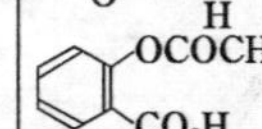	Solid
Cocaine	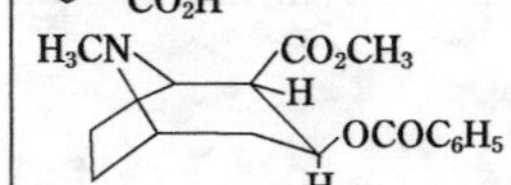	Solid
2,3,7,8-Tetrachlorodibenzo-*p*-Dioxin (TCDD)	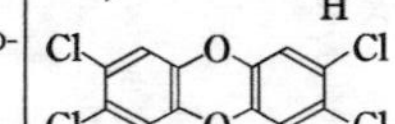	Solid

The state at normal temperatures is shown: S (solid), L (liquid), G (gas). Also, "M" indicates the melting point, and "B" indicates the boiling point in degrees Celsius. (Source: Norihiro Tokitoh)

Macromolecules

Properties of macromolecules

General-purpose polymers

Name / Chemical formula		Usage	Name / Chemical formula		Usage
Polyethylene	$-(CH_2CH_2)_n-$	Industrial fiber, molding materials, insulating coating and films	Nylon 66	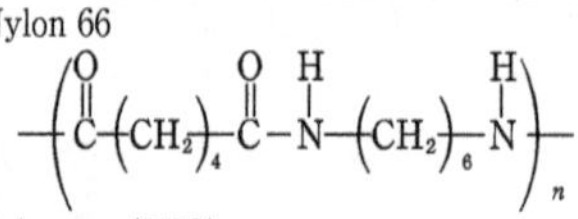	Molding materials and fibers
Polystyrene	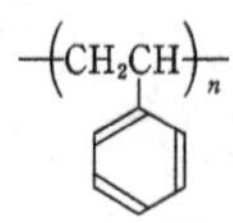	General products and forming materials	Polyester (PET)	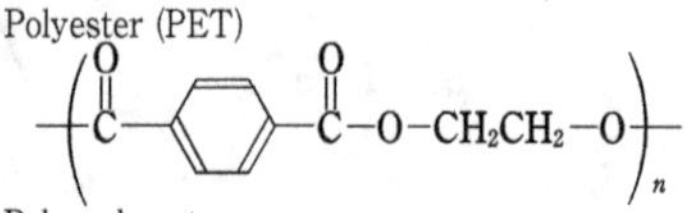	Molding materials, films, and fibers
Poly(methyl methacrylate)	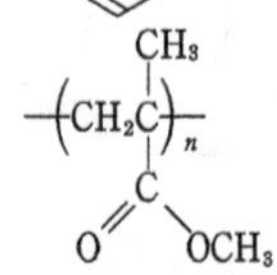	Biomedical materials, optical components and films	Polycarbonate	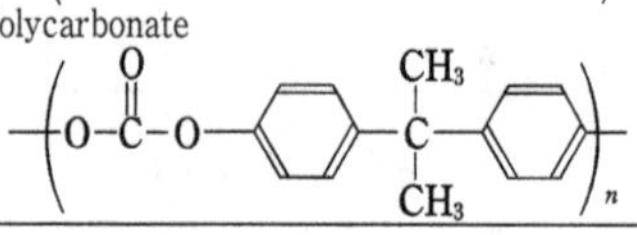	Electrical and automobile components

High-strength polymers

Name	Chemical formula	Elasticity modulus (GPa)
Polyethylene	$-(CH_2CH_2)_n-$	220–230
Poly(*p*-phenylene terephthalamide)	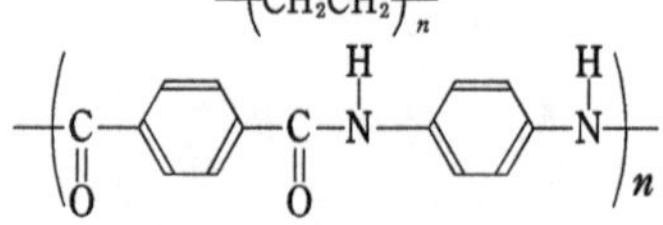	125–186
Poly(*p*-phenylene benzobisthiazole)	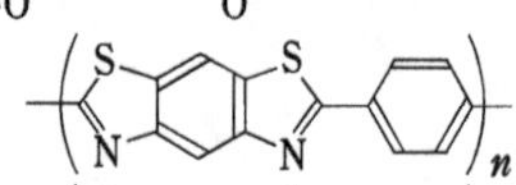	330
Poly(*p*-phenylene benzobisoxazole)	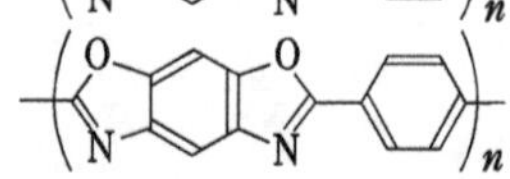	480

Conductive polymers

Polyacetylene

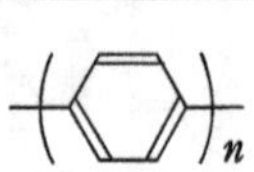

Poly(*p*-phenylene)

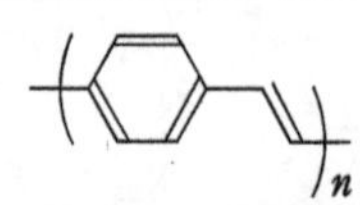

Poly(*p*-phenylene vinylene)

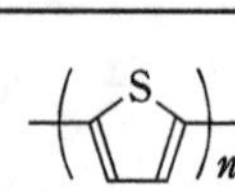

Polythiophene

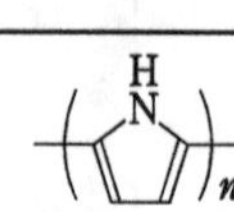

Polypyrrole

Ion-exchange resins

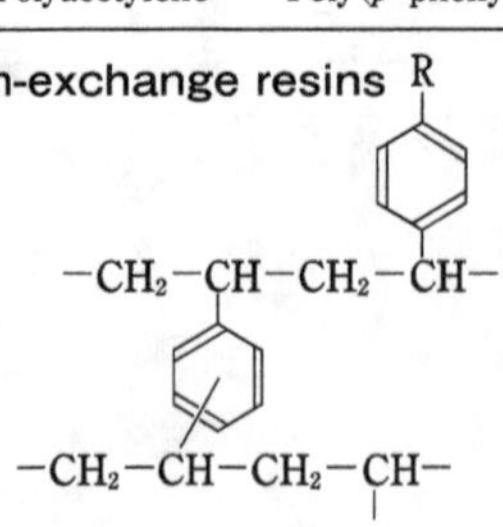

Classification		Example of ion-exchange group (R)
Cation-exchange resin	Strong acid	$\sim SO_3H$
	Weak acid	$\sim COOH$
Anionic-exchange resin	Strong base	$\sim N^+R_1R_2R_3OH^-$
	Weak base	$\sim NH_2$, $\sim NHR_1$, $\sim NR_1R_2$
Chelate resin		$\sim N(CH_2COOH)_2$, $\sim NH(CH_2CH_2NH)_nH$

(Source: Kentaro Tanaka)

Biodegradable polymers

Microbial polymer

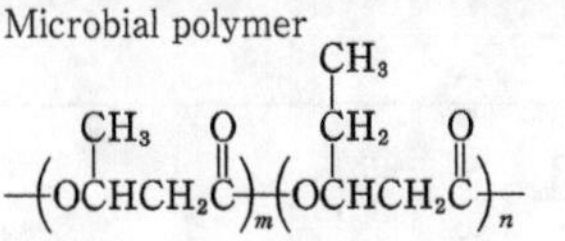

Poly(3-hydroxybutyric acid)-poly(3-hydroxyvalerate) copolymer

Synthetic polymer

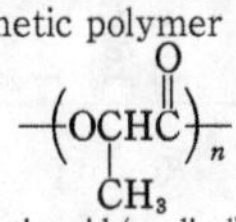

Polylactic acid (medical)

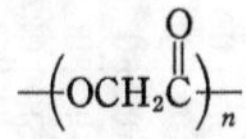

Polyglycolic acid (medical)

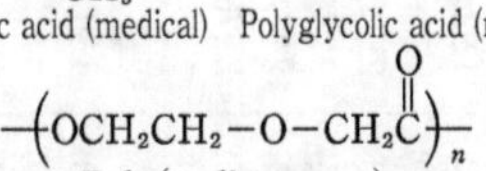

Poly(*p*-dioxanone)

Natural polymer

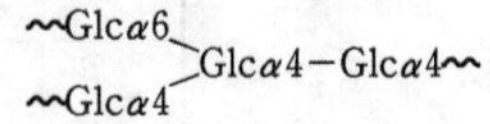

Starch

Structure of macromolecules

Copolymers

A polymer that consists of more than 2 kinds of monomers is called a copolymer.

~ABAABAABAABBBAABBA~

Random copolymer

~ABABABABABABABABAB~

Alternating copolymer

~AAAAAABBBBBBAAAAAA~

Block copolymer

~AAAAAAAAAAAAAAAAA~
B B
B B
B B
B B
B B
B B

Graft copolymer

Stereoregularity of polymers

Stereoregularity (tacticity) indicates the relative stereochemistry of the adjacent chiral centers of the units within a polymer chain. The two adjacent structural units called diad and the three adjacent structural units called triad.

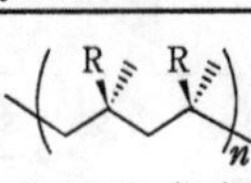

Isotactic diad
(meso diad; m)

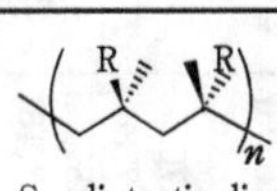

Syndiotactic diad
(racemo diad; r)

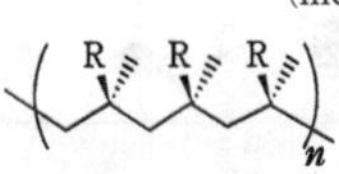

Isotactic triad
(I or mm)

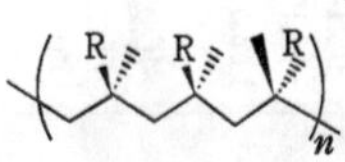

Heterotactic triad
(H or mr)

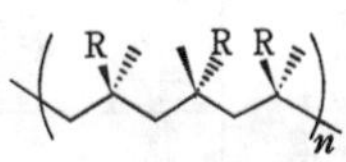

Syndiotactic triad
(S or rr)

Specific Gravities and Boiling Points of Petroleum Products

The following table shows the ranges of boiling point for volatile petroleum products. Although these products are obtained by distillation of crude oil, they are not the distillation range; rather, they are for the sorted, processed, and refined products.

	Specific gravity*2 (15/4℃)	Boiling point range t_v/℃		Specific gravity (15/4℃)	Boiling point range t_v/℃
Crude oil	0.8 -1.00	-	Aviation gasoline	0.70-0.72	45-140
Lpg	0.51-0.56	- -1	Jet fuel	0.75-0.80	55-250
Petroleum ether	0.63-0.68	30- 80	Kerosene	0.78-0.84	150-320
Benzine	0.66-0.68	30-150	Diesel light oil	0.82-0.84	160-400
Ligroin	0.68-0.75	80-120	Light oil	0.80-0.88	200-350
Automotive gasoline	0.71-0.75	30-210	Crude petroleum	0.82-0.96	170-

*1 Mean molecular weight (rough guide): kerosene 100-200, light oil 140-250, crude petroleum: 250+

*2 Ratio of the mass of the sample at 15°C (in vacuum) and the mass of of a unit volume of purified water at 4°C (in vacuum). (Source: Kentaro Tanaka)

Characteristics of Main Organic Solvents

The molecular depression of freezing point and the molecular rise of boiling point are the freezing point and boiling point when 1 mole of solute is dissolved in 1 kg of solvent. The dipole moment is expressed in the unit of Debye = 3.3356×10^{-30}C m. (The density value shown in the table is at 20℃. The refractive index is the value in relation to the sodium D line.) However, values listed in parentheses for the refractive index and relative permittivity are temperatures (℃).

Material name	Molecular formula	Molecular weight	Density	Melting point ℃	Boiling point ℃	Decrease in molecular freezing point ℃	Increase in molecular point ℃	Refractive index	Relative permittivity	Electric dipole moment
Pentane	C_5H_{12}	72.15	0.62638	-129.7	36.1	-	-	1.358 (20)	-	0
Hexane	C_6H_{14}	86.18	0.65937	-95.35	68.8	-	-	1.37506(20)	-	0
Heptane	C_7H_{16}	100.21	0.6837	-90.61	98.4	-	-	1.38777(20)	1.924(20)	0
Octane	C_8H_{18}	114.23	0.7025	-57	125.7	-	-	1.3976 (20)	-	0
Cyclohexane	C_6H_{12}	84.16	0.7786	6.5	80.8	20	-	1.42636(20)	2.023(20)	0
Benzene	C_6H_6	78.12	0.87865	5.49	80.1	5.12	2.57	1.50439(15)	2.284(20)	0
Toluene	C_7H_8	92.14	0.8716	-95	110.8	-	-	1.49985(15)	2.379(25)	0.37(gas)
p-Xylene	C_8H_{10}	106.17	0.861	13.35	138.4	4.3	-	1.496 (15)	2.270(20)	-
Chlorobenzene	C_6H_5Cl	112.56	1.107	-45.2	132	-	-	1.524 (20)	-	-
Methyl alcohol	CH_3OH	32.04	0.7928	-96	64.7	-	0.88	1.32863(20)	32.6 (25)	1.69
Ethyl alcohol	C_2H_5OH	46.07	0.7893	-114.5	78.3	-	1.2	1.36232(20)	24.30 (25)	1.69(gas)
n-Propyl alcohol	C_3H_7OH	60.11	0.8035	-126.5	97.2	-	-	1.38543(20)	20.1 (25)	-
Iso-propyl alcohol	C_3H_7OH	60.11	0.785	-88.5	82.5	-89.5	-	1.37723(20)	-	-
Iso-pentyl alcohol (Isoamyl alcohol)	$C_5H_{11}OH$	88.15	0.8139	-117.2	131.7	-	-	1.40851(15)	-	-
Benzyl alcohol	C_7H_7OH	108.14	1.0415 (25℃)	-15.3	205.7	-	-	1.53987(21.5)	13.1 (20)	-
Phenol (Carbolic acid)	C_6H_5OH	94.11	1.0708	40.8	181.4	7.27	3.6	1.54247(40.6)	9.78 (60)	1.14(gas)
m-Cresol	C_7H_7OH	108.14	1.0336	10.9	200.2	-	-	1.5398 (20)	11.8 (25)	-
Glycerin	$C_3H_8O_3$	92.1	1.2644	18	290	-	-	1.47289(20)	42.5 (25)	-
Diethyl ether	$(C_2H_5)_2O$	74.12	0.71925	-116.3	34.5	-	2.16	1.35555(15)	4.335(20)	1.16(gas)
Tetrahydrofuran	C_4H_8O	72.1	0.8892	-108.5	66	-	-	1.407 (20)	-	1.7
1,4-Dioxane	$C_4H_8O_2$	88.1	1.0329	-39	101.4	-	-	1.4175 (20)	-	0
1,2-Dimethoxyethane (Ethylene glycol dimethyl ether)	$C_4H_{10}O_2$	90.12	0.86285	-58	85	-	-	1.379 (20)	-	-
Diglyme (Diethylene glycol dimethyl ether)	$C_6H_{14}O_3$	134.17	0.9451	-68	162	-	-	1.4097 (20)	-	-

Continued.

Material name	Molecular formula	Molecular weight	Density	Melting point ℃	Boiling point ℃	Decrease in molecular freezing point ℃	Increase in molecular point ℃	Refractive index	Relative permittivity	Electric dipole moment
Acetone	CH_3COCH_3	58.08	0.7908	−94.82	56.5	–	1.725	1.36157(15)	20.7 (25)	2.9
Acetonitrile	CH_3CN	41.05	0.7845	−45	81.6	–	–	1.34604(15)	38.8	–
Propionitrile	C_3H_5N	55.08	0.7818	−91.8	97.2	–	–	1.36585(20)	–	–
Formic acid	HCO_2H	46.03	1.2203	8.6	100.6	2.77	2.4	1.37137(20)	58.5 (16)	1.145
Acetic acid	CH_3CO_2H	60.05	1.0492	16.64	118.1	3.9	3.07	1.37182(20)	6.15	1.73(gas)
Methyl acetate	$CH_3CO_2CH_3$	74.08	0.9337	−98.05	56.3	–	2.06	1.36143(20)	–	1.67(gas)
Ethyl acetate	$CH_3CO_2C_2H_5$	88.11	0.9005	−83.6	76.8	–	2.79	1.37257(20)	–	–
Methylene chloride (Dichloromethane)	CH_2Cl_2	84.94	1.33479	−96.7	−95	–	–	1.4244 (20)	–	–
Chloroform	$CHCl_3$	119.38	1.48945	−63.5	61.2	–	3.88–3.91	1.44858(15)	4.806(20)	1.2
Carbon tertrachloride	CCl_4	153.82	1.5942	−22.6	76.7	–	4.88	1.4604 (20)	–	0
1,2-Dichloroethane	$ClCH_2CH_2Cl$	98.96	1.2569	−35.3	83.5	–	–	1.4443 (20)	–	–
Nitromethane	CH_3NO_2	61.04	1.1322 (25℃)	−29	101.2	–	–	1.38056(22)	–	–
Nitroethane	$C_2H_5NO_2$	75.07	1.052	−50	114	–	–	1.39007(24.3)	–	–
Nitrobenzene	$C_6H_5NO_2$	123.11	1.2037	5.7	210.9	6.9	5.27	1.55261(20)	34.82 (25)	4.21(gas)
Aniline	$C_6H_5NH_2$	93.13	1.0217	−5.98	184.6	5.87	3.69	1.58685(20)	6.89 (20)	1.48(gas)
Pyridine	C_5H_5N	79.1	0.9779 (25℃)	−42	115.5	–	3.01	1.5085 (22)	12.3 (25)	2.15
Morpholine	C_4H_9NO	87.12	1.007	−4.9	128.9	–	–	1.454 (20)	–	1.58
Quinoline	C_9H_7N	129.16	1.0938	−15	240	–	5.72	1.62683(20)	9.00 (25)	2.18
Carbon disulfide	CS_2	76.14	1.2927 (0℃)	−112	46.5	–	2.29	1.63189(15)	–	0
Dimethylsulfoxide (DMSO)	$(CH_3)_2SO$	78.13	1.101	18.45	189	–	–	1.4795 (20)	45	–
N,*N*-Dimethylformamide (DMF)	$(CH_3)_2NCHO$	73.09	0.9445 (25℃)	−61	153	–	–	1.42803(25)	–	–
N,*N*-dimethylacetamide (DMA)	$(CH_3)_2NCOCH_3$	87.12	0.9429 (20℃)	−20	165.5	–	–	1.4358 (25)	–	–
N-Methyl-2-pyrrolidone	C_5H_9NO	99.13	1.028	−24	202 (81–82/ 10 mmHg)	–	–	1.47 (20)	–	–
o-Dichlorobenzene	$C_6H_4Cl_2$	147.00	1.3048	−17.5	180.5	–	–	1.5518 (22)	–	–

(Source: Norihiro Tokitoh)

Biological Material

Amino Acids

(excluding proline)

General type $H_3\overset{+}{N}-\underset{}{CH}(R)-COO^-$

Amino acid	Abbreviation 3-letter	Abbreviation 1-letter	Chemical structural formula(R)	Molecular weight	Isoelectric point
Hydrophobic amino acid					
Glycine	Gly	G	H−	75	5.97
Alanine	Ala	A	CH_3-	89	6.00
* Valine	Val	V	$(CH_3)_2CH-$	117	5.96
* Leucine	Leu	L	$(CH_3)_2CHCH_2-$	131	5.98
* Isoleucine	Ile	I	$CH_3CH_2CH(CH_3)-$	131	6.02
* Methionine	Met	M	$CH_3-S-CH_2CH_2-$	149	5.74
Proline	Pro	P	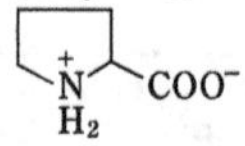	115	6.30
* Phenylalanine	Phe	F	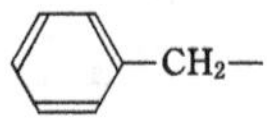	165	5.48
* Tryptophan	Trp	W	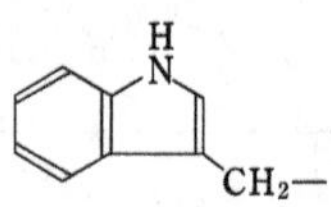	204	5.89
Hydrophilic amino acid					
Neutral amino acid					
Serine	Ser	S	$HO-CH_2-$	105	5.68
* Threonine	Thr	T	$CH_3CH(OH)-$	119	6.16
Cysteine	Cys	C	$HS-CH_2-$	121	5.07
Tyrosine	Tyr	Y	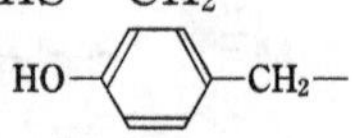	181	5.66
Asparagine	Asn	N	H_2NCOCH_2-	132	5.41
Glutamine	Gln	Q	$H_2NCOCH_2CH_2-$	146	5.65
Acidic amino acid					
Aspartic acid	Asp	D	$^-OOCCH_2-$	133	2.77
Glutamic acid	Glu	E	$^-OOCCH_2CH_2-$	147	3.22
Basic amino acid					
* Lysine	Lys	K	$H_3\overset{+}{N}CH_2CH_2CH_2CH_2-$	146	9.74
Arginine	Arg	R	$H_2NC(=\overset{+}{N}H_2)NHCH_2CH_2CH_2-$	174	10.76
Histidine	His	H	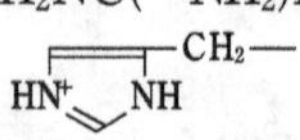	155	7.59

* Essential amino acids for humans (for infatns Arg and His are also needed)

Trace Amino Acid Constituents in Protein Structure

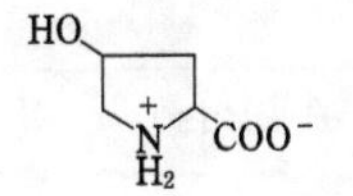

4-Hydroxyproline (Hyp)

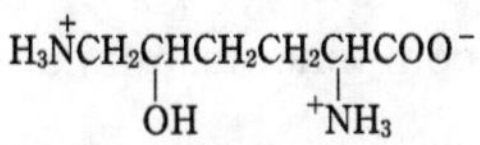

Hydroxylysine (Hyl)

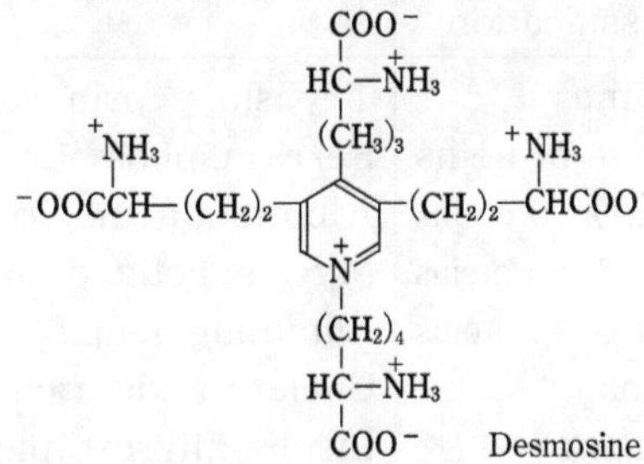

Desmosine

Amino Acids other than Amino Acids in Protein Structure*

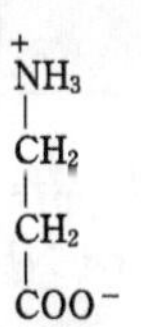

β-alanine

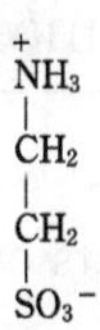

Taurine

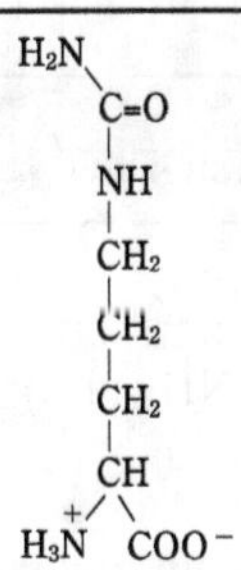

Citrulline

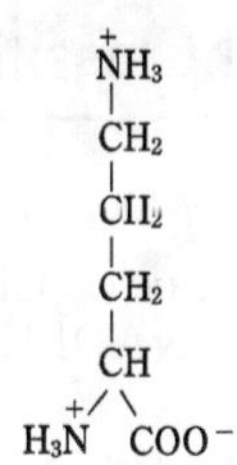

Ornithine

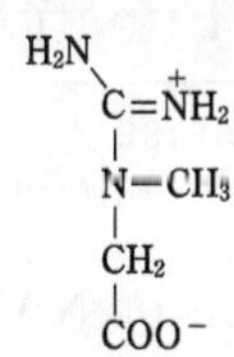

Creatine

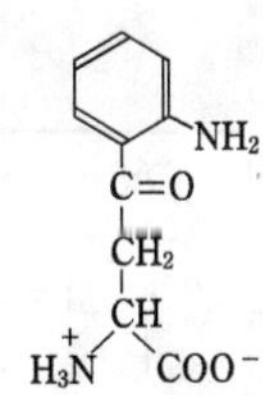

Kynurenine

* For thyroxine, refer to page **494**. For γ-aminobutyric acid, refer to page **493**.

Peptides*

Peptide bond

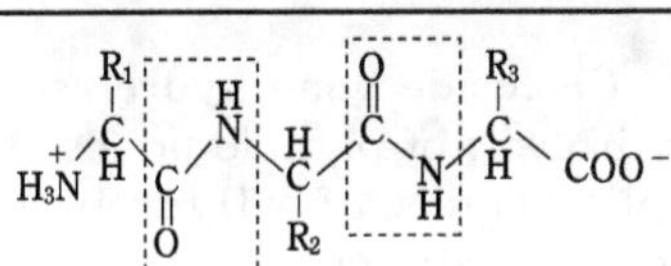

Example R_1 : H–, R_2 : CH_3–, R_3 : HO–CH_2–Glycyl-alanyl-serine (H–$\overset{1}{\text{Gly}}$–$\overset{2}{\text{Ala}}$–$\overset{3}{\text{Ser}}$–NH_2)

S–S (bridge between Cys 1 and Cys 6)

H–$\overset{1}{\text{Cys}}$–Tyr–Ile–Gln–Asn–Cys–Pro–Leu–$\overset{9}{\text{Gly}}$–NH_2

Oxytocin (Human, cow, horse, sheep)

H-$\overset{1}{\text{Ser}}$-Tyr-Ser-Met-Glu-His-Phe-Arg-Trp-$\overset{10}{\text{Gly}}$-Lys-Pro-Val-Gly-Lys-Lys-Arg-Arg-Pro-$\overset{20}{\text{Val}}$-
$\overset{21}{\text{Lys}}$-Val-Tyr-Pro-Asn-Ala-Gly-Glu-Asp-$\overset{30}{\text{Glu}}$-Ser-Ala-Glu-Ala-Phe-Pro-Leu-Glu-$\overset{39}{\text{Phe}}$-OH

Adrenocorticotropic hormone (Corticotropin) (Human)

* For others, refer to page **861**.

Physiological Classification of Proteins

Classification	Example
Enzymes	Trypsin, pepsin, catalase, Na^+, K^+-ATPase
Storage proteins	Ferritin, phosvitin, casein
Transport proteins	Albumin, hemoglobin, α-, β-globulin
Contractile proteins	Myosin, actin, dynein
Protective proteins	Immunoglobulin, complement components, animal lectin
Toxins	Cholera toxin, ricin, cobratoxin
Hormones	Insulin, thyrotropic hormone
Structural proteins	Collagen, histone, keratin, elastin
Regulatory proteins	Hormone receptors, calmodulin, operon repressors

Functional Classification of Enzymes

1 Oxidoreductase

1. 1 It acts on >CHOH, 1. 2 It acts on >C=O, 1. 3 It acts on -CH=CH-, 1. 4 It acts on >CH−NH_2, 1. 5 It acts on >CH−NH-, 1. 6 It acts on NADH, NADPH

2 Transferase

2.1 transfer of groups containing one carbon atom such as methyl and formyl groups, 2.2 transfer of aldehydes and ketones, 2.3 transfer of acyl groups, 2.4 transfer of sugar residues, 2.5 transfer of alkyl groups, 2.6 transfer of amino groups and amides, 2.7 transfer of phosphate groups, 2.8 transfer of groups containing sulfur

3 Hydrolase

3.1 Ester hydrolysis, 3.2 Glycoside bond hydrolysis, 3.3 Peptide bond hydrolysis, 3.4 3.3 Hydrolysis of peptide bonds, 3.4 Hydrolysis of C-N bonds other than peptide bonds, 3.5 Hydrolysis of acid anhydrides, 3.5 Hydrolysis of acid anhydrides

4 Lyase

4. 1 >C=C< It acts on C=C, 4. 2 >It acts on C=O,
4. 3 >It acts on C=N

5 Isomerase

5. 1 Racemization (epimerization), 5. 2 Cis-trans isomerization, 5. 3 Intramolecular redox reaction, 5. 4 Intramolecular rearrangement reaction (mutase), 5. 5 Intramolecular cleavage reaction

6 Synthetase or ligase

6. 1 Formation of C-O bond, 6. 2 Formation of C-S bond,
6. 3 Formation of C-N bond, 6. 4 Formation of C-C bond

Carbohydrates: Monosaccharides

Compound name	Abbreviation	Molecular weight	Chemical structural formula
Neutral sugar (For ribose and deoxyribose, refer to page **485**)			
D-Glyceraldehyde		90	
Dihydroxyacetone		90	
D-Xylose	Xyl	150	
D-Glucose	Glc	180	
D-Fructose	Fru	180	
D-Mannose	Man	180	
D-Galactose	Gal	180	
L-Fucose	Fuc	164	
Sugar acid			
D-Glucuronic acid	GlcUA	194	
L-Iduronic acid	IdoUA	194	
Amino sugar			
D-Glucosamine	$GlcNH_2$	179	
D-Galactosamine	$GalNH_2$	179	
N-Acetylglucosamine	GlcNAc	221	
N-Acetylgalactosamine	GalNAc	221	
N-Acetylmuramic acid	MurNAc	293	
Sialic acid			
N-Acetylneuramic acid	NANA (NeuNAc)	309	
Sugar alcohol			
Glycerol		92	
Inositol		180	

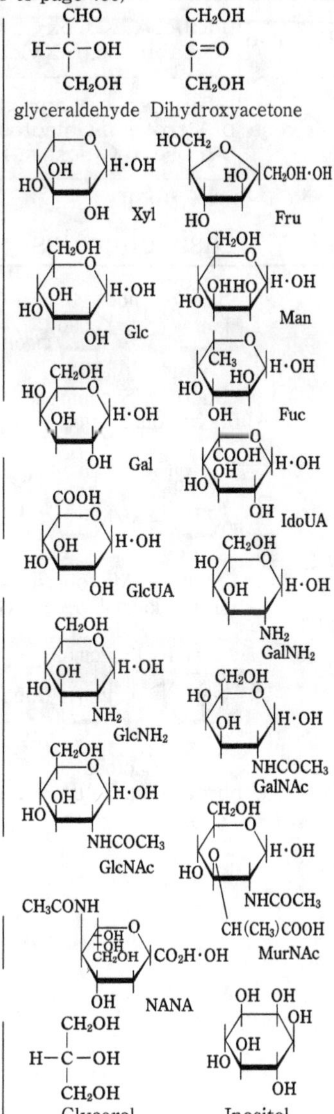

Categories of Monosaccharides

Type	Aldose	Ketose
Triose	D-Glyceraldehyde	Dihydroxyacetone
Tetrose	D-Threose, D-Erythrose	D-Erythrulose
Pentose	D-Ribose, 2-Deoxy-D-ribose*, D-Arabinose, D-Xylose, L-Lyxose,	D-Ribulose, D-Xylulose
Hexose	D-Glucose, D-Mannose, D-Allose, D-Altrose, D-Galactose, D-Talose, D-Idose, D-Gulose, L-Fucose	D-Fructose, L-Sorbose, D-Tagatose, D-psicose

* Deoxy sugar

Disaccharides

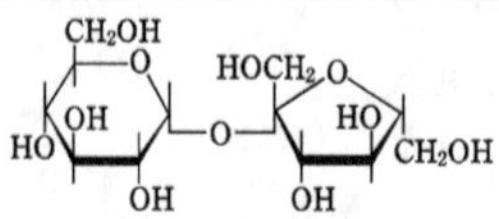

Sucrose (Saccharose)
(Abbreviation: Frufβ2→ 1αGlc)

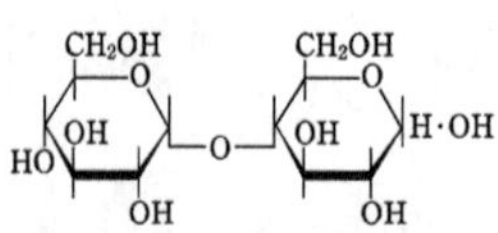

Maltose (Maltobiose)
(Abbreviation: Glcα1 → 4Glc)

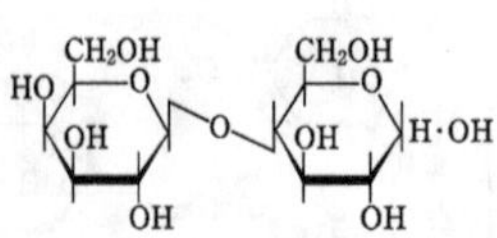

Lactose
(Abbreviation: Galβ1 → 4Glc)

Polysaccharides

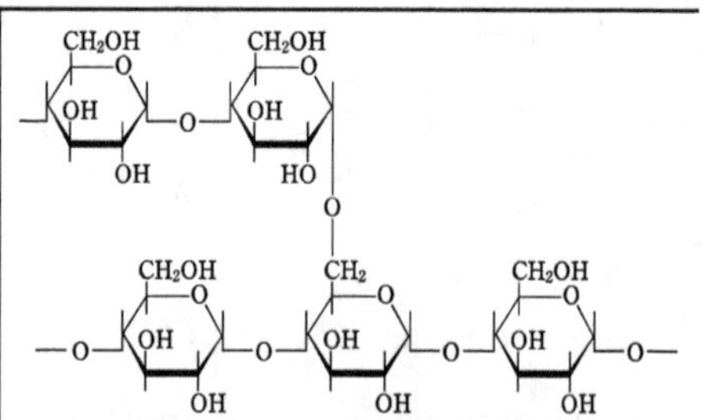

Amylopectin (a component of starch)

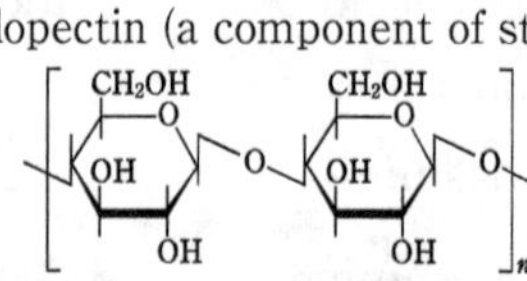

Cellulose

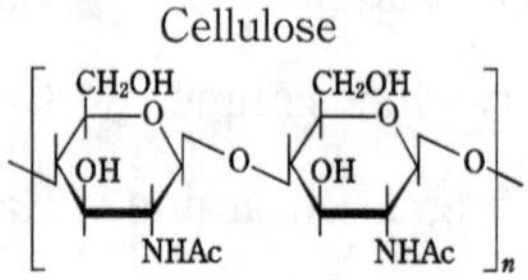

Chitin

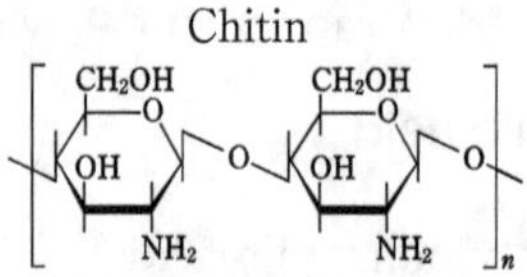

Chitosan

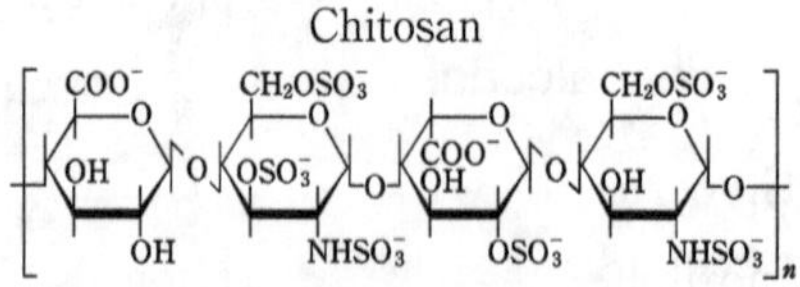

Heparin (one of the partial structures)

Nucleic Acid Constituents

		Name	Abbre-viation	Molecular weight	Chemical structural formula
Sugar		D-Ribose	Rib	150	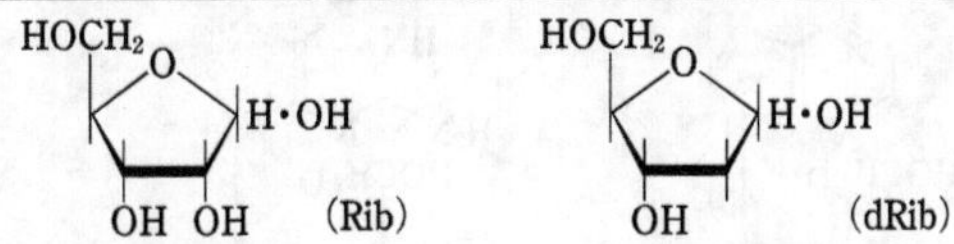 (Rib) (dRib)
		2-Deoxy-D-ribose	dRib	134	
Base	Purine	Adenine	A	135	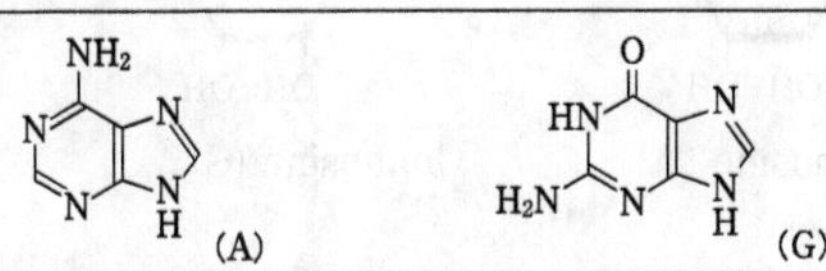 (A) (G)
		Guanine	G	151	
	Pyrimidine	Cytosine	C	111	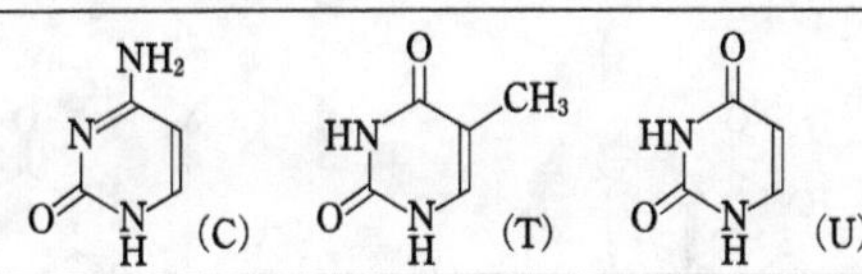 (C) (T) (U)
		Thymine	T	126	
		Uracil	U	112	

Minor Nucleobases

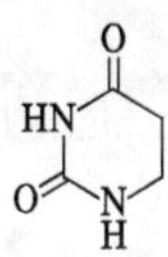

Dihydrouracil

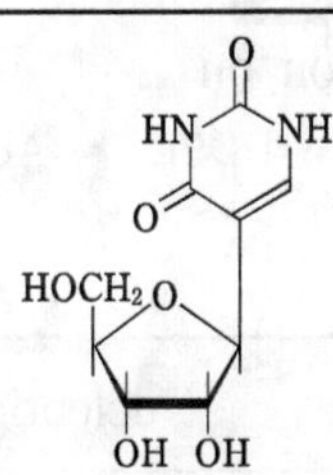

Pseudouridine(Ψ)

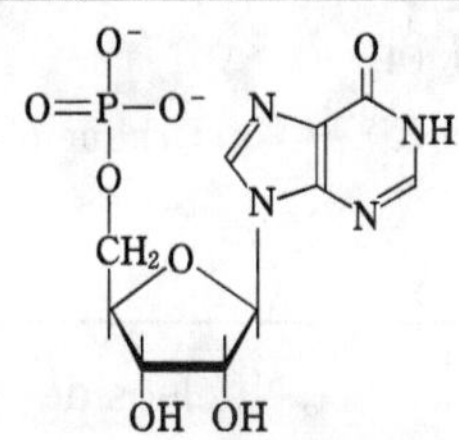

Inosinic acid
(Inosine-5′-monophosphate, IMP)

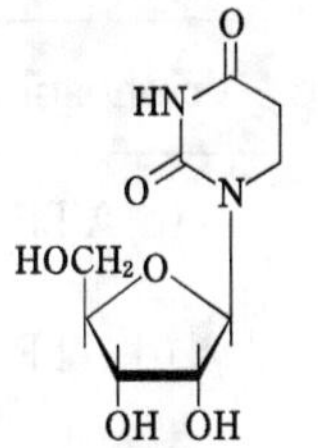

Dihydrouridine(D)

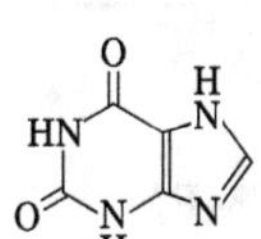

Xanthine

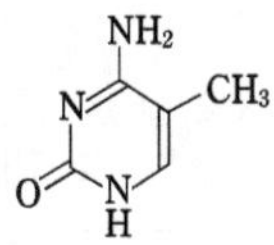

5-Methylcytosine

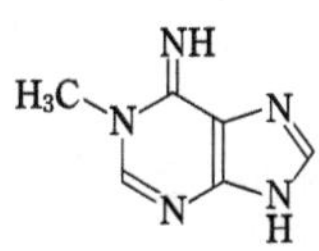

1-Methyladenine

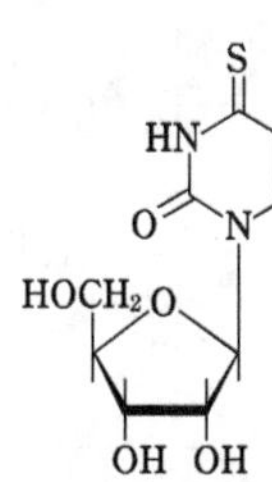

4-Thiouridine(s^4U)

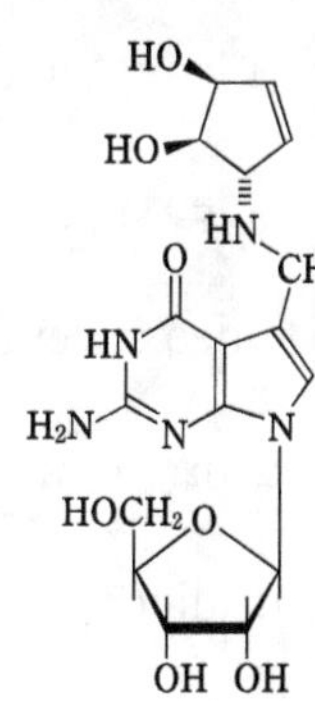

Queuosine(Q)

Nucleosides and Nucleotides

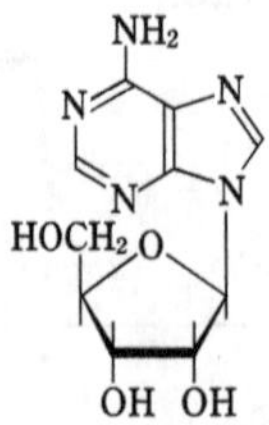

Adenosine (A)

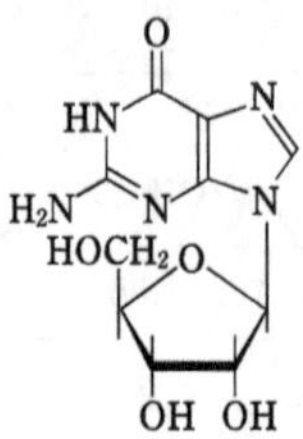

Guanosine (G)

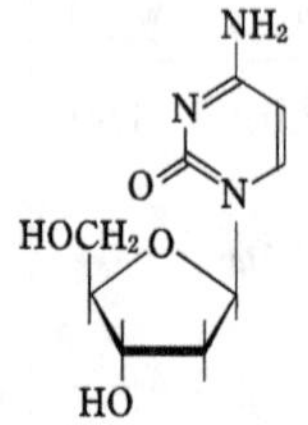

Deoxycytidine (dC)

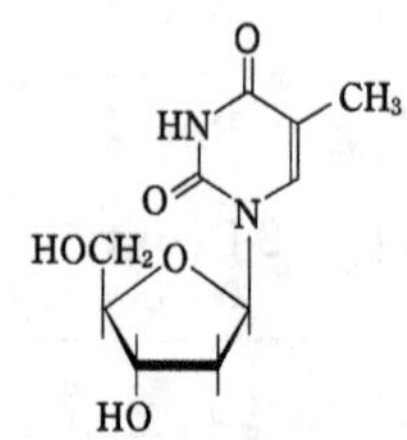

Thymidine (dT)

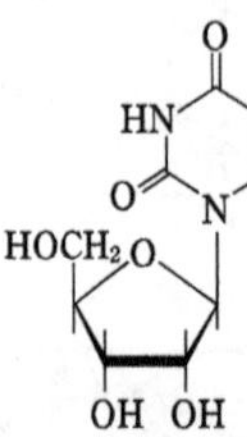

Uridine (U)

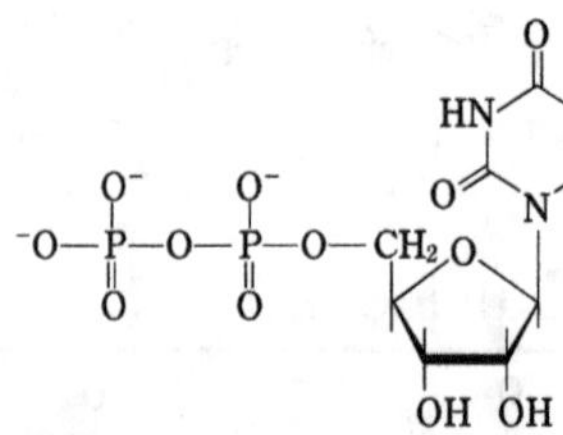

Uridine 5′-diphosphate (UDP)

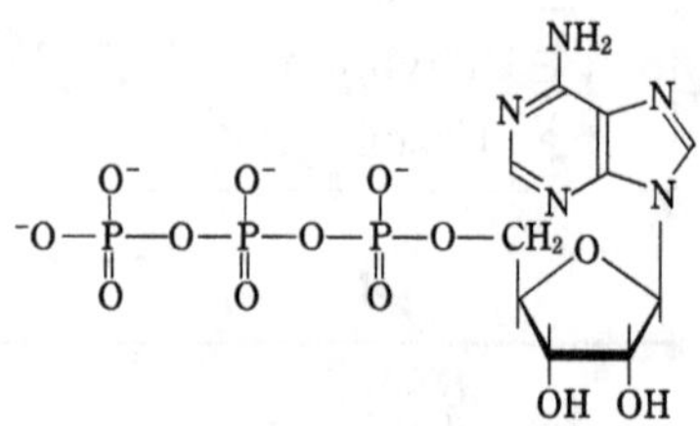

Adenosine 5′-triphosphate (ATP)

Base	Nucleoside	Nucleotide (Abbreviation)		
		Monophosphate	Diphosphate	Triphosphate
A	(deoxy) adenosine	(d)AMP	(d)ADP	(d)ATP
G	(deoxy) guanosine	(d)GMP	(d)GDP	(d)GTP
C	(deoxy) cytidine	(d)CMP	(d)CDP	(d)CTP
T	deoxythymidine	dTMP	dTDP	dTTP
U	uridine	UMP	UDP	UTP

Chemical Structural Formula for DNA

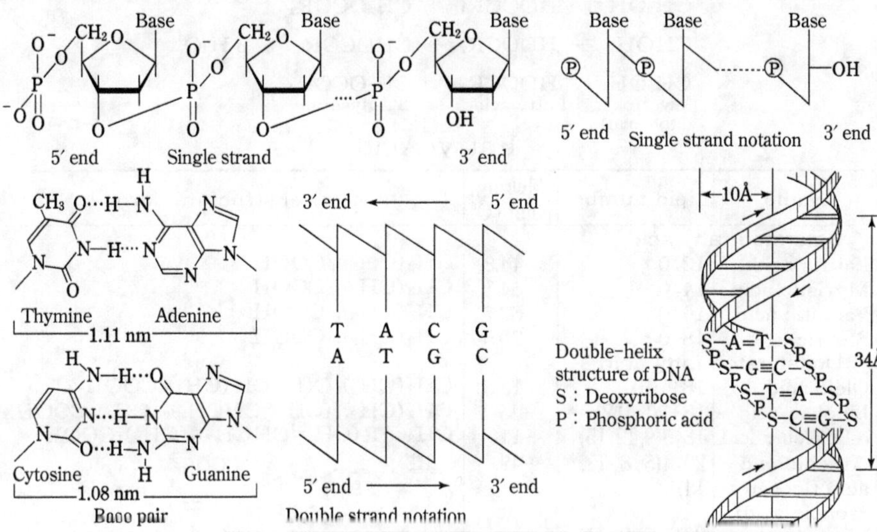

Structure of RNA

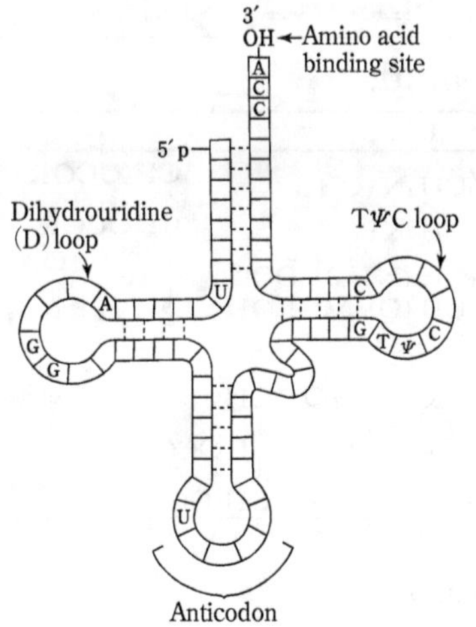

General structure of tRNA molecule

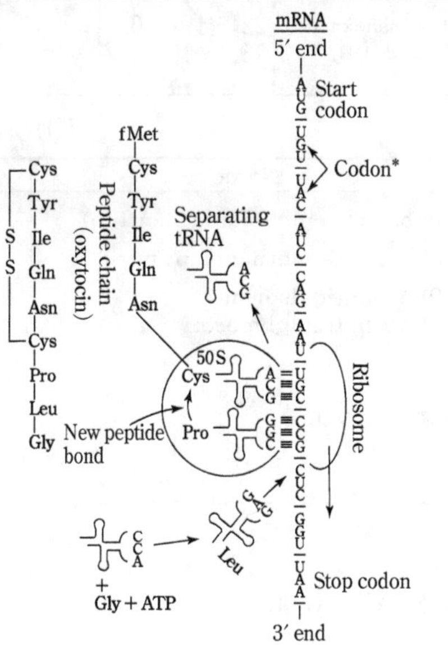

Action of mRNA and tRNA in peptide synthesis

* For the amino acid codon table, refer to page **861**.

Lipids

$$\begin{array}{lclcl} CH_2OH & & HOOCR_1 & & CH_2OCOR_1 \\ | & & & & | \\ CHOH & + & HOOCR_2 & \rightarrow & CHOCOR_2 \quad + \ 3\,H_2O \\ | & & & & | \\ CH_2OH & & HOOCR_3 & & CH_2OCOR_3 \\ \text{Glycerol (glycerin)} & & \text{Fatty acid} & & \text{Triglyceride} \end{array}$$

Fatty Acid

Name	Lipid number	Melting point (℃)	Chemical structural formula
Saturated fatty acid			
Lauric acid	12:0	44.8	$CH_3(CH_2)_{10}COOH$
Myristic acid	14:0	54.1	$CH_3(CH_2)_{12}COOH$
Palmitic acid	16:0	62.7	$CH_3(CH_2)_{14}COOH$
Stearic acid	18:0	70.5	$CH_3(CH_2)_{16}COOH$
Unsaturated fatty acid			
Oleic acid	18:1(9)	13.3	$CH_3(CH_2)_7CH=CH(CH_2)_7COOH$
Linoleic acid	18:2(9, 12)	− 5	$CH_3(CH_2)_4(CH=CHCH_2)_2(CH_2)_6COOH$
α-Linolenic acid	18:3(9,12,15)	−11	$CH_3CH_2(CH=CHCH_2)_3(CH_2)_6COOH$
Arachidonic acid (Icosatetraenoic acid)	20:4(5, 8, 11, 14)	−49	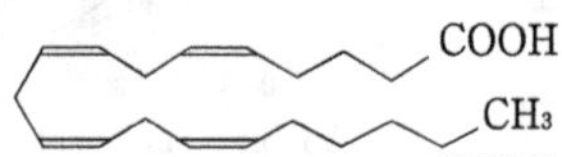
Icosapentaenoic acid (EPA)	20:5(5, 8, 11, 14, 17)		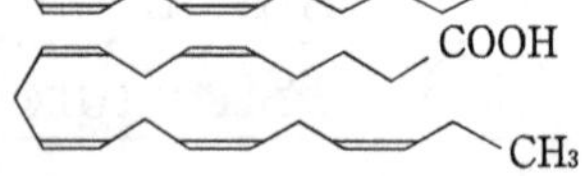
Docosahexaenoic acid (DHA)	22:6(4, 7, 10, 13, 16, 19)		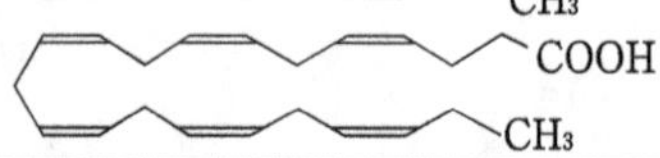

Phospholipids

Name	R_3
Phosphatidylcholine (Lecithin)	$—CH_2CH_2\overset{+}{N}(CH_3)_3$
Phosphatidylethanolamine	$—CH_2CH_2\overset{+}{N}H_3$
Phosphatidylserine	$—CH_2CH(\overset{+}{N}H_3)COO^-$
Phosphatidylglycerol	$—CH_2CH(OH)CH_2OH$
Cardiolipin	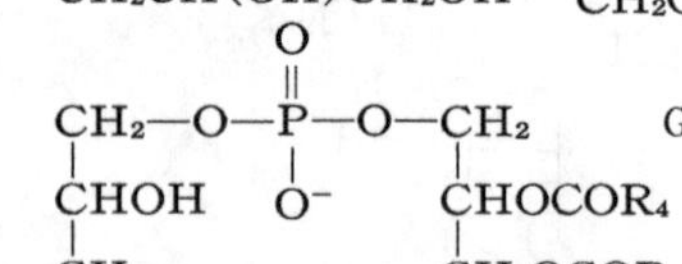
Sphingomyelin	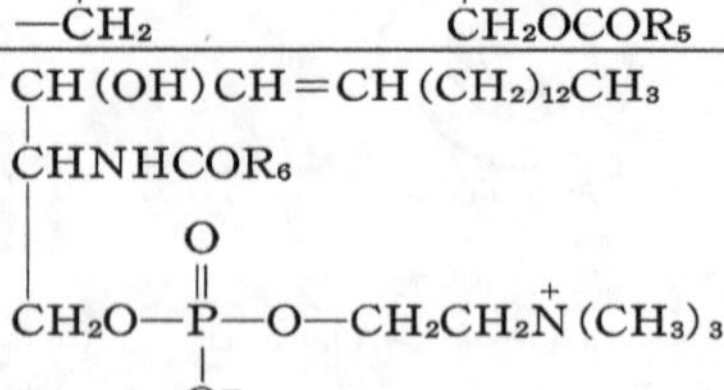

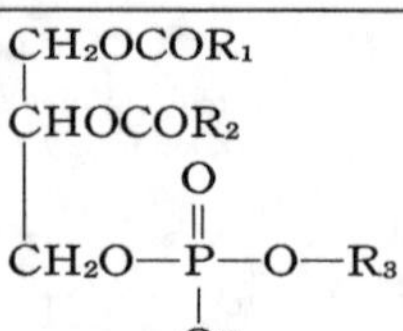

General type

Glycolipids

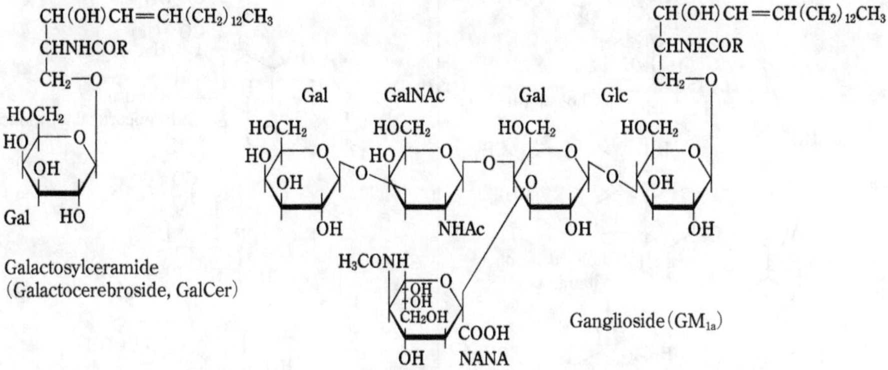

Saturated Hydrocarbons

Chemical name	Molecular formula	Melting point (℃)	Whereabouts
Pristane	$C_{19}H_{40}$	Very low	Basking shark liver
n-Pentakosan	$C_{25}H_{52}$	54	Beeswax
n-Heptadecane	$C_{27}H_{56}$	59	Beeswax
n-Octacosane	$C_{28}H_{58}$	62	Chrysalis oil
n-Nonacosane	$C_{29}H_{60}$	64	Cabbage leaf and beeswax
n-Triacontane	$C_{30}H_{62}$	66	Apple skin
n-Hentriacontane	$C_{31}H_{64}$	66	Candelilla wax
n-Pentatriacontane	$C_{35}H_{72}$	74	Sugarcane leaf

Unsaturated Hydrocarbons

Trivial name	Molecular formula	Whereabouts
α-Farnesene	$C_{15}H_{24}$	Apple epicuticular wax
Muscalure	$C_{23}H_{46}$	House fly sex pheromone
6,9-Heptacosadiene	$C_{27}H_{52}$	Cockroach epidermal, Pheromone (attractant)
Squalene	$C_{30}H_{50}$	Shark liver oil, Human sebum

Steroids

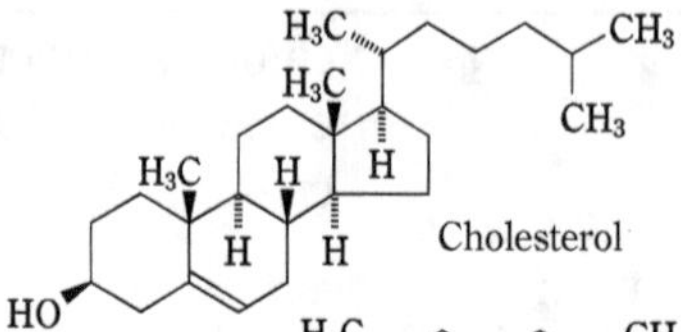

Cholesterol

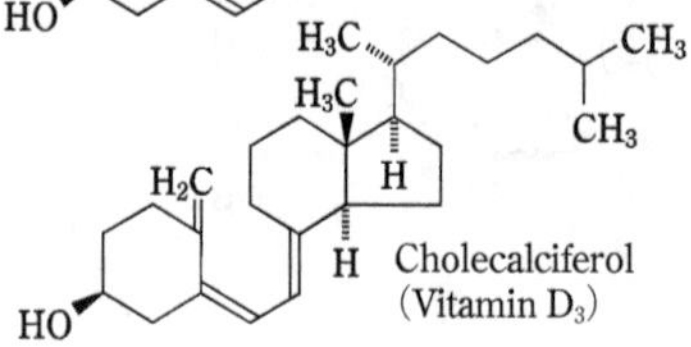

Cholecalciferol (Vitamin D_3)

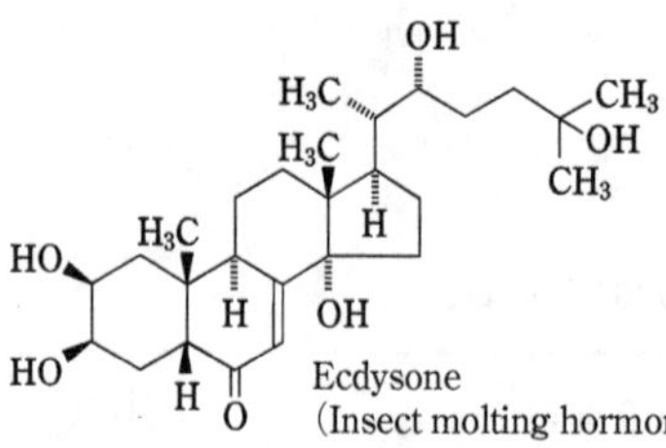

Ecdysone (Insect molting hormone)

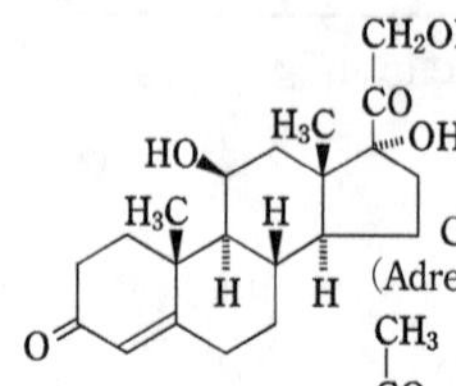

Cortisol (Adrenocortical hormone)

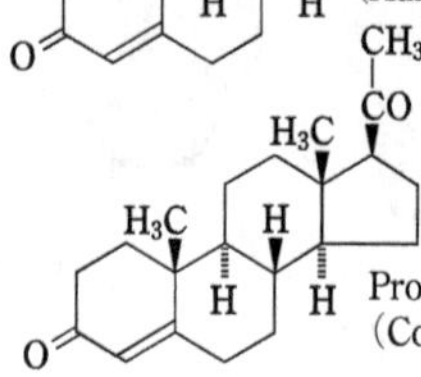

Progesterone (Corpus luteum hormone)

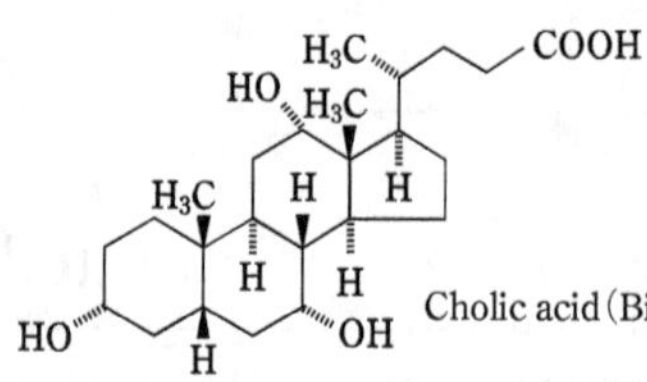

Cholic acid (Bile acid)

Estrogen

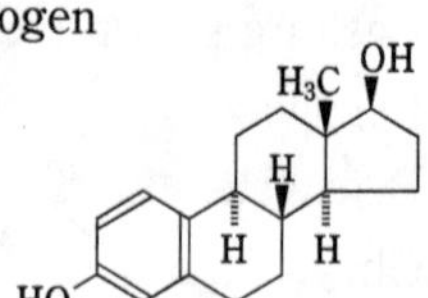

Estradiol

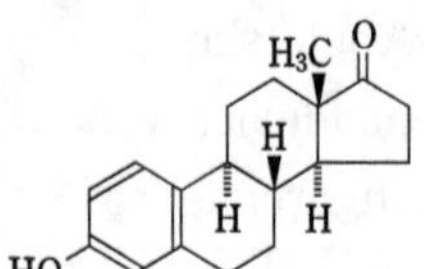

Estrone

Androgen

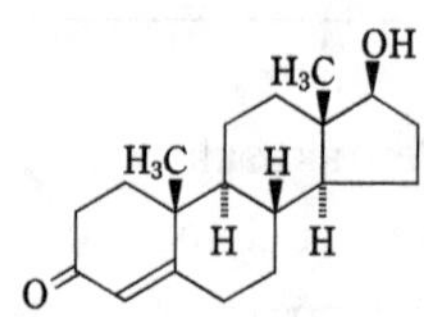

Testosterone

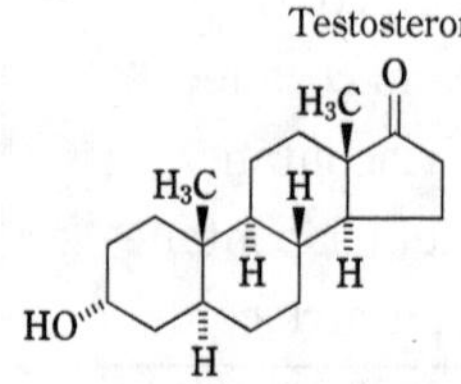

Androsterone

Other lipids

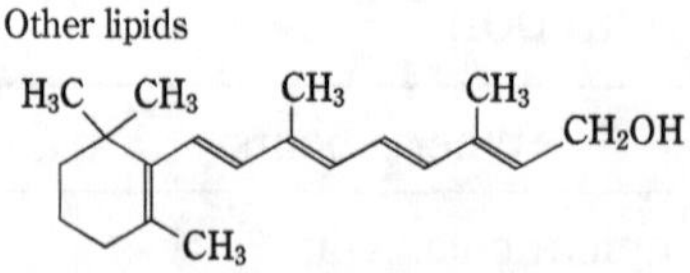

Retinol (Vitamin A_1)

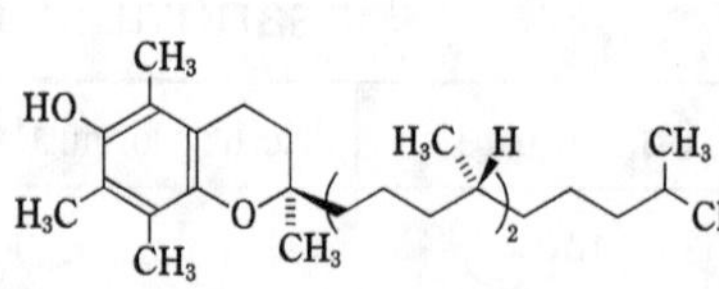

α-Tocopherol (vitamin E)

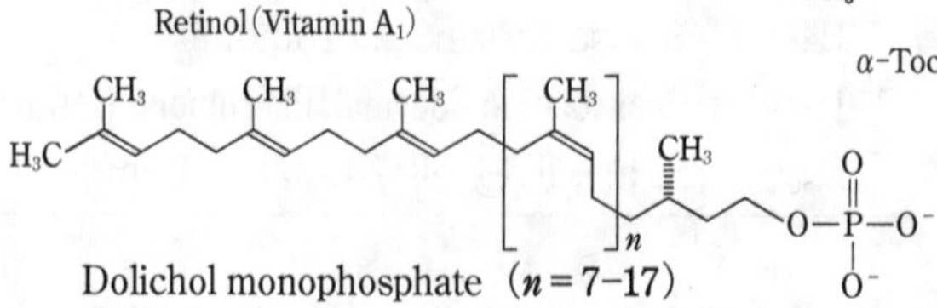

Dolichol monophosphate ($n = 7$–17)

Water-Soluble Vitamins and Coenzymes

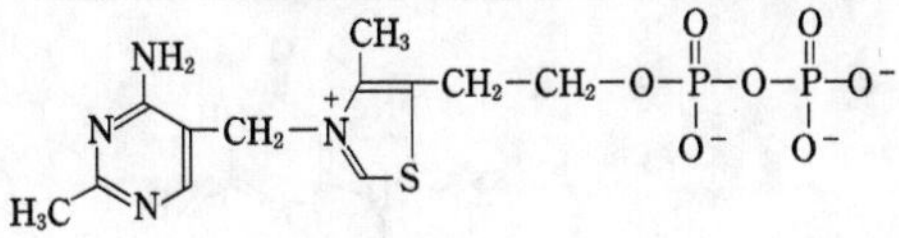

Thiamin (Vitamin B_1)

Thiamin diphosphate

Riboflavin (Vitamin B_2)

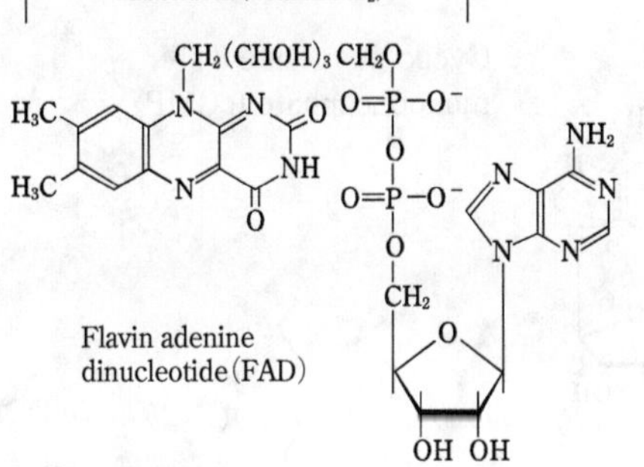

Flavin adenine dinucleotide (FAD)

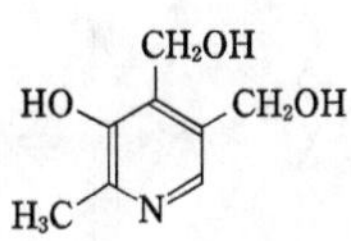

Pyridoxin (Vitamin B_6)

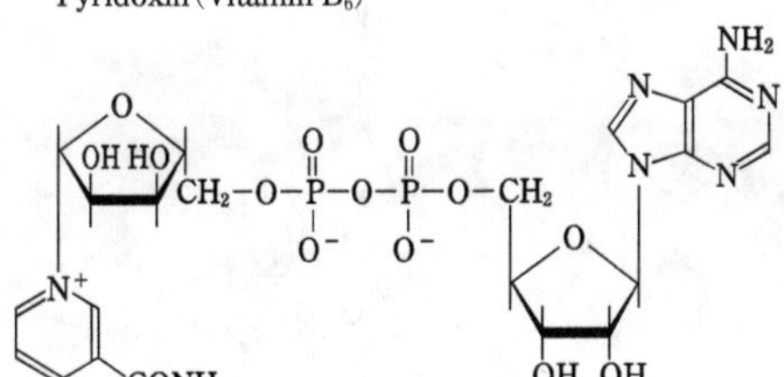

Nicotinamide adenine dinucleotide (NAD^+)

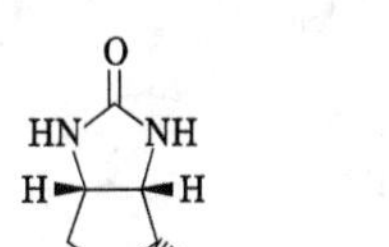

Biotin

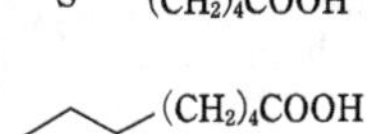

Lipoic acid (Thioctic acid)

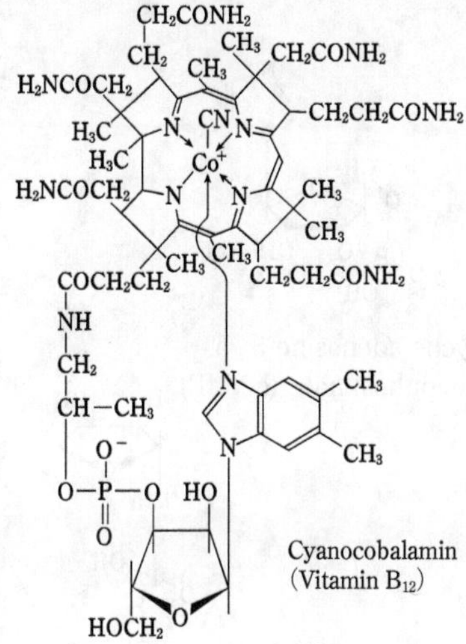

Cyanocobalamin (Vitamin B_{12})

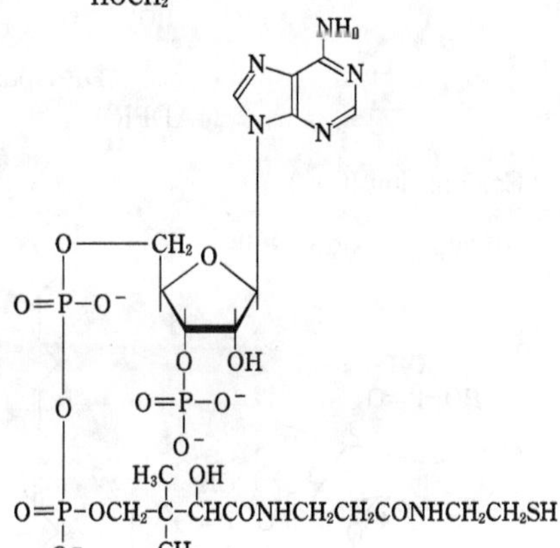

Coenzyme A (CoA)

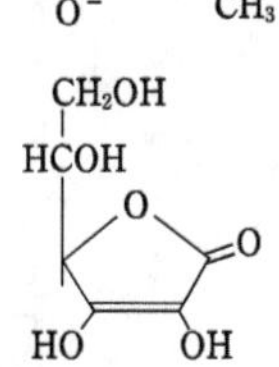

Ascorbic acid (Vitamin C)

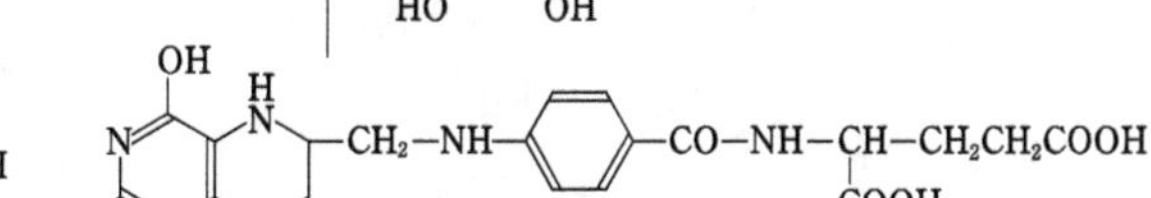

Tetrahydrofolic acid (THF)

(Source : Makoto Sasaki)

Second Messengers

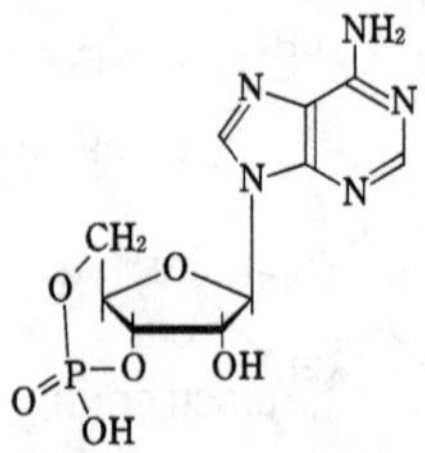

Cyclic adenosine 3′, 5′-monophosphate (cAMP)

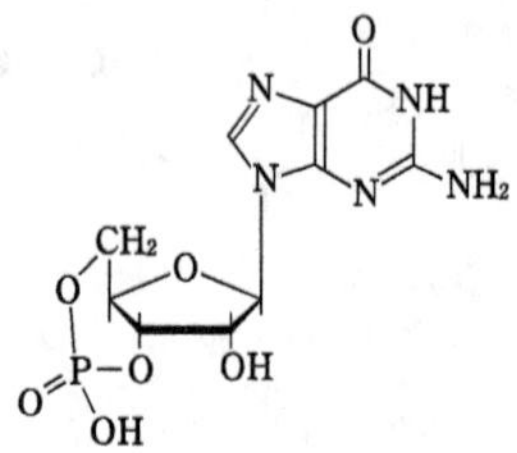

Cyclic guanosine 3′, 5′-monophosphate (cGMP)

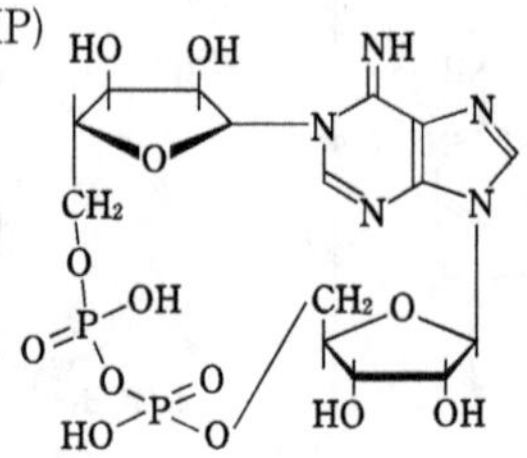

Cyclic ADP ribose (cADPR)

Calcium ion (Ca^{2+})

Nitric monoxide (NO)

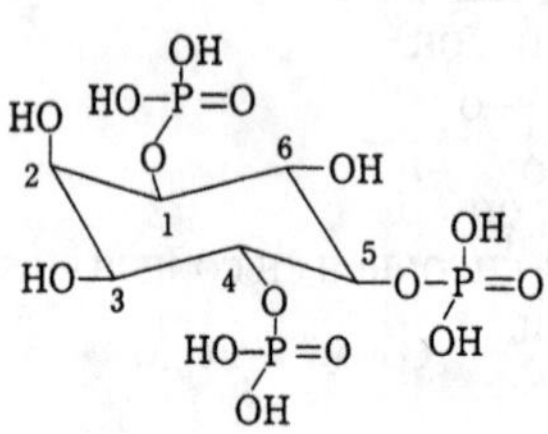

Inositol 1, 4, 5-trisphosphate (IP_3)

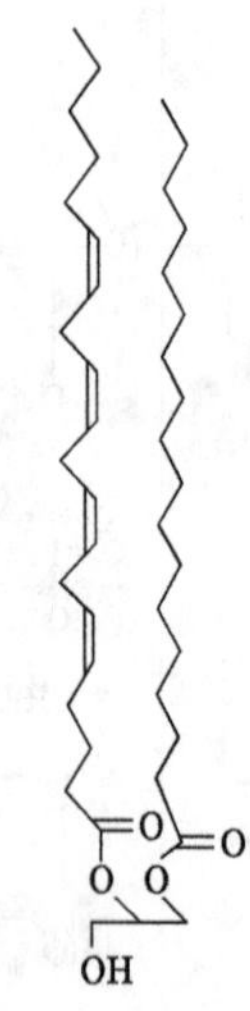

Diacylglycerol (DAG)

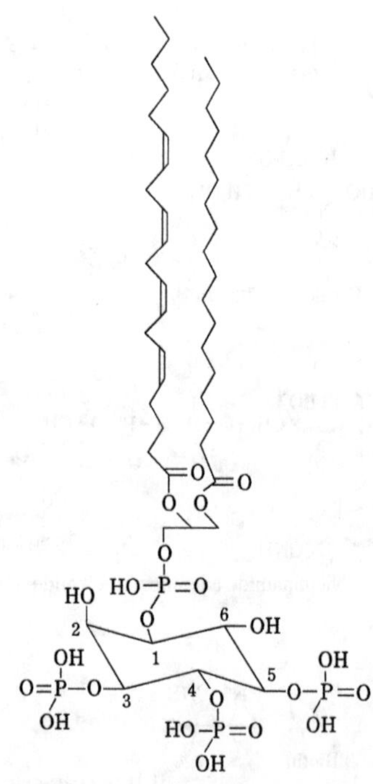

Phosphatidylinositol 3, 4, 5-trisphosphate (PIP_3)

(Source: Moritoshi Sato)

Biologically Active Substances

Neurotransmitters*

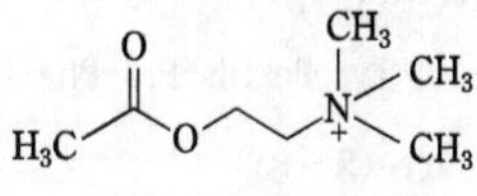

Acetylcholine (ACh)

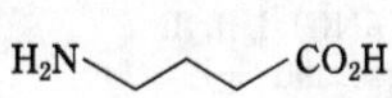

γ-Aminobutyric acid (GABA)

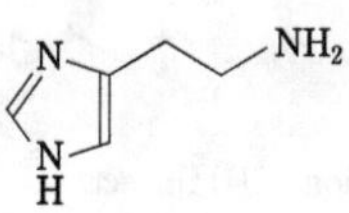

Histamin

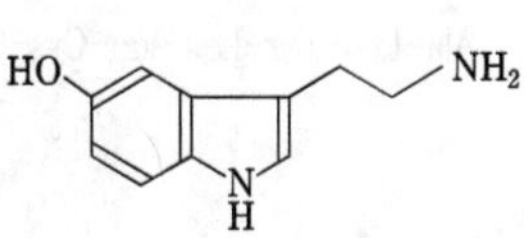

Serotonin (5-Hydroxytryptamine) (5 HT)

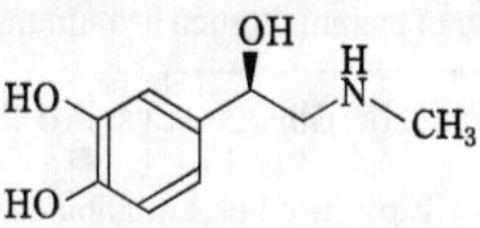

Epinephrine (Adrenalin)

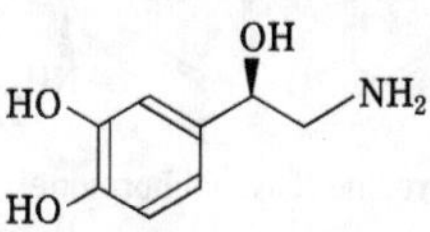

Norepinephrine (Noradrenalin)

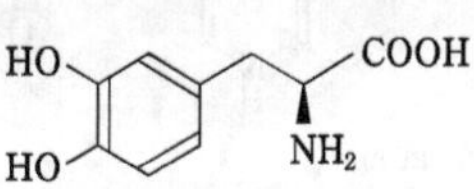

Dopa (3,4-Dihydroxyphenyl-alanine)

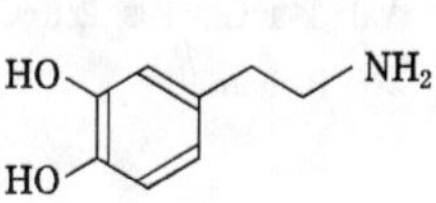

Dopamine

H-Tyr[1]-Gly-Gly-Phe-Met[5]-Thr-Ser-Glu-Lys-Ser[10]-Gln-Thr-Pro-Leu-Val[15]-Thr-Leu-Phe[18]······Lys[28]-Lys-Gly-Gln[31]-OH

β-endorphin (human)

* For glutamic acid, refer to page **480**.

Plant Hormones

$H_2C=CH_2$

Ethylene

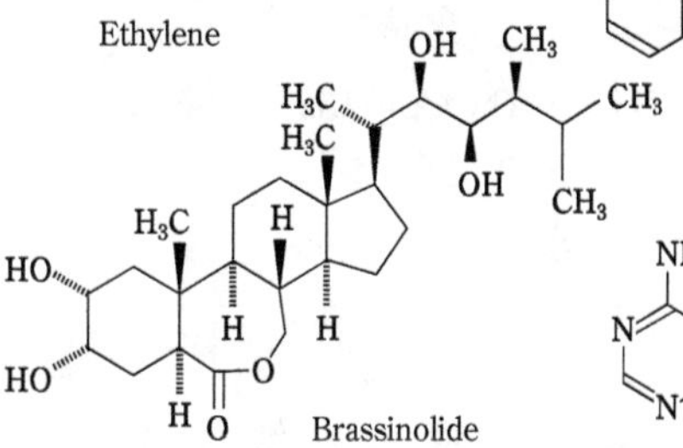

Brassinolide

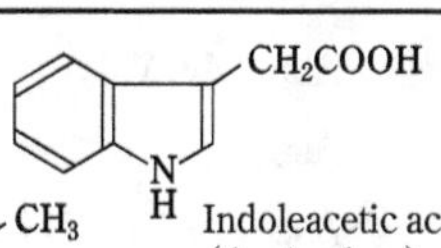

Indoleacetic acid (Auxin class)

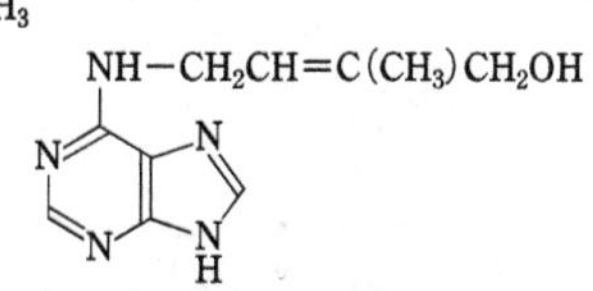

Zeatin (Cytokinin class)

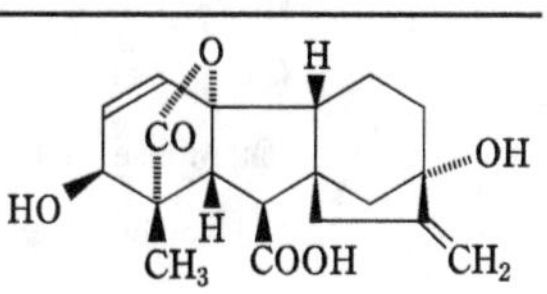

Gibberellin A_3

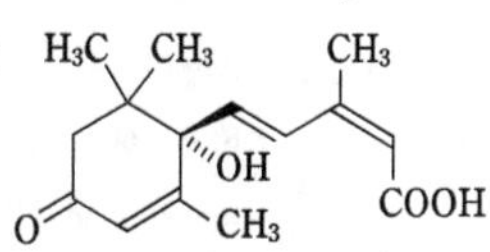

Abscisic acid

Animal Hormones*

H-Cys-Tyr-Phe-Gln-Asn-Cys-Pro-Arg-Gly-NH_2 [disulfide bond between Cys 1 and Cys 6]
(residues numbered 1 2 3 4 5 6 7 8 9)

Vasopressin (AVP)
(Majority of mammals such as humans, cows, etc.)

H-Cys-Tyr-Ile-Gln-Asn-Cys-Pro-Lys-Gly-NH_2 [disulfide bond between Cys 1 and Cys 6]
(residues numbered 1 2 3 4 5 6 7 8 9)

Vasotocin (Birds, reptiles, amphibians, and fishes)

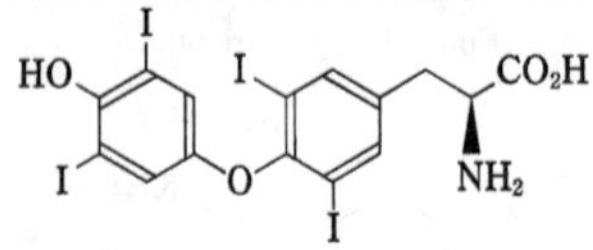

Thyroxine (Thyroid hormone)

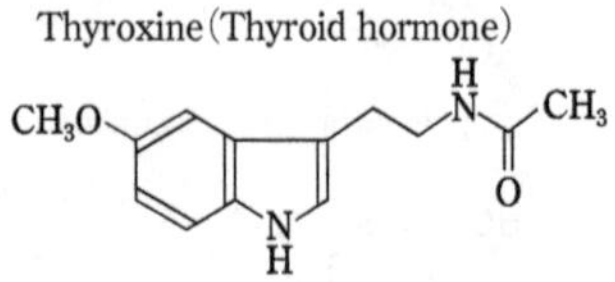

Melatonin

ANG I (ANG-(1~10))
H-Asp-Arg-Val-Tyr-Ile-His-Pro-Phe-His-Leu-OH (residues 1–10)

ANG II (ANG-(1~8))
H-Asp-Arg-Val-Tyr-Ile-His-Pro-Phe-OH (residues 1–8)

ANG III (ANG-(2~8))
H-Arg-Val-Tyr-Ile-His-Pro-Phe-OH (residues 2–8)

Angiotensin (ANG) I, II, III
(Human, horse, and pig)

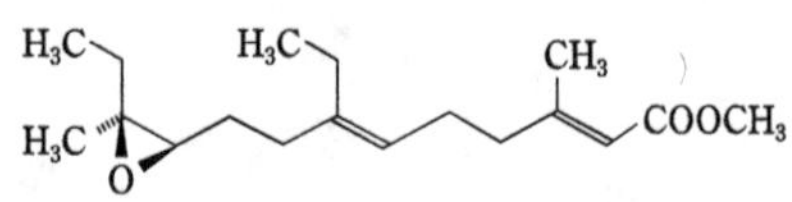

Juvenile hormone (JH) (Insect)

B chain:
H-Phe(1)-Val-Asn-Gln-His-Leu(5)-Cys-Gly-Ser-His(10)-Leu-Val-Glu-Ala-Leu(15)-Tyr-Leu-Val-Cys-Gly(20)-Glu-Arg-Gly-Phe-Phe(25)-Tyr-Thr-Pro-Lys-Thr(30)-OH

A chain (N terminal):
H-Gly(1)-Ile-Val-Glu-Gln(5)-Cys-Cys-Thr-Ser-Ile(10)-Cys-Ser-Leu-Tyr-Gln(15)-Leu-Glu-Asn-Tyr-Cys(20)-Asn(21)-OH

[Disulfide bonds: B7 Cys–S–S–A7 Cys; B19 Cys–S–S–A20 Cys; A6 Cys–S–S–A11 Cys]

Insulin (Human)

Species difference of insulin amino-acid sequence

A chain	Cow	-Ala(8)-····-Val(10)-
	Horse	-Gly(9)-
	Goat, Sheep	-Ala(8)-Gly(9)-Val(10)-
	Rat, Mouse (I, II)	-Asp(4)-
	Guinea pig	-Asp(4)-····-Gly(9)-Thr(10)-····-Thr(12)-Arg(13)-His(14)-····-Ser(18)-
B chain	Pig, Dog, Horse, Cow, Goat, Sheep	-Ala(30)-
	Rabbit	-Ser(30)-
	Rat, Mouse (I)	-Lys(3)-····-Pro(9)-····-Ser(30)-
	Rat, Mouse (II)	-Lys(3)-····-Met(29)-Ser(30)-
	Guinea pig	-Ser(3)-Arg(4)-····-Asn(10)-····-Thr(14)-····-Ser(17)-····-Gln(20)-Asp(21)-Asp(22)-····-Ile(27)-····-Asp(30)

* For oxytocin and adrenocorticotropic hormones, refer to page **481**. For steroid hormones, refer to page **490**.

Biologically Active Substances

Neurotransmitters*

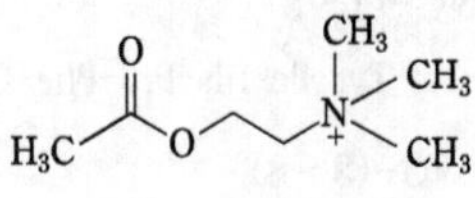
Acetylcholine (ACh)

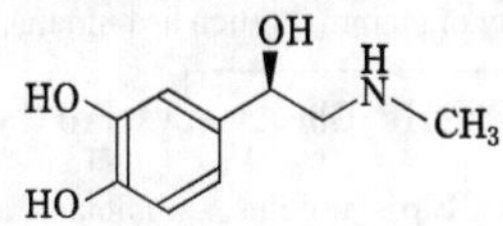
Epinephrine (Adrenalin)

γ-Aminobutyric acid (GABA)

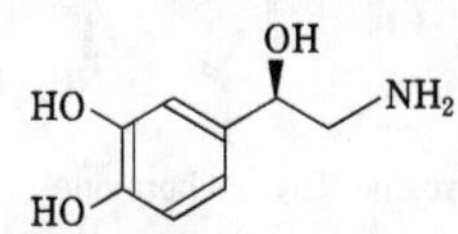
Norepinephrine (Noradrenalin)

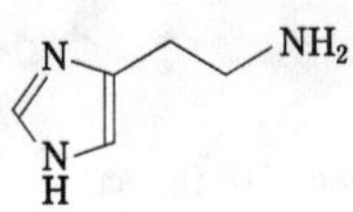
Histamin

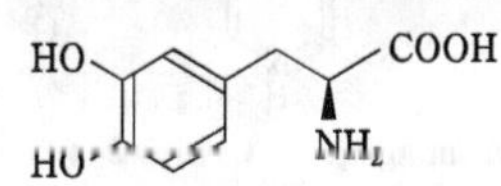
Dopa (3,4-Dihydroxyphenyl-alanine)

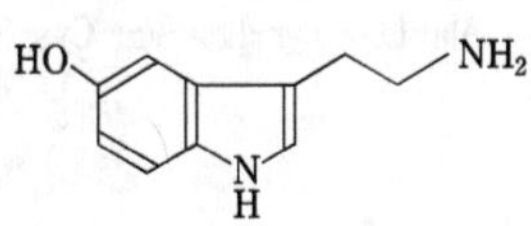
Serotonin (5-Hydroxytryptamine) (5 HT)

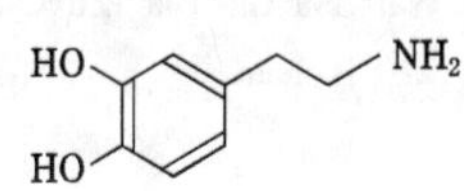
Dopamine

1 5 10 15 18 28 31
H-Tyr-Gly-Gly-Phe-Met-Thr-Ser-Glu-Lys-Ser-Gln-Thr-Pro-Leu-Val-Thr-Leu-Phe······Lys-Lys-Gly-Gln-OH

β-endorphin (human)

* For glutamic acid, refer to page **480**.

Plant Hormones

$H_2C=CH_2$

Ethylene

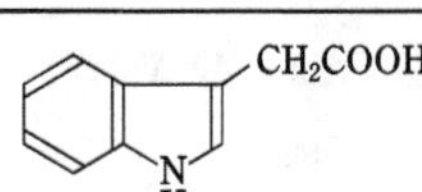
Indoleacetic acid (Auxin class)

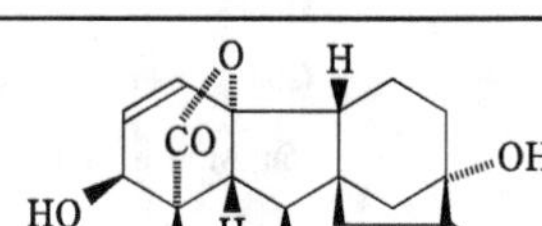
Gibberellin A_3

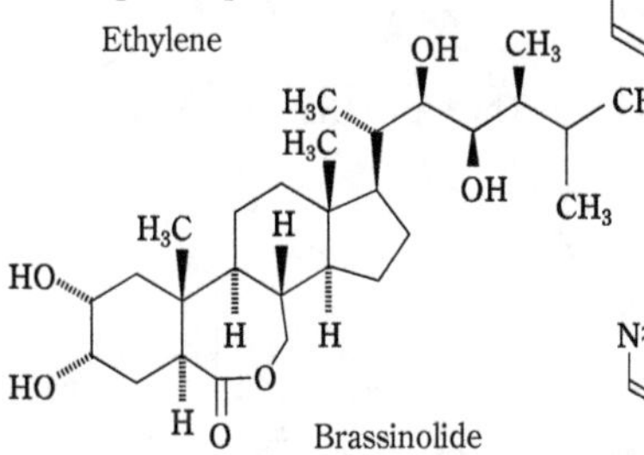
Brassinolide

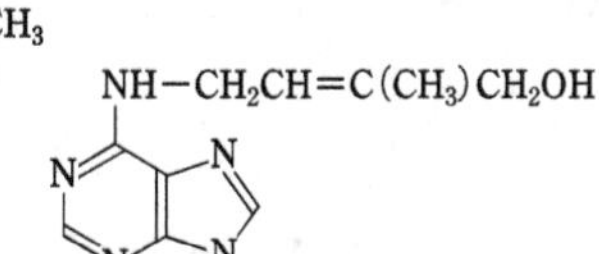
Zeatin (Cytokinin class)

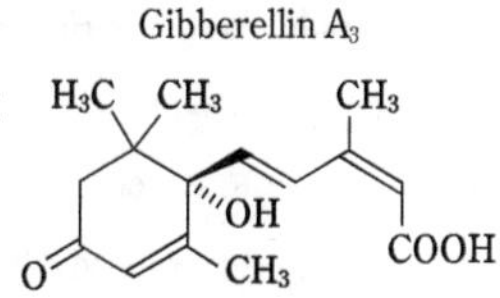
Abscisic acid

Animal Hormones*

H-Cys-Tyr-Phe-Gln-Asn-Cys-Pro-Arg-Gly-NH$_2$ (Cys 1 and Cys 6 bridged)
1 2 3 4 5 6 7 8 9

Vasopressin (AVP)
(Majority of mammals such as humans, cows, etc.)

H-Cys-Tyr-Ile-Gln-Asn-Cys-Pro-Lys-Gly-NH$_2$ (Cys 1 and Cys 6 bridged)
1 2 3 4 5 6 7 8 9

Vasotocin (Birds, reptiles, amphibians, and fishes)

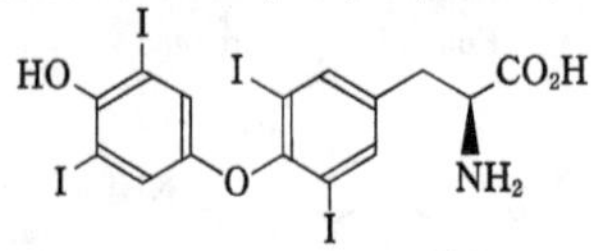

Thyroxine (Thyroid hormone)

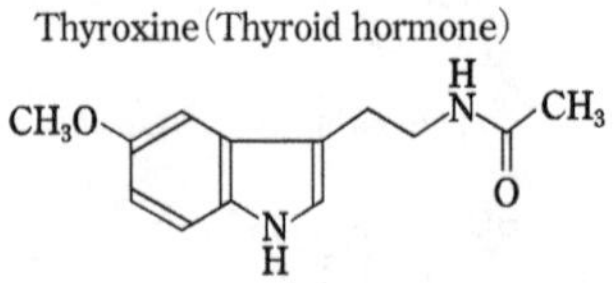

Melatonin

ANG I (ANG-(1~10))

H-Asp-Arg-Val-Tyr-Ile-His-Pro-Phe-His-Leu-OH (residues 1–10)

ANG II (ANG-(1~8))

H-Asp-Arg-Val-Tyr-Ile-His-Pro-Phe-OH (residues 1–8)

ANG III (ANG-(2~8))

H-Arg-Val-Tyr-Ile-His-Pro-Phe-OH (residues 2–8)

Angiotensin (ANG) I, II, III
(Human, horse, and pig)

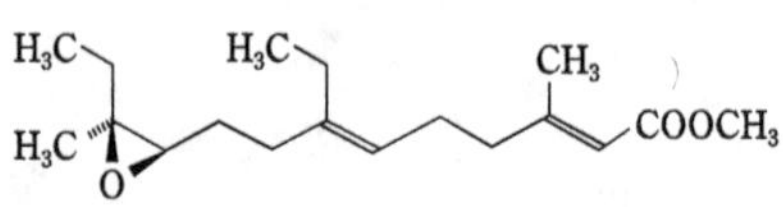

Juvenile hormone (JH) (Insect)

B chain:
H-Phe(1)-Val-Asn-Gln-His(5)-Leu-Cys-Gly-Ser-His(10)-Leu-Val-Glu-Ala-Leu(15)-Tyr-Leu-Val-Cys-Gly(20)-Glu-Arg-Gly-Phe-Phe(25)-Tyr-Thr-Pro-Lys-Thr(30)-OH

A chain (N terminal):
H-Gly(1)-Ile-Val-Glu-Gln(5)-Cys-Cys-Thr-Ser-Ile(10)-Cys-Ser-Leu-Tyr-Gln(15)-Leu-Glu-Asn-Tyr-Cys(20)-Asn(21)-OH

S-S bonds: B7 Cys–A7 Cys; B19 Cys–A20 Cys; A6 Cys–A11 Cys

Insulin (Human)

Species difference of insulin amino-acid sequence

	Species	Sequence
A chain	Cow	-Ala(8)-····-Val(10)-
	Horse	-Gly(9)-
	Goat, Sheep	-Ala(8)-Gly(9)-Val(10)-
	Rat, Mouse (I, II)	-Asp(4)-
	Guinea pig	-Asp(4)-····-Gly(9)-Thr(10)-····-Thr(12)-Arg(13)-His(14)-····-Ser(18)-
B chain	Pig, Dog, Horse, Cow, Goat, Sheep	-Ala(30)-
	Rabbit	-Ser(30)-
	Rat, Mouse (I)	-Lys(3)-····-Pro(9)-····-Ser(30)-
	Rat, Mouse (II)	-Lys(3)-····-Met(29)-Ser(30)-
	Guinea pig	-Ser(3)-Arg(4)-····-Asn(10)-····-Thr(14)-····-Ser(17)-····-Gln(20)-Asp(21)-Asp(22)-····-Ile(27)-····-Asp(30)

* For oxytocin and adrenocorticotropic hormones, refer to page **481**. For steroid hormones, refer to page **490**.

Pheromones

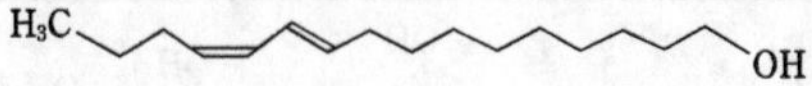

Bombykol (bombycid sex pheromone)

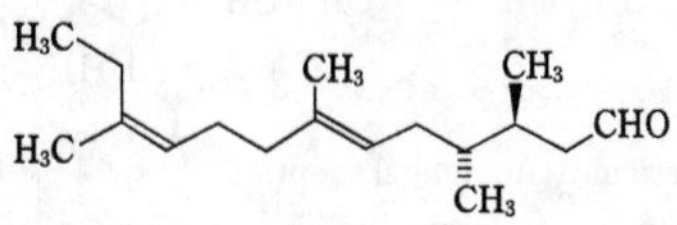

Pharanal (Ant trail phermone)

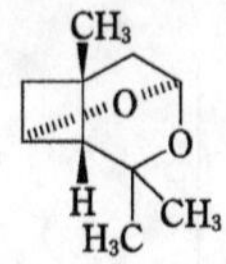

Lineatin
(Bark beetle aggregation pheromone)

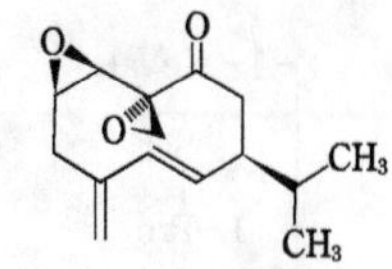

Periplanon-B
(Periplaneta americana sex pheromone)

Autacoids*

H-Arg-Pro-Pro-Gly-Phe-Ser-Pro-Phe-Arg-OH

Bradykinin

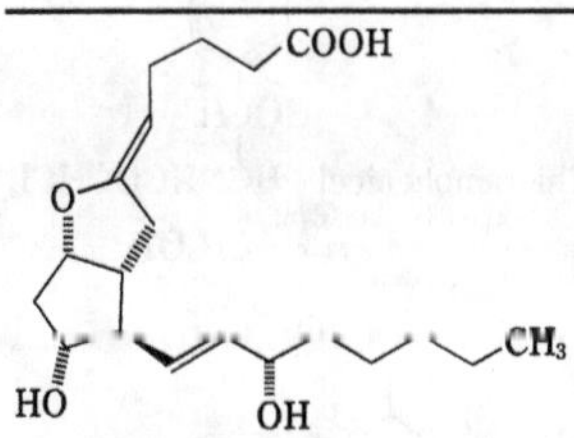

Prostacycline
(PGI_2)

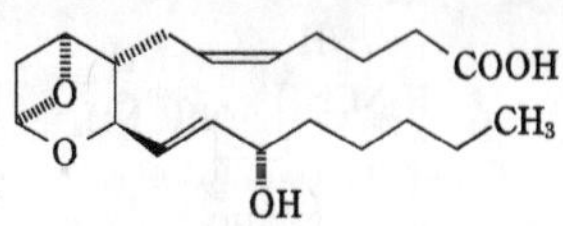

Thromboxane A_2
(TXA_2)

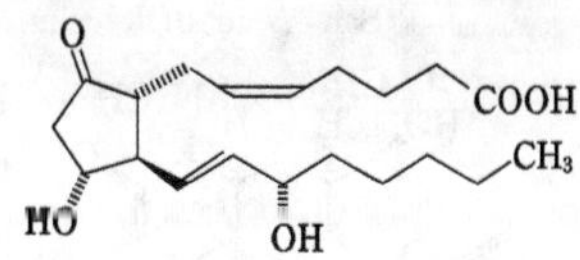

Prostaglandin E_2
(PGE_2)

* For histamine, refer to page **493**. For angiotensin, refer to page **494**.

Pigments

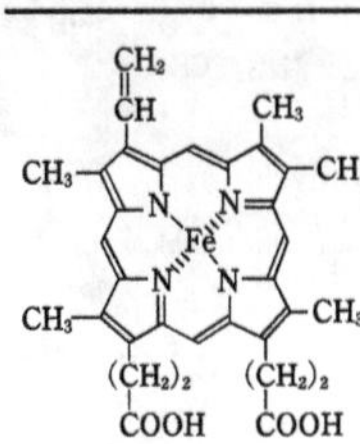

Heme b
(Ferriprotoporphyrin IX)

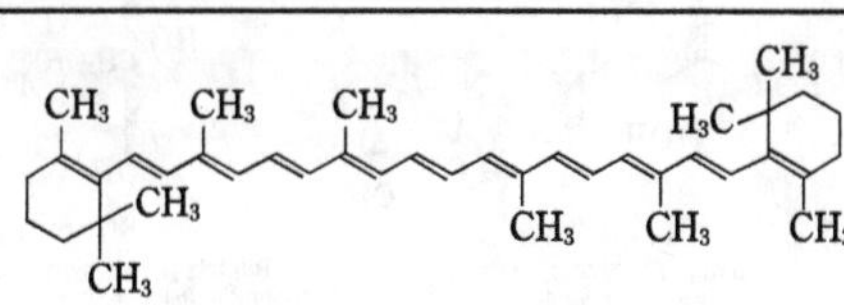

β-Carotene

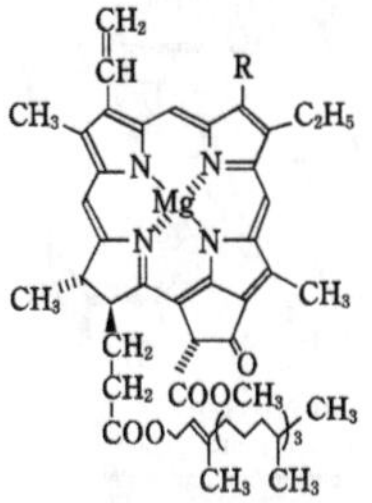

Chlorophyll (*a*, R = CH_3 ; *b*, R = CHO)

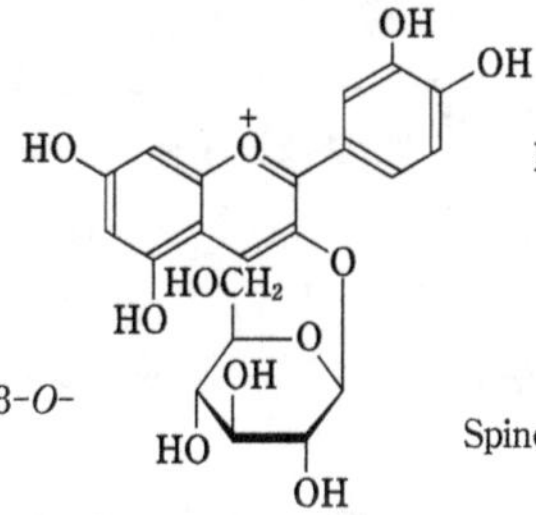

Cyanidin 3-*O*-glucoside

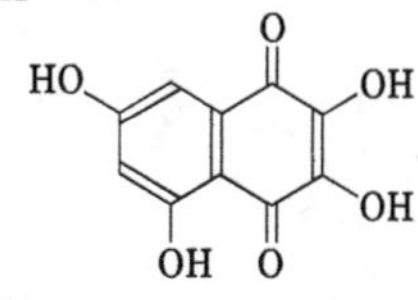

Spinochrome B (Sea urchin dye)

Antibiotics*

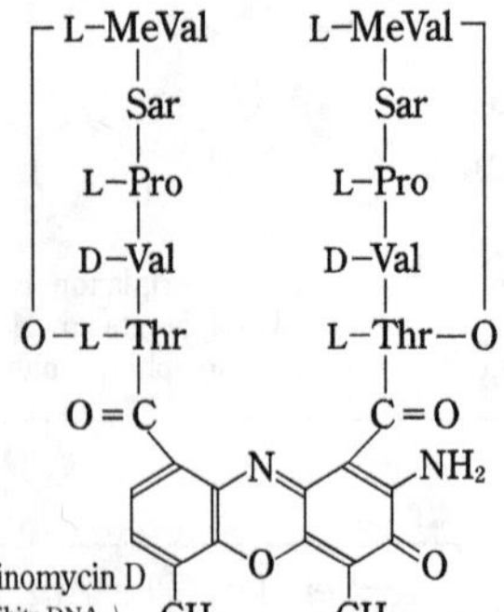

Actinomycin D
(Inhibits DNA dependent RNA synthesis)

Sar : *N*-methylglycine MeVal : *N*-methylvaline

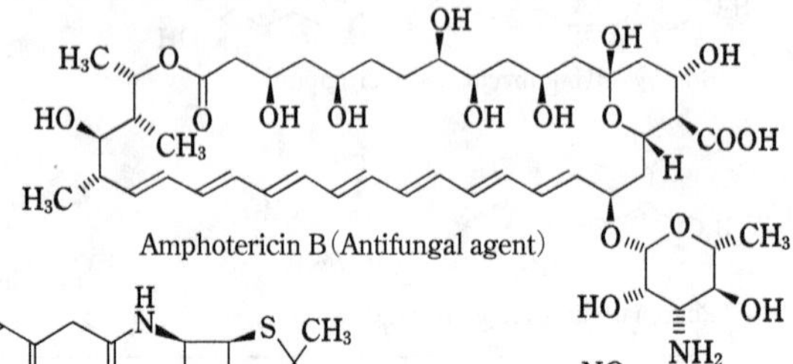

Amphotericin B (Antifungal agent)

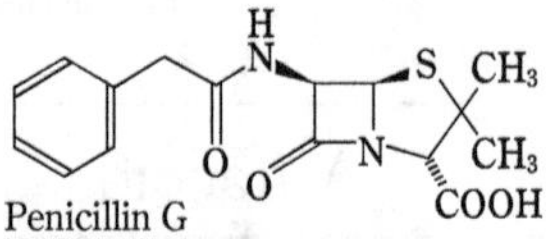

Penicillin G
(Inhibits peptidoglycan synthesis of gram-positive and negative bacteria)

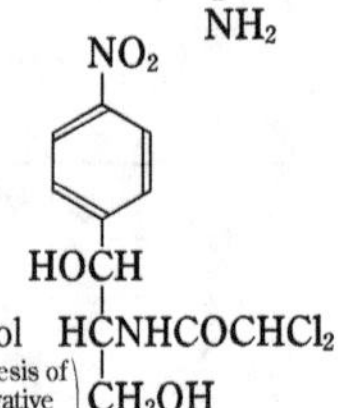

Chloramphenicol
(Inhibits protein synthesis of gram-positive and negative bacteria, Rickettsia)

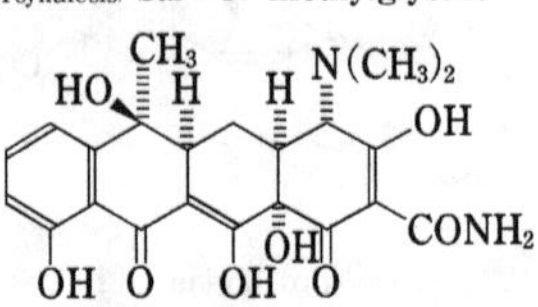

Tetracycline
(Inhibits protein synthesis of gram-positive and negative bacteria, Rickettsia, large viruses, spirochete, and protozoa)

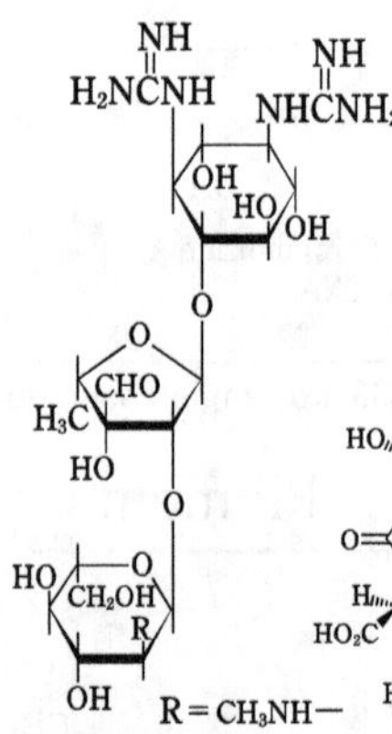

Streptomycin
(Inhibits protein synthesis in gram-positive and negative bacteria and tubercle bacilli)

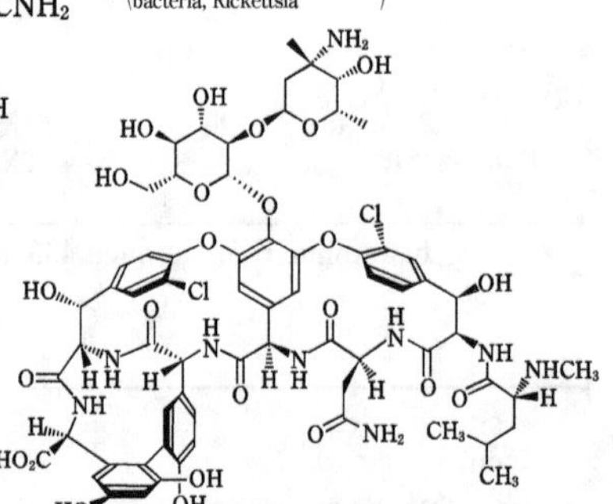

Vancomycin
(Inhibits cell wall peptidoglycan synthesis in gram-positive bacteria)

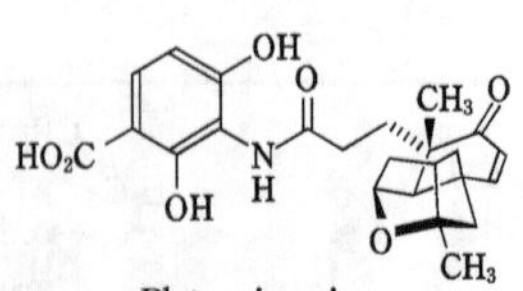

Platensimycin
(Inhibits fatty acid synthesis in gram-positive bacteria)

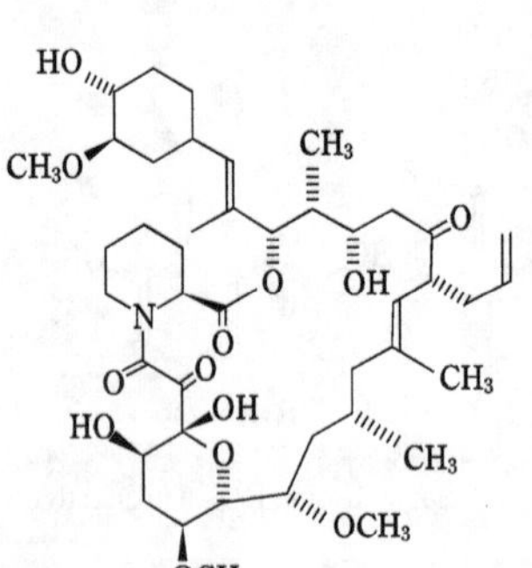

FK506 (Immunosuppressive agent)

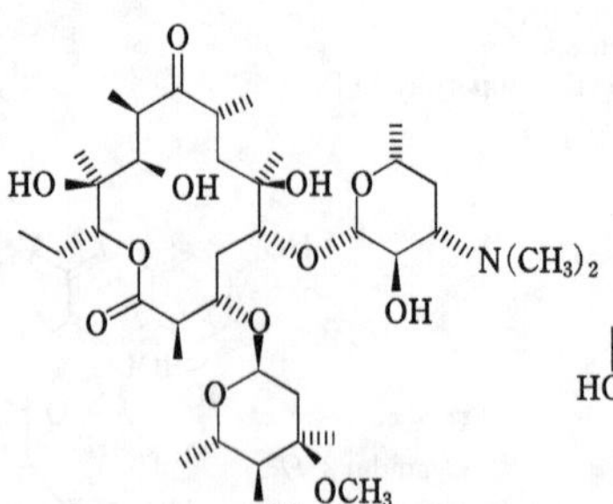

Erythromycin
(Inhibits protein synthesis in gram-positive bacteria)

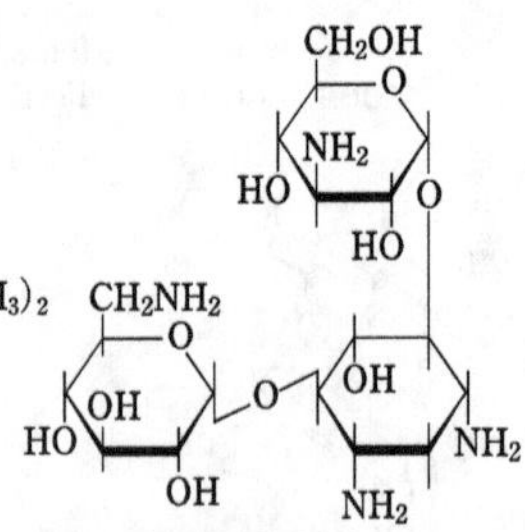

Kanamycin A
(Inhibits protein synthesis of Mycobacterium tuberculosis etc.)

* For cycloheximide, aphidicolin, and puromycin, refer to Rikanenpyo in the 1998 Edition.

Natural Toxins

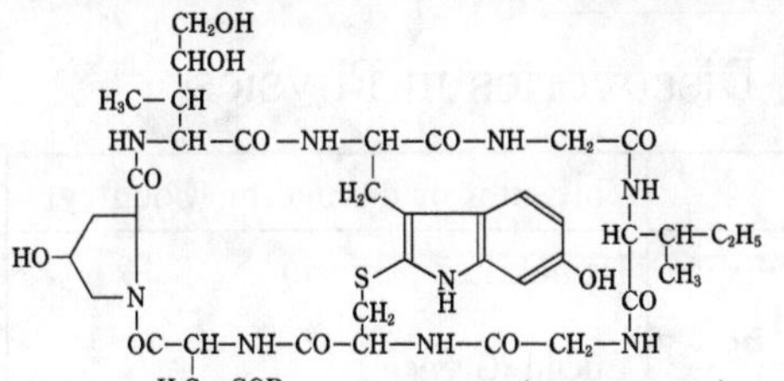

α-Amanitin (Amanita toxin)

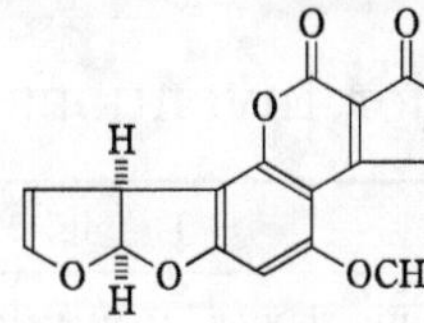

Aflatoxin B_1 (Mycotoxin)

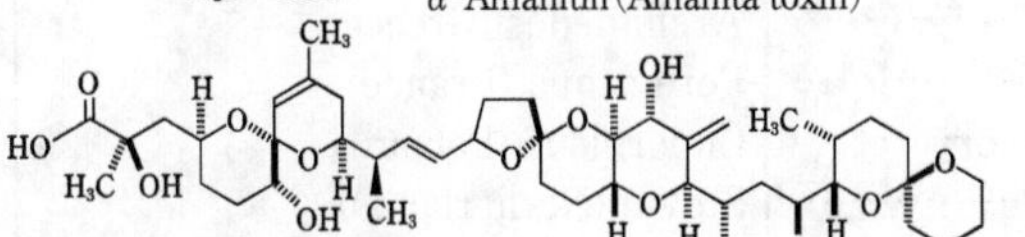

Okadaic acid (Diarrhetic shellfish toxin)

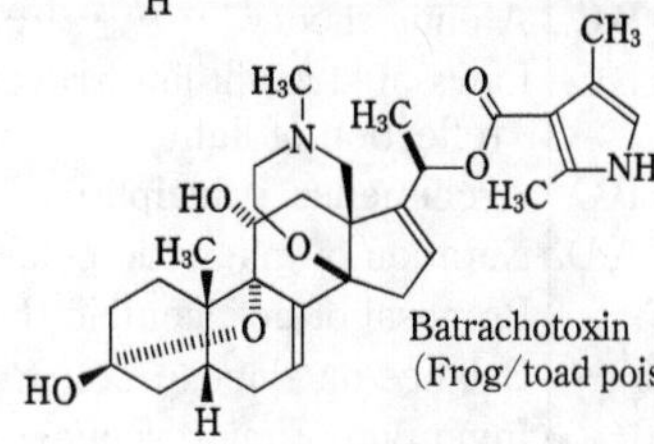

Batrachotoxin (Frog/toad poison)

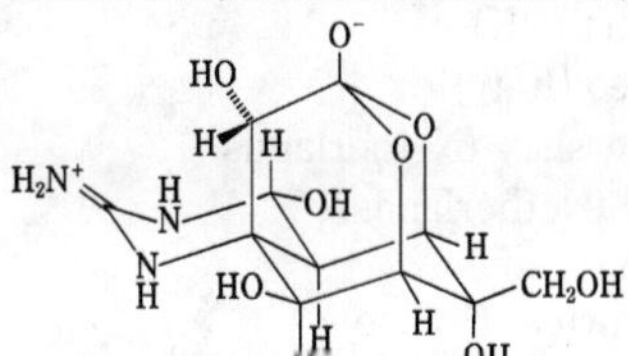

Tetrodotoxin (Pufferfish toxin)

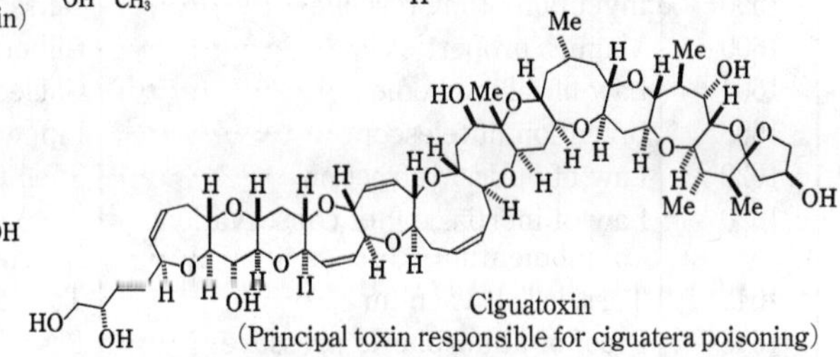

Ciguatoxin (Principal toxin responsible for ciguatera poisoning)

Other Alkaloids

Nicotine

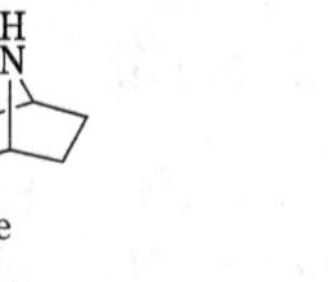

Caffeine

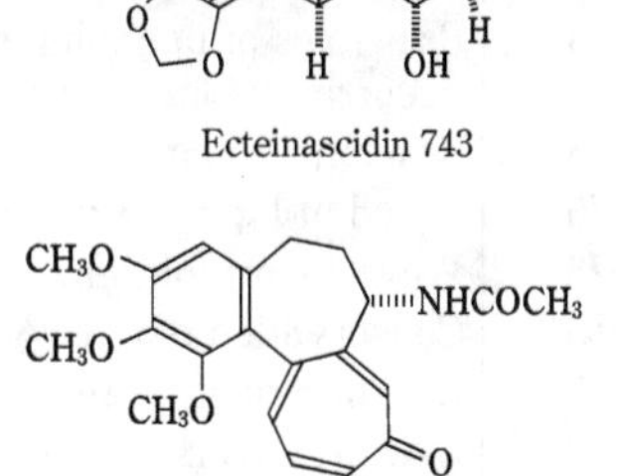

Ecteinascidin 743

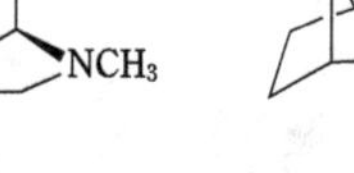

Epibatidine

Morphine

Cocaine

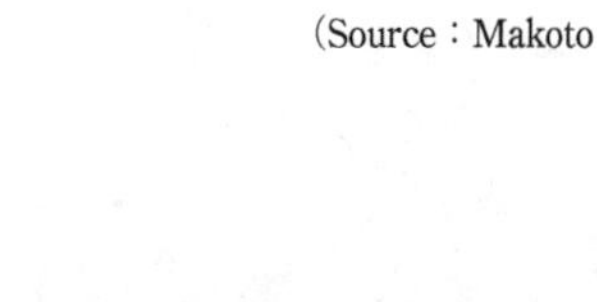

Colchicine

(Source : Makoto Sasaki)

Appendix

Major Inventions and Discoveries in Physics

Year	Details	Inventor or discoverer (Country)
c. 400 B.C.	Atomic theory	Demokritos (Greece)
c. 300 B.C.	Laws of straight-line travel and reflection of light	Euclid (Greece)
c. 250 B.C.	Archimedes' principle	Archimedes (Greece)
1269 A.D.	Notation of magnetic poles	Peregrinus (France)
1543	Proposal of heliocentric theory	Copernicus (Poland)
1586	Balance on slopes	Stevin (Netherlands)
1590	Invention of microscope	Janssen (Netherlands)
1600	Magnet properties	Gilbert (UK)
1604	Law of falling bodies	Galileo (Italy)
1608	Invention of telescope	Lippershey (Netherlands)
c. 1620	Law of optical refraction	Snell (Netherlands)
c. 1620	Law of inertia, law of conservation of momentum	Descartes (France)
1643	Torricellian vacuum	Torricelli (Italy)
1648	Proof of atmospheric pressure	Pascal (France)
c. 1650	Air pump	Guericke (Germany)
1657	Pendulum clock	Huygens (Netherlands)
c. 1660	Diffraction phenomenon of light	Grimaldi (Italy)
1660	Hooke's law	Hooke (UK)
1661	Fermat's principle	Fermat (France)
1662	Boyle's law	Boyle (UK)
1666	Dispersion of light	Newton (UK)
1668	Reflecting telescope	Newton (UK)
1669	Birefringence of calcite	Bartholin (Denmark)
1673	Principles of pendulums and centrifugal force	Huygens (Netherlands)
1675	Newton's rings	Newton (UK)
1676	Speed of light	Cassini (Italy), Romer (Denmark)
1678	Wave theory of light	Huygens (Netherlands)
1686	Conservation of energy	Leibnitz (Germany)
1687	Law of motion	Newton (UK)
1687	Law of universal gravitation	Newton (UK)
1705	Atmospheric engine	Newcomen (UK)

Note: The country name(s) in parentheses roughly indicates the area where the historical event took place or the historical figure lived.

Continued.

Year	Details	Inventor or discoverer (Country)
1720	Thermometer scale (Fahrenheit) ······	Fahrenheit (Germany)
1729	Foundation of actinometry ···········	Bouguer (France)
1729	Conduction of electricity ·············	Gray (UK)
1733	Two types of electricity ················	Du Fay (France)
1738	Bernoulli's principle······················	D. Bernoulli (Switzerland)
1742	Thermometer scale (Celsius)···········	Celsius (Sweden)
1743	d'Alembert's principle·····················	d'Alembert (France)
1744	Principle of least action ················	Maupertuis (France), Euler (Switzerland)
1745	Leyden jar·································	Kleist (Germany)
1746	Leyden jar·································	Musschenbroek (Netherlands)
1747	Hypothesis of electrical fluid ···········	Franklin (USA)
1752	Nature of thunder ·······················	Franklin (USA)
1753	Electrostatic induction ·················	Canton (UK)
1755	Equation of hydromechanics ···········	Euler (Switzerland)
1757	Achromatizing lens ·······················	Dollond (UK)
1760	Rigid-body motion equation ···········	Euler (Switzerland)
1761	Discovery of latent heat and heat capacity ···	Black (UK)
1780	Animal electricity ·······················	Galvani (Italy)
1785	Law of torsion balance, law of friction ···	Coulomb (France)
1785-89	Coulomb's law of magnetism and electricity···	Coulomb (France)
1787	Chladni's diagrams of diaphragms ···	Chladni (Germany)
1788	Analytical Mechanics (book) ···········	Lagrange (France)
1798	Generation of heat from friction ······	Rumford (USA)
1798	Measurement of gravitational constant···	Cavendish (UK)
1799	Voltaic cell·································	Volta (Italy)
1800	Infrared rays ·····························	W. Herschel (UK)
1800	Electrolysis of water ···················	Carlisle (UK), Nicholson (UK)
1801	Ultraviolet rays ·························	Ritter (Germany)
1801	Theory of light interference and three primary colors ·················	Young (UK)
1802	Law of thermal expansion for gas ···	Gay-Lussac (France)
1803	Dalton's law (gas pressure) ···········	Dalton (UK)
1808	Polarization due to reflection ········	Malus (France)
1815	Black line of solar spectrum ···········	Fraunhofer (Germany)
1815	Law of polarization angle ··············	Brewster (UK)
1816-19	Diffraction of light, wave theory of polarized light ························	Fresnel (France)
1816	Discernment for two types of specific heat in gas ······························	Laplace (France)

Major Inventions and Discoveries in Physics Continued.

Year	Details	Inventor or discoverer (Country)
1819	Dulong-Petit law	Dulong (France), Petit (France)
1820	Magnetic action of electric current	Oersted (Denmark)
1820	Ammeter	Schweiger (Germany)
1820	Electromagnets	Arago (France)
1820	Ampere's law	Ampere (France)
1820	Biot-Savart law	Biot (France), Savart (France)
1821	Fundamental equation for elastic body	Navier (France)
1821	Measurement of light wavelength by diffraction grating	Fraunhofer (Germany)
1822	Analytical theory of heat	Fourier (France)
1823	Liquefaction of chlorine	Faraday (UK)
1824	Carnot's principle	Carnot (France)
1826	Ohm's principle	Ohm (Germany)
1827	Brownian motion	Brown (UK)
1831	Electromagnetic induction	Faraday (UK)
1832	Self-induction	Henry (USA)
1833	Faraday's law (electrolysis)	Faraday (UK)
1833-41	Absolute measurement of terrestrial magnetism	Gauss (Germany), Weber (Germany)
1834	Hamilton's principle	Hamilton (Ireland)
1834	Peltier effect	Peltier (France)
1834	Lenz's law	Lenz (Russia)
1835	Coriolis' force	Coriolis (France)
1836	Manufacturing of stationary batteries	Daniel (UK)
1837	Theory of proximity effect for electricity	Faraday (UK)
1840	Law of thermal action for electric current	Joule (UK)
1842	Deviation from Boyle's law	Regnault (France)
1842	Doppler principle	Doppler (Austria)
1842	Conservation of energy, mechanical equivalent of heat	Mayer (Germany)
1843	Wheatstone bridge	Wheatstone (UK)
1843	Mechanical equivalent of heat	Joule (UK)
1845	Faraday effect, diamagnetism	Faraday (UK)
1845	Fluorescence	J. Herschel (UK)
1847	Law of energy conservation	Helmholtz (Germany)
1848	Thermodynamical temperature scale	Kelvin (W. Thomson) (UK)
1849	Experimental measurement of speed of light	Fizeau (France)

Continued.

Year	Details	Inventor or discoverer (Country)
1850	Second law of thermodynamics ······	Clausius (Germany)
1850	Proof of wave theory of light by measuring its speed ····················	Foucault (France)
1851	Second law of thermodynamics ······	Kelvin (UK)
1851	Proof of the earth's rotation ···········	Foucault (France)
1854	Joule-Thomson effect ·····················	Joule (UK), Kelvin (UK)
1855	Eddy current ·····························	Foucault (France)
1856	Five aberrations of lenses ··············	Seidel (Germany)
1858	Scintillation effect and magnetic deflection of cathode rays ···········	Plucker (Germany)
1859	Basics of spectrum analysis ···········	Kirchhoff (Germany), Bunsen (Germany)
1859	Storage battery ·························	Plante (France)
1859	Foundation of kinetic theory of gases···	Maxwell (UK)
1860	Kirchhoff's principle (radiation)········	Kirchhoff (Germany)
1861	Electromagnetic equations··············	Maxwell (UK)
1861	Electromagnetic wave theory of light···	Maxwell (UK)
1862	Theory of Hearing (book) ··············	Helmholtz (Germany)
1863–69	Liquefaction and critical temperature of carbon gas ····························	Andrews (UK)
1865	Principle of entropy increase ········	Clausius (Germany)
1865	Determination of Loschmidt's number ···	Loschmidt (Austria)
1866	Kundt tube experiment ·················	Kundt (Germany)
1869	Straightness of cathode rays···········	Hittorf (Germany)
1873	Equation of state for iquids and gases···	Van der Waals (Netherlands)
1875	Kerr effect·······························	Kerr (UK)
1877	Liquefaction of oxygen ·················	Cailletet (France), Pictet (Switzerland)
1877	Statistical basis for the second law of thermodynamics ······················	Boltzmann (Austria)
1879	Law of similarity for fluid motion······	Reynolds (UK)
1879	Stefan's law (radiation) ·················	Stefan (Austria)
1880	Law of piezoelectric ····················	P. Curie (France)
1881	Hysteresis of magnetism ··············	Warburg (Germany)
1882	Concept of free energy ················	Helmholtz (Germany)
1884	Edison effect (thermionic) ··············	Edison (USA)
1884	Theory for Stefan's law ·················	Boltzmann (Austria)
1885	Formula for hydrogen spectral series···	Balmer (Switzerland)
1886	Anode ray ································	Goldstein (Germany)
1887	Theory of electrolyte solution ········	Arrhenius (Sweden)
1887	Photoelectric effect ······················	Hertz (Germany)

Major Inventions and Discoveries in Physics Continued.

Year	Details	Inventor or discoverer (Country)
1887	Michelson-Morley experiment	Michelson (USA), Morley (USA)
1887	Theory of microscope	Abbe (Germany)
1887	Photography of shock waves	Mach (Austria)
1888	Experimental proof of electromagnetic waves	Hertz (Germany)
1890	Spectral formula	Rydberg (Sweden)
1890	Demonstration of radiation pressure	Lebedev (Russia)
1890	Equivalence of gravitational mass and inertia mass	Eötvös (Hungary)
1892	Contraction hypothesis	Lorentz (Netherlands)
1893	Wien's displacement law	Wien (Germany)
1895	Electromagnetic theory of moving bodies	Lorentz (Netherlands)
1895	X-rays	Röntgen (Germany)
1895	Proof that cathode rays carry a negative charge	Perrin (France)
1895	Curie's law (magnetism)	P. Curie (France)
1896	Radioactivity of uranium	Becquerel (France)
1896	Zeeman effect	Zeeman (Netherlands), Lorentz (Netherlands)
1897	Existence of electrons confirmed	J. J. Thomson (UK)
1897	Wireless telegraphy	Marconi (Italy)
1898	Radium	P. and M. S. Curie (France, Poland)
1900	Radiation theory and quantum of action	Planck (Germany)
1900	Electron theory of metals	Drude (Germany)
1901	Thermal electron	Richardson (UK)
1902	Establishment of statistical mechanics	Gibbs (USA)
1902	Theory for lift of wings	Kutta (Germany)
1903	Decay theory for radioactive elements	Rutherford (New Zealand), Soddy (UK)
1903	Velocity dependence of electron mass	Kaufmann (Germany)
1903	Saturnian model of the atom	H. Nagaoka (Japan)
1905	Special theory of relativity	Einstein (Switzerland, Germany)
1905	Photon hypothesis	Einstein (Switzerland, Germany)
1905	Theory of paramagnetism	Langevin (France)
1906	Third law of thermodynamics	Nernst (Germany)
1906	Theory for specific heat of solids	Einstein (Switzerland, Germany)
1907	Theory for lift of wings	Joukowski (Russia)
1907	Anode ray analysis	J. J. Thomson (UK)
1907	Concept of molecular fields and magnetic domains	Weiss (France)

Continued.

Year	Details	Inventor or discoverer (Country)
1907	Concept for four dimensions of space and time	Minkowski (Russia)
1908	Liquefaction of helium	Kamerlingh Onnes (Netherlands)
1908	Development of counter tube	Rutherford (New Zealand), Geiger (Germany)
1908	Proof for the reality of molecules	Perrin (France)
1908	Proof that α particle = helium nucleus	Rutherford (New Zealand), Lloyds (UK)
1909	Measurement of electron charge using the oil droplet method	Millikan (USA)
1910	Concept of isotopes	Soddy (UK)
1911	Invention of cloud chamber	C. T. R. Wilson (UK)
1911	Superconductivity	Kamerlingh Onnes (Netherlands)
1911	Discovery of the atomic nucleus	Rutherford (New Zealand)
1911-12	Discovery of cosmic rays	Hess (Austria)
1912	Diffraction of X-rays by crystals	Laue (Germany)
1912	Bragg's law	W. L. Bragg (UK)
1913	Quantum theory of atomic structure	N. Bohr (Denmark)
1913	Stark effect	Stark (Germany)
1913	Relation between atomic number and characteristic X-ray	Moseley (UK)
1913	Displacement law for radioactive decay	Soddy (UK), Fajans (Poland)
1914	Energy level of atoms	Frank (Germany), Hertz (Germany)
1915	General theory of relativity	Einstein (Germany)
1915	Fine structure theory for hydrogen spectrum	Sommerfeld (Germany)
1916	Crystal structure analysis by X-ray powder diffraction	Debye (Netherlands), Scherrer (Switzerland)
1916	Theory of X-ray spectrum and atomic value	Kossel (Netherlands)
1918	Correspondence principle	N. Bohr (Denmark)
1919	Isotope research using mass analyzers	Aston (UK)
1919	Atomic nucleus decay caused by α particles	Rutherford (New Zealand)
1919	Barkhausen effect	Barkhausen (Germany)
1921	Magnetic moment of atoms	Stern (Germany), Gerlach (Germany)
1922	Compton effect	Compton (USA)
1923	Concept of matter waves	de Broglie (France)
1924	Bose-Einstein statistics	Bose (India), Einstein (Germany)
1925	Electron spin	Goudsmit (Netherlands), Uhlenbeck (Netherlands)

Major Inventions and Discoveries in Physics Continued.

Year	Details	Inventor or discoverer (Country)
1925	Exclusion principle	Pauli (Austria)
1925	Matrix mechanics	Heisenberg (Germany)
1925	Conservation law in elementary processes	Compton (USA), Simon (USA), Geiger (Germany), Bothe (USA)
1926	Wave mechanics	Schrödinger (Austria)
1926	Fermi statistics	Fermi (Italy)
1927	Principle of uncertainty	Heisenberg (Germany)
1927	Diffraction of electron beams	Davisson (USA), Germer (USA), G. P. Thomson (UK)
1927	Quantum theory of hydrogen covalent bonds	Heitler (Germany), London (USA)
1927	Helium II	Keesom (Netherlands)
1927	Complementary principle	N. Bohr (Denmark)
1928	Theory of ferromagnetism	Heisenberg (Germany)
1928	Quantum theory of metal electrons	Bloch (Switzerland)
1928	Equation of relativistic electrons	Dirac (UK)
1928	Theory of α decay, tunnel effect	Gamow (Russia)
1928	Raman effect	Raman (India)
1929	Relativistic guantum field theory	Heisenberg (Germany), Pauli (Austria)
1929	Cosmic ray shower	Skobelzyn (Russia)
1929	Unified field theory	Einstein (Germany)
1930	Cyclotron	Lawrence (USA), Livingston (USA)
1931	Theory of semiconductors	A. H. Wilson (UK)
1931	Neutrino hypothesis	Pauli (Austria)
1931	Electron microscope	Knorr (Germany), Ruska (Germany)
1931	Reciprocity theorem in irreversible processes	Onsager (USA)
1931	Cosmic radio waves	Jansky (USA)
1932	Neutrons	Chadwick (UK)
1932	Positive electrons	C. Anderson (USA)
1932	Separation of deuterium (heavy hydrogen)	Urey (USA)
1932	Artificial atomic nucleus transmutation by a high-voltage accelerator	Cockcroft (UK), Walton (Ireland)
1933	Theory of β decay	Fermi (Italy)
1933	Perfect diamagnetism in superconductors	Meissner (Germany), Ochsenfeld (Germany)
1934	Artificial radioactivity	J. F. and I. Joliot-Curie (France)
1934	Phase-contrast microscope	Zernike (Netherlands)

Continued.

Year	Details	Inventor or discoverer (Country)
1934	Cherenkov effect	Cherenkov (Russia)
1934	Meson theory	H. Yukawa (Japan)
1937	Cosmic ray μ particle	C. Anderson (USA), Neddermeyer (USA)
1938	Anti-ferromagnetism	Bisset (France) *et al.*
1938	Nuclear fission of uranium	Hahn, Strassmann (Germany)
1939	Explanation of heat source of stars by nuclear reaction	Bethe (Germany)
1939	Magnetic resonance method	Rabi (USA)
1940	Creation of transuranium element (Np)	McMillan (USA), Abelson (USA)
1941	Production of betatrons	Kerst (USA)
1941	Quantum fluid mechanics	Landau (Russia)
1942	Ferroelectrics	Hippel (USA)
1942	Sustaining of fission chain reaction	Fermi (Italy) *et al.*
1944	Second sound in helium II	Peshkov (Russia)
1945	Electron spin resonance method	Zaboisky (Russia)
1945	Synchrotron	McMillan (USA), Veksler (Russia)
1946	Nuclear magnetic resonance	Bloch (Switzerland), Purcell (USA)
1946	Big Bang Theory	Gamow (USA)
1946	^{14}Carbon dating method	Libby (USA)
1947	Existence of π mesons	Powell (UK)
1947	Discovery of V particles	Rochester (UK), Butler (UK)
1947	Lamb shift	Lamb (USA)
1948	Theory of holography	Gabor (Hungary)
1948	Renormalization theory	Tomonaga (Japan), Schwinger (USA)
1948	Artificial production of mesons	Gardner (USA), Lattes (Brazil)
1948	Structural theory for nuclei (magic number)	M. G. Mayer (Germany)
1948	Transistor	Brattain (USA), Bardeen (USA), Shockley (USA)
1948	Ferrimagnetism	Néel (France)
1948	Neutron scattering in solids	Shull (USA)
1949	Direct observation of magnetic domains	Bozorth (USA)
1949	Separated oscillatory field method	Ramsey (USA)
1930–50	Theory of magnetism	Van Vleck (USA)
1949–59	Discovery of new particles (Λ, Σ, Ξ, K)	Powell (UK), C. Anderson (USA) *et al.*
1951–53	Collective motion of atomic nuclei	A. Bohr (Denmark), Rainwater (USA), Mottelson (USA)

Major Inventions and Discoveries in Physics Continued.

Year	Details	Inventor or discoverer (Country)
1953	Bubble chamber	Glaser (USA)
1954	Discovery of masers	Townes (USA) *et al.*
1954	Neutron inelastic scattering	Brockhouse (Canada)
1955	Development of anti-protons	Segrè (Italy)
1956	Confirmation of anti-neutrons	Piccioni (Italy) *et al.*
1957	Parity non-conservation	Lee (China), Yang (China), Wu (China)
1957	Theory of superconductivity	Bardeen (USA), Cooper (USA), Schrieffer (USA)
1957	Möβbauer effect	Möβbauer (Germany)
1957	Absence of dispersion in disordered systems	P. Anderson (UK)
1957	Theory of irreversible processes	R. Kubo (Japan)
1958	Tunnel effect in semiconductors	L. Esaki (Japan)
1959	Verification of electron anti-neutrinos	Reines (USA)
1960	Development of lasers	Maiman (USA)
1960	Excited state of elementary particles	Research group at University of California (USA)
1960	Tunnel effect in superconductors	Giaever (Norway)
1960	Neutrino beam	Lederman (USA), Schwartz (USA), Steinberger (USA)
1960	Spontaneous symmetry breaking	Y. Nambu (Japan, USA)
1961	Production of superconductor magnets	Kunzler (USA)
1962	Two types of of neutrino	Group at Brookhaven National Laboratory (USA)
1962	Tunnel junction for superconductors	Josephson (UK)
1962	X-ray stars	Rossi (USA) *et al.*
1963	Quantum theory of optical coherence	Glauber (USA)
1964	Quark theory	Gell-Mann (USA) *et al.*
1964	Theory of electrical resistance for dilute alloys	J. Kondo (Japan)
1964	Clarification of the origin of mass, prediction of existence of Higgs particle	Englert (Belgium), Higgs (UK)
1965	3K cosmic background radiation	Penzias (USA), R. W. Wilson (USA)
1966	Communication using optical fibers	Kao (China), Hockham (UK)
1967	Pulsar	Bell, Hewish (UK)
1967	Unified theory	Weinberg (USA), Salam (Pakistan)
1968	Multi-wire proportional counter	Charpak (France)
1968	Deep inelastic scattering of electrons (verification of quark)	Friedman (USA), Kendall (USA), Taylor (Canada)
1969–73	Amorphous electron theory	Mott (UK)
1970	Invention of CCD sensor	Boyle (Canada), Smith (USA)

Continued.

Year	Details	Inventor or discoverer (Country)
1971	Clarification of quantum structure for weak electro-interaction ········	't Hooft (Netherlands), Veltman (Netherlands)
1972	^{3}Superfluidity of He ····················	Lee (USA), Richardson (USA), Osheroff (USA)
1973	Prediction of 3rd-generation quark and lepton ······························	M. Kobayashi (Japan), T. Maskawa (Japan), Cabibbo (Italy)
1974	Discovery of particles (J/ψ) ···········	Richter, Ting (USA)
1974	Discovery of new type of pulsar ······	Hulse, Taylor (USA)
1975	Discovery of τ particle ················	Perl (USA)
1980	Discovery of quantum hall effect ······	Klitzing (Germany) *et al.*
1981	Scanning tunneling microscope ······	Binnig (Germany), Rohrer (Switzerland)
1982	Discovery of quantum fluids which possess the excited state of fractional charge ···	Störmer (Germany), Tsui (China)
1983	Discovery of W and Z particles ······	Rubbia (Switzerland) *et al.*
1985	Use of lasers to cool and capture atoms···	Chu (USA), Philipps (USA)
1985	Chirped pulse amplification ···········	Mourou (France), Strickland (Canada)
1986	Discovery of oxide superconductors ···	Bednorz (Germany), Müller (Switzerland)
1987	Development of optical tweezer ······	Ashkin (USA)
1989	Discovery of semiconductors which emit blue light when excited by electric current (blue light-emitting diode) ······	I. Akasaki (Japan), H. Amano (Japan)
1993	Development of high-luminance blue light-emitting diode ····················	S. Nakamura (Japan)
1995	Achievement of Bose-Einstein condensate ···	Cornell (USA), Wieman (USA)
1998	Discovery of accelerating expansion of the universe ························	Perlmutter (USA), Schmidt (USA), Riess (USA)
1998-2001	Discovery of neutrion oscillation ······	T. Kajita (Japan), McDonald (Canada)
1999-2000	Optical comb technology ··············	Hall (USA), Hensch (Germany)
2012	Discovery of Higgs particle ···········	European Organization for Nuclear Research (CERN)
2015	Observation of gravitational wave ···	Weiss (USA), Barish (USA), Thorne (USA)

Major Inventions and Discoveries in Chemistry

Year	Details	Inventor or discoverer (Country)
1250	Arsenic ································	Magnus (Germany)
1661	Criticism of the concept of element (the theory of the tria prima, the theory of four elements) ······················	Boyle (UK)
1662	Boyle's law ······························	Boyle (UK)
1664	Qualitative analysis ······················	Boyle (UK)
1669	Phosphorus ······························	Brand (Germany)
1670	Phlogiston theory························	Becher (Germany), Stahl (Germany)
1735	Cobalt····································	Brandt (Sweden)
1746	Zinc······································	Marggraf (Germany)
1748	Platinum ································	Ulloa (Spain)
1751	Nickel····································	Cronstedt (Sweden)
1754	Discovery of carbon dioxide ··········	Black (UK)
1764	Development of asbestos cloth ········	Hiraga Gennai (Japan)
1766	Hydrogen ································	Cavendish (UK)
1769	Tartaric acid·····························	Scheele (Sweden)
1772	Nitrogen··································	Rutherford (UK)
1772	Oxygen ···································	Scheele (Sweden)
1774	Law of conservation of mass··········	Lavoisier (France)
1774	Chlorine ·································	Scheele (Sweden)
1774	Manganese ······························	Gahn (Sweden)
1774	Oxygen, sulfurous acid gas, ammonia gas···	Priestley (UK)
1776	Carbon monoxide························	Lassone (France)
1777	Explanation of combustion·············	Lavoisier (France)
1777	Hydrogen sulfide ························	Scheele (Sweden)
1781	Molybdenum·····························	Hjelm (Sweden)
1783	Tellurium ································	Müller von Reichenstein (Hungary)
1783	Tungsten ·································	Juan & Fausto Elhuyar (Spain)
1789	Element table ···························	Lavoisier (France)
1789	Zirconium, uranium······················	Klaproth (Germany)
1791	Titanium ·································	Gregor (UK)
1791	Soda production method ··············	Leblanc (France)
1794	Yttrium (first rare earth elements) ···	Gadolin (Finland)
1797	Chrome ·································	Vauquelin (France)
1798	Beryllium ·······························	Vauquelin (France)
1799	Voltaic pile ·····························	Volta (Italiana)
1799	Bleaching powder ······················	Tennant (UK)
1799	Law of definite proportions, glucose	Proust (France)
1801	Niobium·································	Hatchett (UK)

Continued.

Year	Details	Inventor or discoverer (Country)
1802	Tantalum	Ekeberg (Sweden)
1802	Gay-Lussac's law	Gay-Lussac (France)
1803	Law of partial pressure, law of multiple proportions, atom hypothesis	Dalton (UK)
1803	Henry's law	Henry (UK)
1803	Cerium	Berzelius (Sweden), Hisinger (Sweden), Klaproth (Germany)
1803	Osmium, iridium	Tennant (UK)
1803	Palladium, rhodium	Wollaston (UK)
1805	Morphine (first alkaloid)	Serturner (Germany)
1807	Potassium, sodium	Davy (UK)
1808	Law of combining gas volumes	Gay-Lussac (France)
1808	Magnesium, calcium, strontium, barium	Davy (UK)
1808	Boron	Davy (UK), Gay-Lussac (France), Thénard (France)
1811	Molecular hypothesis	Avogadro (Italy)
1812	Iodine	Courtois (France)
1815	Safety lamp	Davy (UK)
1817	Lithium	Arfvedson (Sweden)
1817	Selenium	Berzelius (Sweden)
1817	Cadmium	Strohmeyer (Germany)
1818	Hydrogen peroxide	Thenard (France)
1819	Isomorphic law	Mitscherlich (Germany)
1819	Law of atomic heat	Dulong (France), Petit (France)
1820	Quinine	Peltier (France), Caventou (France)
1824	Silicon	Berzelius (Sweden)
1824	Portland cement	Aspdin (UK)
1825	Benzene	Faraday (UK)
1826	Bromine	Balard (France)
1826	Aniline (from indigo)	Unverdorben (Germany)
1827	Aluminium	Wöhler (Germany)
1827	Synthesis of organic metal compound (platinum-ethylene complex)	Zeise (Germany)
1828	Synthesis of urea	Wöhler (Germany)
1828	Thorium	Berzelius (Sweden)
1830	Vanadium	Sefström (Sweden)
1831	Law of molecular heat	Neumann (Germany)
1831	Chloroform	Liebig (Germany)
1832	Law of gas dispersion	Graham (UK)
1833	Law of electrolysis	Faraday (UK)

Major Inventions and Discoveries in Chemistry Continued.

Year	Details	Inventor or discoverer (Country)
1834	Aniline and carbolic acid (from coal tartar)···	Runge (Germany)
1836	Electroplating method ·················	Jacobi (Germany)
1836	Acetylene ································	Davy (UK)
1839	Lanthanum ······························	Mosander (Sweden)
1839	Discovery of latent images, invention of photograph technology ···········	Daguerre (France), Talbot (UK)
1839	Vulcanization method for rubber······	Goodyear (USA)
1840	Hess' law ································	Hess (Russia)
1840	Ozone·····································	Schönbein (Germany)
1842	Classification of organic compounds ···	Gerhardt (France)
1843	Terbium, erbium ························	Mosander (Sweden)
1844	Ruthenium································	Claus (Russia)
1845	Red phosphorus ························	Schlatter (Austria)
1845	Gun cotton································	Schönbein (Germany)
1850	Theory of ether generation ···········	Williamson (UK)
1855	Bessemer process ······················	Bessemer (UK)
1856	Mauveine (first aniline dye) ···········	Perkin (UK)
1858	Carbon four valence theory ···········	Kekule (Germany)
1859	Atomic spectrum analysis method ···	Bunsen (Germany), Kirchhoff (Germany)
1859	Lead storage battery ···················	Plante (France)
1859	Synthesis of acetylene ················	Berthelot (France)
1860	Caesium···································	Bunsen (Germany), Kirchhoff (Germany)
1861	Rubidium ································	Bunsen (Germany), Kirchhoff (Germany)
1861	Thallium ·································	Crookes (UK), Lamy (France)
1861	Colloid ···································	Graham (UK)
1862	Ammonia soda method ·················	Solvay (Belgium)
1863	Indium ···································	Reich (Germany), Richter (Germany)
1864	Law of mass action ·····················	Guldberg (Norway), Waage (Norway)
1865	Structural formula of benzene ········	Kekule (Germany)
1866	Dynamite ································	Nobel (Sweden)
1867	Formalin ·································	Hoffman (Germany)
1868	Helium ···································	Lockyer (UK), Frankland (UK)
1869	Periodic Table of the Elements ······	Mendeleev (Russia)
1869	Celluloid··································	Hyatt (USA)
1874	Asymmetric carbon atom theory······	Van't Hoff (Netherlands), Le Bel (France)
1875	Gallium ··································	Boisbaudran (France)
1876	Phase rule ·······························	Gibbs (USA)
1877	Liquefaction of oxygen, nitrogen, methane, and carbon monoxide······	Cailletet (France), Pictet (Switzerland)
1878	Holmium, thulium ······················	Cleve (Sweden)

Continued.

Year	Details	Inventor or discoverer (Country)
1878	Ytterbium	Marignac (Switzerland)
1879	Scandium	Nilson (Sweden)
1879	Samarium	Boisbaudran (France)
1880	Gadolinium	Marignac (Switzerland)
1880	Synthesis of indigo	Baeyer (Germany)
1883	Kjeldahl nitrogen analysis method	Kjeldahl (Denmark)
1884	Gas mantle	Welsbach (Austria)
1884	Synthesis of saccharides	Fischer (Germany)
1884	Le Chatelier's principle	Le Chatelier (France), Braun (Germany)
1885	Praseodymium, neodymium	Welsbach (Austria)
1885	Ephedrine	N. Nagai (Japan)
1886	Fluorine	Moissan (France)
1886	Germanium	Winkler (Germany)
1886	Dysprosium	Boisbaudran (France)
1886	Law of depression of vapor pressure	Raoult (France)
1886-87	Dry element battery	S. Yai (Japan), Gassner (Germany), Hellcene (Denmark)
1887	Theory of dilute solution	van't Hoff (Netherlands)
1887	Ionization theory for electrolyte solution	Arrhenius (Sweden)
1888	Ostwald's dilution law	Ostwald (Germany)
1889	Electrolytic solutional tension (Nernst's formula for single electrode potential)	Nernst (Germany)
1891	Carborundum	Acheson (USA)
1892	Indicator theory	Ostwald (Germany)
1893	Coordination theory	Werner (Switzerland)
1894	Argon	Rayleigh (UK), Ramsay (UK)
1894	Air liquefier	Linde (Germany)
1895	Helium	Ramsay (UK)
1896	Walden inversion	Walden (Germany)
1898	Radium, polonium	Pierre and Marie Curie (France, Poland)
1898	Krypton, neon, xenon	Ramsay (UK), Travers (UK)
1898	Liquefaction of hydrogen	Dewar (UK)
1899	Actinium	Debierne (France)
1900	Radon	Dorn (Germany)
1900	Discovery of triphenylmethyl radical	Gomberg (USA)
1901	Europium	Demarçay (France)
1901	Solubility product	Nernst (Germany)
1901	Grignard reagent	Grignard (France)
1901	Crystalline adrenaline	J. Takamine (Japan)
1901	Synthesis of camphor	Komppa (Finland)

Major Inventions and Discoveries in Chemistry Continued.

Year	Details	Inventor or discoverer (Country)
1905	Thermite method	Goldschmidt (Norway)
1905	Method for fixation of airborne nitrogen	Birkeland (Norway), Eyde (Germany)
1905	Hydrogenation method using nickel catalyst	Sabatier (France)
1906	pH glass electrode	Kramer (UK), Haber (Germany)
1906	Third law of thermodynamics	Nernst (Germany)
1906	Method for industrial synthesis of ammonia	Haber (Germany), Bosch (Germany)
1907	Lutetium	Urbain (France)
1907	Lacquer research	T. Majima (Japan)
1908	Liquefaction of helium	Onnes (Netherlands)
1908	Flavoring monosodium glutamate	K. Ikeda (Japan)
1909	Bakelite	Baekeland (USA)
1909	Salvarsan	Ehrlich (Germany), S. Hata (Japan)
1910	Oryzanin (Vitamin B)	U. Suzuki (Japan)
1912	Law of photochemical equivalence	Einstein (Germany)
1912	Dipole moment of molecules	Debye (Netherlands)
1912	Vitamin C	Holst (Norway), Frölich (Norway)
1912	Borane (borohydride)	Stock (Germany)
1913	Tracer method	Hevesy (Hungary)
1913	Displacement law for radioactive elements	Soddy (UK), Fajans (Poland)
1913	Vitamin A	McCollum (USA), Davis (USA)
1913	Liquefaction of coal	Bergius (Germany)
1915	The isolation of thyroxine (A hormone of the thyroid grand)	Kendall (USA)
1916	Organic trace analysis method	Pregl (Austria)
1916	Octet theory for chemical bonding	Lewis (USA)
1916	Powder X-ray diffractometry	Debye (Netherlands), Scherrer (Switzerland)
1916	Adsorption isotherm	Langmuir (USA)
1918	Protactinium	Hahn (Germany), Meitner (Austria)
1919	Mass spectrometry	Aston (UK)
1920	Proposal for the existence of polymers (conceptual)	Staudinger (Germany)
1922	Petroleum cracking	Egloff (USA)
1922	Discovery of insulin	Banting (Canada), Macleod (UK)
1923	Strong electrolyte theory	Debye (Netherlands), Huckel (Germany)
1923	Hafnium	Coster (Netherlands), Hevesy (Hungary)
1924	Ultra-centrifuge	Svedberg (Sweden), Rinde (Sweden)

Continued.

Year	Details	Inventor or discoverer (Country)
1925	Rhenium	Noddack (Germany), Tacke (Germany), Berg (Germany)
1925	Polarography	Heyrovský (Czechoslovakia), M. Shikata (Japan)
1927	Vitamin E	Evans (UK)
1927	Quantum theory of covalent bonds	Heitler (Germany), London (USA)
1928	Raman effect	Raman (India)
1928	Diene synthesis	Diels (Germany), Alder (Germany)
1928	Resonance theory	Pauling (USA)
1928	Penicillin	Fleming (UK)
1929	Synthesis of hemin	Fischer (Germany)
1929	Isotopes of oxygen	Giauque (France), Johnston (UK)
1929	Structural formula for follicular hormone (female hormone)	Butenandt (Germany), Doisy (USA)
1931	Vitamin D	Windaus (Germany) *et al.*
1931	Synthetic rubber	Carothers (USA), Williams (USA)
1932	Molecular orbital method	Hund (Germany), Mulliken (USA)
1932	Heavy hydrogen and heavy water	Urey (USA)
1933	Structural formula for Vitamin A	Karrer (Switzerland)
1934	Synthesis of androsterone (first male hormone)	Butenandt (Germany)
1934	Structural formula for progesterone (female hormone)	Butenandt (Germany), Slotta (Germany)
1934	Artificial radioactivity	Jean Frédéric and Irene Joliot-Curie (France)
1934	Covalent bond radius	Pauling (USA)
1934	Rotational isomer	S. Mizushima (Japan)
1935	Structural formula for Vitamin D_2	Windaus (Germany)
1935	Structural formula for Vitamin B_2	Karrer (Switzerland), Kuhn (Germany)
1935	Synthetic fiber nylon	Carothers (USA)
1936	Neutron activation analysis	Hevesy (Hungary)
1936	Structural determination and synthesis for Vitamin B_1	Williams (USA), Windaus (Germany)
1937	Technetium	Segre (Italy), Perrier (Italy)
1937	Separation of serum protein components by using electrophoresis	Tiselius (Sweden)
1937	Synthesis of Vitamin A	Kuhn (Germany)
1938	Pesticide DDT	Müller (Switzerland)
1938	Nuclear fission of uranium	Hahn, Strassmann (Germany)
1939	Structural determination and synthesis for Vitamin K	Karrer (Switzerland) *et al.*
1939	Francium	Perey (France)

Major Inventions and Discoveries in Chemistry Continued.

Year	Details	Inventor or discoverer (Country)
1940	Astatine	Segre (Italy), Corson (USA)
1940	Neptunium	McMillan (USA) *et al.*
1940	Plutonium	Seaborg (USA) *et al.*
1944	Partition chromatography	Martin (UK), Synge (UK)
1944	Americium	Seaborg (USA) *et al.*
1944	Curium	Seaborg (USA) *et al.*
1944	Streptomycin	Waksman (USA)
1945	Promethium	Marinsky (USA), Glendinin (USA), Coryell (USA)
1946	Age dating using ^{14}C	Libby (USA)
1947	Flash photolysis method	Norrish (UK), Porter (UK)
1947	Tropolone seven-membered ring compound	T. Nozoe (Japan)
1947	Chloromycetin	Burkholder (USA)
1949	Synthesis of adenosine triphosphate (ATP)	Todd (UK)
1949	Reppe reaction	Reppe (Germany)
1949	Berkelium	Seaborg (USA) *et al.*
1950	Californium	Seaborg (USA) *et al.*
1951	Complete synthesis of cholesterol	Woodward (USA), Robinson (UK)
1951	Synthesis of ferrocene	Pauson (UK), Miller (USA)
1952	Frontier electron theory	K. Fukui (Japan)
1952	Einsteinium	Ghiorso (USA) *et al.*
1952	Zone melting	Pfann (USA)
1953	Structural determination for DNA	Watson (USA), Crick (UK)
1953	Fermium	Ghiorso (USA) *et al.*
1953	Low-pressure olefin polymerization method	Ziegler (Germany), Natta (Italy)
1954	Discovery of organic semiconductors	H. Akamatsu (Japan), H. Inokuchi (Japan), Y. Matsunaga (Japan)
1954	Wittig reaction	Wittig (Germany)
1955	Chemical relaxation method	Eigen (Germany)
1955	Structural determination for Vitamin B_{12}	Hodgkin (UK)
1955	Determination of the amino acid sequence for insulin	Sanger (UK)
1955	Artificial diamond	Hall (USA) *et al.*
1955	Mendelevium	Ghiorso (USA) *et al.*
1956	Hydroboration reaction	Brown (USA)
1956	X-ray photoelectron spectroscopy	Siegbahn (Sweden)
1958	Nobelium	Seaborg (USA) *et al.*

Continued.

Year	Details	Inventor or discoverer (Country)
1958	Invention of lasers	Schawlow (USA), Townes (USA)
1958	Enzymatic synthesis of DNA	Kornberg (USA) *et al.*
1958	Moessbauer spectroscopy	Moessbauer (Germany)
1959	Discovery of sex pheromones	Butenandt (Germany)
1959	Radioimmunoassay	Yalow (USA), Berson (USA)
1961	Lawrencium	Ghiorso (USA) *et al.*
1962	Synthesis of xenon compound	Bartlett (UK)
1962	Discovery of green fluorescence protein (GFP)	O. Shimomura (Japan)
1964	Chemical laser	Pimentel (USA)
1965	Woodward-Hoffmann rules	Woodward (USA), Hoffmann (USA)
1967	Synthesis of crown ether	Pedersen (USA)
1969	Rutherfordium	Flerov (Russia) *et al.*, Ghiorso (USA) *et al.*
1969	Identification of pentavalent carbon	Olah (USA)
1970	Dubnium	Flerov (Russia) *et al.*, Ghiorso (USA) *et al.*
1972	Heck reaction (phenylation of olefin caused by Pd)	Heck (USA)
1973	Total synthesis of vitamin B_{12}	Woodward (USA), Eschenmoser (Switzerland)
1974	Seaborgium	Ghiorso (USA) *et al.*
1977	Method for determining the sequencing of nucleic acid	Maxam, Gilbert (USA)
1977	Discovery of conductive polymers	H. Shirakawa (Japan), Heeger (USA), MacDiarmid (USA)
1977	Negishi cross-coupling	E. Negishi (Japan)
1979	Suzuki cross-coupling	A. Suzuki (Japan), N. Miyaura (Japan)
1981	Bohrium	Armbruster (Germany), Münzenberg (Germany) *et al.*
1982	Meitnerium	Armbruster (Germany), Münzenberg (Germany) *et al.*
1982	Quasicrystals	Shechtman (Israel)
1983	Invention of scanning tunneling microscope	Binnig (Switzerland), Rohrer (Switzerland)
1984	Hassium	Armbruster (Germany), Münzenberg (Germany) *et al.*
1984	Electrospray ionization mass spectrometry	M. Yamashita (Japan), Fenn (USA)
1985	PCR (polymerase chain reaction) method for DNA	Mullis (USA)
1985	Discovery of fullerene C_{60}	Curl (USA), Kroto (UK), Smalley (USA)

Major Inventions and Discoveries in Chemistry Continued.

Year	Details	Inventor or discoverer (Country)
1985	Establishment of basic concepts for lithium-ion batteries …………………	A Yoshino (Japan)
1986	High-temperature superconductors…	Bednorz (Switzerland), Müller (Switzerland)
1986	3D structure of photosynthetic bacteria reaction center………………………	Deisenhofer (Germany), Michel (Germany), Huber (Germany)
1987	Femtosecond chemical reaction observation method …………………	Zewail (Egypt)
1988	Matrix Assisted Laser Desorption / Ionization Matter Analysis ………	K. Tanaka (Japan) *et al.*
1989	Synthesis of palytoxin…………………	Y. Kishi (Japan)
1991	Discovery of carbon nanotube ………	S. Iijima (Japan)
1994	Darmstadtium ………………………	Hofmann (Germany), Armbruster (Germany) *et al.*
1994	Roentgenium………………………	Hofmann (Germany), Armbruster (Germany) *et al.*
1995	Asymmetric hydrogenation …………	R. Noyori (Japan)
1996	Copernicium 112 ……………………	Hofmann (Germany) *et al.*
1998	Flerovium 114 ………………………	Organessian (Russia) *et al.*
2000	Livermorium 116 ……………………	Organessian (Russia) *et al.*
2004	Graphene …………………………	Geim (Russia), Novoselov (Russia)
2004	Nihonium 113 ………………………	K. Morita (Japan) *et al.*
2004	Moscovium 115………………………	Organessian (Russia) *et al.*
2006	Oganeson 118 ………………………	Organessian (Russia) *et al.*
2010	Tennessine 117………………………	Organessian (Russia) *et al.*

The elements (carbon, sulfur, iron, copper, silver, tin, antimony, gold, mercury, and lead) not listed in this time line were all known from the B.C. era. Although bismuth is known to have been discovered around 1500, it is not known who discovered the element.

Earth Science Section

Geography

Latest Values Related to the Shape and Size of the Earth

The following are the latest values related to the shape and size of the earth, as announced at a meeting of the Third Special Committee of the International Association of Geodesy (IAG) which was held in Birmingham, England in August 1999 *1.

Velocity of light in vacuum $c = 299\,792\,458\ \mathrm{m \cdot s^{-1}}$

Newtonian gravitational constant $G = (6672.59 \pm 0.30) \times 10^{-14}\ \mathrm{m^3 \cdot s^{-2} \cdot kg^{-1}}$

Mean angular velocity of the Earth rotation $\omega = 7\,292\,115 \times 10^{-11}\ \mathrm{rad \cdot s^{-1}}$

Geocentric gravitational constant including the atmosphere

$$GM = (398\,600\,441.8 \pm 0.8) \times 10^{6}\ \mathrm{m^3 \cdot s^{-2}}$$

Dynamic shape coefficient $J_2 = (1\,082\,626.7 \pm 0.1) \times 10^{-9}$ (tide-free system) *2

$J_2 = (1\,082\,635.9 \pm 0.1) \times 10^{-9}$ (zero tide system)

Equatorial radius of the earth $a = (6\,378\,136.59 \pm 0.10)$ m (tide-free system)

$a = (6\,378\,136.62 \pm 0.10)$ m (zero-tide system)

Regular gravity in equator (zero-tide system)

$$\gamma_e = (978\,032.78 \pm 0.2) \times 10^{-5}\ \mathrm{m \cdot s^{-2}}$$

*1 Although these values were announced as the latest values related to the shape and size of the earth, the following values (Geodetic Reference System 1980) are used when utilizing the equatorial radius and flattening (tide-free system) of the earth as reference values.

Equatorial radius of the earth $a = 6\,378\,137$ m

Flattening (f) $1/f = 298.257\,222\,101$

*2 In order to accurately define the shape and size of the earth, it is necessary to consider changes in the shape of the earth due to the tides. This is because the shape of the earth is constantly changing due to the tidal force of the sun and moon. There are three different systems to handle the permanent tide:

1) "tide-free system" where permanent deformation caused by the tides is completely omitted,
2) "zero-tide system" where tide-generating potential is omitted but its indirect effect is retained, and
3) "mean-tide system" where both direct and indirect effects of permanent deformation caused by the tides are included.

Flattening (f) $1/f = 298.257\,65 \pm 0.000\,01$ (tide-free system)

$1/f = 298.256\,42 \pm 0.000\,01$ (zero-tide system)

Potential on geoid $W_0 = (62\,636\,855.611 \pm 0.5)\ \mathrm{m^2 \cdot s^{-2}}$

Tri-axial parameters

Flattening of the equatorial plane (f_1) $1/f_1 = 91\,026 \pm 10$

The direction of long axis of equator side oval

$\lambda_1 = 14.9291 \pm 0.0010°$ (West longitude)

Values for Earth's main moments of inertia (a_0 corresponds to the value of Earth's equatorial radius: 6 378 137 m)

(zero-tide system)

$$(C-A)/(Ma_0^2) = (1086.267 \pm 0.001) \times 10^{-6}$$

$$(C-B)/(Ma_0^2) = (1079.005 \pm 0.001) \times 10^{-6}$$

$$(B-A)/(Ma_0^2) = (\quad 7.262 \pm 0.004) \times 10^{-6}$$

$$C-A = (2.6398 \pm 0.0001) \times 10^{35}\ \mathrm{kg \cdot m^2}$$

$$C-B = (2.6221 \pm 0.0001) \times 10^{35}\ \mathrm{kg \cdot m^2}$$

$$B-A = (1.765 \pm 0.001) \times 10^{33}\ \mathrm{kg \cdot m^2}$$

$$C/(Ma_0^2) = (330\,701 \pm 2) \times 10^{-6}$$

$$A/(Ma_0^2) = (329\,615 \pm 2) \times 10^{-6}$$

$$B/(Ma_0^2) = (329\,622 \pm 2) \times 10^{-6}$$

Varieties of Earth Ellipsoids

Ellipsoid		Equatorial radius (a ; m)	Reciprocal flattening ($1/f$)	The main countries that use it
Bessel	1841	6377397.155	299.152813	
Revised Clarke	1880	8249.145	293.4663	Various African Countries
Krasovsky	1940	8245	298.3	Russia
Everest	1956	7301.243	300.8017	Republic of India
Australia international (IAU-65)	1965	8160	298.25	
South America 1969	1969	8160	.25	Each country of South America
IAG-67	1967	8160	.247167	
WGS 72	1972	8135	.26	
IAU-76	1976	8140	.257	
Geodetic Reference System 1980 (GRS80)	1979	8137	.257222101	USA, each country of Europe, and Japan
WGS 84	1997	8137	.257223563	

Geometric Quantities for Earth Ellipsoid

Latitude (φ)	Length of arc along 1″ of longitude (l_1)	Length of arc along 1″ of latitude (l_2)	Length of meridian arc from the equator to each circle of latitude (l_3)	Distance from the equator to each circle of latitude in the Mercator map projection (l_4)	Radius of the circle drawn by standard circle of latitude in equidistant conic projection (l_5)	Area of the latitude band between each circle of latitude and the equator (S)	Difference from geocentric latitude based on earth ellipsoid ($\varphi-\varphi'$)
°	m	m	km	km	km	km²	′ ″
0	30.922	30.715	0.000	0.000	∞	0.00	0 0.0
5	30.805	30.717	552.885	553.584	72 904.293	22 128 968.97	1 59.9
10	30.455	30.724	1 105.855	1 111.475	36 175.864	44 093 962.58	3 56.2
15	29.875	30.736	1 658.990	1 678.148	23 808.870	65 731 954.78	5 45.4
20	29.069	30.751	2 212.366	2 258.424	17 530.653	86 881 830.45	7 24.1
25	28.042	30.770	2 766.054	2 857.693	13 686.143	107 385 370.66	8 49.5
30	26.802	30.792	3 320.113	3 482.189	11 056.513	127 088 269.98	9 58.9
35	25.358	30.817	3 874.593	4 139.373	9 118.971	145 841 190.83	10 50.2
40	23.721	30.843	4 429.529	4 838.471	7 611.702	163 500 855.30	11 21.8
45	21.902	30.870	4 984.944	5 591.296	6 388.838	179 931 169.81	11 32.7
50	19.915	30.897	5 540.847	6 413.525	5 362.436	195 004 372.17	11 22.6
55	17.776	30.923	6 097.230	7 326.838	4 476.084	208 602 185.81	10 51.7
60	15.500	30.948	6 654.073	8 362.699	3 691.698	220 616 960.34	10 0.9
65	13.104	30.970	7 211.339	9 569.603	2 982.385	230 952 773.99	8 51.8
70	10.607	30.989	7 768.981	11 028.514	2 328.344	239 526 470.19	7 26.4
75	8.028	31.005	8 326.938	12 890.914	1 714.379	246 268 599.54	5 47.4
80	5.387	31.017	8 885.140	15 496.571	1 128.306	251 124 238.10	3 57.7
85	2.704	31.024	9 443.510	19 929.239	559.878	254 053 655.66	2 0.7
90	0.000	31.026	10 001.966	∞	0.000	255 032 810.86	0 0.0

The above table is calculated based on the Geodetic Reference System 1980 parameters given in page **517** (＊1).
The general formulas used to calculate these values are shown on the next page.

Formulae for Earth Ellipsoids

Taking the Equatorial radius to be a, Flattening f, Eccentricity $e=\sqrt{f(2-f)}$, Latitude φ,

Meridional radius of curvature $M_\varphi=\dfrac{a(1-e^2)}{(1-e^2\sin^2\varphi)^{3/2}}$, Prime vertical radius of curvature $N_\varphi=\dfrac{a}{\sqrt{1-e^2\sin^2\varphi}}$,

$n=f/(2-f)$, $\varepsilon_i=3n/2i-n$,

$$l_1=\frac{\pi N_\varphi\cos\varphi}{648000},\ l_2\cong\frac{\pi M_\varphi}{648000},$$

$$l_3=\int_0^\varphi M_\theta\,\mathrm{d}\theta=a^2\left(1+\frac{\mathrm{d}^2}{\mathrm{d}\varphi^2}\right)\int_0^\varphi\frac{\mathrm{d}\theta}{N_\theta}$$

$$=\frac{a}{1+n}\left(1+\frac{\mathrm{d}^2}{\mathrm{d}\varphi^2}\right)\int_0^\varphi\sqrt{1+2n\cos2\theta+n^2}\,\mathrm{d}\theta$$

$$=\frac{a}{1+n}\sum_{j=0}^{\infty}\left(\prod_{k=1}^{j}\varepsilon_k\right)^2\left\{\varphi+\sum_{l=1}^{2j}\left(\frac{1}{l}-4l\right)\sin2l\varphi\prod_{m=1}^{l}\varepsilon_{j+(-1)^m\lfloor m/2\rfloor}^{(-1)^m}\right\}*$$

$$\approx\frac{a}{1+n}\left\{\left(1+\frac{n^2}{4}+\frac{n^4}{64}\right)\varphi-\frac{3}{2}\left(n-\frac{n^3}{8}\right)\sin2\varphi\right.$$

$$\left.+\frac{15}{16}\left(n^2-\frac{n^4}{4}\right)\sin4\varphi-\frac{35}{48}n^3\sin6\varphi+\frac{315}{512}n^4\sin8\varphi\right\},**$$

$$l_4=a\int_0^\varphi\frac{M_\theta}{N_\theta\cos\theta}\mathrm{d}\theta=a\{\tanh^{-1}\sin\varphi-e\tanh^{-1}(e\sin\varphi)\},\ l_5=N_\varphi\cot\varphi,$$

$$S=2\pi\int_0^\varphi M_\theta N_\theta\cos\theta\mathrm{d}\theta=\pi a^2\left(\frac{1}{e}-e\right)\left\{\frac{e\sin\varphi}{1-(e\sin\varphi)^2}+\tanh^{-1}(e\sin\varphi)\right\}$$

* Source: Kawase, K. (2009)

** Source: Helmert, F. R. (1880); equivalent to values in * up to $j=2$.

With the formula for l_3, $\lfloor x\rfloor$ represents the greatest integer less than or equal to x. The φ of each item shown in the linear expression for φ uses the value expressed in radians. For operations using the summation symbol (Σ) and product symbol (Π), when the range forming the summation and product is empty, the applicable sum is agreed to be "0" and the applicable product to be "1". Although the ** formula is sufficient for calculation of table values on the previous page, for the precise derivation of the Gauss-Krüger projection coordinate conversion formula utilised in Japan, it is necessary to use higher-order terms for n.

Geocentric Cartesian coordinate value at the point of Latitude φ and longitude λ, ellipsoidal height h:

$$\begin{pmatrix}X\\Y\\Z\end{pmatrix}=\begin{pmatrix}(N_\varphi+h)\cos\varphi\cos\lambda\\(N_\varphi+h)\cos\varphi\sin\lambda\\ [N_\varphi(1-e^2)+h]\sin\varphi\end{pmatrix}$$

Geocentric latitude φ' : $\tan\varphi'=\left(1-\dfrac{e^2}{1+h/N_\varphi}\right)\tan\varphi$

In particular, for the case $h=0$, $\varphi-\varphi'=\tan^{-1}\left(\dfrac{e^2\sin\varphi\cos\varphi}{1-e^2\sin^2\varphi}\right)$

Global Potential Coefficients

The gravitational potential at a point geocentric distance r, geocentric latitude φ', and longitude λ (east longitude is positive) can be expressed in the form of W.

$$W = \frac{GM}{r}\left[1 + \sum_{n=2}^{\infty}\sum_{m=0}^{n}\left(\frac{a}{r}\right)^n \overline{P}_n^m(\sin\varphi')(C_{nm}\cos m\lambda + S_{nm}\sin m\lambda)\right] + \frac{1}{2}\omega^2 r^2\cos^2\varphi'$$

a is the equatorial radius of the earth, GM is the geocentric gravitational constant, and $\overline{P}_n^m(\mu)$ is the fully normalized associated Legendre function. $P_n^m(\mu)$ can be defined from the regular associated Legendre function $P_n^m(\mu)$.

$$\overline{P}_n^m(\mu) = \sqrt{(2-\delta_{0m})(2n+1)\frac{(n-m)!}{(n+m)!}}\,P_n^m(\mu),\ \delta_{0m} = \begin{cases} 1 \cdots\cdots m = 0 \\ 0 \cdots\cdots m \neq 0 \end{cases}$$

If $W = W_0$ is inserted here, (r, φ', λ) becomes the geoid point, and the geoid shape is given by the formula listed above.

Value for coefficient C_{nm}, S_{nm}

($n \leqq 6$, $m \leqq 6$, rounded to the 9th decimal place)

(unit: 10^{-6})

n	m	C_{nm}	S_{nm}	n	m	C_{nm}	S_{nm}
2	0	-484.165 143 79	0.000 000 00	5	0	0.068 670 29	0.000 000 00
2	1	-0.000 206 62	0.001 384 41	5	1	-0.062 921 19	-0.094 369 81
2	2	2.439 383 57	-1.400 273 70	5	2	0.652 078 04	-0.323 353 19
3	0	0.957 161 21	0.000 000 00	5	3	-0.451 847 15	-0.214 955 41
3	1	2.030 462 01	0.248 200 42	5	4	-0.295 328 76	0.049 807 06
3	2	0.904 787 89	-0.619 005 48	5	5	0.174 811 80	-0.669 379 94
3	3	0.721 321 76	1.414 349 26	6	0	-0.149 953 93	0.000 000 00
4	0	0.539 965 87	0.000 000 00	6	1	-0.075 921 01	0.026 512 26
4	1	-0.536 157 39	-0.473 567 35	6	2	0.048 648 89	-0.373 789 32
4	2	0.350 501 62	0.662 480 03	6	3	0.057 245 16	0.008 952 01
4	3	0.990 856 77	-0.200 956 72	6	4	-0.086 023 79	-0.471 425 57
4	4	-0.188 519 63	0.308 803 88	6	5	-0.267 166 42	-0.536 493 15
				6	6	0.009 470 69	-0.237 382 35

Source: EGM2008 (Pavlis et al., 2012) Also, the long term tides conform to the tide-free system and the Geoid height distribution world map in page **730** is based on this model.

Surface Area and Ratio for Land and Sea at each Global Latitude Band

Latitude	Area (10^6 km^2)		Percentage (%)	
	Land	Ocean	Land	Ocean
90° -80° N	0.407	3.502	10	90
80 -70 〃	3.494	8.103	30	70
70 -60 〃	13.352	5.558	71	29
60 -50 〃	14.636	10.977	57	43
50 -40 〃	16.457	15.046	52	48
40 -30 〃	15.622	20.790	43	57
30 -20 〃	15.113	25.093	38	62
20 -10 〃	11.249	31.538	26	74
10° N- 0°	10.039	34.055	23	77
90° N- 0°	100.370	154.663	39.4	60.6
0° -10° S	10.399	33.695	24	76
10 -20 〃	9.433	33.355	22	78
20 -30 〃	9.314	30.893	23	77
30 -40 〃	4.146	32.266	11	89
40 -50 〃	0.991	30.512	3	97
50 -60 〃	0.216	25.396	1	99
60 -70 〃	1.601	17.309	8	92
70 -80 〃	7.295	4.302	63	37
80 -90 〃	3.477	0.431	89	11
0° -90° S	46.874	208.159	18.4	81.6
90° N-90° S	147.244	362.822	28.9	71.1

Land : Ocean = 1 : 2.46

The above table uses the second edition of the Total Spherical Map of the Earth (land coverage) and is calculated to conform to the GRS-80 elipsoid. (Refer to page **519** for a table of Various Earth elipsoids.)

* Source: Geospatial Information Authority of Japan

Japan Geodetic Datum 2011

In conjunction with revision of the Survey Act, from April 1, 2002, the Japanese geodetic coordinate system was transferred to the Japan Geodetic Datum 2000 which is the World Geodetic System. Afterwards, it was transferred to the Japan Geodetic Datum 2011 due to diastrophism caused by the 2011 off the Pacific coast of Tohoku Earthquake. The numerical values for the origin of the Japanese horizontal datum and the Japanese datum of leveling were also revised. Various related factors are listed below.

	Region	Standard coordinate system in the earth	Earth ellipsoid
Japan Geodetic Datum 2011	Aomori pref., Iwate pref., Miyagi pref., Akita pref., Yamagata pref., Fukushima pref., Ibaraki pref., Tochigi pref., Gunma pref., Saitama pref., Chiba pref., Tokyo Metropolitan (islands not included), Kanagawa pref., Niigata pref., Toyama pref., Ishikawa pref., Fukui pref., Yamanashi pref., Nagano pref., Gifu pref.	International Terrestrial Reference Frame 2008 (ITRF2008)	Geodetic Reference System 1980 (GRS80)
	Region other than the above	ITRF94	

Origin of the Japanese Horizontal Control Network (Tokyo Minato ward Azabudai 2-18-1)

Latitude and longitude λ ; 139°44′28″. 8869 E φ : 35°39′29″. 1572 N

Geocentric rectangular coordinate value $X = -3\,959\,340.203$ m $Y = 3\,352\,854.274$ m

$Z = 3\,697\,471.413$ m

Origin for azimuth A : 32°20′46′. 209 (azimuth value of the vector from the origin of the Japan horizontal datum to the cross on the metal mark on the Tsukuba Very Long Baseline Interferometry (VLBI) ground marker, projected onto the earth ellipsoid, measured clockwise from true north)

Origin of the Japanese Vertical Control Network (Tokyo Chiyoda ward Nagata-cho 1-1-2)

H : 24.3900 m (Based on the average sea surface level in Tokyo Bay.)

Glossary of Generic Terms

In Earth Science section, Japanese geographical names are generally spelled in the Roman alphabet. This glossary lists generic terms and adjective elements often used for geographical names in Japan. The terms mainly based on the standardized names by Geospatial Information Authority of Japan.

Usual spelling in English	Romanized Japanese	Usual spelling in English	Romanized Japanese
Islands	Shoto, Gunto, Retto	Plateau	Daichi
Island	To, Shima, Jima	Basin	Bonchi
Peninsula	Hanto	Plain	Heiya, Genya
River	Kawa, Gawa	Mountain Range	Sanmyaku
Lake	Ko	Mountains	Sanchi, Kochi, Renzan
Highland	Kogen	Mountain	Yama, San, Zan, Take, Dake
Hills	Kyuryo		

Largest Islands

Island	Location	* Area (10^2 km^2)	
		A	B
Greenland	Denmark	21 756	21 756
New Guinea [Irian]	Indonesia, Papua New Guinea	7 719	8 085
Borneo [Kalimantan]	Indonesia, Malaysia, Brunei	7 366	7 456
Madagascar	Madagascar	5 903	5 870
Baffin	Canada[1)]	5 122	5 075
Sumatra	Indonesia	4 338	4 736
Honshu	Japan	2 302	2 274
Great Britain	England	2 178	2 185
Victoria	Canada	2 173	2 173
Ellesmere	Canada	1 962	1 962
Sulawesi [Celebes]	Indonesia	1 794	1 892
South Island	New Zealand	1 505	1 512
Java	Indonesia	1 261	1 322
Cuba	Cuba	1 145	1 109
North Island	New Zealand	1 143	1 158
Newfoundland	Canada	1 107	1 089
Luzon	Philippines	1 057	1 047
Iceland	Iceland	1 028	1 028
Mindanao	Philippines	956	946
Ireland	Ireland, UK	821	830
Hokkaido	Japan	784	781
Hispaniola [Haiti]	Haiti, Dominican	772	762
Sakhalin [Sakhalin] (Sakhalin)	Russia	770	764
Tasmania	Australia	679	678
Ceylon [Sri Lanka]	Sri Lanka	656	656
Novaja Zemlja (North Island)	Russia	482	471
Tierra del Fuego	Argentina, Chile	480	470
Alexandra I	Antarctica	432	-
Kyushu	Japan	420	366
Spitzbergen	Norway[2)]	395	378
New Britain	Papua New Guinea	337	-
Novaja Zemlja (South Island)	Russia	332	332
Taiwan	Taiwan	360	359
Hainan	China	356	-
Vancouver	Canada	331	-
Timor	Indonesia	330	-
Sicilia	Italy	255	254
Sardinia	Italy	238	241
Shikoku	Japan	188	183
Halmahera [Dilolo]	Indonesia	180	-
New Caledonia	France	161	-
Okt'abr'skojRevol'ucii (October Revolution Island)	Russia[3)]	145	-
Flores	Indonesia	143	-
Sumbawa	Indonesia	133	-

Continued.

Island	Location	* Area (10^2 km^2)	
		A	B
Samar	Philippines	133	-
New Ireland	Papua New Guinea	130	-
Negros	Philippines	127	-
Kotel'ny	Russia[5)]	120	-
Bol'ševik	Russia[4)]	118	-
Palawan	Philippines	117	-
Panay	Philippines	115	-
Jamaica	Jamaica	115	-
Banka	Indonesia	113	-
Sumba	Indonesia	111	-
Hawaii	USA	104	-
Cape Breton	Canada[6)]	104	-
Viti Levu	Fiji	103	-

*1 Area over 10 000 km^2

*2 A is the value from Geogr. Taschenbuch 1951–52 ed., and B is the value from The Times Comprehensive Atlas of the World, 14th ed., 2014. However, it is not possible to guarantee that the new value of B for the published fiscal-year will be more accurate than A.

1) The 12 large islands in Northern Canada, such as Victoria Island and Ellesmere Island are excluded.

2) Main island of the Svalbard Islands

3, 4) Islands in Severnaya Zemlya

5) Main island of the Novosibirsk Archipelago Area of Faddeyevsky (which is connected with Kotelny by samd bar is not included)

6) Island south of Newfoundland

Major Islands of Japan (over 100 km^2)

Island	Location	km^2	Island	Location	km^2
Honshu		227 941	Iriomote Jima	Okinawa	290
Hokkaido		77 984	Tokunoshima	Kagoshima	248
Kyushu		36 783	Shikotan To	Hokkaido	248
Shikoku		18 297	Dogo	Shimane	242
Etorofu To	Hokkaido	3 167	Amakusa-Kamishima	Kumamoto	226
Kunashiri To	Hokkaido	1 489	Ishigaki Jima	Okinawa	222
Okinawa Jima	Okinawa	1 207	Rishiri To	Hokkaido	182
Sado Shima	Niigata	855	Nakadori Jima	Nagasaki	168
Amami-Oshima	Kagoshima	712	Hirado Shima	Nagasaki	163
Tsushima	Nagasaki	696	Miyako Jima	Okinawa	159
Awaji Shima	Hyogo	593	Shodo Shima	Kagawa	153
Amakusa-Shimoshima	Kumamoto	575	Okushiri To	Hokkaido	143
Yaku Shima	Kagoshima	504	Iki Shima	Nagasaki	135
Tanegashima	Kagoshima	444	Yashiro Jima	Yamaguchi	128
Fukue Jima	Nagasaki	326			

Area is the area of the main island only.

Source: Geospatial Information Authority of Japan "Statistical reports on the land area by prefectures and municipalities in Japan" (2020).

Major High Mountains

Mountain	Location		Altitude (m)
	Mountain/Island	Country/Region	
Asia			
Everest [Chomolungma, Sagarmatha]	Himalayas	China, Nepal	8 848
Godwin Austen [K2, chogori]	Karakorum	(Kashmir Shinjang)	8 611
Kangchenjunga	Himalayas	India, Nepal	8 586
Lhotse	Himalayas	China, Nepal	8 516
Makalu	Himalayas	China, Nepal	8 463
Cho Oyu	Himalayas	China, Nepal	8 201
Dhaulagiri I	Himalayas	Nepal	8 167
Manaslu	Himalayas	Nepal	8 163
Nanga Parbat	Himalayas	(Kashmir)	8 126
Annapurna I	Himalayas	Nepal	8 091
Gasherbrum I	Karakorum	(Kashmir)	8 068
Xixabangma Feng	Himalayas	China	8 027
Tirich Mir	Hindu Kush	Pakistan	7 690
Kongur	Pamir	China	7 649
Minya Konka [Gongshan]	Southeastern Kunlun	China	7 556
Muztagata	Pamir	China	7 509
Qullai Ismoili Somoni	Pamir	Tajikistan	7 495
Jengish Chokusu [Pik Pobedy]	Tian Shan	Kyrgyz, China	7 439
Hantengri	Tian Shan	China	6 995
Muztag	Kunlun central	China	6 973
Damavand*	Eruburuzu	Iran	5 671
El'brus*	Caucasus	Russia	5 642
Gora Dykh-Tau	Caucasus	Russia	5 204
Shkhara	Caucasus	Russia, Georgia	5 201
Büyük Ağri Daği [Ararat]*		Turkey	5 165
Kazbek	Caucasus	Russia, Georgia	5 047
Klyuchevskaya*	Kamchatka	Russia	4 853
Zard Kuh	Zagros	Iran	4 548
Kinabalu	Borneo (Kalimantan)	Malaysia	4 095
Yu Shan	Taiwan island	Taiwan	3 950
Kerinci*	Sumatra	Indonesia	3 805
Taibai Shan	Ch'in Ling	China	3 767
Hadūr Shu'ayb	Arabian Peninsula	Yemen	3 760
Rinjani*	Lombok	Indonesia	3 726
Munkus Sardyk	Sayan	Russia, Mongolia	3 491
Victoria	Arakan	Myanmar	3 053
Apo*	Mindanao	Philippines	2 954
Pidurutalagala	Ceylon	Sri Lanka	2 524

Continued.

Mountain	Location		Altitude (m)
	Mountain/Island	Country/Region	
Mayon*	Luzon	Philippines	2 421
Europe			
Mont Blanc	Alps	France, Italy	4 810
Monte Rosa [Dufour Spitze]	Alps	Italy, Switzerland	4 634
Dom	Alps	Switzerland	4 545
Weisshorn	Alps	Switzerland	4 505
Matterhorn	Alps	Italy, Switzerland	4 478
Jungfrau	Alps	Switzerland	4 158
Piz Bernina	Eastern Alps	Italy, Switzerland	4 049
Grossglockner	Eastern Alps	Austria	3 798
Mulhacén	Sierra Nevada	Spain	3 482
Aneto	Pyrenees	Spain	3 404
Etna*	Sicily	Italy	3 323
Olimbos	Píndhou Greek	Píndhou Greek	2 917
Corno	Apennines	Italy	2 912
MonteCinto	Corsica France	Corsica France	2 706
Moldoveanu	Carpathian	Romania	2 544
Durmitor	Dinaric Alps	Montenegro	2 522
Glitterinden	Scandinavia	Norway	2 465
Hvannadalshnukur*	Iceland Island	Iceland	2 119
Narodnaya	Ural	Russia	1 895
Ben Nevis	Scotland	UK	1 344
Africa			
Kilimanjaro [Kibo Peak]*		Tanzania	5 892
Kenya [kirignagger]*		Kenya	5 199
Stanley [Margherita Peak]		Uganda, Congo	5 110
Meru*		Tanzania	4 565
Ras Dashan	Ethiopia	Ethiopia	4 533
Karisimbi*		Rwanda	4 510
Toubkal	Atlas	Morocco	4 167
Cameroun*		Cameroon	4 095
Pico del Teide*	Canary Islands	Spain	3 718
Thabana Ntlenyana	Drakensberg	Lesotho	3 482
Emi Koussi*		Chad	3 415
Jebel Marra		Sudan	3 088
Tahat	Ahaggar [Hoggar]	Algeria	2 918
Maromokotro	Tsaratanana	Madagascar	2 876

Major High Mountains Continued.

Mountain	Location		Altitude (m)
	Mountain/Island	Country/Region	
	North America		
Denali [McKinley]	Alaska	USA (Alaska)	6 194
Logan	St. Elias	Canada	5 959
Orizaba [Citlaltepetl]*	Transversal	Mexico	5 675
St. Elias	St. Elias	Canada, USA (Alaska)	5 489
Popocatepetl*	Transversal	Mexico	5 465
Foraker	Alaska	USA (Alaska)	5 303
Ixtaccihuatl*	Transversal	Mexico	5 280
Lucania	St. Elias	Canada	5 226
Blackburn	Wrangell	USA (Alaska)	4 996
Whitney	Sierra Nevada	USA	4 418
Elbert	Rocky	USA	4 398
Harvard	Rocky	USA	4 395
Rainier*	Cascade	USA	4 392
Blanca Peak	Rocky	USA	4 364
Shasta*	Cascade	USA	4 317
Pikes Peak	Rocky	USA	4 301
Kennedy	St. Elias	Canada	4 237
Tajamulco*		Guatemala	4 210
Waddington	Coast	Canada	4 042
Robson	Rocky	Canada	3 954
Chirripó		Costa Rica	3 819
Baru*		Panama	3 475
Hood*	Cascade	USA	3 427
Baker*	Cascade	USA	3 285
Lassen Peak*	Cascade	USA	3 187
Pico Duarte	Isupaniora island	Dominica	3 175
Keele Peak	Mackenzie	Canada	2 972
Katmai*	Aleutian	USA (Alaska)	2 047
Mitchell	Appalachian	USA	2 037
	South America		
Aconcagua	Andes	Argentina	6 959
Ojos del Salado*	Andes	Argentina, Chile	6 908
Bonete	Andes	Argentina	6 872
Pissis	Andes	Argentina	6 858
Tupungato*	Andes	Argentina, Chile	6 800
Mercedario	Andes	Argentina	6 770
Huascaran	Western Andes	Peru	6 768

Continued.

Mountain	Location		Altitude (m)
	Mountain/Island	Country/Region	
Llullaillaco*	Andes	Argentina, Chile	6 723
Sajama*	Western Andes	Bolivia	6 542
Illampu	Eastern Andes	Bolivia	6 485
Coropuna*	Western Andes	Peru	6 425
Illimani	Eastern Andes	Bolivia	6 402
Auzangate	Eastern Andes	Peru	6 394
Chimborazo*	Andes	Ecuador	6 310
Salccantay	Eastern Andes	Peru	6 271
Cotopaxi*	Andes	Ecuador	5 911
El Misti*	Eastern Andes	Peru	5 822
Cristobal Colon	Central Andes	Columbia	5 775
Maipo*	Andes	Argentina, Chile	5 290
Bolivar	Eastern Andes	Venezuela	5 007
Fitz Roy	Andes	Argentina, Chile	3 375
Bandeiras	Mantiqueira	Brazil	2 890
Roraima	Guiana	Guyana, Venezuela	2 810
Oceania, Antarctica			
Jaya [Carstensz]	New Guinea Island	Indonesia	4 884
Vinson Massif	Antarctica	–	4 897
Trikora	New Guinea Island	Indonesia	4 730
Mandala	New Guinea Island	Indonesia	4 700
Yamin	New Guinea Island	Indonesia	4 595
Wilhelm	New Guinea Island	Papua New Guinea	4 509
Mauna Kea*	Hawaii Island	USA	4 205
Giluwe	New Guinea Island	Papua New Guinea	4 088
Erebus*	Antarctica	–	3 794
Aoraki [Cook, Aorangi]	New Zealand	New Zealand	3 724
Haleakala*	Maui Island	USA	2 935
Ruapehu*	New Zealand North Island	New Zealand	2 797
Balbi*	Bougainville Island	Papua New Guinea	2 685
Orohena*	Tahiti Island	France	2 241
Kosciusko	Snowy	Australia	2 229

Volcanoes are indicated by an asterisk (*), alternate names are listed in parenthes.

Altitudes are based on The Times Atlas of the World, 2014. However, because the reliability of the values of locations in developing countries is low, there are many cases where contradictory values have been published. Region boundaries are based on regional geological features. Volcanoes (*) are based on Volcanoes of the World. However Kilimanjaro, Kenya, Tupungato, Sajama, and Orohena are from The Times Comprehensive Atlas of the World, 14th ed. 2014.

Major Mountains of Japan

(volcanoes are indicated by an asterisk (*))

Mountain	Location	Altitude (m)
Hokkaido		
* Shiretoko Dake	Shiretoko, Akan	1 254
* Io Zan	Shiretoko, Akan	1 562
* Rausu Dake	Shiretoko, Akan	1 661
* Shari Dake	Shiretoko, Akan	1 547
* Meakan Dake	Shiretoko, Akan	1 499
Rebun Dake	Rebun, Rishiri	490
* Rishiri Zan [Rishiri-Fuji]	Rebun, Rishiri	1 721
Teshio Dake	Kitami Sanchi	1 558
* Taisetsu Zan<Asahi Dake>	Ishikari Sanchi	2 291
* Tokachi Dake	Ishikari Sanchi	2 077
* Nipesotsu Yama	Ishikari Sanchi	2 013
Pisshiri Zan	Teshio Sanchi	1 032
Shokanbetsu Dake	Mashike Sanchi	1 492
Ashibetsu Dake	Yubari Sanchi	1 726
Yubari Dake	Yubari Sanchi	1 668
Poroshiri Dake	Hidaka Sanmyaku	2 052
Kamuiekuuchikaushi Yama	Hidaka Sanmyaku	1 979
Kamui Dake	Hidaka Sanmyaku	1 600
* Teine Yama	Shikotsu, Touya, Shakotan	1 023
Yoichi Dake	Shikotsu, Touya, Shakotan	1 488
Eniwa Dake	Shikotsu, Touya, Shakotan	1 320
* Yotei Zan [Ezo-Fuji]	Shikotsu, Touya, Shakotan	1 898
* Showa-shin Zan	Shikotsu, Touya, Shakotan	398
* Usu Zan<Ousu>	Shikotsu, Touya, Shakotan	733
Kariba Yama	Oshima Hanto	1 520
* Komagatake (Hokkaido)<Kengamine>	Oshima Hanto	1 131
Hakodate Yama	Oshima Hanto	334
Daisengen Dake	Oshima Hanto	1 072
Kamui Yama	Okushiri To	584
* Era Dake	Oshima-Oshima	732
Honshu		
* Hiuchi Dake	Shimokita Hanto	781
Oritsume Dake	Kitakami Kochi	852
Himekami San	Kitakami Kochi	1 124
Hayachine San	Kitakami Kochi	1 917
Goyo Zan	Kitakami Kochi	1 351
Kinka San	Kitakami Kochi	444
* Hakkoda San<Odake>	Northern Ou Sanmyaku	1 585
* Hachimantai	Northern Ou Sanmyaku	1 613
* Iwate San	Northern Ou Sanmyaku	2 038
* Komagatake (Akita)<Oname Dake>	Northern Ou Sanmyaku	1 637
* Iwaki San	Shirakami Sanchi	1 625
Shirakami Dake	Shirakami Sanchi	1 235

Continued.

Mountain	Location	Altitude (m)
Taihei Zan	Dewa Sanchi	1 170
* Chokai San<Shin Zan>	Dewa Sanchi	2 236
Honzan	Oga Hanto	715
Waga Dake	Middle Ou Sanmyaku	1 439
* Kurikoma Yama [Sukawa Dake]	Middle Ou Sanmyaku	1 626
Kamuro San	Middle Ou Sanmyaku	1 365
* Funagata Yama [Gosho Zan]	Southern Ou Sanmyaku	1 500
* Zao Zan (Zao San)<Kumano Dake>	Southern Ou Sanmyaku	1 841
* Nishi-Azuma Yama	Southern Ou Sanmyaku	2 035
* Adatara Yama<Tetsu Zan>	Southern Ou Sanmyaku	1 709
* Bandai San	Southern Ou Sanmyaku	1 816
Haguro San	Asahi Sanchi	414
* Gassan	Asahi Sanchi	1 984
Asahi Dake<O-Asahi Dake>	Asahi Sanchi	1 871
Iide San	Iide Sanchi	2 105
Dainichi Dake	Iide Sanchi	2 128
Ryozen	Abukuma Kochi	825
Otakine Yama	Abukuma Kochi	1 192
Yamizo San	Yamizo San, Tsukuba San	1 022
Tsukuba San	Yamizo San, Tsukuba San	877
* Nasu Dake<Chausu Dake>	Nasu, Nikko	1 915
Osabi Yama	Nasu, Nikko	1 908
* Nyoho San	Nasu, Nikko	2 483
* Nantai San	Nasu, Nikko	2 486
* Shirane San (Nikko)	Nasu, Nikko	2 578
* Akagi San<Kurobi San>	Nasu, Nikko	1 828
Taishaku San	Minamiaizu, Oze	2 060
Komagatake (Aizu)	Minamiaizu, Oze	2 133
* Hiuchigatake<Shibayasugura>	Minamiaizu, Oze	2 356
Shibutsu San	Minamiaizu, Oze	2 228
* Sumon Dake	Echigo Sanmyaku	1 537
Komagatake (Uonuma)	Echigo Sanmyaku	2 003
Hakkai San<Nyudo Dake>	Echigo Sanmyaku	1 778
Nakano Dake	Echigo Sanmyaku	2 085
Makihata Yama	Echigo Sanmyaku	1 967
Tanigawa Dake<Shigekura Dake>	Echigo Sanmyaku	1 978
Sennokura Yama	Echigo Sanmyaku	2 026
* Naeba San	Naeba San, Shirane San, Asama Yama	2 145
Iwasuge Yama<Ura-Iwasuge Yama>	Naeba San, Shirane San, Asama Yama	2 341
Yokote Yama	Naeba San, Shirane San, Asama Yama	2 307
* Motoshirane San	Naeba San, Shirane San, Asama Yama	2 171
* Azumaya San	Naeba San, Shirane San, Asama Yama	2 354
* Asama Yama	Naeba San, Shirane San, Asama Yama	2 568
* Haruna San<Kamongatake >	Naeba San, Shirane San, Asama Yama	1 449
Myogi San<Soma Dake>	Kanto Sanchi	1 104
Arafune Yama	Kanto Sanchi	1 423
Ryogami San	Kanto Sanchi	1 723

Major Mountains of Japan Continued.

Mountain	Location	Altitude (m)
Sanpo Yama	Kanto Sanchi	2 483
Kobushigatake	Kanto Sanchi	2 475
Kinpu San	Kanto Sanchi	2 599
Kokushigatake	Kanto Sanchi	2 592
Kita-Oku-Senjo Dake	Kanto Sanchi	2 601
*Kayagatake	Kanto Sanchi	1 704
Buko Zan	Kanto Sanchi	1 304
Kumotori Yama	Kanto Sanchi	2 017
Karamatsuo Yama	Kanto Sanchi	2 109
Daibosatsurei	Kanto Sanchi	2 057
Mito San	Kanto Sanchi	1 531
Jinba San [Jinba San]	Kanto Sanchi	855
Takao San	Kanto Sanchi	599
Kiyosumi Yama<Myoken Yama>	Boso, Miura	377
Kano Zan	Boso, Miura	379
Atago Yama (Boso)	Boso, Miura	408
Ogusu Yama	Boso, Miura	241
Oyama (Tanzawa)	Tanzawa Sanchi	1 252
Tanzawa San<Hirugatake>	Tanzawa Sanchi	1 673
Mishotai Yama	Tanzawa Sanchi	1 681
Mitsutoge Yama	Fuji San and its surroundings	1 785
*Fuji San<Kengamine>	Fuji San and its surroundings	3 776
*Ashitaka Yama<Echizen Dake>	Fuji San and its surroundings	1 504
*Hakone Yama<Kami Yama>	Hakone Yama, Izu Hanto	1 438
*Amagi San<Banzaburo Dake>	Hakone Yama, Izu Hanto	1 406
*Mihara Yama<Mihara-Shin Zan>	Izu Shoto (Oshima)	758
*Miyatsuka Yama	Izu Shoto (Toshima)	508
*Tenjo San	Izu Shoto (Kozu shima)	572
*Oyama (Miyake Jima)	Izu Shoto (Miyake Jima)	775
*Oyama (Mikura Jima)	Izu Shoto (Mikura Jima)	851
*Nishi Yama [Hachijo-Fuji]	Izu Shoto (Hachijo Jima)	854
*Io Yama	Izu Shoto (Torishima)	394
Chuo Zan	Ogasawara Shoto (Chichijima)	320
Chibusa Yama	Ogasawara Shoto (Hahajima)	463
Sakakigamine	Ogasawara Shoto (Kita-Io To)	792
*Suribachi Yama [Paipu Yama]	Ogasawara Shoto (Io To)	170
Minami-Io To	Ogasawara Shoto (Minami-Io To)	916
Kinpoku San	Sado	1 172
Yahiko Yama	Higashi-Kubiki Kyuryo	634
Yone Yama	Higashi-Kubiki Kyuryo	993
*Yake Yama (Niigata)	Myoko San and its surroundings	2 400
Hiuchi Yama	Myoko San and its surroundings	2 462
*Myoko San	Myoko San and its surroundings	2 454
Takatsuma Yama	Myoko San and its surroundings	2 353
Togakushi Yama	Myoko San and its surroundings	1 904
Hijiri Yama	Chikuma Sanchi	1 447
Utsukushigahara<Ogato>	Chikuma Sanchi	2 034

Continued.

Mountain	Location	Altitude (m)
* Kirigamine<Kuruma Yama>	Kirigamine-Yatsugatake	1 925
* Tateshina Yama	Kirigamine-Yatsugatake	2 531
Yoko Dake (Chinoshi, Sakuhomachi)	Kirigamine-Yatsugatake	2 480
* Aka Dake	Kirigamine-Yatsugatake	2 899
Amida Dake	Kirigamine-Yatsugatake	2 805
Kurohime Yama	Northern Hida Sanmyaku	1 221
Ko-Renge San	Northern Hida Sanmyaku	2 766
Shirouma Dake	Northern Hida Sanmyaku	2 932
Asahi Dake	Northern Hida Sanmyaku	2 867
Shakushi Dake	Northern Hida Sanmyaku	2 812
Yarigatake	Northern Hida Sanmyaku	2 903
Goryu Dake	Northern Hida Sanmyaku	2 814
Kashima-Yarigatake	Northern Hida Sanmyaku	2 889
Harinoki Dake	Northern Hida Sanmyaku	2 821
Renge Dake	Northern Hida Sanmyaku	2 799
Mitsudake	Northern Hida Sanmyaku	2 845
Noguchigoro Dake	Northern Hida Sanmyaku	2 924
Akaushi Dake	Northern Hida Sanmyaku	2 864
Suisho Dake [Kuro Dake]	Northern Hida Sanmyaku	2 986
Jii Dake	Northern Hida Sanmyaku	2 825
* Washiba Dake	Northern Hida Sanmyaku	2 924
Mitsumatarenge Dake	Northern Hida Sanmyaku	2 841
Tsurugi Dake	Northern Hida Sanmyaku	2 999
Bessan (Hida)	Northern Hida Sanmyaku	2 880
Masago Dake	Northern Hida Sanmyaku	2 861
Tate Yama<Onanji Yama>	Northern Hida Sanmyaku	3 015
Ryuo Dake	Northern Hida Sanmyaku	2 872
Yakushi Dake (Hida)	Northern Hida Sanmyaku	2 926
Kurobegoro Dake [Nakanomata Dake]	Northern Hida Sanmyaku	2 840
Sugoroku Dake	Northern Hida Sanmyaku	2 860
Nukedo Dake	Northern Hida Sanmyaku	2 813
Kasagatake	Northern Hida Sanmyaku	2 898
O-Tensho Dake (Dai-Tenjo Dake)	Southern Hida Sanmyaku	2 922
Higashi-Tenjo Dake	Southern Hida Sanmyaku	2 814
Jonen Dake	Southern Hida Sanmyaku	2 857
Yarigatake	Southern Hida Sanmyaku	3 180
Obami Dake	Southern Hida Sanmyaku	3 101
Nakadake	Southern Hida Sanmyaku	3 084
Minami Dake	Southern Hida Sanmyaku	3 033
Kita-Hotaka Dake	Southern Hida Sanmyaku	3 106
Karasawa Dake	Southern Hida Sanmyaku	3 110
Oku-Hotaka Dake	Southern Hida Sanmyaku	3 190
Mae-Hotaka Dake	Southern Hida Sanmyaku	3 090
Nishi-Hotaka Dake	Southern Hida Sanmyaku	2 909
* Yakedake	Southern Hida Sanmyaku	2 455
* Norikura Dake<Kengamine>	Southern Hida Sanmyaku	3 026
* Ontakesan<Kengamine>	Ontake San and its surroundings	3 067

Major Mountains of Japan Continued.

Mountain	Location	Altitude (m)
Komagatake (Kiso)	Kiso Sanmyaku	2 956
Sannosawa Dake	Kiso Sanmyaku	2 847
Utsugi Dake	Kiso Sanmyaku	2 864
Minami-Komagatake	Kiso Sanmyaku	2 841
Ena San	Kiso Sanmyaku	2 191
Nyugasa Yama	Northern Akashi Sanmyaku	1 955
Komagatake (Kai)	Northern Akashi Sanmyaku	2 967
Asayomine	Northern Akashi Sanmyaku	2 799
Kannongadake	Northern Akashi Sanmyaku	2 841
Yakushigadake	Northern Akashi Sanmyaku	2 780
Senjogatake	Northern Akashi Sanmyaku	3 033
Kitadake	Northern Akashi Sanmyaku	3 193
Ainodake	Northern Akashi Sanmyaku	3 190
Notori Dake<Nishi-Notori Dake>	Northern Akashi Sanmyaku	3 051
Hirogochi Dake	Northern Akashi Sanmyaku	2 895
Shiomi Dake	Northern Akashi Sanmyaku	3 052
Komori Dake	Northern Akashi Sanmyaku	2 865
Kogochi Dake	Southern Akashi Sanmyaku	2 802
Higashi Dake [Warusawa Dake]	Southern Akashi Sanmyaku	3 141
Arakawa Dake<Nakadake>	Southern Akashi Sanmyaku	3 084
Akaishi Dake	Southern Akashi Sanmyaku	3 121
Osawa Dake	Southern Akashi Sanmyaku	2 820
Nakamori Maruyama	Southern Akashi Sanmyaku	2 807
Usagi Dake	Southern Akashi Sanmyaku	2 818
Hijiri Dake<Mae-Hijiri Dake>	Southern Akashi Sanmyaku	3 013
Kamikochi Dake	Southern Akashi Sanmyaku	2 803
Tekari Dake	Southern Akashi Sanmyaku	2 592
Minobu San	Minobu Sanchi	1 153
Yanbushi	Minobu Sanchi	2 013
Kuno Zan	Minobu Sanchi	216
Chausu Yama	Mino, Mikawa Kogen	1 416
Horaiji San	Mino, Mikawa Kogen	695
Sanage Yama	Mino, Mikawa Kogen	629
Horyu Zan	Noto Hanto	471
Hodatsu San	Hodatsu Kyuryo	637
Sarugababa Yama	Hida Kochi	1 875
Io Zen<Oku-Io Zen>	Hakusan Sanchi	939
* Hakusan<Gozengamine>	Hakusan Sanchi	2 702
Bessan (Hakusan)	Hakusan Sanchi	2 399
Nogo-Hakusan[Gongensan]	Etsumi, Ibuki Sanchi	1 617
Ibuki Yama	Etsumi, Ibuki Sanchi	1 377
Oike Dake	Suzuka, Nunobiki Sanchi	1 247
Gozaisho Yama	Suzuka, Nunobiki Sanchi	1 212
Amagoi Dake	Suzuka, Nunobiki Sanchi	1 238
Bunagadake	Around Biwako	1 214
Minako Yama	Around Biwako	971
Hiei Zan	Around Biwako	848

Continued.

Mountain	Location	Altitude (m)
Kasagi Yama	Kasagi Sanchi	324
Miminashi Yama	Nara Bonchi	139
Amanokagu Yama	Nara Bonchi	152
Unebi Yama	Nara Bonchi	199
Ikoma Yama	Ikoma, Kongo, Izumi Sanchi	642
Shigi San	Ikoma, Kongo, Izumi Sanchi	437
Nijyo San<Odake>	Ikoma, Kongo, Izumi Sanchi	517
Katsuragi San	Ikoma, Kongo, Izumi Sanchi	959
Kongo Zan	Ikoma, Kongo, Izumi Sanchi	1 125
Takami Yama	Takami Sanchi	1 248
Kunimi Yama	Eastern Kii Sanchi	1 419
Odaigahara Zan<Hinodegatake>	Eastern Kii Sanchi	1 695
Hakkyogadake	Eastern Kii Sanchi	1 915
Gomadan Zan	Western Kii Sanchi	1 372
Nachi San<Eboshi Yama>	Western Kii Sanchi	910
Hyakurigadake	Tanbakochi	931
Atago Yama (Tanba)	Tanbakochi	924
Rokko San	Rokko Sanchi	931
Yuzuruha San	Awaji Shima	608
Oginosen	Eastern Chugoku Sanchi	1 310
Hyonosen [Suganosen]	Eastern Chugoku Sanchi	1 510
Ushiro Yama	Eastern Chugoku Sanchi	1 344
Ohira Yama (Kibi)	Eastern Kibi Kogen	698
Daimanji San	Oki	608
Takuhi Yama	Oki	452
Hiruzen<Kami-Hiruzen>	Middle Chugoku Sanchi	1 202
*Daisen<Kengamine>	Middle Chugoku Sanchi	1 729
Dogo Yama	Middle Chugoku Sanchi	1 271
Hiba Yama<Tate-Eboshi Yama>	Middle Chugoku Sanchi	1 299
*Sanbe San<Osanbe San>	Middle Chugoku Sanchi	1 126
Takanosu Zan	Western Kibi Kogen	922
Kan Zan	Western Chugoku Sanchi	863
Osorakan Zan	Western Chugoku Sanchi	1 346
Kanmuri Yama	Western Chugoku Sanchi	1 339
Jakuchi San	Western Chugoku Sanchi	1 337
Kenso Zan<Hoshigajo Yama>	Setonaikai (Shodo Shima)	816
Washigato Zan	Setonaikai (Omi Shima)	436
Misen	Setonaikai (Itsuku Shima)	535
Shikoku		
Iino Yama [Sanuki-Fuji]	Sanuki Sanchi and its surroundings	422
Ryuo Zan	Sanuki Sanchi and its surroundings	1 060
Higashi-Sanpogamori	Takanawa Sanchi	1 233
Tsurugi San	Eastern Shikoku Sanchi	1 955
Miune, Sanrei	Eastern Shikoku Sanchi	1 894
Ishizuchi San<Tengu Dake>	Western Shikoku Sanchi	1 982
Ninomori	Western Shikoku Sanchi	1 930
Kasatori Yama	Western Shikoku Sanchi	1 562

Major Mountains of Japan Continued.

Mountain	Location	Altitude (m)
Kyushu		
Hiko San	Tsukushi Sanchi	1 199
Sefuri San	Tsukushi Sanchi	1 055
*Tsurumi Dake	Aso, Kujyu and its surroundings	1 375
*Yufu Dake<Bungo Fuji>	Aso, Kujyu and its surroundings	1 583
*Kuju Renzan<Nakadake>	Aso, Kujyu and its surroundings	1 791
*Aso San<Takadake>	Aso, Kujyu and its surroundings	1 592
Shaka Dake	Aso, Kujyu and its surroundings	1 231
*Tara Dake<Kyogadake>	Western Saga, Nagasaki, Shimabara	1 076
*Unzen Dake<Heisei-Shinzan>	Western Saga, Nagasaki, Shimabara	1 483
Yatate Yama	Tsushima	648
Sanno Zan<O Dake>	Goto Retto	439
Tetegatake	Goto Retto	460
Sobo San	Kyushu Sanchi	1 756
Kunimi Dake	Kyushu Sanchi	1 739
*Kirishima Yama<Karakuni Dake>	Southern Kyushu	1 700
*On Take (Sakura Jima)<Kitadake>	(Sakura Jima)	1 117
*Kaimon Dake	Southern Kyushu	924
Kura Dake	Amakusa Shoto	682
Kado Yama	Amakusa Shoto	526
Otake	Koshikishima Retto	604
Nanseishoto		
Amamegakura	Osumi Shoto (Tanegashima)	238
*Io Dake (Satsuma-Io Jima)	Osumi Shoto (Satsuma-Io Jima)	704
Yagura Dake	Osumi Shoto (Kuroshima)	620
*Furudake (Kuchinoerabu Jima)	Osumi Shoto (Kuchinoerabu Jima)	657
Miyanoura Dake	Osumi Shoto (Yaku Shima)	1 936
Nagata Dake	Osumi Shoto (Yaku Shima)	1 886
*On Take(Nakanoshima)	Tokara Retto (Nakanoshima)	979
Otake	Tokara Retto (Gajya Jima)	497
*Otake (Suwanose Jima)	Tokara Retto (Suwanose Jima)	796
*Mitake (Akuseki Jima)	Tokara Retto (Akuseki Jima)	584
Yuwan Dake	Amami Gunto (Amami-Oshima)	694
Inokawa Dake	Amami Gunto (Tokunoshima)	645
Oyama (Okinoerabu Jima)	Amami Gunto (Okinoerabu Jima)	240
Yonaha Dake	Okinawa Jima	503
Omoto Dake	Yaeyama Retto (Ishigaki Jima)	526
Komi Dake	Yaeyama Retto (Iriomote Jima)	469
Urabu Dake	Yaeyama Retto (Yonaguni Jima)	231

Based primarily on Geospatial Information authority of Japan, "Nihon no sangaku hyoko ichiran —1003 San—". [ ]: Alternate name, < >: highest peak, ( ): location.

Height of Snow Line

Figure 1

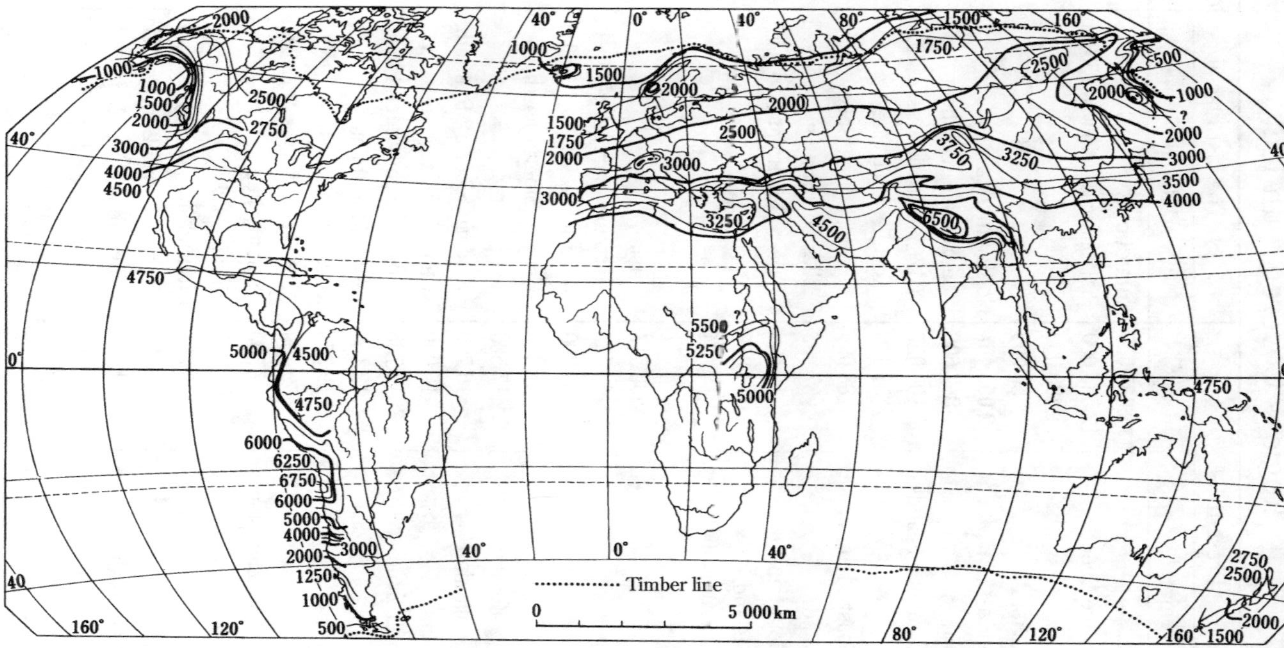

The snow line height given here is the climatic snow line height. If a glacier exists at the location in question, the snow line height given here indicates the border between the accumulation zone and ablation zone for that glacier (unit: m). Source: K. Hermes: Der Verlauf der Schneegrenze. Geograhpisches Taschenbuch, 58–71 (1964).

Regions Covered with Ice

Present		The last glacial period	
Region	km^2	Region	10^6 km^2
Antarctic	**13 985 000**	**Antarctic**	**14.51**
Antarctic Ice Sheet	12 535 000	(Maximum ice thickness 4 500 m)	
Ice shelf	1 400 000		
Other glaciers	50 000		
North America	**2 056 467**	**North America**	**17.19**
Greenland	1 802 600	Laurentide Ice Sheet	11.18
Queen Elizabeth Islands	109 057	(Maximum ice thickness 4 960 m)	
Alaska	51 476	Korujiera Ice Sheet	2.20
Pacific Coast Mountains	38 099	Greenland Ice Sheet	2.16
Baffin Island	37 903	Ellesmere, Baffin Ice Sheet	1.56
Rocky Mountains	12 428	Other Mountains	0.09
Bairotto Island	4 869		
Labrador Peninsula	24		
Mexico	11		
Eurasia	**260 517**	**Eurasia**	**9.01**
Spitsbergen	68 425	Northern Europe Ice Sheet	3.66
Himalayas	33 200	(Maximum ice thickness 3 390 m)	
Novaya Zemlya	24 300	Central Siberia northern Ice Sheet	1.32
Severnaya Zemlya	17 500		
Nan Shan, Kunlun	16 700	Eastern Siberia Mountains	1.24
Karakorum	16 000	Central Asia Mountains	0.87
Franz Josef Land	13 735	Novaya Zemlya Ice Floor	0.43
Iceland	12 173	Altai Mountains	0.32
Arai, Pamir	9 375	Ural Mountains	0.24
Tibet	9 100	Kamchatka, Koryak Mountains	0.19
Southwestern headwaters region of the Salween River	7 500	Spitsbergen	0.16
		Trans-Baikal	0.14
Hindu Kush	6 200	Iceland and Jan Mayen Island	0.12
Tenshan	6 190	East Asia and Central Asia	0.11
Trans-Himalayan	4 000	Novosibirskiye Islands and Wrangel Island	0.09
Scandinavia	3 800		
Alps	3 200	Franz Josef Land	0.05

Continued.

Present		The last glacial period	
Region	km²	Region	10⁶ km²
Caucasus	1 805	Alps Mountains and the surrounding	0.03
Ladakh, Deosai, Rupshu Mountains	1 700	Siberia around	0.02
Kunlun southern and eastern Mountains	1 400	Caucasus and Asia Minor	0.01
		Faroe Islands	0.01
Dzungar Mountains	956		
Kamchatka Mountains	866		
Koljak Mountains	650		
Altai Mountains	646		
Novosibirskie Islands	398		
Suntar-Khayata Mountains	206		
Chersky Mountains	162		
Jan Mayen Island	117		
Virunga Mountains	50		
Elburz Mountains and the Turkish Mountains	50		
Sayan eastern Mountains east	32		
Ural Mountains	28		
Verkhoyansk Mountains	23		
Kodar Mountains	15		
Pyrenees Mountains	15		
South America	**26 500**	**South America**	**0.83**
Oceania	**1 015**	**Oceania and Africa**	**0.06**
New Zealand South Island	1 000		
New Guinea	15		
Africa	**12**		
Total	**16 329 511**	**Total**	**41.60**
(11.0% of the present land)		(27.9% of the present land)	

Source: J. Marcinek: Gletscher der Erde, (1984) Leipzig. The assumed maximum ice thicknesses are based on G. H. Denton & T. J. Hughes (ed.): The Last Great Ice Sheets, (1981) New York.

The areas in the last glacial period show the values when the glaciers spread maximumly although the values have not been established. Even the present area of Antarctic region does not fixed: 13 799 000 km^2 has also been reported.

Major Rivers

River	Drainage area (10^3 km^2)	Length (km)	Whereabouts in Estuary (Country and oceanic name, etc.)
Amazon	7 050	6 516	Brazil · Atlantic Ocean
Madeira		3 200	
Congo [Zaire]	3 700	4 667	Congo · Atlantic Ocean
Nile	3 349	6 695	Egypt · Mediterranean Sea
Mississippi-Missouri	3 250	5 969	United State · Gulf of Mexico
Mississippi		3 765	
Missouri		4 086	
La Plata-Parana	3 100	4 500	Argentina - Uruguay · Atlantic Ocean
Paraguay		2 600	(Parana River tributary)
Ob-Irtysh	2 990	5 568	Russia · Ob Bay
Jenisej-Angara [Yenisei]	2 580	5 550	Russian · Kara Sea
Lena	2 490	4 400	Russia Teputesu Sea
Changjiang [Yangtze]	1 959	6 380	China · East China Sea
Niger	1 890	4 184	Nigeria · Gulf of Guinea
Amur [Heilongjiang]	1 855	4 416	Russia · Mamiya Strait
Mackenzie	1 805	4 241	Canada · Beaufort Sea
Ganges · Brahmaputra	1 621		Bangladesh · Bay of Bengal
Brahmaputra		2 840	
Ganges		2 510	
St. Lawrence	1 463	3 058	Canada · Gulf of St. Lawrence
Volga	1 380	3 688	Russia · Caspian Sea
Zambezi	1 330	2 736	Mozambique · Mozambique Channel
Indas	1 166	3 180	Pakistan · Arabian Sea
Nelson-Saskatchewan	1 150	2 570	Canada · Lake Winnipeg
Orange	1 020	2 100	South Africa · Atlantic Ocean
Hwang Ho [Yellow River]	980	5 464	China · Bohai Sea
Orinoco	945	2 500	Venezuela · Atlantic Ocean
Murray-Darling	1 058	3 672	Australia · Great Australian Bight
Yukon	855	3 185	United State · Bering Sea
Donau [Danube]	815	2 850	Romania · Black Sea
Mekong	810	4 425	Vietnam · South China Sea

Rivers with a basin area of at least 800 000 km^2; alternate names are listed in parentheses
Based primarily on The Times Comprehensive Atlas of the World, 14th ed., 2014.

Largest Rivers

River	Estuary (Confluence river)	Length (km)	Order of the world
	Africa		
Nile	Mediterranean Sea	6 695	1
Congo (Zaire)	South Atlantic Ocean	4 667	8
Niger	Gulf of Guinea	4 184	17
Zambezi	Mozambique channel	2 736	43
Kasai	Congo River	2 153	64
Orange	South Atlantic Ocean	2 100	67
White Nile(al-Bahr-al-Abyad)	Nile	2 084	69
Lualaba	Congo River	1 800	90
Limpopo	Mozambique channel	1 800	90
Jubba (Juba)	Indian Ocean	1 658	99
Sénégal	South Atlantic Ocean	1 641	100
	North America		
Mississippi-Missouri-Red Rock	Gulf of Mexico	5 969	4
Mackenzie-Peace-Finlay	Beaufort Sea	4 241	16
Missouri	Mississippi River	3 767	20
Missouri-Red Rock	Mississippi River	4 086	19
Mississippi	Gulf of Mexico	3 765	21
Yukon-Nisutlin	Bering Sea	3 185	28
St. Lawrence-Great Lakes	Gulf of St. Lawrence	3 058	31
Rio Grande	Gulf of Mexico	3 057	32
Yukon	Bering Sea	3 018	33
Nelson-Saskatchewan	Hudson Bay	2 570	47
Arkansas	Mississippi River	2 348	56
Colorado	Califerania Bay	2 333	57
Ohio-Allegheny	Mississippi River	2 102	66
Red	Mississippi River	2 044	70
Columbia	North Pacific Ocean	2 000	73
Saskatchewan	Lake Winnipeg	1 939	79
Peace	Slave River	1 923	82
Snake	Columbia River	1 670	98
	South America		
Amazon	South Atlantic Ocean	6 516	2
La Plata-Paraná	South Atlantic Ocean	4 500	9
Madeira-Mamoré-Guaporé	Amazon	3 350	25
Jurua	Amazon	3 283	26
Purus	Amazon	3 211	27
São Francisco	South Atlantic Ocean	2 900	35
Japurá (Caquetá)	Amazon	2 816	39
Tocantins	South Atlantic Ocean	2 750	41
Ucayali-Apurimac	Amazon	2 738	42

Rivers with a length in the top 100, listed by continent

Largest Rivers Continued.

River	Estuary (Confluence river)	Length (km)	Order of the world
Araguaia	Tocantins River	2 627	45
Paraguay	Parana River	2 600	46
Orinoco	South Atlantic Ocean	2 500	51
Pilcomayo	Paraguay River	2 500	51
Negro (Guainia)	Amazon	2 253	60
Xingu	Amazon	2 100	67
Tapajos-Teles Pires	Amazon	1 992	74
Mamoré	Madeira River	1 931	80
Marañón-Huallage	Amazon	1 905	83
Guaporé (Iténez)	Mamore River	1 749	94
Parnaiba	South Atlantic Ocean	1 700	96
Madre de Dios	Beni River	1 700	96
Asia			
Chang Jiang	East China Sea	6 380	3
Ob-Irtysh	Ob Bay	5 568	5
Yenisey-Baykal-Selenga	Kara Sea	5 550	6
Huang Ho (Yellow River)	Bo Hai Sea	5 464	7
Amur-Argun	Sea of Okhotsk	4 444	10
Mekong	South China Sea	4 425	11
Amur	Mamiya Strait	4 416	12
Lena	Lapte Sea	4 400	13
Ob-Katun	Ob Bay	4 338	14
Irtysh-Chorny Irtysh	Ob River	4 248	15
Yenisey	Kara Sea	4 090	18
Ob	Ob Bay	3 701	22
Indus	Arabian Sea	3 180	29
Syrdarya-Arabelsu	Aral Sea	3 078	30
Nizhnyaya Tunguska	Yenisei River	2 989	34
Brahmaputra	Jamuna River	2 840	38
Euphrates	Shatt al-Arab	2 800	40
Vilyuy	Lena River	2 650	44
Amu Darya-Pyandzh	Aral Sea	2 540	48
Kolyma-Kulu	East Siberian Sea	2 513	49
Ganges	Padma River	2 510	50
Ishim	Irtysh River	2 450	53
Salween	Gulf of Martaban	2 400	55
Olenyok	Lapte Sea	2 292	58
Aldan	Lena River	2 273	59
Syrdarya	Aral Sea	2 212	61
Zhu Jiang (Pearl)-Xi Jiang	South China Sea	2 197	63
Kolyma	East Siberian Sea	2 129	65
Tarim	Lop Nur	2 030	71
Chulym-Belyy Lyus	Ob River	2 023	72
Irrawaddy	Andaman Sea	1 992	74
Vitim-Vitimkan	Lena River	1 978	76
Indigirka-Khastakh	East Siberian Sea	1 977	77

Continued.

River	Estuary (Confluence river)	Length (km)	Order of the world
Xi(Hsi) Jiang	South China Sea	1 957	78
Songhua Jiang (Sungari)	Amur River	1 927	81
Tigris	Shatt al-Arab	1 900	84
Podkamennaya Tunguska	Yenisei River	1 865	86
Vitim	Lena River	1 837	87
Chulym	Ob River	1 799	92
Angara	Yenisei River	1 779	93
Indigirka	East Siberian Sea	1 726	95
	Europe		
Volga	Caspian Sea	3 688	23
Donau (Danube)	Black Sea	2 850	36
Ural	Caspian Sea	2 428	54
Dnepr	Black Sea	2 200	62
Don	Sea of Azov	1 870	85
Pechora	Barents Sea	1 809	88
Kama	Volga	1 805	89
	Oceania		
Murray-Darling	Great-Australian Bight	3 672	24
Darling	Murray River	2 844	37

Based primarily on The Times Comprehensive Atlas of the World, 14th ed., 2014 and The Water Encyclopedia, 2nd ed., Lewis Publishers, 1990. In cases where the name of the river changes from up stream to down stream, or between the main river and branches, this chart shows the length of the section with the indicated name.

Largest Rivers of Japan

River	Drainage area (km^2)	River trunk flow path extension* (km)	Observation station	Upstream region area of observation station (km^2)	Flowing quantity in 2017** (m^3/s)			Observation period
					Period average	Max.	Min.	
Tone Gawa	16 840	322	Kurihashi	8 588	250	7 900	68	1918-
Ishikari Gawa	14 330	268	Inou	3 379	120	520	45	1953-
Shinano Gawa	11 900	367	Ojiya	9 719	570	6 300	170	1942-
Kitakami Gawa	10 150	249	Tome	7 869	-	3 000	-	1952-
Kiso Gawa	9 100	227	Inuyama	4 684	250	3 900	-	1951-
Tokachi Gawa	9 010	156	Obihiro	2 678	-	990	-	1951-
Yodo Gawa	8 240	75	Hirakata	7 281	-	6 500	-	1952-
Agano Gawa	7 710	210	Maoroshi	6 997	510	5 500	110	1951-
Mogami Gawa	7 040	229	Takaya	6 271	410	2 100	120	1959-
Teshio Gawa	5 590	256	Bifuka bridge	2 899	130	770	31	1957-
Abukuma Gawa	5 400	239	Akutsu	1 865	50	2 100	14	1951-
Tenryu Gawa	5 090	213	Kashima	4 880	190	3 600	67	1939-
Omono Gawa	4 710	133	Tsubakigawa	4 035	290	3 900	90	1938-
Yoneshiro Gawa	4 100	136	Takanosu	2 109	120	1 800	27	1957-
Fuji Gawa	3 990	128	Shimizubata	2 179	(50)	(1 300)	(20)	1952-
Gonokawa	3 900	194	Ozekiyama	1 981	82	2 100	23	1956-

Largest Rivers of Japan Continued.

River	Drainage area (km²)	River trunk flow path extension* (km)	Observation station	Upstream region area of observation station (km²)	Flowing quantity in 2017** (m³/s)			Observation period
					Period average	Max.	Min.	
Yoshino Gawa	3 750	194	Ikeda	2 074	80	4 500	19	1954–
Naka Gawa	3 270	150	Noguchi	2 181	70	2 300	19	1949–
Arakawa	2 940	173	Yorii	905	25	–	–	1938–
Kuzuryu Gawa	2 930	116	Nakatsuno	1 240	–	2 100	–	1952–
Chikugo Gawa	2 860	143	Senoshita	2 295	110	4 700	28	1950–
Jintsu Gawa	2 720	120	Jintsu bridge	2 688	220	3 100	85	1958–
Takahashi Gawa	2 670	111	Hiwa	1 986	59	2 100	15	1963–
Iwaki Gawa	2 540	102	Goshogawara	1 740	90	650	19	1953–
Hii Kawa	2 540	153	Kamishima	895	40	1 200	6	1984–
Kushiro Gawa	2 510	154	Shibecha	895	28	170	21	1956–
Shingu Gawa	2 360	183	Oga	2 251	130	–	19	1951–
Oyodo Gawa	2 230	107	Kashiwada	2 126	390	7 300	210	1961–
Shimanto Gawa	2 186	196	Gudo	1 808	160	2 700	36	1952–
Yoshii Gawa	2 110	133	Tsuze	1 675	73	3 500	19	1986–
Mabechi Gawa	2 050	142	Kenyoshi	1 751	60	1 100	16	1963–
Tokoro Gawa	1 930	120	Kitami	1 394	23	500	7	1954–
Yura Gawa	1 880	146	Fukuchiyama	1 344	67	4 300	8	1953–
Kuma Gawa	1 880	115	Yokoishi	1 856	120	2 500	28	1968–
Yahagi Gawa	1 830	117	Iwatsu	1 356	36	2 100	4	1939–
Gokase Gawa	1 820	106	Miwa	1 044	60	3 800	14	1949–
Asahi Kawa	1 810	142	Makiyama	1 587	59	1 300	15	1965–
Kinokawa	1 750	136	Funato 2	1 558	51	6 100	0	1952–
Ota Gawa	1 710	103	Yaguchi 1	1 527	74	1 600	13	1970–
Shiribetsu Gawa	1 640	126	Nagoma	1 402	58	500	20	1965–
Sendai Gawa	1 600	137	Kurano bridge	1 153	(110)	–	–	1986–
Niyodo Gawa	1 560	124	Ino	1 463	90	6 700	14	1957–
Kuji Gawa	1 490	124	Yamagata	898	18	1 300	4	1958–
Ono Gawa	1 465	107	Shirataki bridge	1 381	67	10 000	6	1950–
Abashiri Gawa	1 380	115	Bihoro	824	14	110	1	1955–
Saru Gawa	1 350	104	Biratori	1 253	44	250	11	1966–
Oi Gawa	1 280	168	Kanza	1 160	–	(1 900)	(3)	1957–
Mukawa	1 270	135	Mukawa	1 228	35	270	6	1952–
Tama Gawa	1 240	138	Ishihara	1 040	33	–	–	1951–
Hijikawa	1 210	103	Ozu	984	40	1 900	10	1956–
Shokawa	1 180	115	Daimon	1 120	52	1 900	8	1955–
Nakagawa	874	125	Furusho	765	60	2 800	5	1956–

Based on material of the Water and Disaster Management Bureau at the Ministry of Land, Infrastructure, Transport and Tourism

(1) Applies to Class A Rivers which have a basin area of at least 2 000 km and a main river channel extension of at least 100 km, and rivers for which flow rate data can be continually acquired. Flow rate values are current as of 2016.

(2) Other * Channel extensions which have a large flow rate and which approximately match main river courses in Japan.

** Unknown data is indicated by "–." Flow rate values are calculated by substituting the observed water level into relational expression for water level and flow rate. The first two digits of the calculated values are shown. These values are preliminary values and there is the possibility that they may change in the future.

Note) Flowing quantity in ( ) are previous data. 2014; Oi gawa (Kanza) 2016; Fuji Gawa (Shimizubata), and Sendai Gawa (Kurano bridge)

Chemical Composition of River, Lake and Marsh Water in Japan (mg/L)

River, Lake, Marsh	Observation point	pH	Evaporation residue	Float	Alkalinity $CaCO_3$	Na^+	K^+	Ca^{2+}	Mg^{2+}	Cl^-	SO_4^{2-}	SiO_2	Fe	P	NO_3-N	NH_4-N	Protein-N
Tone Gawa	Ibaraki, Toride	7.1	97.9	18.5	29.6	5.6	1.5	13.3	3.3	5.3	20.2	23.0	0.14	0.04	0.28	0.03	0.07
Ishikari Gawa	Hokkaido, Bibai	6.5	96.0	167.4	22.6	7.6	1.5	9.4	2.5	5.0	15.5	20.6	0.70	0.01	0.48	0.12	0.02
Shinano Gawa	Niigata, Nagaoka	6.4	111	24	18.9	1.4	0.7	10.2	2.1	4.4	18.7	9.3	0.09	0.002	0.25	0.12	-
Yoneshiro Gawa	Akita, Takanosu[1)]	6.5	103.2	11.6	11.4	8.4	0.9	8.3	2.8	9.1	26.0	23.0	0.18	0.004	0.16	0.03	0.07
Naka Gawa	Tochigi, Kuroiso[2)]	7.0	128.7	4.0	16.9	6.7	1.3	15.8	4.3	8.2	44.9	35.4	0.04	0.004	0.07	0.08	0.04
Chikugo Gawa	Fukuoka, Mii-gun, Ozeki-mura[3)]	7.3	90.4	15.5	21.4	7.8	2.5	9.2	2.3	4.4	7.9	43.4	0.14	0.02	0.17	0.04	-
Yura Gawa	Kyoto, Fukuchiyama	7.1	48.0	25.5	20.3	4.9	0.7	6.4	1.8	5.7	4.8	12.6	0.02	0.004	0.25	0.04	0.05
Kinokawa Gawa	Wakayama, Wakayama	7.1	65.0	10.4	37.3	5.2	0.9	12.9	1.3	4.4	6.6	14.4	0.03	0	0.09	0.05	0.06
Yoshii Gawa	Okayama, Saidaiji[4)]	6.9	53.0	10.4	23.4	5.7	1.0	7.4	1.9	6.2	11.9	15.1	0.13	0	0.11	0.03	0.13
Sugawa Gawa	Gunma, Naganohara	2.5	591.8	51.5	0	13.4	6.3	33.6	2.5	134.8	280.0	61.5	4.71	0.60	0.33	0.18	0.04
Biwako	Shiga, the center of Biwako	7.4	47.6	22.0	25.2	5.1	0.8	8.5	2.7	3.8	3.3	1.6	0.01	0.01	0.20	0.02	-
Kasumi Gaura	Ibaraki, Niihari-gun, Kamiotsu-mura[5)]	7.1	143.7	16.6	43.4	16.7	3.7	16.6	5.9	20.1	36.7	13.8	0.37	0.05	0.48	0.13	0.11
Yamanaka Ko	Yamanashi	7.3	51.2	6.9	38.6	3.3	1.2	8.3	3.4	0.8	1.6	9.3	0.05	0.001	0.02	0.02	-

Mainly converted using values measured by Jun Kobayashi

* Acidic river. 1) Present Kitaakita City, 2) Present Nasushiobara City, 3) Present Mii-gun, Tachiarai-machi, 4) Present Okayama City, 5) Present Tsuchiura City

Major Lakes

Lake/Marsh	Location	Morphogenesis	Surface area (10^3 km^2)	
			A	B
Caspian Sea*	Eurasia	Tectonic	374.000	371.000
L. superior	North America	Glacial	82.367	82.100
L. victoria	Central Africa	Tectonic	68.800	68.870
Aral'skoye More*	Central Asia	Tectonic	×64.100	17.158
L. Huron	North America	Glacial	59.570	59.600
L. Michigan	North America	Glacial	58.016	57.800
L. Tanganyika	Eastern Africa	Tectonic	32.000	32.600
Ozero Baykal	Siberia	Tectonic	31.500	30.500
L. Great Bears	Northern Canada	Glacial	31.153	31.328
L. Great Slave	Northern Canada	Glacial	28.568	28.568
L. Eries	North America	Glacial	25.821	25.700
L. Winnipeg	Canada	Glacial	23.750	24.387
L. Nyasa	Eastern Africa	Tectonic	22.490**	29.500
L. Chad	Northern Central Africa	Tectonic	×20.900	-
L. Ontario	North America	Glacial	19.009	18.960
Oz. Balkhash*	Central Asia	Tectonic	×18.200	17.400
Lakozhskoye Oz	Western Russia	Glacial	18.135	18.390
Lde Maracaibo	Venezuela	Tectonic	13.010	-
Logoa de pato	Brazil	Lagoon	10.140	-
Oz. Onezhskoye	Western Russia	Glacial	9.890	9.600
L. Erye*	Australia	Tectonic	×9.690	8.900
L. Rudolf*	Kenya	Tectonic	8.660	-
LagoTiticaca	Western South-Africa	Tectonic	8.372	8.340
L. Nicaragua	Nicaragua	Tectonic	8.150	8.150
L. Athabasca	Canada	Glacial	7.935	-
L. Reindeer	Canada	Glacial	6.390	-
Oz. Issyk-kul*	Central Asia	Tectonic	6.236	6.200
L. Urmia*	Iran	Tectonic	×5.800	5.585
L. Vanern	Sweden	Glacial	5.648	-
L. Nettilling	Canada	Glacial	5.530	-
L. Winnipegosis	Canada	Glacial	5.375	-
L. Albert	Eastern Africa	Tectonic	5.300	-
L. Kariba	Southeastern Africa	Synthetic	5.400	-
Dongting-hu	China, Hunan	Tectonic	×2.740	-
Tonle Sap	Cambodia	Dammed	×2.450	-
Tai-hu	China, Jiangsu	Dammed	2.428	-
L. Malaren	Sweden	Tectonic, Glacial	1.140	-
L. Songkhla	Thailand	Lagoon	1.082	-
Dead Sea*	Western West-Asia	Tectonic	×1.020	-
Laguna de Bay	Philippines	Lagoon	×0.900	-
L. Taupo	New Zealand	Tectonic, Volcanic	0.616	-
L. Chilwa*	Southeastern Africa	Tectonic	×0.600	-
L. Balaton	Hungary	Tectonic	0.593	-
Lac Lemans	Alps Mountain in the west feet of mt	Glacial	0.584	-
Bodensee	Alps Mountain in the north feet of mt	Glacial	0.539	-
L. Tahoe	USA	Glacial	0.499	-
Skadarsko Jez	Balkans	Tectonic	0.372	-
Lago Maggiore	Alps Mountain in the south feet of mt	Glacial	0.213	-
L. Tiberia	Israel	Tectonic	0.170	-
Bung Boraphet	Thailand	Synthetic	0.106	-
L. Washington	USA	Glacial	0.088	-
L. Sibaya	South Africa	Lagoon	×0.078	-
Zurichsee	Switzerland	Glacial	0.065	-
Loch Ness	Scotland	Tectonic	0.056	-
L. Mendota	USA	Glacial	0.039	-
Dong-hu	China, Hubei	Dammed	0.028	-
Tjeukemeer	Netherlands	Synthetic	0.021	-
Windermere	England	Glacial	0.015	-

(1) Natural lakes with an area over 5 000 km^2 and other major lakes are included. Among the areas, the values under A are based on primary Sources 1-3. Values under B are from The Times Comprehensive Atlas of the World, 14th ed., 2014.

(2) * indicates saline lakes (lakes and marshes within the contental basin with high salinity, brackish lakes adjacent to the ocean are excluded.)

(3) Generally speaking, in the case of lakes in inland arid regions and midstream/downstream from large rivers, there is significant expansion/contraction of lake area and rise/fall of water level depending on the season and year. × denotes lakes which undergo extreme changes in area and water level. The values shown in the table are for a certain point of time.

** The value from Source 3 is shown. 6 400 × 10^3 km^2 according to Source 2.

and Marshes

Altitude (m)	Shoreline (km)	Max. depth (m)	Mean depth (m)	Catchment area (km³)	Transparency (m)	Source
-28	6 000	1 025	(209)	78 200	–	3
183	4 768	406	148	12 221	0–15	1, 2(88)
1 134	3 440	84	40	2 750	0–2	1, 2(88)
53	(2 300)	68	(15)	1 020	–	3
176	5 088	228	53	3 535	12–14	1, 2(88)
176	2 656	281	84	4 871	2–12	1, 2(88)
773	1 900	1 471	572	17 800	5–19	1, 2(88)
456	2 000	1 741	740	23 000	5–23	1, 2(88)
186	2 719	446	72	2 236	10–30	2(90)
156	(2 200)	625	(73)	2 088	–	3
174	1 369	64	18	458	2–4	1, 2(88)
217	1 750	36	12	284	0.4–2	1, 2(88)
500	245	706	292	8 400	13–23	2(89)
282	800	10	8	72	0–1	1, 2(88)
75	1 161	244	86	1 638	2–6	1, 2(88)
343	2 385	26	6	106	1–12	2(93)
5	1 570	230	51	908	2–5	2(91)
1	(900)	60	(21)	280	–	3
1	–	5	(2)	20	–	3
35	(1 600)	120	30	280	3–4	2(91)
-9	1 718	6	3	30	–	2(90)
427	(900)	73	–	251	–	3
3 812	1 125	281	107	893	5–11	1, 2(88)
32	(450)	70	(13)	108	–	3
213	(900)	124	(26)	204	–	3
337	(960)	219	–	96	–	3
1 606	688	668	270	1 738	13–20	2(93)
1 275	(340)	10	(8)	45	–	3
44	1 940	106	27	153	5	2(89)
30	–	1	–	1	–	3
254	(1 100)	12	(3)	16	–	3
615	(520)	58	25	280	2–6	2(89)
485	2 164	78	31	160	3–6	2(88)
34	–	31	7	18	0	1, 2(88)
5	–	12	(4)	10	–	3
3	–	3	2	4.3	0–1	1, 2(88)
0	1 410	61	12	14	2–3	1, 2(89)
0	–	2	1	1.6	1	1, 2(88)
-400	–	426	184	188	–	3
2	220	7	3	3.2	0–1	1, 2(88)
357	153	164	91	60	11–20	1, 2(88)
622	200	3	1	1.8	0	2(88)
105	236	12	3	1.9	1	1, 2(88)
372	167	310	153	89	2–15	1, 2(88)
400	255	252	(100)	49	3–15	1, 2(90)
1 897	120	505	313	375	28	1, 2(88)
5	207	8	5	1.9	4–18	1, 2(88)
194	170	370	177	37	3–22	1, 2(88)
-209	53	43	26	4	3	1, 2(88)
24	63	6	2	0.3	1–2	1, 2(88)
0	–	65	33	2.9	3–7	1, 2(88)
23	127	43	13	1	–	1, 2(88)
406	–	136	51	3.3	3–8	1, 2(88)
16	86	230	132	7.5	4–5	1, 2(88)
850	35	25	12	0.5	1–9	1, 2(88)
21	92	5	2	0.06	1–3	1, 2(88)
-1	25	5	2	0.04	0.3–1	1, 2(88)
39	17	64	21	0.3	–	1, 2(88)

(4) Numbers in parenthesis under Perimeter and average water depth are reference values.

(5) Sources: 1 Shiga Prefecture Lake Biwa Environmental Research Institute and National Institute for Research Advancement (1984): Data Book of World Lake Environments (Japanese Version), 2 Shiga Prefecture Lake Biwa Environmental Research Institute and International Lake Environment Committee Foundation (1988, 1989, 1990, 1991, 1993): Data Book of World Lake Environments, 3 C. E. Herdendorf (1990): Distribution of the World's Large Lakes. For Source 2, the year is shown in parentheses.

Major Lakes and Marshes of Japan

Lake/Marsh	Administrative divisions*	Origin	Estuarine/ Fresh water	Area (km^2)	Altitude (m)	Surroundings length (km)	Max. depth (m)	Mean water depth (m)	Completely freezes over?	Lake type	Trans-parency (m)
Biwako	Shiga	Fault	Fresh	669.3	85	241	103.8	41.2	No	Mesotrophic	6.0
Kasumigaura	Ibaraki	Sea mark	Fresh	168.1	0	120	11.9	3.4	No	Eutrophic	0.6
Saroma Ko	Hokkaido (Okhotsk)	Sea mark	Estuarine	151.6	0	87	19.6	8.7	Yes	Eutrophic	9.4
Inawashiro Ko	Fukushima	Fault	Fresh	103.2	514	50	93.5	51.5	No	Acidic	6.1
Nakaumi	Shimane, Tottori	Sea mark	Estuarine	85.7	0	105	17.1	5.4	No	Eutrophic	5.5
Kussharo Ko	Hokkaido (Kushiro)	Caldera	Fresh	79.5	121	57	117.5	28.4	Yes	Acidic	6.0
Shinji Ko	Shimane	Sea mark	Estuarine	79.2	0	47	6.0	4.5	No	Eutrophic	1.0
Shikotsu Ko	Hokkaido (Ishikari)	Caldera	Fresh	78.5	248	40	360.1	265.4	No	Oligotrophic	17.5
Toya Ko	Hokkaido (Iburi)	Caldera	Fresh	70.7	84	50	179.7	117.0	No	Oligotrophic	10.0
Hamana Ko	Shizuoka	Sea mark	Estuarine	64.9	0	114	13.1	4.8	No	Mesotrophic	1.3
Ogawara Ko	Aomori	Sea mark	Estuarine	62.0	0	47	26.5	10.5	Yes	Mesotrophic	3.2
Towada Ko	Aomori, Akita	Caldera	Fresh	61.1	400	46	326.8	71.0	No	Oligotrophic	9.0
Furen Ko	Hokkaido (Nemuro)	Sea mark	Estuarine	59.0	0	94	13.0	1.0	Yes	Oligotrophic	4.0
Notoro Ko	Hokkaido (Okhotsk)	Sea mark	Estuarine	58.2	0	33	23.1	8.6	Yes	Eutrophic	5.5
Kitaura	Ibaraki	Sea mark	Fresh	35.0	0	64	10.0	4.5	No	Eutrophic	0.6
Akkeshi Ko	Hokkaido (Kushiro)	Sea mark	Estuarine	32.3	0	25	11.0	–	No	Mesotrophic	1.3
Abashiri Ko	Hokkaido (Okhotsk)	Sea mark	Estuarine	32.3	0	39	16.3	6.1	Yes	Eutrophic	1.4
Hachirogata Choseichi	Akita	Sea mark	Fresh	27.8	1	35	11.3	–	Yes	Eutrophic	1.3
Tazawa Ko	Akita	Caldera	Fresh	25.8	249	20	423.4	280.0	–	Acidic	4.0
Mashu Ko	Hokkaido (Kushiro)	Caldera	Fresh	19.2	351	20	211.4	137.5	Yes	Oligotrophic	28.0
Jusan Ko	Aomori	Sea mark	Estuarine	17.8	0	28	1.5	–	No	Mesotrophic	1.0
Kutcharo Ko	Hokkaido (Soya)	Sea mark	Fresh	13.4	0	30	3.3	1.0	Yes	Eutrophic	2.2
Akan Ko	Hokkaido (Kushiro)	Caldera	Fresh	13.3	420	26	44.8	17.8	Yes	Eutrophic	5.0
Suwa Ko	Nagano	Fault	Fresh	12.8	759	17	7.6	4.6	Yes	Eutrophic	0.5
Chuzenji Ko	Tochigi	Dammed	Fresh	11.9	1 269	22	163.0	94.6	No	Oligotrophic	9.0
Ikeda Ko	Kagoshima	Caldera	Fresh	10.9	66	15	233.0	125.5	No	Mesotrophic	6.5
Hibara Ko	Fukushima	Dammed	Fresh	10.9	822	38	30.5	12.0	Yes	Mesotrophic	4.5
Inba Numa	Chiba	Dammed	Fresh	9.4	2	44	4.8	1.7	No	Eutrophic	0.8
Hinuma	Ibaraki	Sea mark	Estuarine	9.3	0	20	3.0	2.1	No	Eutrophic	0.6
Tofutsu Ko	Hokkaido (Okhotsk)	Sea mark	Estuarine	8.2	1	27	2.4	1.1	Yes	Eutrophic	0.8
Mangoku Ura	Miyagi	Sea mark	Estuarine	7.3	0	22	17.5	–	No	Eutrophic	3.0
Kumihama Wan	Kyoto	Sea mark	Estuarine	7.2	0	23	20.6	–	No	Mesotrophic	3.0
Ashinoko	Kanagawa	Caldera	Fresh	7.0	725	19	40.6	25.0	No	Mesotrophic	7.5
Koyama Ike	Tottori	Sea mark	Estuarine	7.0	0	18	6.5	2.8	–	Eutrophic	1.0
Yamanaka Ko	Yamanashi	Dammed	Fresh	6.6	981	14	13.3	9.4	Yes	Mesotrophic	5.5
Toro Ko	Hokkaido (Kushiro)	Sea mark	Fresh	6.3	6	18	6.9	3.1	Yes	Eutrophic	1.1
Matsukawa Ura	Fukushima	Others	Estuarine	6.2	0	23	5.5	–	No	Eutrophic	1.2
Sotonasaka Ura	Ibaraki, Chiba	Sea mark	Estuarine	5.9	0	12	23.3	–	No	Eutrophic	0.6
Onneto	Hokkaido (Nemuro)	Sea mark	Estuarine	5.7	0	14	7.3	1.2	Yes	Oligotrophic	1.7
Kawaguchi Ko	Yamanashi	Dammed	Fresh	5.5	831	18	14.6	9.3	Yes	Eutrophic	5.2
Takahoko Numa	Aomori	Sea mark	Fresh	5.4	0	22	7.0	2.7	Yes	Eutrophic	1.5
Inohana Ko	Shizuoka	Sea mark	Estuarine	5.4	0	14	16.1	4.6	No	Mesotrophic	1.0

Onuma	Hokkaido (Oshima)	Dammed	Fresh	5.3	129	21	11.6	5.9	Yes	Eutrophic	2.5
Naganuma	Miyagi	Dammed	Fresh	4.9	8	12	3.0	1.5	–	Eutrophic	0.6
Kamo Ko	Niigata	Sea mark	Estuarine	4.9	0	17	9.0	5.2	–	Eutrophic	5.4
Komuke Ko	Hokkaido (Okhotsk)	Sea mark	Estuarine	4.8	1	23	5.3	1.2	Yes	Eutrophic	1.4
Asokai	Kyoto	Sea mark	Estuarine	4.8	0	16	13.0	8.4	No	Mesotrophic	1.7
Motosu Ko	Yamanashi	Dammed	Fresh	4.7	900	11	121.6	67.9	No	Oligotrophic	11.2
Kuttara Ko	Hokkaido (Iburi)	Caldera	Fresh	4.7	258	8	148.0	105.1	Yes	Oligotrophic	22.0
Onuma	Hokkaido (Soya)	Sea mark	Fresh	4.7	1	10	2.2	1.6	Yes	Saprophytic	–
Nojiri Ko	Nagano	Others	Fresh	4.5	657	14	38.3	20.8	Yes	Oligotrophic	5.4
Yudo Numa	Hokkaido(Tokachi)	Sea mark	Estuarine	4.4	5	17	3.5	1.3	Yes	Dystrophic	1.8
Kahoku Gata	Ishikawa	Sea mark	Estuarine	4.2	0	25	4.8	2.0	No	Eutrophic	0.6
Suigetsu Ko	Fukui	Others	Estuarine	4.2	0	11	33.7	–	No	Eutrophic	2.2
Togo Ike	Tottori	Sea mark	Estuarine	4.1	0	13	3.6	2.1	–	Eutrophic	0.9
Tega Numa	Chiba	Dammed	Fresh	4.0	3	37	3.8	0.9	No	Eutrophic	0.4
Shikaribetsu Ko	Hokkaido(Tokachi)	Dammed	Fresh	3.6	810	13.8	98.5	57.1	Yes	Eutrophic	9.8
Akimoto Ko	Fukushima	Dammed	Fresh	3.5	736	20	36.0	12.8	Yes	Mesotrophic	3.3
Numazawa Ko	Fukushima	Caldera	Fresh	3.0	474	7.5	96.0	60.4	No	Oligotrophic	9.0
Panketo	Hokkaido (Kushiro)	Dammed	Fresh	2.9	450	12	54.0	23.9	Yes	Oligotrophic	14.0
Saiko	Yamanashi	Dammed	Fresh	2.1	900	9.6	71.5	38.5	No	Oligotrophic	6.5
Aoki Ko	Nagano	Fault	Fresh	1.7	822	6.5	58.0	29.0	No	Oligotrophic	9.8
Unagi Ike	Kagoshima	Volcano	Fresh	1.2	122	4.2	55.8	34.8	No	Mesotrophic	7.6
Hiruga Ko	Fukui	Sea mark	Estuarine	0.9	0	4.0	39.4	14.3	No	Oligotrophic	8.0
Sugenuma	Gunma	Dammed	Fresh	0.8	1 731	6.5	75.0	38.1	Yes	Oligotrophic	13.2
Miike	Miyazaki	Volcano	Fresh	0.7	305	3.9	93.5	57.7	No	Oligotrophic	3.1
Marunuma	Gunma	Dammed	Fresh	0.5	1 428	3.5	47.0	31.1	Yes	Mesotrophic	9.3
Otori Ike	Yamagata	Dammed	Fresh	0.4	966	3.2	68.0	30.1	Yes	Oligotrophic	4.8
Penketo	Hokkaido (Kushiro)	Dammed	Fresh	0.3	520	3.9	39.4	14.0	Yes	Oligotrophic	11.3
Ichinome Gata	Akita	Volcano	Fresh	0.3	87	2.0	42.0	–	–	Mesotrophic	2.5
Otaira Numa	Fukushima	Dammed	Fresh	0.2	458	2.5	35.5	13.3	Yes	Mesotrophic	2.0
Numayama Onuma	Yamagata	Dammed	Fresh	0.1	410	2.0	31.7	11.4	Yes	Mesotrophic	3.7
Bogaike	Niigata	Dammed	Fresh	0.1	460	1.4	33.1	15.0	No	Oligotrophic	5.8
Yanakubo Ike	Nagano	Dammed	Fresh	0.1	630	2.2	38.0	19.2	Yes	Mesotrophic	2.4
Yugama	Gunma	Volcano	Fresh	0.1	2 033	0.8	30.0	12.5	–	Acidic	–

* Integrated Development and Promotion Bureau and Development and Promotion Bureau

(1) As a general rule, lakes with an area of at least 4 km^2 or a maximum depth of at least 30 m are on the table lists. In addition to this table, as lakes with more than 4 km^2 area, Tofutsuko in Kunashiri Islands and Urunonbetsuko in Etorofu Islands, belonging to Hokkaido Government Nemuro Subprefectural Bureau are included.

(2) "Environment Investiation Report" (Environment Agency Nature Conservation Bureau, 1993), "Stastical reports on the land area by prefectures and municipalities in Japan" (Geospatial Information Authority of Japan, 2020), and "Lake Charts" (Geospatial Information Authority of Japan)

(3) The transparency degree is indicated by the depth at which a white disc with a diameter of 25 to 30 cm cannot be seen from the water surface. This value varies significantly depending on seasonal and environmental changes. Numbers listed in the table above are the latest values. The largest values measured in the past are listed below.

Lake	Transparency (m)	Observation year	Observer	Lake	Transparency (m)	Observation year	Observer	Lake	Transparency (m)	Observation year	Observer
Mashu Ko	41.6	1931 VIII	Hokkaido Fisheries Research Inst.	Inawashiro Ko	27.5	1930 VII	N. Yoshimura	Shikotsu Ko	25.0	1926 V	Hokkaido Fisheries Research Inst.
Tazawa Ko	30.0	1926 III	Akita Pref. Fisheries Research Inst.	Ikeda Ko	26.8	1929 V	D. Miyaji	Kuttara Ko	24.3	1916 VI	S. Tanakadate, S. Kokubo

Major Basins*

Basin	Altitude at lowest point or shoreline (m)	Area of the part below sea level (10 km²)	Whereabouts
Dead Sea Region	-423[1]	} 383	Israel, Jordan
Lake Tiberias 〔Galilee〕 Region	-212[2]		Israel, Syria
Lake Assal Region	-174	8	Djibouti
Turfan 〔Rukuchin〕 basin	-154	500	China (Shin Chan wiggle)
Kattara (El Qattara) Depression	-133	4 400	Egypt
Caspian Sea Region	-132[3]	73 600	Russia, Kazakhstan, Iran
Kobaru Depression, Lake Asale	-116	196	Ethiopia
Death Valley	-86	25	USA (California)
Imperial Valley, Lake Salton	-71	-	USA, Mexico
Rayan Wadi	-60	39	Egypt
Fayyun Depression, Lake Qarun	-45	66	Egypt
Lake Enriquillo Region	-40	-	Dominica
Siwah Oasis	-32	11	Egypt
Lake Chott Melrhir Region	-24	713	Algeria
Natrun Wadi	-23	-	Egypt
Lake Chottel Gharsa Region	-17	162	Tunisia
Lake Eyre Region	-12	-	Australia

* Land below Sealevel. Land reclaimed artificially and back marshes of large rivers, etc. are not included. Most are in primarily arid regions. Other names are shown in 〔 〕. Lake levels fluctuate so differing values are shown.

The deepest points at the bottoms of the lakes are 1) -826 m, 2) -252 m, 3) -1023 m.

Major Deserts

Desert	Continent (Address)	Area (10^4 km²)
Sahara	Africa (Northern Africa)	907
Australian	Australia (Midwestern Australia)	337
Great Victoria		(65)
Great Sandy		(40)
Simpson		(15)
Arabian	Asia (Arabian Peninsula)	246
Turkestan	Asia (Central Asia)	194
Karakum		(35)
Kyzylkum		(30)
North American	North American (Southwestern America)	130
Great Basin		(49)
Chihuahuan		(45)
Sonoran		(31)

Desert	Continent (Address)	Area (10^4 km²)
Gobi	Asia (Mongolia, Northeastern China)	130
Patagonian	South America (Southern Argentina)	67
Thar (Great Indian)	Asia (India, Pakistan)	60
kalahari	Africa (South Africa)	57
Namib		(14)
Takla Makan	Asia (Northwestern China)	52
Iranian	Asia (Iran)	39
Kavir		(26)
Lut		(5)
Atacama	South America (Chile, Peru)	36

Desert names and areas are from The Encyclopedia Americana, International Edition, Vol. 9, 1987.

Regional desert names and areas (numbers in parentheses) are from The Encyclopedia Britannica, 15th ed., 1990.

Nationwide Horizontal Crustal Deformation Field Observed by GNSS (Apr. 2019-Apr. 2020)

By observing radio waves from GNSS (Global Navigation Satellite Systems) satellites, it is possible to accurately monitor continuous changes in position of GNSS CORSs (GNSS Continuously Operating Reference Stations). The figure shows horizontal crustal deformation obtained from the observation data during a one-year period (April 2019 to April 2020) at GNSS CORSs distributed all over Japan by the Geospatial Information Authority of Japan. The starting points of arrows shows the location of the observation points on the map. The relative positional change of each station is shown by the direction and size of the arrows, with respect to the station "Fukue" in Goto City, Nagasaki Prefecture.

Postseismic deformation after the 2011 off the Pacific coast of Tohoku Earthquake which occurred on March 11, 2011 can be seen across a broad area, especially in eastern Japan. Furthermore, Crustal deformation associated with volcanic activity on Ioto Island can be seen. Deformation in other areas indicates secular crustal deformation associated with plate motions.

This figure clearly shows the current state of crustal deformation in Japan. The latest results of crustal deformation can be obtained from the Geospatial Information Authority of Japan via the internet (https://mekira.gsi.go.jp/).

Figure 2 ☆ Fixed point (with respect to): "Fukue" station (Goto City, Nagasaki Pref.)

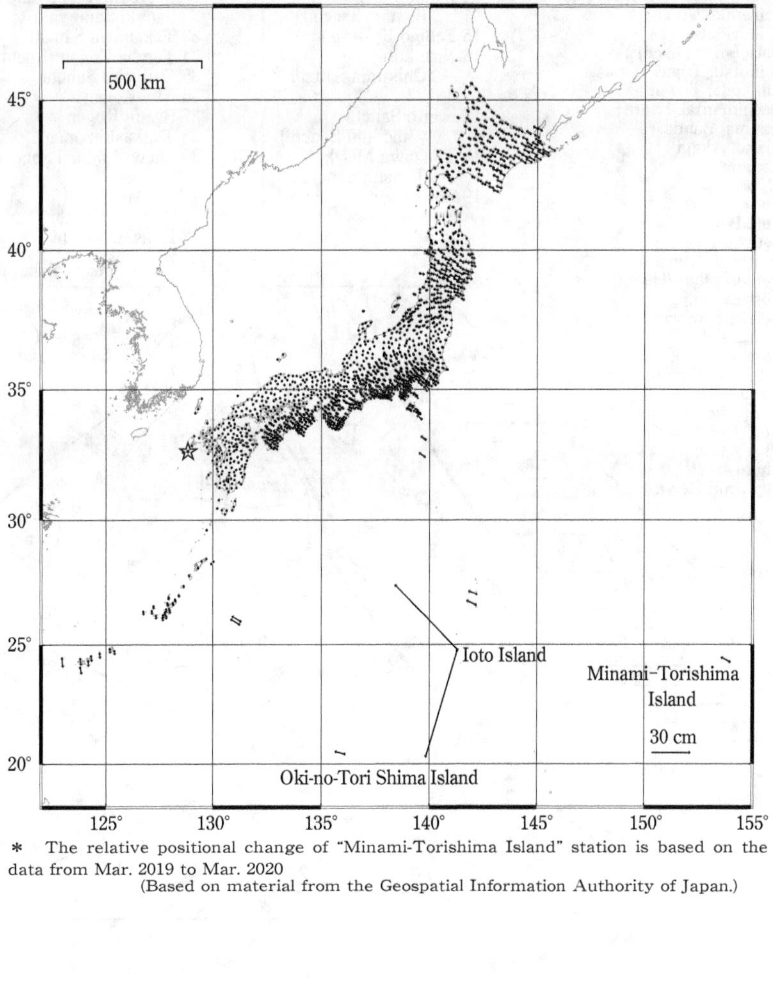

* The relative positional change of "Minami-Torishima Island" station is based on the data from Mar. 2019 to Mar. 2020

(Based on material from the Geospatial Information Authority of Japan.)

Land Classification of Japan

Major landform regions

- A_1 Hokkaido main portion Inner zone (Kuril arc within the arc)
- A_2 Hokkaido main Outer zone (Kuril arc outside the arc)
- B_1 Northeast Japan arc Inner arc (Inner zone)
- B_2 Northeastern Japan Kogaiko (Outer zone)
- C_1 Izu-Ogasawara Inner arc (Inner zone)
- C_2 Izu-Ogasawara Outer arc (Outer zone)
- D_1 Southwest Japan arc Inner zone
- D_2 Southwest Japan arc Outer zone
- DC_1 Central Japan west zone (Central Mountains)
- DC_2 Central Japan Higashitai (Kanto)
- E_1 Ryukyu arc within the arc (Inner zone)
- E_2 Ryukyu Kogaiko (Outer zone)

The main volcanoes and volcanic area

① Shiretoko, Akan
② Shikaribetsu
③ Daisetsu, Tokachi
④ Shikotsu, Toya
⑤ Hakkoda, Towada
⑥ Hachimantai, Iwate
⑦ Azuma, Bandai
⑧ Nikko, Akagi
⑨ Asama
⑩ Myoko
⑪ Yatsugatake
⑫ Fuji, Izu
⑬ Central Kyushu
⑭ Southern Kyushu

The main plain (Heiya)

a Konsen Heiya
b Tokachi Heiya
c Ishikari Heiya
d Kamikita Heiya
e Kitakami Bonchi, Sendai Heiya
f Kanto Heiya
g Echigo Heiya (Niigata Heiya)
h Nobi Heiya
i Osaka Heiya
j Chikushi Heiya
k Miyazaki Heiya

Major mountains (Sanchi)

1 Kitami Sanchi
2 Teshio Sanchi
3 Shiranuka Kyuryo
4 Hidaka Sanmyaku
5 Yubari Sanchi
} A

6 Kitakami Sanchi (Kochi)
7 Abukuma Sanchi (Kochi)
8 Yamizo Sanchi
9 Ashio Sanchi
} B_2

10 Oshima Sanchi
11 Ou Sanmyaku
12 Dewa Sanchi (Kyuryo)
13 Asahi Iide Sanchi
14 Joetsu Sanchi (Echigo Sanchi)
15 Echigo Kyuryo
16 Nishikubiki Chikuma Sanchi
} B_1

17 Kanto Sanchi (Chichibu Sanchi)
18 Tanzawa Misaka Tenshu Sanchi
19 Boso Miura Kyuryo
} DC_2

20 Hida Sanmyaku
21 Kiso Sanmyaku
22 Akaishi Sanmyaku
23 Noto Kyuryo
24 Hida Kogen (Kochi)
25 Minomikawa Kogen
26 Kaga Mino Sanchi (Ryohaku Sanchi)
27 Ibuki Sanchi
} DC_1

28 Suzuka Nunobiki Sanmyaku
29 Takami Sanchi
30 Shigaraki Yamato Kogen (Kasagi Sanchi)
31 Ikoma Kongo Izumi Sanmyaku
32 Sanuki Sanmyaku
33 Takanawa Sanchi
34 Tamba Kogen (Kochi)
35 Chugoku Sanchi
36 Kibi Kogen
37 Iwami Kogen
38 Tsukushi Sanchi
39 Minou Chikuhi Sanchi
} D_1

40 Kii Sanchi
41 Shikoku Sanchi
42 Kyushu Sanchi
} D_2

Source: Sohei Kaizuka

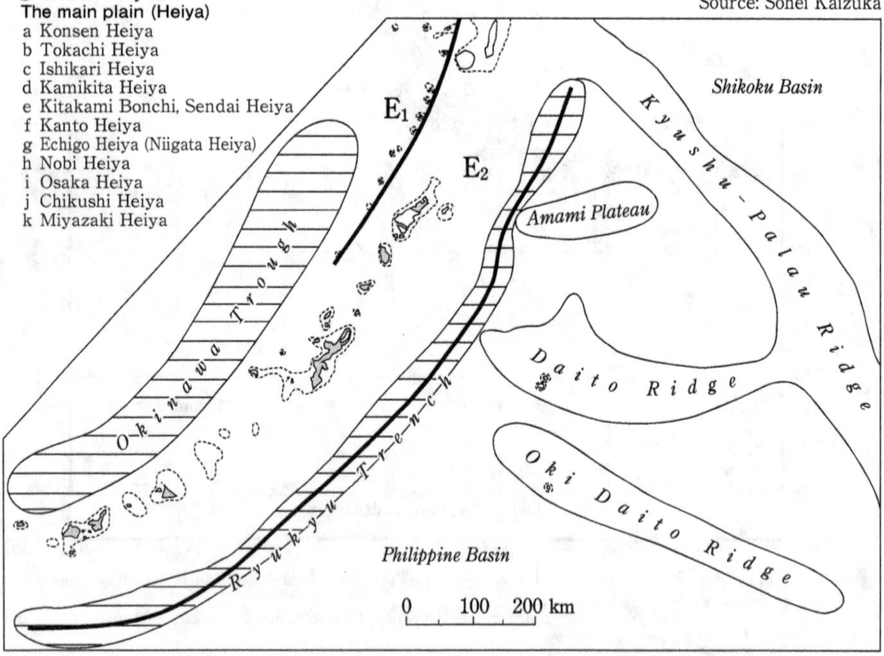

Figure 3

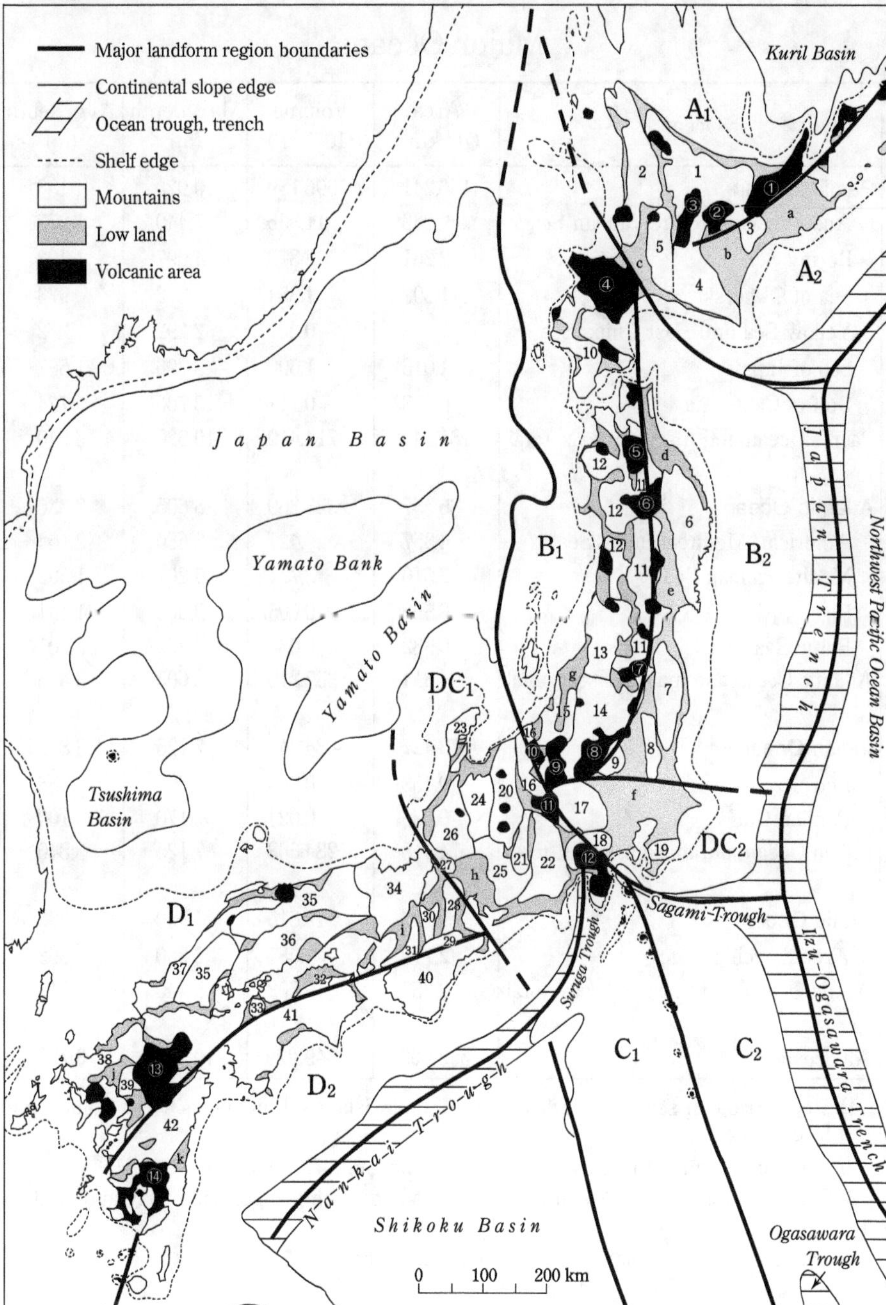
Major landform region boundaries
Continental slope edge
Ocean trough, trench
Shelf edge
Mountains
Low land
Volcanic area
Kuril Basin
Japan Basin
Yamato Bank
Yamato Basin
Tsushima Basin
Japan Trench
Northwest Pacific Ocean Basin
Sagami Trough
Suruga Trough
Izu-Ogasawara Trench
Nankai Trough
Shikoku Basin
Ogasawara Trough
0 100 200 km

Major Oceans

Ocean	Area ($10^6 km^2$)	Volume ($10^6 km^3$)	Max. depth (m)	Avg. depth (m)
Pacific Ocean	166.241	696.189	10 920	4 188
Australasian Mediterranean Sea[1)]	9.082	11.366	7 440	1 252
Bering Sea	2.261	3.373	4 097	1 492
Sea of Okhotsk	1.392	1.354	3 372	973
Yellow Sea and East China Sea	1.202	0.327	2 292	272
Sea of Japan	1.013	1.690	3 796	1 667
Gulf of California	0.153	0.111	3 700	724
Pacific Ocean and marginal seas, total	181.344	714.410	10 920	3 940
Atlantic Ocean	86.557	323.369	8 605	3 736
American Mediterranean Sea[2)]	4.357	9.427	7 680	2 164
Mediterranean Sea	2.510	3.771	5 267	1 502
Black Sea	0.508	0.605	2 200	1 191
Baltic Sea	0.382	0.038	459	101
Atlantic Ocean and marginal seas, total	94.314	337.210	8 605	3 575
Indian Ocean	73.427	284.340	7 125	3 872
Red Sea	0.453	0.244	2 300	538
Persian Gulf	0.238	0.024	170	100
Indian Ocean and marginal seas, total	74.118	284.608	7 125	3 840
Arctic Ocean	9.485	12.615	5 440	1 330
Arctic Archipelago[3)]	2.772	1.087	2 360	392
Arctic Ocean and marginal seas, total	12.257	13.702	5 440	1 117
All oceans	**362.033**	**1 349.929**	**10 920**	**3 729**

1) Mediterranean sea surrounded by the Sunda Islands, Philippines, New Guinea, and Australia.
2) Generic name for the Caribbean Sea and the Gulf of Mexico.
3) Generic name for the North-western Canadian Archipelago, Baffin Bay and Hudson Bay.

Sources: Menard and Smith (1966), Robert L. Fisher (1993)

Frequency Distribution of Depth in Oceans

Range of depth	Pacific Ocean	Atlantic Ocean	Indian Ocean	All oceans
m m	%	%	%	%
0– 200	5.6	8.7	4.1	7.5
200–1 000	3.4	5.9	2.9	4.4
1 000–2 000	3.9	5.2	3.6	4.4
2 000–3 000	7.2	9.6	10.0	8.5
3 000–4 000	20.9	18.9	25.0	20.9
4 000–5 000	32.5	30.4	36.3	31.7
5 000–6 000	24.7	20.6	16.8	21.2
6 000–7 000	1.6	0.6	1.2	1.2
7 000 or more	0.2	Less than 0.1	Less than 0.1	0.1

Source: Menard and Smith (1966)

Major Ocean Trenches

Trench	Max. depth (m)	Coordinates of the deepest point		Total length (km)	Avg. width (km)	Location in Figure 4
Kuril-Kamchatka Trench	9 550	44° 09′ N,	150° 30′ E	2 200	120	①
Japan Trench	8 058	36° 05′ N,	142° 46′ E	800	100	②
Izu-Ogasawara Trench	9 780	29° 28′ N,	142° 42′ E	850	90	③
Mariana Trench	10 920	11° 22′ N,	142° 36′ E	2 550	70	④
Yap (West Caroline) Trench	8 946	10° 30′ N,	138° 41′ E	700	40	⑤
Palau Trench	8 054	7° 52′ N,	134° 57′ E	400	40	⑥
Nansei-Shoto (Ryukyu) Trench	7 480	24° 52′ N,	128° 02′ E	1 350	60	⑦
Philippine Trench	10 057	10° 38′.5N,	126° 36′ E	1 400	60	⑧
Vityaz (East Melanesia) Trench	6 150	10° 27′ S,	170° 17′ E	550	60	⑨
New Britain Trench	8 940	6° 19′ S,	153° 45′ E	1 100	50	⑩
South Solomon (San Cristobal) Trench	8 322	11° 16′ S,	163° 02′ E	800	40	⑪
North New Hebrides (Santa Cruz) Trench	9 175	12° 28′ S,	165° 51′ E	500	70	⑫
South New Hebrides Trench	7 570	20° 37′ S,	168° 37′ E	1 200	50	⑬
Tonga Trench	10 800	23° 15′ S,	174° 44′ W	1 400	55	⑭
Kermadec Trench	10 047	31° 53′ S,	177° 21′ W	1 500	60	⑮
Aleutian Trench	7 679	50° 51′ N,	177° 11′ E	3 700	50	⑯
Middle America Trench	6 662	14° 02′ N,	93° 39′ W	2 800	40	⑰
Peru Trench	6 262	9° 20′ S,	80° 42′ W	1 800	100	⑱
Peru-Chile Trench	8 170	23° 28′ S,	71° 21′ W	3 400	100	⑲
Puerto Rico Trench	8 605	19° 55′ N,	65° 27′ W	1 500	120	⑳
South Sandwich Trench	8 325	55° 43′ S,	25° 56′ W	1 450	90	㉑
Sunda (Java) Trench	7 125	10° 19′ S,	109° 59′ E	4 500	80	㉒

Source: Robert L. Fisher (1993), GEBCO Gazetteer of Undersea Feature Names (IHO-IOC Publication B-8), and information from JHOD (Hydrographic and Oceanographic Department, Japan Coast Guard) and JODC (Japan Oceanographic Data Center).

Figure 4

Distribution of Ocean Trenches

Dotted lines indicate the plate boundaries which are referred to in the NUVEL-1A model (DeMets *et al.*, 1994) The two-letter codes are plate name abbreviations.

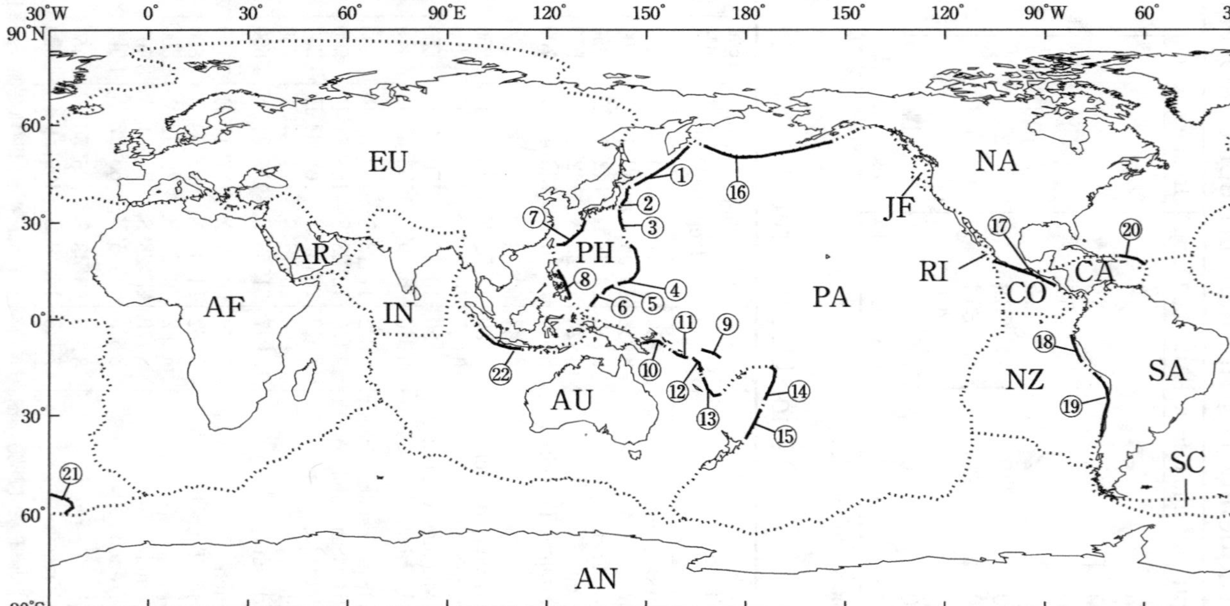

AF: African plate, AN: Antarctic plate, AR: Arabian plate, AU: Australian plate, CA: Caribbean plate, CO: Cocos plate, EU: Eurasian plate, IN: Indian plate, JF: Juan de Fuca plate, NA: North American plate, NZ: Nazca plate, PA: Pacific plate, PH: Philippine Sea plate, RI: Rivera plate, SA: South American plate, SC: Scotia plate.

Figure 5

Submarine Topography of the Western Pacific Ocean

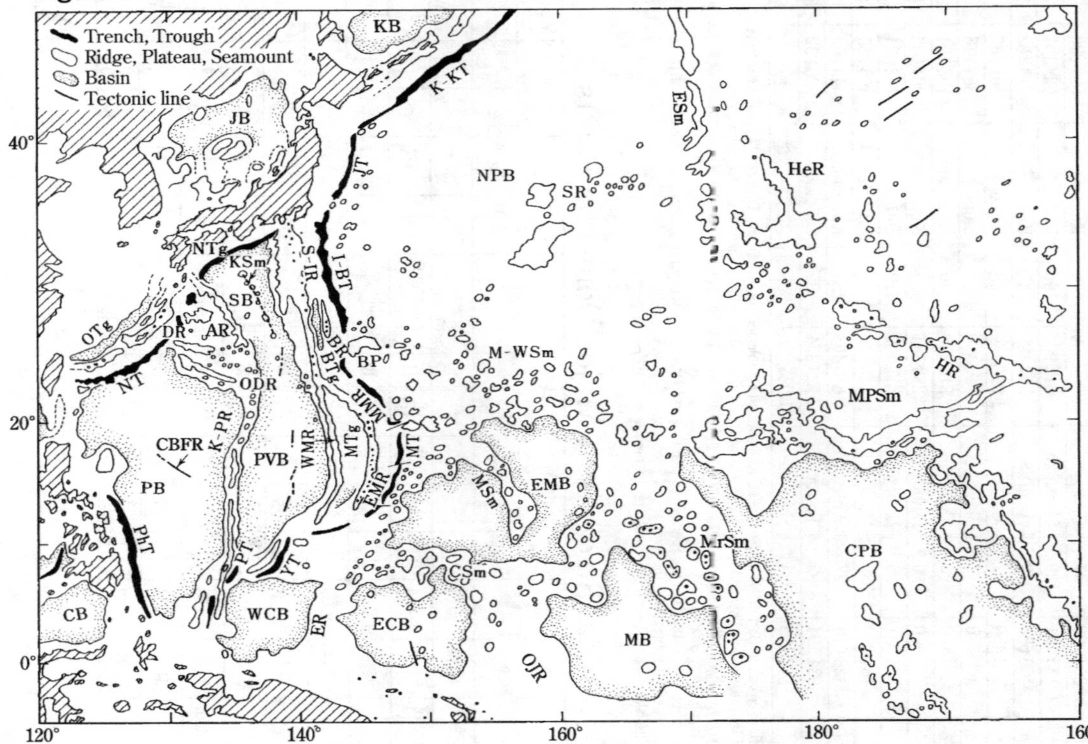

NPB Northwest Pacific Basin
CPB Central Pacific Basin
EMB East Mariana Basin
MB Melanesian Basin
ECB East Caroline Basin
WCB West Caroline Basin
PVB Parece Vela Basin [West Mariana Basin]
PB Philippine Basin (West Philippine Basin)
CB Celebes Basin
SB Shikoku Basin
JB Japan Basin
KB Kuril Basin
OTg Okinawa Trough
BTg Ogasawara Trough
MTg Mariana Trough
K-KT Kuril-Kamchatka Trench
JT Japan Trench
I-BT Izu-Ogasawara Trench
MT Mariana Trench
YT Yap Trench
PT Palau Trench
NT Nansei-Shoto Trench
PhT Philippine Trench
HeR Hess Rise
HR Hawaiian Ridge
ESm Emperor Seamount Chain
SR Shatsky Rise
M-WSm Marcus-Wake Seamounts Group
MPSm Mid-Pacific Seamounts
MrSm Marshall Seamounts
MSm Magellan Seamounts
CSm Caroline Seamounts
OJR Ontong-Java Rise (Solomon Rise)
ER Eauripik Rise
S-IR Shichito-Iojima Ridge
WMR West Mariana Ridge
MMR Middle Mariana Ridge
EMR East Mariana Ridge
K-PR Kyushu-Palau Ridge
DR Daito Ridge
ODR Oki-Daito Ridge
KSm Kinan Seamount Chain
BP Ogasawara Plateau
AR Amami Rise (Amami Plateau)
CBFR CBF Rift
NTg Nankai Trough
BR Ogasawara Ridge

Modified from the figure by Kazuo Kobayashi, Takahiro Sato and Nobuhiro Isezaki (1971). The attached undersea feature names are referred to in the General Bathymetric Chart of the Oceans (GEBCO). Another name is shown in square brakets, and old names are in parentheses.

Coastline Length and Number of Islands

Prefecture	Coastline length (m)	Number of islands	Prefecture	Coastline length (m)	Number of islands	Prefecture	Coastline length (m)	Number of islands
Hokkaido	4 401 737	508	Ishikawa	583 476	110	Okayama	542 752	87 (2)
Aomori	752 300	114	Fukui	414 763	58	Hiroshima	1 129 208	142 (3)
Iwate	708 490	286	Yamanashi	0	0	Yamaguchi	1 503 182	249 (1)
Miyagi	828 246	311	Nagano	0	0	Tokushima	392 348	88 (2)
Akita	264 220	47	Gifu	0	0	Kagawa	699 300	112 (2)
Yamagata	135 421	29	Shizuoka	512 904	106	Ehime	1 633 141	270 (2)
Fukushima	166 550	13	Aichi	597 199	41	Kochi	713 162	159 (2)
Ibaraki	192 450	7	Mie	1 091 474	233	Fukuoka	661 587	62
Tochigi	0	0	Shiga	0	0	Saga	364 918	55
Gunma	0	0	Kyoto	315 235	49	Nagasaki	4 196 104	971
Saitama	0	0	Osaka	233 384	0	Kumamoto	1 084 596	178
Chiba	533 381	95	Hyogo	847 720	110	Oita	772 874	109
Tokyo	761 254	330	Nara	0	0	Miyazaki	405 955	179
Kanagawa	428 618	27	Wakayama	650 463	253	Kagoshima	2 643 000	605
Niigata	634 594	92	Tottori	129 334	35	Okinawa	2 026 852	363
Toyama	147 115	3	Shimane	1 026 786	369	Hokkaido, Honshu, Shikoku, Kyushu		4

Total length of coastline: 35 126 093 m (Source : Costal statistics published by Water and Disaster Management Bureau, MLIT). Number of Japanese islands: 6852. Area of territorial waters: approx. 430 000 km^2 (including internal waters).

Area of Exclusive Economic Zone: approx 4 050 000 km^2 (Source: Material provided by the Hydrographic and Oceanographic Department, Japan Coast Guard).

Note) 1 Numbers in ( ) show the number of the islands across the two prefectures and the total number of Japanese islands is recorded by subtracting the overlapped 7 locations of these islands.

2 The islands, which says here, indicate the perimeter of 0.1 km or more.

Tidal Streams at Major Domestic Points

No.	Point	Location		Mean spring tide period		Max. spring tide period	
		North lat. (° ′)	East long. (° ′)	Direction	Velocity (knot)	Direction	Velocity (knot)
1	East entrance of Tsugaru Kaikyo*	41 37	140 56	. . . . . .	. . . . . .	E W	7 1
2	Entrance of Tokyo Bay	35 17	139 44	NW SE	1.3 1.3	NW SE	1.9 1.9
3	Irako Suido	34 34	137 00	NW SE	1.9 2.1	NW SE	2.3 2.4
4	Naruto Kaikyo	34 14	134 39	N S	8.4 8.6	N S	10.5 9.9
5	Tomogashima Suido	34 17	134 59	N S	2.5 2.5	N S	3.6 3.3
6	Akashi Kaikyo	34 37	135 01	WNW ESE	5.0 4.6	WNW ESE	6.8 5.2
7	Harimanada	34 43	134 30	W E	0.5 0.5	W E	0.6 0.6
8	Bisan Seto	34 25	133 57	W E	2.6 2.4	W E	3.1 2.7
9	Onomichi Suido	34 24	133 12	ENE WSW	2.3 2.3	ENE WSW	2.7 2.5
10	Nagasaki Seto	34 17	133 11	ESE WNW	2.0 2.0	ESE WNW	2.8 2.2
11	Kurushima Kaikyo	34 07	133 00	S N	7.3 7.3	S N	10.3 9.0
12	Obatake Seto	33 58	132 11	E W	6.1 6.4	E W	7.3 6.9
13	Tsurushima Suido	33 58	132 43	NE SW	2.5 2.6	NE SW	3.3 3.0
14	Kanmon Kaikyo	33 58	130 58	SW NE	7.5 7.5	SW NE	10.0 9.7
15	Hayasui Seto	33 18	131 58	N S	5.2 3.9	N S	6.3 4.0
16	Hirado Seto	33 23	129 34	NE SW	4.7 3.9	NE SW	5.4 5.1
17	Hayasaki Seto	32 35	130 10	ESE WNW	5.6 5.6	ESE WNW	6.2 6.8
18	Entrance of Kagoshima Bay	31 12	130 41	NNE SSW	0.5 0.5	NNE SSW	0.9 0.9

1 knot = 1852 m/h ≒ 0.51 m/s

Source: Hydrographic and Oceanographic Department, Japan Coast Guard. (Upper line – Flood Strem, Lower line – Ebb Strem)

* Because Tsugaru Strait is largely influenced by the ocean current toward from west to east and the diurnal inequality in the tidal current is also larger, the valves of tidal strems are described only at the maximum spring tide.

Tides

The diurnal tide area(with large diurnal inequality)

Place	MHHW	MLHW	MHLW	MLLW	MSL (Z0)
Japan					
Sea of Okhotsk					
Esashi	0.9	0.7	0.6	0.2	0.57
Monbetsu	1.1	0.8	0.7	0.3	0.71
Abashiri	1.1	0.7	0.6	0.3	0.68
Pacific Ocean					
Hokkaido					
Kushiro	1.3	1.1	0.9	0.3	0.87
Tomakomai	1.3	1.1	0.8	0.3	0.88
Muroran	1.4	1.2	0.9	0.4	0.95
Hakodate	0.9	0.7	0.5	0.2	0.57
Honshu East Coast					
Hachinohe	1.2	1.1	0.8	0.3	0.85
Miyako	1.2	1.1	0.8	0.3	0.83
Kamaishi	1.3	1.1	0.8	0.3	0.86
Sendai	1.4	1.2	0.9	0.3	0.93
Onahama	1.2	1.1	0.8	0.3	0.84
Kashima	1.3	1.2	0.8	0.3	0.89
Choshi Gyoko	1.3	1.2	0.9	0.3	0.90
Katsuura	1.3	1.2	0.8	0.3	0.90
Honshu South Coast					
Chiba	1.7	1.7	1.0	0.4	1.20
Tokyo	1.7	1.7	1.0	0.4	1.20
Yokohama	1.6	1.6	1.0	0.4	1.15
Yokosuka	1.6	1.5	1.0	0.4	1.10
Shimoda	1.4	1.4	0.9	0.3	1.01
Shimizu	1.4	1.4	0.8	0.3	0.95

Place	MHHW	MLHW	MHLW	MLLW	MSL (Z0)
Kushimoto	1.5	1.5	0.8	0.4	1.05
Wakayama	1.6	1.6	0.9	0.4	1.11
Izu Islands					
Hachijo Island (Kaminato)	1.2	1.1	0.7	0.3	0.81
Shikoku					
Kochi	1.6	1.6	0.8	0.3	1.08
Uwajima	1.9	1.9	1.0	0.4	1.30
Kyushu East Coast					
Oita	2.0	1.8	1.0	0.4	1.30
Hososhima	1.6	1.6	0.8	0.3	1.06
Aburatsu	1.7	1.7	0.9	0.4	1.16
Japan Sea					
Hokkaido					
Wakkanai*2	0.3			0.1	0.18
Otaru	0.2	0.2	0.2	0.0	0.16
Honshu					
Akita	0.3	0.2	0.2	0.1	0.19
Niigata (West Port)	0.3	0.2	0.2	0.1	0.17
Toyama	0.3	0.3	0.2	0.1	0.22
Tsuruga	0.3	0.2	0.2	0.1	0.18
Maizuru (Section 2)	0.3	0.3	0.2	0.1	0.19
Sakai	0.3	0.3	0.2	0.1	0.17
Hamada	0.5	0.3	0.2	0.2	0.29

Place	MHHW	MLHW	MHLW	MLLW	MSL (Z0)
Setonaikai Sea					
Osaka	1.3	1.3	0.9	0.4	0.95
Kobe	1.3	1.2	1.0	0.4	0.95
Himeji (Shikama)	1.4	1.1	0.8	0.3	0.90
Takamatsu	2.3	1.9	1.0	0.4	1.40
Asia					
Korsakov	1.2	0.8	0.7	0.3	0.73
Vladiostok	0.5	0.5	0.4	0.3	0.43
Hong Kong	2.2	1.7	1.1	0.7	1.39
Keelung	0.9	0.7	0.6	0.2	0.58
Kaohsiung	0.9	0.6	0.5	0.3	0.60
Manila	1.0	0.5	0.3	0.0	0.49
Jakarta*	0.9			0.3	0.60
Aden	2.0	1.8	1.3	0.5	1.39
South America					
Buenos Aires	1.1	1.0	0.6	0.5	0.79
Cape Horn	2.2	1.7	0.9	0.5	1.35
Valparaiso	1.5	1.2	0.5	0.4	0.91
Oceania					
Honolulu	0.6	0.3	0.1	0.0	0.24
Sydney	1.6	1.3	0.6	0.4	0.97
Melbourne	1.0	0.6	0.5	0.2	0.59

The tidal heights refer to the chart datum.
- MHHW : The mean of the higher of the two daily high water.
- MHLW : The mean of the higher of the two daily low water.
- MS(Z0) : Height of mean sea level above datum level.
- MLHW : The mean of the lower of the two daily high water.
- MLLW : The mean of the lower of the two daily low water.

*The tide is diurnal.

The semi-diurnal tide area

Place	MHWS	MHWN	MLWN	MLWS	MSL (Z0)
Japan					
Pacific Ocean					
Honshu South Coast					
Mikawa	2.3	1.7	1.0	0.4	1.35
Nagoya	2.4	1.7	1.1	0.5	1.40
Yokkaichi	2.2	1.6	1.0	0.4	1.30
Toba	2.0	1.5	0.9	0.4	1.20
Nansei Islands					
Naze	2.0	1.5	0.8	0.3	1.15
Naha	2.0	1.5	0.8	0.4	1.18
Ishigaki	1.7	1.3	0.8	0.4	1.07
Setonaikai Sea					
Sakaide	2.9	2.3	1.1	0.5	1.70
Mizushima	3.3	2.6	1.3	0.5	1.90
Fukuyama	3.6	2.8	1.4	0.6	2.10
Onomichi	3.5	2.7	1.3	0.5	2.00
Hiroshima	3.4	2.6	1.4	0.6	2.00
Matsuyama	3.3	2.5	1.3	0.5	1.90
Tokuyama	3.1	2.3	1.3	0.5	1.80
Moji	2.3	1.7	0.9	0.3	1.30
Tsushima Strait					
Yahata	1.4	1.0	0.6	0.2	0.80
Hakata	1.9	1.4	0.8	0.3	1.10
Karatsu	2.0	1.5	0.9	0.3	1.18
Izuhara	1.7	1.2	0.7	0.2	0.93

Place	MHWS	MHWN	MLWN	MLWS	MSL (Z0)
East China Sea					
Sasebo	2.8	2.1	1.2	0.5	1.65
Nagasaki (Matsugae)	2.8	2.1	1.2	0.5	1.64
Fukue	2.7	2.0	1.2	0.5	1.59
Suminoe	5.1	3.7	1.9	0.5	2.80
Miike	4.8	3.5	1.8	0.5	2.65
Misumi	3.9	2.9	1.5	0.5	2.21
Yatsushiro	3.8	2.8	1.5	0.5	2.15
Minamata	3.4	2.5	1.4	0.5	1.95
Akune	2.7	2.0	1.2	0.5	1.60
Kagoshima	2.7	2.0	1.1	0.4	1.55
Asia					
Busan	1.2	0.9	0.4	0.1	0.65
Gunsan	6.6	4.9	2.2	0.6	3.62
Incheon	8.6	6.4	2.9	0.4	4.64
Dalian	2.9	2.3	0.9	0.3	1.63
Tanggu	3.5	3.4	2.0	0.7	2.41
Qingdao	4.0	3.2	1.5	0.6	2.39
Wusong	3.4	2.6	1.4	1.0	2.02
Singapore	2.9	2.2	1.3	0.6	1.80
Kolkata	4.8	3.3	1.2	0.6	2.40
Chennai	1.1	0.8	0.4	0.1	0.65
Colombo	0.7	0.5	0.3	0.1	0.38
Mumbai	4.4	3.3	1.9	0.8	2.51
South America					
Rio de Janeiro	1.2	0.9	0.4	0.1	0.69

Place	MHWS	MHWN	MLWN	MLWS	MSL (Z0)
Europe					
Naples	0.4	0.3	0.1	0.1	0.20
Marseille	0.4	0.4	0.3	0.2	0.30
Lisbon	3.8	3.0	1.5	0.6	2.20
Antwerp	6.2	5.2	1.1	0.5	3.17
Hamburg	4.2	3.8	0.3	0.1	2.29
Copenhagen	0.1	0.0	0.0	-0.1	0.00
Oslo	0.8	0.8	0.6	0.5	0.66
London	7.1	5.9	1.3	0.5	3.67
Southampton	4.5	3.7	1.8	0.5	2.91
Liverpool	9.4	7.5	3.2	1.1	5.25
Africa					
Port Said	0.6	0.5	0.4	0.3	0.45
Suez	1.9	1.6	0.7	0.4	1.14
Cape Town	1.7	1.3	0.7	0.2	0.98
North America					
Bay of Fundy	7.9	6.9	1.8	0.8	4.42
Quebec	5.2	3.9	1.0	0.3	2.62
Boston	3.1	2.7	0.4	0.0	1.55
New York	1.6	1.3	0.2	0.0	0.77
Oceania					
Fiji(Suva)	1.8	1.6	0.6	0.4	1.15
Auckland	3.3	2.8	0.9	0.4	1.89
Wellington	1.8	1.4	0.7	0.4	1.11
Darwin	7.0	5.1	3.3	1.4	4.19

- MHWS : The mean of the high water at spring tide.
- MLWN : The mean of the low water at neap tide.
- MHWN : The mean of the high water at neap tide.
- MLWS : The mean of the low water at spring tide.

(Source : Materials provided by "Hydrographic and Oceanographic Department, Japan Coast Guard" and "UK Hydrographic office")

Cotidal and Co-Amplitude Chart for Semidiurnal Tide in the Oceans near Japan (M_2 Tide)

Figure 6

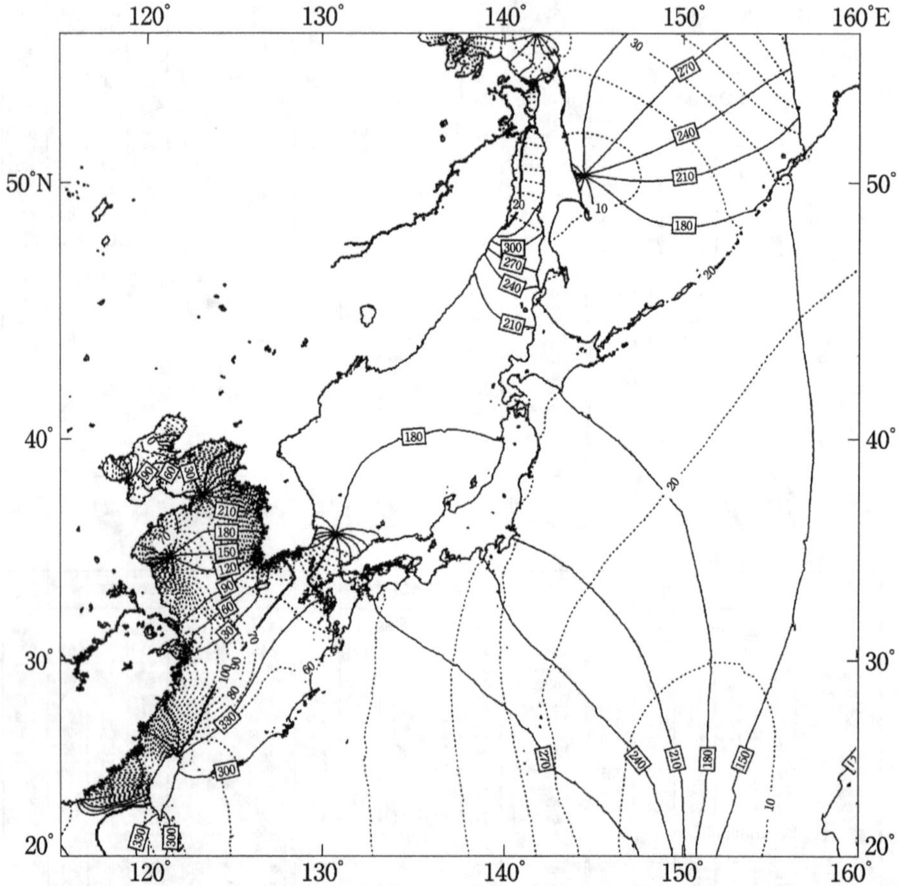

Solid lines are the cotidal lines connecting points where the tidal hour occurred at the same time. The time delay between the passing of the meridian of 0° longitude by the moon and the high tide is indicated in angles. Dotted lines are the co-amplitude contours. (Unit: cm)

(Koji Matsumoto, 2000)

Ocean Current around Japan

Figure 7

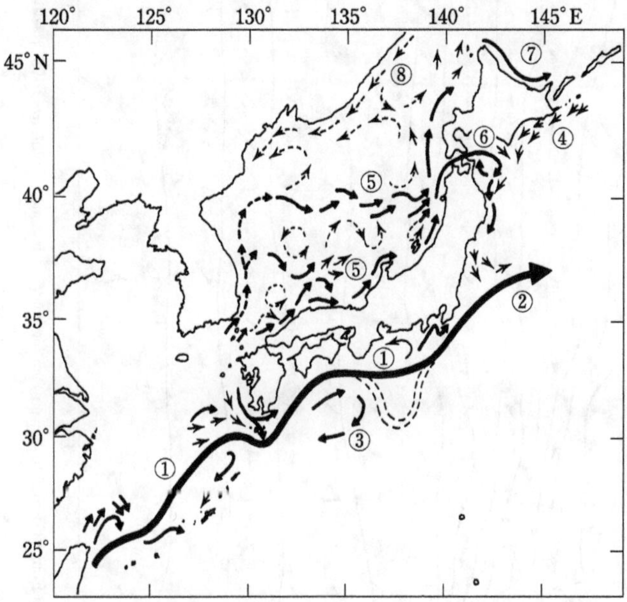

Source: Japan Oceanographic Data Center

This figure is a schematic of ocean current in summer. In winter, the extension of the Tsugaru Warm Current to the east becomes smaller and at the same time, Oyashio becomes stronger and goes down south remarkably. Furthermore, the Soya Warm Current disappears in the winter season.

①Kuroshio (2–3 knots), ② Kuroshio Extension (2–3 knots), ③Kuroshio Countercurrent (around 0.5 knots), ④Oyashio (summer: 0.2–0.5 knots, winter: 0.5–1 knots), ⑤Tsushima Warm Current (summer: around 1 knot, winter: around 0.5 knots), ⑥Tsugaru Warm Current (summer: 1–2 knots, winter: around 0.5 knots; possesses a speed of approx. 2 knots in the Tsugaru Straits), ⑦Soya Warm Current (summer: around 2 knots), ⑧Liman Current.

Changes in Kuroshio Path

Figure 8

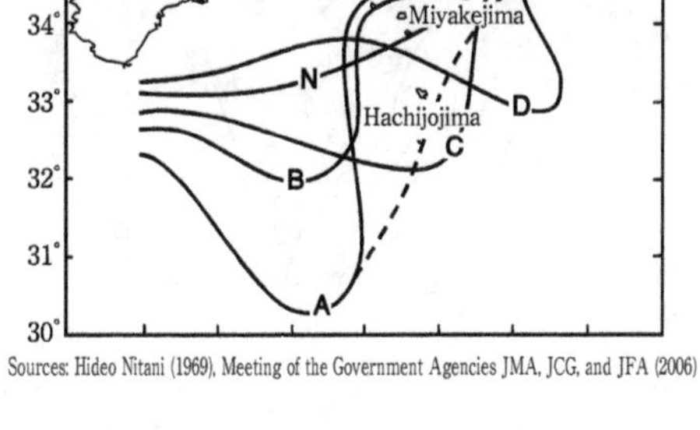

Sources: Hideo Nitani (1969), Meeting of the Government Agencies JMA, JCG, and JFA (2006)

The Kuroshio flow northward, from the eastern shore of Kyushu to the south of Shikoku. In the area spanning from the Kii Peninsula to the Boso Offing, the Kuroshio Current often changes over a period of several weeks or months. The diagram on the left shows representative patterns.

Type-A: At the Kii Peninsula and Enshunada Offing, the current meanders significantly to the south until south of 32° N. Afterwards, there is a pattern in which the current passes north of Hachijo Island (typical large meander channel), and a pattern in which it passes south of Hachijo Island (non-typical large meander channel).

Type-B: medium and small scale meandering of the current in the Enshunada Offing (nearshore non-large-meander).

Type-C: medium and small scale meandering exists in the ocean around the Izu Islands (channel is south of Hachijo Island) (offshore non-large-meander)

Type-D: medium and small scale meandering of the current in the Boso Offing (nearshore non-large-meander).

Type-N: Almost straight (eastern current) along the southern coast of Honshu (nearshore non-large-meander).

Ocean Current (January to March) and Sea Surface Temperature (February) around Japan

Figure 9

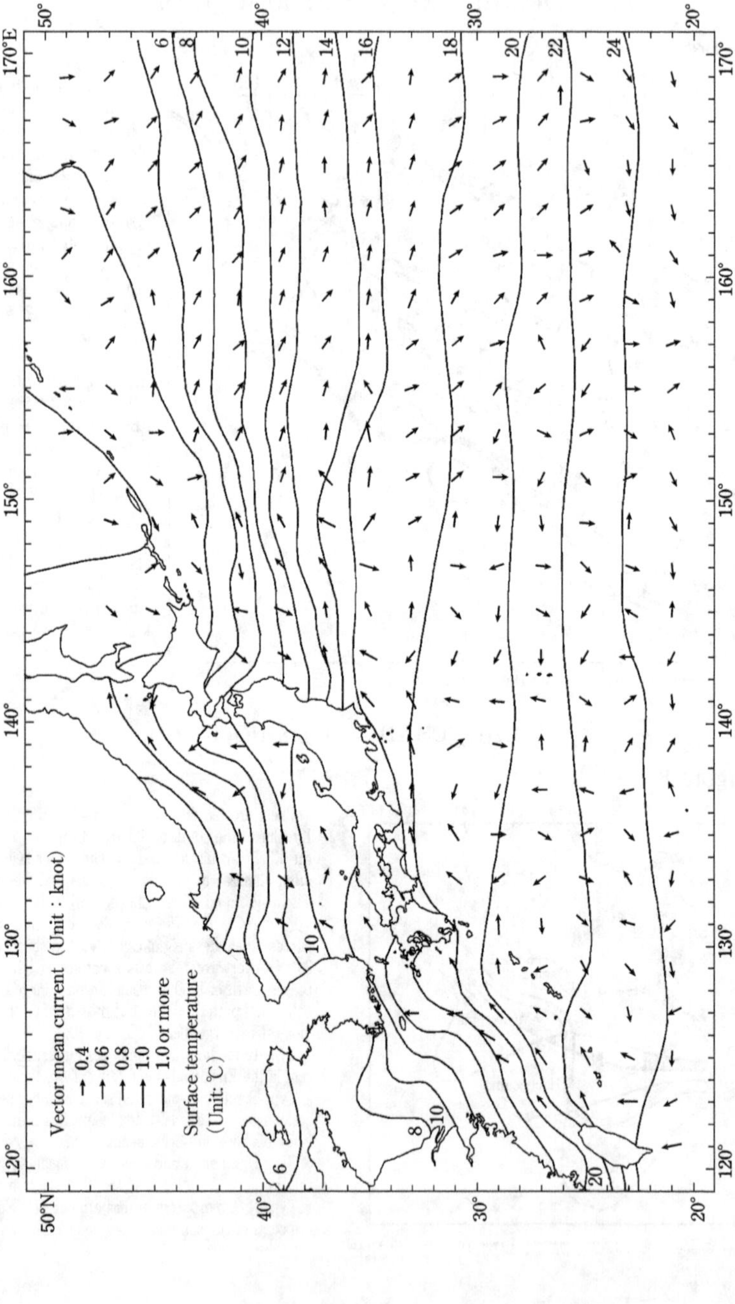

Source: Japan Oceanographic Data Center. (Current is a mean value from 1854 to 2010. Sea surface temperature is a mean value from 1981 to 2010.)

Ocean Current (July to September) and Sea Surface Temperature (August) around Japan

Figure 10

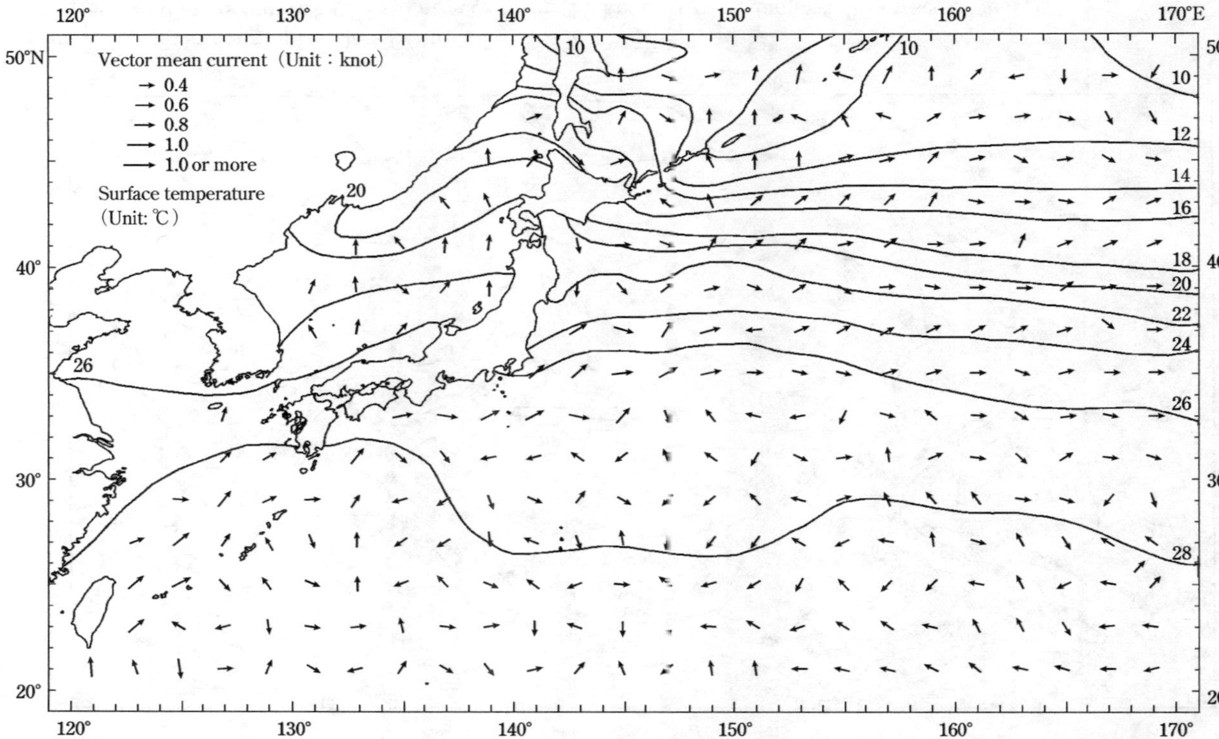

Source: Japan Oceanographic Data Center. (Current is a mean value from 1854 to 2010. Sea surface temperature is a mean value from 1981 to 2010.)

Figure 11

Ocean Current (February)

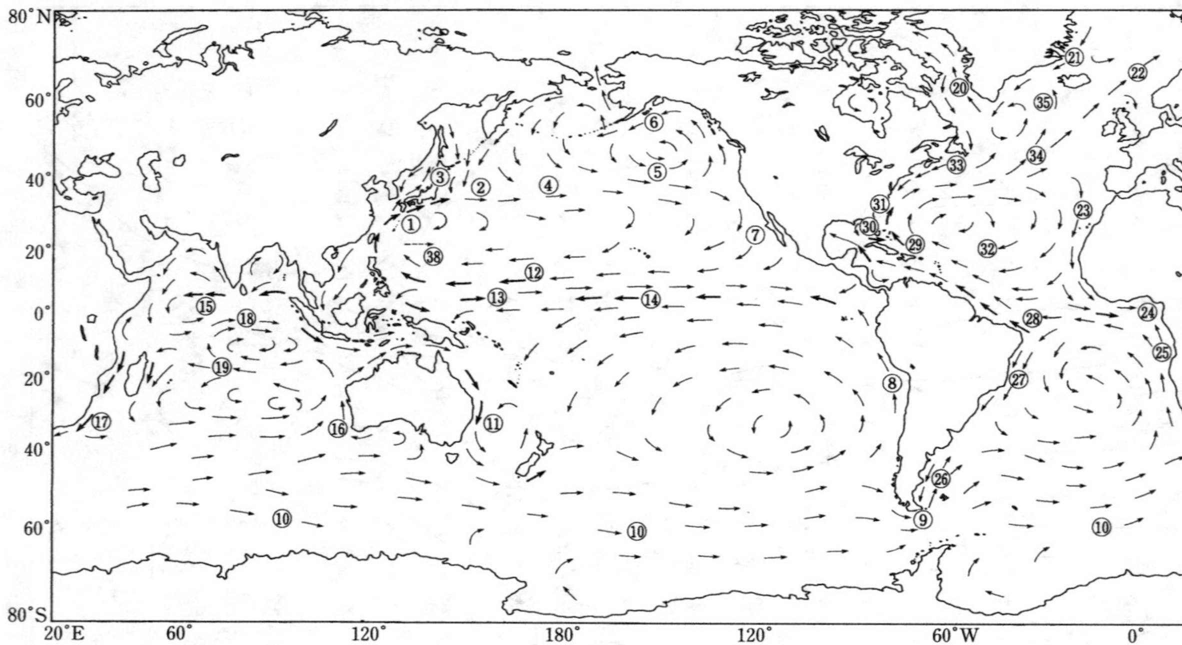

Source: Japan Oceanographic Data Center.

1) ← 0.5–1.0 knots (approx. 0.25–0.5 m/sec), ← 1.0 knots or more (approx. 0.5 m/sec or more)

2) ㊳ is the weak eastward flow which is said to exist as the Subtropical Countercurrent.

3) ㊴ in Figure 12, denotes the equatorial jet which appears as the eastward flow of 3 knots or more at the inter-monsoon period.

Figure 12

Ocean Current (August)

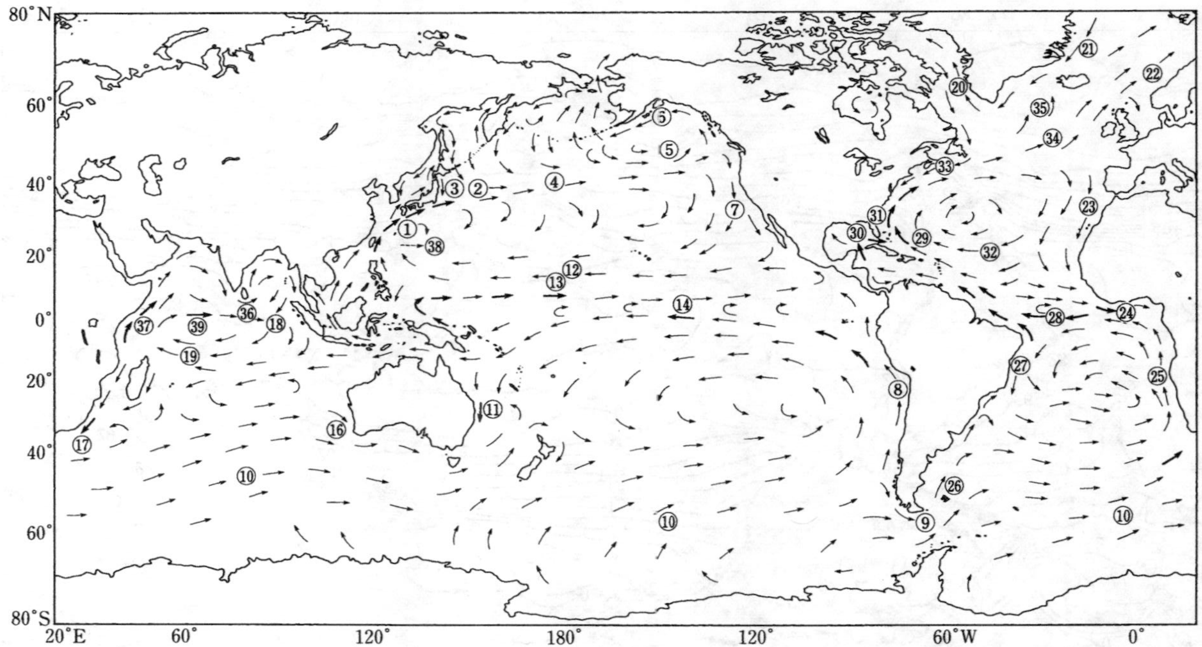

Major Ocean Currents

① Kuroshio
② Kuroshio Extension
③ Oyashio
④ North Pacific Current
⑤ Aleutian Current
⑥ Alaska Current
⑦ California Current
⑧ Peru Current
⑨ Cape Horn Current
⑩ Antarctic Circumpolar Current
⑪ East Australian Current
⑫ North Equatorial Current
⑬ Equatorial Countercurrent
⑭ South Equatorial Current
⑮ Northeast Monsoon Current
⑯ West Australia Current
⑰ Mozambique Current
⑱ Equatorial Countercurrent
⑲ South Equatorial Current
⑳ West Greenland Current
㉑ East Greenland Current
㉒ Norway Current
㉓ Canary Current
㉔ Guinea Current
㉕ Benguela Current
㉖ Falkland Current
㉗ Brazil Current
㉘ South Equatorial Current
㉙ Antilles Current
㉚ Florida Current
㉛ Gulf Stream
㉜ North Equatorial Current
㉝ Labrador Current
㉞ North Atlantic Current
㉟ Irminger Current
㊱ Southwest Monsoon Current
㊲ Somali Current
㊳ Subtropical Countercurrent
㊴ Equatorial Jet

Figure 13

Sea Surface Temperature (February)

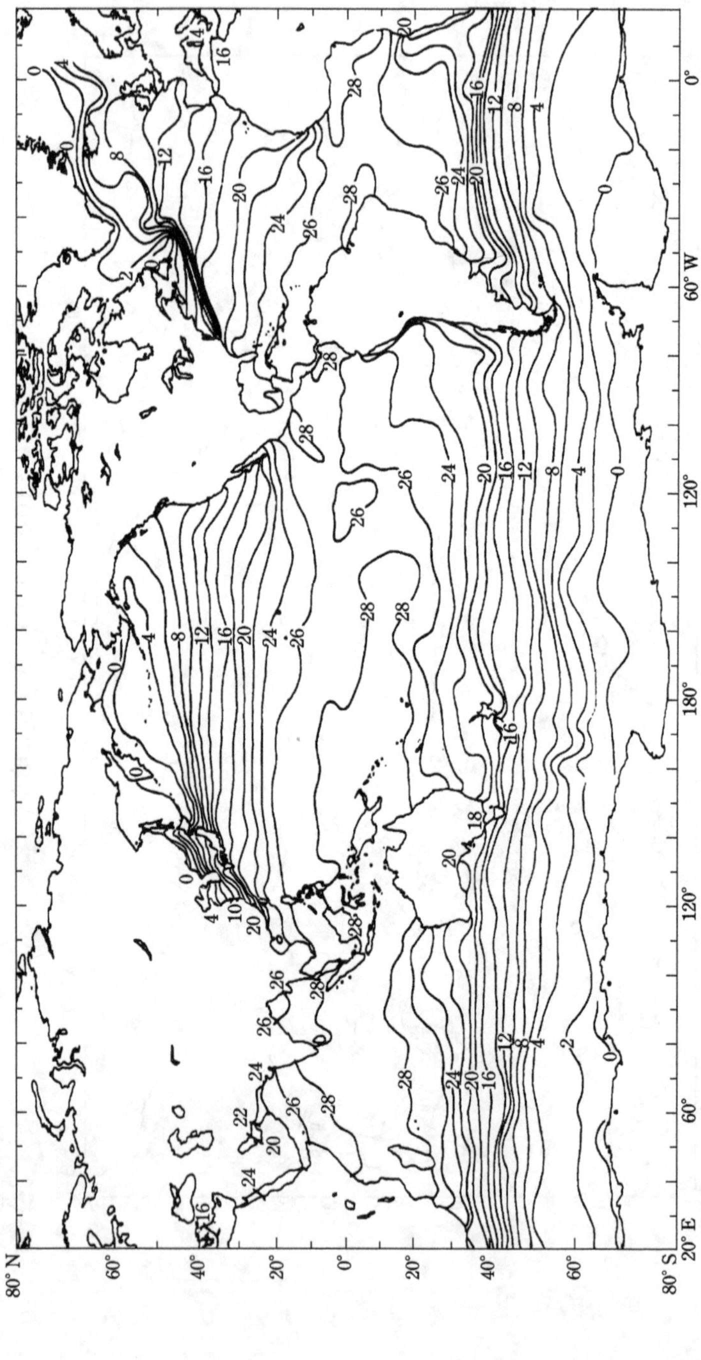

Source: Japan Oceanographic Data Center. The unit is °C. As for the oceans around Japan, refer to pages **562**, **563**.

Figure 14

Sea Surface Temperature (August)

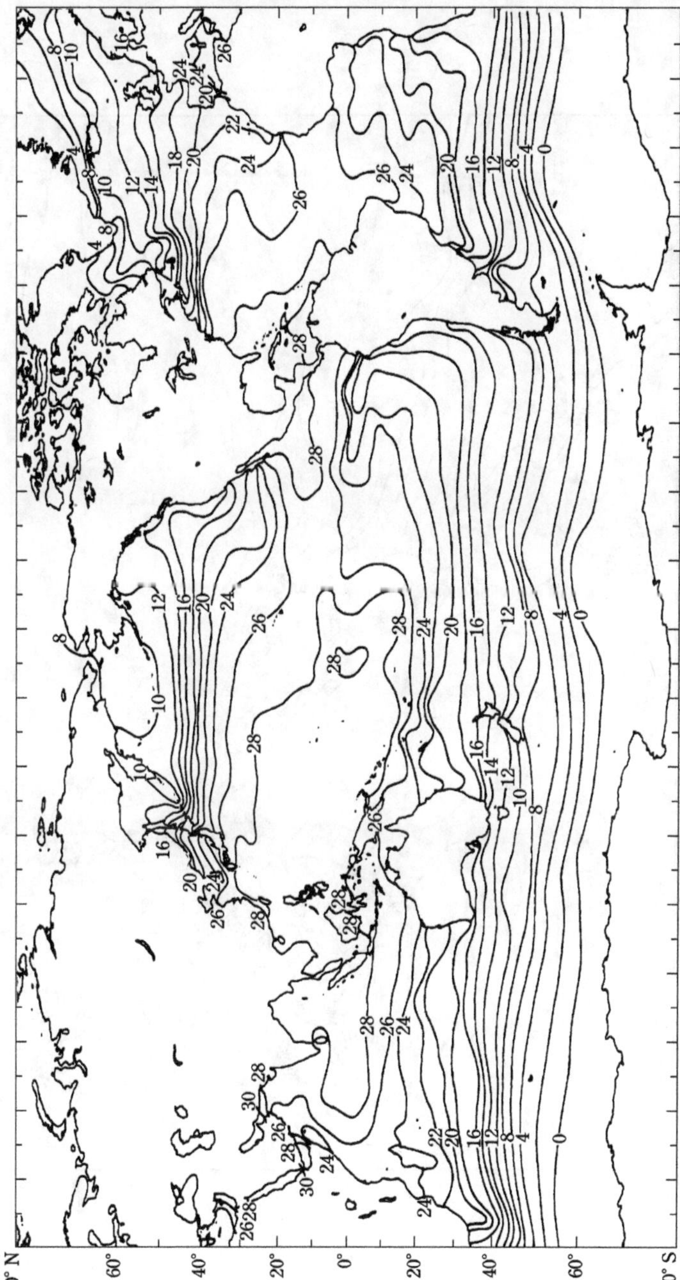

Source: Japan Oceanographic Data Center.

Figure 15 Sea Surface Salinity (PSU) around Japan (January to March)

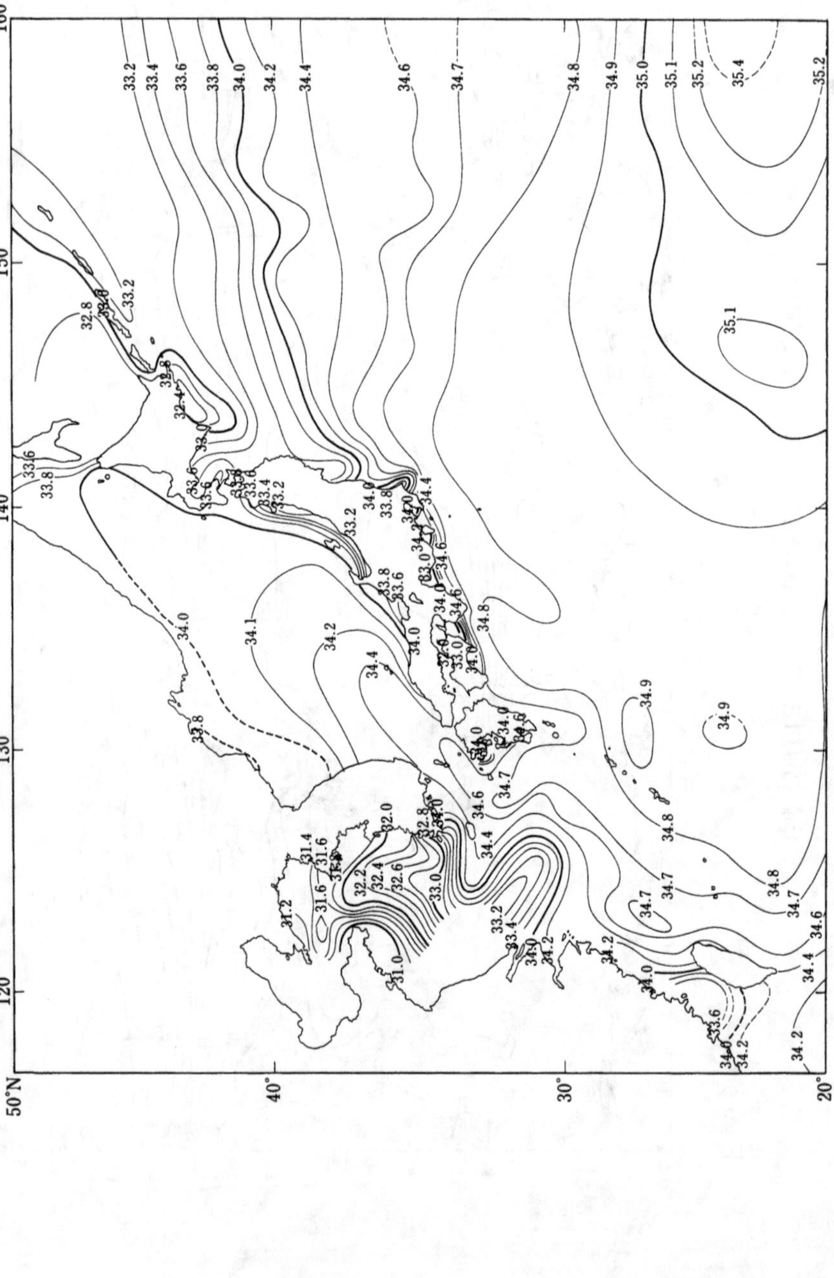

Source: Japan Oceanographic Data Center. (PSU is known as "*Practical Salinity Unit 1978*", unitless)

Figure 16 Sea Surface Salinity (PSU) around Japan (July to September)

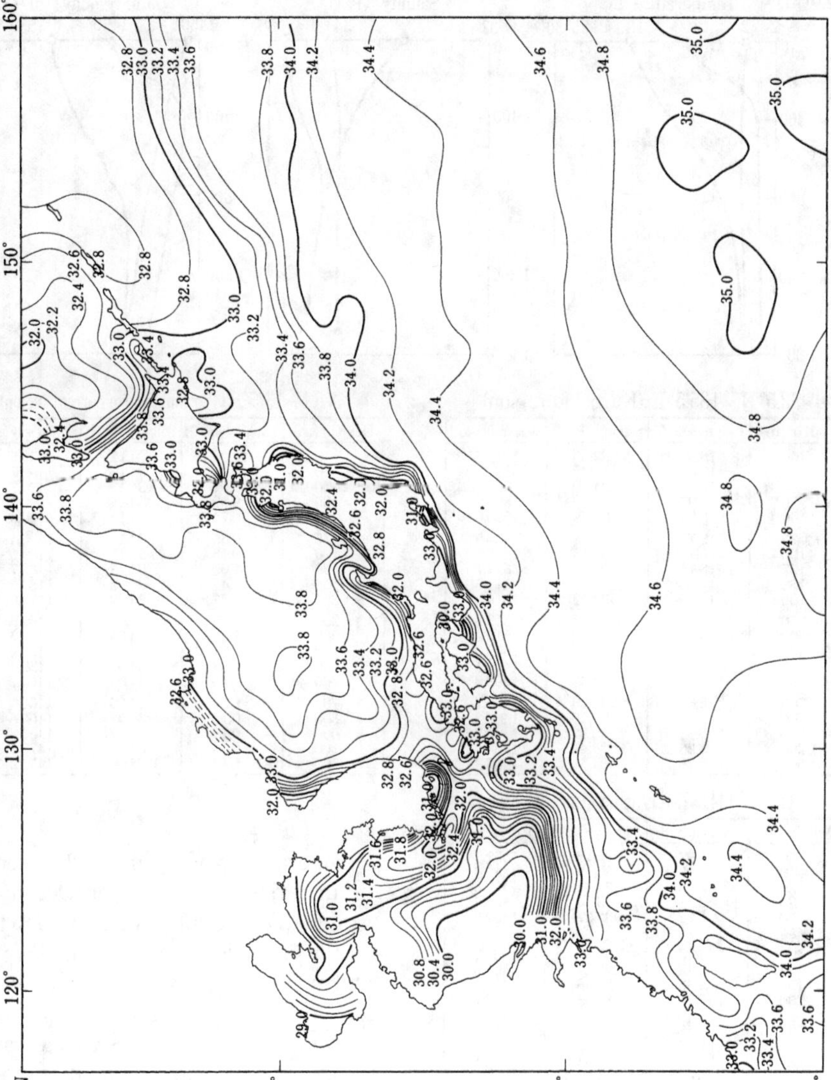

Source: Japan Oceanographic Data Center.

Vertical Profiles of Temperature, Salinity and Sound Velocity in the Oceans around Japan (February)

Figure 17

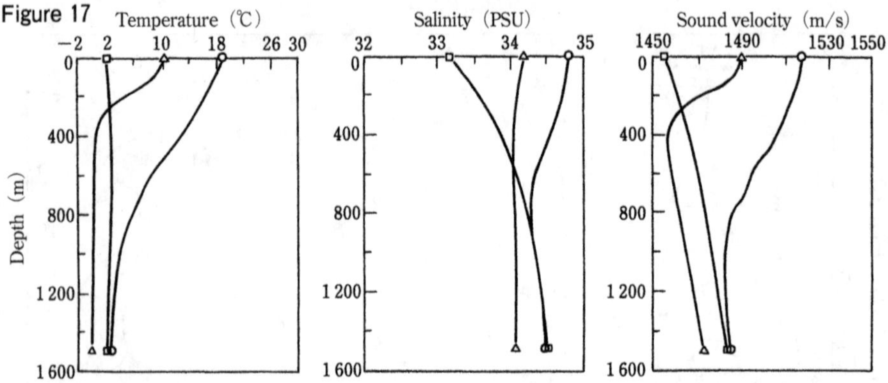

○ 32.5°N, 135.5°E (Kuroshio region)

Depth (m)	Temperature (℃)	Salinity (PSU)	Sound velocity (m/s)
0	18.85	34.80	1519.7
10	19.01	34.79	1520.3
20	18.99	34.79	1520.4
30	18.97	34.79	1520.5
50	18.89	34.79	1520.6
75	18.71	34.78	1520.5
100	18.47	34.78	1520.2
125	18.15	34.77	1519.7
150	17.70	34.76	1518.8
200	16.66	34.72	1516.5
250	15.53	34.68	1513.8
300	14.30	34.62	1510.7
400	12.23	34.51	1505.3
500	10.02	34.39	1499.1
600	8.10	34.32	1493.6
700	6.65	34.28	1489.7
800	5.36	34.29	1486.2
1 000	3.91	34.37	1483.7
1 200	3.26	34.44	1484.4
1 500	2.72	34.51	1487.3

△ 37.5°N, 134.5°E (Tsushima warm current region)

Depth (m)	Temperature (℃)	Salinity (PSU)	Sound velocity (m/s)
0	10.00	34.17	1490.5
10	10.04	34.15	1490.7
20	9.99	34.15	1490.7
30	9.89	34.14	1490.5
50	9.74	34.14	1490.3
75	9.65	34.14	1490.4
100	9.15	34.14	1489.0
125	7.62	34.13	1483.7
150	7.41	34.12	1483.3
200	4.77	34.09	1473.6
250	2.73	34.07	1465.8
300	1.44	34.05	1460.9
400	0.61	34.05	1458.9
500	0.37	34.06	1459.4
600	0.28	34.06	1460.7
700	0.23	34.06	1462.1
800	0.18	34.07	1463.6
1 000	0.15	34.07	1466.8
1 200	0.14	34.07	1470.1
1 500	0.14	34.08	1475.2

□ 41.5°N, 144.5°E (Oyashio current region)

Depth (m)	Temperature (℃)	Salinity (PSU)	Sound velocity (m/s)
0	1.97	33.16	1457.2
10	1.95	33.18	1457.3
20	2.03	33.21	1457.8
30	2.11	33.24	1458.4
50	2.22	33.27	1459.2
75	2.28	33.30	1459.9
100	2.23	33.35	1460.2
125	2.43	33.44	1461.6
150	2.33	33.46	1461.6
200	2.37	33.55	1462.7
250	2.49	33.63	1464.2
300	2.62	33.72	1465.7
400	2.87	33.88	1468.6
500	2.96	34.00	1470.8
600	2.98	34.10	1472.7
700	3.00	34.15	1474.5
800	2.91	34.23	1475.9
1 000	2.76	34.34	1478.8
1 200	2.56	34.44	1481.4
1 500	2.29	34.53	1485.5

Temperature and salinity are based on material from the Japan Oceanographic Data Center. PSU is known as the "Practical Salinity Unit 1978" and unitless. The salinity is equivalent to the amount of solid substances contained in 1 kg of seawater as expressed in grams (but it is not absolutely accurate). Sound velocity was calculated using the UNESCO formula (Chen and Millero formula).

Vertical Profiles of Temperature, Salinity and Sound Velocity in the Oceans around Japan (August)

Figure 18

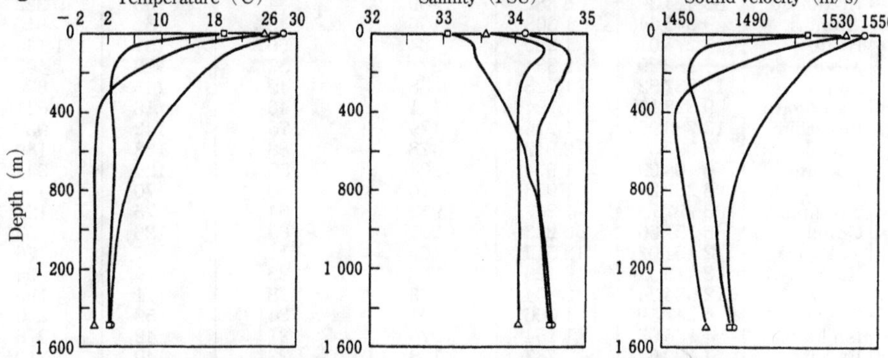

○ 32.5°N, 135.5°E (Kuroshio region)

Depth (m)	Temperature (℃)	Salinity (PSU)	Sound velocity (m/s)
0	28.14	34.11	1542.4
10	27.90	34.14	1542.1
20	27.15	34.25	1540.7
30	26.22	34.34	1538.8
50	24.03	34.51	1534.0
75	21.81	34.66	1528.9
100	20.08	34.73	1524.7
125	18.25	34.72	1520.0
150	17.76	34.72	1518.9
200	16.19	34.67	1515.0
250	14.70	34.60	1511.1
300	13.40	34.55	1507.6
400	11.13	34.44	1501.5
500	8.96	34.35	1495.2
600	7.21	34.32	1490.2
700	6.13	34.32	1487.7
800	5.05	34.32	1485.0
1 000	3.84	34.40	1483.5
1 200	3.15	34.47	1484.0
1 500	2.59	34.54	1486.8

△ 37.5°N, 134.5°E (Tsushima warm current region)

Depth (m)	Temperature (℃)	Salinity (PSU)	Sound velocity (m/s)
0	25.29	33.60	1535.3
10	24.36	33.64	1533.2
20	21.72	33.80	1526.9
30	19.05	34.06	1519.9
50	15.62	34.28	1510.3
75	13.33	34.40	1503.5
100	11.47	34.40	1497.6
125	9.58	34.36	1491.3
150	8.49	34.25	1487.5
200	5.78	34.15	1477.7
250	3.50	34.09	1469.1
300	1.99	34.07	1463.4
400	0.61	34.05	1458.9
500	0.35	34.05	1459.3
600	0.23	34.05	1460.5
700	0.21	34.06	1462.0
800	0.17	34.06	1463.5
1 000	0.20	34.07	1467.0
1 200	0.12	34.06	1470.0
1 500	0.14	34.07	1475.2

□ 41.5°N, 144.5°E (Oyashio current region)

Depth (m)	Temperature (℃)	Salinity (PSU)	Sound velocity (m/s)
0	19.20	33.05	1518.7
10	18.07	33.05	1515.6
20	12.99	33.20	1500.0
30	9.63	33.28	1488.5
50	5.85	33.38	1474.6
75	4.55	33.41	1469.7
100	3.76	33.41	1466.8
125	3.68	33.45	1467.0
150	3.16	33.47	1465.2
200	2.66	33.56	1464.0
250	2.62	33.64	1464.8
300	2.70	33.72	1466.0
400	2.94	33.88	1468.9
500	3.08	34.00	1471.4
600	3.13	34.13	1473.4
700	3.07	34.16	1474.9
800	2.95	34.27	1476.2
1 000	2.72	34.37	1478.7
1 200	2.46	34.42	1481.0
1 500	2.25	34.51	1485.3

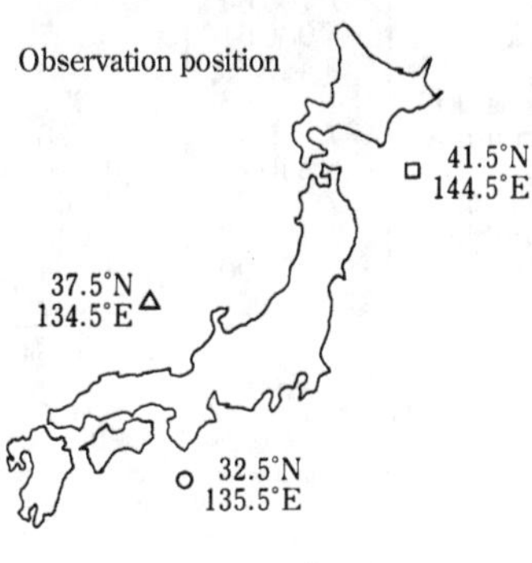

Area, Population, and Land Use by Japanese Prefecture

Prefecture	Area[1] (km^2)	Population[2] (1 000 persons)	Population density[3] (per 1 km^2)	Number of farm household[4] (1 000 households)	Cultivated land area[5] (1 000 ha)	Forest land area[6] (1 000 ha)
Hokkaido	83 424.39	5 382	69	44	1 145	5 322
Aomori	9 645.64	1 308	136	45	152	616
Iwate	15 275.01	1 280	84	66	151	1 144
Miyagi	7 282.29*	2 334	321	52	128	407
Akita	11 637.52	1 023	88	49	148	820
Yamagata	9 323.15*	1 124	121	46	118	641
Fukushima	13 783.90	1 914	139	75	142	936
Ibaraki	6 097.39	2 917	478	88	168	189
Tochigi	6 408.09	1 974	308	55	124	341
Gunma	6 362.28	1 973	310	50	70	406
Saitama	3 797.75*	7 267	1 913	64	75	121
Chiba	5 157.60*	6 223	1 207	63	126	157
Tokyo	2 194.07*	13 515	6 169	11	7	76
Kanagawa	2 416.32	9 126	3 778	25	19	94
Niigata	12 584.24*	2 304	183	78	171	799
Toyama	4 247.59*	1 066	251	24	59	240
Ishikawa	4 186.05	1 154	276	21	42	276
Fukui	4 190.52	787	188	23	40	310
Yamanashi	4 465.27*	835	187	33	24	347
Nagano	13 561.56*	2 099	155	105	107	1 023
Gifu	10 621.29*	2 032	191	61	56	839
Shizuoka	7 777.35*	3 700	476	61	66	491
Aichi	5 173.06*	7 483	1 447	74	76	218
Mie	5 774.45*	1 816	315	43	59	371
Shiga	4 017.38*	1 413	352	29	52	203
Kyoto	4 612.20	2 610	566	31	31	342
Osaka	1 905.29	8 839	4 640	24	13	57
Hyogo	8 400.94	5 535	659	81	74	561
Nara	3 690.94	1 364	370	26	21	283
Wakayama	4 724.65	964	204	30	33	361
Tottori	3 507.14	573	164	28	35	257
Shimane	6 708.27	694	104	34	37	520
Okayama	7 114.33*	1 922	270	63	66	484
Hiroshima	8 479.64	2 844	335	57	55	609
Yamaguchi	6 112.53	1 405	230	36	48	437
Tokushima	4 146.75	756	182	31	29	312
Kagawa	1 876.78*	976	520	35	31	87
Ehime	5 676.16	1 385	244	42	49	399
Kochi	7 103.64	728	103	25	28	592
Fukuoka	4 986.51*	5 102	1 023	53	83	222
Saga	2 440.69	833	341	22	52	110
Nagasaki	4 130.96	1 377	333	34	47	241
Kumamoto	7 409.46*	1 786	241	58	112	448
Oita	6 340.76*	1 166	184	39	56	448
Miyazaki	7 735.33*	1 104	143	38	67	587
Kagoshima	9 187.08*	1 648	179	64	119	582
Okinawa	2 281.00	1 434	628	20	38	105
Total	377 975.21	127 095	341	2 155	4 444	24 433

1) Source: Geospatial Information Authority of Japan "Statistical reports on the land area by prefectures and municipalities in Japan" (2020). The areas marked with *, which contain boundaries not determined between the municipalities, are based on the area values in the "Nationawide Municipality Handbook" (2013 edition) distributed to each prefecture by the Ministry of Internal Affairs and Comuncaitions, Local Administration Bureau, Municipal Merger Promotioion Division, as the approximate value.

2), 3) Source: Ministry of Internal Affairs and Communications Japan, Statistics Bureau "2015 Population Census" (As of October 1, 2015). Note that the population density of Hokkaido is calculated excluding the areas of the Habomai gunto, Shikotan to, Kunashiri to, and Etorofu to.

4), 5), 6) Source: Ministry of Agriculture, Forestry and Fisheries Japan, Statistics Department "The 90th Statistical Yearbook of Ministry of Agriculture, Forestry and Fisheries Japan" 4) is current as of February 1, 2015, 5) is current as of July 15, 2014, and 6) is current as of February 1, 2015.

Area and Population of Major Japanese Cities

City	Prefecture	Area[1] (km^2)	Population[2] (10 000 persons)	City	Prefecture	Area[1] (km^2)	Population[2] (10 000 persons)
Sapporo	Hokkaido	1 121	195.9	Takasaki	Gunma	*459	37.3
Hakodate		678	25.5	Kiryu		274	11.0
Otaru		244	11.4	Isesaki		139	21.3
Asahikawa		748	33.4	Ota		176	22.4
Kushiro		1 363	16.8	Saitama	Saitama	217	131.4
Obihiro		619	16.6	Kawagoe		109	35.3
Kitami		1 427	11.7	Kumagaya		160	19.7
Tomakomai		*562	17.1	Kawaguchi		62	60.7
Ebetsu		187	12.0	Tokorozawa		72	34.4
Aomori	Aomori	825	28.1	Kazo		133	11.3
Hirosaki		524	17.0	Kasukabe		66	23.4
Hachinohe		306	22.8	Sayama		49	15.1
Morioka	Iwate	886	28.8	Konosu		67	11.8
Ichinoseki		1 256	11.5	Fukaya		138	14.3
Oshu		*993	11.6	Ageo		46	22.9
Sendai	Miyagi	786	106.4	Soka		27	25.0
Ishinomaki		555	14.3	Koshigaya		60	34.5
Osaki		797	12.9	Toda		18	14.1
Akita	Akita	906	30.7	Iruma		45	14.8
Yamagata	Yamagata	*381	24.5	Asaka		18	14.2
Tsuruoka		*1 312	12.6	Niiza		23	16.6
Sakata		603	10.1	Kuki		82	15.3
Fukushima	Fukushima	768	27.7	Fujimi		20	11.2
Aizuwakamatsu		*383	11.8	Misato		*30	14.3
Koriyama		757	32.3	Sakado		41	10.1
Iwaki		1 232	32.2	Fujimino		15	11.4
Mito	Ibaraki	*217	27.2	Chiba	Chiba	272	97.3
Hitachi		226	17.8	Ichikawa		*57	49.0
Tsuchiura		123	14.2	Funabashi		86	64.3
Koga		124	14.3	Kisarazu		139	13.6
Toride		70	10.7	Matsudo		61	49.8
Tsukuba		284	23.8	Noda		104	15.4
Hitachinaka		100	15.9	Narita		214	13.3
Chikusei		205	10.4	Sakura		104	17.5
Utsunomiya	Tochigi	417	52.2	Narashino		21	17.4
Ashikaga		178	14.7	Kashiwa		115	42.5
Tochigi		332	16.0	Ichihara		368	27.5
Sano		356	11.8	Nagareyama		35	19.5
Oyama		172	16.8	Yachiyo		51	20.0
Nasushiobara		593	11.7	Abiko		43	13.2
Maebashi	Gunma	312	33.6	Kamagaya		21	11.0

Cities with a population of 100 000 or more as of January 1, 2020. Area is current as of October 1, 2017.

1) Source: Geospatial Information Authority of Japan "Statistical reports on the land area by prefectures and municipalities in Japan" (2020). The areas marked with *, which contain boundaries not determined between the municipalities, are based on the area values in the "Nationawide Municipality Handbook" (2013 edition) distributed to each prefecture by the Ministry of Internal Affairs and Comuncaitions, Local Administration Bureau, Municipal Merger Promotioion Division, as the approximate value.

2) Source: Ministry of Internal Affairs and communications Japan, Local Administration Bureau "the basic resident register population".

Area and Population of Major Japanese Cities Continued.

City	Prefecture	Area[1] (km^2)	Population[2] (10 000 persons)	City	Prefecture	Area[1] (km^2)	Population[2] (10 000 persons)
Urayasu	Chiba	*17	17.0	Ueda	Nagano	552	15.7
Inzai		124	10.4	Iida		659	10.1
Tokyo ku-area	Tokyo	*628	957.1	Gifu	Gifu	204	40.9
Hachioji		186	56.2	Ogaki		207	16.1
Tachikawa		24	18.4	Tajimi		*91	11.0
Musashino		11	14.7	Kakamigahara		88	14.8
Mitaka		16	18.8	Kani		88	10.2
Ome		103	13.3	Shizuoka	Shizuoka	*1 412	69.8
Fuchu		29	26.0	Hamamatsu		*1 558	80.3
Akishima		17	11.3	Numazu		187	19.5
Chofu		22	23.7	Mishima		62	10.9
Machida		72	42.9	Fujinomiya		*389	13.2
Koganei		11	12.2	Fuji		245	25.3
Kodaira		21	19.5	Iwata		*163	17.0
Hino		28	18.6	Yaizu		70	13.9
Higashimurayama		17	15.1	Kakegawa		266	11.8
Kokubunji		11	12.5	Fujieda		194	14.5
Higashikurume		13	11.7	Nagoya	Aichi	*327	230.2
Tama		21	14.9	Toyohashi		262	37.7
Nishitokyo		16	20.5	Okazaki		387	38.8
Yokohama	Kanagawa	438	375.5	Ichinomiya		114	38.5
Kawasaki		143	151.4	Seto		111	13.0
Sagamihara		329	71.8	Handa		47	12.0
Yokosuka		101	40.1	Kasugai		93	31.1
Hiratsuka		*68	25.7	Toyokawa		161	18.7
Kamakura		40	17.6	Kariya		50	15.3
Fujisawa		70	43.6	Toyota		918	42.5
Odawara		114	19.1	Anjo		86	19.0
Chigasaki		*36	24.4	Nishio		161	17.2
Hadano		104	16.1	Konan		30	10.1
Atsugi		94	22.4	Komaki		63	15.3
Yamato		27	23.9	Inazawa		79	13.7
Isehara		56	10.0	Tokai		43	11.5
Ebina		27	13.4	Tsu	Mie	711	27.8
Zama		18	13.2	Yokkaichi		206	31.2
Niigata	Niigata	726	78.8	Ise		208	12.5
Nagaoka		*891	26.9	Matsusaka		624	16.3
Joetsu		974	19.1	Kuwana		137	14.2
Toyama	Toyama	*1 242	41.6	Suzuka		194	20.0
Takaoka		210	17.0	Otsu	Shiga	465	34.4
Kanazawa	Ishikawa	469	45.2	Hikone		197	11.3
Komatsu		371	10.8	Nagahama		681	11.8
Hakusan		755	11.4	Kusatsu		68	13.5
Fukui	Fukui	536	26.3	Higashiomi		388	11.4
Kofu	Yamanashi	212	18.8	Kyoto	Kyoto	828	141.0
Nagano	Nagano	835	37.6	Uji		68	18.6
Matsumoto		978	23.9	Osaka	Osaka	*225	273.0

Continued.

City	Prefecture	Area[1] (km²)	Population[2] (10 000 persons)	City	Prefecture	Area[1] (km²)	Population[2] (10 000 persons)
Sakai	Osaka	150	83.5	Fukuyama	Hiroshima	518	46.9
Kishiwada		73	19.4	Higashihiroshima		635	18.9
Toyonaka		*36	40.8	Hatsukaichi		489	11.7
Ikeda		22	10.4	Shimonoseki	Yamaguchi	716	26.1
Suita		36	37.4	Ube		287	16.4
Takatsuki		105	35.2	Yamaguchi		1 023	19.2
Moriguchi		13	14.4	Hofu		189	11.6
Hirakata		65	40.1	Iwakuni		874	13.4
Ibaraki		76	28.2	Shunan		656	14.2
Yao		42	26.6	Tokushima	Tokushima	191	25.3
Izumisano		57	10.0	Takamatsu	Kagawa	375	42.7
Tondabayashi		40	11.1	Marugame		112	11.3
Neyagawa		25	23.2	Matsuyama	Ehime	429	51.1
Kawachinagano		110	10.5	Imabari		419	15.8
Matsubara		17	12.0	Niihama		235	11.9
Daito		18	12.0	Saijo		510	10.9
Izumi		85	18.6	Kochi	Kochi	309	32.8
Minoh		48	13.8	Kitakyushu	Fukuoka	492	95.1
Habikino		26	11.1	Fukuoka		343	155.4
Kadoma		12	12.2	Ōmuta		81	11.4
Higashiosaka		62	48.9	Kurume		230	30.5
Kobe	Hyogo	*557	153.4	Iizuka		*214	12.8
Himeji		534	53.6	Chikushino		88	10.4
Amagasaki		51	46.3	Kasuga		14	11.3
Akashi		49	30.4	Onojo		27	10.1
Nishinomiya		*100	48.4	Itoshima		216	10.2
Itami		25	20.4	Saga	Saga	*432	23.2
Kakogawa		138	26.4	Karatsu		488	12.1
Takarazuka		*102	23.4	Nagasaki	Nagasaki	406	41.6
Kawanishi		53	15.7	Sasebo		426	25.0
Sanda		210	11.2	Isahaya		342	13.7
Nara	Nara	277	35.6	Kumamoto	Kumamoto	390	73.4
Kashihara		40	12.2	Yatsushiro		681	12.7
Ikoma		53	11.9	Oita	Oita	502	47.8
Wakayama	Wakayama	209	36.7	Beppu		*125	11.7
Tottori	Tottori	765	18.7	Miyazaki	Miyazaki	644	40.3
Yonago		132	14.8	Miyakonojo		*653	16.5
Matsue	Shimane	573	20.2	Nobeoka		868	12.2
Izumo		624	17.5	Kagoshima	Kagoshima	548	60.2
Okayama	Okayama	790[3]	70.9	Kanoya		448	10.3
Kurashiki		356	48.2	Kirishima		603	12.5
Tsuyama		506	10.1	Naha	Okinawa	40	32.2
Hiroshima	Hiroshima	907	119.6	Urasoe		19	11.5
Kure		353	22.2	Okinawa		50	14.3
Onomichi		285	13.6	Uruma		87	12.4

3) The area of Kojima ko is not included in the area of Okayama City, Okayama Prefecture.

Distance between

	Hokkaido	Aomori	Iwate	Miyagi	Akita	Yamagata	Fukushima	Ibaraki	Tochigi	Gunma	Saitama	Chiba
Aomori	253.8											
Iwate	373.6	129.3										
Miyagi	534.0	284.0	161.1									
Akita	385.9	134.2	90.1	174.2								
Yamagata	542.1	288.7	176.2	44.6	165.6							
Fukushima	594.8	342.1	224.9	67.7	220.8	55.2						
Ibaraki	750.4	498.3	378.3	217.2	376.1	210.9	156.3					
Tochigi	732.3	478.6	365.6	208.3	350.5	190.7	141.3	56.3				
Gunma	766.3	513.4	410.9	263.0	380.5	235.5	195.9	124.5	76.2			
Saitama	813.4	559.6	447.0	288.9	430.5	272.1	222.4	89.7	81.4	79.5		
Chiba	834.8	581.9	463.9	303.1	456.6	293.3	240.0	86.8	108.8	129.6	51.2	
Tokyo	831.0	577.3	463.8	304.9	448.7	289.4	238.9	99.3	98.8	96.4	19.0	40.2
Kanagawa	858.2	604.5	490.8	331.7	475.8	316.5	265.9	122.9	125.9	117.1	45.4	47.0
Niigata	606.0	356.5	272.4	167.2	222.3	123.4	128.3	214.5	166.8	167.7	233.7	273.2
Toyama	790.4	551.4	480.2	367.8	420.3	327.3	311.8	292.4	239.4	169.0	237.9	288.6
Ishikawa	823.9	590.2	525.8	419.2	461.6	377.9	364.5	343.6	291.6	219.4	283.9	333.5
Fukui	892.7	659.1	592.4	479.9	529.9	440.0	422.0	381.1	333.5	257.8	310.0	356.2
Yamanashi	855.6	603.5	502.9	354.5	469.9	327.6	287.0	185.3	155.0	92.1	100.0	141.0
Nagano	761.7	513.8	427.2	298.1	379.8	261.5	236.8	205.9	152.6	83.9	158.7	209.9
Gifu	940.4	698.3	618.2	488.8	565.5	453.3	425.3	352.5	313.5	238.5	270.1	309.5
Shizuoka	933.6	681.3	579.1	427.8	547.8	403.1	360.1	240.5	222.5	168.5	150.8	172.9
Aichi	955.4	711.0	626.9	492.8	577.5	459.0	427.9	345.1	309.7	236.6	259.8	296.0
Mie	1015.5	772.1	688.6	553.9	638.8	520.5	488.7	399.3	367.2	295.8	311.8	343.3
Shiga	1012.1	774.8	700.6	575.6	643.6	539.0	512.9	440.2	402.2	327.3	356.1	392.6
Kyoto	1015.1	778.7	705.8	582.2	647.9	545.3	519.9	449.1	410.6	335.5	365.4	402.4
Osaka	1057.9	821.7	748.1	622.8	690.7	586.4	559.7	483.1	446.9	372.6	397.4	431.6
Hyogo	1071.6	838.1	767.9	645.9	708.3	608.7	583.8	511.3	473.8	399.0	426.3	461.4
Nara	1045.4	806.7	729.9	601.5	674.9	566.0	537.8	457.0	422.1	348.6	370.7	404.0
Wakayama	1118.1	882.1	808.0	681.0	751.1	645.2	617.3	534.4	500.8	427.8	447.1	477.9
Tottori	1038.7	820.2	767.7	666.0	697.5	624.7	610.2	567.9	522.2	446.1	491.3	533.6
Shimane	1104.7	897.8	855.4	763.0	780.6	720.6	709.8	674.4	627.6	551.8	598.8	641.4
Okayama	1132.0	909.1	850.3	739.1	783.7	699.6	679.9	619.3	578.7	502.9	536.5	573.5
Hiroshima	1232.6	1020.6	971.2	868.5	899.9	827.6	811.3	757.0	715.2	639.1	674.9	712.1
Yamaguchi	1306.1	1100.9	1056.6	958.2	983.0	916.8	902.2	850.6	808.4	732.2	768.7	806.0
Tokushima	1159.9	927.8	857.9	734.6	798.2	697.8	671.7	592.4	557.8	484.1	505.4	536.7
Kagawa	1157.2	930.9	868.0	752.1	803.9	713.6	691.3	622.9	584.7	509.5	538.0	572.6
Ehime	1267.0	1047.8	991.0	879.6	923.6	840.3	819.8	753.0	714.9	639.5	667.8	701.5
Kochi	1255.8	1028.6	963.4	843.3	900.8	805.9	781.1	702.8	668.4	594.5	615.5	645.8
Fukuoka	1417.1	1215.4	1172.9	1074.9	1098.7	1033.5	1018.5	964.2	923.0	846.9	881.1	917.0
Saga	1454.9	1251.1	1206.4	1105.5	1133.3	1064.6	1048.2	989.4	949.5	873.6	905.1	939.6
Nagasaki	1523.4	1318.9	1272.9	1169.9	1200.4	1129.4	1111.8	1048.9	1010.3	934.8	963.6	996.6
Kumamoto	1470.5	1259.7	1208.5	1100.6	1138.5	1060.9	1041.2	973.2	936.1	861.0	887.1	918.9
Oita	1382.7	1168.4	1115.0	1005.6	1046.0	966.1	946.0	878.3	840.9	765.7	792.5	824.9
Miyazaki	1514.9	1293.5	1231.7	1112.4	1167.4	1075.0	1049.8	965.9	934.3	861.5	877.3	903.6
Kagoshima	1592.3	1374.7	1316.2	1199.6	1250.1	1161.7	1137.6	1055.9	1023.6	950.4	967.5	994.2
Okinawa	2243.8	2019.4	1951.0	1821.7	1891.2	1787.4	1756.4	1652.0	1629.6	1562.0	1562.4	1578.3

	Kagoshima	Miyazaki	Oita	Kumamoto	Nagasaki	Saga	Fukuoka	Kochi	Ehime	Kagawa	Tokushima	Yamaguchi
												285.2
											56.5	237.5
										130.2	167.6	125.5
									77.5	98.7	110.5	202.8
								289.0	219.2	344.8	386.7	116.7
							41.1	302.6	238.4	367.2	405.4	150.3
						68.5	108.0	353.4	295.9	426.0	460.2	218.2
					82.0	65.7	95.5	273.8	221.7	351.5	382.2	169.0
				95.4	171.9	122.4	118.4	182.0	126.3	256.2	288.4	105.9
			148.2	116.7	173.6	182.2	210.3	269.1	248.2	363.7	378.1	252.3
		90.8	210.8	137.4	147.1	188.9	227.3	356.5	326.9	448.6	466.4	303.5
Okinawa	655.7	729.1	866.5	787.0	755.0	820.0	861.1	990.8	977.3	1088.7	1093.4	956.1

	North lat.	East long.		North lat.	East long.		North lat.	East long.		North lat.	East long.
	° ′ ″	° ′ ″		° ′ ″	° ′ ″		° ′ ″	° ′ ″		° ′ ″	° ′ ″
Hokkaido	43 03 51	141 20 49	Fukushima	37 45 00	140 28 04	Tokyo	35 41 22	139 41 30	Yamanashi	35 39 50	138 34 06
Aomori	40 49 28	140 44 24	Ibaraki	36 20 29	140 26 48	Kanagawa	35 26 52	139 38 33	Nagano	36 39 05	138 10 52
Iwate	39 42 13	141 09 09	Tochigi	36 33 57	139 53 01	Niigata	37 54 08	139 01 25	Gifu	35 23 28	136 43 20
Miyagi	38 16 08	140 52 19	Gunma	36 23 28	139 03 39	Toyama	36 41 43	137 12 41	Shizuoka	34 58 37	138 22 59
Akita	39 43 07	140 06 09	Saitama	35 51 25	139 38 56	Ishikawa	36 35 40	136 37 32	Aichi	35 10 49	136 54 24
Yamagata	38 14 26	140 21 48	Chiba	35 36 17	140 07 24	Fukui	36 03 55	136 13 19	Mie	34 43 49	136 30 31

Prefectural Capitals

1. The locations of the prefectural offices are current as of April 1, 2020.
2. Distances are the geodesic length caculated on an ellipsoid (GRS80)
3. In units of km.

* Source: Geospatial Information Authority of Japan

Tokyo	Kanagawa	Niigata	Toyama	Ishikawa	Fukui	Yamanashi	Nagano	Gifu	Shizuoka	Aichi	
27.2											
252.7	277.9										
249.4	259.1	209.2									
293.6	300.2	257.5	53.6								
316.1	316.8	322.1	113.0	69.0							
101.7	100.3	251.7	167.3	203.1	216.5						
172.8	187.5	157.7	86.8	139.3	187.5	115.0					
271.3	265.3	346.4	151.3	133.8	87.4	170.1	192.0				
142.8	126.0	329.6	218.1	239.7	230.2	78.1	186.7	158.1			
259.1	250.5	356.6	170.3	159.0	116.1	160.2	199.7	28.8	136.5		
308.7	296.6	418.2	227.1	207.1	150.4	214.3	261.4	75.9	173.6	61.8	Mie
355.7	347.1	428.2	223.4	189.2	122.0	256.1	277.6	88.8	229.6	96.7	Shiga
365.3	356.9	433.5	227.6	191.4	123.3	265.4	284.1	97.1	239.9	106.4	Kyoto
395.9	385.4	475.7	270.4	234.2	165.7	298.1	324.9	134.7	263.9	138.0	Osaka
425.2	415.3	495.8	288.3	248.3	179.3	326.7	347.8	160.5	294.4	166.5	Hyogo
368.7	357.6	457.7	255.5	223.7	157.1	271.8	304.6	112.8	235.5	112.4	Nara
443.9	431.1	535.7	330.8	294.2	225.5	349.2	383.7	192.2	306.5	191.3	Wakayama
494.6	490.4	503.6	298.6	246.8	189.9	392.8	377.3	225.8	381.7	245.2	Tottori
602.3	598.3	597.9	398.5	345.4	294.2	500.6	480.3	333.5	488.5	352.1	Shimane
536.6	527.9	581.5	372.6	324.6	259.7	436.5	443.3	266.9	408.4	277.6	Okayama
675.1	666.5	707.0	500.6	449.8	389.3	574.8	576.0	404.9	546.5	416.2	Hiroshima
769.0	760.3	795.3	590.7	539.0	480.5	668.7	667.7	498.7	640.1	510.1	Yamaguchi
502.5	489.9	585.7	378.3	337.6	268.7	407.1	436.6	246.7	365.3	248.2	Tokushima
536.7	526.3	597.5	388.4	342.7	275.0	438.3	454.6	271.3	404.0	278.2	Kagawa
666.1	655.0	722.4	513.5	465.5	400.6	568.3	583.0	401.4	531.5	408.1	Ehime
612.2	598.9	691.9	483.2	439.3	371.3	517.5	545.2	356.6	473.6	358.7	Kochi
880.6	870.8	912.1	707.3	655.7	596.7	781.1	783.7	611.8	748.7	621.5	Fukuoka
904.0	893.2	943.7	737.7	686.7	626.2	805.4	812.5	637.0	769.8	645.3	Saga
961.7	949.9	1009.1	802.1	751.7	690.1	864.2	875.6	697.1	825.5	704.0	Nagasaki
884.6	872.1	941.9	733.7	684.4	620.9	788.1	804.5	622.6	747.1	628.2	Kumamoto
790.2	778.2	847.5	639.0	590.2	526.1	693.3	709.3	527.4	653.6	533.2	Oita
872.3	856.7	960.6	751.7	706.5	639.2	781.4	814.2	624.7	730.7	624.9	Miyazaki
962.7	947.3	1046.1	836.9	790.5	724.1	871.2	901.5	713.0	821.3	714.1	Kagoshima
1553.6	1533.2	1678.6	1471.8	1429.7	1361.3	1474.6	1526.3	1334.3	1411.6	1328.9	Okinawa

Hiroshima	Okayama	Shimane	Tottori	Wakayama	Nara	Hyogo	Osaka	Kyoto	Shiga	Mie	
										66.0	Shiga
									10.5	76.0	Kyoto
								43.0	47.5	90.7	Osaka
							30.9	63.9	71.7	121.5	Hyogo
						59.5	28.7	38.0	35.6	62.1	Nara
					79.6	51.6	60.5	103.4	107.6	135.3	Wakayama
				165.2	171.4	124.7	147.9	148.1	158.3	224.0	Tottori
			107.8	237.9	268.3	212.9	241.5	251.2	261.7	325.8	Shimane
		120.8	97.4	123.1	173.9	114.4	145.3	171.2	180.9	235.9	Okayama
	138.6	131.0	203.7	250.0	311.3	252.1	282.8	309.8	319.5	373.4	Hiroshima
93.9	232.5	203.1	292.2	340.6	404.6	345.7	376.2	403.7	413.4	466.7	Yamaguchi
196.9	87.6	208.4	162.2	58.8	135.8	90.0	112.0	152.6	159.0	193.8	Tokushima
145.8	37.0	154.9	130.3	104.3	168.7	111.7	140.9	174.2	182.8	230.4	Kagawa
67.7	140.9	182.8	228.5	225.8	297.5	241.7	270.4	304.3	313.0	358.4	Ehime
135.7	127.8	216.7	225.2	168.4	246.3	197.4	222.0	261.2	268.3	303.7	Kochi
207.9	344.8	318.2	408.9	444.5	513.4	455.6	485.5	515.7	525.0	575.1	Fukuoka
237.0	370.8	353.3	440.2	463.9	535.6	478.8	508.1	540.2	549.3	596.7	Saga
302.1	432.6	421.2	505.8	519.0	593.0	537.6	566.1	600.0	608.8	653.3	Nagasaki
239.1	361.5	365.9	441.1	441.0	516.4	462.4	490.2	525.6	533.9	575.9	Kumamoto
150.5	266.5	280.9	348.5	347.2	421.8	367.2	395.2	430.3	438.7	481.7	Oita
292.1	384.4	422.8	476.2	433.7	513.0	466.5	490.1	530.1	536.9	567.3	Miyazaki
361.3	466.5	491.8	555.0	522.8	601.9	553.8	578.3	617.7	624.8	656.9	Kagoshima
1016.7	1112.1	1147.4	1205.2	1143.1	1221.8	1183.5	1202.9	1245.2	1250.5	1267.7	Okinawa

	North lat.	East long.		North lat.	East long.		North lat.	East long.		North lat.	East long.
	° ′ ″	° ′ ″		° ′ ″	° ′ ″		° ′ ″	° ′ ″		° ′ ″	° ′ ″
Shiga	35 00 16	135 52 06	Tottori	35 30 13	134 14 18	Kagawa	34 20 25	134 02 36	Kumamoto	32 47 23	130 44 30
Kyoto	35 01 17	135 45 20	Shimane	35 28 20	133 03 02	Ehime	33 50 30	132 45 58	Oita	33 14 17	131 36 45
Osaka	34 41 11	135 31 12	Okayama	34 39 42	133 56 06	Kochi	33 33 35	133 31 52	Miyazaki	31 54 40	131 25 26
Hyogo	34 41 29	135 10 59	Hiroshima	34 23 47	132 27 34	Fukuoka	33 36 23	130 25 05	Kagoshima	31 33 37	130 33 29
Nara	34 41 07	135 49 58	Yamaguchi	34 11 09	131 28 17	Saga	33 14 58	130 17 56	Okinawa	26 12 45	127 40 52
Wakayama	34 13 34	135 10 03	Tokushima	34 03 57	134 33 34	Nagasaki	32 45 00	129 52 02			

Distance between Main Capitals

Capital	Seoul	Beijing	Manila	Hanoi	Bangkok	Singapore	Jakarta	New Delhi	Moscow	Tehran
Tokyo	1 160	2 104	2 997	3 675	4 610	5 317	5 776	5 857	7 502	7 683
Ankara	7 765	6 848	8 828	7 117	7 142	8 305	9 097	4 225	1 793	1 699
Brasilia	17 541	16 932	18 837	17 205	16 633	16 541	16 327	14 245	11 165	11 853
Buenos Aires	19 429	19 265	17 787	17 866	16 885	15 889	15 237	15 800	13 461	13 778
Cairo	8 504	7 557	9 192	7 438	7 279	8 270	8 979	4 436	2 899	1 985
Canberra	8 390	8 987	6 273	7 734	7 469	6 210	5 401	10 345	14 477	12 805
Lagos	12 419	11 471	12 767	11 044	10 615	11 160	11 573	8 094	6 247	5 866
Lima	16 311	16 648	18 061	18 966	19 702	18 810	17 947	16 784	12 641	14 244
London	8 882	8 160	10 752	9 250	9 544	10 860	11 712	6 724	2 506	4 410
Mexico	12 071	12 478	14 237	14 766	15 760	16 623	16 862	14 679	10 740	13 172
Moscow	6 626	5 809	8 269	6 744	7 070	8 426	9 298	4 349	0	2 468
Nairobi	10 115	9 216	9 427	7 898	7 218	7 467	7 788	5 428	6 323	4 363
New Delhi	4 699	3 788	4 763	3 007	2 917	4 142	4 988	0	4 349	2 546
Ottawa	10 537	10 475	13 152	12 645	13 442	14 831	15 648	11 364	7 179	9 578
Paris	8 990	8 236	10 761	9 213	9 457	10 743	11 581	6 601	2 492	4 224
Beijing	958	0	2 845	2 323	3 291	4 465	5 199	3 788	5 809	5 614
Pretoria	12 448	11 658	10 986	9 854	8 969	8 646	8 583	7 979	9 074	7 225
Rome	8 991	8 144	10 409	8 747	8 842	10 030	10 823	5 929	2 378	3 424
Singapore	4 666	4 465	2 392	2 196	1 427	0	887	4 142	8 426	6 607
Washington	11 190	11 170	13 797	13 364	14 172	15 555	16 360	12 071	7 842	10 204

Capital	Paris	Rome	Ottawa	Tripoli	Madrid	Alger	Washington	Nairobi	Mexico	Havana
Tokyo	9 738	9 881	10 342	10 600	10 789	10 823	10 925	11 266	11 319	12 134
Ankara	2 605	1 727	8 183	1 925	3 092	2 617	8 747	4 582	11 774	10 335
Brasilia	8 707	8 892	7 334	8 436	7 720	7 901	6 770	9 415	6 827	5 710
Buenos Aires	11 029	11 135	9 031	10 570	10 024	10 149	8 359	10 416	7 366	6 872
Cairo	3 215	2 135	8 877	1 743	3 355	2 716	9 370	3 518	12 392	10 806
Canberra	16 924	16 222	16 105	15 990	17 580	16 985	15 943	11 941	13 173	14 897
Lagos	4 703	4 029	8 650	3 099	3 826	3 359	8 736	3 812	11 084	9 304
Lima	10 246	10 858	6 365	10 754	9 504	9 943	5 639	12 581	4 240	3 935
London	341	1 434	5 379	2 332	1 264	1 654	5 915	6 805	8 947	7 504
Mexico	9 213	10 260	3 603	10 795	9 083	9 773	3 033	14 834	0	1 789
Moscow	2 492	2 378	7 179	3 164	3 446	3 338	7 842	6 323	10 740	9 602
Nairobi	6 471	5 374	11 854	4 526	6 177	5 470	12 150	0	14 834	13 045
New Delhi	6 601	5 929	11 364	6 054	7 288	6 841	12 071	5 428	14 679	13 879
Ottawa	5 664	6 747	0	7 414	5 708	6 418	733	11 854	3 603	2 545
Paris	0	1 108	5 664	1 990	1 054	1 344	6 180	6 471	9 213	7 730
Beijing	8 236	8 144	10 475	8 735	9 243	9 128	11 170	9 216	12 478	12 758
Pretoria	8 655	7 662	13 053	6 682	8 032	7 406	13 024	2 862	14 602	13 065
Rome	1 108	0	6 747	1 001	1 365	991	7 235	5 374	10 260	8 711
Singapore	10 743	10 030	14 831	10 000	11 396	10 889	15 555	7 467	16 623	17 222
Washington	6 180	7 235	733	7 825	6 106	6 810	0	12 150	3 033	1 819

Distances are based on G.L. Fitzpatrick and M.J. Modlin (1986): Direct-Line Distances

(North and South Pole also included)

(Unit : km)

Canberra	Oslo	Riyadh	Ankara	Reykjavik	Berlin	Jerusalem	Wellington	London	Cairo	Capital
7 924	8 428	8 715	8 787	8 819	8 942	9 171	9 246	9 585	9 587	Tokyo
14 497	2 704	2 135	0	4 415	2 040	931	16 823	2 839	1 106	Ankara
14 069	9 891	11 190	10 361	9 137	9 576	10 296	12 315	8 773	9 875	Brasilia
11 751	12 227	12 855	12 472	11 407	11 890	12 236	10 001	11 105	11 811	Buenos Aires
14 269	3 657	1 642	1 106	5 279	2 891	426	16 525	3 513	0	Cairo
0	15 973	12 633	14 497	16 744	16 067	13 992	2 326	16 984	14 269	Canberra
15 280	5 964	5 035	4 736	6 715	5 189	4 336	16 052	5 005	3 915	Lagos
12 865	11 034	13 968	12 559	9 633	11 092	12 811	10 603	10 162	12 430	Lima
16 984	1 157	4 952	2 839	1 896	934	3 615	18 807	0	3 513	London
13 173	9 213	13 899	11 774	7 462	9 746	12 552	11 092	8 947	12 392	Mexico
14 477	1 647	3 535	1 793	3 318	1 612	2 671	16 543	2 506	2 899	Moscow
11 941	7 154	3 060	4 582	8 677	6 353	3 662	13 677	6 805	3 518	Nairobi
10 345	5 996	3 059	4 225	7 608	5 791	4 032	12 642	6 724	4 436	New Delhi
16 105	5 616	10 315	8 183	3 871	6 147	8 993	14 473	5 379	8 877	Ottawa
16 924	1 344	4 692	2 605	2 237	880	3 339	18 981	341	3 215	Paris
8 987	7 041	6 613	6 848	7 902	7 375	7 135	10 759	8 160	7 557	Peking
10 840	9 627	5 921	7 288	10 872	8 789	6 409	11 826	8 994	6 184	Pretoria
16 222	2 008	3 004	1 727	3 308	1 182	2 310	18 549	1 434	2 135	Rome
6 210	10 057	6 656	8 305	11 522	9 927	7 924	8 524	10 860	8 270	Singapore
15 943	6 250	10 867	8 747	4 524	6 729	9 519	14 073	5 915	9 370	Washington

Lagos	Pretoria	Lima	Santiago	Brasilia	Buenos Aires	Monte-video		North Pole	South Pole	Capital
13 496	13 511	15 493	17 234	17 672	18 365	18 575		6 049	13 953	Tokyo
4 736	7 288	12 559	13 361	10 361	12 472	12 318		5 582	14 426	Ankara
6 165	7 899	3 173	3 009	0	2 336	2 271		11 749	8 258	Brasilia
7 912	8 146	3 127	1 135	2 336	0	210		13 834	6 173	Buenos Aires
3 915	6 184	12 430	12 800	9 875	11 811	11 638		6 678	13 330	Cairo
15 280	10 840	12 865	11 339	14 069	11 751	11 800		13 912	6 089	Canberra
0	4 458	9 133	8 945	6 165	7 912	7 733		9 291	10 717	Lagos
9 133	10 920	0	2 458	3 173	3 127	3 292		11 335	8 670	Lima
5 005	8 994	10 162	11 651	8 773	11 105	11 021		4 296	15 712	London
11 084	14 602	4 240	6 585	6 827	7 366	7 531		7 856	12 148	Mexico
6 247	9 074	12 641	14 116	11 165	13 461	13 349		3 823	16 185	Moscow
3 812	2 862	12 581	11 550	9 415	10 416	10 207		10 146	9 862	Nairobi
8 094	7 979	16 784	16 926	14 245	15 800	15 595		6 838	13 168	New Delhi
8 650	13 053	6 365	8 749	7 334	9 031	9 108		4 971	15 034	Ottawa
4 703	8 655	10 246	11 628	8 707	11 029	10 935		4 589	15 419	Paris
11 471	11 658	16 648	19 057	16 932	19 265	19 154		5 579	14 424	Peking
4 458	0	10 920	9 232	7 899	8 146	7 939		12 853	7 155	Pretoria
4 029	7 662	10 858	11 894	8 892	11 135	11 010		5 363	14 645	Rome
11 160	8 646	18 810	16 399	16 541	15 889	15 754		9 860	10 144	Singapore
8 736	13 024	5 639	8 036	6 770	8 359	8 446		5 696	14 309	Washington

International The Scarecrow Press., Inc, Metuchen, N.J. and London.

Major Explorations and Discoveries in Geography

Year	Region discovered or explored	Discoverer or explorer (Country)
334-324 B.C.	Expedition to Persia and India	Alexander III (Alexander the Great) (Greece)
230 B.C.	Circumference of the earth measured in Egypt	Eratosthenes (Greece)
819 A.D.	Circumference of the earth measured in Arabia	Islamic King Al-Ma'mun
c. 1000	Traveled to the North American continent via Greenland and Newfoundland; named the new land "Vinland"	Leif (Scandinavia)
1271-95	Voyage to Central Asia, China, and India	Marco Polo (Italy)
1322	Doctrine established for the earth as a sphere	Doctrine established for the earth as a sphere Mandeville (England)
c. 1487	Followed the West African coast and rounded the Cape of Good Hope (theories for 1486-87 and 1487-88)	Bartolomeu Dias (Portugal)
1492	Discovery of the West Indies (San Salvador, Cuba)	Columbus (Italy); first voyage
1493	Border of Spain and Portugal established in Atlantic	Ocean Pope Alexander VI (Italy)
1493-1504	Discovery of Jamaica, Orinoco Flow, Honduras, Trinidad, etc.	Columbus (Italy); second, third, fourth voyages
1497-98	Rounded the Cape of Good Hope (Africa) and reached India	Vasco da Gama (Portugal)
1497	Reached Newfoundland	Giovanni Caboto (Italy)
1499	Exploration of South American coast in Venezuela	Amerigo Vespucci (Italy)
1499-1500	Reached the mouth of the Amazon River	Pinzón (Portugal)
1500	Brazil discovered en route to India	Alvares Cabral (Portugal)
1501	Sailed southward along the South American coastline and declared the landmass as a continent	Amerigo Vespucci (Italy)
1513	Reached the Pacific Ocean via the Panama Strait	Balboa (Spain)
1516	Discovery of the River Plate in South America	Juan de Solis (Spain)
1519-21	Traveled around South America, crossed the Pacific Ocean, discovered the Mariana Islands, and killed in the Philippines	Magellan (Portugal)
1522	The *Victoria* on which Magellan had embarked circumnavigated the world	Juan Elcano (Spain)
1522	Measured the circumference of the earth between Paris and Amiens	Fernel (France)
1542	Discovered the Palau Islands and made the Philippines a Spanish territory	Villalobos (Spain)
1569	Discovered the Mercator projection method	Mercator (Netherlands)
1577-79	Circumnavigated the world and explored the North American west coast	Drake (England)
1579	Invaded Siberia and occupied the Irtysh River in Sybil	Yermak (Russia)
1606	First passage through the Torres Strait	Luis Torres (Spain)
1606	Discovered the Gulf of Carpentaria	*Duyfken* (Netherlands)
1622	Discovered the southwest corner of Australia	*Leeuwin* (Netherlands)

Continued.

Year	Region discovered or explored	Discoverer or explorer (Country)
1630	Designated the meridian which passes through Ferro Island (Canary Islands) as the prime meridian	
1639	Crossed Siberia and reached the eastern coast	Kopylov (Russia)
1642	Discovered Tasmania and New Zealand	Tasman (Netherlands)
1642	Explored the Heilong River Basin and reached the mouth of the river	Poyarkov (Russia)
1648	Reached the mouth of the Anadyr River via the Bering Strait becoming the first Russian to recognize the northeastern tip of Asia	Dezhnev (Russia)
1675	Explored the Ogasawara Islands per the orders of the Tokugawa Shogunate	Ichizaemon Shimaya (Japan)
1682	Sailed the Mississippi River	La Salle (France)
1685	Sent a mission to explore the southern islands of Japan. The ship never returned.	Mitsukuni Tokugawa (Japan)
1687	Proposed that the earth is a short-axis rotational sphere in Principia Book III	Newton (England)
1687–88	Sent the ship Kaifumaru to explore the eastern side of the Northern Islands of Japan, Ishikari River, the Tatar Region, etc.	Mitsukuni Tokugawa (Japan)
1696	Russians reached Kamchatka	
1718	Jesuits created a map of the empire for the Kangxi Emperor of the Qing Dynasty	
1735–41	The earth was confirmed to be a short-axis rotational sphere through surveying by the French in Lapland and Peru	
1741	Passed through the Bering Strait	Bering (Denmark)
1768–71	Reached New Zealand, passed through the Cook Strait, and discovered the east coast of Australia	Cook (England), first voyage
1776–79	Explored the west coast of North America and discovered Hawaii	Cook (England), third voyage
1785–88	Surveyed Northeast Asia, Japan, and the Sakhalin Coast	La Pérouse (France)
1786	Explored Kuril and Sakhalin	Tokunai Mogami (Japan)
1799–1804	Explored Central America and South America	Freiherr von Humboldt (Germany)
1800	Explored Etorofu Island	Juzo Kondo (Japan)
1800–18	Surveyed and drew Maps of Japan's Coastal Area	Tadataka Ino (Japan)
1808	Explored Sakhalin and Siberia, discovered that Sakhalin is an island	Rinzo Mamiya (Japan)
1821	Completed the Maps of Japan's Coastal Areas and Collection of Land-Survey Data	Tadataka Ino (Japan)
1828–31	Explored the Darling River and Murray River in Australia	Sturt (England)
1840–42	Reached south latitude 78°10′. Discovered Victoria Land and the volcanoes Mount Erebus/Mount Terror	James Ross (England)
1849–73	Explored the Zambezi River and crossed South Africa	Livingstone (England)
1850–54	Found a northwestern nautical route and reached northeastern Canada from Alaska	McClure (England)

Major Explorations and Discoveries in Geography Continued.

Year	Region discovered or explored	Discoverer or explorer (Country)
1860	Crossed Australia from south to north	Burke (England)
1868–72	Explored and surveyed the China Sea	Richthofen (Germany)
1871–88	Conducted four explorations of Central Asia	Przhevalsky (Russia)
1872–76	Conducted a nautical circumnavigation and survey of the world on the *Challenger*	John Murray (England)
1876–89	Explored Africa	Henry Stanley (England)
1878–79	Found a northeastern nautical route and reached Japan from Scandinavia	Nordenskiold (Sweden)
1878–80	Explored Qinling, Sichuan, Nanshan, and Yunnan	Szechenyi, Loczy (Austria)
1893–97	Explored Turkistan, Tibet, and Xinjiang	Sven Hedin (Sweden)
1895	Reached a north latitude of 86°4′	Nansen (Norway)
1903	Reached a south latitude of 82°17′	Robert Scott (England)
1903–04	Explored Central and Northern China	Willis (USA) Carnegie Expedition
1904–15	Surveyed the Caspian Sea	Knipovich (Russia)
1908	Reached a south latitude of 88°23′	Ernest Shackleton (England)
1909	Reached the North Pole	Robert Peary (USA)
1910–12	On a South Pole Expedition, landed at the Bay of Whales and reached a south latitude of 80°5′	Nobu Shirase (Japan)
1911	Reached the South Pole	Amundsen (Norway)
1921–23	Mt. Everest Expedition	Royal Geographical Society Expedition (England)
1926	Reached the North Pole by airplane from Spitsbergen (Svalbard)	Richard Byrd (USA)
1926	Crossed the North Pole and Alaska by airship from Spitsbergen	Amundsen (Norway)
1928–30	Circumnavigated the world by the *Dana*	Schmidt (Denmark)
1929	Circumnavigated the world in the *Graf Zeppelin*	Eckener (Germany)
1929–30	From Little America base to the South Pole by airplane	Byrd (USA)
1929–31	Surveyed the Southeast Asia ocean by the *Snellius* Bart	van Lier (Netherlands)
1931–35	Survey of Silin-Gol and Ulan-Chap	East Asia Archaeology Society (Japan)
1932	Surveyed the Northern sea route by the *Sibiryakov*	Schmidt (Soviet Union)
1933	Surveyed Rehe Province (China)	Shigeyasu Tokunaga (Japan)
1933–	Conducted a multiple ship survey of the Pacific Ocean	Marine Product Survey Ships (Japan)
1933	Aerial exploration of Mount Everest	Houston Expedition, Blacker (England)
1935	Elevation of 22 066 m reached by the *Explorer II* high-altitude balloon	Anderson, Stevens (USA)
1937	Conducting surveying and photographic measurement of Mt. Nanga Parbat	Troll, Finsterwalder, Wien (Germany)
1937–38	Surveyed the Arctic ocean	Papanin (Soviet Union)
1947	Explored the continent of Antarctica	Byrd (USA)
1947–48	Oceanographic expedition around the world with the *Albartross*	Petterson (Sweden)

Continued.

Year	Region discovered or explored	Discoverer or explorer (Country)
1949–	Conducted precise observation of west Pacific Ocean by the *Vityaz*	Zenkevich *et al.* (Soviet Union)
1950–52	Circumnavigated the world and conducted deep-sea exploration by the *Galathea*	Bruun (Denmark)
1950	Summited Annapurna Massif	French Expedition (led by Herzog)
1953	Summited Mt. Everest	Hillary (New Zealand), Tenzing (Nepal), expedition led by Hunt (England)
1953	Summited Mt. Nanga Parbat	German/Austrian Expedition
1954	Summited K2.	Italian Expedition
1954	*Bathyscaphe* submerged to an ocean depth of 4 050 m	Houot (France)
1955	Summited Mt. Kanchenjunga	Band, Brown, expedition led by Evans (England)
1955	Summited Mt. Makalu	French Expedition (led by Jean Franco)
1956	Summited Mt. Manaslu	Japanese expedition (led by Yuko Maki)
1956–58	Explored Antarctica as part of the International Geophysical Year Project	Expedition by multiple countries including Japan
1957	Succeeded in launching the first satellite, *Sputnik 1*	(Soviet Union)
1959	Photographed the far side of the moon. Interplanetary station	(Soviet Union)
1961	Manned satellite, *Vostok 1*	Gagarin (Soviet Union)
1964	Large-scale photograph of the surface of the moon by the *Ranger 7*	(USA)
1965	Spacewalk	Alexey Leonov (Soviet Union)
1966	Succeeded in landing observation equipment on the surface of the moon by the *Luna 9*	(Soviet Union)
1969	Mankind lands on the moon for the first time. Samples taken of moon rocks and set up measurement equipment such as seismographs. *Apollo 11*	Armstrong, Aldrin (and Collins) (USA)
1970	First soft landing on Venus. *Venus 7*	(Soviet Union)
1971	First soft landing on Mars. *Mars 3*	(Soviet Union)
1975	First photograph taken of land surface on Venus. *Venus 9, 10*	(Soviet Union)
1976	Photographed the land surface of Mars; analyzed weather and soil quality at landing point. *Viking 1, 2*	(USA)
1979	Approached Jupiter and photographed the land surface. *Voyager 1*	(USA)
1979	Approached and photographed Saturn. *Pioneer 11*	(USA)
1986	Approached and photographed Uranus, *Voyager 2*	(USA)
1989	Approached and photographed Neptune, *Voyager 2*	(USA)
1998	Surveyed the resources and structure of the moon	Lunar Prospector (USA)
2005	First soft landing on Titan, a satellite orbiting Saturn. *Cassini*	(USA)

Geology and Mineralogy

Elemental Abundances

Atomic number	Element	unit	A	B	C	D	Atomic number	Element	unit	A	B	C	D
1	H	ppm	21 015	-	359.4	-	44	Ru	ppb	692	5	-	0.34
2	He	ppb	9.17	-	-	-	45	Rh	ppb	141	0.9	-	-
3	Li	ppm	1.46	1.6	4.3	24	46	Pd	ppb	588	3.9	-	0.52
4	Be	ppb	25.2	68	-	2 100	47	Ag	ppb	201	8	-	53
5	B	ppb	713	300	-	17 000	48	Cd	ppb	675	40	-	90
6	C	ppm	35 180	120	-	-	49	In	ppb	78.8	11	-	56
7	N	ppm	2 940	2	-	83	50	Sn	ppb	1 680	130	1 100	2 100
8	O	%	45.82	44.3	44.5	47.5	51	Sb	ppb	152	5.5	10	400
9	F	ppm	60.6	25	-	557	52	Te	ppb	2 330	12	-	-
10	Ne	ppb	0.18	-	-	-	53	I	ppb	480	10	-	1 400
11	Na	ppm	5 010	2 670	16 172	24 259	52	Xe	ppb	0.174	-	-	-
12	Mg	%	9.587	22.8	5.87	1.50	55	Cs	ppb	185	21	2 500	4 900
13	Al	ppm	8 500	23 552	88 649	81 505	56	Ba	ppb	2 310	6 600	6 300	628 000
14	Si	%	10.65	21.0	23.1	29.7	57	La	ppb	232	648	2 500	31 000
15	P	ppm	920	90	414.6	654.6	58	Ce	ppb	621	1 675	7 500	63 000
16	S	%	5.41	0.025	-	0.0621	59	Pr	ppb	92.8	254	1 320	7 100
17	Cl	ppm	704	17	-	294	60	Nd	ppb	457	1 250	7 300	27 000
18	Ar	ppb	1.33	-	-	-	62	Sm	ppb	145	406	2 630	4 700
19	K	ppm	530	240	540	23 244	63	Eu	ppb	54.6	154	1 020	1 000
20	Ca	ppm	9 070	25 300	89 336	25 657	64	Gd	ppb	198	544	3 680	4 000
21	Sc	ppm	5.83	16.2	41.4	14	65	Tb	ppb	35.6	99	670	700
22	Ti	ppm	440	1 205	5 394	3 836	66	Dy	ppb	238	674	4 550	3 900
23	V	ppm	55.7	82	-	97	67	Ho	ppb	56.2	149	1 010	830
24	Cr	ppm	2 590	2 625	-	92	68	Er	ppb	162	438	2 970	2 300
25	Mn	ppm	1 910	1 045	-	774.5	69	Tm	ppb	23.7	68	456	300
26	Fe	%	18.28	6.26	6.26	3.92	70	Yb	ppb	163	441	3 050	2 000
27	Co	ppm	502	105	47.07	17.3	71	Lu	ppb	23.7	67.5	455	310
28	Ni	%	1.064	0.196	0.01495	0.0047	72	Hf	ppb	115	283	2 050	5 300
29	Cu	ppm	127	30	74.4	28	73	Ta	ppb	14.4	37	132	900
30	Zn	ppm	310	55	-	67	74	W	ppb	89	29	10	1 900
31	Ga	ppm	9.51	4	-	17.5	75	Re	ppb	37	0.28	-	0.198
32	Ge	ppm	33.2	1.1	-	1.4	76	Os	ppb	486	3.4	-	0.031
33	As	ppm	1.73	0.05	-	4.8	77	Ir	ppb	470	3.2	-	0.022
34	Se	ppm	19.7	0.075	-	0.09	78	Pt	ppb	1 004	7.1	-	0.5
35	Br	ppm	3.43	0.05	-	1.6	79	Au	ppb	146	1	-	1.5
36	Kr	ppb	0.0522	-	-	-	80	Hg	ppb	314	10	-	50
37	Rb	ppm	2.13	0.6	0.56	82	81	Tl	ppb	143	3.5	1.4	900
38	Sr	ppm	7.74	19.9	90	320	82	Pb	ppb	2 560	150	300	17 000
39	Y	ppm	1.53	4.3	28	21	83	Bi	ppb	110	2.5	-	160
40	Zr	ppm	3.96	10.5	74	193	90	Th	ppb	30.9	79.5	120	10 500
41	Nb	ppb	265	658	2 330	12 000	92	U	ppb	8.4	20.3	47	2 700
42	Mo	ppb	1 020	50	310	1 100							

A: Average composition of CI chondrite (Lodders, 2003)
B: Composition of the Earth, excluding the core (silicate portion) (values estimated from CI chondrite and the composition of the upper mantle; McDonough and Sun, 1995)
C: Composition of the oceanic crust (composition of N-type mid-ocean ridge basalt; Sun and McDonough, 1989; Workman and Hart, 2005; Hofmann, 1988; Jambon and Zimmermann, 1990)
D: Average composition of the upper continental crust

Major Element Compositions of the Crust and Mantle

Oxides	Mantle	Uppermost mantle	Oceanic crust	Lower continental crust	Middle continental crust	Upper continental crust	Average of the continental crust
	A	B	C	D	E	F	G
SiO_2	44.9	44.20	49.51	53.40	63.50	66.6	60.6
TiO_2	0.20	0.13	0.90	0.82	0.69	0.64	0.72
Al_2O_3	4.5	2.05	16.75	16.90	15.00	15.40	15.90
Cr_2O_3	0.38	0.44	0.07	–	–	–	–
FeO	8.05	8.29	8.05	8.57	6.02	5.04	6.71
NiO	0.25	0.28	–	–	–	–	–
MnO	0.14	0.13	0.14	0.10	0.10	0.10	0.10
MgO	37.80	42.21	9.74	7.24	3.59	2.5	4.7
CaO	3.54	1.92	12.50	9.59	5.25	3.59	6.41
Na_2O	0.36	0.27	2.18	2.65	3.39	3.27	3.07
K_2O	0.03	0.06	0.07	0.61	2.30	2.80	1.81
P_2O_5	0.02	0.03	0.10	0.10	0.15	0.15	0.13
Total	100.2	100.0	100.0	100.0	100.0	100.1	100.1

A: Chemical composition of the primitive mantle (McDonough and Sun, 1995)
B: Average values of the uppermost mantle peridotite (Maalφe and Aoki, 1977)
C: Chemical composition of the N-type mid-ocean ridge basalt (Presnall and Hoover, 1987)
D: Chemcial composition of the lower continental crust (Rudnick and Gao, 2005)
E: Chemcial composition of the middle continental crust (Rudnick and Gao, 2005)
F: Chemcial composition of the upper continental crust (Rudnick and Gao, 2005)
G: Average values of the continental crust (Rudnick and Gao, 2005)

Chemical Compositions of Main Igneous Rocks

Oxides	Komatiites	Alkali basalt	Flood basalt	Oceanic island basalt	Mid-ocean ridge basalt	Island arc basalt	Calc-alkaline andesite	Calc-alkali dacite	Calc-alkali rhyolite	Granite
	A	B	C	D	E	F	G	H	I	J
SiO_2	45.8	45.4	50.01	50.51	50.68	51.9	59.2	67.2	75.2	72.2
TiO_2	0.30	3.00	1.00	2.63	1.49	0.80	0.70	0.50	0.20	0.30
Al_2O_3	7.30	14.7	17.08	13.45	15.60	16.0	17.1	16.2	13.5	14.6
Cr_2O_3	0.20	–	–	–	–	–	–	–	–	–
Fe_2O_2	–	4.10	–	1.78	–	–	2.90	2.00	1.00	–
FeO	11.2	9.20	10.01	9.59	9.85	9.56	4.20	1.80	1.10	2.40
MnO	–	–	0.14	0.17	–	0.17	–	–	–	–
MgO	26.1	7.80	7.84	7.41	7.69	6.77	3.70	1.50	0.50	1.00
CaO	7.60	10.5	11.01	11.18	11.44	11.8	7.10	3.80	1.60	1.70
Na_2O	0.70	3.00	2.44	2.28	2.66	2.42	3.20	4.30	4.20	2.90
K_2O	0.10	1.00	0.27	0.49	0.17	0.44	1.30	2.10	2.70	4.50
P_2O_5	–	–	0.19	0.28	0.12	0.11	0.20	0.20	0.10	–
Total	99.3	98.7	99.99	99.77	99.70	100.0	99.6	99.6	100.1	99.6

A: Arndt *et al.* (1977), B: Macdonald (1968), C, D: Basaltic Volcanism Study Project (1981), E: Melson *et al.* (1976), F: Jakes & White, (1972), G, H, I: Ewart (1982), J: Phillips *et al.* (1981).

Mean Chemical Compositions of Main Sedimentary Rocks

Oxides	Argillaceous rock (average of 277 samples, Wedepohl)	Arenaceous rock (average of 253 samples, Clarke)	Limestone (average of 345 samples, Clarke)	Deep sea mud (average of 87 samples, Clarke, Goldberg, and Arrhenius)
SiO_2	58.9	78.7	5.2	52.8
TiO_2	0.77	0.25	0.07	0.8
Al_2O_3	16.7	4.8	0.8	16.4
Fe_2O_3	2.8	1.1	} 0.5	} 8.8
FeO	3.7	0.3		
MnO	0.1	0.01	0.05	0.8
MgO	2.6	1.2	7.9	3.4
CaO	2.2	5.5	42.6	3.2
Na_2O	1.6	0.5	0.05	2.1
K_2O	3.6	1.3	0.3	2.8
H_2O^+	} 5.0	1.3	0.6	} 6.2
H_2O^-		0.3	0.2	
P_2O_5	0.16	0.04	0.09	0.15
CO_2	1.3	5.0	41.6	2.1

Classification of Meteorites

<table>
<tr><th colspan="3">Classification</th><th>Group</th><th colspan="3">Classification</th><th>Group</th></tr>
<tr><td colspan="2" rowspan="15">Chondrite</td><td rowspan="8">Carbonaceous (C)</td><td>CI</td><td rowspan="16">Non-chondrite</td><td rowspan="12">Achondrite</td><td rowspan="3">HED</td><td>Eucrite</td></tr>
<tr><td>CM</td><td>Howardite</td></tr>
<tr><td>CR</td><td>Diogenite</td></tr>
<tr><td>CO</td><td rowspan="3">Martian meteorites (SNC)</td><td>Shergottites</td></tr>
<tr><td>CV</td><td>Nakhlites</td></tr>
<tr><td>CK</td><td>Chassignite</td></tr>
<tr><td>CH</td><td rowspan="2">Lunar meteorites</td><td rowspan="2">Lunar (moon)</td></tr>
<tr><td>CB</td></tr>
<tr><td rowspan="3">Ordinary (O)</td><td>H</td><td rowspan="4">Others</td><td>Angrite</td></tr>
<tr><td>L</td><td>Brachinite (BRA)</td></tr>
<tr><td>LL</td><td>Ureilite (URE)</td></tr>
<tr><td rowspan="2">Estatite (E)</td><td>EH</td><td>Aubrite (AUB)</td></tr>
<tr><td>EL</td><td rowspan="2">Stony-iron meteorite</td><td colspan="2" rowspan="2">Pallasite
Mesosiderite</td></tr>
<tr><td rowspan="2">Others</td><td>K</td></tr>
<tr><td>R</td><td rowspan="2">Iron meteorite</td><td>Magmatic</td><td>IC, IIAB, IIC, IID, IIF, IIIAB, IIIE, IIIF, IVA, IVB</td></tr>
<tr><td>Non-chondrite</td><td>Primitive achondrite</td><td colspan="2">Acapulcoite (ACA)
Lodranite (LOD)
Winonaite (WIN)</td><td>Non-magmatic</td><td>IAB, IIE, IIICD</td></tr>
</table>

Ages of Meteorites

Meteorite classification	Type	Meteorite name	Material	Isotope system (** is model age)	Absolute dating (million years) or time after CAI formation (million years)*	An event in which the isotope system is closed	References
Chondrite	CV3	Allende	CAI (SJ101)	Pb-Pb	4567.18 ± 0.50	CAI formation	1
		Efremovka	CAI 3 pieces	Pb-Pb	4567.30 ± 0.16	CAI formation	2
		Allende, NWA5697	Chondrule 5 pieces		From 4567.32 ± 0.42 to 4564.71 ± 0.30	Chondrule formation	
		Allende	Chondrule	Pb-Pb**	4565.32 ± 0.81	Chondrule formation	3
			Chondrule and matrix	Hf-W**	2.2 ± 0.8*	Chondrule formation	4
	L, LL3.0-3.2	NWA5206, NWA8276, MET96503, MET00452, MET00526, NWA7936, QUE97008	Chondrule 31 pieces	Al-Mg**	From −1.76 to −2.92*	Chondrule formation	5
	L3	NWA5697	Chondrule 13 pieces	Pb-Pb	From 4567.61 ± 0.54 to 4563.64 ± 0.51	Chondrule formation	2, 6
	CR	NWA6043, NWA7655	Chondrule 7 pieces		From 4567.26 ± 0.37 to 4563.24 ± 0.62		
		NWA801, Acfer097, NWA1180, GRA06100	Metal, Silicate, Chondrule	Hf-W	−3.6 ± 0.6*	Chondrule formation	7
		QAR99177, MH00426	Chondrule 5 pieces (Mg#>99)	Al-Mg**	From −2.2 to −2.8*	Chondrule formation	8
		QAR99177, MH00426	Chondrule 7 pieces (94<Mg#<99)		From >2.9 to >3.7*		
	CBa	Gjuba	Chondrule 4 pieces	Pb-Pb**	4562.49 ± 0.21	Chondrule formation	9
	H6 or L6	4 meteorite	Phosphate	Pb-Pb**	From ~4510 to ~4500	Thermal metamorphism	10
	H, L, LL, Rock type4-6	15 meteorite	Silicate	Hf-W**	From −3 to −4*(Petrology type 4)	Thermal metamorphism	11
					From −6 to −13*(Petrology type 5)		
					From −9 to −12*(Petrology type 6)		

Meteorite classification	Type	Meteorite name	Material	Isotope system	Absolute dating	Event	References
Iron meteorite	Magma type (IC, IIAB, IIIAB, IIIE, IVA)	10 pieces	Whole rock	Hf-W**	From −0.3 to −1.8*	Core formation in ordinary and E-chondrite objects	12
	Non-magma type (IIC, IID, IIF, IIIF, IVB)	9 pieces			From −2.2 to −2.8*	Core formation in C chondrite-like objects	12
	IIE	10 pieces	Whole rock	Hf-W**	Focused on the 4's of −4-5, −10, −15, −27*	Local metal separation in the parent body	13
	IAB	16 pieces	Whole rock	Hf-W**	3.4 ± 0.7* (Main group)	Core formation or impact in the parent body	14
					From −0.3 to −3* (High Au Subgroup)		
					−5* (Low Au Subgroup)		

Ages of Meteorites Continued.

Meteorite classification	Type	Meteorite name	Material	Isotope system (** is model age)	Absolute dating (million years) or time after CAI formation (million years)*	An event in which the isotope system is closed	References
Eucrite	Basaltic	Average of 5 meteorites	Zircon	Pb-Pb	4554 ± 7	Crystallization	15
				U-Pb	4552 ± 9		
	Basaltic	5 meteorite	Whole rock	Hf-W*	−4, −11, −22* 4's	Magmatic activity	16
	Cumulates	3 meteorite	Whole rock		38 ± 21*	Crystallization	
	Basaltic	6 pieces	Zircon	Hf-W**	From 4565.0 ± 0.9 to 4532 (−11/+6)	Magmatic activity	17
Unique achondrite	Basaltic	NWA2976	Whole rock+pyroxene	Pb-Pb	4562.89 ± 0.59	Crystallization	18
				Al-Mg**	4563.10 ± 0.38		
		Ibitira	Whole rock+pyroxene	Pb-Pb	4556.75 ± 0.57	Final equilibration in thermal metamorphism	19
	Cumulates	NWA7325	Pyroxene	U-Pb**	4563.4 ± 2.6	Crystallization	20
			Plagioclase, pyroxene, olivine, whole rock	Al-Mg**	4563.09 ± 0.26		
Angrite	Angra dos Reis		pyroxene	U-Pb	4557.65 ± 0.13	Crystallization	21
	D'Orbigny		Whole rock+pyroxene		4564.42 ± 0.12		
	LEW 86010		Pyroxene		4558.55 ± 0.15		
	Sahara99555		Whole rock	Pb-Pb	4564.86 ± 0.38	Crystallization	22
Acapulcoite	Acapluco		Whole rock+mineral	Pb-Pb	4555.9 ± 0.6	Final equilibration	23
Ureilite	29 pieces		Whole rock	Hf-W**	3.3 ± 0.7*	Melt separation	24
Martian meteorites	Depleted sherghati	NWA1195	Acid-dissolving component+insoluble component	Pb-Pb	4319 ± 47	Equilibrium with magma ocean	25
	Enriched and intermediate sherghati	NWA480, NWA1068, RBT04262			4092 ± 74	Mantle equilibration	
	Nakhlites and Chassignites	Y-000593, MIL03346, Chassigny			1330 ± 140	Mantle partial melting	
	Sherghati	NWA7635	Whole rock	Sm-Nd	2403 ± 140	Crystallization	26
	ALH84001		Impact glass+ Pyroxene	Pb-Pb	4089 ± 73	Crystallization	27

1: Amelin et al. (2010), 2: Connelly et al. (2012), 3: Connelly & Bizzarro (2009), 4: Budde et al. (2016), 5: Pape et al. (2019), 6: Bollard et al. (2017), 7: Budde et al. (2018), 8: Tenner et al. (2019), 9: Bollard et al. (2015), 10: Blackburn et al. (2017), 11: Hellmann et al. (2019), 12: Kruijer et al. (2017), 13: Kruijer & Klein (2019), 14: Worsham et al. (2019), 15: Misawa et al. (2005), 16: Touboul et al. (2015), 17: Roszjar et al. (2016), 18: Bouvier et al. (2011), 19: Iizuka et al. (2014), 20: Koefoed et al. (2016), 21: Amelin (2008a), 22: Amelin (2008b), 23: Gopel & Manhes (2010), 24: Budde et al. (2015), 25: Bouvier et al. (2009), 26: Lapen et al. (2017), 27: Bellucci et al. (2015).

−: A value that does not have an accurate error but contains some uncertainty. Among the Pb-Pb ages, those that are not model ages Measurement of 235 U/238 U ratio, by Isocron, or Concordia. For the Hf-W system, the Allende CAI value is adopted assuming the homogeneity of the 182 Hf/180 Hf ratio of the early solar system. For the Al-Mg system, the Allende CAI value is adopted assuming the homogeneity of the 26 Al/27 Al ratio in the early solar system.

Mg#: 100 × Mg/(Mg+Fe 2+)

Chemical Compositions (wt%) of Meteorites (Chondrites)

Oxides/ Element	EH Indarch[1)]	EL Hvitis[2)]	H Forest City[1)]	L Modoc[1)]	LL Näs[1)]	CI Orgueil[2)]	CM Murray[2)]	CO Lance[2)]	CV Allende[3)]
SiO_2	35.26	41.53	37.07	39.29	40.96	22.56	28.69	33.23	34.19
TiO_2	0.06	–	0.15	0.12	0.18	0.07	0.09	0.13	0.16
Al_2O_3	1.45	1.55	2.09	2.49	2.23	1.65	2.19	2.93	3.28
Cr_2O_3	0.47	0.56	0.54	0.55	0.59	0.36	0.44	0.49	0.53
FeO	–	–	9.89	14.96	18.90	11.39	21.08	24.80	26.84
MnO	0.25	0.34	0.28	0.33	0.35	0.19	0.21	0.20	0.20
MgO	17.48	23.23	23.62	24.78	25.70	15.81	19.77	23.54	24.62
CaO	0.95	0.74	1.75	1.62	1.62	1.22	1.92	2.64	2.64
Na_2O	1.01	1.26	0.99	0.93	0.84	0.74	0.22	0.58	0.43
K_2O	0.11	0.32	0.07	0.10	0.12	0.07	0.04	0.14	0.03
P_2O_5	0.52	–	0.34	0.30	0.20	0.28	0.32	0.32	0.25
Fe	24.13	20.04	16.21	6.68	1.46	0.00	0.00	2.19	0.13
Ni	1.83	1.96	1.65	1.30	0.99	0.00	0.00	1.50	0.29
Co	0.08	0.07	0.10	0.08	0.06	0.00	0.00	0.07	0.01
P	–	0.08	–	–	–	–	–	–	–
FeS	14.20	7.27	5.21	6.46	6.36	15.07	7.67	6.49	4.11
NiS									1.73
CoS									0.08
C	0.43	–		0.18		3.10	2.78	0.46	–
H_2O	1.17	–	0.39		0.13	19.89	12.42	1.40	–
Other volatiles		–				6.96	0.62	–	–
Total	99.40	99.81	100.34	100.17	100.69	100.65	100.04	101.11	99.80

1) Naoki Onuma, Hitoshi Mizutani (Ed) "Earth in the Solar System" Iwanami Earth Science Series, Vol 13, p. 97, Iwanami Shoten, Publishers (1978). However, the sulfur in C chondrite is present as sulfides, sulfates, Fe-S-O, etc. Here, all sulfur is represented by FeS. 2) Wiik (1956), 3) Jarosewich (1987).

Chemical Compositions (wt%) of Meteorites (Achondrites)

Oxides/ Element	Diogenite[1] Bununu	Hawardite[2] Yurtuk	Eucrite[3] Juvinas	Ureilite[4] Havero	Aubrite[4] Norton County	Shergottite[5] Shergotty	Nakhlites[5] Nakhla	Chassignite[5] Chassigny
SiO_2	48.67	49.45	49.32	40.25	54.40	50.35	48.23	37.00
TiO_2	0.11	-	0.68	0.07	0.06	0.85	0.29	0.07
Al_2O_3	8.87	9.66	12.64	0.26	0.61	7.03	1.45	0.36
Cr_2O_3	0.56	0.04	0.30	0.08	0.07	0.31	0.42	0.83
FeO	16.04	15.61	18.49	14.18	0.00	19.34	20.64	27.45
MnO	0.53	0.72	0.53	0.37	0.16	0.54	0.54	0.53
MgO	14.20	17.40	6.83	38.96	40.72	9.27	12.47	32.83
CaO	6.77	6.39	10.32	0.10	0.66	9.58	15.08	1.99
Na_2O	0.34	0.31	0.42	0.04	0.13	-	0.42	0.15
K_2O	0.04	0.08	0.05	0.01	0.04	0.18	0.10	0.13
P_2O_5	-	0.01	0.09	-	0.02	0.47	0.12	0.04
Fe	1.01		0.04	3.61	0.73			
Ni	0.06			0.12	0.04			
Co				0.01				
FeS	0.96		0.53	0.52	1.37			
C				1.81				
H_2O	1.65		0.05		0.36			
Total	99.81	99.67	100.29	100.39	99.74	97.92	99.76	101.28

1) Mason (1967), 2) Urei and Craig (1953), 3) Wiik (1956), 4) Duke and Silver (1967),
5) McCarthy *et al.* (1974).

Oxygen Isotopes in Meteorites

Figure 19 Oxygen isotopes in meteorites (from Clayton (2005))

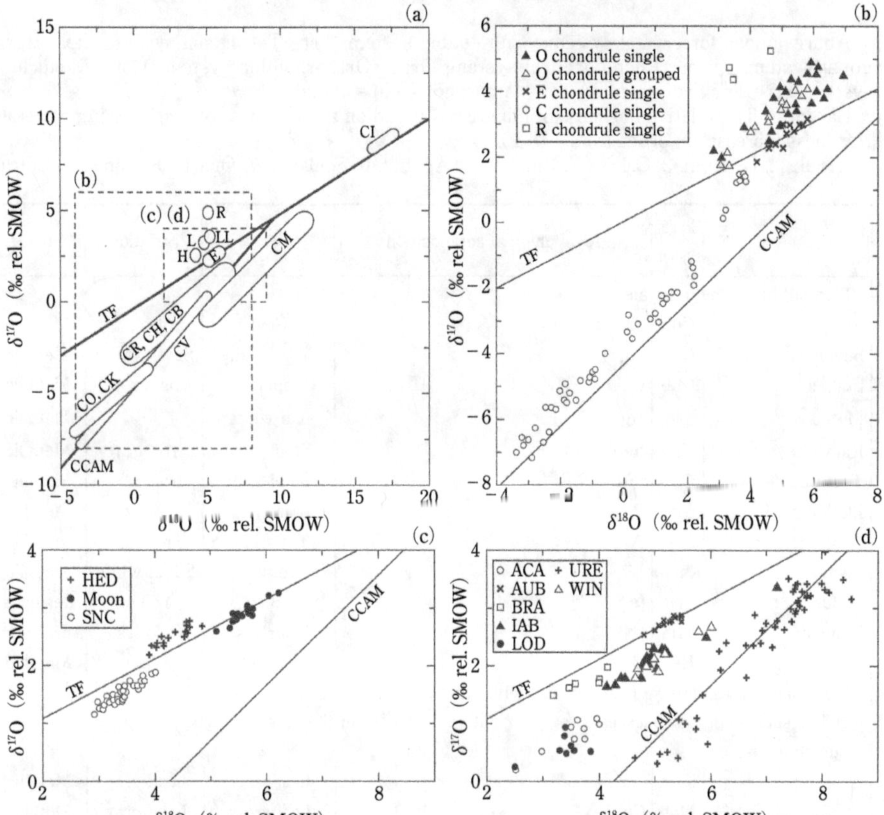

(a) Schematic diagram of the whole-rock oxygen isotopic composition regions of major meteorite groups. The horizontal and vertical axes show the amount of ^{18}O (horizontal axis) and ^{17}O (vertical axis) compared to ^{16}O with respect to SMOW (standard mean ocean water) ($\delta^{17,18}O = ((^{17,18}O/^{16}O)$ sample$/(^{17,18}O/^{16}O)$ SMOW-1$) \times 1000$). F is the earth's oxygen isotope fractionation line, and CCAM (carbonaceous chondrite anhydrous mineral line) is the carbonaceous chondrite anhydrous mineral line. These lines are also shown as reference lines in figures (b)-(d).

(b) Oxygen isotopic composition of chondrules in chondrite meteorites, where single is one chondrule and grouped is the average of multiple chondrules analyzed.

(c) Whole-rock oxygen isotopic compositions of meteorites of differentiated bodies among ray chondrites.

(d) Whole-rock oxygen isotopic composition of meteorites with undifferentiated composition among eichondrites.

For all figures, see "Classification of meteorites" for abbreviations of meteorite class names.v

Major

Abbreviations for crystal systems: Cub = cubic system, Tet = Tetragonal system, Hex = Hexagonal system, Rhom = Rhombohedral system, Orth = Orthorhombic system, Mon = Monoclinic system, Tri = Triclinic system, Amor = Amorphous, Liq = Liquid.

Hardness: The relative hardness of minerals is based on the hardness of the following minerals. Hardness increases in order from 1 to 10.

1. Talc, 2. Gypsum, 3. Calcite, 4. Fluorite, 5. Apatite, 6. Feldspar, 7. Quartz, 8. Topaz, 9. Corun-

Mineral	Crystal system	Ideal chemial composition	Color	Striation	Luster
Element and alloy minerals					
Gold	Cub	Au	Gold	Gold	Metallic
Silver	Cub	Ag	Silver-white	Silver-white	Metallic
Copper	Cub	Cu	Copper red	Copper red	Metallic
Mercury	Liq	Hg	Tin white	-	Metallic
Kamacite	Cub	α-(Fe, Ni)	Steel-gray	Iron gray	Metallic
Taenite	Cub	γ-(Fe, Ni)	Steel-gray	Iron gray	Metallic
Platinum	Cub	Pt	Steel-gray	Gray	Metallic
Arsenic	Rhom	As	Tin white	Tin white	Metallic
Bismuth	Rhom	Bi	Silver-white	Silver-white	Metallic
Sulfur	Orth	α-S	Yellow	Yellow	Resinous
Diamond	Cub	C	Colorless[3)]	Colorless	Adamantine
Graphite	Hex	C	Black	Black	Metallic

1) In many cases, natural gold contains silver and the specific gravity varies greatly.

2) The specific gravity increases together with an increase in the nickel content.

Mineral	Crystal system	Ideal chemial composition	Color	Striation	Luster
Sulfide minerals					
Acanthite	Mon	Ag_2S	Lead-gray	Lead-gray	Metallic
Chalcocite	Mon	Cu_2S	Lead-gray	Lead-gray	Metallic
Bornite	Orth	Cu_5FeS_4	Red-brown[4)]	Gray-black	Metallic
Pentlandite	Cub	$(Fe, Ni)_9S_8$	Brass	Brownish	Metallic
Galena	Cub	PbS	Lead-gray	Lead-gray	Metallic
Alabandite	Cub	MnS	Iron-black	Green-brown	Submetallic
Sphalerite	Cub	(Zn, Fe)S	Brown-Black[5)]	Brown	Adamantine
Wurtzite	Hex	ZnS	Brownish	Brown	Submetallic
Greenockite	Hex	CdS	Yellow-Red	Yellow	Resinous
Pyrrhotite	Hex, Mon	$Fe_{1-x}S$ ($x = 0 - 0.2$)	Brass	Gray-black	Metallic
Covellite	Hex	CuS	Indigo	Gray-black	Metallic
Cinnabar	Rhom	HgS	Carmine	Crimson	Adamantine
Millerite	Rhom	NiS	Brass	Green-black	Metallic
Realgar	Mon	AsS	Red	Red-orange	Resinous

Minerals

dum, 10. Diamond

The space group uses Hermann-Mauguin notation. The lattice constants are: a, b, c (Å), α, β, and γ (degrees), and number of chemical formula (Z) included in the unit lattice.

Unless otherwise indicated, the sources are Dana's New Mineralogy 8th Ed., John Wiley & Sons, Gaines, *et al.* (1997).

Cleavage	Hardness	Specific gravity	Space group	Lattice parameters						
				a	b	c	α	β	γ	Z
None	2.5-3	15.2-19.3[1)]	*Fm3m*	4.079						4
None	2.5-3	10.5	*Fm3m*	4.085						4
None	2.5-3	8.9	*Fm3m*	3.615						4
		13.6								
{001}	4	7.9	*Im3m*	2.859						2
None	5-5.5	7.8-8.2[2)]	*Fm3m*	3.596						4
None	4.5-5	21.5	*Fm3m*	3.924						4
{0001}	3.5	5.7	$R\bar{3}m$	3.76		10.56				6
{0001}	2-2.5	9.8	$R\bar{3}m$	4.546		11.860				6
{001} etc.	1.5-2.5	2.1	*Fddd*	10.45	12.84	24.46				128
{111}	10	3.6	*Fd3m*	3.567						8
{0001}	1-2	2.2	$P6_3/mmc$	2.463		6.714				4

3) Although gray and yellow are the prevalent colors, there are many variations including blue, green, purple, orange, red, blackish-red, and black.

None	2-2.5	7.3	$P2_1/n$	4.229	6.931	7.862		99.61		4
{110}	2.5-3	5.8	$P2_1/c$	15.235	11.885	13.496		116.26		48
None	3	5.1	*Pbca*	10.950	21.862	10.950				16
None	3.5-4	4.9-5.2	*Fm3m*	10.042						4
{100}	2.5	7.6	*Fm3m*	5.94						4
{100}	3.5-4	4	*Fm3m*	5.22						4
{110}	3.5-4	3.9-4.1	$F\bar{4}3m$	5.406						4
{11$\bar{2}$0}	3.5-4	4	$P6_3mc$	3.82		6.26				2
{11$\bar{2}$2}	3-3.5	4.8	$P6_3mc$	4.136		6.713				2
None	3.5-4.5	4.7	*A2/a*	12.811	6.870	11.885		117.3		32
{0001}	1.5-2	4.7	$P6_3/mmc$	3.793 8		16.341				6
{10$\bar{1}$0}	2-2.5	8.1	$P3_121$	4.149		9.595				3
{10$\bar{1}$1} etc.	3-3.5	5.4	*R3m*	9.620		3.149				9
{010}	1.5-2	3.5	$P2_1/n$	9.324	13.534	6.585		106.43		16

Major Minerals

Mineral	Crystal system	Ideal chemial composition	Color	Striation	Luster
Chalcopyrite	Tet	$CuFeS_2$	Brass	Green-black	Metallic
Stannite	Tet	Cu_2FeSnS_4	Steel-gray	Black	Metallic
Orpiment	Mon	As_2S_3	Bitter yellow-orange	Light Yellow	Resinous
Stibnite	Orth	Sb_2S_3	Lead-gray	Lead-gray	Metallic
Bismuthinite	Orth	Bi_2S_3	Lead-gray	Lead-gray	Metallic
Pyrite	Cub	FeS_2	Brass	Green-black	Metallic
Marcasite	Orth	FeS_2	Brass	Gray-black	Metallic
Cobaltite	Orth	CoAsS	Silver-white	Gray-black	Metallic
Arsenopyrite	Mon	FeAsS	Silver-white	Gray-black	Metallic
Molybdenite	Hex	MoS_2	Lead-gray	Lead-gray	Metallic
Polybasite	Mon	$(Ag, Cu)_{16}Sb_2S_{11}$	Iron-black	Black	Metallic
Tetrahedrite	Cub	$(Cu, Fe)_{12}Sb_4S_{13}$	Steel-gray	Black	Metallic
Pyrargyrite	Rhom	Ag_3SbS_3	Dark-red	Dark-red	Adamantine
Jamesonite	Mon	$Pb_4FeSb_6S_{14}$	Gray-black	Gray-black	Metallic
4) Changes to a blueish-purple color in the air. 5) When lacking iron, it is amber-colored.					
Oxide minerals					
Cuprite	Cub	Cu_2O	Dark-red	Dark-red	Adamantine
Periclase	Cub	MgO	White	White	Vitreous
Manganosite	Cub	MnO	Green	Brown	Vitreous
Corundum	Rhom	Al_2O_3	Gray–Blue[7]	Colorless	Metallic
Hematite	Rhom	Fe_2O_3	Iron-black	RedBrown	Metallic
Perovskite	Orth	$CaTiO_3$	Brown	Gray-black	Adamantine
Ilmenite	Rhom	$FeTiO_3$	Black	Black	Metallic
Rutile	Tet	TiO_2	Red-brown	Brown	Adamantine
Pyrolusite	Tet	MnO_2	Black	Black	Metallic
Cassiterite	Tet	SnO_2	Brown	Light-brown	Adamantine
Stishovite	Tet	SiO_2	Colorless	Colorless	Vitreous
Anatase	Tet	TiO_2	Brown, Blue	Colorless, Light-yellow	Adamantine
Brookite	Orth	TiO_2	Brown, Black	Colorless, Gray	Adamantine
Uraninite	Cub	UO_2	Black	Black	Submetallic
Spinel	Cub	$MgAl_2O_4$	Colorless[9]	White	Vitreous
Magnetite	Cub	$Fe^{2+}Fe^{3+}{}_2O_4$	Black	Black	Metallic
Chromite	Cub	$FeCr_2O_4$	Black	Black-brown	Metallic
Chrysoberyl	Orth	$BeAl_2O_4$	Green, Yellow	White	Vitreous
Cryptomelane	Mon	KMn_8O_{16}	Black	Black	Submetallic
Diaspore	Orth	α–AlO(OH)	White	White	Vitreous

Continued.

Cleavage	Hardness	Specific gravity	Space group	Lattice parameters						
				a	b	c	α	β	γ	Z
None	3.5-4	4.2	$I\bar{4}2d$	5.289		10.423				4
None	4	4.4	$I\bar{4}2m$	5.453		10.747				2
{010}	1.5-2	3.5	$P2_1/n$	11.49	9.59	4.25		90.45		4
{010}	2	4.6	$Pbnm$	11.229	11.310	3.839				4
{010}	2	6.8	$Pbnm$	11.149	11.304	3.981				4
None	6-6.5	5	$Pa3$	5.417						4
{101}	6-6.5	4.9	$Pnnm$	4.436	5.414	3.381				2
{001}	5.5	6.3	$Pca2_1$	5.582						4
{001}	5.5-6	6.1	$P2_1/c$	5.744	5.675	5.785		112.3		4
{0001}	1-1.5	4.7	$P6_3mmc$	3.16		12.295				2
None	2-3	6.1	$C2/m$	26.17	15.11	23.89		90		16
None	3-4	4.6-5.1[6]	$I\bar{4}3m$	10.327						2
{10$\bar{1}$1}	2.5	5.8	$R3c$	11.047		8.719				6
{001}	2.5	5.7	$P2_1/a$	15.65	19.03	4.03		91.8		2

6) When containing a large amount of arsenic, the specific gravity decreases.

Cleavage	Hardness	Specific gravity	Space group	a	b	c	α	β	γ	Z
None	3.5-4	6.1	$Pn3m$	4.270						2
{001}	5.5	3.58	$Fm3m$	4.211						4
{100}	5.5	5.4	$Fm3m$	4.445						4
None	9	4	$R\bar{3}c$	4.758		12.991				6
None	5.5	5.3	$R\bar{3}c$	5.036		13.749				6
None	5.5	4	$Pnma$	5.441	7.644	5.381				4
None	5.5	4.7	$R\bar{3}$	5.088		14.093				6
{110}	6-6.5	4.2	$P4_2/mnm$	4.593		2.959				2
{110}	2-6[8]	5	$P4_2/mnm$	4.400		2.874				2
None	6.5	7	$P4_2/mnm$	4.738		3.188				2
None	8	4.3	$P4_2/mnm$	4.178		2.666				2
{001}, {011}	5.5-6	3.9	$I4_1/amd$	3.785		9.514				4
None	5.5-6	4.1	$Pbca$	9.174	5.456	5.138				8
None	5-6	7.5-9.7	$Fm3m$	5.468						4
None	7.5-8	3.6	$Fd3m$	8.083						8
None	5.5-6	5.2	$Fd3m$	8.396						8
None	5.5	4.7	$Fd3m$	8.379						8
{110}	8.5	3.8	$Pnma$	9.404	5.476	4.427				4
None	5-6	4.3	$I2/m$	9.956	2.871	9.706		90.95		1
{010}	6.5-7	3.4	$Pnma$	4.396	9.426	2.844				4

Major Minerals

Mineral	Crystal system	Ideal chemial composition	Color	Striation	Luster
Goethite	Mon	α-FeO(OH)	Brown	Yellow-brown	Adamantine
Manganite	Mon	γ-MnO(OH)	Steel-gray	Black-brown	Submetallic
Brucite	Rhom	$Mg(OH)_2$	White	White	Pearly
Gibbsite	Mon	γ-$Al(OH)_3$	White	White	Vitreous
7) Other colors include yellow and red (ruby). 8) When earthy, the hardness falls to as low as level 2.					
Halogenides					
Halite	Cub	NaCl	Colorless[10]	White	Vitreous
Sylvite	Cub	KCl	Colorless	White	Vitreous
Chlorargyrite	Cub	AgCl	Colorless, Gray-brown	White	Resinous
Fluorite	Cub	CaF_2	Colorless[11]	White	Vitreous
Atacamite	Orth	$Cu_2(OH)_3Cl$	Green	Green	Adamantine
Cryolite	Mon	Na_3AlF_6	Colorless	White	Vitreous
10) Other colors include pale yellow, red, blue and purple.					
Carbonates					
Calcite	Rhom	$CaCO_3$	Colorless, White, Light-yellow	White	Vitreous
Magnesite	Rhom	$MgCO_3$	Colorless, White	White	Vitreous
Siderite	Rhom	$FeCO_3$	Yellow-brown	White	Vitreous
Rhodochrosite	Rhom	$MnCO_3$	Gray, Crimson	White	Vitreous
Aragonite	Orth	$CaCO_3$	Colorless, White	White	Vitreous
Strontianite	Orth	$SrCO_3$	Colorless, White, Gray	White	Vitreous
Cerussite	Orth	$PbCO_3$	White	White	Adamantine
Dolomite	Rhom	$CaMg(CO_3)_2$	Colorless, White, Gray	White	Vitreous
Azurite	Mon	$Cu_3(CO_3)_2(OH)_2$	Blue	Blue	Vitreous
Malachite	Mon	$Cu_2(CO_3)(OH)_2$	Green	Light-green	Adamantine
Nitrates					
Nitratine	Rhom	$NaNO_3$	Colorless, White	White	Vitreous
Niter	Orth	KNO_3	Colorless, White	White	Vitreous
Borates					
Kotoite	Orth	$Mg_3(BO_3)_2$	Colorless, White	White	Vitreous
Borax	Mon	$Na_2B_4O_5(OH)_4 \cdot 8H_2O$	Colorless	White	Vitreous
Sulfates					
Thenardite	Orth	Na_2SO_4	Colorless, White	White	Vitreous
Barite	Orth	$BaSO_4$	Colorless, White[12]	White	Vitreous
Celestine	Orth	$SrSO_4$	Colorless, Light-blue	White	Vitreous
Anhydrite	Orth	$CaSO_4$	Colorless, White[13]	White	Vitreous
Gypsum	Mon	$CaSO_4 \cdot 2H_2O$	Colorless, White	White	Subvitrious
Chalcanthite	Tri	$CuSO_4 \cdot 5H_2O$	Blue	White	Vitreous

Continued.

Cleavage	Hardness	Specific gravity	Space group	Lattice parameters						
				a	b	c	α	β	γ	Z
{010}	5.5	4.3	$Pnma$	4.62	9.95	3.01				4
{010}	4	4.3	$P2_1/c$	8.98	5.28	5.71		90		8
{0001}	2.5	2.4	$P\bar{3}m1$	3.147		4.769				1
{001}	3	2.4	$P2_1/c$	8.684	5.078	9.736		94.53		8

9) Almost all colors are known.

Cleavage	Hardness	Specific gravity	Space group	a	b	c	α	β	γ	Z
{100}	2	2.2	$Fm3m$	5.640						4
{100}	2	2	$Fm3m$	6.93						4
None	2.5	5.6	$Fm3m$	5.549						4
{111}	4	3.2	$Fm3m$	5.463						4
{010}	3.5	3.8	$Pnam$	6.030	6.865	9.120				4
None	2.5	3	$P2_1/n$	5.402	5.596	7.756		90.28		2

11) Other colors include purple, green, blue, yellow, pink and red.

Cleavage	Hardness	Specific gravity	Space group	a	b	c	α	β	γ	Z
{10$\bar{1}$1}	3	2.7	$R\bar{3}c$	4.989 6		17.061				6
{10$\bar{1}$1}	4	3	$R\bar{3}c$	4.633 3		15.013				6
{10$\bar{1}$1}	4	4	$R\bar{3}c$	4.691 6		15.379 6				6
{10$\bar{1}$1}	4	3.7	$R\bar{3}c$	4.768 2		15.635 4				6
{010}	4	2.9	$Pmcn$	4.959 8	7.964 1	5.737 9				4
{110}	3.5	3.8	$Pmcn$	5.090	8.358	5.997				4
{110}, {201}	3.5	6.6	$Pmcn$	5.180	8.492	6.134				4
{10$\bar{1}$1}	4	2.9	$R\bar{3}$	4.806 9		16.003 4				3
{011}	4	3.8	$C2_1/c$	5.00	5.85	10.35		92.33		2
{201}	3.5-4	4.1	$C2_1/a$	9.48	12.03	3.21		98		4
{0112}	2	2.3	$R\bar{3}c$	5.070		16.829				6
{001}, {010}	2	2.1	$Pmcn$	5.414	9.164	6.431				4
{110}	6.5	3.1	$Pnmn$	5.401	8.422	4.507				2
{100}	2.5	1.7	$A2/a$	12.219	10.665	11.884		106.64		4
{010}	2.5	2.7	$Fddd$	9.829	12.302	5.868				8
{001}	3-3.5	4.5	$Pnma$	8.884	5.456	7.157				4
{001}, {210}	3-3.5	4	$Pnma$	8.371	5.355	6.870				4
{010}, {100}	3.5	3	$Cmcm$	6.993	6.995	6.245				4
{010}	2	2.3	$C2/c$	5.678	15.20	6.52		118.43		4
None	2.5	2.3	$P\bar{1}$	6.110	10.673	5.95	97.58	107.17	77.55	2

Major Minerals

Mineral	Crystal system	Ideal chemial composition	Color	Striation	Luster
Brochanite	Mon	$Cu_4(SO_4)(OH)_6$	Green	Aqua	Vitreous
Linarite	Mon	$PbCu(SO_4)(OH)_2$	Blue	Light-blue	Vitreous
Alunite	Rhom	$KAl_3(SO_4)_2(OH)_6$	White, Pink	White	Vitreous
Jarosite	Rhom	$KFe_3(SO_4)_2(OH)_6$	Yellow-brown	Yellow-brown	Subadamantine
12) Other colors include yellow, blackish-red, red, and blue. 13) Other colors include blue, purple, and red.					
Phosphates, Arsenates, Vanadates					
Monazite-(Ce)	Mon	$CePO_4$	Yellow-brown	White, Light-brown	Resinous
Xenotime-(Y)	Orth	YPO_4	Yellow-brown, Red-brown	Light-brown	Vitreous
Autunite	Tet	$Ca(UO_2)_2(PO_4)_2 \cdot 10\text{-}12H_2O$	Yellow, Light-green	Light-yellow	Vitreous
Vivianite	Mon	$Fe_3(PO_4)_2 \cdot 8H_2O$	Colorless[14)]	White[14)]	Vitreous
Scorodite	Orth	$FeAsO_4 \cdot 2H_2O$	Gray-green, Brown	White, Gray	Vitreous
Fluorapatite	Hex	$Ca_5(PO_4)_3F$	Colorless, White[15)]	White	Vitreous
Pyromorphite	Hex	$Pb_5(PO_4)_3Cl$	Green, Light-brown	White	Resinous
Vanadinite	Hex	$Pb_5(VO_4)_3Cl$	Yellow, Orange	Yellow	Resinous
Turquoise	Tri	$CuAl_6(PO_4)_4(HO)_8 \cdot 4H_2O$	Blue, Blue-green	White	Vitreous
14) When exposed to light, it turns blue. 15) Other colors include green, blue, yellow, blackish-red, and red.					
Tungstates					
Ferberite	Mon	$FeWO_4$	Black	Black	Submetallic
Scheelite	Tet	$CaWO_4$	White, Yellow	White	Adamantine
Silicates					
Forsterite	Orth	Mg_2SiO_4	White, Yellow-green	White	Vitreous
Fayalite	Orth	Fe_2SiO_4	Brown, Black	Light-brown	Vitreous
Pyrope	Cub	$Mg_3Al_2(SiO_4)_3$	Red	White	Vitreous
Almandine	Cub	$Fe_3Al_2(SiO_4)_3$	Red, Black-brown	White	Vitreous
Spessartine	Cub	$Mn_3Al_2(SiO_4)_3$	Yellow, Orange	White	Vitreous
Andradite	Cub	$Ca_3Fe_2(SiO_4)_3$	Brown-red, Yellow-green	White	Vitreous
Grossular	Cub	$Ca_3Al_2(SiO_4)_3$	White, Yellow, Orange	White	Vitreous
Uvarovite	Cub	$Ca_3Cr_2(SiO_4)_3$	Dark-green, Rusty-green	White	Vitreous
Zircon	Tet	$ZrSiO_4$	Brown, Orange, Green	White	Adamantine
Sillimanite	Orth	$Al_2O(SiO_4)$	White	White	Silky
Andalusite	Orth	$Al_2O(SiO_4)$	White, Crimson, Green	White	Vitreous
Kyanite	Tri	$Al_2O(SiO_4)$	Blue, Green, Gray	White	Vitreous
Staurolite	Mon	$(Fe, Mg)_4Al_{17}O_{13}[(Si, Al)O_4]_8(OH)_3$	Brown	Gray	Vitreous
Topaz	Orth	$Al_2SiO_4(F, OH)_2$	Colorless[17)]	White	Vitreous

Continued.

Cleavage	Hardness	Specific gravity	Space group	Lattice parameters						
				a	b	c	α	β	γ	Z
{100}	4	4	$P2_1/a$	13.08	9.85	6.02		103.37		4
None	2.5	5.4	$P2_1/m$	9.691	5.650	4.687		102.65		2
{0001}	4	2.8	$R\bar{3}m$	6.970		12.27				3
{0001}	3.5	3.3	$R\bar{3}m$	7.304		17.268				3
{100}	5	4.6	$P2_1/n$	6.761	6.966	6.45		103.47		4
{100}	4-5	4.8	$I4_1/amd$	6.878		6.036				4
{001}	2.5	3.2	$I4/mmm$	7.009		20.736				2
{010}	2	2.7	$C2/m$	10.086	13.441	4.703		104.27		2
None	4	3.3	$Pcab$	9.996	10.278	8.937				8
None	5	3.2	$P6_3/m$	9.367		6.884				2
None	3.5	7	$P6_3/m$	9.976		7.351				2
None	2.5-3	6.0	$P6_3/m$	10.317		7.338				2
{001}	5.5	2.8	$P\bar{1}$	7.690	9.950	7.490	110.57	115.40	68.40	1
{010}	4.5	7.5	$P2/c$	4.72	5.70	4.96		90		2
{101}	5	6.1	$I4_1/a$	5.242		11.372				4
None	7	3.2	$Pbnm$	4.7540	10.197 1	5.980 6				4[28]
None	6.5	4.4	$Pbnm$	4.8211	10.477 9	6.088 9				4[28]
None	7.5	3.7	$Ia3d$	11.455						8
None	7.5	4.3	$Ia3d$	11.526						8
None	7.5	4.2	$Ia3d$	11.63						8
None	7	3.9	$Ia3d$	12.059						8
None	7	3.6	$Ia3d$	11.851						8
None	6.5-7	3.85	$Ia3d$	11.999						8
None	7.5	4.7	$I4_1/amd$	6.604		5.979				4
{010}	7	3.2	$Pnma$	7.484	7.672	5.77				4
{110}	7	3.2	$Pnnm$	7.798	7.903	5.556				4
{100}, {010}	4-7.5[16]	3.6	$P\bar{1}$	7.126	7.852	5.572	89.99	101.11	106.03	4
{010}	7-7.5	3.7-3.8	$C2/m$	7.870	16.623	5.661		90.12		1
{001}	8	3.5	$Pbnm$	4.651 2	8.795	8.399 3				4

Major Minerals

Mineral	Crystal system	Ideal chemial composition	Color	Striation	Luster
Titanite	Mon	$CaTiO(SiO_4)$	Yellow, Orange, Brown	Light-brown	Adamantine
Gehlenite	Tet	$Ca_2Al(SiAlO_7)$	White, Gray, Yellow	White	Vitreous
Ferroaxinite	Tri	$Ca_2FeAl_2(BSi_4O_{15})(OH)$	Gray-brown, Purple	White	Vitreous
Lawsonite	Orth	$CaAl_2(Si_2O_7)(OH)_2 \cdot H_2O$	White, Light-gray-blue	White	Vitreous
hemimorphite	Orth	$Zn_4(Si_2O_7)(OH)_2 \cdot H_2O$	Colorless, White	White	Vitreous
Clinozoisite	Mon	$Ca_2Al_3(SiO_4)(Si_2O_7)O(OH)$	Gray, Light-brown	White	Vitreous
Epidote	Mon	$Ca_2Fe^{3+}Al_2(SiO_4)(Si_2O_7)O(OH)$	Yellow-green, Dark-green	Light-green	Vitreous
Piemontite	Mon	$Ca_2Mn^{3+}Al_2(SiO_4)(Si_2O_7)O(OH)$	Crimson, Red-brown	Crimson	Vitreous
Zoisite	Orth	$Ca_2Al_3(SiO_4)(Si_2O_7)O(OH)$	Gray[18)]	White	Vitreous
Vesuvianite	Tet	$Ca_{19}(Al, Fe, Mg)_{13}(SiO_4)_{10}(Si_2O_7)_4(OH, F, O)_{10}$	Yellow, Green, Brown	White	Vitreous
Beryl	Hex	$Be_3Al_2Si_6O_{18}$	Light-green[19)]	White	Vitreous
Cordierite	Orth	$Mg_2Al_3(AlSi_5O_{18})$	Gray-blue, Blue-green	White	Vitreous
Schorl	Rhom	$NaFe_3Al_6(BO_3)_3Si_6O_{18}(OH, F)_4$	Black, Black-brown	Light-brown	Vitreous
Osumilite	Hex	$(K, Na)(Fe, Mg)_2(Al, Fe)_3(Al_2Si_{10}O_{30}) \cdot H_2O$	Deep-blue	White	Vitreous
Enstatite	Orth	$Mg_2Si_2O_6$	Yellow-green, Brown	Light-gray-brown	Vitreous
Diopside	Mon	$CaMgSi_2O_6$	White, Yellow-green	White	Vitreous
Hedenbergite	Mon	$CaFe^{2+}Si_2O_6$	Black-Green-Brownish-green	Light-green	Vitreous
Augite	Mon	$(Ca, Mg, Fe, Al, Ti)_2(Si, Al)_2O_6$	Dark-green, Dark-brown	Gray	Vitreous
Jadeite	Mon	$NaAlSi_2O_6$	White, Light-green	White	Vitreous
Aegirine	Mon	$NaFeSi_2O_6$	Green, Brown	Light-green, Yellow	Vitreous
Wollastonite-1A	Tri	$CaSiO_3$	White	White	Silky
Cummingtonite	Mon	$(Mg, Fe)_7Si_8O_{22}(OH)_2$	Brown, Green	White	Vitreous
Anthophyllite	Orth	$(Mg, Fe)_7Si_8O_{22}(OH)_2$	Gray	White	Vitreous
Actinolite	Mon	$Ca_2(Mg, Fe)_5Si_8O_{22}(OH)_2$	Green, Dark-green	White	Vitreous
Hornblende	Mon	$Ca_2(Mg, Fe)_4Al(AlSi_7O_{22})(OH)_2$	Green-black	White	Vitreous
Glaucophane	Mon	$Na_2Mg_3Al_2Si_8O_{22}(OH)_2$	Charcoal-gray, Purple-blue	Gray	Vitreous
Kolinite	Tri	$Al_4Si_4O_{10}(OH)_8$	White	White	Dull, Pearly
Antigorite	Mon	$Mg_6Si_4O_{10}(OH)_8$	Dark-green	White	Resinous
Pyrophyllite-1A	Mon, Tri	$Al_2Si_4O_{10}(OH)_2$	White, Light-green	White	Pearly
Talc					
Muscovite-2M$_1$	Mon	$KAl_2(AlSi_3O_{10})(OH, F)_2$	Colorless, White, Yellow	White	Pearly
Phlogopite-1M	Mon	$KMg_3(AlSi_3O_{10})(OH, F)_2$	Yellow, Brown	White, Light-brown	Pearly

Continued.

Cleavage	Hardness	Specific gravity	Space group	Lattice parameters						
				a	*b*	*c*	*α*	*β*	*γ*	*Z*
{110}	5.5	3.5	*A*2/*a*	7.066	8.705	6.561		113.93		4
{001}	5.5	3	$P\bar{4}2_1m$	7.677		5.059				2
{100}	7	3.3	$P\bar{1}$	9.46	9.2	7.15	102.7	98.1	88.2	2
{100}, {010}	6	3.1	*Ccmm*	8.795	5.847	13.142				4
{110}, {101}	4.5–5	3.5	*Imm*2	8.367	10.730	5.115				2
{001}, {100}	6.5	3.3	$P2_1/m$	8.879	5.583	10.149		115.5		2
{001}, {100}	6.5	3.4	$P2_1/m$	8.914	5.640	10.162		115.4		2
{001}	6	3.4	$P2_1/m$	8.843	5.664	10.150		115.25		2
{100}	6.5	3.3	*Pnma*	16.212	5.559	10.036				4
None	5–6.5	3.4	*P*4/*nnc*	15.65		11.8				2
None	7.5	2.7	*P*6/*mcc*	9.208		9.175				2
None	7	2.6	*Cccm*	17.088	9.734	9.359				4
None	7.5	3.1	*R*3*m*	15.992		7.172				3
None	7	2.6	*P*6/*mcc*	10.13		14.31				2
{210}	5.5	3.2	*Pbca*	18.235	8.818	5.179				8
{110}	6	3.3	*C*2/*c*	9.739	8.913	5.253		106.02		4
{110}	6	3.65	*C*2/*c*	9.852	9.031	5.242		104.84		4
{110}	5.5–6	3.2–3.6	*C*2/*c*	9.707	8.858	5.274		106.52		4
{110}	7	3.3	*C*2/*c*	9.418	8.562	5.219		107.58		4
{110}	6	3.6	*C*2/*c*	9.658	8.795	5.295		107.42		4
{100}, {001}	4.5	2.9	$P\bar{1}$	7.926	7.320	7.067 5	90.05	95.22	103.43	6
{110}	6	3.5	$P2_1/m$	9.583	18.091	5.315		102.6		2
{110}	6	3.1	*Pnma*	18.554	18.026	5.282				4
{110}	6	3.1	*C*2/*m*	9.891	18.200	5.305		104.64		2[28)]
{110}	5–6	3.0–3.5	*C*2/*m*	9.9	18.1	5.31		105		2
{110}	6	3.1	*C*2/*m*	9.531	17.759	5.303		103.59		2
{001}	1–2	2.6	*P*1	5.15	8.95	7.39	91.8	105	90	1[28)]
{001}	2.5–4	2.4–2.8	*Pm*	43.4	9.24	7.26		91.4		8
{001}	1.5	2.9	$C\bar{1}$	5.16	8.966	9.347	91.18	100.46	89.64	2
{001}	2.5–3.5	2.9	*C*2/*c*	5.199	9.027	20.106		95.78		4
{001}	2–2.5	2.8–3.0	*C*2/*m*	5.308	9.190	10.155		100.08		2

Major Minerals

Mineral	Crystal system	Ideal chemial composition	Color	Striation	Luster
Clinochlore	Mon	$Mg_3(Mg, Al)_3(Si, Al)_4O_{10}(OH)_8$	Green, Dark-green	Light-green	Greasy, dull, vitreous
Prehnite	Orth	$Ca_2Al(AlSi_3O_{10})(OH)_2$	White, Light-green	White	Vitreous
Cristobalite	Tet	SiO_2	White	White	Vitreous
Tridymite	Mon	SiO_2	White	White	Vitreous
α-Quartz	Rhom	SiO_2	Colorless[20]	White	Vitreous
Coesite	Mon	SiO_2	White	White	Vitreous
Opal	Amor	$SiO_2 \cdot nH_2O$	White[21]	White	Vitreous
Orthoclase	Mon	$KAlSi_3O_8$	White[22]	White	Vitreous
Sanidine	Mon	$(K, Na)AlSi_3O_8$	White	White	Vitreous
Microcline	Tri	$KAlSi_3O_8$	White[23]	White	Vitreous
Low-albite	Tri	$NaAlSi_3O_8$	White[24]	White	Vitreous
Anorthite	Tri	$CaAl_2Si_2O_8$	White[25]	White	Vitreous
Nepheline	Hex	$(Na, K)AlSiO_4$	White, Yellow	White	Vitreous
Leucite (low T type)	Tet	$KAlSi_2O_6$	White	White	Vitreous
Analcime	Cub[26]	$Na(AlSi_2)O_6 \cdot H_2O$	Colorless, White	White	Vitreous
Laumontite	Mon	$Ca(Al_2Si_4)O_{12} \cdot 4H_2O$	White	White	Vitreous
Heulandite	Mon	$(Ca, Na_2, K_2)_{4.5}(Al_9Si_{27})O_{72} \cdot 24H_2O$	Colorless, White, Yellow	White	Vitreous
Natrolite	Orth	$Na_2(Al_2Si_3)O_{10} \cdot 2H_2O$	Colorless, White	White	Vitreous

16) The hardness changes depending on the orientation of crystals. 7 to 7.5 for {010} surfaces, 4 to 5 for {100} blue, and green. 19) Other colors include blue (aqua marine), bright green (emerald), yellow, pink, and red. 20) light. 22) Other colors include pink, yellow, and green. 23) Other colors include blueish-green (amazonite) and red. 26) Other crystal systems include tetragonal, orthorhombic, monoclinic, and triclinic systems. 27) The

Classification of Minerals According

Classification	Structural outline	Si : O	Example of mineral
Nesosilicate	Isolated tetrahedra SiO*	1 4	Olivine, garnet, sillimanite
Sorosilicate	Two tetrahedra sharing an oxygen	2 7	Gehlenite, hemimorphite, lawsonite
Cyclosilicate	Closed ring of tetrahedra, 2 oxygens sharing	1 3	Beryl, cordierite, tourmaline
Inosilicate	Single chain of tetrahedra sharing 2 oxygens	1 3	Augite, wollastonite

* A tetrahedron with an oxygen at each point and a silicon at the center.

Continued.

Cleavage	Hardness	Specific gravity	Space group	Lattice parameters						
				a	*b*	*c*	*α*	*β*	*γ*	*Z*
{001}	2-3	2.3-2.9	$C2/m$	3.350	9.267	14.27		96.35		2[27]
{001}	6-6.5	2.9	$P2cm$	4.65	5.49	18.52				2
None	6.5	2.3	$P4_12_12$	4.969		6.926				4
None	7	2.3	Cc	18.494	4.991	23.758		105.79		48
None	7	2.7	$P6_422$	4.999	5.457					3
None	7.5	3	$C2/c$	7.135	12.372	7.174		120.36		16
None	6.5	2.1								
{001}, {010}	6	2.6	$C2/m$	8.56	12.94	7.21		116.3		4
{001}, {010}	6	2.6	$C2/m$	8.60	13.04	7.18		116.0		4
{001}, {010}	6	2.6	$C\bar{1}$	8.53	12.95	7.19	90.4	116.0	87.80	4
{001}, {010}	6	2.6	$C\bar{1}$	8.14	12.79	7.16	94.3	116.6	87.7	4
{001}, {010}	6	2.8	$P\bar{1}$	8.18	12.87	14.18	93.2	115.9	91.1	8
None	5.5-6	2.6-2.7	$P6_3$	9.989		0.300				8
{110}	5.5-6	2.46	$I4_1/a$	13.09		13.75				16
None	5-5.5	2.3	$Ia3d$	13.73						16[27]
{010}, {110}	4	2.3	$C2/m$	14.9	13.2	7.55		110.5		4
{010}	3.5-4	2.1-2.3	$C2/m$	17.6	17.9	7.45		116.3		1
{110}	5.5	2.3	$Fdd2$	18.3	18.55	6.58				8

surfaces. 17) Other colors include yellow, blackish-red, pink, and blue. 18) Other colors include yellow, pink, Other colors include purple (amethyst), yellow, and black. 21) Sometimes emits amber- and rainbow-colored pink. 24) Other colors include yellow, pale blue, pale green, pink and red. 25) Other colors include yellow and lattice constant was taken from the Handbook of Mineralogy, Mineral Data Publishing (1995).

to Structure of Silicate Minerals

Classification	Structural outline	Si : O		Example of mineral
Inosilicate	Double chains of tetrahedra sharing 2 or 3 oxygens	4	11	Amphibole
Phyllosilicate	Tetrahedra aligned in a plane with 3 oxygens shared	2	5	Mica, chlorite, kaolinite
Tectosilicate	Tetrahedra arrayed in a 3D structure with 4 oxygens shared	1	2	Quartz, feldspar, zeolite

Optical Properties of Major Rock-Forming Minerals

Mineral	Color under open nicol	Refractive index	Optic character and optic axial angle	Optic orientation
Fluorite	Colorless	1.433–1.435	Isotropic	
Analcime	Colorless	1.479–1.493	Isotropic	
Spinel	Colorless, Variable light color	1.719	Isotropic	
Grossular	Colorless	1.734	Isotropic	
Almandine	Colorless, Variable light color	1.830	Isotropic	
Andradite	Colorless, Light-yellow	1.887	Isotropic	
Volcanic glass	Colorless	1.49–1.60	Isotropic	
Quartz	Colorless	ω1.544, ε1.553	Uniaxial(+)	
Osumilite	Colorless–Light-blue	ω1.546, ε1.550	Uniaxial(+)	
Brucite	Colorless	ω1.559, ε1.590	Uniaxial(+)	
Zircon	Colorless–Light-brown	ω1.92–1.96, ε1.97–2.02	Uniaxial(+)	
Rutile	Red-brown–Yellow-brown	ω2.61, ε2.89	Uniaxial(+)	
Cristobalite	Colorless	ω1.487, ε1.484	Uniaxial(–)	
Fluorapatite	Colorless	ω1.633–1.650, ε1.629–1.646	Uniaxial(–)	
Calcite	Colorless	ω1.658, ε1.486	Uniaxial(–)	
Schorl	Blue-black, Dark-gray, Blue, Yellow	ω1.658–1.698, ε1.633–1.675	Uniaxial(–)	
Dolomite	Colorless	ω1.679, ε1.500	Uniaxial(–)	
Corundum	Colorless, Light-blue, Light-red	ω1.767–1.772, ε1.759–1.762	Uniaxial(–)	
Goethite	Brown	ω2.40–2.52, ε2.26–2.28	Uniaxial(–)	
Tridymite	Colorless	α1.469–1.479, γ1.473–1.483	Biaxial(+) : 35°–90°	$a=Y$, $b=X$, $c=Z$
Gypsum	Colorless	α1.519–1.521, γ1.529–1.531	Biaxial(+) : ~58°	$b=Y$, $c\wedge Z=$ ~–52° $a\wedge X=$ ~–15°
Low-albite	Colorless	α1.527–1.533, γ1.538–1.543	Biaxial(+) : 77°–84°	
Cordierite	Colorless–Light-blue	α1.522–1.560, γ1.527–1.578	Biaxial(+) : 90°–75°	$a=Y$, $b=Z$, $c=X$
Gibbsite	Colorless	α1.573–1.576, γ1.594–1.596	Biaxial(+) : 0°–40°	$b=X$, $c\wedge Z=+21°$, $a\wedge Y=25.5°$
Anhydrite	Colorless	α1.570, γ1.614	Biaxial(+) : ~43°	$a=Y$, $b=X$, $c=Z$
Clinochlore	Colorless–Light-yellow–Light-green	α1.57–1.59, γ1.58–1.61	Biaxial(+) : 15°–45°	b = Y, {001} $\wedge$ Z = ~0°–3°

Continued.

Mineral	Color under open nicol	Refractive index	Optic character and optic axial angle	Optic orientation
Coesite	Colorless	α1.593-1.599, γ1.597-1.604	Biaxial(+) : 54°-64°	
Topaz	Colorless	α1.606-1.630, γ1.616-1.638	Biaxial(+) : 48°-68°	$a=X$, $b=Y$, $c=Z$
Prehnite	Colorless	α1.610-1.637, γ1.632-1.673	Biaxial(+) : 60°-70°	$a=X$, $b=Y$, $c=Z$
Anthophyllite	Colorless-Light-brown, Light-green	α1.588-1.663, γ1.613-1.683	Biaxial(+) : 79°-90°	$a=X$, $b=Y$, $c=Z$
Sillimanite	Colorless	α1.653-1.661, γ1.669-1.684	Biaxial(+) : 20°-30°	$a=X$, $b=Y$, $c=Z$
Cummingtonite	Colorless-Light-brown, Light-green	α1.628-1.658, γ1.652-1.692	Biaxial(+) : 70°-90°	$b=Y$, $c\wedge Z=\sim-21^\circ--16^\circ$, $a\wedge X=\sim-9^\circ--3^\circ$
Forsterite	Colorless	α1.635-1.653, γ1.670-1.690	Biaxial(+) : 82°-90°	$a=Z$, $b=X$, $c=Y$
Enstatite	Colorless	α1.654-1.664, γ1.665-1.675	Biaxial(+) : 35°-90°	$a=X$, $b=Y$, $c=Z$
Hornblende	Light-green, Yellow-brown-Blue-green, Brown	α1.610-1.700, γ1.630-1.730	Biaxial(+) : 85°-90°	$b=Y$, $c\wedge Z=-30^\circ--40^\circ$, $a\wedge X=+3^\circ--19^\circ$
Lawsonite	Colorless	α1.665, γ1.684-1.686	Biaxial(+) : 76°-87°	$a=X$, $b=Y$, $c=Z$
Diopside	Colorless, Light-green	α1.664-1.672, γ1.694-1.702	Biaxial(+) : 58°	$b=Y$, $c\wedge Z=-38^\circ--39^\circ$
Zoisite	Colorless	α1.685-1.707, γ1.697-1.725	Biaxial(+) : 0°-60°	$a=Z$, $b=Y$, $c=X$
Augite	Colorless, Light-green, Light-brown-purple	α1.671-1.735, γ1.703-1.761	Biaxial(+) : 25°-60°	$b=Y$, $c\wedge Z=-35^\circ--50^\circ$, $a\wedge X=-20^\circ--35^\circ$
Clinozoisite	Colorless	α1.703-1.715, γ1.709-1.734	Biaxial(+) : 14°-90°	$b=Y$, $c\wedge X=-85^\circ-0^\circ$, $a\wedge Z=-60^\circ-+25^\circ$
Staurolite	Colorless-Yellow-Yellow-brown	α1.736-1.747, γ1.745-1.762	Biaxial(+) : 80°-90°	$a=Y$, $b=X$, $c=Z$
Piemontite	Light-yellow, Pink-Red-purple	α1.725-1.794, γ1.750-1.832	Biaxial(+) : 50°-86°	$b=Y$, $c\wedge X=+2^\circ-+9^\circ$, $a\wedge Z=+27^\circ-+35^\circ$
Monazite	Colorless-Yellow-brown	α1.774-1.800, γ1.828-1.851	Biaxial(+) : 6°-19°	$b=X$, $c\wedge Z=-2^\circ--7^\circ$

Optical Properties of Major Rock-Forming Minerals Continued.

Mineral	Color under open nicol	Refractive index	Optic character and optic axial angle	Optic orientation
Monazite				$a\wedge Y = +12° - +7°$
Titanite	Light-gray-Yellow-brown	α1.840-1.950, γ1.943-2.11	Biaxial(+) : 17°-56°	$b = Y$, $c\wedge Z = -36° - -51°$, $a\wedge X = -6° - -21°$
Microcline	None	α1.514-1.516, γ1.521-1.522	Biaxial(-) : 66°-68°	
Orthoclase	None	α1.518-1.520, γ1.522-1.525	Biaxial(-) : 35°-50°	$b = Z$, $c\wedge Y = -13° - -21°$, $a\wedge X = +14° - +6°$
Sanidine	None	α1.518-1.524, γ1.522-1.530	Biaxial(-) : 18°-42°	$b = Z$, $c\wedge Y = \sim -21°$, $a\wedge X = \sim +5°$
Antigorite	None-Light-green	α1.54-1.58, γ1.55-1.59	Biaxial(-) : -20°	$b = Y$, $c\wedge X = -1°$
Anorthite	None	α1.573-1.577, γ1.585-1.590	Biaxial(-) : 77°-78°	
Talc	None	α1.538-1.550, γ1.565-1.600	Biaxial(-) : 0°-30°	$b = Y$, a-Z, $c\wedge X = \sim -10°$
Muscovite	None	α1.552-1.574, γ1.587-1.610	Biaxial(-) : 30°-47°	$b = Z$, $c\wedge X = 0° - -5°$
Phlogopite	None-Light-yellow-brown	α1.530-1.590, γ1.558-1.637	Biaxial(-) : 0°-15°	$b = Y$, $c\wedge X = -10° - -5°$, $a\wedge Z = 0° - +5°$
Glaucophane	Yellow-Purple-blue	α1.594-1.647, γ1.618-1.663	Biaxial(-) : 50°-0°	$b = Y$, $c\wedge Z = -6° - -9°$, $a\wedge X = +8° - +5°$
Andalusite	None-Pink	α1.629-1.640, γ1.638-1.650	Biaxial(-) : 71°-86°	$a = Z$, $b = Y$, $c = X$
Actinolite	Light-yellow-Light-green	α1.616-1.669, γ1.640-1.688	Biaxial(-) : 84°-75°	$b = Y$, $c\wedge Z = -17° - -13°$, $a\wedge X = -2° - +2°$
Aragonite	None	α1.530, γ1.685	Biaxial(-) : 18°	$a = Y$, $b = Z$, $c = X$
Epidote	None-Light-yellow-green	α1.715-1.751, γ1.734-1.797	Biaxial(-) : 90°-64°	$b = Y$, $c\wedge X = 0° - +15°$, $a\wedge Z = +25° - +40°$
Aegirine	Yellow-Pale-green	α1.722-1.776, γ1.758-1.836	Biaxial(-) : 58°-90°	$b = Y$, $c\wedge X = -10° - +12°$, $a\wedge Z = +28° - +6°$

Major Igneous Rocks (Volcanic Rocks)

(a) Classification of volcanic rocks*

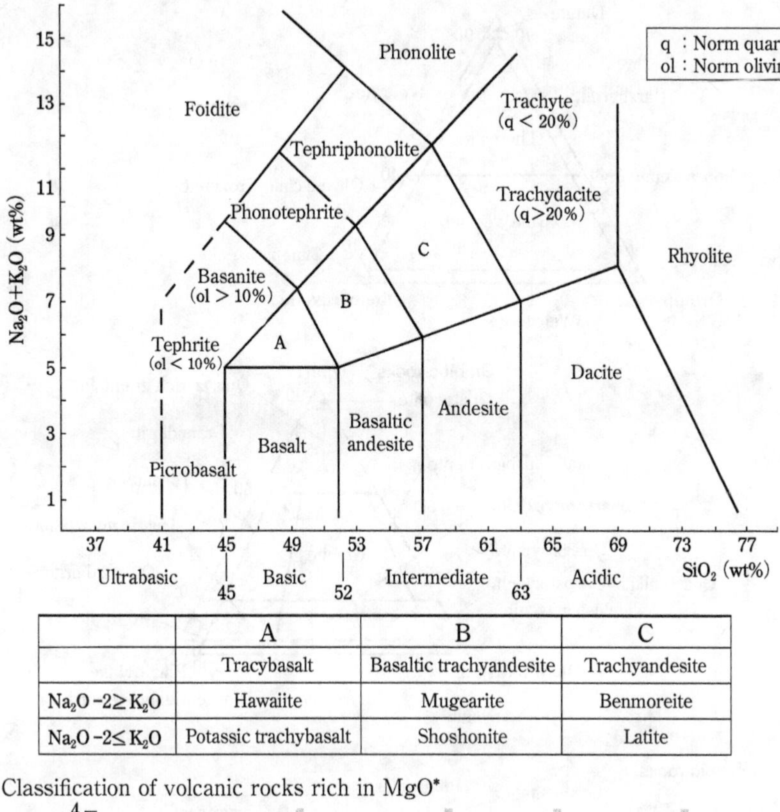

	A	B	C
	Tracybasalt	Basaltic trachyandesite	Trachyandesite
$Na_2O - 2 \geq K_2O$	Hawaiite	Mugearite	Benmoreite
$Na_2O - 2 \leq K_2O$	Potassic trachybasalt	Shoshonite	Latite

(b) Classification of volcanic rocks rich in MgO*

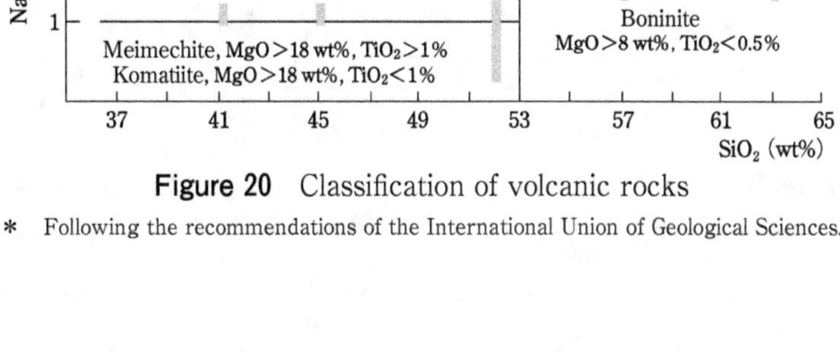

Figure 20 Classification of volcanic rocks

* Following the recommendations of the International Union of Geological Sciences.

Major Igneous Rocks (Plutonic Rocks)

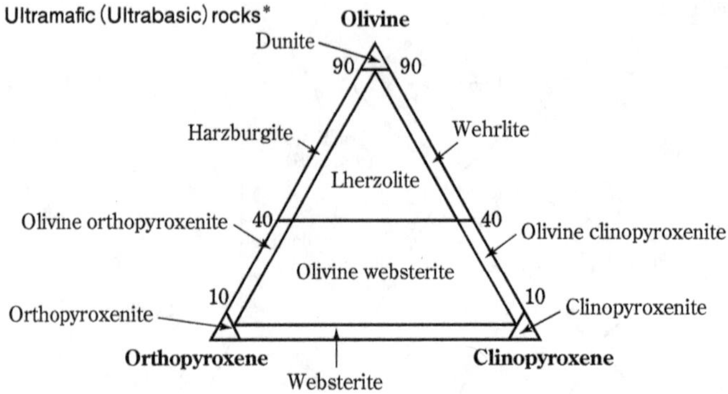

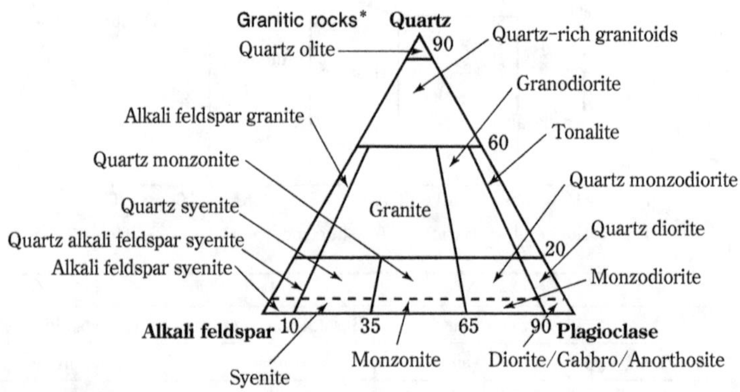

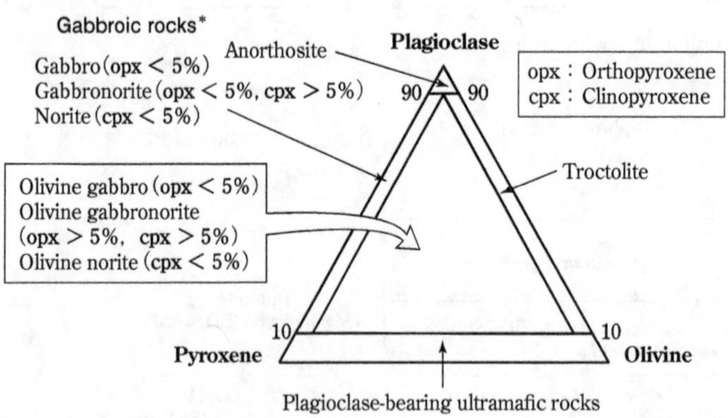

Figure 21 Classification of ultramafic rocks, granites and gabbros

* Following the recommendations of the International Union of Geological Sciences.

Major Sedimentary Rocks

Major constituents	Rock name	Subdivisions	
Clastic rocks Quartz Feldspar Clast Clay minerals, etc.	Clastic rocks	Classification according to grain size Conglomerate Sandstone Mudstone or Shale	Classification according to relative abundances of constituent particles Quartz sandstone Arkose Lithic sandstone
Biogenic particles	Biogenic rocks		
Calcareous fossils Coral, shellfish, foraminifera, etc.	Limestone	Reef limestone Calcareous clastic rocks Dolomite: Various origines, such as metasomatism of limestone, chemcial percipitation, etc.	
Siliceous fossils Diatoms, radiolaria, sponges, etc.	Siliceous rocks	Siliceous ooze → Porcellanite → Chert (diagenetic modification)	
Carbonized plant fragments	Coal	Peat → Brown coal → Bituminous coal → Anthracite (Diagenetic modification)	
Chemcial percipitate Carbonate, silic acid, other salts, iron-manganese oxide	Chemical-sedimentary rocks Evaporite Phosphate rocks Iron-manganese ore	Halide, Sulfate P_2O_5> 20% (early diagenesis) Banded iron ore: limited to Precambrian; Oolitic hematite: Phanerozoic Manganese nodule	
Volcaniclastic materials	Volcaniclastic rocks	Classification according to grain size Volcanic breccia Tuff breccia Lapilli tuff Tuff	

Diameter of Constituent Particles and Clastic Rocks*

ϕ scale†	Grain diameter (mm)	Particle		Most common clastic rocks included in the left particles	
		Gravel	Boulder	Conglomerate	Boulder conglomerate
-8	256				
-7			Cobble		Cobble conglomerate
-6	64				
-5					
-4			Pebble		Pebble conglomerate
-3					
-2	4				
			Granule		Granule conglomerate
-1	2				
		Sand	Very coarse sand	Sandstone	Very coarse sandstone
0	1				
			Coarse sand		Coarse sandstone
1	$\frac{1}{2}$				
			Medium grained sand		Medium sandstone
2	$\frac{1}{4}$				
			Fine sand		Fine sandstone
3	$\frac{1}{8}$				
			Very fine sand		Very fine sandstone
4	$\frac{1}{16}$				
5		Mud particles		Mudstone	
6			Silt particle		Siltstone
7					
8	$\frac{1}{256}$				
9			Clay particle		Claystone
10					

* Constituent particles of volcaniclastic rocks larger than 64 mm are called volcanic blocks, those 2–64 mm are called lapilli, and those under 2 mm are called volcanic ash. Sources: Fisher (1961), Wentworth (1922).

† ϕ is a dimensionless parameter expressing particle size, $\phi = -\log_2 d$, where d is the paricle size in mm.

Ages of Granitic Rocks in Japan

Region or Rock complex name	K-Ar, Ar-Ar ages; Biotite, muscovite (million years)	Rb-Sr mineral ages; Biotite (million years)	U-Pb ages; Zircon (million years)	CHIME, MPB ages; Monazite, uraninite, etc. (million years)
Hidaka	17-19, 35-43		19, 37	
Kitakami	100-129	106-125	117	115
Hikami	109-235	109, 113	411-442	240-260, 340-410
Abukuma	90-110	99-111	100-118	
Hitachi			490-505	
Iide, Asahi mountain range	56-65, 78-94			
Nihonkokuyama	46-52		63-67	
Ashio	48-68, 86-91		104	99
Tsukuba	60-65	58-60		67
Kinshozan	252			
Hiki, Shimonita	62-66			
Sawairi	91-92		94	
Tanzawa	5-8		4-9	
Kofu	4-14		10-16	
Kurobe River (Northern Japan Alps)	0.6-5.1		0.8-10	
Takitani (Northern Japan Alps)	1-2	1-2	1.4	
Hida	167-187	183-186, 226	190-201, 229-256	190-204, 230-240
Sanin-Shirakawa Belt (Chubu)	41-60	63	65	
(Kinki)	56-70	56-68		
(Chugoku)	31-80	60-66	33	
(Kitakyushu)	88-96, 99	83-105	97-106	
Sanyo-Naegi Belt (Chubu)	65-75	72-77	71	64-67
(Kinki)	66-81, 96-97	72-81, 93-101		
(Chugoku)	76-103	80-88		85-87
(Kitakyushu)	85-95	104-128		
Ryoke Belt (Chubu)	64-76	59-64, 74-76	71-86	76-95
(Kinki)	69-88	72-89	75-87	80-86
(Chugoku, Shikoku)	80-90	76, 89-100	86-96	89-95
Kurosegawa Belt	364-425	250-440		
Yatsushiro			446-468	
Anan			472	
Wakayama			446	
Mitaki	365-395		440-442	
Usuki River			292	
Hikawa River			503	
Maizuru			243-292, 405-442	466
Maana			113-117	
Todai	204-244			
Higo	100-103	104-106	108-113	
Outer Zone (Kinki)	14-15			
(Shikoku)	13		13	
(Kyushu)	13-14			
Amami oshima Island	17, 110			
Nansei Islands	17-19, 59			

Based on Imaoka *et al.* (2011), Kon and Takagi (2012), Nakajima *et al.* (1990), Sakashima *et al.* (2003), Suzuki and Adachi (1998), Tani *et al.* (2010), and other sources.

Ages of Metamorphic Rocks in Japan

Region or Rock complex name	K-Ar, Ar-Ar ages; Biotite, muscovite, amphibole (million years)	Rb-Sr mineral ages; Biotite, muscovite (million years)	U-Pb ages; Zircon (million years)	CHIME ages; Monazite, etc. (million years)
Low-Intermediate pressure type				
Hidaka Belt	16-18, 24, 31-33		17-24	
Abukuma Belt	92-123	87-121		
Hida Belt	174-196	157-257		249-251
Ryoke Belt				
Chubu	59-72	57-66	87	98-102
Kinki			90	97
Chugoku		93-100	97	99-101
Higo	72, 98-110	100-102	110-117	108-123
High pressure type				
Kamuikotan Belt	72-135			
Sanbagawa Belt				
Kanto	68-110			
Chubu	58-72	58-70		
Kinki	62-97	69-73		
Shikoku	63-90, 112-129		86, 110-132	
Sangun-Renge Belt	259-336, 378-469	278-323, 362-404		
Sangun-Suo Belt	154-195, 206-227, 245-253	174-227		
Other major metamorphic bodies				
Yaguki	306	293		
Hitachi	96-123		507, 510	
Tsukuba	57-62			
Kamiasou Conglomerate	860-1660	1560-1740		
Maizuru	221-264	244-325		
Osayama	273-327		472	
Okidogo	171, 179	183		250
Nishisonogi	60-88	92		
Nomo Peninsula	153-254, 457-480, 590			
Amakusa	73-94		114	

Based on Nakajima (1992), Suzuki & Adachi (1998), Sakashima *et al.* (2003), Kemp *et al.* (2007), Nishimura (2009), etc.

Geological Chronological Table

Figure 22

(Unit: million years)

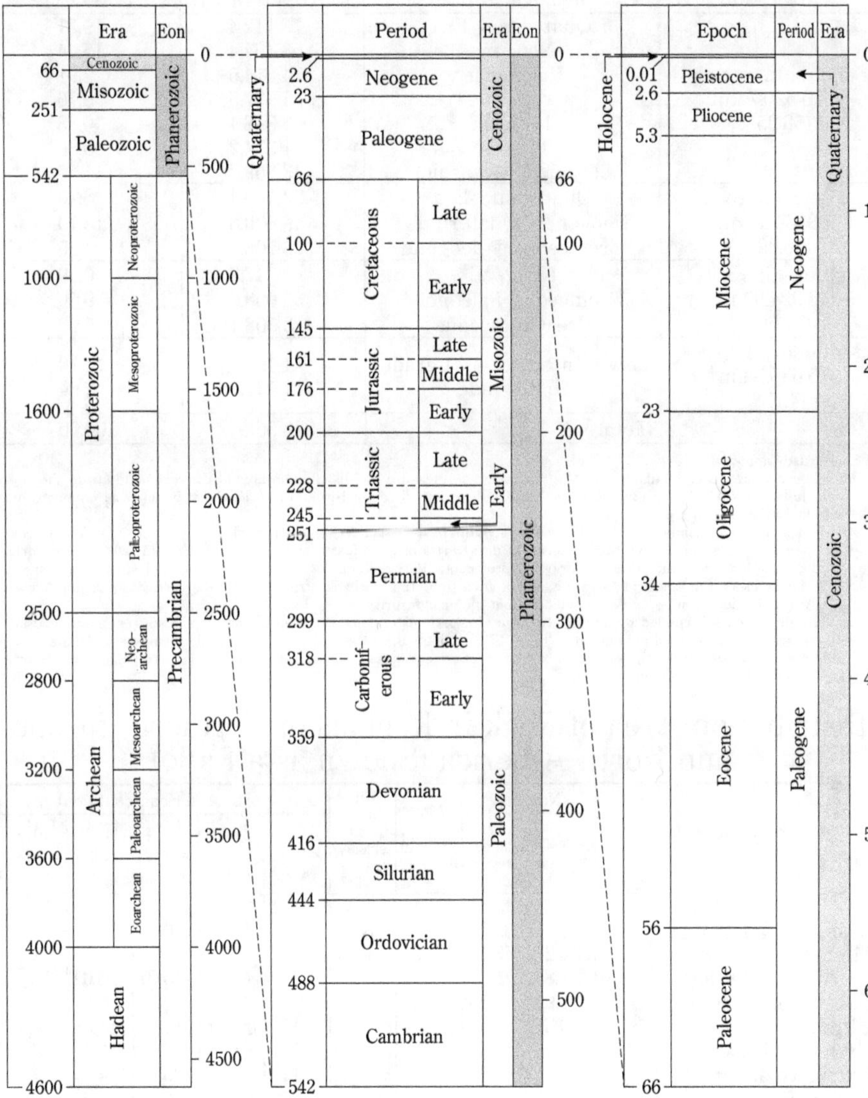

Information for the Phanerozoic Eon is based on the 2009 International Commission on Stratigraphy (2009 International Chronostratigraphic Chart). The decay constant of Steiger & Jager (1977) is also used. The divisions of the Precambrian Era are based on Ogg et al. (2016).

* For a long time, the Paleogene and Neogene were collectively called the Tertiary, but in 2009 the Tertiary was officially abolished, and the Cenozoic was the Paleogene and Neogene. It was divided into the Tertiary and Quaternary.

Distribution Area of Rock Types in the Japanese Archipelago*1

Rock kind	Geological time	Area (km^2)	Ratio (%)
Sedimentary rock*2 220 520.7 km^2 (58.25%)	Quaternary Era	73 112.4	19.31
	Neogene	58 454.4	15.44
	Paleogene	14 814.5	3.91
	Cretaceous	25 577.5	6.76
	Jurassic	3 318.4	0.88
	Triassic	1 374.2	0.36
	Paleozoic (-Mesozoic)*3	43 869.3	11.59
Volcanic rock 98 477.2 km^2 (26.02%)	Quaternary Era	33 168.4	8.76
	Neogene (-Quaternary)	49 600.8	13.10
	Cretaceous-Paleogene	15 708.0	4.15
Plutonic rock 43 924.8 km^2 (11.60%)	Neogene	3 747.8	0.99
	Cretaceous-Paleogene	36 668.6	9.69
	Pre-Cretaceous	3 508.4	0.93
Metamorphic rock 15 639.7 km^2 (4.13%)	Low (-Intermediate) grade,	4 492.7	1.19
	High grade	11 147.0	2.94
Total		378 562.4†	100.00

Source: Isoyama *et al.* (1984)

*1 These values were tabulated by using a digitizer to measure the distribution area of the rocks and strata shown in the Japan Geological Map (2nd Edition, 1:1 000 000 scale, edited by Hirokawa *et al.*, (1978) which was issued by the Geological Survey of Japan.

*2 Volcanic rocks (mainly basalt) in the marine stratum of the ealier Neogene Period are included in sedimentary rocks.

*3 In recent years, there have been a series of discoveries of microfossils such as conodonts from the Triassic Period and radiolarians from the Jurassic Period to Cretaceous Period. As a result, the majority of the stratum which was previously called the Palaeozoic strata is now known to be the Mesozoic strata (mainly the Jurassic systems). Accordingly, these values will be revised significantly in the near future.

† The total area determined by adding the area of lakes and swamps in Japan to this area is approximately 600 km^2 (0.2%) larger than the total area of Japan (page **572**) as defined by the Geospatial Information Authority of Japan. This difference is through to be caused by accumulation of uncertainties in measurement on maps, extrapolation of the original geological maps, and map drawing.

Distribution Area of Rock in Regions of Northern Europe and North America (Shown as a Ratio)

	Norway		Finland	Baltic and ukraine shields (igneous rocks)	North American continent				
	Caledonia belt	Pre-cambrian (west of Oslo Rift)			34-25 $\times 10^8$yr rocks	25-18 $\times 10^8$yr rocks	18-14 $\times 10^8$yr rocks	14-10 $\times 10^8$yr rocks	10-0 $\times 10^8$yr rocks
	%	%	%	%	%	%	%	%	%
Granite, granodiorite	2.4	33.2	52.5	84	76	70	70	66	24
Migmatite	36.2	48.0*	21.8						
Diorite, quartz diorite	1.8				2	1	0.01	0.01	16
Gabbro, diabase	7.9	2.6	8.2	13					
Ultrabasic rock	0.5				0.1	minor	minor	minor	0.1
Extrusive basic rock					12	6	3	3	5
Extrusive acid rock					0.1	0.5	20	4	4
Sedimentary rock and their metamorphic rock	51.2	12.0	17.5		5	18	2	20	46
Others		4.2		3	4	4	5	6	5

* Originally known as mixed gneisses. Sources: Barth (1961), Sederholm (1925), Ronov and Yaroshevsky (1969), Engel (1963)

Density (ρ), Elastic Wave Velocity (V_p, V_s), and Poisson's Ratio for Rocks

After Rudnick & Fountain (1994). V_p is a pressure or irrotational wave. V_s is a shear or torsional wave. The values for V_p and V_s were measured at a pressure of 6×10^8 Pa. The unit of density is 10^3 kg m^{-3} = g cm^{-3}. σ is the standard deviation. N is the number of samples for which the average value was calculated.

	Density	V_p (km/s)	V_s (km/s)	Poisson ratio	SiO_2 (wt%)
Middle crustal rocks (Amphibolite facies)					
Felsic gneiss	2.733	6.329	3.633	0.254	69.88
σ	0.063	0.146	0.104	0.016	2.81
N	17	17	17	17	13
Pelitic gneiss	2.799	6.506	3.638	0.271	63.77
σ	0.075	0.132	0.136	0.023	4.65
N	15	15	15	15	13
Mafic gneiss	3.043	7.039	3.964	0.267	48.55
σ	0.053	0.17	0.154	0.016	1.69
N	12	12	12	12	12
Lower crustal rocks (Granulite facies)					
Felsic granulite	2.718	6.526	3.714	0.26	67.52
σ	0.065	0.125	0.088	0.015	2.98
N	15	15	15	15	12
Intermediate granulite	2.836	6.661	3.626	0.288	57.56
σ	0.086	0.122	0.151	0.026	3.85
N	7	7	7	7	7
Pelitic granulite	3.038	7.069	3.98	0.264	54.39
σ	0.108	0.337	0.125	0.022	6.34
N	13	13	13	13	13
Mafic granulite	3.031	7.209	3.97	0.282	47.74
σ	0.096	0.189	0.127	0.014	1.91
N	27	27	27	27	21
Anorthosite	2.752	6.949	3.728	0.298	53.69
σ	0.036	0.047	0.047	0.009	0.84
N	5	5	5	5	3
Upper mantle rocks					
Eclogite	3.445	8.127	4.583	0.266	47.26
σ	0.096	0.205	0.151	0.018	2.00
N	12	12	12	12	12
Peridotite	3.297	8.185	4.681	0.256	42.75
σ	0.026	0.196	0.127	0.021	2.77
N	14	14	14	14	7

Strength, Elastic Modulus, and Deformation/Failure for Rocks Occurring in Japan

Rock/Locality Sample number and symbol	CP (MPa)	Strength (MPa)	EM (GPa)	Remarks
Kitakami granite				
Iwate, Towa	0.0	149	23.2	Si: 66.4
PR01, KMG	49.0	442	54.9	*t*: 37.8
	98.1	632	65.7	*f*: 43.8
Ryouke granite				
Ehime, Kikuma	0.0	125	20.2	Si: 65.9
PR08, KER	49.0	461	37.4	*d*: 2.58
	98.1	586	60.9	*t*: 40.1
	147.1	672	72.7	*f*: 41.2
Chichibu diorite				
Saitama, Otaki	0.0	220	29.8	Si: 61.0
PR10, KCG	49.0	369	29.8	*t*: 38.5
	98.1	747	58.9	*f*: 45.5
Murotomisaki gabbro				
Kochi, Muroto	0.0	186	28.4	Si: 45.0
PR13, HYD	49.0	408	44.6	*t*: 54.6
	98.1	554	61.8	*f*: 36.8
Hidaka belt peridotite				
Hokkaido, Samani	0.0	207	37.1	Si: 43.4
Horoman	49.0	615	78.1	*d*: 3.31
PR16, PB	98.1	837	123	*t*: 54.5
	147.1	880	150	*f*: 45.6
Abukuma serpentine				
Fukushima, Koriyama	0.0	189	20.4	Si: 41.2
PR18, HYF	49.0	245	31.1	*t*: 62.2
	98.1	368	44.3	*f*: 23.6
Nohi rhyolite				
Gifu, Kashimo	0.0	282	11.9	Si: 70.9
VR06, NKB	49.0	514	29.4	*t*: 62.5
	98.1	702	46.8	*f*: 43.0
Nishiyama layer andesite				
#KanbaraGS-1	0.0	94.2	13.2	*n*: 10.7
Depth 2 313m	49.0	261	16.6	*d* 2.26
SW0523, XP	98.1	491	21.5	*t*: 28.8
	147.1	561	20.4	*f*: 37.3
Matsuura basalt				
Nagasaki, Iki	0.0	215	59.7	Si: 50.5
VR12, T1L	49.0	654	58.4	*t*: 53.9
	98.1	842	59.1	*f*: 47.2

Rock/Locality Sample number and symbol	CP (MPa)	Strength (MPa)	EM (GPa)	Faiture mode	Deformation mode	Remarks
Enbetsu layer siltstone						
Hokkaido Enbetsu	0.0	6.87	0.677	B	W	*n*: 71.1
SR0102, EH	4.9	10.5	0.128	T	F	*t*: 2.79
	9.8	11.1	0.108	T	F	*f*: 11.6
Nishiyama layer mudstone						
#kanbara GS-1	0.0	150	9.26	VB	W	*n*: 11.1
Depth 3 514 m, XT	49.0	292	12.4	VB	W	*t*: 58.3
	98.1	382	12.9	B	S	*f*: 25.9
Teradomari layer sandstone						
#Mishima	0.0	139	9.62	VB	W	*n*: 5.9
Depth 4 880 m	98.1	299	12.9	B	S	*t*: 47.7
SW0537, OMSD	196.1	338	12.5	B-T	S	*f*: 22.6
	245.2	470	12.2	T	S	
Miura group zushi layer siltstone						
Kanagawa, Zushi	0.0	12.2	1.08	VB	W	
SR0911, BM	2.0	13.9	1.27	B	S	*n*: 40.2
	4.9	18.8*	0.834	T	S-N	*d*: 1.69
	9.8	12.4*	0.412	D	F	*t*: 6.53
	19.6	14.6*	0.254	D	F	*f*: 2.4
	29.4	16.5*	0.309	VD	F	
Hayama layer group tuff						
Kanagawa, Hayama	0.0	91.7	5.38	VB	W	*n*: 36.3
SR0924, ZW	49.0	178	8.8	B	S	*t*: 26.6
	98.1	289	6.42	T	S	*f*: 29.3
Nishisonogi layer group maze layer sandstone						
Nagasaki, Nishisonogi	0.0	113	14.5	VB	W	*n*: 6.20
Nagasaki, Oshima	49.0	292	24.0	B	S	*d*: 2.66
SR1411, XC	98.1	396	33.0	B	S	*t*: 49.8
	147.1	423	58.7	T	N	*f*: 28.0
Setogawa layer group sandstone						
Shizuoka, Hiruido	0.0	291	32.4	VB	W	*n*: 1.61
SR1006, SZG	49.0	654	40.3	VB	S	*t*: 71.4
	98.1	835	41.8	B	S	*f*: 44.4
Chichibu type adduct sandstone						
Shizuoka, Sakuma	0.0	314	27.9	VB	W	*n*: 0.53
SR1016, SZTE	49.0	571	41.4	B	S	*d*: 2.65
	98.1	693	37.9	B	S	*t*: 80.6
	147.1	844	67.5	B	-	*f*: 39.8

CP: Confining pressure, EM: Elastic modulus

Source: Kazuo Hoshino, Hirokazu Kato, Deep physical Rock Properties Data Editorial Board "Handbook of mechanical properties of the Japanese rocks under high confining pressure" National Institute of Advanced Industrial Science and Technology, Geological Survey of Japan (2001)

#: drilling site, Strength: differential stress at which rocks fracture. *: differential stress at which the strain reached 5% for experiments without clear failure points. VB: very brittle, B: brittle, T: transitional, D: ductile, VD: very ductile, W: wedge, S: simple shearing, N: network, F: flow. Si: SiO content (wt%), *n*: porosity (%), *d*: desnity (10^3 kg/m^3), *t*: shear strength (MPa), *f*: angle of friction (degrees)

Volcanoes

Major Volcanoes

Map number	Volcano	Latitude longitude	Location (Country)	Altitude (m)	Main eruption type ABCDEFGHIJKLM	Remarks
1	Sheveluch	56° 39′ N 161° 22′ E	Kamchatka Peninsula (Russia)	3283	××××××××-----	Since 1980, the dome has been repeatedly grown and destroyed, pyroclastic flows have occurred, 1652; large eruption of VEI=5
2	Klyuchevskoy	56 3 N 160 38 E	Kamchatka Peninsula (Russia)	4754	×××××××----×-	Highest peak on peninsula; Kamchatka Volcano Observatory
3	Bezymianny	55 58 N 160 36 E	Kamchatka Peninsula (Russia)	2882	××××××××----×	In 1956, the first eruption since the beginning of history (VEI=5), the top was blasted by a large explosion, the elevation dropped 185 m, and the volcano is still active
4	Tolbachik	55 50 N 160 20 E	Kamchatka Peninsula (Russia)	3611	×××××---×----	Frequent eruptions since 18th century; 1975-76 eruption is the largest in recorded history for the Kamchatka region
5	Karymsky	54 3 N 159 27 E	Kamchatka Peninsula (Russia)	1513	××××××××-----	Frequent eruptions since recordkeeping started (1771)
6	Ksudach	51 51 N 157 34 E	Kamchatka Peninsula (Russia)	1079	×-×--×××--×--	1907 large eruption of VEI=5
7	Kurile Lake	51 27 N 157 7 E	Kamchatka Peninsula (Russia)	81	×-×--×××--×--	BC6440 large eruption of VEI=7
8	Alaid	50 52 N 155 34 E	Chishima Islands, Allied Corp. Island (Russia)	2285	×××××-×------	1933-34; Taketomi-jima (117 m) formed from the ocean floor; connected to the main island by a sand bar
9	Kharimkotan	49 7 N 154 30 E	Kuril Islands (Russia)	1145	×-×-××-×----×	1933; large eruption of VEI=5
10	Sarychev Peak	48 6 N 153 12 E	Kuril Islands (Russia)	1496	×-×-××-×-----	2009; eruption of VEI=4
11	Changbaishan	41 59 N 128 5 E	China, North Korea	2744	×-×-?××---×--	942; large eruption of VEI=7, ash fall in Tohoku and Hokkaido areas
12	Farallon de Pajaros (Uracas)	20 33 N 144 54 E	Mariana Islands	337	×××-×----×---	Strombolian eruption; also known as Pajaros Island (old list: Uracas)
13	Pagan	18 8 N 145 48 E	Mariana Islands	570	×××××××------	One of the island that is most greatly active in a volcano of Mariana Islands. It erupts every from several years to ten years after 1981
14	Pinatubo	15 8 N 120 21 E	Luzon Island (Philippines)	1486	××××-×××--×--	1991; great eruption for the first time in 600 years (VEI=6); pyroclastic flow, a mud flood occur, the mountaintop part cave-in. The smoke reached height more than 20 km.
15	Taal	14 0 N 120 60 E	Luzon Island (Philippines)	311	××××××----×-×	1911 1,400 killed; 1965 190 killed; eruption in 1977
16	Mayon	13 15 N 123 41 E	Luzon Island (Philippines)	2462	××××××××-----	1814 1,200 killed; conical volcano
17	Kanlaon	10 25 N 123 8 E	Negros Island (Philippines)	2435	×-×××-×------	Frequent small explosions since 1866
18	Hibok-Hibok (Catarman)	9 12 N 124 40 E	Camiguin Island (Philippines)	1552	?××?××××--?--	1951 500 killed
19	Parker	6 7 N 124 54 E	Mindanao (Philippines)	1824	×-×--××------	1641; large eruption of VEI=5
20	Awu	3 41 N 125 27 E	Sangihe Island (Indonesia)	1318	×-××-×××--×-×	1711 3,200 killed; 1812 1,000 killed; 1856 2,800 killed; 1892 1,500 killed
21	Karangetang	2 47 N 125 24 E	Sangihe Island (Indonesia)	1797	××××××××-----	erupts frequently, erupts in 2018
22	Tangkoko-Duasudara	1 31 N 125 11 E	Sulawesi Island (Indonesia)	1334	×××-×--×--×--	1680; large eruption of VEI=5
23	Soputan	1 7 N 124 44 E	Sulawesi Island (Indonesia)	1785	××××××××----×	About 30 eruptions since record keeping started (1785)
24	Gamkonora	1 23 N 127 32 E	Halmahera Island (Indonesia)	1635	×-×××-------×	1673; large eruption of VEI=5
25	Makian	0 19 N 127 24 E	Halmahera Island (Indonesia)	1357	××××?×××----?	2000 killed by mud flows from the eruption in 1760

Major Volcanoes Continued.

Map number	Volcano	Latitude longitude	Location (Country)	Altitude (m)	Main eruption type ABCDEFGHIJKLM	Remarks
26	Sinabung	3° 10′ N 98° 24′ E	Sumatra (Indonesia)	2460	--x----------	Activity resumed in 2010 after a hiatus of over 1000 years. Pyroclastic flow on February 1, 2014; 16 killed
27	Marapi	0 23 S 100 28 E	Sumatra (Indonesia)	2885	x-xxx-x?-----	Frequent eruptions, short activity period. 2000; 10 eruptions since, 2018; eruption
28	Krakatau	6 6 S 105 25 E	Sunda Strait (Indonesia)	813	xxxxxxx--xx-x	The original Krakatau Island was obliterated by a huge explosion in 1883, forming an underwater caldera, 36 000 killed, the existing Anak Krakatau Island was created in 1930, December 22, 2018 eruption caused part of the mountain to collapse, triggering a tsunami, 429 killed
29	Papandayan	7 19 S 107 44 E	Java Island (Indonesia)	2665	x-xx--x------	1772 2,957 killed
30	Galunggung	7 15 S 108 3 E	Java Island (Indonesia)	2168	x-xxxxxx----x	4,011 killed by 1822 eruption; 80 thousand refugees during 1982-83 eruption
31	Slamet	7 15 S 109 12 E	Java Island (Indonesia)	3428	xxx?x--------	Frequent eruptions since 1772
32	Merapi	7 32 S 110 27 E	Java Island (Indonesia)	2910	xxxxxxxx-----	1006 thousands killed; 1672 3,000 killed; 1930 1,400 killed; 1966 64 killed; 2010 322 killed 1586 10,000 killed; 1872 200 killed; 1919 5,100 killed; 1966
33	Kelut	7 56 S 112 18 E	Java Island (Indonesia)	1731	x-xx-xxx--x--	1586 10,000 killed; 1872 200 killed; 1919 5,100 killed; 1966 282 killed; 2007 disappeared due to eruption in February, 2014
34	Semeru	8 6 S 112 55 E	Java Island (Indonesia)	3657	xxxxxxxx-----	Highest peak on island; effectively continuous eruption; 2000 2 vulcanologists killed
35	Bromo (Tengger Caldera)	7 57 S 112 57 E	Java Island (Indonesia)	2329	x-xxxx----x--	Bromo is Sukoria hill in the ten garfish caldera, it erupts frequently
36	Raung	8 7 S 114 3 E	Java Island (Indonesia)	3260	x-xxxxx---x--	1593; large eruption of VEI=5
37	Ijen	8 3 S 114 15 E	Java Island (Indonesia)	2769	x-xx-xx---x--	In the inside of the crater lake digging of sulfur, it is the fatal accident of the miners by the volcanic gas spout in 1976, 1989
38	Agung	8 21 S 115 30 E	Bali (Indonesia)	2997	x-x-xxx------	Also known as "Bali's Paramount" ; 1963 2,000 killed
39	Rinjani	8 25 S 116 28 E	Kepulauan Sunda Kecil Pulau Lombok (Indonesia)	3726	xxx-xxxx--x--	1257; large eruption of VEI=7
40	Tambora	8 15 S 118 0 E	Sumbawa Island (Indonesia)	2850	x-x-xx-x-----	1815 world's largest eruption; mountain body destroyed; ejecta gross volume 150 km 3; 92,000 killed (including those killed by starvation)
41	Manam	4 5 S 145 2 E	New Guinea Island (Papua New Guinea)	1807	x-xxxx-------	Approximately 10,000 people have been taking refuge off-island from erutpion continuing since 2004
42	Long Island	5 21 S 147 7 E	Papua New Guinea	1280	xxxxxxx---x--	1660; large eruption of VEI=6
43	Lamington	8 57 S 148 9 E	New Guinea Island (Papua New Guinea)	1680	x-x--xxx-----	1951; 1st recorded eruption; 3,000 killed (the most in the world since WWII); summit destroyed; altitude decreased by 600 m
44	Langila	5 32 S 148 25 E	New Britain Island (Papua New Guinea)	1330	xxxxxx-x-----	Of the three main craters, the second crater at the north-eastern end has been erupting since the 19th century
45	Rabaul	4 16 S 152 2 E	New Britain Island (Papua New Guinea)	688	xxxxxxxx-xx-x	1937 505 killed; September, 1994 Tavuruvur volcano and the submarine volcano in Simpson Bay erupted almost simultaneously
46	Billy Mitchell	6 6 S 155 13 E	Solomon Island (Papua New Guinea)	1544	x-x--x-------	1580; large eruption of VEI=6
47	Yasur	19 32 S 169 27 E	Vanuatu	361	x-xx----x---x	Active since 1744, ongoing eruption
48	Savai'i (Matavanu)	13 37 S 172 32 W	Savai Island (Samoan Islands)	1858 (708)	xxx-x---x---x	Fissure eruption (old list: Matavanu Matavanu)
49	Fonuafo'ou	20 19 S 175 25 W	Tonga Island	-17	xxx------x---	Since prehistoric times (from 1787), new volcanic islands have frequently appeared due to eruptions (old list: Falcon Island)
50	White Island	37 31 S 177 11 E	New Zealand North Island	321	x-xx-xx------	1826; record of the first eruption, frequent phreatic eruptions since then, phreatic eruption in December 2019; 21 tourists killed
51	Tarawera (Okataina)	38 7 S 176 30 E	New Zealand North Island	1111	-xxxxxxx--x--	1886 fissure eruption (VEI=5); 100 killed

Continued.

Map number	Volcano	Latitude longitude	Location (Country)	Altitude (m)	Main eruption type ABCDEFGHIJKLM	Remarks
52	Ngrauruhoe (Tongariro)	39° 9′ S 175° 38′ E	New Zealand North Island	1978	××××××××-----	Active; small stratovolcano
53	Ruapehu	39 17 S 175 34 E	New Zealand North Island	2797	××××××××--×--	Highest peak on island; Tongariro National Park; 1953 151 killed
54	Erebus	77 32 S 167 10 E	Ross Island (Antarctica)	3794	×××-×---×----	1841; Discovered by British Ross Corps.; mountain name derived from the name of the exploratin vessel; lava lake has existed continuously since 1972
55	Cerro Hudson	45 54 S 72 58 W	Chile	1905	×××-×××----×-	1991 largest exposion in 20 years (VEI=5); plume altitude 12 km
56	Chaiten	42 50 S 72 39 W	Chile	1122	×-×--××××----	2006; large eruption of VEI=4, 1640; VEI=4, 3100 B.C. and 7750 B.C.; large eruption of VEI=5
57	Calbuco	41 20 S 72 37 W	Chile	1974	×-××××××---×-	1893 1st eruption since recordkeeping started (large explosion)
58	Puyehue-Cordon Caulle	40 35 S 72 7 W	Chile	2236	×××-××-×-----	2011; large eruption of VEI=5
59	Villarrica	39 25 S 71 56 W	Chile	2847	×××××××-×--×-	1948-49 36? Killed; 1964 22 killed; 1971-72 15 killed
60	Cerro Azul	35 39 S 70 46 W	Chile	3788	-×××××-------	1932; large eruption of VEI=5
61	San Jose	33 47 S 69 54 W	Chile・Argentina	6070	×-×××--------	Also known as Maipo, but there is another volcano with the same name
62	Cerro Blanco (Robledo)	26 47 S 67 46 W	Argentina	4670	-------------	2300 B.C.; large eruption of VEI=7
63	Lascar	23 22 S 67 44 W	Chile	5592	×-××××-×-----	1848 1st recorded eruption
64	Huaynaputina	16 36 S 70 51 W	Peru	4850	×××--××------	1600; large eruption of VEI=6
65	Misti, El	16 18 S 71 25 W	Peru	5822	×-××-××------	Typical cone-shaped volcano
66	Sierra Negra (Isabela)	0 50 S 91 10 W	Galápagos Islands (Ecuador)	1124	×××××---×----	1911, 48, 53, 54, 57, 63, 79; eruption, and also 2005, 18 eruption
67	Fernandina	0 22 S 91 33 W	Galápagos Islands (Ecuador)	1476	××××××--×----	More than 20 eruption since 1813
68	Sangay	2 0 S 78 20 W	Ecuador	5286	×××-××-×-----	1628, 1728, 1934; eruption of VEI = 3, frequent eruptions since 2011
69	Tungurahua	1 28 S 78 27 W	Ecuador	5023	×-×××××-?----	Increased activity since 1999, 25,000 people evacuated in 1999, still active
70	Cotopaxi	0 41 S 78 26 W	Ecuador	5911	×××-××××-----	Frequent eruptions from 1532 to the mid 20th century
71	Guagua Pichincha	0 10 S 78 36 W	Ecuador	4784	×-××-×××-----	Pliny-type eruptions in the 16–17th centuries
72	Reventador	0 5 S 77 39 W	Ecuador	3562	×××-××××-----	2002; explosive eruption
73	Galeras	1 13 N 77 22 W	Colombia	4276	××××××××-----	1993 eruption while volcano was being investigated; 9 killed, 6 injured
74	Nevado del Ruiz	4 54 N 75 19 W	Colombia	5279	××××?××?---×-	1595 pyroclastic flow and mud flow; 1985 pyroclastic flow and mud flow; Almelo on the eastern foot completely destroyed, approximately 24,000 killed
75	Soufriere St. Vincent	13 20 N 61 11 W	West Indies (St. Vincent Island)	1220	××××××××--×-×	1902; 1,351 killed (crater lake burst)
76	Pelee	14 49 N 61 10 W	West Indies Martinique (France)	1394	××××-×××--×-×	1902 28,000+1,000 killed (most in the 20th century) across the entire St. Pierre city area, It was called Montagne Pelee once

Major Volcanoes Continued.

Map number	Volcano	Latitude longitude	Location (Country)	Altitude (m)	Main eruption type ABCDEFGHIJKLM	Remarks
77	Soufriere Guadeloupe	16° 3′ N 61° 40′ W	West Indies Isla Guadalupe (France)	1467	xxxxxxxx-----	Dozens of erup tions since recordkeeping started (1400–)
78	Soufriere Hills	16 43 N 62 11 W	West Indies (Montserrat (UK))	915	x-xx-xxx----x	Repeated growth and collapse of lava dome at the summit since 1995; frequent pyroclastic flow; continued until 2013
79	Irazu	9 59 N 83 51 W	Costa Rica	3432	xxxx-xxx-----	30 killed in 1963-64 eruption
80	Poas	10 12 N 84 14 W	Costa Rica	2708	xxxxxx-x--x--	1989; molten sulfur ponds appeared in the crater lake after the eruption
81	Arenal	10 28 N 84 42 W	Costa Rica	1670	xxxxxxxx-----	1968 1st recorded eruption; 76 killed
82	Concepcion	11 32 N 85 37 W	Nicaragua	1700	xxxxxx-x-----	Strombolian eruption
83	Masaya	11 59 N 86 10 W	Nicaragua	635	xxxxxx--x-x--	Basaltic Pliny-type eruption
84	Cerro Negro	12 30 N 86 42 W	Nicaragua	728	xxxxxxxx-----	Formed by 1850 eruption
85	San Miguel	13 26 N 88 16 W	El Salvador	2130	xxxxx?-------	2018; eruption
86	Izalco	13 49 N 89 38 W	El Salvador	1950	xxx-xx-------	1770; formed as a parasitic volcano of the Santa Ana volcano (2,181 m); growth continues; current relative height 650 m
87	Fuego	14 28 N 90 53 W	Guatemala	3763	xxx-xxxx-----	Most active in Guatemala. During June 2018; eruption caused pyroclastic flow and mudslide, 113 killed, 329 missing
88	Santa Maria	14 45 N 91 33 W	Guatemala	3745	-xxxxxxx--x--	1902 large eruption; total ejecta 5.5 km^3
89	El Chichon	17 22 N 93 14 W	Mexico	1150	xxxxxxxx-----	Was a small unnamed volcano, until large erutpion in March-May, 1982; pumice fall; pyroclastic flow; lahar; 1,700 killed
90	Popocatepetl	19 1 N 98 37 W	Mexico	5393	x-xxxxxx-----	Also known as "smoking mountain" ; 2nd highest peak in nation; rises 60 km to the southeast of Mexico City
91	Parictin (Michoac-an-Guanajuato)	19 51 N 101 45 W	Mexico	3860	xxxxxxx------	1943-52; formed in wheatfield; relative height 400 m or more; famous as a new volcano
92	Colima	19 31 N 103 37 W	Mexico	3850	xxxxxxxx-----	Most frequent eruptions in the nation since record-keeping started (1576-)
93	Mono-Inyo Craters	37 48 N 119 2 W	California (USA)	2796	-xxxxxxx-----	Obsidian lava flow
94	Lassen Peak (Lassen Volcanic Center)	40 30 N 121 30 W	California (USA)	3187	xxxxxxxx-----	1914-15 large eruption; national park
95	Shasta	41 25 N 122 12 W	California (USA)	4317	xxx-xxxx-----	History era eruption in 1786, The northwestern foot of the mountain has a clear flowing mountain terrain
96	Crater Lake	42 56 N 122 7 W	State of Oregon (USA)	2487	xxx-xx-x--x--	Caldera formed during eruption 6800 years ago (VEI = 7). One of the largest in the Holocene
97	Yellowstone	44 26 N 110 40 W	Washington (USA)	2805	x--x---------	Caldera eruptions occur every 650 000 years, numerous geysers in the caldera
98	St. Helens	46 12 N 122 11 W	Washington (USA)	2549	xxxxxxxx---x-	1980 large explosion; summit missing due to landslip; lava dome growth began September, 2004; growth stopped 2008
99	Rainier	46 51 N 121 46 W	Washington (USA)	4392	xxxxxx------x	National park; locals of Japanese descent call it "Tacoma Fuji"
100	Spurr	61 18 N 152 15 W	Alaska (USA)	3374	xxxx-xx---x--	1953 1st recorded eruption; smoke height 23 km; maximum width 50 km
101	Redoubt	60 29 N 152 45 W	Alaska (USA)	3108	x-xx-xxx----x	1989 largest eruption in 12 years; smoke heigth 12 km; aircraft engine temoprarily stops inflight due to volcanic ash
102	Augustine	59 22 N 153 26 W	Alaska (USA)	1252	x-xxxxxx----x	2005 eruptions restarted after 19 years
103	Katmai	58 17 N 154 58 W	Alaska (USA)	2047	x-xx?--------	1912 large eruption; ejecta gross volume 21 km^3; caused approximately 4 km caldera collapse

Continued.

Map number	Volcano	Latitude longitude	Location (Country)	Altitude (m)	Main eruption type ABCDEFGHIJKLM	Remarks
104	Trident	58° 14′ N 155° 6′ W	Alaska (USA)	1864	-xx-x-xx-----	1953, 62-63 large eruptions
105	Novarupta	58 16 N 155 9 W	Alaska (USA)	841	x-xx-xx------	1912; large eruptions of VEI=6
106	Pavlof	55 25 N 161 54 W	Alaska (USA)	2493	xxx?xxx------	More than 30 eruptions since the 18th century
107	Shishaldin	54 45 N 163 58 W	Aleutian Islands (USA)	2857	xxxxx-x------	Highest peak in the Aleutian Islands. Erupted on average every 6 years since 1775. Over 40 eruptions
108	Bogoslof	53 56 N 168 2 W	Aleutian Islands (USA)	150	--x-x--x-x---	Since 1796, erupted 8 times, island expanded during 2016-17 eruption
109	Okmok	53 26 N 168 8 W	Aleutian Islands (USA)	1073	xxx-xxx-----x	2008; eruptions of VEI=4
110	Great Sitkin	52 5 N 176 8 W	Aleutian Islands (USA)	1740	x-x-x--x-----	1933, 45, 49, 50, 74; eruption, and also 2018, 19; eruption
111	Mauna Loa	19 29 N 155 36 W	Hawaii (USA)	4170	xxx-x----x--x	Frequent eruptions, many large fissure eruptions, lava fountains form a "curtain of fire". Lava flows cause havoc
112	Kilauea	19 25 N 155 17 W	Hawaii (USA)	1222	xxxxxx--xxx--	Frequent eruptions, many large fissure eruptions, lava fountains form a "curtain of fire". Lava flows cause havoc. The settlement of Kalapana in 1990 and Kapoho in 2018 were obliterated by lava
113	Piton de la Fournaise	21 15 S 55 42 E	Reunion Island (Region of France in the Indian Ocean)	2632	xxxxxx-xx----	East of Madagascar; lava flows almost every year
114	Kilimanjaro	3 4 S 37 21 E	Tanzania	5895	xxx-x---x----	Kibo Peak in the center of the mountain is the highest peak. At the top of Kibo Peak is a caldera with a diameter of 2.5 km and its two craters
115	Ol Doinyo Lengai	2 46 S 35 55 E	Tanzania	2962	x-xxx-x-x----	1960, 88, 93 carbonatite lava flows
116	Nyiragongo	1 31 S 29 15 E	Democratic Republic of Congo	3470	xxxxx---x----	Virunga volcanic cluster; north from Lake Kivu; lava flow entered into Goma during January 2002 eruption
117	Nyamuragira	1 24 S 29 12 E	Democratic Republic of Congo	3058	xxxxx---x-x-x	Next volcano north after Nyiragongo; most frequent eruptions in Africa; 1994 lava lake for 1st time in 11 years
118	Teleki (The Barrier)	2 19 N 36 34 E	Kenya	1032	xxx-x--------	With Andrew pyroclastic cone nearby, it is sometimes fissure emission
119	Erta Ale	13 36 N 40 40 E	Ethiopia	613	xx?-x---x----	lava lake has existed continuously since 1967
120	Cameroon	4 12 N 9 10 E	Cameroon	4095	xxxxx-x-x----	Isolated peak, the highest peak in Midwest Africa
121	NyosLake (Oku Volcanic Field)	6 15 N 10 30 E	Cameroon	3011	x-xx-x----?--	Lake Nios 1986 carbon dioxide plume killed 1 746 people and 3 500 livestock from carbon dioxide poisoning or asphyxiation
122	Ararat	39 47 N 44 18 E	Turkey	5165	-xx--xx------	1 900 people died in pyroclastic flows in 1840 (VEI =3)
123	Santorini	36 24 N 25 24 E	Greece	367	xxxxxxxx-x--x	1610 B.C.; large eruption (VEI=7), volcanic island generation several after history on record (197 B.C.-)
124	Campi Flegrei	40 50 N 14 8 E	Italy	458	x-xxxx-xxx---	Geothermally active; earthquake and remarkable diastrophism; residents have repeatedly taken refuge from earthquakes and remarkable diastrophism
125	Vesuvius	40 49 N 14 26 E	Italy	1281	xxxxxxx-----x	Year 79 many killed; 1631 18,000 killed; most active European volcano; world's oldest volcano observatory
126	Stromboli	38 47 N 15 13 E	Lipari (Italy)	924	xxx-xxx--x--x	Effectively continuous eruptions for the past 24 centuries; Strombolian eruptions; also known as "Light house of the Medilerranean"
127	Vulcano	38 24 N 14 58 E	Lipari (Italy)	500	xxxxxxxx-x--x	Blast furnace of the god Vulcan in Greek mythology; vulcanian eruption
128	Etna	37 45 N 14 60 E	Sicily (Italy)	3295	xxxxxxxx----x	Frequent eruptions since 693 B.C.; 1169 15,000 killed; 1669 10,000 killed; because direction changed; 1983, 1992 lava flow blasted

Major Volcanoes Continued.

Map number	Volcano	Latitude longitude	Location (Country)	Altitude (m)	Main eruption type ABCDEFGHIJKLM	Remarks
129	Beerenberg (Jan Mayen)	71° 5′ N 8° 9′ W	Jan Mayen (Norway)	2197	×××××-×--×-×-	1971 large explosion; plume height 15,000 m
130	Krafla	65 43 N 16 44 W	Iceland	800	×××××---×----	Frequent fissure eruptions
131	Askja	65 2 N 16 47 W	Iceland	1080	××××××--×-×--	Magmatic activity from basalt arrived at day site; 3,500 killed in 1631 eruption
132	Snaefellsjokull	64 48 N 23 47 W	Iceland	1446	×××-×--------	Volcano of the entrance of the Jules Verne bottom of the earth exploration
133	Laki (Lakagigar) (Grimsvotn)	64 25 N 17 19 W	Iceland	1719	×××××××---××-	1783 25 long lava plateau generated by fissure eruption; ejecta gross volume 15 km 3; 9,350 killed; 1996 eruption under the Gr?msv?th glacier was in this vicinity
134	Hekla	63 59 N 19 40 W	Iceland	1490	-××-×××-----×	27 km long caldera chain; many fissure eruptions; 1300 600 killed; 1947 plume height 27,000 m
135	Katla	63 38 N 19 5 W	Iceland	1490	×××-×-×----×-	1625, 1721, 55; large eruption (It is all VEI=5), 1918; eruption of VEI=4
136	Eyjafjallajokull	63 38 N 19 38 W	Iceland	1651	×××-×-×----×-	An air boundary European by the eruption of 2010 by volcanic ashes for six days for the closure
137	Surtsey (Vestmannaeyjar)	64 18 N 21 37 W	Iceland	174	××××××--××---	1963-67 formed 30 km south of Iceland
138	Heimaey (Vestmannaeyjar)	63 25 N 20 16 W	Iceland	283	××××××--××---	1973; eruption, change a duct of it by mass drainage of the seawater because of the lava flow approached the port
139	Fayal	38 36 N 28 44 W	Azores Islands (Portgal)	1043	××××××---×---	1957-58 new island Capelinhos (150 m) formed; was connected to main island, but there has been severe wave erosion
140	Pico	38 28 N 28 24 W	Azores Islands (Portgal)	2351	-××-×----×---	Five times eruption, the latest eruption are 1963 after the 16th century
141	Furnas (Sao Miguel)	37 46 N 25 19 W	Sao Miguel Island (Portgal)	805	×-×--×××-----	191 killed in 1630 eruption
142	Tenerife	28 16 N 16 38 W	Canary Island (Spain)	3715	×××-×--×-----	Also known as Teide; the largest, highest mountain in the archipelago
143	El Hierro	27 44 N 18 2 W	Canary Island (Spain)	1500	×××-×--------	From October, 2011 to May, 2012; bottom of the sea eruption
144	Fogo	14 57 N 24 21 W	Cape Verde Islands (Cape Verde)	2829	×××××--------	Approximately 10 eruptions since recordkeeping starte (1500)

Eruption type = eruption phenomenon prone to occurring, A: Summit eruption, B: Flank eruption (including fissure eruption), C: Normal volcanic explosion, D: Phreatic explosion, E: Lava flow, F: Pyroclastic flow, G: Lahar, H: Lava dome, I: Lava lake, J: Submarine eruption, K: subaqueous (lake) eruption, L: Eruption under glacier, M: Eruption tsunami, Strombolian eruption = an explosive eruption in which a spray of lava with relatively low viscosity or volcanic projectiles are periodically propelled from the crater. Vulcanian eruption = lava boulders with high viscosity or lava projectiles are propelled from the crater and a large amount of ash cloud is emitted. more explosive than a Strombolian eruption. Carbonatite = Lava or igneous rock which is mainly composed of carbonate such as Ca, Na, or K.

As a general rule, the name and height of volcanoes was taken from the International Association of Volcanology and Chemistry of the Earth's Interior (IAVCEI): Catalog of the Active Volcanoes and the Smithsonian Institution: Volcanoes of the World (second edition.1994).

Figure 23

Major volcanoes

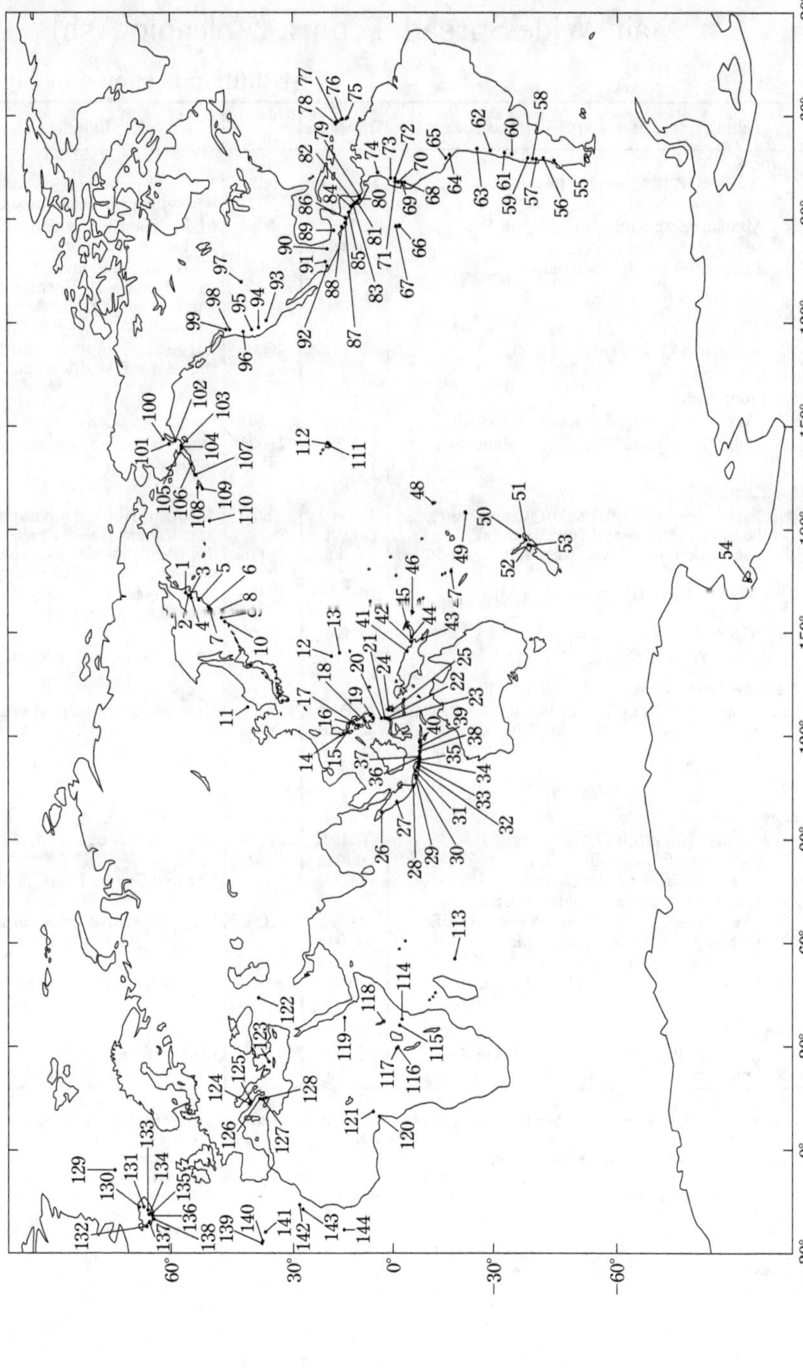

Main Wide-Spread Tephra (Volcanic Ash)

Distribution is shown in Figure 23.

Tephra name and source volcano region	Eruption age	Total volume (km^3)	Remarks
Europe			
① Avelino (Z-1): Vesuvius (Italy)	Pre-17th century	-40	Last major eruption before the one which buried Pompeii (AD 79)
② Minoan (Z-2): Santorini (Aegean Sea)	Pre-16-17th century	>40	Major eruption which devastated Minoan civilization
③ Laacher See (Eifel, Germany)	12.5ka	16	Distributed tephra from Eifel to as far away as Bornholm island, Denmark and Italy. Archeologically important indicator tephra
④ Campania (Y-5) Campi Flegrei	36-37ka	500	Large-scale tephra distributed in the eastern Mediterranean
Southern Asia			
⑤ Tambora (Sumbawa Island, Indonesia)	AD1815	150	Largest eruption in recorded history
⑥ Toba: Toba Lake (Sumatra, Indonesia)	70ka	2500–3000	One of the world's largest eruptions
East Asia			
⑦ Paektu Tomakomai: Paektu (Changbai range, boarder of China and North Korea)	10-11th century	50	Also distributed in northern Japan. Covers Heian Era ruins
⑧ Kikai Akahoya: Kikai (Kagoshima, Japan)	7.3ka	>170	Also widely distributed in western Japan. Devistated the culuture at the end of the early Jomon Era.
⑨ Aira Tn: Aira (Kagoshima, Japan)	26-29ka	>450	Most widely distributed coverage in Japan. Indicator tephra for the Paleolithic, latest glacial period
⑩ Aso 4: Aso (Kumamoto, Japan)	85-90ka	>600	
⑪ Toya: Toya (Hokkaido, Japan)	112-115ka	>150	Widely distributed in northern Japan
New Zealand			
⑫ Taupo: Lake Taupo (North Island, New Zealand)	1.85ka	120	New Zealand's largest eruption
⑬ Kawakawa/Oruanui: Lake Taupo (North Island, New Zealand)	26.5ka	>750	Large eruption involving a lake
⑭ Rotoiti/Rotoehu: Okataina (North Island, New Zealand)	64ka	240	
North America			
⑮ Katmai: Novarupta (Alaska, USA)	AD1912	31-35	One of the largest eruptions of the 20th century
⑯ Old Crow: Emmons Lake (Alaska, USA)	140ka	>300	
⑰ Mazama: Crater Lake (Cascades, USA)	7.63ka	78	Formed Crater Lake Caldera
⑱ Rockland: Breccia (California, USA)	614ka	>80	
⑲ Lava Creek: Yellowstone (Wyoming, USA)	660ka	≧1000*	Distributed over most of the USA
⑳ Bishop: Long Valley (California, USA)	759ka	500*	
㉑ Mesa Falls: Yellowstone (Idaho, USA)	1270ka	≧280*	
㉒ Huckleberry Ridge: Yellowstone I (Wyoming, USA)	2060ka	≧2500*	One of the world's largest tephras
Central America			
㉓ Roseau: Micotrin (Dominican Republic)	30ka	>25	Distributed in 3 seas: the Pacific Ocean, Caribbean Sea, and Gulf of Mexico
㉔ Los Chocoyos: Atitlan Ⅲ (Guatemala)	84ka	280	

* Value converted into the volume of magma. ka: 1 000 years ago

The term tephra (Greek for "ash") or pyroclastic material is used to refer to volcanic eruption substances which erupt from the crater as individual solids. Depending on its particle size, tephra is classified as volcanic block, lapilli, or volcanic ash (particles of less than 2 mm). On the other hand, the term "volcanic ash" is used in the same broad meaning as tephra, regardless of particle size.

Volcanic activity which emits tephra is more explosive than eruptions which emit lava. In most cases, tephra eruptions are caused by the activity of magma which contains a large amount of gas and a high level of silicate content, or by magma coming into contact with water. The ash cloud which is emitted into the sky above the crater is blown by the wind, thus creating tephra fall which is a type of fall deposit. Conversely, when ash cloud collapse, A pyroclastic flows composed of tephra and gas occurs. Both tephra fall and pyroclastic flow deposit spread over an extremely large area in a very short period of time. Figure 23 shows the distribution of tephra which is distributed widely throughout prominent regions of the world. All of these cases are the product of giant explosive eruptions occurring at a low frequency.

Hiroshi Machida and Fusao Arai: "Atlas of tephra in and around Japan, revised edition" University of Tokyo Press (2003)

Main Wide-Spread Tephra

Figure 24 Representative wide-spread tephra (marked by bold line, numbers correspond to the numbers of tephra listed in 104; Shaded area indicate distribution of coarse-grained tephra)

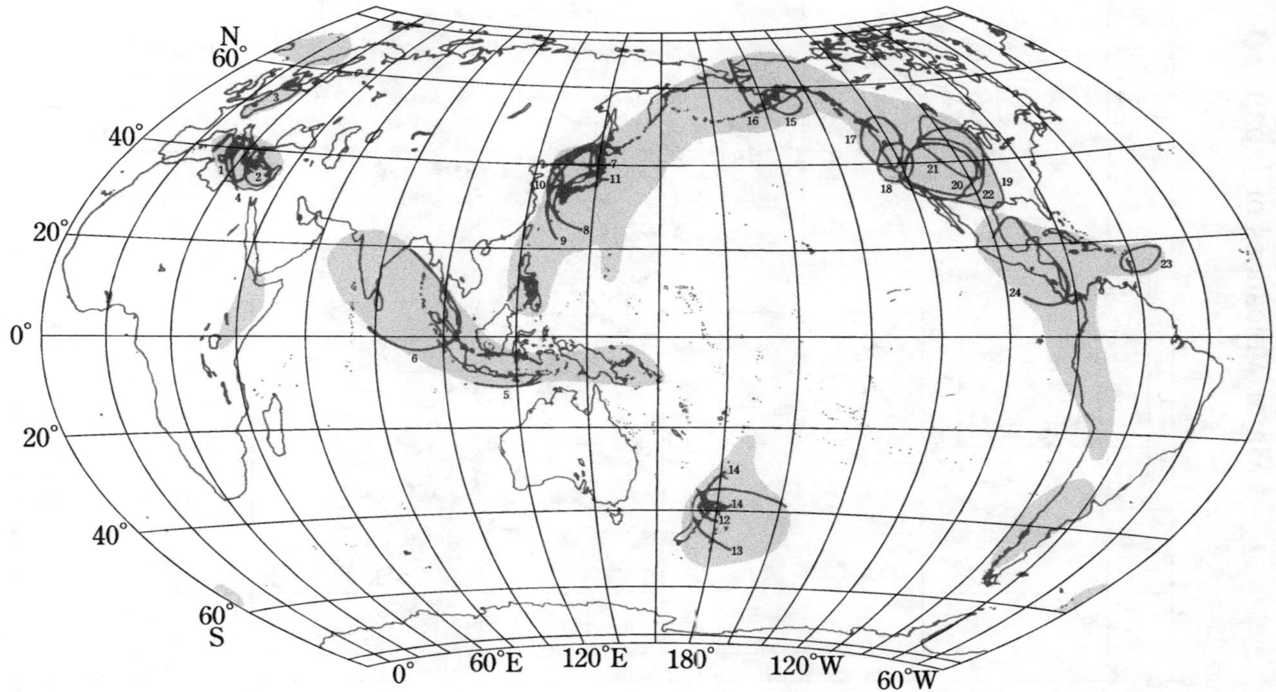

Machida, H. & Arai, F., Atlas of tephra in and around Japan, revised edition, Univ. of Tokyo Press, 2003

Major Volcanoes of Japan

Figure 25

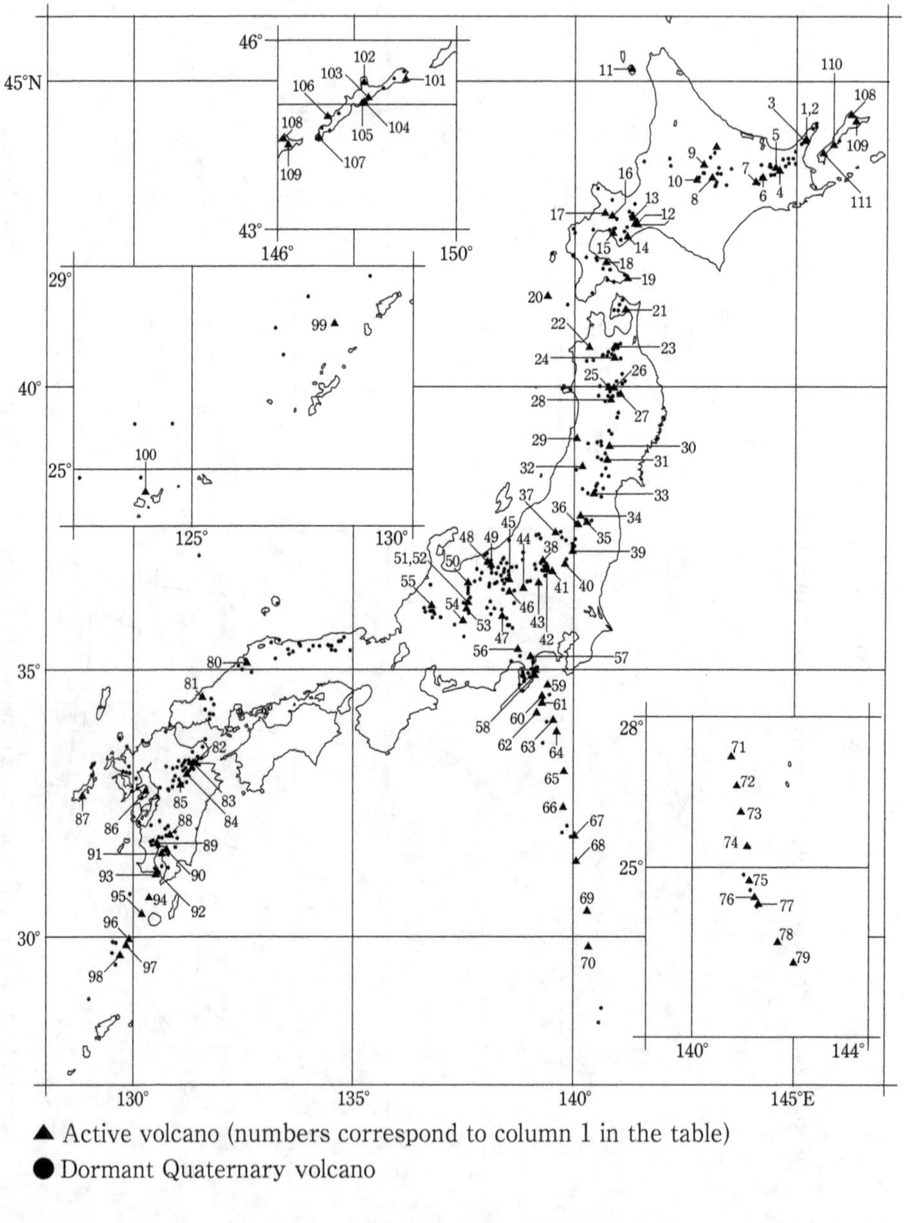

▲ Active volcano (numbers correspond to column 1 in the table)

● Dormant Quaternary volcano

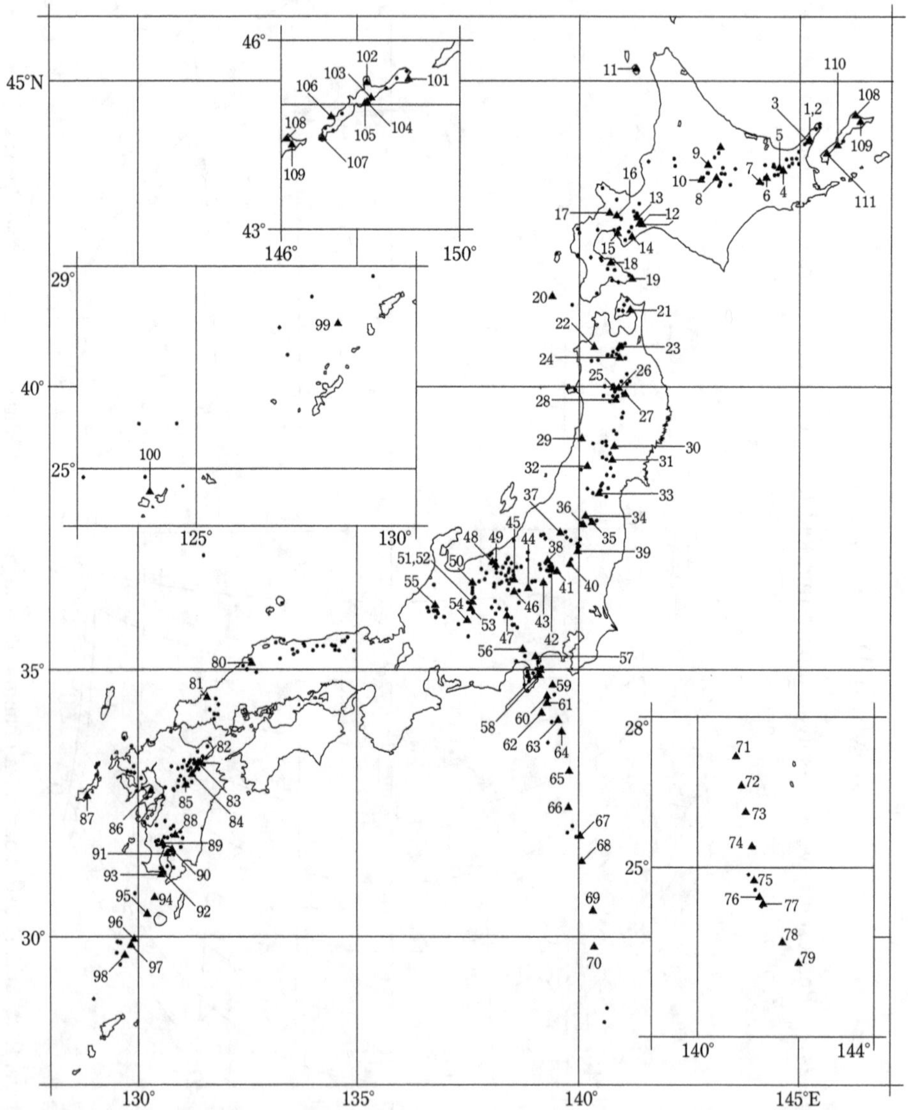

Major Volcanoes of Japan

1) Volcanoes which have erupted within the past 10 000 years are indicated by a bold line encircling the number (same number as Figure 24)
2) Notation is according to page **530**. Volcanoes whose name has been indented one space compose the volcano which is listed above (flank volcanoes, central cones, etc.).
3) Names enclosed in square brackets are the highest peak of the volcano. Names enclosed in angled brackets are aliases.
4) The components of volcanoes are taken from sources including the Japan Quaternary Volcano Catalog (The Volcanological Society of Japan) and the Japan Quaternary Volcano (National Institute of Advanced Industrial Science and Technology). C = Caldera, T = large pyroclastic flow, S = Stratovolcano (a conical volcano formed by alternating layers of lava flow and pyroclastic rock), D = Lava dome (volcano where a large outflow of high-viscosity lava has formed into a dome), L = Lava flow, P = Pyroclastic cone (a small conical volcano formed by repeated layers of pyroclastic material). Ma = Maar (a basin created by an explosive eruption when magma came into contact with groundwater in a near-surface shallow), N = Volcanic neck (lava which had filled the volcanic vent is exposed on the ground surface due to corrosion of the edifice), M = Submarine volcano, U = Unknown form
5) Main rocks: b = Basalt, a = Andesite, d = Dacite, r = Rhyolite, o = Other alkalic rock, etc.

Number	Volcano name	Location	Altitude (m)	Component	Major rock
1	Kamui Dake	Etorofuto	1 323	S	a
2	Moyoro Dake	Etorofuto	1 124	T-C, S, D	b, a, d
3	Shibetoro	Etorofuto	853	T-C, S	-
4	Rucharu	Etorofuto	437	S	a, d
5	Sashiusu Yama	Etorofuto	1 128	S, D	a
6	Chirippu Yama	Etorofuto	1 582	S, D	b, a
7	Odamoi Yama	Etorofuto	1 208	C, S	a
8	Etorofu Yakeyama	Etorofuto	1 147	S, D	a
9	Nishi-Hitokappuyama	Etorofuto	1 629	S, D	a
10	Etorofu Atosa Dake	Etorofuto	1 209	S, P	b, a
11	Urumonbetsu	Etorofuto	908	C, S	a, d
12	Moikeshi	Etorofuto	512	T-C, S	b, a, d
13	Berutarube San	Etorofuto	1 221	S	a, d
14	Ruruidake	Kunashirito	1 481	S	a, d
15	Iwayama	Kunashirito	1 187	S, D	a
16	Chacha Dake	Kunashirito	1 772	C, S	b, a
17	Rausu Yama	Kunashirito	882	C, S, D	a, d
18	Tomari Yama	Kunashirito	535	T-C, S, D	a, d

Major Volcanoes of Japan Continued.

Number	Volcano name	Location	Altitude (m)	Component	Major rock
19	Shiretoko Dake	Abashiri	1 254	S	a
(20)	Shiretoko Iouzan	Abashiri	1 562	S	a
(21)	Rausudake	Abashiri, Nemuro	1 661	S, D	a, d
(21′)	Tencho Zan	Abashiri, Nemuro	1 046	L	a
22	Onnebetsu	Abashiri, Nemuro	1 330	S	a
23	Unabetsu Dake	Abashiri, Nemuro	1 419	S	b, a
24	Musa Dake	Nemuro	1 005	S, D	a, d
25	Samakkenupuri Yama	Abashiri, Nemuro	1 062	S	a
26	Shari Dake	Abashiri	1 547	S, D	b, a, d
(27)	Mashu				
	Kamui-nupuri	Kushiro	857	S, D	a, d, r
	Mashu caldera	Abashiri, Kushiro, Nemuro	-	T-C	d
(28)	Kussharo				
	Atosanupuri	Kushiro	508	S, D	a, d
	Kussharo (Kutcharo) Nakajima	Kushiro	355	S, D	d, r
	Kussharo (Kutcharo) Caldera	Kushiro	-	T-C	d, r
(29)	Akan				
	Meakan Dake ⟨Ponmachineshiri⟩	Kushiro	1 499	S	b, a, d
	Oakan Dake	Kushiro	1 370	S, D	a, d
	Akan Caldera	Kushiro	-	T-C	d, r
30	Shiitokoro Yama	Abashiri	1 232	L	a, d
31	Kumaneshiri Dake	Tokachi	1 586	S, D	a, d
(32)	Maruyama	Tokachi	1 692	S, D	a, d
33	Nipesotsu	Tokachi	2 013	S, D	a, d
34	Tokachi Eboshi Volcano Group	Tokachi	1 291	D	a, d
35	Shikaribetsu Volcano Group	Tokachi	1 401	D	a, d
36	Niseikaushuppe	Kamikawa	1 883	S	b, a, d
(37)	Taisetsuzan [Asahidake]	Kamikawa	2 291	S, D	a, d
38	Chubetsudake	Kamikawa	1 963	S	a
39	Tomuraushi	Kamikawa,Tokachi	2 141	S, D	a, d
(40)	Tokachidake	Kamikawa,Tokachi	2 077	S, D	b, a, d, r
(41)	Rishiri Zan	Soya (Rishirito)	1 721	S, D	b, a, d
42	Sapporo Dake	Ishikari	1 293	L	a
43	Soranuma Dake	Ishikari	1 251	L	a
(44)	Shikotsu				
	Tarumae (Tarumai) San	Ishikari, Iburi	1 041	S, D	a
	Fuppushi Dake	Ishikari	1 102	S, D	a, d

Continued.

Number	Volcano name	Location	Altitude (m)	Component	Major rock
	Eniwa Dake	Ishikari	1 320	S, D	a, d
	Shikotsu Caldera	Ishikari	-	T-C	b, a, d, r
45	Horohoro・Tokusyunbestu	Iburi	1 322	S	a
46	Orofure-Raiba	Iburi	1 231	S	a
47	Kuttara 〈Shihorei〉	Iburi	549	C, S	b, a, d, r
48	Washibetsu Dake	Iburi	911	S, D	a
49	Akaigawa Caldera	Shiribeshi	725	C, S	a, d, r
50	Yotei Zan	Shiribeshi	1 898	S, D	a, d
51	Shiribetsu Dake	Shiribeshi	1 107	D	a
52	Toya				
	Syowa-shin Zan	Iburi	398	D	d
	Usu zan	Iburi	733	S, D	b, a, d, r
	Toya Nakajima	Iburi	455	D	b, a, d
	Toya Caldera	Iburi	-	T-C	d, r
53	Konbudake	Shiribeshi, Iburi	1 045	S, D	a
54	Niseko-Raiden Volcanic Group	Shiribeshi	1 308	S, D	
55	Kariba Yama	Hiyama, Shiribeshi	1 520	S	b, a, d
56	Katsuma Yama	Hiyama (Okushirito)	427	U	r
57	Nigorigawa Caldera	Oshima	-	T-C	a, d
58	Hokkaido Komagatake	Oshima	1 131	S	a
59	Yokotsu Dake	Oshima	1 167	S, D	a
60	Esan Maruyama	Oshima	691	S	a
61	Esan	Oshima	618	S, D	a
62	Zenikame	Oshima	-	T-C	a, d
63	Hakodate Yama	Oshima	334	S	d
64	Shiriuchi	Oshima	855	S, D	a, d
65	Oshima Oshima	Oshima (Oshima)	732	S	b, a
66	Oshima Kojima	Oshima (Kojima)	282	S	a
67	Mutsu Hiuchi Dake	Aomori	781	S, D	a, d
68	Osore Zan	Aomori	878	S	a, d
69	Hakkoda				
	Hakkodasan [Otake]	Aomori	1 585	S, D	b, a, d
	Hakkoda Caldera	Aomori	-	T-C	d, r
70	Iwaki San	Aomori	1 625	S, D	a, d
71	Okiura Caldera 〈Kenashi Yama〉	Aomori	985	T-C, S	b, a, d, r
72	Towada	Aomori, Akita	690	T-C, S, D	b, a, d
73	Tashiro Dake	Akita	1 178	S	a

Major Volcanoes of Japan Continued.

Number	Volcano name	Location	Altitude (m)	Component	Major rock
74	Daira Komagatake	Akita	1 158	S	a
75	West of Kyuroku Island	Coast of Akita	-3 320	M-U	a, d
76	Nanashigure Yama	Iwate	1 063	S, D	a, d
77	Megata	Akita	178	Ma	b, a, d
78	Toga	Akita	-	Ma	r
79	Kanpu Zan	Akita	355	S	b, a
80	Moriyoshi Zan	Akita	1 454	S, D	b, a, d
81	Akita Yakeyama	Akita	1 366	S, D	a, d, r
82	Hachimantai	Akita, Iwate	1 613	S, D	b, a, d
83	Tamagawa Caldera	Akita	1 300	T-C	d, r
84	Iwate San	Iwate	2 038	S, D	b, a
85	Nyuto-Takakura	Iwate	1 478	S, D	b, a
86	Zarumori Yama	Akita, Iwate	1 541	S, D	b, a, d
87	Akita Komagatake	Akita	1 637	C, S	b, a
88	Kayo Dake	Akita	1 254	S, D	b, a
89	Tazawa Ko Caldera	Akita	-	C	-
90	Yakeishi Dake	Iwate	1 547	S, D	b, a, d
91	Chokai San	Akita, Yamagata	2 236	C, S, D	b, a
92	Obinai Yama	Akita	1 004	S	a, d
93	Takamatsu Dake	Akita	1 348	D	d
94	Kurikoma Yama	Akita, Iwate, Miyagi	1 626	S	a
95	Mukaimachi Caldera	Yamagata	-	T-C	a, d
96	Onikoube Caldera	Miyagi	-	T-C	a, d, r
97	Naruko Caldera	Miyagi	461	T-C, D	d, r
98	Hijiori Caldera	Yamagata	-	T-C, D	d
99	Gassan	Yamagata	1 984	S	a, d
100	Funagata Yama	Miyagi, Yamagata	1 500	S	b, a
101	Shirataka Yama	Yamagata	994	S, D	a, d
102	Adachi	Miyagi	-	P	d
103	Ganto Yama	Miyagi, Yamagata	1 486	S	a, d
104	Kamuro Dake	Miyagi, Yamagata	1 356	S	b, a, d
105	Daito Dake	Miyagi, Yamagata	1 365	S, D	b, a, d
106	Sankichi-Hayama	Yamagata	687	S	a
107	Zao Zan (Zao San)	Miyagi, Yamagata	1 841	S, D, P	b, a, d
108	Aoso Yama	Miyagi	799	C, S, D	a, d
109	Azuma Yama	Fukushima, Yamagata	2 035	S, P	a, d
110	Adatara Yama	Fukushima	1 709	S, D	b, a

Continued.

Number	Volcano name	Location	Altitude (m)	Component	Major rock
(111)	Bandai San	Fukushima	1 816	C, S	a, d
112	Nekomagatake	Fukushima	1 404	C, S	a, d
(113)	Numazawa	Fukushima	835	T-C, D	a, d
114	Sumon Dake	Niigata	1 537	S	a, d
115	Asakusa Dake	Niigata, Fukushima	1 585	S	b, a, d
116	Shirakawa	Fukushima	-	T-C	a, d
(117)	Nasudake [Chausudake]	Tochigi, Fukushima	1 915	S, D	b, a, d
(118)	Takahara Yama	Tochigi	1 795	S, D	b, a, d
119	Nyohosan	Tochigi	2 483	S, D	a, d
120	Nantaisan	Tochigi	2 486	S	b, a, d
121	Nikko Volcanic Group [Omanagosan]	Tochigi	2 376	D	a, d, r
(122)	Nikko Shirane San	Gunma, Tochigi	2 578	S, D	a, d
(123)	Hiuchigatake	Fukushima	2 356	S	a, d
124	Ayame Daira	Gunma	1 969	S	a
125	Numanokami Yama	Gunma	1 541	S	b, a
126	Kesamaru Yama	Gunma, Tochigi	1 961	S	b, a
127	Sukai San	Gunma, Tochigi	2 144	S	b, a, d
128	Joshu Hotaka Yama	Gunma	2 158	S	b, a, d
(129)	Akagi San	Gunma	1 828	C, S, D	a, d, r
130	Komochi Yama	Gunma	1 296	S, D	a, d
131	Onoko Yama	Gunma	1 208	S	b, a
(132)	Haruna San	Gunma	1 449	C, S, D	b, a, d
133	Hanamagari Yama	Gunma, Nagano	1 655	S	a
134	Masugata Yama	Niigata	748	L	a
135	Iiji San	Niigata	1 111	S	a, d
136	Naeba San	Nagano, Niigata	2 145	C, S	a, d
137	Torikabuto Yama	Nagano	2 038	S	a, d
138	Kenashi Yama	Nagano	1 650	S	b, a, d
139	Takayashiro Yama	Nagano	1 351	S, D, N	a, d
140	Shiga	Nagano	2 041	L, P	a, d
(141)	Kusatsu Shirane San	Gunma	2 171	T, S, P, Ma	a, d
142	Omeshi Dake	Gunma, Nagano	2 160	S	a
143	Azumaya San	Gunma, Nagano	2 354	C, S, D	a, d
(144)	Asama Yama	Gunma, Nagano	2 568	C, S, D	a, d, r
145	Eboshidake	Nagano	2 066	S, D	a, d
146	Minakami Yama	Nagano	659	D	a

Major Volcanoes of Japan Continued.

Number	Volcano name	Location	Altitude (m)	Component	Major rock
147	Motodori Yama	Nagano	744	D	a
148	Madarao Yama	Nagano	1 382	S	a, d
149	Iizuna Yama	Nagano	1 917	S, D	b, a, d
150	Kurohime Yama	Nagano	2 053	S, D	b, a, d
(151)	Myoko San	Niigata	2 454	C, S, D	b, a, d
(152)	Niigata Yakeyama	Niigata	2 400	S, D	a, d
153	Shirouma Oike	Nagano	2 469	S, D	a, d
(154)	Tateyama	Toyama	2 621	C, T, S	a, d
155	Kaminoroka	Toyama	-	S	a, d
156	Washiba-Kumonotaira	Toyama, Nagano	2 825	S	b, a, d
157	Momisawa Dake	Gifu, Nagano	2 755	T, N	r
158	Hotaka Dake	Gifu, Nagano	3 190	T-C	a, d
(159)	Yakedake	Gifu, Nagano	2 455	D	a, d
(159′)	Akandanayama	Gifu, Nagano	2 109	S, D	a, d
(160)	Norikura Dake	Gifu, Nagano	3 026	S, D	a, d
161	Kamitakara	Gifu	1 831	T	d, r
162	Tomuro Yama	Ishikawa	548	D	a, d
(163)	Hakusan	Ishikawa, Gifu	2 702	S	a, d
164	Kyogadake	Fukui, Ishikawa	1 625	S	a, d
165	Gankyoji Yama	Fukui, Gifu	1 691	S	a
166	Ryohaku Maruyama	Gifu	1 786	S	a
167	Dainichigatake	Gifu	1 709	S	a, d
168	Bishamon Dake	Gifu, Fukui	1 385	S, D	a, d
169	Eboshi-Washigatake	Gifu	1 671	S	a
170	Yugamine	Gifu	1 067	D	r
(171)	Ontake San	Gifu, Nagano	3 067	S	b, a, d, r
172	Ueno [Sakashita]	Gifu	606	L	b, a
173	Kirigamine	Nagano	1 925	S, D	a, d, r
174	Tateshinayama 〈Suwa-Fuji〉	Nagano	2 531	S, D	b, a, d, r
(175)	Yatsugatake				
	Yokodake	Nagano	2 480	S, D	b, a
	Akadake	Nagano, Yamanashi	2 899	S, D	b, a
176	Kayagatake	Yamanashi	1 704	S, D	a, d
(177)	Fuji San	Shizuoka, Yamanashi	3 776	S, P	b, a
178	Ashitaka Yama	Shizuoka	1 504	S, D	b, a, d
(179)	Hakoneyama [Kamiyama]	Kanagawa, Shizuoka	1 438	C, S, D	b, a, d, r
180	Yugawara [Kurakakeyama]	Kanagawa, Shizuoka	1 004	S	a, b
181	Taga [Kurotake]	Shizuoka	798	S	b, a

Continued.

Number	Volcano name	Location	Altitude (m)	Component	Major rock
182	Usami	Shizuoka	592	S	b, a
183	Izu-Tobu Volcanoes	Shizuoka	580	L, D, P, M-P	b, a, d, r
184	Amagi San	Shizuoka	1 406	S	a
185	Daruma Yama	Shizuoka	982	C, S	a
186	Tanaba Yama	Shizuoka	753	S	b, a, d
187	Jaishi Yama	Shizuoka	520	L	a
188	Nanzaki	Shizuoka	106	L, P	b, o
189	Izu Oshima	Tokyo (Oshima)	758	C, S, P	b, a
190	Toshima	Tokyo (Toshima)	508	S	b, a
191	Udonejima	Tokyo (Udonejima)	209	S	b, a
192	Niijima (Miyatsukayama)	Tokyo (Niijima)	432	P, D	b, r
192′	Shikinejima	Tokyo (Shikinejima)	109	P, D	b, r
193	Kozushima [Tenjosan]	Tokyo (Kozushima)	572	P, D	r
194	Miyakejima	Tokyo (Miyakejima)	775	C, S, P	b, a
195	Onoharajima	Tokyo (Onoharajima)	114	N	a
196	Mikurajima	Tokyo (Mikurajima)	851	S, D	b, a
197	Inanbajima	Tokyo (Inanbajima)	74	D	a
198	Hachijojima				
	Higashiyama 〈Miharayama〉	Tokyo (Hachijojima)	701	C, S, P	b, a, d
	Nishiyama 〈Hachijo-Fuji〉	Tokyo (Hachijojima)	854	C, S, P	b, a
199	Hachijokojima	Tokyo (Hachijokojima)	617	S	a
200	Aogashima	Tokyo (Aogashima)	423	C, S, P	b, a
201	East Aogashima Caldera	Tokyo (Near Aogashima)	-	M-C	a, d
202	Myojinsho Caldera				
	Myojinsho	Tokyo (South of Izu Islands)	-50	M-S	a, d
	Beyonesu (Bayonnase) Rocks	Tokyo (South of Izu Islands)	11	M-C	b
203	Sumisujima	Tokyo (Near Sumisujima)	136	M-C	b
204	Izu Torishima	Tokyo (Torishima)	394	S	b, a
205	Sofugan	Tokyo (South of Izu Islands)	99	S	b
206	Nishinoshima	Tokyo (Nishinoshima)	160	S	a
207	Kaikata Seamount	Tokyo (Ogasawara)	-447	M-C	a, b
208	Kaitoku Seamount	Tokyo (Ogasawara)	-97	M-S	b, a, d
209	Kita Io To	Tokyo (Kitaioto)	792	S, M-U	b, a
210	Funka Asane	Tokyo (Ogasawara)	-14	M-S	b, a
211	Kaiseikaikyu [Kaiseinishinoba]	Tokyo (Ogasawara)	-187	M-U	
212	Io To	Tokyo (Ioto)	170	S, P	o
213	Kita-Fukutokutai	Tokyo (Minamiioto)	-55	M-U	d
214	Fukutoku-Oka-no-Ba	Tokyo (Ogasawara)	-25	M-U	a

Major Volcanoes of Japan Continued.

Number	Volcano name	Location	Altitude (m)	Component	Major rock
215	Minami Io To	Tokyo (Minamiioto)	916	S	b, o
216	Minami-Hiyoshi Seamount	Tokyo (South of Ogasawara Shoto)	-84	M-S	b, a
217	Nikko Seamount	Tokyo (South of Ogasawara Shoto)	-392	M-S	b, a
218	Takarayama	Kyoto, Hyogo	350	L, P	b
219	Genbudo	Hyogo	259	L, P	b
220	Kamisano	Hyogo	-	L, P	b, a
221	Mesaka	Hyogo	-	L	b
222	Kannabeyama	Hyogo	469	L, P	b
223	Kebioka, Kazurahata	Hyogo	920	L, P	b, a
224	Oginosen	Tottori	1 310	L, P	b, a
225	Okidogo 〈Misaki〉	Shimane	-	L	b
226	Hiruzen (Kamihiruzen)	Tottori, Okayama	1 202	L	a, d
227	Daisen	Tottori	1 729	S, D, P	a, d
228	Yokota Volcanic Group	Shimane, Tottori	-	L, P	b
229	Noro	Shimane	172	L	b
230	Daikonjima	Shimane	42	L, P	b
231	Sanbe San	Shimane	1 126	T-C, D, P	a, d
232	Oe Takayama	Shimane	808	D	d
233	Abu Volcanoes	Yamaguchi	571	L, P	b
234	Aono Yama Volcano Group				
	Shikumagadake	Yamaguchi	504	D	a
	Sengokudake	Yamaguchi	630	D	a
	Tokuyama Mitake San	Yamaguchi	790	D	a, d
	Aonoyama	Shimane	907	D	a, d
235	Mutsurejima	Yamaguchi	104	L	b
236	Kurose	Fukuoka	8	U	b
237	Himeshima	Oita	266	D, P	d, r
238	Futago San	Oita	720	L, D	a, d
239	Tsurumi Dake	Oita	1 375	S, D	a, d
240	Yufu Dake	Oita	1 583	S, D	a, d
241	Haneyama Volcanic Group	Oita	1 140	L	a, d, r
242	Hohi [Kuenohirayama]	Oita	1 289	L, D, P	a, d
243	Kuju San	Oita	1 791	S, D	b, a, d, r
244	Aso				
	Aso San	Kumamoto	1 592	S	b, a, d, r
	Aso Caldera	Kumamoto	-	T-C	a, d, r
245	Kinbo (Kinpo) Zan	Kumamoto	665	S, D	a, d
246	Ojikajima Volcano Group	Nagasaki (Goto Retto)	111	L, P	b

Continued.

Number	Volcano name	Location	Altitude (m)	Component	Major rock
(247)	Fukue Volcanoes	Nagasaki (Goto Retto)	315	L, P	b
248	Tara Dake	Saga, Nagasaki	1 076	S, D	b, a, d
(249)	Unzen Dake	Nagasaki	1 483	S, D	b, a, d
250	Sendai	Kagoshima	677	L	b, a
251	Satsuma Maruyama	Kagoshima	218	D	d
252	Imuta	Kagoshima	509	D	a, d
253	Kakuto Caldera	Kagoshima	-	T-C	d, r
(254)	Kirishima Yama				
	Karakunidake	Kagoshima, Miyazaki	1 700	S	b, a, d
	Shinmoedake	Kagoshima, Miyazaki	1 421	S	b, a, d
	Takachihonomine	Kagoshima, Miyazaki	1 574	S	b, a, d
(255)	Yonemaru-Sumiyoshi Ike	Kagoshima	15	L, P, Ma	b
(256)	Aira, Sakurajima				
	Sakurajima	Kagoshima	1 117	S	a, d
	Aira Caldera	Kagoshima	-	T-C	r
	Wakamiko Caldera	Kagoshima	-	M-C	
(257)	Ata, Kaimon				
	Kaimon Dake	Kagoshima	924	S, D	b, a
	Nabeshimadake	Kagoshima	256	D	d
	Ikeda	Kagoshima	3	C	r
	Unagiike	Kagoshima	-	Ma	d
	Ata Caldera	Kagoshima	-	T-C	d, r
(258)	Kikai				
	Kikai Caldera	Kagoshima (Near Osumi Islands)	-	T-C	b, r, a
	Satsuma-Iojima	Kagoshima (Iojima)	704	S, D	b, a, d, r
259	Kuroshima	Kagoshima (Kuroshima)	622	S	a
(260)	Kuchinoerabujima	Kagoshima (Kuchinoerabujima)	657	S	a, d
(261)	Kuchinoshima	Kagoshima (Kuchinoshima)	628	S, D	a, d
(262)	Nakanoshima	Kagoshima (Nakanoshima)	979	S	a, d
(263)	Suwanosejima	Kagoshima (Suwanosejima)	796	S	a
264	Akusekijima	Kagoshima (Akusekijima)	584	S	a
265	Yokoatejima	Kagoshima (Yokoatejima)	495	S	a
(266)	Io-Torishima	Okinawa (iotorishima)	212	S, D	a
(267)	Submarine Volcano NNE of Iriomotejima	Okinawa	-	M-U	r
268	Kubashima	Senkakushoto	117	S	b

Active Volcanoes of Japan

Volcano name	Volcano name
Moyoro Dake	Asama Yama
Chirippu San	Yokodake
Sashiusu Dake	Niigata Yakeyama
Odamoi San	Myokosan
Etorofu Yakeyama	Midagahara
Etorofu Atosa Dake	Yakedake
Berutarube San	Akandana Yama
Rurui Dake	Norikura Dake
Chacha Dake	Ontake San
Rausu San	Hakusan
Tomari Yama	Fuji San
Shiretoko Io Zan	Hakone Yama
Rausu Dake	Izu Tobu Volcanoes
Tencho Zan	Izu Oshima
Mashu	Toshima
Atosanupuri	Niijima
Oakan Dake	Kozushima
Meakan Dake	Miyakejima
Maruyama	Mikurajima
Taisetsu Zan	Hachijojima
Tokachi Dake	Aogashima
Rishiri Zan	Beyonesu Rocks
Tarumae San	Sumisujima
Eniwa Dake	Izu Torishima
Kuttara	Sofugan
Usu Zan	Nishinoshima
Yotei San	Kaikata Seamount
Niseko	Kaitoku Seamount
Hokkaido Komagatake	Funka Asane
Esan	Io to
Oshima Oshima	Kita Fukutokutai
Osore Zan	Fukutoku Okanoba
Iwaki San	Minami Hiyoshi Seamount
Hakkoda San	Nikko Seamount
Towada	Sanbe San
Akita Yakeyama	Abu Volcanoes
Hachimantai	Tsurumi Dake and Garandake
Iwate San	Yufu Dake
Akita Komagatake	Kuju San
Chokai San	Aso San
Kurikoma Yama	Unzen Dake
Naruko	Fukue Volcanoes
Hijiori	Kirishima Yama
Zao Zan	Yonemaru-Sumiyoshi Ike
Azuma Yama	Wakamiko
Adatara Yama	Sakurajima
Bandai San	Ikeda and Yamagawa
Numazawa	Kaimon Dake
Hiuchigatake	Satsuma Iojima
Nasu Dake	Kuchinoerabujima
Takahara Yama	Kuchinoshima
Nantai San	Nakanoshima
Nikko Shiranesan	Suwanosejima
Akagi San	Io Torishima
Haruna San	Submarine Volcano NNE of Iriomotejima
Kusatsu Shirane San	

Volcano names are from the "National Catalogue of the Active Volcanoes in Japan (4th ed)," Japan Meteorological Agency.

Volcanoes of Japan which Erupted in These 70 Years

Year	Number of volcanoes	Volcano name
1950	5	Azuma Yama, Asama Yama, Aso San, Sakurajima, Suwanosejima
1951	7	Tarumae San, Azuma Yama, Izu Oshima, Sashiusu Dake, Akita Yakeyama, Suwanosejima, Asama Yama
1952	4	Tokachi Dake, Asama Yama, Bayonnaise Rocks, Suwanosejima
1953	8	Tarumae San, Nasu Dake, Asama Yama, Izu Oshima, Bayonnaise Rocks, Kitafukutoku-tai, Aso San, Suwanosejima
1954	8	Tokachi Dake, Tarumae San, Izu Oshima, Bayonnaise Rocks, Kitafukutoku-tai, Aso San, Sakurajima, Suwanosejima
1955	7	Meakan Dake, Tarumae San, Asama Yama, Izuo Shima, Bayonnaise Retsugan, Aso San, Sakurajima
1956	6	Tokachi Dake, Meakan Dake, Asama Yama, Aso San, Sakurajima, Suwanosejima
1957	7	Meakan Dake, Akita Yakeyama, Izu Oshima, Io To, Aso San, Sakurajima, Suwanosejima
1958	7	Tokachi Dake, Meakan Dake, Asama Yama, Izu Oshima, Aso San, Sakurajima, Suwanosejima
1959	7	Moyoro Dake, Tokachi Dake, Meakan Dake, Aso San, Kirishimayama Shinmoedake, Sakurajima, Suwanosejima
1960	7	Meakan Dake, Nasu Dake, Izu Oshima, Bayonnaise Rocks, Io Torishima, Sakurajima, Suwanosejima
1961	6	Tokachi Dake, Asama Yama, Izu Oshima, Aso San, Sakurajima, Suwanosejima
1962	8	Tokachi Dake, Meakan Dake, Yakedake, Izu Oshima, Miyakejima, Aso San, Sakurajima, Suwanosejima
1963	6	Nasu dake, Niigata Yakeyama, Izu Oshima, Aso San, Sakurajima, Suwanosejima
1964	5	Meakan Dake, Izu Oshima, Aso San, Sakurajima, Suwanosejima
1965	6	Meakan Dake, Asama Yama, Izu Oshima, Aso San, Sakurajima, Suwanosejima
1966	6	Meakan Dake, Izu Oshima, Aso San, Sakurajima, Kuchinoerabujima, Suwanosejima
1967	5	Io Torishima, Io To, Aso San, Sakurajima, Suwanosejima
1968	6	Etorofu Yakeyama, Izu Oshima, Io Torishima, Sakurajima, Kuchinoerabujima, Suwanosejima
1969	5	Io To, Aso San, Sakurajima, Kuchinoerabujima, Suwanosejima
1970	7	Etorofu Yakeyama, Akita Komagatake, Izu Oshima, Bayonnaise Rocks, Aso San, Sakurajima, Suwanosejima
1971	5	Akita Komagatake, Kirishima Yama, Sakurajima, Kuchinoerabujima, Suwanosejima
1972	3	Aso San, Sakurajima, Suwanosejima
1973	8	Etorofu Yakeyama, Chacha Dake, Asama Yama, Izu Oshima, Nishinoshima, Sakurajima, Kuchinoerabujima, Suwanosejima
1974	8	Chacha Dake, Niigata Yakeyama, Nishinoshima, fukutoku Okanoba, Aso San, Sakurajima, Kuchinoerabujima, Suwanosejima
1975	8	Chacha Dake, Chokai Zan, Io To, Minamihiyoshi Seamount, Aso San, Sakurajima, Kuchinoerabujima, Suwanosejima
1976	6	Azuma Yama, Kusatsu Shirane San, Minami Hiyoshi Seamount, Aso San, Sakurajima, Suwanosejima
1977	4	Usu Zan, Aso San, Sakurajima, Suwanosejima
1978	6	Chacha Dake, Usu Zan, Tarumae San, Io To, Sakurajima, Suwanosejima
1979	6	Tarumae San, Ontake San, Nikko Seamount, Aso San, Sakurajima, Suwanosejima
1980	5	Io To, Aso San, Sakurajima, Kuchinoerabujima, Suwanosejima
1981	4	Chacha Dake, Tarumae San, Sakurajima, Suwanosejima
1982	4	Kusatsu Shirane San, Io To, Sakurajima, Suwanosejima
1983	6	Kusatsu Shirane San, Asama Yama, Niigata Yakeyama, Miyakejima, Sakurajima, Suwanosejima
1984	4	Kaitoku Seamount, Aso San, Sakurajima, Suwanosejima

Volcanoes of Japan which Erupted in These 70 Years Continued.

Year	Number of volcanoes	Volcano name
1985	4	Tokachi Dake, Aso San, Sakurajima, Suwanosejima
1986	4	Izu Oshima, Fukutoku Okanoba, Sakurajima, Suwanosejima
1987	5	Izu Oshima, Fukutoku Okanoba, Aso San, Sakurajima, Suwanosejima
1988	6	Meakan Dake, Tokachi Dake, Izu Oshima, Aso San, Sakurajima, Suwanosejima
1989	6	Etorofu Yakeyama, Tokachi Dake, Izu Tobu Volcanoes, Izu Oshima, Sakurajima, Suwanosejima
1990	5	Asama Yama, Aso San, Unzen Dake, Sakurajima, Suwanosejima
1991	7	Ontake San, Izu Oshima, Aso San, Unzen Dake, Kirishima Shinmoe Dake, Sakurajima, Suwanosejima
1992	6	Fukutoku Okanoba, Aso San, Unzen Dake, Kirishima Shinmoe Dake, Sakurajima, Suwanosejima
1993	5	Io To, Aso San, Unzen Dake, Sakurajima, Suwanosejima
1994	5	Io To, Aso San, Unzen Dake, Sakurajima, Suwanosejima
1995	6	Yakedake, Kuju San, Aso San, Unzen Dake, Sakurajima, Suwanosejima
1996	6	Meakan Dake, Hokkaido Komagatake, Akita Yakeyama, Unzen Dake, Sakurajima, Suwanosejima
1997	3	Niigata Yakeyama, Sakurajima, Suwanosejima
1998	6	Meakan Dake, Hokkaido Komagatake, Niigata Yakeyama, Izu Torishima, Satsuma Iojima, Suwanosejima
1999	5	Moyoro Dake, Io To, Sakurajima, Satsuma Iojima, Suwanosejima
2000	6	Usu Zan, Hokkaido Komagatake, Miyakejima, Sakurajima, Satsuma Iojima, Suwanosejima
2001	6	Usu Zan, Miyakejima, Io To, Sakurajima, Satsuma Iojima, Suwanosejima
2002	5	Miyakejima, Izu Torishima, Sakurajima, Satsuma Iojima, Suwanosejima
2003	6	Asama Yama, Miyakejima, Aso San, Sakurajima, Satsuma Iojima, Suwanosejima
2004	8	Tokachi Dake, Asama Yama, Miyakejima, Io To, Aso San, Sakurajima, Satsuma Iojima, Suwanosejima
2005	5	Meakan Dake, Io To, Fukutoku Okanoba, Sakurajima, Suwanosejima
2006	4	Miyakejima, Fukutoku Okanoba, Sakurajima, Suwanosejima
2007	5	Ontake San, Miyakejima, Fukutoku Okanoba, Sakurajima, Suwanosejima
2008	6	Meakan Dake, Asama Yama, Miyakejima, Kirishima Shinmoe Dake, Sakurajima, Suwanosejima
2009	5	Asama Yama, Miyakejima, Aso San, Sakurajima, Suwanosejima
2010	5	Miyakejima, Fukutoku Okanoba, Kirishima Shinmoe Dake, Sakurajima, Suwanosejima
2011	5	Io To, Aso San, Kirishima Shinmoe Dake, Sakurajima, Suwanosejima
2012	4	Etorofu Yakeyama, Io To, Sakurajima, Suwanosejima
2013	7	Etorofu Yakeyama, Miyakejima, Nishinoshima, Io To, Sakurajima, Satsuma Iojima, Suwanosejima
2014	6	Ontake San, Nishinoshima, Aso San, Sakurajima, Kuchinoerabujima, Suwanosejima
2015	8	Asama Yama, Hakone Yama, Nishinoshima, Io To, Aso San, Sakurajima, Kuchinoerabujima, Suwanosejima
2016	5	Niigata Yakeyama, Io to, Aso san, Sakurajima, Suwanosejima
2017	4	Nishinoshima, Kirishima, Shinmoe Dake, Sakura Jima, Suwanose Jima
2018	8	Kusatsu Shirane San, Nishinoshima, Io To, Kirishima Io Yama, Kirishima Shinmoe Dake, Sakurajima, Kuchinoerabujima, Suwanosejima
2019	7	Asama Yama, Nishinoshima, Aso San, Sakurajima, Satsuma Iojima, Kuchinoerabujima, Suwanosejima

Records of Eruptions on Active Volcanoes of Japan

1) Names and numbers correspond to "Major Japanese Volcanoes."
2) A symbol to the right of the volcano name indicates the frequently observed eruption type at the volcano.
   ◯ Summit eruption ♂ flank volcanism ↑ Normal volcanic explosion (Strombolian, Vulkanian, Plinian, etc.) ∨ Steam explosion ⇒ Lava flow → Pyroclastic flow ∼ Volcanic mudflow ∩ Lava dome M submarine eruption m Lake bed eruption ∽ Eruption tsunami
3) For the case of repeated eruptions in the same century, the eruption age only the final 2 digits of the last eruption are written. The dates of the eruptions are inidicated in the western (Gregorian) calendar. Roman numerals indicate the month and Aribic numerals indicate the day. ▲ inidcated ongoing eruptions.
4) The eruption record follows the Japan Meteorological Agency's "National Catalogue of the Active Volcanoes in Japan (4th ed)."

2. Moyoro Dake ↑
1778 or 1779, 1883, 1946, 58, 99

5. Sashiusu Yama ↑ ?
1951 ?

6. Chirippu Yama ↑ ∨
1843, 60

7. Odamoi San
No eruption record

8. Etorofu Yakeyama ↑
1968, 70, 73, 89, 2012, 13

10. Etorofu Atosa Dake ↑ →
1812, 1932 ?

13. Berutarube Yama ↑ ?
1812 ?

14. Rurui Dake
No eruption record

16. Chacha Dake ◯ ↑
1812, 1973–75, 78, 81

17. Rausu Yama ♂ ↑
1880, 1900 ?

18. Tomari Yama ↑
1948

20. Shiretokoio Zan ◯∨ ?
1857–58 (Sulfur discharge), 76, 89–90 (Sulfur discharge), 1935–36 (Approximately 200,000 tons sulfur discharge)

21. Rausudake ↑
Evidence for multiple eruptions starting from 2200–2300 years ago

21′. Tencho Zan ∨
No eruption record

27. Mashu ↑ ∨
No eruption record; occasional earthquake activity

28. Atosanupuri ↑ ∨
No eruption record; occasional earthquake activity; in the strictest sense, Atsusanupuri (Iosan) is solfatara activity

29′. Oakan Dake ↑
No eruption record

29. Meakan Dake ◯∨
1955 (V–VI; forests and arable land damaged; X) –60, 62, 64–66, 88, 96, 98, 2006, 08

Records of Eruptions on Active Volcanoes of Japan Continued.

32. Maru Yama ∨

1898 (Steam explosions)

37. Taisetsu Zan

No eruption record, but there are signs of steam explosions starting from 260 years ago

40. Tokachidake ○♂↑→∼

1857, 87, 1925–26 (V24; normal volcanic explosions; volcanic mudflow; 2 villages buried, 144 kiled, approximately 200 wounded, forests and arable land damaged) -28, 52, 54, 56, 58, 59, 61, 62 (VI29–30; normal volcanic explosions; 5 killed; 11 wounded; sulfur mine facilities, forests, and arable land damaged), 85, 88–89, 2004

41. Rishiri Zan ↑

No eruption record

44. Tarumae San ○↑∩

1667, 1739, 1804–17 (Normal volcanic explosions; many killed and wounded (?)), 67, 74 (Normal volcanic explosions; lava dome collapsed), 83, 85–87, 94, 1909 (Normal volcanic explosions; lava dome reformed), 17–21, 23, 26, 28, 33, 36, 44, 51, 53–55, 78–79, 81

44′. Eniwa Dake ↑∨

Signs of steam explosions since the 17th century; currently gas activity in the bottom of the caldera on the east side of the summit

47. Kuttara ↑∨

No eruption record; signs of steam explosions since approximately 200 years ago

50. Youtei Zan ↑

No eruption record

52. Usuzan ○♂↑→∼∩

1663 (Summit eruptions; normal volcanic explosions; lava dome "Kousu" formed; 5 killed; homes, forests, and arable land damaged), 1769, 1822 (Summit eruptions; normal volcanic explosions; pyroclastic flow; 1 village destroyed; 50–103 killed; 53 wounded; domesticated animals, forests, and arable land damaged), 53 (Summit eruptions; normal volcanic explosions; lava dome "Ousu" formed), 1910 (VII–XI; flank volcanism; normal volcanic explosions; volcanic mudflow; Meijishinzan formed; 1 killed; homes, forests, and arable land damaged), 44 (VI)–45 (IX); flank volcanism; normal volcanic explosions; lava dome; Syowashinzan formed; 1 killed; homes, forests, and arable land damaged, 77 (VIII; summit eruptions; normal volcanic explosions; ejecta gross volume 89 million m^3; mud flow; ash and sand fallout; 3 dead or missing due to ground movements; forests, arable land, and homes damaged), 2000–01(III31–; flank volcanism; normal volcanic explosions; mud flow; ash and sand fallout; homes, forests, roads, and arable land damaged due to ground movements)

54. Niseko ↑∨

No eruption record

58. Hokkaido Komagatake ○↑→∼∽

1640 (Normal volcanic explosions; eruption tsunami; more than 700 drowned), 94, 1765?, 1856 (Normal volcanic explosions; 1 village burned; 19–27 killed), 88, 1905 (Normal volcanic explosions; volcanic mudflow), 19, 23–24, 29 (VI17–IX6; normal volcanic explosions; pyroclastic flow; ejecta gross volume 500 million m^3; 2 killed; 4 wounded; 1915 homes damaged; damage to domesticated animals, forests, and arable land), 37, 42, 96, 98, 2000

61. Esan ○↑∼

1846, 74

65. Oshima Oshima ○↑∽

1741 (VIII23; normal volcanic explosions; eruption tsunami; 1467 drowned along Hokkaido coast), 42, 59

68. Osore Zan ↑∨

No eruption record; occasional earthquake activity

69. Hakkodasan

Steam explosions 1 time in 13–14th centuries, 2 times in 15–17th centuries; occasional earthquake activity

Continued.

70. Iwaki San ○↑∨
1782–83, 1845

72. Towada ↑∨
915

81. Akita Yakeyama ○∨～
807 ? , Sometime between 1300–1460, 1678, 1867, 87, 90, 1929, 48–49, 51, 57, 97

82. Hachimantai
No eruption record

84. Iwate San ○♂↑∨⇒～
1686 (Normal volcanic explosions; lava flows; volcanic mudflow; homes, forests, and arable land damaged), 1732 (Flank volcanism; normal volcanic explosions; lava flows Yakehashiri lava flows), 1919, 98 (Frequent magmatic earthquakes, thermal anomalies)

87. Akita Komagatake ○↑∨⇒
Before 915, 1890–91, 1932, 70–71 (Summit eruptions; normal volcanic explosions; lava flows)

91. Chokai Zan ○↑∨～∩
708–15, 810–23, 30, 71 (Normal volcanic explosions; volcanic mudflow), 939, 1659–63, 1740 (Normal volcanic explosions; volcanic mudflow) –47, 1801 (Summit eruptions; normal volcanic explosions; lava dome; Kyowadake (Shinzan) formed; 8 killed) –04, 21, 34 ? , 1974 (III1–V28; summit eruptions; steam explosions; volcanic mudflow)

94. Kurikoma Yama ○↑～
1744, 1944

97. Naruko ∨
837

98. Hijiori ∨ ?
No eruption record

107. Zao Zan (Zao San) ○∨～m
773, Sometime between 8–13th centuries, 1183, 1227, 30, Sometime between 12–15th centuries, 1620, 22, 23–24, 30, 41, 68–70, 94, 1794, 96, 1804, 06, 09, 21, 22, 30–31, 33, 67 (Lake bed eruption; volcanic mudflow; 3 killed), 73, 94, 95 (Steam explosions; lake bed eruption; volcanic mudflow), 1918–23 (Okama boiled; gas blowout), 40

109. Azuma Yama ○↑ or ∨
1331, 1711, 1893 (Normal volcanic explosions and steam explosions; 2 investigators killed in the line of duty by VI7 eruption) –95, 1950, 52 ? , 77
Note) Since recordkeeping started, all eruptions at Issaikyozan

110. Adatara Yama ○↑ or ∨
1899, 1900 (Normal volcanic explosions and steam explosions; VII; 72 killed; 10 wounded; sulfur mine facilities, forests, and arable land damaged)

111. Bandai San ○∨～
806, 1888 (VII15; summit eruptions; steam explosions; volcanic mudflow; mountain body exploded; ejecta gross volume 1.2 km^3; various villages buried; 461 killed; homes, forests, and arable land damaged; Lake Hibara etc. formed)

113. Numazawa ↑∨
No eruption record

117. Numazawa ○∨～
1408–10 (Summit eruptions; normal volcanic explosions; volcanic mudflow; 180 killed or more), 1846, 81, 1953, 60, 63

118. Takahara Yama ↑∨
No eruption record

122. Nikko Shirane San ∨
1649, 1872–73, 89

123. Hiuchigatake
1544. Oikedake lava dome formed approximately 500 years ago

Records of Eruptions on Active Volcanoes of Japan Continued.

129. Akagi San ∨

1251 ?

132. Haruna San

No eruption record; evidence for 3 eruptions in the 5-6th centuries

141. Kusatsu Shirane San ○∨～m

1805, 82, 97 (Normal volcanic explosions; sulfur mine facilities destroyed), 1900, 02 (Normal volcanic explosions; sulfur mine facilities destroyed), 05, 25, 27-28, 32 (Steam explosions; volcanic mudflow; sulfur mine facilities destroyed ; 2 killed; 7 wounded), 37-39, 40-41, 42, 58 or 59, 76, 82, 83

144. Asamayama ○↑⇒→～

685, 1108 (Normal volcanic explosions; volcanic mudflow; lava flows; forests and arable land damaged), 28, 1281 ?, 1527-28 ?, 32 (Normal volcanic explosions; volcanic mudflow; homes and roads damaged), 34, 82, 90, 91, 95 ?, 96 (Normal volcanic explosions; many killed), 97, 1600, 05, 09, 44-45, 47-48 (Normal volcanic explosions; volcanic mudflow; homes washed away) -52, 55-61, 69, 95 ?, 1703-04, 06, 08-11, 13, 17-21 (Normal volcanic explosions; 5 killed) -23, 28-29, 32-33, 52, 54 (Normal volcanic explosions; eruption; forests and arable land damaged), 76-77, 83 (V9-VIII5; normal volcanic explosions; pyroclastic flow; volcanic mudflow; lava flows; Onioshidashi; ejecta gross weight 200 million m^3; 1151 killed; 1182 homes destroyed, washed away or burned; forests and arable land damaged; altered environment), 1803 (Normal volcanic explosions; houses collapsed), 1815, 67, 69, 75, 79, 89, 94, 99, 1900, 01, 02, 04, 07-09 (Normal volcanic explosions; damage from air shock and falling ash) -11 (Normal volcanic explosions; extensive damage from lava bombs, falling ash, and air shock; many killed or wounded) -13 (Normal volcanic explosions; 1 killed; 1 wounded) -14, 15-17, 19-20 (Normal volcanic explosions; wildfires caused by lava bombs; 1 home burned) -22, 24, 27-28 (Normal volcanic explosions; 1 house set fire by lava bomb) -30 (Normal volcanic explosions; wildfires; 6 killed by lava bombs) -31 (Normal volcanic explosions) -32, 34-35 (Normal volcanic explosions; wildfires caused by lava bombs; damage from falling ash and air shock) -36 (Normal volcanic explosions; 1 killed by lava bomb) -42, 44-47 (VIII14; normal volcanic explosions; 9 killed by lava bombs; wildfires), 49-50 (IX23; 1 killed by lava bomb; 1 wounded; damage from falling ash and air shock) -55, 58 (Normal volcanic explosions; damage from air shock) -59, 61 (VIII18; normal volcanic explosions; 1 killed; forests and arable land damaged), 65, 73, 82-83, 90, 2003, 04, 08, 09, 15

151. Myoko San

Between 430-690

152. Niigata Yakeyama ○♂∨～

887, 989, 1361, 1773, 1852-54, 1949, 62, 63, 74 (VII28; Steam explosions; volcanic mudflow; fissure eruption; 3 killed by lava bombs), 83, 97-98

154. Tateyama (Midagahara) ↑

1836

159. Yake Dake ○♂↑∨～

630, 685, 1270, 1440, 60, 1570, 1746, 1907-14, 15 (VI; Steam explosions; volcanic mudflow; Lake Taisho formed), 16, 19, 22-27, 29-32, 35, 39, 62 (Steam explosions; volcanic mudflow; 2 killed by lava bombs) -63

159′. Akandanayama ↑

No eruption record

160. Norikura Dake ↑∨

No eruption record

163. Hakusan ○↑∨～

1042, 1177 ?, 1239 ?, 1547, 48 ?, 54-56 (Small scale pyroclastic flow; lava bombs; temple destroyed), 79 (Volcanic mudflow; lava bombs; temple destroyed), 1659

171. Ontake San

1979 (X28; steam explosions at summit; volcanic ash fallout), 1991, 2007, 2014 (IX27, phreatic explosion

Continued.

near summit crates, low-temperature pyroclastic flow, fallout tephra, 58 killed by ballistic ejecta, 5 missing)

175. Yoko Dake ↑

No eruption record

177. Fuji San ○♂↑⇒

781, 800–02 (Normal volcanic explosions; Ashigara-road buried under falling ash and sand; Hakone-road open), 826 or 827, 64–66 (Normal volcanic explosions; lava flows "Aokigahara lava flows" divided the lake "Senoumi" into 2 parts: "Saiko" and "shojiko"; houses buried), 70 ?, 937, 52 ?, 93 ?, 99, 1017 ?, 33 (Normal volcanic explosions), 83, 1427 ?, 35 or 36, 1511, 1707 (XII16; normal volcanic explosions; ejecta gross volume 700 million m^3; falling sand and ash 5 cm deep in Kawasaki, 90 km to the east; extensive damage)

179. Hakoneyama

About the latter half of the 12th century to the 13th century Steam explosions 3 times (near Owakudani). Frequent earthquake activity, 2015 (VI 29–30, tiny phreatic explosion at Owakudani crater)

183. Izu Tobu Volcanic Group

Frequent earthquake activity, 1989VII13; submarine eruption off the coast of Ito City; Teishi-knoll (–81 m) formed

189. Izu Oshima ○♂↑∨⇒

680 ?, 84 ?, 5–8th century, 8th century, 9th century, 838, 10–11 century, 11–12 century or 1112, 12th century, 13th century, 14th century, 1338, 15th century, 1552, 1600 or 01 spring, 12 or 13 spring, 23, 34, 36 or 37, 37–38, 84, 90 (Lava flows; extended to northeast coast; "Large eruption of Teikyo"), 95, 1777–92 (Lava flows; extended to northeast and southwest coast; "Large eruption of Anei"), 1803, 21, 22–24, 46, 70, 76–77, 87–1909, 12–14 (Lava flows), 15, 19, 22–23, 33–34, 35, 38–40, 50–51 (VII16–VI28; lava flows; Mihara-Desert buried; ejecta gross volume 30 million m^3), 53–54 (Lava flows), 55–56 (X13; normal volcanic explosions; 1 killed; 53 wounded), 58–60, 62–70, 74, 86 (XI15–XII18; normal volcanic explosions; lava flows; accompanied north slope fissure eruption; ejecta gross volume 40 million m^3; entire population of island took refuge off-island), 87–88, 90

190. Toshima ↑

No eruption record

192. Niijima ↑→∼∩

Sometime between 838–86, 86–87 (Mukai Yama formed), Frequent earthquake activity

193. Kozushima ↑→∼∩

838 (Tenjosan formed), Frequent earthquake activity

194. Miyakejima ○♂↑⇒M

Sometime between 832, 50, 86–1154, 1085, 1154, 1469, 1535, 95, 1643 (Lava flows; 1 village destroyed; forests and arable land damaged), 1712 (Lava flows; many houses buried by mudslides), 63–69 (Normal volcanic explosions; homes, forests, and arable land damaged), 1811 (Normal volcanic explosions; lava flows ?), 35 (Lava flows), 74 (VII; normal volcanic explosions; lava flows; ejecta gross volume 16 million m^3; 45 eaves destroyed in 1 village; 1 killed), 1940 (VII12–VIII8; normal volcanic explosions; lava flows; ejecta gross volume 19 million m^3; 11 killed; 20 wounded; 62 homes destroyed; forests, arable land, and fisheries damaged), 62 (VIII24–25; normal volcanic explosions; lava flows; ejecta gross volume 10 million m^3; 5 houses burned; forests and arable land damaged; residents took refuge outside of Gakudo Island), 83 (X3–4; normal volcanic explosions; magma eruption; lava flows; magma steam explosions along the coast; ejecta gross volume 12 million m^3; approximately 400 homes buried), 2000 [VI27–; submarine eruption; normal volcanic explosions; steam explosions; magma steam eruptions (sea surface discoloration); mud flows produced; sinking caldera formed at summit; fallout tephra; long time, high volume production of poisonous volcanic gasses; all residents take refuge off-island (2005 A.D.II evaculation order lifted)] –06, 08, 09, 10, 13

Records of Eruptions on Active Volcanoes of Japan Continued.

196. Mikurajima ↑

No eruption record

198. Hachijojima ○♂↑M

1487, 1518 (Normal volcanic explosions; forests and arable land damaged) -23, 1605(Normal volcanic explosions; forests and arable land damaged), 06 (Submarine eruption; volcanic island formed)

200. Aogashima ○↑∼m

1652?, 70-80, 1780-85(Normal volcanic explosions; lava flows; pyroclastic cone formed in caldera; approximately 140 killed; almost all homes burned; remaining 108 residents took refuge on Hachijyo Island; was uninhabited island for approximately 40 years after that)

202. Bayonnaise Rocks ○↑∩M

1869, 70, 71, 96 (Submarine eruption; new island formed), 1906, 15, 34, 46 (Submarine eruption; new island formed), 52 (IX17)-53 (IX); submarine eruption; new island "Myojinsho" formed; 52 (IX24) (observation ship Daigo-Kaiyomaru accident; all 31 crew onboard killed in the line of duty), 54, 55, 60, 70, 71, 79, 80, 82, 83, 86, 87, 88

203. Sumisujima

1870 (Submarine eruption; new island formed), 1916 (Submarine eruption; near the west edge of island)

204. Izu Tori Shima ○↑∨M

1902 (VIII; Summit eruptions; steam explosions; submarine eruption; central cone disappeared in explosion; all 125 island residents killed), 39 (VIII18-XII; normal volcanic explosions; lava flows; central cone reformed; entire island population evacuated), 98, 2002

205. Sofugan

No eruption record

206. Nishinoshima ↑⇒M

1973 (Submarine eruption; normal volcanic explosions; new island discovered in VIII1) -74 (Submarine eruption; normal volcanic explosions; lava flows; new island's area reached approximately 238,000 m^2; connected to old island), 2013 (Submarine eruption; normal volcanic explosions; new island discovered in IX20; lava flows; new island's area reached approximately 160,000 m^2; connected to old island) -15 (Normal volcanic explosions; lava flows; ▲)

207. Kaikata Seamount

No eruption record

208. Kaitoku Seamount ↑M

1543?, 1984 (III-VI; submarine eruption)

210. Funka Asane ↑M

1780, 1880, 1930-45

Note) Also known as "Kita Iojima"

212. Io To ○∨

1889 or 90 (Steam explosions), 1922 (Steam explosions), 35 (Steam explosions; in the Chidori Airport guide slope), 44 (Steam explosions), 57 (Steam explosions; central part of island), 67-68 (Steam explosions), 69, 75, 78, 80, 82, 93-94, 99 (Steam explosions), 2001, 04 (Steam explosions), 11, 12, 13, 15

213. Fukutoku Okanoba ↑M

1904-05 (Seabed explosion; new 145 m altitude island formed), 14 (Submarine eruption; new 300 m altitude island formed), 73-74, 86, 92-93, 2005-07, 10, Frequent seawater discoloration

214. Kita Fukutokutai ↑

1947-59

216. Minamihiyoshi Seamount ↑

1975, 76

217. Nikko Seamount ↑

1979 Seawater discoloration

231. Sanbe San ↑∨

No eruption record

Continued.

233. Abu Volcanio Group

No eruption record

239. Tsurumi Dake and Garan Dake ∨

771, 867

240. Yufu Dake ↑

No eruption record

243. Kuju San ∨

1675 ? (Sulfur discharge ?), 1738 ? , 1995-96

244. Aso San ○↑∽

Since recordkeeping started, all eruptions have been Strombolian eruptions of the central cone Nakadake (1323 m). 553 ? (Oldest recorded eruption in Japan; 14th year of the Kinmeitennou Era), 864, 67, 1239-40, 65, 69, 71-74 (Arable land damaged), 81, 86, 1305, 24, 31-33, 35 (Temple damaged), 40, 43, 75-76, 87, 1434, 38, 73-74, 85 (Pyroclastic cone formed), 1505-06, 22, 33, 42, 58-59, 62, 73, 74, 76, 82-83, 84 (Arable land devastated), 87 (Pyroclastic cone formed), 92 (Pyroclastic cone formed), 98, 1611-13, 20, 31, 37, 49, 75, 83, 91, 1709, 65, 72-88, 1804, 06, 14-16 (1 killed; arable land devastated), 26-28, 30 (Pyroclastic cone "Asamayama" formed; arable land devastated) -32, 35, 37, 38, 54, 56, 72 (Many killed) -74, 84, 94, 97, 1906-08, 10, 11-12, 16, 18-20, 23, 26-30 (Forests and arable land devastated), 32-33 (13 wounded; forests and arable land devastated), 34-37, 39-51, 53 (IV27; normal volcanic explosions; 6 killed; 90 wounded or more) -58 (VI24; 12 killed; 28 wounded; tourist facilities destroyed; forests and arable land devastated) -64, 65 (X31; tourist facilities destroyed), 66-68, 69-73, 74-77, 78, 79 (IX6; 3 killed; 11 wounded), 80, 84, 85, 88, 89, 90, 91, 92, 93, 94, 95, 2003, 04, 05, 09, 11, 14-15 (phreatomagmatic eruption, continuous ash emission), 16 (X8, phreatomagmatic eruption)

247. Fukue Volcanio Group

No eruption record

249. Unzen Dake ○♂↑∨⇒→∼∩

1663 (Normal volcanic explosions; "Furuyake Lava" lava flows), 1792 (Normal volcanic explosions; "Shinyake" lava flows; strong tremors; landslip; volcanic mudflow; eruption tsunami in Ariake Sea; approximately 15,000 dead; worst Japanese volcanic disaster on record), 1990-96 (Normal volcanic explosions; lava dome; pyroclastic flow; volcanic mudflow; 1st eruption in 198 years; lava dome formed; many pyroclastic flows produced; 44 dead or missing; homes lost; forests and arable land damaged)

254. Kirishima Yama ○↑∨→∼m

742, 788×, sometime between 10th and 11th centuries, 1112 (Shinto shrine burned), 67, 1235×, sometime between 1350 and 1350×, sometime between 14th and 15th centuries×, around 1350×, sometime between 1350 and 1500×, sometime between 16th and 17th centuries×, 1554-55×, 66×, 74×, 76-78×, 87×, 88×, sometime between 1958 and 1600×, 13×-14×, 15×-16×, 17×-18×, 20×, 28 ?, 37-38, 50×, sometime between 1650 and 1700×, 59×-64×, 77×-78×, 1706×(Shinto shrine burned), 16△-17△ (5 killed; 31 wounded; shrines and temples burned; homes burned; 600 destroyed; damage to forests, arable land, cattle, and horses), 68 (Mountian body collapsed at mountain in Republic of Korea), 71×-72×, 1822 △, 32 ?, 80×, 87×-89×, 91×, 93×-95×(4 killed; 22 homes lost) -96× (2 killed) -97× (Arable land damaged) -1900×(2 killed; 3 wounded), 03×, 13×-14×, 23× (1 killed), 59△ (Forests and arable land damaged), 68-Ebino Earthquake, 71 (Steam explosions), 91-92△, 2008△, 10△, 11△ (No eruptions after September 7), 17△

Note) "×" indicates Takachiho-Ohachi caldera; "△" indicates Shinmoedake caldera; unmarked are of unknown eruption point or other points

255. Yonemaru-Sumiyoshi Ike

No eruption record

256. Sakurajima ○♂↑∨⇒→∼M∽

708 ?, 16 ?, 17 ?, 64 (Nabeyama appeared; Nagasaki-

Records of Eruptions on Active Volcanoes of Japan Continued.

bana Lava (Seto Lava) outflow), 66, around 1200, 1468, 71–76 (Normal volcanic explosions; lava flows; many cases of humans and animals being killed or injured, domesticated animals being buried, etc.), 78 ?, 1642, 78, 1706, 42, 49, 56, 79 ("Large eruption of Anei"; normal volcanic explosions; lava flows; submarine eruption, new island "Moejima" formed; tsunami; 153 killed; homes and arable land damaged) –82 (Normal volcanic explosions; eruption tsunami; 15 killed), 83, 85, 90–91, 92, 94, 97, 99, 1860, 99, 1914 (I12–IV; "Large eruption of Taisho"; normal volcanic explosions; lava flows; connected to Osumi peninsula; ejecta gross volume 2 billion m^3; remarkable earthquake rumbling; various villages burned; 58 killed; 112 wounded; approximately 2140 homes burned), 35, 38–45, 46 (I–XI; normal volcanic explosions; lava flows; ejecta gross volume 100 million m^3; 1 killed; villages buried or burned; homes, forests, and arable land damaged), 48, 50, 54, 55 (X13; magma explosion; 1 killed; 9 wounded; forests and arable land damaged) –56 Shifted to normal volcanic explosions) –64 (Normal volcanic explosions; 8 wounded) –73 (Normal volcanic explosions; 1 wounded) –74 (Normal volcanic explosions; forests, arable land, and homes damaged; VI17; 3 killed in lahar; VIII9; 5 killed in secondary volcanic mudflow) –2006 (VI; 1st eruption in 59 years from Syowa Crater) –15 (VIII15–16, drastic ground deformation due to magma intrusion) –17 ▲

256′. Wakamiko Caldera ↑

No eruption record

257. Ikeda, Yamakawa ↑ V

No eruption record

257. Kaimon Dake ↑

874, 85

258. Satsuma Iojima ↑ M

15–16 centuries, 1934 (Submarine eruption; new island formed) –35 [Submarine eruption; new island reformed; modern 50 m altitude island (Syowa Iojima)], 98–99, 2000–04, 13

260. Kuchinoerabujima ○♂↑V∼

1841, 1931 (Steam explosions; volcanic mudflow; 8 killed; 26 wounded; 1 village burned; forests and arable land damaged), 33–34, 45, 66 (Steam explosions; 3 wounded), 68–69, 72–74, 76, 80, 2014 (VIII3, phreatic explosion, low temperature pyroclastic flow), 15 (V29–, phreatic explosion, low-temperature pyroclastic flow, all the resident evacuated from the island)

261. Kuchinoshima ↑

No eruption record

262. Nakanoshima ○V

1914 (Ontake)

263. Suwanosejima ○↑⇒

1813 (Normal volcanic explosions; lava flows; entire island population moved off island; became uninhabited island for approximately 70 years), 77, 84 (Normal volcanic explosions; lava flows) –85, 89, 1921–22, 25, 38, 40, 49–54, 56–95, 97–2017 ▲

Note) Strombolian eruption ongoing

266. Io Torishima ○↑

1664, 1796, 1829, 55, 68, 1903 (Entire island population moved temporarily to Kumejima), 59 (Entire island population moved; became uninhabited island), 67, 68

267. Submarine Volcano NNE of Iriomotejima ↑ M

1924 (X31; submarine eruption; large amounts of pumice drifted carried on Black Current; washed up on most of the coasts in the Japanese archipelago)

Main Late-Quaternary Wide-Spread Tephra of Japan

Volcanic tephra name	(Abbreviation)	Kind	Eruption age
Hokkaido			
Tarumae a	(Ta-a)	pfa	1739AD
Tarumae b	(Ta-b)	pfa	1667AD
Usu b	(Us-b)	pfa	1663AD
Komagatake d	(Ko-d)	pfa, afa	1640AD
*Paektu Tomakomai	(B-Tm)	afa	10C
Mashu b	(Ma-b)	pfa	1ka
Mashu f	(Ma-f)	pfl	7.3-8ka
Mashu g-j	(Ma-g-j)	pfa	
Tarumae d	(Ta-d)	pfa	8-9ka
Mashu l	(Ma-l)	pfa	>14ka
Rishiri Wankonozawa	(Rs-Wn)	pfa	<15ka
Eniwa a	(En-a)	pfa	19-21ka
*Kutcharo Shoro	(Kc-Sr)	pfl, afa	35-40ka
Shikotsu pyroclastic flow	(Spfl)	pfl	40-45ka
*Shikotsu first	(Spfa-1)	pfa	
Zenikame Menagawa	(Z-M)	pfa	>45ka
Kuttara 1st	(Kt-1)	pfa	>43ka
Kuttara 6th	(Kt-6)	pfa	75-85ka
*Toya	(Toya)	afa	112-115ka
Toya pyroclastic flow	(Toya pfl)	pfl	
*Kutcharo Haboro	(Kc-Hb)	afa	115-120ka
Kutcharo 4	(Kc-4)	pfl	
Tohoku			
Towada a	(To-a)	pfa, afa	915AD
Numazawa 1	(Nm-1)	pfa	5ka
Towada Chuseri	(To-Cu)	pfa	6ka
Towada south	(To-Nb)	pfa	8.6ka
Hijiori Obanazawa	(Hj)	pfa	11-12ka
Towada Hachinohe pyroclastic flow	(To-Hpfl)	pfl	15ka
Towada Hachinohe	(To-HP)	pfa	
Akitakoma Yanagisawa	(Ak-Y)	pfa	11.0-11.9ka
Naruko Yanagisawa	(Nr-Y)	pfl, afa	41-63ka
Adachi Medeshima	(Ac-Md)	pfa	90-100ka
Kanto, Chubu			
Asama Tenmei	(As-A)	pfa	1783AD
Fuji Hoei	(F-Ho)	sfa	1707AD
Asama Tennin	(As-B)	pfa	1108AD
Kozushima Tenjo San	(Iz-Kt)	afa	838AD
Haruna Futatsudake Shibukawa	(Hr-FA)	pfa	Early 6C
Amagi Kawagotaira	(Kg)	pfa	3.1ka
Nantai Imaichi, Nanahonsakura	(Nt-I・S)	pfa	14-15ka
Asama Kusatsu	(As-K)	pfa	15-16.5ka
Asama Itahana yellow	(As-YP)	pfa	
Asama Shiraito	(As-Sr)	pfa	15-20ka
Akagi Kanuma	(Ag-KP)	pfa	>45ka
Hakone New period pyroclastic flow	(Hk-Tpfl)	pfl	60-65ka
Hakone Tokyo	(Hk-TP)	pfa	
Tateyama E	(Tt-E)	pfa	60-75ka
Hakone Obaradai	(Hk-OP)	pfa	80-85ka
*Ontake 1st	(On-Pm1)	pfa	100ka
Tateyama D	(Tt-D)	pfa	120-130ka
Kinki, Chugoku			
*Ulleung Oki	(U-Oki)	pfa	10.7ka
*Daisen Kurayoshi	(DKP)	pfa	>55ka
Daisen Namadake	(DNP)	pfa	>80ka
*Sanbe Kitsugi	(SK)	pfa	110-115ka
Daisen Matsue	(DMP)	pfa	<130ka
Kyushu			
Sakurajima Taisho	(Sz-1)	pfa	1914AD
Sakurajima Bunmei	(Sz-3)	pfa	1471AD
Ikadako	(Ik)	pfa	6.4ka
*Kikai Akahoya	(K-Ah)	afa	7.3ka
〃 Koya pyroclastic flow	(K-Kypfl)	pfl	
〃 Koya	(K-KyP)	pfa	
Sakurajima Satsuma	(Sz-S)	pfa	12.8ka
Kirishima Kobayashi	(Kr-Kb)	pfa	<16ka
*Aira Tn	(AT)	afa	26-29ka
〃 Nitto	(A-Ito)	pfl	
〃 Osumi	(A-Os)	pfa	
Kuju 1st	(Kj-P1)	pfa	50ka
*Aso 4	(Aso-4)	afa	85-90ka
〃 pyroclastic flow	(Aso-4pfl)	pfl	
Aira Fukuyama	(A-Fk)	pfa	90ka
* Kikai Kuzuhara	(K-Tz)	afa	95ka
Ata	(Ata)	afa	105-110ka
Ata pyroclastic flow	(Ata pfl)	pfl	

Note) *: A summary of the distribution is given in Geo-Science Chart 26. pfa: fallout pumice, sfa: fallout scoria, afa: fallout ash, pfl: pyroclastic flow. ka: 1 000 years ago, AD: Anno Domini, C: Century

Source: Hiroshi Machida, Fusao Arai: "Atlas of the Tephira in and around Japan" Revised Edition, Univeristy of Tokyo Press (2003)

Late-Quaternary Wide-Spread Tephra of the Japanese Archipelago and Surrounding Areas

Figure 24

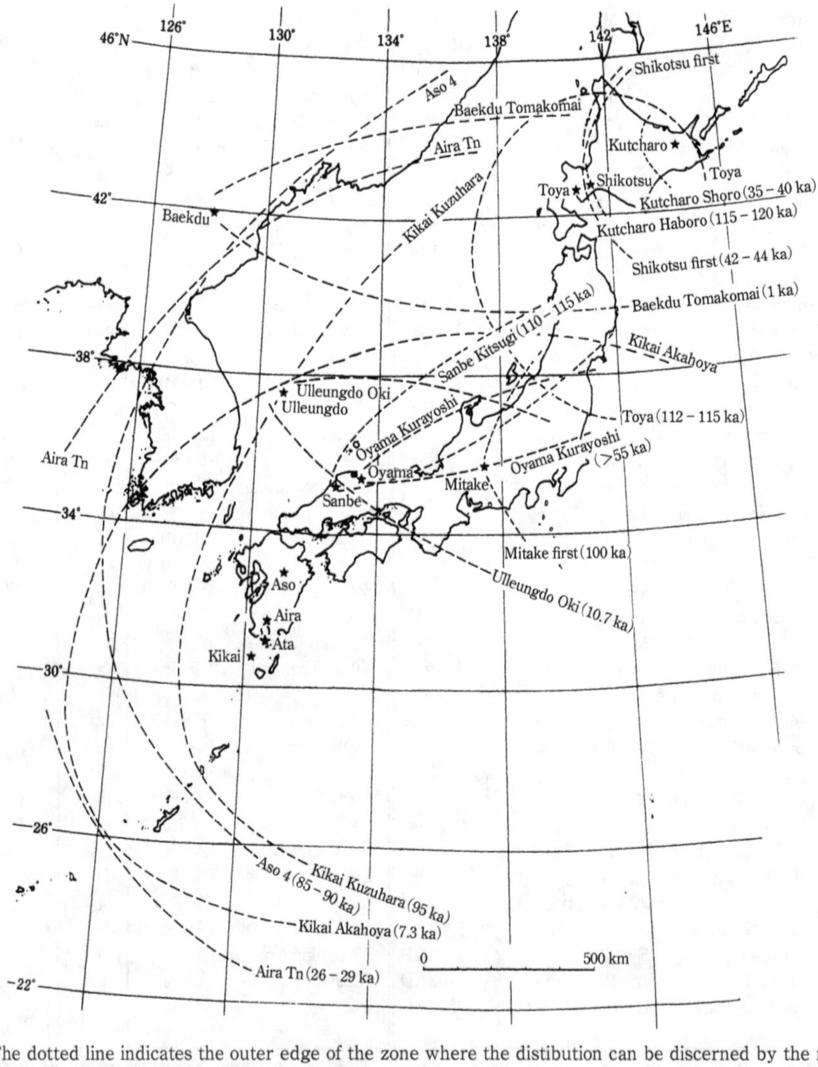

The dotted line indicates the outer edge of the zone where the distibution can be discerned by the naked eye. The numbers in parentheses indicate the eruption age (ka: 1 000 years ago).
Hiroshi Machida, Fusao Arai: "Atlas of the Tephira in and around Japan" Revised Edition, Univeristy of Tokyo Press (2003)

Volcanic Disasters Resulting in at least 1 000 Deaths (Since 1700)

Volcano name	Location	Volcanic Explosivity Index	Year	The main cause of death: Pyroclastic flow	Lahar	Debris avalanche	Tsunami	Famine	Total deaths
Awu	Sangihe Island (Indonesia)	3	1711	3 000					3 000
Oshima Oshima	Japan	4	1741				2 000		2 000
Makian	Halmahera Island (Indonesia)	4	1760		2 000				2 000
Papandayan	Java Island (Indonesia)	3	1772			2 951			2 951
Laki(Grumsboton)	Iceland	4	1783					9 350	9 350
Asamayama	Japan	4	1783		1 400				1 400
Unzen	Japan	2	1792				14 792		14 792
Tambora	Sumbawa island(Indonesia)	7	1812	12 000				49 000	61 000
Galunggung	Java Island (Indonesia)	5	1822	4 011					4 011
Aratoto	Turkey	3	1840	1 900					1 900
Awu	Sangihe Island (Indonesia)	3	1856		2 806				2 806
Krakatau	Sunda Strait (Indonesia)	6	1883	2 000			36 000		38 000
Ritter Island	Papua New Guinea	2	1888				1 650		1 650
Awu	Sunda Strait (Indonesia)	3	1892		1 250				1 250
Perret(former Mont Pelée)	The West Indies Martinique island (French colony)	4	1902	28 000					28 000
Santa María	Guatemala	6	1902	2 000	2 000				4 000
Soufrière Saint Vincent	West Indies Saint Vincent	4	1902		1 351				1 351
Perret(former Mont Pelée)	The West Indies Martinique island (French colony)	4	1902	1 350					1 350
Taal	Luzon (Philippines)	3	1911	1 335					1 335
Kelud	Java Island (Indonesia)	4	1919		5 110				5 110
Merapi	Java Island (Indonesia)	3	1930		1 369				1 369
Lamington	New Guinea (Papua New Guinea)	4	1951	2 942					2 942
Agung	Bali (Indonesia)	5	1963	1 000					1 000
El Chichón	Mexico	5	1982	2 000					2 000
Nevado del Ruiz	Colombia	3	1985		24 000				24 000
Lake Nyos	Cameroon	NA	1986	(1 746 death by carbon dioxide)					1 746
San Cristóbal	Nicaragua	2	1999		1 274				1 274
Mayon	Luzon(hilippines)	1	2006		1 266				1 266

Source: Brown et al. (2017)

Recent Main Eruptions and Magma Eruptive Volume

Volcano name	Location	Year	Amount of magma ejected (kg)
Laki	Iceland	1783	3.4×10^{13}
Mt. Asama	Japan	1783	4×10^{11}
Tambora	Sumbawa island(Indonesia)	1815	2.4×10^{14}
Krakatau	Sunda Strait (Indonesia)	1883	2.7×10^{13}
Santa María	Guatemala	1902	($5.5\ km^3$)*
Katmai	Alaska	1912	$1.2\text{–}1.5\times10^{13}$
Sakurajima	Japan	1914	3×10^{12}
Agung	Bali Island (Indonesia)	1963	2.4×10^{12}
Surutsuei	Iceland	1963	2.8×10^{12}
Heimaey	Iceland	1973	3.6×10^{11}
Soufriere	West Indies, St. Vincent Island	1979	4.0×10^{10}
Saint Helens	Washington (USA)	1980	6.5×10^{11}
Kurafura	Iceland	1981	5.6×10^{10}
El Chichón	Mexico	1982	5.0×10^{11}
Pinatubo	Philippines	1991	5.0×10^{12}

Based on Palis and Sigurdsson (1989), etc. and supplemented.
Source: Sapper (1972)

Size of Eruption

Volcanic explosivity index

The Volcanic Explosivity Index (VEI) is the most widely-used index for relative measure of the explosiveness of volcanic eruption. This index is used in the Global Volcanism Program of the Smithsonian Institution. The index classifies the size of the eruption into different levels based on the ejection amount of explosive eruptions. There are 9 levels (0–8) for eruptions with ejecta from less than 10^4 m^3 to more than 10^{12} m^3. The index increases by one level for every ten-fold increase in the ejection amount. In addition to the ejection amount, factors such as the effusion rate and volcanic plume height are also used as approximations for determining the index.

VEI	0	1	2	3	4	5	6	7	8
Description	Non-explosive eruption	Small scale	Medium scale	Moderately large scale	Large scale	Very large scale			
Tephra volume (m^3)	1×10^4	1×10^6	1×10^7	1×10^8	1×10^9	1×10^{10}	1×10^{11}	1×10^{12}	
Column height (km)									
Above crater	<0.1	0.1–1	1–5						
Above sea level				3–15	10–25	>25 →			
Type of eruption	← Hawaiian → (0–1)	←Strombolian→ (1–2)	← Vulcanian → (2–4)	← Plinian → (3–6)		← Ultra Plinian → (5–8)			

Source: Newhall and Self (1982)

Eruption magnitude and eruption intensity

The VEI is defined in relation to explosive eruptions and cannot be applied to the effusive eruptions. Therefore, there has been proposal of an "eruption magnitude" index which can be applied to eruptions consisting mainly of lava flow, as well as an "eruption intensity" index which indicates the force of eruption. The eruption magnitude index is designed to produce a value which is consistent with the VEI for explosive eruptions.

$$\text{Eruption magnitude} = \log_{10}(\text{amount of ejecta, kg}) - 7$$

$$\text{Eruption strength index} = \log_{10}(\text{effusion rate, kg/s}) + 3$$

Magnitude and Intensity of Explosive Eruptions

Eruption (Location)	Total erupted mass	Thermal energy (J)	Seismic energy (J)	Peak eruption plume height (km)	Peak eruption rate (kg/s)	Magnitude	Intensity
Toba (about 75,000 years ago) (Indonesia)	7×10^{15}	7×10^{21}	-	-	-	8.8	-
Tambora 1815 (Indonesia)	2×10^{14}	2×10^{20}	-	43	2.8×10^{8}	7.3	11.4
Taupo (For the era about 180 years) (New Zealand)	8×10^{13}	8×10^{19}	-	51	1.1×10^{9}	6.9	12.0
Katmai 1912 (Alaska)	3×10^{13}	3×10^{19}	1.6×10^{16}	25	1×10^{8}	6.5	11.0
Krakatau 1883 (Indonesia)	3×10^{13}	3×10^{19}	-	25	ca.5×10^{7}	6.5	10.7
Santa Maria 1902 (Guatemala)	2×10^{13}	2×10^{19}	-	34	1.7×10^{8}	6.3	11.2
Pinatubo 1991 (Philippines)	1.1×10^{13}	10^{19}	6.3×10^{13}	35	4×10^{8}	6.0	11.6
Vesuvius 79 (Italy)	6×10^{12}	6×10^{18}	-	32	1.5×10^{8}	5.8	11.2
Bezymianny 1956 (Russia)	10^{12}	10^{18}	-	36	2.2×10^{8}	5.3	11.3
Saint Helens 1980 (USA)	1.3×10^{12}	10^{18}	2×10^{13}	19	2×10^{7}	4.8	10.3
Saint Augustine (USA)	6×10^{10}	8×10^{16}	-	12	7×10^{6}	3.8	9.8
Saint Augustine (USA)	1×10^{10}	1.5×10^{16}	-	8	1×10^{6}	3	9
Saint Augustine (USA)	4×10^{8}	5×10^{14}	-	4	7×10^{4}	1.6	7.8
Saint Augustine (USA)	3×10^{7}	4×10^{13}	-	1.8	3 600	0.5	6.6
Saint Augustine (USA)	1×10^{7}	1.5×10^{13}	-	0.8	130	0	5.1

Source: Pyle (2000)

Magnitude and Intensity of Non-Explosive Eruptions

Eruption (Location)	Total erupted mass	Thermal energy (J)	Seismic energy (J)	Peak eruption plume height (km)	Mean eruption rate (kg/s)	Magni-tude	Maximum intensity
Laki1783 (Iceland)	3×10^{13}	5×10^{20}	-	2.4×10^{7}	1.4×10^{6}	6.5	10.4
Mauna Loa 1950.6 (USA)	10^{12}	2×10^{18}	-	7×10^{6}	5×10^{5}	5	9.8
Sakurajima 1914 Nishiyama belly	5×10^{11}	7×10^{17}	-	5×10^{6}	1.5×10^{5}	4.7	9.7
Lonquimay 1988-1990 (Chile)	5×10^{11}	7×10^{17}	10^{13}	1.4×10^{5}	1.4×10^{4}	4.7	8.1
Etna 1991-1993 (Italy)	5×10^{11}	7×10^{17}	-	6×10^{4}	10^{4}	4.7	7.8
Hekla 1991 (Iceland)	3×10^{11}	5×10^{17}	-	4×10^{6}	7×10^{5}	4.5	9.6
Etna 1792 (Italy)	2×10^{11}	3×10^{17}	-	3×10^{5}	6 000	4.3	8.5
Nyiragongo 1977 (Congo)	4×10^{10}	6×10^{16}	-	-	10^{7}	3.6	>10
Kilauea 1965.3 (USA)	3×10^{10}	5×10^{16}	-	10^{6}	4×10^{4}	3.5	9
Kilauea 1991 (USA)	8×10^{9}	1×10^{16}	-	6 000	3 400	2.9	6.8
Ol Doinyo Lengai 1988 (Tanzania)	2×10^{5}	1.5×10^{11}	-	20	4	-1.7	4.3

Source: Pyle (2000)

Solubility of H_2O in Magma

Magma is a fluid created when a variety of silicate compounds are melted at high temperatures. However, a considerable amount of gas components (volatile constituents) such as H_2O and CO_2 are dissolved into magma which exists underground. For each level of pressure, the table shows the approximate maximum amount of H_2O which can be dissolved in the 3 main types of magma. In many cases, actual magma contains an H_2O amount which is smaller than these values. However, since the surrounding pressure decreases as magma rises, volatile components such as H_2O ultimately reach their saturation point, transform into bubbles, and separate from the magma. At this time, the overall volume increases (swells) rapidly. This is the driving principle behind volcanic explosion.

Solubility of H_2O Determined via Experiment

Pressure (MPa)	Basalt (Colombian plateau basalt)	Andesite (Mt. Hood lava)	Rhyolite (Harding pegmatite)
	1 100℃	1 100℃	660–720℃
50	(1.6)	(2.7)	(3.0)
100	3.1	4.5	4.6
200	4.6	6.0	6.5
300	5.9	7.4	8.3
500	8.5	9.8	10.9

Sources: Hamilton *et al.* (1964), Burnham *et al.* (1962)

Solubility of H_2O Determined by Model Calculation

Pressure (MPa)	Basalt (JB2)		Andesite (JA2)		Rhyolite (JR2)	
	1 100℃	1 200℃	1 100℃	1 200℃	850℃	1 100℃
50	1.4	1.3	1.5	1.5	2.3	2.1
100	2.4	2.3	2.6	2.6	3.6	3.4
200	4.0	4.0	4.4	4.4	5.2	5.4
300	5.3	5.4	5.8	5.9	6.3	7.1
400	6.6	6.7	7.2	7.3	7.4	8.5
500	7.9	8.0	8.5	8.7	9.5	10.0

Calculated through the Papale (1997) method. The rock numbers are the National Institute of Advanced Industrial Science and Technology, Geological Survey of Japan standard material numbers

Chemical Composition of Volcanic Gas

Although the volatile components which have dissolved into magma ultimately separate from the magma and become volcanic gas, in the case of volcanoes which cause explosive eruptions, the volcanic gas emitted from the magma cannot be directly collected. Instead, it is collected from fumaroles after the eruption has subsided. Therefore, there is the possibility that the chemical composition has changed from the time that the gas separated from magma until it reached the surface. Conversely, volcanic gas collected from lava lakes is thought to show the composition of the volatile components which separated from the magma.

Volcano	(Collection year)	Temperature (℃)	Density (mole%)							
			H_2O	CO_2	SO_2	H_2S	HCl	H_2	CO	N_2
(a) Lava lake gas and lava flow gas										
1 Kilauea (Hawaii)	1918–1919	1 200	37.1	48.9	11.9	-	0.08	0.49	1.50	-
2 Kilauea (Hawaii)	1983	1 120	83.4	2.78	11.1	1.02	0.1	1.54	0.09	-
3 Mauna Loa (Hawaii)		1 100	73.4	4.15	21.0	0.56	0.16	0.48	0.16	-
4 Nyiragongo (Congo)	1959	1 020	42.5	41.5	4.5	-	-	0.75	2.4	8.3
5 Surtsey (Iceland)	1965	1 127	86.2	6.47	1.8	-	0.40	4.70	0.36	0.07
6 Erta Ale (Ethiopia)	1974	1 134	79.4	10.4	6.5	-	0.42	1.49	0.46	0.18
7 Etna (Italy)	1976	1 000	81.0	1.9	15.1	-	-	4.1	<0.05	<2.3
8 Tolbachik (Russia)	1976	1 135	97.4	0.3	0.5	0.3	0.5	1.0	-	<0.2
(b) Fumarole gas										
9 Momotombo (Nicaragua)	1978	800	93.0	4.7	1.2	0.02	0.4	0.6	<0.02	<0.1
10 Merapi (Indonesia)	1978	900	94.0	4.3	0.5	0.5	0.2	0.5	<0.01	<0.1
11 Krakatau (Indonesia)	1980	700	99.0	0.25	0.7	6×10^{-4}	-	0.02	3×10^{-4}	<0.2
12 Saint Helens (USA)	1981	660	98.9	0.9	0.07	0.10	0.4	0.03	$<2\times10^{-3}$	<0.1
13 White island 1971 (New Zealand)	1971	650	79.6	13.9	4.82	1.51	0.11	0.16	-	-
14 Satsuma Ioto (Japan)	1961	745	97.8	0.34	0.92	0.06	0.57	<0.24	-	-
15 Satsuma Ioto (Japan)	1967	570	98.1	0.47	0.82	0.05	0.49	0.07	1×10^{-4}	4×10^{-3}
16 Nasu (Japan)	1964	365	99.7	0.11	0.03	0.12	0.03	0.02	-	0.02
17 Syowa-shinzan (Japan)	1954	800	98.0	1.2	0.043	4×10^{-4}	0.05	0.63	3×10^{-3}	0.06
18 Usu (Japan)	1979	663	96.0	2.64	0.22	0.54	0.16	0.34	5×10^{-3}	0.06

Source: Minoru Kusabe and Osamu Matsubaya (1986)

Temperature and Viscosity Coefficient from Actual Measurement of Lava

In the majority of cases, the temperature is measured using an optical thermometer. However, a thermocouple is used in some cases. Normally, the viscosity coefficient is calculated from the speed of lava flowing down channels for which the cross-sectional shape and slope is known.

In most cases, basaltic lava is 1 100–1 200℃, andesitic lava is 1 000–1 100℃, and dacitic lava is less than 1 000℃. However, there are cases in which the temperature is much higher due to the combustion of volcanic gas. Basaltic lava has the lowest viscosity (approximately 10 Pa s). However, the viscosity increases rapidly when cooled. Generally speaking, the viscosity increases when the lava composition changes from andesitic to dacitic.

Volcano	Eruption year	Lava	Temperature (℃)	Coefficient of viscosity logs η(Pa-)
Miyakejima	1940	Basalt	1 000	5
Mt. Miharayama (Izu Oshima Island)	1950	Basalt	950–1 100	5–6
Mt. Miharayama (Izu Oshima Island)	1951	Basalt	1 150–1 200	2
Sakurajima	1946	Andesite	856–1 000	6–8
Akita Komagatake	1970	Andesite	1 090	-
Mt. Syowa-shinzan	1945	Dacite	900–1 000	8–10
Paricutin	1945–46	Basaltic andesite	1 070	4–5
Etna	1957	Basalt	1 110–1 120	3
Etna	1971	Basalt	1 050–1 100	1–2
Nyiragongo	1959	Basanite	1 180	2
Maunaloa	1950	Basalt	950–1 100	2–3
Kilauea (Iki, Kilauea)	1959	Basalt	1 120–1 190	-
Kilauea (Makaopuhi)	1965	Basalt	1 135	2
Kilauea (E. Rift)	1955	Basalt	1 100	2–3
Lunar surface lava (synthetic)	(Apollo 11) *	Basalt (Fe, Ti rich)	1 395	0

* Based on experiments

Formula for Calculating the Viscosity of Magma (Pa s)

$$\log \eta = \log \eta_{Tg} - \frac{A\,(T - T_g)}{B + T - T_g}$$

Here:

$$\begin{cases} A = 17.9,\ B = 26S^2 + 42S + 114 \\ T_g = 48S^2 - 251S + 1248,\ \log \eta_{Tg} = 12.8 \\ S = \sum \dfrac{Y_i}{1 - Y_{SiO_2}}\ (s_i^{\circ}\ Y_{SiO_2}) \end{cases}$$

Y_i= mole % of oxide; but for Al_2O_3 and Fe_2O_3 the molar ratios were doubled and then converted to 100%.

T = absolute temperature (K)

Oxide component	SiO_2	TiO_2	Al_2O_3	Fe_2O_3	FeO	MnO	MgO	CaO	Na_2O	K_2O	H_2O
s_i°	0	4.5	6.7	3.4	3.4	3.4	3.4	4.5	2.8	2.8	2.0

Note) The formula listed above is used to calculate liquid viscosity. The effective viscosity (η_{eff}) when phenocrysts or bubbles are contained is given by $\eta_{eff} = \eta(1-135\varphi)^{-2.5}$. η is the liquid viscosity and φ is the volume fraction of phenocrysts or bubbles.

According to Goto (2009). s_i° values are from Shaw (1972).

Formula for Calculating the Density of Magma

$$\rho = \frac{\Sigma X_i\, M_i}{V_{liq}\ (T,\ P,\ X)}$$

Here:

X_i = oxide molar fraction

Molar mass of M = oxide (kg/mol)

V_{liq}V (T,P,X) = molar volume of magma given by the following equation

$$V_{liq}\ (T,\ P,\ X) = \Sigma X_i\ [\overline{V}_{i,1673K} + (\partial \overline{V}_i / \partial T)_P (T - 1673\ K) + (\partial \overline{V}_i / \partial P)_T P]$$

Here: $\overline{V}_{i,1673K}$ = the oxide's partial molar volume (at 1673 K)

Oxide component	$\overline{V}_{i,1673K}$ (10^{-6} m^3/mol)	$(\partial \overline{V}_i / \partial T)_P$ (10^{-9} m^3/mol K)	$(\partial \overline{V}_i / \partial P)$ (10^{-6} m^3/mol GPa)
SiO_2	26.86 ± 0.3	0.0	−1.89 ± 0.02
TiO_2	23.16 ± 0.26	7.24 ± 0.46	−2.31 ± 0.06
Al_2O_3	37.42 ± 0.09	0.0	−2.26 ± 0.09
Fe_2O_3	42.13 ± 0.28	9.09 ± 3.49	−2.53 ± 0.09
FeO	13.65 ± 0.15	2.92 ± 1.62	−0.45 ± 0.03
MgO	11.69 ± 0.08	3.27 ± 0.17	0.27 ± 0.07
CaO	16.53 ± 0.06	3.74 ± 0.12	0.34 ± 0.05
Na_2O	28.88 ± 0.06	7.68 ± 0.10	−2.40 ± 0.05
K_2O	45.07 ± 0.09	12.08 ± 0.20	−6.75 ± 0.14
Li_2O	16.85 ± 0.15	5.25 ± 0.81	−1.02 ± 0.06
H_2O	26.27 ± 0.5	9.46 ± 0.83	−3.15 ± 0.61
CO_2	33.0 ± 1.0	0.0	0.0

Based on summary by Spera (2000). Original data from Lange & Carmichael (1990), Lange (1997), Ochs & Lange (1997).

Earthquakes

Official Statements Related to Earthquakes

1. JMA seismic intensity scale (1996)

Seismic intensity	Meter reading
0	0–0.4
1	0.5–1.4
2	1.5–2.4
3	2.5–3.4
4	3.5–4.4
5 Lower	4.5–4.9
5 Upper	5.0–5.4
6 Lower	5.5–5.9
6 Upper	6.0–6.4
7	6.5–

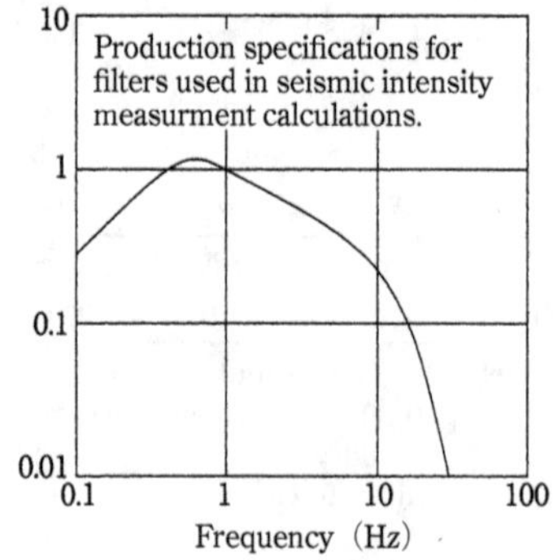

2. Explanatory table for the JMA

Seismic intensity	Human perception, reactions	Indoor situation	Outdoor situation
0	Imperceptible to people, but recorded by seismometers.	–	–
1	Felt slightly by some people keeping quiet in buildings.	–	–
2	Felt by many people keeping quiet in buildings. Some people may be awoken.	Hanging objects such as lamps swing slightly.	–
3	Felt by most people in buildings. Felt by some people walking. Many people are awoken.	Dishes in cupboards may rattle.	Electric wires swing slightly.
4	Most people are startled. Felt by most people walking. Most people are awoken.	Hanging objects such as lamps swing significantly, and dishes in cupboards rattle. Unstable ornaments may fall.	Electric wires swing significantly. Those driving vehicles may notice the tremor.
5 Lower	Many people are frightened and feel the need to hold onto something stable.	Hanging objects such as lamps swing violently. Dishes in cupboards and items on bookshelves may fall. Many unstable ornaments fall. Unsecured furniture may move, and unstable furniture may topple over.	In some cases, windows may break and fall. People notice electricity poles moving. Roads may sustain damage.
5 Upper	Many people find it hard to move; walking is difficult without holding onto something stable.	Dishes in cupboards and items on bookshelves are more likely to fall. TVs may fall from their stands, and unsecured furniture may topple over.	Windows may break and fall, unreinforced concrete-block walls may collapse, poorly installed vending machines may topple over, automobiles may stop due to the difficulty of continued movement.
6 Lower	It is difficult to remain standing.	Many unsecured furniture moves and may topple over. Doors may become wedged shut.	Wall tiles and windows may sustain damage and fall.
6 Upper	It is impossible to remain standing or move without crawling. People may be thrown through the air.	Most unsecured furniture moves, and is more likely to topple over.	Wall tiles and windows are more likely to break and fall. Most unreinforced concrete-block walls collapse.
7		Most unsecured furniture moves and topples over, or may even be thrown through the air.	Wall tiles and windows are even more likely to break and fall. Reinforced concrete-block walls may collapse.

"Instrumental seismic intensity" is an index representing the strength of seismic motion. It is calculated by rounding I as calculated using the following formula to three decimal places and then truncating at the second decimal place.

$$I = 2\log(a_0) + 0.94$$

a_0 is the maximum value which satisfies $\int w(t,a)dt \geqq 0.3$. t represents time (s), a is the parameter (cm/s^2) related to the degree of acceleration for seismic motion, and the interval of integration is the time for during which seismic motion continues. $w(t,a)$ is a function for which "0" is used when $v(t) < a$ and "1" is used when $v(t) \geqq a$. $v(t)$ is the vector-synthesized value (cm/s^2) after multiplying each degree of acceleration for the three orthogonal components of seismic motion by the filter. (From JMA bulletins and homepage)

Filter type and calculation

Filter type	Formula
Filter to show frequency effects	$(1/f)^{1/2}$
Low-pass Filter	$(1+0.694y^2+0.241y^4+0.0557y^6+0.009664y^8+0.00134y^{10}+0.000155y^{12})^{-1/2}$
High-pass Filter	$\{1-\exp[-(f/0.5)^3]\}^{1/2}$

(note) The simbol, f is frequency (Hz) of the seismic ground motion, and y is $f/10$.

seismic intensity scale (2009)

<table>
<tr><th rowspan="2">Seismic intensity</th><th colspan="2">Wooden houses</th><th colspan="2">Reinforced-concrete buildings</th><th rowspan="2">Situation of ground</th><th rowspan="2">Situation of slopes, etc.</th></tr>
<tr><th>High earthquake</th><th>Low earthquake</th><th>High earthquake</th><th>Low earthquake</th></tr>
<tr><td>5 Lower</td><td>-</td><td>Slight cracks may form in walls.</td><td>-</td><td>-</td><td rowspan="2">Small cracks may form and liquefaction may occur.</td><td rowspan="2">Rock falls and landslips may occur.</td></tr>
<tr><td>5 Upper</td><td>-</td><td>Cracks may form in walls.</td><td>-</td><td>Cracks may form in walls, crossbeams and pillars.</td></tr>
<tr><td>6 Lower</td><td>Slight cracks may form in walls.</td><td>Cracks are more likely to form in walls. Large cracks may form in walls. Tiles may fall, and buildings may lean or collapse.</td><td>Cracks may form in walls, crossbeams and pillars.</td><td>Cracks are more likely to form in walls, crossbeams and pillars.</td><td>Cracks may form.</td><td>Landslips and landslides may occur.</td></tr>
<tr><td>6 Upper</td><td>Cracks may form in walls.</td><td>Large cracks are more likely to form in walls. Buildings are more likely to lean or collapse.</td><td>Cracks are more likely to form in walls, crossbeams and pillars.</td><td>Slippage and X-shaped cracks may be seen in walls, crossbeams and pillars. Pillars at ground level or on intermediate floors may disintegrate, and buildings may collapse.</td><td rowspan="2">Large cracks may form.</td><td rowspan="2">Landslips are more likely to occur; large landslides and massif collapses may be seen.</td></tr>
<tr><td>7</td><td>Cracks are more likely to form in walls. Buildings may lean in some cases.</td><td>Buildings are even more likely to lean or collapse.</td><td>Cracks are even more likely to form in walls, crossbeams and pillars. Ground level or intermediate floors may sustain significant damage. Buildings may lean in some cases.</td><td>Slippage and X-shaped cracks are more likely to be seen in walls, crossbeams and pillars. Pillars at ground level or on intermediate floors are more likely to disintegrate, and buildings are more likely to collapse.</td></tr>
</table>

* The "Impact on Lifelines and Infrastructure" and the "Impact on Large-Scale Structures" were omitted.

Seismic intensity represents the strength of seismic motion. Seismic intensity is measured using a seismograph. The Explanatory Table for the JMA Seismic Intensity Scale illustrates the type of phenomena and damage which may actually occur in surrounding areas when a certain seismic intensity is measured. The following cautions must be taken when using this table.

(1) The seismic intensity announced by the JMA is the value measured using a seismograph. It is not determined from the phenomena listed in this table.

(2) Even for the same seismic intensity, damage may vary depending on the condition of affected buildings and structures, and on the characteristics of the seismic motion. This table lists phenomena which normally occur when a certain seismic intensity is measured. In some cases, the actual damage may be greater than or less than the listed damage.

(3) Seismic motion is significantly influenced by the ground condition and terrain. Seismic intensity is the value measured at the point where the seismograph is placed. The actual seismic intensity may vary by location even within the same municipality. Seismic intensity is normally measured on the earth's surface. Generally speaking, larger tremors occur in the upper floors of mid-to-high-rise buildings.

(4) Long-period seismic waves occur in large-scale earthquakes. Accordingly, even though the seismic intensity is relatively small in distant locations, phenomena unique to tremors (damage to elevators, oil tank sloshing, etc.) may occur.

(5) This table was created based mainly on damage caused by earthquakes in recent years. The content of this table may be subject to change in the future if it no longer matches actual conditions due to the acquirement of new cases, improvements in the earthquake-resistance of structures, etc.

3. Calculation of magnitude (M)

1) Initial definition (C. F. Richter, 1935)

$$M = \log A + \log B$$

A is the maximum single amplitude (unit: μm) from the recording paper of the Wood-Anderson seismograph (natural period: 0.8 s, damping constant: 0.8, magnification; 2 800, horizontal motion). $\log B$ is the function of the epicentral distance Δ (unit: km) as shown below.

Δ	50	100	150	200	250	300	350	400	450	500	550	600
$\log B$	−0.37	0	0.29	0.53	0.79	1.02	1.26	1.46	1.62	1.74	1.84	1.94

$\log B$ is an item for compensating for attenuation of the seismic waves. It reaches 0 at a standard epicentral distance of 100 km. This definition was for shallow near-field earthquakes in California, and is sometimes referred to as "local magnitude M_L." The main point of the magnitude proposed by Richter is that it uses the common logarithm for amplitude. This has served as the basis for a number of subsequent definitions.

2) Surface wave magnitude (B. Gutenberg, 1945)

$$Ms = \log A + \log B$$

A is the maximum single amplitude (a synthesis of two horizontal motion components; unit: μm) for ground motion of surface waves around 20 seconds of the period. $\log B$ is the function for the epicentral distance Δ (unit: degrees). The following formula is used to express $\log B$

when Δ is within the range of 15 to 130.

$$\log B = 1.656 \log \Delta + 1.818$$

Currently, the following formula by Vaněk *et al.* (1962) is commonly used internationally as the surface wave magnitude.

$$M_S = \log(A/T) + 1.66 \log \Delta + 3.3$$

T is the period (unit: s) of surface waves. This formula is used when the epicentral distance is within a range of 20 to 160. The amplitude of vertical motion is also frequently used.

3) Body-wave magnitude (B. Gutenberg, 1945)

$$m_B = \log(A/T) + Q(\Delta, h)$$

A is the maximum single amplitude (unit: μm) for ground motion of main motion of the body-wave (P, PP, S). T is the cycle (unit: s) of the body-wave. Q is the function of the epicentral distance Δ and the epicenter depth h. The vertical movement and horizontal motion of each body-wave is shown in a complex graph. This graph was subsequently revised (Gutenberg-Richter, 1956) and is still widely used even today. The seismographs used in the era when Gutenberg and Richter determined m_B had a relatively long period and T was several seconds. Recently, the body-wave magnitude is determined using short-period seismographs. Therefore, T is around one second and is not equivalent to m_B even when using the same formula. Therefore, it is sometimes written as m_b in order to clarify this difference.

4) JMA magnitude (revised on September 25, 2003)

The following formula is used to calculate M for as many observation points as possible. The average value is then taken. Priority is given to average values in the order of a), b), c). (H is the epicenter depth.)

a) $M = \log A + 1.73 \log \Delta - 0.83$

b) $M = \log A_h + \beta_D(\Delta, H) + C_D$

c) $M = \log A_z/0.85 + \beta_V(\Delta, H) + C_V$

A_h is the maximum single amplitude (synthesis of 2 horizontal motion components, cycle of less than 5 seconds, unit: μm) for ground motion as measured by a medium-cycle displacement seismograph. A is the value observed by the meteorological office from among A_h. A_z is the maximum velocity amplitude (vertical motion, unit: 10^{-3} cm/s) for ground motion as measured by a short-cycle velocity seismograph. Δ is the epicentral distance (unit: km). C_D and C_V are compensation items depending on the type of seismograph and installation condition. The distance attenuation items β_D (Δ, H) and β_V (Δ, H) are shown in the table. This definition is sometimes written as M_J.

5) Magnitude using duration of seismic motion

$$M = C_1 \log(F - P) + C_2 \Delta + C_3$$

$F-P$ is the duration from the start of the seismic record (P) until the finish (F). The constants C_1, C_2, and C_3 are values which differ depending on the observation point and characteristics of the seismograph. This formula makes it possible to determine M even for earthquakes where the seismic motion is so great that the record is shaken off and the maximum amplitude cannot be read. Generally speaking, C_2 is small, so M can be estimated even if the epicenter is not known.

6) Magnitude using seismic intensity

Kawasumi (1943) defined the epicentral distance of 100 km as magnitude M_k. There are several formulas which show the relation between M_k and normal M. In addition, for seismic intensity exceeding a certain level, there are methods which use factors such as the regional area and radius in which the earthquake can be felt. For example, the following is a frequently used formula which uses the regional area S_5 (unit: km^2) for seismic intensity of 5 or greater.

$$M = \log S_5 + 3.2 \quad \text{(Muramatsu, 1969)}$$

These methods make it possible to calculate M even for eras in which there was no measurement using seismographs and for rural earthquakes.

7) Moment magnitude (Kanamori, 1977)

$$M_w = (\log M_0 - 9.1)/1.5$$

M_0 is the seismic moment (unit: N m) which is calculated from the seismic wave amplitude, crust fluctuation, etc. It is an amount which expresses the size of an earthquake in terms of fault motion. When the epicenter fault surface area is S, the average displacement amount is D, and the modulus of rigidity where the earthquake occurred is μ, the seismic moment can be calculated using the following formula.

$$M_0 = \mu DS$$

M_s and m_b are determined using seismographs with limited periods. Therefore, the values plateau at approximately 8.5 and 7, respectively. However, this phenomenon does not occur for moment magnitude.

4. Relationship between size of earthquake and energy

(Gutenberg and Richter, 1956; Utsu, 2001)

$$\log E = 4.8 + 1.5\, M_s$$

E is the energy emitted as seismic waves (unit: J)

5. Relationship between size of earthquake and frequency of occurrence

(Gutenberg and Richter, 1944)

For earthquakes which occurred for a certain period of time in a certain region:

$$\log n(M) = a - bM$$

$n(M)$ is the number of earthquakes with a magnitude of M. In actuality, an appropriate value is selected for dM and the number of earthquakes from magnitude M to M+dM is used. The coefficient b is often called the "b value." In addition to possessing regional characteristics, the b value is said to vary according to foreshocks, aftershocks, and earthquake swarms.

The following table lists the number of earthquakes for which the JMA announced a magnitude of at least M5. The listed earthquakes occurred from 1961 to 1999 and in a range of 25 to 48 degrees north latitude and 125 to 150 degrees east longitude. When creating a graph in which the horizontal axis is M and the vertical axis is the logarithm for the number of earthquakes, it can be seen that the Gutenberg-Richter relationship is well supported.

M	5.0	5.1	5.2	5.3	5.4	5.5	5.6	5.7	5.8	5.9
Number of earthquakes	632	581	469	379	306	285	217	216	160	126
M	6.0	6.1	6.2	6.3	6.4	6.5	6.6	6.7	6.8	6.9
Number of earthquakes	109	80	63	48	40	34	33	23	15	14
M	7.0	7.1	7.2	7.3	7.4	7.5	7.6	7.7	7.8	7.9
Number of earthquakes	15	12	7	2	3	4	3	3	4	2
M	8.0	8.1								
Number of earthquakes	0	2								

6. Relationship between the occurrence frequency and maximum amplitude of seismic motion recorded at a certain observation point (Ishimoto and Iida, 1939)

$$n = ka^{-m}$$

n: frequency of earthquakes with a, maximum amplitude. k: constant.
m: constant Approximately 1.8 to 2.2 for normal earthquakes and aftershocks.

7. Damage grades

The damage grades listed below were defined by Utsu (1982, Bulletin of the Earthquake Research Institute, 57, 401–463).

1: Very small damage such as cracking of walls and ground (excluding ground fissures in special areas such as volcanoes, etc.)
2: Small damage such as damage to buildings and roads.
3: Multiple deaths or total collapse of multiple buildings (however, not reaching grade 4)
4: At least 20 deaths or the total collapse of at least 1 000 buildings (however, not reaching grade 5)
5: At least 200 deaths or the total collapse of at least 10 000 buildings (however, not reaching grade 6)
6: At least 2 000 deaths or the total collapse of at least 100 000 buildings (however, not reaching grade 7)
7: At least 20 000 deaths or the total collapse of at least 1 000 000 buildings

8. Size of tsunami

The Imamura-Iida Tsunami Magnitude Scale (Iida, K., 1958, J. Earth Sci., Nagoya Univ., 6, 101–112) is as follows:

Size class	Height of tsunami	Damage level
-1	< 50 cm	No damage
0	-1 m	Very little damage
1	-2 m	Damage to coast and ships
2	4–6 m	Some inland damage or human losses
3	10–20 m	Notable damage along ≥ 400 km of coastline
4	30 m <	Notable damage along ≥ 500 km of coastline

9. Various phase symbols appearing in seismic records

P: Vertical wave, compressional wave
S : Horizontal wave, torsional wave
K: Vertical waves passing through the core
c : Waves reflected by the core surface
I : Vertical waves passing through the inner core
J : Horizontal waves passing through the inner core

P: P waves which reach the observation point directly from the epicenter.

S: S waves which reach the direct observation point from the epicenter.

PP, PPP: P waves which reach the observation point after reflecting on the earth's surface one or two times.
(Similar symbols) SS, SSS

PS: P waves which are reflected on the earth's surface one time and change into S waves before reaching the observation point.
(Similar symbols) SP, PSS, SPP, etc.

PcP: P waves which are reflected on the core surface and reach the observation point as P waves.
(Similar symbols) ScS, ScP, PcS

PKP: P waves which are refracted, enter the core, pass through the core as P waves, are refracted and leave the core, and reach the observation point as P waves. Also written as P′.
(Similar symbols) SKS, PKS, SKP, PKKP, etc.

PKPPKP: PKP waves which are reflected on the earth's surface and once again reach the observation point as PKP waves.
(Similar symbols) SKSSKS, SKPPKP, etc.

Figure 25 Propagation path of the seismic waves

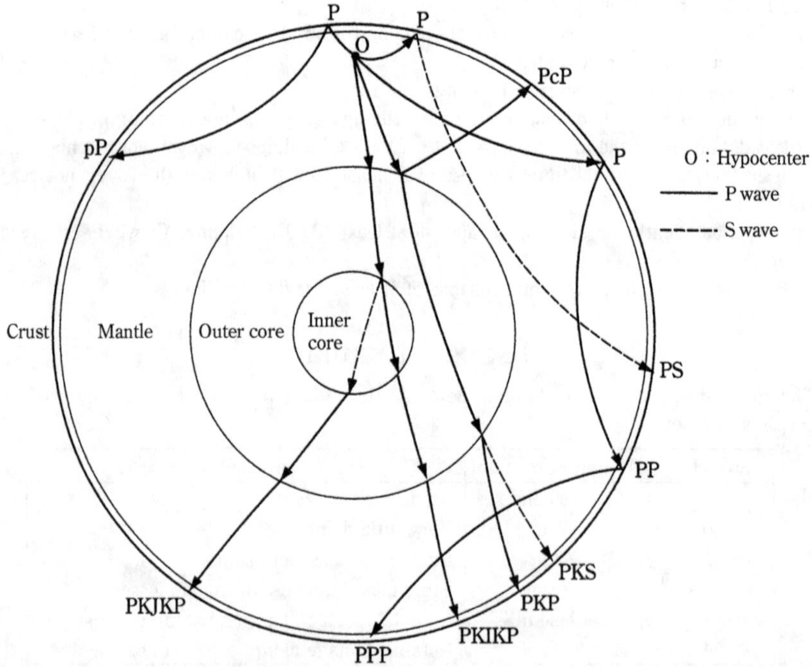

PKIKP: PKP waves which reach the inner core and pass through as P waves.
PKJKP: PKP waves which reach the inner core and pass through as S waves.
pP: The first p represents P waves which were emitted upwards from a horizontal plane at the epicenter.
(Similar symbols) sP, pS, sS, pPP, nPS, etc.
L: Long-period waves, surface waves, Lq: Love waves, Lr: Rayleigh waves

The travel-time curve in Figure 32 is a graph of the time required for these waves to reach a point of the epicenter distance: $\Delta°$.

10. Determining the epicenter distance

(Refer to page **520** for the geocentric latitude)

φ_0, λ_0: geocentric latitude and longitude of the epicenter

φ, λ: geocentric latitude and longitude of the observation point

Δ: corner epicenter distance

$$\cos \Delta = \sin \varphi_0 \sin \varphi + \cos \varphi_0 \cos \varphi \cos (\lambda_0 - \lambda)$$

During actual calculations, when

$$A = \cos \varphi_0 \cos \lambda_0, \quad B = \cos \varphi_0 \sin \lambda_0, \quad C = \sin \varphi_0$$
$$a = \cos \varphi \cos \lambda, \quad b = \cos \varphi \sin \lambda, \quad c = \sin \varphi,$$

then $\cos \Delta = aA + bB + cC$

11. Earth's internal structure

According to Preliminary Reference Earth Model (PREM), Dziewonski and Anderson (1981). α and β are P wave and S wave velocities, ρ is density, and Q μ and Q κ are damping parameters.

Depth(km)	α(km/s)	β(km/s)	ρ(g/cm^3)	Q_μ	Q_κ
0.0	1.45	0.00	1.02	0.0	57 823.0
3.0	1.45	0.00	1.02	0.0	57 823.0
3.0	5.80	3.20	2.60	600.0	57 823.0
15.0	5.80	3.20	2.60	600.0	57 823.0
15.0	6.80	3.90	2.90	600.0	57 823.0
24.4	6.80	3.90	2.90	600.0	57 823.0
24.4	8.11	4.49	3.38	600.0	57 823.0
71.0	8.08	4.47	3.38	600.0	57 823.0
80.0	8.08	4.47	3.37	600.0	57 823.0
80.0	8.08	4.47	3.37	80.0	57 823.0
171.0	8.02	4.44	3.36	80.0	57 823.0
220.0	7.99	4.42	3.36	80.0	57 823.0
220.0	8.56	4.64	3.44	143.0	57 823.0
271.0	8.66	4.68	3.47	143.0	57 823.0
371.0	8.85	4.75	3.53	143.0	57 823.0
400.0	8.91	4.77	3.54	143.0	57 823.0
400.0	9.13	4.93	3.72	143.0	57 823.0
471.0	9.50	5.14	3.81	143.0	57 823.0

Earth's internal structure Continued.

Depth(km)	α(km/s)	β(km/s)	ρ(g/cm^3)	Q_μ	Q_κ
571.0	10.01	5.43	3.94	143.0	57 823.0
600.0	10.16	5.52	3.98	143.0	57 823.0
670.0	10.27	5.57	3.99	143.0	57 823.0
670.0	10.75	5.95	4.38	312.0	57 823.0
771.0	11.07	6.24	4.44	312.0	57 823.0
871.0	11.24	6.31	4.50	312.0	57 823.0
971.0	11.42	6.38	4.56	312.0	57 823.0
1 071.0	11.58	6.44	4.62	312.0	57 823.0
1 171.0	11.73	6.50	4.68	312.0	57 823.0
1 271.0	11.88	6.56	4.73	312.0	57 823.0
1 371.0	12.02	6.62	4.79	312.0	57 823.0
1 471.0	12.16	6.67	4.84	312.0	57 823.0
1 571.0	12.29	6.73	4.90	312.0	57 823.0
1 671.0	12.42	6.78	4.95	312.0	57 823.0
1 771.0	12.54	6.83	5.00	312.0	57 823.0
1 871.0	12.67	6.87	5.05	312.0	57 823.0
1 971.0	12.78	6.92	5.11	312.0	57 823.0
2 071.0	12.90	6.97	5.16	312.0	57 823.0
2 171.0	13.02	7.01	5.21	312.0	57 823.0
2 271.0	13.13	7.06	5.26	312.0	57 823.0
2 371.0	13.25	7.10	5.31	312.0	57 823.0
2 471.0	13.36	7.14	5.36	312.0	57 823.0
2 571.0	13.48	7.19	5.41	312.0	57 823.0
2 671.0	13.60	7.23	5.46	312.0	57 823.0
2 741.0	13.68	7.27	5.49	312.0	57 823.0
2 771.0	13.69	7.27	5.51	312.0	57 823.0
2 871.0	13.71	7.26	5.56	312.0	57 823.0
2 891.0	13.72	7.26	5.57	312.0	57 823.0
2 891.0	8.06	0.00	9.90	0.0	57 823.0
2 971.0	8.20	0.00	10.03	0.0	57 823.0
3 071.0	8.36	0.00	10.18	0.0	57 823.0
3 171.0	8.51	0.00	10.33	0.0	57 823.0
3 271.0	8.66	0.00	10.47	0.0	57 823.0
3 371.0	8.80	0.00	10.60	0.0	57 823.0
3 471.0	8.93	0.00	10.73	0.0	57 823.0
3 571.0	9.05	0.00	10.85	0.0	57 823.0
3 671.0	9.17	0.00	10.97	0.0	57 823.0
3 771.0	9.28	0.00	11.08	0.0	57 823.0
3 871.0	9.38	0.00	11.19	0.0	57 823.0
3 971.0	9.48	0.00	11.29	0.0	57 823.0
4 071.0	9.58	0.00	11.39	0.0	57 823.0
4 171.0	9.67	0.00	11.48	0.0	57 823.0
4 271.0	9.75	0.00	11.57	0.0	57 823.0
4 371.0	9.84	0.00	11.65	0.0	57 823.0
4 471.0	9.91	0.00	11.73	0.0	57 823.0
4 571.0	9.99	0.00	11.81	0.0	57 823.0
4 671.0	10.06	0.00	11.88	0.0	57 823.0
4 771.0	10.12	0.00	11.95	0.0	57 823.0

Earth's internal structure Continued.

Depth(km)	α(km/s)	β(km/s)	ρ(g/cm^3)	Q_μ	Q_κ
4 871.0	10.19	0.00	12.01	0.0	57 823.0
4 971.0	10.25	0.00	12.07	0.0	57 823.0
5 071.0	10.31	0.00	12.12	0.0	57 823.0
5 149.5	10.36	0.00	12.17	0.0	57 823.0
5 149.5	11.03	3.50	12.76	84.6	1 327.7
5 171.0	11.04	3.51	12.77	84.6	1 327.7
5 271.0	11.07	3.54	12.82	84.6	1 327.7
5 371.0	11.11	3.56	12.87	84.6	1 327.7
5 471.0	11.14	3.58	12.91	84.6	1 327.7
5 571.0	11.16	3.60	12.95	84.6	1 327.7
5 671.0	11.19	3.61	12.98	84.6	1 327.7
5 771.0	11.21	3.63	13.01	84.6	1 327.7
5 871.0	11.22	3.64	13.03	84.6	1 327.7
5 971.0	11.24	3.65	13.05	84.6	1 327.7
6 071.0	11.25	3.66	13.07	84.6	1 327.7
6 171.0	11.26	3.66	13.08	84.6	1 327.7
6 271.0	11.26	3.67	13.09	84.6	1 327.7
6 371.0	11.26	3.67	13.09	84.6	1 327.7

Figure 26 Internal structure of the earth (PREM)

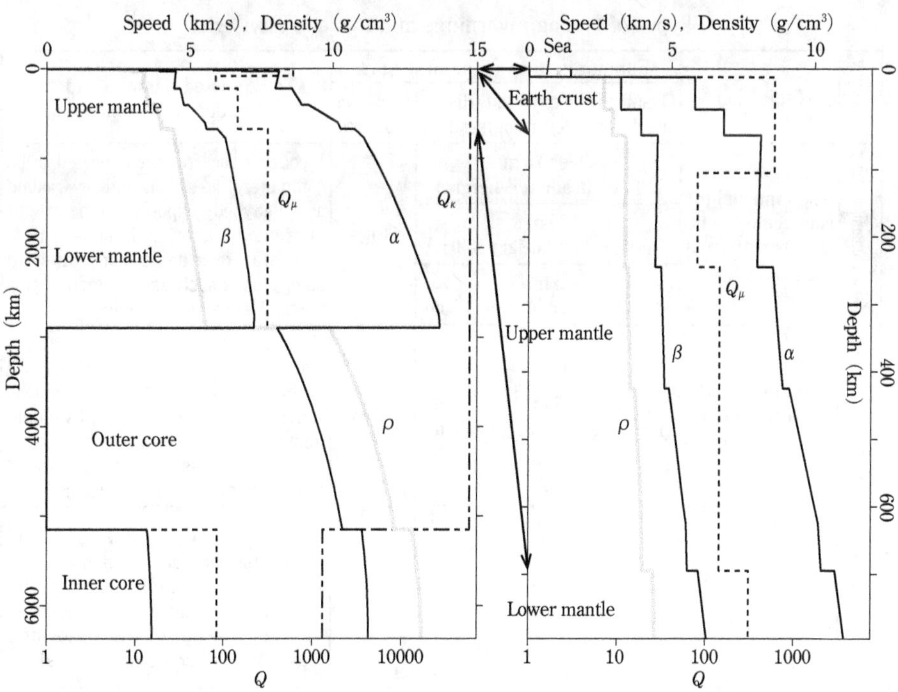

Tsunami Warning, Tsunami Advisory, Tsunami Information

When a tsunami capable of inflicting damage is predicted, the JMA has always issued a tsunami warning or tsunami advisory for each tsunami forecast area (listed on the following page). However, this system did not function effectively during the Tohoku-Pacific Ocean Earthquake, and there were cases in which tsunami damage was increased. In response, revision has been made and the following system was put into operation from March 2013. The forecast for tsunami height is made using the size of the earthquake as estimated based on the JMA magnitude M_J. However, in the case of major earthquakes with a scale where the moment magnitude exceeds 8, the estimate for earthquake size using M_J is the minimum assessment. As a result, the forecast for tsunami height is also the minimum assessment in some cases. Therefore, a method for monitoring the size estimate was implemented. When a minimum assessment is detected, the maximum size predicted for the ocean area where the earthquake occurred is used. Furthermore, in addition to conventional GPS wave meters, measurement values taken by cable seabed pressure gauges are now also used. Initially, by using tsunami warnings (listed below) which contain qualitative expressions such as "major," a state of emergency is conveyed regarding the height of tsunamis associated with major earthquakes which are casted as described above. Afterwards, once the size of the earthquake has been accurately calculated, the tsunami warning is updated and a numerical value (quantitative value) for the tsunami height is announced. When a tsunami warning or tsunami advisory is announced, in addition to the height of the tsunami, other tsunami information such as the predicted arrival time, time of high tide, and tsunami observation information is also provided.

Types of tsunami warnings and tsunami advisories

<table>
<tr><th rowspan="2">Category</th><th rowspan="2">Indication</th><th colspan="2">Estimated max. tsunami height</th><th rowspan="2">Expected damage and action to be taken</th></tr>
<tr><th>Quantitative expression (Height prediction category of tsunami)</th><th>Qualitative expression</th></tr>
<tr><td rowspan="3">Major tsunami warning</td><td rowspan="3">Tsunami height is expected to be greater than 3 meters.</td><td>Over 10 m
(10 m $<$ predicted height)</td><td rowspan="3">Huge</td><td rowspan="3">Wooden structures are expected to be completely destroyed and/or washed away; anybody exposed will be caught in tsunami currents. Evacuate from coastal or river areas immediately to safer places such as high ground or a tsunami evacuation building.</td></tr>
<tr><td>10 m
(5 m $<$ predicted height $\leqq$ 10 m)</td></tr>
<tr><td>5 m
(3 m $<$ predicted height $\leqq$ 5 m)</td></tr>
<tr><td>Tsunami warning</td><td>Tsunami height is expected to be up to 3 meters.</td><td>3 m
(1 m $<$ predicted height $\leqq$ 3 m)</td><td>High</td><td>Tsunami waves will hit, causing damage to low-lying areas. Buildings will be flooded. Anybody exposed will be caught in tsunami currents. Evacuate from coastal or river areas immediately to safer places such as high ground or a tsunami evacuation building.</td></tr>
<tr><td>Tsunami advisory</td><td>Tsunami height is expected to be up to 1 meter.</td><td>1 m
(0.2 m $\leqq$ predicted height $\leqq$ 1 m)</td><td>(None)</td><td>Anybody exposed will be caught in strong tsunami currents in the sea. Fish farming facilities will be washed away and small vessels may capsize. Get out of the water and leave coastal areas immediately.</td></tr>
</table>

Figure 27 Tsunami forecast area

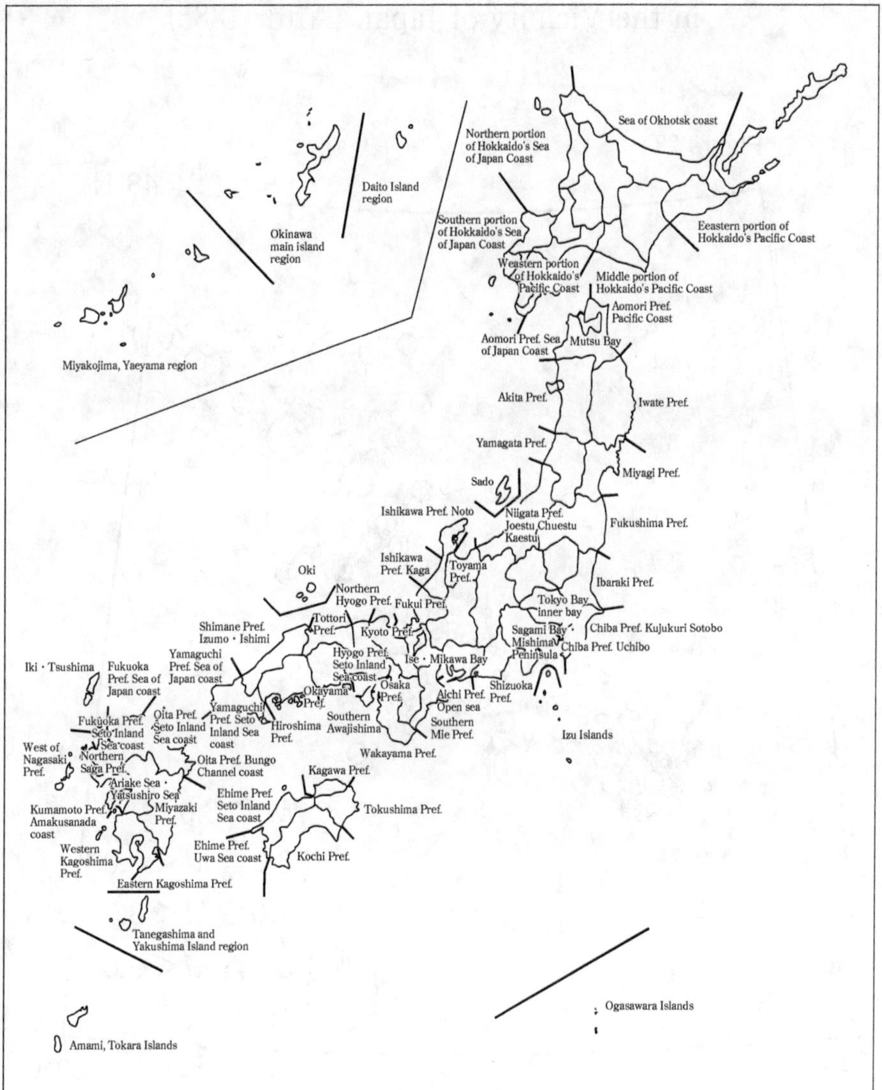

Epicenters of Main Damaging Earthquakes in the Vicinity of Japan (After 1885)

Figure 28

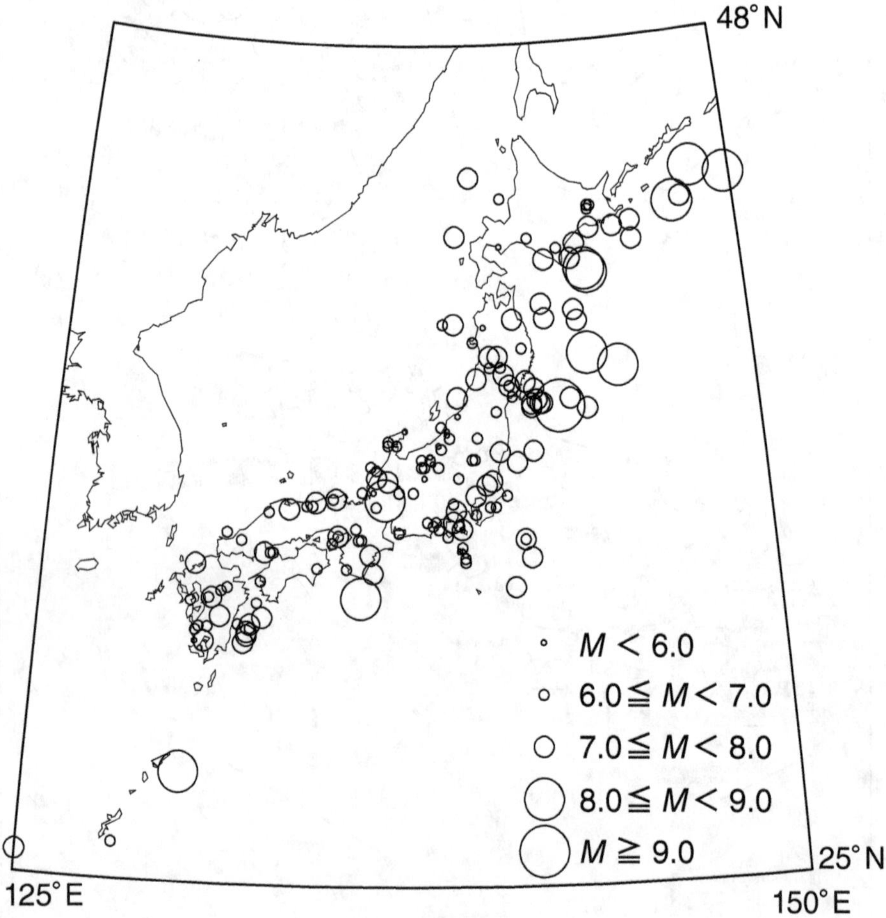

Main Major Earthquakes and Damaging Earthquakes

Starting from the 11th century, we selected earthquakes which had a magnitude (mainly M_s; in rare cases M_s, m_b, etc.) of at least 7.8 or which caused the deaths of at least 1,000 people. The earthquakes were selected from the Table of the World's Damaging Earthquake by Tokuji Utsu. Starting from 2003, earthquakes were selected from Major Earthquakes of the World as listed in this report. Data for earthquakes which occurred long ago was collected from numerous reference literature. It must be noted that, in some cases, further review is required regarding the reliability of this data. The listed date is local time (Universal Time starting from 2003). The Julian calendar is used until 1582 and the Gregorian calendar is used afterwards. For earthquakes in this period, there is a difference ranging from several to ten days. Negative values for latitude and longitude are south latitude and west longitude respectively. In the Region/Damage column, "I" is the maximum seismic intensity equivalent to the revised Mercalli intensity scale, years marked with an "H" are from the Islamic (Hijri) calendar, and "(A)" denotes data taken from additional material. M_w gives priority to USGS. The symbols Mw, Ms, mB, and mb represent M_w, M_s, m_B, and m_b.

Year	Date	Latitude	Longitude	M	Region and Damage
1007		31.°00	47.°40		Iraq: Dijla (Kijla) (Tigris-Ktesiphon); I=8; 10,000 Dead
1008	4	27.68	52.37	6.5	Iran: Siraf; 10,000 Dead (Other Reports) Many Dead (1008/4/11-5/9)
1008	4 27	34.60	47.40	7.0	Iran: Dinavar I=10-11; 16,000 Dead
1038	1 9	38.40	112.90	7.3	Yamanishi [Xinzhou earthquake]; I=10; 32,300 Dead; 10 years of aftershocks (1/24?)
1042	8 21	35.10	38.90	7.2	Syria: Palmyra; 50,000 Dead (first day of 434H)
1042	11 4	38.12	46.33	7.3	Iran: Tabriz; I=10; 40,000 Dead (Other Reports) M7.6/50,000 Dead/I=11?.
1057	3 24	39.70	116.30	6.8	Beijing (south side): I=9; 25,000 Dead (Other Reports) 10,000 Dead/10,000's Dead (5/27?)
1067	11 12	23.60	116.50	6.8	Guangdong [Chaozhou earthquake]; I=9; Extermely high number Dead (Before 11/6, 11/16?)
1068	3 18	28.50	36.70	7.0	Saudi Arabia: Tabuk/Israel: Elat; 20,000 Dead (Other Reports) 25,000 Dead
1068	8 14	38.50	116.50	6.5	Hebei [Hejian earthquake]; I=8; 10,000 Dead (Other Reports) 10,000's Dead (Including Flooding); Heavy Rain
1086		37.00	15.30		Italy: Syracuse; 1,000's Dead
1096	12 11	34.00	137.50	8.3	Eicho 1/11/24: Tokaido, Kinki
1099	2 16	33.00	135.50	8.2	Kowa 1/1/24: Nankaido, Kinki
1101		36.00	59.00	6.5	Iran: Khorasan; I=9-10; 60,000 Dead
1102	1 15	37.70	112.40	6.5	Shanxi; I=8; 1,000 Dead (Other Reports) Many Dead
1114	8 10	37.50	37.80	8.0	Turkey: Antakya (Antioch); Many Dead
1119	2	44.90	124.80	6.8	Heilongjiang; I=9; 7,103 Dead (February-March)
1137	9 19	37.00	38.00	7.4	Iraq/Syria: Jazirah, Mosul; I=10; 230,000 Dead (First day of 552H)
1139		40.30	46.30	6.8	Azerbaijan: Kiapas; 30,000 Dead (Same as 9/30?)
1139	9 30	40.30	46.20	7.7	Azerbaijan: Kirovabad (Ganza); I=9-10; 230,000 Dead; Earthquake lake (9/25?)
1143	4	38.50	106.30	6.5	Ningxia [Yinchuan earthquake]; I=8; 5,000 Dead (Other Reports) 10,000's of humans and animals Dead or Wounded/Many Dead

Main Major Earthquakes and Damaging Earthquakes Continued.

Year	Date	Latitude	Longitude	*M*	Region and Damage
1151		38.70°	-9.10°		Portugal: Lisbon; 1,000's Dead
1154	9 11	14.10	44.10		Yemen: Ibb, San'a, Aden; I=8; 1,345 Dead (Other Reports) 300 Dead/Extremely high number Dead
1155		34.50	36.50		Turkey: Antakya/Syria: Latakia, Damascus; 2,000 Dead (winter)
1157	2 13	36.30	37.30		Syria: Aleppo, Damascus; Extremely high number Dead (Other Reports) 8,000 Dead/80,000 Dead (First day of 532H)
1158		34.50	36.50		Syria: Malatia, Tripoli, Damascus, Aleppo/Lebanon; 20,000 Dead
1168		39.70	39.50	6.0	Turkey: Erzincan; I=9-10; 12,000 Dead (1965-1967?)
1169	2 4	37.50	15.30	6.6	Italy: Catania (Sicily); I=10; 15,000 Dead (Other Reports) 14,000 Dead/16,000 Dead
1170	6 29	35.90	36.40	7.9	Syria to Israel: Latakia, Tripoli, Tyre, Baalbek; Extremely high number Dead
1179	4 29	36.50	44.20	7.1	Iraq: Arbil (Irbil); I=9-10; Extremely high number Dead
1183		34.50	36.50		Turkey: Antakya/Syria: Damascus/Lebanon: Tripoli; 20,000 Dead
1202	5 20	33.50	36.00	7.5	Jordan/Lebanon/Syria/Israel; 1,000's Dead; Dislocation
1208		42.00	60.00	6.1	Turkmenistan/Uzbekistan: Gurgandzhe; I=8-9; 2,000 Dead
1208		36.05	59.22	7.3	Iran: Neyshabur (Nishapur); I=9; 10,000 Dead (605H) (1209?)
1209	12 6	36.00	111.80	6.5	Shanxi [Fushan earthquake]; I=8-9; 2,500 Dead (Other Reports) 2,000-3,000 Dead/1,000's Dead
1219	6 2	35.60	106.20	6.5	Ningxia [Guyuan earthquake]; I=8-9; 10,000's Dead (5/21?)
1222	12 25	45.48	10.23	6.8	Italy: Basso Bresciano; I=9; 2,800 Dead (Other Reports) 12,000 Dead/I=11
1227		43.60	5.30		France: Lambesc; 5,000 Dead (winter)
1248		41.00	13.00		Italy: Sabaudia; 9,000 Dead
1254	4 28	40.20	38.30	7.2	Turkey; I=10; 16,000 Dead (10/11?)
1254	10 11	39.90	39.00		Turkey: West of Erzincan; 15,000 Dead (Other Reports) 56,000 Dead; Dislocation
1268		37.50	35.50		Turkey: Cilicia, Kozan; 60,000 Dead (Other Reports) 50,000 Dead (9/10?)
1268		39.80	40.00	7.2	Turkey: Erzincan; I=9-10; 15,000 Dead
1270	10 7	36.25	58.75	7.1	Iran: Neyshabur; I=10; 10,000 Dead
1290	9 27	41.60	119.30	6.8	Inner Mongolia [Ningcheng earthquake]; I=9; 7220 Dead (Other Reports) 100,000's Dead or Wounded/100,000 Dead
1293	5 20	35.20	139.50	7.5	Einin 1/4/13: Kamakura, echigo? 1,000's Dead (Other Reports) over 30,000 Dead
1303	8 8	34.00	28.00	7.6	Egypt: Faiyum, Alexandria, Acre/Syria/Greece (Crete); 10,000 Dead
1303	9 17	36.30	111.70	8.0	Shanxi [Hongdong-Chengzhi earthquake]; I=11; 200,000 Dead (Other Reports) 475,800 Dead
1303	12	36.10	29.00	8.0	Greece: Crete, Peploponnesus, Rhodes
1304	8 8	36.30	27.30	8.0	Greece: Rhodes, Crete; I=11; (1303/8/8?)
1305	5 3	39.80	113.10	6.5	Shanxi [Datong earthquake]; I=8-9; 2,000 Dead
1306	9 12	35.90	106.10	6.5	Ningxia [Guyuan earthquake]; I=8-9; 5,000 Dead
1328	12 4	42.85	13.02	6.5	Italy: Norcia, Le Preci; I=10; 5,000 Dead (Other Reports) 2,000 Dead/4,000 Dead (12/1?)
1336	10 21	34.49	59.88	7.6	Iran: Jizd, Khwaf; 30,000 Dead (Other Reports) 20,000-30,000 (11,000 Dead from epidemic after quake)
1343		37.43	37.40	7.6	Turkey; I=9; 5,700 Dead (Same as below?)

Continued.

Year	Date	Latitude	Longitude	*M*	Region and Damage
		°	°		
1343	1 1	36.30	37.60		Turkey: Membij; I=8-9; 5,700 Dead
1348	1 25	46.37	13.58	6.9	Austria: Villach/Italy/Slovenia; I=9; 10,000 Dead (Other Reports) 5,000 Dead/40,000 Dead
1349	9 9	41.48	14.07	6.7	Italy: Sant'Elia Fiumerapido; I=10; 2,500 Dead (Other Reports) 800 Dead/1,000's Dead
1352	12 25	43.48	12.15	5.7	Italy: San Sepolcro; I=8-9; 2,000 Dead (Other Reports) 3,000 Dead/I=10-11
1361	7 7	41.23	15.45	6.3	Italy: Ascoli Satriano; I=10; 4,000 Dead
1361	7 26	33.00	135.00	8.4	Shohei 16/6/24; Kinki (middle south), Shikoku; Many Dead
1389	2 6	36.25	58.75	7.3	Iran: Neyshabur; I=10; Extremely high number Dead (Only 100 Survivors?; 1/30-2/27?)
1405	11 23	36.25	58.75	7.2	Iran: Neyshabur; I=10; 30,000 Dead
1411	9 29	30.10	90.50	8.0	Xizang; I=10-11; Many Dead; Dislocation; Landslide; Quake before the night of 9/28 (early morning)
1440		28.42	53.08	6.9	Iran: Karzin, Qir; 10,000 Dead (844H)
1444		38.60	42.20		Turkey: Nemrut Mountains; 30,000 Dead (poisonous fumes?)
1456	12 5	41.30	14.72	7.1	Italy: Campania, Basilicata; I=11; 30,000 Dead (Other Reports) 12,000-100,000 Dead
1457		31.20	34.20		Syria/Israel/Jordan? 32,000 Dead
1458	4	39.90	40.40	7.6	Turkey: Erzincan, Erzurum; I=10; 32,000 Dead (Other Reports) 30,000 Dead
1471		-16.30	-71.00	8.0	Peru; I=8-9
1481		30.00	30.10		Egypt: Cairo/Palestine/Israel/Syria; I=7; 30,000 Dead (Same as below?)
1481	3	39.90	40.40	7.7	Turkey: Erzurum; I=10-11; 3,000 Dead
1482		39.90	40.40	7.5	Turkey: Ergincan, Erzurum; I=10; 30,000 Dead
1487	8 10	34.40	108.90	6.3	Shaanxi [Lintong earthquake]; I=8; 1,900 Dead
1491	10	38.30	26.00		Greece: Kos (Cos); 5,000 Dead
1498		37.18	55.10	6.5	Iran: Old Gorgan, Jorjan; I=9-10; 1,000 Dead (903H)
1498	9 11	34.00	138.00	8.3	Meio 7/8/25: Tokadou · Kii, etc.; 10,000's Dead
1500	1 4	24.90	103.10	7.0	Yunnan [Yiliang earthquake]; I>=9; 10,000's Dead (Other Reports) M>=7.0
1505	7	634.70	69.20		Afghanistan: Kabul, Bala, Hicar, Pangan; I=9-10; Extremely high number Dead
1509	9 14	41.00	28.80	7.4	Turkey: Tsurlu, Istanbul; I=10-11; 13,000 Dead (Other Reports) 1,000 Dead; Mw7.2 (9/10?)
1513		-17.20	-72.30	8.7	Peru I=8
1515	6 17	26.60	100.70	7.8	Yunnan [Heqing earthquake]; I=10; 1,000's Dead (Other Reports) 100's Dead?
1522	10 22	38.00	-25.00		Azores: S. Miguel; I=10; 5,000 Dead
1531	1 26	39.00	-8.00		Portugal: Lisbon/Spain/Morocco; 30,000 Dead (Other Reports) Many Dead/ 1,000 Dead
1543		-19.00	-70.50	8.0	Chile; I=10
1549	2 16	33.70	60.00	6.7	Iran: Qayin; 3,000 Dead
1551	1 28	38.40	-9.10		Portugal: Lisbon; I=9; 2,000 Dead
1556	1 23	34.50	109.70	8.3	Shaanxi [Shaanxi earthquake]; I=11; 830,000 Dead; Names of the Dead are known
1561	7 25	37.50	106.20	7.3	Ningxia [Zhongning earthquake]; I=9-10; 5,000 Dead (Other Reports) 1,000 Dead/Many Dead

Main Major Earthquakes and Damaging Earthquakes Continued.

Year Date	Latitude	Longitude	M	Region and Damage
1562 10 28	-38.70°	-73.15°	8.0	Chile: Santiago, Arauco; I=11; Many Dead
1566	3.00	-77.30	7.8	Colombia; I=7
1570 2 8	-36.75	-73.00	8.3	Chile: Concepcion; I=11; Fatalities reported; (Other Reports) over 2,000 Dead
1574	34.00	51.40		Iran: Fin, Kashan; 1,200 Dead; (autumn) (982H)
1575 12 16	-39.80	-73.20	8.5	Chile: Valdivia, Santiago; I=10; 1,500 Dead (Other Reports) 120 Dead
1582 1 22	-16.60	-71.60	8.2	Peru: Arequipa; I=10; Many Dead (Other Reports) 30 Dead/40 Dead, Mw7.5
1582 8 15	-12.20	-77.60	7.8	Peru; I=7
1584 6 17	40.00	39.00	6.6	Turkey: Erzincan; I=9; 15,000 Dead (Death Toll Uncertain)
1586 1 18	36.00	136.90	7.8	Tensho 13/11/29: Chubu, Kinki; 1,000 Dead (Other Reports) 35.3°/136.6°/M7.9
1586 7 10	-12.30	-77.70	8.1	Peru: Lima, Callao; I=10; Many Dead (Other Reports) 20 Dead; Strong Foreshocks; Mw8.1
1588 8 9	24.00	102.80	7.0	Yunnan [Tonghai earthquake]; I>=9; 1,000 Dead (Other Reports) Many Dead
1593 9	27.70	54.30	6.5	Iran: Lar; 3,000 Dead (1001H summer evening)
1596 9 5	34.65	135.60	7.5	Keicho 1/7/13: Kinki (Fushimi, Osaka, Sakai, Hyogo) 1,000's Dead (Other Reports) 34.9°, 135.65°
1600 2 19	-16.80	-70.90	7.9	Peru: Arequipa, Volcan Huaynaputina; I=11
1604 11 24	-17.90	-70.90	8.4	Peru: Arequipa, Camana/Chile: Arica; I=10-11; Many Dead; Mw8.7
1605 2 3	33.50	138.50	7.9	Keicho 9/12/16: Tokai, Nankai, Saikai-shodo [Keisho earthquake] Tunami Earthquake? 1,000's Dead
1605 7 13	20.00	110.50	7.5	Hainan [Qiongshan earthquake]; I=11; 1,000's Dead
1606 11 30	23.60	102.80	6.3	Yunnan [Jianshui earthquake]; I=9; 1,000's Dead
1611 9 27	37.60	139.80	6.9	Keicho 16/8/21: Aizu; 3,700 Dead; Earthquake Lake
1611 12 2	39.00	144.00	8.1	Keicho 16/10/28: Sanriku, Hokkado (eastern coast); 5,000 Dead (Other Reports) 1,783 Dead/3,000 Dead
1618 5 26	18.90	72.90	6.9	India: Bombay (earthquake + rain storm?); I=9?. 2,000 Dead
1618 8 25	46.30	9.50		Italy-Switzerland border: Piuro; landslide; 1,200 Dead
1619 11 30	18.50	121.60	8.0	Philippines (Luzon): Cagayan, Isabela; I=10; Many Dead
1622 10 25	36.50	106.30	7.0	Ningxia [Guyuan earthquake]; I=9-10; 12,000 Dead (Other Reports) 6,000 Dead/15,000 Dead
1626 6 28	39.40	114.20	7.0	Shanxi [Lingqiu earthquake]; I=9; 5,200 Dead (Other Reports) 10,000's Dead? (5,200 in Lingqiu alone)
1627 7 30	41.73	15.35	6.8	Italy: San Severo, Lesina, Apricina; I=10-12; 5,000 Dead
1638 3 27	39.05	16.28	6.9	Italy: Calabria; I=11; 30,000 Dead (Other Reports) 1,580 Dead/9,571 Dead/19,000 Dead
1640 2	-1.70	-78.60		Ecuador; 5,000 Dead
1641 2 5	37.90	46.10	6.8	Iran: Tabriz, Dehkharqan; I=9; 12,613 Dead (Other Reports) 1,200 Dead/30,000 Dead
1645 11 30	15.60	121.20	7.9	Philippines (Luzon): Manila, Cagayan; 600 Dead (Other Reports) 500 Dead/3,000 Dead
1647 5 13	-14.00	-76.50	7.9	Peru
1647 5 14	-33.40	-70.60	8.5	Chile: Santiago; I=11; 2,000 Dead (Other Reports) 1,000 Dead
1648 4 1	38.30	43.50	6.7	Turkey: Van, Hayotsdzor; I=9-10; 2,000 Dead

Continued.

Year	Date	Latitude	Longitude	*M*	Region and Damage
		°	°		
1650	3 31	-13.50	-71.70	8.1	Peru: Cuzco, Lima; I=10
1652	7 13	25.20	100.60	7.0	Yunnan [Midu earthquake]; I=9 strong; 3,000 Dead
1653	2 23	38.30	27.10		Turkey: Izmir; I=10; 15,000 Dead (Other Reports) 2,500 Dead/3,000 Dead/8,000 Dead
1654	7 21	34.30	105.50	8.0	Gansu [Tianshui earthquake]; I=11; 31,000 Dead (Other Reports) 12,000 Dead (night)
1657	3 15	-36.83	-73.03	8.0	Chile: Concepcion, Chillan, Santiago; I=11; Many Dead
1659	11 5	38.70	16.25	6.4	Italy: Panaia, Polia (C.Calabria); I=10; 2,035 Dead
1660		40.00	41.30	6.5	Turkey: Erzurum; I=9; 1,500 Dead
1664	5 12	-14.10	-75.85	7.8	Peru: Ica, Piseo; I=10-11; 400 Dead (Other Reports) 15 Dead/300 Dead, Mw7.5
1666	2 1	37.10	138.20	6.8	Kanbun 5/12/27: western Niigata pref. [Takada earthquake]; Fire; 1,500 Dead (Other Reports) 138.3°
1667				6.9	Iran: Shirvan, Shamkha; 12,000 Dead (Other Reports) 8,000 Dead (Same as 11/18?)
1667	4 6	42.60	18.10	7.2	Croatia: Dubrovnik (Rugusa); I=10; 5,000 Dead
1668	1 4	40.50	48.50	7.0	Azerbaijan: Shemakha; I=11; 80,000 Dead (Same as below?, last year 12/17?)
1668	1 14	41.00	48.00	7.8	Azerbaijan; I=10; 80,000 Dead
1668	7 25	34.80	118.50	8.5	Santo; I>=11; 47,615 Dead (Other Reports) over 33,000 Dead/50,000 Dead
1668	8 17	40.50	35.00	8.0	Turkey: Anatolia (Bolu to Erzincan); 8,000 Dead (Other Reports) 18,000 Dead; Dislocation
1669	1 14	40.60	48.60	5.7	Azerbaijan: Shemakha; I=9; 7,000 Dead (Other Reports) 6,000-7,000 Dead (1668 A.D.?)
1673	7 30	36.35	59.27	7.1	Iran: Meshed, Neyshabur (Khorasan) 5,600 Dead (Other Reports) 4,000 Dead
1674	2 12	-3.50	128.20		Indonesia (Moluccas): Amboina; 2,342 Dead (Other Reports) 79 Dead from Quake/2,244 Dead from Flooding
1677	4 13	41.00	143.00	7.9	Enpo 5/3/12: Sanriku; Fatalities Reported (Similar to Tokachi 1968 A.D.?)
1677	11 4	35.00	141.50	8.0	Enpo 5/10/9: Fukushima prefecture · Kanto (eastern coast) · Hachijiojima Tsunami; Earthquake; 540 Dead
1678	6 18	-12.30	-77.80	7.9	Peru: Lima, Salinas-Huaura, Callao, Chancay; I=9; Fatalities Reported; Mw7.9
1679	6 4	40.20	44.70	6.4	Armenia: Garnii, Yerevan, Dvina; I=9-10; 7,600 Dead
1679	9 2	40.00	117.00	8.0	Hebei [Sanhe-Pinggu earthquake]; I=11; 45,500 Dead (Other Reports) 10,000's Dead; Dislocation
1680	9 9	25.00	101.60	6.8	Yunnan [Chuxiong earthquake]; I=9; 2,700 Dead
1683	11 22	38.70	112.70	7.0	Shanxi [Yuanping earthquake]; I=9; 8,220 Dead (Other Reports) at least 1,000 Dead/Many Dead
1687	10 20	-15.20	-75.90	8.2	Peru: Lima, Callao, Ica; I=10; 5,000 Dead (Other Reports) 200 Dead/600 Dead; (10/21?); Mw8.4
1688	6 5	41.28	14.57	6.6	Italy: Campania; I=11; 10,000 Dead (Other Reports) 3,311 Dead/8,000 Dead
1688	7 10	38.40	26.90	7.0	Turkey: Izmir (Smyrna); I=10; 17,500 Dead
1692	6 7	17.80	-76.70		Jamaica: [Port Royal EQ]; 3,000 Dead (Other Reports) 2,000 Dead
1693	1 11	37.13	15.02	7.4	Italy: Sicily [Catania EQ]; I=11; 54,000 Dead (Other Reports) 18,000 Dead/93,000 Dead
1693	4 25	23.00	115.30	7.8	Guangdong; I=6
1694	9 8	40.87	15.40	6.8	Italy: Basilicata; I=10; Death 4,820 (Other Reports) 4,057 Dead/6,500 Dead/15,000 Dead/I=11

Main Major Earthquakes and Damaging Earthquakes Continued.

Year	Date	Latitude	Longitude	*M*	Region and Damage
1695	5 18	36.00°	111.50°	7.8	Shanxi [Linfen earthquake]; I=10; 52,600 Dead (Other Reports) 30,000 Dead/10,000's Dead/100,000 Dead?
1697	2 25	16.70	-99.20	7.8	Mexico: Acapulco, San Marcos, Mexico City (2/25-26)
1699	7 14	-11.80	-77.50	7.8	Peru: Lima
1700	1 26			9.0	USA (Oregon/Washington) Cascadia; Subduction Zone
1703	1 14	42.70	13.07	6.7	Italy: Norcia, L'Aquila; I=11; 9,761 Dead (Other Reports) 10,000 Dead/40,000 Dead
1703	12 31	34.70	139.80	8.1	Genroku 16/11/23: southern Kanto [Genroku earthquake]; Fire; 10,000 Dead
1706	11 3	42.08	14.08	6.7	Italy: Maiella; I=10-11; 2,400 Dead (Other Reports) 1,000 Dead/15,000 Dead
1707		40.40	43.00		Turkey: Kars; 1,000's Dead (Other Reports) Many Dead
1707	10 28	33.00	136.00	8.6	Hoei 4/10/4: western Japan [Hoei earthquake]; Largest in Japanese History; 5,000 Dead (Other Reports) 20,000 Dead?
1709	10 14	37.40	105.30	7.5	Ningxia [Zhongwei earthquake]; I=9-10; 2,000 Dead
1713	2 26	25.60	103.30	6.8	Yunnan [Xundian earthquake]; I=9; 1,000's Dead (Other Reports) 2,100 Dead/2,300 Dead
1715	5	36.50	2.60		Algeria: Algiers; 20,000 Dead (During 1715-1717 A.D. same thing, 20,000 Dead, happened 5 times?)
1715	8 5	36.50	2.60		Algeria: Algiers; 20,000 Dead
1716	2 3	36.90	2.90		Algeria: Algiers; I=10; 20,000 Dead (Damage includes the one in May)
1716	2 6	-17.20	-71.20	8.8	Peru: Torata; I=9; Many Dead
1716	2 11	-13.70	-76.05	8.6	Peru; I=9-10
1716	5	36.50	2.60		Algeria: Algiers; 20,000 Dead
1717	8 5	36.90	2.90		Algeria: Algiers; 20,000 Dead
1718	6 19	35.00	105.20	7.5	Gansu [Tongwei-Gangu earthquake]; I=10; 75,000 Dead (Other Reports) 41,000's Dead
1719	5 25	40.70	29.80	7.0	Turkey: Istanbul, Izmit; I=9-10; 1,000 Dead
1721	4 26	37.94	46.66	7.4	Iran: Tabriz; I=11; 40,000 Dead (Other Reports) 8,000 Dead/10,000 Dead/100,000 Dead/250,000 Dead; Dislocations
1721	9	23.00	120.20	6.0	Taiwan (Tainan); I=8; 1,000's Dead (1/5?)
1725	1 7	-9.20	-79.30	7.5	Peru: Trujillo, Ancash; I=9; 1,500 Dead from Mudslides (1/25?) Mw7.5
1725	2 1	56.50	118.50	8.2	Russia (Baikal region): Velikoe; I=11
1727	11 18	38.00	46.20	7.2	Iran: Tabriz; I=8; 77,000 Dead (same as 1721/4/26?)
1730	7 8	-33.05	-71.63	8.7	Chile: Valparaiso, Santiago, Concepcion; I=11; Fatalities Reported
1732	3 27	40.90	14.80		Italy: Avellino; 2,000 Dead (same as 11/29?)
1732	11 29	41.08	15.05	6.6	Italy: Irpinia; I=10-11; 1,942 Dead (Other Reports) 600 Dead/1,500 Dead
1733	8 2	26.30	103.10	7.8	Yunnan [Dongchuan earthquake]; I=10; 1,200 Dead (Other Reports) Dozens Dead/Many Dead
1737	10 17	51.10	158.00	8.3	Russia: Kamchatka; I=10; Fatalities Reported; 25-50 m Giant Tsunami
1737	11 4	55.50	163.00	7.8	Russia: Kamchatka; I=10
1739	1 3	38.80	106.50	8.0	Ningxia [Pingluo-Yinchuan earthquake]; I=10 strong; 50,000 Dead; Dislocation
1741	8 29	41.60	139.40		Kanpo 1/7/19: Oshima Peninsula etc. [Oshima tsunami]; 2,000 Dead; Korea Tsunami (Oshima eruption)

Continued.

Year	Date	Latitude	Longitude	M	Region and Damage
1746	10 29	-12.00°	-77.20°	8.4	Peru: Lima, Callao; I=10; 18,000 Dead (Other Reports) 4,800 Dead/4,941 Dead/7,141 Dead, Mw 8.6
1749	3 25	39.50	-0.40		Spain: Coast of Valencia Prov.; I=10; 5,000 Dead
1750	6 7	36.30	22.80	7.0	Greece: Kythera, Morea, Cerigo; I=10; 2,000 Dead
1751	3 25	-36.90	-73.00	8.5	Chile: Concepcion; 25 Dead (same as 5/25?, M excessive?)
1751	5 21	37.10	138.20	7.2	Houreki 1/4/26: Niigata, northren Nagano [Takada earthquake] 1,541 Dead (Other Reports) 37.3°, 138.25°
1751	5 25	-36.80	-71.60	8.5	Chile: Concepcion, Chillan, Talca, and Tutuben; I=9-11; 65 Dead
1751	5 25	26.50	99.90	6.8	Yunnan [Jianchuan earthquake]; I=9; 1,050 Dead (Other Reports) 940 Dead/3,000 Dead?
1752	7 21	35.50	35.50	7.0	Syria: Latakia (Laodicea)/Lebanon: Tripoli; 20,000 Dead
1754	9	30.00	32.00		Egypt: Grand Cairo [Tanta EQ]; 1,000's Dead (Other Reports) 40,000 Dead
1755	6 7	33.97	51.35	5.9	Iran: Kashan, Tabriz; 1,200 Dead (Other Reports) 40,000 Dead (daytime)
1755	11 1	36.70	-10.00	8.5	Portugal/Spain/Morocco [Lisbon EQ]; 62,000 Dead (Other Reports) 55,000 Dead, Mw 8.5
1755	11 19	34.10	-5.30		Morocco: Mequinez (Mekenes), Fes; I=10; 3,000 Dead
1757	2 22	-0.93	-78.61	7.0	Ecuador: Latacunga, Cotopaxi, Tungurahua; I=9; 1,000 Dead
1757	4 15	34.00	-6.50		Morocco: Sale, Cap Cantin; 3,000 Dead
1759	10 30	33.10	35.60	6.6	Syria: Baalbek, Damascus/Israel: Safad; 2,000 Dead (Other Reports) 20,000 Dead
1759	11 25	33.70	35.90	7.4	Lebanon/Syria: Baalbec/Israel: Safad; 30,000 Dead (Other Reports) 10,000-40,000 Dead; Dislocation
1763	12 30	24.20	102.80	6.5	Yunnan [Tonghai earthquake]; I=8; 1,000 Dead (800 Dead Swept out to Sea, Many other Dead)
1765	9 2	34.80	105.00	6.5	Gansu [Wushan-Gangu earthquake]; I=8 strong; 2,068 Dead (Other Reports) 1,189 Dead
1766	3 8	40.73	140.59	7.3	Meiwa 3/1/28: Aomori, Hirosaki [Tsugaru earthquake]; Fire; 1,335 Dead
1769	8 29	33.00	132.10	7.8	Meiwa 6/7/28: Oita Prefecture, Miyazaki Prefecture, Kumamoto Prefecture; Fatalities Reported
1771	4 24	24.00	124.30	7.4	Syobokuo 20 (Meiwa 8)/3/10: Yaeyama/Miyako Islands [Yaeyama tsunami] 12,000 Dead
1773	6 3	14.60	-91.20		Guatemala: St. Jago (Santiago); 20,000 Dead (Other Reports) 5,000-8,000 Families Dead
1778	12 15	33.97	51.35	6.2	Iran: Kashan; 8,000 Dead (Other Reports) 30,000 Dead
1780		34.00	58.00	6.5	Iran: Khurasan; 3,000 Dead (1194H)
1780	1 8	38.12	46.29	7.4	Iran: Tabriz; I=11; 50,000 Dead (Other Reports) 200,000 Dead; Dislocation; 2/12 and over 20 Aftershock Damage
1783	2 5	38.38	15.97	6.9	Italy: [Calabria EQ]; I=11; Fire; 35,000 Dead (Other Reports) 29,500 Dead/40,000 Dead/50,000 Dead
1784	5 13	-16.50	-72.00	8.0	Peru: Arequipa, Camana; I=10; 400 Dead (Other Reports) 54 Dead/300 Dead, Mw8.4
1784	7 18	39.50	39.70	7.6	Turkey: Erzincan, Erzurum; 5,000 Dead (Other Reports) 5,000 Wounded; Dislocation (7/23?)
1786	6 1	29.90	102.00	7.8	Sichuan [Kangding-Luding earthquake]; I>=10; River blockage; 446 Dead (Other Reports) 442 + Many Dead

Main Major Earthquakes and Damaging Earthquakes Continued.

Year	Date	Latitude	Longitude	M	Region and Damage
1786	6 10	29.40°	102.20°	7.0	Sichuan (Luding); Partial Collapse of Dam across Large River; 10,000's Dead from Flooding; M>=7.0
1787		19.00	-66.00	8.0	Puerto Rico; (5/12?)
1787	3 28	16.50	-98.50	8.2	Mexico: San Marcos, Teotitlan, etc.; 11 Dead (3/28-4/3)
1789	5 29	39.00	40.00	7.0	Turkey: Palu; I=8-9; 51,000 Dead
1789	6 7	24.20	102.90	7.0	Yunnan (Huaning); I=9 strong; 1,000 Dead (Other Reports) 300 Dead (Swept out to Sea) + Many Other Dead
1790	10 9	35.70	0.60		Algeria: Oran; I=10; 3,000 Dead
1792	5 21	32.80	130.30	6.4	Kansei 4/4/1: Shimabara Peninsula [Shimabara-taihen (Higo-meiwaku)]; 15,000 Dead; Bisan Collapse
1792	8 22	54.00	162.00	8.4	Russia: Kamchatka; I=11
1793	2 17	38.50	144.50	8.2	Kansei 5/1/7: Miyagi Prefecture, Iwate Prefecture, Fukushima Prefecture; 39 Dead (Other Reports) 720 Dead
1796	4 26	35.70	36.00	6.6	Syria: Latakia (Ladikije); I=8; 1,500 Dead
1797	2 4	-1.67	-78.64	8.3	Ecuador: Riobamba, Quito; 40,000 Dead (Other Reports) 6,000 Dead/6,406 Dead/16,000 Dead
1797	12 4	10.50	-64.50		Venezuela: Cumana, Cariaco; 16,000 Dead (Damage includes 12/14?)
1799	8 27	23.80	102.40	7.0	Yunnan [Shiping earthquake]; I=9; 2,251 Dead (Other Reports) 2,030 Dead
1805	7 26	41.50	14.47	6.6	Italy: Molise; I=10-11; 5,573 Dead (Other Reports) 6,000 Dead/6,573 Dead
1810	2 16	35.65	25.00	7.8	Greece: Iraklion (Crete); I=10; 2,000 Dead
1811	12 16	36.60	-89.60	8.0	USA. (Missouri): [New Madrid EQ]; I=12; 1 Dead (1st time)
1812	1 23	36.60	-89.60	8.0	USA: [New Madrid EQ]; I=12; (2nd time)
1812	2 7	36.60	-89.60	8.0	USA: [New Madrid EQ]; I=12; Dislocation (secondary) (3rd time)
1812	3 8	43.70	83.50	8.0	Xinjiang [Yining earthquake]; I=11; Many Dead (Other Reports) 58 Dead; Dislocation
1812	3 26	10.60	-66.90	6.3	Venezuela: Caracas; I=9; 20,000 Dead (Other Reports) 10,000 Dead/18,000 Dead/26,000 Dead/40,000 Dead
1815	10 23	34.80	111.20	6.8	Shanxi [Pinglu earthquake]; I=9; 13,090 Dead (Other Reports) <30,000 Dead/37,000 Dead (10/21?)
1815	11 27	-8.00	115.20		Indonesia (Bali); 10,253 Dead
1816	12 8	31.40	100.70	7.5	Sichuan [Luhuo earthquake]; I=10; 2,945 Dead (Other Reports) 2,854 Dead
1819	4 11	-27.35	-70.35	8.3	Chile: Copiapo; I=10
1819	6 16	23.30	70.00	8.3	India: [Cutch EQ]; 1,440 Dead (Other Reports) 1,543 Dead/2,000 Dead/3,000 Dead; Dislocations; Mw7.5
1821	7 10	-16.40	-72.30	8.2	Peru: Camana, Ocona, Caraveli, Arequipa; I=7; 162 Dead
1822	8 13	36.70	36.90	7.4	Syria: Aleppo/Turkey: Antakya; I=10; 20,000 Dead (Other Reports) 30,000-60,000 Dead; Dislocations
1822	9 5	35.00	36.00		Syria: Aleppo, Damascus/Turkey; 22,000 Dead (Damage includes 8/13?)
1822	11 20	-33.10	-71.60	8.5	Chile: Copiapo to Vladivia; I=11; Many Dead; Coastal Upheaval
1825	3 2	36.30	2.50		Algeria: Blida; I=10; 7,000 Dead (Other Reports) 5,000 Dead
1826		-45.80	166.50	8.0	New Zealand: Fjordland (Extermely Large M?)
1827	9 24	31.60	74.30		India: Punjab/Pakistan: Lahore; I=8-9; 1,000 Dead (9/26?)

Continued.

Year	Date	Latitude	Longitude	*M*	Region and Damage
1828	6 6	34.20°	74.50°		India (Kashmir): Srinagar; I=9–10; 1,000 Dead
1828	8 9	40.70	48.40	5.7	Azerbaijan: Shemakha; I=8; 8,000 Dead (Other Reports) I=10
1828	12 18	37.60	138.90	6.9	Bunsei 11/11/12: Niigata Prefecture [Sanjo earthquake]: Fire; 1,681 Dead (Other Reports) M7.2–7.3
1830	6 12	36.40	114.30	7.5	Hebei [Cixian earthquake]; I=10; 1,000's Dead (5,485 in Cixian) (Other Reports) 7,000 Dead/10,000 Dead?
1832	1 22	36.50	71.00	7.4	Afghanistan/Tajikistan/Pakistan: Hindu Kush; I=9; Extremely high number Dead
1833	8 26	28.30	85.50	8.0	Xizang/Nepal; I>=10; Many Dead (Other Reports) 414 Dead (Nepal)
1833	9 6	25.00	103.00	8.0	Yunnan [Songming earthquake]; I>=10; 6,707 Dead (Other Reports) 67,000 Dead
1833	11 24	-3.50	100.00	8.7	Indonesia (Sumatra): Padang, Benkelen, Indrapura
1835	2 20	-36.80	-73.00	8.1	Chile: Concepcion, Valparaiso; I=11; Hawaii Tsunami; Fatalities Reported
1837	1 1	33.00	35.50	7.0	Israel: Safad/Syria/Lebanon; 5,700 Dead (Other Reports) 2,000–3,000 Dead/5,000 Dead
1837	11 7	-39.80	-73.20	8.0	Chile: Valdivia, Concepcion; I=10; 60 Dead
1838		32.00	34.50		Israel: Jaffa; 3,000 Dead
1840	6 20	39.50	43.00		Turkey: Mt. Ararat; 1,500 Dead (Damage is the one on 7/2?)
1840	7 2	39.50	43.90	7.4	Turkey: Balikgolu/Armenia: Nakhichevan; 1,000 Dead (Other Reports) 2,063 Dead
1841	5 17	52.50	159.50	8.4	Russia: Kamchatka; I=10–11; Hawaii Tsunami
1842	5 7	19.70	-72.80		Dominica: St. Domingo, Santiago/Haiti; 4,500 Dead (Other Reports) 200 Dead/3,000 Dead
1843	2 8	16.50	-61.00	7.8	Lesser Antilles: Guadaloupe, Antigua, Montserrat; 5,000 Dead
1843	4 18	38.60	44.80	5.9	Iran: Khoy; 1,000 Dead
1843	4 25	43.00	147.00	8.0	Tenpo 14/3/26: Kushiro・Nemuro; 46 Dead (According to Russ., Minami-Chishima earthquake M8.2)
1844	5 12	33.62	51.36	6.4	Iran: Qohrud, Kashan; 1,500 Dead (5/10, 5/11?)
1844	5 13	37.50	47.97	6.9	Iran: Miyaneh, Garmrud; I=9; Extremely high number Dead
1845	4 7	16.60	-99.20	7.9	Mexico: Acapulco, Puebla, Tlaxacalla, Veracruz
1846	6 28	-14.00	-76.80	7.8	Peru; I=6
1847	5 8	36.70	138.20	7.4	Kouka 4/3/24: Northern Shinano [Zenkoji earthquake]; 8,174 Dead (Other Reports) 8,304 Dead; Dislocations/Flooding
1848	12 3	24.10	120.50	6.8	Taiwan [Changhua earthquake]; I=9; 2,021 Dead (Other Reports) 1,000 Dead/1,200 Dead
1850	9 12	27.80	102.30	7.5	Sichuan [Xichang earthquake]; I=10; 23,860 Dead (Other Reports) 20,652 Dead
1851	6	36.78	58.50	6.9	Iran: Sarvelayat, Quchan, Ma'dan; I=9; 2,000 Dead (Again in 1852/2?)
1851	8 14	40.95	15.67	6.3	Italy: Melfi; I=10; 1,000 Dead (Other Reports) 62 Dead/671 Dead/700 Dead/14,000 Dead?
1851	10 12	40.70	19.70	6.6	Albania: Vlore (Avlona), Berat, Elbasan; I=10–11; 2,000 Dead
1852	2 22	37.10	58.40	5.8	Iran: Quchan; I=9; 2,000 Dead (Same as 1851/6?)
1853	5 5	29.60	52.50	6.2	Iran: Shiraz; 9,000 Dead (Other Reports) 12,000 Dead (Damage includes foreshocks and aftershocks)
1854	4 16	13.80	-89.20	6.6	El Salvador: San Salvador; 1,000 Dead (Foreshocks on 4/14?)

Main Major Earthquakes and Damaging Earthquakes Continued.

Year	Date	Latitude	Longitude	M	Region and Damage
1854	7 9	34.75°	136.00°	7.3	Ansei 1/6/15: Iga, Ise, Yamato [Igaueno earthquake]; 1,600 Dead
1854	12 23	34.00	137.80	8.4	Ansei 1/11/4: Tokai, Nankaido [Ansei-Tokai earthquake]; Fire; 2,000 Dead
1854	12 24	33.00	135.00	8.4	Ansei 1/11/5: Kinai, Nankaido[Ansei-Nankai earthquake]; Fire; 1,000's Dead
1855	1 23	-41.00	175.00	8.0	New Zealand: [Wellington EQ]; Coastal Upheavals; Dislocations
1855	2 28	40.10	28.60	7.1	Turkey: Bursa, Tayabas; I=10; 1,900 Dead (Other Reports) 300 Dead/700 Dead/Small number Dead
1855	4 29	40.20	29.10	6.7	Turkey: Bursa; 1,300 Dead (Same as 4/11?)
1855	11 11	35.65	139.80	6.9	Ansei 2/10/2: Edo [Ansei-Edo earthquake] 7,444 Dead (Other Reports) 10,000 Dead?
1856	3 2	3.50	125.50		Indonesia: Great Sangir (Volcanic); 3,000 Dead
1856	10 12	35.50	26.00	8.3	Greece: Crete, Rhodes/Egypt; I=11; 20 Dead (Other Reports) 10 Dead/538 Dead
1857	12 16	40.35	15.85	7.0	Italy: Basilicata; I=11; 10,939 Dead (Other Reports) 13,488 Dead/19,000 Dead/24,000 Dead?
1859	3 22	-0.30	-78.50	6.3	Ecuador: Quito, Pichincha; I=8; 5,000 Dead
1859	6 2	40.00	41.00	6.4	Turkey: Erzurum; I=9-10; 2,000 Dead (Other Reports) 15,000 Dead
1861	2 16	0.50	97.50	8.4	Indonesia (Sumatra): Lagundi, Padang, Batu I., Nias I.; Extremely high number Dead
1861	3 20	-32.90	-68.90	7.0	Argentina: Mendoza; I=9; 18,000 Dead (Other Reports) 7,000 Dead
1862	6 7	23.30	120.20	6.8	Taiwan [Chaiyi earthquake]; I=9; 1,000's Dead (Other Reports) 500 Dead
1863	4 22	36.40	27.70	7.8	Greece: Rhodes, Kos (Cos); I=10-11; Fatalities Reported
1863	12 30	38.13	48.52	6.1	Iran: Ardabil, Nir; I=8-9; 1,000 Dead (Other Reports) 500 Dead
1868	4 2	19.00	-155.50	8.0	USA: South coast of Hawaii [Kau EQ]; I=10; 81 Dead (Other Reports) 77 Dead
1868	8 13	-18.50	-71.00	8.5	Chile/Peru; I=11; 25,000 Dead (Other Reports) 2,000 Dead/11,500 Dead/40,000 Dead; Mw8.8
1868	8 16	0.31	-78.18	7.7	Ecuador/Colombia: Quayaquil, Ibarra; I=11; 70,000 Dead (Other Reports) 40,000 Dead
1870	4 11	30.00	99.10	7.3	Sichuan [Batang earthquake]; I=10; 2,298 Dead (Other Reports) 100 Dead/Many Dead
1870	5 11	15.80	-96.70	7.9	Mexico: Puebla, Oaxaca, Ejulla; 24 Dead (5/11, 12, and 16)
1871	12 23	37.40	58.40	7.2	Iran: Quchan; I=9; 2,000 Dead (Other Reports) I=7-8/M5.6
1872	1 6	37.10	58.40	6.3	Iran: Quchan; I=8-9; 4,000 Dead (Other Reports) Together with 12/23, 30,000 Dead
1872	4 3	36.40	36.50	7.2	Turkey: Antakya/Syria: Aleppo; I=10; 1,800 Dead (4/2?)
1875	3 28	-21.00	167.00	8.0	New Caledonia/Loyalty Is.: Lifu; Fatalities Reported
1875	5 3	38.10	30.00	6.7	Turkey: Civril, Dinar; I=9; 2,000 Dead (Other Reports) 1,300 Dead/1,000's Dead
1875	5 5	38.10	30.20		Turkey; 1,300 Dead (Includes 5/3 damage?)
1875	5 18	7.90	-72.50	7.3	Colombia: Cucuta; 16,000 Dead (Other Reports) 1,000 Dead
1877	5 10	-19.60	-70.20	8.3	Chile: Iquique, Tarapaca; I=11: Many Dead (5 Dead in Hawaii)
1879	3 22	37.80	47.85	6.7	Iran: Buzqush, Garmrud, Ardabil; I=10; 2,000 Dead (Other Reports) over 3,200 Dead
1879	7 1	33.20	104.70	8.0	Gansu[Wudu earthquake]; I=11; 29,480 Dead (Other Reports) 9,881 Dead (Wudu), 10,792 Dead (Wenxian)
1881	4 3	38.30	26.20	6.5	Greece: Khios (Chios); I=9; 7,866 Dead (Other Reports) 3,541 Dead/4,000 Dead/4,181, I=11

Continued.

Year	Date	Latitude °	Longitude °	M	Region and Damage
1883	7 28	40.75	13.88	5.6	Italy: Casamicciola [Ischia EQ]; 2,333 Dead (Other Reports) 3,100 Dead/3,300 Dead
1883	8 27	-5.80	106.30		Indonesia: Krakatoa; Volcanic Explosion/Earthquake/ Tsunami; 10,000's Dead (Other Reports) =<36,000 Dead
1885	5 30	33.50	75.00		India: Srinagar, Sopur; I=8; 3,000 Dead (Other Reports) 300 Dead/2,000 Dead
1887	6 8	43.10	76.80	7.3	Kyrgyzstan: Alma-Ata (Verny); I=9-10; Landslide; 1,800 Dead
1887	12 16	23.70	102.50	7.0	Yunnan [Shiping earthquake]; I=9 strong; 2,256 Dead (Other Reports) 200 Dead/At least 2,700 Dead
1889	7 11	43.20	78.70	8.3	Kazakhstan: Chilik, Alma Ata; I=10
1891	10 28	35.60	136.60	8.0	Western Gifu Prefecture [Nobi Earthquake]; Fire; 7,273 Dead; Dislocations
1893	3 31	38.30	38.50	7.0	Turkey: Malatya; 1,500 Dead (Other Reports) 400 Dead/469 Dead (Same as 3/2?)
1893	11 17	37.12	58.40	7.1	Iran: Quchan; I=9-10; 10,000 Dead; D=150/18,000
1894	3 22	42.50	146.00	7.9	Nemuro offshore; 1 Dead; Ms8.1
1895	1 17	37.08	58.40	6.8	Iran: Quchan; I=9; 1,000 Dead (Other Reports) 180 Dead/700 Dead/11,000 Dead
1896	1 4	37.70	48.30	6.7	Iran: Sangabad-Khalkhal; I=9-10; 1,100 Dead (Other Reports) 1,600 Dead
1896	6 15	39.50	144.00	8.2	Iwate prefecture offshore [Sanriku earthquake, tsunami] 22,000 Dead; Damage in Hawaii; Ms7.2, Mw8.0
1897	1 10	26.90	56.00	6.4	Iran: Qeshm Is. 1,600 Dead (Other Reports) 750 Dead/10,000 Dead?
1897	6 12	26.00	91.00	8.3	India: [Assam EQ] 1,500 Dead (Other Reports) 1425 Dead/1600 Dead, Dislocations, Ms8.0
1897	9 20	6.00	122.00	7.9	Philippines (Mindanao): Dapitan; Ms7.4; (Other Reports) Ms8.1?
1897	9 21	7.10	122.10	8.7	Philippines (Mindanao/Sulu/Jolo); I=10; 100 Dead; Ms7.5 (Other Reports) Ms8.2?
1897	10 18	12.00	126.00	8.1	Philippines (Samar/Visayas); I=9; Ms7.3 (Other Reports) Ms7.7?
1897	10 20	12.00	126.00	7.9	Philippines: Off Samar; Ms7.1 (Other Reports) Ms7.7?
1899	9 3	60.00	-142.00	8.4	USA (Alaska): Yakutat Bay; 0 Dead; I=11; Ms7.9
1899	9 10	60.00	-140.00	8.6	USA (Alaska): [Yakutat Bay EQ]; I=11; Upheaval; Ms8.0
1899	9 20	37.90	28.80	6.9	Turkey: Aydin, Nazilli, Buldan; I=9; 1,117 Dead; Dislocations; Ms6.9
1899	9 29	-3.00	128.50	7.4	Indonesia (Moluccas): Ceram (south coast); 3,864 Dead; Ms7.1
1900	7 29	-10.00	165.00	8.1	Solomon Is.: Santa Cruz Is.; Ms7.6
1900	10 9	60.00	-142.00	8.3	USA (Alaska): Chugacha Mountains; I=7-8; 0 Dead; Ms7.7
1901	8 9	-22.00	170.00	8.4	Loyalty Is.; Ms7.9
1902	4 19	14.93	-91.50	7.5	Guatemala: Quezaltenango, S. Marcos; I=9; 2,000 Dead (Other Reports) 1,500 Dead; Ms7.5
1902	8 22	39.80	76.20	8.1	Kyrgyzstan/China: Kashgar; I=9-10; Many Dead (Other Reports) 200 Dead/2,500 Dead; Ms7.7
1902	12 16	40.80	72.30	6.4	Uzbekistan: Andizhan; I=9; 4,725 Dead (Other Reports) 700 Dead/4,500 Dead
1903	4 28	39.14	42.65	7.0	Turkey: Malazgirt; I=9; 3,560 Dead (Other Reports) 600 Dead/2,200 Dead/6,000 Dead (also 5/9 and 12)
1903	5 28	40.90	42.80	5.4	Turkey: Ardahan, Varginis, Cardahli; I=7-8; 1,000 Dead
1903	8 11	36.00	23.00	7.9	Greece: Kythera; I=10-11; 2 Dead (Other Reports) Many Dead

Main Major Earthquakes and Damaging Earthquakes Continued.

Year	Date	Latitude	Longitude	M	Region and Damage
1905	4 4	33.00°	76.00°	8.6	India: [Kangra EQ]: 20,000 Dead (Other Reports) 10,600 Dead/18,815 Dead (M is overestimate) Ms7.5, Mw7.8
1905	7 23	49.30	96.20	8.2	Mongolia/Russia: Bolna; I=10; 320 km Dislocation; Ms7.7
1906	1 31	1.00	-81.50	8.8	Ecuador/Colombia; I=9; 1,000 Dead (Other Reports) 400 Dead/600 Dead; Ms8.2, Mw8.8
1906	3 17	23.60	120.50	6.8	Taiwan[Chaiyi earthquake]; 1,258 Dead; Dislocations
1906	4 18	37.70	-122.50	8.3	USA: [San Francisco EQ]; Fire; I=11; 700 Dead (Other Reports) 3,000 Dead? Dislocation; Ms7.8
1906	8 17	51.00	179.00	8.3	USA: Rat Is. (Aleutian Is.); Ms7.8
1906	8 17	-33.00	-72.00	8.4	Chile: Valparaiso; I=11; 3,760 Dead (Other Reports) 1,500 Dead/20,000 Dead; Ms8.1
1907	1 4	2.00	96.30	7.8	Indonesia (Sumatra): Gunung Sitoli (Nias); 400 Dead; Ms7.5
1907	1 14	18.20	-76.70	6.5	Jamaica: Kingston, Port Royal; 1,000 Dead (Other Reports) 1,700 Dead
1907	10 21	38.50	67.90	7.4	Uzbekistan/Tajikistan: Karatag; 15,000 Dead (Other Reports) 12,000 Dead; Ms7.2
1908	3 26	18.00	-99.00	8.1	Mexico: Chilapa, Chilpancingo (M is overestimate); mB7.5
1908	12 28	38.15	15.68	7.1	Italy: [Messina EQ]; I=11; 82,000 Dead (Other Reports) 75,000 Dead/110,000 Dead/200,000 Dead; Ms7.0
1909	1 23	33.40	49.10	7.3	Iran: Silakor; 5,500 Dead (Other Reports) 6,000–8,000 Dead; Dislocation; Ms7.0
1909	2 22	-18.00	-179.00	7.9	Fiji; mB7.6
1909	7 30	17.00	-100.50	7.8	Mexico: Acapulco, Guerrero; Ms7.3
1910	4 12	25.50	122.50	7.8	Taiwan; 20 Dead; mB7.6
1910	6 16	-19.00	169.50	8.6	Vanuatu: New Hebrides Is. (M is overestimate); mB7.9
1910	11 9	-15.00	166.00	7.9	Vanuatu: New Hebrides Is.; mB7.5
1911	1 3	42.90	76.90	8.2	Kazakhstan: Alma-Ata, Kemin; I=10–11; 450 Dead; Dislocation; Ms7.8
1911	6 7	19.70	-103.70	7.9	Mexico: Jalisco, Mexico City; 1,300 Dead (Other Reports) 45 Dead; Ms7.7
1911	6 15	28.00	130.00	8.0	Near Amami-oshima [Kikaijima earthquake]; 12 Dead; mB8.1
1911	7 12	9.00	126.00	7.8	Philippines (Mindanao); I=10; Ms7.5
1911	8 16	7.00	137.00	8.1	Caroline Is.; mB7.7
1912	5 23	21.00	97.00	8.0	Myanmar/Thai: Taunggyi; Fatalities Reported; Ms7.7
1912	7 24	-5.62	-80.41	8.1	Peru: Huancabamba, Cajamarca/Ecuador; I=10; Many Dead; Ms7.0
1912	8 9	40.75	27.20	7.4	Turkey: Saros-Marmara; I=10; 2,836 Dead (Other Reports) 216 Dead/1,950 Dead; Ms7.6
1912	11 19	19.90	-99.80	7.8	Mexico: Acambay, Tixmadeje, Timilpan; Many Dead; Dislocation; mB6.9
1913	3 14	4.50	126.50	8.3	Indonesia (Sangihe); I=9; Fatalities Reported (111 Person Village buried); Ms7.9
1913	8 6	-15.80	-73.50	7.8	Peru: Chuquibamba; I=10; Fatalities Reported; Ms7.8
1913	10 14	-19.50	169.00	8.1	Vanuatu: New Hebrides Is.; mB7.6
1913	12 21	24.10	102.40	7.0	Yunnan [Eshan earthquake]; I=9; 1,314 Dead (Other Reports) 942 Dead/At least 1,900 Dead; Ms7.2
1914	1 30	-33.30	-66.30	8.2	Argentina: San Luis Province (M is overestimate); Ms7.5
1914	5 26	-2.00	137.00	7.9	Indonesia (Irian Jaya): Japen Is. Fatalities Reported; Ms8.0

Continued.

Year Date	Latitude	Longitude	*M*	Region and Damage
1914 6 25	-4.00	102.50	8.1	Indonesia (Sumatra): Benkulen; Many Dead (Other Reports) 11 Dead/20 Dead Ms7.6
1914 10 3	37.82	30.27	7.0	Turkey: Burdur; I=9–10; 4,000 Dead (Other Reports) Fatalities Reported/300 Dead; Ms7.1
1914 11 24	22.00	143.00	8.7	Marianas; mB7.9
1915 1 13	41.98	13.65	7.0	Italy: [Avezzano EQ]; I=11; 32,610 Dead; (Other Reports) 30,000 Dead/35,000 Dead; Dislocation; Ms6.9
1915 5 1	47.00	155.00	8.1	Russia: Kuriles; Ms8.0
1915 10 2	40.50	-117.50	7.8	USA (Nev.): [Pleasant Valley EQ]; I=11; Dislocation; Ms7.7
1916 1 1	-4.00	154.00	7.9	Papua New Guinea: New Ireland; Ms7.8
1916 1 13	-3.00	135.50	8.1	Indonesia (Irian Jaya); Ms7.7
1917 1 20	-8.30	115.00		Indonesia (Bali); 1,300 Dead
1917 1 30	55.20	164.50	8.1	Russia: Kamchatka; I=11; Ms7.8
1917 5 1	-29.00	-177.00	8.6	New Zealand (Kermadec Is.); (M is overestimate) Ms7.9
1917 5 31	54.50	-160.00	7.9	USA: Alaska Pen.; Ms7.9; Mw7.4
1917 6 26	-15.50	-173.00	8.7	Tonga/Samoa; Ms8.4
1917 7 31	28.00	104.00	6.8	Yunnan [Daguan earthquake]; I=9; 1,879 Dead; Ms7.5
1918 5 20	-28.50	-71.50	7.9	Chile; mB7.6
1918 8 15	5.50	123.00	8.5	Philippines (Mindanao): Cotabato; 50 Dead (Other Reports) 102 Dead; Ms8.0
1918 9 7	45.60	151.10	8.2	Russia: Kuriles; 24 Dead; Minor Damage in Hawaii; Ms8.2
1918 11 8	44.90	151.40	7.9	Russia: Kuriles; Ms7.7
1918 11 18	-7.00	129.00	8.1	Indonesia: Banda Sea; (M is Overestimate) mB7.5
1918 12 4	-26.00	-71.00	7.8	Chile: Copiapo, Caldera; Ms7.6
1918 12 18	-27.30	-70.50	8.2	Chile: Copiapo; (M is overestimate, or Earthquake is same as above?)
1919 1 1	-19.50	-176.50	8.3	Tonga; mB7.7
1919 4 30	-19.00	-172.50	8.4	Tonga; Ms8.2
1919 5 6	-5.00	154.00	7.9	Papua New Guinea: New Ireland; Fatalities Reported; Ms7.9
1920 6 5	23.50	122.70	8.0	Taiwan [Hualien earthquake]; 5 Dead; Ms8.0
1920 9 20	-20.00	168.00	8.3	Vanuatu: New Hebrides Is. Ms7.9
1920 12 16	36.70	104.90	8.5	Ningxia [Haiyuan earthquake]; I=12; 235,502 Dead (Other Reports) 220,000 Dead/246,000 Dead; Dislocation; Ms8.6
1921 11 15	36.50	70.50	8.1	Afghanistan (Hindu Kush); mB7.6
1921 12 18	-2.50	-71.00	7.9	Colombia-Peru border; mB7.5
1922 11 11	-28.50	-70.00	8.5	Chile: Atacama; 1,000 Dead (Other Reports) 600 Dead; Ms8.3; Mw8.5
1923 2 3	53.00	161.00	8.5	Russia: Kamchatka; I=11; Fatalities Reported (Hawaii 6 Dead from Tsunami) Ms8.3
1923 5 25	35.30	59.20	5.5	Iran: Kaj Derakht and Torbat; 2,219 Dead (Other Reports) 300 Dead/770 Dead/More than 5,000 Dead
1923 6 14	31.30	100.80	5.8	Sichuan (Kangding); I=7; 1,300 Dead
1923 9 1	35.33	139.14	7.9	Southeastern Kanto [Kanto earthquake]; Fire; 105,000 Dead; Dislocation; Ms8.2; Mw7.9

Main Major Earthquakes and Damaging Earthquakes Continued.

Year Date	Latitude	Longitude	*M*	Region and Damage
1924 4 14	6.50°	126.50°	8.3	Philippines (Mindanao): Davao Bay; I=9; Ms8.3
1924 6 26	-56.00	157.50	8.3	Tasman Sea, Macqurie Is. (M is overestimate); Ms7.7
1925 3 16	25.70	100.40	7.0	Yunnan [Dali earthquake]; I=9; 5,808 Dead (Other Reports) 3,600 Dead/5,847 Dead; Ms7.0
1926 6 26	36.50	27.50	8.0	Greece: Rhodes; I=11; 110 Dead; mB7.7
1926 10 3	-49.00	161.00	7.9	Tasman Sea, N of Macquarie Is.; Ms7.5
1926 10 26	-3.20	138.50	7.9	Indonesia (Irian Jaya); Ms7.6
1927 3 7	35.63	134.93	7.3	northern Kyoto [(north)Tango earthquakes]; (Fire); 2,925 Dead; Dislocation; Ms7.6; Mw7.1
1927 5 23	37.70	102.20	8.0	Gansu [Gulang earthquake]; I=11; 41,419 Dead (Other Reports) 4,000 Dead/35,495 Dead/80,000 Dead; Dislocation; Ms7.9
1928 3 9	-2.50	88.50	8.1	Indian Ocean: [Ninetyeast Ridge EQ]; Ms7.7
1928 6 17	16.20	-98.00	7.9	Mexico: Oaxaca; Ms7.8; Mw8.0
1928 12 1	-35.00	-72.00	8.0	Chile: Talca; I=10; 225 Dead (Other Reports) 218 Dead; Ms8.0
1929 5 1	37.80	57.80	7.2	Iran: Kopet-Dagh; 3,257 Dead (Other Reports) 3,800 Dead/5,803 Dead; Dislocation; Ms7.2
1929 6 27	-54.00	-29.50	8.3	Atlantic Ocean; (M is overestimate) Ms7.7
1930 5 6	38.00	44.70	7.3	Iran: Salmas; I=9-10; 2,514 Dead (Other Reports) 1,360 Dead/3,000 Dead; Dislocation; Ms7.2
1930 7 23	41.05	15.37	6.7	Italy: Vulture (Irpinia); I=10; 1,404 Dead (Other Reports) 1,883 Dead/3,000 Dead
1931 1 15	16.00	-96.70	7.9	Mexico: Oaxaca; I=10; 68 Dead (Other Reports) 114 Dead; Ms7.8
1931 2 2	-39.50	176.90	7.9	New Zealand: [Hawke's Bay EQ]; 256 Dead (Other Reports) 225 Dead/285 Dead; Ms7.8
1931 3 31	12.15	-86.28	6.0	Nicaragua: [Managua EQ]; Fire; 1,000 Dead (Other Reports) 2,000 Dead/2,450 Dead; Dislocation
1931 8 11	47.10	89.80	8.0	Xinjiang [Fuyun earthquake]; I=11; 10,000 Dead (Other Reports) 300 Dead; Dislocation; Ms7.9; Mw8.0
1931 10 3	-10.50	161.70	7.9	Solomon Is. Death 50; Ms7.9
1932 5 14	0.50	126.00	8.3	Indonesia: Molucca Passage; 5 Dead (Other Reports) 6 Dead/Many Dead; Ms8.0
1932 5 26	-25.50	179.20	7.9	Fiji Basin; mB7.5
1932 6 3	19.50	-104.30	8.1	Mexico: Guadalajara, Colima, Zamora; I=11; 60 Dead; Ms8.2; Mw8.1
1932 6 18	19.50	-104.30	7.9	Mexico: Guadalajara, Colima; 52 Dead (Other Reports) 30 Dead; Ms7.8
1933 3 3	39.23	144.52	8.1	Iwate offshore [Sanriku offshore eqrthquake]; 3,064 Dead; Ms8.5; Mw8.4
1933 8 25	31.90	103.40	7.5	Sichuan; 6,865 Dead; Earthquake; Lake Collapse; Flood; 2,500 Dead; Ms7.5
1934 1 15	26.50	86.50	8.3	India/Nepal: [Bihar-Nepal EQ]; 10,700 Dead (Other Reports) 7,253 Dead; Ms8.3; Mw8.2
1934 2 14	17.50	119.00	7.9	Philippines (Luzon); Ms7.6
1934 7 18	-11.70	166.50	8.1	Santa Cruz Is.; Ms8.1
1935 4 21	24.30	120.80	7.1	Taiwan [Miaoli earthquake/Hsinchu-Taichung earthquake]; 3,276 Dead; Dislocation; Ms7.1
1935 5 30	29.50	66.80	7.5	Pakistan: [Quetta EQ]; 60,000 Dead (Other Reports) 25,000 Dead/30,000 Dead; Ms7.6

Continued.

Year Date	Latitude	Longitude	*M*	Region and Damage
1935 9 20	-3.50°	141.70°	7.9	Papua New Guinea; Ms7.9
1935 12 28	0.00	98.20	8.1	Indonesia (Sumatra); I=8; Ms7.7
1937 4 16	-21.50	-177.00	8.1	Tonga; mB7.5
1937 8 1	35.40	115.10	7.0	Shandong [Heze earthquake]; I=9; 3,833 Dead (Other Reports) Many Dead/3350 Dead
1938 2 1	-5.20	130.50	8.5	Indonesia: Banda Sea; I=7; Ms8.2; Mw8.5
1938 5 19	-1.00	120.00	7.9	Indonesia (Sulawesi): Dongala, Mambara; 8 Dead (Other Reports) Many Dead; Ms7.6
1938 11 10	55.50	-158.00	8.7	USA (Alaska); Ms8.3; Mw8.3
1939 1 25	-36.25	-72.25	7.8	Chile: [Chillan EQ]; I=10; 28,000 Dead (Other Reports) 30,000 Dead; Ms7.8
1939 1 30	-6.50	155.50	7.8	Papua New Guinea/Solomon Is.; Ms7.8
1939 4 30	-10.50	158.50	8.1	Solomon Is.; 12 Dead; Ms8.0
1939 12 21	0.00	123.00	8.6	Indonesia (Sulawesi): Minahassa Pen.; mB7.8
1939 12 26	40.10	38.20	7.8	Turkey: [Erzincan EQ]; I=11-12; 32,968 Dead (Other Reports) 45,000 Dead; Dislocation; Ms7.8
1940 5 24	-12.50	-77.00	8.0	Peru: Lima, Callao; I=10; 250 Dead (Other Reports) 179 Dead; Ms7.9; Mw8.2
1940 11 10	45.80	26.80	7.3	Romania: Vrancea Region, Bucharest; 1,000 Dead (Other Reports) 350 Dead; mB7.3
1941 1 11	16.60	43.30	5.8	Yemen; I=9; 1,200 Dead (Number includes damage from 2/4 and 2/23 aftershocks)
1941 6 26	12.50	92.50	8.1	India: Andaman, Colombo, Madras, Calcutta; Fatalities Reported (Other Reports) 5,000 Dead Ms7.7
1941 11 25	37.50	-18.50	8.4	N. Atlantic Ocean; Ms8.2
1942 4 8	13.50	121.00	7.8	Philippines (Mindanao); Ms7.5
1942 5 14	-0.70	-81.50	7.9	Peru/Ecuador; I=9; 200 Dead (Other Reports) Many Dead; Ms7.9; Mw8.2
1942 8 6	13.90	-90.90	7.9	Guatemala: Tecpan, Totonicapan, San Marcos; 38 Dead (Other Reports) 9 Dead; Ms7.9
1942 8 24	-15.00	-76.00	8.1	Peru: Nazca; I=10; 30 Dead (Other Reports) 22 Dead; Ms8.2
1942 11 10	-49.50	32.00	8.3	Indian Ocean: Prince Edward Is.; Ms7.9
1942 12 20	40.70	36.80	7.3	Turkey: Niksar, Erbaa; I=9; 3,000 Dead (Other Reports) 1,000 Dead/1,100 Dead; Dislocation; Ms7.3
1943 4 6	-30.75	-72.00	7.9	Chile: Illapel, Salamanca; 18 Dead (Other Reports) 11 Dead; Ms7.9; Mw8.2
1943 5 25	7.50	128.00	7.9	Philippines: Off E. Mindanao; Ms7.7
1943 7 23	-9.50	110.00	8.1	Indonesia (Java): Jogyakarta; 213 Dead; mB7.6
1943 7 29	19.20	-67.50	7.9	Puerto Rico: San Juan, Mona Passage; Ms7.7
1943 9 6	-53.00	159.00	7.9	Macquarie Is. region; Ms7.7
1943 9 10	35.47	134.19	7.2	Eastern Tottori Prefecture [Tottori earthquake]; Fire; 1,083 Dead; Dislocation; Ms7.4; Mw7.0
1943 11 26	41.00	34.00	7.6	Turkey: Ladik; 4,020 Dead (Other Reports) 2,824 Dead/2,900 Dead/5,000 Dead; Dislocation; Ms7.6
1944 1 15	-31.50	-68.60	7.4	Argentina: San Juan; 8,000 Dead (Other Reports) 2,900 Dead/5,000 Dead/5,600 Dead; Dislocation; Ms7.2

Main Major Earthquakes and Damaging Earthquakes Continued.

Year Date	Latitude	Longitude	M	Region and Damage
1944 2 1	41.50°	32.50°	7.6	Turkey: Gerede (Bolu); 3,959 Dead (Other Reports) 2,381 Dead/2,790 Dead/5,000 Dead; Dislocation; Ms7.4
1944 12 7	33.57	136.18	7.9	Mie offshore [Tonankai earthquake]; 1,251 Dead; Ms8.0; Mw8.1
1945 1 13	34.70	137.11	6.8	Southern Aichi Prefecture [Mikawa earthquake]; 2306 Dead; (Other Reports) 1961 Dead; Dislocation; Ms6.8; Mw6.6
1945 11 27	25.00	63.50	8.0	Iran/Pakistan: Makran coast; 300 Dead (Other Reports) 4,000 Dead; Ms8.0
1945 12 28	-6.00	150.00	7.8	Papua New Guinea: New Britain Is.; Ms7.7
1946 5 31	40.00	41.50	6.0	Turkey: Ustukran; I=7–8; 1,300 Dead (Other Reports) 839 Dead
1946 8 4	19.20	-69.00	8.1	Dominica/Puerto Rico; 100 Dead; Ms8.0
1946 8 8	19.50	-69.50	7.9	Dominica/Puerto Rico; Ms7.6
1946 9 12	23.50	96.00	7.8	Myanmar; Ms7.8
1946 9 29	-4.50	153.50	7.8	Papua New Guinea: New Ireland region; Ms7.7
1946 11 10	-8.50	-77.50	7.3	Peru: Ancash; I=11; 1,400 Dead (Other Reports) 800 Dead/1,500 Dead/1,613 Dead; Dislocation; Ms7.3
1946 12 21	32.93	135.85	8.0	Ki Peninsula offshore [Nankai earthquake]; Fire; 1,330 Dead; Ms8.2; Mw8.1
1947 7 29	28.50	94.00	7.9	India/Bhutan; Ms7.5
1948 1 24	10.50	122.00	8.3	Philippines: Sulu Sea, Iloilo, Jaro; 72 Dead; Ms8.2
1948 6 28	36.17	136.30	7.1	Northern Fukui Prefecture [Fukui earthquake]; Fire; 3,769 Dead (Other Reports) 3,895 Dead/5,390 Dead; Ms7.3; Mw7.0
1948 9 8	-21.00	-174.00	7.9	Tonga; Ms7.8
1948 10 5	37.70	58.70	7.3	Turkmenistan: [Ashkhabad EQ]; 19,800 Dead (Other Reports) 10,000 Dead/10,000's Dead/110,000 Dead; Ms7.3
1949 7 10	39.20	70.80	7.4	Tajikistan: [Khait EQ]; I=9–10; 12,000 Dead (Other Reports) 1,000's Dead 20,000; Ms7.5
1949 8 5	-1.50	-78.25	6.8	Ecuador: [Pelileo EQ]; I=11; 6,000 Dead (Other Reports) 5,050 Dead; Ms6.4
1949 8 22	53.70	-133.20	8.1	Canada: [Queen Charlotte Is. EQ]; Dislocation; Ms8.1
1949 12 17	-54.00	-71.00	7.8	Chile: Punta Arenas; 1 Dead; Ms7.7
1949 12 17	-54.00	-71.00	7.8	Chile; 3 Dead; Ms7.7
1950 2 28	46.00	144.00	7.9	Russia: Off Sakhalin; mB7.5
1950 8 15	28.40	96.70	8.6	Xizang [Chayu earthquake]; I=11; 3,300 Dead (Other Reports) 2,486 Dead/Fatalities Reported in India; Ms8.6; Mw8.6
1950 10 5	10.00	-85.70	7.9	Costa Rica: Nicoya; Ms7.7
1950 11 2	-6.50	129.50	8.1	Indonesia: Banda Sea; mB7.4
1950 12 2	-18.20	167.50	8.1	Vanuatu: New Hebrides Is.; Ms7.2; mB7.6
1950 12 9	-23.50	-67.50	8.0	Chile/Argentina; I=7; 4 Dead (Other Reports) 1 Dead; mB7.7
1950 12 14	-19.20	-175.70	7.9	Fiji; mB7.5
1951 11 18	31.10	91.40	8.0	Xizang [Dangxiong earthquake]; Dislocation; Ms8.0
1951 12 8	-34.00	57.00	7.9	Indian Ocean; Ms7.4
1952 3 4	41.80	144.13	8.2	Tokachi offshore [Tokachi offshore earthquake]; 33 Dead; Ms8.3; Mw8.1
1952 3 19	9.50	127.20	7.9	Philippines: Off Mindanao; I=6; Ms7.6

Continued.

Year	Date	Latitude	Longitude	M	Region and Damage
1952	11 4	52.30°	161.00°	9.0	Russia [Kamchatka EQ]; I=11; Many Dead; Ms8.2; Mw9.0
1953	3 18	40.02	27.53	7.4	Turkey: Onon; 1,103 Dead (Other Reports) 224 Dead/265 Dead/1,070 Dead Ms7.2
1953	12 12	-3.40	-80.60	7.8	Peru/Ecuador: Tumbes; I=9; 7 Dead; Ms7.4
1954	9 9	36.31	1.47	6.7	Algeria: El Asnam (Orleanville); 1,409 Dead (Other Reports) 1,243 Dead; Dislocation
1955	2 27	-28.00	-175.50	7.8	New Zealand (Kermadec Is.); Ms7.7
1957	3 9	51.30	-175.80	8.6	USA: Andreanof Is. (Aleutian Is.); 0 Dead; I=8; M is from USGS; Ms8.1; Mw9.1
1957	7 2	36.10	52.70	7.1	Iran: Mazandaran; 1,200 Dead (Other Reports) 1,100 Dead/1,500 Dead/2,000 Dead; Ms7.0
1957	11 29	-21.00	-66.00	7.8	Bolivia/Chile; mB7.4
1957	12 4	45.20	99.40	8.3	Mongolia: [Gobi-Altai EQ]; Many Dead (Other Reports) 30 Dead/1,200 Dead; Dislocation; Ms8.0; Mw8.1
1957	12 13	34.50	48.00	7.2	Iran: Farsinaj; 2,000 Dead; (Other Reports) 1,130 Dead/1,200 Dead/1,392 Dead/2,500 Dead; Dislocation; Ms6.8
1958	1 19	1.37	-79.34	7.8	Ecuador: Esmeraldas/Colombia; I=9; 20 Dead; Ms7.3
1958	7 9	58.60	-137.10	7.9	USA (Alaska): Lituya Bay; 5 Dead (Other Reports) 3 Dead/ 6 Dead; Dislocation; Ms7.9; Mw8.2
1958	11 7	44.38	148.58	8.1	Etorofu offshore [Etorofu offshore earthquake]; 0 Dead; Ms8.1; Mw8.3
1960	2 29	30.45	-9.62	5.7	Morocco: [Agadir EQ]; I=10; 13,100 Dead (Other Reports) 12,000 Dead/14,000 Dead/15,000 Dead
1960	5 22	-39.50	-74.50	9.5	Chile: [Chilean EQ]; 5,700 Dead (Other Reports) 1,743 Dead/ 2,231 Dead; Ms8.5; Mw9.5
1962	9 1	35.60	49.90	7.2	Iran: Buyin-Zahra [Qazvin EQ]; 12,225 Dead (Other Reports) 10,000 Dead; Dislocation; Ms6.9
1963	7 26	42.00	21.40	6.1	Macedonia: [Skopje EQ]; I=9-10; 1,070 Dead (Other Reports) 1,200 Dead
1963	10 13	44.80	149.50	8.5	Russia (Kuriles): Urup Is.; I=9; Ms8.1; Mw8.5
1963	11 4	-6.80	129.60	8.2	Indonesia: Banda Sea; mB7.8
1964	3 27	61.00	-147.80	9.2	USA: [Alaska EQ]; I=9-10; 131 Dead; Diastrophism; Ms8.4; Mw9.2
1965	2 4	51.30	178.60	8.7	USA: Aleutian Is. [Rat Island EQ]; Ms8.2; Mw8.7
1965	8 23	16.30	-95.80	7.8	Mexico: Oaxaca; 5 Dead (Other Reports) 6 Dead; Ms7.6
1966	3 22	37.50	115.10	7.2	Hebei [Ningjin earthquake]; I=10; 8,064 Dead; (Incudes damage from 3/8); Ms7.1
1966	6 15	-10.40	160.90	7.8	Solomon Is.; 0 Dead; Ms7.7
1966	8 19	39.15	41.55	6.8	Turkey: Varto; I=9; 2,517 Dead (Other Reports) 2,394 Dead/ 2,964 Dead/over 3,000 Dead; Dislocation
1966	12 28	-25.51	-70.74	7.8	Chile: Taltal; I=8; 3 Dead; Ms7.7
1968	2 12	-5.50	153.20	7.8	Papua New Guinea: New Ireland; Ms7.1
1968	5 16	40.73	143.58	7.9	Aomori prefcture eastward offshore [Tokachi offshore earthquake]; 52 Dead; Ms8.1; Mw8.2
1968	8 31	33.97	59.02	7.3	Iran: [Dasht-i Biyaz EQ]; 15,000 Dead (Other Reports) 11,588 Dead/12,100 Dead; Dislocation; Ms7.1

Main Major Earthquakes and Damaging Earthquakes Continued.

Year Date	Latitude	Longitude	M	Region and Damage
1969 2 28	36.01°	−10.57°	8.0	Atlantic Ocean/Portugal/Morocco; 13 Dead (Other Reports) 7 Dead/23 Dead; Ms7.8
1969 8 12	43.44	147.82	7.8	Syakotan Island offshore [Hokkaido eastward offshore earthquake]; 0 Dead; (Other Reports) 42.70°, 147.62°; Ms7.8; Mw8.2
1970 1 5	24.20	102.70	7.8	Yunnan [Tonghai earthquake]; I=10 Strong; 15,621 Dead; Dislocation; Ms7.3
1970 3 28	39.20	29.50	7.1	Turkey: Gediz; I=9; 1,086 Dead (Other Reports) 1,098 Dead; Dislocation
1970 5 31	−9.36	−78.87	7.8	Peru: [Peruvian EQ]; I=10; 66,794 Dead (Other Reports) 54,000 Dead/50,000–70,000 Dead; Ms7.5; Mw7.9
1970 12 15	55.90	163.40	7.8	Russia: Kamchatka; I=10
1971 1 10	−3.13	139.70	8.1	Indonesia (Irian Jaya): Djajapura, Sentani; Ms7.9
1971 7 14	−5.47	153.88	7.9	Papua New Guinea: New Britain Is.; I=7; 2 Dead; Ms7.8
1971 7 26	−4.94	153.17	7.9	Papua New Guinea: New Ireland; Ms7.7
1971 12 15	55.90	163.20	7.8	Russia: Kamchatka; I=10; Ms7.5
1972 1 25	22.60	122.30	8.0	Taiwan eastern offshore earthquake; 1 Dead; Ms7.4
1972 4 10	28.40	52.80	6.8	Iran: [Ghir (Qir) EQ]; 5,010 Dead (Other Reports) 5,374 Dead/5,400 Dead; Dislocation; Ms6.7
1972 12 2	6.47	126.60	7.8	Philippines (Mindanao); mB7.4
1972 12 23	12.15	−86.27	6.3	Nicaragua: [Managua EQ]; 10,000 Dead (Other Reports) 5,000 Dead/6,000 Dead/8,000 Dead/11,000 Dead; Dislocation; Ms6.0
1973 2 6	31.30	100.70	7.6	Sichuan [Luhuo earthquake]; I=10; 2,199 Dead (Other Reports) 2,175 Dead; Dislocation; Ms7.2
1973 12 28	−14.46	166.60	7.8	Vanuatu: Espiritu Santo, Luganvillle; Ms7.3
1974 5 11	28.20	104.10	7.1	Yunnan [Daguan-Yongshan earthquake]; I=9; 1,541 Dead (Other Reports) 1,423 Dead/1,641 Dead/20,000 Dead
1974 12 28	35.05	72.87	6.2	Pakistan: Patan; 5,300 Dead (Other Reports) 700 Dead/900 Dead; Ms6.0
1975 2 4	40.70	122.80	7.3	Liaoning [Haicheng earthquake]; I=9 strong; 1,328 Dead; Predicted; Dislocation; Ms7.2
1975 5 26	36.00	−17.65	8.1	Azores; Ms7.8
1975 7 20	−6.59	155.05	7.9	Papua New Guinea: Bougainville Is.; I=8; Ms7.6
1975 9 6	38.50	40.70	6.7	Turkey: Lice; I=9; 2,370 Dead (Other Reports) 2,100 Dead/2,385 Dead/3,000 Dead; Dislocation
1975 10 11	−24.89	−175.12	7.8	Tonga; Ms7.7
1975 12 26	−16.26	−172.47	7.8	Tonga; I=5; Ms7.5
1976 1 14	−29.20	−177.80	7.8	New Zealand (Kermadec Is.); Ms7.7
1976 1 14	−28.43	−177.66	8.2	New Zealand (Kermadec Is.); Ms7.9
1976 2 4	15.16	−89.18	7.5	Guatemala: [Guatemala EQ]; I=9; 22,870 Dead; Dislocation; Ms7.5
1976 6 25	−4.60	140.09	7.1	Indonesia (Irian Jaya); 6,000 Dead (Other Reports) 422 Dead 5,000–9,000 Dead; Ms6.9
1976 7 28	39.40	118.00	7.8	Hebei [Tangshan earthquake]; I=11; 242,800 Dead; Dislocation; Ms7.8
1976 8 16	6.26	124.02	7.9	Philippines (Mindanao); I=10; 8,000 Dead (Other Reports) 6,500 Dead/3792 Dead; Ms7.8; Mw8.1
1976 10 29	−4.52	139.92	7.2	Indonesia (Irian Jaya); I=8; 6,000 Dead (Other Reports) 108 Dead/133 Dead

Continued.

Year Date	Latitude	Longitude	M	Region and Damage
1976 11 24	39.12°	44.03°	7.3	Turkey: Muradiye/Iran; 3,900 Dead; (Other Reports) 3,626 Dead/3,840 Dead/10,000 Dead; Dislocation; Ms7.1
1977 3 4	45.77	26.76	7.2	Romania: Vrancea, Bucharest; 1,581 Dead (Other Reports) 1,387 Dead; mB7.1; Mw7.5
1977 8 19	-11.09	118.46	7.9	Indonesia: [Sumbawa Is. EQ]; 189 Dead; Normal Fault Type; Ms8.1; Mw8.3
1978 9 16	33.39	57.43	7.4	Iran: [Tabas EQ]; 18,220 Dead (Other Reports) 15,000 Dead/ over 20,000 Dead; Dislocation; Ms7.2; Mw7.3
1979 9 12	-1.68	136.04	7.9	Indonesia (Irian Jaya): Yapan; 15 Dead; Ms7.7; Mw7.5
1979 12 12	1.60	-79.36	7.9	Colombia: Port-Tumaco-Buenaventura area; 600 Dead; Ms7.6; Mw8.1
1980 7 17	-12.53	165.92	7.9	Solomon Is.: Santa Cruz Is.; 0 Dead; Ms7.7; Mw7.7
1980 10 10	36.20	1.35	7.3	Algeria: El Asnam; 3,500 Dead (Other Reports) 2,590–5,000 Dead; Dislocation; Ms7.1; Mw7.1
1980 11 23	40.91	15.37	6.7	Italy: [Irpinia EQ]; 2,483 Dead (Other Reports) 2,735 Dead/ 3,105 Dead/4,680 Dead; Dislocation; Mw6.9
1981 1 19	-4.58	139.23	6.7	Indonesia (Irian Jaya): Jayawijaya Mts.; 1,300 Dead (Other Reports); 261 Dead; Mw6.6
1981 6 11	29.91	57.71	6.7	Iran: Kerman Province, Golbaft; 3,000 Dead; Dislocation; Mw6.6
1981 7 28	30.01	57.79	7.1	Iran: Kerman Province; 1,500 Dead (Other Reports) 8,000 Dead; Dislocation; Mw7.2
1982 12 13	14.70	44.38	6.0	Yemen Arab Republic: Dhamar; 2,800 Dead (Other Reports) 1,900 Dead/3,000 Dead; Dislocation; Mw6.2
1983 10 30	40.33	42.19	6.9	Turkey: [Narman-Horasan EQ]; I=8; 1,400 Dead (Other Reports) 1,155 Dead/1,300–2,000 Dead; Mw6.6
1984 3 6	29.34	138.92	7.6	Near Torishima deeper earthquake; Shock death 1; Mw7.4
1985 3 3	-33.14	-71.87	7.8	Chile: San Antonio, Valparaiso; 177 Dead; Mw7.9
1985 9 19	18.19	-102.53	8.1	Mexico: [Michoacan EQ]; Large Damage in Mexico City; 9,500 Dead; Mw8.0
1986 10 10	13.83	-89.12	5.4	El Salvador: San Salvador; 1,500 Dead (Other Reports) over 1,000 Dead; Dislocation; Mw5.7
1986 10 20	-28.12	-176.37	8.2	New Zealand (Kermadec Is.); Ms8.1; Mw7.7
1986 11 14	23.90	121.57	7.8	Taiwan; 13 Dead (Other Reports) 15 Dead/16 Dead; Mw7.3
1987 3 6	0.15	-77.82	6.9	Ecuador/Colombia: Quito, Tulcan, Riobamba; 5,000 Dead (Other Reports) 1,000 Dead; Mw7.1
1988 8 20	26.76	86.62	6.6	Nepal/India: Bihar Prov.; 1,450 Dead (Other Reports) 721 Dead (Nep) + 280 Dead (Bih); Mw6.8
1988 12 7	40.99	44.19	6.8	Armenia: Spitak, Leninakan/E.Turkey; 25,000 Dead; Dislocation; Mw6.7
1989 5 23	-52.34	160.57	8.2	Macquarie Ridge (Between New Zealand and Antarctica); 0 Dead; Mw8.0
1990 6 20	36.96	49.41	7.7	Iran: Manjil, Rudbar/Azerbaijan; 35,000 Dead (Other Reports) 30,000–50,000 Dead; Dislocation; Mw7.4
1990 7 16	15.68	121.17	7.8	Philippines (Luzon): Baguio, Cabanatuan; 2,430 Dead; Dislocation; Mw7.7
1991 10 19	30.78	78.77	7.0	India: [Uttarkashi EQ]; 2,000 Dead (Other Reports) 768 Dead; Mw6.8
1992 12 12	-8.48	121.90	7.5	Indonesia (Flores): Maumere, Babi Is.; 1,740 Dead (Other Reports) 2,080 Dead; Mw7.7
1993 1 15	42.92	144.36	7.8	Kushiro offshore [Kushiro offshore earthquake] slightly deeper earthquake; 2 Dead; Mw7.6

Main Major Earthquakes and Damaging Earthquakes Continued.

Year	Date	Latitude	Longitude	M	Region and Damage
1993	7 12	42.78°	139.18°	7.8	Okushiri Island offshore [Hokkado southern-west offshore earthquake]; 230 Dead; Russia (3 Missing)/Korea; Mw7.7
1993	8 8	12.98	144.80	8.0	Marianas: Guam; 0 Dead; Mw7.8
1993	9 29	18.07	76.45	6.2	India: Latur-Osmanabad area; 9,748 Dead (Other Reports) 7,601 Dead; Mw6.2
1994	6 9	-13.84	-67.55	8.2	Bolivia/Peru: [Bolivian Deep EQ]; 10 Dead; Mw8.2
1994	10 4	43.37	147.68	8.2	Syakotan Island offshore [Hokkaido eastern offshore earthquake]; Shock; 1 Dead; Mw8.3
1995	1 17	34.59	135.04	7.3	[The Southern Hyogo prefecture earthquake in 1995]; Great Hanshin-Awaji earthquake; 6,434 Dead: Dislocation; Ms6.8; Mw6.9
1995	4 7	-15.20	-173.53	8.0	Tonga; 0 Dead; Mw7.4
1995	5 27	52.63	142.83	7.5	Russia (Sakhalin): Neftegorsk; 1,989 Dead (Other Reports) 1,841 Dead; Dislocation; Mw7.1
1995	8 16	-5.80	154.18	7.8	Solomon Is.; 0 Dead; Mw7.7
1995	12 3	44.66	149.30	7.9	Etorofuto coast; 0 Dead; Mj7.2; Ms7.9; Mw7.9
1996	2 17	-0.89	136.95	8.1	Indonesia (Irian Jaya): Biak, Supiori; 166 Dead; Mw8.2
1997	2 28	38.08	48.05	6.1	Iran: Ardebil; 1,100 Dead (Other Reports) 965 Dead; Mw6.1
1997	4 21	-12.58	166.68	7.9	Santa Cruz Is.; 0 Dead; Mw7.7
1997	5 10	33.83	59.81	7.3	Iran: Qayen, Birjand; 1,572 Dead (Other Reports) 1,640 Dead; Dislocation; Mw7.2
1997	11 8	35.07	87.32	7.9	Xizang; 0 Dead; Mw7.6
1998	2 4	37.05	70.09	6.1	Afghanistan: Rostaq area; 2,323 Dead; Mw5.9
1998	3 25	-62.88	149.53	8.0	Antarctica: Balleny Is. region; 0 Dead; Mw8.1
1998	5 30	37.11	70.11	6.9	Afghanistan: Badakhshan, Takhar Provinces; 4,000 Dead; Mw6.6
1998	7 17	-2.96	141.93	7.1	Papua New Guinea: Sissano; 2,700 Dead; Mw7.0
1999	1 25	4.46	-75.72	5.7	Colombia: Armenia, Calarca, Pereira [Quindio EQ]; 1,900 Dead; Mw6.2
1999	8 17	40.75	29.86	7.8	Turkey: [Kocaeli EQ, Izmit EQ]; 17,118 Dead; Many missing? Dislocation; Mw7.5
1999	9 21	23.77	120.98	7.7	Taiwan [Jiji earthquake]; 2,413 Dead; Dislocation; Mw7.7
2000	6 4	-4.72	102.09	8.0	Indonesia (Sumatra): Bengkulu; 103 Dead (Other Reports) 90 Dead; Mw7.9
2000	6 18	-13.80	97.45	7.8	Indian Ocean; 0 Dead; Mw7.6
2000	11 16	-3.98	152.17	8.2	Papua New Guinea: New Ireland; 2 Dead; Mw8.0
2000	11 16	-5.23	153.10	7.8	Papua New Guinea: New Ireland; 0 Dead; Mw7.6
2000	11 17	-5.50	151.78	8.0	Papua New Guinea: New Britain Is.; 0 Dead; Mw7.5
2001	1 13	13.05	-88.66	7.8	El Salvador/Guatemala; Many landslides (Responsible for majority of the dead); 852 Dead; Mw7.7
2001	1 26	23.42	70.23	8.0	India: Bhuj, Bhachau, Anjar (Gujarat)/Pakistan; 20,023 Dead; Mw7.7
2001	6 23	-16.26	-73.64	8.2	Peru: Arequipa-Camana-Tacna area/Chile: Arica; 139 Dead; Mw8.4
2001	11 14	35.95	90.54	8.0	Xizang, Qinghai [Kokoxili EQ]; 0 Dead; Dislocation; Mw7.8
2002	3 25	36.06	69.32	6.2	Afghanistan: Baghlan Province (Nahrin); 1,000 Dead; Mw6.1
2002	11 3	63.52	-147.44	7.9	USA: Alaska; 0 Dead; Dislocation; Ms8.5; Mw7.9
2003	1 20	-10.49	160.77	7.8	Solomon Is.; Mw7.3
2003	5 21	36.96	3.63	6.9	Algeria: Algiers-Boumerdes-Dellys-Thenia area; 2,266 Dead; Mw6.8

Continued.

Year Date	Latitude	Longitude	*M*	Region and Damage
2003 9 25	41.82°	143.91°	8.1	Tokachi [Tokachi offshore earthquake]; Mj8.0; Mw8.3
2003 12 26	29.00	58.31	6.8	Iran: Bam; 43,200 Dead; Mw6.6
2004 12 26	3.30	95.98	9.1	Indonesia: Sumatra; Large Tsunami; 227,898 Dead or Missing; M is from USGS; Mw9.0
2005 3 28	2.09	97.11	8.4	Indonesia: Sumatra; Tsunami; More than 1,303; Mw8.6
2005 10 8	34.54	73.59	7.7	Pakistan: Kashmir; More than 86,000 Dead; Mw7.6
2006 5 3	-20.19	-174.12	7.8	Tonga; Mw8.0
2006 5 26	-7.96	110.45	6.2	Indonesia (Java): Bantul-Yogyakarta; More than 5,749 Dead; Mw6.3
2006 11 15	46.59	153.27	7.8	Kuril; Mw8.3
2007 1 13	46.24	154.52	8.2	Kuril; Mw8.1
2007 4 1	-8.47	157.04	7.9	Solomon Is.; Tsunami; 52 Dead; Mw8.1
2007 8 15	-13.39	-76.60	7.9	Peru; More than 514 Dead; Mw8.0
2007 9 12	-4.44	101.37	8.5	Indonesia: Sumatra; 25 Dead; Mw8.5
2007 9 12	-2.63	100.84	8.1	Indonesia: Kepulauan-Mentawai; Mw7.9
2007 12 9	-26.00	-177.51	7.8	Fiji; M is Mw
2008 5 12	31.00	103.32	8.1	Sichuan; I=11; 69,227 Dead; Dislocation; Mw7.9
2009 4 6	42.33	13.33	6.3	Italy: [L'Aquila EQ]; I = 7; 309 Dead; M is Mw
2009 9 29	-15.49	-172.10	8.1	Samoa Is.; Tsunami; More than 192 Dead; Mw8.1
2009 9 30	-0.72	99.87	7.1	Indonesia (Sumatra): Padang; More than 1,117 Dead; M is mb; Mw7.5
2009 10 7	-12.52	166.38	7.9	Santa Cruz Is.; Tsunami; Mw7.8
2010 1 12	18.44	-72.57	7.3	Haiti: Port-au-Prince; 316,000 Dead; Mw7.0
2010 2 27	-36.12	-72.90	8.5	Chile: Bio-Bio offshore; Japanese Tsunami; I=9; More than 521 Dead; Mw8.8
2010 4 6	2.38	97.05	7.9	Indonesia: Sumatra; Tsunami; 0 Dead; Mw7.8
2010 4 13	33.17	96.55	7.0	Qinghai; More than 2,220 Dead; Mw6.9
2010 10 25	-3.49	100.08	7.3	Indonesia: South-Sumatra; Tsunami; More than 445 Dead; Mw7.8
2011 3 11	38.30	142.37	9.0	Sanriku offshore [The 2011 off the Pacific coast of Tohoku earthquake]; Large Tsunami; More than 21,000 Dead or Missing; Mw9.1
2011 3 11	36.28	141.11	7.9	Largest aftershock of the above; M is Mw
2012 4 11	2.33	93.06	8.6	Indonesia: Sumatra offshore; Small Tsunami; More than 10 Dead; M is Mw
2012 4 11	0.80	92.46	8.2	Largest Aftershock of the Above; M is Mw
2012 10 28	52.79	-132.10	7.8	Canada: Small Tsunami in Pacific Ocean; M is Mw
2013 2 6	-10.80	165.11	8.0	Santa Cruz Is.; Tsunami; 10 Dead; M is Mw
2013 5 24	54.89	153.22	8.3	The Okhotsk: Depth 600 km; M is Mw
2014 4 1	-19.61	-70.77	8.2	Chilie: Northern Coast; M is Mw
2014 6 23	51.83	178.75	7.9	Aleutian Is. (Rat Is.); M is Mw
2015 4 25	28.23	84.73	7.8	Nepal: [Gorkha EQ]; I = 8; More than 9,164 Dead or Missing; Avalanche; M is Mw
2015 5 30	27.84	140.49	7.8	Ogasawara Islands; Depth 664 km; M is Mw
2016 2 5	22.94	120.64	6.4	Taiwan [Meinong earthquake]; I=7; More than 117 Dead; M is Mw
2016 3 2	-4.95	94.33	7.8	Indonesia (Sumatra): I=5; M is Mw

Main Major Earthquakes and Damaging Earthquakes Continued.

Year	Date	Latitude	Longitude	M	Region and Damage
2016	4 6	0.38°	-79.92°	7.8	Ecuador: I=8; More then 677 Dead or Missing; Small Tsunami; M is Mw
2016	4 15	32.79	130.75	7.0	Kumamoto Prefecture [Kumamoto earthquake]: 50 Dead; M is Mw
2016	8 24	42.72	13.19	6.2	Italy: [Central Italy EQs] More then 297 Dead; M is Mw ; October 30 North
2016	9 3	36.43	-96.93	5.8	USA: Olahoma Triggered earthquake; M is from USGS; Mw5.7
2016	11 13	-42.74	173.05	7.8	New Zealand: [Kaikoura EQ]; 2 Dead; M is Mw
2016	12 8	-10.68	161.33	7.8	Solomon Is.: M is Mw
2016	12 17	-4.51	153.52	7.9	Papua New Guinea:New Ireland region; M is Mw
2017	1 22	-6.25	155.17	7.9	Solomon Is.: 3 M is Mw
2017	9 8	15.02	-96.93	8.2	Mexico: Chiapas offshore; More than 96 Dead; M is Mw
2018	6 17	34.83	135.64	5.5	Osaka prefecture: 6 Dead; M is Mw
2018	9 5	42.69	141.93	6.6	Hokkaido[Hokkaido Eastern Iburi earthquake]; 43Dead; Large-scale power outages; M is Mw
2018	9 6	-18.47	179.35	7.9	Fiji: M is Mw
2018	9 28	-0.26	119.85	7.5	Indonesia: Sulawesi; More than 2077 Dead; M is Mw
2019	5 16	-5.81	-75.27	8.0	Peru: Loreto endorheic; Depth 123km; 2 Dead; M is Mw
2019	7 6	35.77	-117.6	7.1	USA: California[Ridgecrest EQ]; 50 Completely destroyed; Fault; M is Mw, Mw6.5 earthquake 34 hours ago.

(As of December, 2019)

Main Events in Seismology

Year	Event	Inventor/Discoverer (Nationality)
1872	Earthquake observed in Japan using a seismograph	Verbeck (Netherlands) Knipping (Germany)
1880	The Seismological Society of Japan founded after the Yokohama Earthquake (world's first seismological society)	
1880	Earthquake observed using a modern seismograph with pendulum application	Ewing (England)
1884	Started collection of earthquake reports from throughout Japan by the Tokyo District Meteorological Observatory Established Japan's first four-class seismic intensity scale	Sekiya (Japan)
1885	Theory of Rayleigh waves (types of surface waves)	Rayleigh (England)
1892	Earthquake Preparedness Survey Committee started in response to the Nobi Earthquake which occurred the prior year	
1894	Formula related to the decrease in number of aftershocks	Omori (Japan)
1899	Relationship between the duration of preliminary tremors and epicentral distance	Omori (Japan)
1907	Analysis of sub-surface/underground structure using travel-time curve	Herglotz (Germany)
1907	Development of electromagnetic seismometer	Galitzin (Russia)
1910	Discovery of Mohorovicic discontinuity	Mohorovicic (Croatia)

Continued.

Year	Event	Inventor/Discoverer (Nationality)
1911	Theory of Love waves (types of surface waves)	Love (England)
1911	Elastic rebound theory related to occurrence of earthquakes	Reid (USA)
1913	Estimated the radius of the earth's core	Gutenberg (Germany)
1916	Implemented the design seismic coefficient (structural theory for earthquake resistance of buildings)	Sano (Japan)
1917	Research for the regularity of initial distribution for earthquakes	Shida (Japan)
1919	Theory of structural vibration and seismic resistance	Mononobe (Japan)
1925	After completing the survey for the 1923 Kanto Earthquake, the Earthquake Preparedness Survey Committee was abolished and the Earthquake Research Institute was established	
1926	Proposed the existence of a low-velocity layer in the mantle	Gutenberg (Germany)
1927	Confirmed the existence of deep-depth earthquakes	Wadachi (Japan)
1929	Estimated the crust structure near Japan	Matsuzawa (Japan)
1931-	Creation of travel-time table	Jeffreys (England)
1933	Travel-time table for P-waves near Japan	Wadachi, Sagisaka, Masuda (Japan)
1935	Proposed the magnitude scale for earthquakes	Richter (USA)
1936	Proposed the existence of an inner core	Lehmann (Denmark)
1949	Systematic earthquake catalog	Gutenberg (Germany), Richter (USA)
1949	Based on the Fukui Earthquake in the previous year, grade 7 added to the seismic intensity scale	Central Meteorological Observatory (Japan)
1951	Completion of the "Dai Nihon Jishin Shiryo" (Japan Earthquake Materials)	Musha (Japan)
1952	Started issuing warnings for tsunami in Japan	
1960	Started observation of the earth's free oscillations in response to the Chile Earthquake	
1961	Started the establishment of the Worldwide Standard Seismograph Network	
1963	Elasticity theory of dislocation	Maruyama (Japan)
1964	Research for the moving epicentral model	Haskell (USA)
1966	Proposed the concept of seismic moment	Keiichi Aki (Japan)
1967	Seismic model for sinking plates	Utsu (Japan), Oliver, Isacks (USA)
1970	Established the International Seismological Centre	
1973,5	Discovered the double seismic zone	Tsumura, Umino, Hasegawa (Japan)
1976	Seismic wave tomography	Aki (Japan)
1977	Proposed the concept of moment magnitude	Kanamori (Japan)
1986	Started the Global Seismographic Network	IRIS, USGS
1995	In response to the South Hyogo Earthquake, established the Headquarters for Earthquake Research Promotion and started the fundamental observation network	
1996	Instrumental seismic intensity implemented at JMA	

Distribution Map for Shallow Earthquakes ($M \geqq 4.0$, depth = 100 km or less, 1991 to 2010)

Figure 29

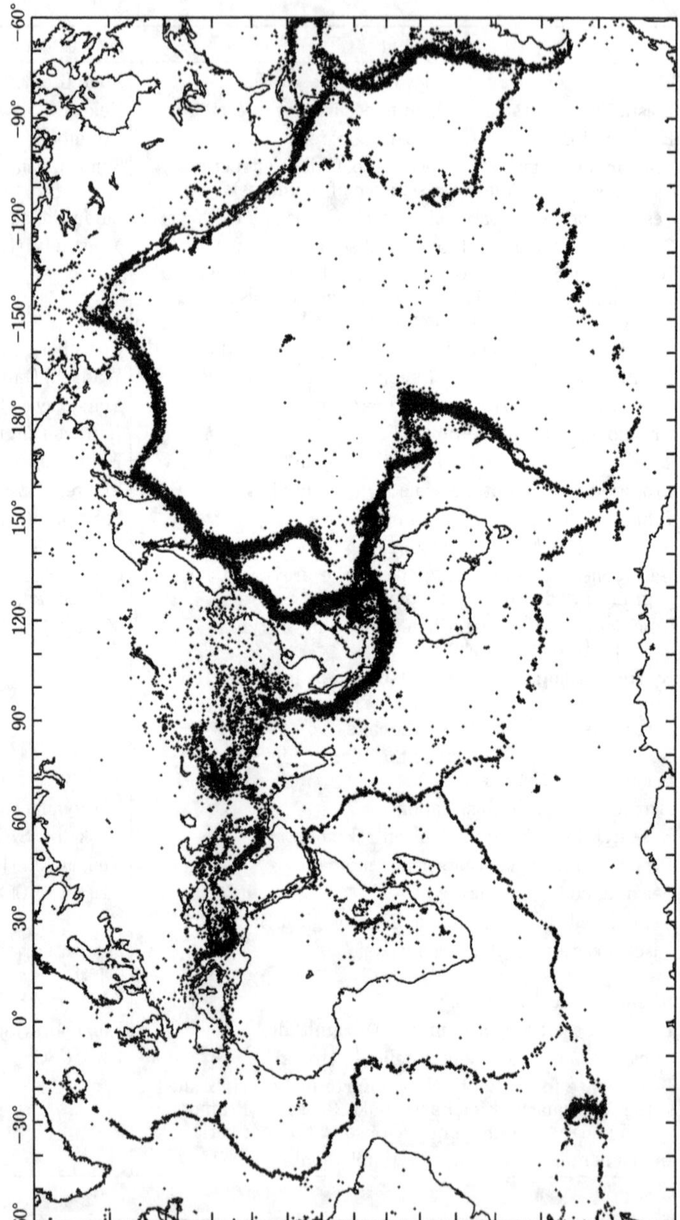

(Based on material from the International Seismological Center, ISC)

* It can be shown that shallow earthquakes mainly occur in belt-shaped regions along plate borders (refer to gray line shown in Figure 31).

Distribution Map for Plate Borders and Deep Earthquakes ($M \geqq 4.0$, depth = 100 km or more, 1991 to 2010)

Figure 30

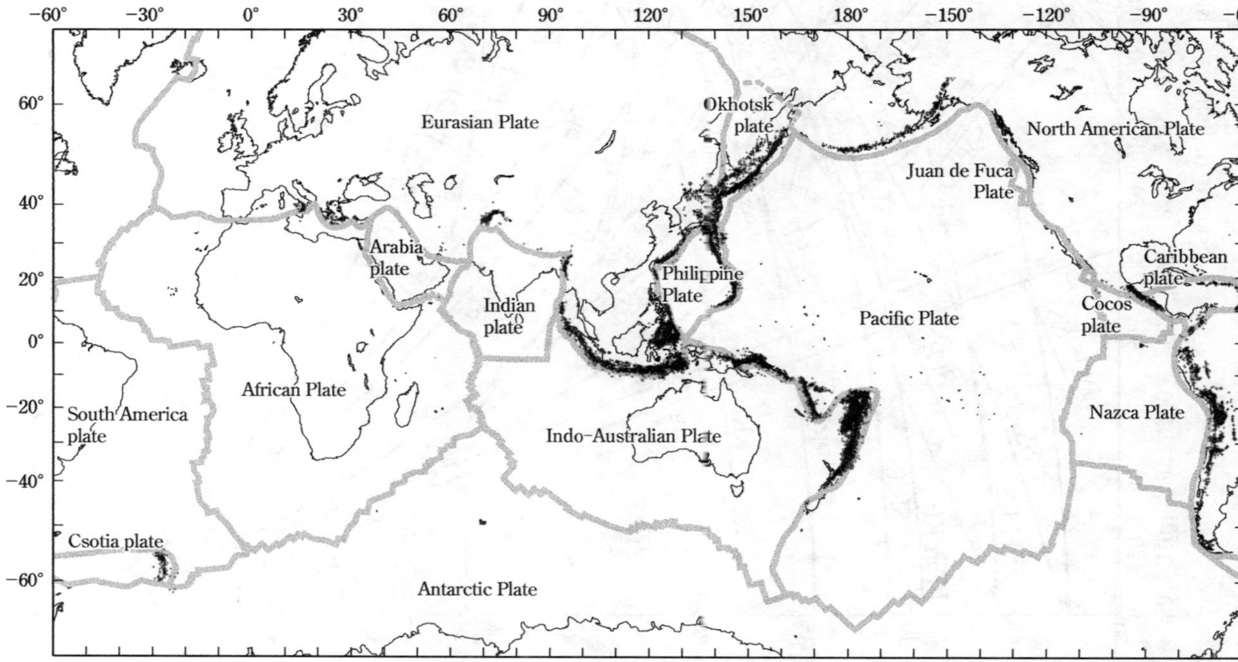

(Based on material from the International Seismological Center, ISC)

* Deep earthquakes only occur in certain regions such as the Pacific rim and the Indonesian coast. The location of deep earthquakes is slightly removed from the plate borders (gray line) because the plates are sinking diagonally.

Logical Travel-Time Curve

Figure 31 (Source: H. Jeffreys & K. E. Bullen)

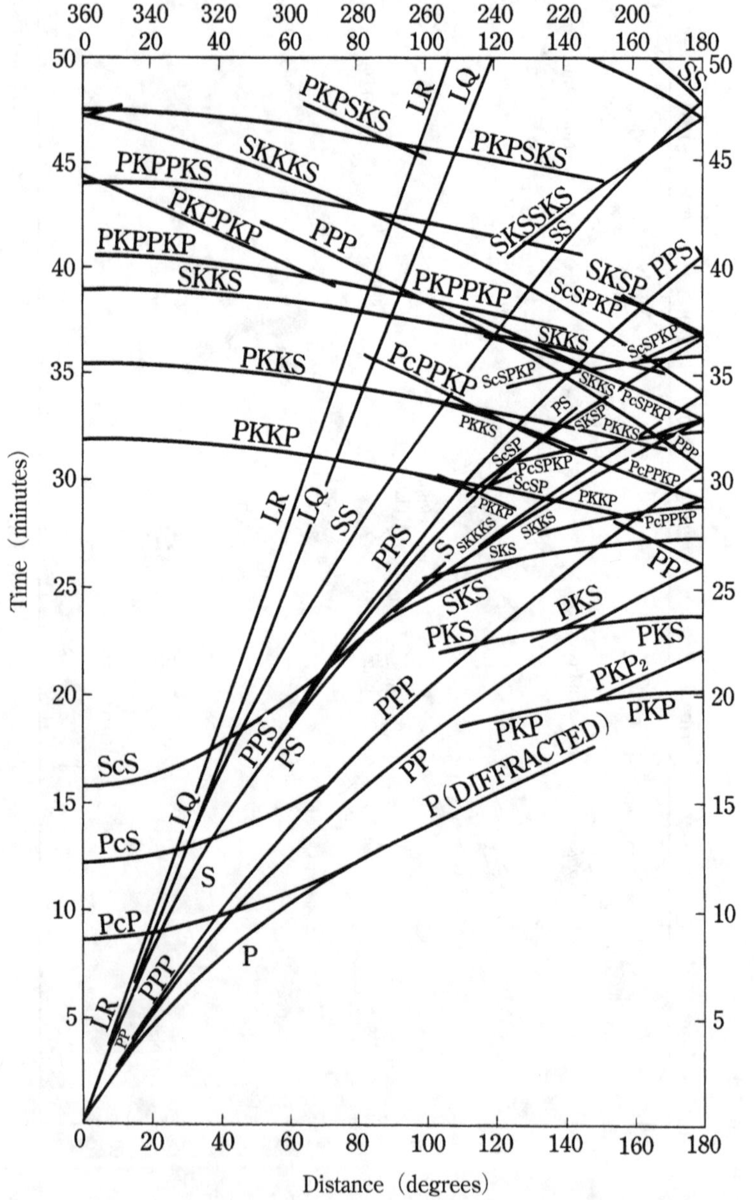

Travel-Time Curve based on Observed Data

Based on the catalogs of the International Seismological Centre (ISC), we plotted more than 190 000 travel-time points for earthquakes with a magnitude of at least 5 and a depth of until 50 km which occurred throughout the world in 2010. When comparing to the logical travel-time curve shown in Figure 32, it can be seen what kinds of phases are actually observed with great frequency.

Figure 32

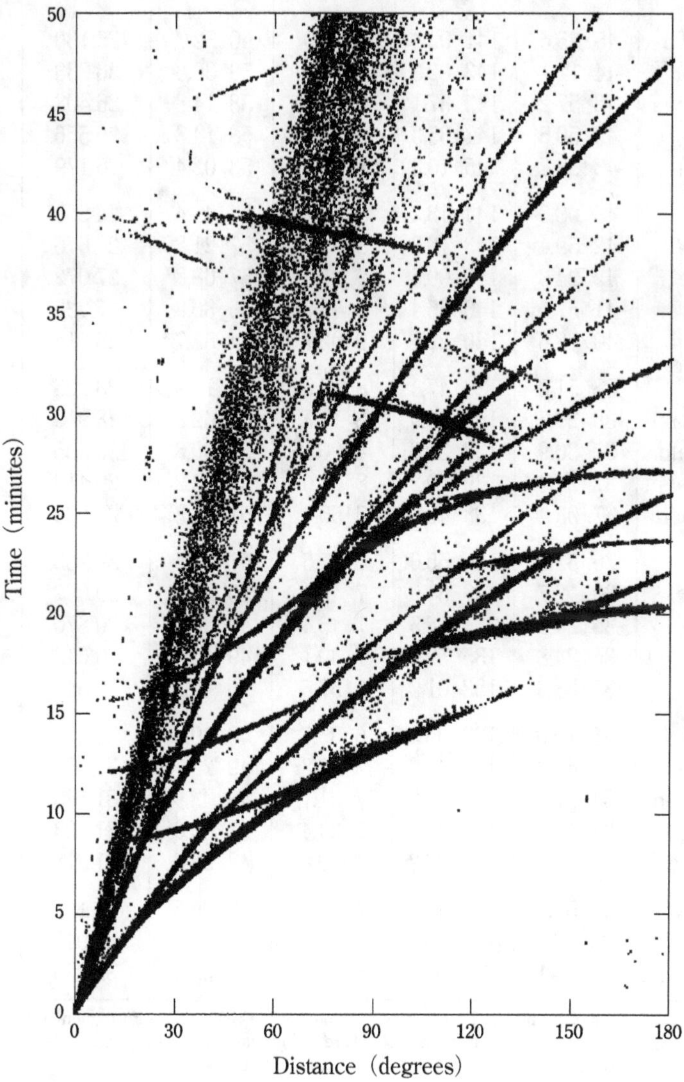

Geomagnetism and Gravity

All Japan Geomagnetic Element (Values for 2015.0)

Magnetic point	Latitude (N)	Longitude (E)	Declination (W)	Inclination	Horizontal component	Total magnetic force
	° ′	° ′	° ′	° ′	nT	nT
Rebun To	45 25.6	141 02.4	8 55.0	60 32.7	25 130	51 104
Takinoue	44 12.1	143 02.3	9 34.4	58 30.8	26 233	50 226
Asahikawa	43 57.2	142 37.1	9 50.7	58 14.8	26 408	50 181
Rumoi	43 50.5	141 35.3	9 20.4	58 14.8	26 556	50 461
Shibetsu	43 38.2	145 10.0	8 44.9	58 02.4	26 179	49 457
Kushiro	43 08.5	144 13.2	8 25.1	57 49.4	26 350	49 480
Obihiro	42 59.4	143 19.8	8 53.5	57 21.8	26 676	49 464
Imakane	42 24.3	140 00.5	9 37.5	57 08.8	27 072	49 903
Kayabe	41 57.6	140 55.4	8 49.7	56 36.8	27 222	49 468
Odate	40 18.6	140 22.1	9 02.1	54 52.7	28 014	48 693
Sakata	38 51.0	139 47.4	8 18.5	53 13.9	28 926	48 324
Ishinomaki	38 29.5	141 10.9	8 18.3	53 02.8	28 910	48 089
Izumozaki	37 30.9	138 39.6	8 06.1	52 00.8	29 505	47 939
Wakamatsu	37 22.4	139 49.2	8 01.6	51 30.2	29 413	47 252
Tokamachi	37 05.4	138 46.1	7 55.0	51 24.8	29 825	47 819
Himi	36 52.2	136 55.0	8 07.5	51 16.8	29 985	47 936
Tomioka	36 12.7	138 54.9	7 40.4	50 25.3	29 987	47 065
Tottori	35 24.9	134 18.5	7 45.4	49 58.4	30 870	47 998
Imazu	35 24.8	135 59.2	7 39.1	49 44.8	30 612	47 375
Matsue	35 23.3	133 01.2	7 57.5	50 47.7	30 548	48 328
Shimizu	35 10.0	138 25.3	6 35.5	48 49.9	30 652	46 564
Okayama	34 46.8	133 47.2	7 42.8	49 22.9	31 154	47 854
Yamaguchi	34 17.6	131 25.2	7 21.1	49 03.9	31 548	48 149
Awaji	34 17.3	134 40.6	7 24.4	48 33.0	31 203	47 137
Kawanoe	33 56.8	133 39.2	7 19.8	48 20.8	31 378	47 212
Nagasaki	32 51.5	129 45.0	7 00.0	47 35.5	32 113	47 618
Nakamura	32 48.7	132 45.5	7 00.6	47 01.4	31 947	46 864
Tanegashima	30 44.1	131 03.9	6 18.7	44 23.4	32 926	46 076

According to results of geomagnetic survey carried out by the Geospatial Information Authority of Japan.
Latitude (N), longitude (E), and declination (W) are values for north latitude, east longitude, and west declination, respectively.

Figure 33 Geomagnetic charts of Japan (Declination) (Values for 2015.0)

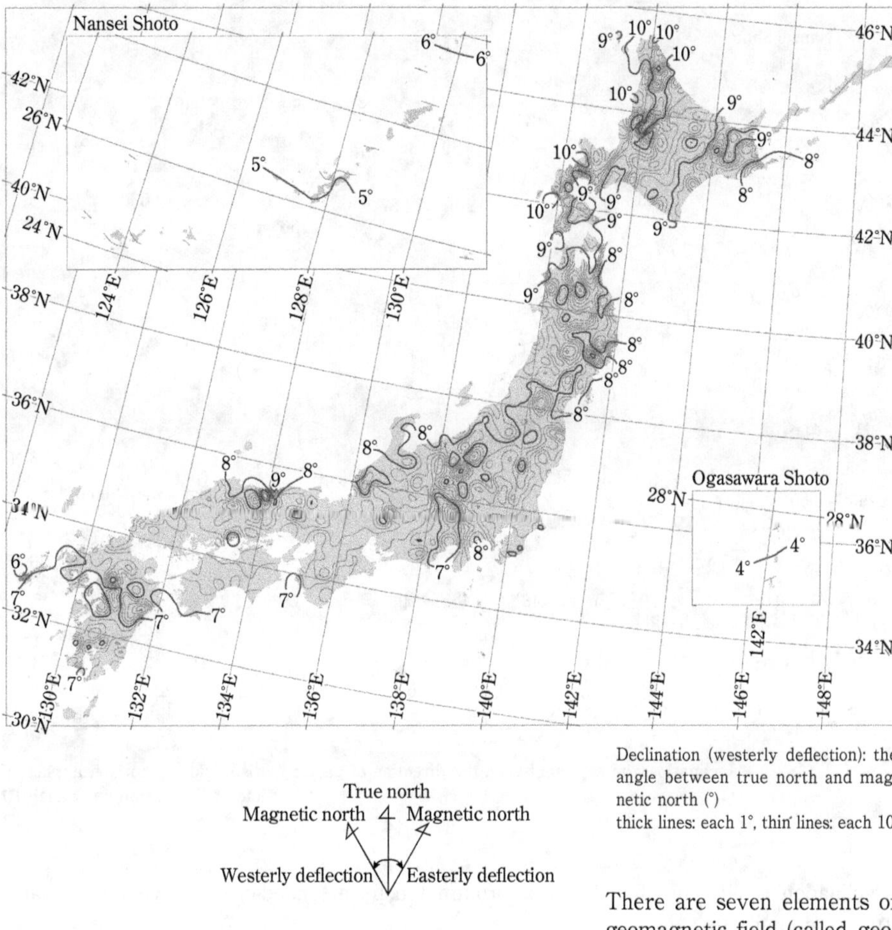

Declination (westerly deflection): the angle between true north and magnetic north (°)
thick lines: each 1°, thin lines: each 10′

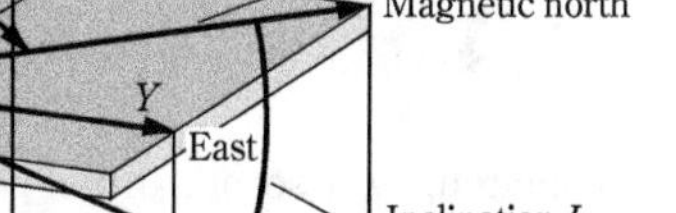

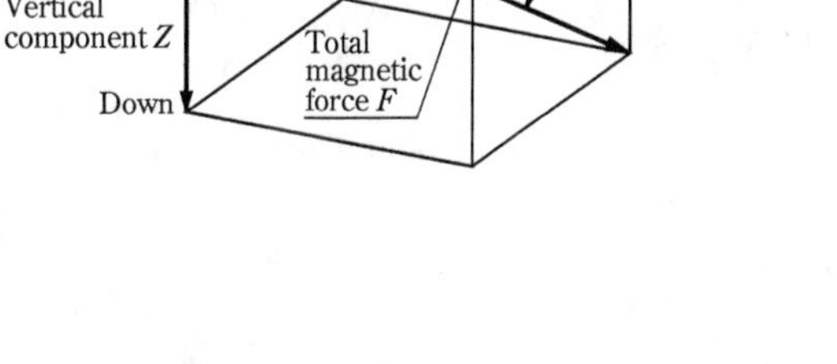

There are seven elements of geomagnetic field (called geomagnetic components): total magnetic force (F), northward component (X), eastward component (Y), vertical component (Z), horizontal component (H), declination (the angle measured eastward from the true north, D), and inclination (the angle measured downward from the horizontal plane, I).

Figure 34 Geomagnetic charts of Japan (Inclination) (Values for 2015.0)

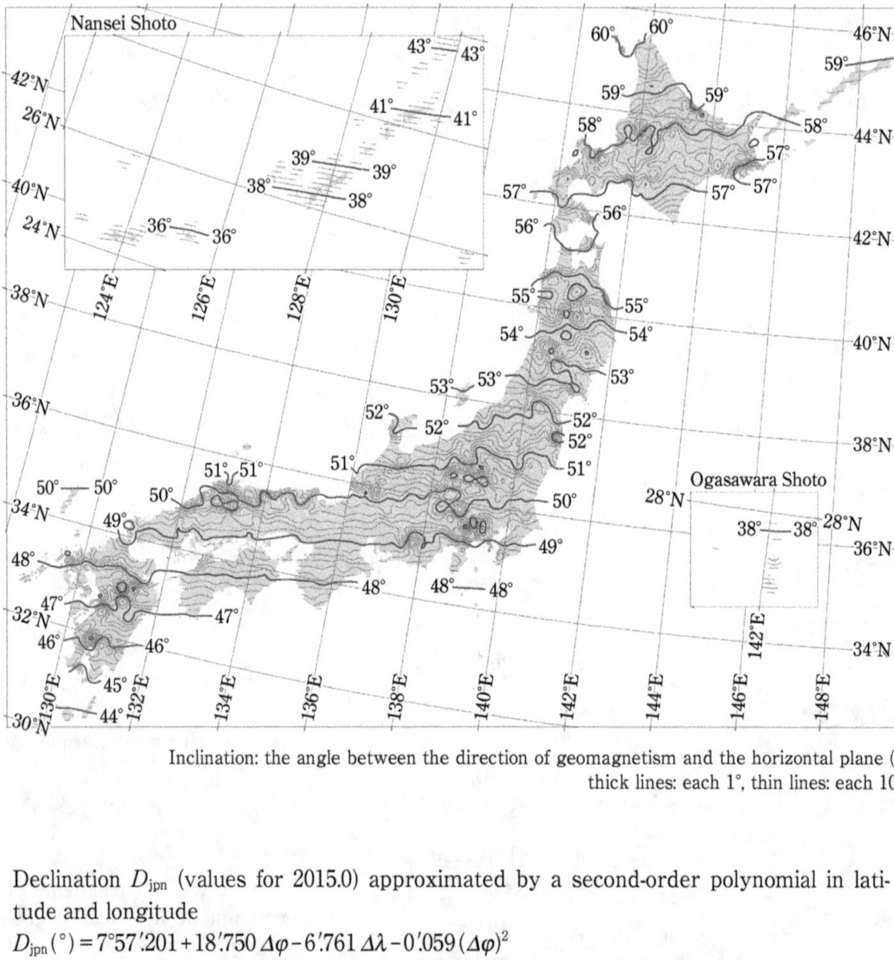

Inclination: the angle between the direction of geomagnetism and the horizontal plane (°)
thick lines: each 1°, thin lines: each 10′

Declination D_{jpn} (values for 2015.0) approximated by a second-order polynomial in latitude and longitude

$$D_{\text{jpn}}(°) = 7°57'.201 + 18'.750\,\Delta\varphi - 6'.761\,\Delta\lambda - 0'.059\,(\Delta\varphi)^2 - 0'.014\,\Delta\varphi\Delta\lambda - 0'.579\,(\Delta\lambda)^2$$

$\Delta\varphi = \varphi - 37°\text{N}$, $\Delta\lambda = \lambda - 138°\text{E}$ (φ is latitude, λ is longitude, in units of degrees)

Inclination I (values for 2015.0) approximated by a second-order polynomial in latitude and longitude

$$I(°) = 51°23'.425 + 72'.639\,\Delta\varphi - 8'.364\,\Delta\lambda - 0'.951\,(\Delta\varphi)^2 - 0'.212\,\Delta\varphi\Delta\lambda + 0'.580\,(\Delta\lambda)^2$$

$\Delta\varphi = \varphi - 37°\text{N}$, $\Delta\lambda = \lambda - 138°\text{E}$ (φ is latitude, λ is longitude, in units of degrees)

Figure 35 Geomagnetic charts of Japan (Horizontal component) (Values for 2015.0)

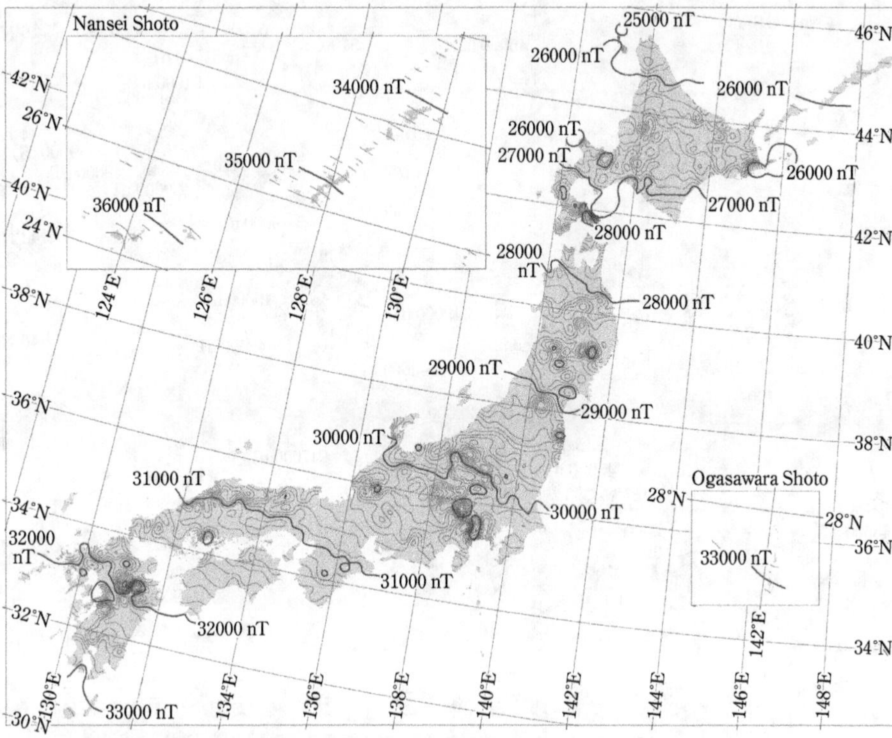

Horizontal component: magnitude of geomagnetic field in a horizontal plane (nT)
thick lines: each 1000 nT, thin lines: each 100 nT

Horizontal component H (values for 2015.0) approximated by a second-order polynomial in latitude and longitude.

$$H\,(\text{nT}) = 29777.000 - 432.944\,\Delta\varphi - 79.882\,\Delta\lambda - 6.464\,(\Delta\varphi)^2 + 8.428\,\Delta\varphi\Delta\lambda - 5.109\,(\Delta\lambda)^2$$

$\Delta\varphi = \varphi - 37°\text{N}$, $\Delta\lambda = \lambda - 138°\text{E}$ (φ is latitude, λ is longitude, in units of degrees)

Note) For the horizontal component of geomagnetic field, a character 'H' is traditionally used. This is different from the H which is used in the Physics and Chemistry section to indicate the magnetic field.

Figure 36 Geomagnetic charts of Japan (Total magnetic force) (Values for 2015.0)

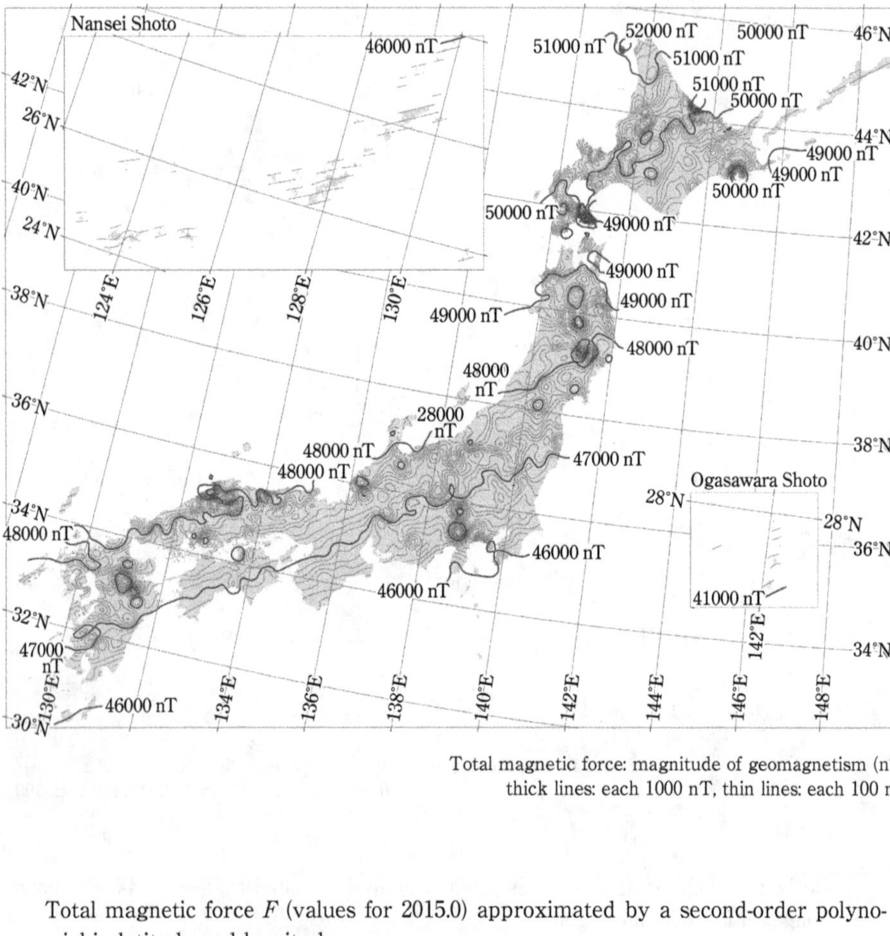

Total magnetic force: magnitude of geomagnetism (nT)
thick lines: each 1000 nT, thin lines: each 100 nT

Total magnetic force F (values for 2015.0) approximated by a second-order polynomial in latitude and longitude

$$F\,(\text{nT}) = 47700.818 + 550.932\,\Delta\varphi - 259.776\,\Delta\lambda - 2.131\,(\Delta\varphi)^2 - 2.992\,\Delta\varphi\Delta\lambda + 4.879\,(\Delta\lambda)^2$$

$\Delta\varphi = \varphi - 37°\text{N}$, $\Delta\lambda = \lambda - 138°\text{E}$ (φ is latitude, λ is longitude, in units of degrees)

Geomagnetic Elements in Different Regions

Location	Area	Latitude	Longitude	Year	Declination	Inclination	Horizontal component	Total magnetic force
Asia and the Pacific(60°E–165°W)								
		° ′	° ′		° ′	° ′	nT	nT
Dixon	Siberia	73 33N	80 34E	2016	30 20.8E	+84 43.3	5406	58768
Tixie Bay	Siberia	71 35N	129 00E	2016	17 54.8W	+83 3.6	7234	59866
Novosibirsk	Siberia	55 02N	82 54E	2018	8 11.5E	+74 18.2	16205	59896
Irkutsk	Siberia	52 10N	104 27E	2018	3 52.6W	+72 14.6	18431	60436
Manzhouli	China	49 36N	117 24E	2017	9 34.9W	+68 41.2	21171	58246
Memambetsu	Japan	43 55N	144 11E	2018	9 3.7W	+58 25.7	26153	49953
Alma At	Kazakhstan	43 15N	76 55E	2015	5 4.2E	+63 15.8	24751	55015
Beijing	China	40 18N	116 12E	2017	8 5.8W	+59 15.2	28215	55191
Kakioka	Japan	36 14N	140 11E	2018	7 36.9W	+50 3.6	30012	46748
Kanoya	Japan	31 25N	130 53E	2018	6 45.5W	+45 31.0	32714	46688
Wuhan	China	30 32N	114 34E	2017	4 20.8W	+46 52.6	33946	49660
Chichijima	Ogasawara Islands	27 06N	142 11E	2018	4 18.5W	+37 32.0	32766	41320
Phu Thuy	Vietnam	21 02N	105 57E	2017	1 30.8W	+31 3.9	38826	45326
Alibag	India	18 38N	72 52E	2018	0 22.7E	+27 51.8	38203	43213
Tondano	Indonesia	1 17N	124 57E	2017	0 10.4E	-13 1.6	39323	40361
Kakadu	Australia North	12 41S	132 28E	2018	2 57.9E	-39 47.9	35489	46192
Apia	Samoa Island	13 48S	171 47W	2018	12 2.9E	-31 2.8	33278	38842
Gnangara	Australia West	31 47S	115 57E	2012	1 51.1W	-66 8.2	23529	58160
Canberra	Australia East	35 19S	149 22E	2018	12 35.3E	-65 52.0	23719	58012
Eyrewell	New Zealand	43 29S	172 24E	2018	23 49.4E	-68 33.5	20993	57427
Port Aux Francais	Indian Ocean	49 21S	70 16E	2013	55 27.2W	-68 39.7	17862	49089
Macquarie island	Macquarie Island	54 30S	158 57E	2018	32 22.1E	-78 33.7	12697	64032
Dumont d'Urville	Antarctic	66 40S	140 01E	2013	169 22.9E	-88 29.5	1823	69285
Mawson	Antarctic	67 36S	62 53E	2018	68 44.1W	-67 52.4	18542	49227
Vostok	Antarctic	78 27S	106 52E	2018	124 35.8W	-76 42.3	13640	59312
Europe and Africa(30°W–60°E)								
Hornsund	Svalbard Island	77 00N	15 33E	2018	8 40.6E	+81 45.9	7830	54662
Tromso	Norway	69 40N	18 57E	2017	6 40.3E	+78 18.8	10851	53566
Sodankyla	Finland	67 22N	26 38E	2017	11 51.8E	+77 28.0	11487	52940
Leirvogur	Iceland	64 11N	21 42W	2019	10 46.8W	+75 46.7	12744	51874
Dombas	Norway	62 04N	9 07E	2017	2 16.8E	+74 2.8	14098	51292
Lerwick	Scotland	60 08N	1 11W	2019	0 54.5W	+72 49.7	15064	51022
Uppsala	Sweden	59 54N	17 21E	2018	6 3.6E	+72 50.5	15193	51500
Arti	Russia	56 26N	58 34E	2018	13 40.1E	+73 32.5	15992	56449
Eskdalemuir	Scotland	55 19N	3 12W	2019	1 49.2W	+69 18.4	17592	49784
Hel	Poland	54 36N	18 49E	2018	5 15.7E	+69 37.4	17570	50463
Dourbes	Belgium	50 06N	4 36E	2019	1 29.9E	+65 32.9	20164	48716

Geomagnetic Elements in Different Regions Continued.

Location	Area	Latitude	Longitude	Year	Declination	Inclination	Horizontal component	Total magnetic force
Europe and Africa(30°W–60°E)								
		° ′	° ′		° ′	° ′	nT	nT
Niemegk	Germany	52 04N	12 41E	2018	3 45.8E	+67 31.6	18907	49464
Hartland	England	51 00N	4 29W	2019	1 31.9W	+65 58.5	19809	48653
Chambon-La-Foret	France	48 01N	2 16E	2019	0 54.1E	+63 41.9	21259	47978
Ebro	Spain	40 57N	0 20E	2018	0 32.2E	+56 4.1	25281	45291
Iznik	Turkey	40 30N	29 44E	2018	5 27.1E	+58 4.8	25179	47621
Coimbra	Portugal	40 13N	8 25W	2017	2 20.9W	+54 39.4	25710	44446
San Fernando	Spain	36 28N	6 12W	2019	1 0.2W	+50 3.8	27670	43105
Bar Gyora	Israel	31 43N	35 05E	2018	4 24.7E	+47 56.7	30071	44892
Eilat	Israel	29 40N	34 57E	2015	4 4.9E	+44 23.1	31165	43608
Guimar	Canary Islands	28 19N	16 26W	2017	7 15.6W	+39 4.0	27882	35912
M'Bour	Senegal	14 23N	16 58W	2018	6 59.5W	+ 6 41.1	32570	32793
Bangui	Central Africa	4 20N	18 34E	2010	0 10.2E	−16 18.6	32034	33377
Tsumeb	Namibia	19 12S	17 35E	2018	8 48.8W	−61 0.8	14194	29289
Maputo	Mozambique	25 55S	32 35E	2018	19 39.8W	−60 38.8	14484	29547
Hermanus	South Africa	34 25S	19 14E	2017	25 48.9W	−65 22.1	10625	25493
Syowa Station	Antarctic	69 00S	39 35E	2019	51 21.1W	−63 18.6	19279	42923
North and South America(165°W–30°W)								
Thule	Greenland	77 29N	69 10W	2018	44 41.3W	+85 46.6	4146	56297
Resolute Bay	Canada	74 41N	94 54W	2018	20 31.3W	+87 8.4	2870	57542
Barrow	Alaska	71 19N	156 37W	2018	13 43.9E	+80 54.2	9069	57358
Godhavn	Greenland	69 15N	53 32W	2018	32 18.3W	+80 50.8	8975	56424
College	Alaska	64 52N	147 52W	2018	17 23.7E	+77 9.8	12575	56600
Baker Lake	Canada	64 19N	96 01W	2017	2 51.8W	+83 16.9	6859	58629
Sitka	Alaska	57 04N	135 20W	2018	18 46.2E	+73 25.9	15806	55432
Meanook	Canada	54 37N	113 21W	2018	14 31.7E	+75 43.8	14070	57082
St. John's	Canada	47 36N	52 41W	2018	17 48.1W	+66 54.3	19990	50962
Fredericksburg	U.S.A	38 12N	77 22W	2018	10 33.2W	+65 2.8	21500	50961
Tucson	U.S.A	32 10N	110 44W	2018	9 6.7E	+59 4.6	24233	47155
Honolulu	Hawaii	21 19N	158 00W	2018	9 43.8E	+37 47.9	27302	34552
San Juan	Puerto Rico	18 07N	66 09W	2016	12 52.6W	+43 14.3	27002	37065
Fuquene	Columbia	5 28N	73 44W	2014	7 17.9W	+29 37.6	27340	31452
Tatuoca	Brazil	1 12S	48 31W	2018	20 13.0W	− 2 17.7	26423	26444
Huancayo	Peru	12 02S	75 19W	2017	3 5.7W	− 0 37.4	24829	24830
Vassouras	Brazil	22 24S	43 39W	2018	22 44.9W	−39 28.8	17975	23289
Pilar	Argentina	31 40S	63 53W	2018	5 46.0W	−34 27.2	18747	22736

Past data from each measurement location are available at http://www.geomag.bgs.ac.uk/data service/data/annual means.shtml.
Geomagnetic elements at a given point are calculated with a model, IGRF (http://wdc.kugi.kyoto-u.ac.jp/igrf/point/index-j.html)

International Geomagnetic Reference Field Expansion Coefficient (2020.0, IGRF-13)

n	m	g_n^m (nT)	$\dot{g}_n^m$ (nT/year)	h_n^m (nT)	$\dot{h}_n^m$ (nT/year)	n	m	g_n^m (nT)	$\dot{g}_n^m$ (nT/year)	h_n^m (nT)	$\dot{h}_n^m$ (nT/year)	n	m	g_n^m (nT)	h_n^m (nT)
1	0	-29404.8	5.7			8	0	23.7	0.0			11	5	0.3	0.6
1	1	-1450.9	7.4	4652.5	-25.9	8	1	9.7	0.1	8.4	-0.2	11	6	-0.7	-0.2
2	0	-2499.6	-11.0			8	2	-17.6	-0.1	-15.3	0.6	11	7	-0.1	-1.7
2	1	2982.0	-7.0	-2991.6	-30.2	8	3	-0.5	0.4	12.8	-0.2	11	8	1.4	-1.6
2	2	1677.0	-2.1	-734.6	-22.4	8	4	-21.1	-0.1	-11.7	0.5	11	9	-0.6	-3.0
3	0	1363.2	2.2			8	5	15.3	0.4	14.9	-0.3	11	10	0.2	-2.0
3	1	-2381.2	-5.9	-82.1	6.0	8	6	13.7	0.3	3.6	-0.4	11	11	3.1	-2.6
3	2	1236.2	3.1	241.9	-1.1	8	7	-16.5	-0.1	-6.9	0.5	12	0	-2.0	
3	3	525.7	-12.0	-543.4	0.5	8	8	-0.3	0.4	2.8	0.0	12	1	-0.1	-1.2
4	0	903.0	-1.2			9	0	5.0				12	2	0.5	0.5
4	1	809.5	-1.6	281.9	-0.1	9	1	8.4		-23.4		12	3	1.3	1.4
4	2	86.3	-5.9	-158.4	6.5	9	2	2.9		11.0		12	4	-1.2	-1.8
4	3	-309.4	5.2	199.7	3.6	9	3	-1.5		9.8		12	5	0.7	0.1
4	4	48.0	-5.1	-349.7	-5.0	9	4	-1.1		-5.1		12	6	0.3	0.8
5	0	-234.3	-0.3			9	5	-13.2		-6.3		12	7	0.5	-0.2
5	1	363.2	0.5	47.7	0.0	9	6	1.1		7.8		12	8	-0.3	0.6
5	2	187.8	-0.6	208.3	2.5	9	7	8.8		0.4		12	9	-0.5	0.2
5	3	-140.7	0.2	-121.2	-0.6	9	8	-9.3		-1.4		12	10	0.1	-0.9
5	4	-151.2	1.3	32.3	3.0	9	9	-11.9		9.6		12	11	-1.1	0.0
5	5	13.5	0.9	98.9	0.3	10	0	-1.9				12	12	-0.3	0.5
6	0	66.0	0.5			10	1	-6.2		3.4		13	0	0.1	
6	1	65.5	-0.3	-19.1	0.0	10	2	-0.1		-0.2		13	1	-0.9	-0.9
6	2	72.9	0.4	25.1	-1.6	10	3	1.7		3.6		13	2	0.5	0.6
6	3	-121.5	1.3	52.8	-1.3	10	4	-0.9		4.8		13	3	0.7	1.4
6	4	-36.2	-1.4	-64.5	0.8	10	5	0.7		-8.6		13	4	-0.3	-0.4
6	5	13.5	0.0	8.9	0.0	10	6	-0.9		-0.1		13	5	0.8	-1.3
6	6	-64.7	0.9	68.1	1.0	10	7	1.9		-4.3		13	6	0.0	-0.1
7	0	80.6	-0.1			10	8	1.4		-3.4		13	7	0.8	0.3
7	1	-76.7	-0.2	-51.5	0.6	10	9	-2.4		-0.1		13	8	0.0	-0.1
7	2	-8.2	0.0	-16.9	0.6	10	10	-3.8		-8.8		13	9	0.4	0.5
7	3	56.5	0.7	2.2	-0.8	11	0	3.0				13	10	0.1	0.5
7	4	15.8	0.1	23.5	-0.2	11	1	-1.4		0.0		13	11	0.5	-0.4
7	5	6.4	-0.5	-2.2	-1.1	11	2	-2.5		2.5		13	12	-0.5	-0.4
7	6	-7.2	-0.8	-27.2	0.1	11	3	2.3		-0.6		13	13	-0.4	-0.6
7	7	9.8	0.8	-1.8	0.3	11	4	-0.9		-0.4					

IGRF is an abbreviation for International Geomagnetic Reference Field. IGRF-13 refers to the 13th Generation model. For conventional naming, IGRF-13 uses model coefficient groups which are sets of the model coefficients for each 5-year period from 1900 for the latest model coefficient of IGRF 2020. Models from 2000 use degrees up to 13. For details, refer to http://www.ngdc.noaa.gov/IAGA/vmod/igrf.html.

Geomagnetic components at epoch 2020.0 are given with following equations.

$$X(\text{northward}) = \frac{1}{r}\frac{\partial V}{\partial \theta},\quad Y(\text{eastward}) = \frac{-1}{r\sin\theta}\frac{\partial V}{\partial \lambda},\quad Z(\text{downward}) = \frac{\partial V}{\partial r}$$

Where $V = a\sum_{n=1}^{13}\sum_{m=0}^{n}\left(\frac{a}{r}\right)^{n+1}[g_n^m \cos m\lambda + h_n^m \sin m\lambda]P_n^m(\mu)$

$$P_n^m(\mu) = \frac{1}{2^n\, n!}\left[\frac{\varepsilon_m (n-m)!(1-\mu^2)^m}{(n+m)!}\right]^{1/2}\frac{d^{m+n}(\mu^2-1)^n}{d\mu^{m+n}}$$

$(\mu = \cos\theta,\ \varepsilon_m = 1,\ m = 0,\ \varepsilon_m = 2,\ m \geqq 1,\ a = 6371.2\ \text{km})$

The coefficients at epoch 't' are expressed as follow if $t_0 = 2020.0$

$$g_n^m(t) = g_n^m(t_0) + (t - t_0)\dot{g}_n^m(t_0),\quad h_n^m(t) = h_n^m(t_0) + (t - t_0)\dot{h}_n^m(t_0)$$

Here, $\dot{g}_n^m$ and $\dot{h}_n^m$ are the rate of secular variation (nT/year) of the coefficients.

Figure 37

Declination (IGRF-13)

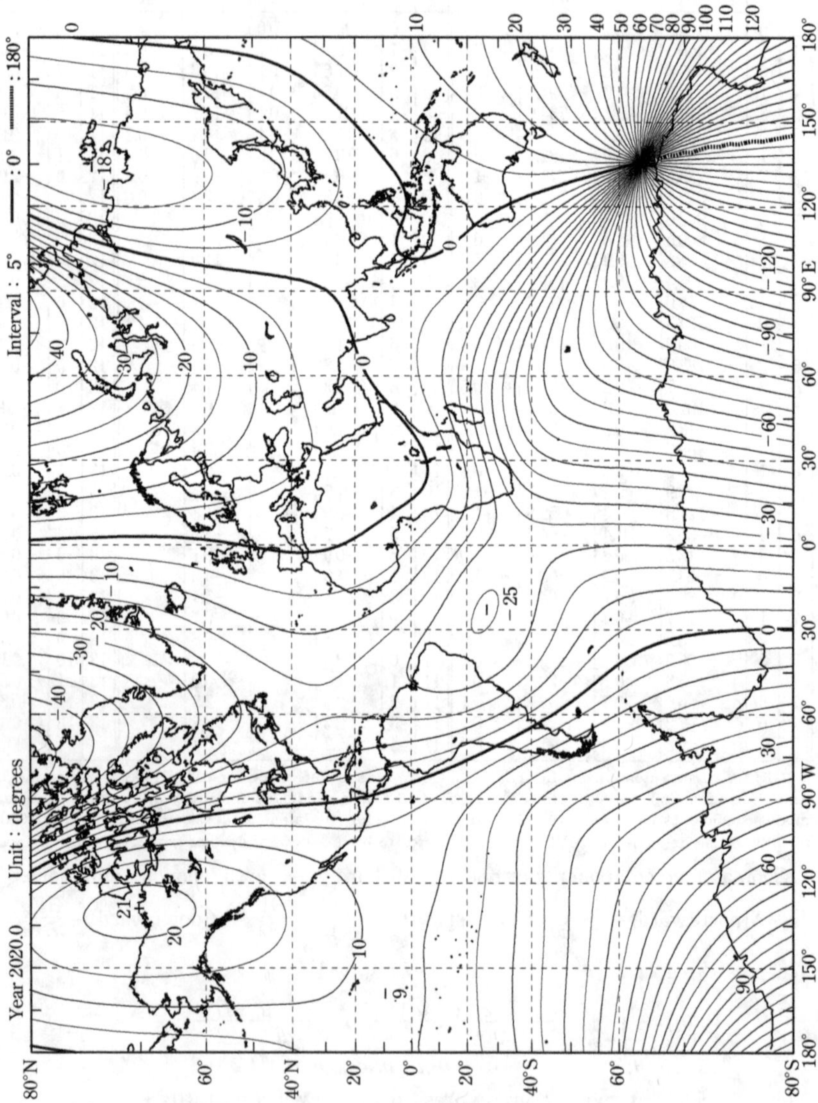

Figure 38 Declination of polar regions (IGRF-13)

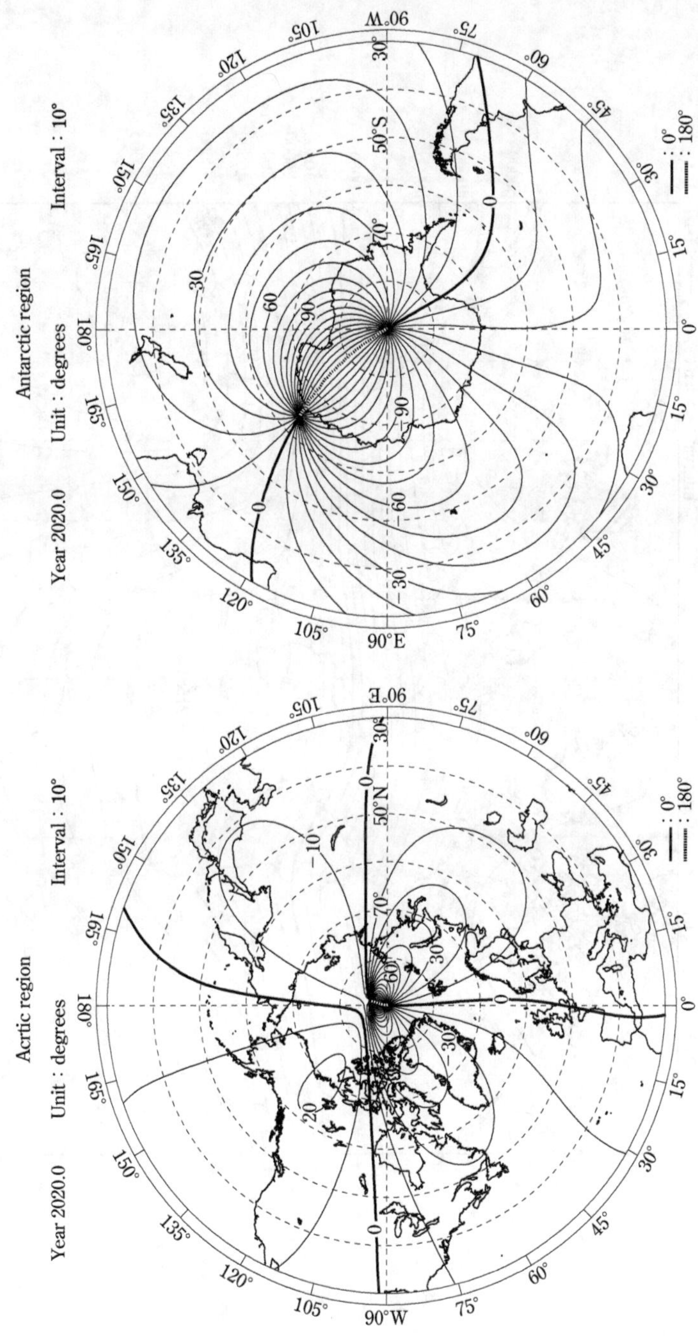

Figure 39

Vertical component (IGRF-13)

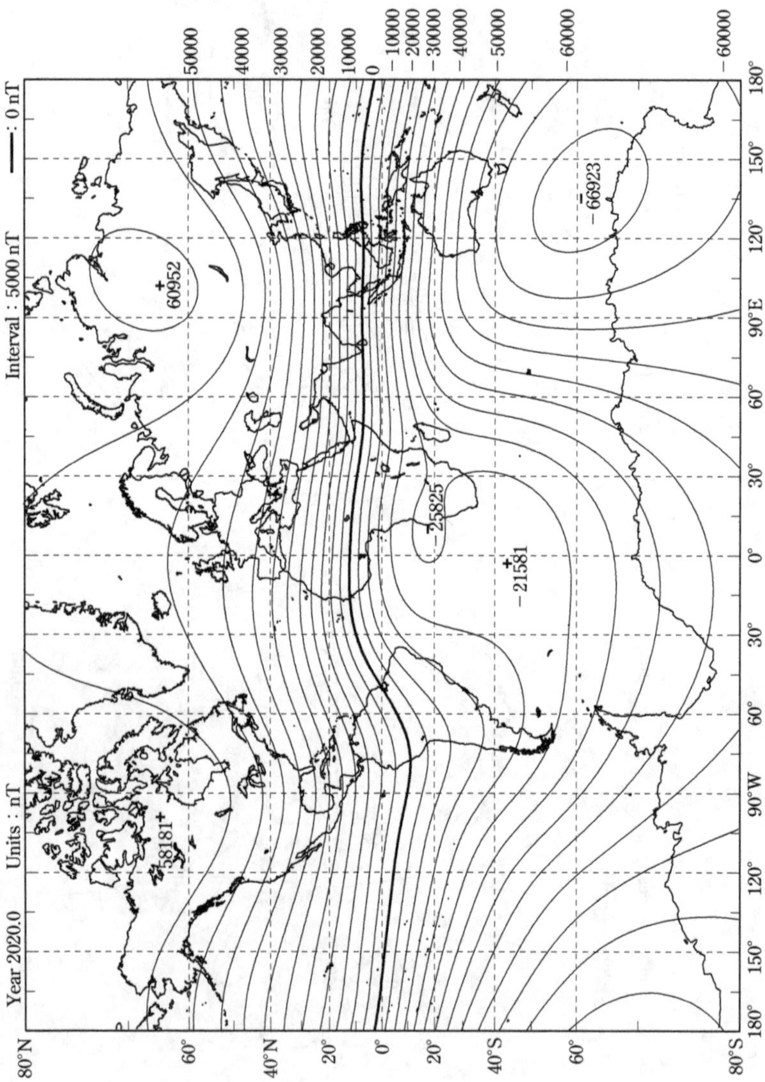

Figure 40

Vertical component of polar regions (IGRF-13)

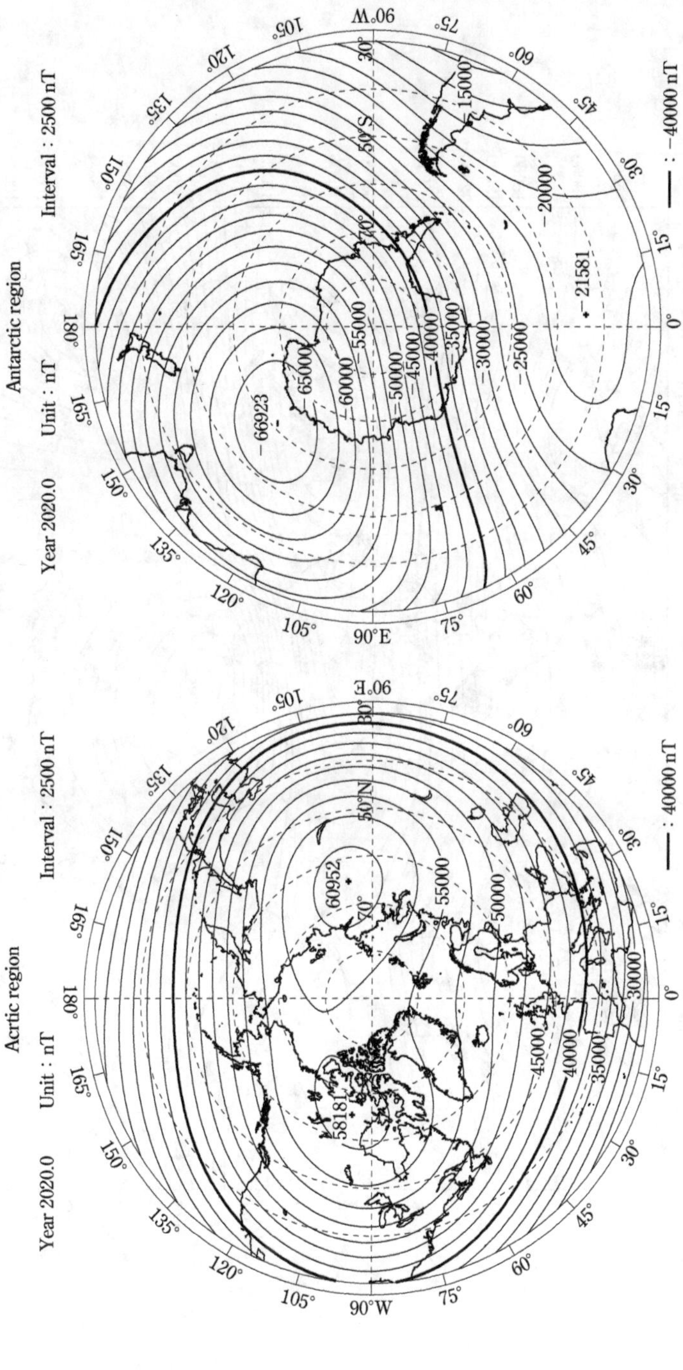

Figure 41

Horizontal component (IGRF-13)

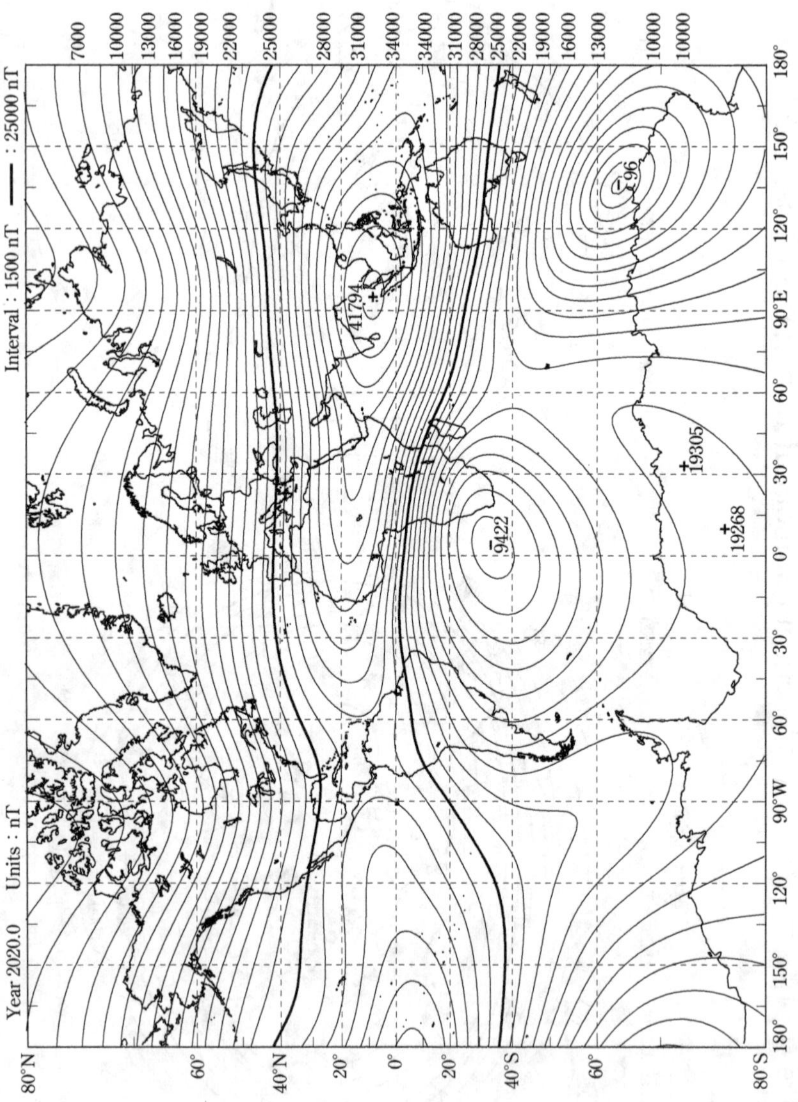

Figure 42 Horizontal component of polar regions (IGRF-13)

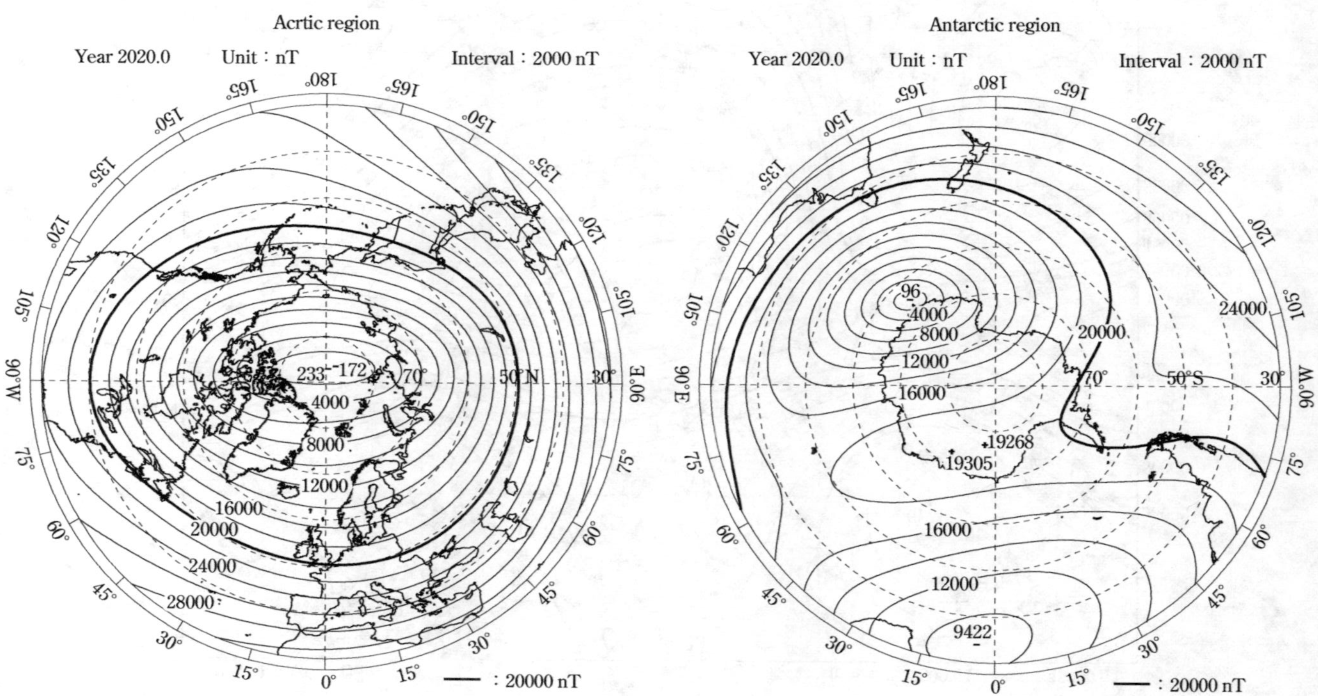

Figure 43

Total magnetic force (IGRF-13)

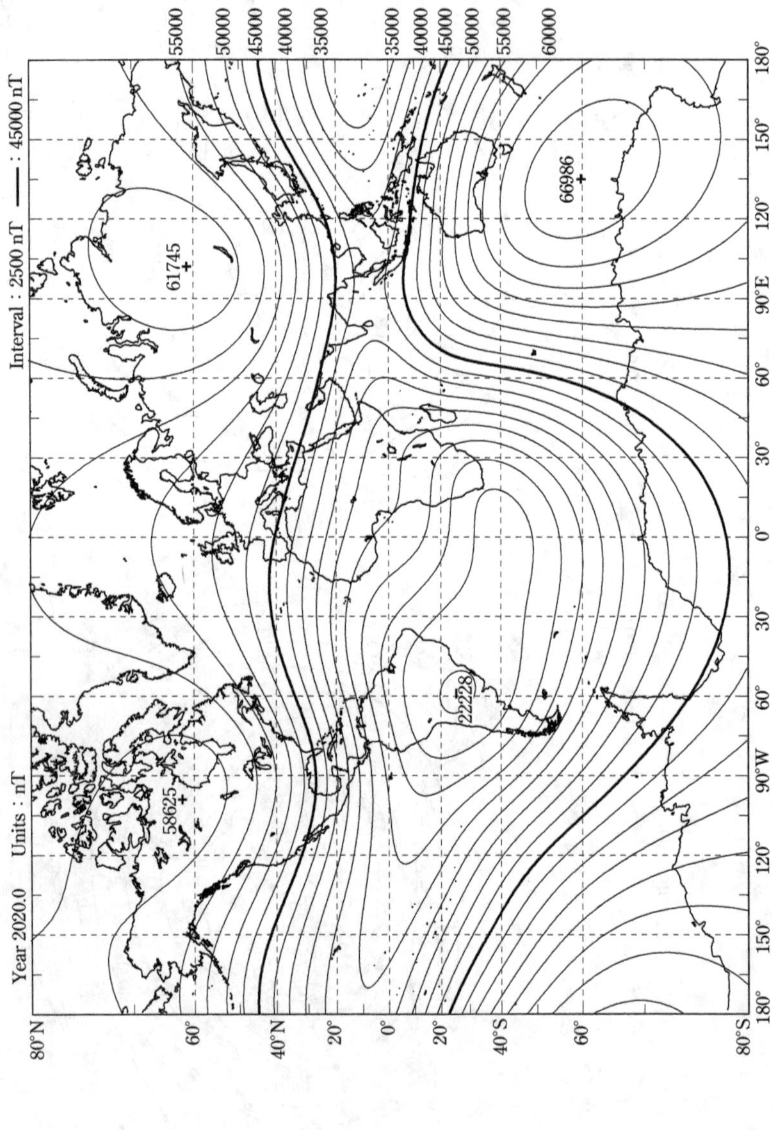

Figure 44

Total magnetic force of polar regions (IGRF-13)

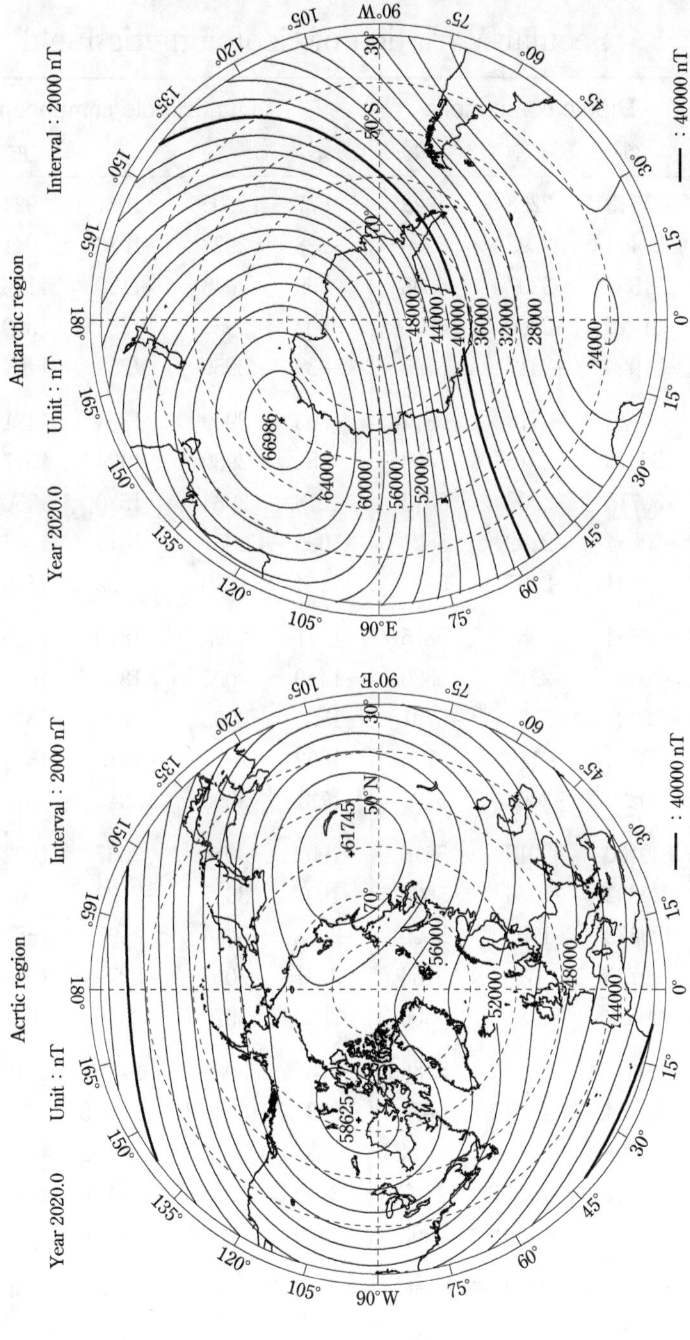

Secular Variation of Geomagnetic Field

Year	Dipole component			Quadrapole component				
	g_1^0	g_1^1	h_1^1	g_2^0	g_2^1	h_2^1	g_2^2	h_2^2
1900	-31543	-2298	5922	-677	2905	-1061	924	1121
1905	-31464	-2298	5909	-728	2928	-1086	1041	1065
1910	-31354	-2297	5898	-769	2948	-1128	1176	1000
1915	-31212	-2306	5875	-802	2956	-1191	1309	917
1920	-31060	-2317	5845	-839	2959	-1259	1407	823
1925	-30926	-2318	5817	-893	2969	-1334	1471	728
1930	-30805	-2316	5808	-951	2980	-1424	1517	644
1935	-30715	-2306	5812	-1018	2984	-1520	1550	586
1940	-30654	-2292	5821	-1006	2981	-1614	1566	528
1945	-30594	-2285	5810	-1244	2990	-1702	1578	477
1950	-30554	-2250	5815	-1341	2998	-1810	1576	381
1955	-30500	-2215	5820	-1440	3003	-1898	1581	291
1960	-30421	-2169	5791	-1555	3002	-1967	1590	206
1965	-30334	-2119	5776	-1662	2997	-2016	1594	114
1970	-30220	-2068	5737	-1781	3000	-2047	1611	25
1975	-30100	-2013	5675	-1902	3010	-2067	1632	-68
1980	-29992	-1956	5604	-1997	3027	-2129	1663	-200
1985	-29873	-1905	5500	-2072	3044	-2197	1687	-306
1990	-29775	-1848	5406	-2131	3059	-2279	1686	-373
1995	-29692	-1784	5306	-2200	3070	-2366	1681	-413
2000	-29619.4	-1728.2	5186.1	-2267.7	3068.4	-2481.6	1670.9	-458.0
2005	-29554.6	-1669.1	5078.0	-2337.2	3047.7	-2594.5	1657.8	-515.4
2010	-29496.6	-1586.4	4944.3	-2396.1	3026.3	-2708.5	1668.2	-575.7
2015	-29442.0	-1501.0	4797.1	-2445.1	3012.9	-2845.6	1676.7	-641.9
2020	-29404.8	-1450.9	4652.5	-2499.6	2982.0	-2991.6	1677.0	-734.6

Refer to page **701** for the International geomagnetic reference field expansion coefficients (g_n^m, h_n^m).

Geomagnetic Variation for the Past 2000 Years in Japan

Variations in the geomagnetic direction (declination, inclination) and geomagnetic intensity in Japan for the past 2000 years. Mainly estimated using archaeological resources and historical lava taken mainly from west Japan.

Figure 45

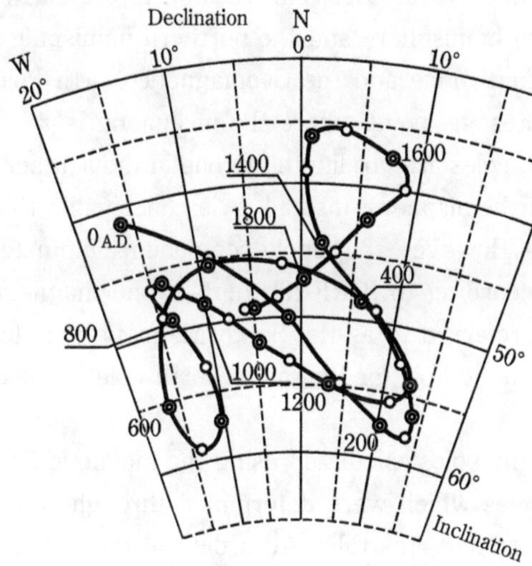

Changes in geomagnetic direction
The horizontal axis represents the declination (*D*) and the vertical axis represents the inclination (*I*). The numbers next to the curve are years (Gregorian calendar). The direction is converted to a reference point, Osaka. The difference in Declination with east Japan (Tokyo) is approximately two degrees. The inclination increases at higher latitudes. There is a difference of about 10 degrees between Kagoshima and Hokkaido.

Source: Kimio Hirooka, Quaternary Research (Daiyonki-kenkyu) Vol. 14, Iss. 4, 200–203, 1977

Figure 46

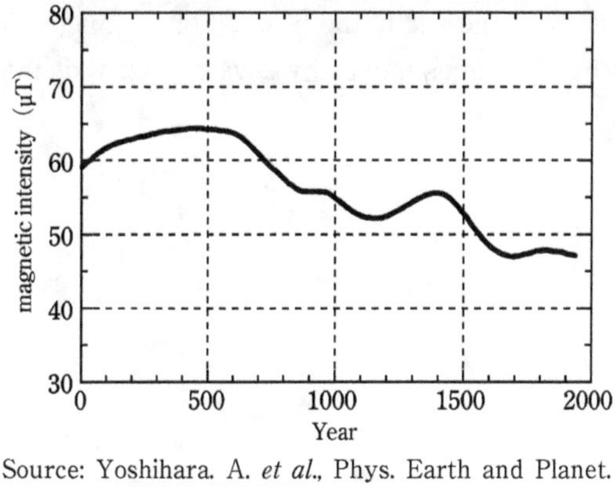

Changes in geomagnetic direction
The vertical axis represents the magnetic field intensity and the horizontal axis is the year (Gregorian calendar). The magnetic field intensity becomes stronger at higher latitudes. There is a difference of about 4 μT (=4000 nT) between Kagoshima and Hokkaido.

Source: Yoshihara. A. *et al.*, Phys. Earth and Planet. Sci. Lett., 210, 219–231, 2003.

Geomagnetic Pole and Magnetic Pole

A geomagnetic pole is the intersection of the earth's surface and the axis of a magnetic dipole when geomagnetism is approximated by a magnetic field created by hypothetically placing the magnetic dipole at the center of the earth. Geomagnetic poles are also called geomagnetic dipole poles and magnetic axis poles. There are geomagnetic poles positioned symmetrically in relation to the earth's center at one location in the southern hemisphere and the northern hemisphere. These are called the Geomagnetic South Pole and the Geomagnetic North Pole, respectively. A magnetic pole indicates the point where the inclination is $\pm 90°$. According to this definition, magnetic poles are not limited to one in the southern hemisphere and one in the northern hemisphere; instead, it is conceivable that there could be many magnetic poles. However, the dipole component dominates in current geomagnetic pole, and according to IGRF-13, there is one magnetic pole in each hemisphere. These are referred to as the South Magnetic Pole (or Magnetic South Pole) and North Magnetic Pole (or Magnetic North Pole), respectively.

For magnetic poles, in addition to the poles calculated using the magnetic field model of the IGRF, there are also poles which were determined through actual measurement by institutions in Canada and Australia. According to those actual measurements, as of 2001, the Magnetic North Pole is at a north latitude of 81.3° and a west longitude of 110.8°. The Magnetic South Pole is at a south latitude of 64.7° and an east longitude of 138.0°. However, due to fluctuations in the magnetic field which arise mainly on the exterior of the earth, the position of the magnetic poles as determined by measurement sometimes moves by several dozen kilometers.

Geographic latitude/longitude and magnetic dipole moment of the geomagnetic poles and magnetic poles (1900 to 2020)

Year	North geomagnetic pole		South geomagnetic pole		North magnetic pole		South magnetic pole		Magnetic dipole moment (10^{22} Am^2)
	Latitude	Longitude	Latitude	Longitude	Latitude	Longitude	Latitude	Longitude	
1900	78.7° N	68.8° W	78.7° S	111.2° E	70.5° N	96.2° W	71.7° S	148.3° E	8.32
1905	78.7 N	68.7 W	78.7 S	111.3 E	70.7 N	96.5 W	71.5 S	148.5 E	8.30
1910	78.7 N	68.7 W	78.7 S	111.3 E	70.8 N	96.7 W	71.2 S	148.6 E	8.27
1915	78.6 N	68.6 W	78.6 S	111.4 E	71.0 N	97.0 W	70.8 S	148.5 E	8.24
1920	78.6 N	68.4 W	78.6 S	111.6 E	71.3 N	97.4 W	70.4 S	148.2 E	8.20
1925	78.6 N	68.3 W	78.6 S	111.7 E	71.8 N	98.0 W	70.0 S	147.6 E	8.16
1930	78.6 N	68.3 W	78.6 S	111.7 E	72.3 N	98.7 W	69.5 S	146.8 E	8.13
1935	78.6 N	68.4 W	78.6 S	111.6 E	72.8 N	99.3 W	69.1 S	145.8 E	8.11
1940	78.5 N	68.5 W	78.5 S	111.5 E	73.3 N	99.9 W	68.6 S	144.6 E	8.09
1945	78.5 N	68.5 W	78.5 S	111.5 E	73.9 N	100.2 W	68.2 S	144.4 E	8.08
1950	78.5 N	68.8 W	78.5 S	111.2 E	74.6 N	100.9 W	67.9 S	143.5 E	8.06
1955	78.5 N	69.2 W	78.5 S	110.8 E	75.2 N	101.4 W	67.2 S	141.5 E	8.05
1960	78.6 N	69.5 W	78.6 S	110.5 E	75.3 N	101.0 W	66.7 S	140.2 E	8.03
1965	78.6 N	69.9 W	78.6 S	110.1 E	75.6 N	101.3 W	66.3 S	139.5 E	8.00
1970	78.7 N	70.2 W	78.7 S	109.8 E	75.9 N	101.0 W	66.0 S	139.4 E	7.97
1975	78.8 N	70.5 W	78.8 S	109.5 E	76.2 N	100.6 W	65.7 S	139.5 E	7.94
1980	78.9 N	70.8 W	78.9 S	109.2 E	76.9 N	101.7 W	65.4 S	139.3 E	7.91
1985	79.0 N	70.9 W	79.0 S	109.1 E	77.4 N	102.6 W	65.1 S	139.1 E	7.87
1990	79.2 N	71.1 W	79.2 S	108.9 E	78.1 N	103.7 W	64.9 S	138.9 E	7.84
1995	79.4 N	71.4 W	79.4 S	108.6 E	79.0 N	105.3 W	64.8 S	138.7 E	7.81
2000	79.6 N	71.6 W	79.6 S	108.4 E	81.0 N	109.6 W	64.7 S	138.3 E	7.79
2005	79.8 N	71.8 W	79.8 S	108.2 E	83.2 N	118.2 W	64.5 S	137.8 E	7.77
2010	80.1 N	72.2 W	80.1 S	107.8 E	85.0 N	132.8 W	64.4 S	137.3 E	7.75
2015	80.4 N	72.6 W	80.4 S	107.4 E	86.3 N	160.0 W	64.3 S	136.6 E	7.72
2020	80.7 N	72.7 W	80.7 S	107.3 E	86.5 N	162.9 E	64.1 S	135.9 E	7.71

Based on IGRF-13.

The white circles in the figure show the position of the geomagnetic poles from 1900 to 2025 according to the international geomagnetic reference field (IGRF-13). In 2020, the Geomagnetic North Pole was located near the northwestern tip of Greenland, at a north latitude of 80.7° and a west longitude of 72.7°. The Geomagnetic South Pole was located at a south latitude of 80.7° and an east longitude of 107.3° in the Antarctic continent. The geomagnetic poles are moving mainly due to slow changes in the main magnetic field arising from the earth's inner core. This phenomenon is known as "geomagnetic secular variation." The black circles in the figure represent the location of the magnetic poles from 1900 to 2025 according to IGRF-13. Geomagnetic field contains non-dipole components. This causes the position of the magnetic poles to be quite different from that of the geomagnetic poles. Similar to the geomagnetic poles, the magnetic poles move due to geomagnetic secular variation. Much attention is given to how the speed of this movement is faster than that of the geomagnetic poles, as well as to how the motion of the Magnetic North Pole has become noticeably faster in recent years.

Figure 47

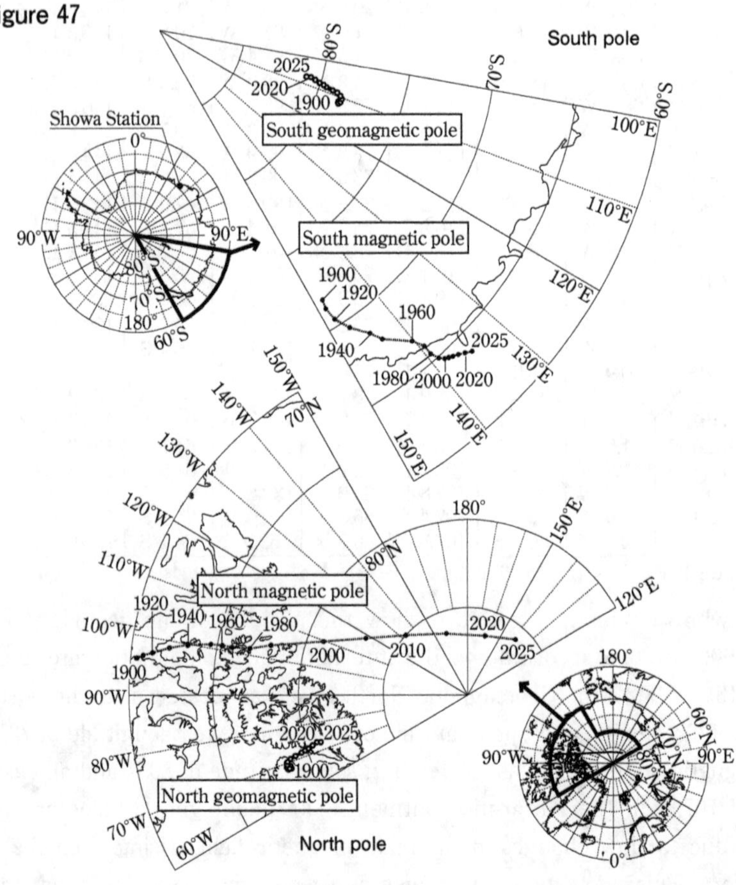

History of Geomagnetic Polarity Reversals

The polarity of geomagnetic field during last 170 million years is shown as a figure with a table. The data are obtained from the rock which forms ocean bottom plate. The columns 'N' indicate 'Normal polarity' which is the same with current polarity and 'R' means 'Reverse polarity' opposite to present one. The numerical values in the figure indicate the age from present in a million years unit. The corresponding period is indicated with black. In the table, the range of age with 'N' polarity and its name (chron) are shown as well as the name of reverse polarity period just before the 'N' period. That is, the reverse polarity period covers from the initial epoch of 'N' period and the last epoch of previous 'N' period. For example, (C1r. 1r) indicates the period with reverse polarity for 0.781–0.988 million years ago.

Figure 48

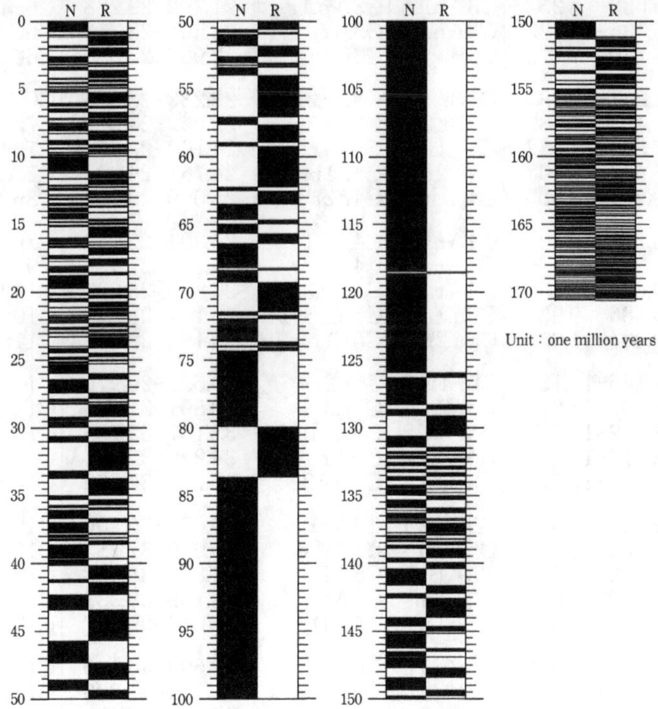

Source: Ogg, J. G., "Geomagnetic Polarity Time Scale" in "The Geological Time Scale 2012", Amsterdam (Elsevier), 85–113, 2012.

Periods of Normal Polarity

Age range (Ma)	Polarity name (Chrons)	Reverse polarity chrons	Age range (Ma)	Polarity name (Chrons)	Reverse polarity chrons
(Top age) (Base age)			(Top age) (Base age)		
0.000- 0.781	C1n	C1r.1r	16.303-16.472	C5Cn.2n	C5Cn.2r
0.988- 1.072	C1r.1n	C1r.2r	16.543-16.721	C5Cn.3n	C5Cr
1.173- 1.185	C1r.2n	C1r.3r	17.235-17.533	C5Dn	C5Dr.1r
1.778- 1.945	C2n	C2r.1r	17.717-17.740	C5Dr.1n	C5Dr.2r
2.128- 2.148	C2r.1n	C2r.2r	18.056-18.524	C5En	C5Er
2.581- 3.032	C2An.1n	C2An.1r	18.748-19.722	C6n	C6r
3.116- 3.207	C2An.2n	C2An.2r	20.040-20.213	C6An.1n	C6An.1r
3.330- 3.596	C2An.3n	C2Ar	20.439-20.709	C6An.2n	C6Ar
4.187- 4.300	C3n.1n	C3n.1r	21.083-21.159	C6AAn	C6AAr.1r
4.493- 4.631	C3n.2n	C3n.2r	21.403-21.483	C6AAr.1n	C6AAr.2r
4.799- 4.896	C3n.3n	C3n.3r	21.659-21.688	C6AAr.2n	C6AAr.3r
4.997- 5.235	C3n.4n	C3r	21.767-21.936	C6Bn.1n	C6Bn.1r
6.033- 6.252	C3An.1n	C3An.1r	21.992-22.268	C6Bn.2n	C6Br
6.436- 6.733	C3An.2n	C3Ar	22.564-22.754	C6Cn.1n	C6Cn.1r
7.140- 7.212	C3Bn	C3Br.1r	22.902-23.030	C6Cn.2n	C6Cn.2r
7.251- 7.285	C3Br.1n	C3Br.2r	23.233-23.295	C6Cn.3n	C6Cr
7.454- 7.489	C3Br.2n	C3Br.3r	23.962-24.000	C7n.1n	C7n.1r
7.528- 7.642	C4n.1n	C4n.1r	24.109-24.474	C7n.2n	C7r
7.695- 8.108	C4n.2n	C4r.1r	24.761-24.984	C7An	C7Ar
8.254- 8.300	C4r.1n	C4r.2r	25.099-25.264	C8n.1n	C8n.1r
8.771- 9.105	C4An	C4Ar.1r	25.304-25.987	C8n.2n	C8r
9.311- 9.426	C4Ar.1n	C4Ar.2r	26.420-27.439	C9n	C9r
9.647- 9.721	C4Ar.2n	C4Ar.3r	27.859-28.087	C10n.1n	C10n.1r
9.786- 9.937	C5n.1n	C5n.1r	28.141-28.278	C10n.2n	C10r
9.984-11.056	C5n.2n	C5r.1r	29.183-29.477	C11n.1n	C11n.1r
11.146-11.188	C5r.1n	C5r.2r	29.527-29.970	C11n.2n	C11r
11.592-11.657	C5r.2n	C5r.3r	30.591-31.034	C12n	C12r
12.049-12.174	C5An.1n	C5An.1r	33.157-33.705	C13n	C13r
12.272-12.474	C5An.2n	C5Ar.1r	34.999-35.294	C15n	C15r
12.735-12.770	C5Ar.1n	C5Ar.2r	35.706-35.892	C16n.1n	C16n.1r
12.829-12.887	C5Ar.2n	C5Ar.3r	36.051-36.700	C16n.2n	C16r
13.032-13.183	C5AAn	C5AAr	36.969-37.753	C17n.1n	C17n.1r
13.363-13.608	C5ABn	C5ABr	37.872-38.093	C17n.2n	C17n.2r
13.739-14.070	C5ACn	C5ACr	38.159-38.333	C17n.3n	C17r
14.163-14.609	C5ADn	C5ADr	38.615-39.627	C18n.1n	C18n.1r
14.775-14.870	C5Bn.1n	C5Bn.1r	39.698-40.145	C18n.2n	C18r
15.032-15.160	C5Bn.2n	C5Br	41.154-41.390	C19n	C19r
15.974-16.268	C5Cn.1n	C5Cn.1r	42.301-43.432	C20n	C20r

Ma: one million years

Continued.

Age range (Ma)	Polarity name (Chrons)	Reverse polarity chrons	Age range (Ma)	Polarity name (Chrons)	Reverse polarity chrons
(Top age) (Base age)			(Top age) (Base age)		
45.724- 47.349	C21n	C21r	138.25-138.42	M13n	M13r
48.566- 49.344	C22n	C22r	138.66-138.91	M14n	M14r
50.628- 50.835	C23n.1n	C23n.1r	139.59-139.94	M15n	M15r
50.961- 51.833	C23n.2n	C23r	140.42-141.64	M16n	M16r
52.620- 53.074	C24n.1n	C24n.1r	142.22-142.57	M17n	M17r
53.199- 53.274	C24n.2n	C24n.2r	144.00-144.64	M18n	M18r
53.416- 53.983	C24n.3n	C24r	145.01-145.14	M19n.1n	M19n.1r
57.101- 57.656	C25n	C25r	145.19-146.28	M19n.2n	M19r
58.959- 59.237	C26n	C26r	146.54-146.90	M20n.1n	M20n.1r
62.221- 62.517	C27n	C27r	146.96-147.72	M20n.2n	M20r
63.494- 64.667	C28n	C28r	148.44-149.35	M21n	M21r
64.958- 65.688	C29n	C29r	149.80-151.12	M22n.1n	M22n.1r
66.398- 68.196	C30n	C30r	151.17-151.21	M22n.2n	M22n.2r
68.369- 69.269	C31n	C31r	151.25-151.32	M22n.3n	M22r
71.449- 71.689	C32n.1n	C32n.1r	151.92-152.06	M22An	M22Ar
71.939- 73.649	C32n.2n	C32r.1r	152.25-152.64	M23n	M23r.1r
73.949- 74.049	C32r.1n	C32r.2r	152.93-152.96	M23r.1n	M23r.2r
74.309- 79.900	C33n	C33r	153.62-153.87	M24n	M24r.1r
83.64 -118.50	C34n	M-1r	154.35-154.38	M24r.1n	M24r.2r
118.60 -125.93	C34n	M0r	154.59-154.73	M24An	M24Ar
126.30 -128.32	M1n	M1r	155.01-155.38	M24Bn	M24Br
128.66 -129.11	M3n	M3r	155.54-155.85	M25n	M25r
130.60 -131.43	M5n	M5r	156.04-156.16	M25An.1n	M25An.1r
131.74 -131.92	M6n	M6r	156.22-156.29	M25An.2n	M25An.2r
132.04 -132.27	M7n	M7r	156.39-156.42	M25An.3n	M25Ar
132.55 -132.80	M8n	M8r	156.56-156.69	M26n.1n	M26n.1r
133.05 -133.30	M9n	M9r	156.75-156.78	M26n.2n	M26n.2r
133.58 -133.88	M10n	M10r	156.84-156.94	M26n.3n	M26n.3r
134.22 -134.48	M10Nn.1n	M10Nn.1r	157.04-157.06	M26n.4n	M26r
134.51 -134.76	M10Nn.2n	M10Nn.2r	157.25-157.40	M27n	M27r
134.78 -135.00	M10Nn.3n	M10Nr	157.57-157.87	M28n	M28r
135.32 -135.92	M11n	M11r.1r	158.02-158.12	M28An	M28Ar
136.11 -136.13	M11r.1n	M11r.2r	158.36-158.41	M28Bn	M28Br
136.29 -136.69	M11An.1n	M11An.1r	158.50-158.60	M28Cn	M28Cr
136.74 -136.80	M11An.2n	M11Ar	158.69-158.78	M28Dn	M28Dr
136.87 -137.06	M12n	M12r.1r	158.89-159.04	M29n.1n	M29n.1r
137.73 -137.80	M12r.1n	M12r.2r	159.07-159.14	M29n.2n	M29r
137.94 -138.17	M12An	M12Ar	159.38-159.43	M29An	M29Ar

Periods of Normal Polarity Continued.

Age range (Ma)	Polarity name (Chrons)	Reverse polarity chrons	Age range (Ma)	Polarity name (Chrons)	Reverse polarity chrons
(Top age) (Base age)			(Top age) (Base age)		
159.49–159.62	M30n	M30r	166.40–166.47	M39n.7n	M39n.7r
159.78–159.88	M30An	M30Ar	166.51–166.61	M39n.8n	M39r
159.91–160.07	M31n.1n	M31n.1r	166.68–166.72	M40n.1n	M40n.1r
160.15–160.18	M31n.2n	M31n.2r	166.89–166.96	M40n.2n	M40n.2r
160.22–160.26	M31n.3n	M31r	167.05–167.11	M40n.3n	M40n.3r
160.32–160.34	M32n.1n	M32n.1r	167.33–167.36	M40n.4n	M40r
160.37–160.47	M32n.2n	M32n.2r	167.43–167.51	M41n.1n	M41n.1r
160.52–160.55	M32n.3n	M32r	167.69–167.75	M41n.2n	M41n.2r
160.63–160.93	M33n	M33r	167.85–167.91	M41n.3n	M41n.3r
161.06–161.14	M33An	M33Ar	168.01–168.03	M41n.4n	M41r
161.21–161.28	M33Bn	M33Br	168.14–168.24	M42n.1n	M42n.1r
161.39–161.43	M33Cn.1n	M33Cn.1r	168.34–168.36	M42n.2n	M42n.2r
161.49–161.61	M33Cn.2n	M33Cr	168.40–168.45	M42n.3n	M42n.3r
161.81–161.89	M34n.1n	M34n.1r	168.49–168.51	M42n.4n	M42n.4r
161.96–162.01	M34n.2n	M34n.2r	168.53–168.57	M42n.5n	M42n.5r
162.04–162.06	M34n.3n	M34n.3r	168.59–168.61	M42n.6n	M42n.6r
162.12–162.15	M34An	M34Ar	168.63–168.65	M42n.7n	M42n.7r
162.28–162.37	M34Bn.1n	M34Bn.1r	168.68–168.70	M42n.8n	M42n.8r
162.43–162.46	M34Bn.2n	M34Br	168.74–168.80	M42n.9n	M42n.9r
162.49–162.55	M35n	M35r	168.92–168.95	M42n.10n	M42n.10r
162.70–162.80	M36n.1n	M36n.1r	169.12–169.14	M42n.11n	M42r
162.86–162.90	M36An	M36Ar	169.23–169.25	M43n.1n	M43n.1r
162.92–162.95	M36Bn	M36Br	169.38–169.47	M43n.2n	M43n.2r
163.10–163.16	M36Cn	M36Cr	169.55–169.60	M43n.3n	M43n.3r
163.29–163.53	M37n.1n	M37n.1r	169.64–169.68	M43n.4n	M43n.4r
163.65–163.76	M37n.2n	M37r	169.74–169.81	M43n.5n	M43r
163.85–163.98	M38n.1n	M38n.1r	169.89–169.93	M44n.1n	M44n.1r
164.03–164.18	M38n.2n	M38n.2r	169.97–170.00	M44n.2n	M44n.2r
164.22–164.30	M38n.3n	M38n.3r	170.05–170.08	M44n.3n	M44n.3r
164.41–164.63	M38n.4n	M38n.4r	170.21–170.23	M44n.4n	M44n.4r
164.69–164.85	M38n.5n	M38r	170.28–170.30	M44n.5n	M44n.5r
164.91–165.08	M39n.1n	M39n.1r	170.37–170.38	M44n.6n	M44n.6r
165.20–165.34	M39n.2n	M39n.2r	170.45–170.45	M44n.7n	M44n.7r
165.40–165.53	M39n.3n	M39n.3r	170.49–170.55	M44n.8n	M44n.8r
165.65–165.80	M39n.4n	M39n.4r	170.57–170.61	M44n.9n	M44r
165.96–166.08	M39n.5n	M39n.5r	170.64–170.67	M45n	M45r
166.21–166.30	M39n.6n	M39n.6r			

Solar Wind and Earth's Magnetosphere

Overall, the sun constantly emits energy of 3.9×10^{26} W. Solar energy can be broadly divided into three categories. The majority of solar energy falls into the first category of infrared rays and visible light which is known as sunlight. The second category is types of X-rays and ultraviolet rays. The majority of this energy is absorbed above the stratosphere of the earth's atmosphere. The third category of solar energy is the emission of high-energy particles and plasma in all directions. This flow of plasma is called "solar wind." It flows out from the corona, which reaches temperatures from 1 million to 10 million K.

Solar wind

The amount of solar wind energy is almost equivalent to X-rays and ultraviolet rays. This amount only accounts for one millionth of sunlight. Although referred to as "wind," there are only about 10 charged particles per 1 cubic centimeter inside of solar wind. Even the ultimate vacuum which can be created by modern engineering technology has several thousand molecules per 1 cubic centimeter. This means that solar wind is an extremely diluted non-collision plasma. Despite its massive gravity, the sun is not able to stop the high-temperature corona gas. In other words, solar wind shakes off the sun's gravity and flies into space. The speed of solar wind is normally about 400 to 500 km per second near the earth.

Magnetosphere

When solar wind nears the earth, it is exposed to interference by the earth's magnetic force. The solar wind flows as if it were enveloping the earth. Furthermore, a magnetopause is created where the solar wind pressure and the geomagnetic force achieve balance. Solar wind is unable to intrude toward the earth past the magnetopause. The result is a comet-shaped space called "magnetosphere." As shown in the figure below, when viewed from the vast universe, the sphere of influence for the geomagnetic field most likely appears as a cavity which has been opened in solar wind. Similar to a comet, the earth develops a long tail-shaped magnetic field.

Electric current is generated when the plasma moves across the line of magnetic force. This phenomenon is known as magnetohydrodynamic (MHD) dynamo. Solar wind and the earth's magnetic field from a huge natural power plant with an output that reaches several trillion watts. A portion of this electromotive force causes a discharge phenomenon at the earth's high latitudes which is the source of the aurora. Solar wind also pulls the sun's magnetic field along with it. This interacts with the earth's magnetic field to cause a variety of phenomena in the magnetosphere and ionosphere.

The magnetosphere is composed of several unique plasma regions known as the plasmasphere, plasma sheet, magnetotail lobe, etc. There is also high-energy radiation belt which surrounds the earth in a donut shape.

Figure 49 Solar wind and earth's magnetosphere (Geomagnetic field's region of influence)

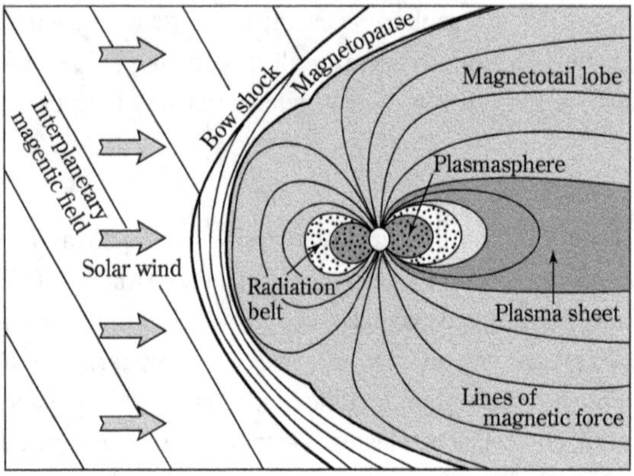

Source: W Baumjohann

Aurora

The aurora which colors the night skies of polar regions is a phenomenon when oxygen atoms, nitrogen molecules, nitrogen molecule ions, etc. in the ionosphere emit light when excited by electrons and protons which fall along the magnetic line of force from the magnetosphere; in other words, from the region where the magnetic field of the earth is confined by the solar wind. Although the height of the luminous layer varies depending on the atoms/molecules which are emitting light and on the precipitating electron/proton energy, it is usually from 90 to 300 km.

(1) Aurora belt

The location with the high probability of aurora occurrence (the aurora belt) is the belt-shaped region surrounding the pole at the geomagnetic latitude of 65° to 70° (see Figure 51). The line labeled as "100" in the figure means that an aurora can be seen in 100 nights on an average in 1 year at that location. Near Japan, a line labeled "0.1" runs through northern Hokkaido. This means that an aurora can be seen an average of 0.1 nights in 1 year; in other words, once every 10 years. This aurora distribution chart changes significantly due to solar activity. By using the latest high-sensitivity equipment, it is possible to observe an aurora in Hokkaido several times a year, when solar activity is high.

(2) Representative spectrums

Oxygen atoms: wavelength 557.7 nm (green), 630.0 to 636.3 nm (red)

Nitrogen molecules: wavelength 646.9 to 687.5 nm (red)

Nitrogen molecule ions: wavelength 391.4 nm, 427.8 nm (blue)

(3) Brightness

The brightness (luminosity) of the aurora is defined by the following four levels, based on the brightness of the oxygen green lines. I: brightness of the Milky Way 0.0003 lx, II; brightness of cirrus clouds when illuminated by the moon 0.003 lx, III: brightness of cumulus clouds when illuminated by the moon 0.03 lx, IV: brightness of the full moon 0.3 lx.

Even in aurora belt where an aurora can be seen almost every night, the auroras which are usually seen are limited to the brightness of level I or II, appearing as a faint arch stretching from east to west. Auroras with a brightness of level III or greater are seen during the development stage of aurora storm (as a technical term "aurora substorm"). Aurora storm is a phenomenon in which bright light suddenly moves violently around midnight. This creates a curtain-shaped aurora or a variety of other shapes which color the entire sky. This vivid display of the aurora lasts for about 5 to 20 minutes.

(Color photography of aurora: When using a lens with an aperture of about F1.2 in equivalence with ISO 200–400, from 3 to 30 seconds of exposure is required.)

Figure 50 Global distribution of aurora appearance probability

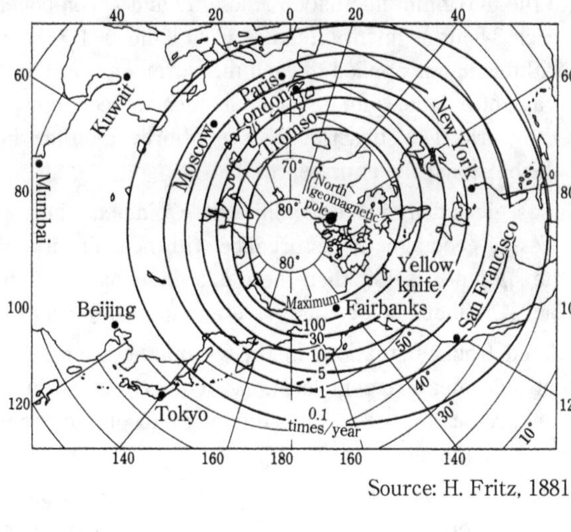

Source: H. Fritz, 1881

(4) Planetary and satellite Auroras

Auroras have been confirmed on planets and a satellite other than Earth which have atmospheres and magnetic fields.

	Mercury	Venus	Earth	Mars	Jupiter	Saturn	Uranus	Neptune	Ganymede
Magnetic moment (Am^2)	2.8×10^{19}	$<10^{18}$	7.7×10^{22}	$<10^{19}$	1.5×10^{27}	4.7×10^{25}	3.7×10^{24}	1.9×10^{24}	1.3×10^{20}
Atmosphere	×	○	○	○	○	○	○	○	dilute
Aurora	×	△	○	×	○	○	○	○	○

○ : exists, × : does not exist, or is very weak, △ : not well understood

Source: M. Galand and S. Chakrabarti, Auroral Processes in the Solar System, in *Atmospheres in the Solar System*, Geophysical Monograph 130, American Geophysical Union, 2002.

Geomagnetic Activity Indices

The geomagnetic activity indices (also known as the geomagnetic disturbance indices) are used to express the overall activity of the geomagnetic field. The merit of using these kind of indices to express the fluctuation of the geomagnetic field is that it possible to quickly obtain a rough quantitative estimate of complex phenomena. However, each of the indices possesses unique physical meaning, and caution is required during use. Reference: P. N. Mayaud, *Derivation, Meaning, and Use of Geomagnetic Indices*, Geophysical Monograph 22, American Geophysical Union, Washington, D. C., 1980.

Frequently used geomagnetic activity indices are listed below.

Kp index: K index is expressed at each measurement location using a 10-level scale (0 to 9) in relation to the maximum fluctuation among H and D components (or X and Y components) for each 3-hour segment (in Greenwich Time, 0–3, 3–6, ⋯, 21–24). The scale is semi-logarithmic and is applied to each measurement location. The Kp index is the weighted average of K indices for 13 locations (near a geomagnetic latitude of 50 degrees) around the world. The subscript "p" is an abbreviation for the German term "planetansche," which means "the entire earth."

Ap index: This index gives daily values. Unlike the Kp index, which is semi-logarithmic, the Ap index expresses geomagnetic disturbance with near linear scale (nT, nanotesla). For example, if the Ap index is 100, then there is a disturbance of about 100 nT at the mid-latitude regions. Ap^* index is a running average for 1 day every 3 hours.

Dst index: Calculated from fluctuations in the geomagnetic horizontal component acquired at 4 low-latitude measurement locations at nearly equal intervals of longitude (unit: nT). Mainly expresses the strength of the magnetospheric equatorial ring current which develops during a magnetic storm.

AE index: Calculated from fluctuations in the geomagnetic horizontal component at 12 locations along in the aurora belt (unit: nT). Mainly expresses the strength of aurora jet current density.

Large Magnetic Storms in Recent Years ($Ap^* \geqq 100$)

Year	Date	Date	Max. Ap^*	Min. *Dst*	Year	Date	Date	Max. Ap^*	Min. *Dst*
1991	3 23 to	3 26	164	-298	2000	8 10 to	8 12	129	-234
1991	6 3	6 6	196	-223	2000	9 17	9 18	103	-201
1991	6 8	6 13	149	-140	2000	10 3	10 5	126	-181
1991	7 7	7 9	129	-194	2001	3 30	3 31	192	-387
1991	7 12	7 14	149	-183	2001	4 10	4 12	124	-271
1991	10 26	10 29	157	-254	2001	10 21	10 22	105	-187
1991	10 30	11 2	128	-196	2001	11 5	11 6	142	-292
1991	11 7	11 9	180	-354	2001	11 23	11 24	104	-202
1992	5 9	5 11	194	-288	2002	9 30	10 2	103	-176
1993	4 4	4 5	116	-165	2003	5 28	5 30	129	-144
1994	2 20	2 22	130	-144	2003	8 17	8 18	108	-148
1994	4 1	4 4	101	-111	2003	10 28	10 31	252	-383
1994	4 16	4 17	114	-201	2003	11 19	11 21	171	-422
1995	4 6	4 7	100	-149	2004	7 24	7 27	195	-170
1998	5 1	5 4	120	-205	2004	11 7	11 10	206	-374
1998	8 26	8 27	144	-155	2005	8 23	8 24	105	-184
1998	9 24	9 25	127	-207	2005	9 10	9 13	102	-139
2000	4 5	4 7	137	-292	2006	12 14	12 15	120	-162
2000	5 23	5 24	101	-147	2015	3 16	3 18	117	-223*
2000	7 13	7 16	192	-300	2015	6 21	6 23	110	-204*
					2017	9 7	9 8	1[illegible]4	1[illegible]*

Period for $Ap^* > 40$. * indicates provisional values.

Top 50 Magnetic Storms since 1932

Rank	Date	Max. Ap^*	Min. *Dst*	Rank	Date	Max. Ap^*	Min. *Dst*
1	1941 9 18	312		26	1949 5 12	197	
2	1960 11 12	294	-339	27	1991 6 5	196	-223
3	1989 3 13	286	-589	28	1946 3 25	195	
4	1940 3 24	278			2004 7 26	195	-170
5	1960 10 6	258	-287	30	1992 5 10	194	-288
6	1959 7 15	252	-429	31	2000 7 15	192	-300
7	2003 10 29	252	-383		2001 3 31	192	-387
8	1960 3 31	251	-327	33	1956 4 26	186	
9	1967 5 25	242	-387		1957 9 22	186	-303
10	1982 7 13	230	-325		1959 3 26	186	-234
11	1986 2 8	229	-307	36	1959 7 17	184	-183
12	1940 3 29	226		37	1957 6 30	182	-218
13	1972 8 4	224	-125	38	1991 11 8	180	-354
14	1941 7 5	222		39	1959 8 16	174	-181
15	1957 9 4	222	-324		1960 4 30	174	-325
16	1946 3 28	216		41	1966 9 3	173	-189
	1958 7 8	216	-330	42	1958 9 4	171	-302
18	1946 9 22	215			2003 11 20	171	-422
19	1941 3 1	212		44	1951 9 25	169	
	1946 7 26	212		45	1946 4 23	168	
21	2004 11 9	206	-374	46	1956 5 16	165	
22	1950 8 19	204		47	1949 10 15	164	
23	1982 9 5	202	-289		1991 3 24	164	-298
24	1946 2 7	199		49	1970 3 8	162	-284
	1958 2 11	199	-426		1989 10 20	162	-268

Dst before 1956 has not yet bee derived.

Figure 51

Geomagnetic Activity Index (Kp index) June 2019 to June 2020

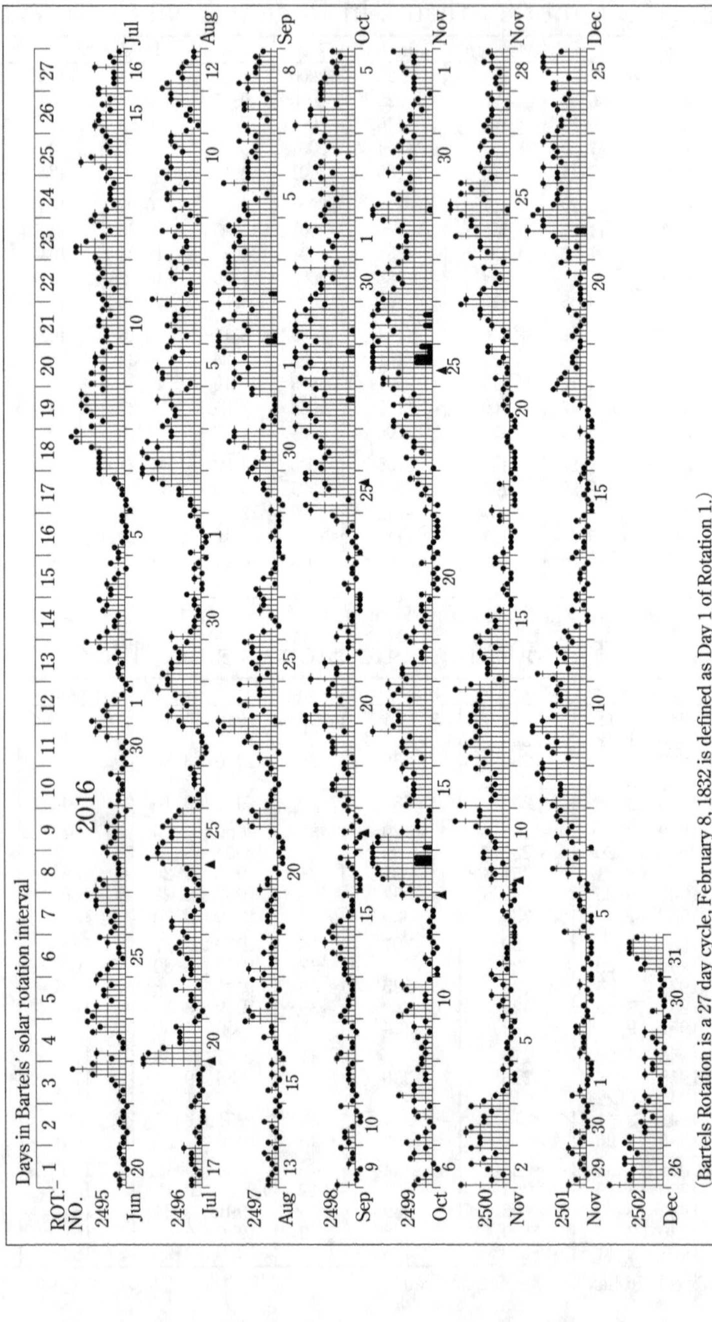

(Bartels Rotation is a 27 day cycle, February 8, 1832 is defined as Day 1 of Rotation 1.)

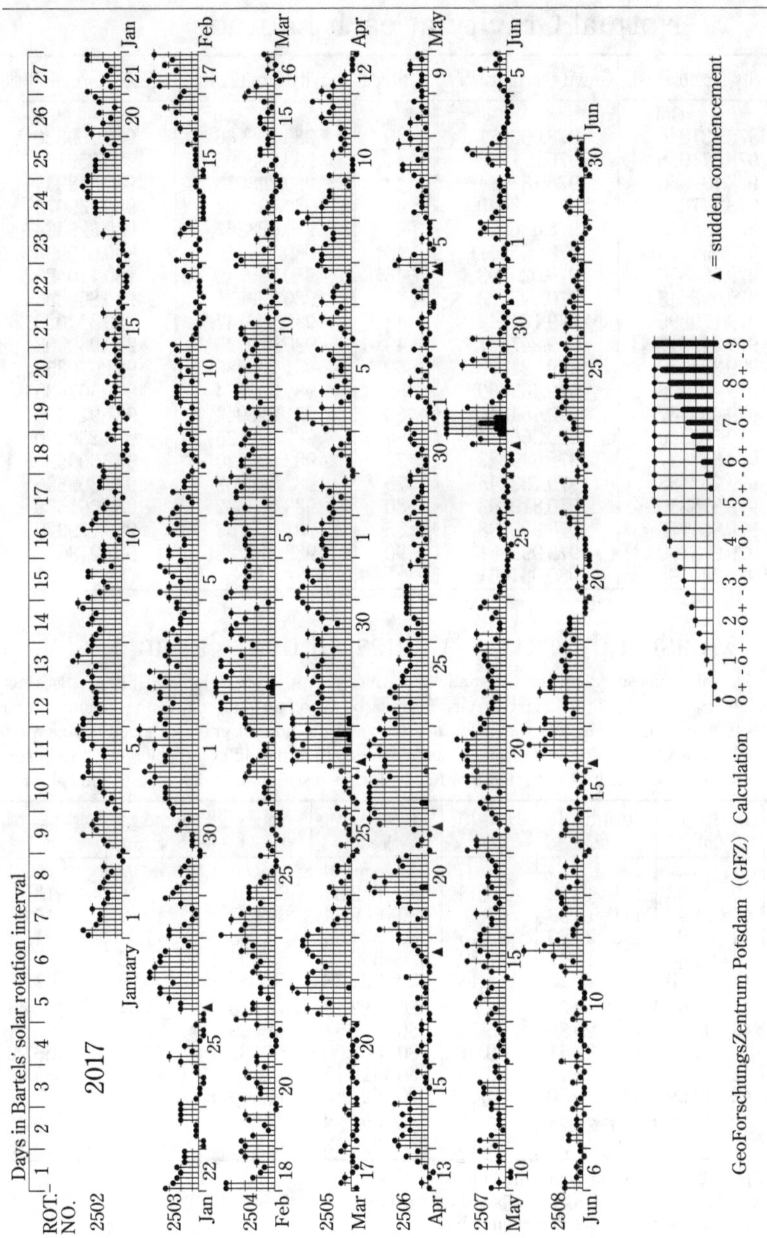
Days in Bartels' solar rotation interval
ROT.-NO.
2017
GeoForschungsZentrum Potsdam (GFZ) Calculation
▲ = sudden commencement

Normal Gravity at each Latitude

Latitude	Gravity formula 1967	Gravity formula 1980	Latitude	Gravity formula 1967	Gravity formula 1980
°	Gal	Gal	°	Gal	Gal
0	978.031846	978.032677	40	980.168966	980.169830
5	978.071066	978.071898	41	980.258295	980.259160
10	978.187550	978.188384	42	980.348078	980.348944
15	978.377803	978.378640	43	980.438204	980.439072
20	978.636113	978.636954	44	980.528565	980.529434
25	978.954716	978.955561	45	980.619050	980.619920
26	979.024856	979.025703	46	980.709549	980.710420
27	979.096945	979.097793	47	980.799951	980.800824
28	979.170895	979.171744	48	980.890147	980.891022
29	979.246617	979.247467	49	980.980026	980.980902
30	979.324019	979.324870	50	981.069480	981.070357
31	979.403008	979.403860	55	981.506554	981.507438
32	979.483487	979.484341	60	981.916949	981.917839
33	979.565360	979.566215	65	982.288125	982.289020
34	979.648527	979.649383	70	982.608720	982.609620
35	979.732888	979.733745	75	982.868902	982.869806
36	979.818339	979.819198	80	983.060682	983.061588
37	979.904778	979.905638	85	983.178162	983.179070
38	979.992100	979.992961	90	983.217728	983.218637
39	980.080198	980.081061			

Measured Gravity Values all over Japan

g is the gravity value measured by the Geospatial Information Authority of Japan (GSI). γ_{1980} is a normal gravity value calculated by the normal gravity formula based on an ellipsoid of the geodetic reference system 1980. A free air anomaly is the difference between the normal gravity value and the value corrected by hight at each measurement point. Bouguer anomaly is the difference between the normal gravity value and the value which is corrected by effects of mass distribution in addition to the height.

Location	Latitude (φ)	Longitude (λ)	Height (H)	Measured gravity value (g)	Normal gravity (γ_{1980})	Free-air anomaly F	Bouguer anomaly B
	° ′ ″	° ′ ″	m	mGal	mGal	mGal	mGal
Wakkanai	45 24 55	141 40 42	3.05	980 642.54	657.51	−13.16	−13.25
Rishiri	45 14 47	141 13 53	69	980 669.64	642.22	49.57	45.09
Nayoro	44 21 53	142 27 40	95.45	980 574.04	562.43	41.93	31.73
Abashiri	44 01 06	144 16 47	37.39	980 589.10	531.09	70.42	66.72
Asahikawa	43 45 28	142 22 22	114.48	980 531.38	507.52	60.05	47.52
Nemuro	43 22 01	145 48 04	13.10	980 682.18	472.20	4.04	213.44
Sapporo	43 04 20	141 20 30	15.21	980 477.54	445.60	37.50	36.31
Kushiro	42 59 10	144 22 41	−0.91	980 603.00	437.82	165.77	165.92
Obihiro	42 55 21	143 12 44	38.95	980 418.15	432.08	−1.04	−5.28
Chitose	42 47 09	141 40 50	24.40	980 421.55	419.74	10.21	7.57
Oshamanbe	42 30 30	140 22 24	5.62	980 421.88	394.73	29.75	29.40
Hakodate	41 49 34	140 44 52	43.43	980 399.28	333.30	80.25	75.96

(1) Refer to page **296** for further about the unit 'Gal'.
(2) The latitude, longitude and height are based on the Japan Geodetic Datum 2011.
(3) Height (H) in integer value are determined by maps, etc.
(4) Numbers with * mark, when upper 3-digits of measured gravity value (g) are '980', upper 3-digits of normal gravity (γ_{1980}) are '979'. Similarly, upper 3-digits of g are '979', upper 3-digits of γ_{1980} are '978'.

Continued.

Location	Latitude (φ)	Longitude (λ)	Height (H)	Measured gravity value (g)	Normal gravity (γ_{1980})	Free-air anomaly F	Bouguer anomaly B
	° ′ ″	° ′ ″	m	mGal	mGal	mGal	mGal
Mutsu	41 18 03	141 12 48	18.57	980 357.31	286.12	5.73	76.11
Aomori	40 49 19	140 46 07	3.37	980 311.07	243.22	69.76	69.79
Misawa	40 40 35	141 22 34	45.40	980 302.68	230.20	87.36	82.34
Hirosaki	40 35 17	140 28 25	50.90	980 261.21	222.31	55.48	50.30
Morioka	39 41 55	141 09 57	153.83	980 189.62	143.01	94.94	79.37
Miyako	39 38 50	141 57 56	45.25	980 270.40	138.44	146.79	142.85
Ofunato	39 03 55	141 42 52	36.70	980 210.67	86.85	136.02	134.53
Sendai	38 15 07	140 50 38	130.99	980 065.09	15.08	91.29	77.41
Akita	39 43 46	140 08 12	27.87	980 175.73	145.74	39.46	36.62
Shinjo	38 45 27	140 18 45	100.74	980 060.08	59.63	32.40	21.81
Iwaki	36 56 52	140 54 12	4.15	980 008.49	901.11*	109.53	109.21
Tsukuba	36 06 14	140 05 13	21.64	979 951.03	828.14	130.44	128.06
Takasaki	36 23 43	139 01 00	132.10	979 819.38	853.26	7.75	-6.28
Kawagoe	35 53 25	139 31 36	8	979 844.94	809.77	38.51	37.65
Choshi	35 44 23	140 51 29	20.09	979 866.88	796.86	77.09	74.91
Kano Zan	35 15 19	139 57 22	351.24	979 690.82	755.45	44.60	8.90
Haneda	35 32 57	139 47 03	−2	979 759.54	780.54	−20.75	−20.51
Chichi Jima	27 05 32	142 11 28	2.51	979 439.62	104.54	336.72	336.78
Hakone	35 14 38	139 03 35	426	979 709.24	754.49	87.04	46.09
Aburatsubo	35 09 37	139 36 56	4.67	979 774.65	747.37	29.59	29.23
Aikawa	38 01 17	138 14 24	5	980 076.67	995.57*	83.51	84.10
Agano	37 49 20	139 13 01	8.00	979 976.83	977.38	2.47	2.20
Nagaoka	37 25 26	138 46 36	58.97	979 931.45	942.55	7.96	1.71
Toyama	36 42 35	137 12 09	9.31	979 867.42	880.45	−9.29	−9.71
Kanazawa	36 32 45	136 42 29	106	979 841.65	866.26	8.96	−1.90
Fukui	36 03 20	136 13 22	8.96	979 838.12	823.97	17.79	17.15
Kofu	35 40 03	138 33 14	273.39	979 705.90	790.67	0.44	−26.92
Fujiyoshida	35 27 13	138 45 45	1029.39	979 566.26	772.37	112.33	0.10
Matsushiro	36 32 38	138 12 11	409.17	979 774.08	866.10	35.08	−2.00
Iida	35 30 04	137 49 57	467.04	979 666.98	776.44	35.49	−13.20
Gifu	35 24 02	136 45 45	12.08	979 745.81	767.85	−17.44	−18.65
Shizuoka	34 58 34	138 24 13	14.45	979 741.61	731.73	15.21	14.10
Hamamatsu	34 45 14	137 42 42	46	979 738.32	712.89	40.50	35.43
Omae zaki	34 36 40	138 12 21	41.76	979 741.08	700.81	54.08	49.47
Nagoya	35 09 18	136 58 08	42.22	979 733.37	746.92	0.35	−3.97
Tsu	34 44 04	136 31 12	−1.26	979 714.99	711.23	4.24	4.55
Owase	34 03 38	136 11 54	14.86	979 716.28	654.47	67.27	69.56
Maizuru	35 27 02	135 19 03	2.72	979 794.90	772.12	24.49	25.41
Kyoto	35 01 50	135 46 59	59.79	979 707.68	736.34	−9.35	−15.11
Itami	34 47 31	135 26 22	15.42	979 703.46	716.10	−7.01	−8.51
Himeji	34 50 25	134 37 21	40.06	979 722.72	720.19	15.76	11.75
Wakayama	34 13 46	135 09 52	13.70	979 689.27	668.64	25.73	24.36
Kushimoto	33 31 13	135 50 11	25.67	979 735.39	609.32	134.86	133.03
Tottori	35 29 17	134 14 18	8	979 790.56	775.32	18.58	18.28
Matsue	35 29 13	133 03 59	8.38	979 794.85	775.23	23.08	22.41
Okayama	34 39 39	133 54 59	−1	979 711.45	705.00	7.01	7.22
Hiroshima	34 22 20	132 27 57	0.95	979 658.59	680.65	−20.90	−20.61
Hagi	34 26 23	131 25 00	16.17	979 685.74	686.35	4.99	4.18
Shimonoseki	33 56 56	130 55 36	0.12	979 675.28	645.10	31.09	31.19
Naruto	34 10 19	134 36 18	0.75	979 670.86	663.80	8.16	8.21

Measured Gravity Values all over Japan Continued.

Location	Latitude (φ)	Longitude (λ)	Height (H)	Measured gravity value (g)	Normal gravity (γ_{1980})	Free-air anomaly F	Bouguer anomaly B
	° ′ ″	° ′ ″	m	mGal	mGal	mGal	mGal
Takamatsu	34 19 06	134 03 16	9	979 698.81	676.12	26.34	25.50
Ehime	33 51 04	132 46 30	30.84	979 597.73	636.93	-28.81	-31.24
Kochi	33 33 25	133 32 01	-0.69	979 625.62	612.39	13.89	14.95
Muroto	33 14 53	134 10 41	10.14	979 669.85	586.73	87.12	88.03
Ashizuri	32 44 09	132 58 33	125.27	979 589.70	544.45	84.77	73.18
Fukuoka	33 35 55	130 22 35	31.51	979 628.56	615.84	23.31	20.30
Tsushima	34 08 09	129 12 24	384.82	979 625.39	660.77	84.21	47.87
Iki	33 44 34	129 44 05	91.38	979 631.78	627.86	32.98	24.06
Nagasaki	32 44 01	129 52 06	23.70	979 588.02	544.27	51.93	50.12
Fukue	32 43 03	128 45 25	72.19	979 564.45	542.96	44.63	40.11
Kumamoto	32 49 01	130 43 40	22.81	979 551.64	551.12	8.43	6.26
Oita	33 14 11	131 37 10	5.10	979 541.83	585.75	-41.48	-41.75
Nobeoka	32 37 01	131 34 38	165.37	979 465.62	534.69	-17.19	-31.14
Miyazaki	31 56 18	131 24 51	9.36	979 429.49	479.35	-46.10	-47.02
Kagoshima	31 33 19	130 32 55	4.58	979 471.21	448.37	25.12	24.86
Naze	28 22 47	129 29 45	3.70	979 250.33	200.29	52.05	52.65
Naha	26 12 27	127 41 13	21.09	979 095.92	40.50	62.80	60.49
Miyako Jima	24 47 41	125 16 41	38.74	978 997.66	941.41	69.08	64.85
Ishigaki Jima	24 20 12	124 09 53	6.67	979 006.02	910.15*	98.80	98.09
Iriomote Jima	24 17 03	123 52 54	14.02	979 012.23	906.60*	110.83	109.53

International Gravity Standardization Net 1971

The International Gravity Standardization Net 1971 (IGSN71) was adopted by the International Union of Geodesy and Geophysics (IUGG) and the American Geophysical Union (AGU) in 1971. It is currently the most accurate set of gravity values covering almost all over the world.

IGSN71 is constructed based on the following measurements.

1) 10 absolute gravity measurements by the free-fall method conducted at eight sites such as Sevres (France), Teddington (England), and Gaithersburg (USA).
2) Approximately 1 200 relative measurements using six types of gravity pendulums such as Gulf (USA), Cambridge (U.K.), GSI (Japan).
3) Relative measurements of approximately 12 000 long-distance or international points using the LaCoste relative gravimeters, and measurements of approximately 11 700 supplementary points within the same city using various gravimeters.

Based on the above, the least-square method was conducted in order to determine the gravity values of 1 854 points in total in 494 cities throughout the world. An accuracy of the values is more than 0.1 mGal over the earth.

The IUGG and the AGU adopted the new Geodetic Reference System 1980 in 1979 (refer to page **517**). According to it, the normal gravity formula based on the Geodetic Reference System 1980 ellipsoid is described as follows.

$$\gamma_{1980} = 978.032\ 677\ 15 \times \frac{1 + 0.001\ 931\ 851\ 353 \times \sin^2\varphi}{\sqrt{1 - 0.006\ 694\ 380\ 022\ 90 \times \sin^2\varphi}}\ \text{Gal}$$

Japan Gravity Standardization Net 2016

GSI established Japan Gravity Standardization Net 1975 (JGSN75) in 1976 which consisted of 39 gravity points (in 11 cities) contained in IGSN71 and former domestic gravity standard network which had been already constructed by GSI at that time. JGSN75 offered the gravity network with about 0.1 mGal accuracy and since then gravity values were decided based on JGSN75 in Japan.

However, while about 40 years passed, differences between JGSN75 and real gravity values became larger in some areas due to crustal deformations. In addition, improvements of gravimeters required more precise and reliable standard gravity values.

Thus GSI prepared the new gravity standard based on latest gravity measurement data and released it as JGSN2016 in March 2017. Thanks to using FG5 absolute gravimeters which were validated their consistency with the international standards by annual domestic comparison campaign, JGSN2016 realised one order better precision than JGSN75 with 0.006 mGal and 0.019 mGal at the fundamental and primary gravity points, respectively.

Geoid Height Distribution in Japan

Figure 52

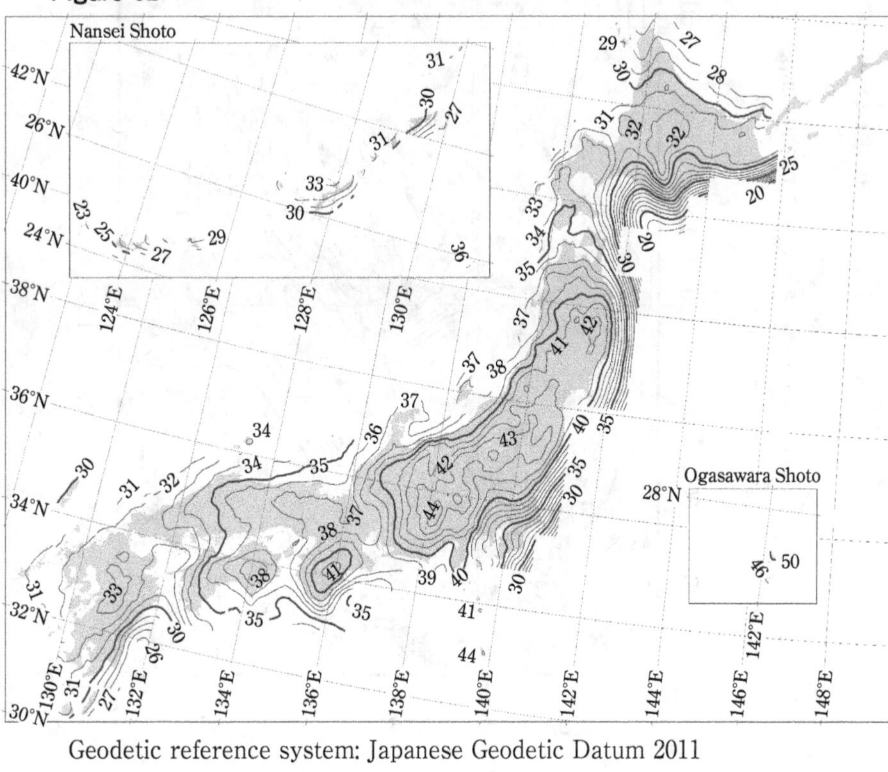

Geodetic reference system: Japanese Geodetic Datum 2011

Geoid Height Distribution

Figure 53
With respect to the WGS-84 coordinate system and ellipsoid, Unit: m, contour interval: 10 m, Dotted lines indicate negative values of the geoid height.

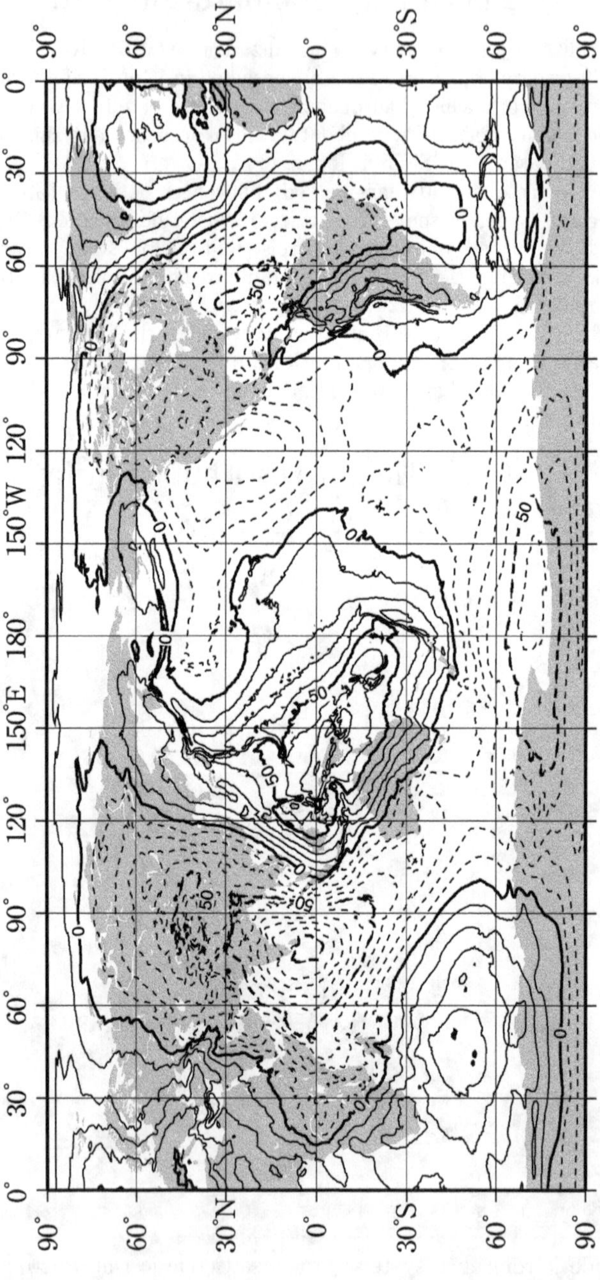

Source: EGM2008 (Pavlis *et al.*, 2012)

Ionosphere

Figure 54 Height profile of the Earth's upper atmosphere properties

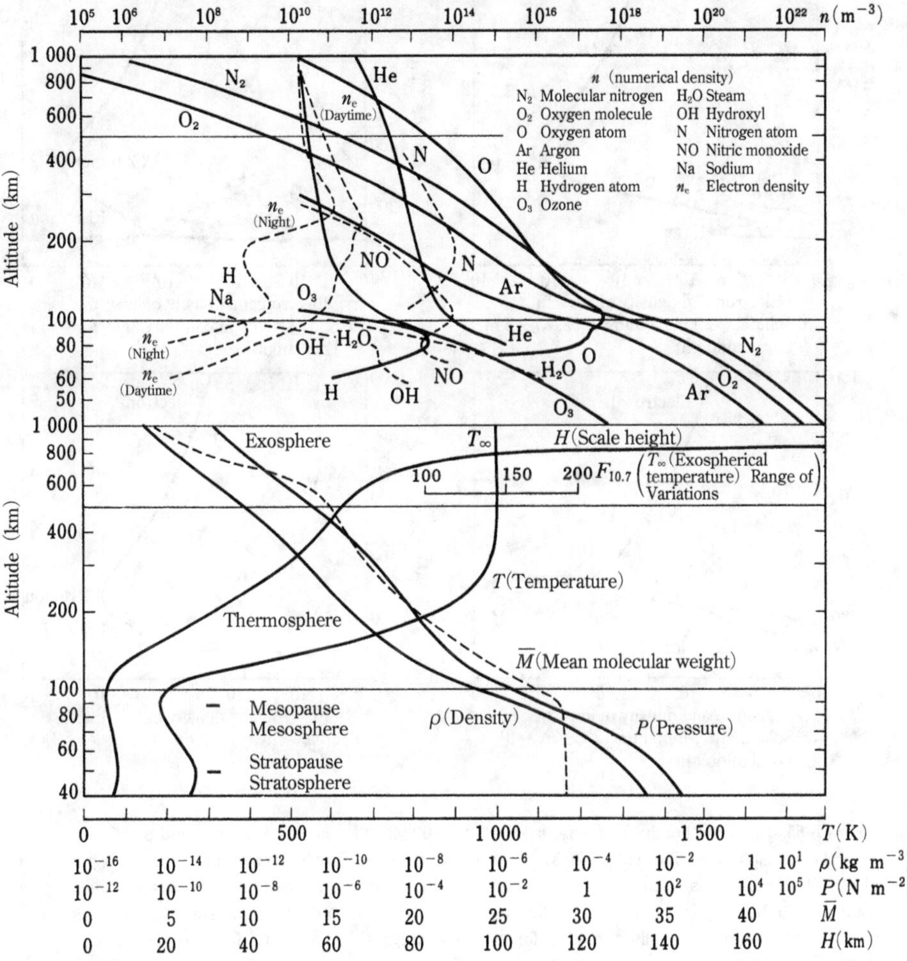

Figure 54 shows the height profile for average physical properties of the Earth's atmosphere (altitude of 40 km to 1 000 km). This information was taken from the COSPAR International Reference Atmosphere 1972 (CIRA) established by the Committee on Space Research (COSPAR). Here, "average" refers to the average at a latitude of 30° in a medium-degree solar activity cycle (solar radio intensity index for a wavelength of 10.7 cm: $F_{10.7}$ = 145). The number density for the atmospheric composition H_2O, OH, N, NO, Na and n_e as shown in the top row of Figure 55 was taken from sources other than CIRA.

Figure 55 Height profile of electron/ion density in the ionosphere above Tokyo (March) (IRI-2016)

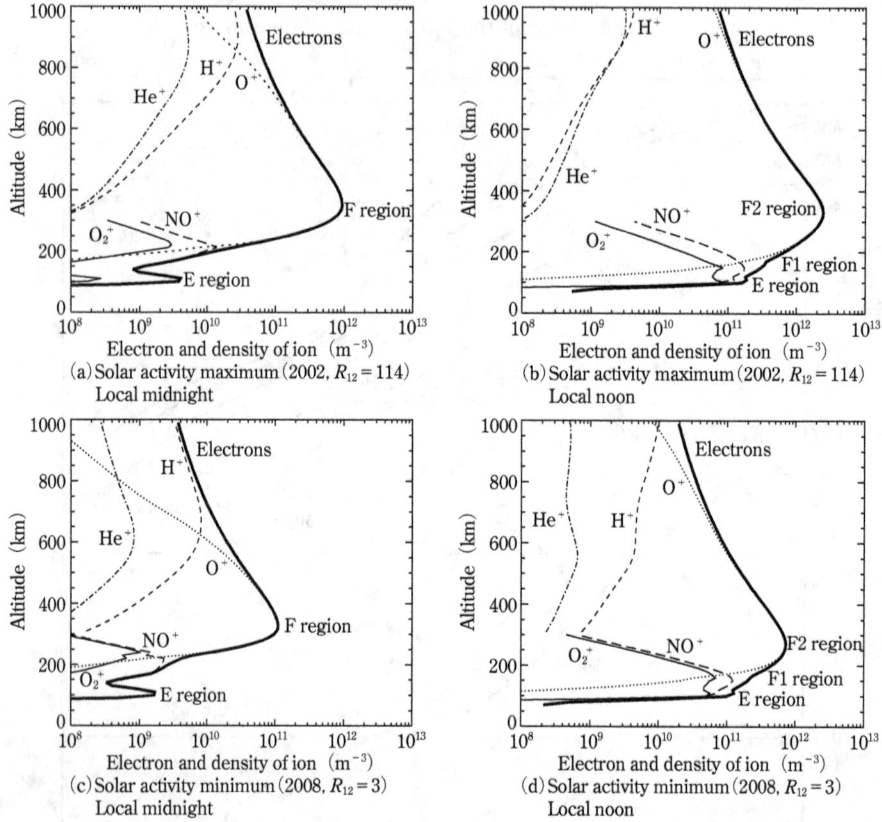

(a) Solar activity maximum (2002, $R_{12}=114$) Local midnight

(b) Solar activity maximum (2002, $R_{12}=114$) Local noon

(c) Solar activity minimum (2008, $R_{12}=3$) Local midnight

(d) Solar activity minimum (2008, $R_{12}=3$) Local noon

Figure 55 shows the height profile (altitude: 80 to 1 000 km) for electron density and positive ion density in the ionosphere above Tokyo in March according to the International Reference Ionosphere (IRI) model. The two graphs at the top ((a) nighttime, (b) daytime) are for the solar maximum, while the two graphs at the bottom ((c) nighttime, (d) daytime) are for the solar minimum. O^+ stands for oxygen ion, H^+ for hydrogen ion, He^+ for helium ion, O_2^+ for molecular oxygen ion, and NO^+ for nitrogen oxide ion. At altitudes of 80 km and higher, the negative ion density is small and the total of the positive ion density is equivalent to the electron density. Starting from low altitude, the ionosphere is divided into layers knowns as D layer (not shown in the figure due to low density), E layer, and F layer (during the daytime, divided in F1 and F2 layer). Furthermore, the Es layer (Sporadic E layer), a layer with high electron/ion density, often suddenly appears in the E layer.

* An ionosphere model established by the IRI Working Committee which consists of the International Union of Radio Science (URSI) and COSPAR.

Radio Wave Observation Records for the Ionosphere (Ionogram)

Figure 56

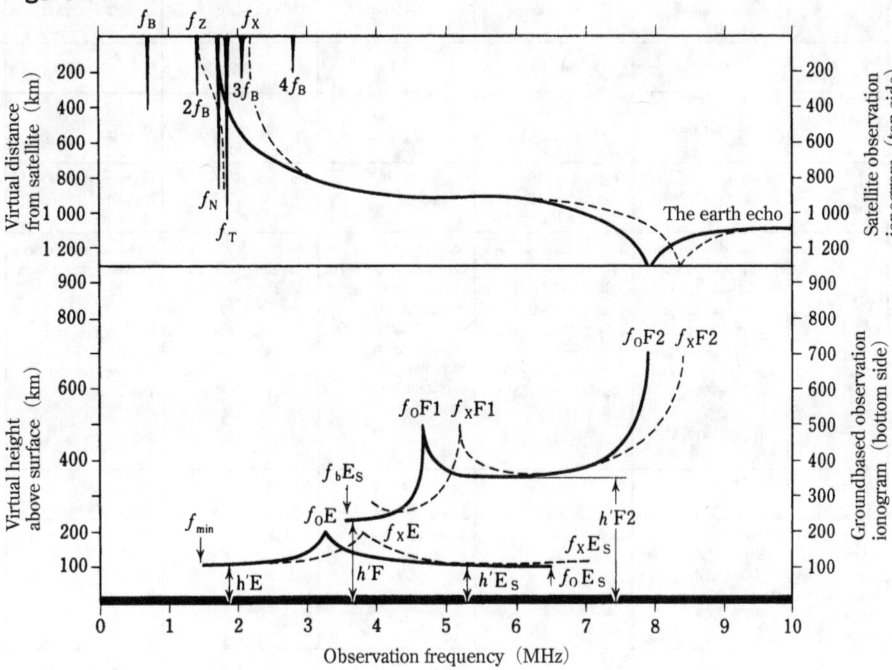

When pulse radio waves are emitted upwards from the Earth's surface, it is possible to receive waves reflected from the ionosphere on the Earth's surface. The reflection points' virtual height (h') is obtained from the delay time of the reflected radio waves.

The bottom panel of Figure 56 shows a record of the virtual height for reflection points in relation to the observed frequency for reflected waves obtained through ground measurement (bottom-side ionogram). From this, it is possible to observe the ionosphere structure for the bottom side from the altitude of maximum electron density of F layer and F2 layer near an altitude of 300 km (refer to Figure 55). Similarly, using observation equipment mounted on a satellite, it is possible to observe the ionosphere for the top side from the altitude of maximum electron density. The top panel of Figure 56 shows a record of the perceived distance for reflection points in relation to the observed frequency obtained from satellite observation. The top-side ionogram receives reflection echoes (earth echoes) from the Earth's surface at high frequencies, which exceed the critical frequency of the F layer.

Characteristics of the Ionosphere

Characteristics \ Name	D region	E region	Es layer	F1 region	F2/F region
Distribution altitude (km)	60-90	90-130	95-130	130-210	210-1000
Peak altitude (km)	no clear peak	Daytime -105 At sunrise-120	95-130 Variable	160-180 Manifests only in daytime	250-400 Always present
Peak electron density (m^{-3})	Daytime 10^9-10^{10} Nighttime 10^8-10^9	Daytime 10^{10}-10^{11} Nighttime 10^9-10^{10} (Refer to Figure 64)	10^{11}-10^{12}	Daytime 10^{11}-5×10^{11}	Daytime 4×10^{11}-4×10^{12} Nighttime 10^{11}-5×10^{11} (Refer to Figures 66-69)
Main ionization source	The sun ultraviolet rays (L_α and wavelength 121.6 nm) X rays of the sun (wavelength 1 nm or less) X rays of aurora (wavelength 1 nm or less) Cosmic rays (1 GeV or more) Solar proton (100 MeV -1 GeV)	The sun ultraviolet rays (wavelength 102.7 nm-80 nm) X rays of the sun (Wavelength 10 nm-1 nm) Aurora particle (Proton 10 keV-1 MeV and electron 1 keV-30 keV)	Same as E layer. However, in the mid to low latitude Es layer, ionization of metal atoms (Na, Mg, Si, Fe, etc.) also contributes as a source of important metallic ions M^+.	Solar ultraviolet rays (wavelength 80 nm-15 nm) Polar region low energy particles (electrons under 1 keV)	Solar ultraviolet rays (wavelength 80 nm-15 nm) Polar region low energy particles (electrons under 1 keV)
1st ions generated	N_2^+, O_2^+, NO^+, O^+	N_2^+, O_2^+, O^+, M^+	Same as E region	N_2^+, O_2^+, O^+	N_2^+, O_2^+, O^+
Primary ions of layers formed through ionization or chemical reactions	Cation NO^+, O_2^+, $H_3O^+\cdot(H_2O)_n$ Anion $NO_3^-\cdot(H_2O)_n$	Cation NO^+, O_2^+, M^+	Same as E region	Cation NO^+, O_2^+	Cation O^+
Stratification factors	The D region is formed as an equilibrium state of production and loss by ionization and chemical reactions of positive and negative ions and electrons.	The E region is formed as an equilibrium state of production and loss by ionization and chemical reactions of positive ions and electrons.	The Es layer is formed as an equilibrium state of production and loss by ionization and chemical reactions, as well as accumulation and divergence by the motion of ions and electrons. Polar region type: density perturbation of ions and electrons due to auroral electrojet currents. Mid- and low latitude type: conversion of metal ions and electrons due to atmospheric winds. Magnetic equator type: density perturbation of ions electrons due to equatorial electrojet currents.	The F1 region is formed as an equilibrium state of production and loss by ionization and chemical reactions of positive ions and electrons.	In addition to the production and loss effects due to the ionization and chemical reactions of positive ions and electrons, the accumulation and divergence effects by the gravitational diffusion motion of ions and electrons are important for the formation of the F2 region.

Continued.

Name / Characteristics	D region	E region	Es layer	F1 region	F2/F region
Diurnal variation	Electron density in the dayside depends on the solar zenith angle because the main source of ionization is solar UV and X-ray.	Electron density in the dayside depends on the solar zenith angle because the main source of ionization is solar UV and X-ray. (Refer to Figure 59–64)	Polar region type: occur during nighttime. mid-latitude type: occur in the daytime to evening. Equator type: occur during daytime.	Electron density in the dayside depends on the solar zenith angle because the main source of ionization is solar UV and X-ray. (Refer to Figure 59)	The electron density is high during the daytime and low during the nighttime but it depends on the ion and electron dynamics (Refer to Figure 60, 66–69).
Seasonal variation	For the same solar zenith angle, the winter anomaly phenomenon in which the electron density increases can be seen in high- to mid-latitudes more often in the winter than in the summer.	Throughout all seasons, the electron density changes in accordance to the solar zenith angle. (Refer to Figure 64)	Low-latitude type: Es layer is highly dependent on the season, with an extremely high rate of occurrence in the summer (May to August in the northern hemisphere, November to February in the southern hemisphere).	Stratification becomes clearly apparent in summer. Generally speaking in winter, layers do not appear often. (Refer to Figure 61)	The winter anomaly phenomenon in which the electron density during the daytime is higher in the winter than in the summer can be seen at high- to mid-latitudes. During both the daytime and nighttime, a biannual variation can be seen in which the electron density reaches a maximum in spring and autumn. (Refer to Figure 61)
Solar activity dependence	The electron density is dependent on solar activity. The ratio of electron density for the period of maximum solar activity and the period of minimum solar activity is about 1.5.	Almost the same as the D region. (Refer to Figure 62)	No particular solar activity dependence is observed.	The ratio of peak electron density for the period of maximum solar activity and the period of minimum solar activity is about 2. The layer appears clearly during the period of minimum solar activity. (Refer to Figure 62)	The ratio of peak electron density for the period of maximum solar activity and the period of minimum solar activity is about 4. (Refer to Figure 62)
Geographical distribution	According to the solar zenith angle, the electron density during the daytime is higher at lower latitudes. However, the winter anomaly appears at high- to mid-latitudes. The impact of particle ionization is strong in polar regions.	According to the solar zenith angle, the electron density during the daytime is higher at lower latitudes. (Refer to Figure 63) The impact of aurora particle ionization is strong in polar regions.	The appearance characteristics vary for the polar regions, mid- to low-latitudes, and magnetic equator. Mid- to low-latitudes: The appearance rate for Es layer is highest in the Far East region near Japan. (Refer to Figure 65)	According to the solar zenith angle, the electron density during the daytime is higher at lower latitudes. There is an impact from low-energy particle ionization in polar regions.	The Equatorial Ionization Anomaly (EIA) phenomenon is characterized by two peak electron density regions at 15° magnetic latitude on either side of the dip equator. This phenomenon can be seen from the daytime until early evening. (Refer to Figure 66–69) The impact of the magnetosphere is strong in polar regions.

Characteristics of the Ionosphere Continued.

Name / Characteristics	D region	E region	Es layer	F1 region	F2/F region
Impact on radio wave propagation	Acts as a reflective layer for VLF (Very Low Frequency) band radio waves. Acts as an absorbent layer for MF (Mid Frequency) and HF (High Frequency) band radio waves. Reflective echoes are not normally observed during normal radio wave observation. (Refer to Figure 56)	During the daytime, acts as a reflective layer for MF and HF band radio waves.	In particular, the Es layer is the cause of FM radio and analog television cross-interference due to over-the-horizon anomalous propagation of VHF band radio waves.	Acts as a reflective layer for MF and HF band radio waves.	Acts as a reflective layer for HF band radio waves. Disturbance in electron density causes spread F and VHF band radio waves to scatter and leads to radio wave scintillation of satellite signal. (Refer to Figure 70)

Isomagnetic Latitude Lines Displayed in Geographical Coordinates**

Figure 57

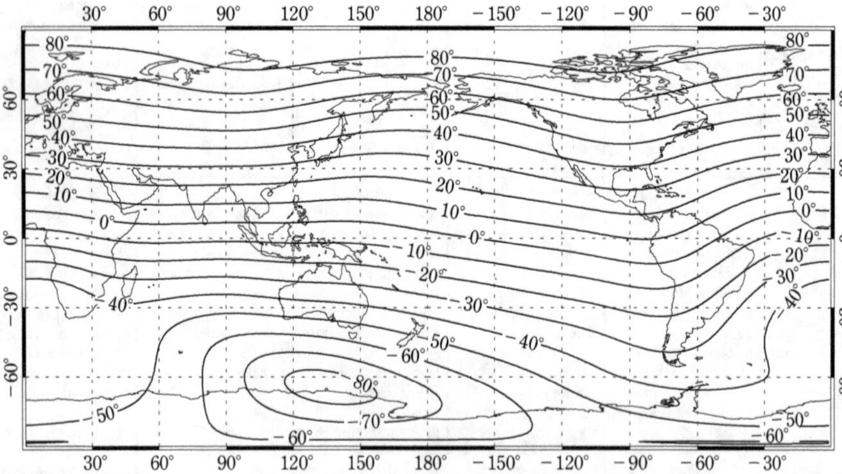

** Based on the 2020 Earth magnetic field model calculated from the International Geomagnetic Reference Field (IGRF-12).

Main Ionospheric Disturbances

Name / Characteristics	Summary	Regional distribution	Altitude region	Temporal variation	Effects on radio propagation
Sudden Ionospheric Disturbance (SID)	A phenomenon in which the X-rays and ultraviolet rays emitted during a solar flare cause abnormal ionization that leads a sudden increase in the electron density.	During the daytime, the electron density increases in accordance to the solar zenith angle.	D, E, and F1 regions	Increases immediately after the flare and continues for several minutes to several hours.	Short waves are absorbed, there are slight fluctuations in frequency, there is a VLF phase anomaly, etc.
Polar Cap Absorption (PCA)	A phenomenon in which the polar cap undergoes abnormal ionization and there is an increase in the electron density and radio wave absorption. Caused by the solar protons emitted during solar flares.	Occurs in polar cap regions (magnetic longitude of 70° or greater) (Refer to Figure 57)	D region	Occurs about 30 minutes to several hours after the flare and continues for several days.	Short waves are absorbed, there are slight fluctuations in frequency, there is a VLF phase anomaly, etc.
Plasma bubble	A phenomenon in which electron density undergoes localized decreases after sunset in magnetic equator regions. Possesses a structure which extends north and south along the magnetic field line.	Spreads from the magnetic equator to the equator anomaly zone (reaches a maximum around +/-15° magnetic latitude) (Refer to Figure 57)	F region	There is seasonal and longitudinal dependence. Occurs frequently in spring and autumn in Southeast Asia.	VHF abnormal propagation, amplitude scintillation of satellite radio waves*, and ionospheric delay occurs due to scattering.
Posi-tive-phase ionospheric storm	A phenomenon in which the electron density decreases to a level much higher relative to the quite time due to a geomagnetic storm. At mid-latitudes, it occurs due to neutral winds blowing toward the equator, a rise in the ionosphere caused by the electric field of the ionosphere origin, or both.	Occurs throughout the world.	F region	Continues for about several hours –1 day.	The ionospheric delay of satellite radio waves* increases.
Nega-tive-phase ionospheric storm	A phenomenon in which the electron density decreases to a level much higher relative to the quite time due to a geomagnetic storm. Occurs due to changes in the atmospheric composition ratio (increase of N_2 / O) caused by aurora activity.	Appears strongly at higher latitudes. In the summer hemisphere, it reaches low-latitude regions. In the winter hemisphere, it is limited to high-latitude regions.	F region	Occurs several hours to 1 day after the start of the geomagnetic storm. Continues for about 1–3 days.	The ionospheric delay of satellite radio waves* decreases. The reflective frequency of short waves decreases.
Traveling Ionospheric Disturbances (TIDs)	A phenomenon in which electron density variations move in a wave-like structure. It is classified into large-scale TIDs (horizontal wavelength of at least 1 000 km) which are propagated from the aurora belt in conjunction with geomagnetic storms, and midium-scale TIDs horizontal (wavelength of a few hundred kilometers) which are unrelat-ed to geomagnetic activity.	Occurs throughout the world.	F region	Large-scale TID: Period of 30–120 minutes. Mid-scale TID: Period of 15–60 minutes. The propagation characteristics differ in the daytime and nighttime. Also, there is seasonal dependence and longitudinal dependence.	The ionospheric delay of satellite radio waves* changes cyclically.

* Effect increases as frequency decreases.

Specific Frequencies Obtained through Radio Wave Observation of the Ionosphere

The ground-based observation ionogram (refer to the bottom panel of Figure 56) shows the ordinary (O) mode reflection echo (bold line) and the extraordinary (X) mode reflection echo (thin dotted line) for radio waves. This occurs because the ionosphere becomes a doubly-refracting medium due to the Earth's magnetic field, decomposing a linearly polarized radio wave into two modes of propagation. The specific frequencies (critical frequencies) f_OE, f_OEs, f_OF1, and f_OF2 (O mode) and f_XE, f_XEs, f_XF1, and f_XF2 (X mode) are observed in correspondence to the maximum electron density for each ionosphere range (E, Es, F1, F2). When the O mode critical frequency is f_o(MHz) and the X mode critical frequency is f_x(MHz), the relationship $f_O^2 = f_X^2 - f_X \cdot f_B$ applies to both. f_B(MHz) is the electron gyrofrequency f_B(MHz) $= 280 \times 10^4$ B (tesla) (refer to Figure 58). The ionosphere maximum electron density N_m(m^{-3}) is given by N_m(m^{-3}) $= 1.24 \times 10^{10} \cdot [f_O(\text{MHz})]^2$.

The satellite observation ionogram (refer to the top panel of Figure 56) shows the ordinary (O) mode reflection echo (bold line) and the extraordinary (X, Z) mode reflection echo (thin dotted line). Also, the resonance spike (vertical bold line) in the vicinity of the satellite is observed in the frequency nf_B ($n = 1, 2, \cdots\cdots\cdots$), f_N, f_T $(= \sqrt{f_N^2 + f_B^2})$.

Distribution for Electron Gyrofrequency f_B(MHz)* at an Altitude of 200 km

Figure 58

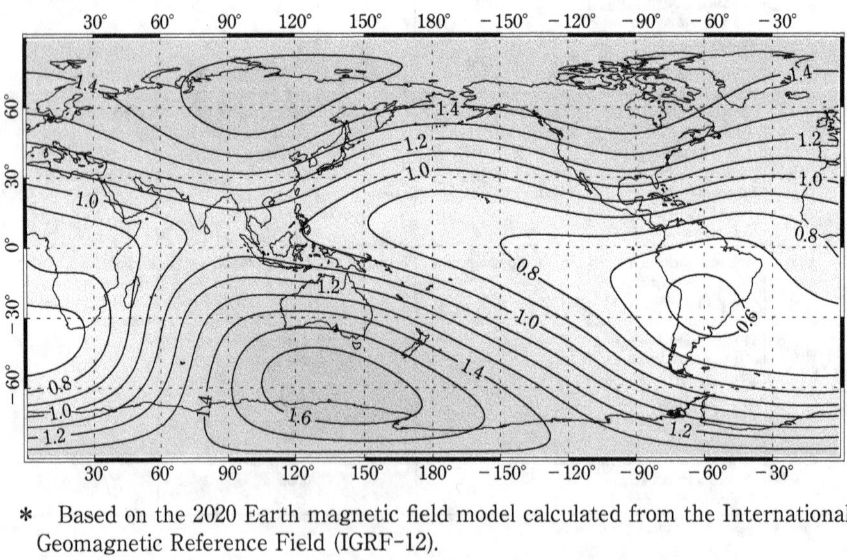

* Based on the 2020 Earth magnetic field model calculated from the International Geomagnetic Reference Field (IGRF-12).

Diurnal Variations for Specific Frequencies, etc.

Figure 59

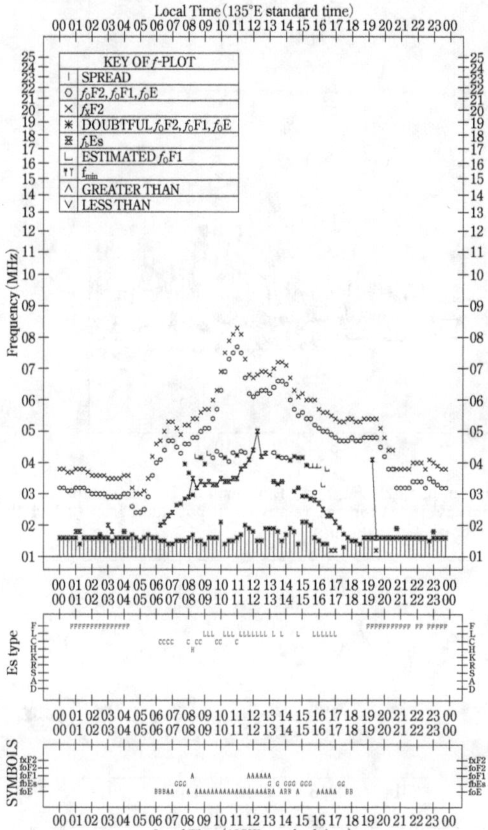

Figure 59 shows the diurnal variations for f_XF2, f_OF2, f_OF1, f_OE, f_bEs, and f_{min} on March 21, 2020 as obtained from ionosphere measurement in Tokyo (Kokubunji) and examples of the f plot which express the Es layer type and the reception echo conditions. The diurnal variation characteristics for f_OF2 as seen in Figure 60 change according to the region of the measurement location.

Example of f_OF2 Diurnal Variation Characteristics of Monthly Averages at each Measurement Location

Figure 60

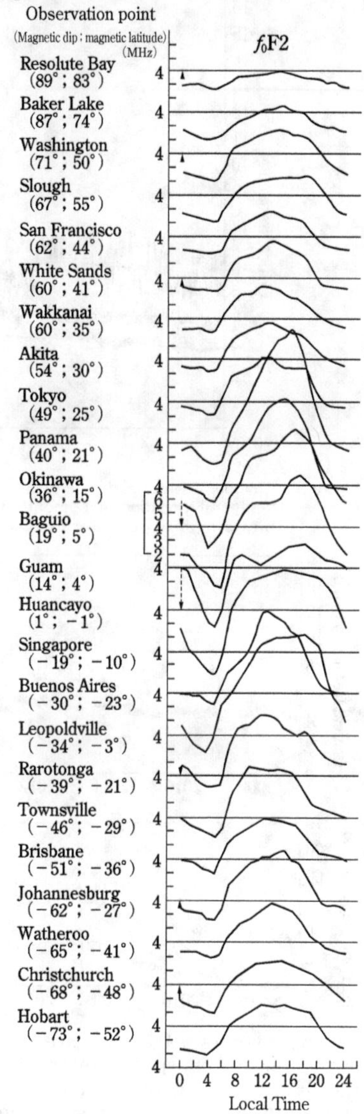

Figure 61 Annual day-to-day variation for f_oF2, f_oF1, and f_oE at noon in Tokyo (Kokubunji) (2019)

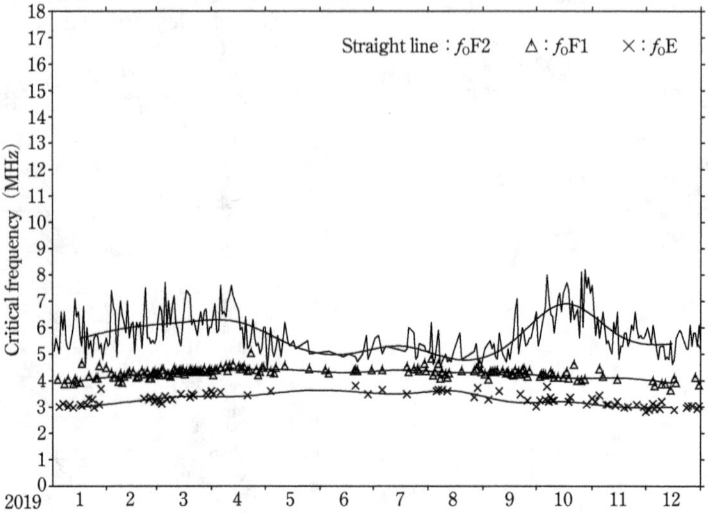

Figure 62 Temporal variation in the monthly median of f_oF2, f_oF1, and f_oE at noon in Tokyo (Kokubunji) and temporal variation in 12-month moving averages for the relative sunspot number (R_{12}) and the wavelength 10.7 cm solar radio wave intensity index (ϕ_{12}).

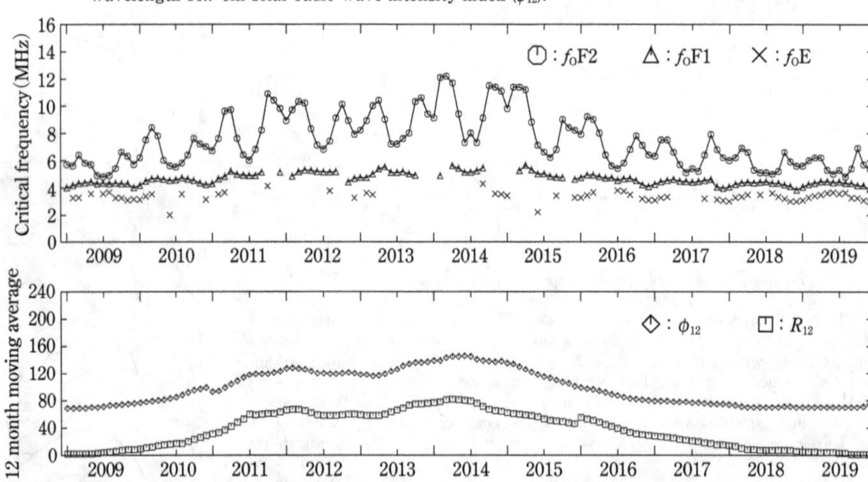

The parameters ϕ_{12} and R_{12} are excerpted from the fundamental indices for radio propagation in the ionosphere, International Telecommunication Union (ITU). R_{12} is based on the data provided by WDC-SILSO, Royal Observatory, Belgium. Since WDC-SILSO started to use revised sun spot number data since 2015, the value of R_{12} after December 2014 becomes bigger than before. As a reference, the estimated value R^*_{12} after December 2014 calculated by NGDC, Boulder US was also shown in this figure.

Short-Wave Fadeout* Observed in Japan (2017–2019)

Occurrence time (UT)		Solar flare class
Year/Month/Day	Hour/Minute	
2017 7 14	01 : 45 – 02 : 45	M2.4/1N
2017 9 5	04 : 45 – 05 : 00	M3.2

* A Sudden Ionospheric Disturbances (SID) phenomenon caused by sudden increase of ionospheric density in D region with increasing X-ray and Ultraviolet during solar flares. During Short-Wave Fadeout, HF radio wave is attenuated in the ionosphere.

Monthly Variation of Occurrence of Sporadic E Layer

Figure 63 Monthly variation of Sporadic E (Es) layer occurrence in 2019 observed at Hokkaido (Wakkanai), Tokyo (Kokubunji), Kagoshima (Yamagawa) and Okinawa (Ogimi).

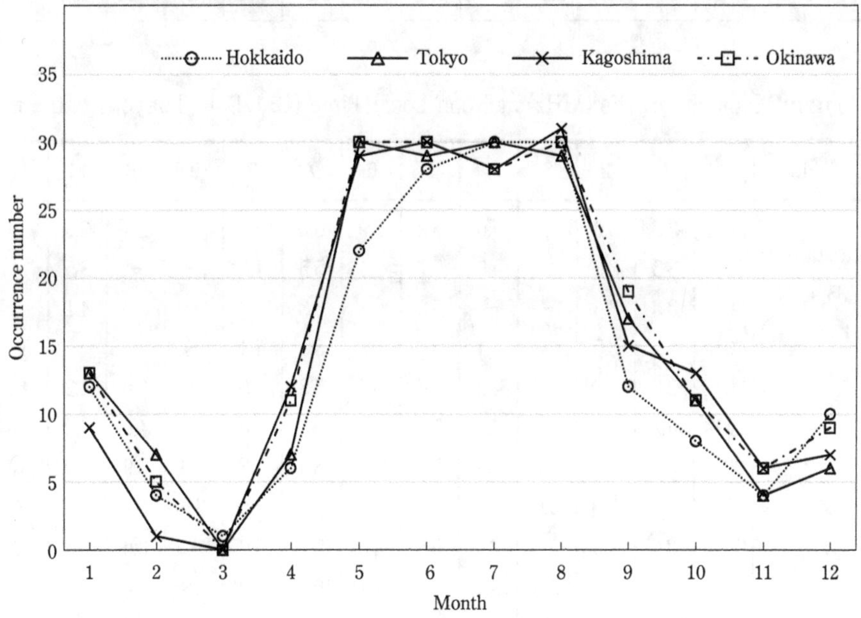

Monthly Medians for f_0E (MHz) at Noon Local Time (12 LT) in Tokyo (Kokubunji)

Station	Mo / Yr	1	2	3	4	5	6	7	8	9	10	11	12
Tokyo (Kokubunji)	2009	-	3.24	3.28	-	3.56	-	3.56	3.66	3.24	3.24	3.08	3.16
35°43′N	2010	3.12	3.36	3.52	-	-	2.00	-	3.54	-	-	-	3.12
139°29′E	2011	-	3.52	3.64	-	-	-	-	-	-	4.12	-	-
	2012	-	-	-	-	-	-	-	3.76	-	-	-	3.24
	2013	-	3.60	3.48	-	-	-	-	-	-	-	-	-
	2014	-	-	-	-	-	-	-	-	4.28	-	3.56	3.52
	2015	3.42	-	-	-	-	2.20	-	-	3.42	-	-	3.28
	2016	3.28	3.44	3.64	-	-	-	3.80	3.74	3.48	-	3.18	3.08
	2017	3.08	3.26	3.32	-	-	-	-	-	3.20	-	3.12	3.06
	2018	3.00	3.28	3.34	3.48	-	3.48	-	3.60	3.34	3.20	3.00	3.00
	2019	3.08	3.28	3.40	3.48	3.60	3.64	3.56	3.62	3.28	3.24	3.08	3.00

Monthly Medians for f_0Es (MHz) at Noon Local Time (12 LT) in Tokyo (Kokubunji)

Station	Mo / Yr	1	2	3	4	5	6	7	8	9	10	11	12
Tokyo (Kokubunji)	2009	3.3	3.0	3.3	4.0	6.2	7.9	6.2	6.0	-	3.4	3.6	3.4
35°43′N	2010	3.5	3.8	-	4.0	5.7	7.1	5.5	4.6	-	-	3.5	-
139°29′E	2011	3.3	3.6	-	-	5.2	6.8	6.3	5.0	-	-	4.1	-
	2012	-	3.5	-	4.1	5.0	6.0	5.1	4.4	-	-	-	3.8
	2013	3.8	3.8	-	4.2	5.6	5.6	5.8	4.5	-	3.8	4.2	4.1
	2014	4.1	-	-	4.4	5.5	7.1	5.5	4.6	4.2	4.0	4.0	3.8
	2015	-	4.0	4.1	4.4	5.1	6.2	8.4	5.3	3.9	3.9	3.9	3.6
	2016	4.0	3.9	3.6	4.1	4.9	6.7	6.3	4.5	3.9	3.7	3.5	-
	2017	3.6	4.0	-	3.9	5.3	8.3	5.7	5.0	-	3.6	3.5	3.7
	2018	3.6	3.6	3.6	3.9	6.3	7.1	9.3	4.8	3.8	3.8	3.4	3.3
	2019	3.4	3.5	-	4.0	5.6	6.6	5.8	4.8	3.8	4.1	3.6	3.6

Note) On pages **742–744**, blanks indicate that data not acquired, ( ) indicate the value is uncertain, - indicates the value cannot be expressed.

Monthly Medians for f_oF2 (MHz) at Midnight Local Time (0 LT)

Station	Yr \ Mo	1	2	3	4	5	6	7	8	9	10	11	12
Tokyo	2009	2.6	2.9	3.3	3.6	3.6	3.3	3.0	2.8	3.4	3.4	3.0	2.8
(Kokubunji)	2010	2.8	3.4	3.9	4.7	4.8	4.8	4.7	4.2	3.9	3.6	3.3	3.0
35°43′N	2011	2.9	3.3	4.4	5.9	6.2	6.3	5.3	5.4	5.3	4.9	3.6	3.4
139°29′E	2012	3.4	3.8	4.8	6.2	7.0	7.3	6.4	5.8	5.6	4.3	3.6	3.1
	2013	3.2	3.7	5.0	6.6	7.6	7.2	6.8	6.0	5.3	4.5	3.6	3.5
	2014	3.5	4.4	6.8	7.4	7.9	6.8	7.3	6.4	5.8	5.2	4.0	3.7
	2015	3.5	4.2	5.2	7.0	7.1	7.4	6.0	5.2	4.4	4.0	3.5	3.2
	2016	3.3	3.7	4.3	4.9	5.7	5.2	5.0	4.4	4.1	3.6	3.2	2.8
	2017	2.9	3.3	3.5	4.4	4.3	4.5	3.8	4.0	3.8	3.7	3.2	3.0
	2018	2.7	3.2	3.6	3.8	4.0	4.4	3.8	3.5	3.4	3.2	3.0	2.9
	2019	2.8	3.0	3.2	3.8	4.0	4.1	3.1	3.4	3.2	3.2	3.0	2.7
Chumphon	2009	2.7	2.5	(4.9)	4.5	2.5	2.4	(2.6)	2.5	3.3	4.2	4.0	3.2
10°43′N	2010	3.3	6.0	7.0	6.0	3.9	2.7	3.6	4.9	6.0	7.7	(6.9)	–
99°22′E	2011	–	–	(8.9)	–	–	–	–	–	–	–	–	–
	2012	–	8.0	9.6	(11.3)	–	–	(8.4)	8.2	10.2	7.6	8.2	6.4
	2013	6.7	7.8	8.8	10.5	10.6	6.8	7.5	8.7	9.6	9.6	8.3	8.2
	2014	7.5	10.2	10.6	11.9	(10.7)	–	–	9.2	10.3	9.8	8.9	7.8
	2015	7.4	9.2	11.2	(11.8)	–	–	–	–	–	–	–	–
	2016	(6.7)	8.2	8.8	8.6	6.1	4.8	4.5	–	–	–	3.9	–
	2017			6.15									
	2018			4.5	4.4	3.2	3.0	2.7	(2.9)	(2.7)			
	2019	(3.8)	4.0	5.0	4.5	3.5	(2.9)	(2.9)	2.5	2.8	4.3		(2.8)
Syowa Station, Antarctica	2009	3.4	2.6	2.0	2.3	3.0	2.4	2.2	2.6	2.7	2.3	3.4	4.0
	2010	4.0	2.8	2.1	3.0	2.3	1.7	2.2	2.8	2.3	2.0	3.2	3.8
69°00′S	2011	3.0	2.7	2.1	2.7	2.8	3.1	2.8	2.8	2.4	3.1	5.4	5.2
39°35′E	2012	4.6	3.2	2.4	3.6	3.1	2.8	3.6	3.1	2.2	3.4	4.2	4.2
	2013	5.0	3.3	3.2	1.6	2.8	2.1	2.6	2.2	2.0	3.3	4.4	5.4
	2014	5.0	3.7	2.5	2.1	2.4	3.0	2.0	3.5	2.0	3.1	4.2	4.8
	2015	4.4	3.4	3.3	3.6	2.4	2.0	2.8	1.6	2.5	2.9	3.7	4.1
	2016	4.1	3.5	2.7	2.1	3.7	1.6	2.4	2.5	1.6	2.4	3.9	3.7
	2017	3.3	3.1	1.8	3.1	3.2	2.4	2.6	2.8	2.1	2.2	3.9	3.8
	2018	3.6	2.6	1.7	2.6	2.2				2.1	2.8	3.6	3.5
	2019	3.3	2.1	1.4	4.4	2.2	1.8	3.4	2.8	1.4	1.8	3.4	3.6

Monthly Medians for f_oF2 (MHz) at Noon Local Time (12 LT)

Station	Yr \ Mo	1	2	3	4	5	6	7	8	9	10	11	12
Tokyo (Kokubunji) 35°43′N 139°29′E	2009	5.7	5.6	6.4	5.8	5.7	4.9	4.8	4.9	5.4	6.6	6.3	5.7
	2010	6.2	7.5	8.4	7.8	6.0	5.6	5.5	5.8	6.4	7.6	7.2	7.0
	2011	6.7	7.6	9.6	9.7	7.6	6.4	6.0	6.8	8.2	10.9	10.4	9.8
	2012	8.9	9.7	10.3	10.2	8.3	7.1	6.8	7.4	9.1	10.1	8.9	7.9
	2013	8.2	8.9	10.0	10.4	9.0	7.2	7.2	7.6	8.0	10.3	10.6	9.4
	2014	9.1	12.1	12.2	11.7	9.4	7.3	8.0	7.3	9.1	11.5	11.4	11.1
	2015	9.8	11.4	11.4	11.2	8.8	7.1	6.6	6.2	6.8	9.0	8.4	8.2
	2016	7.9	9.2	9.0	8.0	6.4	5.6	5.4	5.8	6.8	7.8	7.1	6.4
	2017	6.3	7.5	7.5	6.6	5.7	5.1	5.4	5.2	6.4	7.9	6.8	6.2
	2018	6.0	6.2	6.9	6.6	5.3	5.1	5.1	5.0	5.2	6.6	5.9	5.6
	2019	5.6	6.0	6.2	6.2	5.3	5.0	5.3	4.8	5.4	6.9	5.7	5.4
Chumphon 10°43′N 99°22′E	2009	5.9	5.4	(6.3)	6.6	6.0	5.5	5.6	5.3	5.7	6.2	7.1	6.1
	2010	6.6	6.7	7.2	7.2	7.2	5.6	5.8	6.7	6.8	7.4	(7.1)	
	2011			(7.4)									
	2012		8.5	9.0	(9.6)				8.3	8.8	9.4	9.0	8.6
	2013	8.4	8.2	8.9	9.2	9.4	8.5	8.2	8.4	8.0	9.3	9.7	10.0
	2014	8.9	9.8	10.2	10.5	9.3			8.8	9.9	10.3	10.6	10.4
	2015	9.6	10.6	9.8	9.6								
	2016	(8.7)	8.8	8.4	7.8	7.6	6.6	6.4				7.7	
	2017			6.55									
	2018			7.0	6.6	6.4	6.0	5.2	(5.8)				
	2019	(5.5)	6.2	5.9	6.4	6.2	5.5	(5.3)	5.4	5.8	6.0		(6.0)
Syowa Station, Antarctica 69°00′S 39°35′E	2009	4.8	4.8	4.5	5.0	4.4	3.2	3.3	3.9	4.4	5.2	5.3	5.7
	2010	5.8	5.9	6.2	5.4	5.0	3.4	3.8	5.4	5.5	5.4	5.8	5.8
	2011	5.8	5.8	6.7	7.4	6.4	4.2	4.5	5.8	6.7	8.0	8.1	8.2
	2012	6.6	7.2	6.8	6.7	6.9	4.8	5.6	6.6	8.2	8.0	7.3	7.1
	2013	7.6	6.4	7.2	8.0	7.4	5.0	5.2	6.4	7.0	8.2	7.8	8.2
	2014	7.4	8.0	9.8	9.4	6.9	5.5	5.7	6.6	7.1	7.7	6.7	6.4
	2015	6.6	7.2	7.2	8.2	7.0	5.5	5.2	5.8	6.0	6.8	6.6	6.4
	2016	6.2	6.0	6.2	5.5	5.3	4.1	4.4	5.1	5.4	5.3	5.4	5.2
	2017	4.9	4.8	4.6	4.6	4.6	3.7	3.6	4.4	5.0	4.8	5.0	5.4
	2018	4.8	4.7	4.3	4.5	4.6	3.4	3.6	4.1	4.2	4.5	4.4	4.6
	2019	4.4	4.0	4.1	4.5	4.2	3.2	3.2	4.0	4.3	4.2	4.6	4.8

Distribution Map for f_oE during the Solar Activity Minimum Period* (MHz)

Figure 64

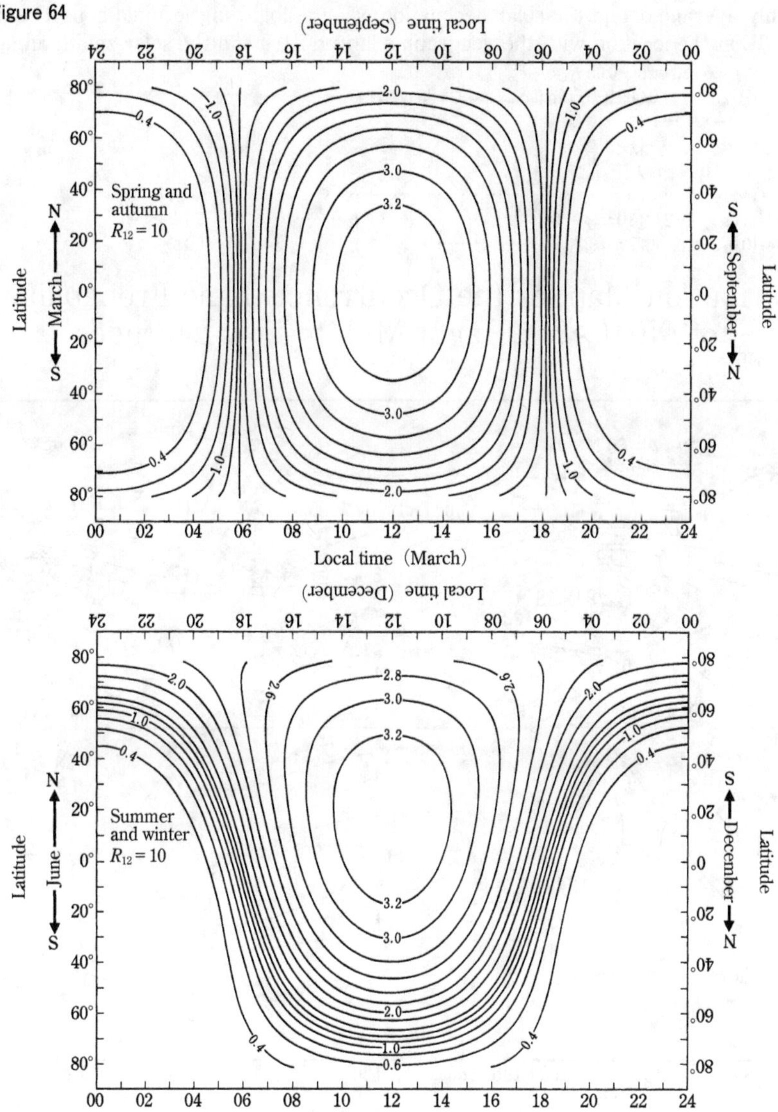

* The relative number of sunspots 12-month moving average R_{12}=10. This is the Sun seen at 10.7 cm wavelengths. The monthly average of the radio strength index corresponds to $\phi \approx 70$.

The f_oE world distribution map model shown in Figure 64 is in accordance with CCIR Rep. 340-4 and conforms to the following. Assuming the relative sunspot number 12-month moving average R_{12}, the wavelength 10.7 cm solar radio wave intensity index monthly average ϕ, and the solar declination δ(°), the following formula is used to determine f_oE at the location with the geographic latitude λ(°) and the solar zenith angle:

$$\phi = R_{12} + 46 + 23\exp(-0.05\,R_{12})$$

$$f_o\mathrm{E(MHz)} = [\{1+0.0094(\phi-66)\}\,(\cos N)^m\,(92+35\cos\lambda)\,D]^{0.25}$$

$$\begin{cases} N(°) = \lambda-\delta, & \cdots\cdots |\lambda-\delta|<80 \\ N(°) = 80, & \cdots\cdots |\lambda-\delta|\geqq 80 \end{cases}$$

$$m = 0.11 - 0.49\cos\lambda$$

$$\begin{cases} D = (\cos\chi)^{1.2} \cdots\cdots & \chi \leq 73 \\ D = [\cos\{\chi - 6.27\times10^{-13}(\chi-50)^{8.20}\}]^{1.2} \cdots\cdots & 73<\chi<90 \\ D = (0.072)^{1.2}\exp(25.2-0.28\chi) \cdots\cdots & 90 \leq \chi \end{cases}$$

Distribution Map for the Occurrence Time Probability (%) of f_oEs (>7 MHz) at Mid- to Low-Latitudes*

Figure 65

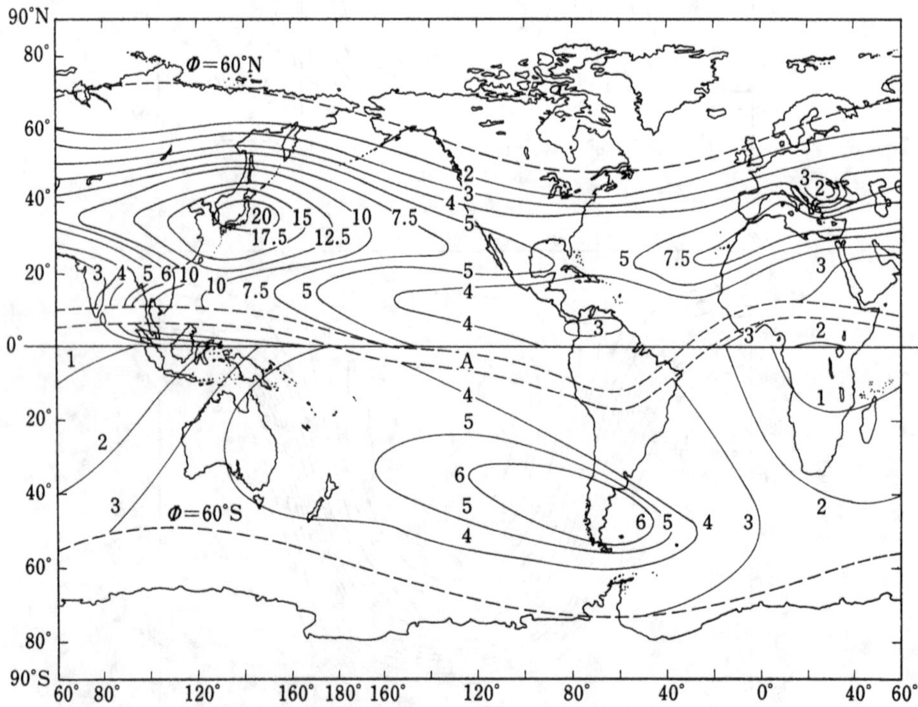

* According to CCIR Rep. 534-1, the time occurrence probability for northern hemisphere summer (May to August) and southern hemisphere summer (November to February) is shown here. The magnetic equator zone (A) and high-latitude regions (magnetic latitude $\Phi \geqq 60°$) are omitted.

f_oF2 Distribution Map for Spring (March 2019) and Summer (June 2019) (MHz)

Figure 66

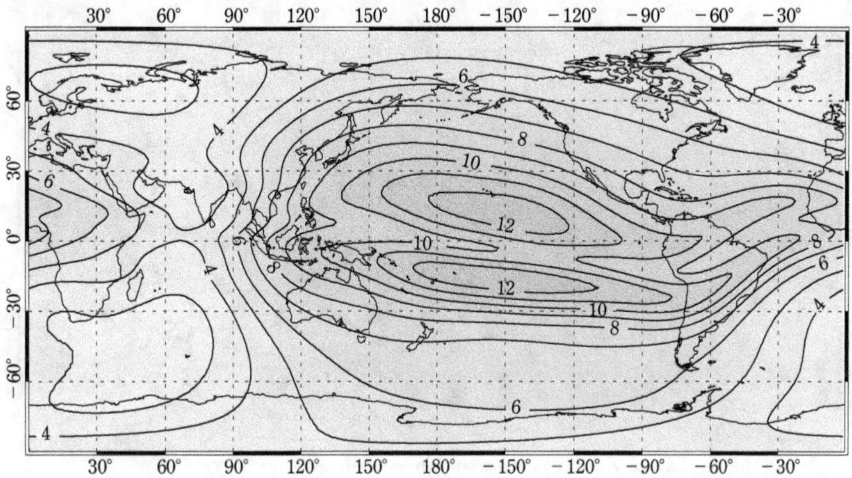

Spring (March 2019, UTC 0:00, 12-month smoothed relative sunspot number $R_{12} = 16$)

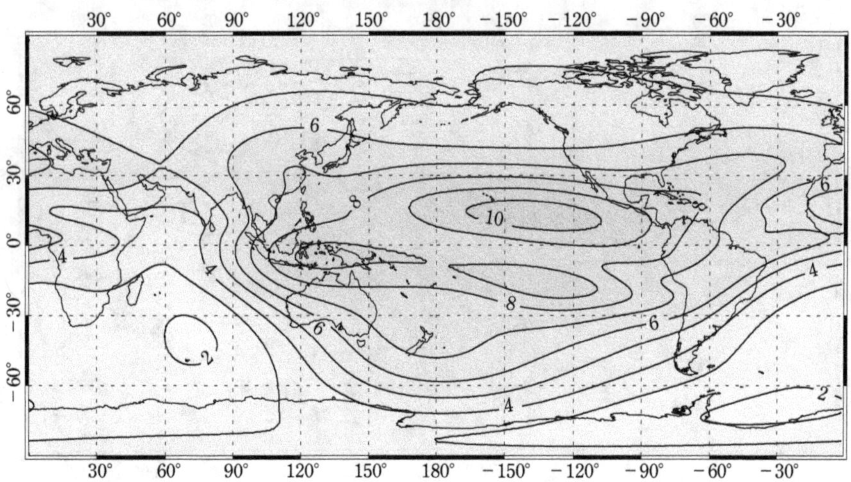

Summer (June 2019, UTC 0:00, 12-month smoothed relative sunspot number $R_{12} = 14$)

* Based on the International Reference Ionosphere (IRI-2016)

f_oF2 Distribution Map for Autumn (September 2019) and Winter (December 2019) (MHz)*

Figure 67

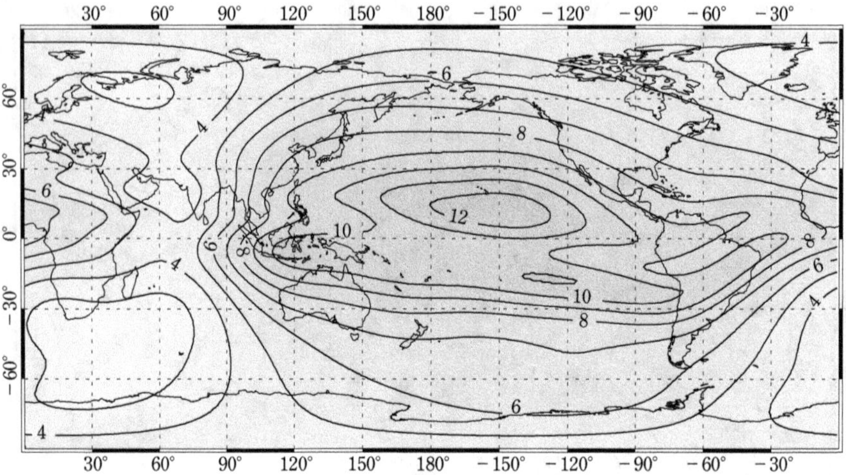

Autumn (September 2019, UTC 0:00, 12-month smoothed relative sunspot number $R_{12} = 12$)

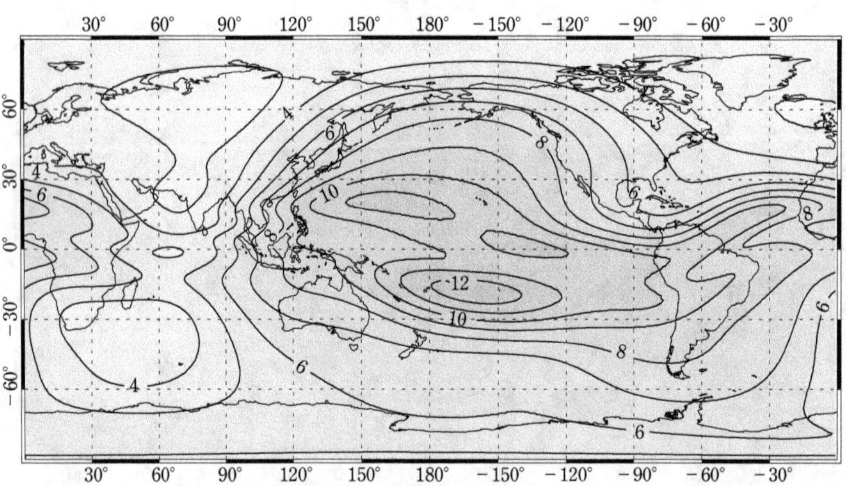

Winter (December 2019, UTC 0:00, 12-month smoothed relative sunspot number $R_{12} = 10$)

* Based on the International Reference Ionosphere (IRI-2016)

Ionosphere Total Electron Content (10^{16} electrons/m^2) Distribution Map for Spring (March 2019) and Summer (June 2019)*

Figure 68

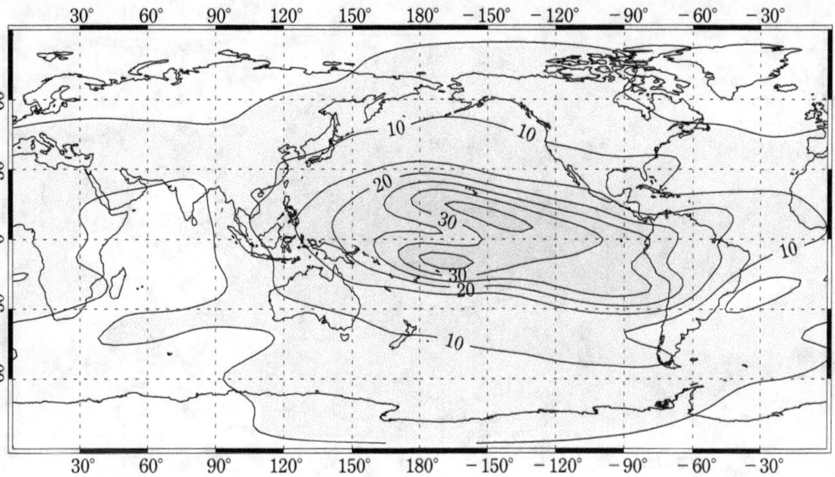

Spring (March 2019 average, UTC 0:00 – 2:00)

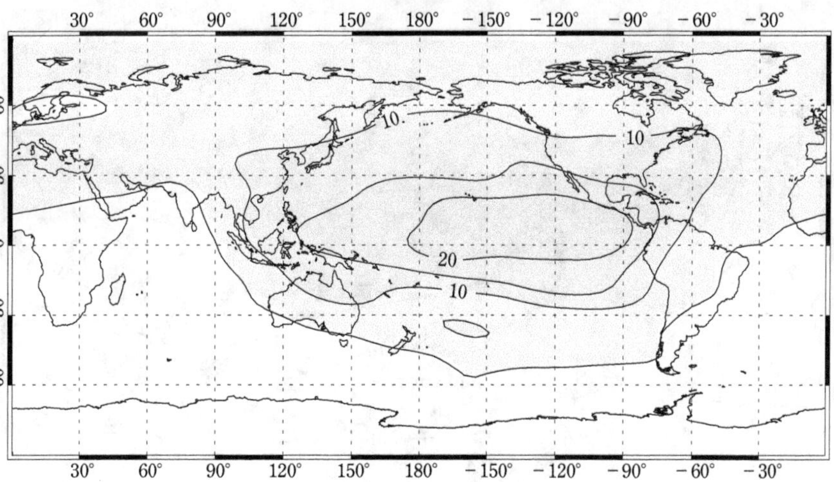

Summer (June 2019 average, UTC 0:00 – 2:00)

* Based on the total electron number observation values for the ionosphere made using GNSS receiver network provided by the International GNSS Service.

Ionosphere Total Electron Content (10^{16} electrons/m^2) Distribution Map for Autumn (September 2019) and Winter (December 2019)*

Figure 69

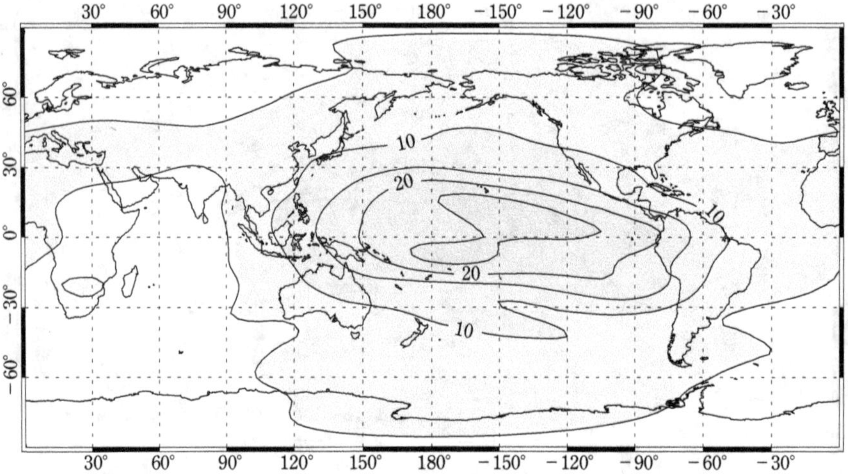

Autumn (September, 2019 average, UTC 0:00−2:00)

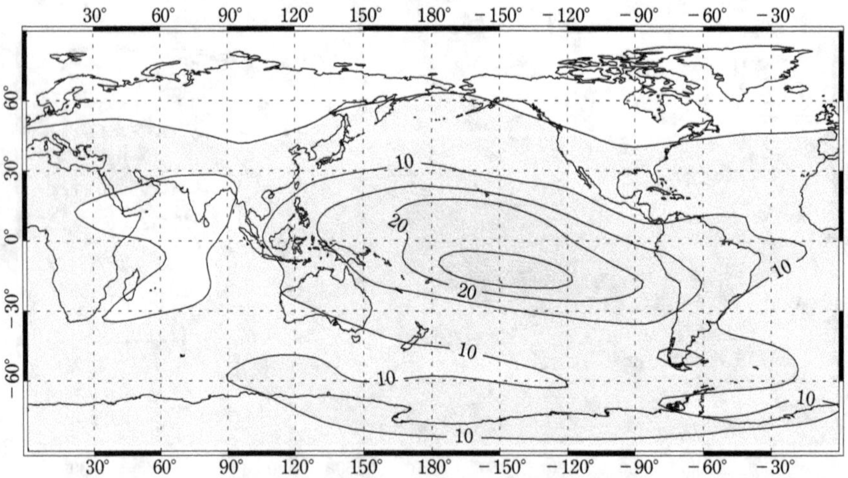

Winter (December, 2019 average, UTC 0:00−2:00)

* Based on the total electron number observation values for the ionosphere made using GNSS receiver network provided by the International GNSS Service.

Distribution Map for Top-side Spread F Occurrence Probability (%)*

Figure 70 August–December 1978

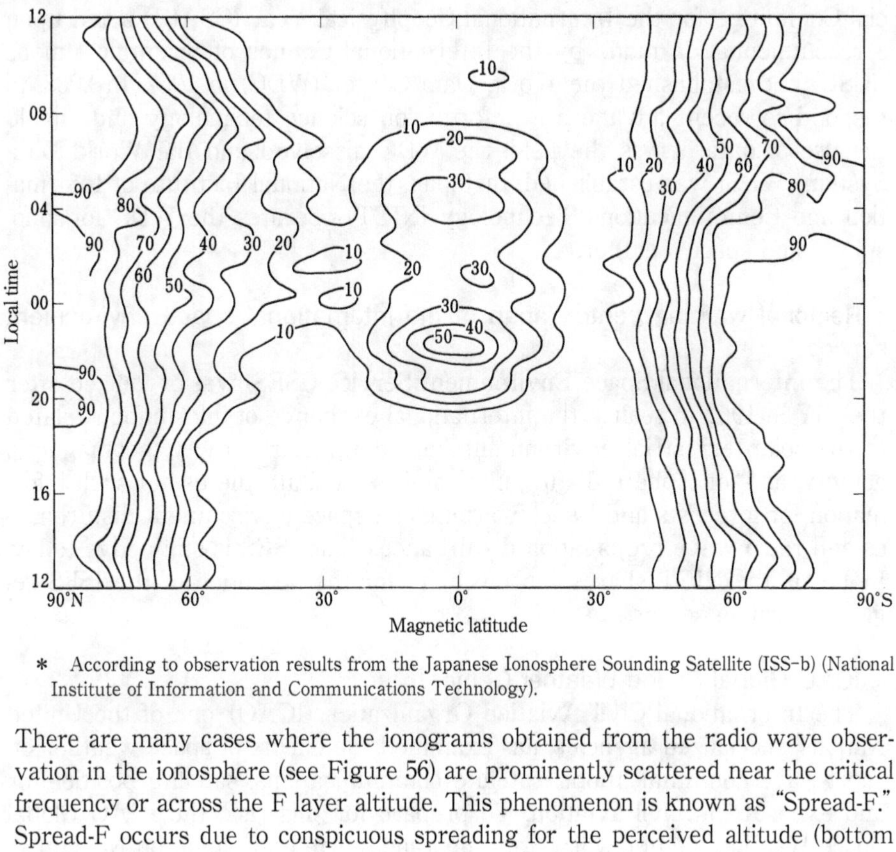

* According to observation results from the Japanese Ionosphere Sounding Satellite (ISS–b) (National Institute of Information and Communications Technology).

There are many cases where the ionograms obtained from the radio wave observation in the ionosphere (see Figure 56) are prominently scattered near the critical frequency or across the F layer altitude. This phenomenon is known as "Spread-F." Spread-F occurs due to conspicuous spreading for the perceived altitude (bottom side) or perceived distance (top side) of the reflection wave, or critical frequency. This results from the observed radio waves being scattered due to a disturbance in the spatial distribution for the electron density of a small scale with wavelengths of about 10 m to 300 m for observed radio waves in the F region. Figure 70 shows the global distribution for the Spread-F occurrence probability (%) determined from multiple top-side ionograms observed by the Ionosphere Sounding Satellite (ISS–b). The Spread-F occurrence probability increases at higher-latitude regions with a magnetic latitude of at least 60°, and in magnetic equator regions during the day.

WDC for ionosphere and space weather

During the International Geophysical Year (IGY: 1957 to 1958), to collect, exchange, and share for use by researchers the material from ionosphere observation locations throughout the world (approx. 250 locations), the Special Committee for the International Geophysical Year (CSAGI) acted upon a recommendation made by the International Council of Scientific Unions (ICSU) and established the World Data Center (WDC) in 1957. In 2008, to respond to recent advances in information science technology and enable the use of data across the field, the WDC dissolved and the World Data System (WDS) was established. In Japan, the National Institute of Information and Communications Technology (NICT) operates the WDC for ionosphere and space weather.

Regional warning center japan of the international space environment service

The International Space Environment Service (ISES) was organized after the IGY in 1962 to realize the international exchange of information related to the solar-terrestrial environment (including solar activity, geomagnetic activity, and ionosphere disturbance) and to facilitate the use of such information for forecasts and issues warnings on space environment disturbances and radio wave propagation disturbances. The ISES is still active today. In Japan, the NICT shares responsibility for the operation of the ISES regional warning center.

ICAO Global Space Weather Center

The International Civil Aviation Organization (ICAO), one of the United Nations' specialized agencies, has examined the effects of space weather on short-wave communications, satellite communications, satellite positioning, and exposure in civil aviation. To prepare for this risk, the ICAO Global Space Weather Center was established and started operation on November 7, 2019. NICT has been approved as a ICAO Global Space Weather Center, and provides commercial airlines with current status and forecast information up to 24 hours later.

Space weather forecast center

The NICT provides space weather forecast at the following URL.

https://www.nict.go.jp (link from the NICT website)

Life Science Section

Shapes and Systems of Organisms

Basic Patterns of Animals

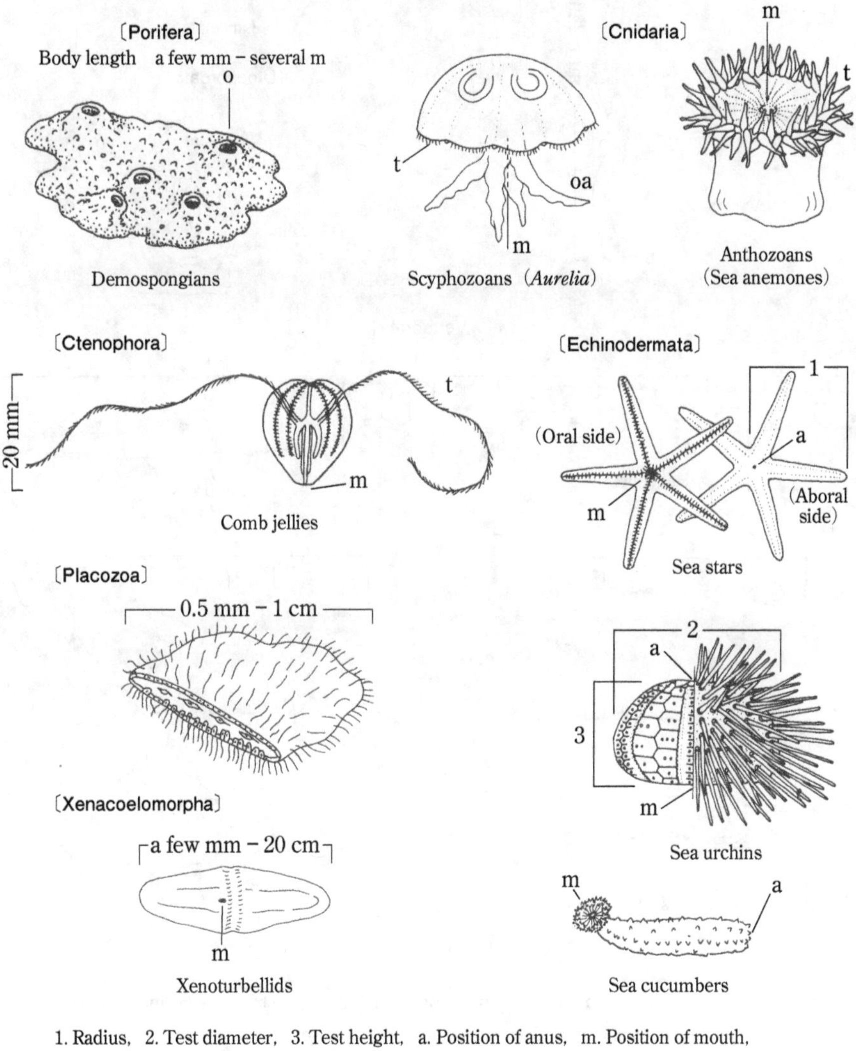

1. Radius, 2. Test diameter, 3. Test height, a. Position of anus, m. Position of mouth, oa. Oral arm, t. Tentacles

Continued.

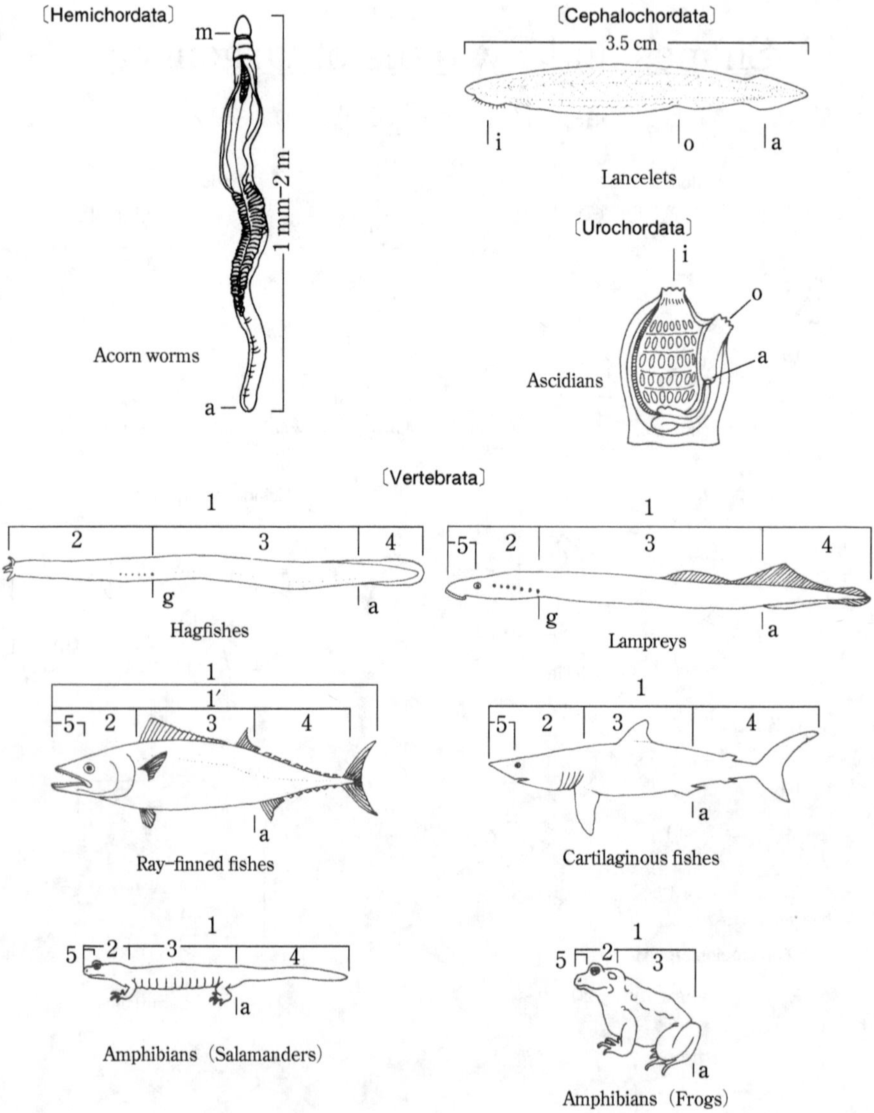

1. Total length, 1′. Standard length, 2. Head, 3. Body, 4. Tail, 5. Rostrum, a. Position of anus, g. Gill pore, i. Incurrent siphon, m. Position of mouth, o. Excurrent siphon

Continued.

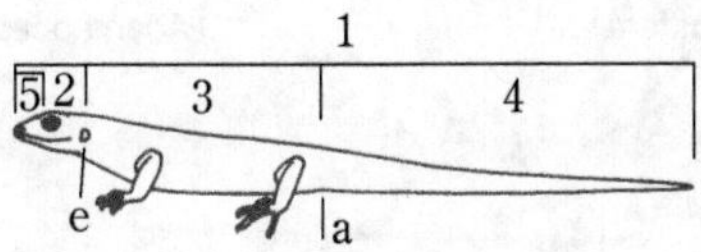

Reptiles (Lizards)

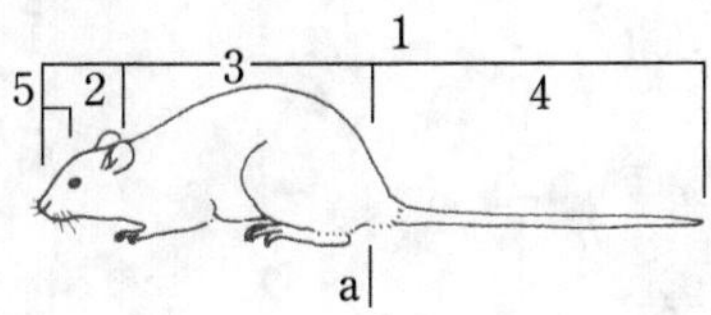

Mammals

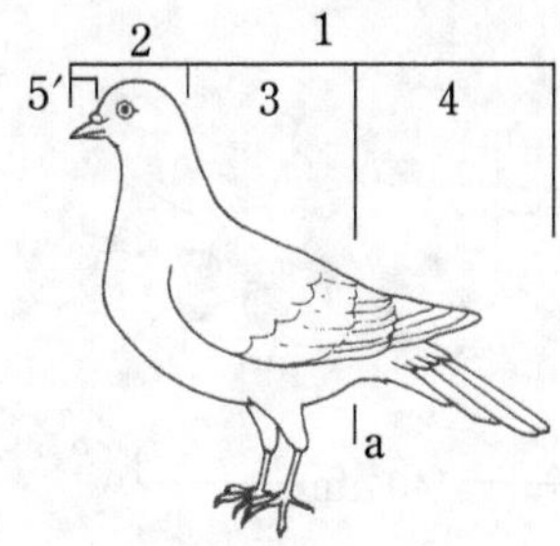

Birds

〔Chaetognatha〕

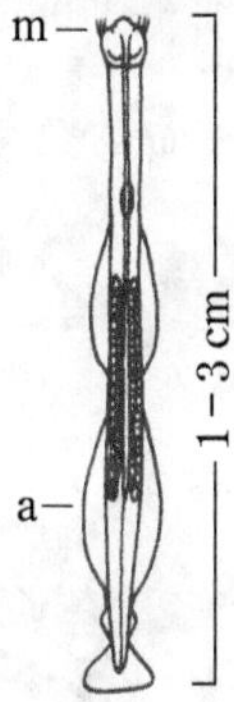

(Arrow worms)

〔Dicyemida〕

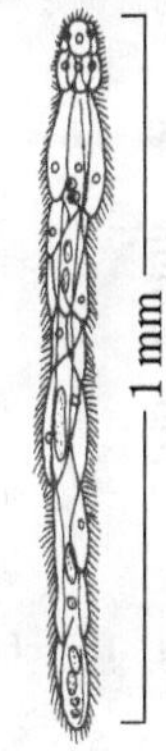

(Dicyemids)

〔Orthonectida〕

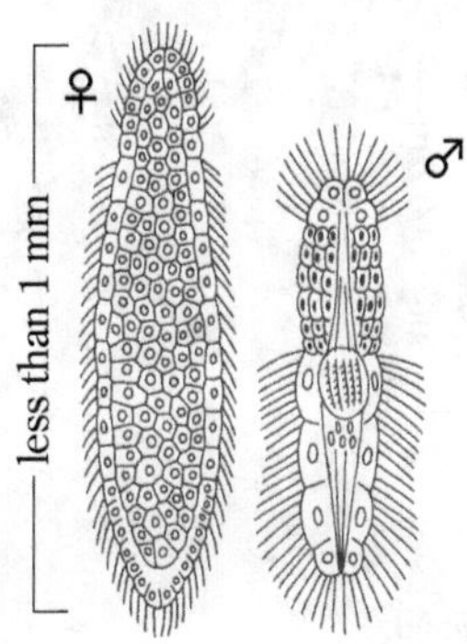

〔Platyhelminthes〕

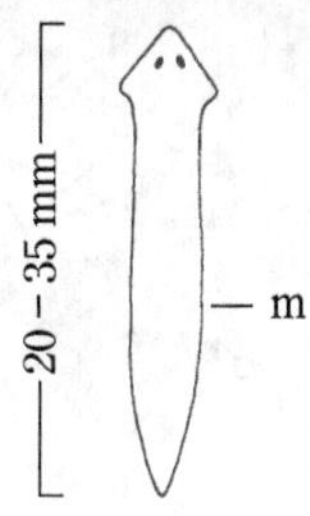

Tricladids
(*Planaria*)

〔Gastrotricha〕

Chaetonotus nodicaudus

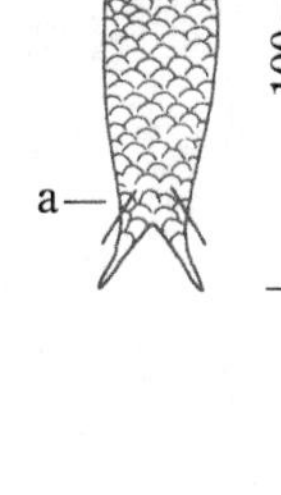

〔Gnathostomulida〕

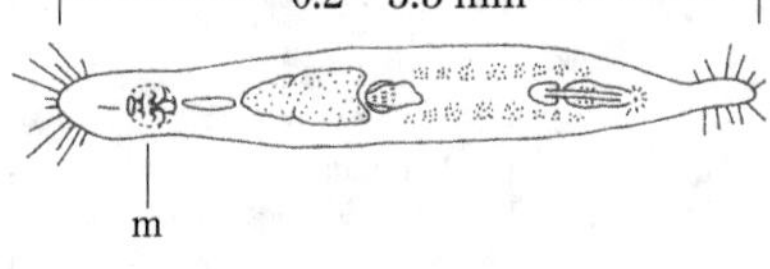

1. Total length, 2. Head, 3. Body, 4. Tail, 5. Rostrum, 5′. Beak,
a. Position of anus or cloacal opening, e. Tympanic membrane,
m. Position of mouth

Continued.

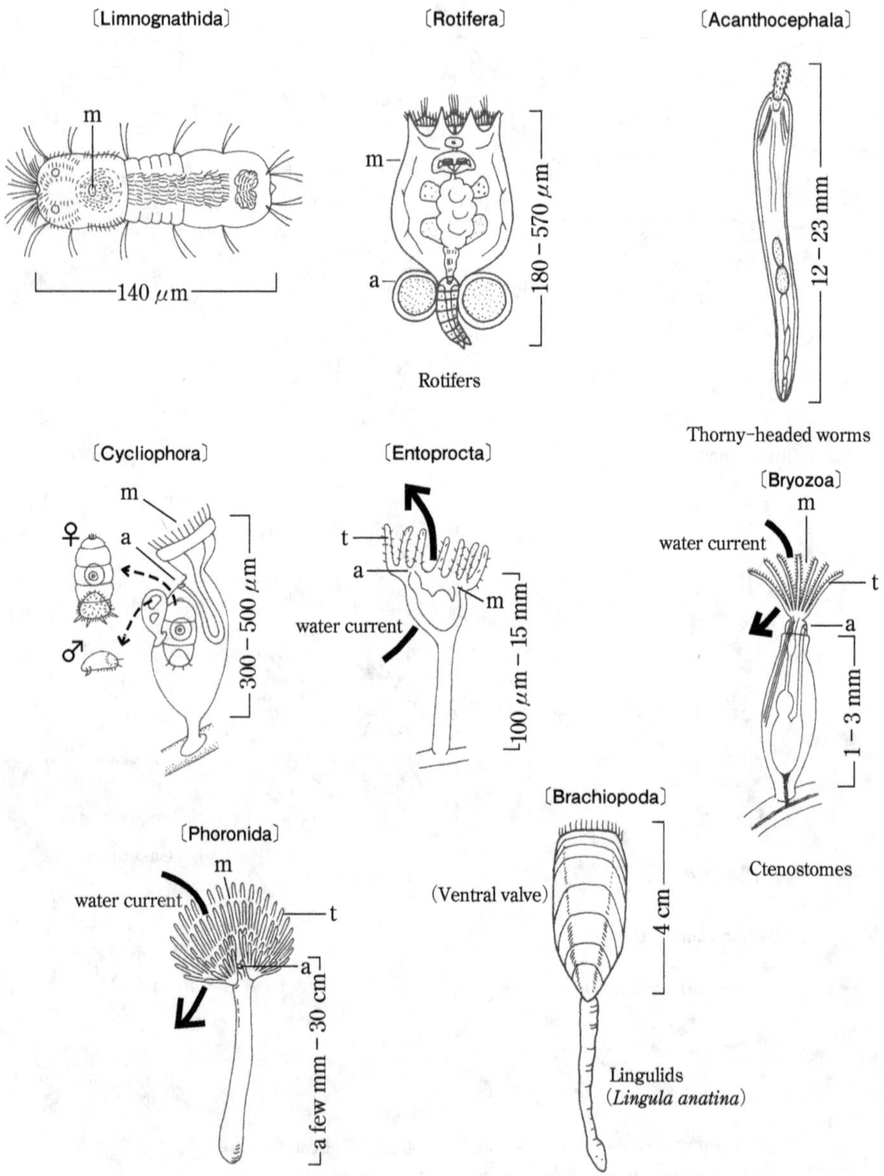

a. Position of anus, m. Position of mouth, t. Tentacles

Continued.

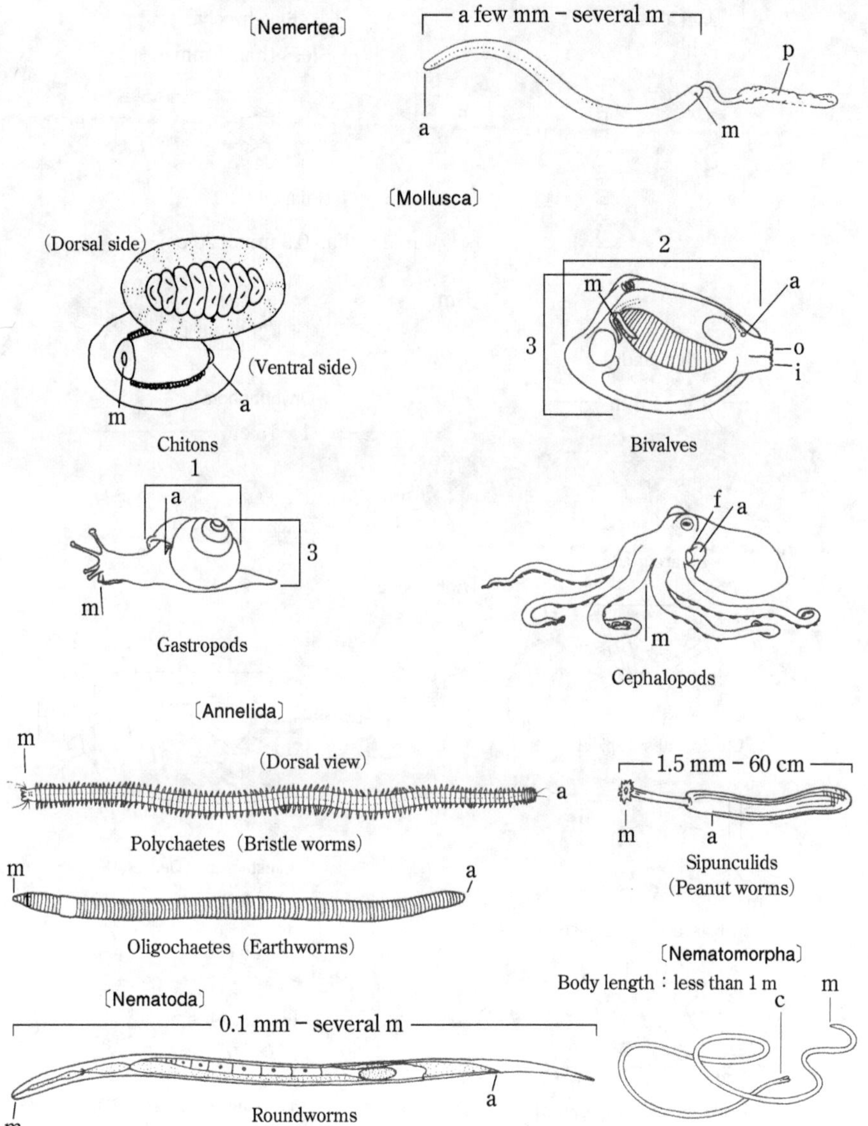

1. Shell diameter, 2. Shell length, 3. Shell height, a. Position of anus, c. Cloacal opening, f. Funnel, i. Incurrent siphon, m. Position of mouth, o. Excurrent siphon, p. Proboscis

Continued.

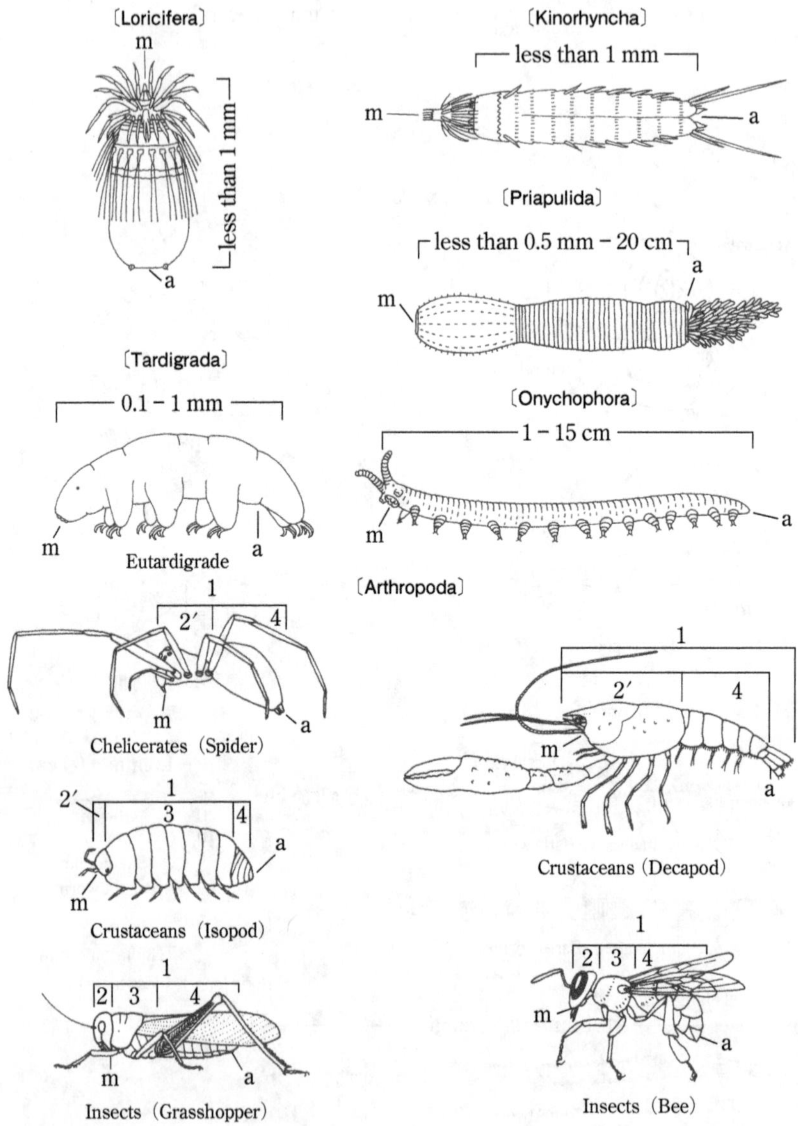

1. Total length, 2. Head, 2′. Cephalothorax, 3. Thorax, 4. Abdomen, a. Position of anus, m. Position of mouth

Basic Shapes of Fungi

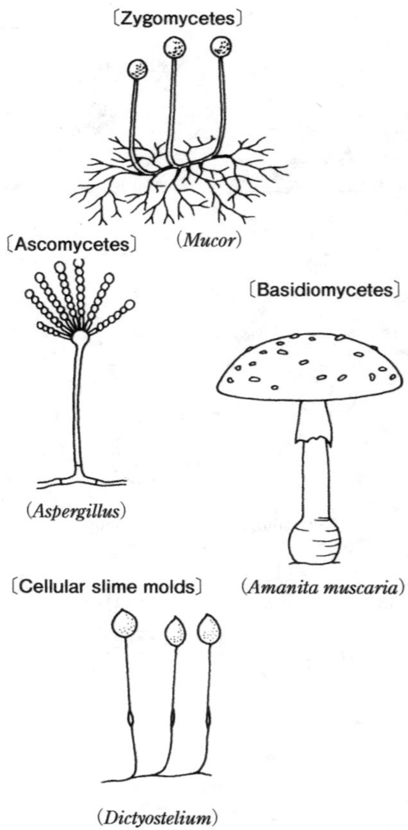

Life Cycles

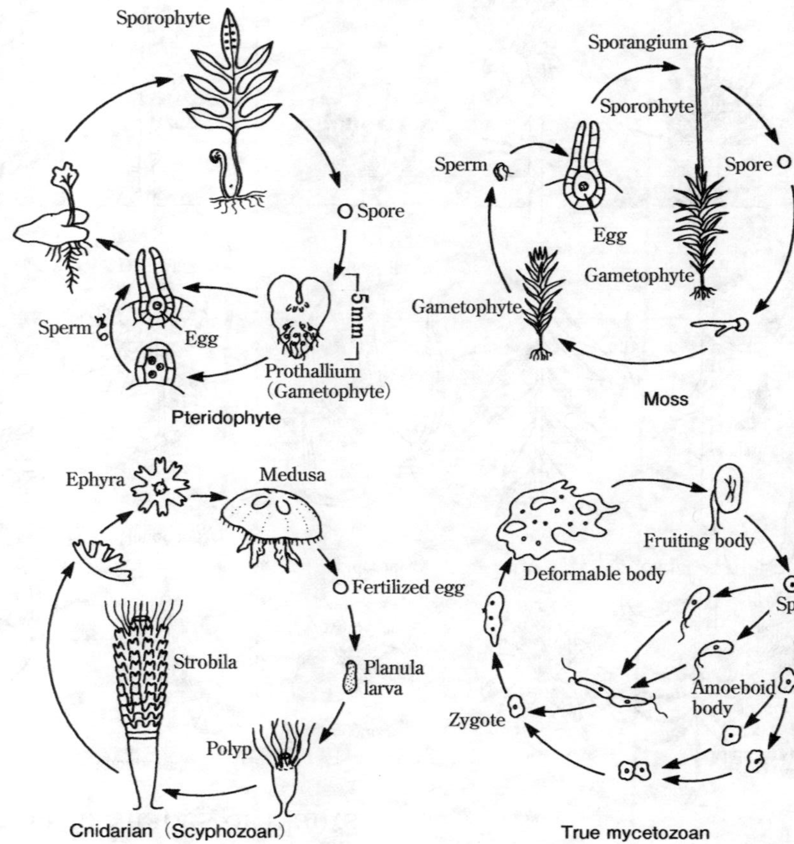

Basic Shapes of Plants

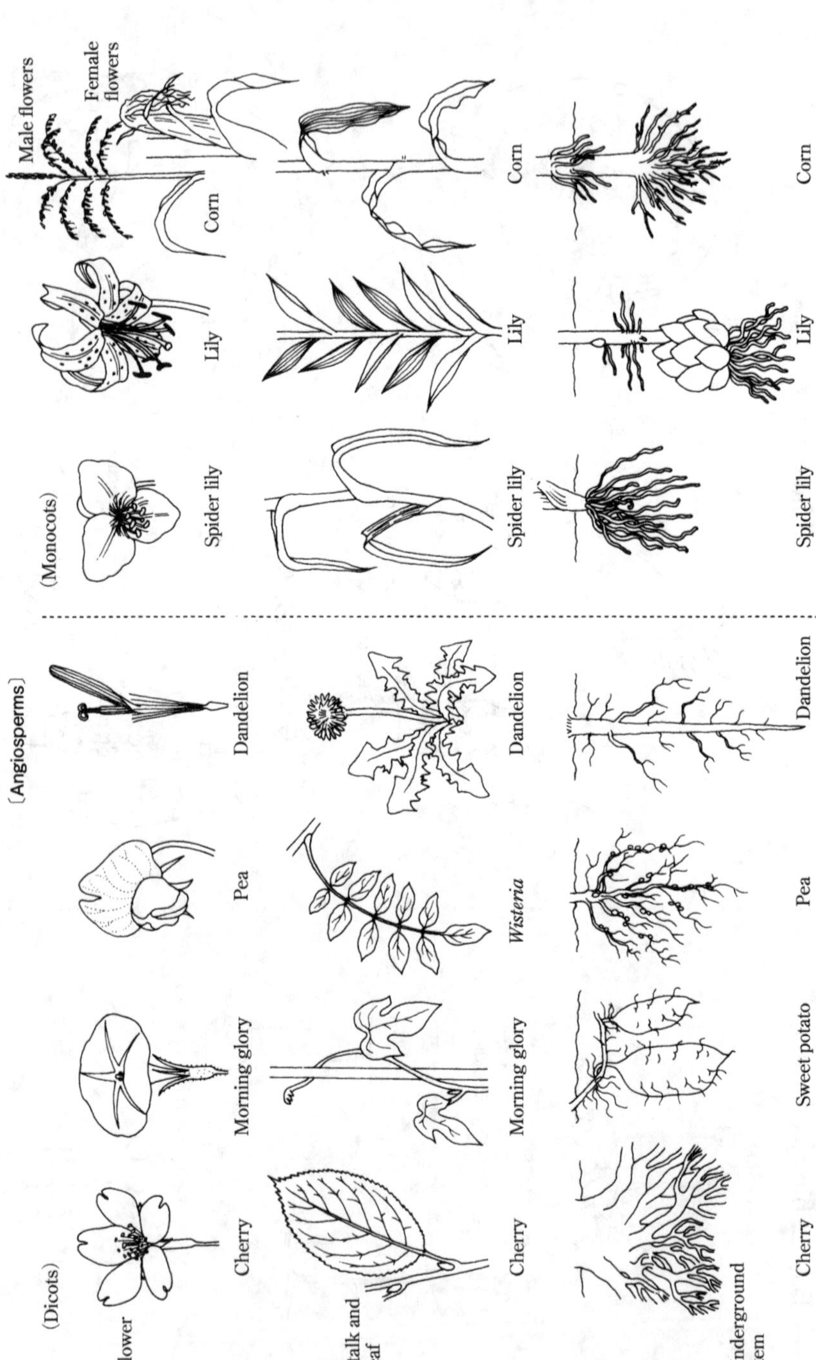

Continued.

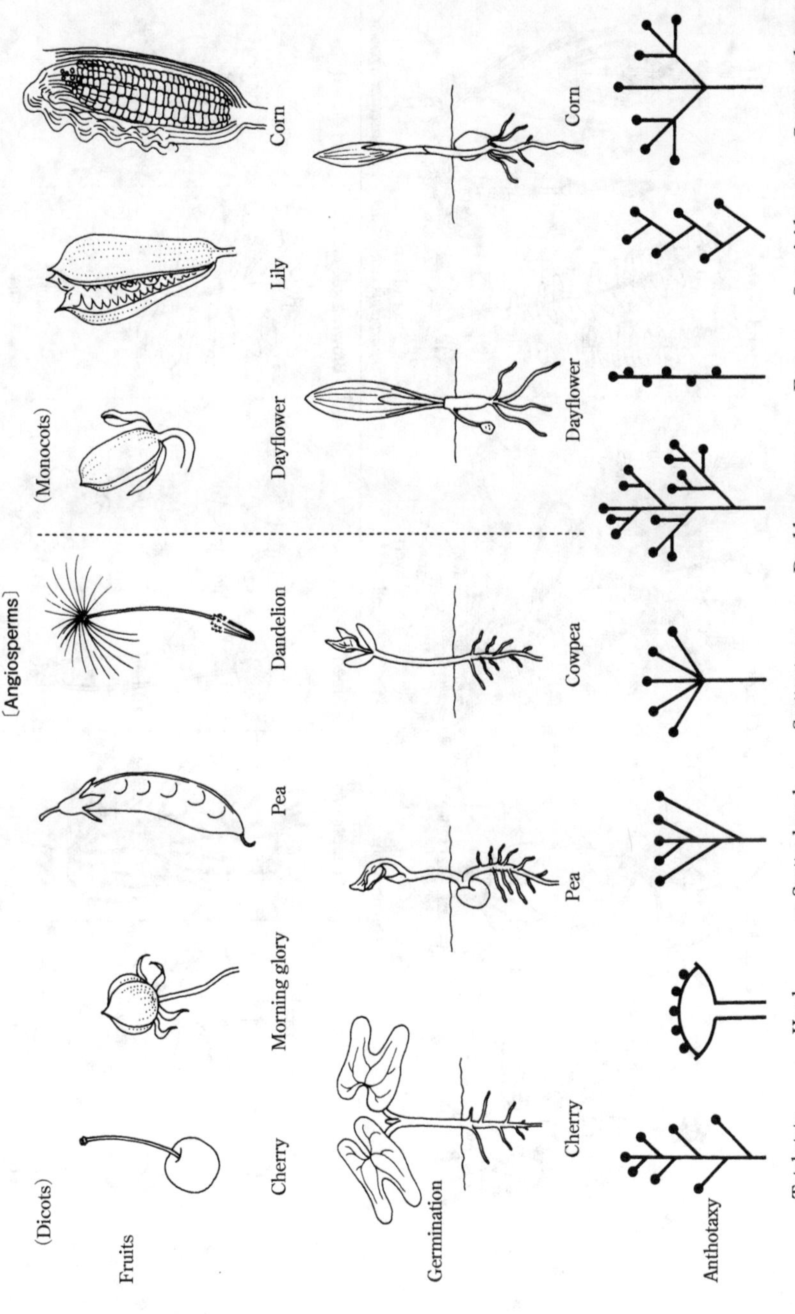
〔Angiosperms〕
(Dicots)
(Monocots)
Fruits
Cherry
Morning glory
Pea
Dandelion
Dayflower
Lily
Corn
Germination
Cherry
Pea
Cowpea
Dayflower
Corn
Anthotaxy
Total state
Head
Scatter bunch
Scatter type
Double racemose
Ear
Scorpioid cyme
Compound cyme

Continued.

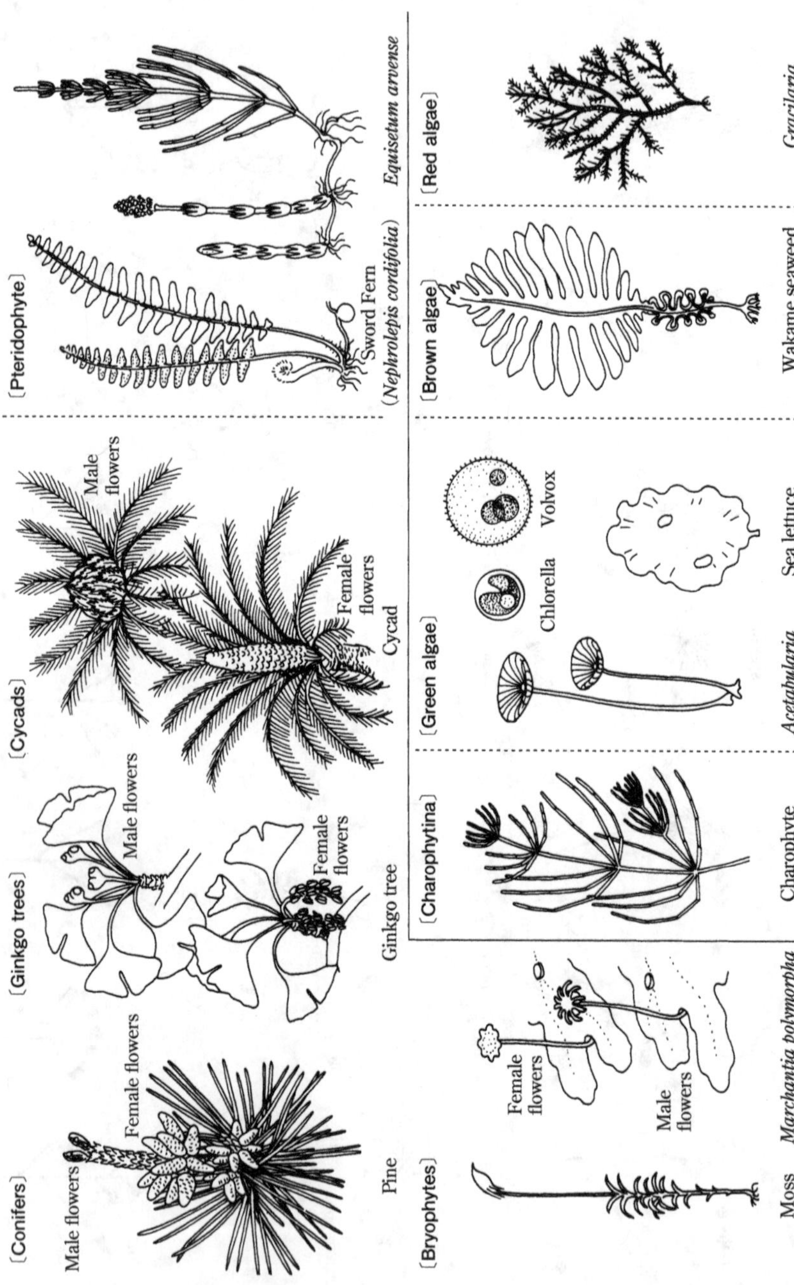

Basic Shapes of Protists (Algae)

Diagrammatic Classification of Organisms

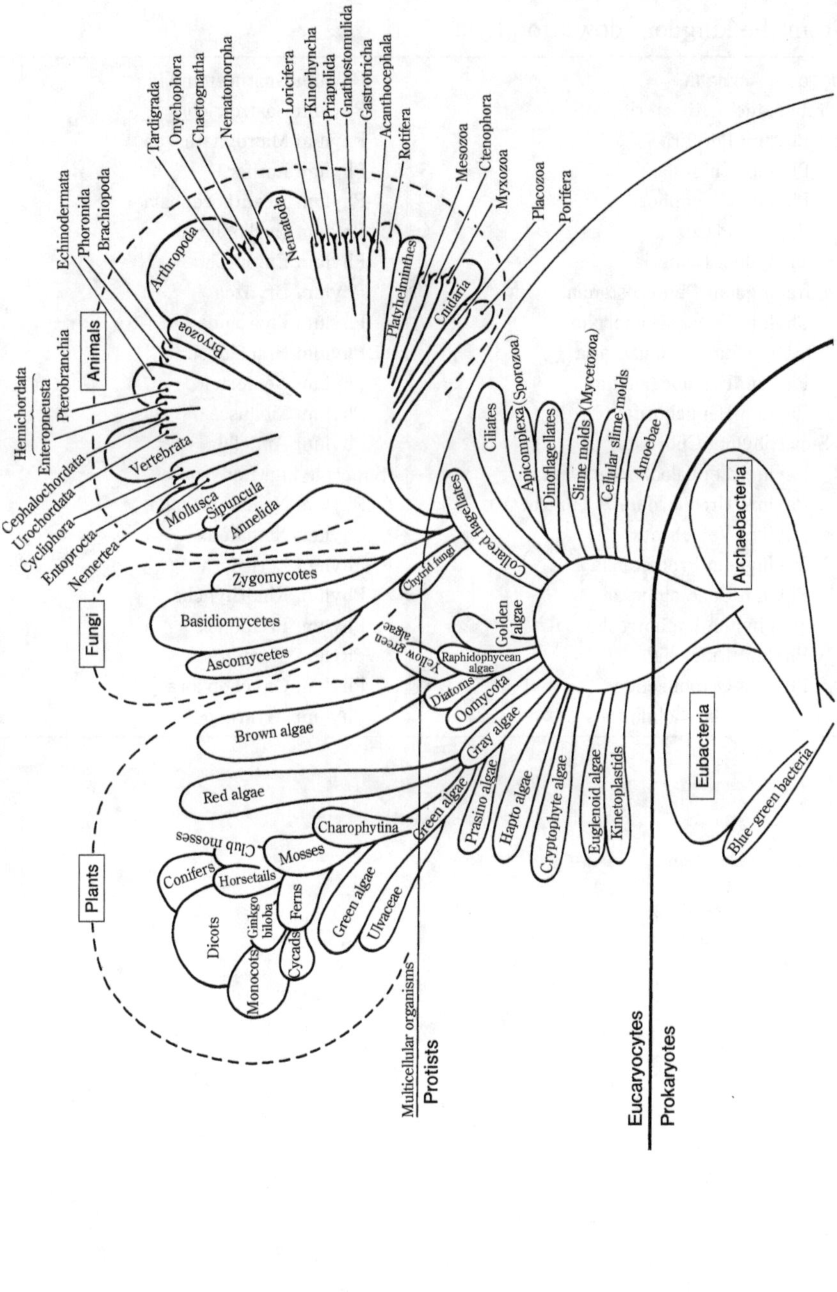

(Source: Shunsuke Mawatari, Takeo Horiguchi. The animal genealogies are also based in part on Zrzavý *et al.*, 1998)

Animal Classification

From the kingdom down to phyla

Kingdom Animalia
- Subkingdom Abilateria
  - Phylum Porifera
  - Phylum Cnidaria
  - Phylum Ctenophora
  - Phylum Placozoa
- Subkingdom Bilateria
- Infrakingdom Deuterostomia
  - Phylum Xenacoelomorpha
- Superphylum Ambulacraria
  - Phylum Echinodermata
  - Phylum Hemichordata
- Superphylum Chordata
  - Phylum Cephalochordata
  - Phylum Urochordata(=Tunicata)
  - Phylum Vertebrata
- Infrakingdom Protostomia
  - Phylum Chaetognatha
- Superphylum Lophotrochozoa
  - Phylum Dicyemida
  - Phylum Orthonectida
  - Phylum Platyhelminthes
  - Phylum Gnathostomulida
  - Phylum Gastrotricha
  - Phylum Micrognathozoa
  - Phylum Rotifera
  - Phylum Acanthocephala
  - Phylum Cycliophora
  - Phylum Entoprocta
  - Phylum Bryozoa
  - Phylum Phoronida
  - Phylum Brachiopoda
  - Phylum Nemertea
  - Phylum Mollusca
  - Phylum Annelida
- Superphylum Ecdysozoa
  - Phylum Nematomorpha
  - Phylum Nematoda
  - Phylum Loricifera
  - Phylum Kinorhyncha
  - Phylum Priapulida
  - Phylum Tardigrada
  - Phylum Onychophora
  - Phylum Arthropoda

From phyla down to orders Continued.

Phylum Porifera

Class Homoscleromorpha
Order Homosclerophorida
Class Calcarea
Subclass Calcinea
Order Clathrinida
Subclass Calcaronea
Order Baerida
Order Leucosolenida
Order Lithonida
Class Hexactinellida
Subclass Amphidiscophora
Order Amphidiscosida
Subclass Hexasterophora
Order Lychniscosida
Order Lyssacinosida
Order Scept ulophora
Class Demospongiae
Subclass Verongimorpha
Order Chondrillida
Order Chondrosiida
Order Verongiida
Subclass Keratosa
Order Dictyoceratida
Order Dendroceratida
Subclass Heteroscleromorpha
Order Agelasida
Order Axinellida
Order Biemnida
Order Bubarida
Order Clionaida
Order Desmacellida
Order Haplosclerida
Order Merliida
Order Poecilosclerida
Order Polymastiida
Order Scopalinida
Order Sphaerocladina
Order Spongillida
Order Suberitida
Order Tethyida
Order Tetractinellida
Order Trachycladida

Phylum Cnidaria

Class Hydrozoa ⟨About 2 700 species⟩
Order Anthoathecata
Order Leptothecata
Order Siphonophorae
Order Limnomedusae
Order Trachymedusae
Order Narcomedusae
Order Actinulida
Class Cubozoa ⟨About 20 species⟩
Order Carybdeida
Order Chirodropida
Class Staurozoa ⟨About 50 species⟩
Order Stauromedusae
Class Scyphozoa ⟨About 200 species⟩
Order Coronatae
Order Semaeostomae
Order Rhizostomae
Class Anthozoa ⟨About 4 700 species⟩
Subclass Octocorallia
Order Alcyonacea
Order Helioporacea
Order Pennatulacea
Subclass Hexacorallia
Order Actiniaria
Order Antipatharia
Order Corallimorpharia
Order Scleractinia
Order Zoantharia
Subclass Ceriantharia
Order Penicillaria
Order Spirularia
Myxozoa (Class unknown)
Malacosporea
Myxosporea

Phylum Ctenophora

Class Tentaculata
Order Cestida
Order Ganeshida
Order Lobata
Order Talassocalycida
Order Cydippida
Order Platyctenida

The fossil groups from phyla down to sub phyla are listed.

Animal Classification Continued.

Class Nuda
Order Beroida

Phylum Placozoa
(Without class and order)

Phylum Xenacoelomorpha
Class Xenoturbellida
Order Xenoturbellida
Class Acoelomorpha
Order Acoela
Order Nemertodermatida

Phylum Echinodermata
〈About 7 300 species〉
Class Crinoidea 〈About 640 species〉
Order Isocrinida
Order Comatulida
Order Hyocrinida
Order Cyrticrinida
Class Asteroidea 〈About 1 800 species〉
Order Brisingia
Order Forcipuladida
Order Velatida
Order Spinulosida
Order Paxillosida
Order Valvatida
Order Peripodida
Class Ophiuroidea 〈About 2 100 species〉
Order Euryophiurida
Order Euryalida
Order Ophiurida
Order Ophintegrida
Order Ophioscolecida
Order Ophiacanthida
Order Ophioleucida
Order Amphilepidida
Class Echinoidea 〈About 1 000 species〉
Order Cidaroida
Order Echinothurioida
Order Micropygoida
Order Diadematoida
Order Aspidodiadematoida
Order Pedinoida
Order Salenioida
Order Stomopneustoida
Order Arbacioida
Order Camarodonta
Order Echinoneoida
Order Cassiduloida
Order Echinolampadoida
Order Clypeasteroida
Order Holasteroida
Order Spatangoida
Class Holothuroidea 〈About 1 700 species〉
Order Apodida
Order Elapsipodida
Order Apidochirotida
Order Dendrochirotida
Order Molpadida

Phylum Hemichordata
Class Enteropneusta 〈About 110 species〉
(Without order)
Class Pterobranchia 〈About 20 species〉
(Without order)

Phylum Cephalochordata 〈35 species〉
Class Leptocaridia
Order Branchiostomatida

Phylum Urochordata 〈About 2 940 species〉
Class Appendicularia 〈About 70 species〉
Class Thaliacea 〈About 70 species〉
Subclass Pyrosomata
(Without order)
Subclass Mysomata
Order Doliolida 〈About 20 species〉
Order Salpida 〈About 40 species〉
Class Ascidiacea 〈About 2 800 species〉
Order Enterogona
Order Pleurogona

Phylum Vertebrata
Superclass Pisces
Class Myxini
Order Mixiniformes 〈80 species〉
Class Petromyzontida
Order Petromyzontiformes 〈38 species〉
Class Chondrichthyes
Subclass Holocephali
Order Chimaeriformes 〈33 species〉
Subclass Elasmobranchii
Order Heterodontiformes 〈8 species〉
Order Orectolobiformes 〈20 species〉
Order Lamniformes 〈15 species〉

Continued.

Order Carcharhiniformes ⟨335 species⟩
Order Hexanchiformes ⟨5 species⟩
Order Echinorhiniformes ⟨2 species⟩
Order Squaliformes ⟨97 species⟩
Order Squatiniformes ⟨15 species⟩
Order Pristiophoriformes ⟨5 species⟩
Order Torpediniformes ⟨59 species⟩
Order Pristiformes ⟨7 species⟩
Order Rajiformes ⟨285 species⟩
Order Myliobatiformes ⟨183 species⟩

Class Actinopterygii

Subclass Cladistia

Order Polypteriformes ⟨16 species⟩

Subclass Chondrostei

Order Acipenseriformes ⟨27 species⟩

Subclass Neopterygii

Order Lepisosteiformes ⟨7 species⟩
Order Amiiformes ⟨1 species⟩

Infraclass Teleostei

Super legion Teleocephala
Legion Elopomorpha
Order Elopiformes ⟨8 species⟩
Order Albuliformes ⟨30 species⟩
Order Notacanthiformes ⟨25 species⟩
Order Anguilliformes ⟨820 species⟩
Legion Osteoglossomorpha
Order Hiodontiformes ⟨2 species⟩
Order Osteoglossiformes ⟨218 species⟩
Legion Otomorpha (= Otocephala)
Order Clupeiformes ⟨370 species⟩
Order Gonorhynchiformes ⟨37 species⟩
Order Cypriniformes ⟨3 300 species⟩
Order Characiformes ⟨1680 species⟩
Order Siluriformes ⟨2 870 species⟩
Order Gymnotiformes ⟨140 species⟩
Legion Euteleostemorpha (= Euteleostei)
Order Argentiniformes ⟨202 species⟩
Order Salmoniformes ⟨164 species⟩
Order Ateleopodiformes ⟨12 species⟩
Order Stomiiformes ⟨400 species⟩
Order Aulopiformes ⟨240 species⟩
Order Myctophiformes ⟨250 species⟩
Order Lampridiformes ⟨21 species⟩
Order Polymixiiformes ⟨10 species⟩
Order Percopsiformes ⟨9 species⟩
Order Gadiformes ⟨560 species⟩
Order Stephanoberyciformes ⟨75 species⟩
Order Zeiformes ⟨32 species⟩
Order Beryciformes ⟨150 species⟩
Order Elassomatiformes ⟨6 species⟩
Order Synbranchiformes ⟨100 species⟩
Order Gasterosteiformes ⟨280 species⟩
Order Mugiliformes ⟨72 species⟩
Order Atheriniformes ⟨320 species⟩
Order Beloniforme ⟨230 species⟩
Order Cyprinodontiformes ⟨1 020 species⟩
Order Ophidiiformes ⟨390 species⟩
Order Batrachoidiformes ⟨80 species⟩
Order Scorpaeniformes ⟨950 species⟩
Order Perciformes ⟨2 630 species⟩
Order Carangiformes ⟨152 species⟩
Order Cottiformes ⟨1 120 species⟩
Order Dactylopteriformes ⟨7 species⟩
Order Labriformes ⟨2 300 species⟩
Order Notothenii formes ⟨112 species⟩
Order Trachiniformes ⟨240 species⟩
Order Pholidichthyiformes ⟨2 species⟩
Order Blenniiformes ⟨820 species⟩
Order Icosteiformes ⟨1 species⟩
Order Gobiesociformes ⟨340 species⟩
Order Gobiiformes ⟨2 250 species⟩
Order Acanthuriformes ⟨130 species⟩
Order Scombriformes ⟨150 species⟩
Order Stromatiformes ⟨70 species⟩
Order Anabantiformes ⟨120 species⟩
Order Caproiformes ⟨10 species⟩
Order Pleuronectiformes ⟨680 species⟩
Order Lophiiformes ⟨320 species⟩
Order Tetraodontiformes ⟨410 species⟩

Class Sarcopterygii

Subclass Coelacanthimorpha

Order Coelacanthiformes ⟨2 species⟩

Subclass Dipnotetrapodomorpha

Order Ceratodontiformes ⟨6 species⟩

Superclass Tetrapoda

Class Amphibia

Subclass Lissamphibia

Order Gymnophiona ⟨206 species⟩
Order Caudata ⟨692 species⟩
Order Anura ⟨6 500 species⟩

Class Reptilia

Subclass Diapsida

Superorder Lepidosauria
Order Sphenodontia ⟨1 species⟩
Order Squamata ⟨Sauria 6 972 species, Serpentes 3 879 species, Amphisbaenia

Animal Classification Continued.

201 species〉
Infraclass Archosauromorpha
Superorder Testudinata
Order Testudines 〈361 species〉
Superorder Arcosauria
Order Crocodilia 〈26 species〉
Class Aves
Infraclass Neornithes
Superorder Palaeognathae
Order Struthioniformes 〈2 species〉
Order Rheiformes 〈2 species〉
Order Apterygiformes 〈5 species〉
Order Casuariiformes 〈4 species〉
Order Tinamiformes 〈46 species〉
Superorder Neognathae
Order Galliformes 〈300 species〉
Order Anseriformes 〈178 species〉
Order Caprimulgiformes 〈122 species〉
Order Apodiformes 〈480 species〉
Order Musophagiformes 〈23 species〉
Order Otidiformes 〈26 species〉
Order Cuculiformes 〈149 species〉
Order Mesitornithiformes 〈3 species〉
Order Pteroclidiformes 〈16 species〉
Order Columbiformes 〈344 species〉
Order Gruiformes 〈189 species〉
Order Podicipediformes 〈23 species〉
Order Phoenicopteriformes 〈6 species〉
Order Charadriiformes 〈387 species〉
Order Eurypygiformes 〈2 species〉
Order Phaethontiformes 〈3 species〉
Order Gaviiformes 〈5 species〉
Order Sphenisciformes 〈18 species〉
Order Procellariiformes 〈147 species〉
Order Ciconiiformes 〈19 species〉
Order Suliformes 〈61 species〉
Order Pelecaniformes 〈118 species〉
Order Opisthocomiformes 〈1 species〉
Order Accipitriformes 〈266 species〉
Order Strigiformes 〈248 species〉
Order Coliiformes 〈6 species〉
Order Leptosomiformes 〈1 species〉
Order Trogoniformes 〈43 species〉
Order Bucerotiformes 〈74 species〉
Order Coraciiformes 〈178 species〉
Order Piciformes 〈445 species〉
Order Cariamiformes 〈2 species〉
Order Falconiformes 〈66 species〉
Order Psittaciformes 〈398 species〉
Order Passeriformes 〈6 533 species〉
Class Mammalia
Subclass Australosphenida
Order Monotremata 〈5 species〉
Subclass Boreosphenida
Infraclass Metatheria
Order Didelphimorphia 〈87 species〉
Order Paucituberculata 〈6 species〉
Order Microbiotheria 〈1 species〉
Order Notoryctemorphia 〈2 species〉
Order Dasyuromorphia 〈71 species〉
Order Peramelemorphia 〈21 species〉
Order Diprotodontia 〈143 species〉
Infraclass Eutheria
Superorder Afrotheria
Order Afrosoricida 〈51 species〉
Order Macroscelidea 〈15 species〉
Order Tubulidentata 〈1 species〉
Order Hyaracoidea 〈4 species〉
Order Sirenia 〈5 species〉
Order Proboscidea 〈3 species〉
Superorder Xenarthra
Order Cingulata 〈21 species〉
Order Pilosa 〈10 species〉
Superorder Euarchontoglires
Order Rodentia 〈2 277 species〉
Order Lagomorpha 〈92 species〉
Order Scandentia 〈20 species〉
Order Dermoptera 〈2 species〉
Order Primates 〈376 species〉
Superorder Laurasiatheria
Order Cetacea 〈84 species〉
Order Artiodactyla 〈240 species〉
Order Perissodactyla 〈17 species〉
Order Carnivora 〈286 species〉
Order Chiroptera 〈1 116 species〉
Order Pholidota 〈8 species〉
Order Eulipotyphla 〈452 species〉

Phylum Chaetognatha 〈About 130 species〉
Class Sagittoidea
Order Phragmophora
Order Aphragmophora

Phylum Dicyemida
Class Dicyemida
Order Dicyemida 〈3 families include 8

Continued.

genera with 150 species, of which 4 genera and 70 species occur in Japan.〉

Phylum Orthonectida

Class Orthonectida
Order Orthonectida 〈1 family include 5 genera with 21 species, of which 1 species live in Japan〉

Phylum Platyhelminthes

Class Catenulida
Order Catenulida
Class Rhabditophora
Order Macrostomida
Order Prorhynchida
Order Polycladida
Order Gnosonesimida
Order Rhabdocoela
Order Proseriata
Order Fecampiida
Order Prolecithophora
Order Tricladida
Order Bothrioplanida
Order Monogenea
Order Cestoda
Order Trematoda

Phylum Gnathostomulida

Order Filospermoidea
Order Bursovaginoidea

Phylum Gastrotricha

Order Macrodasyida 〈258 species〉
Order Chaetonotida 〈422 species〉

Phylum Micrognathozoa

Order Limnognathida

Phylum Rotifera

Class Pararotatoria
Order Seisonacea
Class Eurotatoria
Subclass Bdelloidea
Order Bdelloida
Subclass Monogononta
Order Ploima
Order Flosculariacea
Order Collothecaceae

Phylum Acanthocephala

Class Archiacanthocephala
Order Apororhynchida
Order Gigantorhynchida
Order Moniliformida
Order Oligacanthorhynchida
Class Eoacanthocephala
Order Gyracanthocephala
Order Neoechinorhynchida
Class Palaeacanthocephala
Order Echinorhynchida
Order Heteramorphida
Order Polymorphida
Class Polyacanthocephala
Order Polyacanthorhynchida

Phylum Cycliophora

Class Eucycliophora
Order Symbiida

Phylum Entoprocta(Phylum Kamptozoa)

(Without class and order)

Phylum Bryozoa(Phylum Ectoprocta)
〈About 6 000 recent species〉

Class Phylactolaemata
Order Plumatellida
Class Stenolaemata
Order Cyclostomata
Class Gymnolaemata
Order Ctenostomata
Order Cheilostomata

Phylum Phoronida
〈About 15 recent species〉

(Without class and order)

Phylum Brachiopoda
〈About 380 recent and 30 000 fossil species〉

Subphylum Linguliformea
Class Lingulata
Order Lingulida
Subphylum Craniiformea
Class Craniata
Order Craniida
Subphylum Rhynchonelliformea
Class Rhynchonellata
Order Rhynchonellida

Animal Classification Continued.

Order Terebratulida
Order Thecideida

Phylum Nemertea

Class Palaeonemertea
Class Pilidiophora
Class Hoplonemertea
Order Monostilifera
Order Polysitlifera

Phylum Mollusca

Class Caudofoveata (=Chaetodermomorpha)
(Without orders)
Class Solenogastres (=Neomeniomorpha)
Order Pholidoskepia
Order Neomeniamorpha
Order Sterrofustia
Order Cavibelonia
Class Polyplacophora
Subclass Neoloricata
Order Lepidopleurida
Order Chitonida
Class Monoplacophora
Order Tryblidiida
Class Bivalvia
Subclass Protobranchia
Order Solemyoida
Order Nuculoida
Order Nuculanoida
Subclass Pteriomorphia
Order Arcoida
Order Mytiloida
Order Pterioida
Order Limoida
Order Pectinoida
Order Ostreoida
Subclass Palaeoheterodonta
Order Trigonioida
Order Unionoida
Subclass Heterodonta
Infraclass Archiheterodonta
Order Carditoida
Infraclass Euheterodonta
Order Anomalodesmata
Order Myoida
Order Veneroida
Class Scaphopoda
Order Dentalida
Order Gadilida
Class Gastropoda
Subclass Eogastropoda
Superorder Patellogastropoda
(Without orders)
Subclass Orthogastropoda
Superorder Cocculiniformia
(Without orders)
Superorder Vetigastropoda
(Without orders)
Superorder Neritimorpha
(Without orders)
Superorder Caenogastropoda
Order Architaenioglossa
Order Sorbeoconcha
Superorder Heterobranchia
Order Heterostropha
(Note : Because it is not a monophyletic group, the name Heterostropha tends not tole used recently, and that case is named generally with "lower Heterobranchia".)
Order Opisthobranchia
Order Pulmonata
Class Cephalopoda
Subclass Nautiloidea
Order Nautiloida
Subclass Coleoidea
Superorder Decapodiformes
Order Spirulida
Order Sepiida
Order Sepiolida
Order Teuthida
Superorder Octopodiformes
Order Octopoda
Order Vampyromorpha

Phylum Annelida 〈About 16 500 species〉

Class Scolecida
Order Cossurida
Order Orbiniida
Order Opheliida
Order Capitellida
Class Aciculata
Order Amphinomida
Order Eunicida
Order Nerillida
Order Phyllodocida
Order Spintherida

Continued.

Class Canalipalpata
Order Oweniida
Order Sabellida [including Siboglinidae formerly classified in phylum Pogonophora]
Order Terebellida
Order Cirratulida
Order Flabelligerida
Order Sternaspida
Order Spionida
Order Chaetopterida
Order Magelonida
Class Clitellata
Subclass Oligochaeta
Order Randiellida
Order Tubificida
Order Lumbriculida
Infraclass Diplotesticulata
Superorder Haplotaxidea
Order Haplotaxida
Superorder Metagynophora
Order Moniligastrida
Order Opisthopora
Suborder Alluroidina
Suborder Crassiclitellata
Subclass Hirudinea
Infraclass Branchiobdellida
Infraclass Acanthobdellida
Infraclass Hirudinida
Order Rhynchobdellida
Order Arhynchobdellida
Class Myzostomida
Order Proboscidea
Order Pharyngidea
Class Echiura
Order Echiuroinea
Order Xenopneusta
Class Sipuncula
Subclass Sipunculidea
Order Sipunculiformes
Order Golfingiiformes
Subclass Phascolosomatidea
Order Phascolosomatiformes
Order Aspidosiphoniformes
Class unknown (Class incertae sedis)
Order Aeolosomatida
Order Polygordiida
Order Dinophilida
Order Psammodrilida
Order Diurodrilida
Order Parergodrilida
Order Protodrilida

Phylum Nematomorpha

Order Nectonematoida
Order Gordioida

Phylum Nematoda

Class Enoplea
Subclass Enoplia
Superorder Enoplica
Order Enoplida
Order Ironida
Order Tripyloidida
Order Alaimida
Order Trefusiida
Superorder Rhaptothyreica
Order Rhaptothyreida
Subclass Oncholaimia
Superorder Oncholaimica
Order Oncholaimida
Subclass Triplonchia
Superorder Triplonchica
Order Triplonchida
Order Tripylida
Class Dorylaimea
Subclass Bathyodontia
Superorder Mononchica
Order Mononchida
Order Bathyodontida
Order Mermithida
Subclass Dorylaimia
Superorder Dorylaimica
Order Dorylaimida
Subclass Trichocephalia
Superorder Trichocephalica
Order Trichocephalida
Order Dioctophymatida
Order Marimermithida
Order Muspiceida
Class Chromadorea
Subclass Chromadoria
Superorder Chromadorica
Order Chromadorida
Order Desmodorida
Order Desmoscolecida

Animal Classification Continued.

Order Selachinematida
Subclass Plectia
Superorder Plectica
Order Plectida
Order Leptolaimida
Superorder Monhysterica
Order Monhysterida
Superorder Rhabditica
Order Rhabditida
Order Spirurida
Order Diplogasterida
Order Drilonematida
Order Panagrolaimida
Superorder Teratocephalica
Order Teratocephalida

Phylum Loricifera 〈36 species〉
Order Nanaloricida

Phylum Kinorhyncha 〈About 250 species〉
Class Allomalorhagida (Newly applied name)
(Without orders)
Class Cyclorhagida
Order Echinorhagata (Newly applied name)
Order Kentrorhagata (Newly applied name)
Order Xenosomata (Newly applied name)

Phylum Priapulida 〈22 species〉
Order Priapulimorpha
Order Seticoronaria

Phylum Tardigrada
Class Heterotardigrada
Order Arthrotardigrada
Order Echiniscoidea
Class Mesotardigrada
Order Thermozodia
Class Eutardigrada
Order Apochela
Order Parachela

Phylum Onychophora
Class Onychophora
Order Euonychophora

Phylum Arthropoda
Subphylum Chelicerata
Class Pycnogonida 〈About 1 200 species〉
Order Pantopoda
Class Merostomata
Order Xiphosura
〈4 species, of which 1 species lives in Japan〉
Class Arachnida
Order Ricinulei 〈About 60 species〉
Order Acari
〈About 48 000 species, of which about 1 900 species live in Japan〉
Order Opiliones
〈About 6 500 species, of which about 80 species live in Japan〉
Order Pseudoscorpiones
〈About 3 400 species, of which about 60 species live in Japan〉
Order Solifugae 〈About 1 000 species〉
Order Scorpiones
〈About 1 500 species, of which 2 species live in Japan〉
Order Uropygi
〈About 110 species, of which 2 species live in Japan〉
Order Palpigradi
〈About 80 species, of which 1 species live in Japan〉
Order Schizomida
〈About 260 species, of which 4 species live in Japan〉
Order Amblypygi 〈About 130 species〉
Order Araneae
〈About 45 000 species, of which about 1 600 species live in Japan〉
Subphylum Myriapoda
Superclass Opisthogoneata
Class Chilopoda
Subclass Pleurostigmophora
Order Geophilomorpha
Order Scolopendromorpha
Order Lithobiomorpha
Order Craterostigmomorpha
Subclass Notostigmophora
Order Scutigeromorpha
Superclass Progoneata
Class Symphyla
Order Scolopendrellida
Class Pauropoda
Order Hexamerocerata

Continued.

Order Tetramerocerata	Order Mystacocaridida
Class Diplopoda	**Superclass Multicrustacea**
Subclass Penicillata	**Class Copepoda**
Order Polyxenida	**Infraclass Progymnoplea**
Subclass Chilognatha	Order Platycopioida
Infraclass Pentazonia	**Infraclass Neocopepoda**
Superorder Oniscomorpha	Superorder Gymnoplea
Order Glomerida	Order Calanoida
Order Sphaerotheriida	Superorder Podoplea
Superorder Limacomorpha	Order Cyclopoida
Order Glomeridesmida	Order Gelyelloida
Infraclass Helminthomorpha	Order Harpacticoida
Order Siphoniulida	Order Misophrioida
Infraclass Colobognatha	Order Monstrilloida
Order Polyzoniida	Order Mormonilloida
Order Platydesmida	Order Poecilostomatoida
Order Siphonophorida	Order Siphonostomatoida
Order Siphonocriptida	Superorder Rhabdomoplea
Infraclass Eugnatha	Order Thaumatopsylloida
Superorder Nematophora	**Class Malacostraca**
Order Chordeumatida	**Subclass Phyllocarida**
Order Stemmiulida	Order Leptostraca
Order Callipodida	**Subclass Hoplocarida**
Superorder Merocheta	Order Stomatopoda
Order Polydesmida	**Subclass Eumalacostraca**
Superorder Juliformia	Superorder Syncarida
Order Spirostreptida	Order Anaspidacea
Order Spirobolida	Order Bathynellacea
Order Julida	Superorder Eucarida
Subphylum Crustacea	Order Decapoda
Superclass Oligostraca	Order Euphausiacea
Class Ichthyostraca	Superorder Peracarida
Subclass Branchiura	Order Amphipoda
Order Arguloida	Order Bochusacea
Subclass Pentastomida	Order Cumacea
Order Cephalobaenida	Order Ingolfiellida
Order Porocephalida	Order Isopoda
Order Raillietiellida	Order Lophogastrida
Order Reighardiida	Order Mictacea
Class Ostracoda	Order Mysida
Subclass Podocopa	Order Spelaeogriphacea
Order Palaeocopidae	Order Stygiomysida
Order Platycopida	Order Tanaidacea
Order Podocopida	Order Thermosbaenacea
Subclass Myodocopa	**Class Tantulocarida**
Order Halocyprida	(Without order)
Order Myodocopida	**Class Thecostraca**
Class Mystacocarida	**Subclass Ascothoracida**

Animal Classification Continued.

Order Dendrogastrida
Order Laurida
Subclass Facetotecta
(Without order)
Subclass Cirripedia
Infraclass Acrothoracica
Order Cryptophialida
Order Lithoglyptida
Infraclass Rhizocephala
(Without order)
Infraclass Thoracida
Superorder Phosphatothoracica
Order Iblomorpha
Superorder Thoracicalcarea
Order Balanomorpha
Order Calanticomorpha
Order Pollicipomorpha
Order Scalpellomorpha
Order Verrucomorpha
Superclass Allotriocarida
(Including Subphylum Hexapoda)
Class Branchiopoda
Subphylum Phyllopoda
Order Diplostraca
Order Notostraca
Subclass Sarsostraca
Order Anostraca
Class Cephalocarida
Order Brachypoda
Class Remipedia
Order Nectiopoda
Subclass Hexapoda [Insecta (sensu lato)]
Class Entognatha (Possibly a praphyletic group)
Subclass Ellipura
Order Protura
Order Collembola
Subclass Diplura
Order Diplura
Class Insecta (Sensu stricto) [**Class Ectognatha**]
Subclass Archaeognatha [**Class Monocondylia**]
Order Archaeognatha
Subclass Dicondylia
Infraclass Zygentoma
Order Thysanura
Infraclass Pterygota
Section Palaeoptera
Order Ephemeroptera
Order Odonata
Section Neoptera
Subsection Polyneoptera
Order Plecoptera
Order Dermaptera
Order Zoraptera
Order Embioptera
Order Phasmatodea
Order Orthoptera
Superorder Xenonomia
Order Grylloblattodea
Order Mantophasmatodea
Superorder Dictyoptera
Order Mantodea
Order Blattodea (Including termites)
Eumetabola
Subsection Paraneoptera
Order Psocodea
(Psocoptera+Phthiraptera)
Superorder Condylognatha
Order Thysanoptera
Order Hemiptera
Subsection Holometabola
[Endoperygota]
Superorder Hymenopterida
Order Hymenoptera
Superorder Neuropterida
Order Raphidioptera
Order Megaloptera
Order Neuroptera
Superorder Coleopterida
Order Coleoptera
Order Strepsiptera
Superorder Mecopterida
Antliophora
Order Mecoptera
Order Siphonaptera
Order Diptera
Amphiesmenoptera
Order Trichoptera
Order Lepidoptera

Plant Classification

Phylum Angiospermae

Class Magnoliopsida
(dicotyledonous plants)

Subclass Magnoliidae

- Order Magnoliales
  - Winteraceae
  - Degeneriaceae
  - Himantandraceae
  - Eupomatiaceae
  - Austrobaileyaceae
  - Magnoliaceae
  - Lactoridaceae
  - Annonaceae
  - Myristicaceae
  - Canellaceae
- Order Laurales
  - Amborellaceae
  - Trimeniaceae
  - Monimiaceae
  - Gomortegaceae
  - Calycanthaceae
  - Idiospermaceae
  - Lauraceae
  - Hernandiaceae
- Order Piperales
  - Chloranthaceae
  - Saururaceae
  - Piperaceae
- Order Aristolochiales
  - Aristolochiaceae
- Order Illiciales
  - Illiciaceae
  - Schisandraceae
- Order Nymphaeales
  - Nelumbonaceae
  - Nymphaeaceae
  - Barclayaceae
  - Cabombaceae
  - Ceratophyllaceae

Subclass Ranunculidae

- Order Ranunculales
  - Ranunculaceae
  - Circaeasteraceae
  - Berberidaceae
  - Sargentodoxaceae
  - Lardizabalaceae
  - Menispermaceae
  - Coriariaceae
  - Sabiaceae
- Order Papaverales
  - Papaveraceae
  - Fumariaceae

Subclass Hamamelidae

- Order Trochodendrales
  - Tetracentraceae
  - Trochodendraceae
- Order Hamamelidales
  - Cercidiphyllaceae
  - Eupteleaceae
  - Platanaceae
  - Hamamelidaceae
  - Myrothamnaceae
- Order Daphniphyllales
  - Daphniphyllaceae
- Order Didymelales
  - Didymelaceae
- Order Eucommiales
  - Eucommiaceae
- Order Urticales
  - Barbeyaceae
  - Ulmaceae
  - Cannabaceae
  - Moraceae
  - Cecropiaceae
  - Urticaceae
- Order Leitneriales
  - Leitneriaceae
- Order Juglandales
  - Rhoipteleaceae
  - Juglandaceae
- Order Myricales
  - Myricaceae
- Order Fagales
  - Balanopaceae

The classification system for terrestrial plants (angiosperms, gymnosperms, pteridophytes, bryophytes) is based on "Biodiversity series 2 Shokubutsu no tayosei to keito" (M. Kato ed., Shokabo Publishing, 1997). The recent classification system for angiosperms is now adopting classifications which reflect the results of molecular phylogenetic analysis (known as APG systems). For details, refer to APGIII (2016, Bot. J. Linn. Soc. 181: 1–20).

Plant Classification Continued.

Fagaceae
Betulaceae
Order Casuarinales
Casuarinaceae
Subclass Caryophyllidae
Order Caryophyllales
Phytolaccaceae
Achatocarpaceae
Nyctaginaceae
Aizoaceae
Didiereaceae
Cactaceae
Chenopodiaceae
Amaranthaceae
Portulacaceae
Basellaceae
Molluginaceae
Caryophyllaceae
Order Polygonales
Polygonaceae
Order Plumbaginales
Plumbaginaceae
Subclass Dilleniidae
Order Dilleniales
Dilleniaceae
Paeoniaceae
Order Theales
Ochnaceae
Sphaerosepalaceae
Sarcolaenaceae
Dipterocarpaceae
Caryocaraceae
Theaceae
Actinidiaceae
Scytopetalaceae
Pentaphylacaceae
Tetrameristaceae
Pellicieraceae
Oncothecaceae
Marcgraviaceae
Quiinaceae
Elatinaceae
Paracryphiaceae
Medusagynaceae
Clusiaceae
Order Malvales
Elaeocarpaceae
Tiliaceae
Sterculiaceae
Bombacaceae
Malvaceae
Order Lecythidales
Lecythidaceae
Order Nepenthales
Sarraceniaceae
Nepenthaceae
Droseraceae
Order Violales
Flacourtiaceae
Peridiscaceae
Bixaceae
Cistaceae
Huaceae
Lacistemaceae
Scyphostegiaceae
Stachyuraceae
Violaceae
Tamaricaceae
Frankeniaceae
Dioncophyllaceae
Ancistrocladaceae
Turneraceae
Malesherbiaceae
Passifloraceae
Achariaceae
Caricaceae
Fouquieriaceae
Hoplestigmataceae
Cucurbitaceae
Datiscaceae
Begoniaceae
Loasaceae
Order Salicales
Salicaceae
Order Capparales
Tovariaceae
Capparaceae
Brassicaceae
Moringaceae
Resedaceae
Order Batales
Gyrostemonaceae
Bataceae
Order Ericales
Cyrillaceae
Clethraceae

Continued.

- Grubbiaceae
- Empetraceae
- Epacridaceae
- Ericaceae
- Pyrolaceae
- Monotropaceae

Order Diapensiales
- Diapensiaceae

Order Ebenales
- Sapotaceae
- Ebenaceae
- Styracaceae
- Lissocarpaceae
- Symplocaceae

Order Primulales
- Theophrastaceae
- Myrsinaceae
- Primulaceae

Subclass Rosidae

Order Rosales
- Brunelliaceae
- Connaraceae
- Eucryphiaceae
- Cunoniaceae
- Davidsoniaceae
- Dialypetalanthaceae
- Pittosporaceae
- Byblidaceae
- Hydrangeaceae
- Columelliaceae
- Grossulariaceae
- Greyiaceae
- Bruniaceae
- Anisophylleaceae
- Alseuosmiaceae
- Crassulaceae
- Cephalotaceae
- Saxifragaceae
- Rosaceae
- Neuradaceae
- Crossosomataceae
- Chrysobalanaceae
- Surianaceae
- Rhabdodendraceae

Order Fabales
- Mimosaceae
- Caesalpiniaceae
- Fabaceae

Order Proteales
- Elaeagnaceae
- Proteaceae

Order Podostemales
- Podostemaceae

Order Haloragales
- Haloragaceae
- Gunneraceae

Order Myrtales
- Sonneratiaceae
- Lythraceae
- Penaeaceae
- Crypteroniaceae
- Thymelaeaceae
- Trapaceae
- Myrtaceae
- Punicaceae
- Onagraceae
- Oliniaceae
- Melastomataceae
- Combretaceae

Order Rhizophorales
- Rhizophoraceae

Order Cornales
- Alangiaceae
- Nyssaceae
- Cornaceae
- Garryaceae

Order Santalales
- Medusandraceae
- Dipentodontaceae
- Olacaceae
- Opiliaceae
- Santalaceae
- Misodendraceae
- Loranthaceae
- Viscaceae
- Eremolepidaceae
- Balanophoraceae

Order Rafflesiales
- Hydnoraceae
- Mitrastemonaceae
- Rafflesiaceae

Order Celastrales
- Geissolomataceae
- Celastraceae
- Hippocrateaceae
- Stackhousiaceae

Plant Classification	Continued.
Salvadoraceae	Geraniaceae
Aquifoliaceae	Limnanthaceae
Icacinaceae	Tropaeolaceae
Aextoxicaceae	Balsaminaceae
Cardiopteridaceae	Order Apiales
Corynocarpaceae	Araliaceae
Dichapetalaceae	Apiaceae
Order Euphorbiales	**Subclass Asteridae**
Buxaceae	Order Gentianales
Simmondsiaceae	Loganiaceae
Pandaceae	Retziaceae
Euphorbiaceae	Gentianaceae
Order Rhamnales	Saccifoliaceae
Rhamnaceae	Apocynaceae
Leeaceae	Asclepiadaceae
Vitaceae	Order Solanales
Order Linales	Duckeodendraceae
Erythroxylaceae	Nolanaceae
Humiriaceae	Solanaceae
Ixonanthaceae	Convolvulaceae
Hugoniaceae	Cuscutaceae
Linaceae	Menyanthaceae
Order Polygalales	Polemoniaceae
Malpighiaceae	Hydrophyllaceae
Vochysiaceae	Order Lamiales
Trigoniaceae	Lennoaceae
Tremandraceae	Boraginaceae
Polygalaceae	Verbenaceae
Xanthophyllaceae	Lamiaceae
Krameriaceae	Order Callitrichales
Order Sapindales	Hippuridaceae
Staphyleaceae	Callitrichaceae
Melianthaceae	Hydrostachyaceae
Bretschneideraceae	Order Plantaginales
Akaniaceae	Plantaginaceae
Sapindaceae	Order Scrophulariales
Hippocastanaceae	Buddlejaceae
Aceraceae	Oleaceae
Burseraceae	Scrophulariaceae
Anacardiaceae	Globulariaceae
Julianiaceae	Myoporaceae
Simaroubaceae	Orobanchaceae
Cneoraceae	Gesneriaceae
Meliaceae	Acanthaceae
Rutaceae	Pedaliaceae
Zygophyllaceae	Bignoniaceae
Order Geraniales	Mendonciaceae
Oxalidaceae	Lentibulariaceae

Continued.

Order Campanulales
Pentaphragmataceae
Sphenocleaceae
Campanulaceae
Stylidiaceae
Donatiaceae
Brunoniaceae
Goodeniaceae
Order Rubiales
Rubiaceae
Theligonaceae
Order Dipsacales
Caprifoliaceae
Adoxaceae
Valerianaceae
Dipsacaceae
Order Calycerales
Calyceraceae
Order Asterales
Asteraceae
Class Liliopsida (Monocotyledonous plants)
Subclass Alismatidae
Order Alismatales
Butomaceae
Limnocharitaceae
Alismataceae
Order Hydrocharitales
Hydrocharitaceae
Order Najadales
Aponogetonaceae
Scheuchzeriaceae
Juncaginaceae
Potamogetonaceae
Ruppiaceae
Najadaceae
Zannichelliaceae
Posidoniaceae
Cymodoceaceae
Zosteraceae
Order Triuridales
Petrosaviaceae
Triuridaceae
Subclass Arecidae
Order Arecales
Arecaceae
Order Cyclanthales
Cyclanthaceae
Order Pandanales
Pandanaceae
Order Arales
Araceae
Lemnaceae
Subclass Commelinidae
Order Commelinales
Rapateaceae
Xyridaceae
Mayacaceae
Commelinaceae
Order Eriocaulales
Eriocaulaceae
Order Restionales
Flagellariaceae
Joinvilleaceae
Restionaceae
Centrolepidaceae
Order Juncales
Juncaceae
Thurniaceae
Order Cyperales
Cyperaceae
Poaceae
Order Hydatellales
Hydatellaceae
Order Typhales
Sparganiaceae
Typhaceae
Subclass Zingiberidae
Order Bromeliales
Bromeliaceae
Order Zingiberales
Strelitziaceae
Heliconiaceae
Musaceae
Lowiaceae
Zingiberaceae
Costaceae
Cannaceae
Marantaceae
Subclass Liliidae
Order Liliales
Philydraceae
Pontederiaceae
Haemodoraceae
Cyanastraceae
Liliaceae
Iridaceae

Plant Classification Continued.

Velloziaceae
Aloaceae
Agavaceae
Xanthorrhoeaceae
Hanguanaceae
Taccaceae
Stemonaceae
Smilacaceae
Dioscoreaceae
Order Orchidales
Geosiridaceae
Burmanniaceae
Corsiaceae
Orchidaceae

Phylum Gymnospermae

Class Gnetopsida
Order Gnetales
Gnetaceae
Ephedraceae
Welwitschiaceae
Class Ginkgopsida
Order Ginkgoales
Ginkgoaceae
Class Cycadopsida
Order Cycadales
Cycadaceae
Boweniaceae
Stangeriaceae
Zamiaceae
Class Coniferopsida
Order Coniferales
Araucariaceae
Cephalotaxaceae
Cupressaceae
Pinaceae
Podocarpaceae
Sciadopityaceae
Taxaceae
Taxodiaceae

Phylum Pteridophyta

Class Psilotopsida
Order Psilotales
Psilotaceae
Tmesipteridaceae
Class Lycopsida
Order Licopodiales
Licopodiaceae
Order Selaginellales
Selaginellaceae
Order Isoetales
Isoetaceae
Class Equisetopsida
Order Equisetales
Equisetaceae
Class Filicopsida
Order Marattiales
Marattiaceae
Order Ophioglossales
Ophioglossaceae
Order Filicales
Osmundaceae
Plagiogyriaceae
Gleicheniaceae
Stromatopteridaceae
Schizaeaceae
Hymenophyllaceae
Loxsomataceae
Matoniaceae
Hymenophyllopsidaceae
Cyatheaceae
Dicksoniaceae
Dennstaedtiaceae
Lindsaeaceae
Davalliaceae
Oleandraceae
Adiantaceae
Vittariaceae
Pteridaceae
Aspleniaceae
Blechnaceae
Lomariopsidaceae
Dryopteridaceae
Thelypteridaceae
Woodsiaceae
Dipteridaceae
Cheiropleuriaceae
Polypodiaceae
Grammitidaceae
Marsileaceae
Salviniaceae
Azollaceae

Phylum Bryophyta

Class Bryopsida
Subclass Sphagnidae
Order Sphagnales

Continued.

Sphagnaceae
Subclass Andreaeidae
Order Andreaeales
Andreaeaceae
Subclass Takakiidae
Order Takakiales
Takakiaceae
Order Andreaeobryales
Andreaeobryaceae
Subclass Bryidae
Order Tetraphidales
Tetraphidaceae
Order Buxbaumiales
Buxbaumiaceae
Diphysciaceae
Order Polytrichales
Polytrichaceae
Dawsoniaceae
Order Fissidentales
Fissidentaceae
Order Archidiales
Archidiaceae
Order Dicranales
Ditrichaceae
Viridivelleraceae
Bryoxiphiaceae
Seligeriaceae
Dicranaceae
Dicnemonaceae
Pleurophascaceae
Leucobryaceae
Order Pottiales
Calymperaceae
Encalyptaceae
Pottiaceae
Bryobartramiaceae
Order Grimmiales
Grimmiaceae
Ptychomitriaceae
Erpodiaceae
Order Funariales
Gigaspermaceae
Disceliaceae
Ephemeraceae
Funariaceae
Pseudoditrichaceae
Splachnaceae
Order Schistostegales
Schistostegaceae
Mitteniaceae
Order Bryales
Bryaceae
Leptostomataceae
Mniaceae
Phyllodrepaniaceae
Eustichiaceae
Sorapillaceae
Calomniaceae
Rhizogoniaceae
Hypnodendraceae
Aulacomniaceae
Meesiaceae
Catoscopiaceae
Bartramiaceae
Spiridentaceae
Timmiaceae
Order Orthotrichales
Rhachitheciaceae
Microtheciellaceae
Orthotrichaceae
Helicophyllaceae
Order Isobryales
Racopilaceae
Fontinalaceae
Wardiaceae
Hydropogonaceae
Climaciaceae
Pleuroziopsidaceae
Hedwigiaceae
Cryphaeaceae
Leucodontaceae
Cyrtopodaceae
Ptychomniaceae
Lepyrodontaceae
Prionodontaceae
Rutenbergiaceae
Trachypodaceae
Pterobryaceae
Meteoriaceae
Phyllogoniaceae
Neckeraceae
Lembophyllaceae
Echinodiaceae
Order Hookeriales
Ephemeropsaceae
Callicostaceae

Plant Classification Continued.

Hookeriaceae
Daltoniaceae
Symphyodontaceae
Leucomiaceae
Hypopterygiaceae
Order Hypnobryales
Theliaceae
Fabroniaceae
Leskeaceae
Regmatodontaceae
Thuidiaceae
Amblystegiaceae
Brachytheciaceae
Entodontaceae
Plagiotheciaceae
Sematophyllaceae
Hypnaceae
Rhytidiaceae
Hylocomiaceae

Class Hepaticopsida

Subclass Jungermannidae

Order Calobryales
Haplomitriaceae
Order Jungermanniales
Vetaformaceae
Herbertaceae
Pseudolepicoleaceae
Trichotemnomaceae
Trichocoleaceae
Lepicoleaceae
Lepidoziaceae
Phycolepidoziaceae
Calypogeiaceae
Adelanthaceae
Cephaloziaceae
Cephaloziellaceae
Jackiellaceae
Antheliaceae
Jungermanniaceae
Mesoptychiaceae
Gymnomitriaceae
Scapaniaceae
Geocalycaceae
Plagiochilaceae
Chonecoleaceae
Arnelliaceae
Acrobolbaceae
Schistochilaceae
Perssoniellaceae
Balantiopsaceae
Gyrothyraceae
Pleuroziaceae
Radulaceae
Ptilidiaceae
Chaetophyllopsaceae
Lepidolaenaceae
Porellaceae
Goebeliellaceae
Frullaniaceae
Jubulaceae
Lejeuneaceae
Order Metzgeriales
Treubiaceae
Fossombroniaceae
Phyllothalliaceae
Pelliaceae
Allisoniaceae
Makinoaceae
Pallaviciniaceae
Blasiaceae
Hymenophytaceae
Aneuraceae
Vandiemeniaceae
Metzgeriaceae
Mizutaniaceae

Subclass Marchantiidae

Order Monocleales
Monocleaceae
Order Sphaerocarpales
Riellaceae
Sphaerocarpaceae
Monocarpaceae
Order Marchantiales
Corsiniaceae
Targioniaceae
Lunulariaceae
Wiesnerellaceae
Conocephalaceae
Aytoniaceae
Cleveaceae
Exormothecaceae
Marchantiaceae
Monosoleniaceae
Oxymitraceae
Ricciaceae

Continued.

Class Anthocerotopsida Anthocerotaceae	Notothyladaceae

Algae Classification

<u>Phylum Cyanophyta</u> **Class Cyanophyceae** Order Chroococcales Order Chamaesiphonales Order Pleurocapsales Order Stigonematales Order Nostocales <u>Phylum Prochlorophyta</u> **Class Prochlorophyceae** Order Prochlorales <u>Phylum Glaucophyta</u> **Class Glaucophyceae** Order Cyanophorales Order Glaucocystales <u>Phylum Rhodophyta</u> **Class Cyanidiophyceae** Order Cyanidiales **Class Rhodellophyceae** Order Rhodellales Order Dixoniellales **Class Porphyridiophyceae** Order Porphyridales **Class Stylonematophyceae** Order Stylonema **Class Compsopogonophyceae** Order Compsopogonales Order Erythropeltidales Order Rhodochaetales **Class Bangiophyceae** Order Bangiales **Class Florideophyceae** **Subclass Hildenbrandiophycidae** Order Hildenbrandiales **Subclass Corallinophycidae** Order Rhodogorgonales Order Sporolithales Order Corallinales **Subclass Nemaliophycidae** Order Nemaliales	Order Batrachospermales Order Thoreales Order Balliales Order Acrochaetiales Order Colaconematales Order Palmariales Order Balbianiales Order Rhodachlyales **Subclass Ahnfeltiophycidae** Order Ahnfeltiales Order Pihiellales **Subclass Rhodymeniophycidae** Order Bonnemaisoniales Order Gelidiales Order Acrosymphytales Order Gigartinales Order Peyssonneliales Order Plocamiales Order Nemastomatales Order Gracilariales Order Halymeniales Order Sebdeniales Order Rhodymeniales Order Ceramiales <u>Phylum Chlorophyta</u> **Class Palmophyllophyceae** Order Palmophyllales Order Prasinoccales **Class Mamiellophyceae** Order Mamiellales Order Dolichomastigales Order Monomastigales **Class Chlorodendrophyceae** Order Chlorodendrales **Class Pyramimonadophyceae** Order Pyramimonadales Order Palmophyllales Order Pseudoscourfieldiales Order Prasinococcales

(Source: T. Horiguchi, 2014)

Algae Classification Continued.

Class Nephroselmidophyceae
Order Nephroselmidales
Class Pedinophyceae
Order Pedinomonadales
Class Picocystophyceae
Order Picocystales
Class Chloropicophyceae
Order Chloropicorales
Class Chlorophyceae
Order Volvocales
Order Sphaeropleales
Order Oedogoniales
Order Chaetophorales
Order Chaetopeltidales
Class Trebouxiophyceae
Order Chlorellales
Order Microthamniales
Order Trebouxiales
Order Prasiolales
Class Ulvophyceae
Order Ulvales
Order Cladophorales
Order Caulerpales
Order Dasycladales
Order Ulotrichales
Order Oltmannsiellopsidales
Order Trentepohliales
Class Mesostigmatophyceae
Order Mesostigmatales
Class Charophyceae
Order Chlorokybales
Order Klebsormidiales
Order Zygnematales
Order Coleochaetales
Order Charales

Phylum Cryptophyta

Class Cryptophyceae
Order Cryptomonadales
Order Pyrenomonadales
Order Goniomonadales

Phylum Heterokontophyta

Class Crysophyceae
Order Chrysocapsales
Order Ochromonadales
Order Rhizochrysidales
Order Hydrurales
Order Hibberdiales
Order Chrysosphaerales
Class Phaeophyceae
Order Discosporangiales
Order Ectocarpales
Order Ralfsiales
Order Syringodermatales
Order Sphacelariales
Order Onslowiales
Order Dictyotales
Order Ishigeales
Order Cutleriales
Order Desmarestiales
Order Sporochnales
Order Laminariales
Order Tilopteridales
Order Nemodermatales
Order Fucales
Order Durvillaeales
Order Ascoseirales
Class Raphidophyceae
Order Raphidomonadales
Class Eustigmatophyceae
Order Eustigmatales
Class Xanthophyceae
Order Heterochloridales
Order Rhizochloridales
Order Heterogloeales
Order Mischococcales
Order Tribonematales
Order Vaucheriales
Class Synchromophyceae
Order Synchromales
Class Pinguiophyceae
Order Pinguiochrysidales
Class Schizocladiophyceae
Order Chrysomeridales
Order Schizocladiales
Class Chrysomerophyceae
Order Chrysomeridales
Class Aurearenophyceae
Order Aurearenales
Class Synurophyceae
Order Synurales
Class Bolidophyceae
Order Bolidomonadales
Class Bacillariophyceae
Subclass Thalassiosirophycidae
Order Thalassiosirales

Continued.

Subclass Coscinodiscophycidae
Order Chrysanthemodiscales
Order Melosirales
Order Paraliales
Order Aulacoseirales
Order Orthoseirales
Order Coscinodiscales
Order Ethmodiscales
Order Stictocyclales
Order Asterolampales
Order Arachnoidiscales
Order Stictodiscales
Subclass Biddulphiophycidae
Order Triceratiales
Order Biddulphiales
Order Hemiaulales
Order Anaulales
Subclass Lithodesmiphycidae
Order Lithodesmiales
Subclass Corethrophycidae
Order Corethrales
Subclass Cymatosirophysidae
Order Cymatosirales
Subclass Rhizosoleniophycidae
Order Rhizosoleniales
Subclass Chaetocerotophycidae
Order Chaetocerotales
Order Leptocylindales
Subclass Fragilariophycidae
Order Fragilariales
Order Tabellariales
Order Licmophorales
Order Rhaphoneidales
Order Ardissoneales
Order Toxariales
Order Thalassionematales
Order Rhabdonematales
Order Striatellales
Order Cyclophorales
Order Climacospheniales
Order Protoraphidiales
Subclass Eunotiophycidae
Order Eunotiales
Subclass Bacillariophycidae
Order Lyrellales
Order Mastogloiales
Order Dictyoneidales
Order Cymbellales
Order Achnanthales
Order Naviculales
Order Thalassiophysales
Order Bacillariales
Order Rhopalodiales
Order Surirellales
Class Dictyochophyceae
Order Pedinellales
Order Dictyochales
Order Rhizochromulinales
Order Florenciellales
Class Pelagophyceae
Order Pelagomonadales

Phylum Division Haptophyta

Class Haptophyceae
Subclass Prymnesiophycidae
Order Isochrysidales
Order Prymnesiales
Order Coccolithales
Subclass Pavlovophycidae
Order Pavlovales

Phylum Division Euglenophyta

Class Euglenophyceae
Order Eutreptiales
Order Euglenales
Order Rhabdomonadales
Order Heteronematales
Order Euglenamorphales

Phylum Chlorarachniophyta

Class Chlorarachniophyceae
Order Chlorarachniales

Phylum Dinophyta

Class Dinophyceae
Order Prorocentrales
Order Dinophysales
Order Gymnodiniales
Order Suessiales
Order Ptychodiscales
Order Blastodiniales
Order Gonyaulacales
Order Peridiniales
Order Phytodiniales
Order Thoracosphaerales
Order Haplozoonales
Class Syndinea
Order Syndiniales

Algae Classification Continued.

Class Noctiluciphyceae	Order Noctilucales

Fungi Classification

Fungus

Phylum Aphelidiomycota

Class Aphelidiomycetes
Order Aphelidiales

Phylum Ascomycota

Class Archaeorhizomycetes
Order Archaeorhizomycetales
Class Arthoniomycetes
Order Arthoniales
Order Lichenostigmatales
Class Candelariomycetes
Order Candelariales
Class Coniocybomycetes
Order Coniocybales
Class Dothideomycetes
Order Abrothallales
Order Acrospermales
Order Asterinales
Order Botryosphaeriales
Order Capnodiales
Order Catinellales
Order Cladoriellales
Order Collemopsidiales
Order Dothideales
Order Dyfrolomycetales
Order Eremithallales
Order Eremomycetales
Order Gloniales
Order Hysteriales
Order Jahnulales
Order Kirschsteiniotheliales
Order Lembosinales
Order Lichenotheliales
Order Microthyriales
Order Minutisphaerales
Order Monoblastiales
Order Murramarangomycetales
Order Muyocopronales
Order Myriangiales
Order Mytilinidiales
Order Natipusillales
Order Parmulariales
Order Patellariales
Order Phaeotrichales
Order Pleosporales
Order Stigmatodiscales
Order Strigulales
Order Superstratomycetales
Order Trypetheliales
Order Tubeufiales
Order Valsariales
Order Venturiales
Order Zeloasperisporiales
Class Eurotiomycetes
Order Arachnomycetales
Order Chaetothyriales
Order Coryneliales
Order Eurotiales
Order Mycocaliciales
Order Onygenales
Order Phaeomoniellales
Order Pyrenulales
Order Sclerococcales
Order Verrucariales
Class Geoglossomycetes
Order Geoglossales
Class Laboulbeniomycetes
Order Herpomycetales
Order Laboulbeniales
Order Pyxidiophorales
Class Lecanoromycetes
Order Acarosporales
Order Baeomycetales
Order Caliciales
Order Graphidales
Order Gyalectales
Order Lecanorales
Order Lecideales
Order Leprocaulales
Order Micropeltidales
Order Ostropales
Order Peltigerales
Order Pertusariales

Continued.

Order Rhizocarpales
Order Sarrameanales
Order Schaereriales
Order Sporastatiales
Order Teloschistales
Order Thelenellales
Order Turquoiseomycetales
Order Umbilicariales

Class Leotiomycetes
Order Chaetomellales
Order Cyttariales
Order Helotiales
Order Lahmiales
Order Lauriomycetales
Order Leotiales
Order Lichinodiales
Order Marthamycetales
Order Medeolariales
Order Micraspidales
Order Phacidiales
Order Rhytismatales
Order Thelebolales

Class Lichinomycetes
Order Lichinales

Class Neolectomycetes
Order Neolectales

Class Orbiliomycetes
Order Orbiliales

Class Pezizomycetes
Order Pezizales

Class Pneumocystomycetes
Order Pneumocystidales

Class Saccharomycetes
Order Saccharomycetales

Class Schizosaccharomycetes
Order Schizosaccharomycetales

Class Sordariomycetes
Order Amphisphaeriales
Order Amplistromatales
Order Annulatascales
Order Atractosporales
Order Boliniales
Order Calosphaeriales
Order Cephalothecales
Order Chaetosphaeriales
Order Coniochaetales
Order Conioscyphales
Order Coronophorales
Order Delonicicolales
Order Diaporthales
Order Distoseptisporales
Order Falcocladiales
Order Fuscosporellales
Order Glomerellales
Order Hypocreales
Order Jobellisiales
Order Koralionastetales
Order Lulworthiales
Order Magnaporthales
Order Meliolales
Order Microascales
Order Myrmecridiales
Order Ophiostomatales
Order Pararamichloridiales
Order Parasympodiellales
Order Phomatosporales
Order Phyllachorales
Order Pisorisporiales
Order Pleurotheciales
Order Pseudodactylariales
Order Savoryellales
Order Sordariales
Order Spathulosporales
Order Sporidesmiales
Order Tirisporellales
Order Togniniales
Order Torpedosporales
Order Tracyllalales
Order Vermiculariopsiellales
Order Xenospadicoidales
Order Xylariales

Class Taphrinomycetes
Order Taphrinales

Class Xylobotryomycetes
Order Xylobotryales

Class Xylonomycetes
Order Symbiotaphrinales
Order Xylonales

Phylum Basidiobolomycota

Class Basidiobolomycetes
Order Basidiobolales

Phylum Basidiomycota

Class Agaricomycetes
Order Agaricales
Order Amylocorticiales

Fungi Classification Continued.

Order Atheliales
Order Auriculariales
Order Boletales
Order Cantharellales
Order Corticiales
Order Geastrales
Order Gloeophyllales
Order Gomphales
Order Hymenochaetales
Order Hysterangiales
Order Jaapiales
Order Lepidostromatales
Order Phallales
Order Polyporales
Order Russulales
Order Sebacinales
Order Stereopsidales
Order Thelephorales
Order Trechisporales
Order Tremellodendropsidales

Class Agaricostilbomycetes
Order Agaricostilbales

Class Atractiellomycetes
Order Atractiellales

Class Bartheletiomycetes
Order Bartheletiales

Class Classiculomycetes
Order Classiculales

Class Cryptomycocolacomycetes
Order Cryptomycocolacales

Class Cystobasidiomycetes
Order Buckleyzymales
Order Cystobasidiales
Order Erythrobasidiales
Order Naohideales
Order Sakaguchiales

Class Dacrymycetes
Order Dacrymycetales
Order Unilacrymales

Class Exobasidiomycetes
Order Ceraceosorales
Order Doassansiales
Order Entylomatales
Order Exobasidiales
Order Georgefischeriales
Order Golubeviales
Order Microstromatales
Order Robbauerales
Order Tilletiales

Class Malasseziomycetes
Order Malasseziales

Class Microbotryomycetes
Order Heterogastridiales
Order Kriegeriales
Order Leucosporidiales
Order Microbotryales
Order Sporidiobolales

Class Mixiomycetes
Order Mixiales

Class Moniliellomycetes
Order Moniliellales

Class Pucciniomycetes
Order Helicobasidiales
Order Pachnocybales
Order Platygloeales
Order Pucciniales
Order Septobasidiales

Class Spiculogloeomycetes
Order Spiculogloeales

Class Tremellomycetes
Order Cystofilobasidiales
Order Filobasidiales
Order Holtermanniales
Order Tremellales
Order Trichosporonales

Class Tritirachiomycetes
Order Tritirachiales

Class Ustilaginomycetes
Order Uleiellales
Order Urocystidales
Order Ustilaginales
Order Violaceomycetales

Class Wallemiomycetes
Order Geminibasidiales
Order Wallemiales

Phylum Blastocladiomycota

Class Blastocladiomycetes
Order Blastocladiales
Order Callimastigales
Order Catenomycetales

Class Physodermatomycetes
Order Physodermatales

Phylum Calcarisporiellomycota

Class Calcarisporiellomycetes
Order Calcarisporiellales

Continued.

Phylum Caulochytriomycota	Phylum Kickxellomycota
Class Caulochytriomycetes	Class Asellariomycetes
Order Caulochytriales	Order Asellarialles
Phylum Chytridiomycota	Class Barbatosporomycetes
Class Chytridiomycetes	Order Barbatosporales
Order Chytridiales	Class Dimargaritomycetes
Order Nephridiophagales	Order Dimargaritales
Order Polyphagales	Class Harpellomycetes
Order Saccopodiales	Order Harpellales
Class Cladochytriomycetes	Class Kickxellomycetes
Order Cladochytriales	Order Kickxellales
Class Lobulomycetes	Class Ramicandelaberomycetes
Order Lobulomycetales	Order Ramicandelaberales
Class Mesochytriomycetes	Phylum Monoblepharomycota
Order Gromochytriales	Class Hyaloraphidiomycetes
Order Mesochytriales	Order Hyaloraphidiales
Class Polychytriomycetes	Class Monoblepharidomycetes
Order Polychytriales	Order Monoblepharidales
Class Rhizophlyctidomycetes	Class Sanchytriomycetes
Order Rhizophlyctidales	Order Sanchytriales
Class Rhizophydiomycetes	Phylum Mortierellomycota
Order Rhizophydiales	Class Mortierellomycetes
Class Spizellomycetes	Order Mortierellales
Order Spizellomycetales	Phylum Mucoromycota
Class Synchytriomycetes	Class Endogonomycetes
Order Synchytriales	Order Endogonales
Phylum Entomophthoromycota	Class Mucoromycetes
Class Entomophthoromycetes	Order Mucorales
Order Entomophthorales	Class Umbelopsidomycetes
Class Neozygitomycetes	Order Umbelopsidales
Order Neozygitales	Phylum Neocallimastigomycota
Phylum Entorrhizomycota	Class Neocallimastigomycetes
Class Entorrhizomycetes	Order Neocallimastigales
Order Entorrhizales	Phylum Olpidiomycota
Order Talbotiomycetales	Class Olpidiomycetes
Phylum Glomeromycota	Order Olpidiales
Class Archaeosporomycetes	Phylum Rozellomycota
Order Archaeosporales	Class Microsporidea
Class Glomeromycetes	Order Amblyosporida
Order Diversisporales	Order Glugeida
Order Gigasporales	Order Neopereziida
Order Glomerales	Order Nosematida
Class Paraglomeromycetes	Order Ovavesiculida
Order Paraglomerales	Class Rudimicrosporea
	Order Metchnikovellida

Fungi Classification Continued.

Phylum Zoopagomycota	**Oomycetes**
Class Zoopagomycetes	**Phylum Oomycota**
Order Zoopagales	**Class Oomycetes**
Myxomycota	Order Albuginales
Phylum Eumycetozoa	Order Haptoglossales
Class Dictyosteliomycetes	Order Lagenismatales
Order Acytosteliales	Order Leptomitales
Order Dictyosteliales	Order Myzocytiopsidales
Class Ceratiomyxomycetes	Order Olpidiopsidales
Order Ceratiomyxales	Order Peronosporales
Class Myxomycetes	Order Pythiales
Order Cribrariales	Order Rhipidiales
Order Reticulariales	Order Salilagenidiales
Order Liceales	Order Saprolegniales
Order Trichiales	**Phylum Hyphocytriomycota**
Order Echinosteliopsidales	**Class Hyphochytriomycetes**
Order Echinosteliales	Order Hyphochytriales
Order Clastodermatales	**Phylum Labyrinthulomycota**
Order Meridermatales	**Class Labyrintulomycetes**
Order Stemonitidales	Order Labyrinthulales
Order Physarales	Order Thraustochytriales

The development of a phylogenetic classification system for fungi is currently underway, with revisions being made every year. Kirk et al. (2008) is a standard system up to now, but it is a bit old. However, on the other hand, there is no system that has been proposed that can be accepted by most people. This book follows the system of Wijayawardene et al. (2020), which is the most comprehensive to date. The main body of the system is a reorganization of the taxonomic system of fungi based on molecular phylogenetic findings, but the details are still controversial. In addition, this book covers the deformomycetes (Eumycetozoa), sakage fungi, and ovomycetes, which were previously considered fungi but are now known to be different groups of organisms. Transforming fungi The system of higher-level taxa was based on Wijayawardene et al. Phylum Saccharomyces cerevisiae Chytridiomycota and Oomycota belonging to Chromista are not treated in Wijayawardene et al. (2020), so we followed the system of Katsumoto (2010).

Taxa listed in Katsumoto (2010) are described in Japanese. The taxa not listed in Katsumoto (2010) are described in kana based on the Latin reading of the type genus.

In addition, some fungi are known to lichenize by symbiosis with algae. Lichenizing the major taxonomic groups are marked with an asterisk (*).

Kirk et al. Dictionary of the Fungi. 10 th edition. CABI. (2008)

K. Katsumoto: Fungi of Japan, Kanto Branch, The Mycological Society of Japan (2010).

Wijayawardene et al.: Outline of Fungi and fungus-like taxa. Mycosphere 11: 1060-1456 (2020).

(Tsuyoshi Hosoya et al., 2020)

Prokaryote Classification

Domain Archaea	Order Acidilobales
[5 phyla; 10 Classes; 19 Orders]	Order Cenarchaeales
Phylum Crenarchaeota	Order Desulfurococcales
Class Thermoprotei	Order Fervidicoccales

Continued.

Order Sulfolobales
Order Thermoproteales

Phylum Euryarchaeota

Class Archaeoglobi
Order Archaeoglobales
Class Halobacteria
Order Halobacteriales
Order Haloferacales
Order Natrialbales
Class Methanobacteria
Order Methanobacteriales
Class Methanococci
Order Methanococcales
Class Methanomicrobia
Order Methanocellales
Order Methanomicrobiales
Order Methanosarcinales
Class Methanopyri
Order Methanopyrales
Class Thermococci
Order Thermococcales
Class Thermoplasmata
Order Thermoplasmatales

Phylum Korarchaeota

Phylum Nanoarchaeota

Phylum Thaumarchaeota

Class Nitrososphaeria
Order Nitrososphaerales

Domain Bacteria
[29 phyla; 63 Classes; 145 Orders]

Phylum Acidobacteria

Class Acidobacteria
Order Acidobacteriales
Class Blastocatellia
Order Blastocatellales
Class Holophagae
Order Acanthopleuribacterales
Order Holophagales

Phylum Actinobacteria

Class Acidimicrobia
Order Acidimicrobiales
Class Actinobacteria
Order Acidothermales
Order Actinomycetales
Order Bifidobacteriales
Order Frankiales
Order Geodermatophilales
Order Kineosporiales
Order Micrococcales
Order Nakamurellales
Class Coriobacteria
Order Coriobacteriales
Class Nitriliruptoria
Order Egibacterales
Order Nitriliruptorales
Order Euzebyales
Class Rubrobacteria
Order Gaiellales
Order Rubrobacterales
Order Solirubrobacterales
Class Thermoleophilia
Order Thermoleophilales

Phylum Aquificae

Class Aquificae
Order Aquificales

Phylum Armatimonadetes

Class Armatimonadia
Order Armatimonadales
Order Fimbriimonadales

Phylum Bacteroidetes

Class Bacteroidia
Order Bacteroidales
Class Cytophagia
Order Cytophagales
Class Flavobacteriia
Order Flavobacteriales
Class Sphingobacteriia
Order Sphingobacteriales

Phylum Caldiserica

Class Caldisericia
Order Caldisericales

Phylum Chlamydiae

Class Chlamydiae
Order Chlamydiales

Phylum Chlorobi

Class Chlorobea
Order Chlorobiales
Class Ignavibacteria
Order Ignavibacteriales

Prokaryote Classification Continued.

Phylum Chloroflexi

Class Anaerolineae
Order Anaerolineales
Class Ardenticatenia
Order Ardenticatenales
Class Caldilineae
Order Caldilineales
Class Chloroflexia
Order Chloroflexales
Order Herpetosiphonales
Order Kallotenuales
Class Dehalococcoidia
Order Dehalococcoidales
Class Ktedonobacteria
Order Ktedonobacterales
Order Thermogemmatisporales
Class Thermoflexia
Order Thermoflexales
Class Thermomicrobia
Order Sphaerobacterales
Order Thermomicrobiales

Phylum Chrysiogenetes

Class Chrysiogenetes
Order Chrysiogenales

Phylum Cyanobacteria

Order Chroococcales
Order Oscillatoriales
Order Prochlorales

Phylum Deferribacteres

Class Deferribacteres
Order Deferribacterales

Phylum Deinococcus-Thermus

Class Deinococci
Order Deinococcales
Order Thermales

Phylum Dictyoglomi

Class Dictyoglomia
Order Dictyoglomales

Phylum Elusimicrobia

Class Elusimicrobia
Order Elusimicrobiales

Phylum Fibrobacteres

Class Chitinivibrionia
Order Chitinivibrionales
Class Fibrobacteria
Order Fibrobacterales

Phylum Firmicutes

Class Bacilli
Order Bacillales
Order Lactobacillales
Class Clostridia
Order Clostridiales
Order Halanaerobiales
Order Natranaerobiales
Order Thermoanaerobacterales
Class Erysipelotrichia
Order Erysipelotrichales
Class Limnochordia
Order Limnochordales
Class Negativicutes
Order Acidaminococcales
Order Selenomonadales
Order Veillonellales
Class Thermolithobacteria
Order Thermolithobacterales
Class Tissierellia
Order Tissierellales

Phylum Fusobacteria

Class Fusobacteriia
Order Fusobacteriales

Phylum Gemmatimonadetes

Class Gemmatimonadetes
Order Gemmatimonadales

Phylum Lentisphaerae

Class Lentisphaeria
Order Lentisphaerales
Order Victivallales
Class Oligosphaeria
Order Oligosphaerales

Phylum Nitrospirae

Class Nitrospira
Order Nitrospirales

Phylum Planctomycetes

Class Planctomycea
Order Planctomycetales
Class Phycisphaerae
Order Phycisphaerales
Order Tepidisphaerales

Continued.

Phylum Proteobacteria

Class Zetaproteobacteria
- Order Mariprofundales

Class Alphaproteobacteria
- Order Parvularculales
- Order Caulobacterales
- Order Kiloniellales
- Order Kordiimonadales
- Order Magnetococcales
- Order Rhizobiales
- Order Rhodobacterales
- Order Rhodospirillales
- Order Rickettsiales
- Order Sneathiellales
- Order Sphingomonadales

Class Betaproteobacteria
- Order Procabacteriales
- Order Burkholderiales
- Order Hydrogenophilales
- Order Methylophilales
- Order Neisseriales
- Order Nitrosomonadales
- Order Rhodocyclales
- Order Sulfuricellales

Class Deltaproteobacteria
- Order Bdellovibrionales
- Order Desulfarculales
- Order Desulfobacterales
- Order Desulfovibrionales
- Order Desulfurellales
- Order Desulfuromonadales
- Order Myxococcales
- Order Syntrophobacterales

Class Epsilonproteobacteria
- Order Campylobacterales
- Order Nautiliales

Class Gammaproteobacteria
- Order Enterobacteriales
- Order Vibrionales
- Order Acidiferrobacterales
- Order Acidithiobacillales
- Order Aeromonadales
- Order Alteromonadales
- Order Arenicellales
- Order Cardiobacteriales
- Order Cellvibrionales
- Order Chromatiales
- Order Legionellales
- Order Lysobacterales
- Order Methylococcales
- Order Nevskiales
- Order Oceanospirillales
- Order Orbales
- Order Pasteurellales
- Order Pseudomonadales
- Order Thiotrichales

Class Oligoflexia
- Order Oligoflexales

Phylum Spirochaetes

Class Spirochaetes
- Order Brachyspirales
- Order Brevinematales
- Order Leptospirales
- Order Spirochaetales

Phylum Synergistetes

Class Synergistia
- Order Synergistales

Phylum Tenericutes

Class Mollicutes
- Order Acholeplasmatales
- Order Anaeroplasmatales
- Order Entomoplasmatales
- Order Haloplasmatales
- Order Mycoplasmatales

Phylum Thermodesulfobacteria

Class Thermodesulfobacteria
- Order Thermodesulfobacteriales

Phylum Thermotogae

Class Thermotogae
- Order Kosmotogales
- Order Mesoaciditogales
- Order Petrotogales
- Order Thermotogales

Phylum Verrucomicrobia

Class Spartobacteria

Class Opitutae
- Order Opitutales
- Order Puniceicoccales

Class Verrucomicrobiae
- Order Verrucomicrobiales

Prokaryote Classification based on "Seibutsugaku jiten", Tokyo Kagaku Dojin (2010).

Virus Classification

Double-stranded DNA (dsDNA)
Ackermannviridae
Herelleviridae
Myoviridae
Podoviridae
Siphoviridae
Alloherpesviridae
Herpesviridae
Malacoherpesviridae
Lipothrixviridae
Rudiviridae
Adenoviridae
Ampullaviridae
Ascoviridae
Asfarviridae
Baculoviridae
Bicaudaviridae
Clavaviridae
Corticoviridae
Fuselloviridae
Globuloviridae
Guttaviridae
Hytrosaviridae
Iridoviridae
Lavidaviridae
Marseilleviridae
Mimiviridae
Nimaviridae
Nudiviridae
Ovaliviridae
Papillomaviridae
Phycodnaviridae
Plasmaviridae
Polydnaviridae
Polyomaviridae
Portogloboviridae
Poxviridae
Sphaerolipoviridae
Tectiviridae
Tristromaviridae
Turriviridae

Double-stranded DNA (dsDNA), Single-stranded DNA (ssDNA)
Pleolipoviridae

Double-stranded DNA reverse transcription (dsDNA-RT)
Caulimoviridae
Hepadnaviridae

Double-chain RNA (dsRNA)
Amalgaviridae
Birnaviridae
Chrysoviridae
Cystoviridae
Megabirnaviridae
Partitiviridae
Picobirnaviridae
Quadriviridae
Reoviridae
Totiviridae

Single-chain DNA (ssDNA)
Alphasatellitidae
Bacilladnaviridae
Bidnaviridae
Genomoviridae
Smacoviridae
Tolecusatellitidae

Single-stranded DNA, -sense (ssDNA (−))
Anelloviridae

Single-stranded DNA, +sense (ssDNA (+))
Inoviridae
Microviridae
Nanoviridae
Spiraviridae

Single-stranded ambisense *DNA (ssDNA (+/−))
Circoviridae
Geminiviridae
Parvoviridae

Single-stranded RNA (ssRNA)
Avsunviroidae
Pospiviroidae

Single-stranded RNA, -sense (ssRNA (−))
Qinviridae
Aspiviridae
Chuviridae
Artoviridae
Bornaviridae
Filoviridae
Lispiviridae
Mymonaviridae
Nyamiviridae
Paramyxoviridae
Pneumoviridae
Rhabdoviridae

Continued.

Sunviridae	Astroviridae
Xinmoviridae	Barnaviridae
Yueviridae	Benyviridae
Cruliviridae	Botourmiaviridae
Fimoviridae	Bromoviridae
Hantaviridae	Caliciviridae
Leishbuviridae	Carmotetraviridae
Mypoviridae	Closteroviridae
Nairoviridae	Endornaviridae
Peribunyaviridae	Flaviviridae
Phasmaviridae	Hepeviridae
Wupedeviridae	Hypoviridae
Amnoonviridae	Kitaviridae
Orthomyxoviridae	Leviviridae
Single-stranded RNA, +sense (ssRNA (+))	Luteoviridae
Abyssoviridae	Matonaviridae
Arteriviridae	Narnaviridae
Coronaviridae	Nodaviridae
Medioniviridae	Permutotetraviridae
Mesoniviridae	Potyviridae
Mononiviridae	Sarthroviridae
Euroniviridae	Solemoviridae
Roniviridae	Solinviviridae
Tobaniviridae	Togaviridae
Dicistroviridae	Tombusviridae
Iflaviridae	Virgaviridae
Marnaviridae	**Single-stranded ambisense RNA (ssRNA (+/−))**
Picornaviridae	Arenaviridae
Polycipiviridae	Phenuiviridae
Secoviridae	Tospoviridae
Alphaflexiviridae	**Single-stranded RNA reverse transcription (ssRNA-RT)**
Betaflexiviridae	Belpaoviridae
Deltaflexiviridae	Metaviridae
Gammaflexiviridae	Pseudoviridae
Tymoviridae	Retroviridae
Alphatetraviridae	
Alvernaviridae	

ssDNA: single-stranded DNA, dsDNA: double-stranded DNA, dsRNA : double-strandedRNA, RT: Reverse transcribing, (+): positive sense, (−): negative sense, (+/−): ambisense

* Ambisense: Having both positive and negative strand regions (Reference: Virology for Life Science, edited by Shimotohno and Seya) The virus classification table is based on the classification system of the International Classification Committee ICTV2018b (https://talk.ictvonline.org/taxonomy/p/taxonomy_releases).

(Keizo Nagasaki, Yuji Tomaru, 2020)

Reproduction, Development, Growth

Reproduction of Mammals

Species (scientific name)	Sexual maturity ♀	Breeding season	Estrous cycle	Pregnancy period	Number in litter
Humans (*Homo sapiens*)	11-16 y	year round	28.3 d	253-303 d	1
Chimpanzee (*Pan troglodytes*)	8-11 y	year round	36 (33-38) d	216-260 d	1
Rhesus monkeys (*Macaca mulatta*)	3-4 y	year round	28 (23-33) d	144-197 d	1
Japanese monkey (*Macaca fuscata*)	3-4 y	Oct.-Apr.	28 d	144-197 d	1
Myotis (*Myotis lucifugus*)	1st Summer	Spring-Autumn	–[1]	50-60 d	1
European production shrew (*Sorex araneus*)	2 y	Mar.-Sep.	–	13-19 d	7
Guinea pig (*Cavia procellus*)	55-70 d	Year round	16-19 d	68 (58-75) d	3 (1-8)
Mouse (*Mus musculus*)	35 d	Year round	4 d	19-31 d	6 (1-12)
Rat (*Rattus norvegicus*)	40-60 d	Year round	4-5 d	21 d	6-9
Gray squirrel (*Sciurus carolinensis*)	1-2 y	Dec.-Aug.	–	44 d	4 (1-6)
European rabbit (*Oryctolagus cuniculus*)	5.5-8.5 mo	Year round	–[1]	31 (30-35) d	8 (1-13)
Dogs (*Canis familiaris*)	6-8 mo	Spring-Autumn	9 d	63 (53-71) d	7 (1-12)
Cat (*Felis catus*)	6-15 mo	Feb.-Jul.	15-28 d[1]	63 (52-69) d	4
Cow (*Bos Taurus*)	6-10 mo	Year round	14-23 d	284 (210-335) d	1
Pig (*Sus scrofa*)	7 (5-8) mo	Year round	18-24 d	114 (101-130) d	9 (6-15)
Sheep (*Ovis aries*)	7-8 mo	Autumn-late autumn	14-20 d	151 (144-152) d	1-4
Goat (*Capra hircus*)	8 mo	Winter-Autumn	21 d	151 (135-160) d	1-5
Horse (*Equus caballus*)	15-18 mo	Spring	21-23 d	333 d	1
Ass (*Equus asinus*)	1 y	Mar.-Aug.	2-7 d	365 d	1

1) Post-copulatory ovulation d: days, mo: months, y: years Source: E. S. E. Hafez, *et al.*, 1972

Growth Rate of Embryos and Fetuses (Mammals, Birds)

Stages of growth	Human	Rhesus macaque	Rat	Hamster	European rabbit	Pig	Sheep	Chicken	Sparrow
2 Cell	38 h	24 h	24 h	16 h	8 h	30 h	30 h	3 h	–
4 Cell	48 h	36 h	48 h	40 h	11 h	40 h	34 h	3.25 h	–
Implantation	6.5 d	9 d	6 d	4.5 d	7 d	–	10 d	/	/
Primitive streak	19 d	19 d	8.5 d	6.5-7 d	6.5 d	11 d	13 d	1.5 d	1.5 d
13-20 metamere	27 d	25 d	10.5 d	8 d	9 d	16 d	17 d	3 d	2.5 d
Tail bud formation	29 d	26 d	11.5 d	8.5 d	9.5 d	17 d	18 d	3.25 d	3.25 d
Embryo period end	36 d	28 d	12.5 d	9 d	10 d	20 d	21 d	5 d	5 d
Embryo formation completion	60 d	40 d	16.5 d	13.5 d	14 d	35 d	32 d	8.5 d	8 d
Eyelids fuse	70 d	48 d	18 d	–	19 d	50 d	42 d	13 d	11 d
Eyelids open	140 d	–	38 d	–	42 d	90 d	84 d	21 d	20 d
Birth (hatching)	267 d	164 d	22 d	16 d	32 d	112 d	150 d	22 d	14 d

h: hours, d: days Source: E. Witschi, 1972

Reproduction of Reptiles

Species (scientific name)	Sexual maturity	Spawning season	Hatching days
Green chameleon (*Anolis carolinensis*)	1 y	Apr.-Aug.	42-49 d
Marine iguana (*Amblyrhynchus cristatus*)	-	Dec.-Jan.	112 d
Skink (*Eumeces fasciatus*)	2 y	May	27-47 d
Lacertid (*Takydromus tachydromoides*)	1 y	May-Sep.	42 d
Striped snake (*Elaphe obsoleta*)	4 y	Apr.-May	-60 d
Indian python (*Python molurus*)	5 y	Dec.-Feb.	57-66 d
North American pit viper (*Agkistrodon contortrix*)	3 y	May	-142 d[1)]
Mississippi alligator (*Alligator mississippiensis*)	5-10 y	Jan.-Sep.	56-66 d

1) Ovoviviparous, d: days, y: years Source: H. S. Fitch, *et al.*, 1972

Reproduction of Amphibian

Species (scientific name)	Sexual maturity	Breeding season	Number in spawn	Developmental period	Larva period
Tiger salamander (*Ambystoma tigrinum*)	1 y	Dec.-Jun.	1-110	12-40 d	61-118 d
North American newt (*Triturus viridescens*)	2-4 y	Apr.-Jun.	200-376	10-35 d	60-120 d
Bullfrog (*Rana catesbeiana*)	3 y	Feb.-Nov.	6 000-20 000	2-15.5 d	365-1 095 d
Northern leopard frog (*Rana pipiens*)	3 y	Mar.-Aug.	2 000-6 500	4-20 d	60-84 d
American toad (*Bufo americanus*)	2 y	Feb.-Jun.	2 000-20 603	1-385 d	22-68 d
Tree frog (*Hyla regilla*)	2 y	Jan.-Jul.	500-1 500	6-14 d	50-90 d
African clawed frog					
(*Xenopus laevis*)	1.5 y	Year round	500-15 000	1.5-7 d	28-84 d
(*Xenopus tropicalis*)	5 mo	Year round	1 000-5 000	0.5-2 d	10-20 d

d: days, mo: months, y: years Source: L. E. Brown, & J. L. Vial, 1972

Reproduction of Fish: Spawning Season and Number of Eggs

Species (scientific name)	Spawning season	Number of eggs	Species (scientific name)	Spawning season	Number of eggs
Sturgeon (*Acipenser fulvescens*)	Spring-Summer	128 000-1 000 000	Yellowfin tuna (*Thunnus albacares*)	Spring-Summer	500 000-8 500 000
Herring (*Clupea harengus*)	Spring-Autumn	20 000-40 000	Swordfish (*Xiphias gladius*)	Spring-Summer	16 000 000
Rainbow trout (*Oncorhynchus mikiss = Salmo gairdneri*)	Spring	1 590-3 459	Halibut (*Hippoglossus stenolepis*)	Winter	176 381-3 402 931
Brook Trout (*Salvelinus fontinalis*)	Autumn	18-7 000	Ocean sunfish (*Mola mola*)	-	300 000 000
Chinook salmon (*Oncorhynchus tshawytscha*)	Spring-Autumn	1 718-11 012	Hammerhead shark (*Sphyrna zygaena*)	Summer	29-37[3)]
Goldfish (crucian carp) (*Carassius auratus*)	Spring-Summer	3 000-14 000	Northern dogfish (*Squalus acanthias*)	Autumn-Winter	3-14[3)]
Carp (*Cyprinus carpio*)	Spring-Summer	36 000-7 000 000	Skate (*Raja binoculata*)	Year round	2-7
Cod (*Gadus morhua*)	Spring, Autumn-Winter	3 000 000-9 000 000	Shorttail stingray (*Dasyatis dipterura*)	-	3-5[3)]
Three-spined stickleback (*Gasterosteus aculeatus*)	Spring	50-500[1)]	Lamprey (*Lampetra lamottei*)	Spring-Summer	1 085-3 648
Seahorse (*Hippocampus hudsonius*)	Spring-Summer	150-200[2)]	Hagfish (*Myxine glutinosa*)	Year round	19-30

1) ♂ protect egg nest, 2) ♂ incubation pouch, 3) Ovoviviparous Source: P. E. Porter, *et. al.*, 1972

Growth Rate of Embryos and Fetuses (Amphibians)

Stages of growth	Tokyo daruma pond frog	Hoptoad	African clawed frog	Newt	Tiger salamander
2 cells	2.25 h	3.2 h	1.5 h	6 h	6 h
4 cells	3.08 h	4 h	2 h	9 h	8 h
8 cells	3.92 h	5 h	2.25 h	12 h	10 h
16 cells	4.75 h	6.2 h	2.75 h	15 h	13 h
Morula	7.75 h	8 h	3.5 h	18 h	24 h
Blastosphere	10.5 h	10.5 h	4 h	1 d 5 h	1.5 d
Blastopore appears	22 h	1 d 4 h	9 h	1 d 20 h	3.5 d
Gastrula	1 d 3 h	1 d 8 h	11.75 h	2 d 10 h	4 d
Yolk plug	1 d 6 h	1 d 10 h	12.5 h	3 d	5.5 d
Neural plate	1 d 22 h	2 d	14.75 h	3 d 16 h	7.5 d
Neural tube	2 d 8 h	2 d 9 h	20.75 h	5 d 4 h	8 d
Tail bud	2 d 16 h	2 d 16 h	1 d 11 h	5 d 10 h	12 d
Branchial bud	3 d 16 h	3 d 13 h	2 d	10 d	22 d
Hatching	4 d 12 h	4 d 14 h	3 d	20 d	41 d
Gill cover	8 d 1 h	6 d 10 h	4 d 2 h	-	-
Posterior limb bud	10 d 14 h	6 d 10 h	7 d 12 h	-	-
Tail degeneration beginning	71 d	75 d	42 d	/	/
Foreleg appearance	71 d	70 d	43 d	12 d	49 d
Transformation	79 d	82 d	56 d	-	-
Water temperature (℃)	19*	18	22-24	18	10

* After 11th day 21-29℃. h: hours, d: days

Sources: M. Ichikawa, *et al.*, 1966; P. D. Nieuwkoop, *et al.*, 1967; H. Iwasawa, *et al.*, 1980

Developmental Speed of Embryos (Fish)

Stages of growth	Japanese killifish	Rainbow trout	Brook trout	Red sea bream	Greenling
2 cells	1.5 h	9-10 h	8-10 h	1.5 h	4-4.5 h
4 cells	2.5 h	16 h	10-12 h	2 h	6-7 h
8 cells	3.5 h	24 h	18-20 h	2.5 h	8-9 h
16 cells	4.3 h	24-36 h	24 h	3 h	10-12 h
Morula	6.5 h	2-3.5 d	1-2 d	8 h	20 h
Early blastula	8.5 h	3.5-5.5 d	2-6 d	10 h	24 h
Late blastocyst	17 h	-	-	14 h	-
Gastrula initial	19 h	4.5-6.5 d	4-8 d	16 h	1 d 6 h
Gastrula latter term	23 h	-	-	18 h	-
Neurulae	1 d 3 h	7-7.5 d	9-10 d	20 h	2 d
Optic vesicle differentiation	1 d 10 h	7-9 d	12 d	1 d	3 d
3 metameres	1 d 14 h	-	-	-	4 d
10-12 metameres	2 d 8 h	7-9 d	12 d	-	-
18 metameres	2 d 20 h	9-10 d	-	-	-
22-25 metameres	3 d 6 h	10-11 d	13-13.5 d	1 d 16 h	-
Mouth and pinna pectorale	4 d	17-22 d	25 d	2 d	15 d
Pinna pectorale movement	6 d	22-25 d	37 d	2 d	16 d
Lower jaw differentiation	8 d	25-27 d	42 d	3 d	-
Hatching	11 d	33 d	53 d	2 d 2 h	23 d
Water temperature (℃)	23	10	10	17	11-15

h: hours, d: days

Sources: K. Matsui, 1949; W. W. Ballard, 1972; Y. Suehiro, 1974

Development/Growth Period of Insects

Species (scientific name)	Number of eggs	Generation	Larva	Chrysalis	Imago
Firebrat (*Thermobia domestica*)	50	12-13 d	60-120 d	/*	1-2.5 y
Migratory Grasshopper (*Melanoplus sanguinipes*)	300-400	90-120 d	40-60 d	/*	30-60 d
German cockroach (*Blattella germanica*)	218-267	17 d	40-42 d	/*	80-232 d
American cockroach (*Periplaneta americana*)	200-1 000	32-58 d	285-616 d	/*	102-558 d
Green peach aphid (*Myzus persicae*)	–	120 d	9-12 d	/*	10-49 d
Silkworm (*Bombyx mori*)	300-500	9-12 d	21-25 d	14-21 d	2-3 d
Small cabbage white butterfly (*Pieris rapae*)	200-500	7 d	14 d	7-14 d	–
Northern house mosquito (*Culex pipiens*)	200-1 000	1-2 d	5-8 d	2-3 d	50-70 d
Fruit fly (*Drosophila melanogaster*)	100	<1 d	3-11 d	2-8 d	14 d
Housefly (*Musca domestica*)	75-720	1-3 d	4-10 d	4-18 d	10-50 d
Black blow fly (*Phormia regina*)	100-500	1-2 d	4-15 d	3-13 d	10-40 d
Cat flea (*Ctenocephalides felis*)	200-450	2-4 d	8-24 d	5-7 d	50-200 d
Meal worm (*Tenebrio molitor*)	276	12-16 d	120-350 d	18-20 d	60-90 d
Boll weevil (*Anthonomus grandis*)	20-400	3-5 d	7-12 d	3-5 d	1-250 d
Honey bee (*Apis mellifera*)	600 000[1)]	3 d	7 d	11 d	35-40 d[2)]

* Incomplete metamorphosis. 1) Queen, 2) Worker bee. d: days, y: years. (Source: C. H. Schmidt, *et al.*, 1972)

Annual Number of Generations for Insects

Species (scientific name)	Winter form	Number of generations/breeding season	Species (scientific name)	Winter form	Number of generations/breeding season
Firebrat	–	3-7	Northern house mosquito	Imago	2-18
Black gloss cricket (*Acheta domesticus*)	Egg	1	Tiger mosquito	All forms*	3-12
Migratory grasshopper	Egg	1-2	Fruit fly	Larva, imago	5-6
German cockroach	All forms*	1-4	Housefly	All forms*	4-18
American cockroach	All forms*	≦1	Black blow fly	Imago	2-4
Green peach aphid	Egg	14	Cat flea	Chrysalis	≒10
Body lice (*Pediculus* sp.)	All forms*	10-12	Hoshi ladybug	Imago	1-3
Saturniid (*Platysamia cecropia*)	Chrysalis	1	Flour beetle	–	4-5
Small cabbage white butterfly	Chrysalis	3-6	Meal worm	Larva	1
			Boll weevil	Imago	3-7

* Breeds throught the year (Source: C. H. Schmidt, *et al.*, 1972)

Trends in the Mean Height, Body Weight (Japanese)

Era / Age	1970		1980		1990		2000		2010		2017		2018	
	cm	kg	cm	kg	cm	kg	cm	kg	cm	kg	cm	kg	cm	kg
Male														
1	79.9	10.8	79.9	10.8	80.0	10.9	80.8	11.2	-	-	78.6	10.5	79.7	10.4
2	89.1	12.9	89.2	12.9	89.7	13.1	90.1	13.0	-	-	89.4	12.8	88.6	12.6
3	95.8	14.6	96.7	14.9	96.5	14.8	96.3	14.9	-	-	95.8	14.5	96.0	14.2
4	103.3	16.5	102.4	16.4	104.0	16.5	103.4	16.1	-	-	102.3	15.9	101.2	15.4
5	108.0	18.0	110.1	18.8	110.3	19.2	110.9	19.4	-	-	108.3	18.0	110.0	18.2
6	113.9	19.9	115.3	21.1	115.9	21.2	117.6	22.4	117.1	20.9	116.9	21.2	115.6	20.4
7	119.7	22.5	120.6	23.3	122.5	24.3	121.8	24.2	120.8	23.6	121.3	24.1	122.0	24.2
8	124.7	25.0	126.8	26.6	127.7	26.8	128.5	28.0	129.4	27.2	127.7	27.0	127.8	26.6
9	129.6	27.6	131.5	28.8	132.3	29.4	132.5	30.6	132.7	30.8	134.9	29.4	131.8	29.1
10	134.7	30.4	136.8	32.4	138.6	33.8	138.7	33.5	138.5	32.9	137.4	34.2	138.4	33.8
11	139.7	39.5	142.8	36.7	144.2	38.3	143.6	37.8	143.8	36.2	144.4	38.9	145.7	38.8
12	146.9	38.5	149.4	40.9	152.1	43.3	150.0	42.6	151.9	44.5	150.8	41.5	153.1	43.4
13	153.1	43.8	156.7	46.5	157.5	48.3	159.7	49.3	159.6	46.9	158.0	45.6	160.3	50.4
14	159.3	49.2	162.7	52.2	163.1	52.3	165.1	56.2	164.4	52.6	165.4	53.1	165.2	51.0
15	163.6	53.1	166.9	57.3	167.7	58.5	166.7	56.0	169.0	56.7	168.4	57.9	168.0	57.9
16	165.9	57.1	167.3	58.0	168.0	59.5	169.4	59.9	168.5	57.7	170.4	61.0	173.9	62.6
17	166.7	57.8	169.7	61.6	170.9	62.2	170.9	59.1	170.9	60.9	171.9	61.3	169.2	57.3
18	167.8	58.6	169.3	59.2	170.5	62.7	170.9	62.4	172.5	62.5	168.9	63.6	170.0	61.1
19	167.5	58.0	169.5	60.6	170.2	61.9	171.6	61.1	169.0	59.4	170.7	63.0	174.0	64.3
20	165.9	58.1	170.4	61.0	171.3	63.7	170.4	65.1	171.0	61.8	171.7	64.4	169.1	61.4
21	166.4	59.1	169.5	60.1	170.9	63.8	173.1	65.3	170.2	60.7	171.9	67.3	172.5	65.1
22	166.5	58.8	170.0	60.8	170.9	64.4	170.4	61.9	170.1	69.5	173.7	67.0	172.6	64.0
23	165.9	58.4	169.2	61.0	170.4	62.3	171.2	64.9	171.9	65.7	171.5	73.3	169.4	62.8
24	166.3	58.9	167.1	59.0	171.6	63.8	171.0	63.0	171.5	62.9	171.8	65.0	171.5	63.9
25	165.0	58.3	167.6	60.6	170.8	64.3	170.7	67.0	169.6	63.5	170.9	67.6	173.5	70.3
26–29	165.2	59.5	167.2	61.2	170.4	66.0	171.0	66.8	170.4	66.5	171.0	69.5	171.7	66.0
30–39	163.5	59.6	166.1	62.4	168.7	65.3	170.6	68.2	171.5	69.6	171.2	71.0	172.1	71.0
40–49	162.0	58.6	163.2	61.5	166.4	64.7	168.8	67.2	170.6	70.4	171.2	71.3	171.3	71.1
50–59	159.9	56.7	161.1	58.7	162.8	61.7	165.5	64.6	168.0	68.2	170.2	69.2	170.8	70.4
60–69	157.9	54.2	159.0	55.8	161.0	58.7	163.0	62.5	165.3	64.6	167.3	67.2	167.2	67.1
70–	155.7	51.0	156.9	52.6	158.3	55.2	159.5	57.5	160.9	59.9	162.5	61.7	162.7	62.7
Female														
1	78.7	10.3	79.5	10.5	78.3	10.1	78.9	10.1	-	-	76.8	9.9	77.2	10.2
2	87.6	12.2	88.2	12.5	89.0	12.5	88.3	12.4	-	-	87.9	12.2	87.5	12.3
3	94.4	14.0	95.2	14.2	95.2	14.2	96.7	14.3	-	-	93.8	13.9	96.1	14.6
4	102.1	16.1	102.2	16.4	102.9	16.6	102.8	16.4	-	-	102.8	16.0	102.3	15.9
5	107.0	17.5	108.2	17.8	108.5	18.2	109.1	18.5	-	-	109.1	17.9	109.3	18.1
6	113.0	19.7	114.5	20.3	115.4	20.9	115.0	20.8	115.1	20.1	115.3	20.8	114.8	20.3
7	118.3	21.9	120.2	22.6	121.3	23.8	121.2	23.2	121.8	23.5	122.0	23.3	119.6	22.3
8	123.9	24.3	125.4	25.3	126.8	26.6	126.1	26.1	128.4	26.5	126.7	24.7	125.7	26.0
9	129.6	27.2	131.4	28.3	131.5	28.5	133.6	30.9	135.2	30.8	134.2	29.8	134.4	29.7
10	135.7	30.8	137.8	32.7	139.7	34.1	138.9	34.3	139.2	33.1	138.3	31.7	140.6	33.7
11	142.1	35.4	144.5	36.6	145.2	37.6	146.0	38.5	145.3	36.8	148.1	39.4	146.6	37.1
12	147.7	40.3	149.3	41.0	152.2	43.9	153.0	45.3	150.8	40.8	149.5	41.8	150.1	41.9
13	151.8	44.9	154.0	46.1	154.5	45.9	153.9	46.3	155.7	46.2	154.7	45.6	154.6	47.2
14	153.9	48.3	154.8	48.8	156.8	48.7	156.9	49.0	156.1	48.4	157.4	49.0	154.9	48.7
15	154.3	49.2	156.5	51.0	157.4	50.0	157.1	50.2	156.6	51.0	157.2	51.6	158.2	49.5
16	155.2	51.2	156.7	52.0	157.1	52.2	158.8	52.2	158.0	51.0	157.7	50.2	156.6	49.9
17	155.2	51.6	156.6	51.5	157.8	52.3	157.4	50.8	158.0	51.9	159.1	51.6	154.8	47.2
18	154.6	51.0	155.7	50.9	157.5	52.0	157.7	50.7	157.7	56.3	155.5	51.0	157.4	50.1
19	154.4	51.3	157.1	51.7	157.9	50.6	157.7	49.2	159.2	55.2	158.0	52.2	156.6	51.2
20	154.5	51.0	155.5	50.7	157.1	51.7	158.2	52.5	159.5	55.0	154.9	49.9	157.0	50.7
21	154.4	50.8	155.9	51.8	157.9	51.3	156.8	48.7	156.7	49.4	158.9	52.4	157.1	52.2
22	153.8	50.0	155.9	50.8	156.8	51.4	158.0	51.1	157.2	50.3	158.3	50.2	158.8	52.3
23	153.9	49.9	154.8	51.5	157.5	50.8	157.9	50.8	154.8	48.4	158.3	50.7	159.3	54.6
24	154.0	50.9	155.4	50.8	158.4	50.4	157.8	50.6	159.1	49.8	157.5	50.0	157.5	50.9
25	153.3	49.7	155.2	52.6	157.8	52.3	157.5	49.1	157.7	50.2	157.9	50.3	160.1	53.3
26–29	152.9	50.3	154.3	51.0	157.2	51.9	158.1	51.8	158.7	51.8	157.7	52.5	159.4	54.0
30–39	151.9	51.5	153.4	51.9	155.7	52.7	157.6	53.3	158.3	54.0	158.6	54.4	158.3	53.4
40–49	150.6	52.2	151.8	53.5	153.4	53.6	155.8	55.0	157.8	54.7	158.2	55.8	158.7	55.8
50–59	148.4	50.8	149.5	52.0	151.4	53.7	153.2	54.0	155.2	54.1	156.7	55.4	157.0	55.2
60–69	145.6	48.0	147.3	49.8	148.6	51.9	150.6	53.6	151.8	53.4	153.9	54.7	153.9	54.2
70–	141.9	44.1	142.6	45.8	144.3	47.3	146.1	49.3	147.7	50.4	148.9	51.0	149.0	51.3

Source: National Institute of Health and Nutrition, The National Health and Nutrition Survey in Japan, 2017, The data for 1970: Ministry of Health and Welfare, Syowa 61 kokumineiyo no genjo.

Lifespans of Vertebrates

Common name (*Species*)	Longevity	Common name (*Species*)	Longevity	Common name (*Species*)	Longevity
Human (*Homo sapiens*)	122.5	(*Rhinolophus ferrumequinum*)	30.5	Red junglefowl (chicken) (*Gallus gallus*)	30
Chimpanzee (*Pan troglodytes*)	59.4	Eastern European hedgehog		Barn swallow (*Hirundo rustica*)	16
Orangutan (*Pongo pygmaeus*)	59	(*Erinaceus concolor*)	7	House sparrow (*Passer domesticus*)	23
Gorilla (*Gorilla gorilla*)	60.1	Domestic cat (*Felis catus*)	30	King penguin (*Aptenodytes patagonicus*)	26
Japanese macaque (*Macaca fuscata*)	38.5	Lion (*Panthera leo*)	27	American alligator	
Tree shrew (*Tupaia glis*)	12.4	Tiger (*Panthera tigris*)	26.3	(*Alligator mississippiensis*)	73.1
Old world rabbit (*Oryctolagus cuniculus*)	9	Giant panda (*Ailuropoda melanoleuca*)	36.8	Galapagos tortoise (*Geochelone nigra*)	177
House mouse (*Mus musculus*)	4	Lesser panda (*Ailurus fulgens*)	19	Tuatara (*Sphenodon punctatus*)	90
Norway rat (*Rattus norvegicus*)	3.8	Polar bear (*Ursus maritimus*)	43.8	Olm (*Proteus anguinus*)	102
Naked mole-rat (*Heterocephalus glaber*)	31	Brown bear and grizzly bear (*Ursus arctos*)	40	Axolotl (*Ambystoma mexicanum*)	17
Golden hamster (*Mesocricetus auratus*)	3.9	Domestic dog (*Canis familiaris*)	24	Japanese firebelly newt (*Cynops pyrrhogaster*)	25
Guinea pig (*Cavia porcellus*)	12	Gray wolf (*Canis lupus*)	20.6	Oriental firebelly toad (*Bombina orientalis*)	15.8
Domestic cattle (*Bos taurus*)	20	Red fox (*Vulpes vulpes*)	21.3	Common European toad (*Bufo bufo*)	40
Wild boar (*Sus scrofa*)	27	Giant anteater (*Myrmecophaga tridactyla*)	31	Tokay gecko (*Gekko gecko*)	23.5
Domestic goat (*Capra hircus*)	20.8	African elephant (*Loxodonta africana*)	65	Lake sturgeon (*Acipenser fulvescens*)	152
Domestic sheep or mouflon (*Ovis aries*)	22.8	Dugong (*Dugong dugon*)	73	European eel (*Anguilla anguilla*)	88
Giraffe (*Giraffa camelopardalis*)	39.5	Red kangaroo (*Macropus rufus*)	25	Rougheye rockfish (*Sebastes aleutianus*)	205
Horse (*Equus caballus*)	57	Virginia opossum (*Didelphis virginiana*)	6.6	Common carp (*Cyprinus carpio*)	70-100*
Sika deer (*Cervus nippon*)	26.3	Duck-billed platypus		Whale shark (*Rhincodon typus*)	54
Japanese mole (*Mogera wogura*)	3.2	(*Ornithorhynchus anatinus*)	22.6	Chum salmon (*Oncorhynchus keta*)	7
Bowhead whale (*Balaena mysticetus*)	211	Cockatiel (*Nymphicus hollandicus*)	50	Sea lamprey (*Petromyzon marinus*)	9
Killer whale (*Orcinus orca*)	90	Mew gull (*Larus canus*)	33.7	Bigmouth buffalo (*Ictiobus cyprinellus*)	112
Blue whale (*Balaenoptera musculus*)	110	Ural owl (*Strix uralensis*)	23.8	Greenland Shark (*Somniosus microcephalus*)	512
Greater horseshoe bat		Rock dove (*Columba livia*)	35		

* Hanako, the world's oldest koi carp, who died in 1977 at the age of 226 years (L. Barton, the Guardian, 2007 Apr 12., All Japan Nishikigoi Promotion Association; http://jnpa.info/zen/hanako.htm).
References: The Animal Ageing and Longevity Database (http://genomics.senescence.info/species/), "Suuchi de miru seibutsugaku", Maruzen publishing, 2012.

Lifespan of Invertebrates

Common name (*Species*)	Longevity	Common name (*Species*)	Longevity	Common name (*Species*)	Longevity
Lancelet/amphioxus		Silk warm (*Bombyx mori*)	1-2 mo	Cuttlefish (*Sepia officinalis*)	5 y
(*Branchiostoma belcheri*)	3 y	House fly (*Musca domestica*)	76 d	Ikechougai (*Hyriopsis schlegelii*)	100 y
Red sea urchin	200 y	Body lice (*Pediculus humanus*)	30 d	Escargot (*Helix pomatia*)	>18 y
(*Mesocentrotus franciscanus*)		Silk spider (*Lycosa tarantula*)	15 y	Ocean quahog clam	
Starfish (*Acanthaster planci*)	5 y	Emperor scorpion		(*Arctica islandica*)	507 y
Fruit fly		(*Pandinus imperator*)	5-8 y	Oyster (*Ostrea edulis* (?))	12 y
(*Drosophila melanogaster*)	0.3 y	American lobster		Planaria (*Dugesia* sp.)	14 mo
Honey bee (*Apis mellifera*)		(*Homarus americanus*)	100 y	Tapeworm (*Cestoda* sp.)	35 y
(Queen)	8 y	Horseshoe crab	>20 y	Roundworm	
(Worker)	0.2-0.4 y	(*Limulus polyphemus*)	(20-40 y)	(*Caenorhabditis elegans*)	10 d
Termites		Lithobius (*Lithobius* sp.)	5-6 y	Ascarid (*Ascaris* sp.)	5 y
(*Coptotermes formosanus*)	25 y	Earthworm (*Lumbricina* sp.)	10 y	Paramecium	
Squinting bush brown		Leech (*Hirudinea* sp.)	27 y	(*Paramecium caudatum*)	7-8 mo
(*Bicyclus anynana*)	0.5 y	Lamellibranchia tube worm			
Earwig (*Anisolabis maritima*)	7 y	(*Lamellibrachia luymesi*)	250 y		

d: days, mo: months, y: years
References: AnAge; The Animal Ageing and Longevity Database, Suuchi de miru seibutsugaku, Maruzen publishing (2012), Centers for Disease Control and Prevention HP, University of Michigan MUSEUM OF ZOOLOGY Animal Diversity Web, National Wildlife Federation HP, Boetius A, *PLoS Biol* **3** (3): e102 (2005), Nature Methods, **10**, 665-670 (2013), The 85th annual Meeting of The Zoological Society of Japan: Abstract, etc.

Lifespans of Laboratory Animals (Comparison between Ad Libitum and Restricted Feeding)

Species		Averaged life span of after restricted feeding (Maximum lifespan)		
Common name (*Species*)	Strain[*1]	None (Free feeding)	middle (25% reduction)	high (40–50% reduction)
Ciliates Tokophyra (*Tokophyra infusionum*)		7 days	10 days	13 days
Rotifers (*Philodina acticornis*)		35 days	45 days	55 days
Nematode (*Caenorhabditis elegans*)		15–16 days	-	26 days
Daphnia (*Daphnia longispina*)		30 (46) days	-	40 (48) days
Drosophila (*Drosophila melanogaster*)		17–19 days	-	29 days
Silkworm (*Bombyx mori*)			effective	
Bowl and doily spider (*Frontinella communis*)		42.3 days	63.9 days	81.3 days
Common European Limpet (*Patella vulgata*)		2.5 years	-	16 years
Guppy (♀) (*Poecilia reticulata*)		600 days	-	800 days
Mouse[*2] (*Mus musculus*)	A/J	21 (27) weeks	-	23 (30) weeks
	C57BL/6	21 (27) weeks	-	24 (32) weeks
	C57BL/6 (Start 6 weeks)	26 (31) weeks	-	29 (37) weeks
	C57BL/6 (Start 25 weeks)	29 (-) weeks	-	29 (-) weeks
	(A/J x C57BL) F1	24 (32) weeks	-	28 (37) weeks
	C3H	15 (17) weeks	-	23 (31) weeks
	NZB	15 (20) weeks	-	17 (25) weeks
Rat[*2] (*Rattus norvegicus*)	Wistar	25 (35) weeks	-	30 (38) weeks
	Wistar (Start 12 weeks)	31 (-) weeks	-	35 (-) weeks
	Sprague-Dawley	23 (33) weeks	-	36 (43) weeks
	Fischer 344	23 (29) weeks	-	32 (43) weeks
	Holtzman	20 (24) weeks	-	24 (28) weeks
	Long Evans (Start 6 weeks)	30–33 (40) weeks	-	37 (43) weeks
Hamster (Start 12 weeks) (*Mesocricetus auratus* (?))		23 (28) weeks	-	30 (33) weeks
Dog (Start 6 month) (*Canis lupus familiaris*)	Labrador retrievers	11.2 years	13 years	-

Note: As regard to the effects of dietary restriction on primates, the research groups of University of Wisconsin and National Institute of Aging published their longitude results using rhesus monkeys (Macaca mulatta), but the results of the parallel studies were controversial. However, the direct comparison between both data confirmed that dietary restricion is also effective on rhesus monkeys (Reference: J. A. Mattison, *et al.*, Nature Commun. 8: 14063, 2017).

*1 Strains without a remark; Food restriction (or Restricted feeding) was started of ten weaning (3 weeks of ten birth).

*2 Since diatory restriction protocols have typically increased median lifespan by 14–45% in rats (53 reports) and by 4–27% in mice (72 reports) from 1934–2012 with significant funnel plot asymmetry in rats (WR. Swindell, Ageing Res. Rev. 11: 254–270, 2012.), only one representative was selected here for each species or strain.

Sources: J. A. Mattison, *et al.* (2017), WR. Swindell (2012), R. Weindruch & RL. Walford (1988), A. I. Lansing (1947), SN. Austad (1989), DF. Lawler, *et al.* (2008).

Trends in Population, Live Births and Deaths by Sex (Japanese)

Year[1)]	Male			Female		
	Population[2)]	Live births	Deaths	Population	Live births	Deaths
	×1000	×100	×100	×1000	×100	×100
1900	22 051	72 8	46 4	21 796	69 3	44 7
1905	23 421	73 6	50 5	23 199	71 7	49 9
1910	24 650	87 3	53 5	24 534	84 0	52 9
1915	26 465	91 8	55 6	26 287	88 1	53 8
1920	28 044	103 5	72 1	27 919	99 0	70 1
1925	30 013	106 1	62 1	29 724	102 5	58 9
1930	32 390	107 0	60 4	32 060	101 6	56 7
1935	34 734	112 3	60 4	34 520	106 8	55 8
1940	35 387	108 4	61 5	36 546	103 2	57 1
1945	33 894	–	–	38 104	–	–
1950	40 812	120 3	46 7	42 388	113 4	43 8
1955	43 861	89 0	36 5	45 415	84 1	32 8
1960	45 878	82 5	37 8	47 541	78 1	32 9
1965	48 244	93 5	37 9	50 031	88 8	32 2
1970	50 601	100 0	38 8	52 519	93 4	32 5
1975	54 725	97 9	37 8	56 527	92 2	32 4
1980	57 201	81 1	39 1	59 119	76 5	33 2
1981	57 654	78 7	38 9	59 551	74 3	33 2
1982	58 053	77 8	38 5	59 955	73 8	32 6
1983	58 435	77 5	40 1	60 352	73 3	33 9
1984	58 793	76 5	40 2	60 730	72 5	33 8
1985	59 044	73 5	40 8	61 222	69 6	34 5
1986	59 438	71 1	40 7	61 508	67 2	34 4
1987	59 723	69 2	40 8	61 811	65 4	34 3
1988	59 964	67 5	42 8	62 062	63 9	36 5
1989	60 171	64 1	42 7	62 289	60 6	36 1
1990	60 249	62 7	44 4	62 472	59 5	37 7
1991	60 425	62 9	45 0	62 677	59 5	37 9
1992	60 597	62 2	46 6	62 879	58 7	39 1
1993	60 730	61 0	47 6	63 057	57 8	40 2
1994	60 839	63 6	47 6	63 230	60 2	40 0
1995	60 919	60 9	50 1	63 380	57 9	42 1
1996	61 115	62 0	48 9	63 594	58 7	40 8
1997	61 210	61 1	49 8	63 753	58 1	41 6
1998	61 311	61 7	51 2	63 941	58 6	42 4
1999	61 358	60 5	53 5	64 074	57 3	44 7
2000	61 488	61 2	52 6	64 125	57 8	43 6
2001	61 595	60 1	52 9	64 313	57 0	44 2
2002	61 591	59 3	53 5	64 417	56 1	44 7
2003	61 620	57 7	55 2	64 520	54 7	46 3
2004	61 597	57 0	55 7	64 579	54 1	47 2
2005	61 618	54 5	58 5	64 587	51 7	49 9
2006	61 568	56 0	58 1	64 586	53 2	50 3
2007	61 511	56 0	59 3	64 574	53 0	51 6
2008	61 424	56 0	60 9	64 523	53 2	53 4
2009	61 339	54 9	60 9	64 481	52 1	53 3
2010	61 571	55 1	63 4	64 810	52 1	56 3
2011	61 453	53 8	65 7	64 727	51 3	59 7
2012	61 328	53 2	65 6	64 630	50 5	60 1
2013	61 186	52 8	65 9	64 518	50 2	61 0
2014	61 041	51 6	66 0	64 391	48 8	61 3
2015	61 023	51 5	66 7	64 297	49 0	62 4
2016	60 867	50 2	67 5	64 153	47 5	63 3
2017	60 676	48 4	69 1	63 973	46 2	65 0
2018	60 455	47 1	70 0	63 763	44 8	66 3

Source: Ministry of Health, Labour and Welfare, Vital Statistics.

Note: 1) Excluding Okinawa prefecture for 1947–1972.
2) Figures for 1900–1919 exclude foreigners.
Figures for 1920–1966 include foreigners.
Figures for 1967–2016 exclude foreigners.

Trends in Population by Sex and Age Groups (Japanese)

Age \ Year	1960	1970	1980	1990	2000	2005	2010	2015	2016	2017	2018
Male											
0–4	4 013	4 483	4 337	3 317	3 002	2 841	2 689	2 528	2 504	2 475	2 439
5–9	4 702	4 141	5 109	3 810	3 066	3 024	2 842	2 699	2 689	2 659	2 621
10–14	5 620	3 976	4 564	4 358	3 335	3 071	3 014	2 855	2 799	2 756	2 733
15–19	4 678	4 538	4 195	5 108	3 809	3 355	3 096	3 074	3 058	3 030	2 980
20–24	4 125	5 280	3 932	4 438	4 255	3 689	3 228	3 015	3 026	3 039	3 063
25–29	4 095	4 491	4 513	4 036	4 894	4 119	3 643	3 210	3 127	3 063	3 011
30–34	3 747	4 159	5 388	3 892	4 366	4 866	4 180	3 653	3 579	3 503	3 408
35–39	2 763	4 103	4 569	4 500	4 035	4 347	4 927	4 191	4 033	3 907	3 805
40–44	2 274	3 647	4 138	5 333	3 883	4 021	4 382	4 922	4 854	4 716	4 538
45–49	2 257	2 657	4 017	4 472	4 436	3 838	4 015	4 365	4 626	4 716	4 821
50–54	2 041	2 140	3 531	3 990	5 186	4 362	3 807	3 982	3 919	4 048	4 151
55–59	1 802	2 029	2 494	3 782	4 275	5 065	4 297	3 750	3 724	3 749	3 779
60–64	1 438	1 746	1 933	3 234	3 740	4 149	4 937	4 181	3 990	3 818	3 715
65–69	1 027	1 393	1 734	2 189	3 353	3 543	3 934	4 699	4 947	4 773	4 507
70–74	694	958	1 312	1 557	2 667	3 041	3 235	3 609	3 436	3 611	3 851
75–79	377	531	846	1 197	1 621	2 257	2 593	2 807	2 895	2 997	3 089
80–84	169	241	417	678	913	1 221	1 700	2 010	2 090	2 150	2 187
85–89	48	71	138	276	477	555	747	1 065	1 117	1 171	1 222
90–94	8	17	33	81	176	211	243	336	377	416	451
95–99	–	–	–	–	–	41	56	63	67	70	76
100–	–	–	–	–	–	4	6	8	9	9	9
Female											
0–4	3 832	4 264	4 121	3 152	2 858	2 706	2 565	2 415	2 390	2 361	2 323
5–9	4 502	3 959	4 858	3 627	2 919	2 875	2 708	2 569	2 559	2 532	2 499
10–14	5 397	3 823	4 336	4 138	3 172	2 920	2 870	2 718	2 667	2 627	2 606
15–19	4 631	4 460	4 020	4 860	3 625	3 169	2 932	2 904	2 893	2 867	2 823
20–24	4 193	5 315	3 852	4 284	4 045	3 504	3 076	2 869	2 872	2 882	2 900
25–29	4 115	4 547	4 464	3 940	4 732	3 979	3 512	3 083	3 002	2 936	2 882
30–34	3 771	4 169	5 320	3 821	4 243	4 726	4 034	3 532	3 456	3 379	3 283
35–39	3 275	4 068	4 582	4 446	3 943	4 246	4 761	4 047	3 896	3 776	3 678
40–44	2 745	3 658	4 158	5 284	3 823	3 948	4 269	4 764	4 696	4 564	4 388
45–49	2 560	3 183	4 041	4 518	4 409	3 813	3 951	4 254	4 502	4 583	4 685
50–54	2 161	2 637	3 639	4 078	5 205	4 382	3 801	3 927	3 863	3 982	4 078
55–59	1 839	2 373	3 088	3 932	4 424	5 159	4 360	3 770	3 737	3 754	3 775
60–64	1 494	1 964	2 510	3 512	3 972	4 378	5 118	4 308	4 105	3 919	3 807
65–69	1 133	1 580	2 213	2 902	3 739	3 880	4 296	5 011	5 277	5 095	4 808
70–74	870	1 169	1 700	2 253	3 223	3 594	3 752	4 143	3 937	4 100	4 344
75–79	578	735	1 185	1 818	2 518	3 004	3 379	3 523	3 605	3 714	3 815
80–84	314	408	674	1 153	1 696	2 188	2 663	3 002	3 076	3 127	3 143
85–89	108	158	271	557	1 054	1 294	1 699	2 084	2 150	2 217	2 283
90–94	24	48	86	208	524	630	783	1 024	1 098	1 163	1 221
95–99	–	–	–	–	–	170	242	297	316	334	362
100–	–	–	–	–	–	22	38	53	57	59	60

Unit: 1000 people Source: Ministry of Health, Labour and Welfare, Vital Statistics.

The values listed in 90–94 in 2000 and before are actually the values for ≥90.

Trends in Deaths by Sex and Age Groups (Japanese)

Age \ Year	1950	1960	1970	1975	1980	1985	1990	1995	2000	2005	2010	2015	2016	2017	2018
Male															
0-4	1183	362	191	149	94	60	45	39	29	23	19	15	14	13	13
5-9	106	48	24	20	17	12	8	8	4	4	3	3	2	2	2
10-14	50	33	17	13	10	10	8	7	5	4	4	3	3	3	3
15-19	107	62	50	35	30	32	32	24	17	12	9	8	8	8	7
20-24	187	88	68	49	34	34	35	36	29	23	20	15	15	15	14
25-29	159	94	65	56	41	32	29	32	33	29	24	18	17	15	15
30-34	125	88	72	61	55	42	33	33	37	39	32	23	22	22	21
35-39	141	81	103	81	73	71	55	44	46	49	49	35	33	31	30
40-44	157	93	128	130	105	102	98	82	68	68	66	62	58	55	52
45-49	191	142	133	167	176	151	142	156	131	106	96	87	89	89	88
50-54	233	208	170	170	223	243	202	219	241	195	146	128	125	123	126
55-59	287	302	267	220	230	307	329	305	318	342	271	195	191	185	183
60-64	350	382	381	331	292	309	427	472	422	434	462	361	335	311	300
65-69	410	441	521	461	439	382	427	598	610	553	575	614	651	642	599
70-74	424	484	583	577	572	551	517	609	764	802	735	769	725	743	778
75-79	306	428	521	567	639	656	693	685	739	993	1027	970	971	996	1012
80-84	170	294	364	414	511	591	679	779	735	895	1198	1268	1280	1299	1296
85-89	63	117	166	207	264	347	456	565	627	701	899	1208	1245	1298	1334
90-94	17	30	52	67	98	140	218	250	308	426	492	646	708	784	838
95-99	-	-	-	-	-	-	-	57	76	128	178	199	213	228	238
100-	-	-	-	-	-	-	-	6	10	17	29	37	40	41	40
Female															
0-4	1046	285	138	110	69	48	35	31	23	18	15	12	13	12	11
5-9	92	34	14	12	10	6	5	5	3	2	2	2	2	1	2
10-14	52	23	10	8	6	6	5	5	3	2	2	2	2	2	2
15-19	106	37	20	13	11	10	11	9	7	6	5	4	4	4	4
20-24	172	57	36	25	14	13	13	14	12	11	8	6	6	6	6
25-29	170	64	39	33	22	16	14	14	15	13	10	8	8	7	7
30-34	139	68	43	37	32	25	18	18	18	20	17	12	11	11	10
35-39	140	74	58	47	42	40	31	24	24	26	27	19	19	17	17
40-44	137	81	76	68	56	57	55	46	36	34	35	36	34	33	31
45-49	148	115	99	93	85	76	75	85	66	52	50	49	50	51	52
50-54	171	143	126	124	118	115	101	110	117	94	74	69	70	67	69
55-59	197	185	178	153	150	148	146	142	141	153	122	93	93	90	91
60-64	263	237	239	222	197	200	200	211	185	189	199	161	148	138	133
65-69	348	302	331	305	298	265	273	293	281	256	256	269	284	282	264
70-74	417	409	439	432	422	409	381	415	401	406	368	374	353	349	367
75-79	364	476	494	531	562	557	582	569	571	600	604	565	559	562	571
80-84	257	414	471	511	592	644	716	799	735	847	915	957	958	963	955
85-89	119	212	304	349	409	516	655	779	863	953	1174	1354	1360	1403	1433
90-94	40	76	138	171	226	293	449	473	601	850	1028	1326	1386	1450	1509
95-99	-	-	-	-	-	-	-	142	216	377	575	708	747	812	854
100-	-	-	-	-	-	-	-	21	38	78	147	211	224	237	243

Unit: 100 people. Source: Ministry of Health, Labour and Welfare, Vital Statistics.
1950–1970 Excluding Okinawa Prefecture
The values listed in 90–94 in 1950–1990 are actually the values for ≥90.

Cell, Tissue, Organ

Fine Structure of Eukaryotic Cell

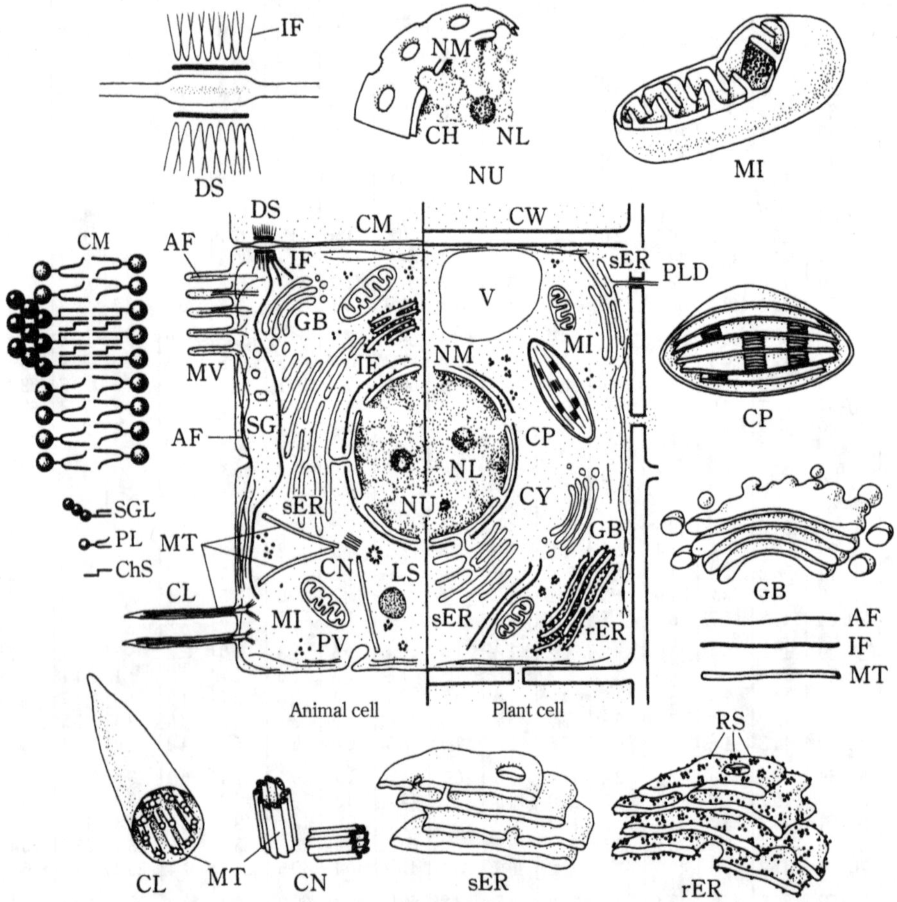

AF: actin filament, CH: chromatin, ChS: cholesterol, CL: cilium, CM: cell membrane, CN: central body, CP: chloroplast, CW: cell wall, CY: cytoplasm, DS: desmosome, GB: Golgi body, IF: intermediate filament, LS: lysosome, MI: mitochondrion, MT: microtubule, MV: microvillus, NL: nucleolus, NM: nuclear membrane, NU: nucleus, PL: phospholipid, PLD: plasmodesmata, PV: pinosome, rER: rough endoplamic reticulum, RS: ribosome, sER: smooth endoplamic reticulum, SG: secretion granule, SGL: sphingoglycolipid, V: vacuole

Structural Elements of Cells and their Functions

Elements	Size	Function
Cell membrane	7.5 nm (thickness)	Trasportation of material between the inside and outside, stimulant reception
Cytoplasm		
Cytosol		Intermediary metabolism
Cytoskeleton		
Microtubule	25 nm (diameter)	Formation of flagella and cilia, cell division, formation of the cell wall
Intermediate filament	10 nm (diameter)	Reticulation structure in cell and intercellular junction
Microfilament	7 nm (diameter)	Contractile apparatus and cell movement
Organelle		
Chloroplast	4.0–8.0 μm (diameter)	Photosynthesis, lipid synthesis
Peroxisome	0.5–1.5 μm (diameter)	Metabolizing glycolic acid and degradation of peroxide
Mitochondrion	0.5–0.8 μm (diameter)	Aerobic respiration, formation of ATP
Golgi body	0.2–5.5 μm (length)	Formation of cell plate and cell wall, formation of secretory organelles
Endoplasmic reticulum (ER)	4–7 nm (thickness)	Product transportation, lipid metabolism
Lysosome	0.4 μm (diameter)	Digestion of waste in cell
Ribosome	15–20 nm (diameter)	Protein synthesis
Nucleus	5.0–25 μm (diameter)	DNA synthesis (genetic code, replication)
Nucleolus	1–4 μm (diameter)	Synthesis of rRNA

Comparison of Structure for Prokaryotic Cells and Eukaryotic Cells

Feature	Prokaryotic cell	Eukaryotic cell
Species	Bacteria and archaea	Protist, fungi, plant, and animal
Cell size	Normally diameter ≤ 2 μm	Normally diamter 2 μm ≥ d ≥ 100 μm
Nuclear membrane	Absent	Present
Cell membrane	Sterol absent or irregular	Sterol present
Cell wall	Peptidoglycan (genuine bacillus) Protein (archaea)	Cellulose (algae and plant) Chitin (fungi)
Intracellular membrane	Mesosome Thylakoid (cyanobacteria)	Endoplasmic reticulum, Golgi body, lysosome, micro body, vacuole (plant)
Ribosome	Subsidence coefficient 70S, molecular weight 2.7×10^6 D	Subsidence coefficient 80S, molecular weight 4×10^6 D
DNA	Mostly circular Does not bind to histone	Linear Binds to histone
Extranuclear DNA	Plasmid	Mitochondria and chloroplast
Nucleoli	Absent	Present
Flagellum	Composed of 1 fiber consisting of flagellion	Composed of microtubules consisting of tubulin*
Other fibers	Pilus composed of pilin (not moveable)	Cilia composed of microtubule (moveable)
Cytoskeleton	Absent	Present
Cell division	Amitosis	Mitosis
Centriole	Absent	Present*
Protoplasmic streaming	Absent	Present
Phagocytosis, exocytosis	Absent	Present

* Most spermatophyta lack flagella, cilia, and centrioles.

Source: K. Ishihara *et al.*, ed, "Seibutsugaku data daihyakkajiten (jo)" Asakura Publishing Co., Ltd. (2002).

Classification of Organelles by Membrane

Kind of organelle	Organelle
Single membrane system organelle	Golgi body, endoplasmic reticulum, lysosome, peroxisome, vacuole, and nucleus*
Double membrane system cell organelles	Mitochondrion and plastid (chloroplast, white body, colored plastid)

* Nuclear membrane is a double membrane, but it is one where a single film forms a saccate structure.

Types of Animal Tissue Cells

Tissue cell		Cell type
Epithelium tissue cell	By configuration, form	Cells of squamous, cuboidal, stratified, ciliated, and transitional epithelia
	By function	Protective, absorptive, glandular, neural, germinal. Corneal epithelial cell
Endothelial cell		Single-layer flat epithelial cell of vertebrate blood vessel, lymphatic vessel, and heart lumen wall
Cells of connective and supportive tissues		Fibroblast, mast cell, histiocyte, mesenchymal cell, reticular cell, testicular interstitial cell (Leydig cell), ovarian interstitial cell, fat globules, chromatophore, chondrocyte, osteocyte, smooth muscle cell, striated muscle cell, cardiac muscle cell
Nerve tissue cell		Neuron (all nerve types: nonpolar, unipolar, bipolar, pseudo-unipolar, multipolar neurons), glial cell
Blood cell		Neutrophilic leukocyte, basophilic leukocyte, eosinophilic leukocyte, monocyte, lymphocyte, red blood corpuscle, blood plasma cell, platelet
Stem cell		Epidermal basal cell, germinal stem cell, myeloblasts, intestinal epithelial stem cell, reticular cell, neural stem cell

Types and Differentiation Potency of Stem Cells (Animals)

Type	Stem cells and their progenitors	Differentiation potential
Stem cells found in living bodies	Umbilical cord stem cells, marrow stem cells, connective tissue (mesenchyme system) stem cells	Very broad potential
	Stem cells within organs (small intestine, skin, muscle, bone, liver, nerve, gonad)	Limited potential
Artificially induced stem cells	Embryonic stem (ES) cells (obtained by culturing mammalian blastocyst inner cell mass)	Able to differentiate into almost any kind of cell
	Induced pluripotent stem (iPS) cells (obtained by indroducing multiple genes into somatic cells)	Able to differentiate into almost any kind of cell
	Embryonic germ (EG) cells (obtained by culturing primordial germ cells)	Able to differentiate into almost any kind of cell

Types of Tissues and Cells based on Form (Plants)

Tissue system	Tissues and cells
Epidermal system	Epidermal cells, guard cells, trichome (hair, scale, papilla, root hair, glandular hair, hydathodal hair)
Fundamental tissue system	Parenchyma (palisadetissue, spongy parenchyma), Collenchyma, sclerenchyma (Sclereids, fiber)
Vascular system	Vessel element, vessel, tracheid, xylem fiber, xylem parenchyma, phloem element, companion cell, phloem fiber, phloem parenchyma, bundle sheath

Types of Organs, Tissues and Cells based on Function (Plants)

Structure	Organ, organization, and cell kind
Growth and differentiation structure	Apical meristem (shoot apical meristem, root apical meristem), lateral meristem (cambium, cork cambium), intercalary meristem
Protecive structure	Periderm, trichome, bud scale, stipule, leaf needle, testa, epidermal cell
Supportive structure	Collenchyma, sclerenchyma (fiber etc.), vessel element, tracheid
Absorptive structure	Parasitic root, absorptive hair, root hair
Assimilative structure	Mesophyll tissue (palisade, spongy tissue)
Conductive structure	Vessel element, vessel, tracheid, sieve tube, companion cell, sieve element
Storage structure	Storage parenchyma (potato tuber, scaly leaf of onion, sweet potato root etc.), water-storage tissue
Aerenchymal structure	Tissue that form large intercellular spaces (air roots, etc., floating leaf of aquatic plants), stomata
Secretory structure	Secretory hair, nectary, hydathod, secretory duct (resin canal, gum canal), laticiferous vessel, secretory cell (oil cell, mucilage cell, tannin cell)
Structures for movement	Pulvinus, stomata, insectivorous leaf (Venus' flytrap etc.)
Reproductive structure	Sexual reproduction: Flower, sepal, petal, stamen (filament, anther, pollen), pistil (ovary, ovule, embryo sac, egg cell), pollen tube, sperm cell, seed, asexual reproduction: adventitious bud, tuber (potato etc.), underground stem (iris etc.), and scaly bud (tiger lily etc.), etc.

(Modification from N. Hara, 1984)

Size of Cells

Organisms groups	Cell kind	Diameter or major axis (μm)	Organisms groups	Cell kind	Diameter or major axis (μm)
Genuine bacillus	Mycoplasma	0.2-1	Eukaryote	Mussel	70
	Escherichia coli	(2-5) × (0.5-1)		Crisia	18
	Staphylococcus	1		Sea urchins	90-110
	Streptococcus faecalis	2		Acorn worm	330-420
	Pseudomonas	2-3		Mouse	60-70
	Synechocystis	1.5-2		Dasyure	240
	Bacillus subtilis	2.5		Ostrich	110 000
	Spirochete	Several μm ×		Porbeagle	220 000
		many hundreds of μm (length)		Dog, Goat, Human	140
Archaebacteria				[Plant cells]	
(methane bacterium)	*Methanobrevibacter*	3		Arabidopsis thaliana	
(thermoacidophile)	*Sulfolobus*	1		Shoot apical meristem cells	5-10
Eukaryote	[Unicellular organism]			Tradescantia	
	Giardia lamblia	(9-20) × (5-15)		Stamen hair cell	(150-200) × (90-100)
	Budding yeast (*Saccharomyces*)	3-8		Onion	
	Paramecium	200-300 (length)		Epidermic cell	(300-400) × (30-40)
	Euglena	38 (length)		Avena sativa	
	Chlorella	5-10		Mesophyll cell	(20-30) × (50-70)
	Heterosigma	10		Human	
	Pinnularia	100-120 (length)		Hepatocyte	20-35
	Chlamydomonas	10		Cartilage cell	3-30
	Amoeba dysenteriae	15-60		Nerve cell	4-135
	Heliozoa	150-270		Striated myocytes	1 000-40 000
	Egg cell			Smooth muscle	
	Ragworm	100		Gut wall	200-250
	Buccinum	1 000-1 700		Wall of small blood vessel	12-15
	Scallop	60-65			

Weight of Human Organs/Tissues and Age-Related Change of Weight Organ Index

Organs/Tissues		Number of cases[1]	Mean weight (g)	Weight organ index[2] of each age group 20-40	60-69	70-79	80-89	90-99	100-[3]
Sex		**Male**							
Brain		1 531	1 416	28.3	30.3	30.6	30.9	29.7	34.0
Pituitary gland		508	0.55	-	-	-	-	-	-
Thyroid		1 183	18.1	-	-	-	-	-	-
Thymus		237	29.4	-	-	-	-	-	-
Heart		1 524	373	5.89	7.78	7.92	7.96	7.25	9.48
Liver		1 500	1 487	27.4	25.5	24.3	21.3	21.7	20.1
Spleen		1 515	97.9	2.18	2.36	1.88	1.68	1.65	1.59
Adrenal gland	(left)	1 486	6.96	-	-	-	-	-	-
	(right)	1 493	6.12	-	-	-	-	-	-
Kidney	(left)	1 565	159	2.63	3.30	3.33	3.01	2.47	2.72
	(right)	1 556	150	2.52	3.21	2.92	2.78	2.41	2.34
Sex		**Female**							
Brain		567	1 278	29.5	32.6	31.7	33.9	33.6	36.5
Pituitary gland		185	0.69	-	-	-	-	-	-
Thyroid		438	14.7	-	-	-	-	-	-
Thymus		91	18.5	-	-	-	-	-	-
Heart		567	297	5.58	7.74	8.55	8.92	9.23	9.60
Liver		554	1 291	28.0	25.8	26.0	23.4	21.5	19.8
Spleen		576	88.6	2.43	2.00	1.94	1.56	1.20	1.35
Adrenal gland	(left)	550	5.63	-	-	-	-	-	-
	(right)	537	4.89	-	-	-	-	-	-
Kidney	(left)	583	131	2.63	3.11	3.02	2.93	2.86	2.67
	(right)	585	123	2.48	3.02	3.01	2.85	2.68	2.43

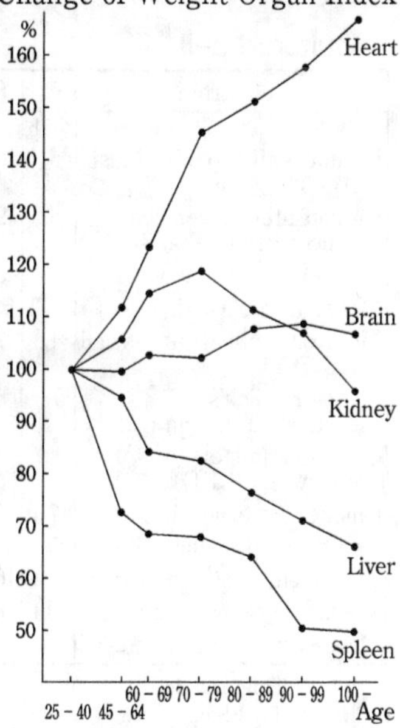

Age-related change of weight ratio between each organ and the body (weight organ index) for females.

1) Mean organs/tissues weight for the 41-60 year-old group (Japanese). Weights and Sizes of Internal Organs in Forensic Autopsy Cases (2009-2013).
2) Ratio of organ weight (g) per body weight (kg), average value (Japanese). Inoue, T., Otsu, S.: Acta Pathol Jpn, **37**: 343-359, 1987.
3) Weight organ index calculated by mean value of each organ (Japanese). Sawabe *et al.*: Pathol Int 56: 315-323, 2006.
4) Values for the 25-40-year-old group are taken as 100 percent.

Cell Cycle

Duration of the cell cycle

Normal cell (*in vivo*)

Cell type	G_1	S	G_2	M	c*
	h	h	h	h	h
Human colon epithelial cell	15	20	3	1	39
Mouse colon epithelial cell	18.5	7.2	1.7	-	27.4
Mouse small intestine epithelial cell	9	7.5	1.5	1	19
Mouse duodenum intestine epithelial cell	5.9	6.9	0.7	1.8	15.3
Mouse follicle granulosa cell	8.0	6.8	1.8	-	16.3
Mouse hair follicle cell	3.4	7.0	1.0	0.5	11.4
Mouse skin epidermal cell	87	11.6	2.0	-	101
Mouse bone marrow erythroblast	0.5	5.5	1.3	-	7.3
Goldfish intestine epithelial cell	5	9	1	2	17
Onion root-tip cell	10	7	3	5	25
Spiderwort root-tip cell	1-4	10.5	2.5-3	3	17-22
Sunflower root-tip cell	3.5	4	1.4	1.6	10.5

(Source: Mendelsohn, M. L., 1976)

Cultured cell

Cell strain	G_1	S	G_2	M	c*
	h	h	h	h	h
Human embryo fibroblast (WI-38)	6	6	4	0.8	17
Human uterine cervical cancer cell (HeLaS3)	8	9.5	3	0.7	21.2
Human kidney epithelial cell (T)	14	8	5	0.8	27
Mouse fibroblast (L-60)	9-11	6-7	2-4	1	20
Mouse lymphoid tumor cell (L5178Y)	1.8	7.3	1.2	0.5	8-11
Mouse Ehrlich's ascites carcinoma cell	5-7	8.5	3.8	1	19
Chinese hamster ovary cell (CHO)	4.7	4.1	2.8	0.8	12.4
Chicken embryo fibroblast	7.5	6	4.5	1	19.5
Newt iris epithelial cell	25	36	6	1.4	69
Human embryonic stem cell[1]	3	10	3 (2+1)		16
Monkey embryonic stem cell[2]	2.4	8.4	4.2		15
Mouse embryonic stem cell[3]	2	6	2		10

* c : Time required for complete cycle
(Except for ES cells, based on Han, A., 1976)
1) Becker *et al.*, J. Cell. Physiol., 2006.
2) Fluckiger *et al.*, Stem Cells, 2006.
3) Savatier *et al.*, Oncogene, 1994. Stead *et al.*, Oncogene, 2002.
iPS cells may be referred to ES cells. (There are differences based on the origin of the cells.)

Cell cycle phases and checkpoints

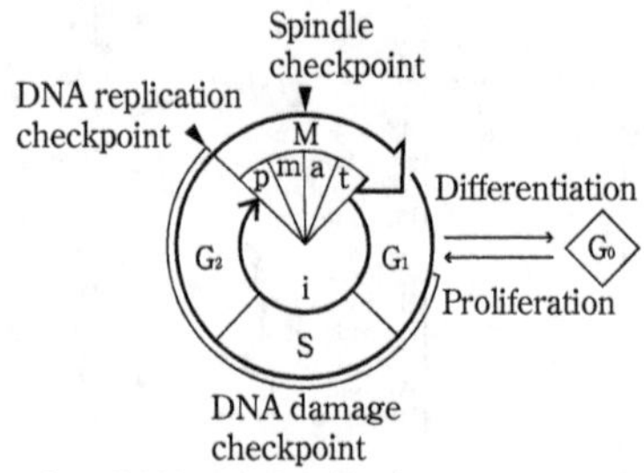

S : DNA synthesis phase
G_1 : Gap 1 phase
(Between M phase and S phase)
G_2 : Gap 2 phase
(Between S phase and M phase)
G_0 : Gap 0 phase (Resting phase)
M : Mitotic phase
p : Prophase
m : Metaphase, a : Anaphase
t : Telophase, i : Interphase

Cell cycle regulators in higher eukaryotic cells (Cell cycle engine)

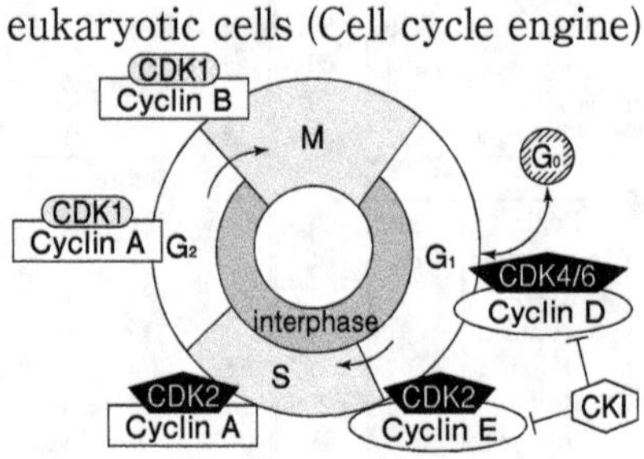

Principle of CDK Regulation

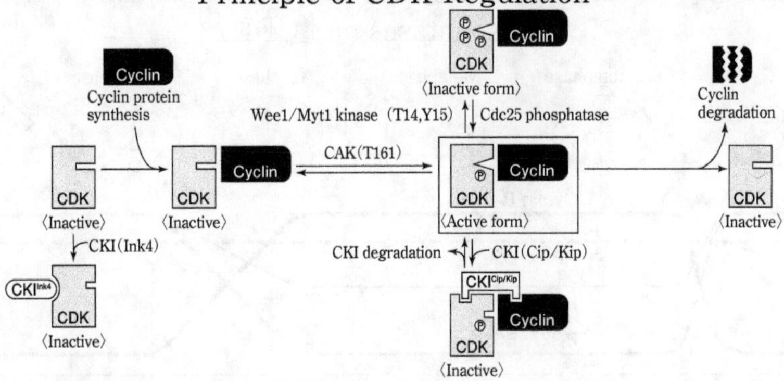

(Based on Martin-Castellanos, C, & Moreno, S., Trends Cell Biol., 1997)

CDK*1, Cyclin, CKI*2 in Yeasts and Vertebrates

CDK	Complexed cyclin	Function	Physiologically associated CKI
Vertebrates			
CDK1 (=Cdc2)	Cyclin A1, A2, B1, B2, (B3, B4, B5)	G_2/M phase transition, M phase progression	-
CDK2	Cyclin A1, A2	S phase progression	$p21^{Cip1}$
CDK2	Cyclin E1, E2	G_1/S phase transition	$p21^{Cip1}$, $p27^{Kip1}$
CDK3	Cyclin C, (A1, A2, E1, E2)	G_0/G_1 phase transition	-
CDK4	Cyclin D1, D2, D3	G_0/G_1 phase transition, G_1 phase progression	$p15^{Ink4b}$, $p16^{Ink4a}$, $p21^{Cip1}$, $p27^{Kip1}$
CDK5	(p35, Different from cyclin)	(Expressed in neuronal cells)	-
CDK6	Cyclin D1, D2, D3	G_0/G_1 phase transition, G_1 phase progression	$p15^{Ink4b}$, $p16^{Ink4a}$, $p21^{Cip1}$, $p27^{Kip1}$
CDK7	Cyclin H	CDK activation throughout the cell cycle	-
CDK8	Cyclin C	(RNA polymerase II activation)	
CDK9	Cyclin T1, T2, K	(RNA polymerase II activation)	
Yeasts			
CDC28 (Budding yeast)	CLN1, 2, 3	G_1 phase (start) progression	FAR1
CDC28 (Budding yeast)	CLB5, 6	G_1/S phase transition	SIC1
CDC28 (Budding yeast)	CLB3, 4	G_2/M phase transition	-
CDC28 (Budding yeast)	CLB1, 2	G_2/M phase transition, M phase progression	-
cdc2 (Fission yeast)	cdc13	G_2/M phase transition, M phase progression	-
cdc2 (Fission yeast)	cig1	M phase control	-
cdc2 (Fission yeast)	cig2	G_1 phase progression	rum1

*1 CDK (cyclin-dependent kinase), *2 CKI (CDK inhibitor)

(Scource: Malumbres, M. & Barbacid, M., Trends Biochem. Sci., 2005)

Mutually Reversed Activity Profiles of Antagonizing Mitotic Phosphatases and CDK1

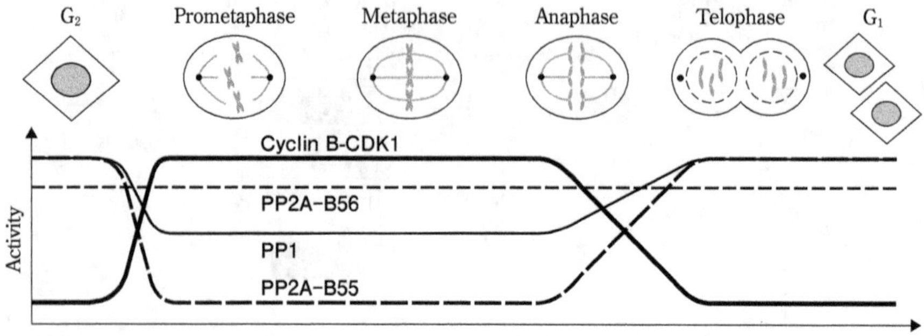

PP: protein phosphatase

(Scource: Nilsson, J., J. Cell Biol., vol. 218, 395-409., 2019)

Checkpoint Signaling

	DNA damage and replication checkpoints[1]			Spindle checkpoint[2]
	Mammal	Fission yeast	Budding yeast	
Sensor	ATR	Rad3	Mec1	Mad1-Mad2
	ATM	Tel1	Tel1	
Adaptor	Claspin	Mrc1	Mrc1	MCC (Mitotic Checkpoint Complex)
(Mediator)	BRCA1/BRCT	Crb2/Rhp1	Rad9	(C-Mad2-CDC20-BubR1-Bub3 complex)
Effector	Chk1 Chk2	Chk1 Cds1	Chk1 Rad53	APC/C (anaphase-promoting complex/cyclosome)
Target	Cdc25A, Cdc25C, p53	Cdc25, Wee1	Pds1, Cdc5, Swi6	securin, cyclin B

1) Morgan, D. O., The Cell Cycle: Principle of Control, Oxford Univ. Press, 2006.
2) London, N. & Biggins, S., Nat. Rev. Mol. Cell Biol., 2014.

Cell Growth Factor, Cytokine

An extremely small amount of a certain group of hormomimetic proteins which show bioactivity via ligand-specific receptors on the cell surface exists widely throughout multicellular organism. These are known by the general term cell growth factor (cell proliferation factor) / cytokine. In the table, alternative names are listed in parentheses. The molecular weight, number of amino acid sequences, and genetic locus (*e.g.* 4q 25–q 27) is for human beings.

Name	Molecular structure, genetic locus (human)	Target cell, function, clinical application
EGF epidermal growth factor (urogastrone)	Composed of 53 amino acids. prepro type; composed of 1217 amino acids. proEGF; appears on the cell surface, possessing growth factor activity. Possesses three distinct disulfide bonds within the molecule. Also seen in many other proteins, known as an "EGF motif." 4 q 25–q 27	Found in many bodily fluids and secretions. Stimulates the proliferation of many cells in bones, vascular endothelial cells, fibroblasts, keratinocytes, etc. Controls the secretion of stomach acid. Medical treatment for the epithelium. Eye opening.
TGF-α transforming growth factor-α	prepro type; composed from of amino acids. Possesses 30–40% homology with EGF. proTGF-α shows the juxtacrine proliferative activity. 2 p 11–p 13	The action of TGF-α is via the EGF receptor. Stimulates proliferation of hepatocytes, mammary epithelial cells, epithelial cells, and fibroblasts. Involved in the action of angiogenic factors and psoriasis vulgaris.
HB-EGF heparin-binding EGF-like growth factor (diphtheria toxin receptor)	Synthesized as a prepro type composed of 208 amino acids. Through processing, becomes a free-form type composed of 75–87 amino acids. Although proHB-EGF is a diphtheria toxin receptor, it also shows juxtacrine proliferative activity. 5 q 23	proHB-EGF acts as a receptor for diphtheria toxin. Stimulates the proliferation of vascular smooth muscle cell, fibroblasts, hepatocytes, and keratinocytes. Required for the formation of heart valves. Involved in arteriosclerosis and healing of wounds.
AR amphiregurin	Synthesized as a prepro type which possesses a transmembrane region. Free-form types are composed of 78–84 amino acids. Similar factors which possess the EGF motif are SDGF, β-cellulin (BTC), epiregulin, and epigen.	Bonds to EGF receptors and propagates signals. Controls proliferation for a portion of human cancer-derived cell lines. Stimulates proliferation of kidney cell lines. Involved in the formation of tubular structures in mammary glands.
NRG 1 neuregulin 1 (NDF/HRG/ARIA/GGF)	Divided into 6 subgroups. There are more than 15 types of isoforms, all of which possess the EGF motif. 8 p 11–p 12	Bonds with ErbB 3 and ErbB 4. NRG 1 deficit causes heart hypoplasia, failure to differentiate into Schwann cells, and failure to form synapses. Similar molecules include NRG 2, 3, 4, 5, and 6.
PDGF A, B platelet-derived growth factor A, B	The molecular weight of homodimer and heterodimer of A chains and B chains is approximately 30 kDa. PDGF-BB is on the cell surface as a membrane bind of 24 kDa. A; 7 p 22, B; 22 q 12.3–13.1	Exists in the α-granule of platelets. Mainly stimulates the migration and differentiation of mesenchymal cells. Many mesoderm-derived cells produce each isoform. Involved in the migration and proliferation of pericytes for small blood vessels, and in the development and differentiation of alveolar line fibroblasts. Also a factor which stimulates proliferation of smooth muscle and causes arteriosclerosis. Required for the differentiation of smooth muscle.
PDGF C, D platelet-derived growth factor C, D	Secreted outside of cells as a latent substance, is severed by protease and becomes active. A homodimer. C; 4 q 32, D; 11 q 22.3	PDGF-CC bonds with α acceptors and PDGF-DD bonds with β acceptors. PDGF-C appears on smooth muscle cells and PDGF-D appears on fibroblasts. Stimulates proliferation of smooth muscle cells and vascularization.

Continued.

Name	Molecular structure, genetic locus (human)	Target cell, function, clinical application
VEGF-A vascular endothelial growth factor-A (vascular permiability factor)	Four sub-types exist due to differences in splicing.; $VEGF_{121}$, $VEGF_{165}$, $VEGF_{189}$, $VEGF_{206}$. $VEGF_{165}$; most abundant. Consisted of 2 subunits, approximately 38 kDa. high similarity to PDGF. 6 p 12–p 21	Shows hyperpermeability activity and growth-stimulation activity for vascular endothelial cells. Stimulates in vivo vascularization. Stimulates migration of vascular endothelial cells and expression of plasminogen activators. Expected to serve as an anti-VEGF neutralizing antibody in anti-cancer drugs. VEGF is induced via low oxygen and low glucose. Members of the VEGF family are PIGF (placenta growth factor), VEGF-B, V-EGF-E. and VEGF-C, VEGF-D those acting on lymphoid vessel synthesis.
FGF-1 fibroblast growth factor-1 (aFGF)	Composed of 154 amino acids (18 kDa). PI 5.6, Does not possess signal peptides. 5 q 31–q 33	Stimulates the proliferation, migration and lumen formation of vascular endothelial cells. Improves the healing of skin wounds, myocardial ischemia, and blood circulation failure in lower limbs.
FGF-2 fibroblast growth factor-2 (bFGF)	In addition to a molecular weight of 18 kDa, there are also isoforms with a high molecular weight of 22–24 kDa. (HMW FGF-2). PI 9.6, Does not possess signal peptides. 4 q 26–q 27	Bonds with the same receptors of FGF-1 and shows similar action. Exists more widely than FGF-1. In particular, during the stimulation of blood vessel induction for cancer tissue, is secreted from cancer cells and takes action. Bone and cartilage formation.
FGF-3 - -25 fibroblast growth factor-3 - -25	Identified as oncogene products (FGF-3 to -6), proliferation factors of epithelial cells (FGF-7), glial cell factors (FGF-9). Each FGF is composed of 150–200 amino acid residues. Inside of the amino acid array, there is a core region which shows high homology. Some FGF possess signal peptides (FGF-3- -8, -17- -19, -21- -23, etc.), while others do not (FGF-11- -14).	Stimulates the proliferation and differentiation of fibroblasts and many other cells. FGF-7 is the same as KGF (keratinocyte growth factor), FGF-8 as AIGF (androgen induced growth factor), and FGF-9 as GAF (glia activating factor). FGF-23; acts on mineral metabolism. FGF-24, -25; control zebrafish fin formation.
IGF-I insulin-like growth factor-I	Composed of 70 amino acids. Composes the same family as insulin. 12 q	Mainly produced in the liver through control of growth hormones. Proliferation and differentiation of bone tissue, application to degenerative disease in nerve cells, control of apoptosis.
IGF-II insulin-like growth factor-II	Composed of 67 amino acids. Member of the same family as insulin. 11p (close proximity to insulin genes)	Large amounts of secretion during the prenatal period. The ICF-II gene is an imprint gene. Genes derived from the mother are inactive. IGF-II-producing tumors are the cause of hypoglycemia.
HGF hepatocyte growth factor (scatter factor)	Synthesized as a prepro body composed from 728 amino acids. Ultimately becomes an active HGF composed of double strands. The α-chain with a molecular weight of 62 kDa is composed of a hairpin domain and four kringle domains. The β-chain with a molecular weight of 32 -34 kDa possesses a structure like serine protease. Variants NK 1 and NK 2 exist due to differences in splicing. 7 q 21	Shows stimulation of proliferation, migration and morphogenesis in a wide range of cells, particularly epithelial cells. Involved in cell proliferation and formation of nerve systems in mature individuals. Expected to serve as medication for liver diseases and medication involved in vascularization.
NGF nerve growth factor	Dimer composed of 118 amino acids. The same family contains NGF 2, neurotrophin 3–7, and BDNA.	NGF acts on limited neurons such as peripheral sympathetic nerves and sensory nerves. Neurotrophin 3–7 and BDNA belong to the same family and act upon different neurons respectively. Attracting attention for its relation to Alzheimer-type dementia.
midkine	Molecular weight 13 kDa. 11 p 11.2 PTN (pleiotrophin) is in the same family.	Possesses neurotrophic factor activity, hyperfibrinolysis, and angiogenic activity.
angiopoietin-1	Molecular weight is 70–75 kDa. Angiopoietin -2 to -4 is similar.	During angiogenesis, mobilizes pericytes and vascular smooth muscle cells.

Cell Growth Factor, Cytokine Continued.

Name	Molecular structure, genetic locus (human)	Target cell, function, clinical application
TGF-β transforming growth factor-β	Protein with a molecular weight of 12.5 kDa forms the dimer. Three types of sub-unit isoforms are known. They exist as homozygotes and heterozygotes.	Suppresses the proliferation of many cells such as epithelial cells, endothelial cells, and hematopoietic cells. Stimulates the accumulation of extracellular matrix, and enhances production of protease inhibitor factors. Stimulates and suppresses the production of growth factors and cytokines. Has an extremely wide variety of effects on cells.
activin	A dimer of the inhibin β-sub-units (24 kDa). There are four types of β-sub-units.	Mesodermic induction in early embryos of *Xenopus* laevis. Determines the asymmetry of organs.
GDNF glial cell line-derived neurotrophic factor	Composed of 134 amino acids. Belongs to the TGF-β super family. 5 p 13 The same family also contains NTN (neurturin), PSP (persephin), etc.	Stimulates the existence of dopaminergic neurons, motor nerves, sympathetic nerves, etc.
FS follistatin	A single-chain peptide with a molecular weight of 35 kDa. There are two splicing forms: FS_{315} and FS_{288}. FLRG is a member of this family.	Inhibits bonding to activin receptors with a 1:2 activin-molar ratio.
BMP bone morphogenetic protein (DVR)	BMP composes the largest sub-family (more than 15 types) within the TGF-β super family. Protein dimer composed of 110–150 amino acids. 20 p 12	Involved in the development, growth and remodeling of bones. Application to the fields of dentistry and orthopedics.
MIS Mullerian inhibiting substance	Homodimer of subunit with a molecular weight of 70 kDa. Homology with TGF-β. 19 p 13.3	Produced from the sertoli cells of testicles and atrophies the Müllerian ducts.
myostatin (GDF-8, MSTN)	Belongs to the TGF-β family. Exists in dimers.	Strong control factor for skeletal muscle formation. Potential as a medication for muscular dystrophy.
TNF-α tumor necrosis factor-α	There is a membrane-bound type (28 kDa) and soluble type (16 kDa). The same family contains approximately 20 other types such as LT (lymphotoxin). Chromosome 6	Broadly involved in inflammation and immunity. Suppresses injury and proliferation of tumor cells. Suppresses proliferation of vascular endothelial cells, stimulates the production of adhesion factors (VCAM, etc.).
Fas Ligand (Fas L)	Membrane protein with a molecular weight of 40 kDa. Belongs to the TNF family.	Eliminates the peripheral activated T cells. Induces apoptosis for many cells.
IFN-α/β interferon-α/β	There are more than 15 isoforms for α-type. β-type is singular. Both are in 9p 22. Composed of 165–166 amino acids.	Involved in viral infection and other biological defense mechanisms. Has a wide range of effects including suppression of cell proliferation, anti-tumor effects, and activation of macrophages. Medication for hepatitis B, chronic hepatitis C, etc.
IFN-γ interferon-γ	Composed of 143 amino acids. IFN-α belongs a different family. 12 q 15	Produced by T cells. Anti-virus effects, suppression of cell proliferation, anti-tumor effects, activation of macrophages, regulation of immune response. Treatment for chronic hepatitis C and stomach cancer.
IL-1 interleukin-1 (endogenous pyrogen)	There are two types: IL-1α and IL-1β. Molecular weight 31 kDa. 2 q 13	Produced by lymphocytes, microglia, endothelial cells, etc. Induction of inflammatory response, regulation of immune response, regulation of hematopoietic function, cell proliferation. Application to radiation therapy.
IL-2 interleukin-2 (TCGF)	Glycoprotein with a molecular weight of 15–30 kDa and composed of 133 amino acids.	Produced by activated T cells. Proliferation of T cells, proliferation of B cells and enhancement of antibody production, proliferation of NK cells.
IL-3 interleukin-3 (pluripotent colony stimulating factor)	Glycoprotein with a molecular weight of 25 kDa and composed of 152 amino acids. 5 q	Produced by activated T cells. Stimulates proliferation of blood cells. Stimulates the existence, proliferation and differentiation of pluripotent hematopoietic stem cells.
IL-4 interleukin-4 (BCGF)	Glycoprotein with a molecular weight of 20 kDa. 5 q 23.3–q 31.2	Produced by activated T cells and obesity cells. Involved in the proliferation of many blood cells and regulation of the immune system. Treatment of infections, autoimmune disorders, and cancer.

Name	Molecular structure, genetic locus (human)	Target cell, function, clinical application
IL-5 interleukin-5 (TRF)	Homodimer with a molecular weight of 50–60 kDa.	Produced by activated T cells and obesity cells. Stimulates the proliferation and differentiation of eosinophils, and superoxide production. Treatment of bronchial asthma.
IL-6 interleukin-6	Glycoprotein with a molecular weight of 21–28 kDa and composed of 184 amino acids. Shares receptors with IL-11, LIF, etc. 7 p 21	Proliferation factor of B cells and plasmacytes. Induces the increase of platelet. Stimulates the differentiate of nerve cells and bone resorption.
IL-7 interleukin-7 (LP-1)	Composed of 152 amino acids. 8 q 12–q 13	Secreted from stromal cells. Proliferation of B cell precursor cells. Induces activity of cytotoxic T cells and LAK.
IL-8 interleukin-8 (NCF)	Has a molecular weight of 8 kDa and a strong heparin binding. Member of the chemokine family. 4 q 12–q 13.	Shows migration activity for neutrophils, lymphocytes, etc. Vascularization and enhancement of adhesion to vascular endothelial cells of neutrophils.
IL-9 interleukin-9 (TCGF Ⅲ)	Molecular weight 40 kDa. 5 q 31–q 35	Stimulates proliferation of T cells. IL-9 genes match the genetic locus of bronchial asthma.
IL-10 interleukin-10	Homodimer protein with a molecular weight of 35–40 kDa. Exists in chromosome 1.	Suppresses production of IFN-γ and suppresses macrophages. T cell proliferation, anti-inflammatory agents.
IL-11 interleukin-11	Has a molecular weight of 23 kDa and is composed of 23–178 amino acids. 19 q 13.3–q 13.4	Induces proliferation of hematopoietic stem cells. Stimulates production of megakaryocyte and blood platelets. Effects erythroblasts and stimulates hematopoiesis.
IL-12 interleukin-12	Heterodimer with a molecular weight of 35 kDa and 40 kDa. 3 q 12–q 13.2, 5 q 31–q 33	Produced from B cells and macrophages. Regulates the functions of T cells and NK cells.
IL-13 interleukin-13	Molecular weight 10 kDa.	Regulates the action of B cells and other immune cells. Also effects endothelial cells, keratinocytes, and fibroblasts.
IL-14 interleukin-14	Molecular weight 53 kDa. Two types, IL-14 α and IL-14 β.	Stimulates the growth of B cells.
IL-15 interleukin-15	Glycoprotein with a molecular weight of 14–15 kDa. 4 q 31	Effects T cells, NK cells, B cells, etc.
IL-16 interleukin-16	Two members, neuronal IL-16 (141 kDa), leukocyte IL-16 (67 kDa).	Stimulates T cell mignation
IL-17 interleukin-17	Protein dimer composed of 155 amino acids. 6 subtypes. 2 q 31	Produced by activated T cells. Effects synovial membrane cells and induces a large amount of cytokine production.
IL-18 interleukin-18	Molecular weight 18 kDa. IL-33, IL-36, IL-37 are in this family.	IFN-γ production. Enhances cellular cytotoxicity.
IL-19 interleukin-19	A related cytokine of IL-10. IL-20, -22, -24, and -26 belong to this family. IL-19 is composed of 176 amino acids.	Monocytes are inducted by the stimulation of LPS and GM-CSF. Produces TNF-α and IL-6, and induces apoptosis.
IL-21 interleukin-21	Glycoprotein with a molecular weight of 15 kDa. 4 q 26–q 27	Produced by activated T cells. Stimulates the differentiation and activation of NK cells.
IL-23 interleukin-23	A heterodimer with IL-12 p 40 and p 19 which is similar to IL-6. IL-35 is a member of this family.	Induces IFN-γ production in dendritic cells and memory T cells.
IL-25 interleukin-25	Belongs to the same family as IL-17. Composed of 177 amino acids. Forms homodimers. 14 q 11.2	Appears in the prostate, trachea and brain. Induces the appearance of various types of cytokine and increase the number of eosinophils. Involved in allergic bronchial asthma.
IL-26 interleukin-26	Belongs to the IL-10 family (formerly named AK 155). 12 q 15 Forms homodimers.	Appears strongly in transformant human T cells in the cancer virus of squirrel monkeys.
IL-27 interleukin-27	Composed of p 28 and EB 13.	Effects T helper cells.

Continued.

Name	Molecular structure, genetic locus (human)	Target cell, function, clinical application
IL-28, IL-29 interleukin-28, -29	A new member of type I interferon. Called IFNsλ. Chromosome 19.	Anti-virus effect.
IL-31 – -33 interleukin-31 – -33	IL-32; 16 p 13.3.	IL-31 is involved in Th2 response. Although IL-32 and -33 have not been sufficiently investigated, they appear to be involved in the inflammatory response to allergies and autoimmune disorders.
IL-34 interleukin-34	Shares same receptors with M-CSF (CSF-1).	Stimulates the growth and differentiation of monocytes, dendritic cells and macrophages.
IL-35 interleukin-35	Synthesized by Breg, a subpopulation of B cells.	Functions in autoimmune diseases.
IL-36 interleukin-36	Highly expressed in epithelial tissues and myeloid cells. a novel cytokine.	A proximal role in the lung innate mucosal immunity during bacterial pneumonia.
IL-37 interleukin-37	Generated by macrophages. A novel cytokine of IL-1 family. Binds IL-18 receptor α	Anti-inflammatory and inhibitor of innate immunity.
IL-38 interleukin-38	Molecular weight 16.9 kDa (precursor) 2p13 IL-1 part of the family.	Expressed in heart, placenta, fetal liver, B cells, etc. It binds to IL-IR and IL-36Ra, promotes the production of inflammatory cytokines, and is involved in various chronic diseases.
IL-39 interleukin-39	IL-23 p 19 and Ebi 3 (Epstein-Barr virus-Induced 3) heterodimer.	It is produced by macrophages in response to inflammation and is involved in innate and acquired immunity.
IL-40 interleukin-40	Chromosome 17 ~27 kDa (No association with other cytokines. Reported in 2017)	Expressed by bone marrow, fetal liver, activated B cells, etc. Regulation of many immune processes through B cells.
IL-41 interleukin-41	Molecular weight 28 kDa 17q25.3	A Novel (2019) Cytokine Also Marked as Metrnl/IL-41. It is expressed in high concentrations in psoriatic skin. It is secreted by fibroblasts, activated macrophages, and mucosa of the articular capsule and synovium.
LIF leukemia inhibitory factor	Glycoprotein with a molecular weight of 38–62 kDa and composed of 180 amino acids.	Suppresses the differentiation of embryonic stem cells (ES cells) and retains multi potency of differentiation. Supports the research of the biology on ES cells and iPS cells.
GM-CSF granulocyte-macrophage colony stimulating factor	Glycoprotein with a molecular weight of 23 kDa and composed of 127 amino acids.	Differentiation and enhancement of myeloid cells. Stimulates the proliferation and migration of vascular endothelial cells. Treatment of aplastic anaemias.
G-CSF granulocyte-colony stimulating factor	Composed from 204 amino acids. Homology with 17q 11–q 22 and IL-6	Produced by monocyte macrophages and vascular endothelial cells. Proliferation and differentiation hyperactivity for neutrophil progenitor cells and neutrophil system cells. Medicine of aplastic anaemias and AIDS.
EPO erythropoietin	Glucose with a molecular weight of 12 kDa bonds with protein of a molecular weight of 18 kDa. 7 q 21–q 22	Produced by the kidney in adults. The appearance of EPO is heightened by the decrease of oxygen partial pressure. Has an important role in differentiation of red blood cells. Medicine for renal anemia.
CSF-1 colony stimulating factor-1 (M-CSF)	Molecular weight is 70–90 kDa. 5 q 33.1	Differentiation of microphages. Involved in the existence and proliferation of nerve cells.
SCF/SLF stem cell factor/ steel factor	Dimer formation with a molecular weight of 53 kDa.	Stimulates proliferation of megakaryocytic stem cells. Proliferates the primordial germ cells and melanocytes.
TPO thrombopoietin	Composed of 332 amino acids. 3 q 26.33–q 27.	Stimulates the stimulatory activity of megakaryocyte colonies and the production of platelets.

(K. Kaji, 2020)

Number of Chromosomes (Vertebrate)

Species	Diploid	Haploid	Sex chromosome	Species	Diploid	Haploid	Sex chromosome
Human (*Homo sapiens*)	46	23♂	♂XY	Turtle (*Clemmys japonica*)	52	–	–
Chimpanzee (*Pan troglodytes*)	48	24♂	♂XY	Loggerhead turtle (*Caretta caretta*)	56	–	♂ZZ, ♀ZW
Gorilla (*Gorilla gorilla*)	48	24♂	♂XY	Spiny-tailed monitor lizard (*Varanus acanthurus*)	40	–	♂ZZ, ♀ZW
Rhesus macaque (*Macaca mulatta*)	42	21♂	♂XY	Japanese grass lizard (*Takydromus tachydromoides*)	40	–	♂ZZ, ♀ZW
Japan squirrel (*Sciurus lis*)	40	20♂	♂XY	Japanese ratsnake (*Elaphe climacophora*)	36	–	♂ZZ, ♀ZW
Japan flying squirrel (*Pteromys momonga*)	38	19♂	♂XY	Tiger keelback (*Rhabdophis tigrinus*)	36	–	♂ZZ, ♀ZW
Guinea pig (*Cavia porcellus*)	64	32♂	♂XY	Pit viper (*Agkistrodon halys*)	36	–	♂ZZ, ♀ZW
Mouse (*Mus musculus*)	40	20♂	♂XY	Leafhopper rattlesnake (*Crotalus cerastes*)	36	–	♂ZZ, ♀ZW
Rat (*Rattus norvegicus*)	42	21♂	♂XY				
Pack rat (*Neotoma floridane*)	52	26♂	♂XY	Tokyo salamander (*Hynobius tokyoensis*)	56	–	♂ZZ, ♀ZW
Rabbit (*Oryctolagus cuniculus*)	44	22♂	♂XY	Smooth newt (*Triturus vulgaris*)	24	–	♂XY, ♀XX
Pika (*Ochotona hyperborea*)	40	20♂	♂XY	Japan brown frog (*Rana japonica*)	26	13♂	♂XY, ♀XX
Dog (*Canis familiaris*)	78	39♂	♂XY	European toad (*Buf bufo*)	22	–	♂ZZ, ♀ZW
Cat (*Felis catus*)	38	19♂	♂XY	Tree frog (*Hyla arborea japonica*)	24	–	♂XY, ♀XX
Cow (*Bos taurus*)	60	30♂	♂XY	African clawed frog (*Xenopus laevis*)	36	–	♂ZZ, ♀ZW
Pig (*Sus scrofa domestica*)	38	19♂	♂XY				
Sheep (*Ovis aries*)	54	27♂	♂XY	Salmon (*Oncorhynchus keta*)	74	–	–
Goat (*Capra hircus*)	60	30♂	♂XY	Rainbow trout (*Oncorhynchus mikiss* (=*Salmo gairdneri*))	58–60	–	–
Horse (*Equus caballus*)	64	32♂	♂XY	Brookie (*Salvelinus fontinalis*)	84	–	–
Indian muntjac (*Muntiacus muntjak*)	7 6	3♂ 3♀	♂X_1X_2Y, ♀XX	Goldfish (Asian carp) (*Carassius auratus*)	100	–	–
Eastern grey kangaroo (*Macropus gigantea*)	22	11♂	♂XY	Carp (*Cyprinus carpio*)	100	–	–
Goose (*Anser anser*)	ca. 80	–	♂ZZ, ♀ZW	Japanese killfish (Medaka) (*Oryzias latipes*)	48	–	♂XY, ♀XX
Duck (Domestic) (*Anas platyrhynchos*)	80	–	♂ZZ, ♀ZW	Zebra fish (*Danio rerio*)	50	–	–
Quail (*Coturnix coturnix*)	78	39♂	–	European eel (*Anguilla anguilla*)	36	–	–
Chicken (*Gallus domesticus*)	78	–	♂ZZ, ♀ZW	Eel (*Anguilla japonica*)	38	–	–
Mongolian pheasant (*Phasianus colchicus*)	ca. 80, ca. 82	–	♂ZZ, ♀ZW	Fresh-water catfish (*Silurus asotus*)	58	–	–
Turkey (*Meleagris gallopavo*)	ca. 80	–	♂ZZ, ♀ZW	Grass puffer (*Takifugu niphobles*)	44	–	–
Sea gull (*Larus canus*)	ca. 66	–	♂ZZ, ♀ZW	Pacific spiny dogfish (*Squalus suckleyi*)	62	31♂	–
Rock pigeon (*Columba livia*)	80	–	♂ZZ, ♀ZW	Bigeyed skate (*Raja meerdervoortii*)	104	52♂	–
Sparrow (*Passer montanas*)	ca. 78	–	♂ZZ, ♀ZW	Far eastern brook lamprey (*Entosphenus reissneri*)	165-174	–	–
Canary (*Serinus canaria*)	ca. 80	–	♂ZZ, ♀ZW	Hagfish (*Myxine glutinosa*)	28	44♂	–

(Sources: Makino, S., 1972; King, M., 1990; Olmo, E., 1980; NCBI, 2017. Compiled by: Yoshida, M., 2005)

Number of Chromosomes (Invertebrate)

Species	Diploid	Haploid	Species	Diploid	Haploid
Sea squirt (*Ciona intestinalis*)	18	–	Green drake mayfly (*Ephemera danica*)	11	–[1]
Lancelet (*Amphioxus lanceolatum*)	24	12♀	Oriental migratory locust (*Locusta migratoria*)	23	11, 12♂
Forbes' sea star (*Asterias forbesi*)	36	18♂	German cockroach (*Blattella germanica*)	23, 24	11, 12♂[1]
Purple-spined sea urchin (*Arbacia punctulata*)	ca.40	–	Silkworm (*Bombyx mori*)	56	28♂
Sand dollar (*Echinarachnius parma*)	52	–	Northern House Mosquito (*Culex pipiens*)	6	3♂
Sea cucumber (*Stichopus regalis*)	28–36	16–18♂	Fruit fly (*Drosophia melanogaster*)	8	4
Hard tick (*Ixodes ricinus*)	28	–	Housefly (*Musca domestica*)	12	6♂
Domestic house spider (*Tegenaria domestica*)	43	23♂	Black blow fly (*Phormia regina*)	12	6♂[2]
Horseshoe crab (*Tachypleustridentatus*)	26	13♂	Japanese beetle (*Popillia japonica*)	18	9♂
Brine shrimp (*Artemia salina*)	42	21♀	Honeybee (*Apis mellifera*)	32♀	16♂
Water flea (*Daphnia magna*)	20	10♀	Great pond snail (*Radix japonica*)	36	18♂
Barnacle (*Lepas anatifera*)	26	13♂ 13♀	Roundworm (*Ascaris lumbricoides*)	43, 48	19, 24♂ 24♀
Prawn (*Marsupenaeus japonicus*)	92	46♂	Horsehair worm (*Gordius tolosanus*)	4	2♀
American crayfish (*Cambarus clarkii*)	200	100♂	Liver fluke (*Fasciola hepatica*)	12	6♀
King crab (*Paralithodes camtschaticus*)	208	104♂	European planaria (*Planaria torva*)	16	8♀
River crab (*Potamon dehaanii*)	82	41♂	Knotted thread hydroid (*Obelia geniculata*)	34	17♀

1) Sex chromosome ♂ XO, 2) Sex chromosome ♂XY

(Sources: Makino, S., 1972; O'Brien, S. J., 1993. Compiled by Yoshida, M., 1998)

Number of Chromosomes (Plants (1))

Species	Diploid	Species	Diploid
Sugar beet (*Beta unlgaris*)	18	Cucumber (*Cucumis sativus*)	14
Cabbage (*Brassica oleracea*)	18	Western pumpkin (*Cucurbita pepo*)	40
Apple (*Malus domestica*)	34	Sunflower (*Helianthus annuns*)	34
Peach (*Prunus persica*)	16	Soybean (*Glycine max*)	40
Plum (*Prunus demestica*)	48	Pea (*Pisum sativum*)	14;28
Pear (*Pirus communis*)	34;51	Red clover (*Trifolium pratense*)	14;28;56
Lemon (*Citrus limon*)	18;36	Catalpa fruit (*Catalpa speciosa*)	40
Olive (*Olea europaea*)	46	Wild oat (*Avena sativa*)	42
Holly (*Ilex aquifolium*)	40	Rice (*Oryza sativa*)	24
Grape (*Vitis vinifera*)	38;76	Sugarcane (*Saccharum officinarum*)	80
Carrot (*Daucus carota*)	18	Bread wheat (*Triticum aestivum*)	42
Sweet potato (*Ipomea batatas*)	90	Maize (*Zea mays*)	20
Tomato (*Lycopersicon esculentum*)	24;48	Coconut (*Cocos nucifera*)	32
Tabaco (*Nicotiana tabacum*)	48	North American spider lily (*Tradescantia paludosa*)	12
Potato (*Solanum tuberosum*)	48	Onion (*Allium cepa*)	16;32
Cayenne pepper (*Capsicum frutescens*)	24	Asparagus (*Asparagus officinalis*)	20;40
Snapdragon (*Antirrhinum majus*)	16;32	Ginko (*Ginkgo biloba*)	24
Watermelon (*Citrullus vulgaris*)	22;33	Thale cress (*Arabidopsis thaliana*)	10
Banana (*Musa sapientum*)	22;33	Petunia (*Petunia hybrida*)	14

(Sources: Riley, H. P., *et al.*, 1972; O'Brien, S. J., 1993. Compiled by Yoshida, M., 1998)

Number of Chromosomes (Plants (2) and Fungus)

Species	Haploid	Species	Haploid
American maidenhair (*Adiantum pedatum*)	58	Fucus (*Fucus vesiculosus*)	10;32
Holly fern (*Cyrtomium falcatum*)	123	Brown algae (*Laminaria digitata*)	27–31
Hart's-tongue (*Phyllitis scolopendrium*)	72	Laver (*Porphyra umbilicalis*)	5
Horsetail (*Equisetum arvense*)	216	Mushroom (*Agaricus campestris*)	12
Broom forkmoss (*Dicranum scoparium*)	12	Shaggy ink cap (*Coprinus comatus*)	14
Common hair cap moss (*Polytrichum commune*)	7	Bracket fungus (*Fomes annosus*)	7
Liverwort (*Marchantia polymorpha*)	9	Yeast (*Saccharomyces cerevisiae*)	16
Crystalwort (*Riccia fluitans*)	8	Aspergillus (*Aspergillus nidulans*)	8
Club moss (*Lycopodium clavatum*)	68	Blue mold (*Penicillum expansum*)	4–5
Cracked lichen (*Acarospora fuscata*)	3	Neurospora (*Neurospora crassa*)	7

(Sources: Storok, R. *et al.*, 1972; O'Brien, S. J., 1993; NCBI, 2017. Compiled by Yoshida, M., 1998)

List of Genome Size*

Virus, bacterium, mitochondrion, and chloroplast		Size (kb)
Virus	SV40	5
	Influenza	13
	T4 phage	168
Bacterium	Bacillus (*Bacillus subtilis*)	413
	Coliform bacterium (*Escherichia coli*)	464
Mitochondrium	Yeast	120
	Rice	490
	Paramecium	41
	Sea urchin	15
	Mouse	17
Chloroplast	Euglena	150
	Maize	140

Organism	Size (Mb)
Aspergillus (*Aspergillus oryzae*)	36.5
Red bread mold (*Neurospora crassa*)	40.7
Cryptococcus (*Cryptococcus neoformans*)	18.4
Baker's yeast (*Saccharomyces cerevisiae*)	12.1
Dictyostelium (*Dictyostelium discoideum*)	34.2
Slim mold (*Physarum polycephalum*)	205.1
Diatom (*Phaeodactylum tricornutum*)	27.4
Brown algae (*Saccharina japonica*)	543.4
Green algae (*Chlamydomonas reinhardtii*)	120.4
Red algae (*Cyanidioschyzon merolae*)	16.5
Chlorella (*Chlorella vulgaris*)	37.3
Liverwort (*Marchantia polymorpha*)	205.7
Physcomitrella (*Physcomitrella patens*)	477.9
Selaginella (*Selaginella moellendorfii*)	212.5
Pine (*Pinus taed*)	22 103.6
Rice (*Oryza sativa*)	368.3
Maize (*Zea mays*)	2 263.9
Thale cress (*Arabidopsis thaliana*)	116.8
Potato (*Solanum tulberosum*)	705.9
Tomato (*Solanum lycopersicum*)	760.0
Peach (*Prunus persica*)	212.7
White oak (*Quercus lobata*)	759.2
Bread wheat (*Triticum aestivum*)	4 568.2
Common sunflower (*Helianthus annuus*)	3 027.8
Radish (*Raphanus sativus*)	392.7
Quinoa (*Chenopodium quinoa*)	1 210.1
Peanut (*Arachis ipaensis*)	1 353.5
Buckweat (*Fagopyrum esculentum*)	1 177.6
Fig (*Ficus carica*)	247.0
Fava bean (*Vicia faba*)	80.3
Wine grape (*Vitis vinifera*)	4 486.1

Organism	Size (Mb)
Paramecium (*Paramecium tetraurelia*)	72.0
Anaerobic parasitic amoebozoa (*Entamoeba histolytica*)	20.8
Malaria parasite (*Plasmodium falciparum*)	24.6
Crenarchaeote (*Cenarchaeum symbiosum*)	2.0
Nematode (*Caenorhabditis elegans*)	101.1
Parasitic filarial nematode (*Brugia malayi*)	93.6
Blood fluke (*Schistosoma japonicum*)	402.7
Silkworm (*Bombyx mori*)	393.6
African malaria mosquito (*Anopheles gambiae*)	250.7
Fruit fly (*Drosophila melanogaster*)	143.7
Fellow fever mosquito (*Aedes aegypti*)	1 383.9
Honey bee (*Apis mellifera*)	246.9
Red flour beetle (*Tribolium castaneum*)	165.9
Tsetse fly (*Glossina morsitans*)	355.5
Starlet sea anemone (*Nematostella vectensis*)	356.6
Sea urchin (*Strongylocentrotus purpuratus*)	990.9
Florida lancelet (*Branchiostoma floridae*)	521.8
Sea squirt (*Ciona intestinalis*)	115.9
Spotted green pufferfish (*Tetraodon nigroviridis*)	342.4
Pufferfish (*Takifugu rubripes*)	391.4
Zebrafish (*Danio rerio*)	1 427.2
Medaka (*Oryzias latipes*)	766.2
African clawed frog (*Xenopus laevis*)	2 718.4
Elephant shark (*Callorhinchus milii*)	974.4
Chicken (*Gallus gallus*)	1 230.2
Platypus (*Ornithorhynchus anatinus*)	1 995.6
Tammar wallaby (*Macropus eugenii*)	3 075.1
Gray short-tailed opossum (*Monodelphis domestica*)	3 598.4
Koala (*Phascolarctos cinereus*)	3 192.5
Tasmanian devil (*Sarcophilus harrisii*)	3 203.5
Mouse (*Mus musculus*)	2 671.1
Rat (*Rattus norvegicus*)	2 743.3
Rabbit (*Oryctolagus cuniculus*)	2 737.4
Pig (*Sus scrofa*)	2 455.8
Domestic cat (*Felis catus*)	2 900.8
Dog (*Canis familiaris*)	2 254.6
Cow (*Bos taurus*)	2 697.5
Horse (*Equus caballus*)	2 377.5
Rhesus monkey (*Macaca mulatta*)	3 033.6
Olive baboon (*Papio anubis*)	2 948.4
Sumatran orangutan (*Pongo abelii*)	3 441.2
Gorilla (*Gorilla gorilla*)	3 071.8
Chimpanzee (*Pan troglodytes*)	2 982.5
Bonobo (*Pan paniscus*)	3 286.6
Human (*Homo sapiens*)	2 996.4

* Adapted from NCBI 2017.

Genes, Immunity

Mendel's Law

Mendel's law

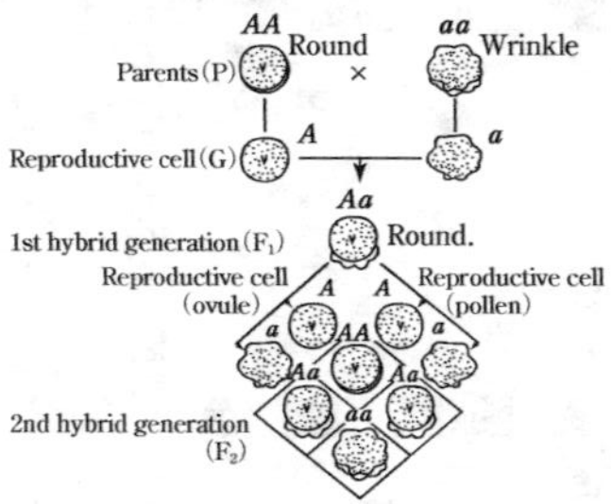

Round : Wrinkle = 3 : 1

Mendelian dominant/recessive and law of segregation

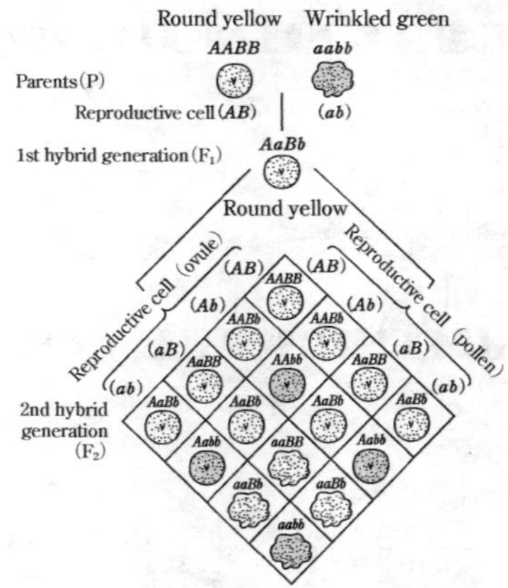

round yellow : round green : wrinkled yellow : wrinkled green = 9 : 3 : 3 : 1

Mendel's law of independent assortment

7 allelic traits using Mendelian

Characteristics	Total # of F_2	Dominant		Recessive		F_2 segregation ratio
Shape of seed	7 324	Round	5 474	Wrinkle	1 850	2.96 : 1.00
Color of cotyledons	8 023	Yellow	6 022	Green	2 001	3.01 : 1.00
Color of testa	929	Gray-brown	705	White	224	3.15 : 1.00
Shape of mature pod	1 181	Swelling and large	882	shrinkage and small	299	2.95 : 1.00
Color of immature pod	580	Green	428	Yellow	152	2.82 : 1.00
Flower attachment	858	Axial	651	Terminal	207	3.14 : 1.00
Stalk height	1 064	High (2 m)	787	Low (25–50 cm)	277	2.84 : 1.00

Mendel (1866) used the seven pairs of allelic traits of peas shown in the table to conduct a crossing experiment. He succeeded in discovering the following three important laws of genetics.

(1) Law of dominance When crossing peas with round seeds and peas with wrinkled, angular peas, the first generation of the hybrid (F_1) are all round. No wrinkled peas appear. In this way, dominant traits and recessive traits appeared in all seven pairs of contrasting traits of peas. When expressing this phenomenon as genotypes, the round seeds are *AA* and the wrinkled seeds are *aa*. The generation F_1 obtained through crossing is *A*. *A* is dominant over *a*.

(2) Law of segregation When crossing F_1 seeds which were obtained in the experiment above, the second hybrid generation (F_2) showed round seeds and wrinkled seeds in a ratio of 3 : 1. When expressed as genotypes, in generation F_2, the ratio is *AA* : *Aa* : *Aa* : *aa* = 1 : 1 : 1 : 1. *A* has dominance over *a* , so the resulting phenotype of round (A–) seeds to wrinkled seeds (*aa*) has a ratio of 3 : 1.

(3) Law of independence This low applies to two or more different allelic traits. For example, the crossing between peas with round seeds and yellow cotyledons (AABB) and ones with wrinkled, angular seeds and green cotyledons (aabb) yields F_1 hybrids with round seeds and yellow cotyledons (AaBb). The crossing between the F_1 hybrids produces F_2 hybrids with round seeds and yellow cotyledons (A–B–), ones with round seeds and green cotyledons (A–bb), ones with wrinkled, angular seeds and yellow cotyledons (aaB–), and ones with wrinkled, angular seeds and green cotyledons (aabb), the ratio of which is 9 : 3 : 3 : 1.

Since the 7 traits shown in the left table are located on different chromosomes, respectively, this low can be applied to them. When two or more traits are located on the same chromosome (linkage), the ratio is distorted depends on the distance between/among their loci (segregation distortion). Linkage is widely applied to uncover the gene responsible for a human disease by using disease phenotype and SNPs discovered by next generation sequencer (linkage analysis).

Genetic Map for Human

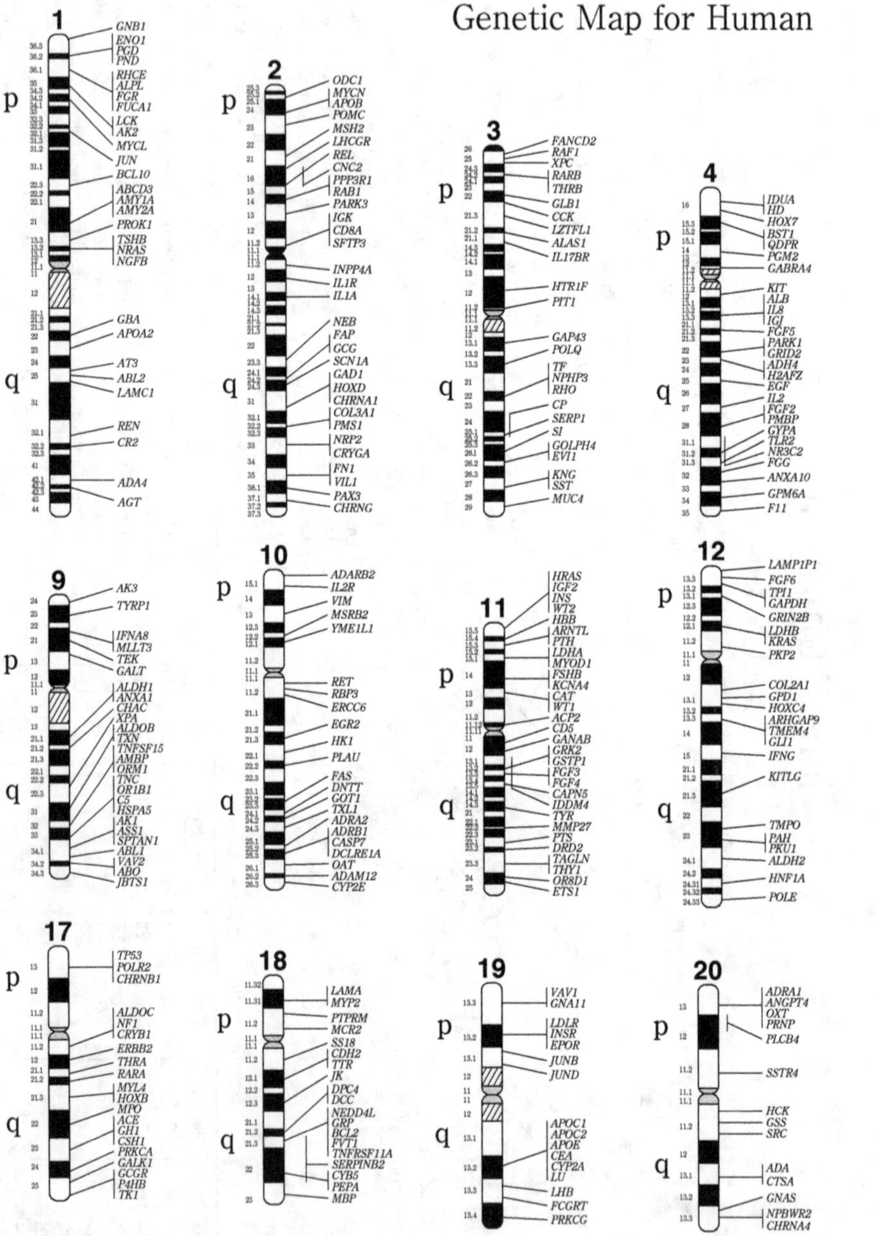

Source: NCBI Map Viewer (2020), HUGO Gene Nomenclature committee (2020)

Chromosomes (Schematic)

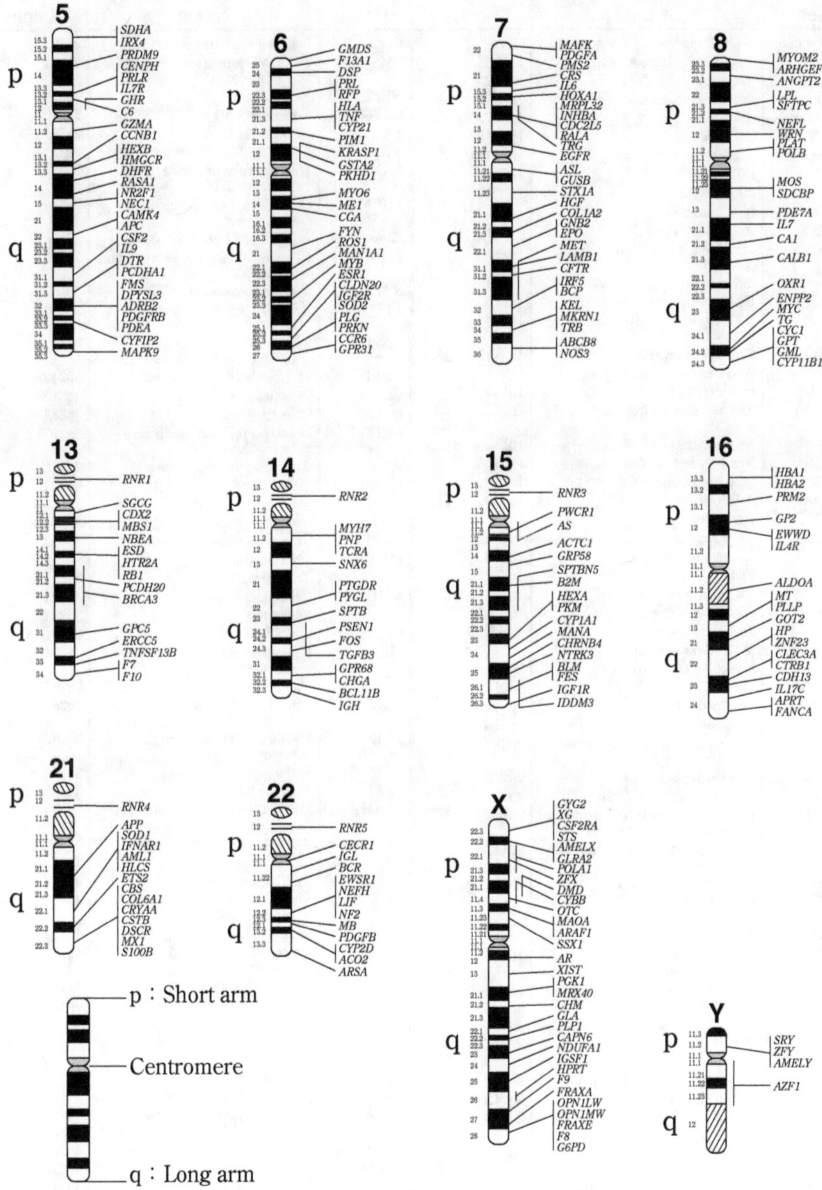

Position indication schematic view

Human Genetic Map

Gene	Name	Locus
Chromosome 1 (2610 number of the genes)		
GNB1	Guanine binding protein	1p36.33
ENO1	Enolase 1	1p36.23
PGD	Phosphogluconate dehydrogenase	1p36.22
PND	Natriuretic peptide A	1p36.22
RHCE	Rh blood type	1p36.11
ALPL	Alkaline phosphatase	1p36.12
FGR	FGR oncogene	1p36.11
FUCA1	α-L-fucosidase 1	1p36.11
LCK	LCK oncogene	1p35.2
AK2	Adenylate kinase 2	1p35.1
MYCL	MYCL oncogene (lung cancer origin)	1p34.2
JUN	*Jun* oncogene	1p32.1
BCL10	B cell lymphoma	1p22.3
ABCD3	ATP binding cassette subfamily D member 3	1p21.3
AMY1A	Amylase (saliva)	1p21.1
AMY2A	Amylase (pancreatic juice)	1p21.1
PROK1	prokineticin 1	1p13.3
TSHB	Thyroid stimulating hormone β chain	1p13.2
NRAS	NRAS oncogene	1p13.2
NGFB	Nerve growth factor	1p13.2
GBA	Glucosylceramidase β	1q22
APOA2	Apolipoprotein	1q23.3
AT3	Anti-thrombin III type	1q25.1
ABL2	*abl* oncogene 2	1q25.2
LAMC1	Laminin subunit γ 1	1q25.3
REN	Renin	1q32.1
CR2	EB virus receptor-2	1q32.2
AD4	Alzheimer's disease 4	1q42.13
AGT	Angiotensinogen	1q42.2
Chromosome 2 (1748 number of the genes)		
ODC1	Ornithine decarboxylase 1	2p25.1
MYCN	MYCN oncogene	2p24.3
APOB	Apolipoprotein	2p24.1
POMC	Propiomelanocortin	2p23.3
MSH2	Hereditary non-polyposis colon cancer	2p21
LHCGR	Luteinizing hormone	2p16.3
REL	*rel* oncogene	2p16.1
CNC2	Carney complex type 2 (lentiginosis and multiple neoplasia)	2p16
PPP3R1	Protein phosphatase 3 regulatory subunit B	2p14
RAB1	ras family oncogene-1	2p14
PARK3	Parkinson's disease 3 (associated lewy body)	2p13
IGK	Immunoglobulin *K* genes	2p11.2
CD8A	CD8A antigen-CD8A	2p11.2
SFTP3	Alveolus protein	2p11.2
INPP4A	Inositol polyphosphate 4-phosphate type 1A	2q11.2
IL1R	Interleukin-1 receptor type 1	2q12.1
IL1A	Interleukin-1a	2q14.1
NEB	Neburin	2q23.3
FAP	Fibroblast activation protein-α	2q24.2
GCG	Glucagon	2q24.2
SCN1A	Sodium voltage-gated α subunit 1	2q24.3
GAD1	Glutamate decarbxylase 1	2q31.1
HOXD	Homeobox D cluster antisense RNA1	2q31.1
CHRNA1	Cholinergic receptor nicotinic α1 subunit	2q31.1
COL3A1	Collagen type III α1	2q32.2
PMS1	Mismatch repair system component	2q32.2
NRP2	Neuropilin 2	2q33.3
CRYGA	Crystallin γ A	2q33.3
FN1	Fibronectin 1	2q35
VIL1	Villin 1	2q35
PAX3	Wallenberg syndrome-related type 1 (Paired box 3)	2q36.1
CHRNG	Cholinergic receptor nicotinic, γ subunit	2q37.1
Chromosome 3 (1381 number of the genes)		
FANCD2	Fanconi anemia complementation group D2	3p25.3
RAF1	*RAF*-1 proto-oncogene	3p25.2
XPC	Xeroderma pigmentosum group C	3p25.1
RARB	Retonikku acid receptor β (microphthalmia factor regions)	3p24.2
THRB	Thyroid hormone receptor β	3p24.2
GLB1	Galactosidase β1	3p22.3
CCK	Cholecystokinin	3p22.1
LZTFL1	Leucine zipper transcription factor-like-1	3p21.31
ALAS1	Aminolevulinate synthase 1	3p21.2
IL17BR	Interleukin-17 receptor β	3p21.1
HTR1F	Serotonin receptor	3p12
PIT1	Growth hormone factor-1	3p11.2
GAP43	Growth-associated protein 1	3q13.31
POLQ	DNA polymerase θ	3q13.33
TF	Transferrin	3q22.1
NPHP3	Cystic kidney disease	3q22.1
RHO	Rhodopsin (rhodopsin)	3q22.1
CP	Ceruloplasmin	3q24-q25.1
SERP1	Stress-associated endoplasmic reticulum protein-1	3q25.1
SI	Sucrose isomaltase	3q26.1
GOLPH4	Golgi integral membrane protein 4	3q26.2
EVI1	Myelocytic leukemia factor regions	3q26.2
KNG	Kininogen 2	3q27.3
SST	Somatostatin	3q27.3
MUC4	Mucin-4 (tracheobronchitis factor regions)	3q29
Chromosome 4 (1024 number of the genes)		
IDUA	Iduronidase, α-L-type	4p16.3
HD	Huntington disease	4p16.3
HOX7	Homeobox gene-7 (ectodermal heteromorphism)	4p16.2
BST1	Bone marrow cell antigen-1	4p15.32
QDPR	Quinoid-dihydropteridine reductase	4p15.32
PGM2	Phosphoglucomutase-2	4p14
GABRA4	γ-aminobutyric acid type A receptor subunit, α4	4p12
KIT	*KIT* oncogene	4q12
ALB	Albumin	4q13.3
IL8	Interleukin-8	4q13.3
IGJ	Immunoglobulin polypeptide-J-	4q13.3
FGF5	Fibroblast growth factor-5	4q21.21
PARK1	Parkinson's disease-1	4q22.1
GRID2	Glutamate receptor δ-2 subunit 2	4q22
ADH4	Alcohol dehydrogenase 4	4q23
H2AFZ	H2A histone-Z	4q23
EGF	Epidermal growth factor	4q25

Continued.

Gene	Name	Locus
IL2	Interleukin-2	4q27
FGF2	Fibroblast growth factor-2	4q28.1
PMBP	Progesterone receptor membrane component 2	4q28.2
GYPA	Glycophorin ? A (MNS blood type)	4q31.21
TLR2	Toll-like protein receptor-2	4q31.3
NR3C2	Nuclear acceptor-3	4q31
FGG	Fibrinogen-γ polypeptide	4q32.1
ANXA10	Annexin A10	4q32.3
GPM6A	M6A glycoprotein	4q34.2
F11	Blood coagulation factor XI	4q35.2
Chromosome 5 (1190 number of the genes)		
SDHA	Succinate dehydrogenase	5p15.33
IRX4	Iroquois homeobox gene 4	5p15.33
PRDM9	PR domain containing-9	5p14.2
CENPH	Centromeric protein H	5q13.2
PRLR	Prolactin receptor	5p13.2
IL7R	Interleukin receptor-7	5p13.2
GHR	Growth hormone receptor	5p13.1-p12
C6	Complement component 6	5p13.1
GZMA	Granzyme	5q11.2
CCNB1	Cyclin B1	5q13.2
HEXB	Hexosaminidase B	5q13.3
HMGCR	hydroxy-methylgutarylcoenzyme A reductase	5q13.3
DHFR	Dihydrofolate reductase	5q14.1
RASA1	RASp21 protein activity body	5q14.3
NR2F1	Nuclear receptor subfamily 2 group F member 1	5q15
NEC1	Neuroendocrine convertase-1	5q15
CAMK4	Calmodulin-dependent protein kinase IV	5q22.1
APC	Adenomatous polyposis coli	5q22.2
CSF2	Colony stimulating factor-2	5q31.1
IL9	Interleukin-9	5q31.1
DTR	Diphtheria toxin receptor	5q31.3
PCDHA1	Protocadherin-α1	5q31.3
FMS	*FMS* oncogene	5q32
DPYSL3	Dihydropyrimidine like-3	5q32
ADRB2	β-2-adrenergic receptor	5q32
PDGFRB	Platelet-derived growth fact	5q32
PDEA	Phosphodiesterase-6A	5q32
CYFIP2	Cytoplasmic FMR1 interacting protein 2	5q33.3
MAPK9	Division-activated protein kinase-9	5q35.3
Chromosome 6 (1394 number of the genes)		
GMDS	GDP mannose 4,6-dehydrogenase	6p25.3
F13A1	Blood clotting factor XIII,A1 polypeptide	6p25.1
DSP	Desmoplakin	6p24.3
PRL	Prolactin	6p22.3
RFP	*Ret* ringfinger protein	6p22.1
HLA	Major histocompatibility complex	6p21.33
TNF	Tumor necrosis factor	6p21.33
CYP21	Cytochrome P450, subfamily XXI	6p21.33
PIM1	*pin*-1 oncogene	6p21.2
KRASP1	KRAS cancer gene	6p12.1
GSTA2	Glutathione S-transferase α2	6p12.2
PKHD1	Fibrocystin / polyductin	6p12.3-p12.2
MYO6	Myosin VI	6q14.1
ME1	Malic enzyme-1	6q14.2
CGA	Chorionic gonadotropin hormone	6q14.3
FYN	*fyn* oncogene	6q21

Gene	Name	Locus
ROS1	*ros* oncogene	6q22.1
MAN1A1	Mannosidase α-1A	6q22.31
MYB	*myb* oncogene	6q23.3
ESR1	Estrogen-receptor 1	6q25.1
CLDN20	Claudin 20	6q25.3
IGF2R	Insulin-like growth factor 2 receptor	6q25.3
SOD2	Superoxide dismutas-2	6q25.3
PLG	Plasminogen	6q26
PRKN	Parkinson disease-2	6q26
CCR6	C-C motif Chemokine receptor 6	6q27
GPR31	G protein-coupled receptor 31	6q27
Chromosome 7 (1378 number of the genes)		
MAFK	Fascia fibrosarcoma	7p22.3
PDGFA	Platelet-derived growth factor-α polypeptide	7p22.3
PMS2	Mismatch repair system component	7p22.1
CRS	Craniosynostosis	7p21.1
IL6	Interleukin-6	7p15.3
HOXA1	Homeobox A1	7p15.2
MRPL32	Mitochondrial ribosomal protein L32	7p14.1
INHBA	Inhibin subunit βA	7p14.1
CDC2L5	Cyclin subunit βA	7p14.1
RALA	*ral* oncogene	7p14.1
TRG	T cell receptor γ chain gene family	7p14
EGFR	Epidermal growth factor receptor	7p11.2
ASL	Arginosuccinate lyase	7q11.21
GUSB	Glucuronidase, β	7q11.21
STX1A	Syntaxin 1A	7q11.23
HGF	Hepatic growth factor	7q21.11
COL1A2	Collagen polypeptide 1, α2	7q21.3
GNB2	G protein subunit β2	7q22.1
EPO	Erythropoietin	7q22.1
MET	*met* oncogene	7q31
LAMB1	Laminin subunit β2	7q31.1
CFTR	Cystic fibrosis transmembrane conductance regulator	7q31.2
IRF5	Interferon regulatory factor-5	7q32.1
BCP	Opsin receptor 1	7q32.1
KEL	Kell blood type	7q34
MKRN1	Makorin ring finger protein 1	7q34
TRB	T cell receptor β chain gene family	7q34
ABCB8	ATP-binding cassette B subfamily 8	7q36.1
NOS3	Nitric oxide synthase 3	7q36.1
Chromosome 8 (927 number of the genes)		
MYOM2	Miyomeshin-2	8p23.3
ARHGEF10	Rho guanine nucleotide exchange factor 10	8p23.3
ANGPT2	Angiopoietin-2	8p23.1
LPL	Lipoprotein lipase	8p21.3
SFTPC	Surfactant protein C	8p21.3
NEFL	Neurofilament-L	8p21.2
WRN	Werner's syndrome	8p12
PLAT	Plasminogen activator	8p11.21
POLB	DNA polymerase β	8p11.21
MOS	*mos* oncogene	8q12.1
SDCBP	Syndecan binding protein	8q12.1
PDE7A	Phosphodiesterase 7A	8q13.1
IL7	Interleukin-7	8q21.13
CA1	Carbonic anhydrase 1	8q21.2

Human Genetic Map Continued.

Gene	Name	Locus
CALB1	Calbindin-1	8q21.3
OXR1	Oxidation resistance-1	8q23.1
ENPP2	Autotaxin connection enzyme 2	8q24.12
MYC	*myc* oncogene	8q24.21
TG	Thyroglobulin (thyroid hormone-related protein)	8q24.22
CYC1	Cytochrome c-1	8q24.3
GPT	Glutamate pyruvate transaminase	8q24.3
GML	GPI like proteins	8q24.3
CYP11B1	Cytochrome P450-XIB1	8q24.3
Chromosome 9 (1076 number of the genes)		
AK3	Adenylate kinase 3	9p24.1
TYRP1	Tyrosinase-related protein 1	9p23
IFNA8	Interferon α8	9p21.3
MLLT3	Myeloid / lymphoid leukemia	9p21.3
TEK	TEK receptor tyrosine kinase	9p21.2
GALT	Galactose-1-phosphate uridyl-transferase	9p13.3
ALDH1	Aldehyde dehydrogenase 1A1	9q21.13
ANXA1	Annexin A1	9q21.13
CHAC	Chorein	9q21.2
XPA	Xeroderma pigmentosum group A	9q22.33
ALDOB	Fructose-bisphosphate B	9q31.1
TXN	Thioredoxin	9q31.3
TNFSF15	Tumor necrosis factor 15	9q32
AMBP	α-1-microglobulin / bikunin precursor	9q32
ORM1	Orosomucoid 1	9q32
TNC	Tenascin C	9q33.1
OR1B1	Olfactory receptor family 1 subfamily B member 1	9q33.2
C5	Complement component-5	9q33.2
HSPA5	Heat shock protein family A5	9q33.3
AK1	Adenylate kinase 1	9q34.11
ASS1	Arginosuccinat synthetase 1	9q34.11
SPTAN1	α spectrin	9q34.11
ABL1	*abl* oncogene	9q34.12
VAV2	*VAV2* oncogene	9q34.2
ABO	ABO blood group	9q34.2
JBTS1	Joubert syndrome 1 (Cerebellum growth failure)	9q34.3
Chromosome 10 (983 number of the genes)		
ADARB2	Adenosine deaminase RNA specific B2	10p15.3
IL2R	Interleukin-2 receptor subunit α	10p15.1
VIM	Vimentin	10p13
MSRB2	Methionine sulfoxide reductase B2	10p12.2
YME1L1	Mitochondrial escape 1-like 1	10p12.1
RET	*ret* oncogene	10q11.21
RBP3	Retinol-binding protein-3	10q11.22
ERCC6	ERCC excision repair 6 (chromatin remodeling factor)	10q11.23
EGR2	Early growth response-2	10q21.3
HK1	Hexokinase-1	10q22.1
PLAU	Urokinase	10q22.2
FAS	*fas* oncogene antigen	10q23.31
DNTT	DNA nucleotidylexotransferase	10q24.1
GOT1	Glutamic-oxaloacetic transaminase-1	10q24.2
TLX1	T cell leukemia homeobox 1	10q24.31
ADRA2	α-2A adrenergic receptor	10q25.2
ADRB1	β-1 adrenergic receptor	10q25.3
CASP7	Caspase 7	10q25.3
DCLRE1A	DNA cross-link repair 1A	10q25.3
OAT	Ornithine aminotransferase	10q26.13
ADAM12	ADAM metallopeptidase domain 12	10q26.2
CYP2E	Cytochrome P450-IIE1	10q26.3
Chromosome 11 (1692 number of the genes)		
HRAS	*Ha-ras* oncogene	11p15.5
IGF2	Insulin-like growth factor 2	11p15.5
INS	Insulin	11p15.5
WT2	Wilms' tumor 2	11p15.5
HBB	β-hemoglobin	11p15.4
ARNTL	Aryl hydrocarbon receptor	11p15.3
PTH	Parathyroid hormone	11p15.3
LDHA	Lactate dehydrogenase A	11p15.1
MYOD1	Myogenic differentiation 1	11p15.1
FSHB	Follicle stimulating hormone subunit β	11p14.1
KCNA4	Potassium ion channel-A4	11p14.1
CAT	Catalase	11p13
WT1	Wilms' tumor 1	11p13
ACP2	Acid phosphatase-2	11p11.2
CD5	CD5 molecule	11q12.2
GANAB	Glucosidase II α unit	11q12.3
GRK2	G protein-coupled receptor kinase 2	11q13.2
GSTP1	Glutathione S-transferase π 1	11q13.2
FGF3	Fibroblast growth factor-3	11q13.3
FGF4	Fibroblast growth factor-4	11q13.3
CAPN5	Calpain 5	11q13.5
IDDM4	Insulin-dependent diabetes mellitus 4	11q13
TYR	Tyrosinase	11q14.3
MMP27	Macrophage metalloproteinase 27	11q22.2
PTS	Pyruvoyltetrahydropterin synthase	11q23.1
DRD2	Dopamine receptor-D2	11q23.2
TAGLN	Transgelin	11q23.3
THY1	Thy-1 T-cell antigen	11q23.3
OR8D1	Olfactory receptor family 8 subfamily D member 1	11q24.2
ETS1	*Est* oncogene	11q24.3
Chromosome 12 (1268 number of the genes)		
LAMP1P1	Lysosome-associated membrane protein-1-pseudogene 1	12p13.3
FGF6	Fibroblast growth factor 6	12p13.32
TPI1	Triosephosphate isomerase-1	12p13.31
GAPDH	Glyceraldehyde-3-phosphate dehydrogenase	12p13.31
GRIN2B	Glutamate receptor-2B	12p13.1
LDHB	Lactate dehydrogenase B	12p12.1
KRAS	KRAS oncogene	12p12.1
PKP2	Plakophilin 2	12p11.21
COL2A1	Collagen II, α1	12q13.11
GPD1	Glycerol-3-phosphate dehydrogenase	12q13.12
HOXC4	Homeobox C4	12q13.13
ARHGAP9	RhoGTP GTPase-activating protein-9	12q13.3
TMEM4	Transmembrane protein-4	12q13.3
GLI1	GLI family zinc finger 1 (associated glioma)	12q13.3
IFNG	γ-interferon	12q15
KITLG	KIT blood type	12q21.32
TMPO	Thymopoietin	12q23.1
PAH	Phenylene alanine hydroxylase	12q23.2
PKU1	Phenylketonuria	12q23.2
ALDH2	Aldehyde dehydrogenase 2	12q24.12

Continued.

Gene	Name	Locus
HNF1A	HNF1 homeobox A (hepatocyte nucleus factor)	12q24.31
POLE	DNA Polymerase ε	12q24.33
Chromosome 13 (496 number of the genes)		
RNR1	Ribosomal RNA	13p12
SGCG	Sarcoglycan γ	13q12.12
CDX2	Caudal type Homeobox transcription factor-2	13q12.2
MBS1	Moebius syndrome 1	13q12.2
NBEA	Neurobeachin	13q13.3
ESD	Esterase D	13q14.2
HTR2A	Serotonin receptor 2A	13q14.2
RB1	RB1 (retinoblastoma)oncogene	13q14.2
PCDH20	Protocadherins 20	13q21.2
BRCA3	Breast cancer type 3	13q21
GPC5	Glypican 5	13q31.3
ERCC5	Excision repair fifth complementation	13q33.1
TNFSF13B	Tumor necrosis factor ligand superfamily, member 13B	13q33.3
F7	Blood coagulation factor VII	13q34
F10	Blood coagulation factor X	13q34
Chromosome 14 (1173 number of the genes)		
RNR2	Ribosome RNA2	14p12
MYH7	Myosin heavy chain 7	14q11.2
PNP	Purine nucleoside phosphorylase	14q11.2
TCRA		14q11.2
SNX6	Sorting nexin 6	14q13.1
PTGDR	Prostaglandin D2 receptor	14q22.1
PYGL	Glycogen phosphorylase L	14q22.1
SPTB	β spectrin	14q23.3
PSEN1	T cell receptor α chain	14q24.2
FOS	*fso* oncogene	14q24.3
TGFB3	Transforming growth factor β3	14q24
GPR68	G protein-coupled receptor 68	14q32.11
CHGA	Chromogranin-A	14q32.12
BCL11B	B-cell leukemia/lymphoma 11B	14q32.2
IGH	Immunoglobulin heavy chain complex	14q32.33
Chromosome 15 (906 number of the genes)		
RNR3	Ribosome RNA3	15p12
PWCR1	Prader-Willy syndrome	15q11.2
AS	Angelman syndrome	15q11.2
ACTC1	Actin α (cardiac muscle)	15q14
GRP58	Glucose-regulated protein (58kDa)	15q15.3
SPTBN5	Spectrin β, non-erythrocytic 5	15q21
B2M	β-2-microglobulin	15q21.1
HEXA	Hexaminidase α	15q23
PKM	Pyruvate kinase M1/M2	15q23
CYP1A1	Cytochrome P450-I-A1	15q24.1
MANA	α-mannosidase	15q24.2
CHRNB4	Cholinergic receptors β4 subunit	15q25.1
NTRK3	Neuro trophic tyrosine kinase receptor-3	15q25.3
BLM	Bloom's syndrome	15q26.1
FES	*fes* oncogene	15q26.1
IGF1R	Insulin-like growth factor-1 receptor	15q26.3
IDDM3	Insulin-dependent diabetes mellitus 3	15q26
Chromosome 16 (906 number of the genes)		
HBA1	Hemoglobin α1 chain	16p13.3
HBA2	Hemoglobin α2 chain	16p13.3
PRM2	Protamine P 2	16p13.13
GP2	Glycoprotein 2	16p12.3
EWWD	Earwax (wet type / dry type)	16p12.1
IL4R	Interleukin-4 receptor	16p12.1
ALDOA	Aldolase A	16p11.2
MT	Metallothionein	16q13
PLLP	Plasmolipin	16q13
GOT2	Glutamic oxaloacetic transferase 2	16q21
HP	Haptoglobin	16q22.2
ZNF23	Zinc finger protein 23	16q22.2
CLEC3A	C-type lectin 3A	16q23.1
CTRB1	Chymotrypsinogen B1	16q23.1
CDH13	Cadherin 13	16q23.3
IL17C	Interleukin-17C	16q24.2
APRT	Adenine phosphoribosyltransferase	16q24.3
FANCA	Fanconi anemia complementation group A	16q24.3
Chromosome 17 (906 number of the genes)		
TP53	Tumor protein p53	17p13.1
POLR2	RNA polymerase II subunit A	17p13.1
CHRNB1	Cholinergic receptor-β1	17p13.1
ALDOC	Aldolase C	17q11.2
NF1	Neurofibromatosis type 1	17q11.2
CRYB1	Crystallin βA1	17q11.2
ERBB2	erb-B2 oncogene	17q12
THRA	Thyroid hormone receptor α1	17q21.1
RARA	Retinoic acid receptor alpha	17q21.2
MYL4	Myosin L chain ? -4	17q21.32
HOXB	Homeobox-antisense RNA 1	17q21.32
MPO	Myeloperoxidase	17q22
ACE	Angiotensin I converting enzyme	17q23.3
GH1	Growth hormone 1	17q23.3
CSH1	Colin Soviet Matoma Motoropin hormone 1	17q23.3
PRKCA	Protein kinase Ca	17q24.2
GALK1	Galactokinase 1	17q25.1
GCGR	Glucagon receptor	17q25.3
P4HB	Prolyl 4-hydroxylase subunit β	17q25.3
TK1	Thymidine kinase 1	17q25.3
Chromosome 18 (906 number of the genes)		
LAMA	Laminin α	18p11.31
MYP2	Myopia-2	18p11.31
PTPRM	Protein tyrosine phosphatase, receptor type M	18p11.23
MC2R	Melanocortin 2 receptor	18p11.2
SS18	Synovial sarcoma chromosomal translocation	18q11.2
CDH2	Cadherin-2	18q12.1
TTR	Transthyretin (Hereditary amyloidosis)	18q12.1
JK	Kidd blood type	18q12.3
DPC4	Pancreatic cancer factor regions	18q21.2
DCC	Netrin receptor (colorectal cancer factor regions)	18q21.2
NEDD4L	Ubiquitin-protein ligase (associated respiratory system disorder)	18q21.3
GRP	Gastrin-releasing protein	18q21.32
BCL2	B-cell leukemia / Lymphoma-related BCL2 gene	18q21.33
FVT1	Follicular lymphoma factor regions	18q21.33
TNFRSF11A	Tumor necrosis factor receptor superfamily 11A	18q21.33
SERPINB2	Serpin B2	18q21.33-q22.1
CYB5	Cytochrome b5 type A	18q22.3

Human Genetic Map Continued.

Gene	Name	Locus
PEPA	Peptidase-A	18q22.3
MBP	Myelin basic protein	18q23
Chromosome 19 (1592 number of the genes)		
VAV1	*Vav*1 oncogene	19p13.3
GNA11	Guanine nucleotide binding protein 11	19p13.3
LDLR	Low density lipoprotein receptor	19p13.2
INSR	Insulin receptor	19p13.2
EPOR	Erythropoietin receptor	19p13.2
JUNB	*Jun*B oncogene	19p13.13
JUND	*Jun*D oncogene	19p13.11
APOC1	Apolipoprotein C-I	19q13.32
APOC2	Apolipoprotein C-II	19q13.32
APOE	Apolipoprotein E (Alzheimer's disease2)	19q13.32
CEA	Carcinoembryonic antigen	19q13.32
CYP2A	Cytochrome P450-IIA7	19q13.32
LU	Lutheran blood type	19q13.32
LHB	luteinizing hormone subunit β	19q13.3
FCGRT	Fc fragment of IgG receptor and transporter	19q13.33
PRKCG	Protein kinase C γ	19q13.42
Chromosome 20 (710 number of the genes)		
ADRA1	Adrenergic receptor, α1D	20p13
ANGPT4	Angiopoietin-4	20p13
OXT	Oxytocin	20p13
PRNP	Prion protein	20p13
PLCB4	Phospholipase-Cβ-4	20p12.3-p12.2
SSTR4	Somatostatin receptor 4	20p11.21
HCK	Hematopoietic cell kinase	20q11.21
GSS	Glutathione synthetase	20q11.22
SRC	*src* oncogene	20q11.23
ADA	Adenosine deaminase	20q13.12
CTSA	Cathepsin A	20q13.12
GNAS	Guanine nucleotide-binding protein	20q13.32
NPBWR2	Neuropeptides B and W receptor 2	20q13.33
CHRNA4	Cholinergic receptor nicotinic alpha α4	20q13.33
Chromosome 21 (337 number of the genes)		
RNR4	Ribosome RNA4	21p12
APP	β-amyloid precursor protein	21q21.3
SOD1	Superoxide dismutase 1	21q22.11
IFNAR1	Interferon α, β receptor subunit 1	21q22.11
AML1	Acute myeloid leukemia	21q22.12
HLCS	Holocarboxylase synthase	21q22.13
ETS2	*est* oncogene-2	21q22.2
CBS	Cystathionine β synthase	21q22.3
COL6A1	Collagen type 4, α-1	21q22.3
CRYAA	Crystallin αA	21q22.3
CSTB	Cystatin B	21q22.3
DSCR	Down syndrome chromosome region	21q22.3
MX1	Mixovirus resistance-1	21q22.3
S100B	S100 calcium-binding protein B	21q22.3
Chromosome 22 (701 number of the genes)		
RNR5	Ribosome RNA5	22p12
CECR1	Cat eye syndrome chromosome region, candidate 1	22q11.1
IGL	Immunoglobulin λ light chain	22q11.2
BCR	Chromosomal breakpoint (chronic myelogenous leukemia / acute lymphatic leukemia)	22q11.23
EWSR1	Ewing sarcoma 1	22q12.2
NEFH	Neurofilament heavy protein	22q12.2
LIF	Leukemia inhibitory factor	22q12.2
NF2	Neurofibromatosis type 2 (nervefibroma symptom type2)	22q12.2
MB	Myoglobin	22q12.3
PDGFB	Platelet-derived growth factor subunit B	22q13.1
CYP2D	Cytochrome P450-IID7	22q13.2
ACO2	Aconitase-2	22q13.2
ARSA	Arylsulfatase A	22q13.33
X chromosome (701 number of the genes)		
GYG2	Glycogenin 2	Xp22.33
XG	Xg blood type	Xp22.33
CSF2RA	Colony-stimulating factor-2 receptor, α	Xp22.32
STS	Steroid sulfatase	Xp22.31
AMELX	Amelogenin (Amelogenesis imperfecta)	Xp22.2
GLRA2	Glycine receptor α-2	Xp22.2
POLA1	DNA polymerase α-1	Xp22.11-p21.3
ZFX	Zinc finger protein	Xp22.11
DMD	Dystrophin (Duchenne muscular dystrophy)	Xp21.2-p21.1
CYBB	Cytochrome b-245 β chain	Xp21.1-p11.4
OTC	Ornithine transcarbamylase deficiency	Xp11.4
MAOA	Monoamine oxidase A	Xp11.3
ARAF1	*ref 1* oncogene	Xp11.3
SSX1	Synovial sarcoma 1	Xp11.23
AR	Androgen receptor	Xq12
XIST	X chromosome controlling element	Xq13.2
PGK1	Phosphoglycerate kinase 1	Xq21.1
MRX40	Mental retardation syndrome (X-linked)	Xq21.1
CHM	Choroideremia (night—blindness factor regions)	Xq21.2
GLA	α-galactosidase	Xq22.1
PLP1	Proteolipid protein 1 (Pelizaeus-Merzbacher disease)	Xq22.2
CAPN6	Calpain 6	Xq23
NDUFA1	Ubiquinone oxidoreductase subunit 1α (mitochondrial disease factor regions)	Xq24
IGSF1	Immunoglobulin superfamily 1	Xq25
HPRT	Hypoxanthine phosphoribosyl-transferase 1	Xq26.2-q26.3
F9	Blood coagulation factor IX (Hemophilia A)	Xq27.1
FRAXA	Fragile site A (fragile X syndrome)	Xq27.3
OPN1LW	Long wave sensitive (abnormal color sensation factor regions)	Xq28
OPN1MW	Medium wave sensitive (abnormal color sensation factor regions)	Xq28
FRAXE	Fragile site E (fragile X syndrome)	Xq28
F8	Blood coagulation factor VIII (Hemophilia B)	Xq28
G6PD	Glucose-6-phosphate dehydroge-nase	Xq28
Y chromosome (255 number of the genes)		
SRY	Sex-determining region Y	Yp11.2
ZFY	Zinc finger protein Y	Yp11.2
AMELY	Amelogenins Y	Yp11.2
AZF1	Azoospermia factor 1 (azoosper-mia factor regions)	Yq11

Structure of Immunoglobulin (Human)

The antibody is also known as immunoglobulin. Based on small differences in the H chain constant region, divided into the five classes of IgA, IgD, IgE, IgG, and IgM.

(Antibody binds the antigenic determinant with the complementarity determining region (CDR)
In the C_{H2} region, there is bonding site for the complement component 1 (→complement activation)
In the C_{H3} region, there is bonding site for the Fc receptor of the neutrophil (→promotes phagocytosis))

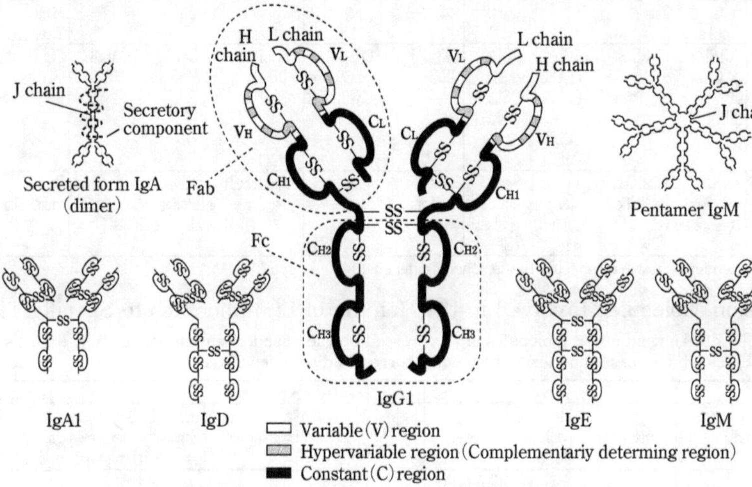

Adhesion Molecules Related to Extravasation of Leukocytes on Leukocytes and Vascular Endothelial Cells

In order for leukocytes to migrate to the point of infiltration by microorganisms, etc., it is necessary for the leukocytes to migrate from blood to outside of blood vessels.

Leukocytes within blood flow roll along the blood vessel wall and attach. After firm adherence, they pass through between endothelial cells and migrate outside of blood vessels. The expression of adhesion molecules on vascular endothelial cells heightens in inflamed areas, and the migration of leukocytes from those areas are facilitated.

	Process								
	Rolling			Tethering			Diapedesis		
Adhesion molecules on surface of white blood corpuscles	LECAM-1 (L selectin)	Sugar chain (CD15s)	CD44	LFA-1	Mac-1	VLA-4	LFA-1	CD31	CD99
	↕	↕	↕	↕	↕	↕	↕	↕	↕
Adhesion molecules on surface of vascular endothelial cells	Sugar chain (Gly-CAM-1)	LECAM-2 (E selectin)	Hyaluronic acid	ICAM-1	RAGE	VCAM-1	JAM-A	CD31	CD99

Integrin family: VLA-4 (CD49d/CD29), LFA-1 (CD11a/CD18), LPAM-1, Mac-1 (CD11b/CD18)

Selectin family: L selectin (LECAM-1, CD62L), E selectin (LECAM-2, CD62E)

Immunoglobulin family: LFA-2 (CD2), CD4, CD8, LFA-3 (CD58), ICAM-1 (CD54), VCAM-1 (CD106), CD28, CD80, NCAM (CD56), CD86, CD152, JAM (CD321-323), CD31

Link protein family: CD44

Sialomucin family: Gly-CAM-1, MadCAM-1, Sialyl Lewis X (CD15s)

TNF family: CD154

Cadherin family: E cadherin

(Source: J. Yata, 2018)

Chemokines Involved in the Homing of Lymphocytes, etc.

The migration of each lymphocytes to the destined area is determined by the type of chemokines produced at that area and expression of the corresponding receptors on the lymphocytes.

Cell	T cell				B cell		Blood plasma cell			Dendritic cell	
Chemokine receptor	CCR7	CXCR5	CCR4 CCR10	CCR6 CCR9	CXCR5	CCR7 CXCR4	CXCR4	CCR9	CCR10	CCR7	CCR6
Homing part	Lymphatic tissue T cell area	Lymphoid follicle	Skin	Intestinal tract	Lymphoid follicle	Peyer's patch	Lymph node medulla, Spleen red pulp, Bone marrow	Lamina propria	Bronchus	Lymphatic tissue, T cell area	Small intestines, Skin
Chemokine	CCL21 (SLC) CCL19 (ELC)	CXCL13 (BLC)	CCL17 (TARC) CCL22 (MDC) CCL 28 (MEC)	CCL20 (LARC) CCL25 (TECK)	CXCL13 (BLC)	CCL21 (SLC) CXCL12 (SDF-1)	CXCL12 (SDF-1)	CCL25 (TECK)	CCL28 (MEC)	CCL21 (SLC)	CCL20 (LARC)
Chemokine producing cells	Vascular endothelium, Dendritic cell	Follicular dendritic cell	Vascular endothelium, Epidermic cell	Small intestinal epithelial cell	Follicular dendritic cell	Vascular endothelium	Stromal cells	Small intestinal epithelial cell	Bronchial epithelial cell	Vascular endothelium	Epithelial cell, Epidermic cell

Homing: tendency towards specific tissue, Chemokine: chemotactic cytokine (Source: J. Yata, 2018)

Adhesion Molecules Involved in the Homing of Lymphocytes to Specific Tissues

Mutually different adhesion molecules are expressed on the vascular endothelial cells of each tissues. Lymphocytes which possess adhesion molecules corresponding to each tissue attach to that area.

		T cell			B cell				Blood plasma cell	
Homing part		Lymph node	Intestinal lymphoid tissue	Skin	Spleen marginal zone		Lymph node	Intestinal lymphoid tissue	Bone marrow, Spleen red pulp, Lymph node medulla	Intestinal lymphoid tissue
Expression bonding molecule	Lymphocyte	L selectin	LPAM-1	CLA derived substance	LFA-1	VLA 4	L selectin	LPAM-1	VLA 4	LPAM-1
		↕	↕	↕	↕	↕	↕	↕	↕	↕
	Vascular endothelial cells	Gly-CAM-1	MadCAM-1	E selectin	ICAM-1	VCAM-1	Gly-CAM-1	MadCAM-1	VCAM-1	MadCAM-1

VLA4: $\alpha_4\beta_1$ integrin, LPAM/-1: $\alpha_4\beta_7$ integrin, LFA-1: $\alpha_L\beta_2$ integrin (Source: J. Yata, 2018)

The Receptors of the Macrophages and Dendritic Cells, for Recognizing Pathogen-Associated Molecular Patterns (PAMPs)

Macrophages and dendritic cells use receptors which recognize molecules existing universally in pathogens to capture and endocytose pathogens, or to cause activation leading to processing of the pathogens. Binding partner molecules of pathogens are shown in parentheses.

Lectin	Toll receptor (TLR)	Others
Dectin-1 (zymosan, β-glucan)	TLR 1 (lipoprotein)	NOD 1 (peptidoglycan derived material)
DC-SIGN (fucose, mannose)	TLR 2 (lipoarabinomannan, lipoteichoic acid, peptidoglycan, zymosan)	NOD 2 (peptidoglycan derived material)
DCIR (fucose, mannose)	TLR 3 (double stranded RNA)	RIG-I (double stranded RNA)
Langerin (mannose, zymosan)	TLR 4 (lipopolysaccharide, fibronectin)	Mda5 (double stranded RNA)
BDCA2 (mannose)	TLR 5 (flagellin)	DAI (double stranded DNA)
MGL-1 (galactose)	TLR 6 (lipopeptide, lipoteichoic acid, zymosan)	AIM 2 (double stranded DNA)
Mannose receptor (mannose)	TLR 7 (single stranded RNA)	Scavenger receptors (SR) (lipopolysaccharide, lipoprotein)
Siglec (sialic acid)	TLR 8 (single stranded RNA)	Peptide glycan recognition protein (PGRP) (peptido glycan)
DEC205 (mannose, sialic acid)	TLR 9 (CpG array of DNA)	LRR protein (CD 14) (lipopolysaccharide, lipoteichoic acid)
Mincle (trehalose, mannose)	TLR 11 (protozoal profilin-like protein)	NLRC4/Ipaf (flagellin)
		Naip-5 (flagellin)
		NLRP3/NALP3 (microbial nucleic acid, muramyldipeptide)

* Exists in cytoplasm, ** Exists in endosomes, others exist on cell surface. (Source: J. Yata, 2018)

Cytokines and Surface Interacting Molecules in the B Cell-Help by T Cells

For many antigens, T cell-help is required for B cells to produce antibodies. This occurs through the action of cytokines produced by the T cells and the action of T cell molecule to the corresponding molecule of the B cells.

Cytokines produced by T cells		IL-2 IL-4 IL-5 IL-10 IL-13 IL-21	IL-2 IL-4 IL-5 IL-6 IL-10 IL-11 IL-13 Interferon γ TGE-β
T cell surface molecules	CD154 ICOS ↕ ↕	↓	↓
B cell surface molecules	CD40 B7RP ↓ ↓ Activation Activation	Activation (Proliferation)	Differentiation into antibody forming cell (IL-5, IL-6, IL-10) Immunoglobulin class switch (IL-4 : IgG1, IgE Interferon γ : IgG1 (Human), IgG2a (Mouse) IL-13 : IgE TGE-β : IgA)

~~~ expressed by activated cells. (Source: J. Yata, 2018)

Chemokines and Migration of Leukocytes to Infiltration Points of Microorganisms or Foreign Matter

Chemokine is a cytokine which possesses a shared molecular structure. It mainly shows leukocyte chemotactic activity.

Classified by the positional relationship of the first 2 cysteines from the N-Terminal as CXCL (17 types), CCL (28 types), CL (2 types), and CX_3CL (1 type).

By reacting to microorganisms, local tissue cells (fibroblasts, vascular endothelial cells, epidermal cells, epithelial cells, macrophages, etc.) produce chemokines and gather leukocytes. The gathered leukocytes process microorganisms, and they also produce chemokines and gather other leukocytes at the same time.

Type of leukocyte	Major chemokines	Type of leukocyte	Major chemokines
Neutrophil (Phagocytosis, bacteriocidal activity, proinflammatric action)	CXCL8(IL-8/NAP-1) CXCL2(MIP-2α) CXCL3(MIP-2β) CXCL1(NAP-3)	T cell	CCL8(MCP-2) CCL19(ELC) CCL21(SLC)
Monocyte (Phagocytosis, bacteriocidal activity, digestion of foreign body, proinflammatric action)	CCL2(MCP-1) CCL8(MCP-2) CCL7(MCP-3) CCL3(MIP-1α) CCL4(MIP-1β) CCL5(RANTES) CXCL10(IP-10)	B cell (antibody production)	CXCL9(Mig) CXCL10(IP-10) CXCL13(BLC) CXCL14(BRAK) CCL19(ELC) CCL21(SLC)
		Eosinophil (Helminth elimination, inflammation induction)	CCL8(MCP-2) CCL11(Eotaxin-1) CCL24(Eotaxin-2) CCL26(Eotaxin-3) CCL5(RANTES)
T cell (Macrophage activatin, destruction of infected cells, antibody production assistance)	CXCL9(Mig) CXCL10(IP-10) CXCL11(I-TAC) CCL2(MCP-1) CCL3(MIP-1α) CCL4(MIP-1β) CCL5(RANTES) CCL7(MCP-3)	Basophil (pro-inflammatic action)	CXCL8(IL-8/NAP-1) CCL2(MCP-1) CCL3(MIP-1α) CCL5(RANTES) CCL11(Eotaxin-1)

Alternate names are given in ( ) after the official names (Source: J. Yata 2018)
~~~

Effector Molecules in the Interaction between T Cells and Antigen-Presenting Cells (APC)

T cells react to the complex of antigenic peptides and MHC molecules on antigen-presenting cells by antigen receptors. In addition, T cells and APC are further activated by the interaction of the molecules on both cells and by the cytokine which is produced by partner cells. The T cells are activated and start work.

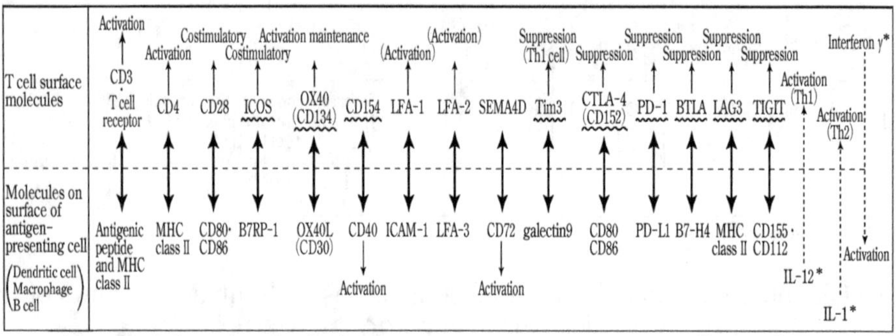

～ (expressed through activation)
* Cytokine

(Source: J. Yata, 2018)

Types, Differentiation Pathway and Factions of Immunocompetent Cells

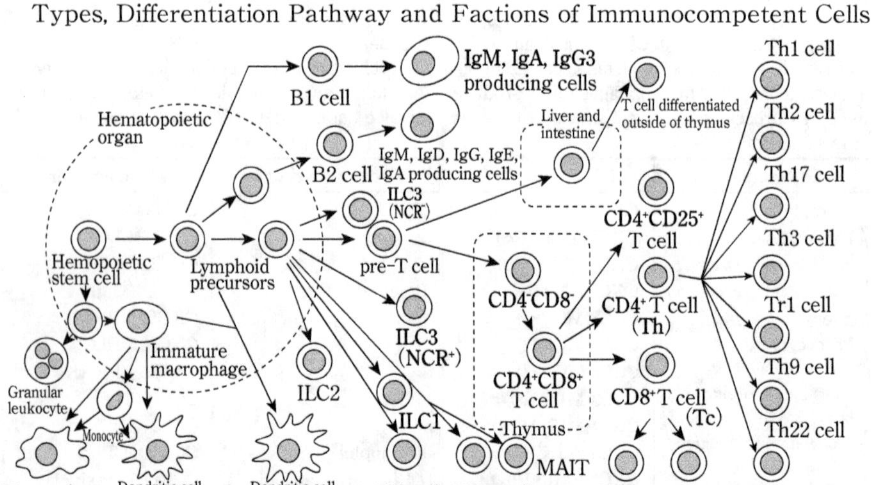

Th (helper T cell) : Assistance of antibody production and activation of macrophage
Tc (cytotoxic T cell) : Destruction of transplanted cells, infected cell, and tumor cell
Th1 cells : production of interferon γ, etc. ; macrophage and Tc activation
Th2 cells : production of IL-4, etc. ; help of antibody production
Th3 cells : TGF-β production ; immunosuppression
Th9 cells : IL-9 production, mast cell activation
Th17 cells : IL-17 production, pro-inflammatory
Th22 cells : IL-22 production, epithelial support
Tr1 cells : IL-10 production, immunosuppression
T_{FH} cells : IL-21 production, B cell help in lymphoid follicle
cDC : Classical dendritic cell
pDC : Plasmacytoid dendritic cell

$CD4^+ CD25^+$ T cells (regulatory T-cells) : suppression of other T cells
NK cells : destruction of virus-infected cells and tumor cells
NKT cells : tumor cell destruction, immunomodulatory
Dendritic cell : Antigen presentation to (T cell)
Macrophage : Antigen presentation, phagocytisis of foreign bodies digestion, and bacteriocidal
ILC2 (NH cell, Ih cell, nuo-cyte) : IL-5, IL-13 production
ILC3 (NCR^+) : IL-22 production, epithelial support
ILC1 : Interferon γ production
ILC3 (NCR^-) : IL-17 production
(ILC : innate lymphoid cells)
Granulocyte : neutrophils ; phagocytosis and killing of bacteria
MAIT (mucosal associated invariant T cell) : cytokine production

Distinctive Surface Molecules and Functions for each Subpopulation Lymphocyte

Lymphocytes include B cells, T cells, NK cells, NKT cells, and ILC. T cells are divided into $CD4^+$ and $CD8^+$ for classification. Furthermore, based on differences of the cytokines produced, $CD4^+$ T cells are divided into Th1 cells (interferon-γ, TNF-β, and IL-2 production), Th2 cells (IL-4, IL-5, IL-6, IL-9, IL-10, and IL-13 production), Th3 cells (TGF-β production), Tr1 cells (IL-10 production), and Th17 cells (IL-17 production). $CD4^+$ $CD25^+$ T cells suppress the function of other T cells and act as regulatory T cells. T cells prior to the reaction with anitgens are called "naive T cells." T cells after the reaction are called "memory T cells." These different types of T cells can be distinguished by differences of the molecules existing on the cell surface.

Matter	B cell	T cell	NK cells	NKT cell	ILC
Surface molecule	Immunoglobulin, CD19, CD20, CD21 (complement receptor)	CD3 (T cell receptor) CD4 : Helper CXCR3, CCR5 : Th1 CCR4, CCR3 : Th2 CD8 : Killer CD45RA : Naive CD45RO : Memory CD4·CD25 : Regulatory	CD56, CD16	Specific T cell receptor (Mouse $V\alpha14V\beta8.2$ Human $V\alpha24V\beta11$) NK receptor (Mouse NK 1.1 Human NKR-P1)	
Alloted function	Immunoglobulin (antibody) production	Th1 cell: Activation of macrophage and induction of killer T cell Th2 cell: Assistance of antibody production Th3 cell: Function control of other T cell, macrophage, and dendritic cell Tr1 cell: Function control of other T cell, macrophage, and dendritic cell $CD4^+$ $CD25^+$ T cell: Function control of other T cells Th17 cells: Mobilization of neutrophilic leukocytes causing inflammation $CD8^+$ T cells: Perforin granzymes are discharaged and injure virus-infected cells, tramsplanted cells, and tumor cells.	Perforin and granzymes are discharaged and injure virus-infected cells, tramsplanted cells, and tumor cells. Interferon γ is produced, and activats macrophage, dendritic cells, and Th1 cells	Interferon γ or IL-4 is produced and controll the recation of T cells. (Interferon γ activates Th1 cells, macrophage, and dendritic cells. IL-4 induces Th2 cells.) IL-17 production which activates neutrophils Perforin, granzymes are produced and injure tumor cells.	ILC1: Interferon γ production and macrophage activation ILC2: IL-13 production, supporting antibody production ILC17 (NCR^- ILC3): IL-17 production, neutrophil activation ILC22 (NCR^+ ILC3): IL-22 production, induces antibacterial peptide production from epitheliuim and supports proliferation of epithelial cells
Cell division inducer (mitogen)	PWM, LPS	PHA, ConA, PWM			

CD: CD (Cluster of differentiation) display antigens CXCR: receptor for CXC chemokine, CCR: receptor for CC chemokines, IL: interleukins, ILC: innate lymphoid cells, PHA: phytohemagglutinin, ConA: concanavalin A, PWM: pokeweed mitogen, LPS: lipopolysaccharide, TNF: tumor necrosis factors (Source: J. Yata, 2016)

Physiology

Photoreaction of Net Photosynthetic Rate for Individual Leaves in Optimal Temperature and Atmosphere CO_2

Plant group	Light compensation point I_c (μmol photon m^{-2} s^{-1})	Saturation light intensity (μmol photon m^{-2} s^{-1})	Plant group	Light compensation point I_c (μmol photon m^{-2} s^{-1})	Saturation light intensity (μmol photon m^{-2} s^{-1})
Terrestrial plant			**Woody plant**		
Flowering herb			Evergreen broadleaf tree located outside the tropics		
C_4 plant	20–50	>1 500	Sun leaf	10–30	600–1 000
Desert plant	>1 500		Shade leaf	2–10	100–300
Agricultural C_3 plant	20–40	1 000–1 500	Conifer		
Sun plant	20–40	1 000–1 500	Sun leaf	30–40	800–1 100
Spring geophyte	10–20	300–1 000	Shade leaf	2–10	150–200
Cryptozoic plant	5–10	100–200(400)	Pteridophyte		
			growing in well lit locations	About 50	400–600(800)
Woody plant			growing in shaded locations	1–5	50–150
Tropical forest tree			Moss	5–20	150–300
Sun leaf	15–25	(400)600–1 500	Lichen	50–150	300–600
Shade leaf	5–10	200–300	Hydrophyte		
Young tree	2–5	50–150	Plankton algae		200–500
Deciduous broad-leaved tree and deciduous shrub			Seaweed in tidal zones	5–8	200–500
Sun leaf	20–50(100)	600–>1 000	Deep water algae	2	150–400
Shade leaf	10–15	200–500	Submersed vascular plant	8–20(30)	(60)100–200(400)

(Source: Larcher W. 1994)

Type of Photosynthetic Pigment and Species†

Species	Chlorophyl						Phycobilin		Major carotenoids	
	a	*b*	*c*	*d*	Bacteri-ochloro-phyll	Chlorobium chlorophyll	Phyco-erythrin	Phycocy-anin	Carotene	Xanthophyll
Higher plant, ferneries, and moss plants	+	+	–	–	–	–	–	–	β-carotene	Lutein Viola xanthine Neo-xanthine
Green algae	+	+	–	–	–	–	–	–	β-carotene	Lutein Viola xanthine Neo-xanthine
Brown algae	+	–	+	–	–	–	–	–	β-carotene	fucoxanthin
Red seaweeds	+	–	–	+	–	–	+	+	α-carotene β-carotene	Lutein Zeaxanthin
Euglena	+	+	–	–	–	–	–	–	β-carotene	Viola xanthine Diadinoxanthin Neo-xanthine
Brown flagellum algae	+	–	+	–	–	–	+	+	α-carotene	alloxantin
Diatoms	+	–	+	–	–	–	–	–	β-carotene	Fucoxanthin Fucoxanthin
Yellow flagellum algae	+	–	+	–	–	–	–	–	β-carotene	Fucoxanthin Diadinoxanthin
Blue-green algae (cyanobacterium)	+	–	–	–	–	–	+	+	β-carotene	Echinenone Myxoxanthophyll
Purple sulfur bacteria	–	–	–	–	+	–	–	–		spiro-xanthine
Purple red non-sulfur bacteria	–	–	–	–	+	–	–	–	Lycopene	spiro-xanthine
Green sulfur bacteria	–	–	–	–	+	+	–	–	Chlorobactin γ-carotene	Hydro-chlorobactin

† Source: Seki Shimizu "Doubutsuseirigaku 3rd ed." Shokabo Co. Ltd. (1982)

Absorption Spectrum of Chlorophyll

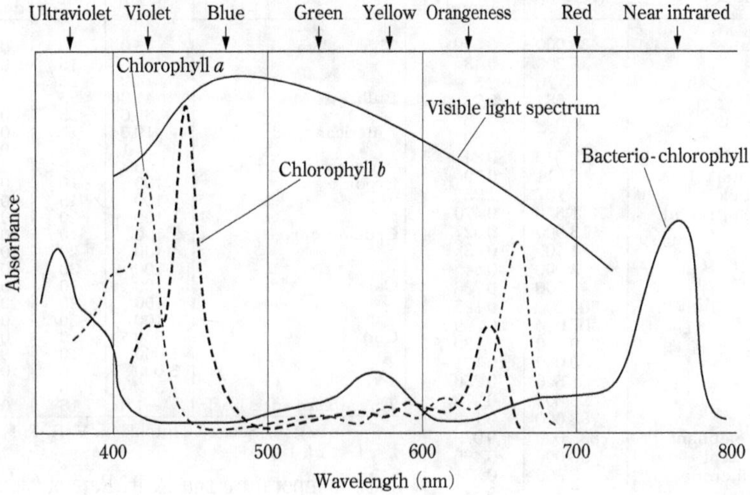

Based on data archived in Tatsuo Sugiyama (Supervisor) : Biochemistry, molecular biology of the plant, society publication center (2005)

Absorption Spectrum of Other Photosynthesis Pigments

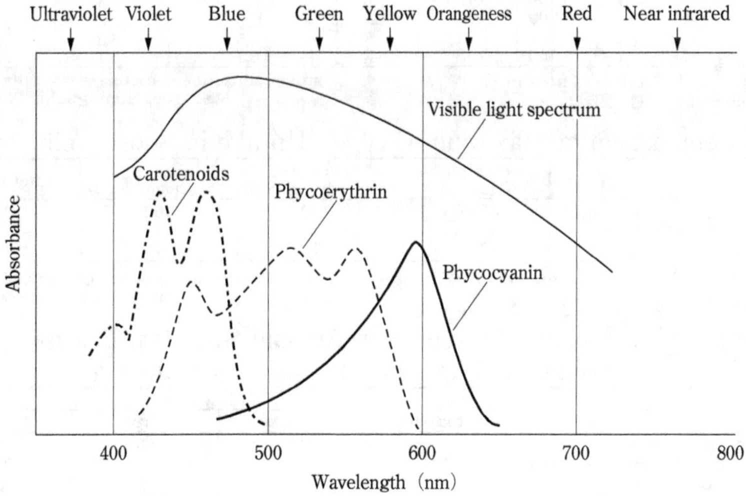

Based on data archived in Tatsuo Sugiyama (Supervisor) : Biochemistry, molecular biology of the plant, society publication center (2005)

Oxygen Consumption (Mammals, Birds)

Species	Weight	Oxygen demand*
	g	
Chimpanzee	38 000	0.249
Rhesus	3 200	0.43
Short-tailed shrew	20	3.5
Hedgehog	684	0.73
Guinea pig	500	0.78
Mouse	35.7	1.59
Rat	545	0.84
Gray squirrel	543.1	0.79
Chipmuck	107	1.25
European rabbit	1 520	0.470
Dog	13 900	0.327
Fox	4 400	0.43
Cat	3 000	0.44
Mink	700	0.738
Cow	500 000	0.125
Camel	407 000	0.099
Pig ♂	250 000	0.130
♀	250 000	0.121
Sheep	51 500	0.256
Goat	20 600	0.272
Horse	703 000	0.145
Indian elephant	3 833 000	0.07
Harbor seal	33 000	0.496
Sandhill crane	3 890	0.37
Goose	3 300	0.58
	5 000	0.49
	5 890	0.40
Brant goose (summer)	1 130	0.85
Brant goose (winter)	1 168	0.71
Gray heron	1 880	0.56
Duck	1 870	0.73
American kestrel	108	1.37
Quail	97	2.06
Chicken	2 000	0.423
Turkey	3 700	0.43
Rock pigeon	266	1.1
Wood pigeon	150	0.98

* mL/g/h (Source: Hudson, J. W. & Lasiewski, R. C., 1974)

Oxygen Consumption (Amphibians, Fish)

Species	Weight	Temp.	Oxygen demand*
	g	℃	
Tiger salamander	15.0	10	0.0595
	16.5	15	0.0918
	14.9	25	0.1052
Bullfrog	47.3	15	0.0571
	35.0	25	0.0430
American toad	19.0	15	0.1101
	21.1	25	0.1217
African clawed frog	63.3	15	0.0338
Brookie	100	10	0.035
	100	15	0.075
	100	20	0.103
Crucian carp	100	10	0.011
	100	20	0.021
	100	30	0.050
Carp	100	10	0.012
	100	20	0.035
	100	30	0.070
Cod	500	10	0.052
	1 000	10	0.048
	3 000	10	0.042
	500	15	0.064
Three-spined stickleback	1.0	18	0.300

* mL/g/h (Source: Hutchison, V. H. & Fry, F. E. J. *et al.* 1974)

Body Temperature and Heart Rate of Chipmunks

Sex	Rearing temperature	Temperature	Ventricular rate
	℃	℃	/min
♂	24	38.8	493.5
♀	6.5	33.0	399.0
♂	24	39.2	485.0
♀	6.5	35.8	476.7
♀	6.5	16.8	244.4**
♂	6.5	12.9	199.0**
♂	6.5	12.3	195.4**
♀	6.5	9.4	96.2**
♀	6.5	8.6	77.5**
♀	6.5	8.3	68.4**
♂	6.5	8.1	63.4**

* Tamias sibricus asiaticus, ** Hibernating (Source: Nomaguchi, T. A., 1982)

Comparison of Maximum Oxygen Uptake in Sport Animals

	Sports player (Human)	Racehorse (Thoroughbred)	Dog race dog (Greyhound kind)	Camel for race
Maximum oxygen intake* (mL/kg/min)	69–85	160	100	51
Run speed (m/sec)	10–11	19	16.6	10–11

* V_{O_2max} (Source: A. Amada, 1998)[oxygen (mL) consumed per minute per gram of body weight]

Respiration Function, Heat Production Amount, and Heat Release Amount for Horse (700 kg) when Standing and Moving

	Walking style			
	Stopped	Walk	Trot	Canter
Run speed (m/min)	0	115	240	340
Pulmonary ventilation (L/min)	212	288	599	948
Oxygen consumption (L/min)	2.5	12.2	22.5	37.8
Heat produced (kJ/min)	50	170	325	550
Heat emitted (kJ/min)	50	146	276	468

Each valuse is the amount for 1 minute. (Taken from A. Amada, 1998)

Heart Rate, Respiration Index, Blood Amount and Red Blood Cell Properties

Species (measurement condition)	Ventricular rate	Respiratory rate	Respiratory volume	Amount of blood	Volume of red blood corpuscle	Red blood corpuscle diameter
	/min	/min	L/min	mL/kg	μm³	μm
Human ♂	64	10.1-13.1	5.8-10.3	71-78	87	6.5-8.0
♀	69	10.4-13.0	4.0-5.1	65-71	87	6.5-8.0
Rhesus macaque	192	31-52	0.31-1.41	44.3-66.6	-	7.4
Guinea pig	280-300	69-104	0.10-0.38	67.0-92.4	71-83	7.0-7.5
Mouse	600-655	84-230	0.011-0.036	-	48-51	6.0
Rat	362	66-114	0.05-0.101	57.5-69.9	57-65	6.0-7.5
European rabbit	205-220	26.7-36.8	0.170-0.188	44.0-70.0	60-68	6.5-7.5
Dog	50-130	11-37	3.3-7.4	86.2	59-68	6.2-8.0
Cat (lying and awaking)	110-140	20-40	0.68-1.36	47.3-65.7	51-63	5.0-7.0
Cow (dairy, standing)	45	26-35	82-109	52.4-60.6	47-54	5.9
Pig (23-27 kg in weight)	60-86	32-58	4.8-8.7	61-68	59-63	-
Sheep	60-80	15.7-23.6	5.95-7.69	59.7-73.8	30-32	4.8
Goat	60-100	19	5.7	56.8-89.4	19.3	-
Horse (thoroughbred, one year old or more)	23-46	10-14	60-72	56.0-118.1	-	5.5
African elephant (under anesthetic)	50 (38-62)	10 (7-13)	-	-	-	-
Goose ♂	80-144	20	-	-	-	-
♀	80-144	40	-	-	-	-
Duck ♂	120-200	42	-	30	-	-
♀	120-200	110	-	25	-	-
Chicken ♂	150-400	17	0.777	22	120-137	11.2×6.8
♀	150-400	27	0.766	16	120-137	11.2×6.8
Turkey ♂	93	28	-	-	-	15.5×7.5
♀	93	49	-	-	-	15.5×7.5
Rock pigeon	120-360	25-30	-	49	-	11.5-14.5 ×5.7-8.5
House sparrow	640-910	81-112	-	-	-	-
Canary	570-840	96-120	-	-	-	-
Mississippi alligator	-	-	-	10.4-13.8	450.0	23.2×12.1
Two-toed amphiuma*	-	-	-	-	13 200-14 513	78×46
Bullfrog	-	-	-	-	625-716	24.8×15.3
Crucian carp	36-40	-	-	-	-	8.8×13.4
Carp	40-78	-	-	-	278-340	-
Eel	48-56	-	-	-	141-170	13.0×8.0
Cod	48-60	-	-	-	159-201	12.2×9.0
Northern dogfish	40-50	-	-	-	820.0	22.7×15.2
Skate	16-50	-	-	-	646-910	23.7×14.4
Hagfish	-	-	-	-	1 470-1 560	24.6×18.3

* *Amphiuma means.*

Unless otherwise noted, values are taken at rest.

(Sources: A. R. Dawe *et al.*, V. Kruta *et al.*, J. J. Gleysteen *et al.*, M. Reynolds, P. D. Altland *et al.*, 1974, M. J. Swenson, 1984, D. J. Heard *et al.*, 1988, etc.)

Reference Values for Blood Constituents (Human)

Age	Red blood cell count ($10^6/\mu L$)		Age	Hematocrit (%)		Age	Hemoglobin (g/dL)		Age	White blood cell count ($10^3/\mu L$)	
	Male	Female		Male	Female		Male	Female		Male	Female
Adult	4.35-5.55	3.86-4.92	Adult	40.7-50.1	35.1-44.4	Adult	13.7-16.8	11.6-14.8	Adult	3.3-8.6	3.3-8.6

Kiyoshi Ichihara *et al.*, : Collaborative derivation of reference intervals for major clinical laboratory tests in Japan. Ann Clin Biochem, 0004563215608875, first published on September 11, 2015.

Reference Values for Blood Serum and Plasma Constituents

Most of the values apply to adults unless the age is specified.

Total protein (Biuret method, g/dL)		Albumin (BCG method, Modified BCP method, g/dL)		Thymol turbidity test (TTT) (U)		Zinc sulfate turbidity test (ZTT) (U)	
Male	Female	Male	Female	Male	Female	Male	Female
6.6-8.1[*2]	6.6-8.1[*2]	4.1-5.1[*2]	4.1-5.1[*2]	<5.0	<4.6	1.6-13.2	3.2-13.5
Fasting blood sugar (mg/dL)		**Total cholesterol** (enzyme method, mg/dL)		**HDL cholesterol** (Homogeneous method, mg/dL)		**LDL cholesterol** (Homogeneous method, mg/dL)	
Male	Female	Male	Female	Male	Female	Male	Female
73-109[*2]	73-109[*2]	142-248[*2]	142-248[*2]	38-90[*2]	48-103[*2]	65-163[*2]	65-163[*2]
Triglyceride (neutral fat) (enzyme method, mg/dL)		**Non-esterified fatty acid** (enzyme method, mEq/L)		**Phospholipid** (enzyme method, mg/dL)		**Hemoglobin A1c** (HPLC method, Immunization method, enzyme method, %)	
Male	Female	Male	Female	Male	Female	Male	Female
40-234[*2]	30-117[*2]	0.14-0.85	0.14-0.85	152-303	157-289	4.9-6.0	4.9-6.0
AST (GOT) (JSCC transferable method, U/L)		**ALT (GPT)** (JSCC transferable method, U/L)		**LDH (LD)** (JSCC transferable method, U/L)		**ALP (Alkaline phosphatase)** (JSCC transferable method, U/L)	
Male	Female	Male	Female	Male	Female	Male	Female
13-30[*2]	13-30[*2]	10-42[*2]	7-23[*2]	124-222[*2]	124-222[*2]	106-322[*2]	106-322[*2]
γGT (JSCC transferable method, U/L)		**Total bilirubin** (enzyme method, mg/dL)		**Cholinesterase (ChE)** (JSCC transferable method, U/L)		**Amylase** (JSCC transferable method, U/L)	
Male	Female	Male	Female	Male	Female	Male	Female
13-64[*2]	9-32[*2]	0.4-1.5[*2]	0.4-1.5[*2]	240-486[*2]	201-421[*2]	44-132[*2]	44-132[*2]
Creatine kinase (CK) (JSCC transferable method, U/L)		**Urea nitrogen (BUN, UN)** (Urease-UV method, mg/dL)		**Creatinine (Cr)** (enzyme method, mg/dL)		**Uric acid (UA)** (Uricase-POD method, mg/dL)	
Male	Female	Male	Female	Male	Female	Male	Female
59-248[*2]	41-153[*2]	8-20[*2]	8-20[*2]	0.65-1.07	0.46-0.79	3.7-7.8[*2]	2.6-5.5[*2]
Sodium (Serum) (Ion-selective electrode method, mEq/L)		**Potassium (Serum)** (Ion-selective electrode method, mEq/L)		**Serum Cl** (Ion-selective electrode method, mEq/L)		**Calcium (Serum)** (colorimetric method, mg/dL)	
Male	Female	Male	Female	Male	Female	Male	Female
138-145[*2]	138-145[*2]	3.6-4.8[*2]	3.6-4.8[*2]	101-108[*2]	101-108[*2]	8.8-10.1[*2]	8.8-10.1[*2]

Continued.

Inorganic phosphorus (Serum) (Molybdate direct method, enzyme method, mg/dL)		**Magnesium (Serum)** (Xylidylblue method, mg/dL)		**Iron (Serum)** (Bathophenanthroline direct method, Nitroso-PSAP method, μg/dL)		**Total iron-binding capacity (TIBC) (Serum)** (Bathophenanthroline direct method, Nitroso-PSAP method, μg/dL)	
Male	Female	Male	Female	Male	Female	Male	Female
2.7-4.6[*2]	2.7-4.6[*2]	1.9-2.5	1.9-2.4	40-188[*2]	40-188[*2]	253-365	246-410
Copper (Serum) (Atomic absorption spectrometry, Colorimetric method, μg/L)		**CRP (C-reactive protein)** (Latex agglutination turbidimetric immunoassay, Turbidimetric immunoassay, Nephelometric immunoassay, mg/dL)		**α_1-antitrypsin** (TIA, Nephelometry, mg/dL)		**Haptoglobin** (TIA, Nephelometry, mg/dL)	
Male	Female	Male	Female	Male	Female	Male	Female
67-123	65-126	0.00-0.14[*2]	0.00-0.14[*2]	94-150[*1]	94-150[*1]	17-152[*1]	17-152[*1]
α_1-acid glycoprotein (TIA, Nephelometry, mg/dL)		**Ceruloplasmin** (TIA, Nephelometry, mg/dL)		**Transferrin** (TIA, Nephelometry, mg/dL)		**Transthyretin (Prealbumin)** (TIA, Nephelometry, mg/dL)	
Male	Female	Male	Female	Male	Female	Male	Female
45-98[*1]	39-86[*1]	21-37[*1]	21-37[*1]	190-300[*1]	200-340[*1]	23-42[*1]	22-34[*1]
α_2-macroglobulin (TIA, Nephelometry, mg/dL)		**Immunoglobulin G (IgG)** (Latex agglutination immunoassay, TIA, Nephelometry, mg/dL)		**Immunoglobulin A (IgA)** (Latex agglutination immunoassay, TIA, Nephelometry, mg/dL)		**Immunoglobulin M (IgM)** (Latex agglutination immunoassay, TIA, Nephelometry, mg/dL)	
Male	Female	Male	Female	Male	Female	Male	Female
100-200	130-250	861-1747[*2]	861-1747[*2]	93-393[*2]	93-393[*2]	33-183[*2]	50-269[*2]
Immunoglobulin D (IgD) (Latex agglutination immunoassay, mg/dL)		**Nonspecific immunoglobulin E (IgE)** (EIA, U/mL)		**Complement C3** (Latex agglutination immunoassay, TIA, Nephelometry, mg/dL)		**Complement C4** (Latex agglutination immunoassay, TIA, Nephelometry, mg/dL)	
Male	Female	Male	Female	Male	Female	Male	Female
<9	<9	<170	<170	73-138[*2]	73-138[*2]	11-31[*2]	11-31[*2]
Total triiodothyronine (T_3) (RIA, ng/mL)		**Free triiodothyronine (FT_3)** (RIA, pg/mL)		**Total thyroxine (T_4)** (RIA, μg/dL)		**Free thyroxine (FT_4)** (RIA, ng/dL)	
Male	Female	Male	Female	Male	Female	Male	Female
0.8-1.6	0.8-1.6	0.9-1.7	0.9-1.7	6.1-12.4	6.1-12.4	0.9-1.7	0.9-1.7
Carcinoembryonic antigens (CEA) (EIA, ng/mL)		**CA19-9** (EIA, U/mL)		**CA125** (EIA, U/mL)		**Nerve peculiar enolase (NSE)**[*3] (RIA, ng/mL)	
Male	Female	Male	Female	Male	Female	Male	Female
<5	<5	<37	<37	<35	<35	<10	<10

Reference Values for Blood Serum and Plasma Constituents Continued.

Age	Growth hormone (GH) (IRMA, ng/mL)		Age	Follicle-stimulating hormone (FSH) (TR-FIA method, mIU/mL)		Age	Luteinizing hormone (LH) (TR-FIA method, mIU/mL)	
	Male	Female		Male	Female		Male	Female
1 mo.			1 mo.			1 mo.		
6 mo.			6 mo.			6 mo.		
1	0.17-31.3	0.18-24.3	1	0.18- 2.06	1.02- 8.44	1	0.04-0.77	0.02- 0.87
5	0.10-23.2	0.15-21.9	5	0.20- 2.34	0.27- 3.84	5	0.02-0.36	0.02- 0.52
10	0.08-20.8	0.21-26.6	10	0.48- 4.66	0.75- 7.04	10	0.07-1.83	0.04- 5.74
15	0.08-19.9	0.17-24.2	15	0.84- 7.02	1.36-10.1	15	0.45-5.32	0.21-15.1
Adult	<0.42	0.66-3.68	Adult	1.0 -10.5	Early follicular phase 1.9 - 9.5 Ovulation period peak 3.1 -10.6 Luteal phase 0.5 - 6.4	Adult	1.0 -8.4	Early follicular phase 1.6 - 9.3 Ovulation period peak 13.8 -71.8 Luteal phase 0.5 -12.8

Age	Thyrotropic hormone (TSH) (Immunoassay, μU/mL)		Age	High sensitivity parathyroid hormone (M-PTH) (Double antibody radioimmunoassay, pg/mL)		Age	Aldosterone (Ald) (RIA method, pg/mL)	
	Male	Female		Male	Female		Male	Female
			1 mo.			1 mo.		
			6 mo.			6 mo.		
			1	110-430	110-430	1	16.4-409	14.5-524
			5	140-510	140-510	5	-280	-353
			10			10	13.8-374	-387
			15			15	15.6-399	12.9-489
Adult	0.35-4.94	0.35-4.94	Adult	160-520	160-520	Adult	35.7-240	35.7-240

Age	Hydroxy 11-corticosteroid (11-OHCS) (fluoroscopic method, μg/dL)		Age	Testosterone (T) (RIA method, ng/dL)		Age	Estradiol (E_2) (RIA method, pg/mL)	
	Male	Female		Male	Female		Male	Female
1 mo.			1 mo.			1 mo.		
6 mo.			6 mo.			6 mo.		
1	6.0-37.6	6.7-41.1	1			1		
5	5.8-36.9	6.4-40.0	5			5		
10	6.1-38.1	6.5-40.3	10	- 317	-44.3	10		
15	5.7-36.4	6.8-41.7	15	125- 767	9.4-76.2	15	-51.2	12.3-170
Adult	7.0-23.0	7.0-23.0	Adult	250-1 100	10 -60	Adult	20-59	Early follicular phase 11 - 82 Later follicular phase 52 -230 Ovulation period peak 120 -390 Luteal phase 9 -230

Age	Progesterone (P_4) (RIA method, ng/mL)		Age	Ferritin (FER) (Latex agglutination, ng/mL)		Age	Alpha fetoprotein (AFP) (IRMA method, ng/mL)	
	Male	Female		Male	Female		Male	Female
1 mo.								
6 mo.								
1	-0.6							
5	-0.6							
10	-0.8							
15	-0.9							
Adult	<0.7	Proliferation phase <1.7 Ovulatory phase <1.9 Luteal phase 0.2-31.6	Adult	24.3-166.1	6.4-144.4	Adult	<10	<10

A. Ohkubo, N. Nara, F. Mashige, S. Ohkubo, 2005., Shonikijunchikenkyuhan hen: nihonjinshoni no rinsyokensakijunchi, Nihonko-syueiseikyokai (1996)

*1 Rinsyobyouritokusyu 101 1996 kesseitanpaku 13koumoku no nihonseijinkijunhani (NCCLS Document C28-P oyobi kokusaikijyunhin CRM 470 niyoru 7 shisetsu no kyodosagyo no seiseki).

*2 Kiyoshi Ichihara *et al.*, : Collaborative derivation of reference intervals for major clinical laboratory tests in Japan. Ann Clin Biochem, 0004563215608875, first published on September 11, 2015.

*3 Tumor marker

Trends in Average Intake of Energy and Nutrients*1

Item (unit) / Era	1975	1980	1985	1990	1995	2000	2005	2010	2014	2015	2016	2017	2018
Energy (kcal)	2 188	2 084	2 088	2 026	2 042	1 948	1 904	1 849	1 863	1 889	1 865	1 897	1 900
Protein													
Total (g)	80.0	77.9	79.0	78.7	81.5	77.7	71.1	67.3	67.7	69.1	68.5	69.4	70.4
Meat, poultry, fish, eggs and dairy products (g)	38.9	39.2	40.1	41.4	44.4	41.7	38.3	36.0	36.3	37.3	37.4	37.8	38.9
Fat													
Total (g)	52.0	52.4	56.9	56.9	59.9	57.4	53.9	53.7	55.0	57.0	57.2	59.0	60.4
Meat, poultry, fish, eggs and dairy products (g)	27.4	27.2	27.6	27.5	29.8	28.8	27.3	27.1	27.7	28.7	29.1	30.0	31.8
Carbohydrate (g)	337	313	298	287	280	266	267	258	257	258	253	255.4	251.2
Calcium (mg)	550	535	553	531	585	547	539	503	497	517	502	514	505
Iron (mg)	13.4	13.1	10.8	11.1	11.8	11.3	8.0	7.4	7.4	7.6	7.4	7.5	7.5
Salt*2 (g)	14.0	13.0	12.1	12.5	13.2	12.3	11.0	10.2	9.7	9.7	9.6	9.5	9.7
Vitamin A (IU)	1 602	1 576	2 188	2 567	2 840	2 654	-	-	-	-	-	-	-
(μgRE)*3	-	-	-	-	-	-	604	529	514	534	524	519	518
Vitamin B_1 (mg)	1.11	1.16	1.34	1.23	1.22	1.17	0.87	0.83	0.83	0.86	0.86	0.87	0.90
Vitamin B_2 (mg)	0.96	1.01	1.25	1.33	1.47	1.40	1.18	1.13	1.12	1.17	1.15	1.18	1.16
Vitamin C (mg)	117	107	128	120	135	128	106	90	94	98	89	94	95

*1 Per capita per day, total, *2 Sodium conversion, *3 RE: Retinol Equivalent

(Source: The National Health and Nutrition Survey in Japan, 2018 – Ministry of Health, Labour and Welfare)

Trends in Average Intake of Food Groups*

Item (unit) / Era	1975	1980	1985	1990	1995	2000	2005	2010	2014	2015	2016	2017	2018
Grains													
Rice and rice products	248.3	225.8	216.1	197.9	167.9	160.4	343.9	332.0	325.0	318.3	310.8	308.0	308.5
Wheat and wheat products	90.2	91.8	91.3	84.8	93.7	94.3	99.3	100.1	101.9	102.6	100.7	103.6	97.3
Potatoes	60.9	63.4	63.2	65.3	68.9	64.7	59.1	53.3	52.9	50.9	53.8	52.7	51.0
Fats and oils	15.8	16.9	17.7	17.6	17.3	16.4	10.4	10.1	10.5	10.8	10.9	11.3	11.0
Legumes	70.0	65.4	66.6	68.5	70.0	70.2	59.3	55.3	59.4	60.3	58.6	62.8	62.9
Vegetables													
Green, Yellow-vegetables	48.2	51.0	73.9	77.2	94.0	95.9	94.4	87.9	88.2	94.4	84.5	83.9	82.9
Others	189.9	192.3	178.1	162.8	184.4	180.1	185.3	180.0	192.2	187.6	181.5	170.8	164.5
Fruits	193.5	155.2	140.6	124.8	133.0	117.4	125.7	101.7	105.2	107.6	98.9	105.0	96.7
Mushrooms	8.6	8.1	9.7	10.3	11.8	14.1	16.2	16.8	15.8	15.7	16.0	16.1	16.0
Seaweeds	4.9	5.1	5.6	6.1	5.3	5.5	14.3	11.0	9.6	10.0	10.9	9.9	8.5
Sugar and sweetener	14.6	12.0	11.2	10.6	9.9	9.3	7.0	6.7	6.3	6.6	6.5	6.8	6.4
Seasoning and beverages													
Beverages } Seasoning and spices (combined)	119.7	109.4	113.4	137.4	190.2	182.3							
Beverages							601.6	598.5	597.9	788.7	605.1	623.4	628.6
Seasoning and spices							92.8	87.0	80.3	85.7	93.5	86.5	60.7
Confectionery	29.0	25.0	22.8	20.3	26.8	22.2	25.3	25.1	26.4	26.7	26.3	26.8	26.1
Fish and shellfish	94.0	92.5	90.0	95.3	96.9	92.0	84.0	72.5	69.4	69.0	65.6	64.4	65.1
Meat and poultry	64.2	67.9	71.7	71.2	82.3	78.2	80.2	82.5	89.1	91.0	95.5	98.5	104.5
Eggs	41.5	37.7	40.3	42.3	42.1	39.7	34.2	34.8	34.8	35.5	35.6	37.6	41.1
Milk and dairy products	103.6	115.2	116.7	130.1	144.5	127.6	125.1	117.3	121.0	132.2	131.8	135.7	128.8
Dietary supplements and food for specified health uses	-	-	-	-	-	-	11.8	12.3	-	-	-	-	-

* Per capita per day, total. Units: grams

Note: Because the classifications changed in 2001, recent data can not be related to data before 2000.

(Source: The National Health and Nutrition Survey in Japan, 2018 – Ministry of Health, Labour and Welfare)

Electrolyte Concentration and Resting Potential of Nerves, Muscles

Animal	Tissue	Intracellular K^+	Intracellular Na^+	Resting potential
		mM	mM	mV
Human	Muscle	140–160	26	-70–-87
Rat	Muscle	74–165	16	-73
Frog	Muscle	138	22	-88–-95
American cockroach	Muscle	112	46	–-60
	Nerve	140–182	84–103	-77
Lobster	Muscle	155	104	–-80
Barnacle (*Balanus nubilis*)	Muscle	168	81	-67
Squid	Nerve	400	50	-63–-72
Cuttlefish	Nerve	268	32	-62
Garden snail	Nerve	86–147	–	-33–-62
Roundworm	Muscle	99	49	-33

(Source: Caldwell, P. C., 1973)

Ion Flux of Nerves, Muscles*

Animal, tissue	At rest ($pM/cm^2/s$)						Flux increment during excitation (pM/cm^2 excitation)		
	Na^+		K^+		Cl^-		Na^+	K^+	Cl^-
	Internal	External	Internal	External	Internal	External	Internal	External	Internal
Common frog, Semitendinous muscle	3.5	3.7	5.4	8.8	–	10	–	–	–
Leopard frog, Sartorius muscle	6	–	–	9.6	–	–	15.6	9.6	–
Lobster, Giant nerve fiber	8	4	15	21	11.5	9.8	6.5	5.3	0.2
Barnacle, Depressor muscle	48	39	28	60	144	143	–	–	–
Isolated muscle	52	–	–	108	144	143	–	–	–
Squid, Giant nerve fiber	28	18	34	–	22.8	20.8	5.7	8.5	0.005
Cuttlefish, Giant nerve fiber	32	39	17	58	–	–	–	–	–
Sea hare, Giant neuron	–	5.8	–	6–8	5–37	–	–	–	–

* Amount of ions transported through unit area of membrane in unit time.

(Source: DeWeer, P. *et al.*, 1976)

Membrane Potential

Animal	Cell	Potential	Animal	Cell	Potential
		mV			mV
Human	Red blood cell	-8.0	Mouse	Immature ovum	-8.3
	Fibroblast	-75–-70		Mature ovum	+1.9
	Pulmonic cells	-21.1		Fertilized ovum	-9.2
	Embryo lungs cell	-14.7–-8.5		2 cell stage	-10.7
Mouse	Fibroblast	-0.6		Morula	-12.8
Rat	Hepatocyte	-53–-50		Blastocyst	-12.9
Dog	Hepatocyte	-40	*Xenopus laevis*	Fertilized ovum	-6.5
Chicken	Embryonic hepatocyte	-45–-8		2 cell stage	-19.0
Cattle	Sperm	-6.1		16 cell stage	-27.0
Sea urchin	Unfertilized ovum	-14–-6		Morula	-31.0
	Fertilized ovum	-70–-60		Blastula	-35.0

(Source: Chowdhury, T. K. *et al.* 1976)

Neurotoxins and Action Mechanisms

Toxin	Found in	LD_{50} (mg/kg)	Action site	Action mechanism
Tetrodotoxin	Pufferfish, etc.	0.01 (mouse oral administration)	Axon	Na channel close
Ciguatoxin (ciguatoxin)	Red snapper, Giant moray	0.45 (mouse intraperitoneal injection)	Axon	Na channel open
Batrachotoxin	Dendrobatidae	0.002(?)	Axon	Na channel open
α-scorpion venom	Scorpion	0.15–0.01(?)	Axon	Delays Na channel inactivation
α-latro toxins	*Latrodeclus hasselti*, etc.	0.55 (mouse intravenous injection)	Presynaptic membrane	Excessive transmitter release
Sea anemone toxin	Sea anemone	–	Axon	Delays Na channel inactivation
Palytoxin	Zoanthids	0.00015 (mouse intravenous injection)	Axon	Na and Ca channels open
α-bungarotoxin	Krait	0.3 (mouse hypodermic injection)	Skeletal muscle postsynaptic membrane	Binds to ACh nicotinic receptor
Conotoxin ω		0.013(mouse?)	Axons terminal	Inhibits voltage dependent Ca channel
〃 α	Cone shell	–	Postsynaptic membrane	Inhibition of Ach receptors
〃 μ		–	Muscle	Na channel close
Saxitoxin	Mussel	0.091–0.727 (Various animal, oral administration)	Axon	Na channel close
Joro spider toxin (JSTX)	*Nephila* etc.	–	Postsynaptic membrane	Inhibition of Glu receptor
Botulinus toxin	*Clostridium botulinum*	0.00000032(presumption)	Axons terminal	Suspension of ACh discharge
Curare (*d*-tubocurarine)	Vine etc.	3.2 (mouse intraperitoneal injection)	Postsynaptic membrane	Binds to Ach receptor
Grayanotoxin	Pieris etc.	0.2(?)	Axon	Na channel open
Atropine	Belladonna	200(?)	Postsynaptic membrane	Inhibition of Ach receptors
Aconitine	Aconite	0.31 (mouse hypodermic injection)	Axon	Na channel open
Picrotoxin	*Sinomenium*, etc.	7.2 (mouse intraperitoneal injection)	Central nerve	Compete with GABA
Strychnine	Loganiaceae	16 (rat oral administration)	Postsynaptic membrane	Inhibition of Gly receptor

– : No data, ? : Dispensing method or animal used unknown.

Source: Curtis D. Klaassen, Mary O. Amdur, John Doull (Supervisors)/Fukuda, H. *et al.* (Official Translation) Toxicology II, Dobunshoin (1988)

Conduction Velocity of Nerve Fibers

Animal, nerve fiber*	Diameter	Velocity	Temperature
	μm	m/s	℃
Cat, motor fiber (thick)*	15-20	80-110	37
Ibid (thin)*	2-6	10-30	35
Frog, motor fiber (thick)*	15-20	30-40	24
Ibid (thin)*	4-8	7-15	24
Carp, Mauthner cell*	55-65	55-63	20-25
Procambarus clarkii, medial giant fiber†	<u>100-250</u>	15-20	20
Periplaneta americana, giant fiber†	10-40	9-12	-
Loligo pealii, giant fiber†	260-520	18-35	23
Earthworms (*Lumbricus*), medial giant fiber†	<u>50-90</u>	15-45	22
Aurelia, neural network†	<u>6-12</u>	0.5	-

* myelinated nerve fibers, † unmyelinated nerve fibers. Myelinated nerve fibers have a faster conduction velocity. If the properties of the nerve cell membrane and cell contents are the same, thicker nerve fibers have a faster conduction velocity. Underline sections indicate gradual change for the fiber diameter in that range.

Source: Bullock, T. H. and Horridge, G. A.: Structure and Function of the Nervous Systems of Invertebrates. Vol.1, Freeman & Comp., 1965

Biogenic Amines in the Brain

Animal	Dopamine	Norepinephrine	Epinephrine	Octopamine	Serotonin
Dog	80	130	30	–	180
Rat	70	150	40	26	410
American cockroach	1 070	370	n.d.	3 450	3 150
Desert locust	870	110	n.d.	2 430	1 470
Oriental migratory locust	1 310	240	n.d.	–	2 340
Field black cricket	1 058	–	n.d.	2 019	2 440
Crayfish	200	120	n.d.	540	150
Sea hare	2 380	–	n.d.	350	2 210

Units indicate ng per g of dry mass. n.d. indicates no detection was possible. – indicates no analysis performed.
Source: Tominaga, Y. ed.: "Studying Insect Brains", Nagao, T., p. 161, Kyoritsu Shuppan, 1995.

Auditory Range of Animals

	Lowest frequency (kHz)	Highest frequency (kHz)		Lowest frequency (kHz)	Highest frequency (kHz)
Vertebrate			**Insects**		
Human	0.031	17.6	Oriental migratory locust (*Locusta migratoria*)	0.6	45
Japanese monkey	0.028	34.5	Longhorned Grasshopper (*Conocephalus saltator*)	6	100
Pig	0.042	40.5	House cricket (*Acheta domesticus*)	2	18
Goat	0.78	37	European Cicadas (*Cicada orni*)	2	15
Mouse	2.3	85.5	Oriental tabacco budworm (*Heliothis zea*)	2	200
Pigeon	0.1	5.8	Lacewing (*Chysopa carnea*)	13	120
Red-eared slider	0.068	0.84			
Bullfrog	0.1	2.5			

The auditory range of vertebrates shows the audible frequency range when the sound pressure is 60 dB. The auditory range of insects shows values measured using tympanic organs.

Sources: Heffner, H. E. and Heffner R. S.: Hearing Range of Laboratory Animals. J. Am. Assoc. Anim. Sci., **46**(1): 20–22, 2007; Frazier, J. L.: Nervous system: sensory system. In: Fundamentals of Insect Physiology. ed. Blum M. S., p. 287–356, John Wiley & Sons, 1985

For the frequencies of vocal music and musical instruments, refer to page **371**.

Number of Auditory Hair Cells in Vertebrate Inner Ear

Animals		Number of hair cell (One cochlear)
Bullfrog	Basilar papilla	60
	Amphibian papilla	600
Chameleon		50
Tokay gecko		1 600
Turtle, Snake		Several 100
Crocodile		11 000–13 000
Owl		12 000
Dorphin		16 000–17 000
Cat	Inner hair cell	2 600
	Outer hair cell	9 900
Human	Inner hair cell	3 500
	Outer hair cell	12 000–20 000

Compiled base on; Wever, E. G: The evolution of vertebrate hearing. "Handbook of sensory physiology" Vol V/1 Auditory System. (Keidel, W. D *et al.* ed.), 1974; Schuknecht, H. F.: Neuroanatomical correlates of auditory sensitivity and pitch discrimination in the cat. "Neural mechanisms of the auditory and vestibular systems" (Rasmussen, G. L. *et al.* ed.), Thomas, Springfield. 1960; Suga, N.: Audition "Physiology" (Iriki, M. and Toyama, K. ed.) Bunkodo, 1986.

Neurotransmitters and Receptors

Class	Transmitter	Receptor	Antagonist	Postsyuaptic potential	Relating matter
Amino acids	Glutamic acid	Ion channel type NMDA type AMPA type Kainate type G protein-coupled type mGluR1–R8	 Kynurenic acid Kynurenic acid	Slow components of fast EPSP Fast components of fast EPSP	Long-term potentiation, Memory and learning Long-term potentiation, Memory and learning Long-term suppression, memory and learning
	γ-aminobutyric acid (GABA)	Ion channel type $GABA_A$ G protein-coupled type $GABA_B$	Bicuculin, Picrotoxin Phaclofen	fast IPSP slowIPSP	Distributed in the central nervous system of vertebrates and invertebrates
	Glycine	Ion channel type	Strychnine	fast IPSP	Distributed in spinal cord renshaw cells
Amines	Adrenaline (epinephrine) Noradrenaline (norepinephrine)	G protein-coupled type $\alpha_{1,2}$ $\beta_{1\text{-}3}$	Prazosin (α_1) Propranolol ($\beta_{1,2}$)		Excess: manic state, Shortage: Loss of motivation
	Dopamine	G protein-coupled type $D_{1\text{-}5}$	Sulpiride (D_2)		Excessive: Schizophrenia and Tourette's syndrome Lack: Mental processes, motor deterioration, and Parkinson's disease
	Histamine	G protein-coupled type $H_{1\text{-}3}$ (H_3 : Autoreceptors)			Arousal/pain threshold adjustment, itching, Vasodilation
	Octopamine	G protein-coupled type			Distributed in central nervous system of invertebrates. Memory and learning in insects
	Serotonin (5-hydroxytryptamine: 5-HT)	Ion channel type $5\text{–}HT_3$ G protein-coupled type $5\text{–}HT_1$ $5\text{–}HT_2$ $5\text{–}HT_4$ $5\text{–}HT_{5\text{-}7}$	Ondansetron Pindolol	slow EPSP	Shortage: depression and panic attack Regulates vomiting.
	Melatonin (*N*-acetyl-5-methoxytryptamine)	G protein-coupled type MT1			Regulates sleep, circadian clock
Peptides	Substance P	G protein-coupled type NKR_1			A kind of tachykinin. Pain transmitter
	Enkephalin	G protein-coupled type μ/δ–Opioid			A kind of opioid peptide. Morphine-like action, net decrease of intestinal movement
	β endorphin	G protein-coupled type κ-opioid			A kind of opioid peptide, morphine-like action. Hypoalgesic effect
	Vasopressin	G protein-coupled type V_{1a} V_{1b} V_2			Promotes reabsorption of water. Blood pressure increase
	Proctolin	G protein-coupled type			Stripe shrinkage in invertebrate
	Oxytocin	G protein-coupled type			Smooth muscle contraction
Others	Acetylcholine	Ion channel type nicotinic muscle type and neuron type G protein-coupled type (muscarinic) M1–M5	*d*-tubocurarine *d*-Neurotoxin Atropine	EPP, fast EPSP slow EPSP, slow IPSP	Muscle type receptors decrase in the myasthenia gravis.
	Nitric monoxide				Vasorelaxing
	Carbon monoxide				Inhibition of oxygen transport of hemoglobin
	Adenosine	G protein-coupled type A_1 A_{2A} A_{2B} A_3	Caffeine and theophylline		
	ATP	Ion channel type P_{2X} G protein-coupled type P_{2Y}		fast EPSP	Coexistence neurotransmitter

Compiled based on "Jintaikinoseirigaku [kaitei 5 ed.]" Nankodo, (2009); "Hyojunseirigaku [7]" Igaku-Shoin, (2009); etc.

Insect Neuropeptides

Group (Family)	Examples of members	Related peptides	Immunocyto-chemistry
Adipokinetic and hypotrehalosaemic peptides			
AKH/RPCH family	AKH I-II	glucagon, corazonin	NC, NSC, CCI
Myotropins			
Proctolin	proctolin	-	NC, NSC, PT
Insect kinins	leucokinins I-VIII	-	NC, NSC, PT
Insect tachykinins	locustatachykinins I-IV	tachykinins	NC, PC, PT
Pyrokinins/myotropins	leucopyrokinin	MRCH, PBAN, LomMT	NC, NSC
Sulfakinins	drosufakinins I-II	gastrin/CCK, FaRPs	NC, NSC
FMRFamide-related	calliFMRFamide	sulfakinins	NC, NSC, PT, PC
Cardioactive peptides	Lom-CAP, Mas-CAP_{2a}	crustacean cardioactive peptide	NC, NSC, PT
Myoinhibitory peptide	Lom-MIP	*Aplysia* APGWamide	-
Accessory gland myotropin	Lom-AG-MT I-II	-	NC, NSC, PC
Corazonin-related	corazonin	hypertrehalosemic hormone	NC, NSC
Diuretic hormones			
Vasopressinlike	F1, F2	vasopressin, conopressin	NC/NSC
Diureric hormones	Mas-DP	urotensin I, CRF	NSC
Other groups			
Pigment-dispersing peptides	PDF, PDH	-	NC, NSC, PT
PBAN/MRCH	Hez-PBAN	pyrokinins, Lom MT	NC, NSC
Diapause hormone	Bom-DP	PBAN/MRCH, pyrokinins	-
Neuroparsins	neuroparsins I-III	neurophysins	NSC
Eclosion hormones	Mas-EH	-	NSC
PTTHs	Bombyxin (4K PTTH)	insulin	NSC
	22K PTTH	-	NSC
Allatotropins	Mas-AT	-	-
Allatostatins	allatostatins A1-4	Met-enkephalinRGL	NC, NSC, RT

NC, neurons; *NSC*, neurosecretory cells; *RT*, peripheral tergets innervated; *CCI*, corpora cardiaca intrinsic cells; *PC*, secretory cells in intestine or other tissues. The insect neuropeptide terminology is according to Raina and Gäde (1989) and the neuropeptide abbreviations are: *AKH*, adipokinetic hormone; *AT*, allatotropin; *CCAP*, crustacean cardioactive peptide; *CCK*, cholecystokinin; *CRF*, corticotropin-releasing factor; *DH* diapause hormone; *DP*, diuretic peptide; *EH*, eclosion hormone; *MIP* myoinhibitory peptide; *MRCH*, melanization and reddish coloration hormone; *MT*, myotropin; *PBAN*, pheromone biosynthesis activating neuropeptide; *PTTH*, prothoracicotropic hormone; *RPCH*, red pigment concentrating hormone

(Source: Nässel, D. R.: Cell Tissue Res., **273** (1): 1–29, 1993)

Mean Value for Systolic (Maximum) Blood Pressure (Japanese)

Age \ Era	2000	2005	2008	2009	2010	2011	2012	2013	2014	2015	2016	2017	2018
Male													
Total	135.7	135.8	135.6	135.6	135.9	135.7	134.6	135.3	135.3	133.8	134.3	135.2	134.7
20-29	120.7	121.1	121.0	118.1	120.1	119.6	119.2	119.9	120.6	117.5	120.4	116.6	116.8
30-39	123.8	122.0	122.0	122.8	123.1	124.5	122.4	121.7	123.1	123.3	121.9	119.5	120.9
40-49	130.2	128.3	130.6	127.8	127.2	126.4	126.5	126.1	125.7	125.7	126.8	128.3	126.0
50-59	137.3	133.7	135.9	136.8	135.4	134.8	135.7	134.9	134.5	133.1	134.7	133.4	134.1
60-69	142.1	139.9	139.7	140.8	139.7	139.6	138.7	138.1	138.7	137.9	137.6	138.2	138.8
70-	146.1	145.3	140.2	142.5	142.9	142.6	140.7	141.5	139.8	137.8	138.5	140.9	140.3
Female													
Total	129.6	130.5	129.2	128.3	129.2	128.6	127.3	129.5	128.7	127.2	127.3	128.9	127.9
20-29	108.6	106.9	108.2	108.1	107.8	106.0	107.3	108.2	108.4	106.9	109.4	108.9	109.5
30-39	113.3	111.8	110.9	110.6	111.5	111.8	111.1	111.6	110.2	109.0	110.0	110.8	110.2
40-49	123.4	122.0	120.7	118.5	118.6	119.3	117.9	119.1	118.7	118.1	116.3	118.4	116.5
50-59	132.5	131.8	130.5	128.8	129.0	128.7	128.1	126.5	125.4	127.2	126.2	125.1	126.9
60-69	140.2	137.7	134.5	134.9	137.6	134.1	134.3	137.1	135.0	132.5	132.8	134.7	132.2
70-	144.7	144.0	140.0	140.5	140.4	140.8	138.5	140.1	138.5	137.8	137.2	138.4	138.5

Units: mmHg
(Source: The National Health and Nutrition Survey in Japan, 2018 – Ministry of Health, Labour and Welfare)
Note: Excluding pregnant women and people taking medication to lower blood pressure. Measurment taken twice and results averaged.

Mean Value for Diastolic (Minimum) Blood Pressure (Japanese)

Age \ Era	2000	2005	2008	2009	2010	2011	2012	2013	2014	2015	2016	2017	2018
Male													
Total	82.0	81.9	82.1	82.3	82.0	81.6	81.2	81.7	81.1	81.3	80.9	82.3	82.3
20-29	74.5	74.6	74.2	73.4	76.0	74.3	73.6	74.1	73.0	72.3	74.3	74.5	74.1
30-39	78.5	78.1	80.0	77.8	79.7	80.2	79.2	78.0	78.8	79.0	78.9	79.1	79.6
40-49	83.9	83.4	84.7	84.5	83.6	82.4	82.3	82.8	81.9	84.3	82.6	83.9	84.3
50-59	85.3	84.6	86.2	86.2	86.3	85.8	86.1	86.9	85.9	85.1	86.0	86.4	86.8
60-69	84.1	83.6	83.5	84.7	83.8	83.6	83.2	84.0	83.4	84.0	82.9	85.0	85.1
70-	79.9	81.5	79.9	80.0	79.0	79.4	78.7	79.9	78.4	78.5	78.3	80.3	79.8
Female													
Total	77.3	77.9	77.4	76.8	77.1	76.9	76.5	77.6	76.5	76.6	76.1	77.6	77.4
20-29	66.1	65.8	68.1	67.6	67.1	66.1	67.2	66.8	68.9	68.0	68.1	68.4	69.1
30-39	71.1	70.8	71.1	70.2	70.6	71.2	70.6	70.9	69.3	69.2	70.3	71.5	70.4
40-49	76.9	77.3	76.6	75.0	75.7	75.9	75.0	76.4	76.1	75.4	73.5	76.0	74.9
50-59	81.4	81.5	81.4	80.3	80.0	80.4	79.8	79.6	78.1	79.8	78.8	79.0	80.6
60-69	81.7	81.6	80.3	80.1	81.3	80.3	79.8	81.6	79.9	79.7	79.2	81.1	80.2
70-	78.5	79.1	77.2	77.6	77.3	77.8	77.2	78.4	76.4	76.6	76.6	78.0	78.2

Units: mmHg
(Source: The National Health and Nutrition Survey in Japan, 2018 – Ministry of Health, Labour and Welfare)
Note: Excluding pregnant women and people taking medication to lower blood pressure. Measurment taken twice and results averaged.

Blood Pressure for Various Animals

Species name	Diastole*	Systole**
	mmHg	mmHg
Chimpanzee	80 (76- 85)	136 (125-147)
Rhesus macaque	127 (112-152)	159 (137-188)
Guinea pig	47 (16- 90)	77 (28-140)
Mouse	81 (67- 90)	113 (95-125)
Rat	91 (58-145)	129 (88-184)
European rabbit	80 (60- 90)	110 (95-130)
Dolphin	118 (111-130)	152 (142-160)
Dog	56 (43- 66)	112 (95-136)
Cat	123 (90-145)	171 (135-200)
Spotted seal	105	150
Cow	110	160
Giraffe	160	260
Dromedary	169	209
Pig	108 (98-120)	169 (144-185)
Sheep	93 (75-103)	123 (104-135)
Goat	84 (76- 90)	120 (112-126)
Ass	103	171
Horse (thoroughbred)	99 (77-121)	142 (118-166)
African elephant	-	106 (87-125)
Red kangaroo	79 ± 18	122 ± 23
Virginia Opossum	135	175
Duck (Peking duck)	134	180
Bobwhite quail	136	150
Quail	152	158
Chicken	61	207
Turkey	204	302
Rock pigeon	105	135
Robin	80	118 (110-125)
Canary	154 (150-160)	175 (110-250)
Starling	130 (100-160)	180 (150-210)
Skink (*Tiliqua*)	10	14
Redbelly turtle (*Pseudemys*)	32	42
Bullfrog	21 (18- 24)	32 (28- 36)
Northern leopard frog	21 (16- 26)	31 (21- 36)
Rainbow trout	32 (31- 33)	40 (39- 41)
Chinook salmon	46 (16- 76)	80 (50-110)
Red trout	38	44
Cod	18	29
Largemouth bass	40	50
Spotted shark	19 (17- 21)	26 (22- 30)
Skate	7	16

* Lowest blood pressure ** Highest blood pressure.

(Sources: Lindsay, H. A., *et al.*, Prosser, C. L., 1973, Swenson, M. J., 1984, Heard, D. J., *et al.*, 1988)

Water Content of Vertebrates

Animal kind	Whole quantity	In cell	Outside cell	Blood
Goat	76%	40%	36%	9.9%
Mississippi alligator	73	58	15	5.1
Rat snake	70	52	17	6.0
Bullfrog	79	57	22	5.3
Carp	71	56	15	3.0
Sea carp	71	57	14	2.2
Dogfish	71	58	13	6.8
Lamprey	76	52	24	8.5

* Body weight ratio

Amount of Body Surface Moisture Evaporation

Animal kind	Evaporation*	Animal kind	Evaporation*
Human (non-perspiration)	48	Tsetse fly	13
		March fly	900
Rat	46	American cockroach	49
Iguana	10		
Frog	300	Schitocerca gregaria	22
Salamander	600		
Helix aspersa (active)	870	Grain mite	2
Helix aspersa (resting)	39	Hard tick	0.8
Mealworm	6	Earthworm	400

* Room temperature, μg/h/cm_2 body surface/mmHg evaporation pressure

Existence pH

pH	Medium	Living Species
0.0	Sulfur and salt	Sulphur bacteria
1.0–3.0	Organism	Wrinkled Mushrooms (Basidiomycota)
3.2–4.6	Peaty soil	Sphagnum
3.4	Vinegar	Vinegar ell, acetic bacteria
5.2	Fermented milk	Lactic acid bacterium
4.5–8.5	Soil	Plants composing forests and uncultivated land
5.2–9.0	Rhizic water	Moss plant (peaty soil)
6.8–8.6	River water	River water flora and fauna
7.8–8.6	Seawater	Marine products flora and fauna
10.5	Organism medium	Dermatophytes (Microsporum, Epidermophyton, Trichophyton)
9.3–11.1	Organism medium	Blue mold, Aspergillus

Optimal pH for Plant Growth

Species name	pH
Apple	5.0–6.5
Peach	6.0–7.5
Pear	6.0–7.5
Japanese chestnut	4.5–6.5
Camellia	4.5–6.0
Grape	6.0–8.0
Carrot	5.5–7.0
Sweet potato	5.0–6.5
Tomato	5.5–7.5
Potato	5.4–6.7
Cucumber	5.5–7.0
Soybean	6.0–7.5
Pease	6.0–8.0
Rice	5.0–6.5
Wheat	5.5–7.5
Corn	5.5–7.5
Pineapple	4.5–6.0
Onion	6.0–7.5
Asparagus	6.0–8.0
Fir	4.5–6.5
Hemlock fir	4.5–6.0
Pineapple	4.5–6.5
Larch	4.5–7.5
Ginko	5.5–7.0

(Source: C. D. Welch, *et al.*, 1973)

Ion Concentration for Marine Animal Body Fluids[1)]

Animal kind	Na^+	Mg^{2+}	Ca^{2+}	K^+	Cl^-	SO_4^{2-}	Amount of body fluid protein[2)]
Hagfish (*Myxine*)	537	18.0	5.9	9.1	542	6.3	67
Sea urchin (*Echinus*)	474	53.5	10.6	10.1	557	28.7	0.3
Japanese dwarf squid (*Loligo*)	456	55.4	10.6	22.2	578	8.1	150
Mussel (*Mytilus*)	474	52.6	11.9	12.0	553	28.9	1.6
Swimming crab (*Carcinus*)	531	19.5	13.3	12.3	557	16.5	60
Spiny crab (*Maja*)	488	44.1	13.6	12.4	554	14.5	–
Lobster (*Nephrops*)	541	9.3	11.9	7.8	552	19.8	33
Sea lice (*Ligia*)	566	20.2	34.9	13.3	629	4.0	–
Sea mouse (*Aphrodita*)	476	54.6	10.5	10.5	557	26.5	0.2
Moon jellyfish (*Aurelia*)	474	53.0	10.0	10.7	580	15.8	0.7
Seawater	478.3	54.5	10.5	10.1	558.4	28.8	/

1) mM/kg bodily fluid, 2) g/L bodily fluid

(Source: K., Schmidt-Nielsen, 1984)

Influence of Light on Plant Growth

Species name	Wavelength (nm)	Energy level (erg/cm^2)	Exposure time
Promotion of seed germination			
Lettuce	660	24 000	4 m
Swinecress[1)]	660	6 000	8 m
Peppergrass	650	1 400 000	8 m
Inhibition of germination			
Lettuce	735	70 000	4 m
Swinecress[1)]	735	180 000	8 m
Peppergrass	735	30 000	8 m
Tomato	740	1 500	4 m
Henbit	730	2	Continuity
Promotes stem elongation			
Common bean	735	7 000	5 m
Inhibits stem elongation			
Common bean	660	1 500 000	2 m
Lettuce	430	500	15 h
〃	730	1 100	8 h
Barley	660	130 000	1.5 m

Species name	Wavelength (nm)	Energy level (erg/cm^2)	Exposure time
Promotes flowering			
Morning glory	650	320 000	2 m
Cocklebur[2)]	735	300 000	4 m
Henbane	650	15 000	6 m
Barley	650	60 000	2 m
Inhibits flowering			
Morning glory	740	1 200 000	2 m
Cocklebur[2)]	630	40 000	5 m
Pigweed[3)]	630	2 000	4 m
〃	750	7 500	90 m
Soybean	630	50 000	30 s
Pigment synthesis (* Anthocyanin synthesis)			
Turnip*	725	100 000 000	8 h
Cabbage*	690	50 000 000	4 h
Tomato	640	6 000	20 s
Sorghum*	470	30 000 000	4 h
Apple*[4)]	650	260 000 000	16 h

1) *Lepidium densiflorum*, 2) *Xanthium pensylvanicum*, 3) *Chenopodium ruburum*, 4) *Males* sp.

(Source: Downs, R. J., 1973)

Photoconvertible Photosensor Protein of Plants, Phytochrome

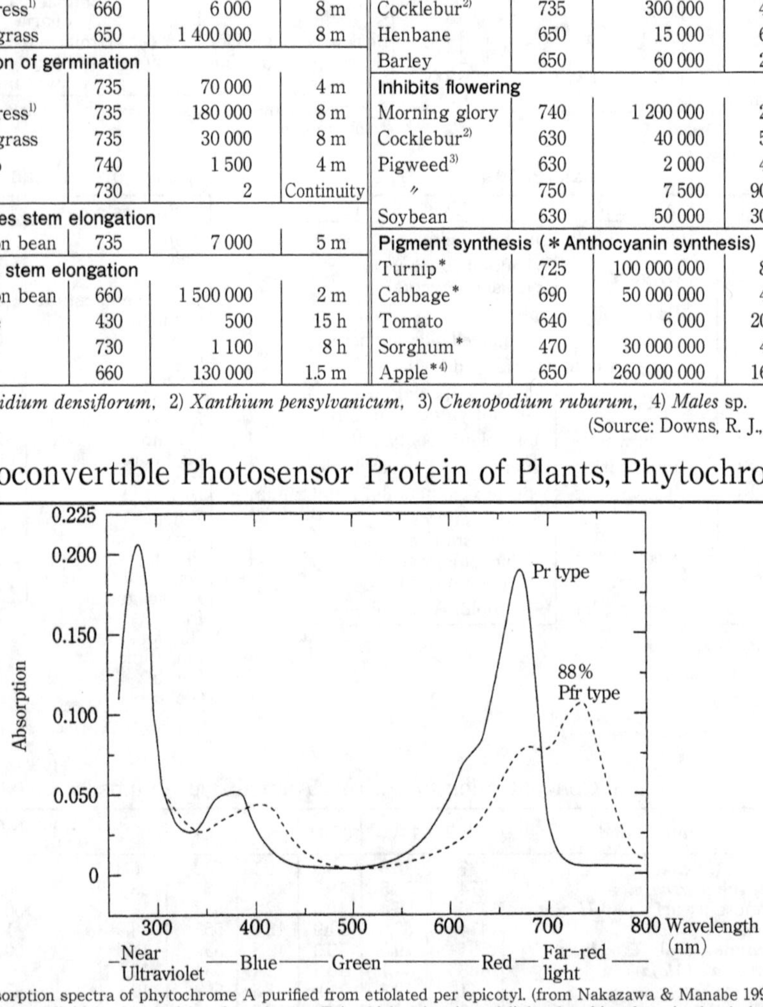

Absorption spectra of phytochrome A purified from etiolated per epicotyl. (from Nakazawa & Manabe 1991.) When red light is absorbed by phytochrome (Pr) which exists in red light absorbing form in the dark, approximately 88% of the phytochrome undergoes reversible photoconversion into far-red light absorbing form (Pfr). The spectroscopic properties are somilar in phytochrome B. Phytochrome molecules are synthesized as Pr form in the cytoplasm. However, they migrate into the nucleus due to the photoconversion and regulate the transcription of some phytochrome regulated genes.

Refer to page **373** for more information about light wavelength and color.

Photoreceptor Pigments for Photomorphogenesis*1 in Plants

Response	Feature of reaction	Photoreceptor protein
Low fluence response (LFR)	Reversible red/far-red reaction	Phytochrome B*2
High irradiance response (FR-HIR)	Strong far-red light	Phytochrome A*2
Very low fluence response (VLFR)	Saturates by very weak light	Phytochrome A*2
Blue-near ultraviolet light	Embryo axis extension control etc	Cryptochrome*3
Reactions	Phototropism and chloroplast movement	Phototropin*3

*1 Induction of seed germination, suppression of extension growth of stem, stimulation of leaf growth, greening, etc. due to light.

*2 Reversible photoconversion between Pr and Pfr, two types which absorb red light and far-red light due to chromoproteins which possess ring-opening tetrapyrroles (refer to figure on previous page).

*3 Flavoprotein

Nitrogen Fixation Organisms

A. Non-symbiotic

1. Bacteria
   - a. Aerobic microorganism
     Azotobacter, Beijerinckia, Azotmonas, Azotococcus
   - b. Facultative anaerobic bacteria
     Klebsiella, Bacillus
   - c. Anaerobic bacteria
     - (i)Non-phototropic bacteria
       Clostridium, Desulfovibrio
     - (ii)Phototropic bacteria
       Rhodospirillum, Chromatium, Chlorobium
2. Blue-green bacteria (Cyanobacterium)
   Nostoc, Anabaena, Calothrix

B. Symbiont

Host	Symbiotic bacteria
1. Symbiosis of leguminous plant and nodule bacteria	
Pea, Kidney bean, Soybean, White clover, Alfalfa, Lupin	*Rhizobium* (Rhizobium)
2. Symbiosis with nonleguminous plant	
Alder (Angiosperm)	*Frankia* (Ray fungus)
Coriaria (〃)	*Streptomyces* (〃)
Gunnera (〃)	*Nostoc*
Macrozamia (Cycad plant)	*Nostoc, Anabaena*
Azolla (Fern)	*Anabaena*
Blasia pusilla (Moss)	*Nostoc*

Metabolism and Biosynthesis System

Glycolysis and Fermentation ([C*n*] *n* = carbon number)

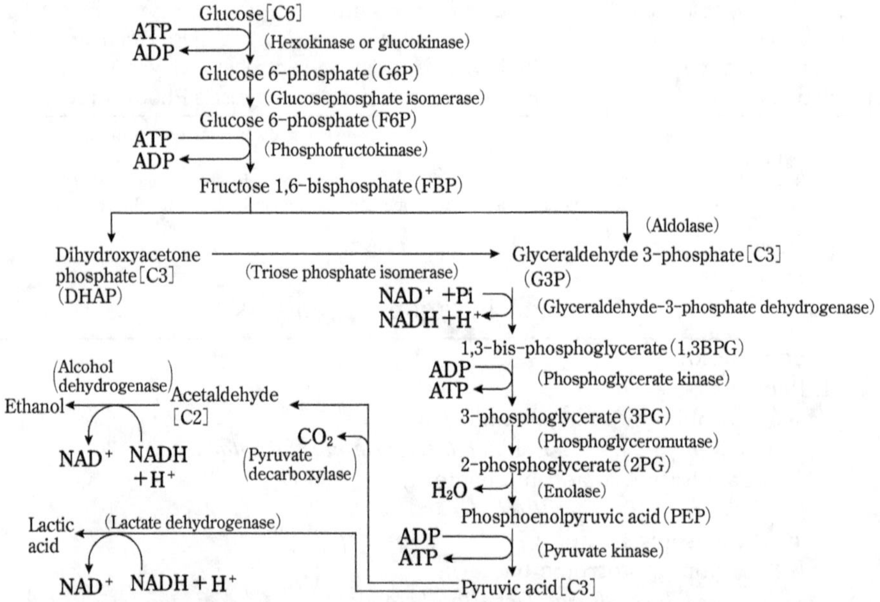

Chemical Structural Formula of Intermediate Metabolites of Glycolysis

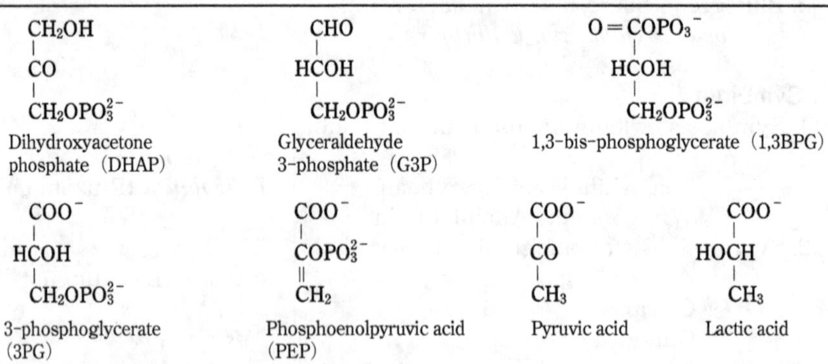

For the structure of ATP and NAD, refer to pages **486**, **491**.

Pentose Phosphate Pathway

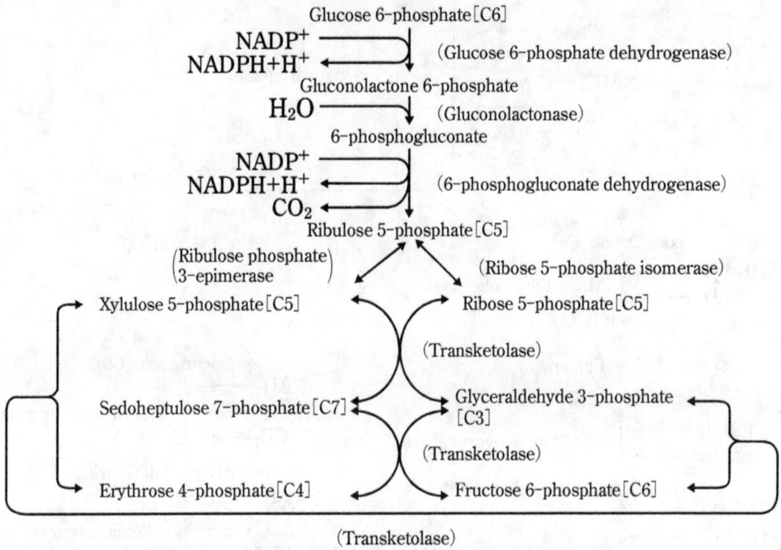

Chemical Structural Formula of Intermediate Metabolites of Pentose Phosphate Pathway

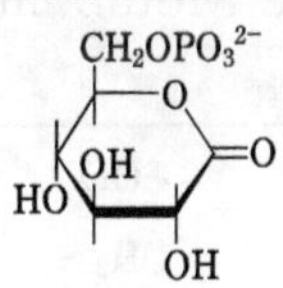

Gluconolactone 6-phosphate

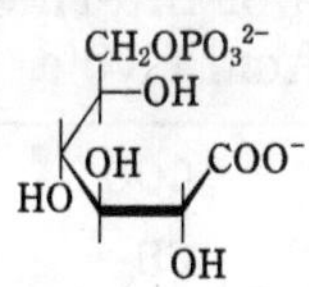

6-phosphogluconate

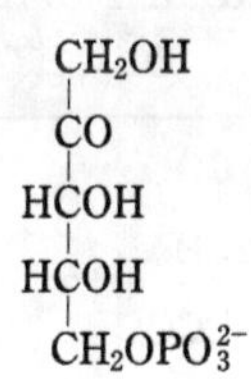

Ribulose 5-phosphate

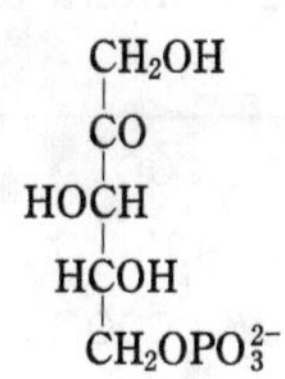

Xylulose 5-phosphate

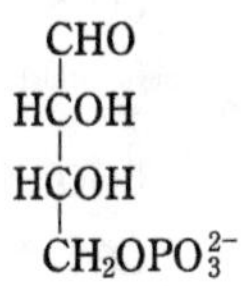

Erythrose 4-phosphate

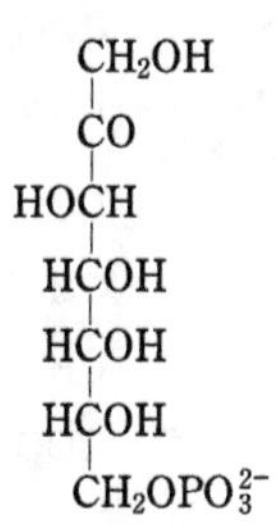

Sedoheptulose 7-phosphate

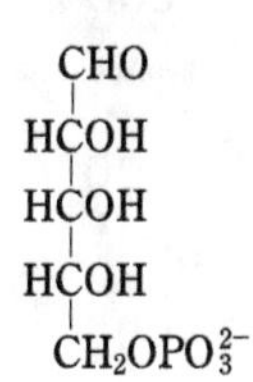

Ribose 5-phosphate

Citric Acid (Tricarboxylic Acid, Krebs) Cycle

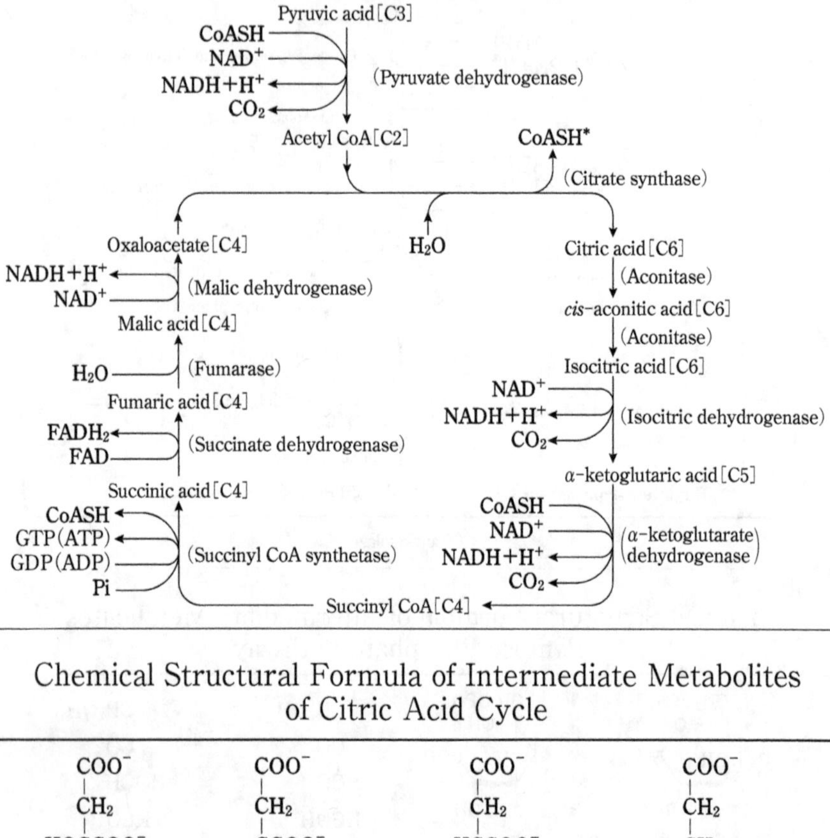

Chemical Structural Formula of Intermediate Metabolites of Citric Acid Cycle

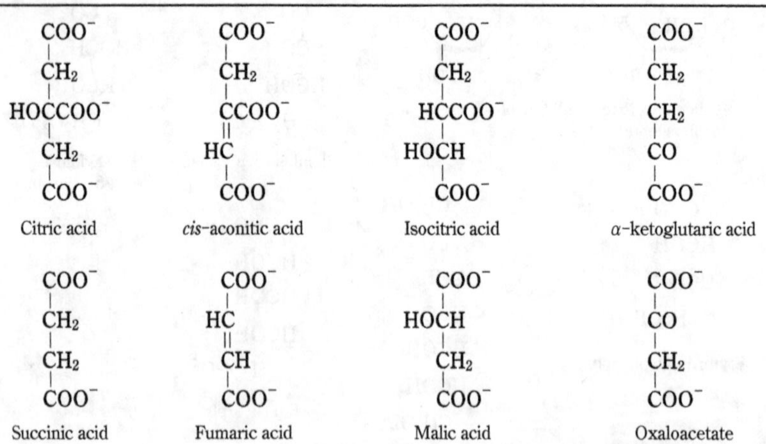

* For the structure of coenzyme A(CoA), NAD, and FAD, refer to page **491**.

Electron Transport Chain for Oxygen Respiration in the Mitochondrial Inner Membrane

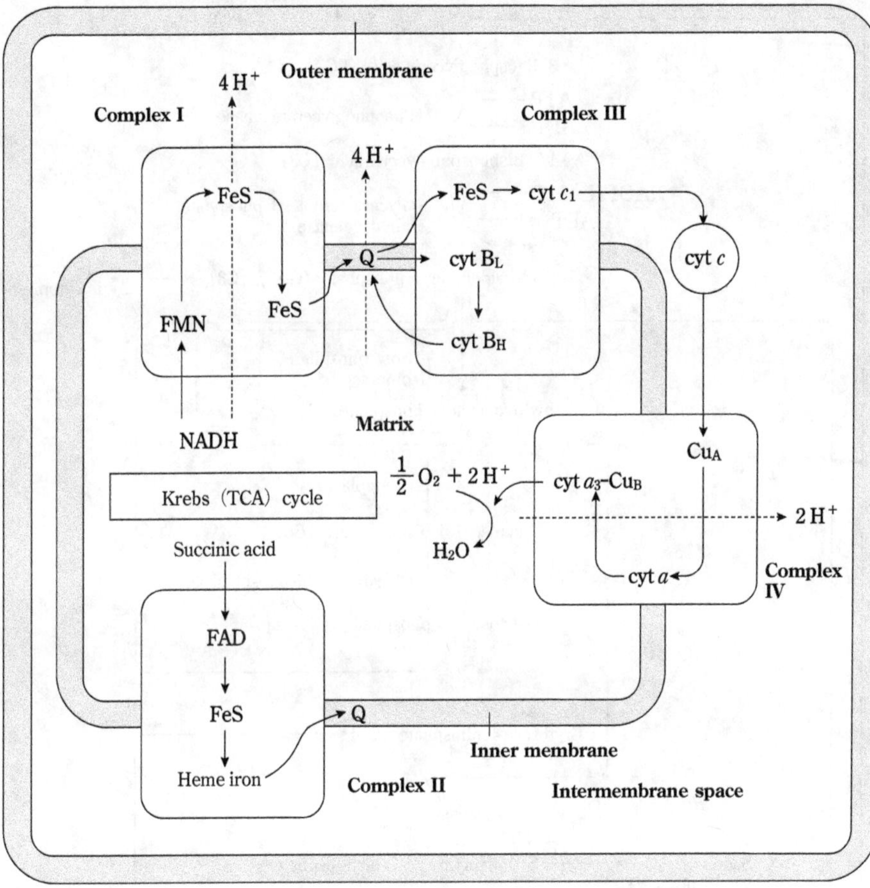

The four types of transmembrane protein complexes listed above exist in the mitochondrial inner membrane. The reduced form Q made from complex II also passes through complex III and transmits electrons in complex IV. Although not shown in the figure, ATP synthase complexes also exist in the inner membrane. These complexes use H^+ gradient stored in the space between the outer and inner membranes due to electron transmission to synthesized ATP from ADP and inorganic phosphate.

The arrows in the figure show the flow of electrons, while the dotted arrows show the movement of H^+ (protons). Q (coenzymeQ, CoQ) shows ubiquinone which exists in the membrane. Among constituents which form the complex, FMN and FAD indicate flavin, FeS indicates an iron-sulfur center, and cyt indicates cytochrome. Among complex IV, Cu_A indicates a copper binuclear center, andcyt a_3-Cu_B indicates a cytochrome a_3-Cu_B binuclear complex.

Calvin Cycle in the Stroma of Chloroplasts

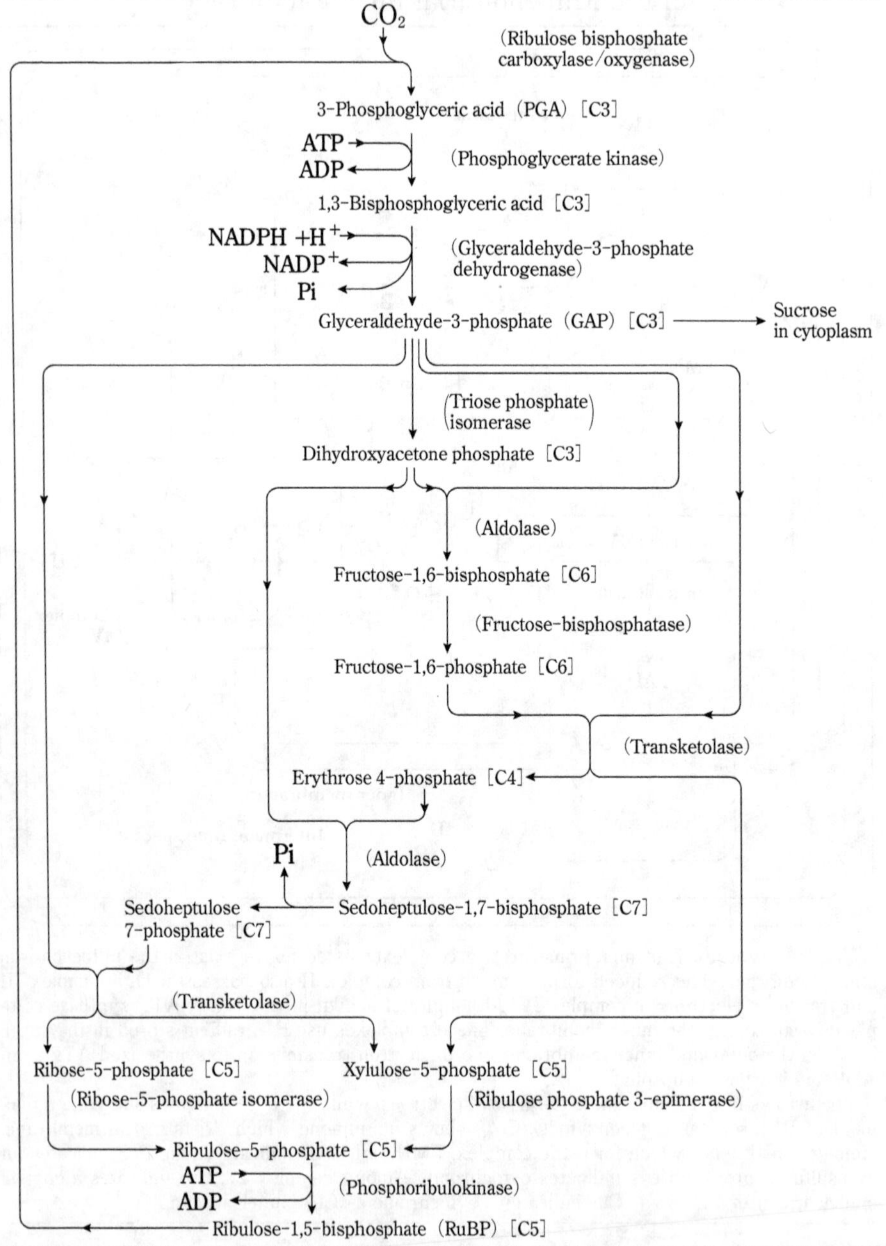

Carbon Dioxide Fixation of C4 Plants (NADP Malic Enzyme Type)

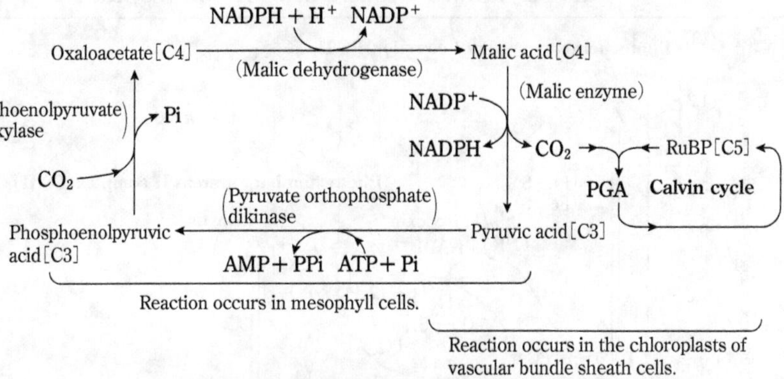

Photorespiration Pathway

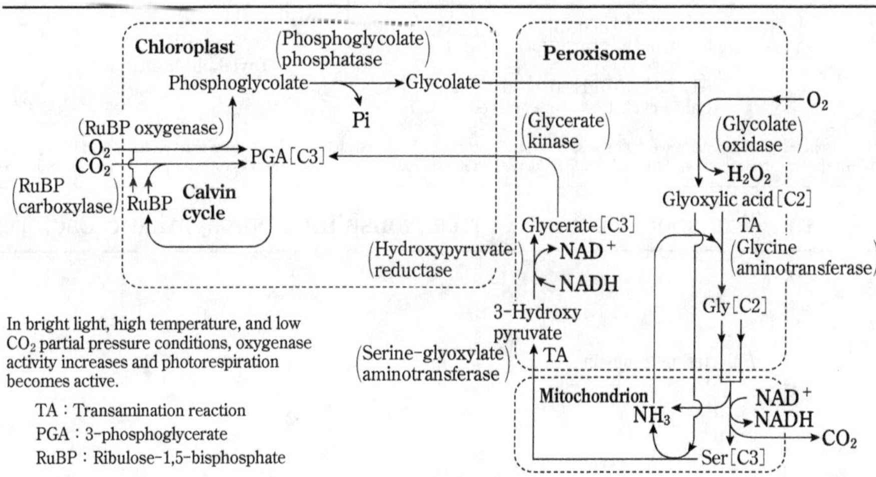

In bright light, high temperature, and low CO_2 partial pressure conditions, oxygenase activity increases and photorespiration becomes active.

TA : Transamination reaction
PGA : 3-phosphoglycerate
RuBP : Ribulose-1,5-bisphosphate

Chemical Structural Formula of Intermediate Metabolites of Photorespiration

Phosphoglycolate	Glycolate	3-Hydroxy pyruvate
$CH_2OPO_3^{2-}$ – COO^-	CH_2OH – COO^-	CH_2OH – CO – COO^-

Electron Transport Chain of Photosynthesis in Thylakoid Membrane of Chloroplast

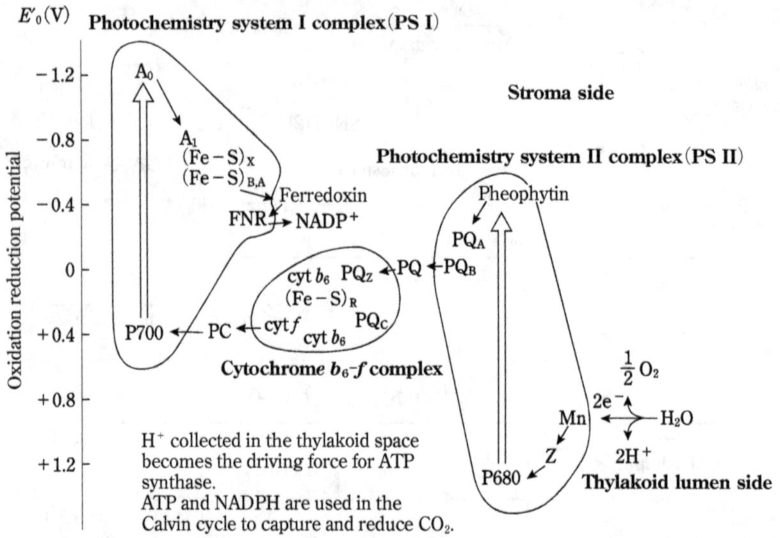

Mn : Manganese, Z : Reaction center tyrosyl residue, cyt : Cytochrome, PQ : Plastoquinone, Fe–S : Iron–sulfur center, PC : Plastocyanin, A_0 : The first electronic receptor, FNR : Fd–NADP Reductase, P680 and P700 are the reaction center chlorophyll a for PS II and PS I, respectively

Electron Transport Chain of Purple Nonsulfur Photosynthetic Bacteria

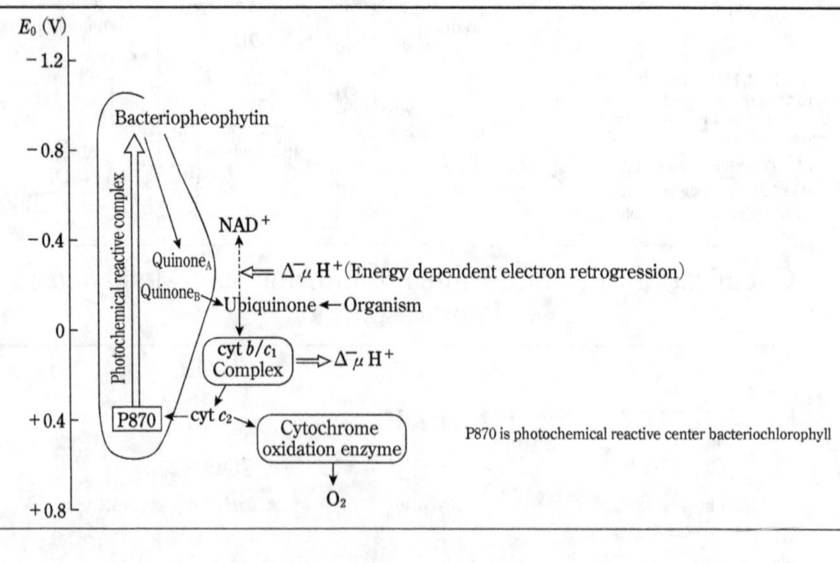

Nitrogen Metabolism

Symbiotic nitrogen fixation

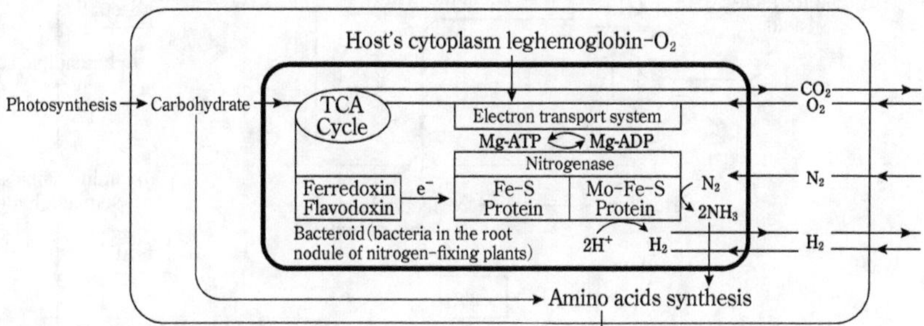

Nitric acid reduction and electron transport

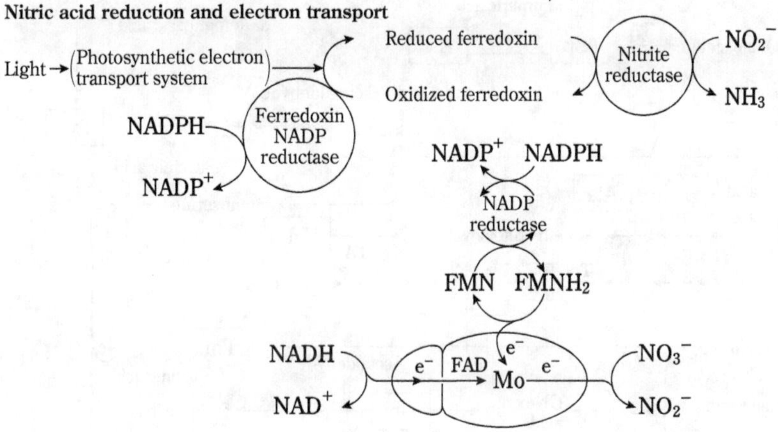

Nitrite oxidation and ATP synthesis

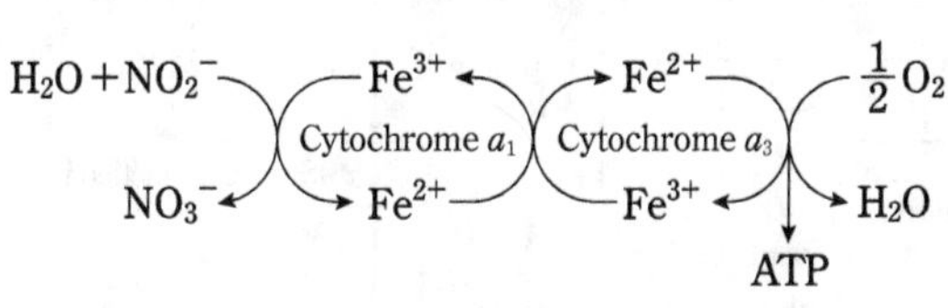

Nitrate reduction (denitrification type)

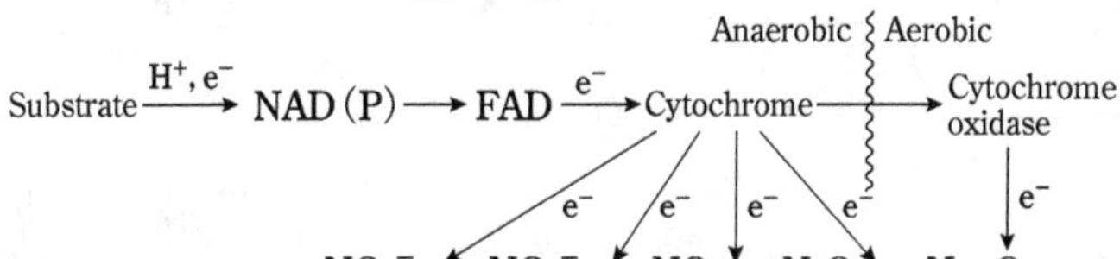

Biosynthesis of Amino Acids*

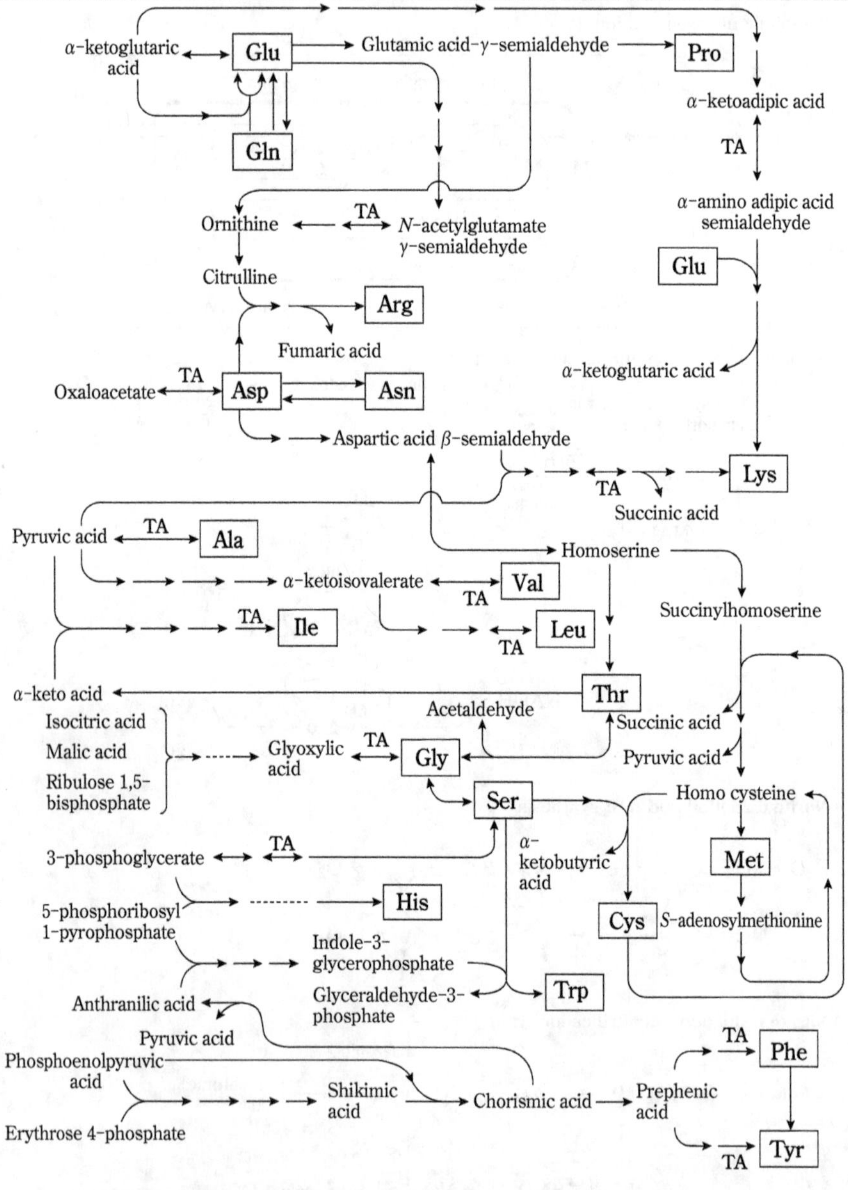

TA : Transamination reaction

* Refer to page **480** for the 3 letter and 1 letter abbreviations for the amino acids.

Chemical Structural Formula of Metabolic Intermediates for Biosynthesis of Amino Acids

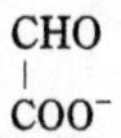
Glyoxylic acid

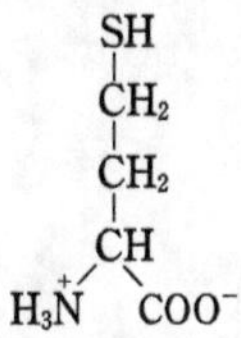
Homo cysteine

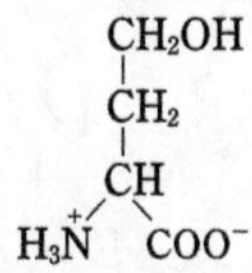
Homoserine

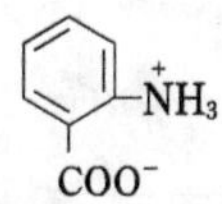
Anthranilic acid

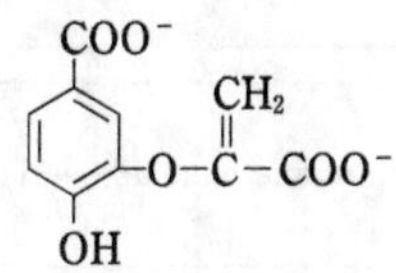
Chorismic acid

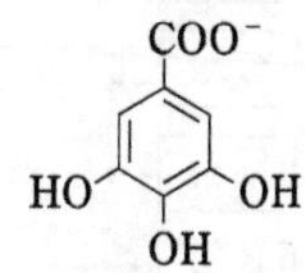
Shikimic acid

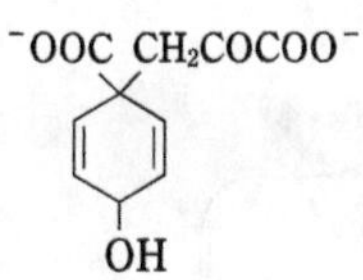
Prephenic acid

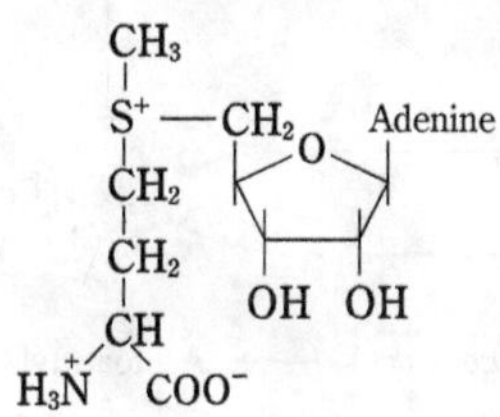
S-adenosylmethionine

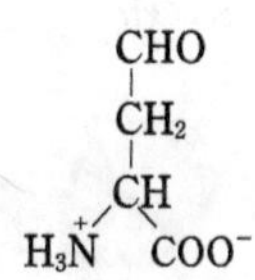
Aspartic acid-β-semialdehyde

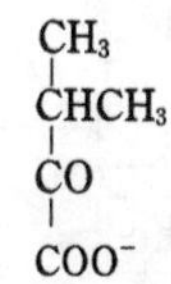
α-ketoisovalerate

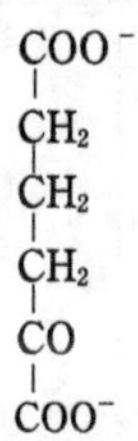
α-ketoadipic acid

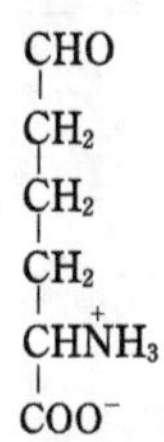
α-amino adipic acid semialdehyde

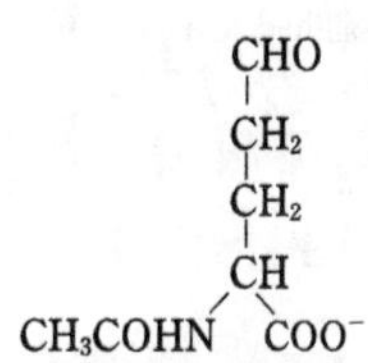
N-acetylglutamate γ-semialdehyde

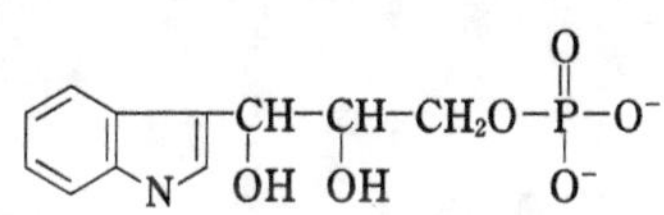
Indole-3-glycerophosphate

Genetic Code (Genetic Codon Table of Amino Acids)

1st letter	2nd letter: U		C		A		G		3rd letter
U	UUU	Phe	UCU	Ser	UAU	Tyr	UGU	Cys	U
U	UUC		UCC		UAC		UGC		C
U	UUA	Leu	UCA		UAA	*	UGA	*	A
U	UUG		UCG		UAG		UGG	Trp	G
C	CUU		CCU	Pro	CAU	His	CGU	Arg	U
C	CUC		CCC		CAC		CGC		C
C	CUA		CCA		CAA	Gln	CGA		A
C	CUG		CCG		CAG		CGG		G
A	AUU	Ile	ACU	Thr	AAU	Asn	AGU	Ser	U
A	AUC		ACC		AAC		AGC		C
A	AUA		ACA		AAA	Lys	AGA	Arg	A
A	AUG	Met	ACG		AAG		AGG		G
G	GUU	Val	GCU	Ala	GAU	Asp	GGU	Gly	U
G	GUC		GCC		GAC		GGC		C
G	GUA		GCA		GAA	Glu	GGA		A
G	GUG		GCG		GAG		GGG		G

AUG: Initiator codons; UAA, UAG, UGA: Termination condon (*)

* Refer to page **480** for the 3 letter and 1 letter abbreviations for the amino acids.

Amino Acid Breakdown and TCA Pathways

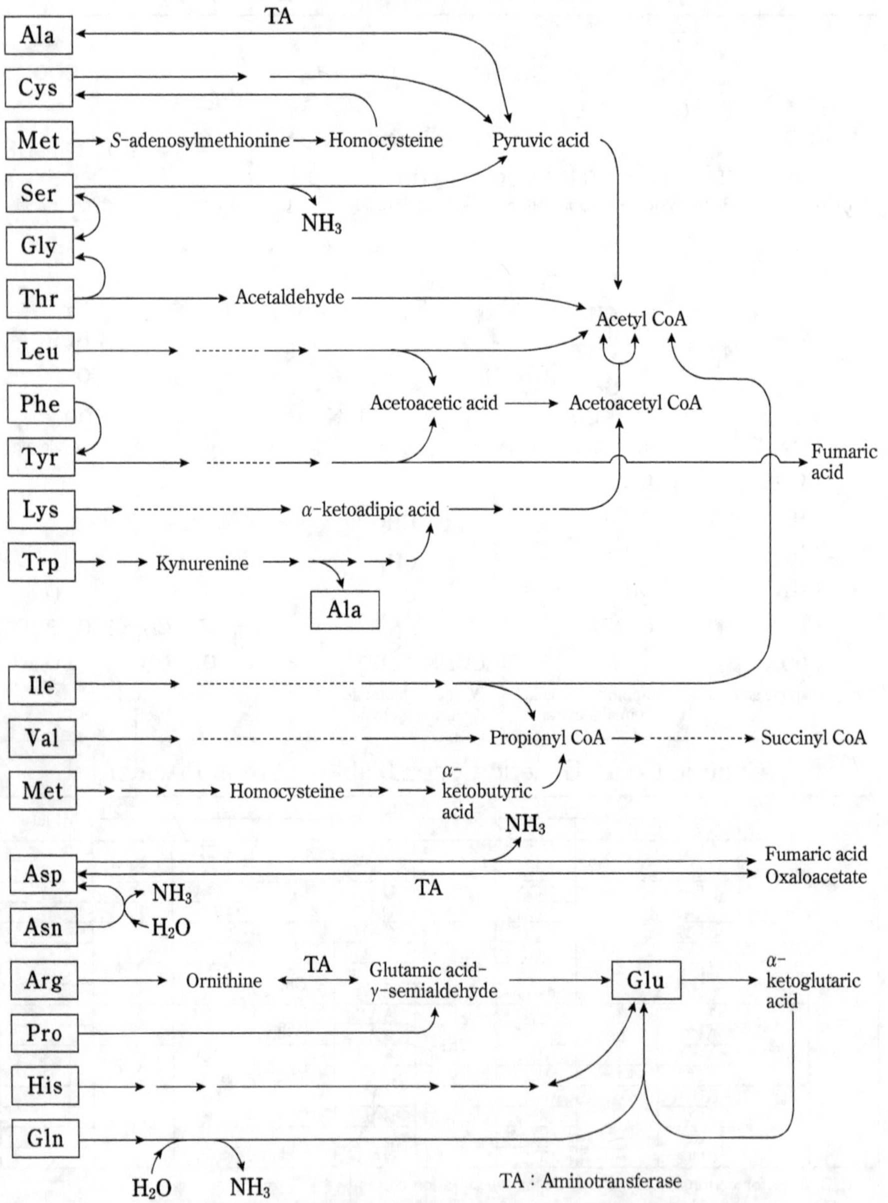

* Refer to page **480** for the 3 letter and 1 letter abbreviations for the amino acids.

Urea Pathway

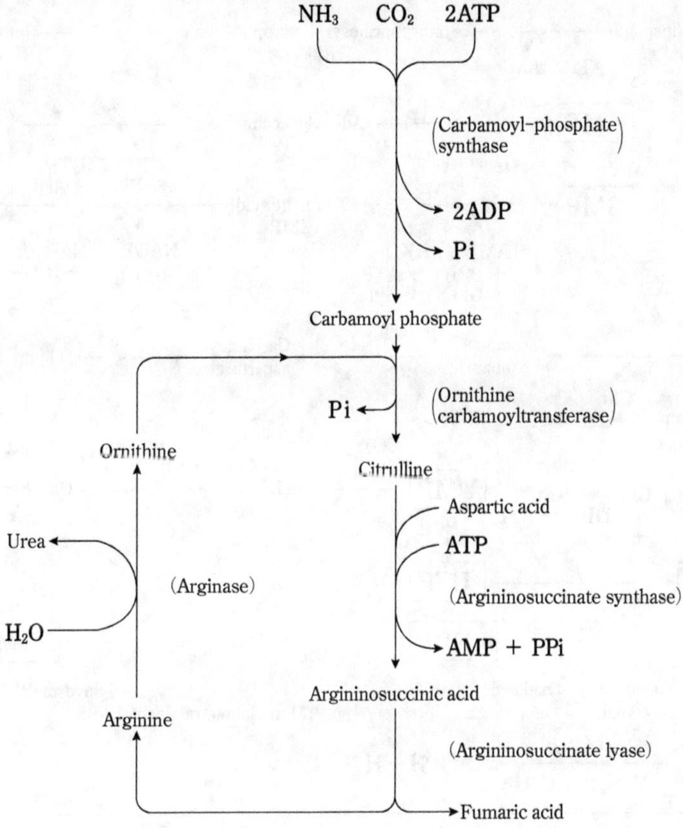

Chemical Structural Formula of Intermediate Metabolites of Urea Pathways

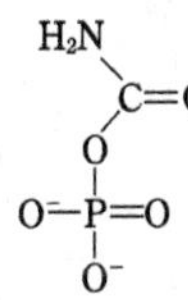

Carbamoyl phosphate

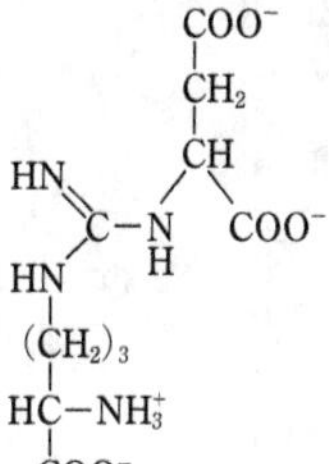

Arginosuccinic acid

Biosynthesis of Nucleotide

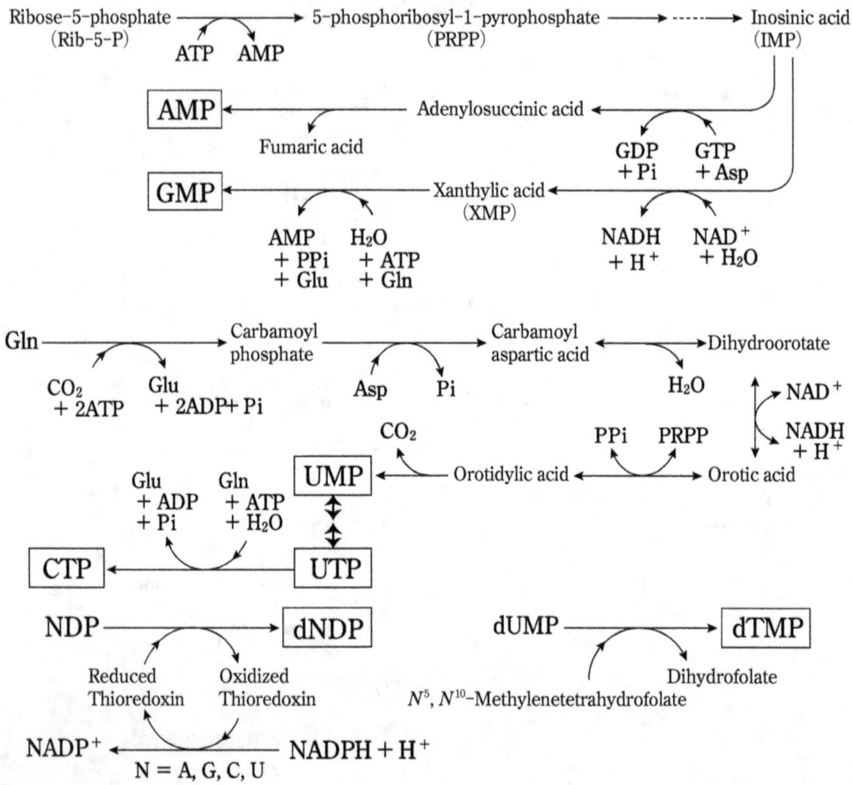

Intermediate Metabolite of Nucleotide Biosynthesis

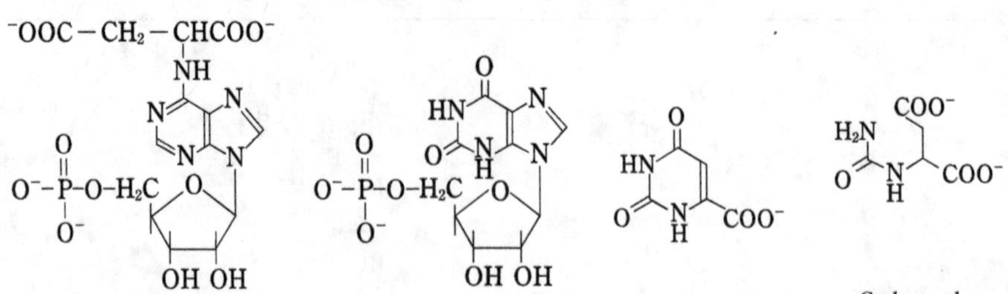

Adenylosuccinic acid

Xanthylic acid

Orotic acid

Carbamoyl aspartic acid

Breakdown of Nucleotide

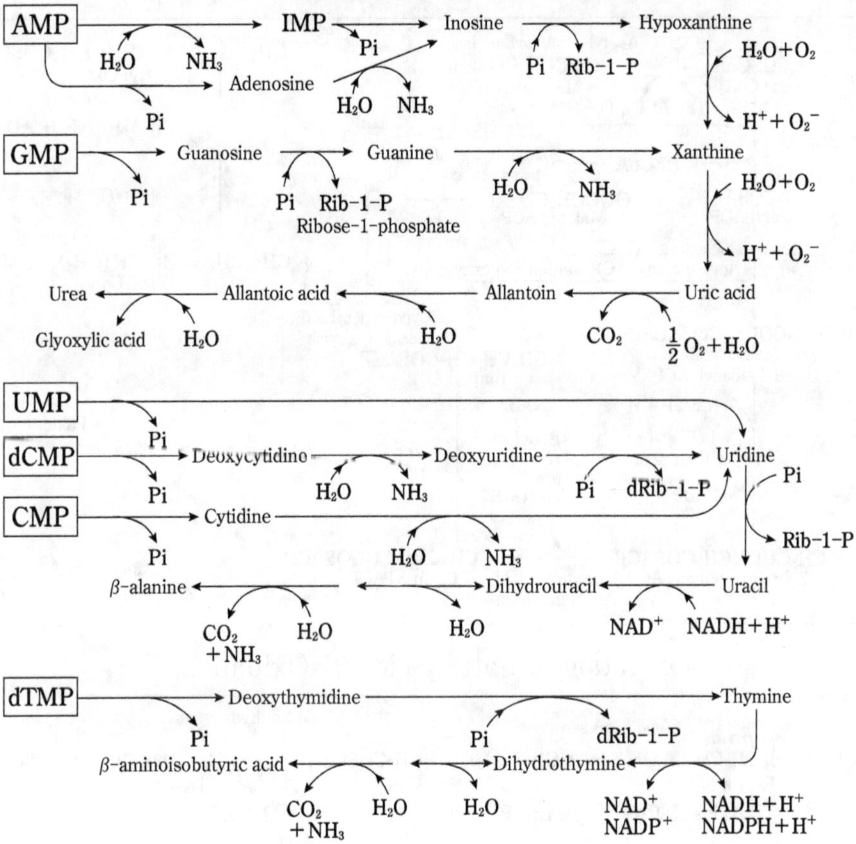

Intermediate Metabolite of Nucleotide Breakdown

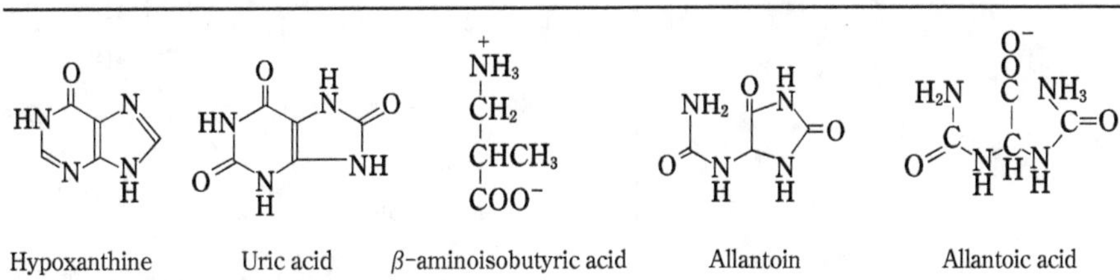

Biosynthesis Pathway of Saturated Fatty Acid and Unsaturated Fatty Acid

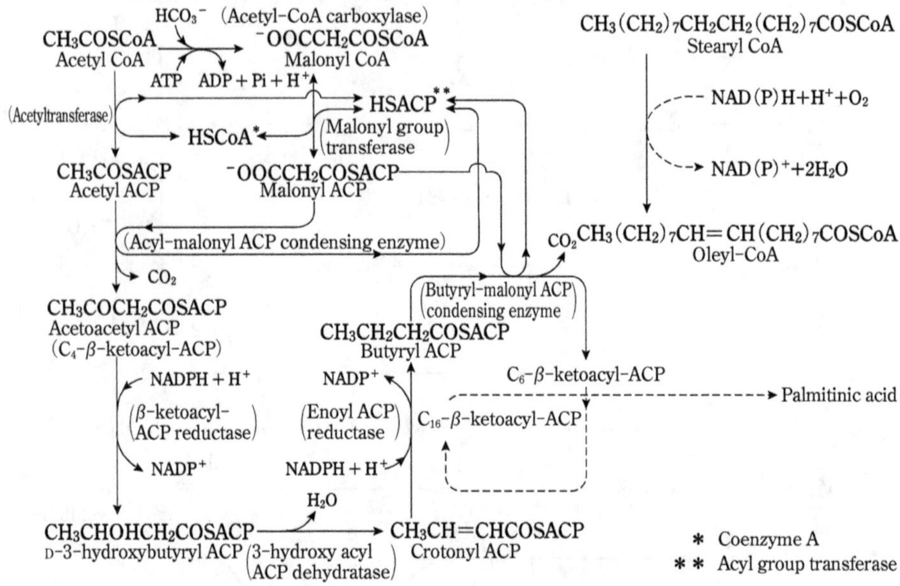

Oxidation of Fatty Acids (β-Oxidation) *

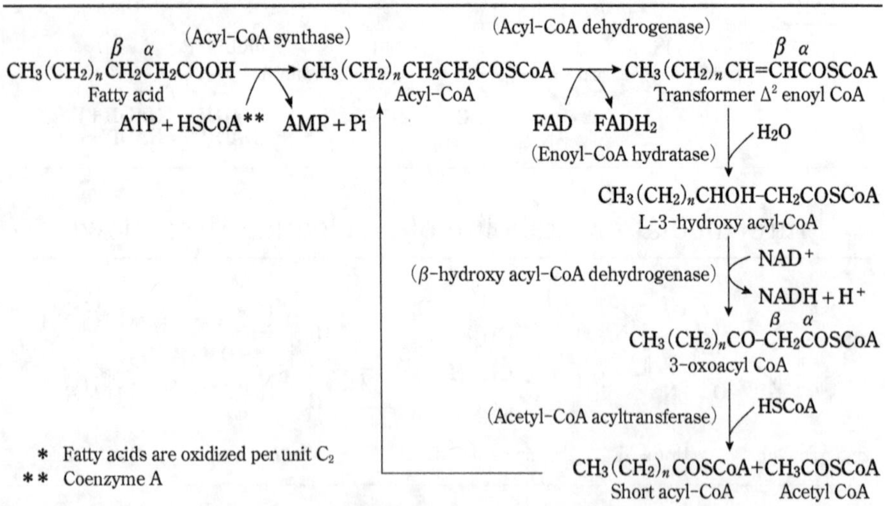

Reciprocal Interactions between Major Metabolic Pathways and Biosynthesis Pathways*

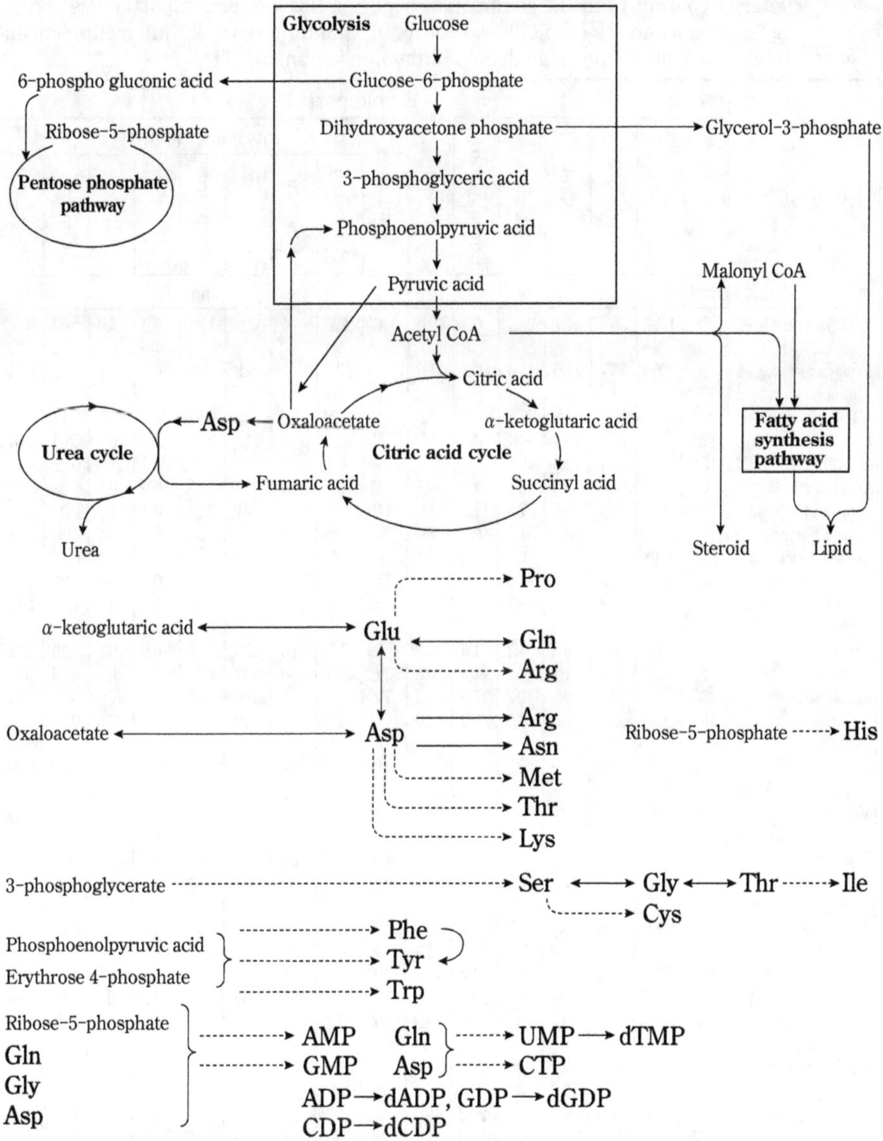

* Reciprocal interactions among glycolysis, pentose phosphate pathway, citric acid cycle, urea cycle, steroidogenesis, lipid synthesis, amino-acids synthesis, and nucleotide synthesis pathway

* Refer to page **480** for the 3 letter and 1 letter abbreviations for the amino acids.

Food Ingredients and Energy (Calorie)

Shows number of grams for each ingredient per 100 g of edible portions of food products. For energy content of foods, combustion heat per 100 g of edible portions is shown as kcal (kilocalories) and kJ (kilojoules). (Source: Kagakugijutsu·gakujutsushingikaishigenchosabunkakai publication "Nihonshokuhinhyojunseibun 2015.")

Food name	Edible part															
	Energy		Moisture	Protein	Lipid	Carbohydrate	Mineral	Inorganic quality								
								Sodium	Potassium	Calcium	Magnesium	Phosphorus	Iron	Zinc	Copper	Manganese
	kcal	kJ	g					mg								
Beef (Japanese beef, chuck)	286	1197	58.8	17.7	22.3	0.3	0.9	47	280	4	19	150	0.9	4.9	0.07	0
Pork (large species, chuck)	216	904	65.7	18.5	14.6	0.2	1.0	53	320	4	21	180	0.5	2.7	0.09	0.01
Mackerel pike (fresh)	297	1241	57.7	17.6	23.6	0.1	0.9	130	190	26	26	170	1.3	0.8	0.12	0.02
Pacific bluefin tuna (lean)	125	523	70.4	26.4	1.4	0.1	1.7	49	380	5	45	270	1.1	0.4	0.04	0.01
Egg (whole egg, fresh)	151	632	76.1	12.3	10.3	0.3	1.0	140	130	51	11	180	1.8	1.3	0.08	0.02
Ordinary milk	67	280	87.4	3.3	3.8	4.8	0.7	41	150	110	10	93	0.02	0.4	0.01	Tr
Evaporated milk	144	602	72.5	6.8	7.9	11.2	1.6	140	330	270	21	210	0.2	1.0	0.02	-
Salted butter	745	3117	16.2	0.6	81.0	0.2	2.0	750	28	15	2	15	0.1	0.1	Tr	0
Polished rice	358	1498	14.9	6.1	0.9	77.6	0.4	1	89	5	23	95	0.8	1.4	0.22	0.81
Flour (soft, type 1)	367	1535	14.0	8.3	1.5	75.8	0.4	Tr	110	20	12	60	0.5	0.3	0.08	0.43
Soybean (whole grain, Japanese, dry)	422	1765	12.4	33.8	19.7	29.5	4.7	1	1900	180	220	490	6.8	3.1	1.07	2.51
Firm tofu	72	301	86.8	6.6	4.2	1.6	0.8	59	140	86	130	110	0.9	0.6	0.15	0.38
Banana (fresh)	86	360	75.4	1.1	0.2	22.5	0.8	Tr	360	6	32	27	0.3	0.2	0.09	0.26
Apple (fresh)	57	240	84.1	0.1	0.2	15.5	0.2	Tr	120	3	3	12	0.1	Tr	0.05	0.02
Japanese chestnut (fresh)	164	686	58.8	2.8	0.5	36.9	1.0	1	420	23	40	70	0.8	0.5	0.32	3.27
Radish (root, skin, fresh)	18	75	94.6	0.5	0.1	4.1	0.6	19	230	24	10	18	0.2	0.2	0.02	0.04
Carrot (root, skin, fresh)	39	164	89.1	0.7	0.2	9.3	0.8	28	300	28	10	26	0.2	0.2	0.05	0.12
Shiitake mushrooms (dry)	182	761	9.7	19.3	3.7	63.4	3.9	6	2100	10	110	310	1.7	2.3	0.5	0.87
Mayonnaise (egg yolk type)	703	2941	16.2	1.5	72.3	4.5	2.0	690	18	9	2	33	0.3	0.2	0.02	0.02
Rice miso (sweet)	217	908	42.6	9.7	3.0	37.9	6.8	2400	340	80	32	130	3.4	0.9	0.22	-
Bean miso	217	908	44.9	17.2	10.5	14.5	12.9	4300	930	150	130	250	6.8	2.0	0.66	-
Curry roux	512	2142	3.0	6.5	34.1	44.7	11.7	4200	320	90	31	110	3.5	0.5	0.13	0.58
Standard soy sauce	71	297	67.1	7.7	0	10.1	15.1	5700	390	29	65	160	1.7	0.9	0.01	-
Light colored soy sauce	54	226	69.7	5.7	0	7.8	16.8	6300	320	24	50	130	1.1	0.6	0.01	-
Grain vinegar	25	105	93.3	0.1	0	2.4	Tr	6	4	2	1	2	Tr	0.1	Tr	-
Rice vinegar	46	192	87.9	0.2	0	7.4	0.1	12	16	2	6	15	0.1	0.2	Tr	-
Superfine sugar	384	1607	0.8	(0)	(0)	99.2	0	1	2	1	Tr	Tr	Tr	0	0.01	-
Starch syrup	328	1372	15.0	(0)	(0)	85.0	Tr	Tr	0	Tr	0	1	0.1	0	Tr	0.01

(0) : estimate, Tr : trace ammounts, - : not determined

Appendix

Main Achievements in Life Science*

Era	Matter	Researcher (native country)
c. 400 B.C.	Four element theory of body fluid ·················	Hippokrates (Greece)
c. 340 B.C.	Classification of animals, systematization of the living world, embryology ····················	Aristotle (Greece)
c. 320 B.C.	Natural history of plants ····························	Theophrastos (Greece)
c. 50	Natural history··	Plinius (Rome)
c. 180	Medical research, particularly anatomy ···········	Galenos (Greece)
c. 1000	Encyclopedia of Greek and Arabian medicine ···	Avicenna (Arabia)
c. 1500	Human anatomy and comparative anatomy ·····	Leonardo da Vinci (Italy)
c. 1530	Establishment of medical chemistry ··············	Paracersus (Switzerland)
1543	Human anatomy ··	Vesalius (Belgium)
1561	Human anatomy including the female reproductive organs ································	Fallopius (Italy)
1604	Discovery of venous valve, embryology for chicken embryo································	Fabricius (Italy)
1628	Theory of blood circulation ·························	Harvey (UK)
1637	Mechanistic view of life ································	Descartes (France)
1665	Observation of cells (cork) using microscope ···	Hooke (UK)
1668	Refutation of spontaneous generation theory ···	Redi (Italy)
1674	Observation of microorganisms using microscope ··	Leeuwenhoek (Netherlands)
1682	Plant anatomy ··	Grew (UK)
1682	Species concept ···	Ray (UK)
1694	Discovery of stamen and pistil ······················	Camerarius (Germany)
1700	Study of plant taxonomy ······························	Tournefort (France)
1727	Plant statics ···	Hales (UK)
1734	Anatomy of insects, digestive function of stomach acid··	Reaumur (France)
1735	Foundation of biological classification ············	Linnaeus (Sweden)
1744	Experimental research in regeneration ···········	Trembley (Switzerland)
1749	Natural history···	Buffon (France)
1758	Foundation of animal nomenclature (10th edition of "Systema Naturae")··············	Linnaeus (Sweden)

* Refer to Major inventions and discoveries in chemistry (pages **508–516**).

(Yasugi, S., 2016)

Main Achievements in Life Science Continued.

Era	Matter	Researcher (native country)
1759	Assertion of epigenesis	Wolff (Germany)
1762	Observation of parthenogenesis	Bonnet (France)
1765	Refutation of spontaneous generation for microorganisms	Spallanzani (Italy)
1774	Research on plant respiration and oxygen	Priestley (UK)
1779	Carbon dioxide assimilation in plants	Ingenhousz (Netherlands)
1780	Discovery of animal electricity	Galvani (Italy)
1796	Founding of vaccination method	Jenner (UK)
1805	Plant geography	Humboldt (Germany)
1809	Lamarckian view of evolution	Lamarck (France)
1812	Comparative anatomy and catastrophism	Cuvier (France)
1816	Vertebrate neurophysiology	Magendie (France)
1817	Germ layer theory in development	Pander (Russia)
1822	Founding of teratology	Geoffroy Saint-Hilaire (France)
1826	Research in sensory physiology	Müller (Germany)
1827	Classification and organ research of higher plants	de Candolle (Switzerland)
1831	Discovery of cell nuclei	Brown (UK)
1837	Establishment of germ layer theory	Baer (Germany)
1837	Fermentation and catalyst reaction	Berzelius (Sweden)
1838	Cell theory for plants	Schleiden (Germany)
1839	Establishment of cell theory	Schwann (Germany)
1848	Physiology of biological electricity	Du Bois-Reymond (Germany)
1852	Three primary color theory for color vision	Helmholtz (Germany)
1858	Systematization of cytopathology	Virchow (Germany)
1858	Natural selection theory for evolution	Darwin, Wallace (UK)
1859	Publishing of "On the Origin of Species"	Darwin (UK)
1860	Research in alcohol fermentation	Pasteur (France)
1861	Refutation of spontaneous generation theory	Pasteur (France)
1863	Man's place in nature	T. H. Huxley (UK)
1865	Methodology for experimental medicine	Bernard (France)
1865	Mendel's laws of genetics	Mendel (Austria)
1866	Recaptulation theory for ontogeny and phylogeny	Haeckel (Germany)
1869	Discovery of nuclein (DNA)	Miescher (Switzerland)
1873	Staining method for nerve cells	Golgi (Italy)

Continued.

Era	Matter	Researcher (native country)
1876	Culturing of *Bacillus anthracis* ······	Koch (Germany)
1879	Natural history of insects ······	Fabre (France)
1882	Isolation of *Mycobacterium tuberculosis* ······	Koch (Germany)
1883	Phagocytosis by leukocytes ······	Metchnikoff (Russia)
1889	Culturing of *Clostridium tetani* ······	Kitazato (Japan)
1891	Discovery of Java man ······	Dubois (Netherlands)
1892	Continuity of germ plasm ······	Weismann (Germany)
1892	Foundation of developmental mechanics ······	Roux (Germany)
1893	Research on neurons ······	Ramón y Cajal (Spain)
1894	Culturing of *Yersinia pestis* ······	Kitazato (Japan), Yersin (Germany)
1897	Research on fermentation in cell-free systems ···	Buchner (Germany)
1898	Discovery of *Shigella dysenteriae* ······	Shiga (Japan)
1900	Rediscovery of Mendel's laws ······	De Vries (Netherlands) *et al.*
1901	Mutation theory ······	De Vries (Netherlands)
1901	Discovery of ABO blood types ······	Landsteiner (Austria)
1902	Discovery of secretin ······	Bayliss, Starling (UK)
1903	Research on conditioned reflex ······	Pavlov (Russia)
1904	Research on genetic linkage ······	Bateson (UK)
1907	Tissue culture methods ······	Harrison (USA)
1910	Genetics of *Drosophila* ······	Morgan (USA)
1910	Discovery of chlorophyll ······	Willstätter (Germany)
1910	Research on syphilitic spirochete ······	Noguchi (Japan)
1912	Proposal of the term "vitamin" ······	Funk (USA)
1912	Necessity of vitamins ······	Hopkins (UK)
1913	First chromosome map ······	Sturtevant (USA)
1916	Research on plant community succession ······	Clements (USA)
1919	Artificial induction of cancer ······	Yamagiwa (Japan)
1921	Proof of chemical transmission for nerves ······	Loewi (Germany)
1922	Discovery of insulin ······	Banting (Canada), Macleod (UK)
1924	Discovery of embryo organizer ······	Spemann (Germany)
1924	Research on respiratory pathway ······	Warburg (Germany)
1926	Gene theory ······	Morgan (USA)
1926	Principle of "all or nothing" for receptors ······	Adrian (UK)
1927	Animal ecology ······	Elton (UK)

Main Achievements in Life Science Continued.

Era	Matter	Researcher (native country)
1927	Creation of artificial mutation	Muller (USA)
1927	Instinctual behavior of honey bees	Frisch (Austria)
1928	Discovery of auxin	Went (Netherlands)
1929	Discovery of ATP	Lohmann (Germany)
1929	Penicillin	Fleming (UK)
1930	Genetic basis of natural selection	Fisher (UK), Wright (USA)
1930	Crystallization of pepsin	Northrop (USA)
1931	Research on the chromosomes of wheat	Kihara (Japan)
1932	Organismal theory of life	Bertalanffy (Austria)
1932	Concept of homeostasis	Canonn (USA)
1933	Research on glycolytic pathway	Embden (Germany), Meyerhof (Germany), Parnas (Germany)
1935	Imprinting phenomenon	Lorentz (Austria)
1935	Crystallization of tobacco mosaic virus	Stanley (USA)
1936	Coacervate theory for the origin of life	Oparin (Russia)
1937	Thermodynamics of muscle activity	Hill (UK)
1942	Life history of phages	Delbrück (Germany)
1942	Contraction of actomyosin via ATP	Szent-Györgyi (Hungary)
1942	General theory of evolution	J. S. Huxley (UK), Simpson (USA) *et al.*
1943	Proposal of epigenetics	Waddington (UK)
1944	Transformation of pneumococcus	Avery (Canada) *et al.*
1945	One-gene one-enzyme hypothesis	Beadle (USA)
1948	Isolation of mitochondria	Palade (Romania)
1948	Proposal of cybernetics	Wiener (USA)
1950	Research on base content of nucleicacid	Chargaff (Austria)
1951	α-Helix structure of protein	Pauling (USA), Corey (USA)
1953	Sodium theory for nerve excitement	Hodgkin (UK), A. F. Huxley (UK)
1953	Double-helix structure of DNA	Watson (USA), Crick (UK)
1954	Sliding theory of muscle contraction	H. E. Huxley (UK) *et al.*
1955	Enzymatic synthesis of RNA	Ochoa (Spain)
1956	Isolation of nerve growth factor	Levi-Montalcini (Italy), Cohen (USA)
1957	CO_2 fixation pathway of photosynthesis	Calvin (USA)

Continued.

Era	Matter	Researcher (native country)
1957	Discovery of cyclic AMP	Sutherland (USA), Rall (USA)
1957	Clone selection theory of immunity	Burnet (Australia)
1958	Formation of individual from single cell of carrot	Steward (UK)
1961	Operon theory of gene regulation	Jacob (France), Monod (France), Lwoff (France)
1962	Chemical properties of pheromones	Butenandt (Germany)
1962	Creation of clone frog	Gurdon (UK)
1963	Primary structure of antibody molecules	Edelman (USA)
1965	Deciphering of genetic code	Khorana (India) *et al.*
1967	Symbiotic origin theory for mitochondria and chloroplasts	Margulis (USA)
1968	Neutral theory of molecular evolution	Kimura (Japan)
1970	Discovery of reverse transcriptase	Temin (USA), Baltimore (USA)
1972	Fluid mosaic model for biological membrane	Singer (USA), Nicolson (USA)
1972	Punctuated equilibrium theory of evolution	Gould (USA), Eldridge (USA)
1973	Development of gene manipulation technology	Cohen (USA), Boyer (USA) *et al.*
1975	Asilomar Conference on gene manipulation	Berg (USA) *et al.*
1975	Monoclonal antibody production method	Kohler (Germany), Milstein (Argentina)
1975	Sociobiology	Wilson (USA)
1977	Determination of entire base sequence for phage	Sanger (UK)
1977	Structure of antibody genes	Tonegawa (Japan) *et al.*
1979	Detection of cellular oncogene	Bishop (USA), Varmus (USA) *et al.*
1981	RNA origin theory of life	Altman (Canada), Cech (USA)
1981	Establishment of embryonic stem cells (ES cells)	Evans (USA) *et al.*
1982	Transgenic animals	Brinstar (USA) *et al.*
1982	Discovery of prion	Prusiner (USA)
1983	Discovery of homeobox	Gehring (Switzerland) *et al.*

Main Achievements in Life Science Continued.

Era	Matter	Researcher (native country)
1983	Isolation of HIV as the cause of AIDS …………	Montagnier (France), Barré-Sinoussi (France)
1986	Discovery of the tumor suppressor gene *Rb*……	Weinberg (USA) *et al.*
1987	Discovery of dystrophin in relation to muscular dystrophy …………………………	Kunkel (USA) *et al.*
1989	Discovery of the tumor suppressor gene *p53* …	Levine (USA) *et al.*
1991	Discovery of sex-determining gene *Sry* in animals ………………………………………	Koopman (UK) *et al.*
1991	ABC model of flowering…………………………	Meyerowitz (USA)
1996	Production of the clone sheep "Dolly" ……………	Wilmut (UK) *et al.*
2000	Draft of human genome sequence ………………	International Human Genome Sequencing Consortium
2000	Discovery of Orrorin man fossils from approx. 6 million years ago ……………………………	Pickford (France) *et al.*
2004	Genome analysis of the primeval red algae Schyzon ………………………………	Kuroiwa (Japan) *et al.*
2006	Establishment of mouse iPS cell …………………	Yamanaka (Japan) *et al.*
2007	Establishment of human iPS cell …………………	Yamanaka (Japan) *et al.*
2013	Genome edditing with CRISPR/Cas 9……………	Doudna (USA), Charpentier (France)

Environmental Science Section

Climate Change and Global Warming

Energy Budget in the Climate System: Greenhouse Effect

The energy source for the Earth's climate system is radiation from the sun. Around 30% of solar energy arriving at the top of the earth's atmosphere is reflected directly back into outer space. The remains are absorbed mainly by the earth's surface and partially in the atmosphere. On average, the earth radiates nearly the same amount of energy to outer space as the absorbed energy.

Significant amounts of thermal emissions from the earth's surface are absorbed by water vapor and greenhouse gases such as carbon dioxide in the atmosphere, reflected back and re-absorbed by the earth's surface in a phenomenon known as the greenhouse effect. According to estimates, the average temperature of the Earth's surface would be about −19℃ without this effect, with it, the actual value is about 14℃. Increased emission of greenhouse gases due to human activity enhances the greenhouse effect, which causes global warming.

Global mean energy budget under present-day climate conditions

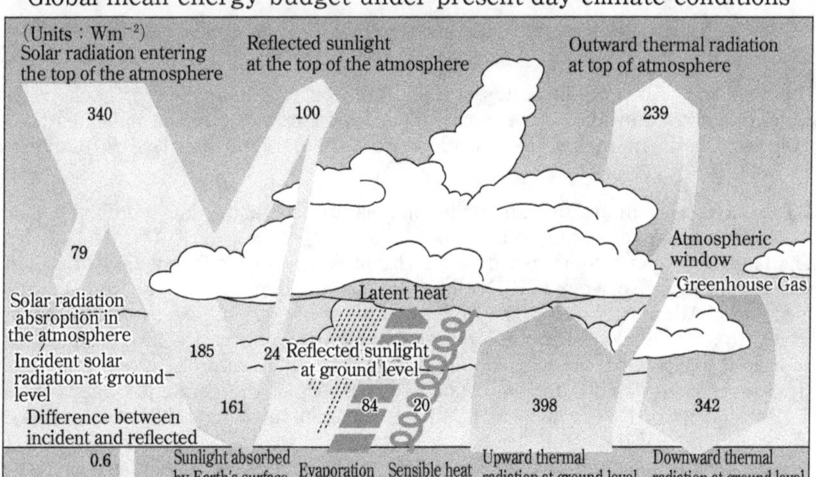

Source: IPCC Fifth Assessment Report (2013)

Annual Anomalies of Global Average Surface Temperature (1891–2019)

The annual anomaly of the global average surface temperature in 2019 (i.e., the combined average of the near-surface air temperature over land and the sea surface temperature (SST)) was +0.43°C above the 1981–2010 average (+0.79℃ above the 20th century average), which was the second highest since 1891. On a longer time scale, it is virtually certain that the global average surface temperature has risen at a rate of about +0.74°C per century. In particular, the warmest years have been often observed since the mid-1990s.

Annual anomalies in surface temperature

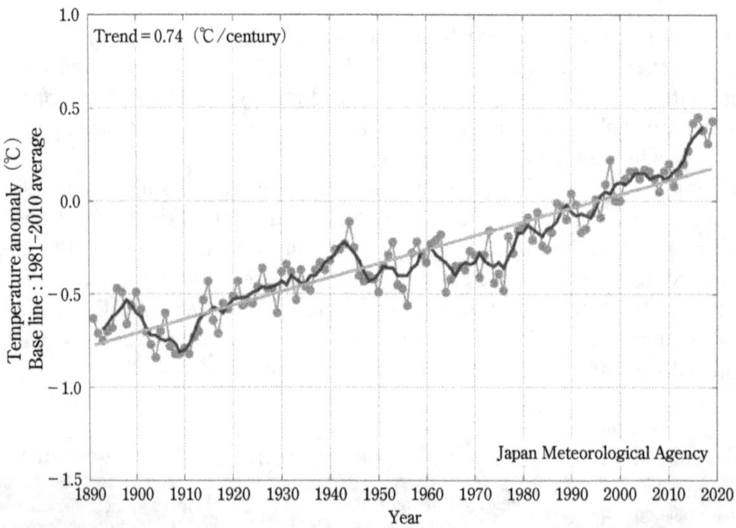

The thin line with dots indicates surface temperature anomalies for each year. The bold line indicates the five-year running mean, and the straight line indicates the long-term linear trend. Anomalies are deviations from the baseline (the 1981–2010 average).

(data used)

Global surface temperature anomalies are estimated using data combined not only over land but also over ocean areas.

The land part of the combined data for the period before 2000 consists of Global Historical Climatology Network (GHCN) information provided by the National Oceanic and Atmospheric Administration (NOAA). The number of stations used for analysis ranges from 300 to 3,900, while that for the period after 2001 consists of CLIMAT messages archived at the Japan Meteorological Agency (JMA). The number of stations used for analysis ranges from 1,000 to 1,300. The oceanic part of the combined data consists of SST data consist of JMA's historical SST data known as COBE-SST, which are 1x1 degree resolution GPV data from 1891.

For details on the calculation method, refer to the section entitled "Data and Analysis Method" under "Global Warming" on the Tokyo Climate Center's website (https://ds.data.jma.go.jp/tcc/tcc/index.html).

Annual Surface Temperature Anomalies in Japan (1898–2019)

The mean surface temperature in Japan for 2019 is estimated to have been +0.92°C above the 1981–2010 average (+1.53°C above the 20th century average), which is the warmest on record since 1898. On a longer time scale, it is virtually certain that the annual mean surface temperature over Japan has risen at a rate of about 1.24°C per century. In particular, the warmest years have been often observed since the 1990s.

Annual surface temperature anomalies in Japan

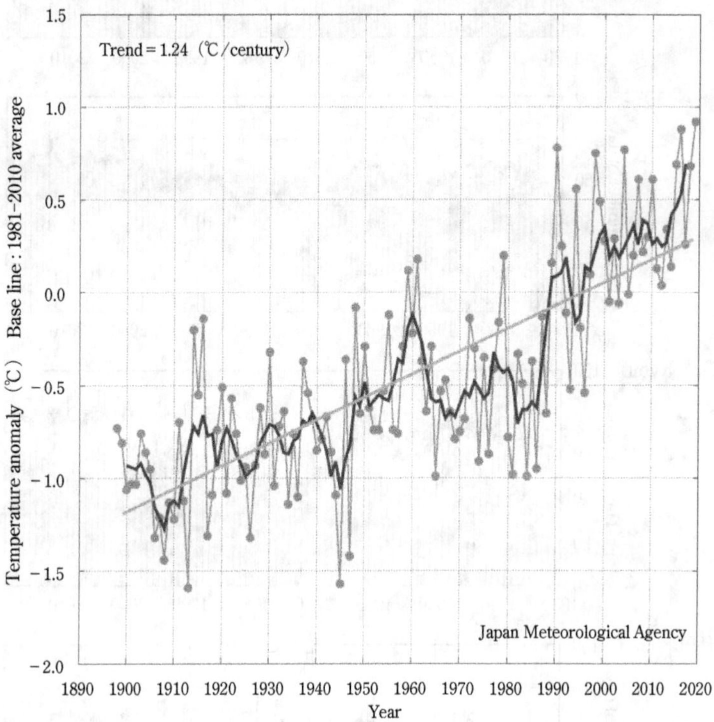

The thin line with dots indicates surface temperature anomalies for each year. The bold line indicates the five-year running mean, and the straight line indicates the long-term linear trend. Anomalies are deviations from the baseline (the 1981–2010 average).

(data used)

Surface temperature anomalies in Japan are estimated using data from 15 observation stations (Abashiri, Nemuro, Suttsu, Yamagata, Ishinomaki, Fushiki, Iida*, Choshi, Sakai, Hamada, Hikone, Tadotsu, Miyazaki*, Naze and Ishigakijima) which have continuous long-term records of temperature data from 1898 onward. They are considered to be affected to a lesser extent by local urbanization and are not biased to a particular region.

(*Miyazaki and Iida were moved in May 2000 and May 2002, respectively, and their temperatures have been adjusted to eliminate the influence of the relocation.)

Annual Number of Days with Maximum Temperatures of ≥ 30℃ (1931–2019)

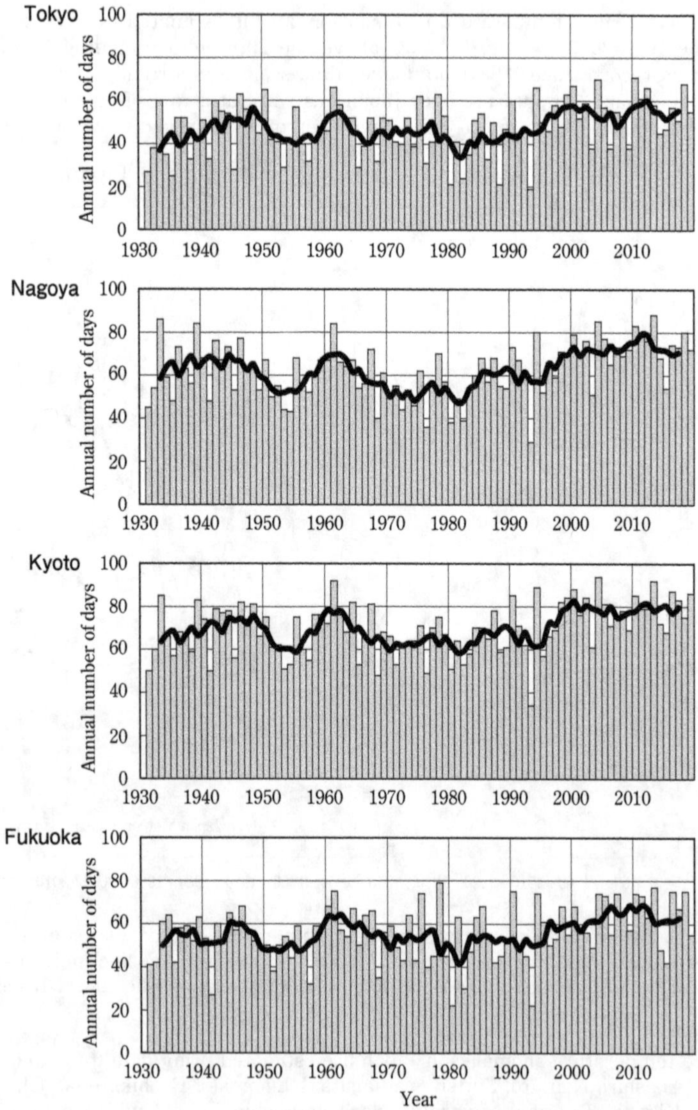

The bars indicate the values for each year. The solid line indicates the five-year running mean.

Tokyo was relocated on 2 December 2014.

Annual Number of Days with Minimum Temperatures of ≥ 25℃ (1931-2019)

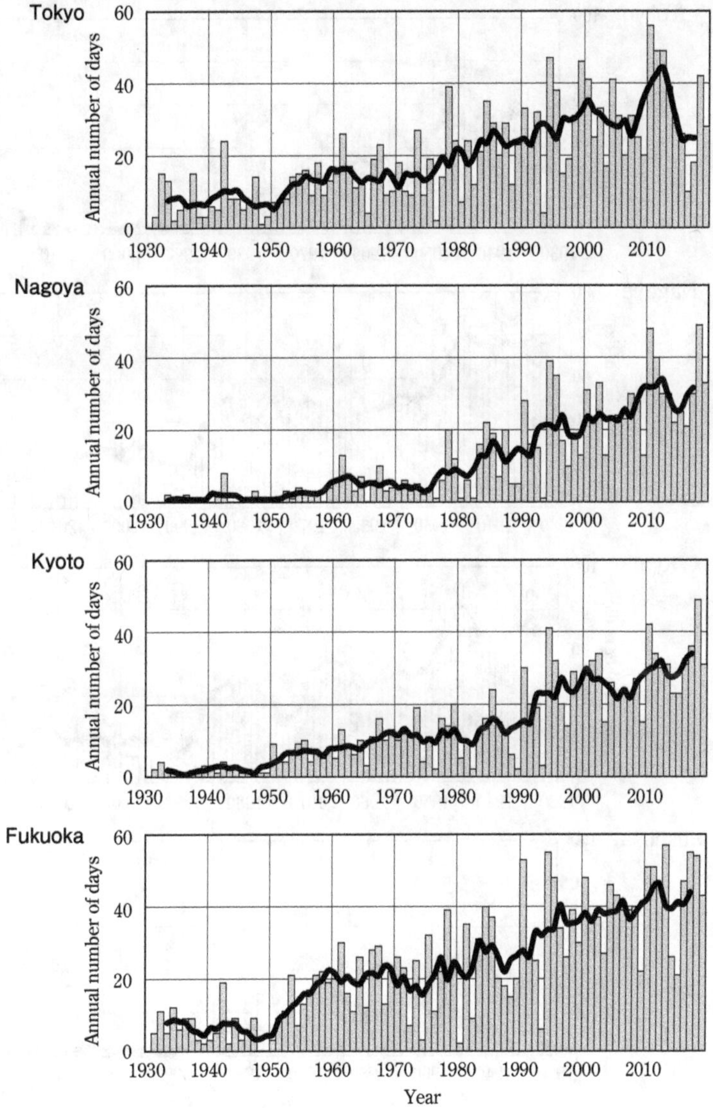

The bars indicate the values for each year. The solid line indicates the five-year running mean.

Tokyo was relocated on 2 December 2014.

Annual Number of Days with Minimum Temperatures of < 0℃ (1931–2019)

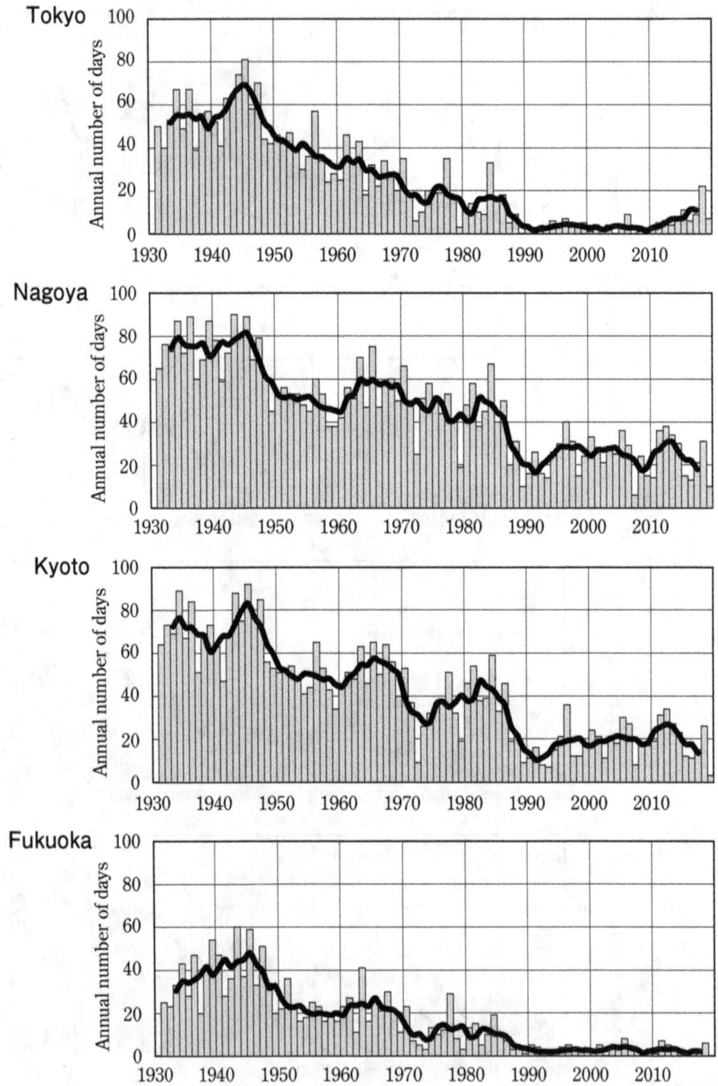

The bars indicate the values for each year. The solid line indicates the five-year running mean.

Tokyo was relocated on 2 December 2014.

First Reported Dates of Cherry Blossom Flowering (1953-2020)

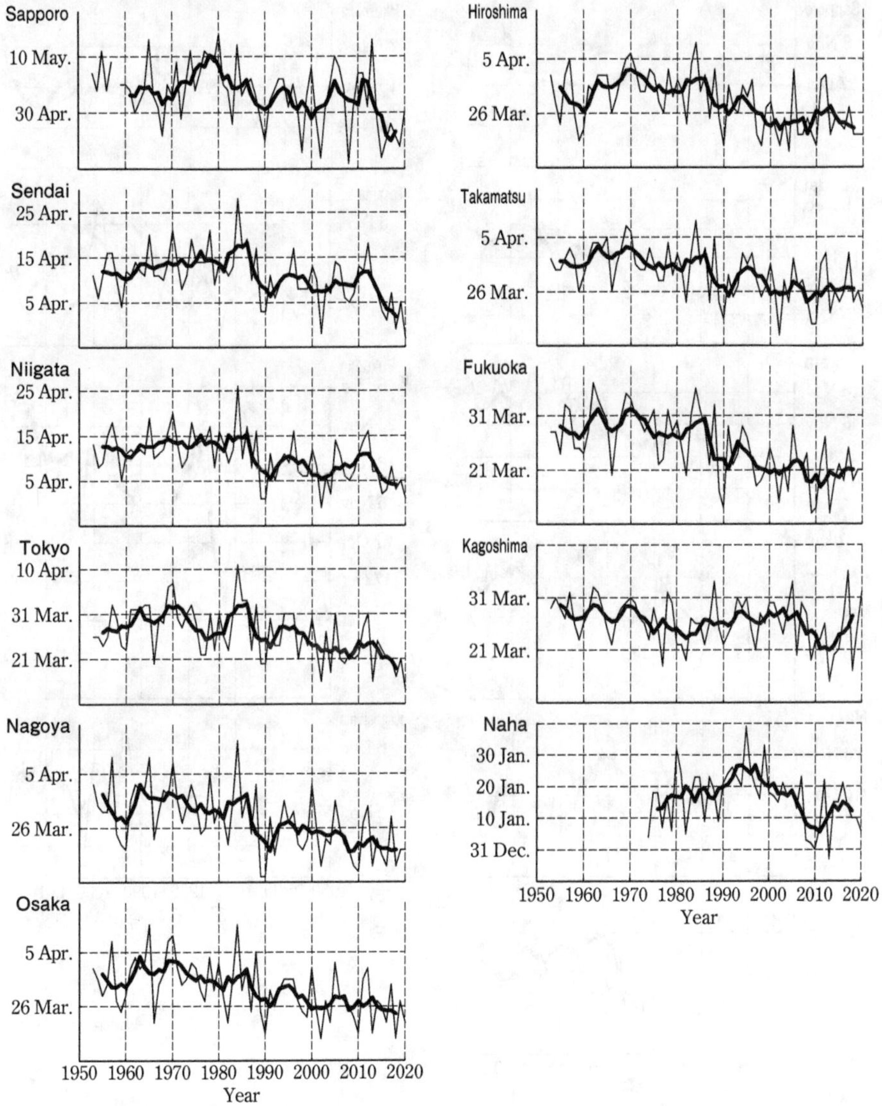

Thin lines indicate the first reported dates of flowering. Thick lines indicate five-year running means.
Cherry blossoms are "Prunus × yedoensis" (Except for Naha data, which are "Prunus campanulata").
Source: Phenological observations implemented by Japan Meteorological Agency

First Reported Dates of Gingko Leaf Color Change (1953-2019)

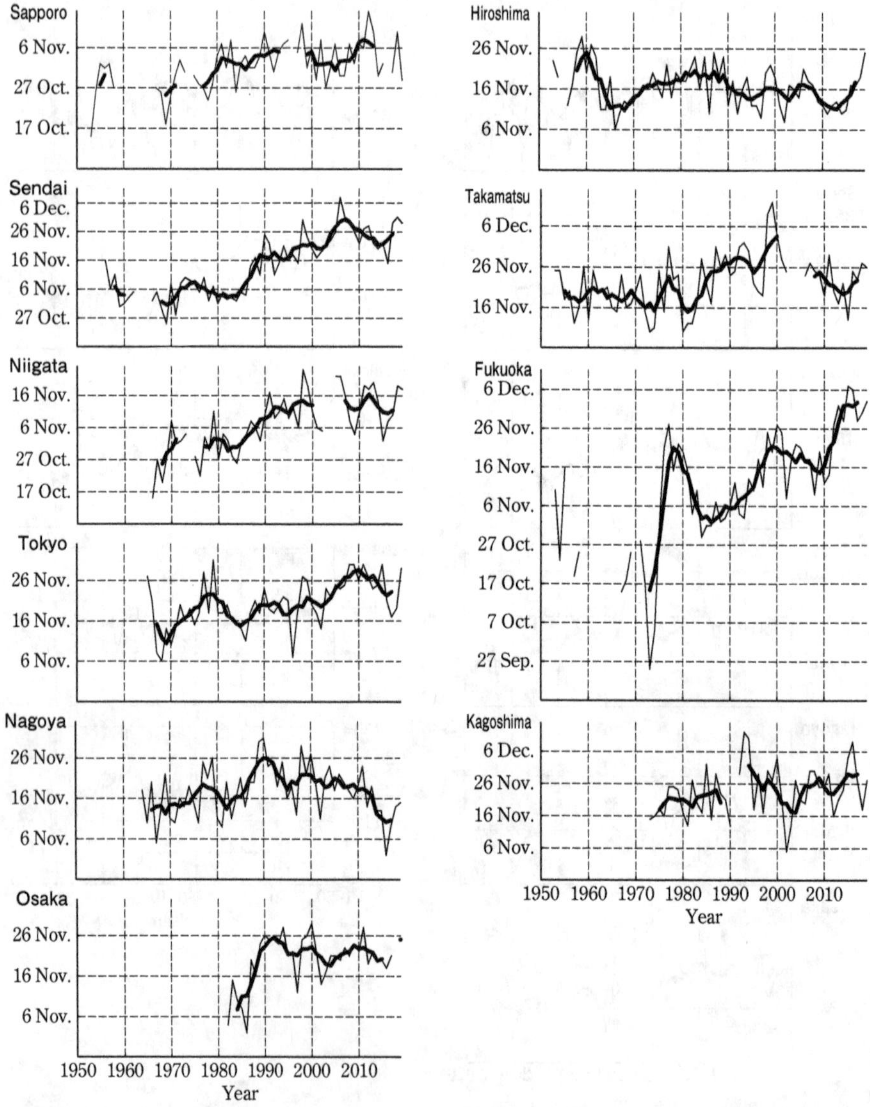

Thin lines indicate the first reported dates of leaf color change. Thick lines indicate five-year running means.
Gingko leaf color change is not observed at Naha.
Source: Phenological observations implemented by Japan Meteorological Agency

Long-Term Trend of Sea Surface Temperature around Japan (1)

The figure shows increase rates (℃/century) of area-averaged annual mean sea surface temperatures (SSTs) around Japan from 1900 to 2019 statistically evaluated using in-situ data observed by marine vessels, etc. The areas are defined based on similarities in SST changes. The increase rates of these areas are generally higher than the corresponding value for the global ocean. SSTs in the northeastern part of the Sea of Japan (marked with [#] in the figure) exhibit no statistical long-term trend.

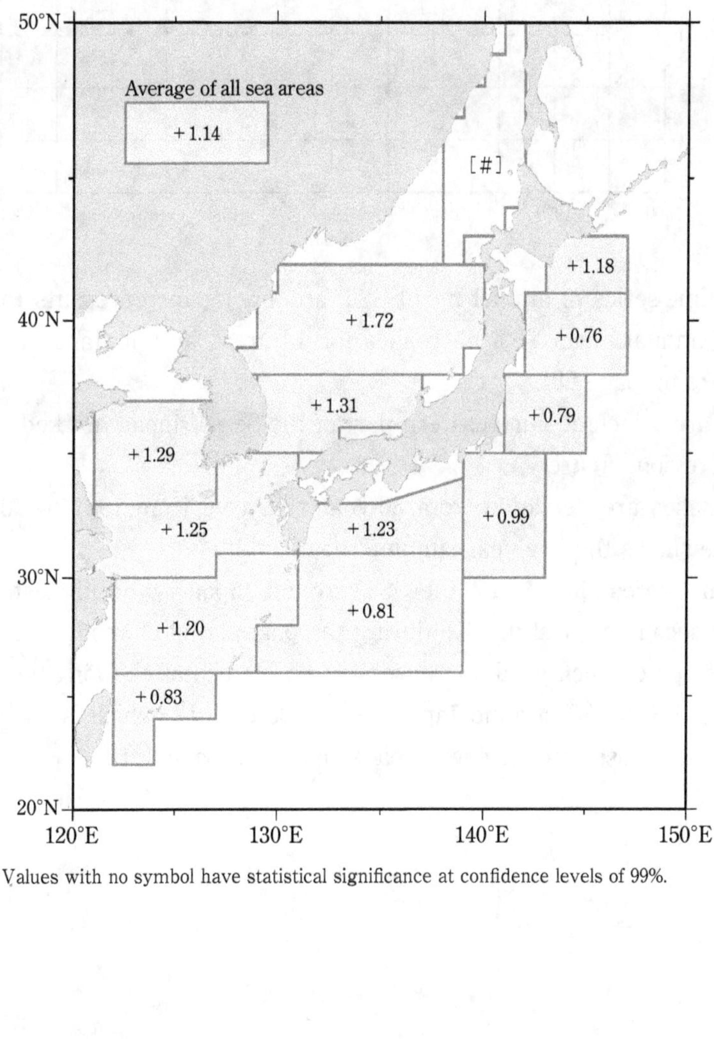

Values with no symbol have statistical significance at confidence levels of 99%.

Long-Term Trend of Sea Surface Temperature around Japan (2)

The figure shows a long-term trend of area-averaged annual mean sea surface temperatures (SSTs) around Japan from 1900 to 2019 statistically evaluated using in-situ data observed by marine vessels, etc.

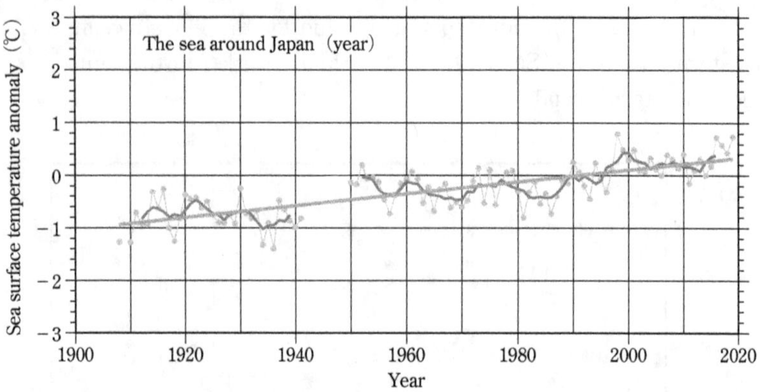

The time-series of annual mean SST around Japan represents the average for annual mean SST anomalies for 13 areas around Japan shown in the figure on page **881**.
The 13 areas include northeastern part of the Sea of Japan marked with [#] in the previous figure.

Anomalies are deviation from 30-year averages from 1981 to 2010 (the bold line shows the five-year running mean).
The time series shows that the SST around Japan is the highest in 1998 and the second highest in 2019 during the past period of about a century.

The slope of thick straight regression line indicates the long-term trend of annual mean SST around Japan from 1900 to 2019, which is +1.14℃ per century (Increase rate for each area is shown on page **881**).

El Niño/La Niña Events

An El Niño event is a phenomenon in which sea surface temperatures (SSTs) are higher than normal across a wide area from the center of the equatorial Pacific to the region off the coast of Peru for a period of between half a year and 1.5 years. In contrast, a La Niña event is a phenomenon in which SSTs are lower than normal in the same area. JMA recognizes the occurrence of an El Niño event when the five-month running mean of SST deviations from the climatological mean (based on the latest sliding 30-year period) averaged over the El Niño monitoring region (5°N–5°S, 150°W–90°W) remains +0.5℃ or above for a period of six months or more. Similarly, a La Niña event is recognized when the corresponding figure is -0.5℃ or below for the same area/period.

Historical El Niño/La Niña events (since 1949)

El Niño events	La Niña events
	summer 1949–summer 1950
spring 1951–winter 1951/52	
spring 1953–autumn 1953	spring 1954–winter 1955/56
spring 1957–spring 1958	
summer 1963–winter 1963/64	spring 1964–winter 1964/65
spring 1965–winter 1965/66	autumn 1967–spring 1968
autumn 1968–winter 1969/70	spring 1970–winter 1971/72
spring 1972–spring 1973	summer 1973–spring 1974
	spring 1975–spring 1976
summer 1976–spring 1977	
spring 1982–summer 1983	summer 1984–autumn 1985
autumn 1986–winter 1987/88	spring 1988–spring 1989
spring 1991–summer 1992	summer 1995–winter 1995/96
spring 1997–spring 1998	summer 1998–spring 2000
summer 2002–winter 2002/03	autumn 2005–spring 2006
	spring 2007–spring 2008
summer 2009–spring 2010	summer 2010–spring 2011
summer 2014–spring 2016	autumn 2017–spring 2018
autumn 2018–spring 2019	

El Niño monitoring region (5°N–5°S, 150°W–90°W) (hatched rectangle)

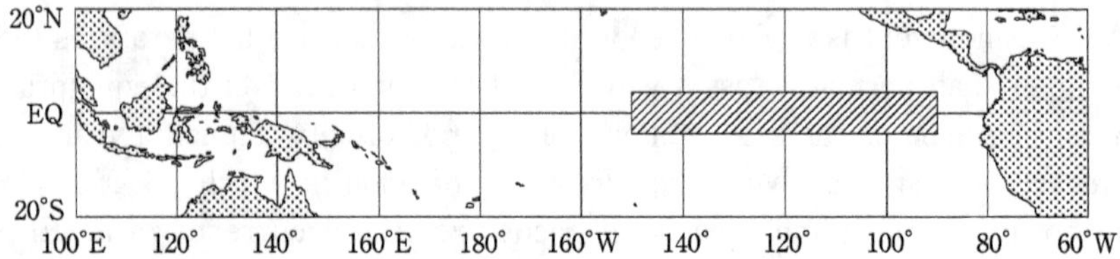

SST deviations (℃) from the climatological mean based on the latest sliding 30-year period for the El Niño monitoring region

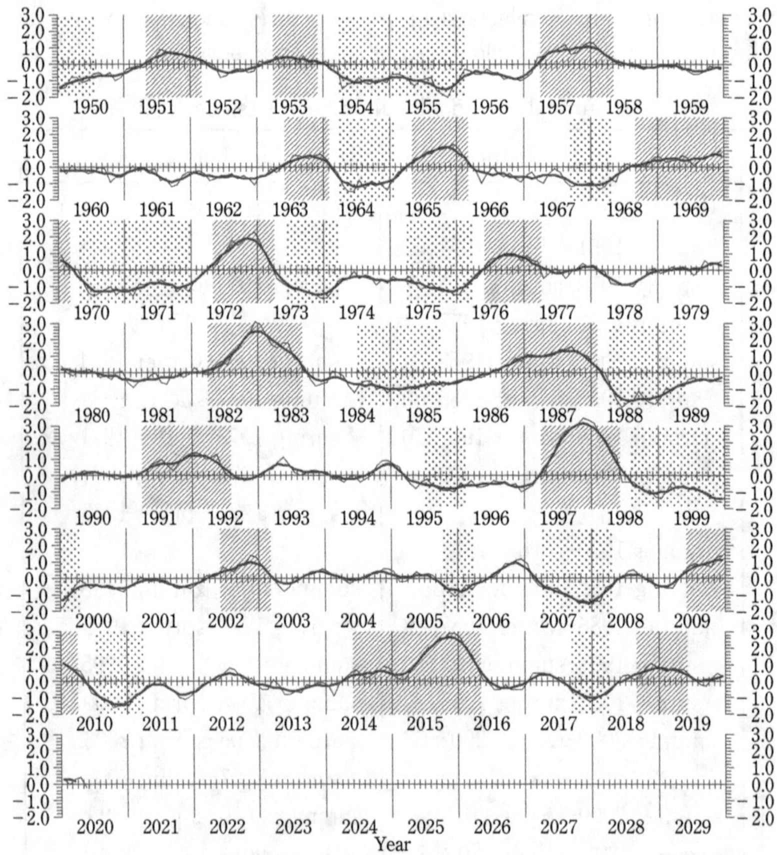

Thin lines indicate a monthly mean value, and smooth thick curves indicate a five-month running mean. Hatched area denotes El Niño periods, and dotted area denotes La Niña periods.

The standard value is the mean of each month until the previous year for the past 30 years.

Sea Ice Extent in the Sea of Okhotsk

The figure shows a time-series representation of sea ice extent in the Sea of Okhotsk. The bars indicate sea ice extents at five-day intervals from December in the previous year to May from 1971 to 2020. The line indicates six-month mean (in this case seasonal mean) sea ice extent for each year.

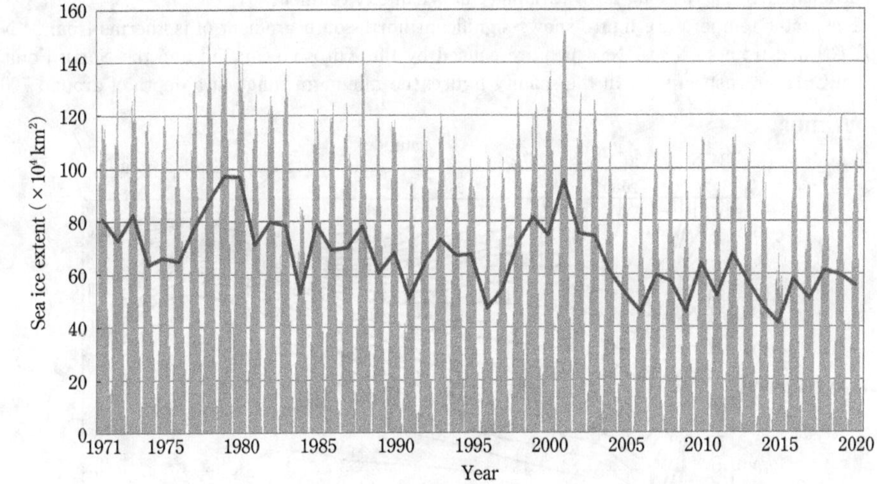

Annual Mean Sea Levels around Japan

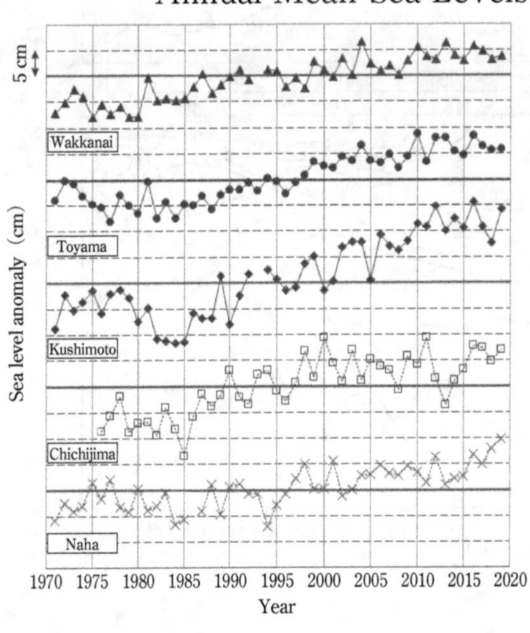

The figure shows time-series representations of annual mean sea level anomalies at five tide gauge stations in Japanese coastal areas, obtained using the 1981–2010 averages as the normals, but each anomaly includes the extent by crustal movement. Each of the bold horizontal lines shows the normal for each station. One division on the vertical axis corresponds to 5 cm.

No marker indicates missing data.

Vertical Sections of Water Temperature and Salinity along the 137°E Line

The figures show vertical sections of the 30-year average water temperature and salinity from the south of Japan (34°N) to the equatorial region (3°N) along the 137°E meridian. JMA has carried out hydrographic observation along the 137°E section using research vessels since 1967 in order to monitor the long-term variability of ocean environment.

The water temperature figure shows significant north-south gradient of isotherms from 34°N to 31°N and from 17°N to 7°N, which are caused by the Kuroshio Current and the North Equatorial Current, respectively. In the salinity figure, the minimum range at a depth of around 700

Winter

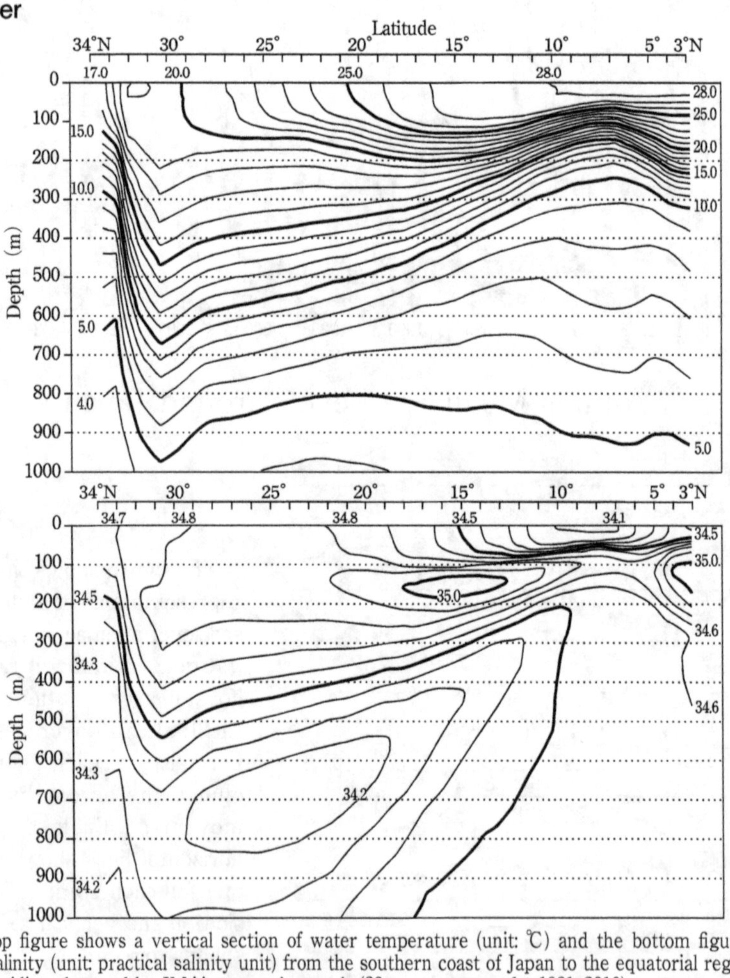

The top figure shows a vertical section of water temperature (unit: ℃) and the bottom figure shows that of salinity (unit: practical salinity unit) from the southern coast of Japan to the equatorial region along 137°E meridian observed by JMA's research vessels (30-year averages for 1981–2010).

m near 25°N corresponds to the low-salinity water mass known as the North Pacific Intermediate Water. This range is thought to be formed by the mixture of the Kuroshio Current and the Oyashio Current in the east of Japan. There are major seasonal fluctuations in water temperature and salinity for the surface layer from the sea surface to a water depth of about 100 m, particularly at latitudes higher than about 15°N.

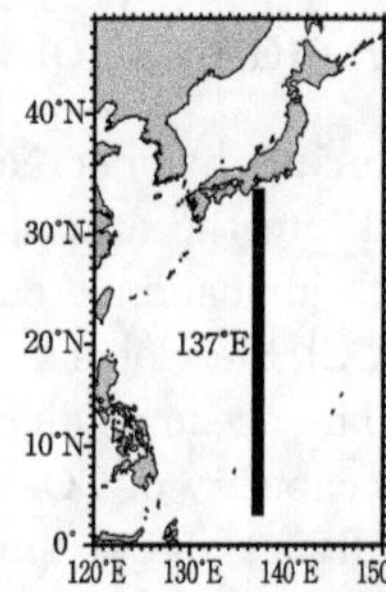

JMA's repeat hydrographic line along 137°E

Summer

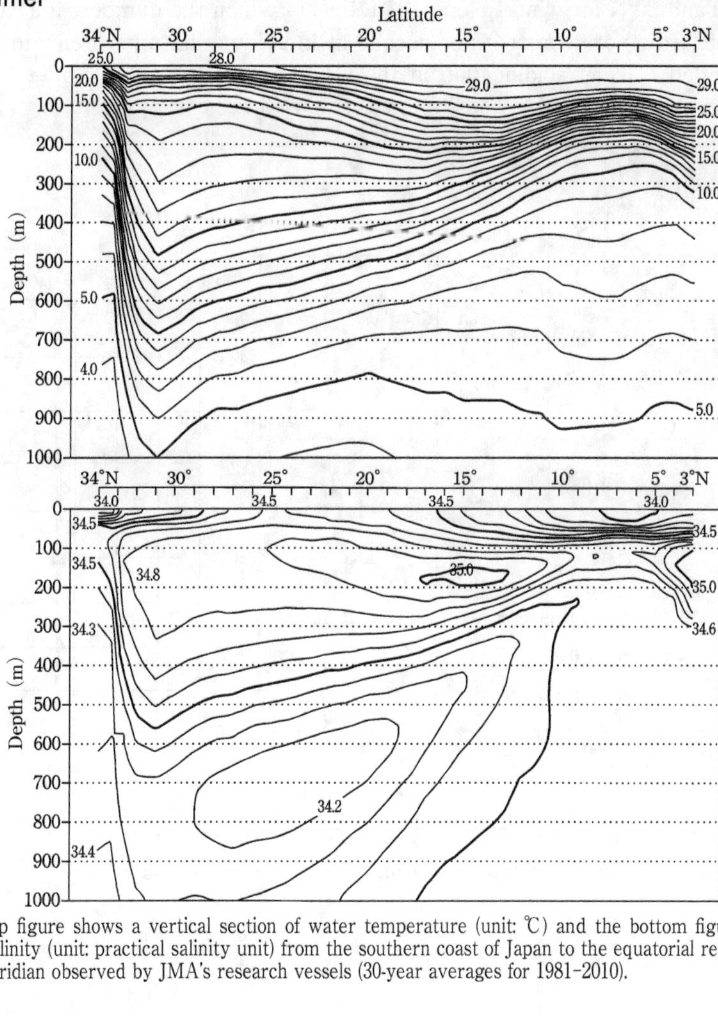

The top figure shows a vertical section of water temperature (unit: ℃) and the bottom figure shows that of salinity (unit: practical salinity unit) from the southern coast of Japan to the equatorial region along 137°E meridian observed by JMA's research vessels (30-year averages for 1981–2010).

Ocean Acidification in the Western North Pacific

Surface seawater pH has decreased as a result of ocean uptake of carbon dioxide (CO_2) emitted into the atmosphere since the beginning of the industrial era, known as ocean acidification. According to the Intergovernmental Panel on Climate Change (IPCC), the global averaged decrease in pH during the period is estimated to be 0.1. The ocean acidification is a particular issue of concern because it limits the ocean's capacity of CO_2 uptake from the atmosphere and affects marine ecosystems. The figure shows time-series representation of surface seawater pH at 10, 20 and 30°N for winter along 137°E line, in which the numbers indicate rates of change (rates of decrease) per decade at individual latitudes. Refer to the section entitled "Ocean acidification in the western North Pacific" under "Oceanic carbon cycle" on the JMA website (https://www.jma.go.jp/index.html).

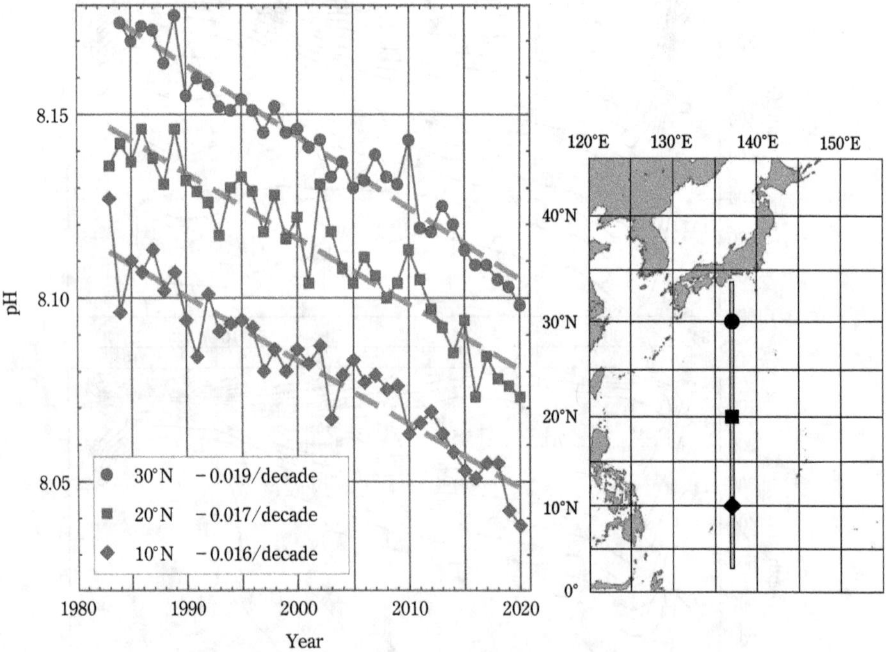

Time-series representation of surface seawater pH at 10, 20 and 30°N at 137°E for winter (left), and JMA's repeat hydrographic line at 137°E (right).

Greenhouse Gases

The Fifth Assessment Report of the Intergovernmental Panel on Climate Change (IPCC) designates the reduction targets in the Kyoto Protocol including carbon dioxide (CO_2), methane (CH_4), nitrous oxide (N_2O), hydrofluorocarbons (HFCs), perfluorocarbons (PFCs) and sulfur hexafluoride (SF_6), and ozone-depleting substances such as chlorofluorocarbons (CFCs) and hydrochlorofluorocarbons (HCFCs), as well-mixed greenhouse gases that are long-lived in the atmosphere. Among these gases, CO_2, CH_4 and N_2O have the greatest impact to global warming. The report also lists tropospheric ozone (O_3) as a short-lived gas with a significant greenhouse effect. Carbon monoxide (CO) is not a greenhouse gas in itself, it affects concentrations of other greenhouse gases, and brings about a greenhouse effect.

Concentrations of greenhouse gases are indicated with mole fractions as parts per million (ppm), parts per billion (ppb) and parts per trillion (ppt). Concentration data are subject to revision depending on changes of the concentration scale of reference gases, etc. For details on greenhouse gases, refer to the section entitled "Greenhouse Gases" under "Atmospheric Environment" - "Climate and Ocean" on the JMA website (https://www.jma.go.jp/jma/indexe.html).

Examples of greenhouse gases

	CO_2 Carbon dioxide	CH_4 Methane	N_2O Nitrous oxide	CFC-11 Chlorofluorocarbon 11	HFC-23 Hydrofluorocarbon	SF_6 Sulfur hexafluoride	CF_4 Carbon tetrafluoride
Pre-industrial concentrations	278 ± 2 ppm	722 ± 25 ppb	270 ± 7 ppb	0	0	0	34.7 ± 0.2 ppt
Concentrations in 2011	391 ± 0.2 ppm	1 803 ± 2 ppb	324.2 ± 0.1 ppb	238 ± 0.8 ppt	24.0 ± 0.3 ppt	7.28 ± 0.03 ppt	79.0 ± 0.1 ppt
Rate of concentration change*1	2.0 ppm/year	4.8 ppb/year	0.8 ppb/year	-2.2 ppt/year	0.9 ppt/year	0.3 ppt/year	0.7 ppt/year
Atmospheric lifetime (years)*2	–	12.4	121	45	222	3200	50 000

Created based on the IPCC Fifth Assessment Report (2013).

*1 The rate of concentration change is the average value for the period from 2005 to 2011.

*2 Atmospheric lifetime: response time (the time scale characterizing the decay of an instantaneous pulse input into the atmosphere) is listed for methane and nitrous oxide, while turnover times (the ratio of the mass of a gaseous compound in the atmosphere and the total rate of removal from the atmosphere) is listed for the other greenhouse gases.

Carbon Dioxide

Of all greenhouse gases, carbon dioxide (CO_2) is the most significant contributor to global warming. According to IPCC, CO_2 accounts for about 64% of total increase in radiative forcing due to well-mixed greenhouse gases since the pre-industrial period. Since oceans are large reservoirs of CO_2, it is also important to monitor oceanic CO_2 concentrations as well as atmospheric concentrations. The first figure shows latitude distribution of atmospheric CO_2 concentrations analyzed using data from fixed stations and some ships around the globe. The second figure shows monthly mean atomospheric CO_2 concentrations observed at JMA stations, Ryori (Ofunato City, Iwate Prefecture), Minamitorishima (Ogasawara Village, Tokyo) and Yonagunijima (Yonaguni Town, Okinawa Prefecture). The third figure shows changes in atmospheric and oceanic CO_2 concentrations observed by JMA research vessels along the 137°E line in the western North Pacific.The table shows annual mean atmospheric CO_2 concentrations at JMA stations.

Latitude distribution of atmospheric CO_2 concentrations

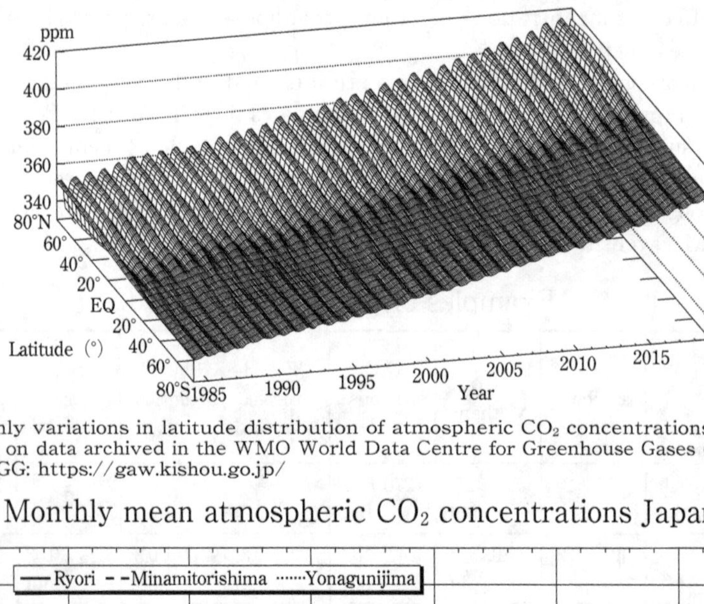

Monthly variations in latitude distribution of atmospheric CO_2 concentrations
Based on data archived in the WMO World Data Centre for Greenhouse Gases (WDCGG).
WDCGG: https://gaw.kishou.go.jp/

Monthly mean atmospheric CO_2 concentrations Japan

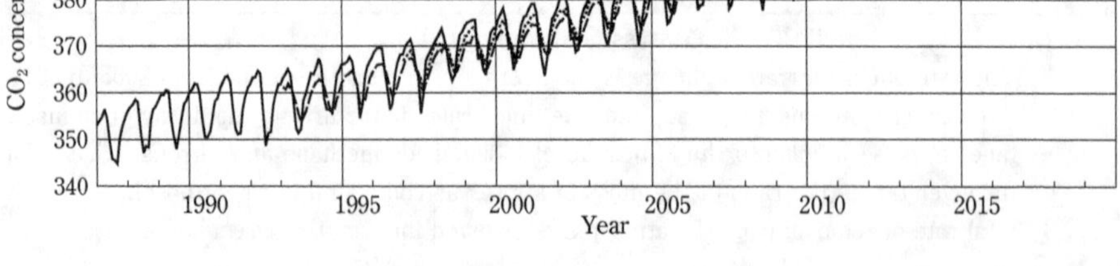

Annual mean atmospheric CO_2 concentrations in Japan (ppm)

Station	1987	1988	1989	1990	1991	1992	1993	1994	1995	1996	1997	1998	1999	2000	2001	2002	2003
Ryori	351.4	354.2	356.1	356.9	358.3	358.6	359.5	361.9	363.8	365.3	366.5	369.5	371.3	372.8	373.5	376.0	378.8
Minamitorishima							358.3*	359.9	361.7	363.5	364.7	367.4	369.1	370.4	371.9	374.0	376.8
Yonagunijima											365.7	368.8	370.7	372.0	373.6	375.7	378.4

Station	2004	2005	2006	2007	2008	2009	2010	2011	2012	2013	2014	2015	2016	2017	2018	2019
Ryori	380.5	382.6	385.4	386.6	388.5	389.8	393.6	394.4*	397.2	399.6	401.3	403.4	407.2	409.3	412.0	(414.0)
Minamitorishima	378.4	380.8	383.8*	384.7	386.6	388.0	390.5	392.8	395.0	397.5	399.5	401.5	404.9	407.7	409.4	(412.2)
Yonagunijima	380.2	382.7	384.8	386.4	388.0	389.4	392.7	394.4	397.0	399.5	401.8	403.9*	407.1	409.5	411.7	(414.8)

Values marked with an asterisk were calculated from 11 or fewer monthly mean values. Values in parentheses are preliminary values prior to the analysis of concentration drift of the standard gases after use.

Please use the latest data because the annual mean values are subject to revision depending on changes of the concentration scale of reference gases, etc.

Annual changes in atmospheric (a) and oceanic (b) carbon dioxide (CO_2) concentrations (ppm) observed by JMA research vessels in the western North Pacific

(Averaged between 7°N and 33°N for winter (January–March) from 1984 to 2020)

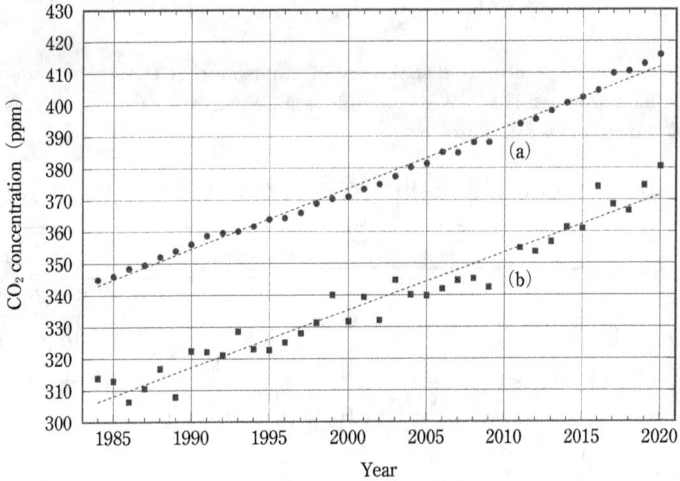

The dashed lines in the figure show mean growth rates of atmospheric and oceanic CO_2 concentrations, which were +1.9 ppm/year and +1.8 ppm/year respectively. The values may be revised backward to the past based on the change of the global standard for CO_2 concentration.

Methane

According to IPCC, methane (CH_4) accounts for about 17% of total increase in radiative forcing due to well-mixed greenhouse gases since the pre-industrial period. It is the most significant greenhouse gas after CO_2. The first figure shows latitude distribution of atmospheric CH_4 concentrations analyzed using data from fixed stations and some ships around the globe. The second figure shows monthly mean atmospheric CH_4 concentrations observed at JMA stations, Ryori (Ofunato City, Iwate Prefecture), Minamitorishima (Ogasawara Village, Tokyo) and Yonagunijima (Yonaguni Town, Okinawa Prefecture). The table shows annual mean atmospheric CH_4 concentrations at JMA stations.

Latitude distribution of atmospheric CH_4 concentrations

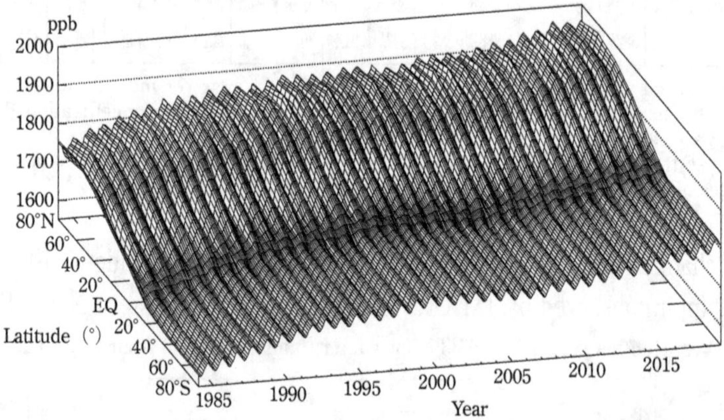

Monthly variations in latitude distribution of atmospheric CH_4 concentrations
Based on data archived in the WMO World Data Centre for Greenhouse Gases (WDCGG).
WDCGG: https://gaw.kishou.go.jp/

Monthly mean atmospheric CH_4 concentrations in Japan

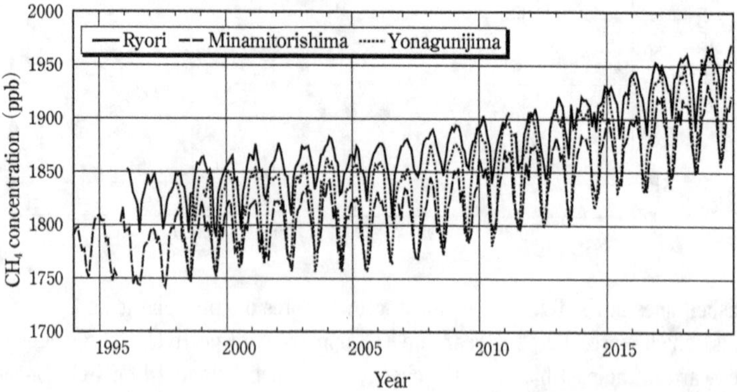

Annual mean atmospheric CH_4 concentrations in Japan (ppb)

Station	1994	1995	1996	1997	1998	1999	2000	2001	2002	2003	2004	2005	2006
Ryori	—	—	1831*	1829	1837	1840	1845	1853	1852	1862	1860	1857	1858
Minamitorishima	1784	1784*	1770	1780	1787	1790	1796	1797	1799	1808	1806	1798	1804*
Yonagunijima					1806	1809*	1816	1825	1817	1825	1824	1823	1824
Station	2007	2008	2009	2010	2011	2012	2013	2014	2015	2016	2017	2018	2019
Ryori	1868	1875	1878	1881	1884*	1886	1901	1912	1919	1929	1939	1941	(1954)
Minamitorishima	1804	1813	1822	1832	1837	1848	1850	1863	1874	1875	1889	1892	(1902)
Yonagunijima	1824	1840	1851	1852	1860	1869	1872	1879	1891*	1897	1905	1916	(1928)

Values marked with an asterisk were calculated from 11 or fewer monthly mean values. Dashes (-) indicate that data were unavailable due to equipment failure, etc. Values in parentheses are preliminary values prior to the analysis of concentration drift of the standard gases after use.

Please use the latest data because the annual mean values are subject to revision depending on changes of the concentration scale of reference gases, etc.

Nitrous Oxide

Nitrous oxide (N_2O) is colorless, and extremely stable in the troposphere. Its radiative forcing per unit mass is estimated to be about 265 times greater than that of CO_2. According to IPCC, N_2O accounts for about 6% of total increase in radiative forcing due to well-mixed greenhouse gases since the pre-industrial period. The first figure shows latitude distribution of atmospheric N_2O concentrations analyzed using data from fixed stations around the globe. The second figure shows monthly mean atmospheric N_2O concentrations observed at Ryori (Ofunato City, Iwate Prefecture). The replacement of the observation system in early 2004 improved monitoring precision and reduced the scale of fluctuations in observed values.

Latitude distribution of atmospheric N_2O concentrations

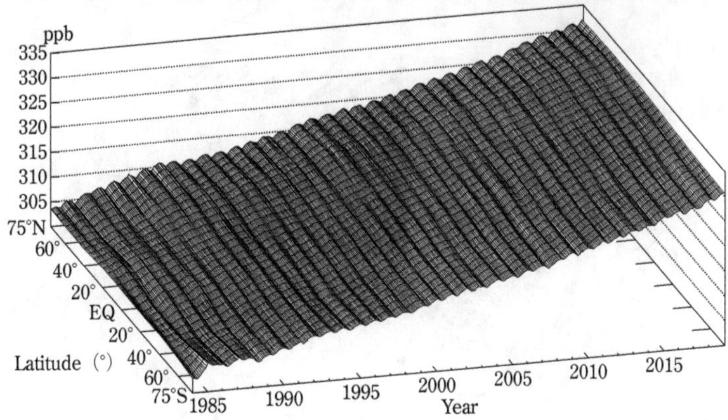

Monthly variations in latitude distribution of atmospheric N_2O concentrations
Based on data archived in the WMO World Data Centre for Greenhouse Gases (WDCGG).
WDCGG: https://gaw.kishou.go.jp/

Monthly mean atmospheric N_2O concentrations at Ryori, Japan

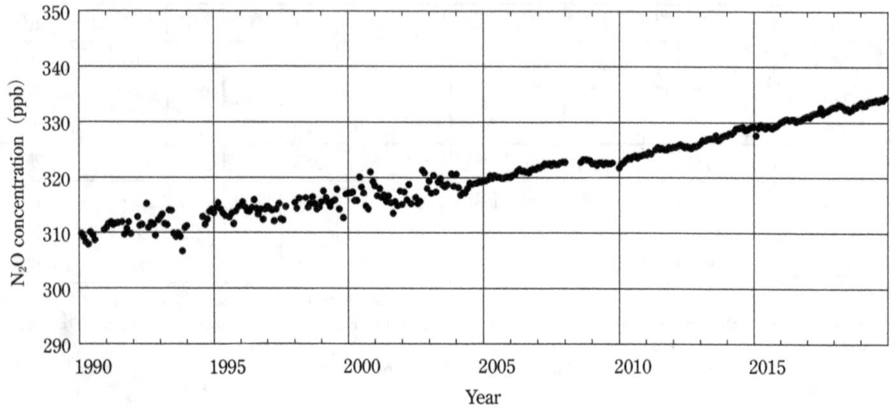

Carbon Monoxide

Carbon monoxide (CO) is not a greenhouse gas in itself because it absorbs hardly any infrared radiation from the ground surface. Through chemical reaction with hydroxyl (OH) radicals, however, it can affect the concentrations of many greenhouse gases such as methane, halocarbons and tropospheric ozone. The first figure shows latitude distribution of atmospheric CO concentrations analyzed using data from fixed stations and some ships around the globe. The second figure shows monthly mean atmospheric CO concentrations observed at JMA stations, Ryori (Ofunato City, Iwate Prefecture), Minamitorishima (Ogasawara Village, Tokyo) and Yonagunijima (Yonaguni Town, Okinawa Prefecture). The table shows annual mean atmospheric CO concentrations at JMA stations.

Latitude distribution of atmospheric CO concentrations

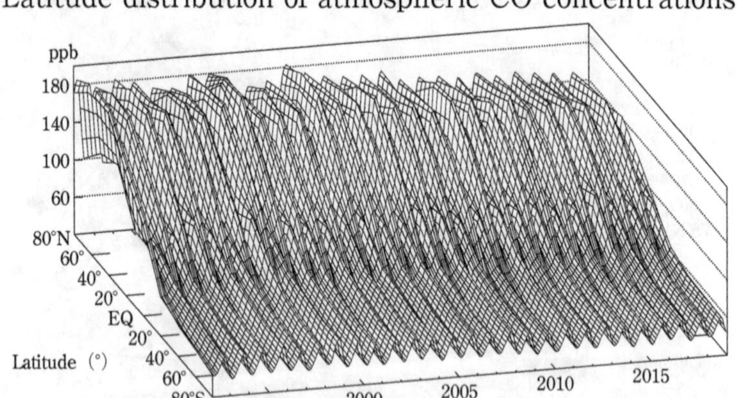

Monthly variations in latitude distribution of atmospheric CO concentrations
Based on data archived in the WMO World Data Centre for Greenhouse Gases (WDCGG).
WDCGG: https://gaw.kishou.go.jp/

Monthly mean atmospheric CO concentrations in Japan

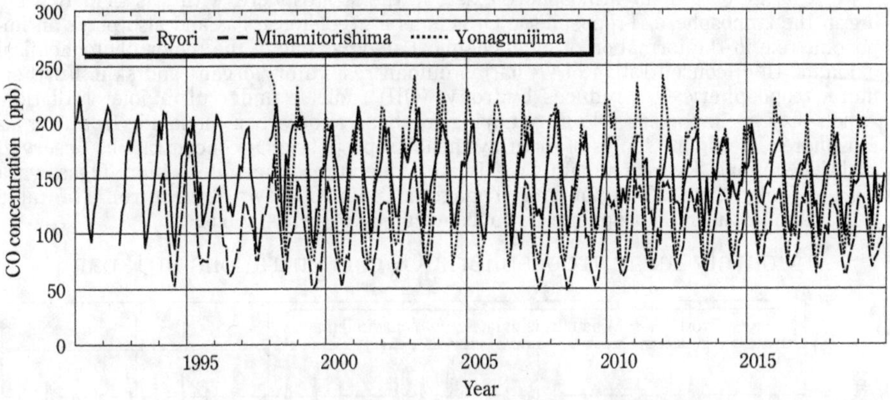

Annual mean atmospheric CO concentrations in Japan (ppb)

Station	1991	1992	1993	1994	1995	1996	1997	1998	1999	2000	2001	2002	2003	2004	2005
Ryori	161	170*	152	158	158	157	141*	177	154	159	155	159	162	156	146
Minamitorishima				106	118	94	108	111	89	93	94	99	112	—	—
Yonagunijima								161	145	142	146	156	154	152	151
Station	**2006**	**2007**	**2008**	**2009**	**2010**	**2011**	**2012**	**2013**	**2014**	**2015**	**2016**	**2017**	**2018**	**2019**	
Ryori	152	153	156	149	158	149*	149	148	156	154	149	146	147	(144)	
Minamitorishima	—	93	88	88	97	101	108	101	105	105	100	99	97	(100)	
Yonagunijima	152	150	144	148	150	143	161	149	144	146*	141	133	137	(145)	

Values marked with an asterisk were calculated from 11 or fewer monthly mean values. Dashes (-) indicate that data were unavailable due to equipment failure, etc. Values in parentheses are preliminary values prior to the analysis of concentration drift of the standard gases after use.

Data at Minamitorishima are missing from January 2004 to October 2006 due to equipment trouble and typhoon damage. Therefore, no annual mean values were calculated from 2004 to 2006.

Please use the latest data because the annual mean values are subject to revision depending on changes of the concentration scale of reference gases, etc.

Tropospheric Ozone

Most ozone (O_3) in the atmosphere exists in the stratosphere, with less than 10% being in the troposphere. Tropospheric O_3 acts as a greenhouse gas and also plays an important role in the formation of photochemical smog as one of the major photochemical oxidants. High concentration of O_3 harms human respiratory organs and skin. Furthermore, tropospheric O_3 produces hydroxyl (OH) radicals under ultraviolet radiation, which play an important role in removing methane and carbon monoxide from the atmosphere. The figure shows of monthly mean tropospheric O_3 concentrations observed at JMA stations, Ryori (Ofunato City, Iwate Prefecture), Minamitorishima (Ogasawara Village, Tokyo) and Yonagunijima (Yonaguni Town, Okinawa Prefecture). The table shows annual mean tropospheric O_3 concentrations at JMA stations.

Monthly mean tropospheric O_3 concentrations in Japan

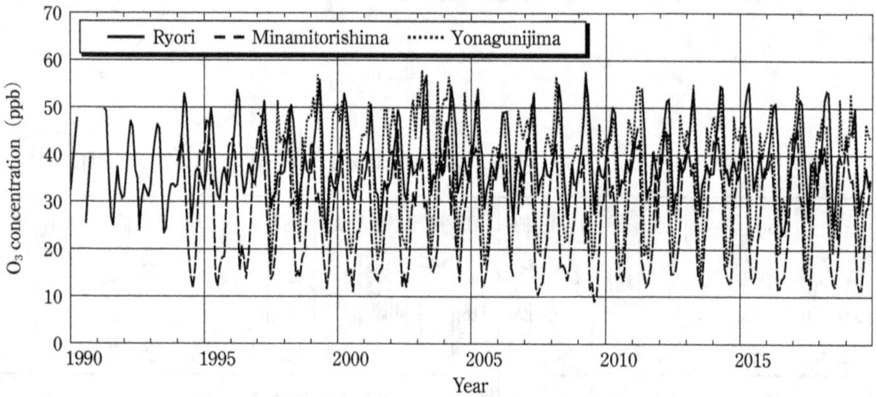

Annual mean tropospheric O_3 concentrations in Japan (ppb)

Station	1990	1991	1992	1993	1994	1995	1996	1997	1998	1999	2000	2001	2002	2003	2004
Ryori	37.2*	35.4*	35.6	34.8	38.2	37.3	39.6	38.3	39.3	38.6	39.1*	35.5	38.6	41.1	41.0
Minamitorishima					30.0	29.6	28.8	31.1	28.3	27.2	25.9	26.0	27.2	28.3	30.6
Yonagunijima								40.7	39.3	41.4	38.4	39.3	39.0	43.4	42.1
Station	**2005**	**2006**	**2007**	**2008**	**2009**	**2010**	**2011**	**2012**	**2013**	**2014**	**2015**	**2016**	**2017**	**2018**	**2019**
Ryori	39.3	39.4	39.8	38.7	40.6	39.9	39.8*	39.1	39.3	40.9	41.1	36.8	40.1	40.6*	(37.5*)
Minamitorishima	29.1	30.4*	25.4	25.8	23.6	26.7	27.2	29.9	28.8	29.8	28.3	25.4	28.8	28.1	(27.0)
Yonagunijima	35.7	38.7	38.2	38.4	38.9	36.6	37.5	38.2	38.0	38.7	37.7	35.5	37.4	37.4	(39.0)

Values marked with an asterisk were calculated from 11 or fewer monthly mean values. Values in parentheses are preliminary values prior to the analysis of drift of ozone monitor outputs.

Please use the latest data because the annual mean values are subject to revision depending on changes of the concentration scale, etc.

Aerosols

In the atmosphere, particulate matters called "aerosol" with radii ranging from approximately 0.001 μm to 10 μm in solid phase, liquid phase and mixed phase are suspending aside from cloud droplets. The composition of aerosols is extremely variable; they include materials such as sulfate generated from anthropogenic and natural gases through particulate conversion, sea salt from seawater spindrift that has dried in the atmosphere, Aeolian dust (also known as "Kosa" in Japanese) brown up by winds, and soot from fossil fuel/biomass combustion (black carbon and organic carbon). Large volcanic eruptions eject large amounts of volcanic smoke and gas into the stratosphere, thus causing a large amount of aerosols to stay in the stratosphere over the long term.

Aerosols directly change the Earth's radiative balance through scattering and absorbing solar radiation. Moreover, they indirectly change the radiative balance by contributing to the formation of clouds as condensation nuclei and ice nuclei. Aerosols are considered to have a significant impact on the climate through these processes. The distribution of aerosols fluctuates greatly depending on time and region, and it is difficult to fully assess their impact on the climate.

Aerosol Optical Depth

JMA monitors aerosol optical depth (AOD), which denotes the total amount of aerosols between the surface and the top of the atmosphere, with different wavelengths of sunlight at three stations in Japan.

Optical depth is a measure of the extinction of solar radiation. The value n means that the light entering the atmosphere perpendicularly is attenuated to $1/e^n$ of its initial intensity at the Earth's surface.

Furthermore, it is empirically known that AOD at a wavelength of λ is proportional to $\lambda^{-\alpha}$. This α is called the Ångström exponent. A larger Ångström exponent indicates a relatively larger amount of aerosols with smaller particle sizes.

Time series of monthly mean aerosol optical depth and Ångström exponent in Japan

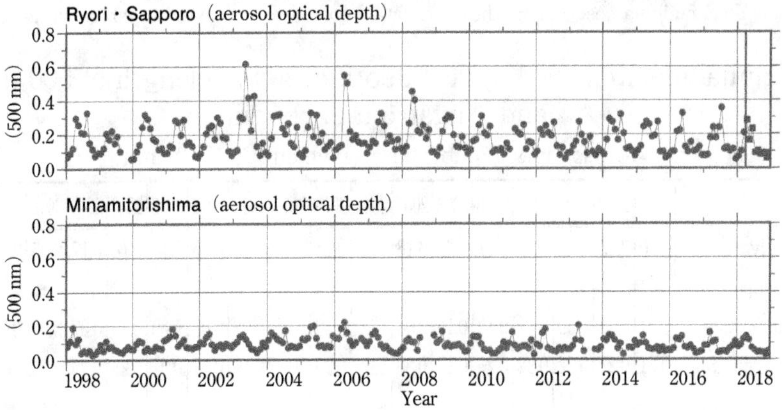

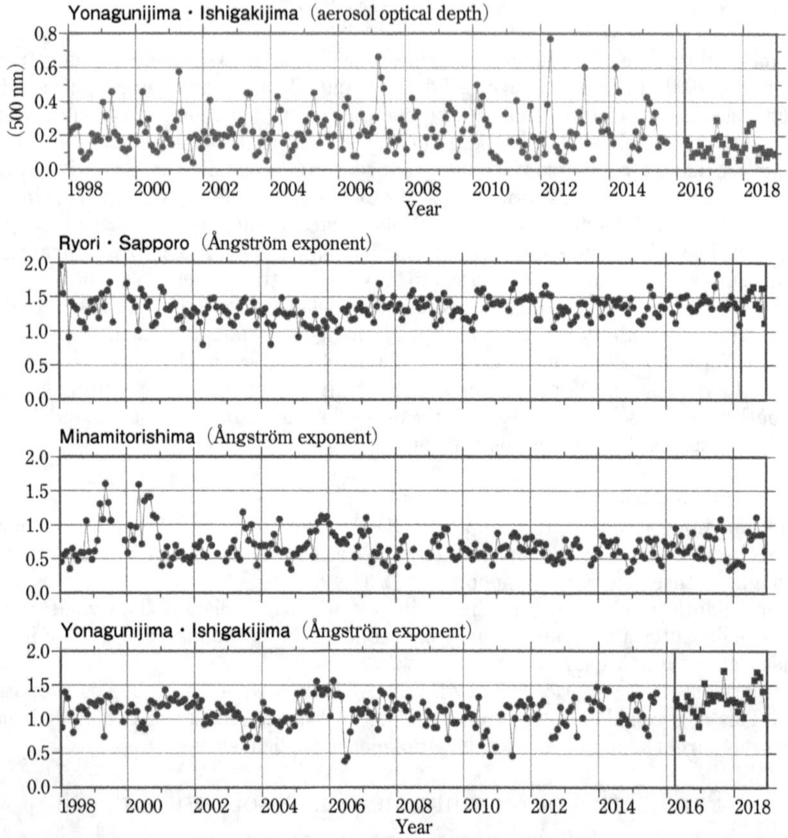

For the Ryori observations, the observation point was rellocated to Sapporo in April 2018. For the Yonagunijima observation, the observation point was rellocated to Ishigakijima in April 2016.

Annual mean aerosol optical depth at a wavelength of 500 nm and Ångström exponent

Aerosol optical depth at a wavelength of 500 nm

Station	2001	2002	2003	2004	2005	2006	2007	2008	2009	2010	2011	2012	2013	2014	2015	2016	2017	2018
Ryori	0.16	0.18	0.24	0.21	0.17	0.21	0.18	0.22	0.18	0.17	0.16	0.16	0.14	0.18	0.17	0.15	0.15	0.13
Sapporo																		0.13
Minamitorishima	0.11	0.09	0.09	0.11	0.12	0.13	0.08	0.10	0.09	0.08	0.08	0.08	0.09	0.10	0.07	0.08	0.08	0.07
Yonagunijima	0.21	0.21	0.22	0.22	0.26	0.23	0.30	0.23	0.25	0.24	0.19	0.21	0.27	0.25	0.25			
Ishigakijima																0.12	0.14	0.14

Ångström exponent

Station	2001	2002	2003	2004	2005	2006	2007	2008	2009	2010	2011	2012	2013	2014	2015	2016	2017	2018
Ryori	1.31	1.27	1.27	1.21	1.11	1.22	1.39	1.38	1.33	1.35	1.47	1.37	1.30	1.38	1.32	1.39	1.47	1.29
Sapporo																		1.44
Minamitorishima	0.54	0.64	0.73	0.65	0.83	0.79	0.61	0.62	0.71	0.59	0.72	0.60	0.63	0.62	0.63	0.69	0.69	0.70
Yonagunijima	1.26	1.12	0.94	1.01	1.32	1.05	1.18	1.13	1.03	0.89	1.09	1.02	1.15	1.17	1.17			
Ishigakijima																1.12	1.34	1.33

Atmospheric Turbidity Coefficient

The atmospheric turbidity coefficient indicates the ratio of the atmospheric optical depth affected by aerosols, water vapor and gases in the atmosphere to that uninfluenced by constituents other than air molecules such as oxygen and nitrogen in the atmosphere. Larger values indicate greater amounts of turbid matter in the air. JMA conducts solar and infrared radiation measurement at five stations in Japan (Sapporo, Tsukuba, Fukuoka, Ishigakijima and Minamitorishima). The atmospheric turbidity coefficient is calculated from direct solar radiation amounts. The figure shows interannual variations in the turbidity coefficient. Increases of the turbidity coefficient and decreases of direct solar radiation, which are considered to be induced by the increase of stratospheric aerosols, can be seen for several years after some large-scale volcanic eruptions.

Atmospheric turbidity coefficient

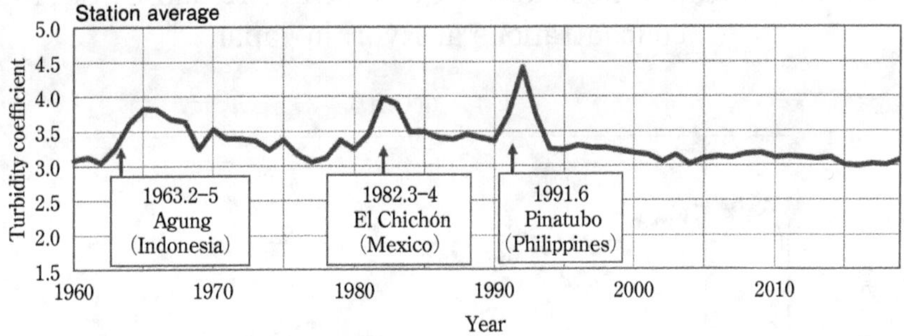

The figure shows a time-series of annual mean atmospheric turbidity coefficients averaged for five stations in Japan. To eliminate the influence of short-term variations affected by water vapor variations and some tropospheric aerosol events such as Kosa (Aeolian dust), the annual mean atmospheric turbidity coefficient is calculated using the minimum turbidity coefficient for each month. Arrows indicate large-scale volcanic eruptions.

Ozone Layer

Ozone-Depleting Substances

Ozone depletion caused by anthropogenic substances has become a concern since the late twentieth century. It is known that ozone depletion leads to an increase in harmful ultraviolet radiation and has impact on human health and ecosystem. Therefore, in order to protect the ozone layer, the "Vienna Convention for the Protection of the Ozone Layer" was adopted in 1985 and the "Montreal Protocol on Substances that Deplete the Ozone Layer" in 1987. Based on the international agreements, production and transportation of ozone-depleting substances are under control. The figures show monthly mean atmospheric concentrations of CFC-11 (CCl_3F), CFC-12 (CCl_2F_2) and CFC-113 (CCl_2FCClF_2) observed at Ryori (Ofunato City, Iwate Prefecture). Concentrations are expressed with the mole fraction of ozone-depleting substances contained in the atmosphere. 'ppt' stands for parts per trillion (10^{-12}).

Monthly mean atmospheric CFC-11, CFC-12 and CFC-113 concentrations at Ryori in Japan

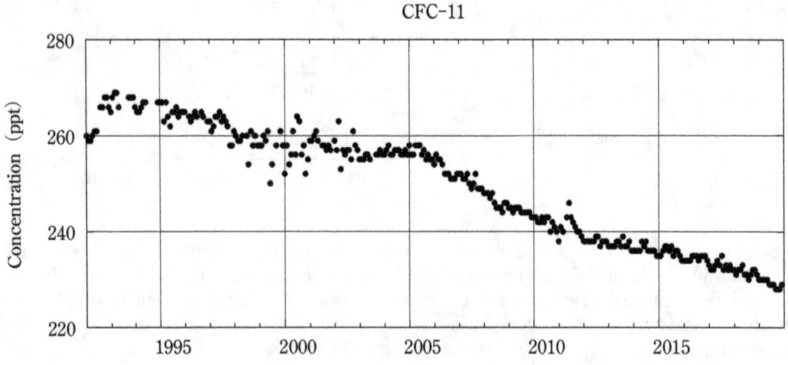

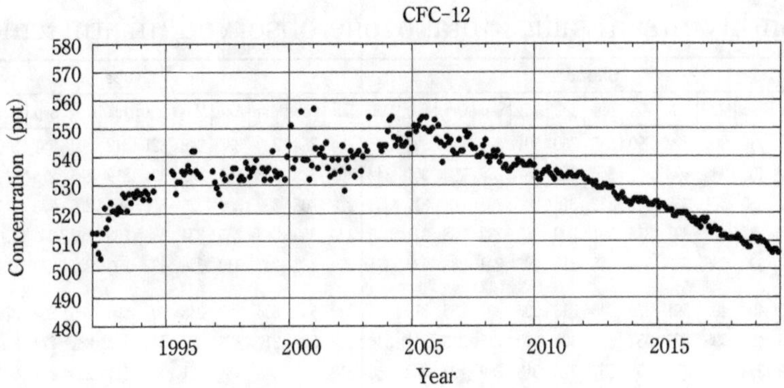

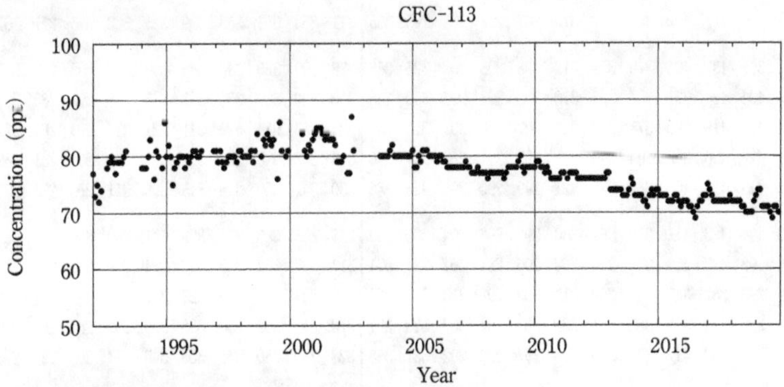

The figures show monthly mean atmospheric concentrations of CFC-11, CFC-12 and CFC-113 at Ryori. Refer to the "Halocarbons" under "Greenhouse gases"-"Atmospheric Environment" on the JMA website (https://www.data.jma.go.jp/ghg/kanshi/ghgp/cfcs_e.html).

Total Ozone in Japan

The tables show monthly mean total ozone observed at four JMA stations. Total ozone refers to the total amount of ozone contained in a vertical column in the atmosphere above the ground extending from the Earth's surface to the upper edge of the atmosphere. It is expressed as the thickness of a layer of pure ozone at standard temperature and pressure conditions of 273.15 K and 101.325 kPa. The unit is m atm-cm (milli-atmosphere centimeter) or DU (Dobson Unit). The values are subject to revision depending on recalibration of measurement instruments, etc.

Monthly mean values total ozone observed (m atm-cm) (1)

Year	Sapporo												Tsukuba											
	Jan.	Feb.	Mar.	Apr.	May	Jun.	Jul.	Aug.	Sep.	Oct.	Nov.	Dec.	Jan.	Feb.	Mar.	Apr.	May	Jun.	Jul.	Aug.	Sep.	Oct.	Nov.	Dec.
1991	401	437	394	402	390	355	330	301	301	295	323	335	334	353	339	338	349	321	312	294	281	270	270	291
1992	361	402	385	398	371	346	325	287	288	294	301	342	305	338	326	355	362	334	319	285	278	269	258	281
1993	352	360	371	369	347	332	316	290	298	301	317	353	278	308	329	339	325	317	305	286	275	271	267	281
1994	394	387	417	391	375	351	302	285	282	288	311	335	320	350	363	335	334	338	298	292	286	270	274	271
1995	389	390	386	382	354	345	311	297	294	288	334	371	299	339	347	319	326	332	293	287	273	264	290	314
1996	389	408	409	391	376	343	313	296	301	299	319	350	322	341	336	353	345	307	296	285	280	274	253	280
1997	385	401	395	375	352	344	308	290	286	313	311	344	312	322	317	323	323	314	297	281	274	274	284	296
1998	399	394	417	374	358	357	323	313	289	297	342	356	312	345	348	334	321	330	311	299	281	271	284	279
1999	403	411	392	398	391	348	316	285	287	310	336	375	302	330	312	348	343	319	300	288	281	282	288	305
2000	385	419	417	402	373	347	310	287	281	301	295	355	311	353	351	358	339	325	302	293	277	267	255	282
2001	417	397	437	404	375	359	312	298	295	302	325	366	329	309	365	346	345	326	299	298	282	279	282	286
2002	377	404	414	371	361	363	310	302	298	295	340	368	302	320	324	330	337	345	288	292	285	277	290	311
2003	408	411	418	380	371	342	337	307	300	307	310	337	335	357	364	329	341	321	318	293	284	274	263	286
2004	395	412	420	395	364	340	311	305	291	284	307	360	314	303	345	338	320	315	301	295	281	269	268	287
2005	413	413	420	405	382	355	321	299	290	291	326	385	330	342	350	349	345	326	311	296	278	273	277	310
2006	400	396	433	411	370	362	330	293	296	304	330	359	325	303	350	367	329	341	309	300	296	280	296	291
2007	394	398	437	423	377	356	328	296	281	301	324	371	317	337	352	363	353	334	309	295	274	270	272	301
2008	389	396	382	387	369	350	317	302	297	313	336	339	309	321	347	346	323	327	307	299	290	279	281	279
2009	380	396	424	398	368	354	330	303	304	307	320	371	312	315	352	350	338	329	302	287	286	277	275	307
2010	408	423	430	427	390	365	328	297	299	292	331	358	323	343	340	358	366	345	314	293	291	273	289	297
2011	410	402	416	396	370	354	305	297	291	303	324	374	316	348	365	352	324	318	292	287	274	271	283	303
2012	388	408	398	403	374	359	320	295	284	293	336	382	309	305	332	354	345	328	304	288	283	270	287	302
2013	392	421	404	411	377	350	324	311	299	283	339	375	305	320	325	344	348	325	308	301	284	262	292	317
2014	389	394	410	399	377	348	324	302	314	306	324	371	320	339	350	354	341	336	307	287	290	271	274	314
2015	402	426	414	384	381	374	324	305	307	321	326	357	320	357	361	336	337	336	299	293	295	278	282	291
2016	394	415	405	390	363	364	315	290	291	300	326	358	309	326	341	333	331	320	302	288	275	259	269	285
2017	399	393	416	399	371	356	318	298	311	291	326	387	312	334	362	356	342	338	304	291	291	267	280	309
2018	405	459	400	401	372	347	302	296	295	303	318	363	322	359	347	358	345	328	295	282	280	268	273	274
2019	384	409	420	408	380	368	323	293	288	298	322	362	295	293	334	350	347	339	310	283	270	269	278	291

Monthly mean values total ozone observed (m atm-cm) (2)

Year	Naha												Syowa station (Antarctica)											
	Jan.	Feb.	Mar.	Apr.	May	Jun.	Jul.	Aug.	Sep.	Oct.	Nov.	Dec.	Jan.	Feb.	Mar.	Apr.	May	Jun.	Jul.	Aug.	Sep.	Oct.	Nov.	Dec.
1991	240	255	264	285	297	286	280	272	261	263	249	247	306	291	295	281	219	295	309	249	211	299	265	318
1992	244	259	265	286	304	289	274	271	258	253	233	225	301	309	288	287	295	315	306	259	187	164	226	306
1993	222	245	251	279	284	276	266	263	259	254	238	227	301	310	308	306	301	324	305	247	200	192	229	292
1994	247	252	274	283	293	280	279	269	269	257	236	219	301	269	256	256	246	263	242	202	174	202	294	300
1995	223	242	267	265	268	275	272	269	266	257	248	245	292	287	269	284	280	287	264	228	193	166	248	253
1996	248	267	273	291	292	283	276	271	265	254	233	227	288	275	269	262	245	272	265	233	187	156	216	305
1997	244	237	254	277	279	289	274	263	267	265	247	242	295	292	279	272	296	270	299	247	232	164	303	290
1998	243	270	267	286	295	284	281	278	270	257	245	228	288	296	285	273	253	278	272	237	172	181	208	251
1999	227	246	247	274	288	275	276	274	267	262	255	246	279	276	278	274	252	285	268	223	203	171	198	227
2000	249	266	277	288	295	285	284	275	279	256	234	244	291	286	270	270	270	263	254	215	185	204	232	316
2001	247	249	275	294	298	284	283	283	267	261	249	229	303	291	289	278	271	284	252	232	172	158	192	284
2002	240	243	256	285	288	287	275	272	272	269	250	253	303	284	289	281	307	274	-	258	241	318	333	327
2003	253	263	282	291	297	296	284	279	271	270	239	236	311	295	276	267	258	257	253	224	165	159	253	296
2004	248	244	267	285	290	296	287	276	272	270	255	246	301	290	282	275	274	282	292	260	222	191	254	309
2005	262	260	288	291	297	288	280	274	269	257	237	245	291	295	276	250	278	284	243	218	173	194	241	311
2006	241	252	268	286	293	286	281	279	278	266	258	243	296	287	292	293	300	263	287	216	171	137	192	262
2007	246	275	273	303	302	293	283	280	272	262	252	245	294	296	291	268	300	286	233	204	180	170	259	268
2008	242	249	274	284	291	284	283	281	272	265	244	238	303	300	283	279	268	304	256	237	172	177	209	260
2009	243	247	271	294	301	292	281	276	273	263	244	248	298	292	278	256	260	253	265	217	187	161	343	308
2010	249	265	275	293	304	290	284	276	270	260	255	246	303	288	273	283	278	285	276	244	203	183	224	287
2011	253	267	278	295	283	283	277	276	274	265	251	245	297	280	272	275	294	291	297	223	182	213	217	257
2012	241	258	264	297	304	287	284	275	279	265	254	241	302	285	288	263	244	277	270	240	216	240	334	327
2013	236	242	265	291	295	290	290	284	278	269	254	253	301	293	291	261	292	271	295	225	241	237	308	311
2014	255	267	281	295	297	290	280	274	273	265	253	255	301	310	286	289	304	275	280	238	195	180	259	300
2015	250	258	276	294	292	287	278	272	276	262	246	242	303	292	284	280	257	284	275	233	203	168	219	247
2016	245	255	274	280	288	279	275	278	262	242	237	234	289	279	276	260	303	277	275	253	198	195	289	320
2017	238	258	268	292	293	279	278	271	264	258	244	250	299	299	281	279	281	281	281	251	235	224	266	323
2018	252	262	281	302	296	284	283	271	266	263	239	223	305	296	278	295	281	285	266	230	177	186	231	320
2019	229	226	251	285	285	286	275	270	258	266	248	238	297	295	290	288	240	246	260	250	296	259	342	319

"–" denotes that there are no data to calculate a monthly mean due to rough weather, etc.

Global Distribution of Total Ozone (m atm-cm) in March

The figures of global distribution of total ozone were created by JMA based on data measured by the Ozone Monitoring Instrument (OMI) on the Aura satellite and the Total Ozone Mapping Spectrometer (TOMS) on the Earth Probe satellite of the National Aeronautics and Space Administration (NASA). As representative distributions, the figures show distributions in March and September averaged for the period from 1997 to 2006.

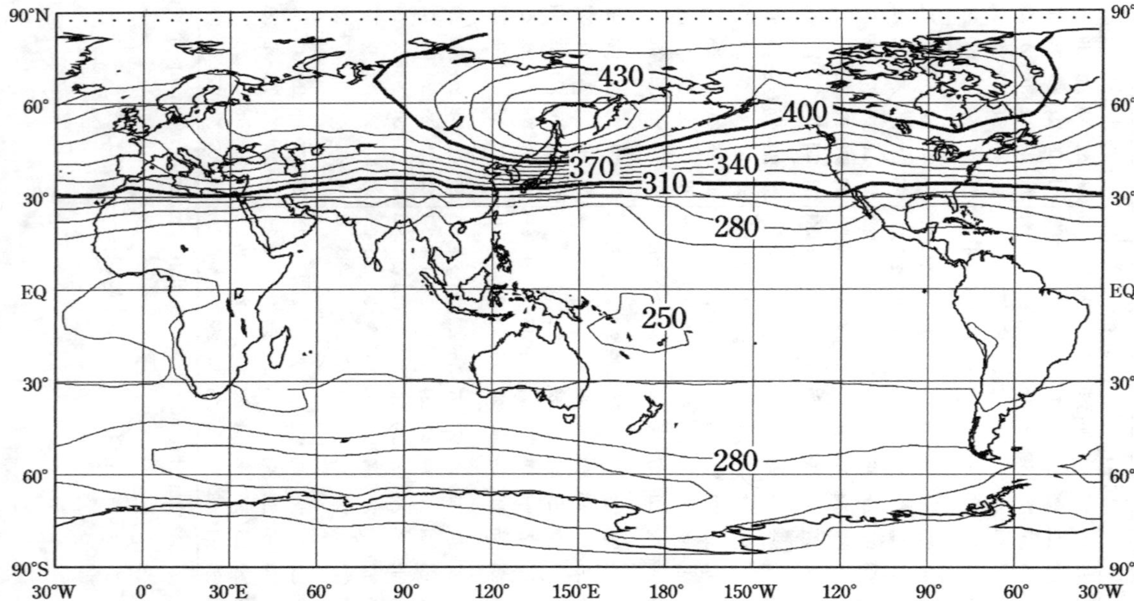

Average for 1997–2006
Contour interval: 15 m atm-cm

Global Distribution of Total Ozone (m atm-cm) in September

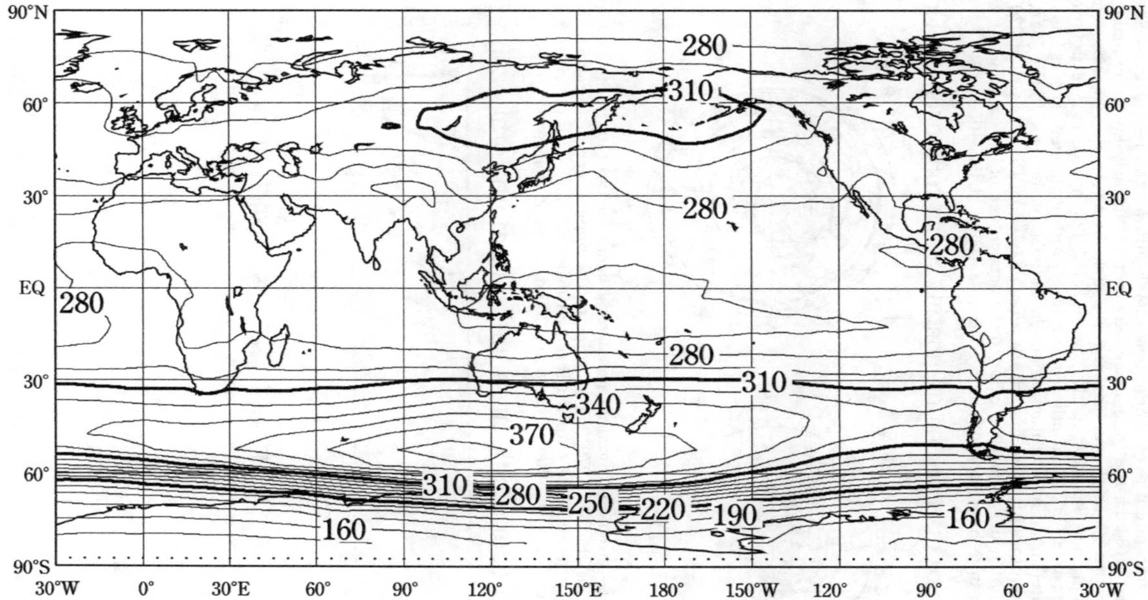

Average for 1997–2006*

Contour interval: 15 m atm-cm

* JMA uses the average for the period from 1994 to 2008 as a "reference value" for expressing fluctuations in the ozone volume and ultraviolet ray volume. During this period, the decreasing trend for the global average ozone volume stopped and the ozone volume remained stable at a low level. For ozone volume measured via satellite observation, the "reference value" is used as the aggregate annual average value for the period from 1997 to 2006 due to reasons such as a portion of data not existing.

Ozone Hole

From the early 1980s, a phenomenon began to appear in which the total ozone fell to a particularly low level in the atmosphere over Antarctica, mainly during the period from September to November. The region in which the total ozone falls to a particularly low level mainly in the atmosphere over Antarctica gives the appearance of a hole being opened in the ozone layer. Consequently, this phenomenon is known as the "ozone hole." The appearance of the ozone hole is attributed to an increase in CFCs and other ozone-depleting substances. The status of ozone depletion can be estimated from annual fluctuations in the size of the ozone hole. JMA defines the ozone hole as the area in which the total ozone amount is equal to or less than 220 m atm-cm (a value which was not widely observed prior to the occurrence of the ozone hole). JMA calculates the area of this range based on data derived from satellite observation.

Southern hemisphere distribution of monthly mean total ozone in October

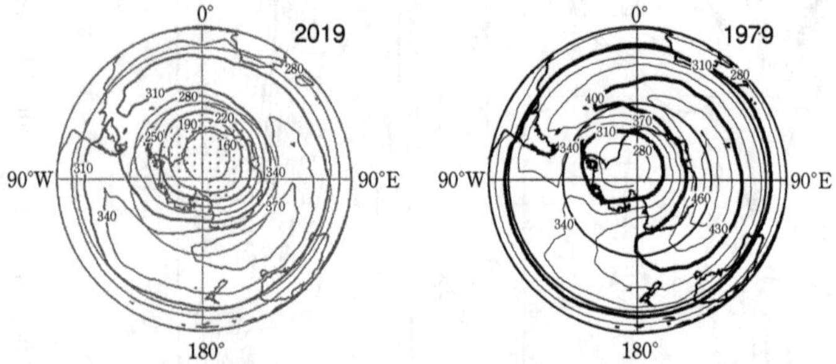

The figures show distributions of monthly mean total ozone in the Southern Hemisphere in October 2019 (left) and in October 1979 (right), prior to the appearance of the ozone hole. The contour interval is 30 m atm-cm. The map was created using satellite data provided by NASA. The dotted area indicates the area where values are 220 m atm-cm or less.

Maximum ozone hole area and minimum total ozone

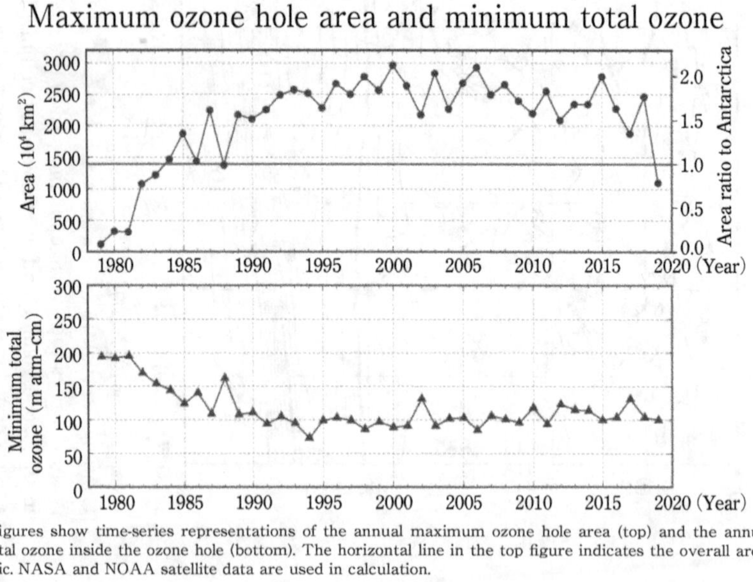

The figures show time-series representations of the annual maximum ozone hole area (top) and the annual minimum total ozone inside the ozone hole (bottom). The horizontal line in the top figure indicates the overall area of the Antarctic. NASA and NOAA satellite data are used in calculation.

UV Index

Solar ultraviolet (UV) radiation is categorized by wavelength into three ranges, UV-A (315–400 nm (nm: nanometer, one-billionth of a meter)), UV-B (280–315 nm) and UV-C (100–280 nm). UV-C is absorbed by oxygen and ozone in the atmosphere before reaching the earth's surface. However, not all of UV-A and UV-B are absorbed in the atmosphere and a portion of these ranges reaches the ground. In particular, UV-B is known to cause cataracts and skin cancer. Acting out of concern that ozone depletion increases the amount of UV-B reaching the earth's surface, organizations such as the World Health Organization (WHO) have recommended the implementation of UV protection measures utilizing the UV Index. The UV Index expresses the strength of UV radiation for the purpose of providing an easy-to-understand measure for expressing the extent to which UV radiation affect the human body. The UV Index is defined as follows by the "Global Solar UV Index: A Practical Guide," which was published by organizations including WHO.

$$I_{\mathrm{CIE}} = \int_{250\mathrm{nm}}^{400\mathrm{nm}} E_\lambda S_{\mathrm{er}} d\lambda \qquad S_{\mathrm{er}} = \begin{cases} 1.0 & (250\ \mathrm{nm} < \lambda < 298\ \mathrm{nm}) \\ 10^{0.094\,(298-\lambda)} & (298\ \mathrm{nm} \leq \lambda \leq 328\ \mathrm{nm}) \\ 10^{0.015\,(139-\lambda)} & (328\ \mathrm{nm} < \lambda < 400\ \mathrm{nm}) \end{cases}$$

E_λ is the solar spectral irradiance expressed in $\mathrm{mW/(m^2 \cdot nm)}$ at wavelength λ. S_{er} is the erythemal reference action spectrum characterizing the relative impact of radiation on skin at each wavelength, which was defined by the International Commission on Illumination (CIE). The erythemal UV irradiance (CIE UV irradiance) I_{CIE} is obtained by integrating E_λ weighted by S_{er} over wavelength. The UV Index (I_{UV}) is obtained by dividing I_{CIE} by 25 $\mathrm{mW/m^2}$.

$$I_{\mathrm{UV}} = I_{\mathrm{CIE}}/25$$

In the Global Solar UV Index: A Practical Guide, it is suggested to avoid being outside during midday hours for UV Index of 8 or greater. The table shows appearance rate (%) for a UV Index of 8 or greater at Tsukuba.

Appearance rate (%) for a UV Index of 8 or higher

Created by hourly observational data for 1990–2019

	9	10	11	12	13	14	15
Tsukuba							
Jan.	0	0	0	0	0	0	0
Feb.	0	0	0	0	0	0	0
Mar.	0	0	0	0	0	0	0
Apr.	0	0	0	1	0	0	0
May	0	1	9	9	1	0	0
Jun.	0	2	11	14	6	0	0
Jul.	0	9	27	33	20	0	0
Aug.	0	5	25	30	14	0	0
Sep.	0	0	4	4	0	0	0
Oct.	0	0	0	0	0	0	0
Nov.	0	0	0	0	0	0	0
Dec.	0	0	0	0	0	0	0

Atmospheric Pollution

Brightness Distribution for the Nighttime Sky in Japan

The brightness of the nighttime sky is an important criterion for measuring atmospheric pollution such as light pollution from the earth's surface and dust which scatter light in the atmosphere. Although it is difficult to directly confirm the condition of the earth's atmosphere by eye, information on the brightness distribution of the sky can be obtained through a measurement method which involves identifying the stars which can be seen.

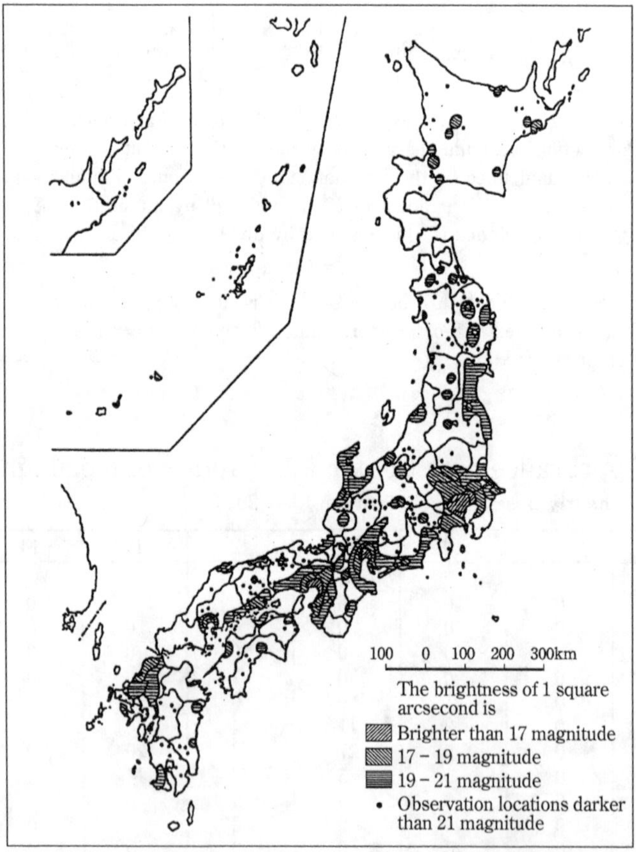

(Based on "Quality of the Environment in Japan 1996" introduction" p. 488; Master Drawing by Hiroki Kosai.)

Kosa (Aeolian Dust)

Kosa (Aeolian dust) is a meteorological phenomenon in which fine dust is blown up to an altitude of several thousand meters by cyclonic or other wind systems from deserts or cropland in semi-arid areas of the Asian continent, and is transported over long distances by westerly winds, resulting in haze or dustfall in downstream areas. It makes the sky yellow and hazy. In Japan, Kosa represents a traffic hazard and causes damage to laundry and vehicle bodies.

Kosa observation is performed by sight at 11 meteorological stations in Japan (as of 31 October, 2020). When a phenomenon can be clearly determined as Kosa from weather conditions, it is recorded regardless of visibility range.

Annual number of days when any station observed Kosa based on the 11 stations in Japan

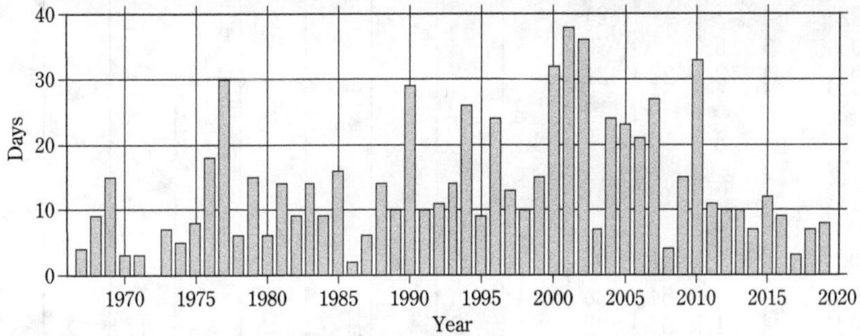

Annual total number of stations observing Kosa based on the 11 stations in Japan

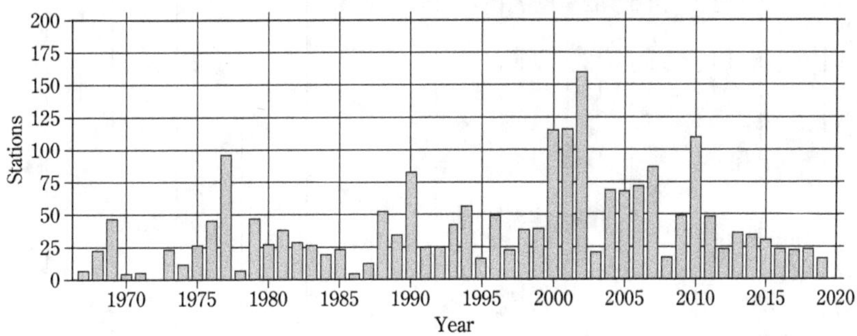

Monthly total number of stations observing Kosa based on the 11 stations in Japan

Normals are averages for 1981–2010.

Year	Jan.	Feb.	Mar.	Apr.	May	Jun.	Jul.	Aug.	Sep.	Oct.	Nov.	Dec.	Totals
1981			12	4	22								38
1982			1		19	1						8	29
1983	1	1	1	18	4	1							26
1984		1	17	1									19
1985		3	5	9	6								23
1986			4										4
1987	2	1	8	1									12
1988	1	1		48	2								52
1989			6	28									34
1990	1	10	36	31	4							1	83
1991		1	3	2	19								25
1992				24						1			25
1993		6	1	30	5								42
1994		13	22	21									56
1995			8	6	2								16
1996	1	18	9	8	12						1		49
1997			8	14									22
1998			17	21									38
1999	13	8	9	9									39
2000			41	60	14								115
2001	2		64	36	14								116
2002		1	57	82		3					17		160
2003			7	14									21
2004		5	29	30	5								69
2005		9	2	38	3						16		68
2006			16	54	2								72
2007		4	14	32	37								87
2008			17										17
2009		18	21	2	1					2		5	49
2010			25	15	35						24	11	110
2011			3	3	42								48
2012			3	17	2							1	23
2013	2		28	1						5			36
2014					29	5							34
2015		11	5	4	4	6							30
2016				16	7								23
2017					22								22
2018			2	21									23
2019				8	1					5	2		16
Normals	0.7	3.3	15.3	21.3	6.9	0.2	0	0	0	0.1	1.9	0.8	50.5

Acid Rain (Acidity of Precipitation)

In recent years, the deposition of acidic substances from the atmosphere to the ground has become a concern, as it adversely influences the environments of rivers, soil and plants. The major acidic substances in the atmosphere are sulfuric acid and nitric acid. They are generated from sulfur dioxide and nitrogen oxides, which are emitted into the atmosphere as a result of fossil fuel combustion, through reactions including photochemical processes. Samples are collected of rain which contains these substances, and hydrogen ion concentration ($pH = -\log [H^+]$) of the samples is measured. A low pH indicates high acidity. The figure shows annual mean acidity (pH) of precipitation at Ryori (Ofunato City, Iwate Prefecture) and Minamitorishima (Ogasawara Village, Tokyo).

Annual mean acidity (pH) of precipitation at Ryori and Minamitorishima in Japan

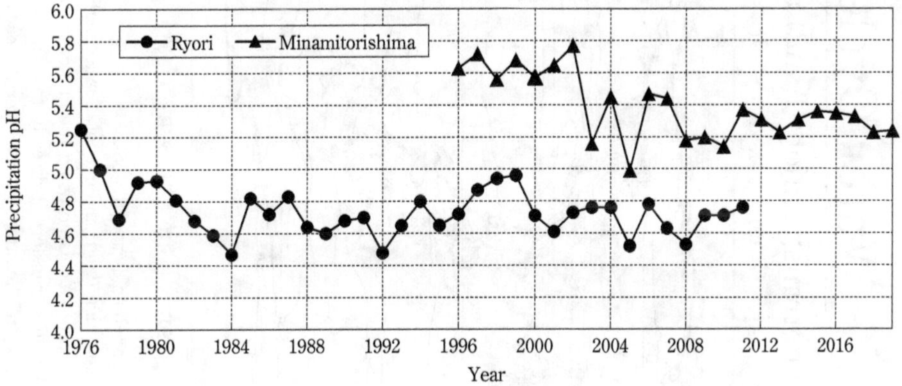

The figure shows of annual mean precipitation pH at Ryori and Minamitorishima. The annual mean value is weighted by the amount of precipitation.

One reason for the low pH values observed in 2003 and 2005 at Minamitorishima was an influx of volcanic gas from the eruption of Mt. Anatahan in the Northern Mariana Islands.

Ryori ended the monitoring at the end of 2011.

Refer to the "Acid rain" under "Atmospheric Environment" on the JMA website (https://www.data.jma.go.jp/gmd/env/acid/en/index.html).

Water Cycle

Global Annual Evapotranspiration (Land) and Evaporation (Ocean)

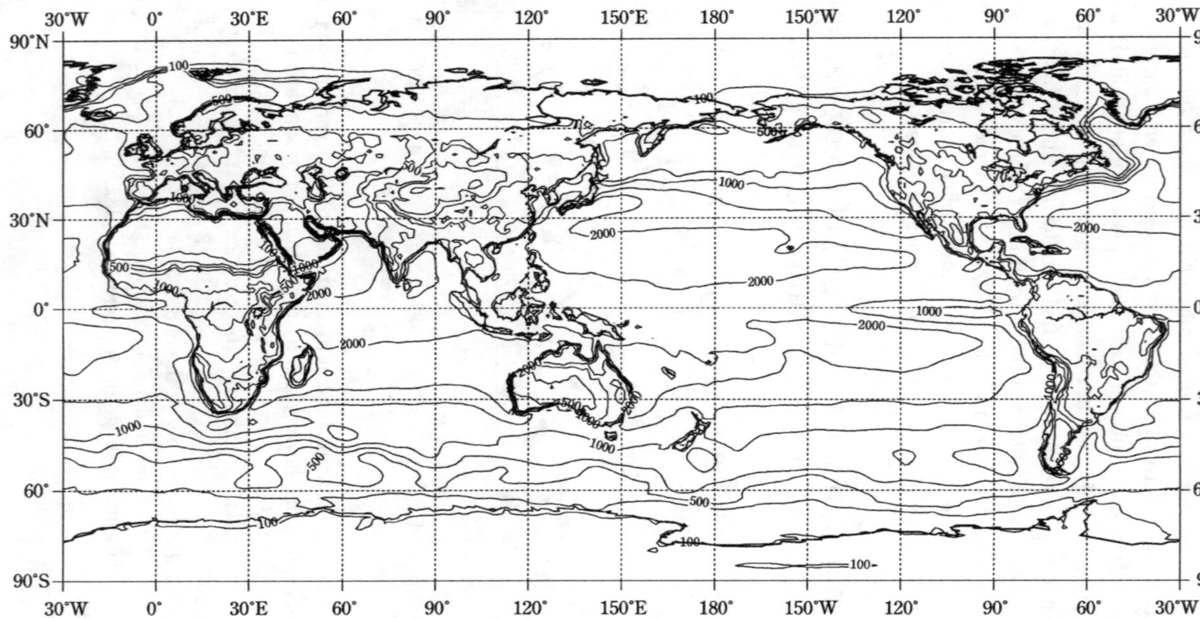

(Unit of contour lines: mm/year)

Based on the results of the Japanese 55-year Reanalysis (JRA-55).

The figure shows annual total evapotranspiration (land) and evaporation (ocean) averaged for the period from 1981 to 2010. They are accumulated values of 6-hour forecasts that were calculated using the numerical weather prediction model of JRA-55. Evapotranspiration is the sum of evaporation (amount of water transferred directly from the Earth's surface to the atmosphere) and transpiration (amount of water transferred from plant leaves to the atmosphere). Over the oceans, only evaporation occurs.

Global Annual Precipitation

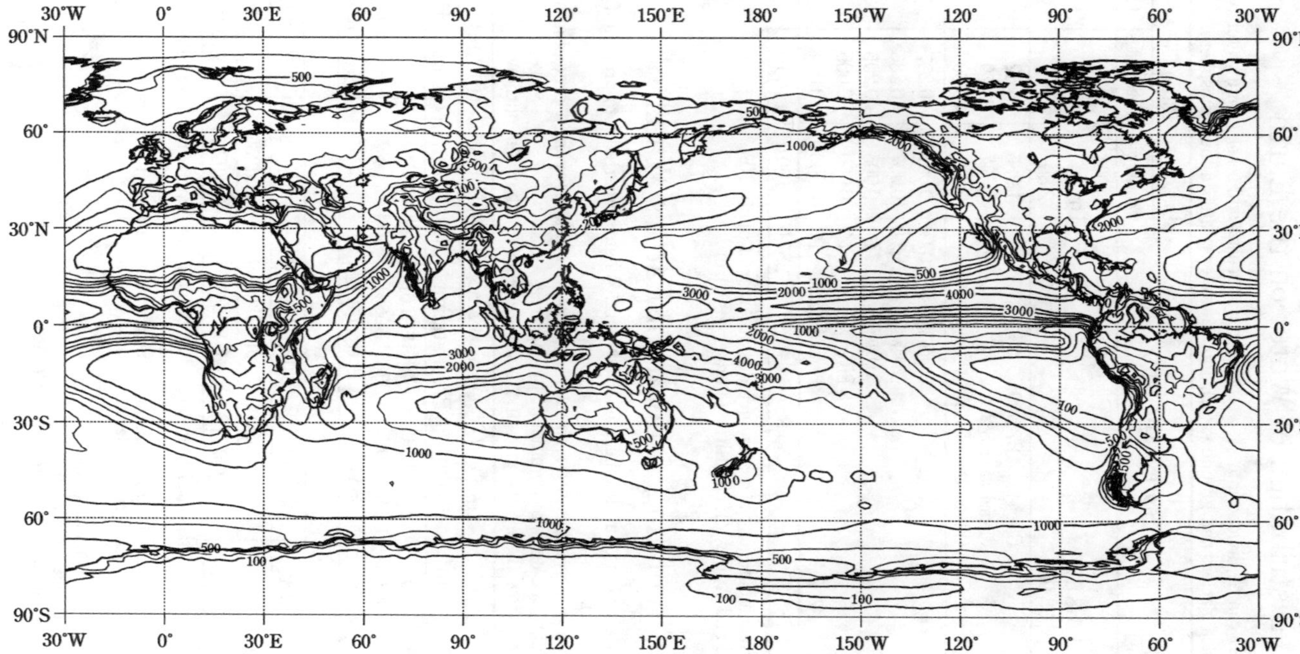

(Unit of contour lines: mm/year)

Based on the results of the Japanese 55-year Reanalysis (JRA-55).

The figure shows annual precipitations averaged for the period from 1981 to 2010. They are accumulated values of 6-hour forecasts that were calculated using the numerical weather prediction model of JRA-55. These values may be greater than estimates using satellite observations over some areas of tropical oceans by about 10%.

Distribution of Water on the Earth

The earth's water can be divided into seawater (water in oceans and seas), inland water (water in land regions; includes glaciers and groundwater), and rainwater (water in the atmosphere).

	Storage (10^3 km^3)			Ratio (%)		
				Compared to total	Inland waters	Others
Rain water	13			0.001		
Seawater	1 348 850			97.4		
Inland water	35 987			2.60		
Details of inland water		Glaciers 27 500		1.986	76.42	
		Ground water 8 200		0.592	22.79	
		Others 287		0.021	0.80	
Details of others			Saline lakes 107.0	0.007 7	0.297	37.28
			Freshwater lakes 103.0	0.007 4	0.286	35.89
			Soil water 74.0	0.005 3	0.206	25.78
			River water 1.7	0.000 1	0.005	0.59
			Plants and animals 1.3	0.000 1	0.004	0.45
Total	1 384 850	35 987	287.0	100.0	100.0	100.0

The storage for each category of water includes an error of about 10%.
Sources: Speidel and Agnew (1988), Hiroo Ohmori (1993), etc.

Circulating Amount of Earth's Water

The storage of water in the atmosphere is 13 000 km^3. When converted into precipitation, this is equivalent to 25 mm.

The value shown in parentheses for circulating amount is the converted precipitation (mm) calculated by dividing the circulating amount by the area.

Unmarked values show the inflow amount. Negative values (–) show the outflow.

	Area (10^6 km^2)	Storage (10^3 km^3)	Annual amount of circulation (10^3 km^3/year)		
			Precipitation	Evaporation	From land to ocean
Atmosphere	510	13	496[972.6]	–496[972.6]	
Ocean	361	1 348 850	385[1 066]	–425[1 177]	40[111]*1
Land	149	35 987	111[745]	–71[477]	–40[268]*2
Details of Land					
Glacier region	18	27 500	3.02[169]	–1.04[58]	–1.98[111]
Internal drainage area	33	125.6	7.59[230]	–7.59[230]	0.00
External drainage area	98	8 360	100.41[1 024]	–62.41[637]	–38.00[387]

Internal drainage: River drainage which does not flow to an ocean. External drainage: River drainage which flows to an ocean.

Land area evaporation is the total of moisture which returns directly to the atmosphere from the earth's surface and moisture which returns to the atmosphere via the leaves of plants. (= evapotranspiration).

*1 Converted to precipitation in oceans.
*2 Converted to precipitation in land areas.
Sources: Speidel and Agnew (1988), Hiroo Ohmori (1993), etc.

Residence Time by Body of Water

Residence time (also called "transit time", "time reguired" and "retention time") is the time until water which has entered a body of water leaves that body of water. It is calculated by dividing the storage by the circulating amount.

	Storage (10^3 km^3)	Amount of circulation* (10^3 km^3/year)	Residence time (Years)	Residence time (Days)		Storage (10^3 km^3)	Amount of circulation* (10^3 km^3/year)	Residence time (Years)	Residence time (Days)
Rain water	13.0	496.00	0.03	10	Freshwater lake	103.0	24.00	4.29	1 566
Seawater	1 348 850.0	425.00	3 173.8	1 158 424	Salina	107.0	7.59	14.10	5 146
Glacier	27 500.0	3.02	9 106.0	3 323 675	Ground water	8 200.0	14.00	585.7	213 786
River water	1.7	24.00	0.07	26	Soil water	74.0	84.00	0.88	322

Residence time also indicates the time required for all of the water in a body of water to be replaced.

* The circulating amount for river water is the amount which flows from land into the ocean in one year (excluding converging amounts of groundwater).

The circulating amount of freshwater lakes is the same as the circulating amount of river water.

The circulating amount of saltwater lakes is the same as the precipitation (=evaporation amount) of inland drainage areas.

The circulating amount for groundwater is the total of amounts converging into rivers (12×10^3 km^3) and the amount which flows directly into the ocean as groundwater (2×10^3 km^3).

The circulating amount of soil water is the evaporation amount from land areas (70×10^3 km^3) and the amount which becomes groundwater after passing through the soil (14×10^3 km^3).

Sources: Speidel and Agnew (1988), Hiroo Ohmori (1993), etc.

Water Circulation of Oceans

	Area (10^6 km^2)	Storage (10^3 km^3)	Amount of circulation (10^3 km^3/year)			Increase and/or decrease
			Precipitation	Inflow from rivers	Evaporation	
Pacific Ocean	181	695 000	228.50 [1 262]	12.20 [67]	213.00 [1 177]	27.70 [153]
Indian Ocean	74	295 000	81.00 [1 095]	5.70 [77]	100.50 [1 358]	-13.80 [-187]
Atlantic Ocean	94	350 000	74.65 [794]	19.40 [211]	111.00 [1 181]	-16.95 [-180]
Arctic Ocean	12	8 850	0.85 [70.8]	2.70 [225]	0.50 [417]	3.05 [254]
Total	361	1 348 850	385 [1 066]	40 [111]	425 [1 177]	0

Increase/decrease = (precipitation amount + amount of inflow from rivers) − amount of evaporation

The value shown in parentheses for circulating amount is the converted precipitation (mm) calculated by dividing the circulating amount by the area.

Sources: Speidel and Agnew (1988), Hiroo Ohmori (1993), etc.

Annual Inflow of Water from Land to Ocean (unit: km^3)

	Europe	Asia	Africa	North America	South America	Australia	Antarctic	Greenland	Total
Pacific Ocean	0	6 530	0	2 000	1 400	1 820	450	0	12 200
Indian Ocean	0	3 600	760	0	0	590	750	0	5 700
Atlantic Ocean	2 470	0	2 820	3 620	9 690	0	600	200*	19 400
Arctic Ocean	40	2 340	0	320	0	0	0	0*	2 700
Total	2 510	12 470	3 580	5 940	11 090	2 410	1 800	200	40 000

* As for the inflow from Greenland, there are calnculated values of 23 km^3 into Arctic Ocean and 231 km^3 into Atlantic Ocean (Steger et al, 2017)

Sources: Speidel and Agnew (1988), Hiroo Ohmori (1993), etc.

Arid Regions

The area of arid regions differs depending on the calculation method used for the extent of drying. In the majority of cases, Thornthwaite's moisture index is used. Defined broadly, all regions ranging from extremely arid regions to semi-arid regions are deserts and account for one-third of all land on the earth. Broken down further, extremely arid regions (hyperarid deserts) account for 7% of all land, arid regions (deserts) account for 11%, and semi-arid regions (semi-deserts) for 15%. Extremely arid regions and arid regions have severe climates. There is little plant growth and almost no agriculture is possible. Prairies and sparse woodlands are found throughout semi-arid regions. These regions are marginal in terms of agriculture and livestock, but serve as the main production area for the world's wheat. These regions are easily effected by climate change and human impact. Therefore, serious desertification is going on.

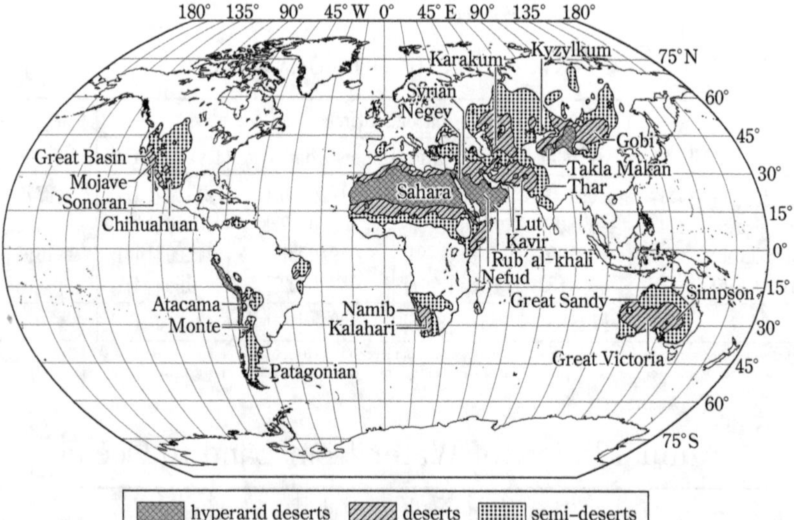

Source: "Desert" in the Encyclopaedia Britannica, based on UNEP/GRID: "Status of Desertification and Plementation of the UN Plan of Action to Combat Desertification" (1992).

Aquatic Environments

Transparency of Water (Secchi Disk Depth)

Transparency (Secchi disk depth) is an index of water clarity or turbidity. The transparency is expressed as the depth at which a white (or black-and-white) disc with a diameter ~30 cm disappers from view when it is lowered into the water. High transparency indicates clear water, while low transparency indicates turbid or colored water.

Sea area		Transparency (m)	Date of observation	Investigator or source
Open ocean				
Pacific Ocean	Taiwan waters	60.0		Yoshimura (1976) 'Limnology (enlarged ed.)'
	Equatorial Pacific Ocean	17.0–18.0	1996/12	Tokyo Univ. of Fisheries
	Off Sanriku coast	6.5–15.0	1997/5	Ocean Res. Inst., The Univ. of Tokyo (KT-97-5)
	Northwest Pacific Ocean	7.5–20.0	1997/7	Ocean Res. Inst., The Univ. of Tokyo (KT-97-2)
	Northeast Pacific Ocean	16.0–22.0	1997/8	Ocean Res. Inst., The Univ. of Tokyo (KT-97-2)
East China Sea	Off Okinawa Main Island	34.0	1996/7/20	Tokyo Univ. of Fisheries
	Off Kyushu	11.5–15.0	1996/7/20	Tokyo Univ. of Fisheries
Japan Sea	Yamato Bank	27.0	1996/7/28	Tokyo Univ. of Fisheries
	Off the Noto Peninsula	27.0	1996/7/29	Tokyo Univ. of Fisheries
	Off Otaru	23.0	1996/7/31	Tokyo Univ. of Fisheries
Bering Sea	Central Bering Sea	12.5–15.0	1997/7	Ocean Res. Inst., The Univ. of Tokyo (KH-97-2)
	Eastern Bering Sea	20.0	1997/7/29	Ocean Res. Inst., The Univ. of Tokyo (KH-97-2)
Atlantic Ocean	Sargasso Sea	66.0		Yoshimura (1976) 'Limnology (enlarged ed.)'
	Western Sargasso sea	48.0	1979/6	Broenkow
	Georges Bank	12.0	1979/6	Broenkow
Indian Ocean	Equatorial Indian Ocean	21.0–30.4	1994/1	Tokyo Univ. of Fisheries
	Southeast Indian Ocean	24.5–29.8	1994/1	Tokyo Univ. of Fisheries
Arabian Sea	Gulf of Oman	13.5	1994/1/4	Tokyo Univ. of Fisheries
	Off India	20.0–27.0	1994/1	Tokyo Univ. of Fisheries
Southern Ocean	Kerguelen Plateau	13.0	2003/1/10	Antarctic Ocean research during the 9th cruise of T/S Umitaka Maru
	Off the French station (Dumont d'Urville Station), Antarctica	17.0	2003/2/3	Antarctic Ocean research during the 9th cruise of T/S Umitaka Maru
	Off the Antarctic coast (Australian Sector)	11.0–20.0	2003/2	Antarctic Ocean research during the 9th cruise of T/S Umitaka Maru
	Off the Antarctic peninsula	8.0–11.0	1987/12/24	The 5th Artarctic Ocean survey by R/V Kaiyo Maru
Costal region				
Tokyo Bay	Off Chiba west coast	3.2	2014 (mean)	Environmental Research Center, Chiba Pref.
	Head (Stn. 3)	3.5	2000 (mean)	Tokyo Univ. of Fisheries
		3.0	2000 (mean)	
	Mouth (Stn. 6)	9.7	2000 (mean)	Tokyo Univ. of Fisheries
		9.0	2012 (mean)	

Continued.

Sea area		Transparency (m)	Date of observation	Investigator or source
Coastal region (continued.)				
	Outside (Stn. 11)	14.7	2000 (mean)	Tokyo Univ. of Fisheries
		15.3	2005 (mean)	
Sagami Bay	Central part (Stn. S3)	16.8	2000 (mean)	Tokyo Univ. of Fisheries
		13.7	2005 (mean)	
Suruga Bay	Western part	5.0–10.0	1966–1980	Coastal Oceanography of Japanese Islands
	Eastern part	15.0–20.0		
Enshunada		1.0–16.5	1976–1982 (mean)	Coastal Oceanography of Japanese Islands
Mikawa Bay		2.0–4.0	1952–1971 (mean)	Coastal Oceanography of Japanese Islands
Ise Bay		4.0–8.0	1952–1971 (mean)	Coastal Oceanography of Japanese Islands
Osaka Bay	Northern part	3.4	2000 (mean)	Wide-area assessment of comprehensive water quality, The Ministry of Environment
	Off north side of Kansai International Airport (A-6)	6.7	2011 (mean)	Department of Environment, Agriculture, Forestry and Fisheries, Osaka Pref.
Setonaikai Sea	Harimanada	9.5	2015 (mean)	Akashiwo Research Institute of Kagawa Pref.
Ariake Sea	Near coast	0.5–3.0	1990–2000	Nakata and Nonaka (2003)
	Central part	2.0–4.0		
	Mouth of a bay	6.0–12.0		
Omura Bay	Central part	5.9–6.6	1979–2002 (mean)	Nagasaki Prefectural Institute of Public Health and Environmental Sciences
	Off Omura city	3.9–4.7		
Shiraho	Coral reef	2.9–16.7*	2000/6	Omija, Mitsumoto
Baltic Sea	Central part	8.1–10.0	1900–1999 (mean)	T. Aarup
	Entrance	6.5–8.6		
	Off the Elba river mouth	0.6–2.8		
Adriatic Sea	Northern part	7.0–14.0	1979–1985 (mean)	M. Morović
	Southern part	14.0–25.0		
California	Off San Diego	7.0–25.0	1976	B. Kimor
Lake and river				
Hokkaido	L. Mashu	41.6	1931/8/31	Hokkaido Fisheries Research Institutes
		18.0	2002/8/23	National Institute for Environmental Studies
		22.6	2013/5/26	
	L. Shikotsu	17.5	1991/8	Nature Conservation Bureau, Environment Agency
	L. Kuttara	22.0	1991/8	Nature Conservation Bureau, Environment Agency
	L. Toya	23.5	1938/8/26	Tanakadate
		9.0–17.5	1993–1994	Nakano & Ban (2003)
	L. Akan	5.0	1991/8	Nature Conservation Bureau, Environment Agency
	L. Kussharo	6.0	1991/8	Nature Conservation Bureau, Environment Agency
Aomori	L. Ogawara	3.2	1991/8	Nature Conservation Bureau, Environment Agency
	L. Towada	20.5	1930/9/7	Hoshino
		6–14	1990–1996	Takamura *et al.* (1999)
Akita	L. Tazawa	4.0	1991/8	Nature Conservation Bureau, Environment Agency
Fukushima	L. Inawashiro	6.1	1991/8	Nature Conservation Bureau, Environment Agency
Ibaraki	Center of L. Kasumigaura	0.96	1980 (mean)	National Inst. for Environmental Studies
		0.80	2014 (mean)	

Transparency of Water Continued.

Sea area		Transparency (m)	Date of observation	Investigator or source
Lake and river (continued.)				
Tochigi	L. Chuzenji	9.0	1991/8	Nature Conservation Bureau, Environment Agency
Niigata	L. Kamo	4.0-5.0	1997-2001	K. Kanzo, Niigata Prefectural Ryotsu High School Science Club *et al.* (2003)
Kanagawa	L. Ahinoko	7.5	1991/8	Nature Conservation Bureau, Environment Agency
Yamanashi	L. Kawaguchi	5.2	1991/7	Nature Conservation Bureau, Environment Agency
	L. Yamanaka	5.5	1991/7	Nature Conservation Bureau, Environment Agency
	L. Motosu	11.2	1991/8	Nature Conservation Bureau, Environment Agency
Nagano	L. Kizaki	4.3	1991/8	Nature Conservation Bureau, Environment Agency
	Center of L. Suwa	1.15	2006	Institute of Mountain Science, Shinshu University
Shizuoka	Center of L. Hamana	3.3	2014 (mean)	Living Environment Dision Environment Protection Bureau Shizuika Pref.
Shiga	North basin of L. Biwa	4.8-6	1990-2000 (mean)	Lake Biwa Environmental Research Institute, Shiga Pref.
		5.8	2012 (mean)	
	South basin of L. Biwa	1.5-2.0	1990-2000 (mean)	Lake Biwa Environmental Research Institute, Shiga Pref.
		2.2	2012 (mean)	
Shimane	Center of L. Shinji	1.3-2.1	2003	E. Goto *et al.*
	Center of L. Nakaumi	2.6	2015 (mean)	Yonago city
Kagoshima	L. Ikeda	6.5	1991/8	Nature Conservation Bureau, Environment Agency
Siberia, Russia	L. Baikal	40.5	1911/4	Schostakowitsch
	L. Baikal (near Bol'shie Koty)	23.0	2003/8/4	Jung *et al.*
Oregon, U.S.	L. Crater	39.0*1	1969/8	Larson
California, U.S.	L. Tahoe	23.7	2014	UC Davis
Outer Mongolia	L. Kossogol	24.6	1903/7	Jelpatjewski
Switzerland	L. Geneva (Lac Léman)	9.0-11.3	1889-1891	Forel
Germany	L. Walchensee	25.0	1903/3 1905/5/19	V. Aufsess
	L. Bodensee (L. Constance)	0.8-16.0	1905/5/19	Straite & Hälbich
Great Lakes	L. Ontario	2.0-4.5		Bukata
	L. Erie	0.9-2.8	1984-1992 (mean)	Holland
	L. Huron	9-15		Bukata
	L. Superior	11-20		Bukata
Kochi	Shimanto R.	7.4*2	1998-1999 (mean value in winter)	Horiuchi et al.
		5.5*2	1998-1999 (mean value in summer)	Horiuchi et al.
U. S.	Mississippi R.	1.3		Broenkow

*1 Transparency using a Secchi disk with 20 cm diameter.
*2 Transparency in a horizontal direction.

In open ocean, transparency (D in m) can be related to diffuse attenuation coefficient of irradiance (K in m^{-1}) by the equation, $KD = 1.7$ (Poole and Atkins, 1929). Different relationships have been also found in coastal region, enclosed bays, lakes, and marshes.

Falkowski and Wilson (1992) showed that transparency (Z in m) can be empirically related to chlorophyll concentration (Chl in mg m^{-3}) in open ocean waters by the equation $Chl = 457\ Z^{-2.37}$.

The National Oceanographic Data Center (NODC) and the Japan Oceanographic Data Center (JODC) archive and distibute a huge number of transparency observations in the world oceans and in adjacent seas of Japan, respectively.

Concentration of Elements Dissolved in Seawater

Atomic number	Element	Dissolved species	Vertical distribution	Concentration in the Atlantic Ocean		Concentration in the Pacific Ocean		Global average concentration (ng/L)	Gross weight in all oceans (t)
				Surface-layer water (nM)	Deep water*1 (nM)	Surface-layer water (nM)	Deep water*1 (nM)		
1	H	H_2O	c	1.11×10^{11}	1.11×10^{11}	1.11×10^{11}	1.11×10^{11}	1.11×10^{11}	1.6×10^{17}
2	He	gas	c	1.9	1.9	1.9	1.9	7.6	1.1×10^{7}
3	Li	Li^+	c	2.6×10^{4}	2.6×10^{4}	2.6×10^{4}	2.6×10^{4}	1.8×10^{5}	2.5×10^{11}
4	Be	$BeOH^+$	s+n	0.01	0.02	0.004	0.025	0.2	3×10^{5}
5	B	$B(OH)_3$	c	4.2×10^{5}	4.2×10^{5}	4.2×10^{5}	4.2×0^{5}	4.5×10^{6}	6.3×10^{12}
6	C	HCO_3^-	n	2.0×10^{6}	2.2×10^{6}	2.0×10^{6}	2.4×10^{6}	2.8×10^{7}	3.9×10^{13}
7	N	N_2 gas	c	5.9×10^{5}	5.9×10^{5}	5.9×10^{5}	5.9×10^{5}	8.3×10^{6}	1.2×10^{13}
7	N	NO_3^-	n	5	2.0×10^{4}	5	4.0×10^{4}	4.2×10^{5}	5.9×10^{11}
8	O	O_2 gas	Inverse of n	2.0×10^{5}	2.5×10^{5}	2.0×10^{5}	1.5×10^{5}	3.2×10^{6}	4.5×10^{12}
9	F	F^+	c	6.8×10^{4}	6.8×10^{4}	6.8×10^{4}	6.8×10^{4}	1.3×10^{6}	1.8×10^{12}
10	Ne	gas	c	7.9	7.9	7.9	7.9	1.6×10^{2}	2.2×10^{8}
11	Na	Na^+	c	4.70×10^{8}	4.70×10^{8}	4.70×10^{8}	4.70×10^{8}	1.08×10^{10}	1.5×10^{16}
12	Mg	Mg^{2+}	c	5.3×10^{7}	5.3×10^{7}	5.3×10^{7}	5.3×10^{7}	1.3×10^{9}	1.8×10^{15}
13	Al	$Al(OH)_3$	s+n	37	20	5	0.5	300	4×10^{8}
14	Si	H_4SiO_4	n	1 000	3.0×10^{4}	1 000	1.5×10	2.50×10^{6}	3.5×10^{12}
15	P	$NaHPO_4^-$	n	50	1 400	50	2 800	6.5×10^{4}	9.1×10^{10}
16	S	SO_4^{2-}	c	2.80×10^{7}	2.80×10^{7}	2.80×10^{7}	2.80×10^{7}	8.98×10^{8}	1.3×10^{15}
17	Cl	Cl^-	c	5.47×10^{8}	5.47×10^{8}	5.47×10^{8}	5.47×10^{8}	1.94×10^{10}	2.7×10^{16}
18	Ar	gas	c	1.6×10^{4}	1.6×10^{4}	1.6×10^{4}	1.6×10^{4}	6.2×10^{5}	8.7×10^{11}
19	K	K^+	c	1.0×10^{7}	1.0×10^{7}	1.0×10^{7}	1.0×10^{7}	3.9×10^{8}	5.5×10^{14}
20	Ca	Ca^{2+}	Nearly c	1.0×10^{7}	1.0×10^{7}	1.0×10^{7}	1.0×10^{7}	4.1×10^{8}	5.7×10^{14}
21	Sc	$Sc(OH)_3$	s+n	0.014	0.020	0.008	0.018	0.8	1×10^{6}
22	Ti	$Ti(OH)_4$	s+n	0.06	0.3	0.005	0.2	10	1×10^{7}
23	V	$NaVO_4^-$	Nearly c	35	35	35	37	1.8×10^{3}	2.6×10^{9}
24	Cr	CrO_4^{2-}(Ⅵ)	n	3.5	4.5	3	5	250	4×10^{8}
24	Cr	$Cr(OH)_3$(Ⅲ)	s			0.2	0.05	2	3×10^{6}
25	Mn	Mn^{2+}	s	2	0.5	2	0.5	30	4×10^{7}
26	Fe	$Fe(OH)_3$	s+n	0.2	1	0.2	1	60	8×10^{7}
27	Co	Co^{2+}	s+n			0.12	0.02	1	2×10^{6}
28	Ni	Ni^{2+}	n	2	7	2	10	500	7×10^{8}
29	Cu	Cu^{2+}	s+n	1.3	2	1.3	4.5	200	3×10^{8}
30	Zn	Zn^{2+}	n	0.8	1.6	0.8	8.2	300	5×10^{8}
31	Ga	$Ga(OH)_4^-$	s+n	0.045	0.030	0.020	0.025	2	2×10^{6}
32	Ge	H_4GeO_4	n	0.001	0.020	0.005	0.10	5	7×10^{6}
33	As	$HAsO_4^{2-}$(Ⅴ)	n	20	21	20	24	2×10^{3}	2×10^{9}
33	As	$As(OH)_3$(Ⅲ)	s			0.3	0.07	5	7×10^{6}
34	Se	SeO_4^{2-}(Ⅵ)	n	0.5	1.0	0.5	1.3	90	1×10^{8}
34	Se	SeO_3^{2-}(Ⅳ)	n	0.03	0.5	0.07	0.9	55	8×10^{7}
35	Br	Br^-	c	8.4×10^{5}	8.4×10^{5}	8.4×10^{5}	8.4×10^{5}	6.7×10^{7}	9.4×10^{13}
36	Kr	gas	c	3.7	3.7	3.7	3.7	310	4.3×10^{8}
37	Rb	Rb^+	c	1 400	1 400	1 400	1 400	1.2×10^{5}	1.7×10^{11}
38	Sr	Sr^{2+}	c	8.9×10^{4}	9.0×10^{4}	8.9×10^{4}	9.0×10^{4}	7.8×10^{6}	1.1×10^{13}
39	Y	YCO_3^+	s+n			0.09	0.3	27	4×10^{7}
40	Zr	$Zr(OH)_4$	s+n	0.02	0.2	0.02	0.3	20	3×10^{7}
41	Nb	$Nb(OH)_6^-$	s+n	0.001	0.002	0.001	0.004	0.3	4×10^{5}
42	Mo	MoO_4^{2-}	c	105	105	105	105	1.0×10^{4}	1.4×10^{10}
44	Ru	RuO_4^-	?			$<5\times10^{-5}$		<0.005	
45	Rh	$Rh(OH)_3$	n			4×10^{-4}	1×10^{-3}	0.08	1×10^{5}

Concentration of Elements Dissolved in Seawater Continued.

Atomic number	Element	Dissolved species	Vertical distribution	Concentration in the Atlantic Ocean		Concentration in the Pacific Ocean		Global average concentration (ng/L)	Gross weight in all oceans (t)
				Surface-layer water (nM)	Deep water*1 (nM)	Surface-layer water (nM)	Deep water*1 (nM)		
46	Pd	$PdCl_4^{2-}$	n			2×10^{-4}	6×10^{-4}	0.06	9×10^{4}
47	Ag	$AgCl_2^-$	n	7×10^{-4}	7×10^{-3}	0.001	0.04	2	3×10^{6}
48	Cd	$CdCl_2$	n	0.01	0.35	0.01	1.0	70	1×10^{8}
49	In	$In(OH)_3$	s	5×10^{-4}	2×10^{-4}	2×10^{-4}	5×10^{-5}	0.01	2×10^{4}
50	Sn	$SnO(OH)_3^-$	s	2.5×10^{-2}	0.005			0.5	7×10^{5}
51	Sb	$Sb(OH)_6^-$	Nearly c	1	1	1.5	2	200	3×10^{8}
52	Te	TeO_3^{2-}(Ⅵ)	s	9×10^{-4}	4×10^{-4}	0.001	4×10^{-4}	0.05	7×10^{4}
52	Te	$TeO(OH)_3^-$(Ⅳ)	s	4×10^{-4}	2×10^{-4}	5×10^{-4}	1×10^{-4}	0.02	3×10^{4}
53	I	IO_3^-(Ⅴ)	Nearly c	410	450	350	460	5.8×10^{4}	8.1×10^{10}
53	I	I^-(−Ⅰ)	s	0.035	0	0.1	0	1	1×10^{6}
54	Xe	gas	c	0.50	0.50	0.50	0.50	66	9.2×10^{7}
55	Cs	Cs^+	c	2.3	2.3	2.3	2.3	310	4.3×10^{8}
56	Ba	Ba^{2+}	n	35	70	35	150	15 000	2.1×10^{10}
57	La	$LaCO_3^+$	s+n	0.013	0.028	0.019	0.051	6	8×10^{6}
58	Ce	$Ce(OH)_4$	s	0.066	0.019	0.011	0.004	1	1×10^{6}
59	Pr	$PrCO_3^+$	s+n	3.0×10^{-3}	5.0×10^{-3}	3.2×10^{-3}	7.3×10^{-3}	0.8	1×10^{6}
60	Nd	$NdCO_3^+$	s+n	0.013	0.023	0.013	0.034	3	4×10^{6}
62	Sm	$SmCO_3^+$	s+n	2.7×10^{-3}	4.4×10^{-3}	2.7×10^{-3}	6.8×10^{-3}	0.6	8×10^{5}
63	Eu	$EuCO_3^+$	s+n	6×10^{-4}	0.0010	7×10^{-3}	0.0018	0.2	3×10^{5}
64	Gd	$GdCO_3^+$	s+n	3.4×10^{-3}	6.1×10^{-3}	4.0×10^{-3}	0.010	1	1×10^{6}
65	Tb	$TbCO_3^+$	s+n	7×10^{-4}	1.0×10^{-3}	5×10^{-4}	1.6×10^{-3}	0.2	3×10^{5}
66	Dy	$DyCO_3^+$	s+n	0.005	6.1×10^{-3}	3.0×10^{-3}	1.2×10^{-2}	1	1×10^{6}
67	Ho	$HoCO_3^+$	s+n	1.5×10^{-3}	1.8×10^{-3}	1.0×10^{-3}	3.6×10^{-3}	0.4	6×10^{5}
68	Er	$ErCO_3^+$	s+n	3.6×10^{-3}	5.3×10^{-3}	2.0×10^{-3}	1.0×10^{-2}	1	1×10^{6}
69	Tm	$TmCO_3^+$	s+n	8×10^{-3}	1.0×10^{-3}	4×10^{-4}	2.0×10^{-3}	0.2	3×10^{5}
70	Yb	$YbCO_3^+$	s+n	3.0×10^{-3}	4.5×10^{-3}	2.2×10^{-3}	0.013	1	1×10^{6}
71	Lu	$LuCO_3^+$	s+n	8×10^{-4}	1.2×10^{-3}	4×10^{-4}	2.4×10^{-3}	0.3	4×10^{5}
72	Hf	$Hf(OH)_4$	s+n	1×10^{-4}	5×10^{-4}	2×10^{-4}	5×10^{-4}	0.09	1×10^{5}
73	Ta	$Ta(OH)_5$	s+n	2×10^{-5}	5×10^{-5}	2×10^{-5}	3×10^{-5}	0.01	2×10^{4}
74	W	WO_4^{2-}	Nearly c	0.05	0.05	0.05	0.05	9	1×10^{7}
75	Re	ReO_4^-	c	0.04	0.04	0.04	0.04	8	1×10^{7}
76	Os	$H_3OsO_6^-$	?			3×10^{-5}	4×10^{-5}	6×10^{-3}	8×10^{3}
77	Ir	$IrCl_6^{3-}$	?			5×10^{-7}		1×10^{-4}	1×10^{2}
78	Pt	$PtCl_4^{2-}$	s+n			2×10^{-4}	3×10^{-4}	0.05	7×10^{4}
79	Au	$AuCl_2^-$	s+n			6×10^{-5}	2×10^{-4}	0.02	3×10^{4}
80	Hg	$HgCl_4^{2-}$	s+n	2.5×10^{-3}	2.5×10^{-3}	0.002	0.001	0.2	3×10^{5}
81	Tl	Tl+	c	0.07	0.07	0.07	0.07	14	2.0×10^{7}
82	Pb	$PbCO_3$	s	0.15	0.02	0.06	5×10^{-3}	20	2×10^{7}
83	Bi	$Bi(OH)_3$	s	2.5×10^{-4}		2×10^{-4}	5×10^{-5}	0.02	3×10^{4}
90	Th	$Th(OH)_4$	s			5×10^{-5}	1.5×10^{-4}	0.02	3×10^{4}
92	U	$UO_2(CO_3)_3^{4-}$	c	13.5	13.5	13.5	13.5	3 200	4.5×10^{9}

*1 Seawater at a depth greater than 1 000 m.

*2 c: Conservative type. Constant concentration from the surface layer to the deep layer. n: Nutrient type. Concentration is low in the surface layer and high in the deep layer. s: Scavenged type. Concentration is high in the surface layer and low in the deep layer.

Li Y.-H.: "Compendium of Geochemistry", p.304-307, Table Ⅶ-1, Princeton University Press (2000).

Nozaki, Y.: "Encyclopedia of Ocean Sciences" (eds. Steele, J. H., Thorpe, S. A. & Turekian, K. K.), p.840-845, Table 1, Academic Press (2001).

Eutrophication of Water

Loading Values of Nitrogen (N kg/m^2 year) and Phosphorus (P kg/m^2 year) in European and North American Wetlands

	Period	TN	TIN	TP	TIP	Surface area (km^2)	Mean water depth (m)	Renewal rate (year)
Central Europe								
Aegeri See	1947-1951		2.40	0.16	0.035	7.24	49	7.9
Pfaffiker See	1950-1953	14.70		1.36	0.615	3.31	18	1.5
Greifen See	1950-1953		31.1	1.56	0.825	8.56	19	2.0
Turler See	1952-1953		7.50	0.30	0.139	0.48	14	2.1
Zurich See	1952-1953	26.20		1.32	0.73	67.3	50	1.25
Hallwiler See	1956-1957	12.80	5.25	0.56	0.26	10.3	28	3.9
Baldegger See	1957-1959		20.70	1.75	1.17	5.25	34	4.5
Lac Leman	1959-1961		9.50	0.70		581.4	155	12.0
Boden See	1960	20.8		4.07	0.28	476.0	100	4.0
Zeller See	1962	4.8		1.20		4.55	37	2.7
Northern Europe								
Vatteru	1965	0.65		0.065		1300		
Vanern	1965	1.50		0.15		5560	21-28	60
Hjalmaren	1965	2.00		0.20				
Malaren	1965	8.00		0.70				
Norrirken	1961-1962	66.4		3.04		2.66	5.4	
Lough Neagh	1965	23.9		0.535		388	12	1
North America								
Washington	1957	31.4		1.34	0.77	87.6	33	3.2
Mendota	1942-1944	3.10	2.24	0.174	0.065	39.4	12	12.0
Menona	1942-1944		9.10	2.14	0.95	14.0	7.8	1.2
Waubesa	1942-1944		48.7	9.93	7.13	8.3	4.8	0.3
Kegonsa	1942-1944	14.65	18.1	6.64	4.24	12.7	4.6	0.35
Koshkonong	1959-1960		10.1					
Mosses	1963-1964	7.0		0.9		27.5	5.6	
Sebasticook	1964-1965	6.7		0.21		17.35	6.0	
Tahoe	1961-1962	0.23		0.04		497	303	700
Geist Reservoir	1963-1964		49.5	3.10		7.30		

TN: Total Nitrogen, TIN: Total Inorganic Nitrogen, TP: Total Phosphorus, TIP: Total Inorganic Phosphorus.

Based on the Science and Technology Agency, Institute of Resource's translation of "OECD Water Resource Managment Research Report" (1971) with modifications.

Water Quality in Wide-Area Enclosed Seas (COD)

(Unit: mg/L)

Sea area	Number of points*1	1994	1997	2000	2002	2004	2006	2008	2010	2012	2014	2016	2017	2018
Tokyo Bay	49	3.3	2.9	2.9	3.0	2.8	2.7	2.6	2.8	2.7	2.8	2.7	3.0	2.9
Ise Bay	32	3.1	3.4	3.5	3.0	3.0	3.3	3.4	3.1	2.8	3.2	2.9	3.3	3.4
Osaka Bay	28	2.9	2.8	2.6	2.8	2.9	2.7	2.8	2.8	2.7	2.4	2.4	2.4	2.7
Setonaikai Sea*2	426	2.0	2.0	1.9	2.0	2.1	2.1	2.0	1.9	1.9	2.0	2.0	1.9	2.0
Ariake Sea	34	–	–	2.4	1.9	2.1	1.8	1.8	1.9	2.3	1.8	1.8	1.7	1.8
Yatsushiro Sea	29	–	–	2.1	1.6	1.8	1.9	1.6	1.8	1.9	1.7	1.7	1.8	1.7

Water Quality in Wide-Area Enclosed Seas (TN)

(Unit: mg/L)

Sea area	Number of points*1	1995	1997	2000	2002	2004	2006	2008	2010	2012	2014	2016	2017	2018
Tokyo Bay	32	0.94	0.93	0.92	0.85	0.80	0.70	0.81	0.73	0.79	0.61	0.59	0.67	0.59
Ise Bay	33	–	0.48	0.45	0.40	0.46	0.41	0.41	0.35	0.36	0.36	0.36	0.37	0.35
Osaka Bay	23	0.73	0.52	0.57	0.49	0.44	0.38	0.41	0.37	0.35	0.34	0.32	0.31	0.31
Setonaikai Sea*2	280	–	0.27	0.27	0.23	0.25	0.23	0.21	0.19	0.19	0.20	0.19	0.20	0.20
Ariake Sea	31	–	–	0.37	0.29	0.34	0.38	0.31	0.27	0.33	0.29	0.33	0.27	0.29
Yatsushiro Sea	14	–	–	0.27	0.20	0.19	0.20	0.19	0.15	0.16	0.18	0.18	0.16	0.17

Water Quality in Wide-Area Enclosed Seas (TP)

(Unit: mg/L)

Sea area	Number of points*1	1995	1997	2000	2002	2004	2006	2008	2010	2012	2014	2016	2017	2018
Tokyo Bay	22	0.073	0.074	0.070	0.065	0.059	0.066	0.069	0.060	0.066	0.055	0.051	0.061	0.052
Ise Bay	33	–	0.051	0.044	0.042	0.045	0.050	0.042	0.038	0.043	0.040	0.040	0.041	0.038
Osaka Bay	23	0.050	0.047	0.046	0.040	0.045	0.037	0.043	0.040	0.038	0.038	0.036	0.033	0.032
Setonaikai Sea*2	280	–	0.025	0.023	0.022	0.024	0.023	0.022	0.021	0.021	0.021	0.022	0.022	0.022
Ariake Sea	31	–	–	0.049	0.036	0.045	0.047	0.046	0.043	0.045	0.049	0.049	0.041	0.042
Yatsushiro Sea	14	–	–	0.025	0.020	0.019	0.024	0.024	0.022	0.021	0.025	0.023	0.021	0.025

Annual Averages for all Enrivonmental Reference Points. COD: Chemical Oxygen Demand, TN: Total Nitrogen, TP: Total Phosphorus.

＊1 "Locations" is the total number of Environmental Standard points. There are some differences between the numbers of locations in each ocean region from one fiscal year to another.

＊2 Setonaikai Sea shows the average value for the ocean region excluding Osaka Bay.

Compiled based on the Ministry of the Environment, Environment Managment Bureau "Water Quality Survey of Public Water Areas".

Load of COD, TN, and TP on Enclosed Seas

(unit: t/day)

		1979	1984	1989	1994	1999	2004	2009	2014
Tokyo Bay	COD	477	413	355	286	247	211	183	163
	TN	364	333	319	280	254	208	185	170
	TP	41.2	30.2	25.9	23.0	21.1	15.3	12.9	12.3
Ise Bay	COD	307	286	272	246	221	186	158	141
	TN	188	185	168	161	143	129	118	110
	TP	24.4	20.4	18.8	17.3	15.2	10.8	9.0	8.2
Setonaikai Sea	COD	1012	900	838	746	672	561	468	404
	TN	666	639	656	697	596	476	433	390
	TP	62.9	47.0	42.7	41.1	40.4	30.6	28.0	24.6

The Ministry of the Environment: "Environmental Statistics Report" (2017).

Generation Load Amount of COD, TN, and TP on Specified Lakes and Marshes (L. Kasumigaura, L. Teganuma, L. Suwa, L. Biwa)

(unit: kg/day)

		1985	1990	1995	2000	2005	2010	2015
L. Kasumigaura	COD	33 151	31 877	30 292	28 328	24 385	26 876	23 750
	TN	15 020	14 932	15 351	14 966	12 743	14 251	13 057
	TP	939	900	975	910	642	685	656
L. Teganuma	COD	6 102	6 286	5 459	4 331	3 525	3 013	2 862
	TN	2 730	2 748	2 357	1 786	1 403	1 261	1 198
	TP	324	358	275	221	146	132	123
L. Suwa*	COD	4 824	5 413	4 629	4 183	3 787	3 303	3 403
	TN	1 929	1 642	1 318	1 201	1 088	1 072	1 227
	TP	247	226	111	94	80	56	58
L. Biwa	COD	55 554	54 336	51 778	43 379	36 870	34 030	34 609
	TN	18 000	17 826	17 816	16 723	14 771	13 972	14 664
	TP	1 232	1 175	1 144	941	755	648	634

* For L. Suwa, the values are from 1991, 1996, 2001, 2006, 2011 and 2016.

L. Kasumigaura: Kasumigaura ni kakawaru kosho suishitsu hozen keikaku. Heisei 28–32nendo (the 7th), L. Teganuma: Teganuma ni kakawaru kosho suishitsu hozen keikaku. Heisei 28–32nendo (the 7th), L. Suwa: Suwako ni kakawaru kosho suishitsu hozen keikaku. Heisei 28–32nendo (the 7th), L. Biwa: Biwako ni kakawaru kosho suishitsu hozen keikaku. Heisei 28–32nendo (the 7th).

BOD of Rivers

(unit: mg/L)

River name	Country	1980	1985	1990	1995	2000	2005	2010
Kum R.	Korea	-	-	-	4.5	3.5	3.7	-
Naktong R.		-	-	-	3.8	3.0	2.6	2.4
Han R.		-	-	3.4	3.8	2.7	3.1	3.2
Yongsan R.		-	-	-	2.6	2.1	5.3	4.3
Gedeizu R.	Turkey	2.4	2.3	10.6	-	3.7	-	(5.2)
Sakarya R.		2.0	3.6	2.7	4.1	3.1	(3.2)	(4.0)
Mississippi R.	USA	1.7	1.2	1.9	1.2	1.5	1.9	(4.9)
Grijalva R.	Mexico	4.3	1.5	2.2	2.0	1.8	2.2	4.3
Bravo R.		3.1	2.5	3.6	3.1	2.2	(3.0)	8.3
Lerma R.		8.0	(2.6)	13.5	(9.6)	30.0	4.2	10.6
Clair R.	Ireland	-	1.7	1.3	1.9	1.5	0.6	0.6
Barrow R.		-	1.7	1.7	2.5	1.5	1.4	1.2
Blackwater R.		-	1.7	2.8	1.9	2.0	2.1	1.4
Boyne R.		-	1.7	1.7	1.8	2.1	1.5	1.7
R. Clyde	England	4.1	3.2	3.5	2.9	2.3	-	-
R. Severn		2.6	1.7	2.8	2.4	(1.9)	-	-
R. Thames		2.7	2.4	2.9	1.8	1.7	(5.2)	-
Inn R.	Austria	2.2	-	1.4	2.8	0.8	0.6	0.8
Danube R.		3.3	-	3.8	3.0	1.2	0.4	-
Rhine R.	Netherlands	3.2	2.3	1.6	1.9	-	-	-
Ebro R.	Spain	3.3	4.3	4.0	9.2	8.1	2.3	-
Guadalquivir R.		11.8	8.8	9.8	77.1	5.9	3.9	-
Guadiana R.		2.7	1.6	4.6	11.0	3.7	3.5	-
Duero R.		2.1	2.7	3.0	5.4	3.1	2.5	-
Odra R.	Czech	12.3	10.1	5.9	7.1	5.8	4.3	3.5
Morava R.		7.8	7.8	7.9	4.6	2.9	3.0	-
Labe R.		8.5	6.6	6.8	3.7	3.9	2.9	2.8
Guzen'ou R.	Denmark	2.5	2.5	2.8	2.4	1.9	1.6	1.2
Scan Roh R.		8.1	5.5	2.3	(2.3)	1.2	0.9	1.0
Danube R.	Germany	3.1	3.2	2.8	2.7	2.1	1.7	1.6
Seine R.	France	6.4	4.3	5.6	4.4	3.2	-	1.1
Rhone R.		7.8	5.0	1.4	1.3	2.0	-	1.0
Loire R.		6.4	6.0	7.0	4.0	4.3	-	1.7
Oder R.	Poland	5.9	4.6	7.0	4.5	5.2	5.5	3.1
Wisła R.		3.7	5.6	6.0	4.2	4.7	4.3	2.9

BOD: Biochemical Oxygen Demand

* The investigation location is the farthest-downstream boarder or location without tidal influences.

Values in ( ) are observed values from the year before or after.

Compiled based on the Ministry of Internal Affairs and Communications Statistics Bureau, Statistics Laboratory (2002), and "OECD Environmental Data Compendium" (2004, 2009).

Supplemented with OECD Environment Statistics/Lake and river quality (2016).

BOD of Rivers in Japan

(Unit: mg/L)

River name/Measurement point	1980	1985	1990	1995	2000	2005	2010	2012	2014	2015	2016	2017	2018
Ishikari R./Ishikari-ohashi Bridge	1.5	1.5	1.2	1.3	1.0	0.9	0.8	0.9	0.9	0.9	0.9	0.8	1.0
Tokachi R./Moiwa Bridge	1.5	1.2	0.8	0.9	1.5	1.4	1.0	1.1	1.4	1.4	1.3	1.4	1.6
Kitakami R./Chitose Bridge	1.7	1.4	1.4	1.1	1.0	1.2	0.9	1.4	1.0	1.0	1.2	1.0	1.0
Mogami R./Ryouu Bridge	1.3	0.9	0.9	1.0	0.8	0.9	0.7	0.7	0.7	0.7	0.8	0.6	0.7
Tone R./Tone-oseki	1.8	1.8	1.4	1.5	1.5	1.4	1.2	1.1	0.9	0.8	0.8	0.7	1.1
Sumida R./Ryōgoku Bridge	5.1	3.6	2.7	2.8	2.6	2.1	2.5	2.1	1.8	2.4	2.0	2.1	1.7
Tama R./Chofu water intake point	6.7	4.7	4.6	3.8	2.0	1.5	1.1	1.3	1.8	1.3	1.2	1.2	1.0
Shinano R./Heisei-ohashi Bridge	–	1.6	2.2	1.3	1.0	1.0	1.3	1.4	1.1	1.2	1.3	1.1	0.9
Jinzu R./Oginoura Bridge	1.2	1.4	1.3	1.5	1.7	1.4	1.1	0.9	1.0	1.0	1.0	0.7	1.6
Chikuma R./Tategahana Bridge	3.4	2.1	2.4	1.9	1.4	1.1	0.9	1.2	1.2	1.2	1.1	1.2	1.5
Tenryu R./Kashima Bridge	–	0.7	0.7	0.6	0.6	0.6	<0.5	0.5	0.5	0.6	0.5	0.8	0.9
Kiso R./Nobi-ohashi Bridge	0.8	0.9	0.8	0.9	0.6	0.6	0.7	0.9	0.5	0.6	0.8	0.6	0.7
Seta R./Karahashi Bridge	1.8	1.7	1.2	1.4	0.9	1.0	0.8	1.0	0.7	0.6	0.9	0.7	0.9
Yodo R./Hirakata-ohashi Bridge	3.3	3.4	2.5	2.3	1.5	1.3	1.0	1.2	0.9	1.0	1.0	1.1	1.0
Kinokawa R./Funato	1.3	1.9	2.1	1.7	1.8	1.8	1.2	1.0	1.2	0.7	0.7	0.9	0.7
Ota R./Asahi Bridge	1.8	1.7	2.0	3.4	1.6	1.3	1.6	1.5	1.6	1.2	1.1	1.6	1.1
Gōnokawa R./Egawa Bridge	0.6	0.9	1.4	1.4	0.8	0.6	0.5	0.5	0.7	0.6	0.6	0.6	0.6
Yoshino R./Takase Bridge	0.9	0.8	0.7	1.0	0.7	0.8	0.5	0.5	0.5	0.6	0.6	0.6	0.6
Chikugo R./Senoshita	1.9	2.2	1.7	1.5	1.5	1.4	1.7	1.4	1.0	1.0	0.8	0.9	1.1
Oyodo R./Aioi Bridge	1.3	1.3	2.7	1.0	1.0	0.8	0.7	0.8	0.6	0.7	0.7	0.8	0.8
Hija R./pump station	5.1	3.6	3.0	1.8	2.6	1.6	0.7	1.1	0.9	1.6	1.0	2.0	1.4

Compiled based on the Ministry of the Environment: Mizukankyo sogosite

Water Quality of Specified Lakes and Marshes in Japan (COD)

(Annual average values for total environmental reference points, Unit: mg/L)

Lake/wetlands	Type*	1980	1985	1990	1995	2000	2005	2010	2012	2014	2015	2016	2017	2018
Kamafusa dam	AA	2.1	1.9	3.8	2.3	1.9	2.3	2.5	2.2	2.7	2.8	2.3	2.1	2.2
L. Hachirogata	A	5.9	5.7	–	6.9	8.5	7.5	7.5	8.5	7.0	7.9	8.0	6.5	7.4
L. Kasumigaura Nishiura	A	9.3	8.1	7.8	9.0	7.6	7.6	8.2	7.5	6.6	7.8	6.8	6.9	6.7
Kitaura	A	7.6	8.6	7.3	7.4	9.2	7.7	9.1	8.3	7.5	8.9	7.8	8.4	8.4
Hitachitone R.	A	8.7	8.1	7.6	8.1	8.3	7.4	9.2	8.0	7.3	8.3	7.2	7.5	7.6
L. Inbanuma	A	11	11	9.2	12	10	8.1	8.9	11	11	11	11	11	12
L. Teganuma	B	23	24	18	25	14	8.2	8.9	9.6	7.6	8.1	8.6	8.6	9.2
L. Suwa	A	7.8	5.1	6.9	5.1	6.0	5.7	4.5	4.9	5.0	4.7	4.4	5.2	4.7
L. Nojiri	AA	2.0	1.7	1.4	1.4	1.8	1.6	1.9	2.0	2.1	1.9	2.1	2.2	2.0
L. Biwa North Lake	AA	2.2	2.1	2.3	2.5	2.6	2.6	2.6	2.6	2.4	2.5	2.6	2.6	2.4
South Lake	AA	3.1	3.0	3.3	3.1	3.2	3.2	3.7	3.7	3.1	3.2	3.3	3.3	3.4
L. Nakaumi	A	4.0	3.6	4.2	4.6	5.0	4.2	3.8	3.6	3.4	3.7	3.7	3.5	3.6
L. Shinji	A	4.0	3.5	4.8	3.9	4.5	4.5	5.1	5.3	4.1	4.3	4.4	4.4	4.6
L. Kojima	B	8.6	10	10	11	8.2	7.5	7.6	6.9	7.3	7.0	7.1	7.5	7.9

*1 Lakes and marshes designated as requiring comprehensive measures to maintain water quality environmental standards based on Act on Special Measures concerning Conservation of Lake Water Quality.

*2 Environemtnal Standard Values (AA: 1 mg/L, A: 3 mg/L, B: 5 mg/L)

Sources: The Ministry of the Environment: "Results Water Quality Survey of Public Water Areas", National Institute for Environmental Studies: Kankyosuchi Database.

Water Quality of Specified Lakes and Marshes in Japan (TN)

(Annual averages for all enrivonmental reference points, Units: mg/L)

Lake/wetlands	Type*	1985	1990	1995	2000	2005	2010	2012	2014	2015	2016	2017	2018
Kamafusa dam	–		0.60	0.50	0.63	0.61	0.59	0.49	0.52	0.57	0.40	0.46	0.40
L. Hachirogata	IV				1.1	1.1	1.0	1.5	0.89	0.99	1.1	1.2	1.2
L. Kasumigaura Nishiura	III		1.08	0.96	1.0	1.1	1.3	1.0	1.2	1.1	1.1	1.0	0.88
Kitaura	III		0.83	0.71	0.95	1.1	1.6	1.2	1.4	1.2	1.3	1.2	1.3
Hitachitone R.	III		0.85	0.85	0.95	1.0	1.1	0.91	1.1	0.89	0.92	0.86	0.96
L. Inbanuma	III	2.1	2.3	2.1	2.2	2.9	2.9	2.6	2.5	2.4	2.6	2.3	2.2
L. Teganuma	V	5.3	4.3	5.3	3.2	2.8	2.5	2.3	2.2	2.1	2.2	2.1	2.1
L. Suwa	IV	1.1	1.4	0.81	0.95	0.69	0.76	0.81	0.85	0.82	0.80	0.87	0.62
L. Nojiri	–		0.18	0.21	0.12	0.11	0.09	0.13	0.11	0.13	0.13	0.11	0.12
L. Biwa North Lake	II		0.31	0.32	0.29	0.30	0.24	0.26	0.25	0.24	0.23	0.21	0.20
South Lake	II	0.36	0.49	0.42	0.39	0.36	0.28	0.30	0.25	0.24	0.25	0.23	0.32
L. Nakaumi	III		0.56	0.49	0.61	0.42	0.46	0.50	0.45	0.40	0.44	0.41	0.41
L. Shinji	III		0.50	0.54	0.56	0.54	0.59	0.63	0.49	0.44	0.45	0.47	0.44
L. Kojima	V		1.8	2.0	1.6	1.3	1.2	1.2	1.1	1.1	1.2	1.5	1.2

* Type Standard Values (mg/L, less than) : I : 0.1, II : 0.2, III : 0.4, IV : 0.6, V : 1.0.
Sources: The Ministry of the environment: "Result Walter Quality of Public Water Areas", National Institute for Environmental Studies: Kankyosuchi Database.

Water Quality of Specified Lakes and Marshes in Japan (TP)

(Annual averages for all enrivonmental reference points, Units: mg/L)

Lake/wetlands	Type*	1985	1990	1995	2000	2005	2010	2012	2014	2015	2016	2017	2018
Kamafusa dam	–		0.015	0.014	0.015	0.019	0.019	0.015	0.017	0.022	0.017	0.018	0.018
L. Hachirogata	IV				0.067	0.082	0.074	0.10	0.066	0.075	0.066	0.073	0.072
L. Kasumigaura Nishiura	III		0.061	0.10	0.12	0.10	0.090	0.090	0.085	0.090	0.089	0.086	0.084
Kitaura	III		0.057	0.093	0.12	0.092	0.13	0.090	0.096	0.11	0.11	0.11	0.12
Hitachitone R.	III		0.061	0.082	0.080	0.093	0.10	0.080	0.092	0.090	0.082	0.088	0.093
L. Inbanuma	III	0.081	0.10	0.14	0.12	0.11	0.14	0.16	0.14	0.13	0.14	0.14	0.16
L. Teganuma	V	0.55	0.44	0.51	0.26	0.17	0.16	0.18	0.13	0.13	0.15	0.15	0.16
L. Suwa	IV	0.083	0.13	0.064	0.051	0.053	0.042	0.046	0.048	0.049	0.042	0.052	0.042
L. Nojiri	–		0.006	0.006	0.005	0.005	0.006	0.005	0.009	0.006	0.006	0.005	0.005
L. Biwa North Lake	II		0.010	0.008	0.006	0.007	0.007	0.007	0.008	0.007	0.008	0.006	0.005
South Lake	II	0.020	0.033	0.021	0.020	0.018	0.016	0.014	0.012	0.012	0.013	0.014	0.017
L. Nakaumi	III		0.054	0.049	0.063	0.039	0.045	0.056	0.038	0.035	0.041	0.040	0.040
L. Shinji	III		0.036	0.037	0.047	0.039	0.064	0.11	0.040	0.035	0.040	0.045	0.048
L. Kojima	V		0.23	0.20	0.19	0.19	0.19	0.19	0.16	0.17	0.18	0.18	0.16

* Type Standard Values (mg/L, less than) : I : 0.005, II : 0.01, III : 0.03, IV : 0.05, V : 0.10.
Sources: The Ministry of the environment: "Result Walter Quality of Public Water Areas", National Institute for Environmental Studies: Kankyosuchi Database.

Water Quality of Lakes and Marshes (TN)

Lake/wetlands	Country	1980	1985	1990	1995	2000	2005	2010
Ontario	Canada	0.48	0.54	0.54	-	(1.91)	1.92	1.86
Superior		-	0.42	1.48	-	(1.60)	1.68	(1.70)
Chapala	Mexico	0.21	0.59	0.15	(0.23)	0.20	(2.58)	1.92
Chairel		-	0.17	-	0.11	0.31	(1.08)	0.38
Chunchonho	Korea	-	1.05	0.60	1.53	1.43	1.26	1.29
Chungjuho		-	(1.44)	0.62	1.75	2.27	2.23	2.10
Taupo	New Zealand	-	-	-	0.09	0.07	0.07	0.08
Mondsee	Austria	0.48	0.56	0.62	0.57	-	-	-
Ossiacher See		-	0.30	0.33	0.47	-	-	-
Wallersee		-	-	-	1.10	-	-	-
Zeller See		-	-	-	0.28	-	-	-
Arreso	Denmark	-	4.08	3.50	3.39	2.61	2.62	1.63
Fureso		-	-	0.97	0.91	0.82	0.72	0.66
Pääjärvi	Finland	0.91	0.91	0.93	0.88	0.89	1.07	1.01
Pääjänne		0.54	0.56	0.62	0.50	0.78	1.01	0.83
Parentis-Biscarrosse	France	-	-	0.86	1.00	1.10	-	-
Lac d'Annecy		-	-	0.07	0.27	-	-	-
Bodensee	Germany	0.87	0.92	0.96	1.01	0.78	-	-
Balaton	Hungary	0.93	0.78	0.78	0.70	0.79	0.93	-
Velencei		-	-	3.63	1.83	1.66	2.10	-
Ennell	Ireland	-	-	0.26	0.05	0.07	1.59	-
Owel		-	-	-	0.01	-	1.05	-
Maggiore	Italy	0.91	0.90	0.99	0.82	0.82	-	-
Como		-	0.96	0.96	0.88	0.88	1.07	0.93
Garda		-	0.43	0.41	0.32	0.03	1.17	0.35
Orta		9.90	7.66	4.71	2.70	1.97	-	(1.72)
Remerschen	Luxembourg	-	-	0.29	0.02	(0.23)	(0.45)	-
Ijsselmeer	Netherlands	4.40	4.16	3.88	3.55	3.31	2.83	2.81
Ketermee		5.43	5.89	5.30	4.44	3.57	4.00	3.21
Mjoesa	Norway	0.41	0.45	0.38	0.49	0.49	1.34	-
Randsfjorden		0.51	-	0.54	0.54	0.39	0.48	-
Jasien Pólnocny	Poland	-	-	-	-	0.92	0.84	-
Wuksniki		-	-	-	-	0.57	0.74	-
Castero de Bode	Portugal	-	0.46	0.97	0.24	0.72	-	0.65
Valdecanas	Spain	-	-	2.12	0.37	(0.45)	-	3.63
Alcántara		2.86	-	0.91	0.11	(0.13)	-	0.73
Hjälmaren	Sweden	0.71	0.80	0.67	0.79	0.65	-	-
Mälaren		0.61	0.81	0.58	0.68	0.66	0.49	0.55
Vänern		0.87	0.92	0.79	0.80	0.82	0.78	0.60
Vättern		0.60	0.65	0.69	0.69	0.73	0.72	(0.65)
Lac Léman	Switzerland	0.66	0.73	0.69	0.67	0.68	-	-
Constance		0.93	1.05	1.19	1.21	0.97	-	0.97
Sapanca	Turkey	0.94	0.62	(0.37)	0.17	-	-	-
Altinapa		1.55	0.55	(1.81)	1.76	(1.64)	-	-
Lough Neagh	England	0.48	0.48	0.77	0.42	0.40	-	-
Lomond		0.30	0.29	0.13	0.39	(0.37)	-	-
Bewl Water		0.91	0.77	1.12	0.56	(1.29)	(3.35)	-

Average annual values, Unit: mg/L

* Values in ( ) are observed values from the year before or after.

Compiled based on "OECD Environmental Data Compendium" (2009).

Supplemented with OECD Environment Statistics/Lake and river quality (2016).

Water Quality of Lakes and Marshes (TP)

Lake/wetlands	Country	1980	1985	1990	1995	2000	2005	2010
Ontario	Canada	0.015	0.011	0.010	0.008	(0.008)	0.007	0.006
Superior		–	0.003	0.003	(0.003)	(0.003)	0.003	(0.002)
Chapala	Mexico	0.280	0.730	0.240	(0.320)	0.570	0.660	0.550
Chairel		–	0.010	0.040	0.020	0.120	0.080	0.060
Chunchonho	Korea	–	0.036	0.014	0.136	0.015	0.023	0.029
Chungjuho		–	(0.012)	0.044	0.023	0.025	0.022	0.017
Taupo	New Zealand	–	–	–	0.004	0.007	0.006	0.006
Mondsee	Austria	0.025	0.014	0.009	0.008	0.008	0.008	0.007
Ossiacher See		0.012	0.013	0.014	0.009	0.011	0.013	0.010
Wallersee		0.030	0.025	0.027	0.016	0.016	0.014	0.022
Zeller See		0.018	0.010	0.011	0.009	0.006	0.005	0.004
Arreso	Denmark	–	1.113	0.514	0.406	0.194	0.175	0.089
Fureso		–	–	0.169	0.174	0.097	0.063	0.087
Pääjärvi	Finland	0.026	0.030	0.027	0.025	0.024	0.028	0.026
Pääjänne		0.017	0.017	0.014	0.011	0.013	0.017	0.010
Parentis-Biscarrosse	France	–	–	0.084	0.091	0.086	–	0.373
Lac d'Annecy		–	–	0.010	0.008	0.006	–	–
Bodensee	Germany	–	0.038	0.021	0.017	0.011	(0.009)	0.007
Balaton	Hungary	0.010	0.020	0.036	0.076	0.086	0.041	–
Velencei		–	0.163	0.087	0.072	0.065	0.079	–
Ennell	Ireland	0.029	0.032	0.017	0.020	0.015	0.023	–
Owel		0.020	0.015	0.010	0.011	0.010	0.011	0.012
Sheelin		0.049	0.023	0.013	0.032	(0.019)	0.025	0.024
Maggiore	Italy	0.036	0.019	0.015	0.009	0.011	–	–
Como		0.078	0.052	0.047	0.038	0.039	0.024	0.027
Garda		0.020	0.011	0.015	0.017	0.018	0.024	0.017
Orta		0.004	0.006	0.004	0.004	0.005	(0.005)	–
Remerschen	Luxembourg	–	0.500	0.600	0.100	(0.030)	(0.036)	–
Ijsselmeer	Netherlands	0.350	0.286	0.177	0.139	0.141	0.111	0.045
Ketermee		0.480	0.480	0.247	0.190	0.153	0.144	0.151
Mjoesa	Norway	0.009	0.007	0.007	0.005	0.004	0.026	–
Randsfjorden		0.004	–	0.004	0.005	0.006	0.003	–
Jasien Pólnocny	Poland	–	–	–	–	0.047	0.061	–
Wuksniki		–	–	–	–	0.040	0.027	–
Castero de Bode	Portugal	0.150	0.110	0.022	0.012	0.035	(0.029)	(0.025)
Valdecanas	Spain	2.60	1.457	1.478	0.828	0.555	–	0.152
Alcántara		0.428	0.141	0.251	0.193	–	–	0.172
Hjälmaren	Sweden	0.042	0.044	0.046	0.062	0.051	–	–
Mälaren		0.028	0.024	0.025	0.021	0.024	0.018	0.023
Vänern		0.014	0.009	0.009	0.008	0.006	0.007	0.006
Vättern		0.009	0.007	0.007	0.006	0.003	0.005	(0.005)
Lac Léman	Switzerland	0.083	0.073	0.055	0.041	0.036	(0.030)	–
Constance		0.083	0.066	0.039	0.024	0.014	(0.010)	0.007
Sapanca	Turkey	0.030	0.030	0.030	0.040	–	–	–
Altinapa		0.020	0.150	0.110	0.110	–	–	–
Lough Neagh	England	0.108	0.115	0.096	0.120	0.145	–	(0.144)
Lomond		0.009	0.009	0.019	0.009	–	–	–
Bewl Water		–	0.023	0.081	0.030	–	–	0.048

Average annual values, Unit: mg/L

* Values in ( ) are observed values from the year before or after.

Compiled based on "OECD Environmental Data Compendium" (2009).

Supplemented with OECD Environment Statistics/Lake and river quality (2016).

Vertical Ecological Classification of Oceans, Seas, Lakes and Marshes

Oceans/Seas

Benthic environment

As listed below, further divided moving from the coast to the offing.

1. Supralittoral zone: Zone under the influence of waves and spray
2. Littoral zone: Coastal zone between the high-water line of high tide and the low-water line of low tide
3. Sublittoral zone: From the low-water line of low tide to a depth of approximately 200 m
4. Bathyal zone: From a depth of 200 m to 2 000 m
5. Abyssal zone: From a depth of 2 000 m to 6 000 m
6. Hadal zone: Deeper than 6 000 m

Pelagic environment

1. Epipelagic zone: From the sea surface to a depth of 200 m. Sea areas with a depth shallower than 200 m are called "Neritic zones." Sea areas with a depth deeper than 200 m are called "Oceanic zones."
2. Mesopelagic zone: From a depth of 200 m to 1 000 m
3. Bathypelagic zone: From a depth of 1 000 m to 4 000 m
4. Abyssopelagic zone: From a depth of 4 000 m to 6 000 m
5. Hadopelagic zone: Deeper than 6 000 m

Areas which are shallower than the photosynthesis compensation depth of plants (approximately 1% of the sea surface light intensity) are called Euphotic zones. Deeper areas are divided in Disphotic zones, which are the limit where plants can detect light, and Aphotic zones, where there is no light. Euphotic zones and Disphotic zones are referred to together as Photic zones.

Friendrich, H.: "Marine Biology", Sidgwick & Jackson (1965).

Lalli, C. M. & Parsons, T. R. (translated by Fumitake Seki): "Biological Oceanography", published by Kodansha Scientific (1996).

Lakes/Marshes

Benthic environment

Epilittoral zone: Zone which is not impacted by water.

Supralittoral zone: Zone which is about the water surface and is impacted by splashing water

Littoral zone—Further divided into the following 3 sub-classifications.

1. Eulittoral zone: Zone which is between the maximum water level and minimum water level depending on the season.
2. Infralittoral zone: Zone which is below the water surface for the entire year and in which there is an overgrowth of high flowering plants. Normally classified from top to bottom as emergent vegetation, floating vegetation, and submerged vegetation.
3. Littoriprofundal zone: A transition zone in which plants are scattered. In deep lakes and marshes, nearly equivalent to the depth of the thermocline.

Profundal zone: Deeper than the phytoplankton photosynthesis compensation depth. The Profundal zone does not exist in shallow lakes and marshes.

Pelagic environment

Limnetic zone: A deeper zone adjacent to the coastal zone. Ranges from the water surface to the phytoplankton photosynthesis compensation depth. The Limnetic zone does not exist in shallow lakes and marshes.

In lakes and marshes where thermocline occurs in the summer, the zone from the top of the thermocline to the water surface is called the "epilimnion", and the zone from the bottom of the thermocline to the bottom of the lake is called the "hypolimnion". The thermocline is also called the "metalimnion."

Hutchinson, G. E.: "A Treatise on Limnology", John Wiley & Sons Inc. (1967).

Yoshimura, S. "Limnology," Sanseido Co., Ltd (1937).

Size Classifications for Pelagic Organisms

Taxonomic groups of plankton (vertical) and classification according to size (horizontal)

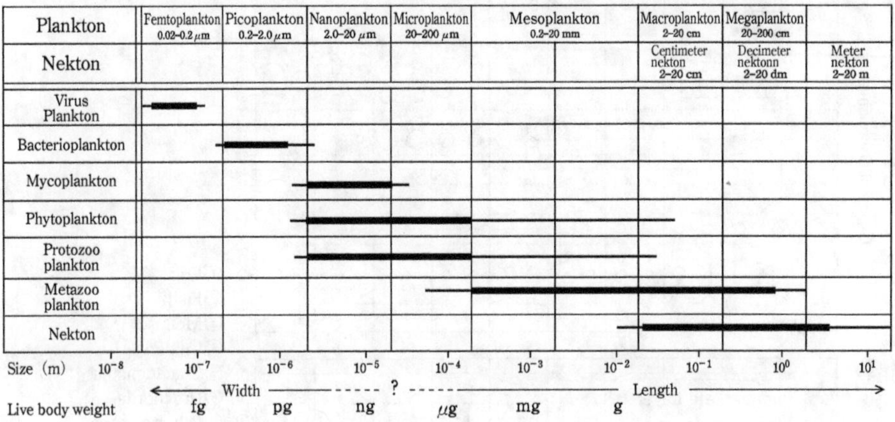

Sieburth, J. M., *et al.*: Pelagic ecosystem structure: Heterotrophic compartments of the plankton and their relationship to plankton size fractions. Limnol. Oceanogr. 23: 1256–1263 (1978).

Vertical distribution of total plankton biomass and its major toxonomic composition

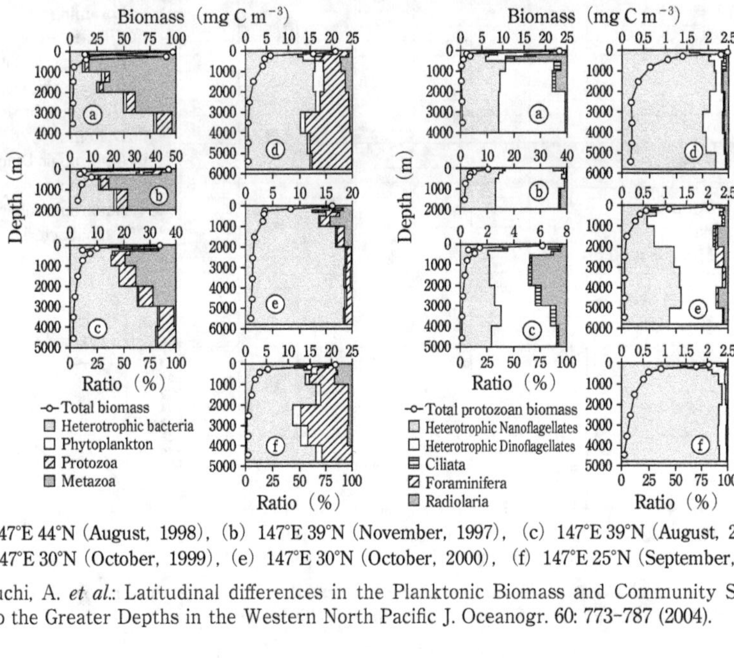

(a) 147°E 44°N (August, 1998), (b) 147°E 39°N (November, 1997), (c) 147°E 39°N (August, 2001), (d) 147°E 30°N (October, 1999), (e) 147°E 30°N (October, 2000), (f) 147°E 25°N (September, 1999)

Yamaguchi, A. *et al.*: Latitudinal differences in the Planktonic Biomass and Community Structure Down to the Greater Depths in the Western North Pacific J. Oceanogr. 60: 773–787 (2004).

Size and taxonomic abundance of plankton carbon biomass in surface, intermediate, and deep layers (November, 1997 in 147°E and 39°N)

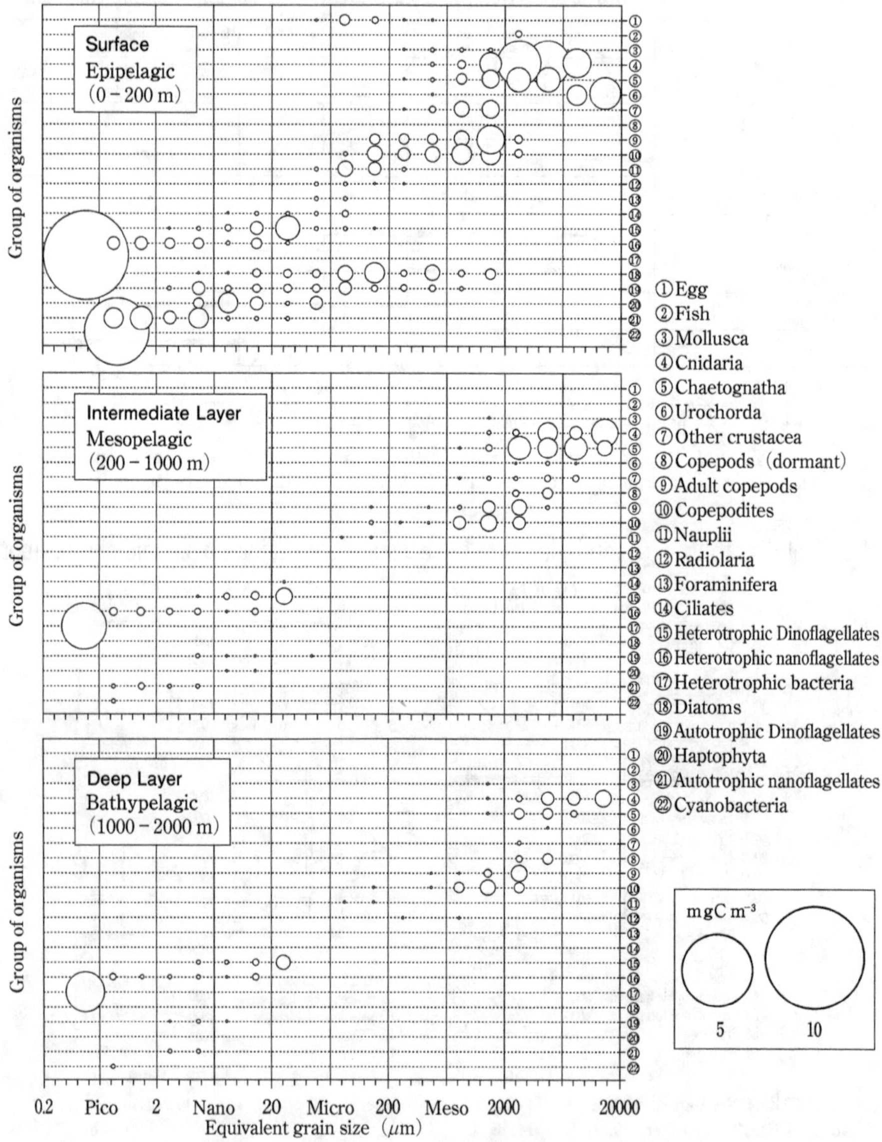

Yamaguchi, A. *et al.* "Vertical distribution of plankton community in the western North Pacific Ocean (WEST-COSMIC)" The Plankton Society of Japan, **47**: 144–156 (2000).

Characteristic Distribution of Organism Types in Fresh Water

Vertical distribution of phytoplankton in offings of lakes and marshes

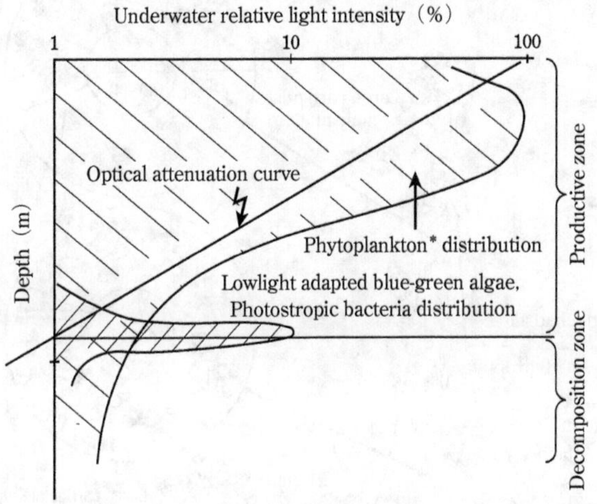

* Phytoplankton is the amount of chlorophyll *a*, displayed as a relative value with 100 as the maximum value for vertical distribution. Using a layer with relative light intensity of about 1% as the border, divided into the production zone (top layer) and decomposition zone (bottom layer).

Plant distribution in coastal areas of lakes/marshes

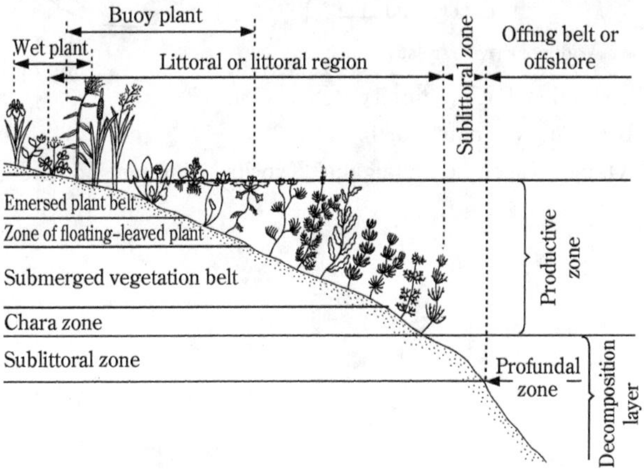

Geographical distribution of benthos and aquatic insects in rivers

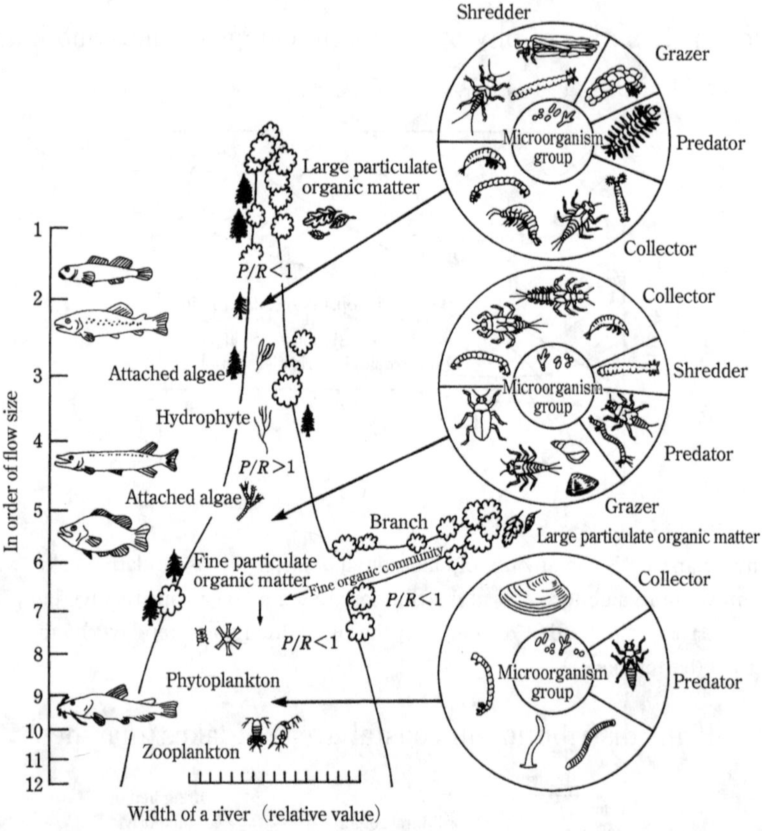

P: Gross Production *R*: Community Respiration
Compiled based on Vannote (1980).
Source: T. Okino "Kasen no Seitaigaku" Kyoritsu Shuppan Co., Ltd. (2002).

Amount of Fish Caught on Sea Surface by Year and Main Type of Fish*1

Species \ Era	1960	1965	1970	1975	1980	1985	1990	1995	2000	2005	2010	2015	2017	2018	2019
Tuna	390	430	291	311	378	391	293	332	286	239	208	190	169	165	163
Bluefin tuna	66	56	44	41	49	30	14	11	17	19	10	8	10	8	10
Skipjacks	94	167	232	274	377	339	325	336	369	399	331	264	227	260	242
Skipjack tuna	79	136	203	259	354	315	301	309	341	370	303	248	219	248	233
Salmon masses	147	146	118	159	123	203	223	282	179	246	180	140	72	96	60
Salmonids	86	79	79	110	100	173	209	257	154	229	165	136	69	84	56
Herrings	15	50	97	67	11	9	2	4	2	9	3	5	9	12	15
Spotlined sardine	78	9	17	526	2 198	3 866	3 678	661	150	28	70	311	500	522	535
Round herring	49	29	24	44	38	30	50	48	24	35	50	98	72	55	61
Anchovy	349	406	365	245	151	206	311	252	381	349	351	169	146	111	133
Agitations	596	560	269	235	145	225	331	385	282	214	185	167	165	135	113
Mackerals	351	669	1 302	1 318	1 301	773	273	470	346	620	492	530	518	542	445
Mackerel pike	287	231	93	222	187	246	308	274	216	234	207	116	84	129	46
Yellowtails	41	44	55	38	42	33	52	62	77	55	107	123	118	100	109
Olive flounder	6	7	7	7	7	8	6	8	8	6	8	8	7	7	7
Flatfish	503	209	288	341	282	206	72	76	71	54	49	41	47	41	40
Codfish	447	781	2 464	2 770	1 649	1 650	930	395	351	243	306	230	174	178	207
Alaska pollack	380	691	2 347	2 677	1 552	1 532	871	339	300	194	251	180	129	128	154
Sea breams	45	40	38	29	28	26	25	27	24	25	25	25	25	16	25
Prawns	62	68	56	69	51	53	43	36	29	24	19	16	17	15	13
Crabs	64	64	90	76	78	100	61	57	42	34	32	29	26	24	22
Shelfish	296	293	321	280	338	355	418	412	405	380	407	292	284	350	386
Littleneck clams	102	121	142	122	127	133	71	49	36	34	27	14	7	8	8
Scallop	14	6	16	30	83	118	230	275	304	287	327	234	236	305	339
Squids	542	499	519	538	687	531	565	547	624	330	267	167	103	84	75
Octopuses	58	78	96	74	46	40	55	52	47	55	42	33	35	36	35
Seaweed	286	253	212	231	183	184	208	151	119	105	97	94	70	79	67
Seaweeds	140	127	111	158	125	133	132	121	94	79	74	72	46	56	47

*1 Unit: 1000 t.
(Ministry of Agriculture, Forestry and Fisheries Statistics Department "Census of Fisheries" [Japanese Version])

Amount of Fish Caught on Rivers, Lakes and Marshes by Year and Main Type of Fish*

Species \ Era	1960	1970	1975	1980	1985	1990	1995	2000	2005	2010	2014	2015	2016	2017	2018
Salmons	1 465	1 750	6 603	9 207	10 694	15 284	17 736	12 326	16 269	12 580	10 212	12 330	7 471	5 802	6 696
Rainbow trout	259	604	698	725	504	552	597	536	328	…	…	…	…	…	…
Rockfish	156	202	192	259	271	412	413	496	369	…	…	…	…	…	…
Pond smelt	4 761	2 782	3 054	3 181	3 907	2 535	2 077	2 124	1 937	1 967	1 242	1 417	1 181	943	1 146
Sweetfish	6 860	9 879	13 951	14 723	14 492	17 795	13 700	11 172	7 149	3 422	2 395	2 407	2 390	2 168	2 140
Carp	2 205	4 043	6 699	8 479	7 830	6 302	4 896	4 079	1 484	401	258	227	220	213	210
Crucian	7 870	10 443	9 890	10 066	7 987	5 853	4 286	3 423	2 021	778	596	555	534	512	456
Eel	2 871	2 726	2 202	1 936	1 526	1 128	899	765	484	245	112	70	71	71	69
Corbicula	23 178	56 144	47 035	41 491	30 839	37 017	26 938	19 295	13 455	11 189	9 804	9 819	9 580	9 868	9 646
Prawns	979	3 277	6 840	5 846	4 783	3 305	2 717	1 676	1 035	676	409	372	360	364	409
Other types of salmon and trout										307	252	237	311	269	205

* Unit: t. (Ministry of Agriculture, Forestry and Fisheries Statistics Department "Census of Fisheries" [Japanese Version])
Note 1: The hauls published for 2000 and ealier are for all Japanese lakes and wetlands. The 2005 hauls are for the 106 major rivers and 24 major wetlands, and 2010–2012 hauls are for the 108 major rivers and 24 major wetlands.
Note 2: Starting from 2006, the river and wetland hauls are the amount of fish caught in commercial fishing only; the amounts caught by sports fishermen are not included. Also starting from 2006, "Rainbow Trout Rockfish" also includes "other varieties of salmon and trout."

Amount of Fish Caught by Year, River/Lake/Marsh, and Type of Fish*

Species	Era / Wate system	1970	1975	1980	1985	1990	1995	2000	2005	2010	2014	2015	2016	2017	2018
Salmon	Tokachi R.	623	779	436	600	480	608	563	1 136	x	x	x	x	x	x
	Ishikari R.	1	-	437	493	1 296	1 708	692	993	341	314	672	379	374	320
	Kitakami R.	15	28	50	146	146	174	187	251	338	318	207	232	261	191
	Abukuma R.	9	23	18	16	34	25	14	17	33	51	60	28	34	44
	Mogami R.	8	10	12	11	19	28	23	41	17	40	60	27	33	27
	Kuji (Iwate) R.	0	5	48	39	94	69	-	110	x	x	x	x	x	x
	Kuji (Fukushima, Ibaraki) R.	10	5	11	91	28	10	6	17	x	x	x	x	x	x
	Naka R.	44	9	33	84	73	112	73	212	173	104	99	57	47	40
	Tone R.	17	24	4	11	5	15	7	49	10	14	13	7	3	5
	Agano R.	17	34	33	19	17	21	32	61	59	123	134	81	120	44
	Shinano R.	27	44	83	50	90	70	57	88	63	68	114	74	95	40
	Jinzu R.	10	16	32	44	52	48	58	62	39	41	30	21	21	13
	Kuzuryu R.	0	1	0	4	4	5	-	-	-	—	-	—	—	—
Pond smelt	Ishikari R.	113	118	89	72	164	110	101	52	62	12	86	81	58	75
	Chikugo R.	-	6	20	12	12	14	7	5	4	6	6	6	4	4
	L. Abashiri	142	294	282	360	222	234	333	264	x	x	x	x	x	x
	L. Ogawara	252	532	771	467	534	638	656	512	x	x	x	x	x	x
	L. Hachirogata	302	468	440	382	83	249	242	300	x	x	x	x	x	x
	L. Kasumigaura	557	440	46	857	312	169	19	78	499	199	247	159	83	92
	L. Kitaura	216	129	353	234	151	68	32	108	21	43	26	18	9	6
	L. Inbanuma	-	42	18	14	5	5	3	2	x	x	x	x	x	x
	L. Suwa	307	400	384	100	86	90	46	20	x	x	x	x	x	x
	L. Shinji	405	34	115	180	290	5	2	1	x	x	x	x	x	x
Sweetfish	Naka R.	448	410	509	352	814	1 493	457	1 031	684	418	343	362	364	384
	Tone R.	1 062	857	1 157	857	801	398	249	181	7	0	0	0	0	0
	Sagami R.	133	131	278	405	457	501	490	268	318	363	372	380	373	360
	Jinzu R.	153	221	260	243	183	190	114	96	85	89	67	79	79	70
	Kuzuryu R.	190	353	506	529	571	253	253	195	31	15	15	11	11	11
	Tenryu R.	241	498	477	716	758	202	291	162	15	8	17	6	9	10
	Yahagi R.	72	108	165	312	271	204	102	76	1	2	2	2	2	2
	Kiso R.	220	259	320	277	430	430	277	222	43	36	35	33	29	24
	Ibi R.	206	210	240	188	168	140	131	59	40	19	2	2	2	1
	Kinokawa R.	139	364	406	543	578	305	301	156	1	1	2	2	3	2
	Yodo R.	157	211	218	344	396	308	234	173	14	11	4	4	4	4
	Gōnokawa R.	432	591	430	150	343	156	191	123	52	30	33	27	25	22
	Takahashi R.	171	272	125	193	249	145	132	46	19	14	11	9	9	9
	Yoshino R.	262	436	201	92	208	257	325	61	45	33	65	62	50	20
	Shimanto R.	176	1 603	846	877	926	373	266	222	20	16	25	16	27	16
	Kuma R.	110	222	450	478	213	317	432	229	x	x	x	x	x	x
	L. Biwa	678	891	1 345	965	1 832	1 258	953	390	681	408	476	461	279	336
Corbicula	Naka R.	1 518	3 434	2 111	2 012	1 614	2 067	1 269	861	1 047	533	271	433	591	555
	Tone R.	37 955	18 151	14 908	5 064	3 142	6 588	1 418	15	5	1	1	1	1	0
	Ibi R.	162	258	282	708	601	337	165	260	220	264	212	126	130	97
	Yoshino R.	215	153	227	156	118	164	185	76	44	17	14	12	9	17
	Chikugo R.	337	75	747	812	647	453	258	207	158	53	50	48	35	32
	L. Abashiri	403	397	443	518	671	782	732	803	x	x	x	x	x	x
	L. Jusan	2 969	1 296	1 079	1 371	1 747	2 363	2 747	1 642	x	x	x	x	x	x
	L. Ogawara	120	420	1 748	2 800	3 615	2 349	2 496	1 534	x	x	x	x	x	x
	L. Hachirogata	548	567	93	108	10 750	58	3	1	x	x	x	x	x	x
	L. Hinuma	1 365	2 719	2 660	2 390	2 376	1 183	605	412	x	x	x	x	x	x
	L. Kitaura	3 372	1 155	458	106	-	-	-	-	-	—	-	—	—	—
	L. Biwa	1 725	992	700	313	211	113	80	161	41	43	36	51	53	58
	L. Shinji	4 191	15 597	14 300	12 320	9 100	8 400	7 500	6 100	x	x	x	x	x	x

Cultivated yield is excluded. * Unit: t. (Ministry of Agriculture, Forestry and Fisheries Statistics Department "Census of Fisheries" [Japanese Version])
"x" indicates that the statistics are not made public to protect the confidential information of indivdiuals, corporations, or other groups.

Breeding Seasons for Main Experimental Marine Product Invertebrate in Japan

Observation region / Species names	Akkeshi[1] (Hokkaido)	Asamushi[2] (Aomori)	Sado (Niigata)	Noto[3] (Ishikawa)	Tateyama[4] (Chiba)	Misaki[5] (Kanagawa)	Shimoda (Shizuoka)	Sugashima (Mie)	Shirahama[6] (Wakayama)	Oki (shimane)	Ushimado[7] (Okayama)	Mukojima (Hiroshima)	Usa[8] (Kochi)	Nakajima (Ehime)	Aizu[9] (Kumamoto)	Amakusa (Kumamoto)	Sesoko[10] (Okinawa)
Purple sea urchin (*Heliocidaris crassispina*)			Early July–Late Aug.	Late June–Late Aug.	Early June–Late Aug.	June–July	Early June–Early Oct.	Early July–Late Aug.	Early June–Late Sep.	Mid June–Mid Sep.	June	Early July–Mid Sep.	Early Aug.–Mid Sep.	Early June–Mid Aug.	Late June–Mid July	Late May–Late Aug.	
Northern sea urchin (*Mesocentrotus nudus*)		Late Sep.–Late Oct.	Mid Sep.–Late Oct.		Early Sep.–Mid Oct.	Oct.–Dec.											
Elegant sea urchin (*Hemicentrotus pulcherrimus*)		Mid Dec.–Early Apr.	Early Jan.–Mid Apr.	Early Jan.–Mid Mar.	Late Jan.–Late Mar.	Jan.–Mar.	Late Dec.–Early May	Late Dec.–Mid Mar.	Early Jan.–Early Apr.	Early Jan.–Mid Apr.	Jan.–Mar.	Early Jan.–Late Mar.		Mid Jan.–Mid Apr.	Early Jan.–Mid Mar.	Early Jan.–Late Mar.	
Red sea urchin (*Pseudocentrotus depressus*)				Late Nov.–Mid Dec.	Late Oct.–Late Dec.	Oct.–Dec.	Late Oct.–Late Jan.	Late Oct.–Late Dec.	Mid Oct.–Late Dec.	Late Nov.–Mid Feb.	Early Oct.–Early Dec.			Early Oct.–Mid Dec.			
– (*Temnopleurus toreumaticus*)		Late June–Late Aug.		Early June–Late Aug.	Late June–Early Sep.			Late June–Mid Sep.	Mid June–Early Oct.		July–Aug.	Late May–Early Aug.	Late May–Early July		Late June–Mid July	Mid June–Late Aug.	
Flower sea urchin (*toxopneustes pileolus*)					Early June–Early Aug.				Early July–Late Aug.				Late June–Late July				Mid Sep.–Early Jan.
Cake urchin (*Clypeaster japonicus*)				Late June–Early July	Mid June–Early Aug.	May–July	Late May–Mid Sep.		Early June–Early Sep.				Mid May–Late June				
Sand dollar (*Astriclypeus mannii*)					Early Aug.–Late Aug.	July–Aug.								Late Apr.–Late June			
Northern Pacific seastar (*Asterias amurensis*)	Early May–Early July	Mid Mar.–Mid Apr.			Early Dec.–Mid Apr.	Jan.–Mar.					Feb.–Mar.			Mid Dec.–Late Mar.		Early June–Late July	
Bat star (*Patiria pectinifera*)		Sep.		Early May–Late Oct.	Early May–Mid July	Apr.–June		Early May–Late Sep.		Early May–Late July	Dec.			Early June–Mid Sep.			
Japanese oyster (*Magallana gigas*)	Early Aug.–Late Oct.	Early July–Mid Aug.	Late June–Late Aug.			June–Aug.	Mid May–Mid Oct.	Late June–Early Sep.				Early July–Late Sep.			Late July–Aug.		
Japanese spiny oyster (*Saccostrea kegaki*)				Late May–Late Sep.	Early July–Late Aug.	July–Aug.	Early May–1Early Jan.	Mid June–Early Sep.	Early July–Mid Aug.		Mid June–Early Aug.	Early July–Late Sep.				Mid June–Late Aug.	
Blue Mussel (*Mytilus galloprovincialis*)		Early Mar.–Early Apr.			Early Dec.–Late Mar.	Dec.–Apr.						Early Dec.–Late Mar.					
Sea squir (*Styela plicata*)		Apr., Mid Nov.–Early Dec.	Late July–Late Aug.				Early July–Late Sep.	Mid June–Late Aug.	Mid Mar.–Late Nov.							Early June–Late Aug.	

1) *Strongylocentrotus intermedius* (Mid May–Mid August), 2) *Strongylocentrotus intermedius* (Mid March–Mid April), *Temnopleurus hardwickii* (Mid June–Early August), *Glyptocidaris crenularis* (March), *Echinocardium cordatum* (Late May–Mid June), *A. manii* (Mid September–Early October), *Apostichopus japonicus* (Mid May–Mid June), *Distolasterias nipon* (Mid March–Early April), *Cinoa savignyi* (Mid May–Mid July, Mid October–1Mid January), *Halocynthia roretzi* (Late October–Late December), *Mizuhopecten yessoensis* (Mid February–Mid March), *Mytilisepta virgata* (August), *Chlamys farreri nipponensis* (Late July–Late August), *Aplysia kurodai* (July, November), *Aurelia aurita* (Early July–Late August), *Spirocodon saltatrix* (May), 3) *Ophioplocus japonicus* (Mid July–Mid August), *Peronella japonica* (Mid June–Mid July), *Brissus agassizii* (Mid June–Mid July), *Balanoglossus misakiensis* (Mid July–Late August), 4) *Astropecten polyacanthus* (Late June–Late July), *A. japonicus* (Early February–Mid March), *Holothuria moebii* (Early July–Mid August), *Halichondria okadai* (Late July–Early September), *Carybdea brevipedalia* (Mid August–Early September), 5) *Antedon serrata* (January–June), *Anneissia japonica* (October–November), *P. japonica* (July–August), 6) *Echinometra* sp. (June–October), *Mespilia globulus* (July–August), *Holothuria leucospilota* (July–August), *Cinoa savignyi* (July), *Spirobranchus kraussii* (July–August), 7) *Cinoa intestinalis* (June–July), *Scaphechinus mirabilis* (November), *A. japonicus* (April–May), *Eupentacta quinquesemita* (December), *Ophiothrix exiqua* (June), *Amphioctopus fangsiao* (February–March), *Octopus vulgaris* (March–April, August–September), *Sepiella japonica* (May–June), *Sepia esculenta* (May–June), *Sepia lycidas* (June–July), *Sepioteuthis lessoniana* (June), *Cellana nigrolincata* (July–September), *Aplysia kurodai* (June), *Dendrodoris rubra* (June–July), *Pleurobranchaea japonica* (June), *Balanus amphitrite* (May–September), *Siphonosoma cumanense* (June–August), *Phascolosoma scolops* (June–August), *Planocera multitentaculata* (May–June), *Notoplana humilis* (February–April), *Bolinopsis mikado* (August), *Chrysaora pacifica* (May–June), *Cyanea nozakii* (August–September), *A. aurita* (June–July), *Nemopsis dofleini* (April–May), *S. saltatrix* (March–April), 8) *Echinometra mathaei* (Early July–Late July), 9) *Chaetopterus cautus* (August), *Herdmania momus* (Mid July–Mid August), *Branchiostoma japonicum* (Mid June–arly July), *Austruca lactea* (Early July, Early August), 10) *Austruca perplexa* (Early May–Late August), *Tripneustes gratilla* (Mid September–Early January), *E. mathaei* (Early July–Late August), *Diadema setosum* (Mid June–Mid August), *Acanthopleura loochooama* (August–Early September), *Acanthopleura gemmata* (August–Early September), *Acanthopleum tenuispinosa* (August–Early September)

(Based on Michibata *et al.*, 2001, with modifications.)

Invertebrate Spawning Season

Family name (Scientific name)	Habitat	Observation site	Sexual maturity	Full length	Oviposition period	Number spawned
Asteriidae (*Asterias amurensis*)	Japanese coast	Mutsu Bay	-	mm 55	Mar.-May	-
Asteriidae (*Asterias forbesi*)	Mexico-Maine (USA)	Long Island (USA)	1-2 y	60-210	July-Oct.	Thousands
Asteriidae (*Asterias rubens*)	North Atlantic coasts	Plymouth (United Kingdom)	-	50-130	Apr.-May	Thousands
Asteriidae (*Asterias vulgaris*)	Cape Cod (USA)-Labrador Peninsula (Canada)	Prince Edward (Canada)	1 y	20	May-June	-
Arbaciidae (*Arbacia punctulata*)	Gulf of Mexico-Cape Cod (USA)	Woods Hole (USA)	1 y	30-50	Early summer	Tens of thousands
Ishnochitonidae (*Ischnochiton magdalenensis*)	Mexico-California (USA)	-	2 y	35-36	-	57 970
Haliotidae (*Haliotis rufescens*)	California (USA)	-	6 y	100	Feb.-Apr.	100 000-2 500 000
Littorinidae (*Littorina littorea*)	New Jersey (USA)-Labrador Peninsula (Canada)	Nova Scotia (Canada)	1 y	20-25	May-Aug.	5 000
Muricidae (*Urosalpinx cinerea*)	Georgia (USA)-Canada	North Carolina (USA)	15 mo	16.5-29.6	May-Sep.	252
Melongenidae (*Busycon carica*)	Mexico-Cape Cod	North Carolina (USA)	-	30-210	Mar.-June, Aug.-Sep.	4 000 -6 000
Lymnaeidae (*Lymnaea stagnalis*)	Worldwide	Wisconsin (USA)	4-14 mo	50-60	July-Oct.	6 000
Helicidae (*Helix pomatia*)	USA and Europe	-	33-39 mo	7	May-Aug.	40-200
Mytilidae (*Mytilus edulis*)	Worldwide	Nova Scotia (Canada)	1-2 y	45-55	June-Sep.	5 000 000 -12 000 000
Ostreidae (*Crassostrea virginica*)	Texas (USA)-Canada	Long Island (USA)	1 y	25-50	June-Aug.	500 000 -1 000 000
Veneridae (*Mercenaria mercenaria*)	Yucatan Peninsula (Mexico)-Nova Scotia (Canada)	New Jersey (USA)	1-2 y	50-70	June-Aug.	About 1 000 000
Mactridae (*Spisula solidissima*)	Cape Hatteras (USA)-Canada	New Jersey (USA)	1.5-2 y	40-70	July, Oct.	Thousands
Myidae (*Mya arenaria*)	Cape Hatteras (USA)-Labrador peninsula (Canada)	Maine (USA)	1-2 y	13-19	June-Oct.	1 000 000 -4 000 000
Loliginidae (*Loligo pealei*)	Caribbean-Nova Scotia (Canada)	Woods Hole (USA)	1-2 y	150-200	June-Sep.	5 000 -12 000
Limulidae (*Limulus polyphemus*)	Yucatan Peninsula (Mexico)-Nova Scotia (Canada)	Delaware Bay (USA)	9-11 y	♂178-258 ♀243-251	May-June	3 000
Daphniidae (*Daphnia longispina*)	North America, Europe, Asia	Florida (USA)	75-86 h	♂1.2 ♀1-9	Excluding winter	28(4-35)
Penaeidae (*Penaeus setiferus*)	Gulf of Mexico-New Jersey (USA)	Georgia (USA)	1 y	♂130-170 ♀135-190	Mar.-Sep.	500 000 -1 000 000
Nephropidae (*Homarus americanus*)	North Carolina (USA)-Newfoundland (Canada)	In North Atlantic	4-5 y	♂170-600 ♀180-400	July-Sep.	8 500
Lithodidae (*Paralithodes camtschatica*)	North Pacific Ocean	Kojak Island (Alaska)	5 y	100	Mar.-May	150 000 -400 000
Portunidae (*Callinectes sapidus*)	Uruguay-Nova Scotia (Canada)	Chesapeake Bay (USA)	13 mo	♂135-215 ♀134-185	July-Aug.	1 750 000

(Based on Merrill, A. S. *et al.*, 1972, with modifications.)

Land Environments

Distribution of the world's biomes

A biome is a group of organisms and their habitats formed mainly by climate and topography. There are many different classifications of biomes, but they all share one thing in common: climatic conditions are the most important.

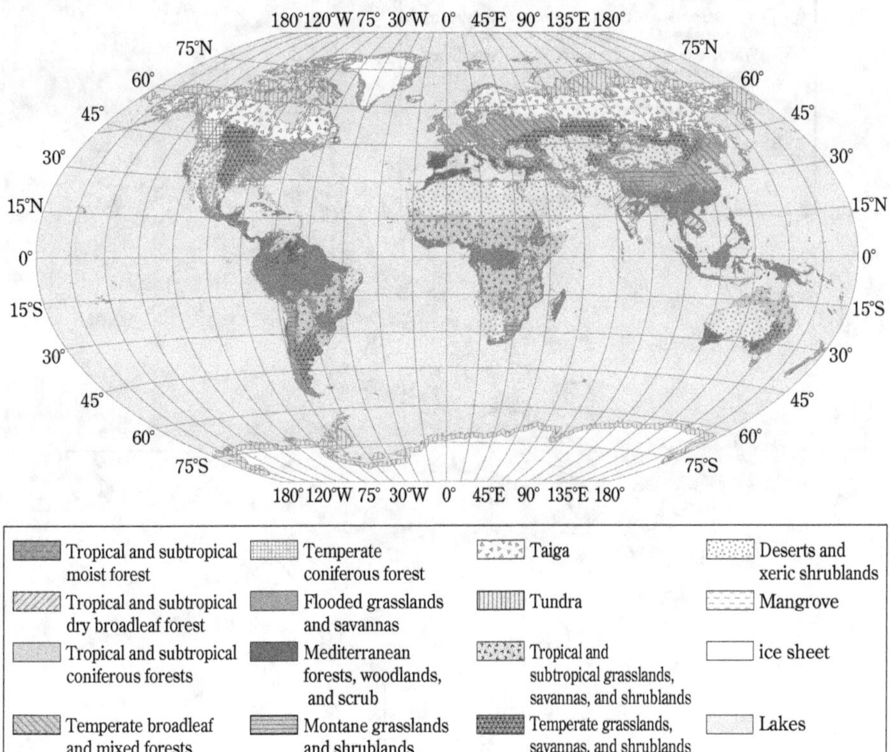

WWF: "Terrestrial ecoregion of the world" (2012)

Vegetation Distribution in Japan

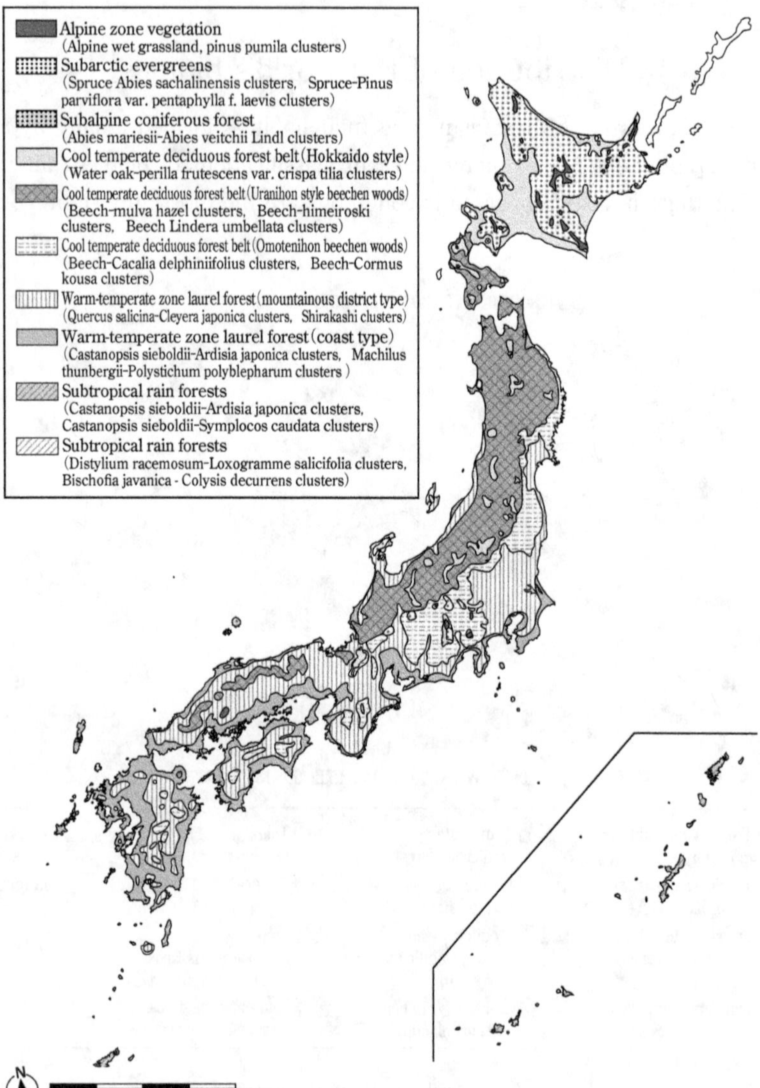

Source: Miyawaki (1975)

W. Osawa, *et al.* "Kiso seibutsu handbook" p. 134, Iwanami Shoten, Publishers (1980).

Number of Living Species

Currently, there are approximately 1.9 million known (scientific name has been assigned) living species on earth. About half of these living species are insects. However, there are still many species on the earth which are unknown to mankind. Although there are various estimations for the total number of living species on earth, the majority of estimates range from 10 million to 100 million. The estimates listed in the following table are somewhat conservative.

Living species numbers of known species and presumed numbers of species

Groups of organisms	Known species	Presumed species*
Animal	1 424 153	6 836 330
Fungi	98 998	1 500 000
Plant	310 129	390 800
Protist	53 915	1 200 500
Prokaryote	10 307	1 400 000
Total	1 897 502	11 327 630

Invertebrate numbers of known species and presumed number of species

Invertebrate	Known species	Presumed species*
Hemichordates	108	110
Echinoderm	7 003	14 000
Insect kind	1 000 000	5 000 000
Arachnidae	102 248	600 000
Pycnogonids	1 340	-
Myriapod	16 072	90 000
Crustacean	47 000	150 000
Peripatus	165	220
Hexapoda	9 048	52 000
Molluscous	85 000	200 000
Annelid	16 763	30 000
Roundworm	25 000	500 000
Acanthocephala	1 150	1 500
Flatworm	20 000	80 000
Cnidarian	9 795	-
Poriferan	6 000	18 000
Others	12 673	20 000
Total	1 359 365	6 755 830

Plant numbers of known species and presumed numbers of species

Plant	Known species	Presumed species*
Moss plant	16 236	22 750
Pteridophyte	12 000	15 000
Gymnosperm	1 021	1 050
Angiosperm	268 600	352 000
Green alga and red seaweed	12 272	-
Total	310 129	390 800

* Number excluding unknown organism groups.

Chordata numbers of known species and presumed numbers of species

Chordata	Known species	Presumed species*
Mammal	5 487	5 500
Birds	9 990	10 000
Reptiles	8 734	10 000
Amphibians	6 515	15 000
Fishes	31 153	40 000
Jawless vertebrate	116	Uncertainty
Lancelets	33	Uncertainty
Urochord animal	2 760	Uncertainty
Total	64 788	80 500

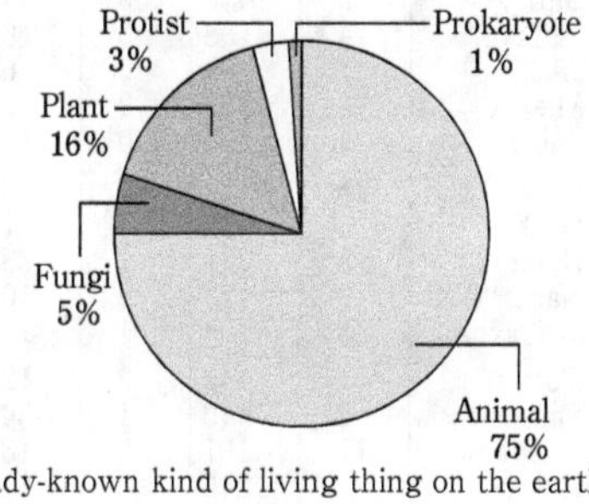

Already-known kind of living thing on the earth

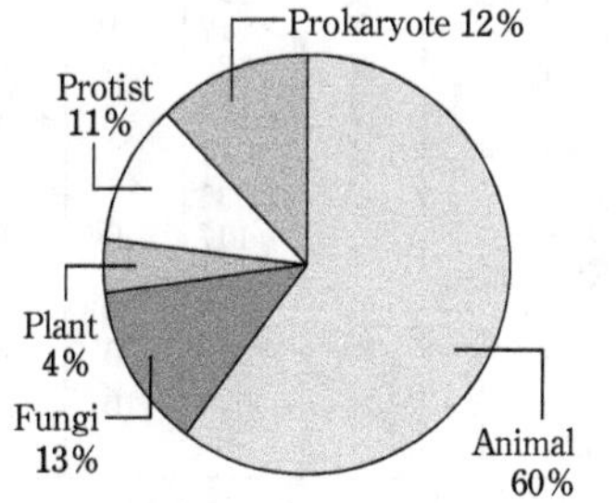

Presumed numbers of living species in the world

Chapman 2009, "Numbers of Living Species in Australia and the World", 2nd edition.

Area of Broad-Leaved Forests by Year and Region in Japan*

Region	1960		1970		1980		1990		2000		2005	
	Planted forest	Natural forest	Planted forest	Natural forest	Planted forest	Natural forest	Planted forest	Natural forest	Planted forest	Natural forest	Planted forest	Natural forest
Hokkaido	27	3 499	27	3 240	30	3 155	43	3 052	47	2 767	41	2 757
Aomori	1	346	4	320	1	358	1	268	2	261	2	263
Iwate	6	753	6	705	3	653	3	548	5	542	5	541
Miyagi	7	287	3	244	2	208	2	189	3	185	3	181
Akita	5	475	4	456	4	424	4	383	9	381	11	381
Yamagata	1	511	1	477	1	436	1	425	2	424	3	421
Fukushima	14	673	7	615	6	543	11	516	9	511	16	509
Ibaraki	15	58	7	60	2	57	2	62	2	61	2	61
Tochigi	16	226	1	192	3	161	3	150	2	145	4	145
Gunma	6	233	5	206	5	193	8	185	7	184	10	184
Saitama	5	69	0[1)]	61	0[8)]	56	0[12)]	52	0[14)]	51	0[16)]	45
Chiba	13	51	2	63	1	64	1	70	1	77	1	75
Tokyo	8	38	8	37	3	38	2	38	2	37	2	37
Kanagawa	8	50	3	50	1	45	1	52	1	50	1	49
Niigata	1	624	2	574	3	549	4	533	3	547	4	546
Toyama	1	173	0[2)]	177	1	164	1	158	1	154	1	153
Ishikawa	2	193	1	184	2	167	2	152	2	147	1	131
Fukui	2	216	1	213	1	182	1	179	1	170	2	169
Yamanashi	7	180	1	159	1	136	3	130	4	126	4	128
Nagano	3	417	2	406	1	373	1	362	2	354	2	360
Gifu	2	482	3	447	4	392	5	365	4	356	11	348
Shizuoka	14	188	3	160	4	146	4	142	5	147	5	146
Aichi	1	69	0[3)]	62	0[9)]	56	1	55	1	58	1	57
Mie	5	157	2	134	1	119	1	120	2	120	2	119
Shiga	1	109	0[4)]	104	0[10)]	87	0[13)]	77	1	76	1	75
Kyoto	5	172	1	161	0[11)]	148	1	139	0[15)]	136	1	133
Osaka	3	14	4	13	2	12	2	13	2	13	2	13
Hyogo	5	271	2	252	3	218	4	214	4	214	4	221
Nara	9	112	5	106	3	97	3	90	2	93	2	92
Wakayama	3	163	1	143	2	121	2	127	1	126	1	126
Tottori	1	143	2	123	1	105	1	97	2	97	3	98
Shimane	2	340	1	340	1	293	1	265	2	265	3	265
Okayama	11	197	2	187	2	167	3	159	3	159	3	158
Hiroshima	3	246	0[5)]	215	1	202	2	191	4	213	5	212
Yamaguchi	3	171	1	177	1	159	2	161	3	167	4	166
Tokushima	3	133	1	120	1	100	3	95	3	99	3	99
Kagawa	2	17	1	20	1	20	1	29	1	35	1	37
Ehime	21	130	0[6)]	114	1	95	3	112	3	113	4	113
Kochi	1	288	1	238	4	168	7	176	7	180	9	179
Fukuoka	8	62	4	52	4	49	4	51	4	51	4	50
Saga	3	35	1	30	1	26	1	27	1	26	1	26
Nagasaki	3	147	0[7)]	137	1	123	1	120	2	123	2	122
Kumamoto	3	184	3	154	6	137	9	143	9	141	9	142
Oita	25	155	4	163	8	155	11	156	11	165	12	163
Miyazaki	18	308	16	252	16	214	22	198	23	205	25	201
Kagoshima	13	275	10	242	8	225	13	219	15	230	17	231
Okinawa	-	-	-	-	4	76	4	74	4	71	5	72

* Unit: 1000 ha (Source: Ministry of Agriculture, Forestry and Fisheries "REPORT ON RESULTS OF 2010 WORLD CENSUS OF AGRICULTGURE AND FORESTRY IN JAPAN")

1) 45 ha, 2) 367 ha, 3) 192 ha, 4) 396 ha, 5) 413 ha, 6) 245 ha, 7) 435 ha, 8) 178 ha, 9) 425 ha, 10) 312 ha, 11) 191 ha, 12) 206 ha, 13) 400 ha, 14) 182 ha, 15) 398 ha, 16) 218 ha

Area of Coniferous Forests by Year and Region in Japan*

Region	1960		1970		1980		1990		2000		2005	
	Planted forest	Natural forest	Planted forest	Natural forest	Planted forest	Natural forest	Planted forest	Natural forest	Planted forest	Natural forest	Planted forest	Natural forest
Hokkaido	528	1 104	877	964	1 341	655	1 467	677	1 477	891	1 465	823
Aomori	140	117	175	103	240	25	266	81	269	82	269	78
Iwate	190	72	291	69	424	50	490	70	500	67	499	66
Miyagi	109	17	148	15	187	10	200	15	199	14	198	14
Akita	209	52	273	34	354	28	400	26	400	24	398	24
Yamagata	112	11	139	13	171	18	180	19	181	17	180	19
Fukushima	170	59	234	71	307	68	329	66	334	65	329	65
Ibaraki	121	14	134	11	137	7	119	3	115	3	114	2
Tochigi	104	32	133	41	147	39	154	36	155	33	153	33
Gunma	122	40	153	29	173	28	177	28	176	28	172	28
Saitama	44	15	54	14	58	10	60	10	60	9	59	9
Chiba	92	3	98	0[1)]	85	0[2)]	78	1	62	0[3)]	61	0[4)]
Tokyo	27	2	32	3	33	2	33	2	33	2	33	2
Kanagawa	36	2	40	1	37	1	35	1	36	1	35	1
Niigata	103	24	119	24	141	24	158	22	161	17	160	17
Toyama	28	19	33	15	44	16	49	16	52	17	52	17
Ishikawa	51	29	64	22	81	21	94	19	100	18	100	33
Fukui	56	20	66	15	107	10	115	10	123	9	123	9
Yamanashi	84	58	107	49	143	46	148	45	150	45	149	45
Nagano	289	240	349	214	419	192	436	184	442	188	442	180
Gifu	195	141	257	114	329	103	361	95	371	93	376	88
Shizuoka	236	48	271	43	283	42	283	41	279	38	277	38
Aichi	112	39	135	27	141	22	144	18	141	15	140	14
Mie	179	32	218	22	234	18	234	16	231	14	230	14
Shiga	38	57	46	51	67	45	79	40	82	38	82	37
Kyoto	69	86	89	81	111	76	123	71	128	67	128	65
Osaka	27	19	25	21	25	17	25	14	26	14	26	13
Hyogo	136	145	173	139	213	125	231	106	237	92	235	85
Nara	128	31	145	24	161	18	171	14	171	13	170	13
Wakayama	161	28	191	20	217	20	220	11	220	10	219	10
Tottori	63	27	93	23	125	20	135	17	137	13	137	11
Shimane	83	56	108	44	164	45	194	42	203	35	203	35
Okayama	87	148	118	146	165	135	190	120	194	112	194	111
Hiroshima	74	271	122	259	158	244	182	230	190	192	193	189
Yamaguchi	81	151	127	106	168	88	183	71	188	57	190	56
Tokushima	126	36	151	26	182	22	190	17	191	13	190	13
Kagawa	24	48	23	43	26	38	27	26	26	19	26	17
Ehime	180	55	207	62	232	58	243	27	244	25	243	25
Kochi	203	48	289	32	381	24	379	20	381	16	380	16
Fukuoka	136	12	138	14	142	9	141	6	139	5	138	5
Saga	57	4	62	3	68	1	71	1	72	1	72	1
Nagasaki	67	13	75	10	95	5	103	4	103	3	103	3
Kumamoto	201	5	232	10	267	7	276	7	277	6	274	6
Oita	174	22	211	17	229	14	235	11	227	7	228	7
Miyazaki	196	26	265	22	326	14	340	13	335	9	331	11
Kagoshima	198	37	252	27	287	26	294	22	292	20	285	22
Okinawa	-	-	-	-	8	13	8	12	8	11	8	12

* Unit: 1000 ha (Source: Ministry of Agriculture, Forestry and Fisheries "REPORT ON RESULTS OF 2010 WORLD CENSUS OF AGRICULTGURE AND FORESTRY IN JAPAN")
1) 187 ha, 2) 211 ha, 3) 187 ha, 4) 195 ha

Important Wetlands in Japan

Primary Wetland Type (●)	Number of Locations
High-Moor	67
Intermediate-Moor	41
Low-Moor	55
Snow-covered grass lands	22
River	128
Freshwater lake	72
Brackish water lake	31
Brackish water area	39
Mudflat	105
Salt marsh	27
Seaweed bed	119
sandy beache	20
Shallow sea area	27
Coral reef	30
Mangrove wetland	29
Paddy field	34
Fallow paddy field	12
Reservoir	62
Waterway	27
Spring water	55
Quellmoor	25
Other wetlands	102
Total	1129

* Because there are some wetlands with combination types, the total is not exactly 633.

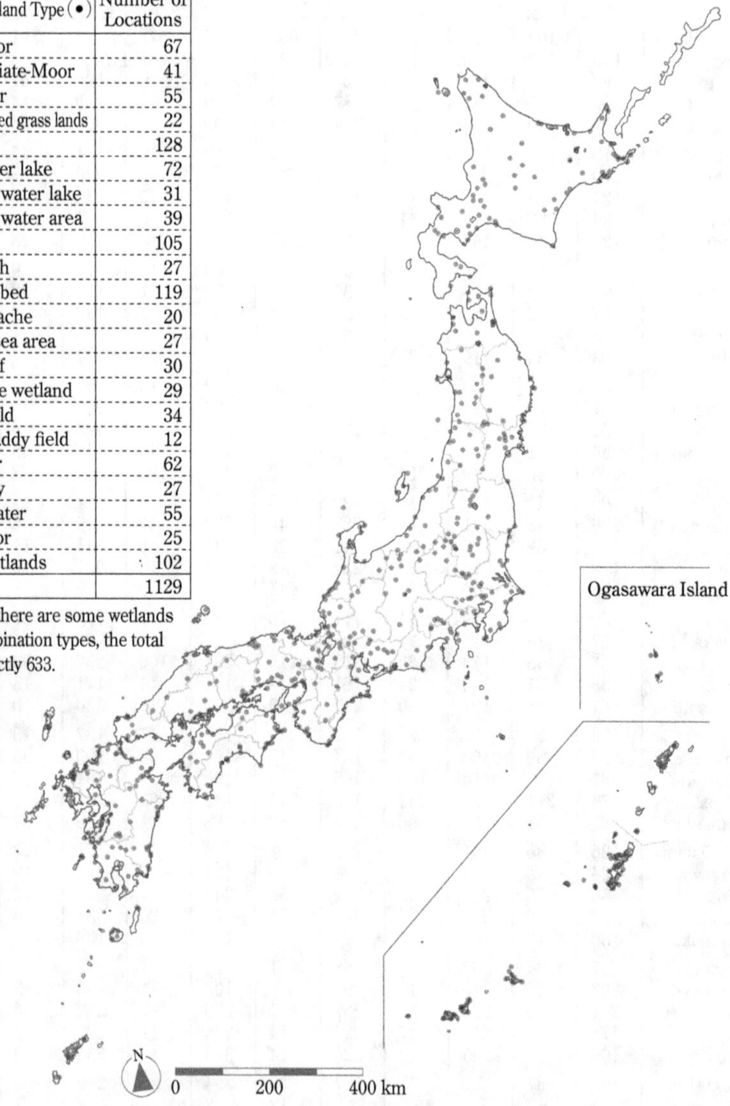

Minsitry of the Environment, Nature Conservation Bureau, "Nihon no juyoshicchi" (2016). http://www.env.go.jp/nature/import_wetland/index.htm

Observed Number of Swans/Geese/Ducks and Sandpipers/Plovers

The maximum number of individuals per species from 2004 to 2008 at each of 1000 observation sites monitored by the Ministry of the Environment (MOE) was extracted and 20 sites from the top sites are displayed in a bar graph. This graph shows species for which the total observed number of individuals was at least 10 000 (excluding whooper swans). As reference, the graph shows the 1% reference value which is used in registration Criterion 6 of the Ramsar Convention. The broken line graph shows the increase/decrease in the total number of observed individuals when 2005 is set as "1."

Whooper swan

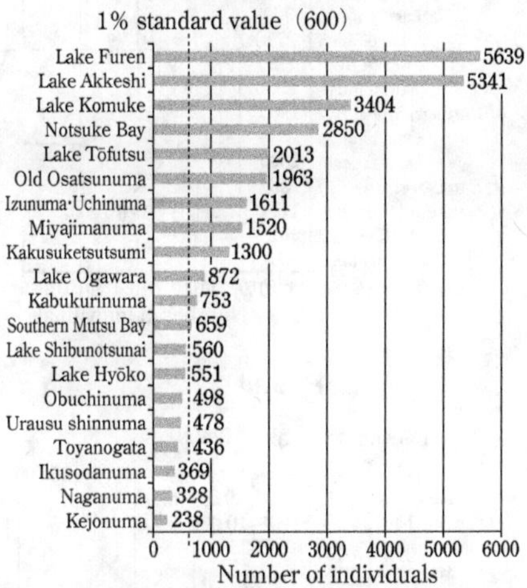

Tundra swan

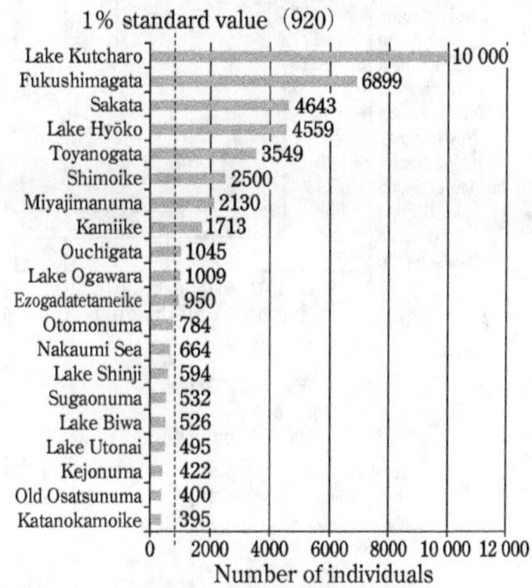

White-fronted goose

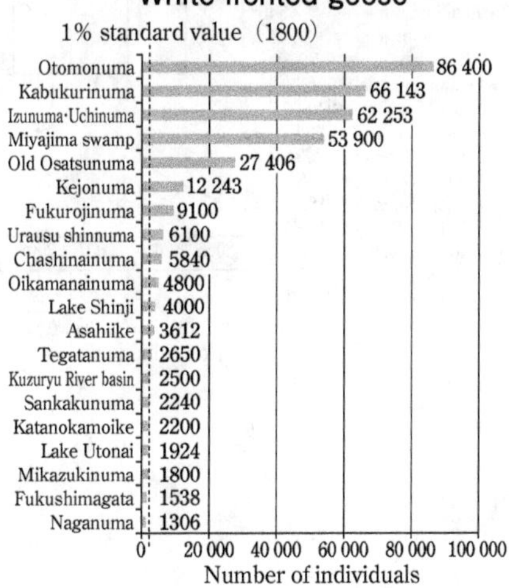

Mallard

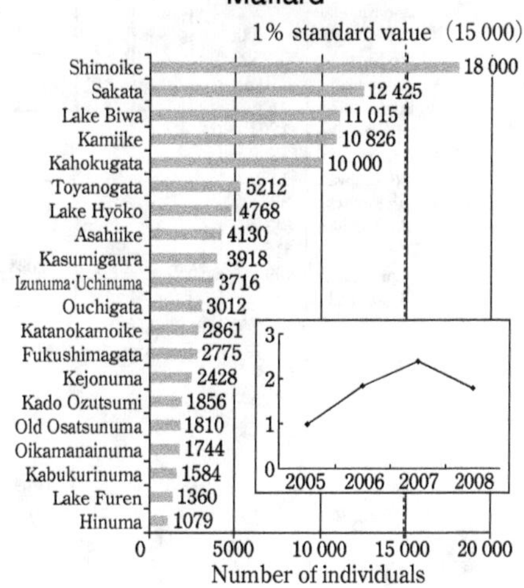

Teal

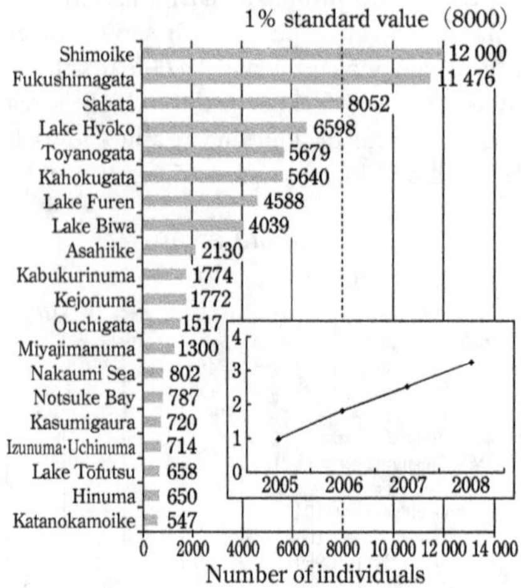

Scaup duck

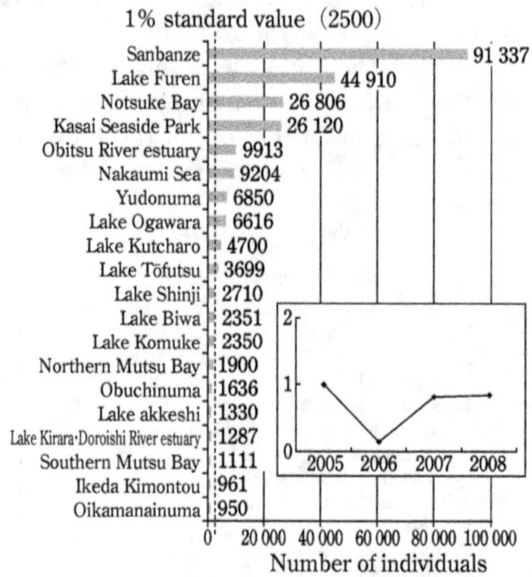

Widgeon

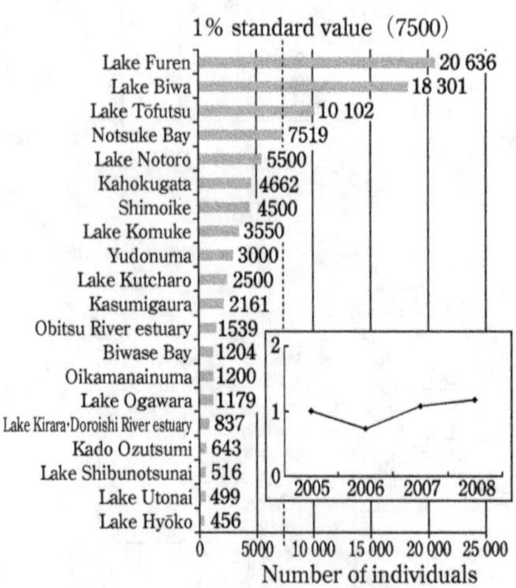

Pintail

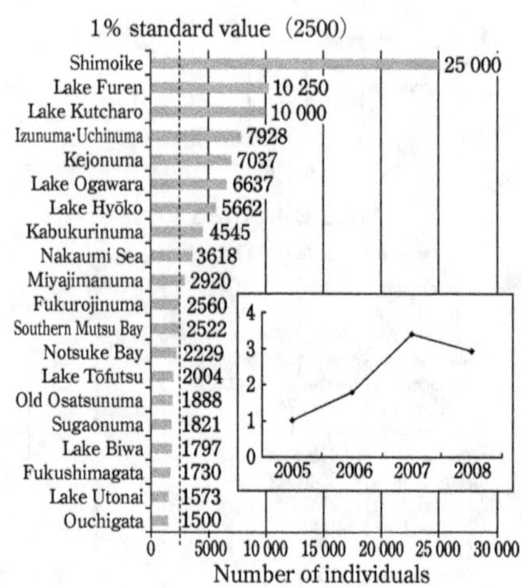

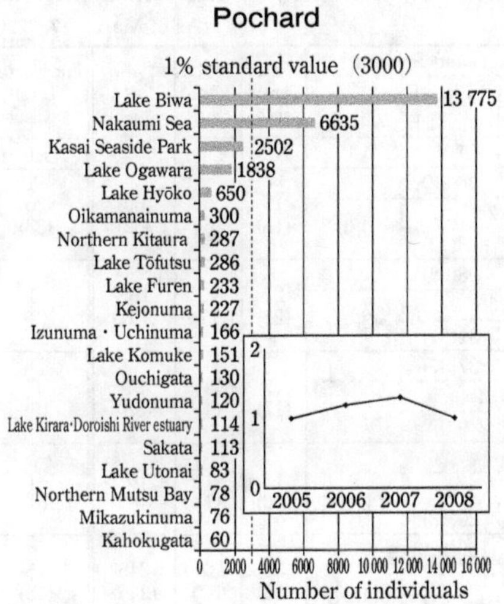

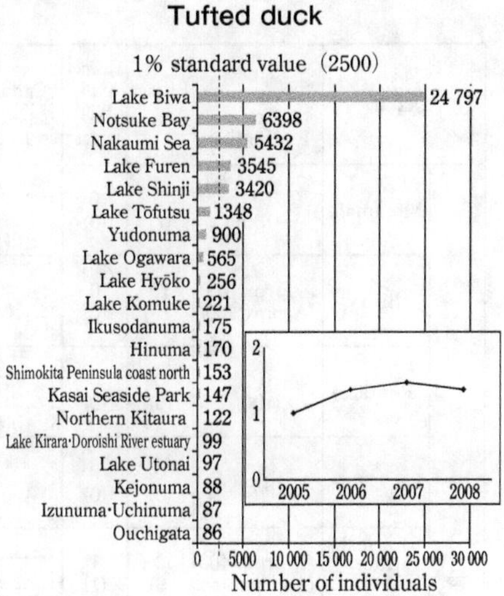

Ministry of the Environment: "Biodiversity Monitoring Program (Monitoring-Site 1000) Summary of the Goose and Duck Survey Takend During the 1st Term" (2009).
URL : http://www.biodic.go.jp/moni1000/index.html

Species of Japanese Wildlife in Danger of Extinction

(Numbers of species listed in Red List 2020 by Ministry of the Environment)

(As of March 27, 2020.)

	Taxon	Species Targeted for Evaluation	Extinct EX	Extinct in the Wild EW	Threatened Species (total)	Endangered Class I	Class I A CR	Class I B EN	Endangered Class II VU	Near Threatened NT	Data Deficient DD	Total of listed Species	Endangered Local Population LP
Fauna	Mammals	160 (160)	7 (7)	0 (0)	34(33)	25(24)	12(12)	13(12)	9(9)	17 (18)	5 (5)	63 (63)	26 (23)
Fauna	Birds	Approx. 700 (Approx. 700)	15 (15)	0 (0)	98(98)	55(55)	24(24)	31(31)	43(43)	22 (21)	17 (17)	152 (151)	2 (2)
Fauna	Reptiles	100 (100)	0 (0)	0 (0)	37(37)	14(14)	5(5)	9(9)	23(23)	17 (17)	3 (4)	57 (58)	5 (5)
Fauna	Amphibians	91 (76)	0 (0)	0 (0)	47(29)	25(17)	5(4)	20(13)	22(12)	19 (22)	1 (1)	67 (52)	0 (0)
Fauna	Brackish water/ freshwater fishes	Approx. 400 (Approx. 400)	3 (3)	1 (1)	169(169)	125(125)	71(71)	54(54)	44(44)	35 (35)	37 (37)	245 (245)	15 (15)
Fauna	Insects	Approx. 32 000 (Approx. 32 000)	4 (4)	0 (0)	367(363)	182(177)	75(71)	107(106)	185(186)	351 (350)	153 (153)	875 (870)	2 (2)
Fauna	Shellfishes	Approx. 3 200 (Approx. 3 200)	19 (19)	0 (0)	629(616)	301(288)	39(33)	28(16)	328(328)	440 (445)	89 (89)	1 177 (1 169)	13 (13)
Fauna	Other invertebrate	Approx. 5 300 (Approx. 5 300)	1 (0)	0 (0)	65(65)	22(22)	0(0)	2(2)	43(43)	42 (42)	44 (44)	152 (151)	0 (0)
Fauna	Subtotal of Fauna		49 (48)	1 (1)	1 446(1 410)	749(722)			697(688)	943 (950)	349 (350)	2 787 (2 759)	63 (60)
Flora	Vascular plants	Approx. 7 000 (Approx. 7 000)	28 (28)	11 (11)	1 790(1 786)	1 049(1 045)	529(525)	520(520)	741(741)	297 (297)	37 (37)	2 163 (2 159)	0 (0)
Flora	Bryophytes	Approx. 1 800 (Approx. 1 800)	0 (0)	0 (0)	240(241)	137(138)			103(103)	21 (21)	21 (21)	282 (283)	0 (0)
Flora	Algae	Approx. 3 000* (Approx. 3 000)	4 (4)	1 (1)	116(116)	95(95)			21(21)	41 (41)	40 (40)	202 (202)	0 (0)
Flora	Lichens	Approx. 1 600 (Approx. 1 600)	4 (4)	0 (0)	63(61)	43(41)	2(0)	0(0)	20(20)	41 (41)	46 (46)	154 (152)	0 (0)
Flora	Fungi	Approx. 3 000* (Approx. 3 000)	25 (26)	1 (1)	61(62)	37(39)	0(0)	1(0)	24(23)	21 (21)	51 (50)	159 (160)	0 (0)
Flora	Subtotal of flora		61 (62)	13 (13)	2 270(2 266)	1 361(1 358)			909(908)	421 (421)	195 (194)	2 961 (2 956)	0 (0)
Total of thirteen taxonomic groups			110 (110)	14 (14)	3 616(3 676)	2 110(2 080)			1 606(1 596)	1 364 (1 371)	544 (544)	5 748 (5 715)	63 (60)

(Numbers of species listed in Red List of marine species)

(As of March 27, 2020.)

<table>
<tr><th rowspan="3">Taxon</th><th rowspan="3">Species Targeted for Evaluation</th><th rowspan="2">Extinct</th><th rowspan="2">Extinct in the Wild</th><th colspan="3">Threatened Species</th><th rowspan="2">Near Threat-ened</th><th rowspan="2">Data Defi-cient</th><th rowspan="3">Total of listed Species</th><th rowspan="2">Endangered Local Population</th></tr>
<tr><th colspan="2">Endangered Class I
Class I A / Class I B</th><th>Endangered Class Ⅱ</th></tr>
<tr><th>EX</th><th>EW</th><th>CR</th><th>EN</th><th>VU</th><th>NT</th><th>DD</th><th>LP</th></tr>
<tr><td rowspan="2">Fishes</td><td rowspan="2">Approx. 3 900</td><td rowspan="2">0</td><td rowspan="2">0</td><td colspan="3">16</td><td rowspan="2">89</td><td rowspan="2">112</td><td rowspan="2">217</td><td rowspan="2">2</td></tr>
<tr><td>8</td><td>6</td><td>2</td></tr>
<tr><td rowspan="2">Hermatypic corals</td><td rowspan="2">Approx. 690</td><td rowspan="2">1</td><td rowspan="2">0</td><td colspan="3">6</td><td rowspan="2">7</td><td rowspan="2">1</td><td rowspan="2">15</td><td rowspan="2">0</td></tr>
<tr><td>0</td><td>1</td><td>5</td></tr>
<tr><td rowspan="2">Crustaceans</td><td rowspan="2">Approx. 3 000</td><td rowspan="2">0</td><td rowspan="2">0</td><td colspan="3">30</td><td rowspan="2">43</td><td rowspan="2">98</td><td rowspan="2">171</td><td rowspan="2">2</td></tr>
<tr><td>8</td><td>11</td><td>11</td></tr>
<tr><td rowspan="2">Cephalopoda</td><td rowspan="2">Approx. 230</td><td rowspan="2">0</td><td rowspan="2">0</td><td colspan="3">0</td><td rowspan="2">3</td><td rowspan="2">0</td><td rowspan="2">3</td><td rowspan="2">0</td></tr>
<tr><td>0</td><td>0</td><td>0</td></tr>
<tr><td rowspan="2">Other invertebrates</td><td rowspan="2">Approx. 2 300</td><td rowspan="2">0</td><td rowspan="2">0</td><td colspan="3">4</td><td rowspan="2">20</td><td rowspan="2">13</td><td rowspan="2">37</td><td rowspan="2">1</td></tr>
<tr><td>1</td><td>2</td><td>1</td></tr>
<tr><td colspan="2" rowspan="2">Total</td><td rowspan="2">1</td><td rowspan="2">0</td><td colspan="3">56</td><td rowspan="2">162</td><td rowspan="2">224</td><td rowspan="2">443</td><td rowspan="2">5</td></tr>
<tr><td>17</td><td>20</td><td>19</td></tr>
</table>

Annotation: Numbers of species listed in Red List 2020 by Ministry of the Environment

(1) The numbers in parentheses in the table indicate the number of species (including subspecies and variants for plants, and varieties for algae) in the Red List 2019 (published in 2019). LP is the number of target populations.

(2) For some species of shellfish, other invertebrates, lichens and fungi, the evaluation are conducted by divided Endangered Class I into Class I A (CR) and Class IB (EN).

* The number of species excluding those that cannot be evaluated by the naked eye.

The categories are considered as follows:

Extinct [EX]: Species that are likely to already be extinct

Extinct in the Wild [EW]: Species that exist only in captivity or as a naturalized population outside its natural habitat

Endangered Class I (Critically Endangered+Endangered) [CR+EN]: Species that are threatened to extinction

Endangered Class I A (Critically Endangered) [CR]: Species that are facing an extremely high risk of extinction in the Wild in the near future

Endangered Class I B (Endangered) [EN]: Species that are facing a high risk of extinction in the Wild in the near future

Endangered Class Ⅱ (Vulnerable) [VU]: Species with and increasing risk of extinction

Near Threatened [NT]: Species that are not currently endangered, but may possibly qualify for “endangered” status with changes in their habitat conditions

Data Deficient [DD]: Species with data insufficient for adequate evaluation

Endangered Local Population [LP]: Species with a population isolated regionally, and face a high risk of extinction

List of Regulated Living Organisms under the Invasive Alien Species Act

Vertebrate

Class	Order	Family	Genus	Invasive Alien Species (IAS)	Uncategorized Alien Species (UAS)
Mammalia	Marsupialia	Didelphidae	*Didelphis*	None	Any species of the genus *Didelphis*
			All other genera of Didelphidae	None	None
		Phalangeridae	*Trichosurus*	Brushtail possum (*T. vulpecula*)	Any species of the family Phalangeridae excluding IAS
			All other genera of Phalangeridae	None	
	Insectivora	Erinaceidae	*Erinaceus*	Any species of the genus *Erinaceus*	None
			Atelerix *Hemiechinus* *Mesechinus*	None	Any species of the genera *Atelerix*, *Hemiechinus* and *Mesechinus* excluding *Atelerix albiventris*
	Primates	Cercopithecidae	*Macaca*	Taiwan macaque (*M. cyclopis*)	Any species of the genus *Macaca* excluding Japanese macaque (*M. fuscata*) and IAS
				Crab-eating macaque (*M. fascicularis*)	
				Rhesus macaque (*M. mulatta*)	
				Taiwan macaque × Japanese macaque *M. cyclopis* × *M. fuscata*	Any living hybrid organisms of species of the genus *Macaca* excluding Taiwan macaque × Japanese macaque, Rhesus macaque × Japanese macaque
				Rhesus macaque × Japanese macaque *M. mulatta* × *M. fuscata*	
	Rodentia	Agoutidae	All genera of Agoutidae	None	None
		Capromyidae	All genera of Capromyidae	None	None
		Dinomyidae	All genera of Dinomyidae	None	None
		Myicastoridae	*Myocastor*	Coypu or Nutria (*M. coypus*)	None
		Sciuridae	*Callosciurus*	Pallas's squirrel or Taiwan squirrel (*C. erythraeus*)	Any species of the genus Callosciurus excluding Pallas's squirrel (*C. erythraeus*) and Finlayson's squirrel (*C. finlaysonii*)
				Finlayson's Squirrel (*C. finlaysonii*)	
			Pteromys	Russian (or Siberian) flying squirrel (*P. volans*) excluding Japanese subspecies	None
			Sciurus	Gray squirrel (*S. carolinensis*)	Any species of the genus Sciurus excluding Japanese squirrel (*S. lis*) and red squirrel (*S. vulgaris*), IAS
				Eurasian red squirrel (*S. vulgaris*) excluding Japanese subspecies (*S. vulgaris orientis*)	
			All other genera of Sciuridae	None	None
		Muridae	*Ondratra*	Muskrat (*O. zibethicus*)	None
	Carnivora	Procyonidae	*Procyon*	Raccoon (*P. lotor*)	None
				Crab-eating raccoon (*P. cancrivorus*)	None
		Mustelidae	*Mustela*	American mink (*M. vison*)	Any species of genus *Mustela* excluding Ermine (*M. erminea*), Japanese Weasel (*M. itatsi*), Least Weasel (*M. itatsi*), ferret (*M. putoriusfuro*), Siberian weasel (*M. sibilica*), American mink
		Herpestidae	*Herpestes*	Small Indian mongoose (*H. auropunctatus*)	Any species of the family Herpestidae excluding IAS, any species of genus *Suricata*
				Javan mongoose (*H. javanicus*)	
			Mungos	Banded mongoose (*M. mungo*)	
			All other genera of Herpestidae	None	

Mammalia	Artiodactyla	Cervidae	*Axis*	All species of the genus *Axis*	None
			Cervus	All species of the genus *Cervus* excluding *C. nippon centralis* *C. nippon keramae* *C. nippon mageshimae* *C. nippon nippon* *C. nippon pulchellus* *C. nippon yakushimae* *C. nippon yesoensis*	None
			Dama	All species of the genus *Dama*	None
			Elaphurus	Pere David's deer (*E. davidianus*)	None
			Muntiacus	Reeves's muntjac (*M. reevesi*)	Any species of the genus *Muntiacus* excluding IAS
Aves	Anseriformes	Anatidae	*Branta*	Canada goose (*B. canadensis*)	Any species of the genus *Branta* excluding Canada goose (*B. canadensis*) Aleutian Canada Goose (*B. huhutchinsii. leucopareia*), Cackling Canada Goose (*Branta h. minima*), Brent goose (*Branta h. bernicla*)
	Passeriformes	Pycnonotidae	*Pycnonotus*	Red-vented Bulbul (*P. cafer*)	None
		Timaliidae	*Garrulax*	Laughing thrushes (*G. canorus*)	Any species of the family Timaliidae excluding IAS
				Moustached laughingthrush (*G. cineraceus*)	
				White-browed laughingthrush (*G. sannio*)	
				Masked laughingthrush (*G. perspicillatus*)	
			Leiothrix	Red-billed mesia (*L. lutea*)	
			All other genera of Timaliidae	None	
Reptilia	Testudinata	Chelydridae	*Chelydra*	Snapping turtle (*C. serpentina*)	None
			All other genera of Chelydridae	None	None
		Geoemydidae	*Mauremys*	*M. sinensis*	None
				M. sinensis × *M. japonic*	None
				M. sinensis × *M. mutica*	None
				M. sinensis × *M. eevesi*	None
	Squamata	Agamidae	*Japalura*	*J. swinhonis*	None
		Iguanidae (Polychrotidae)	*Anolis*	Bueycito anole (*A. allogus*)	Any species of the genus *Anolis* (and/or *Norops*) excluding *A. allogus*, *A. alutaceus*, Twig anole (*A. angusticeps*), Green anole (*A. carolinensis*), Knight anole (*A. equestris*), *A. garmanni*, *A. homolechis* and Brown anole (*A. sagrei*)
				Blue-eyed grass-bush anole, Monte Verde Anole (*A. alutaceus*)	
				A. angusticeps	
				Green anole (*A. carolinensis*)	
				Knight anole (*A. equestris*)	
				Garman anole (*A. garmanni*)	
				Habana Anole, Cuban white-fanned anole (*A. homolechis*)	
				Brown anole (*A. sagrei*)	
			Norops	None	

List of Regulated Living Organisms under the Invasive Alien Species Act

Continued.

Class	Order	Family	Genus	Invasive Alien Species (IAS)	Uncategorized Alien Species (UAS)
Reptilia	Squamata	Colubridae	*Boiga*	Green cat snake (*B. cyanea*)	Any species of the genus *Boiga* excluding IAS
				Dog-toothed cat snake (*B. cynodon*)	
				Gold-ringed cat snake, Mangrove snake (*B. dendrophila*)	
				Brown tree snake (*B. irregularis*)	
				Black-heade Cat snake (*B. nigriceps*)	
			Psammodynastes	None	None
			Elaphe	Taiwan beauty snake (*E. taeniura friesi*)	Any subspecies of *E. taeniura* excluding *E. taeniura schmackeri* and *E. taeniura friesi*
		Viperidae	*Protobothrops*	Taiwan pit vipers (*P. mucrosquamatus*)	Any species of the genus *Protobothrops* excluding *P. elegans*, *P. flavoviridis*, *P. mucrosquamatus* and *P. tokarensis*
			Bothrops	None	None
Amphibia	Anura	Bufonidae	*Bufo*	Great plains toad (*B. cognatus*)	Any species of the genus *Bufo* excluding Great Plains toad (*B. cognatus*), Green toad (*B. marinus*) *B. debilis*, *B. gargarizans miyakonis*, Smooth sided toad (*B. guttatus*), *B. japonicus*, Cane toad (*B. marinus*), *B. melanostictus*, *B. paracnemis*, Red-spotted toad (*B. punctatus*), Oak toad (*B. quercicus*), Texas toad (*B. speciosus*), *B. terrestris*, *B. torrenticola*, South american common toad (*B. typhonius*), *B. valliceps*, and *B. viridis*
				Spotted toad (*B. guttatus*)	
				Cane toad (*B. marinus*)	
				B. melanostictus	
				Red-spotted toad (*B. punctatus*)	
				Oak toad (*B. quercicus*)	
				Texas toad (*B. speciosus*)	
				South American common toad (*B. typhonius*)	
		Hylidae	*Osteopilus*	Cuban treefrog (*O. septentrionalis*)	Any species of the genus Osteopilus excluding Cuban treefrog (*O. septentrionalis*)
		Leptodactylidae	*Eleutherodactylus*	Puerto Rican coqui (*E. coqui*)	None
				E. johnstonei	
				Greenhouse Frog (*E. planirostris*)	
		Microhylidae	*Kaloula*	*K. pulchra*	None
		Ranidae	*Rana*	Bullfrog (*R. catesbeiana*)	Green frog (*R. clamitans*), Pig frog (*R. grylio*), River frog (*R. heckscheri*), Florida bog frog (*R. okaloosae*), Mink frog (*R. septentrionalis*), and Carpenter frog (*R. virgatipes*)
		Rhacorhoridae	*Polypedates*	Asian tree frog (*P. leucomystax*)	Any species of the genus *Polypedates* excluding *P. leucomystax*
Osteichthyes	Lepisosteiformes	Lepisosteidae	*Atractosteus*	Any member of the family Lepisosteidae	None
			Lepisosteus	Any living hybrid organisms of species of the family Lepisosteidae	

Osteichthyes	Cypriniformes	Cyprinidae	*Acheilognathus*	*A. macropterus*	None
	Siluriformes	Bagridae	*Tachysurus*	*T. fulvidraco*	None
		Ictaluridae	*Ameiurus*	*A. nebulosus*	Any species of the genus *Ameiurus* excluding *A. nebulosus*
			Ictalurus	Channel catfish (*I. punctatus*)	Any species of the genus *Ictalurus* excluding *I. punctatus*
			Pylodictis	*P. olivaris*	None
		Siluridae	*Silurus*	*S. glanis*	None
	Esociformes	Esocidae	*Esox*	Any species of the family Esocidae	None
				Any living hybrid organisms of species of family Esocidae	
	Cyprinodontiformes	Poeciliidae	*Gambusia*	Western mosquito fish (*G. affinis*)	None
				G. holbrooki	
	Perciformes (Percoidei)	Centrarchidae	*Lepomis*	Bluegill (*L. macrochirus*)	Any species of the family Centrarchidae excluding largemouth bass (*M. salmoides*), smallmouth bass (*M. dolomieu*), and bluegill (*L. macrochirus*)
			Micropterus	Smallmouth bass (*M. dolomieu*)	
				Largemouth bass (*M. salmoides*)	
			All other genera of Centrarchidae	None	
		Gobiidae	*Neogobius*	*N. melanostomus*	None
		Latidae	*Lates*	*L. niloticus*	None
			All genera of Latidae	None	None
		Moronidae	All genera of Moronidae	None	
			Morone	*M. americana*	Any species of the family Moronidae excluding *M. saxatilis*, *M. chrysops* and *M. americana*
				White bass (*M. chrysops*)	
				Striped bass (*M. saxatilis*)	
				Striped bass × White bass (*M. chrysops* × *M. saxatillis*)	Any living hybrid organisms of species of the family Moronidae excluding IAS
		Nandidae	All genera of Nandidae	None	None
		Percichthyidae	*Gadopsis*	None	Any species of the genera *Gadopsis*, *Maccullochella*, *Macquaria* and *Percichthys* excluding Murray cod (*Maccullochella peelii*) and Golden perch (*Macquaria ambigua*)
			Maccullochella	None	
			Macquaria	None	
			Percichthys	None	
		Percidae	*Gymnocephalus*	*G. cernuus*	Any species of the genus *Gymnocephalus* excluding *G. cernuus*
			Perca	Eurasian perch (*P. fluviatilis*)	Any species of the genera *Gymnocephalus*, *Perca*, *Sander* (*Stizostedion*), and *Zingel* excluding pikeperch (*S. licioperca*) and Eurasian perch (*P. fluviatilis*)
			Sander (*Stizostedion*)	Pikeperch (*S. lucioperca*)	
			Zingel	None	
		Sinipercidae	*Siniperca*	Mandarin fish (*S. chuatsi*)	Any species of the genus *Siniperca* excluding *S. chuatsi* and *S. sherzeri*
				Mandarin fish (*S. scherzeri*)	
Insecta	Lepidoptera	Nymphalidae	*Hestina*	*Hestina assimilis* (excluding *H. assimilis shirakii*)	None
	Coleoptera	Cerambycidae	*Aromia*	*Aromia bungii*	None

List of Regulated Living Organisms under the Invasive Alien Species Act

Continued.

Class	Order	Family	Genus	Invasive Alien Species (IAS)	Uncategorized Alien Species (UAS)
Insecta	Coleoptera	Lucanidae	*Neolucanus*	*Neolucanus angulatus*	None
				Neolucanus baladeva	
				Neolucanus giganteus	
				Neolucanus katsuraorum	
				Neolucanus maedai	
				Neolucanus maximus	
				Neolucanus perarmatus	
				Neolucanus saundersii	
				Neolucanus tanakai	
				Neolucanus waterhousei	
		Scarabaeidae	*Cheirotonus*	Any species of the genus *Cheirotonus* excluding Yanbaru Long-armed scarab (*C. jambar*)	None
			Euchirus	Any species of the genus *Euchirus*	None
			Propomacrus	Any species of the genus *Propomacrus*	None
	Hymenoptera	Apidae	*Bombus*	Large earth bumblebee (*B. terrestris*)	Any species of the genus *Bombus* excluding *B. ardens ardens* *B. ardens sakagamii* *B. ardens tsushimanus* *B. beaticola beaticola* *B. beaticola moshkarareppus* *B. beaticola shikotanensis* *B. consobrinus wittenburgi* *B. deuteronymus deuteronymus* *B. deuteronymus maruhanabachi* *B. diversus diversus* *B. diversus tersatus* *B. florilegus* *B. honshuensis honshuensis* *B. honshuensis tkalcui* *B. hypnorum koropokkrus* *B. hypocrita hupocrita* *B. hypocrita sapporensis* *B. ignitus* *B. oceanicus* *B. pseudobaicalensis* *B. schrencki albidopleuralis* *B. schrencki konakovi* *B. schrencki kuwayamai* *B. terrestris* *B. ussurensis* *B. yezoensis*
		Formicidae	*Linepithema*	Argentine ant or Tropical fire ant (*L. humile*)	None
			Solenopsis	Red imported fire ant (*S. invicta*)	None
				Fire ant (*S. geminata*)	None
			Wasmannia	Little fire ant (*W. auropunctata*)	None
		Vespidae	*Vespa*	Asian black hornet (*Vespa velutina*)	None

Crustacea	Decapoda	Astacidae	*Astacus*	Any species of the genus *Astacus*	Any species of the family Astacidae excluding the genus *Astacus* and signal crayfish (*P. leniusculus*)
			Atlantoastacus *Austropotamobius* *Caspiastacus* *Pacifastacus*	Signal crayfish (*P. leniusculus*)	
		Cambaridae	*Orconectes*	Rusty crayfish (*O. rusticus*)	Any species of the family Cambaridae excluding *O. rusticus*, zarigani (*Cambaroides japonicus*) and red swamp crayfish (*Procambarus clarkii*)
			All genera of Cambaridae	None	None
		Parastacidae	*Cherax*	Any species of the genus *Cherax*	Any species of the family Parastacidae excluding the genus *Cherax*
			All genera of Parastacidae	None	None
		Varunidae	*Eriocheir*	Any species of the genus *Eriocheir* excluding Japanese mitten crab (*E. japonica*)	None
Arachnid	Scorpiones	Buthidae	All genera of Buthidae	Any species of the family Buthidae	None
	Araneae	Hexathelidae	*Atrax*	Any species of the genus *Atrax*	None
			Hadronyche	Any species of the genus *Hadronyche*	None
		Loxoscelidae	*Loxosceles*	*L. gaucho*	None
				L. laeta	None
				L. reclusa	None
		Theridiidae	*Latrodectus*	Any spiecies of the genus *Latrodectus* excluding *L. elegans*	None
Mollusca	Mytiloida	Mytilidae	*Limnoperna*	Any species of the genus *Limnoperna*	None
	Veneroida	Dreissenidae	*Dreissena*	Quagga mussel (*D. bugensis*)	None
				Zebra mussel (*D. polymorpha*)	None
	Stylommatoph ora	Haplotrematidae	*Ancotrema* *Haplotrema*	None	Any species of the family Haplotrematidae
		Oleacinidae	All genera of Oleacinidae	None	Any species of the family Oleacinidae
		Rhytididae	All genera of Rhytididae	None	Any species of the family Rhytididae
		Spiraxidae	*Euglandina*	Cannibal snail (*E. rosea*)	Any species of the family Spiraxidae excluding *E. rosea*
			All genera of Spiraxidae	None	
		Streptaxidae	All genera of Streptaxidae	None	Any species of the family Streptaxidae excluding the genus *Sinoennea* and *Indoennea bicolor*
		Subulinidae	All genera of Subulinidae	None	Any species of the family Subulinidae excluding *Allopeas brevispirum* *A. clavulinum kyotoense* *A. gracilis* *A. heudei* *A. javanicum* *A. mauritianum obesispira* *A. pyrgula* *A. satsumense* *Rumina decollata* *Subulina octona*
Platyhelminthes	Tricladida	Rhynchodemidae	*Platydemus*	Predatory flatworm (*P. manokwari*)	None

List of Regulated Living Organisms under the Invasive Alien Species Act

2. Plant Kingdom

Class	Order	Family	Genus	Invasive Alien Species (IAS)	Uncategorized Alien Species (UAS)
Tracheophyte	Sympetalae	Compositae	*Coreopsis*	Lanceleaf tickseed (excluding cut flowers) (*C. lanceolata*)	None
			Gymnocoronis	Senegal tea plant (*G. spilanthoides*)	None
			Mikania	*M. micrantha*	None
			Rudbeckia	Cutleaf coneflower (excluding cut flowers) (*R. laciniata*)	None
			Senecio	Madagascar ragwort (*S. madagascariensis*)	None
		Scrophulariaceae	*Veronica*	Water speedwell (excluding cut flowers) (*V. anagallis-aquatica*)	None
	Caryophyllales	Amaranthaceae	*Alternanthera*	Alligatorweed (*A. philoxeroides*)	None
		Apiaceae	*Hydrocotyle*	Floating marshpennywort or Pennywort (*H. ranunculoides*)	*H. bonariensis* *H. umbellata*
		Cucurbitaceae	*Sicyos*	Bur cucumber (*S. angulatus*)	None
		Haloragaceae	*Myriophyllum*	Parrotfeather (*M. aquaticum*)	None
		Onagraceae	*Ludwigia*	Large-flower primrose-willow (*L. grandiflora*)	None
	Liliopsida	Poaceae	*Ammophila*	*A. arenaria*	None
			Spartina	Any species of the genus *Spartina*	None
		Araceae	*Pistia*	Water lettuce (*P. stratiotes*)	None
	Pteridophyta	Azollaceae	*Azolla*	Water fern (*A. cristata*)	None
	Caryophyllales	Droseraceae	*Drosera*	*D. intermedia*	None

Usage Status of Farmland

Pastures and farmland in each region

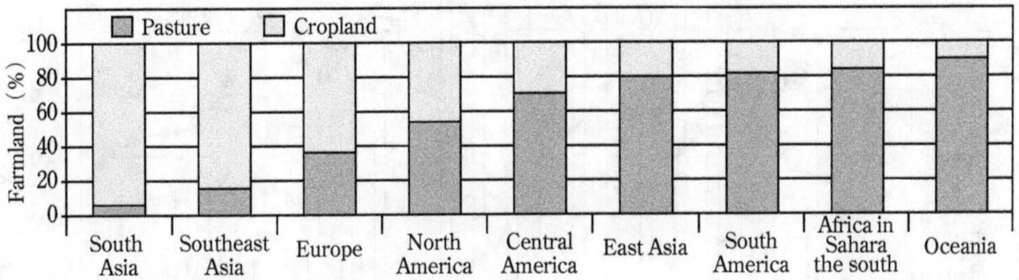

World Resources Institute : "World Resources 2000–2001" (2002).

Rice Production

Harvest amount and harvest area for rice in major countries

Country	1995		2000		2005		2010		2015		2018	
	Production	Area harvested	Production	Area harvested	Production	Area harvested	Production	Area harvested	Production	Area harvested	Production	Area harvested
China*	187 298	31 107	189 814	30 301	182 055	29 116	197 212	30 117	213 724	31 036	214 079	30 461
India	115 440	42 800	127 465	44 712	137 690	43 660	143 963	42 862	156 540	43 390	172 580	44 500
Indonesia	49 744	11 439	51 898	11 793	54 151	11 839	66 469	13 253	75 398	14 117	83 037	15 995
Bangladesh	26 399	9 952	37 628	10 801	39 796	10 524	50 061	11 529	51 805	11 381	56 417	11 910
Vietnam	24 964	6 766	32 530	7 666	35 833	7 329	40 006	7 489	45 091	7 829	44 046	7 571
Thailand	22 015	9 113	25 844	9 891	30 648	10 225	35 703	11 932	27 702	9 718	32 192	10 407
Myanmar	17 670	6 033	20 987	6 302	27 246	7 384	32 065	8 011	26 210	6 769	25 418	6 706
Philippines	10 541	3 759	12 389	4 038	14 603	4 070	15 772	4 354	18 150	4 656	19 066	4 800
Brazil	11 226	4 374	11 135	3 665	13 193	3 916	11 236	2 722	12 301	2 138	11 749	1 861
Pakistan	5 950	2 162	7 204	2 377	8 321	2 621	7 235	2 365	10 202	2 739	10 803	2 810
Cambodia	3 448	1 924	4 026	1 903	5 986	2 415	8 245	2 777	9 335	2 789	10 647	2 982
Nigeria	7 887	1 252	8 658	1 230	10 108	1 361	11 027	1 463	8 725	1 042	10 170	1 180
Japan	13 435	2 118	11 863	1 770	11 342	1 706	10 604	1 628	9 986	1 506	9 728	1 470
USA	2 920	1 796	3 298	2 199	3 567	2 494	4 473	2 433	6 256	3 122	6 809	3 346
Korea	6 389	1 056	7 197	1 072	6 435	980	5 811	892	5 771	799	5 195	738
Nepal	3 579	1 497	4 216	1 560	4 290	1 542	4 024	1 481	4 789	1 425	5 152	1 470
Madagascar	2 450	1 150	2 480	1 209	3 392	1 250	4 738	1 307	3 722	908	4 030	928
Sri Lanka	2 810	890	2 860	832	3 246	915	4 301	1 060	3 783	1 005	3 930	1 041
World Total	547 162	149 579	598 668	154 002	634 225	155 266	701 139	161 700	745 905	162 630	782 000	167 133

Statistics for Rice Paddy Conversin, Production: 1 000 t, Harvested area: 1 000 ha

* Includes Hong Kong, Macau, and Taiwan

Source: FAO: "FAOSTAT-Production-Crops" (August 14, 2020).

Planted Area and Harvest Amount for Rice in Japan

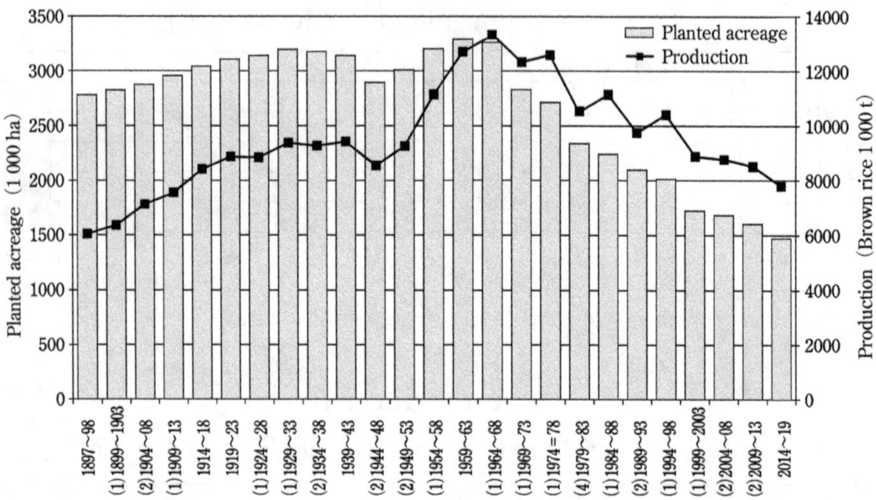

The number in parentheses shows the number of years for which crop failure occurred in the 5-year period.
Chart based on the Ministry of Agriculture, Forestry and Fisheries Secretariat Statistics Division: "Agriculture, Forestry and Fisheries Statistics" (2015).

Amount of Major Crops

Harvest area and harvest amount for major crops by region

Crops	World		Africa		North Central America		South America		Asia		Europe		Russian Federation		Oceania	
	P	AH	P	AH	P	AH	P	AH	P	AH	P	AH	P	AH	P	AH
Cereals (Total)	29 629	7 281	2 026	1 249	5 672	805	2 073	446	14 503	3 429	4 986	1 175	1 098	420	349	168
Wheat	7 340	2 143	293	102	860	265	271	91	3 282	970	2 421	606	721	265	213	110
Barley	1 414	479	85	50	127	35	65	17	210	101	831	234	170	79	96	42
Rye	113	41	1	1	5	2	1	1	15	5	91	33	19	10	0	0
Oats	231	98	2	1	44	14	21	8	16	7	135	59	47	27	13	9
Maize	11 476	1 937	789	387	4 374	435	1 399	266	3 616	673	1 286	170	114	24	6	1
Rice (paddy)	7 820	1 671	332	142	114	15	261	43	7 054	1 461	40	6	10	2	6	1
Potatoes	3 682	176	260	19	290	6	173	9	1 886	93	1 052	48	224	13	17	0
Sweet potatoes	919	81	260	46	13	1	15	1	607	30	1	0	—	—	9	2
Soybeans	3 487	1 249	36	26	1 313	384	1 711	571	306	211	121	57	40	27	1	0
Tobacco (unmanufactured)	61	34	7	6	3	2	9	4	39	20	2	1	0	0	0	0
Sugar cane	19 070	263	949	15	1 526	18	8 416	114	7 519	103	23	0	0	0	351	5
Sugar beet	2 749	48	150	3	306	5	24	0	417	7	1 851	33	421	11	—	—

Area harvested (AH): 100 000 ha, Production (P): 100 000 t
Aggregate, may include official, semi-official, estimated or calculated data
Source: FAO: "FAOSTAT-Production-Crops" (August 14, 2020).

Materials Cycle

Biomass and Net Annual Primary Production of the Biosphere

Ecosystem	Area ($10^{12}m^2$)	Net primary production (10^{15}gC/year)
Continental		
Tropical wet & moist forest	10.4	8.3
Tropical dry forest	7.7	4.8
Temperate forest	9.2	6
Boreal forest	15	6.4
Tropical woodland & savanna	24.6	11.1
Temperate steppe	15.1	4.9
Desert	18.2	1.4
Tundra	11	1.4
Wetland	2.9	3.8
Cultivated land	15.9	12.1
Rock & ice	15.2	0
Total	145.2	60.2
Marine		
Open ocean	326	42
Coastal zone	36	9
Upwelling area	0.36	0.15
Total	362.36	51.15

Continental: Houghton and Skole (1990).
Marine: Martin *et al.* (1987).

Ecosystem	Net primary production (10^{15}gC/year)
Continental	
Tropical rain forests	17.8
Broadleaf deciduous forests	1.5
Mixed Broadleaf & needleleaf forests	3.1
Savannas	16.8
Needleleaf evergreen forests	3.1
Needleleaf deciduous forests	1.4
Perennial grasslands	2.4
Broadleaf shrubs with bare soil	1
Tundra	0.8
Desert	0.5
Cultivation	8
Total	56.4
Marine	
Trade Winds Domain (tropical & subtropical)	13
Westerly Winds Domain (temperate)	16.3
Polar Domain	6.4
Coastal Domain	10.7
Salt marshes, estuaries & macrophytes	1.2
Coral Reefs	0.7
Total	48.3

Geider *et al.* (2001).

Biomass of Plants and Animals, and Related Dimensions of the Biosphere

Ecosystem	Area (10^{12} m^2)	Plant biomass (kg dry weight/m^2)	Plant biomass (10^9 dry weight ton)	Animal biomass (10^6 dry weight ton)
Continental				
Tropical rain forest	17.0	45	765	330
Tropical seasonal forest	7.5	35	260	90
Temperate evergreen forest	5.0	35	175	50
Temperate deciduous forest	7.0	30	210	110
Boreal forest	12.0	20	240	57
Woodland & shrubland	8.5	6	50	45
Savanna	15.0	4	60	220
Temperate glassland	9.0	1.6	14	60
Tundra & alpine	8.0	0.6	5	3.5
Desert scrub	18.0	0.7	13	8
Extreme desert	24.0	0.02	0.5	0.02
Cultivated land	14.0	1	14	6
Swamps & marsh	2.0	15	30	20
Lakes & stream	2.0	0.02	0.05	10
Total continental	149	12.3	1 835	1 010
Marine				
Open ocean	332.0	0.003	1.0	800
Upwelling zones	0.4	0.02	0.008	4
Continental shelf	26.6	0.01	0.27	160
Algal beds & reefs	0.6	2	1.2	12
Estuaries	1.4	1	1.4	21
Total marine	361	0.01	3.878	997

Whittaker and Likens (1973), and Woodwell and Pecan (1973).

Global Carbon Cycle in the Biosphere

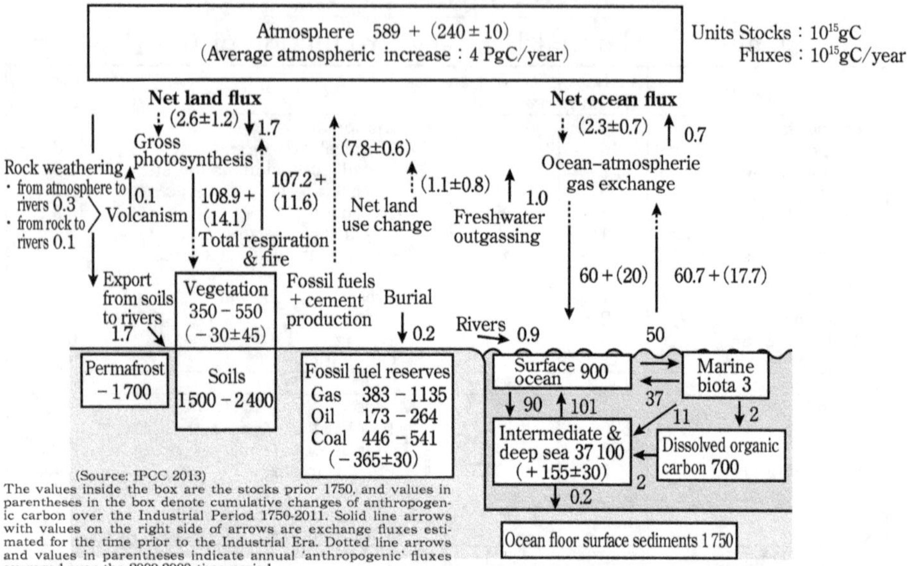

(Source: IPCC 2013)
The values inside the box are the stocks prior 1750, and values in parentheses in the box denote cumulative changes of anthropogenic carbon over the Industrial Period 1750-2011. Solid line arrows with values on the right side of arrows are exchange fluxes estimated for the time prior to the Industrial Era. Dotted line arrows and values in parentheses indicate annual 'anthropogenic' fluxes averaged over the 2000-2009 time period.

Global Nitrogen Cycle in the Biosphere

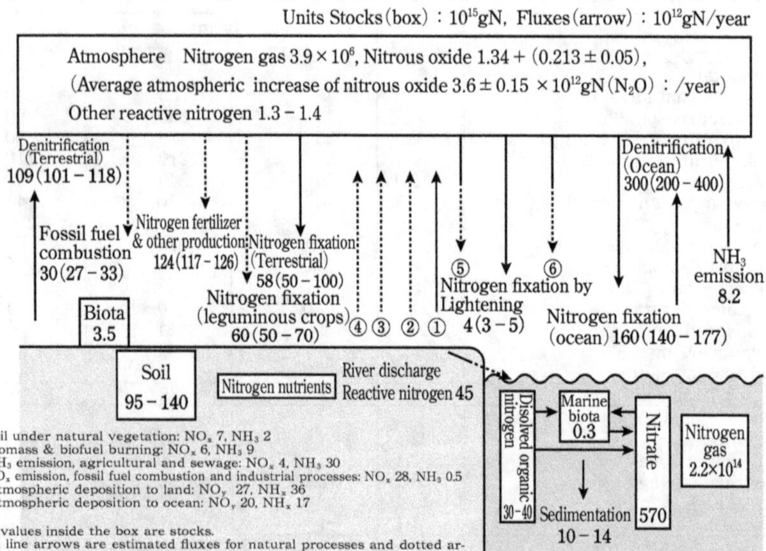

① Soil under natural vegetation: NO_x 7, NH_3 2
② Biomass & biofuel burning: NO_x 6, NH_3 9
③ NH_3 emission, agricultural and sewage: NO_x 4, NH_3 30
④ NO_x emission, fossil fuel combustion and industrial processes: NO_x 28, NH_3 0.5
⑤ Atmospheric deposition to land: NO_y 27, NH_x 36
⑥ Atmospheric deposition to ocean: NO_y 20, NH_x 17

The values inside the box are stocks.
Solid line arrows are estimated fluxes for natural processes and dotted arrows for anthropogenic activities.
(Fluxes are from IPCC 2013, and stocks are based on Galloway *et al.* 2004)

Global Sulfur Cycle in the Biosphere

Units Stocks : 10^{12}gS
Fluxes : 10^{12}gS/year

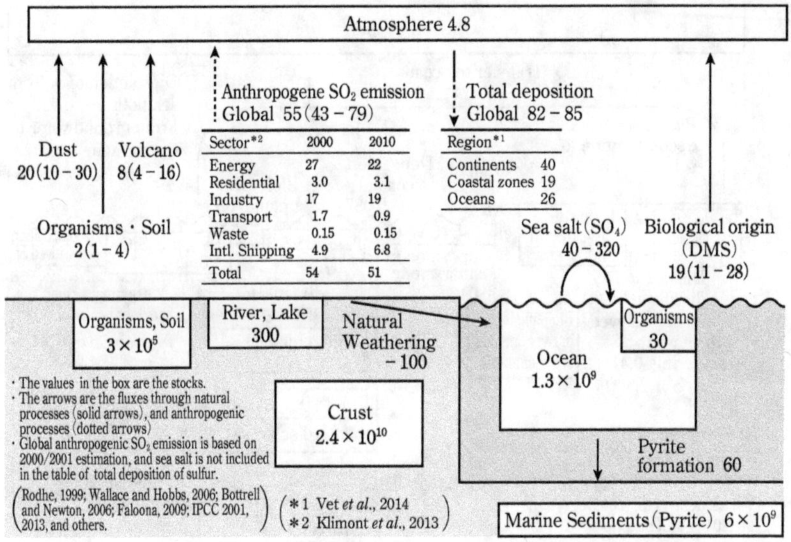

Global Phosphorus Cycle in the Biosphere

Units Stocks (box) TgP
Fluxes (arrow) TgP/year

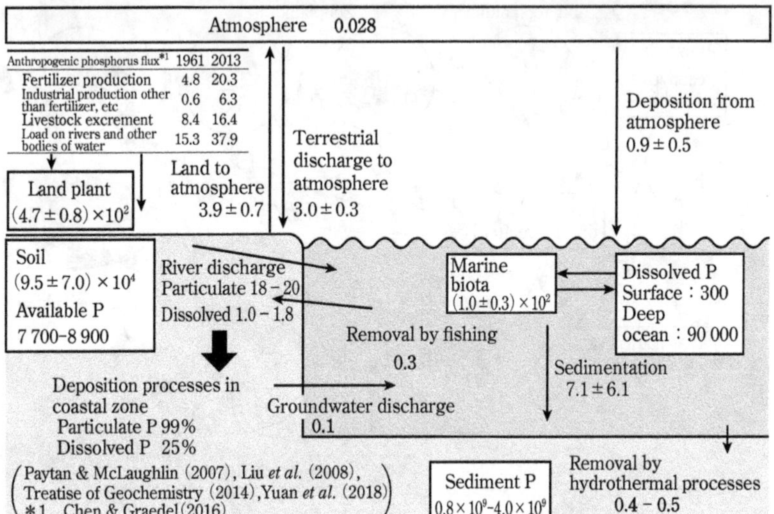

Global Silica Cycle in the Biosphere

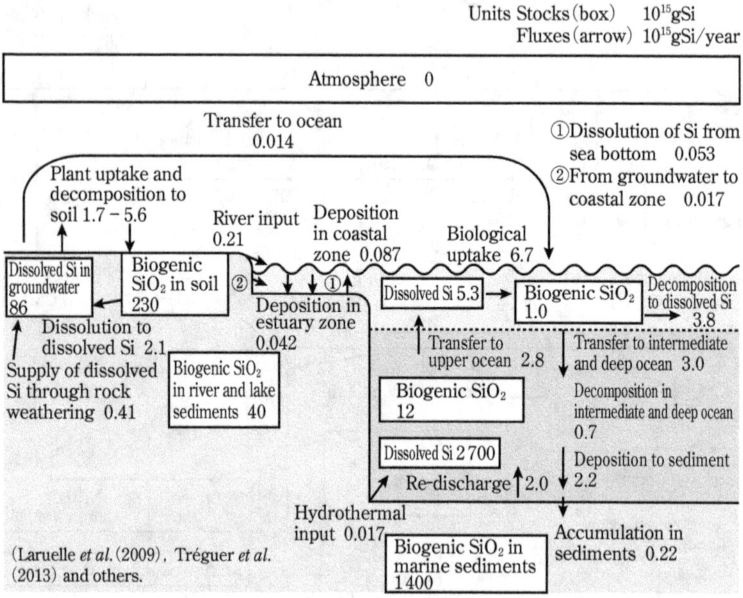

Two Planktonic Food Chains in Aquatic Environments

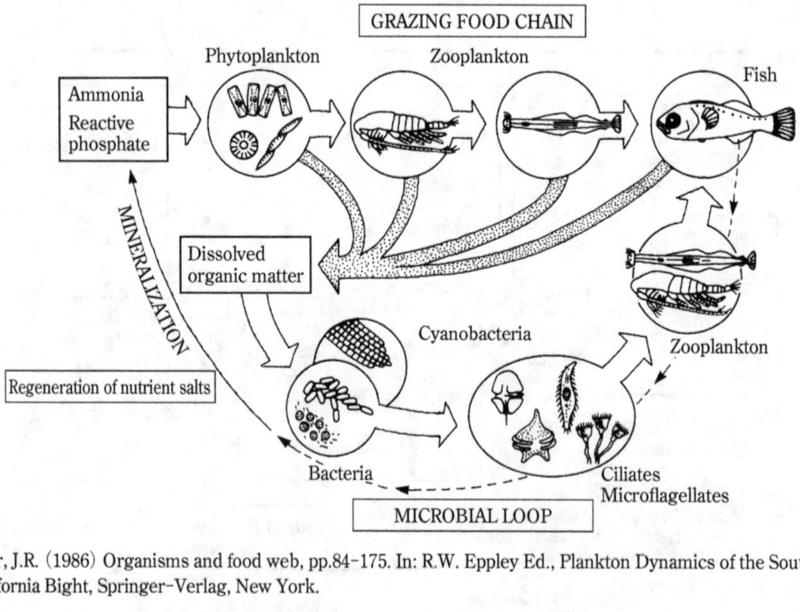

Beer, J.R. (1986) Organisms and food web, pp.84–175. In: R.W. Eppley Ed., Plankton Dynamics of the Southern California Bight, Springer–Verlag, New York.

Chemical Substances which are Contained in the Atmosphere, Seas, and Earth's Crust and which Influence River Water Quality

Chemcial substances in air (Unit : $10^{-9}g/m^3$)

	Sodium Na^+	Potassium K^+	Ammonia NH_4^+	Chlorine Cl^-	Nitric acid NO_3^-	Sulfuric acid SO_4^{2-}
Free atmosphere	51.00	13.00	8.00	69.00	3.00	22.00
ratio (%)	26.2	6.7	4.1	35.4	1.5	11.3
Near ground (forest)	162.00	20.00	105.00	74.00	13.00	256.00
ratio (%)	18.9	2.3	12.3	8.6	1.5	29.9

	Methyl sulphate $CH_3SO_4^-$	Phosphoric acid PO_4^{3-}	Fluorine F^-	Oxalate $C_2O_4^{2-}$	Formic acid $HCOO^-$	Acetic acid CH_3COO^-	Total
Free atmosphere	0.91	20.00	0.74	4.00	1.00	2.00	194.65
ratio (%)	0.5	10.3	0.4	2.1	0.5	1.0	100.0
Near ground (forest)	27.00	157.00	5.10	25.00	0.50	11.00	855.60
ratio (%)	3.2	18.3	0.6	2.9	0.1	1.3	100.0

Chemicals in Ocean Water (Unit : mg/kg)

	Na^+	K^+	Mg^{2+}	Ca^{2+}	Cl^-	SO_4^{2-}	HCO_3^-	Total
Chemcial Abundance ratios	10 762.0	399.0	1 293.0	411.0	19 353.0	2 709.0	142.0	35 069.0
(%)	30.7	1.1	3.7	1.2	55.2	7.7	0.4	100.0

Chemicals in Earth's Crust (Unit : %)

	SiO_2	Na_2O	K_2O	MgO	CaO	FeO	Al_2O_3	TiO_2	Total
Poldervaart (1955)	55.2	2.9	1.9	5.2	8.8	8.6	15.3	1.6	99.6
Taylor (1964)	60.5	3.2	2.5	3.9	5.8	7.2	15.6	1.0	99.7
Clarke & Washington (1924)	59.1	3.7	3.1	3.5	5.1	6.8	15.2	1.0	97.5

Sources: Ono *et al.* (1975), Smith (1976), Summerfield (1991), Hiroo Ohmori (1993)

Chemical Substances in River Water and Sources of Substance Supply

	SiO_2	Na^+	K^+	Mg^{2+}	Ca^{2+}	Cl^-	SO_4^{2-}	HCO_3^-	Total
Chemcial Abundance ratios	10.4	7.4	1.4	3.6	13.5	9.6	8.8	52.1	106.8
(%)	9.7	6.9	1.3	3.4	12.6	9.0	8.2	48.8	100.0
Quantity supplied from atmosphere	0.0	0.0	0.0	0.0	0.0	0.0	0.0	29.7	29.7
Quantity supplied from Earth's crust	10.4	3.5	0.2	3.1	13.2	2.7	7.1	22.4	62.6
Quantity supplied from rain water	0.0	3.9	1.2	0.5	0.3	6.9	1.7	0.0	14.5
Total	10.4	7.4	1.4	3.6	13.5	9.6	8.8	52.1	106.8

Unit: mg/kg = ppm

Sources: Smith (1976), Summerfield (1991), Hiroo Ohmori (1993)

Supplied Amount and Periods for Supply of Chemical Substances from Rivers to Seas*

Chemicals in Seawater	Na^+	K^+	Mg^{2+}	Ca^{2+}	Cl^-	SO_4^{2-}	HCO_3^-	Total
Density of dissolved chemical in seawater (mg/kg)	10 762.0	399.0	1 293.0	411.0	19 353.0	2 709.0	142.0	35 069.0
Amount of dissolved chemical (Gt)	14 819 274	549 423	1 780 461	565 947	26 649 081	3 730 293	195 534	48 290 013
Annual inflow amount from rivers (Kt)	177 600	33 600	86 400	324 000	230 400	211 200	1 250 400	2 313 600
Years of supply from rivers (millions of years)	83.4	16.4	20.6	1.7	115.7	17.7	0.2	20.9

* Calculated from (amount of dissolved chemical)/(annual inflow from rivers).
Total quantity of seawater: 1 377 000 trillion m^3 (trillion t)
Annual riverwater flow: 24 trillion m^3 (trillion t)
Annual sediment load washed out of land: 20 billion tons
Sources: Smith (1976), Summerfield (1991), Hiroo Ohmori (1993).

Erosion Rate for the Major Rivers

River	Region and country	Erosion rate	River	Region and country	Erosion rate
Colder and arid region (pleistocene permanent frozen ground region)			**Warm dryness region (half dried and dry region)**		
Mackenzie R.	Northern Canada	30	Ural R.	Southern Kazakhstan	13
Slave R.	Northern Canada	12	Tigris R.	Iraq	0.05
Saskatchewan R.	Northern Canada	5	Euphrates R.	Iraq	6
Yukon R.	Alaska	16	Indus R.	Pakistan	182
Injigiruka R.	Northeastern Russia	14	Yellow R.	Northern China	1 160
Yana R.	Northeastern Russia	6	White R.	Southwestern America	59
Lena R.	Northeastern Russia	18	Orange R.	Botswana	56
Colder and arid region (forest area mid latitude continental)			Murray R.	Southeastern Australia	12
Amur R.	Southeastern Russia	24	**Tropical arid region**		
Yenisei R.	North central Russia	15	Nile R.	Egypt	13
Ob R.	North central Russia	29	Blue Nile R.	Sudan	15
Northern Dvina R.	Northwestern Russia	18	Shari R.	Chad	3
Volga R.	Southern Russia	21	Rogon R.	Chad	13
Dnieper R.	Ukraine	8	Zambezi R.	Mozambique	30
Saint Lawrence R.	Southeastern Canada	18	Parana R.	Argentina	32
Missouri R.	Central America	55	Uruguay R.	Uruguay	14
Ohio R.	Central America	85	**Tropical humid region**		
Columbia R.	Northwestern America	14	Congo R.	Zaire River	22
Warm humid region (mid latitude coastal woodland)			Sonkoi R.	Northern Vietnam	431
West Dovina R.	Western Russia	12	Mekong R.	Southern Vietnam	174
Thames R.	United Kingdom	17	Chao Phraya R.	Thailand	42
Rhine R.	Netherlands	32	Eyawadei R.	Myanmar	328
Seine R.	France	18	Damodar R.	Eastern India	560
Savannah R.	Southeastern America	50	Mahanaji R.	Eastern India	187
Mississippi R.	Southern America	59	Orinoco R.	Venezuela	36
Yangtze R.	Mid-eastern China	216	Amazon R.	Brazil	58

Erosion rate: Amount of sediment delivered from river basins (unit: $m^3/km^2/year$)
Even in the same region, the erosion rate varies significantly due to the topography and climate of the basin.
Furthermore, in many cases, measurement is performed using floating substances at the mouths of rivers.
Sources: Ohmori (1983), Summerfield (1991), Hiroo Ohmori (1993), etc.

Chemical Substances and Radiation

Emissions Amount of Greenhouse Gases

Fossil-fuel and cement production-emissions by fuel type (from 1901)

Year	Total	Natural gas	Oil	Coal	CM	Gas flaring	per capita	Year	Total	Natural gas	Oil	Coal	CM	Gas flaring	per capita
1901	552	4	18	531	0	0		1958	2 330	192	731	1 336	36	35	0.80
1902	566	4	19	543	0	0		1959	2 454	206	789	1 382	40	36	0.83
1903	617	4	20	593	0	0		1960	2 569	227	849	1 410	43	39	0.85
1904	624	4	23	597	0	0		1961	2 580	240	904	1 349	45	42	0.84
1905	663	5	23	636	0	0		1962	2 686	263	980	1 351	49	44	0.86
1906	707	5	23	680	0	0		1963	2 833	286	1 052	1 396	51	47	0.89
1907	784	5	28	750	0	0		1964	2 995	316	1 137	1 435	57	51	0.92
1908	750	5	30	714	0	0		1965	3 130	337	1 219	1 460	59	55	0.94
1909	785	6	32	747	0	0		1966	3 288	364	1 323	1 478	63	60	0.97
1910	819	7	34	778	0	0		1967	3 393	392	1 423	1 448	65	66	0.98
1911	836	7	36	792	0	0		1968	3 566	424	1 551	1 448	70	73	1.01
1912	879	8	37	834	0	0		1969	3 780	467	1 673	1 486	74	80	1.05
1913	943	8	41	895	0	0		1970	4 053	493	1 839	1 556	78	87	1.10
1914	850	8	42	800	0	0		1971	4 208	530	1 947	1 559	84	88	1.12
1915	838	9	45	784	0	0		1972	4 376	560	2 057	1 576	89	95	1.14
1916	901	10	48	842	0	0		1973	4 614	588	2 241	1 581	95	110	1.18
1917	955	11	54	891	0	0		1974	4 623	597	2 245	1 579	96	107	1.16
1918	936	10	53	873	0	0		1975	4 596	604	2 132	1 673	95	92	1.13
1919	806	10	61	735	0	0		1976	4 864	630	2 314	1 710	103	108	1.17
1920	932	11	78	843	0	0		1977	5 016	650	2 398	1 756	108	104	1.19
1921	803	10	84	709	0	0		1978	5 074	680	2 392	1 780	116	106	1.18
1922	845	11	94	740	0	0		1979	5 357	721	2 544	1 875	119	98	1.23
1923	970	14	111	845	0	0		1980	5 301	737	2 422	1 935	120	86	1.19
1924	963	16	110	836	0	0		1981	5 138	755	2 289	1 908	121	65	1.14
1925	975	17	116	842	0	0		1982	5 094	738	2 196	1 976	121	64	1.11
1926	983	19	119	846	0	0		1983	5 075	739	2 176	1 977	125	58	1.08
1927	1 062	21	136	905	0	0		1984	5 258	807	2199	2 074	128	51	1.10
1928	1 065	23	143	890	10	0		1985	5 417	835	2 186	2 216	131	49	1.12
1929	1 145	28	160	947	10	0		1986	5 583	830	2 293	2 277	137	46	1.13
1930	1 053	28	152	862	10	0		1987	5 725	892	2 306	2 339	143	44	1.14
1931	940	25	147	759	8	0		1988	5 936	935	2 412	2 387	152	50	1.16
1932	847	24	141	675	7	0		1989	6 066	982	2 459	2 428	156	41	1.16
1933	893	25	154	708	7	0		1990	6 074	1 026	2 492	2 359	157	40	1.14
1934	973	28	162	775	8	0		1991	6 142	1 051	2 601	2 284	161	45	1.14
1935	1 027	30	176	811	9	0		1992	6 078	1 085	2499	2 290	167	36	1.11
1936	1 130	34	192	893	11	0		1993	6 070	1 117	2 515	2 225	176	37	1.09
1937	1 209	38	219	941	11	0		1994	6 174	1 133	2 539	2 278	186	39	1.09
1938	1 142	37	214	880	12	0		1995	6 305	1 151	2 560	2 359	197	39	1.1
1939	1 192	38	222	918	13	0		1996	6 448	1 198	2 626	2 382	203	40	1.11
1940	1 299	42	229	1 017	11	0		1997	6 556	1 197	2 701	2 409	209	40	1.11
1941	1 334	42	236	1 043	12	0		1998	6 576	1 224	2 763	2 343	209	36	1.1
1942	1 342	45	222	1 063	11	0		1999	6 561	1 258	2 741	2 310	217	35	1.08
1943	1 391	50	239	1 092	10	0		2000	6 733	1 289	2 845	2 327	226	46	1.1
1944	1 383	54	275	1 047	7	0		2001	6 893	1 316	2 848	2 445	237	47	1.11
1945	1 160	59	275	820	7	0		2002	6 994	1 342	2 832	2 518	252	49	1.11
1946	1 238	61	292	875	10	0		2003	7 376	1 397	2 958	2 695	276	48	1.16
1947	1 392	67	322	992	12	0		2004	7 743	1 443	3 043	2 906	298	54	1.2
1948	1 469	76	364	1 015	14	0		2005	8 042	1 485	3 068	3 108	320	60	1.23
1949	1 419	81	362	960	16	0		2006	8 336	1 534	3 091	3 293	356	62	1.26
1950	1 630	97	423	1 070	18	23	0.65	2007	8 503	1 562	3 071	3 422	382	66	1.27
1951	1 767	115	479	1 129	20	24	0.69	2008	8 776	1 630	3 103	3 587	388	69	1.3
1952	1 795	124	504	1 119	22	26	0.69	2009	8 697	1 584	3 042	3 590	415	66	1.27
1953	1 841	131	533	1 125	24	27	0.69	2010	9 128	1 696	3 107	3 812	446	67	1.32
1954	1 865	138	557	1 116	27	27	0.69	2011	9 503	1 756	3 134	4 055	494	64	1.36
1955	2 042	150	625	1 208	30	31	0.74	2012	9 673	1 783	3 200	4 106	519	65	1.36
1956	2 177	161	679	1 273	32	32	0.77	2013	9 773	1 806	3 220	4 126	554	68	1.36
1957	2 270	178	714	1 309	34	35	0.79	2014	9 855	1 823	3 280	4 117	568	68	1.36

Unit: Million tonnes of carbon per year (MtC/yr), except the per capita emissions which are in tonnes of carbon per person per year (tC/person/yr). CM: Cement manufacturing.

Source: Boden, T.A., G. Marland, and R.J. Andres. 2017. Global, Regional, and National Fossil-Fuel CO2 Emissions. Carbon Dioxide Information Analysis Center, Oak Ridge National Laboratory, U.S. Department of Energy, Oak Ridge, Tenn, U.S.A.

Annual Global Fossil-Fuel CO_2 Emissions

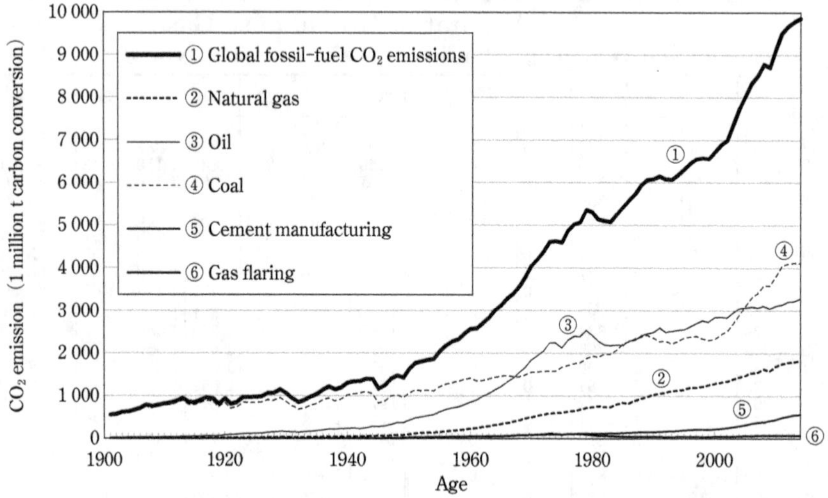

Total emissions amount of greenhouse gases in Japan *1

	GWP*2	1990	1991	1992	1993	1994	1995	1996	1997	1998	1999	2000	2001	2002
Carbon dioxide (CO_2)	1	1 164	1 175	1 185	1 177	1 232	1 245	1 257	1 250	1 210	1 246	1 269	1 254	1 283
Methane (CH_4)	25	44.4	43.3	44.1	40.0	43.4	41.9	40.7	40.0	38.1	38.0	38.0	37.1	36.4
Nitrous oxide (N_2O)	298	31.9	31.6	31.8	31.6	32.9	33.2	34.3	35.1	33.5	27.4	29.9	26.3	25.7
Hydrofluorocarbon (HFCs)	HFC-134a e.g. 1 430	15.9	17.3	17.8	18.1	21.1	25.2	24.6	24.4	23.7	24.4	22.9	19.5	16.2
Par fluorocarbon (PFCs)	PFC-14 : e.g. 7 390	6.5	7.5	7.6	10.9	13.4	17.6	18.3	20.0	16.6	13.1	11.9	9.9	9.2
Sulphur hexafluoride (SF_6)	22 800	12.9	14.2	15.6	15.7	15.0	16.4	17.0	14.5	13.2	9.2	7.0	6.1	5.7
Nitrogen Trifluoride (NF_3)	17 200	0.03	0.03	0.03	0.04	0.08	0.20	0.19	0.17	0.19	0.32	0.29	0.29	0.37
Total		1 275	1 289	1 302	1 294	1 358	1 379	1 392	1 384	1 335	1 359	1 379	1 353	1 377

	2003	2004	2005	2006	2007	2008	2009	2010	2011	2012	2013	2014	2015	2016	2017	2018
Carbon dioxide (CO_2)	1 291	1 286	1 293	1 270	1 306	1 235	1 165	1 216	1 266	1 308	1 317	1 267	1 225	1 205	1 190	1 138
Methane (CH_4)	35.0	36.0	35.8	35.3	35.5	35.2	34.3	34.8	33.8	32.9	32.5	31.9	31.1	30.7	30.2	29.9
Nitrous oxide (N_2O)	25.6	25.4	25.0	24.8	24.2	23.4	22.7	22.2	21.8	21.5	21.5	21.1	20.7	20.2	20.4	20.0
Hydrofluorocarbon (HFCs)	16.2	12.4	12.8	14.6	16.7	19.3	20.9	23.3	26.1	29.4	32.1	35.8	39.3	42.6	44.9	47.0
Par fluorocarbon (PFCs)	8.9	9.2	8.6	9.0	7.9	5.7	4.0	4.2	3.8	3.4	3.3	3.4	3.3	3.4	3.5	3.5
Sulphur hexafluoride (SF_6)	5.4	5.3	5.0	5.2	4.7	4.2	2.4	2.4	2.2	2.2	2.1	2.0	2.1	2.2	2.1	2.0
Nitrogen Trifluoride (NF_3)	0.42	0.49	1.5	1.4	1.6	1.5	1.4	1.5	1.8	1.5	1.6	1.1	0.57	0.63	0.45	0.28
Total	1 383	1 375	1 382	1 360	1 396	1 324	1 251	1 305	1 356	1 399	1 410	1 361	1 322	1 305	1 291	1 240

*1 Converted to 1 million tons CO_2. *2 Global Warming Index

Materials form the Ministry of the Environment: "Japan's National Greenhouse Gas Emissions in Fiscal Year 2018 (Final Figures)" (2020)

Pesticides

Amount of main pesticides produced in Japan

Agricultural chemicals name	Unit	2006 Aguri-cultural FY	2007	2008	2009	2010	2011	2012
Insecticide								
Acephate granules	t	7 100	6 143	5 399	6 140	4 388	4 241	4 070
Ethylthiometon granules	t	5 027	3 731	4 125	3 672	2 297	2 074	2 412
Ethofenprox dusting powder	t	2 872	3 083	3 134	2 726	2 020	2 245	1 859
Oxamyl granules	t	1 651	1 959	2 007	2 263	1 923	2 103	2 093
Cartap granules	t	1 793	2 216	1 693	2 302	1 414	1 941	1 418
Clothianidin dusting powder	t	1 887	2 533	1 628	1 426	1 216	1 166	987
Chloropicrin fumigation treatment	kL	8 857	9 441	9 238	9 411	6 580	8 081	7 379
Chlorpyrifos granules	t	741	903	1 018	1 175	954	1 110	947
Dinotefuran dusting powder	t	2 224	1 782	2 734	2 336	1 764	1 724	1 687
Methyl bromide fumigation treatment	kL	1 656	1 391	1 092	945	716	740	730
Diazinon granules	t	5 716	5 108	5 377	5 686	4 889	5 626	5 330
Tefluthrin granules	t	1 823	2 160	1 844	1 908	1 986	1 787	2 097
Benfuracarb granules	t	1 223	1 049	1 103	974	976	627	1 274
Fosthiazate granules	t	6 256	6 109	4 784	6 753	6 897	6 810	7 445
Machine oil emulsion	kL	6 739	5 930	7 319	6 014	5 550	5 825	5 199
D-D treatment	kL	10 009	9 262	9 783	9 103	9 351	9 996	6 253
DEP dusting powder	t	1 924	1 185	1 389	1 387	1 336	1 382	678
Fungicide								
Lime sulphur mixture	kL	5 808	6 103	5 978	5 909	5 357	5 140	5 267
Dazomet powder granules	t	2 967	3 071	3 233	3 633	3 249	3 416	3 359
Copper wettable powder	t	4 093	4 267	4 393	4 113	3 277	3 786	3 636
Pyroxylon granules	t	2 791	2 603	1 434	1 048	1 089	988	691
Fluazinam dusting powder	t	4 063	4 322	4 071	4 017	4 398	4 801	4 208
Flusulfamide dusting powder	t	4 167	4 215	4 187	3 917	3 749	3 499	3 383
Probenazole granules	t	2 551	2 503	2 606	2 242	2 104	1 783	1 516
Mancozeb wettable powder	t	3 562	2 675	2 983	2 833	2 882	2 336	2 388
Combination insecticide-fugicide								
Fipronil - Probenazole granules	t	1 642	1 604	1 714	1 392	2 187	1 094	400
Permethrin - myclobutanil solutions	kL	1 508	1 553	1 594	1 610	-	660	723
Herbicide								
Isouron-DBN-DCMU granules	t	991	1 415	1 426	585	98	9	-
Chlorate granules	t	2 849	2 949	2 767	2 720	2 360	2 148	2 290
Glyphosate isopropylamine salt solution	kL	6 459	6 917	7 360	8 178	8 084	9 241	9 754
Potassium salt of glyphosate solution	kL	455	3 243	7 303	5 125	4 235	4 813	6 832
Glufosinate solution	kL	2 402	2 601	2 442	1 889	1 586	1 932	1 594
Diquat-paraquat solution	kL	2 744	2 517	3 131	2 676	1 644	1 801	1 932
Cyhalofop-butyl -dimethametryn - pyrozosulfurn Ethyl - pretilachlor granules	t	1 506	1 423	1 138	678	762	803	663
Simetryn - molinate - MCPB granules	t	1 097	963	1 318	535	1 253	517	877
Trifluralin granules	t	2 035	2 906	2 632	2 946	2 140	2 276	2 222
Pretilachlor granules	t	1 635	1 329	1 371	1 264	1 107	1 293	1 065
Promasil granules	t	1 780	2 193	1 789	1 759	1 648	1 515	1 621
Bentazone granules	t	1 833	1 397	2 271	2 305	1 564	1 803	1 420
DBN granules	t	2 735	3 017	2 111	3 429	2 689	2 626	2 335
MCPP solution	kL	290	2 420	1 731	501	533	514	416

The pesticide year is from October of the previous year to September of this year.
Source: Japan Plant Protection Association "Agricultural Chemcials Handbook" (2013)

Japanese imports/exports of insecticides, fungicides, antiseptics, herbicides, etc.

Age	Insecticide		Fungicide		Herbicides*		Disinfectant	
	Amount	Price	Amount	Price	Amount	Price	Amount	Price
Import								
2005	9 695	9 511	10 698	10 027	20 374	12 084	1 359	459
2008	12 302	11 745	10 831	9 603	26 205	18 832	2 797	753
2009	12 749	12 008	12 211	11 223	23 735	16 483	7 131	2 248
2010	12 695	12 113	11 033	9 685	21 291	11 641	4 865	1 460
2011	16 226	14 375	10 212	10 038	21 712	11 359	7 883	1 930
2012	14 065	11 893	11 898	10 266	24 289	12 365	9 698	2 310
2013	15 863	15 410	13 066	11 557	25 992	15 649	11 846	3 185
2014	16 169	15 706	12 103	11 844	27 591	16 477	13 035	4 617
Export								
2005	4 548	13 116	7 632	8 614	3 134	10 695	538	256
2008	5 054	18 034	4 863	7 531	4 073	10 831	306	206
2009	4 476	13 677	4 523	6 771	4 456	10 962	216	185
2010	5 363	15 946	5 472	7 587	4 601	9 719	405	216
2011	5 115	15 528	6 316	8 161	4 605	9 971	743	436
2012	5 556	14 026	5 592	8 215	3 710	8 033	659	428
2013	5 483	16 128	5 739	9 089	3 408	8 363	715	636
2014	6 071	17 959	6 128	10 815	3 820	11 541	803	1 292

Weight: tons; Monetary value: 1 million yen

* Includes germination inhibitors and plant growth regulators.

Insecticides, fungicides, antiseptics, herbicides, etc. include substances intended for uses other than agriculture.

Based on material from Ministry of Finance Customs and Tariff Bureau "Trade Statistics of Japan" Ministry of Agriculture, Forestry and Fisheries Secretariat Statistics Division: "Agriculture, Forestry and Fisheries Statistics" (2015)

Occupational Exposure Limits for Chemical Substances

Occupational exposure limit is the concentration of a chemical substance in air which is determined as not having a negative effect on the health of the majority of workers during a job without the use of protective respiratory equipment. It is assumed that workers engage in work for approximately 8 hours a day and 40 hours a week under a moderate work-load. Exposure which exceeds the occupational exposure limit should be avoided even in the case of short-period exposure or light labor intensity.

The values in this report are the values recommended by the Japan Society for Occupational Health (2020). These values are for when the applicable substance exists alone in the air. When exposed to multiple substances, it is not acceptable to make judgements based only on the occupational exposure limit of each individual substance. Furthermore, when using the values listed below, consideration must be given to conditions such as labor intensity and temperature conditions. When these conditions are added into consideration, there is the possibility that the impact of hazardous substances on health will increase.

Characteristics of Occupational Exposure Limits and Instruction for Users (excerpt)

- Occupational exposure limits (OELs) should be applied by individuals well-trained and experienced in occupational health.
- OELs cannot be applied in cases where exposure duration or work intensity exceeds the prerequisite conditions for setting an OEL.
- OELs are set based on various information obtained from experiences in industries and experiments on humans and animals. However, the quantity and quality of information used in setting OELs is not always the same.
- Types of health effects considered in setting OELs depend on the substances involved; an explicit health impairment provides the basis for OELs in certain substances, while health effects such as discomfort, irritation or central nervous system suppressive effects afford the basis in others. Thus, OELs cannot be used simply as a relative scale of toxicity.
- Due to the variance in individual susceptibilities, discomfort, deterioration of pre-existing ill health or occupational disease may be induced at levels of exposure below the OELs, even though the chances of this should be remote.
- Because OELs do not represent a definitive borderline between safe and hazardous conditions, it is not correct to conclude that working environments above OEL are the direct and sole cause of health impairment in workers, or vice versa.
- OELs cannot be applied as reference values in non-occupational environments.

Occupational Exposure Limits for Chemical Substances

Substance name	Chemical formula	Occupational Exposure Limits		Proposed fiscal year
		ppm	mg/m^3	
Acetaldehyde	CH_3CHO	50*	90*	'90
Acetic acid	CH_3COOH	10	25	'78
Acetic anhydride	$(CH_3CO)_2O$	5*	21*	'90
Acetone	CH_3COCH_3	200	470	'72
Acrylamide (skin)	CH_2=$CHCONH_2$	–	0.1	'04
Acrylaldehyde	CH_2=$CHCHO$	0.1	0.23	'73
Acrylonitrile (skin)	CH_2=$CHCN$	2	4.3	'88
Allyl alcohol (skin)	CH_2=$CHCH_2OH$	1	2.4	'78
2-Aminoethanol	$H_2NCH_2CH_2OH$	3	7.5	'65
Ammonia	NH_3	25	17	'79
Aniline (skin)	$C_6H_5NH_2$	1	3.8	'88
o-Anisidine (skin)	$H_3COC_6H_4NH_2$	0.1	0.5	'96
p-Anisidine (skin)	$H_3COC_6H_4NH_2$	0.1	0.5	'96
Antimony and compounds (as Sb, except Stibine)	Sb	–	0.1	('13)
Arsenic and compounds (as As)	As	Excessive cancer risk		'00
Arsine	AsH_3	0.01 0.1*	0.032 0.32*	'92
Asbestos		Excessive cancer risk		'00
Atrazine	$C_8H_{14}ClN_5$	–	2	'15
Benomyl	$C_{14}H_{18}N_4O_3$	–	1	'18
Benzene (skin)	C_6H_6	Excessive cancer risk		'97
Benzyl alcohol	C_7H_8O	–	25	'19
Beryllium and compounds (as Be)	Be	–	0.002	'63
Boron trifluoride	BF_3	0.3	0.83	'79
Bromine	Br_2	0.1	0.65	'64
Bromoform	$CHBr_3$	1	10.3	'97
1-Bromopropane	$Br-CH_2CH_2CH_3$	0.5	2.5	'12
2-Bromo-propane (skin)	$CH_3CHBrCH_3$	1	5	'99
Buprofezin	$C_{16}H_{23}N_3OS$	–	2	'90
Butane (all isomers)	C_4H_{10}	500	1200	'88
2-Butanol	$CH_3CH(OH)CH_2CH_3$	100	300	'87
1-Butanol (skin)	$CH_3CH_2CH_2CH_2OH$	50*	150*	'87
Butyl acetate	$CH_3COO(CH_2)_3CH_3$	100	475	'94
t-Butyl alcohol	$(CH_3)_3COH$	50	150	'87
Butylamine (skin)	$CH_3CH_2CH_2CH_2NH_2$	5*	15*	('94)
n-Butyl-2,3-epoxypropylether	$C_7H_{14}O_2$	0.25	1.3	'16
Cadmium and compounds (as Cd)	Cd	–	0.05	'76
Calcium cyanide (as CN) (skin)	$Ca(CN)_2$	–	5*	'01
Carbaryl (skin)	$C_{12}H_{11}NO_2$	–	5	'89
Carbon dioxide	CO_2	5000	9000	'74
Carbon disulfide (skin)	CS_2	1	3.13	'15
Carbon monoxide	CO	50	57	'71
Carbon tetrachloride (skin)	CCl_4	5	31	'91
Chlorine	Cl_2	0.5*	1.5*	'99
Chlorobenzene	C_6H_5Cl	10	46	'93
Chlorodifluoromethane	$CHClF_2$	1000	3500	'87
Chloroethane	C_2H_5Cl	100	260	'93
Chloroform	$CHCl_3$	3	14.7	'05
Chloromethane	CH_3Cl	50	100	'84
Chloromethyl methyl ether (technical grade)	CH_3OCH_2Cl	No numerical value		'92
Chloropicrin	Cl_3CNO_2	0.1	0.67	'68
Chromium and compounds (as Cr)	Cr			'89
Chromium Metal		–	0.5	

Continued.

Substance name	Chemical formula	Occupational Exposure Limits		Proposed fiscal year
		ppm	mg/m^3	
Chromium (Ⅲ) compounds		–	0.5	
Chromium (Ⅵ) compounds		–	0.05	
Certain Chromium (Ⅵ) compounds		–	0.01	
Cobalt and compounds (as Co)	Co	–	0.05	'92
Cresol (all isomers) (skin)	$C_6H_4CH_3(OH)$	5	22	'86
Cumene	C_9H_{12}	10	–	'19
Cyclohexane	C_6H_{12}	150	520	'70
Cyclohexanol	$C_6H_{11}OH$	25	102	'70
Cyclohexanone	$C_6H_{10}O$	25	100	'70
Di-2-ethylhexyl phthalate	$C_{24}H_{38}O_4$	–	5	'95
Diazinon (skin)	$C_{12}H_{21}N_2O_3PS$	–	0.1	'89
Diborane	B_2H_6	0.01	0.012	'96
Dibutyl phthalate	$C_6H_4(COOC_4H_9)_2$	–	5	'96
2,2'-Dichloro ethyl ether (skin)	$(ClCH_2CH_2)_2O$	15	88	'67
2,2-Dichloro-1,1,1-trifluoroethane	CF_3CHCl_2	10	62	'00
1,4-Dichloro-2-butene	$C_4H_6Cl_2$	0.002	–	'15
3,3'-Dichloro-4,4'-diaminodiphenylmethane (MBOCA) (skin)	$CH_2(C_6H_3NH_2Cl)_2$	–	0.005	'12
o-Dichlorobenzene	$C_6H_4Cl_2$	25	150	'94
p-Dichlorobenzene	$C_6H_4Cl_2$	10	60	'98
Dichlorodifluoromethane	CCl_2F_2	500	2500	'87
1,1-Dichloroethane	Cl_2CHCH_3	100	400	'93
1,2-Dichloroethane	$ClCH_2CH_2Cl$	10	40	'84
1,2-Dichloroethylene	$ClCH{=}CHCl$	150	590	'70
2,4-Dichlorophenoxyacetic acid (skin)	$C_8H_6Cl_2O_3$	–	2	'19
Dichloromethane (skin)	CH_2Cl_2	50	170	'99
1,2-Dichloropropane	$C_3H_6Cl_2$	1	4.6	'13
Di (2-ethylhexyl) phthalate	$C_{24}H_{38}O_4$	–	5	'95
Diethyl phthalate	$C_6H_4(COOC_2H_5)_2$	–	5	'95
Diethylamine	$(C_2H_5)_2NH$	2	4	('17)
Dimethyl sulfate (skin)	$(CH_3)_2SO_2$	0.1	0.52	'80
N, N-Dimethylacetamide (skin)	$(CH_3)_2NCOCH_3$	10	36	'90
Dimethylamine	$(CH_3)_2NH$	2	3.7	'16
N, N-Dimethylaniline (skin)	$C_6H_5N(CH_3)_2$	5	25	'93
N, N-Dimethylformamide (DMF) (skin)	$(CH_3)_2NCHO$	10	30	'74
1,2-Dinitrobenzene (skin)	$C_6H_4(NO_2)_2$	0.15	1	'94
1,3-Dinitrobenzene (skin)	$C_6H_4(NO_2)_2$	0.15	1	'94
1,4-Dinitrobenzene (skin)	$C_6H_4(NO_2)_2$	0.15	1	'94
1,4-Dioxane (skin)	$C_4H_8O_2$	1	3.6	'15
Diphenylmethane-4,4'-diisocyanate (MDI)	$CH_2(C_6H_4NCO)_2$	–	0.05	'93
Disulphur dichloride	S_2Cl_2	1*	5.5*	'76
Ethofenprox	$C_{25}H_{28}O_3$	–	3	'95
Ethyl acetate	$CH_3COOC_2H_5$	200	720	'95
Ethyl ether	$(C_2H_5)_2O$	400	1200	('97)
Ethylamine	$C_2H_5NH_2$	10	18	'79
Ethylbenzene	$C_6H_5C_2H_5$	50	217	'01
Ethylene diamine (skin)	$N_2NCH_2CH_2NH_2$	10	25	'91
Ethylene glycol monobutyl ether (skin)	$C_6H_{14}O_2$	20	97	'17
Ethylene glycol monoethyl ether (skin)	$C_2H_5OCH_2CH_2OH$	5	18	'85
Ethylene glycol monoethyl ether acetate (skin)	$C_2H_5OCH_2CH_2OCOCH_3$	5	27	'85
Ethylene glycol monomethyl ether (skin)	$CH_3OCH_2CH_2OH$	0.1	0.31	'09
Ethylene glycol monomethyl ether acetate (skin)	$CH_3OCH_2CH_2OCOCH_3$	0.1	0.48	'09
Ethylene oxide	C_2H_4O	1	1.8	'90
Ethylenimine (skin)	C_2H_5N	0.5	0.88	('90)

Occupational Exposure Limits for Chemical Substances Continued.

Substance name	Chemical formula	Occupational Exposure Limits		Proposed fiscal year
		ppm	mg/m^3	
2-Ethyl-1-hexanol	$C_8H_{18}O$	1	5.4	'17
Ethylidene norbornane	C_9H_{12}	2	10	'18
Fenitrothion (skin)	$C_9H_{12}NO_5PS$	–	1	'81
Fenobucarb (skin)	$C_{12}H_{17}NO_2$	–	5	'89
Fenthion (skin)	$C_{10}H_{15}O_3PS_2$	–	0.2	'89
Flutolanil	$C_{17}H_{16}NO_2F_3$	–	10	'90
Formaldehyde	HCHO	0.1 0.2*	0.12 0.24*	'07
Formic acid	HCOOH	5	9.4	'78
Fthalide	$C_8H_2Cl_4O_2$	–	10	'90
Furfural (skin)	$C_5H_4O_2$	2.5	9.8	('89)
Furfuryl alcohol	$C_4H_3OCH_2OH$	5	20	'78
Gasoline		100[b]	300[b]	'85
Glutaraldehyde	$OHC(CH_2)_3CHO$	0.03*	–	'06
Heptane	$CH_3(CH_2)_5CH_3$	200	820	'88
Hexachlorobutadiene (skin)	C_4Cl_6	0.01	0.12	'13
Hexane (skin)	$CH_3(CH_2)_4CH_3$	40	140	'85
Hexane-1,6-diisocyanate	$OCN(CH_2)_6NCO$	0.005	0.034	'95
Hydrazine (anhydrous) and hydrazine hydrate (skin)	H_2NNH_2 and $H_4N_2H_2O$	0.1	0.13 and 0.21	'98
Hydrogen chloride	HCl	2*	3.0*	'14
Hydrogen cyanide (skin)	HCN	5	5.5	'90
Hydrogen fluoride	HF	3*	2.5*	'00
Hydrogen selenide	SeH_2	0.05	0.17	'63
Hydrogen sulfide	H_2S	5	7	'01
Indium and compound	In	Theoretical tolerable quantity in serum of living organisms	3 μg/L	'07
Iodine	I_2	0.1	1	'68
Isobutyl alcohol	$(CH_3)_2CHCH_2OH$	50	150	'87
Isopentyl alcohol	$(CH_3)_2CHCH_2OH$	100	360	'66
Isoprene	C_5H_8	3	8.4	'17
Isopropyl acetate	$CH_3COOCH(CH_3)_2$	100		'17
Isopropyl alcohol	$CH_3CH(OH)CH_3$	400*	980*	'87
Isoprothiolane	$C_{12}H_{18}O_4S_2$	–	5	'93
Lead and compounds (as Pb, except alkyl lead compounds)	Pb	–	0.03	'16
Lithium hydroxide	LiOH		1	'95
Malathion (skin)	$C_{10}H_{16}O_6PS_2$	–	10	'89
Maleic anhydride	$C_4H_2O_3$	0.1	0.4	('15)
Maleic anhydride	$C_4H_2O_3$	0.2*	0.8*	
Manganese and compounds (as Mn, except organic compounds)	Mn	–	0.2	'08
Man-made mineral fiber**				
Ceramic fibers, Micro glass fibers		–	–	'03
Continuous filament glass fibers		1 (fiber/mL)		'03
Glass wool fibers, Rock wool fibers, Slag wool fibers		1 (fiber/mL)		'03
Mepronil	$C_{17}H_{19}NO_2$	–	5	'90
Mercury lithium	LiOH	–	1	'95
Mercury vapor	Hg	–	0.025	'98
Methacrylic acid	$C_4H_6O_2$	2	7.0	'12
Methacrylic acid 2,3-epoxypropyl (glycidyl methacrylate) (skin)	$C_7H_{10}O_3$	0.01	0.06	'18
Methanol (skin)	CH_3OH	200	260	'63
Methyl acetate	CH_3COOCH_3	200	610	'63
Methyl acrylate	$C_4H_6O_2$	2	7	'04

Continued.

Substance name	Chemical formula	Occupational Exposure Limits		Proposed fiscal year
		ppm	mg/m^3	
Methyl bromide (skin)	CH_3Br	1	3.89	'03
Methyl cyclohexanone (skin)	$CH_3C_6H_9O$	50	230	'87
Methyl ethyl ketone	$C_2H_5COCH_3$	200	590	'64
Methyl isobutyl ketone	$CH_3COCH_2CH(CH_3)_2$	50	200	'84
Methyl methacrylate	$C_5H_8O_2$		8.3	'12
N-methyl-2-pyrrolidone (skin)	C_5H_9NO	1	4	'02
Methylamine	CH_3NH_2	10	13	'79
Methylcyclohexane	$CH_3C_6H_{11}$	400	1600	'86
Methylcyclohexanol	$CH_3C_6H_{10}OH$	50	230	'80
4,4'-Methylene dianiline (skin)	$CH_2(C_6H_4NH_2)_2$	–	0.4	'95
Methyl-*n*-butyl ketone (skin)	$CH_3CO(CH_2)_3CH_3$	5	20	'84
Methyltetrahydrophthalic anhydride	$C_9H_{10}O_3$	0.007 0.015*	0.05 0.1*	'02
Mineral oil mist		–	3	'77
Nickel	Ni	–	1	'11
Nickel carbonyl	$Ni(CO)_4$	0.001	0.007	'66
Nickel compounds (total dusts) (as Ni)				
Nickel compounds, not soluble		–	0.1	'11
Nickel compounds, soluble		–	0.01	'11
Nitric acid	HNO_3	2	5.2	'82
p-Nitro chlorobenzene (skin)	$C_6H_4ClNO_2$	0.1	0.64	'89
Nitro glycol (skin)	$O_2NOCH_2CH_2ONO_2$	0.05	0.31	'86
p-Nitroaniline (skin)	$H_2NC_6H_4NO_2$	–	3	'95
Nitrobenzene (skin)	$C_6H_5NO_2$	1	5	('88)
Nitrogen dioxide	NO_2	(pending)		'61
Nitroglycerin (skin)	$(O_2NOCH_2)_2CHONO_2$	0.05*	0.46*	'86
Nonane	$CH_3(CH_2)_7CH_3$	200	1050	'89
Octane	$CH_3(CH_2)_6CH_3$	300	1400	'89
Oil mist, mineral		–	3	'77
Ozone	O_3	0.1	0.2	'63
Parathion (skin)	$(C_2H_5O)_2PSOC_6H_4NO_2$	–	0.1	('80)
Pentachlorophenol (skin)	C_6Cl_5OH	–	0.5	('89)
Perfluorooctanoic acid	$C_8HF_{15}O_2$		0.005	'08
Pentane	$CH_3(CH_2)_3CH_3$	300	880	'87
Pentyl acetate, All isomers	$CH_3COO(CH_2)_2CH(CH_3)_2$	50 100*	266.3 532.5*	'08
Phenol (skin)	C_6H_5OH	5	19	'78
m-Phenylenediamine	$C_6H_4(NH_2)_2$	–	0.1	'99
o-Phenylenediamine	$C_6H_4(NH_2)_2$	–	0.1	'99
p-Phenylenediamine	$C_6H_4(NH_2)_2$	–	0.1	'97
Phosgene	$COCl_2$	0.1	0.4	'69
Phosphine	PH_3	0.3*	0.42*	'98
Phosphoric acid	H_3PO_4	–	1	('90)
Phosphorus (yellow)	P_4	–	0.1	('88)
Phosphorus pentachloride	PCl_5	0.1	0.85	'89
Phosphorus trichloride	PCl_3	0.2	1.1	'89
Phthalic anhydride	$C_8H_4O_3$	0.33*	2*	'98
o-Phthalodinitrile (skin)	$C_6H_4(CN)_2$	–	0.01	'09
Platinum (soluble salts (as Pt)	Pt	–	0.001	'00
Polychlorinated biphenyls (skin)	$C_{12}H_{(10-n)}Cl_n$	–	0.01	'06
Potassium cyanide (as CN) (skin)	KCN	–	5*	'01
Potassium hydroxide	KOH	–	2*	'78
Propyl acetate	$CH_3COO(CH_2)_2CH_3$	200	830	'70

Occupational Exposure Limits for Chemical Substances Continued.

Substance name	Chemical formula	Occupational Exposure Limits ppm	Occupational Exposure Limits mg/m^3	Proposed fiscal year
Propyleneimine (skin)	C_3H_7N	0.2	0.45	'17
Pyridaphenthion (skin)	$C_{14}H_{17}N_2O_4PS$	–	0.2	'89
Rhodium (soluble compound, as a Rh)	Rh	–	0.001	'07
Selenium and compounds (as Se, except SeH_2 and SeF_6)	Se	–	0.1	'00
Silane	SiH_4	100*	130*	'93
Silver and silver compounds (as Ag)	Ag	–	0.01	'91
Sodium cyanide (as CN) (skin)	NaCN	–	5*	'01
Sodium hydraoxide	NaOH	–	2*	'78
Styrene (skin)	$C_6H_5CH{=}CH_2$	20	85	'99
Sulfur dioxide	SO_2	(pending)		'61
Sulfuric acid	H_2SO_4	–	1*	'00
Sulfur monochloride	Cl_2S_2	1	5.5	'76
Tetrachlorethylene (skin)	$Cl_2C{=}CCl_2$	(pending)		'72
1,1,2,2-Tetrachloroethane (skin)	$Cl_2CHCHCl_2$	1	6.9	'84
Tetraethoxysilane	$Si(OC_2H_5)_4$	10	85	'91
Tetraethyl lead (as Pb) (skin)	$Pb(C_2H_5)_4$	–	0.075	'65
Tetrahydrofuran	C_4H_8O	50	148	'15
Tetramethoxysilane	$Si(OCH_3)_4$	1	6	'91
Thiuram	$C_6H_{12}N_2S_4$	–	0.1	'08
Titanium dioxide (nano-particles)	TiO_2	–	0.3	'13
Toluene (skin)	$C_6H_5CH_3$	50	188	'94
Toluene diisocyanates (TDI)	$C_6H_3CH_3(NCO)_2$	0.005 0.02*	0.035 0.14*	'92
o-Toluidine (skin)	$CH_3C_6H_4NH_2$	1	4.4	'91
Trichlorethylene	$Cl_2C{=}CHCl$	25	135	'15
Trichlorhon (skin)	$C_4H_8Cl_3O_4P$	–	0.2	'10
1,1,2-Trichloro-1,2,2-trifluoroethane	$Cl_2FCCClF_2$	500	3800	'87
1,1,1-Trichloroethane	Cl_3CCH_3	200	1100	'74
1,1,2-Trichloroethane (skin)	Cl^2CHCH_2Cl	10	55	('78)
Trichlorofluoromethane	CCl_3F	1000*	5600*	'87
Tricyclazole	$C_9H_7N_3S$	–	3	'90
Trimellitic anhydride (skin)	$C_9H_4O_5$	– –	0.0005 0.004*	'15
1,2,3-Trimethylbenzene	$C_6H_3(CH_3)_3$	25	120	'84
1,2,4-Trimethylbenzene	$C_6H_3(CH_3)_3$	25	120	'84
1,3,5-Trimethylbenzene	$C_6H_3(CH_3)_3$	25	120	'84
Trinitrotoluene (all isomers) (skin)	$C_6H_2CH_3(NO_2)_3$	–	0.1	'93
Turpentine		50	280	'91
Vanadium compounds				
Ferro vanadium dust	FeVdust	–	1	'68
Vanadium pentoxide	V_2O_5	–	0.05	'03
Xylene (all isomers and their mixtures)	$C_6H_4(CH_3)_2$	50	217	'01
Zinc oxide fume	ZnO	(pending)		'69

ppm: parts of vapors and gases per million of substance in air by volume at 25℃ and atmospheric pressure (760 torr, 1 013 hPa)

( ) in the year of proposal column indicates that revision was done in the year without change of the occupational exposure level value.

* Occupational Exposure Limit-Ceiling; exposure concentration must be kept below this level.

** Fibers longer than 5 μm and with an aspect ratio equal to or greater than 3:1 as determined by the membrane filter method at 400 × magnification phase contrast illumination.

[a] Exposure concentration should be kept below a detectable limit though occupational exposure limit is set at 2.5 ppm provisionally.

[b] Occupational exposure level for gasoline is 300 mg/m^3, and an average molecular weight is assumed to be 72.5 for conversion to ppm unit.

Main Chemical Substances and Manufacturing Process Judged to be Human Carcinogens

Substance	Chemical formula [Compositional formula]
Acetaldehyde associated with consumption of alcoholic beverages	C_2H_4O
Acheson process, occupational exposure associated with	
Acid mists, strong inorganic[4)]	
Aflatoxins	$C_{17}H_{12}O_6$
Aluminium production[4)]	Al
4-Aminobiphenyl	$C_{12}H_{11}N$
Areca nut, Betel nut or Betel quid with tobacco	
Aristolochic acid, plants containing	$C_{17}H_{11}NO_7$
Arsenic and inorganic arsenic compounds[2)]	
Asbestos (all forms, including actinolite, amosite, anthophyllite, chrysotile, crocidolite, tremolite)	
Auramine production[4)]	$C_{17}H_{22}ClN_3$
Azathioprine	$C_3H_2N_2(CH_3)NO_2SC_5H_2N_4$
Benzene	C_6H_6
Benzidine	$C_{12}H_{12}N_2$
Benzo[*a*]pyrene	$C_{20}H_{12}$
Beryllium and beryllium compounds[2)]	Be
Bis(chloromethyl)ether; chloromethyl methyl ether (technical-grade)	$ClCH_2OCH_2Cl$, $ClCH_2OCH_3$
Busulfan	$C_6H_{14}O_6S_2$
1,3-Butadiene	$CH_2=CHCH=CH_2$
Cadmium and cadmium compounds[2)]	Cd
Chlorambucil	$(ClCH_2CH_2)_2NC_6H_4CH_2CH_2CH_2COOH$
Chlornaphazine	C_2H_5ClO
Chromium (VI) compounds[2)]	
Coal gasification[4)]	
Coal, indoor emissions from household combustion of	
Coal-tar distillation[3)]	
Coke production[4)]	
Cyclophosphamide	$C_3ONHP(=O)N(CH_2CH_2Cl)_2 \cdot H_2O$
Cyclosporine	$C_{62}H_{111}N_{11}O_{12}$
1,2-Dichloropropane	$C_3H_6Cl_2$
Diethylstilbestrol	$HOC_6H_4C(C_2H_5)C(C_2H_5)C_6H_4OH$
Engine exhaust, diesel	
Epstein-Barr virus	
Erionite	$Al_4CaKNaO_{33}Si_{12.12}H_2O$
Estrogen therapy, postmenopausal	$C_{18}H_{24}O_2$
Estrogen-progestogen menopausal therapy (combined)	$C_{18}H_{24}O_2$, $C_{21}H_{30}O_2$

Main Chemical Substances and Manufacturing Process Judged to be Human Carcinogens Continued.

Substance	Chemical formula [Compositional formula]
Ethanol in alcoholic beverages	C_2H_6O
Ethylene oxide	C_2H_4O
Etoposide in combination with cisplatin and bleomycin	$C_{29}H_{32}O_{13}$, $Cl_2H_6N_2Pt$, $C_{55}H_{84}N_{17}O_{21}S_3$
Fluoro-edenite fibrous amphibole	$NaCa_2Mg_5(Si_7Al)O_{22}F_2$
Formaldehyde	HCHO
Haematite mining (underground)	Fe_2O_3
Helicobacter pylori (infection with)	
Hepatitis B and C virus (chronic infection with)	
Human immunodeficiency virus type 1 (infection with)	
Human papillomavirus types 16, 18, 31, 33, 35, 39, 45, 51, 52, 56, 58, 59	
Human T-cell lymphotropic virus type I	
Iron and steel founding (occupational exposure during)[4]	
Isopropyl alcohol manufacture using strong acids[4]	C_3H_8O, H_2SO_4
Kaposi sarcoma herpesvirus	
Leather dust[3]	
Lindane (see also Hexachlorocyclohexanes)	$C_6H_6Cl_6$
Magenta production[4]	
Melphalan	$HOOCCH(NH_2)CH_2C_6H_4N(CH_2CH_2Cl)_2$
Methoxsalen (8-methoxypsoralen) plus ultraviolet A radiation	$C_{11}H_6O_3$
4,4'-Methylenebis(2-chloroaniline) (MOCA)	$CH_2(C_6H_3ClNH_2)_2$
Mineral oils, untreated or mildly treated[3]	
MOPP and other combined chemotherapy including alkylating agents	
2-Naphthylamine	$C_{10}H_7NH_2$
Neutron radiation	
Nickel compounds[2]	
Opisthorchis viverrini (infection with)	
Outdoor air pollution, particulate matter in	
Painter (occupational exposure as a)[4]	
2,3,4,7,8-Pentachlorodibenzofuran, 3,3',4,4',5-Pentachlorobiphenyl (PCB-126)	$C_{12}H_3Cl_5O$, $C_{12}H_5Cl_5$
Pentachlorophenol (see also Polychlorophenols)	C_6HCl_5O
Phenacetin, analgesic mixtures containing	$CH_3CONHC_6H_4OCH_2CH_3$
Plutonium	Pu
Processed meat (consumption of)	
Radioactive substances:	
X- and Gamma-Radiation	
Radionuclides, alpha- and beta-particle-emitting, internally deposited	
Radioiodines, including iodine-131	

Continued.

Substance	Chemical formula [Compositional formula]
Fission products, including strontium-90	
Ionizing radiation (all types)	
Radium-224, -226, -228 and its decay products	
Radon-222 and its decay products	
Phosphorus-32, as phosphate	
Rubber manufacturing industry[4)]	
Salted fish, Chinese-style	
Schistosoma haematobium (infection with)	
Semustine [1-(2-Chloroethyl)-3-(4-methylcyclohexyl)-1-nitrosourea, Methyl-CCNU]	$C_{10}H_{18}ClN_3O_2$
Shale oils[3)]	
Silica dust, crystalline, in the form of quartz or cristobalite[4)]	
Solar radiation	
Soot (as found in occupational exposure of chimney sweeps)[3)]	
Sulfur mustard	$C_4H_8Cl_2S$
Tamoxifen	$(CH_3)_2NCH_2CH_2OC_6H_4(C_6H_5)C=C(C_6H_5)C_2H_5$
2,3,7,8-Tetrachlorodibenzo-*para*-dioxin	$C_{12}H_4Cl_4O_2$
Thiotepa	$(CH_2CH_2N)_3P=S$
Thorium-232 and its decay products	
Tobacco smoking[3)]	
Ingredient of tobacco smoking:	
N-Nitrosonornicotine (NNN) and 4-(*N*-Nitrosomethylamino)-1-(3-pyridyl)-1-butanone (NNK)[3)]	
o-Toluidine	$CH_3C_6H_4NH_2$
Treosulfan	$CH_3SO_2OCH_2CH(OH)CH(OH)CH_2OSO_2CH_3$
Trichloroethylene	$Cl_2C=CHCl_2$
Ultraviolet radiation (wavelengths 100–400 nm, encompassing UVA, UVB, and UVC)	
Vinyl chloride[1)]	$H_2C=CHCl$
Welding fumes	
Wood dust	

Based on evaluations (vol. 1 to 116) of the WHO/International Agency for Research on Cancer (WHO/IARCC) and recommendations for allowable concentrations by the Japan Society for Occupational Health (JOH) (2015).

1) Evaluation for manufacturing processes and other exposure environments.
2) Refers to vinyl chloride and vinylchloride monomer.
3) An evaluation of the entire chemical substance group. In some cases, sufficient evaluation may not have been conducted for the carcinogenicity of individual substances.
4) Evaluation for mixtures.

Persistent Organic Pollutants (POPs) and Survey on Chemical Substance Environmental Conditions

In order to prevent global environmental pollution caused by persistent organic pollutants (POPs) such as PCB, DDT, dioxin and other substances with a high rate of environmental residue, actions by only some countries are not sufficient. Instead, international cooperation is required in order to eliminate and reduce POPs. Accordingly, the Stockholm Convention on Persistent Organic Pollutants (POPs Convention) was conducted in May 2001. Japan conducted a treaty in August 2002 and the treaty was enacted in May 2004.

The POPs Convention requires each party to (1) prohibit and/or eliminate the intentional manufacturing and use of POPs, (2) reduce or eliminate releases from unintentionally produced POPs, (3) conduct appropriate management and processing of stockpiles and wasters which include POPs, (4) develop and implement domestic plans for executing the duties of the convention, (5) and to take other related measures (measures to prevent the manufacturing and use of new POPs; surveys, research, monitoring, exchange of information, and education related to POPs; technical and financial support for developing nations).

Currently, the 32 substances listed below are targeted by the POPs Convention.

Substances subject to the POPs Convention

Chemical	Annex
Aldrin	Pesticide
Dieldrin	Pesticide
Endrin	Pesticide
Chlordane	Pesticide
Heptachlor	Pesticide
Toxaphene	Pesticide
Mirex	Pesticide
Hexachlorobenzene (HCB)	Pesticide/Unintentional Production
Polychlorinated biphenyls (PCB)	industrial chemical/Unintentional Production
Chlordecone	Pesticide
Pentachlorobenzene (PeCB)	Pesticide/Unintentional Production
Tetrapentabromodiphenyl ether	Industrial chemical
Hexaheptabromodiphenyl ether	Industrial chemical
Hexabromobiphenyl	Industrial chemical
Lindane	Pesticide
α-hexachlorocyclohexane	Pesticide
β-hexachlorocyclohexane	Pesticide
Hexabromocyclododecanes	Industrial chemical
Endosulfan	Pesticide
DDT	Pesticide
Perfluorooctane sulfonic acid (PFOS), its salts, perfluorooctane sulfonyl fluoride (PFOSF)	Water and oil repellant/Surfactant/Industrial chemical
Polychlorinated dibenzo-p-dioxins (PCDD)	Pesticide/Unintentional Production
Polychlorinated dibenzofurans (PCDF)	Pesticide/Unintentional Production
Polychlorinated naphthalene	Preservative/Unintentional Production
Hexachlorobutadiene	Solvent
Pentachlorophenol and its salts and esters	Agricultural chemicals
Decabromodiphenyl ether (DecaBDE)	Flame retardant
Short-chain chlorinated paraffins (SCCPs)	Flame retardant/Metal working oil
Dicofol	Insecticide
Perfluorooctanoic acid (PFOA), its salts, related substances	Fluoropolymer processing material/Surfactant

In addition, Article 16 of the POPs Convention stipulates that long-term continuous monitoring, analysis and evaluation shall be conducted to assess the effectiveness of the Convention. In response to this, Therefor, in order to ascertain the actual status of residues in the environment, for substances covered by the POPs Convention other than dioxins for which detailed monitoring is conducted separately, surveys of actual residues in air, water, quality, and organisms have been conducted in Japan.

Results of 2018 POPs Monitoring Survey

Surveyed media / Target chemicals	Surface Water (pg/L)	Bottom Sediment (pg/g-dry)	Wildlife (pg/g-wet)			Air (pg/m^3)
			Shellfish	Fish	Birds	warm season
	Detection range (detected frequency)	Detection range (detected frequency)	Detection range (detected frequency)	Detection range (detected frequency)	Detection range (detected frequency)	Detection range (detected frequency)
Polychlorinated biphenyls (PCBs) (total)	tr (11)~2 600 (47/47)	nd~720 000 (58/61)	740~12 000 (3/3)	1 200~280 000 (18/18)	85 000~130 000 (2/2)	20~750 (37/37)
Hexachlorobenzene (HCB)	4.0~380 (47/47)	3.1~8 900 (61/61)	14~28 (3/3)	25~900 (18/18)	2 600~3 100 (2/2)	72~140 (37/37)
Aldrin	–	nd~270 (50/61)	–	–	–	–
Dieldrin	–	nd~860 (60/61)	–	–	–	–
Endrin	–	nd~7 500 (48/61)	–	–	–	–
Dichlorodiphenyltrichloroethanes (DDTs)						
p,p'-DDT	–	–	32~280 (3/3)	tr (2)~4 800 (18/18)	29~63 (2/2)	0.15~14 (37/37)
p,p'-DDE	–	–	150~2 200 (3/3)	290~16 000 (18/18)	22 000~290 000 (2/2)	0.31~49 (37/37)
p,p'-DDD	–	–	17~830 (3/3)	40~3 100 (18/18)	210~260 (2/2)	nd~0.72 (36/37)
o,p'-DDT	–	–	10~120 (3/3)	tr (1.1)~1 500 (18/18)	nd~tr (2.5) (1/2)	0.08~6.3 (37/37)
o,p'-DDE	–	–	tr (2)~250 (3/3)	nd~2 000 (17/18)	tr (1) (2/2)	tr (0.04)~1.2 (37/37)
o,p'-DDD	–	–	4.9~720 (3/3)	nd~1 100 (17/18)	3.7~9.9 (2/2)	nd~0.38 (36/37)
Chlordanes						
cis-Chlordane	–	–	–	–	–	–
trans-Chlordane	–	–	–	–	–	–
Oxychlordane	–	–	–	–	–	–
cis-Nonachlor	–	–	–	–	–	–
trans-Nonachlor	–	–	–	–	–	–
Heptachlors						
Heptachlor	–	–	–	–	–	–
cis-Heptachlor epoxide	–	–	–	–	–	–
trans-Heptachlor epoxide	–	–	–	–	–	–
Toxaphenes						
Parlar-26	nd~5 (7/47)	nd (0/61)	nd~tr (15) (2/3)	nd~280 (12/18)	53~54 (2/2)	nd~tr (0.3) (12/37)
Parlar-50	nd~tr (2) (1/47)	nd~tr (3) (1/61)	nd~17 (2/3)	nd~300 (16/18)	tr (11)~tr (13) (2/2)	nd~tr (0.2) (2/37)
Parlar-62	nd (0/47)	nd~tr (20) (1/61)	nd (0/3)	nd~150 (3/18)	nd (0/2)	nd (0/37)
Mirex	nd~1.0 (3/47)	nd~240 (44/61)	1.8~20 (3/3)	1.9~70 (18/18)	47~260 (2/2)	0.05~0.20 (37/37)
Hexachlorocyclohexanes (HCHs)						
α-HCH	–	–	–	–	–	–
β-HCH	–	–	–	–	–	–
γ-HCH (Lindane)	–	–	–	–	–	–
δ-HCH	–	–	–	–	–	–
Chlordecone	–	–	–	–	–	–
Hexabromobiphenyls	–	–	–	–	–	–
Polybrominated diphenyl ethers (PBDEs) (Br_4–Br_{10})						
Tetrabromodiphenyl ethers	nd~72 (22/47)	nd~3 100 (43/61)	26~68 (3/3)	tr (13)~440 (18/18)	280~310 (2/2)	0.05~3.9 (37/37)

Results of 2018 POPs Monitoring Survey Continued.

Surveyed media / Target chemicals	Surface Water (pg/L)	Bottom Sediment (pg/g-dry)	Wildlife (pg/g-wet)			Air (pg/m^3)
			Shellfish	Fish	Birds	warm season
	Detection range (detected frequency)	Detection range (detected frequency)	Detection range (detected frequency)	Detection range (detected frequency)	Detection range (detected frequency)	Detection range (detected frequency)
Polybrominated diphenyl ethers (PBDEs) (Br$_4$–Br$_{10}$)						
Pentabromodiphenyl ethers	nd~110 (13/47)	nd~2 800 (53/61)	tr (5)~23 (3/3)	nd~100 (17/18)	140~240 (2/2)	nd~4.1 (18/37)
Hexabromodiphenyl ethers	nd~54 (15/47)	nd~1 300 (52/61)	nd~34 (2/3)	nd~190 (17/18)	330~1 300 (2/2)	nd~1.5 (9/37)
Heptabromodiphenyl ethers	nd~65 (3/47)	nd~1 900 (46/61)	nd~tr(10) (1/3)	nd~58 (11/18)	110~480 (2/2)	nd~1.3 (16/37)
Octabromodiphenyl ethers	nd~69 (35/47)	nd~5 500 (57/61)	nd (0/3)	nd~74 (8/18)	61~580 (2/2)	nd~1.3 (34/37)
Nonabromodiphenyl ethers	nd~170 (46/47)	nd~56 000 (60/61)	nd (0/3)	nd (0/18)	46~53 (2/2)	nd~3 (31/37)
Decabromodiphenyl ether (c-DecaBDE)	12~2 700 (47/47)	tr (14)~520 000 (61/61)	nd (0/3)	nd~tr (110) (2/18)	tr (90)~500 (2/2)	nd~19 (31/37)
Perfluorooctane sulfonic acid (PFOS)	nd~4 100 (42/47)	nd~700 (55/61)	–	–	–	–
Perfluorooctanoic acid (PFOA)	160~28 000 (47/47)	nd~190 (58/61)	–	–	–	–
Pentachlorobenzene (PeCB)	2.7~320 (47/47)	1.2~3 400 (61/61)	tr (5)~tr (13) (3/3)	nd~70 (15/18)	280~480 (2/2)	30~100 (37/37)
Endosulfans						
α-Endosulfan	nd~tr (50) (1/47)	nd~30 (21/61)	–	–	–	–
β-Endosulfan	nd~tr (20) (3/47)	nd~41 (11/61)	–	–	–	–
1,2,5,6,9,10-Hexabromocyclododecanes (HBCDDs)						
α-1,2,5,6,9,10-Hexabromocyclododecane	–	–	76~270 (3/3)	nd~530 (17/18)	590~610 (2/2)	–
β-1,2,5,6,9,10-Hexabromocyclododecane	–	–	nd (0/3)	nd (0/18)	nd (0/2)	–
γ-1,2,5,6,9,10-Hexabromocyclododecane	–	–	nd~46 (2/3)	nd~130 (10/18)	nd (0/2)	–
δ-1,2,5,6,9,10-Hexabromocyclododecane	–	–	–	–	–	–
ε-1,2,5,6,9,10-Hexabromocyclododecane	–	–	–	–	–	–
Polychlorinated naphthalenes (total)	nd~260 (39/47)	9.9~34 000 (61/61)	tr (13)~700 (3/3)	nd~520 (16/18)	220~250 (2/2)	5.3~590 (37/37)
Hexachloro-1,3-butadiene	–	–	–	–	–	150~8 500 (37/37)
Pentachlorophenol and its salt, Pentachlorophenol ethers						
Pentachlorophenol	nd~4 400 (44/47)	nd~3 900 (59/61)	tr (10)~30 (3/3)	nd~80 (13/18)	180~1 200 (2/2)	0.9~30 (37/37)
Pentachloroanisole	nd~230 (30/47)	nd~160 (53/61)	tr (2)~21 (3/3)	nd~73 (16/18)	11~20 (2/2)	4.6~110 (37/37)
Short-Chain Chlorinated Paraffins (SCCPs)						
Chlorinated decanes	nd~1 600 (8/47)	nd~7 000 (7/61)	nd~tr (400) (2/3)	nd~tr (800) (1/18)	nd~tr (600) (1/2)	tr (130)~1 700 (37/37)
Chlorinated undecans	nd~3 500 (6/47)	nd~tr (13 000) (7/61)	nd (0/3)	nd~tr(700) (1/18)	nd (0/2)	tr (100)~2 600 (37/37)
Chlorinated dodecanes	nd~3 000 (16/47)	nd~38 000 (28/61)	nd (0/3)	nd (0/18)	nd (0/2)	tr (60)~880 (37/37)
Chlorinated tridecanes	nd~11 000 (18/47)	nd~36 000 (24/61)	nd (0/3)	nd (0/18)	nd (0/2)	nd~470 (26/37)
dicofol	–	–	nd~30 (1/3)	nd~280 (9/18)	nd (0/2)	–
Perfluorohexanesulfonic acid (PFHxS)	nd~2 600 (44/47)	nd~27 (15/61)	–	–	–	–

nd: Less than the method detection limit (MDL) tr ( ): Value equal to or greater than the method detection limit and less than the method quantification limit (MQL). – : Medium excluded from the survey. The range is shown in a specimen base and the detection frequency is shown in a location base. Therefore, there are some cases in which the range begins from nd even when detected at all locations.

Ministry of the Environment, Environmental Health Department, Environmental Health and Safety Division: "Investigation into the Actual Conditions of the Chemical Environment for FY 2019 –Chemicals in the Environment–" (2020).

http://www.env.go.jp/chemi/kurohon/2019/index.html.

Annual Exposure Effective Dose per Person for Natural Radiation Sources

Exposure source		Annual effective dose (mSv) Average	Typical range
Cosmic ray	Direct ionization and photo components	0.28(0.30)	(In the range from sealevel altitude to high altitude regions) 0.3–1.0
	Neutron component	0.10(0.08)	
Cosmic rays generation radionuclide		0.01(0.01)	
Total of cosmic rays and generation radio nuclide		0.39	
Radiation from outer Space	Outdoor	0.07(0.07)	(Depends on chemical composition of soil and construction materials) 0.3–0.6
	Room	0.41(0.39)	
Total of outdoor and indoor		0.48	
Inhalation exposure	Uranium and thrium series	0.006(0.01)	(Depends on the indoor density of radon gas) 0.2–10
	Radon(^{222}Rn)	1.15(1.2)	
	Thoron(^{220}Rn)	0.10(0.07)	
Total inhalation exposure		1.26	
Ingestion exposure	^{40}K	0.17(0.17)	(Depends on the elemental composition of the radionuclides in food and water) 0.2–0.8
	Uranium and thrium series	0.12(0.06)	
Total ingestion exposure		0.29	
Total		2.4	1–10

Based on the Report of the 2000 General Assembly of United Nations Scientific Committee con the Effects of Atomic Radiation translation by the Publication Company (Ltd.) in the National Institute of Radilogical Sciences. Published in March 2002. Figures in parentheses are previous results.

High Natural Radiation Background Regions

Country	Region	Regional features	Population estimate	Air absorbed dose rate (nGy/h)*
Brazil	Guarapari	Monazite sand, coastal areas	73 000	90–170 (street) 90–90 000 (beach)
China	Yangjiang city Guangdong Province	Monazite particle	80 000	370 averages
India	Kerala Chennai	Monazite sand, coastal areas 200 km in length and 0.5 km in width	100 000	200–4 000 1800 averages
Iran	Ramsar Mahasrat	Pond	2 000	70–17 000 800–4 000
Italy	Campania province Orviet	Volcanic soil	5 600 000 21 000	200 averages 560 averages

* Including cosmic rays and terrestrial radiation.
Based on the Report of the 2000 General Assembly of United Nations Scientific Committee con the Effects of Atomic Radiation translation by the Publication Company (Ltd.) in the National Institute of Radilogical Sciences. Published in March 2002. Figures in parentheses are previous results.

Median Lethal Dose ($LD_{50/30}$) by Species for One-Time Whole-Body Exposure to X, γ Radiation

Animal kind	Approximate median lethal dose (Gy)	Animal kind	Approximate median lethal dose (Gy)
Medaka	20–25	Rabbit	7–9
Mouse	5–7	Ape	5–6
Rat	7–8	Dog	2–3
Squirrel	7	Pig	2–3
Hamster	7–8	Sheep	2
Cavy (guinea pig)	4–5	Donkey	2–3
		Human	4–5 (estimation)

*Median lethal dose ($LD_{50/30}$): Dose which will be lethal to 50% of the population within 30 days after exposure. The majority of population deaths in this case are due to failure of hematopoietic organs and is therefore referred to as hematopoietic death. Strain differences exist even within the same species. For human beings, this dose is expressed as $LD_{50/30}$. However, conditions vary when receiving medical treatment.

Appearance of Biological Effects due to Radiation

The frequency of biological effects due to radiation differs depending on what biological effects are measured. **a-c** is called the linear non-threshold hypothesis. This linear non-threshold hypothesis is formed for the frequency of radiation-induced mutations and the frequency of radiation-induced chromosome abnormalities during the experimental use of bacteria, mammalian cell cultures, and mice populations. Generally, investigation is conducted for single (acute) irradiation dosage in the level of Gy. In many cases, biological effect is analogized for low doses. In many cases, this linear non-threshold hypothesis is used internationally in anticipation of safety in relation to the effects of radiation on the human body. The reason that the effect value is not 0 when the dose is 0 is to express the frequency of naturally-occurring effects. **a-b** is a phenomenon in which the effect is larger than the direct relationship for low doses (**b**). This is called the bystander effect. This is examined through survival rate experiments and experiments on the frequency of sudden mutation for mammalian cell cultures. In some cases, an effect on the organs of mice populations is observed. **d** is known as the low-dose rate effect. When dividing the radiation and irradiation at the same dose, or when continually irradiating at a low dose, the effect observed during single (acute) irradiation decreases. This phenomenon occurs because organism possesses the ability (DNA repair ability) to lessen the damage caused by radiation. $\mathbf{e_1}-\mathbf{e_2}$ is called the threshold hypothesis. It is recognized as a safe dose region ($\mathbf{e_1}$, threshold) where biological effects do not appear at low doses. Experiments for radiation-induced cancer in mice are based on this hypothesis. **f**-$\mathbf{e_2}$ is called the hormesis hypothesis. In a low dose range, it is even lower than the biological effect caused by natural radiation, and is considered to be good for health. Reports for the results of experiments conducted on paramecium and drosophila support this hypothesis. Radium hot springs are another example of this hypothesis. In some cases, there is confusion between hormesis and the fact that a resistance to high radiation doses has been observed in mice populations and mammalian cell cultures which were pre-exposed to low doses of radiation or low dose rate radiation (radioadaptive response). **g** is a phenomenon observed in mutant strains of radiosensitivity and in cultured cells of radiosensitive patients in human beings. The DNA repair ability has decreased due to radiation. Therefore, the biological effect of radiation increases even at low doses. Furthermore, there are some cases in which there is a high naturally-occurring frequency even when not irradiated with radiation (**h**). This is thought to occur due to naturally-occurring DNA damage.

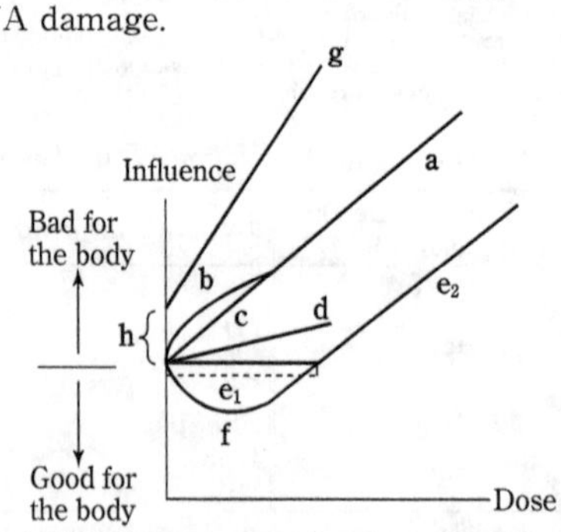

Takeo Onishi (Supervisor): Biological Effect by low dose and low dose rate irradiation, Aipricom (2003); National Institute of Radiological Sciences, Academic Publishing Center (2007); Dictionary of the body and light, Asakura Publishing Co. Ltd. (2010).

Method for Calculating the Equivalent Dose and Effective Dose for Radiation on the Human Body

The Cabinet Order for Enforcement of the Radiation Hazard Prevention Act was revised due to incorporation of recommendations from the International Commission on Radiological Protection (ICRP Publication 60) and was enacted on April 1, 2001. The cabinet order defines new dose limits (values which must not be exceeded for the equivalent dose and effective dose of the population through managed action) for labor exposure and public exposure.

Due to differences in the quality of radiation, the extent to which living organisms are effected by the same absorbed dose of radiation may differ. In consideration of this fact, the equivalent dose H_T (unit: J/kg, sievert(Sv)) for a certain tissue (T) is calculated by multiplying the average absorbed dose $D_{T,R}$ by the radiation weighting coefficient w_R. A value from 1 to 20 is used for the radiation weighting coefficient depending on the type and energy of the radiation.

The effective dose E (unit: J/kg, sievert(Sv)) is calculated by multiplying the equivalent dose H_T on each tissue and organ by the tissue weighting coefficient w_T and then tabulating the result for each tissue and organ. The tissue weighting factor expresses the relative risk for cancer in all tissues and organs when the human body is exposed to average radiation. It also includes hereditaty disease for the gonads.

H: Equivalent dose to tissue T
D: Average absorbed dose of radiation type R to tissue T
w: Weighting factor for radiation type R

$$H_T = \sum_R w_R D_{T,R}$$

E: Effective dose
w: Weighing factor for tissue T

$$E = \sum_T w_T H_T$$

Summary of the Effect of Human Health by Radiation

Radiation amount	Effect on individual	Results for exposed group
Extremely low dose: Approx. 10 mSv or less (effective dose)	No acute effects. Extremely small increase in the risk of cancer.	Even in large exposed groups, no increase in the cancerogenesis rate is observed.
Low dose: Up to approx. 100 mSv (effective dose)	No acute effects. Afterwards, the risk of cancer increases by less than 1%.	In the case of large exposed groups (approx. 100 000 or more), there is the possibility of observing an increase in the cancerogenesis rate.
Medium dose: Up to approx. 1000 mSv (acute whole-body dose)	Possibility of nausea and vomiting; small decrease in bone marrow function. Afterwards, the risk of cancer increases by approx. 10%.	In the case of exposed groups of at least several hundred individuals, an increase in the cancerogenesis rate will most likely be observed.
High dose: Approx. 1000 mSv or more (acute whole-body dose)	Nausea will always occur. Myelodysplastic syndrome may occur in some cases. At acute whole-body doses exceeding approx. 4000 mSv, high risk of death unless medical treatment is received. Significant increase in the risk of cancer.	Increase in the cancerogenesis rate will be observed.

Publication 96, Annuals of the ICRP (International Commission on Radiological Protection), Vol 35, No.1 (2005) translation publication by the Japan Radioisotope Association (inc.) in April 2011.

Human Effect by Radiation (mSv)

<table>
<tr><th colspan="2">Partial-body exposure</th><th colspan="3">Whole-body exposure</th></tr>
<tr><td>10 000</td><td>Radiation exposure to skin/ Acute ulcer</td><td>7 000–10 000</td><td colspan="2">100% of people die</td></tr>
<tr><td>5 000</td><td>Radiation exposure to skin/ Erythema</td><td>3 000–5 000</td><td colspan="2">50% of people die/30 days*</td></tr>
<tr><td>5 000</td><td>Radiation exposure to crystalline lens/Cataract</td><td>1 000</td><td colspan="2">Nausea and vomiting (10% of people)</td></tr>
<tr><td>3 000</td><td>Radiation exposure to skin/ Loss of hair</td><td>500</td><td colspan="2">Decrease in lymphocyte of peripheral blood</td></tr>
<tr><td>2 500–6 000</td><td>Radiation exposure to gonad/Permanent sterility</td><td></td><td colspan="2"></td></tr>
<tr><td>500–2 000</td><td>Radiation exposure to crystalline lens/Lens opacity</td><td></td><td colspan="2"></td></tr>
<tr><td></td><td></td><td>6.9</td><td>Whole body CT scanning/1 time</td><td rowspan="3">No clinical symptom</td></tr>
<tr><td></td><td></td><td>0.6</td><td>Mass radiography of the stomach</td></tr>
<tr><td></td><td></td><td>0.05</td><td>Mass radiography of the chest</td></tr>
</table>

* It is different in medical environment.

Avoidable Radiation Dose that is Urged for Public Protection

Countermeasures	Avoidable irradiation dose (because countermeasures are generally optimization.)
Sheltering	-10 mSv/2 days (Effective dose)
Temporary sheltering	-50 mSv/1 week (Effective dose)
Preventive administration by iodine agent	-100 mSv
Migration	1 000 mSv or 100 mSv/first year (Effective dose)

Publication 96, Annuals of the ICRP (International Commission on Radiological Protection), Vol 35, No.1 (2005) translation publication by the Japan Radioisotope Association (inc.) in April 2011.

Symptom to Appear Immediately after Whole-Body Exposure in Human

Symptom or influence part	Whole body irradiation (mSv)				
	Slight (1000-2000)	Moderate (2000-4000)	Severe (4000-6000)	Quite severe (6000-8000)	Lethal (>8000)
Vomiting Expression time Incidence (%)	 After 2 hours 10-50	 1-2 hours 70-90	 within 1 hour 100	 within 30 min 100	 within 10 min 100
Diarrhea Expression time Incidence (%)	None - -	None - -	Slight 3-8 hours <10	Severe 1-3 hours >10	Severe within 1 hour about 100
Headache Expression time Incidence (%)	Some - -	Slight - -	Moderate 4-24 hours 50	Severe 3-4 hours 80	Severe 1-2 hours 80-90
Consciousness Expression time Incidence (%)	No effect - -	No effect - -	No effect - -	May change - -	Unconscious Several seconds/several minutes continuation 100 (>50 Sv)
Body temperature Expression time Incidence (%)	Normal - -	Increase 1-3 hours 10-80	High fever 1-2 hours 80-100	High fever <1 hour 100	High fever <1 hour 100
Medical countermeasure	Medical examination of an outpatient	Observation in the general hospital. If necessary, treatment in special hospital	Treatment in special hospital	Treatment in special hospital	Relieving medication (Only symptomatic treatment)

Publication 96, Annuals of the ICRP (International Commission on Radiological Protection), Vol 35, No.1 (2005) translation publication by the Japan Radioisotope Association (inc.) in April 2011.

Evidence for the Important Stage of Acute Radiation Symptom after Whole-Body Exposure in Human

Symptom or part of the influence	Acute exposure dose of whole-body irradiation (mSv)				
	Slight (1000-2000)	Moderate (2000-4000)	Severe (4000-6000)	Quite severe (6000-8000)	Lethal (>8000)
Expression time	>30 days	18-28 days	8-18 days	<7 days	<3 days
Lymphocyte (10^9/L)	0.8-1.5	0.5-0.8	0.3-0.5	0.1-1.3	0.0-0.1
Platelet (10^9/L)	60-100 10-25%	30-60 25-40%	25-35 40-80%	15-25 60-80%	<20 80-100%
Clinical symptom	Feebleness, weakness	Fever, infectious, hemorrhage, weakness, loss of hair	High fever, infectious, hemorrhage, loss of hair	High fever, diarrhea, vomiting, vertigo, disorientation, hypotension	High fever, diarrhea, consciousness disorder
Lethality (%) Expression time	0 -	0-50 6-8 weeks	20-70 4-8 weeks	50-100 1-2 weeks	100 1-2 weeks
Medical countermeasure	Prevention	Special countermeasures for prevention from 14-20 days; Quarantine from 10-20 days	Special countermeasures for prevention from 7-10 days; Quarantine from the beginning	Special countermeasures for prevention from first day; Quarantine from the beginning	Only symptomatic treatment

Publication 96, Annuals of the ICRP (International Commission on Radiological Protection), Vol 35, No.1 (2005) translation publication by the Japan Radioisotope Association (inc.) in April 2011.

Material Balance on a National Level (Resource Productivity*[1], Recycling Rate*[2], Final Disposal Amount)

FY	GDP (Gross Domestic Product) (Trillion yen)	DMI (Direct Material Input) (Billion t)	GDP/DMI Resource productivity (10 000 yen/t)	Amount recycled (million t)	Recycling rate (%)	Industrial waste Final disporsal amount (1 000 t/year)	Municipal waste Final disposal amount (1 000 t/year)	Waste Final disposal amount (1 000 t/year)
1980	287.4	18.9	15.2	170	8.3	68 000	19 715	87 715
1985	355.1	17.4	20.4	155	8.2	91 000	16 048	107 048
1990	453.6	21.8	20.8	175	7.4	89 000	16 809	105 809
1992	467.5	20.4	23.0	177	8.0	89 000	15 296	104 296
1993	465.3	19.9	23.4	170	7.9	84 000	14 959	98 959
1994	426.6	20.2	21.1	180	8.2	80 000	14 142	94 142
1995	441.4	20.1	21.9	193	8.7	69 000	13 602	82 602
1996	453.5	20.2	22.4	196	8.8	68 000	13 093	81 093
1997	453.6	19.7	23.1	192	8.9	67 000	12 008	79 008
1998	449.9	18.2	24.7	188	9.4	58 000	11 350	69 350
1999	452.9	18.3	24.8	195	9.7	50 000	10 869	60 869
2000	464.3	19.3	24.1	213	10.0	45 000	10 510	55 510
2001	461.5	19.4	23.8	207	9.6	42 000	9 949	51 949
2002	465.7	18.7	24.9	211	10.1	40 000	9 030	49 030
2003	475.4	17.6	26.9	222	11.2	30 440	8 452	38 892
2004	482.6	17.1	28.2	226	11.7	25 827	8 093	33 920
2005	492.7	16.5	29.9	228	12.1	24 229	7 341	31 570
2006	499.6	15.9	31.4	228	12.5	21 799	6 809	28 608
2007	505.5	15.6	32.4	243	13.5	20 143	6 349	26 492
2008	488.0	14.9	32.7	245	14.1	16 701	5 531	22 232
2009	477.5	13.1	36.5	229	14.9	13 591	5 072	18 663
2010	492.8	13.6	36.1	246	15.3	14 255	4 837	19 092
2011	495.1	13.3	37.1	238	15.1	12 439	4 821	17 260
2012	499.6	13.6	36.7	244	15.2	13 102	4 648	17 750
2013	512.7	14.1	36.5	269	16.1	11 721	4 538	16 259
2014	510.4	13.9	37.8	261	15.8	10 399	4 302	14 701
2015	517.4	13.6	38.1	251	15.6	10 085	4 165	14 250
2016	522.0	13.2	39.6	240	15.4	9 894	3 980	13 874
2017	531.7							
2018	533.7							

*1 Resource productivity = GDP/(amount of natural resource investment). Because the GDP benchmark years are different, data for years upto 1993 and data for years from 1994 can not be directly compared.

*2 Recycling ratio = amount recycled/ (amount recycled + Amount of natural resource investment)

The GDP for 1980–1993 is the chain-linking method (benchmark year 2000). The GDP since 1994 is the fixed-base method (benchmark year 2011).

DMI, Amount recycled (FY 1980–1989): Materials from the Ministry of the Environment (issued on May 29, 2003)

DMI, Amount recycled (FY 1990–): Ministry of the Environment: "Annual Report on Environmental Statics" (issued each year).

Amount of industiral waste ultimately disposed of: Ministry of Health and Welfare, Ministry of the Environment: "Sangyohaikibutsu no haishutsu oyobi shorijokyo ni tsuite" (issued each year).

Amount of municipal waste ultimately disposed of: Ministry of Health and Welfare, Ministry of the Environment: "Nihon no haikibutsushori" (issued each year).

Recycling Rate for each Product Group

Year	Paper and Paperboard Domestic consumption (1 000 t)	Recovery rate of used paper*1 (%)	Fiber raw material consumption total for paper manufacture (1 000 t)	Utilization rate of recovered paper*2 (%)			Glass bottle			
				For paper	For paper-board	Total	Production (1 000 t)	Amount of cullet used*3 (1 000 t)	Cullet availability rate*4 (%)	Cullet use rate*5 (%)
1990	28 227	49.7	28 399	25.2	85.8	51.5	2 610		47.9	
1993	28 153	51.1	28 011	26.5	86.7	53.0	2 351		55.5	
1994	28 838	51.7	28 618	26.7	86.8	53.3	2 440		55.6	
1995	30 015	51.6	29 593	26.7	87.7	53.4	2 233		61.3	
1996	30 735	51.3	29 943	27.2	87.8	53.6	2 210		65.0	
1997	31 161	53.1	30 814	27.1	88.3	54.0	2 160		67.4	
1998	29 931	55.3	29 761	29.2	89.0	54.9	1 975		73.9	
1999	30 541	55.9	30 394	30.7	89.3	56.1	1 906		78.6	
2000	31 758	57.7	31 648	32.1	89.5	57.0	1 820	1 416	77.8	
2001	31 072	61.5	30 910	33.8	90.3	58.0	1 738	1 425	82.0	
2002	30 646	65.4	30 752	36.2	91.1	59.6	1 689	1 408	83.3	
2003	30 930	66.1	30 560	36.5	92.3	60.2	1 561	1 410	90.3	
2004	31 377	68.5	30 957	37.1	92.4	60.4	1 554	1 409	90.7	
2005	31 382	71.1	31 057	37.4	92.6	60.3	1 501	1 370	91.3	
2006	31 541	72.4	31 225	38.1	92.7	60.6	1 472	1 383	94.5	71.4
2007	31 304	74.5	31 651	40.1	92.4	61.4	1 433	1 368	95.6	72.7
2008	30 303	75.1	30 954	40.5	92.8	61.9	1 387	1 343	96.7	74.2
2009	27 194	79.7	26 797	42.1	92.7	63.1	1 330	1 297	-	74.2
2010	27 752	78.2	27 850	40.5	92.8	62.5	1 337	1 295	-	73.4
2011	27 663	77.9	27 106	39.6	92.8	63.0	1 342	1 284	-	73.3
2012	27 230	79.9	26 501	41.1	93.0	63.7	1 281	1 285	-	75.9
2013	27 203	80.4	26 661	40.9	93.3	63.9	1 287	1 274	-	74.8
2014	26 917	80.8	26 918	40.3	93.2	63.9	1 257	1 230	-	74.4
2015	26 314	81.3	26 589	40.2	93.5	64.3	1 246	1 228	-	75.9
2016	26 132	81.3	26 712	39.2	93.8	64.2	1 237	1 211		75.4
2017	26 023	80.9	26 859	37.9	93.8	64.1	1195	1189	-	75.1
2018	25 332	81.6	26 512	37.3	93.4	64.3	1156	1160	-	74.7
2019	24 901	79.5	25 823	36.0	93.4	64.0				

*1 Recovery rate of used paper = Amount of waste paper collected / Paper and Paperboard Domestic consumption

*2 Utilization rate of recovered paper = Domestic consumption of waste paper / Fiber raw material consumption total for paper manufacture

Paper Recycling Promotion Center (as of 28 July 2020)

*3 "Cullet" refers to finely-crushed glass bottles, etc.

*4 Cullet availability rate = Amount of cullet used / Amount of glass bottles produced

*5 Cullet use rate = Amount of cullet used / Total dissolution volume

The total dissolution volume is the total volume of raw materials (virgin raw materials + cullet) which were melted in order to produce glass bottles.

Glass Bottle 3R Promotion Association (as of 28 July 2020)

Recycling Rate for each Product Group Continued.

Year	Steel can		Aluminum can				PET bottle					
	Consumption weight (production weight) (1 000 t)	Recycling rate*6 (%)	Consumption: Number of cans (Billion cans)	Consumption: Weight (1 000 t)	Recycling rate*7 (%)	CAN TO CAN rate*8	Volume of sold PET bottles*8 (1 000 t)	Domestic recycling*10 (1 000 t)	Overseas recycling*10 (1 000 t)	Municipality collection rate*11 (%)	Collection rate*12 (%)	Recycling rate*13 (%)
1990	1 459	44.8	91.5	161	42.6							
1993	1 360	61.0	117.8	201	57.8		123.8			0.4		
1994	1 476	69.8	148.5	248	61.1		150.3			0.9		
1995	1 421	73.8	159.2	265	65.7	45.6	142.1			1.8		
1996	1 422	77.3	163.9	271	70.2	71.2	172.9			2.9		
1997	1 351	79.6	165.6	275	72.6	73.3	218.8			9.8		
1998	1 285	82.5	166.5	271	74.4	79.0	281.9			16.9		
1999	1 269	82.9	169.6	276	78.5	75.8	332.2			22.8		
2000	1 215	84.2	167.5	266	80.6	74.5	361.9			34.5		
2001	1 055	85.2	174.4	283	82.8	67.8	402.7			40.1	44.0	
2002	949	86.1	177.8	292	83.1	70.3	412.6			45.6	53.4	
2003	911	87.5	177.4	297	81.8	63.7	436.6			48.5	61.0	
2004	908	87.1	185.2	303	86.1	61.7	513.7			46.4	62.3	
2005	868	88.7	184.3	302	91.7	57.3	529.8			47.6	61.7	
2006	832	88.1	183.6	299	90.9	62.1	543.8	234	175	49.3	66.3	75.1
2007	834	85.1	185.2	301	92.7	62.7	573.2	235	229	49.4	69.2	81.2
2008	772	88.5	184.3	299	87.3	66.8	571.4	233	238	49.7	77.9	82.2
2009	699	89.1	182.4	293	93.4	62.5	564.2	245	263	50.9	77.5	89.9
2010	684	89.4	185.6	296	92.6	68.3	594.7	242	256	49.6	72.1	83.5
2011	682	90.4	188.1	298	92.5	64.5	604.0	265	253	–	79.6	85.8
2012	664	90.8	191.2	301	94.7	66.7	582.9	254	241	–	90.4	85.0
2013	611	92.9	194.0	304	83.8	68.4	578.7	258	239		91.4	85.8
2014	571	92.0	201.6	313	87.4	63.4	569.3	271	253		93.5	82.6
2015	486	92.9	222.0	332	77.1	74.7	563.0	262	227		92.4	86.9
2016	463	93.9	223.8	341	76.1	62.8	596.1	279	221		88.8	83.9
2017	451	93.4	219.3	336	75.1	67.3	587.4	298	201		92.2	84.8
2018	439	92.0	216.6	331	72.4	71.4	625.5	334	195			

*6 (1998 and earlier) Recycling rate = used weight of steel can scraps / weight of steel cans produced
(1999 and later) Recycling rate = weight of recycled steel cans / weight of steel cans consumed
Japan Steel Can Recycling Association (as of 28 July 2020)

*7 Aluminum recycling rate = volume reused / weight consumed

*8 Can-to-Can rate = weight of can material / weight reused
Japan Aluminum Can Recycling Association (as of 28 July 2020)

*9 Weight of PET bottles only (excludes the cap and label)
(2004 and earlier) Designated PET bottle (soft drinks, soy sauce, other types) production volume. Does not include other PET bottles (detergents, shampoo, edible oils, condiments, cosmetics, medical products, etc.)
(2005 and later) Designated PET bottle sales volume

*10 Recycled materials excludes the cap and label

*11 Weight including the cap and label
(2004 and earlier) Municipality collection rate = volume collected by municipality / designated PET bottle production volume
(2005 and later) Municipality collection rate = volume collected by municipality / designated PET bottle sales volume

*12 Weight including the cap and label. Collection rate including business-related collection volume.
(2004 and earlier) Includes the molding loss during bottle manufacturing in business-related collection activities. The denominator is the designated PET bottle production volume.
(2005 and later) The "business-related collection volume of designated PET bottles which have been used," excluding the molding loss during bottle manufacturing in business-related collection activities. The denominator is the designated PET bottle sales volume.

*13 Recycling rate = (Domestic recycling + Overseas recycling) / designated PET bottle sales volume
The Council for PET Bottle Recycling (as of 28 July 2020)

Generated Volume of General Waste

FY	Total amount of solid waste generated (1 000 t)							Total waste produced per person per day (g)	Amount of waste incinerated (1 000 t)	Amount recycled (1 000 t)	Amount ultimately disposed (1 000 t)
	Amount bulk collected	Amount processed by private households	Amount collected by municipalities	Amount transported commercially	Homes	Businesses	Total generated				
1975		3 987	27 916	10 262			42 165	1 033	19 939	–	21 017
1980		2 425	32 015	9 496			43 935	1 025	25 090	–	19 715
1985	564	1 919	35 383	6 147			43 449	982	29 335	1 619	16 048
1990	986	1 171	42 495	6 776			50 443	1 119	36 676	2 669	16 809
1995	2 318	788	44 100	5 806			50 694	1 105	39 495	5 100	13 602
1998	2 521	511	44 771	6 313	35 994	17 612	51 595	1 118	41 417	6 491	11 350
1999	2 604	352	45 736	5 359	36 220	17 478	51 446	1 114	41 623	7 032	10 869
2000	2 765	293	46 695	5 373	36 844	17 990	52 362	1 132	42 149	7 860	10 514
2001	2 837	253	46 528	5 316	37 381	17 300	52 097	1 124	42 280	8 246	9 949
2002	2 807	218	46 202	5 190	37 118	17 081	51 610	1 111	42 016	8 638	9 030
2003	2 829	165	46 044	5 398	37 321	16 950	51 607	1 106	42 012	9 157	8 452
2004	2 919	130	45 114	5 343	36 838	16 538	50 587	1 086	40 986	9 400	8 093
2005	2 996	92	44 633	5 090	36 471	16 249	52 720*1	1 131*1	40 267	10 026	7 328
2006	3 058	74	44 155	4 810	36 220	15 804	52 024	1 115	39 914	10 204	6 809
2007	3 049	56	42 629	5 138	35 724	15 092	50 816	1 089	38 737	10 305	6 349
2008	2 926	45	40 946	4 234	34 104	14 003	48 106	1 033	37 233	9 776	5 531
2009	2 792	31	39 616	3 845	32 974	13 278	46 252	994	35 989	9 502	5 072
2010	2 729	28	38 827	3 803	32 385	12 974	45 359	976	35 254	9 446	4 837
2011	2 682	37	39 025	3 724	32 385	13 045	45 430	976	35 419	9 375	4 821
2012	2 646	21	38 890	3 697	32 137	13 097	45 234	964	35 407	9 263	4 648
2013	2 583	19	38 546	3 745	31 757	13 117	44 874	958	35 146	9 268	4 538
2014	2 503	36	38 095	3 718	31 242	13 075	44 317	947	34 859	9 129	4 302
2015	2 394	22	37 867	3 720	30 935	13 046	43 981	939	34 813	9 001	4 165
2016	2 270	28	37 245	3 654	30 182	12 988	43 170	925	34 293	8 792	3 980
2017	2 172	13	37 092	3 630	29 880	13 014	42 894	920	34 181	8 683	3 859
2018	2 044	25	36 929	3 743	29 673	13 043	42 716	918	34 052	8 529	3 835

MHLW and MOE: "Nihon no haikibutsushori" (yearly edition)
Based on the "Ippanhaikibutsushorijigyo jittaichosa" conducted by the MOE (previously the MHLW).
The generated volume of household trash includes group collection volumes.
"Total waste produced per person per day" is "Total amount of solid waste generated" divided by total population and 365 (or 366).
For 2004 and earlier, the "total generated volume of waste" includes the household disposal volume but excludes the group collection volume.
For 2005 and later, the "total generated volume of waste" excludes the household disposal volume but includes the group collection volume. The total generated volume of waste according to the conventional definition was 49.828 million tons in fiscal 2005.
Data for fiscal 2010 are tabulated values which exclude Minami-Sanriku Town (Miyagi Prefecture).
For data from fiscal 2011 and later, the top value excludes waste disposal related to disasters, while the bottom value includes said waste. For 2010 and earlier, values include waste disposal related to disasters.
For 2012 and later, total population is included foreign population.

Generated Volume of Industrial Waste

(Unit: 1 000 tons)

Kind / FY	Cinder	Dirt	Wasted oil	Spent pickling solution	Waste alkali	Waste plastics	Waste-paper	Waste wood	Waste textile	Flora and fauna residue	Animal solid waste	Rubber rubbish	Waste metal	Cullet, concrete rubbish, and pottery rubbish	Slag	Toles and pebbles (Construction Scrap)	Animal's excreta	Animal carcass	Corbicula dust	Other	total
1980	1 797	88 190	2 419	10 219	6 090	2 232	1 624	6 628	101	4 323	–	92	13 111	2 297	60 567	30 007	49 629	62	11 731	1 199	292 000
1985	2 409	112 821	3 672	4 320	923	2 816	1 472	8 058	98	2 207	–	78	8 877	3 910	41 649	48 948	62 462	96	6 224	1 230	312 000
1990	2 678	171 450	3 471	2 674	1 547	4 334	1 193	6 573	99	3 543	–	94	8 533	5 295	42 507	54 798	77 208	28	7 491	1 228	395 000
1995	3 258	185 508	3 173	4 441	2 020	6 253	1 897	7 161	84	3 961	–	87	6 482	6 067	24 242	58 460	72 996	145	7 578	–	394 000
2000	1 892	189 181	3 248	2 938	1 563	5 790	2 156	5 511	76	4 052	–	44	8 096	4 797	16 448	58 829	90 489	163	10 765	–	406 000
2004	1 935	188 306	3 310	2 738	2 039	5 939	1 756	5 959	75	3 393	119	47	10 039	5 473	21 192	62 497	87 686	186	14 466	–	417 000
2005	1 857	187 688	3 471	2 477	2 079	6 052	1 748	5 951	93	3 117	97	55	10 947	4 555	26 186	60 562	87 204	196	17 342	–	422 000
2006	1 969	185 327	3 406	5 405	2 561	6 094	1 664	5 852	80	3 008	104	48	11 004	4 922	21 288	60 823	87 573	234	17 135	–	418 000
2007	2 028	185 305	3 610	5 662	2 777	6 428	1 466	5 971	75	3 066	78	62	11 461	5 183	20 715	60 900	87 476	197	16 964	–	419 000
2008	2 053	176 114	3 617	2 721	2 648	6 445	1 383	6 262	73	3 194	124	41	8 766	6 174	18 440	61 189	87 698	168	16 550	–	404 000
2009	1 821	173 629	3 048	2 542	1 867	5 665	1 265	6 294	69	2 888	113	27	7 830	5 411	14 109	58 921	88 162	161	15 923	–	390 000
2010	1 835	169 885	3 251	2 483	2 563	6 185	1 153	6 121	79	2 902	126	32	7 246	6 031	16 006	58 264	84 847	156	16 823	–	386 000
2011	1 836	166 132	3 118	2 752	1 889	5 710	1 118	6 233	79	2 754	84	32	7 242	6 361	15 493	59 839	84 459	172	15 903	–	381 000
2012	1 869	164 638	3 212	2 595	1 778	5 691	1 020	6 229	68	2 572	70	34	7 267	6 083	16 398	58 887	85 434	153	15 138	–	379 000
2013	1 833	164 169	2 912	2 778	2 243	6 120	896	6 991	89	2 603	97	26	7 815	6 468	16 761	63 233	82 626	125	16 911	–	385 000
2014	2 046	168 821	3 044	3 191	2 306	6 509	985	7 487	103	2 706	83	28	9 284	8 267	14 563	64 394	81 416	126	17 479	–	393 000
2015	1 912	169 318	2 953	2 826	2 677	6 823	938	7 248	90	2 557	92	23	8 647	7 348	15 161	64 212	80 512	112	17 736	–	391 000
2016	1 967	167 316	3 049	2 740	2 348	6 836	988	7 098	120	2 604	81	36	8 221	8 002	14 089	63 587	80 465	114	17 373	–	387 000
2017	1 876	170 695	2 869	2 609	2 392	6 456	935	7 413	88	2 429	59	16	8 008	8 109	15 011	59 773	77 894	124	16 788	–	384 000

Ministry of Health and Welfare, Ministry of the Environment: "Sangyohaikibutsu no haishutsu oyobi shorijokyo ni tsuite" (issued each year).

Recycle Use, Intermediate Treatment Loss, and Final Disposal for Industrial Waste

(Unit: 1 million tons)

FY	Total recycle use	Reduced disposal	Total final disposal	Discharge	Fy	Total recycle use	Reduced disposal	Total final disposal	Discharge	FY	Total recycle use	Reduced disposal	Total final disposal	Discharge
1980	124	100	68	292	2002	182	172	40	393	2010	205	167	14	386
1985	129	92	91	312	2003	201	180	30	412	2011	200	169	12	381
1990	151	155	89	395	2004	214	177	26	417	2012	208	158	13	379
1995	147	178	69	394	2005	219	179	24	422	2013	205	168	12	385
1996	150	187	68	405	2006	215	182	22	418	2014	210	173	10	393
1996*	181	185	60	426	2007	219	180	20	419	2015	208	174	10	391
2000	184	177	45	406	2008	217	170	17	404	2016	204	173	10	387
2001	183	175	42	400	2009	207	169	14	390	2017	200	174	10	384

* Based on Basic Policies for Dioxins (passed at the Ministerial Conference on Dioxins), shows the generated volume for fiscal 1996 in terms of the "Haikibutsu no genryoka no mokuhyoryo" (set by the Japanese government on 28 September 1999), for which the government set fiscal 2010 as the target fiscal year. The aforementioned calculation conditions were also used for the generated volume from fiscal 1997 and later.

MHLW and MOE: "Sangyohaikibutsu no haishutsu oyobi shorijokyo ni tsuite" (yearly edition)

Appendix

Nobel Laureates/Reason

This list follows the descriptions on the Official Web Site of the Nobel Prize (http://nobelprize.org)
The entries are formatted: Year, Name (Nationality); Reason for the Award
Year: Entries are listed in the order of the year of award. Cases where the actual decision was made the following year are noted between the year and the names.
Names: All names except the surname are abbreviated as initials.
Nationality: Laureates citizenship. References to Germany and West Germany follow the conventions of the Nobel Foundation.
Reason: Objects of discovery, invention, or improvement are specified, where they are clearly mentioned. Wording may have been slightly modified in some cases.

Prizes in Physics

Year	Name (Nationality)	Reason for the Award
1901	W. C. Röntgen (Germany)	Discovery of Röntgen Rays (X-rays)
1902	H. A Lorentz (Netherlands), P. Zeeman (Netherlands)	Research into the influence of magnetism upon radiation phenomena
1903	A. H. Becquerel (France)	Discovery of radioactivity
	M. Curie (France), P. Curie (Poland, France)	Research into radiation phenomena
1904	Lord Rayleigh (J. W. Strutt) (UK)	Research into the density of gases, and discovery of argon
1905	P. E. A. von Lenard (Germany)	Research into cathode rays (beams of electrons)
1906	J. J. Thomson (UK)	Theoretical and experimental research into electrical conduction in gasses
1907	A. A. Michelson (USA)	Design of a precise interferometer, spectroscopy conducted with it, and research used to establish the meter
1908	G. Lippmann (France)	Method based on interferometry for photographing natural colors
1909	G. Marconi (Italy), K. F. Braun (Germany)	Contributions to the development of wireless communications
1910	J. D. van der Waals (Netherlands)	Gas-liquid equation of state
1911	W. Wien (Germany)	Discovery of laws governing thermal radiation
1912	N. G. Dalén (Sweden)	Invention of automatic regulators for the gas accumulators used for lighting lighthouses
1913	H. Kamerlingh Onnes (Netherlands)	Research into cryogenics leading to the production of liquid helium
1914	M. von Laue (Germany)	Discovery of X-ray diffraction in crystals
1915	W. H. Bragg (UK), W. L. Bragg (UK)	Analysis of crystal structure by using X-rays
1916	No recipient	
1917	C. G. Barkla (UK) [awarded the following year]	Discovery of X-ray radiation characteristic of each element
1918	M. K. E. L Planck (Germany) [awarded the following year]	Discovery of energy quanta
1919	J. Stark (Germany)	Discovery of the Doppler Effect in canal rays (beams of positive ions) and discovery of the Stark Effect
1920	C. E. Guillaume (Switzerland)	Contributions to precision measurements rendered by his discovery of invar

Year	Name (Nationality)	Reason for the Award
1921	A. Einstein (Switzerland, Germany) [awarded the following year]	Contributions to theoretical physics, in particular discovery of the laws governing the photoelectric effect
1922	N. H. D Bohr (Denmark)	Research into atomic structure and atomic radiation
1923	R. A. Millikan (USA)	Research into the elementary electric charge and the photoelectric effect
1924	K. M. G. Siegbahn (Sweden) [awarded the next year]	Research into X-ray spectroscopy
1925	J. Franck (Germany) [awarded the next year], G. L. Hertz (Germany) [awarded the next year]	Discovery of the laws governing the impact of an electrons upon an atom
1926	J. B. Perrin (France)	Research into the discontinuous structure of matter, in particular the discovery of sedimentation equilibrium
1927	A. H. Compton (USA)	Discovery of Compton Scattering
	C. T. R. Wilson (UK)	Visualization of the tracks of electrically charged particles by condensation of vapour
1928	O. W. Richardson (UK) [awarded the next year]	Research into the thermionic phenomenon, in particular the discovery of the Richardson Effect
1929	L.-V. P. R. de Broglie (France)	Discovery of the wave nature of electrons
1930	C. V. Raman (India)	Research into light scattering, and discovery of Raman Scattering
1931	No recipient	
1932	W. K. Heisenberg (Germany) [awarded the next year]	Creation of quantum mechanics, in particular the discovery of the allotropic forms of hydrogen
1933	E. Schrödinger (Austria), P. A. M. Dirac (UK)	Discovery of new forms of atomic theory
1934	No recipient	
1935	J. Chadwick (Great Britain)	Discovery of the neutron
1936	V. F. Hess (Austria)	Discovery of cosmic radiation
	C. D. Anderson (USA)	Discovery of the positron
1937	C. J. Davisson (USA), G. P. Thomson (UK)	Discovery of electron diffraction in crystals
1938	E. Fermi (Italy)	Discovery of new radioactive elements made through neutron bombardment and discovery of nuclear reactions induced by slow neutrons
1939	E. O. Lawrence (USA)	Invention and development of the cyclotron and research into artificial radioactive elements
1940	No recipient	
1941	No recipient	
1942	No recipient	
1943	O. Stern (USA) [awarded the next year]	Development of the molecular ray method and discovery of the magnetic moment of a proton
1944	I. I. Rabi (USA)	Resonance method for recording the maginetic properties of atomic nuclei
1945	W. Pauli (Austria)	Discovery of the (Pauli) Exclusion Principle
1946	P. W. Bridgman (USA)	Invention of a device for achieving extremely high pressures and discoveries in high pressure physics
1947	E. V. Appleton (UK)	Research into the upper atmosphere, in particular discovery of the Appleton Layer
1948	P. M. S. Blackett (UK)	Development of the Wilson Cloud Chamber and discoveries in nuclear and cosmic-ray physics
1949	H. Yukawa (Japan)	Prediction of the existence of mesons based on the study of nuclear forces
1950	C. F. Powell (UK)	Development of photographic method for researching nuclear processes and discoveries about mesons
1951	J. D. Cockcroft (UK), E. T. S. Walton (Ireland)	Pioneering work on the transmutation of the atomic nuclei using particle accelerators
1952	F. Bloch (USA), E. M. Purcell (USA)	Development of a new, precise nuclear magnetic resonance method and discoveries made with it
1953	F. Zernike (Netherlands)	Invention of the phase contrast microscope
1954	M. Born (UK)	Fundamental research into quantum mechanics, in particular statistical interpretation of the wavefunction
	W. Bothe (Germany)	Coincidence method and discoveries made with it
1955	W. E. Lamb (USA)	Discoveries related to the fine structure of the hydrogen spectrum

Year	Name (Nationality)	Reason for the Award
1955	P. Kusch (USA)	Precision measurements of the magnetic moment of the electron
1956	W. B. Shockley (USA), J. Bardeen (USA), W. H. Brattain (USA)	Semiconductor research, discovery of the transistor effect
1957	C. N. Yang (China), T. D. Lee (China)	Research into the laws of parity
1958	P. A. Cherenkov (USSR), I. M. Frank (USSR), I. Y. Tamm (USSR)	Discovery and interpretation of the Cherenkov Effect
1959	E. G. Segrè (USA), O. Chamberlain (USA)	Discovery of the antiproton
1960	D. A. Glaser (USA)	Invention of the bubble chamber
1961	R. Hofstadter (USA)	Research into electron scattering by atomic nuclei and related discoveries concerning nucleon structure
	R. L. Mössbauer (Germany)	Research related to γ-ray resonant absorption and discovery of the Mössbauer Effect
1962	L. D. Landau (USSR)	Condensed matter physics, in particular theoretical research related to liquid helium
1963	E. P. Wigner (USA)	Contributions to the theory of the atomic nucleus and the elementary particle and discovery and application of the symmetry principles
	M. Goeppert Mayer (USA), J. H. D. Jensen (Germany)	Discoveries related to nuclear shell structure
1964	C. H. Townes (USA), N. G. Basov (USSR), A. M. Prokhorov (USSR)	Fundamental research into quantum electronics and the invention of lasers and masers
1965	S. Tomonaga (Japan), J. Schwinger (USA), R. P. Feynman (USA)	Fundamental research into quantum electrodynamics
1966	A. Kastler (France)	Discovery and development of an optical method for studying Hertzian resonances (electron energy sublevels)
1967	H. A. Bethe (USA)	Theory of nuclear reactions, in particular discoveries related to the energy sources of stars
1968	L. W. Alvarez (USA)	Contributions to elementary particle physics, in particular the discovery of resonance states made with the hydrogen bubble chamber
1969	M. Gell-Mann (USA)	Discoveries related to the classification and interactions of elementary particles
1970	H. O. G. Alfvén (Sweden)	Fundamental research into magnetohydrodynamics and its application to plasma physics
	L. E. F. Néel (France)	Fundamental research into antiferromagnetism and ferrimagnetism
1971	D. Gabor (Great Britain)	Invention and development of holography
1972	J. Bardeen (USA), L. N. Cooper (USA), J. R. Schrieffer (USA)	BCS theory of superconductivity
1973	L. Esaki (Japan)	Discovery of tunneling phenomena in semiconductors
	I. Giaever (USA)	Discovery of tunneling phenomena in superconductors
	B. D. Josephson (UK)	Theoretical prediction of the Josephson (tunneling) Effect
1974	M. Ryle (UK)	Invention of aperture synthesis technique
	A. Hewish (UK)	Discovery of pulsars
1975	A. N. Bohr (Denmark), B. R. Mottelson (Denmark), L. J. Rainwater (USA)	Discovery of the relation between collective motion and particle motion in atomic nuclei, and development of the theory of structure of atomic nuclei
1976	B. Richter (USA), S. C. C. Ting (USA)	Discovery of a new kind of heavy elementary particle (J/ψ meson)
1977	P. W. Anderson (USA), N. F. Mott (UK), J. H. van Vleck (USA)	Theoretical research into the electronic structure of magnetic and disordered systems
1978	P. L. Kapitsa (USSR)	Inventions and discoveries related to low temperature physics
	A. A. Penzias (USA), R. W. Wilson (USA)	Discovery of cosmic microwave background radiation
1979	S. L. Glashow (USA), A. Salam (Pakistan), S. Weinberg (USA)	Theory unifying the nuclear weak force and the electromagnetic force, in particular the prediction of weak neutral current
1980	J. W. Cronin (USA), V. L. Fitch (USA)	Discovery of symmetry breaking in the decay of neutral K-mesons
1981	N. Bloembergen (USA), A. L. Schawlow (USA)	Contributions to the development of laser spectroscopy
	K. M. Siegbahn (Sweden)	Contributions to the development of high-resolution electron spectroscopy

Year	Name (Nationality)	Reason for the Award
1982	K. G. Wilson (USA)	Theory for critical phenomena related to phase transitions
1983	S. Chandrasekhar (USA)	Theoretical research into the physical processes responsible for the structure and evolution of stars
	W. A. Fowler (USA)	Research into nuclear reactions important for the formation of elements in the Universe
1984	C. Rubbia (Italy), S. van der Meer (Netherlands)	Contributions to the discovery of the W and Z particles which mediate nuclear weak interactions
1985	K. von Klitzing (West Germany)	Discovery of the quantum Hall Effect
1986	E. Ruska (West Germany)	Contributions to electron optics, in particular the development of the electron microscope
	G. Binnig (West Germany), H. Rohrer (Switzerland)	Design of the scanning tunneling electron microscope
1987	J. G. Bednorz (West Germany), K. A. Müller (Switzerland)	Discovery of superconductivity in ceramic materials
1988	L. M. Lederman (USA), M. Schwartz (USA), J. Steinberger (USA)	Neutrino beam method and proving the doublet (spin-1/2) structure of the leptons through the discovery of the muon neutrino
1989	N. F. Ramsey (USA)	Invention of the separated oscillatory fields method and its application to atomic clocks, particularly those based on hydrogen masers
	H. G. Dehmelt (USA), W. Paul (West Germany)	Development of ion trapping technology
1990	J. I. Friedman (USA), H. W. Kendall (USA), R. E. Taylor (Canada)	Research into deep inelastic scattering of electrons by protons and deuterium nuclei, contributions to the quark model
1991	P.-G. de Gennes (France)	Discovery that the methods for studying order phenomena in simple systems can be generalized to more complex forms of matter, in particular liquid crystals and polymers
1992	G. Charpak (France)	Invention and development of particle detectors, in particular the multiwire proportional chamber
1993	R. A. Hulse (USA), J. H. Taylor Jr. (USA)	Discovery of a new kind of pulsar with new potential for the study of gravity
1994	B. N. Brockhouse (Canada), C. G. Shull (USA)	Development of the neutron scattering method and neutron spectroscopy for condensed matter research
1995	M. L. Perl (USA)	Contributions to lepton physics, in particular the discovery of the τ lepton
	F. Reines (USA)	Contributions to lepton physics, in particular the detection of the neutrino
1996	D. M. Lee (USA), D. D. Osheroff (USA), R. C. Richardson (USA)	Discovery of helium-3 superfluid
1997	S. Chu (USA), C. Cohen-Tannoudji (France), W. D. Phillips (USA)	Laser cooling and trapping of atoms
1998	R. B. Laughlin (USA), H. L. Störmer (Germany), D. C. Tsui (USA)	Discovery of a new form of quantum fluid with fractionally charged excitations
1999	G. 't Hooft (Netherlands), M. J. G. Veltman (Netherlands)	Clarification of the quantum nature of electromagnetic-weak interactions
2000	Z. I. Alferov (Russia), H. Kroemer (Germany)	Development of the semiconductor heterostructure used in high-speed electronics and optoelectronics
	J. S. Kilby (USA)	Invention of the integrated circuit
2001	E.A. Cornell (USA), W. Ketterle (Germany), C.E. Wieman (USA)	Bose-Einstein Condensation in dilute gases of alkali atoms
2002	R. Davis Jr. (USA), M. Koshiba (Japan)	Contributions to astrophysics through the detection of cosmic neutrinos
	R. Gaiacconi (USA)	Contributions to astrophysics through the discovery of cosmic X-ray sources
2003	A. A. Abrikosov (Russia, USA), V. L. Ginzburg (Russia), A. J. Leggett (USA)	Contributions to the theory of superconductivity and superfluidity
2004	D. J. Gross (USA), H. D. Politzer (USA), F. Wilczek (USA)	Discovery of asymptotic freedom in the theory of the nuclear strong interaction
2005	R. J. Glauber (USA)	Quantum theory for optical coherence
	J. L. Hall (USA), T. W. Hänsch (Germany)	Precision spectroscopy using lasers, in particular the optical comb method
2006	J. C. Mather (USA), G. F. Smoot (USA)	Discovery of the blackbody form and anisotropy of the cosmic microwave background radiation

Year	Name (Nationality)	Reason for the Award
2007	A. Fert (France), P. Grünberg (Germany)	Discovery of Giant Magentoresistance
2008	Y. Nambu (USA)	Discovery of the mechanism for spontaneous symmetry breaking in elementary particle and nuclear physics
	M. Kobayashi (Japan), T. Maskawa (Japan)	Discovery of the origin of symmetry breaking which predicts the number of generations of quarks
2009	C. K. Kao (UK, USA)	Achieving transmission through optical fibers
	W. S. Boyle (Canada, USA), G. E. Smith (USA)	Invention of CCD sensors
2010	A. Geim (Netherlands), K. Novoselov (UK, Russia)	Experiments related to graphene
2011	S. Perlmutter (USA), B. P. Schmidt (USA, Australia), A. G. Riess (USA)	Discovery of the accelerating expansion of the Universe through observations of distant supernovae
2012	S. Haroche (France), D. J. Wineland (USA)	Measurement and manipulation of discrete quantum systems
2013	F. Englert (Belgium), P. W. Higgs (UK)	Theoretical predictions related to the origin of mass (Higgs Field, Higgs Boson)
2014	I. Akasaki (Japan), H. Amano (Japan), S. Nakamura (USA)	Invention of efficient blue light-emitting diodes
2015	T. Kajita (Japan), A. B. McDonald (Canada)	Discovery of neutrino oscillations, indicating that neutrinos have non-zero mass
2016	D. J. Thouless (UK, USA), F. D. M. Haldane (UK, USA), J. M. Kosterlitz (UK, USA)	Theoretical discoveries of topological phase transitions and topological phases of matter
2017	R. W. Weiss (USA), B. C. Barish (USA), K. S. Thorne (USA)	Observation of gravitational waves
2018	A. Ashkin (USA)	Optical tweezers and their application to biological systems
	G. Mourou (France), D. Strickland (Canada)	Method of generating high-intensity, ultra-short optical pulses
2019	J. Peebles (Canada, USA)	Theoretical discoveries in physical cosmology
	M. Mayor (Switzerland), D. Queloz (Switzerland)	Discovery of an exoplanet orbiting a solar-type star
2020	R. Penrose (UK)	Discovery that black hole formation is a robust prediction of the general theory of relativity
	R. Genzel (Germany), A. Ghez (USA)	Discovery of a supermassive compact object at the centre of our galaxy
2021	S. Manabe (USA), K. Hasselmann (Germany)	Physical modelling of Earth' s climate, quantifying variability and reliably predicting global warming
	G. Parisi (Italy)	Discovery of the interplay of disorder and fluctuations in physical systems from atomic to planetary scales

Prizes in Chemistry

Year	Name (Nationality)	Reason for the Award
1901	J. H. van't Hoff (Netherlands)	Discovery of the laws governing chemical dynamics and osmotic pressure
1902	H. E. Fischer (Germany)	Research into sugar and purine syntheses
1903	S. A. Arrhenius (Sweden)	Electrolytic theory of dissociation (acid-alkali formulation)
1904	W. Ramsay (UK)	Discovery of inert gasses and determination of their positions in the periodic system (table)
1905	J. F. W. A. von Baeyer (Germany)	Contributions to organic chemistry and the chemical industry through research into organic dyes and hydroaromatic compounds
1906	H. Moissan (France)	Separation and study of fluorine and scientific applications of the Moissan Electric Furnace
1907	E. Buchner (Germany)	Research into organic chemistry and the discovery of cell-free fermentation
1908	E. Rutherford (Great Britain)	Elemental decay and the chemistry of radioactive substances
1909	W. Ostwald (Germany)	Catalyst research and the fundamental principles of chemical equilibrium and reaction rates

Year	Name (Nationality)	Reason for the Award
1910	O. Wallach (Germany)	Research in the field of alicyclic compounds
1911	M. S. Curie (France, Polant)	Discovery of radium and polonium, the refinement of radium, and research into its characteristics and compounds
1912	V. Grignard (France)	Discovery of Grignard Reagent
1912	P. Sabatier (France)	Development of a method to hydrogenate organic compounds using minute metal particles
1913	A. Werner (Switzerland)	Research related to the bonds between atoms in molecules
1914	T. W. Richards (USA) [awarded the next year]	Accurate determination of the atomic weights of many elements
1915	R. M. Willstätter (Germany)	Research into plant pigments, in particular chlorophyll
1916	No recipient	
1917	No recipient	
1918	F. Haber (Germany) [awarded the next year]	Synthesis of ammonia from its elements
1919	No recipient	
1920	W. H. Nernst (Germany)	Research into thermochemistry
1921	F. Soddy (UK) [awarded the next year]	Research into radioactive substances, especially the origin and characteristics of isotopes
1922	F. W. Aston (UK)	Discovery of non-radioactive isotopes, discovery of the integer law
1923	F. Pregl (Austria)	Invention of micro-analysis method for organic substances
1924	No recipient	
1925	R. A. Zsigmondy (Germany) [awarded the next year]	Proof of the heterogeneous nature of colloidal suspensions and the development of colloid suspension research methodology
1926	T. Svedberg (Sweden)	Research into disperse systems
1927	H. O. Wieland (Germany) [awarded the next year]	Research into the structure of bile acid and related substances
1928	A. O. R. Windaus (Germany)	Research into the structure of sterols and their relation to vitamins
1929	A. Harden (UK), H. K. A. S. von Euler-Chelpin (Sweden)	Research into sugar fermentation and fermentation enzymes
1930	H. Fischer (Germany)	Research into hemin and chlorophyll, in particular the production of hemin
1931	C. Bosch (Germany), F. Bergius (Germany)	Invention and refinement of high-pressure techniques in chemistry
1932	I. Langmuir (USA)	Research and discoveries related to surface chemistry
1933	No recipient	
1934	H. C. Urey (USA)	Discovery of deuterium
1935	F. and I. Joliot-Curie (France)	Production of new radioactive elements
1936	P. J. W. Debye (Netherlands)	Research into dipole moments and research into the diffraction of X-rays and electrons in gas
1937	W. N. Haworth (UK)	Research into carbohydrates and vitamin C
	P. Karrer (Switzerland)	Research into carotenoids, flavins, and vitamins A and B_2
1938	R. Kuhn (Germany) [awarded the next year]	Research into carotenoids and vitamins
	[Kunh refused the award at the behest of the German government; but later he did accept the diploma and medal.]	
1939	A. F. J. Butenandt (Germany)	Research into sex hormones
	[Butenandt refused the award at the behest of the German government; but later he did accept the diploma and medal.]	
1939	L. Ruzicka (Switzerland)	Research into polymethylenes and higher terpenes
1940	No recipient	
1941	No recipient	
1942	No recipient	
1943	G. de Hevesy (Hungary) [awarded the next year]	Method using isotopes to trace chemical reactions
1944	O. Hahn (Germany) [awarded the next year]	Discovery of heavy element fission
1945	A. I. Virtanen (Finland)	Research and discoveries in agricultural and nutritional chemistry, in particular fodder preservation methods
1946	J. B. Sumner (USA)	Discovery of enzyme crystallization
	J. H. Northrop (USA), W. M. Stanley (USA)	Preparation of pure enzymes and virus proteins
1947	R. Robinson (Great Britain)	Research into plant products, in particular alkaloid research

Year	Name (Nationality)	Reason for the Award
1948	A. W. K. Tiselius (Sweden)	Research into electrophoresis and adsorption analysis, in particular discoveries related to the nature of serum proteins
1949	W. F. Giauque (USA)	Contributions to chemical thermodynamics, in particular research into the behavior of substances at cryogenic temperatures
1950	O. P. H. Diels (Germany), K. Alder (Germany)	Discovery and development of diene synthesis
1951	E. M. McMillan (USA), G. T. Seaborg (USA)	Discoveries in the chemistry of the transuranium elements
1952	A. J. P. Martin (UK), R. L. M. Synge (UK)	Invention of partition chromatography
1953	H. Staudinger (Germany)	Discoveries in polymer chemistry
1954	L. C. Pauling (USA)	Research into the nature of chemical bonds clarifying the structure of complex substances
1955	V. du Vigneaud (USA)	Research into biochemically important sulfur compounds, in particular the synthesis of a polypeptide hormone
1956	C. N. Hinshelwood (UK), N. N. Semenov (USSR)	Research into the mechanisms of chemical reactions
1957	A. R. Todd (UK)	Research into nucleotides and nucleotide coenzymes
1958	F. Sanger (UK)	Research into the structure of proteins, in particular insulin
1959	J. Heyrovsky (Czechoslovakia)	Discovery and refinement of polarography
1960	W. F. Libby (USA)	Carbon 14 dating technique used in archaeology, geology, geophysics, etc.
1961	M. Calvin (USA)	Research into the carbon absorption (photosynthesis) technique used by plants
1962	M. F. Perutz (UK), J. C. Kendrew (UK)	Research into the structures of globular proteins
1963	K. Ziegler (Germany), G. Natta (Italy)	Discoveries in high polymer chemistry and technology
1964	D. C. Hodgkin (UK)	Determination of the structures of important biochemical substances through X-ray diffraction
1965	R. B. Woodward (USA)	Achievements in organic synthesis
1966	R. S. Mulliken (USA)	Research into chemical bonds and electronic structure of molecules by the molecular orbital method
1967	M. Eigen (West Germany), R. G. W. Norrish (UK), G. Porter (UK)	Research into extremely fast chemical reactions through the use of very short pulses of energy
1968	L. Onsager (USA)	Discovery of Onsager reciprocal relations which contributed to the thermodynamics of irreversible processes
1969	D. H. R. Barton (UK), O. Hassel (Norway)	Development of the concept of conformation and its application to chemistry
1970	L. F. Leloir (Argentina)	Discovery of sugar nucleotides and their role in carbohydrate biosyhtheses
1971	G. Herzberg (Canada)	Electronic and geometric structure of molecules, in particular free radicals
1972	C. B. Anfinsen (USA)	Ribonucleic research, in particular into the relation between amino acid sequencing and protein structure
	S. Moore (USA), W. H. Stein (West Germany)	Relation between the chemical structure of the active centers of ribonuclease molecules and their catalytic function
1973	E.O. Fischer (West Germany), G. Wilkinson (UK)	Chemistry of organometallic sandwich compounds
1974	P. J. Flory (USA)	Theoretical and experimental work related to polymers
1975	J. W. Cornforth (UK)	Stereochemistry of enzyme-catalyzed reactions
	V. Prelog (Switzerland)	Stereochemistry of organic molecules and reactions
1976	W. N. Lipscomb (USA)	Research into the structure of boranes and chemical bonds
1977	I. Prigogine (Belgium)	Theory of non-equilibrium thermodynamics, in particular dissipative structures
1978	P. D. Mitchell (UK)	Energy transfer in living bodies, chemiosmotic theory
1979	H. C. Brown (USA)	Application of compounds containing boron to organic synthesis
	G. Wittig (West Germany)	Application of compounds containing phosphorus to organic synthesis
1980	P. Berg (USA)	Research into nucleic acid biochemistry, in particular recombinant-DNA
	W. Gilbert (USA), F. Sanger (UK)	Determination of the base sequences in nucleic acids
1981	K. Fukui (Japan), R. Hoffmann (USA)	Theories related to the process of chemical reactions
1982	A. Klug (UK)	Development of crystallographic electronic spectroscopy, and clarification of the structure of nucleic acid-protein complexes

Year	Name (Nationality)	Reason for the Award
1983	H. Taube (USA)	Electron transfer reactions, in particular those occurring in metal complexes
1984	R. B. Merrifield (USA)	Method for chemical synthesis on solid substrates
1985	H. A. Hauptman (USA), J. Karle (USA)	Methods for direct determination of crystalline structure
1986	D. R. Herschbach (USA), Y. T. Lee (USA), J. C. Polanyi (Canada)	Dynamics of chemical elementary reactions
1987	D. J. Cram (USA), J.-M. Lehn (France), C. J. Pedersen (USA)	Development and application of molecules with highly selective structure-specific interactions
1988	J. Deisenhofer (West Germany), R. Huber (West Germany), H. Michel (West Germany)	Determination of the 3-D structure of a photosynthesis reaction center
1989	S. Altman (USA, Canada), T. R. Cech (USA)	Discovery of the catalytic characteristics of RNA
1990	E. J. Corey (USA)	Theory and methodology of organic synthesis
1991	R. R. Ernst (Switzerland)	High resolution nuclear magnetic resonance spectroscopy
1992	R. A. Marcus (USA)	Theory for electron transfer reactions in chemical systems
1993	K. B. Mullis (USA)	DNA chemistry, in particular the polymerase chain reaction method
	M. Smith (Canada)	DNA chemistry, in particular site-directed mutagenesis based on oligonucleotides
1994	G. A. Olah (USA)	Contributions to carbocation chemistry
1995	P. J. Crutzen (Netherlands), M. J. Molina (USA), F. S. Rowland (USA)	Atmospheric chemistry, in particular research into the formation of the ozone hole
1996	R. F. Curl Jr. (USA), H. W. Kroto (UK), R. E. Smalley (USA)	Discovery of Fullerene (Buckyballs)
1997	P. D. Boyer (USA), J. E. Walker (UK)	Explanation of the enzyme mechanism of ATP synthesis
	J. C. Skou (Denmark)	Discovery of ion transport enzymes
1998	W. Kohn (USA)	Development of density-functional theory
	J. A. Pople (UK)	Application of computers to quantum chemistry
1999	A. H. Zewail (Egypt, USA)	Research into transition states by applying femtosecond spectroscopy to chemical reactions
2000	A. J. Heeger (USA), A. G. MacDiarmid (USA), H. Shirakawa (Japan)	Discovery and development of conductive polymers
2001	W. S. Knowles (USA), R. Noyori (Japan)	Chirally catalyzed hydrogenation reactions
	K. B. Sharpless (USA)	Chirally catalyzed oxidation reactions
2002	J. B. Fenn (USA), K. Tanaka (Japan)	Development of the soft desorption ionization method for mass spectroscopy of organic macromolecules
	K. Wüthrich (Switzerland)	Development of nuclear magnetic resonance spectroscopy for determining the structure of biological macromolecules in solution
2003	P. Agre (USA)	Discoveries of cell membrane water channels
	R. MacKinnon (USA)	Research into structure and mechanics of ion channel
2004	A. Ciechanover (Israel), A. Hershko (Israel), I. Rose (USA)	Discovery of the protein decomposition governed by ubiquitin
2005	Y. Chauvin (France), R.H. Grubbs (USA), R. R. Schrock (USA)	Development of the metathesis method of organic synthesis
2006	R. D. Kornberg (USA)	Research into the molecular foundation of transcription of genetic information in eukaryotes
2007	G. Ertl (Germany)	Research into chemical reactions on solid surfaces
2008	O. Shimomura (Japan), M. Chalfie (USA), R. Y. Tsien (USA)	Discovery and development of green fluorescent proteins (GFP)
2009	V. Ramakrishnan (Israel), T.A. Steitz (UK), A. E. Yonath (USA)	Research into ribosome structure and function
2010	R. F. Heck (USA), E. Negishi (Japan), A. Suzuki (Japan)	Cross coupling utilizing palladium catalyses in organic synthesis
2011	D. Shechtman (Israel)	Discovery of quasicrystals
2012	R. J. Lefkowitz (USA), B. K. Kobilka (USA)	Research into G-protein-coupled receptors
2013	M. Karplus (USA, Austria), M. Levitt (USA, UK, Israel), A. Warshel (USA, Israel)	Development of multiscale models for complex chemical systems
2014	E. Betzig (USA), S. W. Hell (Germany), W. E. Moerner (USA)	Development of super-resolved fluorescence microscopy
2015	T. Lindahl (Sweden), P. Modrich (USA), A. Sancar (Turkey, USA)	Research into the mechanics of DNA repair

Year	Name (Nationality)	Reason for the Award
2016	J.-P. Sauvage (France), J. F. Stoddart (UK, USA), B. L. Feringa (Netherlands)	Design and synthesis of molecular machines
2017	J. Dubochet (Switzerland), J. Frank (Germany, USA), R. Henderson (UK)	Development of cryo-electron microscopy for the high-resolution structure determination of biomolecules in solution
2018	F. H. Arnold (USA)	Directed evolution of enzymes
	G. P. Smith (USA), G. P. Winter (UK)	Phage display of peptides and antibodies
2019	J. B. Goodenough (USA), M. S. Whittingham (UK, USA), A. Yoshino (Japan)	Development of lithium-ion batteries
2020	E. Charpentier (France), J. A. Doudna (USA)	Development of a method for genome editing
2021	B. List (Germany), D. W.C. MacMillan (UK, USA)	Development of asymmetric organocatalysis

Prizes in Physiology or Medicine

Year	Name (Nationality)	Reason for the Award
1901	E. A. von Behring (Germany)	Research into serum therapy, in particular its applications to diphtheria
1902	R. Ross (UK)	Research into malaria, clarification of infection vectors
1903	N. R. Finsen (Denmark)	Concentrated light irradiation treatment, in particular its application to the disease lupus vulgaris
1904	I. P. Pavlov (Russia)	Physiology of digestion
1905	R. Koch (Germany)	Research into tuberculosis
1906	C. Golgi (Italy), S. Ramón y Cajal (Spain)	Research into the structure of the nervous system
1907	C. L. A. Laveran (France)	Research into the role of protozoa in causing diseases
1908	I. I. Mechnikov (Russia), P. Ehrlich (Germany)	Research into immunity
1909	E. T. Kocher (Switzerland)	Thyroid physiology, pathology, and surgical procedures
1910	A. Kossel (Germany)	Cellular chemistry and research into proteins, including nucleic substances
1911	A. Gullstrand (Sweden)	Dioptrics (refraction) of the eyeball
1912	A. Carrel (France)	Research into vascular structure and the transplantation of blood vessels and internal organs
1913	C. R. Richet (France)	Research into anaphylaxis
1914	R. Bárány (Austria)	Physiology and pathology of the inner ear
1915	No recipient	
1916	No recipient	
1917	No recipient	
1918	No recipient	
1919	J. Bordet (Belgium) [awarded the next year]	Discoveries related to immunity
1920	S. A. S. Krogh (Denmark)	Discovery of the mechanism regulating capillary blood flow
1921	No recipient	
1922	A. V. Hill (UK) [awarded the next year]	Discoveries related to heat generation in muscles
1922	O. F. Meyerhof (Germany) [awarded the next year]	Discovery of the relation between the consumption of oxygen and the metabolism of lactic acid
1923	F. G. Banting (UK), J. J. R. Macleod (Canada)	Discovery of insulin
1924	W. Einthoven (Netherlands)	Discovery of the mechanism of electrocardiogram
1925	No recipient	
1926	J. A. G. Fibiger (Denmark) [awarded the next year]	Discovery of Spiroptera carcinoma (mistakenly believed to be the cause of tumors)
1927	J. Wagner-Jauregg (Austria)	Discovery of the therapeutic benefits of a malaria inoculation to treat dementia paralytica
1928	C. J. H. Nicolle (France)	Research into typhus
1929	C. Eijkman (Netherlands)	Discovery of the antineuritic vitamin
	F. G. Hopkins (UK)	Discovery of the growth-stimulating vitamin
1930	K. Landsteiner (Austria)	Discovery of human blood types
1931	O. H. Warburg (Germany)	Discovery of the nature and functioning of the (yellow) respiratory enzyme
1932	C. S. Sherrington (UK), E. D. Adrian (UK)	Discoveries related to the function of neurons
1933	T. H. Morgan (USA)	Discovery of the role of chromosomes in heredity

Year	Name (Nationality)	Reason for the Award
1934	G. H. Whipple (USA), G. R. Minot (USA), W. P. Murphy (USA)	Discoveries related to liver treatments for anemia
1935	H. Spemann (Germany)	Discovery of embryonic induction
1936	H. H. Dale (UK), O. Loewi (Austria)	Discoveries related to the chemical transmission of neural signals
1937	A. von Szent-Györgyi Nagyrápolt (Hungary)	Discoveries related to metabolism, with special reference to vitamin C and the catalysis of fumaric acid
1938	C. J. F. Heymans (Belgium) [awarded the next year]	Discovery of the role of the sinus and aortic mechanisms in regulating respiration
1939	G. Domagk (Germany)	Discovery of the antibacterial properties of prontosil
[Domagk refused the award at the behest of the German government; but later he did accept the diploma and medal.]		
1940	No recipient	
1941	No recipient	
1942	No recipient	
1943	H. C. P. Dam (Denmark) [awarded the next year]	Discovery of vitamin K
	E. A. Doisy (USA) [awarded the next year]	Discovery of the chemistry of vitamin K
1944	J. Erlanger (USA), H. S. Gasser (USA)	Discoveries related to the highly differentiated functions of individual nerve fibers
1945	A. Fleming (UK), E. B. Chain (UK), H. W. Florey (UK)	Discovery of penicillin and its use in treating infectious diseases
1946	H. J. Muller (USA)	Discovery of genetic mutations resulting from X-ray irradiation
1947	C. F. and G. T. Cori (USA)	Discovery of the glycogen catalytic conversion process
	B. A. Houssay (Argentina)	Discovery of the role of the anterior pituitary lobe hormones in sugar metabolism
1948	P. H. Müller (Switzerland)	Discovery that DDT acts as a strong contact poison against several types of arthropods
1949	W. R. Hess (Switzerland)	Discovery of the role of the interbrain in coordinating internal organs
	A. C. A. F. E. Moniz (Portugal)	Discovery of the effectiveness of prefrontal lobotomy in treating certain mental disorders
1950	E. C. Kendall (USA), T. Reichstein (Switzerland), P. S. Hench (USA)	Discoveries related to adrenocortical hormones, their structures, and their biological effects
1951	M. Theiler (South Africa)	Discoveries related to yellow fever and how to combat it
1952	S. A. Waksman (USA)	Discovery of streptomycin, the first antibiotic effective against tuberculosis
1953	H. A. Krebs (UK)	Discovery of the citric acid cycle (Krebs Cycle)
	F. A. Lipmann (USA)	Discovery of coenzyme A and its importance in metabolism
1954	J. F. Enders (USA), T. H. Weller (USA), F. C. Robbins (USA)	Discovery of the capability of poliomyelitis viruses (polio) to grow in various types of tissue cultures
1955	A. H. T. Theorell (Sweden)	Research related to the nature and functional mechanisms of oxidation enzymes
1956	A. F. Cournand (USA), W. Forssmann (West Germany), D. W. Richards (USA)	Discoveries related to inserting catheters into the heart and the progression of circulatory disease
1957	D. Bovet (Italy)	Discoveries related to synthetic compounds which inhibit the functions of bodily substances, especially the vascular system and skeletal muscles
1958	G. W. Beadle (USA), E. L. Tatum (USA)	Discovery that genes act by regulating specific chemical reactions
	J. Lederberg (USA)	Discoveries related to genetic recombination and the genetic material of bacteria
1959	S. Ochoa (USA), A. Kornberg (USA)	Discovery of the mechanicsm of the biological production of ribonucleic acid and deoxyribonucleic acid
1960	F. M. Burnet (Australia), P. B. Medawar (UK)	Discovery of acquired immunity
1961	G. von Békésy (USA)	Discovery of the physical stimulation mechanism in the cochlea (portion of the inner ear)
1962	F. H. C. Crick (UK), J. D. Watson (USA), M. H. F. Wilkins (UK)	Discovery of the molecular structure of DNA and its importance for the transmission of genetic information

Year	Name (Nationality)	Reason for the Award
1963	J. C. Eccles (Australia), A. L. Hodgkin (UK), A. F. Huxley (UK)	Discoveries related to ionic stimulating/inhibiting mechanisms in the peripheral and central sections of nerve cell membranes
1964	K. Bloch (USA), F. Lynen (West Germany)	Discoveries related to the mechanism and regulation of cholesterol and fatty acids metabolism
1965	F. Jacob (France), A. Lwoff (France), J. Monod (France)	Discoveries related to genetic control of enzyme and virus production
1966	P. Rous (USA)	Discovery of tumor inducing viruses
	C. B. Huggins (USA)	Discoveries related to hormone treatment for prostate cancer
1967	R. Granit (Sweden), H. K. Hartline (USA), G. Wald (USA)	Discoveries related to the physical and chemical processes of sight in the eye
1968	R. W. Holley (USA), H. G. Khorana (USA), M. W. Nirenberg (USA)	Deciphering the genetic code and its function in protein synthesis
1969	M. Delbrück (USA), A. D. Hershey (USA), S. E. Luria (USA)	Discoveries related to the genetic structure and reproduction method of viruses
1970	B. Katz (UK), U. von Euler (Sweden), J. Axelrod (USA)	Discoveries related to humoral transmitters at nerve terminals and their mechanisms of storage, release, and inactivation
1971	E. W. Sutherland, Jr. (USA)	Discoveries related to the functional mechanisms of hormones
1972	G. M. Edelman (USA), R. R. Porter (UK)	Discoveries related to the chemical structure of antibodies
1973	K. von Frisch (West Germany), K. Lorenz (Austria), N. Tinbergen (UK)	Discoveries related to the organization and invoking of individual and group behavior patterns
1974	A. Claude (Belgium), C. de Duve (Belgium), G. E. Palade (USA)	Discoveries related to the structure and function of cells
1975	D. Baltimore (USA), R. Dulbecco (USA), H. M. Temin (USA)	Discoveries related to the interaction between tumor viruses and cellular genetic material
1976	B. S. Blumberg (USA), D. C. Gajdusek (USA)	Discoveries related to a new mechanism of generation and diffusion of infectious diseases
1977	R. Guillemin (USA), A. V. Schally (USA)	Discoveries related to peptide production in the brain
	R. Yalow (USA)	Development of peptide radioimmunoassays (using antibodies to measure the amounts of specific materials)
1978	W. Arber (Switzerland), D. Nathans (USA), H. O. Smith (USA)	Discovery of restriction enzymes and their applications to molecular genetics (slicing DNA strands)
1979	A. M. Cormack (USA), G. N. Hounsfield (UK)	Development of the computer assisted tomography (CAT) scan
1980	B. Benacerraf (USA), J. Dausset (France), G. D. Snell (USA)	Discoveries related to genetically determined markers on the surface of cells which govern immune reactions
1981	R. W. Sperry (USA)	Discoveries related to functional specialization of the cerebral hemispheres
	D. H. Hubel (USA), T. N. Wiesel (Sweden)	Discoveries related to information processing in the visual cortex
1982	S. K. Bergström (Sweden), B. I. Samuelsson (Sweden), J. R. Vane (UK)	Discoveries related to prostaglandins and related biologically active substances
1983	B. McClintock (USA)	Discovery of mobile genetic elements
1984	N. K. Jerne (Denmark), G. J. F. Köhler (West Germany), C. Milstein (UK, Argentina)	Theories related to development and regulation of the immune system and the discovery of the production principle of monoclonal antibodies
1985	M. S. Brown (USA), J. L. Goldstein (USA)	Discoveries related to the regulation of cholesterol metabolism
1986	S. Cohen (USA), R. Levi-Montalcini (Italy, USA)	Discovery of growth factors
1987	S. Tonegawa (Japan)	Discovery of the genetic basis for antibody diversity
1988	J. W. Black (UK), G. B. Elion (USA), G. H. Hitchings (USA)	Discovery of important principles of drug treatment
1989	J. M. Bishop (USA), H. E. Varmus (USA)	Discovery of the cellular origins of retrovirus oncogene
1990	J. E. Murray (USA), E. D. Thomas (USA)	Discoveries related to organ and cell transplants to treat human diseases
1991	E. Neher (Germany), B. Sakmann (Germany)	Discoveries related to the function of single ion channels in cells

Year	Name (Nationality)	Reason for the Award
1992	E. H. Fischer (USA), E. G. Krebs (USA)	Discoveries related to reversible protein phosphorylation as a biological regulatory mechanism
1993	R. J. Roberts (USA), P. A. Sharp (USA)	Discovery of split genes
1994	A. B. Gilman (USA), M. Rodbell (USA)	Discovery of G-proteins and their role in signal transduction in cells
1995	E. B. Lewis (USA), C. Nüsslein-Volhard (Germany), E. F. Wieschaus (USA)	Discoveries related to genetic regulation of early embryonic development
1996	P. C. Doherty (Australia), R. M. Zinkernagel (Switzerland)	Discoveries related to the selectiveness of cell-mediated immune defenses
1997	S. B. Prusiner (USA)	Discovery of prions, a new biological principle of infection
1998	R. F. Furchgott (USA), L. J. Ignarro (USA), F. Murad (USA)	Discovery that nitric oxide works as a cardiovascular signal
1999	G. Blobel (USA)	Discovery that inherent signals govern protein transportation and distribution within cells
2000	A. Carlsson (Sweden), P. Greengard (USA), E. R. Kandel (USA)	Discoveries related to signal transduction in the nervous system
2001	L. H. Hartwell (USA), T. Hunt (UK), P. M. Nurse (UK)	Discovery of important regulators of the cell cycle
2002	S. Brenner (UK), H. R. Horvitz (USA), J. E. Sulston (UK)	Discoveries related to genetic regulation of organ development and programed cell death
2003	P. C. Lauterbur (USA), P. Mansfield (UK)	Discoveries related to nuclear magnetic resonance imaging (MRI)
2004	R. Axel (USA), L. B. Buck (USA)	Discovery of smell receptors and organization of the olfactory system
2005	B. J. Marshall (Australia), J. R. Warren (Australia)	Discovery of Helicobacter pylori
2006	A. Z. Fire (USA), C. C. Mello (USA)	Discovery of RNA interference (gene silencing)
2007	M. R. Capecchi (USA), M. J. Evans (USA), O. Smithies (UK)	Discovery of the principles related to introducing specific genetic modifications in mice by employing embryonic stem cells
2008	H. zur Hausen (Germany)	Discovery of human papilloma virus inducing cervical cancer
2008	F. Barré-Sinoussi (France), L. Montagnier (France)	Discovery of human immunodeficiency virus (HIV)
2009	E. H. Blackburn (USA), C. W. Greider (USA), J. W. Szostak (USA)	Discovery of telomeres and enzyme telomerase mechanisms for chromosomal defense
2010	R. G. Edwards (UK)	Development of in vitro fertilization
2011	B. A. Beutler (USA), J. A. Hoffmann (France)	Discoveries related to activation of innate immunity
	R. M. Steinman (USA)	Discovery of dendritic cells and their role in adaptive immunity
2012	J. B. Gurdon (UK), S. Yamanaka (Japan)	Discovery that mature cells can be turned back into pluripotent (stem cells)
2013	J. E. Rothman (USA), R. W. Schekman (USA), T. C. Südhof (Germany, USA)	Discovery of the vesicule traffic regulation mechanism
2014	J. O'Keefe (USA, UK), M.-B. Moser (Norway), E. I. Moser (Norway)	Discovery of cells composing the positional system in the brain
2015	W. C. Campbell (Ireland, USA), S. Ōmura (Japan)	Discoveries related to a treatment for roundworm infections
2015	Y. Tu (China)	Discoveries related to a treatment for malaria
2016	Y. Ohsumi (Japan)	Discoveries of mechanisms for autophagy
2017	J. C. Hall (USA), M. Rosbash (USA), M. W. Young (USA)	Discoveries of molecular mechanisms controlling the circadian rhythm
2018	J. P. Allison (USA), T. Honjo (Japan)	Discovery of cancer therapy by inhibition of negative immune regulation
2019	W. G. Kaelin Jr (USA), P. J. Ratcliffe (UK), G. L. Semenza (USA)	Discoveries of how cells sense and adapt to oxygen availability
2020	H. J. Alter (USA), M. Houghton (UK), C. M. Rice (USA)	Discovery of Hepatitis C virus
2021	D. Julius (USA), A. Patapoutian (USA)	Discoveries of receptors for temperature and touch

Index

H

www.ingramcontent.com/pod-product-compliance
Lightning Source LLC
LaVergne TN
LVHW010636300926
849161LV00005B/958